Element	Symbol	Atomic Number	Relative Atomic Mass[a,b]	Element	Symbol	Atomic Number	Relative Atomic Mass[a,b]	Element	Symbol	Atomic Number	Relative Atomic Mass[a,b]
actinium	Ac	89	(227)	hassium	Hs	108	(265)	radium	Ra	88	(226)
aluminum	Al	13	26.981538	helium	He	2	4.002602	radon	Rn	86	(222)
americium	Am	95	(243)	holmium	Ho	67	164.93032	rhenium	Re	75	186.207
antimony	Sb	51	121.760	hydrogen	H	1	1.00794	rhodium	Rh	45	102.90550
argon	Ar	18	39.948	indium	In	49	114.818	rubidium	Rb	37	85.4678
arsenic	As	33	74.92160	iodine	I	53	126.90447	ruthenium	Ru	44	101.07
astatine	At	85	(210)	iridium	Ir	77	192.217	rutherfordium	Rf	104	(261)
barium	Ba	56	137.327	iron	Fe	26	55.845	samarium	Sm	62	150.36
berkelium	Bk	97	(247)	krypton	Kr	36	83.80	scandium	Sc	21	44.955910
beryllium	Be	4	9.012182	lanthanum	La	57	138.9055	seaborgium	Sg	106	(263)
bismuth	Bi	83	208.98038	lawrencium	Lr	103	(262)	selenium	Se	34	78.96
bohrium	Bh	107	(262)	lead	Pb	82	207.2	silicon	Si	14	28.0855
boron	B	5	10.811	lithium	Li	3	6.941	silver	Ag	47	107.8682
bromine	Br	35	79.904	lutetium	Lu	71	174.967	sodium	Na	11	22.989770
cadmium	Cd	48	112.411	magnesium	Mg	12	24.3050	strontium	Sr	38	87.62
calcium	Ca	20	40.078	manganese	Mn	25	54.938049	sulfur	S	16	32.066
californium	Cf	98	(251)	meitnerium	Mt	109	(266)	tantalum	Ta	73	180.9479
carbon	C	6	12.0107	mendelevium	Md	101	(258)	technetium	Tc	43	(98)
cerium	Ce	58	140.116	mercury	Hg	80	200.59	tellurium	Te	52	127.60
cesium	Cs	55	132.90545	molybdenum	Mo	42	95.94	terbium	Tb	65	158.92534
chlorine	Cl	17	35.4527	neodymium	Nd	60	144.24	thallium	Tl	81	204.3833
chromium	Cr	24	51.9961	neon	Ne	10	20.1797	thorium	Th	90	232.0381
cobalt	Co	27	58.933200	neptunium	Np	93	(237)	thulium	Tm	69	168.93421
copper	Cu	29	63.546	nickel	Ni	28	58.6934	tin	Sn	50	118.710
curium	Cm	96	(247)	niobium	Nb	41	92.90638	titanium	Ti	22	47.867
dubnium	Db	105	(262)	nitrogen	N	7	14.00674	tungsten	W	74	183.84
dysprosium	Dy	66	162.50	nobelium	No	102	(259)	uranium	U	92	238.0289
einsteinium	Es	99	(252)	osmium	Os	76	190.23	vanadium	V	23	50.9415
erbium	Er	68	167.26	oxygen	O	8	15.9994	xenon	Xe	54	131.29
europium	Eu	63	151.964	palladium	Pd	46	106.42	ytterbium	Yb	70	173.04
fermium	Fm	100	(257)	phosphorus	P	15	30.973761	yttrium	Y	39	88.90585
fluorine	F	9	18.9984032	platinum	Pt	78	195.078	zinc	Zn	30	65.39
francium	Fr	87	(223)	plutonium	Pu	94	(244)	zirconium	Zr	40	91.224
gadolinium	Gd	64	157.25	polonium	Po	84	(209)	[110]		110	(269)
gallium	Ga	31	69.723	potassium	K	19	39.0983	[111]		111	(272)
germanium	Ge	32	72.61	praseodymium	Pr	59	140.90765	[112]		112	(277)
gold	Au	79	196.96655	promethium	Pm	61	(145)				
hafnium	Hf	72	178.49	protactinium	Pa	91	231.03588				

[a]Source: Commission on Atomic Weights and Isotopic Abundances, International Union of Pure and Applied Chemistry, *Pure and Applied Chemistry*, Vol. 70, 237–257 (1998). © 1998 IUPAC
[b]Values in parentheses give the mass number of the radioactive isotope with the longest half-life.

Student Media Package

Chemistry: The Science in Context is supplemented by an outstanding media package designed to insure your success in introductory chemistry. Free with every new copy of the textbook, the Student CD-ROM offers comprehensive materials for review, practice, and learning assessment of each chapter's contents. Central to the Student CD-ROM are the over 100 compelling tutorials that use animation and interactivity to reinforce conceptual understanding and develop quantitative skills. The Student CD-ROM requires no installation, browser tune-ups, or plug-in updates. Just drop it in your disc drive and go!

A list of the tutorials follows.

Section 1.3	Light Diffraction	pp. 12–18		* Section 10.2	Unit Cell	pp. 472–476
* Section 1.3	Electromagnetic Radiation	pp. 18–19		Section 10.3	Crystal Packing	pp. 476–484
Section 1.3	Doppler Effect	pp. 20–23		Section 10.5	Allotropes of Carbon	pp. 489–490
* Section 1.4	Dimensional Analysis	pp. 23–24		* Section 10.7	Crystal Field Splitting	pp. 494–502
* Section 1.4	Significant Figures	pp. 24–29		Section 11.2	State Functions and Path Functions	pp. 513–515
* Section 1.4	Scientific Notation	pp. 30		* Section 11.4	PV Work	pp. 518–523
Section 1.5	Big Bang	pp. 31–35		* Section 11.5	Internal Energy	pp. 523–524
* Section 1.6	Temperature Conversion	pp. 35–37		* Section 11.6	Heating Curves	pp. 524–531
Section 2.1	Fusion of Hydrogen	pp. 55–57		* Section 11.7	Estimating Enthalpy Changes	pp. 531–535
Section 2.3	Modes of Radioactive Decay	pp. 62–67		* Section 11.8	Calorimetry	pp. 535–537
Section 2.3	Synthesis of Elements	pp. 62–67		* Section 11.11	Hess's Law	pp. 542–545
* Section 2.3	Balancing Nuclear Reactions	pp. 62–67		* Section 12.1	Fractional Distillation	pp. 558–562
* Section 2.8	Half-life	pp. 89–93		* Section 12.2	Structure of Cyclohexane	pp. 562–571
Section 3.2	Light Emission and Absorption	pp. 106–109		Section 12.2	Cyclohexane in 3-D	pp. 562–571
Section 3.2	Rutherford Experiment	pp. 106–109		Section 12.3	Structure of Benzene	pp. 571–573
* Section 3.4	Bohr Model of the Atom	pp. 112–118		* Section 12.5	Formation of Sucrose	pp. 576–586
* Section 3.5	De Broglie Wavelength	pp. 118–122		* Section 13.1	Enthalpy	pp. 612–616
* Section 3.7	Quantum Numbers	pp. 123–125		Section 13.1	Dissolution of Ammonium Nitrate	pp. 612–616
* Section 3.10	Electron Configuration	pp. 132–143		* Section 13.2–3	Entropy	pp. 616–622
Section 4.3	NaCl Reaction	pp. 166–172		* Section 13.2	Molecular Motion	pp. 616–621
* Section 4.4	Avogadro's Number	pp. 173–180		* Section 13.4	Gibbs Free Energy	pp. 622–629
* Section 4.5	Balancing Equations	pp. 181–184		* Section 13.5	Chiral Centers	pp. 630–645
Section 4.7	Carbon Cycle	pp. 190–193		Section 13.5	Chirality	pp. 630–645
* Section 4.8	Limiting Reactant	pp. 193–198		Section 13.8	Condensation of Biological Polymers	pp. 655–661
* Section 4.8	Percent Composition	pp. 193–198		* Section 14.2	Reaction Rate	pp. 673–679
* Section 5.2	Molarity	pp. 214–220		* Section 14.3	Reaction Order	pp. 679–697
Section 5.4	Osmotic Pressure	pp. 224–229		* Section 14.4	Reaction Mechanisms	pp. 697–704
* Section 5.5	Boiling and Freezing Points	pp. 230–236		* Section 14.5	Arrhenius Equation	p. 704–712
Section 5.7	Saturated Solutions	pp. 257–263		* Section 14.4–6	Collision Theory	pp. 697–716
Section 6.1	Chemistry of the Upper Atmosphere	pp. 285–286		* Section 15.1	Equilibrium	pp. 732–736
Section 6.3	Bonding	pp. 289–292		Section 15.3	Equilibrium in the Gas Phase	pp. 743–747
* Section 6.3	Lewis Dot Structures	pp. 289–292		* Section 15.4	Equilibrium and Thermodynamics	pp. 747–750
* Section 6.5	Periodic Table	pp. 293–297		* Section 15.5	Le Châtelier's Principle	pp. 750–756
Section 6.6	Resonance	pp. 303–305		* Section 15.7	Solving Equilibrium Problems	pp. 756–762
* Section 6.9	Molecular Orbitals	pp. 316–324		* Section 16.1	Acid Rain	pp. 782–790
Section 7.1	Greenhouse Effect	pp. 334–340		* Section 16.2	Acid-Base Ionization	pp. 790–797
Section 7.2	Vibrational Modes	pp. 340–343		* Section 16.3	Acid Strength and Molecular Structure	pp. 797–801
* Section 7.3	Expanded Valence Shells	pp. 343–347				
Section 7.6	VSEPR Model	pp. 354–363		* Section 16.4	pH Scale	pp. 801–805
* Section 7.7	Hybridization	pp. 363–371		* Section 16.4	Self-Ionization of Water	pp. 801–805
* Section 7.8	Partial Charges and Bond Dipoles	pp. 371–375		* Section 16.6	Buffers	pp. 811–816
* Section 8.3	Ideal Gas Law	pp. 400–403		* Section 16.7–8	Strong Acid and Strong Base Titrations	pp. 816–824
* Section 8.5	Dalton's Law	pp. 406–410				
* Section 8.6	Henry's Law	pp. 410–417		* Section 16.8	Titrations of Weak Acids	pp. 817–824
* Section 8.7	Molecular Speed	pp. 417–422		* Section 17.1	Zinc-Copper Cell	pp. 843–847
* Section 9.2	Lattice Energy	pp. 439–444		* Section 17.2	Free Energy	pp. 847–850
* Section 9.2–5	Intermolecular Forces	pp. 439–450		* Section 17.4	Cell Potential	pp. 856–863
* Section 9.6	Raoult's Law	pp. 450–453		* Section 17.3	Ni-Cd Battery	pp. 850–856
* Section 9.7	Phase Diagrams	pp. 454–456		* Section 17.8	Fuel Cell	pp. 878–883
Section 9.8	Hydrogen Bonding in Water	pp. 457–467		Section 18.2	Superconductors	pp. 921–926
Section 9.8	Capillary Action	pp. 457–461		* Section 18.4	Polymers	pp. 936–940
* Section 10.1	X-ray Diffraction	pp. 470–472		Section 18.4	Fiber Strength and Elasticity	pp. 929–940

(*includes quantitative exercises)

Thomas R. Gilbert

Rein V. Kirss

Geoffrey Davies

NORTHEASTERN UNIVERSITY

W. W. Norton & Company
NEW YORK • LONDON

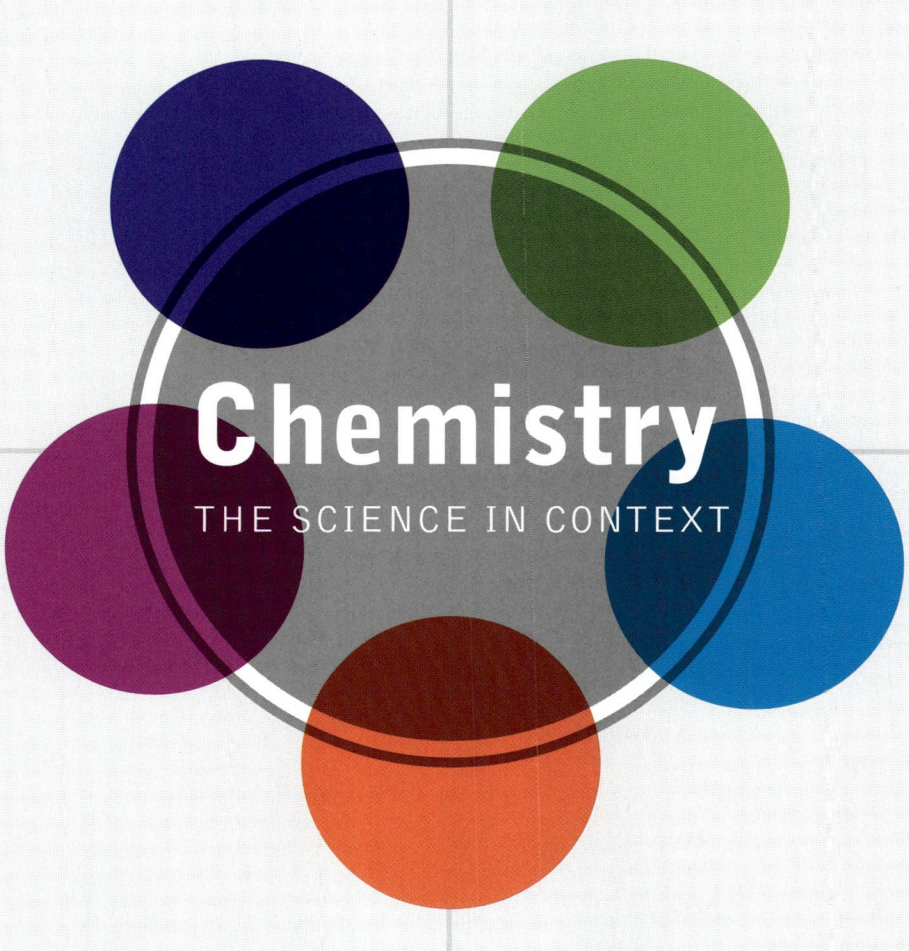

Copyright © 2004 by W. W. Norton & Company, Inc.

All rights reserved.
PRINTED IN THE UNITED STATES OF AMERICA
First Edition

The text of this book is composed in Photina with the display set in Bell Gothic
Composition by UG / GGS Information Services, Inc.
Manufacturing by RR Donnelley & Sons Company
Editor: Vanessa Drake-Johnson
Project Editor: Christopher Miragliotta
Photo Researcher: Penni Zivian
Assistant Editor: Sarah Chamberlin
Director of Manufacturing—College: Roy Tedoff
Managing Editors—College: Jane Carter, Marian Johnson
Book Designer: Rubina Yeh
Layout Artist: Paul Lacy
Cover Illustration: Digital Art/Corbis

Library of Congress Cataloging-in-Publication Data

Gilbert, Thomas R.
   Chemistry : the science in context / Thomas R. Gilbert, Rein V. Kirss, Geoffrey Davies.
     p. cm.
   Includes bibliographical references and index.
   ISBN 0-393-97531-2
   1. Chemistry. I. Gilbert, Thomas R. II. Kirss, Rein V. III. Davies, Geoffrey, 1942– IV. Title.

   QD33.2 .G55 2002
   547—dc21

                                                                                   2002069643

W. W. Norton & Company, Inc. 500 Fifth Avenue, New York, NY 10110
www.wwnorton.com

W. W. Norton & Company Ltd., Castle House, 75/76 Wells Street, London W1T 3QT

1 2 3 4 5 6 7 8 9 0

W. W. Norton & Company has been independent since its founding in 1923, when William Warder Norton and Mary D. Herter Norton first published lectures delivered at the People's Institute, the adult education division of New York City's Cooper Union. The Nortons soon expanded their program beyond the Institute, publishing books by celebrated academics from America and abroad. By midcentury, the two major pillars of Norton's publishing program—trade books and college texts—were firmly established. In the 1950s, the Norton family transferred control of the company to its employees, and today—with a staff of four hundred and a comparable number of trade, college, and professional titles published each year—W. W. Norton & Company stands as the largest and oldest publishing house owned wholly by its employees.

To our families

About the Authors

Thomas R. Gilbert has a B.S. in chemistry from Clarkson College and a Ph.D. in analytical chemistry from M.I.T. After 10 years with the Research Department of the New England Aquarium in Boston, he joined the faculty of Northeastern University, where he is currently Associate Professor of Chemistry and Education and Associate Director for Academic Affairs of the School of Education. His research interests are in environmental analytical chemistry and science education. He teaches general and analytical chemistry for majors other than chemistry, as well as graduate courses in analytical and environmental chemistry; he also conducts professional-development workshops and preservice courses in science education for pre-K–12 teachers.

Rein V. Kirss received both a B.S. in chemistry and a B.A. in history in 1981 as well as an M.A. in chemistry in 1982 from SUNY Buffalo. He received his Ph.D. in inorganic chemistry from the University of Wisconsin-Madison in 1986, where the seeds for this textbook were undoubtedly planted. After 2 years of postdoctoral study at the University of Rochester, he spent a year at Advanced Technology Materials, Inc., before returning to academics at Northeastern University in 1989. He is an associate professor of chemistry with an active research interest in organometallic chemistry. In addition to this textbook, he is coauthor with Thomas Gilbert of *Why Chemistry?*

Geoffrey Davies has B.Sc., Ph.D., and D.Sc. degrees in chemistry from Birmingham University, England. He joined the faculty at Northeastern University in 1971 after postdoctoral research on the kinetics of very rapid reactions at Brandeis University, Brookhaven National Laboratory, and the University of Kent at Canterbury. He is now Matthews Distinguished University Professor at Northeastern. His research group has explored experimental and theoretical redox chemistry, alternative fuels, transmetalation reactions, tunable metal-zeolite catalysts and, most recently, the chemistry of humic substances, the essential brown animal and plant metabolites in sediments, soils, and water. He edits a column on experiential and study-abroad education in the *Journal of Chemical Education* and a book series on humic substances. He is a Fellow of the Royal Society of Chemistry and was awarded Northeastern's Excellence in Teaching Award in 1981, 1993, and 1999.

Brief Contents

1. **Matter** AND ITS ORIGINS 1
2. **Nuclear Chemistry** AND THE ORIGINS OF THE ELEMENTS 54
3. **Electrons** AND ELECTROMAGNETIC RADIATION 104
4. **Stoichiometry** AND THE FORMATION OF EARTH 160
5. **Solution Chemistry** AND THE HYDROSPHERE 208
6. **Chemical Bonding** AND ATMOSPHERIC MOLECULES 284
7. **Molecular Shape** AND THE GREENHOUSE EFFECT 332
8. **Properties of Gases** AND THE AIR THAT WE BREATHE 386
9. **Intermolecular Forces and Liquids** WATER, NATURE'S UNIVERSAL SOLVENT 436
10. **The Solid State** A MOLECULAR VIEW OF GEMS AND MINERALS 466
11. **Thermochemistry** AND THE QUEST FOR ENERGY 510
12. **Energy** AND ORGANIC CHEMISTRY 556
13. **Entropy and Free Energy** AND FUELING THE HUMAN ENGINE 610
14. **Chemical Kinetics** AND AIR POLLUTION 670
15. **Chemical Equilibrium** AND WHY SMOG PERSISTS 730
16. **Equilibrium in the Aqueous Phase** AND ACID RAIN 780
17. **Electrochemistry** AND ELECTRICAL ENERGY 842
18. **Materials Chemistry** PAST, PRESENT, AND FUTURE 892

Contents

PREFACE — xxi
ACKNOWLEDGMENTS — xxvii
TO THE STUDENT — xxix

1 Matter and Its Origins — 1

1.1 Classes and Properties of Matter — 2

1.2 Creation of Matter — 9
- The Scientific Method — 9
- The Big Bang — 10

1.3 Light Waves — 11
- The Electromagnetic Spectrum — 12
- Wavelength, Frequency, and Energy — 18
- Shifting Wavelengths and the Doppler Effect — 20

1.4 Measurements in Scientific Studies — 23
- SI Units — 23
- Significant Figures — 24
- Precision and Accuracy — 30

1.5 Matter and Energy — 31
- Radioactive Decay — 31
- Radioactive-Decay Rates — 33
- The Formation of Nuclei — 34

1.6 Temperature Scales and a Cooling Universe — 35
- Then Came Atoms — 37
- Cold Microwaves — 42
- Continuum Radiation — 42

Chapter Review — 46
Questions and Problems — 49

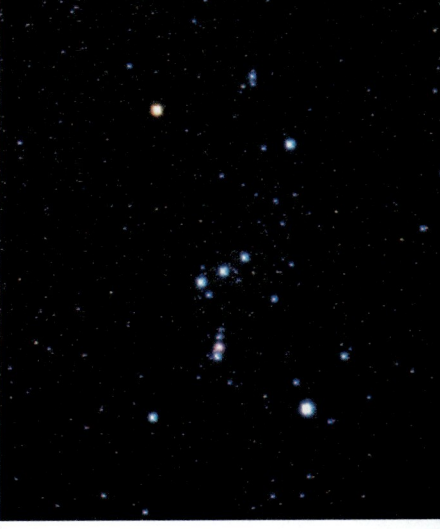

CONTENTS

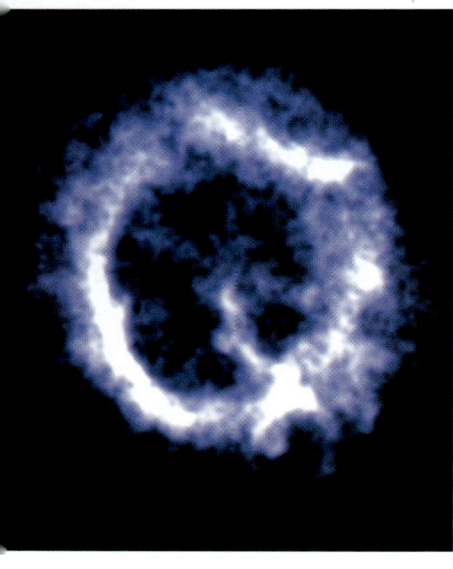

2 Nuclear Chemistry and the Origins of the Elements — 54

2.1	Hydrogen Fusion	55
2.2	Nuclear Binding Energies	58
2.3	Neutron Capture and Modes of Radioactive Decay	62
2.4	Supernova: Radiochemistry of the Heaviest Elements	67
2.5	Artificial Isotopes and Elements	74
2.6	Measuring Radioactivity	78
2.7	Biological Effects of Radiation	80
	Ionizing Radiation and Living Matter	80
	Radiation Dosage	81
	Assessing the Risks of Radiation	83
	Radiation Therapy	86
	BOX: THE CHEMISTRY OF RADON, RADIUM, AND URANIUM	86
	Medical Imaging with Radionuclides	87
2.8	Radiochemical Dating	89
2.9	The Composition of the Universe	93
	Chapter Review	95
	Questions and Problems	97

3 Electrons and Electromagnetic Radiation — 104

3.1	The Fraunhofer Lines	105
3.2	Electrons in Atoms	106
3.3	Particles of Light	110
3.4	The Bohr Model of the Hydrogen Atom	112
3.5	Matter Waves	118
3.6	Schrödinger's Wave Equation	122
3.7	Quantum Numbers	123
3.8	Shapes and Sizes of Atomic Orbitals	126
3.9	Spinning Electrons	131
3.10	The Periodic Table and Filling in the Orbitals	132
	BOX: THE CHEMISTRY OF THE NOBLE GASES	134
3.11	More Evidence for the Existence of Atomic Orbitals	144
	Ionization Energies	144
	X-Ray Photoelectron Spectroscopy	146

3.12	The Uncertainty Principle	148
	Chapter Review	151
	Questions and Problems	155

4 Stoichiometry and the Formation of Earth — 160

4.1	The Composition of Earth	162
4.2	The Composition of Compounds	164
4.3	Naming Compounds	166
	Binary Molecular Compounds	166
	Binary Ionic Compounds	167
	Binary Compounds of Transition Metals	169
	Polyatomic Ions	169
4.4	Chemical Reactions and the Mole	173
	BOX: THE CHEMISTRY OF THE ALKALI METALS	178
4.5	Completing and Balancing Chemical Equations	181
4.6	Percent Composition and Empirical Formulas	184
4.7	Stoichiometric Calculations and the Carbon Cycle	190
4.8	Limiting Reactants and Percent Yields	193
	Chapter Review	198
	Questions and Problems	200

5 Solution Chemistry and the Hydrosphere — 208

5.1	Earth: The Water Planet	209
5.2	Solution Concentration and Molarity	214
5.3	Electrolytes and Nonelectrolytes	221
5.4	Colligative Properties of Solutions	224
	Osmosis and Osmotic Pressure	224
	Boiling-Point Elevation and Freezing-Point Depression	230
	The van't Hoff Factor	236
	Measuring Molar Mass	238
5.5	Introduction to Oxidation–Reduction Processes	241
	Oxidation Numbers	244
	Balancing Redox Reactions	246
5.6	Acid–Base Reactions and Net Ionic Equations	250
	BOX: THE CHEMISTRY OF THE ALKALINE EARTH METALS	254
5.7	Precipitation Reactions	257
5.8	Ion Exchange	263

5.9	Titrations	266
5.10	Colloids	269
	Chapter Review	271
	Questions and Problems	275

6 Chemical Bonding and Atmospheric Molecules — 284

6.1	Introduction	285
6.2	Electron Sharing	286
6.3	Lewis Structures	289
6.4	Unequal Sharing	292
6.5	Electronegativity and Other Periodic Properties of the Elements	293
6.6	More Lewis Structures	298
	The Structure of Ozone	299
	Resonance	303
	Holes in the Ozone Layer	305
	BOX: THE CHEMISTRY OF THE HALOGENS	307
6.7	Choosing between Lewis Structures: Formal Charges	311
6.8	Electron Diffraction, Bond Lengths, and Predictions Confirmed	315
6.9	Molecular-Orbital Theory	316
	The Molecular Orbitals of H_2	316
	Nitrogen and Oxygen	318
	Ultraviolet and Visible Spectra and Auroras	322
	Chapter Review	324
	Questions and Problems	327

7 Molecular Shape and the Greenhouse Effect — 332

7.1	Bond Vibration and Climate Change	334
7.2	Infrared Spectropscopy	340
7.3	Exceptions to the Octet Rule	343
	BOX: THE CHEMISTRY OF OXYGEN AND THE GROUP 6A ELEMENTS	348
7.4	The Molecular-Orbital Diagram of Nitric Oxide	351
7.5	Electron-Spin Resonance: Locating Unpaired Electrons	352

7.6	**Molecular Shape: The VSEPR Model**	**354**
	Tetrahedra of Electrons	355
	Triangles of Electrons	357
	Linear Molecular Geometry	358
	Shapes of Expanded-Octet Molecules	359
	Summary	362
7.7	**Valence-Bond Theory**	**363**
	Hybrid Orbitals	364
	Hybrid Orbitals for Beryllium and Boron	368
7.8	**Polar Bonds and Polar Molecules**	**371**
7.9	**Molecular Vibration and Infrared Absorption**	**376**
	Chapter Review	376
	Questions and Problems	379

8 Properties of Gases and the Air That We Breathe   386

8.1	Introduction	388
8.2	The Atmosphere: A Molecular View	392
	Boyle's Law	392
	The Combined Gas Law	396
8.3	The Ideal-Gas Law	400
8.4	Gas Density	403
8.5	Dalton's Law and Mixtures of Gases	406
8.6	Henry's Law and the Solubility of Gases	410
	BOX: THE CHEMISTRY OF THE GROUP 5A ELEMENTS	414
8.7	The Kinetic Molecular Theory of Gases and Graham's Law	417
8.8	Real Gases	422
	Chapter Review	425
	Questions and Problems	428

9 Intermolecular Forces and Liquids: Water, Nature's Universal Solvent   436

9.1	Sea Spray and the States of Matter	437
9.2	Ion–Ion Interactions and Lattice Energy	439
9.3	Interactions of Polar Molecules	445
9.4	Dispersion Forces	446
9.5	Polarity and Solubility	447

CONTENTS

	9.6	**Vapor Pressure**	**450**
		Vapor Pressure of Solutions: a Molecular View	451
		Vapor Pressure and Solute Concentration	452
		Vapor Pressure and Temperature	452
	9.7	**Phase Diagrams: Intermolecular Forces at Work**	**454**
	9.8	**The Remarkable Behavior of Water**	**457**
		Chapter Review	461
		Questions and Problems	463

10 The Solid State: A Molecular View of Gems and Minerals — 466

10.1	**Crystal Lattices**	**467**
	Crystalline versus Amorphous	468
	X-Ray Diffraction	469
10.2	**The Unit Cell**	**472**
10.3	**Packing Efficiency**	**476**
	Cubic Closest Packing	477
	Simple Cubic Packing	481
	Hexagonal Closest Packing	482
10.4	**Network Solids: The Many Forms of Silica**	**484**
	Orthosilicates	486
	Metasilicates	488
10.5	**Allotropes of Carbon and Sulfur**	**489**
10.6	**Metallic Bonds and Structures**	**491**
	BOX: THE CHEMISTRY OF THE GROUP 4A ELEMENTS	492
10.7	**Gemstones: An Introduction to Crystal Field Theory**	**494**
	Crystal Field Splitting Energy	495
	Magnetic Properties	500
	Chapter Review	502
	Questions and Problems	504

11 Thermochemistry and the Quest for Energy — 510

11.1	**An Historical Perspective**	**511**
11.2	**Energy: Some Definitions**	**513**
11.3	**Natural Gas**	**515**

11.4	Combustion and Energy Transfer	518
11.5	Enthalpy (*H*) and Enthalpy Changes (Δ*H*)	523
11.6	Heating Curves and Heat Capacity	524
	Hot Soup on a Cold Day	525
	Cold Drinks on a Hot Day	528
11.7	Estimating Δ*H* from Average Bond Energies	531
11.8	Calorimetry: Measuring Heats of Combustion	535
11.9	Enthalpies of Formation and Reaction	537
11.10	Fuel Values	540
11.11	Hess's Law	542
	Chapter Review	546
	Questions and Problems	549

12 Energy and Organic Chemistry — 556

12.1	Petroleum Refining: Fractional Distillation and Raoult's Law	558
12.2	Alkanes in Gasoline and Structural Isomerism	562
	Cycloalkanes	564
	Structural Isomerism and Octane Ratings	566
	Rules for Naming Alkanes	567
12.3	Aromatic Hydrocarbons	571
12.4	Alcohols, Ethers, and Reformulated Gasoline	573
12.5	Carbohydrates	576
	Molecular Structures of Glucose and Other Sugars	577
	Condensation Reactions	582
	Starch and Cellulose	583
12.6	More Fuels from Biomass	586
	Carboxylic Acids	587
	Amines	588
12.7	Coal	590
12.8	Hydrogen As Fuel	591
12.9	Combustion Analysis and Elemental Composition	593
12.10	Alkanes and Alkynes	596
	Chapter Review	600
	Questions and Problems	603

xvi CONTENTS

13 Entropy and Free Energy and Fueling the Human Engine — 610

- 13.1 Enthalpies of Solution — 612
- 13.2 Entropy and Why Endothermic Processes Take Place — 616
- 13.3 Entropy Calculations — 621
- 13.4 Free Energy — 622
  - Connecting ΔH and ΔS — 624
  - The Meaning of Free Energy — 628
- 13.5 Fueling the Human Engine — 630
  - Carbohydrates Revisited — 630
  - Amino Acids and Proteins — 631
  - Stereoisomerism — 635
  - BOX: THE CHEMISTRY OF GROUP 5B — 640
  - Lipids — 642
- 13.6 The Energy Values of Carbohydrates, Fats, and Proteins — 645
- 13.7 Driving the Human Engine — 648
  - BOX: THE CHEMISTRY OF GROUP 7B — 650
- 13.8 DNA and Making Proteins — 655
- Chapter Review — 661
- Questions and Problems — 664

14 Chemical Kinetics and Air Pollution — 670

- 14.1 Photochemical Smog — 671
- 14.2 Reaction Rates — 673
  - Average Reaction Rates and the Formation of NO — 675
  - Instantaneous Reaction Rates and the Formation of NO_2 — 677
- 14.3 Effect of Concentration on Reaction Rate — 679
  - Reaction Order and Initial Rates — 679
  - The Single-Experiment Approach — 684
  - Second-Order Reactions — 690
- 14.4 Reaction Mechanisms — 697
  - BOX: THE CHEMISTRY OF THE GROUP 8B METALS — 701
- 14.5 Reaction Rates, Temperature, and the Arrhenius Equation — 704
- 14.6 Catalysis — 712
- Chapter Review — 717
- Questions and Problems — 719

CONTENTS

15 Chemical Equilibrium and Why Smog Persists — 730

- 15.1 Achieving Equilibrium — 732
- 15.2 Equilibrium Constants and Reaction Quotients — 736
  - Reactions in Reverse — 739
  - K and Q for Combined Equations — 740
  - Multiplying a Chemical Equation by a Constant — 742
- 15.3 Equilibrium in the Gas Phase and K_p — 743
- 15.4 K, Q, and ΔG — 747
- 15.5 Le Châtelier's Principle — 750
  - BOX: THE CHEMISTRY OF AMMONIA — 754
- 15.6 The Role of Catalysts — 756
- 15.7 Calculations Based on K — 757
- 15.8 Changing K with Changing Temperature — 763
- 15.9 Heterogeneous Equilibria — 767
- Chapter Review — 769
- Questions and Problems — 772

16 Equilibrium in the Aqueous Phase and Acid Rain — 780

- 16.1 Acid Rain and Acid Strength — 782
  - Weak and Strong Acids — 782
  - Diprotic Acids — 786
- 16.2 Acids and Bases: A Molecular View — 790
  - Acids in Water — 790
  - Bases in Water — 792
  - Lewis Acids and Bases — 794
  - Conjugate Pairs — 795
- 16.3 Acid Strength and Molecular Structure — 797
  - BOX: THE CHEMISTRY OF TWO STRONG ACIDS: SULFURIC AND NITRIC ACIDS — 798
- 16.4 The Concept of pH — 801
  - The pH Scale — 803
  - The pH of "Natural" and Acid Rain — 803
  - pOH — 805
- 16.5 The pH of Solutions of Acidic and Basic Salts — 806
- 16.6 Buffer Solutions and the pH of Natural Waters — 811
- 16.7 Acid–Base Indicators — 816

CONTENTS

16.8	Acid–Base Titrations	817
16.9	Solubilities of Minerals and Other Compounds	824
16.10	Complex Ions	824
	Complexation and Solubility	828
	Metal Complexes in Biomolecules	832
	Chapter Review	834
	Questions and Problems	836

17 Electrochemistry and Electrical Energy 842

17.1	Voltaic Cells	843
17.2	Voltage and Free Energy	847
17.3	The Chemistries of Some Common Batteries	850
	Dry Cells	850
	Alkaline Batteries	852
	Nickel–Cadmium Batteries	853
17.4	Standard Potentials and Batteries for Laptops	856
	Cell Potentials	856
	A Reference Point: The Standard Hydrogen Electrode	857
	Nickel–Metal Hydride Batteries	860
	Lithium-Ion Batteries	862
17.5	The Effect of Concentration on Potential	863
	BOX: THE CHEMISTRY OF THE GROUP 2B ELEMENTS	864
17.6	Quantities of Reactants and Battery Power	871
17.7	Electrolytic Cells and Recharging Batteries	873
17.8	"Low Emission" Vehicles and More Voltaic Devices	878
	Hybrid Vehicles	879
	Fuel Cells	879
	Photochemical Cells	881
	Biochemical Fuel Cells	882
	Chapter Review	883
	Questions and Problems	885

18 Materials Chemistry: Past, Present, and Future 892

18.1	Metals	894
	The Age of Copper	894
	BOX: THE CHEMISTRY OF THE GROUP 1B ELEMENTS	898
	The Bronze Age	901
	The Iron Revolution	904
	Aluminum Alloys: Lightweight and High-Performance	909

	BOX: THE CHEMISTRY OF GROUP 3A ELEMENTS: BORON, ALUMINUM, GALLIUM, INDIUM, AND THALLIUM	910
	BOX: THE CHEMISTRY OF GROUP 4B METALS: Ti, Zr, AND Hf	916
18.2	**Ceramics**	**918**
	Made of Clay	918
	Making Ceramics	919
	Superconducting Ceramics	921
	BOX: THE CHEMISTRY OF GROUP 3B AND THE LANTHANIDES	923
18.3	**Semiconductors**	**926**
18.4	**Fibers for Clothing and Other Uses**	**929**
	Natural Fibers	929
	Synthetic Polymers from Condensation Reactions	933
	Synthetic Polymers from Addition Reactions	936
18.5	**The Scientific Method Revisited**	**940**
	Chapter Review	**941**
	Questions and Problems	**942**
	APPENDIX 1: MATHEMATICAL PROCEDURES	A-1
	APPENDIX 2: SI UNITS AND CONVERSION FACTORS	A-5
	APPENDIX 3: THE ELEMENTS AND THEIR PROPERTIES	A-7
	APPENDIX 4: CHEMICAL BONDS AND THERMODYNAMIC DATA	A-15
	APPENDIX 5: EQUILIBRIUM CONSTANTS	A-22
	APPENDIX 6: STANDARD REDUCTION POTENTIALS	A-27
	PHOTO CREDITS	A-30
	GLOSSARY	A-33
	ANSWERS TO IN-CHAPTER QUESTIONS AND PROBLEMS	A-45
	ANSWERS TO SELECTED END-OF-CHAPTER QUESTIONS AND PROBLEMS	A-57
	INDEX	A-96

Descriptive Chemistry Boxes

The Chemistry of Radon, Radium, and Uranium **86**
The Chemistry of the Noble Gases **134**
The Chemistry of the Alkali Metals **178**
The Chemistry of the Alkaline Earth Metals **254**
The Chemistry of the Halogens **307**
The Chemistry of Oxygen and the Group 6A Elements **348**
The Chemistry of the Group 5A Elements **414**
The Chemistry of the Group 4A Elements **492**
The Chemistry of Group 5B **640**
The Chemistry of Group 7B **650**
The Chemistry of Group 6B **658**
The Chemistry of the Group 8B Metals **701**
The Chemistry of Ammonia **754**
The Chemistry of Two Strong Acids: Sulfuric and Nitric Acids **798**
The Chemistry of the Group 2B Elements **864**
The Chemistry of the Group 1B Elements **898**
The Chemistry of Group 3A Elements: Boron, Aluminum, Gallium, Indium, and Thallium **910**
The Chemistry of Group 4B Metals: Ti, Zr, and Hf **916**
The Chemistry of Group 3B and the Lanthanides **923**

Preface

Understanding chemical principles leads to a better understanding of a diverse range of issues, from environmental concerns to health-care choices to the geopolitics of resource use. Although few of the students who take general chemistry are planning careers in the chemical sciences, many are interested in these and other issues that are connected to chemistry. The challenge facing those of us who teach general chemistry is how to effectively make these connections. It is important that we do because students truly learn chemical priniciples only when they construct their own meaning of them, and that construction process requires a disire to construct as well as familiar foundation material on which to build. Introducing chemical principles in discussions of issues drawn from the physical and social sciences can provide both the desire to create meaning and the foundation on which to create it. It also delivers the message that chemistry truly is the central science. Many of us base our teaching on this principle. *Chemistry: The Science in Context* was written to support that teaching.

This book evolved from a supplement called *Why Chemistry?*, which we wrote in the early 1990s to connect the chemical principles in a standard general chemistry book to contemporary issues. After a number of years of teaching with this combination of materials, we decided that our students would be best served by a single textbook that seamlessly weaves the themes and issues in *Why Chemistry?* into the main narrative of a general chemistry textbook.

Thus, *Chemistry: The Science in Context* presents innovation within a familiar framework. The breadth of coverage and level of detail are appropriate for a general chemistry audience of science, pre-med, and engineering majors, but its blend of chemical principles with an interesting story line is distinctive. The book is innovative in several ways:

It is contextual.
Chemistry: The Science in Context brings chemistry alive by introducing chemical principles within the context of issues and topics that interest students, incorporating ideas and applications from fields as diverse as biology, nutrition, environmental sciences, astronomy, and geology. Chemical principles are presented as they are needed to explain the atomic and molecular foundations of the topics under discussion. This approach makes chemistry both more interesting and more relevant to other courses in which the students are enrolled, illustrating that chemistry is truly the central science. This innovative book is both engaging

and comprehensive, and it is adaptable to a wide range of courses and syllabi. Sample exercises and end-of-chapter problems include many conceptual and open-ended questions, some of them quite challenging, but they are interspersed with traditional quantitative problems. By providing this balance of conceptual questions and quantitative problem solving, we achieve the mathematical rigor of a traditional course while enhancing the students' understanding of important concepts.

It makes connections to the other sciences.
Chemistry is inherently interesting to most chemistry professors but seems boring or irrelevant to many students because few textbooks do an effective job of building intellectual bridges to biology, environmental sciences, and other areas that interest them. *Chemistry: The Science in Context* builds these bridges by incorporating themes that span the sciences. The first chapter, for example, defines terms and introduces data-manipulation techniques—unit conversions, significant figures, and so forth—in a discussion of the origins of the universe and topics drawn from cosmology and particle physics. These connections and others show the centrality of chemistry among the sciences. They also exemplify hot areas of current scientific inquiry. As former American Chemical Society president Ron Breslow and many others have pointed out, the greatest growth in chemical research is taking place at its intellectual borders: in molecular biology and the synthesis and characterization of advanced materials. These themes should therefore receive more than parenthetical treatment in general chemistry textbooks. Connections to other sciences are interwoven throughout the chapters, reinforcing concepts and themes that span the sciences. In addition, there is a capstone chapter on materials chemistry, and biochemistry is woven into several chapters.

Its contextual threads provide a more coherent view of chemistry.
General chemistry is often presented as a grab bag of seemingly unconnected topics. This textbook addresses that problem in several ways. The book has an overarching theme that might be called the "evolution of matter": from the formation of subatomic particles, to nuclei of progressively bigger mass, to atomic structure, to theories of molecular bonding, and finally to the formation of natural and synthetic polymers. Important connections between topics are highlighted with a Connections icon and marginal annotation. Among the organizational improvements is the presentation of atomic structure within the context of a discussion of how compounds are formed and why elements have the properties that they do. A unique section on energy presents a cohesive view of thermochemistry and thermodynamics. Descriptive chemistry is integrated into each chapter in "The Chemistry of..." boxes that relate to the chemical and contextual topics. A more complete description of the contents of each chapter appears in "How This Book Is Organized."

It emphasizes the spirit of inquiry that is a hallmark of science.
Chemistry: The Science in Context fosters a spirit of inquiry by emphasizing the importance of developing good questions and by describing how scientists figure things out and how our understanding of the chemical principles has evolved through time. Many textbooks present chemistry as a collection of facts and problem-solving algorithms to be learned or at least memorized. Throughout our book, we try to explain how our understanding of the properties of matter has evolved. Students are frequently asked why a statement is true or why an outcome was observed. As they engage in these intellectual exercises, they begin to acquire a conceptual understanding of chemical principles and a fascination for science (and chemistry) as a way of understanding how nature works and why chemical reactions take place. The emphasis on inquiry is highlighted in a magazine-style format. This feature helps students build the skills to move away from exclusive reliance on memorization and toward conceptual understanding.

PEDAGOGICAL FEATURES

The text incorporates several teaching aids that help students identify important chemical ideas and make connections between them. Sample Exercises lead students through solutions in a conversational and interactive way, helping them build confidence in their own problem-solving skills, which they can test in the paired Practice Exercises. Concept Tests probe students' understanding of chemical principles, and both conceptual and quantitative exercises are featured in extensive end-of-chapter problem sets. Details of the pedagogical features appear in the section titled "How to Use This Book."

HOW THIS BOOK IS ORGANIZED

Descriptive chemistry is integrated into each chapter in "The Chemistry of..." boxes, which relate to the chemical and contextual topics. See page xx for a list of these boxes.

Chapter 1 Matter and Its Origins An examination of the theories for the creation of the Universe is the context for introducing basic concepts about matter, atoms, and the scientific method. The properties and energy of electromagnetic radiation (light) are introduced as needed to evaluate the supporting evidence for the Big Bang theory. The importance and limitations (significant figures, accuracy, and precision) of measurements are described as well as the common units and their interconversion.

Chapter 2 Nuclear Chemistry and the Origins of the Elements This optional chapter offers a unique opportunity to grab students' interest early in the course because it is new material for most students. Fusion, fission, and the energies associated with these processes are used to explain the origins of the elements. Nuclear chemistry is often front-page news and offers many opportunities to relate nuclear chemistry to medical and other technologies.

Chapter 3 Electrons and Electromagnetic Radiation This chapter examines the historical development of the nuclear and electronic structure of atoms. The properties of light introduced in Chapter 1 are used to trace the evolution of electronic structure from the Bohr model to the quantum mechanical picture. The chapter is constructed in a fashion that allows it to be taught after Chapters 4 and 5.

Chapter 4 Stoichiometry and the Formation of Earth Chapter 4 begins to assemble atoms of elements into compounds in the context of how the solar system and Earth were formed some 4.5 billion years ago. The nomenclature of binary compounds and those with polyatomic anions, determination of percent composition, and empirical formulas are illustrated by using compounds found in Earth's core, mantle, and crust (lithosphere). Stoichiometric relations in chemical reactions and limiting reagents are introduced through a discussion of chemistry of Earth's primitive atmosphere.

Chapter 5 Solution Chemistry and the Hydrosphere Earth's oceans and other bodies of water (hydrosphere) serve as the context for solution chemistry. The composition of seawater and fresh water is used to introduce concentration units. The properties of electrolytes, their effect on the colligative properties of solution, and solution stoichiometry (titrations) are presented by using environmental themes. The classification of chemical reactions (acid–base, redox, precipitation, and ion exchange) in solution is discussed in the context of chemical weathering and other reactions that have altered the composition of Earth's surface and that control the composition of natural waters.

Chapter 6 Chemical Bonding and Atmospheric Molecules Small molecules in Earth's upper atmosphere are the context for the introduction of covalent bonding, Lewis structures, resonance, and molecular-orbital theory. Ozone and carbon dioxide are featured prominently because of students' familiarity with issues of air quality and depletion of the ozone layer. Electron diffraction is used to provide experimental evidence for the molecular structures presented in this chapter.

Chapter 7 Molecular Shape and the Greenhouse Effect As a continuation of the theme of Chapter 6, trace molecules in the lower atmosphere (nitrogen and sulfur oxides) are used to examine Lewis structures that do not follow the octet rule. Evidence for odd-electron molecules is presented in an optional section of ESR spectroscopy. Concerns about global warming on this planet create the need to explain why carbon dioxide is such a potent greenhouse gas in an optional section on IR spectroscopy. The explanation also includes coverage of polarity, molecular geometry, (VSEPR) valence-bond theory, hybrid orbitals, and permanent dipole moments.

Chapter 8 Properties of Gases and the Air That We Breathe Continuing the discussion of atmospheric compounds, we turn to the properties of gases on the molecular level. The ideal-gas law and Boyle's, Charles's, Dalton's, and Graham's laws are presented in the context of the air that we breathe and the changes that a mountaineer experiences as a function of elevation or a diver experiences as a function of depth. The solubility of gases is relevant in both situations. Kinetic molecular theory leads us to the behavior of real gases and a consideration of the interactions between molecules.

Chapter 9 Intermolecular Forces and Liquids: Water, Nature's Universal Solvent We begin by considering the intermolecular forces responsible for deviations from ideal-gas behavior and for the formation of liquids. Water, nature's universal solvent, is a natural context for this discussion. We also examine the role of intermolecular forces on physical properties (boiling point, vapor pressure, viscosity, and capillary action) of water and seawater, a solution first encountered in Chapter 5. The chapter ends with an introduction to phase diagrams, setting the stage for the investigation of solids.

Chapter 10 The Solid State: A Molecular View of Gems and Minerals The solid-state structures (unit cells and close-packed structures) of ionic and covalent compounds (network solids) are described by using examples drawn from earlier chapters. Evaporated seawater produces binary ionic compounds, whereas Earth's crust contains more-complicated structures. The origin, location, and occupancy of holes in close-packed structures are described. The band theory of bonding in solid materials is presented, as well as other theories of metallic bonding. This chapter also examines the structure and color (crystal field theory) of gemstones as examples of valuable materials obtained from the crust.

Chapter 11 Thermochemistry and the Quest for Energy Students are introduced to thermochemistry by examining the combustion of familiar fuels: natural gas and propane. This chapter covers the origin of the heat of combustion (bond energies), heats of formation, the flow of heat to and from a system, enthalpy, and work. Calorimetry and Hess's law are presented as methods of determining the enthalpies of reaction. The calculation of fuel values appears late in the chapter. A connection is made to the problems associated with combustion (CO_2) mentioned in Chapters 6 and 7, and the possibility of using alternate fuels is discussed.

Chapter 12 Energy and Organic Chemistry This introductory organic chemistry chapter builds on the topics introduced in Chapter 11 in the context of issues related to the availability and environmental impact of gasoline consumption. The structures (isomerism) and nomenclature of alkanes, alkenes, alkynes, and aromatic hydrocarbons present in hydrocarbon fuels are described. The structure and function of carbohydrates and cellulose in wood and wood products (alcohols, methane from the biomass) also are explored. Hydrogen and hydrogen storage are discussed as a possible solution to energy needs.

Chapter 13 Entropy and Free Energy and Fueling the Human Engine This chapter begins with the question of why endothermic reactions take place as a way of introducing the concepts of entropy, free energy, and the connection between enthalpy, entropy, temperature, and free energy. A consideration of the thermodynamics of solution allows for a review of intermolecular forces, whereas a discussion of proteins and fats reviews the role of Hess's law in determining how carbohydrates, proteins, and fats are catabolized and in determining the free energy of these processes.

Chapter 14 Chemical Kinetics and Urban Air Pollution This chapter continues the discussion of the environmental impacts of fossil-fuel combustion begun in Chapter 11. With smog formation as the context, the kinetics of nitrogen oxide formation are examined. With a careful choice of reactions, average and instantaneous rates are contrasted. Both the method of initial rates and the integrated rate laws are introduced. The role of mechanism on the rate law is presented. The effect of temperature and catalysts on reaction rates is described, and the potential for heterogeneous catalysis to reduce smog formation is discussed.

Chapter 15 Chemical Equilibrium and Why Smog Persists The reactions responsible for smog formation reach equilibrium, a convenient context for introducing the concepts of equilibrium, equilibrium constants, and the related calculations without disrupting the story begun in Chapter 14. Le Châtelier's principle is applied to important reactions in the troposphere. We make the connection between equilibrium constants and free energy, which is possible only by the introduction of ΔG two chapters earlier.

Chapter 16 Equilibrium in the Aqueous Phase and Acid Rain Shifting our focus from gas-phase nitrogen oxides to sulfur oxides, acid–base equilibrium is introduced in the context of acid precipitation. The calculation of pH for strong acid solutions (SO_2 forming sulfuric acid) and weak acids (CO_2 forming carbonic acid) is directly related to combustion processes described just a few chapters earlier. The effects of acid precipitation allow the introduction of buffers and solubility products (K_{sp}) for crust minerals. Titrations, first introduced in Chapter 5, are revisited. Weak bases present in the environment serve to introduce base equilibria.

Chapter 17 Electrochemistry and Electrical Energy In this chapter, electrochemistry is presented as a solution to the problems associated with hydrocarbon combustion. The oxidation–reduction reactions that take place in batteries are discussed in the context of electric vehicles and portable electronic devices. The amount of electricity available (EMF) and the effects of concentration (Nernst equation) and temperature on the cell potential are described. The chapter ends with a section of fuel-cell technology, bringing to a close the story begun in Chapter 11 on energy use in industrial societies.

Chapter 18 Materials Chemistry: Past, Present, and Future The observation that humans have relied on metals, ceramics, and fibers for millenia serves to organize a discussion of materials. The Copper and Bronze Ages form the context for an introduction to metallurgy, followed by a description of the iron and steel revolution. The structures and properties of alloys are presented as they appear in this historical sequence. A description of ceramic materials from clays to alumina to ceramic superconductors follows a similar historical progression. A discussion of the structure of natural fibers (silk, wool, and cotton) reviews the structures of proteins presented in Chapter 13. Synthetic clothing fibers (nylon and polypropylene) serve as the context for discussing both addition and chain-polymerization reactions.

SUPPLEMENTS

For Instructors

Instructor's Resource Manual This thoughtfully developed resource helps teachers to integrate a multidisciplinary and contextual narrative approach into existing lectures and to apply this approach in helping to develop students' conceptual understanding and quantitative problem-solving skills. This manual includes lecture outlines and objectives, ideas for classroom demonstrations and discussion, a guide to available media resources, and more.

Transparencies Contains 275 color acetates.

Norton Media Library CD-ROM This collection of PowerPoint slides includes editable lecture outlines, all of the drawings from the textbook, selected photographs, and animations suitable for classroom display. Available on adoption.

Norton Resource Library This online resource offers Web-ready materials for your WebCT, Blackboard, or personal course page. Contents include test and quiz questions, lecture outlines, images from the textbook, animations, and elements from the student Web site. Available on adoption.

Test Item File The Test Item File includes approximately 2000 multiple-choice questions.

Computerized Test Item File This resource includes questions from the Test Item File and the Study Guide. Available for Macintosh and Windows.

Solutions Manual Worked solutions to all of the end of chapter problems are available for adopters. This item is not for sale. A separate student solutions manual contains solutions for the even-numbered end-of-chapter problems.

For Students

Student Web site (www.wwnorton.com/chemistry). Developed specifically for *Chemistry: The Science in Context*, this student resource includes helpful review materials, self-tests, animations, and assignable exercises that emphasize key concepts and help students build problem-solving skills.

Student CD-ROM An enhanced version of the Student Web site is also available in a convenient offline format.

Solutions Manual, Student Version Worked solutions to the even-numbered end-of-chapter problems are available for students.

Study Guide This resource includes helpful review materials, problem-solving tutorials, worked problems, additional exercises and solutions, and multiple-choice self-tests.

Acknowledgments

We are very grateful for the assistance of many people in taking this book from the concept stage to the shelf. First and foremost, we wish to thank our wives, Carla Verschoor (Wellesley College) and Beth Gilbert, for their patience, support, and suggestions during the writing of this text. We thank our children, Michael, Alex, and Emily Verschoor-Kirss; Michael, Christina, and Victoria Gilbert; and Warwick, Russell, and Claire Davies, for tolerating our absence and for helping us keep this project in perspective.

We also wish to thank those who encouraged us throughout our lives and throughout this project, including Dr. Voldemar Kirss (1912–1999), Professor O. T. Beachley, Jr. (SUNY Buffalo), Mr. Donald A. Mills, and Mr. Walter E. Guild. Their influence started us on the road to an academic career in chemistry. We also thank Mr. J. B. Baileff, Professor Kenneth Kustin (Brandeis University), and Dr. Norman Sutin (Brookhaven National Laboratory).

The publisher, W. W. Norton, has been incredibly supportive and generous in its investment in this project. We are particularly grateful to the editorial staff. Joseph Wisnovsky helped lay out the basic framework of the book and provided valuable suggestions for content, and Vanessa Drake-Johnson refined the framework and set it on track to its final form. Both have been sources of encouragement and have provided numerous innovative ideas that have strengthened the book. Sarah Chamberlin and Erin O'Brien coordinated details, large and small, and kept many aspects of the project running smoothly at the same time. Rubina Yeh created an inspired design for the book. Our project editor, Christopher Miragliotta, managed the herculean task of overseeing the production of this vast project. Penni Zivian hunted high and low to find the photographs that we requested. Many thanks must go to Jane Carter, Marian Johnson, and Roy Tedoff for keeping the project on track in the face of many obstacles.

We thank Precision Graphics, of Champaign, Illinois, for producing overwhelming volumes of high-quality illustrations very rapidly.

We are also grateful to the many people who worked to develop our ancillary program. This sprawling effort was ably led by April Lange, and we thank her for her creativity and conscientious attention to detail. We had the good fortune to work with a team of educators and programmers at Science Technologies, who crafted the outstanding animated tutorials contained on the Student CD-ROM and Web site. We particularly appreciate the efforts of James Caras, who worked tirelessly on this project. Stephanie Myers at Augusta State College wrote a study guide loaded with patient and helpful advice. John Goodwin of Coastal Carolina University and Brian Gilbert of Linfield College brought creative energy and experience to the development of our Test-Item File. Through their careful scrutiny of the problems and writing of solutions, our Northeastern Univesity colleague Ed Witten, along with Dan Durfey of the Naval Academy Preparatory School, helped us to refine the problem sets and strengthen the text. We were fortunate to draw on the experience of many individuals in creating our Instructor's Resource Manual: Alexander Pines of University of California at Berkeley, David Laws of The Lawrenceville School, Sandra Laursen of University of Colorado at Boulder, Sharon Anthony of The Evergreen State College, Tricia Ferrett of Carleton College, George Lisensky of Beloit College, and Heather Mernitz of Tufts University.

This book has benefited greatly from the care and thought that our many reviewers, listed on the next page, gave to their readings of our earlier drafts, both in written reviews and at focus groups. Their responses, questions, and concerns illuminated the task at hand and forced us to be clearer, more careful, and more creative. Several reviewers merit special mention for reading large sections of the text and providing invaluable guidance in the overall development of the text. We are particularly grateful to Professors Mortimer Hoffman, Kenneth Robertson, and David Laws for their thoughtful and detailed comments.

R. Allendoefer	State University of New York, Buffalo
Jeffrey Appling	Clemson University
Robert Balahura	University of Guelph (Canada)
Robert Bateman	University of Southern Mississippi
Eric Bittner	University of Houston
David Blauch	Davidson College
Robert Boggess	Radford University
Simon Bott	University of Houston
Michael Bradley	Valparaiso University
Robert Burk	Carleton University (Canada)
Julia Burdge	Florida Atlantic University
Kevin Cantrell	University of Portland
Nancy Carpenter	University of Minnesota, Morris
Tim Champion	Johnson C. Smith University
Penelope Codding	University of Victoria (Canada)
Jeffrey Coffer	Texas Christian University
Renee Cole	Central Missouri State University
Brian Coppola	University of Michigan
Richard Cordell	Heidelberg College
Dwaine Eubanks	Clemson University
Lucy Eubanks	Clemson University
Matt Fisher	St. Vincent College
Richard Foust	Northern Arizona University
David Frank	California State University, Fresno
Cynthia Friend	Harvard University
Jack Gill	Texas Woman's University
Frank Gomez	California State University, Los Angeles
John Goodwin	Coastal Carolina University
Tom Greenbowe	Iowa State University
Stan Grenda	University of Nevada
Todd Hamilton	Adrian College
Robert Hanson	St. Olaf College
David Harris	University of California, Santa Barbara
Holly Ann Harris	Creighton University
Donald Harriss	University of Minnesota, Duluth
C. Alton Hassell	Baylor University
Dale Hawley	Kansas State University
Vicki Hess	Indiana Wesleyan University
Donna Hobbs	Augusta State University
Tamera Jahnke	Southern Missouri State University
Kevin Johnson	Pacific University
Martha Joseph	Westminster College
John Krenos	Rutgers University
C. Krishnan	State University of New York, Stony Brook
Jerry Lokensgard	Lawrence University
Sue Nurrenbern	Purdue University
Jung Oh	Kansas State University, Salinas
Robert Orwoll	College of William and Mary
Robert Pribush	Butler University
Gordon Purser	University of Tulsa
Barbara Sawrey	University of California, Santa Barbara
Truman Schwartz	Macalester College
Estel Sprague	University of Cincinatti
Steven Strauss	Colorado State University
Mark Sulkes	Tulane University
Keith Symcox	University of Tulsa
Brian Tissue	Virginia Technological University
William Vining	University of Massachusetts
Andrew Vreugdenhill	Trent University (Canada)
Charles Wilkie	Marquette University

Thomas R. Gilbert
Rein V. Kirss
Geoffrey Davies
Northeastern University, Boston

To the Student

HOW TO USE THIS BOOK

Here you will find a description of the major pedagogical features that have been carefully constructed to help you get the most out of your study of chemistry. A familiarity with the uses of these features will greatly aid you as you read and study from this book. We hope you will find them useful.

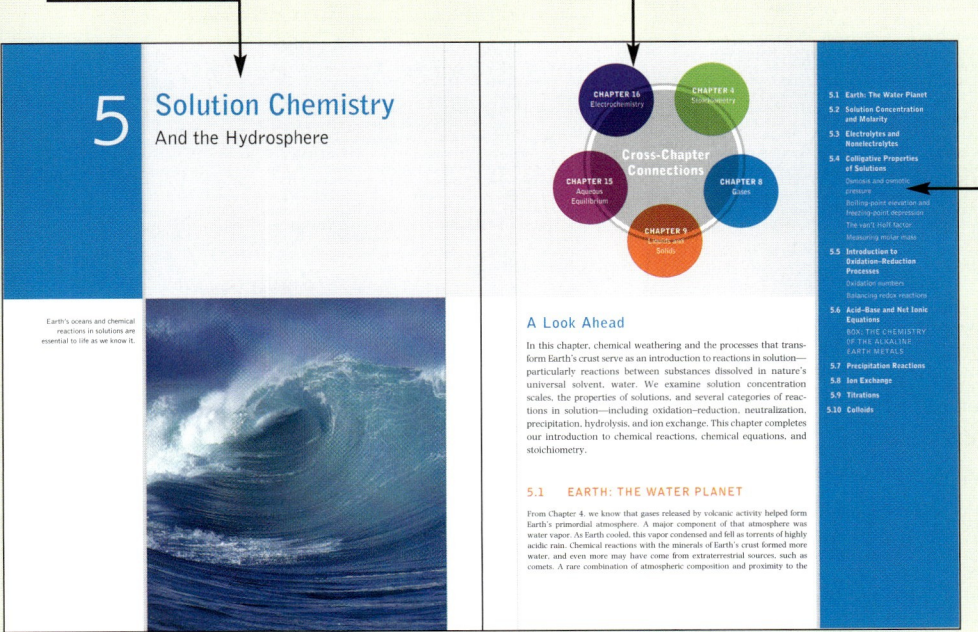

Chapter Title Each chapter has a title and a subtitle, emphasizing the chemistry, on the one hand, and the "contextual" theme on the other.

Connections Connections between important chemistry topics are illustrated schematically at the beginning of each chapter. Within the text, icons are paired with marginal notations to reinforce these connections.

Chapter Outline and "A Look Ahead" Each chapter begins with an outline that presents the organization of the chapter and a section called "A Look Ahead" that previews the major chemistry topics and contextual themes that are introduced in each chapter.

Extensive Illustration Program
Clear and informative drawings illustrate important concepts, and many of the photographs are annotated to clearly illustrate their relevance to chemistry.

What a Chemist Sees
Visualization at the molecular level is crucial to understanding chemistry. "What a Chemist Sees" drawings dramatically illustrate the molecular perspective, pairing a microscopic view with a macroscopic view.

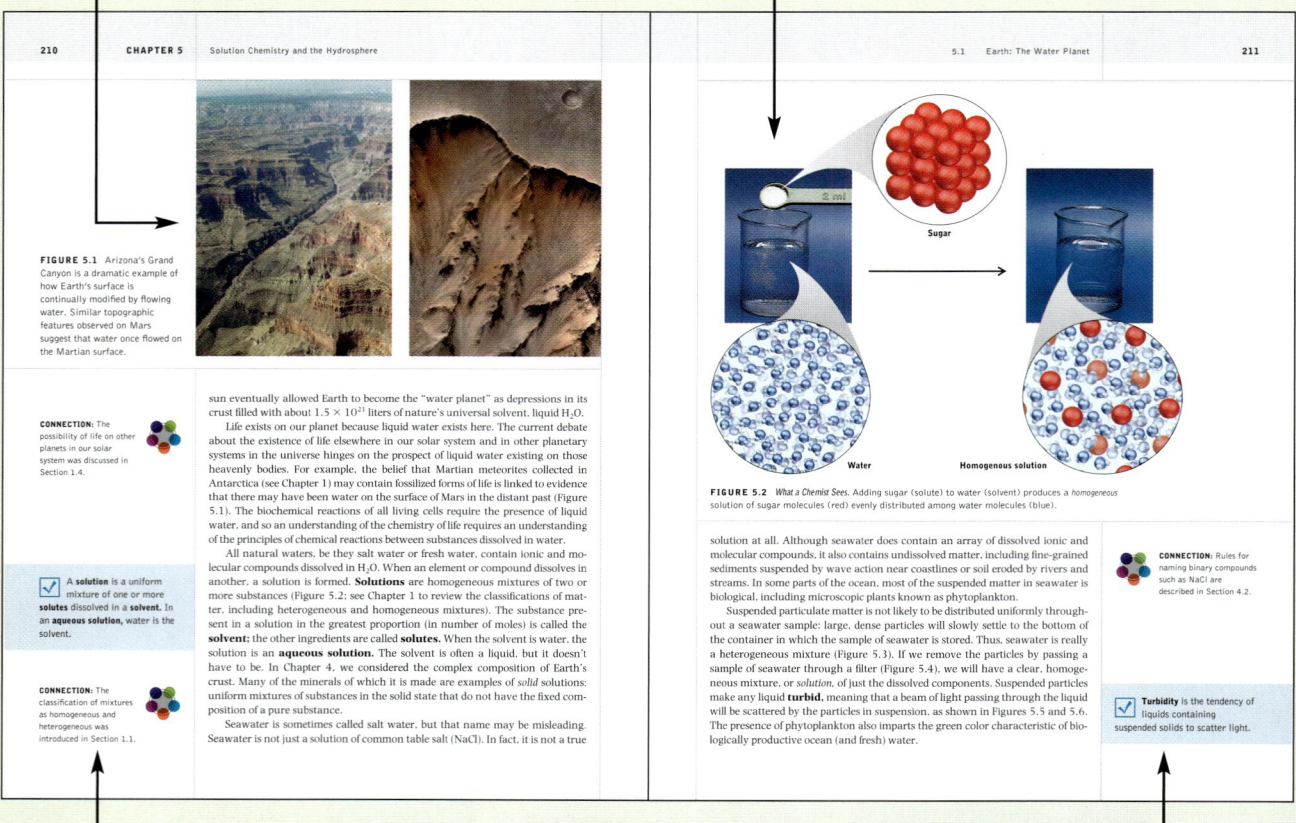

Connections
Icons are paired with marginal annotations to reinforce cross-chapter connections.

Chemistry Checkpoints
Marginal notations emphasize key chemical concepts and terms. These checkpoints reinforce the importance of these concepts and terms and facilitate review of the chapter. All key terms are defined here, as well as in the glossary at the end of the book.

TO THE STUDENT

Highlighted Questions This book attempts to foster a spirit of inquiry. The authors frequently pose a question that leads into a discussion of why or how we know something. Some of these questions are visually highlighted in magazine-style boxes.

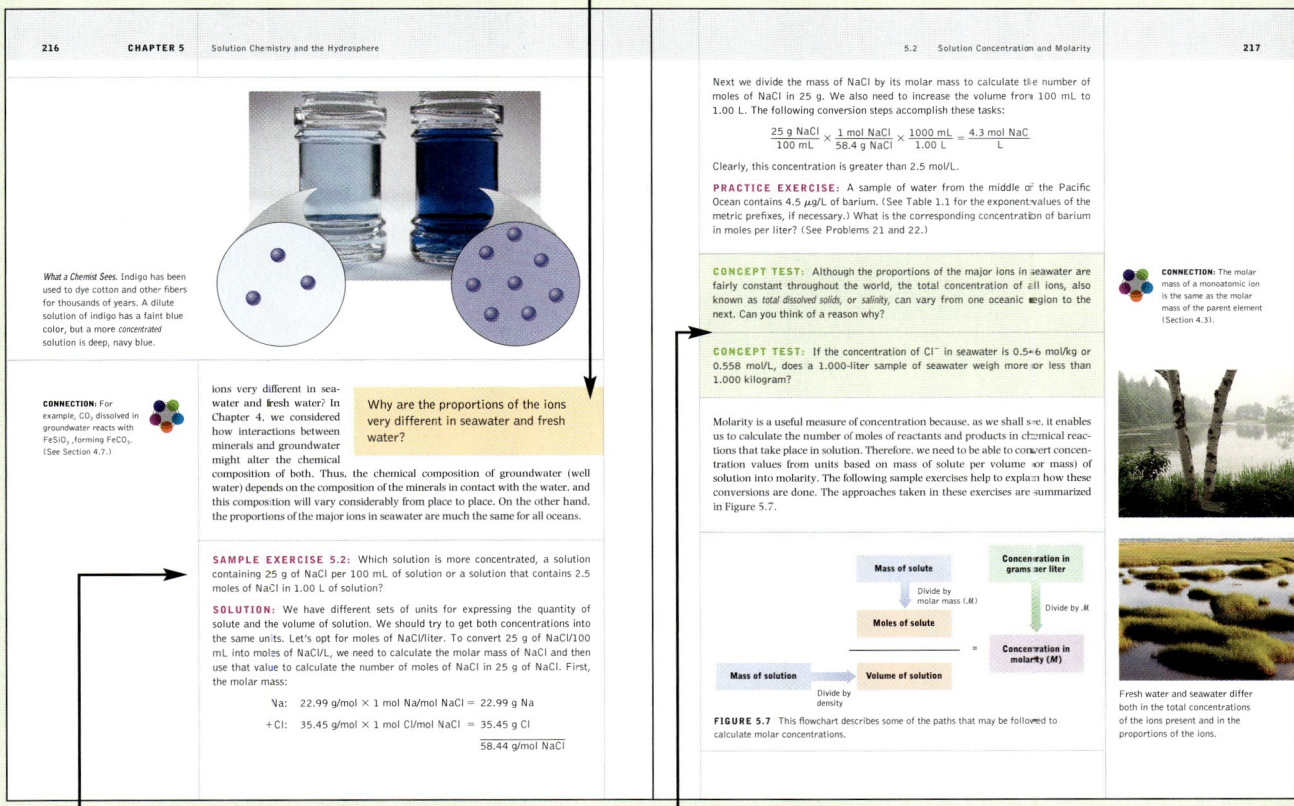

Sample Exercise and Practice Exercise Pairs The ability to solve problems is an important skill for students. Sample exercises illustrate how to solve many different kinds of problems. Each sample problem is followed by a practice problem, to enable you to practice newly learned skills. Sample Exercise and Practice Exercise pairs also direct you to similar problems in each end-of-chapter problem set.

Concept Test Questions Each chapter includes questions that test conceptual understanding. When appropriate, brief answers to these problems are provided at the back of the book.

"The Chemistry of ..." Boxes

Descriptive chemistry is integrated into the text so that the chemistry of different periodic groups can be presented within the context of relevant chemical topics. A list of boxes appears on page xx.

THE CHEMISTRY OF THE ALKALINE EARTH METALS

The elements in Group 2A of the periodic table are called the alkaline earth metals. Like their Group 1A neighbors, the Group 2A elements are very reactive, and none are found in nature as the free metal. All of the isotopes of radium (Ra) are radioactive and were discussed in Chapter 2. Calcium (Ca) and magnesium (Mg) are the fifth and sixth most abundant elements in Earth's crust: Ca makes up 3.6% of the crust and magnesium about 2.5%, by weight. Both elements are widely found as their carbonates, $CaCO_3$ and $MgCO_3$, in a variety of minerals such as chalk, limestone, and marble. Calcium is also found as its hydrated sulfate in gypsum ($CaSO_4 \cdot 2 H_2O$), and Mg is present as its hydroxide, $Mg(OH)_2$, in brucite. The Dolomite Range in Italy takes its name from the mineral dolomite, which is a mixture of the Ca and Mg carbonates.

With the exception of beryllium, all of the alkaline earth metals form compounds in which their atoms have lost the two s electrons in their outermost shells, forming ions with +2 charges. Beryllium behaves differently, sharing electrons rather than losing them, which leads to the formation of covalent rather than ionic Be compounds (Chapters 6 and 7).

Metallic alkaline earth elements are typically prepared by the electrolysis of molten chloride salts, which may be obtained by the evaporation of seawater. Magnesium metal forms homogeneous mixtures with aluminum and titanium, forming strong, yet lightweight, metals that are used in, for example, airplanes and automobiles. The other alkaline earth metals are more likely to be used in compounds rather than as pure elements. Gypsum is a widely used building material (the principal ingredient in "wall board" and plaster), and minerals rich in calcium phosphate are the raw materials for phosphate fertilizers. Among the commonly encountered compounds of Mg are its hydrated sulfate ($MgSO_4 \cdot 7 H_2O$), also known as Epsom salts, and aqueous suspensions of $Mg(OH)_2$, called milk of magnesia, which is widely used in antacids and laxatives. The most common barium mineral is barite, mostly in the form of $BaSO_4$. It is used to adjust the density of the fluids called drilling muds that are used in drilling for oil, and suspensions of $BaSO_4$ are ingested by patients suffering from intestinal disorders. When the suspension reaches the intestines, X-ray images are taken of these organs, which are filled with X-ray-absorbing ions of Ba.

Oxides of the Group 2A elements can be readily prepared by heating the respective carbonates. The reaction for calcium carbonate may be written as follows

$$CaCO_3(s) \longrightarrow CaO(s) + CO_2(g)$$
(limestone)   (lime)

Calcium oxide, also known as lime, is used in agriculture to reduce the acidity of soils. It is also used in water treatment and as an absorber of sulfur dioxide in the smokestacks of coal-burning factories and power plants.

$$CaO(s) + SO_2(g) \longrightarrow CaSO_3(s)$$

The alkaline earth metals react directly with nitrogen to form nitrides:

$$3 Mg(s) + N_2(g) \longrightarrow Mg_3N_2(s)$$

Nitrides belong to a class of materials known as ceramics and will be discussed in Chapter 18. Alkali earth metal nitrides are largely ionic compounds and so include the N^{3-} anion in their structures.

The Dolomite Range in Italy is rich in calcium and magnesium carbonate minerals known as dolomite.

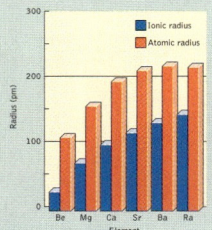

The atomic and ionic radii of the Group 2A elements increase down the group reflecting the increase in atomic number. An increase in atomic number corresponds to population of higher energy orbitals which are found further from the nucleus.

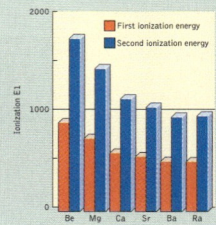

The ionization energies of the Group 2A elements are larger than for the Group 1A elements as a result of their smaller size and greater effective nuclear charge. The decrease in ionization energy down the group reflects the shielding of the outer electrons from the nuclear charge by the inner electrons. The larger second ionization energies reflect the greater attraction of the electrons in the monocation to the nucleus.

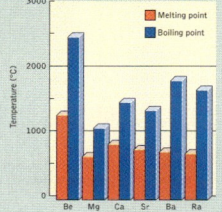

As with the Group 1A elements, there is a general trend toward lower boiling and melting points down the group. This is explained by weaker metallic bonds (Chapter 9) between the elements as size increases; however, anomalies arise for the melting point of Mg. Like lithium, the small beryllium has a particularly high boiling point compared to the other members of its group. The trend in boiling points for the other Group 2A elements does not follow a smooth periodic trend.

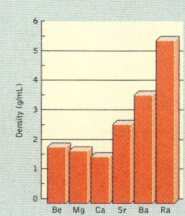

The densities of the Group 2A elements follow a general trend toward increasing density down the group. The trend is not smooth, with the densities of magnesium and calcium being slightly less than the density of beryllium.

Chapter Review Extensive chapter-review sections ensure that you will be able to identify and master the important chemical ideas of each chapter. These review sections are divided into several subsections, including Summary, Key Terms, Key Skills and Concepts, and Key Equations and Relations.

218 Chapter 14 The Chemical Cold Pack, Free Energy and Intermolecular Forces

CHAPTER REVIEW

Summary

SECTIONS 4.1 AND 4.2

Two or more elements can combine in definite proportions to form **compounds** described by **chemical formulas** which give the relative number of atoms of each element present in one **formula unit** of the compound. Compounds can be either ionic or molecular. Ionic compounds are neutral overall, so the charges of the cations and anions in the compound's formula must balance.

SECTION 4.3

A **mole** of particles, especially atoms, ions or molecules, is a collection of Avogadro's number (6.022x1023) of them. Most non-metal oxides form acids on reaction with water. The **molar mass** of a compound is the sum of the molar masses of the constituent elements each multiplied by a factor equal to the subscript of that element in the formula of the compound. **Chemical reactions** between elements and compounds can lead to new compounds. **Chemical equations** are used to describe the conversion of **reactants** to **products**.

SECTION 4.4

Chemical equations must be **balanced**: equal numbers of atoms of each element must appear on both sides of the equation. The coefficients in balanced chemical equations describe the **stoichiometry** or **mole ratios** of reactants and products. . Mole ratios can be used to obtain quantitative information from chemical equations, such as calculating a mass of product from a given mass of reactant. Combustion is a reaction in which an element or compund burns in an atmosphere containing oxygen. The products of complete combustion of products containing carbon and hydrogen are H_2O and CO_2.

SECTION 4.5

A compound can be described with its **percent composition** by mass of each element.

SECTION 4.6

Analysis of the percent composition of a sample allows calculation of the **empirical formula** or simplest whole number ratio of the elements. The molecular formula can be derived if the compound's molar mass is known.

SECTION 4.7

In cases where multiple reactants are involved, one of the reactants may act as a **limiting reactant**, dictating the maximum amount of product that can result from the reaction. The maximum amount of product formed from a given amount of reactant is the **theoretical yield**. For many reactions the actual yield is less than the theoretical yield; the ratio of the two is called the **percent yield**.

KEY TERMS

compound	**percent composition**	**Avogadro's number**
chemical formula	**empirical formula**	**product**
ionic bonds	**chemical equations**	**limiting reagent**
stoichiometry	**reactant**	**limiting reagent**
heterogeneous mixture	**covalent bonds**	**theoretical yield**
homogeneous mixture	**pure substance**	
mol ratio	**formula weight**	
molar mass	**mol**	

Key Terms 219

Key Skills and Concepts

SECTIONS 4.1 AND 4.2

Predict the formulas of binary ionic compounds from their names.
Give the names of binary compounds from their formulas. Binary compounds names end in ...ide.
Relate chemical names to subscripts and formulas. The names of simple molecular compounds reflect the subscripts in their formulas because more than one combination of their elements is possible.
Know the names and formulas of common cations and anions and the naming relationship between anions and their corresponding acids
Know how to derive binary chemical formulas from common stable cations and anions.
Relate the names of the oxoanions to their corresponding acids.

Example	-ite	Example	-ous
$NaNO_2$	sodium nitrite	HNO_2	nitrous acid
Na_2SO_3	sodium sulfite	H_2SO_3	sulfurous acid

Example	-ate	Example	-ic
$NaNO_3$	sodium nitrate	HNO_3	nitric acid
Na_2SO_4	sodium sulfate	H_2SO_4	sulfuric acid

Name transition metal compounds from their formulas and vice versa.

SECTION 4.3

Balance chemical equations involving known reactants and products
Understand the concept of the mole.
Know that the molar mass is the mass in grams of 1 mole of a substance.
Calculate the molar mass from a formula
Convert from molecules to moles and vice versa
Convert from grams of a compound to moles using the molar mass, and vice versa

SECTION 4.4

Calculate the yields of products of chemical reactions using the mole ratios from the balanced chemical equation.
Understand that during combustion a substance reacts with oxygen, O_2, generally producing heat and light.

SECTION 4.5

Calculate the percent composition of a substance from its formula

SECTION 4.6

Calculate the empirical formula of a substance from its percent composition from chemical analysis
Calculate the molecular formula of a substance from its molar mass and empirical formula

SECTION 4.7

Understand strategies for finding the limiting reactant.
Calculate the theoretical yield of a reaction with information on the reactant(s)
Understand why the actual yield of a reaction can not exceed 100% and often is considerably less.
Be able to calculate the percentage yield of the product(s) of a chemical reaction.

Key Equations and Relationships

SECTIONS 4.1 AND 4.2

Balanced chemical equations have the same number of atoms of each element on both sides of the reaction arrow.

SECTION 4.3

Convert from molecules to moles.

Number of molecules x 1 mol/6.022 x 10^{23} molecules = number of moles

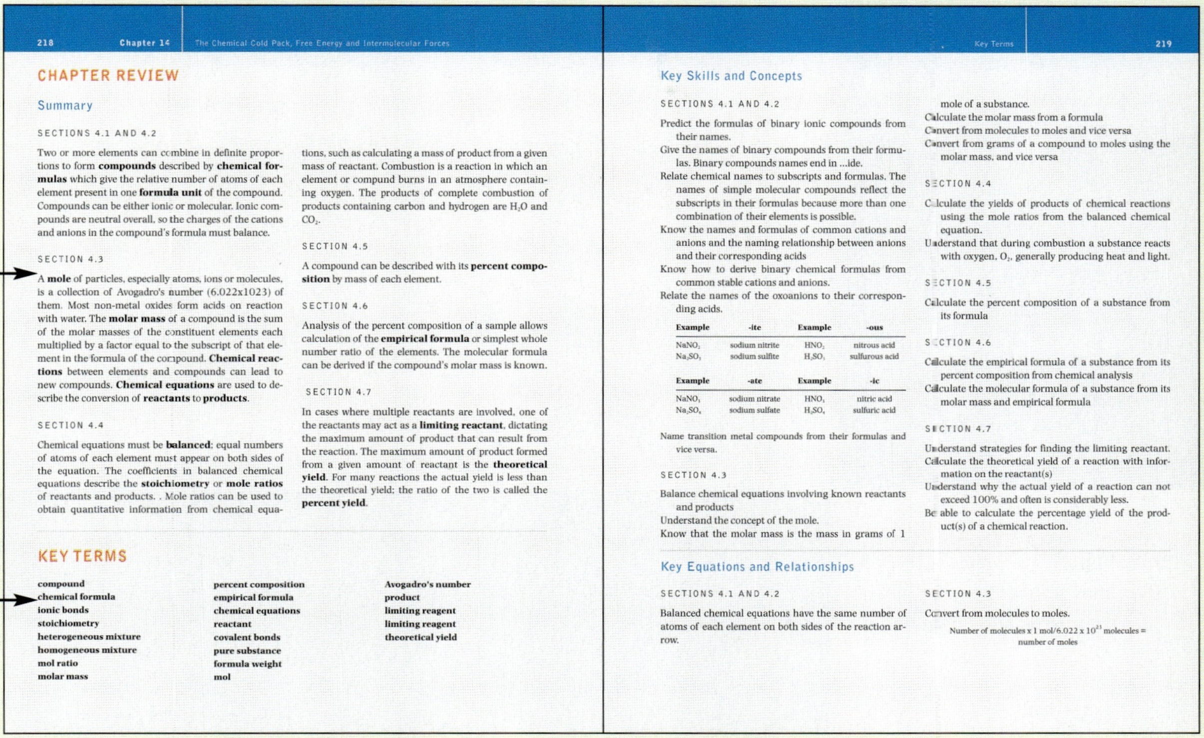

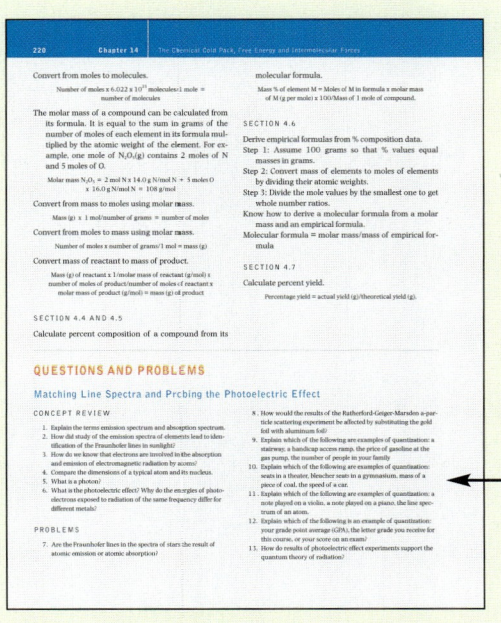

Chapter Problems Each chapter concludes with an extensive set of practice problems, grouped into two sections: "Concept Questions" and "Problems." Brief answers to all Concept Questions and half of the Problems can be found at the end of the book. Particularly challenging problems are identified with an asterisk.

220 Chapter 14 The Chemical Cold Pack, Free Energy and Intermolecular Forces

Convert from moles to molecules.

Number of moles x 6.022 x 10^{23} molecules/1 mole = number of molecules

The molar mass of a compound can be calculated from its formula. It is equal to the sum in grams of the number of moles of each element in its formula multiplied by the atomic weight of the element. For example, one mole of $N_2O_5(g)$ contains 2 moles of N and 5 moles of O.

Molar mass N_2O_5 = 2 mol N x 14.0 g N/mol N + 5 moles O x 16.0 g N/mol N = 108 g/mol

Convert from mass to moles using molar mass.

Mass (g) x 1 mol/number of grams = number of moles

Convert from moles to mass using molar mass.

Number of moles x number of grams/1 mol = mass (g)

Convert mass of reactant to mass of product.

Mass (g) of reactant x 1/molar mass of reactant (g/mol) x number of moles of product/number of moles of reactant x molar mass of product (g/mol) = mass (g) of product

SECTION 4.4 AND 4.5

Calculate percent composition of a compound from its molecular formula.

Mass % of element M = Moles of M in formula x molar mass of M (g per mole) x 100/Mass of 1 mole of compound.

SECTION 4.6

Derive empirical formulas from % composition data.
Step 1: Assume 100 grams so that % values equal masses in grams.
Step 2: Convert mass of elements to moles of elements by dividing their atomic weights.
Step 3: Divide the mole values by the smallest one to get whole number ratios.
Know how to derive a molecular formula from a molar mass and an empirical formula.
Molecular formula = molar mass/mass of empirical formula

SECTION 4.7

Calculate percent yield.

Percentage yield = actual yield (g)/theoretical yield (g).

QUESTIONS AND PROBLEMS

Matching Line Spectra and Prcbing the Photoelectric Effect

CONCEPT REVIEW

1. Explain the terms emission spectrum and absorption spectrum.
2. How did study of the emission spectra of elements lead to identification of the Fraunhofer lines in sunlight?
3. How do we know that electrons are involved in the absorption and emission of electromagnetic radiation by atoms?
4. Compare the dimensions of a typical atom and its nucleus.
5. What is a photon?
6. What is the photoelectric effect? Why do the energies of photoelectrons exposed to radiation of the same frequency differ for different metals?

PROBLEMS

7. Are the Fraunhofer lines in the spectra of stars the result of atomic emission or atomic absorption?
8. How would the results of the Rutherford-Geiger-Marsden a-particle scattering experiment be affected by substituting the gold foil with aluminum foil?
9. Explain which of the following are examples of quantization: a stairway, a handicap access ramp, the price of gasoline at the gas pump, the number of people in your family
10. Explain which of the following are examples of quantization: seats in a theater, bleacher seats in a gymnasium, mass of a piece of coal, the speed of a car.
11. Explain which of the following are examples of quantization: a note played on a violin, a note played on a piano, the line spectrum of an atom.
12. Explain which of the following is an example of quantization: your grade point average (GPA), the letter grade you receive for this course, or your score on an exam?
13. How do results of photoelectric effect experiments support the quantum theory of radiation?

1 Matter
And Its Origins

In this chapter, we probe a few of the many connections between chemistry and cosmology.

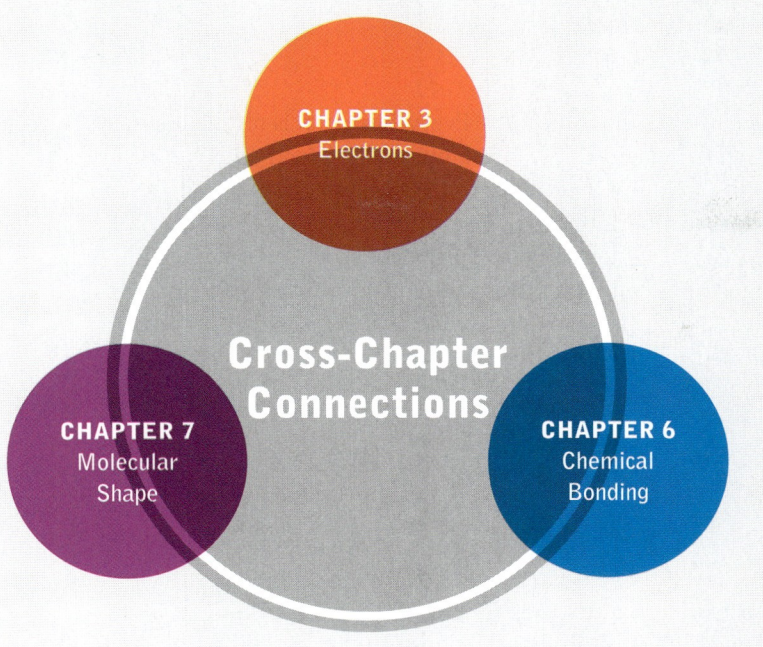

A Look Ahead

Chapter 1 begins with a discussion of the physical world and a brief look at the forms of matter, their characteristic properties, and the transformations in matter that take place in chemical reactions. Then we consider current theories of the creation of matter, which leads to an examination of the scientific method: how scientists interpret the results of observations and experiments, and develop theories that explain them. We address the inherent uncertainties in measured quantities, and how to express those uncertainties. An examination of the interactions between atoms and electromagnetic radiation leads to a first look at the structure of atoms and their nuclei.

1.1 **Classes and Properties of Matter**

1.2 **Creation of Matter**
   The Scientific Method
   The Big Bang

1.3 **Light Waves**
   The Electromagnetic Spectrum
   Wavelength, Frequency, and Energy
   Shifting Wavelengths and the Doppler Effect

1.4 **Measurements in Scientific Studies**
   SI Units
   Significant Figures
   Precision and Accuracy

1.5 **Matter and Energy**
   Radioactive Decay
   Radioactive-Decay Rates
   The Formation of Nuclei

1.6 **Temperature Scales and a Cooling Universe**
   Then Came Atoms
   Cold Microwaves
   Continuum Radiation

CHAPTER 1 Matter and Its Origins

> **Chemistry** is the science of matter: its composition, structure, and properties.
>
> An **atom** is the smallest particle of an element that retains the characteristics of the element.
>
> A **pure substance** is a particular kind of matter with well-defined properties and a fixed chemical composition. An **element** is a pure substance composed of only one kind of atom.
>
> A **physical property** is a characteristic of a substance that can be observed without changing it into another substance.
>
> **Density** is the ratio of the mass of an object to its volume.
>
> A **compound** is a substance composed of two or more elements combined together in fixed proportions.
>
> A **chemical property** is a characteristic of a substance that can be observed only through a chemical reaction that includes the substance.

Consider the world around us. Everything that is physically real—the air that we breathe, the earth that we walk on, and the food and water that we need to survive—are examples of matter. **Chemistry** is the science of matter. Chemists study the composition, structure, and properties of matter, as well as the changes that take place and the energy that is produced or consumed when matter is changed in chemical reactions. Chemistry provides an understanding of natural processes and has led to the synthesis of new forms of matter that have greatly affected the way in which we live and the planet on which we live.

1.1 CLASSES AND PROPERTIES OF MATTER

Chemists view matter and its properties on an atomic level. Today they even have the ability to produce images of matter at the atomic level (Figure 1.1). Chemists know, for example, that a nugget of gold consists of an enormous number of **atoms** of the element gold. **Elements** are the simplest of **pure substances** because they consist of only one kind of atom (Figure 1.2).

There are three classes of elements: metals, nonmetals, and metalloids. They are distinguished by differences in their physical and chemical properties. The **physical properties** of an element can be observed without changing it into something else. These properties include luster, hardness, color, the capacity to be hammered into thin sheets (malleability) or drawn out into wires (ductility), and the ability to conduct electricity.

Pure gold (Au), for example, has a distinctive yellow color and metallic luster. It is relatively soft compared with other metals and very **dense**—a property that can be exploited by prospectors, as shown in Figure 1.3. Gold is also one of the few elements found in nature as a free element. Most others are found chemically bonded to other elements in the form of **compounds**. A lack of reactivity is a distinctive **chemical property** of gold. To observe a chemical property of a substance, the substance must undergo a chemical reaction. Such a reaction, the rate at which it proceeds, the identities of the other reacting substances, and the identities of the products formed serve to define the chemical properties of a sub-

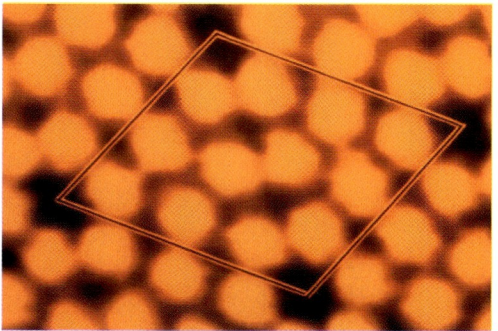

FIGURE 1.1 Since the 1980s, scientists have been able to observe individual atoms by using a scanning tunneling microscope. In this photograph of silicon atoms, each fuzzy sphere represents one silicon atom with a radius of 117 picometers (pm), or 117 trillionths of a meter.

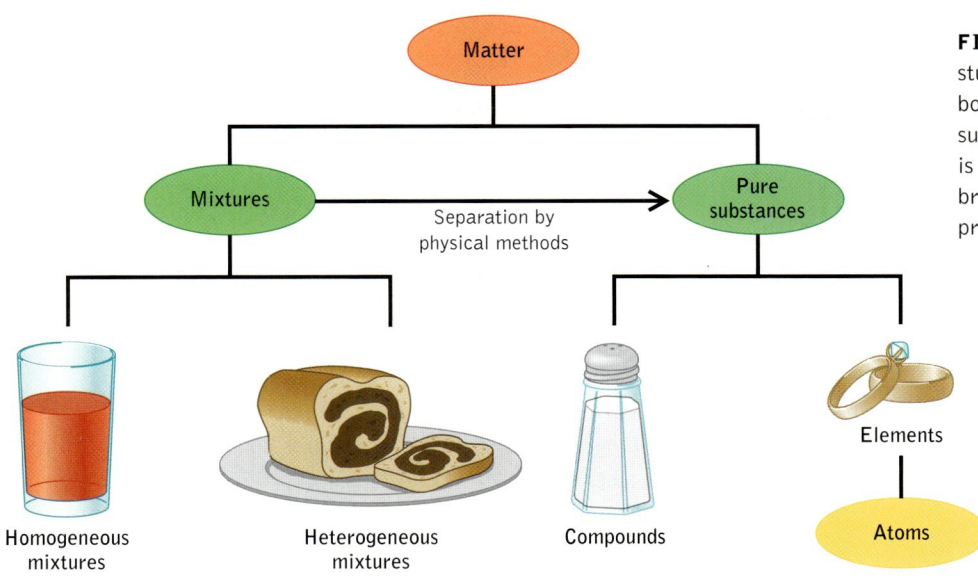

FIGURE 1.2 Chemistry is the study of matter. Matter appears both as mixtures and as pure substances, but the simplest of all is an element, which cannot be broken down further by chemical processes.

stance. The following list illustrates some of the differences and similarities in the physical and chemical properties of the elements:

- **Metals,** such as gold, have a metallic luster and are malleable, ductile, and good conductors of electricity. Metals form compounds with oxygen, which, when the compounds dissolve in water, produce alkaline solutions.
- **Nonmetals** do not conduct electricity and are neither malleable nor ductile. Nonmetal compounds formed with oxygen produce acids when they dissolve in water.
- **Metalloids** have many of the physical properties of metals, but they have the chemical properties of nonmetals.

CONNECTION: The metallurgy and properties of gold are discussed in Chapter 18.

FIGURE 1.3 Prospectors take advantage of the high density of gold when they "pan" for gold in sediments from riverbeds. Silt, sand, and small pebbles are made mostly of compounds of silicon, aluminum, and oxygen. These compounds are only about a third as dense as gold and can be washed away with swirling water in the pan, leaving particles of gold behind.

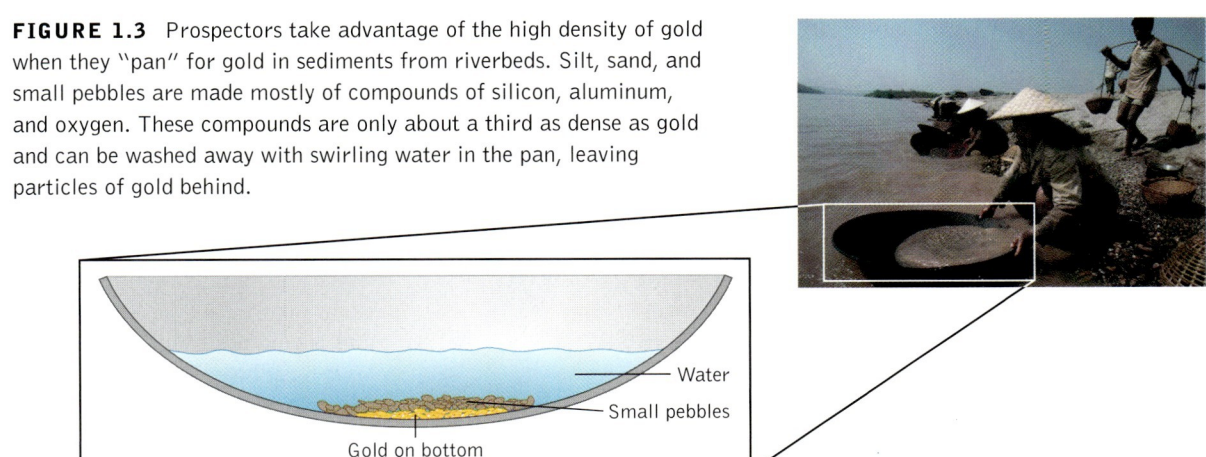

All the known elements appear in the periodic table on the inside front cover of this book. Most of the elements are metals. Hydrogen and the 16 elements in green on the right-hand side of the table are nonmetals, and the elements in yellow adjacent to the "staircase" line separating metals and nonmetals are metalloids. You should become familiar with the names and symbols of the first 20 elements and the common metals that appear in the middle of the table.

CONCEPT TEST: Consider the following properties of gold. Which of them are chemical properties and which are physical properties?

(a) Gold metal can be dissolved by reacting it with a mixture of nitric and hydrochloric acids known as *aqua regia*.
(b) Gold melts at 1063°C.
(c) Gold can be hammered into sheets (called gold leaf) that are so thin that light passes through them.
(d) Gold can be recovered from gold ore by treating the ore with a solution containing cyanide. Tiny particles of gold in the ore react with and dissolve in the cyanide solution.

> ✓ A **molecular compound** consists of atoms combined together in molecules. A **molecule** is a collection of atoms bonded together.
>
> A **molecular formula** is the exact formula of a molecular compound expressing the types of atoms and the number of atoms of each type in a molecule of the compound.

Consider another element: hydrogen (H). Like most elements, hydrogen is not found in the world around us as the pure element. Instead it is found in hydrogen-containing compounds. The most familiar of these compounds is water. Water is a **molecular compound.** This means that it consists of **molecules,** which are collections of atoms bonded together. The **molecular formula** of water, H_2O, tells us that its molecules consist of two hydrogen atoms and one oxygen atom bonded together.

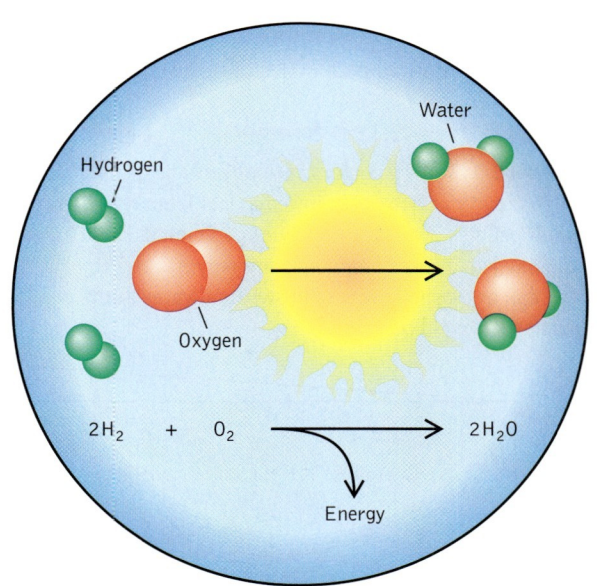

FIGURE 1.4 Hydrogen (H_2) reacts with oxygen (O_2), producing H_2O and releasing energy. This reaction supplies the energy that propels the U.S. space shuttles into space. During liftoff, each of the five main engines of the shuttle consumes about 40 tons of hydrogen and oxygen (stored in the large brown tank under the shuttle) each minute.

The physical and chemical properties of water are very different from those of the elements that combine to form it, as shown in Figure 1.4. Water is a colorless, odorless liquid at room temperature, whereas hydrogen and oxygen are colorless, odorless gases. Water expands when it freezes; hydrogen and oxygen do not. Oxygen and hydrogen, like most elements that are gases at room temperature, exist as two-atom, or diatomic, molecules: O_2 and H_2. Oxygen supports combustion reactions, whereas hydrogen is a highly flammable fuel. Water, as we know, neither supports combustion nor is flammable. Indeed, it is widely used to put out fires. These observations about the differences in the properties of O_2, H_2, and H_2O apply to all substances that undergo chemical changes: the products of chemical reactions are different substances from the reactants, with different physical and chemical properties.

There are very few pure substances, either elements or compounds, in nature. Rather, the world around us, and we ourselves, are complex **mixtures** of substances in which the different substances retain their identities. Mixtures are either homogeneous (of uniform composition) or heterogeneous (not of uniform composition). Examples of pure substances and mixtures are shown in Figure 1.5. Many scientists and engineers are engaged in separating these mixtures to isolate, for example, a target compound from myriad others. These separations are usually based on differences in the physical properties of the ingredients of the mixture, such as the extent to which they dissolve in water or other liquids or the temperatures at which they melt or boil.

Consider, for example, the water that we drink. It is not *pure* water. Drinking water that is drawn from a lake or river or pumped from a well may contain substances dissolved in it and fine particles of soil, sediment, and other suspended matter that do not dissolve in water. These particles can be separated from the water by allowing them to settle (assuming they are large enough and more dense than water) or they can be separated by **filtration.** The water is either

CONNECTION: The composition of seawater will be discussed in detail in Chapter 5.

A **mixture** contains more than one pure substance.

Filtration is a process for separating particles suspended in a liquid from that liquid by passing the mixture through a medium that retains the particles.

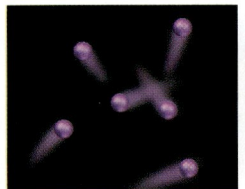

A Atoms of helium

B Molecules of carbon dioxide

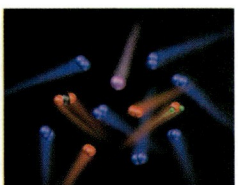

C Mixture of gases in air

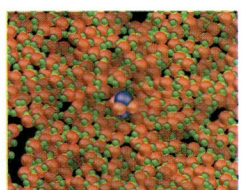

D Particles suspended in water

FIGURE 1.5 All matter is made up of either pure substances (of which there are few in nature) or mixtures of substances. The examples shown here for each class are: (A) an element, helium (He) (the gas used to fill the Goodyear blimps); (B) a compound, carbon dioxide (CO_2) (the gas used in many fire extinguishers); (C) a homogeneous mixture (such as the filtered, compressed air in a scuba tank, which is a mixture of nitrogen, oxygen, and lesser amounts of other elements and compounds); (D) a heterogeneous mixture, which contains suspended particles (such as the Missouri River, nicknamed "Big Muddy" because of the large quantities of soil and sediment suspended in it.)

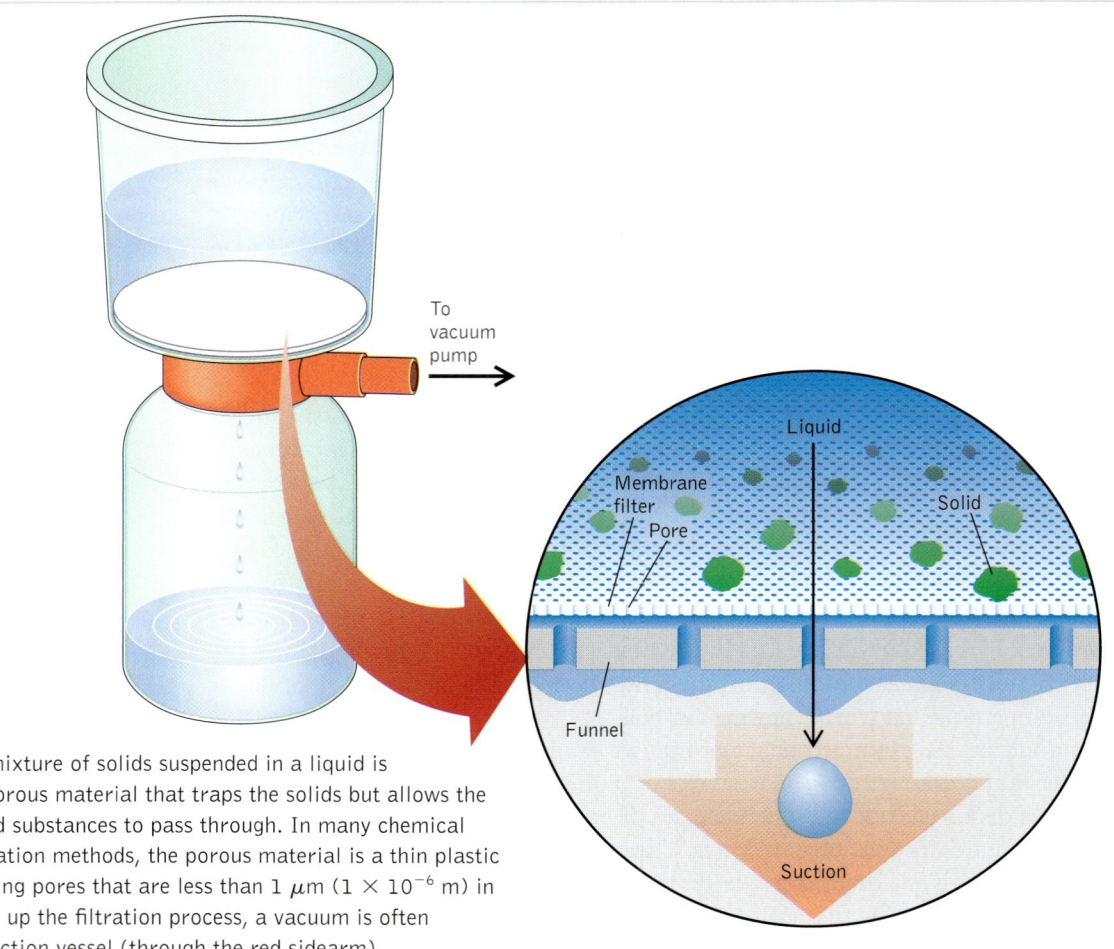

FIGURE 1.6 A mixture of solids suspended in a liquid is drawn through a porous material that traps the solids but allows the liquid and dissolved substances to pass through. In many chemical and biological filtration methods, the porous material is a thin plastic membrane containing pores that are less than 1 μm (1×10^{-6} m) in diameter. To speed up the filtration process, a vacuum is often applied to the collection vessel (through the red sidearm).

> ☑ An **ionic compound** consists of oppositely charged atoms and groups of atoms.
>
> An **ion** is an electrically charged atom or group of atoms. Negative ions, called **anions**, are made of atoms that have gained electrons; **cations** are positively charged atoms that have lost electrons.
>
> **Distillation** is a separation technique in which the more volatile components of a mixture are vaporized and then condensed, separating them from the less volatile components.

pumped through or pulled through (Figure 1.6) a porous membrane or other material that allows the water and substances dissolved in it to pass but that retains particles, including some too small for us to see. Muddy water is an example of a heterogeneous mixture. It is heterogeneous because the suspended solids that make it muddy are not likely to be distributed uniformly throughout the water. If these solids are removed by, for example, filtration, then the filtered water contains only dissolved substances.

The identities and quantities of the dissolved substances in the water that we drink vary, depending on the geology of the water's source. These substances include **ionic compounds** made of oppositely charged **ions**, which are atoms or groups of atoms with positive or negative electrical charges. Ions with positive charges are called **cations**; ions with negative charges are called **anions**. In arid regions, the concentrations of these compounds may be so high that the water is not fit to drink. One way to render such water drinkable is **distillation**—a process in which the mineral-rich water is heated to a temperature at which

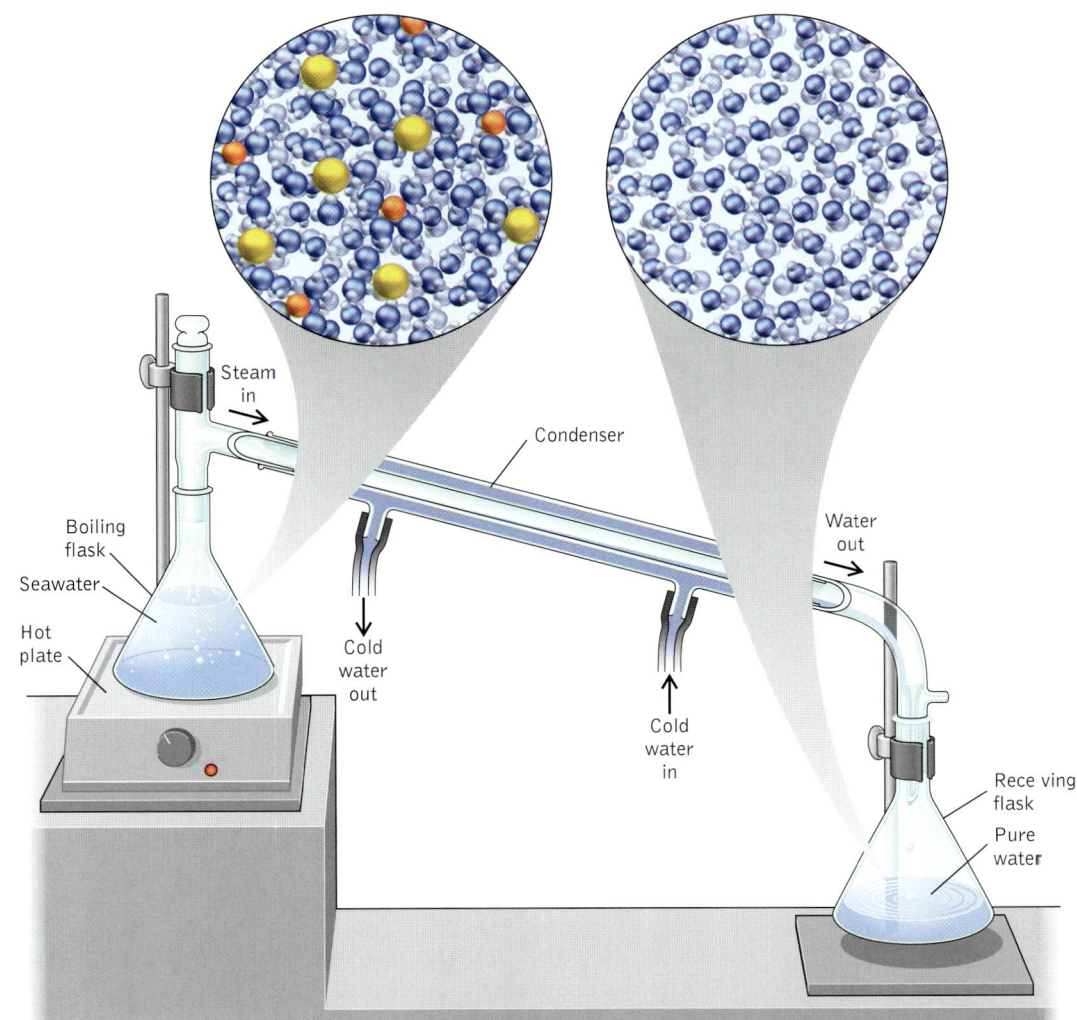

FIGURE 1.7 *What a Chemist Sees.* We can obtain pure water from seawater by heating seawater to the temperature at which it boils (its boiling point) in the flask on the left. Water vapor released from the boiling seawater passes into the condenser where it condenses into pure liquid water and is collected. Sea salts (mostly sodium chloride) are not so easily vaporized and remain in the flask on the left. Distillation is used to obtain drinking water from seawater in those regions where fresh water is scarce but sources of energy are abundant, such as the area around the Arabian (Persian) Gulf.

water is vaporized. Water vapor flows through a cooling condenser where it is converted back into liquid water and collected as purified "distillate." The minerals remain behind in the boiling mixture (Figure 1.7) because they are much less **volatile** than water.

The air that we breathe is not a pure substance either. It is a homogeneous mixture of gaseous elements, principally nitrogen (N_2), oxygen (O_2), and argon (Ar), and many compounds, the most abundant of which are carbon dioxide

✓ **Volatile substances** are either gases or easily vaporized liquids or solids.

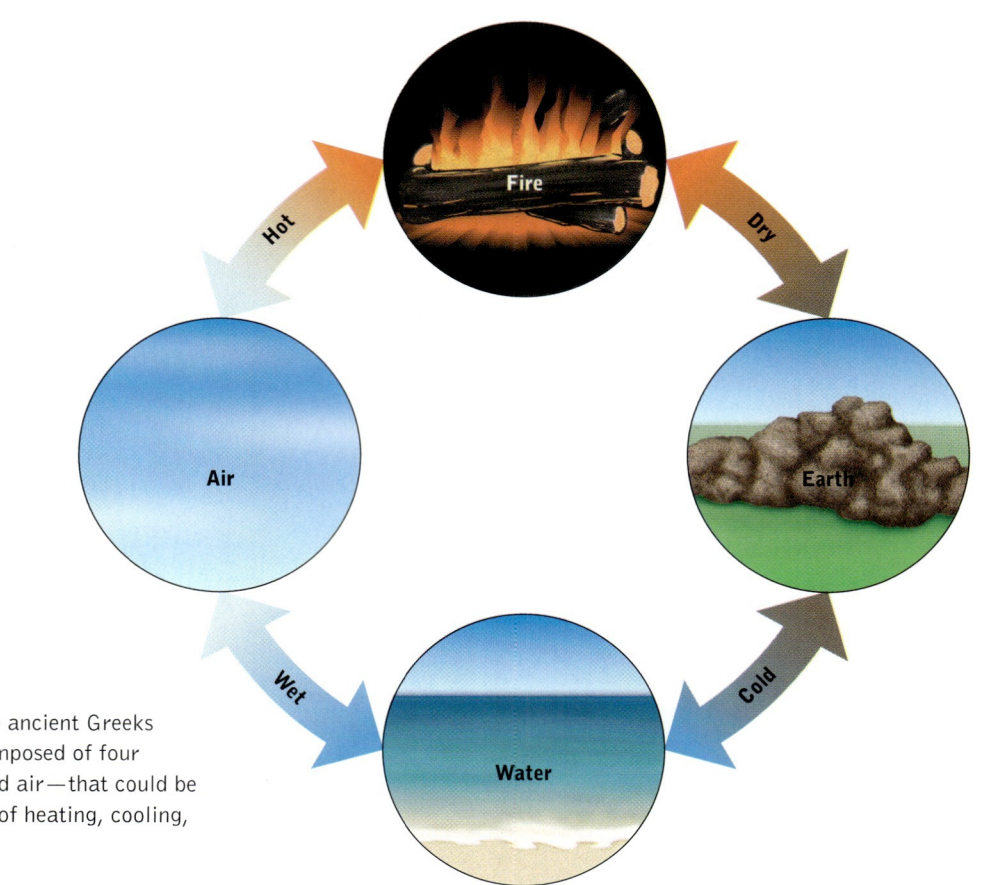

FIGURE 1.8 Aristotle and the ancient Greeks believed that everything was composed of four elements—fire, earth, water, and air—that could be interconverted by the processes of heating, cooling, wetting, or drying.

CONNECTION: Chapter 8 addresses the properties of gases in the atmosphere.

(CO_2) and water vapor. Note that nitrogen gas and oxygen gas consist of diatomic molecules, whereas argon exists as free atoms. We can prove that air is a mixture by cooling it to extremely low temperatures. As the temperature of air drops, water vapor will condense as liquid water or form crystals of solid ice. At lower temperatures, CO_2 becomes solid "dry" ice. At much lower temperatures, argon, then oxygen, and finally nitrogen condense into liquids.

Our ability to fractionate air into its component elements by cooling it proves that air is a mixture. The philosopher scientists of ancient Greece did not have the needed refrigeration systems, and so they could not know that air is a mixture. They believed that air and water were two of four basic elements that combined to form all the matter found in nature (Figure 1.8). The theory of the four elements remained a cornerstone of chemistry until the eighteenth century.

CONCEPT TEST: An element is the simplest of pure substances, with a defined set of properties and chemical composition. It cannot be separated into other elements. With these ideas in mind, try to explain why air and water were thought to be elements for so many years.

1.2 CREATION OF MATTER

What is the universe made of? How was it created? What forces control it? Humans have sought answers to such questions since we evolved as a species. The questions may be timeless, but our tools for exploring them have changed dramatically through the ages. Throughout history, different cultures have held the belief that one or more supernatural beings created an ordered world out of vast emptiness. People living in ancient Greece called the emptiness *Chaos* and believed that from this emptiness emerged *Gaia*, the first Supreme Being (also known as Mother Earth), who gave birth to *Uranus* (Father Sky). The opening verses of the book of Genesis describe a dark world that "was without form and void" and from which God created light (energy) and then the heavens and Earth (matter). Similar ideas of order created from primordial chaos through divine intervention are part of Asian, African, and Native American cultures (Figure 1.9).

> What is the universe made of? How was it created? What forces control it?

Today, many explanations of natural phenomena are based on an approach to understanding that we call *science*. Based on careful observation and thinking about the meaning of those observations, science is about formulating explanations of why things happen and then testing those explanations to see if they always work.

FIGURE 1.9 For centuries, cultures have passed on stories about the creation of the universe. Native American cultures left pictographs on canyon walls in the desert Southwest; Europeans painted elegant pictures of creation in the Bible.

The scientific method

Modern science evolved in the late Renaissance during a time when economic and social stability gave people an opportunity to look more closely at the workings of nature and to question old beliefs. By the early seventeenth century, the English philosopher Francis Bacon had published his *Novum Organum* in which he described the creation of knowledge and understanding through observation, experiments, and careful reflection. We call this approach to understanding the world around us the **scientific method** of inquiry (Figure 1.10). Careful observation leads to a tentative explanation for why a phenomenon occurs. This explanation is called a **hypothesis.** Further testing and observation may support the hypothesis or perhaps require that it be modified so that it adequately explains *all* of the information available. A hypothesis that is supported by the results of extensive testing and study is considered a scientific **theory.**

Galileo Galilei was perhaps the first modern scientist. Galileo used the newly invented telescope to observe the motions of the sun and planets relative to Earth. His inescapable (and correct) conclusion was that planets revolve around the sun. This view of our solar system was not new: it had been developed by Greek astronomers in the third century B.C. and was reinforced by the observations of Copernicus 18 centuries later. However, the view conflicted with a popular seventeenth-century interpretation of the creation described in Genesis: that the sun and planets revolved around Earth. Galileo was forced to appear before an Inquisition (religious tribunal) in 1633 and was convicted of:

CONNECTION: Theories about the structure of atoms are described in Chapter 3.

The **scientific method** is an approach to acquiring knowledge based on careful observation of phenomena.

A **hypothesis** is a tentative explanation for an observation or a series of observations. A **theory** is a general explanation of a widely observed phenomenon that has been extensively tested.

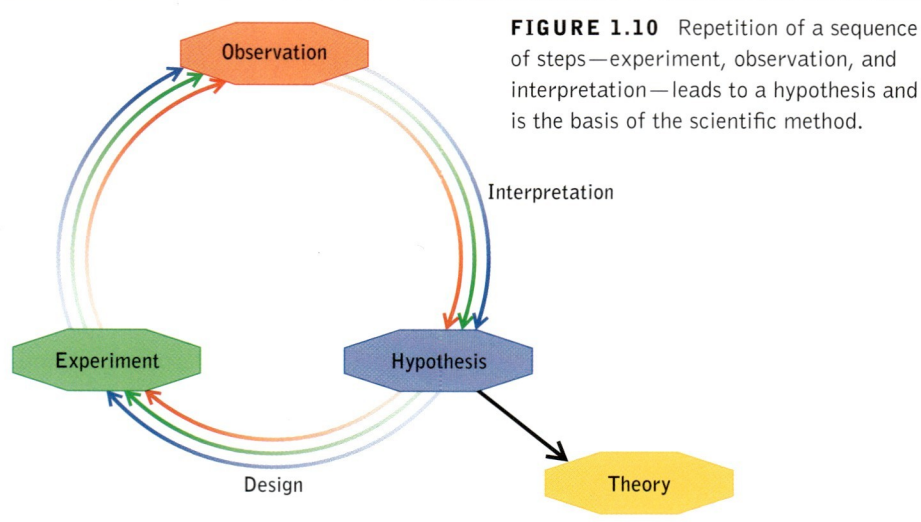

FIGURE 1.10 Repetition of a sequence of steps—experiment, observation, and interpretation—leads to a hypothesis and is the basis of the scientific method.

European science advanced in the seventeenth century with the work of Bacon, Galileo, and Copernicus, among others. Their contributions to our understanding of the universe came from application of the scientific method.

believing and holding the doctrines—false and contrary to the Holy and Divine Scriptures—that the sun is the center of the world, and that it does not move from east to west, and that the earth does move and is not the center of the world

Galileo was forced to disavow the results of his observations. Shortly thereafter his health failed, and his observations and experiments ended when he became completely blind in 1636.

Tensions between religious beliefs and scientific theories continue to this day. This chapter describes a theory of how the universe may have begun some 10 billion to 20 billion years ago. The theory was proposed early in the twentieth century and has been under close scrutiny ever since. It makes an interesting case study of the scientific method in action. Keep in mind that this theory is constantly under review and revision. It is very much a work in progress. Indeed, there have been significant changes in it just since the authors began to write this book. No student reading the following pages should believe that they represent the final word on how the universe came to be. To do so is to miss the dynamic nature of science: there are inevitably more discoveries to be made, serving as new sources of insight into the workings of nature.

The Big Bang

The universe contains billions of galaxies containing billions of billions of stars. All of those stars and galaxies attract, and are attracted to, their celestial neighbors by gravity. So why has the universe not collapsed on itself? What force keeps the stars and galaxies away from one another? In 1927

> So why has the universe not collapsed on itself?

Belgian astronomer Georges Lemaître proposed an intriguing explanation for why the universe has not collapsed. He proposed that the universe formed in the explosion of a cosmic "egg." According to his explanation, the universe was not in danger of collapsing, because it was actually *expanding*.

Lemaître's explanation is a good example of a scientific hypothesis. It was based on a growing body of evidence that the universe was indeed expanding. Much of this evidence came from the research of American astronomer Edwin Powell Hubble (for whom the Hubble Space Telescope was named). In the 1920s Hubble discovered galaxies outside our Milky Way and found that: (1) these other galaxies were moving away from our own and (2) the speeds with which these galaxies were receding were proportional to the distances to them—the galaxies farthest away were moving away the fastest.

The significance of Hubble's discoveries can be understood if we think about what might happen if time ran backward. What if the motion of the galaxies since the beginning of time were somehow recorded on videotape and we played it in reverse? The other galaxies would then be moving toward ours (and toward one another), with the ones farthest away closing in the fastest. Near the beginning of the tape, everything in the universe would approach the same point at the same time. As all the matter in the universe was squeezed into a smaller and smaller volume, its density would become enormous and it would get very hot. To understand why, you might remember what happened if you ever used a hand-operated pump to inflate a bicycle tire. With every stroke of the pump, air is compressed as it is forced into the tire. After several strokes, the barrel of the tire pump starts to feel warm. With a lot of pumping, it may get too hot to hold. Some of the energy used to force air into the tire heats up the air in the pump. When our tape of history is played backward, the force of gravity that compresses the galaxies together also produces temperatures so high that the matter that we know could not exist. At the start of the tape, there would be no matter at all, only energy in a very small space of unimaginably high temperature. Many scientists believe that this moment was the starting point for the universe and all matter in it: the beginning of space and time. It is the moment that has come to be known as the "Big Bang."

CONCEPT TEST: If the volume of the universe is expanding and if the mass of the universe is not changing, is the density of universe: (a) increasing, (b) decreasing, or (c) constant?

1.3 LIGHT WAVES

How did Hubble discover that the galaxies farthest away are moving away the fastest? To understand his method, we must first understand some basic properties of the energy that stars emit into space. Visible sunlight and its rainbow of colors provide a useful model for describing the energy of the stars. In the seventeenth century, Sir Isaac Newton (he was born in 1642, the year Galileo died) discovered that rainbows are a blend of the different colors of light emitted by the sun. Newton used glass prisms to separate sunlight into its different colors and to

The rainbow of colors in sunlight is the result of Earth's atmosphere acting like a prism.

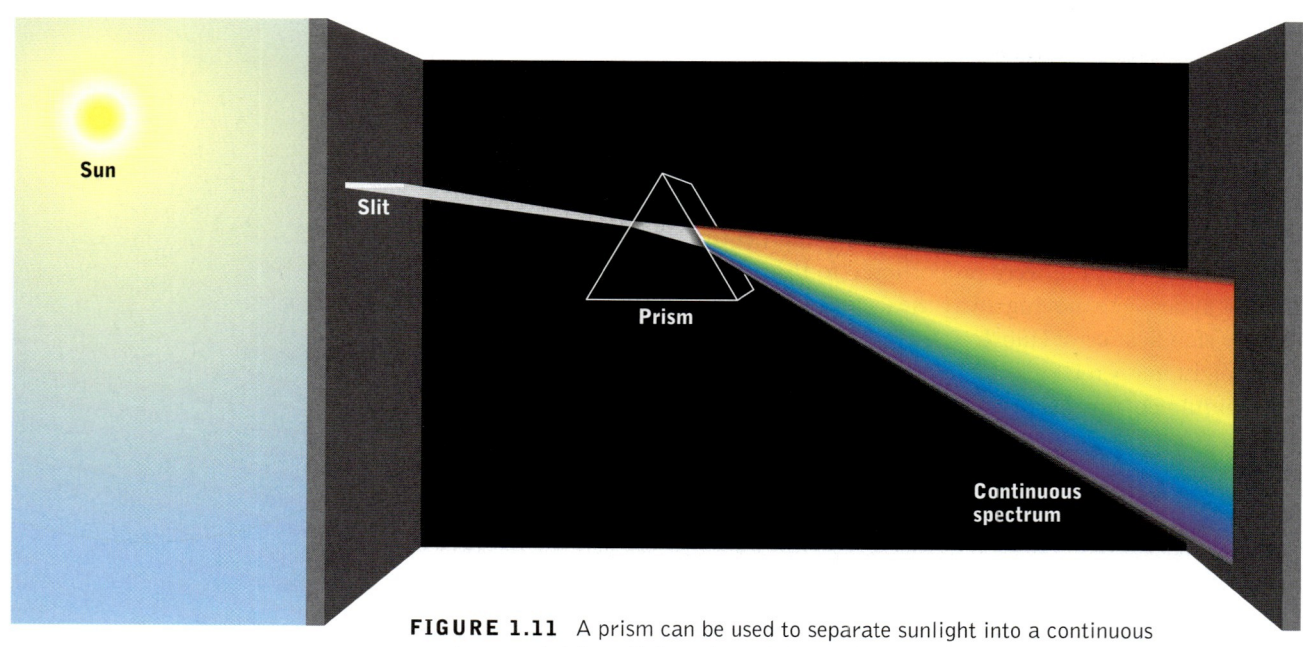

FIGURE 1.11 A prism can be used to separate sunlight into a continuous spectrum containing all the colors of the rainbow. Raindrops also bend light in this way, a process known as refraction, forming real rainbows.

study the **spectrum** of the sun. The separation process is called **refraction** because the path of a beam of light entering a prism (or raindrop) is bent, or *refracted*. Different colors of light are bent through different angles, as shown in Figure 1.11: violet is bent the most; red is bent the least; and the others are in between.

The sun emits all the colors of visible light and so is a source of **continuum radiation**. An incandescent light bulb also emits a continuum of radiation. However, some devices emit only one or a few colors of light. These devices include the lasers that emit red light and that are used in optical pointers, compact-disk (CD) players, and bar-code readers. Such devices are said to be *line* sources of light. We will explore the nature of the spectra emitted by continuum sources later in this chapter, and we will consider the origins of the distinctive colors of line sources of radiation in Chapter 3.

The electromagnetic spectrum

Visible light constitutes a small range of a much larger domain of energy emitted by the sun and stars that we call **electromagnetic radiation** (Figure 1.12). How do we know that the sun's spectrum includes forms of energy that we cannot see? Proof of their existence dates to the early nineteenth century. In 1800, William Herschel discovered that the solar spectrum contains a component that is refracted by prisms even less than the red end of visible light. He detected this

> ☑ The distribution of the radiant energy that an object or a substance emits or absorbs over different wavelengths is its **spectrum**.
>
> **Refraction** is the change in direction of a beam of electromagnetic radiation as it passes from one medium into another.
>
> **Continuum radiation** contains electromagnetic radiation of all wavelengths within a wavelength range.
>
> **Electromagnetic radiation** is radiant (*electromagnetic*) energy including light and ultraviolet, infrared, X-ray, γ ray, and other forms of radiation.

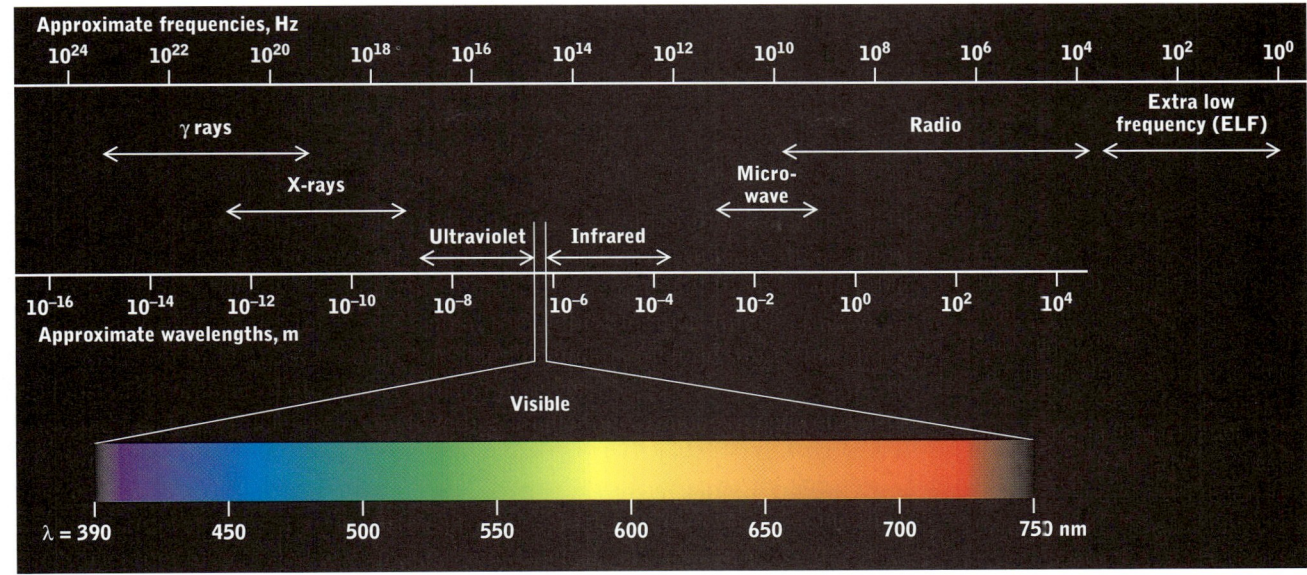

FIGURE 1.12 The visible-light region between about 390 and 760 nm is a tiny fraction of the electromagnetic spectrum that can be separated and detected by using instruments called spectrometers. Note the inverse relation between frequency and wavelength: frequencies increase from right to left, but wavelengths increase from left to right.

component with a thermometer. This "warm" region of the electromagnetic spectrum is the one we call **infrared**. In 1801, German physicist Johann Wilhelm Ritter found that the solar spectrum contains components that are refracted even more than violet light. He discovered these **ultraviolet** (UV) rays when he observed that they made the compound silver chloride, which is white, turn black. The chemical reaction that produced this color change is similar to the reaction that produces images on photographic film. It is also used in diagnostic X-ray imaging and can be used to monitor a person's exposure to the gamma (γ) rays given off by radioactive substances.

> How do we know that the sun's spectrum includes forms of energy that we cannot see?

You may be wondering why light and these other forms of radiant energy are called "electromagnetic." What do electricity and magnetism have to do with light? The connections between them were developed in the 1860s by James Clerk Maxwell, a Scot working in London. Maxwell proposed that light and all the other forms of electromagnetic radiation travel through space or transparent objects as oscillating electric and magnetic fields, as depicted by the arrows in Figure 1.13. The oscillations are described mathematically by sine functions. The oscillating fields have a wavelike appearance similar to the waves produced by the wind blowing across a body of water or by a boat moving through it.

☑ **Ultraviolet radiation** has shorter wavelengths than those of visible light; **infrared radiation** has longer wavelengths.

CONNECTION: Electromagnetic radiation interacts with matter in ways that will be discussed in Chapters 3, 6, and 7.

FIGURE 1.13 Light and the other forms of electromagnetic radiation are believed to travel through space as oscillating electrical and magnetic fields. The strengths of these fields, represented by the arrows in this illustration, oscillate in planes at right angles. Although wavelike properties of light were recognized by Sir Isaac Newton, the mathematical connections between the propagation of light and the oscillating motion of electrical and magnetic fields were largely the work of physicist James Clerk Maxwell (1831–1879).

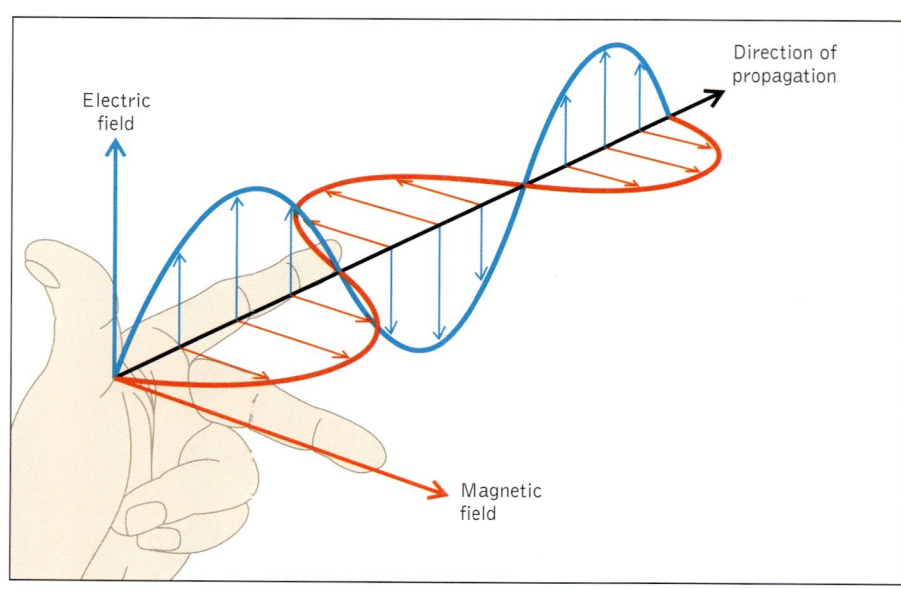

✓ **Wavelength (λ)** is the distance from one wave crest to the next; **frequency (ν)** is the rate at which wave crests pass a point per unit time.

Like waves of water, electromagnetic rays have characteristic **wavelengths**—that is, the distance from one wave crest to the next—that are represented by the Greek letter λ (lambda), as shown in Figure 1.14. All forms of electromagnetic radiation travel through a vacuum (and outer space) at the same velocity: the speed of light (c), the value of which is 2.998×10^8 m s^{-1}. Because all electromagnetic radiation travels with the same velocity, the longer the distance between waves, the fewer the waves that will pass through space per unit time. This waves-per-unit-time property is called **frequency** and is represented by the

FIGURE 1.14 The distance between consecutive wave crests is wavelength. The top wave has a longer wavelength than that of the bottom one. If both sets of waves travel with the same velocity, fewer crests of the top waves will pass a given point per unit time. This crests-per-unit-time property of waves is called frequency. Note that longer wavelength means lower frequency. A third property of waves is amplitude—that is, the height of their crests (which equals the depth of their troughs).

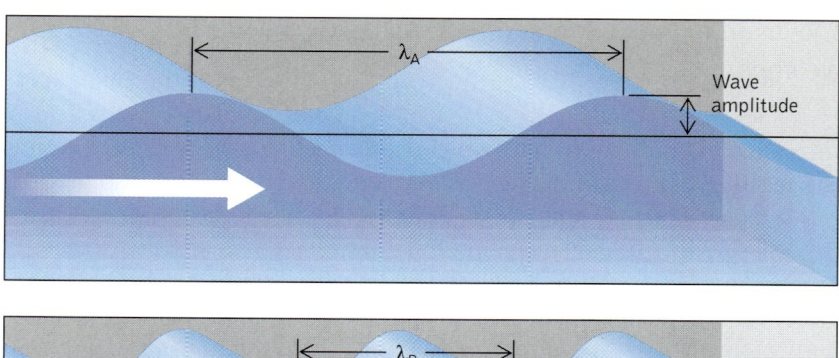
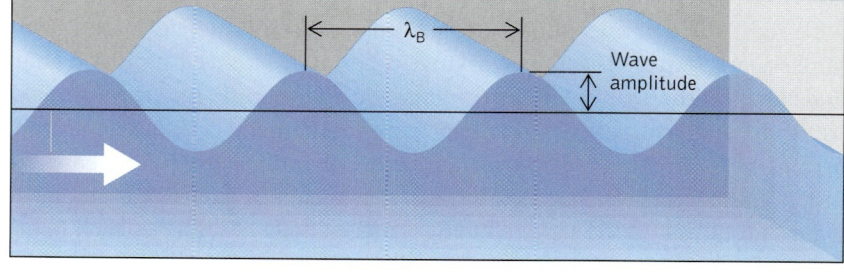

Greek letter ν (nu). Another property of a wave is the height of its crests (or depth of its troughs). This property is called **amplitude**.

The inverse relation between wavelength and frequency of electromagnetic rays can be expressed mathematically as follows:

$$c = \lambda \nu \qquad (1.1)$$

If the product of frequency times wavelength is a constant, then wavelength must decrease as frequency increases, and vice versa. This relation between wavelength and frequency makes sense if we consider the units on these parameters. Wavelength is the distance from one wave to the next, and frequency is the number of waves passing a point per unit time. Multiplying λ by ν results in a parameter with the units of distance/time; that is, a velocity equal to the speed of light:

$$\text{distance/wave} \times \text{waves/time} = \text{distance/time}$$

As a rule we do not bother to express the "wave" part of the units used for frequency and wavelength. Thus, frequencies are expressed simply in units of "per time" such as *per second*, or s^{-1}, and wavelengths are expressed simply in units of distance. For wavelengths of visible light, the most convenient unit of distance is the nanometer (nm): 1 nm = 10^{-9} m (see Table 1.1 for a list of the prefixes commonly used to express distances that are small fractions of a standard length, such as a meter). See Figure 1.12 for a view of the ranges of frequencies and wavelengths that make up the electromagnetic spectrum.

 The **amplitude** of a wave is the height of the wave.

SAMPLE EXERCISE 1.1: What is the wavelength in nanometers (nm) of yellow-orange light that has a frequency of 5.09×10^{14} s^{-1}?

SOLUTION: We can rearrange Equation 1.1 to convert frequency into wavelength:

$$\lambda = \frac{c}{\nu}$$

$$= \frac{2.998 \times 10^8 \text{ m/s}^{-1}}{5.09 \times 10^{14} \text{ s}^{-1}}$$

$$= 5.89 \times 10^{-7} \text{ m}$$

To convert 5.89×10^{-7} meters into nanometers, we start with the definition of a nanometer (see Table 1.1):

$$1 \text{ nm} = 10^{-9} \text{ m}$$

Using this definition as a conversion factor, we have

$$5.89 \times 10^{-7} \text{ m} \times \frac{1 \text{ nm}}{10^{-9} \text{ m}} = 589 \text{ nm}$$

PRACTICE EXERCISE: The laser light used in bar-code scanners has a wavelength of 785 nm. What is its frequency? (See Problems 25–28.)

TABLE 1.1 Commonly Used Prefixes for SI Units

Prefix		Value	
Name	Symbol	Numerical	Exponential
giga	G	1,000,000,000	10^9
mega	M	1,000,000	10^6
kilo	k	1,000	10^3
hecto	h	100	10^2
deka	da	10	10^1
deci	d	0.1	10^{-1}
centi	c	0.01	10^{-2}
milli	m	0.001	10^{-3}
micro	μ	0.000001	10^{-6}
nano	n	0.000000001	10^{-9}
pico	p	0.000000000001	10^{-12}
femto	f	0.000000000000001	10^{-15}

☑ **Diffraction** is the bending of electromagnetic rays around the edge of an object or as they pass through minute openings, forming circular waves.

Newton and his contemporaries knew that there was another way to "bend" light besides refraction. This other way, known as **diffraction,** was extensively studied by another Englishman, Thomas Young (1773–1829). We can reproduce one of Young's key experiments on diffraction in the following way. Suppose a distant source of monochromatic light illuminates a barrier that has two tiny holes in it, as shown in Figure 1.15. Light passing through the holes falls on a screen behind the barrier. If the widths of the two holes are large, then we will simply observe an image of each of them on the screen, which is what Newton would have predicted. Newton believed that light consists of tiny particles of energy. Those particles that hit the barrier would be absorbed or reflected by it; those that passed through the holes would travel straight ahead to the screen on the other side.

However, if the widths of the two holes are very small (Young used pinholes for his experiment), then a pattern of alternating light and dark images appears on the screen, as shown in Figure 1.15. Young explained this pattern this way: the tiny holes behave as two new sources of the original beam of light. Light from each hole radiates outward in a circular pattern much like the spreading ripples produced when a pebble is dropped into a pool of water (Figure 1.16). In Young's experiment, there were two "pebbles" (holes) and thus two sets of waves. As they spread out, the two sets overlap each other. Their undulating waves are superimposed one upon the other, producing light on the screen if the waves match up crest to crest and trough to trough. We call this coincidence **constructive interference.**

☑ Two sets of waves undergo **constructive interference** when their crests and troughs coincide.

For the two waves to constructively interfere with each other, either the distance that each beam of light travels from its hole to the screen must be the same or the difference in distance traveled by the two beams must be some multiple of

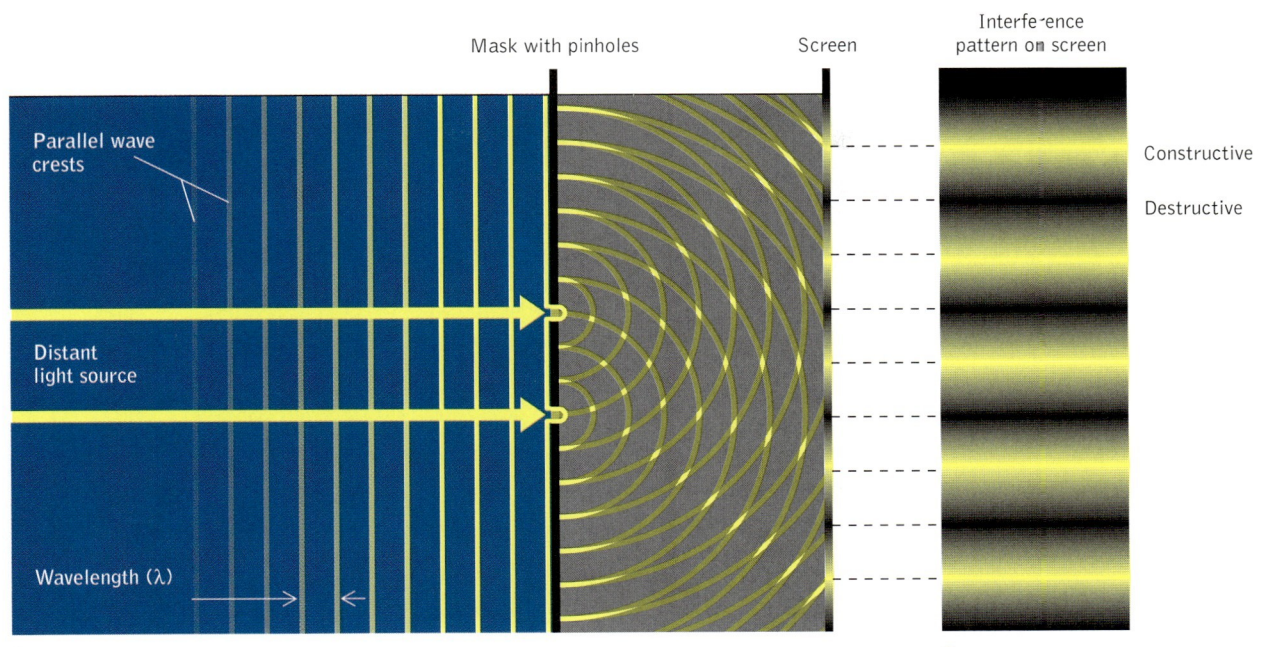

FIGURE 1.15 A. Waves of light approach a barrier and pass through twin pinholes in the barrier. The vertical parallel lines represent the crests of these waves. As they pass through the holes, the waves are bent around the edges of the holes, radiating out from the holes in two circular wave patterns. The circular waves overlap. When two crests overlap, they enhance each other (this is called *constructive* interference). When a crest from one pattern overlaps a trough of the other, the two crests cancel each other (this is called *destructive* interference). B. Constructive interference results in bright spots on the screen; destructive interference results in dark spots on the screen.

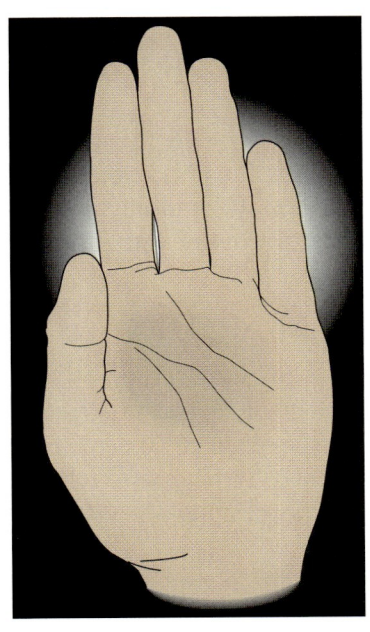

FIGURE 1.16. Pebbles dropped into a pond generate waves that produce an interference pattern. As the concentric circles of waves expand and intersect, regions of constructive and destructive interference arise. The same is true for light waves.

Hold your hand up in front of a light with palm facing you. Slowly bring fingers together until a tiny crack remains. You should see one or more dark lines in the narrow gap that remains between your fingers, representing the interference pattern generated by diffraction.

 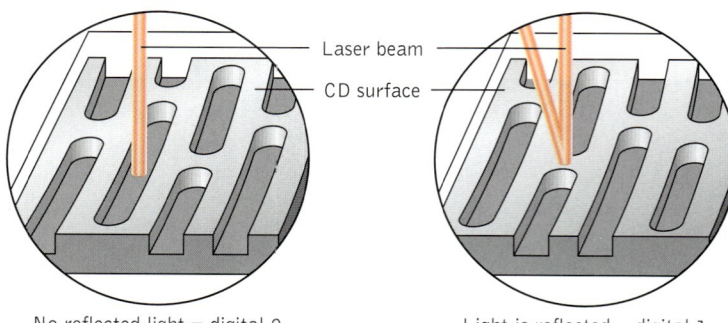

FIGURE 1.17 The surface of a CD contains circles of tiny craters, as shown to the right in the magnified views of the surface. The etched surface does not reflect light, but the unetched surface does. The tiny spaces between adjacent circles are similar in size to the wavelengths of visible light. White light reflecting off the unetched spaces in adjacent circles may interfere with each other. We see rainbows of colors of light that have undergone constructive interference.

> Two sets of waves undergo **destructive interference** when they do not coincide.

the wavelength of the light. If the difference in distance traveled is not a multiple of λ, then there is a mismatch in the two waves, such as when the crests of one wave overlap the troughs of the other. This mismatch causes **destructive interference** and results in a dark area on the screen.

The diffraction of white light produces rainbows of color. You may have seen this phenomenon in the colorful patterns produced by light reflected off the surface of a compact disk (Figure 1.17). Each "groove" of the disk becomes a secondary source of reflected light. Adjacent grooves will produce an interference pattern that creates constructive interference for a particular wavelength, depending on the angle produced by the primary source of light, the surface of the disk, and your eye. You see the color corresponding to that wavelength from that part of the disk.

Wavelength, frequency, and energy

The scales in Figure 1.12 show that the energies of different types of electromagnetic radiation increase as their frequencies increase. Mathematically, we can express the proportionality between energy (E) and frequency with a simple equation:

$$E = h\nu \quad (1.2)$$

where the constant of proportionality, h, is the Planck constant and has the value 6.626×10^{-34} J·s. The letter "J" is the abbreviation for joule, the standard unit of energy (Table 1.2) and "s" represents time in seconds. By combining Equations 1.1 and 1.2, we obtain an equation that defines the inverse relation between the energy of electromagnetic radiation and wavelength:

$$E = hc/\lambda \quad (1.3)$$

> The energies of different types of electromagnetic radiation are proportional to their frequencies and are inversely proportional to their wavelengths.

The Planck constant is named after German physicist Max Planck (1858–1947). We will consider some of Planck's important contributions to science in

TABLE 1.2 SI Base Units

Quantity or Dimension	Unit Name	Unit Abbreviation
Mass	kilogram	kg
Length	meter	m
Temperature	kelvin	K
Time	second	s
Energy	joule	J
Electrical current	ampere	A
Amount of a substance	mole	mol
Luminosity	candela	cd

Chapter 3. For now, simply note that Planck's constant is a very small number and so the energy values calculated using it are small. These tiny packets of energy, called **photons,** represent individual building blocks of radiation just as atoms represent individual building blocks of matter. They fit into Sir Isaac Newton's "particle" view of light. However, the wavelike properties of light revealed by the interference patterns observed by Young and seen in colorful reflections of light from the surfaces of CDs offer another perspective. We must conclude that electromagnetic radiation can exhibit *both* wavelike *and* particle-like properties. In Chapter 3 we will see that this same duality applies to the behavior of tiny particles of matter as well.

> ✓ A **photon** is the smallest "building block" of electromagnetic radiation, such as that emitted or absorbed by a single atom.
>
> Electromagnetic radiation exhibits both wavelike and particle-like properties.

SAMPLE EXERCISE 1.2: What is the energy of a photon of red light with a wavelength of 656 nm?

SOLUTION: We use Equation 1.3 to calculate the energy of electromagnetic radiation:

$$E = hc/\lambda$$

$$E = \frac{(6.626 \times 10^{-34} \text{ J} \cdot \text{s})\left(2.998 \times 10^8 \frac{\text{m}}{\text{s}}\right)}{(656 \text{ nm})\left(\frac{10^{-9}\text{m}}{\text{nm}}\right)}$$

$$= 3.03 \times 10^{-19} \text{ J}$$

wavelength (λ)

$$\downarrow E = \frac{hc}{\lambda}$$

energy (E)

This amount of energy is extremely small. To understand why, keep in mind this relation: a photon is a particle of radiant energy in the same way that an atom is a particle of matter.

PRACTICE EXERCISE: There are instruments used in chemical analysis that discriminate between individual photons of energy on the basis of their energies. Suppose such an instrument is adjusted so that it detects photons with 1.00×10^{-16} J of energy. What wavelength of radiation would be detected? (See Problems 29 and 30.)

FIGURE 1.18 The pattern of lines missing in the light emitted by the sun is the same in light emitted by other stars, but the pattern may be shifted to longer wavelengths. By measuring the amount of these shifts, astronomers can determine how rapidly a star or galaxy of stars is moving away from Earth.

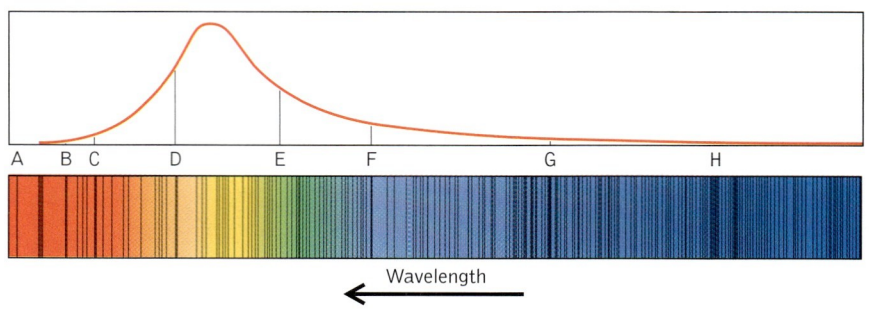

Shifting wavelengths and the Doppler effect

In 1800, English scientist William Hyde Wollaston (1766–1828) used carefully ground prisms to study the spectrum of the sun and discovered that some of the sun's spectrum was missing. There were narrow dark lines in it (Figure 1.18) where little light was transmitted. Using even better quality prisms of his own making, German physicist and optical craftsman Joseph von Fraunhofer (1787–1826) was able to resolve and map the wavelengths of hundreds of these missing lines in the solar spectrum. He labeled the most prominent of them with letters of the alphabet. Fraunhofer's lines are important in estimating the speeds of recession of other galaxies because the same pattern of lines is missing from the light produced by distant stars. Hubble discovered, however, that the wavelengths of the lines in the spectra of starlight are not exactly the same as in sunlight. In starlight from distant galaxies, the pattern of lines is the same but the lines themselves are all shifted to longer wavelengths. Because the color of the longest wavelength of visible light is red, scientists call this shift a *redshift*.

✓ Fraunhofer lines in the spectra of distant stars and galaxies are Doppler shifted to longer wavelengths. This shift is caused by the rapid rates at which these sources are moving away from Earth.

CONNECTION: The missing lines in the spectrum of the sun helped scientists decipher the arrangement of electrons in atoms (Chapter 3).

How do these redshifts occur? To answer this question, let's consider a source of light that emits only a single frequency. Now suppose that this source and a viewer were moving rapidly away from each other. Under these conditions, fewer wave crests of the light from the source would reach the viewer per unit time and would result in the perception that the source produced a lower frequency than it actually did. Such motion-induced shifts in frequency and wavelengths are examples of the **Doppler effect.** This phenomenon is the same as that which produces the higher pitch (frequency) that we hear when the source of sound, such as a train whistle or the horn of a truck, approaches us, and the apparent lowering of frequency as it passes by.

> How do these redshifts occur?

✓ The **Doppler effect** is the shift in perceived frequency of waves caused by movement of their source away from (*redshift*), or toward (*blueshift*), the observer.

The magnitude of the redshift in starlight is expressed by the following ratio:

$$\frac{(\nu - \nu')}{\nu}$$

where ν' is the perceived frequency of the source and ν is its natural (unshifted) frequency. The magnitude of a redshift is equal to the ratio of the velocity (**v**) at which the source and observer are moving apart to the speed of light (*c*):

$$\frac{(\nu - \nu')}{\nu'} = \frac{v}{c} \tag{1.4}$$

Note that a redshift of, for example, 10% means that the velocity with which a galaxy and Earth are moving apart must be 10% of the speed of light, or more than 30,000 km/s. Actually, some galaxies are moving away from us much more quickly than that.

SAMPLE EXERCISE 1.3: Suppose the frequencies of the Fraunhofer lines in light from a distant galaxy are 25% less than the corresponding lines in sunlight. How rapidly are Earth and this galaxy moving apart?

SOLUTION: The problem states that the Fraunhofer lines have a redshift of 25%. Therefore:

$$\frac{(\nu - \nu')}{\nu'} = 25\%, \text{ or } 0.25$$

Inserting 0.25 into Equation 1.4, we have:

$$\frac{v}{c} = 0.25$$

$$v = 0.25c$$

redshift
$$\left(\frac{\nu' - \nu}{\nu}\right)$$

$$\downarrow \quad v = c\frac{\nu' - \nu}{\nu}$$

recession speed
(v)

Thus, the galaxy and Earth must be moving away from each other at 0.25 times the speed of light, or:

$$0.25 \times 2.998 \times 10^8 \text{ m/s} = 7.5 \times 10^7 \text{ m/s}$$

PRACTICE EXERCISE: Suppose an express train traveling at 54 m/s blows its whistle as it approaches a highway crossing. Atmospheric conditions are such that the speed of sound is 330 m/s. Will the pitch of the train whistle seem higher or lower to a person standing at the crossing than to the engineer on the train? Express the shift to higher or lower pitch as a percentage of the actual pitch of the whistle. (See Problems 31–32.)

The requirement built into Equation 1.4 that recession speeds be significant fractions of the speed of light presents a major challenge when scientists analyze the light from nearby galaxies. If these galaxies are still close by, they must not be moving away very rapidly, and so their redshifts will be tiny and hard to measure accurately. On the other hand, the farthest galaxies have speeds of recession that are easy to measure, but astronomers have difficulty estimating the distances to them. These uncertainties affect how well we can know how long the universe has been expanding and, therefore, how long ago the expansion began.

SAMPLE EXERCISE 1.4: The galaxy known to astronomers as 3C-295 is estimated to be 5 billion light-years from Earth. The frequencies of the starlight from this galaxy are redshifted 46.1%.
 a. Calculate how rapidly galaxy 3C-295 and Earth are moving apart.

b. On the basis of your answer to part a and the distance to galaxy 3C-295, calculate the time interval during which Earth and this galaxy have been moving apart.

Given: A light-year = 9.4601×10^{15} m.

SOLUTION: Edwin Hubble discovered that the distance to galaxies are proportional to their recession speeds. His discovery provides us with a way to calculate the age of the universe if we assume that all the galaxies began their journeys through space at the same time: the moment of creation. Dividing the distance to a galaxy by its recession speed provides a measure of the time during which that galaxy and ours (the Milky Way) have been moving apart.

a. We can use the size of the redshift to calculate the galaxy's recession speed. If the frequencies of the starlight from 3C-295 are redshifted 46.1%, then, according to Equation 1.4,

$$(\nu - \nu')/\nu = v/c = 46.1\%, \text{ or } 0.461$$

therefore:

$$v = 0.461c$$

To calculate the value of v, we should consider using as the value for c the distance that light travels in a year (9.4601×10^{15} m) rather than the distance that it travels in a second (2.998×10^{8} m). The choice is a matter of convenience: it makes more sense for us to calculate the age of the universe in years than in seconds. So, using $c = 9.4601 \times 10^{15}$ m/yr, we calculate a recession speed in meters per year:

$$v = 0.461 \times 9.4601 \times 10^{15} \text{ m/yr} = 4.361 \times 10^{15} \text{ m/yr}$$

b. The distance to galaxy 3C-295 is 5 billion (10^{9}) light-years. Because light travels 9.4601×10^{15} m in one year it must travel

$$(5 \times 10^{9}) \times 9.4601 \times 10^{15} \text{ m} = 4.7 \times 10^{25} \text{ m}$$

in 5 billion years, which is the distance to galaxy 3C-295.

The distance between two objects that are moving apart should be equal to the time during which they have been moving apart multiplied by the rate at which they are moving apart; that is,

$$\text{distance} = \text{velocity} \times \text{time}$$

or

$$d = v \times t$$

Solving for time (t), we have

$$t = \frac{d}{v}$$

$$= 4.7 \times 10^{25} \text{ m}/4.36 \times 10^{15} \text{ m/yr}$$

$$= 1.1 \times 10^{10} \text{ yr}$$

This value falls within the range of recent estimates of the age of the universe, which are between 10 billion and 20 billion years. As astronomers develop more sophisticated techniques for estimating distances to other galaxies, they will be better able to calculate the true age of the universe. (See Problems 39–42.)

CONCEPT TEST: The value for the speed of light was used twice in the preceding calculation of the age of the universe. Look at the calculations closely. Can you think of a simpler way to use the distance to 3C-295 in light-years (5×10^9) and its redshift (46.1%) in a one-step calculation of the age of the universe?

1.4 MEASUREMENTS IN SCIENTIFIC STUDIES

Accurate measurements are crucial to testing scientific theories such as those accounting for the creation of the universe. They are also essential to understanding the physical and chemical properties of matter and to advancing our understanding of natural phenomena. Inevitably, the rise of scientific inquiry in the seventeenth and eighteenth centuries brought about a heightened awareness of the need for accurate measurements and for expressing the results of observations and measurements in ways that were understandable to others.

✓ Accurate measurements and standardized units of measurement, such as SI units, are essential to scientific inquiry.

SI units

In March 1791, the French Academy of Sciences proposed that the basis for a standard unit of length be 1/10,000,000 of Earth's quadrant—that is, the distance from the North Pole to the equator along a meridian of longitude. The members of the academy knew that the meridian passing through Paris also passed through Dunkirk on the English Channel and near Barcelona, Spain. Astronomers were able to accurately determine the latitudes of these two cities (they were just over one-tenth of a quadrant apart), and teams of surveyors set out to precisely measure the distance between them. Their work led to the adoption of a length corresponding 39.38 inches as the standard meter (after the Greek *metron*, which means "measure") in 1794.

Later these French scientists settled on a decimal-based system for designating lengths that were multiples or fractions of a meter (m). They chose Greek prefixes for the names of units of distance much greater than a meter, such as *deca* and *kilo* for lengths 10 and 1000 times as great, and they chose Latin prefixes for much smaller lengths, such as *centi* and *milli* for lengths that were 1/100 and 1/1000 of a meter (see Table 1.1). Eventually, these decimal prefixes carried over to the names of standard units for other dimensions. These dimensions included the weight of a cubic centimeter of water at the temperature of its maximum density, which was the basis for the *gram* (g), and the volume corresponding to 1 cubic decimeter came to be known as the liter (L). In July 1799, a platinum (Pt) rod 1 meter long and a platinum block weighing 1 kilogram were placed in the French National Archives to serve as the legal standards for length and mass. These objects served as references for what came to be known as the metric system for expressing measured quantities.

CONCEPT TEST: Iron (Fe), unlike gold and platinum, combines with oxygen from the atmosphere, forming Fe_2O_3 (a component of rust). How would rust formation affect the mass of an iron object, assuming the rust adhered to the surface of the object?

Since 1960 scientists have, by international agreement, used a modern version of the metric system: *Système International d'Unités*, commonly abbreviated SI. The base SI units are listed in Table 1.2. Units for other quantities are derived from these base units. For example, the standard unit for volume is the cubic meter (m^3) and the standard unit for speed is meters per second (m/s).

The SI unit of length is the same as in the original metric system: the meter. However, the needs of modern science require that the length of the meter, as well as the dimensions of other SI units, be known or defined by quantities that are much more constant than, for example, the length of a rod of platinum. Two such parameters are the speed of light (c) and time. Thus, the meter has been redefined as the distance traveled by the light from a helium–neon laser in 1/299,772,458th of a second.

Significant figures

If light travels exactly 1 meter in 1/299,772,458th of a second, then it must travel 299,772,458 meters in 1 second. Starting with this distance, let's calculate how far (in meters) light travels in 1 year and check the value of the distance in a light-year (9.4601×10^{15}m) that we used in our calculation of the age of the universe.

If we knew the number of seconds in a year, we could simply multiply that number by 299,772,458 m/s to calculate the distance that light travels in a year. Assuming we don't know that number, we will need to work up to it by using units with which we are familiar. Among these units are 60 seconds = 1 minute, 60 minutes = 1 hour, 24 hours = 1 day, and 365.25 days = 1 year. Let's arrange these equalities as a series of fractions. Each fraction represents unity because the numerator and denominator represent equivalent quantities, and so multiplying a quantity by one or more of these fractions does not change its value, just the units on it. We need to be sure the fractions are written right side up. For example, if the unit that we are changing appears in the numerator of our starting value, it must appear in the denominator of a conversion factor. Factors for determining how far light travels in 1 year are given in the following calculation. Notice how the units of time change from seconds to minutes to hours to days to years in succeeding steps. All units of time except years cancel out and the final value has the dimension meters per year.

$$\text{Step:} \qquad\qquad 1 \qquad\qquad 2 \qquad\qquad 3 \qquad\qquad 4$$

$$\frac{299{,}772{,}458 \text{ m}}{\text{s}} \times \frac{60 \text{ s}}{\text{min}} \times \frac{60 \text{ min}}{\text{hr}} \times \frac{24 \text{ hr}}{\text{d}} \times \frac{365.25 \text{ d}}{\text{yr}}$$

$$= \frac{9{,}460{,}099{,}320{,}581{,}000 \text{ m}}{\text{yr}}$$

There are two problems with the result of this calculation. For one thing, it has 16 digits in it, which is more than it should have. To understand why 16 digits are too many, we must address the concept of significant figures. The number of significant figures in a measured value or in the result of a calculation based on one or more measured values tells us how well we know the value. For example, say that we were to determine the mass of a penny on the two balances shown in Figure 1.19. The balance on the left can be used to weigh objects weighing as much as 500 grams to the nearest 0.01 g; the one on the right can be used to weigh as much as 50 grams to the nearest 0.0001 g. According to the balance on the left, our penny weighs 2.53 g; according to the balance on the right, it weighs 2.5271 g. The mass of the penny can be determined more precisely when we use the balance on the right. Much more is indicated by the greater number of significant figures in this balance's reading. The value obtained with the left balance has three significant figures; the one on the right has five. As a general rule, all of the digits from the leftmost nonzero digit to the rightmost nonzero digit in a number are considered significant. The location of the decimal place does not affect the number of significant figures.

How many significant figures are in 9,460,099,320,581,000 m/yr? Including all of those from the first nonzero digit (the first 9) to the last (the 1), there are at least 13. The last three zeroes may or may not be significant: they could just be serving to fix the decimal place. How many significant figures should there be? Recall that this value was calculated from the speed of light and several conversion factors. Factors such as 60 seconds per minute and 60 minutes per hour are considered exact numbers, like 12 eggs in a dozen. We assume that there is no uncertainty in their values, and so they do not affect the number of significant figures in our result. On the other hand, the number of days in a year is not exact. The decimal part of 365.25 represents the extra day in every fourth (leap)

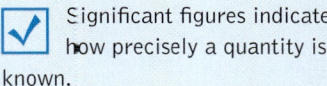 Significant figures indicate how precisely a quantity is known.

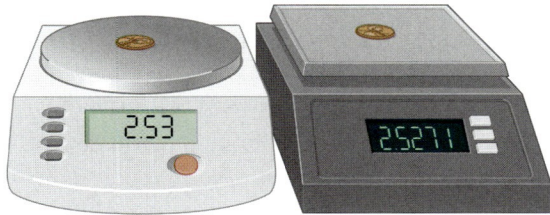

FIGURE 1.19 The mass of a penny can be measured with a variable number of decimal places, depending on the balance used. The total number of digits in the number corresponds to the number of significant figures in the number.

year. Even so, the calendar is not in exact agreement with the revolution of Earth around the sun, which is why we skip a leap year about once every 100 years. We would have to include this correction and others in our days-to-years conversion factor if we wanted to obtain a value with more than five significant figures.

Because we know the number of days in a year to only five significant figures, we can know the result of our multiplication of conversion factors to only five significant figures. The principle behind this limitation is that a calculated value can be known only as well as the least well known value used in its calculation. This principle is a variation of the "weak link" philosophy that states, "a chain is only as strong as its weakest link." In our calculation, 365.25 days per year is the weak link. Rewriting the calculated value so that it has only five significant figures, we get 9,460,100,000,000,000 m/yr. All of the zeros to the right of the 1 are not significant. They serve only to set the decimal place, which is a task we can accomplish using exponential notation. Thus, the distance light travels in a year is 9.4601×10^{15} m.

We apply the weak-link principle in a similar fashion to calculations that require division. Suppose, for example, we wished to determine whether a small nugget of gold in a mineral collection was pure gold. To find out, we might determine the mass and volume of the nugget and then calculate its density. If that density matched that of gold (19.3 g/mL), chances would be good that the nugget is pure (or nearly so). Suppose we found that the mass of the nugget is 4.72 g and that its volume is 0.25 mL. What would be the density of the nugget expressed with the appropriate number of significant figures?

To calculate density (d), we divide mass (m) by volume (V):

$$d = \frac{m}{V} = \frac{4.72 \text{ g}}{0.25 \text{ mL}} = 18.9 \text{ g/mL}$$

This result appears to be slightly less than that of pure gold, and so we might suspect the presence of a lower-density impurity. However, we need to answer the question, "How well do we know the density of the nugget?" There are three significant figures in the value of the mass but only two in the value of volume. Therefore, according to the weak-link principle, we can have only two significant figures in the final answer. We must round off 18.9 to 19. g/mL. The density of pure gold, expressed to two significant figures, is also 19. g/mL; so the nugget may well be "pure" gold.

CONCEPT TEST: Did the preceding determination of the density of the nugget prove that it was pure gold or did it fail to prove that the nugget was impure? Explain the difference between these two conclusions and explain why you prefer one over the other.

SAMPLE EXERCISE 1.5: The meteorite pictured in Figure 1.20 was found in Antarctica and is believed to have been part of a debris field formed by the impact of an asteroid on the surface of Mars more than a billion years ago. It has a mass of 1.90 kg. Calculate the mass of the meteorite in ounces by using the

FIGURE 1.20 Antarctica is a profitable place to search for meteorites, given the absence of vegetation. The meteorite ALH84001 is believed to have originated on Mars. An asteroid impact led to its ejection from that planet, and it eventually landed on Earth.

unit-conversion information in Table 1.3 on the next page. Be sure to express your answer with the appropriate number of significant figures.

SOLUTION: There are exactly 16 ounces in a pound, and there are 453.6 grams in a pound.

We also need to convert kilograms into grams by using the factor 1000 grams in a kilogram. Setting up the appropriate conversion factors, we have

Step: 1   2   3

$$1.90 \text{ kg} \times \frac{1000 \text{ g}}{\text{kg}} \times \frac{1 \text{ lb}}{453.6 \text{ g}} \times \frac{16 \text{ oz}}{\text{lb}} = 67.0194004 \text{ oz}$$

In this calculation, the starting mass (1.90 kg) has three significant figures and the number of grams (453.6) in a pound has four. We know the exact number of grams in a kilogram (1000) and the exact number of ounces in a pound (16). Thus, the weak link in this calculation is the three significant figures in the starting value, and the calculated mass in ounces can have only three significant figures. Therefore we must round off the calculated answer to:

67.0 oz

Note that the zero to the right of the decimal place in 67.0 is significant. It is there to indicate that we know the value to three significant figures and not, for example, to only two, which would be implied if we had rounded off the answer to 67.

Note about rounding off: When rounding off a value, round "up" or round "down," depending on whether the next digit is greater or less than 5. The question remains of how to handle cases in which the next digit *is* 5. A good rule to follow in those cases is to always round to the nearest *even number* when the first nonsignificant digit is a 5. For example, if only three significant figures are allowed, we round off 45.45 to 45.4, but we round off 45.55 to 45.6.

PRACTICE EXERCISE: The meteorite in the preceding Sample Exercise would have to have been flying away from the Martian surface at a velocity of at

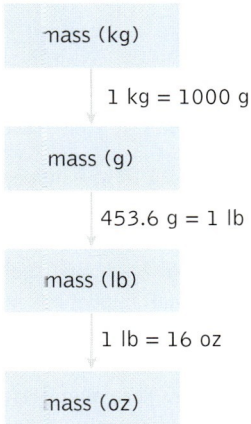

least 5.4 km/s to escape the gravity of Mars and make its way to Earth. What is this minimum velocity in miles per hour? (See Problems 35–38.)

TABLE 1.3 Conversion Factors for SI and Other Commonly Used Units

Quantity or Dimension	Equivalent Units
Mass	1 kg = 2.205 pounds (lb) or 1 lb = 0.4536 kg or 453.6 g
	1 g = 0.03527 ounce (oz) or 1 oz = 28.35 g
Length	1 m = 1.094 yards (yd) or 1 yd = 0.9144 m
	1 m = 39.37 inches (in) or 1 foot (ft) = 0.3048 m
	1 in = 2.54 cm (exactly)
	1 km = 0.62 miles (mi) or 1 mi = 1.63 km
Volume	1 m^3 = 35.3 ft^3 or 1 ft^3 = 0.0283 m^3
	1 m^3 = 1000 liters (L) (exactly)
	1 L = 0.2642 gallon (gal) or 1 gal = 3.785 L
	1 L = 1.057 quarts (qt) or 1 qt = 0.9464 L

One final note on the weak-link principle for significant figures: the principle also applies to calculations requiring addition and subtraction but in a different way from that for multiplication and division. Let's revisit the determination of the density of a nugget of gold. How might we determine the volume of such an object? One approach is to measure the volume of water that the nugget displaces (assuming that it is more dense than water, which gold certainly is). Suppose we half fill a 100-mL graduated cylinder with water so that it contains exactly 50. mL as shown in Figure 1.21. Then we place a chunk of shiny yellow mineral into the graduated cylinder, being careful not to splash any water out, and discover that the level of the water has increased to 58. mL. What is the volume of the mineral? Clearly, its volume is the volume of the water that it displaced, which is indicated by the increase in the level of water from 50. to 58. mL. That increase is

$$58. - 50. = 8. \text{ mL}$$

The result has only one significant figure. Yet the two volumes used to calculate it had two significant figures. Shouldn't the answer also have two? The answer is no, because the initial and final volumes were known to the nearest milliliter, and so we can know the difference between them to only the nearest milliliter. The location of the decimal point in values that are added or subtracted *is* important in setting the number of significant digits in the answer.

Consider one more example. Suppose we decide to weigh a roll of pennies with the balance on the left in Figure 1.19 to determine whether there are exactly 50 pennies in the roll. The roll of pennies should weigh

$$50 \text{ pennies} \times 2.53 \text{ g/penny} = 126.5 \text{ g}$$

plus the weight of the paper wrapper. The "tare" feature on the balance enables us to correct for the weight of the wrapper, and we find that the weight of just the

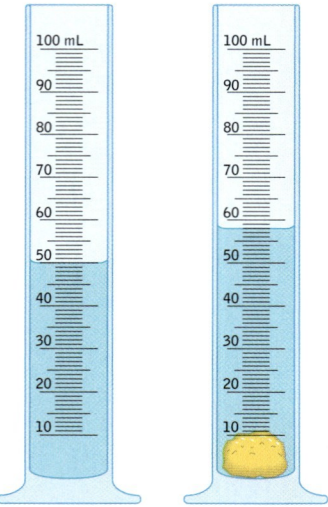

FIGURE 1.21 The volume of a small, irregularly shaped object can be determined by measuring the volume of water that it displaces. A small nugget believed to be pure gold is placed in a graduated cylinder containing 50. mL of water. The volume rises to 58. mL, demonstrating that the volume of the nugget is 8. mL.

pennies is 124.01 g. We conclude that the roll has only 49 pennies; so we add one more. Now suppose the penny that we add is one that we know weighs 2.5271 g. What is the total weight of all 50 pennies to the appropriate number of significant figures? Adding the weigh of the 50th penny to the other 49, we have

$$\begin{array}{r} 124.01 \text{ g} \\ + 2.5271 \text{ g} \\ \hline 126.5371 \text{ g} \end{array}$$

The total mass of the first 49 pennies has two decimal places, and the mass of the 50th has four decimal places. Therefore, we can know the sum of the two to only two decimal places, and so we must round off the total weight to 126.54 g.

SAMPLE EXERCISE 1.6: A shiny yellow object has a mass of 30.01 g. Its volume is determined by placing it in a 100-mL graduated cylinder partly filled with water. The volume of the water before adding the mineral is 56. mL. The volume after the mineral was added is 62. mL.
a. What is the density of the object to the appropriate number of significant figures?
b. Could the object be made of gold?

SOLUTION: a. Density is calculated by dividing the mass of an object by its volume. The volume of the object is the volume of water that it displaced; that is, the difference in the volume of the contents of the graduated cylinder before and after the object was added. Calculating this difference, we have

$$\begin{array}{r} 62. \text{ mL} \\ - 56. \text{ mL} \\ \hline 6. \text{ mL} \end{array}$$

Because the initial and final volumes are known to only the nearest milliliter, we can know the difference between them to only the nearest milliliter.

To calculate density, we divide mass by volume:

$$d = \frac{m}{V} \times \frac{30.01 \text{ g}}{6. \text{ mL}} = 5.00 \text{ g/mL}$$

Because we know the volume of the object to only one significant figure, we can know its density to only one significant figure, and so we must round off the result to 5. g/mL.
b. The density of gold is 19.3 g/mL; so the object cannot be made of gold. It may well be made of "fool's gold," the common name for a mineral that geologists call iron pyrite. The density of iron pyrite is 5.0 g/mL.

PRACTICE EXERCISE: Express the results of the following calculation to the appropriate number of significant figures:

$$\frac{0.391 \times 0.0821 \times (273 + 25)}{8.401}$$

(See Problems 45–48.)

 Significant-figure rules:

1. In multiplication and division, the result is expressed with the same number of significant figures as the measured value with the fewest number of significant figures.

2. In addition and subtraction, the result can have no more digits to the right of the decimal point than the measured value with the fewest number of digits to the right of the decimal.

> **Precision** is the repeatability of a measurement. **Accuracy** is the agreement between an experimental value and the true value.

Precision and accuracy

Two terms—**precision** and **accuracy**—are used (and sometimes misused) to describe how well a measured quantity or a value calculated from a measured quantity is known. Precision indicates how *repeatable* a measurement is. Suppose that we used the balance on the left in Figure 1.19 to weigh a penny over and over again, and suppose that, every time we did so, the result was the same: 2.53 grams. These results tell us that our determinations of the mass of the penny were precisely 2.53 g. Said another way, the results were precise to the nearest 0.01 g.

Now suppose that we weigh the same penny using the balance on the right in Figure 1.19 and obtained the following results for five measurements:

Measurement	Mass (g)
1	2.5270
2	2.5371
3	2.5371
4	2.5371
5	2.5372

Note that there is a small variability in the last decimal place. Such variability is not unusual when using a balance that can weigh to the nearest 0.0001 g. Particles of dust landing on the pan of the balance, vibration of the laboratory bench under it, or the transfer of moisture from our fingers to the penny as we pick it up and place it on the pan could all produce a change in mass of 0.0001 g or more. (Using tweezers or rubber gloves to handle objects to be weighed to the nearest 0.0001 g would be a good idea.) One way to express the precision of these results is to use the range between the highest and lowest values: 2.5270 to 2.5272. Range can also be expressed by using the average value (2.5271 g) and the range above (+0.0001) and below (−0.0001) the average that includes all the observed results. A convenient way to express the observed range is 2.5271 ± 0.0001, where the symbol ± means "plus or minus" the value that follows it.

While precision relates to the repeatability of a measurement, accuracy relates to how close the measured value is to the true value. Using our example of the penny, suppose the true mass of the penny was 2.5267 g. The results obtained by using the balance on the left in Figure 1.19 would have been precise *and* accurate to the nearest 0.01 g, because 2.5267 rounded

> How can we be sure that the results of a measurement are accurate?

off to the nearest 0.01 g is 2.53 g. However, the average result obtained with the balance on the right (2.5271 g) is 0.0004 g too heavy. Thus, the measurements with the balance on the right may have had a precision of ±0.0001 g, but they were not accurate to within ±0.0001 g of the true value. Another view of the difference between accuracy and precision is presented in Figure 1.22.

How can we be sure that the results of a measurement are accurate? The accuracy of a balance could be checked by weighing objects of known masses.

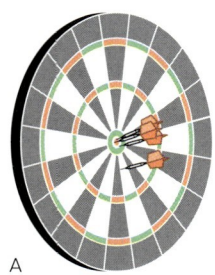 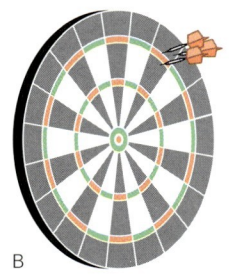 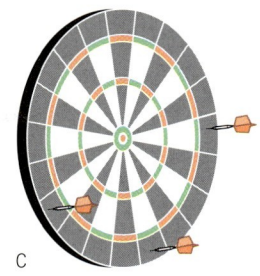

A   B   C

FIGURE 1.22 Accuracy refers to the proximity of a result to the true value. On a target, the closer to the bull's-eye the dart lands, the more accurate the throw is. Precision refers to the proximity of several results to one another. On the target, proximity is represented by three darts very close to each other. Target A shows good accuracy and good precision, target B shows good precision but poor accuracy, and target C shows neither good accuracy nor good precision.

A thermometer could be calibrated by using it to measure the temperatures of substances that have known temperatures. A mixture of ice and pure water, for example, should have a temperature of 0.0°C, and the normal boiling point of water is 100.0°C at exactly 1.00 atm of atmospheric pressure. A measurement that is validated by calibration with an accepted standard material should be accurate.

1.5 MATTER AND ENERGY

Cosmologists who study the dynamics of the universe believe that, just after the Big Bang, the energy released began to be transformed into matter. The relation between energy and matter is a simple one, first described by Albert Einstein (1879–1955) in his well-known equation:

$$E = mc^2 \qquad (1.5)$$

where E is the amount of energy, m is the mass of the corresponding amount of matter, and c is the speed of light.

What sort of matter was created just after the Big Bang? Cosmologists believe that, for about a millisecond, matter in the rapidly expanding universe consisted of a class of extremely tiny particles called quarks. We will not consider the nature of quarks here; we leave that to our colleagues in physics. For now, we note that individual quarks are not discernible in nature today; they exist by themselves only at extremely high temperatures.

CONNECTION: In Chapter 2, we will use Einstein's equation to explain the formation of the elements in the periodic table.

Radioactive decay

Within a millisecond of the Big Bang, the expanding universe had "cooled" enough—its temperature was still billions of degrees—to allow groups of quarks to be drawn together under the influence of the strongest of the four basic

| | The **basic forces** include *gravity* (the attraction between bodies based on their masses), *electromagnetism* (the force of attraction between oppositely charged particles or between the north and south poles of magnets), and the *strong nuclear force*, which is extremely strong but which operates over only very small distances. There is also a *weak nuclear force* that is stronger than gravity but not as strong as electromagnetism.

Subatomic particles include uncharged *neutrons* and positively charged *protons*, which are of similar mass and form the nuclei of atoms, and negatively charged *electrons*, which are of much smaller mass.

Radioactive decay is the spontaneous disintegration of unstable particles accompanied by the release of radiation.

Mass number (A) is a scale in which the masses of a proton and neutron are 1 and an electron is zero.

The **atomic mass unit** (amu) is widely used to express the relative masses of atoms and subatomic particles.

---

forces of nature, known as the strong nuclear force. When these quarks combined, they formed neutrons. Free neutrons are not stable; instead they spontaneously disintegrate in a process known as **radioactive decay.** Each neutron disintegrates into a proton, which has a positive electrical charge and most of the mass of the neutron, and an electron, with a negative charge and very little mass (Table 1.4). We can write this process with symbols as follows:

$$_0^1 n \rightarrow {}_1^1 p + {}_{-1}^0 e \qquad (1.6)$$

where "n" represents the neutron, "p" the proton, and "e" the electron. The subscripts preceding the symbols show the electrical charges of the particles (see Table 1.4); the superscripts represent their masses expressed in **mass numbers** (A). Though the mass number of an electron is zero, an electron is a particle of matter and so must have some mass. In Table 1.4, which gives a more precise view of the charges and masses of these particles, the mass values are expressed in **atomic mass units** (amu) and in grams. As you can see, these particles are tiny indeed.

Equation 1.6 or any equation describing a nuclear reaction must be balanced. In other words, the electrical charges denoted by the subscripts and the masses denoted by the superscripts on the left side of the reaction arrow must equal the sum of the electrical charges and masses on the right side. A check of the superscripts in Equation 1.6 reveals that they do indeed add up:

$$1 = 1 + 0$$

(left side) = (right side)

and so do the subscripts:

$$0 = 1 + (-1)$$

(left side) = (right side)

Although Equation 1.6 is complete in regard to the masses of the particles, it is not quite complete from the perspective of energy. When a neutron decays into a proton and an electron, a packet of energy, called a neutrino, is released. The existence of neutrinos was first proposed in the 1930s and, for most of the time

TABLE 1.4 Some Properties of the Neutron, Proton, and Electron

	Mass		Charge	
Particle	Atomic Mass Units (amu)	Grams (g)	Relative Value	Coulombs*
Neutron	1.00867	1.67494×10^{-24}	0	0
Proton	1.00728	1.67263×10^{-24}	+1	$+1.602 \times 10^{-19}$
Electron	5.48580×10^{-4}	9.10939×10^{-28}	−1	-1.602×10^{-19}

*The coulomb is the SI unit of electrical charge. When a current of 1 ampere (see Table 1.2) passes through a conductor for 1 second, the quantity of electrical charge that moves past any point in the conductor equals 1 coulomb.

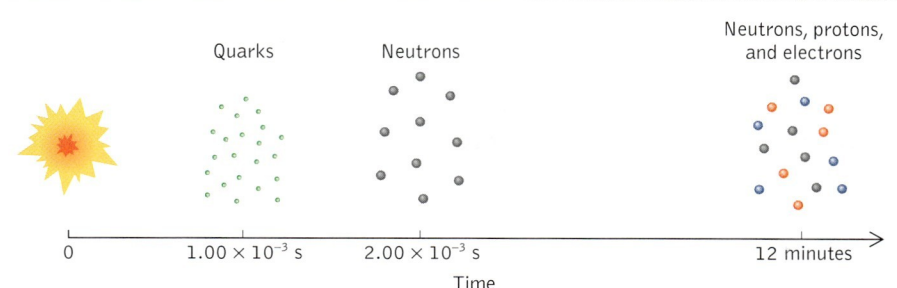

We can follow the aftermath of the Big Bang through time. The Big Bang created quarks, which began to combine into neutrons after 1 millisecond. Neutrons slowly decay into protons and electrons. It takes 12 minutes for half of the neutrons to convert into protons and electrons.

since then, they were thought to be massless particles of energy. In recent years, results of studies of the behavior of neutrinos suggest that they do have a slight mass, but only about a billionth that of a neutron and not enough to affect the mass balance in Equation 1.6. We will revisit the notion that particles can have energy-like as well as mass-like properties in Chapter 3.

 CONNECTION: Theories about the structure of atoms are described in Chapter 3.

Radioactive-decay rates

The disintegration of neutrons is a rapid process but not exactly instantaneous. Suppose that, in a tiny bit of newly created space, 1 million neutrons formed within a second of the Big Bang. After 12 minutes, half of them would have disintegrated into protons, electrons, and neutrinos, leaving half a million still intact. In the next 12 minutes, half of the remaining half million would have disintegrated, leaving only a quarter million left. After the next 12 minutes, half of the remaining quarter million would have decayed, leaving one-eighth of a million neutrons still intact. This decay process in which half of the remaining neutrons disintegrate every 12 minutes is shown in the graph in Figure 1.23. As you can see, there should have been few free neutrons left within hours of the Big Bang. Actually, within half an hour, essentially all of the neutrons in the universe had either decayed or, as we will see, met a different fate.

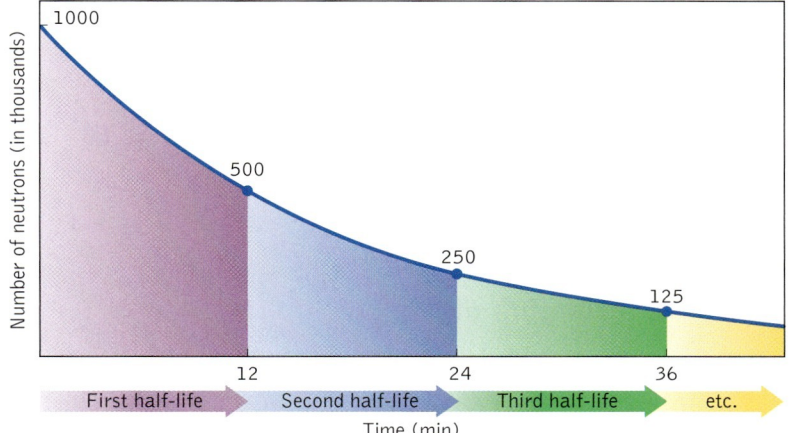

FIGURE 1.23 Neutrons and other radioactive particles decay at a constant rate, or half-life. The half-life is the time required for half of the neutrons present at any time to decay into protons and electrons. The half-life for neutrons is 12 minutes.

> **Half-life** is the interval over which half of a quantity of radioactive substance decays.

The 12-minute interval is a property characteristic of the decay of neutrons. Other radioactive species decay at different rates. We call the time interval in which half of any radioactive substance decays its **half-life.** The faster the decay process, the shorter the half-life. If we know the half-life of a radioactive substance, as well as the quantity of it at the start, we can calculate how much of it should be left at some later time. The mathematical relation between the initial amount (A_0, or the amount at zero time) and the amount left at some later time (A_t) can be written as follows:

$$\frac{A_t}{A_0} = 0.5^n \tag{1.7}$$

where n is the number of half-lives that has elapsed. If $n = 1$, then $A_t/A_0 = 0.5^1 = 0.5$, and half of the starting amount is left. If $n = 2$, $A_t/A_0 = 0.5^2 = 0.25$, and one-quarter of the starting amount is left. However, elapsed time may not be an exact multiple of the half-life. What if we wished to know the fraction of an initial quantity of free neutrons remaining after only 10 minutes? We could begin by calculating the number of half-lives that has elapsed:

$$10 \text{ minutes} \times \frac{\text{one half-life}}{12 \text{ minutes}} = 0.833 \text{ half-life}$$

Using this value for n in Equation 1.7, we have

$$A_t/A_0 = 0.5^n = 0.5^{0.833} = 0.561, \text{ or } 56\% \text{ left (to two significant figures)}$$

CONNECTION: The equation describing the decay of neutrons can be applied to other nuclear (Chapter 2) and chemical (Chapter 14) reactions.

SAMPLE EXERCISE 1.7: Of a population of 1 million free neutrons, how many would remain after 2.0 minutes?

SOLUTION: Neutrons undergo radioactive decay and have a half-life of 12 minutes. In 2.0 minutes, there are 2.0/12. = 0.1667 half-lives. This value is used as the n in Equation 1.7. The number of neutrons to begin with (A_0) is 1,000,000 and the number after 2.0 minutes can be calculated as follows:

$$\frac{A_t}{A_0} = 0.5^n$$

or

$$A_t = A_0\, 0.5^n = 1,000,000 \times 0.5^{0.1667}$$
$$= 890,878 = 8.9 \times 10^5 \text{ (to two significant figures)}$$

PRACTICE EXERCISE: Of a population of 1 million free neutrons, how many would remain after 3.0 minutes? (See Problems 61 and 62.)

time (t)

$$n = \frac{t}{t_{1/2}}$$

number of half-lives (n)

number of neutrons initially (A_0)

$$A_t = A_0(0.5)^n$$

number of neutrons at time t (A_t)

The formation of nuclei

The graph in Figure 1.23 indicates that, within several hours of the Big Bang, just about all of the neutrons in the universe should have disintegrated. Actually, the population of free neutrons decreased even faster than that. Some did decay, but others, starting about 100 seconds after the Big Bang when the universe had cooled to "only" 10^9 degrees, formed stable aggregates of neutrons and protons. As temperature dropped, so did the velocities with which protons and neutrons were moving. This decrease in velocity allowed the strong nuclear force to produce clusters of protons and neutrons. Eventually, these clusters would become the nuclei of atoms. Within these clusters, individual protons and neutrons are called **nucleons**. If one proton and one neutron combine they make a **deuteron** (d) with a charge of $+1$ (from the charge on the proton) and a mass number (which is the sum of the number of protons and neutrons) of 2:

$$^1_1 p + ^1_0 n \rightarrow ^2_1 d \tag{1.8}$$

When two deuterons fuse together, they produce an α (alpha) particle:

$$2\,^2_1 d \rightarrow ^4_2 \alpha \tag{1.9}$$

An α particle has four nucleons—two protons and two neutrons—and so has a charge of $+2$ and a mass number of 4.

According to the Big Bang theory, the universe continued to expand and to cool, but its composition did not change much for more than a hundred thousand years. During that time, it remained a collection of hot, electrically charged particles: protons, α particles, and electrons. Such a collection of high-temperature ions is called a plasma. In today's universe, plasmas exist in the cores of our sun and the other stars.

> The protons and neutrons in atomic nuclei are called, collectively, **nucleons.** They are held together in nuclei by the strong nuclear force.
>
> A **deuteron** contains one proton and one neutron.
>
> The mass number (A) of a nucleus is the sum of the number of neutrons and protons in it.

1.6 TEMPERATURE SCALES AND A COOLING UNIVERSE

After a hundred thousand years, the temperature of the expanding universe had decreased to 10^4 degrees. At this point in our trip through time, we need to be precise on what kind of temperatures we are referring to. There are several temperature scales in common use. In the United States, the Fahrenheit scale is the most popular, although American weather forecasters may give temperatures in both Fahrenheit and Celsius (Centigrade) degrees. The latter is the most widely used scale in other countries.

The two scales differ in two ways, as shown in Figure 1.24. First of all, their zero points are different. Zero degrees Celsius (0°C) is the temperature at which water freezes, but that temperature is 32 degrees on the Fahrenheit scale (32°F). The other difference is in the size of the temperature change corresponding to a degree. The difference between water's freezing and boiling points is $212 - 32 = 180$ degrees on the Fahrenheit scale but only 100 degrees on the Celsius scale. This difference means that a Celsius degree is bigger than a Fahrenheit degree: 1.8 times (or 9/5) as big. To convert temperatures from Fahrenheit into Celsius, we need to account for the differences in zero point and in degree size. The following equation allows us to do both:

$$°C = \frac{5}{9}(°F - 32) \tag{1.10}$$

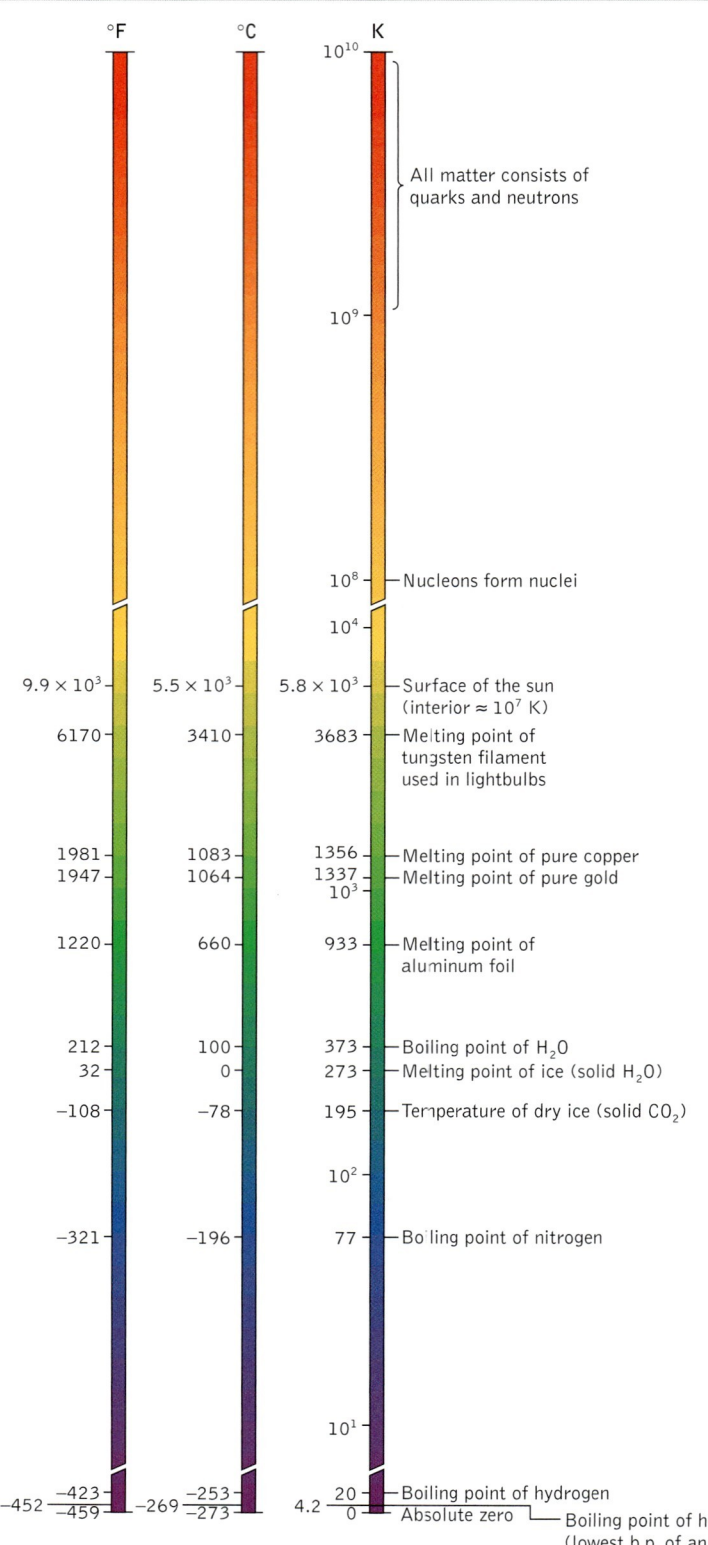

FIGURE 1.24 Three temperature scales are commonly used today, although the Fahrenheit (°F) scale is rarely used in chemistry. Fortunately, we can readily convert between the scales.

SAMPLE EXERCISE 1.8: A thermometer is calibrated in °C and °F. At what point do the scales overlap? That is, at what temperature are the values on the Celsius and Fahrenheit scales the same?

SOLUTION: The logical starting place for this exercise is Equation 1.10. We wish to calculate the temperature at which °C = °F. At that temperature, the left side of Equation 1.10 is equal to °F as well as °C. Substituting °F for °C and solving for °F, we have

$$°F = \frac{5}{9}(°F - 32)$$

$$9°F = 5(°F - 32)$$

$$= 5°F - 160$$

$$4°F = -160$$

$$°F = -40$$

Thus, the two scales cross over at −40°. Above that temperature, °F values are greater than °C values; below it, °F values are smaller than °C values.

PRACTICE EXERCISE: Rearrange the terms in Equation 1.10 to create an equation for converting Celsius temperatures into Fahrenheit temperatures. (See Problems 73–78.)

The base SI unit of temperature (see Table 1.2) is neither Celsius nor Fahrenheit degrees; rather it is the **kelvin** (K). The zero point on the Kelvin scale is not related to the freezing of a particular substance such as water but rather is the coldest temperature that can theoretically exist. This coldest temperature is called **absolute zero.** It is the same as −273.15°C. No one has ever been able to chill matter to absolute zero, although scientists studying the structure of matter have cooled samples to microkelvin temperatures. The zero point of the Kelvin scale differs from that of the Celsius scale, but their increments (degrees) are the same size. Thus, the conversion from Celsius degrees into kelvins is simply a matter of adding 273.15 to account for the difference in zero points:

$$K = °C + 273.15 \qquad (1.11)$$

> ✓ The **Kelvin scale** is an absolute temperature scale (**0 K, absolute zero,** is the lowest temperature possible). The kelvin is the SI unit of temperature.

Then came atoms

As the universe cooled below 10^4 K, a new era began. It was marked by another level of particle aggregation and the formation of electrically neutral particles that we call atoms. At temperatures above 10^4 K, positively charged aggregates of nucleons may have been attracted to negatively charged electrons, but their energies were too great and their speeds were too high for them to develop long-term relations. Below 10^4 K, however, the electrostatic forces that attract opposite charges and repel like charges began to prevail. Electrons approaching positive aggregates did not merge with them and form clusters of nucleons and electrons (as was believed at the beginning of the twentieth century). Instead, the electrons swirled around the aggregates, forming atoms with positive nuclei at their centers and surrounding spaces in which the electrons moved.

When one electron occupies the space around one proton, an atom of hydrogen is created. An atom containing an electron and a deuteron also is an atom of hydrogen, but this form of hydrogen is called deuterium. Hydrogen and deuterium atoms, which differ only by the number of neutrons in their nuclei, are called **isotopes** (Table 1.5). The term **nuclide** is used to refer to an atom of a particular isotope of an element.

Hydrogen has three isotopes, as shown in Figure 1.25. The third isotope has a nucleus with *two* neutrons and a proton and is called tritium. We distinguish between nuclides by using symbols with superscripts that give each mass number—as shown for the three isotopes of hydrogen in Table 1.5. The symbols may also contain subscripts that indicate the number of protons, though these subscripts are redundant because the number of protons in the nucleus of an atom defines which element's atom it is. They can be useful, however, when we are balancing a nuclear reaction. They also represent an atom's **atomic number** (Z). Any atom with just one proton (Z = 1) is, by definition, a hydrogen atom. If it has more than one proton, it must be an atom of one of the other elements arranged in the periodic table of the elements (Figure 1.26). For example, an atom with 20 protons is, by definition, a calcium (Ca) atom. The atomic numbers of the elements appear at the top of their respective boxes in the table. Because the number of protons must be a whole number, all the atomic numbers in the boxes are whole numbers.

The value at the bottom of each box in the periodic table of the elements (see inside front cover of book) is the average mass of the atoms of that element. To understand the meaning of "average," consider the information in Table 1.5. The right-hand column gives the natural abundances of the hydrogen isotopes in the universe today. As you can see, the vast majority of them have just a proton and no neutrons in their nuclei. Only 15 of every 100,000 have one neutron, and nearly none have two. The reason for there being so little tritium is that the combination of two neutrons and only one proton is not stable. Like free neutrons, tritium nuclei are radioactive. They decay with a half-life of 12.26 years. Appendix 3 contains a selection of stable isotopes of the elements.

> **Isotopes** are atoms of an element that have different numbers of neutrons and so different mass numbers.
>
> A **nuclide** is an atom of a specific isotope of an element.
>
> The **atomic number** of an element equals the number of protons in each of its atoms.

FIGURE 1.25 Hydrogen, deuterium, and tritium are isotopes of the same element. They contain one proton (pink), but a different number of neutrons (gray). The region occupied by the electrons is shown in blue (not to scale).

TABLE 1.5 The Isotopes of Hydrogen

Name	Symbol	Mass (amu)*	Relative Abundance
Hydrogen	$^{1}_{1}H$	1.00782505	0.99985
Deuterium	$^{2}_{1}H$	2.0141079	0.00015
Tritium	$^{3}_{1}H$	3.01602930	Insignificant

*The mass numbers of these isotopes and their actual masses rounded off to the nearest atomic mass unit (amu) are the same value. This equality is true for all the isotopes of all the elements.

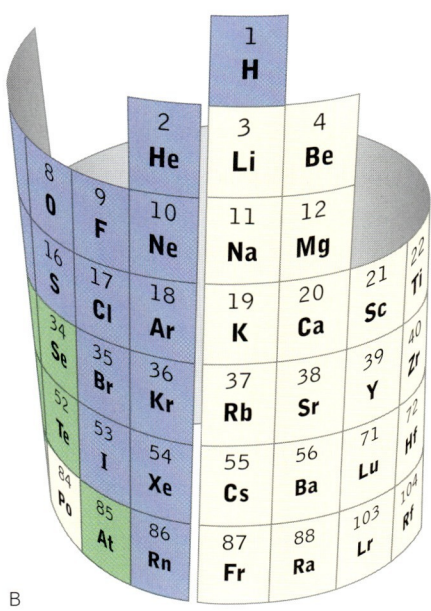

FIGURE 1.26 A. In the periodic table of the elements, the elements are arranged in order of increasing atomic number and according to their chemical properties. Being able to interpret this arrangement is a critical skill in gaining an understanding of chemistry. B. In this view of the periodic table, the sides have been joined together to illustrate the sequence of the elements from one row to the next and the connections between elements on the left and right sides of the table. For example, the chemical properties of potassium ($Z = 19$) are closely linked to the fact that its atoms have one more proton and one more electron than those of argon ($Z = 18$).

The information in the last two columns of Table 1.5 can be used to calculate the average mass of all hydrogen atoms. This calculation will yield a weighted average, which means that we do not simply average the mass values of the two stable isotopes (and get an average of 1.5 amu). Instead, we calculate an average that reflects the relative abundances of the isotopes. To do the calculation, we multiply the values in the last two columns of Table 1.5 and then add up these products:

$$\text{average atomic mass of hydrogen} = (1.007825 \times 0.99985) + (2.0141079 \times 0.00015)$$

$$= 1.0079 \text{ amu (to five significant figures)}$$

This average is not much bigger than the mass of the lightest isotope, ^1_1H. This result makes sense given that the vast majority of all hydrogen atoms *are* ^1_1H.

SAMPLE EXERCISE 1.9: The precious metal platinum has five isotopes with the following natural abundances:

Symbol	Mass	Relative abundance
^{192}Pt	191.96	0.008
^{194}Pt	193.96	0.329
^{195}Pt	194.96	0.338
^{196}Pt	195.96	0.253
^{198}Pt	197.97	0.072

Using these data, calculate the average atomic mass of platinum.

SOLUTION: The atomic mass of an element is the weighted average of the masses of its isotopes. To calculate the weighted average, we need to sum the products of the mass of each isotope times its abundance:

$$(191.96)(0.008) + (193.96)(0.329) + (194.96)(0.338) + (195.96)(0.253) + (197.97)(0.072) = 195.1$$

PRACTICE EXERCISE: There are only two stable isotopes of silver (Ag): ^{107}Ag has a mass of 106.90; ^{109}Ag has a mass of 108.90. If the atomic mass of silver is 107.87, what is the percent abundance of each of the isotopes? Hint: Let x be the relative abundance of one of the isotopes. Then $(1 - x)$ will be the relative abundance of the other. (See problems 67–70.)

> **Percent composition** is the percentage by mass of each component in a mixture or each element in a compound.

Cosmologists believe, on the basis of the stability of nuclei, that α particles made up 24% of the mass in the universe when atoms began to form about 10^4 years after the Big Bang. When two electrons orbit an α particle, they form a neutral helium atom. And so, by the time the universe was 10^4 years old, essentially all of the matter in it consisted of just two elements: hydrogen (76% of the mass of the universe) and helium (24%). Said another way, the **percent composition** of

   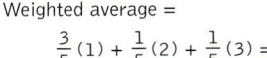 

Weighted average =

$$\frac{3}{5}(1) + \frac{1}{5}(2) + \frac{1}{5}(3) = 1.6$$

Average mass =

$$\frac{8}{5} = 1.6$$

The average mass of five spheres, three with a mass of one, and one each with a mass of two and three is 1.6. Similarly, the weighted average mass of an element is calculated from the abundance and mass of each isotope.

the early universe was 76% hydrogen and 24% helium. Today, hydrogen and helium still make up more than 99% of the mass of the observable universe. Moreover, the 76:24 mass ratio of hydrogen to helium predicted by the theory of the Big Bang is close to the hydrogen:helium ratio that astronomers observe in stars as they form today.

If the ratio of helium to hydrogen in the early universe was 24:76 by mass, what was the ratio of helium to hydrogen in terms of numbers of atoms of each element? To answer this question, we need to keep in mind that the mass of an atom of helium (4.003 amu) is nearly four times the average mass of a hydrogen atom (1.008 amu).

If we let x be the number of hydrogen atoms in the universe and y the number of helium atoms, then the total mass of each is $1.008x$ and $4.003y$. Applying these expressions to the given ratio of the mass of helium to that of hydrogen, we have

$$\frac{4.003y}{1.008x} = \frac{24}{76}$$

Cross multiplying, we get:

$$304.228y = 24.192x$$

Solving for y/x, we have

$$y/x = 0.0795$$

Thus, there were nearly 8 atoms of helium for every 100 atoms of hydrogen in the universe 10^4 years after the Big Bang.

If the element-building process had ended at this stage, there would be only two elements: hydrogen and helium gas. What processes led to the formation of the other elements? Cosmologists believe that the others did not form until stars and galaxies had formed. However, galaxies could have formed from an expanding cloud of hot hydrogen and helium only if

> What processes led to the formation of the other elements?

these gases did not expand and cool uniformly. Instead the expansion must have produced eddies and swirls of matter of different densities. A cluster of high density with more matter (and more mass) per unit volume would have a larger gravitational field and so would attract more matter into itself. Cycles of more matter → more gravity → more matter could have produced clusters of hot gases that eventually formed the first galaxies and stars.

Cold microwaves

Until the 1990s, there was no direct proof of clustering in the early universe. On the contrary, there was evidence that the universe had expanded smoothly and uniformly. The most compelling evidence of this sort was (and is) microwaves (see Figure 1.12). Today, a very low level of microwave radiation, called cosmic background radiation, permeates the universe. This radiation was discovered in the early 1960s. The story of its discovery provides a view of how major advances in science may happen: through careful observation, through communication between scientists sometimes with different research interests, and with the understanding that sometimes scientists discover things they weren't even looking for.

In the early 1960s, the United States launched the first communication satellites called *Echo* and *Telstar*. These early satellites were basically reflective spheres designed to bounce microwave signals from transmitters to receivers back on Earth. An antenna designed to receive such signals had been built at a Bell Laboratories research facility in New Jersey. Two scientists, Robert W. Wilson and Arno A. Penzias, were working to improve its reception when they encountered a problem. No matter where they directed their antenna it picked up background "noise" much like the static from a radio. They assumed that the noise came from their antenna or the instruments connected to it. At one point they suspected its source to be a pair of pigeons roosting on the antenna and coating parts of it with their droppings but when the droppings were cleaned up, the problem persisted.

Then, in 1964 Robert Dicke, a theoretical physicist at Princeton University, presented a seminar on the Big Bang theory, sharing his ideas about a cosmic microwave background with colleagues at Johns Hopkins University in Baltimore. Through a series of telephone calls between scientists at Johns Hopkins, the Carnegie Mellon Institute in Pittsburgh, the Massachusetts Institute of Technology in Cambridge, and Bell Labs in New Jersey, the theory of a cosmic background and the observations of Penzias and Wilson came together. The two scientists realized that their antenna was not malfunctioning. It was actually detecting a faint echo of the Big Bang. In recognition of their discovery, Penzias and Wilson were awarded the Nobel Prize in physics.

Continuum radiation

How do cosmic microwaves relate to the Big Bang? To understand the connection, we need to examine the relation between the spectra emitted by hot objects, such as the sun, and their temperatures.

Sources of electromagnetic radiation, such as our sun or the incandescent light bulbs that illuminate our homes at night, produce rainbows of colors. The radiant energies produced by these two sources are plotted against wavelength

> ✓ Cosmic background refers to microwave radiation found throughout the universe and believed to be an echo of the Big Bang. It was predicted to exist but was later discovered by accident in the 1960s.

Robert Dicke predicted the existence of a cosmic microwave background. His prediction was confirmed by Robert Wilson and Arthur Penzias.

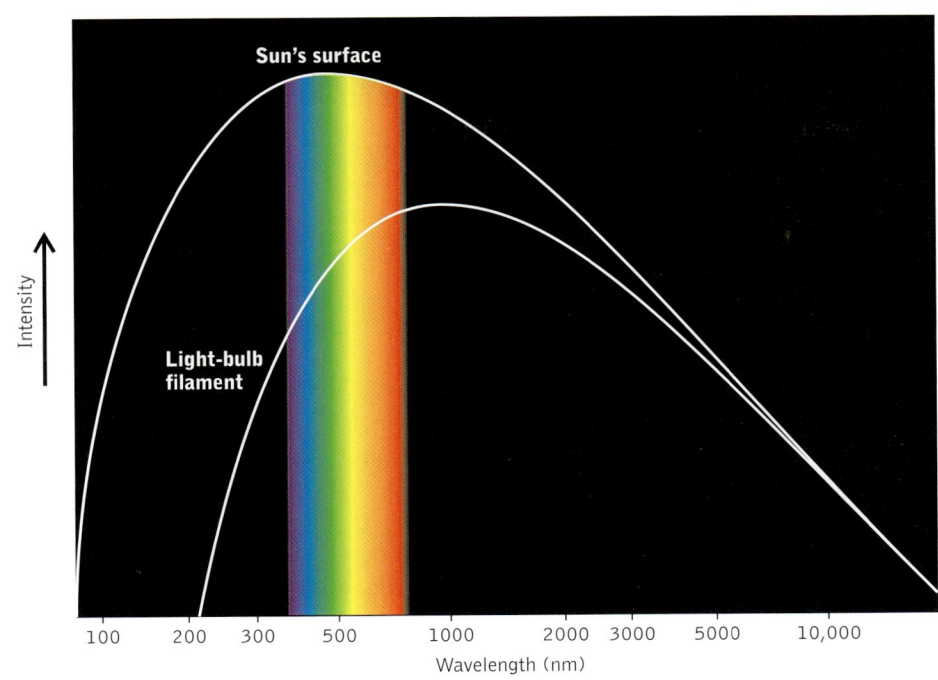

FIGURE 1.27 Both the sun and the 3000-K filament used in a light bulb radiate throughout the visible region; so both sources are "white hot." However, there are proportionately fewer of the short (blue and violet) wavelengths in the filament's spectrum, which is why photographs of white objects illuminated by incandescent light bulbs have a reddish yellow tint.

in Figure 1.27. Note the broad spectral profiles of the sun and light-bulb filament. They emit light *throughout* the visible region and beyond. As mentioned earlier, such broad spectral profiles are referred to as *continuum* radiation. Also note that the tops of the two curves are not at the same wavelength. The profile for the filament reaches a maximum at a longer wavelength and has proportionately more red than blue light. You may have seen evidence of this profile in color photographs taken at night, with illumination only from incandescent lights. The colors in such photographs do not look natural: white objects appear yellow and the faces of people with pale skin usually have a ruddy complexion. These effects are due to the fact that most photographic film is manufactured to accurately reproduce the colors of *sunlit* objects.

> How do cosmic microwaves relate to the Big Bang?

Why do light bulbs and the sun produce different spectral profiles? At 5700 K, the surface of the sun is nearly twice as hot as the filament inside a switched-on light bulb. The hotter a continuum source is, the more radiant it is, and the shorter the wavelength corresponding to the maximum wavelength (λ_{max}) in its spectral profile. A simple equation relates λ_{max} to temperature (T):

$$\lambda_{max} = \frac{2.897 \times 10^6 \text{ nm}}{T} \quad (1.12)$$

where T is in kelvins. We can use this equation to trace the shifting spectrum of the universe as it cooled and expanded after the Big Bang.

> ✓ The hotter a continuum source is, the shorter the wavelengths of radiation that it emits.

SAMPLE EXERCISE 1.10: Protons and neutrons combined to form the nuclei of the lightest elements as the universe cooled below 10^9 K. What form of electromagnetic radiation was released by these processes?

SOLUTION: To answer this question, let's turn it around and ask how much energy might be needed to break apart an atomic nucleus into its component nucleons. Presumably, the energy corresponding to a temperature of 10^9 K is needed. If this energy came from a continuum source, we could calculate λ_{max} directly from temperature as follows:

$$\lambda_{max} = \frac{(2.897 \times 10^6)}{T}$$

$$= \frac{(2.897 \times 10^6)}{(1 \times 10^9)}$$

$$= 3 \times 10^{-3} \text{ nm, or } 3 \times 10^{-12} \text{ m}$$

A check of the electromagnetic spectrum in Figure 1.12 tells us that this wavelength is in the γ-ray region. It is not surprising, then, that the electromagnetic radiation given off by nuclear reactions consists of γ rays.

PRACTICE EXERCISE: What wavelength of radiation is emitted by an object with a temperature of only 2.7 K? (See Problems 81 and 82.)

A **plasma** is a hot, extensively ionized gas.

We noted at the beginning of this section that the temperature of the universe (essentially a hydrogen–helium **plasma**) had fallen to about 10^4 K, by a hundred thousand (10^5) years after the Big Bang. If we assume that the plasma was a continuum source of electromagnetic radiation, this temperature corresponded to a λ_{max} of 3×10^2 nm. The continuum emission of the plasma would have spanned the ultraviolet, visible, and infrared regions. In other words, the universe at a hundred thousand years, with a temperature almost twice that of the surface of the sun, was *white-hot*.

Continued expansion and cooling allowed neutral atoms of hydrogen and helium to form. At first, the lifetimes of these atoms in a white-hot universe would have been brief. If a hydrogen atom encountered radiation with a wavelength much shorter than 100 nm, the energy of the radiation could break up the atom into a free electron and a proton. This process is an example of **ionization**. This term applies to any reaction in which a neutral atom (or molecule) gives up an electron and forms a positively charged ion. When the energy needed to separate the electron comes from a photon of electromagnetic radiation, the process may also be called **photoionization**.

Ionization is the process by which an atom or a molecule releases an electron and forms a positively charged ion. When the energy to produce ionization comes from electromagnetic radiation, the process is called **photoionization**.

When a free electron combines with a proton to form a hydrogen atom, the two particles give up their independence and some freedom of motion, and they neutralize each other's electrical charges. The energy given up in this process is emitted as electromagnetic radiation. Thus, the early universe was marked by a close coupling between matter and electromagnetic radiation: atoms absorbed high-energy radiation and were ionized; ions and electrons recombined to form atoms and emitted high-energy radiation. Atoms did not stay intact for very long, and radiation did not travel very far before being absorbed.

CONCEPT TEST: It takes much more energy to ionize helium than it does to ionize hydrogen. Can you think of a reason why?

With further expansion and cooling, the white-hot universe lost much of its radiance, and there was no longer enough energy available to ionize hydrogen. As a result, electromagnetic radiation was able to travel with fewer interruptions caused by interactions with matter.

For billions of years, the universe has continued to expand, and its interstellar space has continued to cool (Figure 1.28). During this time, the spectrum of space has shifted through the ultraviolet, visible, and infrared regions. Theoreticians predicted as long ago as 1948 that, if the Big Bang theory were true, the temperature of interstellar space should have dropped to about 2.7 K. A continuum source even at this extremely low temperature would still produce a faint "glow," but with a λ_{max} in the microwave region—the very microwaves detected by Penzias and Wilson.

The discovery of cosmic microwaves simultaneously reinforced the Big Bang theory and created an inconsistency in it. Recall that Wilson and Penzias detected these microwaves no matter where they pointed their antenna. If the afterglow from the Big Bang really was uniform throughout the galaxies, then how could the galaxies have formed in the first place? Scientists thought that clustering in the early universe should be observable in small differences in the cosmic background. Differences in intensity of as little as 1 part in 10,000, or 0.01%, might represent enough heterogeneity to account for galaxy formation. Microwave antennae on Earth could not pick up such tiny differences, because of atmospheric interferences. The only way to find them, if they did exist, was from an antenna in space.

In late 1989, the United States launched the *Cosmic Background Explorer* (COBE) satellite to obtain spectra of interstellar cosmic microwaves. In April 1992, the results of measurements taken by the COBE satellite (Figure 1.29)

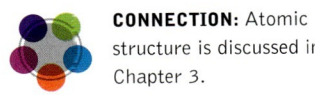

CONNECTION: Atomic structure is discussed in Chapter 3.

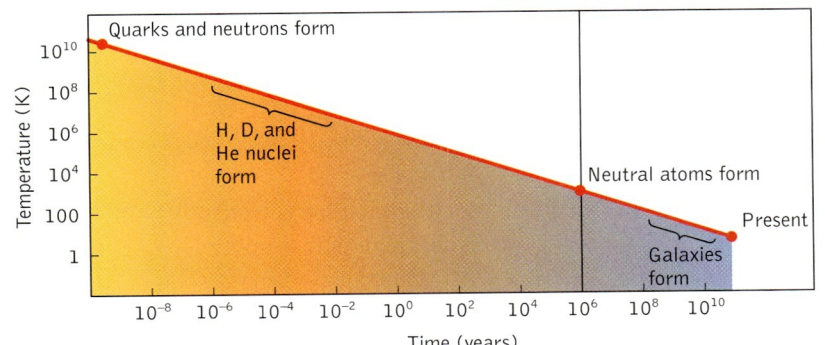

FIGURE 1.28 The early universe was extremely hot, too hot in fact for neutral (uncharged) atoms to exist. Only nuclides survived these conditions. As the universe cooled through millions of years, atoms and, eventually, stars and galaxies began to form.

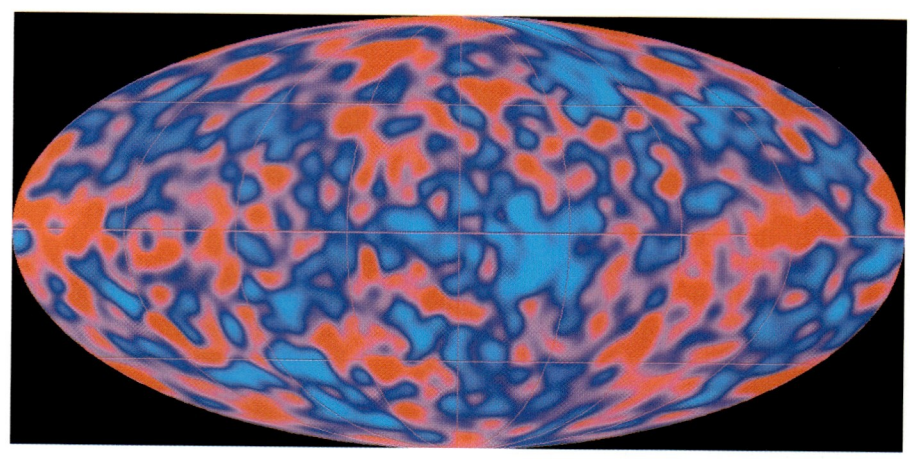

FIGURE 1.29 The map of the cosmic microwave background is a 360-degree image of the sky made by collecting microwave signals for a year from the COBE satellite's sensitive microwave telescopes. The map shows the plane of the Milky Way galaxy horizontally with the center of our galaxy at its center. The colors represent temperature variations, with red indicating regions that are 0.01% warmer and blue indicating regions that are 0.01% cooler than the average temperature of 2.7 K.

were published in newspapers and magazines throughout the world. At a news conference in Washington, D.C., the lead scientist on the COBE project called the map a "fossil of creation." In his excitement he added, "If you're religious, it's like looking at God."

COBE's measurements and those obtained more recently from newer satellites show that, indeed, the universe did not expand and cool uniformly. There were blobs and ripples in the microwaves perhaps due to galaxy "seed" clusters that formed as the universe cooled to less than 10^4 K. Thus, scientists believe they have discovered a record of the next step in the creation of matter as we know it today. In this next phase, the first galaxies and stars formed and, within them, the remaining natural elements were made.

CHAPTER REVIEW

Summary

SECTION 1.1

We have begun our study of chemistry by looking for the connections between the substances, compounds, elements, atoms, molecules, and ions that make up our everyday existence and what is known about the universe, how it probably was formed, and how it is changing. We have learned some distinctions between physical and chemical properties, as well as between physical and chemical changes, and that elements are classified as metals, nonmetals, and metalloids. Filtration and distillation are major methods of separating mixtures to give pure substances.

SECTION 1.2

Our observations and conclusions are guided by the scientific method, which is rooted in logic and depends on the interpretation of measurements to formulate theories.

SECTION 1.3

Electromagnetic radiation is transmitted in the form of waves with characteristic amplitude, wavelength, and frequency. The transmission speed of all electromagnetic energy is the speed of light. Waves have the property of constructive and destructive interference, depending on whether their amplitudes are in or out of phase. This interference is the origin of a kind of bending of light called diffraction. Particles of electromagnetic radiation are called photons, the building blocks of light whose energy is given by $E = h\nu = hc/\lambda$, where h is Planck's constant, ν is the frequency, c is the speed of light, and λ is the wavelength.

The Doppler effect is the shift in frequency of electromagnetic radiation (or sound) caused by movement of its source either toward (blueshift) or away from (redshift) an observer.

SECTION 1.4

The results of scientific measurements are given in SI units. An example is the meter, the unit of length that is about 10% longer than a yard. Data are converted from one unit into another with the use of unit conversion factors. Prefixes indicate the fraction or multiple of a measurement unit: 1 km is 10^3 m and 1 mm is 10^{-3} m. Significant figures indicate the precision of a measurement. When we combine or manipulate numbers, the least-precise number determines the precision (number of significant figures) in the result. Accuracy indicates how close a measurement is to the true value represented by a standard. Density is the ratio of the mass of an object to its volume.

SECTION 1.5

Energy and matter are related by $E = mc^2$, where E is energy and m is mass. A free neutron, ^1_0n (with zero electrical charge), decays with a half-life of 12 seconds to give a positively charged proton, ^1_1p, and a negatively charged electron, $^{\;\;0}_{-1}e$.

SECTION 1.6

Temperature, a measure of the intensity of heat, is expressed by using related Fahrenheit, Celsius, and Kelvin (the SI unit) scales. As the early universe cooled, positively charged protons, zero-charged neutrons, and negatively charged electrons combined to form atoms such as hydrogen and helium. All the atoms of an element contain the same number of protons, called the element's atomic number. But atoms of the same element can contain different numbers of neutrons. The nuclei of atoms that differ in this way are called nuclides, and the whole atoms are called isotopes. The percentages of the isotopes of an element determine its average atomic mass.

Key Terms

absolute zero (p. 37)
accuracy (p. 30)
amplitude (p. 15)
anion (p. 6)
atom (p. 2)
atomic mass unit (p. 32)
atomic number (p. 38)
basic forces of nature (p. 32)
cation (p. 6)
chemical property (p. 2)
chemistry (p. 2)
compound (p. 2)
continuum radiation (p. 12)
constructive interference (p. 16)
density (p. 2)
destructive interference (p. 16)
deuteron (p. 35)
diffraction (p. 16)
distillation (p. 6)

Doppler effect (p. 20)
electromagnetic radiation (p. 12)
element (p. 2)
filtration (p. 5)
frequency (p. 14)
half-life (p. 34)
hypothesis (p. 9)
infrared (p. 13)
ion (p. 6)
ionic compound (p. 6)
ionization (p. 44)
isotope (p. 38)
kelvin (p. 37)
mass number (p. 35)
metal (p. 3)
metalloid (p. 3)
mixture (p. 5)
molecular compound (p. 4)
molecular formula (p. 4)

molecule (p. 4)
nonmetal (p. 3)
nucleon (p. 35)
nuclide (p. 38)
percent composition (p. 40)
photoionization (p. 44)
photon (p. 19)
physical property (p. 2)
plasma (p. 44)
precision (p. 30)
pure substance (p. 2)
radioactive decay (p. 32)
refraction (p. 12)
scientific method (p. 9)
spectrum (p. 12)
theory (p. 9)
ultraviolet (p. 13)
volatile (p. 7)
wavelength (p. 14)

Key Skills and Concepts

SECTION 1.1

Distinguish between mixtures and pure substances, between compounds and elements, and between molecules, atoms, and ions.

Understand density as an example of a physical property of a sample.

Distinguish between physical and chemical properties and between physical and chemical changes.

Describe the distinguishing properties of metals, nonmetals, and metalloids.

Understand the concepts of a molecule and a molecular formula (e.g., CO_2).

Describe two methods of separating mixtures.

Distinguish between an atom and an ion and between a cation and an anion.

SECTION 1.2

Explain the difference between a theory and a hypothesis.

SECTION 1.3

Describe different spectral regions of the electromagnetic spectrum.

Illustrate the amplitude, frequency, speed, and wavelength of a wave.

Explain the constructive and destructive interference of waves.

Describe the relation between the energy of a photon and its frequency.

Explain how the redshifted frequency of a retreating object can be used to estimate its velocity.

SECTION 1.4

Name the SI units for length, mass, temperature, time, energy, and electrical charge and their commonly used prefixes.

Understand the importance of significant figures in working with data, and how they are taken into account when manipulating scientific data (addition, subtraction, multiplication, and division).

Be able to convert from one set of units into another by using the concept of unit conversion factors.

Clearly distinguish between precision and accuracy.

SECTION 1.5

Understand the concepts of radioactive decay and radioactive half-life.

Know the nuclear symbols for a neutron, a proton, and an electron, and give their relative masses and relative charges.

Balance equations for simple nuclear processes.

SECTION 1.6

Be able to interconvert a Fahrenheit, Celsius, and absolute (Kelvin) temperature.

Differentiate between the nuclides and the isotopes of an element.

Differentiate between the atomic number and the mass number of a nuclide or an isotope.

Be able to calculate the atomic weight of an element from isotope abundance data.

Calculate the temperature of a hot object from the wavelength of maximum emission in its spectrum.

Explain the concepts of ionization and photoionization.

Key Equations and Relations

SECTION 1.3

To relate the speed (c) of a light wave to its wavelength (λ) and frequency (ν):

$$c = \lambda \nu \quad (1.1)$$

To relate the energy (E) of a photon to its frequency (ν) and wavelength (λ), where h is Planck's constant (a universal constant):

$$E = h\nu = hc/\lambda \quad (1.2 \text{ and } 1.3)$$

To relate the perceived frequency of a source (ν') to its natural frequency (ν) and the speed of light (c):

$$(\nu - \nu')/\nu = v/c \quad (1.4)$$

SECTION 1.4

To relate the density (d) of an object to mass (m) and volume (V):

$$d = m/V$$

SECTION 1.5

To relate the initial amount A_0 of a radioactive nuclide to A_t remaining after the number of half-lives n for the decay process have elapsed:

$$A_t/A_0 = 0.5^n \quad (1.7)$$

SECTION 1.6

To convert from a temperature measured in Fahrenheit (°F) units into the corresponding Celsius temperature (°C):

$$°C = 5/9(°F - 32) \quad (1.10)$$

To convert from a temperature measured in Celsius units into an absolute (Kelvin) temperature:

$$K = °C + 273.15 \quad (1.11)$$

To calculate the average atomic mass x of an element X that has isotopes X_1, X_2, X_3, and so on:

$$x = f_1 m_1 + f_2 m_2 + \cdots$$

where f_1 is the fractional abundance (measured to be, say, 0.45) of isotope 1 and m_1 is its atomic mass, and so on.

To relate the maximum wavelength (λ_{max}, in nanometers) in the spectrum of a hot object to its temperature (T):

$$\lambda_{max} = 2.897 \times 10^6 \text{ nm}/T \quad (1.12)$$

QUESTIONS AND PROBLEMS

Matter

CONCEPT REVIEW

1. Give an example of each of the following: (a) matter; (b) an element; (c) a compound; (d) a metal; (e) a nonmetal; (f) a cation; (g) an anion.
2. Distinguish between a substance, a compound, and an element.
3. List four physical properties of a typical metal.
4. Explain how filtration can be used to separate salt from sand.
5. Explain how distillation can be used to desalinate seawater.
6. Which of the following processes entails a chemical change? (a) distillation; (b) combustion; (c) filtration; (d) condensation.

PROBLEMS

7. Indicate which of the following is a pure substance: (a) iron (Fe); (b) gasoline; (c) cow's milk; (d) sodium chloride (NaCl).
8. Indicate which of the following is a pure substance: (a) dry ice; (b) distilled water; (c) helium; (d) filtered seawater.
9. Indicate whether each of the following properties is a physical or a chemical property of sodium (Na).
    a. Its density is greater than that of kerosene and less than that of water.
    b. It has a lower melting point than those of most metals.
    c. It is an excellent conductor of heat and electricity.
    d. It is soft and can be easily cut with a knife.
    e. Freshly cut sodium is shiny, but it rapidly tarnishes in contact with air.
    f. It reacts very vigorously with water to form hydrogen gas (H_2) and sodium hydroxide (NaOH).
10. Indicate whether each of the following is a physical or chemical property of hydrogen gas (H_2).
    a. At room temperature, its density is less than that of any other gas.
    b. It reacts vigorously with oxygen (O_2) to form water.
    c. Liquefied H_2 boils at a very low temperature ($-253°C$).
    d. H_2 gas does not conduct electricity.
11. Indicate which of the following is a heterogeneous mixture: (a) iron (Fe); (b) gasoline; (c) cow's milk; (d) sodium chloride (NaCl); (e) sterile saline (a solution of NaCl in water); (f) orange juice; (g) nitrogen gas (N_2); (h) ammonia gas (NH_3).
12. Indicate which of the following is a homogeneous mixture: (a) copper; (b) rainwater; (c) cow's milk; (d) sodium chloride (NaCl); (e) filtered air; (f) apple juice; (g) hydrogen gas (H_2).
13. How would you separate ethanol (boiling point, 80°C) from water (boiling point, 100°C). Sketch the equipment used for the separation
14. How would you separate crude oil from sand? Sketch the equipment used for the separation.

The Scientific Method

CONCEPT REVIEW

15. What kinds of information are needed to formulate a hypothesis?
16. How does a hypothesis become a theory?
17. Is it possible to prove that a scientific theory is wrong?
18. Why is the theory that matter consists of atoms universally accepted?
19. Explain how Hubble's observations led to the belief that ours is an expanding universe.
20. How would you convince someone that Earth is round?

Waves

CONCEPT REVIEW

21. When X-ray images are taken of your teeth and gums in your dentist's office, your body is covered with a lead shield. Explain the need for this precaution.
22. Explain with a sketch why long-wavelength waves have lower frequencies than those of short-wavelength waves.
23. Ultraviolet radiation causes skin damage that may lead to cancer, but exposure to infrared radiation does not seem to cause skin cancer. Why do you think this is so?
24. How do the wave properties of radiation account for the phenomenon of diffraction?
25. A neon light emits radiation of $\lambda = 616$ nm. Calculate the frequency of this radiation.

PROBLEMS

26. A communications network for submarines that is based on ultralong wavelength (126 km) signals has been proposed. What is the frequency of these signals?
27. FM radio stations broadcast at different frequencies. Calculate the wavelengths corresponding to the broadcast frequencies of the following radio stations: (a) KKNB (Lincoln, NE), 104.1 MHz; (b) WFNX (Boston, MA), 101.7 MHz; (c) KRTX (Houston, TX), 100.7 MHz.
28. Which radiation has the longer wavelength: radio waves from an AM radio station broadcasting at 680 kHz or infrared radiation emitted by the surface of Earth ($\lambda = 15$ μm)?
29. Which radiation has the lower frequency: radio waves from an AM radio station broadcasting at 1090 kHz or the green light ($\lambda = 550$ nm) from an LED (light-emitting diode) on a stereo system?
30. Which radiation has the higher energy: red light from fireworks ($\lambda = 671$ nm) or signals transmitted by a cellular telephone ($\nu = 10^{11}$ s^{-1})?
31. A physicist was ticketed for going through a red light ($\lambda = 700$ nm). In court, the physicist claimed that the Doppler effect made the light seem green ($\lambda = 550$ nm) at the speed at which he had been traveling. At what speed would the physicist need to have been traveling to see this wavelength shift?
32. A space probe equipped with a microwave transmitter is traveling away from Earth at 6.7×10^5 m/s. Will the apparent wavelength of its transmissions be redshifted by more than 0.1%?

Measurements and Unit Conversions

CONCEPT REVIEW

33. List the SI base units of mass, length, temperature, time, and amount of substances.
34. Which is a larger quantity: 110 μg or 1.00 mg?

PROBLEMS

35. The speed of light in a vacuum is 2.9979×10^8 m/s. Calculate the speed of light in miles per hour.
36. Women runners must run a marathon race (26.2 miles) in less than 3 hours and 40 minutes to qualify for the Boston Marathon. To qualify, what must a woman's average speed be (a) in miles per hour and (b) in meters per second?
37. An Olympic "mile" is actually 1500 m. What percentage is an Olympic mile of a real mile (5280 feet)? (1 inch = 2.54 centimeters)
38. The price of a popular soft drink is $1.00 for 24 oz or

$0.75 for 0.50 L. Which is a better buy? (1.000 L = 1.057 qt; 1 qt = 32 oz)

39. If the price of gasoline is $1.25/gal in the United States and $0.35/L (in U.S. currency) in Canada, in which country is gasoline cheaper? (1.000 L = 1.057 qt; 1 gal = 4 qt)

40. The level of water in a rectangular swimming pool needs to be lowered 6.0 inches. If the pool is 40. ft long and 16. ft wide and the water is pumped out at a rate of 5.2 gal/min, how long will the pumping take? (1 ft^3 = 62.4 gal)

41. If a wheelchair-marathon racer moving at 13.1 miles/hour expends energy at a rate of 665 Calories/hour, how much energy in Calories would be required to complete a marathon race (26.2 miles) at this pace?

42. A sports utility vehicle has an average mileage rating of 18 miles/gallon. How many gallons of gasoline are needed for a 389-mile trip?

43. A single strand of natural silk may be as long as 4.0×10^3 m. Convert this length into feet. Convert it into miles.

44. The Ford GT40 sports cars of the 1960s were powered by engines with 427. cubic inches of displacement. A new generation of GT40s are to be powered by 5.4-liter engines. How much bigger (%) were the old engines?

45. The density of magnesium is 1.74 g/cm^3. What is the mass of a magnesium block that measures 2.5 cm × 3.5 cm × 1.5 cm?

46. Osmium has a density of 22.59 g/cm^3, making it the most dense element known. What is the mass of an osmium block that measures 6.5 cm × 9.0 cm × 3.25 cm? Do you think you could lift it with one hand?

47. The average density of Earth is 5.5 g/cm^3. The mass of Venus is 81.5% of Earth's mass, and the volume of Venus is 88% of Earth's volume. What is the density of Venus?

48. Earth has a mass of 6.0×10^{27} g and an average density of 5.5 g/cm^3.
    a. What is the volume of Earth in cubic kilometers?
    b. Geologists sometimes express the "natural" density of Earth after doing a calculation that corrects for "gravitational squeezing." Should the natural density be more or less than the observed value calculated in part *a*?

49. A small plastic cube is 1.2×10^{-5} km on a side and has a mass of 1.10×10^{-3} kg. Water has a density of 1.00 g/cm^3. Will the cube float on water?

50. The sun is estimated to have a mass of 2×10^{36} kg. Assume it to be a sphere of average radius 6.96×10^5 km and calculate the average density of the sun in units of grams per cubic centimeter (the volume of a sphere is $4/3 (\pi r^3)$.

51. Diamonds are measured in carats, where 1 carat = 0.200 g. The density of diamond is 3.51 g/cm^3. What is the volume of a 5.0-carat diamond?

52. If the concentration of mercury in the water of a polluted lake is 0.33 μg per liter of water, what is the total mass (kg) of mercury in the lake? The lake has a surface area of 100 square miles and an average depth of 20 feet.

53. The following cartoon applies accuracy and precision to the measurement of body weight.
    a. Give definitions of accuracy and precision.
    b. Is the lawyer using the terms correctly?
    c. Is it possible to be "precisely accurate"?
    d. What does the sign "Accurate Weight to the nearest pound" say about the uncertainty in the measurements?

54. Three different analytical techniques were used to determine the quantity of sodium in a Mars Milky Way Dark candy bar. Each technique was used to analyze five portions of the same candy bar with the following results expressed in milligrams of sodium per candy bar:

Technique 1	Technique 2	Technique 3
109	110	114
111	115	115
110	120	116
109	116	115
110	113	115

The actual quantity of sodium in the candy bar was

115 mg. Comment on the accuracy and precision of the three techniques.
55. Which of the following quantities has four significant figures?
    a. the number of eggs in a dozen
    b. 0.08206
    c. 8.314
    d. 5400.
    e. 5.4×10^3
    f. 5400.0
56. Which of the following numbers have just three significant figures?
    a. 7.02
    b. 6.452
    c. 302
    d. 0.0821
    e. 6.02×10^{23}
    f. 12.77
    g. 3.43
57. Perform each of the following calculations and express the answer with the correct number of significant figures:
    a. $0.6274 \times 1.00 \times 10^3/[2.205 \times (2.54)^3] =$
    b. $6 \times 10^{-18} \times (1.00 \times 10^3) \times 17.4 =$
    c. $(4.00 \times 58.69)/(6.02 \times 10^{23} \times 6.84) =$
    d. $[(26.0 \times 60.0) + 43.53]/(1.000 \times 10^4) =$
58. Perform each of the following calculations and express the answer with the correct number of significant figures:
    a. $[(12 \times 60.0) + 55.3]/(5.000 \times 10^3) =$
    b. $(2.00 \times 183.9 \text{ g/mole})/[6.02 \times 10^{23} \text{ atoms} \times (1.61 \times 10^{-8})^3] =$
    c. $0.8161/[2.205 \times (2.54)^3] =$
    d. $(9.00 \times 60.0) + (50.0 \times 60.0) + (3.00 \times 10^1) =$

Matter and Energy

CONCEPT REVIEW

59. Write the symbols for (a) an electron; (b) a proton; (c) a neutron; (d) a deuteron; (e) an α particle.
60. 100.0 g of a radioactive element is decaying. How many grams of the element are left after five half-lives?

PROBLEMS

61. A radioactive isotope of the element cobalt (^{60}Co) is used in hospital radiation-therapy units. It has a half-life of 5.26 yr. If a hospital installs a new ^{60}Co source on January 1, how much of it will remain at the end of May (180 days later)?
62. Radioactive radon-222 decays with a loss of one α particle. The half-life of the process is 3.82 days. What percentage of the radon in a sealed vial would remain after 1 week (7.0 days)?

Average Atomic Mass

CONCEPT REVIEW

63. What is meant by a *weighted average*?
64. Explain the difference between an isotope and a nuclide.
65. Explain how isotope abundances are related to atomic masses.
66. Boron, lithium, nitrogen, and neon each have two stable isotopes. In which of the following pairs of isotopes is the heavier isotope more abundant?
    a. ^{10}B or ^{11}B (average mass, 10.81 amu)
    b. ^{6}Li or ^{7}Li (average mass, 6.941 amu)
    c. ^{14}N or ^{15}N (average mass, 14.01 amu)
    d. ^{20}Ne or ^{22}Ne (average mass, 20.18 amu)

PROBLEMS

67. Naturally occurring copper contains a mixture of 69.09% copper-63 (62.9298 amu) and 30.91% copper-65 (64.9278 amu). What is the average atomic mass of copper?
68. Naturally occurring chlorine consists of two isotopes: ^{35}Cl, 75.5%, and ^{37}Cl, 24.5%. Calculate the average atomic mass of chlorine.
69. Naturally occurring sulfur consists of four isotopes: ^{32}S (31.97207 amu, 95.0%); ^{33}S (32.97146 amu, 0.76%); ^{34}S (33.96786 amu, 4.22%); and ^{36}S (35.96709 amu, 0.014%). Calculate the average atomic mass of sulfur in atomic mass units.

70. The 1997 mission to Mars included a small robot, the *Sojourner*, with the capacity to analyze the composition of rocks found on the "Red Planet." Magnesium oxide from a boulder dubbed "Barnacle Bill" was analyzed and found to have the following isotopic composition. All of the oxygen is oxygen-16 with an exact mass of 15.9948. Show that the average mass of magnesium is 24.32.

Exact mass of MgO	Abundance (%)
39.9872	78.70
40.9886	10.13
41.9846	11.17

71. From the following table of abundances and masses of the three naturally occurring argon isotopes, calculate the exact mass of ^{40}Ar.

Isotope	Exact mass (amu)	Abundance (%)
^{36}Ar	35.96755	0.337
^{38}Ar	37.96272	0.063
^{40}Ar	?	99.60
Average	39.948	

72. From the following table of abundances and masses of five naturally occurring titanium isotopes, calculate the exact mass of ^{48}Ti.

Isotope	Exact mass (amu)	Abundance (%)
^{46}Ti	45.95263	7.93
^{47}Ti	46.9518	7.28
^{48}Ti	?	73.94
^{49}Ti	46.94787	5.51
^{50}Ti	49.9448	5.34
Average	47.90	

Temperature Scales

CONCEPT REVIEW

73. How is the temperature of a solid related to the wavelengths of electromagnetic radiation emitted by the solid?
74. What is meant by an *absolute* temperature scale?

PROBLEMS

75. Liquid helium boils at 4 K. What is the boiling point of helium in °C?
76. Liquid hydrogen boils at −253°C. What is the boiling point of H_2 in kelvins (K)?
77. A person has a fever of 102.5°F. What is this temperature in °C?
78. At what temperature do the Fahrenheit and Celsius scales meet (i.e., °F = °C)?
79. The lowest temperature measured on Earth is −128.6°F recorded at Vostok, Antarctica in July 1983. What is this temperature in °C and in kelvins?
80. The highest temperature ever recorded in the United States is 134°F at Greenland Ranch, Death Valley, California, on July 13, 1913. What is this temperature in °C and in kelvins?
81. The discovery of new "high temperature" superconducting materials in the mid-1980s spurred a race to prepare the material with the highest superconducting temperature. The critical temperatures (T_c) of $YBa_2Cu_3O_7$, Nb_3Ge, and $HgBa_2CaCu_2O_6$ are 92 K, −250°C, and −221.3°F, respectively. Convert these temperatures into a single temperature scale, and determine which superconductor has the highest T_c.
82. The boiling point of O_2 is −183°C; the boiling point of N_2 is 77 K. As air is cooled, which gas condenses first?
83. Calculate λ_{max} for: (a) the human body (about 37°C); (b) a pot of boiling water (100°C); (c) an automobile engine (2000 K); (d) a hydrogen bomb (10^7 K).
84. Calculate λ_{max} for: (a) a block of ice at 0.0°C; (b) a metal container filled with liquid nitrogen at 77 K; (c) liquid iron at 1350°C.

2 Nuclear Chemistry
And the Origins of the Elements

Supernova explosions have fascinated and frightened humans for centuries. Scientists believe that these massive explosions are sources of and distribution systems for most of the elements in the universe.

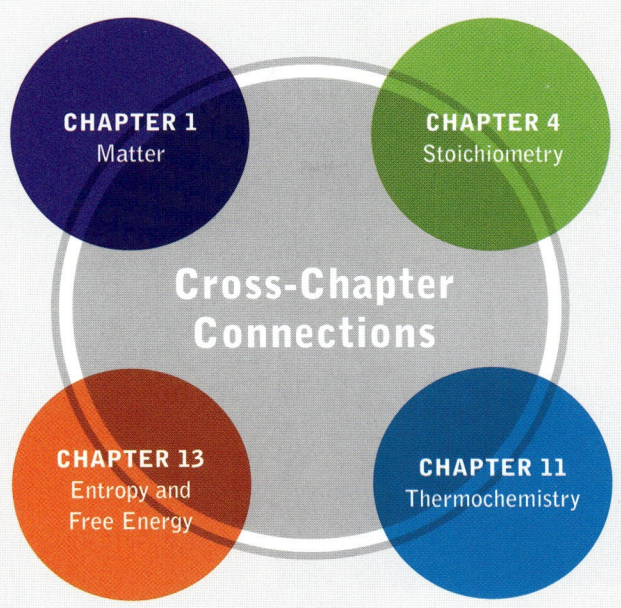

2.1	**Hydrogen Fusion**
2.2	**Nuclear Binding Energies**
2.3	**Neutron Capture and Modes of Radioactive Decay**
2.4	**Supernova: Radiochemistry of the Biggest Elements**
2.5	**Artificial Isotopes and Elements**
2.6	**Measuring Radioactivity**
2.7	**Biological Effects of Radiation**
	Ionizing radiation and living matter
	Radiation dosage
	Assessing the risks of radiation
	Radiation therapy
	BOX: THE CHEMISTRY OF RADON, RADIUM, AND URANIUM
	Medical imaging with radionuclides
2.8	**Radiochemical Dating**
2.9	**The Composition of the Universe**

A Look Ahead

In this chapter we examine the synthesis of the elements by fusion and other radiochemical processes in the cores of giant stars and in the laboratory. Then we consider the biological effects of nuclear radiation and the use of radioisotopes in medical imaging and treatment, as well as in determining the ages of natural and human-made objects.

2.1   HYDROGEN FUSION

Many scientists believe that, within a million years of the Big Bang, galaxies began to form. High-density, high-temperature clusters of hydrogen and helium became the birthing zones of stars: a process that still continues to create new generations of stars, as captured by the cameras of the Hubble Space Telescope (Figure 2.1). The energy of these stars, including our sun, is derived from the fusion of hydrogen nuclei (protons). Hydrogen fusion consists of a series of nuclear reactions beginning with the fusing of hydrogen nuclei ($^{1}_{1}\text{H}$) and the formation of deuterons—that is, the nuclei of deuterium ($^{2}_{1}\text{H}$) atoms. Let's try to write this nuclear reaction in equation form, starting with two hydrogen nuclei:

$$^{1}_{1}\text{H} + ^{1}_{1}\text{H} \longrightarrow ^{2}_{1}\text{H}$$

> The symbols of atomic isotopes are often used in writing nuclear equations even when the reactions take place at such high temperatures that the atoms are partly or completely ionized.

FIGURE 2.1 It was at least a million years after the Big Bang before the gases had cooled enough to allow star formation. Stars become the factories for elements beyond hydrogen, helium, and their isotopes. The star cluster NGC 2264 is one of the youngest star clusters known.

Comparing the masses (denoted by the superscripts—see Section 1.5) and the electrical charges on the three particles, we discover that the masses are in balance (1 + 1 = 2) but the sum of the charges of the particles on the two sides of the reaction arrow are not (1 + 1 ≠ 1). This inequality means that the equation is not balanced. To make it so, we need to add a product with a charge of +1 but with an insignificant mass. Such a particle exists. It is called a **positron.** It is represented by the symbol $^{0}_{1}\text{e}$, which means that it has the mass of an electron but that it has a positive charge. Adding a positron to the preceding reaction, we have a balanced equation:

> A **positron** is a particle with the mass of an electron but with a positive charge.

$$^{1}_{1}\text{H} + ^{1}_{1}\text{H} \longrightarrow ^{2}_{1}\text{H} + ^{0}_{1}\text{e}$$

or

$$2\,^{1}_{1}\text{H} \longrightarrow ^{2}_{1}\text{H} + ^{0}_{1}\text{e} \tag{2.1}$$

In the second stage of hydrogen fusion, a deuteron formed in the first reaction fuses with another proton to form the nucleus of a helium-3 ($^{3}_{2}\text{He}$) atom:

$$^{1}_{1}\text{H} + ^{2}_{1}\text{H} \longrightarrow ^{3}_{2}\text{He} \tag{2.2}$$

Finally, fusion of two helium-3 atoms produces a nucleus of $^{4}_{2}\text{He}$ (α particle) and two other protons:

$$^{3}_{2}\text{He} + ^{3}_{2}\text{He} \longrightarrow ^{4}_{2}\text{He} + ^{1}_{1}\text{H} + ^{1}_{1}\text{H} \tag{2.3}$$

or

$$2\,^{3}_{2}\text{He} \longrightarrow ^{4}_{2}\text{He} + 2\,^{1}_{1}\text{H}$$

> A substance that is formed in one step of a multistep process and consumed in a later step is an **intermediate.**

Deuterium and helium-3 nuclei are **intermediates** in the hydrogen-fusion process because they are made in one step but are then consumed in another. To gain a perspective on the overall process in which helium is formed from hydrogen fusion, we can combine Equations 2.1, 2.2, and 2.3. We do so by adding the content of the left-hand sides of the three equations together and then adding the content of the right-hand sides together. Then we cancel out the particles that appear on both the left- and right-hand sides of the overall reaction equation. It turns out that we cannot simply add the three equations together, because we need two ^3_2He particles in Equation 2.3, but only one ^3_2He particle is made according to Equation 2.2. To get around this mismatch, we multiply Equation 2.2 by two. Doing so requires that we also multiply Equation 2.1 by two because we need two ^2_1H particles from Equation 2.1 to use in Equation 2.2. Doubling Equations 2.1 and 2.2 and summing them with Equation 2.3:

$$2 \times [2\,^1_1\text{H} \longrightarrow\,^2_1\text{H} + \,^0_1\text{e}]$$
$$+2 \times [^1_1\text{H} + \,^2_1\text{H} \longrightarrow\,^3_2\text{He}]$$
$$+2\,^3_2\text{He} \longrightarrow\,^4_2\text{He} + 2\,^1_1\text{H}$$
$$6\,^1_1\text{H} + 2\,^2_1\text{H} + 2\,^3_2\text{He} \longrightarrow\,^4_2\text{He} + 2\,^2_1\text{H} + 2\,^3_2\text{He} + 2\,^1_1\text{H} + 2\,^0_1\text{e}$$

Simplifying by canceling out the terms that are common to both sides of the combined equation, we have the overall reaction shown in Figure 2.2:

$$4\,^1_1\text{H} \longrightarrow\,^4_2\text{He} + 2\,^0_1\text{e} \tag{2.4}$$

The positrons formed in Reaction 2.4 belong to a group of subatomic particles that are the charge opposites of those typically found in atoms. In addition to the electron with a positive charge, there is a proton with a negative charge. These charge opposites are particles of **antimatter.** Subatomic particles and their antimatter opposites are mortal enemies. If a pair of them collide, they instantly annihilate each other. In their mutual destruction, they cease to exist as matter, and all of their collective mass is released as energy in an amount predicted by Einstein's equation (see Equation 1.5):

$$E = mc^2$$

where m is the sum of the masses of the two particles. Clearly, matter and antimatter cannot exist in the same place for very long. The positrons produced in hydrogen fusion are rapidly annihilated by the high density of free electrons that also must be present. The sole product of the reaction is energy in the form of two γ rays:

$$^0_1\text{e} + \,^0_{-1}\text{e} \longrightarrow 2\,\gamma \tag{2.5}$$

Hydrogen fusion (Figure 2.2) releases enough energy to sustain a star the size of our sun for billions of years. However, the nuclear furnaces in giant stars many times the size of the sun are much hotter, and consume hydrogen at much higher rates. When the core temperatures of these stars reach 2×10^8 K, helium nuclei can fuse together. This temperature is higher than the 4×10^7 K needed to initiate hydrogen fusion. Why? Because the electrostatic repulsion between α particles with their $+2$ charges is greater than the repulsion between protons with charges of only $+1$. To overcome their greater repulsion toward each other, α particles must be moving toward each other at very high velocities. Very high velocities are achieved at very high temperatures—a concept that we will revisit in later chapters of this book.

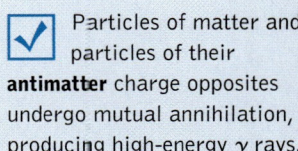

Particles of matter and particles of their **antimatter** charge opposites undergo mutual annihilation, producing high-energy γ rays.

CONNECTION: Einstein's equation also describes the energy-to-mass conversion that may have accompanied the creation of the universe (Section 1.5).

CONNECTION: The link between velocity and temperature is also used to account for the properties of gases (Chapter 8).

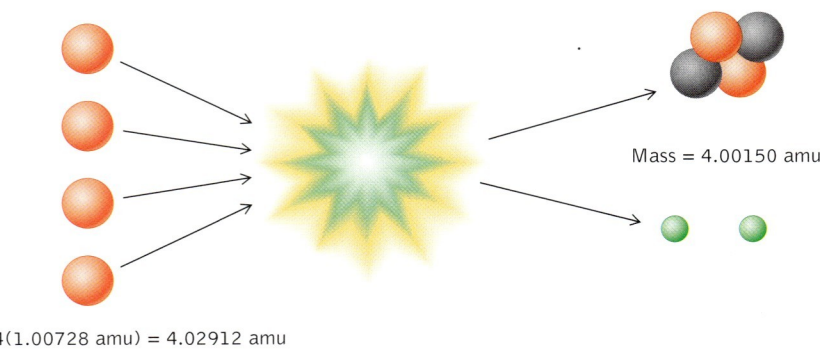

FIGURE 2.2 The fusion of four hydrogen nuclei (or protons—the orange particles) produces a helium nucleus that contains two protons and two neutrons (gray particles) and emits two positrons (green particles). The difference in mass between four protons and the sum of the masses of a helium nucleus and two positrons is released as energy. The amount of energy (E) released is given by $E = mc^2$, where m is the difference in mass and c is the speed of light. The two positrons encounter electrons and are converted into more energy in an annihilation reaction.

2.2 NUCLEAR BINDING ENERGIES

An important result of helium fusion is the synthesis of heavier nuclei. However, there are some significant barriers to the nucleus-building process. If two helium nuclei, $^4_2$He, fuse together, they produce a nuclear particle with an atomic number of $2 + 2 = 4$ and a mass number of $4 + 4 = 8$. This nuclear particle is the nucleus of an isotope of beryllium: $^8_4$Be. It turns out that beryllium-8 is extremely unstable and quickly falls apart into two α particles in a process with a half-life of less than 10^{-15} s. The instability of $^8_4$Be and other isotopes that might have formed from the fusion of hydrogen and helium nuclei is one reason why the early universe was, and the universe of today is, made up mostly of only the two lightest elements.

The relative stabilities of different nuclei are related to their masses and to the energy that binds their nucleons together. In the 1930s, scientists discovered that the mass of a stable nucleus is always less than the sum of the masses of the individual nucleons that make it up. This difference is called the **mass defect** (Δm) of a nucleus: the greater the mass defect, the stronger the energy that binds the nucleons together. To calculate this **binding energy** (E), we simply insert the value of the mass defect for that nucleus into $E = mc^2$ (in this case, $E = (\Delta m)c^2$) and solve for E. As noted in Section 1.4, the SI unit for energy is the joule (J), and the mathematical connection between a joule of energy and Δmc^2 is the following equality:

$$1\,\text{J} = 1\,\text{kg (m/s)}^2$$

The value of c (2.998×10^8 m/s) is usually expressed in meters per second; so calculating binding energies in joules is relatively straightforward as long as Δm is in kilograms.

> ☑ The stability of a nucleus is proportional to its **binding energy** (E), which can be calculated by using the equation $E = (\Delta m)c^2$, where Δm is the **mass defect** of the nucleus.

CONNECTION: The prefixes for SI units are listed in Table 1.1, and the SI units for energy, mass, and distance are given in Table 1.2.

Consider the nucleus of the helium-4 atom. It consists of two neutrons and two protons and has a mass of 6.64465×10^{-24} g. The mass of a neutron (see Table 1.4) is 1.67494×10^{-24} g, and the mass of a proton is 1.67263×10^{-24} g. Summing the individual masses of the nucleons,

$$
\begin{array}{ll}
\text{2 neutrons:} & 2(1.67494 \times 10^{-24} \text{ g}) \\
\text{+ 2 protons:} & 2(1.67263 \times 10^{-24} \text{ g}) \\
\hline
& 6.69513 \times 10^{-24} \text{ g}
\end{array}
$$

and subtracting the mass of the nucleus from this sum,

$$
\begin{array}{ll}
\text{total mass of nucleons:} & 6.69513 \times 10^{-24} \text{ g} \\
- \text{ mass of nucleus:} & 6.64465 \times 10^{-24} \text{ g} \\
\hline
& 0.05048 \times 10^{-24} \text{ g} \\
\text{or} & 5.048 \times 10^{-26} \text{ g} \\
\text{or} & 5.048 \times 10^{-29} \text{ kg}
\end{array}
$$

we obtain the mass defect (Δm) of a helium nucleus, inserting this value into a modified version of Einstein's equation:

$$E = (\Delta m)c^2 \tag{2.6}$$

$$= (5.048 \times 10^{-29} \text{ kg})(2.998 \times 10^8 \text{ m/s})^2$$

$$= 4.537 \times 10^{-12} \text{ kg (m/s)}^2 = 4.537 \times 10^{-12} \text{ J}$$

This quantity may not seem like much, but remember that we are dealing with only one nucleus. To put this quantity in perspective, consider the enormous energy released during the combustion of hydrogen in the main engines of the U.S. space shuttles (see Figure 1.4). The balanced chemical equation for the reaction may be written as follows:

$$H_2(g) + \tfrac{1}{2} O_2(g) \longrightarrow H_2O(l)$$

It turns out that the energy released during the combustion of just one molecule of hydrogen is only about 4.7×10^{-19} J. Thus, the binding energy in the helium nucleus is nearly 10 million times as great as the energy released by a very high energy chemical reaction.

CONNECTION: The energy released when hydrogen burns is described in Figure 1.4 and Chapters 11 and 12.

SAMPLE EXERCISE 2.1: Calculate the binding energy in joules that holds a proton and a neutron together in a deuteron. Given: The speed of light is 2.998×10^8 m/s, and the masses are:

Particle	Mass (g)
deuteron	3.34370×10^{-24}
neutron	1.67494×10^{-24}
proton	1.67263×10^{-24}

CONNECTION: Chemical reactions that release energy are more likely to take place than those that do not release energy (Sections 11.5 and 11.6).

SOLUTION: Equation 2.6 relates the binding energy (E) of a nucleus to its mass defect (Δm):

$$E = (\Delta m)c^2$$

To calculate Δm, we need to find the difference between the sum of the masses of the component nucleons and the mass of the nucleus itself. In this case, there are two nucleons: one neutron and one proton. Therefore:

$$\Delta m = (m_{neutron} + m_{proton}) - m_{deuteron}$$
$$= (1.67494 \times 10^{-24} \text{ g} + 1.67267 \times 10^{-24} \text{ g}) - 3.34370 \times 10^{-24} \text{ g}$$
$$= 0.00391 \times 10^{-24} \text{ g, or } 3.91 \times 10^{-27} \text{ g}$$

A joule of energy $= 1$ kg (m/s)^2; so we need to convert Δm into kilograms:

$$3.91 \times 10^{-27} \text{ g} \times \frac{1 \text{ kg}}{1000 \text{g}} = 3.91 \times 10^{-30} \text{ kg}$$

Inserting this value into $E = (\Delta m)c^2$:

$$E = (3.91 \times 10^{-30} \text{ kg})(2.998 \times 10^8 \text{ m/s})^2$$
$$= 3.51 \times 10^{-13} \text{ kg (m/s)}^2 = 3.51 \times 10^{-13} \text{ J}$$

neutron mass (m_n)   proton mass (m_p)   deuteron mass (m_d)

$\Delta m = (m_n + m_p) - m_d$

mass defect (Δm)

$E = (\Delta m)c^2$

binding energy (E)

PRACTICE EXERCISE: Calculate the binding energy of beryllium-8. The mass of an atom of this isotope is 8.0053 amu (1.32931 × 10^{-23} g). Caution: 1.32931 × 10^{-23} g is the mass of an atom; therefore it includes the mass of the nucleus *and* the masses of four electrons. (See Problems 11 and 12.)

✓ Fusion accounts for the stellar synthesis of all elements with atomic numbers up to $Z = 26$ (iron).

The fusion of nuclei heavier than iron consumes energy.

Within the extremely dense, extremely hot cores of giant stars, collisions between nuclei occur so quickly that even particles with short half-lives, such as $^8_4$Be, may survive long enough to collide and fuse with more $^4_2$He. Such a fusion reaction produces a nucleus with an atomic number of $4 + 2 = 6$ and an atomic mass of $8 + 4 = 12$, which is the nucleus of a carbon-12 atom, $^{12}_6$C. Such a nucleus is stable and may undergo yet another fusion event with $^4_2$He, making the oxygen-16 nucleus, $^{16}_8$O. These fusion reactions between heavier and heavier nuclei with greater and greater positive charges are possible only if the nuclei are moving fast enough. There is evidence that the core temperatures of giant stars can exceed 3×10^9 K—hot enough to allow the synthesis of nuclei with as many as 26 protons. In this way, giant stars create the elements of the periodic table from helium in the outer regions to iron at the highest temperatures in their centers, as shown in Figures 2.3 and 2.4. These elements include those that make up most of the matter in our planet and in ourselves.

Building bigger nuclei by fusing together smaller ones reaches a dead end with the nucleus of iron that has 26 protons and 30 neutrons in it, ^{56}Fe. Why does fusion end with iron? Because fusing lighter nuclei to make heavier ones with as many as 26 protons and 30 neutrons produces nuclei that are more stable than the nuclei that fused together to make them. Nuclear reactions (and chemical reactions) that form more stable products from less-stable reactants also release energy. This energy produces temperatures of billions of kelvins in the cores of stars. However, nuclear stability reaches a maximum with ^{56}Fe and then decreases with increasing atomic number. This trend is shown in the graph in Figure 2.5 in which binding energies *per nucleon* are plotted against mass number. Dividing the binding energy of a nucleus by the total number of nucleons in it provides a useful measure of the stability of that nucleus relative to others with different numbers of nucleons. The data in Figure 2.5 tell us that fusing, for example, an α particle with ^{56}Fe to make ^{60}Ni, would require the *addition* of energy. As we will see throughout this book, reactions that produce energy are often more likely to occur than those that consume energy. Besides, the core of a giant star would not stay hot for very long if its nuclear fuel consumed rather than created energy.

FIGURE 2.3 The fusion of α particles forms, sequentially, beryllium-8, carbon-12, oxygen-16, neon-20, and magnesium-24. Each of the fusion processes after the first one results in the formation of products with slightly less mass than that of the nuclei that fused to make them. This loss of mass is accompanied by an enormous release of energy—the energy that fuels the nuclear furnaces of giant stars.

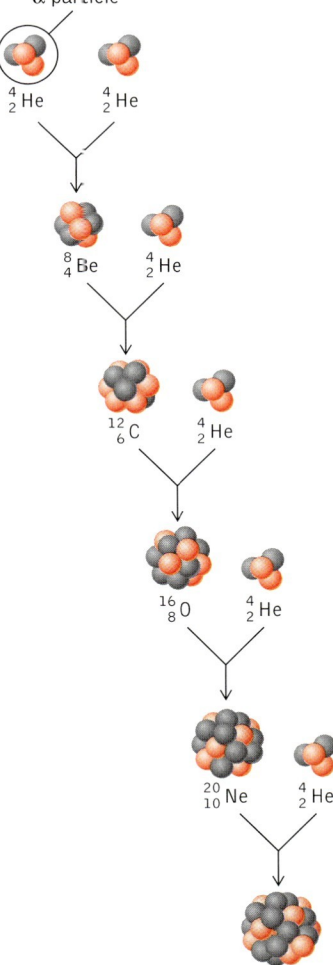

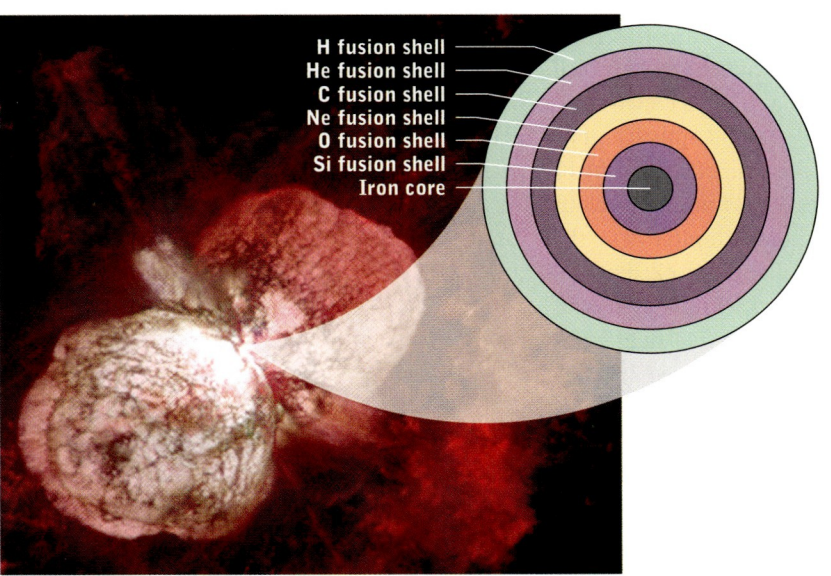

FIGURE 2.4 The star η-Carinae is believed to be evolving toward a supernova explosion. At its center are concentric regions where the fusion of different elements is taking place around an iron core.

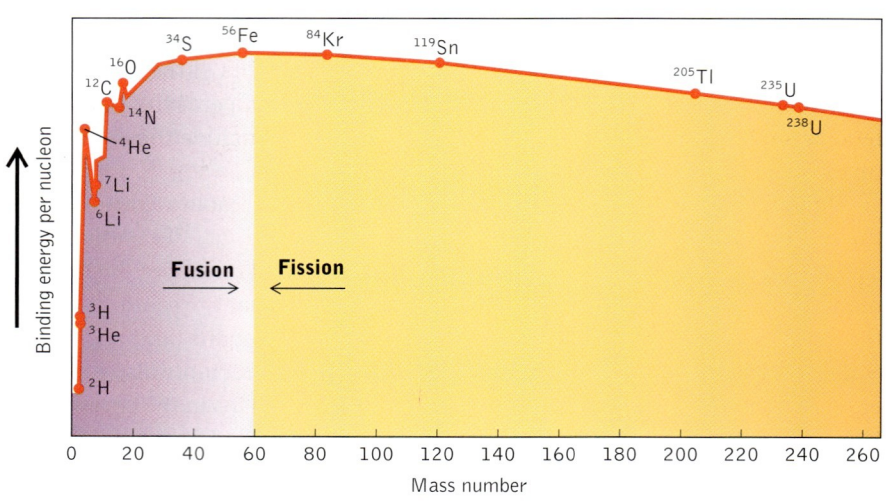

FIGURE 2.5 The stability of a nucleus is indicated by the binding energy. This quantity is best expressed as binding energy per nucleon (proton or neutron). For light elements, fusion processes lead to nuclei with greater binding energy, whereas heavy elements gain binding energy through fission.

2.3 NEUTRON CAPTURE AND MODES OF RADIOACTIVE DECAY

How are nuclei heavier than ^{56}Fe synthesized? The key to their formation lies in giant stars' core temperatures of billions of kelvins. These temperatures are high enough to break up some nuclei, producing free protons and neutrons. Free neutrons create an important pathway for making elements. Because they have no charge, they are not repelled by positively charged nuclei. Thus even slowly moving neutrons, called thermal neutrons, can collide and fuse with atomic nuclei in a process called **neutron capture.**

Neutron capture can produce isotopes with unusually high ratios of neutrons to protons, as shown in Figure 2.6. However, for each element, the number of neutrons that its nucleus can hold is limited. For the light elements (Z ≤ 20), stable nuclei have neutron-to-proton ratios near 1:1. Most helium nuclei have 2 neutrons and 2 protons; most carbon nuclei have 6 of each and a mass of 12. The fact that most carbon nuclei have neutron-to-proton ratios near 1:1 does not mean that other isotopes of carbon with more neutrons cannot form. It means that, if they do, they will likely be radioactive. They will fall apart in nuclear reactions, yielding nuclides with fewer neutrons and greater stability. Carbon, for example, has an isotope with 8 neutrons in its nucleus: carbon-14. It is radioactive and slowly (half-life = 5730 years) undergoes a decay process in which one of its neutrons disintegrates into a proton (which stays in the nucleus) and a high-speed electron, called a beta (β) particle (which escapes to the surroundings), as shown in Figure 2.7. This example of **β decay** is represented in equation form as follows:

$$^{14}_{6}C \longrightarrow {}^{14}_{7}N + {}^{0}_{-1}e \tag{2.7}$$

> ✓ **Neutron capture** is the absorption of a neutron by a nucleus.

> ✓ In **β decay**, a neutron disintegrates into a proton and a high-energy electron called a β particle.

> How are nuclei heavier than ^{56}Fe synthesized?

FIGURE 2.6 Nuclei with atomic numbers greater than 26 can be built up by the absorption of neutrons followed by β decay, as shown here for cadmium. Cadmium-110 absorbs five neutrons (gray particles) before emitting a β particle (blue) and forming indium-115. Indium-115 absorbs one neutron before undergoing β decay to tin-116.

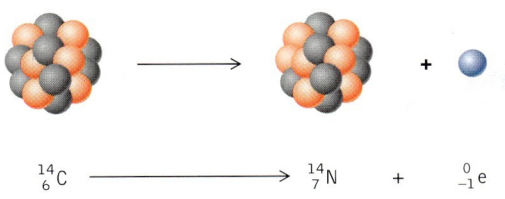

FIGURE 2.7 Carbon-14 undergoes β decay to nitrogen-14.

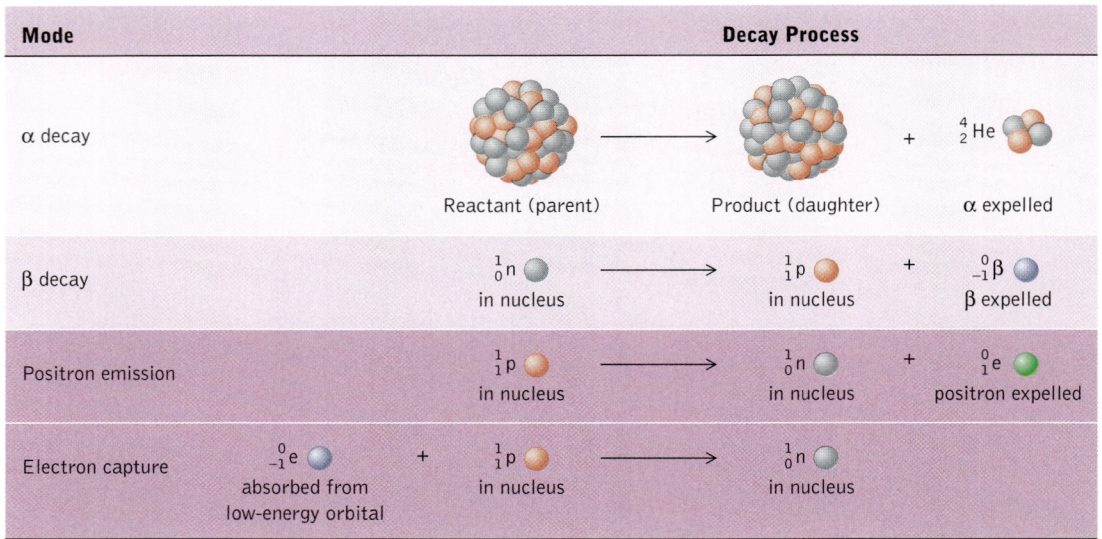

FIGURE 2.8 The paths of radioactive decay include α decay, β decay, positron emission, and electron capture. In all but electron capture, the nucleus expels a particle. For all four modes, the atomic number of the nucleus and the number of neutrons changes.

Beta decay produces a nucleus with one more proton and thus an atomic number that is larger by one unit than the starting nuclide (Figure 2.8).

If a neutron-rich iron nucleus (Z = 26) undergoes β decay, it will produce a nucleus of cobalt, Co (Z = 27). A cobalt nucleus also could absorb neutrons until it has too many and then undergo β decay, producing a nucleus of nickel, Ni (Z = 28). This process of neutron capture and β decay continues in the cores of giant stars, producing nuclei as heavy as bismuth-209 (^{209}Bi), a nuclide with 83 protons and 126 neutrons. Bismuth-209 is the heaviest stable nuclide and the end of the line in the element-building process in giant stars.

Some nuclides are produced neither by the fusion of lighter nuclei nor by neutron capture. They have natural abundances that are hundreds of times lower than those produced by, for example, neutron capture and β decay, and many are radioactive. These nuclides are produced by collisions of lighter nuclei with the high-speed protons that are released in some nuclear reactions. Whereas neutron capture makes nuclides that are neutron rich and are thus likely to undergo β decay, proton capture leads to the formation of radionuclides that are *proton-rich* (or *neutron-poor*) and are therefore likely to undergo decay processes that *increase* their neutron:proton ratio.

One such proton-rich radionuclide is carbon-11, which has six protons but only five neutrons and a half-life of only 20.4 minutes. The proton-to-neutron ratio in carbon-11 can be reduced by a process known as **positron emission**. An equation for this process can be written in the following way:

$$^{11}_{6}\text{C} \longrightarrow {}^{11}_{5}\text{B} + {}^{0}_{1}\text{e} \tag{2.8}$$

> Neutron capture leads to the formation of radioactive nuclides that are neutron rich and undergo β decay.
>
> Proton-rich nuclei undergo **positron emission** or **electron capture**.

SAMPLE EXERCISE 2.2: A neutron-rich nuclide of krypton, Kr ($Z = 36$), undergoes β decay. Which element is produced?

SOLUTION: When a neutron-rich nucleus undergoes β decay, it ejects a high-speed electron from its nucleus, and a neutron becomes a proton. Thus, the mass number of the atom does not change, but the atomic number (Z) increases by one. The element with $Z = 36 + 1 = 37$ is rubidium.

PRACTICE EXERCISE: A proton-rich nuclide of fluorine ($Z = 9$) undergoes positron emission. Which element is produced?

As stated earlier in this chapter, $^{0}_{1}e$ represents a positron. The positrons emitted by carbon-11 nuclei rapidly undergo annihilation reactions with electrons, producing two γ rays, as shown in Equation 2.5. The boron-11 produced by the reaction is a stable isotope: 80.2% of all the boron in nature is boron-11; 19.8% is boron-10.

There is another way in which carbon-11 can increase its neutron:proton ratio and so achieve nuclear stability. Its nucleus can "capture" one of its own electrons—that is, undergo **electron capture.** Remember that six electrons must surround the nucleus of each carbon atom for it to be electrically neutral. One of these electrons may fall into a ^{11}C nucleus, converting one of the protons into a neutron. The nucleus would then have six neutrons and five protons instead of the other way around. It would have been transformed, or **transmuted,** into boron-11.

In this section, we have seen how nuclides that are either neutron rich or neutron poor undergo spontaneous nuclear decay reactions that produce stable nuclei. The graph in Figure 2.9 depicts the combinations of neutrons and protons that form stable isotopes. The band running diagonally through the graph is called the **belt of stability.** Each green square represents a neutron–proton combination that is stable. Among the lighter elements, the most stable neutron:proton ratios are close to 1:1. With increasing atomic number, the neutron:proton ratios increase to about 1.5:1 for the heaviest stable nuclei. Isotopes with neutron–proton combinations shown in orange in Figure 2.9 *do* exist, but not indefinitely. All of them are radioactive and undergo decay reactions that ultimately produce stable isotopes. Nuclides with neutron–proton combinations above the belt of stability are neutron rich and tend to undergo β decay. Nuclides below the belt of stability are proton rich and undergo positron emission or electron capture. The isotopes of carbon listed in Table 2.1 illustrate this trend.

> **Transmutation** is the conversion of the nucleus of an element into the nucleus of another.
>
> The **belt of stability** is a plot of neutrons versus protons in the nuclei of stable and radioactive nuclides.

SAMPLE EXERCISE 2.3: Predict a likely mode of radioactive decay of $^{30}_{15}P$ (see Figure 2.9).

SOLUTION: According to the information in the belt of stability (see Figure 2.9), there is only one stable isotope of phosphorus. It has 15 protons and 16

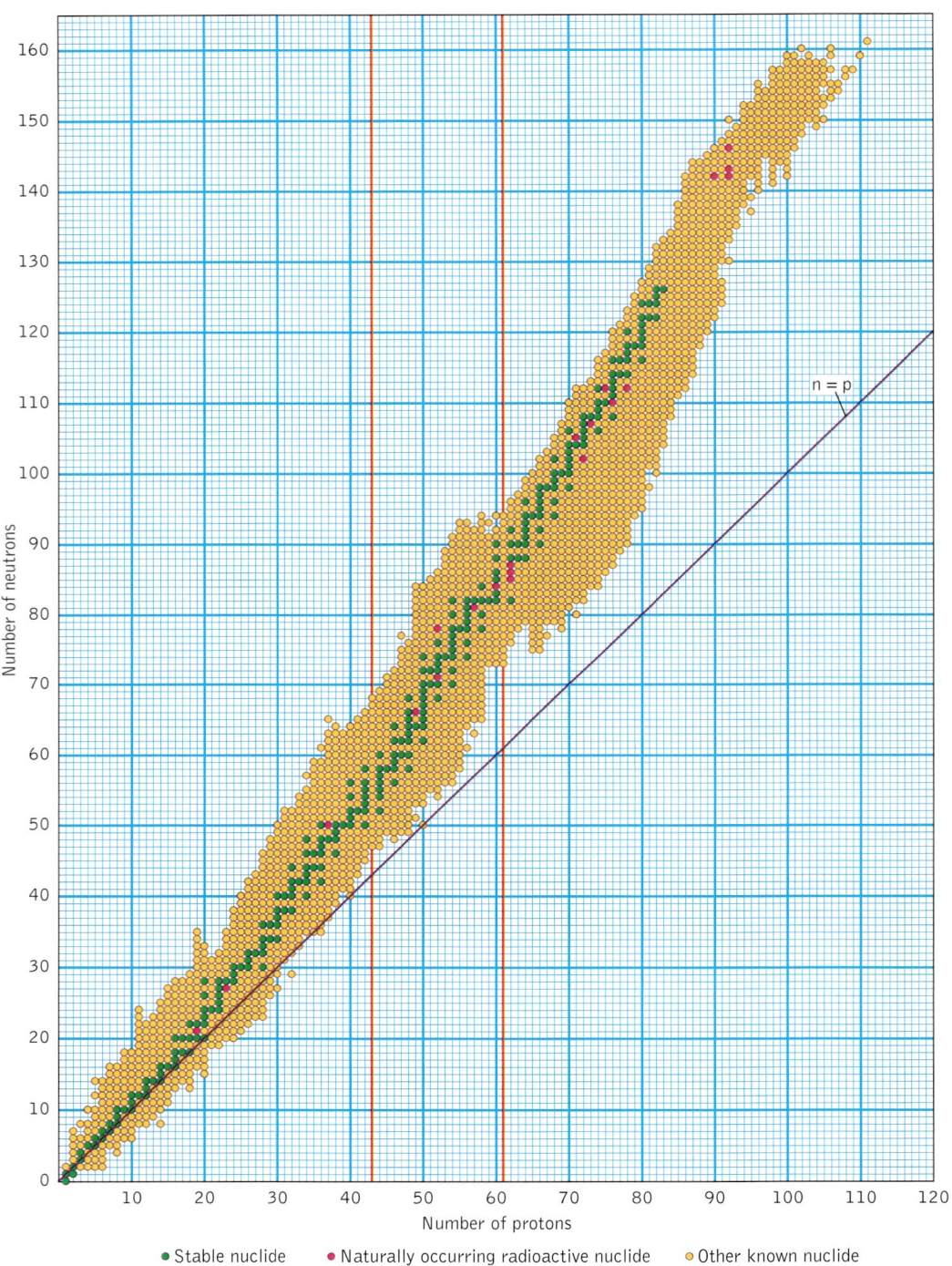

FIGURE 2.9 In this depiction of the belt of stability, the green dots represent stable combinations of protons and neutrons (nuclides). The red dots represent naturally occurring radioactive nuclides, and the orange dots represent man-made radioactive nuclides. Note that there are no stable nuclides for $Z = 43$ (technetium) and $Z = 61$ (promethium). These elements are the only ones among the first 83 not found in nature.

neutrons. Because phosphorus-30 has one neutron fewer than the stable nuclide, it is likely to undergo a radioactive-decay process in which its neutron:proton ratio increases. Two modes of decay that accomplish this increased ratio are positron emission and electron capture. Either one would be a reasonable prediction. Actually, the principal mode of decay of $^{30}_{15}P$ is positron emission.

PRACTICE EXERCISE: Predict the modes of radioactive decay of $^{32}_{15}P$ and $^{18}_{9}F$. (See Problems 27–30.)

TABLE 2.1 Isotopes of Carbon and Their Radioactive-Decay Products

Name	Symbol	Mass (amu)	Mode(s) of Decay	Half-life	Natural Abundance (%)
Carbon-10	$^{10}_{6}C$		Positron emission	19.45 s	
Carbon-11	$^{11}_{6}C$		Positron emission, EC*	20.3 min	
Carbon-12	$^{12}_{6}C$	12.0000		(Stable)	98.89
Carbon-13	$^{13}_{6}C$	13.00335		(Stable)	1.11
Carbon-14	$^{14}_{6}C$		β decay	5730 yr	
Carbon-15	$^{15}_{6}C$		β decay	2.4 s	
Carbon-16	$^{16}_{6}C$		β decay	0.74 s	

*EC, electron capture.

2.4 SUPERNOVA: RADIOCHEMISTRY OF THE HEAVIEST ELEMENTS

Having reached core temperatures of billions of kelvins through the fusion of lighter elements, a giant star's supply of these elements eventually begins to dwindle and the star begins to cool. As it does, it begins a gravity-induced collapse. Its huge mass of elements from Z = 1 to Z = 83 is compressed into a tiny space of immense density and temperature. The result is tremendous compression heating and an enormous explosion in an event called a supernova.

In a supernova, free neutrons collide with nuclei and are absorbed by them so rapidly that neutron-rich, radioactive nuclei do not have a chance to undergo β decay before they absorb even more neutrons. These nuclei become so overloaded with neutrons that additional neutrons smashing into them simply pass right through. These neutron-rich nuclei undergo multiple β-decay events. In so doing, they form nuclei of all the elements found in nature through Z = 94. Even heavier elements may be formed in this way (Figure 2.10), but all are radioactive.

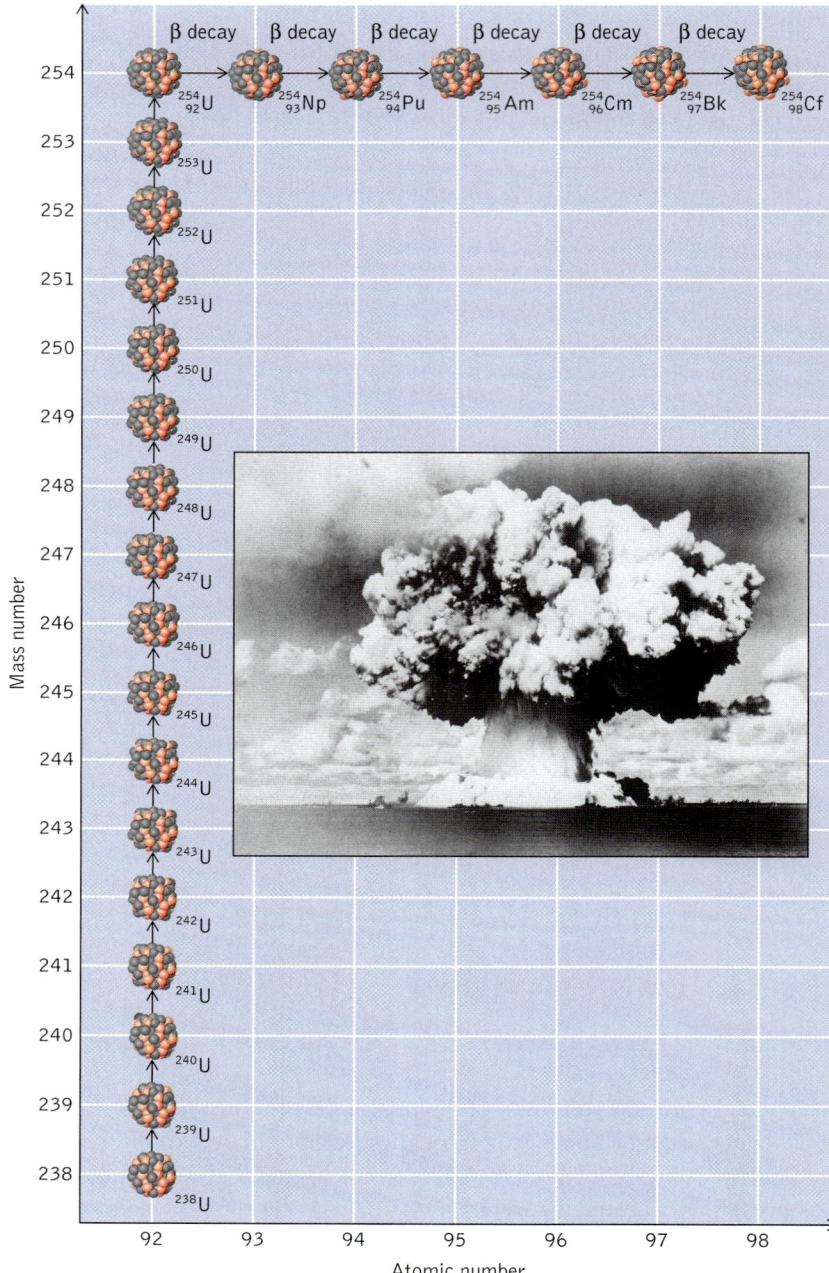

FIGURE 2.10 The high neutron densities of a supernova are simulated in the explosion of a nuclear bomb. In the atomic bomb tests on Bikini Island in 1952, californium-254 (^{254}Cf) was observed among the products. This nuclide is produced by the absorption of 16 neutrons by ^{238}U followed by six β-decay events.

In addition to finishing the job of synthesizing all of the elemental building blocks found in the universe, a supernova serves as its own element-distribution system, blasting its inventory of nuclides throughout its galaxy (Figure 2.11). The legacies of supernovas are found in the elemental composition of later-generation stars (such as our sun) and the planets orbiting these stars, including our own.

FIGURE 2.11 An explosion that occurred in the Cygnus Loop nebula about 15,000 years ago caused this expanding blast wave from a supernova. Emission from oxygen atoms that have lost two electrons is shown in blue. Emission from sulfur atoms that are missing a single electron is shown in red. Light emitted by hydrogen atoms is shown in green.

Neutron capture and β decay yield nuclides with greater and greater numbers of nucleons (see Figure 2.10), but there are limits to nuclear size. Extremely large nuclei tend to be extremely unstable. All nuclides with more than 83 protons ($Z > 83$) are radioactive. Many of their radioactive-decay processes entail the loss of an α particle ($_2^4$He) and so yield nuclides with two fewer protons and two fewer neutrons (Figure 2.12). For example, the most abundant isotope of uranium, ^{238}U, undergoes **α decay**, producing thorium-234 (^{234}Th). This process can be written in equation form as follows:

✓ The loss of a $_2^4$He particle by the nucleus of a heavy element is called **α decay**.

$$_{92}^{238}\text{U} \longrightarrow {}_2^4\text{He} + {}_{90}^{234}\text{Th} \tag{2.9}$$

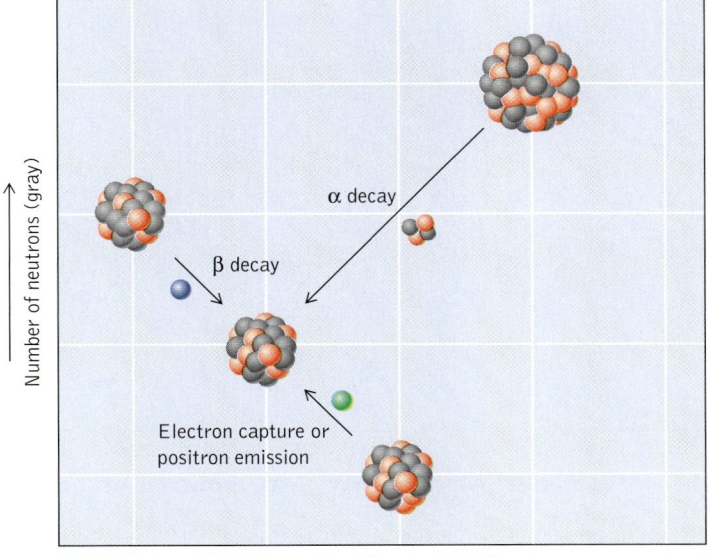

FIGURE 2.12 Radioactive decay results in predictable changes in the number of protons and neutrons in a nucleus. Alpha decay leads to the release of two protons and two neutrons, a decrease of two in the atomic number and four in the mass number. Beta decay leads to an increase of one proton at the expense of a neutron. The atomic number increases, but the mass number is unchanged. In positron emission and electron capture, the number of protons decreases by one and the number of neutrons increases by one. Again, the atomic number decreases, but the mass number remains the same.

The thorium produced in this reaction has an atomic number above 83 and so has no stable isotopes. Therefore, the nuclear disintegration process must continue. Thorium-234 undergoes two β decay steps, producing ^{234}U. In a series of α decay steps ^{234}U turns into thorium-230 (^{230}Th), radium-226 (^{226}Ra), radon-222 (^{222}Rn), polonium-218 (^{218}Po), and finally lead-214 (^{214}Pb). Although some isotopes of lead (Z = 82) are stable, ^{214}Pb is not one of them. Therefore, the radioactive decay series that began with ^{238}U continues with two more β decay steps that produce ^{214}Po, an α decay to ^{210}Pb, two more β decays to ^{210}Po, and finally an α decay to a stable isotope, ^{206}Pb. A graphic view of these decay processes and their products is shown in Figure 2.13. We will return to this series later in a discussion of the hazards posed by environmental radon.

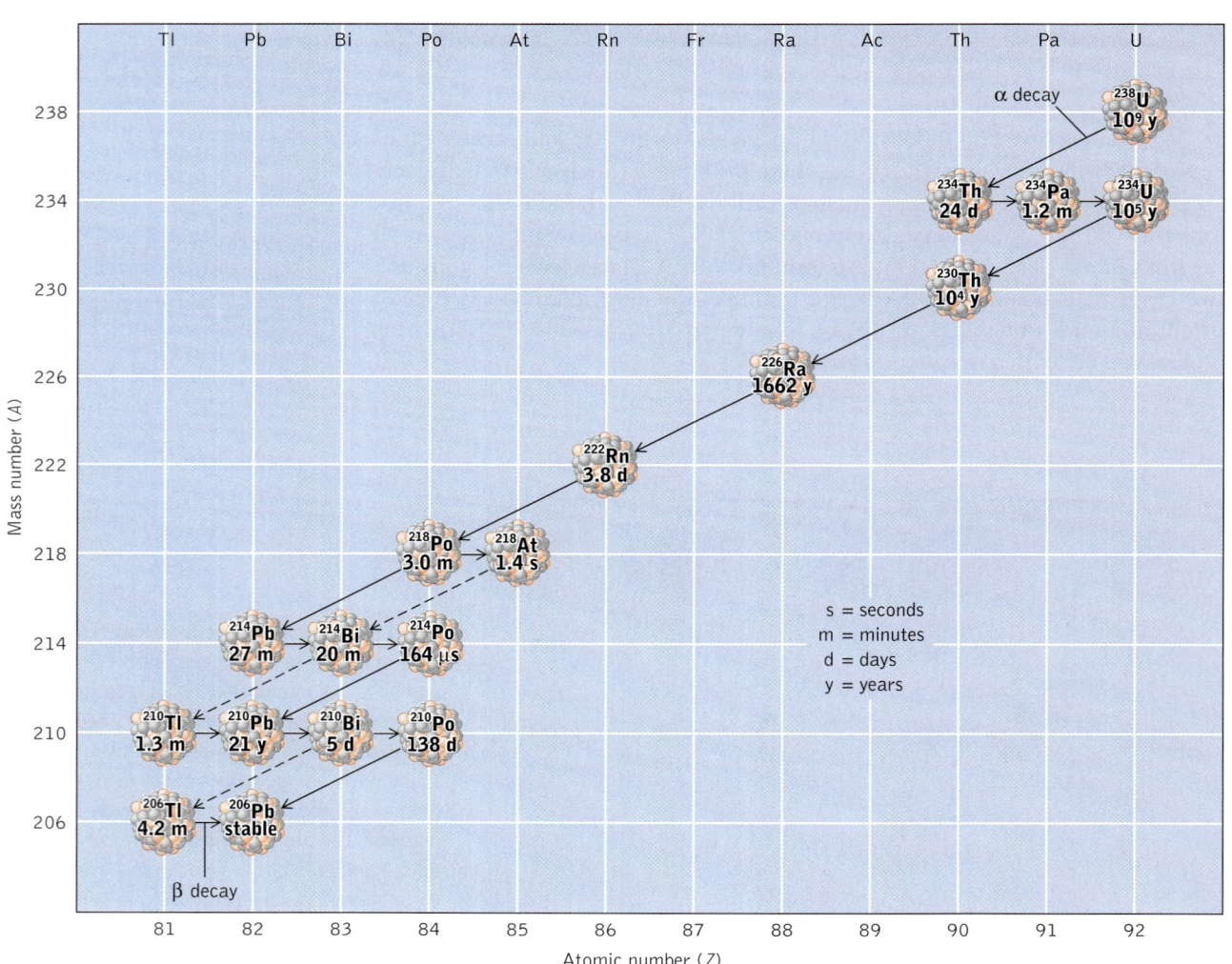

FIGURE 2.13 In this graph of the nuclear disintegration of ^{238}U, the α-decay steps are designated by long arrows and β decay by short ones. Note how β decay produces no change in atomic mass but increases atomic number by one.

There is another process by which some isotopes of uranium and some other heavy nuclides disintegrate. They may split into lighter nuclei; that is, they undergo **nuclear fission**. The most common of the fissionable isotopes is ^{235}U. It makes up 0.72% of the uranium in the universe, but it can be enriched to make the fuel for nuclear reactors (Figure 2.14) and for atomic bombs. There are many ^{235}U fission reactions; here are three of them:

$$1\,{}_{0}^{1}\text{n} + {}_{92}^{235}\text{U} \longrightarrow {}_{56}^{142}\text{Ba} + {}_{36}^{91}\text{Kr} + 3\,{}_{0}^{1}\text{n} \tag{2.10}$$

$$1\,{}_{0}^{1}\text{n} + {}_{92}^{235}\text{U} \longrightarrow {}_{52}^{137}\text{Te} + {}_{40}^{97}\text{Zr} + 2\,{}_{0}^{1}\text{n} \tag{2.11}$$

$$1\,{}_{0}^{1}\text{n} + {}_{92}^{235}\text{U} \longrightarrow {}_{55}^{144}\text{Ba} + {}_{37}^{90}\text{Rb} + 2\,{}_{0}^{1}\text{n} \tag{2.12}$$

In all these reactions, the sums of the masses of the products are less than the sums of the masses of the reactants, and so all these reactions release energy. They also produce more neutrons, which can smash into other ^{235}U nuclei and produce more fission events. This process is called a **chain reaction.** It will proceed as long as there are enough ^{235}U nuclei present to absorb the neutrons being produced. On average, at lease one neutron from each fission decay must cause the fission of another nucleus for the chain reaction to be self-sustaining. The amount of fissionable material needed to sustain a chain reaction is called the **critical mass.** For uranium-235, the critical mass is about 1 kg of the pure isotope. A critical mass of pure uranium-235 can undergo an uncontrolled chain reaction accompanied by an enormous release of energy, such as the energy of the atomic bomb that devastated Hiroshima on August 6, 1945.

The fuel rods in the nuclear reactors used to generate electricity (Figure 2.15) are only about 3 or 4% ^{235}U, and so the nuclide is not present in the critical mass that produces a nuclear explosion. Still, the rate of the chain reaction and the production of nuclear energy in these fuel rods must be regulated. Control rods made of cadmium or boron are lowered between the fuel rods to absorb neutrons and so reduce the number that can produce more ^{235}U fission. Water surrounding the fuel rods transports the heat that they produce to the steam-generation area of the nuclear power plant. Water also acts as a moderator, slowing down the high-speed neutrons produced during fission so that they can be more effectively absorbed by ^{235}U nuclei.

> ✓ In **nuclear fission,** collision with a neutron splits a heavy nucleus into two nuclei of about equal mass.

> ✓ Self-sustaining fission due to the production of neutrons that split other nuclei is a **chain reaction.**

> The mass of a fissionable substance needed to produce a self-sustaining chain reaction is its **critical mass.**

> Energy production from ^{235}U fission in a nuclear reactor is regulated with control rods and a moderator.

SAMPLE EXERCISE 2.4: Strontium-90 is among the radioactive isotopes produced during the fission of uranium-235. complete the following equation describing the fission process that yields ^{90}Sr:

$$_{92}^{235}\text{U} + {}_{0}^{1}\text{n} \longrightarrow {}_{38}^{90}\text{Sr} + 3\,{}_{0}^{1}\text{n} + ?$$

SOLUTION: In the balanced equation for a nuclear reaction, the sum of the superscripts (atomic mass numbers) on the left-hand side equals the sum of the superscripts on the right, and the sum of the subscripts (electrical charges) also must be the same. To balance the mass numbers of the particles, we must solve the following equation for x:

$$235 + 1 = 90 + 3(1) + x$$

$$x = 143$$

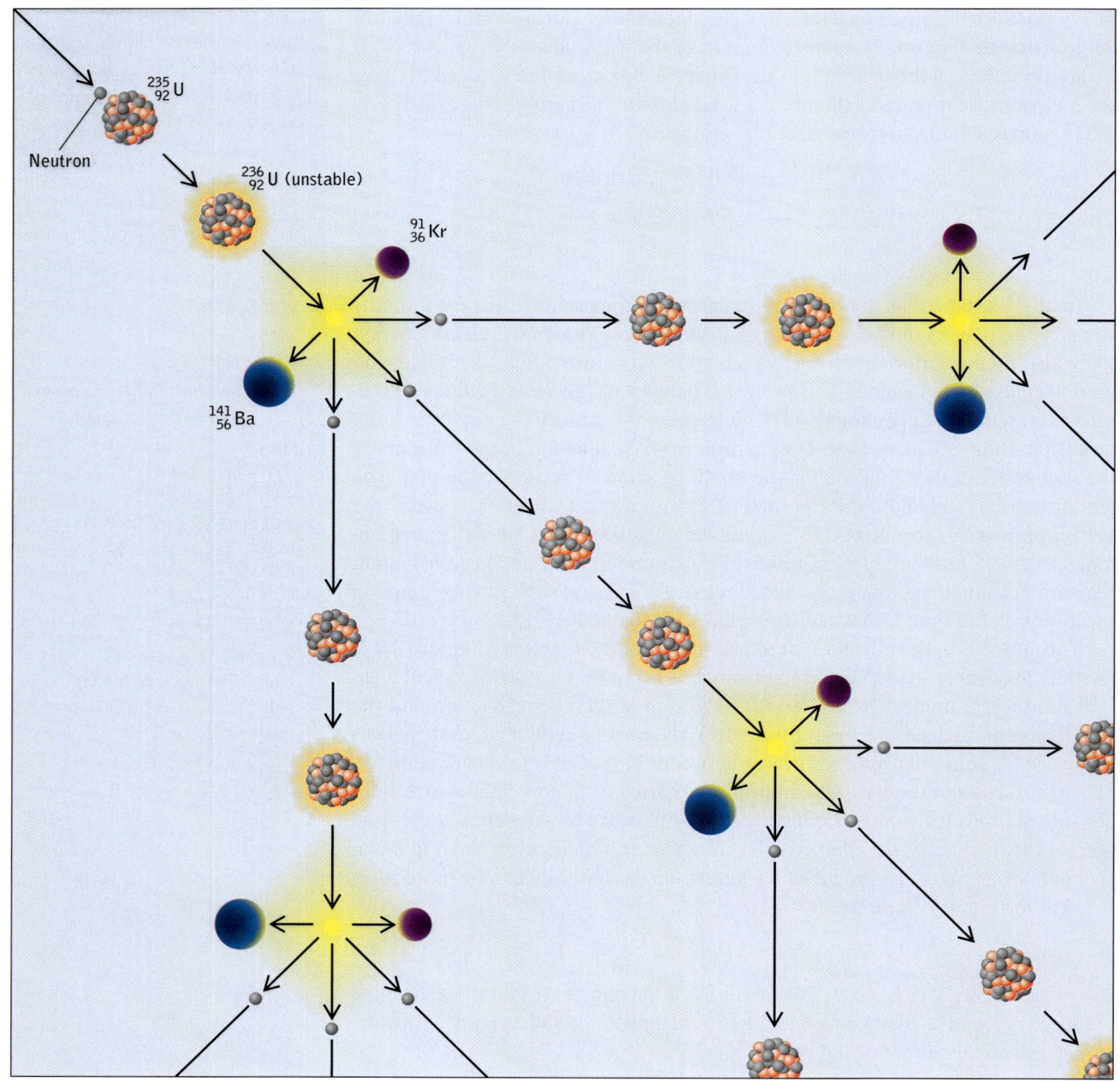

FIGURE 2.14 Each fission event in the chain reaction of ^{235}U begins with the nucleus capturing a neutron. The first such event forms an unstable nucleus of ^{236}U, which splits apart in one of several ways. In the process shown here, it splits into krypton-91 (^{91}Kr), barium-141 (^{141}Ba), and three neutrons. If, on the average, more than one of the three neutrons from each fission event causes the fission of another nucleus, then the process accelerates to an explosive rate in a chain reaction.

To balance the charges of the particles, we must solve this equation for x:

$$92 + 0 = 38 + 3(0) + x$$

$$x = 54$$

A nucleus with 54 protons is the nucleus of an atom of Xe, and so the balanced equation is:

$$^{235}_{92}U + ^{1}_{0}n \longrightarrow ^{90}_{38}Sr + 3^{1}_{0}n + ^{143}_{54}Xe$$

PRACTICE EXERCISE: In another fission reaction that may occur when an atom of ^{235}U absorbs a neutron, the products include 4 neutrons and an atom of antimony-51. What other nuclide is formed in the reaction?

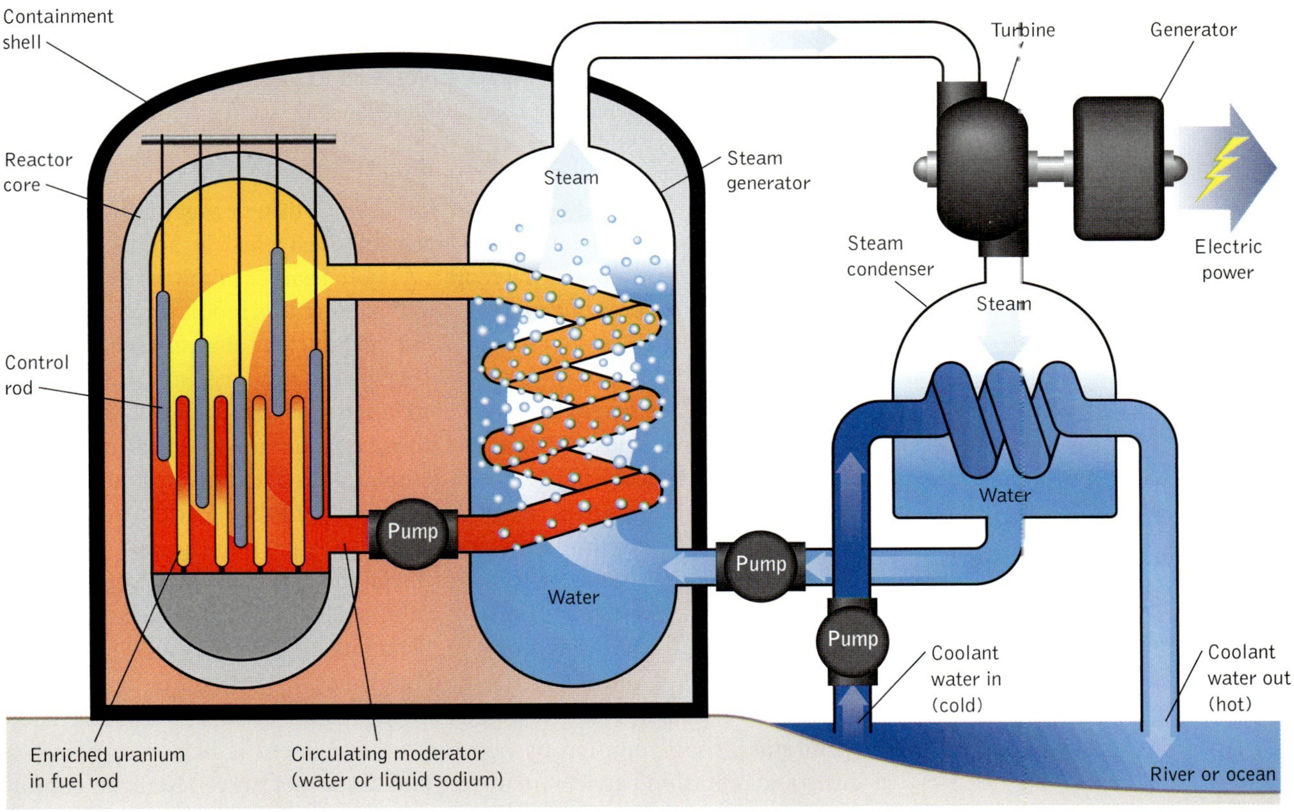

FIGURE 2.15 In a pressurized light-water nuclear power plant (as shown in the photo), the fuel rods contain uranium that has been enriched from 0.7% to 3 or 4% of its fissionable isotope, ^{235}U, so that a fission chain reaction can be sustained. Rods made of boron or cadmium are lowered between the fuel rods to control the rate of the fission by absorbing some of the neutrons that it produces. Water under high pressure flows around the fuel and control rods, removing the heat created during fission and transferring it to a steam generator. The water also acts as a moderator, slowing down the neutrons released by the fission of ^{235}U so that they can be more efficiently captured by other atoms of ^{235}U. Normal, or "light," water, H_2O, both absorbs and slows down neutrons. Heavy water, D_2O, absorbs fewer neutrons than does H_2O. Thus the fuel rods in heavy-water reactors can be made of unenriched uranium.

CONNECTION: Ernest Rutherford also played a key role in deciphering the structures of atoms (Section 3.2).

2.5 ARTIFICIAL ISOTOPES AND ELEMENTS

We have yet to account for those elements in the periodic table with atomic numbers greater than 92. These elements, which are heavier than uranium and are thus called "transuranic" elements, are inherently unstable and all are radioactive. We are able to study and use them, as well as radioactive isotopes of lighter elements, because they were produced not by our stars but by ourselves.

Nuclear scientists have been transmuting elements into others since 1919. In that year, Ernest Rutherford (1871–1937) of Cambridge University reported the nuclear synthesis of oxygen and hydrogen by bombarding nitrogen-14 with α particles:

$$^{14}_{7}\text{N} + ^{4}_{2}\text{He} \longrightarrow ^{17}_{8}\text{O} + ^{1}_{1}\text{H} \tag{2.13}$$

Bombardment of nuclei by α particles became a popular approach for transmuting elements in the 1920s and 1930s. In 1933, Irène and Frédéric Joliot-Curie synthesized the first radionuclide not found in nature, $^{30}_{15}\text{P}$, by bombarding aluminum with α particles in a process that also yielded a free neutron:

$$^{27}_{13}\text{Al} + ^{4}_{2}\text{He} \longrightarrow ^{30}_{15}\text{P} + ^{1}_{0}\text{n} \tag{2.14}$$

To successfully bombard a nucleus with an α particle or any positive ion, the electrostatic repulsion between a positively charged particle and a positively charged target nucleus must be overcome. As noted earlier in the discussion of fusion processes in the cores of giant stars, overcoming this repulsion requires that the bombarding particle be shot toward the target nucleus at a very high velocity. Facilities for accelerating particles to such velocities are large and expensive to build and operate. The **linear accelerator** shown in Figure 2.16 uses an array of alternating electrical fields to accelerate positively charged ions and nuclei in a straight line. Other accelerators use a combination of magnetic and electrical fields to swirl positive particles in spiral pathways until they exit the accelerator, called a **cyclotron,** and smash into target nuclei, as shown in Figure 2.17.

✓ **Linear accelerators** and **cyclotrons** use oscillating electrical fields to accelerate charged particles to velocities high enough to hit target nuclei and initiate nuclear reactions.

Since 1933, more than a thousand artificial radionuclides have been synthesized. Beginning in 1940, the list of artificial nuclides included artificial elements that had never been observed in nature. Elements number 93 and 94 were first produced in 1940 at the University of California, at Berkeley, by bombarding uranium-238 with neutrons. As noted in regard to the synthesis of the elements in giant stars, transmutation by neutron bombardment is helped by a lack of electrostatic repulsion between neutrons and the target nuclei. Neutron capture by uranium-238 initiates a series of reactions leading to the formation of neptunium-239 (Z = 93) and plutonium-239 (Z = 94):

$$^{238}_{92}\text{U} + ^{1}_{0}\text{n} \longrightarrow ^{239}_{92}\text{U} + \gamma \tag{2.15}$$

$$^{239}_{92}\text{U} \longrightarrow ^{239}_{93}\text{Np} + ^{0}_{-1}\text{e} \tag{2.16}$$

$$^{239}_{92}\text{Np} \longrightarrow ^{239}_{93}\text{Pu} + ^{0}_{-1}\text{e} \tag{2.17}$$

Like uranium-235, plutonium-239 is fissionable. In World War II, it was produced in nuclear reactors for use in atomic bombs, including the one dropped on the Japanese city of Nagasaki.

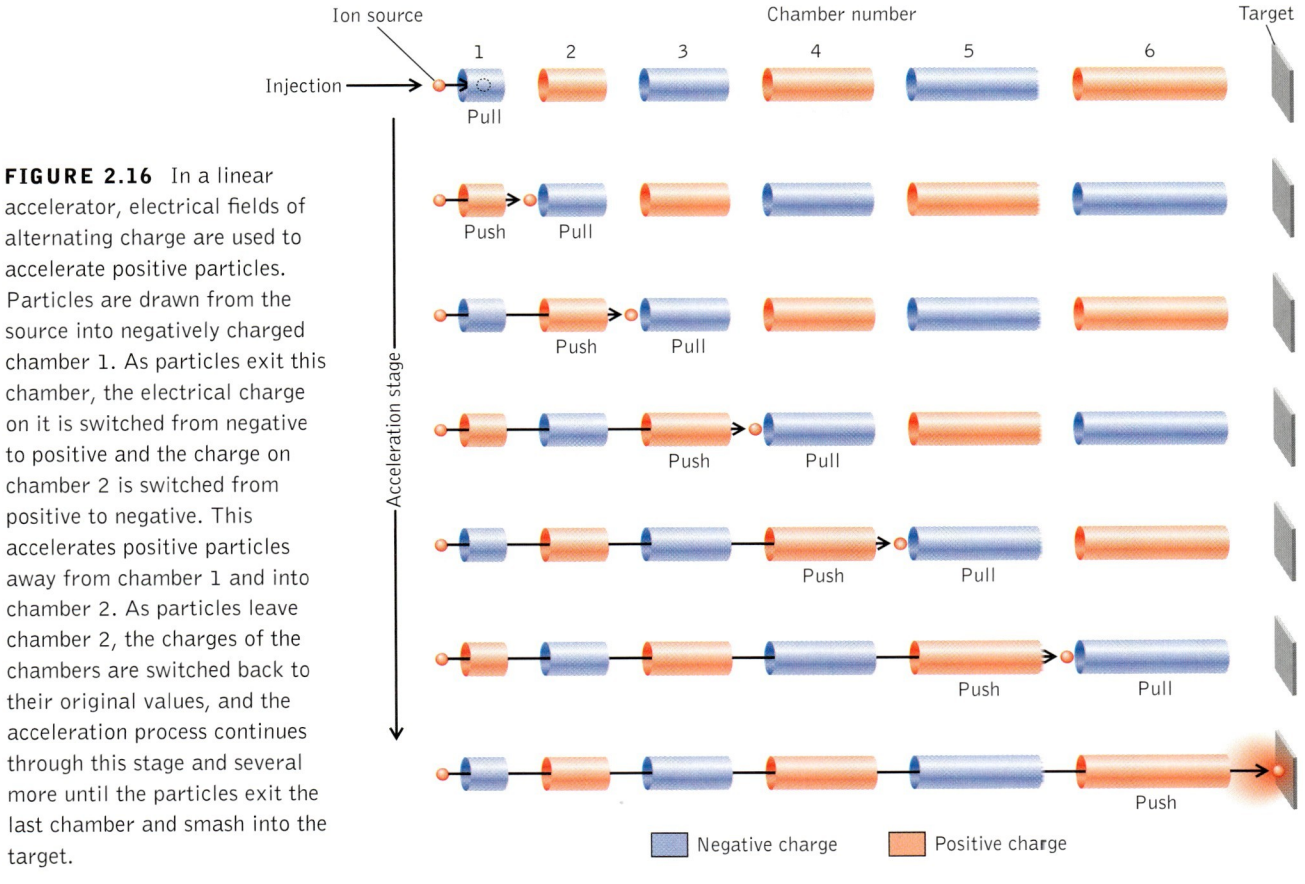

FIGURE 2.16 In a linear accelerator, electrical fields of alternating charge are used to accelerate positive particles. Particles are drawn from the source into negatively charged chamber 1. As particles exit this chamber, the electrical charge on it is switched from negative to positive and the charge on chamber 2 is switched from positive to negative. This accelerates positive particles away from chamber 1 and into chamber 2. As particles leave chamber 2, the charges of the chambers are switched back to their original values, and the acceleration process continues through this stage and several more until the particles exit the last chamber and smash into the target.

In 1952, the first of a new kind of nuclear reactor, called a **breeder reactor**, was used to make plutonium-239 (^{239}Pu) from uranium-238 while also producing energy to make electricity. The fuel in a breeder reactor is a mixture of plutonium-239 and uranium-238. As ^{239}Pu undergoes fission, some of the neutrons that it produces sustain the fission chain reaction; others convert ^{238}U into more plutonium fuel: $^{238}_{92}\text{U} + ^{1}_{0}\text{n} \rightarrow ^{239}_{92}\text{U} + \gamma$. In less than 10 years of operation, a breeder reactor can make enough ^{239}Pu to refuel itself *and* another reactor as well.

Unfortunately, breeder reactors have a down side. Plutonium-239 is one of the most toxic substances known; it is also carcinogenic (it causes cancer). Only about half a kilogram of plutonium-239 would be needed to make a terrorist's atomic bomb, and it has a very long half-life, 2.4×10^4 yr. Understandably, extreme caution and tight security surround the handling of plutonium fuel and the transportation and storage of nuclear wastes containing even small amounts of plutonium. Health and safety matters related to reactor operation and spent fuel disposal are the principal reasons why there are no breeder reactors in nuclear power stations in the United States, although they have been built in at least seven other countries.

> ✓ A **breeder reactor** is a nuclear reactor in which fissionable material is produced during normal reactor operation.

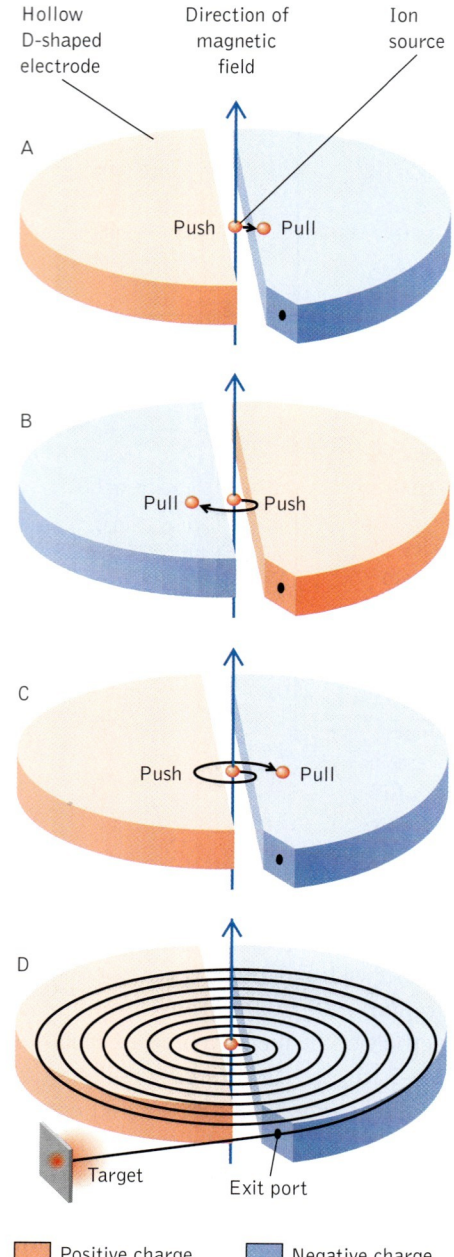

FIGURE 2.17 Particles are introduced near the center of a cyclotron between two D-shaped hollow half-cylinders. The electrical charge on each half-cylinder is opposite that of the other; so positive particles are (A) repelled by the positively charged half-cylinder and attracted to the other. As they enter the negatively charged half-cylinder, the signs of the charges are switched and (B) the particles turn back toward the first half-cylinder. Magnets above and below the half-cylinders (C) bend the path taken by the particles in traveling back and forth between the two. As a result, the particles (D) spiral outward as they pick up speed until they reach the exit and the target.

The leader of the research team at Berkeley that discovered plutonium was American chemist Glenn T. Seaborg (1912–1999). Between 1944 and 1961, Seaborg and his colleagues synthesized elements 95 through 103 by using combinations of neutron and α-particle bombardment of transuranic target nuclei. Unfortunately, radionuclides heavier than $^{249}_{98}$Cf have extremely short half-lives and so cannot be used as target material to make even bigger nuclides through neutron or α-particle bombardment. However, scientists have aimed larger positive ions, such as those in Equations 2.19 and 2.20, at $^{249}_{98}$Cf nuclei to create even heavier elements, including rutherfordium (Z = 104), dubnium (Z = 105), and, in 1974, seaborgium (Z = 106):

$$^{249}_{98}\text{Cf} + ^{12}_{6}\text{C} \longrightarrow ^{257}_{104}\text{Rf} + 4\,^{1}_{0}\text{n} \qquad (2.18)$$

$$^{249}_{98}\text{Cf} + ^{15}_{7}\text{N} \longrightarrow ^{260}_{105}\text{Db} + 4\,^{1}_{0}\text{n} \qquad (2.19)$$

$$^{249}_{98}\text{Cf} + ^{18}_{8}\text{O} \longrightarrow ^{263}_{106}\text{Sg} + 4\,^{1}_{0}\text{n} \qquad (2.20)$$

The names for the synthetic elements are chosen by an organization called the International Union of Pure and Applied Chemistry (IUPAC). Typically, the names selected are those of the places in which the elements were first synthesized, such as Berkeley (Z = 97), California (Z = 98), or of famous scientists, including Albert Einstein (Z = 99) and Enrico Fermi (Z = 100), who made key contributions to our understanding of nuclear chemistry and physics. In August 1997, IUPAC decided to name element 106 after Glenn T. Seaborg. The decision was a surprise because no element had ever been named in honor of a living scientist. That so many scientists supported the name seaborgium is a measure of the high regard in which his colleagues held him. He died in 1999, having had the pleasure of viewing the latest version of the periodic table with his name immortalized on it.

In recent years, scientists at the Gesellschaft für Schwerionenforschung (translation: Society for Heavy Ion Research), or GSI, in Darmstadt, Germany, reported the synthesis of nuclides with as many as 112 protons. To make these "superheavy" elements, the scientists bombard targets of ^{209}Bi or ^{208}Pb with nuclei of medium-weight elements (Table 2.2). The trick to fusing these large positive particles with target nuclei is to control their velocities so that they are moving just fast enough to overcome electrostatic repulsion and fuse with the target. The fusion event is said to be like two nuclear particles "kissing" each other. If they approach each other too slowly, electrostatic repulsion keeps them apart, and no fusion occurs. If the bombarding particle is going too fast, the superheavy nucleus that is formed will be unstable and undergo fission. It is not easy to control this kind of fusion, and many days of bombardment may lead to the synthesis of only one or two, if any, atoms of a superheavy element. Moreover, these new elements are extremely radioactive, with half-lives of milliseconds or less.

> So why bother to make superheavy elements?

Glenn T. Seaborg.

So why bother to make superheavy elements? Although these elements have very short lives, there is both theoretical and now experimental evidence that some isotopes of these elements may be much more stable than their near neighbors in the periodic table. In January 1999, an atom with 114 protons and 175 neutrons was synthesized and lasted for 30 seconds before undergoing a series of α-decay steps that yielded isotopes of elements numbered 112, 110, and 108. These isotopes also had much longer lives than those of the same atomic numbers that had been synthesized by fusing lighter nuclei together. The mere existence of these elements, no matter for how brief a time, can be a source of insight into the nature of nuclear structure and the competition between the force that holds nucleons together and the electrostatic repulsion that drives them apart. Superheavy elements are pieces of a puzzle that someday may tell us whether there is a limit to the size of atoms that make up our world.

TABLE 2.2 Some of the Isotopes of Nine Superheavy Elements Synthesized by Colliding Heavy Nuclei

Bombarding Ion	Target	New Element*	Date Created
^{54}Cr	^{209}Bi	$^{262}_{107}$Bh	February 1981
^{58}Fe	^{208}Pb	$^{265}_{108}$Hs	March 1984
^{58}Fe	^{209}Bi	$^{266}_{109}$Mt	September 1982
^{62}Ni	^{208}Pb	$^{269}_{110}$X	November 1994
^{64}Ni	^{209}Bi	$^{272}_{111}$X	December 1994
^{69}Zn	^{208}Pb	$^{277}_{112}$X	February 1996
^{48}Ca	^{244}Pu	$^{289}_{114}$X	January 1999†
		$^{289}_{116}$X	April 1999‡
^{86}Kr	^{208}Pb	$^{293}_{118}$X	April 1999‡

*No names or symbols have been proposed for elements 110 through 118. The names for elements 107 through 109: bohrium, hassium, and meitnerium, respectively, are based on the recommendations of the scientists at GSI who created them.
†Element was made in Dubna, Russia, by a team of scientists from the Joint Institute for Nuclear Research in Dubna and the Lawrence Livermore National Laboratory, California.
‡Element 118 was believed to have been synthesized at the Lawrence Livermore National Laboratory in 1999. In $<10^{-4}$ s, it underwent α decay, forming element 116. This synthesis had not been reproduced as of December 2001.

2.6 MEASURING RADIOACTIVITY

The French scientist Henri Becquerel (1852–1908) discovered radioactivity in 1896, when he observed that uranium and other substances produce an invisible radiation that fogs photographic film. Two years later, Marie Curie (1867–1934) named the process that produces such radiation *radioactivity*. Photographic film is still widely used to detect and record radioactivity. People who work with radioactive materials must wear badges called radiation dosimeters. The badges contain photographic film that darkens with exposure to radiation: the greater the exposure, the darker the film.

Radioactivity also can be detected with a Geiger counter (named after German physicist Hans Geiger (1882–1945). A Geiger counter detects the common

Marie Curie.

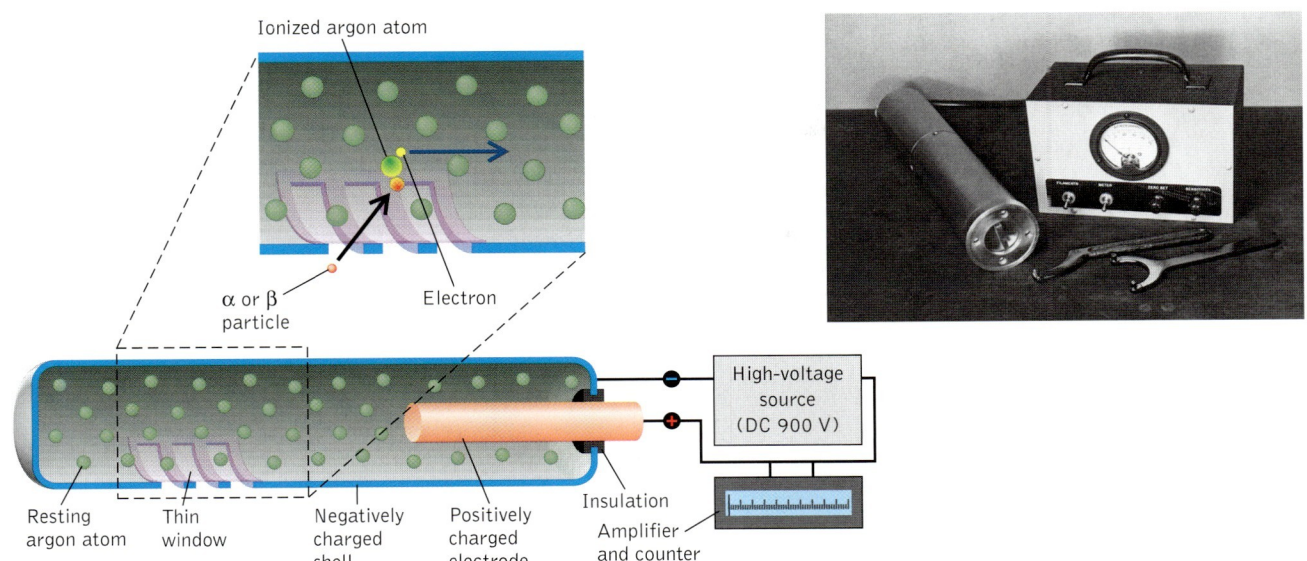

FIGURE 2.18 In a Geiger counter (like the one pictured at the upper right), a particle produced by radioactive decay, such as an α or a β particle, passes through a thin window usually made of beryllium or a plastic film. Inside the tube, the particle collides with atoms of argon gas and ionizes them. The resulting positively charged argon ions migrate toward the negatively charged tube housing and the electrons migrate toward a positive electrode, causing a pulse of current to flow through the tube. The current pulses are amplified and recorded with the use of a meter and a small speaker that produces an audible "click" for each pulse.

products emitted by radioactivity: α and β particles and γ rays, on the basis of their abilities to ionize atoms. A Geiger counter (Figure 2.18) consists of a sealed metal cylinder filled with gas, usually argon, and with a wire in its center. At one end of the cylinder is a very thin window that allows α and β particles and γ rays to enter. Once inside the cylinder, these products of radioactive decay break up argon atoms into free electrons and Ar^+. If an electrical voltage is applied between the cylinder and the central wire, free electrons rapidly migrate toward the positive electrode and argon ions migrate toward the negative electrode. This rapid migration of ions produces a pulse of electrical current whenever radiation enters the cylinder. The current is amplified and read out to a meter and a microphone that makes a clicking sound.

How do we express the quantities, or levels, of radioactivity that Geiger counters detect? One way is based on the number of radioactive decay events per unit time. The SI unit for radioactivity is the becquerel, Bq (Table 2.3), in honor of Henri Becquerel. It equals *one decay event per second*. An older, much bigger, and more commonly used unit of radioactivity is the curie, Ci (1 Ci = 3.70×10^{10} Bq), named in honor of Marie and Pierre Curie (1859–1906). In practice, the levels of radioactivity used in scientific studies and in medical imaging (see Section 2.7) are often less than a millicurie (10^{-3} Ci) and typically in the microcurie

TABLE 2.3 Units for Expressing Quantities of Radiation and Their Effects on Matter

Unit	Parameter	Description
Curie (Ci)	Amount of radioactivity	3.7×10^{10} nuclear disintegrations per second
Becquerel (B)*	Amount of radioactivity	1 disintegration per second
Roentgen (R)	Ionizing intensity of radiation	Amount of ionizing radiation producing 2.1×10^9 charges in 1 cm^3 of air
Rad	Energy of radiation absorbed per mass of matter	0.01 J of ionizing energy per kilogram of matter
Rem	Amount of tissue damage	1 rem = 1 rad × RBE

*SI unit of radioactivity.

(10^{-6} Ci) to nanocurie (10^{-9} Ci) range. When a nuclear reactor at the Chernobyl power station in Ukraine exploded in 1986, at least 20 million curies of radioactivity was released into the atmosphere, causing an increase in background radiation levels throughout the world.

2.7 BIOLOGICAL EFFECTS OF RADIATION

Concern about uncontrolled radiation from nuclear waste and other sources stems from the invisibility of radiation and its ability to cause health problems and even death. On the positive side, radiation is widely used for the diagnosis of disease and for therapy, and there is a renewed interest in its use to kill bacteria in fresh fruit and vegetables and to sterilize materials contaminated by bioterrorists. In this section, we consider how radiation is measured, how its effects are evaluated, and how radiation is applied to benefit human health.

Ionizing radiation and living matter

We have seen that the γ rays and many of the α and β particles produced by nuclear reactions have more than enough energy to tear atoms apart, producing free electrons and positive ions. Therefore, these products of radioactive decay are examples of **ionizing radiation**. The ionization of atoms (and molecules) in living tissue results in tissue damage such as burns, and molecular changes that can lead to radiation sickness, altered cell growth, cancer, or birth defects in offspring. The scientists who first worked with radioactive materials were not aware of these hazards, and some of them suffered for it. Marie Curie died of leukemia caused by her many years of exposure to radiation from radium, polonium, and

> **Ionizing radiation** comprises the high-energy products of radioactive decay that can ionize substances.

other radionuclides. The same disease claimed her daughter Irène Joliot-Curie, who had continued the research of her parents.

Radiation-induced alterations to the biochemical machinery that controls cell growth are most likely to occur in those tissues in which cell-division rates are normally rapid. Such tissues include bone marrow, where billions of white blood cells (called leukocytes) are produced each day to fortify the body's immune system. Molecular damage to bone marrow can lead to leukemia, an uncontrolled production of leukocytes that do not fully mature and so cannot destroy invading pathogens. Eventually, leukemic cells spread throughout the body (metastasize), invading and crowding out the cells of other tissues. Ionizing radiation can also cause molecular alterations to genes and chromosomes in sperm and egg cells that may cause birth defects in offspring.

> ✓ Radiation can break molecules into highly reactive molecular fragments, electrons, and positive ions that cause radiation sickness.

Radiation dosage

We need to distinguish between a *level* of radioactivity, which is the number of radioactive decay events per unit time, and *dose* of radioactivity, which is the *quantity of ionizing radiation absorbed by a unit mass of matter.* The most common unit for expressing dose is the rad (radiation absorbed dose). One rad represents the absorption of 0.01 joule of ionizing radiation energy by 1 kilogram of matter:

$$1 \text{ rad} = 1 \times 10^{-2} \text{ J/kg}$$

A unit related to the rad is the roentgen (R), which is defined as the quantity of ionizing radiation producing 2.1×10^9 units of charge in 1 cm³ of dry air at atmospheric pressure. For γ rays and X-rays, 1 rad is the same as 1 roentgen.

Although rads and roentgens provide measures of the amount of ionizing radiation to which an organism is exposed, they do not indicate the amount of tissue damage. Different products of nuclear reactions affect living tissue differently: a rad of γ rays produces about the same amount of tissue damage as a rad of β particles, but a rad of α particles, which move about 10 times slower than β particles but have nearly 10^4 times the mass, produce 20 times as much tissue damage. Neutrons do more damage than β particles but less than α particles. To account for these differences in relative biological effectiveness (RBE), another unit that takes the type of radiation into account as well as its amount was developed (Figure 2.19). It is called the rem (roentgen equivalent man):

$$1 \text{ rem} = 1 \text{ rad} \times \text{RBE}$$

Beta particles and γ rays have RBE values of 1; the RBE for α particles is 20. These values may lead you to believe that α particles pose the greatest health threat from radioactivity; not exactly. Alpha particles are so big that they have little penetrating power. They are stopped by a sheet of paper, your clothing, or even a layer of dead skin. On the other hand, if you ingest or breathe in an α-particle emitter, tissue damage can be severe. Gamma rays are considered the most dangerous form of radiation emanating from a source outside the body because they have the greatest penetrating power. Table 2.3 summarizes the various units used to express quantities of radiation.

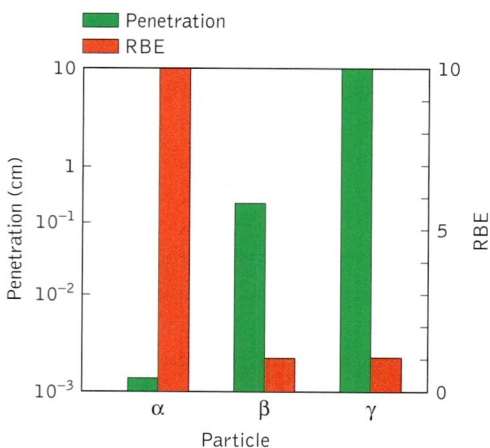

FIGURE 2.19 The biological effect of radiation depends on the relative body effectiveness (an indication of the energy) and the depth to which the radiation can penetrate. Thus the extremely potent α particle does not penetrate very deeply and is easily shielded, whereas lower RBE γ rays penetrate more deeply and are harder to block.

The acute toxic effects of exposure to different single doses of radiation are summarized in Table 2.4. Sadly, most of what is known on the subject has come from the accidental exposure of workers in the nuclear industry and the victims of the atomic bombs dropped on Hiroshima and Nagasaki, Japan. To put the data in Table 2.4 in perspective, the dose from a typical dental X-ray is about 0.5 mrem (5×10^{-4} rem).

Widespread exposure to much higher levels of radiation occurred after an explosion at a nuclear reactor in the Chernobyl power station. Early in the morning of April 26, 1986, workers were testing emergency systems when the reactor overheated and its fuel rods began to melt. The water that was supposed to carry heat away from the reactor vaporized into steam, and the reactor exploded, releasing fuel and reactor materials into the atmosphere. The reactor core contained graphite (carbon) to absorb and slow down neutrons produced by nuclear fission. After the explosion, this graphite caught fire and burned for nearly 2 weeks. The explosion and fire are believed to have released more than 200 times the amount of radioactivity released by the atomic bombs dropped on Hiroshima and Nagasaki combined. Many of the responding firefighters and power-plant workers were exposed to more than 100 rem. At least 30 of them died in the

TABLE 2.4 Acute Effects of Single Whole-Body Doses of Ionizing Radiation

Dose (rem)	Toxic Effect
5–25	No acute effect, possible carcinogenic or mutagenic damage to DNA
25–100	Temporary reduction in white blood cell count
100–200	Radiation sickness: fatigue, vomiting, diarrhea, impaired immune system
200–400	Severe radiation sickness: intestinal bleeding, bone marrow destruction
400–1000	Death, usually through infection, within weeks
>1000	Death within hours

weeks after the accident. Many of the more than 300,000 workers who cleaned up the area around the reactor exhibited symptoms of radiation sickness (see Table 2.4), and at least 5 million people in Ukraine, Belarus, and Russia were exposed to fallout from the accident.

Studies of the biological effects of radiation from the Chernobyl accident uncovered a 200-fold increase in the incidence of thyroid cancer (^{131}I was released in the accident; see page 87) in children in southern Belarus. Children born in this region 8 years after the accident had twice the number of mutations in their DNA as that of other children. The latter results were surprising because no abnormal incidences of birth defects or chromosomal damage had been found among children born to wartime residents of Hiroshima and Nagasaki. Moreover, mutation rates in rodents living in heavily contaminated fields near Chernobyl were thousands of times as great as normal rates. Note, however, that the biomolecular techniques used in the Chernobyl study were much more sensitive than the assessments of mutagenesis among survivors of the atomic bomb attacks. The significance of the Chernobyl results is the subject of considerable debate in the scientific community.

CONNECTION: Chapter 13 examines the structure of DNA and mechanisms of its mutation.

Assessing the risks of radiation

The cloud of radioactivity released from Chernobyl spread rapidly across northern Europe. Within 2 weeks, its radioactivity was detected throughout the entire Northern Hemisphere. The release produced a global increase in human exposure to ionizing radiation estimated to be equivalent to 5 mrem per year. To put this exposure in perspective, we need to consider the amount of ionizing radiation from other sources to which we are typically exposed each year.

Until the mid-1980s, the average American was thought to be exposed to about 200 mrem/yr from natural and artificial sources, most of it coming from the sun, other stars, and radioactive elements that are present naturally in soil, water, and air. The intensity of cosmic (γ) radiation is a function of altitude, because γ rays entering the atmosphere are absorbed by the nuclei of atmospheric gases (Figure 2.20). These nuclei may be broken apart, releasing their component nucleons and creating a population of free neutrons that take part in nuclear chemical reactions in the upper atmosphere that produce, for example, the carbon-14 that is widely used in dating prehistoric artifacts as is discussed in the next section.

This picture of background-radiation dosage changed in the 1980s (Figure 2.21) with growing awareness of the presence of radon in indoor air and in well water. Like all of the elements in the last column of the periodic table, radon is a chemically inert gas. Unlike the others, all of its isotopes are radioactive. The most common isotope, radon-222, is produced by the decay series in which uranium-238 eventually turns into lead-206 (Figure 2.13). Trace amounts of uranium are present in most rocks and soils. Radon gas formed underground percolates toward the surface through the pores in soil and along cracks and fissures in rocks. It can also migrate through cracks and pores in the foundations of buildings and into living spaces.

Destroyed nuclear reactor at Chernobyl, Ukraine.

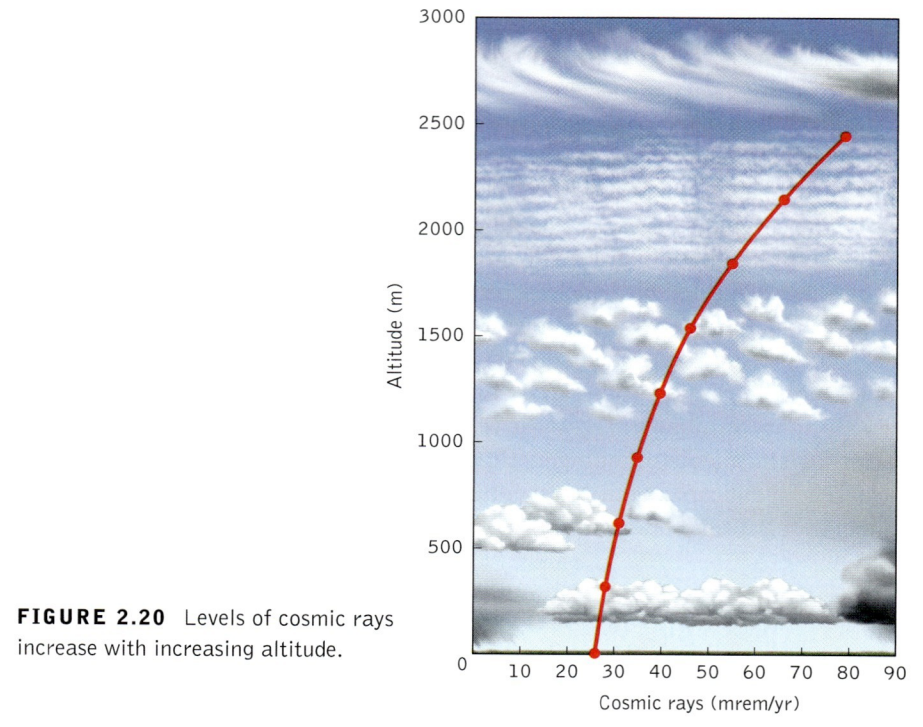

FIGURE 2.20 Levels of cosmic rays increase with increasing altitude.

If you breathe air containing radon and then exhale before it undergoes radioactive decay, no harm is done. However, if radon-222 decays inside the lungs, it emits an α particle and forms an atom of radioactive polonium-218, as shown in Equation 2.21. Polonium is a reactive solid and becomes attached to tissue in the respiratory system, where it will undergo another α-decay reaction, forming lead-214. As we have seen, α particles are the most damaging product of nuclear decay when formed *inside the body*.

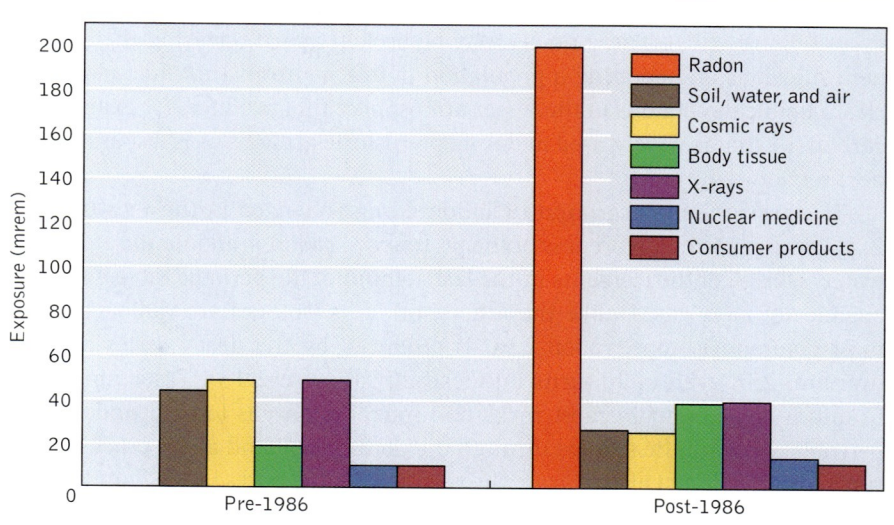

FIGURE 2.21 Estimates of average annual exposure to ionizing radiation in the United States.

$$^{222}_{86}\text{Rn} \longrightarrow {}^{4}_{2}\text{He} + {}^{218}_{84}\text{Po} \qquad t_{1/2} = 3.8 \text{ days} \qquad (2.21)$$

$$^{218}_{84}\text{Po} \longrightarrow {}^{4}_{2}\text{He} + {}^{214}_{82}\text{Pb} \qquad t_{1/2} = 3.1 \text{ minutes} \qquad (2.22)$$

How big a threat does radon pose to human health? Concentrations of indoor radon depend on local geology and how gas tight building basements and foundations are. Although there are variations in every region, different parts of the United States tend to have lower or higher concentrations of radon (Figure 2.22). Still, the air in many buildings contains concentrations of radon in the picocurie-per-liter range (1 pCi = 10^{-12} Ci). How hazardous are such tiny concentrations? There appears to be no simple answer. The U.S. Environmental Protection Agency has established 4 pCi/L as an "action level," meaning that people occupying houses with higher concentrations should take measures to minimize their exposure to radon. Sealing cracks in the foundation is one approach that is not always successful. An expensive alternative entails excavating around the foundation and installing perimeter ductwork and exhaust fans to remove radon before it seeps in.

> How big a threat does radon pose to human health?

The 4-pCi/L action level is based on the results of studies of the incidence of lung cancer in workers in uranium mines. These workers are exposed to much higher concentrations of radon (and radiation from other radionuclides) than those found in homes and other buildings. However, many scientists believe that people exposed to very low levels of radon for many years are as much at risk as miners exposed to high levels of radiation for shorter periods. They assume a linear relation between radon exposure and incidence of lung cancer. On the basis of this dose–response model, an estimated 15,000 Americans die of lung cancer each year because of exposure to indoor radon. This number comprises 10% of all lung-cancer fatalities and 30% of those among nonsmokers.

Is this linear radon-induced-cancer model valid? Perhaps, but some scientists believe that there may be a threshold of exposure below which radon poses no significant threat to public health. The factors that control who gets cancer and who does not are many and complex. More research will determine which of these two models is correct.

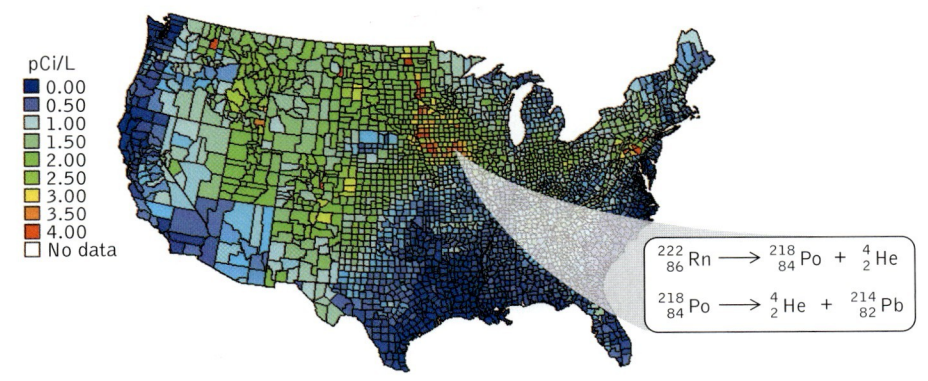

FIGURE 2.22 Dense radon gas released from the radioactive decay of uranium in rocks collects in basements. Radon poses a health hazard because it is easily inhaled and decays by releasing α particles.

Radiation therapy

Because ionizing radiation causes the most damage to those cells that are growing and dividing the fastest, it is also a powerful tool in the fight *against* cancer. Radiation therapy consists of exposing cancerous tissue to γ radiation emitted by radioactive nuclides. Sometimes the radionuclide is encased in a platinum capsule and surgically implanted in a cancerous tumor. The platinum provides a

THE CHEMISTRY OF RADON, RADIUM, AND URANIUM

Uranium is the heaviest naturally occurring element found on Earth. All uranium is radioactive, with only two isotopes of uranium, ^{235}U and ^{238}U, present in nature. In comparison with other elements in the periodic table, uranium is present in approximately the same amount as tin, Sn. Uranium is found primarily in combination with oxygen as the black mineral pitchblende, with the formula U_3O_8, and as a complex mineral uraninite, which has the composition $K_2(UO_2)_2(VO_4)_2 \cdot 3\ H_2O$.

The preparation of uranium metal consists of several steps in which uranium ore is converted into UO_3 and then into UO_2. Conversion of UO_2 into UF_4 followed by reaction with magnesium metal produces silvery uranium metal. Uranium is one of the densest metals (19.04 g/mL) known and reacts with most elements in the periodic table. The most significant use of uranium is in nuclear reactors where the fission of ^{235}U is used to produce energy, as shown in Figure 2.15. Both uranium metal and UO_2 are used as reactor fuel.

The spontaneous α and β decay of both uranium isotopes produces other elements such as thorium, Th, proactinium, Pa, and radium, Ra (as shown in Figure 2.8 and summarized below). Radium-226 is derived from ^{238}U and is typically found in uranium-containing ores. Radium belongs to the same group in the periodic table as magnesium and calcium and has the chemical properties of these elements. For example, radium forms Ra^{2+} ions, which dissolve in water. In the nineteenth century before the hazards of radioactivity were understood, hot springs rich in radium were thought to be useful in curing diseases. Radium was also used in watches in the early twentieth century to provide glowing dials. The radium-containing paint was applied by hand with

small brushes. The painters would often pass the tip of the brush between their lips in the process. The similar chemical properties of Ra^{2+} and Ca^{2+} led to the incorporation of $^{226}Ra^{2+}$ in bones, increasing the risk of cancer.

$^{238}U \rightarrow\ ^{234}Th \rightarrow\ ^{234}Pa \rightarrow\ ^{234}U \rightarrow\ ^{230}Th \rightarrow\ ^{226}Ra \rightarrow\ ^{222}Rn$

Radium-226 decays by the emission of α particles to radon-222, a radioactive gas that seeps through the ground in areas with uranium-containing minerals. The hazards of radon are discussed in Section 2.7. A simple radon detector containing carbon is available for homeowners who wish to test the air in their homes. The radioactive decay of ^{222}Rn leaves solid lead and bismuth on the carbon. Measuring the radiation from ^{214}Pb and ^{214}Bi is representative of the amount of radon present.

Chemically, radon belongs in the last column of the periodic table with helium, neon, argon, krypton, and xenon. We will explore the chemistry of these elements further in Chapter 3.

chemically inert outer layer and acts as a radiation filter, absorbing α and β particles but allowing γ rays to pass into the cancerous tissue. Some radioactive nuclides used in cancer therapy are listed in Table 2.5.

The chemical properties of a nuclide can be exploited to direct it to the site of a tumor. For example, most of the iodine in the body is concentrated in the thyroid gland. An effective therapy against thyroid cancer is the ingestion of potassium iodide (KI) containing radioactive ^{131}I.

Surgically inaccessible cancerous tumors can often be treated with a beam of γ rays from a radiation source outside the body. Unfortunately, γ radiation destroys not only cancerous cells but also the healthy tissue surrounding them. Thus, radiation-therapy patients frequently suffer many of the symptoms of radiation sickness, including nausea and vomiting (the tissues that make up intestinal walls are especially susceptible to radiation-induced damage), fatigue, weakened immune response (from depressed leukocyte production), and hair loss. For this reason, radiologists must carefully control the amount, or dosage, that each patient receives.

Medical imaging with radionuclides

The movement of radionuclides in the body and their accumulation in certain organs provide ways to visualize the circulatory system and to assess organ function. Tiny amounts of a radioactive isotope are used for these studies, together with a much larger amount of a stable isotope, called a carrier, of the same element. For example, the circulatory system can be imaged by injecting a solution of sodium chloride (NaCl) containing a trace amount of ^{24}NaCl into the bloodstream. Circulation of the radioactive sodium chloride can be monitored by measuring the γ rays emitted by ^{24}Na as it decays. In this example, the injected sodium chloride solution is said to have been labeled with a radioactive **tracer**—namely, the chloride of ^{24}Na.

The ideal radioactive tracer for medical imaging is one with a half-life about as long as the time it takes to do the imaging measurements. It should emit moderate-energy γ rays, but no α particles or high-energy β particles that cause

> A radioactive **tracer** is a species that has been labeled with a radioactive isotope of a constituent element. A tracer is administered with a much higher concentration of nonradioactive carrier.

TABLE 2.5 Some Radionuclides Used in Radiation Therapy*

Nuclide	Radiation	Half-Life	Treatment
^{32}P	β	14.3 d	Leukemia therapy
^{60}Co	β, γ	5.3 yr	External cancer therapy
^{90}Sr	β	28 yr	
^{123}I	γ	13.3 hr	Thyroid therapy
^{131}I	β, γ	8.1 d	Thyroid cancer therapy
^{137}Cs	β	30 yr	
^{192}Ir	β, γ	74 d	Surgical implantation

*For a more comprehensive list, see Appendix 3.

tissue damage. Thus the ideal mode of decay is electron capture, as described in Section 2.3.

One widely used nuclide in medical imaging is technetium-99*m*. Technetium has no stable isotopes, and so it has to be synthesized in a particle accelerator just before it is used. The process includes the synthesis of a radioactive isotope of molybdenum, which undergoes β decay to give technetium-99*m*:

$$^{99}_{42}\text{Mo} \longrightarrow {}^{99m}_{43}\text{Tc} + {}^{0}_{-1}\text{e} \tag{2.23}$$

You may be wondering what the *m* in 99*m* means. It means that the nucleus is metastable; that is, the nucleons in it are arranged in such a way that, with time, they will rearrange to form a more stable nuclear structure. In the process, the nucleus emits a γ ray. Because no particles are emitted, there is no change in atomic number or mass number:

$$^{99m}_{43}\text{Tc} \longrightarrow {}^{99}_{43}\text{Tc} + \gamma \tag{2.24}$$

The half-life of the nuclear reaction in Equation 2.24 is 6.0 hours, which is suitable for most radio-imaging studies. The "normal" technetium-99 that is produced is itself radioactive, but it decays extremely slowly (half-life = 2.1×10^5 yr) and does not damage tissue at the concentrations used in medical imaging. Table 2.6 lists some other isotopes used for medical imaging. With the exception of technetium-99*m*, they decay by electron capture, and so all emit only γ rays.

A powerful tool for diagnosing brain function makes use of short-lived radionuclides, such as carbon-11, oxygen-15, and fluorine-18, which undergo positron decay. The diagnostic technique is called positron emission tomography (PET) and was developed by Alfred P. Wolf and Joanna S. Fowler at the Brookhaven National Laboratory in Long Island, New York. In PET imaging, a patient might be administered a solution of glucose (a simple sugar) in which a small fraction of the sugar molecules contain one of the preceding three isotopes. The rate at which glucose is metabolized in various regions of the brain can then be monitored by arrays of detectors surrounding the patient's head. These detectors monitor the production of γ rays from positron–electron annihilation reactions (see Equation 2.5). Computers merge the signals that they produce into three-dimensional images of the brain. Artificial colorizing is used to distinguish regions with different levels of γ-ray production, which represent different rates of glucose metabolism, as shown in Figure 2.23. Unusual patterns in these images can indicate damage from strokes, mental illnesses including schizophrenia, manic depression, and Alzheimer's disease, and even nicotine addiction in tobacco

CONNECTION: The metabolism of glucose is described in Section 13.6.

TABLE 2.6 Some Radionuclides Used for Medical Imaging

Nuclide	Radiation Emitted	Half-Life (hr)	Use
^{99m}Tc	γ	6.0	Bones, circulatory system, various organs
^{67}Ga	γ	78.	Tumors in the brain and other organs
^{201}Tl	γ	73.	Coronary arteries, heart muscle
^{123}I	γ	13.3	Thyroxin production in thyroid gland

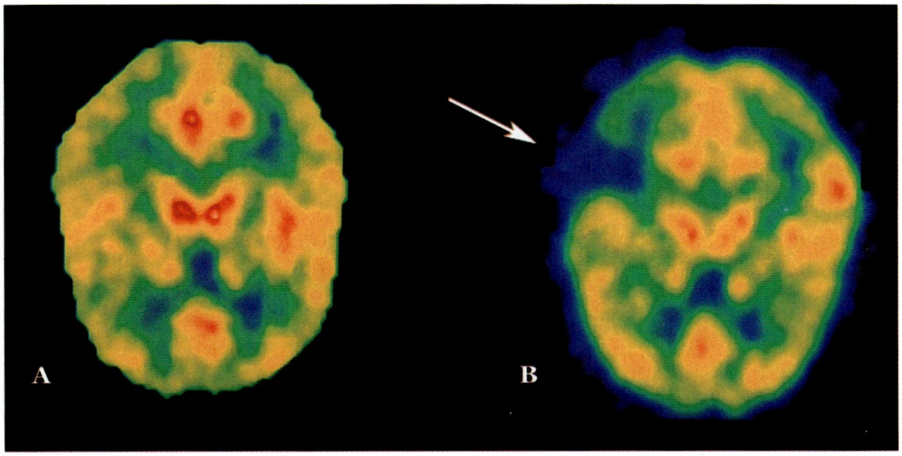

FIGURE 2.23 Positron-emission tomography is used to image soft tissue, such as brain tissue. A. A healthy person's brain. B. Image of brain function in a patient who has had a stroke. The stroke has affected blood flow in the region highlighted by the arrow.

smokers. Because the nuclides used in PET have half-lives of only a few minutes, PET imaging can be done only in locations near the accelerators in which these isotopes are made.

2.8 RADIOCHEMICAL DATING

As noted in Chapter 1, each radioactive nuclide decays at a rate inversely proportional to its half-life ($t_{1/2}$). If we know how long a radionuclide has decayed (t), we can convert this time interval into an equivalent number of half-lives (n):

$$n = t/t_{1/2} \qquad (2.25)$$

and then use Equation 1.7 to calculate from the amount of nuclide present initially (A_0) how much is left at time t (A_t):

$$A_t/A_0 = 0.5^n$$

SAMPLE EXERCISE 2.5: The cobalt-60 used in radiotherapy of cancerous tumors has a half-life of 5.3 years. If an oncology unit installs a new cobalt-60 source each year, what fraction of the initial radiation level produced by one of these sources will remain after a year?

SOLUTION: The amount of radiation produced by the source is directly related to how much radionuclide is present (A). Using Equation 2.25, we find that the number of ^{60}Co half-lives in 1 year is

$$n = t/t_{1/2} = 1.0 \text{ yr}/5.3 \text{ yr} = 0.189$$

The fraction of the radionuclide left after 0.189 half-lives is

$$A_t/A_0 = 0.5^n = 0.5^{0.189} = 0.877$$

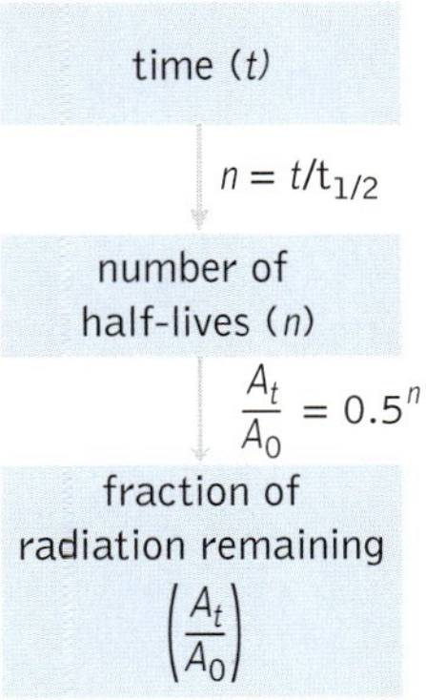

time (t)
↓
$n = t/t_{1/2}$
↓
number of half-lives (n)
↓
$\frac{A_t}{A_0} = 0.5^n$
↓
fraction of radiation remaining
$\left(\frac{A_t}{A_0}\right)$

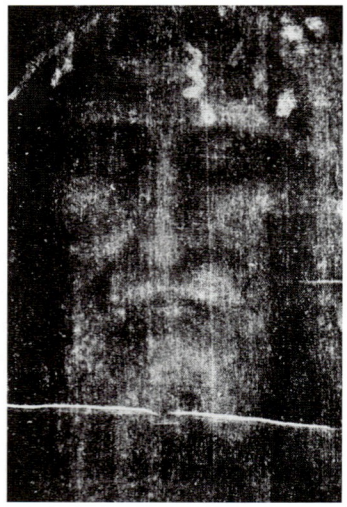

For centuries the Shroud of Turin was believed to be the burial shroud of Jesus Christ. Radiocarbon dating of tiny fragments of the fabric in 1988 indicate that it was woven between A.D. 1260 and A.D. 1390.

Radiocarbon dating is based on the presence of trace amounts of radioactive ^{14}C in the carbon dioxide that plants incorporate into their structures during photosynthesis.

CONNECTION: The carbon cycle is summarized in Section 4.6.

Thus, the radiation level of a source after 1 year will be 0.877, or 88% (to two significant figures) of the initial amount.

PRACTICE EXERCISE: The carbon-11 used in PET imaging has a half-life of only 20.3 minutes. What fraction of the ^{11}C in a sugar solution used to image the brain will remain after 1.00 hour? (See Problems 83 and 84.)

We can do this calculation in a different direction to determine how long a radionuclide has decayed (t) if we know what fraction of it is left (A_t/A_0) and its half-life ($t_{1/2}$). Using Equation 1.7, we calculate how many half-lives (n) have elapsed, and, with Equation 2.25, we calculate time (t). We can also combine these two equations by substituting the right side of Equation 2.25 for n in Equation 1.7:

$$A_t/A_0 = 0.5^{t/t_{1/2}} \tag{2.26}$$

Taking the natural logarithm (ln) of each side of Equation 2.26 to remove the exponent gives us

$$\ln A_t/A_0 = -0.693 t/t_{1/2} \tag{2.27}$$

Rearranging the terms in Equation 2.27 to solve for t [and keeping in mind that $\ln(x/y) = -\ln(y/x)$], we get

$$t = \frac{t_{1/2}}{0.693} \ln \frac{A_0}{A_t} \tag{2.28}$$

In 1947, American chemist Willard Libby (1908–1980) developed a method called **radiocarbon dating** for determining the age of artifacts from prehistory and early civilizations. The method is based on measuring the amount of radioactive carbon-14 in samples derived from plant or animal tissue. The starting point for the method is the upper atmosphere, where, as we have seen, cosmic rays produce free neutrons that can be absorbed by the nuclei of atmospheric gases. In one such nuclear reaction, nitrogen-14 absorbs a neutron and disintegrates into carbon-14 and a proton:

$$^{14}_{7}N + ^{1}_{0}n \longrightarrow ^{14}_{6}C + ^{1}_{1}H \tag{2.29}$$

Neutron-rich carbon-14 undergoes β decay with a half-life of 5730 years (Equation 2.7):

$$^{14}_{6}C \longrightarrow ^{14}_{7}N + ^{0}_{-1}e$$

After carbon-14 is formed in the upper atmosphere, it combines with oxygen to form $^{14}CO_2$. This radioactive carbon dioxide mixes with normal $^{12}CO_2$ and is carried by atmospheric transport processes to Earth's surface where both $^{14}CO_2$ and $^{12}CO_2$ are incorporated into the structures of green plants. The ratio of ^{14}C to ^{12}C in the atmosphere should be constant, assuming the intensity of cosmic rays striking Earth is constant. However, after the two nuclides have been incorporated into plant tissue, such as the trunk of a tree, the $^{14}C/^{12}C$ ratio begins to decline as carbon-14 undergoes radioactive decay. If we could determine the $^{14}C/^{12}C$ ratio of a piece of wood from an ancient building or the charcoal from a

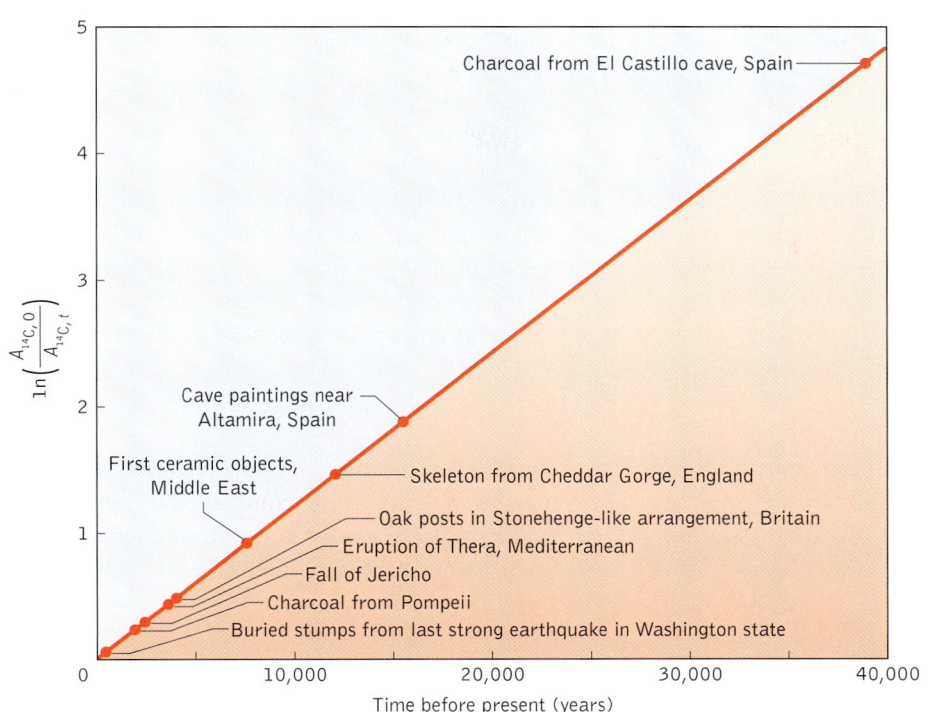

FIGURE 2.24 Radiocarbon dating: a time line of artifacts.

Papyrus was used by Egyptians for thousands of years to make a woven paperlike material. The ^{14}C-to-^{12}C ratio in these materials can be used to date them.

cave-dwelling fire or the papyrus from an early Egyptian scroll and if we knew what the corresponding $^{14}C/^{12}C$ ratio was in these materials originally (presumably the same ratio found in trees and papyrus plants growing today), we could calculate the age of these artifacts. Willard Libby showed how these measurements could be made. (For examples of objects that have been dated by using this technique, see Figure 2.24.) For this discovery and other research in isolating radionuclides, he was awarded the Nobel Prize in chemistry in 1960.

SAMPLE EXERCISE 2.6: A piece of wooden shaft from a harpoon used to hunt seals was found in the remains of an early Inuit encampment in western Alaska. Its $^{14}C/^{12}C$ ratio was found to be 61.9% of the $^{14}C/^{12}C$ ratio in an equal dry mass of the same type of wood from a recently cut tree. The half-life of ^{14}C is 5730 years. When was the harpoon made?

SOLUTION: We will assume that the sample of harpoon shaft and an equal mass of modern wood have the same total carbon content. Thus, the difference in the $^{14}C/^{12}C$ ratio between the two must be due to the age of the spear. We will also assume that the rate of carbon-14 production in the upper atmosphere is the same today as it was when the spear was made. Therefore, the A_0/A_t ratio in Equation 2.28 is 1.00/0.619. Using this information and the half-life of ^{14}C, we have

CONNECTION: Equations expressing the rate of radioactive decay were described in Section 1.5. They also apply to the rates of many chemical reactions (Section 14.3).

fraction of
carbon-14 remaining
$\left(\dfrac{A_t}{A_0}\right)$

$\downarrow$ $t = t_{1/2}/0.693 \ln\left(\dfrac{A_t}{A_0}\right)$

time of decay
(t)

$\downarrow$ date = present year − time of decay

date

$$t = \frac{5730 \text{ yr}}{0.693} \ln \frac{1.00}{0.619}$$

$$t = (8268 \text{ yr}) \ln 1.615 = (8268 \text{ yr}) \cdot 0.4796 = 3960 \text{ yr}$$

to three significant figures.

PRACTICE EXERCISE: The carbon-14 decay rate in papyrus growing along the Nile River in Egypt today is 13.8 counts per minute per gram of carbon. If a papyrus scroll found in an uncovered tomb near the Great Pyramid at Cairo has a carbon-14 decay rate of 7.6 counts per minute, how old is the scroll?

Hint: Decay rate is proportional to the amount of carbon-14 in the sample.
(See Problems 99 and 100.)

The accuracy of radiocarbon dating can be checked by determining the ^{14}C activity in the rings of very old trees such as the bristlecone pines that grow in the American Southwest (Figure 2.25). These checks reveal that radiocarbon dates are accurate to within 10%, with much of this variability due to variations in cosmic-ray production by the sun.

CONCEPT TEST: Why is ^{14}C better suited to dating wooden tools that are several thousand years old than to dating a limestone deposit that formed millions of years ago?

Other isotopes can be used to date other kinds of material. Some nuclides with half-lives of millions to billions of years can be used to estimate the age of rocks, meteorites, and Earth itself (Figure 2.26). Neptunium-237, for example, has a half-life of 2.2×10^6 years. Because none of this isotope is found in nature, Earth must be many times more than 2.2 million years old. On the other hand, uranium-235, with a half-life of 7.1×10^8 years, still makes up 0.72% of the uranium found in nature. This percentage suggests that Earth may be several billion years old. The ratio of ^{238}U to ^{206}Pb in rock samples can be used to obtain a better estimate of the age of Earth or at least its oldest rocks. Recall that ^{238}U undergoes a series of α- and β-decay steps that end with the formation of stable ^{206}Pb. The half-life of the overall process is 4.5×10^9 years. If we assume that radioactive decay of ^{238}U is the only source of ^{206}Pb in a rock sample, then the ratio of the two isotopes should provide a measure of when the rock formed. To be sure that there are no other sources of ^{206}Pb, analytical chemists can check for other isotopes of lead that would have accompanied any of the 206 isotope from another source. If none are present, then the original assumption should be valid. On the basis of their ^{238}U/^{206}Pb ratios, the oldest rocks on Earth formed about 3.8 billion years ago. Geologists believe that it took more than half a billion years for Earth to cool enough for a solid crust to form; so the overall age of Earth is believed to be about 4.5 billion years. This conclusion is confirmed by isotopic analyses of meteorites. Their ^{238}U/^{206}Pb ratios are consistently 50:50, indicating

FIGURE 2.25 Dating artifacts by using carbon-14 relies on knowing the atmospheric concentrations of carbon-14 over time. Ancient living trees such as the bristlecone pines in the American Southwest act as a check of the carbon-14 levels over thousands of years. The ages of the rings can be determined by counting them, and their carbon-14 content can be determined independently.

that half of the ^{238}U in them when they formed during the creation of the solar system has decayed. Because the half-life of ^{238}U is 4.5 billion years, the solar system and the planets must be that old, too.

2.9 THE COMPOSITION OF THE UNIVERSE

We have seen how the Big Bang theory and the synthesis of elements in giant stars may have led to the formation of our physical world. The first generation of giant stars may have formed from an unevenly expanding plasma of hydrogen and helium. These stars could have created the heavier elements found in latter-day stars, including our sun, and the planets that orbit them, including Earth. What is the elemental composition of the universe after some 15 billion years of expansion, cooling, and nuclear synthesis? From spectral analyses of the light produced by the sun and other stars and of the light absorbed by the dust and gases in interstellar space, scientists calculate that 93.4% of the atoms in the universe are hydrogen and 6.5% are helium. That leaves only 0.1% for all the other elements.

> What is the elemental composition of the universe after some 15 billion years of expansion, cooling, and nuclear synthesis?

In Figure 2.27, a logarithmic y-axis is used to plot the vastly different atomic abundances of the elements against their atomic numbers in the universe today. Keep in mind that each unit on the y-axis represents a 10-fold difference in relative abundance. We see a clear trend toward lesser abundance with increasing atomic number, as we would expect if heavier elements formed from lighter ones by nuclear fusion and neutron capture in giant stars. Some elements, including iron and lead, seem unusually abundant. Recall that iron is the heaviest element that can be synthesized through fusion of lighter elements and that the decay series for uranium-238 and other nuclides with Z > 83 end with the formation of lead. Note that, among the lighter elements, those with even atomic numbers (O, Ne, Mg, Si, S, Ar, and Ca) are about 10 times as abundant as the odd-atomic-number elements bracketing them in the chart. This greater abundance makes sense, given the theory that helium nuclei serve as nuclear building blocks in the fusion processes that take place in giant stars (see Section 2.1). We would expect nuclei that are essentially multiples of α particles—that is, $(^4_2He)_n$—to be more abundant than those not formed through the fusion of helium nuclei. Some elements, particularly lithium, beryllium, and boron, have unusually low natural abundances. These three elements are scarce because they are easily transmuted in the cores of giant stars into other, more stable nuclides. Their natural abundances would be even less if it were not for cosmic rays bombarding lighter nuclides in cosmic gases and dust, releasing high-speed protons and neutrons. Boron-11, for example, is formed by the collision of a high-speed proton with carbon-12:

$$^{12}_{6}C + ^{1}_{1}H \longrightarrow ^{11}_{5}B + 2\,^{1}_{1}H$$

FIGURE 2.26 The age of Earth can be estimated from the ratio of ^{238}U to ^{206}Pb, of ^{40}K to ^{40}Ar, or of ^{87}Rb to ^{87}Sr in rocks. For example, the amount of ^{87}Rb present originally equals the sum of the amounts of ^{87}Rb and ^{87}Sr present now. An ^{87}Rb-to-^{87}Sr ratio of 0.154 means that 154 atoms of ^{87}Rb remain for every 1000 atoms of ^{87}Sr.

✓ Hydrogen and helium account for 99.9% of the atoms in the universe.

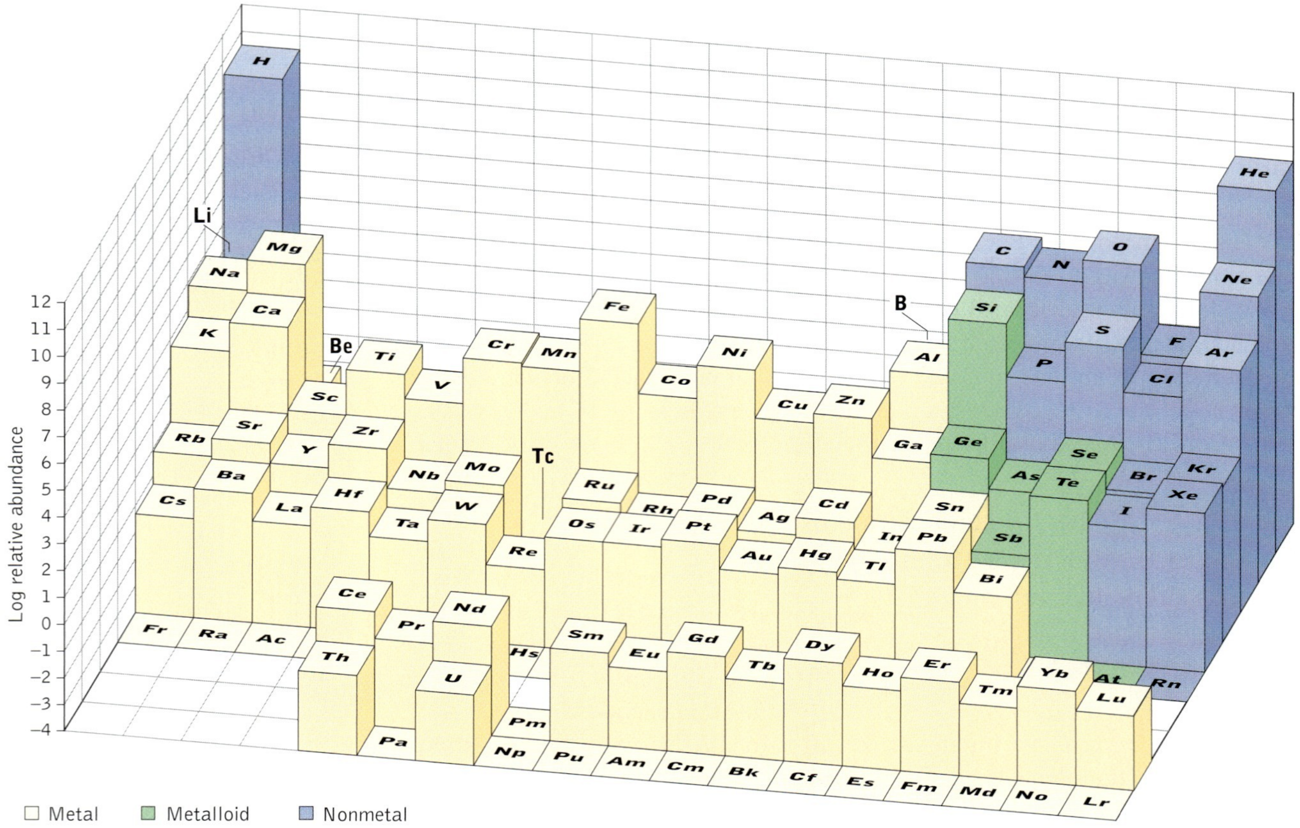

FIGURE 2.27 Atomic abundances of the elements in the universe are plotted against their atomic numbers. Note that each unit on the y-axis corresponds to a 10-fold change in relative abundance. The reference point in the chart is Si = 1,000,000 atoms of silicon, the log of which is 6.

CONCEPT TEST: Can you explain why some elements seem to be missing in Figure 2.9 (at Z = 43, 61, and 84–89)?

CONNECTION: The elemental composition of Earth is discussed in Section 4.1.

A universe rich in hydrogen and helium and poor in everything else may seem at odds with our knowledge of the composition of our own planet, which has a core of molten iron and a solid crust made up mostly of magnesium, silicon, iron, and oxygen. Actually, the elemental composition of Earth is very different from that of the sun and the universe as a whole. Why? The answer is linked to the nuclear and chemical properties of the elements and to the processes by which our planet and the other members of our solar system formed. These matters will be addressed in more detail in Chapter 4.

CHAPTER REVIEW

Summary

SECTION 2.1

The fusion of hydrogen nuclei into helium nuclei releases tremendous amounts of energy that fuels the sun and stars. Hydrogen fusion also releases positrons. A positron has the mass of an electron but a positive charge and is an example of antimatter. A particle of antimatter annihilates the corresponding particle of ordinary matter, releasing γ radiation and a quantity of energy (E) given by Einstein's equation $E = mc^2$, where m is the sum of the masses of the particles of matter and antimatter. Helium fusion requires higher temperatures than hydrogen fusion because of the greater positive charge of helium nuclei.

SECTION 2.2

The fusion of $^{4}_{2}\text{He}$ releases energy and produces nuclides with even mass numbers such as $^{12}_{6}\text{C}$ and $^{16}_{8}\text{O}$. Energy is released because the product nuclei have masses that are less than the sum of the masses of the reactant nuclei. Stellar synthesis of elements by fusion of lighter nuclei stops at Fe because the formation of heavier nuclei *consumes* energy instead of *producing* energy. (Processes that release energy are more likely to take place than those that do not.)

SECTION 2.3

Neutron capture by a nucleus results in a nuclide with higher mass. The neutron-to-proton ratio of a nucleus determines whether the nucleus is stable. Stable nuclides exist in a belt of stability where the neutron-to-proton ratio increases with atomic number, Z. Stable light nuclides have a ratio near 1, whereas heavier nuclides have ratios as high as 1.5. Nuclides outside the belt of stability are radioactive and undergo nuclear decay. Beta decay results from the conversion of a neutron in the nucleus into a proton and a high-energy electron, or β particle. Beta decay produces the nucleus of the next element (e.g., $^{14}_{6}\text{C} \rightarrow {^{14}_{7}\text{N}} + {^{0}_{-1}\beta}$). Neutron capture and β decay in giant stars result in nuclides up to $^{209}_{83}\text{Bi}$, the heaviest stable nuclide. Proton capture results in proton-rich nuclides such as $^{11}_{6}\text{C}$, which converts a proton into a neutron with the release of a positron, as in $^{11}_{6}\text{C} \rightarrow {^{11}_{5}\text{B}} + {^{0}_{-1}\text{e}}$. Another option for proton-rich nuclides is electron capture, which converts a proton into a neutron, increasing the neutron-to-proton ratio and nuclide stability. Beta decay, positron emission, and electron capture result in nuclear transmutations that convert the nucleus of an element into the nucleus of another.

SECTION 2.4

Nuclides that are rich in neutrons are produced in a supernova event. They undergo multiple β decays and produce all the elements with $Z > 26$. Nuclides with more than 83 protons are radioactive. Most decay by emitting $^{4}_{2}\text{He}$, the α particles discussed in Chapter 1 (e.g., $^{238}_{92}\text{U} \rightarrow {^{234}_{90}\text{Th}} + {^{4}_{2}\text{He}}$). This reaction begins a series of α-decay and coupled β-decay reactions in a nuclear decay series that ends with ^{206}Pb. Some nuclides such as ^{235}U undergo nuclear fission induced by reaction with slow neutrons. Processes such as $^{1}_{0}\text{n} + {^{235}_{92}\text{U}} \rightarrow {^{142}_{56}\text{Ba}} + {^{91}_{36}\text{Kr}} + 3\,^{1}_{0}\text{n}$ release energy and create more neutrons than they consume. A critical mass of a fissionable neutron absorber results in a chain reaction that releases huge amounts of energy and a nuclear explosion. The mass of ^{235}U in the fuel rods of a nuclear reactor is subcritical, and control rods moderate the neutron-induced fission reactions. The nuclear energy released converts water into steam for electricity generation.

SECTION 2.5

Artificial nuclides such as elements 95 through 112 are produced by bombarding nuclei with α particles and other nuclei in a linear accelerator or a cyclotron. Plutonium-239 is fissionable: it produces neutrons that convert nonfissionable ^{238}U into more ^{239}Pu. Breeder reactors based on this principle produce more nuclear fuel than they consume. Elements 98 through 112 have very short half-lives, but heavier nuclei may be more stable. Photographic film or electronic detectors such as the Geiger counter measure radioactivity. The ionization of gas molecules in the counter results in a current when a voltage is applied. The becquerel (Bq) is one decay per second and the curie (Ci) is 3.7×10^{10} Bq.

SECTION 2.6

Radiation ionizes molecules and damages cells. Radiation dosage is measured in rads (1 rad is 1×10^{-2} J/kg absorber) and roentgens, R (for γ radiation and X-rays; 1 rad = 1 R). The biological effects of α, β, γ, and X-ray radiation are expressed in rems, with 1 rem = 1 rad × RBE, where RBE is the relative biological effectiveness (RBE = 1 for β particles and γ radiation, and RBE = 20 for α particles). Most environmental radiation comes from radon gas. Alpha decay of ^{222}Rn and ^{218}Po may induce lung cancer. Radiation therapy kills cancer cells but often with significant side effects. Medical imaging utilizes nuclides such as ^{24}Na and ^{99m}Tc that have short half-lives and emit moderate-energy γ rays that are detectable outside the body. Positron emission tomography (PET) uses proton-rich ^{11}C, ^{15}O, and ^{18}F isotopes. Gamma rays from emitted positron annihilation allow PET diagnosis of brain malfunctions and other diseases.

SECTION 2.8

The half-life of a radioactive nuclide is independent of temperature and the initial nuclide amount. This property allows radiation to be used for estimating the ages of objects, including wooden artifacts and Earth itself. Carbon dating is based on different ^{14}C/^{12}C ratios for live and dead matter.

SECTION 2.9

After about 15 billion years of the universe expanding and cooling, hydrogen and helium still account for most matter in the universe. The relatively high abundances of iron (Fe) and lead (Pb) are due to the nuclear stability of ^{26}Fe and to Pb's position at the end of the uranium decay series. Lithium, beryllium, and boron are easily transmuted, which explains their unusually low abundance. Earth's high abundances of elements that contain even numbers of protons in their nuclei such as oxygen (O), magnesium (Mg), silicon (Si), and iron (Fe) are due to their formation in helium fusion processes.

Key Terms

α decay (p. 69)
antimatter (p. 57)
belt of stability (p. 65)
β decay (p. 62)
binding energy (p. 58)
breeder reactor (p. 75)
chain reaction (p. 71)

critical mass (p. 71)
cyclotron (p. 74)
electron capture (p. 64)
ionizing radiation (p. 80)
intermediate (p. 56)
linear accelerator (p. 74)
mass defect (p. 58)

neutron capture (p. 62)
nuclear fission (p. 71)
positron (p. 56)
positron emission (p. 64)
radiocarbon dating (p. 90)
tracer (p. 87)
transmutation (p. 65)

Key Skills and Concepts

SECTION 2.1

Understand the concepts of matter and antimatter.

SECTION 2.2

Understand why energy is produced by the fusion of small nuclei.
Be able to calculate mass defects and nuclear binding energies.

SECTION 2.3

Understand the role of neutron capture in the formation of heavy nuclei.
Understand the process of β decay and be able to predict its products.
Understand the role of proton capture in the formation of nuclei.
Understand the processes of positron emission and electron capture and be able to predict their products.

Understand the principles underlying the belt of stability and be able to predict, on the basis of its relation to the belt of stability, what decay processes a radioactive nuclide is likely to undergo.

SECTION 2.4

Understand the process of α decay and be able to predict its products.
Understand the principles of nuclear fission and chain reactions.

SECTION 2.5

Understand the approach taken in the synthesis of transuranic elements.
Understand the principles underlying the energy production of breeder reactors.

SECTION 2.6

Understand the concept of radioactivity and how it is measured.

SECTION 2.7

Understand the various measures of radiation, including the rad, roentgen, and rem.
Be aware of some of the sources and biological effects of ambient ionizing radiation.
Be aware of some everyday uses of radiation in medical treatment and imaging.

SECTION 2.8

Understand the principles underlying radiochemical dating.
Be able to calculate the ages of objects, given the amount of carbon-14 or other radioactive isotopes present.

SECTION 2.9

Understand why hydrogen and helium are the most abundant elements in the universe.

Key Equations and Relations

SECTION 2.1

Positron annihilation generates γ radiation.

$$^{0}_{1}e + ^{0}_{-1}e \longrightarrow 2\gamma \tag{2.5}$$

Calculate the binding energy (E) of a nucleus from its mass defect (Δm).

$$E = (\Delta m)c^2$$

SECTION 2.2

Convert the units in Equation 1.5 into commonly used energy units, joules.

$$1\,J = 1\,kg\,(m/s)^2$$

SECTION 2.8

Calculate the age of an object (t), for example, by using ^{14}C data.

$$t = \frac{t_{1/2}}{0.693} \ln \frac{A_0}{A_t} \tag{2.28}$$

QUESTIONS AND PROBLEMS

Hydrogen Fusion

CONCEPT REVIEW

1. Give the symbol of each of the following particles: electron, β particle, positron, proton, neutron, α particle, deuteron.
2. Indicate whether each of the following particles is positive, neutral, or negative: electron, β particle, positron, proton, neutron, α particle, deuteron.
3. Arrange the following particles in order of increasing

mass: electron, β particle, positron, proton, neutron, α particle, deuteron.

4. Electromagnetic radiation is emitted when two protons fuse to make a deuteron. In which region of the electromagnetic spectrum is the radiation?

5. Scientists at the Fermi National Accelerator Laboratory in Illinois announced in the fall of 1996 that they had created "antihydrogen." How does antihydrogen differ from hydrogen?

6. Can there be an antineutron?

Calculating Nuclear Binding Energies

CONCEPT REVIEW

7. What do *mass defect* and *nuclear binding energy* mean?
8. Why is energy *released* in a nuclear fusion process when the product is an element preceding iron in the periodic table?

PROBLEMS

9. Calculate the energy and wavelength of radiation released by the annihilation of a proton and an antiproton.
10. Calculate the energy released and the wavelength of the two photons emitted in the annihilation of an electron and a positron.
11. What is the binding energy of ^{60}Ni? The mass of ^{60}Ni is 59.9308 amu.
12. What is the binding energy of ^{50}Ti? The mass of ^{50}Ti is 49.9448 amu.
13. Calculate the energy released by the following fusion reactions, all of which produce ^{28}Si, from the exact masses of the isotopes (^{2}H, 2.0146 amu; ^{4}He, 4.00260 amu; ^{10}B, 10.0129 amu; ^{12}C, 12.000 amu; ^{14}N, 14.00307 amu; ^{24}Mg, 23.98504 amu; ^{28}Si, 27.97693 amu):
    a. ^{14}N + ^{14}N $\longrightarrow$ ^{28}Si
    b. ^{10}B + ^{16}O (15.99491 amu) + ^{2}H $\longrightarrow$ ^{28}Si
    c. ^{16}O + ^{12}C $\longrightarrow$ ^{28}Si
    d. ^{24}Mg + ^{4}He $\longrightarrow$ ^{28}Si
14. Calculate the energy released by the following fusion reactions, all of which produce ^{32}S, from the exact masses of the isotopes (^{4}He, 4.00260 amu; ^{6}Li, 6.01512 amu; ^{12}C, 12.000 amu; ^{14}N, 14.00307 amu; ^{16}O, 15.99491 amu; ^{24}Mg, 23.98504 amu; ^{28}Si, 27.97693 amu; ^{32}S, 31.97207 amu):
    a. ^{16}O + ^{16}O $\longrightarrow$ ^{32}S
    b. ^{28}Si + ^{4}He $\longrightarrow$ ^{32}S
    c. ^{14}N + ^{12}C + ^{6}Li $\longrightarrow$ ^{32}S
    d. ^{24}Mg + 2 ^{4}He $\longrightarrow$ ^{32}S
15. Our sun is a fairly small star, with barely enough mass to fuse hydrogen to give helium. Calculate the nuclear binding energy per nucleon for ^{4}He, given the exact masses of ^{4}He (4.00260 amu), ^{1}H (1.00728 amu), and neutrons (1.00867 amu).
16. Our sun contains carbon even though the sun is too small to manufacture carbon by nuclear fusion.
    a. Explain where the carbon may have come from.
    b. Calculate the binding energy per nucleon for ^{12}C, given the exact masses of ^{12}C (12.000 amu), ^{1}H (1.00728 amu), and neutrons (1.00867 amu).

Neutron Capture and Modes of Radioactive Decay

CONCEPT REVIEW

17. What factor favors neutron capture over proton capture?
18. Explain how β decay produces the nucleus of an element with a higher atomic number.
19. Describe the *belt of stability* and how it can be used to predict the likely decay mode of an unstable nucleus.
20. Compare and contrast positron-emission and electron-capture processes. What happens to the atomic number of a nucleus that undergoes either of these processes?
21. Is the following statement correct? "Elements with atomic numbers between 27 and 83 form by β decay because the neutron-to-proton ratio of the parent isotope is too low."
22. Iodine-137 decays to give xenon-137, which decays to give cesium-137. Which of the following statements is true?
    a. Xenon-137 decays by β decay, but iodine-137 decays by γ decay.

b. Xenon-137 decays by β decay, but iodine-137 decays by α decay.
   c. Both xenon-137 and iodine-137 decay by β decay.

PROBLEMS

23. Which of the following nuclides could not be produced by the capture of one or more neutrons by ^{116}Sn followed by β decay in stars? (a) ^{117}Sn; (b) ^{116}In; (c) ^{121}Sb; (d) ^{122}Te.
24. Write a balanced nuclear equation for:
    a. β emission by ^{28}Mg
    b. α emission by ^{255}Lr
    c. electron capture by ^{129}Cs
    d. positron emission by ^{25}Al
25. If the mass number of an isotope is more than twice the atomic number, is the neutron-to-proton ratio less than, greater than, or equal to 1?
26. In each of the following pairs of isotopes, which isotope has more protons and which one has more neutrons? (a) ^{127}I or ^{131}I; (b) ^{188}Re or ^{188}W; (c) ^{14}N or ^{14}C.
27. Aluminum is found on Earth exclusively as ^{27}Al. However, ^{26}Al is formed in stars. Aluminum-26 decays to give magnesium-26 with a half-life of 7.4×10^5 years. Write an equation describing the decay of ^{26}Al to ^{26}Mg.
28. Which of the products ^{131}Te, ^{131}Xe, and ^{131}Sb results from the decay of ^{131}I?
29. Refer to the belt of stability and predict the modes of decay for the following radioactive isotopes: (a) ^{32}P; (b) ^{10}C; (c) ^{50}Ti; (d) ^{19}Ne; (e) ^{116}Sb.
30. Nine isotopes of sulfur are known with mass numbers ranging from 30 to 38. Five of the nine are radioactive: ^{30}S, ^{31}S, ^{35}S, ^{37}S, and ^{38}S. Which of these isotopes do you expect to decay by β emission?
31. The isotopes ^{56}Co and ^{44}Ti were observed in supernova SN 1987A. Predict the decay pathway for these radioactive isotopes.
32. Predict the decay pathway of the radioactive isotope ^{56}Ni.

Neutron Capture and Nuclear Fission

CONCEPT REVIEW

33. What is the difference between nuclear fusion and nuclear fission?
34. What is meant by the term *fissionable nucleus*?
35. What process initiates nuclear fission of uranium-235?
36. Why does a nuclear explosion occur if the critical mass of uranium-235 is exceeded?
37. How is the rate of energy release controlled in a nuclear reactor?
38. What is a *breeder reactor*?
39. Why are neutrons always by-products of the fission of heavy nuclides? (Hint: Look closely at the neutron-to-proton ratios shown in Figure 2.9.)
40. The origin of the two naturally occurring isotopes of boron, ^{11}B and ^{10}B, are unknown. Both isotopes may be formed from collisions between protons and carbon, oxygen, or nitrogen in the aftermath of supernova explosions. Propose nuclear reactions for the formation of ^{10}B from such collisions with ^{12}C and ^{14}N.

PROBLEMS

41. The fission of uranium produces dozens of isotopes. For each of the following fission reactions, determine the number of protons and neutrons in the missing product and write the symbol for it.
    a. ^{235}U + ^{1}n $\longrightarrow$ ^{96}Zr + ? + 2 ^{1}n
    b. ^{235}U + ^{1}n $\longrightarrow$ ^{99}Nb + ? + 4 ^{1}n
    c. ^{235}U + ^{1}n $\longrightarrow$ ^{90}Rb + ? + 3 ^{1}n
42. For each of the following fission reactions, determine the number of protons and neutrons in the missing product and write the symbol for it.
    a. ^{235}U + ^{1}n $\longrightarrow$ ^{137}I + ? + 2 ^{1}n
    b. ^{235}U + ^{1}n $\longrightarrow$ ^{94}Kr + ? + 2 ^{1}n
    c. ^{235}U + ^{1}n $\longrightarrow$ ^{95}Sr + ? + 2 ^{1}n
43. Complete the following nuclear reactions:
    a. ^{210}Po $\longrightarrow$ ^{206}Pb + ?
    b. ^{3}H $\longrightarrow$ ^{3}He + ?
    c. ^{11}C $\longrightarrow$ ^{11}B + ?
    d. ^{111}In $\longrightarrow$ ^{111}Cd + ?
*44. The presence of uranium-containing ores has made part of the Northern Territory of Australia a battleground between those seeking to mine the uranium and the indigenous aborigine population. An article in *Outside* magazine in March 1999 described the dangers of a proposed mine as follows:

> Thorium 230 becomes radium 226. . . . Radium 226 goes to radon 222. Radon 222, a heavy gas that will flow downhill, goes to polonium 218 when one alpha pops out of the nucleus. . . . Polonium 218 goes to lead 214, lead 214 to bismuth 214, bismuth 214 to polonium 214, and then that goes to lead 210, all within minutes, amid a crackle of alphas and betas and gammas.

a. Write balanced nuclear reactions for the decay of thorium-230 and determine how many "alphas and betas" are produced.
b. Using an appropriate reference such as the *Handbook of Chemistry and Physics* (CRC Press, Boca Raton, FL), find the half-lives for each isotope and comment on the statement that all these processes take place "within minutes."

Making Artificial Elements

CONCEPT REVIEW

45. How are linear accelerators and cyclotrons used to make artificial elements?
46. Why must the velocity of the nuclide that is fired at a target nuclide to form a superheavy element be not too fast and not too slow, but "just right"?

PROBLEMS

47. Complete the following nuclear reactions used in the preparation of isotopes for nuclear medicine:
    a. $^{32}S + {}^1n \longrightarrow ? + {}^1H$
    b. $^{55}Mn + {}^1H \longrightarrow {}^{52}Fe + ?$
    c. $^{75}As + ? \longrightarrow {}^{77}Br$
    d. $^{124}Xe + {}^1n \longrightarrow ? \longrightarrow {}^{125}I + ?$
48. Complete the following nuclear reactions used in the preparation of isotopes for nuclear medicine:
    a. $^6Li + {}^1n \longrightarrow {}^3H + ?$
    b. $^{16}O + {}^3H \longrightarrow {}^{18}F + ?$
    c. $^{56}Fe + ? \longrightarrow {}^{57}Co + {}^1n$
    d. $^{121}Sb + {}^4He \longrightarrow ? + 2\,{}^1n$
49. Complete the following nuclear reactions:
    a. $? \longrightarrow {}^{122}Xe + {}^{\ 0}_{-1}e$
    b. $? + {}^4He \longrightarrow {}^{13}N + {}^1n$
    c. $? + {}^1n \longrightarrow {}^{59}Fe$
    d. $? + {}^1H \longrightarrow {}^{67}Ga + 2\,{}^1n$
50. Seaborgium (Sg, element 106) is prepared by the bombardment of curium-248 with neon-22, which produces two isotopes, ^{265}Sg and ^{266}Sg. Write balanced nuclear reactions for the formation of both isotopes. Are these reactions better described as fusion or fission processes?
51. The synthesis of elements 114 and 118 was reported in early 1999. Bombardment of ^{208}Pb with ^{86}Kr produced $^{293}_{118}$X, whereas bombardment of ^{244}Pu with ^{48}Ca produced a single atom of $^{289}_{114}$XX. Write balanced nuclear reactions for the formation of both elements.
52. Describe how a ^{209}Bi target might be bombarded with subatomic particles to form ^{211}At. Use balanced equations for the required nuclear reactions.

Measuring Radioactivity, Its Biological Effects, Cancer Therapy, and Medical Imaging

CONCEPT REVIEW

53. Describe two ways of detecting and measuring radiation levels.
54. What is the difference between a *level* of radioactivity and a *dose* of radioactivity?
55. What are some of the molecular effects of exposure to radioactivity?
56. Describe the dangers of exposure to radon-222.
57. How does the selection of an isotope for radiotherapy relate to (a) its half-life? (b) its mode of decay? (c) the properties of the products of decay?
58. Describe an example of medical imaging with a radioactive isotope.
59. Explain why radiocarbon dating is reliable only for artifacts and fossils younger than about 50,000 years.
60. Which of the following statements about ^{14}C dating are true?
    a. The amount of ^{14}C in all objects is the same.
    b. Carbon-14 is unstable and is readily lost from the atmosphere.
    c. The ratio of ^{14}C to ^{12}C in the atmosphere is a constant.
    d. Living tissue will absorb ^{12}C but not ^{14}C.
61. Explain why ^{40}K dating ($t_{1/2} = 1.28 \times 10^9$ years) generally works only for fossils or rocks older than 300,000 years.

62. Periodic outbreaks of food poisoning from *E. coli*-contaminated meat have renewed the debate about irradiation as an effective treatment of food. In one newspaper article on the subject, the following statement appeared: "Irradiating food destroys bacteria by breaking apart their molecular structure." How might you improve or expand on this explanation?

PROBLEMS

63. Dental X-rays expose patients to about 0.5 mrem of radiation. Given an RBE of 1 for X-rays, how many rads does 0.5 mrem represent? For a 50-kg person, how much energy does 0.5 mrem correspond to?
64. Some workers responding to the explosion at the Chernobyl nuclear power plant were exposed to 500 rem of radiation, resulting in death for many of them. If the exposure was primarily in the form of γ rays with an energy of 3.3×10^{-14} J and an RBE of 1, how many γ-ray photons did an 80-kg person absorb?
65. In 1991, the Environmental Protection Agency proposed a maximum radon level in drinking water of 300 pCi per liter of water.
    a. How many disintegrations per second does this level correspond to for 1 liter of water?
    b. Using the relation rate of decay = kN, where N = atoms of Rn and $k = 0.693/t_{1/2}$, how many Rn atoms does this level correspond to in 1 liter of water ($t_{1/2, Rn} = 3.8$ d)?
66. Suppose that the air in a house is contaminated with 4.0 pCi/L of radon. How many atoms of Rn does a person breathe in with a 4.0-L lungful of air?
67. Yttrium-90, rhenium-188, dysprosium-165, and bismuth-213 are used in nuclear medicine. Which ones are unlikely to undergo β decay?
68. Samarium-153, phosphorus-32, iodine-131, and radium-223 are used in nuclear medicine. Which ones are unlikely to undergo β decay?
69. In boron neutron-capture therapy (BNCT), a patient is given a compound containing ^{10}B that accumulates inside the cancer tumor. Then the tumor is irradiated with neutrons, which are absorbed by ^{10}B nuclei. The product of neutron capture is an unstable form of ^{11}B that undergoes a decay to ^{7}Li.
    a. Write a balanced nuclear equation for the neutron absorption and α-decay process.
    b. Calculate the energy released by each nucleus of boron-10 that captures a neutron and undergoes α decay, given the following masses of the particles in the process: ^{10}B (10.0129 amu), ^{7}Li (7.01600 amu), ^{4}He (4.00260 amu), and ^{1}n (1.008665 amu).
    c. Why might a nuclide that undergoes α decay be particularly effective for cancer therapy?
70. The absorption of a neutron by ^{11}B produces ^{12}B, which decays by two pathways: α decay and β decay.
    a. Write balanced nuclear reactions for these processes.
    b. Calculate the energy released by each boron-11 nucleus from the following masses: ^{11}B (11.00931 amu), ^{12}C (12.0000 amu), ^{4}He (4.00260 amu), and ^{1}n (1.008665 amu).
71. What percentage of a sample's original radioactivity remains after two half-lives?
72. What percentage of a sample's original radioactivity remains after five half-lives?
73. Balloon angioplasty is a common procedure for unclogging arteries in patients suffering from arteriosclerosis. Iridium-192 therapy is being tested as a treatment to prevent reclogging of the arteries. In the procedure, a thin ribbon containing pellets of ^{192}Ir is threaded into the artery. The half-life of ^{192}Ir is 74 days. How long will it take 99% of the radioactivity from 1.00 mg of ^{192}Ir to disappear?
74. Americium-241 ($t_{1/2} = 433$ yr) is used in smoke detectors. How long will it take for the activity of a sample of ^{241}Am to drop to 1% of its original activity?
75. In a letter to *Science News* magazine, a knowledgable reader responds to a recent article about radiation therapy as follows:

    > First, I believe you mean iridium-192, not iridium-92 (which doesn't exist). Second, iridium-192 not only generates gamma rays, it first emits an energetic electron via beta decay. Characteristic energies of both beta electrons and gamma rays are on the order of 0.3 MeV. However, the range of the beta electron in tissue is on the order of 1,000 microns, while that of the gamma ray is about 10 centimeters.

    a. Why doesn't iridium-92 exist?
    b. Write a balanced nuclear reaction for the β decay of ^{192}Ir.
    c. Comment on whether the reader is correct about the effects of β and γ rays.
76. Francium has chemical properties similar to those of sodium, potassium, and other elements found in the first column of the periodic table. Francium can be produced by bombarding gold with oxygen-18 atoms, which creates francium-210.
    a. Write an equation describing the formation of ^{210}Fr from gold and ^{18}O. Which isotope of gold is needed for the reaction?

b. Predict the products of the radioactive decay of ^{210}Fr.
c. Radiation from which isotope will penetrate more deeply into the body?

77. Predict the most likely mode of decay for the following isotopes used as imaging agents in nuclear medicine: (a) ^{197}Hg (kidney); (b) ^{75}Se (parathyroid gland); (c) ^{113}In (brain); (d) ^{18}F (bone).

78. Predict the most likely mode of decay for the following isotopes used as imaging agents in nuclear medicine: (a) ^{133}Xe (cerebral blood flow); (b) ^{57}Co (tumor detection); (c) ^{51}Cr (red blood cell mass); (d) ^{67}Ga (neoplastic diseases).

79. A 1.00-μg sample of ^{192}Ir was inserted into the artery of a heart patient. After 30 days, 0.756 μg remained. What is the half-life of ^{192}Ir?

80. Dysprosium-165 with an activity of 1100 counts per second was injected into the knee of a patient suffering from rheumatoid arthritis. After 24 hours, the activity had dropped to 1.14 counts per second. Calculate the half-life of ^{165}Dy.

81. Iodine isotopes are used in brain imaging of people suffering from Tourette syndrome. Mammalian brain cells were treated with a solution containing ^{131}I with an initial activity of 108 counts per minute. The cells were removed after 30 days, and the remaining solution was found to have an activity of 4.1 counts per minute. Did the brain cells absorb any ^{131}I ($t_{1/2}$ = 8.1 d)?

82. Immediately after the unclogging of an artery in a patient suffering from arteriosclerosis, 1.75 pg of ^{188}Re ($t_{1/2}$ = 18.7 min) was pumped into the balloon catheter. How many half-lives passed before the amount of ^{188}Re was reduced to 1% of its original value?

83. Mercury-197 has a half-life of 65 hours. A patient with a kidney problem is given 10.0 ng of mercury-197 to help diagnose kidney function. How much mercury-197 will remain after 6 days?

84. Selenium-75 is a β emitter with a half-life of 120 days that is used for radioimaging the pancreas. Approximately how much selenium-75 would remain in a 0.050-pCi sample that has been stored for 1 year (365 days)?

85. Gold-198 is a β emitter used to treat leukemia. Its half-life is 2.7 days. How long does it take a 30-mCi sample of gold-198 to decay until only 3.75 mCi remains?

86. The half-life for the β decay of strontium-90 is 28.8 years. A milk sample is found to contain 10.3 ppm (parts per million) of ^{90}Sr. How many years will pass before the amount of ^{90}Sr drops to 1.0 ppm?

87. Carbon-11 is an isotope used in positron-emission tomography and has a half-life of 20.3 minutes. How long will it take for 99% of the ^{11}C injected into a patient to decay?

88. Technetium-99m is used to image the circulatory system and has a half-life of 6.0 hours. How long will it take for 95% of the ^{99m}Tc that is injected into a patient to decay?

89. Consider the following decay series:

$$A\ (t_{1/2} = 4.5\ s) \longrightarrow B\ (t_{1/2} = 15.0\ d) \longrightarrow C$$

If we start with 10^6 atoms of A, how many atoms of A, B, and C are there after 30 days?

90. Which element in the following series will be present in the greatest amount after 1 year?

$$t_{1/2} = \ ^{214}_{83}Bi \xrightarrow{-\alpha} \ ^{210}_{81}Ti \xrightarrow{-\beta} \ ^{210}_{82}Pb \xrightarrow{-\beta} \ ^{210}_{83}Bi$$
$$\quad\quad\ 20\ min \quad\quad 1.3\ min \quad\quad 20\ yr \quad\quad 5\ d$$

91. Sodium-24 is used to treat leukemia and has a half-life of 15 hours. A patient was injected with a salt solution containing sodium-24. What percentage of the ^{24}Na remained after 48 hours?

92. There was once a plan to store radioactive waste that contained plutonium-239 in the reefs of the Marshall Islands. The planners claimed that the plutonium would be "resonably safe" after 240,000 years. If its half-life is 24,400 years, what percentage of the ^{239}Pu would remain?

93. Rhenium-188 is formed by the decay of tungsten-188.
a. Write a balanced equation describing the decay of tungsten-188 to rhenium-188.
b. Why doesn't tungsten-188 decay by electron capture or positron emission?

94. Rhenium-188, a promising isotope for the treatment of colorectal cancer, decays by β emission. Write a balanced nuclear equation describing the decay of rhenium-188.

95. Radiocarbon dating of charcoal from a cave in Chile was used to establish the earliest date of human habitation in South America as 8700 years ago. What fraction of the ^{14}C present initially remained after 8700 years?

96. Native Americans living along the north coast of Peru for thousands of years have woven ornate twined cotton textiles. One of the oldest samples is believed to have been woven 4800 years ago. Compared with the fibers of cotton plants growing today, what is the ratio of carbon-14 to carbon-12 in the sample?

97. A giant sequoia tree cut down in 1891 in what is now known as Kings Canyon National Park contained 1342 annual growth rings. If samples of the tree were re-

moved for radiocarbon dating today, what would be the ratio of the carbon-14 concentration in the innermost (oldest) ring compared with that in the youngest? Assume that the density and overall carbon content of the wood in the rings is the same.

98. Geologists who study volcanoes can develop historical profiles of previous eruptions by determining the $^{14}C/^{12}C$ ratios of charred remains of plants entrapped in old magma and ash flows. If the uncertainty in determining these ratios is 0.1%, could radiocarbon dating distinguish between debris from the eruptions of Mt. Vesuvius that occurred in a.d. 472 and a.d. 512? (Hint: Calculate the $^{14}C/^{12}C$ ratios for samples from the two dates.)

99. A mammoth tusk containing grooves made by a sharp stone edge (indicating the presence of humans or Neanderthals) was uncovered at an ancient camp site in the Ural Mountains in 2001. The $^{14}C:^{12}C$ ratio in the tusk was only 1.19% of that in modern elephant tusks. How old is the mammoth tusk?

100. The ratio of carbon-14 to carbon-12 in charred bones is 92% of the $^{14}C:^{12}C$ ratio found in living organisms. The half-life of ^{14}C is 5730 years. How old are the bones?

3 Electrons
And Electromagnetic Radiation

During a solar eclipse, the outer layer of the sun, or corona, is clearly visible behind the dark sphere of the moon. Studying sunlight provides clues about the electronic structure of atoms.

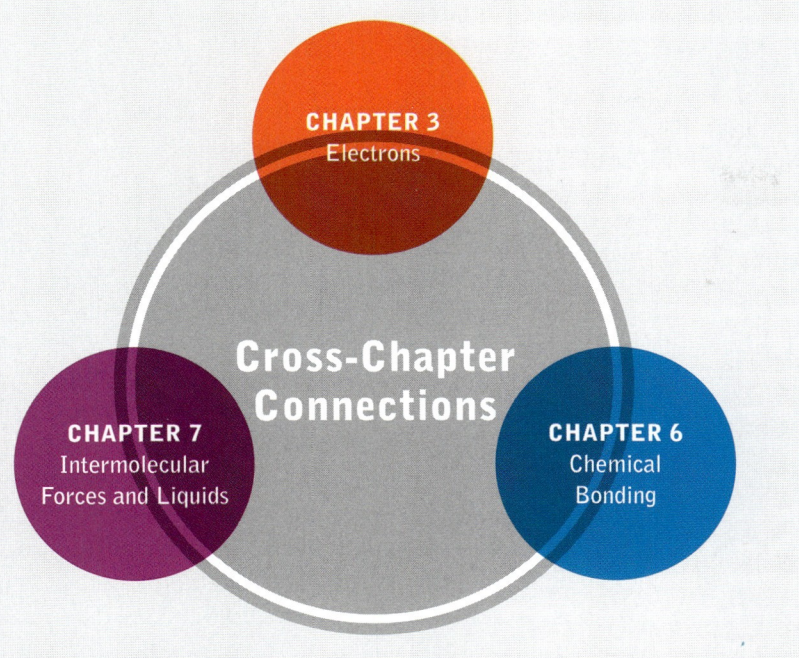

3.1 The Fraunhofer lines
3.2 Electrons in atoms
3.3 Particles of light
3.4 The Bohr model of the hydrogen atom
3.5 Matter waves
3.6 Schrödinger's wave equation
3.7 Quantum numbers
3.8 Shapes and sizes of atomic orbitals
3.9 Spinning electrons
3.10 The periodic table and filling in the orbitals
   BOX: THE CHEMISTRY OF THE NOBLE GASES
3.11 More evidence for the existence of atomic orbitals
   Ionization energies
   X-ray photoelectron spectroscopy
3.12 The uncertainty principle

A Look Ahead

This chapter is about the structure of atoms. We start with the discovery of dark lines in the sun's spectrum (the Fraunhofer lines described in Chapter 1), noting that these lines result from absorption of radiation by atoms and ions, and that these atoms and ions can emit the colors missing from sunlight. We will see how the spectra of atoms are related to atomic energy states that are, in turn, related to how electrons are distributed within atoms. We will also follow the development of quantum theory, which accounts for the spectra and chemical properties of the elements on the basis of the electron configurations of their atoms.

3.1   THE FRAUNHOFER LINES

In Chapter 1, we noted that the sun's visible spectrum contains a number of narrow, dark lines. German physicist and optical craftsman Joseph von Fraunhofer mapped hundreds of these lines in the early 1800s. He assigned letters of the alphabet to the darkest of them. However, Fraunhofer and his contemporaries did not understand *why* the lines existed in the sun's spectrum. That understanding was provided by two other German scientists half a century later.

✓ At high temperatures, atoms of each element emit and absorb electromagnetic radiation with characteristic wavelengths; these are **line spectra**.

CONNECTION: Hubble used the redshift in the Fraunhofer lines to determine that the universe was expanding (Section 1.3).

In the mid–nineteenth century, German chemist Robert Wilhelm Bunsen (1811–1899) and physicist Gustav Robert Kirchoff (1824–1887) extensively studied the **line spectra** emitted by elements being vaporized in the transparent flames of a burner designed by Bunsen. The wavelengths of the emission lines that they obtained were the same as the wavelengths of some of the Fraunhofer lines in the sun's spectrum. For example, Fraunhofer's D line (actually a pair of closely spaced lines at 589.0 and 589.6 nm that were later given numerical subscripts D_1 and D_2) matched the wavelength of orange light produced by hot sodium vapor, as shown in Figure 3.1. The wavelengths of Fraunhofer's H and K lines near 393 nm exactly matched the wavelengths of calcium emission lines. A prominent line in the green region matched that produced by hot magnesium atoms. From these experiments and others came the understanding that each element emits a unique electromagnetic spectrum when the element is heated to a sufficiently high temperature *and* that atoms of each element can absorb electromagnetic radiation of the same characteristic wavelengths. This atomic absorption process casts the shadows of Fraunhofer's lines in the spectrum of the sun.

Some of the Fraunhofer lines could not be accounted for by the spectra of elements known in the mid–nineteenth century. Kirchoff and Bunsen suspected that some elements in the sun had not yet been isolated and identified on Earth. Two of these elements were named after the colors of their prominent Fraunhofer lines: cesium from the Latin *cæsius*, meaning "bluish gray," and rubidium from the Latin *rubidus* (red). The English astronomer Joseph Lockyer proposed that a series of unidentified Fraunhofer lines were due to an unknown element that he named helium after *Helios*, the Greek sun god. It was not until 1895 that Scottish chemist William Ramsay isolated helium from a mineral containing uranium.

CONCEPT TEST: Can you think of a reason why helium might be present in a uranium-bearing mineral?

Kirchoff conducted other experiments that demonstrated the ability of free atoms to emit and absorb electromagnetic radiation at characteristic wavelengths. He discovered that, by passing sunlight through sodium vapor, he could make the Fraunhofer D lines darker. Moreover, when a beam of sunlight passed through a flame glowing bright yellow orange from hot sodium atoms, the D lines disappeared. The sodium in the flame had added back the yellow orange light that sodium atoms in the outer atmosphere of the sun had absorbed (see Figure 3.1).

3.2 ELECTRONS IN ATOMS

Why do the vapors of an element emit and absorb light of characteristic wavelengths? Why are some of the darkest Fraunhofer lines produced not by hydrogen and helium but by sodium, magnesium, iron, and calcium—elements that make up only a tiny fraction of the sun's photosphere (outer region)? To begin to

FIGURE 3.1 *What a Chemist Sees.* The wavelengths of the distinctive yellow orange glow produced by sodium atoms in flames (A) exactly match those of the dark absorption lines produced by ground-state sodium atoms when illuminated by lamps or the sun (B). Thus, the characteristic wavelengths of radiant energy absorbed by sodium atoms exactly match the wavelengths of radiation that hot sodium atoms emit.

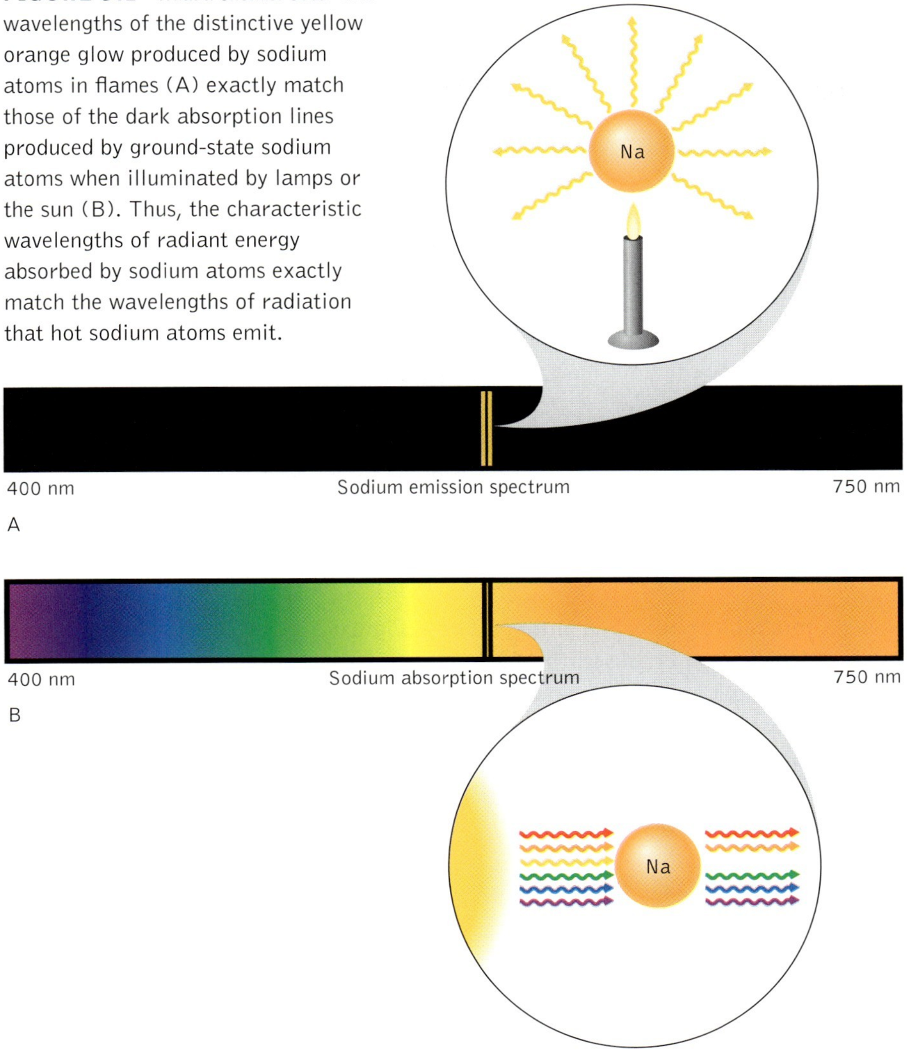

400 nm — Sodium emission spectrum — 750 nm

A

400 nm — Sodium absorption spectrum — 750 nm

B

answer these questions, we need to understand more about the structure of atoms and how their structures relate to their interaction with electromagnetic radiation.

Why do the vapors of an element emit and absorb light of characteristic wavelengths?

At the beginning of the twentieth century, scientists had very different ideas about atomic structure from those we have today. One popular idea was called the "plum pudding" model, which assumed that an atom is a diffuse sphere of positive charge with negatively charged electrons embedded in the

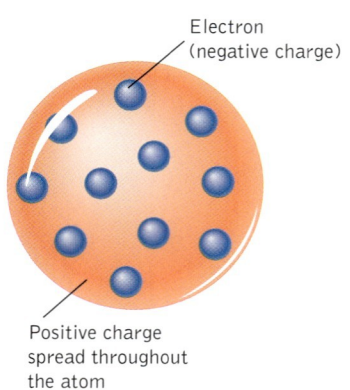

FIGURE 3.2 The plum-pudding model of atomic structure assumed that the positive charge in an atom was evenly distributed throughout its volume and that electrons were embedded within the sphere of positive charge, like plums in a plum pudding.

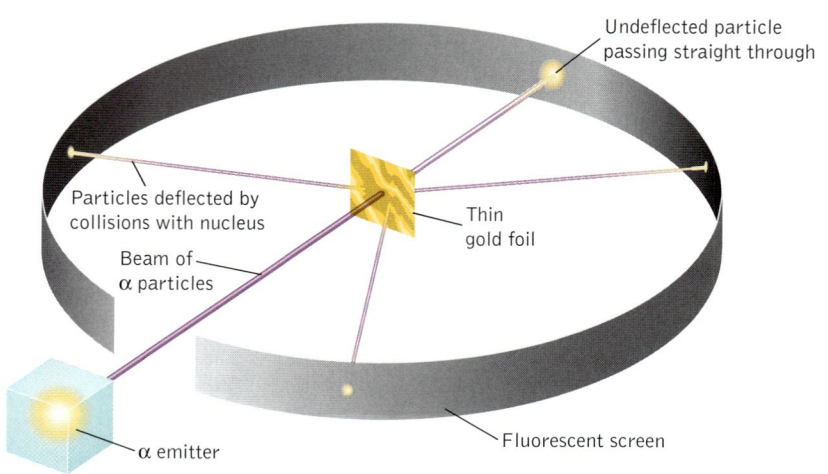

FIGURE 3.3 Rutherford's experiment entailed the scattering of α particles by gold foil. The results, as shown, led to the theory that atoms have small, positively charged nuclei.

sphere, much like plums in a plum pudding (Figure 3.2). In 1907, while working in the laboratory of Ernest Rutherford, two students, Hans Geiger (the Geiger for whom the Geiger counter was named) and Ernest Marsden, conducted an experiment designed to investigate the plum-pudding model. They bombarded a thin foil of gold with α particles and measured how many of the α particles were deflected from their original track. They assumed that most of the α particles would pass straight through the diffuse positive spheres of the gold atoms but that, occasionally, some of the particles would interact with the tiny electrons embedded in these spheres and be deflected slightly. And that is mostly what they observed. However, they also observed something completely unexpected. They found that about 1 in 8000 particles deflected from the foil through an average angle of 90 degrees (Figure 3.3), and a very few bounced right back at the source of the particles. Years later, Rutherford recounted his astonishment at these results:

> It is about as incredible as if you had fired a 15-inch shell at a piece of tissue paper and it came back and hit you.

Rutherford concluded that these large deflections could occur only if the positive α particles encountered large concentrations of positive charge in the foil. Therefore, the positive charge in each atom must be concentrated in a tiny volume (the nucleus) that occupies only about 1/10,000 of the volume occupied by the electrons. Rutherford's model of the atom is shown in Figure 3.4. It replaced the plum-pudding model and is the basis for our current understanding of atomic structure.

As already mentioned, some of the Fraunhofer lines in the sun's spectrum could not be attributed to known elements and were presumed to be produced by elements yet to be discovered. This interpretation worked well for cesium,

☑ The results of Geiger and Marsden's experiments on α-particle scattering showed that a typical atom is about 10,000 times as large as its nucleus.

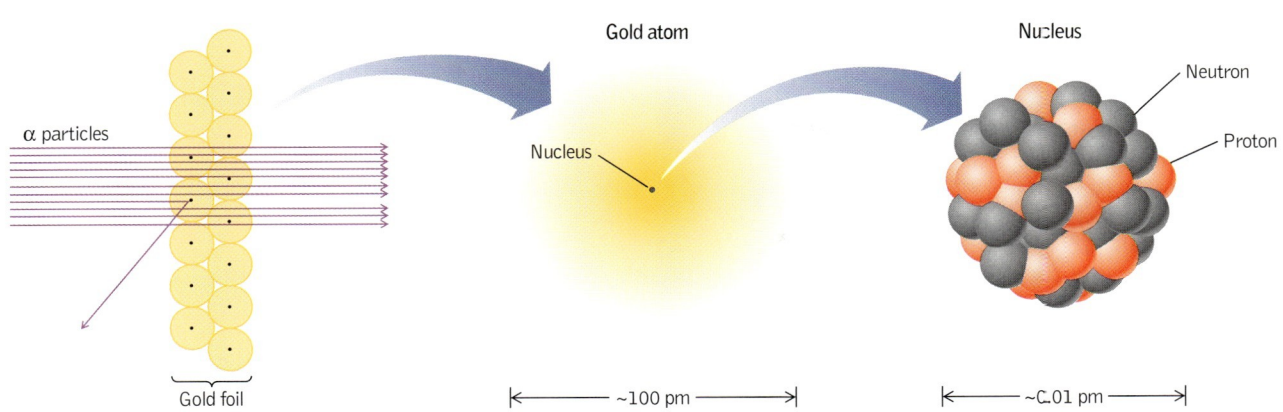

FIGURE 3.4 Rutherford's model of the atom includes a nucleus that is about 1/10,000th the size of the atom. The nucleus would be too small to see if drawn to scale.

rubidium, and helium lines but not so well for others. One series of mystery lines was attributed to the yet-to-be discovered element "coronium," named after the sun's outer atmosphere, the corona. In 1927, American astronomer Ira Bowen discovered that "coronium" lines were actually caused by the absorption of sunlight by oxygen and nitrogen atoms that had lost electrons and so were present in the sun's atmosphere as positively charged ions. Similarly, Fraunhofer's H and K lines (Figure 3.5) are due to the absorption of sunlight not by calcium *atoms* (Ca) but by singly charged calcium *ions* (Ca^+).

If the absorption and emission spectra of free atoms of an element differ from those of free ions of the same element, what does that tell us about the processes responsible for absorption and emission? Because the only differences between atoms and ions of the same element are related to the number of electrons surrounding identical nuclei, the electrons must have a role in producing line spectra. Therefore, the absorption and emission of electromagnetic energy by atoms and ions must be related to increases and decreases in the energies of their electrons.

☑ Atoms absorb and emit electromagnetic radiation as the energies of their electrons change.

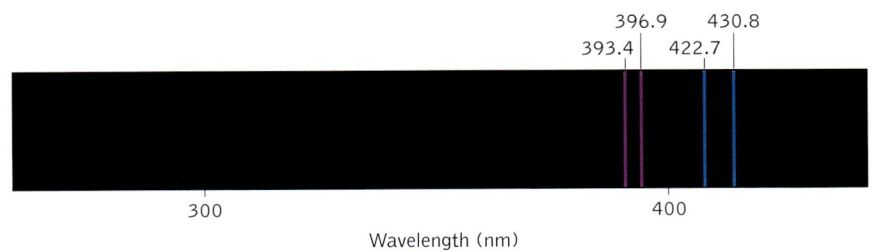

FIGURE 3.5 The line spectra of calcium atoms and calcium ions are different. The lines at 422.7 and 43.8 nm are produced by calcium atoms; the pair at 393.4 and 396.9 nm (Fraunhofer's H and K lines) arise from calcium ions.

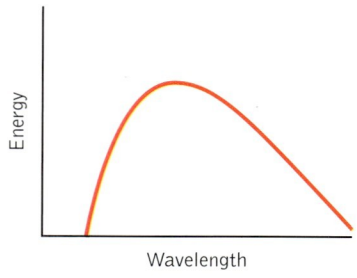

FIGURE 3.6 A glowing filament in an incandescent light bulb emits a continuum of wavelengths. Planck was modeling this type of spectrum when he proposed a particle description of radiation.

3.3 PARTICLES OF LIGHT

In Chapter 1, we noted that both the sun and a light bulb appear to emit all wavelengths of visible light (Figure 3.6). The fact that free atoms and ions can absorb narrow parts of the sun's spectrum raises the question, Is any spectrum *really* continuous? Toward the end of the nineteenth century, many scientists were trying to develop equations that accurately describe the spectra of high-temperature continuum sources of electromagnetic energy. None of the mathematical models agreed with published experimental results. Then, in 1900, German physicist Max Planck developed an equation that did fit the data. But Planck was not satisfied with his equation, because it was only an **empirical** equation. By empirical we mean that the equation agreed with experimental observations, but it had no theoretical basis. Planck set out to derive a theoretical model. He succeeded in doing so but only after he made a critical assumption: that high-temperature continuum radiators do not absorb and release energy continuously but, rather, in tiny chunks or packets of energy. He named such a packet a **quantum** of energy. Albert Einstein would later call the particles of electromagnetic radiation corresponding to these tiny energy packets *photons* (see Section 1.3).

To understand the significance of Planck's assumption, consider the two ways in which you might go from the sidewalk to the entrance of the building in Figure 3.7. If you walked up the steps, there would be discrete heights between the sidewalk and the entrance: the height of each step. You could not stand with both feet at an elevation in between two adjacent steps: there would be nothing to stand on at that height. However, if you walked up the wheelchair ramp, you could stop at any height between the sidewalk and the entrance. The steps and the discrete height changes that they represent are a useful model of Planck's hypothesis that energy is released (as in walking down the steps) or absorbed (walking up the steps) in discrete packets, or quanta, of energy.

> Is any spectrum *really* continuous?

CONNECTION: The continuum spectra of the sun and a light bulb are compared in Section 1.6.

✓ An **empirical** relation accounts for experimental measurements but does not necessarily have a theoretical basis.

The energy (E) of a photon, the smallest **quantum** (quantity) of radiant energy, is given by the equation $E = h\nu$.

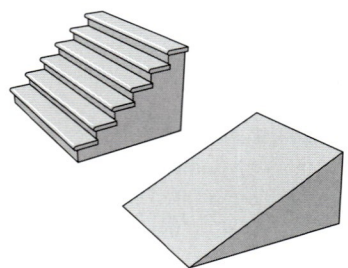

FIGURE 3.7 A flight of stairs exemplifies quantization: there are discrete heights in a flight of stairs. In contrast, the heights on a ramp are not quantized.

CONCEPT TEST: Which of the following quantities vary continuously and which of them vary by discrete values?

- The volume of water that evaporates from a lake each day during a summer heat wave
- The quantity of eggs remaining in a carton in a refrigerator
- The time that you take to get ready for class in the morning
- The number of red lights that one might encounter in driving the length of Fifth Avenue in New York City

Einstein used the concept that electromagnetic radiation consists of tiny particles to explain the **photoelectric effect**, a process by which some metals emit electrons when they are illuminated. According to classical (Newtonian) physics, the production of these **photoelectrons** should be proportional to the intensity of the beam of light striking the metal surface. According to the classical model, the energy of a wave is proportional to its amplitude: the bigger the wave, the more energy it has. But this classical model failed to explain a well-known feature of the photoelectric effect: to produce photoelectrons, the radiation illuminating the metal has to have a minimum, or threshold, *frequency*, not *amplitude*. Thus, a very bright source of low-frequency radiation produced no photoelectrons, and no flow of electrical current in a device such as the phototube shown in Figure 3.8. However, even a dim source of high-frequency radiation produced a little current, and a brighter source of high frequency produced even more current.

Einstein explained these observations by proposing that a photon striking the surface of a metal *transmits all of its energy to a single atom.* If this energy exceeds the energy needed to release an electron from that atom and from the surface of

The **photoelectric effect** is the release of electrons, called **photoelectrons**, from a metal as a result of the absorption of electromagnetic radiation. The minimum frequency of light for releasing photoelectrons depends on the metal. Higher frequencies give electrons with more kinetic energy.

CONNECTION: Although he didn't call them photons, Newton was the first to propose the existence of tiny packets of light (Section 1.3).

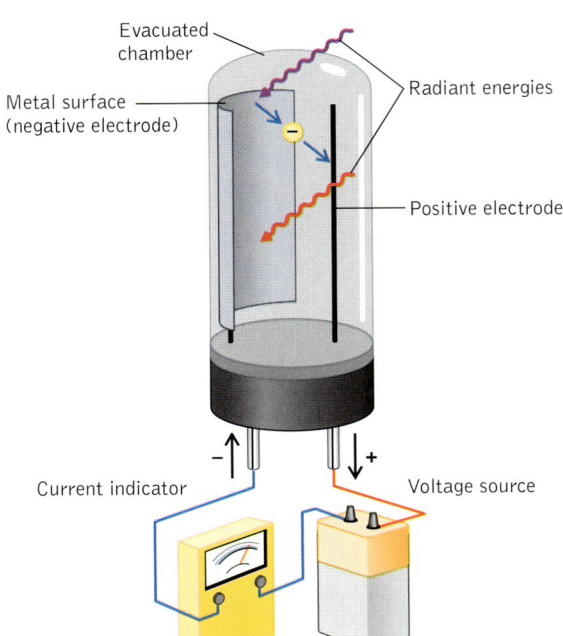

FIGURE 3.8 A device called a phototube includes a metallic electrode with a negative electrical charge. If radiation of high enough energy (shown in violet) illuminates this electrode, electrons "boil off" its surface (the photoelectric effect) and flow toward the positive electrode. This flow of electrons produces an electrical current. The size of the current is proportional to the intensity of the radiation—that is, to the number of photons per unit time striking the negative electrode. Photons that are of much lower energy (shown in red) do not produce the photoelectric effect.

> The **work function** (ϕ) is the amount of energy needed to release a photoelectron from the surface of a metal.

the metal—a quantity of energy called the **work function** (ϕ) of the metal—then the photoelectric effect is observed. A bright source of such radiation produces many photons per unit time, and they will produce many photoelectrons and a large current in a phototube.

SAMPLE EXERCISE 3.1: The work function (ϕ) of silver is 7.59×10^{-19} J. What is the longest wavelength (corresponding to the lowest frequency) of electromagnetic radiation that can eject an electron from the surface of a piece of silver?

SOLUTION: The exercise gives the amount of energy that must be overcome to dislodge an electron from a piece of silver. This amount is the minimum energy (E) that a photon must have to produce the photoelectric effect in silver. Equation 1.3 allows us to calculate the wavelength that corresponds to this amount of energy:

$$E = hc/\lambda$$

or

$$\lambda = hc/E = (6.626 \times 10^{-34} \text{ J} \cdot \text{s})(2.998 \times 10^8 \text{ m/s})/7.59 \times 10^{-19} \text{ J}$$
$$= 2.62 \times 10^{-7} \text{ m} = 262 \text{ nm}$$

energy (E)

$\lambda = \frac{hc}{E}$

wavelength (λ)

The question asks for the "longest" wavelength. Because the energy of a photon is inversely proportional to its wavelength, a maximum wavelength corresponds to a "minimum" energy. In this case, the minimum energy is that of a photon with $\lambda = 262$ nm. A longer wavelength of light would not have enough energy to overcome the work function of silver. A shorter wavelength would have more than enough energy to do so. Any extra energy would become the kinetic energy (see Section 3.11) of the ejected electron.

PRACTICE EXERCISE: The energy required to remove the electron from a hydrogen atom is 2.18×10^{-18} J. What is the wavelength of a photon of radiation with this much energy? (See Problems 11–16.)

3.4 THE BOHR MODEL OF THE HYDROGEN ATOM

The development of the idea that there are discrete energy levels inside atoms was closely followed by the experiments of Geiger and Marsden and the emergence of the modern view of atomic structure. These developments led another of Rutherford's brilliant assistants, Danish physicist Niels Bohr, to propose that the energies of electrons surrounding atomic nuclei also are quantized. In 1913, he published a model of the hydrogen atom based on the notion that the electron in a hydrogen atom must reside in one of an array of discrete energy states, or "allowed" energy levels. Key features of Bohr's hydrogen model (Figure 3.9) are:

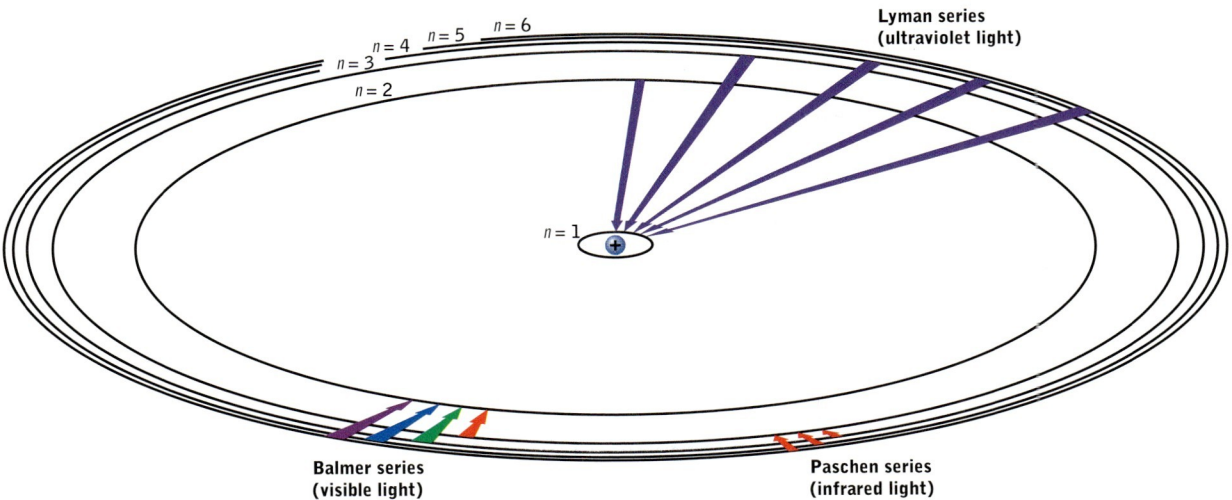

FIGURE 3.9 The Bohr model places electrons in concentric circular orbits with certain energy levels. Each orbit is identified by the principal quantum number n, where $n = 1, 2, 3,$ and so forth. The arrows represent electron transitions between some of these levels. Transitions in dark red correspond to the emission of infrared radiation, and those in violet correspond to the emission of ultraviolet radiation. Bohr's model resembles our solar system.

- The electron in a hydrogen atom swirls around its one-proton nucleus in one of many concentric orbits.
- Each orbit represents a different energy level.
- The electron may jump from one energy level to another by either acquiring or releasing an amount of energy equal to the difference in the energies of the two levels.
- Each orbit is designated by a specific value for n, called a quantum number. The orbit closest to the nucleus has the lowest value for n; that is, $n = 1$. The next closest orbit has an n value of 2, and so on, as shown in Figure 3.9.

In developing his model, Bohr was influenced by the studies of a Swiss high school teacher named Johann Balmer, who had shown in 1885 that the frequencies of the visible lines in the hydrogen spectrum fit a simple equation:

$$\nu = C\left(\frac{1}{2^2} - \frac{1}{n^2}\right) \tag{3.1}$$

where C is a constant equal to 3.29×10^{15} s^{-1} and n is a positive integer greater than 2. Because we usually refer to spectral features by their wavelengths rather than frequencies, let's convert Equation 3.1 into an equivalent expression based on wavelength (actually, 1/wavelength). Starting with Equation 1.1 from Chapter 1, we have

$$c = \lambda \nu$$

CONNECTION: The wavelengths and energies of the electromagnetic spectrum are presented in Section 1.3.

or

$$\frac{1}{\lambda} = \nu/c$$

Substituting Equation 3.1 for ν gives

$$\frac{1}{\lambda} = \frac{C}{c}\left(\frac{1}{2^2} - \frac{1}{n^2}\right)$$

Inserting the numerical values of C and c, we have

$$\frac{1}{\lambda} = \frac{(3.29 \times 10^{15}\,\text{s}^{-1})}{(2.998 \times 10^8\,\text{m/s})}\left(\frac{1}{2^2} - \frac{1}{n^2}\right)$$

$$\frac{1}{\lambda} = \frac{1.097 \times 10^7}{\text{m}}\left(\frac{1}{2^2} - \frac{1}{n^2}\right) \tag{3.2}$$

If we change the units for wavelength from meters to nanometers, we have a more useful version of Equation 3.2 for calculating the wavelengths of the principal lines in the hydrogen spectrum:

$$\frac{1}{\lambda} = \frac{1.097 \times 10^7}{\cancel{\text{m}}}\left(\frac{1}{2^2} - \frac{1}{n^2}\right)\left(\frac{10^{-9}\,\cancel{\text{m}}}{\text{nm}}\right)$$

$$\frac{1}{\lambda} = \frac{1.097 \times 10^{-2}}{\text{nm}}\left(\frac{1}{2^2} - \frac{1}{n^2}\right) \tag{3.3}$$

Before we consider the significance of Balmer's equation and its relation to Bohr's model, we should note that there is more to the spectrum of hydrogen than is visible to the eye. As noted in Chapter 1, the electromagnetic spectrum extends beyond the visible region to both shorter and longer wavelengths; so the spectrum of hydrogen might be expected to have lines in the ultraviolet and infrared as well as visible regions. Indeed it does. In addition to the visible spectral lines, called the Balmer series (Figure 3.10), there are the Lyman series of lines in

CONNECTION: The relation between energy, wavelength, and frequency is discussed in Section 1.3.

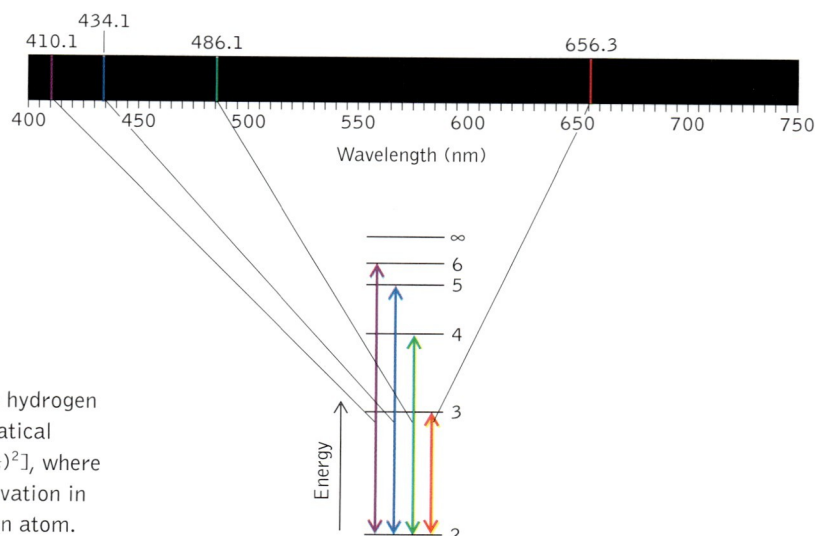

FIGURE 3.10 The four visible lines of the hydrogen emission spectrum fit the following mathematical equation: $\frac{1}{\lambda} = 1.097 \times 10^{-2}\,nm^{-1}[(\frac{1}{2})^2 - (\frac{1}{n})^2]$, where $n = 3, 4, 5,$ or 6. Niels Bohr used this observation in the development of his model of the hydrogen atom.

the ultraviolet and the Paschen series in the infrared, each named after the scientist who first characterized it. An important feature of these other spectral series is that the wavelengths of the lines in them fit Balmer's equation if we replace the "2" in the denominator of the first fraction with a "1" for the Lyman series or with a "3" for the Paschen series (see Figure 3.9). This pattern suggests a general form of Equation 3.3 that should (and does) predict the wavelengths of all the lines in the hydrogen spectrum:

$$\frac{1}{\lambda} = 1.097 \times 10^{-2} \text{ nm}^{-1} \left(\frac{1}{n_1^2} - \frac{1}{n_2^2} \right) \tag{3.4}$$

where n_1 is any positive integer and n_2 is another positive integer larger than n_1.

 The equation

$$\frac{1}{\lambda} = 1.097 \times 10^{-2} \text{ nm}^{-1} \left(\frac{1}{n_1^2} - \frac{1}{n_2^2} \right)$$

accurately predicts the wavelengths of all the lines in the hydrogen spectrum.

SAMPLE EXERCISE 3.2: What is the wavelength of the line in the hydrogen spectrum corresponding to $n_1 = 2$ and $n_2 = 3$?

SOLUTION: Using Equation 3.4 and substituting 2 for n_1 and 3 for n_2, we have

$$\frac{1}{\lambda} = 1.097 \times 10^{-2} \text{ nm}^{-1} \left(\frac{1}{2^2} - \frac{1}{3^2} \right) = 1.097 \times 10^{-2} \text{ nm}^{-1} \left(\frac{1}{4} - \frac{1}{9} \right)$$

$$= 1.097 \times 10^{-2} \text{ nm}^{-1}(0.1389) = 1.524 \times 10^{-3} \text{ nm}^{-1}$$

$$\lambda = 656 \text{ nm}$$

This wavelength is that of the red line in the hydrogen spectrum. It is also the wavelength of one of Fraunhofer's missing lines in the solar spectrum.

PRACTICE EXERCISE: What are the wavelengths of the lines of the hydrogen spectrum corresponding to $n_1 = 2$ and $n_2 = 4, 5,$ and 6? Before calculating them, predict which value of n corresponds to the longest wavelength of the three. (See Problems 29 and 30.)

We can derive an equation for calculating the energies of the photons emitted or absorbed by atomic hydrogen if we combine Equation 3.4 with Equation 1.3 ($E = hc/\lambda$):

$$E = \frac{hc}{\lambda} = (6.626 \times 10^{-34} \text{ J} \cdot \text{s}) \left(\frac{2.998 \times 10^8 \text{ m}}{\text{s}} \right) \left(\frac{10^9 \text{ nm}}{\text{m}} \right)$$

$$\left(\frac{1.097 \times 10^{-2}}{\text{nm}} \right) \left(\frac{1}{n_1^2} - \frac{1}{n_2^2} \right)$$

$$E = (2.18 \times 10^{-18} \text{ J}) \left(\frac{1}{n_1^2} - \frac{1}{n_2^2} \right) \tag{3.5}$$

If we assume that this rock on its geological perch has zero energy, then after it falls from its perch, it will have less energy than it has in this photograph: the further it falls, the more negative its energy will be. Similarly, the closer an electron is to the nucleus of a hydrogen atom, the more negative its energy is.

The Bohr model says that energy values calculated by using Equation 3.5 must equal the changes in energy (ΔE) that the electron in an atom of hydrogen may undergo as it jumps from one orbit (energy level) to another. This equality is achieved when the energy of each orbit is:

$$E_n = (-2.18 \times 10^{-18} \text{ J}) \left(\frac{1}{n^2} \right) \tag{3.6}$$

✓ The Bohr model says that the energy of the electron in a hydrogen atom depends only on quantum number, n. This energy is always negative: the lower the value of n, the more negative the energy of the electron.

According to Equation 3.6, the energy of the electron in a hydrogen atom is always *negative*. A hydrogen atom has its lowest (most negative) energy when its electron is in the $n = 1$ orbit closest to the nucleus:

$$E_1 = (-2.18 \times 10^{-18} \text{ J})\left(\frac{1}{1^2}\right) = -2.18 \times 10^{-18} \text{ J}$$

This level of energy is the **ground state** of the atom. The value of E_n increases (becomes less negative) as n increases, because the value of the fraction $1/n^2$ decreases. As the value of n approaches infinity ($n \to \infty$), E_n approaches zero ($E_n \to 0$). At $n = \infty$, the electron is no longer in the hydrogen atom. In other words, the hydrogen atom has ionized, producing a free electron and leaving behind a proton.

All the energy states above the $n = 1$ ground state are called **excited states**. Bohr proposed that the electron in a hydrogen atom can jump from the ground state to one of the excited states, or from one excited state to another, or from an excited state back to the ground state. But it can make one of these transitions only if it absorbs or emits a quantum of energy that exactly matches the energy difference (ΔE) between the two states. We can express this energy difference in equation form:

$$\Delta E = E_f - E_i \quad (3.7)$$

where E_i is the energy of the initial state (with principal quantum number n_i) and E_f is the energy of the final state (with principal quantum number n_f). Combining Equations 3.6 and 3.7, we get:

$$\Delta E = \left[(-2.18 \times 10^{-18} \text{ J})\left(\frac{1}{n_f^2}\right)\right] - \left[(-2.18 \times 10^{-18} \text{ J})\left(\frac{1}{n_i^2}\right)\right]$$

which can be simplified by combining the two terms in brackets on the right side:

$$\Delta E = \left[(-2.18 \times 10^{-18} \text{ J})\left(\frac{1}{n_f^2} - \frac{1}{n_i^2}\right)\right] \quad (3.8)$$

> The lowest-energy, or most stable, state of an atom is its **ground state**. Any state with energy higher than that of the ground state is called an **excited state**.

SAMPLE EXERCISE 3.3: What amount of energy is required to ionize a ground-state hydrogen atom?

SOLUTION: Ground state means that the atom's electron is in the lowest energy orbit, where $n = 1$. The atom undergoes ionization when n goes to ∞. Using these two values for n_i and n_f, respectively, in Equation 3.8, we have

$$\Delta E = \left[(-2.18 \times 10^{-18} \text{ J})\left(\frac{1}{n_f^2} - \frac{1}{n_i^2}\right)\right]$$

$$= \left[-2.18 \times 10^{-18} \text{ J})\left(\frac{1}{\infty^2} - \frac{1}{1^2}\right)\right] = 2.18 \times 10^{-18} \text{ J}$$

PRACTICE EXERCISE: Calculate the energy required to ionize a hydrogen atom from the $n = 3$ excited state. Before doing the calculation, predict whether this amount of energy is greater than or less than the energy required to ionize a ground-state hydrogen atom. (See Problems 31 and 32.)

Equation 3.8 looks a lot like Equation 3.5. The only difference is the negative sign on the right-hand side of Equation 3.8. The logic of having a negative sign in Equation 3.8 may become clearer if we consider an electron transition that results in a gain of energy, as when a hydrogen atom *absorbs* a photon of energy. The electron undergoes a transition from a lower energy level to a higher one, as represented by each of the arrows pointing upward in Figure 3.10. In each of these transitions n_f is greater than n_i, and the value of $(1/n_f^2) - (1/n_i^2)$ is *negative*. According to Equation 3.8, this negative value makes ΔE positive.

In other words, the atom *gains* energy, the energy of the photon. Such a transition can take place only if the energy of the photon *exactly matches the difference in energy* (ΔE) between the two energy states.

Now let's consider what happens when an atom of hydrogen *emits* a photon. Its electron is initially in one of its excited states and falls to either a lower-energy excited state or the ground state. In either case n_f is less than n_i, which means that ΔE is negative. The quantum of energy lost must match the energy of the photon emitted. Mathematically, the energy of the photon must match the absolute value of ΔE, which is written as $|\Delta E|$:

$$E_{photon} = |\Delta E| \tag{3.9}$$

Before we leave this discussion of the hydrogen spectrum and Bohr's model, we will examine why the Fraunhofer lines are so rich in the spectra of elements other than hydrogen and helium. These other elements, taken together, make up less than 1% of the atoms in the sun, yet they account for most of the prominent absorption lines. The reason is embedded in Balmer's equation for the visible lines of the hydrogen spectrum. We might interpret Equation 3.6 in the following way: to absorb a photon of one of the wavelengths in the Balmer series, a hydrogen atom must first be in the $n = 2$ energy level ($n_i = 2$ in Equation 3.8). But the $n = 2$ energy level is an *excited state*, one level above the ground state. What are the chances that hydrogen atoms on the surface of the sun are in this first excited state instead of the ground state? It turns out that the chances are not good. We can calculate what the chances are by using Equation 3.8 and the energy required to raise a hydrogen atom's electron from the ground state ($n = 1$) to the first excited state ($n = 2$):

$$\Delta E = \left[(-2.18 \times 10^{-18} \text{ J})\left(\frac{1}{n_f^2} - \frac{1}{n_i^2}\right)\right]$$
$$= \left[(-2.18 \times 10^{-18} \text{ J})\left(\frac{1}{2^2} - \frac{1}{1^2}\right)\right] = [(-2.18 \times 10^{-18} \text{ J})(-0.75)]$$
$$= 1.635 \times 10^{-18} \text{ J}$$

Next, we use an equation developed by Austrian physicist Ludwig Boltzmann (1844–1906) that allows us to calculate the ratio of the number of hydrogen atoms in an excited state (N_i) to the number in the ground state (N_0):

$$\frac{N_i}{N_0} = n_i^2 \times e^{-E_i/kT} \tag{3.10}$$

where n_i is the principal quantum number of the excited state, E_i is the energy difference between the ground state and the excited state, T is temperature in

CONNECTION: Molecules also can absorb radiation, forming molecular excited states (Section 6.9).

CONNECTION: Bohr built on the observations of Balmer and others in developing his theory of the hydrogen atom—an example of the dynamics of the scientific method (Section 1.2).

kelvins, and k is Boltzmann's constant $(1.3806 \times 10^{-23}$ J/K). Using the E_i value for $n = 2 (1.635 \times 10^{-18}$ J) and the sun's surface temperature (5700 K), we get:

$$\frac{N_i}{N_0} = n_i^2 \times e^{-E_i/kT} = (2^2) \times e^{-1.635 \times 10^{-18} \text{ J}/(1.3806 \times 10^{-23} \text{ J/K}) 5700 \text{K}} = 4 \times e^{-21.08}$$

$$= 2.8 \times 10^{-9}$$

This result tells us that only about three hydrogen atoms in a billion on the surface of the sun are in the first excited state and, therefore, capable of absorbing radiation corresponding to one of Balmer's lines. On the other hand, *ground-state* sodium atoms produce Fraunhofer's D lines at 589.0 and 589.6 nm and *ground-state* calcium ions produce the H and K lines at 397 and 393 nm, respectively. Most of these atoms and ions are in their ground states at 5700 K, and so, despite their relatively low abundance, they play major roles in creating the line shadows in the sun's spectrum.

As we close this discussion of the Bohr model, we should note that the model has two major limitations:

1. It applies only to hydrogen atoms and those ions, such as He^+, having a single electron. The model does not account for the observed spectra of multielectron elements and ions and so does not provide an accurate view of their electronic structure. The reason is that electrons interact with each other. In addition to electromagnetic repulsion, there is another factor that will be addressed later in this chapter: electrons spin. This spinning creates tiny magnetic fields that interact with the fields produced by other electrons.

2. Although the Bohr model faithfully describes the features of the hydrogen spectrum, it has no theoretical basis. It assumes that an electron swirls around each hydrogen nucleus in a stable orbit; it does not explain *why* it does. According to the laws of classical physics, an orbiting electron moving in the electrical field created by the positive nucleus should radiate energy and eventually spiral into the nucleus. The Bohr model is silent on why this event does not happen. A decade after Bohr published his model of the hydrogen atom, another young scientist provided a theoretical basis for the stability of electron orbits and greatly affected our view of the structure of matter. His approach incorporated yet another significant change in the way in which early-twentieth-century scientists viewed the behavior of atoms and subatomic particles—namely, to think of them not only as particles of matter, but also as waves.

✓ Bohr's empirical atomic model fails for atoms or ions with two or more electrons because it does not account for interactions between electrons.

3.5   MATTER WAVES

In 1905, Einstein published his special theory of relativity. In doing so, he tackled two phenomena that could not be explained by Newtonian physics: (1) the transmission of light through a vacuum, and (2) the constancy of the speed of light, independent of the motion of the source or observer. Einstein postulated that the speed of light is not only a constant, independent of one's frame of reference, but also an ultimate speed that cannot be exceeded. According to his theory, the energy (E) of a "particle" of light depends on its **momentum** (p), which is the product of its mass (m) and its velocity (c), and on its mass at zero velocity called **rest mass** (m_0) as follows:

$$E^2 = p^2c^2 + m_0^2c^4 \quad (3.11)$$

Einstein also proposed that the mass of any particle moving nearly as fast as the speed of light is related to its velocity (v), according to the following formula:

$$m = \frac{m_0}{\left(1 - \frac{v^2}{c^2}\right)^{0.5}} \quad (3.12)$$

As the velocity of a particle approaches the speed of light, the ratio v^2/c^2 approaches 1, making the denominator of Equation 3.12 nearly zero and the value of m nearly infinite. Clearly, matter as we know it, with some definable rest mass, cannot travel at the speed of light, which leads to the conclusion that the only kind of particle that can travel at the speed of light is a particle with zero rest mass. A photon must be such a particle. If we apply Equation 3.11 to a photon with zero mass ($m_0 = 0$), the second term on the right-hand side is zero and the equation simplifies to

$$E^2 = p^2c^2$$

or

$$E = pc \quad (3.13)$$

In 1923, a French graduate student named Louis de Broglie (1892–1977) proposed that the energy defined by Equation 3.13 and the energy of a photon ($E = hc/\lambda$) are really the same energy and therefore equal to each other:

$$E = pc = hc/\lambda$$

or

$$\lambda = h/p$$

If we assume that the momentum of a particle is equal to its mass times its velocity (and if it is a particle of light, then its velocity is c), then $p = mc$. Substituting this expression into $\lambda = h/p$, we have

$$\lambda = h/mc \quad (3.14)$$

Equation 3.14 provides a mathematical link between the wavelike and the particle-like properties of light. But de Broglie did not stop there. If light could behave as a particle, he reasoned, why couldn't a particle in motion, such as an electron, behave as a wave? If that were true, then the electron should have a characteristic wavelength that could be calculated from a modification of Equation 3.14:

> If light could behave as a particle, de Broglie reasoned, why couldn't a particle in motion, such as an electron, behave as a wave?

$$\lambda = h/mv \quad (3.15)$$

where m is the mass of the electron and v is its velocity.

The **momentum** of an object is the product of its mass and its velocity ($p = mv$).

The mass of a particle of light with zero velocity is called its **rest mass**.

CONNECTION: De Broglie's equation relating the mass of an object and its velocity with a wavelength was alluded to in Section 1.5.

The de Broglie equation $\lambda = h/mv$ links the wavelike and particle-like properties of electromagnetic radiation and can be used to calculate wavelike properties of electrons and other particles of matter.

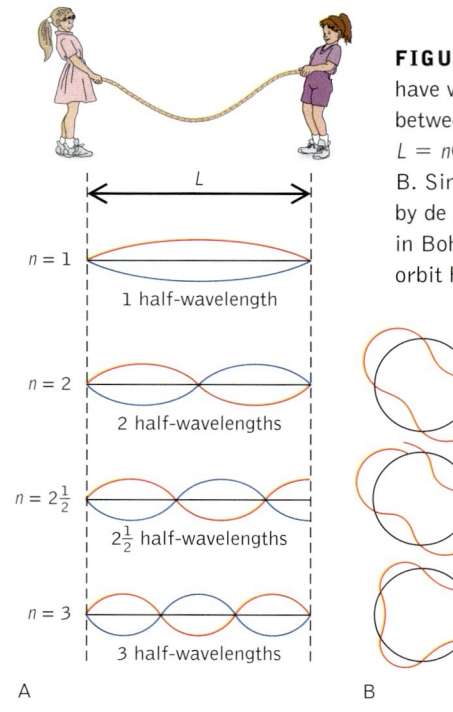

FIGURE 3.11 A. Standing waves in a jump rope have wavelengths that are related to the distance between the ends of the rope (L) by the equation $L = n(\lambda/2)$. In the examples shown, $n = 1, 2,$ and 3. B. Similarly, the circular standing waves proposed by de Broglie account for the stability of the orbits in Bohr's model of the hydrogen atom. Each stable orbit has a circumference equal to $n\lambda$.

A wave that is confined to a given space and has a wavelength that is some multiple of the dimensions of the space is a **standing wave**.

CONNECTION: Constructive and destructive interference patterns are shown in Figure 1.16.

De Broglie applied this reasoning to explain the stability of the orbits in Bohr's model of the hydrogen atom. He proposed that the electron in a hydrogen atom could behave as a wave, and in particular as a circular wave oscillating around the nucleus. To understand the implications of this reasoning, we need to examine what it takes to make a stable circular wave. Consider the motion of a vibrating string on a guitar or violin or the oscillations of a rope set in motion with a flick of the wrist as in a game of jump rope. These patterns of motion can be represented by **standing waves**, the wavelengths of which are related to the length of the string or rope, as shown in Figure 3.11.

The standing-wave pattern for a circular orbit differs slightly from the model of an oscillating rope in that there are no defined stationary ends. Rather, the electron wave oscillates in an endless series of waves, provided the waves are superimposable: crests overlapping crests and troughs overlapping troughs, as in the description of constructive interference in Chapter 1. This requirement is met only if the circumference (cf) of the orbit equals a multiple of the wavelength of the electron:

$$cf = n\lambda \tag{3.16}$$

Equation 3.16 gives a new meaning to Bohr's principal quantum number, n: it represents the number of characteristic wavelengths in an orbit.

De Broglie went on to predict that moving particles much bigger than electrons, such as atomic nuclei, molecules, and even served tennis balls, airplanes in flight, and Earth revolving around the sun, would have characteristic wavelengths that could be calculated by using Equation 3.15. The wavelengths of such large objects must be extremely small, given the tiny size of Planck's constant in the numerator of Equation 3.15. Therefore, we would never notice the wave nature of large objects.

SAMPLE EXERCISE 3.4: The speed of the electron in the ground state of the hydrogen atom is 2.2×10^6 m/s. What is the wavelength of the electron?

SOLUTION: The rest mass of an electron is 9.110×10^{-31} kg. Because its speed is less than one-thousandth that of the speed of light, we will assume that its effective mass (m) is the same as its rest mass. Using this value in Equation 3.16 and the equality 1 joule = 1 kg · m²/s², we have

$$\lambda = h/mv = \frac{6.626 \times 10^{-34} \frac{\text{kg} \cdot \text{m}^2}{\text{s}^2} \cdot \text{s}}{(9.110 \times 10^{-31} \text{ kg})\left(\frac{2.2 \times 10^6 \text{ m}}{\text{s}}\right)}$$

$$= 3.3 \times 10^{-10} \text{ m or } 0.33 \text{ nm}$$

A photon with this very short wavelength would be in the X-ray region.

Microscopes called scanning electron microscopes (SEMs) use beams of electrons to obtain images at resolutions of less than a nanometer (10^{-9} m). They can achieve this resolution because the de Broglie wavelengths of these electrons are about this size. This resolution is hundreds of times better than can be obtained with the best optical microscopes, which rely on visible wavelengths between 400 and 750 nm.

PRACTICE EXERCISE: What is the wavelength of a baseball (mass = 0.143 kg) traveling toward home plate at 100.0 miles per hour? (See Problems 37, 38, and 41.)

A scanning electron microscope uses beams of electrons to obtain images as small as a few nanometers (10^{-9} m).

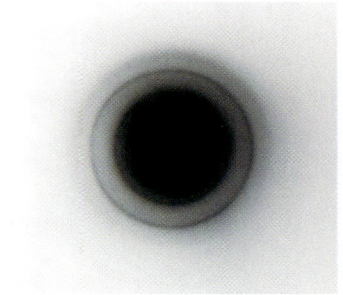

FIGURE 3.12 De Broglie's proposal that particles behave as waves is illustrated by the X-ray and electron-diffraction patterns of aluminum foil. The pattern of concentric rings is characteristic of the diffraction patterns of light waves seen by Young in Chapter 1. X-rays are known to behave as waves, whereas electrons (β particles) definitely behave as particles (they have mass). The observation of similar diffraction patterns from electron and X-ray diffraction supports the idea that small particles such as electrons have wave properties in addition to behaving as particles.

De Broglie's thesis research created a quandary for the graduate faculty at the University of Paris, where he studied. Bohr's model of electrons moving between "allowed" energy levels had been widely criticized as an arbitrary suspension of well-tested physical laws. De Broglie's rationalization of Bohr's model seemed even more outrageous to many scientists. Before the faculty would accept his thesis, they wanted another opinion; so they sent it to Albert Einstein for his review. Einstein wrote back that he found the young Frenchman's work "quite interesting." That was good enough for the faculty: de Broglie's thesis was accepted in 1924 and immediately submitted for publication.

If de Broglie was correct and electrons did behave as waves, then electron beams should undergo diffraction and produce interference patterns like beams of light. In 1925, American scientist Clinton J. Davisson (1881–1958) observed that electrons reflecting off very clean surfaces of nickel crystals produced unexpected patterns not unlike those shown for aluminum in Figure 3.12. He did not offer an explanation for the patterns when he published these results, but other scientists recognized in them the diffraction patterns of electron waves. Also in 1925, British scientist George Thomson (1892–1975) produced

Waves are produced by the diffraction of electrons by an atomic-scale defect in a copper surface.

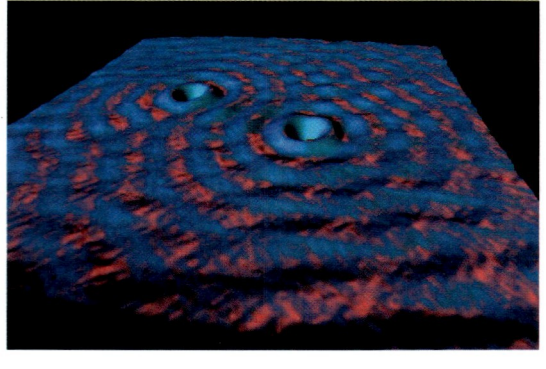

CONNECTION: Electron diffraction can also be used to determine the structures of molecules (Section 6.8).

✓ The wavelike properties of electrons can be observed in the interference patterns produced by beams of electrons.

electron-diffraction patterns by passing a beam of electrons through thin gold foil. Twelve years later, Davisson and Thomson shared the Nobel Prize in Physics for their investigations of the wavelike properties of electrons. In 1989, a team of Japanese scientists produced an interference pattern from an electron beam by using two slits, as Thomas Young had done with a beam of light 189 years earlier.

3.6 SCHRÖDINGER'S WAVE EQUATION

Many of the leading scientists of the 1920s were unwilling to accept the wavelike–particle-like dual nature of electrons proposed by de Broglie until it could be used to accurately predict the features of the hydrogen spectrum. Such use required the development of equations describing the behavior of electron waves. In November 1925, Austrian physicist Erwin Schrödinger (1897–1966) had just presented a seminar on de Broglie's thesis research when a colleague reminded him that a rigorous mathematical basis for electron waves was still needed. Schrödinger must have taken this advice to heart, because, during a most productive Christmas vacation in the Swiss Alps, he developed the mathematical foundation for what has come to be called **wave mechanics** or **quantum mechanics**.

✓ The description of the behavior of particles as waves is called **wave mechanics** or **quantum mechanics,** which is the formulation and solution of **wave equations**. A **wave function** is a solution of a wave equation.

Schrödinger's mathematical description of electron waves is called the Schrödinger **wave equation**. It is a second-order differential equation and will not be discussed in detail in this book. However, you should know that solutions to the Schrödinger wave equation are mathematical expressions called **wave functions** represented by the Greek letter psi (ψ). Wave functions are mathematical descriptions of the motion of electron waves as they vary with location and with time. Wave functions can be simple trigonometric functions, such as sine or cosine waves, or they can be very complex.

Schrödinger published a series of four papers in 1926 describing his wave equation that astounded the physical science community. Not only did his work provide a mathematical basis for de Broglie's electron waves, but also solutions to his wave equation for hydrogen exactly fit its spectral data. In addition, his approach held promise for solutions for multielectron atoms.

 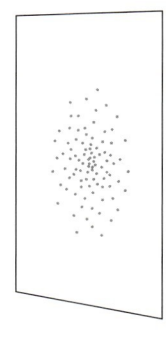

FIGURE 3.13 The probability of finding an ink spot decreases with increasing distance from the center of the pattern in much the way that electron density in the 1s orbital decreases with increasing distance from the nucleus.

What is the physical significance of a wave function? Actually, there is none. However, the square of a wave function, ψ^2, does have physical meaning. Initially, Schrödinger believed that a wave function depicted the smearing of an electron through three-dimensional space. This notion of subdividing what also was a discrete particle was later rejected in favor of the model developed by German physicist Max Born (1882–1970). Less than a year after Schrödinger introduced the world to quantum mechanics, Born developed his interpretation of it, arguing that ψ^2 defines the space, called an **orbital**, in an atom where the probability of finding an electron is high. Born later showed that his interpretation could be used to calculate the probability of a transition between two orbitals, as happens when an atom absorbs or emits a photon. To help you think about ψ^2 as a matter of probability, imagine if you were to spray ink from a perfume sprayer, as shown in Figure 3.13. If you drew a circle encompassing most of the ink spots, you would have identified the region of maximum probability for finding the ink spots.

> ✓ Psi squared (ψ^2) defines the size and shape of **orbitals,** the regions in atoms within which there is a high probability of finding an electron.

3.7 QUANTUM NUMBERS

It is important to distinguish between an *orbital*, a region of space where an electron is likely to be found, and an *orbit* in Bohr's model of the hydrogen atom. Orbitals *are not concentric circles;* they are three-dimensional regions of space with distinctive shapes, orientations, and average distances from the nucleus. Each is a solution to Schrödinger's wave equation and is identified by a unique combination of three integers called **quantum numbers**:

- The **principal quantum number (n)** is a positive integer that indicates the relative size of an orbital or a group of orbitals in an atom. Orbitals with the same value of n are said to be in the same *shell*. Orbitals with larger values of n are farther from the nucleus and, in the hydrogen atom, represent higher energy levels (consistent with Bohr's model of the hydrogen atom). Energy levels of orbitals are more complex in multielectron atoms, but generally they increase with increasing values of n.

- The **angular momentum quantum number (l)** is an integer from zero to $n - 1$. It defines the shape of an orbital. Orbitals with the same values of n and l are said to be in the same *subshell*. Orbitals in the same subshell represent equivalent energy levels. Orbitals with a given value of l are identified with a letter (to keep from getting l and n values mixed up) according to the following scheme:

Value of l:	0	1	2	3	4
Letter identifier:	s	p	d	f	g

- The **magnetic quantum number (m_l)** is an integer with a value from $-l$ to $+l$. It defines the orientation of an orbital in the space around the nucleus of an atom.

> ✓ Each orbital has a unique set of the three quantum numbers n, l, and m_l. For example, a 2s orbital has the quantum numbers $n = 2$, $l = 0$, and $m_l = 0$.

Each subshell has a two-part designation containing the appropriate numerical value of n and a letter designation for l. For example, orbitals with $n = 3$ and $l = 1$ are said to be in the 3p subshell. They are called 3p orbitals. Electrons that are in 3p orbitals are called 3p electrons. How many 3p orbitals are there? We can answer this question by finding all possible values of m_l. Because p orbitals are those for which $l = 1$, they can have m_l values of -1, 0, and $+1$. These three values tell us that there are *three* orbitals in a p subshell. The unique combinations of the three quantum numbers for the *3p orbital* are represented by the contents of the following three columns of integers:

n	3	3	3
l	1	1	1
m_l	-1	0	$+1$

SAMPLE EXERCISE 3.5: (a) What are the designations of all the subshells in the $n = 4$ shell? (b) How many orbitals are in these subshells?

SOLUTION: Let's calculate the possible combinations of l and m_l when $n = 4$. The possible values of l from 0 to $n - 1$ are 0, 1, 2, and 3. These values of l correspond to the subshell designations s, p, d, and f. The appropriate subshell names are 4s, 4p, 4d, and 4f.

There is only one orbital in the 4s subshell. It has the following combination of quantum numbers:

n	4
l	0
m_l	0

The n and l values for the 4p orbitals are 4 and 1, respectively. If $l = 1$, then the possible m_l values are -1, 0, and $+1$. Thus, there are three orbitals in the 4p subshell with the following quantum numbers:

n	4	4	4
l	1	1	1
m_l	-1	0	$+1$

The n and l values for the 4d orbitals are 4 and 2, respectively. When $l = 2$, there are five possible values for m_l, -2, -1, 0, $+1$, and $+2$, and therefore five orbitals with unique combinations of n, l, and m_l:

n	4	4	4	4	4
l	2	2	2	2	2
m_l	-2	-1	0	$+1$	$+2$

Finally, the n and l values for the 4f orbitals are 4 and 3, respectively. When $l = 3$, there are seven possible values for m_l, -3, -2, -1, 0, $+1$, $+2$, and $+3$, and so there are seven orbitals in the 4f subshell:

n	4	4	4	4	4	4	4
l	3	3	3	3	3	3	3
m_l	-3	-2	-1	0	$+1$	$+2$	$+3$

and a grand total of $1 + 3 + 5 + 7 = 16$ orbitals in the $n = 4$ shell.

PRACTICE EXERCISE: How many orbitals are there in the second shell ($n = 2$)? (See Problems 47–56.)

Several trends are worth noting in the quantum numbering system (Figure 3.14):

- There are n subshells in the nth shell: there is one subshell (1s) in the first shell; there are two subshells (2s and 2p) in the second shell; there are three subshells (3s, 3p, and 3d) in the third shell, and so on.
- There are n^2 orbitals in the nth shell: one in the first shell, four in the second, nine in the third, and so on.
- There are $2l + 1$ orbitals in each subshell: one s orbital ($2 \times 0 + 1 = 1$); three p orbitals ($2 \times 1 + 1 = 3$); five d orbitals ($2 \times 2 + 1 = 5$), and so on.

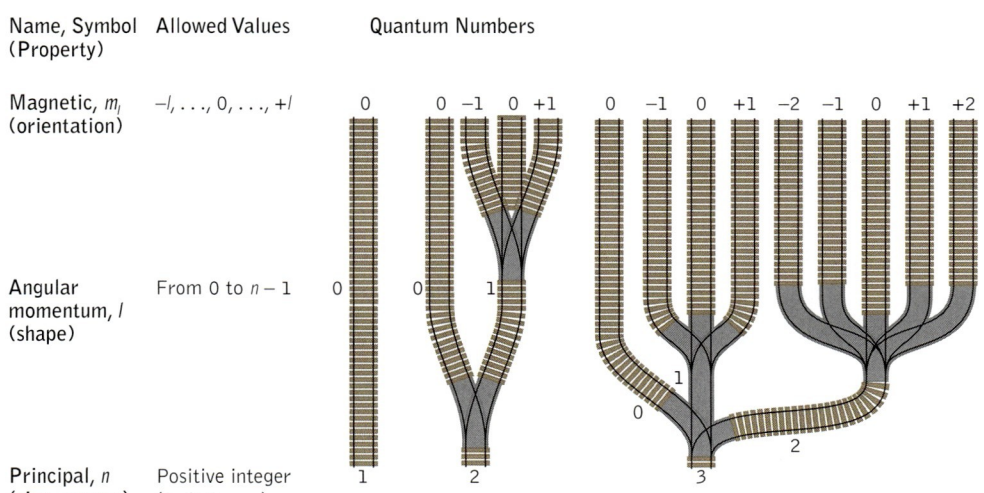

FIGURE 3.14 The quantum numbers are determined by a set of simple rules. The principal quantum number n can have integer values of 1, 2, 3, and so forth. The value of the angular momentum quantum number l depends on n; $l = 0, 1, 2 \ldots (n - 1)$. The magnetic quantum number m_l ranges in value from l to $-l$.

3.8 SHAPES AND SIZES OF ATOMIC ORBITALS

Atomic orbitals are regions of atomic space where electrons are likely to be found. Because they are defined by the square of a mathematical function, ψ^2, they have three-dimensional shapes related to graphical representations of ψ^2. Figure 3.15 provides several such representations of the 1s orbital of the hydrogen atom. In Figure 3.15A, ψ^2 is plotted against distance from the nucleus. Note that the scale of the x-axis is in *pico*meters (10^{-12} m). The plot seems to suggest that the most likely place to find the electron is in the nucleus. We know that can't be true. Is there another interpretation of ψ^2? It turns out that Figure 3.15A is really a plot of probable electron density. To obtain a more useful view of where the electron is likely to be, we need to adjust this profile for the volume term that must be in the denominator of any expression of density.

Figure 3.15B provides a more useful profile of electron location. To understand the meaning of this plot, think of a hydrogen atom as a tiny onion made of many layers all of the same thickness, as shown in Figure 3.15C. What is the probability of finding an electron in one of these layers? A layer very close to the nucleus must have a very small radius (r), and so it must be a small fraction of the total volume of the atom. A layer with a larger radius makes up a much larger fraction of the volume of the atom, because the volumes of these layers increase as a function of r^2. So, even though electron densities are high close to the nucleus, the volumes of the layers there are so small that the chances of the electron being near the center of an atom is extremely low. Farther away from the nucleus, electron densities are lower, but the volumes of the layers are much larger; so the probability of finding the electron in one of them goes up. At even greater distances, layer volumes are very much larger, but ψ^2 (see Figure 3.15A) drops to nearly zero, and so the chances of finding the electron decreases.

The plot in Figure 3.15B shows how increasing layer volume but decreasing electron probability combine to produce a **radial distribution plot**. It is not a plot of ψ^2 versus distance (r), but rather of $4\pi r^2 \psi^2$ versus r. In geometry, $4\pi r^2$ is the formula for the area of a sphere, though here it represents the volume of one of those thin spherical layers described in the preceding paragraph. This makes sense because the volume of a spherical layer that is infinitely thin is approximated by its surface area.

A significant feature of the plot in Figure 3.15B is that it has a maximum corresponding to the distance from the nucleus at which the electron is most likely to be found. The value of r corresponding to the maximum electron density for the 1s (ground state) orbital of hydrogen is 53 picometers, which is the radius predicted by Bohr's model.

Figure 3.15D provides a view of the spherical shape of this and other s orbitals. The volume of the sphere is the volume within which the probability of finding a 1s electron is 90%. This type of depiction is called a *boundary-surface*

> ✓ A **radial distribution plot** is a graphical representation of the probability of finding an electron near the nucleus of an atom.

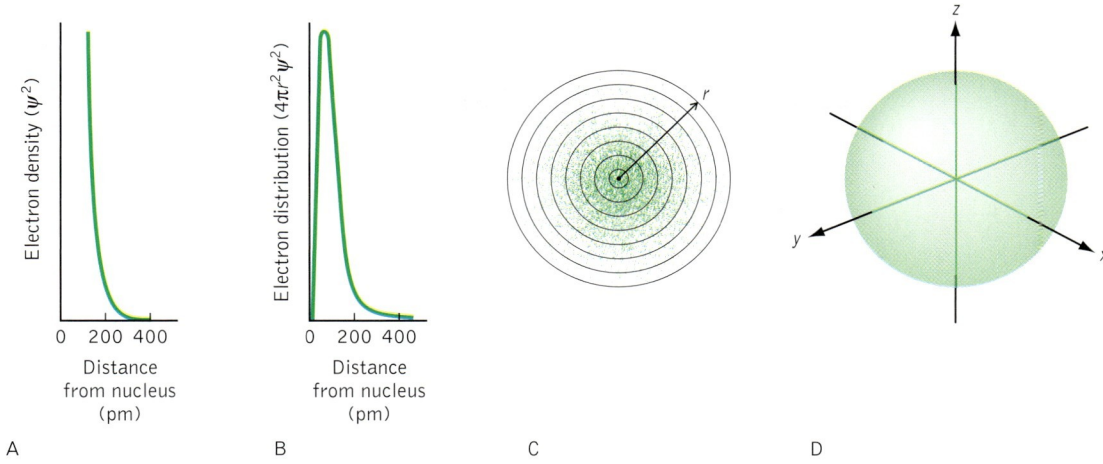

FIGURE 3.15 Probable electron density in the 1s orbital (ground state) of the hydrogen atom can be represented by: A. a plot of electron density (ψ^2) versus distance from the nucleus; B. a radial distribution profile, which indicates the probability of an electron being in a thin spherical shell with the radius r, as shown in C.; and D. a sphere within which the probability of finding a 1s electron is 90%.

representation and is one of the most useful ways to view the relative size, shape, and orientation of an orbital. All s orbitals are spheres and so have no "orientation." For them, boundary surfaces are a useful way to depict relative size.

The sizes of the 1s, 2s, and 3s orbitals are shown in Figure 3.16. Note that their sizes increase with increasing values of the principal quantum number (n). Also note that, in each profile, there is a maximum in electron density close to the nucleus. The heights of these maxima decrease with increasing n, and they are separated from other maxima farther from the nucleus by **nodes** of zero probability. The radial distribution profiles in Figure 3.16 show how electrons in s orbitals, even those with high values of n, can penetrate close to the nucleus. This penetration ability decreases among the different orbitals in the same shell as follows: $s > p > d > f > g$.

Orbital penetration is important when an outer-shell orbital is separated from the nucleus by one or more filled inner shells. Put another way, the **effective nuclear charge** (Z_{eff}) is made smaller by the shielding effect of the inner-shell electrons. An electron that can penetrate through the shield will experience a greater Z_{eff}, lowering its energy (Figure 3.17). For this reason, the orbitals in an energy level fill up in the order s, p, d, f, and g because the degree of penetration of the orbitals is $s > p > d > f > g$ (Figure 3.18).

We need to keep in mind that there are no shielding effects inside a hydrogen atom, because there is only one electron and no others to shield it from the nucleus. Therefore, in a hydrogen atom, all of the orbitals in a given shell should

> ✓ A **node** is a region in an orbital where the probable electron density is zero.
>
> The **effective nuclear charge** (Z_{eff}) is the attractive force toward the nucleus felt by an electron in an atom.

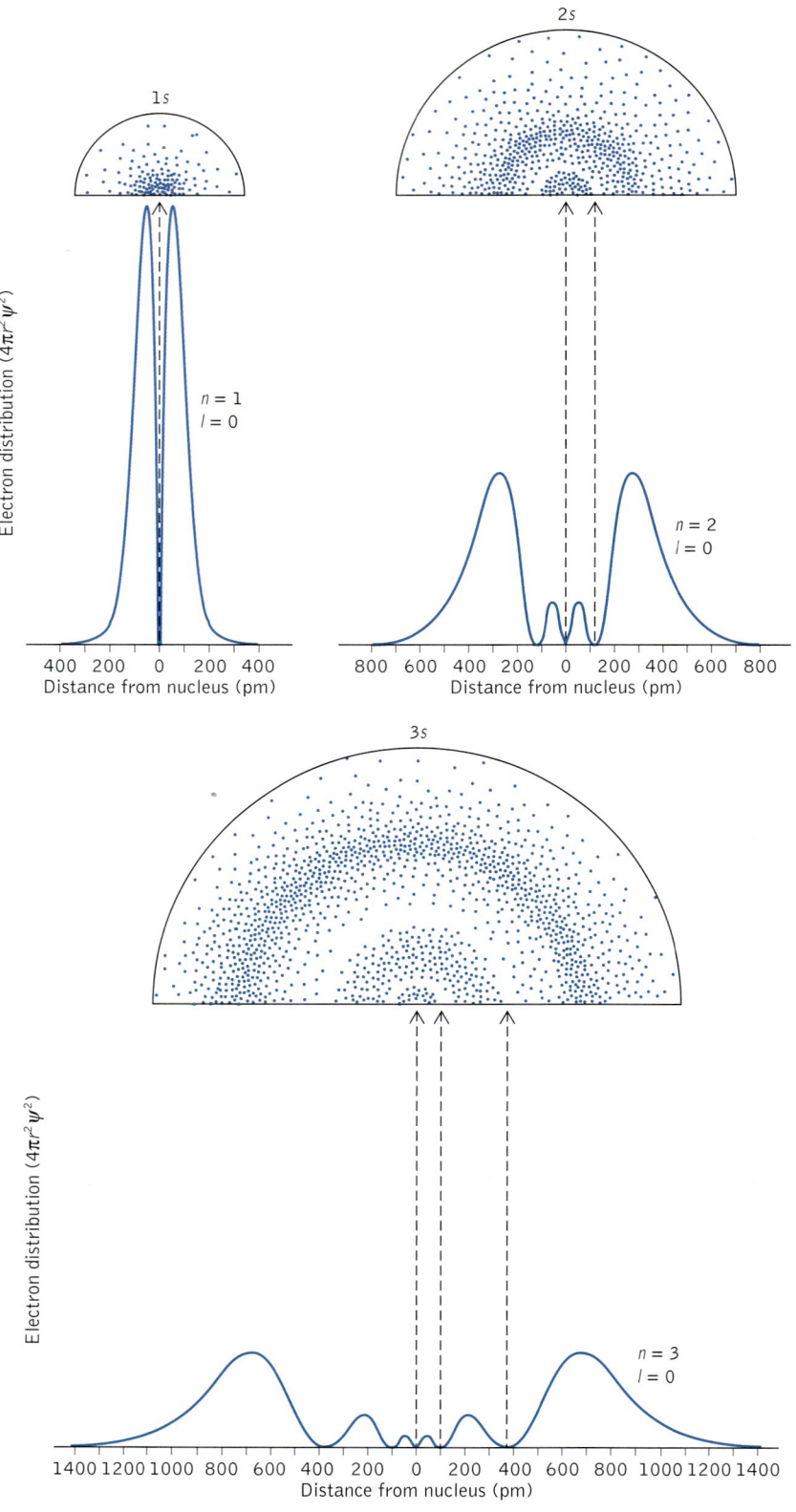

FIGURE 3.16 Electron-distribution profiles for 1s, 2s, and 3s orbitals have 0, 1, and 2 nodes of zero density, respectively. Note the wider distribution with increasing n.

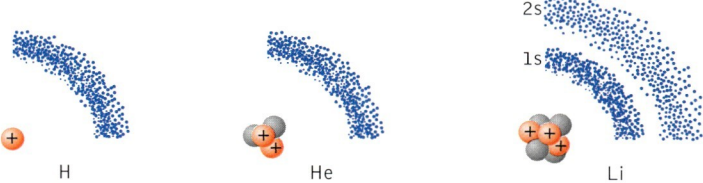

FIGURE 3.17 The effective nuclear charge felt by the 2s electron in lithium is less than the effective nuclear charge in hydrogen or helium for two reasons: the 2s electron is farther from the nucleus and the 1s electrons shield the 2s electron from the positive charge of the nucleus. As in lithium, a second electron in the 1s orbital of helium causes electron–electron repulsions.

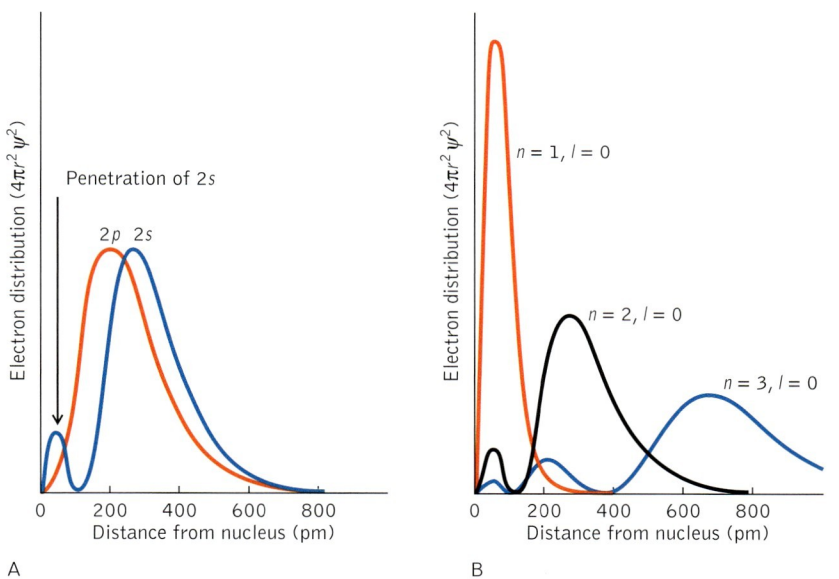

FIGURE 3.18 A. The 2s orbital appears to be farther from the nucleus than the 2p orbital. However, the 2s orbital is of lower energy because it allows 2s electrons to penetrate closer to the nucleus. This penetration means that 2s electrons experience greater nuclear charge and so have lower energy. B. All three of the s orbitals have electron distributions that allow their electrons to penetrate close to the nucleus, but 3s electrons are more likely to be farther away from the nucleus than 2s electrons, which are farther away than 1s electrons.

have the same energy. They are said to be **degenerate**. Only the principal quantum number defines energy levels in a hydrogen atom. Actually, that fact is embedded in the observations of Johann Balmer and in Equation 3.6.

Figures 3.19 and 3.20 are boundary-surface views of p and d orbitals, respectively. All shells with $n \geq 2$ have a subshell containing three p orbitals. Each of these orbitals has two balloon-shaped lobes along one of the three axes and so is designated p_x, p_y, or p_z, depending on which axis the lobes are aligned. The two

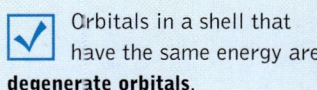

Orbitals in a shell that have the same energy are **degenerate orbitals**.

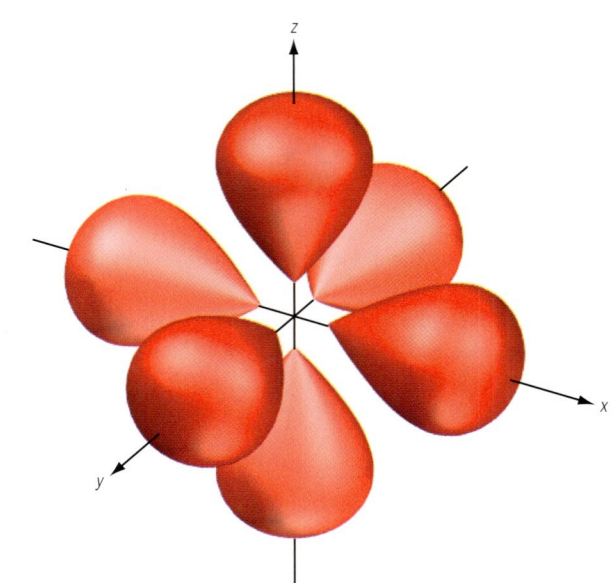

FIGURE 3.19 Boundary-surface views of a set of three p orbitals show the orientation along the x-, y-, and z-axes. The nucleus is located at the origin.

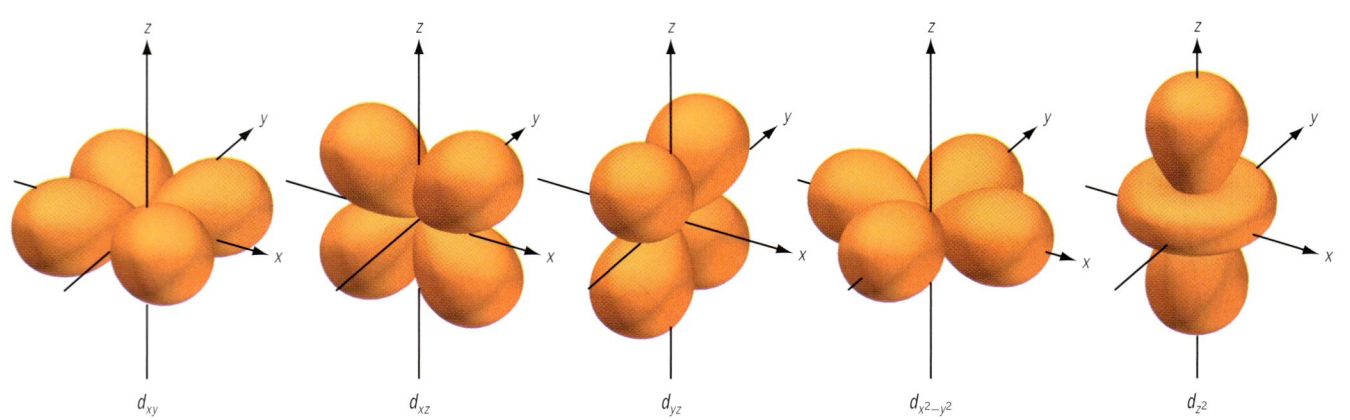

FIGURE 3.20 Boundary-surface views of five d orbitals are projected on x-, y-, and z-axes.

CONNECTION: The shapes of orbitals are key to the formation of chemical bonds and to predicting molecular geometries (Section 7.5).

lobes of a p orbital are sometimes designated with plus and minus signs, indicating that the sign of the wave function defining them can have negative and positive values. Because a node of zero probability separates the two lobes, you may be wondering how an electron gets from one lobe to the other. The best way to understand how it does so is to remember that an electron behaves as a *wave*, and waves have no difficulty passing through nodes. After all, there is a node right in the middle of a simple sine wave separating the positive and negative displacement regions.

The five d orbitals found in shells of $n \geq 3$ have different shapes as well as orientations. Three of them have a cloverleaf array of four lobes that are *between*, rather than along, the x-, y-, and z-axes. These orbitals are designated d_{xy}, d_{xz}, and d_{yz}. The four lobes of the fourth d orbital lie *along* the x- and y-axes. This orbital is designated $d_{x^2-y^2}$. The fifth d orbital has a much different shape, with two major lobes oriented along the z-axis and a donut surrounding the middle. It is designated d_{z^2}. We will not address the shapes of other types of orbitals (f, g, and so on), because they are much less important than s, p, and d to the discussions of chemical bonding in the chapters to come.

CONNECTION: The shape and orientation of the d orbitals influence the colors of many transition-metal compounds (Section 10.7).

3.9 SPINNING ELECTRONS

We began this chapter by noting that the distinctive orange glow produced by hot sodium atoms was produced by two closely spaced emission lines with wavelengths of 589.0 and 589.6 nm. The Schrödinger equation accounts for the spectral features of many elements, but it does not explain the closely spaced pair of yellow orange lines that are found in the spectrum of sodium, the pair of red lines in the hydrogen spectrum (the red line at 656 nm is actually a pair of lines at 656.272 nm and 656.285 nm as shown in Figure 3.21), and other pairs of lines in the spectra of atoms with a single electron in the outermost shell.

In 1925, two students at the University of Leiden in the Netherlands, Samuel Goudsmit (1902–1978) and George Uhlenbeck (1900–1988), proposed that the various pairs of lines, called *doublets*, could be the result of electrons spinning in one of two directions, producing two energy states not accounted for by Schrödinger's equation. Just as Earth and other planets simultaneously revolve around the sun and rotate on their axes, so, too, an electron moving around a nucleus may spin on its axis. Just as planets might spin in one of two directions—west to east (as Earth does) or east to west—an electron can have one of two spin directions. These directions are usually designated "spin up" or "spin down," owing to the fact that a spinning charged particle, such as an electron or the nucleus of an atom, produces a tiny magnetic field as a result of that spin. According to quantum mechanics, the resulting magnetic fields can have two orientations: up or down. Goudsmit and Uhlenbeck proposed a fourth quantum number, called the **spin magnetic quantum number** (m_s), to account for the two spin orientations. The values for m_s are $+\frac{1}{2}$ and $-\frac{1}{2}$ for spin up and spin down, respectively.

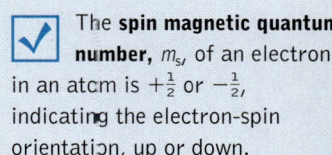

The **spin magnetic quantum number,** m_s, of an electron in an atom is $+\frac{1}{2}$ or $-\frac{1}{2}$, indicating the electron-spin orientation, up or down.

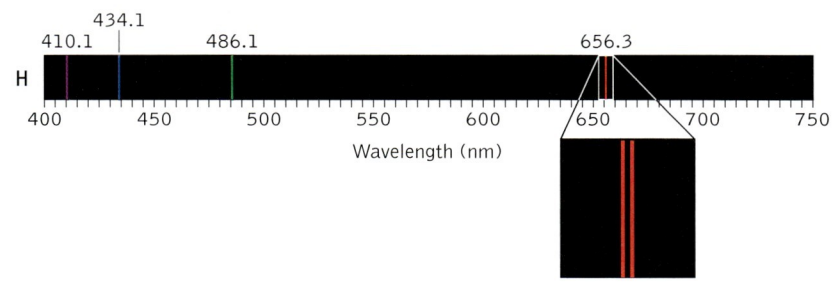

FIGURE 3.21 The Schrödinger equation accounted for most of the atomic spectral features of hydrogen and other elements, but it did not account for the appearance of closely spaced pairs of lines, such as the red lines at 656.272 and 656.285 nm. The two lines are the result of the two spin orientations of electrons.

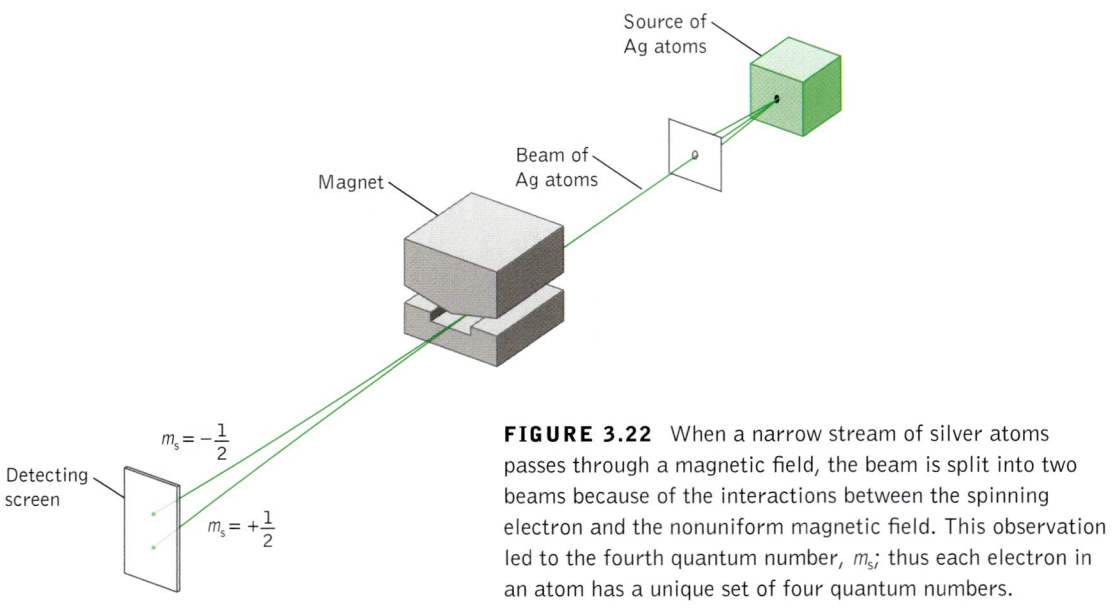

FIGURE 3.22 When a narrow stream of silver atoms passes through a magnetic field, the beam is split into two beams because of the interactions between the spinning electron and the nonuniform magnetic field. This observation led to the fourth quantum number, m_s; thus each electron in an atom has a unique set of four quantum numbers.

The direction of spin of a hydrogen atom's electron (and the two different orientations of the resulting tiny magnetic field) can influence how the atom behaves in a magnetic field. Even before Goudsmit and Uhlenbeck proposed the electron-spin hypothesis, two other scientists, Otto Stern (1888–1969) and Walther Gerlach (1889–1979), observed the effect of electron-spin direction by shooting a beam of silver atoms through a magnetic field such as that produced by the magnets shown in Figure 3.22. The beam was split into two beams because of the two different directions of electron spin on the atoms.

In 1925, Austrian physicist Wolfgang Pauli proposed that *no two electrons in an atom may have the same set of four quantum numbers.* This idea came to be known as the **Pauli exclusion principle**. The three quantum numbers derived from solutions to Schrödinger's wave equation define the orbitals in which electrons are likely to be found. The two allowed values of the spin magnetic quantum number indicate that each orbital can hold just two electrons, one with $m_s = +\frac{1}{2}$ and the other with $m_s = -\frac{1}{2}$. Thus, each electron in an atom has a unique "quantum address" defined by a unique combination of n, l, m_l, and m_s values.

☑ Each orbital can hold as many as two electrons. The two are *paired*, meaning that they are spinning in opposite directions.

Pauli's exclusion principle states that no two electrons in an atom may have the same set of four quantum numbers.

3.10 THE PERIODIC TABLE AND FILLING IN THE ORBITALS

It is time to begin a survey of the arrangement of electrons in the atoms of the elements. We will use the periodic table as a guide, building up the number of electrons per atom as we proceed from one element to the next. In deciding which orbitals contain electrons and which do not, we will use the following rule:

orbitals with the lowest energies are always filled first. The 1s orbital ($n = 1$) is filled first, orbitals in the second shell ($n = 2$) are filled next, and so on.

The orbital-filling process would be very straightforward were it not for the fact that the differences in energy between shells gets smaller as n gets larger (see Figure 3.10). This compression of levels leads to similar energies for orbitals having high values of l in one shell and having low values of l in the next shell. We shall soon see how these similarities affect the arrangement of the elements in the periodic table.

We know that the single electron in a ground-state atom of hydrogen (H) should be in the 1s orbital. Accordingly, the **electron configuration** of hydrogen is $1s^1$, where the superscript indicates that there is one electron in the 1s orbital.

The next element in the periodic table is helium (He). Its atomic number of 2 tells us that there are two protons in its nucleus and therefore two electrons in a helium atom. The principle of filling the lowest-energy orbitals first requires the second electron also to be in the 1s orbital. Therefore, helium has the electron configuration $1s^2$. With two electrons, the 1s orbital is filled to capacity. Because the first shell contains only this single orbital, the first shell is completely filled, too. Its filled shell has major implications for the chemical properties of helium. As will be repeated in the remainder of this chapter, elements with atoms that have filled shells or at least filled s and p subshells in the outermost shells are chemically stable. The lack of chemical reactivity of helium coupled with its low density (about one-seventh that of air) makes it the ideal gas for giving buoyancy to the blimps seen hovering over sporting events throughout the world. Hydrogen gas was once used in these lighter-than-air craft, but that use ended shortly after the historic explosion of the Hindenburg at Lakehurst, New Jersey, in 1937 (Figure 3.23).

 Orbitals with the lowest energies are always filled first.

Electron configuration describes the distribution of electrons among the orbitals of an atom or ion.

FIGURE 3.23 Both hydrogen and helium gases have been used to fill lighter-than-air craft, but the hydrogen-filled Hindenburg ended its last trans-Atlantic flight in flames in Lakehurst, New Jersey, in 1937. Modern blimps get their buoyancy from helium, which is chemically inert.

THE CHEMISTRY OF THE NOBLE GASES

The six elements in the far-right-hand column of the periodic table (Group 8A) are called the noble gases. We encountered several of them in this and earlier chapters. We described and addressed the properties of radon, the only noble gas that has no stable isotopes, in Chapter 2.

Helium is the second element in the periodic table and the second most abundant element in the universe. Its presence in the gases of the sun was detected in 1868 by a set of Fraunhofer lines that did not align with those of the known elements on Earth. These lines were assigned to helium—the name having been derived from the Greek work for the sun, *helios*. The element was isolated on Earth by Sir William Ramsay in 1895.

Within Earth's crust and mantle, helium is formed by the α decay of radioactive elements such as uranium. When α particles acquire two electrons, they form atoms of helium. The gas may remain trapped within minerals of the radioactive nuclides that formed it (Ramsay discovered helium in uranium ore) or it may leak into underground deposits of natural gas. The gas deposits are the source of the helium used to fill balloons and blimps and in other applications in which a gas is needed that is (1) inert and (2) much lighter than air.

Argon was isolated from the atmosphere the year before Ramsay discovered helium. The name argon is from the Greek *argos*, meaning "lazy," in reference to the chemical inertness of the element and the others in Group 8A. Argon is the most abundant of the Group 8A elements, making up about 0.94% (by volume) of the atmosphere, and is the twelfth most abundant element in the universe. Most of the argon in the atmosphere is the product of radioactive decay. It is produced in rocks and minerals when nuclei of radioactive ^{40}K undergo electron-capture decay. The slow rate of the decay process allows scientists to date the age of rocks on the basis of their $^{40}Ar/^{40}K$ ratio. Eventually, the argon leaks into the atmosphere, from which it is isolated by first liquefying the air and then separating Ar from N_2 and O_2 by distillation. Argon is used to provide an inert atmosphere in arc-welding certain metals such as stainless steel. It is also used in light bulbs and other photoelectric devices.

Neon was discovered as an impurity in argon by Ramsay and his coworkers in 1898. It takes its name from the Greek *neos*, meaning "new." Although it is found in some minerals as a result of radioactive decay, its principal industrial source is the atmosphere. How-

ever, its tiny concentration in the atmosphere (0.0018% by volume) means that nearly 100 kg of air must be liquefied to obtain 1 g of neon. When an electrical current is passed through a tube containing neon at low pressure, the gas emits the characteristic red orange light that we associate with neon signs (such as the one shown in the above photo). Electrical discharge tubes filled with the other noble gases emit their own unique combinations of emission lines and characteristic colors.

Krypton (from the Greek *kryptos*, or "hidden") and xenon (from the Greek *xenos*, or "foreign") are even more rare components of the atmosphere. The concentration of krypton is only 1.1 parts per million (ppm) by volume, which means that there are, on average, 1.1 atoms of krypton for every million atoms and molecules of the other ingredients in the atmosphere. The atmospheric concentration of xenon is even smaller: only 86 parts per billion (ppb) by volume, which means that there are 86 atoms of xenon for every billion (10^9) atoms and molecules of the other ingredients in the atmosphere. Because these gases have higher boiling points than do the principal components of the atmosphere, they can be separated by the fractional distillation of liquefied air. These two elements are recovered in the least-volatile fraction.

All of the noble gases except helium have completely filled *s* and *p* orbitals (and eight electrons) in their outer shells. Helium has a filled 1*s* orbital and so only two electrons. The stability of these filled orbitals gives the noble gases their lack of chemical reactivity. Each noble gas has the highest ionization energy of all the elements in its row (period) of the periodic table. For example, it takes about 20% more energy to remove an electron from neon than from fluorine, and 400% more than from lithium. The ionization energies of the six noble gases decrease with increasing atomic number because the increasing numbers of inner-shell electrons shield the outermost electron from the positive charge on the nucleus.

The noble gases do not easily share electrons with other atoms in molecules. However, xenon and krypton do react with fluorine, forming, for example, XeF_2, XeF_4, XeF_6, and KrF_2. The xenon compounds react with water, yielding compounds containing xenon, fluorine, and oxygen.

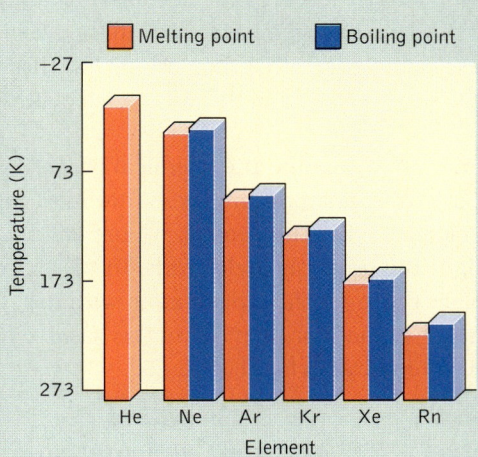

All of the Group 8A elements are gases at room temperature, with boiling and melting points below 0°C. The trend toward higher boiling and melting points with increasing atomic size will be discussed in Chapter 9. As the Group 8A elements increase in size, their van der Waals forces (London dispersion forces) increase.

The densities of the Group 8A elements at 20°C and 1 atmosphere pressure are proportional to their atomic masses.

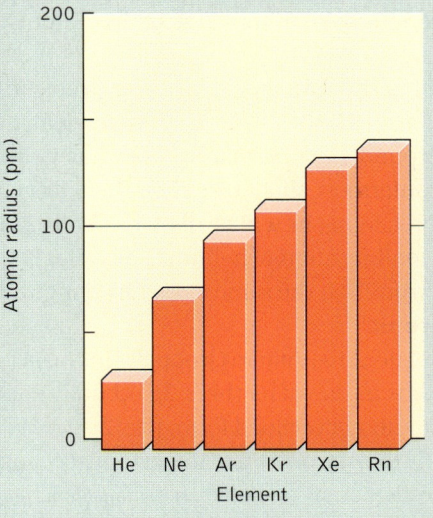

The atomic radii of the Group 8A elements increase down the group, corresponding to the increase in atomic number, which in turn corresponds to populations of higher-energy orbitals found farther from the nucleus.

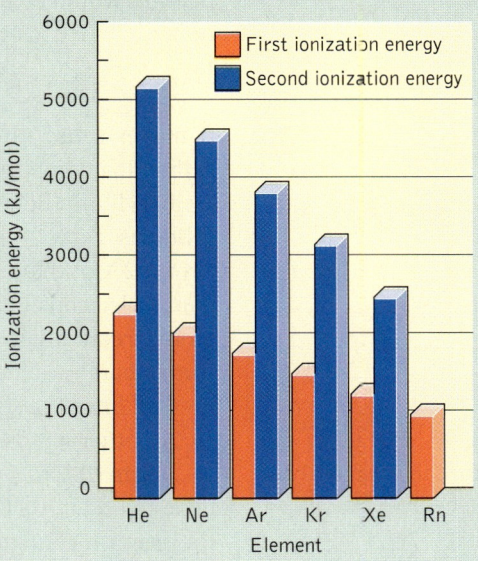

The first ionization energies of the Group 8A elements decrease down the group because the inner electrons shield the outer electrons from the nuclear charge. The larger, second ionization energies are due to the greater amount of energy required to remove an electron from an ion with a positive charge.

> The horizontal rows of the periodic table are called **periods**.

Lithium (Li; Z = 3) is the next element after helium in the periodic table and the first element in the second row (also called the second **period**) of the periodic table. This location is a signal that an atom of lithium has an electron in its second shell. The second shell has four orbitals (one s and three p) and so can hold as many as eight electrons. Not coincidentally, there are eight elements in the second row. Which of the four orbitals contains the third electron of an atom of lithium? As noted in Section 3.8, the energy levels of the subshells in a given shell are related to their shapes and the degree to which electrons in them experience the charge of the nucleus. These factors mean that the s orbital has the lowest energy within a shell and that the single electron in the $n = 2$ shell of lithium is in the $2s$ orbital.

Its single $2s$ electron plays a key role in defining the chemical properties of lithium. In general, the outermost electrons of atoms are those that most influence the chemical behavior of the elements. For this reason, we use *abbreviated* electron configurations that highlight the outer-shell electrons. In an abbreviated electron configuration, we replace the inner shells of an atom with the symbol of the Group 8A element that has that configuration (He in this case), which means that the abbreviated electron configuration for Li is $[He]2s^1$.

Element	Electron Configuration	Abbreviated Electron Configuration
Li	$1s^2 2s^1$	$[He]2s^1$

This nomenclature reinforces the fact that a stable electron configuration, that of helium, would be achieved if a lithium atom lost its $2s$ electron through ionization:

$$Li \longrightarrow Li^+ + e^- \tag{3.17}$$

It makes sense, then, that lithium is found in nature not as free atoms but, rather, in compounds containing Li^+.

Beryllium (Be) is the fourth element in the periodic table. Its electron configuration is $1s^2 2s^2$, or, in abbreviated form, $[He]2s^2$. Like the other elements in Group 2A of the periodic table, the ground-state beryllium atom has two spin-paired electrons in the s orbital of its outermost shell. Like the others, beryllium atoms frequently engage in chemical reactions in which they lose both $2s$ electrons, thereby achieving the electron configuration of helium. Thus, He, Li^+, and Be^{2+} all have the same number of electrons and the same electron configuration: $1s^2$. In other words, they are **isoelectronic**.

> Different atoms or ions having the same number of electrons and the same electron configurations are said to be **isoelectronic**.

Boron (B; Z = 5) comes next, but we need to jump to Group 3A in the periodic table to find it. Its fifth electron should be in one of its three $2p$ orbitals, giving it the electron configuration $1s^2 2s^2 2p^1$, or, in abbreviated form, $[He]2s^2 2p^1$. Which of the three $2p$ orbitals contains the fifth electron is not important, because all three have the same energy. This equivalency is called orbital *degeneracy*, as stated in Section 3.8.

The next element after boron is carbon (C). Its chemical properties are linked to an electron configuration in which there are four electrons in the outermost shell and its abbreviated electron configuration is $[He]2s^2 2p^2$. Note that there are two electrons in *2p* orbitals. Are they both in the same orbital or are they in separate orbitals? Remember that all electrons have a negative charge and so repel each other. Thus they tend to occupy orbitals that allow them to be as far away from each other as possible. Thus, the two *2p* electrons are in separate *2p* orbitals.

Information about the distribution of electrons within a subshell is not contained in conventional configurations or in abbreviated electron configurations such as [He]$2s^2 2p^2$. We need a visual representation such as that provided by **orbital diagrams**. In such a diagram, each orbital is represented by a square. Electrons in an orbital are represented by arrows in the square. An arrow pointed upward represents an electron with spin up and $m_s = +\frac{1}{2}$. An arrow pointed downward represents an electron with the opposite spin orientation, spin down, and $m_s = -\frac{1}{2}$. The orbital diagram for carbon follows:

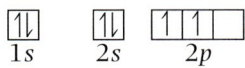

> An **orbital diagram** shows how the electrons in an atom or ion are distributed among its orbitals.

Note that the two *2p* electrons are unpaired, reside in separate orbitals, and have parallel spins. This diagram exemplifies **Hund's rule,** which states that *the lowest-energy electronic configuration of an atom or ion is the one with the maximum number of unpaired electrons (that are allowed by the Pauli exclusion principle) in a subshell, with all unpaired electrons having parallel spins.*

> **Hund's rule** states that the lowest energy state of an atom or ion contains the maximum number of unpaired electrons with parallel spins.

The next element is nitrogen (N; Z = 7). According to Hund's rule, the third *2p* electron resides alone in the third *2p* orbital. Nitrogen has the electron configuration $1s^2 2s^2 2p^3$, or [He]$2s^2 2p^3$, and the electrons are distributed among its orbitals as follows:

As we proceed across the second row to neon (Ne; Z = 10), we continue to fill in the *2p* orbitals. The last three electrons to be added pair up with the first three; so, in an atom of neon, all *2p* orbitals are filled to capacity and so, too, is the second shell. Note that neon is directly below helium in Group 8A. Elements in the same column of the periodic table have similar chemical properties because their atoms have similar electron configurations in their outermost shells. Helium is chemically unreactive (inert) because it has a totally filled first shell. Neon is chemically inert because it has a completely filled outermost (second) shell. The electron configurations of argon (Ar), krypton (Kr), xenon (Xe), and radon (Rn) also have completely filled *s* and *p* orbitals in their outermost shells, and they also are inert gases at room temperature. Only under the most extreme reaction conditions do Ar, Kr, and Xe (but not He and Ne) form chemical bonds. This lack of chemical reactivity led to their being called **noble gases**. The electron configurations for the first 10 elements are shown in Figure 3.24.

> Chemical stability is achieved when an atom or ion has a completely filled set of *s* and *p* orbitals in its outermost shell.

> The elements in Group 8A are chemically stable and are called **noble gases**. Only the largest of them undergo chemical reactions and then only under extreme reaction conditions.

Sodium (Na) follows neon in the periodic table and is the first element in the third row. It is the third element in Group 1A and so has a single electron in its *3s* orbital. Its abbreviated electron configuration is [Ne]$3s^1$ and a similarly abbreviated orbital diagram would be:

[Ne]

> **CONNECTION:** Valence-shell electrons in molecules can also occupy molecular orbitals (Section 6.9).

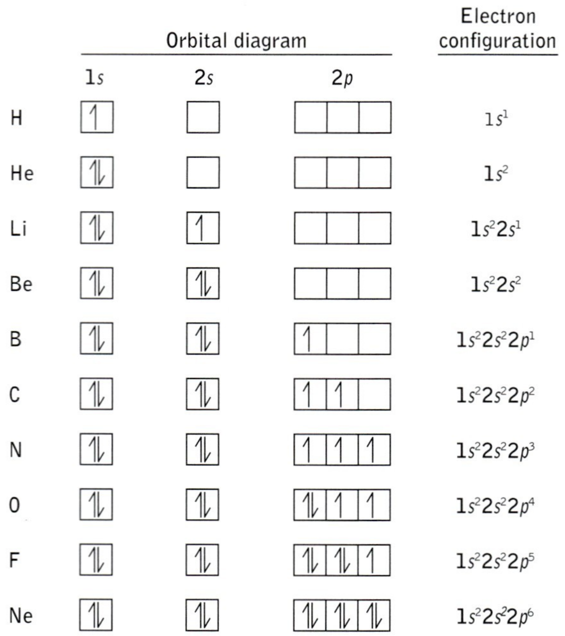

	Orbital diagram			Electron configuration
	1s	2s	2p	
H	↑	☐	☐☐☐	$1s^1$
He	↑↓	☐	☐☐☐	$1s^2$
Li	↑↓	↑	☐☐☐	$1s^2 2s^1$
Be	↑↓	↑↓	☐☐☐	$1s^2 2s^2$
B	↑↓	↑↓	↑☐☐	$1s^2 2s^2 2p^1$
C	↑↓	↑↓	↑↑☐	$1s^2 2s^2 2p^2$
N	↑↓	↑↓	↑↑↑	$1s^2 2s^2 2p^3$
O	↑↓	↑↓	↑↓ ↑ ↑	$1s^2 2s^2 2p^4$
F	↑↓	↑↓	↑↓ ↑↓ ↑	$1s^2 2s^2 2p^5$
Ne	↑↓	↑↓	↑↓ ↑↓ ↑↓	$1s^2 2s^2 2p^6$

FIGURE 3.24 Orbital diagrams and electron configurations of the first 10 elements show that each orbital holds as many as two electrons of opposite spin. The orbitals are filled in order of increasing quantum numbers n and l. A complete list of electron configurations for all elements is given in Appendix 3.

SAMPLE EXERCISE 3.6: Predict the charge of a sodium ion.

SOLUTION: Sodium can achieve the chemical stability associated with the electron configuration of neon if it loses its 3s electron. If sodium loses a single electron, the resulting ion will have 11 protons in its nucleus and only 10 electrons surrounding it, which leaves a net charge on the sodium ion of +1. Na^+ is isoelectronic with Ne and so is expected to be more stable than a sodium atom.

PRACTICE EXERCISE: Draw complete orbital diagrams for the most stable ions formed by an atom of each of the following elements: O, Mg, Al, and Cl. (See Problems 67 and 68.)

Any electron configuration for sodium other than $[Ne]3s^1$, such as $[Ne]3p^1$, would not represent the lowest possible energy level, or the ground state, of sodium atoms. Instead $[Ne]3p^1$ represents an *excited-state* configuration. This particular one is of interest to us in connection with our earlier consideration of Fraunhofer's D lines in the solar spectrum. The electron transition from the ground state of sodium to the $[Ne]3p^1$ excited state can result in the absorption of the yellow orange light missing from the solar spectrum at 589 nm. In a sodium-vapor streetlight, atoms in the $[Ne]3p^1$ excited state spontaneously lose quanta of energy and fall back to the $[Ne]3s^1$ ground state, emitting the same yellow orange light.

The bright red glow of an emergency road flare comes from the line emission by strontium atoms. There are several lines in this spectrum, but the most intense ones are in the red region.

In magnesium (Mg; Z = 12) atoms, the 3s orbital is filled. To achieve the chemical stability of neon's electronic configuration, magnesium must lose its two 3s electrons. As a result, magnesium, like strontium (Sr) and the other elements in Group 2A, exists in nature as +2 ions.

The next six elements from aluminum (Al) to argon (Ar) contain increasing numbers of 3p electrons until all these orbitals are filled. Thus, argon atoms have completely filled outer-shell p orbitals, and argon is chemically inert, as predicted by its position in Group 8A.

After argon comes potassium (K; Z = 19), the Group 1A element of the fourth row. Its position indicates that—like hydrogen, lithium, sodium, and all other elements in this group—it has a single electron in its outermost s orbital. Therefore, we write its abbreviated electron configuration as $[Ar]4s^1$. (Have you noticed that the row number corresponds to the value of n for the outer shells of the atoms in that row?) Potassium is followed by calcium (Ca; Z = 20), which has the abbreviated electron configuration $[Ar]4s^2$. At this point, the 4s orbital is filled, but the 3d orbitals are still empty. Why were they not filled before 4s? We began to answer this question in Section 3.9 when we noted that the differences in energy between shells get smaller as n increases. The smaller differences result in similar energies between orbitals with large values of l in one shell and orbitals with small values of l in the next shell, as shown in Figure 3.25. As a result of such an overlap, the energy of the 4s orbital is slightly lower than that of the 3d orbitals and so the 4s orbital is filled first, followed by the five 3d orbitals.

The next element (Z = 21) in the periodic table is scandium (Sc). It is the first element in the center region of the table populated by elements called **transition metals**. The properties of scandium and its place after calcium in the periodic table indicate that scandium has the abbreviated electron configuration $[Ar]4s^2 3d^1$. This electron-filling order explains the positions of other transition metals in the middle of the periodic table. The atoms of most of these elements contain partly filled d orbitals. Note that the 3d orbitals are filled in the procession of transitions metals in the fourth row. This pattern of filling the d orbitals of shell n in the transition metals of row n + 1 is followed throughout the periodic table: the 4d orbitals are filled in the transition metals of the fifth row, and so on, as shown in Figure 3.26.

☑ Elements in Group 1A form ions with a +1 charge; those in Group 2A form ions with a +2 charge; and those in Group 7A form ions with a −1 charge.

☑ The number of the row in which an element resides corresponds to the value of n for the outer shell of an atom of that element.

☑ A **transition metal** is an element in Groups 3 to 12. Atoms of the transition metals typically have partly filled d orbitals.

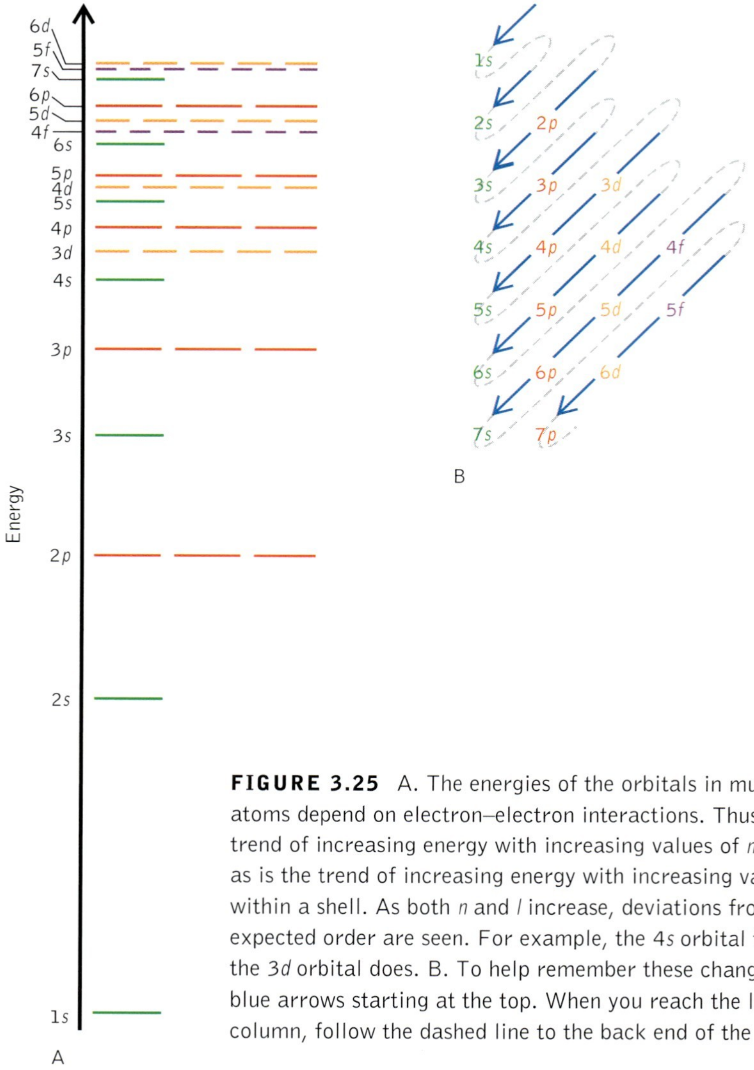

FIGURE 3.25 A. The energies of the orbitals in multielectron atoms depend on electron–electron interactions. Thus, the general trend of increasing energy with increasing values of n is still true, as is the trend of increasing energy with increasing values of l within a shell. As both n and l increase, deviations from the expected order are seen. For example, the 4s orbital fills before the 3d orbital does. B. To help remember these changes, follow the blue arrows starting at the top. When you reach the left-hand column, follow the dashed line to the back end of the next arrow.

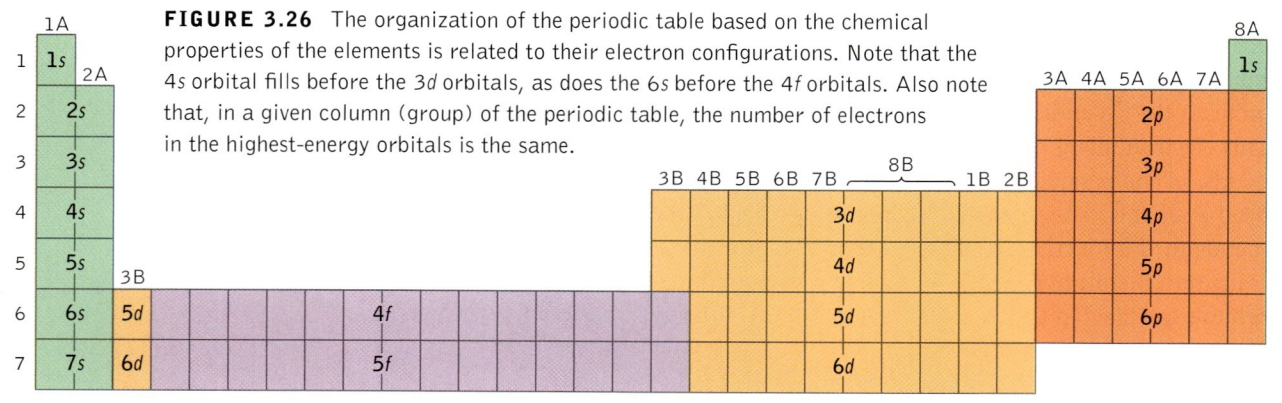

FIGURE 3.26 The organization of the periodic table based on the chemical properties of the elements is related to their electron configurations. Note that the 4s orbital fills before the 3d orbitals, as does the 6s before the 4f orbitals. Also note that, in a given column (group) of the periodic table, the number of electrons in the highest-energy orbitals is the same.

SAMPLE EXERCISE 3.7: Which of the following species are isoelectronic with Ne? Na^+; F^-; O^{2-}; Mg^{2+}.

SOLUTION: The abbreviated electron configuration of Na is $[Ne]3s^1$. Na^+ is a sodium atom that has lost its $3s^1$ electron, and so Na^+ is isoelectronic with Ne.

Fluorine atoms have the electron configuration $1s^2 2s^2 2p^5$. F^- has one more electron, giving it the electron configuration $1s^2 2s^2 2p^6$ and making it isoelectronic with Ne.

Oxygen atoms have the electron configuration $1s^2 2s^2 2p^4$. O^{2-} has two more electrons than an O atom, giving it the electron configuration $1s^2 2s^2 2p^6$ and making it isoelectronic with Ne.

The abbreviated electron configuration of Mg is $[Ne]3s^2$. Mg^{2+} is a magnesium atom that has lost both its $3s$ electrons, and so Mg^{2+} also is isoelectronic with Ne.

In summary, all four species are isoelectronic with Ne.

PRACTICE EXERCISE: Which of the following species is isoelectronic with Ar? K^+; Cl^-; S^{2-}; Sc^{3+}. (See Problems 73 and 74.)

The next element after Sc is titanium (Ti; Z = 22), which has one more d electron, and so its abbreviated electron configuration is $[Ar]4s^2 3d^2$. At this point, you may feel that you can accurately predict the electron configurations of the remaining transition metals in the fourth row. However, because the energies of the $3d$ and $4s$ orbitals are similar, the sequence of d-orbital filling can be a bit unusual. For example, vanadium (V) has the expected configuration $[Ar]4s^2 3d^3$, but the next element, chromium (Cr), has the configuration $[Ar]4s^1 3d^5$. The reason for this difference is that a *half*-filled set of d orbitals is an energetically favored configuration. Apparently, the energy needed to raise a $4s$ electron to a $3d$ orbital is compensated by the stability of five half-filled $3d$ orbitals. As a result, $[Ar]4s^1 3d^5$ is more stable than $[Ar]4s^2 3d^4$. Another variation in the expected filling pattern is observed near the ends of the rows of transition metals. Copper (Cu; Z = 29), for example, has an $[Ar]4s^1 3d^{10}$ configuration instead of $[Ar]4s^2 3d^9$ because a completely filled set of d orbitals also represents a stable electron configuration.

SAMPLE EXERCISE 3.8: Write the abbreviated electron configuration of an atom of silver (Ag; Z = 47).

SOLUTION: Silver is a transition metal in the fifth row of the periodic table. The electrons in its outermost shells surround a core of inner-shell electrons corresponding to the electron configuration of the noble gas at the end of the fourth row: Kr. Therefore, we begin the abbreviated electron configuration with Kr in brackets and then add the 11 additional electrons that take us from 36 to a total of 47. Ordinarily, the first 2 electrons would be in the $5s$ orbital, and the next 9 would be in $4d$ orbitals, giving the abbreviated electron configuration $[Kr]5s^2 4d^9$. However, a completely filled set of d orbitals represents a stable electron configuration; so Ag has a $[Kr]5s^1 3d^{10}$ electron configuration.

PRACTICE EXERCISE: Draw the abbreviated orbital diagram of an atom of rhodium (Rh; $Z = 45$). (See Problems 79–86.)

Before leaving this discussion of filling in the d orbitals of transition-metal atoms, we will examine the electron configurations of the cations made by the loss of these electrons. The electronic configuration of manganese (Mn) is $[Ar]4s^2 3d^5$, whereas the electron configuration of Mn^{2+} is $[Ar]3d^5$. We might have expected Mn^{2+} to be $[Ar]4s^2 3d^3$ from the fact that we put the last two electrons in $3d$ orbitals as we went from vanadium (V; $[Ar]4s^2 3d^3$) to chromium (Cr; $[Ar]4s^1 3d^5$) and then to Mn ($[Ar]4s^2 3d^5$). The configuration $[Ar]4s^2 3d^3$ has three unpaired electrons, but Mn^{2+} spectra indicate that it has five unpaired electrons. Therefore, its configuration must be $[Ar]3d^5$. To form Mn^{2+}, the Mn atom could have lost two $4s$ electrons, or one $4s$ electron and one $3d$ electron, or two $3d$ electrons, but we will never know, because electrons outside atoms have lost their quantum-number labels. The electron configurations of all the $+2$ first-row transition metal ions are $[Ar]3d^n$ and not $[Ar]4s^2 3d^{n-2}$, because in these ions the $3d$ orbitals are lower in energy than the $4s$ orbitals, whereas the reverse is true for the corresponding atoms. Many of the transition metals also lose one or more d electrons. Scandium, for example, forms cations with a $+3$ charge. The chemistry of titanium is dominated by its tendency to lose both of its s *and* both of its d electrons.

After the $4s$ and all the $3d$ orbitals are filled, the $4p$ orbitals are filled for elements with atomic numbers 31 to 36, leaving us with the chemically stable configuration of krypton. The pattern emerging from our orbital-filling exercise can be summarized as follows:

- s orbitals of the shell corresponding to the row number are filled in Groups 1A and 2A.
- d orbitals of a shell one number less than the row number are filled in Groups 1B through 10B.
- p orbitals of the shell corresponding to the row number are filled in Groups 3A through 8A.

We have yet to consider the two rows at the bottom of the periodic table. Elements 58 through 71 are called **lanthanide (rare earth)** elements and those of $Z = 90$ through 103 are called **actinide** elements. These two rows of elements have partly filled $4f$ (lanthanide) and $5f$ (actinide) orbitals. There are 14 elements in each row, reflecting the capacity of the seven orbitals in each $f(l = 3)$ subshell, with m_l values of $-3, -2, -1, 0, +1, +2,$ and $+3$. As you can see from the sequence of atomic numbers, the $4f$ orbitals are not filled until after the $6s$ orbital has been filled. This order of filling is due to the similar energies of $6s$ and $4f$ orbitals. Similarly, the $5f$ orbitals are filled after $7s$ but before $6d$ (see Figure 3.26). The actinides are elements with no stable isotopes, and most of them are not found in nature.

The periodic table is a useful reference for understanding why elements have particular chemical properties and why some groups (also known as *families* of elements because of their related properties) of elements behave simi-

> ✓ Elements 58 through 71 are the **lanthanide** (rare earth) **elements**. Elements 90 through 103 are the **actinide elements**.

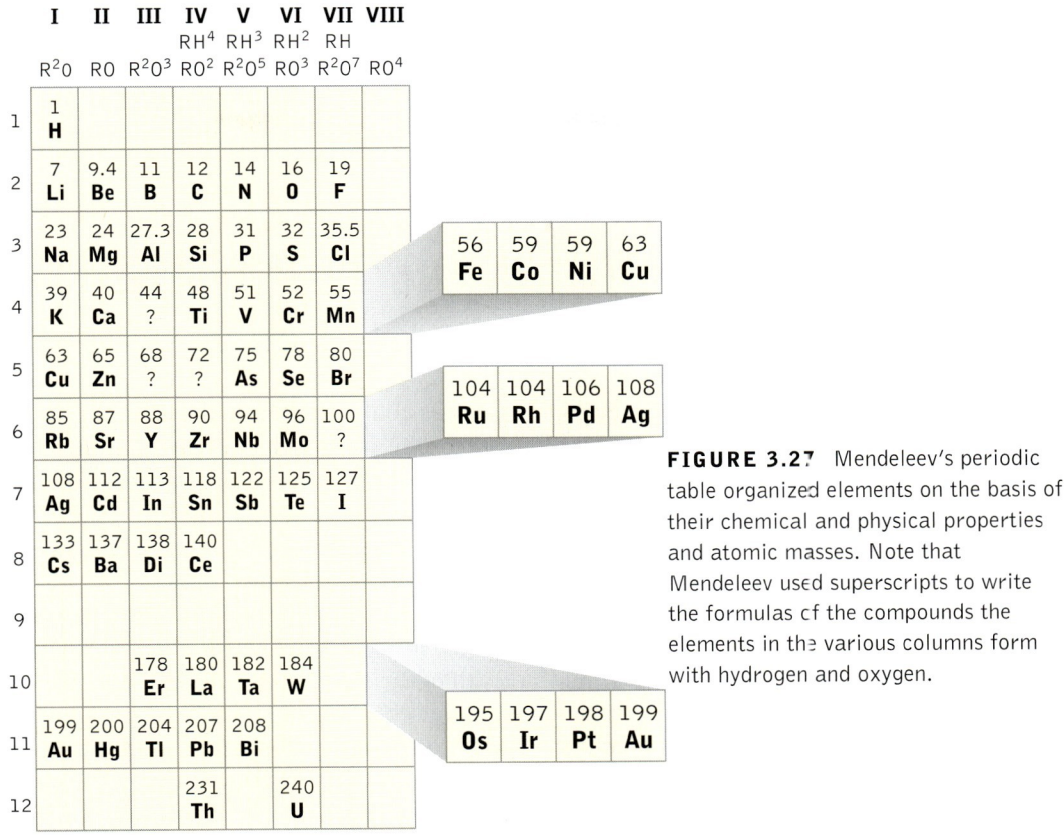

FIGURE 3.27 Mendeleev's periodic table organized elements on the basis of their chemical and physical properties and atomic masses. Note that Mendeleev used superscripts to write the formulas of the compounds the elements in the various columns form with hydrogen and oxygen.

larly. Russian chemist Dimitri Mendeleev developed the first useful periodic table of the elements decades before de Broglie gave us electron waves and Schrödinger pioneered quantum mechanics. Mendeleev's table (Figure 3.27) was based on periodic trends in *the chemical properties of the elements*. What Mendeleev did not know was that *the chemical properties of an element are closely linked to its electron configuration*.

The development of the periodic table in the middle of the nineteenth century was an incredible feat. For one thing, not all of the elements were known in those days, as we saw earlier in this chapter. Being a brilliant and circumspect scientist, Mendeleev knew that there was still much that he did not know; so he left "blanks" in his periodic table to account for yet-to-be-discovered elements. Moreover, he was able to predict the chemical properties of these missing elements on the basis of the positions of the blanks in the table. These insights greatly facilitated their subsequent discovery.

Keep in mind that Mendeleev arranged the elements in his periodic table in order of increasing *atomic mass*, not atomic number. Remember that the concept of an atomic number equaling the number of protons in the nucleus is a product of twentieth-century science and the work of Lord Rutherford and his brilliant students.

CONCEPT TEST: Find the elements in the periodic table that should have been out of place in Mendeleev's version, which was based on atomic mass.

3.11 MORE EVIDENCE FOR THE EXISTENCE OF ATOMIC ORBITALS

Ionization energies

CONNECTION: Ionization energy is important in considering the overall energy that drives the formation of ionic compounds, as described in Section 9.2.

✓ The **ionization energy** is the minimum amount of energy needed to remove an electron from a gaseous atom or ion.

Ionization energies of the elements in a group of the periodic table decrease with increasing atomic number. Within a row, ionization energies increase with increasing atomic number.

In developing electron configurations in the previous section, we followed a theoretical framework for the arrangement of electrons in orbitals that was developed in the early years of the twentieth century. Is there experimental evidence for the existence of orbitals representing different energy levels inside atoms? Indeed there is. It includes the measurements of the energies needed to remove electrons from atoms and their ions, as described below.

Earlier in this chapter, we calculated how much energy is needed to remove an electron from a hydrogen atom, 2.18×10^{-18} J. This **ionization energy** is due to the strength of the electrostatic attraction between the single proton in a hydrogen nucleus and the electron in its $1s$ orbital. How much energy is needed to remove one electron from a helium atom? With two protons in its nucleus and twice the positive charge, the attraction between the electrons and the nucleus should be about twice as much as the force between a hydrogen atom's nucleus and electron. In fact, it *is* nearly twice as much: 3.94×10^{-18} J.

Let's consider the ionization energy of lithium. It has three times the nuclear charge of hydrogen, so you might expect its ionization energy to be much larger. However, our orbital model tells us that the outermost electron in a lithium atom is in a $2s$ orbital, which means that the outermost electron is farther away from the nucleus and of higher energy than is an electron in a $1s$ orbital. Moreover, the completely filled $1s$ orbital of lithium shields the $2s$ electron from the nucleus. As a result, the energy needed to remove this electron ought to be less than that of hydrogen. Actually, it is less than half that of hydrogen: 8.6×10^{-19} J.

Figure 3.28 shows the energies required to remove an electron from the outermost subshells of single gas-phase atoms of the first 20 elements. The trend in ionization energies is consistent with our view of how electrons fill the s and p orbitals of the first three shells. The ionization energies of hydrogen, helium, and lithium have already been discussed. Beryllium has a larger ionization energy than does lithium for the same reason that the removal of a $1s$ electron from helium is more difficult than the removal of a $1s$ electron from hydrogen: the positive charge of a beryllium nucleus is greater (by 4/3) than that of a lithium nucleus.

We might expect the ionization energy of a $2p$ electron in boron to be somewhat less than that of a $2s$ electron in beryllium because the filled $2s$ orbital shields the more distant $2p$ electrons from the nucleus (see Figure 3.18). As

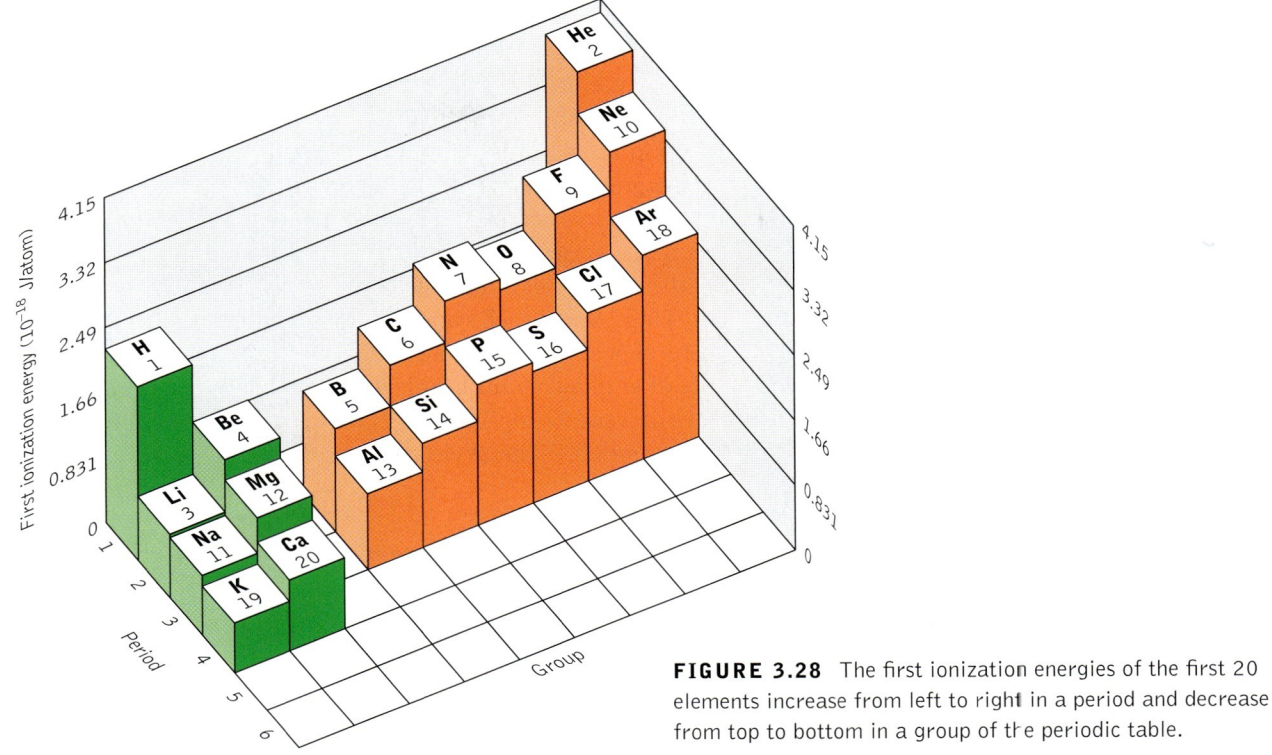

FIGURE 3.28 The first ionization energies of the first 20 elements increase from left to right in a period and decrease from top to bottom in a group of the periodic table.

expected, there is a decrease in ionization energy from beryllium to boron. Carbon and nitrogen nuclei have greater positive charges than does boron, and so removal of one of their $2p$ electrons requires more energy.

Note the small decrease in ionization energy between nitrogen and oxygen. To understand why it happens, think back to the stability associated with half-filled d orbitals in the transition metals. The same pattern of enhanced stability is observed with half-filled sets of p orbitals. When oxygen loses a $2p$ electron, it achieves the slightly more stable configuration of a half-filled set of $2p$ orbitals.

$$\begin{array}{ccc} \boxed{\uparrow\downarrow} & \boxed{\uparrow\downarrow}\ \boxed{\uparrow\downarrow|\uparrow|\uparrow} & \longrightarrow \quad \boxed{\uparrow\downarrow} \quad \boxed{\uparrow\downarrow}\ \boxed{\uparrow|\uparrow|\uparrow} \\ 1s & 2s \quad\quad 2p & \quad\quad 1s \quad\quad 2s \quad\quad 2p \end{array}$$

The relative stability of the product of oxygen ionization means that the ionization process happens more easily and so the energy needed is less.

Neon has a greater ionization energy than those of the elements at its left in the second row because its atoms have the largest numbers of protons in the second row and the greatest effective nuclear charge. Thus, they hold on to their $2p$ electrons the most tightly of all the elements in the second row.

The trends in ionization energies of the elements in the third row of the periodic table are the same as those observed in the second row and for the same reasons. A sodium atom is easily ionized because the completely filled orbitals of the first and second shells shield the 3s electron from the +11 nuclear charge. Loss of the single 3s electrons gives the very stable electron configuration of the noble gas neon. The magnesium nucleus exerts a slightly greater attractive force on its two 3s electrons because its nuclear charge (+12) is slightly higher. The shielding effect of the filled 3s orbital means that the effective nuclear charge exerted on the 3p electron in an aluminum atom is less, and so its ionization energy is less than that of magnesium. The pattern in ionization energies from aluminum to argon is due to the increasing positive charge of the nucleus and its stronger attractive force on the 3p electrons. The break between phosphorus and sulfur is associated with the stability of a half-filled set of 3p orbitals. Finally, we see the shielding effects of filled inner-shell s and p orbitals on the low ionization energies of the 4s electrons of potassium and calcium. Ionization energy decreases in the downward procession of a column in the periodic table. The decreasing ionization energy as we descend Group 1A from Li to Na to K is due to the shielding of the 3s orbital in potassium by the 2s electrons, as shown in Figure 3.28.

CONCEPT TEST: Why do ionization energies decrease with increasing atomic number for the three noble gases: He, Ne, and Ar?

X-ray photoelectron spectroscopy

The discussion of the photoelectric effect in Section 3.3 noted that some metals emit electrons when they are illuminated with electromagnetic radiation. It turns out that the atoms of *any* element will eject electrons if they are illuminated with photons of radiation with enough energy. X-rays have such energies. With wavelengths of a few nanometers or less, X-rays have enough energy to knock out an electron from any atom that they encounter. (Tissue damage caused by such ionization is one reason why people should limit their exposure to the X-rays used in medical imaging.) Moreover, the electron does not have to be one of the outermost electrons of the atom. Even 1s electrons in atoms with large Z values can be dislodged by X-rays. An important feature of these X-ray-induced ionization events is that one photon ejects only one electron. Thus all of the energy of the X-ray is consumed in removing the electron from its atom. If the energy of the X-ray is more than that needed for ionization, the excess energy of the X-ray accelerates the ejected electron to a velocity proportional to the energy difference.

The technique of X-ray photoelectron spectroscopy (XPS) is based on an instrument in which X-rays of the same wavelength (and energy) irradiate a sample, and the velocities of the ejected electrons are measured. From such a velocity (v), the energy of motion, or **kinetic energy** (KE), of an electron can be calculated:

$$KE = \tfrac{1}{2}mv^2 \quad (3.18)$$

where m is the mass of an electron. Equation 3.18 allows us to calculate the kinetic energy of any mass in motion. Because we can determine the energies of the ejected electrons from their measured velocities and we know the initial

✓ In X-ray photoelectron spectroscopy, the kinetic energy of electrons emitted by shining X-rays of fixed wavelength on a material can be used to measure ionization energies.

CONNECTION: The kinetic energies of molecules may be used to explain the properties of gases (Section 8.6).

✓ The energy of motion is called **kinetic energy**.

energy of the X-rays that ejected them, we can take the difference between the two to obtain the energy that had to be overcome to dislodge the electrons. If the sample consists of atoms in the gas phase, then this difference in energy is an ionization energy (E_i) that can be calculated with the use of Equation 3.19.

$$\text{ionization energy} = \text{energy of the X-ray} - \text{kinetic energy of the ejected electron} \quad (3.19)$$

$$E_i = \frac{hc}{\lambda} - KE$$

Consider the data in Table 3.1. The trends in experimental E_i values fit our orbital model of atomic structure extremely well. As we would expect, E_i values within a row increase with increasing atomic number (Z) because the electrostatic attraction between the nucleus and the electrons in a given subshell ought to increase with increasing nuclear charge. Down a column, E_i values decrease for electrons believed to be at greater distances and more shielded from the nucleus. Although not shown in Table 3.1, the relative sizes of the peaks also reinforce our model. For

TABLE 3.1 Ionization Energies (J × 10^{18}) from XPS Spectra of the First 20 Elements

Element	Peak 1 (1s)	Peak 2 (2s)	Peak 3 (2p)	Peak 4 (3s)	Peak 5 (3p)	Peak 6 (4s)
H	2.18					
He	3.94					
Li	10.4	0.86				
Be	19.1	1.50				
B	32.1	2.26	1.32			
C	47.5	2.86	1.81			
N	65.8	4.07	2.32			
O	87.4	5.19	2.18			
F	112.	6.45	2.79			
Ne	140.	7.78	3.46			
Na	173.	11.4	6.10	0.83		
Mg	209.	15.1	8.83	1.23		
Al	251.	20.1	13.0	1.81	0.96	
Si	296.	25.1	17.1	2.43	1.31	
P	346.	31.1	22.4	3.24	1.68	
S	397.	37.7	27.4	3.41	1.66	
Cl	454.	44.6	33.6	4.06	2.08	
Ar	514.	52.4	40.1	4.69	2.53	
K	577.	61.7	48.4	6.53	3.96	0.70
Ca	648.	71.0	56.5	7.73	4.82	0.98

TABLE 3.2 Successive Ionization Energies (J × 10¹⁸) of the First Ten Elements*

Element	Z	E_{i1}	E_{i2}	E_{i3}	E_{i4}	E_{i5}	E_{i6}	E_{i7}	E_{i8}	E_{i9}	E_{i10}
H	1	2.18									
He	2	3.94	8.72								
Li	3	0.86	12.12	20.							
Be	4	1.49	2.92	25.	35.						
B	5	1.33	4.03	6.08	41.	54.					
C	6	1.80	3.90	7.67	10.3	63.	78.				
N	7	2.33	4.75	7.61	12.4	15.6	88.	107.			
O	8	2.18	5.62	8.80	12.4	18.2	22.1	118.	140.		
F	9	2.79	5.60	10.0	14.0	18.3	25.2	29.7	153.	177.	
Ne	10	3.46	6.56	10.2	15.6	20.2	25.3	33.2	38.3	192.	218.

*Values for inner shell (1s) electrons in the elements of the second period are shaded.

example, the relative sizes of the six peaks in the XPS spectrum of potassium (K) are, in order of decreasing E_i value, 2, 2, 6, 2, 6, 1. This pattern is just what we would expect, given potassium's electron configuration: $1s^2 2s^2 2p^6 3s^2 3p^6 4s^1$.

Another perspective on the different energy levels of orbitals is provided by variations in the successive ionization energies of the elements; that is, the different energies required to remove one electron (E_{i1}), then a second (E_{i2}), then a third (E_{i3}), and so on, from an atom of an element. Consider the trend in successive ionization energies for the first 12 elements in Table 3.2. For multielectron atoms, the energy needed to remove a second electron is always greater than that needed to remove the first, because the second, negatively charged electron is being removed from an ion that already has a positive charge. The energy needed to remove a third electron is greater still because it is being removed from an ion with a +2 charge. Superimposed on this trend is a much more dramatic increase in ionization energy when the outer-shell electrons have been removed and the next electron must come from an inner shell. These electrons are clearly held much more tightly by the atom and are not likely to take part in chemical reactions.

3.12 THE UNCERTAINTY PRINCIPLE

To end this chapter, let's return to the momentous advances in chemistry and physics of the first three decades of the twentieth century. Figure 3.29 summarizes and connects some of those advances. They are due to the brilliant work of Einstein, Planck, Rutherford, de Broglie, Schrödinger, Born, and others, which has forever changed scientists' view of the fundamental structure of matter and the universe.

Consensus in the scientific community did not come easily. We have seen how hot sodium atoms in the [Ne]$3p^1$ excited state emit photons of yellow orange

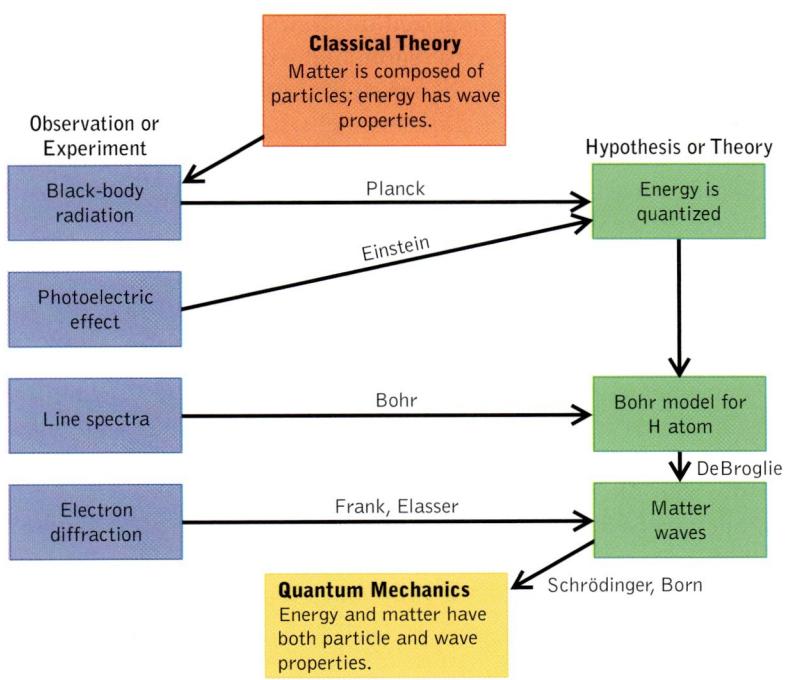

FIGURE 3.29 In Chapter 1, we considered the scientific method of inquiry. The development of quantum theory from classical theory in the first decades of the twentieth century shows the interplay between observations and hypotheses or theories. Follow the arrows to trace the development of modern quantum theory, which assumes that energy and mass are equivalent, showing both particle and wave properties.

light as they fall to the [Ne]$3s^1$ ground state. This phenomenon is called *spontaneous* emission and was a concept that Einstein puzzled over for several years before deciding that the exact moment when spontaneous emission occurs and the direction of the photon emitted could not be predicted exactly. He concluded that quantum theory allows us only to calculate the *probability* of a spontaneous electron transition occurring. The details of the event are left to chance. No force of nature actually *causes* a hot sodium atom to fall to a lower energy level at an exact instant.

This situation is much like the decay of a radioactive nucleus. The half-life of a nuclide tells us how long it will take half of a large number of nuclei to decay, but there is no way to say with certainty when a particular nucleus will decay. How different these processes are from the laws governing the behavior of large objects. If you pick up a pebble and then drop it, it immediately falls to the ground. However, electrons remain in excited states for indeterminate (though usually short) times before falling to ground states. This lack of determinacy bothered Einstein and many of his colleagues. Had they discovered an underlying theme of nature: that some processes could not be described or known with certainty? Are there fundamental limits to how well we can know and understand our world and the events that change it?

> Are there fundamental limits to how well we can know and understand our world and the events that change it?

Many scientists in the early decades of the twentieth century did not care to make such admissions. They preferred the Newtonian view of the world as a place where things happened for a reason and where there were causes and effects. They believed that the more they studied nature with ever more sophisticated tools, the more they would understand why things happened in the way that they did. Soon after Max Born published his probabilistic interpretation of Schrödinger's wave functions in 1926, Einstein wrote a letter to Born:

> Quantum Mechanics is very impressive. But an inner voice tells me that it is not yet the real thing. The theory produces a great deal but hardly brings us closer to the secret of the *Old One*. I am at all events convinced that *He* does not play dice.

Although God may or may not play dice with the universe, there do seem to be limits to what we mere mortals can know about it. In 1927, a young German scientist named Werner Heisenberg (1901–1976) speculated what might happen if we tried to "see" the path followed by an electron around an atom. He proposed to watch the electron with a hypothetical γ-ray microscope. Although such a microscope does not exist, we would need one because only γ rays have wavelengths short enough to match the diminutive size of electrons. However, their short wavelengths mean that γ rays have very high frequencies and so, according to $E = h\nu$, enormous energies. Their energies are so large that, if a γ ray were to strike an electron, it would knock the electron off course. The only way not to affect the electron's motion would be to use a much lower energy, longer-wavelength source of radiation to illuminate it, but then we could not see the tiny electron clearly. And so we have a quantum mechanical dilemma. The only way to clearly observe an electron makes it impossible to know about its motion, or, more precisely, its momentum. Therefore, we can never know both. This conclusion came to be known as **Heisenberg's uncertainty principle**. To Heisenberg, this uncertainty is the essence of quantum mechanics. Its message for us is that there are limits to what we can observe, measure, and therefore know.

When he proposed his uncertainty principle, Heisenberg was working with Niels Bohr at the University of Copenhagen. The two scientists had widely different views about the significance of the uncertainty principle. To Heisenberg, it was a fundamental characteristic of nature; to Bohr, it merely represented a mathematical consequence of the wave–particle duality of electrons. For Bohr, there was no physical meaning to an electron's position and path. The debate between these two gifted scientists was heated at times. Heisenberg later wrote about one particularly emotional debate:

> [A]t the end of the discussion I went alone for a walk in the neighboring park [and] repeated to myself again and again the question: "Can nature possibly be as absurd as it seems . . . ?"

In the first chapter of this book, we examined some of the conflicts that have arisen through the ages between religion and science. We noted that religion provided humankind with answers to questions that were, at the time, unanswerable. Despite all we have learned about the origins and composition of the universe, the behavior of the fundamental particles that make up all matter may never be known, at least not to us.

> ✓ **Heisenberg's uncertainty principle** says that you cannot at the same time determine the position and momentum of an electron in an atom.

CHAPTER REVIEW

Summary

SECTION 3.1

Early evidence of atomic structure came from the Fraunhofer lines in sunlight, which exactly match the line spectra emitted by elements in hot flames. Related species such as Ca and singly charged calcium ions (Ca^+) have different emission and absorption spectra. Atoms and ions change their energies by changing the arrangements of their electrons.

SECTION 3.2

Alpha particle scattering by a gold foil shows that the nucleus is only about 0.01% of the size of a typical atom. A cloud of electrons surrounds the nucleus so that thus, an atom is mostly empty space.

SECTION 3.3

Planck proposed that light energy consists of packets called quanta with $E = h\nu$, where h is Planck's constant and ν is the frequency of a wave. Quanta of light are called photons. Energy and light thus have particle and wave duality. Einstein used the quantum principle to explain the photoelectric effect, in which light of particular frequency ejects electrons with definite kinetic energy from some metals after the work function (ϕ) of the metal has been overcome. A minimum frequency of light is needed to eject any electrons at all.

SECTION 3.4

Bohr's model of the simplest atom (H) consists of concentric circular orbits in which the single electron moves around the nucleus. It predicts the energy levels of the atom $E_n = -(2.18 \times 10^{18} \text{ J})/n^2$, where orbit number $n = 1, 2, 3, \ldots$ in steps of 1 to infinity (∞). E_n is negative—assembling an atom from a nucleus and electrons releases energy. When $n = 1$, the electron in a hydrogen atom is nearest the nucleus in the ground state. In excited states, the electron is farther away from the nucleus in orbits with r values of 2, 3, and so on. The differences in energy between neighboring orbits decrease with increasing n. When $n = \infty$, the electron has left the hydrogen atom, creating a hydrogen ion and a free electron. Bohr's model exactly predicts the absorption and emission spectra of the hydrogen atom, as well as its size (its radius is 53 picometers in the ground state, called the Bohr radius, and increases with $n = 2, 3$, and so on) and the ionization energy needed to remove its electron.

SECTION 3.5

Einstein's theory of relativity predicts that photons (particles of light) must have zero mass. De Broglie found the connection between the wavelength, λ, of a particle and its mass, m: $\lambda = h/mv$, where v is its velocity. The stability of Bohr orbits is due to the electron wave being a standing wave. Bohr's principal quantum number n is the number of characteristic wavelengths in an orbit. The internal energy of an H atom depends on which orbit contains its electron. The Bohr model cannot account for the energy states of atoms with two or more electrons.

SECTION 3.6

Electrons can be diffracted, so they must also have wavelike properties. Schrödinger described the total energy of an atom in a wave equation with solutions called wave functions (ψ). This quantum theory accounts for the spectrum of the H atom and atoms with more electrons. The function ψ^2 gives the probability of finding an electron at a particular point in an atom. Orbitals are three-dimensional regions in an atom, in which an electron exists with a certain probability.

SECTION 3.7

ψ^2 describes an orbital with three quantum numbers that serve as a label for the orbital. Principal quantum number n sets the relative size of an orbital and most of its energy. The angular momentum quantum number l, an integer from zero to $n - 1$, determines the orbital's shape. The magnetic quantum number m_l varies in steps of one from $-l$ to $+l$ and determines the orbital's orientation in space. The letters s, p, d, and f are given to electrons with $l = 0, 1, 2,$ and 3, respectively. No two orbitals in an atom can have the same combination of n, l, and m_l values.

SECTION 3.8

The s orbitals are spherical and increase in size with increasing n. The three dumbbell-shaped orbitals p_x, p_y and p_z lie along the three dimensions of space with a node (zero value) at the nucleus. They also increase in size with increasing n. The five d orbitals consist of d_{xy}, d_{xz}, d_{yz}, $d_{x^2-y^2}$, and d_{z^2}.

Radial distribution functions give the probability of finding an electron in a spherical shell at a distance from the nucleus and are degenerate. For heavier atoms with more electrons, the shielding of the outer electrons by inner ones gives an effective nuclear charge that is less than the actual nuclear charge. Because orbitals penetrate an atom to different degrees and the energy levels get closer with increasing n, the orbital energy order of atoms is $1s < 2s < 2p < 3s < 3p < 4s < 3d$. In an atom that is not exposed to a magnetic field, all orbitals with the same value of n and l (e.g., $3p_x$ and $3p_z$) are degenerate.

SECTION 3.9

Electrons in an atom have a spin quantum number m_s that is either $+\frac{1}{2}$ or $-\frac{1}{2}$. Each orbital can hold two electrons. The Pauli exclusion principle states that no two electrons in an atom can have the same four quantum numbers, which means that two electrons in the same orbital must have opposite spins.

SECTION 3.10

The electron configurations of atoms determine their chemical properties. The lowest-energy orbitals are filled first. The last column in the periodic table is occupied by noble gases, which have limited chemical reactivity. Noble gases are stable because their outermost s and p subshells are completely filled. Elements from Groups 1A, 2A, 5A, 6A, and 7A can achieve noble-gas electron configurations by losing or gaining electrons. Atoms and ions with the same number of electrons are isoelectronic.

Different elements in the same column of the periodic table have the same outer-shell electronic configuration. Orbital diagrams allow us to visualize the electron populations of the orbitals in an atom. Hund's rule says that electrons in an unfilled subshell will be as far from each other as possible and spinning in the same direction. Transition metals are found in the middle of the periodic table and have partly filled d orbitals. The lanthanides (rare earth elements) are metals with Z = 58 through 71. The actinides (Z = 90–103) are radioactive and mostly absent from nature. These families contain partly filled $4f$ (lanthanide) and $5f$ (actinide) orbitals. There are 14 members of each family because there are seven f orbitals, each of which can hold as many as two electrons.

SECTION 3.11

The Bohr model and the Schrödinger wave equation predict the measured ionization energy of the hydrogen atom. The ionization energies of the other elements increase in the upward procession of a column and from left to right in a row of the periodic table because of differences in effective nuclear charge. Small variations in these trends are due to the special stability of half-filled subshells. X-ray photoelectron spectroscopy reinforces our understanding of the energy levels within atoms.

SECTION 3.12

We cannot predict when an electron in an atom in an excited state will fall into a vacant lower-energy orbital or when a single radionuclide will decay. According to the Heisenberg uncertainty principle, we cannot simultaneously determine the location and momentum of an electron in an atom.

Key Terms

actinide element (p. 142)
angular momentum quantum number (p. 124)
degenerate orbital (p. 129)
effective nuclear charge (p. 127)
electron configuration (p. 133)
empirical (p. 110)
excited state (p. 116)
ground state (p. 116)
Heisenberg's uncertainty principle (p. 150)
Hund's rule (p. 137)
ionization energy (p. 144)
isoelectronic (p. 136)
kinetic energy (p. 146)
lanthanide element (p. 142)
line spectra (p. 106)
magnetic exclusion number (p. 124)
momentum (p. 119)
noble gas (p. 137)
node (p. 127)
orbital (p. 123)
orbital diagram (p. 137)
Pauli exclusion principle (p. 132)
period (p. 136)
photoelectric effect (p. 111)
photoelectron (p. 111)
principal quantum number (p. 123)
quantum (p. 110)
quantum number (p. 123)
radial distribution plot (p. 126)
rest mass (p. 119)
spin magnetic quantum number (p. 131)
standing wave (p. 120)
transition metal (p. 139)
wave equation (p. 122)
wave function (p. 122)
wave mechanics (quantum mechanics) (p. 122)
work function (p. 112)

Key Skills and Concepts

SECTION 3.1

Understand how atoms and ions of an element emit and absorb radiation.

SECTION 3.2

Describe how we know from α-particle-scattering data that atoms are mostly empty space.

SECTION 3.3

Understand that light behaves as waves *and* as individual packets called photons with energy $E = h\nu$.

Describe and understand the photoelectric effect and the concept of the work function of a metal.

SECTION 3.4

List the properties of a hydrogen atom that are predicted by the Bohr model.

Explain the concepts of a ground-state and an excited-state hydrogen atom.

Calculate the energies of the hydrogen atom in specified excited states.

Calculate the wavelengths of light emitted and absorbed by a hydrogen atom in its different electronic states (e.g., emission resulting from $n = 3 \rightarrow n = 2$).

SECTION 3.5

Understand the significance of de Broglie's equation linking particle and wave properties.

Sketch a standing wave.

SECTION 3.6

Explain an orbital as a region of space where an electron has a certain probability of being found.

SECTION 3.7

Realize that quantum numbers are necessary labels for electrons in atoms because of the need for an atomic structure.

Understand the possible combinations of the three quantum numbers n, l, and m_l of orbitals in an atom or ion. The fourth quantum number is the spin magnetic quantum number $m_s = +\frac{1}{2}$ or $-\frac{1}{2}$. The principal quantum number n is a positive integer from 1 to ∞. The angular momentum quantum number l is an integer from 0 to $n - 1$.

Value of l:	0	1	2	3	4
Letter identifier:	s	p	d	f	g

The magnetic quantum number m_l is an integer with a value from $-l$ to $+l$.

Predict how the size and energy of an orbital changes with increasing principal quantum number n.

Explain why there are *five d* orbitals.

SECTION 3.8

Describe three ways of illustrating *s*-orbital electron density as a function of radial distance from the nucleus.

Understand the concept of effective nuclear charge and its origins.

Understand the concept of degeneracy (electronic states being degenerate).

Explain the notation $2p_x$, $2p_y$, and $2p_z$ and sketch the three orbitals.

SECTION 3.9

Describe the concept of electron spin and the need for a fourth quantum label for an electron in an atom.

Understand the consequences of the Pauli exclusion principle in predicting electron structures and the exact shape of the periodic table.

SECTION 3.10

Be able to interpret and write electron configurations, including abbreviated electron configurations.

Be able to distinguish between ground-state and excited-state electron configurations.

Understand what is meant by the term *isoelectronic*.

With the use of Hund's rule, be able to predict the number of unpaired electron spins from an electron configuration.

Understand the importance of noble-gas configurations in predicting the stability of isoelectronic species.

Be familiar with patterns of physical and chemical properties of groups and rows predicted by the periodic table.

SECTION 3.11

Explain the trends in ionization energy in the procession across a row and down a column of the periodic table.

SECTION 3.12

Understand the Heisenberg uncertainty principle, which states that, because electrons are so small, it's impossible to measure both their speed and their positions in an atom. Rather, only one of the two can be measured.

Key Equations and Relations

SECTION 3.4

Empirical equation for the visible part of the line spectrum of hydrogen, where C is $3.29 \times 10^{15} \text{s}^{-1}$ and n is a positive integer greater than 2:

$$\nu = C\left(\frac{1}{2^2} - \frac{1}{n^2}\right) \quad (3.1)$$

Empirical equation for the line spectrum of hydrogen, where λ is the wavelength of a line observed in the spectrum and $n_2 > n_1$:

$$\frac{1}{\lambda} = 1.097 \times 10^{-2} \, \text{nm}^{-1}\left(\frac{1}{n_1^2} - \frac{1}{n_2^2}\right) \quad (3.4)$$

The energy states of the electron in a hydrogen atom, where $n \geq 1$:

$$E_n = (-2.18 \times 10^{-18} \, \text{J})\left(\frac{1}{n^2}\right) \quad (3.6)$$

The difference between two energy states in the Bohr model of the hydrogen atom:

$$\Delta E = (-2.18 \times 10^{-18} \, \text{J})\left(\frac{1}{n_f^2} - \frac{1}{n_i^2}\right) \quad (3.8)$$

The Boltzmann distribution, where N_i and N_0 are the numbers of hydrogen atoms in the i excited and ground states, n_i is the principal quantum number of the i excited state, E_i is the energy difference between the ground state and the i excited state, T is the temperature in kelvins, and k is Boltzmann's constant (1.3806×10^{-23} J/K:

$$\frac{N_i}{N_0} = n_i^2 \cdot e^{-E_i/kT} \quad (3.10)$$

SECTION 3.5

The relation of the wavelength λ to the momentum mc and mass m of the particles (photons) of electromagnetic radiation, where c is the speed of light and h is Planck's constant:

$$\lambda = h/mc \quad (3.14)$$

SECTION 3.8

Effective charge Z_{eff} is the positive force of the nucleus felt by an electron in an atom. Intervening electron density decreases this force; so $Z_{eff} < Z$, the true nuclear charge.

SECTION 3.9

The Pauli exclusion principle: no two electrons in an atom can have the same four quantum numbers.

QUESTIONS AND PROBLEMS

Line Spectra and the Photoelectric Effect

CONCEPT REVIEW

1. What is a *photon*?
2. Explain the terms *emission spectrum* and *absorption spectrum*.
3. How did the study of the emission spectra of elements lead to the identification of the Fraunhofer lines in sunlight?
4. Are the Fraunhofer lines in the spectra of stars the result of atomic emission or atomic absorption?
5. Cite evidence that electrons take part in the absorption and emission of electromagnetic radiation by atoms.
6. Why do the energies of photoelectrons from metals exposed to radiation of the same frequency differ for different metals?
7. Compare the dimensions of a typical atom and its nucleus.
8. How would the results of the Rutherford-Geiger-Marsden α-particle-scattering experiment be affected by replacing the gold foil with aluminum foil?

PROBLEMS

9. Which of the following have "quantized" (discrete) values? Explain your selections.
   a. number of eggs remaining in a open carton of eggs
   b. elevation up a handicapped access ramp
   c. elevation up a flight of stairs
   d. speed of an automobile
   e. speed of the electron orbiting the nucleus of a hydrogen atom
10. Which of the following have "quantized" (discrete) values? Explain your selections.
    a. pitch of a note played on a slide trombone
    b. pitch of a note played on a flute
    c. wavelengths of light produced by a hydrogen lamp
    d. wavelengths of light produced by the heating elements in a toaster
    e. wind speed on the top of Mt. Everest
11. Solar energy can be converted into electricity with the use of the photoelectric effect. Could tantalum ($\phi = 6.41 \times 10^{-19}$ J) be used to construct such solar-powered photocells? Assume that most of the electromagnetic energy from the sun is in the visible region near 500 nm.
12. With reference to Problem 11, could tungsten ($\phi = 7.16 \times 10^{-19}$ J) be used to construct solar-powered photocells?
13. When a piece of metal is irradiated with UV radiation ($\lambda = 162$ nm), electrons are ejected with a kinetic energy of 5.34×10^{-19} J. What is the work function of the metal?
14. The first ionization energy for a gas-phase atom of a particular element is 6.24×10^{-19} J. What is the maximum wavelength of electromagnetic radiation that could ionize this atom?
15. Thin layers of potassium ($\phi = 3.68 \times 10^{-19}$ J) and sodium ($\phi = 4.41 \times 10^{-19}$ J) are exposed to radiation of wavelength 300 nm. Which metal emits electrons with the greater velocity? What is the velocity of these electrons?
16. Titanium ($\phi = 6.94 \times 10^{-19}$ J) and silicon ($\phi = 7.77 \times 10^{-19}$ J) surfaces are irradiated with UV radiation of wavelength 250 nm. Which metal emits electrons with the longer wavelength? What is the wavelength of the electrons emitted by the titanium surface?

Hydrogen Atoms

CONCEPT REVIEW

17. Why should hydrogen have the simplest atomic spectrum of all the elements?
18. What role does the principal quantum number n play in the Bohr model of the H atom?
19. Does the energy of light emitted by H atoms depend on the values of n_1 and n_2 or only on the difference between n_1 and n_2?
20. Explain the difference between a ground-state H atom and an excited-state H atom.
21. Explain why Bohr's model works only for single-electron atoms and ions.
22. The Hubble Space Telescope can detect hydrogen atoms in the atmosphere of Jupiter by using the Lyman lines. In what region of the electromagnetic spectrum are the Lyman emission lines observed?
23. Without calculating any wavelength values, predict which of the following transitions in the hydrogen atom is associated with radiation having the shortest wavelength.
    a. from $n = 1$ to $n = 2$
    b. from $n = 2$ to $n = 3$
    c. from $n = 3$ to $n = 4$
    d. from $n = 4$ to $n = 5$
24. Without calculating any frequency values, predict which of the following transitions in the hydrogen atom is associated with radiation having the highest frequency.
    a. from $n = 5$ to $n = 6$
    b. from $n = 6$ to $n = 7$
    c. from $n = 9$ to $n = 11$
    d. from $n = 12$ to $n = 15$
25. In what ways are the emission spectra of H and He^+ alike and in what ways are they different?
26. Sodium fog lamps contain gas-phase sodium atoms and sodium ions. Sodium atoms emit yellow orange light at 589 and 590 nm. Do sodium ions emit the same yellow orange light?
27. Transitions from $n = 2$ to $n = 3, 4, 5,$ or 6 in hydrogen atoms are responsible for some of the Fraunhofer lines in the sun's spectrum. Are there any Fraunhofer lines due to transitions starting from the $n = 3$ state in hydrogen atoms?
28. Are there any visible lines in the atomic emission spectrum of hydrogen due to transitions to the $n = 1$ state?

PROBLEMS

29. The emission lines of one-electron atoms and ions can all be fit to the equation describing the spectrum of the hydrogen atom:

    $$E_n = -(2.18 \times 10^{-18} \text{ J})Z^2(1/n_f^2 - 1/n_i^2)$$

    where Z is the atomic number.
    a. As the value of Z increases, does the wavelength of the photon associated with the transition from $n = 2$ to $n = 1$ increase or decrease?
    b. Can the wavelength associated with the transition from $n = 2$ to $n = 1$ ever be observed in the visible region of the spectrum?
30. With reference to the equation in Problem 29, can transitions from higher-energy states to the $n = 2$ state in He^+ ever produce visible light? If so, for what values of n_i?
*31. Calculate the energy difference between the excited states with $n = 3$ and $n = 2$ in the hydrogen atom.
*32. Calculate the wavelength of light emitted in the transition $n = 3 \rightarrow n = 2$ in the hydrogen atom. In which region of the electromagnetic spectrum does this radiation occur?
*33. Astronomers observing the constellation Bootes, which is 5.00×10^6 light-years from Earth, can see emission from the $n = 4$ to $n = 2$ transition of the H atom. What is the wavelength of this transition as viewed from Earth?
*34. The light from a particular star is redshifted by 23.5% and includes the Balmer series of H emission lines. Calculate the wavelength for the line arising from the transition from $n = 3$ to $n = 2$ as viewed from Earth.

Electrons As Waves

CONCEPT REVIEW

35. Identify the symbols in the de Broglie relation $\lambda = h/mc$ and explain how the relation links the properties of a particle to those of a wave.
36. Explain how the observation of electron diffraction supports the description of electrons as waves.

PROBLEMS

37. The serve of some professional tennis players crosses the net at 120 miles per hour. What is the de Broglie wave-

length of a tennis ball (mass = 56 grams) going at that speed?

38. Calculate the wavelengths of the following objects:
    a. a muon (mass 1.884×10^{-25} g) traveling at 325 m/s
    b. electrons (mass = 9.11×10^{-28} g) in an electron microscope moving at 4.05×10^6 m/s
    c. an 80-kg athlete running a 4-minute mile
    d. Earth (mass = 6.0×10^{27} g) moving through space at 3.0×10^4 m/s

39. Two objects are moving at the same speed. Which (if any) of the following statements about them are true?
    a. The wavelength of the heavier object is longer.
    b. If one object has twice as much mass as the other, its wavelength is one-half the wavelength of the other.
    c. Doubling the mass of one of the objects will have the same effect on its wavelength as does doubling its speed.

40. Which (if any) of the following statements about the frequency of a particle is true?
    a. Heavy, fast-moving objects have smaller frequencies than those of lighter, faster-moving objects.
    b. Only very light objects can have large frequencies.
    c. Doubling the mass of an object and halving its speed result in no change in its frequency.

41. How rapidly would each of the following particles be moving if they all had the same wavelength as a photon of red light (λ = 750 nm)?
    a. an electron of mass 9.10939×10^{-28} g
    b. a proton of mass 1.67262×10^{-24} g
    c. a neutron of mass 1.67493×10^{-24} g
    d. an α particle of mass 6.64×10^{-24} g

42. Do electrons, positrons, and β particles with a velocity that is 90% of the speed of light have the same wavelength?

Labeling Electrons in Atoms

CONCEPT REVIEW

43. How does the concept of an orbit in the Bohr model of the hydrogen atom differ from the concept of an orbital in quantum theory?
44. What properties of an orbital are defined by each of the following quantum numbers? n, l, and m_l
45. How many quantum numbers are needed to identify an orbital?
46. How many quantum numbers are needed to identify an electron in an atom?

PROBLEMS

47. How many orbitals are there in an atom with each of the following principal quantum numbers? 1, 2, 3, 4, and 5
48. How many orbitals are there in an atom with the following combinations of quantum numbers?
    a. $n = 3, l = 2$
    b. $n = 3, l = 1$
    c. $n = 4, l = 2, m_l = 2$
49. What are the possible values of quantum number l when $n = 4$?
50. Which are possible values of m_l when $l = 2$?

51. What set of orbitals corresponds to each of the following sets of quantum numbers?
    a. $n = 2, l = 0$ c. $n = 4, l = 2$
    b. $n = 3, l = 1$ d. $n = 1, l = 0$
52. What set of orbitals corresponds to each of the following sets of quantum numbers?
    a. $n = 2, l = 1$ c. $n = 3, l = 2$
    b. $n = 5, l = 3$ d. $n = 4, l = 3$
53. How many electrons could occupy orbitals with the following quantum numbers?
    a. $n = 2, l = 0$ c. $n = 4, l = 2$
    b. $n = 3, l = 1, m_l = 0$ d. $n = 1, l = 0, m_l = 0$
54. How many electrons could occupy an orbital with the following quantum numbers?
    a. $n = 3, l = 2$ c. $n = 3, l = 0$
    b. $n = 5, l = 4$ d. $n = 4, l = 1, m_l = -1$
55. Which of the following combinations of quantum numbers are allowed?
    a. $n = 1, l = 1, m_l = 0$ c. $n = 1, l = 0, m_l = -1$
    b. $n = 3, l = 0, m_l = 0$ d. $n = 2, l = 1, m_l = 2$
56. Which of the following combinations of quantum numbers are allowed?
    a. $n = 3, l = 2, m_l = 0$ c. $n = 3, l = 0, m_l = 1$
    b. $n = 5, l = 4, m_l = -4$ d. $n = 4, l = 4, m_l = -1$

57. List the following orbitals in order of increasing energy (assume a multielectron atom).
    a. $n = 3, l = 2$
    b. $n = 5, l = 4$
    c. $n = 3, l = 0$
    d. $n = 4, l = 1, m_l = -1$

58. Place the following orbitals in order of increasing energy (assume a multielectron atom).
    a. $n = 2, l = 1$
    b. $n = 5, l = 3$
    c. $n = 3, l = 2$
    d. $n = 4, l = 3$

Electron Configurations

CONCEPT REVIEW

59. Does an orbital that contains no electrons really exist?
60. Describe the shapes of s, p, and d orbitals.
61. Explain the meaning of *effective nuclear charge*.
62. What is meant when two or more orbitals are said to be *degenerate*?
63. Are F^- and Ne isoelectronic?
64. Explain how the electron configuration of an element is predicted by the location of the element in the periodic table.
65. How do we know from the periodic table's structure that the $4s$ orbital is filled before the $3d$ orbital?
66. Do the $4s$ orbitals fill before the $3d$ orbitals because s orbitals always fill before d orbitals, because the $4s$ orbital is lower in energy, or because the $4s$ orbital is higher in energy?

PROBLEMS

67. Give the electron configurations of Li, Li^+, He, F^-, Ne, Na^+, Mg^{2+}, and Al^{3+}. For which of these species can the electron configuration be abbreviated to read $[He]2s^1$?
68. Give the abbreviated electron configurations of the species in Question 67.
69. How are the electron configurations of Ne and Na different?
70. In what way are the electron configurations of H, Li, Na, K, Rb, and Cs similar?
71. According to Hund's rule, how many unpaired electrons are there in the following ground-state atoms and ions? (a) N; (b) O; (c) P^{3-}; (d) Na^+.
72. Identify the atom with electron configuration $[Ar]4s^23d^2$. How many unpaired electrons are there in the ground state of this atom?
73. Explain why the following ions are more stable than their respective parent atoms: K^+, S^{2-}, and I^-.
74. Predict the most stable ionic form of each of the following elements: Al, N, Mg, Cs, S, P, and F.

75. Write the electron configurations of the following species: (a) Na; (b) Cl; (c) Mn; (d) Mn^{2+}.
76. Write the electron configurations of the following species: (a) C; (b) S; (c) Ti; (d) Ti^{4+}.
77. Which of the following electron configurations represent an excited state?
    a. $[He]2s^12p^5$
    b. $[Kr]5s^24d^{10}5p^1$
    c. $[Ar]4s^23d^{10}4p^5$
    d. $[Ne]3s^23p^24s^1$
78. Which of the following electron configurations represent an excited state?
    a. $[Ne]3s^23p^1$
    b. $[Ar]3d^{10}4p^2$
    c. $[Kr]5s^14d^{10}5p^1$
    d. $[Ne]3s^23p^64s^1$
79. How many different elements are represented by the following electron configurations?
    $1s^22s^22p^63s^23p^64s^23d^1$ $[Ar]4s^23d^{10}4p^1$
    $[Ar]4s^24p^1$ $[Ne]3s^23p^24s^24p^1$
80. How many different elements are represented by the following electron configurations?
    $1s^22s^22p^63s^1$ $[He]2s^22p^6$
    $1s^22s^22p^6$ $[Ne]3s^1$
81. Which of the following electron configurations refers to the element phosphorus?
    $1s^22s^22p^63s^23p^4$ $1s^22s^22p^63s^23p^3$
    $[Ar]3s^23p^3$
82. Which of the following electron configurations is that of the element rubidium (Rb)?
    $1s^22s^22p^63s^23p^63d^{10}4s^24p^64d^1$ $[Kr]5s^1$
    $1s^22s^22p^63s^23p^63d^{10}4s^24p^65s^1$ $[Ar]4s^1$
83. How many unpaired electrons are there in ground-state As, Te, Sn, and Ge atoms?
84. How many unpaired electrons are there in ground-state K, Ca, Al, and I atoms?
85. How many unpaired electrons are there in ground-state Ti, Cr, Cu, and Zn atoms?
86. How many unpaired electrons are there in ground-state V, Mn, Fe, and Co atoms?
87. Write the quantum numbers describing the highest-energy electrons in a ground-state atom of the isotope ^{131}I.

Do these electrons in ^{131}I and ^{127}I have the same set of quantum numbers? Explain.

*88. Although no currently known elements contain electrons in g orbitals, these elements may be synthesized some day. What is the atomic number of the first element that would have a g^1 electron configuration in the ground state?

Ionization Energy

CONCEPT REVIEW

89. What is meant by the term *ionization energy*?
90. How do ionization energies change as we go (a) down a group of elements in the periodic table; (b) from left to right across a period of elements?

PROBLEMS

91. The ionization energies of the main-group elements are given in Figure 3.28. Explain the differences in ionization energy between the following pairs of elements: (a) He and Li; (b) Li and Be; (c) Be and B; (d) N and O.
92. Explain why it is more difficult to ionize a fluorine atom than a boron atom.
*93. Albert Einstein did not fully accept the uncertainty principle, remarking that "He [God] does not play dice." What do you think Einstein meant? Niels Bohr allegedly responded to Einstein by saying, "Albert, stop telling God what to do." What do you think Bohr meant?
94. Explain why it is not possible to locate an electron in an atom and at the same time accurately measure its speed and direction.

4 Stoichiometry
And the Formation of Earth

Gases released by volcanic activity probably produced Earth's first atmosphere.

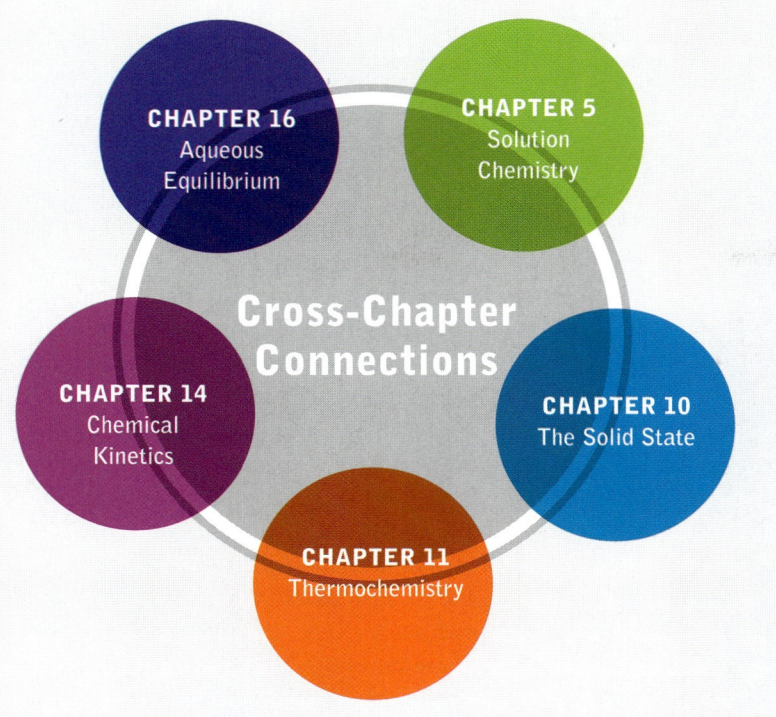

4.1	**The Composition of Earth**
4.2	**The Composition of Compounds**
4.3	**Naming Compounds**
	Binary molecular compounds
	Binary ionic compounds
	Binary compounds of transition metals
	Polyatomic ions
4.4	**Chemical Reactions and the Mole**
	BCX: THE CHEMISTRY OF THE ALKALI METALS
4.5	**Completing and Balancing Chemical Equations**
4.6	**Percent Composition and Empirical Formulas**
4.7	**Stoichiometric Calculations and the Carbon Cycle**
4.8	**Limiting Reactants and Percent Yields**

A Look Ahead

Some of the chemical processes that shaped early Earth and led to the formation of the oceans and atmosphere provide the context for an introduction to chemical reactions and the formation of compounds. We will approach the naming of compounds systematically and then consider the composition and some of the chemical properties of the common minerals of Earth's crust. Chemical reactions among the substances present in the atmosphere and the crust form the contexts for writing balanced chemical equations and for calculating quantities of reactants and products. Balanced chemical equations are also developed for photosynthesis and for energy-producing reactions based on combustion. We then examine how the chemical composition of a substance can be used to determine its formula and examine reactions that do not give the amount of product expected.

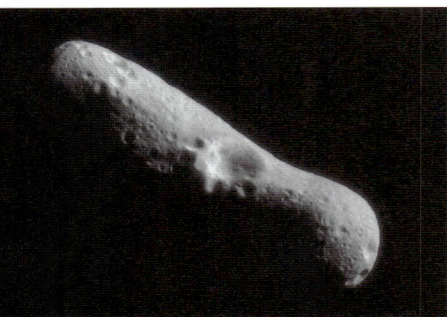

The *NEAR* (Near Earth Asteroid Rendezvous) *Shoemaker* spacecraft landed on this 33-km long, peanut-shaped asteroid, known as 433 Eros, on February 12, 2001. Instruments onboard the spacecraft determined the elemental composition of the asteroid and found that it was made of the same elements that make up our planet, indicating that the asteroid formed at the same time as the planetesimals that aggregated to form Earth and the other terrestrial planets.

4.1   THE COMPOSITION OF EARTH

We began this book with a description of a theory of the origin and evolution of our universe. According to this theory, the matter in our universe that formed from an enormous quantity of primordial energy eventually gathered into galaxies. Nuclear reactions in the biggest stars in these galaxies led to the formation of the elements of the periodic table. Many scientists further believe that about 4.6 billion years ago a rotating disk of matter called a nebula formed at the fringes of the Milky Way galaxy. As the nebula collapsed under the force of its own gravity, it became denser and hotter until nuclear fusion began at its center. The resulting release of energy stopped the collapse, and a protostar that would become our sun was born.

In cooler regions of the solar nebula millions of miles from the sun, some of the swirling matter condensed into structures called planetesimals. As these structures collided and fused with each other, the planets of the solar system formed. The distribution of the elements among the forming planets was not uniform. The lightest and most easily vaporized elements were swept away from the planets closest to the sun by a combination of solar heat and solar wind: high-velocity ions emitted by the sun. This separation of substances based on their volatilities (not unlike the process of distillation discussed in Chapter 1) left the planets closest to the sun—Earth, Mercury, Venus, and Mars—rich in those elements that are not volatile or that form nonvolatile compounds. More easily vaporized elements, including hydrogen and helium, were enriched in the outermost planets, as illustrated in Figure 4.1.

The elemental composition of Earth is dramatically different from the composition of the universe, as shown in Figure 4.2. Note that just 4 elements—iron, oxygen, silicon, and magnesium—make up 92% of all the mass of Earth, and the

FIGURE 4.1 The solar wind has swept away nearly all the atmosphere of Mercury, leaving it, like the other four inner (terrestrial) planets, enriched in the less-volatile elements. On the other hand, Uranus and the outer planets of the solar system are called gas giants because of their high concentrations of volatile substances. Uranus has a small mantle of mostly ice and an atmosphere that is 83% hydrogen and 15% helium by mass.

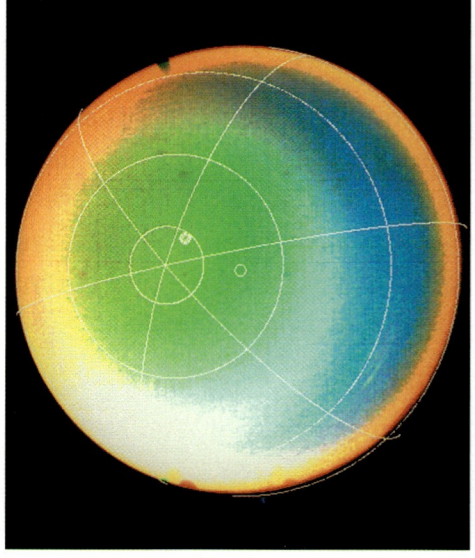

FIGURE 4.2 The composition of Earth is much different from that of the universe, the mass of which is 99% hydrogen and helium.

top 10 make up more than 99%. In contrast, 99% of the matter in the universe is hydrogen and helium.

Early Earth was heavily bombarded by asteroids and other remnants of the solar disk. The energy released by these collisions and by the decay of radioactive elements in the planet's interior made the planet a hot, molten mass. In this state, its elements underwent additional fractionation owing to their different densities and melting points. Dense elements, notably iron and nickel, sank to the center of the planet, forming a molten core that is still there today. A second, less-dense mantle rich in aluminum, magnesium, silicon, and oxygen formed around the core. As time passed and Earth cooled, the mantle was fractionated further, and a solid crust formed from the components of the mantle that were the least dense and had the highest melting points (Figure 4.3).

CONNECTION: The elements present in the solar nebula were probably synthesized by earlier generations of giant stars (Chapter 2).

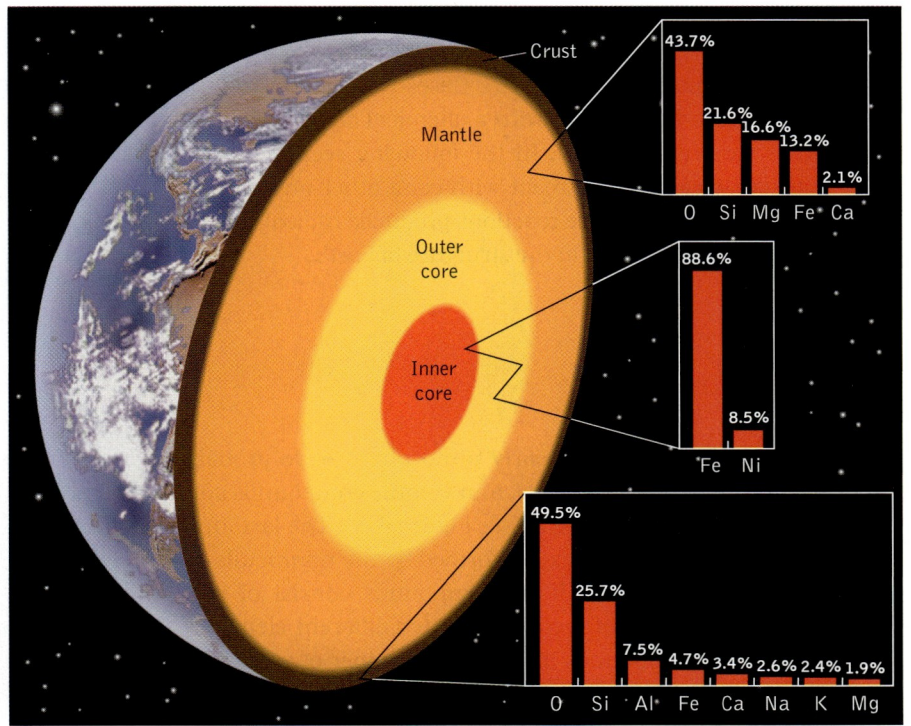

FIGURE 4.3 *What a Chemist Sees.* Earth is composed of a solid inner core consisting mostly of nickel and iron surrounded by a molten outer core of similar composition. A liquid mantle composed mostly of oxygen, silicon, magnesium, and iron lies between the molten core and a relatively thin solid crust.

FIGURE 4.4 Although collisions with large asteroids, such as the one that produced this crater in northern Arizona, are infrequent today, such collisions were much more frequent just after Earth formed. These collisions coupled with widespread volcanic activity prevented the formation of a solid crust for millions of years.

FIGURE 4.5 Volcanic eruptions, such as this one at Mount St. Helens, Washington, on May 18, 1980, release sulfur and nitrogen oxides including SO_2, SO_3, NO, and NO_2. Volcanologists monitor atmospheric concentrations of SO_2 near volcano craters to determine when an eruption is likely to occur.

Earth's early crust was torn by the impact of asteroids (Figure 4.4) and widespread volcanic activity (Figure 4.5). Gases released during volcanic eruptions formed a primitive atmosphere with a chemical composition very different from the air that we breathe today. In the following sections, we will explore how some of these volcanic gases react with each other and with some of the compounds of Earth's crust. These reactions produced radical changes in the composition of the atmosphere and in the structure of Earth.

4.2 THE COMPOSITION OF COMPOUNDS

The volcanic gases of Earth's early atmosphere were probably rich in water vapor (H_2O), carbon dioxide (CO_2), and volatile compounds containing oxygen, including SO_2, SO_3, NO, and NO_2 (see Figure 4.5). The last four compounds are also examples of the volatile oxides formed by the nonmetals on the right-hand side of the periodic table. All six compounds consist of molecules that are formed by combinations of the atoms of two different elements. The molecular formulas of these compounds include the number of atoms of each element in one of their molecules (see Section 1.1). These proportions are an atomic-scale extension of a **scientific law** first published in 1799 by French chemist Joseph Louis Proust (1754–1826). Proust's **law of definite proportions** states that *a compound always contains the same proportions (by mass) of its constituent elements, no matter how the compound was produced.* Thus, the ratio of the mass of carbon to the mass of oxygen in carbon dioxide is always 3.00 grams of C for every 8.00 grams of O no matter whether the CO_2 was produced by the combustion of carbon in coal,

$$C + O_2 \longrightarrow CO_2$$

or the thermal decomposition of limestone ($CaCO_3$),

$$CaCO_3 \xrightarrow{\Delta} CaO + CO_2$$

(We will see how the formula of a compound can be used to predict the ratio of masses of the constituent elements in it, and vice versa, in Section 4.5.)

The four oxides of sulfur and nitrogen described in the preceding paragraph illustrate another property of molecular compounds that was first described by British scientist and mathematician John Dalton (1766–1844). At the beginning of the nineteenth century, Dalton laid the foundation for modern chemistry when he developed his atomic theory of matter. According to this theory, each element is composed of identical atoms of a particular size and mass and having characteristic chemical properties. Dalton also proposed that compounds are formed from atoms of more than one element and that, in a given compound, the ratio of the numbers of atoms of any two constituent elements is either an integer or a simple fraction. Dalton's atomic view of compounds explained Proust's law of definite proportions. It also explained another property of some compounds: if the same two elements (call them X and Y) form more than one compound, then *the ratio of the masses of Y that react with a given mass of X to form any two such compounds is the ratio of two small whole numbers.* This experimentally observed principle has come to be known as the **law of multiple proportions.** Consider SO_2 and SO_3. Suppose we could measure the mass of O that reacts with a given mass of S to form SO_2. Next, we measured the mass of O that reacts with the same mass of S to form SO_3. Then we compared the mass of O in SO_2 with that in SO_3. We would discover that the ratio of the two masses of O is 2 : 3. Similarly, there is twice as much oxygen in NO_2 as there is in NO; that is, the ratio of the mass of O that reacts with a given mass of N to form NO_2 and NO is 2 : 1.

> ✓ A **scientific law** is a statement, often expressed in a mathematical equation, that summarizes a broad range of observations and experimental results.

> The **law of definite proportions** states that a specific chemical compound obtained from any source always contains the same proportions by mass of its elements.

> ✓ The **law of multiple proportions** states that the masses of element Y that combine with a fixed mass of element X to form two or more different compounds are in the ratios of small whole numbers.

SAMPLE EXERCISE 4.1: Predict the ratio of the masses of oxygen that combine with a given mass of nitrogen to form N_2O and N_2O_5.

SOLUTION: There is one atom of O for every two atoms of N in N_2O, but there are five atoms of O for every two atoms of N in N_2O_5. Therefore, the mass ratio of O to N is five times as great in N_2O_5 as in N_2O.

PRACTICE EXERCISE: Predict the ratio of the masses of oxygen that combine with a given mass of iron to form FeO and Fe_2O_3. (See Problems 7 and 8.)

4.3 NAMING COMPOUNDS

Before we continue our consideration of the composition and chemical properties of components of Earth's early atmosphere, we need to establish some rules for how we name these compounds and write their **chemical formulas.** Each compound has a chemical formula and name. The names and formulas of common compounds form a foundation for the language of chemistry.

> A **chemical formula** uses atomic symbols and subscripts to identify the constituent elements in a substance and their proportions.

Binary molecular compounds

Can you name the compounds with the molecular formulas SO_2, SO_3, NO, and NO_2? These compounds are **binary molecular compounds,** meaning that they consist of molecules made up of atoms of only two elements. The method for naming them is fairly straightforward and follows the pattern for assigning the name *carbon dioxide* to CO_2. The starting point is the periodic table of the elements, an extremely valuable source of chemical information if you know how to use it. It is important that you learn how to use the periodic table because a goal of this chapter is for you to be able to

1. name simple chemical compounds,
2. translate the names of compounds into chemical formulas, and
3. translate chemical formulas into names.

> A **binary molecular compound** consists of molecules containing atoms of only two elements.

Like many of the nonmetallic elements on the right-hand side of the periodic table, sulfur and nitrogen combine with oxygen in more than one way. Therefore, we need a naming system that indicates what proportions of the constituent elements are present in these compounds. Thus prefixes such as *mono, di, tri,* and *tetra* indicate the number of atoms of each element in a molecular formula. For example, NO is nitrogen monoxide, whereas NO_2 is nitrogen *di*oxide (we do not bother with the prefix *mono-* for the first element in a name). Similarly, SO_2 is the formula for sulfur dioxide, and SO_3 is the formula for sulfur trioxide. Note that the ending on the oxygen part of these names has been changed to *-ide*. This pattern holds for all binary compounds: the ending on the name of the second element in the chemical formula always ends in *-ide*. The order in which the names of elements are given in the names and formulas of compounds corresponds to the relative positions of the elements in the periodic table: the name of the leftmost element appears first in the name of a binary compound. When two elements are in the same column—for example, sulfur and oxygen—then the name of the lower element appears first.

SAMPLE EXERCISE 4.2: Name the following oxides of nitrogen: N_2O, N_2O_4, N_2O_5.

SOLUTION: N_2O is dinitrogen oxide (also called *nitrous oxide* or laughing gas), N_2O_4 is dinitrogen tetroxide, and N_2O_5 is dinitrogen pentoxide.

PRACTICE EXERCISE: Name the following compounds: MgO, CO, As_2O_3, and PCl_5. (See Problems 13 and 14.)

Binary ionic compounds

A **binary ionic compound** consists of a positively charged ion, or *cation*, formed by a metallic element (from the middle or the left-hand side of the periodic table) and a negatively charged ion, or *anion*, formed by a nonmetal (from the right-hand side). To name a binary ionic compound, we start with the elemental name of the cation and follow it with the name of the anion. The names of single-atom anions are based on the names of the corresponding elements, but the name of the element always ends in *-ide*. Consider the substance that geologists call *halite*, but which most of us know as table salt. Halite deposits may be found underground or on Earth's surface in arid regions where an ancient lake or inland sea evaporated to dryness (Figure 4.6). They are even found in meteorites. The chemical name of halite is sodium chloride; its formula is NaCl. NaCl consists of ordered arrays of sodium cations, Na^+, surrounded by chloride anions, Cl^-, and vice versa, as shown in Figure 4.6.

> **Binary ionic compounds** consist of crystals (three-dimensional arrays with repeating patterns) of alternating *cations* (ions with positive charges) and *anions* (ions with negative charges).

CONNECTION: The charges on the ions formed by single atoms are linked to the electron configurations of the atoms (Section 3.7).

CONCEPT TEST: Why are surface deposits of halite more likely to be found in arid regions, such as the American Southwest, than in humid regions?

Figure 4.7 contains a version of the periodic table showing the charges of the most common single-atom ions formed by some of the elements. Group 1A elements, called **alkali metals**, form ions with a +1 charge. Group 2A elements, or **alkaline earth metals**, exist in nature as ions with a +2 charge. When Group 1A and Group 2A elements combine with those from the right-hand side of the periodic table, such as those in Group 6A or 7A, the right-hand-side atoms gain one (7A) or two (6A) electrons, forming ions with −1 and −2 charges, respectively.

> **Alkali metals** and **alkaline earth metals** belong to Groups 1A and 2A of the periodic table and exist in nature as cations with charges of +1 and +2, respectively.

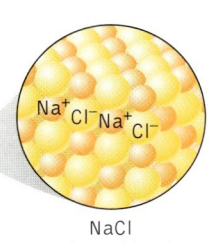

NaCl
Sodium chloride

FIGURE 4.6 *What a Chemist Sees.* The Bonneville salt flats in Utah were once part of ancient Lake Bonneville, which covered 50,000 km². As the lake evaporated, it left behind this extremely flat layer of the salt that was once dissolved in it. Most of the salt is NaCl, which is made of crystals composed of ordered arrays of Na^+ and Cl^- ions. Another part of Lake Bonneville became the Great Salt Lake—a body of water that is eight times as salty as seawater.

Sodium fluoride, NaF, is added to toothpaste and most municipal water supplies to prevent tooth decay. The F⁻ ion in NaF is incorporated into tooth enamel, making the enamel more resistant to acidic substances.

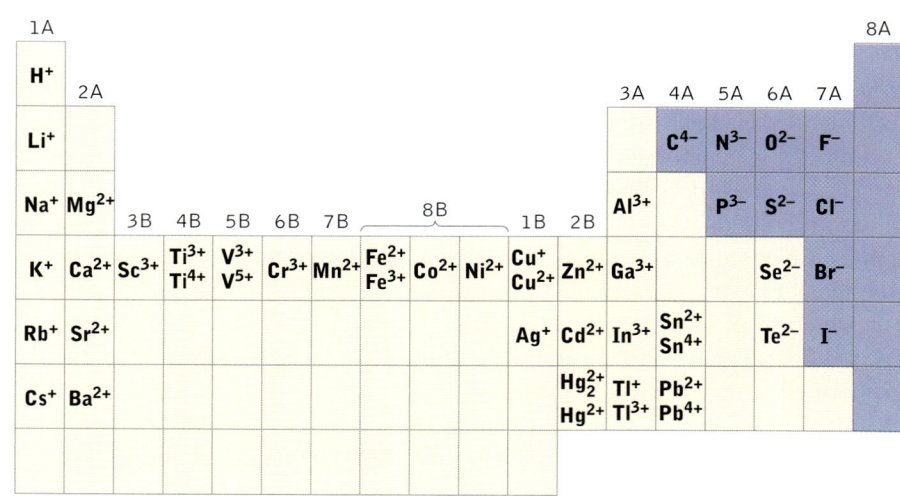

FIGURE 4.7 This version of the periodic table shows the most common charges on the ions derived from some of the elements. The elements in a group typically have ions with the same charges.

Ionic compounds, like all substances, are electrically neutral. (A saltshaker does not have an electrical charge even though it is filled with an ionic compound, NaCl.) Thus the negative and positive charges of the ions in an ionic compound must be in balance. For example, the chemical formula for sodium chloride derives from the fact that, for every Na^+, there must be one Cl^- so that the positive and negative charges balance. Similarly, the chemical formula for calcium sulfide is CaS because the ratio of Ca^{2+} to S^{2-} also should be 1:1. However, the chemical formula for magnesium fluoride is MgF_2 because *twice* as many fluoride ions, F^-, are needed to balance the +2 charges on magnesium ions, Mg^{2+}.

 CONNECTION: The structures of sodium chloride and other ionic compounds are described in Section 10.2.

SAMPLE EXERCISE 4.3: Write the chemical formulas of the following compounds: (a) potassium bromide; (b) calcium oxide; (c) sodium sulfide; (d) magnesium chloride; (e) aluminum oxide.

SOLUTION: To write chemical formulas, we need to accurately predict the charges on the ions. Refer to Figure 4.7; locate the given elements in parts *a* through *e* and note their charges. You should come up with the following ionic charges: K^+, Br^-, Ca^{2+}, O^{2-}, Na^+, S^{2-}, Mg^{2+}, Cl^-, and Al^{3+}. The need to balance these charges in each compound leads to the following chemical formulas: (a) KBr, (b) CaO, (c) Na_2S, (d) $MgCl_2$, and (e) Al_2O_3.

PRACTICE EXERCISE: Write the chemical formulas of the following compounds: (a) strontium chloride; (b) magnesium oxide; (c) sodium fluoride; (d) calcium bromide. (See Problems 15 and 16.)

Binary compounds of transition metals

Some metallic elements, including many of the transition metals in the middle of the periodic table, form cations with different charges. For example, most of the copper found in nature is present as Cu^{2+}; however, some copper compounds contain Cu^+. Thus, the name *copper chloride* could apply to $CuCl_2$ or CuCl. To distinguish between the two compounds, we need a way to identify the charge on the copper ion. One way is to use a Roman numeral after the word copper in the name of the compound. Thus, copper(II) chloride represents the chloride of Cu^{2+}, which is $CuCl_2$. The formula of copper(I) chloride is CuCl. Chemists have for many years also used different names to identify cations of the same element with different charges. For Cu^+ and Cu^{2+}, these names are *cuprous* and *cupric*, respectively. Similarly, the common ions of iron, Fe^{2+} and Fe^{3+}, are called *ferrous* and *ferric*, respectively. Note that, in both these pairs of ions, the name of the ion with the smaller charge ends in *-ous* and the name of the one with the higher charge ends in *-ic*.

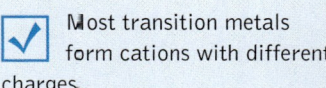
Most transition metals form cations with different charges.

SAMPLE EXERCISE 4.4: Write the chemical formulas of iron(II) sulfide and iron(III) oxide. Write alternative names for these compounds that do not make use of Roman numerals to indicate the charge on iron.

SOLUTION: The Roman numerals indicate the sizes of the positive charges on the iron cations: +2 for iron(II) and +3 for iron(III). The charge on the sulfide ion is −2 (see Figure 4.7). Likewise, the charge on the oxide ion is −2. Therefore, iron(II) sulfide consists of an array of equal numbers of Fe^{2+} and S^{2-} ions and has the chemical formula FeS.

Iron(III) oxide is an array of Fe^{3+} and O^{2-} ions. To balance the different charges on the two ions, we need three O^{2-} ions for every two Fe^{3+} ions. Thus the formula of iron(III) oxide is Fe_2O_3.

The common names for these compounds are ferrous sulfide and ferric oxide. We will encounter this *-ous* or *-ic* convention again in naming other ions and molecules.

PRACTICE EXERCISE: Write the chemical formulas of manganese(II) chloride and manganese(IV) oxide. Propose alternative names for these compounds that do not include Roman numerals. (See Problems 17 and 18.)

Polyatomic ions

Table 4.1 lists the names and chemical formulas of some commonly encountered ions. Note that several of them consist of more than one kind of atom and are called **polyatomic ions.** They are essentially charged molecules that contain a different number of electrons from the sum of the protons in the nuclei of their

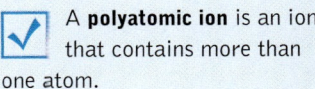
A **polyatomic ion** is an ion that contains more than one atom.

The white cliffs of Dover are an impressive geological formation on the southeastern coast of England. The cliffs are made of chalk, a form of calcium carbonate, $CaCO_3$.

TABLE 4.1 Names and Charges of Some Common Ions

Name	Chemical Formula	Name	Chemical Formula
Acetate	$C_2H_3O_2^-$	Hydride	H^-
Ammonium	NH_4^+	Hydrogen phosphate	HPO_4^{2-}
Azide	N_3^-	Hydroxide	OH^-
Bromide	Br^-	Nitrate	NO_3^-
Carbonate	CO_3^{2-}	Nitride	N^{3-}
Chlorate	ClO_3^-	Nitrite	NO_2^-
Chloride	Cl^-	Oxide	O^{2-}
Chromate	CrO_4^{2-}	Perchlorate	ClO_4^-
Cyanide	CN^-	Permanganate	MnO_4^-
Bicarbonate	HCO_3^-	Peroxide	O_2^{2-}
Bisulfite	HSO_3^-	Phosphate	PO_4^{3-}
Dichromate	$Cr_2O_7^{2-}$	Sulfate	SO_4^{2-}
Dihydrogen phosphate	$H_2PO_4^-$	Sulfide	S^{2-}
Disulfide	S_2^{2-}	Sulfite	SO_3^{2-}
Fluoride	F^-	Thiocyanate	SCN^-

CONNECTION: N and O also form the diatomic cation NO^+ (Section 6.4).

✓ An **oxoanion** is an ion that contains atoms of oxygen and another element.

component atoms. These polyatomic ions with more electrons than protons are negatively charged, and those with fewer are positively charged. For example, the NO_2^- ion, with one nitrogen atom (containing 7 protons) and two oxygen atoms (8 protons each), has a total of 23 protons and one more electron (i.e., 24 electrons), giving it a net charge of -1. On the other hand, the ammonium ion, NH_4^+, has 1 electron fewer than the 11 protons found in one nitrogen atom and four hydrogen atoms, or 10 electrons in all. It turns out that all the stable ions in Table 4.1 contain an *even* number of electrons.

Many polyatomic ions are anions containing oxygen and another element. These polyatomic ions are called **oxoanions.** Each has a name based on the name of the element that appears first in the formula, but the ending is changed to either *-ite* or *-ate*, depending on the number of oxygen atoms in the formula. Thus, SO_4^{2-} is sul*fate*, whereas SO_3^{2-} is sul*fite*. Notice that a sulfate ion has one more oxygen atom than a sulfite ion has. The same rule applies to naming the oxoanions of nitrogen, nitrate (NO_3^-) and nitrite (NO_2^-), and to naming other sets of oxoanions.

SAMPLE EXERCISE 4.5: Write the chemical formulas for the ionic compounds containing the following combinations of cations and anions: (a) Ca^{2+}, O^{2-}; (b) Na^+, SO_4^{2-}; (c) Mg^{2+}, PO_4^{3-}; (d) Al^{3+}, S^{2-}.

SOLUTION: To write the correct formula for an ionic compound, the total positive charge on the cations must equal the total negative charge on the anions. For Ca^{2+} and O^{2-}, CaO exactly balances the cation and anion charges. In Na_2SO_4, two singly charged Na^+ ions are needed to balance the -2 charge of the sulfate anion. To balance the numbers of ions with $+3$ and -2 or $+2$ and -3 charges, use the charge on one ion as the subscript for the other. This approach gives $Mg_3(PO_4)_2$ and Al_2S_3.

PRACTICE EXERCISE: Write the chemical formulas of the compounds formed by the following combinations of ions: (a) Sr^{2+}, NO_3^-; (b) K^+, SO_3^{2-}; (c) barium cations and iodide anions; (d) potassium cations and oxide anions. (See Problems 19 and 20.)

If an element forms more than two kinds of oxoanions, as chlorine and the other Group 7A elements do, then prefixes are used to distinguish them. The oxoanion with the largest number of oxygen atoms gets the prefix *per*, and that with the smallest number of oxygen atoms may have the prefix *hypo* in its name. Applying these rules to the oxoanions of chlorine, we have

ClO^- hypochlorite
ClO_2^- chlorite
ClO_3^- chlorate
ClO_4^- perchlorate

Note that these rules do not necessarily enable us to predict the chemical formula from the name or the charge on the anion. You will need to commit them to memory.

SAMPLE EXERCISE 4.6: What are the names of the following ionic compounds? (a) $CaCO_3$; (b) $NaNO_3$; (c) $MgSO_3$; (d) $PbSO_4$; (e) $KClO_3$; (f) $Ca_3(PO_4)_2$; (g) $NaHCO_3$.

SOLUTION: (a) Calcium carbonate; (b) sodium nitrate; (c) magnesium sulfite; (d) lead sulfate; (e) potassium chlorate; (f) calcium phosphate; (g) sodium hydrogen carbonate (also called sodium bicarbonate—the prefix *bi-* is sometimes used to indicate that a hydrogen ion, H^+, is attached to the oxoanion).

PRACTICE EXERCISE: What are the names of the following ionic compounds? (a) NH_4NO_3; (b) $Mg(ClO_3)_2$; (c) $LiNO_2$; (d) $NaClO$; (e) K_2SeO_4. (See Problems 21–24.)

The active ingredient in household bleach is sodium hypochlorite, NaClO. The hypochlorite oxoanion, ClO^-, is one of four oxoanions formed by chlorine that have the general formula ClO_x^- ($x = 1-4$). The prefix *hypo-* is reserved for the member of this series with the smallest number of oxygen atoms (one).

We need to consider one more naming convention—for those compounds having formulas that begin with hydrogen and end with the formula of an oxoanion. These compounds are **acids.** Their acidic properties are related to the release of hydrogen ions when the compounds dissolve in water. The scheme for naming the acids of oxoanions is shown in Figure 4.8. If the name of the oxoanion ends

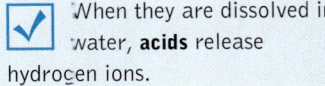

When they are dissolved in water, **acids** release hydrogen ions.

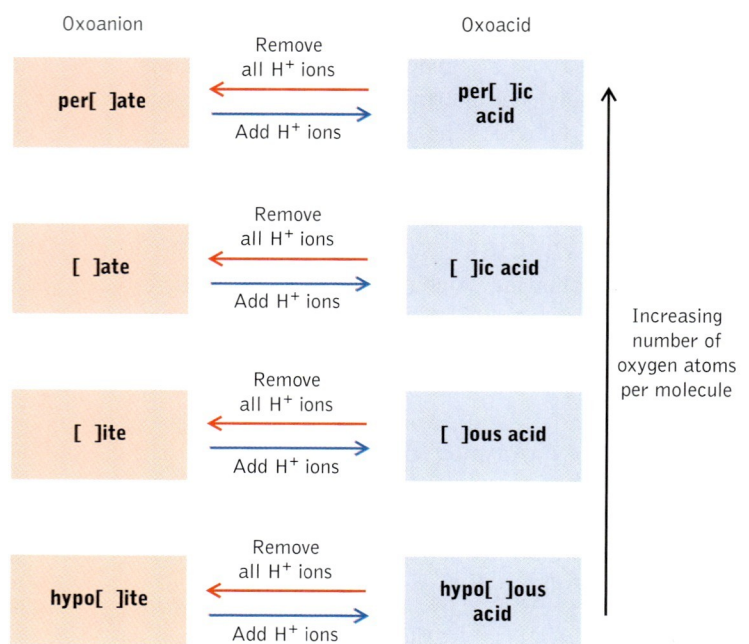

FIGURE 4.8 The names of the acids derived from oxoanions are related to the names of the oxoanions, as shown. An *-ate* suffix becomes an *-ic* ending, whereas an *-ite* ending is replaced by *-ous*. The prefixes *hypo-* and *per-* indicate the smallest and the largest number of oxygen atoms per anion, respectively.

in *-ate*, the name of the corresponding acid ends in *-ic*; if the name of the oxoanion ends in *-ite*, the name of the corresponding acid ends in *-ous*. Thus, Na_2SO_4 is the formula for sodium sulfate, and H_2SO_4 is the formula for sulfuric acid. Similarly, $NaNO_2$ is sodium nitrite, and HNO_2 is nitrous acid.

There are other acids besides those with oxoanions in their structures. Many of them are produced by life processes. For example, fermentation reactions produce *acetic acid*, which has the formula $HC_2H_3O_2$. A molecule of acetic acid in water releases the first hydrogen in its chemical formula as H^+, also forming the *acetate* ion, $C_2H_3O_2^-$.

Another important group of acidic compounds are the binary acids that contain a hydrogen and a halogen atom. This group includes aqueous solutions of HF, HCl, HBr, and HI. The names of these acids take the form *hydro*[halogen base name]*ic acid*. Thus, a solution of hydrogen chloride (HCl) is called *hydrochloric acid*.

SAMPLE EXERCISE 4.7: What are the names of the acids formed by the following oxoanions? (a) SO_3^{2-}; (b) ClO_4^-; (c) $C_2H_3O_2^-$.

SOLUTION: (a) Sulfite, SO_3^{2-}, is the oxoanion in sulfurous acid, H_2SO_3; (b) perchlorate, ClO_4^-, is the oxoanion in perchloric acid, $HClO_4$; (c) acetate $C_2H_3O_2^-$, is the oxoanion in acetic acid, $HC_2H_3O_2$.

PRACTICE EXERCISE: What are the names of the following acids? (a) HClO; (b) $HClO_2$; (c) H_2CO_3. (See Problems 25 and 26.)

4.4 CHEMICAL REACTIONS AND THE MOLE

The primordial Earth described in Section 4.1 was a planet torn by incessant volcanic activity and bombardment by debris from the solar nebula. In time, volcanic gas emissions produced an atmosphere rich in carbon dioxide, water vapor, and oxides of sulfur and nitrogen. Life as we know it could not have survived in such an environment. Besides the lack of O_2, the atmosphere was very corrosive: it contained a number of strong acids that formed when sulfur and nitrogen oxides combined with water. For example, when sulfur trioxide combines with water vapor, sulfuric acid, H_2SO_4, is produced (Figure 4.9). This reaction can be expressed by using the following **chemical equation:**

$$SO_3(g) + H_2O(g) \longrightarrow H_2SO_4(l) \quad (4.1)$$

Reactants   Product

FIGURE 4.9 A molecule of SO_3 and a molecule of H_2O react to produce a molecule of H_2SO_4.

In chemical equations, the chemical formulas of the reacting substances, or **reactants,** appear first, followed by a reaction arrow and the chemical formulas of the substances that are formed, or **products.** Sometimes we use symbols in parentheses to indicate the physical states of reactants and products: (g) for gas, (l) for liquid, and (s) for solid.

Equation 4.1 describes a reaction between molecules of sulfur trioxide and water vapor. In our macroscopic world, we must work with macroscopic quantities, and so we need a "scale up" factor that allows us to convert individual atoms and molecules into numbers of them that can be measured. Such a conversion factor exists. It is the **mole (mol),** the SI base unit for expressing quantities of substances (see Table 1.2): a mole of particles (atoms, ions, or molecules)

CONNECTION: Sulfuric acid is the principal contributor to acid rain in regions of North America and Europe (Chapter 16).

☑ A **chemical equation** describes the proportions of the substances called **reactants** that are consumed during a chemical reaction and the substances called **products** that are formed.

☑ A **mole (mol)** of particles, especially atoms, ions, or molecules, is a quantity equal to **Avogadro's number** ($N_A = 6.022 \times 10^{23}$) of the particles.

is defined by international agreement as the number of particles equal to the number of atoms in exactly 12 grams of carbon-12, which is a very large number of atoms. To four significant figures, it equals 6.022×10^{23} atoms. This number is called **Avogadro's number (N_A)** after the Italian scientist Amedeo Avogadro (1776–1856) whose research enabled scientists to accurately determine the atomic masses of the elements. It is used to convert numbers of particles into moles of particles, and vice versa, as demonstrated in the following sample exercises.

SAMPLE EXERCISE 4.8: The solar wind is made up of ions, mostly protons, flowing out from the sun at 400 km/s. Near Earth, each cubic centimeter of space contains on average six solar-wind ions. How large a volume of space contains a mole of these ions?

SOLUTION: We know that 1 cm³ of space holds six ions. One way to calculate the volume of space that holds one mole of ions is to multiply 1 cm³/6 ions by Avogadro's number:

$$\frac{1 \text{ cm}^3}{6 \text{ ions}} \times \frac{6.022 \times 10^{23} \text{ ions}}{\text{mol}} = 1.0 \times 10^{23} \text{ cm}^3/\text{mol}$$

We can reduce the size of the exponent by expressing this volume in larger-sized units, such as cubic kilometers (km³):

$$1.0 \times 10^{23} \text{ cm}^3/\text{mol} \times (1 \text{ m}/100 \text{ cm})^3 \times (1 \text{ km}/1000 \text{ m})^3 = 1.0 \times 10^8 \text{ km}^3/\text{mol}$$

PRACTICE EXERCISE: An analytical method used to map DNA can detect as few as 20 molecules of DNA fragments. How many moles of DNA fragments does this quantity represent? (See Problems 31–34.)

SAMPLE EXERCISE 4.9: It is not unusual for the air above major U.S. cities to contain as much as 1.0×10^{-9} mol of SO_2 per liter of air. How many molecules of SO_2 are in a liter of such polluted air?

SOLUTION: To convert from moles into particles (atoms, ions, or molecules), we multiply by the number of particles in a mole—that is, by Avogadro's number ($N_A = 6.022 \times 10^{23}$):

$$1.0 \times 10^{-9} \text{ mol } SO_2 \times 6.022 \times 10^{23} \text{ molecules/mol} =$$
$$6 \times 10^{14} \text{ molecules of } SO_2$$

PRACTICE EXERCISE: One milliliter of seawater contains about 2.5×10^{-14} moles of gold. How many atoms of gold are in a milliliter of seawater? (See Problems 35 and 36.)

moles

molecules = moles × (6.022 × 10²³)

molecules

The mole provides an important perspective on the atomic mass values in the periodic table. In preceding chapters, we viewed each of these values as the average mass of the atoms of an element expressed in atomic mass units (amu). Given the definition of a mole, we may also view each mass value in the periodic table as *the mass in grams of a mole of that element.* Thus, the average mass of an atom of helium is 4.003 amu, *and* the mass of a mole of helium (6.022×10^{23} atoms of He) is 4.003 g. In other words, the **molar mass** ($\mathcal{M}$) of helium is 4.003 g/mol. Similarly, there are 6.022×10^{23} atoms of iron in 55.85 grams of iron, and so the molar mass of iron is 55.85 g/mol.

> The **molar mass** ($\mathcal{M}$) of an element is the average atomic mass of the element expressed in grams per mole (g/mol).

SAMPLE EXERCISE 4.10: If the helium balloons sold at an amusement park each contain 1.85 g of He, how many moles of He are in each of them?

SOLUTION: To calculate the number of moles of He in a given mass of the element, we divide by the molar mass of He (4.003 g/mol):

$$1.85 \text{ g}/(4.003 \text{ g/mol}) = 0.462 \text{ mol}$$

PRACTICE EXERCISE: A length of copper wire weighs 4.86 g. How many moles of copper are in the wire? (See Problems 41 and 42.)

SAMPLE EXERCISE 4.11: Each Tums Ultra tablet contains 1.00×10^{-2} moles of calcium (as Ca^{2+} ions). How many milligrams of calcium are in each tablet?

SOLUTION: To convert moles of calcium into an equivalent mass, we multiply by the molar mass of Ca (40.08 g/mol), which gives us a mass in grams. Then we convert that mass into the equivalent mass in milligrams:

$$(1.00 \times 10^{-2} \text{ mol Ca}) \times (40.08 \text{ g/mol}) \times (1000 \text{ mg/ g}) = 401 \text{ mg Ca}$$

PRACTICE EXERCISE: How many grams of gold are there in 0.250 moles of gold? (See Problems 43 and 44.)

For a molecular compound, there are 6.022×10^{23} molecules in a mole of the compound. The corresponding molar mass is the *sum of the molar masses of the elements that make up the compound, each multiplied by the number of moles of that element in a mole of the compound.* Let's calculate the molar mass of a compound that consists of small molecules: SO_3. The molar mass is the sum of the masses of 1 mole of S and 3 moles of O:

> The molar mass of a compound is the sum of the molar masses of the elements in that compound, each multiplied by its subscript in the chemical formula.

$$[(1 \text{ mol S} \times 32.07 \text{ g/mol}) + (3 \text{ mol O} \times 16.00 \text{ g/mol})]/\text{mol SO}_3 =$$
$$80.07 \text{ g/mol SO}_3$$

CONNECTION: Methods of determining the molar mass of an unknown substance are described in Section 5.4.

SAMPLE EXERCISE 4.12: Calculate the molar masses of (a) H_2O; (b) H_2SO_4.

SOLUTION: (a) In 1 mole of H_2O, there are 2 moles of H and 1 mole of O. Therefore, the molar mass of H_2O is

$$(2 \text{ mol H} \times 1.008 \text{ g/mol}) = 2.016 \text{ g H}$$
$$+ (1 \text{ mol O} \times 16.00 \text{ g/mol}) = \underline{16.00 \text{ g O}}$$
$$18.02 \text{ g}$$

or 18.02 g/mol.

(b) In 1 mole of H_2SO_4, there are 2 moles of H, 1 mole of S, and 4 moles of O. Therefore, the molar mass of H_2SO_4 is

$$(2 \text{ mol H} \times 1.008 \text{ g/mol}) = 2.016 \text{ g}$$
$$+ (1 \text{ mol S} \times 32.07 \text{ g/mol}) = 32.07 \text{ g}$$
$$+ (4 \text{ mol O} \times 16.00 \text{ g/mol}) = \underline{64.00 \text{ g}}$$
$$98.09 \text{ g}$$

or 98.09 g/mol.

PRACTICE EXERCISE: Carbon dioxide combines with water, forming carbonic acid, H_2CO_3. Calculate the molar masses of (a) CO_2 and (b) H_2CO_3. (See Problems 46–48.)

SAMPLE EXERCISE 4.13: Suppose 0.360 moles of SO_3 are bubbled through 100.0 mL of water (d = 1.00 g/mL) to form H_2SO_4. How many grams of SO_3 and how many moles of H_2O do the quantities represent?

SOLUTION: To convert the moles of SO_3 to grams of SO_3, we multiply by the molar mass:

$$0.360 \text{ mol } SO_3 \times \frac{80.06 \text{ g } SO_3}{1 \text{ mol } SO_3} = 28.8 \text{ g } SO_3$$

To find the number of moles in 100.0 mL of water, we first need to find the mass of the water using its density:

$$100.0 \text{ mL } H_2O \times \frac{1.00 \text{ g } H_2O}{1 \text{ mL } H_2O} = 100.0 \text{ g } H_2O$$

We then divide by the molar mass of water:

$$100.0 \text{ g } H_2O \times \frac{1 \text{ mol } H_2O}{18.02 \text{ g } H_2O} = 5.55 \text{ mol } H_2O$$

PRACTICE EXERCISE: If 0.246 moles of N_2O_5 are dissolved in water, 31.0 g of HNO_3 are formed, as described in the chemical reaction on page 181. Calculate the mass of 0.246 moles of N_2O_5. How many moles of HNO_3 are in 31.0 g of HNO_3? (See Problems 51–54.)

The concepts of the mole and molar mass provide three more interpretations of Equation 4.1. We can say that

1 *mole* of SO_3 reacts with 1 *mole* of H_2O, producing 1 *mole* of H_2SO_4

or

80.07 *grams* of SO_3 reacts with 18.02 *grams* of H_2O, producing 98.09 *grams* of H_2SO_4

or

6.022×10^{23} *molecules* of SO_3 react with 6.022×10^{23} *molecules* of H_2O, forming 6.022×10^{23} *molecules* of H_2SO_4.

The concept of molar mass applies to ionic as well as molecular compounds. For example, a mole of NaCl contains a mole of Na^+ ions and a mole of Cl^- ions. Therefore, the molar mass of NaCl equals the sum of the molar masses of Na and Cl (electrons have so little mass that their loss or gain does not significantly change the mass of an atom when it forms an ion):

[(1 ~~mol~~ Na × 22.99 g/~~mol~~) × (1 ~~mol~~ Cl × 35.45 g/~~mol~~)]/mol NaCl = 58.44 g/mol NaCl

Note that there are 2 moles of ions within 1 mole of NaCl: 1 mole each of Na^+ and Cl^-. Keep in mind that there are no discrete molecules of NaCl, and so the term "formula unit" is used to indicate one Na^+ ion and one Cl^- ion. Salt crystals consist of ordered arrays of equal numbers of Na^+ and Cl^- ions. Figure 4.10 summarizes the mathematical connections between mass, moles, and particles (atoms, ions, or molecules).

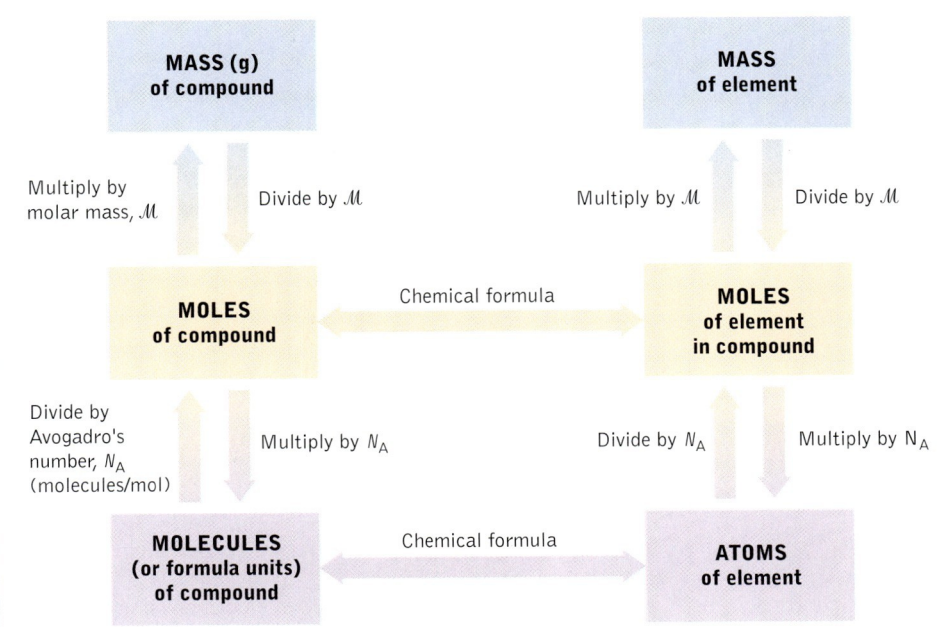

FIGURE 4.10 The mass of a pure substance can be converted into the equivalent number of moles or number of particles (atoms, ions, or molecules), and vice versa.

THE CHEMISTRY OF THE ALKALI METALS

The elements lithium (Li), sodium (Na), potassium (K), rubidium (Rb), and cesium (Cs) are shiny, silvery metals that belong to Group 1A in the periodic table and are known collectively as the alkali metals. A sixth element, francium (Fr), is a short-lived ($t_{1/2}$ = 21.8 minutes) radioactive isotope whose chemical properties are poorly understood. It forms in trace amounts by the decay of actinium-227. Only about 15 g of francium exists at any time on Earth. Of the other five alkali metals, sodium is the most abundant, comprising 2.7% of Earth's crust (see Figure 4.3). Potassium is nearly as abundant as sodium, but the other alkali metals are minor components of the crust with the abundances of 78 μg/g for rubidium, 18 μg/g for lithium, and only 2.6 μg/g for cesium. All the alkali metals are found in nature as +1 cations.

Lithium and sodium metal can be produced by passing electricity through molten lithium chloride and sodium chloride. For example:

$$2\ NaCl(l) \xrightarrow{\text{electrical energy}} 2\ Na(l) + Cl_2(g)$$

Potassium is produced by the reaction of molten potassium chloride with liquid sodium metal at 850°C:

$$Na(l) + KCl(l) \xrightarrow{\text{electrical energy}} K(l) + NaCl(l)$$

Rubidium and cesium are produced in a similar fashion from their molten chlorides and calcium metal.

Alkali metals react rapidly with the halogens (F_2, Cl_2, Br_2, and I_2), producing ionic compounds containing the +1 alkali metal cations and halide anions (F^-, Cl^-, Br^-, I^-).

$$2\ M(s) + X_2(g) \longrightarrow 2\ MX(s)$$

$$M = Li,\ Na,\ K,\ Rb,\ Cs \qquad X = F,\ Cl,\ Br,\ I$$

Alkali metals react with hydrogen, producing metal hydrides (MH) that contain M^+ cations and H^- anions:

$$2\ M(s) + H_2(g) \longrightarrow 2\ MH(s)$$

Alkali metals react with oxygen, yielding the alkali metal oxides (M_2O),

$$4\ M(s) + O_2(g) \longrightarrow 2\ M_2O(s)$$

peroxides (M_2O_2),

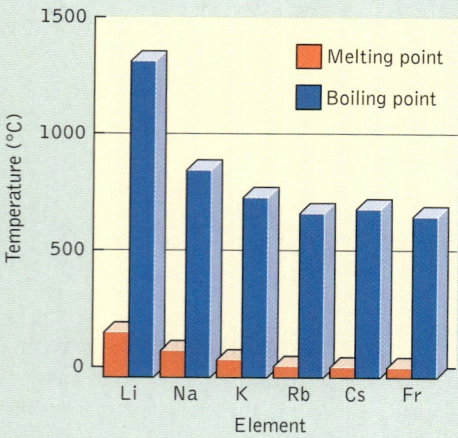

The boiling and melting points of the Group 1A elements tend to decrease with increasing atomic size. This trend corresponds to weaker interactions between atoms of these elements with increasing size.

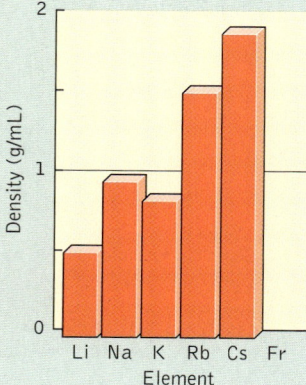

The densities of the Group 1A elements increase with increasing molar masses of the elements.

$$2\ M(s) + O_2(g) \longrightarrow M_2O_2(s)$$

and superoxides (MO_2),

$$M(s) + O_2(g) \longrightarrow MO_2(s)$$

The peroxide ion is a diatomic anion with a -2 charge (O_2^{2-}), whereas the diatomic superoxide ion has a single negative charge (O_2^-).

Potassium superoxide is used in a self-contained breathing apparatus (such as the one shown in the facing photo). A trace of water initiates the flow of oxygen:

$$2\ KO_2(s) + 2\ H_2O(l) \longrightarrow 2\ KOH(s) + O_2(g) + H_2O_2(l)$$

Exhaled carbon dioxide reacts with the superoxide, forming potassium carbonate and additional oxygen:

$$4\ KO_2(s) + 2\ CO_2(g) \longrightarrow 2\ K_2CO_3(s) + 3\ O_2(g)$$

Sodium and lithium peroxides react with CO and CO_2 in a similar fashion, forming carbonates:

$$M_2O_2(s) + CO(g) \longrightarrow M_2CO_3(s)$$
$$2\ M_2O_2(s) + 2\ CO_2(g) \longrightarrow 2\ M_2CO_3(s) + O_2(g)$$

These peroxides are used in submarines and spacecraft to scrub carbon monoxide and carbon dioxide.

Reactions between alkali metals and water produce alkali metal hydroxides and hydrogen gas. The reaction becomes increasingly violent in the procession down the group and sometimes leads to the ignition of the hydrogen. The increase in reactivity down the group is in accord with the trend in ionization energies.

$$M(s) + 2\ H_2O(l) \longrightarrow MOH(aq) + H_2(g)$$

The alkali metals form a wide variety of water-soluble compounds. Among the more common are NaCl (table salt), KOH (drain cleaner), $NaHCO_3$ (baking soda), Na_2CO_3 (in baking powder), KNO_3 (saltpeter used in gunpowder), and Li_2CO_3 (for the treatment of manic depression). For years, sodium bicarbonate was manufactured from sodium chloride by the Solvay process:

$$NH_3(aq) + NaCl(aq) + H_2CO_3(aq) \longrightarrow$$
$$NH_4Cl(aq) + NaHCO_3(s)$$

and converted into sodium carbonate by heating:

$$2\ NaHCO_3(s) \xrightarrow{\Delta} Na_2CO_3(s) + H_2O(g) + CO_2(g)$$

Recently, a more direct, economical, and environmentally benign route to the manufacture of sodium carbonate was developed. It entails heating the mineral trona, $Na_5(CO_3)_2HCO_3 \cdot 2\ H_2O$:

$$Na_5(CO_3)_2HCO_3 \cdot 2\ H_2O(s) \xrightarrow{\Delta}$$
$$5\ Na_2CO_3(s) + CO_2(g) + 3\ H_2O(g)$$

Sodium carbonate is used extensively in the manufacture of glass.

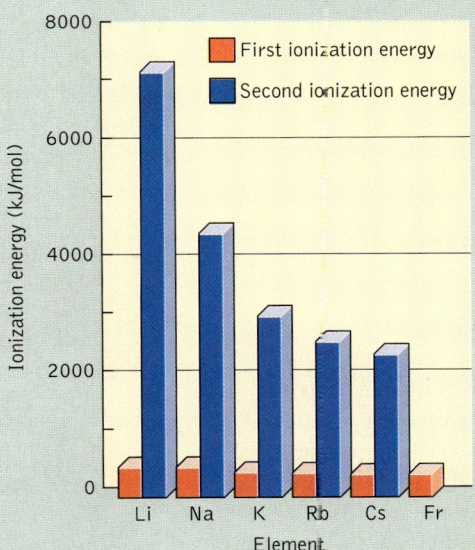

The ionization energies of the Group 1A elements decrease down the group as the inner electrons increasingly shield the outer electrons from the nuclear charge. The Group 1A elements are particularly easy to ionize because the removal of one electron gives a Group 1A element a noble gas configuration. The large second ionization energies reflect the difficulty in ionizing a noble gas.

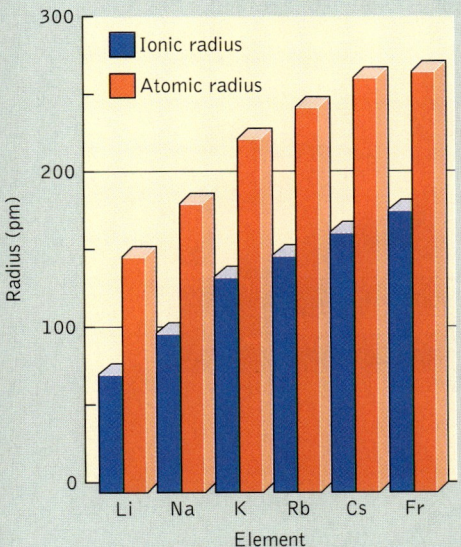

The atomic and ionic radii of the Group 1A elements increase down the group due to the increase in atomic number and in the population of higher-energy orbitals, which are found farther from the nucleus.

Our several interpretations of Equation 4.1 are not limited to only 1 mole of SO_3 reacting with only 1 mole of water. In general, Equation 4.1 tells us that any number of moles of SO_3 (x) will react with an equal number of moles of water vapor (x) to produce x moles of sulfuric acid. This interpretation is valid because the *mole ratio* of SO_3 to H_2O to H_2SO_4 in the reaction equation is 1 : 1 : 1. By multiplying x by the appropriate molar masses, we can determine either how much of one reactant is needed to react with any quantity of another or how much product could be made from any quantity of reactant.

	$SO_3(g)$	+	$H_2O(g)$	$\longrightarrow$	$H_2SO_4(l)$	(4.1)
	1 molecule	+	1 molecule	$\longrightarrow$	1 molecule	
Mole ratios:	1 mol	+	1 mol	$\longrightarrow$	1 mol	
Mass ratios:	80.07 g	+	18.02 g	$\longrightarrow$	98.09 g	
General case:	x mol	+	x mol	$\longrightarrow$	x mol	
or, for masses:	$x(80.07$ g$)$	+	$x(18.02$ g$)$	$\longrightarrow$	$x(98.09)$	

The mole ratios of reactants and products of a chemical equation are referred to collectively as the **stoichiometry** of the reaction. Note that stoichiometric values do not necessarily add up. In Equation 4.1, a total of *two* moles of reactants yields only *one* mole of product. On the other hand, the sum of the masses of the reactants *does* equal the mass of the product (Figure 4.11). This fact illustrates a fundamental relation for all chemical reactions known as the **law of conservation of mass**. This law states that the sum of the masses of the reactants in a chemical reaction equals the sum of the masses of the products. Thus the total number of moles of each element on the left-hand side of the reaction arrow must match the total number of moles of that element on the right-hand side, a key concept in balancing chemical equations to be addressed in Section 4.5.

> **Stoichiometry** refers to the relation between the quantities of reactants and the quantities of products in a chemical reaction.
>
> The **law of conservation of mass** states that the sum of the masses of the reactants of a chemical reaction is equal to the sum of the masses of the products.

FIGURE 4.11 The law of conservation of mass states that the total masses of reactants and products in a chemical reaction must be the same. For example, the combined mass of SO_3 and H_2O equals the mass of the H_2SO_4 that they form (as represented by the masses of the molecular models on the balance). Note that the numbers of each kind of atom are the same on the left-hand and right-hand sides of the balance.

4.5 COMPLETING AND BALANCING CHEMICAL EQUATIONS

Earth's primordial atmosphere contained acidic compounds besides sulfuric acid that were formed by reactions between water vapor and other oxides of nonmetals. Among these other oxides was dinitrogen pentoxide, N_2O_5. Nitric acid is formed when N_2O_5 combines with water as shown in Figure 4.12 and as described in Equation 4.2:

$$N_2O_5(g) + H_2O(g) \longrightarrow 2\ HNO_3(l) \qquad (4.2)$$

CONNECTION: Nitrogen oxides are formed in automobile engines and the atmosphere by processes described in Section 14.1.

Reactants   Product

FIGURE 4.12 Dinitrogen pentoxide (N_2O_5) reacts with water vapor (H_2O) to form 2 moles of nitric acid (HNO_3) for 1 mole of each of the reactants.

Note that, in this reaction, 2 moles of HNO_3 are produced for 1 mole of N_2O_5. This result makes sense because there are 2 moles of N in 1 mole of N_2O_5, whereas there is just 1 mole of N in a mole of HNO_3. Clearly, we need twice as many moles of HNO_3 on the right-hand side of the chemical equation as we have moles of N_2O_5 on the left-hand side if the equation is to be balanced.

Let's consider the reaction of water vapor with another oxide of nitrogen, NO_2. Two different acids may be produced by the reaction: nitric acid, HNO_3, and nitrous acid, HNO_2. To balance a chemical equation for this reaction, we start by writing the known products and reactants on the appropriate sides of the reaction arrow:

$$NO_2(g) + H_2O(g) \longrightarrow HNO_3(l) + HNO_2(l)$$

As written, the equation is not balanced, because there are 2 moles of N on the right-hand side and only 1 mole on the left-hand side. To bring the moles of nitrogen into balance, we add a coefficient of 2 to NO_2:

$$2\ NO_2(g) + H_2O(g) \longrightarrow HNO_3(l) + HNO_2(l) \qquad (4.3)$$

Now the number of moles of N (2), H (2), and O (5) are the same on both sides of the reaction arrow, and the equation is balanced.

SAMPLE EXERCISE 4.14: Write a balanced chemical equation for the reaction between SO_2 and O_2 that forms SO_3.

SOLUTION: Let's start by writing the formulas of the reactants and products in an *unbalanced* chemical equation:

$$SO_2(g) + O_2(g) \longrightarrow SO_3(g)$$

There is 1 mole of S on each side of the reaction arrow as written, but there are 4 moles of O on the left-hand side and only 3 on the right-hand side. Putting a 2 in front of the SO_3 would at least give us an even number of moles of O on the right:

$$SO_2(g) + O_2(g) \longrightarrow 2\,SO_3(g)$$

Now there are 2 moles of S on the right and only 1 on the left; so let's put a 2 in front of the SO_2:

$$2\,SO_2(g) + O_2(g) \longrightarrow 2\,SO_3(g)$$

We now have the same number of moles of S (2) and of O (6) on both sides, and the reaction is balanced.

PRACTICE EXERCISE: Balance the following unbalanced equation, describing the reaction between P_4O_{10} and water that yields phosphoric acid.

$$P_4O_{10}(s) + H_2O(l) \longrightarrow H_3PO_4(l)$$

(See Problems 61–68.)

As already noted, Earth's primordial atmosphere contained little oxygen, certainly not enough to sustain life as we know it. However, there were some sources of oxygen, most probably from the decomposition of water vapor as a result of its absorbing ultraviolet radiation from the sun. If the energy ($h\nu$) of a UV photon matches that of a chemical bond holding the hydrogen and oxygen atoms together in a water molecule, the molecule may be broken apart and fragments may recombine to yield oxygen and hydrogen gas:

CONNECTION: The ability of UV radiation to break the molecular bonds of atmospheric molecules is revisited in Section 6.3.

$$2\,H_2O(g) \xrightarrow{h\nu} 2\,H_2(g) + O_2(g) \qquad (4.4)$$

In Equation 4.4, $h\nu$ is placed above the reaction arrow to indicate that the reaction proceeds as a result of the addition of radiant energy. This photon-induced breaking of the bonds that hold molecules of water together could have produced significant quantities of atmospheric oxygen in Earth's primitive atmosphere, but molecules of O_2 would not have lasted long. In addition to carbon dioxide, the atmosphere contained two other carbon-containing compounds: carbon monoxide (CO) and methane (CH_4). Carbon monoxide readily consumes oxygen to form carbon dioxide,

$$2\,CO(g) + O_2(g) \longrightarrow 2\,CO_2(g) \qquad (4.5)$$

in a reaction that is balanced when there are 2 moles of CO for every mole of O_2.

Methane, too, reacts with oxygen, producing carbon dioxide and water. Let's write a balanced equation for this reaction, starting with the reactants and products:

$$CH_4(g) + O_2(g) \longrightarrow H_2O(g) + CO_2(g)$$

The number of moles of carbon in this equation is balanced, but those of hydrogen and oxygen are not. There are 4 moles of H on the left-hand side but only 2 on the right. A simple way to balance these values is to add a coefficient of 2 to water on the right-hand side:

$$CH_4(g) + O_2(g) \longrightarrow 2\,H_2O(g) + CO_2(g)$$

To balance oxygen, of which there are 4 moles on the right and only 2 on the left, we add a coefficient of 2 to O_2:

$$CH_4(g) + 2\,O_2(g) \longrightarrow 2\,H_2O(g) + CO_2(g) \tag{4.6}$$

Equation 4.6 is now balanced. The approach just taken—balancing the moles of C, then H, and then O—is useful for many other reactions in which oxygen reacts with compounds containing carbon and hydrogen. In our industrialized world, these reactions are often associated with burning natural gas and petroleum-based fuels. Equation 4.6 describes the principal chemical reaction that takes place when natural gas is burned to heat buildings and to heat water. It is an example of an important class of chemical reactions known as **combustion** reactions. We will examine the quantities of energy released by such reactions in Chapter 11.

CONNECTION: Section 11.3 describes the energetics of methane (natural gas) combustion.

Combustion is a reaction in which an element or a compound burns in an atmosphere containing oxygen.

SAMPLE EXERCISE 4.15: The geological deposits that today serve as sources of natural gas consist mostly of methane, but they also contain ethane (C_2H_6) and propane (C_3H_8). Complete and balance a chemical equation describing the combustion of C_2H_6.

SOLUTION: Like methane, ethane combines with oxygen to form carbon dioxide and water. Let's begin by writing the reactants and products for the reaction:

$$C_2H_6(g) + O_2(g) \longrightarrow H_2O(g) + CO_2(g)$$

In accord with the procedure used for methane, we'll first balance the number of carbon atoms on each side of the reaction arrow. The simplest way to do so is to put a coefficient of 2 in front of CO_2:

$$C_2H_6(g) + O_2(g) \longrightarrow H_2O(g) + 2\,CO_2(g)$$

To balance the atoms of hydrogen requires six of them on the right-hand side. Writing the coefficient 3 in front of H_2O balances the total number of hydrogen atoms.

$$C_2H_6(g) + O_2(g) \longrightarrow 3\,H_2O(g) + 2\,CO_2(g)$$

Finally, we must balance oxygen. There are a total of seven oxygen atoms on the right-hand side. This number seems to present a problem in that we can have only even numbers of oxygen atoms on the left-hand side. One solution to this problem is to use the fraction $\frac{7}{2}$ as a coefficient for O_2:

$$C_2H_6(g) + \tfrac{7}{2}\,O_2(g) \longrightarrow 3\,H_2O(g) + 2\,CO_2(g)$$

To avoid using fractions as coefficients, we double all the terms in the equation:

$$2\ C_2H_6(g) + 7\ O_2(g) \longrightarrow 6\ H_2O(g) + 4\ CO_2(g)$$

Either of the last two equations is acceptable, though the fraction-free version is preferred.

PRACTICE EXERCISE: Complete and balance a chemical equation describing the combustion of propane, C_3H_8. (See Problems 69 and 70.)

Nitric, sulfuric, and other acids in Earth's primordial atmosphere eventually fell to the surface of the planet. This process was accelerated as the atmosphere cooled and water vapor condensed as liquid water and fell to Earth as rain. Sulfuric acid (H_2SO_4) and nitric acid (HNO_3) readily dissolve in water, forming aqueous solutions that we designate $H_2SO_4(aq)$ and $HNO_3(aq)$. Reactions of these and other acidic compounds with the minerals of Earth's crust dramatically altered the chemical composition of both, as we will see in Chapter 5.

4.6 PERCENT COMPOSITION AND EMPIRICAL FORMULAS

Some regions of Earth's crust are rich in minerals that have had a significant effect on human development and economic growth. Since the beginning of the industrial revolution, access to geological deposits rich in iron oxides has been key to the development of national economies. The more accessible and the higher the iron content of a particular iron ore, the more valuable it is. Higher iron content in this case means a greater mass of iron in a given mass of ore. Let's consider two iron-containing minerals. One is called wüstite and has the chemical formula FeO. The other is hematite and has the formula Fe_2O_3. Which of them has the higher iron content and what is that content based on % Fe by mass?

A simple way to answer the first question is to examine the Fe/O mole ratios in the two minerals. In FeO, that ratio is obviously 1:1. In Fe_2O_3, the ratio is 2:3, or 1:1.5. Clearly, there is a higher proportion of O and a lower proportion of Fe in Fe_2O_3; so FeO has the higher iron content on a mole basis. Now let's calculate the iron content of FeO on a mass basis. Suppose we have 1 mole of FeO. The molar mass of FeO is the sum of the molar masses of Fe and O:

$$58.56\ \text{g} + 16.00\ \text{g} = 74.56\ \text{g/mol}$$

Of this 74.56 g/mol, Fe accounts for 58.56 g. Therefore, the iron content of FeO is

$$\text{mass of Fe/total mass} \times 100 = 58.56\ \text{g}/74.56\ \text{g} \times 100 = 78.54\%$$

It follows that the O content is

$$100.00 - 78.54 = 21.46\%$$

The Iron Range in northern Minnesota has one of the richest deposits of iron ore in the world and has provided iron ore for steelmaking in the United States for more than a century. The richness of the ore is indicated by the high percent composition of iron in the minerals found there.

These values give the percentage by mass of FeO that is Fe and that which is O. Together, they describe the *percent composition* of FeO. A summary of how to calculate the percent composition of a compound from its chemical formula is given in the flowchart in Figure 4.13.

SAMPLE EXERCISE 4.16: Determine the percent composition of enstatite, Mg_2SiO_4.

SOLUTION: A mole of Mg_2SiO_4 contains 2 moles of Mg, 1 mole of Si, and 4 moles of O. The molar mass of Mg_2SiO_4 is

$$2 \text{ mol Mg} \times \frac{24.31 \text{ g}}{\text{mol}} = 48.62 \text{ g Mg}$$

$$+ 1 \text{ mol Si} \times \frac{28.09 \text{ g}}{\text{mol}} = 28.09 \text{ g Si}$$

$$+ 4 \text{ mol O} \times \frac{16.00 \text{ g}}{\text{mol}} = 64.00 \text{ g O}$$

$$= 140.71 \text{ g/mol}$$

Mass percent values are calculated by dividing the mass of each element in one mole of Mg_2SiO_4 by the molar mass of Mg_2SiO_4:

$$\text{mass \% Mg} = \frac{48.62}{140.71} \times 100 = 34.55\% \text{ Mg}$$

$$\text{mass \% Si} = \frac{28.09}{140.71} \times 100 = 19.96\% \text{ Si}$$

$$\text{mass \% O} = 100.00 - 34.55 - 19.96 = 45.49\% \text{ O}$$

PRACTICE EXERCISE: Earth's crust contains a mineral called forsterite, which has the chemical formula $MgSiO_3$. Determine the percent composition of $MgSiO_3$. (See Problems 75–80.)

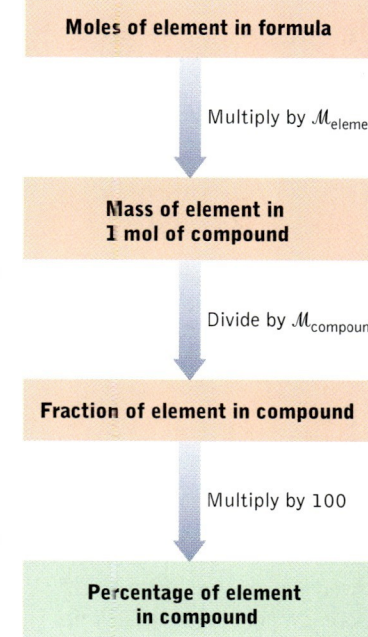

FIGURE 4.13 This chart summarizes the method for calculating the percentage of a constituent element in a compound from the compound's formula.

Sometimes we know the percent composition of a compound from the results of chemical analysis, and we wish to determine its chemical formula. Suppose, for example, that a mineral known to contain only iron and oxygen is carefully analyzed and found to be 72.36% iron by weight. To determine the formula of this mineral, let us convert this percentage into mass in grams by assuming that we have 100.00 g of the sample. Thus:

$$\frac{72.36}{100} \text{ Fe} \times 100.00 \text{ g} = 72.36 \text{ g Fe}$$

If the sample contains only Fe and O, then the O content must be (100.00 − 72.36) = 27.64%, and the mass of oxygen in a 100.00-gram sample must be 27.64 g.

The chemical formula of a compound represents the mole ratio of its constituents; so let's convert the preceding masses into equivalent numbers of moles of Fe and O by dividing them by the molar masses of these elements (their atomic masses from the periodic table expressed in grams):

$$72.36 \text{ g Fe} \times \frac{1 \text{ mol Fe}}{55.847 \text{ g Fe/mol}} = 1.296 \text{ mol Fe}$$

$$27.64 \text{ g O} \times \frac{1 \text{ mol O}}{16.000 \text{ g O/mol}} = 1.728 \text{ mol O}$$

Thus the molar ratio of Fe:O in this sample is 1.296:1.728. We can simplify this ratio by dividing both values by the smaller of the two:

$$\frac{1.296 \text{ mol Fe}}{1.296} = 1.000 \text{ mol Fe}$$

$$\frac{1.728 \text{ mol O}}{1.296} = 1.333 \text{ mol O}$$

The ratio of iron to oxygen is not a simple whole number. However, we could make it so if we first recognized that 1.333 is the decimal equivalent of 4/3 and then multiplied both mole values by 3:

$$3 \times \text{Fe}_1\text{O}_{4/3} = \text{Fe}_3\text{O}_4$$

Clearly, this formula does not match that of wüstite or hematite. It is, in fact, the formula of another iron-containing mineral called magnetite. A summary of the steps just taken to convert the results of an analysis of percent composition into a chemical formula is presented in Figure 4.14.

The chemical formula just derived was based on the result of a chemical analysis. Information obtained through experiments and measurements is called *empirical* information, and a chemical formula so obtained is called an **empirical formula.** Is an empirical formula different from a true chemical or molecular formula? To answer this question, we need to keep in mind that an empirical formula represents the simplest whole-number ratio of the elements in a compound. Although an empirical formula sufficiently describes the chemical formula of most ionic compounds, it may not represent the true molecular formula of a molecular compound; nor does it indicate how the atoms in a molecule are bonded together.

> Is an empirical formula different from a true chemical or molecular formula?

Hematite and magnetite are iron ores with different proportions of iron and oxygen. The "richer" ore is magnetite because it has the higher percentage of iron of the two.

✓ The **empirical formula** of a compound is the simplest whole-number ratio of the elements in the compound. An empirical formula of a molecular compound is not necessarily the same as its true molecular formula.

SAMPLE EXERCISE 4.17: What is the empirical formula of a sample of the mineral dolomite with the following percent composition: 21.734% Ca, 13.181% Mg, 13.027% C, and 52.058% O?

SOLUTION: Let's assume that we have 100.00 g of sample. Then, the percentages given become masses in grams of each of the elements. Converting these

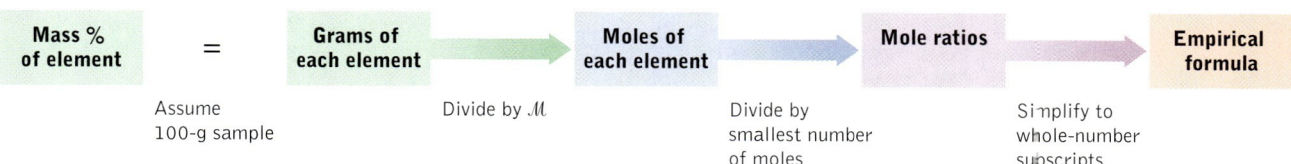

FIGURE 4.14 The empirical formula of a compound can be calculated from percent-composition data for the compound by following the steps in this chart.

masses into the equivalent numbers of moles of the elements, we get

$$21.734 \text{ g Ca} \times \frac{1 \text{ mol Ca}}{40.078 \text{ g Ca}} = 0.5423 \text{ mol Ca}$$

$$13.181 \text{ g Mg} \times \frac{1 \text{ mol Mg}}{24.305 \text{ g Mg}} = 0.5423 \text{ mol Mg}$$

$$13.027 \text{ g C} \times \frac{1 \text{ mol C}}{12.011 \text{ g C}} = 1.0846 \text{ mol C}$$

$$52.058 \text{ g O} \times \frac{1 \text{ mol O}}{15.999 \text{ g O}} = 3.254 \text{ mol O}$$

Next, we divide each number of moles by the smallest value (0.5423) to obtain a simple ratio of the four elements. That ratio, 1 Ca : 1 Mg : 2 C : 6 O, corresponds to the empirical formula $CaMgC_2O_6$. It turns out that dolomite is a carbonate mineral, meaning that the carbon and oxygen atoms in it are bonded together as CO_3^{2-} ions. Therefore, a formula more representative of the true composition of dolomite would be $CaMg(CO_3)_2$. However, elemental analysis alone does not provide information about how the atoms are bonded together, and so we could only guess that carbonate ions were part of dolomite's structure.

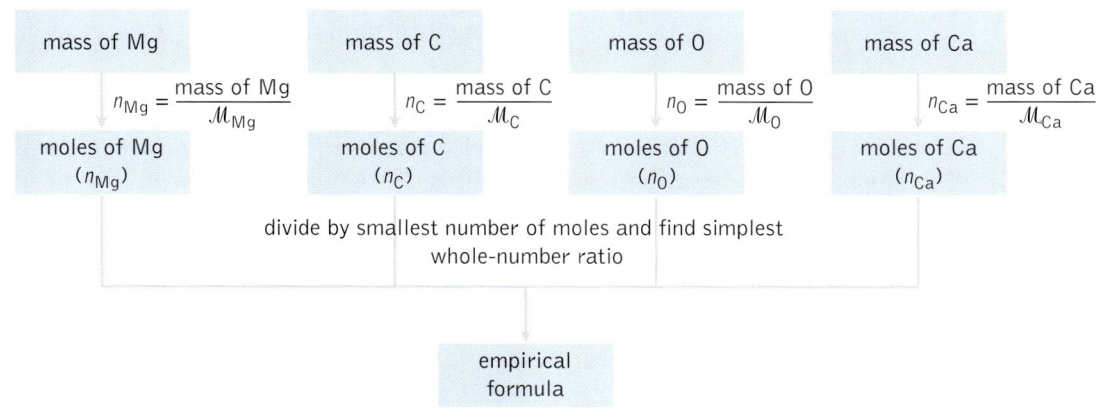

PRACTICE EXERCISE: Determine the empirical formula of a mineral sample with the following composition: 28.59% O, 24.95% Fe, and 46.46% Cr. (See Problems 83 and 84.)

Consider the gas acetylene and its molecular structure in Figure 4.15. It is one of the many volatile organic compounds found in interstellar space and in the comets that sometimes pass near Earth. On Earth, acetylene is widely used as fuel in torches that cut through metal. Its molecular formula is C_2H_2. However, its empirical formula, the simplest mole ratio of carbon to hydrogen in the compound, is simply CH. On the basis of a chemical analysis alone, we could not determine whether an unknown compound is acetylene or some other compound in which the C:H ratio also is 1:1. Such compounds do exist. One of them is the widely used solvent benzene, which has the molecular formula C_6H_6.

Let's calculate the percent composition of C_2H_2 and C_6H_6. The molar masses of C_2H_2 and C_6H_6 are 26.04 and 78.11 g/mol, respectively. Using the method developed earlier in this section, we find that

$$\text{mass \% C in } C_2H_2 = \text{mass of C/total mass} \times 100$$
$$= 24.02 \text{ g}/26.04 \text{ g} \times 100 = 92.24\%$$

$$\text{mass \% C in } C_6H_6 = \text{mass of C/total mass} \times 100$$
$$= 72.06 \text{ g}/78.11 \text{ g} \times 100 = 92.24\%$$

The H content of both compounds is

$$100.00 - 92.24 = 7.76\%$$

Any compound with the composition 92.24% C and 7.76% H will have the same empirical formula. The chemical formulas for acetylene and benzene are related by the formula $(CH)_x$ where x is a whole number; $x = 2$ for acetylene and 6 for benzene. The value of x cannot be obtained from the results of elemental analysis alone. We must also know the molar mass of the compound. The value of x can then be determined from the following relation:

$$x = \frac{\text{molar mass}}{\text{mass of empirical formula}} \quad (4.7)$$

SAMPLE EXERCISE 4.18: Determine the empirical and molecular formulas of eicosene, an insect pheromone that is 85.63% C and 14.37% H. The molar mass of eicosene is 280.54 g/mol.

SOLUTION: Once again, let's assume that we have 100.00 g of sample which means that we have 85.63 g C and 14.37 g H. Converting these masses into the equivalent numbers of moles, we get:

$$85.63 \text{ g C} \times \frac{1 \text{ mol C}}{12.011 \text{ g C}} = 7.129 \text{ mol C}$$

A

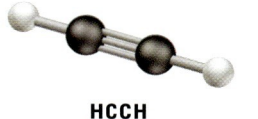

HCCH

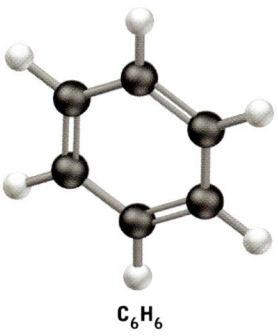

C₆H₆

B

H₂CCH₂

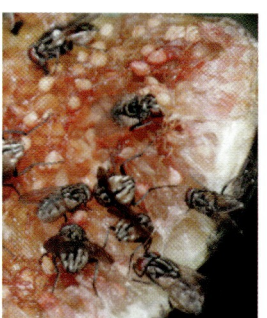

CH₃(CH₂)₁₉CHCH₂

FIGURE 4.15 *What a Chemist Sees.* Compounds can have the same empirical formula but different molecular formulas and different properties. A. Both acetylene (C_2H_2) and benzene (C_6H_6) have the empirical formula CH. However, acetylene is a gas used in welding, and benzene is a liquid used as a solvent in chemistry. B. Ethylene (C_2H_4) is a gas used to ripen fruit and make plastic, whereas muscalure ($C_{23}H_{46}$) is a waxy white solid that acts as a chemical signaling agent (pheromone) for houseflies. Both have the empirical formula CH_2.

$$14.37 \text{ g H} \times \frac{1 \text{ mol H}}{1.0079 \text{ g H}} = 14.26 \text{ mol H}$$

Dividing by the smaller value (7.129 moles of C) we find that the mole ratio of C : H is 1 : 2. Therefore, the empirical formula is CH_2.

To find the molecular formula, we need to determine the number of CH_2 building blocks in a molecule of eicosene. To do so, we determine the value of x in $(CH_2)_x$ by using Equation 4.7:

$$x = \frac{\text{molar mass of eicosene}}{\text{molar mass of } CH_2} = \frac{280.54}{14.027} = 20$$

Thus the molecular formula of eicosene is $(CH_2)_{20} = C_{20}H_{40}$.

PRACTICE EXERCISE: Determine the empirical and molecular formulas of a compound that is 56.36% O and 43.64% P and has a molar mass of 141.94 g/mol. (See Problems 85–88.)

4.7 STOICHIOMETRIC CALCULATIONS AND THE CARBON CYCLE

The composition of Earth's early atmosphere underwent a major change beginning about 600 million years ago with the evolution of green plants and the onset of photosynthesis. The following equation describes this chemical reaction, which takes place in several steps and is driven by the energy contained in visible light.

$$6 \; CO_2(g) + 6 \; H_2O(l) \longrightarrow C_6H_{12}O_6(aq) + 6 \; O_2(g) \tag{4.8}$$

The energy that drives photosynthesis is released in the reverse reaction:

$$C_6H_{12}O_6(aq) + 6 \; O_2(g) \longrightarrow 6 \; CO_2(g) + 6 \; H_2O(l) \tag{4.9}$$

This reaction takes place during combustion; in biological systems, it takes place during respiration in living organisms and during the decay of their tissue after these organisms die.

CONNECTION: The role of sugar (carbohydrates) as fuel for the human body is described in Section 12.5.

Photosynthesis and respiration are nearly in balance in Earth's biosphere. They are key chemical reactions in the carbon cycle, illustrated in Figure 4.16. If photosynthesis and respiration were *exactly* in balance, there would have been no net decrease in atmospheric carbon dioxide or net increase in oxygen in the past 600 million years. However, about 0.01% of the decaying mass of plants and animals (called detritus) is incorporated into sediments and soil when they die and so is shielded from oxygen and conversion back into CO_2. This proportion may not seem like much, but through hundreds of millions of years it has added up to the burial of about 10^{73} kg of carbon and the removal of an equivalent amount of carbon dioxide from the atmosphere. About 10^{28} kg of this buried carbon is in the form of fossil fuels: coal, petroleum, and natural gas.

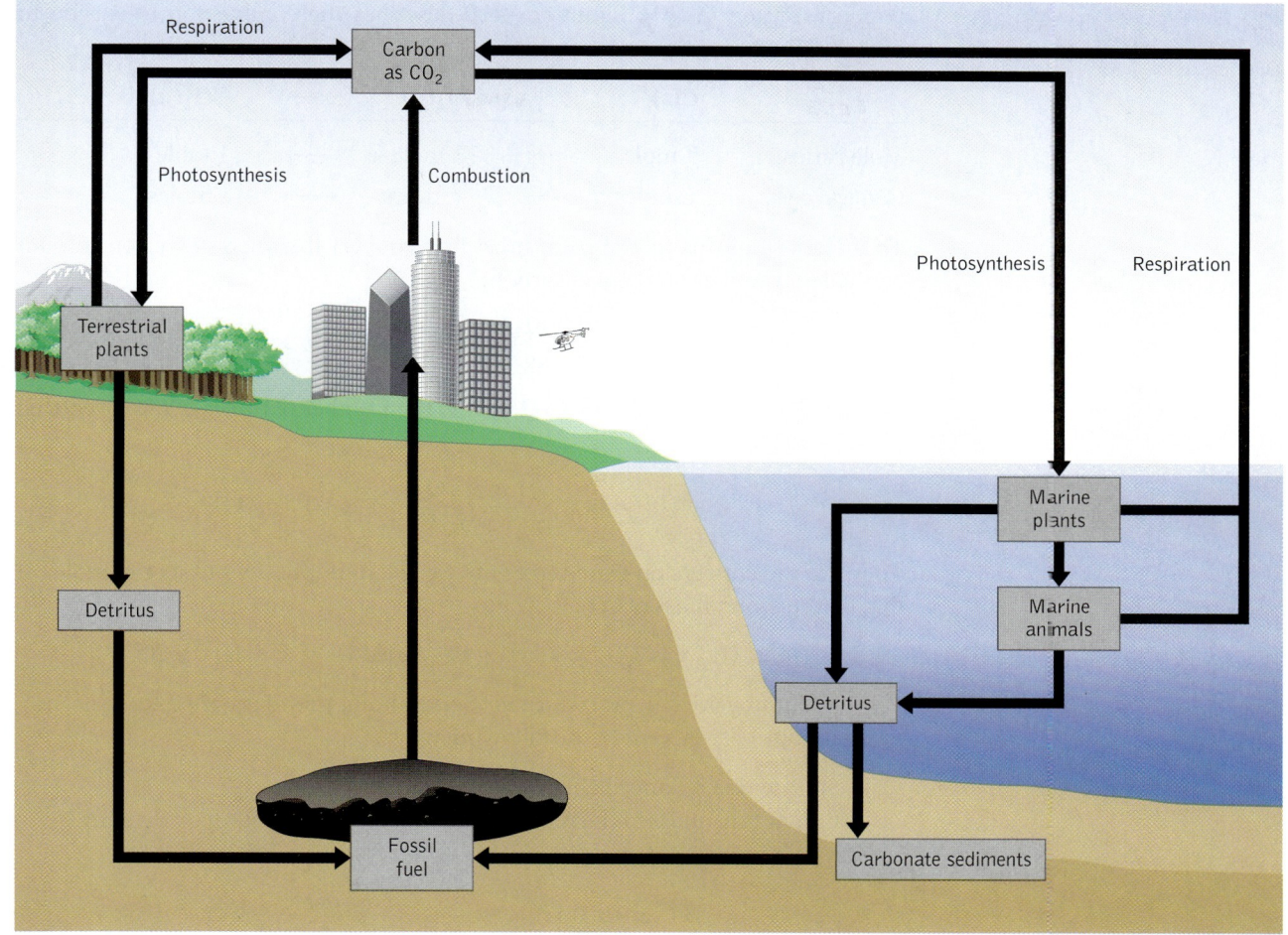

FIGURE 4.16 Carbon is cycled through the environment by many processes. Green plants incorporate CO_2 into their biomass. Some of the plant biomass becomes the biomass of animals. As plants and animals respire, they release CO_2 back into the environment. When they die, the decay of their tissues releases most of their carbon content as CO_2, but about 0.01% is incorporated into carbonate minerals and deposits of coal, petroleum, and natural gas (fossil fuels). Mining and the combustion of fossil fuels for human use may be shifting the natural equilibrium that has controlled the concentration of CO_2 in the atmosphere.

As a result of human activity and the combustion of these fossil fuels, the natural balance that helped limit the concentration of CO_2 in the atmosphere has been altered. Annually, an estimated 5 trillion (5×10^{12}) kilograms of carbon is reintroduced to the atmosphere as CO_2 as a result of fossil-fuel combustion. Deforestation adds another 2×10^{12} kg C/yr. The effects of these additions on global climate have been the subject of considerable debate, and we will examine them in detail in Chapter 7.

If the combustion of fossil fuel adds about 5×10^{12} kg of carbon into the atmosphere as CO_2 each year, what is the actual mass of CO_2 that is added? Clearly,

CONNECTION: Carbon dioxide emissions are linked to global climate change in Section 7.3.

it is more than 5×10^{12} kg because this amount does not take into account the mass of oxygen in CO_2. Equation 4.10 describes the combustion of carbon to carbon dioxide:

$$C(s) \quad + \quad O_2(g) \quad \longrightarrow \quad CO_2(g) \quad (4.10)$$

Mole ratios:	1 mole	+	1 mole	⟶	1 mole
General case:	x moles	+	x moles	⟶	x moles

To make use of this relation, we must first convert the mass of carbon into an equivalent number of moles of carbon:

$$5 \times 10^{12} \; \cancel{\text{kg C}} \times \frac{1000 \text{ g}}{1 \text{ kg}} \times \frac{1 \text{ mol C}}{12.011 \; \cancel{\text{g C}}} = 4 \times 10^{14} \text{ mol C}$$

Because the mole ratio of CO_2 to C is 1:1, the number of moles of CO_2 produced also is 4×10^{14}:

$$4 \times 10^{14} \; \cancel{\text{mol C}} \times \frac{1 \text{ mol } CO_2}{1 \; \cancel{\text{mol C}}} = 4 \times 10^{14} \text{ mol } CO_2$$

To make the conversion from moles of CO_2, we multiply the number of moles of CO_2 by the molar mass of CO_2:

$$4 \times 10^{14} \; \cancel{\text{mol } CO_2} \times 44.01 \text{ g } CO_2 / \cancel{\text{mol } CO_2} = 2 \times 10^{16} \text{ g } CO_2$$

We can summarize the procedure for determining the mass of products from a given amount of reactant in the following way:

- Start with a balanced chemical equation.
- Convert the mass of the reactant (r) into moles of reactant by dividing by its molar mass ($\mathcal{M}_r$):

$$\text{moles of r } (n_r) = \frac{\text{mass of r}}{\text{molar mass of r } (\mathcal{M}_r)}$$

- Use the mole ratio of product to reactant, as defined by the ratio of their coefficients (C_p/C_r) in the balanced chemical equation, to calculate the moles of product:

$$\text{moles of product } (n_p) = \text{moles of reactant } (n_r) \times (C_p/C_r)$$

- Convert moles of product into mass of product by multiplying by its molar mass ($\mathcal{M}_p$):

$$\text{mass of product} = \text{moles of product } (n_p) \times \text{molar mass } (\mathcal{M}_p)$$

This process is summarized in Figure 4.17 for the general case in which we start with the mass of a reactant or product and calculate the equivalent mass of another reactant or product.

SAMPLE EXERCISE 4.19: In 1990, the United States consumed approximately 3.7×10^{15} kg of natural gas (mostly methane) for energy production.

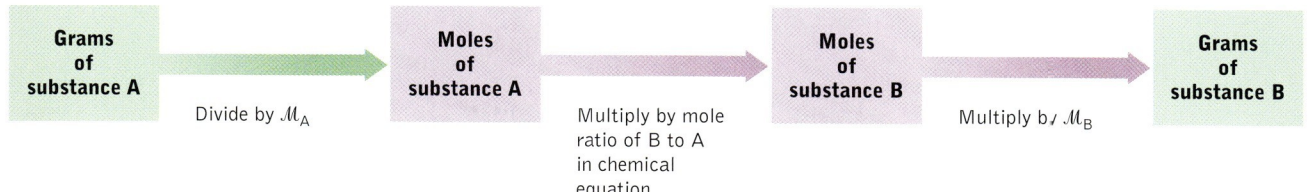

FIGURE 4.17 The masses of reactants and products in a chemical reaction are connected by the stoichiometry of the reaction and by their molar masses. This chart shows how to calculate the mass of one substance in a reaction from the mass of another.

What mass of carbon dioxide would have been produced by the combustion of 3.7×10^{15} kg of CH_4?

SOLUTION: Let's start with the balanced equation for the combustion of methane (Equation 4.6):

$$CH_4(g) + 2\,O_2(g) \longrightarrow 2\,H_2O(g) + CO_2(g)$$

Next, let's convert the mass of reactant into an equivalent number of moles. The mass in this case is 3.7×10^{15} kg of CH_4:

$$3.7 \times 10^{15} \text{ kg CH}_4 \times \frac{1000 \text{ g}}{1 \text{ kg}} \times \frac{1 \text{ mol CH}_4}{16.04 \text{ g CH}_4} = 2.3 \times 10^{17} \text{ mol CH}_4$$

The stoichiometry of the reaction tells us that 1 mole of carbon dioxide is produced for every mole of methane consumed; so 2.3×10^{17} moles of CO_2 will be produced. The equivalent mass of CO_2 is calculated as follows:

$$2.3 \times 10^{17} \text{ mol CO}_2 \times 44.01 \text{ g CO}_2/\text{mol CO}_2 = 1.0 \times 10^{19} \text{ g CO}_2$$

PRACTICE EXERCISE: Disposable lighters burn butane (C_4H_{10}), forming CO_2 and H_2O. Balance the chemical equation for the combustion of butane and determine how much CO_2 is produced from burning 1.00 g of C_4H_{10}. (See Problems 91–102.)

4.8 LIMITING REACTANTS AND PERCENT YIELDS

The production of glucose by photosynthesis plays a key role in regulating the concentrations of O_2 and CO_2 in the atmosphere. Six moles of water and six moles of carbon dioxide are needed to make one mole of glucose:

$$6\,CO_2(g) + 6\,H_2O(l) \longrightarrow C_6H_{12}O_6(aq) + 6\,O_2(g) \qquad (4.8)$$

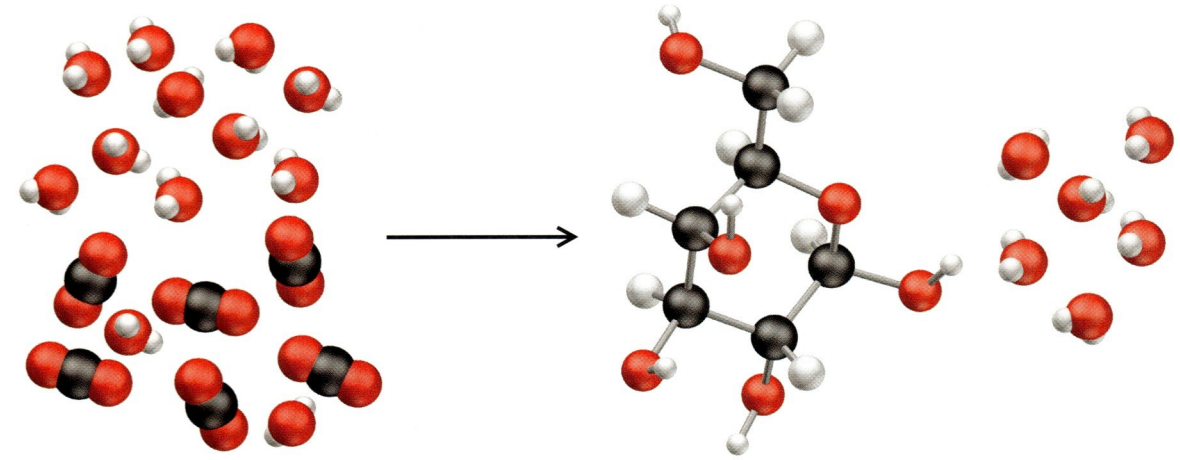

FIGURE 4.18 During photosynthesis, a reaction mixture of 6 molecules of CO_2 and 12 molecules of H_2O can produce only 1 molecule of glucose, $C_6H_{12}O_6$. Six molecules of H_2O remain unreacted, and CO_2 is the limiting reactant.

> A **limiting reactant** is completely consumed in a chemical reaction, thereby limiting the quantity of products that can be made.

What if there were more than six molecules of water for every six molecules of CO_2 (a common occurrence in biological systems)? The photosynthetic production of glucose could continue only until there was no more CO_2 left, leaving the extra molecules of water unreacted (Figure 4.18). In this example, carbon dioxide is the **limiting reactant,** meaning that the extent to which the reaction proceeds is limited by the quantity of CO_2 available for the reaction and not by the quantity of H_2O.

In the preceding example, it was easy to identify CO_2 as the limiting reactant. However, for many reactions, selecting the limiting reactant is more challenging. Let's consider the reaction between acidic precipitation that contains nitric acid and minerals that contain calcium carbonate. The balanced equation for the reaction is

$$2\ HNO_3(aq) + CaCO_3(s) \longrightarrow Ca(NO_3)_2(aq) + CO_2(g) + 6\ H_2O(l) \quad (4.11)$$

Note that the acid is consumed by the reaction, thereby reducing its environmental impact. However, there are regions in the United States and elsewhere where soils and rocks have little $CaCO_3$. This lack of natural neutralizing capacity has allowed acidity levels to rise in lakes and rivers, damaging the aquatic ecosystems that they support. One way to mitigate this problem is to add ground limestone ($CaCO_3$) to the acidified water. Suppose a rain storm adds 50.0 kg of HNO_3 to a large lake. A conservation officer proposes to treat the lake by dispersing the contents of two 50-pound bags of limestone into it. Is 100. pounds of limestone enough to neutralize 50.0 kg of nitric acid?

To answer this question, we need to convert the quantities of the two reactants into equivalent numbers of moles. Starting with limestone ($CaCO_3$, M = 100.1 g/mol), we have:

$$100. \text{ lbs} \times \frac{453.6 \text{ g}}{\text{lbs}} \times \frac{1 \text{ mol}}{100.1 \text{ g}} = 453 \text{ mol CaCO}_3$$

The number of moles of nitric acid ($M = 63.01$ g/mol) added by the storm is

$$50.0 \text{ kg} \times \frac{1000 \text{ g}}{\text{kg}} \times \frac{1 \text{ mol}}{63.01 \text{ g}} = 794 \text{ mol HNO}_3$$

Clearly, there are more moles of HNO_3 in 50.0 kg of HNO_3 than there are moles of $CaCO_3$ in 100. pounds of $CaCO_3$. However, that does not mean that there is not enough $CaCO_3$. We need to consider the stoichiometry of the reaction and the fact that 1 mole of $CaCO_3$ reacts with 2 moles of HNO_3. Thus, 453 moles of $CaCO_3$ will react with twice that many moles of HNO_3, or:

$$453 \text{ mol CaCO}_3 \times \frac{2 \text{ mol HNO}_3}{1 \text{ mol CaCO}_3} = 906 \text{ mol HNO}_3$$

Because the 906 moles of HNO_3 that could be neutralized is more than the 794 moles of HNO_3 added by the storm, HNO_3 is the limiting reactant in this case and there is more than enough $CaCO_3$.

In the preceding example, we identified the limiting reactant by calculating the moles of both reactants, and then used one of those values (for $CaCO_3$) and the stoichiometry of the reaction to calculate how many moles of the other reactant (HNO_3) would be consumed. Because that number was more than the quantity available, HNO_3 was the limiting reactant. Another way to identify the limiting reactant in a reaction system is to calculate the quantity of product that each reactant would make if it were the limiting reactant. The reactant that makes the least product really is the limiting reactant. This approach is taken in the following illustration.

Carbon dioxide dissolved in water is a reactant in many chemical reactions that can alter the composition of minerals. In one of these reactions, iron silicate minerals are transformed into iron carbonates. Consider the following reaction:

$$FeSiO_3(s) + CO_2(aq) + 2 \text{ H}_2O(l) \longrightarrow FeCO_3(s) + H_4SiO_4(aq)$$

A product of the reaction is a water-soluble version of silica, known as silicic acid: H_4SiO_4. The reaction requires one mole of CO_2 and two moles of water for every mole of $FeSiO_3$. Let's consider what would happen if we placed 1.000 g of $FeSiO_3$ in a sealed reaction chamber and reacted it with 1.000 g of rainwater. We will assume that the water has dissolved in it 1.4×10^{-5} g of CO_2. Clearly, not a lot of carbon dioxide is available for the preceding reaction. Under these conditions, carbon dioxide is the limiting reactant.

Now consider a slightly different scenario: the reaction of 1.000 g of $FeSiO_3$ with 1.000 g of H_2O in a large reaction vessel filled with pure carbon dioxide—more than enough CO_2 for the reaction. Under these conditions, how much $FeCO_3$ could be formed? The stoichiometry of the reaction calls for 2 moles of H_2O for every mole of $FeSiO_3$, but we have equal masses of these reactants. So, which is the limiting reactant? To answer this question, let's convert the masses of the two reactants into equivalent numbers of moles:

CONNECTION: Section 5.2 describes more chemical reactions that involve dissolved carbon dioxide.

$$1.000 \text{ g FeSiO}_3 \times \frac{1 \text{ mol FeSiO}_3}{131.94 \text{ g FeSiO}_3} = 0.00758 \text{ mol FeSiO}_3$$

$$1.000 \text{ g H}_2\text{O} \times \frac{1 \text{ mol H}_2\text{O}}{18.02 \text{ g H}_2\text{O}} = 0.0555 \text{ mol H}_2\text{O}$$

The number of moles of $FeCO_3$ that could be produced from either of these quantities of reagents can be found by two parallel calculations:

$$0.00758 \text{ mol FeSiO}_3 \times \frac{1 \text{ mol FeCO}_3}{1 \text{ mol FeSiO}_3} = 0.00758 \text{ mol FeCO}_3$$

$$0.0555 \text{ mol H}_2\text{O} \times \frac{1 \text{ mol FeCO}_3}{2 \text{ mol H}_2\text{O}} = 0.0278 \text{ mol FeCO}_3$$

$FeSiO_3$ is the reactant producing the lesser quantity of product, and so $FeSiO_3$ is the limiting reactant.

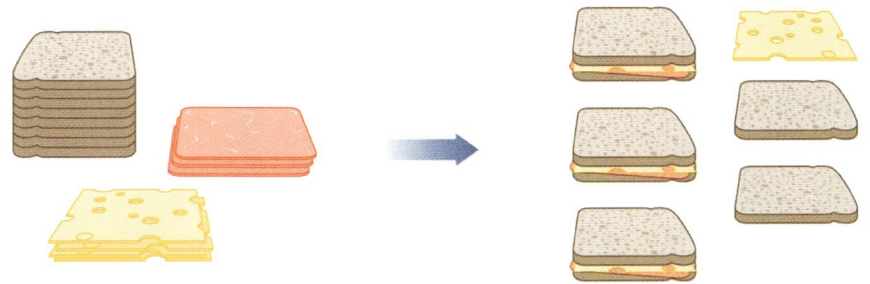

The principle of limiting reactant (*ingredient*) is illustrated by this culinary example. Suppose we wish to prepare ham and cheese sandwiches: each requires two slices of bread and one slice each of ham and cheese. Now suppose we have available eight slices of bread, four slices of cheese, and three slices of ham. How many ham and cheese sandwiches can be prepared? As the illustration shows, only three sandwiches can be prepared; ham in this case is the limiting ingredient, and bread and cheese are left over.

SAMPLE EXERCISE 4.20: In the reaction above between 1.000 g of $FeSiO_3$ and 1.000 g of H_2O in an excess of CO_2, how much water would be left unreacted after all the $FeSiO_3$ has been consumed?

SOLUTION: We determined that 0.00758 mol of $FeCO_3$ would be produced by the consumption of 1.000 g of $FeSiO_3$. Let's determine how much water is needed to produce this quantity of $FeCO_3$. We need to take into account the 2:1 mole ratio of H_2O to $FeCO_3$ in the following reaction:

$$FeSiO_3(s) + CO_2(aq) + 2 \text{ H}_2O(l) \longrightarrow FeCO_3(s) + H_4SiO_4(aq)$$

Therefore the moles of H_2O needed are:

$$0.00758 \text{ mol FeCO}_3 \times \frac{2 \text{ mol H}_2\text{O}}{1 \text{ mol FeCO}_3} = 0.0152 \text{ mol H}_2\text{O}$$

The amount of water left over is the difference between the available water (0.0278 mol) and the amount needed for the reaction (0.0152 mol), or:

$$0.0278 \text{ mol } H_2O - 0.0152 \text{ mol } H_2O = 0.0126 \text{ mol } H_2O$$

Converting moles into grams, we have

$$0.0126 \text{ mol } H_2O \times \frac{18.01 \text{ g } H_2O}{1 \text{ mol } H_2O} = 0.227 \text{ g } H_2O$$

PRACTICE EXERCISE: Geologists believe that Earth's iron-rich core is divided into two zones. The outer zone is estimated to have a mass of 1.85×10^{27} g and is 86% iron, 12% sulfur, and 2% nickel by mass. The sulfur is probably present as iron sulfide (FeS). How much FeS could be present in the outer core? (See Problems 109–114.)

We have just calculated the mass of product that should be formed by a given mass of reactant. This value represents the **theoretical yield** of a chemical reaction. In nature or in the laboratory, the *actual* yield is often less than the theoretical yield for several reasons. The same reactants may undergo different chemical reactions, yielding different sets of products. Sometimes the rate of a reaction is so low that reactants remain unreacted even after an extended reaction time. Other reactions do not go to completion no matter how long they are allowed to run, yielding a mixture of reactants and products of which the composition does not change with time. For these and other reasons, it is useful to distinguish between the theoretical and the actual yields of a chemical reaction and to calculate the **percent yield**:

$$\text{percent yield} = \frac{\text{actual yield (g)}}{\text{theoretical yield (g)}} \times 100$$

> **Percent yield** is the ratio (expressed as a percentage) of the actual amount of product formed in a chemical reaction to the **theoretical yield** that could have formed if the reaction had been complete.

SAMPLE EXERCISE 4.21: The reaction of 0.216 g of N_2O_5 with excess water produces 0.232 g of HNO_3. Calculate the percent yield of nitric acid.

$$N_2O_5(g) + H_2O(l) \longrightarrow 2 \, HNO_3(aq)$$

SOLUTION: To calculate the percent yield, we first need to calculate the theoretical yield of HNO_3 obtained from the complete reaction of 0.216 g of N_2O_5. In a three-step approach, we:

1. convert grams of N_2O_5 into moles of N_2O_5;
2. convert moles of N_2O_5 into moles of HNO_3;
3. convert moles of HNO_3 into grams of HNO_3.

Step: (1)   (2)   (3)

$$0.216 \text{ g } N_2O_5 \times \frac{1 \text{ mol } N_2O_5}{108 \text{ g } N_2O_5} \times \frac{2 \text{ mol } HNO_3}{1 \text{ mol } N_2O_5} \times \frac{63.0 \text{ g } HNO_3}{63.0 \text{ mol } HNO_3} = 0.252 \text{ g } HNO_3$$

CONNECTION: Reactions that don't go to completion reach chemical equilibrium, a concept explored in Chapter 15.

The percent yield is calculated by dividing the actual yield by the theoretical yield and multiplying by 100:

$$\text{percent yield } HNO_3 = \frac{\text{actual yield}}{\text{theoretical yield}} \times 100 = \frac{0.232 \text{ g } HNO_3}{0.252 \text{ g } HNO_3} \times 100 = 92.1\%$$

$$\frac{\text{actual yield}}{\text{theoretical yield}} \times 100 = \text{percent yield}$$

PRACTICE EXERCISE: The combustion of 58 g of butane (C_4H_{10}) yields 158 g of CO_2. What is the percent yield of the reaction?

$$\text{Given: } 2\ C_4H_{10}(g) + 13\ O_2(g) \longrightarrow 10\ H_2O(g) + 8\ CO_2(g)$$

(See Problems 115–118.)

CHAPTER REVIEW

Summary

SECTION 4.2

Two or more elements combine in definite proportions to form compounds. Chemical formulas give the relative number of atoms of each element present in a compound. Compounds can be either ionic or molecular. Ionic compounds are neutral overall; so the charges of the cations and anions in an ionic compound's formula must balance.

SECTION 4.4

A mole of particles, especially atoms, ions, or molecules, is a collection of Avogadro's number (6.022×10^{23}) of them. Most nonmetal oxides form acids on reaction with water. The molar mass of a compound is the sum of the molar masses of the constituent elements, each multiplied by a factor equal to the subscript of that element in the formula of the compound. Chemical reactions between elements and compounds can lead to new compounds.

SECTION 4.5

Chemical equations are used to describe the conversion of reactants into products. Chemical equations must be balanced; an equal number of atoms of each element must appear on both sides of the equation. The coefficients in balanced chemical equations describe the stoichiometry, or mole ratios, of reactants and products. Mole ratios can be used to obtain quantitative information from chemical equations, such as the mass of a product calculated from a given mass of reactant. Combustion is a reaction in which an element or a compound burns in an atmosphere containing oxygen. The products of the complete combustion of products containing carbon and hydrogen are H_2O and CO_2.

SECTION 4.6

The percent composition of a compound describes the proportions of its mass contributed by each of its constituent elements. Elemental analysis of a compound allows the calculation of its empirical formula, which is the simplest whole-number ratio of its elements. The molecular formula can be derived from the empirical formula if the compound's molar mass is known.

SECTION 4.8

In reactions having multiple reactants, one of the reactants may act as a limiting reactant, dictating the maximum amount of product that can result from the reaction. The maximum amount of product formed from a given amount of reactant is the theoretical yield. For many reactions, the actual yield is less than the theoretical yield; the ratio of the two is called the percent yield.

Key Terms

acid (p. 171)
alkali metal (p. 167)
alkaline earth metal (p. 167)
Avogadro's number (p. 173)
binary ionic compound (p. 167)
binary molecular compound (p. 166)
chemical equation (p. 173)
chemical formula (p. 166)

combustion (p. 183)
empirical formula (p. 186)
law of conservation of mass (p. 180)
law of definite proportions (p. 165)
law of multiple proportions (p. 165)
limiting reactant (p. 194)
mole (p. 173)
molar mass (p. 174)

oxoanion (p. 170)
percent yield (p. 198)
polyatomic ion (p. 169)
product (p. 173)
reactant (p. 173)
scientific law (p. 165)
stoichiometry (p. 180)
theoretical yield (p. 198)

Key Skills and Concepts

SECTION 4.3

Predict the chemical formulas of binary compounds from their names.
Give the names of binary compounds from their chemical formulas.
Know the names, formulas, and charges of common cations and anions and the naming relation between anions and their corresponding acids.
Relate the names of the oxoanions to their corresponding acids.
Name transition-metal compounds from their chemical formulas, and vice versa.

SECTION 4.4

Balance chemical equations in which reactants and products are known.
Understand the concept of the mole.

Know that the molar mass is the mass in grams of 1 mole of a substance.
Calculate the molar mass of a compound from its chemical formula.
Convert from molecules into moles, and vice versa.
Convert from grams of a compound into moles by using the molar mass, and vice versa.

SECTION 4.5

Understand that, during combustion, a substance reacts with oxygen, O_2, producing energy.

SECTION 4.6

Calculate the percent composition of a substance from its chemical formula.
Calculate the empirical formula of a substance from its percent composition.

Calculate the molecular formula of a substance from its molar mass and empirical formula.

SECTION 4.7

Predict the quantity of a product formed or the quantity of a reactant consumed in a chemical reaction by using the stoichiometry of the reaction.

SECTION 4.8

Know how to identify the limiting reactant.
Calculate the theoretical yield of a reaction.
Understand why the actual yield of a reaction cannot exceed 100% and often is less.
Be able to calculate the percent yield of the product(s) of a chemical reaction.

Key Equations and Relations

SECTION 4.4

Convert from molecules into moles (n).
$$n = \text{number of molecules} \times \frac{1 \text{ mol}}{6.022 \times 10^{23} \text{ molecules}}$$
Convert from moles into molecules.
$$\text{number of molecules} = n \times \frac{6.022 \times 10^{23} \text{ molecules}}{1 \text{ mol}}$$
Convert from mass into moles by using molar mass ($\mathcal{M}$).
$$n = \text{mass (g)} \times \frac{1}{\mathcal{M}}$$
Convert from moles into mass by using molar mass.
$$\text{mass (g)} = n \times \mathcal{M}$$

SECTION 4.6

Calculate percent composition of a compound from its molecular formula.

$$\text{mass percent of element X} = \frac{\text{moles of X in formula} \times \text{molar mass of X (g/mol)}}{\text{mass of 1 mole of compound}} \times 100$$

Derive a molecular formula from a molar mass and an empirical formula.

$$\text{molecular formula} = \frac{\text{molar mass}}{\text{mass of empirical formula}} \quad (4.7)$$

SECTION 4.6

Calculate percent yield.

$$\text{percent yield} = \frac{\text{actual yield (g)}}{\text{theoretical yield (g)}} \times 100$$

QUESTIONS AND PROBLEMS

The Composition of Earth

CONCEPT REVIEW

1. On the basis of the distribution of the elements in Earth's layers (see Figure 4.3), which of the following substances should be the most dense? $SiO_2(s)$; $Al_2O_3(s)$; $Fe(l)$.
2. On the basis of the compositions and physical states of Earth's various layers, which of the following substances has the highest melting point? Al_2O_3; Fe; Ni; S.
3. The most abundant ion in the solar wind is the proton (H^+). Which of the following ions is the second most abundant? He^+; Fe^+; Na^+; K^+.
*4. The proportions of the elements that make up the asteroid 433Eros are similar to those that make up Earth. Scientists believe that this similarity means that the asteroid and Earth formed in the same period. If the asteroid had formed after Earth had formed layers and a solid crust, perhaps from debris from the collision of an even larger asteroid and Earth, how might the composition of the asteroid be different from what it is? Explain your answer.

The Composition of Compounds

CONCEPT REVIEW

5. Their names suggest that the law of definite proportions and the law of multiple proportions are incompatible. Explain why they are not incompatible.
*6. Explain why Proust's law of definite proportions is classified a scientific *law*, whereas Dalton's view of the atomic structure of matter is classified a scientific *theory*.

PROBLEMS

7. Cobalt forms two sulfides, CoS and Co_2S_3. Predict the ratio of the masses of sulfur that combine with a fixed mass of cobalt to form CoS and Co_2S_3.
8. Lead forms two oxides, PbO and PbO_2. Predict the ratio of the masses of oxygen that combine with a fixed mass of lead to form PbO and PbO_2.

Naming Compounds

CONCEPT REVIEW

9. Consider a mythical element X, which forms only two oxoanions: XO_2^{2-} and XO_3^{2-}. Which of the two would have a name that ends in *-ite?*
10. Concerning the oxoanions in Question 9, would the name of either of them require a prefix such as *hypo* or *per?* Explain why or why not.
11. What is the role of the Roman numerals in the names of the compounds formed by transition metals?
12. Why do the names of the ionic compounds formed by the alkali metals and by the alkaline earth metals not include Roman numerals?

PROBLEMS

13. The following list contains the chemical formulas for eight binary compounds formed by nitrogen and oxygen. Name the compounds.
    a. NO_3
    b. N_2O_5
    c. N_2O_4
    d. NO_2
    e. N_2O_3
    f. NO
    g. N_2O
    h. N_4O
14. More than a dozen binary compounds containing sulfur and oxygen have been identified. Give the chemical formulas of the following six:
    a. sulfur monoxide
    b. sulfur dioxide
    c. sulfur trioxide
    d. disulfur monoxide
    e. hexasulfur monoxide
    f. heptasulfur dioxide
15. Predict the formula and give the name of the binary ionic compound containing:
    a. sodium and sulfur
    b. strontium and chlorine
    c. aluminum and oxygen
    d. lithium and hydrogen
16. Predict the formula and give the name of the binary ionic compound containing:
    a. potassium and bromine
    b. calcuim and hydrogen
    c. lithium and nitrogen
    d. aluminum and chlorine
17. Give the chemical names of the cobalt oxides that have the formulas CoO, Co_2O_3, CoO_2.
18. Give the formula of each of the following copper minerals:
    a. cuprite [copper(I) oxide]
    b. chalcocite [copper(I) sulfide]
    c. covellite [copper(II) sulfide]
19. The most abundant anion in seawater is the chloride ion. Write the formulas for the chlorides and sulfates of the most abundant cations in seawater: sodium, magnesium, calcium, potassium, and strontium.
20. The most abundant cation in seawater is the sodium ion. The evaporation of seawater gives a mixture of ionic compounds containing sodium combined with chloride, sulfate, carbonate, bicarbonate, bromide, fluoride, and borate, $B(OH)_4^-$. Write the chemical formulas of all these compounds.
21. Give the formula and charge of the oxoanion in each of the following compounds:
    a. sodium hypobromite
    b. potassium sulfate
    c. lithium iodate
    d. magnesium nitrite
22. Give the formula and charge of the oxoanion in each of the following compounds:
    a. potassium tellurite
    b. sodium arsenate
    c. calcium selenite
    d. potassium chlorate
23. Give chemical names for the following ionic compounds:
    a. $NiCO_3$
    b. NaCN
    c. $LiHCO_3$
    d. $Ca(ClO)_2$

24. Give names for the following ionic compounds:
    a. $Mg(ClO_4)_2$
    b. NH_4NO_3
    c. $Cu(CH_3CO_2)_2$
    d. K_2SO_3
25. Give the name or chemical formula of each of the following acids:
    a. HF
    b. $HBrO_3$
    c. phosphoric acid
    d. nitrous acid
26. Give the name or chemical formula of each of the following acids:
    a. HBr
    b. HIO_4
    c. selenous acid
    d. hydrocyanic acid

Moles and Molar Mass

CONCEPT REVIEW

27. In principle, we could use the more familiar unit "dozen" in place of "mole" when expressing the quantities of particles (atoms, ions, or molecules) in chemical reactions. What is the disadvantage in doing so?
28. What is the difference between the molar mass of a molecular compound and the mass of one of its molecules?
29. Do molecular compounds containing three atoms always have a molar mass greater than that of molecular compounds containing two atoms?
30. Without calculating their molar masses (though you may consult the periodic table), predict which of the following oxides of nitrogen has the larger molar mass: NO_2 or N_2O.

PROBLEMS

31. Earth's atmosphere contains many volatile compounds that are present in trace amounts. The following quantities of these trace gases were found in a 1.0-mL sample of air. Calculate the number of moles of each compound in the sample.
    a. 4.4×10^{14} molecules of Ne
    b. 4.2×10^{13} molecules of CH_4
    c. 2.5×10^{12} molecules of O_3
    d. 4.9×10^9 molecules of NO_2
32. The following quantities of trace gases were found in a 1.0-mL sample of air. Calculate the number of moles of each compound in the sample.
    a. 1.4×10^{13} molecules of H_2
    b. 1.5×10^{14} molecules of He
    c. 7.7×10^{12} molecules of N_2O
    d. 3.0×10^{12} molecules of CO
33. Memory in computer hard drives and floppy disks is represented by the number of bytes of information that can be stored. How many micromoles (1 μm = 10^{-6} mol) of bytes are there in
    a. a 10-gigabyte hard drive?
    b. a 100-megabyte Zip disk?
    c. a 1.4-kilobyte diskette?
34. Express the following population estimates for the year 2010 in nanomoles (1 nmol = 10^{-9} mol) of people:
    a. United States, 298. million people
    b. China, 1.34 billion people
    c. the world, 6.86 billion people
35. How many atoms of titanium are there in 0.125 moles of
    a. ilmenite, $FeTiO_3$?
    b. titanium(IV) chloride?
    c. Ti_2O_3?
    d. Ti_3O_5?
36. How many atoms of iron are there in 2.5 moles of
    a. wolframite, $FeWO_4$?
    b. pyrite, FeS_2?
    c. magnetite, Fe_3O_4?
    d. hematite, Fe_2O_3?
37. In the following pairs of quantities, which one contains more moles of oxygen?
    a. 1 mole of Al_2O_3 or 1 mole of Fe_2O_3
    b. 1 mole of SiO_2 or 1 mole of N_2O_4
    c. 3 moles of CO or 2 moles of CO_2
38. In the following pairs of quantities, which one contains more moles of oxygen?
    a. 2 moles of N_2O or 1 mole of N_2O_5
    b. 1 mole of NO or 1 mole of calcium nitrate
    c. 2 moles of NO_2 or 1 mole of sodium nitrite
39. Aluminum, silicon, and oxygen form minerals known as aluminosilicates. How many moles of aluminum are in 1.50 moles of
    a. pyrophyllite, $Al_2Si_4O_{10}(OH)_2$?
    b. mica, $KAl_3Si_3O_{10}(OH)_2$?
    c. albite, $NaAlSi_3O_8$?
40. The uranium used for nuclear fuel exists in nature in several minerals. Calculate how many moles of uranium are in 1 mole of
    a. carnotite, $K_2(UO_2)_2(VO_4)_2 \cdot 3\ H_2O$
    b. uranophane, $CaU_2Si_2O_{11} \cdot 7\ H_2O$
    c. autunite, $Ca(UO_2)_2(PO_4)_2 \cdot 10\ H_2O$

41. How many moles of carbon are there in 500.0 grams of carbon?
42. How many moles of gold are there in 2.00 ounces of gold?
43. How many moles of Ca^{2+} ions are in 0.25 moles of calcium titanate, $CaTiO_3$? What is the mass in grams of these Ca^{2+} ions?
44. How many moles of O^{2-} ions are 0.55 moles of aluminum oxide, Al_2O_3? What is the mass in grams of these O^{2-} ions?
45. Calculate the molar masses of the following atmospheric molecules:
    a. SO_2
    b. O_3
    c. CO_2
    d. N_2O_5
46. Determine the molar masses of the following minerals:
    a. rhodanite, $MnSiO_3$
    b. scheelite, $CaWO_4$
    c. ilmenite, $FeTiO_3$
    d. magnesite, $MgCO_3$
47. Calculate the molar masses of the following common flavors in food:
    a. vanilla, $C_8H_8O_3$
    b. oil of cloves, $C_{10}H_{12}O_2$
    c. anise oil, $C_{10}H_{12}O$
    d. oil of cinnamon, C_9H_8O
48. Calculate the molar masses of the following common sweeteners:
    a. sucrose, $C_{12}H_{22}O_{11}$
    b. saccharin, $C_7H_5O_3NS$
    c. aspartame, $C_{14}H_{18}N_2O_5$
    d. fructose, $C_6H_{12}O_6$
49. Suppose pairs of balloons were filled with 10.0 grams of the following pairs of gases. In which balloon in each pair would there be a greater number of particles?
    a. CO_2 or NO
    b. CO_2 or SO_2
    c. O_2 or Ar
50. If you had equal masses of the following pairs of compounds, in which one of the two would there be the greater number of ions?
    a. $NaBr$ or KCl
    b. $NaCl$ or $MgCl_2$
    c. $BaCl_2$ or Li_2CO_3
51. How many moles of SiO_2 are there in a quartz crystal (SiO_2) that weighs 45.2 g?
52. How many moles of $NaCl$ are there in a crystal of halite that weighs 6.82 g?
53. What is the mass of 0.122 mol of $MgCO_3$?
54. What is the volume of 1.00 mol of benzene (C_6H_6) at 20°C? The density of benzene is 0.879 g/mL.
*55. The density of uranium (19.05 g/cm^3) is more than five times as great as that of diamond (C, 3.514 g/cm^3). If you have a cube of each element, 1 cm on a side, which cube contains more atoms?
*56. Aluminum ($d = 2.70$ g/mL) and strontium ($d = 2.54$ g/mL) have nearly the same density. If we manufacture two cubes, each containing 1 mole of one element or the other, which cube will be smaller? What are the dimensions of this cube?

Chemical Equations

CONCEPT REVIEW

57. In a balanced chemical equation, does the number of moles of reactants always equal the number of moles of products?
58. In a balanced chemical equation, does the sum of the coefficients for the reactants always equal the sum of the coefficients of the products?
59. In a balanced chemical equation, must the sum of the masses of all the gaseous reactants always equal the sum of the masses of the gaseous products?
60. In a balanced chemical equation, must the sum of the volumes occupied by the reactants always equal the sum of the volumes occupied by the products?

PROBLEMS

61. Balance the following reactions that are believed to take place in the environment of a "protosun."
    a. $CH_4(g) + H_2O(g) \longrightarrow CO(g) + H_2(g)$
    b. $NH_3(g) \longrightarrow N_2(g) + H_2(g)$
    c. $CO(g) + H_2O(g) \longrightarrow CO_2(g) + H_2(g)$
62. Balance the following reactions, which take place in volcanoes.
    a. $SO_2(g) + O_2(g) \longrightarrow SO_3(g)$
    b. $H_2S(g) + O_2(g) \longrightarrow SO_2(g) + H_2O(g)$
    *c. $H_2S(g) + SO_2(g) \longrightarrow S_8(s) + H_2O(g)$
63. Balance the following chemical reactions, which contribute to weathering of the iron-silicate minerals ferrosillite ($FeSiO_3$), fayallite (Fe_2SiO_4), and greenalite [$Fe_3Si_2O_5(OH)_4$].
    a. $FeSiO_3(s) + H_2O(l) \longrightarrow Fe_3Si_2O_5(OH)_4(s) + H_4SiO_4(aq)$
    b. $Fe_2SiO_4(s) + CO_2(g) + H_2O(l) \longrightarrow FeCO_3(s) + H_4SiO_4(aq)$
    c. $Fe_3Si_2O_5(OH)_4(s) + CO_2(g) + H_2O(l) \longrightarrow FeCO_3(s) + H_4SiO_4(aq)$

64. Copper was one of the earliest metals used by humans because it can be prepared from a wide variety of copper minerals, such as cuprite (Cu_2O), chalcocite (Cu_2S), and malachite [$Cu_2CO_3(OH)_2$]. Balance the following reactions for converting these minerals into copper metal.
    a. $Cu_2O(s) + C(s) \longrightarrow Cu(s) + CO_2(g)$
    b. $Cu_2O(s) + Cu_2S(s) \longrightarrow Cu(s) + SO_2(g)$
    c. $Cu_2CO_3(OH)_2(s) \longrightarrow CuO(s) + CO_2(g) + H_2O(g)$

65. Nitrogen oxides play a key role in smog formation. Balance the following reactions, which contribute to the formation of nitrogen oxides in the lower atmosphere.
    a. $N_2(g) + O_2(g) \longrightarrow NO(g)$
    b. $NO(g) + O_2(g) \longrightarrow NO_2(g)$
    c. $NO(g) + NO_3(g) \longrightarrow NO_2(g)$
    d. $N_2(g) + O_2(g) \longrightarrow N_2O(g)$

66. Some scientists believe that life on Earth may have originated near deep-ocean vents. Balance the following reactions, which are among those taking place near such vents:
    a. $CH_3SH(l) + CO(g) \longrightarrow CH_3CO(SCH_3)(l) + H_2S(g)$
    b. $H_2S(g) + CO(g) \longrightarrow CH_3CO_2H(g) + S_8(s)$

67. Write a balanced chemical equation for each of the following reactions.
    a. Dinitrogen pentoxide reacts with sodium metal to produce sodium nitrate and nitrogen dioxide.
    b. A mixture of nitric acid and nitrous acid is formed when water reacts with dinitrogen tetroxide.
    c. At high pressure, nitrogen monoxide decomposes to dinitrogen monoxide and nitrogen dioxide.

68. Write a balanced chemical equation for each of the following reactions.
    a. Carbon dioxide reacts with carbon to form carbon monoxide.
    b. Potassium reacts with water to give potassium hydroxide and the element hydrogen.
    c. Phosphorus (P_4) burns in air to give diphosphorus pentoxide.

69. Write a balanced chemical equation for the combustion of acetylene, C_2H_2.

70. Write a balanced chemical equation for the combustion of octane (C_8H_{18}).

Percent Composition and Empirical Formulas

CONCEPT REVIEW

71. What is the difference between an empirical and a molecular formula?
72. Do the empirical and molecular formulas of a compound have the same percent composition values?
73. Is the element with the largest atomic mass always the element present in the highest percentage by mass in a compound?
74. Sometimes the composition of a compound is expressed as a "mole percent" or an "atom percent." Are the values of these parameters likely to be the same or different for a given compound?

PROBLEMS

75. Calculate the percent composition of (a) Na_2O; (b) NaOH; (c) $NaHCO_3$; (d) sodium carbonate.
76. Calculate the percent composition of (a) sodium sulfate; (b) dinitrogen tetroxide; (c) strontium nitrate; (d) aluminum sulfide.
77. The following compounds have been detected in outer space. Which of them contains the greatest percentage of carbon by mass?
    a. naphthalene, $C_{10}H_8$
    b. chrysene, $C_{18}H_{12}$
    c. pentacene, $C_{24}H_{12}$
    d. pyrene, $C_{16}H_{10}$

78. Ancient Egyptians used a variety of lead compounds as white pigments in their cosmetics, including PbS, $PbCO_3$, PbCl(OH), and $Pb_2Cl_2CO_3$. Which of these compounds contains the highest percentage of lead?

79. Which of the nitrogen oxides N_2O, NO, N_2O_3, and NO_2 is more than 50% oxygen on a molar basis and which is more than 50% oxygen by mass?

80. Which of the sulfur oxides S_2O, SO, SO_2, and SO_3 is more than 50% oxygen on a molar basis and which is more than 50% oxygen by mass?

81. Carbon in the cosmos exists as carbon atoms and as complex carbon-containing molecules. Do any two of the following compounds, which have been detected in outer space, have the same empirical formula?
    a. naphthalene, $C_{10}H_8$
    b. chrysene, $C_{18}H_{12}$
    c. anthracene, $C_{14}H_{10}$
    d. pyrene, $C_{16}H_{10}$
    e. benzoperylene, $C_{22}H_{12}$
    f. coronene, $C_{24}H_{12}$

82. Which, if any, of the following nitrogen oxides have the same empirical formula?
    a. N_2O b. NO c. NO_2 d. N_2O_2 e. N_2O_4

83. Zircon is a common substitute for diamond in inexpensive jewelry. The percent composition of zircon is 49.76% Zr, 15.32% Si, and the remainder is oxygen. Determine the empirical formula of zircon.

84. A sample of an iron-containing compound is 22.0% Fe,

50.2% oxygen, and 27.8% chlorine by mass. What is the empirical formula of this compound?

85. 2.43 g of magnesium reacts with 1.60 g of oxygen, forming 4.03 g magnesium oxide.
    a. Use these data to calculate the empirical formula of magnesium oxide.
    b. Write a balanced chemical reaction for this reaction.

86. Ferrophosphorus, Fe_2P, reacts with pyrite, FeS_2, producing iron(II) sulfide and a compound that is 27.87% P and 72.13% S by mass and has a molar mass of 444.56 g/mol.
    a. Determine the empirical and molecular formulas of this compound.
    b. Write a balanced chemical equation for this reaction.

87. Asbestos was used for years as an insulating material in buildings until prolonged exposure to asbestos was demonstrated to cause lung cancer. Asbestos is a mineral containing magnesium, silicon, oxygen, and hydrogen. One form of asbestos, chrysolite (520.27 g/mol), has the composition 28.03% magnesium, 21.60% silicon, and 1.16% hydrogen.
    a. Determine the empirical formula of chrysotile.
    b. Determine the molecular formula of chrysotile.

88. A candle flame produces easily seen specks of soot near the edges of the flame, especially when the candle is moved. A piece of glass held over a candle flame will become coated with soot, which is the result of the incomplete combustion of candle wax. Elemental analysis of a compound extracted from a sample of this soot gave these results: 7.74% H and 92.26% C by mass. Calculate the empirical formula of the compound.

Stoichiometry

CONCEPT REVIEW

89. Can you correctly calculate the amounts of products of a reaction without first balancing the reaction equation?

90. There are two ways to write the equation for the combustion of ethane:

    $C_2H_6(g) + \frac{7}{2} O_2(g) \longrightarrow 3 H_2O(g) + 2 CO_2(g)$
    $2 C_2H_6(g) + 7 O_2(g) \longrightarrow 6 H_2O(g) + 4 CO_2(g)$

    Do these two different ways of writing the equation affect the calculation of how much CO_2 is produced from a known quantity of C_2H_6?

PROBLEMS

91. The United Nations Intergovernmental Panel on Climate Change reported in June, 2000, that better management of cropland, grazing land, and forests would reduce the amount of carbon dioxide in the atmosphere by 5.4×10^9 kg of carbon per year.
    a. How many moles of carbon are present in 5.4×10^9 kg of carbon?
    b. How many kilograms of carbon dioxide does this quantity of carbon represent?

92. Energy generation results in the addition of an estimated 27 billion tons of CO_2 to the atmosphere each year.
    a. How many moles of CO_2 does 27 billion tons represent?
    b. How many grams of carbon are in 27 billion tons of CO_2?

93. When $NaHCO_3$ is heated above 270°C, it decomposes to Na_2CO_3, H_2O, and CO_2.
    a. Write a balanced chemical equation for the decomposition reaction.
    b. Calculate the mass of CO_2 produced from the decomposition of 25.0 g of $NaHCO_3$.

94. One of the lead compounds, Pb(OH)Cl, used in ancient Egyptian cosmetics (see Problem 78) was prepared from PbO according to the following ancient recipe:

    $PbO + NaCl + H_2O \longrightarrow Pb(OH)Cl + NaOH$

    How many grams of PbO and how many grams of NaCl would be required to produce 10.0 g of Pb(OH)Cl?

95. The manufacture of aluminum includes the production of cryolite, Na_3AlF_6, from the following reaction:

    $6 HF + 3 NaAlO_2 \longrightarrow Na_3AlF_6 + 3 H_2O + Al_2O_3$

    How much $NaAlO_2$ is required to produce 1.00 kg of Na_3AlF_6?

96. Chromium metal can be produced from the high-temperature reaction of Cr_2O_3 with silicon or aluminum by each of the following reactions:

    $Cr_2O_3 + 2 Al \longrightarrow 2 Cr + Al_2O_3$

    or

    $2 Cr_2O_3 + 3 Si \longrightarrow 4 Cr + 3 SiO_2$

    a. Calculate the number of grams of aluminum required to prepare 400. g of chromium metal by the first reaction.
    b. Calculate the number of grams of silicon required to prepare 400. g of chromium metal by the second reaction.

97. Twenty-five tons of coal that is 3.0% sulfur by mass is burned at an electricity-generating plant. During the combustion of the coal, the sulfur is converted into sulfur dioxide. How many tons of sulfur dioxide are produced?

98. Charcoal (C) and propane (C_3H_8) are used as fuel in backyard grills.
    a. Write balanced chemical equations for the combustion reactions for C and C_3H_8.
    b. How many grams of carbon dioxide are produced from burning 500.0 g of each of the two fuels?

99. The uranium minerals found in nature must be refined and enriched in ^{235}U before the uranium can be used as a fuel in nuclear reactors. One procedure for enriching uranium relies on the reaction of UO_2 with HF to form UF_4, which is then converted into UF_6 by reaction with fluorine:

    (1) $UO_2 + 4\, HF \longrightarrow UF_4 + 2\, H_2O$

    (2) $UF_4 + F_2 \longrightarrow UF_6$

    a. How many kilograms of HF are needed to completely react with 5.00 kg of UO_2?
    b. How much UF_6 can be produced from 850. g of UO_2?

100. The mineral bauxite, which is mostly Al_2O_3, is the starting material for manufacturing aluminum ware. How much aluminum can be produced from 1.00 ton of Al_2O_3?

*101. "Native," or elemental, copper can be found in nature, but most copper is recovered from oxide or sulfide minerals. Chalcopyrite ($CuFeS_2$) is an abundant copper mineral that can be converted into elemental copper.
    a. How much Cu could be produced from 1.00 kg of $CuFeS_2$?
    b. If the yield for the overall conversion process is 82%, how many grams of $CuFeS_2$ would be needed to produce 100. g Cu?

*102. Unlike most metals, gold is found in nature as the pure element. Miners in California in 1849 searched for gold nuggets and gold dust in streambeds, where the denser gold could be easily separated from sand and gravel. However, larger deposits of gold are found in veins of rock and can be separated chemically in a two-step process:

$$4\, Au(s) + 8\, NaCN(aq) + O_2(g) + 2\, H_2O(l) \longrightarrow 4\, NaAu(CN)_2(aq) + 4\, NaOH(aq)$$

$$2\, NaAu(CN)_2(aq) + Zn(s) \longrightarrow 2\, Au(s) + Na_2[Zn(CN)_4](aq)$$

If a 1000-kilogram sample of rock is 0.019% gold by mass, how much Zn is needed to react with the gold extracted from the rock? Assume that the reaction is 100% efficient.

Limiting Reactants and Percent Yields

CONCEPT REVIEW

103. If a reaction vessel contains equal masses of Fe and S, what mass of FeS could theoretically be produced?
    a. the sum of the masses of Fe and S
    b. more than the sum of the masses of Fe and S
    c. less than the sum of the masses of Fe and S

104. A reaction vessel contains equal masses of magnesium metal and oxygen gas. The mixture is ignited, forming MgO. After the reaction has gone to completion, the mass of the MgO is less than the mass of the initial quantities of Mg and O_2. Is this result a violation of the law of conservation of mass? Explain your answer.

105. Explain how the parameters *theoretical* yield and *percent yield* differ.

106. Can the percent yield of a chemical reaction ever exceed 100%?

107. Give two reasons why the observed yield from a chemical reaction is usually less than the theoretical yield.

108. A chemical reaction produces less than the expected amount of product. Is this result a violation of the law of conservation of mass?

PROBLEMS

109. A recipe for 1 cup of Hollandaise sauce calls for $\frac{1}{2}$ cup of butter, $\frac{1}{4}$ cup of hot water, 4 egg yolks, and the juice of a medium-size lemon. How many cups of this sauce can be made from a pound (2 cups) of butter, a dozen eggs, 4 medium-size lemons, and an unlimited supply of hot water?

110. A factory making toy wagons has 13,466 wheels, 3360 handles, and 2400 wagon beds in stock. What maximum number of wagons can the factory make?

111. Potassium superoxide, KO_2, reacts with carbon dioxide to form potassium carbonate and oxygen. This reaction makes potassium superoxide useful in a self-contained breathing apparatus. How much O_2 could be produced from 2.50 g of KO_2 and 4.50 g of CO_2?

$$4 KO_2 + 2 CO_2 \longrightarrow 2 K_2CO_3 + 3 O_2$$

112. A reaction vessel contains 10.0 g of CO and 10.0 g of O_2. How many grams of CO_2 could be produced according to the following reaction?

$$2 CO + O_2 \longrightarrow 2 CO_2$$

113. Ammonia rapidly reacts with hydrogen chloride, making ammonium chloride. Write a balanced chemical equation for the reaction and calculate the number of grams of excess reactant when 3.0 g of NH_3 reacts with 5.0 g of HCl.

114. Sulfur trioxide dissolves in water, producing H_2SO_4. How much sulfuric acid can be produced from 10.0 mL of water ($d = 1.00$ g/mL) and 25.6 g of SO_3?

115. The reaction of 3.0 g of carbon with excess O_2 yields 6.5 g of CO_2. What is the percent yield of this reaction?

116. Baking soda ($NaHCO_3$) can be made in large quantities by the following reaction:

$$NaCl + NH_3 + CO_2 + H_2O \longrightarrow NaHCO_3 + NH_4Cl$$

If 10.0 g of NaCl reacts with excesses of the other reactants and 4.2 g of $NaHCO_3$ is isolated, what is the percent yield of the reaction?

117. Yeast converts glucose ($C_6H_{12}O_6$) into ethanol ($d = 0.789$ g/mL) in a process called fermentation. An unbalanced equation for the reaction can be written as follows:

$$C_6H_{12}O_6(aq) \longrightarrow C_2H_5OH(l) + CO_2(g)$$

a. Write a balanced chemical equation for this fermentation reaction.
b. If 100.0 g of glucose yields 50.0 mL of ethanol, what is the percent yield for the reaction?

118. A 1-liter sample of seawater contains 19.4 g of Cl^-, 10.77 g of Na^+, and 1.29 g of Mg^{2+}.
a. How many moles of each ion are present?
b. If we evaporated the seawater, would there be enough Cl^- present to form the chloride salts of the sodium and magnesium present?

5 Solution Chemistry
And the Hydrosphere

Earth's oceans and chemical reactions in solutions are essential to life as we know it.

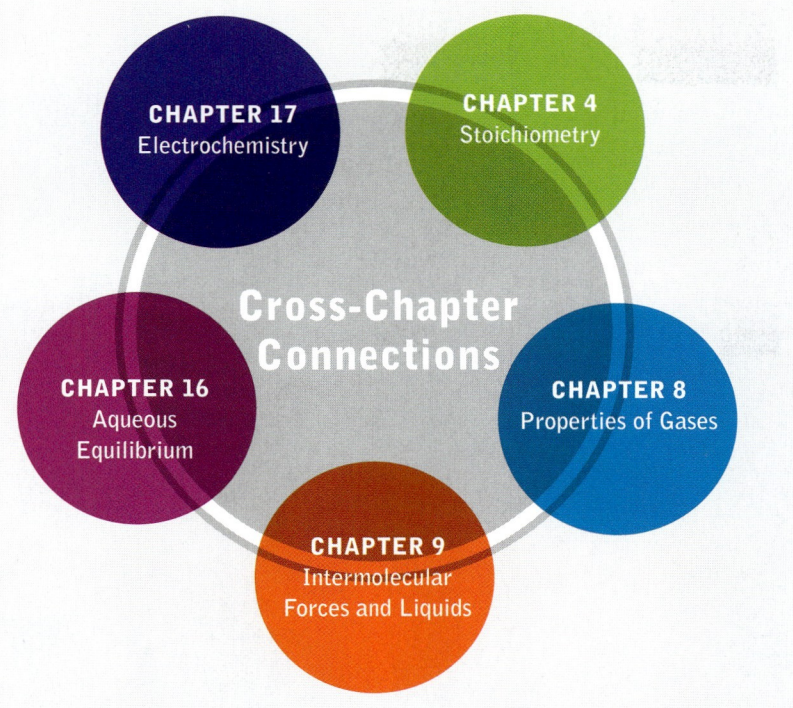

- 5.1 **Earth: The Water Planet**
- 5.2 **Solution Concentration and Molarity**
- 5.3 **Electrolytes and Nonelectrolytes**
- 5.4 **Colligative Properties of Solutions**
  - Osmosis and osmotic pressure
  - Boiling-point elevation and freezing-point depression
  - The van't Hoff factor
  - Measuring molar mass
- 5.5 **Introduction to Oxidation–Reduction Processes**
  - Oxidation numbers
  - Balancing redox reactions
- 5.6 **Acid–Base and Net Ionic Equations**
  - BOX: THE CHEMISTRY OF THE ALKALINE EARTH METALS
- 5.7 **Precipitation Reactions**
- 5.8 **Ion Exchange**
- 5.9 **Titrations**
- 5.10 **Colloids**

A Look Ahead

In this chapter, chemical weathering and the processes that transform Earth's crust serve as an introduction to reactions in solution—particularly reactions between substances dissolved in nature's universal solvent, water. We examine solution concentration scales, the properties of solutions, and several categories of reactions in solution—including oxidation–reduction, neutralization, precipitation, hydrolysis, and ion exchange. This chapter completes our introduction to chemical reactions, chemical equations, and stoichiometry.

5.1 EARTH: THE WATER PLANET

From Chapter 4, we know that gases released by volcanic activity helped form Earth's primordial atmosphere. A major component of that atmosphere was water vapor. As Earth cooled, this vapor condensed and fell as torrents of highly acidic rain. Chemical reactions with the minerals of Earth's crust formed more water, and even more may have come from extraterrestrial sources, such as comets. A rare combination of atmospheric composition and proximity to the

FIGURE 5.1 Arizona's Grand Canyon is a dramatic example of how Earth's surface is continually modified by flowing water. Similar topographic features observed on Mars suggest that water once flowed on the Martian surface.

CONNECTION: The possibility of life on other planets in our solar system was discussed in Section 1.4.

> A **solution** is a uniform mixture of one or more **solutes** dissolved in a **solvent**. In an **aqueous solution**, water is the solvent.

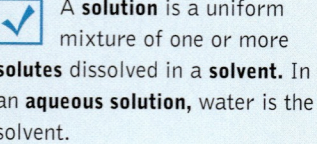

CONNECTION: The classification of mixtures as homogeneous and heterogeneous was introduced in Section 1.1.

sun eventually allowed Earth to become the "water planet" as depressions in its crust filled with about 1.5×10^{21} liters of nature's universal solvent, liquid H_2O.

Life exists on our planet because liquid water exists here. The current debate about the existence of life elsewhere in our solar system and in other planetary systems in the universe hinges on the prospect of liquid water existing on those heavenly bodies. For example, the belief that Martian meteorites collected in Antarctica (see Chapter 1) may contain fossilized forms of life is linked to evidence that there may have been water on the surface of Mars in the distant past (Figure 5.1). The biochemical reactions of all living cells require the presence of liquid water, and so an understanding of the chemistry of life requires an understanding of the principles of chemical reactions between substances dissolved in water.

All natural waters, be they salt water or fresh water, contain ionic and molecular compounds dissolved in H_2O. When an element or compound dissolves in another, a solution is formed. **Solutions** are homogeneous mixtures of two or more substances (Figure 5.2; see Chapter 1 to review the classifications of matter, including heterogeneous and homogeneous mixtures). The substance present in a solution in the greatest proportion (in number of moles) is called the **solvent**; the other ingredients are called **solutes**. When the solvent is water, the solution is an **aqueous solution**. The solvent is often a liquid, but it doesn't have to be. In Chapter 4, we considered the complex composition of Earth's crust. Many of the minerals of which it is made are examples of *solid* solutions: uniform mixtures of substances in the solid state that do not have the fixed composition of a pure substance.

Seawater is sometimes called salt water, but that name may be misleading. Seawater is not just a solution of common table salt (NaCl). In fact, it is not a true

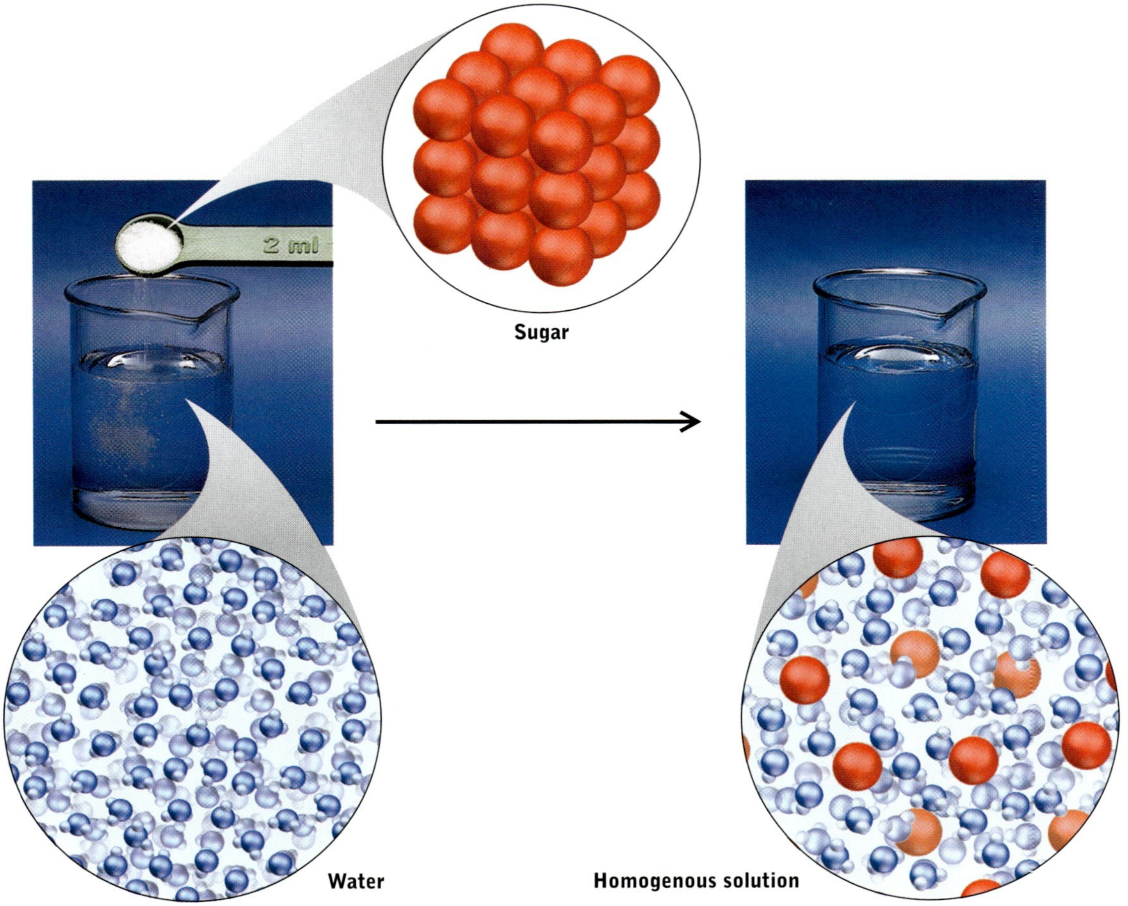

FIGURE 5.2 *What a Chemist Sees.* Adding sugar (solute) to water (solvent) produces a *homogeneous* solution of sugar molecules (red) evenly distributed among water molecules (blue).

solution at all. Although seawater does contain an array of dissolved ionic and molecular compounds, it also contains undissolved matter, including fine-grained sediments suspended by wave action near coastlines or soil eroded by rivers and streams. In some parts of the ocean, most of the suspended matter in seawater is biological, including microscopic plants known as phytoplankton.

Suspended particulate matter is not likely to be distributed uniformly throughout a seawater sample: large, dense particles will slowly settle to the bottom of the container in which the sample of seawater is stored. Thus, seawater is really a heterogeneous mixture (Figure 5.3). If we remove the particles by passing a sample of seawater through a filter (Figure 5.4), we will have a clear, homogeneous mixture, or *solution,* of just the dissolved components. Suspended particles make any liquid **turbid,** meaning that a beam of light passing through the liquid will be scattered by the particles in suspension, as shown in Figures 5.5 and 5.6. The presence of phytoplankton also imparts the green color characteristic of biologically productive ocean (and fresh) water.

CONNECTION: Rules for naming binary compounds such as NaCl are described in Section 4.2.

 Turbidity is the tendency of liquids containing suspended solids to scatter light.

FIGURE 5.3 Coastal seawater is a complex mixture of dissolved solutes, such as NaCl, and suspended particulate matter including sediments and beach sand as well as microscopic plants known as phytoplankton. The presence of suspended matter that is not evenly distributed makes these waters heterogeneous mixtures.

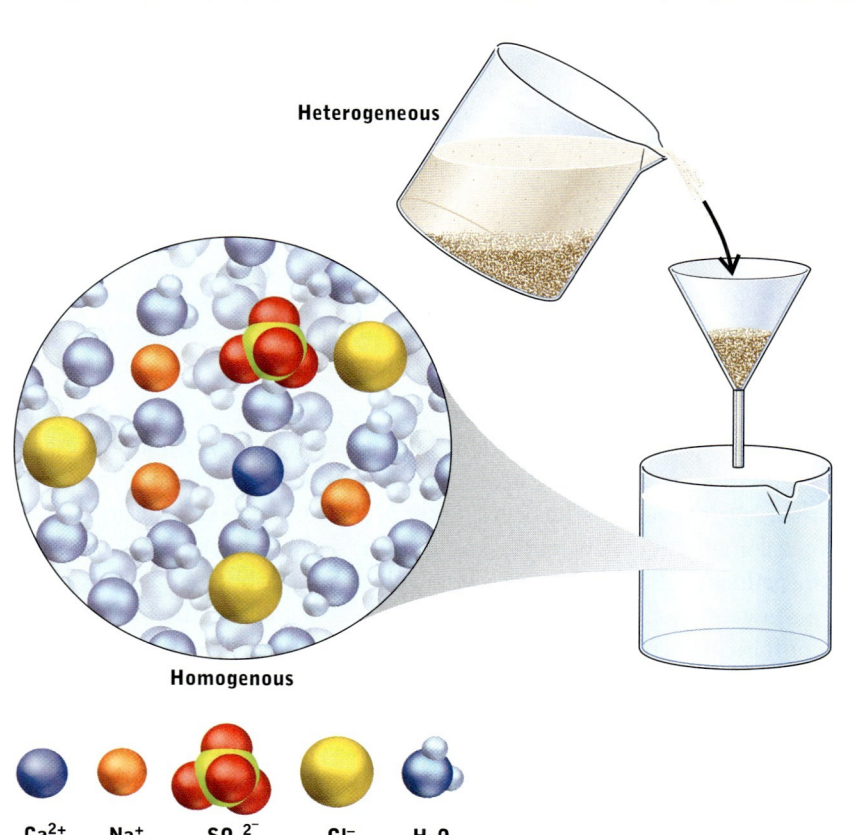

FIGURE 5.4 *What a Chemist Sees.* Suspended solids can be separated from water by filtration. Filtering a sandy mixture of seawater collected in waves breaking on a beach leaves the sand behind and produces a homogeneous solution of ions.

Ca^{2+} Na^+ SO_4^{2-} Cl^- H_2O

FIGURE 5.5 A Secchi disk is used to measure the turbidity of bodies of water. The two-colored wheel is lowered over the side of a vessel until the wheel becomes invisible. Turbidity is evaluated on the basis of the depth of the disk. In the photograph on the left, the disk is clearly visible, whereas, in the photograph on the right, the disk is more difficult to see due to the muddy water.

FIGURE 5.6 The beam of light passing through the two containers of liquid is scattered by particles in suspension in the container on the right; thus, less light passes through turbid suspensions than through clear liquids.

SAMPLE EXERCISE 5.1: Rubbing alcohol is a solution of isopropyl alcohol, C_3H_8O, and water. Which compound is the solvent in a solution of 30.0 g of isopropyl alcohol and 20.0 g of water?

SOLUTION: It is tempting to say that isopropyl alcohol is the solvent because a greater mass of isopropyl alcohol is present. However, our definition of solvent refers to the number of *moles* rather than mass or volume to determine the solvent.
  For the solution in question, the number of moles of isopropyl alcohol is

$$30.0 \text{ g } C_3H_8O \times 1 \text{ mol } C_3H_8O/60.0 \text{ g } C_3H_8O = 0.500 \text{ mol } C_3H_8O$$

and the number of moles of water is

$$20.0 \text{ g } H_2O \times 1 \text{ mol } H_2O/18.0 \text{ g } H_2O = 1.11 \text{ mol } H_2O$$

Water is present in the greatest number of moles; so water is the solvent and isopropyl alcohol is the solute.

PRACTICE EXERCISE: Vinegar is a solution of 5.0 g of $C_2H_4O_2$ in 100 g of water. Which is the solvent and which is the solute? (See Problems 9 and 10.)

CONNECTION: To a chemist, isopropyl alcohol is 2-propanol. The rules that chemists follow for naming alcohols are discussed in Section 12.2.

CONCEPT TEST: Which of the following forms of matter is a solution—that is, a homogeneous mixture?

  a. muddy river water
  b. helium gas
  c. cough syrup
  d. filtered air

5.2 SOLUTION CONCENTRATION AND MOLARITY

Table 5.1 lists the major ions in seawater and in a sample of drinking water from a well in New England. Note that all but one of the ions listed, sulfate (SO_4^{2-}), are single-atom ions of Group 1A, 2A, or 7A elements. The values in Table 5.1 are **concentration** values. Each of them is the quantity of solute in a given mass or volume of solution. The entries in the first column of concentration values for seawater are expressed in g/kg—that is, grams of solute per kilogram of seawater. The corresponding column for well water has the units of mg/kg, meaning milligrams of solute per kilogram of well water. These different units are used because the concentrations of the major ions in fresh water are much lower in the well-water sample than they are in seawater.

Although units based on the mass of solute are widely used, their value can sometimes be misleading. For example, the concentration of chloride in seawater

> **Concentration** is the ratio of the quantity of solute to either the volume or the mass of the solution or solvent in which the solute is dissolved.

TABLE 5.1 Major Ions of Seawater and a Sample of New England Well Water

Ion	Seawater		Well water	
	g/kg	M*	mg/kg	mM
Chloride (Cl^-)	19.354	0.558	8.	0.22
Sodium (Na^+)	10.77	0.479	13.	0.6
Sulfate (SO_4^{2-})	2.712	0.0289	26.	0.26
Magnesium (Mg^{2+})	1.290	0.0543	2.	0.08
Calcium (Ca^{2+})	0.412	0.0105	40.	1.0
Potassium (K^+)	0.399	0.0104	0.2	0.005
Bicarbonate (HCO_3^-)	0.12	0.0020	0.9	0.015

*Surface seawater at 25°C.

(19.35 g/kg) is greater than the concentration of sodium (10.77 g/kg). Do these values mean that there is much more Cl^- in seawater than there is Na^+? Not really. Because the molar mass of chloride (35.45 g/mol) is greater than that of sodium (22.99 g/mol), we would expect the concentration of chloride, expressed on a mass-of-solute basis, to be greater. To check whether seawater is principally a solution of NaCl, we need to express the concentrations of Na and Cl by using *moles* of solute rather than mass. Such units are used in the second columns of Table 5.1. These concentrations represent the number of moles (n) of solute in a volume (V) of one liter of seawater. They are **molar** concentrations. In equation form, the **molarity** (M) of any solution is

$$\text{molarity} = \frac{\text{moles of solute}}{\text{volume of solution in liters}}$$

or

$$M = \frac{n}{V} \qquad (5.1)$$

> ✓ A 1.0-**molar** solution has a **molarity** of 1.0 mole of solute per liter of solution.

Thus, a 1.0-molar (1.0 M) solution is one in which there is 1.0 mole of solute for every liter of solution. If we know the volume of a solution and the molarity of a solute dissolved in it, we can calculate the number of moles of the solute. In such a calculation, it is convenient to rearrange the terms in Equation 5.1 as follows:

$$\text{moles of solute} = \text{molarity} \times \text{volume of solution in liters}$$

or

$$n = M \times V \qquad (5.2)$$

The concentration values in the last column in Table 5.1 are *millimolar* (mM) values, meaning millimoles (10^{-3} mole) of solute per liter of well water. The use of millimoles (mM) fits the relatively small concentrations of the ions in the well-water sample. Even smaller concentrations might be expressed in micromolar, μM (10^{-6} mole of solute per liter of solution), or nanomolar, concentrations, nM (10^{-9} mol/L).

Keep in mind that any concentration value, regardless of its units, is the ratio of the quantity of solute to either the volume or the mass of solution (or solvent). Because a concentration value is a ratio, it does not change whether we have a milliliter of a solution or a million liters of it. On the other hand, the mass of solute in solution is proportional to the volume of solution. Thus, a million liters of a solution contains a billion times as much solute as does only a milliliter of the same solution.

Let's compare the two sets of data in Table 5.1. Notice that, overall, the concentrations of the ions are one or more (usually several more) factors of 10 higher in seawater than in the drinking-water sample, which comes as no surprise to anyone who has tasted the saltiness of seawater. Also notice that the well-water sample is not simply a less concentrated, or more *dilute*, form of seawater. For example, the top two ions in seawater are sodium and chloride, but this is not the case for the sample of well water. Why are the proportions of the

What a Chemist Sees. Indigo has been used to dye cotton and other fibers for thousands of years. A dilute solution of indigo has a faint blue color, but a more *concentrated* solution is deep, navy blue.

CONNECTION: For example, CO_2 dissolved in groundwater reacts with $FeSiO_3$, forming $FeCO_3$. (See Section 4.7.)

ions very different in seawater and fresh water? In Chapter 4, we considered how interactions between minerals and groundwater might alter the chemical composition of both. Thus, the chemical composition of groundwater (well water) depends on the composition of the minerals in contact with the water, and this composition will vary considerably from place to place. On the other hand, the proportions of the major ions in seawater are much the same for all oceans.

> Why are the proportions of the ions very different in seawater and fresh water?

SAMPLE EXERCISE 5.2: Which solution is more concentrated, a solution containing 25 g of NaCl per 100 mL of solution or a solution that contains 2.5 moles of NaCl in 1.00 L of solution?

SOLUTION: We have different sets of units for expressing the quantity of solute and the volume of solution. We should try to get both concentrations into the same units. Let's opt for moles of NaCl/liter. To convert 25 g of NaCl/100 mL into moles of NaCl/L, we need to calculate the molar mass of NaCl and then use that value to calculate the number of moles of NaCl in 25 g of NaCl. First, the molar mass:

Na: 22.99 g/~~mol~~ × 1 ~~mol~~ Na/mol NaCl = 22.99 g Na/mol NaCl

+Cl: 35.45 g/~~mol~~ × 1 ~~mol~~ Cl/mol NaCl = 35.45 g Cl/mol NaCl

_____

58.44 g/mol NaCl

Next we divide the mass of NaCl by its molar mass to calculate the number of moles of NaCl in 25 g. We also need to increase the volume from 100 mL to 1.00 L. The following conversion steps accomplish these tasks:

$$\frac{25 \text{ g NaCl}}{100 \text{ mL}} \times \frac{1 \text{ mol NaCl}}{58.4 \text{ g NaCl}} \times \frac{1000 \text{ mL}}{1.00 \text{ L}} = \frac{4.3 \text{ mol NaCl}}{\text{L}}$$

Clearly, this concentration is greater than 2.5 mol/L.

PRACTICE EXERCISE: A sample of water from the middle of the Pacific Ocean contains 4.5 μg/L of barium. (See Table 1.1 for the exponent values of the metric prefixes, if necessary.) What is the corresponding concentration of barium in moles per liter? (See Problems 21 and 22.)

CONCEPT TEST: Although the proportions of the major ions in seawater are fairly constant throughout the world, the total concentration of all ions, also known as *total dissolved solids*, or *salinity*, can vary from one oceanic region to the next. Can you think of a reason why?

CONNECTION: The molar mass of a monoatomic ion is the same as the molar mass of the parent element (Section 4.3).

CONCEPT TEST: If the concentration of Cl⁻ in seawater is 0.546 mol/kg or 0.558 mol/L, does a 1.000-liter sample of seawater weigh more or less than 1.000 kilogram?

Molarity is a useful measure of concentration because, as we shall see, it enables us to calculate the number of moles of reactants and products in chemical reactions that take place in solution. Therefore, we need to be able to convert concentration values from units based on mass of solute per volume (or mass) of solution into molarity. The following sample exercises help to explain how these conversions are done. The approaches taken in these exercises are summarized in Figure 5.7.

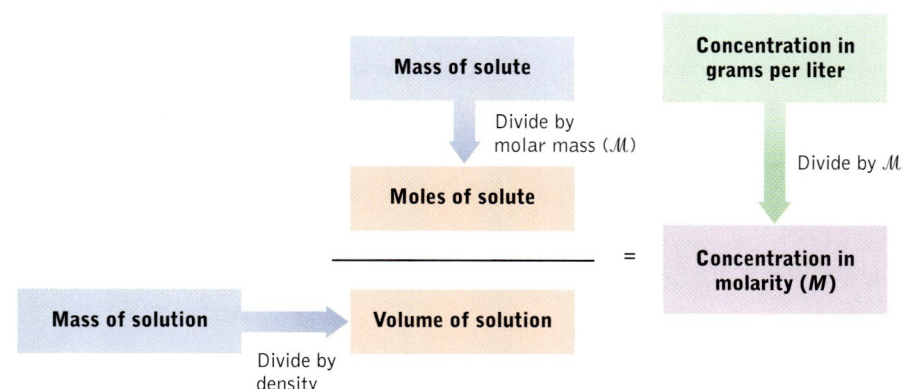

FIGURE 5.7 This flowchart describes some of the paths that may be followed to calculate molar concentrations.

Fresh water and seawater differ both in the total concentrations of the ions present and in the proportions of the ions.

mass NaCl

$\downarrow$ mol = $\dfrac{\text{mass}}{\text{molar mass } \mathcal{M}}$

mol NaCl

$\downarrow$ molarity = $\dfrac{\text{mol NaCl}}{\text{volume (L)}}$

molarity NaCl

SAMPLE EXERCISE 5.3: Calculate the molarity of a solution in which 9.98 g of NaCl is dissolved in a total of 200.0 mL of solution.

SOLUTION: We need to convert the given quantities of solute mass and solution volume into the equivalent number of moles of solute per liter of solution. Let's begin by expressing the given quantities as a concentration—that is, quantity of solute per volume of solution:

$$\dfrac{9.98 \text{ g NaCl}}{200.0 \text{ mL}}$$

We convert the 9.98 g of NaCl into an equivalent number of moles by dividing by the molar mass of NaCl (58.44 g/mol). Then we convert moles per milliliter into moles per liter by multiplying by 1000 mL/L. The overall calculation looks like this:

$$\dfrac{9.98 \text{ g NaCl}}{200.0 \text{ mL}} \times \dfrac{1 \text{ mol NaCl}}{58.44 \text{ g NaCl}} \times \dfrac{1000 \text{ mL}}{1 \text{ L}} = 0.854 \text{ mol/L} = 0.854 \text{ M}$$

PRACTICE EXERCISE: What is the molarity of the Na^+ in Gatorade if a 240-mL serving contains 110 mg of Na^+? (See Problems 11 and 12.)

SAMPLE EXERCISE 5.4: How many grams of magnesium sulfate are needed to prepare 100.0 milliliters of 2.50×10^{-2} M $MgSO_4$?

SOLUTION: We need to convert the given concentration and volume of solution into an equivalent mass of solute. The following equation allows us to calculate the number of moles of solute in a given volume of solution.

$$\text{number of moles } (n) = \text{molarity } (M) \times \text{volume (liters)} \quad (5.2)$$

Using the information given, we have

$$n = \dfrac{2.50 \times 10^{-2} \text{ mol } MgSO_4}{L} \times 100.0 \text{ mL} \times \dfrac{1.000 \text{ L}}{1000 \text{ mL}}$$

$$= 2.50 \times 10^{-3} \text{ mol } MgSO_4$$

Next, we need to convert moles of $MgSO_4$ into grams of $MgSO_4$ by multiplying by the molar mass of $MgSO_4$ (120.37 g/mol):

$$2.50 \times 10^{23} \text{ mol } MgSO_4 \times \dfrac{120.37 \text{ g}}{\text{mol } MgSO_4} = 0.301 \text{ g } MgSO4$$

PRACTICE EXERCISE: How many grams of ammonium nitrate, NH_4NO_3, are required to prepare 400 mL of a 0.0863 M solution? (See Problems 13 and 14.)

SAMPLE EXERCISE 5.5: In the Great Salt Lake in Utah, there are 83.6 g of Na^+ per kilogram of lake water. What is the molarity of Na^+ in the lake if the water has a density of 1.160 g/mL?

5.2 Solution Concentration and Molarity

SOLUTION: In this problem, we are given a concentration based on a mass of solute in a given mass of solution. To convert this concentration into molarity (*moles* of solute per *liter* of solution), we need to convert the mass of solute into an equivalent number of moles, and we need to convert the mass of solution into an equivalent volume in liters. Let's start with the solute: to convert the mass of an element into moles, we need to multiply the mass by 1 over its molar mass. In this case, the mass is in milligrams. Note how the *milli* prefix carries through the calculation, giving us *milli*moles of Na^+:

$$\frac{83.6 \text{ mg } Na^+}{\text{g soln}} \times \frac{1 \text{ mol } Na^+}{22.99 \text{ g } Na^+} = \frac{3.64 \text{ mmol } Na^+}{\text{g soln}}$$

In the next step, we convert grams of water into milliliters by using the density value given:

$$\frac{3.64 \text{ mmol } Na^+}{\text{g soln}} \times \frac{1.160 \text{ g soln}}{1.000 \text{ mL soln}} = \frac{4.22 \text{ mmol } Na^+}{\text{mL soln}}$$

The resulting units are millimoles per milliliter. The *milli* (or 10^{-3}) prefixes in the numerator and the denominator cancel out, yielding moles per liter or molarity (M). Thus the concentration of Na^+ in the Great Salt Lake is 4.22 M (nearly 10 times the concentration of Na^+ in seawater).

PRACTICE EXERCISE: Many municipal drinking-water companies notify their customers when the concentration of Na^+ in their water exceeds 20 mg/L. What is the molar concentration of Na^+ that is equivalent to 20 mg/L? (See Problems 19–22.)

$\dfrac{\text{mass (mg) } Na^+}{\text{mass (g) soln}}$

↓ $\text{mmol} = \dfrac{\text{mass (mg)}}{\mathcal{M}}$

$\dfrac{\text{mmol } Na^+}{\text{mass (g) soln}}$

↓ $V \text{ (mL)} = \dfrac{\text{mass (g)}}{\text{density (g/mL)}}$

$\dfrac{\text{mmol } Na^+}{\text{volume (mL) soln}}$

↓ $M = \dfrac{\text{mmol}}{\text{mL}}$

molarity Na^+

Coastal zones where rivers flow into the sea are places where the concentrations of dissolved ions change dramatically as fresh water mixes with seawater. Before mixing occurs, the fresh water from a river sometimes flows over the top of more-saline, and more-dense, seawater, producing surface waters that are much less concentrated than those beneath. Eventually, wind and wave action mix the layers together. Calculating the concentrations of the major ions in seawater in a coastal area such as a bay in which reasonably well defined volumes of seawater and fresh water mix is relatively easy. The concentrations of the major ions are about 1000 times as high in seawater as in fresh water (see Table 5.1); so their concentrations are proportional to the ratio of seawater to fresh water. Said another way, their concentrations depend on the degree to which the seawater is diluted by the fresh water.

Suppose that a coastal bay has a volume of 5.0 km³ and that, within the bay, on average 1.2 km³ of fresh water mixes with 3.8 km³ of open ocean water in which the concentration of Na^+ is 0.48 M. What is the molarity of Na^+ in the water in the bay? If we assume that the fresh water contributes an insignificant amount of Na^+ to the bay, then the concentration of Na^+ will be equal to the amount of Na^+ provided by the sea divided by the total volume of the bay. The amount of Na^+ provided by the sea could be calculated by using Equation 5.2 if we knew how many liters there are in 3.8 km³. It turns out that there are

1000 liters in a cubic meter and that there are 1000 meters in a kilometer. Therefore, in 3.8 km³ of seawater, there are:

$$3.8 \text{ km}^3 \times \frac{(1000 \text{ m})^3}{\text{km}^3} \times \frac{1000 \text{ L}}{\text{m}^3} = 3.8 \times 10^{12} \text{ L}$$

Using Equation 5.2 to calculate the number of moles of Na^+ in this volume, we get:

$$n = M \times V$$
$$= 0.48 \text{ mol/L} \times 3.8 \times 10^{12} \text{ L}$$
$$= 1.8 \times 10^{12} \text{ mol}$$

To calculate the molar concentration of Na^+ in the bay, we employ Equation 5.1 and divide the moles of Na^+ by the total volume in the bay, which is 5.0×10^{12} L:

$$M = n/V$$
$$= 1.8 \times 10^{12} \text{ mol}/5.0 \times 10^{12} \text{ L}$$
$$= 0.36 \ M$$

The two steps that we just took in this calculation can be combined into one by using an equation that applies to all situations in which an initial volume (V_i) and concentration (M_i) and either a final volume (V_f) or concentration (M_f) are known:

$$V_i \times M_i = V_f \times V_f \tag{5.3}$$

Equation 5.3 works because each side of the equation represents a number of moles and that number does not change, even though the volumes change.

SAMPLE EXERCISE 5.6: To what volume must one dilute 50 mL of an aqueous solution of a green dye to reduce its concentration by 80%?

SOLUTION: Let's start with Equation 5.3:

$$M_{initial} \times V_{initial} = M_{final} \times V_{final}$$

If the concentration is reduced by 80%, or 4/5, then the final concentration must be 1/5 of the initial concentration, that is, $M_{final} = 1/5 \ M_{initial}$. Using this information in Equation 5.3, we have

$$M_{initial} \times V_{initial} = 1/5 \ M_{initial} \times V_{final}$$

or

$$V_{final} = 5 \ V_{initial}$$
$$= 5 \times 50 \text{ mL}$$
$$= 250 \text{ mL}$$

PRACTICE EXERCISE: What volume of a 0.543 M solution of $MgCl_2$ should be diluted to a final volume of 250. mL to prepare a 0.030 M solution of $MgCl_2$? (See Problems 27–30.)

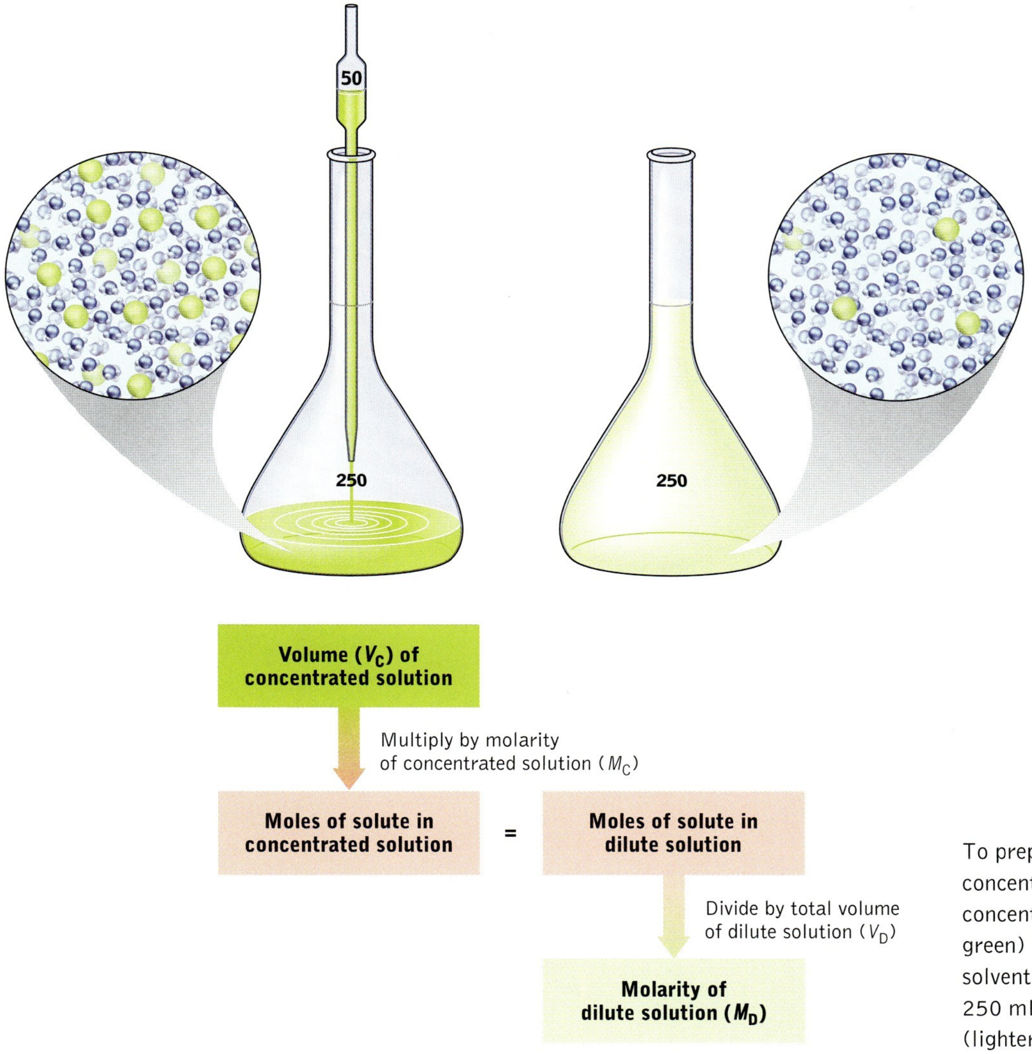

To prepare a fivefold dilution of a concentrated solution, 50 mL of concentrated solution (darker green) is mixed with sufficient solvent (water) to make the 250 mL of the dilute solution (lighter green).

5.3 ELECTROLYTES AND NONELECTROLYTES

Seawater is a much better conductor of electricity than pure water, as shown by the results of the following experiment. Suppose we inserted two copper metal wires into a sample of seawater. Then we connected these wires to a battery and to a small light bulb. We would observe bubbles of gas forming in the solution immediately surrounding the wires, and the light bulb would light as shown in Figure 5.8A. However, if we repeat the experiment with distilled water instead of seawater, the light bulb does not light.

CONNECTION: The details of the dissolution of ionic compounds in water are described in Section 9.2.

What makes seawater a better conductor of electricity than pure water? Consider what happens when a crystal of table salt (NaCl is the major solute in seawater) dissolves in water. A molecular view of the dissolution process is shown in Figure 5.9. We can write a chemical equation for the process in the following way:

$$NaCl(s) \xrightarrow{H_2O} Na^+(aq) + Cl^-(aq)$$

Note how sodium chloride dissociates into its component ions when it dissolves in water. Other soluble ionic compounds behave similarly.

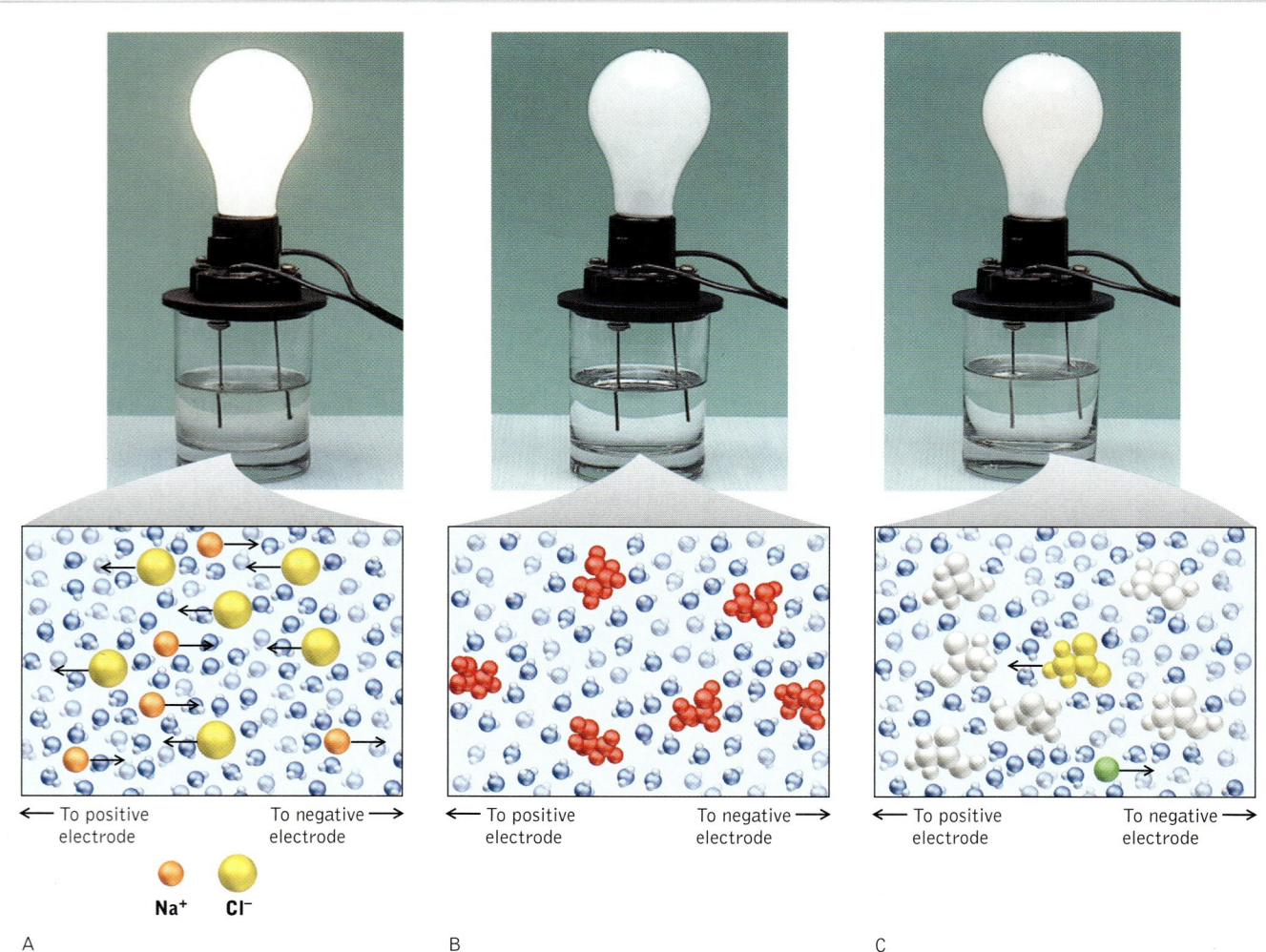

FIGURE 5.8 *What a Chemist Sees.* A. Many of the compounds that are dissolved in seawater, such as NaCl, dissociate completely into ions. Migration of their ions makes seawater a good conductor of electricity. B. Solutions of nonelectrolytes, such as ethanol (C_2H_5OH, shown in red) do not conduct electricity well, because they do not dissociate into ions. C. Compounds such as acetic acid ($C_2H_3O_2H$, shown in white), which partly dissociate into ions ($C_2H_3O_2^-$, shown in yellow, and H^+, shown in green), are weak electrolytes.

5.3 Electrolytes and Nonelectrolytes

How does a solution of Na^+ and Cl^- ions conduct electricity? Electricity flows through a metal wire because electrons surrounding metal atoms are able to migrate from one atom to the next, and each electron carries -1.609×10^{-19} coulombs of electrical charge. However, there are no free electrons in a solution of NaCl. The solution is a good electrical conductor because its *ions* can migrate, taking their electrical charges with them. In the solution of NaCl, Cl^- ions move toward the wire connected to the positively charged (+) pole of the battery, and Na^+ ions move toward the negatively charged (−) wire. This flow of oppositely charged ions in opposite directions produces a net flow of electrical charge and makes the solution a conductor of electricity. A solute that imparts this electrical conductivity to a solution is called an **electrolyte**. The wires or other materials that make the electrical connections between the electrolyte and the outside electrical circuit are called **electrodes** (see Figure 5.8).

> What makes seawater a better conductor of electricity than pure water?

Not all solutes dissociate when they dissolve, at least not completely. If we replace NaCl with a comparable quantity of acetic acid, $C_2H_3O_2H$, the principal ingredient in vinegar, we discover that the light bulb in Figure 5.8C does not glow as brightly. If we dissolve ethanol, C_2H_5OH, or glucose, $C_6H_{12}O_6$, instead of NaCl, the light bulb does not glow at all (see Figure 5.8B). From these observations, we may conclude that acetic acid forms fewer ions than does NaCl when dissolved in water and that ethanol and sugar do not form any ions. Substances such as acetic acid that only partly dissociate into ions in aqueous solution are called **weak electrolytes**. Because ethanol and sugar do not form ions in aqueous solutions, they are considered **nonelectrolytes**.

CONNECTION: Unlike the chemical reactions in batteries (see Chapter 17), a solution of sodium chloride doesn't *produce* electricity; rather, it *conducts* electricity.

An **electrolyte** is a substance that dissociates into ions when it dissolves, enhancing the conductivity of water.

An **electrode** is a conducting solid used to make electrical contact with a liquid or a solution.

A **weak electrolyte** only partly dissociates into ions when it dissolves and only slightly increases the conductivity of water.

A **nonelectrolyte** is a substance that does not enhance the conductivity of water when it dissolves.

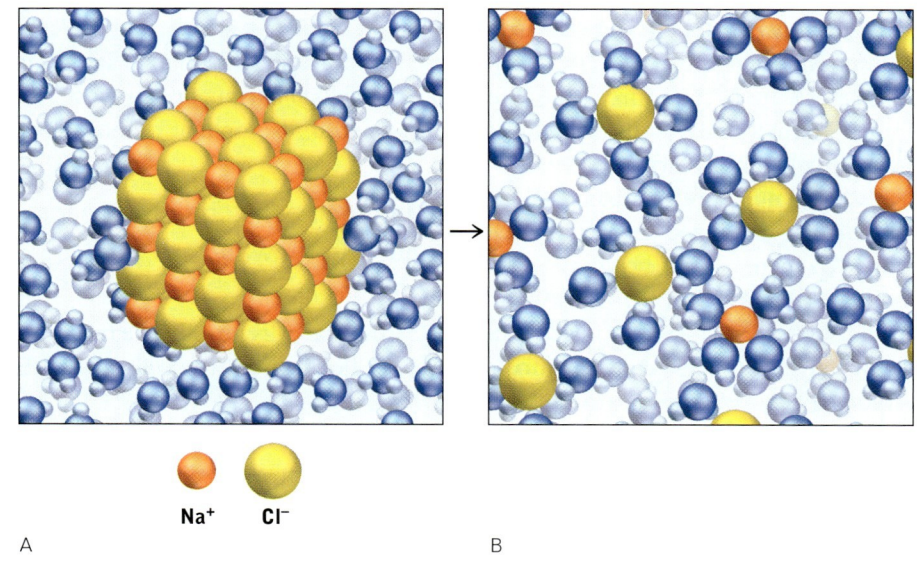

FIGURE 5.9 When a crystal of solid sodium chloride (A) dissolves in water (B), the crystal breaks into Na^+ and Cl^- ions. Each ion is surrounded by a cluster of water molecules.

CONCEPT TEST: Which of the following solutions should be the best conductor of electricity?

1 M acetic acid   1 M NaCl   1 M KBr   1 M CaCl$_2$ 1 M ethanol

5.4 COLLIGATIVE PROPERTIES OF SOLUTIONS

Adding a nonvolatile solute, such as salt, to a pure solvent, such as water, changes the physical properties of the solvent. The arrangement of solvent molecules around solute ions or molecules allows more mass to occupy very little more volume. As a result solutions have greater densities than that of the pure solvent. Other changes take place that are perhaps less obvious. For example, adding a nonvolatile solute lowers the temperature at which the solvent freezes, and it raises the temperature at which the solvent boils. These changes in melting point and boiling point depend directly on the *number of moles* of solute dissolved in a given quantity of solvent and not on the *identity* of the solute. Thus, a mole of NaCl produces the same changes in freezing point and boiling point as does a mole of KNO$_3$, even though the latter has considerably more mass. The presence of dissolved solutes is also responsible for a phenomenon known as osmotic pressure.

Osmosis and osmotic pressure

More than 97% of the water on Earth is seawater. None of it is fit for human consumption. We can't drink seawater, because the total concentration of solutes in seawater is more than three times the total in our blood plasma and in most cells in the human body. Consider what would happen if a sample of human blood were mixed with seawater. The cells suspended in the blood, such as the red blood cells that carry oxygen, would begin to collapse (Figure 5.10B). They collapse because the water inside them flows from the cells into the surrounding seawater in a process called **osmosis.** As water flows out of the cells and their volume decreases, they begin to shrivel up.

Why do blood cells collapse when suspended in seawater?

Why do blood cells collapse when suspended in seawater? For one thing, the membranes that give cells their shape are slightly porous, or **semipermeable,** which means that water molecules can pass through them, but most solutes cannot. When a cell is surrounded by an aqueous solution with a solute concentration differing from that inside the cell, water molecules naturally tend to flow

Osmosis is the process by which solvent passes through a **semipermeable membrane** to balance the concentrations of solutes on both sides of the membrane.

A **semipermeable membrane** blocks the passage of all solution components except the solvent.

5.4 Colligative Properties of Solutions

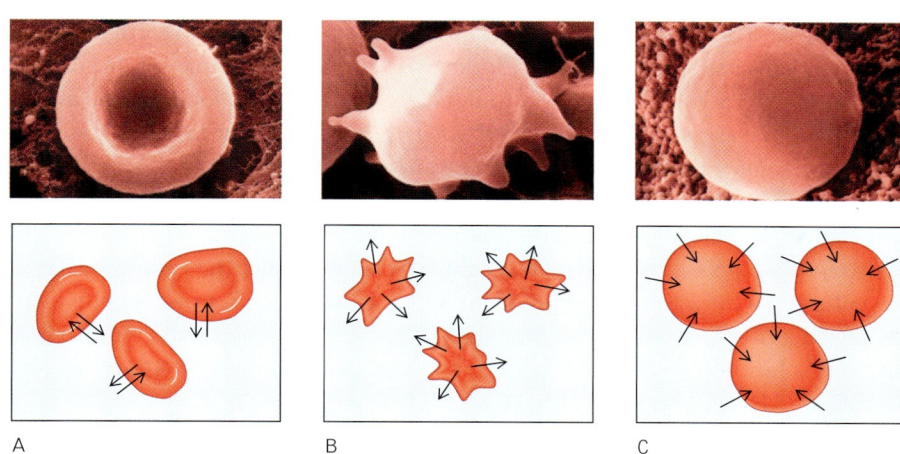

FIGURE 5.10 The membranes of red blood cells are semipermeable. Water can readily pass into and out of these cells to equalize the concentration of solutes inside and outside. A. When red blood cells are bathed in a solution of NaCl with a concentration equal to the concentration of the fluid inside the cell (isotonic), the flow of water into and out of the cell (shown by the arrows) is in balance. B. When the concentration of solutes in the fluid surrounding the cells is higher, water flows from the cells across the membrane until the concentration of dissolved particles inside matches the concentration outside. C. When the cells are immersed in pure water, they expand as additional water enters, diluting the fluid inside the cells.

through the membrane to bring the total solute concentrations inside and outside the cell into balance. A red blood cell bathed in seawater, for example, achieves this balance when nearly three-fourths of the water in it passes through its membrane into the surrounding seawater, causing the cell to shrink and the concentrations of solutes inside it to increase until they match the concentrations of the salts in the surrounding seawater.

During a medical emergency, body fluids may need to be administered intravenously (Figure 5.11). The solutions used in these injections must have the same total solute concentration as that of blood serum. One such solution is *physiological saline*, which is a salt solution that contains 0.92% NaCl. Thus, physiological saline *by weight* contains 0.92 gram of NaCl for every 100 grams of solution. The density of dilute aqueous solutions is close to 1.00 g/mL; so 100 grams has a volume of 100 milliliters, and a concentration of 0.92 g/100 g is nearly the same as 0.92 g/100 mL. Scaling up the mass of solute and volume of solution by 10, we have 9.2 g/L. This value is less than a third the concentration (35 g/L) of sea salts in seawater.

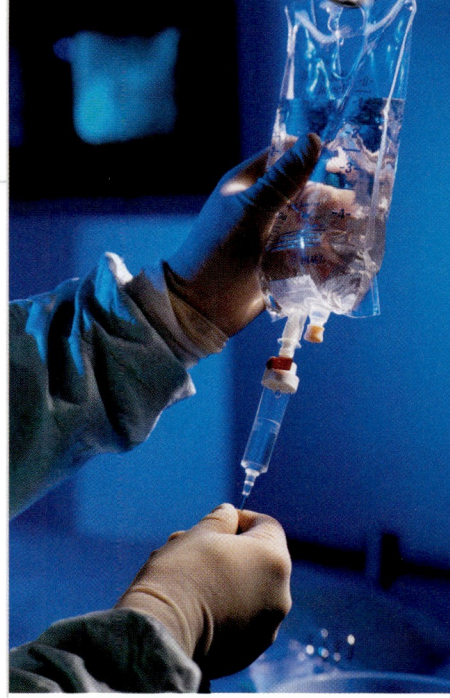

FIGURE 5.11 A solution of physiological saline has a concentration of 0.92 g NaCl/L, or 0.155 M Na^+ and 0.155 M Cl^-. This concentration is equal to the concentration inside the cells of the body.

CONCEPT TEST: Why do cucumbers shrivel when pickled in brine (salt water), and why do wilted flowers revive when placed in fresh water?

Emergency medical technicians may inject trauma patients with a solution called *D5W*. The acronym stands for 5.5% solution of *dextrose* (a form of sugar) in *water*. This use suggests that 5.5% dextrose contains the same solute concentration as 0.92% NaCl. To confirm that it does so, we need to focus on the meaning of the phrase *same solute concentration*. What we really mean is the *same number of solute particles (molecules or ions)* in a given volume of solution. To calculate the number of particles in solution, we need to convert these percent-concentration values into molarity values. In doing so, we assume that percent by weight values for dilute aqueous solutions is essentially the same as percent by volume; so 0.92% NaCl is equivalent to 9.2 g/L, and 5.5% sugar is equivalent to 55 g/L. These assumptions allow us to calculate the molar concentrations of the two solutions by using the molar masses of NaCl, which is 58.44 g/mol, and dextrose ($C_6H_{12}O_6$), which is 180.16 g/mol:

$$\frac{9.2 \text{ g NaCl}}{L} \times \frac{1 \text{ mol NaCl}}{58.44 \text{ g NaCl}} = 0.157 \frac{\text{mol}}{L} = 0.157 \, M$$

$$\frac{55 \text{ g } C_6H_{12}O_6}{L} \times \frac{1 \text{ mol } C_6H_{12}O_6}{180.16 \text{ g } C_6H_{12}O_6} = 0.31 \frac{\text{mol}}{L} = 0.31 \, M$$

A comparison of these two molar concentrations discloses that the molar concentration of dextrose is about twice that of NaCl. How can both match the solute levels in blood? The answer is contained in the dissolution process for NaCl as described in the equation on page 223: when each mole of NaCl dissolves, it yields a mole of Na^+ *and* a mole of Cl^-, or 2 moles of ions. Therefore the total solute concentration in 0.92% NaCl is 0.31 *M*, the same as in D5W.

Water molecules migrate through the membrane of a blood cell or through any semipermeable membrane because a *force* makes it happen—a force derived from the different concentrations of solutes on the two sides of the membrane. When we divide the magnitude of this force (*F*) by the surface area (*A*) of the membrane, we get pressure (*P*):

$$P = F/A \tag{5.4}$$

To stop molecules of water from moving across the membrane, we could apply an opposing pressure equal to *F/A*. This pressure is called **osmotic pressure** (Figure 5.12). The Greek letter pi (π) is used as the symbol for osmotic pressure to distinguish it from other kinds of pressure, such as atmospheric pressure, which we will encounter in Chapter 8.

The magnitude of the osmotic pressure required to stop the net flow of solvent across a membrane separating pure solvent from a solution is proportional to the concentration of the solution expressed in molarity (*M*) of particles (ions or molecules) of solute. It is also proportional to absolute temperature (*T*). We can write a mathematical equation connecting these variables to pressure if we add a constant, *R*, which has the value 0.0821 (L · atm)/(mol · K). The resulting equation relating osmotic pressure to concentration and temperature is

$$\pi = MRT \tag{5.5}$$

Note the units on *R*. They include the inverse of molarity (L/mol) and the inverse

> **Osmotic pressure** (π) is the pressure that must be applied across a semipermeable membrane to stop the flow of solvent from a less-concentrated toward a more-concentrated solution.

CONNECTION: The pressure exerted by gases will be addressed in Section 8.1.

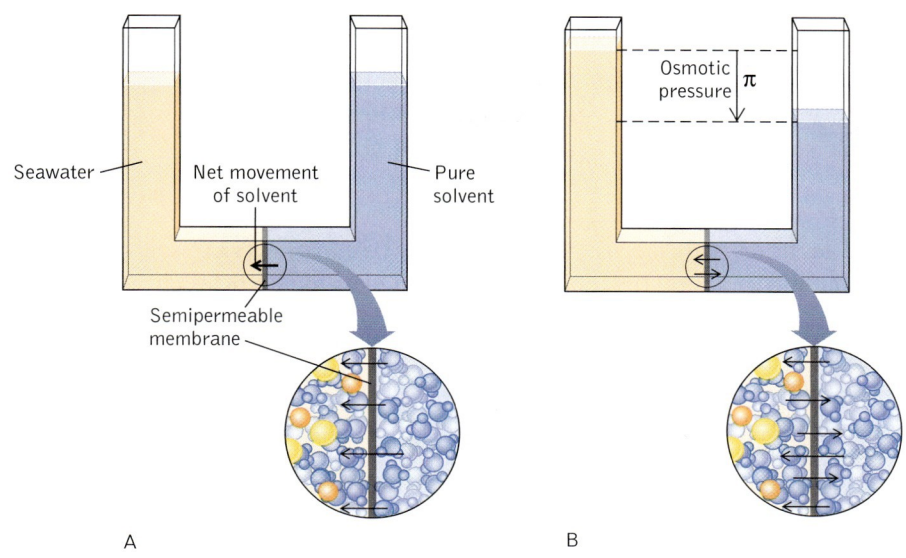

FIGURE 5.12 *What a Chemist Sees.* A. If equal volumes of seawater and pure water are separated by a semipermeable membrane, a net flow of water from the pure-water side to the seawater side is observed. B. The volume of the seawater increases and the volume of the pure water decreases until balanced by pressure produced by the difference in the heights of the fluid levels. This pressure, which pushes water molecules into the pure-water side as fast as they flow toward the seawater side, is called osmotic pressure (π).

of temperature (1/K). Therefore, multiplying $M \times R \times T$ leaves the unit "atm," which stands for pressure in "atmospheres."

Let's apply Equation 5.5 to calculating the osmotic pressure across a semipermeable membrane separating a 0.31 M solution of dextrose from pure water at the temperature of the human body: 37°C. First, we need to convert body temperature from degrees Celsius into kelvins (see Section 1.6):

$$K = °C + 273.15$$
$$= 37°C + 273.15$$
$$= 310$$

Using this kelvin temperature in Equation 5.5, we have

$$\pi = MRT$$
$$= 0.31 \text{ mol/L} \times 0.0821 \text{ L} \cdot \text{atm/mol} \cdot \text{K} \times 310 \text{ K} = 7.9 \text{ atm}$$

A pressure of 7.9 atm is a lot of pressure: it is more than three times the air pressure in a typical automobile tire and about the same as the water pressure experienced by a diver at a depth of 82 meters (268 feet).

CONNECTION: The constant movement of solvent molecules back and forth across the membrane represents a dynamic equilibrium. We will consider this concept in more detail in Chapter 16.

SAMPLE EXERCISE 5.7: Calculate the osmotic pressure across a semipermeable membrane separating pure water from seawater at 25°C.

SOLUTION: Osmotic pressure is related to the total concentration of all particles in solution. Table 5.1 lists the molar concentrations of the most abundant ions in seawater. Summing these concentrations should provide a good estimate of the concentration of all ions in seawater. This sum is 1.14 M. Inserting this value in Equation 5.5 and converting temperature into kelvins, we have

molarity (M)

$\pi = MRT$

osmotic pressure (π)

$$\pi = MRT = 1.14 \text{ mol/L} \times 0.0821 \text{ L} \cdot \text{atm/mol} \cdot \text{K} \times (25 + 273 \text{ K})$$
$$= 27.9 \text{ atm}$$

PRACTICE EXERCISE: Calculate the osmotic pressure across a semipermeable membrane separating seawater (1.14 M total particles) from a solution of normal saline (0.31 M total particles). (See Problems 53 and 54.)

> ✓ A **colligative property** is a property of a solution that depends on the concentration but not the identity of the particles dissolved in it.

Osmotic pressure is referred to as a **colligative property** of a solution, meaning that it depends on the total concentration of solute particles but not on their identity. Thus a 1.0 M solution of NaCl produces the same osmotic pressure as 1.0 M KCl or 1.0 M NaNO$_3$ because all three solutions are 2.0 M in total ions. These solutions have twice the osmotic pressure of a 1.0 M solution of glucose because glucose is a nonelectrolyte and so produces a solution that is only 1.0 M overall in dissolved particles.

As already noted, the osmotic pressure of intravenous solutions must be identical with that of blood plasma. Such solutions are said to be isotonic. Solutions with higher (hypertonic) or lower (hypotonic) concentrations cause dehydration or swelling (edema), respectively, and must never be used in an intravenous injection.

If an opposing pressure can stop osmotic flow, an even greater opposing pressure should *reverse* osmotic flow, forcing solvent to flow from a more concentrated solution to a more dilute solution. A technique called **reverse osmosis** is based on this principle. In a widely used procedure for purifying water, undrinkable water is pumped through semipermeable membranes. Because water molecules can go through the membrane, whereas most solute particles cannot, the water on the high-pressure side of the membrane becomes more concentrated with solute molecules and ions while that which flows through the membrane has much lower solute concentrations (Figure 5.13). This technology is used by some municipal water-supply systems to make somewhat saline, or brackish, water fit to drink and by various industries whose processes require water that is purer than conventional tap water.

> ✓ **Reverse osmosis** is used in the water-purification process in which water is pumped through semipermeable membranes, leaving dissolved impurities behind.

SAMPLE EXERCISE 5.8: What is the reverse osmotic pressure required at 20°C to purify brackish well water containing 0.355 M dissolved solids if the product water is to contain no more than 87 mg of dissolved solids (such as NaCl) per liter?

SOLUTION: The molarity of the more-concentrated solution is 0.355 M. The molarity of the less-concentrated product water must be calculated from the information given. If it is to contain no more than 87 mg of NaCl per liter, the maximum molar concentration of NaCl can be calculated from the molar mass of NaCl:

$$\frac{87 \text{ mg NaCl}}{\text{L}} \times \frac{1.000 \text{ g}}{1000 \text{ mg}} \times \frac{1 \text{ mol NaCl}}{58.44 \text{ g NaCl}} = \frac{1.50 \times 10^{-3} \text{ mol NaCl}}{\text{L}}$$
$$= 1.50 \times 10^{-3} \text{ } M \text{ NaCl}$$

A solution that is 1.50×10^{-3} M NaCl is twice that, or 3.00×10^{-3} M, in total ions. The difference in the concentration of total ions between the two solutions is

$$0.355 \text{ M (brackish water)}$$
$$-0.003 \text{ M (product water)}$$
$$0.352 \text{ M (across membrane)}$$

Calculating the osmotic pressure for this concentration, we have

$$\pi = MRT = \frac{0.352 \cancel{\text{ mol}}}{\cancel{\text{L}}} \times \frac{0.0821 \cancel{\text{L}} \cdot \text{atm}}{\cancel{\text{mol} \cdot \text{K}}} \times (273 + 20) \cancel{\text{K}}$$

$$= 8.47 \text{ atm}$$

Therefore, a pressure greater than 8.47 atm will be required to force water from the sample through a reverse-osmosis membrane.

PRACTICE EXERCISE: Calculate the pressure that must be applied to the brackish water of the preceding Sample Exercise if the maximum dissolved concentration allowed in the product water is twice that of the product water in the Sample Exercise. (See Problems 57 and 58.)

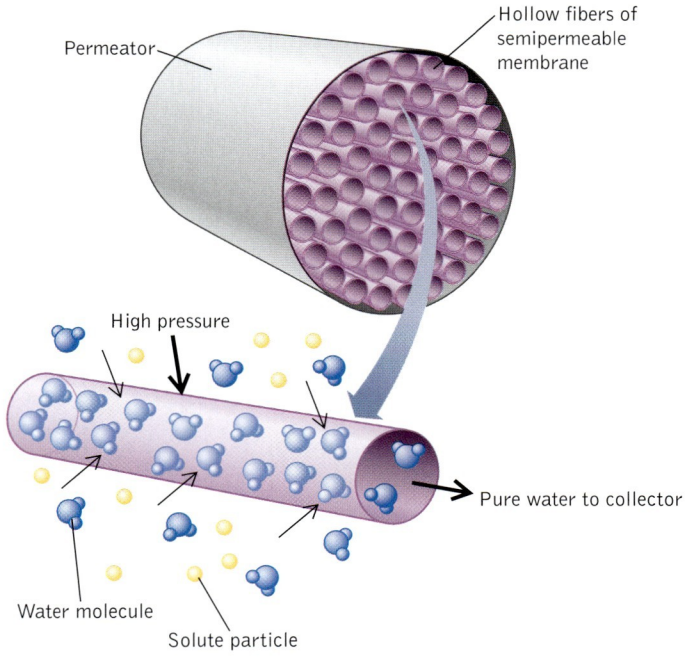

FIGURE 5.13 Seawater can be desalinated by reverse osmosis. In a reverse-osmosis system, seawater flows at a pressure greater than its osmotic pressure around bundles of tubes with semipermeable walls. Water molecules pass from the seawater through the walls into the tubes, leaving dissolved sea salts behind. Pure water flows from the insides of the tubes and is collected.

Boiling-point elevation and freezing-point depression

CONNECTION: The energy required to heat water to its boiling point and vaporize it is considered in Section 11.7.

In several of the desert countries bordering the Persian Gulf, the source of drinking water is the sea. Reverse osmosis is one way to desalinate seawater. Another approach is distillation (Figure 5.14; see Chapter 1). It takes a great deal of energy to heat seawater to its boiling point and convert it into steam, particularly when supplying the water needs of a city or a region of a country. However,

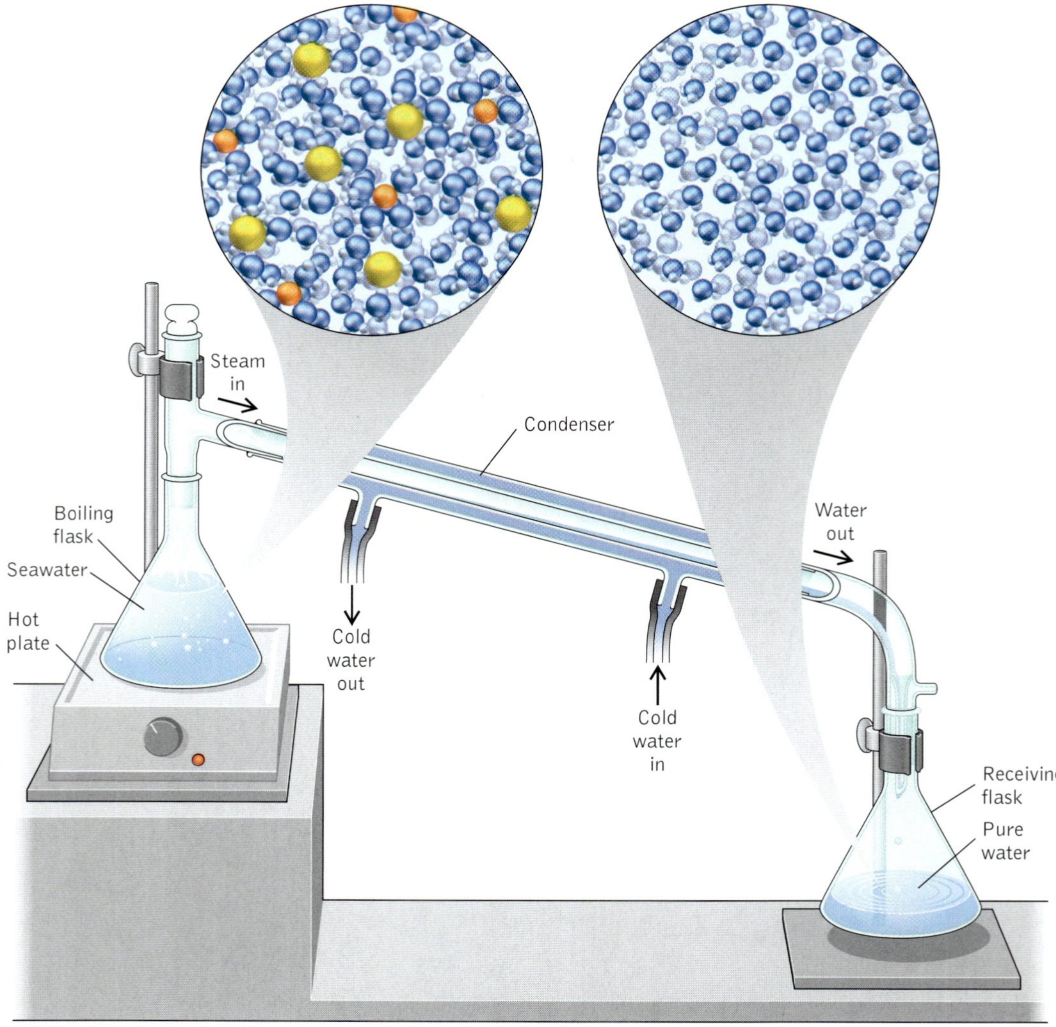

FIGURE 5.14 Pure water can be obtained from seawater by heating seawater to its boiling point (in the flask on the left). Water vapor released from the boiling seawater passes into the condenser where it condenses into pure liquid water and is collected as pure, distilled water.

many of the arid countries bordering the Persian Gulf are energy rich. For these countries, distillation makes economic sense. Still, there are ways to make the distillation process more energy efficient. In one approach, seawater is used to cool the condensers of the distillation apparatus and then, after absorbing the energy of water vapor, the preheated seawater is heated even more to make steam.

A factor to be considered in distilling seawater is that seawater boils at a slightly higher temperature than pure water does. Why is the boiling point of seawater higher than that of pure water? The reason is connected to the ways in which ionic and molecular solutes interact with water when they dissolve, as shown in Figure 5.15. The presence of solute molecules reduces the number of solvent molecules at the surface of a liquid that are free to enter the vapor phase, while not inhibiting vapor molecules from condensing into the liquid. The net effect is to reduce the vaporization process, and so the solvent does not boil at its normal boiling point.

> Why is the boiling point of seawater higher than that of pure water?

 CONNECTION: The boiling point of a solution is linked to its vapor pressure (Section 9.3).

To overcome this effect and achieve a steady boil, more heat must be added and the temperature of the solution must be raised. As you might expect, the greater the concentration of solute, the higher the boiling point of the solution. Note that the identity of the solute is not important, only the concentration of dissolved particles in solution, be they molecules or ions. This dependence on

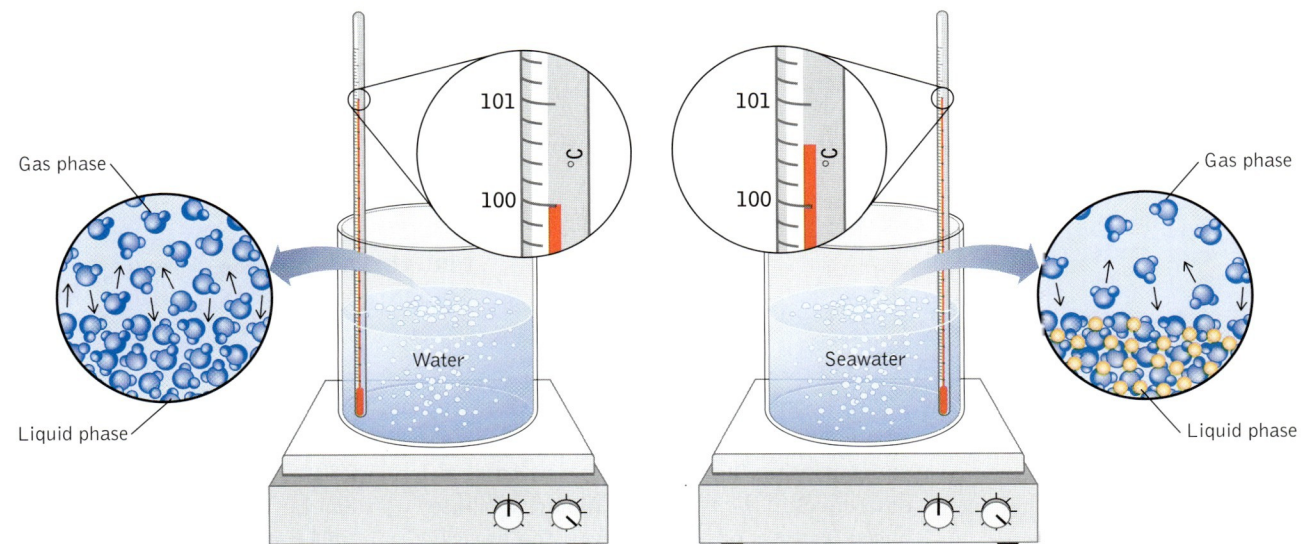

FIGURE 5.15 *What a Chemist Sees.* Seawater (1.14 m particles) boils at 100.6°C, an increase of 0.6 degrees above the boiling point of pure water (100.0°C). At the boiling point, the rates of solvent molecules entering the gas phase and condensing into a liquid are equal. The presence of solute (orange spheres) decreases the rate at which solvent molecules enter the gas phase at a given temperature, requiring a higher temperature to boil the solution (see text).

> The **boiling-point elevation** is the increase in the boiling point of a solvent when particles of solute are dissolved in it.
>
> The **molality** (*m*) of a solution is the number of moles of solute per kilogram of solvent.

concentration alone makes **boiling-point elevation** a colligative property of the solvent. It is described in equation form as follows:

$$\Delta T_b = K_b m \tag{5.6}$$

where K_b is the boiling-point-elevation constant of the solvent (0.52 °C/*m* for water) and *m* represents **molality,** which is the number of *moles of solute per kilogram of solvent.* In equation form,

$$m = \frac{\text{mol solute}}{\text{kg solvent}} \tag{5.7}$$

Molality (*m*) is sometimes confused with molarity (*M*), but note the difference: molarity is the number of moles of solute per *liter* of *solution,* whereas molality is the number of moles of solute per *kilogram* of *solvent.* Figure 5.16 summarizes the calculation of molality.

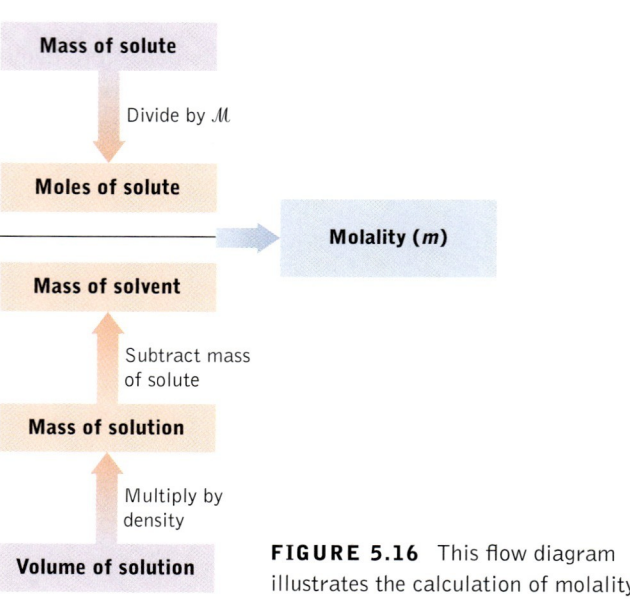

FIGURE 5.16 This flow diagram illustrates the calculation of molality.

To examine the difference between the molarity and molality of solutions, let's calculate the molality of chloride in seawater by starting with the molar concentration of chloride given in Table 5.1: 0.558 *M*. We need to focus on the differences in the two units; molarity is based on *volume* of *solution* and molality is based on *mass* of *solvent*. Given the mass–volume difference, you might assume that the density of the solution is important. And it is. The density of seawater is 1.022 g/mL at 25°C. Scaling up this value a thousand times, we get a density of 1022 g/L:

$$\frac{1.022 \text{ g}}{\text{mL}} \times \frac{1000 \text{ mL}}{\text{L}} = \frac{1022 \text{ g}}{\text{L}}$$

Of this 1022 g, the combined mass of dissolved salts is 35 grams, leaving (1022 − 35) = 987 g (0.987 kg) of pure water in every liter of seawater. Let's use this information to calculate molality from molarity:

$$0.558\ M = \frac{0.558\ \text{mol}}{L} = \frac{0.558\ \text{mol}}{0.987\ \text{kg}} = \frac{0.565\ \text{mol}}{\text{kg}} = 0.565\ m$$

Therefore, the chloride concentration in seawater is 0.558 molar; it is also 0.565 molal.

> **CONCEPT TEST:** The differences between the *molar* and *molal* concentration values for dilute aqueous solutions are small. Why?

SAMPLE EXERCISE 5.9: How many grams of Na_2SO_4 should be added to 250. mL of water to prepare a 0.750 m solution of Na_2SO_4?

SOLUTION: To calculate the mass of Na_2SO_4 that is needed, we must first calculate the number of moles of Na_2SO_4 that are needed. This value can be obtained from Equation 5.7:

$$\text{molality} = \frac{\text{moles of solute}}{\text{mass of solvent (kg)}}$$

Rearranging the terms to solve for moles of solute:

$$\text{moles of solute} = \text{molality} \times \text{mass of solvent (kg)}$$

Inserting the given values into this equation (and converting grams of solvent into kilograms of solvent):

$$\text{moles of solute} = \text{molality} \times \text{mass of solvent (kg)}$$

$$= \frac{0.750\ \text{mol}\ Na_2SO_4}{\text{kg water}} \times 250.\ \text{g water} \times \frac{1\ \text{kg}}{1000\ \text{g}}$$

$$= 0.1875\ \text{mol}\ Na_2SO_4$$

Next, we need to convert moles of Na_2SO_4 into grams of Na_2SO_4 by multiplying by its molar mass:

$$0.1875\ \text{mol}\ Na_2SO_4 \times \frac{142.04\ \text{g}\ Na_2SO_4}{\text{mol}\ Na_2SO_4} = 26.6\ \text{g}\ Na_2SO_4$$

PRACTICE EXERCISE: What is the molality of a solution prepared by dissolving 78.2 g of ethylene glycol, $C_2H_6O_2$, in 1.50 L of water? The density of water is 1.00 g/mL. (See Problems 63 and 64.)

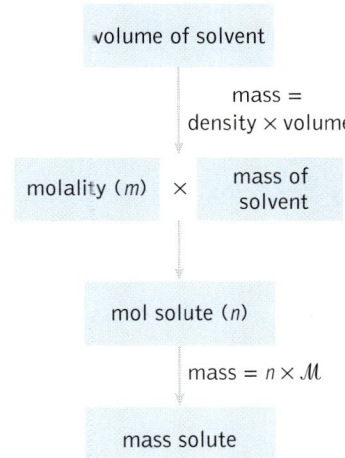

SAMPLE EXERCISE 5.10: What is the increase in boiling point of a sample of seawater versus pure water if the total concentration of dissolved particles in the sample is 1.14 m?

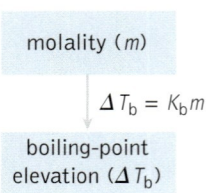

SOLUTION: Equation 5.6 relates boiling-point elevation to concentration:

$$\Delta T_b = K_b m$$
$$= (0.52\ °C/m)(1.14\ m)$$
$$= 0.60°C$$

PRACTICE EXERCISE: When crude oil is pumped out of the ground, it may be accompanied by "formation water" that contains high concentrations of NaCl. If the boiling point of a sample of formation water is 2.3 Celsius degrees above the boiling point of pure water, what is the molality of NaCl in the sample? (See Problems 67 and 68.)

Adding salt to water raises its boiling point. Adding salt to water also lowers its freezing point. You may have seen the effects of this phenomenon in the application of salt to roads and sidewalks in the winter to melt ice and snow. The salty solutions that result have freezing points lower than the normal freezing point of water (0°C). If the lowered freezing point is below the temperature of the ice, the ice melts. A molecular view of freezing-point depression is shown in Figure 5.17.

The magnitude of **freezing-point depression** also is directly proportional to the concentration of dissolved solute. The equation describing this dependence

$$\Delta T_f = K_f m \tag{5.8}$$

looks a lot like Equation 5.6 and is used in much the same way either to predict changes in freezing point from known concentrations or to estimate concentrations from measured changes in freezing-point temperature. The chart in Figure 5.18 summarizes calculations of freezing-point depression and boiling-point elevation.

✓ **Freezing-point depression** is the lowering of the freezing point of a solvent when particles of solute are dissolved in it.

SAMPLE EXERCISE 5.11: The fluid used in automobile cooling systems is prepared by dissolving ethylene glycol, $C_2H_6O_2$, in water. What is the freezing point of radiator fluid prepared by mixing 1.00 gallon of ethylene glycol with 1.00 gallon of water? Given: The density of ethylene glycol is 1.114 g/mL.

SOLUTION: To answer this question, we need to calculate the ΔT_f value for the radiator fluid, which requires converting the given proportion of solute to solvent into molality, or moles of ethylene glycol per kilogram of solvent. We can save steps by recognizing that:

1. the ratio of 1.00 gallon of ethylene glycol to 1.00 gallon of water is equivalent to 1.00 liter of ethylene glycol per liter of water; and that
2. the density of water is 1.000 g/mL or 1.000 kg/L.

Thus, the composition of radiator fluid is 1.00 L of ethylene glycol per kilogram of water.

The following steps convert this concentration into molality by calculating:

1. the volume of solute in milliliters;

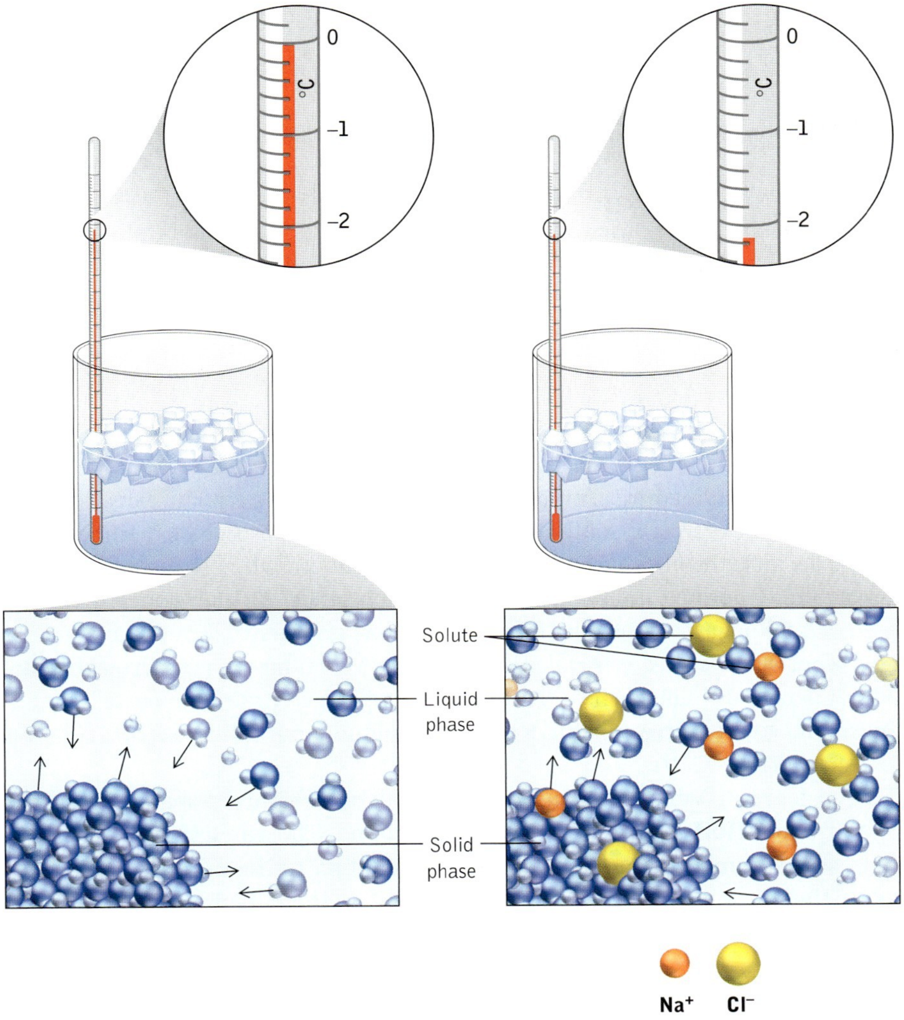

FIGURE 5.17 Ocean water freezes at a lower temperature (−2.15°C) than does fresh water (∼0.00°C) because of the dissolved salts present in seawater. At the freezing point, the number of molecules melting equals the number of molecules freezing. The solutes interfere with the solidification of water, making its freezing point lower.

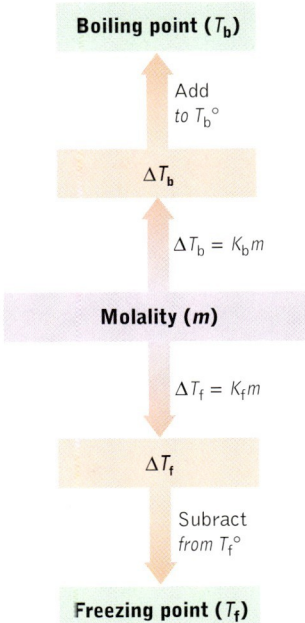

FIGURE 5.18 This flow chart summarizes how to calculate the freezing and boiling points of solutions of known molality.

2. the mass of solute in grams;
3. the moles of solute; and
4. the molality of the solution (moles of solute per kilogram of solvent)

Step: 1.   2.   3.   4.

$$\frac{1.00 \text{ L}}{\text{kg water}} \times \frac{1000 \text{ mL}}{\text{L}} \times \frac{1.114 \text{ g}}{\text{mL}} \times \frac{1 \text{ mol}}{62.07 \text{ g}} = 17.95 \ m$$

This molal concentration can be used to calculate ΔT_f by using Equation 5.8:

$$\Delta T_f = K_f m$$
$$= 1.86 \ °C/m \times 17.95 \ m = 33.4 \ °C$$

Applying this change to the normal freezing point (T_f^0) of water, 0.0 °C, we have the freezing point of coolant (T_f):

$$T_f = T_f^0 - \Delta T_f$$
$$= (0.0 - 33.4)\,°C$$
$$= -33.4\,°C$$

Converting the freezing-point value into the more familiar Fahrenheit scale, we have

$$F = 9/5\,°C + 32 = 9/5(-33.4\,°C) + 32 = -28.1\,°F$$

PRACTICE EXERCISE: What is the boiling point of the automobile radiator fluid in the preceding Sample Exercise? (See Problems 67 and 68.)

The van't Hoff factor

Sometimes calculated freezing-point depressions or boiling-point elevations for concentrated electrolyte solutions are less than those observed experimentally. This is because the positive and negative ions produced when strong electrolytes dissolve may not be totally independent of each other. At high concentrations, positive and negative ions may interact with one another to form ionic clusters. The simplest such cluster is called an **ion pair.** When an ion pair forms in solution, it acts as a single particle. Thus, the overall concentration of particles is reduced and so are freezing-point depression, boiling-point elevation, and osmotic pressure (Figure 5.19).

The degree to which free ions form when an electrolyte dissolves is expressed by the **van't Hoff factor,** (i), for that compound. The van't Hoff factor is the ratio of the number of moles of particles in solution to the number of moles of solute that dissolved. This definition is expressed mathematically as follows:

$$i = \frac{\text{moles of particles in solution}}{\text{moles of solute dissolved}} \quad (5.9)$$

The van't Hoff factor for NaCl in water should be 2, because there are two moles of dissolved sodium and chloride ions for each mole of NaCl. However, it is actually 1.97 for 0.0100 m NaCl and only 1.87 for 0.100 m NaCl. Figure 5.20 summarizes some calculated and observed values for the van't Hoff factor. The equations for calculating osmotic pressure, boiling-point elevation, and freezing-point depression can be modified to account for the formation of ion pairs by including the van't Hoff factor:

$$\pi = i\,MRT$$
$$\Delta T_b = i\,K_b m$$
$$\Delta T_f = i\,K_f m$$

In these modified equations, the molarity and molality terms refer to the concentration of the solute, not the sum of the concentrations of its individual ions. The value of i for a particular solution is evaluated experimentally.

The cation and anion in an **ion pair** move together as a single, neutral cluster in solution.

The **van't Hoff factor** (i) of a solution is the ratio of the moles of particles actually in the solution and the moles of solute that dissolved.

CONNECTION: The clustering of water molecules around the ions in Figure 5.19 is further explored in Section 9.2.

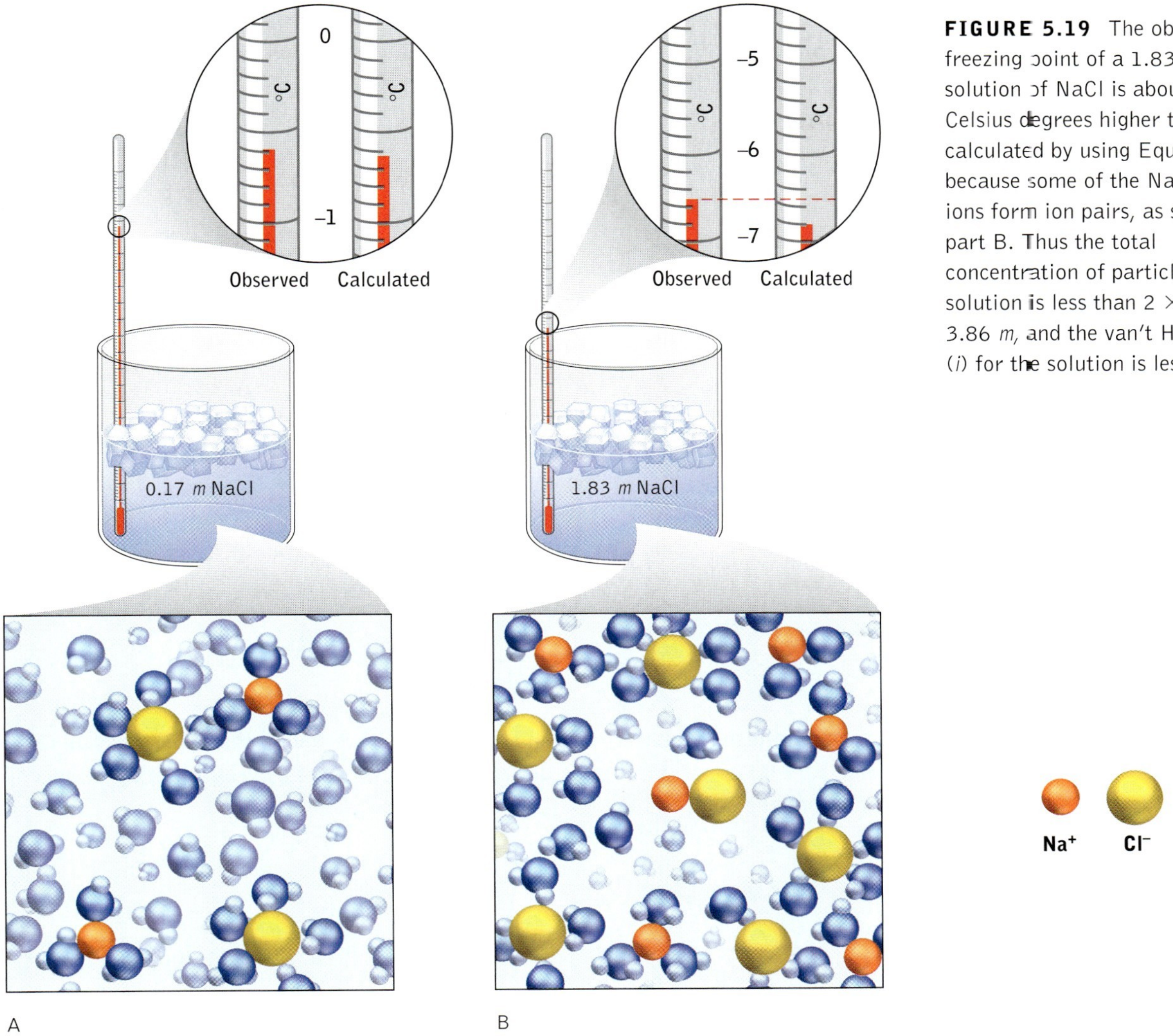

5.4 Colligative Properties of Solutions | 237

FIGURE 5.19 The observed freezing point of a 1.83 m solution of NaCl is about 0.3 Celsius degrees higher than calculated by using Equation 5.6 because some of the Na$^+$ and Cl$^-$ ions form ion pairs, as shown in part B. Thus the total concentration of particles in solution is less than 2 × 1.83, or 3.86 m, and the van't Hoff factor (i) for the solution is less than 2.

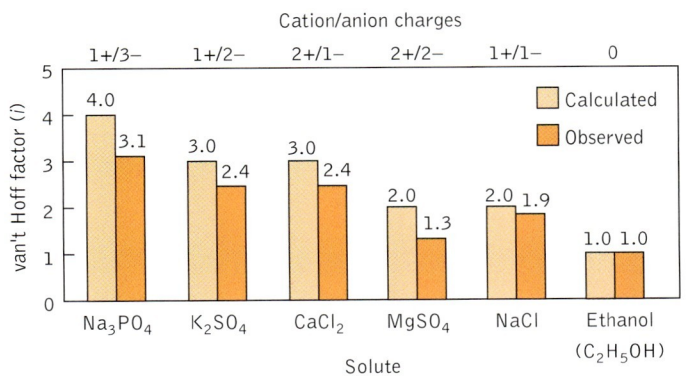

FIGURE 5.20 The vertical bars represent theoretical and experimentally derived values for the van't Hoff factors for 0.1 m solutions of ethanol (a nonelectrolyte) and several electrolytes. The differences between theoretical and experimental values increase with increasing ionic charge.

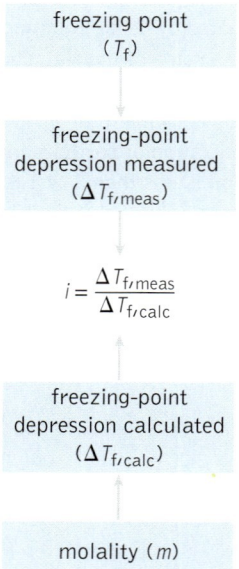

SAMPLE EXERCISE 5.12: The observed freezing point of a 1.83-m aqueous solution of NaCl is -6.50 °C. If we assume the freezing point of pure water to be 0.00 °C, what is the value of the van't Hoff factor for this solution?

SOLUTION: The freezing point of the solution is depressed 6.50 Celsius degrees below the freezing point of the pure solvent; that is,

$$\Delta T_f = 6.50 \text{ °C}$$

We can calculate the molal concentration of solute particles in solution from this value and the freezing-point depression constant (1.86 °C/m) for water by using Equation 5.8. First, let's rearrange the terms to solve for m:

$$\Delta T_f = K_f m$$

or

$$m = \frac{\Delta T_f}{K_f}$$

Using the values of $\Delta T_f/K_f$ given, we have

$$m = \frac{6.50 \text{ °C}}{1.86 \text{ °C}/m}$$

$$= 3.49 \, m$$

To calculate the value of the van't Hoff factor, we divide the experimentally observed molal concentration of dissolved particles (3.49 m) by the molal concentration of the solute (1.83 m):

$$i = 3.49/1.83$$

$$= 1.91$$

DISCUSSION: One mole of NaCl dissociates into a total of 2 moles of ions; so an i value of 1.91 instead of 2.00 indicates that some ion pairing is taking place in solution.

PRACTICE EXERCISE: The van't Hoff factor for a 0.05 m solution of magnesium sulfate is 1.3. What is the freezing point of the solution? (See Problems 77 and 78.)

Measuring molar mass

In Chapter 4 (Section 4.5), we needed information about molar mass to convert the empirical formula of a compound derived from elemental analysis into its molecular formula. Knowing the molar mass ($\mathcal{M}$) of a compound enables us to determine the number of empirical formula units in each of its molecules and, from that number, its molecular formula.

5.4 Colligative Properties of Solutions

The molar mass of a solute can be determined by osmotic-pressure, boiling-point-elevation, or freezing-point-depression measurements. These methods are suitable only for nonelectrolytes, which have a van't Hoff factor of 1.0. Let's see how a freezing-point-depression measurement can be used to find the molar mass of eicosene, a compound encountered in Section 4.5.

SAMPLE EXERCISE 5.13: The freezing point of a solution of 100 mg of eicosene (a nonelectrolyte) in 1.00 g of benzene was lower by 1.87 Celsius degrees than the freezing point of pure benzene. Determine the molar mass of eicosene. (K_f for benzene is 4.90 °C/m.)

SOLUTION: Let's start by calculating the molality of the solution from ΔT_f and Equation 5.8:

$$\Delta T_f = K_f m$$

or

$$m = \Delta T_f / K_f = 1.87 \text{ °C} / 4.90 \text{ °C}/m = 0.382 \ m \text{ eicosene}$$

At this point, we know that the molal concentration of the solution if 0.382.

The second part of the problem requires application of the definition of molality, which is moles of solute per kilogram of solvent.

$$m = \frac{\text{mol solute}}{\text{kg solvent}}$$

We know the molality of the solution (0.382) and the mass of the solvent (1.00 g, or 1.00×10^{-3} kg). With this information, we can calculate the moles of eicosene (solute) in the sample, as follows:

$$0.382 \ m = \frac{\text{mol eicosene}}{1.00 \times 10^{-3} \text{ kg benzene}}$$

$$\text{mol eicosene} = 0.382 \ m \times 1.00 \times 10^{-3} \text{ kg}$$

$$= 3.82 \times 10^{-4}$$

This value is the number of moles of eicosene in 100 mg. The ratio of these two values can be used to calculate molar mass:

$$\text{molar mass} = \frac{\text{mass eicosene}}{\text{moles of eicosene}} = \frac{0.100 \text{ g eicosene}}{3.82 \times 10^{-4} \text{ moles}} = 262 \text{ g/mole}$$

Check Section 4.5 to see how the molar mass of eicosene was used in the calculation of its molecular formula.

PRACTICE EXERCISE: A solution prepared from 360 mg of sugar in 1.00 g of water froze at −3.72 °C. What is the molar mass of this sugar? Assume that it is a nonelectrolyte. (See Problems 81 and 82.)

---

freezing-point depression (ΔT_f)

$$m = \frac{\Delta T_f}{K_f}$$

molality (m)

$$(n_{\text{solute}}) = m \times \text{mass solvent}$$

mol solute (n_{solute})

$$\mathcal{M} = \frac{\text{mass solute}}{n_{\text{solute}}}$$

molar mass ($\mathcal{M}$)

TABLE 5.2 Molal Freezing-Point-Depression and Boiling-Point-Elevation Constants for Selected Solvents

Solvent	Freezing point (°C)	K_f (°C/m)	Boiling point (°C)	K_b (°C/m)
Water (H_2O)	0.0	1.86	100.0	0.52
Benzene (C_6H_6)	5.5	4.90	80.1	2.53
Ethanol (C_2H_5OH)	−114.6	1.99	78.4	1.22
Carbon tetrachloride (CCl_4)	−22.3	29.8	76.8	5.02

Osmotic pressure is the best of the three methods of determining molar mass for water-soluble substances for the following reasons. First, the K_f and K_b parameters of water are much smaller than those of other solvents (Table 5.2). Thus, more-concentrated molal solutions are needed to observe the same ΔT_f or ΔT_b value than with, say, carbon tetrachloride as the solvent. Second, many biomaterials, such as proteins and carbohydrates, are available only in small quantities, and they are often not very soluble in solvents with larger K_f and K_b values than those for water. These biomaterials often have high molar masses, which means that large quantities are needed to give high enough molal concentrations for reliable ΔT_f or ΔT_b measurements. Third, a solute might need to be recovered unchanged for other uses, and so boiling-point-elevation measurements are ruled out for heat-sensitive solutes.

In contrast, very small osmotic pressures can be measured precisely, the equipment can be miniaturized, and the measurements can be made at room temperature. The following Sample Exercise illustrates the use of osmotic pressure to determine the molar mass of a compound.

SAMPLE EXERCISE 5.14: A 47-mg sample of a nonelectrolyte compound isolated from a South African tree was dissolved in water to make 2.50 mL of solution at 25 °C. The measured osmotic pressure of the solution was 0.489 atm. Calculate the molar mass of the compound.

SOLUTION: The osmotic pressure of this noneletrolyte solution at 25°C is 0.489 atm; so

$$\pi = MRT = 0.489$$

Substituting $R = 0.0821$ L · atm/mol · K and $T = 273 + 25 = 298$ K, we have

$$M = \pi/RT = \frac{0.489 \text{ atm}}{\left(0.821 \frac{\text{L} \cdot \text{atm}}{\text{mol} \cdot \text{K}}\right)(298 \text{ K})} = 2.03 \times 10^{-2} \text{ mol/L}$$

We can calculate the number of moles (n) of solute from the molarity (M) of a solution and its volume (V) by using Equation 5.2. Since the volume is in milliliters, the result of the calculation is in millimoles:

$$n = M \times V$$
$$= 2.03 \times 10^{-2} \text{ mol/L} \times 2.50 \text{ mL}$$
$$= 5.08 \times 10^{-2} \text{ mmol}$$

The molar mass (M) of a substance is the mass in grams of one mole of it and so has the units of grams per mol. In this exercise, we know the mass (47 mg) of 5.08×10^{-2} mmol. The ratio of these values gives us the molar mass because the "milli" prefixes cancel out:

$$M = \frac{47 \text{ mg}}{5.08 \times 10^{-2} \text{ mmol}} = \frac{9.3 \times 10^2 \text{ mg}}{\text{mmol}} = \frac{9.3 \times 10^2 \text{ g}}{\text{mol}}$$

PRACTICE EXERCISE: A solution was made by dissolving 5.00 mg of hemoglobin in water to give a final volume of 1.00 mL. The osmotic pressure of this solution was 1.91×10^{-3} atm at 25°C. Calculate the molar mass of hemoglobin. (See Problems 79 and 80.)

5.5 INTRODUCTION TO OXIDATION–REDUCTION PROCESSES

In the discussion of how ionic solutions conduct electricity (Section 5.3), we learned that the migration of ions is a requirement for the flow of electricity through a solution. This ion migration is not the only requirement. When chloride ions reach the surface of the positively charged electrode in the example on page 223, they do not simply accumulate there. Instead, each chloride ion gives up an electron to the electrode, forming a free atom of chlorine. Two such atoms bond together to make a molecule of chlorine gas, the presence of which is indicated by the presence of bubbles on the electrode surface and the acrid smell of chlorine gas just above the sample. We can describe this process by using the following chemical equation:

$$2 \text{ Cl}^-(aq) \longrightarrow \text{Cl}_2(g) + 2 \text{ e}^-$$

where the symbol e^- on the product side represents electrons transferred to the electrode from chloride ions. This reaction, or any reaction in which electrons are lost, is an **oxidation** reaction. In this example, we would say that *chloride has been oxidized to chlorine.*

There is a fundamental rule in chemistry: *if one species in a reaction loses electrons, then some other species must gain electrons* so that overall the number of electrons lost and gained is the same. For the purposes of completing the electrical circuit represented by the battery and salt solution in Figure 5.8, the loss of electrons by Cl⁻ must be balanced by a *gain* of electrons by another species at the opposite electrode. In other words, a **reduction** reaction takes place at the negatively charged electrode. Therefore, the oxidation reaction of chloride to chlorine describes only half of the chemistry as electrical current passes through

> ✓ **Oxidation** is a chemical change accompanied by the loss of electrons.
>
> **Reduction** is a chemical change accompanied by a gain of electrons.

> A **half-reaction** is the oxidation or reduction component of an oxidation–reduction reaction.

a solution of NaCl. Thus, chemists call the conversion of chloride into chlorine an oxidation **half-reaction**. It takes place not by itself, but only when a reduction half-reaction also takes place.

Which species in a solution of NaCl is most likely to be reduced? You might guess Na^+, and indeed we could write a reduction half-reaction for sodium ions that results in the formation of sodium metal:

> **Which species in a solution of NaCl is most likely to be reduced?**

$$Na^+(aq) + e^- \longrightarrow Na(s)$$

If sodium metal were to form at the electrode surface, it would not last long. Like all alkali metals, sodium metal reacts rapidly with water to form sodium and hydroxide (OH^-) ions and hydrogen gas:

$$2\ Na(s) + 2\ H_2O(l) \longrightarrow 2\ Na^+(aq) + 2\ OH^-(aq) + H_2(g)$$

Because sodium metal is not stable in aqueous media, it is likely not to be formed in the first place. Some other ingredient in a solution of NaCl must be reduced at the negatively charged electrode. But which one? The answer lies within the sodium oxidation reaction in water. In the sodium part of the reaction, each atom of sodium loses an electron, forming a sodium ion, Na^+. The remaining part of the reaction is the reduction of water itself. This half-reaction may be written as follows:

$$2\ H_2O(l) + 2\ e^- \longrightarrow 2\ OH^-(aq) + H_2(g)$$

If water were reduced at the negatively charged electrode in the solution of NaCl, we would expect to observe bubbles of hydrogen gas forming at the electrode surface. That is exactly what is observed (Figure 5.21).

CONCEPT TEST: Can you think of a way to test whether the gas formed at the negatively charged electrode in the electrolysis of NaCl is hydrogen gas? What chemical property of hydrogen might be the basis for such a test? (See Section 1.1.)

CONNECTION: The electrolysis of a solution containing metal cations is useful in making thin films of metal, as described in Section 17.7.

We can combine the half-reaction for chlorine oxidation, which takes place at the positively charged electrode, with the reduction half-reaction for water, which takes place at the negatively charged electrode, and write the overall reaction that takes place as electricity flows through a solution of NaCl:

$$2\ Cl^-(aq) + 2H_2O(l) + 2\ e^- \longrightarrow Cl_2(g) + 2\ OH^-(aq) + H_2(g) + 2\ e^-$$

Eliminating the $2\ e^-$ terms that appear on both sides of the reaction arrow, we have

$$2\ Cl^-(aq) + 2H_2O(l) \longrightarrow Cl_2(g) + 2\ OH^-(aq) + H_2(g)$$

Thus, the overall result of the passage of electricity through a solution of NaCl is the generation of hydrogen and chlorine gases and the conversion of the NaCl solution into a solution of NaOH.

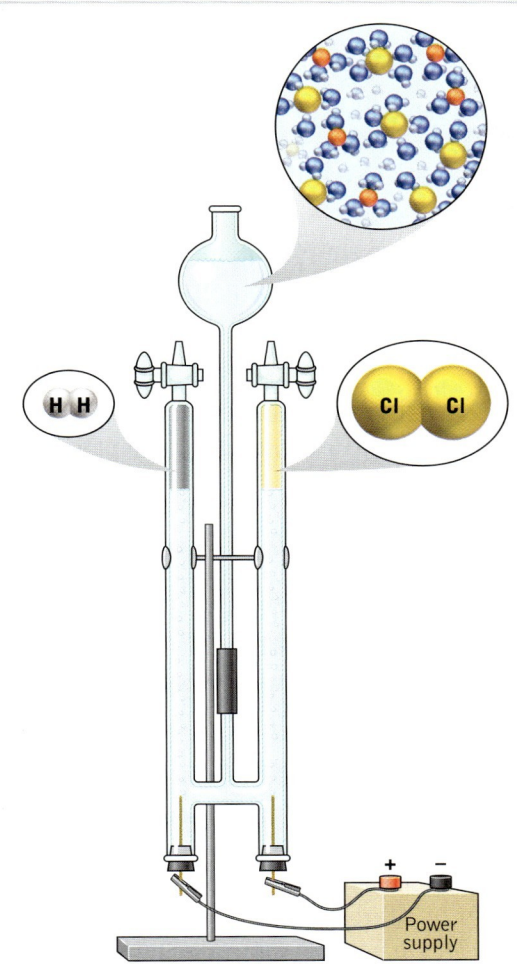

FIGURE 5.21 In the electrolysis of a solution of NaCl, hydrogen gas (H_2) forms at the negatively charged electrode and chlorine gas (Cl_2) forms at the positively charged electrode. In the molecular view (shown at the top), the orange spheres are Na^+ ions and the yellow spheres are Cl^- ions.

The preceding chemical equation is an example of a **net ionic equation.** It provides detailed information on which of the ions in solution were consumed or produced by the reaction. Sodium ions do not appear in this net ionic equation, even though they were present in solution, because sodium ions were not changed by the chemical reaction. They were simply **spectator ions.**

Chemical changes that are caused by the flow of electricity through a solution are examples of **electrolysis** reactions. The reaction vessel consisting of an electrolyte and two electrodes is called an **electrochemical cell.** This particular electrochemical cell, in which the electrodes are connected to a supply of electrical power that drives the chemical changes in the cell is an example of an **electrolytic cell.** There is another important category of electrochemical cells in which chemical reactions at the two electrodes cause electrical current to flow through an external circuit. Such cells are known as **voltaic cells.** Their internal chemistry produces electrical energy. Batteries are familiar examples of voltaic cells. How they work will be described in Chapter 17.

> ✓ A **net ionic equation** is a chemical equation in which the **spectator ions** not changed by the reaction have been eliminated, revealing only those species taking part in the chemical reaction.
>
> **Electrolysis** is a chemical reaction caused by the passage of an electrical current through the reactant.
>
> An **electrochemical cell** is a reaction system in which oxidation and reduction reactions at separate **electrodes** either consume (in an **electrolytic cell**) or produce (in a **voltaic cell**) electrical energy.

A **redox reaction** is an oxidation–reduction reaction.

Oxidation number, or *oxidation state,* is a positive or negative whole number equal to the number of charges that a chemically bonded atom would have if all the bonding electrons were transferred to the more electronegative atom.

Oxidation numbers

Oxidation–reduction, or **redox,** reactions can take place even without the aid of electrical energy and electrodes. Redox reactions can be driven by the chemical properties of substances that readily donate electrons to other substances or that tend to gain electrons for themselves. Consider the reactions of many metals with strong acids. A strip of zinc (Zn), for example, reacts with hydrochloric acid (HCl) with

> How do we know which reactant is being reduced and which is being oxidized in a redox reaction?

the rapid release of hydrogen gas (Figure 5.22). In the process, the zinc dissolves, forming soluble zinc chloride, $ZnCl_2$. We can write a chemical equation for this reaction as follows:

$$Zn(s) + 2\,HCl(aq) \longrightarrow ZnCl_2(aq) + H_2(g)$$

This equation is an example of a redox reaction. How do we know which reactant is being reduced and which is being oxidized in a redox reaction? One way to find out involves the use of **oxidation numbers.** Every atom has an oxidation number. The convention used for assigning oxidation numbers starts with the following rule:

All pure elements have an oxidation number of *zero.*

Thus, the oxidation number of zinc in zinc metal or hydrogen in H_2 gas is zero. In simple binary ionic compounds such as $ZnCl_2$, the oxidation numbers of the constituent elements are simply the electrical charges on their ions. Thus the oxidation numbers of zinc and chlorine in $ZnCl_2$ are $+2$ and -1, respectively.

Just as the sum of the charges on the ions in an ionic compound must add up to zero (all compounds are, after all, neutral), so, too, the sum of the oxidation numbers in any compound must add up to zero. For polyatomic ions, such as oxoanions, the sum of the oxidation numbers must add up to the charge of the ion. The hydroxide ion, for example, has a net charge of -1 because its oxygen atom has an oxidation number of -2 and the hydrogen bonded to it has an oxidation number of $+1$.

Assigning oxidation numbers to elements in molecular compounds is a bit more complicated because the elements do not actually have electrical charges. Instead, their oxidation numbers correspond to the electric charges that they *appear* to have. To determine these apparent charges, we need to use the following guidelines:

- The oxidation number of oxygen is nearly always -2, analogous to the -2 charge on oxide ions in ionic compounds. The principal exception is the peroxide ion, O_2^{2-}, in which each oxygen has a -1 oxidation number. The more common -2 oxidation state is also the more chemically stable of the two. The oxygen atoms in peroxides readily engage in chemical reac-

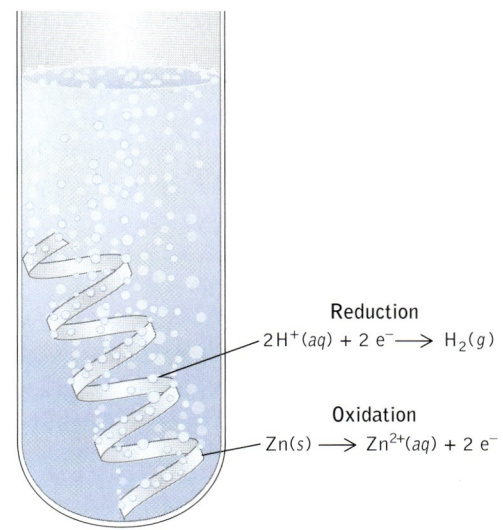

FIGURE 5.22 Zinc metal dissolves in aqueous HCl to give zinc ions and hydrogen gas. Zinc metal is oxidized to Zn^{2+} and H^+ is reduced to H_2 gas.

tions in which they are reduced from the -1 to the -2 oxidation state. If peroxide oxygen is to be reduced, another substance must be oxidized, and so peroxides are said to be effective **oxidizing agents.** In these reactions, the oxidation state of oxygen changes from -1 to -2 as each oxygen atom accepts an electron from the substance being oxidized.

- The oxidation state of hydrogen is usually $+1$. It is so as long as hydrogen is bonded to an element from the right-hand side of the periodic table, including those elements to which it is usually bonded in nature and, in particular, in biological systems. In some synthetic compounds, such as the metal hydrides described in Section 12.7, hydrogen acquires an electron and forms the hydride ion, H^-. Hydride ions are not very stable, especially in contact with water, and readily lose an electron, forming atomic hydrogen and hydrogen gas. Their considerable electron donating power makes hydrides extremely effective **reducing agents.**
- After assigning -2 to oxygen and $+1$ to hydrogen, assign oxidation numbers to the other elements in a compound so that the sum of all oxidation numbers for all the elements in the compound add up to zero. In HCl, for example, the oxidation state of hydrogen is $+1$, which means that the chlorine in HCl must have an oxidation number of -1 to balance hydrogen's $+1$. If we need to assign oxidation numbers to more than one element in this step and if the compound is ionic, we use oxidation numbers for single-atom ions that correspond to the electrical charge that these ions would normally have in binary ionic compounds.

> ☑ In a redox reaction, the **reducing agent** provides the electrons and the **oxidizing agent** accepts them. Thus, the reducing agent is oxidized and the oxidizing agent is reduced.

SAMPLE EXERCISE 15.5: Determine the oxidation state of sulfur in (a) sulfur dioxide, SO_2, (b) sodium sulfide, Na_2S, and (c) calcium sulfate, $CaSO_4$.

SOLUTION:

a. We assign an oxidation number of −2 to oxygen and let the oxidation number of sulfur be x. The sum of the oxidation numbers of the sulfur atom and two oxygen atoms must add up to zero; so $x + 2(-2) = 0$ and $x = +4$.

b. Sodium is present in all its compounds as Na^+, which means that the oxidation number of Na is +1. To balance the oxidation numbers in Na_2S, we can use the following equality, letting the oxidation number of sulfur be x.

$$2(+1) + x = 0; x = -2$$

c. The charge on the calcium ion is always +2, which means that the charge on the sulfate ion (SO_4^{2-}) must be −2. Assigning oxygen an oxidation number of −2 and letting x represent the oxidation number of sulfur, we have the following equality for sulfate ion:

$$x + 4(-2) = -2; x = +6$$

PRACTICE EXERCISE: Determine the oxidation state of nitrogen in (a) NO_2, (b) N_2O, and (c) HNO_3. (See Problems 89 and 90.)

Balancing redox reactions

Redox reactions abound in nature. Many, including photosynthesis and respiration, take part in the production and transfer of energy in living systems. Others take place when materials from reducing environments devoid of molecular oxygen mix with natural waters that contain O_2. Consider, for example, the unique ecosystems that were discovered in the deep oceans near geothermal vents in the 1970s (Figure 5.23). The food chains for the communities growing near these vents start not with photosynthesis but with chemical and biochemical reactions as elements in low oxidation states are oxidized to higher oxidation states. A good example is sulfur. As hot magma intrudes into the water-laden layers of rocks under the seafloor, sulfur may react with water and be liberated from these rocks as volatile hydrogen sulfide, H_2S. What is the oxidation number of sulfur in H_2S? If we use the convention that the oxidation number of hydrogen is +1 when combined with elements from the right-hand side of the periodic table, the oxidation number of sulfur in H_2S must be −2.

As H_2S percolates toward the seafloor, it encounters water containing molecular oxygen and undergoes a series of oxidation reactions leading to the formation of elemental sulfur, S, then thiosulfate, $S_2O_3^{2-}$, and, finally, sulfate, SO_4^{2-}, which is the chemical form of sulfur that is the most stable in seawater. In this series of reactions, the oxidation number of S increases from −2 to 0 to +2 to +6. Let's try writing chemical equations for some of these redox reactions, beginning with the reaction of H_2S with O_2 that yields elemental S. Let's start with a half-reaction that includes molecular oxygen:

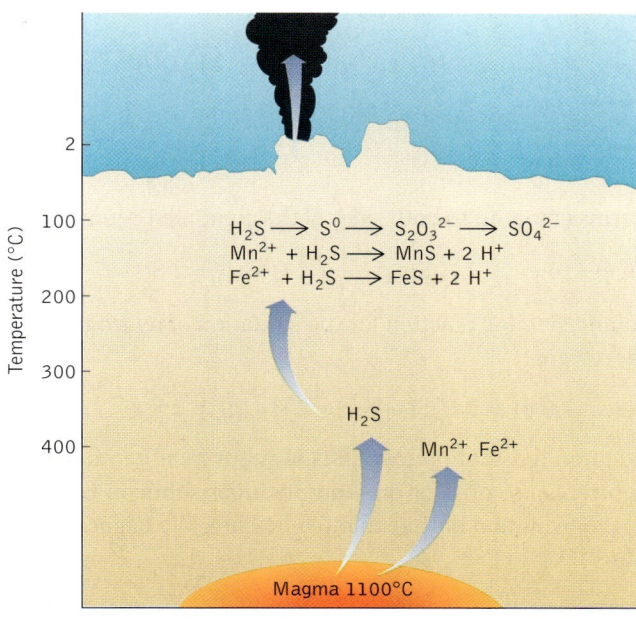

FIGURE 5.23 *What a Chemist Sees.* Geothermal vents at the bottom of the ocean release black clouds of metal sulfides formed by precipitation reactions between H_2S and such metal ions as Fe^{2+} and Mn^{2+}. H_2S mixes with seawater and undergoes a series of oxidation–reduction reactions that yield S^0, $S_2O_3^{2-}$, and SO_4^{2-}.

$$O_2(aq) + 4\,H^+(aq) + 4\,e^- \longrightarrow 2\,H_2O(l)$$

This half-reaction includes H^+, which is appropriate because the reaction takes place in somewhat acidic conditions where there is an excess of H^+ (see Section 4.3). To develop a half-reaction for the oxidation of H_2S to elemental S, we begin with the two species in an unbalanced equation:

$$H_2S(aq) \longrightarrow S(s)$$

Because this equation describes an oxidation half-reaction, one of the products will be electrons. How many electrons will depend on the change in oxidation state for sulfur. Because the change is from -2 to 0, there must be a loss of two electrons for each sulfur atom:

$$H_2S(aq) \longrightarrow S(s) + 2\,e^-$$

We are left with the need to balance two negative charges on the right-hand side of the equation and the need to add two hydrogen atoms to balance the two hydrogen atoms on the left-hand side. Both needs are satisfied if we add two hydrogen ions to the right-hand side, giving us a balanced half-reaction:

$$H_2S(aq) \longrightarrow S(s) + 2\,H^+(aq) + 2\,e^-$$

Now let's combine this oxidation half-reaction with the reduction half-reaction of O_2. We can do so only if we balance the number of electrons gained and lost in the two half-reactions. Four electrons are consumed in the reduction of O_2, but only two electrons are produced in the oxidation of H_2S. Therefore, we will need to multiply the H$_2$S half-reaction by two and then add the two half-reactions together:

$$O_2(aq) + 4\,H^+(aq) + 4\,e^- \longrightarrow 2\,H_2O(l)$$
$$+\ 2[H_2S(aq) \longrightarrow S(s) + 2\,H^+(aq) + 2\,e^-]$$
$$O_2(aq) + 4\,H^+(aq) + 4\,e^- + 2\,H_2S(aq) \longrightarrow 2\,H_2O(l) + 2\,S(s) + 4\,H^+(aq) + 4\,e^-$$

Canceling the terms common to both sides of the combined equation, we have

$$O_2(aq) + \cancel{4H^+(aq)} + \cancel{4e^-} + 2\,H_2S(aq) \longrightarrow 2\,H_2O(l) + 2\,S(s) + \cancel{4H^+(aq)} + \cancel{4e^-}$$

and have the balanced redox reaction for the oxidation of hydrogen sulfide to sulfur by molecular oxygen:

$$O_2(aq) + 2\,H_2S(aq) \longrightarrow 2\,H_2O(l) + 2\,S(s)$$

There are additional oxidation reactions of sulfur near geothermal vents and elsewhere in nature. Let's consider one that includes sulfur in its highest oxidation state (+6): the oxidation of thiosulfate to sulfate. We begin with the following sulfur species:

$$S_2O_3^{2-}(aq) \longrightarrow SO_4^{2-}(aq)$$

Let's first balance the number of sulfur atoms on both sides of the reaction arrow:

$$S_2O_3^{2-}(aq) \longrightarrow 2\,SO_4^{2-}(aq)$$

We then add the appropriate number of electrons as defined by the change in oxidation state of sulfur. The sulfur in sulfate is in the +6 oxidation state, but it is only in the +2 state in thiosulfate. Therefore, each sulfur atom will lose 4 electrons, or 8 electrons for two of them:

$$S_2O_3^{2-}(aq) \longrightarrow 2\,SO_4^{2-}(aq) + 8\,e^-$$

Now we need to balance the 10 extra negative charges on the right-hand side. (Can you see why there are 10?) We could add 10 hydrogen ions, except for one thing. If this final oxidation step takes place in seawater, as it likely does, then it takes place not in an acidic environment, but rather in a slightly basic one. Under these conditions, there will not be an excess of H$^+$ present; rather there will be an excess of OH$^-$, as will be discussed later in this chapter. This situation gives us another way to balance the electrical charges on the electrons: instead of adding 10 H$^+$ to the right-hand side, we can add 10 OH$^-$ to the left-hand side:

$$S_2O_3^{2-}(aq) + 10\,OH^-(aq) \longrightarrow 2\,SO_4^{2-}(aq) + 8\,e^-$$

This addition leaves us with 10 hydrogen atoms on the left-hand side but none on the right-hand side. Adding five molecules of water to the right-hand side solves this problem:

$$S_2O_3^{2-}(aq) + 10\,OH^-(aq) \longrightarrow 2\,SO_4^{2-}(aq) + 5\,H_2O(l) + 8\,e^-$$

If our work has been done correctly, the number of oxygen atoms should be balanced. Counting them up, we find that there are a total of 13 on the left-hand side and 13 on the right-hand side, and we have a balanced half-reaction.

$$2\,S_2^{2-}(s) + 2\,H_2O(l) + 7\,O_2(g) \longrightarrow$$
$$4\,SO_4^{2-}(aq) + 4\,H^+(aq)$$

The mineral pyrite, FeS$_2$, is often found in gold mines and is called fool's gold because of its luster. The S$_2^{2-}$ anion in pyrite is oxidized to SO$_4^{2-}$ by oxygen dissolved in water.

Now we can combine this oxidation half-reaction with the reduction half-reaction of O_2. However, the half-reaction that we used previously

$$O_2(aq) + 4\,H^+(aq) + 4\,e^- \longrightarrow 2\,H_2O(l)$$

assumes an acidic environment. To make it relevant to a slightly basic environment, we can remove the hydrogen ions by adding 4 OH^- to both sides of the reaction arrow.

$$O_2(aq) + 4\,H^+(aq) + 4\,OH^-(aq) + 4\,e^- \longrightarrow 2\,H_2O(l) + 4\,OH^-(aq)$$

This approach works because the following reaction takes place between acids and bases:

$$H^+(aq) + OH^-(aq) \longrightarrow H_2O(l)$$

Thus, the half-reaction becomes

$$O_2(aq) + 4\,H_2O(l) + 4\,e^- \longrightarrow 2\,H_2O(l) + 4\,OH^-(aq)$$

which simplifies to

$$O_2(aq) + 2\,H_2O(l) + 4\,e^- \longrightarrow 4\,OH^-(aq)$$

Combining this half-reaction with that for the oxidation of thiosulfate, we have

$$S_2O_3^{2-}(aq) + 10\,OH^-(aq) \longrightarrow 2\,SO_4^{2-}(aq) + 5\,H_2O(l) + 8\,e^-$$
$$+\ 2\,[O_2(aq) + 2\,H_2O(l) + 4\,e^- \longrightarrow 4\,OH^-(aq)]$$

$$S_2O_3^{2-}(aq) + 10\,OH^-(aq) + 2\,O_2(aq) + 4\,H_2O(l) + 8\,e^- \longrightarrow 2\,SO_4^{2-}(aq)$$
$$+\ 5\,H_2O(l) + 8\,OH^-(aq) + 8\,e^-$$

which simplifies to the following balanced equation:

$$S_2O_3^{2-}(aq) + 2\,OH^-(aq) + 2\,O_2(aq) \longrightarrow 2\,SO_4^{2-}(aq) + H_2O(l)$$

SAMPLE EXERCISE 5.16: In a classical method for determining the concentration of Fe^{2+} in acidic solutions, it is reacted with a solution containing the permanganate ion, MnO_4^-. Starting with the following half-reactions, write a balanced equation describing the reaction between these two ions.

Given: $MnO_4^-(aq) + 8\,H^+(aq) + 5\,e^- \longrightarrow Mn^{2+}(aq) + 2\,H_2O(l)$

$Fe^{2+}(aq) \longrightarrow Fe^{3+}(aq) + e^-$

SOLUTION: Before we can combine these two half-reactions, we need to multiply the iron half-reaction by five so that the number of electrons gained and lost is the same. Then the two half-reactions can be combined.

$$MnO_4^-(aq) + 8\,H^+(aq) + 5\,e^- \longrightarrow Mn^{2+}(aq) + 2\,H_2O(l)$$
$$+\ 5\,[Fe^{2+}(aq) \longrightarrow Fe^{3+}(aq) + e^-]$$

$$5\,Fe^{2+}(aq) + MnO_4^-(aq) + 8\,H^+(aq) \longrightarrow 5\,Fe^{3+}(aq) + Mn^{2+}(aq) + 2\,H_2O(l)$$

PRACTICE EXERCISE: Write a balanced chemical equation describing the oxidation of elemental sulfur, S_8, to thiosulfate, $S_2O_3^{2-}$, by molecular oxygen in a basic solution. (See Problems 97 and 98.)

5.6 ACID–BASE REACTIONS AND NET IONIC EQUATIONS

CONNECTION: The early atmosphere of Earth was rich in CO_2, SO_2, SO_3, and NO_2 (Section 4.3).

As noted in Chapter 4, volcanic activity introduces a variety of gases into the atmosphere, including volatile oxides of sulfur, nitrogen, and other nonmetals. These oxides form compounds (see the following table) that serve as hydrogen ion (H^+) donors when they dissolve in rainwater. On the other hand, Earth's crust contains many metal oxides and other basic compounds that accept hydrogen ions. Acidic compounds falling from the sky in the form of rain engage in a variety of chemical reactions with the basic minerals of Earth's crust. These reactions are examples of an important class of reactions—**acid–base** neutralization reactions—and they play key roles in the chemical transformations of rocks and minerals and in the formation of soil from them that are known as **chemical weathering**.

✓ An **acid** is a substance that can donate H^+ to water and other substances. A **base** is a substance that accepts H^+ and so can produce an excess of OH^- in aqueous solutions.

Chemical weathering refers to chemical reactions of constituents of the atmosphere (often dissolved in rain) with rocks and minerals. As a result parts of the rocks and minerals dissolve.

Formulas of Volatile Nonmetal Oxides and Acids Formed When the Oxides Dissolve in Water

Nonmetal oxide	Acid
SO_2, SO_3	H_2SO_3, H_2SO_4
NO_2	HNO_2, HNO_3
CO_2	H_2CO_3

CONCEPT TEST: Using the naming conventions discussed in Chapter 4 (see Figure 4.8), name the acids in the preceding table.

Some of the acids in the table completely dissociate into ions when they dissolve in water; others do not. Those that do are classified as strong acids and include HNO_3 and H_2SO_4. These acids are so strong that they can produce other strong acids. For example, if HNO_3 and H_2SO_4 encounter Cl^-, the following reactions take place, producing hydrochloric acid (HCl):

$$HNO_3(aq) + Cl^-(aq) \longrightarrow NO_3^-(aq) + HCl(aq)$$

$$H_2SO_4(aq) + Cl^-(aq) \longrightarrow HSO_4^-(aq) + HCl(aq)$$

Note that only one of the two hydrogen atoms in H_2SO_4 was ionized, because H_2SO_4 ionizes in two steps. The first step,

$$H_2SO_4(aq) \longrightarrow HSO_4^-(aq) + H^+(aq)$$

is complete, but the second one is not. The equation for the second step is often drawn with two reaction arrows:

$$HSO_4^-(aq) \rightleftharpoons SO_4^{2-}(aq) + H^+(aq)$$

to indicate that not all of the HSO_4^- forms ions. The fact that the arrows point from right to left as well as from left to right is a signal that the reaction may run in reverse, forming HSO_4^- from SO_4^{2-} and H^+. In fact, the forward reaction and the reverse reaction can happen at the same time: some HSO_4^- ions undergo further ionization, forming SO_4^{2-} and H^+, while, simultaneously, other pairs of SO_4^{2-} and H^+ are combining to make HSO_4^-. The net effect of these reactions running in the opposite directions is a **chemical equilibrium** in which the rates of the forward and reverse reactions are in balance and the concentrations of the reactants and products do not change with time. We will address the matter of reactions at equilibrium in detail in Chapters 15 and 16. For now, we need to keep in mind that weak acids, including H_2SO_3, HNO_2, H_2CO_3, and HSO_4^-, only partly dissociate into ions in solution, whereas strong acids, including HNO_3, and HCl, are completely ionized.

 When a reaction reaches **chemical equilibrium,** the rate of formation of products from reactants is equal to the rate of formation of reactants from products.

How do acids and bases neutralize each other? Let's examine just a few of the many neutralization reactions that may take place when strong acids come in contact with basic minerals. Included in the latter category is a mineral called gibbsite. It is formed in tropical climates from the chemical weathering of other minerals and has the chemical formula $Al(OH)_3$. Consider what would happen if water containing hydrochloric acid (HCl) came in contact with $Al(OH)_3$. HCl is a strong acid and therefore an effective H^+ donor. In solution, HCl is completely ionized:

> **How do acids and bases neutralize each other?**

CONNECTION: Chapter 15 will explore the concept of chemical equilibrium in more depth, but for now it is sufficient to think of it as a balance between forward and reverse reactions.

$$HCl(aq) \longrightarrow H^+(aq) + Cl^-(aq)$$

$Al(OH)_3$ is a basic compound because it contains hydroxide ions (OH^-), and OH^- is an excellent H^+ acceptor. We can write a simple, but significant, equation for reaction between H^+ and OH^-:

$$H^+(aq) + OH^-(aq) \longrightarrow H_2O(l)$$

The important message embedded in this equation is that there must be one mole of hydrogen ion donors for every mole of hydrogen ion acceptors in a balanced neutralization reaction. Let's apply this principle to the reaction between HCl and $Al(OH)_3$. There are three moles of hydroxide ions for every mole of $Al(OH)_3$. Therefore, there must be three moles of H^+ to neutralize them and thus three moles of HCl in the reaction for every one mole of $Al(OH)_3$. Putting this information into equation form, we have

$$3\ HCl(aq) + Al(OH)_3(s) \longrightarrow AlCl_3(aq) + 3\ H_2O(l)$$

Antacids for neutralizing excess stomach acid include milk of magnesia, which is a suspension of $Mg(OH)_2$.

SAMPLE EXERCISE 5.17: Complete and balance the neutralization reaction between H_2SO_4 and the mineral brucite, $Mg(OH)_2$. Assume that H_2SO_4 donates two hydrogen ions per molecule in the reaction.

SOLUTION: One mole of H_2SO_4 will release two moles of H^+ in solution. Each mole of $Mg(OH)_2$ contains two moles of OH^-. Therefore the number of H^+ donors and acceptors will be balanced (at two moles) if one mole of H_2SO_4 reacts with one mole of $Mg(OH)_2$. The resulting equation is

$$H_2SO_4(aq) + Mg(OH)_2(s) \longrightarrow MgSO_4(aq) + 2\,H_2O(l)$$

PRACTICE EXERCISE: Complete and balance the neutralization reaction between a solution of HNO_3 and solid $Mg(OH)_2$. (See Problems 113 and 114.)

Now let's consider a reaction in which a weak acid is a reactant: reactions responsible for the formation of caverns in geological formations made of limestone. Limestone is mostly $CaCO_3$, and the reaction between it and H_2CO_3 can be written as follows:

$$CaCO_3(s) + H_2CO_3(aq) \longrightarrow Ca(HCO_3)_2(aq)$$

The reaction results in the dissolution of the rock and may result in the formation of caverns of mammoth proportions. The product of this reaction is an aqueous solution of calcium hydrogencarbonate, also called calcium *bi*carbonate. If the calcium-rich solution produced by this reaction seeps out of the ground and evaporates, it will likely leave behind residues of limestone as the preceding reaction essentially runs in reverse, driven by the loss of carbon dioxide and water vapor:

$$Ca(HCO_3)_2(aq) \longrightarrow CaCO_3(s) + CO_2(g) + H_2O(g)$$

CONNECTION: The rates of chemical reactions are discussed in Chapter 13.

Like many other reversible chemical reactions, this one appears to go in a particular direction when the rate of the reaction in that direction is much higher than the rate in the opposite direction. For carbonic acid and calcium carbonate, the forward reaction is observed and limestone dissolves when the rate at which it dissolves is greater than the rate at which it re-forms as a result of the reaction running in reverse. When the opposite is true, more solid calcium carbonate forms than dissolves. This happens as calcium-rich water evaporates as it drips from the ceilings of limestone caves. The resulting deposits of $CaCO_3$ may grow into geological structures known as stalactites and stalagmites (Figure 5.24).

There are many other important acid–base reactions in nature. Human activities, particularly the burning of fossil fuels, add to the formation of strong acids in the atmosphere and to the phenomenon known as **acid rain.** We will consider the acid-rain problem in much more detail in Chapter 16. For now, we will focus on some of the methods used to combat the effects of acid rain.

The environmental problems posed by acid rain are connected to the strengths of the acids in acid rain. All rain contains carbonic acid as a result of

Acid rain is highly acidic rain owing to pollutant nonmetal oxides and other acidic compounds dissolved in it.

$$Ca(HCO_3)_2(aq) \longrightarrow CaCO_3(s) + CO_2(g) + H_2O(g)$$

FIGURE 5.24 In Carlsbad Caverns (New Mexico), white stalagmites of limestone grow up from the cavern floor as groundwater containing CaHCO$_3$ drips from the stalactites above them and evaporates, forming solid CaCO$_3$.

The reefs found in tropical seas consist of carbonate minerals, principally CaCO$_3$ and MgCO$_3$. These minerals are formed by corals from Ca^{2+}, Mg^{2+}, and HCO$_3^-$ dissolved in seawater.

Atmospheric sulfuric acid attacks marble (mostly CaCO$_3$) statues. The sulfate products are more soluble than carbonates and are slowly washed away by rain and snow.

the dissolution of atmospheric CO$_2$. But carbonic acid is a weak acid and not completely ionized in solution. On the other hand, sulfuric acid is a strong acid. Its presence in rainwater means that the concentration of hydrogen ions in the rain is much greater than it would be if only carbonic acid were present.

 CONNECTION: Section 16.1 addresses the concept of acid strength and the effects of acid precipitation.

CONCEPT TEST: When ammonia dissolves in water, the following reaction takes place:

$$NH_3(g) + H_2O(l) \longrightarrow NH_4^+(aq) + OH^-(aq)$$

Which reactant is acting as an acid (H$^+$ donor) and which is acting as a base (H$^+$ receptor)?

THE CHEMISTRY OF THE ALKALINE EARTH METALS

The elements in Group 2A of the periodic table are called the alkaline earth metals. Like their Group 1A neighbors, the Group 2A elements are very reactive, and none are found in nature as the free metal. All of the isotopes of radium (Ra) are radioactive and were discussed in Chapter 2. Calcium (Ca) and magnesium (Mg) are the fifth and sixth most abundant elements in Earth's crust: Ca makes up 3.6% of the crust and magnesium about 2.5%, by weight. Both elements are widely found as their carbonates, $CaCO_3$ and $MgCO_3$, in a variety of minerals such as chalk, limestone, and marble. Calcium is also found as its hydrated sulfate in gypsum ($CaSO_4 \cdot 2\ H_2O$), and Mg is present as its hydroxide, $Mg(OH)_2$, in brucite. The Dolomite Range in Italy takes its name from the mineral dolomite, which is a mixture of the Ca and Mg carbonates.

The Dolomite Range in Italy is rich in calcium and magnesium carbonate minerals known as dolomite.

With the exception of beryllium, all of the alkaline earth metals form compounds in which their atoms have lost the two *s* electrons in their outermost shells, forming ions with +2 charges. Beryllium behaves differently, sharing electrons rather than losing them, which leads to the formation of covalent rather than ionic Be compounds (Chapters 6 and 7).

Metallic alkaline earth elements are typically prepared by the electrolysis of molten chloride salts. which may be obtained by the evaporation of seawater. Magnesium metal forms homogeneous mixtures with aluminum and titanium, forming strong, yet lightweight, metals that are used in, for example, airplanes and automobiles. The other alkaline earth metals are more likely to be used in compounds rather than as pure elements. Gypsum is a widely used building material (the principal ingredient in "wall board" and plaster), and minerals rich in calcium phosphate are the raw materials for phosphate fertilizers. Among the commonly encountered compounds of Mg are its hydrated sulfate ($MgSO_4 \cdot 7\ H_2O$), also known as Epsom salts, and aqueous suspensions of $Mg(OH)_2$, called milk of magnesia, which is widely used in antacids and laxatives. The most common barium mineral is barite, mostly in the form of $BaSO_4$. It is used to adjust the density of the fluids called drilling muds that are used in drilling for oil, and suspensions of $BaSO_4$ are ingested by patients suffering from intestinal disorders. When the suspension reaches the intestines, X-ray images are taken of these organs, which are filled with X-ray–absorbing ions of Ba.

Oxides of the Group 2A elements can be readily prepared by heating the respective carbonates. The reaction for calcium carbonate may be written as follows

$$CaCO_3(s) \longrightarrow CaO(s) + CO_2(g)$$
(limestone)   (lime)

Calcium oxide, also known as lime, is used in agriculture to reduce the acidity of soils. It is also used in water treatment and as an absorber of sulfur dioxide in the smokestacks of coal-burning factories and power plants.

$$CaO(s) + SO_2(g) \longrightarrow CaSO_3(s)$$

The alkaline earth metals react directly with nitrogen to form nitrides:

$$3\ Mg(s) + N_2(g) \longrightarrow Mg_3N_2(s)$$

Nitrides belong to a class of materials known as ceramics and will be discussed in Chapter 18. Alkaline earth metal nitrides are largely ionic compounds and so include the N^{3-} anion in their structures.

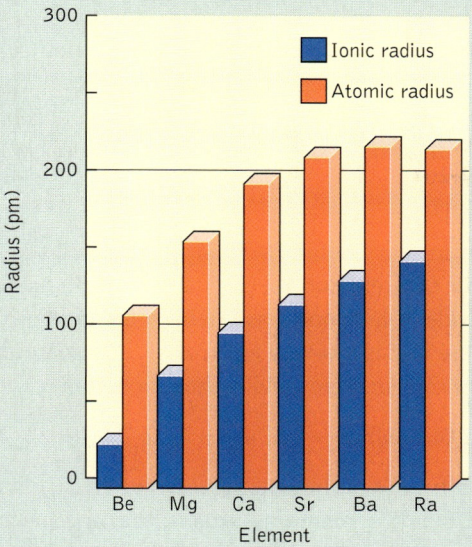

The atomic and ionic radii of the Group 2A elements increase down the group reflecting the increase in atomic number. An increase in atomic number corresponds to population of higher energy orbitals which are found farther from the nucleus.

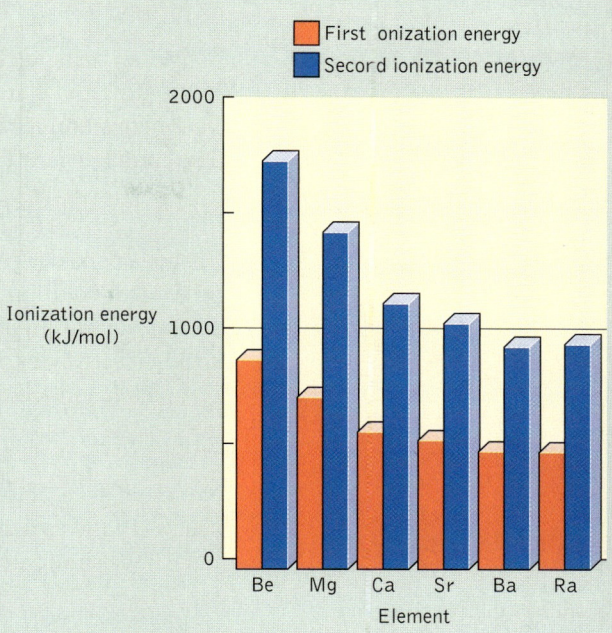

The ionization energies of the Group 2A elements are larger than for the Group 1A elements as a result of their smaller size and greater effective nuclear charge. The decrease in ionization energy down the group reflects the shielding of the outer electrons from the nuclear charge by the inner electrons. The larger second ionization energies reflect the greater attraction of the electrons in the monocation to the nucleus.

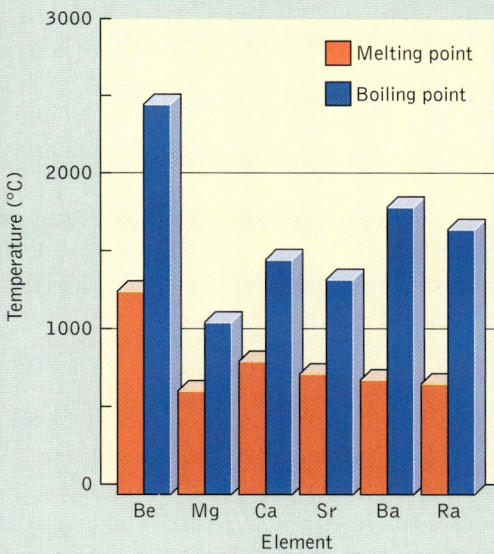

As with the Group 1A elements, there is a general trend toward lower boiling and melting points down the group. This is explained by weaker metallic bonds (Chapter 10) between the elements as size increases; however, anomalies arise for the melting point of Mg. Like lithium, the small beryllium has a particularly high boiling point compared to the other members of its group. The trend in boiling points for the other Group 2A elements does not follow a smooth periodic trend.

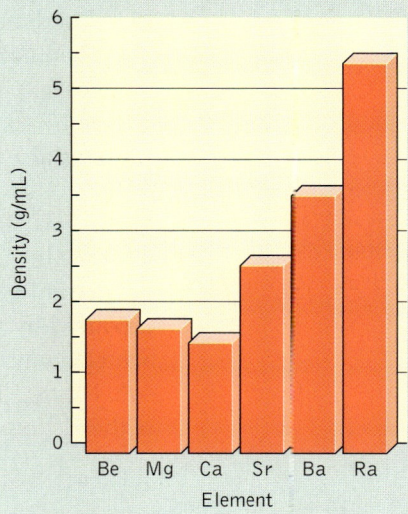

The densities of the Group 2A elements follow a general trend toward increasing density down the group. The trend is not smooth, with the densities of magnesium and calcium being slightly less than the density of beryllium.

When acid precipitation acidifies a public water supply, it can be neutralized by treating the water with potassium hydroxide, KOH. The hydroxides of the alkali metals and alkaline earth elements are all effective hydrogen ion acceptors. The complete neutralization of sulfuric acid by KOH can be written as follows:

$$H_2SO_4(aq) + 2\ KOH(aq) \longrightarrow K_2SO_4(aq) + 2\ H_2O(l)$$

Let's consider the last equation in more detail by writing out all of the ions present in solution before and after the reaction takes place. Sulfuric acid is a strong acid and potassium hydroxide is a strong base, which means both are strong electrolytes. Potassium sulfate also is a strong electrolyte, and so it, too, is completely ionized. Therefore, the complete ionic view of the reaction is

$$2\ H^+(aq) + SO_4^{2-}(aq) + 2\ K^+(aq) + 2\ OH^-(aq) \longrightarrow 2\ K^+(aq) + SO_4^{2-}(aq) + 2\ H_2O(l)$$

Note that several ions appear on both sides of the reaction arrow. These ions are species that were unaffected by the chemical reaction—that is, they are spectator ions. If we eliminated them from the preceding equation, we would be left with

$$2\ H^+(aq) + 2\ OH^-(aq) \longrightarrow 2\ H_2O(l)$$

or simply

$$H^+(aq) + OH^-(aq) \longrightarrow H_2O(l)$$

The preceding equation is the net ionic equation for the neutralization reaction between sulfuric acid and potassium hydroxide. Actually, it could be the net ionic equation for any one of many neutralization reactions between a strong acid and a strong base. In these reactions, hydrogen ions from acids combine with hydroxide ions from bases, producing water. The anion of the acid and the cation of the base are present before and after the reaction takes place and are therefore simply spectator ions.

When acid rain acidifies the water in swimming pools, the compound most often used to neutralize it is sodium carbonate, Na_2CO_3, also known as soda ash. Assuming that the acid in acid rain is principally sulfuric acid, we can write a **molecular equation** for the neutralization reaction as follows:

$$H_2SO_4(aq) + Na_2CO_3(s) \longrightarrow Na_2SO_4(aq) + H_2O(l) + CO_2(g)$$

To write the net ionic equation for this reaction, we need to consider which ions are present before and after the reaction takes place. The ions present are those produced by the ionization of any strong acids or bases and the dissociation of soluble ionic salts. In this case, there is one strong acid, H_2SO_4, and one soluble salt, Na_2SO_4. Modifying the preceding equation to include the ionic components of these compounds, we have

$$2\ H^+(aq) + SO_4^{2-}(aq) + Na_2CO_3(s) \longrightarrow 2\ Na^+(aq) + SO_4^{2-}(aq) + H_2O(l) + CO_2(g)$$

We can simplify this equation by eliminating the ingredients that are present on both sides of the reaction arrow—namely, $SO_4^{2-}(aq)$, which leaves us with the following net ionic equation:

$$2\ H^+(aq) + Na_2CO_3(s) \longrightarrow 2\ Na^+(aq) + H_2O(l) + CO_2(g)$$

An antacid tablet containing calcium carbonate reacts with stomach acid (HCl), producing $CO_2(g)$, water, and $CaCl_2(aq)$.

✓ A **molecular equation** is a chemical equation in which the formulas of all reactants and products are expressed as if they were intact molecules, even when they are not.

SAMPLE EXERCISE 5.18: Write the net ionic equation for the reaction between a solution of nitric acid and solid calcium carbonate.

SOLUTION: Let's begin by writing the molecular equation for the reaction:

$$2\ HNO_3(aq) + CaCO_3(s) \longrightarrow Ca(NO_3)_2(aq) + H_2O(l) + CO_2(g)$$

The equation containing all the ions in the reactants and products looks like this:

$$2\ H^+(aq) + 2\ NO_3^-(aq) + CaCO_3(s) \longrightarrow Ca^{2+}(aq) + 2\ NO_3^-(aq) + H_2O(l) + CO_2(g)$$

Canceling the nitrate ions that are common to both sides, we have the following net ionic equation:

$$2\ H^+(aq) + CaCO_3(s) \longrightarrow H_2O(l) + CO_2(g) + Ca^{2+}(aq)$$

PRACTICE EXERCISE: Write the net ionic equation for the reaction between hydrochloric acid and solid calcium hydroxide. (See Problems 121 and 122.)

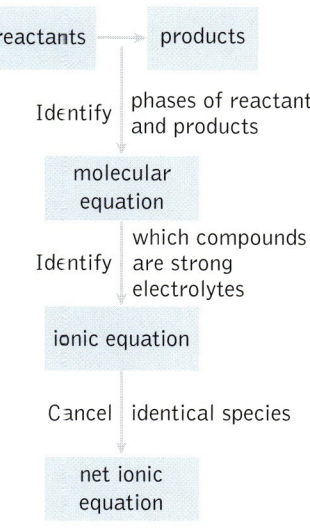

5.7 PRECIPITATION REACTIONS

The reaction that results in the formation of stalactites and stalagmites in caverns

$$Ca(HCO_3)_2(aq) \longrightarrow CaCO_3(s) + CO_2(g) + H_2O(g)$$

is an example of a **precipitation reaction.** These solution reactions take place when reactants that are soluble form one or more products that are much less soluble. As these products form solid phases and fall out of solution, they **precipitate.** *The Handbook of Chemistry and Physics* published by the Chemical Rubber Company (CRC) contains solubility data on thousands of ionic compounds. These solubility values are expressed as the number of grams of each compound that dissolves in 100. cubic centimeters (a cubic centimeter is the same as a milliliter) of water. The solubility of NaCl, for example, is 35.7 g/100 mL at 0°C. This value is about 10 times the salt concentration in seawater. If a solution contains the maximum concentration of solute, it is said to be **saturated** (Figure 5.25). Thus, seawater is **unsaturated** with respect to NaCl (but not necessarily with respect to compounds that are much less soluble than NaCl, such as $CaCO_3$).

Sometimes more solute temporarily dissolves than is predicted by its solubility, creating a **supersaturated** solution. This can happen when the volume of an unsaturated solution is reduced through evaporation or when the temperature of a solution changes. For example, more sugar dissolves in hot water than in cold water, as anyone who has tried to sweeten a glass of iced tea knows. If the temperature of a saturated solution of sugar in water were lowered, the solution could be temporarily supersaturated. Sooner or later the solute in a supersaturated solution will come out of solution. If the solute is a solid, it will form a solid precipitate, as shown in Figure 5.26 on page 259. If the solute is a gas, as is the

☑ In a **precipitation reaction,** a solid product called a precipitate is formed from a reaction in solution.

CONNECTION: The solubility of gases in water is described by Henry's law (Section 8.5).

☑ A **saturated** solution contains the maximum amount of solute that can be dissolved in a given volume or mass of solvent. An **unsaturated** solution contains less than this amount of solute. A **supersaturated** solution temporarily contains more than this amount.

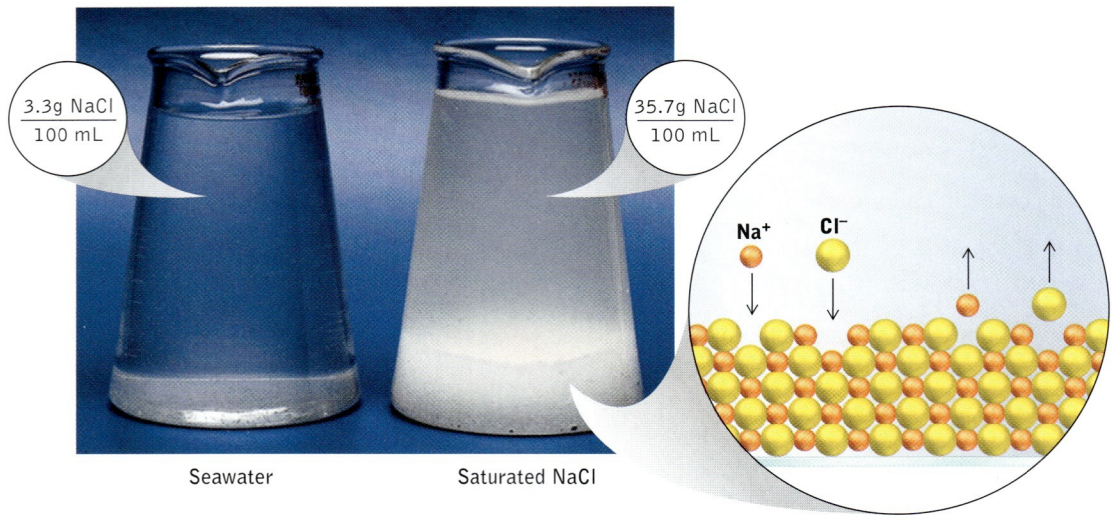

FIGURE 5.25 The seawater in the beaker on the left is a concentrated solution of NaCl, but it is far from a *saturated* solution of NaCl. The solution in the beaker on the right is in equilibrium with solid NaCl and so is a saturated solution of NaCl. The interface between the solution and the undissolved solid is constantly changing. Particles of solute constantly dissolve and precipitate, so the number of particles in the solution remains constant.

TABLE 5.3 Abundances of the Most Common Elements in Earth's Crust

Element	Weight (%)
O	45.5
Si	26.8
Al	8.40
Fe	7.06
Ca	5.3
Mg	3.2
Na	2.3
K	0.90

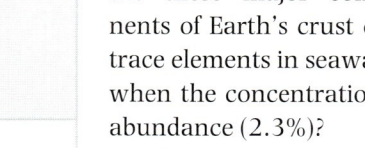

CONNECTION: The solubility trends for ionic compounds are related to the strengths of interactions between the ions and between them and molecules of water (Sections 9.2 and 13.1).

carbon dioxide in carbonated beverages, then bubbles of gas will start to form, just as they do inside a bottle of soft drink when it is opened and the pressure that keeps the CO_2 in solution is released.

A comparison of the composition of ocean water (Table 5.1) with that of Earth's crust (Table 5.3) reveals that some of the most abundant elements in the Earth's crust, including Si, Al, and Fe, are not among the most abundant elements in seawater. Why are these major components of Earth's crust only trace elements in seawater, when the concentration of sodium in seawater (1.1%) is nearly half its crustal abundance (2.3%)?

> Some of the most abundant elements in Earth's crust, including Si, Al, and Fe, are not among the most abundant elements in seawater. Why?

The concentrations of ions in seawater and in bodies of fresh water are related to the chemical reactivities of minerals in soils and rocks and the solubilities (or *in*solubilities) of the compounds in those minerals. Rocks and minerals must be made of compounds of limited solubility or they would not survive as solid rocks for millions of years. However, weathering reactions can chemically change minerals and result in the formation of soluble products. The following equations describe the chemical weathering of two examples of a class of minerals known as feldspars (Figure 5.27):

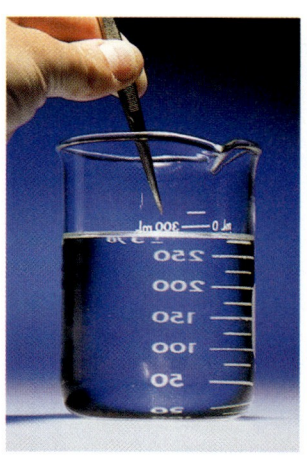
A

B

C

D

FIGURE 5.26 Sodium acetate precipitates from a supersaturated solution when: A. a seed crystal is added; B. the crystal becomes a site for rapid crystal growth as sodium acetate begins to crystallize; and C. and D. crystal growth continues until the solution is no longer supersaturated but is merely saturated with sodium acetate.

$2 \text{ NaAlSi}_3\text{O}_8(s) + 3 \text{ H}_2\text{O}(l) + 2 \text{ CO}_2(aq) \longrightarrow$
$\text{Al}_2\text{Si}_2\text{O}_5(\text{OH})_4(s) + 2 \text{ Na}^+(aq) + 2 \text{ HCO}_3^-(aq) + 4 \text{ SiO}_2(s)$

$2 \text{ KAlSi}_3\text{O}_8(s) + 3 \text{ H}_2\text{O}(l) + 2 \text{ CO}_2(aq) \longrightarrow$
$\text{Al}_2\text{Si}_2\text{O}_5(\text{OH})_4(s) + 2 \text{ K}^+(aq) + 2 \text{ HCO}_3^-(aq) + 4 \text{ SiO}_2(s)$

The products of the reaction include soluble sodium and potassium bicarbonate, as well as two other insoluble substances: kaolinite, $\text{Al}_2\text{Si}_5\text{O}_5(\text{OH})_4$, and silica, SiO_2. Sodium and potassium compounds formed in this way may percolate into the ground dissolved in groundwater or flow into lakes and rivers, and, eventually, the ocean. Note that the aluminum and silicon compounds formed from these reactions are insoluble and so remain behind after the soluble salts have washed away.

What makes some compounds, such as SiO_2, insoluble, whereas others, such as NaCl, readily dissolve in water? The reason is buried with the strength of the chemical bonds that hold them together. We will consider the nature of these bonds in detail in the next few chapters. For now, simply be aware that the solubilities of the geologically important compounds (and inorganic compounds in general) follow a trend. There are some important exceptions to the trend, but you may still find it useful. Most compounds made of cations with a +1 charge and anions with a −1 charge (e.g., NaCl) are soluble in water, including all the common compounds of the Group 1A elements. Most compounds made of +1

> Why are some compounds, such as SiO_2, insoluble, whereas others, such as NaCl, readily dissolve in water?

FIGURE 5.27 Rainwater and groundwater saturated with carbon dioxide are responsible for the weathering of many minerals, including this piece of granite. Chemical weathering has turned one surface of this pink granite green.

FIGURE 5.28 The solubilities of common ionic compounds often follow patterns. For example, lead nitrate (*left*) and lead acetate (*middle*) are soluble, but lead sulfide (*right*) is insoluble in water.

cations and −2 anions, as well as +2 cations and −1 anions, also are soluble. However, many compounds made of +2 cations and −2 anions or of +3 cations and −2 anions have very little solubility in water. This group includes the sulfides (S^{2-}) and oxides of most metals (Figure 5.28). This pattern does have some notable exceptions, such as the compounds formed by the −1 anions of Group 7, called halide ions, and Ag^+, Cu^+, and $Hg_2^{\,+}$. These compounds are insoluble, as are the hydroxides (OH^-) of most transition metals and aluminum. Still, it is generally true that the smaller the charges on the cation and anion of an ionic compound, the more soluble is the compound. Table 5.4 summarizes the solubility rules for common ionic compounds.

TABLE 5.4 Solubility Rules for Common Ionic Compounds

All compounds containing the following ions are soluble:
- Cations: Group 1A ions (alkali metals) and NH_4^+
- Anions: NO_3^- and $C_2H_3O_2^-$ (acetate)

Compounds containing the following anions are soluble except as noted:
- Group 7A ions (halides), except the salts of Ag^+, Cu^+, Hg_2^{2+}, and Pb^{2+}
- SO_4^{2-}, except the salts of Ba^{2+}, Ca^{2+}, Hg_2^{2+}, Pb^{2+}, and Sr^{2+}

All other compounds are insoluble except the following Group 2A (alkaline earth) hydroxides:
- $Ba(OH)_2$, $Ca(OH)_2$, and $Sr(OH)_2$

Chemists make use of the insolubility of ionic compounds to determine the concentrations of their constituent ions in solution. For example, in cold climates, NaCl is widely used to melt ice and snow on roads and sidewalks during the winter (by lowering the freezing point of water as discussed earlier). This salt seeps into the ground and may make its way into wells and reservoirs that serve as drinking-water supplies. To find out whether a water supply has been contaminated by road salt, analytical chemists may determine the concentration of chloride in a sample of it. They do so by adding a solution of silver nitrate, $AgNO_3$, to a known volume of the water sample. Chloride in the sample combines with Ag^+ and forms a precipitate of solid AgCl.

$$Cl^-(aq) + AgNO_3(aq) \longrightarrow AgCl(s) + NO_3^-(aq)$$

This precipitate can be filtered, dried, and weighed. From its mass, the concentration of chloride (and NaCl) in the water sample can be calculated, as illustrated in the following Sample Exercise.

SAMPLE EXERCISE 5.19: Write the net ionic equation for the precipitation reaction that takes place when a solution of $AgNO_3$ is added to a solution of NaCl.

SOLUTION: To begin, let's first write the balanced molecular equation for the reaction. Then we write a *total* ionic equation that includes all of the species present in solution before and after the reaction. The balanced molecular equation is

$$AgNO_3(aq) + NaCl(aq) \longrightarrow AgCl(s) + NaNO_3(aq)$$

The two reactants are both aqueous solutions of soluble ionic compounds. We assume that they are both totally ionized into their component ions. Thus, a solution of $AgNO_3$ is really an equal mixture of $Ag^+(aq)$ and $NO_3^-(aq)$, and a solution of NaCl is an equal mixture of $Na^+(aq)$ and $Cl^-(aq)$. Therefore, the reactants in their ionic forms are

$$Ag^+(aq) + NO_3^-(aq) + Na^+(aq) + Cl^-(aq)$$

The precipitate formed by the reaction is silver chloride, AgCl(s). Still in solution are $Na^+(aq)$ and $NO_3^-(aq)$. Thus, the products of the reaction are

$$AgCl(s) + Na^+(aq) + NO_3^-(aq)$$

Combining the reactant and product sides of the equation into a total ionic equation, we have

$$Ag^+(aq) + NO_3^-(aq) + Na^+(aq) + Cl^-(aq) \longrightarrow AgCl(s) + Na^+(aq) + NO_3^-(aq)$$

We can simplify this equation by eliminating the components that do not change—namely, $Na^+(aq) + NO_3^-(aq)$:

$$Ag^+(aq) + \cancel{NO_3^-(aq)} + \cancel{Na^+(aq)} + Cl^-(aq) \longrightarrow AgCl(s) + \cancel{Na^+(aq)} + \cancel{NO_3^-(aq)}$$

leaving as the net ionic equation,

$$Ag^+(aq) + Cl^-(aq) \longrightarrow AgCl(s)$$

PRACTICE EXERCISE: A white precipitate forms when a solution of Na_2SO_4 is added to a solution of $BaCl_2$.
  a. Identify the insoluble compound that is formed. (Hint: The positive and negative ions with the greatest charges are likely to be the ions that form the insoluble compound.)
  b. Write the net ionic equation for the precipitation reaction. (See Problems 120–124.)

SAMPLE EXERCISE 5.20: To determine the concentration of chloride in a sample of groundwater, a chemist adds 1.0 mL of 1.00 M $AgNO_3$ solution to 100. mL of the sample. The mass of the resulting precipitate of AgCl is 71.7 mg. What is the concentration of chloride in the original sample expressed in milligrams of chloride per liter?

SOLUTION: In this analysis, we assume that essentially all the chloride in the sample is precipitated as AgCl. Our task is to calculate how much chloride is contained within 71.7 mg of AgCl precipitate. The chemical formula, AgCl, tells us that there is one mole of Cl^- in every mole of AgCl. The mass of one mole of Cl^- is the same as the atomic weight of chlorine: 35.45 g. The molar mass of AgCl is the sum of the atomic masses of Ag and Cl, or 143.32 g. Therefore, the mass of Cl^- in 71.7 mg of AgCl is

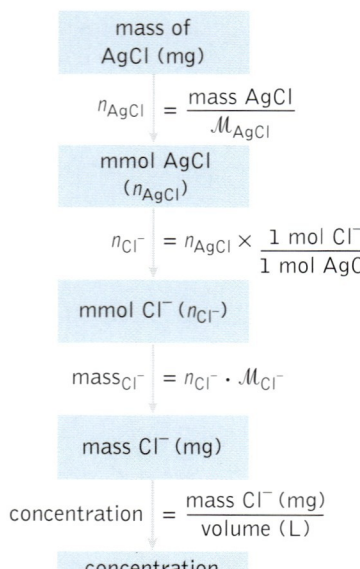

$$71.7 \text{ mg AgCl} \times \frac{1 \text{ mol AgCl}}{143.32 \text{ g AgCl}} \times \frac{1 \text{ mol Cl}^-}{1 \text{ mol AgCl}} \times \frac{35.45 \text{ g Cl}^-}{\text{mol Cl}^-} = 17.73 \text{ mg Cl}^-$$

This 17.73 mg of Cl^- is dissolved in a 100. mL of sample. Therefore, the concentration of Cl^- in milligrams per liter is

$$\frac{17.73 \text{ mg Cl}^-}{100. \text{ mL}} \times \frac{1000 \text{ mL}}{1 \text{ L}} = \frac{177 \text{ mg Cl}^-}{L}$$

Note that the volume and concentration of the $AgNO_3$ solution added to the sample were not used in the calculation, because we assumed that more than enough Ag^+ had been added to precipitate all the Cl^- in the sample. Was this assumption valid? One way to find out is to compare the amount of Ag^+ precipitated as AgCl with the amount of Ag^+ added. The latter can be calculated from the volume and molarity of the $AgNO_3$ solution (0.100 M) and the volume added (1.0 mL):

$$1.0 \text{ mL} \times 1.00 \text{ mol/L} = 1.00 \text{ mmol}$$

Next, we calculate the number of millimoles of Ag^+ in 71.7 mg of AgCl:

$$71.7 \text{ mg AgCl} \times \frac{1 \text{ mol AgCl}}{143.32 \text{ g AgCl}} \times \frac{1 \text{ mol Ag}^+}{1 \text{ mol AgCl}} = 0.500 \text{ mmol Ag}^+$$

So, twice as much Ag^+ (1.00 mmol) was added to the sample than was needed (0.50 mmol) to precipitate the chloride in it.

PRACTICE EXERCISE: The silver compounds of other halides also are insoluble and can be used to precipitate the halides from solution. If the addition of excess Ag^+ to 50.0 mL of a solution containing Br^- results in the precipitation of 22.8 mg of AgBr, what was the original concentration (in milligrams per liter) of Br^- in the sample? (See Problems 133 and 134.)

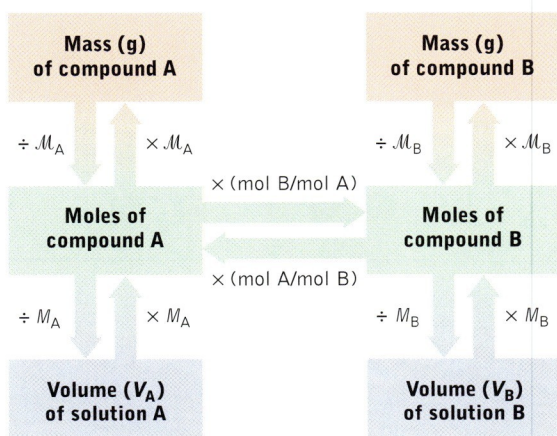

This flowchart shows how the quantities of two dissolved reactants (A and B) are related to the volumes (V_A and V_B) and concentrations (M_A and M_B) of their solutions. The moles of A and B are related to each other by the stoichiometry of the reaction.

5.8 ION EXCHANGE

The structure of kaolinite features two-dimensional sheets made up of either silicon bonded to oxygen or aluminum bonded to oxygen (Figure 5.29). Geologists call minerals with such structures aluminosilicates. They are among the most commonly encountered materials in Earth's crust. The multilayered structure of kaolinite and other clay minerals gives them distinctive physical and chemical properties. Some of these properties are related to imperfections in their molecular structures due to the presence of cations of lower charge in sites normally occupied by aluminum(III). In many clay minerals, for example, Fe^{2+} or Mg^{2+} ions substitute for Al^{3+} in the Al–O layer (Figure 5.30). These substitutions of $+2$ ions for Al^{3+} mean that the structure is deficient in positive charge and so has a negative charge overall. In nature, this charge is balanced by additional positively charged ions, such as Na^+ or K^+, that reside in the gaps between aluminosilicate layers. These ions are not held tightly in place, and so they can be replaced by other positive ions. For example, hydrogen ions in acidic groundwater can take the place of Na^+ ions in kaolinite:

$$\text{kaolinite}^-Na^+(s) + H^+(aq) \longrightarrow \text{kaolinite}^-H^+(s) + Na^+(aq)$$

The H^+-for-Na^+ replacement process is an example of an **ion-exchange reaction**. In this case, the result is the removal of hydrogen ions from solution,

CONNECTION: Kaolinite is a primary ingredient in the manufacture of ceramics described in Section 18.3.

 In an **ion-exchange reaction,** ions on a solid phase are replaced by ions in solution.

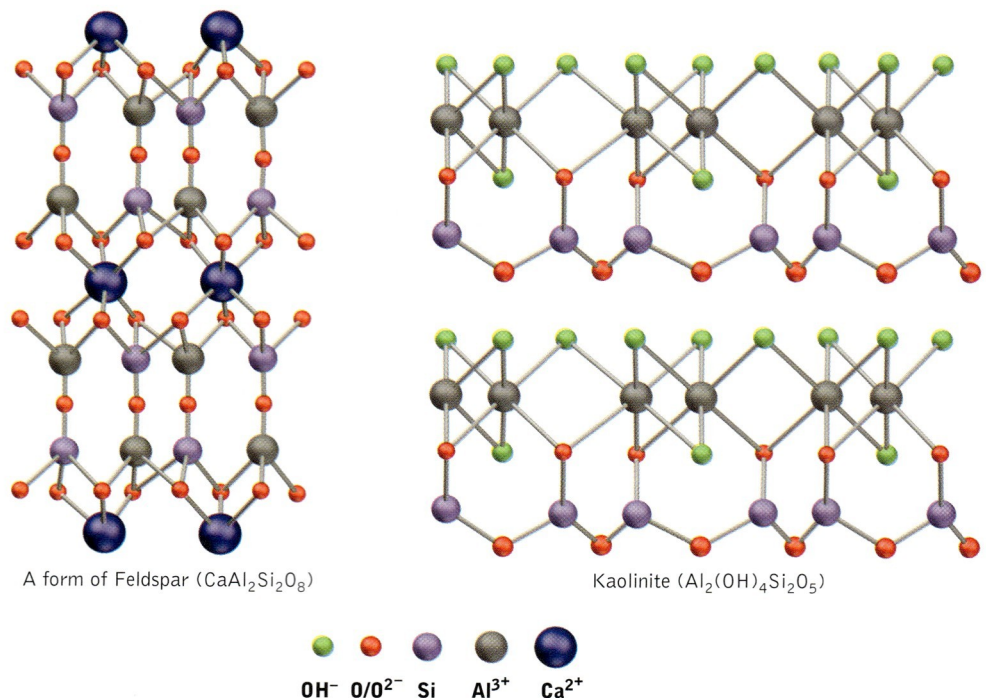

FIGURE 5.29 The structures of kaolinite, a clay mineral, and feldspar, the mineral from which kaolinite forms, feature layers of aluminum and silicon bonded to oxygen.

A form of Feldspar ($CaAl_2Si_2O_8$)

Kaolinite ($Al_2(OH)_4Si_2O_5$)

OH⁻ O/O²⁻ Si Al^{3+} Ca^{2+}

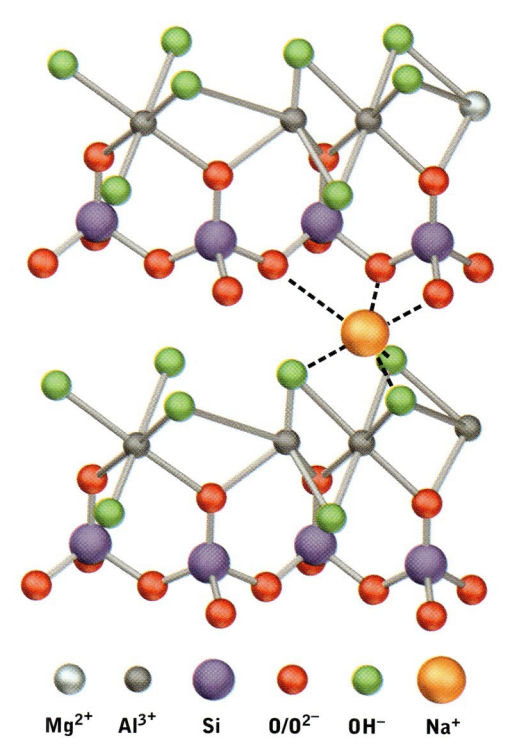

FIGURE 5.30 If an ion with a +2 charge such as Mg^{2+} is incorporated into the kaolinite structure in place of an Al^{3+} ion, then the difference in their charges gives the kaolinite a net negative charge. This negative charge is balanced by the inclusion of a cation such as Na^+ in the spaces between the layers of hydroxide ions.

Mg^{2+} Al^{3+} Si O/O²⁻ OH⁻ Na^+

which also makes it a neutralization reaction. There are many other forms of ion-exchange reactions in nature. Many consist of cation exchange on particles of clay and can control, or even arrest, the movement of positively charged trace metal ions through soils. Thus, they play an important role in defining the environmental fate of potentially toxic metals in groundwater.

Ion exchange is widely used to purify water. The groundwater in many parts of the world is considered "hard" owing to high concentrations of Ca^{2+} and Mg^{2+}. To "soften" the water, it is passed through a treatment system based on ion exchange. The system contains cartridges of porous plastic resin (R) to which have been bonded ion-exchange reaction sites. These sites usually consist of anionic groups bonded to the resin. One type is the *carboxylate* group; its formula is R—COO$^-$. The water surrounding these sites contains excess Na^+ to balance the negative charges on the anionic groups in fresh resin.

As hard water flows through the pores of the resin, calcium and magnesium ions in the water exchange with, or displace, the sodium ions in the pore water. The ion-exchange reaction may be written as follows:

$$2\,Na^+(R\text{—}COO^-)(s) + Ca^{2+}(aq) \longrightarrow Ca^{2+}(R\text{—}COO^-)_2(s) + 2\,Na^+(aq)$$

Hard water that has been softened in this way contains increased concentrations of Na^+, as shown in Figure 5.31. Although increased concentrations of sodium ions may not be a problem for healthy children and adults, people suffering from high blood pressure often must limit their intake of Na^+. They must not drink water softened by this kind of ion-exchange reaction.

Many chemical reactions in solution require the use of deionized water. It is often prepared by distillation or reverse osmosis followed by further purification

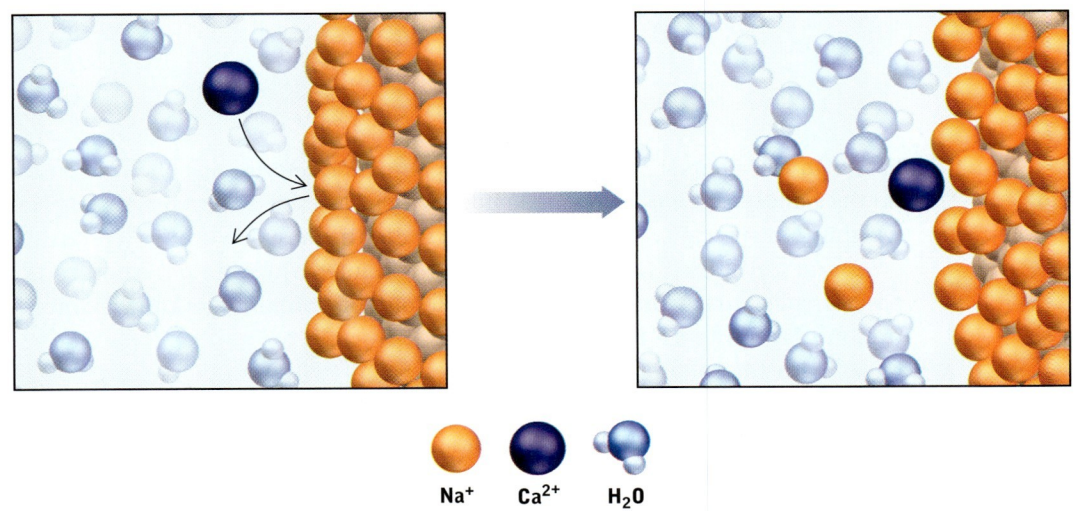

FIGURE 5.31 *What a Chemist Sees.* Ion exchange can be used to remove Ca^{2+} and Mg^{2+} ions from "hard" water. Sodium ions are released in the process.

by ion exchange. These ion-exchange systems include cartridges packed with materials that have both cationic exchange sites in the OH$^-$ form and anionic exchange sites in the H$^+$ form. As water containing small amounts of ions such as Na$^+$ and Cl$^-$ flows past these reaction sites, the following reactions take place:

$$RC^-H^+(s) + Na^+(aq) \longrightarrow RC^-Na^+(s) + H^+(aq)$$

and

$$RA^+OH^-(s) + Cl^-(aq) \longrightarrow RA^+Cl^-(s) + OH^-(aq)$$

where RC$^-$ and RA$^+$ represent cation-exchange and anion-exchange sites on the resin, respectively.

The aqueous products of these ion-exchange reactions, H$^+$ and OH$^-$, react with each other, forming water:

$$H^+(aq) + OH^-(aq) \longrightarrow H_2O(l)$$

Thus, the overall result of the ion-exchange process is the replacement of Na$^+$ and Cl$^-$ in the original water supply with H$_2$O. Similar reactions could be written for other ionic species. The liquid product that flows from the ion-exchange cartridge is essentially pure, *deionized* water. The ion-exchange reaction will proceed as long as there are still reaction sites in the RC$^-$H$^+$ and RA$^+$OH$^-$ forms. The cartridges are replaced when all sites have been consumed by the ion-exchange process (often indicated by a change in the color of the resin).

CONCEPT TEST: What advantage does an ion-exchange device have over distillation in producing deionized water?

5.9 TITRATIONS

Earlier in this chapter, we considered a method for determining the concentration of halides in solution by precipitating them as their silver compounds and weighing the precipitates. That approach is not the only one that we can use to determine an unknown concentration of a solute in a solution. Another approach, called **titration,** requires a second solution containing a known concentration of a substance that reacts with the solution of unknown concentration. If we accurately measure the volume of the second solution, called a **standard solution,** containing just enough reactant to completely react with the unknown solution, we can calculate the concentration of the unknown.

> ✓ In a **titration,** a carefully measured volume of a **standard solution** of known concentration is added to a sample solution. The concentration of the solute in the sample is calculated from the volume of the sample, the concentration of the standard solution, and the volume of the standard solution needed to react with all the solute.

> How do we determine an unknown concentration of a solute in a solution?

Consider the problem of sulfuric acid in the water draining from an abandoned coal mine. It turns out that iron sulfide is often found in coal deposits. When groundwater leaches through iron sulfide, oxygen in the water reacts with the iron sulfide, producing H_2SO_4. Suppose we have a 100-mL sample of drainage water containing an unknown concentration of sulfuric acid. We decide to react the sulfuric acid with a standard solution of 0.00100 M NaOH. The following acid–base reaction takes place:

$$H_2SO_4(aq) + 2\ NaOH(aq) \longrightarrow Na_2SO_4(aq) + 2\ H_2O(l)$$

Now suppose that exactly 22.4 mL of the NaOH solution is required to completely react with the H_2SO_4 in the 100-mL sample. Because we know the concentration of the NaOH solution, we can calculate the number of moles of NaOH consumed in the neutralization reaction:

$$\frac{0.00100\ mol\ NaOH}{\cancel{L}} \times 22.4\ \cancel{mL} \times \frac{1\ \cancel{L}}{1000\ \cancel{mL}} = 2.24 \times 10^{-5}\ mol\ NaOH$$

We know from the stoichiometry of the neutralization reaction that two moles of NaOH are required to neutralize one mole of H_2SO_4; so the number of moles of H_2SO_4 in the sample must be

$$2.24 \times 10^{-5}\ \cancel{mol\ NaOH} \times \frac{1\ mol\ H_2SO_4}{2\ \cancel{mol\ NaOH}} = 1.12 \times 10^{-5}\ mol\ H_2SO_4$$

If the 100-mL sample of mine effluent contains 1.12×10^{-5} moles of H_2SO_4, then the concentration of H_2SO_4 is

$$\frac{1.12 \times 10^{-5}\ mol\ H_2SO_4}{100\ \cancel{mL}} \times \frac{1000\ \cancel{mL}}{L} = 1.12 \times 10^{-4}\ mol/L = 1.12 \times 10^{-4}\ M$$

How did we know that exactly 22.4 mL of standard solution was needed to react with the sulfuric acid in the sample? A setup for doing a titration is illustrated in Figure 5.32. The solution of known concentration is placed in a narrow tube with volume markings called a *buret*. This solution is carefully added to the solution of unknown concentration until reaction is complete. A visual indicator that changes color may be used to signal that the reaction is complete. This stage of the titration is called the **equivalence point.** In our sulfuric acid and sodium hydroxide titration, a dye called phenolphthalein is used as the indicator. Phenolphthalein is colorless in acid solution but changes to pink when even a slight excess of base is present. The calculation of the H_2SO_4 concentration is summarized in Figure 5.32.

CONNECTION: Titrations are used to determine the concentrations of unknown solutions of acids and bases. (Further details are given in Section 16.8.)

✓ In a titration, the **equivalence point** is reached when just enough standard solution has been added to completely react with a solute in the sample. The concentration of the solute is calculated from the volume and concentration of the standard solution and the volume of the sample.

SAMPLE EXERCISE 5.21: To determine the concentration of Cl^- in a sample of drinking water from a well that is suspected of being contaminated by road salt (NaCl), the sample is titrated with a standard solution of Ag^+:

$$Cl^-(aq) + Ag^+(aq) \longrightarrow AgCl(s)$$

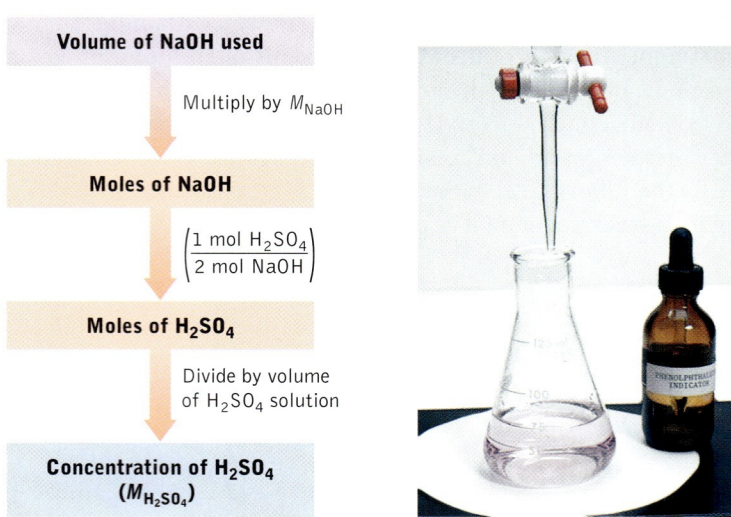

FIGURE 5.32 The concentration of a sulfuric acid solution can be determined through titration with a standard solution of sodium hydroxide. A known volume of the H_2SO_4 solution is placed into the flask. The buret is filled with an aqueous solution of sodium hydroxide of known concentration. A few drops of phenolphthalein indicator solution are added to the flask. Sodium hydroxide is carefully added until the indicator changes color from colorless to pink. The concentration of H_2SO_4 is calculated by using the steps in the flowchart above.

What is the concentration of Cl^- in a 100.0-mL sample if 24.1 mL of 0.100 M Ag^+ is needed to react with all the Cl^- in the sample?

SOLUTION: We know the volume (24.1 mL) and the molarity (0.100 M) of the Ag^+ solution. From this information, we can calculate the number of moles of Ag^+ needed to react with all the Cl^-:

$$\frac{0.100 \text{ mol Ag}^+}{\text{L}} \times 24.1 \text{ mL} \times \frac{1 \text{ L}}{1000 \text{ mL}} = 2.41 \times 10^{-3} \text{ Ag}^+$$

We know from the stoichiometry of the precipitation reaction that one mole of Ag^+ is required to precipitate one mole of Cl^-; so the number of moles of Cl^- in the sample must be

$$2.41 \times 10^{-3} \text{ mol Ag}^+ \times \frac{1 \text{ mol Cl}^-}{1 \text{ mol Ag}^+} = 2.41 \times 10^{-3} \text{ mol Cl}^-$$

If the 100-mL sample contains 2.41×10^{-4} moles of Cl^-, then the concentration of Cl^- is

$$\frac{2.41 \times 10^{-3} \text{ mol Cl}^-}{100 \text{ mL}} \times \frac{1000 \text{ mL}}{\text{L}} = 2.41 \times 10^{-3} \text{ mol/L} = 2.41 \times 10^{-2} \text{ } M$$

PRACTICE EXERCISE: The concentration of SO_4^{2-} in a sample of water can be determined by titrating the sample with Ba^{2+}:

$$SO_4^{2-}(aq) + Ba^{2+}(aq) \longrightarrow BaSO_4(s)$$

What is the concentration of SO_4^{2-} in a 50.0-mL sample if 6.55 mL of 0.00100 M Ba^{2+} is needed to react with all the SO_4^{2-} in the sample? (See Problems 129–132.)

5.10 COLLOIDS

We started this chapter on solution chemistry by observing that seawater is much more than a solution of salt dissolved in water. Seawater is also a heterogeneous mixture of particles that may separate from the solution on standing. On a scale of particle sizes that includes molecules and single-atom ions that are less than a nanometer across, as well as filterable particles that are hundreds of nanometers across or larger, there is an intermediate range of particles from about 1 to 100 nm in diameter that may remain suspended indefinitely in seawater or any fluid (including gases: fog and smoke include particles small enough to remain suspended in the air). These particles are called **colloids**. In natural waters, colloids may include clay minerals, decaying organic matter, or large biomolecules. Colloidal particles scatter light, a phenomenon known as the **Tyndall effect** (Figure 5.33).

In aqueous suspensions, colloidal particles are classified as **hydrophobic** (water fearing) or **hydrophilic** (water loving), depending on their interactions with water molecules. Hydrophilic colloids behave almost like soluble molecules. Hydrophobic colloids tend to aggregate into larger particles and precipitate from the suspension unless they are stabilized by the adsorption of ions onto their

CONNECTION: The sizes of atoms and ions are described in Chapter 9.

Colloids are stable suspensions of particles with diameters ranging from 1 to 100 nm in liquids or gases. Colloids are detected by light-scattering phenomena called the **Tyndall effect**.

Hydrophilic substances are attracted to water and dissolve in it; **hydrophobic** substances are repelled by water and are insoluble in it.

FIGURE 5.33 Smoke and fog are examples of gas-phase colloids consisting of small particles of carbon and droplets of water, respectively, suspended in air. Light scattered by these particles produces the Tyndall effect.

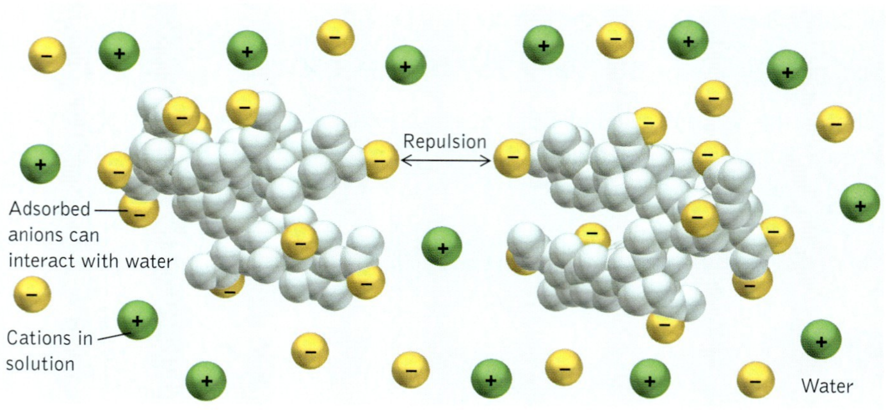

FIGURE 5.34 Colloids owe their stability to the adsorption of ions onto their surfaces. Repulsion between charged particles prevents their aggregation and precipitation.

surfaces. When that happens, repulsions between like-charged colloidal particles prevent their aggregation, as shown in Figure 5.34.

The aggregation of colloidal particles into larger particles provides a way of separating them from a solution. Adding electrolytes to a hydrophobic colloid reduces the electrostatic repulsions between colloidal particles, allowing them to join into larger particles. Heating a hydrophilic colloid also can increase molecular interactions between particles, leading to coagulation and an increase in particle size. Finally, semipermeable membranes such as those used in reverse osmosis can be used to remove colloidal particles. Recall that only solvent can pass through such membranes; so applying an appropriate pressure will separate pure water from colloidal as well as dissolved species.

CHAPTER REVIEW

Summary

SECTION 5.1

Scientists call water the universal solvent because of its ability to dissolve both ionic and molecular solutes. When ionic compounds dissolve in water, they separate into positive cations and negative anions.

SECTION 5.2

Molarity is a concentration scale in which a 1.0 molar solution has 1.0 mole of solute per liter of solution.

SECTION 5.3

When placed in an electrochemical cell, compounds called electrolytes and their aqueous solutions conduct electricity as a result of the migration of the ions that they form toward the electrodes in the cell. Weak electrolytes are compounds that only partly ionize in aqueous solutions, and nonelectrolytes are molecular solutes that do not ionize at all.

SECTION 5.4

Osmosis is a process in which solvent flows from a solution of lower solute concentration through a semipermeable membrane into a solution of higher solute concentration. Like freezing point and boiling point, osmotic pressure is a colligative property of a solution. The concentration scales used for colligative measurements include molarity and molality, which is the number of moles of solute per kilogram of solvent. The more concentrated a solution is, the higher its boiling point, the lower its freezing point, and the higher its osmotic pressure. Colligative properties can be used to measure the molar masses of dissolved solutes.

SECTION 5.5

Chemical equations describing reactions in solution may be molecular equations, which include the complete chemical formulas of all reactants and products, or net ionic equations, which include only those species consumed or produced in a reaction. Those ions not changed by a reaction are called spectator ions. Reactions in which electrons are transferred from one reactant to another are called oxidation–reduction (redox) reactions. The half-reaction components of redox reactions indicate the number of electrons lost or gained and are written as net ionic equations. Electrolysis results in oxidation (loss of electrons) and reduction (gain of electrons) reactions at electrodes. Oxidation numbers are used to differentiate different compounds of an element and to keep track of electrons in redox reactions. Balancing redox reactions requires the same number of electrons to be produced in one half-reaction and consumed in the other.

SECTION 5.6

Electrolytes include acids, bases, and salts. Acids donate hydrogen ions in aqueous solution. Acidic compounds are produced when nonmetal oxides dissolve in water. Bases are hydrogen ion acceptors, and they include the hydroxides, oxides and carbonates of metals. Salts and water are produced when acids and bases react with and neutralize each other.

SECTION 5.7

Other important classes of chemical reactions between solutes include precipitation reactions in which a product of limited solubility is formed.

SECTION 5.8

Water can be deionized and purified by ion-exchange reactions at reactive sites on solid materials.

SECTION 5.9

Titrations are based on reactions between acids and bases and on precipitation reactions. They are used in quantitative chemical analysis to determine the concentrations of solutes in unknown samples.

SECTION 5.10

Microscopic particles called colloids that are suspended in liquids can be detected with light scattering in the Tyndall effect. Colloids can coagulate when salts are added to the liquid or the liquid is heated.

Key Terms

acid (p. 250)
acid rain (p. 252)
aqueous solution (p. 210)
base (p. 250)
boiling-point elevation (p. 232)
chemical equilibrium (p. 251)
chemical weathering (p. 250)
colligative property (p. 228)
colloid (p. 269)
concentration (p. 214)
electrochemical cell (p. 243)
electrode (p. 223)
electrolysis (p. 243)
electrolyte (p. 223)
electrolytic cell (p. 243)
equivalence point (p. 267)
freezing-point depression (p. 234)
half-reaction (p. 242)

hydrophilic (p. 269)
hydrophobic (p. 269)
ion-exchange reaction (p. 263)
ion pair (p. 236)
molality (p. 232)
molar (p. 215)
molarity (p. 215)
molecular equation (p. 256)
net ionic equation (p. 243)
nonelectrolyte (p. 223)
osmosis (p. 224)
osmotic pressure (p. 226)
oxidation (p. 241)
oxidation number (p. 244)
oxidizing agent (p. 245)
precipitate (p. 257)
precipitation reaction (p. 257)
redox reaction (p. 244)

reducing agent (p. 245)
reduction (p. 241)
reverse osmosis (p. 228)
saturated (p. 257)
semipermeable membrane (p. 224)
solute (p. 210)
solution (p. 210)
solvent (p. 210)
spectator ion (p. 243)
standard solution (p. 266)
supersaturated (p. 257)
titration (p. 266)
turbidity (p. 211)
Tyndall effect (p. 269)
unsaturated (p. 257)
van't Hoff factor (p. 236)
voltaic cell (p. 243)
weak electrolyte (p. 223)

Key Skills and Concepts

SECTION 5.1

Understand the importance of water as the universal solvent.
A solution consists of one or more solutes dissolved in a solvent.

SECTION 5.2

Understand that the concentration of a solution is independent of the amount of solution.
Be able to calculate the molarity of a solution from a specified quantity of solute and the solution volume.

Be able to calculate the amount of solute in a specified volume of a solution of known molarity.

Understand the difference between dilute and concentrated solutions and that the density of a very dilute solution is close to the density of the solvent.

Be able to calculate the concentration of a solution made by diluting a more concentrated solution with the solvent.

SECTION 5.3

Explain the terms nonelectrolyte, weak electrolyte, and strong electrolyte.

SECTION 5.4

Describe the apparatus that can be used to measure the osmotic pressure of a solution.

List the factors that affect the osmotic pressure of a solution.

Explain why red blood cells shrivel when placed in seawater or expand when placed in distilled water, and explain the concept of isotonic solutions.

Explain how osmotic pressure measurements can be used to measure the molar mass of a dissolved solute.

Describe how reverse osmosis can be used to desalinate water.

Be able to calculate the molality of a specified solution.

Explain why the boiling point of a solution is higher than that of the solvent, and be able to calculate the boiling point of a specified solution.

Be able to explain why the freezing point of a solution is lower than that of the solvent, and be able to calculate the freezing point of a specified solution.

Describe how freezing-point-depression measurements can be used to measure the molar mass of a dissolved solute.

Explain how the van't Hoff factor accounts for the colligative properties of electrolytes.

SECTION 5.5

Understand the concept of a net ionic equation.

Distinguish between oxidation and reduction half-reactions, and understand that they are complementary.

Be able to find the oxidation number of an element in a compound by using the oxidation number rules.

Be able to combine half-reactions to write balanced redox reactions in acid, neutral, and basic aqueous solutions.

Describe the factors that determine the products of electrolysis experiments.

SECTION 5.6

Define an acid, a base, and a salt.

Know that nonmetal oxides produce acidic solutions when they dissolve in water and that metal oxides produce basic solutions when they dissolve in water.

Give an example of a strong acid, a weak acid, and a weak base.

Be able to write net ionic equations for acid–base reactions between strong acids and bases and between weak acids and bases.

Know that the products of reactions of carbonates with acids are CO_2 and water.

SECTION 5.7

Distinguish between undersaturated, saturated, and supersaturated solutions.

Predict the insolubility of a compound from the solubility rules.

Understand how precipitation reactions can be used for quantitative analysis, and be able to calculate unknown amounts or concentrations of sample components from titration data.

SECTION 5.8

Describe how ion-exchange resins work to deionize water.

SECTION 5.9

Describe a titration setup, and explain why an indicator is needed for most acid–base titrations.

Be able to calculate the unknown concentration of an acid or a base from titration data.

SECTION 5.10

Describe the detection of colloids with the use of the Tyndall effect.

Distinguish between hydrophobic and hydrophilic colloids.

Explain how changes in salt concentration and temperature can cause colloids to coagulate.

Key Equations and Relations

The molar concentration (M) of a solution:

$$\text{molarity } (M) = \frac{\text{number of moles of solute}}{\text{volume of solution in liters}} = \frac{n}{V} \quad (5.1)$$

The osmotic pressure (π) of a solution:

$$\pi = MRT \quad (5.5)$$

where R is the universal gas constant (0.0821 L·atm./mol·K), T is absolute temperature (K), and π is in atmospheres.

Boiling-point elevation (ΔT_b) of a solution:

$$\Delta T_b = K_b m \quad (5.6)$$

where K_b is the boiling-point-elevation constant of the solvent and m is the molal concentration of the solution.

The molal concentration (m) of a solution:

$$\text{molality } (m) = \frac{\text{number of moles of solute}}{\text{mass of solvent in kilograms}} = \frac{n}{\text{kg}} \quad (5.7)$$

Freezing-point depression (ΔT_f) of a solution:

$$\Delta T_f = K_f m \quad (5.8)$$

where K_f is the freezing-point-depression constant of the solvent and m is the molal concentration of the solution.

The van't Hoff factor (i), which expresses the degree to which a solute in a solution dissociates into ions:

$$i = \frac{\text{moles of particles in solution}}{\text{moles of solute dissolved}} \quad (5.9)$$

Correcting the values of colligative properties by using the van't Hoff factor:

a. osmotic pressure

$$\pi = iMRT$$

b. boiling-point elevation

$$\Delta T_b = iK_b m$$

c. freezing-point depression

$$\Delta T_f = iK_f m$$

QUESTIONS AND PROBLEMS

Solutions and Molarity

CONCEPT REVIEW

1. How do we decide which component in a solution is the solvent?
2. Which ingredient is the solvent in a 50:50 (by mass) solution of methanol (CH_3OH) and water?
3. Can a solid ever be a solvent?
4. Is a saturated solution always a concentrated solution? Explain.
5. What is meant by the molarity of a solution?
6. What is the molarity of a solution that contains 1.00 millimole of solute per milliliter of solution?
7. A beaker contains 100 grams of 1.00 M NaCl. If you transfer 50 grams of the solution to another beaker, what is the molarity of the solution remaining in the first beaker?
8. Are the contents of a can of soda a homogeneous or a heterogeneous mixture under the following conditions?: (a) closed can; (b) freshly opened can; (c) can left open for 3 days.

PROBLEMS

9. Calculate the molarity of each of the following solutes in their respective aqueous solutions:
   a. 0.56 mol of $BaCl_2$ in 150. mL of solution
   b. 20. mmol of Na_2CO_3 in 750. mL of solution
   c. 0.325 mol of $C_6H_{12}O_6$ in 100. mL of solution
   d. 1.48 mol of KNO_3 in 250 mL of solution
10. Calculate the molarity of each of the following solutes in their respective aqueous solutions:
    a. 15 mmol of urea (CH_4N_2O) in 225. mL of solution
    b. 14.6 mol of $NH_4C_2H_3O_2$ in 675. mL of solution
    c. 1.94 mol of methanol (CH_3OH) in 1.25 L of solution
    d. 45 mmol of sucrose ($C_{12}H_{24}O_{11}$) in 50.0 mL of solution
11. Calculate the molarity of each of the following ions dissolved in the volume of solution indicated:
    a. 0.33 g Na^+ in 100. mL
    b. 0.38 g Cl^- in 100. mL
    c. 0.46 g SO_4^{2-} in 50. mL
    d. 0.40 g Ca^{2+} in 50. mL
12. Calculate the molarity of each of the following solutes in their respective aqueous solutions:
    a. 64.7 g of LiCl dissolved in enough water to make 350. mL of solution
    b. 29.3 g of $NiSO_4$ dissolved in enough water to make 200. mL of solution
    c. 50.0 g of KCN dissolved in enough water to make 425. mL of solution
    d. 0.155 g. of $AgNO_3$ dissolved in enough water to make 100. mL of solution
13. How many grams of solute are needed to prepare each of the following solutions?
    a. 1.00 L of 0.092 M NaCl
    b. 300. mL of 0.125 M $CuSO_4$
    c. 250. mL of 0.400 M CH_3OH
14. How many grams of solute are needed to prepare each of the following solutions?
    a. 500. mL of 0.250 M KBr
    b. 125. mL of 0.200 M $NaNO_3$
    c. 100. mL of 0.375 M CH_3OH
15. The MacKenzie River in the Canadian Arctic contains, on average, 0.820 mM Ca^{2+}, 0.430 mM Mg^{2+}, 0.300 mM Na^+, 0.0200 mK^+, 0.250 mM Cl^-, 0.380 mM SO_4^{2-}, and 1.82 mM HCO_3^-. What, on average, is the total mass of these ions in 2.75 L of MacKenzie River water?
16. Zinc, copper, lead, and mercury ions are toxic to Atlantic salmon at concentrations of 6.42×10^{-2} mM, 7.16×10^{-3} mM, 0.965 mM, and 5.00×10^{-2} mM, respectively. What are the corresponding concentrations in milligrams per liter?
17. Calculate the number of moles of solute contained in the following volumes of saturated aqueous solutions of four pesticides:
    a. 400 mL of 0.024 M Lindane
    b. 1.65 L of 0.473 mM Dieldrin
    c. 25.8 L of 3.4 μM DDT
    d. 154 L of 27.4 μM Aldrin

18. A sample of crude oil was found to contain 3.13 mM napththalene, 12.0 mM methylnaphthalene, 23.8 mM dimethylnaphthalene, and 14.1 mM trimethylnaphthalene. How many total moles of these four naphthalene compounds are present in 300 mL of the oil?

19. The toxicity of DDT ($C_{14}H_9Cl_5$) led to a ban on its use in the United States in 1972. Measurements of DDT concentrations in groundwater samples from Pennsylvania between 1969 and 1971 gave the following results:

Location	Sample size	Mass of DDT
orchard	250 mL	0.030 μg
residential	1.75 L	0.035 μg
residential after a storm	50 mL	0.57 μg

Express these concentrations in moles per liter.

20. Calculate the molarity of each of the following solutes in their respective solutions:
    a. 5.0 g of acetic acid ($C_2H_3O_2H$) in enough water to make 100 mL of solution
    b. 2.4 g of commercial tile cleaner (NaClO) in enough water to make 100. mL of solution
    c. 5.0 g of household ammonia (NH_3) in enough water to make 100. mL of solution

21. Pesticide concentrations in the Rhine River between Germany and France between 1969 and 1975 averaged 0.55 μg/L of hexachlorobenzene (C_6Cl_6), 0.06 μg/L of Dieldrin ($C_{12}H_8Cl_6O$), and 1.02 μg/L of hexachlorocyclohexane ($C_6H_6Cl_6$). Express these concentrations in moles per liter.

22. Effluent from municipal sewers and mining operations can contain high concentrations of zinc. A sewer pipe discharges effluent that contains 10 mg Zn/L. What is the molar concentration of Zn in the effluent?

23. The concentration of copper(II) sulfate in one brand of soluble plant fertilizer is 0.07% by weight. If a 20.-g sample of this fertilizer is dissolved in 2.0 L of solution, what is the molar concentration of Cu^{2+}?

24. For which of the following compounds is it possible to make a 1.0 M solution?
    a. $CuSO_4 \cdot 5\,H_2O$, solubility = 31.6 g/100 mL
    b. $AgNO_3$, solubility = 122. g/100 mL
    c. $Fe(NO_3)_2 \cdot 6\,H_2O$, solubility = 83.5 g/100 mL
    d. $Ca(OH)_2$, solubility = 0.131 g/mL

25. About 6. × 10^9 g of gold is thought to be dissolved in the oceans of the world. If the total volume of the oceans is 1.5×10^{21} L, what is the average molarity of gold in seawater?

26. The concentration of Mg^{2+} in seawater is about 1.29 g/kg. If the density of seawater is 1.04 g/mL, what is the molarity of Mg^{2+}?

27. Calculate how much water must be added to each of the following solutions to reduce their concentrations to 1.00 μM.
    a. 1.00 mL 0.024 M Lindane, a pesticide
    b. 15.5 mL of 7.16×10^{-3} mM Cu^{2+}
    c. 25.0 μL of 75. μM PO_4^{3-}

28. Calculate the final concentrations of the following aqueous solutions after adding 20.0 mL of distilled water:
    a. 1.00 mL of 0.452 M Na^+
    b. 11.2 mL of 3.4 μM DDT
    c. 0.566 mL of 6.42×10^{-2} mM Zn^{2+}

29. Chemists who analyze samples for trace metals may buy standard solutions that contain 1000. mg/L concentrations of the metals. If a chemist wishes to dilute 1000. mg/L standard to prepare 500.0 mL of a "working" standard that has a concentration of 2.25 mg/L, what volume of the 1000. mg/L standard solution is needed?

30. A puddle of seawater that is caught in a depression formed by some coastal rocks at high tide begins to evaporate on a hot summer day. If the volume of the puddle decreases to 23% of its initial volume, what is the concentration of Na^+ after evaporation? Initial $[Na^+]$ = 0.479 M.

Electrolytes

CONCEPT REVIEW

31. Which of the following solutes increase the conductivity of water? table salt; table sugar; formic acid (HCOOH); methanol (CH_3OH).

32. Which solution should be the better conductor of electricity: $1.0\ M$ NaCl or $1.0\ M$ $MgCl_2$?

PROBLEMS

33. What is the molar concentration of Na^+ ions in $0.025\ M$ solutions of the following sodium salts in water? (a) NaBr; (b) Na_2SO_4; (c) Na_3PO_4

34. What is the total molar concentration of all the ions in $0.025\ M$ solutions of the following salts in water? (a) KCl; (b) $CuSO_4$; (c) $CaCl_2$

Colligative Properties and Molality

CONCEPT REVIEW

35. What effect does dissolving a solute have on the following properties of a solvent? (a) its osmotic pressure; (b) its freezing point; (c) its boiling point.

36. What is a semipermeable membrane?

37. A pure solvent is separated from a solution containing the same solvent by a semipermeable membrane. In which direction does the solvent tend to flow across the membrane, and why?

38. A dilute solution is separated from a more concentrated solution containing the same solvent by a semipermeable membrane. In which direction does the solvent tend to flow across the membrane, and why?

39. How is the osmotic pressure of a solution related to its molar concentration and its temperature?

40. What is reverse osmosis?

41. What role does the van't Hoff factor play in describing colligative properties of solutions?

42. How can osmotic-pressure, freezing-point-depression, and boiling-point-elevation measurements be used to find the molar mass of a solute? Why is it important to know whether the solute is an electrolyte or a nonelectrolyte?

43. Is the following statement true or false? For solutions of the same osmotic pressure and temperature, the molarity of a solution of NaCl will always be less than the molarity of a solution of $CaCl_2$. Explain your answer.

44. Suppose you have $1.00\ M$ aqueous solutions of each of the following solutes: glucose ($C_6H_{12}O_6$), NaCl, and acetic acid ($C_2H_3O_2H$). Which solution has the highest osmotic pressure, and which has the lowest?

45. What is the difference between molarity and molality?

46. As a solution of NaCl becomes more concentrated, does the difference between its molarity and its molality increase or decrease?

47. Which aqueous solution has the lowest freezing point: $0.5\ M$ glucose, $0.5\ M$ NaCl, or $0.5\ M$ $CaCl_2$?

48. Which aqueous solution has the highest boiling point: $0.5\ M$ glucose, $0.5\ M$ NaCl, or $0.5\ M$ $CaCl_2$?

49. Why does seawater have a lower freezing point than freshwater?

50. The thermostat in a refrigerator filled with cans of soft drinks for a party malfunctions and the temperature of the refrigerator drops below 0°C. The contents of the cans of diet soft drinks freeze, rupturing many of the cans and causing an awful mess. However, none of the cans containing regular, nondiet soft drinks rupture. Why?

PROBLEMS BASED ON OSMOTIC PRESSURE

51. The following pairs of aqueous solutions are separated by a semipermeable membrane. In which direction will solvent flow?
a. $A = 1.25\ M$ NaCl, $B = 1.50\ M$ KCl
b. $A = 3.45\ M$ $CaCl_2$, $B = 3.45\ M$ NaBr
c. $A = 4.68\ M$ dextrose, $B = 3.00\ M$ NaCl

52. The following pairs of aqueous solutions are separated by a semipermeable membrane. In which direction will solvent flow?
    a. A = 0.48 M NaCl, B = 5.85 g of NaCl dissolved 1.00 L of solution
    b. A = 100 mL of 0.982 M $CaCl_2$, B = 16 g of NaCl in 100 mL of solution
    c. A = 100 mL of 6.56 mM $MgSO_4$, B = 24 g of $MgCl_2$ in 250 mL of solution
53. Calculate the osmotic pressure of each of the following aqueous solutions at 20°C:
    a. 2.39 M methanol (CH_3OH)
    b. 9.45 mM $MgCl_2$
    c. 40.0 mL of glycerol ($C_3H_8O_3$) in 250. mL of solution (density of glycerol = 1.265 g/mL)
    d. 25. g of $CaCl_2$ in 350 mL of solution
54. Calculate the osmotic pressure of each of the following aqueous solutions at 27°C:
    a. 10.0 g of NaCl in 1.50 L of solution
    b. 10.0 mg/L of $LiNO_3$
    c. 0.222 M glucose
    d. 0.00764 M K_2SO_4
55. Determine the molarity of each of the following solutions from its osmotic pressure at 25°C. Include the van't Hoff factor for the solution when the factor is given.
    a. π = 0.674 atm for a solution of ethanol (C_2H_5OH)
    b. π = 0.0271 atm for a solution of aspirin ($C_9H_8O_4$)
    c. π = 0.605 atm for a solution of $CaCl_2$, i = 2.47
56. Determine the molarity of each of the following solutions from the osmotic pressure at 25°C. Include the van't Hoff factor for the solution when the factor is given.
    a. π = 0.0259 atm for a solution of urea (CH_4N_2O)
    b. π = 1.56 atm for a solution of sucrose ($C_{12}H_{22}O_{11}$)
    c. π = 0.697 atm for a solution of KI, i = 1.90
57. 100. mL of a solution of physiological saline (0.92% NaCl by weight) is diluted by the addition of 250. mL of water. What is the osmotic pressure of the final solution at 37°C? Assume that NaCl dissociates completely into Na^+ and Cl^-.
58. 100. mL of 2.50 mM NaCl is mixed with 80.0 mL of 3.60 mM $MgCl_2$ at 20°C. Calculate the osmotic pressure of each starting solution and that of the mixture, assuming that the volumes are additive and that both salts dissociate completely into their component ions.

PROBLEMS BASED ON MOLALITY

59. Calculate the molality of each of the following solutions:
    a. 0.875 mol of glucose in 1.5 kg of water
    b. 11.5 mmol of acetic acid in 65. g of water
    c. 0.325 mol of baking soda ($NaHCO_3$) in 290. g of water
60. Table 5.1 lists molar concentrations of major ions in seawater. Using a density of 1.022 g/mL for seawater, convert the concentrations into molalities.
61. What mass of the following solutions contains 0.100 mol of solute? (a) 0.334 m NH_4NO_3; (b) 1.24 m ethylene glycol, $C_2H_6O_2$; (c) 5.65 m $CaCl_2$.
62. How many moles of solute are there in the following solutions?
    a. 0.150 m glucose solution made by dissolving the glucose in 100. kg of water
    b. 0.028 m Na_2CrO_4 solution made by dissolving the Na_2CrO_4 in 1000. g of water
    c. 0.100 m urea solution made by dissolving the urea in 500. g of water
63. High concentrations of ammonia (NH_3), nitrite ion (NO_2^-), and nitrate ion (NO_3^-) in water can kill fish. Lethal concentrations of these species for rainbow trout are 1.1 mg/L, 0.40 mg/L, and 1360. mg/L, respectively. Express these concentrations in molality units, assuming a solution density of 1.00 g/mL.
64. The concentrations of six important crustal elements in a river-water sample are: 0.050 mg/kg of Al, 0.040 mg/kg of Fe, 13.4 mg/kg of Ca, 5.2 mg/kg of Na, 1.3 mg/kg of K, and 3.4 mg/kg of Mg. Express each of these concentrations in molality.

PROBLEMS BASED ON BOILING-POINT ELEVATION AND FREEZING-POINT DEPRESSION

65. Cinnamon owes its flavor and odor to cinnamaldehyde (C_9H_8O). Determine the boiling-point elevation of a solution of 100 mg of cinnamaldehyde dissolved in 1.00 g of carbon tetrachloride (K_b = 5.03 °C/m).
66. Determine the boiling-point elevation of a solution of 125. mg of carvone ($C_{10}H_{14}O$, oil of spearmint) dissolved in 1.50 g of carbon disulfide (K_b = 2.34 °C/m).
67. What molality of a nonvolatile, nonelectrolyte solute is needed to lower the melting point of camphor by 1.000 °C (K_f = 39.7 °C)?

68. What molality of a nonvolatile, nonelectrolyte solute is needed to raise the boiling point of water by 7.60 °C ($K_b = 0.52$ °C/m)?
69. Determine the melting point of an aqueous solution containing 186 mg of saccharin ($C_7H_5O_3NS$) in 1.00 mL of water (density = 1.00 g/mL, $K_f = 1.86$ °C/m).
70. Determine the boiling point of an aqueous solution that is 2.50 m ethylene glycol ($C_2H_6O_2$) and 5.00 m methanol (CH_3OH); $K_b = 0.52$ °C. Assume that the boiling point of pure water is 100.00°C.
71. Which one of the following aqueous solutions should have the highest boiling point: 0.0200 m CH_3OH, 0.0125 m KCl, or 0.0100 m $Ca(NO_3)_2$?
72. Which one of the following aqueous solutions should have the lowest freezing point: 0.0500 m $C_6H_{12}O_6$, 0.0300 m KBr, or 0.0150 m Na_2SO_4?
73. Arrange the following aqueous solutions in order of increasing boiling point
    a. 0.06 m $FeCl_3$ ($i = 3.4$)
    b. 0.10 m $MgCl_2$ ($i = 2.7$)
    c. 0.20 m KCl ($i = 1.9$)
74. Arrange the following solutions in order of increasing melting-point depression:
    a. 0.10 m $MgCl_2$ in water, $i = 2.7$, $K_f = 1.86$ °C/m
    b. 0.20 m toluene in diethyl ether, $i = 1.00$, $K_f = 1.79$ °C/m
    c. 0.20 m ethylene glycol in ethanol, $i = 1.00$, $K_f = 1.99$ °C/m
75. $CaCl_2$ is widely used to melt frozen precipitation on sidewalks after a winter storm. Could $CaCl_2$ melt ice at -20°C? Assume that the solubility of $CaCl_2$ at this temperature is 70. g/100. g of H_2O and that the van't Hoff factor for a saturated solution of $CaCl_2$ is 2.5.
76. A mixture of table salt and ice is used to chill the contents of hand-operated ice-cream makers. What is the melting point of a mixture of 2.00 pounds of NaCl and 12.00 pounds of ice if exactly half of the ice melts. Assume that all the NaCl dissolves in the melted ice and that the van't Hoff factor for the resulting solution is 1.44.
77. The freezing points of 0.0935 m ammonium chloride and 0.0378 m ammonium sulfate in water were found to be -0.322°C and -0.173°C, respectively. What are the values of the van't Hoff factor for these salts?
78. The following data were collected for three compounds in aqueous solution. Determine the value of the van't Hoff factor for each salt.

Compound	Concentration	Observed ΔT_f
LiCl	5.0 g/kg	0.410
HCl	5.0 g/kg	0.486
NaCl	5.0 g/kg	0.299

PROBLEMS BASED ON CALCULATING MOLAR MASS

79. A 188 mg sample of a nonelectrolyte isolated from throat lozenges was dissolved in enough water to make 10.0 mL of solution at 25°C. The osmotic pressure of the resulting solution was 4.89 atm. Calculate the molar mass of the compound.
80. An unknown compound (152 mg) was dissolved in water to make 75.0 mL of solution. The solution did not conduct electricity and had an osmotic pressure of 0.328 atm at 27°C. Elemental analysis revealed 78.90% C, 10.59% H, and 10.51% O. Determine the molecular formula of this compound.
81. Eugenol is one of the compounds responsible for the flavor of cloves. A 111-mg sample of eugenol was dissolved in 1.00 g of chloroform ($K_b = 3.63$ °C/m), increasing the boiling point of chloroform by 2.45 Celsius degrees. Calculate eugenol's molar mass. Eugenol is 73.17% C, 7.32% H, and 19.51% O by mass. What is the molecular formula of eugenol?
82. The freezing point of a solution prepared by dissolving 150. mg of caffeine in 10.0 g of camphor is 3.07 Celsius degrees lower than that of pure camphor ($K_f = 39.7$ °C/m). What is the molar mass of caffeine? Elemental analysis of caffeine yields the following results: 49.49% C, 5.15% H, 28.87% N, and the remainder is O. What is the molecular formula of caffeine?

Oxidation–Reduction Reactions

CONCEPT REVIEW

83. What is meant by a *half*-reaction?
84. Electron gain is associated with _____ half-reactions and electron loss is associated with _____ half-reactions.
85. How are the gains or losses of electrons related to changes in oxidation numbers?
86. Describe the processes that take place in the electrolysis of molten NaCl.
87. What is the sum of the oxidation numbers of the atoms in a molecule?
88. What is the sum of the oxidation numbers of all the atoms in each of the following polyatomic ions: OH^-, NH_4^+, SO_4^{2-}, and PO_4^{3-}?

PROBLEMS

89. Give the oxidation number of chlorine in each of the following molecules: (a) hypochlorous acid (HClO); (b) chloric acid ($HClO_3$); (c) perchlorate (ClO_4^-).
90. Give the oxidation number of nitrogen in each of the following molecules: (a) elemental nitrogen (N_2); (b) hydrazine (N_2H_4); (c) ammonium ion (NH_4^+).
91. Balance the following half-reactions by adding the appropriate number of electrons. Which half-reactions are oxidation half-reactions and which are reduction half-reactions?
    a. $Br_2(l) \rightarrow 2\ Br^-(aq)$
    b. $Pb(s) + 2\ Cl^-(aq) \rightarrow PbCl_2(s)$
    c. $O_3(g) + 2\ H^+ \rightarrow O_2(g) + H_2O(l)$
    d. $H_2S(g) \rightarrow S(s) + 2\ H^+(aq)$
92. Balance the following half-reactions by adding the appropriate number of electrons. Which half-reactions are oxidation half-reactions and which are reduction half-reactions?
    a. $Fe^{2+}(aq) \rightarrow Fe^{3+}(aq)$
    b. $AgI(s) \rightarrow Ag(s) + I^-(aq)$
    c. $VO_2^+(aq) + 2\ H^+(aq) \rightarrow VO^{2+} + H_2O(l)$
    d. $I_2(s) + 6\ H_2O(l) \rightarrow 2\ IO_3^-(aq) + 12\ H^+(aq)$
93. Iron is oxidized in a number of chemical-weathering processes. Write a half-reaction for the oxidation of magnetite (Fe_3O_4) to hematite (Fe_2O_3) in acidic groundwater. Add H_2O, H^+, and electrons as needed to balance the half-reaction.
94. The mineral rhodochrosite [manganese(II) carbonate, $MnCO_3$] is a commercially important source of manganese. Write a half-reaction for the oxidation of the manganese in $MnCO_3$ to MnO_2 in neutral groundwater where the principal carbonate species is HCO_3^-. Add H_2O, H^+, and electrons as needed to balance the half-reaction.
95. The following chemical reactions have helped shape Earth's crust. Determine the oxidation numbers of all the elements in the reactants and products, and identify which of the elements are oxidized or reduced.
    a. $3\ SiO_2 + 2\ Fe_3O_4 \rightarrow 3\ Fe_2SiO_4 + O_2$
    b. $SiO_2 + 2\ Fe + O_2 \rightarrow Fe_2SiO_4$
    c. $4\ FeO(s) + O_2(g) + 6\ H_2O(l) \rightarrow 4\ Fe(OH)_3(s)$
96. Determine the oxidation numbers of each of the elements in the following reactions, and identify which of them are oxidized or reduced, if any.
    a. $SiO_2 + 2\ H_2O \rightarrow H_4SiO_4$
    b. $2\ MnCO_3 + O_2 \rightarrow 2\ MnO_2 + 2\ CO_2$
    c. $3\ NO_2 + H_2O \rightarrow 2\ H^+ + 2\ NO_3^- + NO$
97. Combine the following oxidation half-reactions, which are based on common iron minerals, with the half-reaction for the reduction of O_2,
    $$O_2 + 4\ H^+ + 4\ e^- \longrightarrow 2\ H_2O$$
    to develop complete redox reactions.
    a. $2\ FeCO_3 + H_2O \rightarrow Fe_2O_3 + 2\ CO_2 + 2\ H^+ + 2\ e^-$
    b. $3\ FeCO_3 + H_2O \rightarrow Fe_3O_4 + 3\ CO_2 + 2\ H^+ + 2\ e^-$
    c. $2\ Fe_3O_4 + H_2O \rightarrow Fe_2O_3 + 2\ H^+ + 2\ e^-$
98. Uranium is found in Earth's crust as UO_2 and an assortment of compounds containing UO_2^{n+} cations. Add the following pairs of half-reactions to develop overall chemical reactions for converting soluble uranium polyatomic ions into insoluble UO_2.
    a. $6\ H^+ + UO_2(CO_3)_3^{4-} + 2\ e^- \rightarrow UO_2 + 3\ CO_2 + 3\ H_2O$
       $Fe^{2+} + 3\ H_2O \rightarrow Fe(OH)_3 + 3\ H^+ + e^-$
    b. $6\ H^+ + UO_2(CO_3)_3^{4-} + 2\ e^- \rightarrow UO_2 + 3\ CO_2 + 3\ H_2O$
       $HS^- + 4\ H_2O \rightarrow SO_4^{2-} + 9\ H^+ + 8\ e^-$
    c. $2\ e^- + UO_2(HPO_4)_2^{4-} \rightarrow UO_2 + 2\ HPO_4^{2-}$
       $3\ OH^- \rightarrow H_2O + HO_2^- + 2\ e^-$
99. Nitrogen in the hydrosphere is found primarily as ammonium ions, NH_4^+, and nitrate ions, NO_3^-. Complete and balance the following chemical equation describing the oxidation of ammonium ions to nitrate ions in acid solution.
    $$NH_4^+(aq) + O_2(g) \longrightarrow NO_3^-(aq)$$

100. The solubilities of Fe and Mn in freshwater streams are affected by changes in their oxidation states. Complete and balance the following redox reaction in which soluble Mn^{2+} becomes solid MnO_2:
$Fe(OH)^{2+}(aq) + Mn^{2+}(aq) \longrightarrow MnO_2(s) + Fe^{2+}(aq)$

101. The classical method for determining the amount of dissolved oxygen in natural waters requires a series of redox reactions. Balance the following chemical equations under the conditions indicated:
 a. $Mn^{2+} + O_2 \rightarrow MnO_2$ (basic solution)
 b. $MnO_2 + I^- \rightarrow Mn^{2+} + I_2$ (acidic solution)
 c. $I_2 + S_2O_3^{2-} \rightarrow S_4O_6^{2-} + I^-$ (neutral solution)

102. Silver can be extracted from rocks by using cyanide ion. Complete and balance the following reaction for this process.

$Ag(s) + CN^-(aq) + O_2(g) \longrightarrow Ag(CN)_2^-(aq)$ (basic solution)

103. Permanganate ion (MnO_4^-) is used in water purification to remove oxidizable substances. Complete and balance the following reactions for the removal of sulfide, cyanide, and sulfite. Assume that reaction conditions are slightly basic.
 a. $MnO_4^-(aq) + S^{2-}(aq) \rightarrow MnS(s) + S(s)$
 b. $MnO_4^-(aq) + CN^-(aq) \rightarrow CNO^-(aq) + MnO_2(s)$
 c. $MnO_4^-(aq) + SO_3^{2-}(aq) \rightarrow MnO_2(s) + SO_4^{2-}(aq)$

104. Chlorine dioxide (ClO_2) is used as a bactericide in water treatment. Complete and balance the following chemical reactions for the preparation of ClO_2 in slightly acidic solutions.
 a. $ClO_3^-(aq) + SO_2(g) \rightarrow ClO_2(g) + SO_4^{2-}(aq)$
 b. $ClO_3^-(aq) + Cl^-(aq) \rightarrow ClO_2(g) + Cl_2(g)$
 c. $ClO_3^-(aq) + Cl_2(g) \rightarrow ClO_2(g) + O_2(g)$

Acid–Base Neutralization Reactions

CONCEPT REVIEW

105. What name is given to a proton donor?
106. What is the difference between a strong acid and a weak acid?
107. Give the formulas of two strong acids and two weak acids.
108. Why is HSO_4^- a weaker acid than H_2SO_4?
109. What name is given to a proton acceptor?
110. What is the difference between a strong base and a weak base?
111. Give the formulas of two strong bases and two weak bases.
112. Write the net ionic reaction for the neutralization of a strong acid by a strong base.

PROBLEMS

113. Identify the acid and the base and then write the net ionic equation for each of the following acid-base reactions:
 a. $H_2SO_4(aq) + Ca(OH)_2(aq) \rightarrow CaSO_4(s) + 2\ H_2O(l)$
 b. $PbCO_3(s) + H_2SO_4(aq) \rightarrow PbSO_4(s) + CO_2(g) + H_2O(l)$
 c. $Ca(OH)_2(aq) + 2\ CH_3COOH(aq) \rightarrow Ca(CH_3COO)_2(aq) + 2\ H_2O(aq)$

114. Complete and balance each of the following neutralization reactions, name the products, and write the net ionic reactions.
 a. $HBr(aq) + KOH(aq) \rightarrow$
 b. $H_3PO_4(aq) + Ba(OH)_2(aq) \rightarrow$
 c. $Al(OH)_3(s) + HCl(aq) \rightarrow$
 d. $CH_3COOH(aq) + Sr(OH)_2(aq) \rightarrow$

Precipitation Reactions

CONCEPT REVIEW

115. What is the difference between a saturated solution and a supersaturated solution?
116. What are common solubility units?

117. What is a precipitation reaction?
118. A precipitate may appear when two completely clear aqueous solutions are mixed. What circumstances are responsible for this event?

PROBLEMS

119. According to the solubility rules, which of the following compounds have limited solubility in water? barium sulfate, barium hydroxide, lanthanum nitrate, sodium acetate, lead hydroxide, and calcium phosphate.

120. Complete and balance the chemical equations for the precipitation reactions between the following pairs of reactants, and write the net ionic equations:
   a. $Pb(NO_3)_2(aq) + Na_2SO_4(aq) \rightarrow$
   b. $NiCl_2(aq) + NH_4NO_3(aq) \rightarrow$
   c. $FeCl_2(aq) + Na_2S(aq) \rightarrow$
   d. $MgSO_4(aq) + BaCl_2(aq) \rightarrow$

121. The use of lead(II) carbonate and lead(II) hydroxide as pigments in white paint has been discontinued because their solubility increases in acidic solutions and lead is toxic to humans. Using the appropriate net ionic equations, show why lead carbonate and lead hydroxide dissolve in acid.

122. Many homeowners treat their lawns with $CaCO_3(s)$ to reduce the acidity of the soil. Write a net ionic equation for the reaction of $CaCO_3(s)$ with a strong acid.

123. Show with appropriate net ionic reactions how Cr^{3+} and Cd^{2+} can be removed from wastewater by treatment with solutions of sodium hydroxide.

124. An aqueous solution containing Ca^{2+}, Cl^-, NO_3^-, and CO_3^{2-} is allowed to evaporate. Which compound will precipitate first?

125. 10 mL of a 5×10^{-3} M solution of Cl^- is reacted with increasing volumes of a 0.500 M solution of $AgNO_3$. What is the maximum mass of AgCl that precipitates?

126. Calculate the mass of $MgCO_3$ precipitated by mixing 10 mL of a 0.2 M Na_2CO_3 solution with 5 mL of a 0.05 M $Mg(NO_3)_2$ solution.

Ion Exchange

PROBLEMS

127. Explain how a mixture of anion and cation exchangers can be used to deionize water.

128. How is a used cation exchanger regenerated for further use?

Solution Stoichiometry

PROBLEMS

129. How many milliliters of 0.100 M NaOH are required to neutralize the following solutions? (a) 10.0 mL of 0.0500 M HCl; (b) 23.6 mL of 0.126 M HNO_3; (c) 15.8 mL of 0.215 M H_2SO_4.

130. How many milliliters of 0.100 M HNO_3 are required to neutralize the following solutions? (a) 45.0 mL of 0.667 M KOH; (b) 58.5 mL of 0.0100 M $Al(OH)_3$; (c) 34.7 mL of 0.775 M NaOH.

131. The solubility of slaked lime, $Ca(OH)_2$, in water is 0.148 g/100 mL. What volume of 0.00100 M HCl is needed to neutralize 10 mL of a saturated $Ca(OH)_2$ solution?

132. The solubility of magnesium hydroxide, $Mg(OH)_2$, in water is 1.33×10^{-3} g/100 mL. What volume of 0.00100 M HNO_3 is required to neutralize 1.00 L of saturated $Mg(OH)_2$ solution?

133. Highly toxic Cr(VI) can be precipitated from an aqueous solution by bubbling SO_2 through the solution. How many grams of SO_2 are required to remove the Cr(VI) from 100 mL of 0.25 M CrO_4^{2-}?

$$2\,CrO_4^{2-} + 3\,SO_2 + 4\,H^+ \longrightarrow Cr_2(SO_4)_3 + 2\,H_2O$$

134. Fe(II) can be precipitated from a slightly basic aqueous solution by bubbling oxygen through the solution, which oxidizes Fe(II) to insoluble Fe(III):
$$4\,Fe(OH)^+(aq) + 4\,OH^-(aq) + O_2(g) + 2\,H_2O(l) \longrightarrow 4\,Fe(OH)_3(s)$$
How many grams of O_2 are consumed to precipitate all of the iron in 75 mL of 0.090 M Fe(II)?

135. Toxic cyanide ions can be removed from wastewater by adding hypochlorite:
$$2\,CN^-(aq) + 5\,OCl^-(aq) + H_2O(l) \longrightarrow N_2(g) + 2\,HCO_3^-(aq) + 5\,Cl^-(aq)$$
How many liters of 0.125 M OCl^- are required to remove the cyanide in 3.4×10^6 liters of wastewater in which the cyanide concentration is 0.58 mg/L?

136. Iron(II) can be precipitated from neutral solutions by reacting it with MnO_4^-:
$$Mn_4^-(aq) + 3\,Fe(OH)^+(aq) + 2\,OH^-(aq) + 2\,H_2O(l) \longrightarrow MnO_2(s) + 3\,Fe(OH)_3(s)$$
What volume of 0.825 M MnO_4^- is needed to completely react with 50 mL of 0.545 M Fe(II)?

137. How many grams of $PbCO_3$ will dissolve when 1.00 L of 1.00 M H^+ is added to 5.00 g of $PbCO_3$? Given:
$$PbCO_3(s) + 2\,H^+(aq) \longrightarrow Pb^{2+}(aq) + H_2O(l) + CO_2(g)$$

138. Phosphate can be removed from drinking-water supplies by treating the water with $Ca(OH)_2$:
$$5\,Ca(OH)_2(aq) + 3\,PO_4^{3-}(aq) \longrightarrow Ca_5OH(PO_4)_3(s) + 9\,OH^-(aq)$$
How much $Ca(OH)_2$ is required to remove 90% of the PO_4^{3-} from 4.5×10^6 L of drinking water containing 25 mg/L of PO_4^{3-}?

Colloids

CONCEPT REVIEW

139. How can colloids be detected?
140. What are the approximate dimensions of colloidal particles?
141. What are the two main types of colloids?
142. What additions to a colloidal system can lead to coagulation of the colloid?

6 Chemical Bonding
And Atmospheric Molecules

Auroras are spectacular displays produced by the solar wind colliding with Earth's upper atmosphere.

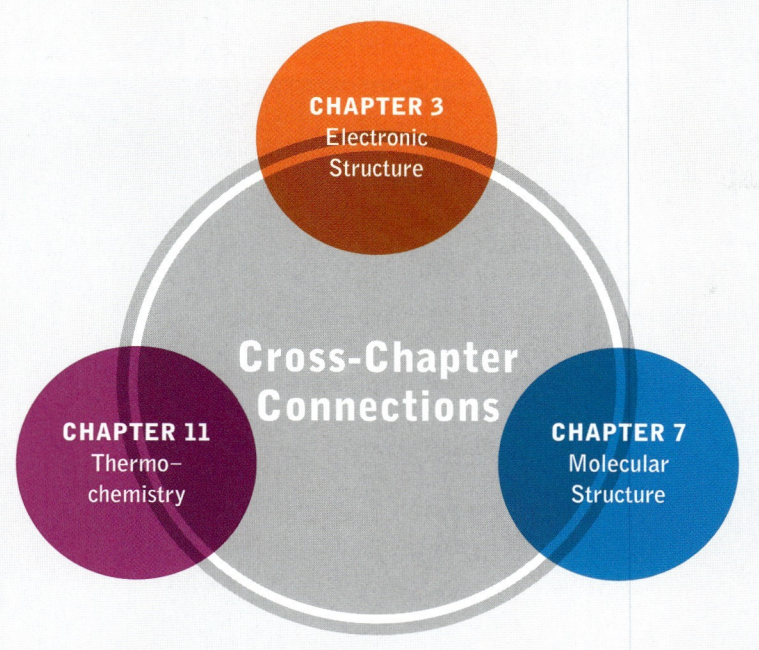

6.1 Introduction
6.2 Electron sharing
6.3 Lewis structures
6.4 Unequal sharing
6.5 Electronegativity and other periodic properties of the elements
6.6 More Lewis structures
    The structure of ozone
    Resonance
    Holes in the ozone layer
6.7 Choosing between Lewis structures: formal charges
6.8 Electron diffraction, bond lengths, and predictions confirmed
6.9 Molecular-orbital theory
    The molecular orbitals of H_2
    Nitrogen and oxygen
    Ultraviolet and visible spectra and auroras

A Look Ahead

The concept of chemical bonding is introduced by using the structure and properties of substances found in the upper atmosphere. The covalent bonding between atoms in molecules arises from the sharing of electrons in the outermost (valence) shells of these atoms. The number of bonds in a molecule and the number of bonds to each atom can often be predicted by using the *octet rule* and by writing Lewis structures. A consideration of formal charges enables us to choose between alternative bonding patterns and to accurately predict the structures of molecules. Molecular-orbital (MO) theory can be used to explain the spectra and magnetic properties of molecules.

6.1 INTRODUCTION

Sometimes the skies at extreme northern and southern latitudes brighten with a shimmering glow known as the *aurora borealis* (northern lights) and the *aurora australis* in the Southern Hemisphere. Auroras form in the upper atmosphere in a layer called the thermosphere that extends from 65 to 500 km above Earth

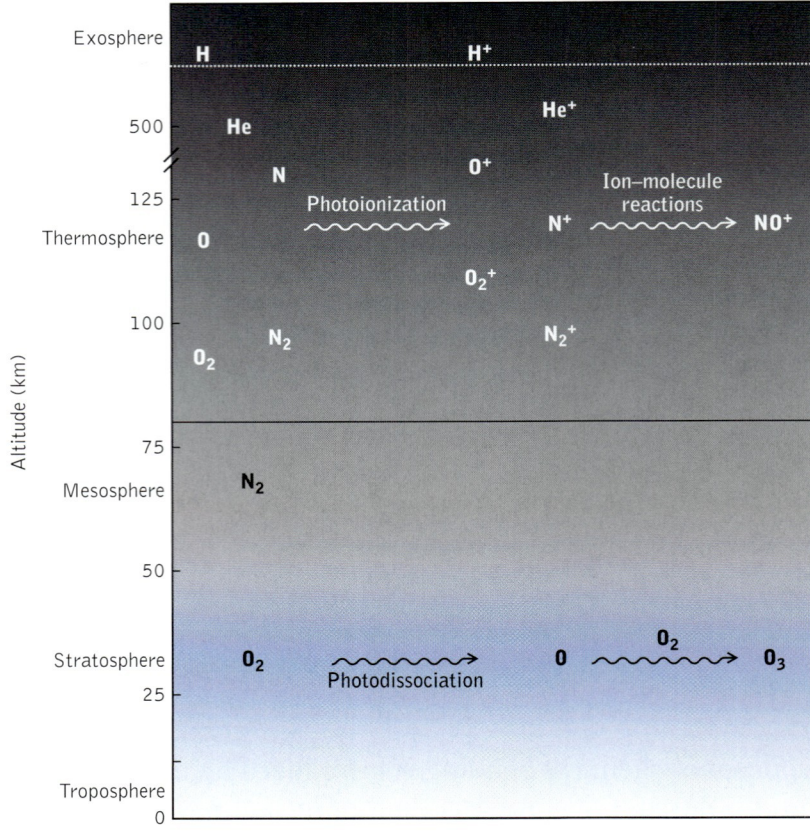

FIGURE 6.1 The major species present in the atmosphere and the photochemical reactions they undergo vary with altitude. Radiation from the sun dissociates atmospheric molecules and ionizes both atoms and molecules. See Figure 6.5 for information on the concentrations of these constituents.

(Figure 6.1). In its upper region, the thermosphere consists mostly of free atoms of helium, hydrogen, and oxygen; at lower altitudes, it is mostly molecular nitrogen, N_2, and molecular oxygen, O_2. Auroras occur when the molecules of these atmospheric gases absorb ultraviolet radiation from the sun or when they collide with the high-velocity electrons and positive particles of the solar wind. These absorption and collision events produce excited-state molecules and ions that emit the characteristic colors of aurora displays as they return to their ground states. To understand how these particles absorb and emit electromagnetic radiation, we need to examine their structures and the way in which the atoms in molecules share their electrons as they form chemical bonds.

6.2 ELECTRON SHARING

At the beginning of the past century, there was a growing body of evidence that electrons are involved in the formation of chemical bonds. Some of this evidence came from the pioneering work of British chemist and physicist Michael Faraday (1791–1867), who found that passing an electrical current through solutions caused chemical reactions to take place (see Section 5.5). By 1900, leading sci-

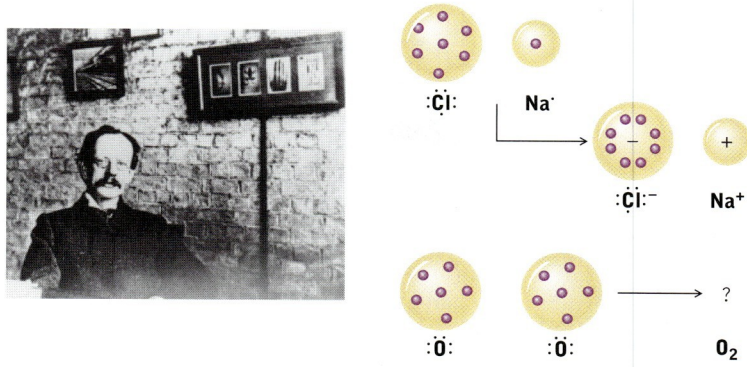

FIGURE 6.2 J. J. Thomson thought that the formation of NaCl from elemental sodium and chlorine entailed electron transfer from sodium atoms to chlorine atoms, as shown with his plum-pudding model of the atom (see Section 3.2). However, this description of bond formation could not explain the bonding in a homonuclear molecule such as O_2.

entists, including J. J. Thomson, the British physicist who discovered the electron, believed that chemical bonds formed when electrons were transferred from one atom to another, as shown in Figure 6.2. This transfer would leave the donor atom with a net positive charge, and the electron acceptor would acquire a negative charge. The two ions would then be "bonded" together by the electrostatic attraction that draws any pair of oppositely charged particles together.

Thomson's model was a good description of how ionic compounds are formed. Compounds such as sodium chloride that consist of positive and negative ions really are held together by the electrostatic attraction between positively charged cations such as Na^+ and negatively charged anions such as Cl^-. However, the electrostatic model did not explain how two atoms of the same element chemically bond together. If all oxygen atoms have the same electron configuration and the same affinity for their electrons, then why should one of them donate any of its electrons to another to form, say, one O^+ cation and one O^- anion? Another model of chemical bonding was needed to explain the existence of O_2, N_2, H_2, and millions of other molecules.

In 1916, American chemist Gilbert N. Lewis proposed that atoms form bonds by sharing their electrons. In the Lewis model, atoms contained:

- positive "kernels" that are unchanged in chemical reactions and
- clouds of electrons around their kernels that Lewis described as "mutually interpenetrable."

By interpenetrable Lewis meant that an electron could simultaneously contribute to the structure of two atoms. Through this sharing process, each atom could acquire enough electrons to mimic the electron configuration of a noble gas. For many atoms, this configuration is marked by the presence of eight electrons in their outermost energy levels, or shells. In Lewis's day, the chemical stability achieved by such an electron configuration was called the *rule of eight*. Today we call it the **octet rule** and associate it with filled *s* and *p* orbitals of an atom's outermost shell.

✓ Electrostatic forces hold cations and anions together in ionic compounds.

 **CONNECTION:** Atomic orbitals are discussed in Sections 3.7 and 3.8.

✓ The **octet rule** refers to the fact that the lighter atoms of the Group 1A–7A elements tend to form bonds that result in their having eight valence electrons.

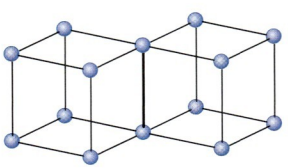

Cubes sharing an edge and one pair of electrons

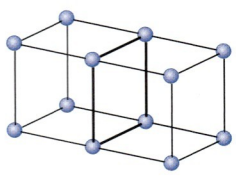

Cubes sharing a face and two pairs of electrons

Gilbert N. Lewis proposed a model based on cubes that shared edges and faces to explain the formation of single and double covalent bonds, such as those in F_2 and O_2. The cube idea was more a teaching tool than a physical model and has long since vanished, but Lewis's idea of shared electrons in covalent bonding remains a fundamental concept in chemistry.

Covalent bonds result from electron sharing by atoms in molecules.

Lewis's theory still dominates our view of the formation of **covalent bonds**—that is, bonds formed by the sharing of electrons. However, this theory was rather vigorously attacked when Lewis proposed it in 1916. One well-known chemist of the time made the following, seemingly logical criticism:

> Saying that each of two atoms can attain closed electron shells by sharing a pair of electrons is equivalent to saying that a husband and wife, by having a total of two dollars in a joint account and each having six dollars in individual bank accounts, have got eight dollars apiece.
>
> Kasimir Fajans, quoted in *The Norton History of Chemistry* by William H. Brock (Norton, 1993), p. 477.

The statement illustrates what can happen when an attempt to explain a phenomenon in the world of subatomic particles collides with familiar observations and behavior patterns of the macroscopic world.

In 1923, Lewis published *Valence and the Structure of Atoms and Molecules* in which he defined valency as the capacity of the atoms of an element to form covalent bonds. The periodic table is a powerful tool for predicting an element's *valency*, or bonding capacity, as illustrated by the following examples:

The outermost electrons, or **valence electrons**, of an atom form chemical bonds.

- Fluorine in Group 7A has seven **valence electrons** per atom and so should need one more to complete its octet. A fluorine atom can complete its octet by sharing one of its own electrons and one belonging to another atom. By sharing this pair of electrons, the two atoms form a single covalent bond. It turns out that fluorine forms only one bond per atom. Fluorine and the other Group 7A elements exist as homonuclear (same atom) diatomic molecules: F_2, Cl_2, Br_2, and I_2. Each of these molecules is held together by a single covalent bond.
- Group 6A elements such as oxygen have six valence electrons per atom and so need two more to complete their octets. Within molecules, oxygen atoms form either two single bonds (as in the two O—H bonds in a molecule of H_2O) or one double bond, such as the O=O bond in a molecule of O_2.

- Group 5A elements such as nitrogen need three more electrons per atom to complete their octets. Within molecules, nitrogen forms either three single bonds (such as the three N—H bonds in ammonia, NH_3), a single and a double bond, or one triple bond, such as the N≡N bond in N_2.
- Group 4A elements need to acquire four more valence-shell electrons per atom to complete their octets. Carbon atoms they form four single bonds such as the four C—H bonds in a molecule of methane, CH_4. In other molecules, they form two single bonds and a double bond such as the two C—H bonds and one C=O bond in formaldehyde, CH_2O; or two double bonds such as the two C=O bonds in a molecule of carbon dioxide, CO_2; or a triple and a single bond such as the C≡N and C—H bonds in hydrogen cyanide, HCN.

6.3 LEWIS STRUCTURES

In developing his theory of chemical bonding, Lewis proposed a system for diagramming the bonding patterns inside molecules, a system that is still widely used today. The starting points for his system are atomic symbols surrounded by dots representing valence electrons. Consider the Lewis symbol for nitrogen, which has the abbreviated electron configuration $[He]2s^2 2p^3$. The valence shell of a nitrogen atom has a pair of electrons in the $2s$ orbital and three unpaired electrons in each of the $2p$ orbitals. The Lewis symbol of N illustrates this distribution of five valence electrons as follows:

The Lewis symbols for all the elements in Groups 1A through 8A of the periodic table are shown in Figure 6.3. In the procession from left to right across a period, note how the Lewis symbol for each element contains one more electron than does the Lewis symbol for the preceding element. Elements in the same group have identical Lewis symbols (and similar chemical properties).

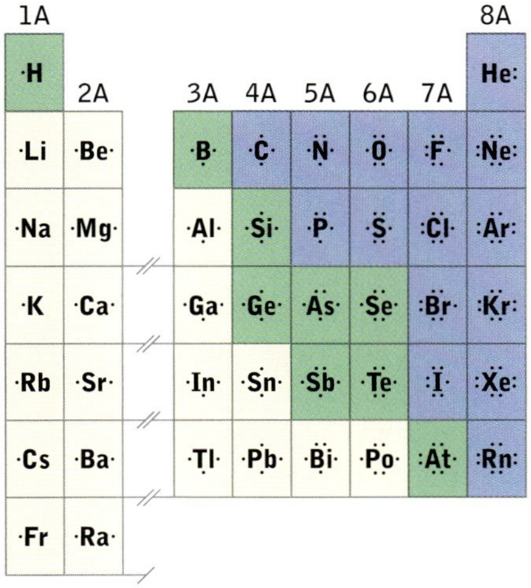

FIGURE 6.3 This abbreviated periodic table shows Lewis symbols for the elements in Groups 1A to 8A. Dots are used to indicate the valence electrons for each element. Notice that the number of dots representing electrons increases from left to right. The Lewis symbols for all the elements in a group contain the same number of dots, indicating that elements in a group have the same number of valence electrons.

FIGURE 6.4 Lewis symbols for molecular nitrogen and oxygen indicate that two pairs of electrons are shared in the O═O double bond in O_2 and three pairs are shared in the N≡N triple bond in N_2. We distinguish between the shared (bonding) electrons and the unshared (nonbonding) electrons by using a straight line for the former and dots for the latter. The overlapping spheres in the space-filling models of O_2 and N_2 also indicate that electrons are shared by covalently bonded atoms.

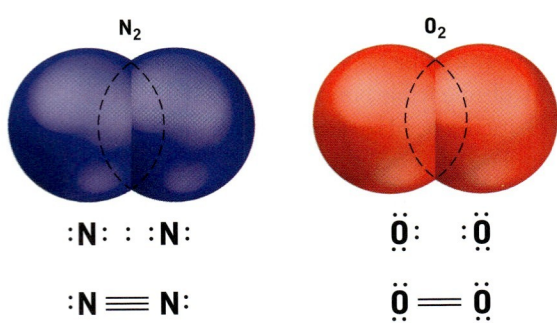

> A **Lewis structure** represents the arrangement of valence electrons in molecules: bonding pairs of electrons are shown as straight lines; nonbonding pairs, or **lone pairs**, are shown as pairs of dots.

CONNECTION: These excited states in molecules are analogous to the excited states in atoms described in Sections 3.10 and 3.11.

> **Molecular ions** are formed when molecules lose or gain electrons.

To examine how Lewis symbols can be used to draw a **Lewis structure** showing the bonding in a molecule, let's consider some of the gases found in the atmosphere. The Lewis structures of the two most abundant gases, N_2 and O_2, are shown in Figure 6.4. As noted in Section 6.2, atoms of oxygen share two pairs of electrons to complete their octets. Sharing two pairs of electrons leads to the formation of a double bond in O_2. Nitrogen atoms must share three pairs of electrons to fill their octets, and so the atoms in N_2 are connected by a triple bond. To distinguish electron pairs that form bonds from those that do not, we show each bonding pair in a Lewis structure as a straight line (Figure 6.4), and the pairs of valence electrons that are not taking part in bond formation, called **lone pairs,** are represented by pairs of dots.

In the thermosphere, bombardment of N_2 and O_2 by high-energy electromagnetic radiation and high-velocity ions from the sun raises the energies of the bonding electrons in these molecules, resulting in excited-state molecules analogous to the excited-state atoms discussed in Chapter 3. If the wavelengths of absorbed radiation are short enough or if the ions of the solar wind are energetic enough, these interactions may result in electron loss and the formation of **molecular ions.** They may also break the bonds holding N_2 and O_2 together, producing free atoms of N and O. Some photons of ultraviolet radiation can simultaneously ionize *and* excite the ions formed from molecular oxygen and nitrogen, producing O_2^{+*} and N_2^{+*}. The asterisks indicate that these molecular ions are in excited states. As a result of these high-energy interactions, the thermosphere contains a variety of nitrogen and oxygen species, including:

> neutral atoms and molecules (N, O, N_2, O_2),
> ionized atoms and molecules (N^+, O^+, N_2^+, O_2^+), and
> excited states of all of the above.

Collisions between the various forms of nitrogen and oxygen also can result in the formation of new chemical bonds—that is, in chemical reactions. One product of these collisions in the upper atmosphere is the molecular ion NO^+. If you were to analyze Earth's atmosphere starting from the fringes of outer space and working your way down to the planet's surface, the first substance that you would detect that was made from different elements would be NO^+ (Figure 6.5). Equations 6.1 through 6.3 describe three ways in which NO^+ might be formed in the thermosphere. Note that, in all three equations, the numbers of atoms of

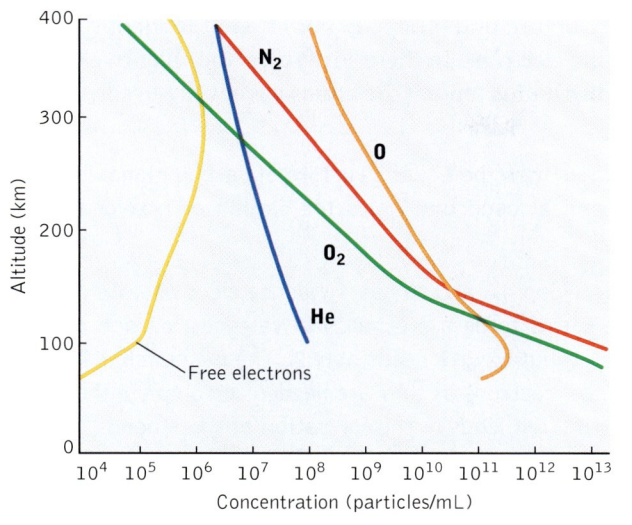
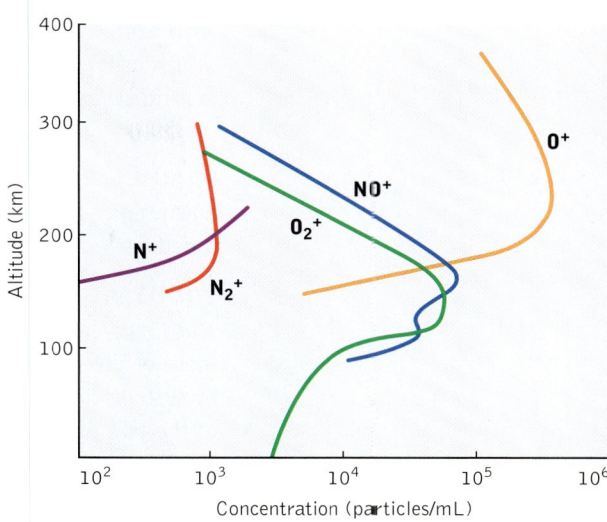

FIGURE 6.5 The concentrations of (A) the principal atoms and molecules of the upper atmosphere and (B) the products of their photoionization vary with altitude. Increasing intensity of ionizing radiation, with increasing altitude (but decreasing concentrations of N_2 and O_2), lead to maxima in the concentrations of ions made of N and O, as shown in (B).

each element are in balance *and* the sum of the electrical charges on both sides is the same.

$$O^{+*} + N_2 \longrightarrow NO^+ + N^* \qquad (6.1)$$
$$N_2^+ + O \longrightarrow NO^+ + N^* \qquad (6.2)$$
$$O_2^+ + N^* \longrightarrow NO^+ + O \qquad (6.3)$$

How are nitrogen and oxygen atoms bonded together in NO^+? To answer this question, we must first determine the number of valence electrons that NO^+ contains. Nitrogen has 5 valence electrons and oxygen has 6. However, the +1 charge on NO^+ means there is 1 valence electron fewer than the total of 11 in N and O. Because there are only 10 valence electrons available, which works out to 5 for each atom, three bonds are needed so that each atom has a filled octet. Thus there must be a triple bond between the N and O. The Lewis structure of NO^+ shows triple-bonded N and O atoms surrounded by brackets to indicate that the species is an ion. The charge on the ion appears as a superscript outside the right-hand bracket.

$$[:N\equiv O:]^+$$

The remaining four electrons in NO^+ are lone pairs: one pair on each of the nitrogen and oxygen atoms. Note that NO^+ and N_2 have the same number of valence electrons—that is, they are isoelectronic—and they have similar Lewis structures. Also note that this Lewis structure and others do not attempt to identify which atoms provide which bonding electrons. In these structures, the valence electrons are indistinguishable and distributed among the atoms as they are needed.

SAMPLE EXERCISE 6.1: Draw the Lewis structure of carbon monoxide (CO) and predict the type of chemical bond that holds the carbon and oxygen atoms together.

SOLUTION: Carbon is in Group 4A and so has 4 valence electrons. As we have seen, oxygen has 6. There seems to be a mismatch in valences: carbon needs 4 more electrons to fill its octet, but oxygen needs only 2. The mismatch is handled by distributing the 10 valence electrons as they are needed to complete the octets of the two atoms. This distribution leads to the formation of three bonds, leaving single lone pairs on C and O:

$$:C\equiv O:$$

PRACTICE EXERCISE: Unlike O_2 and CO, the compound SO is highly unstable. Draw a Lewis structure for SO and predict the type of bond between S and O. (See Problems 25–28.)

6.4 UNEQUAL SHARING

Although the atoms in diatomic molecules of pure elements, such as N_2, O_2, and H_2, share their bonding pairs equally, the same cannot be said for bonds between atoms of different elements that have different effective nuclear charges and so attract bonding pairs of electrons differently.

Gilbert Lewis realized that electron sharing does not necessarily mean *equal* sharing. For example, he knew from the chemical properties of HCl that it is a *polar* molecule, which means that its hydrogen and chloride "ends" are electrically polarized in much the way that the ends of a bar magnet have north and south magnetic poles. He explained the polarity by assuming that the shared electrons and their negative charges are closer to the chlorine end of the molecule than the hydrogen end. The chlorine atom has a larger effective nuclear charge, and so it attracts the electrons that it shares with an atom of hydrogen more than the hydrogen atom does. As a result, the electrons are drawn more closely to the chlorine end of the molecule, which means that the H—Cl bond is a **polar covalent bond**.

Lewis regarded the formation of ions as unequal sharing taken to an extreme. In NaCl, for example, chlorine attracts the bonding electrons that it shares with sodium so much more strongly than sodium does that chlorine effectively acquires the bonding pair, forming Cl^-. By giving away its electron, sodium becomes positively charged Na^+. In the Lewis diagram of NaCl,

$$[Na]^+ \ [:\ddot{Cl}:]^-$$

sodium is shown with no valence electrons, because it lost the only one that it had, and the chloride ion appears with a complete octet, having acquired the

 **Polar covalent bonds** result from unequal electron sharing of bonding pairs of electrons between atoms.

CONNECTION: The polarity of a molecule depends not only on its having polar bonds, but also on its shape (Section 7.6).

electron that sodium lost. The transfer of an electron from a sodium atom to a chlorine atom means that the two atoms take on the electron configurations of neon and argon, respectively.

All alkali and alkaline earth elements tend to lose, rather than share, their valence electrons. However, the cations that they form still satisfy the octet rule because they have the electron configurations of the noble gases that precede the parent elements in the periodic table.

6.5 ELECTRONEGATIVITY AND OTHER PERIODIC PROPERTIES OF THE ELEMENTS

Another American chemist, Linus Pauling (1901–1994), developed a useful concept, called **electronegativity,** for predicting the degree to which bonding pairs of electrons are shared unequally. Electronegativity is a measure of an element's ability to attract electrons in a covalent bond to its end of the bond. Pauling's electronegativity scale is based on the idea that bonds between atoms of different elements are neither 100% covalent (which would represent equal electron sharing) nor 100% ionic (completely unequal electron sharing). Instead, he proposed, heteroatomic compounds are held together by bonds that are somewhere between these two extremes.

> **Electronegativity** is a measure of an element's ability to attract bonding electrons.

Electronegativity is a periodic property of the elements, which means that electronegativity is related to an element's position in the periodic table. Electronegativity values generally increase from left to right and decrease from top to bottom, as shown in Figure 6.6. Thus the most-electronegative elements, including fluorine, oxygen, and nitrogen, are found in the upper right corner of the table, and the least-electronegative elements are the metals in the lower left corner.

Electronegativity values such as those in Figure 6.6 have no units, because they are expressed on a relative scale. However, electronegativity is related to parameters that do have absolute values, such as the sizes of atoms (Figure 6.7) and the ionization energies of the elements (Figure 6.8). For the **main-group elements**—that is, the elements in Groups 1A, 2A, and 3A through 8A—there is a clear inverse relation between atomic size and electronegativity: the smaller the atom, the greater its electronegativity. This relation makes sense. Within a period (row) of elements, those on the right have more protons in their nuclei than do those on the left, but their valence electrons are in the same shell as those on the left. Therefore, the pull on the valence electrons, including shared valence electrons, is greater on the right-hand side than on the left-hand side. Greater attraction to the nucleus draws the valence electrons inward, making the size of the atoms on the right-hand side smaller. Greater attraction also applies to bonding electrons, and so electronegativity also increases from left to right across a period.

> The **main-group** (or **representative**) **elements** are those in Groups 1A, 2A, and 3A through 8A.

Within a group (column) of elements, charges in electronegativity are related to increasing atomic size with increasing atomic number. In the procession

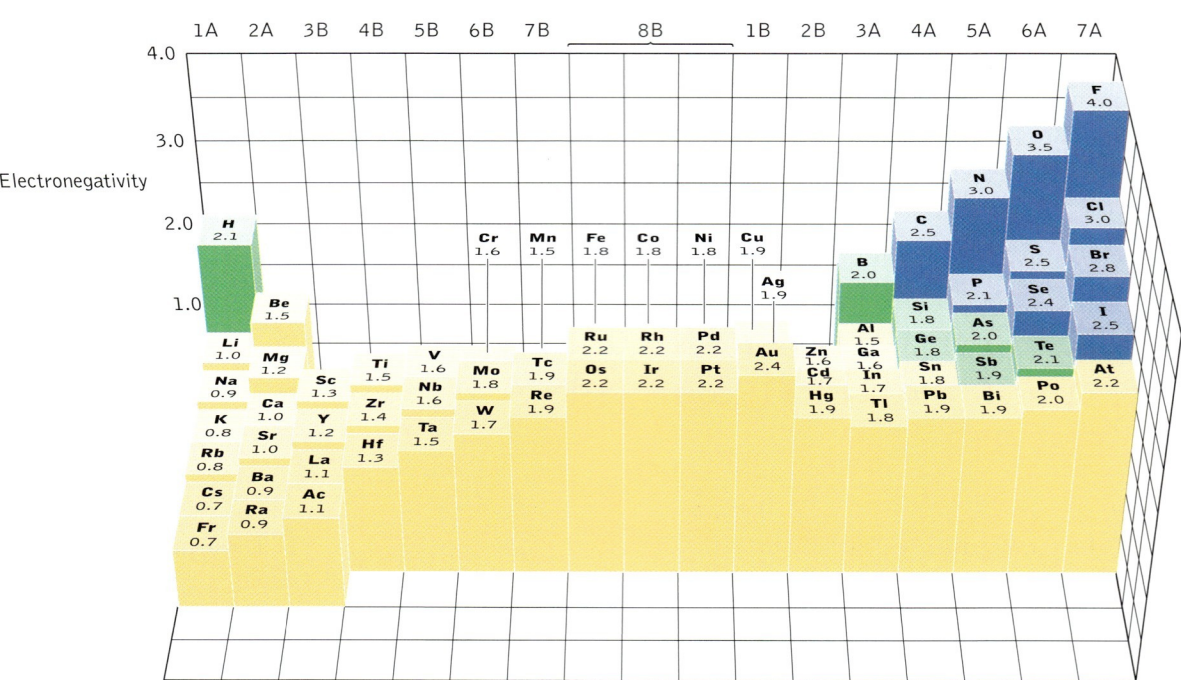

FIGURE 6.6 Electronegativity increases from left to right within rows (periods of elements) of the periodic table and decreases from top to bottom within columns (groups of elements). A greater electronegativity indicates a greater ability of an atom to attract bonding electrons.

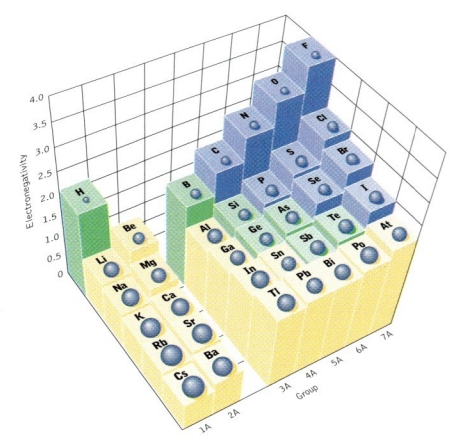

FIGURE 6.7 The sizes of the atoms of the main-group elements increase with increasing atomic number in a group but decrease with increasing atomic number in a row. Electronegativities decrease with increasing atomic number in a group but increase with increasing atomic number in a row. These opposite trends are consistent with the trends in effective nuclear charge, which decreases with increasing atomic number in a group but increases with increasing atomic number in a row.

CONNECTION: Inner-shell electrons shield outer-shell electrons from the positive charge of the nucleus (Sections 3.10 and 3.11).

down a group, the size of the valence orbitals increases as the value of the principal quantum number (n) increases. The valence electrons of those elements toward the bottom of a group are separated and shielded from their nuclei by greater distances and by more layers of inner-shell electrons than are the elements toward the top of the group. This shielding is the basis for the concept of effective nuclear charge (Z_{eff}) discussed in Section 3.10. More shielding leads to less

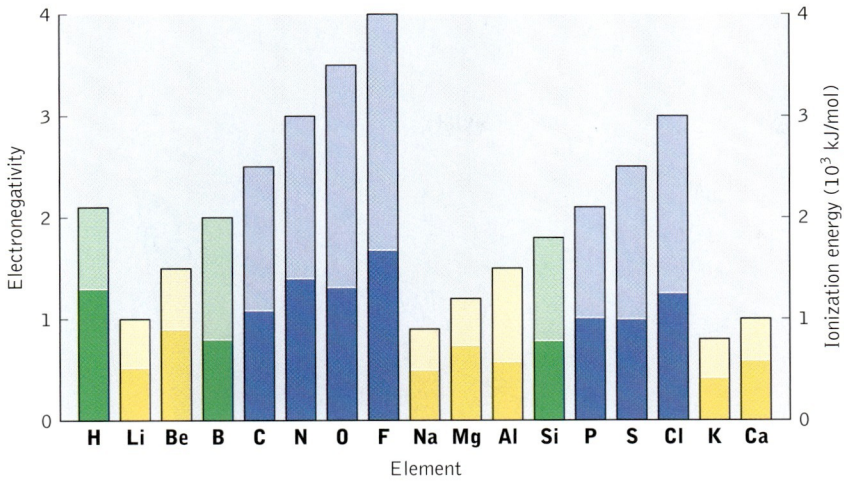

FIGURE 6.8 The trend in ionization energies among the main-group elements follows the trend in their electronegativities. The transparent columns represent electronegativity values.

increase in Z_{eff} with increasing atomic number. More shielding and greater atomic size lead to less pull on bonding pairs of electrons and, therefore, lower electronegativity with increasing atomic number.

The correlation between electronegativity and ionization energy is easy to explain. The same factors that make it more difficult to remove an outer-shell electron from an atom should also make that atom more effective in attracting shared valence electrons. The periodic nature of ionization energies was developed in Section 3.11, where it was noted that

- ionization energies among the elements in a group of the periodic table decrease with increasing atomic number and,
- across a period in the periodic table, ionization energies increase with increasing atomic number.

Electronegativity values follow the same pattern.

Another periodic property that correlates reasonably well with electronegativity is called **electron affinity.** As the term implies, electron affinity is the attraction between atoms and additional electrons. Specifically, the electron affinity of an element is defined as the energy change that occurs when a mole of its atoms in the gaseous state acquires 1 mole of electrons. Suppose a mole of free fluorine atoms acquires a mole of electrons and becomes a mole of fluoride ions:

$$F(g) + e^- \longrightarrow F^-(g)$$

This reaction is accompanied by a considerable release of energy: 328 kJ, as the less-stable fluorine atoms with their 7 valence electrons become more stable, 8-valence-electron fluoride ions. This flow of energy from the reaction system means that the electron affinity for fluorine is -328 kJ. The negative sign means that the product of the reaction, a mole of fluoride ions, has less energy than the reactants: a mole of fluorine atoms and a mole of electrons. Thus, large *negative* values for electron affinity indicate strong attraction for electrons.

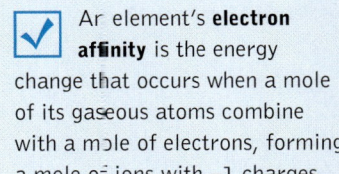

An element's **electron affinity** is the energy change that occurs when a mole of its gaseous atoms combine with a mole of electrons, forming a mole of ions with −1 charges.

CONNECTION: Energy that is released during a reaction is lost by the reaction system and so has a negative sign; energy absorbed has a positive sign (Section 11.5).

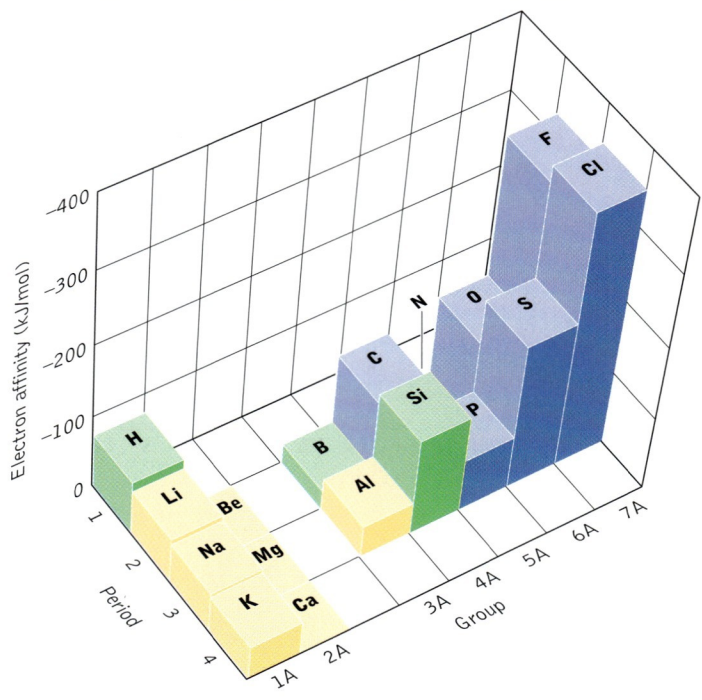

FIGURE 6.9 The main-group elements with atoms that have the highest (most-negative) electron-affinity values tend to be on the right-hand side of the periodic table. Nitrogen and phosphorus have low electron affinities because when one of their atoms acquires an electron it loses the stability of a half-filled set of *p* orbitals.

It makes sense that stronger attraction for an additional electron should correlate with stronger pull on shared pairs of bonding electrons (electronegativity). Thus, the elements with the most-negative values of electron affinity are located on the upper right-hand side of the periodic table (Figure 6.9), as are the elements with the greatest electronegativities. In general, the least-negative electron affinities are among the elements on the left-hand side. Electron affinities for some elements are actually greater than zero, meaning that energy must be added to force their atoms to acquire electrons. For these elements, the addition of an electron produces an electron configuration with less stability and, therefore, higher energy.

CONNECTION: Ground-state electron configurations of atoms are described in Section 3.10.

CONCEPT QUESTION: Do you think the electron-affinity values of the Group 8A elements (noble gases) are as negative as those of the Group 7A elements or less so?

 Electronegativities increase with increasing atomic number across periods but decrease with increasing atomic number down groups of elements in the periodic table.

Electronegativity allows us to determine which end of any covalent bond will be electron rich and which will be electron poor. It also allows us to estimate the covalent versus ionic character of a bond (and a compound) if we assume that ion formation is an extreme case of unequal sharing in a covalent bond. There is no simple rule for all compounds, but in general we can assume that ionic compounds are formed between two elements when the difference in electronegativities between the two is 2.0 or greater. Such differences exist in compounds formed between metallic and nonmetallic elements. For example, calcium oxide

6.5 Electronegativity and Other Periodic Properties of the Elements

is considered an ionic compound because the difference in electronegativity between Ca (1.0) and O (3.5) is 2.5. On the other hand, the bonds formed between different nonmetals are covalent bonds. These bonds are more polar or less polar, depending on the difference in the electronegativities of the elements. For example, an O—Cl bond is less polar than a P—Cl bond because the difference in electronegativity between O and Cl (3.5 − 3.0 = 0.5) is less than the difference in electronegativity between P and Cl (3.0 − 2.1 = 0.9).

SAMPLE EXERCISE 6.2: In carbon monoxide, which end of the C≡O triple bond is electron rich?

SOLUTION: The unequal sharing of bonding pairs of electrons is related to differences in the electronegativities of the elements on the ends of the bond. The electronegativity of oxygen (3.5) is greater than that of carbon (2.5), and so the oxygen end of the C≡O bond is electron rich.

PRACTICE EXERCISE: In hydrogen bromide, which end of the H—Br bond is electron rich? (See Problems 37 and 38.)

SAMPLE EXERCISE 6.3: Rank the bonds formed between the following pairs of elements in order of increasing polarity. Do any of these pairs form an ionic compound? O and C; Cl and Ca; N and S; O and Si.

SOLUTION: Polarity is related to differences in the electronegativities of the elements on the ends of the bond. Calculating these differences for the pairs of elements in this exercise, we get:

O and C: 3.5 − 2.5 = 1.0

Cl and Ca: 3.0 − 1.0 = 2.0

N and S: 3.0 − 2.5 = 0.5

O and Si: 3.5 − 1.8 = 1.7

These differences should be proportional to the polarity of the bonds formed between the pairs of elements. Therefore, ranking them in order of increasing polarity, we have:

1. N and S   2. O and C   3. O and Si   4. Ca and Cl

The difference in electronegativity between Ca and Cl is 2.0, which meets our guideline for calling calcium chloride, $CaCl_2$, an ionic compound. Note: The presence of polar bonds in a molecule may or may not make the molecule itself polar. For example, there are two polar O—H bonds in a molecule of H_2O, and H_2O is indeed a polar molecule. However, there are two polar C═O bonds in CO_2, yet CO_2 is a nonpolar molecule. We will explore why this is so in Chapter 7.

PRACTICE EXERCISE: Which of the following pairs of elements forms the most polar bond? O and S; Be and Cl; N and H; C and F. (See Problems 39 and 40.)

6.6 MORE LEWIS STRUCTURES

Just beneath the thermosphere is a layer called the stratosphere. Some of the molecules in this layer contain more than two atoms. A particularly important *triatomic* (three-atom) molecule in the stratosphere is a molecular form of oxygen called ozone, O_3. Different molecular forms of the same element, such as O_2 and O_3, are called **allotropes.** The allotropes of an element have very different chemical and physical properties. Ozone, for example, is an acrid, pale blue gas that is toxic even at low concentrations, whereas O_2 is a colorless, odorless gas that is essential for life. Their formulas tell us that they must also have different molecular structures.

> **Allotropes** are forms of the same element with different structures and different physical and chemical properties.

Ozone forms in the stratosphere when O_2 absorbs ultraviolet radiation from the sun, breaking the O=O double bond. This dissociation process is a photochemical reaction because it is initiated by photons. Any excess energy of the UV photon above that needed to break the bond increases the kinetic energy of the two free O atoms. The excess energy also can produce excited-state oxygen atoms. These atoms are extremely reactive and rapidly combine with O_2 to give ozone, O_3, as described in Equations 6.5 through 6.7:

$$O_2 \xrightarrow{h\nu} 2\,O^* \tag{6.5}$$

$$O_2 + O^* \longrightarrow O_3^* \tag{6.6}$$

$$O_3^* + M \longrightarrow O_3 + M^* \tag{6.7}$$

The product of Equation 6.6 is an excited-state ozone molecule, as indicated by the asterisk. These excited ozone molecules give up their excess energies and return to the ground state in several ways. One of the most important ways, shown in Equation 6.7, is collision with another molecule "M," the identity of which is not important. Because of the collision, the extra electronic energy in the ozone molecule is transferred to M, creating M*.

Ultraviolet rays are responsible for making stratospheric ozone *and* for destroying it. The products of the photochemical dissociation of ozone are O_2 and an excited-state oxygen atom:

$$O_3 \xrightarrow{h\nu} O_2 + O^* \tag{6.8}$$

Excited-state oxygen atoms return to the ground state by transferring their extra energies to other molecules:

$$O^* + M \longrightarrow O + M^* \tag{6.9}$$

or by emitting light:

$$O^* \longrightarrow O + h\nu \tag{6.10}$$

Free oxygen atoms may combine to form O_2:

$$2\,O \longrightarrow O_2 \tag{6.11}$$

The net effect of Equations 6.8 through 6.10 is the conversion of ozone into O_2 and an oxygen atom:

$$O_3 \xrightarrow{h\nu} O_2 + O$$

If we multiply this equation by two and add it to Equation 6.11, we have the overall equation for the photodecomposition of ozone:

$$2\,O_3 \longrightarrow 2\,O_2 + \cancel{2\,O}$$
$$\underline{+\,\cancel{2\,O} \longrightarrow O_2}$$
$$2\,O_3 \xrightarrow{h\nu} 3\,O_2 \qquad (6.12)$$

In a pristine environment, the conversion of O_3 into O_2 is in equilibrium with the creation of O_3 from O_2 (essentially Equation 6.12 in reverse) so that there is a relatively constant concentration of ozone in the stratosphere. This equilibrium is reached when the rates of forward and reverse reactions are equal. We represent this balancing of reaction rates with a double reaction arrow as shown in Equation 6.13. The equation means that O_2 is formed from O_3 while O_3 is simultaneously formed from O_2.

$$3\,O_2 \underset{}{\overset{h\nu}{\rightleftharpoons}} 2\,O_3 \qquad (6.13)$$

CONNECTION: We will consider chemical equilibrium in much more detail in Chapters 15 and 16.

The structure of ozone

How are the oxygen atoms in ozone bonded together? To answer this question, we need to decide the connectivity, or arrangement, of the atoms in the molecule. Then we must determine how many pairs of electrons must be shared by three atoms of oxygen so that each atom has a complete octet. The following procedure is one that many students find useful for drawing Lewis structures of polyatomic molecules.

> How are the oxygen atoms in ozone bonded together?

1. Connect the atoms in the molecule that are likely to be bonded together with single bonds to create a molecular "skeleton." The molecular skeleton for ozone looks like this:

$$O\!-\!O\!-\!O$$

   If there are two kinds of atoms in the compound, as in CO_2, the less-electronegative element (C in CO_2) is usually the *central* atom (H_2O is an exception to this rule); so CO_2 would have the following skeleton:

$$O\!-\!C\!-\!O$$

2. Count the total number of valence electrons in the molecule. We will represent the number of valence electrons that we *have* with an H. Because ozone consists of three oxygen atoms and each has six valence electrons, the total number of valence electrons (H) = 3 × 6 = 18.

3. Determine the number of valence electrons needed by the atoms in the molecule so that they have noble-gas configurations. For elements with $Z > 5$, this number is eight electrons per atom. (Hydrogen needs only two electrons per atom to mimic the electron configuration of helium. Lithium tends to form +1 ions with the electron configuration of helium, and beryllium and boron may form stable molecules with only four valence electrons around each Be atom and only six around each B atom.) Multiply the number of atoms of each element in the molecule by the number of valence electrons needed by each atom. Add up the valence electrons *needed*, N. For an ozone molecule with three atoms of oxygen, $N = 3 \times 8 = 24$.

4. Calculate the number of available valence electrons that must be *shared* (S) in the molecule through covalent bond formation by using Equation 6.14.

$$S = N - H \quad (6.14)$$

For ozone, the solution to Equation 6.14 is

$$S = 24 - 18 = 6$$

5. Because each bond consists of a pair of shared electrons, the *total* number of bonds in the molecule will be $S/2$. Applying this relation to ozone, we find that 6/2, or three, bonds hold the molecule together.

6. Now add bonds to your original skeleton of the molecule to reach the **S**/2 value. For ozone, we have two options: we could add another bond on the left side of the molecule to create a double bond there:

$$O\!=\!O\!-\!O$$

or we could add one more bond to the right side:

$$O\!-\!O\!=\!O$$

For the moment, let's proceed by using the second structure.

7. When all of the bonds have been drawn, complete the Lewis structure of the molecule by adding nonbonding electrons as needed to complete the octets of all the atoms with $Z > 5$. Remember that hydrogen atoms need only two electrons and so can be connected by only one covalent bond to another atom. Count the total number of valence electrons in the molecule to ensure it is the same as H. Applying step 7 to ozone, we add two pairs of nonbonding electrons to the left-side atom to complete its octet, three pairs to the right-side atom, and one pair to the middle atom. Taking an inventory of all the bonding and nonbonding pairs in our ozone structure, we find that there are nine pairs in all or 18 electrons (as there should be).

$$:\!\ddot{O}\!-\!\ddot{O}\!=\!\ddot{O}\!:$$

Note that we have drawn a bent structure for the molecule. This structure represents the true molecular shape of ozone, which we will explore in Chapter 7. However, Lewis structures are not meant to represent molecular shapes accurately, and an acceptable Lewis structure for ozone could be drawn as follows:

$$:\ddot{O}-\ddot{O}=\ddot{O}$$

SAMPLE EXERCISE 6.4: Draw Lewis structures for water (H_2O) and formaldehyde (CH_2O).

SOLUTION: In each case, we first draw a molecular skeleton for the molecule and then apply the $S = N - H$ rule to determine the number of bonds per molecule. Finally, we will add lone pairs of electrons as needed to complete the octets on those atoms that should have 8 valence electrons. Let's start with water.

An oxygen atom must be in the center of the skeletal structure of H_2O, because hydrogen atoms can have only one bond apiece. Thus:

$$H-O-H$$

Each oxygen atom has 6 valence electrons and each hydrogen atom has 1; so the number of valence electrons that we have is

$$H = 6 + 2(1) = 8$$

Each oxygen atom must have 8 valence electrons and each hydrogen must have 2 electrons to complete their valence shells. Thus the total number of valence electrons needed is

$$N = 8 + 2(2) = 12$$

The number of electrons that must be shared is

$$S = N - H = 12 - 8 = 4$$

and the total number of bonds is $S/2 = 2$. The skeletal structure has two bonds, and so no more are needed. Two lone pairs of electrons are added around the oxygen atom to complete its octet, giving us the following Lewis structure:

$$\overset{:\ddot{O}:}{\underset{H \quad\quad H}{\diagup \quad \diagdown}}$$

Note that the molecule as drawn has a bond angle of 104.5° to represent its true molecular geometry. We will see why H_2O has this geometry in Chapter 7.

To draw the Lewis structure of CH_2O, we must decide which atom lies in the center of the structure. Because carbon is less electronegative than oxygen, we put the carbon atom in the center and connect the other atoms to it with single bonds, producing the following skeletal structure:

$$\begin{array}{c} H-C-O \\ | \\ H \end{array}$$

An oxygen atom has 6 valence electrons, a carbon atom has 4, and each hydrogen atom has 1; so the number of valence electrons in the molecule is

$$H = 6 + 4 + 2(1) = 12$$

Each oxygen and carbon atom must have 8 valence electrons and each hydrogen must have 2 electrons to complete their valence shells. Thus the total number of valence electrons needed is

$$N = 8 + 8 + 2(2) = 20$$

The number of electrons that must be shared is

$$S = N - H = 20 - 12 = 8$$

and the total number of bonds is $S/2 = 4$. The skeletal structure has three bonds. The only place to put a fourth bond is between carbon and oxygen, making a C=O double bond:

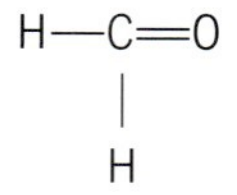

$$\text{H}-\text{C}=\text{O}$$
$$|$$
$$\text{H}$$

Carbon has 4 bonding pairs of electrons in this structure and a complete octet, but oxygen has only 2 bonding pairs and so needs 2 lone pairs to complete its octet, giving us the following Lewis structure:

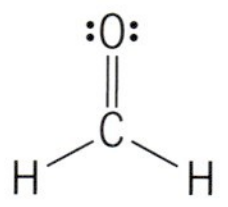

:O:
‖
H–C–H

There are 4 bonding pairs of electrons in this structure and 2 lone pairs for a total of 12 valence electrons, which is the correct number.

PRACTICE EXERCISE: Draw the Lewis structures of chloroform ($CHCl_3$) and hydrogen cyanide (HCN). (See Problems 43–46.)

SAMPLE EXERCISE 6.5: Draw the Lewis structure of the ammonium ion (NH_4^+).

SOLUTION: We can use the same method as in the preceding sample exercise to calculate the number of bonds. However, we must take into account the charge on the ion. The positive charge on the ammonium ion means that it has one electron fewer than a neutral NH_4 molecule would have. The value of H must be decreased by 1 to account for the loss of this electron. Nitrogen atoms each have 5 valence electrons, and the four hydrogen atoms have 1 electron each; so:

$$H = 5 + 4(1) - 1 = 8$$

Calculating the number of bonds follows from the preceding sample exercise:

$$N = 8 + 4(2) = 16$$

$$S = N - H = 16 - 8 = 8$$

$$S/2 = 4$$

Finally, we place square brackets around the Lewis structure of the ion, and add the charge outside the upper right-hand corner:

$$\left[\begin{array}{c} H \\ | \\ H-N-H \\ | \\ H \end{array}\right]^+$$

PRACTICE EXERCISE: Draw the Lewis structure of the hydroxide ion (OH⁻). (See Problems 47 and 48.)

Resonance

The Lewis structure of ozone on page 300 has a double bond on the right side, but we could have drawn it on the left. How do we know where the double bond really is? We don't. Drawing a double bond on one side of the molecule and a single bond on the other creates an uneven distribution of electrons. But there is experimental evidence (see Section 6.8) that both the O–O bonds have identical lengths. If both bonds are equivalent, there should not be a preferred location for the double bond; nor should there be any side-to-side asymmetry. How can the bonds in ozone molecules be distributed evenly? It turns out that no single Lewis structure for ozone describes its bonding. The bonds are neither single bonds nor double bonds; rather, they are something in between. We represent this situation by using two equivalent **resonance structures** connected by a double-ended arrow:

> How can the bonds in ozone molecules be distributed evenly?

:Ö::Ö:⋯:Ö: ↔ :Ö:⋯:Ö::Ö:

The true electron arrangement is an average of both forms. Resonance does *not* mean that the molecule spends half its time in one form and half in the other; rather, its structure is in between these two extremes. Think of resonance as a blending of molecular structures in much the way that a liger is a genetic blend of two different animal species: a lion and a tiger (Figure 6.10).

SAMPLE EXERCISE 6.6: Draw Lewis structures for SO_2, including resonance forms.

SOLUTION: Sulfur is the less-electronegative element and so S will be the center atom. Let's determine the number of valence electrons and the number of shared electrons, as we did for ozone:

$H = 2(6) + 6 = 18$

$N = 3(8) = 24$

$S = N - H = 24 - 18 = 6$

$S/2 = 3$

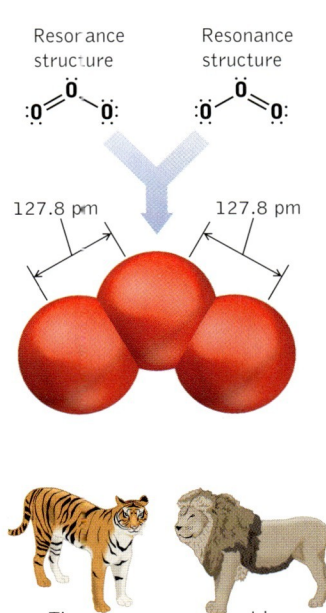

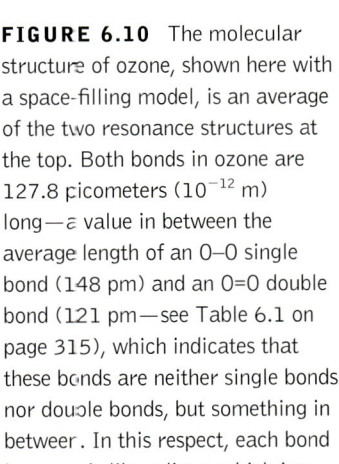

FIGURE 6.10 The molecular structure of ozone, shown here with a space-filling model, is an average of the two resonance structures at the top. Both bonds in ozone are 127.8 picometers (10^{-12} m) long—a value in between the average length of an O–O single bond (148 pm) and an O=O double bond (121 pm—see Table 6.1 on page 315), which indicates that these bonds are neither single bonds nor double bonds, but something in between. In this respect, each bond in ozone is like a liger, which is a hybrid of a lion and a tiger.

✓ A **resonance structure** is one of two or more Lewis structures with the same skeletal structure, but different bonding arrangements.

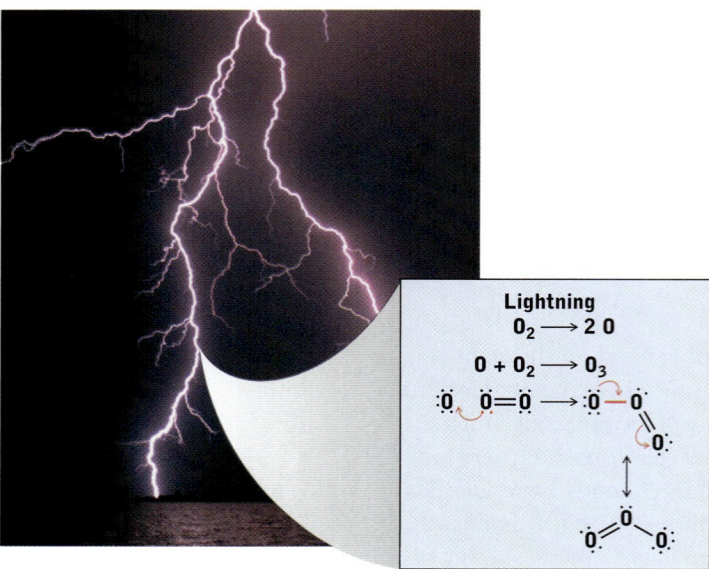

What a Chemist Sees. Lightning strikes contain sufficient energy to break O=O double bonds. The O atoms formed in this fashion collide with O_2 molecules, forming another allotrope of oxygen: ozone (O_3). The curved arrows in the Lewis structures show how the movement of pairs of valence electrons can account for the formation of O_3 and its resonance.

What a Chemist Sees. Hot springs are a natural source of hydrogen sulfide, H_2S, which reacts with O_2 in the air, forming SO_2.

The three bonds in SO_2 can be arranged in two different ways without changing the positions of the atoms. Thus there are two resonance forms of SO_2.

Have you noticed the similarity in the resonance forms of ozone and sulfur dioxide? Sulfur and oxygen are in the same group of the periodic table and have the same number of valence electrons; so we should expect similarities in bonding in O_3 and SO_2.

PRACTICE EXERCISE: Draw Lewis structures for all the resonance forms of SO_3. (See Problems 57–60.)

SAMPLE EXERCISE 6.7: Draw all the Lewis structures for the nitrate ion (NO_3^-).

SOLUTION: As in the preceding sample exercise, we first decide the arrangement of the atoms (nitrogen in the center), the number of available valence electrons, and the number of shared electrons. Because NO_3^- is an anion, we must account for the charge on the ion. For a monoanion, we *add* 1 to the number of valence electrons from the atoms:

$$H = 3(6) + 5 + 1 = 24$$
$$N = 4(8) = 32$$
$$S = N - H = 32 - 24 = 8$$
$$S/2 = 4$$

The four bonds can be arranged in three equivalent ways with one N=O double bond and two N—O single bonds in each resonance form:

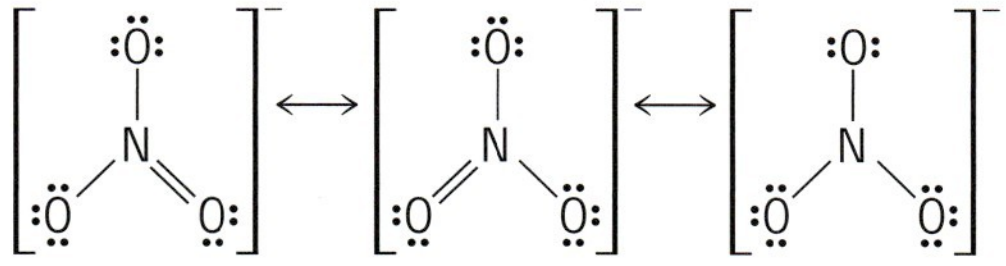

PRACTICE EXERCISE: Draw all the resonance forms of the azide ion (N_3^-) and the nitronium ion (NO_2^+). (See Problems 61 and 62.)

Holes in the ozone layer

Absorption of ultraviolet radiation by stratospheric ozone helps protect life forms on Earth from most of the sun's ultraviolet rays, particularly the wavelengths known as UVB (λ = 280–320 nm) and UVA (λ = 320–400 nm) that cause sunburn, skin aging, and skin cancer. O_2 and N_2 absorb ultraviolet radiation but only at wavelengths shorter than about 180 nm. Stratospheric ozone partly fills a gap in Earth's UV shield by absorbing some of the UVB though none of the UVA radiation.

In 1974, two American scientists, Sherwood Rowland and Mario Molina, predicted significant depletion of stratospheric ozone because of the release of a widely used class of gases known as chlorofluorocarbons (CFCs). Their predictions were later supported by experimental evidence of a thinning of the ozone layer over Antarctica during October (springtime in the Southern Hemisphere). This region of lowered ozone concentrations came to be known as the ozone hole. By 2000, stratospheric ozone concentrations over Antarctica were less than half of what they were in the late 1950s (Figure 6.11), and the ozone hole covered all of Antarctica and parts of South America and Australia (Figure 6.12). Less-severe thinning of stratospheric ozone were observed in the Northern Hemisphere (see Figure 6.11).

What are CFCs, and why do they threaten the ozone layer?

In 1987, an international agreement known as the Montreal Protocol called for reduced production of ozone-depleting compounds. Revisions in 1992 called for an end to the production of most such compounds by 1996. The Montreal Protocol has had a dramatic effect on CFC production and emission into the atmosphere (Figure 6.13). However, these compounds last for many years in the atmosphere, and recovery of the ozone layer may take most of the twenty-first century.

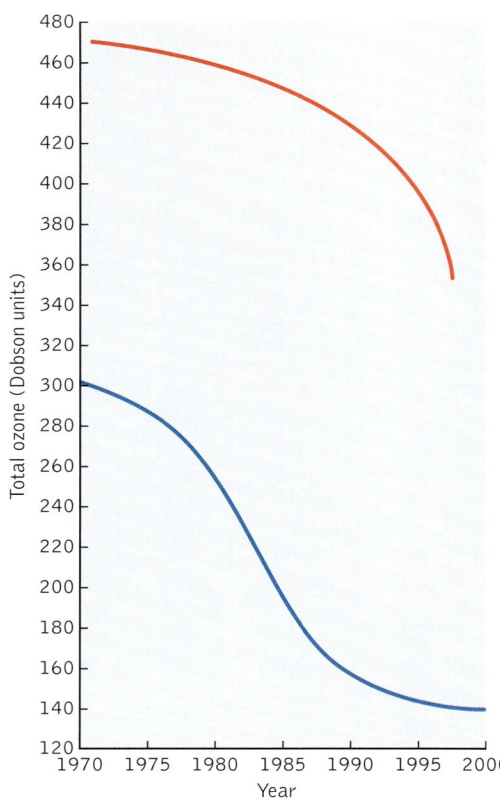

FIGURE 6.11 Springtime concentrations of ozone in the stratosphere over the North Pole (red) and over Halley Bay, Antarctica (blue), decreased in the final quarter of the twentieth century. Ozone concentrations are expressed in Dobson units; 1 Dobson unit is equivalent to a gas pressure of 0.01 mm Hg (see Chapter 8).

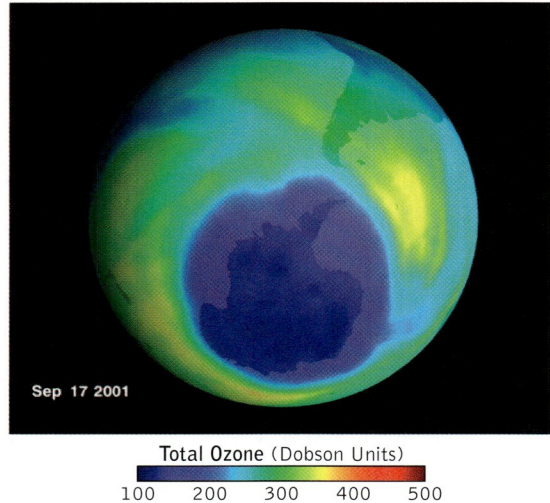

FIGURE 6.12 At its maximum the ozone hole in 2001 was 26 million square kilometers in area (about the size of North America). The hole has remained roughly the same size in recent years and may soon begin to shrink as the atmospheric concentrations of CFCs and other ozone-depleting gases continue to decline.

What are CFCs, and why do they threaten the ozone layer? Chlorofluorocarbons were first manufactured in the 1930s mainly for use as coolants in refrigerators. They are easily liquefied, nontoxic, chemically inert gases and so appeared to be ideal for use as refrigerants. Under the trade name Freon, CFCs were also widely used in air conditioners, as propellants in aerosol cans, to prepare plastic-foam cups and food packaging, and for cleaning microchips and circuit boards.

Chlorofluorocarbons are composed of carbon, fluorine, chlorine, and occasionally hydrogen. Two of the more widely used CFCs were CCl_2F_2 and CCl_3F. Their Lewis structures feature single bonds between the carbon atom and four halogen atoms:

THE CHEMISTRY OF THE HALOGENS

Halogen is derived from Greek *hal* (meaning "salt," as in *hal*ite) and *gen* (meaning "to produce"). The members of Group 7A of the periodic table—fluorine (F), chlorine (Cl), bromine (Br), iodine (I), and radioactive astatine (As)—are called halogens because they all readily form salts with sodium and the other reactive metals.

The halogens are among the most reactive elements in the periodic table. Their chemistry is linked to the presence of seven electrons in the valence shells of their atoms and to the stability that comes with having eight electrons. Thus the halogens are the most effective oxidizing agents (electron acceptors) of all the elements. It also means that halogens are not found as free elements in nature. Instead, they combine with metallic elements in such minerals as fluorite (CaF_2) and halite (NaCl). Fluorine is the most abundant of the halogens: it is 13th on the list of the most abundant elements in Earth's crust, and chlorine is 20th. Igneous rocks are typically 0.06% F, 0.03% Cl, 0.0002% Br, and 0.00003% I. All naturally occurring isotopes of At are radioactive with half-lives of less than 1 minute. Fluorine minerals may be more abundant than those of chlorine, but they are more dispersed in nature, and so fluorine must be recovered from rocks in which it may be only a minor component.

As free elements, halogens are diatomic molecules joined by a single covalent bond: F_2 and Cl_2 are gases at room temperature; Br_2 is a dark red liquid, and I_2 is a dark violet solid. Fluorine is the strongest oxidizing agent known; so it is difficult to produce from fluoride-containing compounds by chemical reactions. However, it can be produced electrochemically by passing an electrical current through molten fluoride compounds. Most of the F_2 produced commercially is used in the purification of uranium for nuclear reactors (Chapter 2). In this process, uranium is converted into volatile hexafluorides. Fissionable (Chapter 2) $^{235}UF_6$ gas can be separated from the more abundant and nonfissionable $^{238}UF_6$ because the slightly lighter $^{235}UF_6$ gas diffuses more quickly (Chapter 8). Diatomic chlorine is produced by the electrolysis of molten NaCl, and sodium metal also is produced. Chlorine is used extensively in the paper industry, as a disinfectant for drinking water, and in manufacturing plastics. Bromine and iodine are produced by reaction of bromide or iodide compounds with chlorine gas. For example:

$$2\ Br^-(aq) + Cl_2(g) \longrightarrow Br_2(l) + Cl^-(aq)$$

Calcium hypochlorite is used to kill bacteria and algae in swimming pools.

Diatomic bromine is used in the manufacture of pesticides, and solutions of I_2 are used in medicine to disinfect wounds and incisions.

Halogens react with water to form a mixture of hydrohalic and hypohalous acids:

$$Cl_2(g) + H_2O(l) \longrightarrow HCl(aq) + HClO(aq)$$

The addition of NaOH to solutions of HOCl (neutralization) produces aqueous NaOCl, which is the active ingredient in household bleach:

$$HOCl(aq) + NaOH(aq) \longrightarrow NaOCl(aq) + H_2O(l)$$

The dissolution of Cl_2 or Br_2 in hot solutions containing OH^- produces ClO_3^- and BrO_3^- ions, respectively. For example:

$$3\ Cl_2(aq) + 6\ OH^-(aq) \longrightarrow$$
$$ClO_3^-(aq) + 5\ Cl^-(aq) + 3\ H_2O(l)$$

Chlorate and chlorite salts are used as bleaching agents in the paper industry. Effluent from paper mills is often quite toxic to aquatic life. Paper manufacturers are experimenting with less environmentally damaging bleaching agents.

The oxidizer in many fireworks is a perchlorate salt.

Sodium chlorate, $NaClO_3$, reacts with SO_2, producing ClO_2, which is reduced by hydrogen peroxide in aqueous NaOH, forming sodium chlorite, $NaClO_2$.

$$2\ NaClO_3(aq) + SO_2(g) \longrightarrow 2\ ClO_2(aq) + Na_2SO_4(aq)$$

$$2\ ClO_2(aq) + H_2O_2(aq) + 2\ NaOH(aq) \longrightarrow$$
$$NaClO_2(aq) + 2\ H_2O(l) + O_2(g)$$

Match heads contain a mixture of sulfur and antimony sulfide, Sb_2S_3, and potassium chlorate, which serves as an oxidizing agent. Chlorates produce oxygen when mixed with traces of manganese(IV) oxide and heated above 70°C. This reaction has been employed on passenger aircraft to provide emergency oxygen.

$$2\ KClO_3(s) \longrightarrow 2\ KCl(s) + 3\ O_2(g)$$

Sodium chlorate can be oxidized to sodium perchlorate by passing an electrical current through a solution of $NaClO_3$:

$$NaClO_3(aq) + H_2O(l) \longrightarrow NaClO_4(aq) + H_2(g)$$

A reaction between $NaClO_4$ and ammonium chloride is used to produce the ammonium perchlorate that is used as a solid propellant in the booster rockets of the U.S. space shuttles.

Mixtures of potassium perchlorate, sulfur, and aluminum provide the white flash and noise in fireworks.

The flash powder used in stage shows is a mixture of magnesium metal and $KClO_4$.

Oxoanions of the other halogens, including bromate, iodate, perbromate, and periodate, have chemistries that are similar to those of the corresponding chlorine oxoanions. The short half-life of astatine isotopes makes working with this element difficult, and less is known about its chemistry.

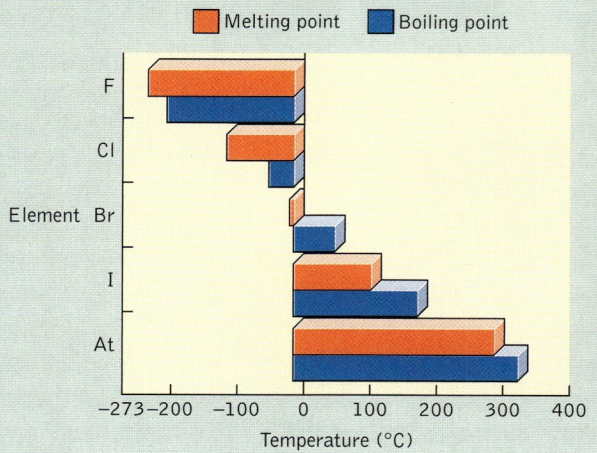

The melting and boiling points of the halogens increase uniformly with increasing molar mass. F_2 and Cl_2 are gases at room temperature; bromine is a liquid and iodine is a solid.

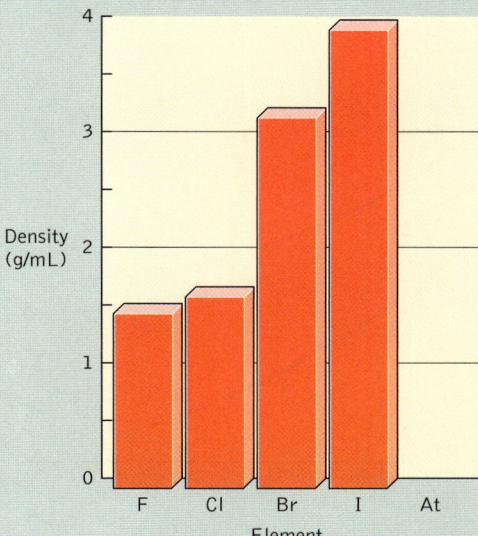

The densities of the Group 7A elements are given for their liquid states. These densities increase with increasing atomic number.

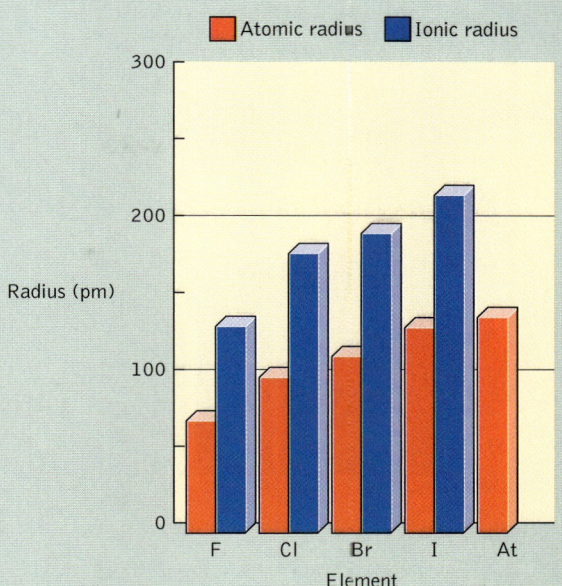

The atomic and ionic radii of the Group 7A elements increase with increasing atomic number. The ionic radius of each halide anion is larger than the radius of its parent atom. Astatine has such a short half-life that its atomic and ionic dimensions are not known.

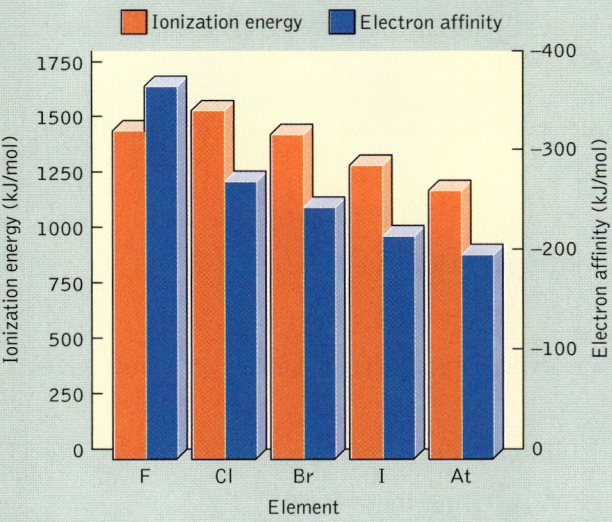

The ionization energies of the group 7A elements are among the highest of all the elements (second only to the noble gases). Ionization energies decrease with increasing atomic number due to greater shielding of the valence electrons from nuclear charge. The electron affinities of all four stable Group 7A elements have large negative values that decrease with increasing atomic number.

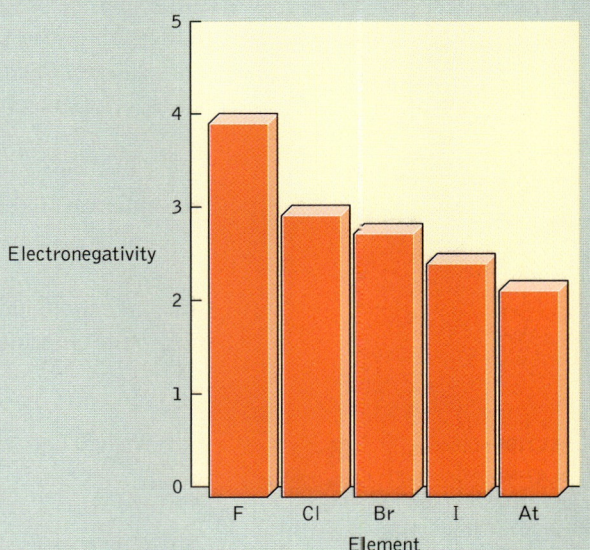

The Group 7A elements are the most electronegative elements in their respective periods, though electronegativities decrease with increasing atomic number due to more shielding of the outer shell electrons.

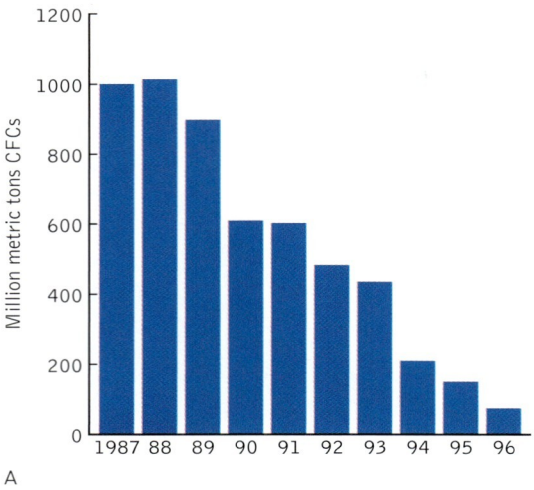
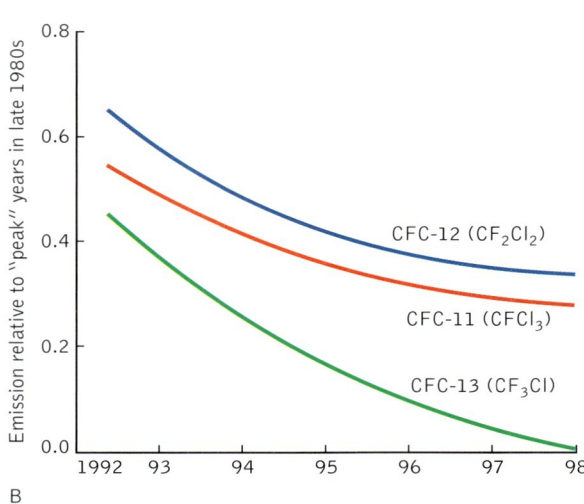

FIGURE 6.13 The worldwide production of CFCs (A) and the release of these compounds into the atmosphere (B) have decreased since the ratification of the Montreal Protocol in 1986. However, the chemical stability of these volatile compounds means that their concentrations in the atmosphere will decrease much more slowly.

The chemical inertness of CFCs means that they persist in the environment and so can be swept up to the stratosphere by wind and air currents in the bottom layer of the atmosphere, called the troposphere (see Figure 6.1). In the stratosphere, CFCs encounter UV rays with enough energy to break their C–Cl bonds, releasing chlorine atoms:

$$CCl_3F \xrightarrow{h\nu} CCl_2F + Cl \qquad (6.15)$$

Free chlorine atoms react with ozone to make chlorine monoxide:

$$Cl + O_3 \longrightarrow ClO + O_2 \qquad (6.16)$$

Chlorine monoxide then reacts with more ozone, making oxygen and regenerating atomic chlorine:

$$ClO + O_3 \longrightarrow Cl + 2\,O_2 \qquad (6.17)$$

If we add Equations 6.16 and 6.17 together, we get

$$Cl + 2\,O_3 + ClO + \longrightarrow ClO + Cl + 3\,O_2 \qquad (6.18)$$

Canceling species that appear on both sides of the reaction arrow,

$$2\,O_3 \longrightarrow 3\,O_2 \qquad (6.19)$$

Equation 6.19 is nearly the same as Equation 6.12. However, the chlorine-introduced destruction of ozone occurs much more rapidly. Also, Cl atoms are not consumed in the overall process. We call a substance that speeds up a chemical

CONNECTION: Chlorine acts as a catalyst, increasing the rate of ozone decomposition (Section 14.6).

Short winter

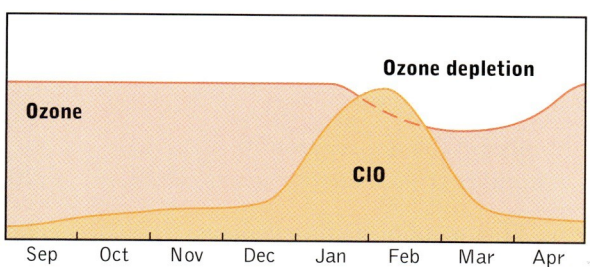

The ozone concentration decreases more sharply after a long, cold winter in the Arctic than after a short winter. Notice how the concentrations of ClO increase to a maximum 1 to 2 months before the minimum ozone concentration. The graphs

Long winter

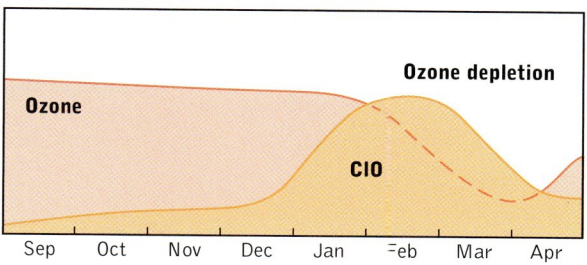

show that ClO levels remain high longer into the spring after a long, cold winter, making more of the ozone-depleting compound available when radiation from the sun returns after the darkness of the polar winter.

process without being consumed a **catalyst**. A single chlorine atom can catalyze the destruction of hundreds to thousands of O_3 molecules before it is destroyed in another reaction. We will return to the important roles that catalysts play in chemistry in Chapter 14.

> A **catalyst** is a substance that increases the rate of a reaction without being consumed by it.

6.7 CHOOSING BETWEEN LEWIS STRUCTURES: FORMAL CHARGES

The greatest variety of atmospheric gases is found in the troposphere, the layer next to Earth. Among these gases are compounds that have been in the news because (1) their concentrations in the troposphere have been increasing and (2) they may be altering global weather patterns. The most abundant of these compounds is CO_2. The capacity of CO_2 and other gases to alter climate is linked to their molecular structures. Let's consider the structure of CO_2. Each molecule of CO_2 is held together by four bonding pairs of electrons. We usually draw them evenly distributed in a Lewis structure with two C=O double bonds:

> Which of the three resonance forms contributes most to the bonding in CO_2?

  CONNECTION: The role of CO_2 in global warming is described in Section 7.1.

$$\ddot{\text{O}}=\text{C}=\ddot{\text{O}}$$

However, we are not limited to symmetric arrangements for the four bonding pairs. We might also draw structures with C—O single and C≡O triple bonds, giving three possible resonance structures for CO_2:

$$:\ddot{\text{O}}-\text{C}\equiv\text{O}: \longleftrightarrow \ddot{\text{O}}=\text{C}=\ddot{\text{O}} \longleftrightarrow :\text{O}\equiv\text{C}-\ddot{\text{O}}:$$

> **Formal charges** are used to assess the relative stability of resonance structures. The formal charge on an atom in a molecule equals the number of valence electrons on the free atom minus the number assigned to the atom in the molecular structure.

Which of the three resonance forms contributes most to the bonding in CO_2? The relative importance of resonance forms can be judged by assigning a **formal charge** to each atom in the molecule. A formal charge is the difference between the number of valence electrons in a free atom and the number assigned to that atom in a Lewis structure:

$$\text{formal charge} = \text{number of valence electrons in an atom} \\ - \text{number of electrons assigned to atom} \quad (6.20)$$

The number of electrons assigned to an atom can be calculated by adding up all the lone pairs of electrons on the atom and half the shared pairs (bonding electrons):

$$\text{number of electrons assigned to atom} = \text{number of nonbonding electrons} \\ + \frac{1}{2}(\text{number of bonding electrons}) \quad (6.21)$$

Structures in which atoms have formal charges that are zero or close to zero have the lowest energy and most stability. In addition, resonance forms with *negative* formal charges on the more *electronegative* elements are preferred.

Let's apply the concept of formal charge to the three resonance forms of CO_2. In the first one ($:\ddot{O}\!\!-\!\!C\!\!\equiv\!\!O:$), the oxygen atom on the left has three lone pairs of electrons and one bonding pair. In calculating its formal charge, we credit this oxygen atom with all six electrons in its three lone pairs plus half of the two electrons in the bonding pair for a total of seven electrons in Equation 6.21,

> **Resonance forms** with negative formal charges on the most-electronegative atoms are preferred.

$$\text{number of electrons assigned to atom} = \text{number of nonbonding electrons} \\ + \frac{1}{2}(\text{number of bonding electrons}) \\ = (3)(2) + \frac{1}{2}(2) = 7$$

Seven is one more than the six valence electrons in a free oxygen atom. According to Equation 6.20, the formal charge on this oxygen atom is,

$$\text{formal charge} = \text{number of valence electrons for atom} \\ - \text{number of electrons assigned to atom} \\ = 6 - 7 = -1$$

Carbon gets half credit for the four bonding pairs of electrons. Because a carbon atom normally has four electrons, the C atom in the first structure has a formal charge of zero. The right-hand oxygen atom in the first structure has one lone pair and gets half credit for the three bonding pairs of electrons. Therefore, it is assigned five electrons—one fewer than the six electrons in a free oxygen atom, and so its formal charge is $+1$. Notice that the sum of the formal charges is zero. A net value of zero makes sense because the sum of the formal charges should equal the actual charge, which, for a neutral molecule, is zero. It would be a positive value for a polyatomic cation and a negative value for a polyatomic anion.

Applying this approach to the other two resonance forms of CO_2, we find that all the atoms in the structure with two C=O double bonds have zero formal charge, but the formal charges on the three atoms in the third structure (:O≡C—Ö:) are, from left to right, +1, 0, and −1, respectively. The structure in the middle with zero formal charge on all its atoms is the preferred one.

Note that, in all three resonance forms of CO_2, there are four bonds on the carbon atom. This number matches carbon's bonding capacity (see page 289). Also note that the formal charge on the carbon atoms in all three structures is zero. This outcome is in accord with an important principal: when the number of bonding pairs of electrons to an atom matches its bonding capacity, the formal charge on the atom is zero. If the number of bonding pairs is one more than the bonding capacity (e.g., O atoms with three bonds), the formal charge is +1, and, if the number of bonding pairs is one fewer than the bonding capacity (e.g., O atoms with one bond), the formal charge is −1.

CONCEPT TEST: Without actually calculating a formal-charge value by using Equations 6.20 and 6.21, predict the formal charge on a nitrogen atom that has three lone pairs of electrons and one bonding pair.

SAMPLE EXERCISE 6.8: The atmospheric concentrations of dinitrogen oxide, N_2O, are very low, but they have been increasing in recent years. Draw the Lewis structure that best describes the bonding in N_2O.

SOLUTION: Let's start by putting a nitrogen atom in the middle (nitrogen is less electronegative than oxygen) and connect the other nitrogen and oxygen atoms to the middle nitrogen atom with single bonds:

$$N—N—O$$

Next, we need to calculate the number of bonds holding the molecule of N_2O together. Each nitrogen atom brings five valence electrons to the molecule, and oxygen brings six; so

$$H = (2 \times 5) + 6 = 16$$

All three atoms require eight valence electrons to achieve an octet; so

$$N = 3 \times 8 = 24$$

Thus the number of electrons to be shared in bond formation is

$$S = N - H = 24 - 16 = 8$$

and we need $S/2 = 8/2$, or 4, bonds.

As with CO_2, there are three ways to distribute the four bonds:

:N̈—N≡O: N̈=N=Ö   :N≡N—Ö:

Unlike the three Lewis structures of CO_2, the first and third structures of N_2O are not equivalent. Which of the three contributes most to the actual structure (and the chemical reactivity) of N_2O?

To answer this question, we need to determine the formal charge on each atom in each resonance form. We could do so by using Equations 6.20 and 6.21 or we could compare the number of bonds to each atom with the atom's bonding capacity. The terminal nitrogen atom in the right-hand structure has three bonds (which matches its bonding capacity), and so it has a formal charge of 0. The middle nitrogen atom in all three structures has four bonds, one more that its bonding capacity, and has a formal charge of +1. Oxygen atoms with two bonds (equaling oxygen's bonding capacity) have a formal charge of 0; those with three bonds have a formal charge of +1; and those with only one bond have a formal charge of −1. The terminal nitrogen in the left-hand structure has a formal charge of −2. The other two atoms in the structure have +1 formal charges; so the formal charge on the terminal nitrogen atom must be −2 to make the overall charge 0. Summarizing these calculations, we have:

$$\overset{-2}{:\ddot{N}}-\overset{+1}{N}\equiv\overset{+1}{O}: \longleftrightarrow \overset{-1}{\ddot{N}}=\overset{+1}{N}=\overset{0}{\ddot{O}} \longleftrightarrow :\overset{0}{N}\equiv\overset{+1}{N}-\overset{-1}{\ddot{O}}:$$

At first glance, the middle and right-hand structures look equivalent in that their atoms have formal charges closest to 0. However, the −1 value is on oxygen in the right-hand structure but on nitrogen in the middle one. Remember that resonance forms with negative formal charges on the more-electronegative elements are preferred. Oxygen is more electronegative than nitrogen; so the right-hand structure with the negative formal charge on oxygen is the most important of the three. In Section 6.8, we will see that measurements of the lengths of the N–O and N–N bonds in N_2O can be used to confirm this prediction.

PRACTICE EXERCISE: Which of the resonance forms of carbonyl sulfide, COS, contributes the most to bonding in this molecule? (See Problems 67–74.)

SAMPLE EXERCISE 6.9: Which of the resonance forms of the nitrate ion (NO_3^-) in the sample exercise on page 304 contribute the most to bonding?

SOLUTION: As in the preceding sample exercise, we first assign formal charges of 0 to all double-bonded oxygen atoms, −1 to all single-bonded oxygen atoms, and +1 to all the nitrogen atoms, because they all have four bonds.

In all three resonance forms, equivalent atoms of each element have the same formal charges. Therefore, all three structures are equivalent. Notice that the sum of the formal charges equals the charge on the molecular ion: $-1 + (-1) + 1 = -1$.

PRACTICE EXERCISE: Which resonance forms of the azide ion (N_3^-) and the nitronium ion (NO_2^+) contribute most to bonding? (See Problems 75–76.)

6.8 ELECTRON DIFFRACTION, BOND LENGTHS, AND PREDICTIONS CONFIRMED

In Chapter 3, we considered how a beam of electrons can produce a diffraction pattern (Figure 6.14) in much the way that a beam of light is diffracted as waves of light undergo constructive and destructive interference. Beams of high-velocity electrons have much shorter wavelengths than those of visible light, typically less than a nanometer. Thus an electron beam diffracted by structures with features that are less than a nanometer apart will produce interference patterns. Gas-phase molecules are such structures. Analysis of the interference patterns produced by diffracted electrons allows scientists to calculate the distances between atoms in molecules. That is, they can determine the lengths of the covalent bonds that hold the atoms together (Table 6.1). This method was used, for example, to determine that the O–O bond lengths in ozone are identical.

Electron-diffraction patterns from N_2O reveal that the N–N distance is 113 pm (10^{-12} m) and the N–O distance is 119 pm. The N–N distance is slightly

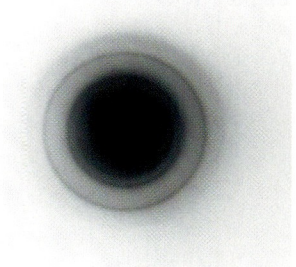

FIGURE 6.14 Electron diffraction by atoms produces interference patterns such as that shown here. Distances between atoms in gas-phase molecules can be determined from analyses of these interference patterns.

TABLE 6.1 Selected Average Covalent Bond Lengths and Average Bond Strengths

Bond	Bond Length (pm)	Bond Strength (kJ/mol)*	Bond	Bond Length (pm)	Bond Strength (kJ/mol)*
C—C	154	348	N≡O	106	678
C=C	134	614	O—O	148	146
C≡C	120	839	O=O	121	495
C—N	143	293	O—H	96	458
C=N	138	615	S—O	151	265
C≡N	116	891	S=O	143	523
C—O	143	358	S—S	204	266
C=O	123	799	S—H	134	347
C≡O	113	1072	H—H	75	432
C—H	110	411	H—F	92	565
C—F	133	453	H—Cl	127	427
C—Cl	177	339	H—Br	141	363
N—H	104	386	H—I	161	295
N—N	147	163	F—F	143	159
N=N	124	418	Cl—Cl	200	243
N≡N	110	941	Br—Br	228	193
N—O	136	201	I—I	266	151
N=O	122	607			

*Bond strength is defined as the energy required to break a mole of bonds. The energetics of bond breaking (and bond formation) are discussed in Section 11.8. For now, just note that increasing bond order (from single to triple bonds) results in shorter bond lengths and greater bond strengths.

CONNECTION: The diffraction of light is described in Section 1.3.

longer than the length of the N≡N triple bond in N_2 (110 pm) but significantly shorter than the N═N double bond in FN═NF (125 pm). A similar comparison of N–O bond lengths leads to the conclusion that the N–O bond in N_2O is between a single and a double bond. These experimental results fit our predictions based on formal-charge calculations that the N–N bond in N_2O has considerable triple-bond character. However, they also show that the resonance form with two double bonds plays a role in defining the molecular bonding in N_2O.

6.9 MOLECULAR-ORBITAL THEORY

CONNECTION: Molecular orbitals are formed by combining atomic orbitals (Sections 3.7 and 3.8).

In the late 1920s, scientists tried to develop mathematical models that would accurately predict the spectra of molecules in the way in which Schrödinger's wave equation had predicted the spectrum of atomic hydrogen. German physicist Friedrich Hund (1896–1997), who developed Hund's rule for filling atomic orbitals, and American chemist Robert S. Mulliken (1896–1986) succeeded in developing such a model. In their approach, they assumed that the wave functions of atomic orbitals combine to create **molecular orbitals (MOs)** in molecules.

Molecular orbitals represent discrete energy states in molecules just as atomic orbitals represent allowed energy states in free atoms. As with atomic orbitals, the lowest-energy MO is filled first and the higher-energy levels are filled as more electrons are added. Electrons in molecules can be raised to higher-energy MOs by absorbing quanta of electromagnetic radiation. When they return to lower-energy MOs, they emit distinctive wavelengths of UV and visible radiation, including some of the shimmering colors in the aurora.

✓ **Molecular orbitals (MOs)** formed by combinations of atomic orbitals are **delocalized** over the entire molecule.

Molecular orbitals differ from atomic orbitals in several important ways. Each MO is associated with an entire molecule, not just a single atom. Thus, molecular orbitals are spread out, or **delocalized** over all the atom centers in a molecule rather than between just two atoms. We will see how MO theory works by examining the simplest molecular compound, hydrogen gas.

The molecular orbitals of H_2

Like N_2 and O_2, hydrogen is a diatomic gas at room temperature and has the molecular formula H_2. According to MO theory, molecules of hydrogen are formed by mixing the 1s orbitals on two hydrogen atoms. Molecular-orbital theory stipulates that mixing two atomic orbitals creates two molecular orbitals representing two different energy states (Figure 6.15). The lower-energy MO is oval in shape and spans the two atomic centers. It is called a **bonding molecular orbital.** Its shape leads to enhanced electron density between atoms. When two electrons occupy a bonding MO, a single covalent bond is formed. When the region of highest density lies along the bond axis, as it does in the bonding MO in H_2, the MO is designated a **sigma (σ) molecule orbital** and the resulting covalent bond is called a **sigma bond.** The σ bonding molecular orbital in H_2 is labeled σ_{1s} in Figure 6.15 because it is formed by mixing two 1s atomic orbitals.

✓ A **sigma (σ) bond** is a covalent bond in which the highest electron density lies along the bond axis.

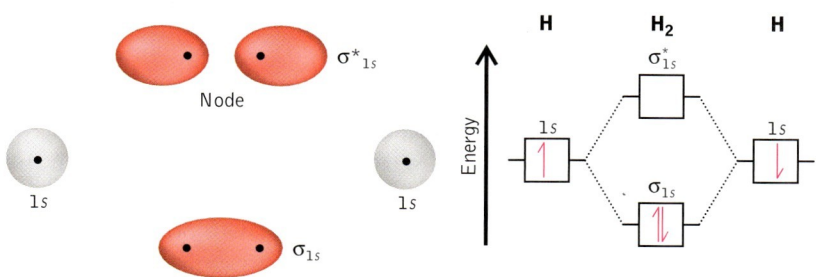

FIGURE 6.15 Mixing 1s orbitals in two hydrogen atoms creates two molecular orbitals: a filled bonding σ_{1s} and an empty antibonding σ_{1s}^*. The shapes of the orbitals are shown on the left, and a molecular-orbital diagram showing the relative energies of the atomic and molecular orbitals is shown on the right.

The second hydrogen molecular orbital is a higher-energy **antibonding molecular orbital** designated σ_{1s}^*. Scientists speak of such an orbital as a "sigma star" orbital to verbally note the presence of the asterisk and to indicate that it is an antibonding orbital. This antibonding orbital has two separate lobes of electron density and a region of no electron density (a node) between the two hydrogen atoms, as shown in Figure 6.15.

Note that two hydrogen *atomic* orbitals combine to form two hydrogen *molecular* orbitals, in accord with a general rule in MO theory: *The total number of molecular orbitals formed in a molecule equals the number of atomic orbitals used in the mixing process.* The σ_{1s} bonding MO is lower in energy than the 1s atomic orbitals by nearly the same amount that the σ_{1s}^* antibonding MO is higher in energy than the atomic orbitals. Therefore, the formation of the two MOs does not significantly change the total energy of the system. However, the valence electrons in H_2 reside in the lower-energy bonding orbital, giving the molecule a lower-energy electron configuration (which is written as σ_{1s}^2) than that of two separate H atoms ($1s^1$). The stability of the σ_{1s}^2 configuration explains why hydrogen gas exists as H_2 molecules rather than as H atoms.

Hydrogen is a diatomic gas, but helium exists as free atoms and not as molecular He_2. Let's use MO theory to explain why. The molecular-orbital diagram for He_2 is similar to that of H_2 except that there are four valence electrons in He_2. Therefore, both the bonding and the antibonding molecular orbitals are fully occupied. The presence of two electrons in the higher-energy σ_{1s}^* antibonding orbital cancels the stability gained from having two electrons in the lower-energy σ_{1s} bonding orbital. Because there is no net gain in stability on formation of He_2 from two separate He atoms, the molecule He_2 does not exist.

SAMPLE EXERCISE 6.10: Develop an MO diagram for the H_2^- ion. Does the diagram represent a stable molecular structure?

SOLUTION: There are three valence electrons in H_2^-. Two electrons occupy the σ_{1s} orbital, and the third electron occupies the σ_{1s}^* orbital. Because there are more bonding than antibonding electrons in H_2^- ions, they have a stable structure.

PRACTICE EXERCISE: Use MO theory to predict the existence of the H_2^+ ion. (See Problem 92.)

 Interaction between two atomic orbitals creates a **bonding molecular orbital** and a higher-energy **antibonding molecular orbital**.

CONNECTION: Atomic orbitals also contain nodes, as discussed in Section 3.8.

 Each molecular orbital, like each atomic orbital, can hold as many as two electrons.

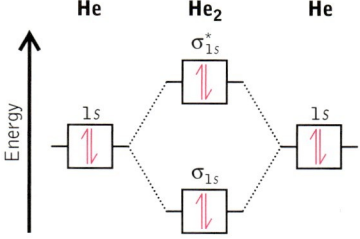

This molecular-orbital diagram for He_2 indicates that there are as many electrons in antibonding orbitals as there are in bonding orbitals. Therefore, the molecule is not stable.

Nitrogen and oxygen

Molecular-orbital diagrams of N_2 and O_2 are more complicated than that of H_2 because of the greater number and variety of atomic orbitals available. Not all combinations of atomic orbitals result in effective bonding. Orbitals with similar energy and shape mix more effectively than do those that have different energies and shapes. Thus, an s orbital more effectively mixes with other s orbitals than with p orbitals. Orbitals of different principal quantum numbers (e.g., $1s$ and $2s$) have different sizes and energies, resulting in less-effective mixing than, say, two $1s$ or two $2s$ orbitals.

Figure 6.16 shows how the $2s$ and $2p$ orbitals from two atoms mix together to form MOs. The different orientations of the $2p_x$, $2p_y$, and $2p_z$ atomic orbitals yield different kinds of MOs. The $2p_z$ orbitals of the bonded nitrogen atoms point toward each other. When they mix, they form a bonding σ_{2p} orbital and an antibonding σ^*_{2p} orbital. The lobes of the $2p_x$ and $2p_y$ atomic orbitals are oriented 90° from the bonding axis. When the $2p_x$ orbital on one atom mixes with the $2p_x$ orbital on the other, they form a set of π (pi) and π^* **molecular orbitals.** When electrons occupy a π orbital, they can form a π **bond.** The electron density in a π bond is greatest above and below the bonding axis (Figure 6.16). Similar mixing can occur between two $2p_y$ orbitals, forming a second set of π and π^* orbitals, which have lobes in front of and behind the bonding axis. If a second pair of electrons occupies this π orbital, then a second π bond can form. Note that mixing two $2p_x$ and two $2p_y$ atomic orbitals creates a total of $2 + 2 = 4$ molecular orbitals. The two bonding MOs are labeled π_{2p}; the two antibonding MOs are labeled π^*_{2p}, as shown in Figure 6.16.

What are the relative energies of σ and π molecular orbitals? In other words, which ones are occupied first? Let's consider the MO diagrams for N_2 and O_2 shown in Figure 6.17. In both molecules, the MOs derived from mixing $2s$ orbitals (σ_{2s} and σ^*_{2s}) are lower in energy than the energy of σ_{2p} for the same reason that a $2s$ atomic orbital is lower in energy than a $2p$. For N_2, the π_{2p} orbitals are lower in energy than the σ_{2p} orbital. Why? To answer this question we need to consider the mixing of the $2s$ orbital of one nitrogen atom with the $2p_x$ orbital of the other nitrogen atom. It turns out that such mixing does occur in N_2 because of the stability (and lower energy) of the half-filled $2p$ orbitals in N atoms, bringing their energy closer to that of the $2s$ orbital. Mixing $2s$ and $2p_z$ orbitals tends to lower the energy of the σ_{2s} orbital (as shown in Figure 6.17A and raise the energy of the σ_{2p} orbital. Therefore, the MOs are filled by the 10 valence electrons in N_2 from the lowest energy to the highest in the following order:

$$\sigma_{2s}, \sigma^*_{2s}, \pi_{2p}, \sigma_{2p}$$

The resulting valence-shell electron configuration is

$$\sigma_{2s}^2 \sigma^{*2}_{2s} \pi^4_{2p} \sigma^2_{2p}$$

> **What are the relative energies of σ and π molecular orbitals?**

CONNECTION: The shapes and orientations of atomic orbitals are described in Section 3.8.

✓ Atomic orbitals with similar energies and shapes mix to give molecular orbitals.

✓ **Pi (π and π^*) molecular orbitals** form by the mixing atomic orbitals that are not oriented along the bonding axis in a molecule. Electrons occupying π orbitals form π **bonds.** In a covalent π bond, electron density is greatest above and below or in front of and behind the bonding axis.

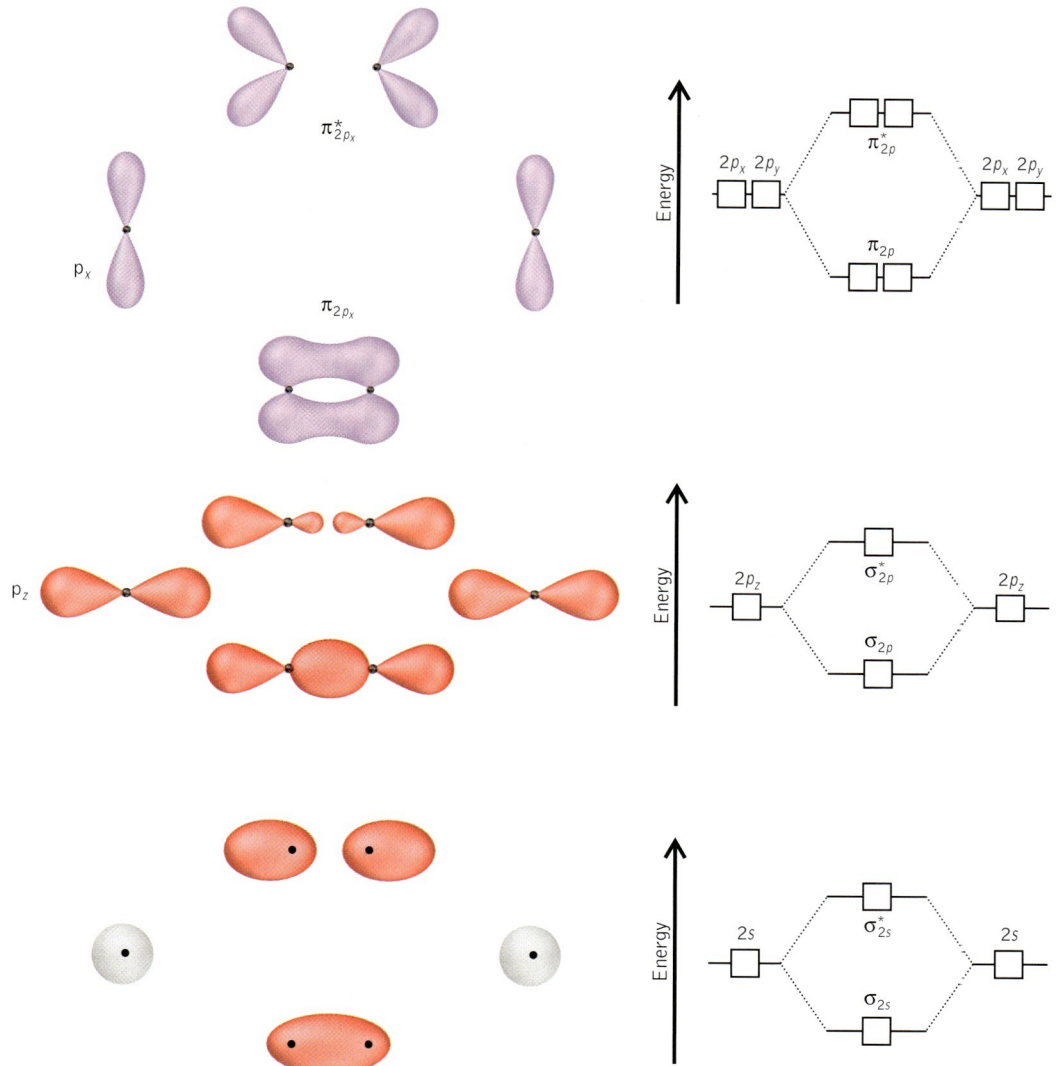

FIGURE 6.16 When 2s and 2p atomic orbitals from two atoms mix, they form eight molecular orbitals. The two 2s orbitals form bonding σ_{2s} and antibonding σ_{2s}^* orbitals. Mixing the two $2p_z$ orbitals that are directed along the axis connecting the two nuclei results in σ_{2p} and σ_{2p}^* molecular orbitals When the pairs of $2p_x$ and $2p_y$ orbitals mix, they form π_{2px} and π_{2py} bonding orbitals, and π_{2px}^* and π_{2py}^* ant bonding orbitals. The shapes of the σ_{2s}, σ_{2s}^*, σ_{2p}, σ_{2p}^*, π_{2px}, and π_{2px}^* orbitals are shown on the left. Not shown are the π_{2py} and π_{2py}^* orbitals. They have the same shape and energy as the π_{2px} and π_{2px}^* orbitals, but are oriented in and out of the plane of the diagram, instead of up and down.

The array of molecular orbitals for O_2 (Figure 6.17B) differs from that for N_2. In molecules of O_2, interaction between the 2s and 2p orbitals of the two oxygen atoms is not as strong as in N_2, because there is a greater energy difference between the 2s and 2p orbitals. As a result, the σ_{2p} orbital in O_2 is lower in energy than the π_{2p} orbital. Molecules of O_2 have two more electrons than do molecules of N_2. These two electrons must occupy π_{2p}^* antibonding orbitals because all the

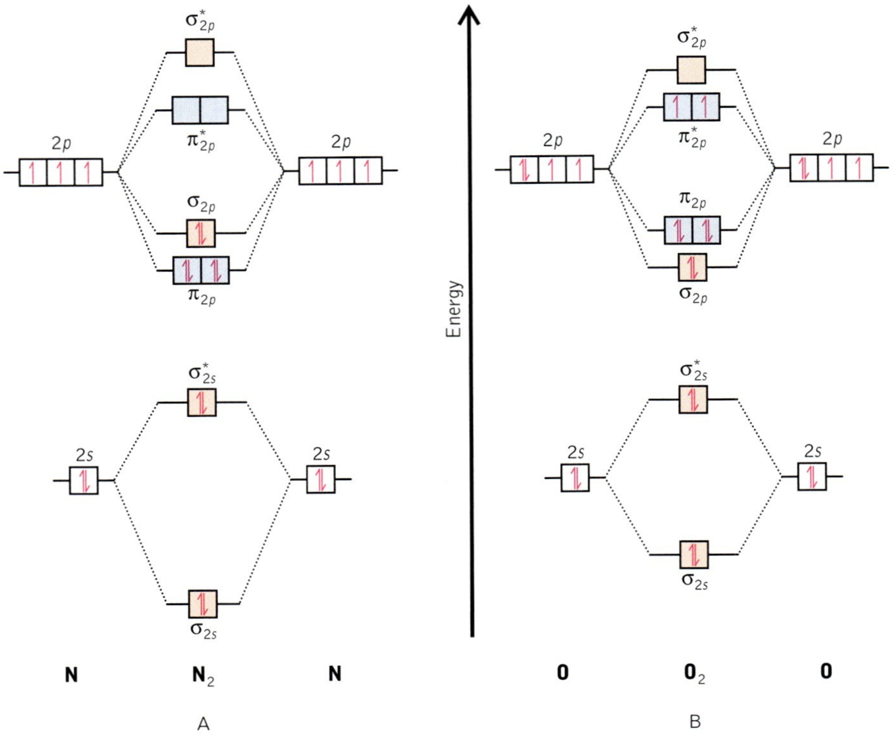

FIGURE 6.17 A. In the molecular-orbital diagram of N_2, the σ_{2p} orbital is higher in energy than the π_{2p} because of mixing of 2s and 2p orbitals. B. In the molecular-orbital diagram of O_2, the σ_{2p} orbital is lower in energy than the π_{2p} because much less 2s and 2p mixing occurs.

✓ Single, double, and triple bonds have **bond orders** of 1, 2, and 3, respectively.

bonding orbitals are filled. There are two energetically equivalent π_{2p}^* antibonding orbitals; so, in accord with Hund's rule, each of them is half filled with one electron each. The resulting electron configuration is

$$\sigma_{2s}^2 \sigma_{2s}^{*2} \sigma_{2p}^2 \pi_{2p}^4 \pi_{2p}^{*2}$$

On the basis of their Lewis structures, we predicted a triple bond in N_2 and a double bond in O_2. Molecular-orbital theory leads us to the same conclusions. In N_2, the four electrons from the two 2s atomic orbitals fill the σ_{2s} bonding orbital and the σ_{2s}^* antibonding orbital, and so there is no net bonding from 2s overlap. However, the six 2p electrons fill three bonding MOs: the two π_{2p} orbitals and the σ_{2p} orbital. These three filled bonding MOs between only two atom centers are equivalent to a triple bond. In O_2, the same three bonding MOs are occupied, but there is one electron in each of the two π_{2p}^* antibonding orbitals. The presence of these two electrons in antibonding MOs offsets the stabilizing effect of six electrons in lower-energy bonding MOs, and so there is a net of $6 - 2 = 4$ electrons in bonding orbitals. Dividing four bonding electrons by two electrons per bond, we calculate that there should be a double bond in O_2. This calculation establishes the number of bonds, or **bond order,** between the oxygen atoms in O_2. In equation form we can define bond order as:

$$\text{bond order} = \frac{1}{2}(\text{number of bonding electrons} - \text{number of antibonding electrons}) \quad (6.22)$$

FIGURE 6.18 Molecular-orbital theory explains the behavior of liquid oxygen in a magnetic field. The unpaired electrons in the π_{2p}^* orbital are attracted to the poles of a magnet. Substances made of molecules with paired electrons do not exhibit this property.

The MO diagram for O_2 predicts that there are two unpaired electrons in each molecule. The Lewis structure for O_2 suggests that all the valence electrons are paired. Which is correct? Experimental evidence can help us decide. Recall that all electrons spin and that their spins produce tiny magnetic fields. When two atoms reside in the same orbital, their spins are paired and the magnetic fields cancel out. If all the electrons in a molecule are paired, the molecule is **diamagnetic,** meaning that it has no electron spin-related magnetic field of its own. As a result, it is repelled by externally applied magnetic fields. Most molecules, including N_2, are diamagnetic, but O_2 is **paramagnetic,** which means that it is attracted to the poles of a magnet, as demonstrated in the experiment shown in Figure 6.18. A molecule is paramagnetic only if it has unpaired electrons in its structure. In this case, MO theory correctly predicts the magnetic properties of O_2, and the Lewis structure does not.

Figure 6.19 contains MO-based electron configurations for the homonuclear (same element) diatomic molecules formed by the elements in the second row of the periodic table. These configurations enable us to make several predictions about the behavior of these molecules. First, Be_2 and Ne_2 do not exist for the same reason that He_2 does not exist: they have as many antibonding electrons as they have bonding electrons and a net bond order of zero. Second, Li_2, B_2, and F_2 have a bond order of 1, whereas C_2 has a bond order of 2. Like O_2, B_2 is paramagnetic, whereas Li_2, C_2, N_2, and F_2 are diamagnetic.

 Diamagnetic atoms, ions, and molecules have no unpaired electrons. **Paramagnetic** atoms, ions, and molecules contain at least one unpaired electron and are attracted by an external magnetic field.

CONNECTION: The magnetic fields generated by spinning electrons are discussed in Section 3.9.

CONCEPT TEST: Based on the molecular-orbital diagrams for the homonuclear diatomic molecules of the second-row elements in Figure 6.19, identify the molecules in which there is 2s–2p mixing.

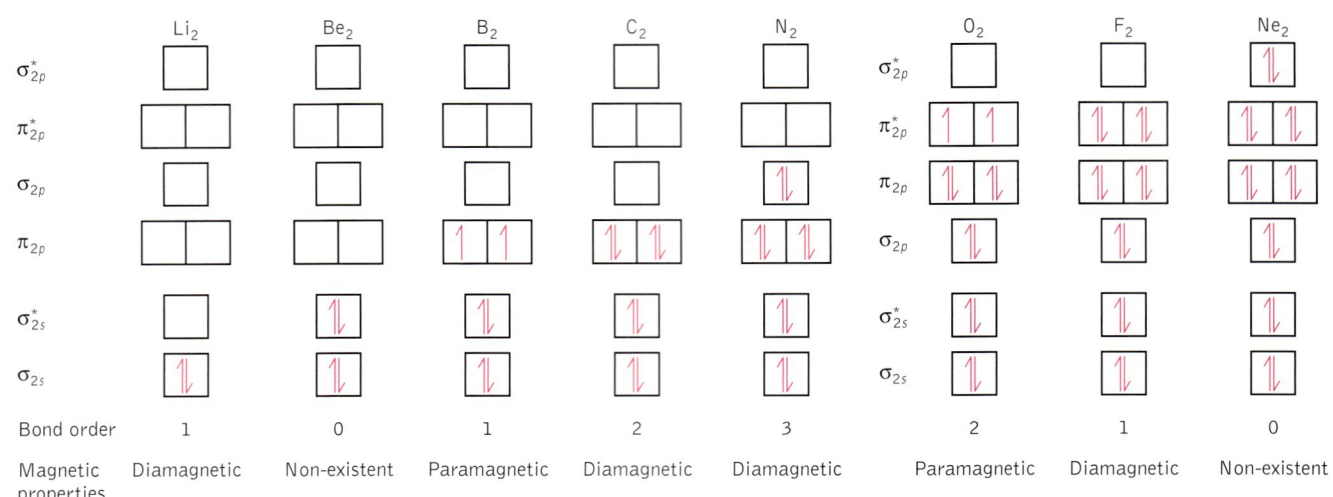

FIGURE 6.19 Molecular-orbital diagrams of the valence electrons and the magnetic properties of the homonuclear diatomic molecules of the second-row elements.

Ultraviolet and visible spectra and auroras

In addition to predicting the magnetic properties of molecules, MO theory is particularly useful for predicting their spectroscopic properties. Table 6.2 lists some of the principal species that contribute to the colors of the aurora. Excited states of two nitrogen-containing species, N_2^{+*} and N_2^*, produce, respectively, the blue violet (391–470 nm) and deep, crimson red (650–680 nm) waves in auroras. The MO diagrams for these excited states are shown in Figure 6.20. Compare the molecular-orbital diagram of N_2^* with that of N_2 in its ground state (see Figure 6.19). Note that an electron has been raised from the σ_{2p} MO in N_2 to a π_{2p}^* orbital in N_2^*, leaving an unpaired σ_{2p} electron behind. The corresponding excited state of the molecular ion N_2^{+*} also has one electron in a π_{2p}^* orbital, but the π_{2p} orbital is empty because the other σ_{2p} electron was lost when the molecule was ionized. As the π_{2p}^* electrons return from their antibonding orbital and excited state to the bonding (ground state) σ_{2p} orbital, the distinctive emissions of N_2^{+*} and N_2^* appear.

✓ In the molecular-orbital diagram of an excited state, an electron occupies an orbital of higher energy than that of an orbital that is empty or partially filled.

TABLE 6.2 Origins of Colors in the Aurora

Wavelength (nm)	Color	Chemical Species
650–680	Deep red	N_2^*
630	Red	O^*
558	Green	O^*
391–470	Blue violet	N_2^{+*}

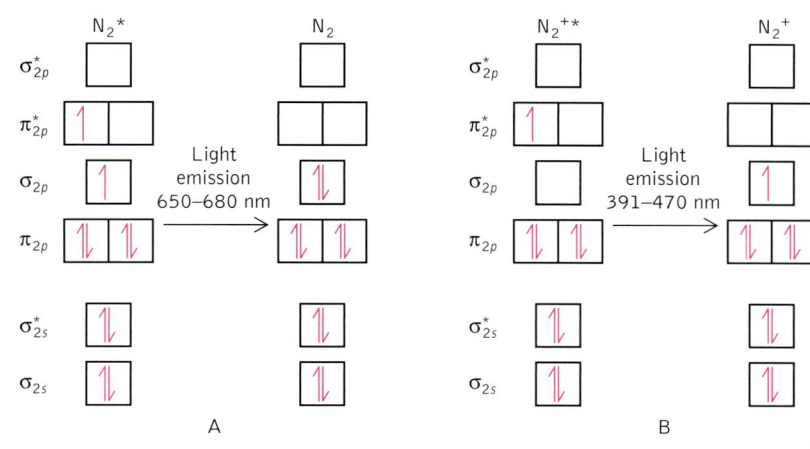

FIGURE 6.20 These molecular orbital diagrams for (A) N_2 and (B) N_2^+, show electronic transitions that produce visible light. Collisions with ions in the solar wind result in the promotion of an electron from the σ_{2p} orbital in N_2 to the π_{2p}^* orbital, creating an excited-state N_2^* molecule. When the electron returns to the ground state, the difference in energy between the states is emitted as light in the 650- to 680-nm range. Higher energies create N_2^{+*} molecular ions, which also have electrons in π_{2p}^* orbitals. When these electrons return to the ground state, light in the wavelength range of 391 to 470 nm is emitted.

SAMPLE EXERCISE 6.11: Determine the bond order in N_2^* and N_2^{+*}.

SOLUTION: Using the MO diagrams in Figure 6.20 and Equation 6.4, we find the bond order for N_2^* to be

(5 bonding electrons − 1 antibonding electron)/2 = 2 (a double bond)

For N_2^{+*}, it is

(4 bonding electrons − 1 antibonding electron)/2
= 1.5 (between a single and a double bond)

PRACTICE EXERCISE: Draw the MO diagram for O_2^+ and a possible MO diagram for its excited state, O_2^{+*}. What are the bond orders of O_2^+ and the excited-state O_2^{+*} that you have drawn? (See Problems 97 and 98.)

Scientists use the spectral "signatures" of molecules and molecular ions to search for them in the cosmos and here on Earth. Sometimes, the emission of spectra of high-energy substances are recorded and analyzed. Other times, the light that a substance absorbs is determined. In the field of analytical chemistry, absorption spectroscopy is a widely used technique in which different wavelengths of ultraviolet, visible, or infrared radiation (see Section 1.3) are passed through a sample and the absorption spectrum of the sample is recorded. The features of the spectrum can be used to determine which molecular species are present. Infrared spectroscopy is particularly useful for such a **qualitative analysis**. The *amount* of light absorbed at one or more characteristic wavelengths can be used to determine *how much* of a molecular substance is present in a sample. Ultraviolet and visible spectroscopy are widely used in a **quantitative analysis**.

Astronomers studying the cosmos also employ absorption spectroscopy, but the sources of light for their measurements are the nuclear furnaces of the stars. The substances absorbing starlight may be free atoms or ions (such as those that produce the Fraunhofer lines in the solar spectrum) or they may be molecules. Substances near stars also may emit distinctive spectra. If they are close enough

 Qualitative chemical analysis tells us what substances are present in a sample; **quantitative analysis** tells us how much of one or more substances is present.

CONNECTION: Infrared spectroscopy is widely used to identify organic compounds (Section 7.2).

FIGURE 6.21 Intense ultraviolet radiation from these four bright stars, called the Trapezium, is absorbed by gaseous molecules in the surrounding Orion nebula, creating excited-state molecules and ions. Radiation from these excited molecules and ions— including H_2, H_2CO, HCN, CO^+, HCO^+, and HOC^+—contribute to the multicolored glow of the nebula.

CONNECTION: The origin of Fraunhofer lines is described in Section 3.1.

to a source of energy, they may emit ultraviolet, visible, and infrared radiation, as happens for instance in the gases of the nebula shown in Figure 6.21. If the source of energy is too distant or too weak, molecules can still emit low-energy radiation such as radio waves and microwaves. In 1970, Arno Penzias and Robert Wilson, the scientists who first detected the cosmic microwave background (see Section 1.6), discovered interstellar carbon monoxide in this way.

Scientists scan the heavens with telescopes capable of analyzing the absorption and emission spectra of distant objects to learn about the chemical composition of the universe and the processes that created it. Data from these instruments enable us to know that the atmosphere of Venus consists mostly of carbon dioxide and other greenhouse gases and that the polar caps of Mars are made of solid carbon dioxide (dry ice). In 1996, scientists found a region of space so distant that the electromagnetic radiation that they detected must have been emitted before the universe was a tenth its current age. In the radiation was the spectral signature of carbon monoxide. This finding is significant because, as stated in Chapter 2, carbon and oxygen can form only in giant stars. Therefore, stars, galaxies, and the elements of the periodic table must have formed within a billion years of the beginning of our universe.

CHAPTER REVIEW

Summary

SECTION 6.1

The colors of the *aurora borealis*, or northern lights, are due to the emission of light by excited-state atoms, molecules and molecular ions.

SECTION 6.2

Electrostatic attraction between oppositely charged ions, such as Na^+ and Cl^-, accounts for the formation of ionic compounds. The atoms in molecules are held together by shared pairs of outermost, or valance, electrons.

SECTION 6.3

The Lewis symbol of an atom shows only its valence electrons. The Lewis structures of molecules and molecular ions show how pairs of valence electrons may be shared in covalent bonds. The lone pairs of electrons on atoms in a molecule do not contribute to bonding. The valency of an atom is the number of electron pairs that it can share. Stable small molecules tend to have an octet of electrons around every atom with Z > 5. Hydrogen requires only two electrons.

SECTION 6.4

Unequal electron sharing between atoms of different elements results in polar covalent bonds.

SECTION 6.5

Unequal sharing of the electrons in covalent bonds results from different electronegativities of the bonded atoms. The electronegativities of elements decrease down a column in the periodic table and increase from left to right in a row. They generally increase with increasing ionization energies and electron affinities and with decreasing atomic size. Thus, fluorine is the most-electronegative element.

SECTION 6.6

Lewis structures with the same molecular skeleton but different electron arrangements are resonance structures.

SECTION 6.7

The most favored structure of a molecule is one in which the formal charges on its atoms are closest to zero, and any negative formal charges are on the more-electronegative atoms.

SECTION 6.8

The bond distances between atoms in molecules and polyatomic ions in substances can be determined from electron-diffraction measurements.

SECTION 6.9

According to molecular-orbital theory, mixing atomic orbitals of atoms produces delocalized molecular orbitals. Mixing two atomic orbitals makes two MOs: one bonding (σ or π) and one antibonding (σ^* or π^*). A bonding MO has lower energy than that of the atomic orbitals that formed it; an antibonding orbital has higher energy. The most stable MOs are formed by mixing atomic orbitals of similar energies and the appropriate orientation. The bond order of a diatomic molecule is half the difference between the number of bonding and the number of antibonding electrons. Molecules absorb energy as electrons shift to higher-energy MOs. Molecules emit a distinctive pattern of radiation when electrons in higher-energy MOs fall into unfilled or partly filled MOs of lower energy.

Key Terms

allotrope (p. 298)
antibonding molecular orbital (p. 317)
bonding molecular orbital (p. 317)
bond order (p. 319)
catalyst (p. 311)
covalent bond (p. 288)
delocalization (p. 316)
diamagnetic (p. 321)
electron affinity (p. 295)

electronegativity (p. 293)
formal charge (p. 312)
Lewis structure (p. 290)
lone pair (p. 290)
main-group elements (p. 293)
molecular ion (p. 290)
molecular orbital (MO) (p. 316)
octet rule (p. 287)
paramagnetic (p. 321)

pi (π) bond (p. 318)
pi (π) molecular orbital (p. 318)
polar covalent bond (p. 292)
qualitative chemical analysis (p. 323)
quantitative chemical analysis (p. 323)
resonance structure (p. 303)
sigma (σ) bond (p. 316)
sigma (σ) molecular orbital (p. 316)
valence electron (p. 288)

Key Skills and Concepts

SECTION 6.1

Know that molecular substances are held together by covalent bonds.

SECTION 6.2

Understand the origin of the octet rule and be able to predict the number of bonds that an atom typically forms.

SECTION 6.3

Be able to draw the Lewis symbols of the main-group (representative) elements.
Understand the difference between bonding pairs and lone pairs of electrons.
Distinguish single, double, and triple covalent bonds.

SECTION 6.4

Recognize that atoms of different elements do not share bonding electrons equally and that unequal sharing leads to bond polarity.

SECTION 6.5

Understand the trends in electronegativity, size, ionization energy, and electron affinity in the columns and rows of the periodic table.
Be able to use electronegativity differences to estimate the polarity of bonds and to identify which end of a bond has higher electron density.

SECTION 6.6

Be able to construct Lewis structures for molecules and molecular ions.
Be able to recognize resonance and to draw resonance structures.

SECTION 6.7

Be able to use formal charges to identify the most-preferred molecular structures.

SECTION 6.8

Know that electron-diffraction data can be used to determine the distances between atoms in substances.
Be able to compare interatomic distances in different molecules to assess the relative importance of contributors to resonance structures.

SECTION 6.9

Understand how atomic orbitals of different atoms mix, forming σ, σ^*, π, and π^* molecular orbitals.
Know that each MO can hold as many as two electrons.
Know that the electrons in a molecule are allocated to MOs of successively higher energy.
Be able to sketch MO diagrams.
Be able to calculate the bond order of a homonuclear diatomic molecule from its MO electron configuration.
Be able to predict whether a diatomic molecule of an element is diamagnetic or paramagnetic from its MO diagram.
Understand that absorption and emission spectroscopy at various wavelengths can be used to identify molecules and molecular ions in a sample.

Key Equations and Relations

SECTION 6.3

The energy (E) of a photon:

$$E = hc/\lambda \qquad (1.3)$$

where λ is wavelength, h is Planck's constant, and c is the speed of light.

SECTIONS 6.4 AND 6.5

Bond polarity is related to the difference in electronegativity between two bonded atoms.

SECTION 6.6

The number of bonds in a molecule or molecular ion for which the octet rule applies:

$$S = N - H \qquad (6.14)$$

where S is the number of electrons that must be shared to build the Lewis structure; N is the total number of valence electrons if each atom in the molecule had a noble-gas electron configuration, and H is the total number of valence electrons in the molecule or ion.

SECTION 6.7

Formal charge of an atom in a Lewis structure:

formal charge = number of valence electrons in the free atom
− number of valence electrons assigned to the atom in the structure

SECTION 6.9

Bond order in a molecular-orbital description of a molecule or an ion:

$$\text{bond order} = \tfrac{1}{2}(\text{number of bonding electrons} - \text{number of antibonding electrons})$$

QUESTIONS AND PROBLEMS

Lewis Symbols

CONCEPT REVIEW

1. Which electrons in an atom are considered the *valence* electrons?
2. Does the number of valence electrons in a neutral atom ever equal the atomic number?
3. Does the number of valence electrons in a neutral atom ever equal the group number?
4. Do all the elements in a group in the periodic table have the same number of valence electrons?

PROBLEMS

5. Draw Lewis symbols for atoms of lithium, magnesium, and aluminum.
6. Draw Lewis symbols for atoms of nitrogen, oxygen, fluorine, and chlorine.
7. Draw the Lewis symbols of Na^+, As^{3-}, Ca^{2+}, and S^{2-}.
8. Draw Lewis symbols for the most stable ion formed by lithium, magnesium, aluminum, and fluorine.
9. Which of the following ions have a complete valence-shell octet? B^{3+}, I^-, Ca^{2+}, or Pb^{2+}.
10. Draw Lewis symbols for Xe, Sr^{2+}, Cl, and Cl^-. How many valence electrons are in each atom or ion?
11. Which group among the main-group elements in the periodic table contains atoms that have
    a. one valence electron?
    b. four valence electrons?
    c. six valence electrons?
12. Which group among the main-group elements in the periodic table contains atoms that have
    a. two valence electrons?
    b. five valence electrons?
    c. three valence electrons?
13. Draw the Lewis symbol of an ion that has
    a. a 1+ charge and one valence electron.
    b. a 3+ charge and no valence electrons.
14. Draw the Lewis symbol of an ion that has
    a. a 1− charge and eight valence electrons.
    b. a 1+ charge and five valence electrons.
15. Which of the following Lewis symbols correctly portrays the most stable ion of magnesium?

    $[Mg\cdot]^+$ $[Mg]^+$ $[\ddot{M}g{:}]^{2+}$ $[Mg{:}]^{2+}$ $[Mg]^{2+}$

16. Which of the following Lewis symbols are correct?

    $\dot{N}\cdot$ $[\ddot{N}]^{2+}$ $[\ddot{N}{:}]^{3-}$ $[{:}\ddot{O}]^{2-}$ $[{:}\ddot{O}]^{2+}$

Lewis Structures of Diatomic Molecules

CONCEPT REVIEW

17. Describe the differences in bonding in *covalent* and *ionic* compounds.
18. Some of his critics described Gilbert N. Lewis's approach to explaining covalent bonding as an exercise in double counting and therefore invalid. Explain the basis for this criticism.
19. Does the octet rule mean that a diatomic molecule has 16 valence electrons?
20. Does the octet rule mean that a diatomic molecule has 8 valence electrons?
21. What is meant by the term *polar covalent* bond?
22. Why are the electrons that bond different elements together not equally shared?

PROBLEMS

23. How many valence electrons does each of the following species contain? (a) BN; (b) HF; (c) OH^-; (d) CN^-.
24. How many valence electrons does each of the following species contain? (a) N_2^+; (b) CS^+ (c) CN; (d) CO.
25. Draw Lewis structures of the following diatomic molecules and ions. (a) CO; (b) O_2; (c) ClO^-; (d) CN^-.
26. Draw Lewis structures of the following diatomic molecules and ions. (a) F_2; (b) NO^+; (c) SO; (d) HI.
27. How many electron pairs are shared in each of the molecules and ions in Problem 25?
28. How many covalent bonds are there in each of the molecules and ions in Problem 26?

Electronegativity, Ionization Energy, and Electron Affinity

CONCEPT REVIEW

29. Which of the following periodic properties can be determined experimentally: electronegativity, ionization energy, or electron affinity?
30. How can we use electronegativity to predict whether a bond between two atoms is likely to be covalent or ionic?
31. How do the electronegativities of the elements change across a period and down a group? Explain, on the basis of atomic structure, why these trends are observed.
32. Why do the ionization energies of the elements increase across a period and decrease down a group?
33. Why do electron affinities of the elements not follow the same trend as that of ionization energy across the second row of the periodic table?
34. Why does atomic size decrease across a period and increase down a group?
35. What does a positive value for electron affinity mean?
36. Is the element with the largest (most negative) electron affinity also the most electronegative?

PROBLEMS

37. Which of the following bonds are polar? C—S, C—O, Cl—Cl, O=O, N—H, C—H. In the bonds that you selected, which atom has the greater electronegativity?
38. Which is the least polar of the following bonds? C—S, C—O, Cl—Cl, O=O, N—H, C—H.
39. Which of the binary compounds formed by the following pairs of elements contain polar covalent bonds, and which of them are considered ionic compounds?
    a. C and S
    b. C and O
    c. Al and Cl
    d. Ca and O
40. Which of the beryllium halides, if any, are considered ionic compounds?

Lewis Structures of Polyatomic Molecules and Ions

CONCEPT REVIEW

41. Why is the bonding pattern in water H—O—H and not H—H—O?
42. Does each atom in a pair that is covalently bonded always contribute the same number of valence electrons to form the bonds between them?

PROBLEMS

43. Chlorofluorocarbons (CFCs) are linked to the depletion of stratospheric ozone. They are also greenhouse gases. Draw Lewis structures for the following CFCs:
    a. CF_2Cl_2 (Freon 12)
    b. Cl_2FCCF_2Cl (Freon 113, containing a C–C bond)
    c. C_2Cl_3F (Freon 1113, containing a C–C bond)
44. The replacement of one halogen in a CFC by hydrogen makes the compound more environmentally "friendly." Draw Lewis structures of the following compounds:
    a. CHF_2Cl (Freon 22)
    b. $CHCl_3$ (chloroform, formerly used as an anesthetic)
    c. CH_2Cl_2 (a common laboratory solvent)
45. Many sulfur-containing organic compounds have characteristically foul odors: butyl thiol, $CH_3CH_2CH_2CH_2SH$, is responsible for the odor of skunks, and rotten eggs smell the way they do because they produce tiny amounts of pungent hydrogen sulfide, H_2S. Draw the Lewis structures of $CH_3CH_2CH_2CH_2SH$ and H_2S.
46. Formic acid, HC(O)OH, is the smallest organic acid. Draw its Lewis structure.
47. Chlorine combines with oxygen in several proportions. Dichlorine monoxide, Cl_2O, is used in the manufacture of bleaching agents. Potassium chlorate, $KClO_3$, is used in oxygen generators aboard aircraft. Draw the Lewis structures of Cl_2O and ClO_3^-.
48. Labels on household cleansers caution against mixing bleach with ammonia because the reaction produces monochloramine (NH_2Cl), hydrazine (N_2H_4), and the hypochlorite ion (OCl^-).

    $NH_3(aq) + OCl^-(aq) \longrightarrow NH_2Cl(aq) + OH^-(aq)$
    $NH_2Cl(aq) + NH_3(aq) + OH^-(aq) \longrightarrow N_2H_4(aq) + H_2O(l)$

    Draw the Lewis structures of monochloramine, hydrazine, and the hypochlorite ion.

Resonance

CONCEPT REVIEW

49. Explain the concept of molecular *resonance*.
50. Does resonance help to stabilize a molecule or an ion?
51. Explain why SO_2 exhibits resonance, whereas H_2S does not.
52. What features do all the resonance forms of a molecule or an ion have in common?
53. What role does resonance play in our understanding of covalent bonding?
54. How do you recognize when resonance forms are needed to account for the bonding in a molecule?
55. Are the following three structures resonance forms of the thiocyanate ion, SCN^-?

    $[:\ddot{S}-N\equiv C:]$ $[:N\equiv C-\ddot{S}:]$ $[:\ddot{N}-S\equiv C:]$

56. Why are the structures below not resonance forms of the molecule S_2O?

    $\ddot{S}=\ddot{S}-\ddot{O}:$ $:\ddot{S}-\ddot{O}=\ddot{S}$

PROBLEMS

57. Draw two Lewis structures showing the resonance that occurs in the cyclic molecule benzene (C_6H_6), which has a skeletal structure that includes a ring of six carbon atoms.
58. Oxygen and nitrogen combine to form a variety of nitrogen oxides, including the following two unstable compounds that have two nitrogen atoms per molecule: N_2O_2 and N_2O_3. Draw Lewis structures for these molecules, including all resonance forms.
59. Oxygen and sulfur combine to form a variety of different sulfur oxides. Some are stable molecules and some, including S_2O_2 and S_2O_3, decompose when they are heated. Draw Lewis structures for these compounds, including all resonance forms.
60. Draw Lewis symbols for hydrocyanic acid, HCNO, including all resonance forms.
61. Draw three Lewis structures showing the resonance that occurs in the carbonate ion (CO_3^{2-}).
62. Nitrogen-fixing bacteria convert urea, $H_2NC(O)NH_2$ into nitrite ions (NO_2^-). Draw Lewis structures for these two species. Include all resonance forms. Hint: There is a C=O double bond in urea.

Formal Charges

CONCEPT REVIEW

63. Describe how formal charges are used to choose between possible molecular structures.
64. How do the electronegativities of elements influence the selection of which Lewis structure is favored?
65. In a molecule containing S and O atoms, is a structure with a negative formal charge on sulfur more likely to contribute to bonding than an alternative structure with a negative formal charge on oxygen?
66. In a cation containing N and O, why do Lewis structures with a formal positive charge on nitrogen contribute more to bonding than do those structures with a formal positive charge on oxygen?

PROBLEMS

67. Nitrogen is the central atom in molecules of nitrous oxide, N_2O. Draw Lewis structures for another possible arrangement: N–O–N. Assign formal charges and suggest a reason why this structure is not likely to be stable.
68. Complete the Lewis structures and assign formal charges to the atoms in five of the resonance forms of SN_4. Indicate which of your structures should be most stable.
69. Hydrogen isocyanide, HNC, has the same elemental composition as HCN, but the H atom in HNC is bonded to nitrogen. Draw a Lewis structure for HNC and assign formal charges to each atom. What are the differences in formal charges between the atoms in the Lewis structures of HCN and HNC?
70. Hydrogen cyanide, HCN, and cyanoacetylene, HC_3N, have been detected in the interstellar regions of space. Draw Lewis structures for these molecules and assign formal charges to each atom. The hydrogen atom is bonded to carbon in both cases.

71. The discovery of polyatomic organic molecules such as cyanamide, H_2NCN, in interstellar space has led some scientists to believe that the molecules from which life began on Earth may have come from space. Draw Lewis structures for cyanamide and select which of them is the preferred structure on the basis of formal charges.

72. Formamide, $HCONH_2$, and methyl formate, HCO_2CH_3, also have been detected in space. Draw Lewis structures for these compounds, based on the following skeletal structures:

73. Nitromethane (CH_3NO_2) reacts with hydrogen cyanide, producing $CNNO_2$ and CH_4

$$HCN + CH_3NO_2 \longrightarrow CNNO_2 + CH_4$$

   a. Draw Lewis structures of CH_3NO_2, showing all resonance forms.
   b. Draw Lewis structures, showing all resonance forms, based on the following two possible skeletal structures for $CNNO_2$, assign formal charges, and predict which structure is more likely to exist.

   c. Are the two structures resonance forms of each other?

*74. AgOCN reacts with Br_2 and NO_2, forming N_2O and CO_2. During the reaction, an unstable molecule with the formula $OCNNO_2$ may form. If two of these molecules collide, they may form OCNNO and O_2CNNO_2.
   a. Draw three of the resonance forms of OCNNO.

   $$O-C-N-N-O$$

   b. On the basis of formal charges, is the Lewis structure for CN_2O_3 with the framework shown on the left more stable than the one on the right?

75. Use formal charges to determine which resonance form of each of the following ions is preferred: CNO^-, NCO^-, and CON^-.

*76. The dinitramide anion, $N(NO_2)_2^-$, was first isolated in 1996. Its skeletal structure is

Complete the Lewis structure of $N(NO_2)^-$. Include any resonance forms and assign formal charges.

Bond Length

CONCEPT REVIEW

77. Do you expect the N–O bond length in the nitrate ion to be the same as in the nitrite ion?
78. Why is the O–O bond length in O_3 not the same as in O_2?
79. Explain why the N–O bond lengths in N_2O_4 (which has an N–N bond) and N_2O are nearly identical: 118 and 119 pm, respectively.
80. Do you expect the S–O bond lengths in sulfite (SO_3^{2-}) and sulfate (SO_4^{2-}) ions to be about the same? Why?

PROBLEMS

81. Rank the following ions in order of increasing N–O bond lengths: NO_2^-, NO^+, and NO_3^-.

82. Estimate the C–O bond lengths in the following ions and molecules (see Table 6.1 for average lengths of C—O, C=O, and C≡O bonds): CO, CO_2, and CO_3^{2-}.

*83. Carbon and oxygen form three oxides, CO, CO_2, and carbon suboxide, C_2O_3. Draw a Lewis structure for C_2O_3 in which the two carbon atoms are bonded to each other, and predict whether the C–O bond lengths in the molecule are equal.

*84. Draw a Lewis structure for the molecule N_4O. Spectroscopic analysis of the molecule reveals that the N–O bond length is 120 pm and that there are three N–N bond lengths: 148, 127, and 115 pm.

Molecular Orbitals

CONCEPT REVIEW

85. Do all σ molecular orbitals result from the mixing of s atomic orbitals?

86. Do all π molecular orbitals result from the mixing of p atomic orbitals?

87. Are s atomic orbitals with different principal quantum numbers (n) as likely to mix and form MOs as s atomic orbitals with the same value of n?

88. Are atomic orbitals with the same principal quantum number (n) but different angular momentum quantum numbers (l) as likely to mix and form MOs as orbitals with the same values of n and l?

PROBLEMS

89. Make a sketch showing how two $1s$ orbitals overlap to form a σ_{1s} bonding molecular orbital and a σ_{1s}^* antibonding molecular orbital.

90. Make a sketch showing how two $2p_y$ orbitals overlap "sideways" to form a π_{2p} bonding molecular orbital and a π_{2p}^* antibonding molecular orbital.

91. Use MO theory to predict the bond orders of the following molecular ions: N_2^+, O_2^+, C_2^{2+}, and Br_2^{2-}. Do you expect any of these species to exist?

92. Diatomic noble-gas molecules, such as He_2 and Ne_2, do not exist. Would removing an electron create molecular ions, such as He_2^+ and Ne_2^+, that are more stable than He_2 and Ne_2?

93. Which of the following molecular ions is expected to have one or more unpaired electrons? N_2^+, O_2^+, C_2^{2+}, Br_2^{2-}.

94. Which of the following molecular ions is expected to have one or more unpaired electrons? O_2^-, O_2^{2-}, N_2^{2-}, or F_2^+.

95. Which of the following anions have electrons in π antibonding orbitals? C_2^{2-}, N_2^{2-}, O_2^{2-}, Br_2^{2-}.

96. Which of the following molecular cations have electrons in π antibonding orbitals? N_2^+, O_2^+, C_2^{2+}, Br_2^{2+}.

97. For which of the following diatomic molecules does the bond order increase with the gain of two electrons, forming the corresponding anion with a -2 charge?
 a. $B_2 + 2\,e^- \rightarrow B_2^{2-}$
 b. $C_2 + 2\,e^- \rightarrow C_2^{2-}$
 c. $N_2 + 2\,e^- \rightarrow N_2^{2-}$
 d. $O_2 + 2\,e^- \rightarrow O_2^{2-}$

98. For which of the following diatomic molecules does the bond order increase with the loss of two electrons, forming the corresponding cation with a $+2$ charge?
 a. $B_2 \rightarrow B_2^{2+} + 2\,e^-$
 b. $C_2 \rightarrow C_2^{2+} + 2\,e^-$
 c. $N_2 \rightarrow N_2^{2+} + 2\,e^-$
 d. $O_2 \rightarrow O_2^{2+} + 2\,e^-$

99. Do the $1+$ cations of homonuclear diatomic molecules of the second-row elements always have shorter bond lengths than those of the corresponding neutral molecules?

100. Do any of the anions of the homonuclear diatomic molecules formed by B, C, N, O, and F have shorter bond lengths than those of the corresponding neutral molecules? Consider only the anions with -1 and -2 charge.

7 Molecular Shape
And the Greenhouse Effect

The primary mission of the *Galileo* spacecraft, launched in October 1989, was to explore the atmosphere and moons of Jupiter, which it is still doing today. The spacecraft passed the inner planets on its journey to Jupiter, taking the picture of Venus shown in Figure 7.1.

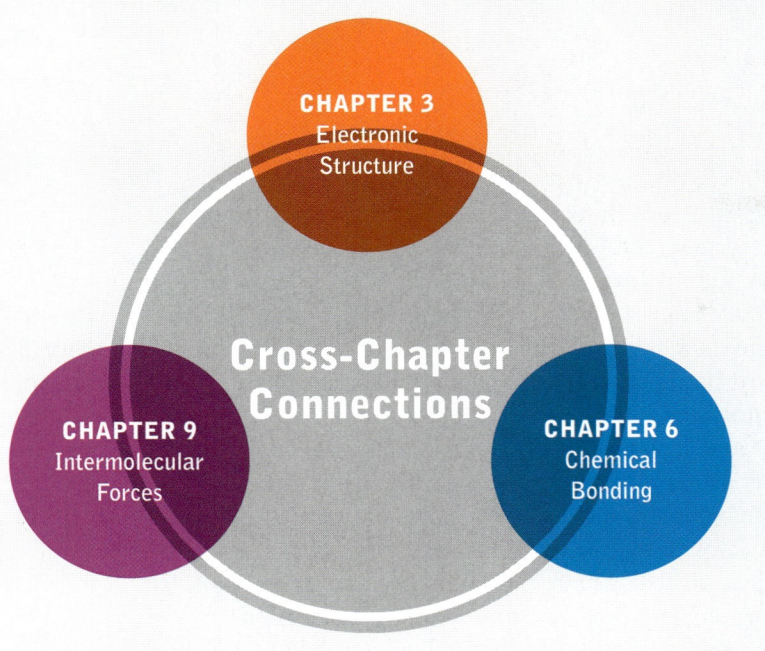

- 7.1 Bond vibration and climate change
- 7.2 Infrared spectroscopy
- 7.3 Exceptions to the octet rule
- 7.4 The molecular-orbital diagram of nitric oxide
- 7.5 Electron-spin resonance: locating unpaired electrons
- 7.6 Molecular shape: the VSEPR model
  - Tetrahedra of electrons
  - Triangles of electrons
  - Linear molecular geometry
  - Shapes of expanded-octet molecules
  - Summary
- 7.7 Valence-bond theory
  - Hybrid orbitals
  - Hybrid orbitals for beryllium and boron
- 7.8 Polar bonds and polar molecules
- 7.9 Molecular vibration and infrared absorption

A Look Ahead

In this chapter, we continue our examination of the structure of atmospheric gases and the bonding in molecules that contain atoms with more or less than eight electrons in their valence shells. We also consider the shapes of molecules and two theories of bond formation and orientation that account for molecular shapes. The combined effect of molecular shape and the alignment of polar bonds within a molecule accounts for the nonpolar character of some molecules and the polar character of others. The vibration of polar bonds is linked to the absorption and emission of infrared radiation, a phenomenon that is explored in the context of global warming and that leads to a brief examination of the analytical capabilities of infrared spectroscopy.

FIGURE 7.1 The surface of Venus, as viewed from the *Galileo* spacecraft, is obscured by clouds of sulfuric acid suspended in an atmosphere that is nearly 100 times as dense as that of Earth and is 97% carbon dioxide. This CO_2 level has created an atmospheric greenhouse around Venus that traps the energy of the sun, contributing to surface temperatures higher than 450°C.

CONNECTION: The temperature of the surface of the sun is about 5700 K. The electromagnetic spectrum that it emits is shown in Figure 1.12.

7.1 BOND VIBRATION AND CLIMATE CHANGE

For years, people have pondered the prospects of life on the other planets of the solar system. Until recently, scientists believed that only Earth had an atmosphere and hydrosphere with temperatures and chemical compositions that allowed life to exist. Many scientists still believe that Earth is the only such planet, though there is some evidence that primitive life forms may have existed on Mars in the distant past. The outer planets of the solar system (Jupiter, Saturn, Uranus, Neptune, and Pluto) have hydrogen and helium atmospheres and are probably frigid and lifeless, though there is some evidence that the frozen exterior of Jupiter's moon Europa may encase a warm interior that contains liquid water, a key ingredient in the evolution of life. Mercury and Venus are far too hot for life as we know it to exist. Venus is similar to Earth in mass, volume, and gravity. It has been visited by more than 20 spacecraft since 1962. These missions confirm that Venus is a hot, dry, inhospitable planet with an atmosphere nearly 100 times as dense as Earth's and made up mostly of carbon dioxide (97% by volume). The clouds that cover much of the planet (Figure 7.1) are made not of water but of sulfuric acid.

Is there a connection between the composition of the Venusian atmosphere and the planet's high surface temperatures? Indeed there is, and it is related to the molecular structure of CO_2. We noted in Chapter 1 that most of the electromagnetic energy emitted by the sun is in the visible region. When this energy is absorbed by the surface of a planet, the surface is warmed. A warm surface radiates electromagnetic energy. The surface temperatures of the planets are much lower than the temperature of the sun; so the wavelengths of electromagnetic energy that they radiate are much longer, in the infrared (IR) region. On Venus, this IR radiation encounters an atmosphere rich in CO_2. Because of its molecular structure, CO_2 is an effective absorber of IR energy. This capability means that the heat leaving the surface of Venus is trapped in its atmosphere, in much the way that the glass of a greenhouse allows sunlight to pass through but absorbs and traps infrared radiation. Earth's atmosphere contains much less CO_2 and other "greenhouse" gases than Venus does, but their concentrations are sufficient to influence global temperatures (Figure 7.2).

> **What makes CO_2 an effective absorber of infrared radiation?**

What makes CO_2 an effective absorber of infrared radiation? The absorption process is related to the fact that molecular bonds are not completely rigid. They behave more like flexible springs than rigid rods, which means that they can undergo a variety of vibrational motions: stretching, bending, and twisting, for example. In CO_2, the average carbon–oxygen bond length is 116 pm, but this dimension changes slightly as the two bonds stretch and compress in unison, a *symmetric-stretch* vibration mode (Figure 7.3). These bonds can also undergo an *asymmetric stretch* as one bond stretches while the other compresses. The mole-

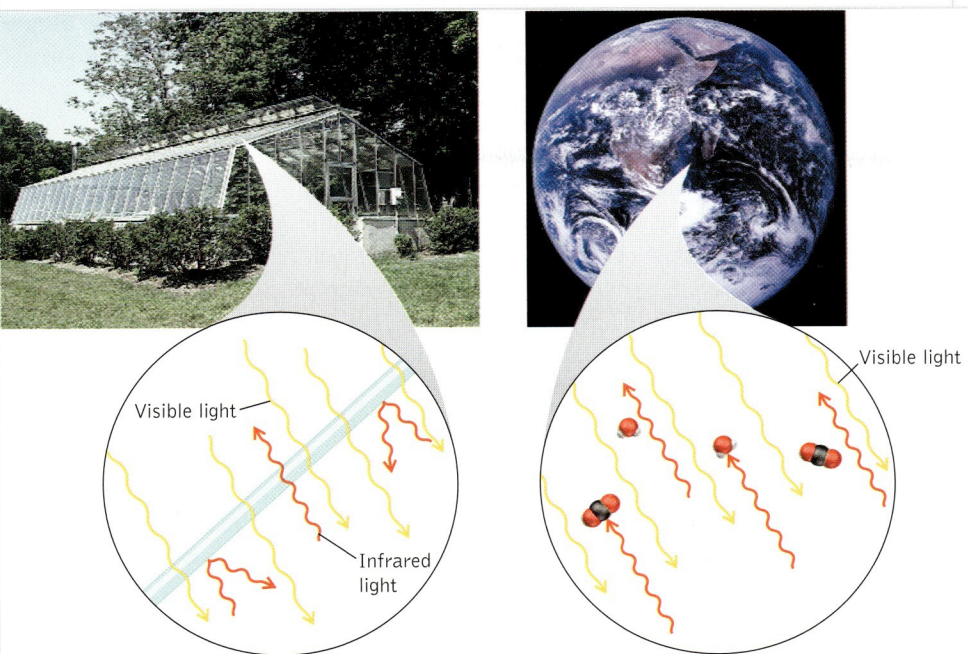

FIGURE 7.2 *What a Chemist Sees.* A. In a greenhouse, sunlight passes through the glass roof and walls, warming the plants and soil inside. These warmed materials radiate heat in the form of infrared radiation, but this radiation is absorbed by the glass, trapping much of it inside the greenhouse. B. Gases, including carbon dioxide and water, in Earth's atmosphere, also are transparent to visible sunlight but absorb the infrared radiation emitted by Earth's surface, trapping a significant percentage of it and increasing global temperatures.

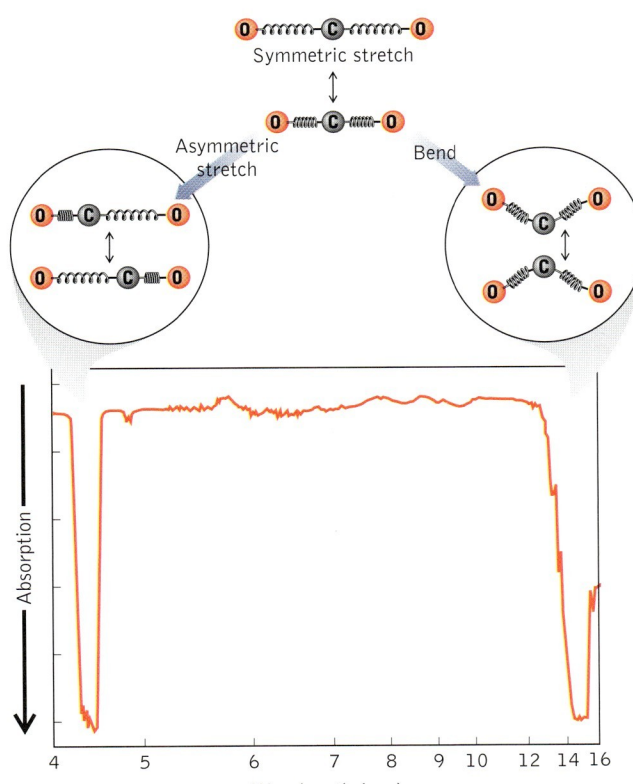

FIGURE 7.3 The infrared spectrum of carbon dioxide includes two absorption bands corresponding to bending and asymmetric stretching of the C=O double bonds. The symmetric-stretching mode does not produce an infrared absorption band for reasons discussed in Section 7.8.

cule can also bend. These motions are linked to infrared radiation because they occur at characteristic frequencies, and these frequencies correspond to the frequencies of infrared radiation.

The frequencies of bond vibrations depend on the force that it takes to distort the bond and on the masses of the atoms at either end. Just as molecules absorb quanta of ultraviolet and visible electromagnetic radiation, forming electronic excited states, molecules may absorb photons of IR radiation as they populate excited vibrational-energy levels. These vibrational levels are quantized, and only IR frequencies (wavelengths) with energies that match the quantized differences in energy between these levels may be absorbed. For example, asymmetric stretching in CO_2 gives rise to an IR absorption band with a wavelength of 4.3 μm. The bending vibration produces a band at 15. μm. However, not all vibrations produce IR absorption bands. For example, the symmetric-stretching vibration in CO_2 does not. We will examine why it does not in Section 7.8.

Carbon dioxide is not the only gas in Earth's atmosphere that absorbs infrared radiation. Among the others is water vapor. The principal vibrational motions for molecules of water are shown in Figure 7.4. All of them absorb IR radiation, including the symmetric O–H stretching vibration that gives rise to a strong IR band at 2.7 μm. Bending vibrations cause the bond angle to change slightly from its average value of 104.5 degrees and result in a characteristic band at 6.2 μm.

The ability of the bonds in CO_2, H_2O, and other atmospheric gases to absorb infrared radiation is important when we consider the emission spectrum

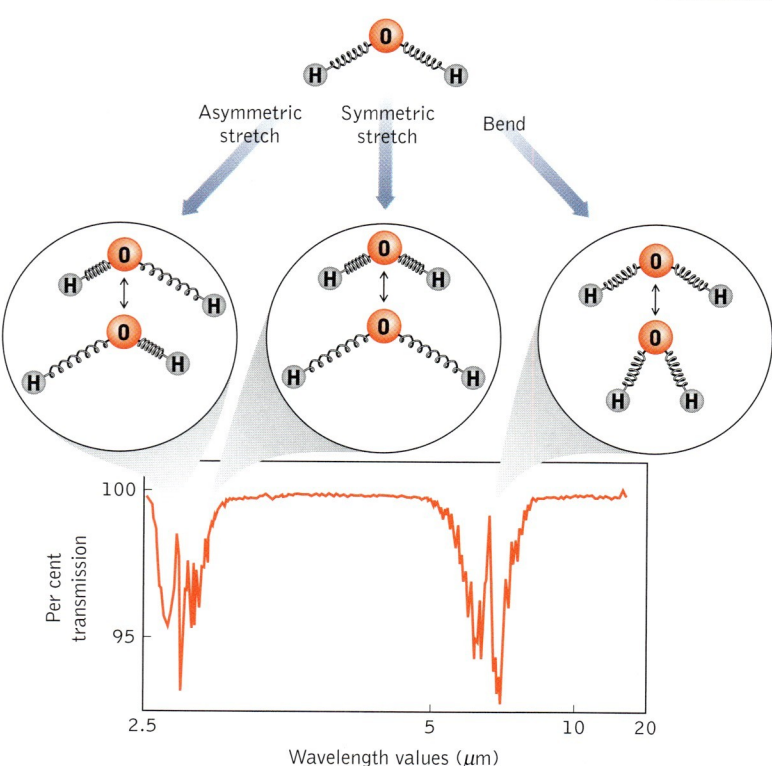

FIGURE 7.4 A water molecule has three vibration modes, all of which are infrared active. The bands on the left are linked to O–H stretching modes; those on the right are the result of O–H bending vibrations.

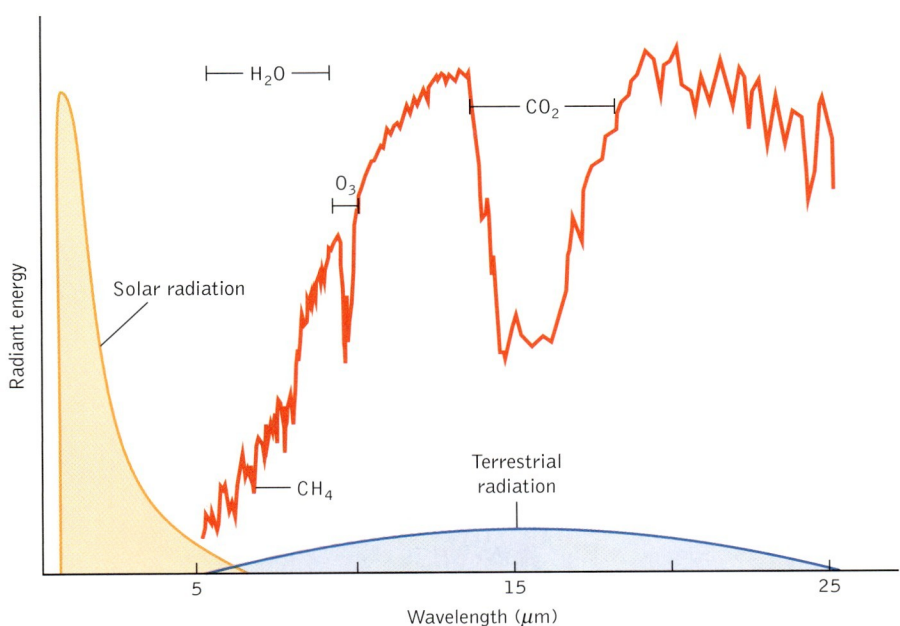

FIGURE 7.5 The spectrum of the sun that reaches Earth is shown on the left, and a magnified view of the infrared (IR) spectrum that the Earth radiates back into space, as viewed by a *Nimbus* satellite, is shown on the right. The IR spectrum includes several broad and many narrow valleys that coincide with the IR absorption bands of atmospheric gases. Note the strong absorption from carbon dioxide in the region of maximum emission at 15 mm.

of Earth's surface (Figure 7.5). Earth's emission reaches a maximum at about 15. μm, which coincides with a strong CO_2 absorption band. The excited vibrational states produced by the absorption process have very short lifetimes. Molecules quickly return to their ground states by releasing energy through collisions with other molecules or by reemitting infrared radiation. Some of the reemitted energy radiates into space, but much of it is reflected back into the planet's atmosphere, warming the atmosphere and the planet below.

The idea that CO_2 acts as a greenhouse gas on Venus was first proposed in 1940 and laid the foundation for one of the great environmental debates—global warming of Earth. Carbon dioxide and water vapor in Earth's atmosphere play an important role in moderating the temperature of our planet. Average annual global surface temperatures are believed to have varied within a narrow range—between 7° and 15°C (280–288 K)—since life began. But small temperature differences have caused significant changes in climatic conditions. The coldest temperatures have triggered periods of extensive glaciation, including the most recent Ice Age, which ended about 10^4 years ago. A 1 degree (Celsius) rise in global temperatures between A.D. 900 and 1200 turned the icy coast of Greenland into farmland and allowed vineyards to flourish in England. This warming was followed by a gradual cooling that began in the early fifteenth century, bringing harsh winters and advancing mountain glaciers. These natural climatic changes

> ✓ Greenhouse gases are transparent to sunlight, but they absorb heat in the form of infrared radiation emitted by Earth's surface, trapping the heat from this infrared radiation in the atmosphere.

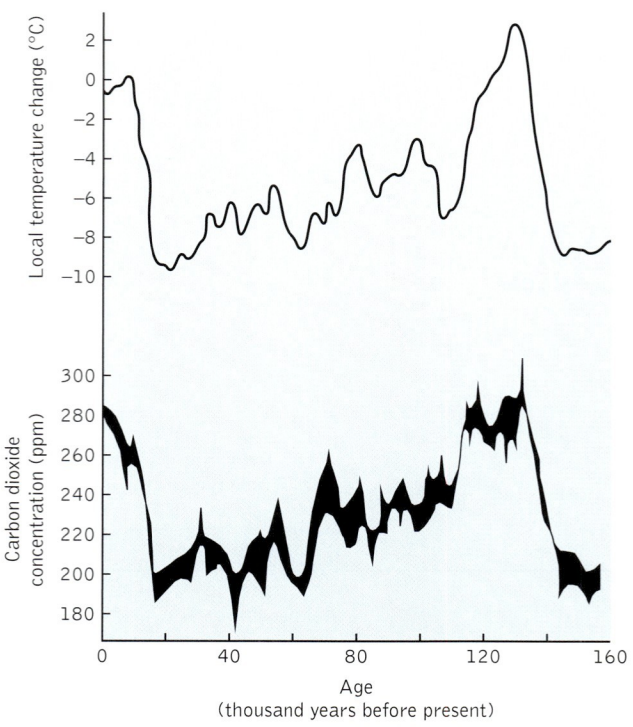

Analyses of ice cores from a site in Antarctica have allowed scientists to estimate temperatures and atmospheric concentrations of carbon dioxide over the past 160,000 years. These data show a strong correlation between temperature and atmospheric concentrations of CO_2. Concentration values are expressed in ppm (parts per million) by volume. These values correspond to microliters of CO_2 per liter of air.

(Figure 7.6) have recently been followed by a significant warming of Earth, which may have been due to the effects of human activity that increased concentrations of atmospheric CO_2 (Figure 7.7) and other greenhouse gases (Figure 7.8).

Climate models predict that continued emission of CO_2 based on the current rate of fossil-fuel consumption will result in a planetary warming from 1.0 to 3.5

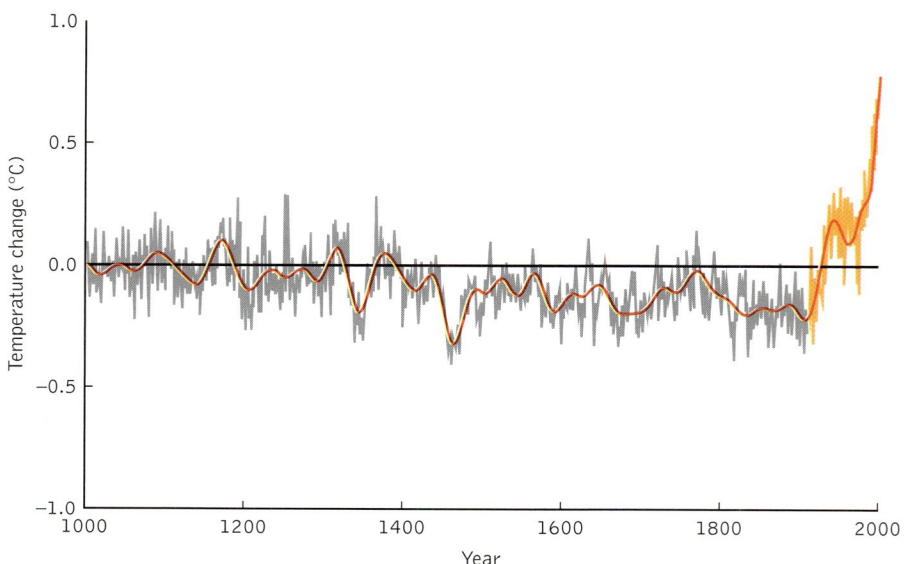

FIGURE 7.6 Changes in average global temperatures between A.D. 1000 and 2000 have been obtained from analyzing tree rings and sediments (shown in gray) and from direct measurements (shown in orange). The red line represents a 40-year running average. Throughout most of the millennium, global temperatures declined, but they began to rise sharply in the twentieth century. Many people believe that the recent increase is the result of human activity that has contributed to increasing concentrations of carbon dioxide and other greenhouse gases in the atmosphere.

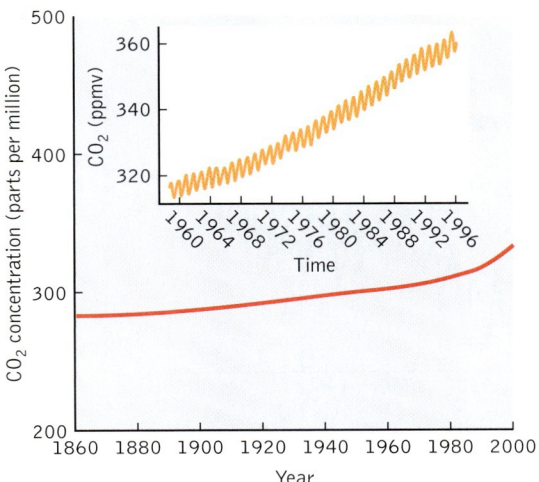

FIGURE 7.7 Trends in atmospheric concentration of carbon dioxide since 1860 and since 1958 (inset) show increases that seem to be most rapid in the last half of the twentieth century.

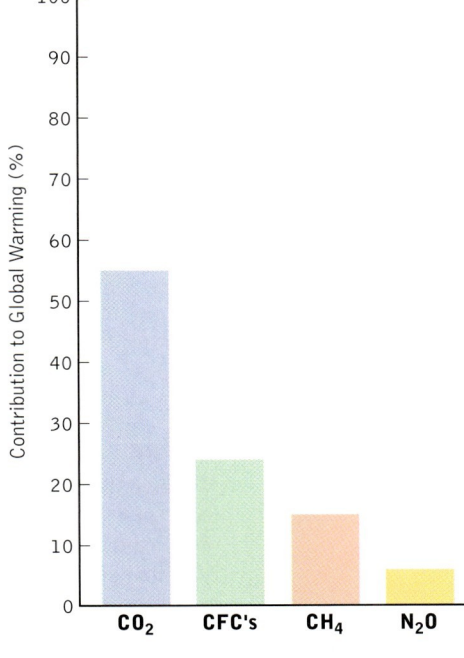

FIGURE 7.8 Human activity has increased atmospheric concentrations of carbon dioxide and several other greenhouse gases. The lengths of the bars in this graph represent the contribution of these gases to global warming.

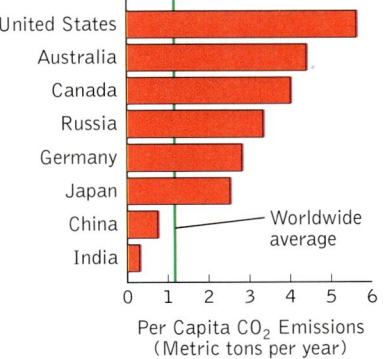

Emission of carbon dioxide from human activity averaged about 1 metric ton per person worldwide (vertical green line) in 1999. Industrialized nations exceeded this worldwide average, with the United States producing about 5.5 metric tons of carbon in the form of CO_2 per capita.

Celsius degrees in the twenty-first century. Some scientists have argued that increased heat from trapped IR radiation will be consumed by the evaporation of water, which could lead to an increase in cloud formation. More clouds would reflect more sunlight back into space and so would moderate the global-warming process. Indeed, computer models that incorporate the scattering of sunlight back into space from clouds and other atmospheric aerosols have done a better job of predicting changes in global temperature than those that have not, as shown in Figure 7.9.

Higher temperatures will also accelerate the rates of biological processes. Biological processes are essentially combinations of chemical reactions, and, as we

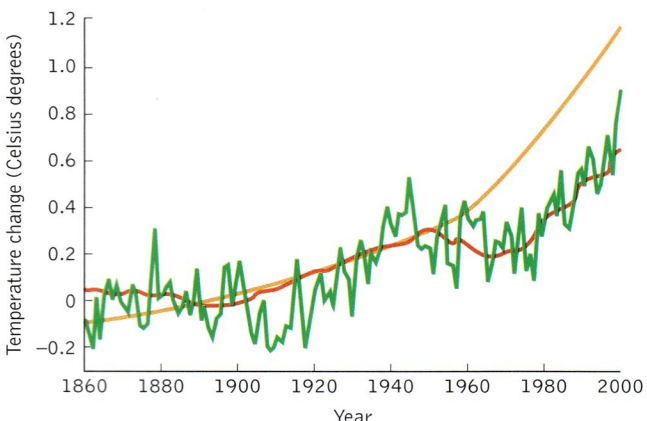

FIGURE 7.9 Accurate models of global warming must include variables in addition to atmospheric concentrations of greenhouse gases. Temperature changes based on increasing concentrations of greenhouse gases alone (orange line) do not track the observed temperature changes (green line) as well as models that account for increased reflection of sunlight from atmospheric aerosols.

shall see in Chapter 14, the rates of chemical reactions increase with increasing temperature. Warmer oceans could stimulate the growth of phytoplankton, microscopic plant life that forms the base of the marine food chain. Like terrestrial green plants, phytoplankton contain chlorophyll and so can convert carbon dioxide and water into glucose and oxygen through photosynthesis:

$$6\ CO_2(g) + 6H_2O(l) \longrightarrow C_6H_{12}O_6(aq) + 6\ O_2(g)$$

Some scientists have even suggested fertilizing the oceans by adding trace nutrients such as iron to stimulate the growth of phytoplankton and reduce atmospheric carbon dioxide.

CONNECTION: The effect of temperature on the rate of chemical reactions is described in Section 14.5.

CONNECTION: Photosynthesis and the carbon cycle are discussed in Section 4.6.

7.2 INFRARED SPECTROSCOPY

The various vibrational modes in the molecules of atmospheric gases and in nearly all molecules give rise to distinctive **infrared spectra** that can be recorded with an instrument called an infrared spectrophotometer (Figure 7.10). Infrared spectroscopy is a powerful analytical tool for identifying molecules and probing their structures. Consider the structures of carbon monoxide, CO, and formaldehyde, H_2CO. The Lewis structure of CO

$$:C\!\equiv\!O:$$

predicts the presence of a C≡O triple bond. The Lewis structure of formaldehyde, a carcinogenic compound found in cigarette smoke, has a C=O double bond:

✓ A compound's **infrared spectrum** can serve as a molecular fingerprint, enabling chemists to detect the compound's presence in a sample and to determine its molecular structure.

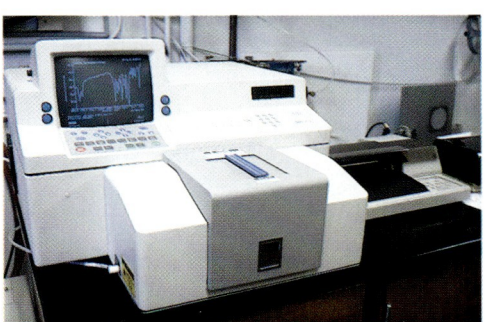

FIGURE 7.10 Infrared spectrophotometers are instruments that record the absorption of electromagnetic radiation at wavelengths between 2.5 and 20.0 μm by transparent solids, liquids, or gases. Special sample holders allow spectra to be recorded for opaque samples.

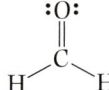

Electron-diffraction measurements of the C–O bond distances in these two molecules indicate that the carbon–oxygen bond distance in CO is shorter than that in H_2CO, indicating a higher bonding order in CO than in H_2CO. The IR spectra of these compounds reinforce this conclusion. The spectrum of CO has a prominent absorption band at 4.4 μm due to stretching of the C≡O triple bond. The same spectral feature in the spectrum of H_2C=O is at 6.2 μm (Figure 7.11). These wavelengths correspond to energies of 4.3×10^{-20} J and 3.2×10^{-20} J, respectively. It is reasonable that more energy is required to stretch a C≡O bond than a C=O bond.

Infrared spectra can also be used to distinguish between the resonance forms of molecules. Consider the spectrum of N_2O in Figure 7.12. Two strong absorp-

CONNECTION: Electron diffraction is used to determine bond distances in gas-phase molecules (Section 6.8).

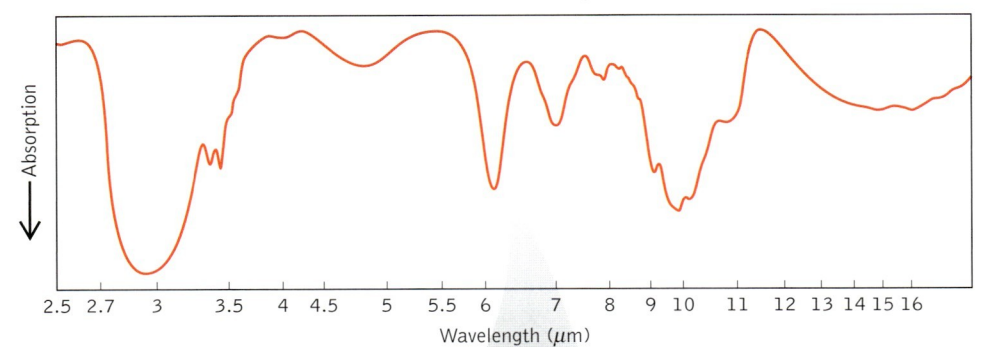

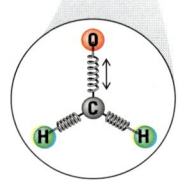

FIGURE 7.11 The IR spectrum of formaldehyde (H_2CO) has a greater number of absorption bands than does the IR specta of carbon dioxide (CO_2) or H_2O (see Figures 7.3 and 7.4). There are more types of bonds and more modes of bond vibration in H_2CO. The C=O stretch at 6.2 μm is characteristic of C=O double bonds in many molecules.

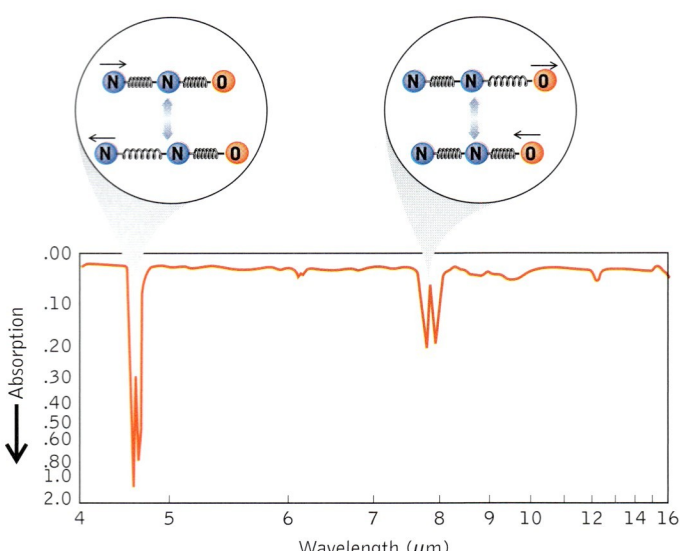

FIGURE 7.12 The IR spectrum of nitrous oxide (N_2O) shows strong absorption bands at 4.5 and 7.8 μm from N–N triple-bond and N–O single-bond stretching vibrations, respectively.

tion bands are observed at 4.5 and 7.8 μm. The band at 7.8 μm is, according to a reference table of IR spectral features, indicative of an N—O single bond. The band at 4.5 μm is closer to the wavelength expected for an N≡N triple bond than it is for an N=N double bond. These results support a Lewis structure for N_2O that has an N≡N triple bond

$$:N≡N—\ddot{\underset{..}{O}}:$$

rather than the resonance form that has an N=N double bond:

$$\ddot{\underset{..}{N}}=N=\ddot{\underset{..}{O}}$$

Thus, IR spectroscopy provides experimental evidence that supports our prediction in Section 6.3 for the preferred structure of N_2O based on formal charges.

SAMPLE EXERCISE 7.1: Describe how the IR spectrum of KSCN might be used to determine which of the resonance forms of SCN^- is the most important.

SOLUTION: Using the $S = N - H$ rule described in Section 6.7, we can determine that there are four bonds in this ion. There are three ways to arrange the four bonds between the three atoms, which leads to three resonance forms (with formal charges assigned):

$$\left[:S≡C—\ddot{\underset{..}{N}}:\right]^{+1\ 0\ -2} \longleftrightarrow \left[\ddot{\underset{..}{S}}=C=\ddot{\underset{..}{N}}\right]^{0\ 0\ -1} \longleftrightarrow \left[:\ddot{\underset{..}{S}}—C≡N:\right]^{-1\ 0\ 0}$$

The resonance form in the middle might be preferred because two atoms have zero formal charges and the negative charge is on the most electronegative element (N). The resonance form on the right with a C–N triple bond and a C–S single bond also might contribute to the true structure of the molecule because the difference in electronegativity between S and N is not very large.

The infrared spectrum of KSCN should contain spectral bands that match up with the wavelengths expected for either S=C and C=N, if the middle form is the dominant form, or S—C and C≡N, if the form on the right is the dominant one. If the S–C and C–N bands in the spectrum of KSCN do not match either of these combinations, it may mean that no one resonance form dominates and the actual structure of SCN^- is an average of the middle and right-hand resonance forms.

PRACTICE EXERCISE: Write Lewis structures for the resonance forms of the cyanate anion (OCN^-). Describe how the IR spectrum of potassium cyanate (KOCN) might be used to determine which of the resonance forms is the most dominant. (See Concept Review Questions 5 and 6.)

7.3 EXCEPTIONS TO THE OCTET RULE

Let's examine the structures and properties of some of the constituents of the atmosphere closest to Earth's surface, the troposphere. We have a head start on this process in that we considered the bonding patterns and molecular orbitals of N_2 and O_2 in Chapter 6. Now we will turn our attention to several gases present in small but variable concentrations in the air that we breathe (Table 7.1). As we shall see, the magnitude and significance of these changes in concentration are currently the subjects of considerable research and a great deal of controversy.

Three of the gases in Table 7.1—O_3, SO_2, and NO_2—are found in a range of concentrations that vary from small to too-small-to-measure. Ozone is present in the lower atmosphere because of natural processes such as lightning or as a result of human activity. Both SO_2 and NO_2 are produced through natural processes, such as volcanic emissions, but they are of concern to environmental scientists because of emissions from fossil-fuel combustion and other human activity.

The combustion of sulfur-containing fuels, particularly coal, releases SO_2 to the atmosphere. Further reaction of sulfur dioxide with oxygen in the atmosphere

TABLE 7.1 Typical Concentrations of Gases in Dry Air

Gas	Percent by Volume	Gas	Percent by Volume
Nitrogen (N_2)	78.084	Methane (CH_4)	0.0002
Oxygen (O_2)	20.948	Krypton (Kr)	0.0001
Argon (Ar)	0.934	Hydrogen (H_2)	0.00005
Carbon dioxide (CO_2)	0.036	Sulfur dioxide (SO_2)	< 0.0001
Neon (Ne)	0.0018	Nitrogen dioxide (NO_2)	< 0.000002
Helium (He)	0.0005	Ozone (O_3)	< 0.000001

CONNECTION: The chemistry of sulfur and nitrogen oxides was introduced in Section 4.3 and will be explored further in Section 16.2.

yields sulfur *trioxide*, SO$_3$. The balanced chemical equation representing this reaction is

$$2\ SO_2(g) + O_2(g) \longrightarrow 2\ SO_3(g)$$

Sulfur trioxide reacts with water vapor, forming sulfuric acid, H$_2$SO$_4$:

$$SO_3(g) + H_2O(g) \longrightarrow H_2SO_4(l)$$

which is a principal component of acidic precipitation in the eastern United States and Europe.

The molecular structures of sulfur dioxide and sulfur trioxide were the subjects of exercises in Chapter 6; so let's start with an examination of the molecular structure of sulfuric acid. Sulfuric acid is a more complicated molecule for us to draw because it has three kinds of atoms. It is an oxoacid—a group that includes nitric acid (HNO$_3$), phosphoric acid (H$_3$PO$_4$), and carbonic acid (H$_2$CO$_3$). To draw the skeletal structure of H$_2$SO$_4$, we start with S at the center because it is less electronegative than oxygen (a convention described in Section 6.6). Hydrogen is even less electronegative than sulfur, but a hydrogen atom cannot be the central atom, because it forms only one bond. The four O atoms are bonded to the central S atom, and each of the two H atoms is bonded to one of the O atoms. This arrangement gives us the following skeletal structure:

$$\begin{array}{c} H \\ \diagup \\ O \\ | \\ O-S-O \\ | \\ O \\ \diagdown \\ H \end{array}$$

Applying the $S = N - H$ rule, we have
- N (valence electrons needed): (5 atoms × 8 electrons each) + (2 hydrogen atoms × 2 electrons each) = 44
- H (number of valence electrons that we have): (5 Group 6A atoms × 6 electrons each) + (2 hydrogen atoms × 1 electron each) = 32
- $S = N - H = 44 - 32 = 12$ electrons, or 6 covalent bonds

The skeletal structure already has six bonds; so all that remains is to complete the octets of the oxygen atoms with lone pairs:

$$\begin{array}{c} \phantom{:\ddot{O}-S-\ddot{O}:}H \\ \phantom{:\ddot{O}-S-\ddot{O}:}\diagup \\ \phantom{:\ddot{O}-S-}:\ddot{O}: \\ \phantom{:\ddot{O}-S-}| \\ :\ddot{O}-S-\ddot{O}: \\ \phantom{:\ddot{O}-S-}| \\ \phantom{:\ddot{O}-S-}:\ddot{O}: \\ \phantom{:\ddot{O}-S-\ddot{O}:}\diagdown \\ \phantom{:\ddot{O}-S-\ddot{O}:}H \end{array}$$

All seems to be in order with this structure. There is considerable symmetry in the molecule and all the S–O bonds are essentially equivalent. However, if we calculate the formal charges on the sulfur and oxygen atoms, we find that sulfur is +2 and the left- and right-side oxygen atoms are −1. In addition, electron-diffraction data reveal that the left- and right-side S–O bonds have lengths (143 pm) that are more consistent with S=O double bonds than S—O single bonds.

These observations support an alternative Lewis structure. What if we created two double bonds to the left- and right-side oxygen atoms by using a lone pair from each atom, as shown below? Then, all formal charges would be zero, and the bond order would agree with the bond-length data.

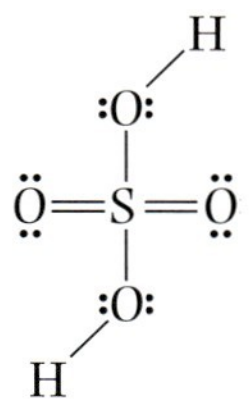

$$\ddot{\text{O}}=\text{S}=\ddot{\text{O}}$$
with :O:—H above and :O:—H below the S

The trouble with this proposal is that the sulfur atom has too many (12) valence electrons! Can a sulfur atom in a molecule ever have more than 8 valence electrons? The answer is yes. Atoms of the elements in the third row and below ($Z > 12$) in the periodic table have the ability to expand their valence shells by using their empty d orbitals. Having this ability does *not* mean that these atoms *always* expand their octets. They tend to do so when

> Can a sulfur atom in a molecule ever have more than 8 valence electrons?

> ✓ Elements with $Z > 12$ expand their octets by using empty d orbitals to form covalent bonds with strongly electronegative elements and when formal charges closer to zero result.

1. they form compounds with strongly electronegative elements, particularly F, O, and Cl, or
2. smaller formal charges are the result.

SAMPLE EXERCISE 7.2: A space probe to Venus detected the compound sulfur hexafluoride, SF_6, in the Venusian atmosphere. This compound is also present in Earth's atmosphere. It is made by reacting sulfur with fluorine and is used as an insulator in electrical transformers. Does sulfur have an expanded octet in SF_6?

SOLUTION: To determine the bonding pattern in SF_6, we first draw a skeletal structure for the molecule:

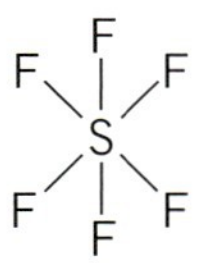

$$SF_6 \text{ skeletal structure with six F atoms around central S}$$

We immediately have a clear indication of an expanded octet on the sulfur atom: there are six bonds emanating from it, which means 12 valence electrons—4 more than could fit into an octet. Electron-diffraction analysis of SF_6 confirms that our skeletal structure is the actual molecular structure. These analyses reveal that SF_6 has six S–F bonds of equal length, 158 pm.

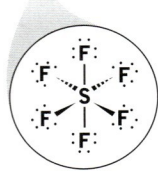

What a Chemist Sees. Sulfur hexafluoride (SF_6) is used in electrical transformers because it is an excellent insulator and chemically unreactive. The sulfur atom in a molecule of SF_6 must have an expanded octet because there are six equivalent S–F bonds. Sulfur hexafluoride has an octahedral molecular geometry (see page 359).

PRACTICE EXERCISE: Reaction of PCl_3 with Cl_2 gives PCl_5, an important industrial chemical. In the gas phase, it exists as PCl_5 molecules. As a solid, it exists as an ionic compound, PCl_4PCl_6. Write Lewis structures for PCl_4^+, PCl_5, and PCl_6^-. In which of the two ions does the phosphorus atom expand its octet? (See Problems 13–22.)

SAMPLE EXERCISE 7.3: Draw a Lewis structure for the phosphate ion (PO_4^{3-}) that minimizes the formal charges on its atoms.

SOLUTION: Let's first draw the skeleton of PO_4^{3-} with the minimum number of bonds connecting all the atoms in it:

$$\begin{array}{c} O \\ | \\ O-P-O \\ | \\ O \end{array}$$

Next, we calculate the theoretical number of bonds in PO_4^{3-} by using our $S = N - H$ rule. We find that $H = 5 + (4 \times 6) + 3 = 32$ and that $N = 5 \times 8 = 40$; so $S = 40 - 32 = 8$ and $S/2 = 4$. Four bonds are the minimum needed to draw the skeleton of a phosphate ion. Adding the remaining valence electrons to complete the octets on all the atoms, we have a complete Lewis structure:

$$\left[\begin{array}{c} :\ddot{O}: \\ | \\ :\ddot{O}-P-\ddot{O}: \\ | \\ :\ddot{O}: \end{array} \right]^{3-}$$

Calculating formal charges, we find that each single-bonded oxygen has a formal charge of -1 and the formal charge on the phosphorus atom is $+1$. Is this the best we can do? Actually, we can reduce the formal charge on phosphorus by one and increase it on an oxygen atom by one if we use one of the lone pairs on that oxygen atom, say the bottom one, to make another PO bond:

$$\left[\begin{array}{c} :\ddot{O}: \\ | \\ :\ddot{O}-P-\ddot{O}: \\ \| \\ :\ddot{O}: \end{array} \right]^{3-}$$

Now there are three single-bond oxygen atoms each with a formal charge of -1, which corresponds to an overall charge on the ion of -3. The final structure puts 10 valence electrons around the phosphorus atom. This suspension of the octet rule is allowed because phosphorus is in the third period and has d orbitals available with which to expand its octet.

PRACTICE EXERCISE: Draw a Lewis structure for the phosphite ion (PO_3^{3-}) that minimizes the formal charges on its atoms. (See Problems 23–30.)

Sulfuric acid is not the only compound with a molecular structure that appears to disobey the octet rule. Consider nitrogen dioxide, NO_2, the compound that gives "brown LA haze" its distinctive color. The formation of NO_2 in urban areas begins inside automobile engines, where high temperatures promote the formation of nitrogen monoxide, NO, from N_2 and O_2:

$$N_2(g) + O_2(g) \longrightarrow 2\ NO(g)$$

NO combines with more oxygen, forming nitrogen dioxide:

$$2\ NO(g) + O_2(g) \longrightarrow 2\ NO_2(g)$$

Unlike most other atmospheric gases, NO and NO_2 are odd-electron molecules. They have 11 and 17 valence electrons, respectively. An odd number of valence electrons, **H,** guarantees an odd number for the calculated value **S.** Therefore the total number of bonds that we calculate by using $S = N - H$ will not be a whole number. For example, in NO, $H = 11$, $N = 16$, and $S/2 = 2.5$. This result suggests that the bond order in NO is 2.5. How can we draw one or more Lewis structures to represent it? A triple bond between nitrogen and oxygen allows us to complete the octets of both atoms but leaves one electron left over. If we draw a double bond between nitrogen and oxygen, there are not enough valence electrons to satisfy the octet rule for each atom. We are left with the realization that, although odd-electron molecules, called *radicals*, exist in the atmosphere (and elsewhere), the atoms in these molecules must violate the octet rule by having fewer than 8 valence electrons.

If we are to draw a Lewis structure for an odd-electron molecule, we need to decide which atom gets the odd number—that is, 7 valence electrons instead of 8. The approach that makes the most sense is to put an octet around the more-electronegative element and to shortchange the less-electronegative element. Therefore, the nitrogen atom in the Lewis structure of NO gets only 7 electrons.

CONNECTION: The reactions that lead to the formation of NO_2 and photochemical smog are discussed in Chapter 14.

What a Chemist Sees. The high temperatures inside vehicle engines promote the reaction between N_2 and O_2, forming an odd-electron compound: nitric oxide (NO). Further reactions between NO and O_2 produce another odd-electron compound, nitrogen dioxide (NO_2).

THE CHEMISTRY OF OXYGEN AND THE GROUP 6A ELEMENTS

Oxygen and the other members of Group 6A, known as chalcogens, have valence-shell electron configurations ns^2np^4. Therefore, the chemical properties of these elements are based on the formation of compounds in which the atoms of Group 6A fill their valence octets by forming structures in which there are two bonds, or a -2 charge, on each atom. Oxygen is the most abundant of the chalcogens and the fourth most abundant element in the universe, behind hydrogen, helium, and neon. On Earth, oxygen makes up 21% of the atmosphere, 46% of the crust, and 89% of the water by mass.

Most of the oxygen in the atmosphere is diatomic O_2 gas, though oxygen is also present, particularly in the stratosphere, as highly reactive triatomic ozone (O_3). Oxygen gas would probably not be in our atmosphere were it not for the evolution of green plants and photosynthesis:

$$6\ CO_2(g) + 6\ H_2O(l) \longrightarrow C_6H_{12}O_6(aq) + 6\ O_2(g)$$

Much of the oxygen in Earth's crust is chemically bonded to silicon (Si) and aluminum (Al) in silicates and aluminosilicates, the structures of which are described in Chapter 10. Eighteenth-century French chemist Antoine Lavoisier was the first scientist to understand that oxygen is an element, and he gave it its name. He was also the first to recognize that oxygen combines with other elements and compounds in combustion reactions.

Sulfur and the other Group 6A elements are solids at room temperature. Deposits of elemental sulfur exist in regions that were once sites of volcanic activity, including Poland, East Texas, Louisiana, and under the Gulf of Mexico. Sulfur deposits form when salt domes in underlying rock are capped by layers of gypsum ($CaSO_4 \cdot 2\ H_2O$). Bacterial reduction of SO_4^{2-} to S^0 produces layers of pure sulfur as thick as 30 m. Bacterial action can also produce odorous and toxic hydrogen sulfide gas (H_2S), the sulfur analog of water. Sulfur is the 15th most abundant element in Earth's crust, making it more abundant than chlorine. At least 14 allotropes of sulfur have been identified; most of them have cyclic structures, as described in Chapter 10.

Iron pyrite, FeS_2, or fool's gold, is a relatively common iron-containing mineral. Its structure consists of Fe^{2+}

Sulfur deposit.

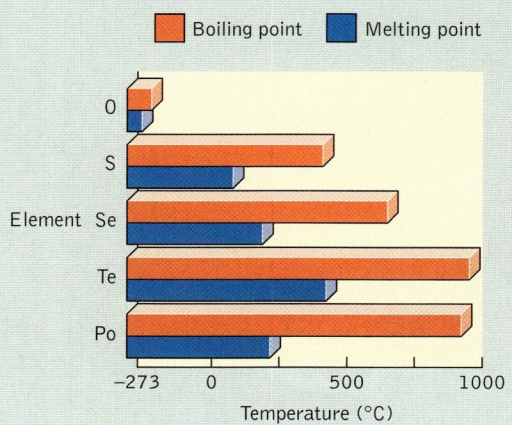

With the exception of oxygen, the Group 6A elements are solids at room temperature. Their melting and boiling points follow a smooth trend toward higher temperatures down the group. Radioactive polonium is the exception.

cations and disulfide (S_2^{2-}) anions. Disulfide ions have the Lewis structure

$$[:\ddot{S}-\ddot{S}:]^{2-}$$

The smelters built to extract metals from sulfide ores, such as iron pyrite, are a source of atmospheric sulfur dioxide, as described in the following chemical equation:

$$4\ FeS_2(s) + 11\ O_2(g) \longrightarrow 2\ Fe_2O_3(s) + 8\ SO_2(g)$$

Sulfur dioxide released into the atmosphere may react with oxygen, forming sulfur trioxide, which in turn combines with water vapor, forming sulfuric acid, as described in Section 7.3 and elsewhere. In many regions of North America and Europe, even more sulfuric acid forms in the atmosphere as a result of the combustion of sulfur-containing fossil fuels. To reduce this environmental problem, scrubbers have been installed in the smokestacks of smelters and power plants to trap SO_2. Many use calcium oxide (lime), which reacts with SO_2 and traps it as solid calcium sulfite:

$$CaO(s) + SO_2(g) \longrightarrow CaSO_3(s)$$

The oxidation of calcium sufite yields calcium sulfate:

$$2\ CaSO_3(s) + O_2(g) \longrightarrow 2\ CaSO_4(s)$$

which is the principal ingredient of gypsum, a material widely used in the building industry.

More sulfuric acid is produced each year for use in industry than any other chemical in the world. In the United States alone, 4.0×10^{10} kg of H_2SO_4 was produced in 2000. It is used in the manufacture of fertiliz-

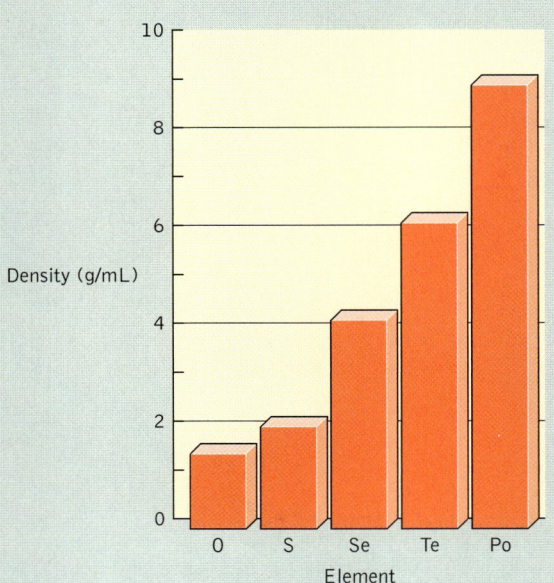

The densities of the solid Group 6A elements increase with increasing atomic number.

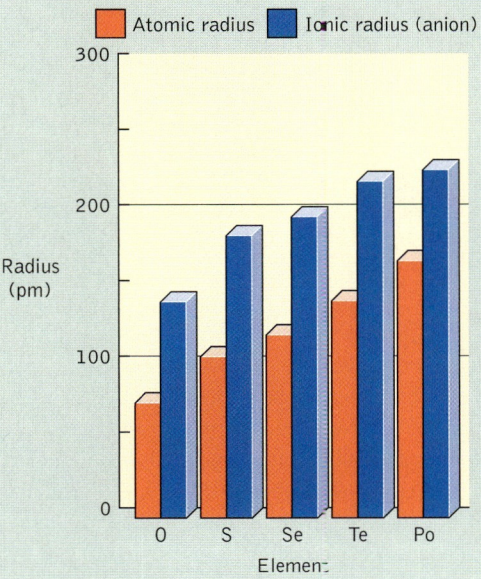

The atomic and ionic (charge = −2) radii of the Group 6A elements increase down the group due to increasing numbers of shells of electrons.

ers, explosives, detergents, paint pigments, dyes, and drugs; in the synthesis of other industrial chemicals; in metallurgy; and to refine petroleum. It is also used as the electrolyte in the lead-acid batteries used to start most vehicles, as will be discussed in Chapter 17.

Selenium (Se) and tellurium (Te) are present as metal selenides and tellurides in sulfur-containing minerals, such as iron(II) sulfide (FeS), lead sulfide (PbS), and copper(II) sulfide (CuS). Selenium and tellurium can be recovered from these minerals by heating them and selectively volatilizing Se and Te or by electrolysis during which Se^{2-} and Te^{2-} are oxidized to the free elements (Se^0 and Te^0). Selenium exists in several allotropes, the most stable of which has a gray metallic luster. Like most metalloids, selenium has the physical properties of metals but the chemical properties of nonmetals. For example, solutions of selenium oxides are acidic and selenic acid (H_2SeO_4) is as strong an acid as sulfuric acid (H_2SO_4). Metal selenides can convert light into electricity, and so they are used in photocells and in the light meters used in photography. Selenium is widely used in xerography. Both selenium and tellurium are important elements in the microelectronics industry.

Polonium (Po), an extremely rare element, is radioactive and is produced by the decay of uranium isotopes. It was first isolated from pitchblende by the Curies, but it takes 25 tons of the ore to harvest just a milligram of Po. Even though Po is a Group 6A element, it has the chemical properties of a metal.

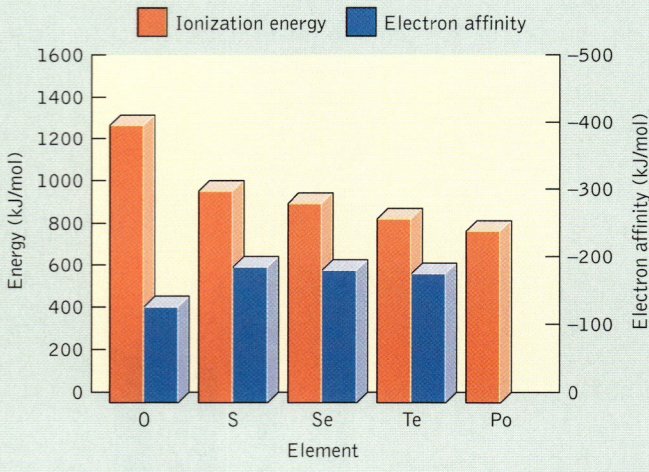

The ionization energies of the Group 6A elements decrease with increasing atomic number, corresponding to greater shielding of the outer electrons from nuclear charge by more layers of inner shell electrons. Compounds containing cations of the heavier Group 6A elements (Se, Te, Po) are known. The addition of an electron to atoms of the Group 6A elements leads to a release of energy and electron affinities less than zero. The addition of a second electron, however, requires energy. The periodic trend toward lower electron affinities down a group holds for the 6A elements other than oxygen, which has the lowest electron affinity of the group.

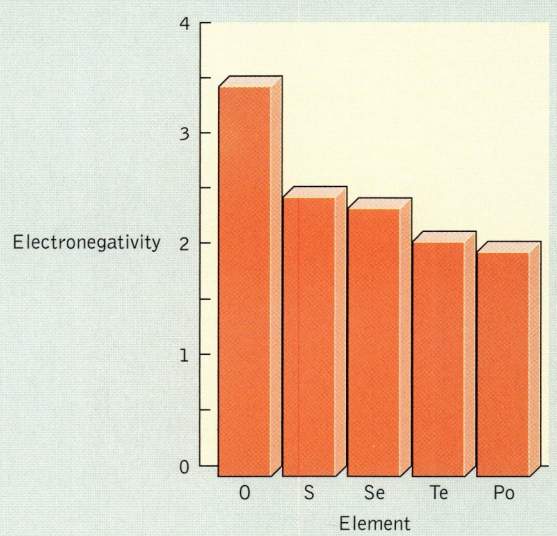

Increased shielding of the outer electrons in the heavier elements leads to the observed trend toward lower electronegativities down the group. The Group 6A elements are more electronegative than elements to their left in the periodic table but less electronegative than the Group 7A elements.

7.4 THE MOLECULAR-ORBITAL DIAGRAM OF NITRIC OXIDE

There is no way to draw a Lewis structure for an odd-electron molecule by following the rules discussed in Chapter 6. Clearly, another model for describing the bonding in NO and other odd-electron compounds would be useful. Such a model was described in Chapter 6; it is molecular-orbital theory. In fact, we developed an MO diagram for one odd-electron species, N_2^+, in the discussion of the electron transitions that create the colors of the aurora.

The MO diagram of nitric oxide is different from those of homonuclear diatomic gases described in Chapter 6. The 2s and 2p orbitals on nitrogen have higher energies than do the same orbitals on oxygen (Figure 7.13). This difference gives the NO molecular-orbital diagram a distorted appearance compared with the MO diagrams of N_2 and O_2 (Section 6.9). Eleven valence-shell electrons lead to a valence electron configuration, σ_{2s}^2, $\sigma^*_{2s}^2$, π_{2p}^4, σ_{2p}^2, and $\pi^*_{2p}^1$. The unpaired electron is in a π^*_{2p} antibonding orbital. If we take half the difference in the number of valence-shell electrons in bonding and antibonding MOs, we calculate a bond order of $\frac{1}{2}(8 - 3) = 2.5$. This value is the same as that obtained in our $S = N - H$ calculation.

Note that the π_{2p} bonding MOs in NO are closer in energy to the 2p orbitals of oxygen and that the antibonding π^*_{2p} MOs are closer in energy to the 2p orbitals of nitrogen. This trend is observed in other heteronuclear diatomic molecules:

CONNECTION: The molecular orbitals of homonuclear diatomic molecules are discussed in Section 6.9.

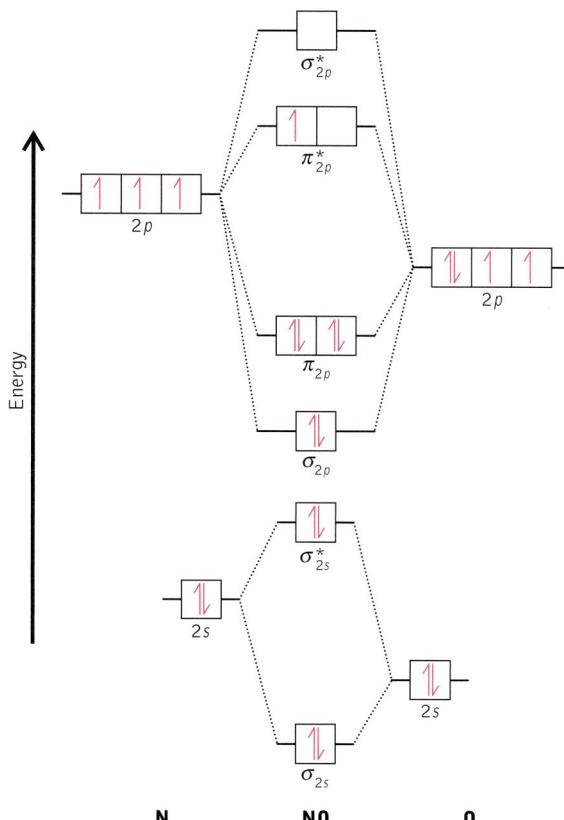

FIGURE 7.13 The molecular-orbital diagram of the odd-electron compound nitric oxide (NO) shows that the odd electron occupies a π^* antibonding orbital that is closer in energy to the atomic orbitals of nitrogen and so has more nitrogen character.

bonding MOs tend to be closer in energy to the atomic orbitals of the more-electronegative element from which they are derived, and the antibonding MOs are closer in energy to the atomic orbitals of the less-electronegative element. The proximity of the nitrogen atomic orbitals to the π^*_{2p} MOs gives these MOs more nitrogen *character*. Greater nitrogen character suggests that the single electron in the π^*_{2p} MOs is more likely to be on nitrogen than on oxygen. This observation is consistent with our earlier prediction that the odd electron in NO is on the nitrogen atom.

Odd-electron molecules tend to be chemically reactive. We have noted the rapid reaction between NO and atmospheric oxygen. Nitric oxide also reacts readily with N_2^+ or O_2^+ in the thermosphere. In these reactions, the NO molecule loses an electron, forming NO^+:

$$N_2^+ + NO \longrightarrow N_2 + NO^+$$

$$O_2^+ + NO \longrightarrow O_2 + NO^+$$

CONNECTION: Other reactions of N_2^+ and O_2^+ are described in Section 6.3.

7.5 ELECTRON-SPIN RESONANCE: LOCATING UNPAIRED ELECTRONS

Among the reactions of NO_2 is one in which NO_2 reacts with itself. Two NO_2 molecules combine by sharing their unpaired electrons and forming a *dimer* of NO_2 that has the formula N_2O_4. This reaction might be predicted from the Lewis structure of NO_2 that shows an unpaired electron on the nitrogen atom. Dimerization leads to pairing of the two unpaired electrons:

$$:\ddot{O}-\dot{N}=\ddot{O} + :\ddot{O}-\dot{N}=\ddot{O} \longrightarrow \text{(structure of } N_2O_4\text{)}$$

✓ Molecules and ions with an odd number of electrons exhibit **electron-spin resonance** (ESR). Electron-spin resonance spectra enable scientists to predict where unpaired electrons are likely to be in a molecule.

Is there experimental evidence that the nitrogen atom in NO_2 has seven valence electrons? The location of an unpaired electron in NO_2 can be determined by using an analytical technique called **electron-spin resonance** (ESR) spectroscopy. In Chapter 3, we noted that the electrons on atoms spin and that electron spin creates small magnetic fields. When two electrons occupy the same orbital, they must spin in opposite directions so that their magnetic fields are opposed to each other and cancel each other. Electron spins in odd-electron molecules, however, cannot cancel each other, and so the molecule generates a small magnetic field. This magnetic field will interact with an applied, external magnetic field (with a strength designated by the letter *H*). In the process, the spins of unpaired electrons in these molecules may either line up in the direction of the applied field or line up opposed to it. These two alignments produce two energy states, as shown in Figure 7.14.

CONNECTION: Section 3.9 describes the experiments that confirmed that electrons in atoms have spin.

The state in which electron spin is aligned with the external magnetic field is lower in energy than the state in which the spin is opposed to the field. Thus most molecules line up so that their spins are aligned with the field, as shown in Figure 7.14B. However, absorption of just the right frequency of electromagnetic

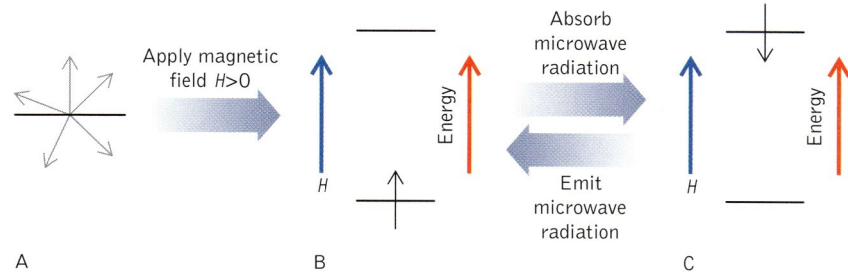

FIGURE 7.14 A. A spinning unpaired electron in a molecule generates a small magnetic field with no preferred orientation. B. When a magnetic field (H) is applied, most of the tiny magnetic fields produced by the spinning unpaired electrons align with the magnetic field. C. Molecules absorb a characteristic frequency of microwave radiation, and the spins of the unpaired electrons flip so that they are opposed to the external field. Molecules emit microwave radiation as their unpaired electrons flip again and they return to the lower energy state.

radiation, with an energy that exactly matches the energy difference between the two states, can "flip" electron spins, flipping the direction of the molecules' magnetic fields so that they are opposed to the external field. Molecules may then emit these quanta of energy and return to the aligned energy level. This flip-flop of electron spins is called electron-spin resonance. Instruments called electron-spin resonance spectrometers (Figure 7.15) detect the wavelengths of electromagnetic energy required for resonance to occur. Because the energy differences are very small, the radiation absorbed is in the long-wavelength microwave region ($\lambda \sim 3$ cm). Unpaired electrons on different atoms produce different ESR spectral patterns. Thus, ESR enables us to determine which atom or atoms in a molecule have unpaired electrons.

CONNECTION: The energy needed to flip the spin on an electron is quantized (Section 3.4).

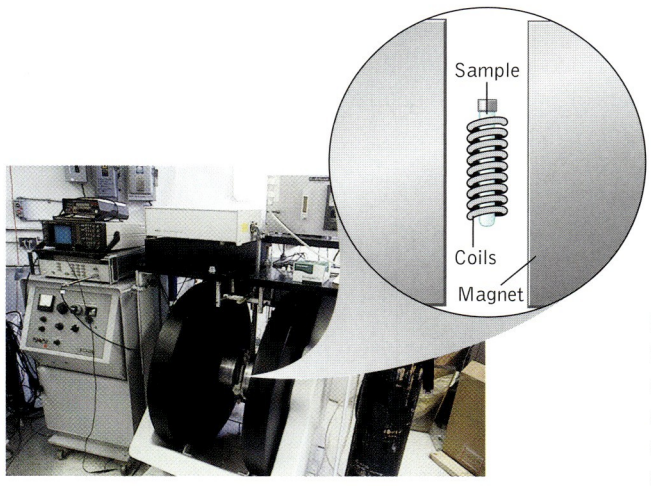

FIGURE 7.15 A sample is placed in the field of the powerful magnet of an ESR spectrometer. Microwave frequencies are applied through the coils surrounding the sample, and those frequencies that produce electron-spin resonance are detected.

FIGURE 7.16 The ESR spectrum of nitrogen dioxide (NO_2) shows three lines of equal intensity. Interaction between the unpaired electron and the odd number of protons in the nitrogen nucleus accounts for the triplet pattern. If the unpaired electron were located on an oxygen atom, a single line would be expected because O has even numbers of protons and neutrons in its nuclei.

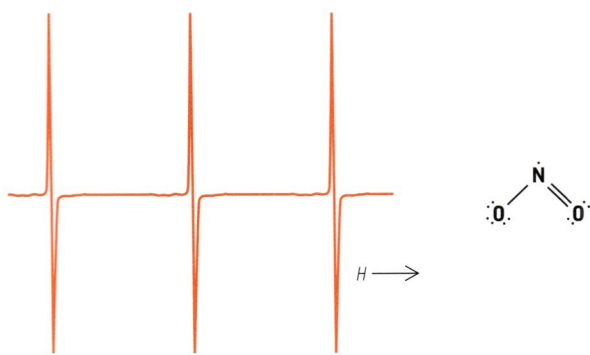

The fact that NO_2 has an ESR spectrum (Figure 7.16) confirms the presence of an unpaired electron somewhere in the molecule. The spectrum is a *triplet* of three closely spaced lines with equal intensity. The presence of these three lines, rather than a single line, tells an experienced ESR spectroscopist that the unpaired electron in the molecule is on nitrogen and not on one of the oxygen atoms.

How does the ESR spectroscopist arrive at this conclusion? The triplet is the result of interactions between the magnetic field of the spinning unpaired electron in NO_2 and the nucleus of the atom on which it resides. This interaction exists because the nucleons in the nuclei of atoms also spin. In an external magnetic field, these spins also align in one of two directions: either with the field or opposed to it. In a nucleus with even numbers of protons and neutrons, these spins may all be paired. Oxygen atoms have such nuclei. This pairing leaves no net spin and no magnetic field to interact with a spinning unpaired electron around the nucleus. Such an electron would produce a simple, one-line ESR spectrum. However, in other nuclei interactions between nuclear magnetic fields and an unpaired electron result in a triplet in the ESR spectrum. Which atom in NO_2 has an odd number of protons in its nucleus? A check of the periodic table tells us that it is nitrogen ($Z = 7$). The ESR spectrum of NO_2 contains such a triplet, which means that the unpaired electron must be on an atom with odd numbers of protons and neutrons, e.g., an atom of nitrogen.

7.6 MOLECULAR SHAPE: THE VSEPR MODEL

We have written the Lewis structures of a number of atmospheric gases, but we have yet to explore how the bonding and nonbonding pairs of electrons are oriented within their molecules. For a given molecule, this orientation helps define the shape of the molecule, which is also called its *molecular geometry*. For example, in Chapter 6, we drew the Lewis structures of O_3 and H_2O with O–O–O and

H–O–H bond angles of 116.5 and 104.5 degrees, respectively, in accord with their true molecular geometry. Why are molecules of ozone and water bent? This question is worth answering because a molecule's bond angles and its molecular geometry, together with unequal sharing of bonding electrons, contribute to the physical and chemical properties of molecular compounds. These properties include a compound's physical state at room temperature, whether the compound dissolves in other molecules or reacts with them, its biological activity, and how it is distributed in the environment.

> Why are molecules of ozone and water bent?

Tetrahedra of electrons

One explanation of why each molecule of water or ozone has an angular (bent) geometry is based on the electrostatic repulsion between electrons—electron pairs in valence-shell orbitals repel other pairs. To minimize these repulsions and produce lower energy orientations of the orbitals, electron pairs arrange themselves around atoms in ways that minimize interactions with each other. One model of chemical bonding is based on achieving minimal interaction by allowing orbitals to be as far away from each other as possible. This model is called the valence shell electron pair repulsion or VSEPR model.

How do you arrange an octet (four pairs) of electrons around a central point—for example, the oxygen atom in H_2O—so that all four pairs are as far apart as possible? The geometric shape that meets this criterion is that of a tetrahedron, as shown at the top of Figure 7.17. If we place the oxygen atom at the center, then its four pairs of valence electrons are directed toward the four corners of the tetrahedron.

In a perfect tetrahedron, we would predict a bond angle between O–H bonds of 109.5 degrees. The bond angle in water is a little less (104.5 degrees), because the two nonbonding, or *lone,* pairs of electrons are not shared with other atoms and so are drawn closer to the nucleus of the oxygen atom. The proximity of these lone pairs to the two bonding pairs means that they exert stronger repulsive forces on the bonding pairs, squeezing them closer together and reducing the bond angle between them slightly.

To find a molecule with a perfect **tetrahedral geometry,** we need look no further than Table 7.1 and the greenhouse gas methane, CH_4. A methane molecule has a carbon atom at the center with single bonds to four hydrogen atoms. Because all four pairs of valence electrons on the carbon atom are bonding pairs and because all the bonds go to the same kind of atom, we have a completely symmetric molecule and tetrahedral molecular geometry (see Figure 7.17).

SAMPLE EXERCISE 7.4: Explain why CH_4 (see Figure 7.17) has a tetrahedral structure rather than the square planar structure shown here:

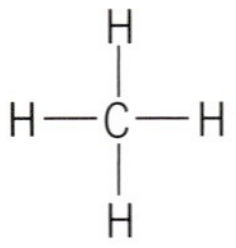

$$\begin{array}{c} H \\ | \\ H-C-H \\ | \\ H \end{array}$$

SOLUTION: The tetrahedral arrangement puts the electron pairs 109.5 degrees apart, whereas the square planar arrangement puts them only 90 degrees apart. Electron–electron repulsions favor the arrangement with the largest bond angles and the greatest distance between electron pairs.

PRACTICE EXERCISE: Two possible arrangements of three electron pairs around an atom are shown here. Which of them is preferred?

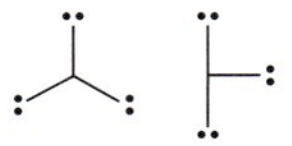

(See Concept Review Questions 57 and 58.)

FIGURE 7.17 The molecular geometries of methane (CH_4), ammonia (NH_3), and water (H_2O) can be rationalized on the basis of minimizing interactions between four pairs of valence electrons. The gray lobes in the upper diagram represent the four regions around the central atom in which these electrons are likely to be so that the repulsion forces between them are minimized. The tetrahedral orientation of these regions gives CH_4 a tetrahedral molecular geometry, NH_3 a trigonal pyramidal geometry, and H_2O an angular (bent) geometry.

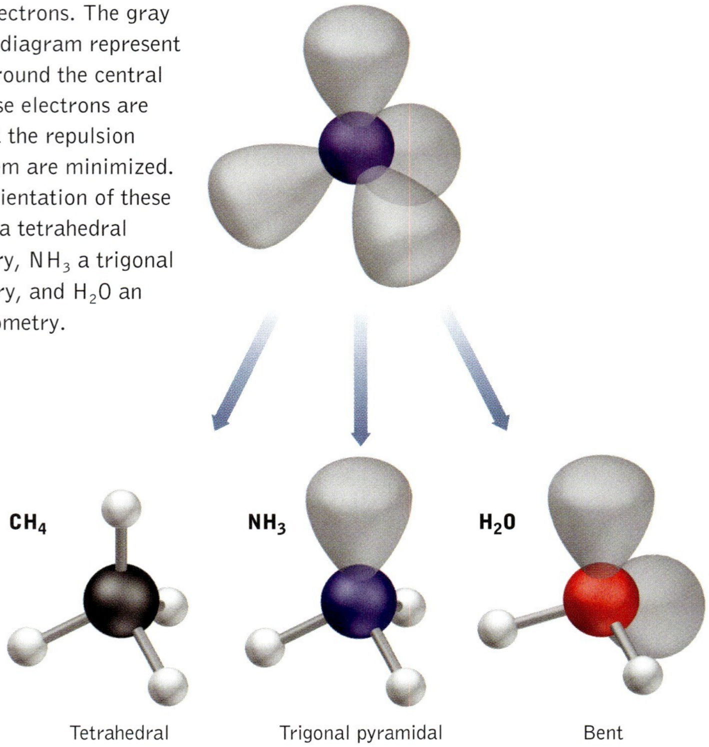

CH_4 — Tetrahedral
NH_3 — Trigonal pyramidal
H_2O — Bent

Ammonia, NH$_3$, may have been an important ingredient in Earth's primordial atmosphere, but it is present at only trace levels in the atmosphere today. Still, it is one of the most widely used chemicals in industry, agriculture, and the home (great for washing windows), and so its molecular structure is worth our attention. The Lewis structure features four independent groups of electrons—one lone pair and three bonding pairs—around nitrogen, as shown in Figure 7.17.

What is the molecular geometry of this structure? If we stay with the assumption that the four groups of electrons will arrange themselves so that they minimize interactions with each other, we again have a tetrahedral orientation of electron groups. Because one of the groups is a lone pair, ammonia does not have a true tetrahedral molecular geometry. Instead, it is a **trigonal pyramid**. As in water, the lone pair in ammonia is closer to the central (nitrogen) atom than are the bonding pairs. Therefore, the lone pair exerts stronger repulsive forces on the bonding pairs, reducing the N–H bond angles from the 109.5 degrees of a true tetrahedron to 107 degrees in its trigonal pyramidal geometry.

✓ A tetrahedral orientation of groups of valence electrons on the central atom of a molecule leads to **tetrahedral, trigonal pyramidal, or angular (bent) molecular geometries,** depending on the numbers of bonding and lone pairs of electrons on the central atom.

Triangles of electrons

The central atoms of many small molecules have four separate pairs of valence electrons, but two or three of these pairs may be grouped together in double or triple bonds. Let's revisit one such molecule: formaldehyde, CH$_2$O. How are the pairs of electrons arranged around the central carbon atom in CH$_2$O? The Lewis structure of formaldehyde

$$\begin{array}{c} \ddot{\text{O}}\text{:} \\ \| \\ \text{C} \\ \diagup \quad \diagdown \\ \text{H} \qquad \text{H} \end{array}$$

suggests that the *three* groups of *four* pairs of electrons might form a triangle having

1. two pairs in the C=O double bond toward the top,
2. one pair in the C—H single bond on the left, and
3. one pair in the C—H single bond on the right.

How can these three groups be arranged so that there is minimal interaction between them? Basic geometry tells us that the best way to achieve this goal is to direct them toward the three corners of a triangle, as we have done. This triangular shape produces bond angles of about 120 degrees and is called a **trigonal planar geometry** (Figure 7.18).

The Lewis structures of the resonance forms of ozone

$$\ddot{\text{O}}=\overset{..}{\text{O}}-\ddot{\text{O}}\text{:} \quad \longleftrightarrow \quad \text{:}\ddot{\text{O}}-\overset{..}{\text{O}}=\ddot{\text{O}}$$

✓ Three groups of electrons on the central atom of a molecule lead to a **trigonal planar** orientation of electron groups, to bond angles of about 120 degrees, and to **trigonal planar or angular geometries,** depending on the numbers of bonding and lone pairs of electrons on the central atom.

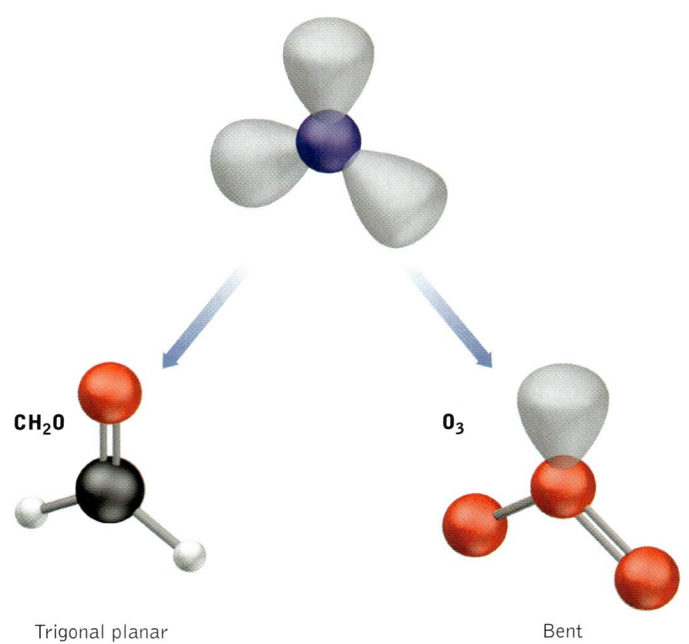

FIGURE 7.18 Repulsions between three groups of electrons around a central atom are minimized when the groups are in regions that have a trigonal planar orientation. This orientation gives formaldehyde (H_2CO) a trigonal planar molecular geometry and ozone (O_3) an angular (bent) geometry. Note that only one of the two resonance forms of O_3 is shown.

also have three groups of electrons around the central (oxygen) atom. As with formaldehyde, the three groups are oriented toward the three corners of a triangle (see Figure 7.18). However, the top corner is occupied by a lone pair. The pairs in the single bond and the double bond are directed toward the corners at the base of the triangle. This triangular arrangement of electrons produces an O–O–O bond angle of 116.5 degrees and is called an **angular geometry**. As with H_2O and NH_3, the lone pair of electrons influences the bond angle formed by the bonding pairs of electrons, but molecular geometry designations are based only on the arrangements of the atoms in the molecule, not on all the groups of electrons. Therefore, we say that ozone has a *trigonal planar* orientation of valence-electron pairs but an *angular* molecular geometry.

CONCEPT TEST: Why is the bond angle in SO_2 slightly less than the 120 degrees predicted for a trigonal planar orientation of valence electrons?

Linear molecular geometry

Let's revisit the molecular structure of carbon dioxide. In Chapter 6, the concept of formal charges was used to establish that CO_2 has the following Lewis structure:

$$\ddot{\text{O}}=\text{C}=\ddot{\text{O}}$$

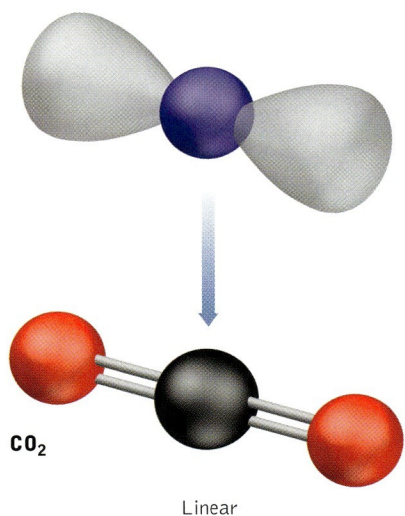

FIGURE 7.19 Repulsions between two groups of electrons around a central atom are minimized when the groups are in regions on opposite sides of the atom. This orientation gives carbon dioxide a linear molecular geometry and a bond angle of 180 degrees.

The molecule was drawn as if it were linear, with a C–O bond angle of 180 degrees. Is the bond angle in CO_2 really 180 degrees? To answer this question, we need to recognize that there are two independent groups of four electrons each around the central carbon atom. To minimize their interactions with each other, the two groups are on opposite sides of the carbon atom. Thus, the linear shape drawn for CO_2 (Figure 7.19) is, according to the VSEPR model, correct.

> Is the bond angle in CO_2 really 180 degrees?

CONCEPT TEST: Why is CO_2 a linear molecule, but SO_2 an angular molecule?

Shapes of expanded-octet molecules

We have yet to examine the geometry of molecules with expanded octets and more than four independent groups of electrons around their central atoms. Consider, for example, sulfur hexafluoride, SF_6. The molecule has six pairs of bonding electrons around its central sulfur atom. The orientation for the six bonding pairs in these molecules that minimizes their interactions is one in which the six pairs are in lobes directed toward the vertices of an eight-sided double pyramid called an octahedron (Figure 7.20). This orientation gives SF_6 an **octahedral geometry** in which all the electron pairs are 90 degrees apart. Phosphorous pentafluoride, PF_5, has five pairs of bonding electrons. The five

> ✓ Expanded octets can lead to **trigonal bipyramidal** or **octahedral orientations** of five or six pairs of valence electrons, respectively, on a central atom in a molecule.

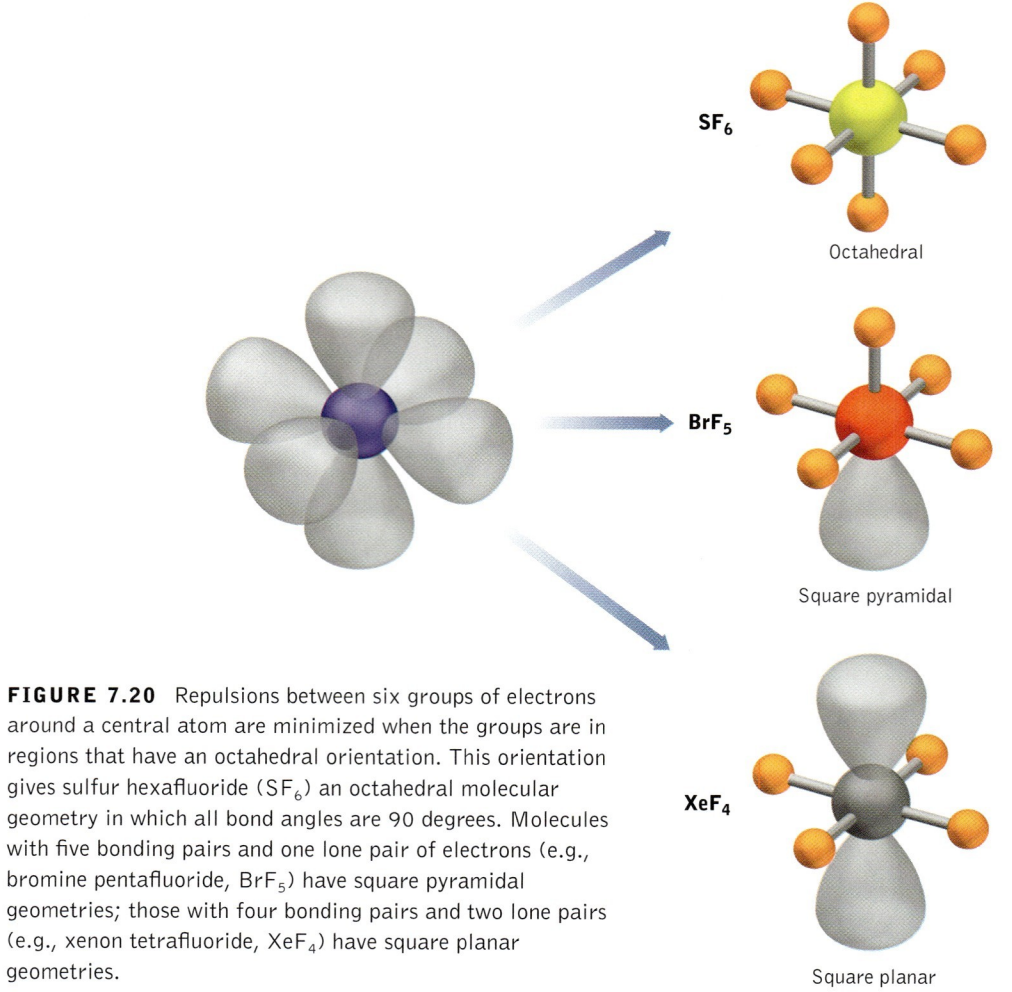

FIGURE 7.20 Repulsions between six groups of electrons around a central atom are minimized when the groups are in regions that have an octahedral orientation. This orientation gives sulfur hexafluoride (SF_6) an octahedral molecular geometry in which all bond angles are 90 degrees. Molecules with five bonding pairs and one lone pair of electrons (e.g., bromine pentafluoride, BrF_5) have square pyramidal geometries; those with four bonding pairs and two lone pairs (e.g., xenon tetrafluoride, XeF_4) have square planar geometries.

pairs can minimize their repulsions by adopting a **trigonal bipyramidal** orientation. Three of the five pairs are 120 degrees apart in the middle, or *equatorial*, plane, and the other two pairs are oriented 90 degrees, or *axial*, to the plane, as shown in Figure 7.21.

In a trigonal bipyramidal orientation of electron pairs, there are a total of three 120-degree interactions between electron pairs and six 90-degree interactions between the two pairs above and below the plane and the three pairs in the plane. These different angles are significant when we try to predict the molecular shapes of molecules in which one or more of the five independent groups of electrons are lone pairs. Consider sulfur tetrafluoride, SF_4. Each molecule has four S–F bonds and a lone pair on sulfur. If the lone pair were located above or below the trigonal plane, the molecular structure would differ from the molecular

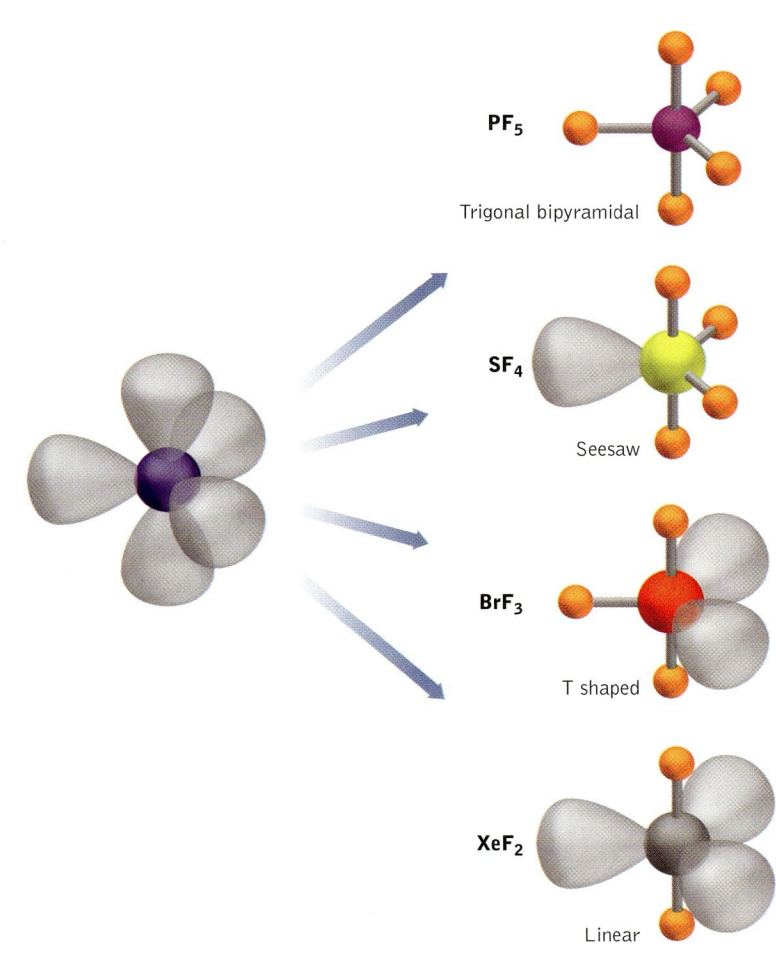

FIGURE 7.21 Repulsions between five groups of electrons around a central atom are minimized when the groups are in regions that have a trigonal bipyramidal orientation. This orientation gives phosphorus pentafluoride (PF_5) a trigonal bipyramidal molecular geometry in which three bonds are in an equatorial plane and the other two bonds are above and below the plane. Molecules with four bonding pairs and one lone pair of electrons (e.g., sulfur tetrafluoride, SF_4) have seesaw geometries; those with three bonding pairs and two lone pairs (e.g., bromine trifluoride, BrF_3) have T-shaped geometries; and those with two bonding pairs and three lone pairs (e.g., xenon difluoride, XeF_2) are linear molecules.

structure in which the lone pair is located in the equatorial plane. It turns out that the one lone pair is indeed located in the equatorial plane, and SF_4 has what is known as a seesaw structure (see Figure 7.21). We can rationalize this location with the argument that it provides less interaction between the lone pair and adjacent bonding pairs. The lone pair in the plane has the most interaction with the two bonding pairs above and below the plane, which are perpendicular (90-degree bond angle) to it. The other two bonding pairs are 120 degrees away and interact much less. However, a lone pair above or below the plane is 90 degrees from all three bonding pairs in the plane. This greater level of interaction results in a less-stable molecular geometry. Similar arguments can be made for all the other molecular geometries based on octahedral (Figure 7.20) and trigonal bipyramidal (Figure 7.21) orientations of valence-electron pairs.

Summary

We have seen how the presence of nonbonding electrons affects the molecular geometry of molecules. We can predict the molecular geometries of simple molecules in which the central atoms obey the octet rule from the number of bonding pairs of electrons and lone pairs of electrons. Preferred geometries are those that allow electrons to get as far away from one another as they can. The approach consists of the following steps:

1. Draw the Lewis structure of the molecule.
2. Count the number of *independent groups of electrons* around the central atom.
   a. Each bond (single, double, or triple) represents one independent group.
   b. Each lone pair is an independent group.
3. Orient the groups of electrons so that inter–group interactions are minimized. These orientations are listed in the second column of Table 7.2.
4. The molecule will have a particular molecular geometry listed for its electron-group orientation in the third column of Table 7.2, depending on the number of lone pairs and bonding pairs of electrons.

SAMPLE EXERCISE 7.5: What are the molecular geometries of CF_2Cl_2 and SeF_4?

SOLUTION: Let's begin by drawing Lewis structures because the geometries of small molecules depend on the number of bonding and nonbonding electron pairs around their central atoms. In a molecule of CF_2Cl_2 there are four single bonds around the central carbon atom, representing four independent groups of electrons. They will be directed to the four corners of a tetrahedron, giving CF_2Cl_2 a *tetrahedral molecular geometry*.

To draw the Lewis structure for SeF_4 we might begin by calculating the number of bonds, assuming that the octet rule is obeyed and that our $S = N - H$ rule applies. There are a total of 34 valence electrons in one atom of Se and four atoms of F. They need 40 valence electrons to satisfy their octets. These values lead to the prediction that there are six electrons to be shared

$$40 - 34 = 6$$

which means that only 3 bonds are formed. Clearly, this is not enough, since at least 4 bonds are needed between the Se atom and four F atoms. This result tells us that the octet rule does not work for this molecule, and that Se must have an expanded octet.

Let's draw the skeleton form of SeF_4 with four single bonds and then add lone pairs until there are a total of 34 valence electrons. These additions require that one atom has five pairs of electrons. It must be Se because Se can have an expanded octet, whereas F cannot. According to Table 7.2, five electron pairs around a central atom are oriented in a trigonal bipyramid arrangement. If one

TABLE 7.2 Electron-Group Orientations and Corresponding Molecular Geometries

Number of Independent Groups	Group Orientation	Number of Bonding Pairs	Corresponding Molecular Geometry	Bond Angle (degrees)	Example
2	Linear	4	Linear	180	CO_2
3	Trigonal planar	4	Trigonal planar	~120	H_2CO
		3	Angular (bent)		O_3
4	Tetrahedral	4	Tetrahedral	~109	CH_4
		3	Trigonal pyramidal		NH_3
		2	Angular (bent)		H_2O
5	Trigonal bipyramidal	5	Trigonal bipyramidal	90, 120	PCl_5
		4	Seesaw		SF_4
		3	T shaped		BrF_3
		2	Linear		XeF_2
6	Octahedral	6	Octahedral	90	SF_6
		5	Square pyramidal		BrF_5
		4	Square planar		XeF_4

of these pairs is a lone pair, the resulting molecular geometry is called a *seesaw geometry*.

PRACTICE EXERCISE: Determine the molecular geometries of NOF (which has a central N atom) and SO_2Cl_2. (See Problems 63–78.)

7.7 VALENCE-BOND THEORY

In Section 7.6, we predicted molecular shapes based on the VSEPR model. Unfortunately, there is an inconsistency between the shapes that we predicted and the shapes and orientations of atomic orbitals described in Chapter 3. Consider methane, CH_4. At its center is a carbon atom. A carbon atom has the electron configuration $[He]2s^22p^2$ and forms four covalent bonds to complete its octet. How can four bonds form from a filled 2s orbital and two partly filled 2p orbitals so that they are equivalent and have bond angles of 109.5 degrees? (Keep in mind that the 2s orbital is spherical and three 2p orbitals, p_x, p_y, and p_z, are 90 degrees from one another.) **Valence-bond theory** provides one way to account for the observed bonding orientation.

Valence-bond theory arose in the late 1920s, largely through the genius and efforts of Linus Pauling. After earning his Ph.D. at the California Institute of Technology, Pauling traveled to Europe in 1926 to study quantum mechanics with Erwin Schrödinger and Niels Bohr. In the years that followed, Pauling developed a theory of molecular bonding based on quantum mechanics. This valence-bond theory assumes that chemical bonding results from the overlap of

> ✓ According to **valence-bond theory,** atomic orbitals can be mixed together to create **hybrid atomic orbitals** that yield observed molecular geometries.

atomic orbitals on different atoms. The theory also assumes that the greater the overlap, the stronger and more stable is the bond that is formed. Pauling believed that orbital overlap would occur as long as the shared electrons had opposite spins, just as two electrons in the same atomic orbital must have opposite spins. Shared electrons in a chemical bond would be attracted to the nuclei of two atoms, not just one atom. This greater attraction would lead to lower energy and greater stability.

In the 1930s, Pauling published a series of seven papers on "The nature of the chemical bond." His book of the same title in 1939 dominated chemists' views of molecule bonding for more than 20 years. In his book, Pauling foresaw the application of valence-bond theory to biomolecules, anticipating his own deciphering of the coiled structures of proteins. In recognition of his many contributions to chemistry and biochemistry, Linus Pauling was awarded the Nobel Prize in chemistry in 1954.

Hybrid orbitals

In valence-bond theory, atomic orbitals of different shapes and energies are mixed together to form equivalent **hybrid atomic orbitals.** Hybrid atomic orbitals differ from molecular orbitals in that molecular orbitals belong to the molecule as a whole. According to molecular-orbital theory, bonding takes place when more electrons occupy bonding MOs than antibonding MOs. In valence-bond theory, hybrid orbitals are associated only with a particular atom in the molecule. The end-on overlap of a hybrid orbital with a hybrid orbital or an unhybridized atomic orbital on a neighboring atom results in the formation of a σ bond. The sideways overlap of unhybridized atomic orbitals produces the π bonds of double and triple bonds.

In methane, for example, the 2s orbital and three 2p orbitals in the valence shell of carbon are, according to valence-bond theory, *hybridized*, forming four equivalent hybrid orbitals, each of which contains one electron (Figure 7.22). Because the ingredients in this set of hybrid orbitals were a single s orbital and three p orbitals, their mixing creates four equivalent **sp^3 hybrid** orbitals. The lobes of these orbitals are directed toward the four corners of a tetrahedron and so are 109.5 degrees apart, as shown in Figure 7.22. Note that the number of hybrid orbitals (four) equals the number of atomic orbitals (four) from which they were created.

CONNECTION: The shapes and orientations of atomic orbitals are discussed in Section 3.8.

✓ The tetrahedral orientation of valence electrons is achieved by the sp^3 **hybridization** of atomic orbitals, which can result in the formation of as many as four σ bonds.

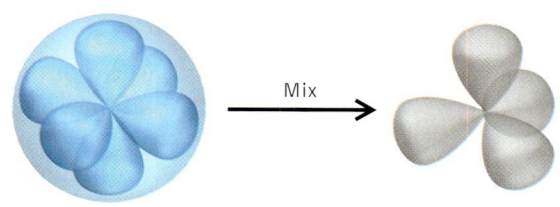

FIGURE 7.22 Four equivalent sp^3 hybrid orbitals are formed by mixing an s orbital and three p orbitals. In the hybridization process for carbon, the four valence electrons are distributed among the 2s and 2p orbitals and then all four are mixed, forming a set of four equivalent sp^3 hybrid orbitals with lobes pointed toward the corners of a tetrahedron.

The end-on overlap of the four hybrid orbitals on carbon with four 1s orbitals from four hydrogen atoms results in the sharing of the four pairs of electrons in these eight atomic orbitals and the formation of four C–H σ bonds in methane (Figure 7.23A). Any molecule with a tetrahedral orientation of its valence electrons has, according to valence-bond theory, a set of four equivalent sp^3 hybrid orbitals. These molecules include those in which one or more hybrid orbitals are filled before any bonding takes place. For example, in ammonia sp^3 hybridization of the valence electrons on nitrogen produces three orbitals that are half filled and one orbital that is completely filled, as shown in Figure 7.23B. The three sp^3 orbitals with one electron each form the three N–H σ bonds in ammonia. The orbital that was already filled contains the lone pair on nitrogen. Similarly, the oxygen atoms in water have, according to valence-bond theory, sp^3 hybrid orbitals in their valence shells, two of which are filled before any bonding takes place and two of which are half filled and available for bond formation. The latter two are the ones that overlap with 1s orbitals from hydrogen atoms and form the two O–H σ bonds in H_2O (Figure 7.23C).

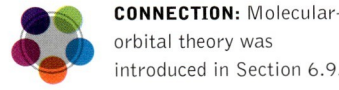

CONNECTION: Molecular-orbital theory was introduced in Section 6.9.

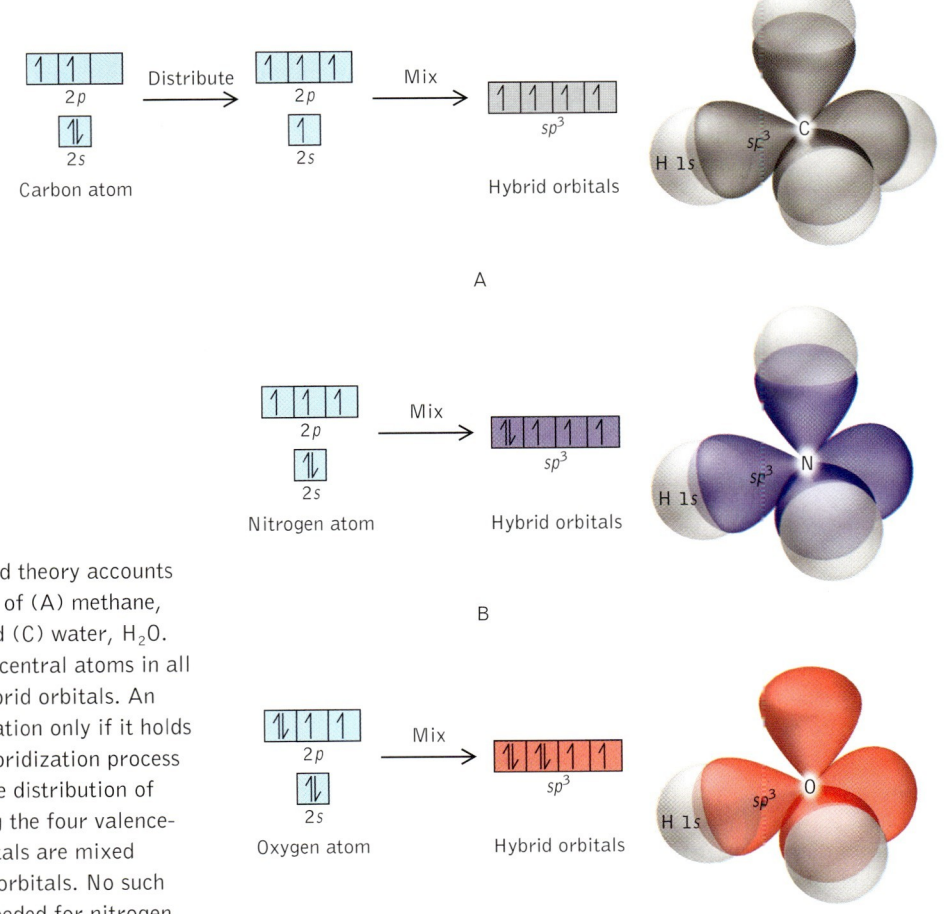

FIGURE 7.23 Valence-bond theory accounts for the molecular geometries of (A) methane, CH_4; (B) ammonia, NH_3, and (C) water, H_2O. The valence electrons on the central atoms in all three molecules are in sp^3 hybrid orbitals. An orbital can undergo hybridization only if it holds at least one electron. The hybridization process for carbon atoms requires the distribution of four valence electrons among the four valence-shell orbitals before the orbitals are mixed together, forming sp^3 hybrid orbitals. No such distribution of electrons is needed for nitrogen and oxygen atoms.

FIGURE 7.24 Three equivalent sp^2 hybrid orbitals are formed by mixing an s orbital and two p orbitals, leaving one p orbital unhybridized. This combination of three hybridized orbitals and one unhybridized p orbital can account for the formation of as many as three σ bonds and one π bond and a trigonal planar orientation of the lobes of these orbitals.

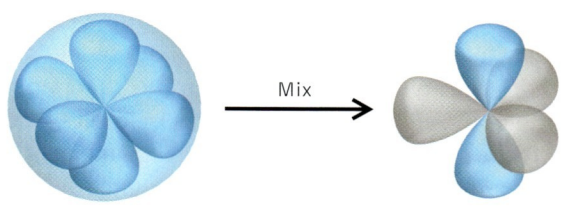

> The trigonal planar orientation of valence electrons is achieved with **sp^2 hybridization,** which can result in the formation of as many as three σ bonds and one π bond.

We need a different kind of mixing, or *hydridization,* of atomic orbitals to account for the trigonal planar geometries of such molecules as H_2CO. Only three σ bonds are needed to connect the atoms in this molecule, and so we need a hybridization scheme that uses only three of the atomic orbitals in the valence shell of carbon. Such a hybridization scheme includes the $2s$ orbital and two of the carbon atom's $2p$ orbitals, leaving the third $2p$ orbital unhybridized. This mixing results in **sp^2 hybridization** (Figure 7.24). The end-on overlap of the three sp^2 hybrid orbitals with $1s$ orbitals on two hydrogen atoms and with a half-filled p orbital on an oxygen atom results in the formation of three σ bonds. The fourth bond is the π bond in the C=O double bond. Sideways overlap of the unhybridized p orbital on carbon with the other half-filled p orbital on oxygen results in the formation of this π bond (Figure 7.25). The σ and π bonds between carbon and oxygen together make a C=O double bond. The trigonal planar molecular geometry of H_2CO is consistent with the geometry of sp^2 hybrid orbitals on the central carbon atom.

CONCEPT TEST: In which of the following molecules does the central atom have sp^3 hybrid orbitals? CCl_4; HCN; SO_2; NH_3.

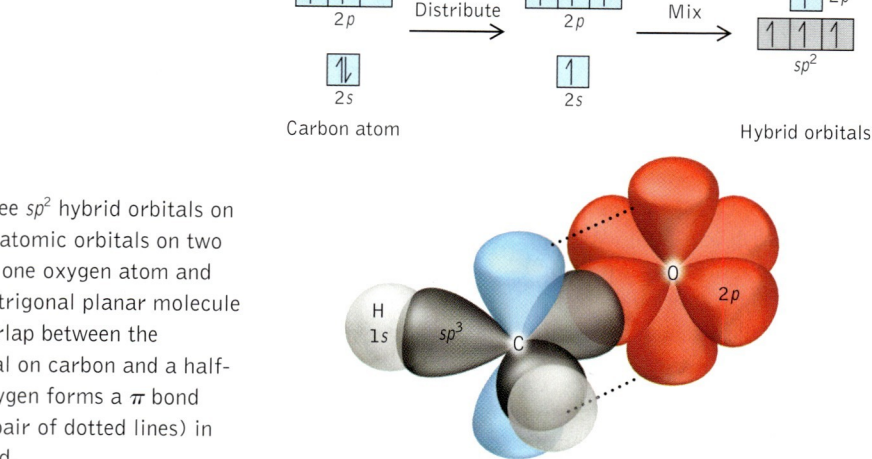

FIGURE 7.25 Three sp^2 hybrid orbitals on carbon overlap with atomic orbitals on two hydrogen atoms and one oxygen atom and form σ bonds in the trigonal planar molecule CH_2O. Sideways overlap between the unhybridized p orbital on carbon and a half-filled p orbital on oxygen forms a π bond (represented by the pair of dotted lines) in the C=O double bond.

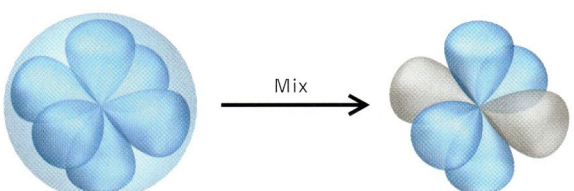

FIGURE 7.26 Linear molecular geometries and bond angles of 180 degrees can be accounted for by *sp* hybridization on central atoms. Two equivalent *sp* hybrid orbitals are formed by mixing one *s* orbital and one *p* orbital, leaving two *p* orbitals unhybridized. This hybridization accounts for the formation of two σ bonds by the end-on overlap of the *sp* orbitals and the formation of as many as two π bonds from the sideways overlap of unhybridized *p* orbitals with *p* orbitals on other atoms.

Let's see how valence-bond theory accounts for the linear shape of CO_2. The Lewis structure of CO_2 tells us that we have two C=O bonds in each molecule. The VSEPR model predicts that the two double bonds will be on opposite sides of the carbon atom so that they are as far away from each other as possible, resulting in a bond angle of 180 degrees. To produce a bond angle of 180 degrees, we need two equivalent hybrid orbitals that have lobes in opposite directions. One way to get two equivalent hybrid orbitals is to blend the 2s orbital with one of the 2p orbitals of a carbon atom, creating two equivalent **sp hybrid** orbitals (Figure 7.26). End-on overlap of these two *sp* hybrid orbitals with *p* orbitals on two oxygen atoms results in two σ bonds with the desired bond angle of 180 degrees. This arrangement leaves the other two 2p orbitals on carbon unhybridized. These 2p orbitals form two π bonds by sideways overlap with a *p* orbital on each of the two oxygen atoms. As a result, a σ and a π bond form between each oxygen atom and the central carbon atom, resulting in the two C=O bonds shown in Figure 7.27.

> ✓ A linear orientation of valence electrons and the formation of as many as two σ bonds and two π bonds are achieved by *sp* **hybridization.**

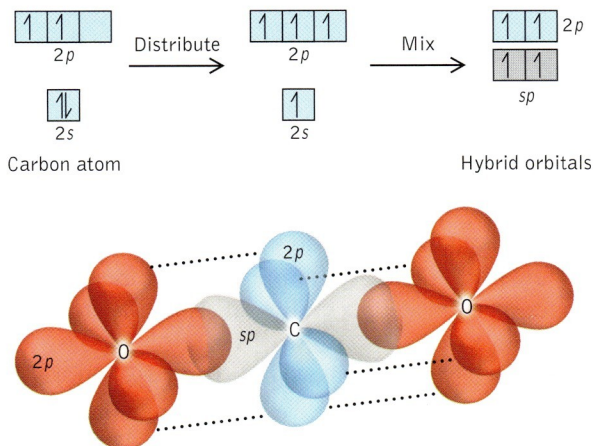

FIGURE 7.27 Two *sp* hybrid orbitals on carbon overlap end-on with *p* orbitals on two oxygen atoms, forming two σ bonds in a molecule of carbon dioxide. Sideways overlap between the two unhybridized *p* orbitals on carbon and half-filled *p* orbitals on the two oxygen atoms creates two π bonds (the two pairs of dotted lines) and two C=O double bonds.

CONCEPT TEST: Resonance forms of CO_2 can be drawn in which the atoms are linked by a C—O single bond and a C≡O triple bond. Could these bonds be formed with *sp* hybrid orbitals on the carbon atom?

Hybrid orbitals for beryllium and boron

Beryllium compounds can have considerable ionic character, but some of them, such as beryllium chloride, $BeCl_2$, have considerable covalent character. How are Be and Cl bonded together in $BeCl_2$ gas? What is the shape of the $BeCl_2$ molecule? Beryllium has only two valence electrons, and chlorine atoms each form only one bond with less-electronegative elements. Therefore, the maximum number of valence electrons that the beryllium atom in $BeCl_2$ can have is four: two of its own and one each from the chlorine atoms with which it forms Be–Cl bonds. Apparently, the small size of the beryllium atom allows beryllium to achieve chemical stability with less than a complete octet of electrons in its valence shell. The best way to orient the two bonding pairs of valence electrons on beryllium so that their interactions with each other are minimized is to place them on opposite sides of the beryllium atom. This geometry can be achieved with *sp* hybridization, as illustrated in Figure 7.26. Thus, valence-bond theory predicts that the bonding in $BeCl_2$ (and the other beryllium halides) is based on *sp* hybrid orbitals on Be.

CONNECTION: The rule of thumb for predicting ionic versus covalent bonding between two elements is described in Section 6.5.

Boron is another element that forms compounds in which its atoms have less than complete octets. Consider boron trifluoride, BF_3. Although the difference in electronegativities of fluorine and boron (4.0 − 2.0 = 2.0) meets our guideline for designating BF_3 an ionic compound, it is not ionic. In this case, the 2.0 guideline does not hold. Actually, boron rarely exists as a B^{3+} ion, and BF_3 is a molecular compound. According to valence-bond theory, three equivalent B–F bonds form as a boron 2*s* orbital blends with two of its 2*p* orbitals, creating three sp^2 orbitals. The three valence-shell electrons on the boron atom half fill each of the hybrid orbitals. The lobes of these three equivalent orbitals lie in a plane and are separated by an angle of 120 degrees. This geometry is consistent with sp^2 hybridization (see Figure 7.24). End-on overlap between the three half-filled boron sp^2 orbitals and the half-filled *p* orbital on each of the three fluorine atoms yields three B–F σ bonds and a molecule of BF_3 with a trigonal planar shape.

Hybridization can also be used to describe the shapes of compounds in which the central atom has more than eight valence electrons. For example, we noted in Section 7.6 that SF_6 has an octahedral geometry. Which six atomic orbitals on sulfur could be hybridized to produce six equivalent orbitals that point to the six corners of an octahedron? Mixing sulfur's 3*s* orbital, three 3*p* orbitals, and two of its 3*d* orbitals gives six equivalent d^2sp^3 hybrid orbitals that have the appropriate orientation (Figure 7.28).

Other molecules, such as phosphorus pentachloride (PCl_5), have a trigonal bipyramidal geometry, and the central phosphorus atom has 10 valence electrons.

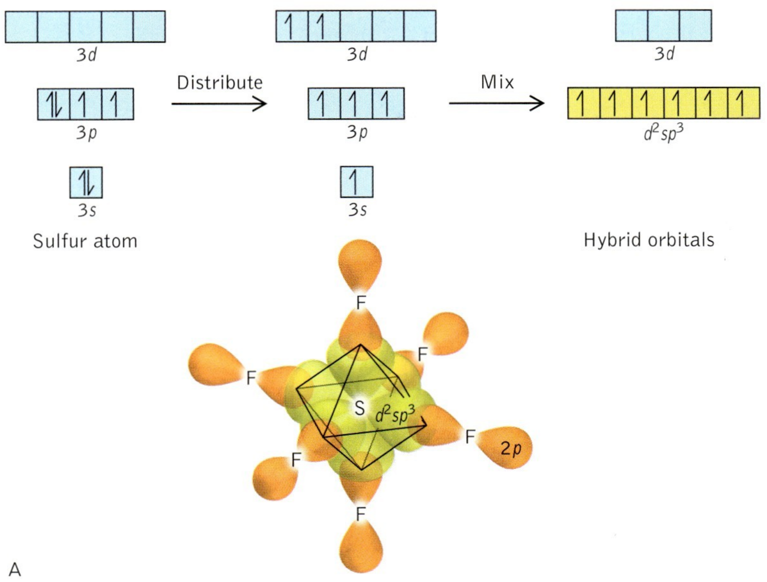

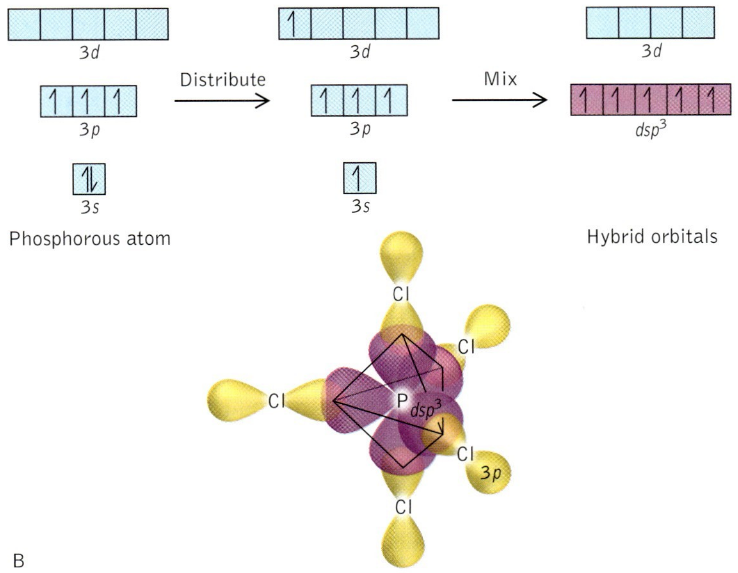

FIGURE 7.28 Molecules with expanded octets require hybridization schemes that include s, p, and d orbitals. A. The geometry of molecules with six pairs of valence electrons (e.g., sulfur hexafluoride, SF_6) can be accounted for by the hybridization of six atomic orbitals on the central atom: an s, three p, and two d orbitals—that is, d^2sp^3 hybridization. B. The geometry of molecules with five pairs of valence electrons (e.g., phosphorus pentachloride, PCl_5) can be accounted for by the hybridization of five atomic orbitals on the central atom: one s, three p, and one d, forming five dsp^3 orbitals.

Hybridization	Orientation of hybrid orbitals	Number of σ bonds	Molecular geometrics
sp		2	Linear
sp^2	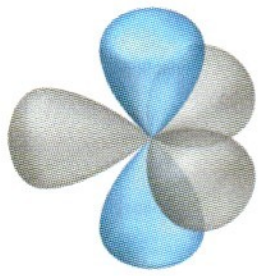	3 2	Trigonal planar Angular
sp^3		4 3 2	Tetrahedral Trigonal pyramidal Bent
dsp^3		5 4 3 2	Trigonal bipyramidal Seesaw T shape Linear
d^2sp^3		6 5 4	Octahedral Square pyramidal Square planar

FIGURE 7.29 The hybridization schemes and the orientations of the orbitals derived from them account for the bonding in small molecules and their geometries.

Which five atomic orbitals could be hybridized to produce five equivalent orbitals that point to the five corners of a trigonal bipyramid? Mixing a 3s orbital, three 3p orbitals, and one 3d orbital yields five equivalent dsp^3 hybrid orbitals with lobes that point toward the corners of a trigonal bipyramid (Figure 7.28). Remember that only elements with Z > 12 expand their octets by using valence-shell d orbitals in bond formation. A summary of the orientation of different hybrid orbitals appears in Figure 7.29.

SAMPLE EXERCISE 7.6: Below are Lewis structures for thiocyanate ion and formaldehyde. What is the hybridization of the carbon atom in the two structures?

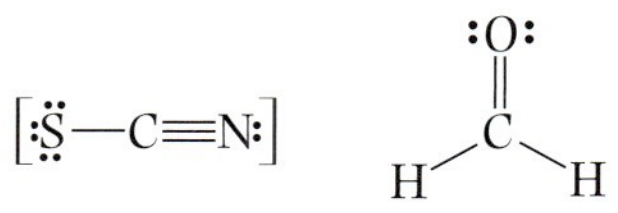

SOLUTION: The carbon atom in the thiocyanate structure forms a single (σ) bond to sulfur and a triple bond (one σ and two π bonds) to nitrogen. This bonding pattern around carbon can be explained by the formation of two sp hybrid orbitals that form the two σ bonds, leaving two unhybridized p orbitals on carbon that form π bonds through sideways overlap with p orbitals on nitrogen.

In formaldehyde there are three σ bonds connecting the carbon atom to two hydrogen atoms and to oxygen, and one π bond that is part of the C=O double bond. This bonding pattern is consistent with a set of threee sp^2 hybrid orbitals on carbon and one unhyridized p orbital for forming the π bond to oxygen.

PRACTICE EXERCISE: Determine the hybridization of the carbon atoms in ethane, C_2H_6, and in carbon disulfide, CS_2. (See Problems 85–96.)

7.8 POLAR BONDS AND POLAR MOLECULES

Of all the minor gases in the atmosphere, H_2O is the only one that exists in nature as a liquid. Why is H_2O a liquid at room temperature, whereas other compounds of comparable molar mass, such as N_2, O_2, and CH_4, are gases? The answer is because water is a polar molecule, but the others are not. This difference in molecular polarities depends on the geometries of these molecules and the polarities of their bonds. For example, the covalent bonds in N_2 and O_2 are nonpolar, making N_2 and O_2 nonpolar molecules. The C=O double bonds in CO_2 are polar bonds because carbon and oxygen have different electronegativities. The carbon end of each bond has a partial positive charge and the oxygen end has a partial negative charge. When two bonded atoms have partial electrical charges of equal but opposite sign, a **bond dipole** is created. Bond dipoles are represented by arrows pointing in the direction of higher density of the bonding electrons. Thus, the heads of these arrows represent partial negative charges and the tails represent partial positive charges, as shown here for carbon dioxide.

$$\overset{\leftharpoonup}{\ddot{O}}=C=\overset{\rightharpoonup}{\ddot{O}}$$

The overall polarity of the molecule can be determined by summing individual bond dipoles. This summing process must take into consideration the geometry of the molecule. The linear geometry of CO_2 means that the bond dipole of one C=O bond is offset by the bond dipole of the other C=O bond so that, overall, CO_2 is nonpolar. Offsetting bond dipoles is the reason why CH_4 also is a nonpolar molecule, as shown in Figure 7.30.

On the other hand, the bent geometry of a water molecule means that the bond dipoles of the two O–H bonds do not offset each other. Instead, there is an overall dipole directed toward the oxygen side of the molecule, as shown in Figure 7.30. The presence of this overall dipole means that water is a polar molecule. This polarity leads to intense interactions between molecules based on

FIGURE 7.30 The overall polarity of a molecule is determined by the magnitude of the individual bond dipoles in its structure and the orientation of these dipoles. In some molecular structures, bond dipoles offset one another and the molecule has a zero dipole moment, which is the case for methane (CH₄), carbon dioxide (CO₂), and sulfur trioxide (SO₃). In other molecules, including water (H₂O), ammonia (NH₃), and hydrogen fluoride (HF), asymmetric orientation of the bond dipoles leads to a permanent dipole moment.

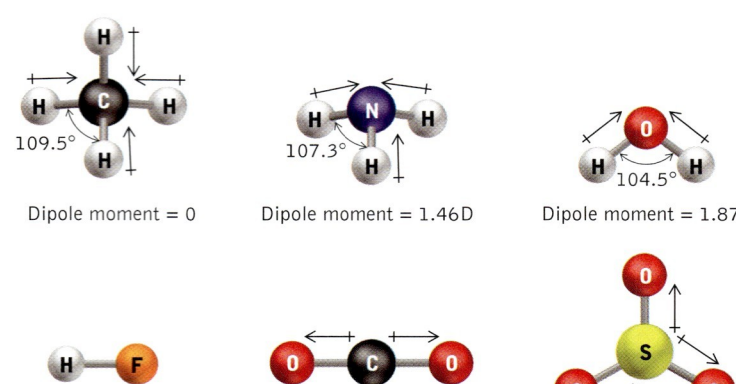

CONNECTION: Some of the remarkable physical properties of H₂O are described in Section 9.5.

attractions between the positive poles of some and the negative poles of others. We shall discuss these interactions in detail in Chapter 9. For now, keep in mind that the remarkable physical properties of water, such as its being a liquid at room temperature, its expanding when it freezes, and it's ability to dissolve ionic compounds, are linked to the polarity of its molecules.

SAMPLE EXERCISE 7.7: SO₃ has a trigonal planar structure with O–S–O bond angles of 120 degrees. Is SO₃ a polar molecule?

SOLUTION: Because there is no preferred orientation of the double bond, it is delocalized over all three bonds in three equivalent resonance structures. Thus, the molecule has three equivalent oxygen atoms distributed equally around the central sulfur atom. Any unequal sharing of bonding electrons in one direction is canceled out by offsetting shifts of bonding electrons in other directions. Therefore, the molecule is nonpolar.

PRACTICE EXERCISE: Is CF₄ a polar molecule? (See Problems 105–108.)

When water or any polar compound is placed in an electrical field created by two charged plates, as shown in Figure 7.31, its molecules align with the field so that their negative poles are oriented toward the positive plate and their positive poles are oriented toward the negative plate. The more polar the molcules are, the more strongly they align with the field. The magnitude of a molecule's polarity is expressed by the value of its **permanent dipole moment** (μ). The value of μ is equal to the magnitude of the partial electrical charges (Q_+ and Q_-) in a molecule times the distance (r) between them:

$$\mu = Qr \tag{7.1}$$

> A molecule has a **permanent dipole moment** when it has an asymmetric orientation of polar bonds.

7.8 Polar Bonds and Polar Molecules

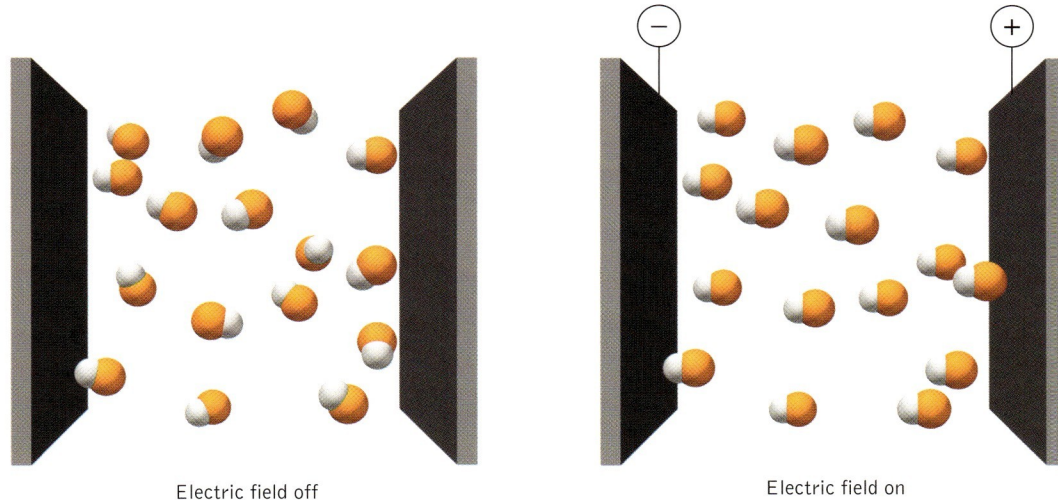

Electric field off   Electric field on

FIGURE 7.31 Polar molecules (such as hydrofluoric acid, HF) adopt a random arrangement of their dipoles in the absence of an electric field but align themselves when an electric field is applied to two metal plates. The negative end of each dipole (in this case, the fluorine atom) is directed toward the positively charged plate.

Dipole moments of molecules are usually expressed in *debyes* (D), where 1 D = 3.336×10^{-30} coulomb-meter (C-m). As noted in Chapter 1 (Table 1.4), a coulomb is the SI unit of electrical charge and the charge on an electron (e) is 1.602×10^{-19} C. Let's use this relation, the dipole moment of water (1.84 D), and the O–H bond distance (95.7 pm) to calculate the magnitude of the partial positive charge ($\delta+$) and the partial negative charge ($\delta-$) on H and O, respectively, in H_2O. To calculate the partial charges on H and O, we need to calculate Q by using Equation 7.1. We know the value for μ, but the O–H bond distance is not the same as r. Keep in mind that water has a bent structure with a H–O–H bond angle of 104.5 degrees. The partial negative charge on a molecule of water resides on the oxygen atom, but there are two sources of positive charge: the two hydrogen atoms. The net effect of their individual positive charges will be a center of positive charge halfway between them, shown by the tail of the arrow in the following diagram:

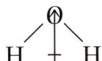

We can calculate the distance between the positive and negative charges (r) by using the O–H bond length (95.7 pm) and some trigonometry. Let's draw a triangle by joining the centers of the three atoms and then draw a perpendicular line from the oxygen to the center of positive charge. This line bisects the H–O–H

7.9 MOLECULAR VIBRATION AND INFRARED ABSORPTION

As noted at the beginning of this chapter, some of the vibration modes that molecular bonds undergo produce absorption bands in their infrared spectra, but other vibrations do not. Why? All of the vibration modes for water molecules shown in Figure 7.4 produce IR absorption bands. They do so because (1) water has a permanent dipole moment and (2) the vibrations slightly alter the bond dipoles and the dipole moment of the molecule.

In CO_2, however, only some of the vibration modes are *infrared active*. Why does the asymmetric stretch in CO_2 produce an IR band, whereas the symmetric stretch does not? The asymmetric stretching motion of the C=O double bonds means that one bond stretches while the other compresses, and so the carbon atom oscillates back and forth between them. This motion means that the value of r in Equation 7.1 increases for one C=O bond dipole as r decreases for the other. As a result, the two bond dipoles no longer offset each other, and the molecule has a *temporary* dipole that fluctuates with the motion of the atoms. This fluctuating dipole can absorb infrared radiation of a frequency that matches its motion.

> Why does the asymmetric stretch in CO_2 produce an IR band, whereas the symmetric stretch does not?

On the other hand, symmetric stretching of the C=O double bonds means that the values of r for the two C=O bonds are always the same, and so the bond dipoles always offset each other. Therefore, the motion does not produce a temporary dipole, and so it cannot absorb electromagnetic radiation.

CONCEPT TEST: CH_4 is a nonpolar molecule, yet it has a distinctive IR spectrum. Describe a vibration mode that could result in an IR absorption band.

CHAPTER REVIEW

Summary

SECTION 7.1

Greenhouse gases in Earth's atmosphere, which include CO_2, H_2O, O_3, and CH_4, transmit visible sunlight but absorb infrared radiation emitted by Earth's surface. The wavelengths of infrared radiation absorbed by these molecules are related to changes in the vibrational energies of their bonds.

SECTION 7.2

Infrared spectroscopy can be used to probe the chemical bonding in the molecules of a substance and to determine whether that substance is present in a sample.

SECTION 7.3

Sulfuric acid, H_2SO_4, contains two S—O—H units and two shorter S=O double bonds, as indicated by electron-diffraction data. Thus bonding pattern indicates that the central sulfur atom has 12 valence electrons. Elements such as sulfur, with $Z > 12$, can expand their octets by using empty valence-shell d orbitals. Other molecules, including NO and NO_2, have odd numbers of valence electrons and so contain atoms with incomplete octets. Atoms of the less-electronegative elements are more likely to have incomplete octets in odd-electron molecules.

SECTION 7.4

The molecular-orbital diagrams of heteronuclear diatomic molecules, such as NO, have bonding orbitals with energies closer to the energies of the atomic orbitals of the more-electronegative element and antibonding orbitals with energies closer to the energies of the atomic orbitals of the less-electronegative element. This pattern in NO gives the antibonding orbitals more N character and suggests that the odd electron in NO resides on the N atom.

SECTION 7.5

Odd-electron molecules exhibit electron-spin resonance, in which the magnetic field of the spinning odd electron lines up with an applied magnetic field. Microwave radiation is absorbed and released as the electron spin "flips" in directions opposed to the applied field or aligned with it, producing electron-spin resonance. Electron-spin resonance allows us to determine where an unpaired electron is in an odd-electron molecule.

SECTION 7.6

According to the VSEPR model, molecules and polyatomic ions adopt structures that keep the shared and lone-pair electrons as far away from one another as possible. Four pairs of electrons are farthest apart when pointed toward the corners of a tetrahedron. This orientation gives CH_4 a tetrahedral geometry, NH_3 a trigonal pyramidal geometry, and H_2O an angular, or bent, geometry. Three groups of electrons are farthest apart when they point toward the corners of a triangle. This orientation gives SO_3 and BF_3 trigonal planar geometries and SO_2 an angular, or bent, geometry. Two groups of electrons are farthest apart when they point in opposite directions. This orientation gives CO_2 a linear geometry. Molecules with expanded octets around the central atom include SF_6 (which is octahedral) and PF_5 (which is a trigonal bipyramid).

SECTION 7.7

Valence-bond theory uses the hybridization of atomic orbitals to explain observed molecular geometries. The hybridization of one s orbital and three p orbitals produces four equivalent sp^3 hybrid orbitals that point toward the four corners of a tetrahedron. Other hybridization schemes explain other molecular geometries: sp hybridization for linear molecules and sp^2 hybridization for those molecules with a trigonal planar orientation of electron groups. The geometries of molecules in which the central atom has an expanded octet are described by dsp^3 hybridization, which gives a trigonal bipyramidal orientation, and d^2sp^3 hybridization, which gives an octahedral orientation of electron groups.

SECTION 7.8

Molecules that contain polar bonds can be polar or nonpolar, depending on their molecular geometry. Polar molecules have a permanent dipole moment, $\mu = Qr$, expressed in debyes (D). Most heteronuclear diatomic molecules are polar. Carbon dioxide has a zero dipole moment because it is a linear molecule, and SO_3 is nonpolar because of its trigonal planar symmetry. Angular molecules, such as SO_2 and H_2O, have permanent dipole moments, as do molecules with different atoms around a central atom, such as $CHCl_3$ and CH_2O.

Key Terms

angular (bent) molecular geometry (p. 357)
bond dipole (p. 371)
electron-spin resonance (p. 352)
hybrid atomic orbital (p. 363)
infrared spectrum (p. 340)
octahedral molecular geometry (p. 359)

permanent dipole moment (p. 372)
sp hybridization (p. 367)
sp^2 hybridization (p. 366)
sp^3 hybridization (p. 364)
tetrahedral molecular geometry (p. 357)
trigonal bipyramidal orientation (p. 359)

trigonal planar molecular geometry (p. 357)
trigonal pyramidal molecular geometry (p. 357)
valence-bond theory (p. 363)

SECTION 7.1

Be able to explain the greenhouse effect in regard to the transmission of visible light and the absorption of infrared radiation by gases in the atmosphere.

Know some of the kinds of vibrations that molecules and polyatomic ions undergo.

Know that the frequencies of bond vibrations correspond to frequencies of infrared radiation.

SECTION 7.2

Understand how the infrared spectrum of a molecule relates to its structure.

SECTION 7.3

Be able to use formal charges to predict whether the atoms of elements with $Z > 12$ have expanded octets in molecular structures.

Be able to draw Lewis structures for odd-electron molecules.

SECTION 7.4

Recognize that NO and NO_2 are odd-electron molecules (radicals) and be able to sketch the molecular-orbital diagrams of NO and NO^+.

SECTION 7.5

Explain the principle of electron-spin resonance (ESR).

Know that ESR can be used to predict which atom in an odd-electron molecule has the odd electron.

SECTION 7.6

Be able to predict the geometry of a molecule or polyatomic ion from its Lewis structure with the VSEPR model.

Be able to sketch and label the major geometries of common small molecules, including those with expanded octets.

Be familiar with the importance of the presence of lone pairs on the central atom in determining a molecule's geometry.

SECTION 7.7

Describe the concept of the hybridization of atomic orbitals.

Understand the meaning of sp, sp^2 sp^3, dsp^3, and d^2sp^3 hybridization and be able to determine the hybridization of the atoms in a molecule with known geometry.

SECTION 7.8

Be able to recognize polar bonds in molecules and polyatomic ions.

Know which is the negative end of a polar bond.

Be able to explain why some molecules with polar bonds are polar molecules, whereas others are nonpolar molecules.

Be able to relate the permanent dipole moment of a molecule to the magnitude of the partial charges within it and the distance between them.

Key Equations and Relations

Relations between the Orientation of Groups of Electrons around the Central Atom of a Molecule, Hybridization, and Molecular Geometry

Number of Electron Groups	Orientation	Hybridization	Molecular Geometries	Example
2	Linear	sp	Linear	$BeCl_2$, CO_2
3	Trigonal planar	sp^2	Trigonal planar	H_2CO, BF_3
			Angular	O_3
4	Tetrahedral	sp^3	Tetrahedral	CH_4
			Trigonal pyramid	NH_3
			Angular	H_2O
5	Trigonal bipyramidal	dsp^3	Trigonal bipyramidal	PF_5
			Seesaw	SF_4
			T shaped	BrF_3
			Linear	XeF_2
6	Octahedral	d^2sp^3	Octahedral	SF_6
			Square pyramidal	BrF_5
			Square planar	XeF_4

The dipole moment of a molecule, μ:

$$\mu = Qr \quad (7.1)$$

where Q is the partial charge at each "end" of the dipole and r is the distance between the ends.

QUESTIONS AND PROBLEMS

Molecular Vibrations

CONCEPT REVIEW

1. A scientist is not sure whether a compound contains a C—N single bond or a C=N double bond. How might IR spectroscopy be used to determine which type of bond is present?
2. Infrared spectrometers aboard the *Mariner* spacecraft were used to analyze the atmosphere of Venus. Were the spectrometers able to distinguish between carbon monoxide and carbon dioxide in the Venusian atmosphere?
3. Why does infrared radiation cause bonds to vibrate but not break (as UV radiation can)?
4. Argon is the third most abundant species in the atmosphere. Why isn't it a greenhouse gas?
5. The N–O bond distances in N_2O_4 (which has an N–N bond) and N_2O are the same. Do you expect N_2O_4 and N_2O to have IR absorption bands arising from N–O stretching vibrations at nearly the same wavelengths?
6. How could IR spectroscopy be used to determine whether S_2O has the skeletal structure S–S–O or S–O–S?

PROBLEMS

7. Rank the following molecules in order of increasing stretching frequencies of the S—O bonds: SO, SO_2, and SO_3.
8. In which species is the C–N stretching frequency higher: CN^- or C_2N_2 (which has a N–C–C–N– skeleton)?
9. Predict the order of N–O stretching frequencies in the following ions: NO^+, NO_2^-, and NO_4^{3-}.
10. Rank the following molecules in order from highest N–N stretching frequency to lowest: $H_2N–NH_2$, N_2, and HNNH.

Exceptions to the Octet Rule

CONCEPT REVIEW

11. Why do C, N, O, and F atoms in covalently bonded molecules and ions have no more than eight valence electrons?

12. Are all odd-electron molecules exceptions to the octet rule?

PROBLEMS

13. In which of the following molecules does the sulfur atom have an expanded octet? (a) SF_6; (b) SF_5; (c) SF_4; (d) SF_2.

14. In which of the following molecules and ions does the phosphorus atom have an expanded octet? (a) PF_6^-; (b) PF_5; (c) PF_3; (d) P_2F_4 (which has a P–P bond).

15. Which of the following molecules and ions contains an atom with an expanded octet? (a) Cl_2; (b) ClF_3; (c) ClF_2^+; (d) ClO^-.

16. Which of the following molecules contains an atom with an expanded octet? (a) XeF_2; (b) $GaCl_3$; (c) ONF_3; (d) SeO_2F_2.

17. How many electrons are there in the covalent bonds surrounding the sulfur atom in the following molecules and ion? (a) SF_4O; (b) SOF_2; (c) SO_3; (d) SF_5^-.

18. How many electrons are there in the covalent bonds surrounding the phosphorus atom in the following molecules and ion? (a) $POCl_3$; (b) H_3PO_4; (c) H_3PO_3; (d) PF_6^-.

19. How many pairs of electrons does xenon share in the following molecules and ions? (a) XeF_2; (b) $XeOF_2$; (c) XeF^+; (d) XeF_5^+; (e) XeO_4.

20. How many pairs of valence electrons do bromine atoms have in the following molecules and ions? (a) BrF; (b) BrF_2^+; (c) BrF_6^+; (d) BrF_2^-; (e) BrF_5.

21. Why is SF_4 a stable gas, but OF_4 does not exist?

22. Why is PF_5 a stable molecule, but NF_5 does not exist?

23. Draw Lewis structures for NOF_3 and POF_3 in which the Group 5A element is the central atom. The oxygen and fluorine atoms are all bonded to the central atom. What differences are there in the types of bonding in these molecules?

24. The phosphate anion, PO_4^{3-}, is common in minerals. The corresponding unstable NO_4^{3-} anion can be prepared by reaction of sodium nitrate with sodium oxide at 300°C. Draw Lewis structures for each anion. What are the differences in bonding between these ions?

25. Dissolving NaF in selenium tetrafluoride (SeF_4) produces $NaSeF_5$. Draw Lewis structures for SeF_4 and SeF_5^-. In which structures does Se have more than eight valence electrons?

26. Reaction between NF_3, F_2, and SbF_3 at 200°C and 100-atm pressure gives the ionic compound NF_4SbF_6:

$$NF_3(g) + 2\,F_2(g) + SbF_3(g) \longrightarrow NF_4SbF_6(s)$$

Draw Lewis structures for the NF_4^+ and SbF_6^- ions in this product.

27. The compound Cl_2O_2 may play a role in ozone depletion in the stratosphere. In the laboratory, reaction of $FClO_2$ with aluminum chloride produces Cl_2O_2 and $AlCl_2F$.

$$FClO_2 + AlCl_3 \longrightarrow Cl_2O_2 + AlFCl_2$$

Draw a Lewis structure of Cl_2O_2 based on the following arrangement of atoms:

$$Cl-Cl{\begin{smallmatrix}O\\O\end{smallmatrix}}$$

Does either of the chlorine atoms in the structure have an expanded octet?

28. Cl_2O_2 decomposes to chlorine and chlorine dioxide:

$$2\,Cl_2O_2 \longrightarrow Cl_2 + 2\,ClO_2$$

Draw the Lewis structure of ClO_2. Which atom in the structure does not have eight valence electrons?

29. Draw a Lewis structure for $AlFCl_2$, the second product in the synthesis of Cl_2O_2 in Problem 27.

*30. Draw Lewis structures for BF_3 and $(CH_3)_2BF$. The B–F distance in both molecules is the same (130 pm). Does this observation support the argument that all the B–F bonds in BF_3 are single bonds?

31. Which of the following chlorine oxides are odd-electron compounds? (a) Cl_2O_7; (b) Cl_2O_6; (c) ClO_4; (d) ClO_3; (e) ClO_2.

32. Which of the following nitrogen oxides are odd-electron compounds? (a) NO; (b) NO_2; (c) NO_3; (d) N_2O_4; (e) N_2O_5.

33. In the following species, which atom is more likely to have an unpaired electron? (a) SO^+; (b) NO; (c) CN; (d) OH.

34. In the following molecules, which atom is more likely to have an unpaired electron? (a) NO_2; (b) CNO; (c) ClO_2; (d) HO_2.

35. Which of the following Lewis structures contributes most to the bonding in CNO?

a. $\cdot\ddot{C}-N\equiv O:$

b. $\cdot\ddot{C}=N=\ddot{O}$

c. $:C\equiv N-\ddot{O}\cdot$

d. $\cdot C\equiv N-\ddot{O}:$

36. Why is the following Lewis structure unlikely to contribute much to the bonding in NCO?

    :N̈—C≡O·

37. A compound with the formula Cl_2O_6 decomposes to a mixture of ClO_2 and ClO_4. Draw two Lewis structures for Cl_2O_6: one with a Cl–Cl bond and one with a Cl–O–Cl arrangement of atoms. Draw a Lewis structure for ClO_2.

    $Cl_2O_6 \longrightarrow ClO_2 + ClO_4$

38. A compound consisting of chlorine and oxygen, Cl_2O_7, decomposes by the following reaction:

    $Cl_2O_7 \longrightarrow ClO_4 + ClO_3$

    a. Draw two Lewis structures for Cl_2O_7: one with a Cl–Cl bond and one with a Cl–O–Cl arrangement of atoms.
    b. Draw a Lewis structure for ClO_3.

39. The odd-electron molecule CN dimerizes to give cyanogen, C_2N_2. Draw a Lewis structure for CN and predict which skeletal arrangement for cyanogen, NCCN or CNNC, is more likely.

40. The odd-electron molecule SN forms S_2N_2, which has a cyclic structure.
    a. Draw a Lewis structure for SN and complete the following possible Lewis structures of S_2N_2.

    ```
    S—N        S—N
    |  |       |  |
    S—N        N—S
    ```

    b. Which of the two is the preferred structure?

*41. The molecular structure of sulfur cyanide trifluoride, SF_3CN, has been studied by electron diffraction and shown to have the following skeletal structure with the indicated bond lengths:

    115.9 pm → N
               |
               C
               | ← 173.6 pm
               S
             / | \
            F | F
               F   160 pm

    Using the observed bond lengths to guide you, complete the Lewis structure of SF_3CN and assign formal charges.

42. Heating phosphorus with sulfur gives P_4S_3, a solid used in the heads of strike-anywhere matches. P_4S_3 has the following Lewis structure framework. Complete this Lewis structure so that each atom has the optimum formal charge.

    ```
          P
         / \
        S   S
         \ /
          S
        / |
       P  |  P
        \ | /
          P
    ```

*43. The heavier Group 6A elements can expand their octets. The $TeOF_6^{2-}$ anion, first prepared in 1993, has a pentagonal bipyramidal geometry. A pentagonal bipyramid is similar to the trigonal bipyramid but has a pentagon rather than a triangle in the middle. Draw a Lewis structure for $TeOF_6^{2-}$, and justify the observed geometry.

44. Sulfur is cycled in the environment through compounds such as dimethylsulfide (CH_3SCH_3), hydrogen sulfide (H_2S), sulfite (SO_3^{2-}), and sulfate (SO_4^{2-}) ions. Draw Lewis structures for these four molecules and ions. Are expanded octets needed to optimize the formal charges for any of these molecules?

Molecular Orbitals of Odd-Electron Molecules

CONCEPT REVIEW

45. Why are the 2p orbitals on an oxygen atom lower in energy than the 2p orbitals on a nitrogen atom?
46. Which MO diagram is more distorted: the one for NO or the one for CO?

PROBLEMS

47. The CN molecule has been detected in interstellar space. Sketch an MO diagram for CN and determine the bond order.
48. The OH radical plays an important role in atmospheric chemistry. Sketch an MO diagram for OH and determine the bond order.
49. The compound PO is believed to be the principal phosphorus-containing molecule in space. It can be prepared in the laboratory by reaction between $POBr_3$ and magnesium metal. Sketch an MO diagram for PO and determine the bond order.
50. The odd-electron compound SN has a short lifetime. Sketch an MO diagram for SN and determine the bond order.

51. The first compound containing a xenon–xenon bond, $Xe_2Sb_4F_{21}$, was reported in 1997. Sketch a Lewis structure and an MO diagram for the Xe_2^+ cation and determine the bond order.

52. Ionized carbon monoxide, CO^+, has been detected in interstellar space. Sketch an MO diagram for CO^+. What type of orbital (σ, π, π^*, or σ^*) contains the highest-energy electrons and what is the bond order?

ESR Spectroscopy

CONCEPT REVIEW

53. Why do odd-electron molecules and ions have electron-spin resonance (ESR) spectra?
54. How might the ESR spectrum of NO_2 help determine whether the odd electron is on nitrogen or oxygen?

PROBLEMS

55. Complete a Lewis structure of $ON(SO_3)_2^{2-}$ that is consistent with a triplet in its ESR spectrum.

56. Can ESR spectroscopy distinguish between the following two Lewis structures of NCO?

Molecular Geometry

CONCEPT REVIEW

57. Is the shape of a molecule determined by repulsions between nuclei or by repulsions between electron pairs?
58. In a molecule of ammonia, why is the repulsion between the lone pair and a bonding pair of electrons on nitrogen greater than the repulsion between two N–H bonding pairs?
59. Why is it important to draw a correct Lewis structure for a molecule before determining its geometry?
60. Do all resonance forms of a molecule have the same molecular geometry?
61. How does VSEPR account for the range of bond angles from about 104 to 180 degrees in triatomic molecules?
62. How is it that SO_3 and BF_3 have different numbers of bonds but the same trigonal planar geometry?

PROBLEMS

63. Determine the molecular geometries of the following molecules: (a) GeH_4; (b) PH_3; (c) H_2S; (d) $CHCl_3$.
64. Determine the molecular geometries of the following molecules and ions: (a) NO_3^-; (b) NO_4^{3-}; (c) S_2O; (d) NF_3.
65. Determine the molecular geometries of the following ions: (a) NH_4^+; (b) CO_3^{2-}; (c) NO_2^-; (d) XeF_5^+.
66. Determine the geometries of the following ions: (a) SCN^-; (b) $CH_3PCl_3^+$ (P is the central atom and this cation contains a C–P bond); (c) ICl_2^+; (d) PO_3^{3-}.
67. Determine the geometries of the following ions and molecules: (a) $S_2O_3^{2-}$; (b) PO_4^{3-}; (c) NO_3; (d) NCO.
68. Determine the geometries of the following molecules: (a) ClO_2; (b) ClO_3; (c) IF_3; (d) SF_4.
69. Which of the following triatomic molecules, O_3, SO_2, and CO_2, have the same molecular geometry?
70. Which of the following species, N_3^-, O_3, and CO_2, have the same molecular geometry?
71. Which of the following molecular ions, SCN^-, CNO^-, and NO_2^-, have the same molecular geometry?
72. Which of the following molecules, N_2O, S_2O, and CO_2, have the same molecular geometry?
73. A number of sulfur oxides not found in Earth's atmosphere have been detected in the atmosphere of Venus, including S_2O and S_2O_2. Draw Lewis structures for S_2O and S_2O_2, and determine their molecular geometries.

74. The structures of NOCl, NO_2Cl, and NO_3Cl were determined in 1995. They have the following skeletal structures. Draw Lewis structures of these three compounds, and predict the bonding geometry at each nitrogen atom.

```
        O             O
        ||            ||
O—N—Cl   O—N—Cl    O—N
                       |
                       O
                        \
                         Cl
```

75. For many years, it was believed that the noble gases could not form covalently bonded compounds. However, xenon may react with fluorine and oxygen. Reaction between xenon tetrafluoride and fluoride ions produces the pentafluoroxenate anion XeF_5^-.

$$XeF_4 + F^- \longrightarrow XeF_5^-$$

Draw Lewis structures for XeF_4 and XeF_5^-, and predict the geometry around xenon in XeF_4. The crystal structure of XeF_5^- compounds indicates pentagonal bipyramidal orientation of valence pairs around Xe. Sketch a pentagonal pyramidal structure.

76. The first compound containing a xenon–sulfur bond was isolated in 1998. Draw a Lewis structure for HXeSH, and determine its molecular geometry.

77. A 1995 terrorist attack by a cult in Japan on the Tokyo subway system with the nerve gas Sarin focused world attention on the dangers of chemical warfare agents. The following structure shows the connectivity of the atoms in the Sarin molecule. Complete the Lewis structure by adding bonds and lone pairs as necessary. Assign formal charges to the P and O atoms, and determine the bonding geometry around P.

```
              O
              ||           CH_3
      CH_3—P—O—C—H
              |           \
              F            CH_3
            Sarin
```

78. Determine the bonding geometry around the nitrogen atom in the following unstable nitrogen oxides: N_2O_2, N_2O_5, and N_2O_3. (N_2O_2 and N_2O_3 have N–N bonds; N_2O_5 does not.)

Valence-Bond Theory

CONCEPT REVIEW

79. Why can't atoms with sp^3 hybridization form π bonds?
80. Which better explains molecular geometry: valence-bond theory or molecular-orbital theory?
81. Which better explains the magnetic properties of a diatomic molecule: valence-bond theory or molecular-orbital theory?
82. Give an example of a hybrid orbital constructed from atomic orbitals with different principal quantum numbers.
83. Why aren't the orbitals on free atoms hybridized?
84. Do the resonance forms of N_2O have the same hybridization at the central N atom?

PROBLEMS

85. What is the hybridization of nitrogen in each of the following ions and molecules? (a) NO_2^+; (b) NO_2^-; (c) N_2O; (d) N_2O_5; (e) N_2O_3.
86. What is the hybridization of sulfur in each of the following molecules? (a) SO; (b) SO_2; (c) S_2O; (d) SO_3.
87. Azides such as sodium azide, NaN_3, are used in automobile airbags as a source of nitrogen gas. Another compound with three nitrogen atoms bonded together is N_3F. What differences are there in the arrangement of the electrons in NaN_3 and N_3F? Is there a difference in the hybridization of the central nitrogen atom?
88. N_3F decomposes to nitrogen and N_2F_2 by the following reaction:

$$2\,N_3F \longrightarrow 2N_2 + N_2F_2$$

N_2F_2 has two possible structures:

```
   :F:              :F:
     \                \
      N=N              N=N
                            \
                             :F:
```

Are the differences between these structures related to differences in the hybridization of nitrogen in N_2F_2? Identify the hybrid orbitals that account for the bonding in N_2F_2. Are they the same as those in acetylene, C_2H_2?

89. How does the hybridization of the sulfur atom change in the series SF_2, SF_4, and SF_6?
90. How does the hybridization of the central atom change in the series CO_2, NO_2, O_3, and ClO_2?
91. What is the hybridization of the central oxygen atom in the disulfate anion, which has the following skeletal structure?

92. Which atomic or hybrid orbitals or both overlap to form the S–O and S–N bonds in the sulfamate anion, which has the following skeletal structure?

93. In Chapter 4, we identified an oxoanion of chlorine, ClO_2^- (chlorite), which is used as a bleaching agent. Do ClO_2^- and SO_2 have the same hybridization of the central atom?

94. Does the chlorine atom in chlorate (ClO_3^-) or perchlorate (ClO_4^-) ion have the same hybridization as the carbon atom in methane (CH_4)?

95. Draw a Lewis structure for Cl_3^+; determine its molecular geometry and the hybridization of the central Cl atom.

96. Synthesis of the first compound of argon was reported in 2000. HArF was made by reacting Ar with HF. Draw a Lewis structure for HArF, and determine the hybridization of Ar in this molecule.

97. The Lewis structure of N_4O, with the skeletal structure O–N–N–N–N contains one N—N single bond, one N=N double bond, and one N≡N triple bond. Is the hybridization of all nitrogen atoms the same?

98. The three nitrogen atoms in the nitramide anion $N(NO_2)_2^-$ are connected, with two oxygen atoms on each terminal nitrogen. Is the hybridization of all of the nitrogen atoms the same?

99. Cyclic structures exist for many compounds of carbon and hydrogen. Describe the hybridization of the carbon atoms in benzene (C_6H_6), cyclobutane (C_4H_8), and cyclobutene (C_4H_6).

100. The trifluorosulfite anion SOF_3^- was isolated in 1999 as the tetramethylammonium salt $(CH_3)_4NSOF_3$. Determine the geometry around sulfur in the SOF_3^- anion and describe the bonding in SOF_3^- according to valence-bond theory.

Molecular Polarity

CONCEPT REVIEW

101. Explain the difference between a polar bond and a polar molecule.
102. Must a polar molecule contain polar covalent bonds? Why?
103. Can a nonpolar molecule contain polar covalent bonds?
104. What does a dipole moment measure?

PROBLEMS

105. The following molecules contain polar covalent bonds. Which of them are polar molecules and which are nonpolar? CCl_4; $CHCl_3$; CO_2; H_2S; SO_2.
106. Photolysis of Cl_2O_2 is thought to produce compounds with the following skeletal structures. Do the two compounds have the same dipole moment?

    Cl—O—Cl—O   Cl—O—O—Cl

107. Which of the following chlorofluorocarbons (CFCs) are polar and which are nonpolar?
    a. Freon 11 ($CFCl_3$)
    b. Freon 12 (CF_2Cl_2)
    c. Freon 113 (Cl_2FCCF_2Cl)
108. Which of the following chlorofluorocarbons (CFCs) are polar and which are nonpolar?
    a. Freon C318 (C_4F_8, cyclic structure)
    b. Freon 1113 (C_2Cl_3F)
    c. $C_2HCl_3F_2$
109. Predict which molecule in each of the following pairs is the more polar.
    a. Freon 13 ($CClF_3$) or Freon 13B1 ($CBrF_3$)
    b. Freon 12 (CF_2Cl_2) or Freon 22 (CHF_2Cl)
    c. Freon 113 (Cl_2FCCF_2Cl) or Freon 114 (ClF_2CCF_2Cl)
110. Which molecule in each of the following pairs is more polar?
    a. NH_3 or PH_3
    b. CCl_2F_2 or CBr_2F_2

111. A series of carbonyl dihalide compounds of formula COX$_2$ (where X = I, Cl, or Br) has been prepared. Place these compounds—COI$_2$, COCl$_2$, and COBr$_2$—in order of increasing polarity. Explain your reasoning.
112. Simple diatomic molecules that have been detected in interstellar space include CO, CS, SiO, SiS, SO, and NO. Arrange these molecules in order of increasing dipole moment based on the location of the constituent elements in the periodic table, and then calculate the electronegativity differences from the data in Figure 6.6.
113. Two compounds with the same formula, S$_2$F$_2$, have been isolated. The following structures show the arrangements of the atoms in these different compounds. Can these two compounds be distinguished by their dipole moments?

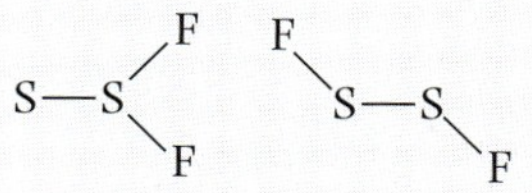

$$\text{S—S} \diagdown^{F}_{F} \qquad ^{F}\diagdown\text{S—S}\diagdown_{F}$$

114. Could you distinguish between the two structures of N$_2$F$_2$ (see Problem 88) by the magnitude of their dipole moments?

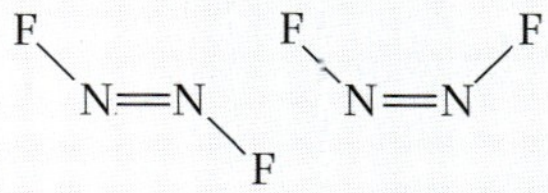

$$^{F}\diagdown\text{N=N}\diagdown_{F} \qquad ^{F}\diagdown\text{N=N}\diagup^{F}$$

115. Given the following data for carbon monoxide, calculate the partial charge (relative to the charge on an electron) on the carbon atom. The length of the C≡O triple bond = 113 pm and the dipole moment of CO = 0.112 D.
*116. Given the following data for sulfur dioxide, calculate the partial charge (relative to the charge on an electron) on the sulfur atom. Length of the S–O bond in SO$_2$ = 142 pm; the dipole moment of SO$_2$ = 1.63 D; O–S–O bond angle = 120 degrees.

8 Properties of Gases
And the Air That We Breathe

The severity of a tropical storm is related to depressed atmospheric pressure at its center. This photograph of Typhoon Odessa was taken by the space shuttle *Discovery* in August 1985, when the maximum winds of the storm were about 90 mph and the central pressure was 40 millibars (mb) lower than its surroundings. In contrast, the central pressure of Hurricane Andrew (see page 392) was 90 mb lower than its surroundings when it hit south Florida with winds as high as 165 mph.

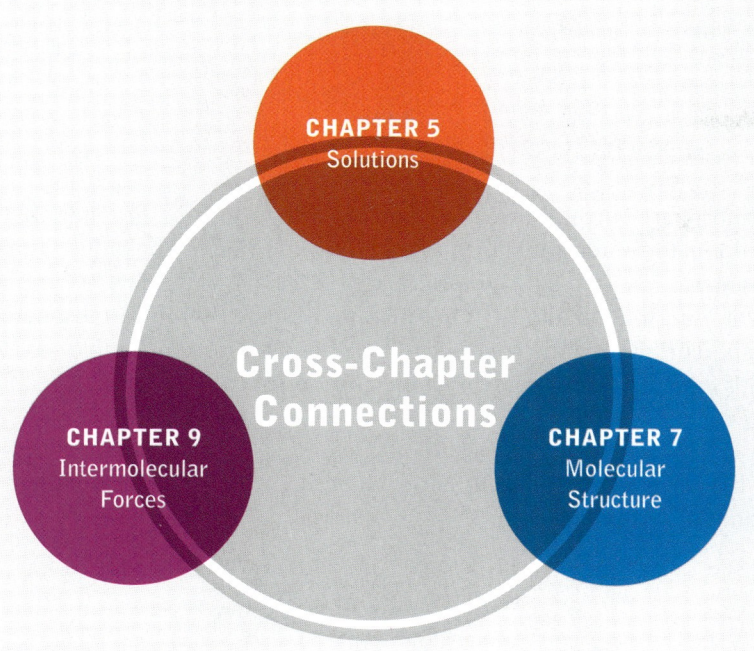

- 8.1 Introduction
- 8.2 The atmosphere: a molecular view
  - Boyle's law
  - The combined gas law
- 8.3 The ideal-gas law
- 8.4 Gas density
- 8.5 Dalton's law and mixtures of gases
- 8.6 Henry's law and the solubility of gases
- 8.7 The kinetic molecular theory of gases and Graham's law
- 8.8 Real gases

A Look Ahead

In this chapter, changes in atmospheric pressure (P) and temperature (T) with increasing altitude are used to introduce the ideal-gas law. We will explore how the number of moles (n) of a gas is proportional to its volume (V) at constant T and P; how the density (d) of a gas is proportional to its molar mass ($\mathcal{M}$); how a gas leaks through tiny openings at a rate inversely proportional to $\mathcal{M}^{1/2}$; and how the solubility of a gas in water and other liquids is proportional to its partial pressure. Then we use the kinetic molecular theory to explain the properties of gases and gas mixtures, including those at high pressures or low temperatures whose behavior is better described by the van der Waals equation than by the ideal gas law.

8.1 INTRODUCTION

The atmosphere surrounding our planet is drawn to it by gravitational forces proportional to the masses of the gases in the atmosphere. The total mass of these gases is about 5×10^{18} kg. The surface area of Earth is about 5×10^{14} m² (Figure 8.1). This information allows us to estimate the force that the atmosphere exerts on Earth's surface and the resulting atmospheric pressure. A simple equation relates this force (F) to the mass (m) of the atmosphere:

$$F = ma \qquad (8.1)$$

where a is acceleration due to gravity, which is 9.81 m/s². Inserting these values into Equation 8.1, we have

$$F = (5 \times 10^{18} \text{ kg})(9.81 \text{ m/s}^2)$$
$$= 5 \times 10^{19} \text{ kg·m/s}^2 = 5 \times 10^{19} \text{ N}$$

where N is the abbreviation for newton, the SI unit of force. The pressure (P) that the atmosphere exerts on the surface of Earth is the result of this force spread out over the area (A) of the surface. In equation form, we have

$$P = \frac{F}{A} \qquad (8.2)$$

$$= 5 \times 10^{19} \text{ N}/5 \times 10^{14} \text{ m}^2$$

$$= 1.0 \times 10^5 \text{ N/m}^2 = 1.0 \times 10^5 \text{ Pa}$$

where Pa is the abbreviation for pascal, the SI unit of pressure, named in honor of French mathematician and physicist Blaise Pascal (1623–1662). He was the first to propose that atmospheric pressure decreases with increasing altitude.

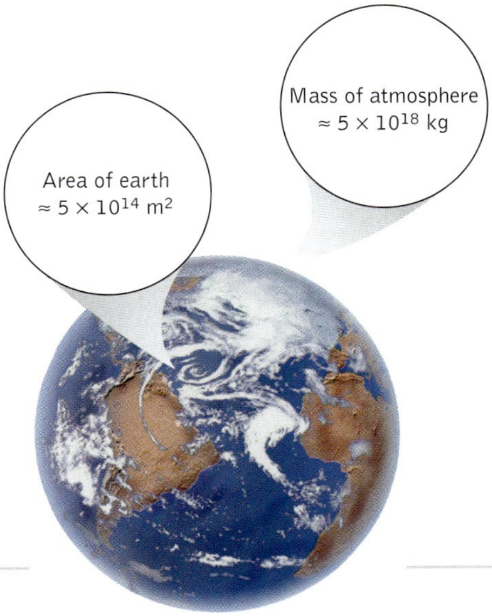

FIGURE 8.1 Atmospheric pressure results from the force exerted by the atmosphere acting on the surface of Earth. The force is the product of the combined mass of atmospheric gases and acceleration due to gravity. Pressure equals this force divided by Earth's surface area.

Mass of atmosphere ≈ 5×10^{18} kg

Area of earth ≈ 5×10^{14} m²

8.1 Introduction

Atmospheric pressure is routinely measured with an instrument called a barometer. A simple, but effective, barometer design is shown in Figure 8.2. It consists of a tube nearly a meter long and closed at one end. The tube is filled with mercury and inverted in a pool of mercury. Gravity pulls the mercury in the tube downward, creating a vacuum at the top of the tube, but atmospheric pressure pushes the mercury in the pool up into the tube. The net effect of these opposing forces is indicated by the height of mercury in the tube, which provides a measure of atmospheric pressure.

CONNECTION: The definition of pressure as force per unit area also applies to osmotic pressure (Section 5.4).

SAMPLE EXERCISE 8.1: Calculate the force exerted by an atmospheric pressure of 1.013×10^5 Pa at the base of the column of mercury in Figure 8.2. The inside diameter of the glass tube is 10.00 mm.

SOLUTION: Pressure is the force per unit area (Equation 8.2):

$$P = \frac{F}{A}$$

Because we know the value of P and wish to calculate F, let's rearrange the terms in Equation 8.2:

$$F = P \times A$$

The pressure at the base of the column is 1.013×10^5 Pa = 1.013×10^5 N/m². The cross-sectional area of the tube equals the area of a circle with a diameter of 10 mm. The area of a circle is πr^2. In this case, the radius is 10.00 mm/2, or 5.00 mm, or 5.00×10^{-3} m; so the area is

$$A = \pi r^2$$
$$= 3.1416 \, (5.00 \times 10^{-3} \text{ m})^2 = 7.85 \times 10^{-5} \text{ m}^2$$

The force exerted across this area is

$$F = P \times A = (1.013 \times 10^5 \text{ N/m}^2)(7.85 \times 10^{-5} \text{ m}^2) = 7.95 \text{ N}$$

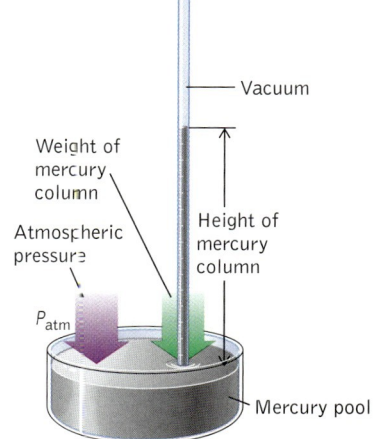

FIGURE 8.2 The height of the mercury column in the barometer designed by Evangelista Torricelli is proportional to atmospheric pressure.

PRACTICE EXERCISE: Calculate the force in newtons exerted by an atmospheric pressure of 1.01×10^5 Pa on a square inch of Earth's surface. (See Problems 9 and 10.)

SAMPLE EXERCISE 8.2: What is the height of the column of mercury in the preceding sample exercise if the density of mercury is 13.58 g/cm³.

SOLUTION: The height of the mercury column will be determined by the balance between the atmospheric pressure pushing up on it and the force of gravity on the mass of the mercury that is pulling it down. The force due to atmospheric pressure pushing up is 7.95 N, or 7.95 kg·m/s². We can use this value and Equation 8.1 to calculate the mass of mercury in the column:

$$F = ma$$

or

$$m = \frac{F}{a}$$

$$= \frac{7.95 \text{ kg} \cdot \text{m/s}^2}{9.81 \text{ m/s}^2}$$

$$= 0.8104 \text{ kg} = 810.4 \text{ g of Hg}$$

We can use the density of mercury to determine what volume is occupied by 810 g of Hg:

$$d = \frac{m}{V}$$

or

$$V = \frac{m}{d}$$

$$= \frac{810.4 \text{ g Hg}}{13.58 \text{ g/cm}^3} = 59.68 \text{ cm}^3$$

This volume of mercury is in the shape of a cylinder with a radius of 5.00 mm (or 0.500 cm). We can calculate the height (h) of the cylinder by using the equation for its volume (V):

$$V = \pi r^2 h$$

or

$$h = \frac{V}{\pi r^2}$$

$$= \frac{59.68 \text{ cm}^3}{3.1416 (0.500 \text{ cm})^2}$$

$$= 76.0 \text{ cm} = 760. \text{ mm Hg}$$

CONCEPT TEST: Would the height of the column of mercury in the preceding sample exercise be the same if atmospheric pressure were the same but the diameter of the tube were different?

Atmospheric pressure varies from place to place and with changing weather conditions. For reference purposes, standard atmospheric pressure, or 1 *atmosphere* (atm), is defined as the pressure that supports a column of mercury 760 mm high. The pressure unit *mm Hg* is also called *torr* in honor of Evangelista Torricelli (1608–1647), the Italian mathematician and physicist who invented the barometer. Thus,

$$1 \text{ atm} = 760 \text{ mm Hg} = 760 \text{ torr}$$

Evang. Torricelli.

Evangelista Torricelli was an accomplished scientist and mathematician; his work contributed to the development of integral calculus. His invention of the barometer was influenced by another famous Italian scientist, Galileo, whom Torricelli served as a secretary and research assistant during the last year of Galileo's life.

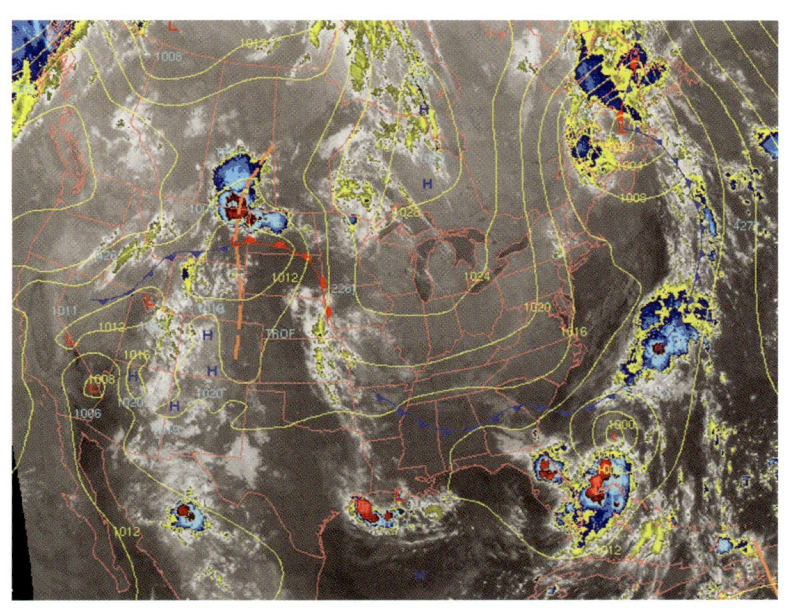

FIGURE 8.3 Small differences in atmospheric pressure are associated with major changes in weather. Adjacent contours of constant pressure (called isobars) on this weather map differ by 4 millibars.

The relation between atmospheres of pressure and pascals is

$$1 \text{ atm} = 101{,}325 \text{ Pa}$$

Clearly, the pascal is a tiny quantity of pressure. In many applications, we find it more convenient to express pressures in kilopascals (1000 Pa):

$$1 \text{ atm} = 101.325 \text{ kPa} = 1.01325 \times 10^2 \text{ kPa}$$

For many years, meteorologists have expressed atmospheric pressure in millibars (mb). The weather maps prepared by the U.S. National Weather Service show changes in atmospheric pressure by using contours of constant pressure that are 4 mb apart (Figure 8.3). There are exactly 10 mb in one kPa; so, at standard atmospheric pressure,

$$1 \text{ atm} = (101.325 \text{ \sout{kPa}}) \left(\frac{10 \text{ mb}}{\text{\sout{kPa}}} \right) = 1013.25 \text{ mb}$$

SAMPLE EXERCISE 8.3: The lowest atmospheric pressure ever recorded at sea level was 25.69 in. Hg. This pressure was measured on October 12, 1979, in the eye of typhoon Tip, northwest of Guam in the Pacific Ocean. What was the pressure in typhoon Tip in (a) mm Hg, (b) atm, and (c) kPa? There are exactly 2.54 cm/in.

SOLUTION: The following conversions provide the desired new units of pressure:

P(inches Hg)

↓ 1 in. = 2.54 cm

P(cm Hg)

↓ 1 cm = 10 mm

P(mm Hg)

↓ 1 atm = 760 mm Hg

P(atm)

↓ 1 atm = 101.325 kPa

P(kPa)

(a) $(25.69 \text{ in. Hg}) \left(\dfrac{2.54 \text{ cm}}{\text{in.}}\right) \left(\dfrac{10 \text{ mm}}{\text{cm}}\right) = 652.5$ mm Hg

(b) $(652.5 \text{ mm Hg}) \left(\dfrac{1 \text{ atm}}{760 \text{ mm Hg}}\right) = 0.8586$ atm

(c) $(0.8586 \text{ atm})(101.325 \text{ kPa/atm}) = 87.00$ kPa

PRACTICE EXERCISE: Atmospheric pressure in the eye of hurricane Andrew was 922 mb when it hit the Florida coast near Miami on August 24, 1992. What was this pressure in (a) kPa, (b) atm, and (c) torr? (See Problems 11 and 12.)

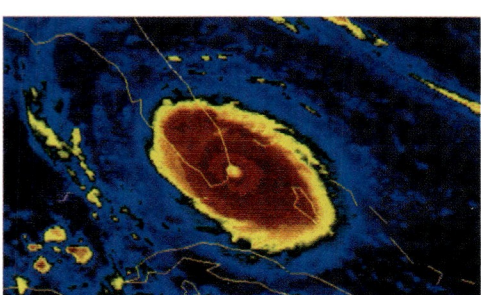

This infrared satellite image of Hurricane Andrew was taken as the eye of the hurricane struck the coast of Florida with wind speeds over 150 miles per hour.

8.2 THE ATMOSPHERE: A MOLECULAR VIEW

Atmospheric pressure decreases with increasing altitude (Figure 8.4). We can explain this phenomenon by noting that pressure is related to the total mass of molecules of gas *above* a given altitude. As altitude increases, the mass of the gases still above it must decrease. Less mass means less pressure.

Boyle's law

Air, like any gas or mixture of gases, is compressible. The weight of the air in the atmosphere (about 1 kg/cm² of Earth's surface area) presses down on the layer of air nearest the surface, squeezing the molecules and atoms of the atmospheric gases and making them occupy less volume. If there are more moles (n) of atmospheric gases per volume (V) nearer the surface, then there is more mass of air in a given volume, and so air is densest near the surface.

Consider the effects of compressing a collection of molecules of gas into less space. The molecules in the gas phase are constantly in random motion, which means that they collide with one another and with any surfaces with which they come in contact. If there are more molecules squeezed into the space, there will be more collisions per unit time (Figure 8.5). These collisions exert a force on

✓ Gas pressure is the result of collisions between molecules of a gas and their surroundings.

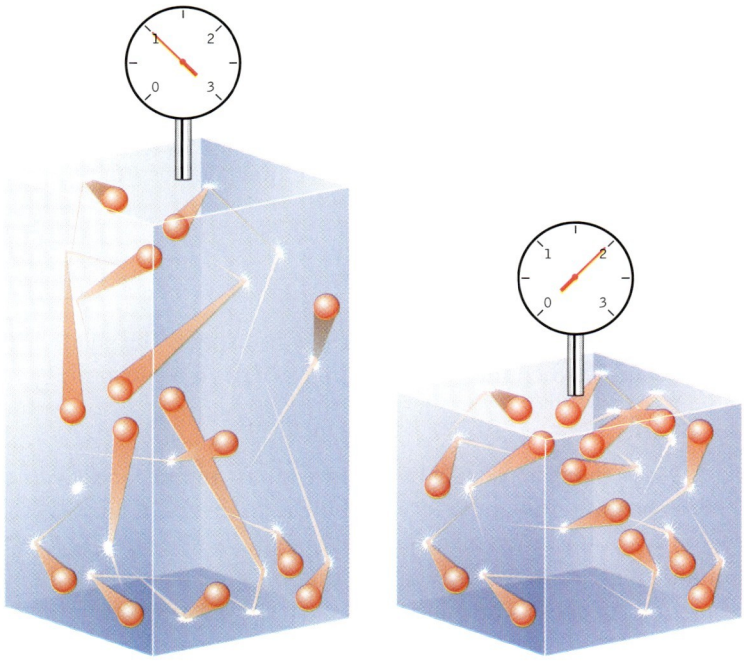

FIGURE 8.4 Atmospheric pressure decreases with increasing altitude as the mass of the column of air above a given area decreases with increasing altitude. One atmosphere of pressure at sea level corresponds to 0.83 atm in Denver and 0.35 atm at the summit of Mt. Everest.

FIGURE 8.5 Gas molecules are in constant random motion, exerting a pressure through collisions with their surroundings. When a quantity of gas is squeezed into half its original volume, the frequency of collisions per unit of surface area increases by a factor of two, and so does the pressure.

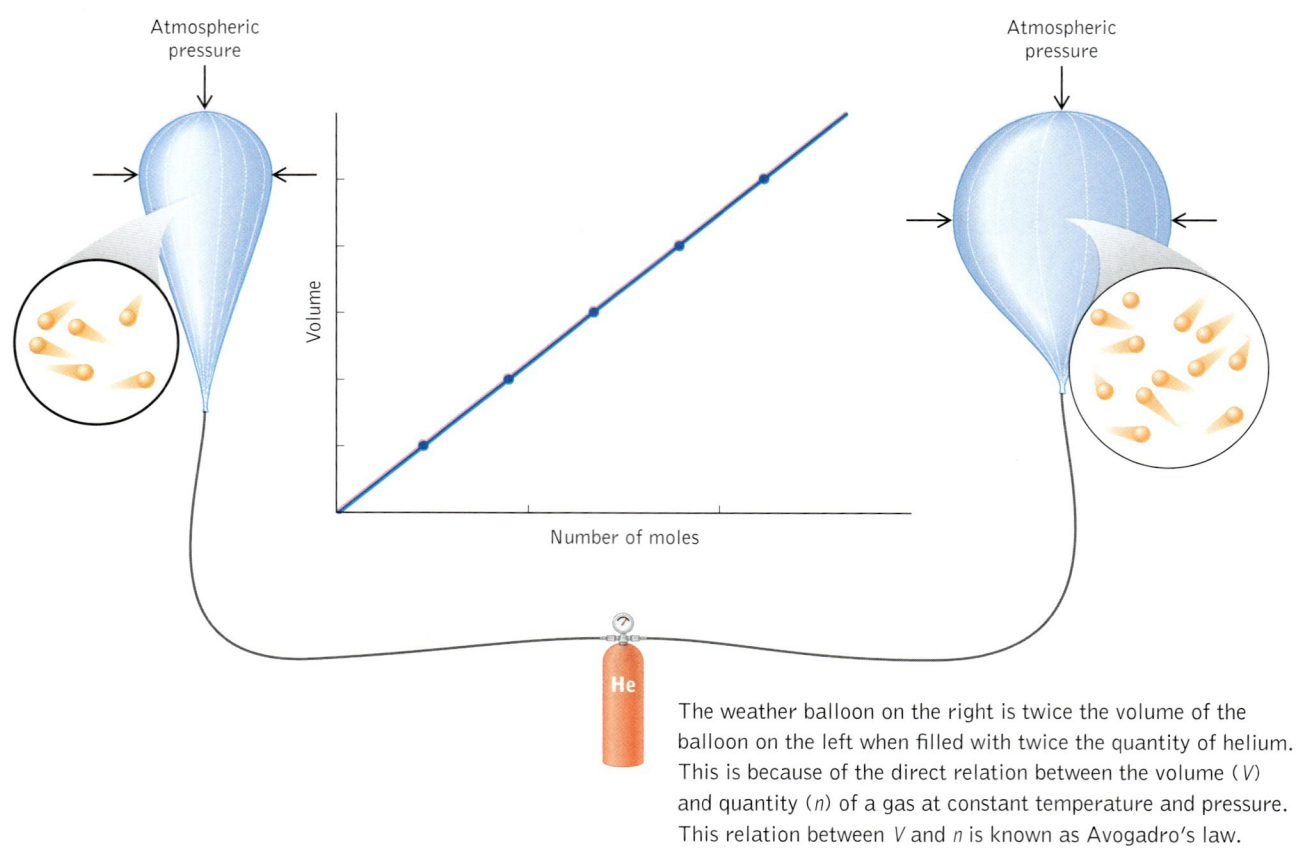

The weather balloon on the right is twice the volume of the balloon on the left when filled with twice the quantity of helium. This is because of the direct relation between the volume (V) and quantity (n) of a gas at constant temperature and pressure. This relation between V and n is known as Avogadro's law.

the pressure sensor. The more frequent the collisions, the greater the force. According to Equation 8.2, dividing the force (F) of the collisions by area (A) yields the pressure (P) measured by a barometer. The magnitude of P should be proportional to the number of molecules in a given volume. Expressing this proportionality mathematically, we have

$$P \propto \frac{n}{V} \tag{8.3}$$

We can rearrange the terms in Equation 8.3 to express the relation between pressure and volume when the number of moles of gas is a constant:

$$n \propto P \times V = \text{constant} \tag{8.4}$$

Equation 8.4 says that the product of pressure times volume for a given quantity (n) of gas is a constant. Put another way, there is an inverse relation between the pressure and the volume of a given quantity of gas, as shown in Figure 8.6. We need to qualify these statements by noting that they are true only if the temperature of the gas is constant. The relation between the pressure of a quantity of gas and its volume at constant temperature was investigated by British chemist Robert Boyle (1627–1691) and is known as **Boyle's law** in his honor.

> ✓ **Boyle's law** states that the product of the pressure times the volume of a gas is constant at constant temperature. In other words, volume is inversely proportional to pressure.

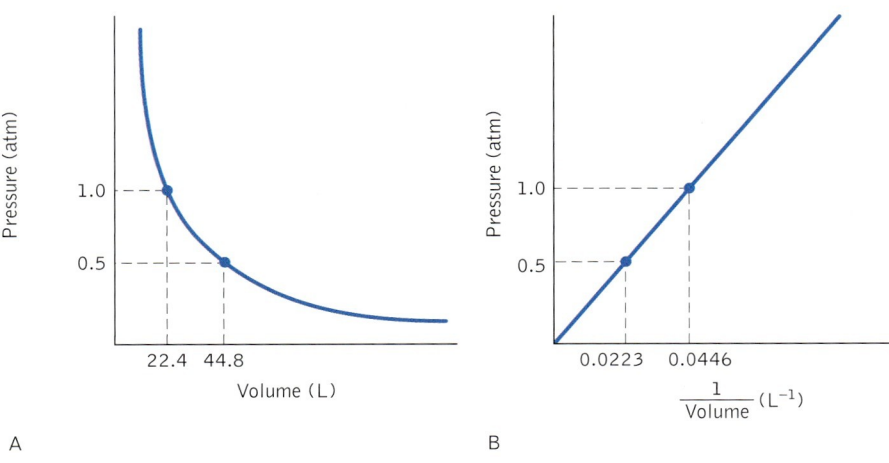

FIGURE 8.6 A. At a constant temperature of 0°C, the pressure of a mole of a gas is inversely proportional to the volume that it occupies. B. This relation means that a plot of P versus 1/V is a straight line.

If $P \times V$ does not change for a given quantity of gas at constant temperature, then any two combinations of pressure and volume will be related as follows:

$$P_1 V_1 = P_2 V_2 \qquad (8.5)$$

Equation 8.5 can be used to calculate the change in volume that would take place at constant temperature when the pressure of a quantity of gas changes or to calculate the change in pressure that is required to compress a volume of gas to a smaller value.

SAMPLE EXERCISE 8.4: A balloon is partly inflated with 5.00 liters of helium at sea level where the atmospheric pressure is 1008 mb. The balloon ascends to an altitude of 3000 meters where the pressure is 855 mb. What is the volume of the helium in the balloon? Assume that the temperature of the gas in the balloon does not change in the ascent.

SOLUTION: Equation 8.5 relates the pressure and volume of a given quantity of gas under two sets of conditions. Let sea level be condition 1, and 3000 m be condition 2. We know the values for P_1, V_1, and P_2, and we need to calculate V_2. Solving Equation 8.5 for V_2 and inserting the values given, we have

$$V_2 = \frac{P_1 V_1}{P_2}$$

$$\frac{(1008 \text{ mb})(5.00 \text{ L})}{(855 \text{ mb})} = 5.89 \text{ L}$$

PRACTICE EXERCISE: A scuba diver exhales 3.5 L of air while swimming at a depth of 20 m where the sum of atmospheric and water pressure is 3.0 atm. By the time the bubbles of air rise to the surface, where the pressure is 1.0 atm, what is their total volume? (See Problems 17–22.)

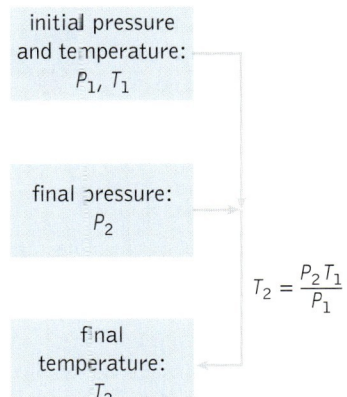

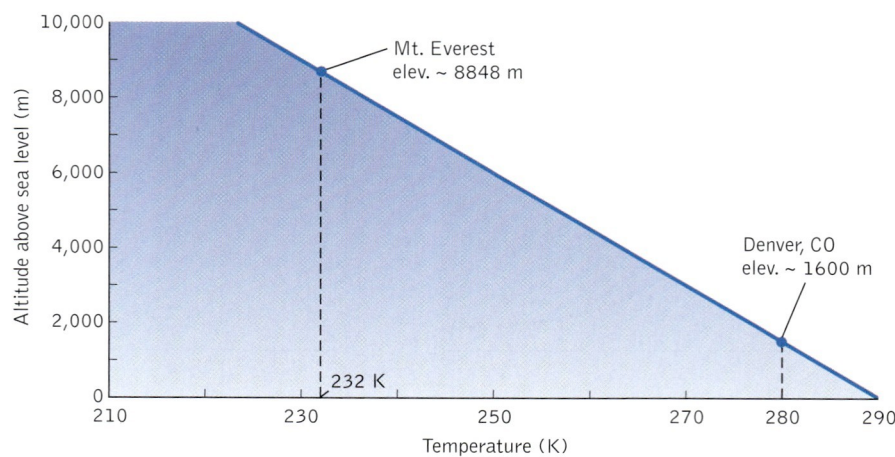

FIGURE 8.7 The temperature in the troposphere decreases with increasing elevation.

The combined gas law

Increasing altitude in the troposphere also brings a decrease in temperature, as shown in Figure 8.7. Why does temperature decrease with increasing altitude? To answer this question, let's revisit the discussion in Chapter 7 about the greenhouse effect and the absorption of infrared radiation (heat) by carbon dioxide, water vapor, and other atmospheric gases. When sunlight warms the surface of our planet, the surface radiates infrared radiation back into the atmosphere. Molecules of greenhouse gases absorb this radiant energy, and we feel the result as a warming of the air. With increasing altitude, there are fewer molecules per unit volume to trap the infrared radiation not already absorbed by the atmosphere below. With fewer molecules to trap less radiant energy, the air feels, and is, colder at higher altitudes.

> Why does temperature decrease with increasing altitude?

Temperature, like pressure, is connected to molecular motion. The speed with which molecules move increases with increasing temperature. For a given number of molecules, collisions at high temperature are more frequent and forceful than collisions at a lower temperature. Because pressure is related to the frequency and force of these collisions, it follows that higher temperatures bring higher pressures, other factors being equal. This linear relation between pressure and temperature for a given quantity of gas at constant volume can be confirmed experimentally, as shown in Figure 8.8. It can also be expressed in equation form, but only if temperature values are converted into an absolute scale, such as Kelvin:

$$P \propto T \tag{8.6}$$

Taken together, Equations 8.3 and 8.6 connect the pressure (P) exerted by n moles of gas in a given volume (V) to the absolute temperature (T) of the gas.

CONNECTION: The molecular basis for the greenhouse effect is discussed in Section 7.1.

✓ The speed of gas molecules increases with increasing temperature, causing higher pressure due to more frequent and forceful collisions between the molecules and their surroundings.

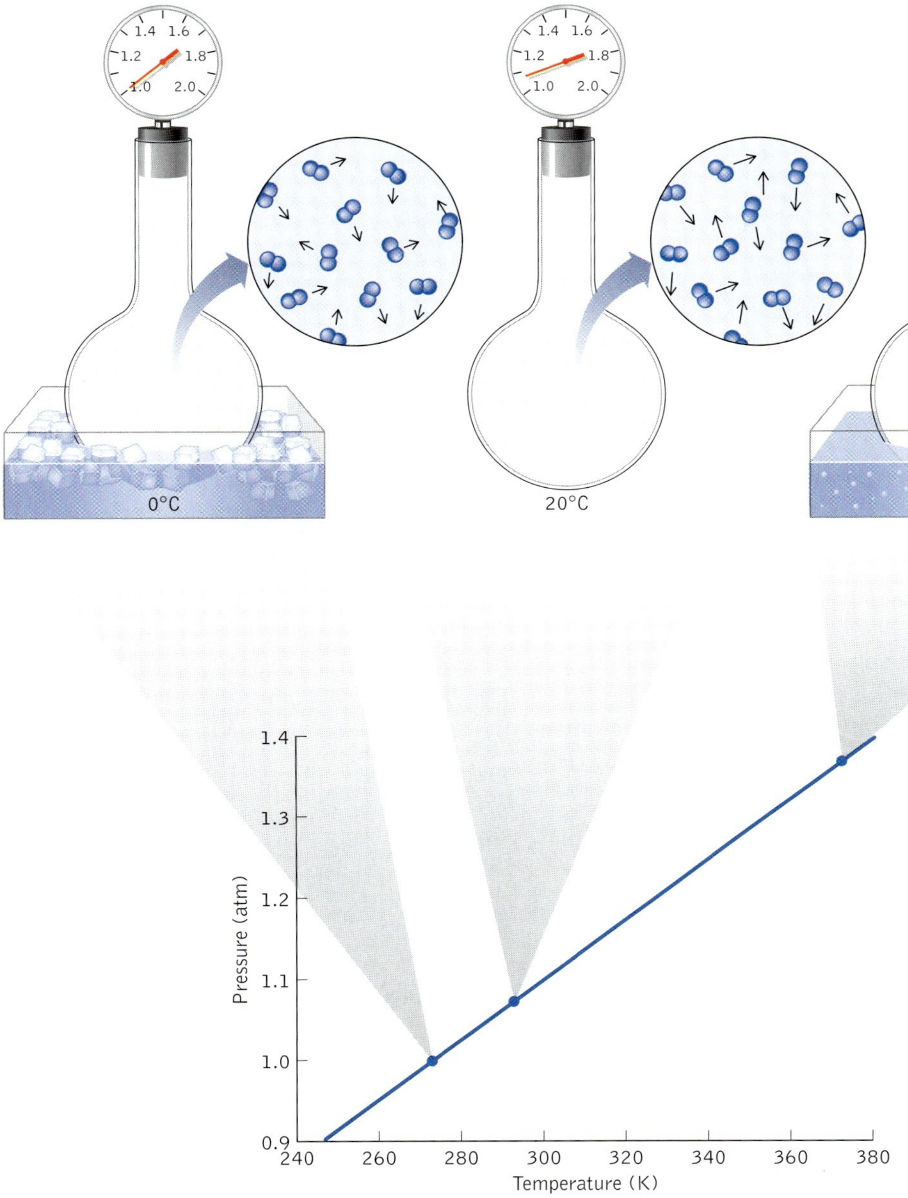

FIGURE 8.8 *What a Chemist Sees.* The pressure of a given quantity of gas is directly proportional to its absolute temperature at constant volume. Each flask has the same volume and number of molecules. The lengths of the arrows represent the relative velocities of the molecules, which increase with increasing temperature, creating more frequent and forceful collisions and thus higher pressures.

These relations can be combined into one expression of proportionality as follows:

$$P \propto T\left(\frac{n}{V}\right)$$

Rearranging the terms, we have

$$PV \propto nT \tag{8.7}$$

We can modify Equation 8.7 to correspond to the situation in which pressure, volume, and temperature change for a given quantity of gas. If n is a constant, then, using the logic that we followed in developing Equation 8.5, we have

$$n \propto \frac{PV}{T} = \text{constant}$$

so that

$$\frac{P_1 V_1}{T_1} = \frac{P_2 V_2}{T_2} \quad (8.8)$$

Equation 8.8, known as the *combined* gas law, can be used to assess the effects of changing pressure, volume, and temperature on a given quantity of a gas.

SAMPLE EXERCISE 8.5: A weather balloon filled with 100. liters of He is launched from sea level ($T = 20°C$, $P = 755$ torr). If the balloon rises to an altitude of 10 km where the temperature is $-52°C$ and atmospheric pressure is 195 torr, what is its volume?

SOLUTION: Because the quantity of gas does not change during the flight of the balloon, pressure, volume, and temperature at any point in the flight (condition 2) are linked to initial pressure, volume, and temperature (condition 1) by Equation 8.8. In this sample exercise, we are to solve for V_2. To do so, let's first rearrange the terms in Equation 8.8:

$$\frac{P_1 V_1}{T_1} = \frac{P_2 V_2}{T_2}$$

or

$$V_2 = \frac{P_1 V_1 T_2}{T_1 P_2}$$

Before inserting the values of P, V, and T from the exercise, we need to convert the given temperatures into kelvins:

$$T_1 = (20 + 273) = 293 \text{ K} \quad \text{and} \quad T_2 = (-52 + 273) = 221 \text{ K}$$

Inserting these values and the others into the expression for V_2, we have

$$V_2 = \frac{(755 \text{ torr})(100 \text{ L})(221 \text{ K})}{(293 \text{ K})(195 \text{ torr})} = 292 \text{ L}$$

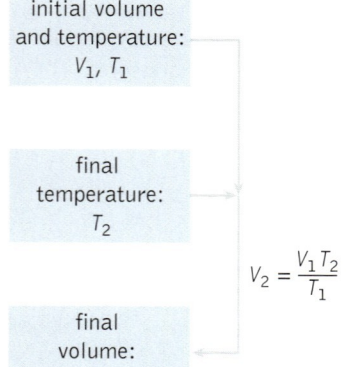

initial volume and temperature: V_1, T_1

final temperature: T_2

$V_2 = \frac{V_1 T_2}{T_1}$

final volume: V_2

DISCUSSION: The volume of the balloon increases by nearly three times as it ascends to 10 km. This result makes sense because atmospheric pressure decreases to less than a third of its sea-level value during the ascent. Volume would have increased more than three times if not for the lower temperature at 10 km.

PRACTICE EXERCISE: The balloon in the preceding sample exercise is designed to continue its ascent to an altitude of 30 km, where it bursts, releasing a package of meteorological instruments that parachute back to Earth and take measurements on the way down. If the atmospheric pressure at 30 km is 28 torr and the temperature is −45°C, what is the volume of the balloon when it bursts? (See Problems 35 and 36.)

SAMPLE EXERCISE 8.6: Labels on aerosol cans caution against incineration because the cans will explode when pressures inside them exceed 3 atm. At what temperature (in Celsius degrees) will the pressure inside an aerosol can, which is 2.20 atm above ambient at 25°C, reach 3.00 atm?

SOLUTION: In this problem, the quantity of gas and its volume are constant. Under these conditions ($V_1 = V_2$), we can simplify Equation 8.8 to

$$\frac{P_1}{T_1} = \frac{P_2}{T_2}$$

or

$$T_2 = \frac{T_1 P_2}{P_1}$$

The starting temperature and the answer are in Celsius degrees, but the equation works only for absolute temperatures. Therefore, we will need to convert from °C into K and back again. Inserting the values given in the exercise and converting T_1 from °C into K, we have

$$T_2 = \frac{(25 + 273) \text{ K } (3.00 \text{ atm})}{2.20 \text{ atm}} = 406 \text{ K}$$

Converting T_2 from K into °C gives us

$$T_2 = (406 - 273) = 133°C$$

PRACTICE EXERCISE: The air pressure in the tires of an automobile are adjusted to 28 psi (pounds per square inch) at a gas station in San Diego, California, where the air temperature is 68°F. The automobile is then driven east on I-8 toward Yuma, Arizona. Along the way, the temperature of the tires reaches 136°F. What is the pressure in the tires then? (Hint: Absolute zero on the Fahrenheit scale is −460°F.) (See Problems 37 and 38.)

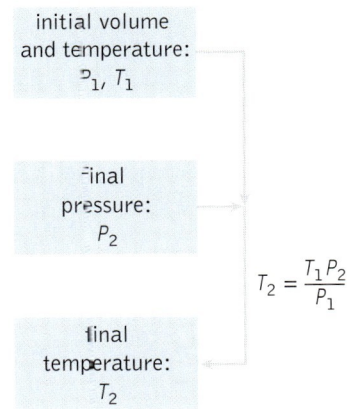

At constant pressure the volume of a given quantity of gas is directly proportional to its temperature (Figure 8.9). This proportionality is also embedded in Equation 8.8. If pressure is constant, then $P_1 = P_2$ and Equation 8.8 simplifies to

$$\frac{V_1}{T_1} = \frac{V_2}{T_2}$$

The linear relation between the volume and the temperature of a fixed quantity of gas at constant pressure was first documented by French scientist Jacques Charles (1746–1823) and the relation has come to be known as **Charles's law.** It helps explain why a balloon inflated with hot air rises. A given mass of hot air occupies more volume than does the same mass of cooler ambient air. Therefore, a balloon

✓ **Charles's law** states that the volume of a gas at constant pressure is proportional to its absolute temperature.

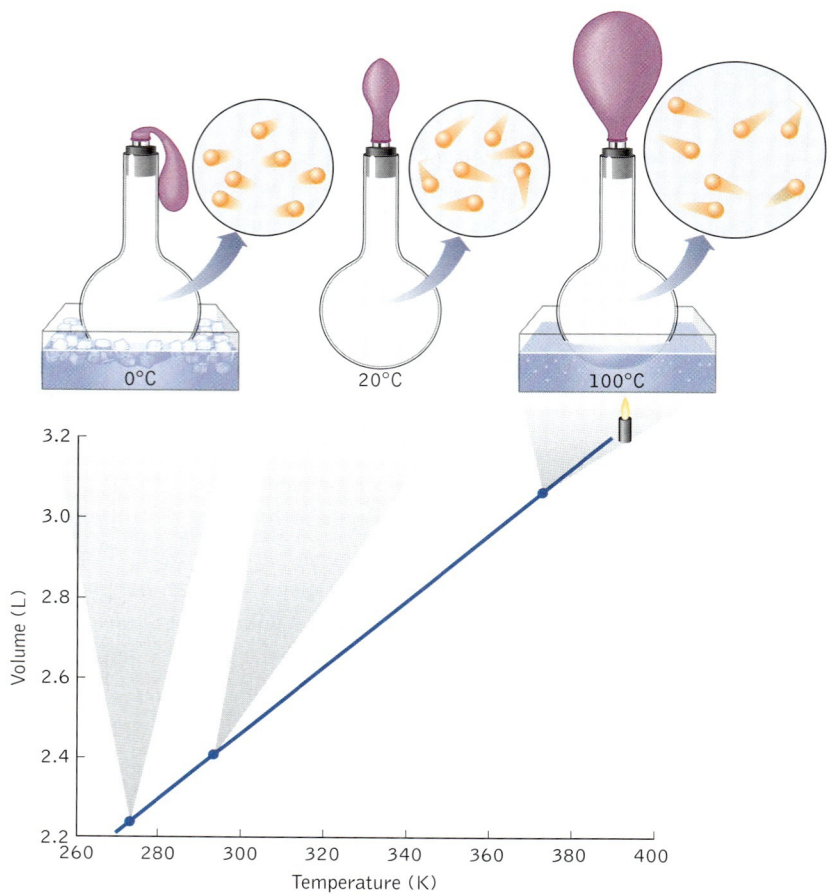

FIGURE 8.9 *What a Chemist Sees.* A deflated balloon attached to a flask inflates as the temperature of the gas inside the flask increases. This behavior is predicted by Charles's law.

filled with hot air displaces a mass of ambient air larger than its own mass, giving the balloon buoyancy. Put another way, the density of a given quantity of gas decreases with increasing temperature. It also decreases with decreasing pressure.

Before ending this discussion of the effects of changing pressure and temperature on the volumes of balloons, let's address the process of inflating a balloon. Experience tells us (and as shown in the figure on page 394) the more gas we add to a balloon, the larger its volume. This relation is described by **Avogadro's law**, which states that the volume (V) of a gas at a given temperature (T) and pressure (P) is proportional to the quantity (n) of the gas.

☑ **Avogadro's law** states that the volume (V) of a gas is proportional to the quantity (n) of the gas at a given temperature (T) and pressure (P).

8.3 THE IDEAL-GAS LAW

We can turn Equation 8.7 into an equality by replacing the proportionality symbol with an equal sign. To do so we must insert the appropriate constant of proportionality. By convention, this constant, called the *gas constant*, is given the symbol R:

$$PV = nRT \tag{8.9}$$

8.3 The Ideal-Gas Law

There are different values for R, depending on the units used for pressure, volume, and temperature (see Table 8.1). For many of our calculations, we will find it convenient to use the first value of R listed: 0.082058 L·atm/mol·K. We can use it in Equation 8.9 only when the quantity of gas is expressed in moles, volume is in liters, pressure is in atmospheres, and temperature is in kelvins.

Equation 8.9 is called the **ideal-gas equation**, and the mathematical expression of the **ideal-gas law**. It describes the relation between pressure, volume, and temperature for any gas, provided it behaves as an *ideal* gas. Most gases exhibit such behavior at the pressures and temperatures typically encountered in the atmosphere. Under these conditions, the molecules of gas have little volume compared to the volume that they occupy. All of the open space between gas molecules makes gases compressible. Gas molecules are also assumed to not interact with one another; rather, they move independently with velocities that are related to the temperature of the gas and, as we shall see, to their masses.

✓ Most gases behave as **ideal gases** at the temperatures and pressures encountered in nature. They obey the **ideal-gas equation**, which expresses the **ideal-gas law** by relating the pressure, volume, and temperature of a given quantity of gas: $PV = nRT$.

TABLE 8.1 Values for the Gas Constant (R)

Units	Value of R
L·atm/(mol·K)	0.08206
kg·m^2/s^2(mol·K)	8.314
J/(mol·K)	8.314
m^3·Pa/(mol·K)	8.314
L·torr/(mol·K)	62.36

SAMPLE EXERCISE 8.7: The bottles of compressed O_2 carried by alpine climbers have an internal volume of 5.9 L. Such a bottle was filled with O_2 to a pressure of 2000. psi at 25.°C. How many moles of O_2 are in the bottle? What is the mass of O_2 in the bottle in grams? Given: 1 atm = 14.7 psi. Assume that O_2 behaves as an ideal gas.

SOLUTION: We are given the pressure, volume, temperature, and identity of an ideal gas and asked to calculate its mass. The ideal-gas equation enables us to use P, V, and T to calculate n, the number of moles O_2. We can use its molar mass (32.00 g/mol) to calculate the mass of O_2 in the bottle. Let's start with the ideal-gas equation and rearrange the terms to solve for n:

$$PV = nRT$$

or

$$n = \frac{PV}{RT}$$

Before using this expression for n, we need to convert pressure into atmospheres and temperature into kelvins.

$$P = 2000.\ \text{psi} \left(\frac{1\ \text{atm}}{14.7\ \text{psi}}\right) = 136.\ \text{atm}$$

$$T = 25.°C + 273. = 298\ K$$

Inserting these values and the volume given in the above expression for n, we have

$$n = \frac{(136.\ \text{atm})(5.9\ \text{L})}{(0.8206\ \text{L}\cdot\text{atm/mol}\cdot\text{K})(298\ \text{K})} = 32.8\ \text{mol}$$

Converting moles into grams is a matter of multiplying by the molar mass:

$$(32.8\ \text{mol})\left(\frac{32.00\ \text{g}}{\text{mol}}\right) = 1050.\ \text{g}$$

Thus, each bottle is filled with about 1 kg, or 2.2 lb, of oxygen.

pressure, volume, and temperature:
V, P, T

$\downarrow\ n = \frac{PV}{RT}$

n_{O_2}

$\downarrow\ $ mass $O_2 = n_{O_2} \mathcal{M}_{O_2}$

mass of O_2

PRACTICE EXERCISE: Starting with the moles of O_2 calculated in the preceding sample exercise, calculate the volume of O_2 that the bottle could deliver to a climber near the top of Mount Everest. Assume the temperature there is $-38°C$ and atmospheric pressure is 0.35 atm. (See Problems 39–46.)

SAMPLE EXERCISE 8.8: Oxygen generators in some airplanes are based on a chemical reaction between sodium chlorate and iron:

$$NaClO_3(s) + Fe(s) \longrightarrow O_2(g) + NaCl(s) + FeO(s)$$

How many grams of $NaClO_3$ would be needed to produce 120 liters of O_2 at 1.00 atm and 20°C?

CONNECTION: For a review of stoichiometry calculations, refer to Section 4.6.

SOLUTION: We may assume that oxygen behaves as an ideal gas under the given temperature and pressure, and so the ideal-gas law applies:

$$PV = nRT$$

Solving for n,

$$n = PV/RT$$

$$n = \frac{(1.00 \text{ atm})(120 \text{ L})}{(0.08206 \text{ L} \cdot \text{atm/mol} \cdot \text{K})(273+20) \text{ K}}$$

$$= 5.0 \text{ mol } O_2$$

To convert moles of O_2 into an equivalent mass of $NaClO_3$, we need to first use the stoichiometry of the reaction to calculate the equivalent number of moles of $NaClO_3$. Then we will convert moles of $NaClO_3$ into grams of $NaClO_3$ by multiplying by the molar mass of $NaClO_3$. Stringing these two steps together, we have

$$(5.0 \text{ mol } O_2) \left(\frac{1 \text{ mol NaClO}_3}{1 \text{ mol } O_2} \right) \left(\frac{106.49 \text{ g NaClO}_3}{\text{mol NaClO}_3} \right) = 5.3 \times 10^2 \text{ g NaClO}_3$$

pressure, temperature, and volume:
P, V, T

$n = \dfrac{PV}{RT}$

n_{O_2}

$n_{NaClO_3} = n_{O_2} \times \dfrac{1 \text{ mol NaClO}_3}{1 \text{ mol } O_2}$

n_{NaClO_3}

mass $NaClO_3 = n_{NaClO_3} \mathcal{M}_{NaClO_3}$

mass of $NaClO_3$

PRACTICE EXERCISE: Automobile air bags inflate during a crash by the rapid generation of nitrogen gas from sodium azide according to the following reaction:

$$2 \text{ NaN}_3 \longrightarrow 2 \text{ Na} + 3 \text{ N}_2$$

How many grams of sodium azide are needed to fill a $40 \times 40 \times 20$ cm bag to a pressure of 1.20 atm and a temperature of 15°C? (See Problems 47–50.)

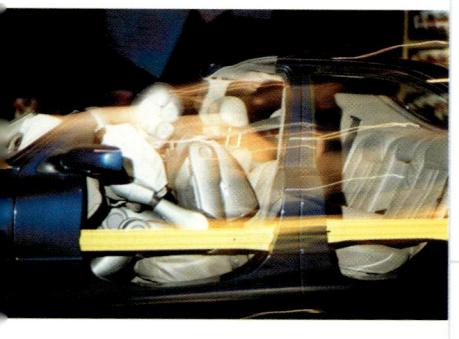

An automobile air bag inflates when solid NaN_3 rapidly decomposes, producing N_2 gas.

A handy reference point in studying the properties of gases is **standard temperature and pressure** (STP), which is 0°C and 1 atm. Another useful reference point is the volume that a mole of ideal gas occupies at STP. We can calculate this volume from the ideal-gas equation:

or
$$PV = nRT$$
$$V = \frac{nRT}{P}$$

and, then, inserting the values of n, P, and T:

$$V = \frac{(1 \text{ mol})(0.08206 \text{ L·atm/mol·K})(273 \text{ K})}{1 \text{ atm}}$$

$$= 22.4 \text{ L}$$

This volume is known as the **molar volume** of an ideal gas. It is a useful reference because many chemical and biochemical processes take place at pressures near 1 atm and at temperatures between 0°C and 40°C. Under these conditions, the volume that a mole of gaseous reactant or product occupies will be no more than about 15% greater than the molar volume.

> ✓ **Standard temperature and pressure** (STP) are 0°C and 1 atmosphere of pressure. The **molar volume** (the volume of 1 mole of an ideal gas) at STP is 22.4 L. Note that the standard state of a gas (discussed in Chapter 11) was once defined as 1 atm, but is now 1 bar or 10^5 Pa (1 atm = 1.013 bar).

8.4 GAS DENSITY

An important feature of molar volume is that it applies to all ideal gases, and, at STP, most gases do behave as ideal gases. Thus the density of a gas at STP can be easily calculated from its molar mass. Carbon dioxide, for example, has a molar mass of 44.0 g/mol. Therefore, the density of CO_2 at STP will be

$$\frac{44.0 \text{ g/mol}}{22.4 \text{ L/mol}} = 1.96 \text{ g/L}$$

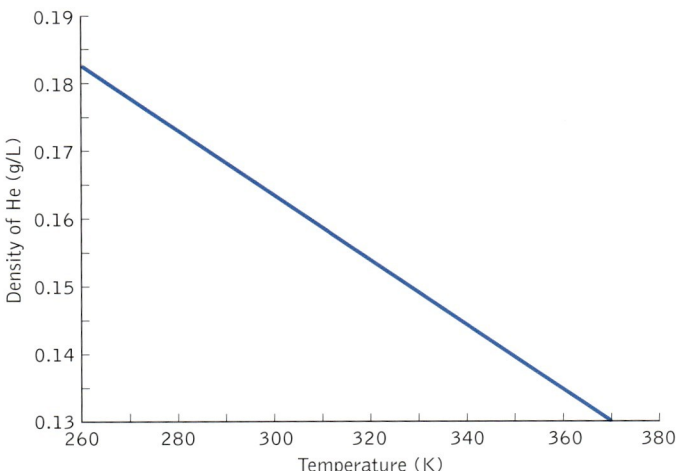

At 1.00 atm of pressure, the density of helium decreases with increasing temperature. Similarly, the density of any ideal gas at constant pressure is inversely proportional to its temperature.

CONCEPT TEST: Which of the following gases is the most dense at STP: CO_2, CH_4, Cl_2, Kr, or C_3H_8?

CONCEPT TEST: Which of the following changes will result in the greater decrease in the density of a gas: (1) decreasing its pressure from 2.00 atm to 1.00 atm or (2) increasing its temperature from 20°C to 40°C?

The density of an ideal gas can be calculated for any combination of temperature and pressure by using the ideal-gas equation. Let's start by rearranging the terms in the equation so that the ratio of amount (n) to volume (V) is on one side:

$$\frac{P}{RT} = \frac{n}{V}$$

> The density of an ideal gas increases with increasing pressure and decreasing temperature.

If we multiply both sides of this equation by the molar mass of the gas ($\mathcal{M}$), then the numerator on the right-hand side will have the units of $\left(\text{mol} \times \frac{\text{g}}{\text{mol}}\right)$, or grams. Mass in grams divided by volume in liters gives us the density of the gas in grams per liter:

$$\frac{\mathcal{M}P}{RT} = \frac{n\mathcal{M}}{V} = d$$

or

$$d = \frac{\mathcal{M}P}{RT} \tag{8.10}$$

SAMPLE EXERCISE 8.9: In August, 1986, 1700 people died from asphyxiation in the Lake Nyos valley, Cameroon, as a result of the lake's massive release of carbon dioxide that blanketed the valley. Calculate the density of CO_2 and of air at 1.00 atm of pressure and 300 K. Assume that air has an average molar mass of 28.8 g/mol.

SOLUTION: Inserting the values P and T and $\mathcal{M}$ for CO_2 and air in Equation 8.10, we have

> pressure, temperature and molar mass:
> $P, T, \mathcal{M}$
> $d = \frac{\mathcal{M}P}{RT}$
> ↓
> density of CO_2

$$d_{CO_2} = \frac{\mathcal{M}P}{RT} = \frac{(44.0 \text{ g/mol})(1.00 \text{ atm})}{(0.08206 \text{ L·atm/mol·K})(300 \text{ K})} = 1.79 \text{ g/L}$$

$$d_{air} = \frac{\mathcal{M}P}{RT} = \frac{(28.8 \text{ g/mol})(1.00 \text{ atm})}{(0.08206 \text{ L·atm/mol·K})(300 \text{ K})} = 1.17 \text{ g/L}$$

DISCUSSION: Carbon dioxide is clearly more dense than air. The release of CO_2 from Lake Nyos killed so many because the dense gas flowed from the lake and over the valley floor, effectively displacing the air (and oxygen) that the inhabitants and their livestock needed to survive.

PRACTICE EXERCISE: Air is a mixture of mostly nitrogen and oxygen. Will a balloon filled with oxygen sink to the floor or float to the ceiling if released in a room full of air? (See Problems 55 and 56.)

We can rearrange the terms in Equation 8.10 to solve for molar mass:

$$\mathcal{M} = \frac{dRT}{P} \tag{8.11}$$

Equations 8.11 can be used to calculate the molar mass of a gas from its density at any temperature and pressure (see the following sample exercise). Gas density can be determined with a specially designed glass bulb of known volume that can be attached to a vacuum pump. The mass of the bulb is determined when it has essentially no gas in it and, again, when it is filled with the test gas. The difference in the masses divided by the internal volume of the bulb is the density of the gas.

 The molar mass of an ideal gas can be calculated from its density: $\mathcal{M} = dRT/P$.

SAMPLE EXERCISE 8.10: Vent pipes at solid waste landfills often emit foul-smelling gases. A sample of such a gas is found to have a density of 0.65 g/L at 25°C and 757 mm Hg. What is the (average) molar mass of the gas?

SOLUTION: Equation 8.11 may be used to calculate molar mass, provided we convert temperature from Celsius degrees into kelvins:

$$T = 25.°C + 273. = 298. \text{ K}$$

and pressure from mm Hg into atm:

$$P = (757. \text{ mm Hg})(1 \text{ atm}/760 \text{ mm Hg}) = 0.996 \text{ atm}$$

Inserting these values into Equation 8.11, we have

$$\mathcal{M} = dRT/P = (0.65 \text{ g/L})(0.08206 \text{ L·atm/mol·K})(298. \text{ K})/$$
$$(0.996 \text{ atm}) = 16.0 \text{ g/mol}$$

pressure, density, and temperature: d, P, T

$$\downarrow$$

$$\mathcal{M} = \frac{dRT}{P}$$

molar mass $\mathcal{M}$

DISCUSSION: We have no proof that the sample consists of only one gaseous substance; so the calculated value may be a weighted average of all the gases coming out of the vent pipe. It happens that a principal ingredient in the gases emitted by decomposing solid waste is methane (CH_4), which has a molar mass of 16.0 g/mol.

PRACTICE EXERCISE: When solutions of hydrochloric acid (HCl) and sodium bicarbonate ($NaHCO_3$) are mixed together, a chemical reaction takes place in which a gas is one of the products. A sample of the gas has a density of 1.81 g/L at 1.00 atm and 23°C. What is the molar mass of the gas? Can you identify the gas? (See Problems 59–62.)

8.5 DALTON'S LAW AND MIXTURES OF GASES

We have noted that air is a mixture of mostly N_2 and O_2 and smaller proportions of a variety of other gases (see Table 7.1). Atmospheric pressure is the sum of the pressures that each of these gases exerts by itself. These individual contributions to total pressure are called **partial pressures.** Atmospheric pressure is the sum of the partial pressures of all of the gases in the air:

$$P_{\text{atmospheric}} = P_{N_2} + P_{O_2} + P_{Ar} + P_{CO_2} + \cdots$$

A similar expression can be written for any mixture of gases:

$$P_{\text{total}} = P_1 + P_2 + P_3 + P_4 + \cdots \quad (8.12)$$

Summing partial pressures to calculate the total pressure of a mixture of gases is known as **Dalton's law** of partial pressures.

You might guess that the most abundant gases in a mixture have the greatest partial pressures and contribute the most to the total pressure of the mixture. The mathematical term used to express the abundance of each component is its **mole fraction** (X), which is the ratio of the moles of component to the total number of moles in the mixture:

$$X_x = \frac{n_x}{n_{\text{total}}} \quad (8.13)$$

CONNECTION: The composition of the atmosphere is described in Section 7.1.

Dalton's law states that the total pressure of a mixture of gases equals the sum of the **partial pressures** of each of the gases in the mixture.

Mole fraction is the ratio of the number of moles of a component of a mixture to the total number of moles of all the components of the mixture.

To see how mole fractions work, consider a portion of the atmosphere that contains 100.0 moles of atmospheric gases. In this 100.0 moles, there are 20.9 moles of O_2 and 78.1 moles of N_2. The mole fraction of O_2 (X_{O_2}) in air is the ratio of the moles of O_2 to the total number of moles in the mixture:

$$X_{O_2} = n_{O_2}/n_{\text{total}} = (20.9 \text{ mol})/(100 \text{ mol}) = 0.209$$

Similarly, the mole fraction of N_2 is

$$X_{N_2} = n_{N_2}/n_{\text{total}} = (78.1 \text{ mol})/(100 \text{ mol} = 0.781$$

CONCEPT TEST: The values for the mole fractions of N_2 and O_2 in the atmosphere are the same as their percentage-by-volume values in Table 7.1. Explain why these values are the same.

The partial pressure of O_2 (P_{O_2}) in air is the product of the mole fraction O_2 times total atmospheric pressure. At a total pressure of 1.00 atm, the partial pressure of O_2 is

$$P_{O_2} = X_{O_2} P_{\text{total}} = (.209)(1.00 \text{ atm}) = 0.209 \text{ atm}$$

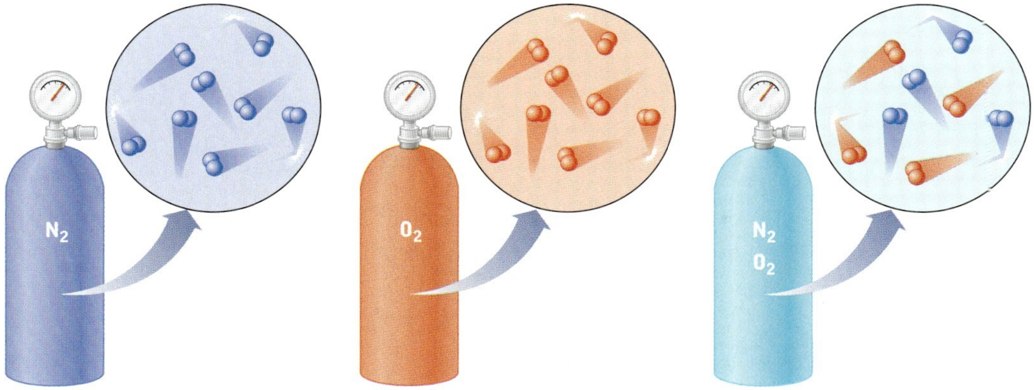

FIGURE 8.10 *What a Chemist Sees.* Pressure is directly proportional to the number of moles of an ideal gas but is independent of the chemical composition of the gas. Three gas tanks of equal volume are filled with equal numbers of moles of N_2, O_2, and a mixture of the two gases. The pressure inside all three tanks is the same.

and the partial pressure of N_2 is

$$P_{N_2} = X_{N_2} P_{total} = (.781)(1.00 \text{ atm}) = 0.781 \text{ atm}$$

The general equation for the partial pressure of a gas in a mixture of gases is

$$P_x = X_x P_{total} \tag{8.13}$$

Thus, the partial pressure of a gas is proportional to the quantity of the gas in a given volume and does not depend on the identity of the gas (See Figure 8.10).

SAMPLE EXERCISE 8.11: Scuba divers intending to descend below 130 feet may breathe a gas mixture called Trimix, which is 11.7% He, 56.2% N_2, and 32.1% O_2 by mass. Calculate the mole fraction of each gas in this mixture.

SOLUTION: To calculate the mole fraction, we need to determine the number of moles of each gas present in a given quantity of the gas mixture. One hundred grams of Trimix will contain 11.7 g of He, 56.2 g of N_2, and 32.1 g of O_2. Using the molar mass of each gas to convert these quantities into moles:

$$11.7 \text{ g He} \left(\frac{1 \text{ mol He}}{4.003 \text{ g He}} \right) = 2.91 \text{ mol He}$$

$$56.2 \text{ g } N_2 \left(\frac{1 \text{ mol } N_2}{28.01 \text{ g } N_2} \right) = 2.01 \text{ mol } N_2$$

$$32.1 \text{ g } O_2 \left(\frac{1 \text{ mol } O_2}{32.00 \text{ g } O_2} \right) = 1.00 \text{ mol } O_2$$

Mole fraction values are calculated using Equation 8.13: $X_x = n_x/n_{total}$. The total number of moles in this exercise is the sum of the moles of He, N_2, and O_2:

$$n_{total} = 2.91 + 2.01 + 1.00 = 5.92 \text{ mol}$$

Thus, X_{He} and X_{N_2} are:

$$X_{He} = \frac{2.91 \text{ mol He}}{5.92 \text{ mol}} = 0.49$$

$$X_{N_2} = \frac{2.01 \text{ mol N}_2}{5.92 \text{ mol}} = 0.34$$

We could solve for X_{O_2} in a similar fashion; however, the sum of the mole fractions of all of the components of a mixture must equal 1:

$$X_{He} + X_{N_2} + X_{O_2} = 1$$

So,

$$X_{O_2} = 1 - (X_{He} + X_{N_2}) = 1 - (0.49 + 0.34) = 0.17$$

PRACTICE EXERCISE: A different gas mixture, 6.11% O_2 and 93.89% He by mass, is known as Heliox and also is used in scuba tanks for descents below 190 feet. Calculate the mole fraction of He and O_2 in this mixture. (See Problems 67 and 68.)

Let's consider the implications of Equation 8.14 in the context of the "thin" air at high altitudes. The mole fraction of oxygen in air is 0.209 and does not change significantly with increasing altitude in the troposphere. However, the total pressure of the atmosphere and, therefore, the partial pressure of oxygen decrease with increasing altitude, as illustrated in the following exercises.

SAMPLE EXERCISE 8.12: Calculate the partial pressure (atm) of O_2 in the air outside an airplane cruising at an altitude of 33,000 feet (about 10 km) where the atmospheric pressure is 190 mm Hg. Assume that the mole fraction of O_2 in air is 0.209.

SOLUTION: Inserting the values given into Equation 8.14, we have

$$P_{O_2} = X_{O_2} P_{total}$$

$$= (0.209)(190. \text{ mm Hg})\left(\frac{1 \text{ atm}}{760 \text{ mm Hg}}\right) = 0.0522 \text{ atm}$$

total pressure, mole fraction:
P, X

$P_{O_2} = X_{O_2} P_{total}$

partial pressure of O_2

PRACTICE EXERCISE: If a scuba diver is working at a depth (about 50 m) at which the total pressure is 5.0 atm, what mole fraction of oxygen in an artificial gas mixture will provide $P_{O_2} = 0.21$ atm in the gas that the diver breathes? (See Problems 69 and 70.)

Dalton's law of partial pressures is used in the laboratory when a gaseous product of a chemical reaction is collected in an inverted vessel filled with water as shown in Figure 8.11. For example, heating potassium chlorate ($KClO_3$)

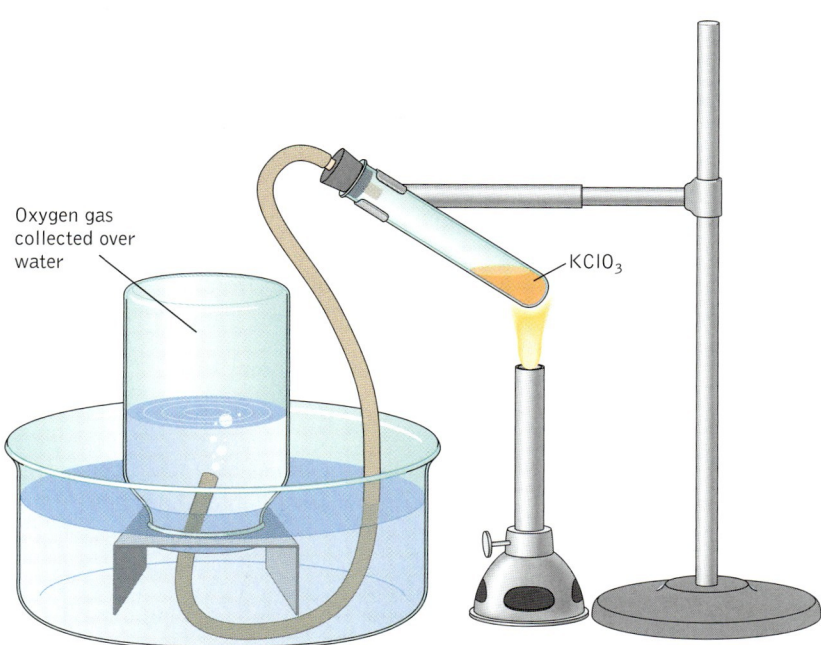

FIGURE 8.11 During the thermal decomposition of $KClO_3$, O_2 is collected by displacing water from an inverted bottle initially filled with water.

causes it to decompose into KCl and O_2. The oxygen gas produced by this reaction can be collected by bubbling the gas into an inverted bottle that is initially filled with water. As the reaction proceeds, O_2 displaces the water in the bottle. When the reaction is complete, the volume of water displaced provides a measure of the volume of O_2 produced. If the temperature of the water and the barometric pressure are known, then the moles of O_2 produced can be calculated by using the ideal-gas law.

The procedure works for oxygen over water or for any gas that neither reacts with nor dissolves in the liquid being displaced. However, one additional step is needed to calculate the moles of O_2 produced. At room temperature, water has a significant vapor pressure (see Section 9.3). Thus the gas collected will be a mixture of mostly O_2 but with some water vapor. The total pressure of the mixture will be

$$P_{total} = P_{atmospheric} = P_{O_2} + P_{H_2O}$$

To calculate the quantity of oxygen produced (n_{O_2}), we must first subtract P_{H_2O} (Table 8.2) from $P_{atmospheric}$ to calculate P_{O_2}. If we know the values of P_{O_2}, T, and V, we can calculate n_{O_2} by using the ideal-gas equation, as illustrated in the following sample exercise.

TABLE 8.2
Vapor Pressure of Water at Several Temperatures

Temperature (°C)	Pressure (mm Hg)
5.	6.5
10.	9.2
15.	12.8
20.	17.5
25.	23.8
30.	31.8
35.	42.2
40.	55.3
45.	71.9
50.	92.5

CONNECTION: Vapor pressure is discussed in more detail in Section 9.3.

SAMPLE EXERCISE 8.13: During the decomposition of $KClO_3$ (see Figure 8.11), 92 mL of O_2 is collected by displacement of water at 25°C. If atmospheric pressure is 756 mm Hg, what mass of O_2 is collected?

SOLUTION: Collecting a gas over water means that we need to account for the presence of water vapor in the collected gas. These considerations lead to the following three steps for solving this problem:

total pressure and vapor pressure of H_2O:
P

$P_{O_2} = P_{total} - P_{H_2O}$

partial pressure of O_2

volume and temperature:
V, T

$n_{O_2} = \dfrac{PV}{RT}$

moles of O_2

mass = $n_{O_2} \mathcal{M}_{O_2}$

mass of O_2

1. Calculate P_{O_2} in the collected gas by subtracting from P_{total} the vapor pressure of water (P_{H_2O}) at 25°C (24 mm Hg; see Table 8.2):

$$P_{O_2} = P_{total} - P_{H_2O} = (756 - 24) \text{ mm Hg}$$
$$= 732 \text{ mm Hg}$$

2. Use the ideal-gas equation to calculate the moles of O_2 (n_{O_2}) produced (we will also need to convert P_{O_2} into atmospheres, V into liters, and T into kelvins):

$$n_{O_2} = \dfrac{PV}{RT} = \dfrac{(732 \text{ mm Hg})\left(\dfrac{1 \text{ atm}}{760 \text{ mm Hg}}\right)(92 \text{ mL})\left(\dfrac{1 \text{ L}}{1000 \text{ mL}}\right)}{\left(\dfrac{0.08206 \text{ L} \cdot \text{atm}}{\text{mol} \cdot \text{K}}\right)(25 + 273) \text{ K}}$$

$$= 3.62 \times 10^{-3} \text{ mol}$$

3. Calculate the mass of O_2 by multiplying n_{O_2} by the molar mass of O_2:

$$n_{O_2} \mathcal{M}_{O_2} = (3.62 \times 10^{-3} \text{ mol})\left(\dfrac{32.00 \text{ g}}{\text{mol}}\right)$$

$$= 0.116 \text{ g}$$

PRACTICE EXERCISE: Electrical energy can be used to separate water into O_2 and H_2. In one demonstration of electrolysis, 27 mL of H_2 was collected over water at 25°C. Atmospheric pressure was 761 mm Hg. How many grams of H_2 were collected? (See Problems 71 and 72.)

8.6 HENRY'S LAW AND THE SOLUBILITY OF GASES

In Section 8.2, we considered the decreases in atmospheric pressure and temperature that occur with increasing altitude in the troposphere. These trends have a major effect on those hardy souls who choose to climb the highest mountains on Earth. Lower partial pressures of oxygen at high altitudes result in low concentrations of oxygen in the blood and tissues of people who visit those altitudes. Low blood oxygen causes climbers to become weak and unable to think clearly—

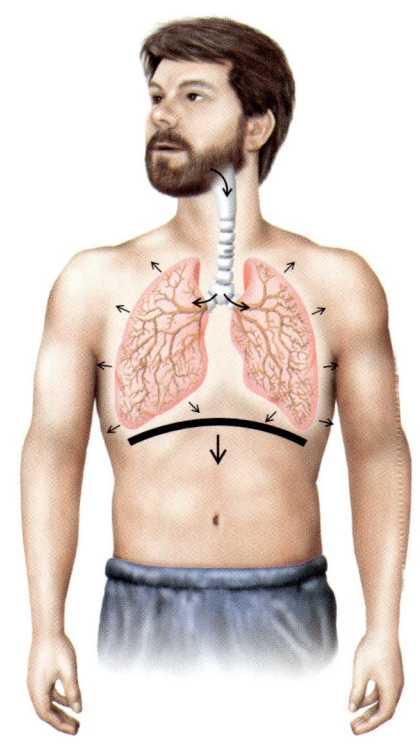

Breathing illustrates the relation between P, V, and n. When you inhale, your rib cage expands and your diaphragm moves down, increasing the volume (V) of your lungs. Increased volume decreases the pressure inside the lungs, in accord with Boyle's law. Decreased pressure allows more air to enter until the pressure inside your lungs matches atmospheric pressure. At high altitudes, atmospheric pressures are low and so is the pressure in the lungs. Therefore the quantity (n) of air (as well as O_2) that enters the lungs is smaller than normal.

symptoms of a condition known as anoxia. Anoxia develops because the solubility of O_2 in our blood and tissues is proportional to its partial pressure (Figure 8.12). This relation applies to all gases and is known as **Henry's law**. In equation form, it is written as follows:

$$C_{gas} = k_H P_{gas} \tag{8.15}$$

where C_{gas} represents the dissolved concentration of gas and k_H is the Henry's-law constant for the gas in that particular solvent. When dissolved concentrations are expressed in molarity, a common set of units for Henry's-law constant is moles per liter-atmosphere. Equivalent values, based on moles of gas per kilogram of liquid, are listed in Table 8.3.

Henry's law may lead you to assume that the concentration of dissolved oxygen in a person's blood is directly proportional to the partial pressure of oxygen and so directly proportional to atmospheric pressure. This assumption would mean that residents of Denver, where the average atmospheric pressures is 0.85 atm, would be forced to live with less blood oxygen than did the residents of New York City or Los Angeles. This is not the case. The reason why is connected to molecular interactions between dissolved oxygen and a substance called hemoglobin. The human body uses hemoglobin, a large iron-containing protein found in red blood cells, to bind oxygen breathed in through the lungs and transport it

> ✓ **Henry's law** states that the solubility of a gas in a liquid is proportional to the partial pressure of the gas.

FIGURE 8.12 Henry's law predicts that the solubility of a gas is directly proportional to its pressure. The higher pressure in the container at the right produces an increase in the quantity of gas dissolved in the liquid.

increase P by increasing n

CONNECTION: The response of the body to living at high elevation is an example of a shift in chemical equilibrium, which will be discussed in Section 15.5.

to cells (only about 3% of the oxygen in blood is actually freely dissolved O_2). Our supply of blood oxygen is related to the concentration of hemoglobin in the blood and to the fraction of the oxygen-binding sites on hemoglobin that are occupied by oxygen as the blood leaves the lungs. This *saturation* of binding sites does depend on the partial pressure of oxygen (Figure 8.13), as well as proper lung function. For most people, breathing air with $P_{O_2} > 85$ mm Hg results in nearly 100% saturation of the oxygen-binding sites of hemoglobin after their blood has circulated through their lungs. Moreover, the body responds to lower oxygen partial pressures by producing more red blood cells and more hemoglobin. This increase in O_2 carriers compensates for the lower partial pressure of O_2.

TABLE 8.3 Henry's-Law Constants for the Solubilities of Some Gases in Water at 20°C

Gas	k_H (mol/L·atm)	k_H (mol/kg·mm Hg)
He	3.5×10^{-4}	4.6×10^{-7}
O_2	1.3×10^{-3}	1.7×10^{-6}
N_2	6.7×10^{-4}	8.8×10^{-7}
CO_2	3.5×10^{-2}	4.6×10^{-5}

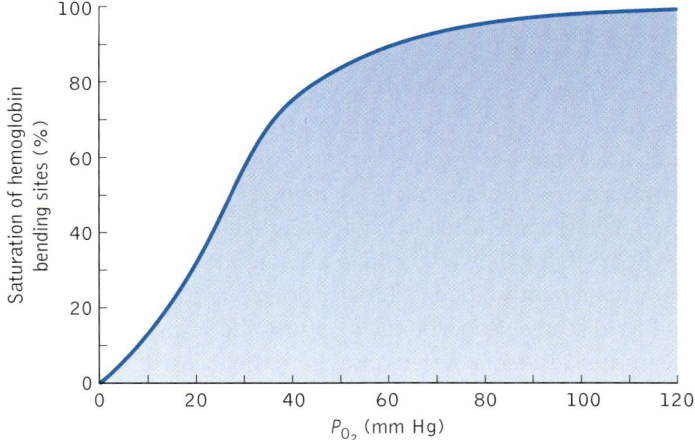

FIGURE 8.13 The proportion of hemoglobin binding sites in blood that carry oxygen increases with increasing partial pressure of O_2.

SAMPLE EXERCISE 8.14: Calculate the solubility of oxygen in water at 1.00 atm of atmospheric pressure and 20°C. The mole fraction of O_2 in air is 0.209.

SOLUTION: Henry's law (Equation 8.15) describes the relation between the solubility of oxygen and its partial pressure. Partial pressure is related to the mole fraction of a gas and the total pressure (Equation 8.14):

$$P_{O_2} = X_{O_2} P_{total} = (0.209)(1.00 \text{ atm}) = 0.209 \text{ atm}$$

Using this value for P_{O_2} and k_H for O_2 in water (see Table 8.3) in Equation 8.15 gives

$$C_{O_2} = k_H P_{O_2} = \left(\frac{1.3 \times 10^{-3} \text{ mol}}{\text{L} \cdot \text{atm}} \right) (0.209 \text{ atm}) = 2.7 \times 10^{-4} \text{ mol/L}$$

pressure of O_2

$C_{O_2} = k_H P_{O_2}$

dissolved concentration of O_2

PRACTICE EXERCISE: Calculate the solubility of oxygen in water at 20°C and an atmospheric pressure of 0.35 atm (a typical pressure at the top of Mt. Everest). The mole fraction of O_2 in air is 0.209. (See Problems 89 and 90.)

Scuba divers must cope with high concentrations of dissolved gases caused by breathing air at the high underwater pressures that they encounter. The pressure around a diver increases by about 1 atm with every 33-foot (10-m) increase in depth. Thus, the air in the lungs of a scuba diver at a depth of 10 m is at twice the pressure (and is twice as dense) as at the surface. Increased pressure increases the solubility of atmospheric gases (Figure 8.14). After diving for a time at depths of 20 m or more, divers should not come quickly to the surface, because

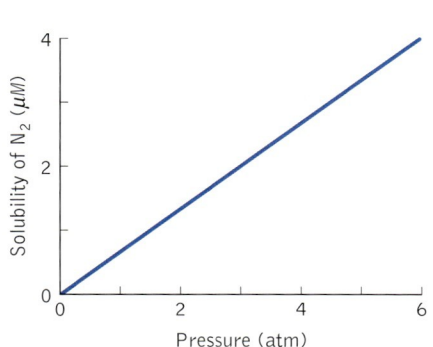

FIGURE 8.14 The pressure of the air in the lungs of a scuba diver increases with increasing depth. Increased pressure means that more N_2 and O_2 dissolve in the diver's blood and tissue.

THE CHEMISTRY OF THE GROUP 5A ELEMENTS

Nitrogen is the most abundant element in the atmosphere, making up 78% of it by volume, and the sixth most abundant element in the universe. However, it is a relatively minor constituent of Earth's crust (only about 0.003% N by mass). Nitrogen exists in the crust mostly in deposits of potassium nitrate, KNO_3, or sodium nitrate, $NaNO_3$. Common names for KNO_3 include saltpeter or niter. The latter name is the source of the name of the element itself: in Greek *nitro* and *gen* mean "niter forming." In living systems, nitrogen makes up 16% (by mass) of proteins.

In the early twentieth century, mined nitrates were widely used in the manufacture of gunpowder and other explosives. For example, a mixture of potassium nitrate, sulfur, and carbon can react exothermically and explosively, producing nitrogen and carbon dioxide. Rapid expansion of these hot gases adds to the explosive character of the reaction:

$$2\ KNO_3(s) + S(s) + 3\ C(s) \longrightarrow$$
$$K_2S(s) + N_2(g) + 3\ CO_2(g)$$

When naval blockades cut off Germany's supplies of KNO_3 in World War I, German scientist Fritz Haber developed a process for making these compounds, starting with the synthesis of ammonia from hydrogen gas and nitrogen from the atmosphere. The Haber process relies on high temperature and pressure to promote the reaction:

$$N_2(g) + 3\ H_2(g) \longrightarrow 2\ NH_3(g)$$

In this reaction the strong N≡N triple bond in N_2 must be broken, which requires an energy investment of 941 kJ/mol. (In nature, bacteria living in the roots of leguminous plants—such as peas, beans, and peanuts—harness biochemical energy to break N≡N triple bonds and form chemical compounds from atmospheric nitrogen.) Today, the Haber process is the principal source of ammonia and the nitrogen compounds derived from ammonia that are widely used in industry and agriculture as described in Chapter 15.

The first step in converting ammonia into other important nitrogen compounds entails its oxidation to NO_2 and H_2O:

$$4\ NH_3(g) + 5\ O_2(g) \longrightarrow 4\ NO(g) + 6\ H_2(g)$$

$$2\ NO(g) + O_2(g) \longrightarrow 2\ NO_2(g)$$

Nitrogen dioxide dissolves in water, producing a mixture of nitrous and nitric acids:

$$2\ NO_2(g) + H_2O(l) \longrightarrow HNO_2(aq) + HNO_3(aq)$$

Heating the mixture converts nitrous acid into nitric acid as steam and NO are given off:

$$3\ HNO_2(aq) \longrightarrow HNO_3(aq) + H_2O(g) + 2\ NO(g)$$

About 75% of nitric acid produced in this way is combined with more ammonia to produce ammonium nitrate:

$$NH_3(g) + HNO_3(l) \longrightarrow NH_4NO_3(s)$$

Ammonium nitrate is an important industrial chemical and the source of water-soluble nitrogen in many formulations of fertilizers.

Just below nitrogen in Group 5A of the periodic table is phosphorus, the 11th most abundant element in Earth's crust. Unlike nitrogen, phosphorus is not found in nature as the free element, though it can be produced from phosphorus minerals known as apatites. These minerals are mostly calcium phosphates with the general formula $Ca_5(PO_4)_3X$ (X = F, Cl, or OH). Reaction of these minerals with silicon dioxide (silica) and carbon at temperatures near 1500°C produces white phosphorus (P_4).

$$2\ Ca_3(PO_4)_2(s) + 6\ SiO_2(s) + 10\ C(s) \xrightarrow{1400-1500°C}$$
$$6\ CaSiO_3(s) + 10\ CO(g) + P_4(s)$$

Heating white phosphorus produces a less-reactive allotrope, red phosphorus. White phosphorus consists of discrete P_4 tetrahedra, whereas red phosphorus contains chains of P_4 tetrahedra, as described in Chapter 10.

Most lush, green lawns require fertilizers that supply nitrogen, phosphorus, and potassium.

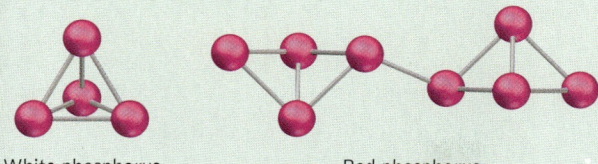

White phosphorus   Red phosphorus

Reaction of red or white phosphorus with oxygen yields the solid oxide P_4O_{10}. When P_4O_{10} dissolves in water, it forms phosphoric acid (H_3PO_4):

$$P_4O_{10}(s) + 6\ H_2O(l) \longrightarrow 4\ H_3PO_4(aq)$$

Phosphoric acid can also be produced by reaction between phosphate (apatite) minerals and sulfuric acid (H_2SO_4) such as

$$Ca_5(PO_4)_3F(s) + 5\ H_2SO_4(l) + 10\ H_2O(l) \longrightarrow$$
$$3\ H_3PO_4(l) + 5\ CaSO_4 \cdot 2\ H_2O(s) + HF(g)$$

Nearly 90% of phosphoric acid is used to produce ammonium phosphate for fertilizer. Other products that contain phosphates include soft drinks, toothpaste, and detergents.

The remaining three elements in Group 5A—arsenic, antimony, and bismuth—were actually discovered before either nitrogen or phosphorus. Arsenic and antimony are metalloids and bismuth is a metal. None of them is found as the free element in nature, but all three can be isolated from the thermal decomposition of minerals such as arsenopyrite (FeAsS), stibnite (Sb_2S_3), or bismite (Bi_2O_3).

$$FeAsS(s) \xrightarrow{\Delta} FeS(s) + As(s)$$
$$Sb_2S_3(s) + 3\ Fe(s) \xrightarrow{\Delta} 2\ Sb(s) + 3\ FeS(s)$$
$$Bi_2O_3(s) + 3\ C(s) \xrightarrow{\Delta} 2\ Bi(s) + 3\ CO(g)$$

All five Group 5A elements form compounds with the formula XH_3. We have discussed the commercial synthesis of NH_3, which is the sixth largest industrial chemical produced in the United States. Phosphine (PH_3), arsine (AsH_3), and stilbine, SbH_3, are used in the microelectronics industry.

Arsenic, antimony, and bismuth all form oxides having the formula X_2O_3. Nitric acid oxidizes arsenic trioxide (As_2O_3) to form arsenic acid (H_3AsO_4), which was widely used in herbicides and wood preservatives until concerns about groundwater pollution and soil contamination around structures built from "pressure treated" wood led to restrictions on these uses. Antimony trioxide (Sb_2O_3) is used primarily as a flame retardant. Bismuth compounds are used in pharmaceuticals and as additives to metals.

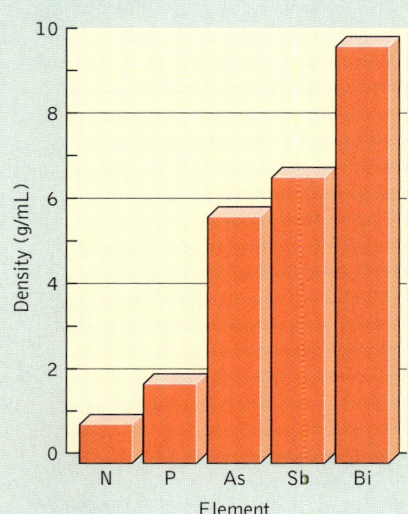

The melting points of the Group 5A elements do not follow the trend found in other groups. Melting points increase with increasing atomic number from nitrogen to arsenic but decrease for antimony and bismuth. The boiling points follow the expected trend; that is, they increase with increasing atomic number.

The densities of the Group 5A elements increase with increasing atomic number.

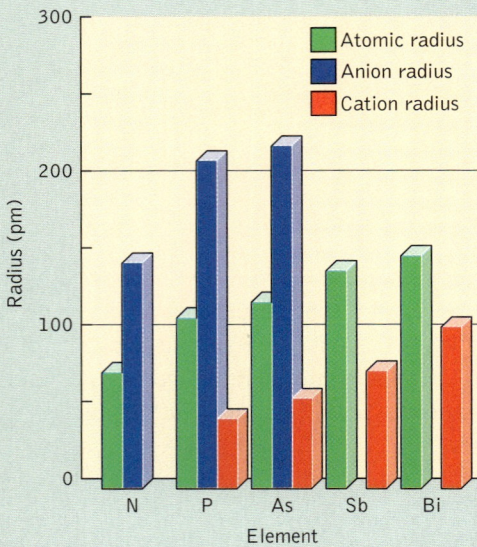

The atomic radii of the Group 5A elements increase with increasing atomic number. The radii for their +3 cations are considered "effective" ionic radii obtained from crystal structures of compounds containing these cations. Nitrogen does not form a stable +3 cation. The radius of the N^{3+} cation has been estimated to be 16 pm, fitting the expected trend. The first three members of the group form −3 anions whose radii are larger than the atomic radii of these elements and increase with increasing atomic number.

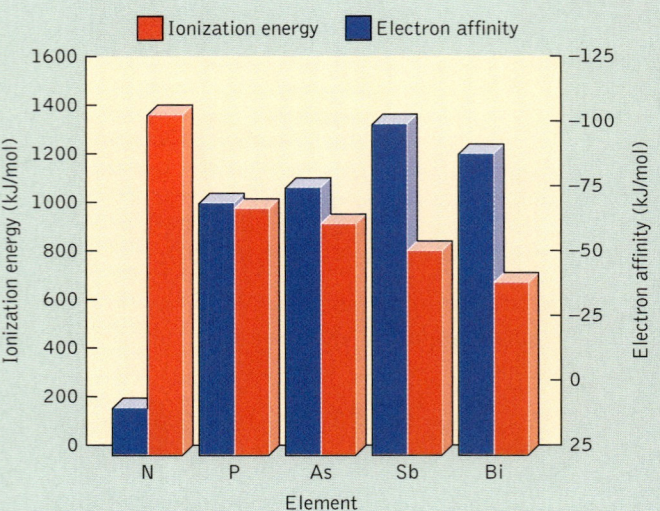

The ionization energies of the Group 5A elements decrease with increasing atomic number, owing to the shielding of the outer electrons from the nuclear charge. The electron affinities of these elements become more negative with increasing atomic number. This trend is different from that of other groups (3A, 4A, and 6A through 8A). Addition of an electron to a Group 5A atom puts a second electron in a half-filled p orbital, resulting in electron–electron repulsion. The sizes of valence p orbitals increases with increasing atomic number, which reduces electron–electron repulsions for the heavier Group 5A elements and results in more-negative (larger) electron affinities.

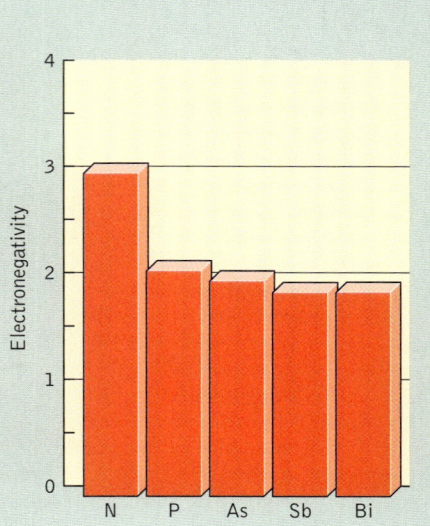

The observed trend toward lower electronegativities for the Group 5A elements as atomic number increases is consistent with the trend for other groups in the periodic table. The Group 5A elements are more electronegative than their neighbors to the left but less electronegative than elements in Groups 6A or 7A.

the decrease in pressure will cause gases dissolved in the blood to rapidly become less soluble and to revert to the gaseous state. The result may be formation of bubbles of N_2 in the blood, blocking capillaries and creating a medical condition, known commonly as "the bends," that is both painful and life threatening. The only cure for the bends is for the diver to repressurize so that the bubbles of N_2 redissolve and then to slowly return to lower pressure so that N_2 can be gradually eliminated from the blood with exhaled breath. To avoid the bends or long decompression times, as well as the toxic effects of high concentrations of N_2 in the blood, divers may fill their scuba tanks with artificial gas mixtures such as those described in the sample and practice exercises on page 407.

8.7 THE KINETIC MOLECULAR THEORY OF GASES AND GRAHAM'S LAW

Gas molecules move constantly and randomly throughout the volume they occupy as shown in Figure 8.5. At a given temperature the molecules of a gas have a range of velocities, the *average* of which is related to the average kinetic energy (K.E.) of the gas. The kinetic energy of a gas increases with increasing temperature and can be calculated by using the following equation:

$$K.E. = \tfrac{1}{2} m u_{rms}^2 \qquad (8.16)$$

where m is the mass of a molecule of the gas and u_{rms} is the **root-mean-square speed** of its molecules.

This vision of independent molecular motion in gases is the basis for the **kinetic molecular theory** of gases. A feature of the kinetic molecular theory is that, in a mixture of gases at a given temperature, the average kinetic energies of the different gases are the same. In a sample of air, for example, the average kinetic energy of N_2 is the same as that of O_2 and of all the other gases present. Because different gases have different molar masses, the heavier molecules in a mixture have lower root-mean-square speeds and the lighter ones move faster as shown in Figure 8.15.

The **root-mean-square speed** (u_{rms}) of a gas is the square root of the average of the squares of the velocities of a collection of its molecules.

The **kinetic molecular theory** states that the average kinetic energy of a gas is directly proportional to its absolute temperature. Molecules of gas at a given temperature move at speeds that are, on average, inversely proportional to the square roots of their masses.

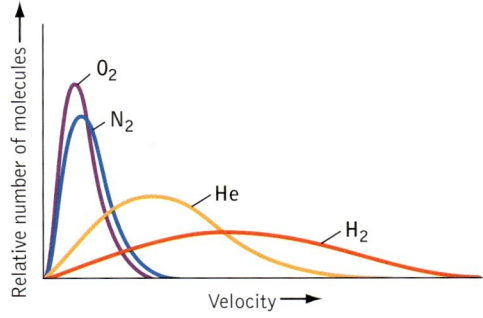

FIGURE 8.15 The speeds of gas atoms or molecules at constant temperature cover a range of values. The most probable, or root-mean-square, speed (u_{rms}) is the maximum in each curve and is inversely proportional to the square root of the mass of the atoms or molecules of each gas.

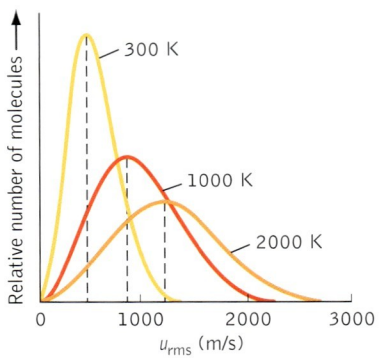

FIGURE 8.16 The root-mean-square speed (u_{rms}) of a gas increases with temperature, as shown for N_2 gas at three temperatures.

The kinetic molecular theory assumes that the average kinetic energies of gases are proportional to their temperatures. Figure 8.16 illustrates the distribution of molecular speeds at three temperatures. We can calculate the root-mean-square speed, u_{rms}, for a gas with molar mass $\mathcal{M}$ at temperature T with the following equation:

$$u_{rms} = \sqrt{\frac{3RT}{\mathcal{M}}} \qquad (8.17)$$

SAMPLE EXERCISE 8.15: Calculate the root-mean-square speed of nitrogen at 300 K (80°F). Express the answer in meters per second and in miles per hour.

SOLUTION: To calculate the root-mean-square speed of a gas, we will need to use a value of the gas constant (R) with the appropriate units. The second value in Table 8.1 is promising because it will yield a velocity in meters per second. If we use this value of R, 8.314 $\frac{\text{kg} \cdot \text{m}^2}{\text{s}^2 \cdot \text{mol} \cdot \text{K}}$, we will need to express molar mass in kilograms. For N_2, this value is 0.0280 kg/mol. Inserting the appropriate values into Equation 8.17, we have

$$u_{rms, N_2} = \sqrt{\frac{3RT}{\mathcal{M}}} = \sqrt{\frac{3\left(8.314\,\frac{\cancel{\text{kg}} \cdot \text{m}^2}{\text{s}^2/(\cancel{\text{mol}} \cdot \cancel{\text{K}})}\right)(300\,\cancel{\text{K}})}{0.0280\,\frac{\cancel{\text{kg}}}{\cancel{\text{mol}}}}} = 517\,\text{m/s}$$

To convert speed from meters per second into miles per hour (mi/hr) requires the use of several conversion factors:

$$1\,\text{mi} = 1.6093\,\text{km} = 1.6093 \times 10^3\,\text{m}$$

and

$$1\,\text{hr} = 3600\,\text{s}$$

$$u_{N_2} = \left(517\,\frac{\text{m}}{\text{s}}\right)\left(\frac{1\,\text{mi}}{1.6093 \times 10^3\,\text{m}}\right)\left(3600\,\frac{\text{s}}{\text{hr}}\right) = 1157\,\text{mi/hr}$$

molar mass, temperature:
$\mathcal{M}$, T

$u_{rms} = (3RT/\mathcal{M})^{1/2}$

root-mean-square speed

8.7 The Kinetic Molecular Theory of Gases and Graham's Law

Molecules of N_2 move rather quickly at room temperature!

PRACTICE EXERCISE: Calculate the root-mean-square speed of helium at 300 K. Express the answer in meters per second and compare it to the root-mean-square speed of nitrogen. (See Problems 103–106.)

Equation 8.17 allows us to compare the relative root-mean-square speeds of two gases at the same temperature. Consider a balloon filled with dry air at 25°C. Most of the molecules in the balloon are N_2 (79%) and O_2 (21%). Because both are in the same space, their temperatures must be the same, which means their average kinetic energies must be the same. Expressing this equality mathmatically, we have

$$K.E._{N_2} = K.E._{O_2}$$

or

$$(\tfrac{1}{2}m_{N_2})(u_{N_2})^2 = (\tfrac{1}{2}m_{O_2})(u_{O_2})^2$$

or, simply,

$$(m_{N_2})(u_{N_2})^2 = (m_{O_2})(u_{O_2})^2$$

Rearranging the terms to express the ratio of the root-mean-square speeds of the two gases in terms of the ratio of their masses, we have

$$\left(\frac{u_{N_2}}{u_{O_2}}\right)^2 = \frac{m_{O_2}}{m_{N_2}}$$

or

$$\frac{u_{N_2}}{u_{O_2}} = \sqrt{\frac{m_{O_2}}{m_{N_2}}} = \sqrt{\frac{32.0}{28.0}} = 1.07$$

Thus, the molecules of N_2 in air are moving on average 7% faster than the molecules of O_2.

We can write a generic form of the preceding equation that applies to any pair of gases (x and y):

$$\frac{u_x}{u_y} = \sqrt{\frac{m_y}{m_x}} \qquad (8.18)$$

The ratio of the root-mean-square speeds of two gases at the same temperature is inversely proportional to the square root of the ratio of their masses. This relation explains two similar properties of gases: their rates of **effusion** and **diffusion**. Effusion is the process by which a gas escapes through a tiny hole, such as an imperfection in a rubber balloon, into a space of lower pressure. Diffusion is the spread of one substance through another, such as the diffusion of the molecules of perfume throughout a room. The rates of both these processes will depend on the average speeds of the molecules of gas. In fact, the equation that relates the rates (r) of effusion of two gases to their molar masses—

$$\frac{r_x}{r_y} = \sqrt{\frac{M_y}{M_x}} \qquad (8.19)$$

> ✓ **Effusion** is the process by which a gas escapes through tiny holes into a space of lower pressure. **Diffusion** is the spread of one substance through another.

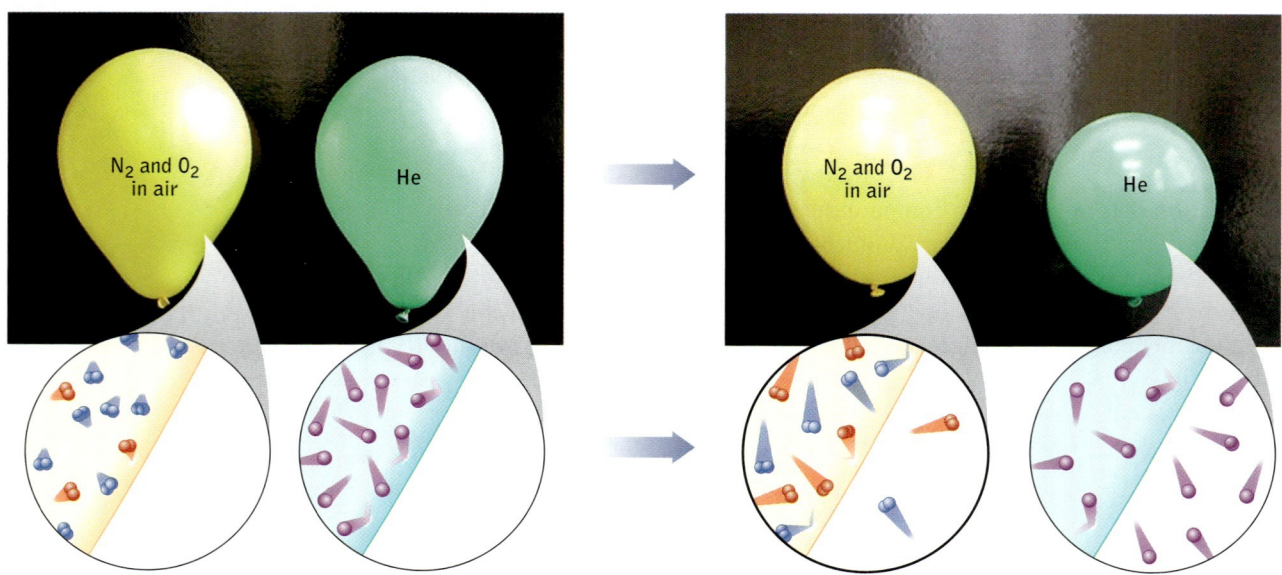

What a Chemist Sees. Two balloons containing equal volumes of air and helium (He) will both decrease in volume over time as gas effuses through small pores in the material of the balloon.

The lighter-mass atoms of helium have greater root-mean-square speeds than those of the molecules of N_2 and O_2 in air and so effuse faster.

☑ **Graham's law of effusion** states that the rate of effusion of a gas is inversely proportional to the square root of its molar mass.

looks a lot like Equation 8.18. Equation 8.19 provides a mathematical expression of a phenomenon first observed by nineteenth-century Scottish chemist Thomas Graham (1805–1869): that the rate of effusion of gases is inversely proportional to the square root of their molar masses. Today, this relation is known as **Graham's law of effusion.**

SAMPLE EXERCISE 8.16: Calculate how much faster than nitrogen will helium effuse through a balloon?

SOLUTION: Equation 8.19 allows us to calculate the relative rates of effusion of two gases on the basis of their molar masses. Inserting the molar mass of N_2 and the atomic mass of He into Equation 8.19, we have

$$\frac{r_{He}}{r_{N_2}} = \sqrt{\frac{\mathcal{M}_{N_2}}{\mathcal{M}_{He}}} = \sqrt{\frac{28.02 \text{ g/mol}}{4.00 \text{ g/mol}}} = 2.65$$

molar masses:
$\mathcal{M}_{N_2}, \mathcal{M}_{He}$

$r_{He}/r_{N_2} = (\mathcal{M}_{N_2}/\mathcal{M}_{He})^{1/2}$

relative rate of effusion

Therefore, He will effuse 2.65 times as rapidly as N_2.

PRACTICE EXERCISE: Helium effuses 3.16 times as fast as another noble gas. Which one? (See Problems 115 and 116.)

Moving gas molecules continuously collide with one another and with their container walls. These collisions do not result in a net transfer of energy to the

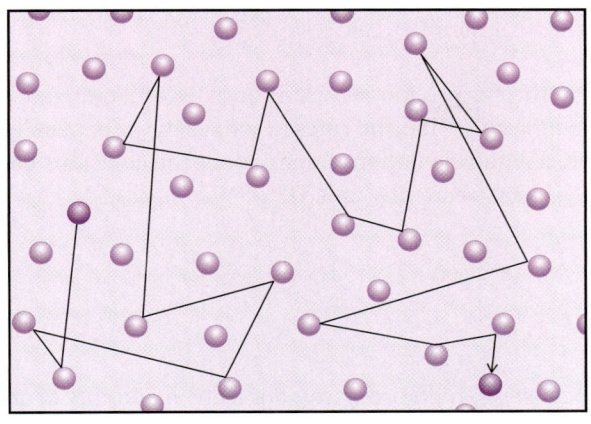

FIGURE 8.17 The path traveled by an atom or a molecule of gas is affected by collisions with other atoms or molecules and with the walls of the container. The result is a highly erratic path. The greater the number of particles, the greater the number of collisions and the more erratic the path.

walls; therefore, the average kinetic energy (and root-mean-square speeds) of gas molecules are not affected by these collisions and remain constant as long as there is no change in temperature. Physical scientists classify such collisions, in which there is no transfer of energy, as *elastic* collisions.

Another parameter that describes molecular motion is **mean free path**, which represents the average distance traveled by a molecule or atom of gas at a given temperature and pressure between collisions. The length of the mean free path depends on the number of gas molecules in a given volume: the more there are, the more likely they are to run into each other and the shorter the mean free path. The mean free path of air ranges from about 60 nm at sea level to 10 cm in the uppermost layer of the atmosphere, called the thermosphere. Collisions between molecules of an ideal gas do not change the value of u, but they do change the directions in which individual molecules are traveling (Figure 8.17). These frequent collisions account for the random motion of gases. They also contribute to the time that it takes for molecules from different containers to diffuse through one another and mix. The rate at which the molecules of a gas diffuse is related to their root-mean-square speeds, which, as we have seen, is a function of temperature and molecular mass.

> The **mean free path** is the average distance traveled by a molecule or atom of a gas between collisions.

CONNECTION: The layers of Earth's atmosphere are described in Section 6.1.

CONCEPT TEST: The mean free path of molecules in air increases with increasing altitude. Why?

If we divide the root-mean-square speed of a molecule by its mean free path, we have a parameter with the units of 1/time, such as s^{-1}. We encountered such a unit in Chapter 1. It is associated with frequency. For the motion of gas molecules, frequency must mean the frequency of molecular collisions. We have calculated in previous exercises in this chapter that, at 300 K (27°C), $u_{N_2} = 517$ m/s. If the mean free path for N_2 in air at sea level is 60 nm, then the frequency with which molecules of N_2 collide with other molecules in air (mostly other molecules of N_2) will be

CONNECTION: Chapter 14 addresses the rates of chemical reactions.

$$\text{collision frequency} = \left(\frac{517 \text{ m/s}}{60 \text{ nm/collision}}\right)\left(\frac{10^9 \text{ nm}}{\text{m}}\right) = 8.6 \times 10^9 \text{ collisions/s}$$

In Chapter 14, we will explore the connection between the rates and energetics of collisions between reactive molecules and the rates at which they chemically react with each other. The preceding calculation suggests that the collisions between gaseous reactants can happen frequently indeed.

8.8 REAL GASES

Let's consider the behavior of gases at high pressures. For a given quantity of an ideal gas, n, the fraction PV/RT should be a constant equal to n. Therefore a plot of PV/RT versus P should give a flat straight line with a y-intercept equal to n (see Figure 8.18). However, plots of PV/RT versus P for CH_4, H_2, and CO_2 at high pressure do not give flat straight lines, as shown in Figure 8.18.

Why don't real gases behave like ideal gases at high pressure? One reason is that ideal-gas behavior assumes that gas molecules have little volume compared with the volume of their container. To test the validity of this assumption, let's calculate the volume of a mole of) O_2 molecules and compare it to the total volume they occupy at STP (22.4L).

> Why don't real gases behave like ideal gases at high pressure?

Oxygen is a linear molecule with an O–O bond distance of 121 pm. An oxygen atom has an atomic radius of about 74 pm. Thus, an oxygen molecule is (approximately) a cylinder of radius 74 pm and a height of 269 pm. The volume of a cylinder is $\pi r^2 h$, where r is the radius of the cylinder and h is its height. Let's cal-

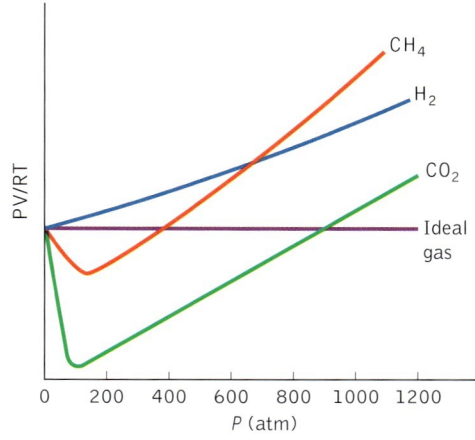

FIGURE 8.18 The quantity PV/RT remains constant over all pressures for a constant amount of an ideal gas but varies significantly for real gases. The deviations from ideal are the result of interactions between gas molecules, as well as the volume occupied by them.

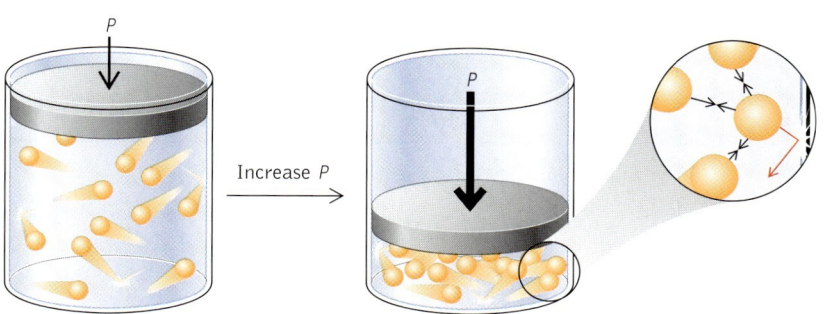

FIGURE 8.19 At high pressures, more of the volume of a gas is occupied by the atoms or molecules of the gas. The greater density of gas particles also results in more interactions between them (shown by the black arrows). These interactions reduce the frequency and force of particle collisions with the walls of the container (red arrow) and so reduce pressure.

culate the volume of our O_2 molecular cylinder in cubic centimeters. This approach is useful because $1 \text{ cm}^3 = 1 \text{ mL} = 10^{-3} \text{ L}$.

$$V = \pi r^2 h = \pi(7.4 \times 10^{-9} \text{ cm})^2(2.69 \times 10^{-8} \text{ cm})$$
$$= 4.63 \times 10^{-24} \text{ cm}^3 = 4.63 \times 10^{-24} \text{ mL}$$

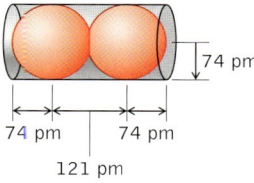

The volume occupied by one oxygen molecule is 4.63×10^{-27} L.
The volume occupied by one mole of oxygen molecules is

$$V_{O_2} = (4.63 \times 10^{-27} \text{ L/molecule})(6.022 \times 10^{23} \text{ molecules/mol})$$
$$= 2.79 \times 10^{-3} \text{ L/mol}$$

The percentage of the total volume occupied by oxygen molecules at STP is

$$\frac{2.79 \times 10^{-3} \text{ L/mol}}{22.4 \text{ L/mol}} \times 100 = 0.0124\%$$

Thus, the volume occupied by O_2 molecules in pure O_2 at STP is indeed negligible. However, at high pressures, more molecules are squeezed into a given volume of space, as shown in Figure 8.19.

Ideal-gas behavior also assumes that there are no interactions between gas molecules. This assumption is not likely to be true for polar molecules, because of the attraction of a positive pole of one molecule to the negative pole of another. The nature of intermolecular forces will be presented in greater detail in Chapter 9. For now it is important to recognize that intermolecular interactions reduce the pressure of a gas by decreasing the force of collisions with the container, as shown in Figure 8.19. These interactions are more likely to occur at high pressure and when molecules are moving slowly, which they do at low temperatures.

The ideal-gas law can be modified to accommodate the behavior of real gases. The **van der Waals equation**

$$(P + n^2a/V^2)(V - nb) = nRT \tag{8.20}$$

 CONNECTION: Strong intermolecular interactions among polar volatile compounds (see Sections 7.7 and 9.2) contribute to their nonideal behavior.

☑ The **van der Waals equation** includes corrections for intermolecular interactions and the volume occupied by particles of gas.

TABLE 8.4 Van der Waals Constants of Selected Gases

Substance	$a\left(\dfrac{L^2 \cdot atm}{mol^2}\right)$	b (L/mol)	Substance	$a\left(\dfrac{L^2 \cdot atm}{mol^2}\right)$	b (L/mol)
He	0.0341	0.02370	CO	1.48	0.03985
Ar	1.34	0.0322	H_2O	5.46	0.0305
H_2	0.244	0.0266	NO	1.34	0.02789
N_2	1.39	0.0391	NO_2	5.28	0.04424
O_2	1.36	0.0318	HCl	3.67	0.04081
CH_4	2.25	0.0428	SO_2	6.71	0.05636
CO_2	3.59	0.0427			

includes terms to correct for the pressure lost through intermolecular interactions (n^2a/V^2) and for the volume occupied by molecules and atoms of gas (nb). The symbols a and b are called van der Waals constants and can be determined experimentally for each gas. Table 8.4 lists the van der Waals constants for common atmospheric gases.

SAMPLE EXERCISE 8.17: Calculate the pressure of 1.00 mol of N_2 in a 1.00-L container at 300 K using the van der Waals equation. Compare the result of this calculation with the pressure of N_2 calculated by using the ideal-gas equation.

SOLUTION: Rearranging the terms in Equation 8.20 to solve for pressure, we have

$$P = \frac{nRT}{V - nb} - \frac{n^2 a}{V^2}$$

Inserting the appropriate values of n, V, T, a, and b given in the problem and in Table 8.4:

$$P_{N_2} = \frac{(1.00 \text{ mol})\left(0.08206 \frac{L \cdot atm}{mol \cdot K}\right)(300 \text{ K})}{1.00 \text{ L} - (1.00 \text{ mol})\left(0.0391 \frac{L}{mol}\right)}$$

$$- \frac{(1.00 \text{ mol})^2 (1.39 \text{ L}^2 \cdot atm/mol^2)}{(1.00 \text{ L})^2}$$

$$= 24.2 \text{ atm}$$

If nitrogen behaved as an ideal gas, then

$$P = \frac{nRT}{V}$$

$$P_{N_2} = \frac{(1.00 \text{ mol}) \left(0.08206 \frac{\text{L} \cdot \text{atm}}{\text{mol} \cdot \text{K}}\right)(300 \text{ K})}{1.00 \text{ L}}$$

$$= 24.6 \text{ atm}$$

Thus, the actual pressure is 0.4 atm lower than the ideal-gas pressure, and the deviation from ideal behavior is $-0.4/24.6 = -0.016$, or -1.6%.

PRACTICE EXERCISE: Which of the following gases would you expect to behave least like an ideal gas under the conditions described in the preceding sample exercise: O_2, N_2, F_2, HF, or Ne? Explain your selection. (See Problems 133 and 134.)

CHAPTER REVIEW

Summary

SECTION 8.1

Atmospheric pressure, measured with a barometer, is due to the weight of the atmosphere pressing on Earth's surface. Atmospheric pressure decreases with altitude. Common units of pressure are torr (mm Hg), atmospheres, and kilopascals.

SECTION 8.2

The volume of a gas sample increases with decreasing pressure at constant temperature (Boyle's law), but it increases with increasing temperature at constant pressure. Increasing the number of moles of gas in a given volume produces a proportionate increase in pressure at constant temperature.

SECTION 8.3

The ideal-gas equation, $PV = nRT$, accurately describes the behavior of most gases under normal conditions. Here, R is the ideal-gas constant, which is the same value for all gases. The molar volume of any gas is 22.4 liters at STP (0 and 1 atm pressure).

SECTION 8.4

Gas densities increase with increasing molar mass and pressure, and they decrease with increasing temperature. The molar mass of a gas can be calculated from its density at a specified temperature and pressure.

SECTION 8.5

In a gas mixture, the contribution of each component to the total gas pressure is called its partial pressure. Dalton's law states that the total pressure of a gas mixture is the sum of the partial pressures of its constituents. Dalton's law allows us to calculate the partial pressure, P_x, of constituent x in a gas mixture from its mole fraction, X_x, and the total pressure. The total pressure of a gas sample collected over a liquid is the sum of the vapor pressure of the liquid and the pressure of the gas itself.

SECTION 8.6

According to Henry's law, the solubility of a gas in a liquid increases with increasing pressure of the gas.

SECTION 8.7

The kinetic molecular theory of gases assumes that (1) gas molecules are in constant, random motion, (2) collisions with the vessel wall are responsible for gas pressure, (3) collisions between molecules and with the vessel wall are elastic, and (4) that the average kinetic energy of the gas molecules is proportional to absolute temperature. The root-mean-square speed of the molecules increases with increasing temperature and decreasing molar mass. The kinetic molecular theory of gases explains Graham's law, which states that the rate of effusion (escape through a pinhole) or diffusion (spreading) of a gas at fixed temperature is inversely proportional to its molar mass. Another parameter of gas behavior is the mean free path, which is the average distance between molecular collisions.

SECTION 8.8

Real gases disobey the ideal-gas law under conditions of high pressure, where their molecules take up a greater proportion of the volume that they occupy, and low temperature, which favors intermolecular interaction. The van der Waals equation is a modified form of the ideal-gas law that accounts for real gas properties.

Key Terms

Avogadro's law (p. 400)
Boyle's law (p. 394)
Charles's law (p. 400)
Dalton's law (p. 406)
diffusion (p. 419)
effusion (p. 419)
gas constant (p. 400)

Graham's law (p. 420)
Henry's law (p. 411)
ideal gas (p. 401)
ideal-gas equation (p. 401)
ideal-gas law (p. 401)
kinetic molecular theory (p. 417)
mean free path (p. 421)

molar volume (p. 403)
mole fraction (p. 406)
partial pressure (p. 406)
root-mean-square speed (p. 417)
standard temperature and pressure (STP) (p. 402)
van der Waals equation (p. 423)

Key Skills and Concepts

SECTION 8.1

Understand how a barometer is used to measure atmospheric pressure.
Understand why atmospheric pressure, density, and temperature decrease with increasing altitude in the troposphere.

SECTION 8.2

Describe how the volume of a gas sample varies with pressure and temperature.

SECTION 8.3

Be able to use the ideal-gas law to describe the behavior of gases.
Know the meaning of STP and understand that all gases have the same molar volume at fixed temperature and pressure.

SECTION 8.4

Be able to calculate the density of a known gas at a specified temperature and pressure.
Be able to calculate the molar mass of a gas from its density under specified conditions.

SECTION 8.5

Know how the partial pressure of a gas in a mixture, its mole fraction, and the total pressure of the gas mixture are related.
Understand why the vapor pressure of the liquid must be subtracted from the total pressure of a gas sample collected over a liquid to calculate the moles of gas collected.

SECTION 8.6

Be able to calculate the solubility of a gas from Henry's law.

SECTION 8.7

Know that the root-mean-square speed of the molecules of a gas is proportional to the square root of its absolute temperature.

Be able to use Graham's law to calculate the relative rates of effusion and diffusion of different gases at fixed temperature.

Be able to use Graham's law to calculate the molar mass of a gas from its rate of effusion or diffusion relative to a known gas at fixed temperature and pressure.

Understand the concepts of mean free path and collision frequency.

SECTION 8.8

Understand why real gases disobey the ideal-gas law at low temperatures and high pressures.

Be able to use the van der Waals equation to allow for the behavior of real gases such as SO_2.

Key Equations and Relations

SECTION 8.1

Force (F):

$$F = ma \quad (8.1)$$

where F is force, m is mass, and a is acceleration due to gravity.

Pressure (P):

$$P = F/A \quad (8.2)$$

where F is force and A is area.

SECTION 8.2

Pressure (P), volume (V), and temperature (T) of a gas sample:

$$\frac{P_1 V_1}{T_1} = \frac{P_2 V_2}{T_2} \quad (8.8)$$

SECTION 8.3

The ideal-gas law:

$$PV = nRT \quad (8.9)$$

where P is pressure (atm), V is volume (liters), n is number of moles of gas, and T is absolute temperature (K). The appropriate value for constant R is 0.08206 L·atm/mol·K.

The molar volume of an ideal gas at STP (1 atm pressure, 0°C) is 22.4 L.

SECTION 8.4

The density, d, of a gas:

$$d = MP/RT \quad (8.10)$$

where P is pressure, T is temperature, and M is the molar mass of the gas. Note that the units of gas density are grams per liter.

The molar mass of a gas:

$$M = dRT/P \quad (8.11)$$

where d is density.

SECTION 8.5

Dalton's law of partial pressure:

$$P_{total} = P_1 + P_2 + P_3 + P_4 + \cdots \quad (8.12)$$

where P_{total} is the total pressure of a gas mixture and P_1, P_2, and so forth, are the pressures of the constituents.

The mole fraction X_1 of constituent 1 in a gas mixture or solution:

$$X_1 = \frac{n_1}{n_{total}}$$

where n_1 is the number of moles of constituent 1 and n_{total} is the total number of moles of constituents in the mixture.

The partial pressure of a gas, P_1:

$$P_1 = X_1 P_{total} \quad (8.13)$$

where χ_1 is the mole fraction of the gas in the mixture and P_{total} is the total pressure of the gas mixture.

The pressure, P_{gas}, of a gas sample collected over a liquid:

$$P_{total} = P_{atmospheric} = P_{gas} + P_{liquid}$$

where P_{liquid} is the vapor pressure of the liquid.

SECTION 8.6

Henry's law for the solubility of a gas (C_{gas}) in a liquid:

$$C_{gas} = k_H P_{gas} \tag{8.14}$$

where k_H is the Henry's-law constant for the gas in that liquid.

SECTION 8.7

The root-mean-square speed (u_{rms}) of the molecules of a gas:

$$u_{rms} = \sqrt{\frac{3RT}{\mathcal{M}}} \tag{8.16}$$

where $\mathcal{M}$ is the molar mass of the gas.

Graham's law of effusion:

$$\frac{r_x}{r_y} = \sqrt{\frac{\mathcal{M}_y}{\mathcal{M}_x}} \tag{8.18}$$

where r_x and r_y are the rates of effusion of gases x and y, respectively, and $\mathcal{M}_x$ and $\mathcal{M}_y$ are their molar masses.

SECTION 8.8

The van der Waals equation for a real gas:

$$(P + n^2 a/V^2)(V - nb) = nRT \tag{8.19}$$

where constants a and b correspond to the effects of intermolecular attraction and molecular volume, respectively, and vary from gas to gas (see Table 8.4).

CONCEPT REVIEW

1. Describe the difference between force and pressure.
2. How did Torricelli's barometer measure atmospheric pressure?
3. What is the relation between *torr* and *atmospheres* of pressure?
4. What is the relation between *millibars* and *pascals* of pressure?
5. Three barometers based on Torricelli's design are constructed by using water (density 1.00 g/mL), ethanol (0.789 g/mL), and mercury (13.546 g/mL). Which barometer contains the tallest column of liquid?
6. In constructing a barometer, what advantage is there in choosing a dense liquid?
7. Why does an ice skater exert more pressure on ice when wearing newly sharpened skates than when wearing skates with dull blades?
8. Why is it easier to travel over deep snow when wearing snowshoes rather than just boots?

PROBLEMS

9. Calculate the downward pressure due to gravity exerted by the bottom face of a 1.00-kg cube of iron that is 5.00 cm on a side.
10. The gold block represented here has a mass of 38.6 g. Calculate the pressure exerted by the block when it is on (a) a square face and (b) a rectangular face.

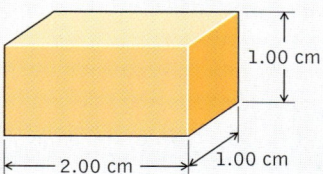

11. Convert the following pressures into atmospheres.
    a. 2.0 kPa   b. 562 mm Hg   c. 4.19×10^5 N/m^2
12. Convert the following pressures into mm Hg (torr).
    a. 0.541 atm   b. 2.8 kPa   c. 3.00×10^4 N/m^2

Connecting P, V, n, and T

CONCEPT REVIEW

13. Why does atmospheric pressure decrease with increasing altitude?

14. What is meant by standard temperature and pressure (STP)? What is the volume of 1 mole of an ideal gas at STP?

15. A quantity of gas is compressed into half its initial volume while it is cooled from 20°C to 10°C. Will the pressure of the gas increase, decrease, or remain the same?

16. If the volume of gasoline vapor and air in an automobile engine cylinder is reduced to one-tenth of its original volume before ignition, by what factor does the pressure in the cylinder increase? (Assume there is no change in temperature.)

PROBLEMS BASED ON CONSTANT n

17. The volume of 1.00 mol of ammonia gas at 1.00 atm of pressure was gradually decreased from 78. mL to 39. mL. What was the final pressure of ammonia if there was no change in temperature?

18. The pressure on a sample of an ideal gas was increased from 715 mm Hg to 3.55 atm at constant temperature. If the initial volume of the gas was 485. mL, what was the final volume of the gas?

19. A scuba diver released a balloon containing 153 L of helium attached to a tray of artifacts at an underwater archeological site. When the balloon reached the surface, it had expanded to a volume of 352 L. If the pressure at the surface was 1.00 atm, what was the pressure at the underwater site? If the pressure increased by 1.0 atm for every 10 meters of depth, at what depth was the diver working?

20. The world record for diving without supplemental air tanks ("breath-hold diving") is about 125 meters: a depth at which the pressure is about 12.5 atm. If a diver's lungs have a volume of 6 L at the surface, what is their volume at 125 m?

21. Use the following data to draw a graph of the volume (V) of H_2 as a function of the reciprocal of pressure (1/P) at 298 K. Would the graph be the same for the same number of moles Ar?

P_{H_2} (mm Hg)	V (L)
100	186
120	155
240	77.5
380	48.9
500	37.2

22. Use the following data to draw a graph of the volume (V) of He as a function of temperature for 1.0 mol of He gas at a constant pressure of 1.00 atm. How would the graph change if the amount of gas were halved?

V (L)	T (K)
7.88	96.
3.94	48.
1.97	24.
0.79	9.6
0.39	4.8

23. A cylinder with a piston contains a sample of gas at 25°C. The piston moves in response to changing pressure inside the cylinder. At what gas temperature would the piston move so that the volume inside the cylinder doubled?

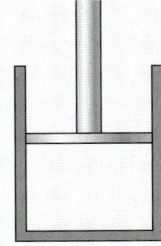

24. The temperature of the gas in Problem 23 is reduced to a temperature at which the volume inside the cylinder decreased by 25% from its initial volume at 25.0°C. What is the new temperature?

25. A 2.68-L sample of gas was warmed from 250 K to a final temperature of 398 K. Assuming no change in pressure, what was the final volume of the gas?

26. A 5.6-L sample of gas was cooled from 78°C to a temperature at which its volume is 4.3 L. What was this temperature? Assume no change in pressure of the gas.

27. Which of the following plots of volume (V) versus temperature (T) at constant pressure is not consistent with the ideal-gas law?

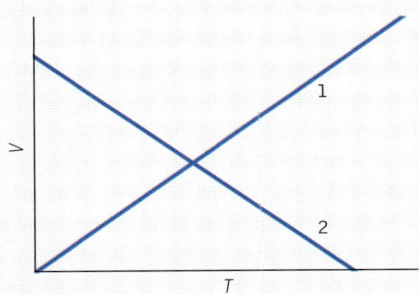

28. Which of the following plots of volume (V) versus pressure (P) at constant temperature is not consistent with the ideal-gas law?

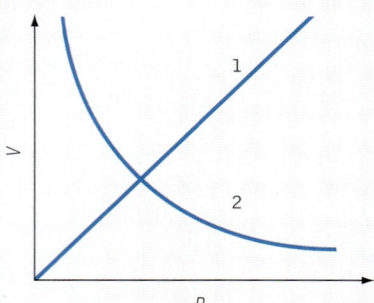

29. Which of the following actions would produce the greatest increase in the volume of a gas sample: (1) lower the pressure from 760 mm Hg to 720 mm Hg at constant temperature or (2) raise the temperature from 10°C to 40°C at constant pressure?

30. Which of the following actions would produce the greatest increase in the volume of a gas sample: (1) double the amount of gas in the sample at constant temperature and pressure or (2) raise the temperature from −44°C to +100°C?

31. What happens to the volume of gas in a cylinder with a movable piston (see sketch in Problem 23) under the following conditions?
    a. Both the absolute temperature and the external pressure on the piston are doubled.
    b. The absolute temperature is halved and the external pressure on the piston is doubled.
    c. The absolute temperature is increased by 75% and the external pressure on the piston increases by 50%.

32. What happens to the pressure of a gas under the following conditions?
    a. The absolute temperature is halved and the volume is doubled.
    b. Both the absolute temperature and the volume are doubled.
    c. The absolute temperature is increased by 75% and the volume decreased by 50%.

33. Which line corresponds to the higher temperature in the following graph?

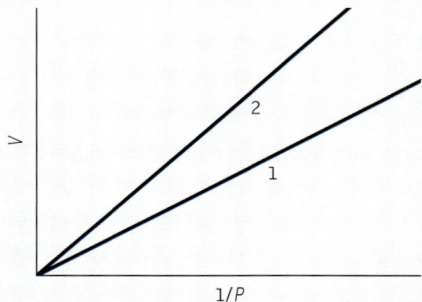

34. Which line in the following graph represents a gas at higher pressure? Is the x-azis an absolute temperature scale?

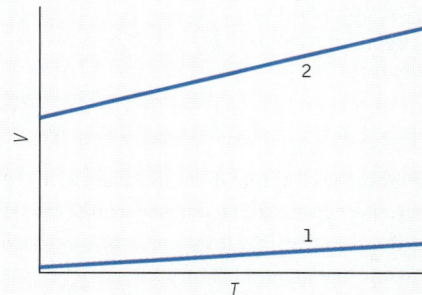

35. A weather balloon with a volume of 200 L is launched at 20°C at sea level where the atmospheric pressure is 1.00 atm. It rises to an altitude of 20,000 m, where atmospheric pressure is 63 mm Hg and the temperature is 210 K. What is the volume of the balloon at 20,000 m?

36. For some reason, a skier decides to ski from the summit of a mountain near Park City, Utah (elevation = 9970 feet, T = −10°C, P = 623 mm Hg), to the base of the mountain (elevation 6920 feet, T = −5°C, P = 688 mm Hg) with a balloon tied to each of her ski poles. If each balloon was filled to a volume of 2.00 L at the summit, what was the volume of each of them at the base?

37. Balloons for a New Year's Eve party in Fargo, North Dakota, are filled to a volume of 2.0 L at a temperature of 22°C and then hung outside where the temperature is −22°C. What is the volume of the balloons after they have cooled to the outside temperature? Assume that the atmospheric pressure inside and outside the house is the same.

38. The air inside a hot-air balloon is heated to 45°C. If the air inside the balloon then cools to 25°C, by what percentage does the volume of the balloon change?

PROBLEMS BASED ON PV = nRT

39. How many moles of air must be in a bicycle tire with a volume of 2.36 L if it has an internal pressure of 6.8 atm at 17.0°C?
40. At what temperature will 1.00 mol of an ideal gas in a 1.00-L container exert a pressure of 1.00 atm?
41. What is the pressure inside a 500-mL cylinder containing 1.00 g of deuterium gas (D_2) at 298 K?
42. What is the volume of 100 g of H_2O vapor at 120°C and 1.00 atm of pressure?
43. Hydrogen holds promise as an "environment friendly" fuel. How many grams of H_2 gas are present in a 50.0-L fuel tank at a pressure of 2850 pounds/inch^2 (psi) at 20°C (1 atm = 14.7 psi)?
44. Students at the University of North Texas and the University of Washington built a car propelled by compressed nitrogen gas. The gas was obtained by boiling liquid nitrogen stored in a 182-L tank. What volume of N_2 is released at 0.927 atm of pressure and 25°C from a tank full of liquid N_2 (density 0.808 g/mL)?
45. Before the development of reliable batteries, miners' lamps burned acetylene produced by the reaction of calcium carbide with water:

$$CaC_2(s) + H_2O(l) \longrightarrow C_2H_2(g) + CaO(s)$$

A lamp uses 1.00 L of acetylene per hour at 1.00 atm pressure and 18°C.
a. How many moles of C_2H_2 are used per hour?
b. How many grams of calcium carbide must be in the lamp for a 4-h shift?

46. Acid precipitation dripping on limestone produces carbon dioxide by the following reaction:

$$CaCO_3(s) + 2\,H^+(aq) \longrightarrow Ca^{2+}(aq) + CO_2(g) + H_2O(l)$$

If 15.0 mL of CO_2 was produced at 25°C and 760 mm Hg,
a. how many moles of CO_2 were produced?
b. how many milligrams of $CaCO_3$ were consumed?

47. Oxygen is generated by the thermal decomposition of potassium chlorate:

$$2\,KClO_3(s) \xrightarrow{\Delta} 2\,KCl(s) + 3\,O_2(g)$$

How much $KClO_3$ is needed to generate 200. L of oxygen at 0.85 atm of pressure at 273 K?

48. Calculate the volume of carbon dioxide at 20°C and 1.00 atm produced from the complete combustion of 1.00 kg of methane, and compare the result with the volume of CO_2 produced from the complete combustion of 1.00 kg of propane (C_3H_8).
49. The CO_2 that builds up in the air of a submerged submarine can be removed by reacting it with sodium peroxide:

$$2\,Na_2O_2(s) + 2\,CO_2(g) \longrightarrow 2\,Na_2CO_3(s) + O_2(g)$$

If a sailor exhales 150 mL of CO_2 per minute at 20°C and 1.02 atm, how much sodium peroxide is needed per sailor in a 24-h period?

50. Self-contained self-rescue breathing devices convert CO_2 into O_2 according to the following reaction:

$$2\,KO_2(s) + CO_2(g) \longrightarrow K_2CO_3(s) + \tfrac{3}{2}O_2(g)$$

How many grams of KO_2 are needed to produce 100. L of O_2 at 20°C and 1.00 atm?

Gas Density

CONCEPT REVIEW

51. Do all gases at the same pressure and temperature have the same density?
52. Birds and sailplanes take advantage of "thermals," which are rising columns of warm air, to gain altitude with less effort than usual. Why does warm air rise?
53. How does the density of a gas sample change when (a) its pressure is increased and (b) its temperature is decreased?
54. How would you measure the density of a gas sample?

PROBLEMS

55. Radon is hazardous because it is easily inhaled and it emits α particles when it undergoes radioactive decay.
a. Calculate the density of radon at 298 K and 1 atm of pressure.
b. Are radon concentrations likely to be greater in the basement or on the top floor of a building?
56. Four balloons, each with a mass of 10.0 g, were inflated to a volume of 20.0 L with either helium, neon, carbon monoxide, or nitrogen monoxide. If the density of air at

25°C and 1.00 atm is 0.00117 g/mL, will any of the balloons float in this air?

57. Which line in the following graph of density versus pressure at constant temperature for methane and nitrogen should be labeled *methane*?

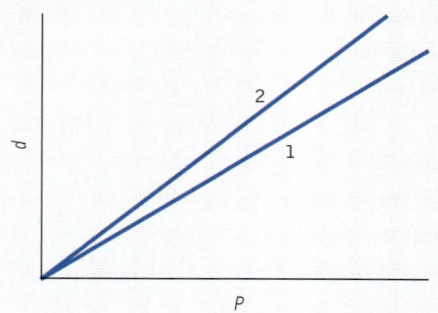

58. Add lines showing the densities of He and NO as a function of pressure to a copy of the graph in Problem 57.

59. A 150.-mL flask contains 0.391 g of a volatile oxide of sulfur. The pressure in the flask is 750 mm Hg and the temperature is 22°C. Is the gas SO_2 or SO_3?

60. A 100.-mL flask contains 0.193 g of a volatile oxide of nitrogen. The pressure in the flask is 760 mm Hg at 17°C. Is the gas NO, NO_2, or N_2O_5?

61. The density of an unknown gas is 1.107 g/L at 300 K and 740 mm Hg. Could this gas be CO or CO_2?

62. The density of a gas containing chlorine and oxygen has a density of 2.875 g/L at 756 mm Hg and 11°C. What is the most likely formula of the gas?

Dalton's Law

CONCEPT REVIEW

63. What is meant by the *partial* pressure of a gas?

64. Can a barometer be used to measure just the partial pressure of oxygen in the atmosphere? Why or why not?

65. Which gas sample has the largest volume at 25°C and 1 atm pressure: (a) 0.500 mol of dry H_2, (b) 0.500 mol of dry N_2, or (c) 0.500 mol of H_2 that has been collected over water?

66. Two identical balloons are filled to the same volume at the same pressure and temperature. One balloon is filled with air and the other with helium. Which balloon contains more particles (atoms and molecules)?

PROBLEMS

67. A gas mixture contains 0.70 mol of N_2, 0.20 mol of H_2, and 0.10 mol of CH_4. What is the mole fraction of H_2 in the mixture?

68. A gas mixture contains 7.0 g of N_2, 2.0 g of H_2, and 16.0 g of CH_4. What is the mole fraction of H_2 in the mixture?

69. Calculate the pressure of the gas mixture and the partial pressure of each constituent gas in Problem 67 if the mixture is in a 10.0-L vessel at 27°C.

70. Calculate the pressure of the gas mixture and the partial pressure of each constituent gas in Problem 68 if the mixture is in a 1.00-L vessel at 0°C.

71. A sample of oxygen was collected over water at 25°C and 1.00 atm pressure. If the total sample volume was 0.480 L, how many moles of O_2 were collected?

72. Water was removed from the O_2 sample in Problem 71. What is the volume of the dried O_2 gas sample at 25°C and 1.00 atm pressure?

73. The following reactions were carried out in sealed containers. Will the total pressure after each reaction is complete be greater than, less than, or equal to the total pressure before the reaction? Assume all reactants and products are gases at the same temperature.
a. $N_2O_5 \rightarrow NO_2 + NO + O_2$
b. $2\,SO_2 + O_2 \rightarrow 2\,SO_3$
c. $C_3H_8 + 5\,O_2 \rightarrow 3\,CO_2 + 4\,H_2O$

74. In each of the following gas-phase reactions, determine whether the total pressure at the end of the reaction (carried out in a sealed rigid vessel) will be greater than, less than, or equal to the total pressure at the beginning. Assume all reactants and products are gases at the same temperature.
a. $H_2 + Cl_2 \rightarrow 2\,HCl$
b. $4\,NH_3 + 5\,O_2 \rightarrow 4\,NO + 6\,H_2O$
c. $2\,NO + O_2 \rightarrow 2\,NO_2$

75. Alpine climbers use pure oxygen near the summits of 8000-m peaks ($P = 0.35$ atm). How much more O_2 is there in a lung full of pure O_2 at this elevation than in a lung full of air at sea level?

76. A scuba diver is at a depth of 50 m, where the pressure is 5.0 atm. What should be the mole fraction of O_2 in the gas mixture the diver breathes to replicate P_{O_2} at sea level?

77. Carbon monoxide at a pressure of 680 mm Hg reacts completely with O_2 at a pressure of 340 mm Hg in a sealed vessel to produce CO_2. What is the final pressure in the flask?

78. Ozone reacts completely with NO, producing NO_2 and O_2. A 10.0-L vessel was filled with 0.280 mol of NO and 0.280 mol of O_3 at 350 K. What is the partial pressure of each product and what is the total pressure in the flask at the end of the reaction?

*79. Ammonia is produced industrially from the reaction of hydrogen with nitrogen under pressure in a sealed reactor. What is the percent decrease in pressure of a sealed reaction vessel during the reaction between 3.60×10^3 mol H_2 and 1.20×10^3 mol N_2 if half of the N_2 is consumed?

*80. A mixture of 0.156 mol of C is reacted with 0.117 mol of O_2 in a sealed, 10.0-L vessel at 500 K, producing a mixture of CO and CO_2. The total pressure is 0.640 atm. What is the partial pressure of CO?

Gas Solubility

CONCEPT REVIEW

81. Why does the solubility of gases in liquids increase with increasing gas pressure?

82. Why are the bends caused by the formation of bubbles of N_2, and not O_2, in the blood?

83. Air is primarily a mixture of nitrogen and oxygen. Is the Henry's-law constant for the solubility of air in water the sum of k_{H,N_2} and k_{H,O_2}? Explain why or why not.

84. Power plants often discharge warm water (used for cooling) into the environment. This thermal pollution affects the survival of species of fish in the vicinity because the solubility of all gases in liquids is lower at higher temperature. Which term (k_H or P) in the Henry's-law equation is influenced by changing temperature?

85. Why is the Henry's-law constant for carbon dioxide, k_{CO_2}, so much larger than that for either O_2 or N_2? (Hint: Consider the chemical reactivity of CO_2 with H_2O described in Section 5.3.)

86. A student observes bubbles while heating water in a beaker to 60°C. What are the gases in the bubbles and why did they form?

PROBLEMS

87. Arterial blood contains about 0.25 g of oxygen per liter at 37°C and standard atmospheric pressure. Express this solubility of O_2 in blood in mol/L · atm.

88. The solubility of O_2 in water is 6.5 mg/L at $P_{atmosphere} = 1.00$ atm and 40°C. Calculate the Henry's-law constant for O_2 at 40°C.

89. Use the information in Problem 87 to calculate the solubility (g/L) of O_2 in the blood of (a) an alpine climber on Mt. Everest ($P_{atm} = 0.35$ atm) and (b) a scuba diver breathing air at a depth of 100 feet ($P = 3.0$ atm).

90. The Henry's-law constant for nitrogen dissolved in blood is 7.0×10^{-4} mol/L·atm at 37°C. Calculate the solubility (g/L) of N_2 in the blood of (a) an alpine climber on Mt. Everest ($P_{atm} = 0.35$ atm) and (b) a scuba diver breathing air at a depth of 100 feet ($P = 3.0$ atm).

91. Use the following graph of solubility of O_2 expressed in liters of O_2 per liter of water versus temperature to calculate the value for k_{H,O_2} at 10°C, 20°C, and 30°C.

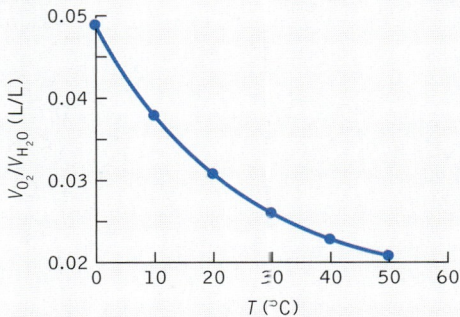

92. On the basis of the data in Problem 91, which has a greater effect on the solubility of oxygen in water: (a) decreasing the temperature from 20°C to 10°C, or (b) raising the pressure from 1.00 atm to 1.25 atm?

Kinetic Theory and Graham's Law

CONCEPT REVIEW

93. What is meant by the *root-mean-square speed* of gas molecules?
94. Why don't all molecules in a sample of air move at exactly the same speed?
95. How does the root-mean-square speed of the molecules in a gas vary with (a) molar mass and (b) temperature?
96. Does pressure affect the root-mean-square speed of the molecules in a gas? Explain your answer.
97. How can Graham's law of effusion be used to determine the molar mass of an unknown gas?
98. Is the ratio of the rates of effusion of two gases the same as the ratio of their root-mean-square speeds?
99. What is the difference between *diffusion* and *effusion*?
100. If gas X diffuses faster in air than gas Y, is gas X also likely to effuse faster than gas Y?

PROBLEMS

101. Rank the gases SO_2, CO_2, and NO_2 in order of increasing root-mean-square speed at 0°C.
102. In a mixture of CH_4, NH_3, and N_2, which gas molecules are, on average, moving fastest?
103. At 286 K, three gases, A, B, and C, have root-mean-square speeds of 360 m/s, 441 m/s, and 472 m/s, respectively. Which gas is O_2?
104. Air is approximately 21% O_2 and 78% N_2. Calculate the root-mean-square speed of each gas at 273 K.
105. Calculate the root-mean-square speed of Ne atoms at the temperature at which their average kinetic energy is 5.18 kJ/mol.
106. Determine the root-mean-square speed of CO_2 molecules that have an average kinetic energy of 4.2×10^{-21} J/molecule.
107. What is the ratio of the root-mean-square speed of D_2 to that of H_2 at constant temperature?
108. The two isotopes of uranium, ^{238}U and ^{235}U, can be separated by diffusion of the corresponding UF_6 gases. What is the ratio of the root-mean-square speed of $^{238}UF_6$ to that of $^{235}UF_6$ at constant temperature?
109. The following graph shows the distribution of molecular speeds of CO_2 and SO_2 molecules at 25°C. Which curve is the profile for SO_2? Which of these profiles should match that of propane (C_3H_8), a common fuel in portable grills?

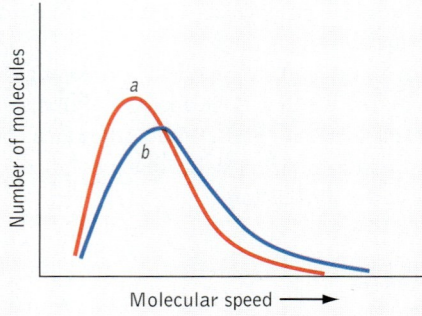

110. How would a graph showing the distribution of molecular speeds of CO_2 at −100°C differ from the graph for CO_2 in Problem 109?
111. Molecular hydrogen effuses four times as fast as gas X at the same temperature. What is the molar mass of gas X?
112. Gas Y effuses half as fast as O_2. What is the molar mass of gas Y?
113. If an unknown gas has one-third the root-mean-square speed of H_2 at 300 K, what is its molar mass?
114. A flask of ammonia is connected to a flask of an unknown acid HX by a 1.00-m glass tube. As the two gases diffuse down the tube, a white ring of NH_4X forms 68.5 cm from the ammonia flask. Identify element X.
115. Studies of photosynthesis in green plants often make use of $^{13}CO_2$. Calculate the relative rates of diffusion of $^{13}CO_2$ and $^{12}CO_2$. Specify which gas diffuses faster.
116. At fixed temperature, how much faster does NO effuse than NO_2?
117. Two balloons were filled with H_2 and He, respectively. The person responsible for filling them neglected to label them. After 24 hours, the volume of both balloons had decreased, but by different amounts. Which balloon contained hydrogen?
118. Compounds sensitive to oxygen are often manipulated in *glove boxes* that contain a pure nitrogen or argon atmosphere. A balloon filled with carbon monoxide was placed in a glove box. After 24 hours, the volume of the balloon was unchanged. Did the glove box contain N_2 or Ar?
119. H_2S is an extremely toxic gas that is easily identified by its "rotten egg" odor. The root-mean-square speed of H_2S at 25°C is 220 m/s. Why does it take the odor of H_2S much longer than 1 second to travel 220 m?
120. The mean free paths of nitrogen and oxygen are about 60 nm at sea level, but increase to 10 cm in the thermosphere. Calculate the frequency of collisions of N_2 and O_2 molecules in the thermosphere (T = 2200°).

Real Gas Behavior

CONCEPT REVIEW

121. Explain why real gases behave nonideally at low temperatures and high pressures?

122. Explain why the van der Waals constant a for water vapor (5.46 L²·atm/mol²) is significantly larger than that for N_2 (1.39 L²·atm/mol²). (Hint: Consider the polarity of the molecules.)

123. Under what conditions is the pressure exerted by a real gas *less* than that predicted for an ideal gas?

124. Explain why the values of the van der Waals constant b of the noble-gas elements increase with atomic number.

125. Explain why the constant a in the van der Waals equation generally increases with the molar mass of the gas.

126. How does the polarity of a molecule influence the value of its a constant in the van der Waals equation?

PROBLEMS

127. The graphs of PV/RT versus P for a mole of CH_4 and He differ in how they deviate from ideal behavior. For which gas is the effect of the volume occupied by the gas molecules more important than the attractive forces between molecules?

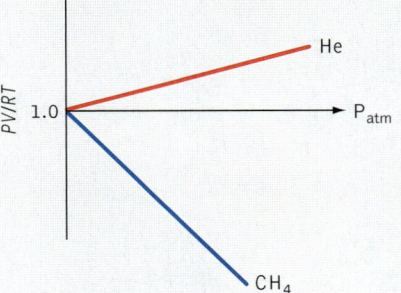

128. Which gas would you predict to have a larger value for the van der Waals constant a, ammonia or methane?

129. At high pressures, real gases do not behave ideally. Use the van der Waals equation and data in the text to calculate the pressure exerted by 40.0 g H_2 at 20°C in a 1.00-L container. Repeat the calculation assuming that the gas behaves like an ideal gas.

130. Calculate the pressure exerted by 1.00 mol of CO_2 in a 1.00-L vessel at 300 K, assuming that the gas behaves ideally. Repeat the calculation by using the van der Waals equation.

9 Intermolecular Forces and Liquids
Water, Nature's Universal Solvent

Ocean waves crashing on a shoreline produce a spray that carries the ions dissolved in the ocean into the air. As the water in the spray evaporates, dry aerosols of sea salt form that are dispersed by the wind over land and sea.

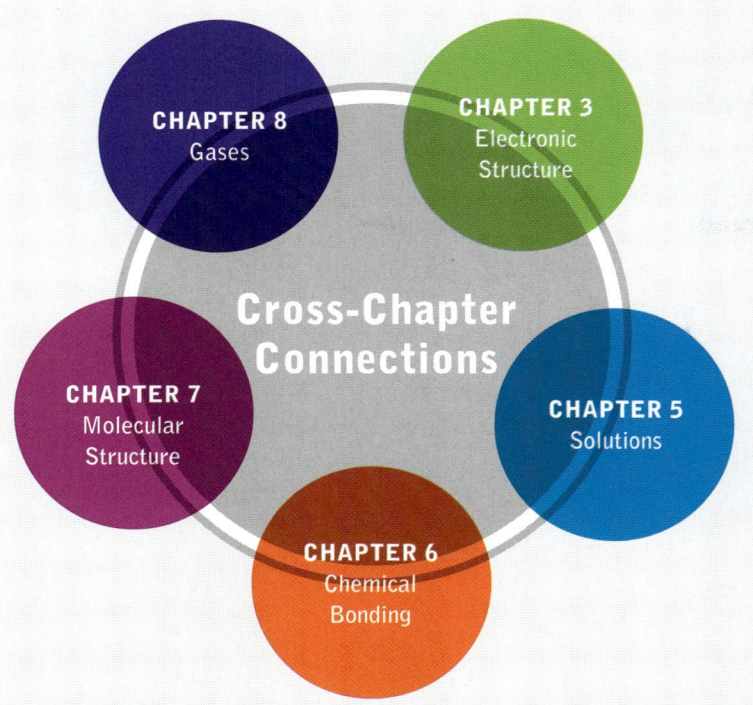

9.1 Sea spray and the states of matter

9.2 Ion–ion interactions and lattice energy

9.3 Interactions of polar molecules

9.4 Dispersion forces

9.5 Polarity and solubility

9.6 Vapor pressure
   Vapor pressure of solutions: a molecular view
   Vapor pressure and solute concentration
   Vapor pressure and temperature

9.7 Phase diagrams: intermolecular forces at work

9.8 The remarkable behavior of water

A Look Ahead

In this chapter, discussion of the solubility of substances in water introduces the different types of intermolecular forces. We examine the forces that attract atoms, ions, and molecules to one another and how the strengths of these forces determine the physical properties of substances. The strengths of intermolecular forces influence which physical state of a substance is stable at a particular combination of temperature and pressure, as shown in its phase diagram. Finally, we examine how hydrogen bonds contribute to some of the remarkable behavior of nature's universal solvent, H_2O.

9.1 SEA SPRAY AND THE STATES OF MATTER

Ocean waves crashing on a rocky shoreline create plumes of sea spray carrying small drops of seawater into the atmosphere, where they evaporate. As seawater turns into water vapor, the concentrations of sea salts, rich in Cl^-, Na^+, SO_4^{2-}, Mg^{2+}, Ca^{2+} and other ions, increase. Eventually, the diminished volume of the drops produces supersaturated solutions of the various sea salts, and they begin to precipitate. Among the first solids to form is $CaSO_4$. Among the last is $NaCl$,

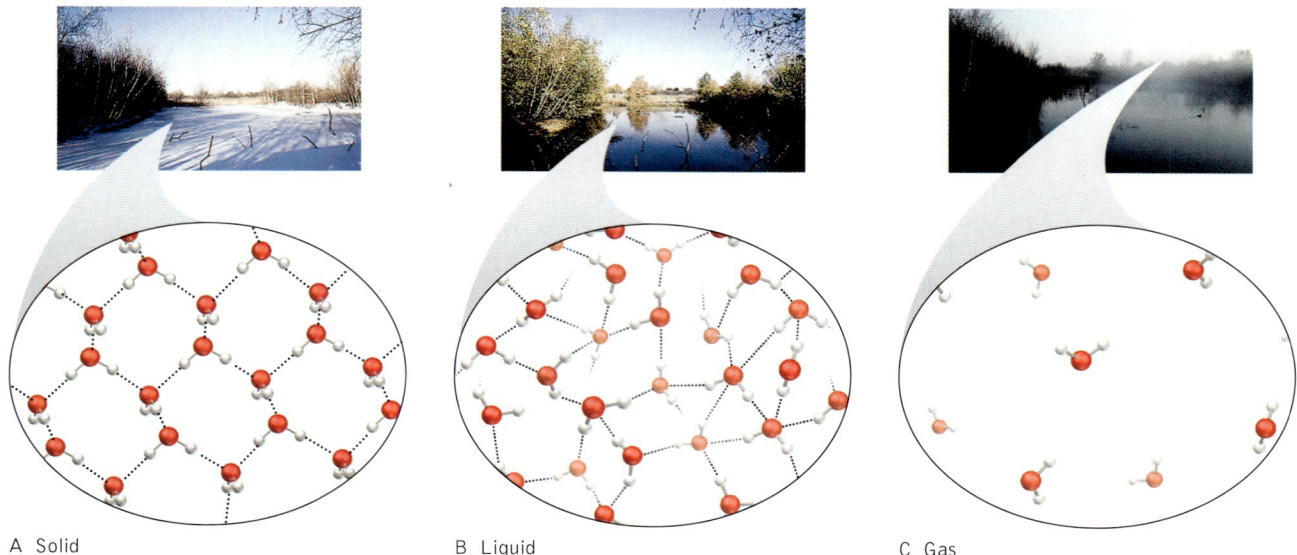

A Solid   B Liquid   C Gas

FIGURE 9.1 Water, like all matter, can exist in three states: (A) solid (ice), (B) liquid (water), or (C) gas (water vapor), depending on temperature and pressure. Water molecules in ice have the greatest number of mutual interactions; there are fewer such interactions in liquid water and hardly any in the gas phase. Water vapor is invisible; however, we can see the clouds and fog that form when atmospheric water vapor condenses into tiny drops of liquid H_2O.

CONNECTION: Table 5.1 lists the concentrations of the principal ions in seawater.

which does not precipitate until 90% of the seawater has evaporated. This sequence takes place even though the concentrations of Ca^{2+} and SO_4^{2-} are much smaller than the concentrations of Na^+ and Cl^-. Why does $CaSO_4$ precipitate before NaCl? Put another way: Why is NaCl more soluble in water than $CaSO_4$? To answer these questions, we need to examine the nature and strengths of the interactions between pairs of ions and between the ions and molecules of water.

> Why is NaCl more soluble in water than $CaSO_4$?

Let's examine the three common states of matter (Figure 9.1) from the perspective of the interactions between the atomic-scale building blocks: atoms, ions, and molecules. We will use the term *intermolecular forces* inclusively to refer to the forces of attraction between all such particles, not just molecules. In solids, there are extensive mutual interactions between adjacent particles that have only limited freedom of motion. In liquids, particles have weaker interactions with their neighbors and more freedom of movement. They are free to tumble over one another, allowing liquids to flow and to take the shape of their containers. In the gas phase, particles have the greatest freedom of motion and the least interaction with one another. Gases are free to expand and occupy the entire volume of their container.

According to this molecular view of the states of matter, the process of melting a solid requires that some intermolecular forces be overcome to achieve the freedom of motion of particles in the liquid phase. The decrease in molecular interaction during vaporization is much greater as particles achieve nearly total independence from one another in the gas phase. It makes sense that a substance made of particles that interact relatively strongly will have high melting and boiling points. Such a substance is likely to be a solid at room temperature. Substances with somewhat weaker intermolecular interactions are more likely to be liquids at room temperature, and those with the least intermolecular interaction are likely to be gases.

CONNECTION: The flows of energy that accompany phase changes are described in Chapter 11.

9.2 ION–ION INTERACTIONS AND LATTICE ENERGY

Ionic compounds are among the substances most likely to be solids at room temperature. They are solids because the attraction between ions of opposite charge is the strongest kind of intermolecular interaction. The strength of ion–ion interactions can be calculated by using **Coulomb's law**, which states that the energy (E) of the interaction between two ions is proportional to the product of the charges of the two ions (Q_1 and Q_2) and inversely proportional to the distance (d) between them:

$$E = 2.31 \times 10^{-19} \, \text{J} \cdot \text{nm} \left(\frac{Q_1 Q_2}{d} \right) \quad (9.1)$$

where the value of the constant of proportionality depends on the charge of an electron and the units of distance. The distance values in Equation 9.1 are calculated by summing the radii of the ions. The ionic radii of many of the main-group elements are given in Figure 9.2. Note the differences between the sizes of the ions and those of their parent atoms. The cations formed by Group 1A and 2A elements are considerably smaller than their parent atoms. This difference in size makes sense because these atoms lose all their valence-shell electrons to form their ions, and loss of their outermost electrons inevitably means a decrease in size. On the other hand, the anions of the elements on the right-hand side of the periodic table are larger than their parent atoms. Additional valence electrons in these anions are subject to more electron–electron repulsion, which results in an increase in the size of the valence shell.

Note that the value of E in Equation 9.1 is negative when Q_1 and Q_2 have opposite signs. These negative values of E make sense because separating oppositely charged ions that are attracted to each other requires the *addition* of energy. (We assume that ions that are completely free from each other have zero energy of interaction.) If we need to add energy to get to zero energy, then we must have started with a negative energy value. The energies calculated with Equation 9.1 are the energies that oppositely charged free ions *lose* when they come together and form an ion pair.

Let's calculate the energy of interaction between a Na$^+$ ion and a Cl$^-$ ion in NaCl. The charges of the ions are +1 and −1, and the distance d between them

CONNECTION: The charge on a monatomic ion is related to the electron configuration of its parent atom (Section 6.2).

☑ **Coulomb's law** states that the energy (E) of the interaction between two ions is proportional to the product of the charges of two ions (Q_1 and Q_2) and inversely proportional to the distance (d) between them.

should be the sum of their ionic radii. According to Figure 9.3, these values are 95 pm for Na^+ and 181 pm for Cl^-, so d is:

$$95 + 181 = 276 \text{ pm, or } 0.276 \text{ nm}$$

Substituting these values into Equation 9.1 and solving for E gives

$$E = 2.31 \times 10^{-19} \text{ J} \cdot \text{nm} \frac{[(+1)(-1)]}{0.276 \text{ nm}}$$

$$= -8.37 \times 10^{-19} \text{ J}$$

The strength of ion–ion interactions affects the **lattice energy** (U) of an ionic compound. Lattice energy is the energy released when free gaseous ions combine to form a mole of a solid ionic compound. The lattice energies of some common binary ionic compounds are given in Table 9.1. The formula for lattice energy (Equation 9.2) resembles the Coulomb's-law equation, except for one difference:

> ✓ The **lattice energy** (U) of an ionic compound is the energy released when a mole of the compound forms from free gaseous ions.

TABLE 9.1 Lattice Energies (U) of Some Common Ionic Compounds

Compound	U (kJ/mol)
LiF	−1047
LiCl	−861
NaCl	−769
KCl	−701
KBr	−671
$MgCl_2$	−2326
MgO	−3795

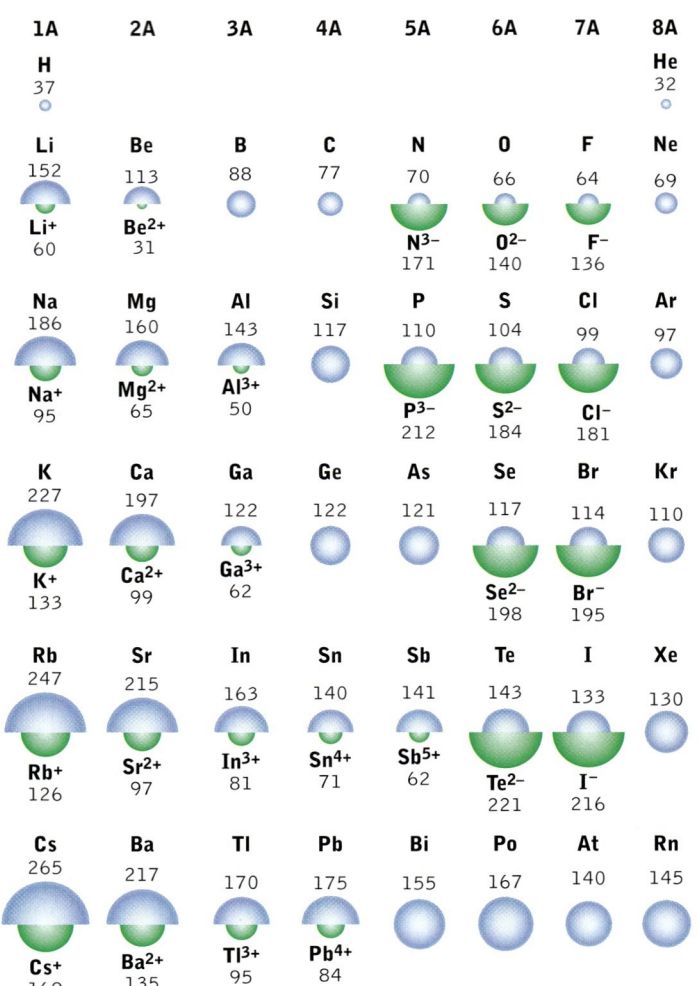

FIGURE 9.2 The radii of the anions formed by the main-group elements are larger than the radii of their parent atoms because additional valence-shell electrons lead to more electron–electron repulsion and an expanded valence shell. The radii of cations are smaller than those of their parent atoms. All values are in picometers (1 pm = 10^{-12} m).

Equation 9.2 includes a constant of proportionality (k) that depends on the structure of the ionic solid.

$$U = k\left(\frac{Q_1 Q_2}{d}\right) \quad (9.2)$$

We will examine the structures of solids in detail in Chapter 10. For now, we need to keep in mind that U values cannot be calculated simply by multiplying the right-hand side of Equation 9.1 by Avogadro's number. Consider NaCl, which has a lattice energy of -769 kJ/mol. The energy needed to separate a mole of Na^+ and Cl^- ions might be calculated in the following way:

$$(-8.37 \times 10^{-19} \text{ J}) \times (6.022 \times 10^{23}) \times (1 \text{ kJ}/1000 \text{ J}) = -504 \text{ kJ}$$

Why is this value different from the actual lattice energy of -786 kJ/mol? The answer lies in the different kinds of ion–ion interactions that take place in the three-dimensional array of ions in NaCl (Figure 9.3). The structure does not contain discrete pairs of ions. Instead, each Na^+ ion is surrounded by *six* Cl^- ions, and each Cl^- ion is surrounded by *six* Na^+ ions. In addition, there are forces of repulsion between ions of the same charge, particularly between the Cl^- ions that occupy adjoining spaces in the crystal. The net effect of these forces of attraction and repulsion is a lattice energy of NaCl that is not the same as that based only on the energy of interaction between a mole of Na^+ and Cl^- ion pairs.

SAMPLE EXERCISE 9.1: List the following ionic compounds in order of decreasing attraction between their ions: CaO, NaF, and CaF_2.

SOLUTION: Coulomb's law states that the strengths of the interactions between the ions in these compounds depend on the charges on the ions and the distances between them. We can calculate the distances between the three pairs of ions by using the ionic-radii data in Figure 9.3. For CaO, the distance is equal to the sum of the radii of Ca^{2+} and O^{2-} = 99 pm + 140 pm = 239 pm. Similar calculations for NaF and CaF_2 give ion–ion distances of 231 pm and 235 pm, respectively. Thus, the ion–ion distances in these three compounds are nearly the same. On the other hand, the combinations of charges on the ions in these compounds are quite different. The products of their charges ($Q_1 \times Q_2$) are -4 for CaO, -1 for NaF, and -2 for CaF_2. Therefore, the compound list in order of decreasing ionic interaction should be CaO, CaF_2, and NaF.

PRACTICE EXERCISE: Arrange the following ionic compounds in order of decreasing energy of attraction between their ions: $CaCl_2$, BaO, and NaCl. (See Problems 19 and 20.)

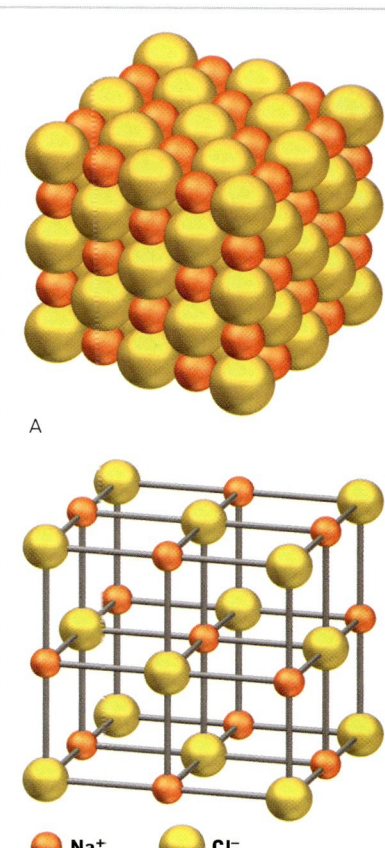

FIGURE 9.3 The crystalline structure of sodium chloride is represented by (A) a space-filling model and (B) a ball-and-stick model. Each Na^+ ion is surrounded by six Cl^- ions, and each Cl^- ion is surrounded by six Na^+ ions. The overall lattice energy of NaCl is the result of electrostatic attraction between Na^+ and Cl^- ions and repulsion between ions of the same charge.

How do differences in lattice energies affect the properties of ionic solids? For one thing, they affect the temperatures at which the solids melt. An ionic structure that is more tightly held together should require more energy and a higher temperature to melt into a liquid of freely flowing ions than does a structure that is less tightly held together—a trend that we observe experimentally. Consider two ionic compounds: LiF ($U = -1047$ kJ/mol) and MgO ($U = -3795$ kJ/mol). The nearly four times greater lattice energy of MgO is manifested in its much higher melting point, 2800°C versus 845°C for LiF, and much higher boiling point, 3600°C versus 1676°C for LiF. The solubilities of ionic compounds in water also are affected by the strengths of ion–ion interactions (as well as other factors that we are about to consider). The greater the force required for water molecules to pull apart the ions in a crystal of solute, the less soluble the solute is likely to be. We might predict that LiF is more soluble than MgO on the basis of the greater lattice energy of MgO, and that prediction would be correct: at 20°C, the solubility of LiF in water is 2.7 g/L and the solubility of MgO is only 0.006 g/L.

> How do differences in lattice energies affect the properties of ionic solids?

SAMPLE EXERCISE 9.2: Rank the following three ionic compounds in order of increasing melting point: LiF, NaCl, and KBr. These compounds have the same structure, and they have the same value of k in Equation 9.2.

SOLUTION: The compound with the greatest (most negative) lattice energy should have the highest melting point. Lattice energy, as given by Equation 9.2,

$$U = k\left(\frac{Q_1 Q_2}{d}\right)$$

is proportional to the product of the charges on the ions and inversely proportional to the distances between them. In this exercise, all of the cations are alkali-element cations and so have a charge of $+1$; all of the anions are halides and have a charge of -1. Therefore, any differences in lattice energy must be related to differences in the distance between ions, which are related to differences in their radii. A check of Figure 9.3 discloses that, within a group of elements, ionic radius increases with atomic number. Therefore, listing the cations in order of increasing size, we have: $Li^+ < Na^+ < K^+$. A similar list of the anions in order of increasing size is $F^- < Cl^- < Br^-$. Combining these list allows us to rank the three compounds in order of increasing distance between their ions. No math is required because, conveniently, LiF is made of the smallest cation and anion of the two groups, KBr is made of the largest, and NaCl is in the middle. Thus, the compounds ranked in order of increasing d values are: LiF < NaCl < KBr. Because U is inversely proportional to d, the rank ordering for these compounds based on increasing lattice energy and thus increasing melting point should be KBr < NaCl < LiF. This predicted order is confirmed by their actual melting points:

Compound	Melting point (°C)
KBr	734
NaCl	801
LiF	845

PRACTICE EXERCISE: Predict which of the following compounds should have the highest melting point: $CaCl_2$, $PbBr_2$, or TiO_2. All three have nearly the same structure. The radius of Ti^{4+} is 68 pm. (See Problems 21 and 22.)

Lattice energies are difficult to measure directly, but they can be calculated from the energy released when a binary ionic compound is formed by reaction of its constituent elements. Consider the reaction between sodium metal and chlorine gas to form NaCl. The reaction produces heat, which can be measured by an increase in the temperature of the reaction mixture. The heat released corresponds to an energy change (ΔE) of -411 kJ/mol NaCl. The negative sign again indicates that energy is lost, or released, by the reaction mixture. This energy change is actually the net result of the energy changes that are associated with a series of five processes (Figure 9.4):

1. The energy needed to vaporize a mole of sodium metal into free sodium atoms
2. The energy needed to break the covalent bonds in half a mole of Cl_2 to form a mole of free Cl atoms
3. The energy needed to ionize a mole of Na atoms, forming a mole of Na^+ ions and a mole of electrons
4. The energy released when a mole of Cl atoms combines with a mole of electrons, which is the electron affinity of Cl (-349 kJ/mol)
5. The energy released when a mole of Na^+ ions combines with a mole of Cl^- ions, forming solid NaCl, which is the lattice energy of NaCl.

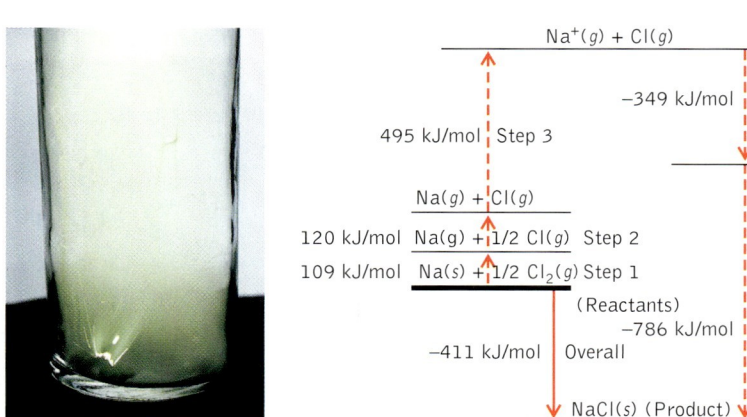

FIGURE 9.4 A. The violent reaction between sodium metal and chlorine gas releases more than 400 kJ per mole of NaCl product. B. The principal reason for this release is the energy lost when free sodium and chloride ions combine to form solid NaCl. This lost energy is the lattice energy of NaCl.

Energy is consumed by the reaction mixture in the first three steps but released in the last two steps:

Step	Process	Energy change (kJ)
1	$Na(s) \rightarrow Na(g)$	+109
2	$\frac{1}{2}Cl_2(g) \rightarrow Cl(g)$	+120
3	$Na(g) \rightarrow Na^+(g) + e^-$	+495
4	$Cl(g) + e^- \rightarrow Cl^-(g)$	−349
5	$Na^+(g) + Cl^-(g) \rightarrow NaCl(s)$	U
Overall:	$Na(s) + \frac{1}{2}Cl_2(g) \rightarrow NaCl(s)$	−411

The overall energy change is the sum of the individual energy changes in the five steps;

$$-411 \text{ kJ} = 109 \text{ kJ} + 120 \text{ kJ} + 495 \text{ kJ} - 349 \text{ kJ} + U$$

and so

$$U = -786 \text{ kJ/mol}$$

SAMPLE EXERCISE 9.3: The energy change in the reaction between a mole of lithium metal and half a mole of fluorine gas is −617 kJ. Calculate the lattice energy of LiF. Given: The energy required to vaporize a mole of Li metal is 161 kJ; the energy required to break half a mole of F–F bonds is 77 kJ; the energy required to ionize a mole of Li vapor is 520 kJ; and the electron affinity of F is −328 kJ/mol.

SOLUTION: The overall energy change for the reaction

$$Li(s) + \tfrac{1}{2}F_2(g) \longrightarrow LiF(s)$$

is equal to the sum of the energies required to vaporize solid Li, break F–F bonds, and ionize Li vapor plus the electron affinity of F and the lattice energy of LiF.

Substituting the given values of these energies gives

$$-617 \text{ kJ} = 161 \text{ kJ} + 77 \text{ kJ} + 520 \text{ kJ} - 328 \text{ kJ} + U$$

and, solving for the lattice energy, we have

$$U = -617 \text{ kJ} - (161 \text{ kJ} + 77 \text{ kJ} + 520 \text{ kJ} - 328 \text{ kJ})$$

$$= -1047 \text{ kJ/mol}$$

PRACTICE EXERCISE: Burning magnesium metal in air produces MgO and a very bright white light, making the reaction popular in fireworks and signaling devices. The energy change that accompanies this reaction is −602 kJ/mol MgO. Calculate the lattice energy of MgO from the following energy changes:

Process	Energy change (kJ)	Process	Energy change (kJ)
$Mg(s) \rightarrow Mg(g)$	150	$Mg(g) \rightarrow Mg^{2+}(g)$	2188
$O_2(g) \rightarrow 2\,O(g)$	499	$O(g) + 2\,e^- \rightarrow O^{2-}(g)$	638

(See Problems 23 and 24.)

9.3 INTERACTIONS OF POLAR MOLECULES

The solubilities of ionic compounds in water are affected by factors in addition to their lattice energies. These other factors include the intermolecular force between an ion and the end of the polar water molecule with the opposite partial charge. This force is an **ion–dipole** intermolecular force. When ionic solids dissolve, ion–dipole forces pull solute ions from the solid into solution. As an ion is pulled away from its solid-state neighbors, it is surrounded by water molecules that form a **sphere of hydration.** These dissolved ions are said to be *hydrated* or, in the general case, *solvated.* When the strengths of the ion–dipole forces in these spheres of hydration are strong enough to overcome the ion–ion interactions of the ionic compound, the compound dissolves. Within a sphere of hydration, the water molecules closest to an ion are oriented so that either their oxygen atoms and negatively charged (−) poles are directed toward a cation (such as Na^+ in Figure 9.5A) or their hydrogen atoms and positively charged (+) poles are directed toward an anion (such as Cl^- in Figure 9.5B). The number of molecules oriented in this way depends on the size of the ion. The number is typically six but can range from four to nine.

Surrounding the water molecules closest to the ions in Figure 9.5 are other water molecules that are less structured than the layer next to the ion but are not completely randomly oriented either. The ordering of these other water molecules is caused by yet another intermolecular force: **dipole–dipole interactions,** which are shown as blue dashed lines in Figure 9.5. Dipole–dipole interactions take place between molecules that have permanent dipole moments. In molecules of water, the partial electrical charges on oxygen and hydrogen atoms result in attractions between a hydrogen atom of one molecule and the oxygen atom of another. Dipole–dipole interactions are not as strong as ion–dipole interactions, because dipole–dipole interactions entail only partial charges caused by unequal sharing of electrons, whereas one of the particles in an

 CONNECTION: The energy changes associated with the dissolution of a compound are discussed in Section 13.1.

 CONNECTION: The solubility trends for ionic compounds given in Section 5.7 are related to the strengths of intermolecular forces.

Hydration is the process by which a dissolved ion or molecule is surrounded by an ordered array of water molecules called a **sphere of hydration.** The hydration of ions is a result of **ion–dipole interactions.**

Attractive forces between polar molecules are called **dipole–dipole interactions.**

 CONNECTION: The solvation of ions is described in Sections 5.3 and 13.1.

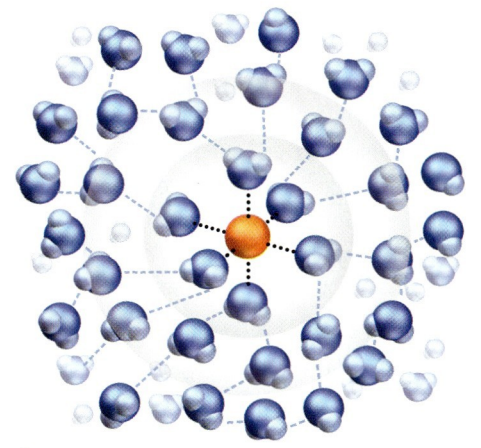
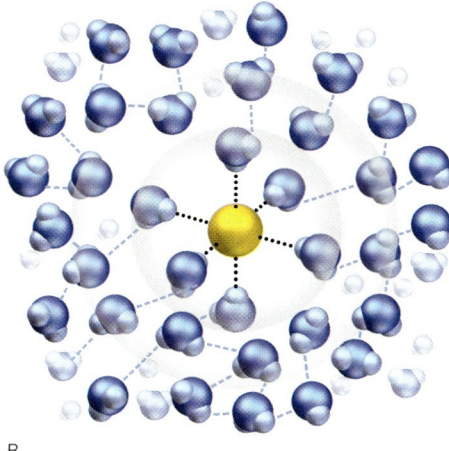

A   B

FIGURE 9.5 When an ionic compound dissolves in water, surrounding water molecules are oriented so that their negative dipoles point toward dissolved cations (A) and their positive dipoles point toward anions (B). Surrounding these water molecules are others that interact with one another through hydrogen bonds (dashed blue lines).

CONNECTION: Dipole moments (Section 7.8) are measures of the polarities of molecules.

✓ A **hydrogen bond** is a particularly strong kind of dipole–dipole interaction between a hydrogen atom bonded to a highly electronegative element (F, O, N) and an atom of one of these elements in another molecule.

CONNECTION: The strengths of covalent bonds are discussed in Section 11.3.

CONNECTION: The role of hydrogen bonding in the structure of DNA is described in Section 13.11.

✓ **London,** or **dispersion, forces** are the intermolecular forces caused by temporary dipoles induced by neighboring atoms or molecules. The strengths of the forces depend on how easily the atoms or molecules can be **polarized.**

ion–dipole interaction has completely lost or gained electrons and has a full positive or negative charge.

CONCEPT TEST: Dimethyl ether, CH_3OCH_3, and acetone, $CH_3C(O)CH_3$, have similar formulas and molar masses. However, the dipole moments of dimethyl ether and acetone are quite different: 1.30 and 2.88 debyes, respectively. Predict which compound has the higher boiling point.

Polar molecules containing O–H and N–H bonds and the molecule HF have large dipole moments and stronger-than-average dipole–dipole interactions. These interactions between partially positive H atoms and partially negative O, N, or F atoms on adjacent molecules are called **hydrogen bonds.** About 1/10th the strength of covalent bonds, hydrogen bonds play a key role in defining the remarkable properties of H_2O described in Section 9.6. They also strongly influence the overall three-dimensional structures of biopolymers such as proteins and DNA.

9.4 DISPERSION FORCES

Are there intermolecular forces between nonpolar molecules? Indeed there are; they are called **dispersion forces** or **London forces** named in honor of German American physicist Fritz London (1900–1954). His work in quantum mechanics, beginning in the 1920s, provided an explanation for the attractions between molecules. These forces are weaker than ion–ion and ion–dipole forces, but they play an important role in defining the physical properties of molecular substances. When two noble-gas atoms or molecules approach one another, mutual repulsion between their electron clouds perturbs the distributions of the electrons in these clouds *inducing* nonuniformity in them and creating *temporary* dipoles, as shown in Figure 9.6. The magnitude of an induced dipole depends on the ease with which the electron cloud surrounding a molecule, ion, or atom can be perturbed, or **polarized.**

CONNECTION: Intermolecular forces are responsible for nonideal behavior in gases, as described in Section 8.7.

SAMPLE EXERCISE 9.4: The van der Waals equation, (See Section 8.6) relates the pressure, volume, and temperature of a quantity of a real gas. It includes a correction term that accounts for interactions between molecules: an^2/V^2. The van der Waals constant *a* for SO_2, 6.71 $L^2 \cdot atm/mol^2$, is nearly twice the value of *a* for CO_2, which is 3.59 $L^2 \cdot atm/mol$. Can you think of a reason why?

SOLUTION: The magnitude of van der Waals constant *a* corresponds to the strength of the intermolecular forces in a gaseous substance. Carbon dioxide molecules are symmetric and nonpolar; so they interact through relatively weak London forces. However, sulfur dioxide molecules have a permanent dipole moment and so are attracted to one another by stronger dipole–dipole forces. These stronger interactions lead to a larger value of *a*.

PRACTICE EXERCISE: Without looking up the values in Chapter 8, place the following substances in order of increasing value of the van der Waals constant a: H_2O, O_2, and CO. (See Problems 27 and 28.)

What factors control how easily the cloud of electrons around an atom or molecule is polarized? To answer this question, let's examine how dispersion forces influence the physical properties of nonpolar compounds, including their melting and boiling points. A comparison of the boiling points of the monatomic noble gases and the nonpolar diatomic halogens (Table 9.2) discloses the same trend for both groups: boiling points increase with increasing molar mass. The molecular interpretation of this trend is that larger clouds of electrons are more easily perturbed and polarized than smaller ones, leading to stronger induced dipoles and stronger intermolecular interactions, which correspond to higher melting points and boiling points.

> ✓ The strengths of dispersion forces are proportional to the sizes of the interacting particles.

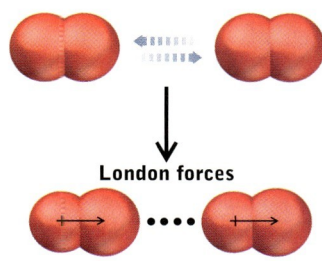

FIGURE 9.6 Nonpolar molecules such as N_2 or O_2 and atoms of the noble gases interact through London (dispersion) forces. As two molecules of O_2 approach each other, repulsions between their electrons induce a dipole in each molecule. London forces are the result of induced-dipole–induced-dipole interactions.

TABLE 9.2 Molar Masses and Boiling Points of the Halogens and Noble Gases

Halogen	$\mathcal{M}$ (g/mol)	Boiling Point (K)	Noble Gas	$\mathcal{M}$ (g/mol)	Boiling Point (K)
			He	4	4
F_2	38	85	Ne	20	27
Cl_2	71	239	Ar	40	87
Br_2	160	333	Kr	84	120
I_2	254	458	Xe	131	165
			Rn	222	211

CONCEPT TEST: Explain, using the concept of intermolecular forces, why CF_4 is a gas at room temperature but CCl_4 is a liquid.

9.5 POLARITY AND SOLUBILITY

We have seen how the solubilities of compounds can be explained in terms of solute–solute and solute–solvent interactions. For example, if ion–ion interactions are particularly strong or ion–dipole interactions in solution are weak, then an ionic solute will have only limited solubility. These considerations help explain the observation made on the first page of this chapter: $CaSO_4$ precipitates before $NaCl$ as seawater evaporates because the ion–ion interactions between Ca^{2+} and SO_4^{2-} are much stronger than the interactions between Na^+ and Cl^-, making $CaSO_4$ less soluble in H_2O than $NaCl$.

Just as ionic compounds dissolve in polar solvents because of strong ion–dipole interactions, polar solutes tend to dissolve in polar solvents because of dipole–dipole interactions between solute and solvent molecules. On the other hand, nonpolar solutes tend not to dissolve in polar solvents because the solvent–solute interactions that promote dissolution are weaker than those that keep solute molecules together in one phase and solvent molecules together in another. Consider the solubilities of three organic compounds with these structures:

$$\underset{\text{2-Propanone}}{H_3C-\overset{\overset{O}{\|}}{C}-CH_3} \quad \underset{\text{2-Butanone}}{H_3C-\overset{\overset{O}{\|}}{C}-CH_2-CH_3} \quad \underset{\text{2-Pentanone}}{H_3C-\overset{\overset{O}{\|}}{C}-CH_2-CH_2-CH_3}$$

All three compounds contain a C=O double bond. Unequal sharing of the electrons in this C=O double bond makes it, as well as the overall molecule, polar. All three compounds are liquids at room temperature. The smallest of them, 2-propanone, also known as acetone, dissolves in water in all proportions. When two liquids dissolve in each other in this way, they are said to be completely **miscible**. The unlimited solubility of acetone in water (and water in acetone) is the result of hydrogen bonding between the hydrogen atoms of water molecules and the oxygen atom of acetone, as shown by the blue dashed lines in Figure 9.7.

> ☑ **Miscible** liquids form homogeneous solutions when mixed in all proportions.

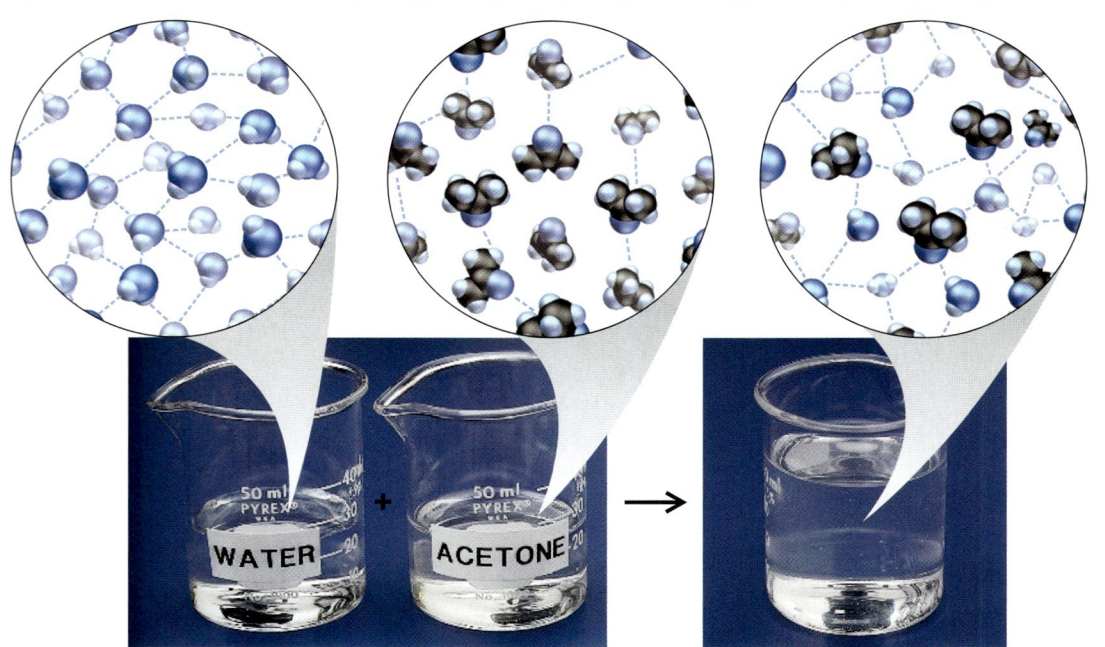

FIGURE 9.7 *What a Chemist Sees.* 2-Propanone (acetone) dissolves in water (and vice versa) because of hydrogen bonds between the hydrogen atoms in water molecules and the oxygen atoms in polar C=O groups in 2-propanone molecules.

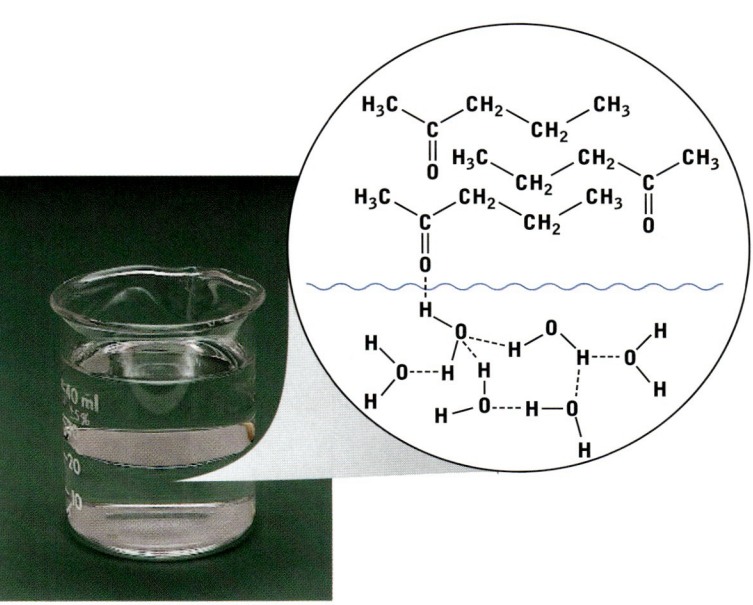

FIGURE 9.8 Water and 2-pentanone are immiscible. Although there is some hydrogen bonding between hydrogen atoms in water molecules and the oxygen atoms in polar C=O groups in 2-pentanone molecules, this hydrogen bonding does not offset the energy needed to overcome the London forces between the nonpolar hydrocarbon chains in 2-pentanone and to break the hydrogen bonds in water.

If we add a —CH_2— group to a molecule of acetone, we have a molecule of 2-butanone. This compound also is quite soluble in water, though it does not have the unlimited solubility of acetone. Adding another —CH_2— group gives us a molecule of 2-pentanone, a liquid that is nearly insoluble in water. In other words, water and 2-pentanone are **immiscible** liquids. Why do seemingly small differences in the molecular structures of these compounds lead to such large differences in their solubility in water? The answer lies in the lengths of the nonpolar chains of carbon and hydrogen atoms attached to the polar C=O groups. Compounds with longer nonpolar chains are less polar overall. In addition, longer nonpolar chains have larger clouds of electrons, which are more easily polarized, leading to stronger dispersion forces. These stronger forces tend to keep nonpolar molecules or nonpolar parts of molecules together, which inhibits their solubility in a polar solvent such as water, as shown in Figure 9.8. Such nonpolar interactions are sometimes called *hydrophobic* (literally, "water fearing") interactions, whereas interactions that promote solubility in water are called *hydrophilic* ("water loving") interactions.

> Do nonpolar solutes dissolve in water?

Do nonpolar solutes dissolve in water? It turns out that they can, but only a little. The limited solubility that they have results from the ability of polar solvent molecules to induce dipoles in nonpolar solutes (Figure 9.9). These **dipole–induced dipole interactions** are weaker than the dipole–dipole forces between molecules of a polar solvent, and they may not be stronger than the London dispersion forces between molecules of a nonpolar solute. Thus, nonpolar molecules

> ✓ The interaction of a polar molecule with a non-polar molecule or an atom may result in a **dipole–induced dipole interaction**.

FIGURE 9.9 Water contains dissolved oxygen. As O_2 and H_2O approach each other, the dipole of water causes polarization of the O=O double bond in the O_2 molecule. The resulting dipole–induced-dipole interaction accounts for the solubility of O_2 in water that allows fish and other aquatic life to survive.

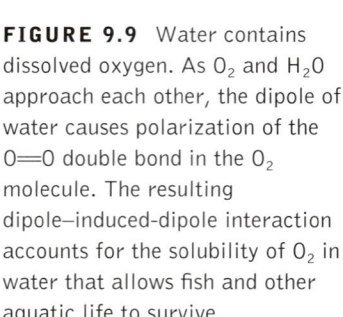
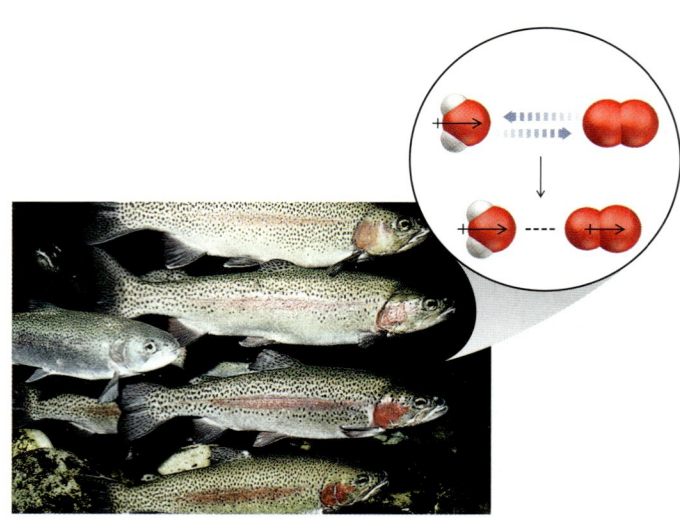

such as O=O are very sparingly soluble in polar solvents such as H_2O. Fortunately for aquatic organisms, the limited solubility of oxygen in water, about 10 mg/L at normal atmospheric pressures, is sufficient to sustain life.

CONCEPT TEST: Which of the noble gases should be the most soluble in water?

CONCEPT TEST: Do you think there can be any intermolecular interaction between ions and nonpolar molecules? How would you describe such an interaction?

9.6 VAPOR PRESSURE

An open beaker of seawater on a laboratory bench will evaporate over time, leaving behind a moist residue of "sea salts." However, if a beaker of seawater and a beaker of distilled water are placed in a sealed chamber, as shown in Figure 9.10, the volume of the seawater does not decrease; it actually *increases* over time. Meanwhile, the volume of distilled water decreases at the same rate at which the volume of the liquid in the beaker that originally contained seawater increases. What causes these changes in volume? The answer is connected to the intermolecular interactions between water molecules and the presence of dissolved ions in seawater.

CONNECTION: Like boiling-point elevation and melting-point depression (Section 5.4), vapor-pressure lowering is a colligative property of a solution.

Vapor pressure of solutions: a molecular view

As water in either of the beakers in Figure 9.10 evaporates, the concentration of water molecules in the air space in the bell jar increases. On a molecular level, evaporation takes place when molecules of water at the surface break from the other molecules of liquid water that surround them and become molecules of water vapor. This process is shown by the blue arrows in Figure 9.10. As the concentration of water vapor increases, the partial pressure of water vapor increases. Eventually this partial pressure stabilizes at a value equal to the **vapor pressure** of water (P_{H_2O}). Under these conditions, the rate at which molecules evaporate from the beakers is balanced by the rate at which molecules of water vapor *condense* into them, as represented by the red arrows in Figure 9.10. If the rates of evaporation and condensation are the same, the liquid levels in the beakers do not change over time.

However, the liquid levels in the two beakers in Figure 9.10 *do* change until there is no distilled water left. Because the molecules of water vapor in the chamber are free to condense into either beaker, we can reasonably assume that the rates of condensation of water vapor into both beakers is the same. Therefore, the transfer of water from one beaker to the other must be due to a higher rate of evaporation of distilled water than that of seawater. This conclusion leads to another one: the vapor pressure of seawater is less than the vapor pressure of distilled water.

☑ The evaporation of a liquid substance produces gas molecules that are responsible for the **vapor pressure** of the liquid.

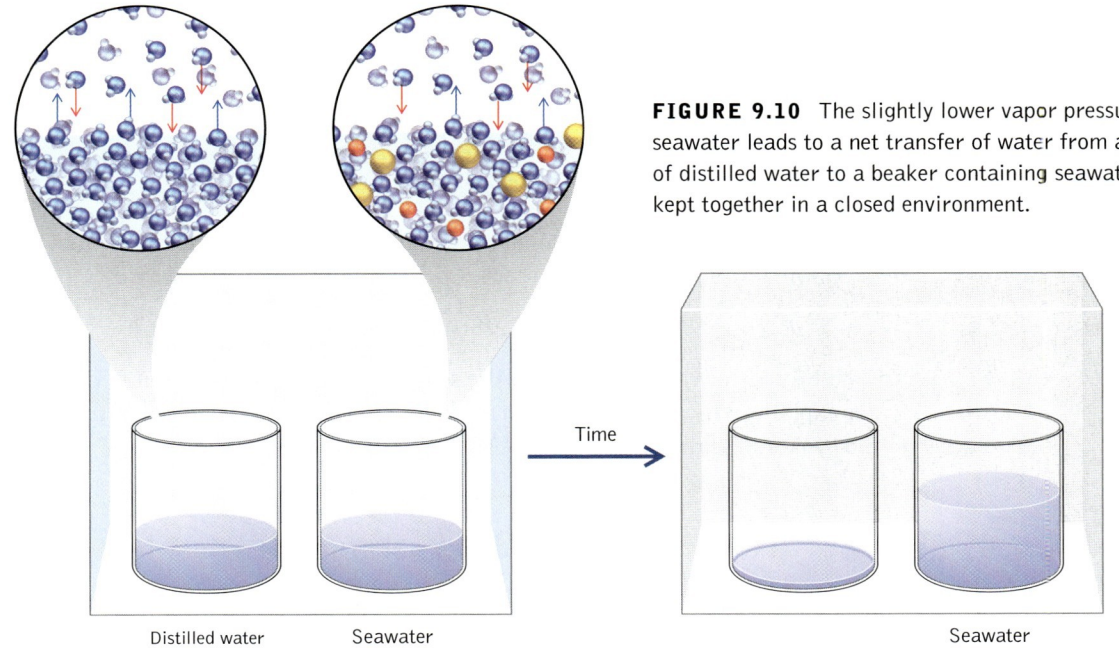

FIGURE 9.10 The slightly lower vapor pressure of seawater leads to a net transfer of water from a beaker of distilled water to a beaker containing seawater when kept together in a closed environment.

> **Raoult's law** states that the vapor pressure of a solution containing nonvolatile solutes is proportional to the mole fraction of the solvent.

CONNECTION: The mole fraction is also used to calculate the partial pressure of a gas in a mixture of gases (Section 8.4).

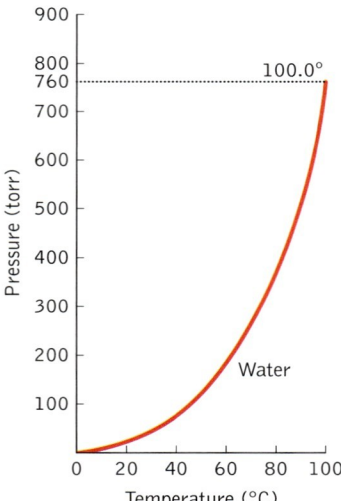

FIGURE 9.11 The vapor pressure of water (or any pure liquid) increases with increasing temperature, which increases the average kinetic energy of the molecules in the liquid.

Vapor pressure and solute concentration

The connection between the vapor pressure of solutions and their concentrations of nonvolatile solutes was studied extensively by French chemist François Marie Raoult (1830–1901). He discovered the following relation between the vapor pressure of a solution and that of the pure solvent, a relation called **Raoult's law:**

$$P_{\text{solution}} = X_{\text{solvent}} \times P_{\text{solvent}} \quad (9.3)$$

where X_{solvent} represents the mole fraction of solvent—that is, the moles of solvent divided by the total number of moles of solvent and solutes:

$$X_{\text{solvent}} = \frac{\text{mol solvent}}{\text{mol solvent} + \text{mol solutes}}$$

Notice that X_{solute} is defined as moles of solute/total moles. The sum of X_{solutes} and X_{solvent} is always equal to one.

Let's reconsider the two beakers in Figure 9.10. At 20°C (293 K), the vapor pressure of water ($P_{\text{H}_2\text{O}}$), produced by evaporation of the distilled water is 0.0231 atm. What is the vapor pressure produced by evaporation of water from seawater if the total concentration of all sea-salt ions is 1.15 m? To calculate this vapor pressure, we first need to convert the molality of the sea-salt ions into the mole fraction of solvent (water) in seawater. In a 1.15-m aqueous solution there are 1.15 moles of ions per kilogram of water. Using the molar mass of water, we can calculate the moles of water in a kilogram (1000 g):

$$\frac{1000 \text{ g H}_2\text{O}}{18.0 \text{ g/mol}} = 55.5 \text{ mol H}_2\text{O}$$

The mole fraction of water (X_{solvent}) in seawater is

$$X_{\text{H}_2\text{O}} = \frac{\text{mol H}_2\text{O}}{\text{total mol}} = \frac{\text{mol H}_2\text{O}}{\text{mol H}_2\text{O} + \text{mol solute}} = \frac{55.6 \text{ mol}}{55.6 + 1.15 \text{ mol}} = 0.980$$

We can use this mole-fraction value to calculate the vapor pressure of the seawater solution:

$$P_{\text{solution}} = X_{\text{H}_2\text{O}} \times P_{\text{H}_2\text{O}} = (0.980)(0.0231 \text{ atm}) = 0.0226 \text{ atm}$$

The slightly lower vapor pressure of the seawater compared with that of distilled water explains why, over time, water is transferred from the beaker of pure water to the beaker of seawater in Figure 9.10.

Vapor pressure and temperature

The vapor pressure of water increases with increasing temperature, as shown in the graph in Figure 9.11. We can explain this phenomenon on a molecular level by applying the principle that molecular motion of molecules increases with increasing temperature. Thus higher temperature means that a greater fraction of the water molecules have enough kinetic energy to break away from the surface and enter the vapor phase. At 100°C, the vapor pressure of pure water is 1 atm, and water boils. At this temperature, essentially *all* the water molecules have enough energy to enter the vapor phase. In the general case, the boiling point of any liquid is the temperature at which its vapor pressure equals atmospheric pressure.

SAMPLE EXERCISE 9.5: The fluid used in automobile cooling systems is prepared by dissolving ethylene glycol ($C_2H_6O_2$, M = 62.07 g/mol) in water (see Section 5.4). What is the vapor pressure of a solution prepared by mixing 1.00 gallon of ethylene glycol (density = 1.114 g/mL) with 1.00 gallon of water (density = 1.000 g/mL) at 100°C? Assume that the mixture obeys Raoult's law.

SOLUTION: To answer this question, we need to calculate the mole fraction of water (X_{water}) in the solution. To do so requires the conversion of the volumes given into equivalent numbers of moles. Let's start with the observation that the ratio of 1.00 gallon of ethylene glycol to 1.00 gallon of water is equivalent to the ratio of 1.00 liter of ethylene glycol to 1.00 liter of water. The number of moles of solute and solvent in 1.00 liter of each can be calculated from their densities and molar masses as follows:

vapor pressure and mole fraction of solvent:
$P_{solvent}$, $X_{solvent}$

$P_{solution} = X_{solvent} P_{solvent}$

vapor pressure of solution

For $C_2H_6O_2$: $1.0 \;\cancel{L} \left(\dfrac{1000 \;\cancel{mL}}{\cancel{L}}\right)\left(\dfrac{1.114 \text{ g}}{\cancel{mL}}\right)\left(\dfrac{1 \text{ mol}}{62.07 \text{ g}}\right) = 17.95 \text{ mol}$

For H_2O: $1.0 \;\cancel{L} \left(\dfrac{1000 \;\cancel{mL}}{\cancel{L}}\right)\left(\dfrac{1.000 \text{ g}}{\cancel{mL}}\right)\left(\dfrac{1 \text{ mol}}{18.02 \text{ g}}\right) = 55.6 \text{ mol}$

The mole fraction of water (X_{H_2O}) is

$$X_{H_2O} = \dfrac{\text{mol } H_2O}{\text{total mol}} = \dfrac{\text{mol } H_2O}{\text{mol } H_2O + \text{mol } C_2H_6O_2} = \dfrac{55.6 \;\cancel{\text{mol}}}{(55.5 + 17.95) \;\cancel{\text{mol}}} = 0.756$$

Because 100°C is the boiling point of water at 1.00 atm of pressure, P_{H_2O} must be 1.00 atm. Using this value and the calculated mole fraction of water in the mixture in Equation 9.3, we have

$$P_{solution} = X_{H_2O} \times P_{H_2O} = (0.756)(1.00 \text{ atm}) = 0.756 \text{ atm}$$

PRACTICE EXERCISE: Glycerol ($C_3H_8O_3$), like ethylene glycol, is a member of a family of organic compounds that have multiple C–O–H groups in their molecular structures and that readily dissolve in water. What is the vapor pressure of water in a 50:50 (by volume) mixture of glycerol (density = 1.261 g/mL) and water at 25°C (P_{H_2O} = 23.8 torr)? (See Problems 37 and 38.)

9.7 PHASE DIAGRAMS: INTERMOLECULAR FORCES AT WORK

The strengths of intermolecular forces control whether a compound or element exists as a solid, liquid, or gas at room temperature. With increasing temperature, substances are more likely to be gases and less likely to be solids. However, temperature is not the only factor that influences the state of a substance. Pressure also plays a significant role. Scientists use **phase diagrams** such as the one for water in Figure 9.12 to show which physical states are stable at various combinations of temperature and pressure.

The phase diagram of water (like those of many pure substances) has three lines separating regions that correspond to the three states of matter. The line separating the solid and liquid states represents a series of melting points: combinations of temperature and pressure at which the solid and liquid states can exist in equilibrium with each other. The line separating liquid and gaseous states is a series of boiling points, and the line separating the solid and gaseous states represents a series of sublimation points. (Recall that sublimation is the direct conversion of a solid into a gas without an intermediate liquid state.)

Notice that the line representing boiling points in Figure 9.12 curves from the lower left to the upper right. As already noted, a liquid boils when it is heated to the temperature at which its vapor pressure equals atmospheric pressure. If atmospheric pressure is increased, then the vapor pressure required to overcome it also must increase, which requires a higher temperature. This phenomenon applies to the pressure cooker, as discussed in Section 5.4. A similar orientation of sublimation points (the solid–gas curve) also makes sense: higher atmospheric pressures make it more difficult for molecules of ice to leave the solid phase and become water vapor.

The orientation of the boiling point and sublimation curves in Figure 9.12 can also be explained by the much lower density of water vapor (and gases, in general) than the densities of liquid water or solid ice. Liquids and solids are hun-

> A **phase diagram** is a graphical presentation of the dependence of the stabilities of the physical states of a substance on temperature and pressure.

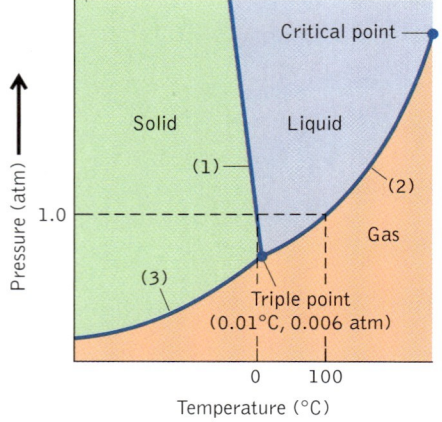

FIGURE 9.12 The phase diagram for water indicates which phase (solid, liquid, or gas) is stable at various combinations of pressure and temperature. The three solid lines represent temperature and pressure combinations at which two phases can coexist and so represent collections of (1) melting points (between solid and liquid), (2) boiling points (between liquid and gas), and (3) sublimation points (between solid and gas). The triple point is the temperature and pressure at which all three phases can co-exist. The line separating liquids and gases ends at the critical point, above which the liquid and gaseous states are indistinguishable.

Boiling points increase with increasing pressure. In a pressure cooker such as the one shown here, the boiling point of water is greater than 100°C because the pressure inside the pot exceeds 1 atm.

dreds to thousands of times more dense as the corresponding gases. Therefore, applying pressure to a gas may force it to change into a state (liquid or solid, depending on temperature) that takes up much less volume, thereby relieving the pressure.

CONCEPT TEST: Why do directions for cooking pasta call for longer cooking times at high altitudes?

The line representing melting points in Figure 9.12 is oriented differently from the other two. Its tilt from lower right to upper left means that the temperatures at which ice melts *decrease* as pressure *increase.* This trend is the opposite of what is observed for almost all other substances. This unusual melting-point dependence on pressure is caused by the fact that water expands when it freezes. Most other substances shrink a little when they freeze. Water expands because hydrogen bonds in the structure of ice create a more open structure than that of liquid water, making ice less dense. Thus, applying enough pressure to ice will force it into a physical state (liquid water) in which it takes up less volume, thereby relieving some of the pressure.

The three lines in Figure 9.12 meet at the **triple point** for water, which is the temperature and pressure at which all three states can co-exist with one another. For water, the triple point is just above the normal melting point at 0.010°C, but at a very low pressure: 0.0060 atm (4.6 torr). The curve separating the liquid and gaseous states ends at a combination of temperature and pressure called the **critical point** because, at that point, the liquid and gaseous states are indistinguishable. This point is reached because thermal expansion at these high temperatures causes the liquid form to become less dense, while higher pressures mean that the gas phase is compressed into a small volume, making it more dense. The densities of the liquid and gaseous states are equal at the critical point so that one cannot be distinguished from the other.

Above its **critical temperature** and **critical pressure,** a substance exists as a **supercritical fluid.** It has most of the physical properties of a gas but has the ability to dissolve other substances as if it were a liquid. This property has considerable importance in the food-processing industry, where supercritical

✓ All three states of a substance are stable at a particular temperature and pressure called its **triple point.**

The liquid and gaseous states of a substance are indistinguishable at its **critical point.**

At temperatures and pressures above its **critical temperature** and **critical pressure,** the liquid and gaseous states of a substance are indistinguishable. Under these conditions, the substance is a **supercritical fluid.**

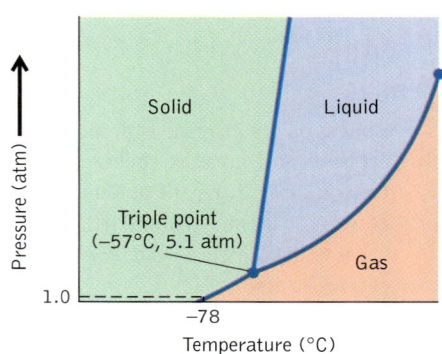

FIGURE 9.13 The phase diagram of carbon dioxide has a triple point that is well above standard pressure (1 atm). The tilt of the line separating the solid and liquid phases indicates that solid CO_2 (unlike solid H_2O) melts at temperatures that increase with increasing pressure.

carbon dioxide is widely used to, for example, decaffeinate coffee and to remove fat from potato chips. The phase diagram for CO_2 is shown in Figure 9.13. Note that the triple point is at $-57°C$ and 5.1 atm, which means that solid dry ice does not melt at room temperature and normal atmospheric pressures; rather, it sublimes directly to CO_2 gas. Liquid CO_2 does not exist in nature except under very high pressures (Figure 9.14). The critical point of CO_2 is at $31.1°C$ and 73.0 atm, which is a pressure readily achieved by compressors in laboratories, factories, and food-processing plants.

CONCEPT TEST: Would you expect supercritical CO_2 to be a better solvent for polar or nonpolar solutes?

SAMPLE EXERCISE 9.6: Describe the phase changes that take place as the pressure on a quantity of water at 0°C is increased from 1 torr to 100 atm.

SOLUTION: At extremely low pressure, water or any substance is likely to be a gas because gases take up the most volume. Thus, water at 0°C and 1 torr is a vapor, as predicted by its phase diagram. If we follow a vertical line starting at 0°C upward toward higher pressure, the vapor will solidify to ice at about 0.005 atm. Increasing the pressure further to 1 atm will melt the ice.

PRACTICE EXERCISE: Describe the phase change that occurs when the pressure of CO_2 at $-60°C$ is increased from 0.02 atm to 70 atm. (See Problems 43 and 44.)

CONCEPT TEST: Propane is a gas at room temperature; its normal boiling point is $-42°C$. Yet the propane in the fuel tanks of barbecue grills is in the liquid state. How can this be?

A

B

FIGURE 9.14 A. At room temperature, solid CO_2 does not melt but rather sublimes. B. Liquid CO_2 forms only under high pressures such as those found deep in the ocean, which scientists are evaluating as a disposal site for CO_2 that might otherwise contribute to global warming.

9.8 THE REMARKABLE BEHAVIOR OF WATER

Water has some remarkable properties. Its melting and boiling points are much higher than those of all other substances with similar molar masses. Ice is less dense than liquid water. Small insects that are more dense than water can walk on it. Water can rise 100 meters and more through the trunks of tall trees. All of these phenomena are related to the strength of the hydrogen bonds that attract water molecules to one another.

The strength of the hydrogen bonds between surface molecules of water accounts for its high **surface tension,** which is the resistance of a liquid to an increase in its surface area. It is also equal to the energy required to move molecules apart so that an object can break through the surface. Hydrogen bonding creates such a high surface tension of water, 7.29×10^{-2} J/m^2, that a steel needle can float on water (Figure 9.15). The same needle would sink in, for example, oil or gasoline.

CONCEPT QUESTION: A cold needle floats on water, but a hot needle sinks. Can you explain why?

Water in a graduated cylinder or pipet forms a concave surface, called a **meniscus.** The curved surface is the result of competing *cohesive forces* (hydrogen bonds between water molecules) and *adhesive forces* (dipole–dipole interactions between water molecules and polar Si–O–H groups on the glass surface). The adhesive forces are strong enough to cause the water to climb upward to increase contact with the container. Water molecules on the surface and next to the container wall have less contact with other water molecules and so are drawn upward by adhesive forces. In very narrow tubes, such as capillaries, ad-

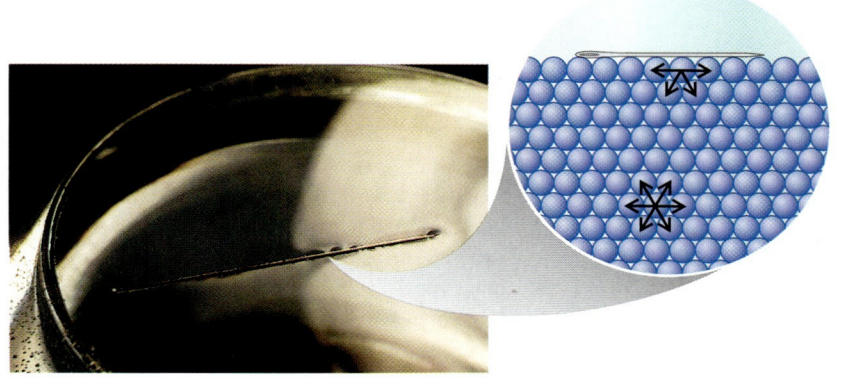

FIGURE 9.15 Unlike intermolecular forces in the interior of a liquid, intermolecular forces, such as hydrogen bonds in water, between the molecules at the liquid's surface are not equal in all directions. The asymmetry of the surface interactions makes it more difficult to move them apart and so creates surface tension. If the magnitude of the surface tension exceeds the downward force due to gravity that is exerted by a small object, such as a needle, the object floats.

Surface tension coupled with a lack of interaction between polar water molecules (blue spheres) and the nonpolar surface of a leaf (gray spheres) is why drops of water bead up on the surface.

hesion to the capillary walls draws the outer layer of water molecules upward, as shown in Figure 9.16. Adjacent water molecules are drawn upward by cohesive interactions with the outermost molecules. If the capillary is narrow enough, the level of water will climb up it, seeming to defy the force of gravity. This **capillary**

☑ **Capillary action** is the rise of a liquid up a narrow tube as a result of adhesive forces between the liquid and the tube and cohesive forces within the liquid.

The **meniscus** is the curved surface of a liquid. The **surface tension** of a liquid is the energy needed to separate the molecules of a unit area of the liquid surface.

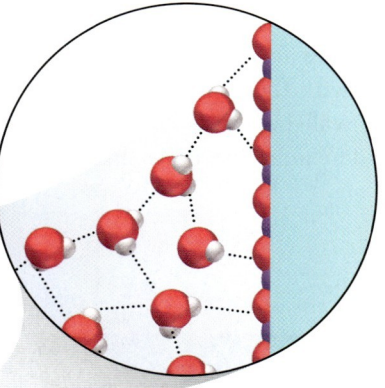

FIGURE 9.16 *What a Chemist Sees.* Hydrogen bonds (dotted lines) between water molecules and exposed surface oxygen atoms in silicon dioxide glass allows water molecules to counteract the force of gravity acting on the bulk of the sample. The result is a meniscus in which the water level is higher near the edge of the tube than in the center.

FIGURE 9.17 Water rises in a thin (capillary) tube of glass and in a stick of celery by capillary action. If the diameter of the tube is narrow enough, water molecules will rise through a combination of adhesive dipole–dipole interactions between water molecules and those of the tube and cohesive hydrogen bonds between water molecules.

action is the process that results in the flow of water upward in the stalks of plants and the trunks of trees (Figure 9.17). The maximum height of the column is reached when the force of gravity balances the adhesive and cohesive forces.

The **viscosity,** or resistance to flow, of liquids is another property related to the strength of intermolecular forces. Among nonpolar compounds such as those found in petroleum products (described in Chapter 12), viscosity increases with increasing molar mass. The higher viscosity of lubricating oil than that of gasoline, for example, is due to stronger dispersion forces between the larger molecules in lubricating oil. Stronger interactions mean that the molecules do not slide past one another as easily as the shorter-chain hydrocarbons in gasoline do, and so the larger molecules do not flow as easily. On the other hand, water is more viscous than gasoline, even though water molecules are much smaller than those of the nonpolar compounds in gasoline. The greater viscosity of water is another of its properties that is directly related to the strong hydrogen bonds between water molecules.

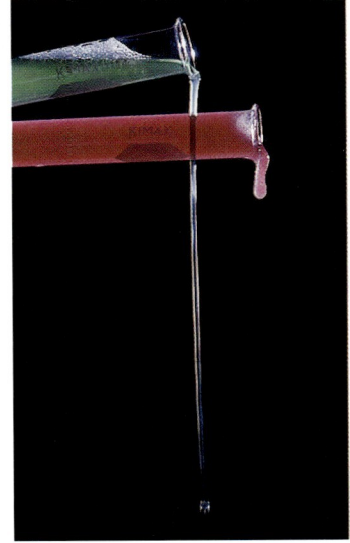

The high viscosity of molasses results from strong intermolecular forces between large, polar molecules.

CONNECTION: The composition of gasoline and structures of the molecules in gasoline are described in Section 12.1.

CONCEPT TEST: Do you think the viscosity of seawater is greater than that of distilled water? Why or why not?

 Viscosity is a measure of the resistance to flow of a fluid (liquid or gas).

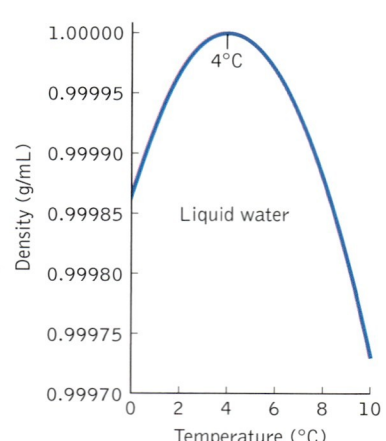

FIGURE 9.18 The density of water increases as it cools to 4°C but then decreases as the temperature approaches 0°C.

As liquid water cools from room temperature to 4°C, its density increases, as shown in Figure 9.18. This pattern is observed for most liquids and solids and for all gases. However, as water is cooled from 4°C to 0°C, it expands and its density *decreases.* As water freezes, its density drops even more, to about 0.92 g/mL for ice, causing ice to float on water (Figure 9.19). This unusual behavior is caused by the formation of a network of hydrogen bonds in ice: two per oxygen atom. With each oxygen atom covalently bonded to two hydrogen atoms and hydrogen bonded to two others, molecules of H_2O form a tetrahedral network in ice (see Figure 9.19). The resulting rigid array of molecules occupies more volume than does the same number of molecules in liquid water, where hydrogen bonds are constantly made and broken as H_2O molecules slip by one another.

This expanded structure of ice plays a crucial role in the ecology of bodies of water in temperate and polar climates. The lower density of ice means that lakes, rivers, and polar oceans freeze from the top down, allowing fish and other aquatic life to overwinter in the waters below. Each fall, surface waters cool first, and their densities increase. The denser, colder water sinks to the bottom, bringing warmer water from the bottom to the surface. This autumnal *turnover* of the water column stirs up dissolved nutrients, including nitrates and phosphates in bottom waters, making them available for photosynthesis in the sunlit surface during the next growing season. When all of the water has reached 4°C, the surface water begins to cool further, and ice eventually forms. The layer of ice effectively shields the 4°C water beneath it in lakes and deep ponds, allowing aquatic organisms to survive. In the spring, the ice melts and the surface water warms to 4°C. At this temperature, the entire column of water has nearly the same temperature and density; dissolved nutrients for plant growth are evenly distributed, and the stage is set for a burst of photosynthesis and biological activity called the spring *bloom.*

Further warming of the surface waters during the summer months creates a warm upper layer separated from colder, denser deep water by a *thermocline,* a

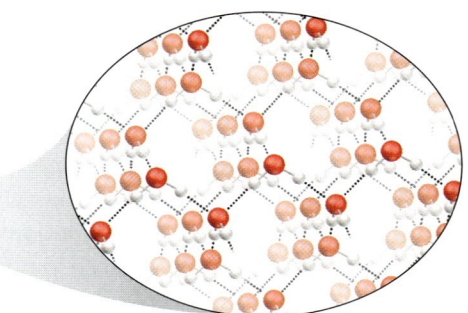

FIGURE 9.19 Changes in the density of water with changing temperature near water's freezing point (see Figure 9.18) coupled with the even lower density of ice (about 0.92 g/mL) means that ice and ice-cold water float on top of 4°C water in lakes and rivers during the winter. In ice, each oxygen atom is in a rigid tetrahedral environment of two O–H covalent bonds and two O···H hydrogen bonds.

sharp change in temperature between the two layers. Biological activity depletes the pool of nutrients above the thermocline as decaying biomass settles into the bottom layer. Consequently, photosynthetic activity drops from its spring maximum during the summer, even though there is much more photosynthetic energy available from the sun. The thermocline persists until the autumnal turnover mixes the water column and the cycle begins anew.

CHAPTER REVIEW

Summary

SECTION 9.1

The strengths of intermolecular interactions determine whether a compound will be a gas, liquid, or solid under normal conditions.

SECTION 9.2

Strong ion–ion interactions hold ionic solids together. The strength of these interactions correlate with high melting and boiling points and low aqueous solubilities among ionic compounds.

SECTION 9.3

Ions interact with water through ion–dipole forces. Dipole–dipole forces are expected between water molecules, but water's properties indicate the existence of particularly strong dipole–dipole interactions called hydrogen bonds. These bonds form between polar molecules containing O–H, N–H, and F–H covalent bonds.

SECTION 9.4

London (dispersion) forces are due to the polarization of atoms and molecules. These interactions are weak compared with ion–ion and ion–dipole interactions. The strongest dispersion forces occur among the largest atoms and molecules.

SECTION 9.5

Polar solutes dissolve in polar solvents when the dipole–dipole interactions between solute and solvent molecules offset the interactions that keep solute or solvent molecules together. The limited solubility of weak nonpolar solutes in polar solvents is a result of dipole–induced-dipole interactions.

SECTION 9.6

During evaporation, molecules at the surface of a liquid overcome intermolecular interactions with neighboring molecules and enter the vapor phase. These vapor-phase molecules are the reason for the liquid's vapor pressure. A greater proportion of liquid molecules enter the vapor phase as the temperature increases, leading to higher vapor pressures. The presence of particles of solute decreases the vapor pressure of a solvent.

SECTION 9.7

The phase diagram of a substance indicates whether it exists as a solid, liquid, gas, or supercritical fluid at a particular pressure and temperature. All three states (solid, liquid, and gas) can exist in equilibrium at the triple point. Above their critical temperatures and critical pressures substances exist as supercritical fluids.

SECTION 9.8

The remarkable properties of water, including its high melting and boiling points, viscosity, surface tension and its capillary action, result from the strength of intermolecular forces.

Key Terms

capillary action (p. 458)
Coulomb's law (p. 439)
critical point (p. 455)
critical pressure (p. 455)
critical temperature (p. 455)
dipole–dipole interaction (p. 455)
dipole–induced dipole interaction (p. 449)
hydration (p. 445)

hydrogen bond (p. 446)
immiscible liquid (p. 449)
ion–dipole interaction (p. 445)
lattice energy (p. 440)
London (dispersion) force (p. 446)
meniscus (p. 457)
miscible liquid (p. 448)
phase diagram (p. 454)

polarization (p. 446)
Raoult's law (p. 452)
sphere of hydration (p. 445)
supercritical fluid (p. 455)
surface tension (p. 457)
triple point (p. 455)
vapor pressure (p. 451)
viscosity (p. 459)

Key Skills and Concepts

SECTION 9.1

Be familiar with the molecular views of the three states of matter.

SECTIONS 9.2–9.5

Be able to distinguish between the major types of intermolecular interactions in solids and liquids: ion–ion, ion–dipole, dipole–dipole, hydrogen bonds, and dispersion forces.
Be able to calculate the energies of ion–ion interactions.
Understand that stronger interaction energies in ionic solids lead to higher melting points and lower solubilities in water.
Understand the relative strengths of the different types of intermolecular interactions.

SECTION 9.6

Understand the concept of the vapor pressure of a liquid and its dependence on temperature.

SECTION 9.7

Recognize which of the three regions in a simple phase diagram correspond to solid, liquid, or gas.
Understand that the boundaries of the regions in a phase diagram of a substance indicate the pressure and temperature conditions at which the substance can co-exist in different states.
Be able to predict how changing pressure affects the melting point or boiling point of a substance from its phase diagram.

SECTION 9.8

Understand the role of hydrogen bonding in defining the physical properties of water.
Know that ice has a lower density than liquid water because of geometrically specific hydrogen bonding in ice.

Key Equations and Relations

SECTION 9.2

Coulomb's law relates the energy of interaction (E) of ions in ionic solids to the charges of the ions (Q_1 and Q_2) and the distance between them (d):

$$E = 2.31 \times 10^{-19} \, \text{J} \cdot \text{nm} \left(\frac{Q_1 Q_2}{d} \right) \quad (9.1)$$

Raoult's law relates the vapor pressure of a solution (P_{solution}) of a nonvolatile solute to the mole fraction of the solvent in the solution (X_{solvent}) and the vapor pressure of the pure solvent (P_{solvent}):

$$P_{\text{solution}} = X_{\text{solvent}} \times P_{\text{solvent}} \quad (9.3)$$

QUESTIONS AND PROBLEMS

Intermolecular Forces

CONCEPT REVIEW

1. In which physical state are there the fewest interactions between atoms, ions, or molecules?
2. Which phase change requires the disruption of the largest number of intermolecular interactions: melting, boiling, or sublimation?
3. Which type of intermolecular force affects all molecules?
4. Can all nonpolar molecules always form hydrogen bonds?
5. Why doesn't methane (CH_4) exhibit hydrogen bonding?
6. Why are dipole–dipole interactions generally weaker than ion–dipole interactions?
7. Why are hydrogen bonds considered a special class of dipole–dipole interactions?
8. Two molecules—one polar, one nonpolar—have similar molar masses. Which one is more likely to have the higher boiling point?
9. Why do the strengths of London (dispersion) forces generally increase with increasing molecular size?
10. Why is $CaSO_4$ less soluble in water than NaCl?
11. Explain why the van der Waals constant a for Ar is greater than it is for He.
12. The van der Waals constant a for CO_2 is 3.59 $L^2 \cdot atm/mol^2$. Would you expect the value of a for CS_2 to be larger or smaller than 3.59 $L^2 \cdot atm/mol^2$?
13. The boiling point of phosphine (PH_3) ($-88°C$) is lower than that of ammonia ($-33°C$) even though PH_3 has twice the molar mass of NH_3. Why?
14. Why does methanol (CH_3OH) boil at a lower temperature than water, even though CH_3OH has the greater molar mass?
15. The dipole moment of CH_2F_2 (1.93 D) is larger than that of CH_2Cl_2 (1.60 D), yet the boiling point of CH_2Cl_2 (40°C) is much higher than that of CH_2F_2 ($-52°C$). Why?
16. How is it that the dipole moment of HCl (1.08 D) is larger than the dipole moment of HBr (0.82 D), yet HBr boils at a higher temperature?
17. Explain why the boiling point of Br_2 (59°C) is lower than that of iodine monochloride, ICl (97°C), even though they have nearly the same molar mass.
18. Explain why the boiling point of pure sodium chloride is much higher than the boiling point of an aqueous solution of sodium chloride.

PROBLEMS

19. Rank the following ionic compounds in order of increasing attraction between their ions: KBr, $SrBr_2$, and CsBr.
20. Rank the following ionic compounds in order of increasing attraction between their ions: BaO, $BaCl_2$, and CaO.
21. How do the melting points for the series of sodium halides NaX (X = F, Cl, Br, I) relate to the atomic number of X?
22. Rank the following ionic compounds in order of increasing melting point: $BaCl_2$, $CaCl_2$, $MgCl_2$, and $SrCl_2$.
23. Calculate the lattice energy of potassium chloride (KCl) from the following data:
    - Ionization energy of K = 425 kJ/mol.
    - Electron affinity of Cl = -349 kJ/mol.
    - Energy to vaporize K = 89 kJ/mol.
    - Cl_2 bond energy = 240 kJ/mol.
    - Energy change for the reaction
    $$K(s) + \tfrac{1}{2} Cl_2(g) \longrightarrow KCl(s)$$
    is -438 kJ/mol KCl.
24. Calculate the lattice energy of sodium oxide (Na_2O) from the following data:
    - Ionization energy of Na = 495 kJ/mol.
    - Electron affinity of O for 2 e^- = 638 kJ/mol.
    - Energy to vaporize Na = 109 kJ/mol.
    - O_2 bond energy = 499 kJ/mol.
    - Energy change for the reaction
    $$2\,Na(s) + \tfrac{1}{2} O_2(g) \longrightarrow Na_2O(s)$$
    is -416 kJ/mol Na_2O.
25. In each of the following pairs of molecules, which compound experiences the stronger London (dispersion) forces? (a) CCl_4 or CF_4; (b) CH_4 or C_3H_8.
26. In which of the following compounds do the molecules experience the strongest dipole–dipole interactions? CF_4, CF_2Cl_2, CCl_4.
27. Which of the following compounds, CO_2, NO_2, SO_2, or H_2S, is expected to have the weakest interactions between its molecules?
28. What kinds of intermolecular forces must be overcome as (a) solid CO_2 sublimes? (b) $CHCl_3$ boils? (c) ice melts?
29. In each of the following pairs of compounds, which compound is likely to be more soluble in water?
    a. CCl_4 or $CHCl_3$ b. CH_3OH or $C_6H_{11}OH$
    c. NaF or MgO

30. In each of the following pairs of compounds, which compound is likely to be more soluble in CCl_4?
   a. Br_2 or NaBr
   b. C_2H_5OH or CH_3OCH_3
   c. CS_2 or KOH

31. Which of the following compounds is likely to be the least soluble in water: NaCl, KI, $Ca(OH)_2$, or CaO?

*32. Which sulfur oxide would you predict to be more soluble in nonpolar solvents: SO_2 or SO_3? (Hint: Draw the Lewis structures of SO_2 and SO_3.)

Vapor Pressure and Raoult's Law

CONCEPT REVIEW

33. Why does the vapor pressure of a liquid increase with increasing temperature?
34. What happens when the vapor pressure of a liquid is equal to or greater than atmospheric pressure?

PROBLEMS

35. Rank the following compounds in order of increasing vapor pressure at 298 K: CH_3CH_2OH, CH_3OCH_3, and $CH_3CH_2CH_3$.
36. Rank the following compounds in order of increasing vapor pressure at 298 K:

37. A solution contains 3.5 mol of water and 1.5 mol of glucose ($C_6H_{12}O_6$). What is the mole fraction of water in this solution? What is the vapor pressure of the solution at 25°C, given that the vapor pressure of pure water at 25°C is 23.8 torr?

38. A solution contains 4.5 mol of water, 0.3 mol of sucrose ($C_{12}H_{22}O_{11}$), and 0.2 mol of glucose. What is the mole fraction of water in this solution? What is the vapor pressure of the solution at 35°C, given that the vapor pressure of pure water at 35°C is 42.2 torr?

Phase Diagrams

CONCEPT REVIEW

39. Which phase of a substance (gas, liquid, or solid) is more likely to be the stable phase: (a) at low temperatures and high pressures; (b) at high temperatures and low pressures?
40. At what temperatures and pressures does a substance behave as a supercritical fluid?
41. Freeze-drying is used to preserve food at low temperature with minimal loss of flavor. Freeze-drying is accomplished by freezing the food and then lowering the pressure with a vacuum pump to sublime the ice. Must the pressure be lower than the pressure at the triple point of H_2O?
42. Solid helium cannot be converted directly into the vapor phase. Does the phase diagram of He have a triple point?

PROBLEMS

43. What phase changes, if any, does liquid water initially at 100°C and 5.0 atm undergo when the pressure on the liquid is reduced to 0.5 atm?
44. What phase changes, if any, does CO_2 initially at −80°C and 8.0 atm undergo when the CO_2 is allowed to warm to −25°C at 5.0 atm?
45. Below what temperature can solid CO_2 (dry ice) be converted into CO_2 gas simply by lowering the pressure?

46. What is the maximum pressure at which solid CO_2 (dry ice) can be converted into CO_2 gas without melting?
47. Predict the phase of water that exists under the following conditions:
    a. 2 atm of pressure at 110°C
    b. 0.5 atm of pressure and 80°C
    c. 6.0×10^{-3} atm of pressure and 0°C
48. Which phase or phases of water exist under the following conditions?
    a. 0.32 atm and 70°C   b. 300 atm and 400°C
    c. 1 atm and 0°C
49. Does the freezing point of water increase or decrease with increasing pressure?
50. Does the sublimation point of water increase or decrease with increasing pressure?
51. A phase diagram of a substance can be sketched from its normal ($P = 1.00$ atm) boiling point, melting point, and triple point. Draw a phase diagram for the compound for which the normal boiling point is 90 K, the normal melting point is 55 K, and the triple point is 54 K and 1.14 torr.
52. Sketch a phase diagram for element X, which has a triple point (152 K, 0.371 atm), a normal ($P = 1.00$ atm) boiling point of 166 K, and a normal melting point of 161 K.
*53. The melting point of compound Y at $P = 1.0$ atm is 3850 K, and its triple point is at 4000 K and 0.5 atm. Does Y expand or contract when it freezes?
*54. The melting point of hydrogen is 14.96 K at $P = 1.00$ atm, and the temperature of its triple point is 13.81 K. Does H_2 expand or contract when it freezes?

The Remarkable Behavior of Water

CONCEPT REVIEW

55. Explain why a needle floats on the surface of water but sinks in a container of methanol (CH_3OH).
56. Explain why water climbs higher in a narrow glass capillary than in a test tube.
57. Explain why different liquids do not reach the same height in capillary tubes of the same diameter.
58. Explain why ice floats on water.
59. Explain why pipes filled with water are in danger of bursting when the temperature drops below 0°C.
60. Predict how changing temperature might affect the capillary action of water. Explain your answer.
61. A hot needle sinks when put in cold water. Would a cold needle float in hot water?
62. Do nonpolar liquids have surface tension?
*63. The meniscus of mercury in a thermometer (shown here) is convex, rather than concave. Explain why.

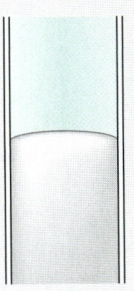

*64. The mercury level in a capillary tube placed in a dish of mercury is actually below the surface of the mercury in the dish. Explain why.

PROBLEMS

65. One of two glass capillaries of the same diameter is placed in a dish of water and the other in a dish of ethanol (CH_3CH_2OH). Which liquid will rise higher in its capillary?
66. Would you expect water to rise to the same height in a tube made of a polyethylene plastic as it does in a glass capillary of the same diameter? The molecular structure of polyethylene is

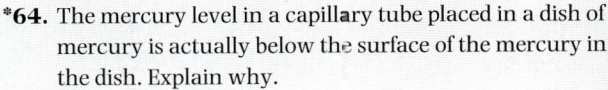

10 The Solid State
A Molecular View of Gems and Minerals

Amethyst is one of the most beautiful (and valuable) forms of quartz. Its distinctive lavender and purple colors are due to the presence of manganese impurities in crystals of silicon dioxide.

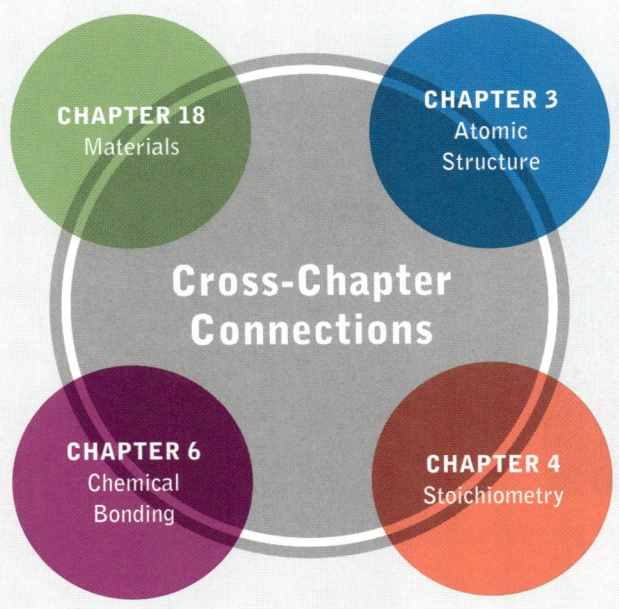

10.1	**Crystal lattices**
	Crystalline versus amorphous
	X-ray diffraction
10.2	**The unit cell**
10.3	**Packing efficiency**
	Cubic closest packing
	Simple cubic packing
	Hexagonal closest packing
10.4	**Network solids: the many forms of silica**
	Orthosilicates
	Metasilicates
10.5	**Allotropes of carbon and sulfur**
10.6	**Metallic bonds and structures**
10.7	**Gemstones: an introduction to crystal field theory**
	Crystal field splitting energy
	Magnetic properties

A Look Ahead

In this chapter, we examine the structures of solids at the atomic level. We will investigate how the arrangement of ions in NaCl and other crystalline compounds can be determined by X-ray diffractometry and is consistent with the densities of these compounds. We will see how the spaces between closely packed atoms or ions can accommodate smaller atoms or ions, and how some elements and compounds form more than one crystalline structure, depending on the pressures and temperatures at which they form.

10.1 CRYSTAL LATTICES

We begin with a familiar ionic compound, table salt (NaCl). A magnified view of salt granules (Figure 10.1) reveals that they consist of tiny cubic crystals. This geometry reflects the structure of table salt on an atomic level: NaCl is a **crystalline solid** made of a cubic array of Na^+ and Cl^- ions tightly packed together so that the distances between their opposite charges are as small as possible.

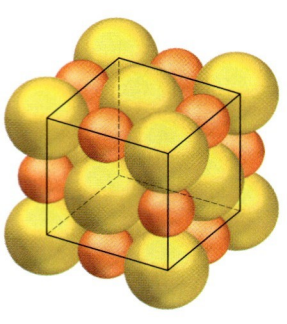

FIGURE 10.1 Sodium chloride is a crystalline solid consisting of an ordered array of Na⁺ and Cl⁻ ions.

CONNECTION: The connection between lattice energy and the stability of ionic solids is discussed in Section 9.2.

CONNECTION: The crystal structure of NaCl, also appears in Figure 4.6 and 9.3.

CONNECTION: Periodic trends in the sizes of atoms and their monoatomic ions are shown in Figure 9.2.

✓ **Crystalline solids** are made of ordered arrays of atoms, ions, or molecules and give characteristic **X-ray diffraction** patterns. **Amorphous solids** have no long-range ordering in their structures.

Smaller distances (d) mean more negative lattice energies (U) according to Equation 9.2:

$$U = k\left(\frac{Q_1 Q_2}{d}\right)$$

and more negative lattice energies mean more-stable structures. The relative sizes of cations and anions, such as Na⁺ and Cl⁻, control how tightly packed they can be. Because the ionic radius of Na⁺ (98 picometers) is smaller than that of Cl⁻ (191 pm), the best fit for these two ions is the one shown in Figure 10.1.

Crystalline versus amorphous

Note that the Na⁺ and Cl⁻ ions in Figure 10.1 are arranged in an orderly three-dimensional pattern. This means that NaCl is a **crystalline solid.** Most ionic compounds form crystalline solids, as do some covalent compounds. Consider, for example, one of the many forms of silica, or silicon dioxide (SiO_2). The most abundant mineral in Earth's crust is *quartz*, a type of SiO_2 that can form impressively large, nearly transparent crystals (Figure 10.2A). The distinctive shape of quartz crystals is linked to the ordered array of silicon and oxygen atoms shown in the atomic view of quartz in Figure 10.2A. However, not all silica-based minerals are crystalline. When lava containing molten SiO_2 flows from a volcano and into a lake or the ocean, it cools quickly. During rapid cooling, Si and O atoms may not have enough time to achieve an ordered structure as the viscous lava solidifies. A product of such rapid solidification is a disordered, noncrystalline **amorphous** form of silica known as volcanic glass, or *obsidian* (Figure 10.2B). Actually, *glass* is a term generally used to describe any solid that either has no crystalline structure or that has only very tiny crystals surrounded by disordered arrays of atoms. This definition applies to the glassware that we use in the laboratory and the drinking glasses that you use at home.

 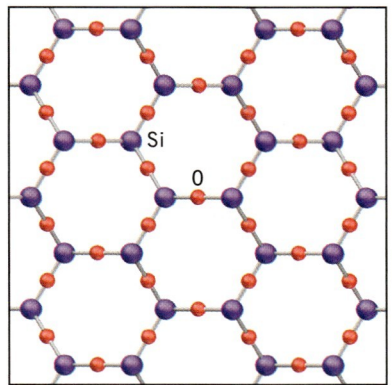

 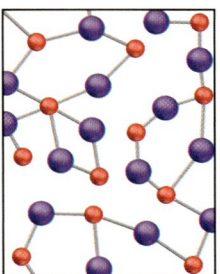

FIGURE 10.2 *What a Chemist Sees.* A. Quartz is a crystalline form of silicon dioxide (SiO_2). B. Obsidian (volcanic glass) contains mostly amorphous silica with random arrangements of silicon and oxygen atoms.

X-ray diffraction

How do we know that NaCl has the structure shown in Figure 10.1? A technique for characterizing the structures of crystalline solids is **X-ray diffractometry** (XRD). X-rays are diffracted when they bounce off the nuclei and electrons of atoms. To understand how X-ray diffraction works, let's consider what happens when a beam of X-rays strikes the surface of a crystal, as shown in Figure 10.3B. The angle between the beam and the surface of the crystal (called the angle of incidence) is represented by the Greek letter theta (θ). Suppose the beam is reflected by adjacent layers of atoms or ions separated by a distance d. If the angle of reflection matches the angle of incidence so that both equal θ, then the total change in direction of the beam of X-rays—that is, the *angle of diffraction*—is 2θ. X-rays that reflect off different layers of atoms or ions in the crystal will travel slightly different distances. The right triangles in Figure 10.3B indicate the greater distance traveled through the crystal by X-rays reflected by the deeper layer. The added

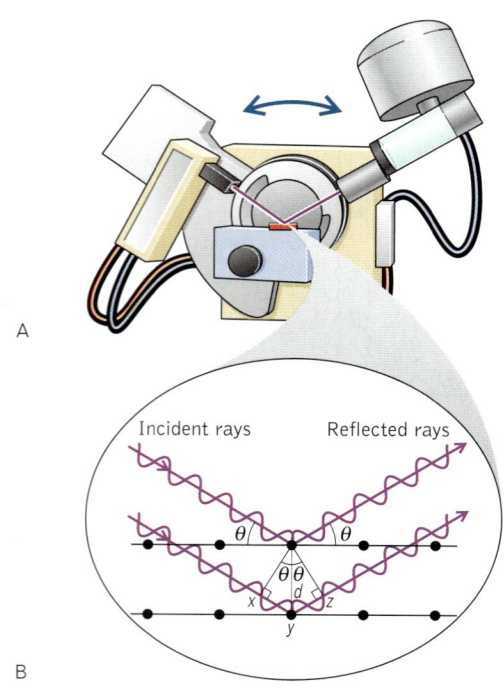

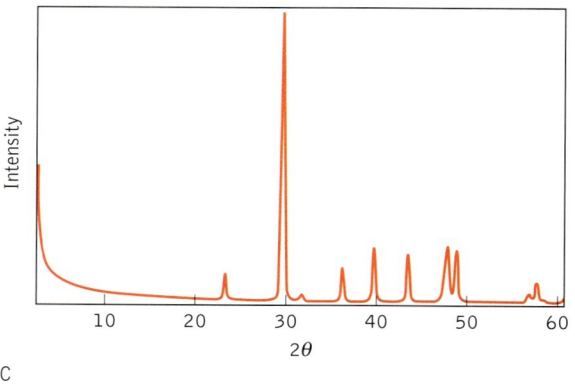

FIGURE 10.3 A. An X-ray diffractometer (XRD) is used to determine the crystal structures of solids. A source of X-rays and a detector are mounted so that they can rotate around the sample. B. A beam of X-rays of wavelength λ is directed at the surface of a sample. The beam will be diffracted by the sample to the detector if the angle of diffraction (2θ) satisfies the Bragg equation: $n\lambda = 2d \sin \theta$, where d is the distance between layers of atoms or ions in crystals of the sample. C. Moving the source and detector around the sample as X-ray signals are recorded produces a scan such as this one for quartz.

CONNECTION: Electron diffraction can be used to determine the structures of gas-phase molecules (Section 6.8).

distance is the sum of the distances $\overline{xy}$ and $\overline{yz}$. The two right triangles incorporating these distances share a hypotenuse equal to the distance d. It turns out that the angles opposite $\overline{xy}$ and $\overline{yz}$ are the same and equal to θ. According to trigonometry, the ratio of either $\overline{xy}$ or $\overline{yz}$ to d is the sine of θ, that is,

$$\frac{\overline{xy}}{d} = \frac{\overline{yz}}{d} = \sin \theta$$

Rearranging these terms, we find that the difference in distance traveled ($\overline{xy} + \overline{yz}$) is

$$d \sin \theta + d \sin \theta = 2d \sin \theta$$

If this difference equals some multiple (n) of the wavelength (λ) of the X-ray beam, the following relation, which is called the **Bragg equation,** is true:

$$n\lambda = 2d \sin \theta \tag{10.1}$$

Under these conditions, the crests and valleys of the two X-rays will be in phase with each other as they emerge from the crystal, as shown in Figure 10.3B.

> The **Bragg equation** relates the angle of diffraction (θ) of X-rays to the spacing (d) of layers of ions or atoms in a crystal.

When waves of electromagnetic radiation are aligned in this way, they undergo constructive interference, which means that a detector aligned θ degrees to the surface of the sample will detect X-rays diffracted by the sample. Moving the source and the detector around the sample as the intensity of diffracted X-rays is recorded produces an XRD scan, such as the one shown for quartz in Figure 10.3C.

Scientists use the values of the angle of diffraction (2θ) that produce peaks in an XRD scan with Equation 10.1 to calculate the distances between atoms or ions in a crystal. With this information, they can identify which crystalline solid is present in a sample and determine its crystal structure.

SAMPLE EXERCISE 10.1: An XRD scan (λ = 154 pm) of NaCl has peaks at 2θ = 15.7, 31.8, 48.4, and 66.2°. What is the distance (d) between layers of ions in a crystal of NaCl that produced this diffraction pattern?

CONNECTION: Constructive and destructive interference of light waves was discussed in Section 1.3.

SOLUTION: Equation 10.1 allows us to calculate the distance d in a crystal structure on the basis of those angles of diffraction associated with constructive interference in a beam of scattered X-rays. To use Equation 10.1, we must first convert the 2θ values into θ values; so θ = 7.85, 15.9, 24.2, and 33.1°. Rearranging the terms in Equation 10.1, we have

$$d = n\lambda/2 \sin \theta$$

The additional unknown in Equation 10.1 is the wavelength multiplier, n. Its value must be evaluated before we can solve for d. The key to determining the value of n is to look for a pattern in the values of θ. Perhaps you noticed that the higher values of θ are approximately two, three, and four times the lowest value, 7.85°. This pattern suggests that the values of n for this set of data might be 1, 2, 3, and 4 for θ = 7.85, 15.9, 24.2, and 33.1°, respectively. Let's use these combinations of n and θ to see if they all give the same value of d.

$$d = (1)(154 \text{ pm})/2 \sin 7.85° = 154 \text{ pm}/2(0.1366) = 564 \text{ pm}$$

$$= (2)(154 \text{ pm})/2 \sin 15.9° = 308 \text{ pm}/2(0.2740) = 562 \text{ pm}$$

$$= (3)(154 \text{ pm})/2 \sin 24.2° = 462 \text{ pm}/2(0.4099) = 564 \text{ pm}$$

$$= (4)(154 \text{ pm})/2 \sin 33.1° = 616 \text{ pm}/2(0.1366) = 564 \text{ pm}$$

Our assumption about the combinations of n and θ appears to be correct, and a single value for d accounts for the observed diffraction pattern. The results of many careful analyses of NaCl crystals put this value at 562.8 pm.

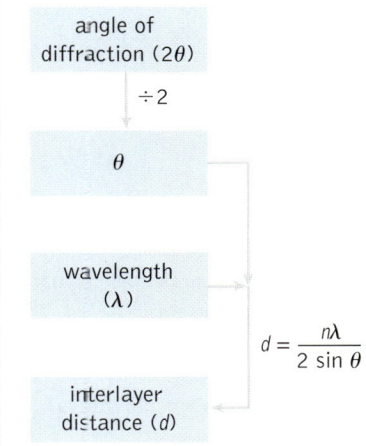

PRACTICE EXERCISE: An X-ray diffraction scan (λ = 71.2 pm) of CsCl has a prominent peak at 2θ = 19.9°. If this peak corresponds to n = 2, what is the spacing (d) between the layers of ions in a crystal of CsCl? What are the corresponding 2θ values for $n = 3$ and 4? (See Problems 7–10.)

10.2 THE UNIT CELL

> **Lattice** refers to the three-dimensional array of particles in a crystalline solid. Each particle occupies a **lattice point** in the array.
>
> The **unit cell** is the basic repeating unit of the arrangement of atoms, ions, or molecules in a crystalline solid.

Figure 10.4 contains detailed views of the structure of NaCl. The Na^+ and Cl^- ions are located at the **lattice points** of a crystalline array, or **lattice.** The lattice continues in all three dimensions, though there is a distinctive and repeating pattern in the structure. This repeating pattern is called a **unit cell.** Ball-and-stick and space-filling models of the unit cell of NaCl are shown in Figure 10.4B and C, respectively. As drawn, the unit cell has eight Cl^- ions (the larger spheres) at its eight corners and one Cl^- ion at the center of each of its six faces. One Na^+ ion is in the center, and one Na^+ ion is along each edge. If we add up these ions, we find that there are 14 Cl^- ions but only 13 Na^+ ions. How can this be when the ratio of the two ions must be exactly 1:1? The apparent (but not real) inconsistency arises because the corner, edge, and face ions belong to more than one unit cell; only the central Na^+ ion belongs entirely to this cell. For example, each corner Cl^- ion is shared by the eight unit cells that meet in each corner, and so each unit cell gets credit for only one-eighth of the corner Cl^- ions. Similarly, the Cl^- ion in the center of each of the six faces is shared by two unit cells, and so each unit cell gets credit for one-half of these Cl^- ions; each Na^+ ion along an edge is shared by four unit cells and so counts as one-fourth. Only the Na^+ in the center belongs completely to one unit cell. Therefore the total number of Cl^- ions in the unit cell of NaCl is

$$8 \text{ corner ions} \times 1/8 = 1$$
$$+ \ 6 \text{ face ions} \quad \times 1/2 = 3$$
$$= 4$$

A. The unit cell of a chessboard is the minimum set of squares that defines the pattern that is repeated over the entire board. That set is a 2 × 2 block of four squares. B. This wrapping paper has a more-complex array. One repeat pattern is outlined. Can you locate others?

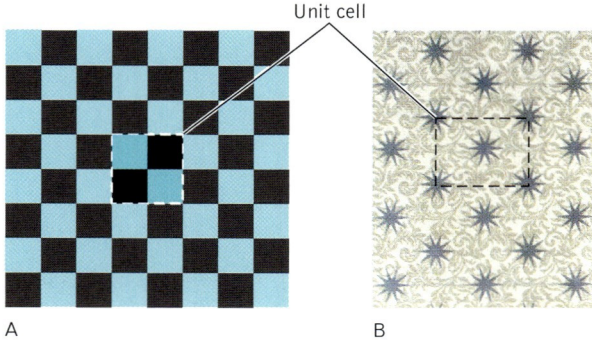

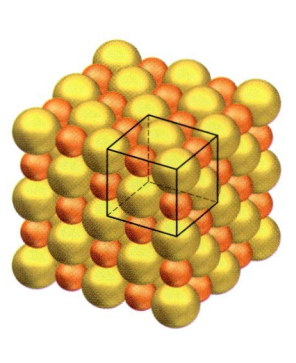
A

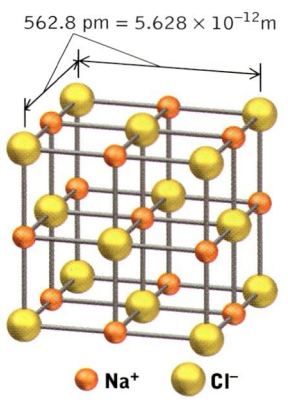
562.8 pm = 5.628 × 10⁻¹²m

● Na⁺   ● Cl⁻

B

C

FIGURE 10.4 A. Within the crystalline structure of NaCl, the simplest repeating unit of Na⁺ and Cl⁻ ions (the cube within the black lines) is called the unit cell. B. A ball-and-stick model of the unit cell of NaCl shows that there are eight Cl⁻ ions (yellow spheres) at the eight corners of the cell and six Cl⁻ ions at the centers of the six faces. There are twelve Na⁺ ions (orange spheres) on the twelve edges of the cell and one Na⁺ ion at the center. This arrangement is called a face-centered cubic (fcc) structure. C. This space-filling model of the NaCl unit cell shows that only one-eighth of each corner Cl⁻ ion is in this cell; only one-fourth of a Na⁺ ion along an edge is in this cell; and only one-half of each face Cl⁻ ion is in this cell. Only the center Na⁺ ion belongs completely to the cell.

and the total number of Na⁺ ions is

$$\begin{aligned}12 \text{ edge ions} \times 1/4 &= 3 \\ + 1 \text{ center ion} \times 1 &= 1 \\ &= 4\end{aligned}$$

Thus, the ratio of Na⁺ to Cl⁻ ions in the unit cell is 4:4 (or 1:1) and is consistent with the chemical formula of NaCl. Table 10.1 summarizes how the ions in different positions around the perimeter of a unit cell are counted in the ion inventory of the cell.

We could have drawn the unit cell of NaCl so that the smaller Na⁺ ions were at the corners of the unit cell. Either way, the unit cell has the shape of a cube, and the ion in each face is the same as that in the corners. This arrangement makes the unit cell for NaCl a **face-centered cubic** (fcc) **unit cell**.

There are other kinds of cubic unit cells. All of them are the building blocks of cubic crystals. One of them is a **body-centered cubic** (bcc) **unit cell** (Figure 10.5A), which has a particle at each corner and the same kind of particle in the middle of the cell. The particles in the corners do not touch one another, but they do touch the center particle in their cell and in seven other cells. Thus, each particle touches eight others, and is said to have a **coordination number** of eight. On the other hand, each Na⁺ ion in an fcc crystal of NaCl touches six Cl⁻ ions and each Cl⁻ ion touches six Na⁺ ions, giving both ions a coordination number of six. A third type of cubic unit cell, the **simple cubic unit cell,** consists of just eight corner particles, as shown in Figure 10.5B. Each corner particle touches six

☑ A **face-centered cubic unit cell** has the same kind of particle at the eight corners of a cube and at the center of each face.

A **body-centered cubic unit cell** has lattice points at the eight corners of a cube and at the center of the cell.

The **coordination number** of a particle in a crystalline structure is the number of surrounding particles that touch it.

A **simple cubic unit cell** has lattice points only at the eight corners of a cube.

TABLE 10.1 Fractional Contributions of Particles to Unit Cells

Location	Contribution to Unit Cell
Center	1
Face	1/2
Edge	1/4
Corner	1/8

Body-centered cubic
A

Simple cubic
B

FIGURE 10.5 Cubic crystals can have unit cells with face-centered cubic structures or with (A) body-centered cubic or (B) simple cubic structures.

others: three in its own unit cell and three in adjoining cells, giving it a coordination number of six. You may be wondering why some solids adopt one kind of unit cell and other solids another. The answer to that question has a lot to do with the relative sizes of the particles in the solid and will be addressed later in this chapter.

SAMPLE EXERCISE 10.2: Lithium chloride (LiCl) crystallizes with the same structure as NaCl except that the chloride anions touch along the face diagonal of the unit cell. If the unit cell edge is 513 pm, what is the radius of a chloride (Cl^-) anion?

SOLUTION: In a face-centered arrangement of Cl^- ions, the face diagonal is equal to four chloride radii, or $4r_{Cl^-}$. Applying the Pythagorean theorem to a right triangle composed of the unit-cell edges and the face diagonal, we have

$$(4r_{Cl^-})^2 = (\text{edge})^2 + (\text{edge})^2 = 2\,(\text{edge})^2$$

Solving for r_{Cl^-} gives

$$4r_{Cl^-} = \sqrt{2}(\text{edge})$$

$$r_{Cl^-} = \frac{\sqrt{2}}{4}(\text{edge})$$

unit-cell edge length

$r = \frac{\sqrt{2}}{4}$ (edge length)

radius (r)

Substituting 513 pm for the unit-cell edge:

$$r_{Cl^-} = \frac{\sqrt{2}}{4}(513 \text{ pm}) = 181 \text{ pm}$$

This value for r_{Cl^-} is the same as that in Figure 9.3. Actually, the ionic radii values in Figure 9.3 are determined from crystal structures.

PRACTICE EXERCISE: A hypothetical compound composed of elements A and B has the empirical formula A_3B_2. The crystal structure is composed of a bcc arrangement of the B atoms, with the B atoms touching along the body diagonal.

The smaller A atoms fit into the voids in the structure. Calculate the radius of B if the unit-cell edge is 665 pm. (See Problems 31 and 32.)

One way to test whether a particular crystal structure is the correct one for a substance is to evaluate whether it accurately predicts the density of the substance. Let's calculate the density of NaCl based on an fcc unit cell. We will sum the masses of the four Na⁺ and Cl⁻ ions in the cell and then divide the sum by the cell volume. Sodium atoms have an average mass of 22.99 amu. The average mass in grams is

$$22.99 \text{ amu} \times \frac{1 \text{ g}}{6.022 \times 10^{23} \text{ amu}} = 3.818 \times 10^{-23} \text{ g}$$

Doing a similar calculation for chlorine, we have

$$35.45 \text{ amu} \times \frac{1 \text{ g}}{6.022 \times 10^{23} \text{ amu}} = 5.887 \times 10^{-23} \text{ g}$$

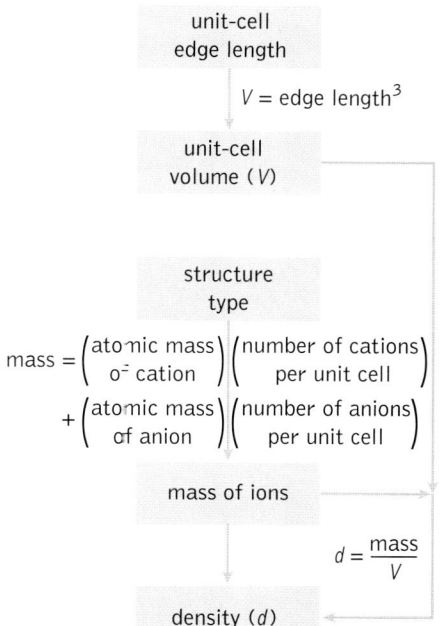

We can use these values as the masses of sodium and chloride ions, respectively, because the difference in the mass between an ion and its parent atom is only the mass of an electron—a tiny quantity compared with the masses of nucleons. The combined mass of the four Na⁺ and Cl⁻ ions in the unit cell is

$$4 \text{ Na atoms} \times 3.818 \times 10^{-23} \text{ g/Na atom} = 1.527 \times 10^{-22} \text{ g}$$

$$+ 4 \text{ Cl atoms} \times 5.887 \times 10^{-23} \text{ g/Cl atom} = 2.355 \times 10^{-22} \text{ g}$$

$$\text{mass of unit cell} = 3.882 \times 10^{-22} \text{ g}$$

The volume of the unit cell can be determined from cell dimensions obtained from X-ray diffraction data. These data indicate that the edge length of the NaCl unit cell is 562.8 pm, or 5.628×10^{-8} cm. The volume (V) of a cube (the unit cell) with this edge length is

$$V = (5.628 \times 10^{-8} \text{ cm})^3 = 1.83 \times 10^{-22} \text{ cm}^3$$

Dividing the sum of the masses of four Na⁺ and Cl⁻ ions by this volume, we have

$$d = \frac{m}{V} = \frac{3.882 \times 10^{-22} \text{ g}}{1.341 \times 10^{-22} \text{ cm}^3} = 2.17 \text{ g/cm}^3$$

The density of NaCl at 25°C is, in fact, 2.16 g/cm³; so our fcc model for NaCl seems to be correct.

All crystals of NaCl have the same shape (cubic), but their sizes can vary from the small granules in salt shakers (Figure 10.1) to single crystals with edge lengths of 5 cm or more. Crystal size depends on the rate at which the crystals form and grow from supersaturated solutions of NaCl. In nature, supersaturation occurs when bodies of saline water evaporate. Slow evaporation leads to the formation of crystals 2–10 mm on an edge, such as the ones shown on the left. These were collected by students from the University of Nebraska, Lincoln from a *salina* (saline pond) on West Caicos Island in the British West Indies, shown on the right.

SAMPLE EXERCISE 10.3: Calculate the edge length of the fcc unit cell of NaCl on the basis of the average radii of Cl^- (181 pm) and Na^+ (98 pm) assuming that the corner Cl^- ions and the Na^+ along the edges touch.

SOLUTION: The view of the unit cell of NaCl in Figure 10.4C shows how the edge of the cell accommodates half of two Cl^- ions and one Na^+ ion. Therefore, the edge distance should equal the sum of twice the ionic radii of Cl^- and Na^+ = (2)(181 pm) + (2)(98 pm) = 558 pm. This value is a good approximation of the value obtained from X-ray diffraction analysis: 562.8 pm.

PRACTICE EXERCISE: Solid KBr also has a face-centered cubic structure. Calculate the approximate edge length of the fcc unit cell of KBr by using the ionic radii data in Figure 9.2, and predict the density of KBr. (See Problems 27–32.)

10.3 PACKING EFFICIENCY

One more property of the unit cell of NaCl deserves our attention: the efficiency with which the ions are packed into the volume of the cell. We have noted that the Na^+ and Cl^- ions are tightly packed along the edges of the unit cell. Adjacent Cl^- ions nearly touch along the diagonals of the cell faces, as shown in Figure 10.4C. This close proximity of the spheres representing Cl^- suggests that the ions are packed in a crystal pattern that comes close to minimizing empty space and

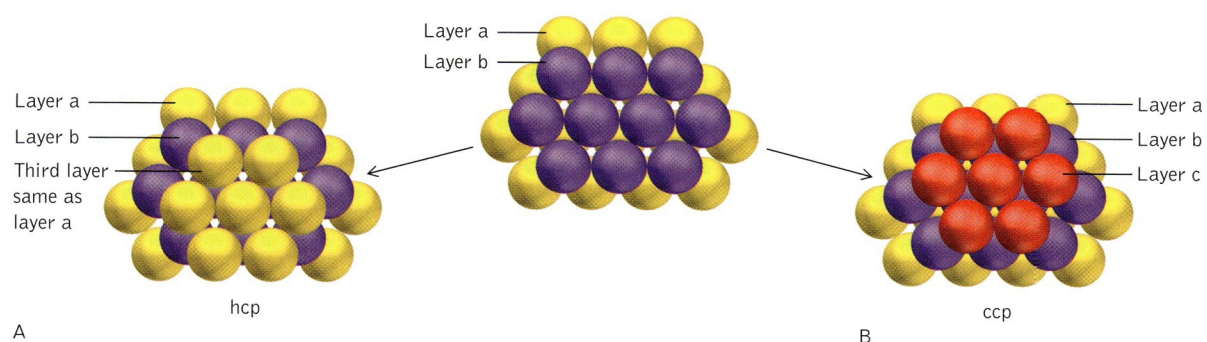

FIGURE 10.6 Two efficient ways to pack solid spheres are (A) hexagonal closest packed (hcp) and (B) cubic closest packed (ccp). The difference is in the arrangement of the third layer. In hcp, the spheres in the third layer are directly above those in the first, giving an *abab*... pattern. In ccp, the third layer is different from either of the two preceding layers, generating an *abcabc*... pattern.

achieving a **closest-packed structure.** In such a structure, each layer of chloride ions is nestled into the spaces between the chloride ions of the layers above and below it.

Cubic closest packing

Figure 10.6 illustrates two closest-packing patterns, starting with two nested layers of spheres and then adding a third. There are two closest-packed options for stacking in the third layer on the top of the second. In the **cubic closest packed** (ccp) stacking pattern, the spheres in the third layer fit snuggly into the spaces between the spheres of the second layer (Figure 10.6B) in such a way that this third layer of spheres is not directly above the spheres in the first layer. This option is shown in Figure 10.6B. When a fourth layer is nestled into the spaces of the third, the spheres of the fourth layer do lie directly above the spheres in the first. The three layers with their distinctive alignments are labeled *a*, *b*, and *c*. Together they form a repeating layering pattern *abcabcabc*... throughout the crystal. This ccp arrangement is one of the two most efficient methods of layering spheres of equal size and is the pattern observed in NaCl. Thus, a face-centered cubic lattice of NaCl contains a cubic closest-packed array of Cl^- ions.

When Na^+ ions are included in the *abc* stack of Cl^-, they go into the spaces, or *holes*, in between the layers. Two kinds of holes are created by *abc* layering. One kind is formed by three spheres of one layer and a single sphere of an adjacent layer, as shown in Figure 10.7A. A smaller sphere located in such a hole will touch the four adjacent larger spheres, giving it a coordination number of four. The four spheres surrounding the hole form a tetrahedron, and so the hole is called a *tetrahedral hole*. The other kind of hole in *abc* layering is located above

> ✓ **Closest packing** is the most efficient stacking of particles in a crystalline solid; 74% of the available volume is occupied.
>
> **Cubic closest packing** is a crystal structure in which the composition of layers of particles in face-centered cubic unit cells has an *abcabcabc*... pattern.

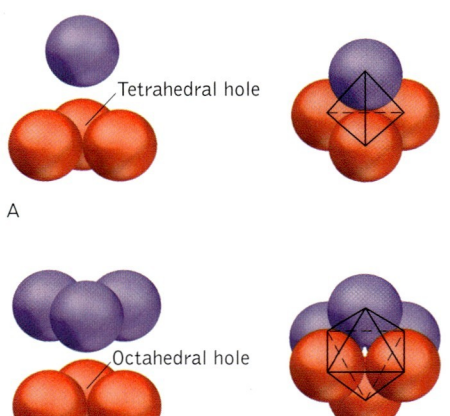

FIGURE 10.7 There are two kinds of holes between layers of closest-packed spheres: (A) tetrahedral holes between three spheres in one layer and a fourth sphere in the layer above and (B) octahedral holes between six spheres in two adjacent layers.

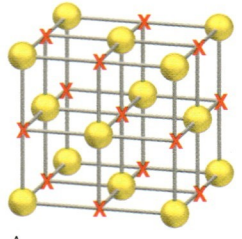

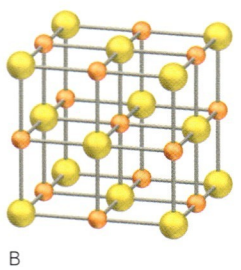

FIGURE 10.8 A. An X marks the location of an octahedral hole in NaCl. B. The unit cell of NaCl consists of a cubic closest-packed array of Cl^- ions with Na^+ ions (smaller spheres) in all of the octahedral holes.

three spheres in one layer and below three spheres of another layer. The two groups of three spheres each are rotated 60 degrees from one another, as shown in Figure 10.7B. The centers of the six large spheres, if connected together, would form an octahedron, and so this kind of hole is called an *octahedral hole*. A small sphere inside an octahedral hole is surrounded by six large spheres, giving it a coordination number of six.

Are Na^+ ions in NaCl in tetrahedral or octahedral holes formed by Cl^- ions? We have already observed that each Na^+ is surrounded by six Cl^- ions. A coordination number of six indicates that sodium ions are located in the octahedral holes of a cubic closest-packed array of chloride ions. A structure in which anions are in a ccp array and cations occupy the octahedral holes is a common crystalline form and is called the *rock-salt* or *NaCl structure*. Each unit cell in the rock-salt structure has one octahedral hole at its center, and each unit cell shares each of 12 other octahedral holes, located along its 12 edges (Figure 10.8A), with three other unit cells. Thus, each unit cell lays claim to $1 + \frac{1}{4}(12) = 4$ octahedral holes, which correspond to the 4 Na^+ ions allocated to each unit cell of NaCl. There are also 8 tetrahedral holes within a unit cell of NaCl. Each of them is formed by one of the eight corner Cl^- ions and three Cl^- ions from the centers of the faces that come together in that corner, as shown in Figure 10.9. In NaCl, these holes are empty, but, as we shall see, they may be occupied in fcc structures of other compounds.

All these holes may seem to constitute a lot of empty space in a structure that is supposed to be closest packed. However, in a ccp array of spheres, the spheres of the lattice-point particles take up 74% of the available space (see Sample Exercise 10.4). The octahedral and tetrahedral holes account for the remaining 26%, and much of that space could be occupied by, for example, smaller ions. The tetrahedral holes are the smaller of the two types. Sodium ions occupy the octahedral holes rather than the tetrahedral holes in a ccp array of Cl^- partly because the tetrahedral holes are too small to accommodate Na^+ ions.

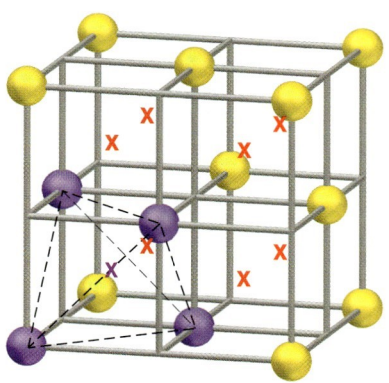

FIGURE 10.9 An X marks the location of one of eight tetrahedral holes in a face-centered cubic structure. Each tetrahedral hole is the empty space between a corner particle and three particles at the centers of the faces that come together at that corner.

 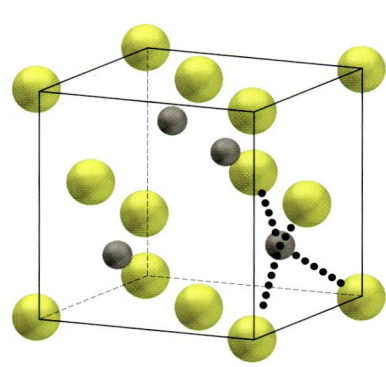

FIGURE 10.10 The crystal structure of the mineral sphalerite is based on a face-centered cubic array of S^{2-} ions with Zn^{2+} ions in four of the eight tetrahedral holes.

In other ionic solids, the cations are small enough to fit into the tetrahedral holes formed by the anions. An example is the mineral sphalerite, also known as zinc blende or zinc sulfide (ZnS; Figure 10.10). The sulfide anions, with an ionic radius of 180 pm, are arranged in a face-centered cubic lattice. The tetrahedral holes of the S^{2-} lattice can accommodate Zn^{2+} cations, which have an ionic radius of only 74 pm. The unit cell contains eight tetrahedral holes, but only four S^{2-} ions. Therefore only four Zn^{2+} ions are needed, and so only half of the tetrahedral holes are occupied. This pattern of half-filled tetrahedral holes in an fcc structure is sometimes called the *sphalerite structure*. Many other compounds, particularly those formed between transition metal cations with a +2 charge and anions of the Group 6A elements with a −2 charge, have sphalerite structures.

The mineral fluorite, CaF_2, has an fcc structure in which Ca^{2+} ions are at the lattice points, and the eight tetrahedral holes per unit cell are filled by F^- ions. Because there are four Ca^{2+} ions in each unit cell, this arrangement satisfies the 1:2 mole ratio of Ca^{2+} to F^- ions. This structure is so common that it, too, has its own name, the *fluorite structure* (Figure 10.11). Other compounds having this structure include SrF_2, $BaCl_2$, CaF_2, and PbF_2. Another group of compounds with

FIGURE 10.11 The crystal structure of the mineral fluorite (CaF_2) consists of a face-centered cubic arrangement of Ca^{2+} ions with F^- anions in all eight of the tetrahedral holes. Thus, there are eight F^- ions within the unit cell—twice as many as the four Ca^{2+} ions assignable to it and consistent with the chemical formula CaF_2.

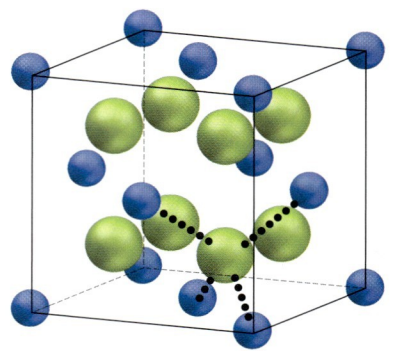

a cation-to-anion mole ratio of 2 : 1 have *antifluorite structures.* In the crystal lattices of these compounds, which include Li_2O and K_2S, the much smaller cations occupy the tetrahedral holes formed by cubic closest packing of the larger anions.

To summarize, the arrangement of ions in a crystalline lattice depends on their relative sizes. If we treat ions as hard spheres of constant radius, then the sizes of the tetrahedral and octahedral holes in a lattice will be directly related to the size of the larger ion, which is often the anion. The ratio of the ionic radius of the cation to that of the anion predicts which holes the cation is more likely to occupy. Table 10.2 contains guidelines for predicting how ions are distributed in these structures. We can check the usefulness of these guidelines by comparing the known structures of NaCl and Li_2O against those predicted by the values in the table. The ionic radius ratio of Na^+ to Cl^- is (98 pm/181 pm) = 0.54. According to the criteria in Table 10.2, sodium ions should be in octahedral holes, which indeed they are. The ionic radius ratio of Li^+ to is O^{2-} is (60 pm/140 pm) = 0.43. On the basis of this ratio, we predict that the smaller Li^+ ions will occupy tetrahedral holes—a prediction confirmed by X-ray diffraction data and the compound's antifluorite structure.

TABLE 10.2 Ion-Size Ratios and the Location of the Smaller Ion in Closest Packed Structures

Packing	Type of Hole for Smaller Ion	Radius Ratio*
hcp or ccp	Tetrahedral	0.22–0.44
hcp or ccp	Octahedral	0.44–0.73
Simple cubic	Cubic	0.73–1.00

*Radius ratios as predictors of solid-state structures have their limitations because ions are not truly solid spheres with a constant radius; the radius of an ion may be different in different compounds, and so these ranges are only approximate.

Simple cubic packing

The criteria in Table 10.2 might lead you to wonder what types of holes accommodate ions that are greater than 0.73 times the size of the larger ion in a binary ionic compound. It turns out that the larger ions cannot have closest-packed structures. One way to accommodate ions of nearly the same size is to stack the larger ions directly above one another. This stacking results in the simple cubic structure shown in Figure 10.5A and an *aaa* packing pattern. Such a pattern is not as efficient as cubic closest packing: the particles in *aaa* layering occupy only 52% of the available space, leaving more space for the holes between *aaa* layers. These larger holes are called *cubic holes*. Consider, for example, the structure of CsCl (Figure 10.12). The ionic radius of Cs^+ (169 pm) is only slightly smaller than that of Cl^- (181 pm). This similarity enables Cs^+ ions to occupy cubic holes inside a simple cubic lattice of Cl^- ions. The unit cell of CsCl could also be drawn with Cs^+ ions in the corners and Cl^- ions in the cubic holes of a Cs^+ simple cubic structure.

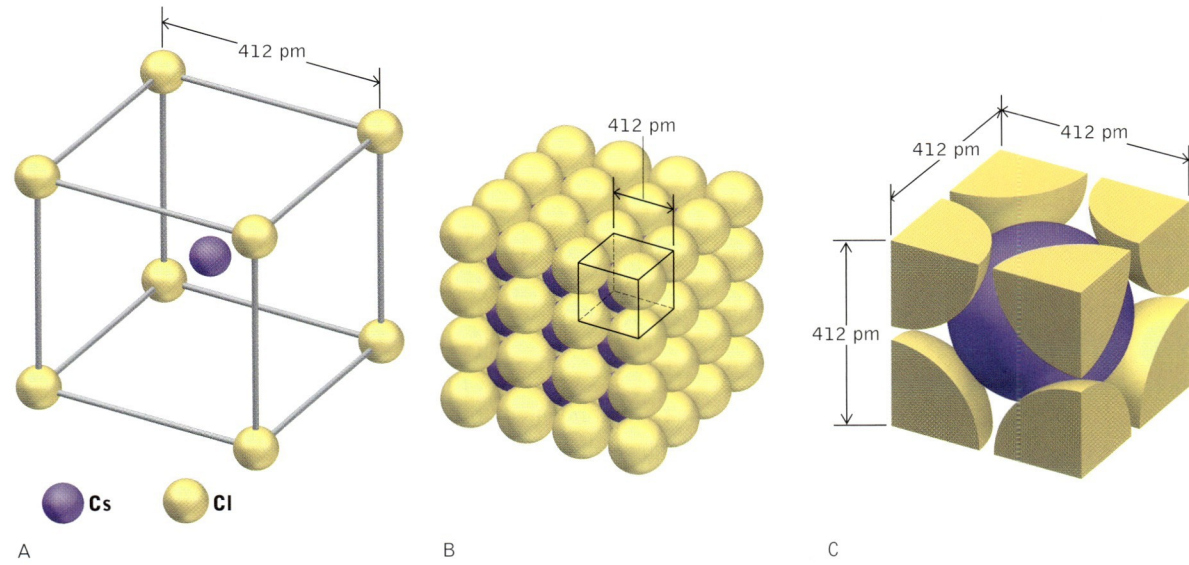

FIGURE 10.12 The crystal structure of the cesium chloride on the watch glass to the right resembles that of a body-centered cubic (bcc) substance, but it is more accurately described as two intertwined simple cubic structures: one of Cs^+ ions and the other of Cl^- ions. Each Cs^+ ion resides in a cubic hole in a Cl^- lattice and touches eight Cl^- ions as shown here. However, the structure could also be drawn with each Cl^- ion residing in a cubic hole in a Cs^+ lattice and touch eight Cs^+ ions.

SAMPLE EXERCISE 10.4: Calculate the percentage of the available space occupied by the chloride ions in lithium chloride, given that LiCl contains a face-centered arrangement of Cl⁻ with radius r_{Cl^-} = 181 pm and a unit-cell edge equal to 513 pm.

SOLUTION: The volume of the unit cell is:

$$V_{unit\ cell} = (edge)^3 = (513\ pm)^3 = 1.35 \times 10^8\ pm^3$$

A face-centered unit cell of LiCl contains 4 Cl⁻ ions. The volume of a sphere is

$$V_{sphere} = \frac{4}{3}\pi r^3$$

Substituting the radius of the chloride ion gives

$$V_{sphere} = \frac{4}{3}\pi(181\ pm)^3 = 2.48 \times 10^7\ pm^3$$

The volume for all four chloride ions is:

$$4 \times (2.48 \times 10^8\ pm^3) = 9.93 \times 10^7\ pm^3.$$

The percentage of the available space occupied by the chloride ions, or the packing efficiency, is calculated from the following relation:

$$packing\ efficiency = \frac{V_{spheres}}{V_{unit\ cell}} \times 100$$

Substituting the appropriate values gives

$$packing\ efficiency = \frac{9.93 \times 10^7\ pm^3}{1.35 \times 10^8\ pm^3} \times 100 = 73.6\%$$

PRACTICE EXERCISE: Calculate the packing efficiency of CsCl, which consists of a simple cubic arrangement of Cl⁻ ions (r_{Cl^-} = 181 pm) with Cs⁺ ions (r_{Cs^+} = 169 pm) in the cubic hole. The unit-cell edge is 412 pm. (See Problems 33 and 34.)

CONCEPT TEST: Do Cs⁺ ions occupy *all* the cubic holes inside a simple cubic lattice of Cl⁻ ions in CsCl?

Hexagonal closest packing

So far, we have considered stacking patterns having either three distinct layers (*abcabcabc* . . .) or only one kind of layer (*aaa* . . .). We have yet to consider an example of the pattern described in Figure 10.5 that has an *ababab* . . . repeating pattern. The name of this pattern is **hexagonal closest packing** (hcp). The corresponding unit cell has the shape of a hexagonal prism, as shown in Figure 10.13A. Hexagonal closest packing is just as efficient a packing pattern as cubic

Hexagonal closest packing is a crystal structure based on a hexagonal unit cell in which the composition of layers of particles has an *ababab* . . . pattern.

closest packing (Figure 10.13B). In both packing patterns, the larger ionic spheres occupy 74% of the available space. Hexagonal closest-packed structures also have the same number of tetrahedral and octahedral holes as do ccp structures. One mineral with the hcp structure is würtzite, which is another crystalline form of ZnS. Its structure consists of an hcp array of S^{2-} ions with Zn^{2+} ions in half of the tetrahedral holes. Perhaps you are wondering how one compound, ZnS, has two crystalline structures: the cubic closest packing of sphalerite

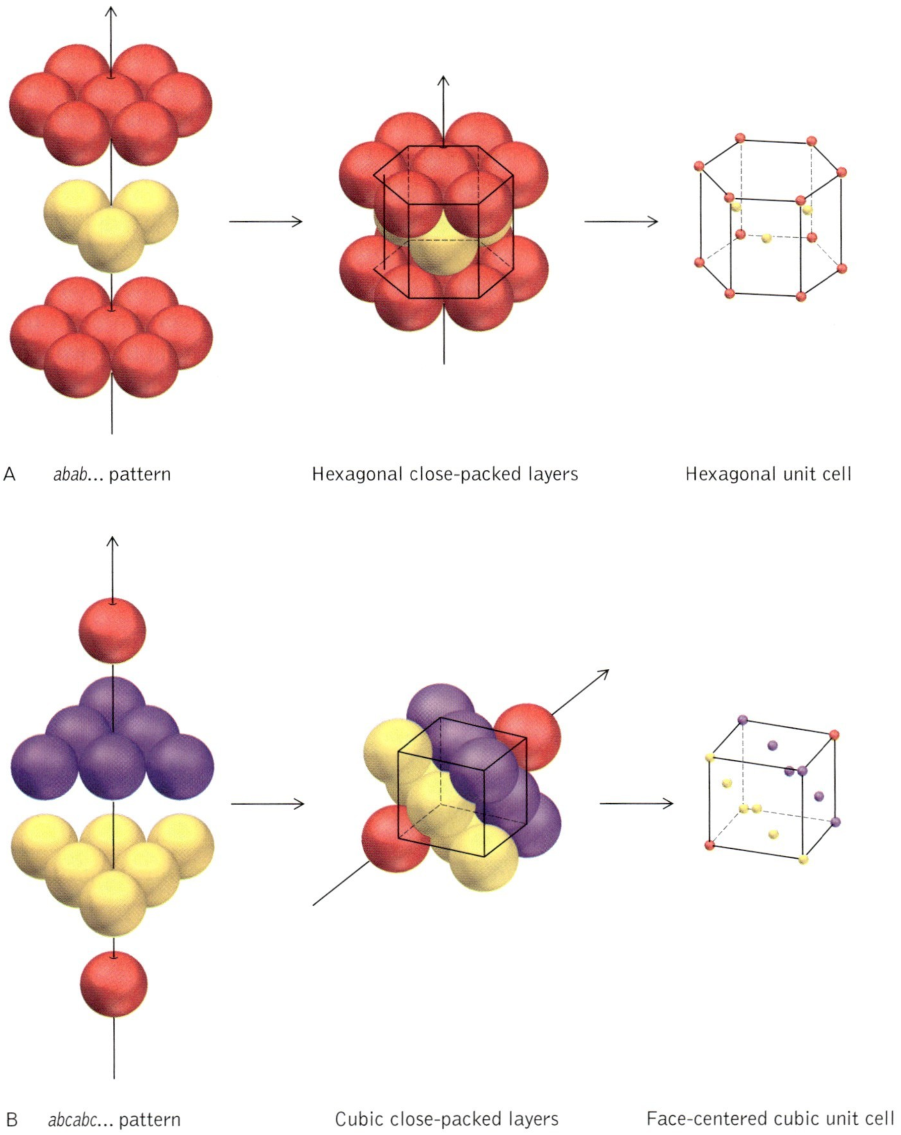

A *abab...* pattern   Hexagonal close-packed layers   Hexagonal unit cell

B *abcabc...* pattern   Cubic close-packed layers   Face-centered cubic unit cell

FIGURE 10.13 Hexagonal closest-packed structures and hexagonal unit cells (A) are as efficiently packed as cubic closest-packed structures (B), which have face-centered cubic unit cells. The fcc unit cell is more easily seen if the layers are tipped about 45 degrees.

Displays of fruit in an open-air market or cannonballs at an eighteenth-century fort illustrate closest-packed arrays of objects.

and the hexagonal closest packing of würtzite. The answer to this question lies in the relative stabilities of the two structures and on the conditions under which they are formed. We will see how these factors affect crystal structures in the next section.

CONCEPT TEST: Does a stack of cannonballs have an *aaa*..., *ababab*..., or *abcabcabc*... packing pattern?

Before ending our consideration of the structures of ionic compounds, we should examine the structure of at least one compound made of polyatomic ions, calcium carbonate ($CaCO_3$). The most stable form of $CaCO_3$ is called *calcite*. It makes up about 15% of marine sediments and Earth's sedimentary rocks and is the principal component of coral reefs, shellfish shells, limestone, and marble. To understand the structure of calcite, let's extend our model of ions as solid spheres to include polyatomic ions, such as CO_3^{2-}. Multiatom CO_3^{2-} anions occupy the lattice points of a ccp structure in $CaCO_3$. Calcium ions occupy the octahedral holes, providing the necessary 1:1 mole ratio of Ca^{2+} to CO_3^{2-} ions. Many other minerals are made of polyatomic ions, including those formed from the two most abundant elements in Earth's crust: silicon and oxygen. The structures of silicates do not fit our sphere model for reasons to be explored in the next section.

CONNECTION: Calcium carbonate is a principal constituent of the white cliffs of Dover, England (Section 4.8), and limestone caves (Section 5.6).

10.4 NETWORK SOLIDS: THE MANY FORMS OF SILICA

As noted in Section 10.1, quartz (SiO_2) is the most common mineral in Earth's crust. Despite its formula, quartz is not made of discrete SiO_2 molecules. The silicon and oxygen atoms in quartz are covalently bonded together in three-dimensional networks of linked Si and O atoms, making the quartz structure an example of a **network solid.**

In all SiO_2 minerals, each silicon atom is bonded to four oxygen atoms, forming oxygen tetrahedra with silicon atoms at their centers. Each oxygen atom is bonded to two silicon atoms, thereby linking the tetrahedra into networks.

A **network solid** is made of a rigid, three-dimensional array of covalently bonded atoms.

What sorts of structures do linked silica tetrahedra have?

Because each corner oxygen atom is bonded to two silicon atoms, each silicon atom gets credit for only half of the four oxygen atoms to which it is bonded—hence the formula SiO_2.

What sort of structures do linked silica tetrahedra have? At least eight different minerals have the chemical composition SiO_2. The members of such a family of substances, with shared chemical composition but different structures and

FIGURE 10.14 The crystal structure of quartz includes hexagonal arrays of silicon and oxygen atoms.

properties, are called **polymorphs.** The quartz group is the most abundant of the silica family of minerals, and quartz itself (see Figure 10.2) is the most abundant member of the group. Other members of the group also have crystal structures in which SiO$_4$ tetrahedra are linked by sharing their oxygen atoms, but the three-dimensional arrays of these tetrahedra vary from one mineral to the next. For example, the mineral tridymite (Figure 10.14) occurs in nature in crystals made of thin, transparent hexagonal plates or scales that are usually twinned together in groups of three (hence the name of the mineral). The shapes of these plates reflects the crystal structure of tridymite, as shown by the hexagonal pattern of SiO$_4$ tetrahedra in one view of its structure in Figure 10.14.

Silica has many other structural options. One of them is in the mineral α-cristobalite—a common mineral in volcanic rocks. In the unit cell of α-cristobalite (see Figure 10.15), four SiO$_4$ tetrahedra reside within what appears to be a face-centered cubic lattice of silicon atoms. There are also β-cristobalite and β-quartz structures that differ from the α-cristobalite and quartz structures described here. Why are there so many forms of SiO$_2$ in nature? The answer is related to the complex phase diagram of SiO$_2$, which has numerous solid phases, unlike H$_2$O and CO$_2$, each of which has just one (see Figures 9.13 and 9.14). These multiple phases mean that, as solid silica minerals form from cooling molten SiO$_2$, they take on different structures, depending on the pressures and temperatures at which they crystallize. Denser forms, such as β-cristobalite and β-quartz, are more likely to form at higher pressures. The presence of these different structures provides geologists with clues about the temperature and pressure under which igneous rocks in a region formed.

> Why are there so many forms of SiO$_2$ in nature?

Polymorphs are compounds with the same molecular formula but with different structures and different properties.

CONNECTION: A phase diagram indicates which forms of a substance are stable at various pressures and temperatures (Section 9.7).

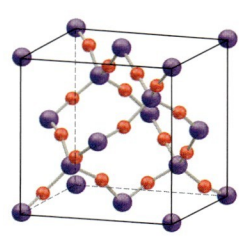

FIGURE 10.15 In the crystal structure of the white cristobalite deposits that formed on these pieces of volcanic glass, four SiO₄ tetrahedra reside within an expanded face-centered cubic unit cell of silicon atoms.

Orthosilicates

Earth's crust also contains ionic minerals that are rich in silicon and oxygen. These minerals, called *silicates*, have some of the tetrahedral structure of pure silica, but not all the oxygen corner atoms take part in Si–O–Si linkages. Some, or all, of these oxygen atoms complete their octets by acquiring extra electrons, giving the structures negative charges. The simplest silicates are isolated tetrahedra in which each corner atom has an extra electron, giving the tetrahedron a -4 charge and the formula SiO_4^{4-} (Figure 10.16A). These silicates are called *orthosilicates* and are found in nature in structures that contain cations with a total of four positive charges for every one SiO_4^{4-}. The most common orthosilicate mineral is *olivine*, the main mineral of Earth's mantle and a principal ingredient in igneous rocks. It is also found in meteorites. For each SiO_4^{4-} ion in olivine's structure, there are two Mg^{2+} or to two Fe^{2+} ions or one of each. The chemical formula of olivine:

$$(Mg, Fe)_2SiO_4$$

indicates the uncertainty in its cationic composition; Mg appears first because Mg^{2+} is present more often than Fe^{2+}. In nature, olivine minerals contain different proportions of the two cations, and so geologists use the following formula:

$$Mg_xFe_{2-x}SiO_4$$

to express the range in their chemical compositions where x can have any value (not just whole-integer values) from 0 to 2. Thus, a particular crystal of olivine could have the formula $Mg_{1.2}Fe_{0.8}SiO_4$ if 60% of the cations in it were Mg^{2+} ions and 40% of them were Fe^{2+} ions. Clearly, olivine does not meet our definition of a pure substance in that it does not have a fixed chemical composition. However, it is classified as a *mineral* because it is a geological material with a chemical composition that falls within a limited range.

10.4 Network Solids: The Many Forms of Silica

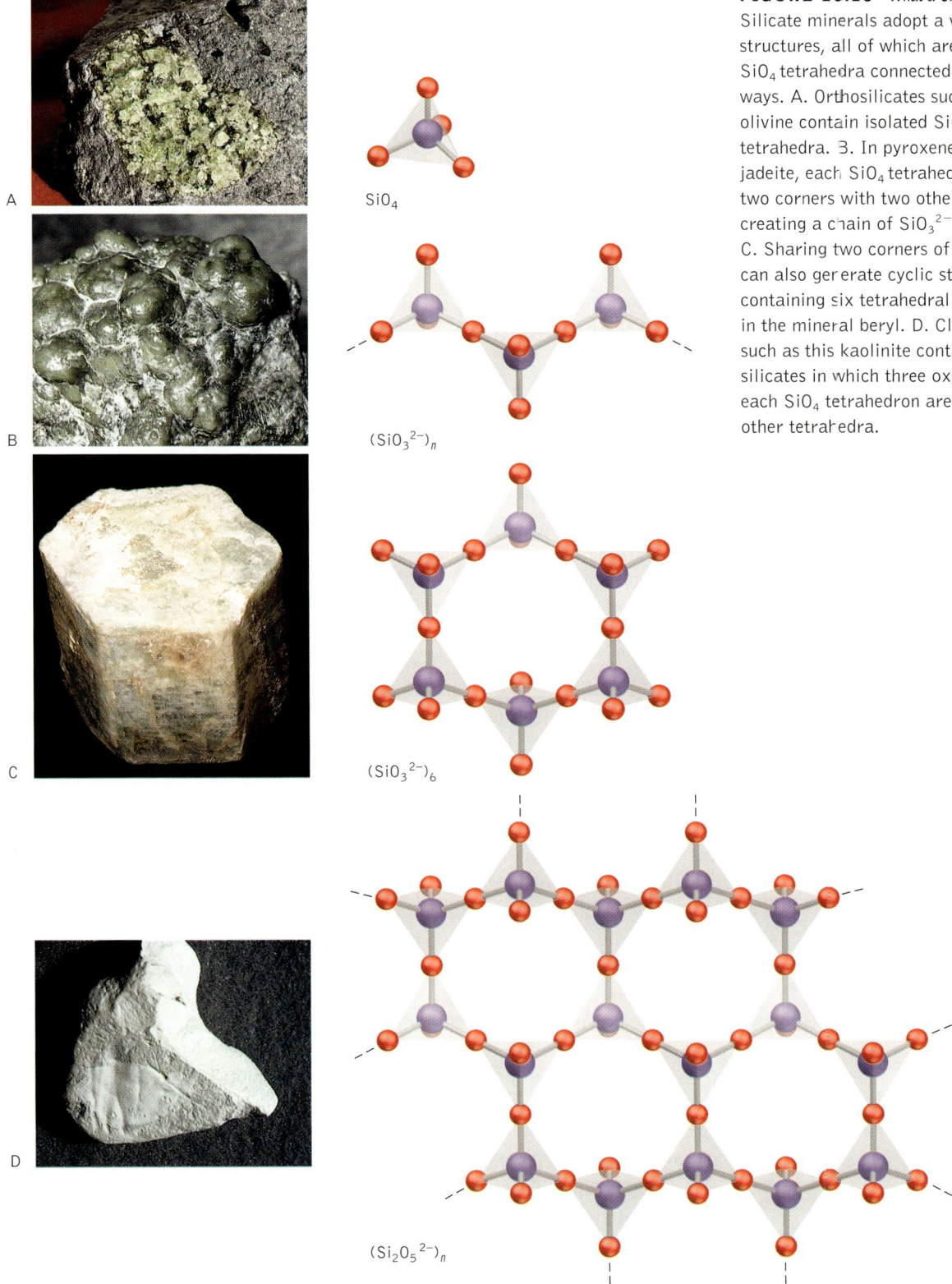

FIGURE 10.16 *What a Chemist Sees.* Silicate minerals adopt a variety of structures, all of which are based on SiO$_4$ tetrahedra connected in different ways. A. Orthosilicates such as this olivine contain isolated SiO$_4$$^{4-}$ tetrahedra. B. In pyroxenes such as this jadeite, each SiO$_4$ tetrahedron shares two corners with two other tetrahedra creating a chain of SiO$_3$$^{2-}$ units. C. Sharing two corners of SiO$_4$ units can also generate cyclic structures containing six tetrahedral SiO$_4$ units as in the mineral beryl. D. Clay minerals such as this kaolinite contain layered silicates in which three oxygen atoms of each SiO$_4$ tetrahedron are shared with other tetrahedra.

How can Fe^{2+} and Mg^{2+} substitute for each other in olivine and other minerals? They have the same charges and similar ionic radii: 66 pm for Mg^{2+} and 74 pm for Fe^{2+}. Ions with the same charge and similar radii can substitute for one another in ionic crystalline lattices without disturbing the structure of the lattice. Some structures can accommodate differences in ionic size as great as 30%.

CONCEPT TEST: Suppose you have two rocks—one composed of olivine rich in iron and the other composed of olivine rich in magnesium. Which one should be denser?

Metasilicates

☑ In most silicates, one or more of the oxygen atoms are shared by two silicon atoms. These metasilicates have chain-, ring-, or sheetlike structures.

The molecular structures of most silicates are two- and three-dimensional arrays of silicate tetrahedra linked together by sharing one or more of their oxygen atoms with other silicate tetrahedra. These structures are called condensed silicates or *meta*silicates. Among the major classes of minerals with such structures are the *pyroxenes*. They consist of chains of silicate tetrahedra in which each tetrahedron shares two of its oxygen atoms with two other silicate tetrahedra, creating Si–O–Si links in the chain, as shown in Figure 10.16B. Adjoining chains are held together by their mutual attraction to Ca^{2+}, Mg^{2+}, or Fe^{2+} ions located in between the chains. The two nonlinking oxygen atoms of each tetrahedron complete their octets by acquiring an additional electron and taking on a negative charge. Thus, each tetrahedron contributes a charge of -2 to the overall charge of the chain.

To derive a chemical formula for pyroxene, we need to account for the fact that each silicate tetrahedron can claim only half ownership of each shared oxygen. Thus, each link in the silicate chain has two oxygen atoms that it does not share (for which it gets full credit) and two that are shared (for which it gets only half credit). The resulting Si:O ratio is one Si atom for every $[2 + 2(\frac{1}{2})] = 3$ O atoms so the formula for each link in the chain is SiO_3^{2-}.

Earth's crust and atmospheric aerosols are rich in silicates with even more intricate two- and three-dimensional arrays of silicate tetrahedra. Some metasilicates are made of rings of tetrahedra (see Figure 10.16C), rather than chains. An example of such a mineral is *beryl* ($Be_3Al_2Si_6O_{18}$), the principal source of the element beryllium. Within the structure of beryl, Al^{3+} ions are located in octahedral holes formed by six oxygen atoms from six surrounding silicate tetrahedra. Small variations in the structure of beryl are created when transition-metal cations such as Fe^{3+} or Cr^{3+} replace Al^{3+} in these holes. These substitutions add color and considerable value to minerals such as beryl, as will be discussed in Section 10.7.

Finally, there are silicates in which three of the four corner oxygen atoms are shared. They consist of sheets of linked hexagons each made of six silicate tetrahedra, as shown in Figure 10.16D. Each tetrahedron gets credit for the one oxygen atom that it does not share plus half of the three others that it does share.

Thus, the basic repeating unit has $[1 + 3(\frac{1}{2})] = 2.5$ O atoms for every Si atom and the formula $SiO_{2.5^-}$. For convenience, we use whole numbers for subscripts when possible. To do so in this case, we multiply $SiO_{2.5^-}$ by two; so the formula for the silicate part of these sheetlike, or layered, structures is $Si_2O_5^{2-}$. Among the more common layered silicates is the clay mineral kaolinite, $Al_2(Si_2O_5)(OH)_4$. It is a member of a class of minerals in which layers of silicate sheets are interspersed with layers containing aluminum ions at their centers instead of silicon atoms. These *aluminosilicates* are among the most common minerals in Earth's crust.

CONNECTION: The structures of silicates and the changes they undergo at high temperature are important in understanding the properties of many ceramic materials (see Section 18.3).

10.5 ALLOTROPES OF CARBON AND SULFUR

The atmosphere above land areas contains particles of dust with chemical compositions similar to those of the soils of those areas. Forest fires and human activity add particles produced by incomplete combustion: soot and smoke. A principal ingredient in these materials is elemental carbon, which can have more than one structure, or allotrope—a term used in Section 6.6 in reference to different structural and molecular forms of oxygen. The most commonly encountered allotrope of carbon is graphite, the form found in pencils, lubricants, and gunpowder, as well as in soot and smoke. Graphite contains sheets of carbon atoms, each of which is connected by overlapping sp^2 hybrid orbitals to three neighboring carbon atoms in a network of six-membered rings (Figure 10.17A). Each C–C bond is 142 pm long. Overlapping unhybridized p orbitals on the carbon atoms form a network of π bonds that are delocalized across the network. Their presence makes graphite an excellent conductor of electricity for reasons we will discuss in the next section.

Diamond is another allotrope of carbon that forms under intense heat (>1700 K) and pressure (>50,000 atm) deep in Earth. In the diamond structure, each carbon atom is bonded by sp^3 hybrid orbitals to four neighboring carbon atoms. The bonding electrons are entirely localized in σ bonds, making

CONNECTION: The different chemical and physical properties of the allotropes of oxygen, O_3 and O_2, are described in Section 6.6.

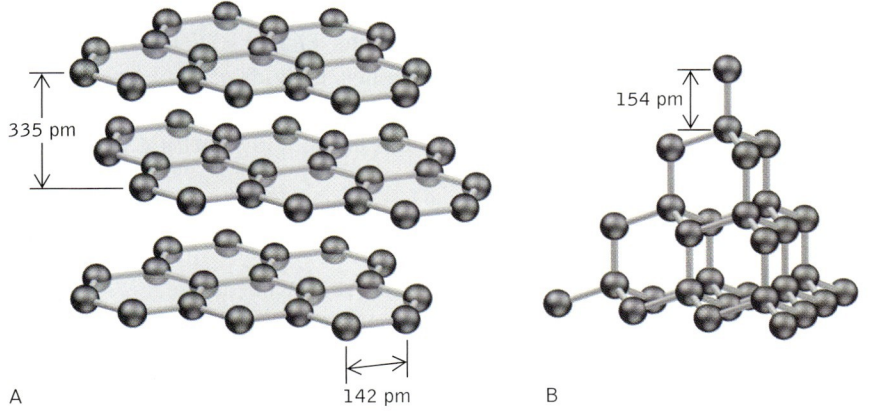

FIGURE 10.17 Carbon crystallizes as (A) graphite, in which carbon atoms connected by σ bonds and delocalized π bonds, or (B) diamonds, which is a network solid in which each carbon atom is connected by σ bonds to four adjacent carbon atoms.

diamonds poor electrical conductors. The network of σ bonds produces an extremely rigid structure.

The C–C bond distance in diamonds is 154 pm. These longer C–C bonds are not as strong in diamond as are the C–C bonds in graphite. Yet diamond is the hardest substance known, and graphite is quite soft. How can we rationalize these differences on the basis of chemical bonding? The key difference between these two allotropes of carbon is in the forces *between* the two-dimensional sheets in graphite. The distance between the sheets is 335 pm, which is much too long to be a covalent bonding distance. Rather, relatively weak London (dispersion) forces (see Section 9.4) hold the sheets to one another, allowing adjacent sheets to slide past each other with ease. This slippage makes graphite soft and a good lubricant. On the other hand, the hardness of diamond makes it an excellent, though expensive, abrasive. Since the 1950s, high temperatures and pressures have been used to artificially transform graphite into "industrial" diamonds. These artificial diamonds are used principally as abrasives and in cutting tools. Most lack the optical clarity of gemstones.

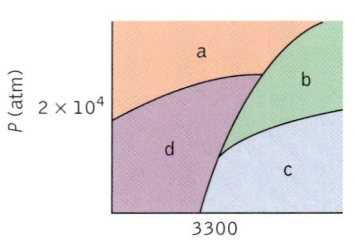

Phase diagram of carbon.

CONCEPT TEST: The phase diagram for carbon to the left shows the temperatures and pressures at which carbon vapor, liquid carbon, and two solid forms, graphite and diamond, are stable. Which region (a, b, c, or d) represents diamond?

A third allotrope of carbon was discovered in the 1980s. Networks of five- and six-atom carbon rings can form spherical structures of 60, 70, or more carbon atoms that look like molecular-sized soccer balls (Figure 10.18). They are called *fullerenes* in recognition of the similarities in their shapes to the geodesic domes designed by American architect R. Buckminster Fuller. Many chemists call them "bucky balls" for the same reason. When discovered, fullerenes were believed to be a form of carbon rarely found in nature. In recent years, analyses of soot and of emission spectra from giant stars have disclosed that fullerenes may be present wherever carbon is found.

Elemental sulfur is another element that exists in nature in one of several structures. All of them are crystalline molecular solids in which the lattice is constructed of S_8 molecules, called cyclo-octasulfur, rather than individual atoms or ions. Cyclo-octasulfur consists of puckered rings (Figure 10.19), each containing eight sulfur atoms covalently bonded together. Only dispersion forces hold one ring to another in solid sulfur. The weakness of these interactions is the reason why the crystalline forms of elemental sulfur are soft, low-melting solids.

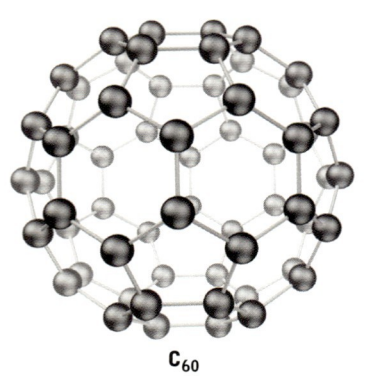

FIGURE 10.18
Buckminsterfullerene (C_{60}) is a molecular solid containing 60 sp^2-hybridized carbon atoms. Both five- and six-membered rings are required to construct the nearly spherically shaped molecule.

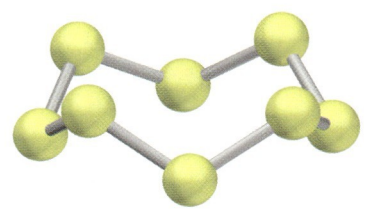

FIGURE 10.19 Sulfur crystallizes as a molecular solid containing puckered eight-member rings of σ bonds formed by overlapping sp^3 hybrid orbitals on adjacent sulfur atoms.

10.6 METALLIC BONDS AND STRUCTURES

Ionic solids are held together by electrostatic forces, and nonmetallic elements and molecular solids are held together by a combination of covalent bonds and intermolecular forces. What, then, is the nature of bonding in metals, in which there are no ions and no readily identifiable covalent bonds? Two models can be used to explain bonding in metals. In one approach, metal atoms are assumed to be immersed in a sea of mobile electrons (Figure 10.20). These electrons are free to redistribute themselves throughout the solid in response, for example, to an applied electrical field, making metals excellent conductors of electricity.

A more sophisticated model of metallic bonding is based on molecular-orbital theory. When the atomic orbitals on two neighboring metal atoms mix, they form sets of high-energy (antibonding) and lower energy (bonding) molecular orbitals (Figure 10.21). When more atoms bond to the first two, more molecular orbitals are created. In all cases, the number of bonding MOs equals the number of antibonding MOs. As more molecular orbitals are created, the differences in their energy levels becomes blurred. In structures with enormous numbers of atoms, the molecular orbitals form a *band* of orbital energies. Because the valence shells of many metal atoms are only partly filled, the corresponding MOs, particularly the higher energy MOs (which fill last) will be empty or only partly occupied. If the energy range of a band of MOs filled with valence electrons (shown in blue in Figure 10.21) is close to the energy range of an unoccupied band of MOs (shown in yellow), bonding MO valence electrons can readily move into the unfilled band. This additional freedom of motion gives the electrons considerable mobility, which imparts the qualities of good thermal and electrical conductivity that we associate with metals. This theory of MO bonds can be used to describe

> What is the nature of bonding in metals, in which there are no ions and no readily identifiable covalent bonds?

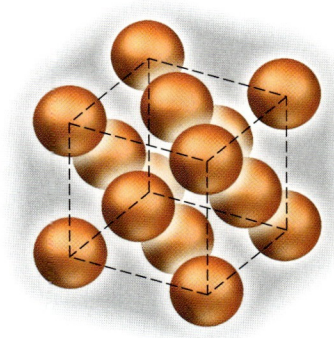

FIGURE 10.20 In one model of metallic bonding metal atoms are immersed in a sea of valence electrons (shown in gray). The electrons are associated with the whole sample rather than with individual specific atoms.

CONNECTION: Molecular-orbital theory was introduced in Section 6.9.

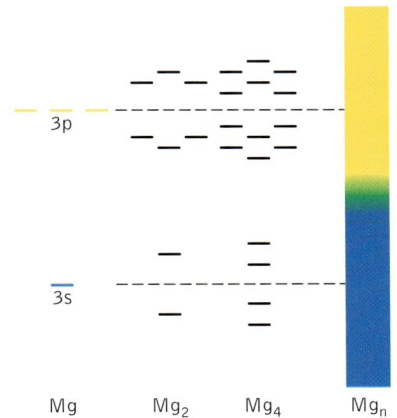

FIGURE 10.21 Band theory of metals and other solids derives from molecular-orbital theory. As the atomic orbitals of an increasing number of atoms overlap, bands of molecular orbitals are formed. Because there are small differences in the energy levels of filled and unfilled bands in many metals, electrons are able to move between bands.

THE CHEMISTRY OF THE GROUP 4A ELEMENTS

Carbon is only a minor component of Earth's crust, ranking 19th in abundance. What carbon there is in the crust exists mostly as magnesium and calcium carbonates and in deposits of petroleum and natural gas. Still, more compounds contain carbon than any other element except hydrogen. Carbon is the central atom of organic and biochemistry, as will be explored in the next unit of this textbook.

Elemental carbon has three allotropes: graphite, diamond, and fullerenes (bucky balls). Red-hot graphite (called coke) is an effective reducing agent. Its reaction with iron ore:

$$3\ C(s) + 2\ Fe_2O_3(s) \longrightarrow 3\ CO_2(g) + 4\ Fe(s)$$

is a preliminary step in the production of steel (Chapter 18). The combustion of carbon compounds in fossil fuels is the principal source of energy in our world. The role of the CO_2 produced by these combustion reactions in global climate change is described in Chapter 6.

The inorganic chemistry of carbon includes reactions with metals to form carbides such as calcium carbide (CaC_2). Miners and cavers used calcium carbide in lamps. The reaction of calcium carbide with water produces acetylene, which burns with a bright white light.

$$CaC_2(s) + H_2O(l) \longrightarrow C_2H_2(g) + Ca(OH)_2(aq)$$
$$2\ C_2H_2(g) + 5\ O_2(g) \longrightarrow 4\ CO_2(g) + 2\ H_2O(g)$$

Carbon dioxide is used in fire extinguishers and as solid, "dry" ice for keeping food and other materials cold (the sublimation point of CO_2 is $-57°C$).

Silicon is the second most abundant element in Earth's crust (behind oxygen) and is present mostly as silica and silicate minerals. Elemental silicon is the basic material of the microelectronics industry. Silicon is produced by the reaction of silicon dioxide with carbon:

$$SiO_2(s) + 2\ C(s) \longrightarrow Si(s) + 2\ CO_2(g)$$

The silicon prepared in this way must be purified before it can be used to make microchips. This purification is done by the reaction of impure silicon with chlorine gas to form $SiCl_4$:

$$Si(s) + 2\ Cl_2(g) \longrightarrow SiCl_4(g)$$

Impurities that do not form volatile chlorides are removed, and the $SiCl_4$ is thermally decomposed back to silicon:

$$SiCl_4(g) \longrightarrow Si(s) + 2\ Cl_2(g)$$

Crystalline silicon has the same structure as that of diamond and is a semiconductor (see Chapter 18).

For years miners and cavers used head lamps fueled by acetylene made by reacting CaC_2 and H_2O.

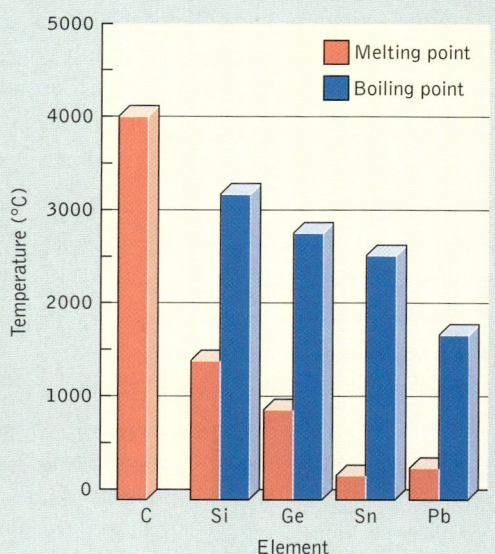

The melting points and boiling points of the Group 4A elements all decrease with increasing atomic number. The boiling point of carbon has not been determined.

The chemistry of germanium, tin, and lead resembles that of silicon. All three form oxides with the formula MO_2 and react with halogens to form tetrahalides, MX_4 (X = F, Cl, Br, I). The dihalides of tin and lead, $SnCl_2$ and $PbCl_2$, also are stable. Both germanium and one of the allotropes of tin, α-tin, have the diamond structure, but β-tin has a more complicated structure. Elemental lead has a face-centered cubic structure. Germanium and α-tin are semiconductors, but β-tin and lead are conductors. Tin is used extensively in metallurgy in combination with other metals. Tin oxides are used to make ceramics, as we will see in Chapter 18. Lead and lead(IV) oxide are the electrode materials in most automobile batteries (Chapter 17).

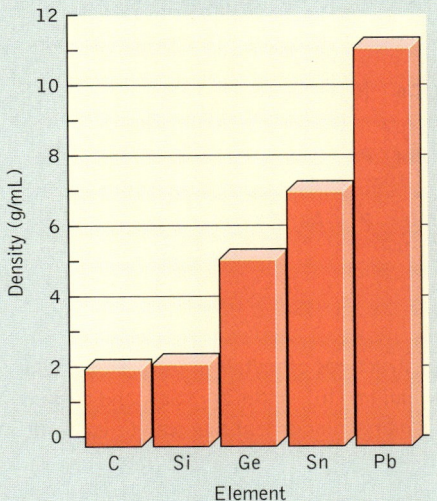

The densities of the Group 4A elements increase with increasing atomic number.

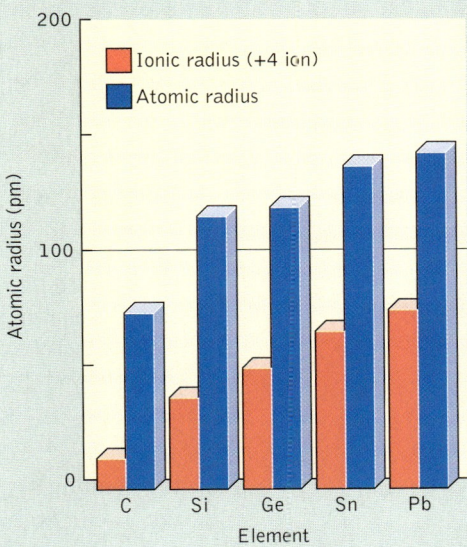

The atomic and ionic radii of the Group 4A elements increase with increasing atomic number.

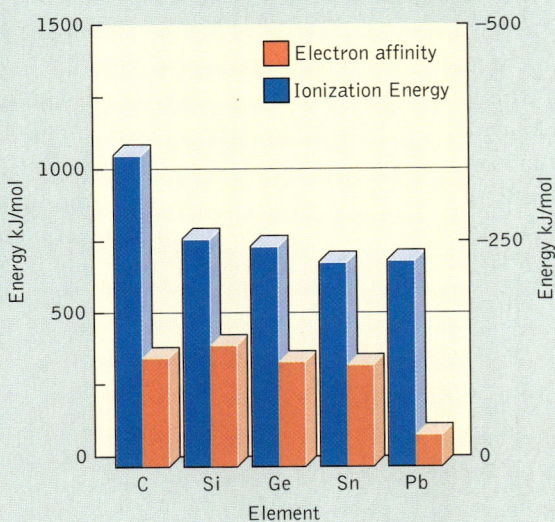

The ionization energies of the Group 4A elements decrease with increasing atomic number. The electron affinities of these elements are smaller than those for the Group 5A–7A nonmetals in the same row. Moderate electron affinities and relatively high ionization energies lead to chemistries based on the formation of covalent bonds.

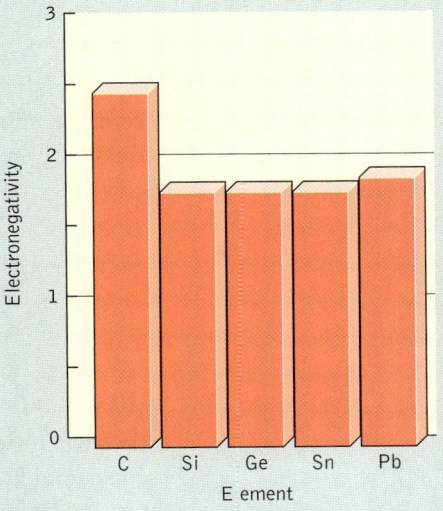

Carbon is the most electronegativite of the Group 4A elements.

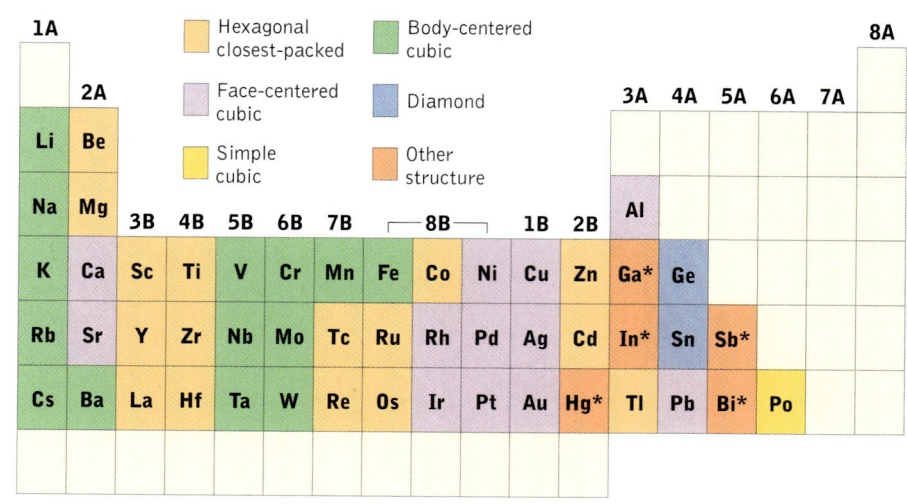

FIGURE 10.22 Summary of the crystal structures of metals in the periodic table. The asterisks on mercury (Hg), gallium (Ga), indium (In), antimony (Sb), and bismuth (Bi) indicate that they have structures that are more complicated than can be easily described in this book.

the bonding in all solids. Those atomic orbitals with the proper energy and orientation can be blended into sets of molecular orbitals and perhaps into bands of molecular orbitals. Whether a solid is a good conductor, or a semiconductor, or an insulator depends on the energy gap between its filled bonding band and its unfilled antibonding band. We will return to this topic in Chapter 18.

The crystal structures of metals follow many of the patterns described in this chapter for ionic compounds. Many transition metals, including the coinage elements gold, silver, nickel, and copper, have cubic closest-packed structures based on face-centered cubic unit cells. A few, including titanium and zinc, form hexagonal closest-packed arrays, and manganese adopts the more open body-centered cubic structure. Figure 10.22 summarizes the crystal structures of the metallic elements.

10.7 GEMSTONES: AN INTRODUCTION TO CRYSTAL FIELD THEORY

We noted in Section 10.3 that the substitution of transition metals for Al^{3+} in the structure of beryl changes the color of the mineral and increases its value considerably. For example, Cr^{3+} substitution in beryl, $Be_3Al_2Si_6O_{18}$, produces the precious gemstone emerald. How does the substitution of Cr^{3+} for Al^{3+} in beryl yield an enormously expensive gemstone? The answer is contained in the crystal structure of beryl and its rings of six SiO_4 tetrahedra (see Figure 10.16C). The oxygen atoms in the beryl structure are nearly closest packed, with silicon occupying a fraction of the tetrahedral holes. Aluminum ions occupy octahedral holes in the structure. The Cr^{3+} ion has the same charge as Al^{3+} and its ionic radius is only slightly larger: 62 pm versus 54 pm. As a result, the substitution of Cr^{3+} for Al^{3+} does not cause significant changes in the structure of the parent beryl mineral. Partial substitution yields the chemical formula for emeralds:

> How does the substitution of Cr^{3+} for Al^{3+} in beryl yield an enormously expensive gemstone?

$$Be_3Cr_xAl_{2-x}Si_6O_{18}$$

Crystal field splitting energy

The distinctive green color of emeralds can be traced to the partly filled d orbitals of Cr^{3+} ions. Aluminum ions are isoelectronic with neon and have no d electrons, but Cr^{3+} ions have the electron configuration $[Ar]3d^3$. Let's focus on the immediate environment around a Cr^{3+} ion in an octahedral hole in beryl. If a Cr^{3+} ion were in the gas phase, all five of its $3d$ orbitals would have the same energy and so would be degenerate (see Section 3.8). Now consider what happens to a Cr^{3+} ion in an octahedral hole in a crystal of emerald as we bring the six octahedral oxygen atoms closer to the Cr^{3+} ion (Figure 10.23). As the atoms approach, there is repulsion between the electrons in the $3d$ orbitals of Cr^{3+} and the lone pairs of electrons on oxygen atoms. This repulsion is greatest for the $3d_{z^2}$ and $3d_{x^2-y^2}$ orbitals of Cr^{3+} because the lobes of these orbitals are pointed directly toward the oxygen atoms in the corners of the octahedron. These repulsions raise the energy of the $3d_{z^2}$ and $3d_{x^2-y^2}$ orbitals above that of $3d_{xy}$, $3d_{xz}$, and $3d_{yz}$ because the lobes of the latter three orbitals do not point toward the corners of the octahedron. This splitting of a set of d orbitals into two subsets with different energies because of interaction with pairs of electrons in orbitals on surrounding atoms is known as **crystal field splitting**, and the difference in energy between the two subsets is called the **crystal field splitting energy** (Δ).

Hund's rule predicts that the three $3d$ electrons in a Cr^{3+} ion in an octahedral field will each occupy one of the lower-energy $3d$ orbitals: $3d_{xy}$, $3d_{xz}$, or $3d_{yz}$, leaving the higher-energy $3d_{z^2}$ or $3d_{x^2-y^2}$ orbitals unoccupied. The energy difference between the two subsets of $3d$ orbitals is symbolized by Δ_o. We use the subscript "o" to indicate that the d-orbital splitting is caused by an *o*ctahedral field of electron repulsions.

Now suppose a photon of electromagnetic radiation with an energy that exactly equals Δ_o strikes a Cr^{3+} ion in an emerald crystal. The ion may absorb this photon as a $3d$ electron jumps from one of the three occupied lower-energy orbitals to one of the two higher-energy orbitals, as shown in Figure 10.24. The wavelength of the absorbed radiation is related to Δ_o by a slightly modified version of Equation 1.3:

$$\Delta_o = hc/\lambda \qquad (10.2)$$

According to **crystal field theory**, the colors and magnetic properties of species containing transition-metal ions are due to the splitting of the energies of the ions' d orbitals caused by interactions with electrons on surrounding atoms.

CONNECTION: Orbital *degeneracy* means that a set of orbitals have the same level of energy (Section 3.8).

CONNECTION: The shape and orientation of the d orbitals are described in Section 3.10.

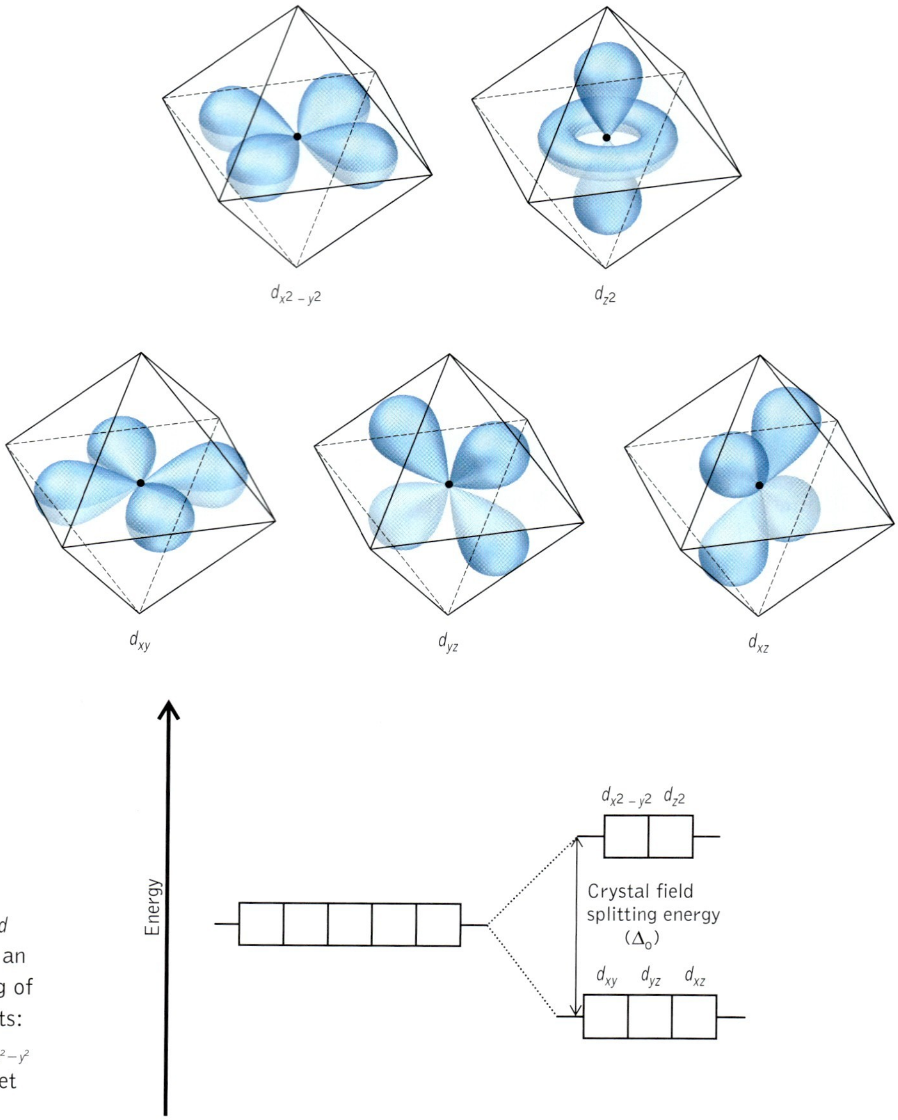

FIGURE 10.23 Interactions between d orbitals on a metal ion and p orbitals in an octahedral geometry lead to the splitting of the energies of the d orbitals into two sets: (1) a higher-energy set containing the $d_{x^2-y^2}$ and d_{z^2} orbitals and (2) a lower-energy set containing the d_{xy}, d_{xz}, and d_{yz} orbitals.

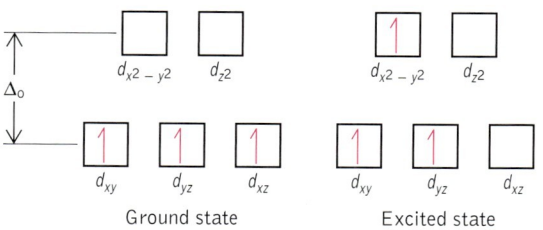

FIGURE 10.24 Absorption of a photon of red-orange light causes a $3d$ electron in a lower-energy orbital of a Cr^{3+} ion in an octahedral field to move to a higher-energy d orbital.

The value of Δ_o for Cr^{3+} in emeralds is 3.24×10^{-19} J. To calculate the corresponding wavelength, we can rearrange the terms in Equation 10.2, insert the value for Δ_o, and solve for λ:

$$\Delta_o = hc/\lambda$$

or

$$\lambda = \frac{hc}{\Delta_o}$$

$$\lambda = \frac{(6.63 \times 10^{-34} \text{ J} \cdot \text{s})(3.00 \times 10^8 \text{ m/s})}{3.24 \times 10^{-19} \text{ J}}$$

$$\lambda = 6.13 \times 10^{-7} \text{ m}$$

Converting λ into nanometers, the unit typically used to express wavelengths of visible light, we have

$$\lambda = (6.13 \times 10^{-7} \text{ m})(10^9 \text{ nm/m}) = 613 \text{ nm}$$

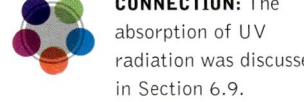

CONNECTION: The absorption of UV radiation was discussed in Section 6.9.

The absorption spectrum of emerald (Figure 10.25) does indeed have an absorption band of red-orange light at 613 nm.

You may be wondering why emeralds that are green absorb light that is red orange. The color that an object appears to have is not the color that it absorbs, but rather the colors that it does *not* absorb—that is, the colors that it reflects. When an object absorbs light that spans yellow, orange, and red wavelengths, as emerald does, it will appear to have the color that is a mix of the other colors of the visible spectrum: violet, blue, and green. In emeralds, another electron transition absorbs some of the violet and blue light; so what we see as the color reflected by emeralds is, no surprise, emerald green.

Emeralds are not the only gems that contain traces of Cr^{3+}. Consider another common component of Earth's crust: alumina, or aluminum oxide (Al_2O_3). The crystal structure of alumina consists of hexagonal closest-packed oxide ions, O^{2-}, within which Al^{3+} ions occupy two-thirds of the octahedral holes. Sometimes a minerals forms in which Cr^{3+} substitutes for some of the Al^{3+} ions in the alumina crystal lattice. The result is a crystal of ruby.

Why are rubies red and emeralds green?

Why are rubies red and emeralds green? Both contain Cr^{3+} in octahedral holes, but their different colors mean that different wavelengths are absorbed as

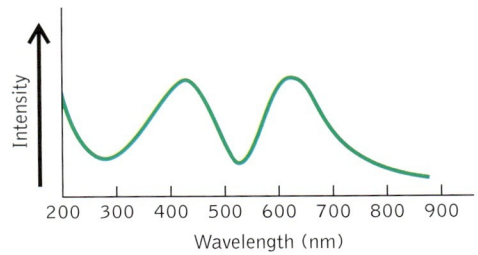

FIGURE 10.25 The absorption spectrum of emerald has two absorption bands in the visible region at 430 and 613 nm.

the electrons from the lower-energy $3d$ orbitals in Cr^{3+} jump to higher-energy $3d$ orbitals. Equation 10.2 tells us that these wavelength differences correspond to differences in the value of Δ_o for Cr^{3+}. We must conclude that the value of Δ_o in rubies is greater than the value of Δ_o in emeralds, and so rubies absorb higher-energy, shorter-wavelength radiation while reflecting ruby red. This conclusion is supported by the results of the following sample exercise.

SAMPLE EXERCISE 10.5: The value for Δ_o for ruby is 3.61×10^{-19} J. What wavelength of visible light do rubies absorb that corresponds to this energy?

SOLUTION:

$$\Delta_o = hc/\lambda$$

Rearranging the equation gives

$$\lambda = hc/\Delta_o$$

$$\lambda = \frac{(6.63 \times 10^{-34} \text{ J} \cdot \text{s})(3.00 \times 10^8 \text{ m/s})(10^9 \text{ nm/m})}{(3.61 \times 10^{-19} \text{ J})} = 551 \text{ nm}$$

which is a wavelength of green light. Rubies have another absorption band near 400 nm; so the light that they reflect is mostly red with some blue and orange. Our eyes interpret the sum of these colors as ruby red.

PRACTICE EXERCISE: Predict the color of the mineral with the formula $Mn_xAl_{2-x}O_3$ if $\Delta_o = 3.85 \times 10^{-19}$ J. (See Problems 67 and 68.)

Magnetite Samples often include large black crystals.

The mineral magnetite, Fe_3O_4, is not considered a gemstone, but samples of it may incorporate beautiful large black crystals. In the structure of magnetite, Fe^{2+} occupies octahedral holes in a cubic closest-packed arrangement of oxide ions, and Fe^{3+} ions are distributed among both octahedral and tetrahedral sites in the lattice. What effect does a tetrahedral geometry have on the d orbitals of a transition-metal cation such as Fe^{3+}? The greatest electron–electron repulsions are between lone pairs on oxygen atoms approaching a set of d orbitals from the four corners of a tetrahedron (Figure 10.26). The greatest repulsions will be experienced by the d_{xy}, d_{xz}, and d_{yz} orbitals. The energy of the d_{z^2} and $d_{x^2-y^2}$ orbitals will be less affected, because their lobes do not point toward the tetrahedral corners. The resulting splitting pattern is just the opposite of that produced in an octahedral geometry. The difference in energies between the two subsets of d orbitals as a result of the tetrahedral interactions is labeled Δ_t. In general, the values of Δ_t are smaller than those of Δ_o.

CONCEPT TEST: The stable oxidation states of Fe are Fe(II) and Fe(III). What is the Fe(II)/Fe(III) ratio in a crystal of Fe_3O_4?

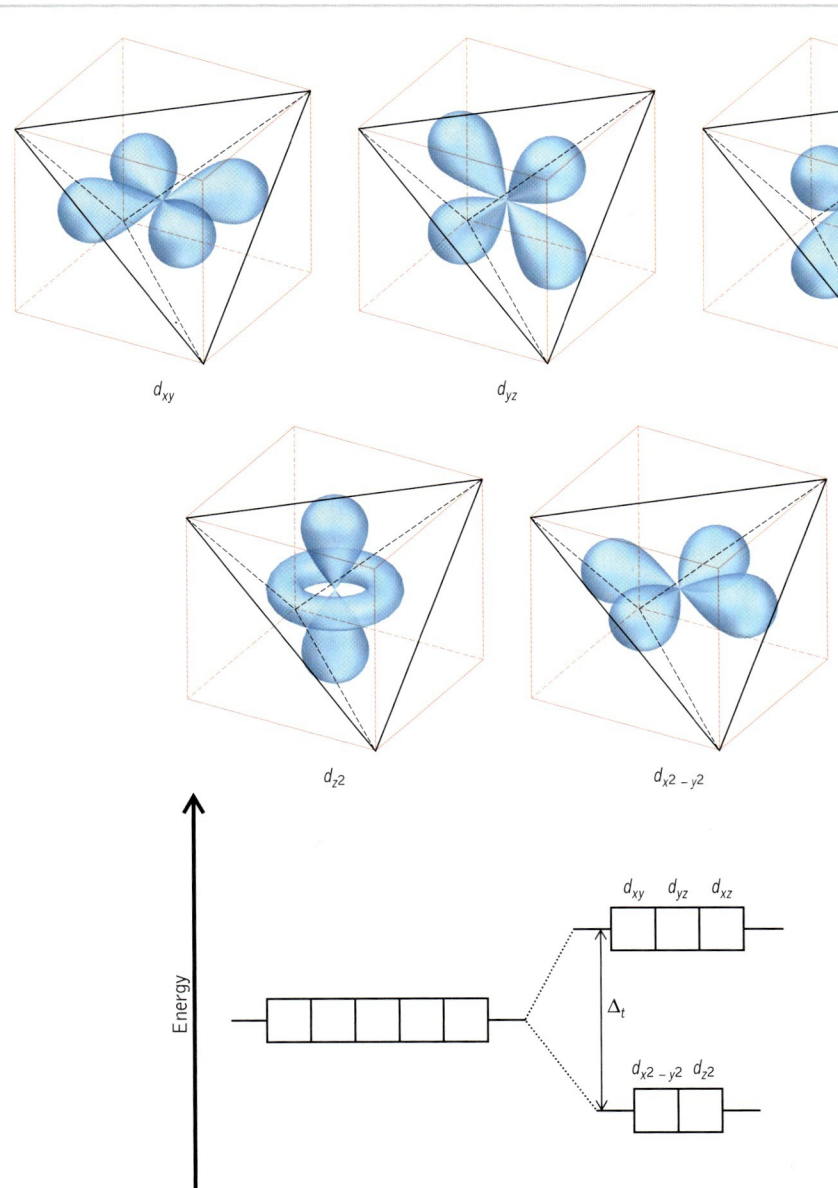

FIGURE 10.26 Interactions between d orbitals on a metal ion and p orbitals in a tetrahedral geometry also lead to the splitting of the energies of the d orbitals into two sets. These sets are the mirror image of those in an octahedral field. One set includes the d_{xy}, d_{xz}, and d_{yz} orbitals, and the other contains the $d_{x^2-y^2}$ and d_{z^2} orbitals. In a tetrahedral field, the d_{xy}, d_{xz}, and d_{yz} orbitals are higher in energy.

Turquoise is a mineral widely used in jewelry crafted by Native Americans in the southwestern United States. The pale blue color of turquoise is the result of d-orbital splitting in Cu^{2+} ions. These ions are located in holes of a complex crystal structure with the following formula:

$$CuAl_6(PO_4)_4(OH)_8 \cdot 4\,H_2O$$

The hole in which each Cu^{2+} ion resides is formed by six oxygen atoms, but the holes are not exactly in an octahedral geometry. The distance from the Cu^{2+} ion

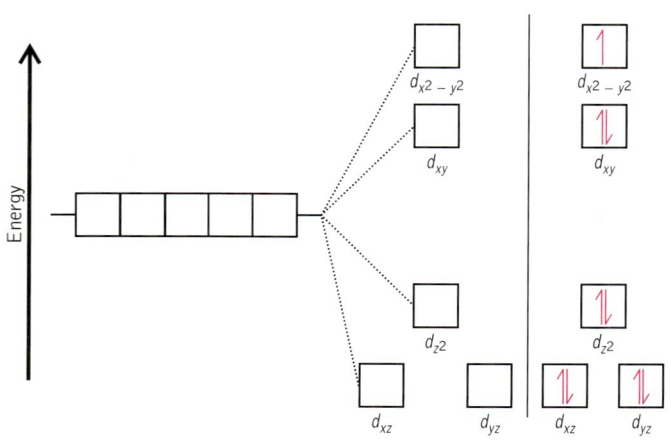

FIGURE 10.27 The splitting of the five 3d orbitals in a square planar field produces several energy levels. The $d_{x^2-y^2}$ and d_{z^2} orbitals are still at higher energy than that of the d_{xz} and d_{yz} orbitals, but the d_{xy} orbital is destabilized by considerable repulsion in a square planar field.

to the oxygen atoms above and below it are much greater than the distance to the four oxygen atoms in the corners of the center plane. As a result, most of the interactions occur between the 3d orbitals on Cu^{2+} and lone pairs on the four nearest oxygen atoms. Thus the environment of the Cu^{2+} ion has a square planar geometry rather than octahedral. A square planar geometry affects the relative energies of the d orbitals, as shown in Figure 10.27. Stronger interactions lead to higher energy for the $d_{x^2-y^2}$ orbital with its lobes directed toward the four corners of the square plane and to somewhat less energy for the d_{xy} orbital because its lobes, though in the xy plane, are directed 45 degrees away from the square planar corners, as also shown in Figure 10.27. Electrons in d orbitals with lobes out of the xy plane undergo less interaction with the lone pairs of electrons on atoms in the corners of the square plane and so have even lower energies. The square planar crystal field splitting, Δ_{sp}, is usually larger than Δ_o. Square planar geometry tends to be limited to the transition metals with nearly filled valence-shell d orbitals, particularly those with d^8 or d^9 electron configurations.

CONNECTION: Square planar geometry was described in Section 7.4.

Magnetic properties

Other transition metal ions occur as impurities in alumina, silica, and aluminosilicate crystal structures. For example, Fe^{3+} may substitute for Al^{3+} in the structure of beryl ($Be_3Al_2Si_6O_{18}$), forming a mineral with the following formula:

$$Be_3Fe_xAl_{2-x}Si_6O_{18}$$

This is the formula of another gemstone, aquamarine. In aquamarine, there are two ways to arrange the five 3d electrons of Fe^{3+} among the five 3d orbitals. The

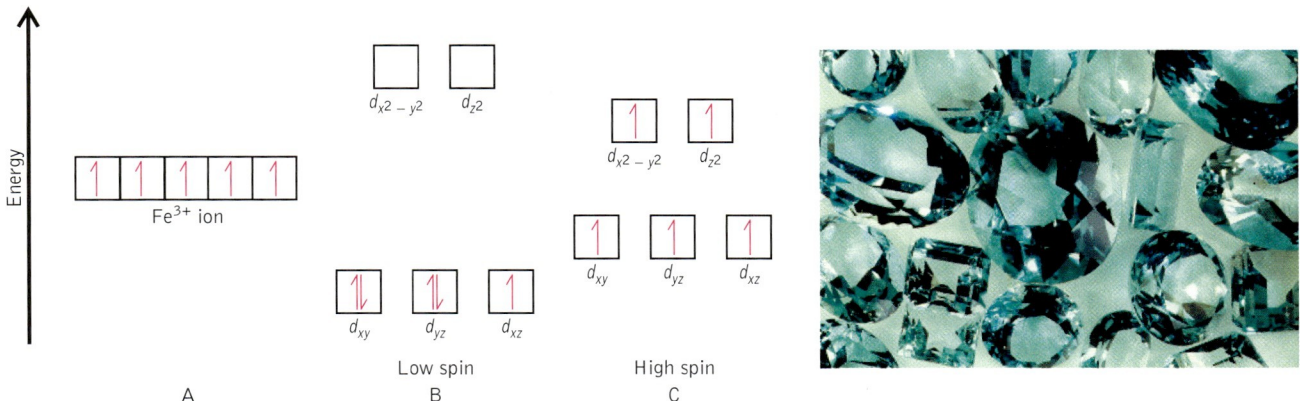

FIGURE 10.28 The ground state of an Fe^{3+} ion has a half-filled set of 3d orbitals as shown in A. In a strong octahedral field the energy of the 3d orbitals may be split enough to produce the low-spin state shown in B. Less splitting produces the high-spin state shown in C. The Fe^{3+} ions in aquamarine are high-spin.

obvious one, in accord with Hund's rule, places a single electron in each orbital, leaving them all unpaired. However, when the value of Δ_o is relatively large, the three lower-energy orbitals are lower in energy than the other two, so all five electrons go into the lower level, leaving the upper two orbitals empty. In this configuration, only one electron is left unpaired. When all five 3d electrons are unpaired, the configuration is called *high spin* because the spin on all five electrons is in the same direction and the resulting magnetic field produced by their spins is maximal. When four of the five electrons are paired and only one is left unpaired, the electron configuration is called *low spin* (Figure 10.28). The two configurations can be distinguished by magnetic measurements. Both configurations are paramagnetic (see Section 6.9) because both have at least one unpaired electron. However, the high-spin state is much more paramagnetic—a property that can be measured with an instrument called a magnetometer.

CONNECTION: Substances made of atoms, ions, or molecules that contain unpaired electrons are paramagnetic (Section 6.9).

SAMPLE EXERCISE 10.6: Not all transition-metal ions can have high-spin and low-spin configurations, because they have either too few (fewer than four) or too many (more than seven) valence-shell d electrons. Determine which of the following ions could have high-spin and low-spin configurations: Mn^{2+}, Mn^{4+}, Co^{3+}, and Cu^{2+}; and draw energy-level diagrams for their 3d orbitals in an octahedral field.

SOLUTION: Mn^{4+}, Mn^{2+}, Co^{3+}, and Cu^{2+} ions have three, five, six, and nine d electrons, respectively. (Remember that transition metals preferentially lose their valence-shell s electrons when they form ions.)

Each of the three 3d electrons in Mn^{4+} must be in a lower-energy orbital. According to Hund's rule, no two of them could be paired in the same orbital if that meant leaving another orbital of the same energy empty. Therefore, there is no low-spin state possible.

Mn^{2+} has a d^5 (five $3d$ electrons) electron configuration. It can exist in high-spin and low-spin states in an octahedral field, as shown in the following energy-level diagram:

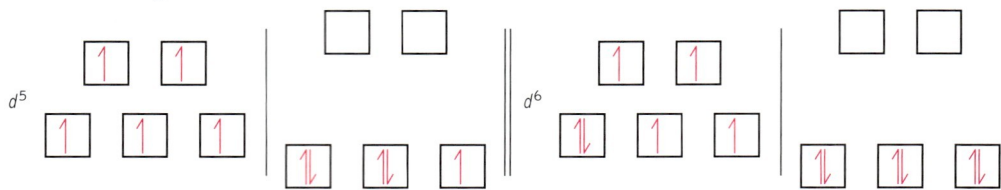

Co^{3+} has a d^6 electron configuration. Distributing the six electrons as evenly as possible (on the left) or all in the lower-energy orbitals (on the right) produces high-spin (four unpaired electrons) and low-spin (no unpaired electrons) states.

Six of the nine $3d$ electrons in Cu^{2+} fill the lower-energy levels. There is no arrangement possible other than a single electron in one of the two high-energy orbitals, and so there can be only one spin state.

PRACTICE EXERCISE: Which, if any, of the spin configurations in the preceding sample exercise are diamagnetic? (See Problems 71 and 72.)

What determines whether a transition-metal ion is in a high-spin or a low-spin configuration? It all depends on the relative energy cost of pairing two electrons in the same lower-energy d orbital compared with the energy cost of promoting one of them to a higher-energy d orbital (Δ_o). If Δ_o is greater than the energy required to pair two electrons in a single d orbital, the result is a low-spin state. If not, then there will be a high-spin distribution. Several factors, including the nature of the atoms surrounding the metal ion, affect the size of Δ_o. For example, Fe^{2+} and Fe^{3+} ions found in most minerals are surrounded by oxygen atoms and are in high-spin states. However, the Fe^{2+} ions in hemoglobin, the oxygen-transport protein in blood (see Section 16.10), are in an octahedral field with four nitrogen atoms in the central plane of the octahedron. The lone pairs of electrons on nitrogen atoms interact more strongly with electrons in the $3d$ orbitals Fe^{2+} ions. The result is more splitting of d orbital energies and Fe^{2+} ions in low-spin states.

CHAPTER REVIEW

Summary

SECTION 10.1

Solids can be crystalline or amorphous. The particles (atoms or ions) in crystalline solids are arranged in ordered arrays, or lattices. X-ray diffraction arises from constructive and destructive interference of X-rays reflecting off different layers of atoms or ions in crystals.

SECTION 10.2

Crystals consist of repeating units called unit cells. The composition of the unit cell of a compound is consistent with the formula of the compound.

SECTION 10.3

Many ionic compounds consist of crystals with closest-packed anions at the lattice points. Closest packing leads to octahedral and tetrahedral holes occupied by cations that tend to be smaller than the anions. The correct crystal structure of a compound accurately predicts its density.

SECTION 10.4

Some nonmetals and many solid covalent compounds consist of vast, covalently bonded networks. Silica consists of SiO_4 tetrahedra, each of which can share as many as three of its oxygen atoms with other tetrahedra, forming Si–O–Si bridges and two- and three-dimensional networks.

SECTION 10.5

The allotropes of carbon are graphite, diamond, and fullerenes. One allotrope of sulfur consists of puckered rings of eight sulfur atoms.

SECTION 10.6

The conductivity of metals can be explained by two models of metallic bonding: (1) atoms in a "sea" of mobile, valence electrons and (2) conduction bands predicted by molecular-orbital theory.

SECTION 10.7

The d orbitals of transition-metal ions that are surrounded by arrays of electron-pair donors undergo crystal field splitting owing to differential interaction with different donors. The promotion of an electron from a lower- to a higher-energy d orbital can result from the absorption of a photon of visible light. Tetrahedral fields produce less d-orbital splitting than do octahedral fields. Cations such as Cu^{2+} (d^9) are always paramagnetic, but ions such as Fe^{2+} (d^6) can be diamagnetic (low spin, with no unpaired electrons) or paramagnetic (high spin, with four unpaired electrons), depending on the crystal field splitting energy.

Key Terms

amorphous solid (p. 469)
body-centered cubic (bcc) unit cell (p. 473)
Bragg equation (p. 471)
closest-packed structure (p. 477)
coordination number (p. 473)
cubic closest packing (ccp) (p. 477)
crystal field splitting (p. 495)
crystal field splitting energy (p. 495)
crystalline solid (p. 469)
face-centered cubic (fcc) unit cell (p. 473)
hexagonal closest packing (hcc) (p. 482)
lattice (p. 472)
lattice point (p. 472)
network solid (p. 484)
polymorph (p. 485)
simple cubic unit cell (p. 474)
unit cell (p. 472)
X-ray diffraction (p. 469)

Key Skills and Concepts

SECTION 10.1

Understand the meaning of crystalline and amorphous solids.
Be able to calculate the distance between the layers in a crystal from its X-ray diffraction pattern.

SECTION 10.2

Be able to sketch simple cubic, face-centered cubic, and body-centered cubic unit cells.

SECTION 10.3

Be able to describe the similarities and differences between cubic closest packing (abcabc . . .), hexagonal closest packing (abab . . .), and simple cubic packing (aaa . . .).
Understand how the relative sizes of the ions in an ionic compound help to predict whether the smaller ions occupy octahedral or tetrahedral holes.

SECTION 10.4

Be able to describe the tetrahedral networks of σ bonds in covalent solids made of atoms with sp^3 hybridization.
Be able to interpret the chemical formula of olivine: $Mg_xFe_{2-x}SiO_4$.

SECTION 10.5

Explain why diamond is hard and an insulator, whereas graphite is soft and slippery and a very good electrical conductor.

SECTION 10.6

Understand how band theory explains why metals conduct electricity.

SECTION 10.7

Explain the origins of the colors of compounds continuing transition-metals.
Be able to calculate the wavelength of the maximum light absorption of an ion in a compound from its crystal field splitting energy.
Understand the terms *high spin* and *low spin* as applied to transition-metal ions.

Key Equations and Relations

The Bragg equation relates the angle of diffraction (2θ) of X-rays to the distance (d) between layers of ions or atoms in a crystal.

$$n\lambda = 2d \sin \theta \qquad (10.1)$$

The difference in energy, Δ_o, between d orbitals split by an octahedral crystal field is

$$\Delta_o = hc/\lambda \qquad (10.2)$$

where λ is the wavelength of the radiation absorbed by the crystal, h is Planck's constant, and c is the speed of light.

QUESTIONS AND PROBLEMS

Structures of Ionic Solids

CONCEPT REVIEW

1. Why doesn't an amorphous solid produce an X-ray diffraction pattern with distinct peaks?
2. Why can't X-ray diffraction be used to determine the structures of compounds in solution?
3. Why are X-rays rather than microwaves chosen for diffraction studies of crystalline solids?
4. The sources used in X-ray diffractometers can be changed. What happens to the values of θ in the Bragg equation when a crystal of NaCl is irradiated with X-rays with a wavelength of 71 pm rather than 154 pm? Why might a crystallographer use different X-ray wavelengths to determine a crystal structure?

PROBLEMS

5. The spacing between the layers of ions in KCl (sylvite) is larger than in NaCl (halite). Which crystal will diffract X-rays of a given wavelength through greater 2θ values?
6. Silver halides are used in black-and-white photography. Which compound would you expect to have the larger distances between ion layers: AgCl or AgBr? Which compound would you expect to diffract X-rays through larger values of 2θ if the same wavelength of X-ray were used?
7. Galena, Illinois, once the home of Ulysses S. Grant, is named for the rich deposits of lead(II) sulfide (PbS) found nearby. When PbS is exposed to X-rays with $\lambda = 71.2$ pm,

strong reflections from a single crystal of PbS are observed at 13.98 and 21.25 degrees. Determine the values of n to which these reflections correspond, and calculate the spacing between the crystal layers.

8. Cobalt(II) oxide is used as a pigment in ceramics. It has the same type of crystal structure as that of NaCl. When cobalt(II) oxide was exposed to X-rays with $\lambda = 154$ pm, reflections were observed at 42.38°, 65.68°, and 92.60°. Determine the values of n to which these reflections correspond, and calculate the spacing between the crystal layers.

9. Pyrophyllite, $Al_2Si_4O_{10}(OH)_2$ is a silicate mineral with a layered structure. The distances between the layers is 1855 pm. What is the smallest angle of diffraction of X-rays with $\lambda = 154$ pm from this solid?

10. Minnesotaite, $Fe_3Si_4O_{10}(OH)_2$, is a silicate mineral with a layered structure of the same type found in pyrophyllite. The distance between the layers in minnesotaite is 1940 ± 10 pm. What is the smallest angle of diffraction of X-rays with $\lambda = 154$ pm from this solid?

Unit Cells

CONCEPT REVIEW

11. Explain why corner, face, and edge atoms or ions in a unit cell contribute to different numbers of unit cells.
12. Which has the greater packing efficiency, a simple cubic or a body-centered cubic unit cell?
13. Predict based on Figure 10.5 which has the greater packing efficiency, a body-centered cubic or a face-centered cubic unit cell?
14. Can $CaCl_2$ have the rock-salt (NaCl) crystal structure?
15. In the crystals of ionic compounds, how do the relative sizes of the ions influence the location of the smaller ions?
16. Explain the difference between cubic closest packing and hexagonal closest packing arrangements of identical spheres.
17. Do you expect the cation:anion ratio for a triangular hole to be larger or smaller than that for a tetrahedral hole?

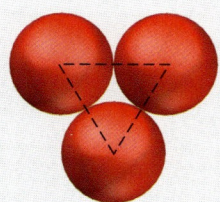

*18. In some books the structure of CsCl is described as body-centered cubic instead of simple cubic. Explain how CsCl might be assigned to both categories?

PROBLEMS

19. Draw a box around the unit cells for the following two patterns (they continue infinitely in two dimensions). How many light squares and how many dark squares are in each unit cell?

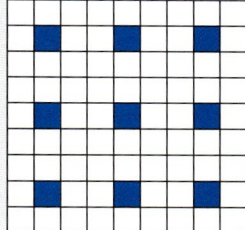 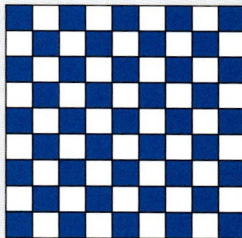

20. Draw a box around the unit cell for the following pattern (it continues indefinitely in three dimensions). If the red circles represent element A and the blue circles represent element B, what is the chemical formula of the compound?

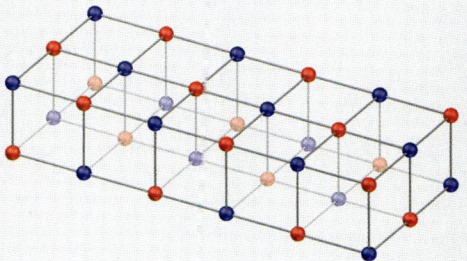

21. How many cations (A and B) and anions (X) are there in the following unit cell?

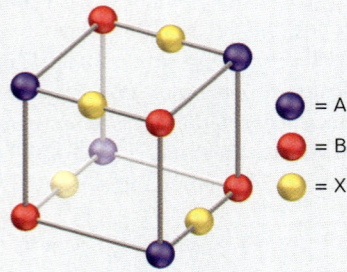

= A
= B
= X

22. How many equivalent atoms of elements A and B are there in the following unit cell? (Spheres labeled A are located at the corners and center of each face. Those within the cell are shaded.)

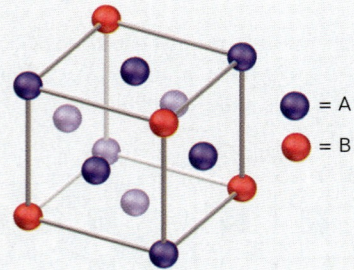

= A
= B

23. What is the chemical formula of the compound whose unit cell is shown here?

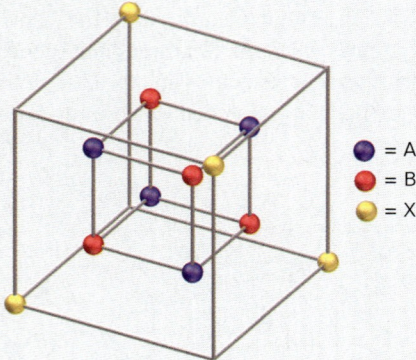

= A
= B
= X

24. What is the chemical formula of the ionic compound whose unit cell is shown here? (A and B are cations, X is an anion.)

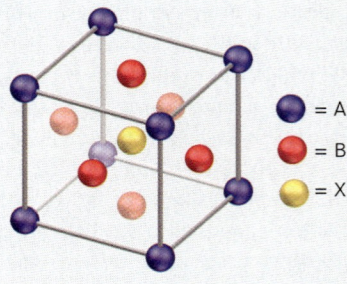

= A
= B
= X

*25. In 2000, magnesium boride (MgB_2) was observed to behave as a superconductor. Is the unit cell of MgB_2 shown here consistent with the formula MgB_2? A boron atom is in the center of the unit cell (shown on the left), which is part of the hexagonal structure shown on the right.

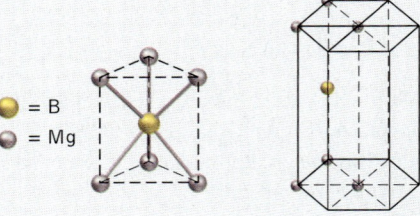

= B
= Mg

*26. The 1987 Nobel Prize in physics was awarded to G. Bednorz and K. A. Muller for their discovery of superconducting ceramic materials such as $YBa_2Cu_3O_7$. The following drawing represents the unit cell of another yttrium-barium-copper oxide. What is the chemical formula of this compound? Eight oxygen atoms must be removed from the drawing to produce the unit cell of $YBa_2Cu_3O_7$. Does it make a difference which oxygen atoms are removed?

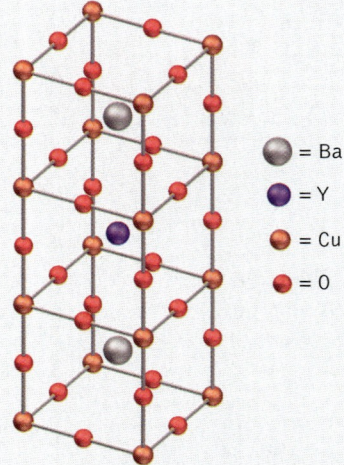

= Ba
= Y
= Cu
= O

27. Copper metal has a density of 8.95 g/cm³. If the radius of copper atoms is 127.8 pm, is the copper unit cell simple cubic, body-centered cubic, or face-centered cubic?
28. The radius of a molybdenum atom is 139 pm, and the density of molybdenum metal is 10.28 g/cm³. Which unit cell is consistent with these data: simple cubic, body-centered cubic, or face-centered cubic?
29. The unit cell of rhenium trioxide (ReO_3) consists of a cube with rhenium atoms at the corners and an oxygen atom on each of the 12 edges. The atoms touch along the edge of the unit cell. The radii of Re and O atoms in ReO_3 are 137 and 74 pm, respectively. Calculate the density of ReO_3.
30. Iron may crystallize in either body-centered cubic or face-centered cubic structures. Calculate the densities of iron in these structures, given that the radius of an iron atom is 124 pm.
31. Europium is one of the lanthanide elements used in television screens. Europium crystallizes in a body-centered structure with a unit cell edge of 240.6 pm. Calculate the radius of a europium atom.
32. Nickel has a face-centered cubic unit cell with an edge length of 350.7 pm. Calculate the radius of a nickel atom.
33. Calculate the packing efficiency in a simple cubic unit cell.
34. Calculate the packing efficiency in body-centered cubic and face-centered cubic unit cells.
35. What is the formula of the compound that crystallizes with barium ions occupying one-half of the cubic holes in a simple cubic arrangement of fluoride ions?
36. What is the formula of the compound that crystallizes with lithium ions occupying all of the tetrahedral holes in a cubic closest-packed arrangement of sulfide ions?
37. What is the formula of the compound that crystallizes with aluminum ions occupying one-half of the octahedral holes and magnesium ions occupying one-eighth of the tetrahedral holes in a cubic closest-packed arrangement of oxide ions?
38. What is the formula of the compound that crystallizes with zinc atoms occupying one-half of the tetrahedral holes in a cubic closest-packed arrangement of sulfur atoms?
39. The 1996 Nobel Prize in chemistry was shared by Harold Kroto, Robert Curl, and Richard Smalley for the discovery of buckminsterfullerene, a molecule that contains 60 carbon atoms (C_{60}) In the solid state, C_{60} can be described as cubic closest-packed spheres (C_{60} is approximately a sphere), with an edge length of 1410 pm. Calculate the density of crystalline C_{60} and the radius of the C_{60} molecule.
40. Under certain conditions, buckminsterfullerene (C_{60}) reacts with alkali metals and forms a compound with the formula M_6C_{60} (M = Na, K) in which the C_{60} molecules have a body-centered cubic arrangement. Calculate the density of a bcc arrangement of C_{60}.
41. In 1989, Stanley Pons and Martin Fleischmann, at the University of Utah, claimed to have achieved the fusion of deuterium (D_2) at room temperature in a palladium electrode. Palladium metal crystallizes in a cubic closest-packed structure with a density of 11.99 g/cm³. Does a D atom (radius = 74.1 pm) fit in either the octahedral or the tetrahedral holes of palladium metal?
42. The products of "cold" fusion of deuterium in Problem 41 are expected to include ^{3}He, radius 32 pm. Could a ^{3}He atom fit in an octahedral hole of Pd?
*43. Buckminsterfullerene (C_{60}) may react with alkali metals to form M_3C_{60} (M = Na, K). The structure of M_3C_{60} contains a cubic closest-packed array of C_{60} spheres with metal ions in the holes. If the radius of the K^+ cation is 138 pm, which type of hole is K^+ most likely to occupy? What fraction of these holes will be occupied?
*44. A number of transition metals, including titanium, zirconium, and hafnium can store hydrogen as metal hydrides. Calculate the size of the tetrahedral and octahedral holes in the ccp structure of these three metals, given that their atomic radii are 147, 160, and 159 pm, respectively. Which metal(s) is (are) most likely to accommodate H atoms (radius 37 pm) with the least distortion?

Network Solids

CONCEPT REVIEW

45. Why is S_8 not a flat octagon?
46. When amorphous red phosphorus is heated at high pressure, it is transformed into the allotrope black phosphorus, which can exist in one of several forms. One of these forms consists of six-atom rings of phosphorus atoms. Why are the six-atom rings in black phosphorus puckered, whereas the six-atom rings in graphite are planar?

47. Selenium exists as Se_8 rings or in a crystalline structure with helical chains of Se atoms. Are these two structures of selenium polymorphs, allotropes, or both? Explain your answer.
48. Iron metal adopts either an fcc or a bcc structure, depending on its temperature. Do these two crystal structures represent allotropes of iron? Explain your answer.

PROBLEMS

49. The distance between atoms in a cubic form of phosphorus is 238 pm. Calculate the density of this form of phosphorus.

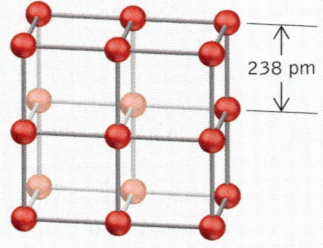

50. Carbon and silicon have allotropes with the diamond structure. The unit-cell-edge lengths of carbon and silicon are 356.7 and 543.1, respectively. The density of diamond is 3.514 g/cm³. What is the density of the diamond allotrope of silicon?
51. Ice is a network solid. However, theory predicts that, under high pressure, ice (solid H_2O) becomes an ionic compound composed of H^+ and O^{2-} ions. The proposed structure of ice under these conditions is a body-centered cubic arrangement of oxygen ions with hydrogen ions in holes. How many equivalent H^+ and O^{2-} ions are in the unit cell? Calculate the density of "ionic" ice, given the radius of O^{2-} (126 pm). Draw a Lewis structure for "ionic" ice.
52. Alistair MacLean's novel *Ice Station Zebra* describes a high-pressure form of ice, called "ice nine." With the assumption that ice nine has a cubic closest-packed arrangement of oxygen atoms with hydrogen atoms in the appropriate holes, what type of hole will accommodate H atoms? What is the density of ice nine, given atomic radii of 32 pm for H and 73 pm for O?
53. Replacement of Al^{3+} ions in kaolinite, $Al_2(OH)_4(Si_2O_5)$, with Mg^{2+} ions yields the mineral antigorite. What is the formula of antigorite?
54. What is the formula of the aluminum silicate mineral phlogopite, obtained by the replacement of Al^{3+} ions in muscovite, $KAl_2(OH)_2(AlSi_3O_{10})$, with Mg^{2+} ions?

Metallic Bonds

CONCEPT REVIEW

55. Explain how both the electron-sea model and band theory can be used to explain the conductivity of a metal such as copper.
56. The melting and boiling points of sodium metal are much lower than those of sodium chloride. What does this difference reveal about the relative strengths of metallic bonds and ionic interactions?

PROBLEMS

57. Some scientists believe that solid hydrogen at high pressure may conduct electricity. Draw a molecular-orbital diagram for hydrogen and show how it transforms into two bands for a very large number of atoms of hydrogen. Is the lower-energy band filled or half-filled?
58. Would you expect solid helium to conduct electricity?

Gemstones and Crystal Field Theory

CONCEPT REVIEW

59. Explain why most first-row transition-metal compounds are colored.
60. Why are the d_{z^2} and $d_{x^2-y^2}$ orbitals in an octahedral field raised in energy relative to the other three d orbitals?
61. Why is the d_{z^2} orbital lower in energy relative to the $d_{x^2-y^2}$ orbital in a square planar crystal field?
62. What determines whether a transition metal ion is in a *high-spin* configuration rather than a *low-spin* configuration?

63. Why are Ni(II) and Cu(II) compounds colored, but most Zn(II) compounds are not?
64. Titanium(IV) oxide (TiO_2) consists of hexagonally close-packed oxide anions, with Ti^{4+} cations occupying half of the octahedral holes. Why is titanium(IV) oxide colorless?

PROBLEMS

65. Which of the following cations can have either a high-spin or a low-spin electron configuration in an octahedral hole: Fe^{2+}, Fe^{3+}, Mn^{4+}, and Cr^{3+}?
66. Which of the following cations can, in principle, have either a high-spin or a low-spin electron configuration in a tetrahedral hole: Co^{2+}, Cr^{3+}, Ni^{2+}, and Zn^{2+}?
67. Two minerals that contain transition-metal ions in octahedral holes absorb blue ($\lambda = 450$ nm) and yellow ($\lambda = 580$ nm) light, respectively. Which compound has the larger crystal field splitting energy Δ_o?
68. A solid containing tetrahedral $CuCl_4^{2-}$ ions strongly absorbs red light ($\lambda = 800$ nm), whereas the corresponding $CoCl_4^{2-}$ compound absorbs orange light ($\lambda = 680$ nm). Which of these anions has the larger crystal field splitting energy Δ_t?
69. Calculate the octahedral crystal field splitting energy Δ_o for the two compounds in Problem 67.
70. Calculate the tetrahedral crystal field splitting energy Δ_t for the $CuCl_4^{2-}$ and $CoCl_4^{2-}$ anions in Problem 68.
71. Give the number of unpaired electrons in the following transition-metal ions in an octahedral crystal field: high-spin Fe^{3+}, Rh^+, and V^{3+} and low-spin Mn^{3+}.
72. Give the number of unpaired electrons in the following transition-metal ions in a tetrahedral crystal field: high-spin Fe^{2+}, Cu^{2+}, and Co^{2+} and low-spin Mn^{3+}.
73. Dissolving cobalt(II) nitrate in water gives a beautiful purple solution containing a cobalt(II) species that has the formula $Co(H_2O)_6^{2+}$ and an octahedral geometry. The ESR spectrum (see Chapter 7) of the solution is consistent with three unpaired electrons in this cobalt(II) complex. When cobalt(II) nitrate is dissolved in aqueous ammonia and oxidized with air, the resulting yellow $Co(NH_3)_6^{3+}$ complex has no unpaired electrons. Which cobalt complex has the larger crystal field splitting energy Δ_o?
74. A solid compound containing Fe(II) in an octahedral crystal field has four unpaired electrons at 298 K. When the compound is cooled to 80 K, the same sample appears to have no unpaired electrons. How do you explain this change in the compound's properties?

11 Thermochemistry
And the Quest for Energy

In recent years the search for energy has taken oil companies to remote regions of our planet. These rigs are drilling for oil and natural gas in the Beaufort Sea off the coast of Canada's Northwest Territories.

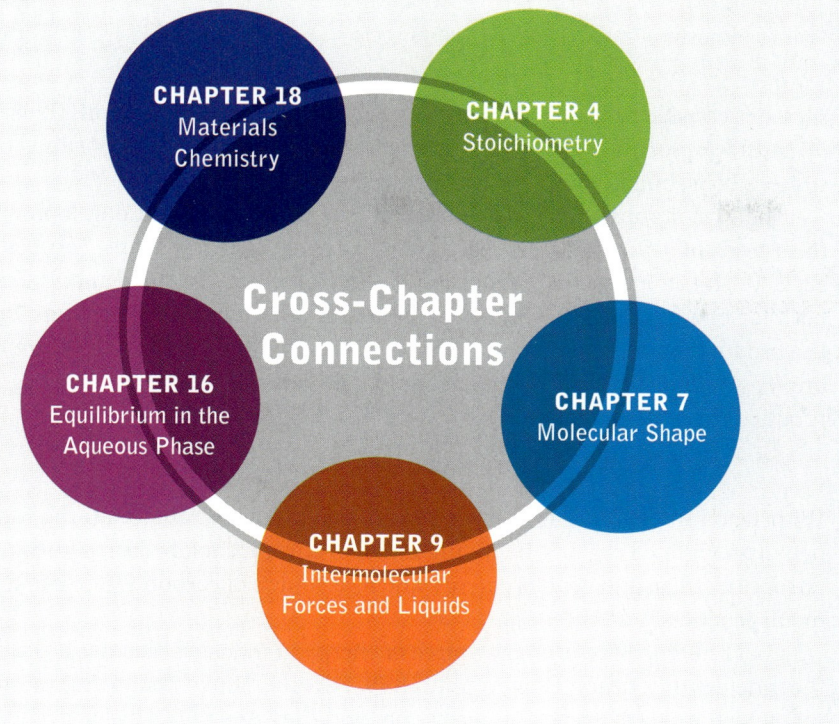

11.1	**An historical perspective**
11.2	**Energy: some definitions**
11.3	**Natural gas**
11.4	**Combustion and energy transfer**
11.5	**Enthalpy (H) and enthalpy change (ΔH)**
11.6	**Heating curves and heat capacity**
	Hot soup on a cold day
	Cold drinks on a hot day
11.7	**Estimating ΔH from average bond energies**
11.8	**Calorimetry: measuring heats of combustion**
11.9	**Enthalpies of formation and reaction**
11.10	**Fuel value**
11.11	**Hess's law**

A Look Ahead

The quest for energy forms the context for an examination of the energy changes that take place in chemical reactions. We explore several methods for estimating or actually measuring the energy released or consumed during chemical reactions. These include calculations based on average bond energies, standard enthalpies of formation, and data obtained from calorimetry experiments. We use the energy released during combustion to evaluate fuel values. Molar heat capacities and heats of fusion (melting) and vaporization or sublimation are used to calculate heat flow as substances change temperature and physical state.

11.1 AN HISTORICAL PERSPECTIVE

Our consideration of the chemical structures and reactivities of compounds found in Earth's crust has yet to include an important class of geological deposits: fossil fuels. These residues of plant and animal matter buried in the crust for millions of years include coal, crude oil, and natural gas. Fossil fuels contain complex mixtures of compounds composed principally of two elements: carbon and hydrogen. The combustion of these compounds plays a major role in satisfying society's considerable appetite for energy. They also serve as the initial reactants, or *feedstocks*, for the synthesis of the materials used to make thousands of consumer goods, building materials, and agricultural products.

The exploitation of Earth's fossil fuels is a relatively recent development. For thousands of years, humans relied on wood and other renewable sources of energy. With the Industrial Revolution came an increase in the demand for wood both for energy and as a source of raw material for chemical reactions, such as the production of metals from metal ores. For example, a heated mixture of iron ore (Fe_2O_3) and charcoal (a source of carbon that is produced by charring wood) produces iron metal:

$$2\ Fe_2O_3(s) + 3\ C(s) \longrightarrow 4\ Fe(s) + 3\ CO_2(g)$$

With wood in demand as fuel and as an important raw material, the forests of heavily populated regions, including much of Europe, rapidly disappeared. And so Europeans turned to coal for energy and as a source of carbon for manufacturing metal products.

Coal remained the primary source of energy in industrialized nations well into the nineteenth century. However, in 1859, Edwin L. Drake drilled the first successful oil well near Oil Creek, Pennsylvania, and the "Age of Oil" began. Since then, heating oil has replaced coal for heating buildings, diesel engines have replaced coal-fired steam engines in factories and train locomotives, and gasoline engines now propel most cars and trucks. In addition, petroleum and natural gas provide raw materials for the petrochemical industry. By 1970, petroleum and natural gas accounted for two-thirds of energy consumption worldwide. Supplies were plentiful and cheap: the price of a 42-gallon barrel of crude oil on the world market was about $1. Then came the 1973 Arab-Israeli war and an oil embargo directed against the United States and other nations. The price of a barrel of crude oil increased by more than 10 times its prewar price. Since then, increasing demand for energy coupled with self-imposed limits on production by the oil-producing and -exporting countries (OPEC) has led to crude oil prices as high as $40 per barrel.

Cost and uncertainties in supply are only part of the motivation for developing alternative fuels. A more fundamental problem is the finite size of fossil-fuel reserves. Petroleum, natural gas, and coal formed millions of years ago from the decay of plant and animal matter in the sediments of bogs, swamps, and coastal regions where limited supplies of oxygen prevented their oxidation to carbon dioxide and water.

Although deposits of these fuels continue to form, they are doing so at a very slow rate. The amount of the sun's energy striking Earth and stored as biomass is estimated to be 10^{19} kJ/yr. This value is 30 times the energy consumed by all the industrialized countries of the world, but only about 0.01% of the biomass is trapped in anaerobic (devoid of oxygen) sediments and wetlands where it is protected from oxidation. Even worse, less than 0.1% of that amount accumulates in deposits that are of sufficient size to be economically recoverable. Thus the rate at which biomass energy is preserved naturally as recoverable fossil fuels is less than 10^{12} kJ/yr, which is a lot less than the 3×10^{17} kJ consumed each year. It is not surprising, then, that many experts believe that the world's supplies of petroleum and natural gas will be exhausted before the middle of the twenty-first century.

CONNECTION: The production of iron and steel is described in Section 18.2.

Edwin L. Drake's successful oil well near Oil Creek, Pennsylvania, in 1859 marked the beginning of the "Age of Oil." Today oil and natural gas supply most of the world's energy needs.

11.2 ENERGY: SOME DEFINITIONS

In this chapter, we will focus on the release of energy through the combustion of natural gas. We will examine the chemical bonding and structures of the compounds in natural gas, and we will see how breaking those bonds during combustion reactions and creating of new ones in the products of combustion relate to the amount of energy produced. These discussions will lead us to examine the fundamentals of **thermochemistry,** the study of how energy in the form of heat is consumed and produced by chemical reactions. We start with a few definitions.

Energy is defined as the capacity to do work or to transfer heat. What do we mean by *work*? In physical science, work (w) is done when a force (F) moves an object through a distance (d). The amount of work done is the product of the force exerted on the object and the distance through which it was exerted, or:

$$w = F \times d \tag{11.1}$$

Consider the connections between work and energy as they relate to skiers ascending a mountain (Figure 11.1). Energy is expended and work is done to get

> ✓ **Thermochemistry** is the study of the heat produced or consumed in chemical reactions.
>
> **Energy** is the capacity to do work or to transfer heat.

FIGURE 11.1 Work is done as skiers ascend to the top of a mountain. The amount of work will be different, depending on whether the skiers (A) ride a chairlift or (B) hike to the top. Either way, a part of the work is stored as potential energy that depends only on the skiers' mass and their positions at the top of the mountain; it is independent of how they got there. This independence of pathway makes potential energy a *state function*.

each skier up the mountain. For example, the work (w) done by a gondola lift on each skier equals the length of the ride on the lift (d) times the force (F) needed to overcome gravity and transport the skier up the mountain. Some of the work done is stored in the mass of the skier as **potential energy.** This potential energy is strictly a function of the skier's position on the mountain. How the skier got there is not important. Had the skier climbed up the mountain rather than riding a lift, the amount of work done to get to the top would have been different. Had the skier climbed the mountain wearing snowshoes rather than ski boots or climbed a different trail, the work done would have been different. Still, the gain in potential energy would be the same in all cases. This independence of the potential energy of a position, or *state*, from how the state was achieved makes potential energy a **state function.** It is independent of the path followed to achieve it.

Consider the conversion of the potential energy of a ski jumper leaving the starting gate (the left photo in Figure 11.2) into kinetic energy as he gains speed down the "in-run" structure prior to take off. At any moment between the start of the run and coming to a stop at the bottom of the hill, the ski jumper's kinetic energy (K.E.) is proportional to the product of the mass (m) of the skier times the square of the skier's speed (u):

$$K.E. = \tfrac{1}{2}mu^2 \qquad (11.2)$$

> The **potential energy** of an object is the energy that it has because of its position.

> A **state function** of a system is a property based solely on its chemical or physical state or both and not on how it acquired those states.

CONCEPT TEST: Two skiers with masses m_1 and m_2 are poised at the starting gate of a downhill course. Do the two skiers have the same or different potential energies? If you think they are different, which skier has more potential energy?

CONCEPT TEST: Two skiers with masses m_1 and m_2 go through the first gates of parallel racecourses at the same time. At that moment, which of them has the greater potential energy, or do they have the same potential energy? If the two are moving at the same speed, which of them has the greater kinetic energy, or do they have the same kinetic energy?

FIGURE 11.2 A. A skier leaving the starting gate of a ski jump has potential energy owing to his position at the top of the "in-run". B. During his run, the jumper's potential energy is converted into kinetic energy (K.E.), which is the energy of motion related to each skier's mass (m) and speed (u) by the equation $K.E. = \tfrac{1}{2}mu^2$.

FIGURE 11.3 Energy can be converted from one form into another. A. Engines in cars and trucks convert the chemical energy in gasoline or diesel oil and oxygen into heat and kinetic energy. B. Solar panels on the Hubble Space Telescope (each nearly 12 m long) convert energy from sunlight into electrical energy. C. A burning candle converts the chemical energy in candle wax and oxygen into heat and light. D. A runner converts the chemical energy in food and oxygen into heat and kinetic energy.

Chemical energy is another form of potential energy. It is the energy stored in the bonds of substances. It is released when those bonds are broken and new bonds are formed in chemical reactions. This chemical energy also is a state function: it depends on the identities and quantities of the reactants; it does not depend on how the substances were made.

According to the **law of conservation of energy,** energy cannot be created or destroyed. However, it can be converted from one form into another, as illustrated in Figure 11.3. In the example of the ski jumper, potential energy at the top of the run becomes kinetic energy during the run. Ski jumping and other forms of physical exercise entail the conversion of chemical energy from food and oxygen into heat and kinetic energy. Similarly, an automobile engine converts the chemical energy in gasoline and oxygen into heat and kinetic energy, and a photovoltaic solar cell converts light energy into electrical energy. In Section 11.4, we will examine the production of energy in the form of heat and light by an important class of chemical reactions known as combustion.

> ✓ The **chemical energy** of a substance is a kind of potential energy, but it is a function of the *composition*, not the *position*, of the substance.
>
> The **law of conservation of energy** states that energy can be neither created nor destroyed.

11.3 NATURAL GAS

Natural gas is the simplest of fossil fuels in regard to chemical composition. It is found in large deposits underground, and as we will see in Chapter 12, can be produced from biomass by bacterial action. The major ingredients in natural gas are presented in Table 11.1. All of them are **hydrocarbons,** meaning that their

> ✓ **Hydrocarbons** are organic compounds composed of only carbon and hydrogen.

> **Alkanes, or saturated hydrocarbons,** are composed of molecules in which all the C–C bonds are single bonds.

Natural gas is a mixture of methane and C_2 to C_4 hydrocarbons. In rural areas where natural gas is not available, liquefied propane is used to fuel stoves and water heaters. Tanks of propane also supply fuel for camp stoves, lanterns, and barbecue grills. Butane is the fuel used in many disposable lighters.

molecules contain only carbon and hydrogen atoms. These particular hydrocarbons have from one to four carbon atoms per molecule and so are called C_1 to C_4 hydrocarbons. They are members of a class of hydrocarbons called **alkanes** or **saturated hydrocarbons.** These terms apply to hydrocarbons with molecular structures in which every carbon atom has four single bonds, each connecting it to another carbon atom or to a hydrogen atom. The molecular formulas for the alkanes in natural gas fit the general formula C_nH_{2n+2}, where n, the number of carbon atoms per molecule, is an integer ≥ 1. The hydrogen:carbon ratio in these compounds $[(2n + 2)/n]$ is as large as it can be, given the limit of no more than four bonds per carbon atom. You might say that these compounds are "saturated" with hydrogen.

Note the similarities in the structures of the compounds listed in Table 11.1. In all cases, the carbon atoms are bonded to no more than two other carbon atoms. In propane and butane, the C–C bonds form a molecular chain in which each link is a carbon atom bonded to two hydrogen atoms:

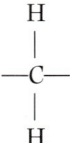

These —CH_2— units are called *methylene groups.* The carbon atom at each end of a chain is bonded to three hydrogen atoms:

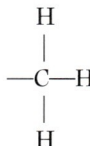

forming a *methyl group.* Add another hydrogen atom to a methyl group and you have methane, CH_4.

In writing the structures of hydrocarbons and other organic compounds, we can show how the atoms are bonded together without drawing complete Lewis structures. Instead, we might write abbreviated, or *condensed,* structural formulas in which the C–H bonds are not drawn. Thus, the Lewis structure of propane simplifies to the condensed structural formula:

$$CH_3—CH_2—CH_3$$

Condensed structural formulas of linear alkanes can be further simplified by also leaving out the C–C bonds. Thus, the condensed structural formula for propane could also be written as

$$CH_3CH_2CH_3$$

An even simpler line structure for propane:

has segments that represent C–C bonds. Each carbon atom is assumed to be bonded to two or three hydrogen atoms.

TABLE 11.1 Alkanes in Natural Gas

Compound	Typical Abundance (% by Volume)	Formula	Lewis Structure	Boiling Point (°C)
Methane	75–90	CH_4	H—C(H)(H)—H	−164
Ethane	5–15	C_2H_6	H—C(H)(H)—C(H)(H)—H	−89
Propane	2–5	C_3H_8	H—C(H)(H)—C(H)(H)—C(H)(H)—H	−42
Butane	<3	C_4H_{10}	H—C(H)(H)—C(H)(H)—C(H)(H)—C(H)(H)—H	0

Lewis structures and condensed structural formulas indicate the patterns of the covalent bonds in these molecules, but they do not necessarily give us accurate descriptions of the structures of the molecules in three dimensions. That description is provided by ball-and-stick and space-filling models, as shown for propane in Figure 11.4. The structures in Figure 11.4 have bond angles between C–C and C–H bonds in propane (and in all alkanes) are about 109.5 degrees. This geometry minimizes repulsion between the four bonding pairs of electrons on carbon atoms and is consistent with sp^3 hybridization of the valence-shell orbitals of carbon atoms.

CONNECTION: Hybridization of atomic orbitals is described in Section 7.7.

FIGURE 11.4 The three-dimensional shape of propane molecules can be represented by ball-and-stick and space-filling models. Ball-and-stick models offer better views of the bond angles in molecules; space-filling models better represent of the proximity and orientation of atoms to one another.

CONCEPT TEST: Table 11.1 lists the boiling points of the C_1 to C_4 alkanes. Explain why the boiling points of these compounds increase with increasing molecular size.

11.4 COMBUSTION AND ENERGY TRANSFER

Natural gas is one of the principal sources of energy in the United States, particularly for heating buildings, heating water, and generating electricity. It has the advantage of burning more cleanly than do other fossil fuels, meaning that it contains fewer impurities, such as sulfur, that contribute to air pollution. Let's consider how energy is released in the combustion of the principal ingredient of natural gas, CH_4. Like all hydrocarbons, methane burns in air, producing CO_2 and H_2O,

CONNECTION: The contribution of sulfur impurities to air pollution is described in Section 16.2.

$$CH_4(g) + 2\,O_2(g) \longrightarrow CO_2(g) + 2\,H_2O(g)$$

and releasing considerable energy in the form of heat and light.

Let's next consider the flow of heat from the flame produced by a burner on a gas stove (Figure 11.5). The flame, which includes the reactants of the combustion reaction (natural gas and oxygen from the air) and the products of combustion, is a thermodynamic **system.** The stove, the pot above the flame, the food in the pot, and everything else in the universe make up the system's **surroundings**. Heat flows from the system to its surroundings. Because energy is conserved, the energy lost by the system must equal the energy gained by its surroundings. The study of how energy is transferred from one form to another and how it flows from systems to surroundings and vice versa is called **thermodynamics.** The idea that the total energy gained or lost by a system must be balanced by the total energy lost or gained by its surroundings is known as the **first law of thermodynamics.**

> A **system** is that part of the universe that is the focus of study. In thermochemistry, the system includes the reaction vessel and its contents. The **surroundings** include everything not part of the system.
>
> **Thermodynamics** is the study of energy and its transformations. The **first law of thermodynamics** states that the energy gained or lost by a system must equal the energy lost or gained by its surroundings.

FIGURE 11.5 Natural gas is a common fuel for cooking food. The flame, including the reactants (natural gas and oxygen from the air) and the products of combustion, may be considered a thermodynamic *system*. Everything else, including the food being cooked, is considered part of the *surroundings*.

A chemical reaction or a physical change of state that produces a net flow of heat from a system to its surroundings is **exothermic**. This flow can be detected by an increase in the temperature of the surroundings. Combustion reactions that release heat and light are examples of exothermic reactions. Chemical reactions and physical changes that absorb heat *from* their surroundings are **endothermic**. For example, ice cubes melting in a glass of water on a hot summer day or perspiration evaporating from your skin are examples of *endothermic* physical changes. The opposite transitions—water freezing into ice or atmospheric water vapor condensing into droplets of liquid water on the outside of an ice-cold drink—require that heat flow from the system to its surroundings; so these processes are *exothermic*. They illustrate an important point: an exothermic process in one direction is endothermic in the reverse direction.

The flow of heat during physical changes in the state of matter is illustrated in Figure 11.6. The arrows pointing upward correspond to changes requiring that heat flow into the system from its surroundings. The arrows pointing downward correspond to changes requiring that heat flow from the system into its surroundings. We use the symbol q to represent heat produced or consumed by a chemical reaction or physical change. If the reaction or process is exothermic, then q is negative, meaning that the system *loses* heat to its surroundings. If the reaction or process is endothermic, then q is positive, indicating that heat is *gained* as it flows into the system from its surroundings. In Figure 11.6, changes of state that are endothermic are represented by the arrows pointing upward: solid ⟶ liquid, liquid ⟶ gas, and solid ⟶ gas. The opposite changes represented by arrows pointing downward are exothermic. To summarize:

$$q < 0 \quad \text{exothermic}$$
$$q > 0 \quad \text{endothermic}$$

> An **exothermic** process is one in which heat from the system flows to its surroundings; in an **endothermic** process, heat is absorbed by the system from its surroundings.

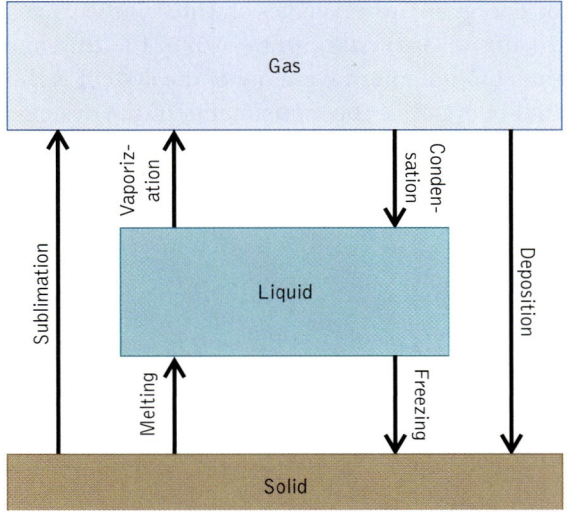

FIGURE 11.6 The three physical states of matter can be transformed from one into another by the application or removal of heat. The arrows pointing upward represent endothermic ($q > 0$) processes. The arrows pointing downward represent exothermic ($q < 0$) processes.

CONCEPT TEST: On a cool autumn morning, dense fog may form over a lake or river still warmed by the past summer's sunshine and high temperatures. Starting with the warm water, describe the flows of heat (and the associated signs of q) during the phase changes that result in the formation of fog, as shown in the above photo.

Consider the flow of energy when a tray of ice cubes is inadvertently left on a kitchen counter. If we focus on the tray and its contents, then they become our thermodynamic system. As the ice cubes warm and start to melt, the hydrogen bonds that hold molecules of H_2O in the ice rigidly in place are broken. These molecules then have more freedom to move, and so they have more kinetic energy. After all the ice has melted, the temperature of the water in the tray slowly rises to room temperature. Higher temperature is also the result of increasing kinetic energy of the molecules of water in the tray. This kinetic energy is part of the overall **internal energy** (E) of the tray-and-water system (Figure 11.7). The absolute internal energy of the system is the sum of the kinetic and potential energies of the components of the system. The values of these energies are difficult to determine on an absolute scale. However, *changes* in internal energy, ΔE, are fairly easy to measure, because a change in a system's physical state or temperature provides a measure of the change in its internal energy.

> **Internal energy** (E) is the sum of the potential and kinetic energies of the constituents of a system. The internal energy of a system is increased by heating the system or by doing work on it. Internal energy decreases as a system loses heat or does work on its surroundings.

CONNECTION: The connection between molecular size and the strength of London dispersion forces is described in Section 9.4.

SAMPLE EXERCISE 11.1: What is the sign of q as (a) a match burns, (b) molten candle wax solidifies, and (c) gasoline evaporates? Which of these processes are endothermic and which are exothermic?

SOLUTION: A burning match (the *system*) gives off heat to its surroundings; so the sign of q is negative. The same is true when molten wax cools to its melting

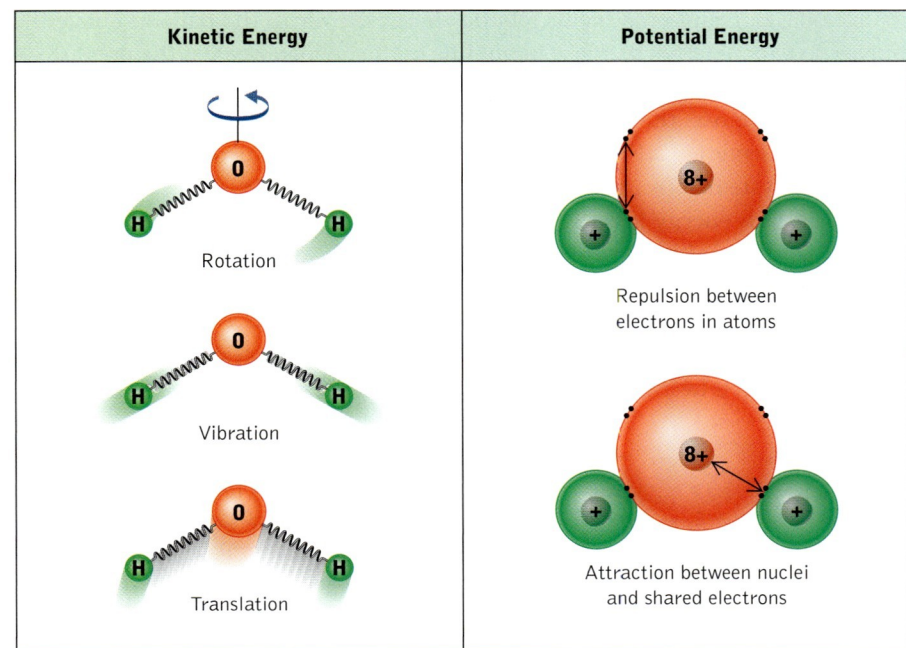

FIGURE 11.7 The internal energy of a water molecule is the sum of its kinetic and potential energies. Some of the contributions to its kinetic energy include rotation and vibration of the molecule as well as its principal source of kinetic energy: its motion through space (translation). Contributions to the potential energy include repulsion between electrons in an atom and the attraction between nuclei and shared pairs of electrons.

point and then solidifies. On the other hand, heat must flow into gasoline when it evaporates. This energy is needed to overcome the London dispersion forces between the hydrocarbon molecules in gasoline in its liquid state.

PRACTICE EXERCISE: Describe the flows of heat during the purification of water by distillation. Consider the boiling water and its vapors to be the system. (See Problems 23–26.)

Doing work on a system is another way to add to its internal energy. For example, compressing a quantity of gas into a smaller volume (such as filling a scuba tank with air) is work done on the gas and causes its temperature to rise. The quantity of energy gained by doing work on a system is given the symbol w. The total increase in the energy of a system is the sum of the work done on it (w) and the heat (q) that flows into it. Mathematically, we write this relation as follows:

$$\Delta E = q + w \tag{11.3}$$

A system undergoing an exothermic reaction can do work on its surrounding. Consider the enormous balloons used by adventurers to fly themselves around the world. These balloons typically have a central chamber filled with helium and a secondary chamber of hot air heated by burners fueled by ethane and propane (Figure 11.8). The burners allow a balloonist to adjust the temperature

Filling scuba tanks requires the compression of air, which means that work is done on the air. This work heats the air so much that the tanks need to be immersed in a tub of water to cool them.

CONNECTION: Section 8.3 includes a description of the behavior of gases in hot-air balloons.

of the air (and its density), thereby controlling the buoyancy of the balloon and its altitude. Burning propane or ethane causes the air in the balloon above the burner to expand. This increase in volume of the balloon against the pressure of the atmosphere outside the balloon is a kind of work known as $P\Delta V$ work. If we define the system as the flame and hot air, then the expansion of the air trapped above the burner is $P\Delta V$ work done by the system on its surroundings. The energy of the system decreases as it performs this $P\Delta V$ work. We can express the relation between energy lost (ΔE) by the system, the heat lost by the system (q), and the work done by the system ($P\Delta V$) in equation form as follows:

$$\Delta E = q - P\Delta V \tag{11.4}$$

The negative sign in front of $P\Delta V$ is needed because, if the system expands (positive ΔV), then it is losing energy (negative ΔE) as it does work on its surroundings.

FIGURE 11.8 Energy released by the combustion of ethane and propane warms the air in hot-air balloons. Higher temperature causes an increase in volume (ΔV) and a decrease in the density of the air, making the balloon buoyant. The product $P\Delta V$ represents the work done by the combustion system on its surroundings.

CONCEPT TEST: Equation 11.1 provides one definition of work: the product of force times distance. Now we have defined the work done by an expanding gas as the product of pressure times changing volume. Show that these definitions are equivalent on the basis of the units of the four parameters.

SAMPLE EXERCISE 11.2: A tank of compressed helium is used to inflate balloons for sale at a carnival. If 100 balloons are each inflated to a final volume of 4.8 liters at 1.01 atm of pressure, how much $P\Delta V$ work was done to inflate the balloons? Express your answer in L · atm and in joules. Assume that the pressure in the balloons was constant during the filling process.

SOLUTION: The total volume change (ΔV) as all the balloons are inflated is

$$100 \text{ balloons} \times 4.8 \text{ L/balloon} = 480 \text{ L}$$

The work (w) to inflate this volume at a pressure of 1.01 atm is

$$w = P\Delta V$$
$$= 1.01 \text{ atm} (480 \text{ L}) = 485 \text{ L} \cdot \text{atm}$$

The SI unit for energy (the ability to do work) is the joule (J). There are several approaches to converting L · atm into J. One of the simplest involves the use of two of the values for the universal gas constant (R) listed in Table 8.1:

$$R = 8.3145 \text{ J/(mol} \cdot \text{K)} = 0.082058 \text{ L} \cdot \text{atm/(mol} \cdot \text{K)}$$

Dividing the first value by the second

$$\frac{8.3145 \text{ J/(mol} \cdot \text{K)}}{0.082058 \text{ L} \cdot \text{atm/(mol} \cdot \text{K)}} = 101.32 \frac{\text{J}}{\text{L} \cdot \text{atm}}$$

gives a handy conversion factor for changing L · atm into J. Using this factor for the work done in filling the balloons, we get

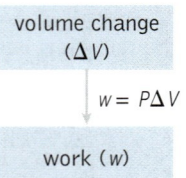

$$w = 485 \, \cancel{L \cdot atm} \left(101.32 \, \frac{J}{\cancel{L \cdot atm}}\right) = 4.91 \times 10^4 \, J$$

PRACTICE EXERCISE: The balloons used in recent attempts to fly around the world use a combination of helium and hot air for buoyancy. The *Spirit of Freedom* (Figure 11.8), which was flown around the world by American Steve Fossett in June and July 2002, contained 550,000 cubic feet of helium. How much $P\Delta V$ work was done to inflate the balloon, assuming 1.00 atm of pressure? Express your answer in joules. Given: 1 m³ = 1000 L = 35.3 ft³. (See Problems 27 and 28.)

11.5 ENTHALPY (H) AND ENTHALPY CHANGES (ΔH)

The physical and chemical changes discussed so far, as well as many others including the biochemical reactions in Chapters 12 and 13, take place at nearly constant (atmospheric) pressure (P). An important thermodynamic variable that relates heat flow into or out of a system during chemical reactions at constant pressure is called **enthalpy** (H). Enthalpy is defined as the sum of a system's internal energy and the product of its pressure and its volume: in equation form, this relation is written as

$$H = E + PV \quad (11.5)$$

> ✓ The **enthalpy** (H) of a system is the sum of its internal energy and the product of the pressure and volume of the system.

Suppose a chemical change takes place at constant pressure. This change is accompanied by a change in internal energy (ΔE) and by work done if the system changes volume (ΔV) at constant pressure (P). These changes produce a change in enthalpy (ΔH), described by inserting Δ in front of the appropriate variables in Equation 11.5, giving us a mathematical definition of enthalpy change (ΔH):

$$\Delta H = \Delta E + P\Delta V \quad (11.6)$$

To see where all this is headed, let's substitute the mathematical expression for ΔE from the right-hand side of Equation 11.4 for the ΔE term in Equation 11.6:

$$\Delta H = \Delta E + P\Delta V$$

$$\Delta H = (q - P\Delta V) + P\Delta V$$

Simplifying this expression, we get

$$\Delta H = q_p \quad (11.7)$$

> ✓ The change in enthalpy (ΔH) of a system undergoing a physical or chemical change at constant pressure is equal to the heat (q_p) lost or gained by the system.

where q_p stands for heat flow at constant pressure. Equation 11.7 tells us that the enthalpy change of a reaction at constant pressure is simply the heat gained or lost by the reaction system. When heat flows from the system to the surroundings, the change is exothermic and ΔH is less than zero. When heat flows from the surroundings into the system, the change is endothermic and ΔH is greater than zero. For example, heat flows into a melting ice cube from its surroundings. This process is endothermic ($\Delta H > 0$). However, heat must be removed from a tray of water to make a tray of ice cubes in a freezer, and so ΔH is less than zero. The changes in enthalpy for the two processes have different signs but, for a given amount of water, they have the same absolute value. For example, $\Delta H = +6.01$ kJ as a mole of ice melts, but $\Delta H = -6.01$ kJ as a mole of water freezes.

SAMPLE EXERCISE 11.3: In between periods of a hockey game, an ice-refinishing machine spreads 850. liters of water across the surface of a hockey rink.

  a. If the "system" is the water, what is the sign of ΔH_{system} as it freezes?
  b. What is the quantity of heat that must be lost by this volume of water at 0°C to completely freeze it if 6.01 kJ of heat must be removed to freeze 1 mole of water? Assume that the density of water is 1.00 g/mL.

SOLUTION: a. We may assume that the freezing process takes place at constant pressure so that the heat lost (negative q) by the water as it freezes equals the enthalpy change. Therefore, the sign of ΔH_{system} also is negative.

  b. To calculate the heat lost, we must convert 850 L into an equivalent number of moles because the connection between the quantity of heat removed and the quantity of water that freezes is 6.01 kJ/mol. The following series of conversion steps gives the quantity of heat that must flow from the water:

$$q = 850 \, \cancel{L} \left(\frac{1000 \, \cancel{mL}}{\cancel{L}} \right) \left(\frac{1.00 \, \cancel{g}}{\cancel{mL}} \right) \left(\frac{1 \, \cancel{mol}}{18.02 \, \cancel{g}} \right) \left(\frac{-6.01 \, kJ}{\cancel{mol}} \right) = -2.83 \times 10^5 \, kJ$$

PRACTICE EXERCISE: The flame in a torch used to cut metal is produced by acetylene burning in an atmosphere of pure oxygen. If the combustion of a mole of acetylene releases 1251 kJ of heat, what volume of acetylene at STP is needed to cut through a piece of steel that requires 5.42×10^4 kJ to cut? (See Problems 39–46).

11.6 HEATING CURVES AND HEAT CAPACITY

Winter hikers and high-altitude mountain climbers use portable stoves fueled by propane or butane to prepare hot meals where the only source of water may be snow. In this section, we examine the flow of heat into water that begins as snow and ends up as steam.

FIGURE 11.9 Sometimes winter hikers rely on melting snow to provide water for cooking and drinking. As heat is added, the temperature of the snow increases in four stages represented by the four line segments: heating snow to its melting point (AB), melting the snow (BC), heating the meltwater to its boiling point (CD), and boiling the water (DE).

Hot soup on a cold day

Let's consider the changes in temperature and changes of state that water undergoes as some hikers set out to prepare hot soup by using dry soup mix and melted snow. Suppose they start with a saucepan filled with snow at 0°F (−18°C). It is placed on top of the flame of a portable stove, and heat begins to flow into the snow. The temperature of the snow immediately begins to rise. If the flame of the stove is steady so that the flow of heat is constant, then the temperature of the snow should follow the series of connected straight lines shown in Figure 11.9. First, heat from the stove warms the snow to its melting point, 0°C. This temperature rise and the accompanying heat flow into the snow are represented by line $\overline{AB}$ in Figure 11.9. When the temperature reaches 0°C, it remains steady as absorbed heat melts the snow, producing a *phase change* that yields liquid water. The melting process, called *fusion*, is represented by the constant temperature (flat) line $\overline{BC}$.

When all the snow has melted, the temperature of the water again rises, along line $\overline{CD}$, until its temperature reaches 100°C. At 100°C, another phase change takes place as liquid water is vaporized to steam with the absorption of a quantity of energy denoted by the length of horizontal line $\overline{DE}$. If all of the liquid water in the pot were converted into steam (to the dismay of the hikers), the temperature of whatever steam remained in the pot would begin to rise again as more heat was added.

The differences in the x-axis coordinates of the points along the curve in Figure 11.9 tell us how much heat is required in each of the preceding steps in the heating process. In the first step, along line $\overline{AB}$, the heat required to raise the temperature of snow (or ice) from $-18°C$ to $0°C$ can be calculated by using the following equation:

$$q = nc_P\Delta T \qquad (11.8)$$

where q represents heat, n is the number of moles, ΔT is the temperature change in Celsius degrees, and c_P is the **molar heat capacity** of ice at constant pressure.

The molar heat capacity of a substance is the amount of heat required to raise the temperature of 1 mole of the substance 1 Celsius degree. Some tables of thermodynamic data list a related parameter called *specific heat*, which is the heat required to raise the temperature of 1 gram of a substance 1 Celsius degree and so has units J/g · °C. Later in this chapter, we will make use of the parameter **heat capacity** (C), which is the quantity of energy required to increase the heat of a particular object by 1 Celsius degree. The heat capacity of a quantity of a pure substance (x) is the product of the quantity in moles times the molar heat capacity of the substance: $C = n_x c_x$ and has units of J/°C.

Let's assume that the hikers decided to cook their meal with 270. g of snow. Dividing the mass of snow by the molar mass of water (18.01 g/mol), we find that there are 15.0 mol of frozen water in the snow. To calculate how much heat is needed to raise the temperature of 15.0 mol of $H_2O(s)$ from $-18°C$ to $0°C$, we use Equation 11.8 and insert the quantity of snow (15.0 mol), the temperature change (18 Celsius degrees), and the molar heat capacity of ice, which is 37.1 J/mol · °C:

$$q = nc_P\Delta T$$
$$= (15.0 \text{ mol})\left(\frac{37.1 \text{ J}}{\text{mol} \cdot °C}\right)(18 \, °C)$$
$$= 10{,}012 \text{ J} = 10.0 \text{ kJ}$$

During the next phase of the heating process, the snow melts, or *fuses*, into liquid water. The heat absorbed during this step, represented by the length of line $\overline{BC}$, can be calculated by using the **molar heat of fusion** (ΔH_{fus}) of water, which is 6.01 kJ/mol, and the following equation:

$$q = n\Delta H_{\text{fus}} \qquad (11.9)$$

Substituting the given information into the equation, we get:

$$q = (15.0 \text{ mol})\left(\frac{6.01 \text{ kJ}}{\text{mol}}\right) = 90.0 \text{ kJ}$$

When all of the snow has melted, the temperature of the water rises from its freezing point to its boiling point of 100°C. During this stage, the relation between temperature and heat absorbed is defined by Equation 11.8 for liquid water, which has a molar heat capacity of 75.3 J/mol · °C.

The **molar heat capacity** of a substance is the amount of heat required to raise the temperature of 1 mole of the substance by 1 Celsius degree or 1 kelvin.

The **heat capacity** (C) of an object (or group of objects) is the quantity of heat required to increase the temperature of the object by 1 Celsius degree.

The **molar heat of fusion** (ΔH_{fus}) is the enthalpy change that takes place when 1 mole of a solid substance melts.

$$q = nc_p \Delta T$$

$$= (15.0 \text{ mol}) \left(\frac{75.3 \text{ J}}{\text{mol} \cdot °C} \right) (100°C)$$

$$= 112{,}950 \text{ J} = 113 \text{ kJ}$$

Note that the c_P values of solid and liquid water are not the same. The c_P value of water vapor is different from the values for liquid or solid water. This trend is common: nearly all substances have different molar heat capacities in their different physical states.

At this point in our story, we assume that our inattentive hiker chefs accidentally leave the boiling water unattended and it is completely vaporized. The conversion of liquid water into steam is depicted by line $\overline{DE}$ in Figure 11.9. The temperature of the water remains at 100°C until enough energy (the length of horizontal line $\overline{DE}$) has been absorbed to vaporize all of its liquid water content. This quantity of energy absorbed is the product of the number of moles of water present and the **molar heat of vaporization** (ΔH_{vap}) of water, which is 40.67 kJ/mol:

$$q = n\Delta H_{vap} \quad (11.10)$$

$$= (15.0 \text{ mol})(40.67 \text{ kJ/mol})$$

$$= 610.0 \text{ kJ}$$

Only after all the water has vaporized can its temperature increase above 100°C along the line above point E, the slope of which is a function of the quantity of steam still in the pot and the molar heat capacity of steam (43.1 J/mol · °C).

The horizontal line ($\overline{DE}$) in Figure 11.9 representing the boiling of water is much longer than the horizontal line ($\overline{BC}$) representing the melting of snow. The relative lengths of these lines indicate that the molar heat of vaporization of water (40.67 kJ/mol) is much greater than the molar heat of fusion of ice (6.01 kJ/mol). Why does it take more energy to boil a mole of water than it does to melt a mole of snow (or ice)? The answer is connected to the number of intermolecular forces that must be overcome in each process. Melting snow requires that some hydrogen bonds between molecules in the crystalline snow lattice be broken, but there are still many hydrogen bonds between molecules of H_2O in liquid water. When water vaporizes, essentially all hydrogen bonds between water molecules are broken. Breaking this greater number of hydrogen bonds requires more energy.

> **Why does it take more energy to boil a mole of water than it does to melt a mole of snow (or ice)?**

CONNECTION: Boiling points of liquids decrease with increasing altitude and decreasing atmospheric pressure are (See Section 9.5).

The **molar heat of vaporization** (ΔH_{vap}) is the enthalpy change that takes place when 1 mole of a liquid substance vaporizes.

CONNECTION: The origin and strength of hydrogen bonds is described in Section 9.3.

SAMPLE EXERCISE 11.4: Calculate the amount of heat required to raise the temperature of one cup (8.0 fluid ounces) of water from 15.° to 100.°C. Given: 1 fluid ounce = 29.6 mL.

SOLUTION: The amount of heat required can be determined by using Equation 11.8 after we have calculated the number of moles of H_2O in 8.0 fluid ounces of water. The fluid ounce is a measure of volume; so we must use the density of water to convert volume into mass. Concurrently, we must convert ounces into milliliters.

We can calculate the number of milliliters in an 8-fluid-ounce cup, the mass of H_2O (the density is 1.00 g/mL), and, from the mass, the number of moles:

volume of H_2O (V_{H_2O})

$mass_{H_2O} = d_{H_2O} V_{H_2O}$

mass of H_2O

$n_{H_2O} = \dfrac{mass_{H_2O}}{\mathcal{M}_{H_2O}}$

moles of H_2O n_{H_2O}

$q = n_{H_2O}\, c_{P, H_2O(l)}$

heat (q)

$$8 \text{ oz } H_2O \left(\dfrac{29.6 \text{ mL}}{1 \text{ oz}}\right)\left(\dfrac{1.00 \text{ g}}{\text{mL}}\right)\left(\dfrac{1 \text{ mol}}{18.01 \text{ g}}\right) = 13.2 \text{ mol } H_2O$$

Using this quantity of H_2O in Equation 11.8 gives

$$q = nc_p \Delta T$$

$$q = (13.2 \text{ mol})\left(\dfrac{75.3 \text{ J}}{\text{mol} \cdot {}^\circ\text{C}}\right)(100 - 15)\, {}^\circ\text{C} = 84{,}287 \text{ J, or } 84.3 \text{ kJ}$$

PRACTICE EXERCISE: Calculate the heat released when 100. g of water vapor at 100.°C condenses to liquid water and then cools to 15.0°C. (See Problems 51 and 52.)

Cold drinks on a hot day

Let's consider another study in heat transfer. Suppose you are throwing a summertime party and you need to chill three cases (72 aluminum cans, each containing 355 mL) of your favorite beverages by placing them in the bottom of a well-insulated cooler and covering them with ice cubes. If the temperature of the ice (sold in 10-pound bags) is −8°C and the temperature of the beverages is 25°C initially, how many bags of ice do you need to chill the cans and their contents to 0°C (as in "ice cold")? Perhaps you have confronted this problem in the past and already have an idea that more than 1 bag, but fewer than 10 will be needed. With a little math we can predict more exactly how much ice will be needed. In doing so, we will assume that whatever heat is absorbed by the ice is lost by the cans of beverage. As ice absorbs heat from the cans, its temperature will increase from −8°C to 0°C, and then it should remain at 0°C until all the ice has melted (see the heating curve in Figure 11.9). We need enough ice so that the last of it melts just as the temperature of the cans of beverage reaches that of the melting ice, 0°C. The cooling process is a little simpler because there is no phase change, only cooling of liquid beverage and solid aluminum cans.

Let's consider the heat lost in the cooling process first. Two materials are to be chilled: 72 aluminum cans and 72 × 355 mL = 25,560 mL of beverage. The beverages are mostly water. The other ingredients are present in such small concentrations that they will not significantly affect our calculation; so we will assume that we need to reduce the temperature of what is essentially 25,560 mL of water by 25°C. We can calculate the amount of heat lost with Equation 11.8 if we first calculate the number of moles of water in 25,560 mL of water:

$$25{,}560 \text{ mL H}_2\text{O} \times \frac{1.00 \text{ g}}{\text{mL}} \times \frac{1 \text{ mol H}_2\text{O}}{18.01 \text{ g H}_2\text{O}} = 1419 \text{ mol}$$

The heat that is lost by this quantity of water as its temperature decreases from 25°C to 0°C is

$$q = nc_p\Delta T$$

$$= (1419 \text{ mol})\left(\frac{75.3 \text{ J}}{\text{mol} \cdot °\text{C}}\right)(-25 \text{ °C})$$

$$= -2.67 \times 10^7 \text{ J}$$

We must also consider the energy needed to lower the temperature of 72 aluminum cans by 25 Celsius degrees. Suppose the average mass of each can is 12.5 g. The molar heat capacity of solid aluminum is 24.4 J/mol · °C and its molar mass is 26.98 g/mol. Using these values in Equation 11.8, we have

$$q = nc_p\Delta T$$

$$= 72 \text{ cans} \left(\frac{12.5 \text{ g Al}}{\text{can}}\right)\left(\frac{1 \text{ mol}}{26.98 \text{ g}}\right)\left(\frac{24.4 \text{ J}}{\text{mol} \cdot °\text{C}}\right)(-25 \text{ °C}) = -2.03 \times 10^4 \text{ J}$$

Thus, the total quantity of energy that must be removed from the cans of beverage is

$$q_{\text{total lost}} = q_{\text{beverage}} + q_{\text{can}} = (-2.67 \times 10^7 - 2.03 \times 10^4) \text{ J}$$

$$= -2.69 \times 10^7 \text{ J} = -2.69 \times 10^4 \text{ kJ}$$

This quantity of heat must be absorbed by the ice as it warms to its melting point and then melts. To calculate how much ice is needed, we will make use of algebra. Let x be the number of moles of ice needed. Then the heat absorbed will be the sum of the heat needed to (1) raise the temperature of x moles of ice from $-8°$ to 0°C and (2) melt x moles of ice. The quantities of heat needed can be calculated with Equation 11.8 for step 1 and Equation 11.9 for step 2:

$$q_{\text{total gained}} = q_1 + q_2$$

$$= nc_p\Delta T + n\Delta H_{\text{fus}}$$

$$= x\left(37.1 \frac{\text{J}}{\text{mol} \cdot °\text{C}}\right)(8 \text{ °C}) + x\left(6.01 \frac{\text{kJ}}{\text{mol}}\right)$$

$$= (x)(6.31 \text{ kJ/mol})$$

The heat lost by the cans of beverage balances the heat gained by the ice:

$$-q_{\text{total lost}} = q_{\text{total gained}}$$

$$2.69 \times 10^4 \text{ kJ} = (x)(6.31 \text{ kJ/mol})$$

$$x = 4.26 \times 10^3 \text{ moles of ice}$$

Converting 4.26×10^3 moles of ice into pounds gives

$$4.26 \times 10^3 \text{ mol} \left(18.01 \frac{\text{g}}{\text{mol}}\right)\left(\frac{1 \text{ lb}}{453.6 \text{ g}}\right) = 75 \text{ pounds of ice}$$

Thus, you will need at least eight 10-pound bags of ice to chill three cases of your favorite beverages. But maybe you already knew that.

CONCEPT TEST: The heat lost by the beverage inside the 72 cans in the preceding illustration was more than 100 times the heat lost by the cans themselves. What factors contributed to this large difference between heat lost by the cans and the heat lost by their contents?

SAMPLE EXERCISE 11.5: Your assignment is to prepare freshly brewed iced tea. If you added 250 g of ice that is initially at $-18°C$ to one cup (237 g) of freshly brewed tea initially at 100°C, what would be the final temperature of the tea? Assume that the mixture is in an insulated container of negligible heat capacity, and that 237 g of tea is, for the purposes of this calculation, the same as 237 g of water.

SOLUTION: Before beginning to solve this problem, we need to think about the changes that take place during the process in which the temperatures of the hot tea and ice cubes approach each other. Three processes (assuming that the ice cubes entirely melt) will consume heat lost by the hot tea:

1. raising the temperature of the ice cubes to their melting point (0°C);
2. melting the ice cubes; and
3. raising the temperature of the melted ice water to the final temperature of the mixture (T_{final})

Let's begin by noting that the heat gained by the ice cubes equals the heat lost by the tea:

$$q_{ice\ cubes} = -q_{tea}$$

The heat lost by the tea as it cools from 100°C to its final temperature (T_{final}) is

$$q_{tea} = nc_p \Delta T_{hot\ water}$$

$$= 237 \text{g} \left(\frac{1 \text{ mol}}{18.01 \text{ g}}\right)\left(\frac{75.3 \text{ J}}{\text{mol} \cdot °C}\right)(T_{final} - 100°C)$$

$$= 991 \text{ J/°C}(T_{final} - 100°C)$$

Let's treat the process of heat transfer from the hot tea to the ice cubes in stages. In stage 1, the ice is warmed to its melting point:

$$q_1 = nc_p \Delta T$$

$$= 250 \text{ g} \left(\frac{1 \text{ mol}}{18.01 \text{ g}}\right)\left(\frac{37.1 \text{ J}}{\text{mol} \cdot °C}\right)(18°C) = 9270 \text{ J}$$

In stage 2, the ice melts, requiring the absorption of the following quantity of heat based on ΔH_{fus}:

$$q_2 = n\Delta H_{fus}$$

$$= 250 \text{ g} \left(\frac{1 \text{ mol}}{18.01 \text{ g}}\right)\left(6.01 \frac{\text{kJ}}{\text{mol}}\right) = 83.4 \text{ kJ}$$

In stage 3, the melted ice water warms to the final temperature ($\Delta T = T_{final} - T_{initial} = T_{final} - 0°C$):

$$q_3 = n_{water} C_{H_2O(l)} \Delta T_{water}$$

$$= 250 \text{ g} \left(\frac{1 \text{ mol}}{18.01 \text{ g}}\right)\left(\frac{75.3 \text{ J}}{\text{mol} \cdot °C}\right)(T_{final} - 0°C)$$

$$= (1045 \text{ J/°C})(T_{final})$$

The sum of the quantities of heat absorbed by the ice during stages 1 through 3 (q_{ice}) must balance the heat lost by the hot tea; so:

$$q_{ice} = -q_{tea}$$

$$9270 \text{ J} + 83.4 \text{ kJ} + (1045 \text{ J/°C})(T_{final}) = -[991 \text{ J/°C})(T_{final} - 100°C)]$$

Rearranging the terms to solve for T_{final} and converting J/°C to kJ/°C, we have

$$(2036 \text{ J/°C}) T_{final} = (2.04 \text{ kJ/°C}) T_{final} = -9.27 \text{ kJ} - 83.4 \text{ kJ} + 99.1 \text{ kJ} = 6.43 \text{ kJ}$$

$$T_{final} = 3°C$$

PRACTICE EXERCISE: Calculate the final temperature of a mixture of 350 g of −18°C ice cubes and a cup (237 g) of 100°C water. Is your calculated answer less than 0°C? If so, did you include the heat lost in freezing a cup of water? (See Problems 55–58.)

11.7 ESTIMATING ΔH FROM AVERAGE BOND ENERGIES

Combustion reactions produce energy in the form of heat and light. Where does the energy of combustion reactions come from? Let's answer this question by considering the molecular changes that take place in the combustion of a mole of propane:

$$C_3H_8(g) + 5 O_2(g) \longrightarrow 3 CO_2(g) + 4 H_2O(g)$$

As the reaction proceeds, the bonds that hold carbon and hydrogen together in a mole of C_3H_8 must be broken. Similarly the bonds in 5 moles of O_2 also must be broken. The free atoms of C, H, and O from the reactants can then recombine to form the bonds in 3 moles of CO_2 and 4 moles of H_2O. Breaking bonds takes energy, but forming them releases energy. Keep in mind an important equality: the quantity of energy needed to break a chemical bond equals the quantity of energy released when the same bond forms. If a chemical reaction is an exothermic reaction, then more energy must be released in forming the bonds of the products than was required to break apart the bonds in the reactants. To explore this point, let's use average bond energies to estimate the change in enthalpy during the combustion of propane.

> **Where does the energy of combustion reactions come from?**

Bond energy is usually expressed in terms of the enthalpy change (ΔH) required to break a mole of the bonds in the gas phase. A list of bond energies appears in Table 11.2. Note that they are *average* bond energies. Bond energy can vary for a given bond, depending on molecular structure. For example,

> ✓ **Bond energy** is the enthalpy change required to break 1 mole of bonds in the gas phase. Average bond energies can be used to estimate the enthalpy change in chemical reactions (ΔH_{rxn}).

TABLE 11.2 Average Bond Energies (ΔH) of Some Covalent Bonds (See also Appendix 4)

Atom	Bond	ΔH (kJ/mol)	Atom	Bond	ΔH (kJ/mol)
H	H—H	436		C—Cl	338
	H—F	565		C—Br	276
	H—Cl	431		C—I	238
	H—Br	366	N	N—N	163
	H—I	299		N=N	409
C	C—C	348		N≡N	946
	C=C	612		N—H	388
	C≡C	838		N—O	157
	C—H	413		N=O	630
	C—N	305	O	O—O	146
	C=N	613		O=O	497
	C≡N	890		O—H	463
	C—O	360	F	F—F	155
	C=O	743	Cl	Cl—Cl	242
	C≡O	1076	Br	Br—Br	193
	C—F	484	I	I—I	151

the average bond energy for a C=O bond is 745 kJ/mol, but the energy of the C=O bonds in CO_2 is 799 kJ/mol. Another view of the variability in bond-energy values for the same type of bond comes from the step-by-step decomposition of CH_4:

Decomposition Step	Energy Needed (kJ/mol)
$CH_4 \longrightarrow CH_3 + H$	435
$CH_3 \longrightarrow CH_2 + H$	453
$CH_2 \longrightarrow CH + H$	425
$CH \longrightarrow C + H$	339
	Total = 1652
	Average = 413

These results tell us that the chemical environment of a bond affects its dissociation energy: breaking the first C—H bond in methane is easier than breaking the second but is more difficult than breaking the third or fourth.

An examination of the structure of C_3H_8 (see Figure 11.4) reveals that, during the combustion of 1 mole of C_3H_8, 2 moles of C—C bonds and 8 moles of C—H bonds must be broken. The number of moles of oxygen consumed is 5; so 5 moles of O=O bonds must also be broken. In the formation of 3 moles of CO_2 and 4 moles of H_2O, 6 moles of C=O bonds and 8 moles of O—H bonds form. The net change in energy resulting from these bonds breaking and forming can be calculated from the following tabulation of data:

Type of Bond	Number of Bonds (mol)	Bond Energy (kJ/mol)	Sign of ΔH in calculation	ΔH (kJ)
C—H	8	413	+	$+(8)(413) = +3304$
C—C	2	348	+	$+(2)(348) = +696$
O=O	5	497	+	$+(5)(497) = +2485$
O—H	8	463	−	$-(8)(463) = -3704$
C=O	6	799	−	$-(6)(799) = -4794$
				$\Delta H_{rxn} = -2013$

If we sum the positive enthalpy changes in bond breaking and the negative enthalpy changes in bond formation, we get the overall enthalpy change for the combustion of a mole of propane (see Fig. 11.10): -2013 kJ. Because this calculation was based on the combustion of 1 mole of fuel, the enthalpy change for the reaction, ΔH_{rxn}, is called the **molar heat of combustion** (ΔH_{comb}) of propane.

Take another look at the bond energies in Table 11.2, and note the differences in bond energies for O—O and O=O bonds. At 497 kJ/mol, the bond energy of the oxygen–oxygen double bond is more than three times the bond energy of the oxygen–oxygen single bond. This correlation between bond order and bond energy is true for other pairs of atoms, too: greater bond order means greater bond energy.

The **molar heat of combustion** is the change in enthalpy that takes place when 1 mole of a substance reacts with oxygen in a combustion reaction.

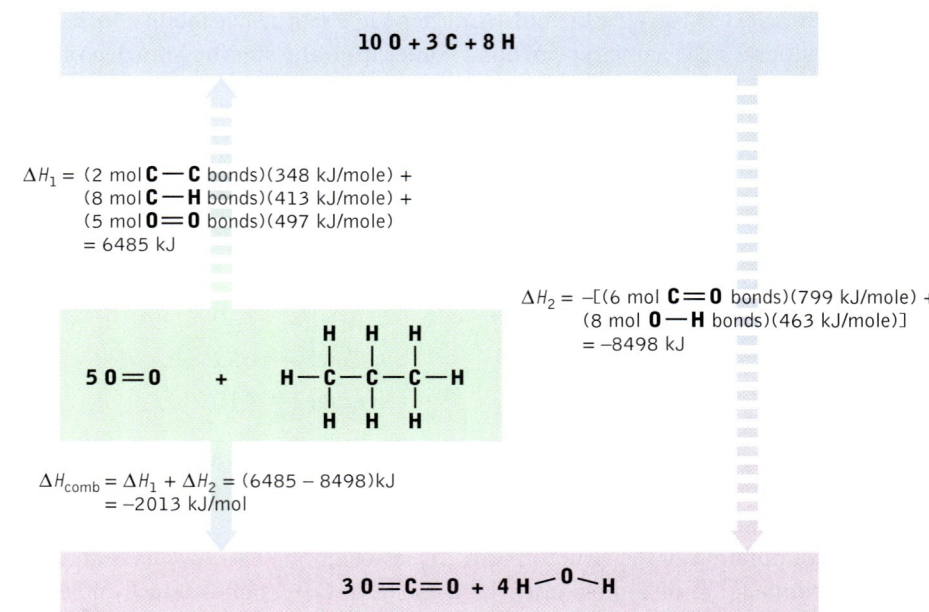

FIGURE 11.10 The combustion of 1 mole of propane requires that 8 moles of C—H bonds, 2 moles of C—C bonds, and 5 moles of O═O bonds be broken. These processes are endothermic, requiring an enthalpy gain ($\Delta H_1 = +6485$ kJ). Formation of 6 moles of C═O and 8 moles of O—H bonds are exothermic processes that result in a loss of enthalpy ($\Delta H_2 = -8498$ kJ) and that lead to an overall exothermic reaction ($\Delta H_{comb} = -2013$ kJ/mol of propane).

The bond energy of the nitrogen–nitrogen triple bond (946 kJ/mol) is one of the largest in Table 11.2. It is more than twice the bond energy of the nitrogen–nitrogen double bond and more than four times that of the nitrogen–nitrogen single bond. The large amount of energy that must be consumed in breaking N≡N bonds is a reason why N_2 participates in few chemical reactions.

SAMPLE EXERCISE 11.6: Estimate the heat of combustion (ΔH_{comb}) of methane gas (CH_4) from the bond energies in Table 11.2. Given:

$$CH_4(g) + 2\, O_2(g) \longrightarrow CO_2(g) + 2\, H_2O(g)$$

SOLUTION: Within a mole of CH_4, there are 4 moles of C—H bonds. There are 2 moles of O═O double bonds in 2 moles of O_2. These bonds must be broken in the course of the reaction. Two moles of C═O double bonds are formed in a mole of CO_2, as are 4 moles of O—H bonds in 2 moles of H_2O. Putting this information in table form along with the appropriate average bond energies from Table 11.2, we have

Type of Bond	Number of Bonds (mol)	Bond Energy (kJ/mol)	Sign of ΔH	ΔH (kJ)
C—H	4	413	+	$+(4)(413) = +1652$
O=O	2	497	+	$+(2)(497) = +994$
O—H	4	463	−	$-(4)(463) = -1852$
C=O	2	799	−	$-(2)(799) = -1598$
				$\Delta H_{comb} = -804$

PRACTICE EXERCISE: Estimate ΔH_{comb} for ethane from the bond energies in Table 11.2 and the following chemical equation:

$$C_2H_6(g) + \tfrac{7}{2} O_2(g) \longrightarrow 2\ CO_2(g) + 3\ H_2O(g)$$

(See Problems 63 and 64.)

11.8 CALORIMETRY: MEASURING HEATS OF COMBUSTION

Average bond energies provide a basis for estimating the chemical energy of a substance and the enthalpy change of a chemical reaction, but there is uncertainty in these calculations due to the variability in bond energies noted in Section 11.7. There is, however, an experimental approach for accurately determining the quantities of heat released by chemical reactions. The technique is called **calorimetry**. When the reactions of interest are combustion reactions, we measure the enthalpy change, or heat of combustion, with a device called a bomb calorimeter (Figure 11.11). The sample is placed in a sealed vessel (called a *bomb*) capable of withstanding high pressures. Oxygen is introduced into the vessel and the mixture is ignited with an electrical spark. As combustion occurs, heat generated by the reaction flows into water surrounding the bomb and an insulated container.

Bomb calorimetry is based on the assumption that whatever heat is lost as reactants form products (the system) in an exothermic chemical reaction will be gained by the reaction system's surroundings (the water and reaction vessel). The heat gained by the surroundings is determined by accurately measuring the temperature of the surroundings before and after the reaction takes place.

Measuring the change in temperature is not the whole story. We also need to know how much heat is required to raise the temperature of the surroundings by, say, 1 Celsius degree. In other words, we need to know the heat capacity (C) of the surroundings. If we know the value of C and if we can accurately measure the change in temperature (ΔT), then we can calculate the quantity of heat (q) that flowed into the surroundings:

$$q = C\Delta T \qquad (11.11)$$

> **Calorimetry** is the measurement of the change in heat that occurs during a physical or chemical change.

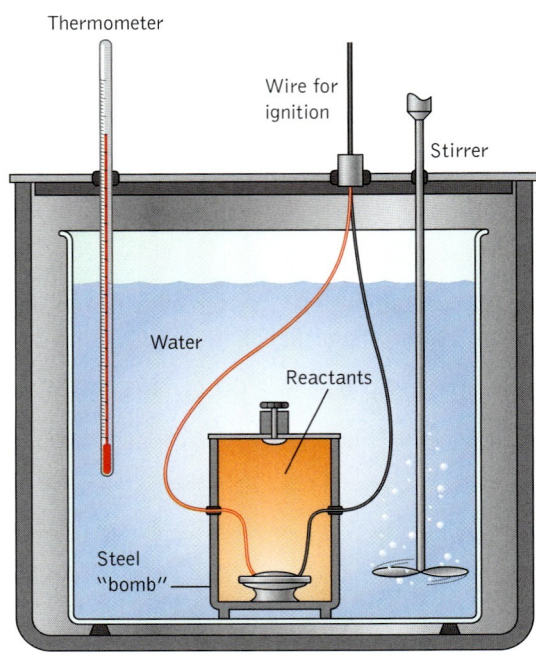

FIGURE 11.11 The combustion of substances in oxygen releases heat that is absorbed by the insulated mass of water surrounding the reaction vessel in a bomb calorimeter. The quantity of heat released is calculated from the temperature change and the heat capacity of the water and its container.

Rearranging the terms in Equation 11.11, we have the following expression for the heat capacity of the surroundings:

$$C = q/\Delta T \qquad (11.12)$$

Equation 11.12 indicates that heat capacity is expressed in units of heat/temperature, or kJ/C°. Equation 11.12 can be used to determine the heat capacity of a bomb calorimeter. To use it, we need to combust a quantity of material that produces a known quantity of heat when it burns. Benzoic acid ($C_7H_6O_2$) is often used for this purpose; the combustion of 1 gram of benzoic acid produces 26.38 kJ of heat. After the heat capacity of the calorimeter has been determined, other substances can be combusted, and the observed increases in temperature can be used to calculate the quantities of heat that their combustion produces. Because there is no change in the volume of the reaction mixture, no $P\Delta V$ work is done, and so, according to Equation 11.4 the heat lost by the system equals the energy lost by the system during the combustion process:

$$q_{system} = \Delta E_{comb}$$

Pressure inside the calorimeter may change during the combustion reaction; so q_{system} is not exactly the same as ΔH_{comb}. However, the pressure effects are often so small that ΔH_{comb} is nearly the same as q_{system} and ΔE_{comb}.

SAMPLE EXERCISE 11.7: What is the heat capacity, C, of a bomb calorimeter if the combustion of 1.000 g of benzoic acid causes the temperature of the calorimeter to rise by 7.248 Celsius degrees?

SOLUTION: The heat capacity of a calorimeter can be calculated by using Equation 11.12 and knowing that the combustion of 1.000 g of benzoic acid produces 26.38 kJ of heat.

$$C = q/\Delta T$$
$$= \frac{26.38 \text{ kJ}}{7.248 \text{ C}°}$$
$$= 3.640 \text{ kJ/C}°$$

PRACTICE EXERCISE: When 0.500 g of a mixture of volatile hydrocarbons was combusted in the bomb calorimeter in the preceding Sample Exercise, the temperature rose 6.76 Celsius degrees. How much energy (in kilojoules) was released during combustion? How much energy would be released by the combustion of 1.000 g of the same mixture? (See Problems 75 and 76.)

heat (*q*), temperature change (ΔT)

$$C = \frac{q}{\Delta T}$$

heat capacity (*C*)

11.9 ENTHALPIES OF FORMATION AND REACTION

As noted in Section 11.4, it is difficult to measure the *absolute* value for the internal energy of a substance. The same is true for the enthalpy of a substance. However, we can establish *relative* enthalpy values that are referenced to a convenient standard. This approach is similar to using the freezing point of water as the zero point for the Celsius temperature scale or to using sea level as the zero point for expressing altitude. The zero point for enthalpy values is the **standard enthalpy of formation** ($\Delta H_f°$), which is defined as the enthalpy change that takes place when 1 mole of a substance in its standard state is formed from its elements in their standard states. The **standard state** of an element is its most stable physical form under standard conditions, such as 25°C (298.15 K, to be more precise) and a pressure of 10^5 Pa (slightly less than 1 atm—see Section 8.2). By definition, $\Delta H_f°$ for a pure element in its most stable form is zero. Because enthalpy change is equivalent to the transfer of heat at constant pressure, $\Delta H_f°$ is frequently called the *standard heat of formation* or, simply, *heat of formation*. About the symbol $\Delta H_f°$: you probably guessed that the subscript "f" represents "formation" but may be wondering about the superscript °. Throughout chemistry, and particularly in thermochemistry, the superscript ° refers to the value of a parameter, such as ΔH_f, under *standard conditions*. Implied in our notion of standard states and standard conditions is the assumption that parameters such as ΔH change with temperature and pressure. That assumption is correct, though the changes are so small that we will ignore them in many of the calculations in this chapter and those that follow.

✓ The **standard enthalpy of formation** ($\Delta H_f°$) of a substance is the enthalpy change that takes place when 1 mole of the substance in its standard state is formed from its elements in their standard states. The **standard state** is the most stable physical form of a substance at standard temperature and pressure.

TABLE 11.3 Standard Enthalpies of Formation at 298 K (See Appendix 4)

Compound	ΔH_f° (kJ/mol)	Compound	ΔH_f° (kJ/mol)
$CH_4(g)$	−74.8	$CO(g)$	−110.5
$C_2H_6(g)$	−84.7	$CO_2(g)$	−393.5
$C_3H_8(g)$	−103.8	$H_2O(g)$	−241.8
$C_4H_{10}(g)$	−125.7	$H_2O(l)$	−285.8
$O_2(g)$	0.0	$H_2(g)$	0.0

Table 11.3 lists some ΔH_f° values for compounds connected to the current discussion on energy from natural gas. A more complete list can be found in Appendix 4. Note that all the standard enthalpies of formation in Table 11.3 are negative, which means that the reactions in which these compounds are formed from their component elements are exothermic. If the reactions are exothermic, then the bonds that hold the atoms together in these molecules have more energy than the bonds that hold atoms of the pure elements together. Stronger bonds make more-stable molecules. In this context, it is reasonable that the compounds that are the products of the combustion reactions (CO_2 and H_2O) have the most negative ΔH_f° values in Table 11.3 and are more stable. Also note that the ΔH_f° values for the C_1 to C_4 hydrocarbons become more negative with increasing molar mass. This trend makes sense because greater molar mass means a larger number of bonds. As more bonds are formed, more energy is released.

When we know standard enthalpies of formation, we can calculate the enthalpy changes that take place in chemical reactions by taking the difference in ΔH_f° between their products and reactants. To see how this approach works, let's recalculate ΔH_{rxn} for the combustion of methane by using the ΔH_f° values from Table 11.3. In this calculation, we make use of the following equation:

$$\Delta H_{rxn}^\circ = \Sigma n \Delta H_{f,products}^\circ - \Sigma m \Delta H_{f,reactants}^\circ \qquad (11.13)$$

which states that ΔH_{rxn}° equals the sum (Σ) of the ΔH_f° values for n moles of products minus the sum of the ΔH_f° values of m moles of reactants in a balanced chemical equation. Thus, we will multiply the value of ΔH_f° for O_2 and H_2O by two before summing them because the coefficient for both compounds is two in the balanced equation for the combustion of methane:

$$CH_4(g) + 2\,O_2(g) \longrightarrow CO_2(g) + 2\,H_2O(g)$$

Inserting ΔH_f° values for the products (CO_2 and H_2O) and the reactants (CH_4 and O_2) and their coefficients from the balanced chemical equation into Equation 11.13, we have

$$\Delta H_{rxn}^\circ = [(1\text{ mol }CO_2)(-393.5\text{ kJ/mol }CO_2) + (2\text{ mol }H_2O)(-241.8\text{ kJ/mol }H_2O)]$$
$$- [(1\text{ mol }CH_4)(-74.8\text{ kJ/mol }CH_4) + (2\text{ mol }O_2)(0.0\text{ kJ/mol }O_2)]$$

$$\Delta H_{rxn}^\circ = [-393.5\text{ kJ} - 483.6\text{ kJ}] - [-74.8\text{ kJ} + 0.0\text{ kJ}] = -802.3\text{ kJ}$$

Calculations of ΔH°_{rxn} from ΔH°_f values can be done for all kinds of chemical reactions. Consider the following reaction between methane and steam, which yields a mixture of hydrogen and carbon monoxide known as *water gas*:

$$CH_4(g) + H_2O(g) \longrightarrow CO(g) + 3\,H_2(g)$$

This reaction is used to synthesize the hydrogen used in the chemical industry to make, for example, ammonia, NH_3. It is also used to make hydrogen fuel for fuel cells, an efficient technology for the generation of electricity discussed in detail in Chapter 17. Inserting the appropriate ΔH°_f values from Table 11.3 and a coefficient of three for H_2 into Equation 11.13, we have

$$\Delta H^\circ_{rxn} = [1\text{ mol CO}](-110.5\text{ kJ/mol CO}) + (3\text{ mol H}_2)(0.0\text{ kJ/mol H}_2)]$$
$$- [1\text{ mol CH}_4)(-74.8\text{ kJ/mol CH}_4) + (1\text{ mol H}_2O)(-241.8\text{ kJ/mol H}_2O)]$$
$$\Delta H^\circ_{rxn} = [-110.5\text{ kJ} + 0.0\text{ kJ}] - [-74.8\text{ kJ} - 241.8\text{ kJ}] = +206.1\text{ kJ}$$

The positive enthalpy change tells us that this reaction is endothermic. Therefore, heat must be added to make the reaction take place. In fact, the *steam-reforming process*, as it is called in the chemical industry, is typically conducted at temperatures near 1000°C.

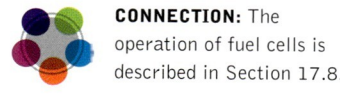

CONNECTION: The operation of fuel cells is described in Section 17.8.

SAMPLE EXERCISE 11.8: Calculate ΔH°_{rxn} for the combustion of propane in air by using the appropriate ΔH°_f values from Table 11.3.

SOLUTION: The balanced equation for the combustion of propane is

$$C_3H_8(g) + 5\,O_2(g) \longrightarrow 3\,CO_2(g) + 4\,H_2O(g)$$

Inserting ΔH°_f values for the products (CO_2 and H_2O) and reactants (C_3H_8 and O_2) from Table 11.3 and the coefficients in the preceding chemical equation into Equation 11.13, we have

$$\Delta H^\circ_{rxn} = [(3\text{ mol CO}_2)(-393.5\text{ kJ/mol CO}_2)$$
$$+ (4\text{ mol H}_2O)(-241.8\text{ kJ/mol H}_2O)]$$
$$- [(1\text{ mol C}_3H_8)(-103.8\text{ kJ/mol C}_3H_8) + (5\text{ mol O}_2)(0.0\text{ kJ/mol O}_2)]$$
$$\Delta H^\circ_{rxn} = [-1180.5\text{ kJ} - 967.2\text{ kJ}] - [-103.8\text{ kJ} + 0.0\text{ kJ}] = -2043.9\text{ kJ}$$

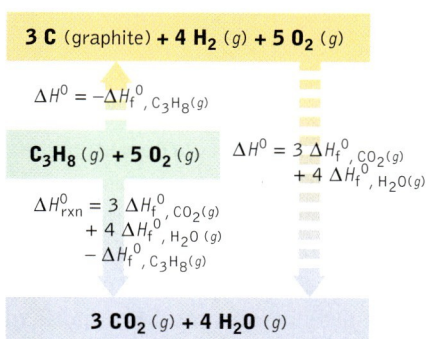

The heat of reaction can be calculated from the difference in the sums of the heats of formation of the products and the sums of the heats of formation of the reactants.

PRACTICE EXERCISE: Calculate ΔH°_{rxn} for the *water-gas shift reaction* in which the carbon monoxide formed by the steam-reforming process is reacted with more steam, producing CO_2 and H_2:

$$CO(g) + H_2O(g) \longrightarrow CO_2(g) + H_2(g)$$

(See Problems 85 and 86.)

11.10 FUEL VALUES

The results of calculations in Section 11.9 tell us that the enthalpy change is much greater and much more heat is released in the combustion of a mole of propane ($\Delta H^\circ_{rxn} = -2043.9$ kJ) than in the combustion of a mole of methane ($\Delta H^\circ_{rxn} = -802.3$ kJ). Does this make propane an inherently better (i.e., higher-energy) fuel? Not necessarily. Expressing ΔH values on a per-mole basis creates an unfair advantage for propane because there are many more bonds and reacting atoms in a mole of propane than there are in a mole of methane. What if we recalculated ΔH for a given mass (say, 1 gram) of the two hydrocarbons? The numbers of moles of CH_4 and C_3H_8 in 1 gram of each compound are the reciprocals of their molar masses:

CONNECTION: Converting grams into moles is discussed in Section 4.3.

$$1 \text{ g } CH_4 \times \frac{1 \text{ mol } CH_4}{16.04 \text{ g } CH_4} = 6.234 \times 10^{-2} \text{ mol } CH_4$$

$$1 \text{ g } C_3H_8 \times \frac{1 \text{ mol } C_3H_8}{44.10 \text{ g } C_3H_8} = 2.268 \times 10^{-2} \text{ mol } C_3H_8$$

To calculate the enthalpy change that takes place when 1 gram of methane or propane burns in air, we multiply the absolute value of ΔH°_{rxn} for each of their combustion reactions by the appropriate numbers of moles:

CH_4: $(6.234 \times 10^{-2} \text{ mol/g})(802.3 \text{ kJ/mol}) = 50.02$ kJ/g

C_3H_8: $(2.268 \times 10^{-2} \text{ mol/g})(2043.9 \text{ kJ/mol}) = 46.35$ kJ/g

These quantities of energy per gram of fuel are called **fuel values.** If we were to perform similar calculations for the other hydrocarbons in natural gas, we would get the fuel values listed in Table 11.4. Note that fuel values decrease with increasing molar mass. Why should this be so? The answer lies in the hydrogen:carbon ratio in these compounds. As the number of carbon atoms per molecule increases, the hydrogen-to-carbon ratios decrease. Less hydrogen content means less fuel value because hydrogen has the smallest atomic mass of all the elements. Thus, there are 12 times as many H atoms in a gram of hydrogen as there are C atoms in a gram of carbon, which means that, gram for gram, hydrogen can form six times as many moles of H_2O as carbon can form moles of CO_2. More energy is released by the formation of a mole of CO_2 (393.5 kJ) than by

The **fuel value** of a substance is the quantity of energy released when 1 gram of the substance undergoes combustion.

11.10 Fuel Values

TABLE 11.4 Fuel Values of the Hydrocarbons of Natural Gas

Compound	Molecular Formula	Fuel Value (kJ/g)*
Methane	CH_4	50.0
Ethane	C_2H_6	47.6
Propane	C_3H_8	46.3
Butane	C_4H_{10}	45.8

*Based on the formation of $H_2O(g)$.

a mole of $H_2O(g)$ (241.8 kJ) but, even if we take this difference into account, hydrogen has, gram for gram, many times the fuel value as that of carbon, as indicated by the values in Table 11.4.

CONCEPT TEST: Without doing any calculations, predict which of the following paired quantities of fuel releases more heat during combustion in air.

1 mole of CH_4 or 1 mole of H_2
1 gram of CH_4 or 1 gram of H_2

SAMPLE EXERCISE 11.9: Calculate how much heat is released during the combustion of 20 pounds of propane (the amount of fuel in the tank of a typical gas barbecue grill), starting with the appropriate fuel value in Table 11.4.

SOLUTION: To calculate the flow of heat (q) to the surroundings, we could start with the fuel value of propane in Table 11.4 and then convert kilojoules per gram into kilojoules per pound and then into kilojoules per 20 pounds:

$$q = \frac{-46.3 \text{ kJ}}{\text{g}} \times \frac{453.6 \text{ g}}{1 \text{ lb}} \times 20 \text{ lb} = 4.2 \; 10^5 \text{ kJ}$$

PRACTICE EXERCISE: Use the fuel values in Table 11.4 to calculate how much heat is released in the combustion of 20 pounds of ethane (C_2H_6) gas, a fuel used on board high-performance hot-air balloons. (See Problems 97 and 98.)

Burning propane releases 5 times as much heat per mole as does burning charcoal but only about 1.4 times as much heat per pound. ¯he combustion of propane also produces less CO_2 per kilo_oule of energy.

SAMPLE EXERCISE 11.10: Oceanographers estimate that there are 11.2 trillion tons of methane in the sediments of Earth's oceans.
a. Starting with the fuel value of methane (50 kJ/g), calculate the energy content of this sedimentary methane.

b. If this methane could be mined, how many years of the world's energy needs would it meet if we continued to consume energy at the current rate of 3.0×10^{17} kJ annually?

SOLUTION: a. Fuel values have units of kilojoules per gram, and so we need to convert tons into grams and then into an equivalent amount of energy. The following mass conversion factors should come in handy:

$$1 \text{ trillion} = 1 \times 10^{12}; \quad 1 \text{ ton} = 2000 \text{ pounds}; \quad 1 \text{ pound} = 453.6 \text{ grams}$$

Putting this information together, we can calculate the enthalpy change for the combustion of 11.2 trillion tons of methane:

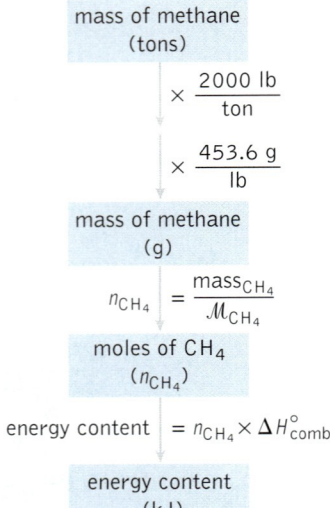

$$11.2 \times 10^{12} \text{ tons CH}_4 \left(\frac{2000 \text{ lb}}{\text{ton}}\right)\left(\frac{453.6 \text{ g}}{1 \text{ lb}}\right)\left(\frac{50.0 \text{ kJ}}{\text{g CH}_4}\right) = 5.1 \times 10^{20} \text{ kJ}$$

Thus, burning 11.2 trillion tons of methane releases 5.1×10^{20} kJ of energy. To grasp how much energy this is, let's divide its absolute value by the current rate of energy consumption worldwide:

$$\frac{5.1 \times 10^{20} \text{ kJ}}{3.0 \times 10^{17} \text{ kJ/yr}} = 1.7 \times 10^{3} \text{ yr}$$

Thus, the methane locked in the ocean's sediments would, if it could be mined, meet our energy needs for more than 1700 years at the present rate of energy consumption.

PRACTICE EXERCISE: How many metric tons (1 metric ton = 1000 kg) of methane would have to be burned to meet the current annual worldwide demand for energy (3.0×10^{17} kJ)? (See Problems 99 and 100.)

11.11 HESS'S LAW

In Section 11.9 we described a process for synthesizing hydrogen gas from methane and steam at high temperatures. There are two steps in the process. The first step is an endothermic reaction between methane and a limited supply of high-temperature steam, producing carbon monoxide and hydrogen gas:

$$\text{Step 1:} \quad CH_4(g) + H_2O(g) \longrightarrow CO(g) + 3\,H_2(g) \qquad \Delta H° = 206 \text{ kJ}$$

This chemical equation, to which we have added the enthalpy change for the reaction, is called a *thermochemical equation*. A similar expression for step 2 describes how the carbon monoxide from the first reaction reacts with more steam, producing carbon dioxide and more hydrogen gas:

$$\text{Step 2:} \quad CO(g) + H_2O(g) \longrightarrow CO_2(g) + H_2(g) \qquad \Delta H° = -41 \text{ kJ}$$

We can write an overall reaction equation that combines steps 1 and 2:

$$CH_4(g) + H_2O(g) + CO(g) + H_2O(g) \longrightarrow CO_2(g) + H_2(g) + CO(g) + 3\,H_2(g)$$

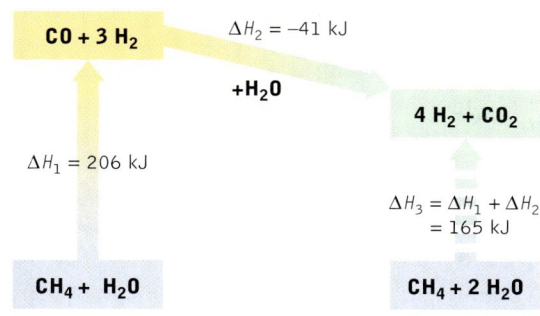

FIGURE 11.12 Hess's law predicts that the enthalpy change for the production of CO_2 and hydrogen from methane and water is the sum of the enthalpies of two reactions: the formation of CO and hydrogen from methane and water and the reaction of CO with water producing CO_2 and H_2.

which simplifies to:

Overall: $CH_4(g) + 2\,H_2O(g) \longrightarrow CO_2(g) + 4\,H_2(g)$ $\Delta H° = ?$

By adding the values of $\Delta H°$ for steps 1 and 2, we can obtain the enthalpy change for the overall reaction. The thermochemical equation for the overall reaction is

Overall: $CH_4(g) + 2\,H_2O(g) \longrightarrow CO_2(g) + 4\,H_2(g)$ $\Delta H° = 165\text{ kJ}$

This calculation of $\Delta H°$ for a step-by-step reaction by summing the enthalpy changes of the steps within it is an application of **Hess's law**. The law, also known as *Hess's law of constant heat of summation*, states that *the enthalpy change for a reaction that is the sum of two or more other reactions is equal to the sum of the enthalpy changes of the constituent reactions.* This relation is shown in graphical form in Figure 11.12.

Now let's consider a slightly more complicated case. When hydrocarbons burn in a limited supply of air, they may not burn completely. One of the reasons why furnaces and hot-water heaters fueled by natural gas need to be properly vented is that incomplete combustion may produce toxic carbon monoxide:

$$2\,CH_4(g) + 3\,O_2(g) \longrightarrow 2\,CO(g) + 4\,H_2O(g)$$

Eventually, CO will combine with O_2 in the atmosphere, forming CO_2:

$$2\,CO(g) + O_2(g) \longrightarrow 2\,CO_2(g)$$

If we combine these two equations, we get the following overall reaction equation:

$$2\,CH_4(g) + 3\,O_2(g) + 2\,CO(g) + O_2(g) \longrightarrow 2\,CO(g) + 4\,H_2O(g) + 2\,CO_2(g)$$

which simplifies to

$$2\,CH_4(g) + 4\,O_2(g) \longrightarrow 2\,CO_2(g) + 4\,H_2O(g)$$

If we divide both sides of this equation by two, we get the combustion reaction discussed earlier in this chapter and the following thermochemical equation:

Overall: $CH_4(g) + 2\,O_2(g) \longrightarrow CO_2(g) + 2\,H_2O(g)$ $\Delta H° = -802\text{ kJ}$

The combustion of CO is a well-studied reaction. The thermochemical equation for this second step in the combustion of CH_4 is:

Step 2: $2\,CO(g) + O_2(g) \longrightarrow 2\,CO_2(g)$ $\Delta H° = -566\text{ kJ}$

> **Hess's law** states that the enthalpy change of a reaction that is the sum of two or more other reactions is equal to the sum of the enthalpy changes of the constituent reactions.

CONNECTION: Oxygen is a limiting reagent in these reactions (Section 4.7).

How can we combine the thermochemical equations for the overall reaction and for step 2 to calculate $\Delta H°$ for the incomplete reaction in step 1?

Step 1: $\quad 2\,CH_4(g) + 3\,O_2(g) \longrightarrow 2\,CO(g) + 4\,H_2O(g) \qquad \Delta H° = ?$

Carbon dioxide is a product in the overall reaction, and in Step 1, but CO_2 does not appear in the equation for step 1. Therefore, our strategy for combining the overall and step 2 equations must involve reversing one of them so that CO_2 is a reactant. Then the CO_2 terms cancel when the two equations are combined. Carbon monoxide is a reactant in step 2 but a product of step 1; so we should choose step 2 as the one to be reversed. When we do, we need to change the sign of $\Delta H°$:

Reverse step 2: $\quad 2\,CO_2(g) \longrightarrow 2\,CO(g) + O_2(g) \qquad \Delta H° = +566\ \text{kJ}$

Before we can combine this equation and the overall equation, we need to multiply the overall expression by two so that the quantities of CO_2 are the same in both equations. If we multiply the chemical equation for the overall reaction by two, we need to multiply its $\Delta H°$ by two as well:

$$2\,CH_4(g) + 4\,O_2(g) \longrightarrow 2\,CO_2(g) + 4\,H_2O(g) \qquad \Delta H° = -1604\ \text{kJ}$$

Now we can combine this equation with the equation for the reverse of step 1, the CO_2 terms cancel, and the sum of the $\Delta H°$ values gives us the enthalpy change for the incomplete combustion of methane:

$$2\,CH_4(g) + 3\,O_2(g) \longrightarrow 2\,CO(g) + 4\,H_2O(g) \qquad \Delta H° = -1038\ \text{kJ}$$

SAMPLE EXERCISE 11.11: Hess's law can be used to calculate energy changes that are difficult to measure directly. For example, CO_2 is the principal product of the combustion of carbon (as in charcoal), but CO can also be produced if oxygen is the limiting reactant. It is difficult to directly measure the enthalpy of combustion for carbon to carbon monoxide

(1) $\quad 2\,C(s) + O_2(g) \longrightarrow 2\,CO(g)$

because burning carbon in a limited supply of oxygen is more likely to give carbon dioxide and soot (unburned carbon) than pure carbon monoxide. From the following thermochemical equations, calculate $\Delta H°$ for reaction 1:

(2) $\quad C(s) + O_2(g) \longrightarrow CO_2(g) \qquad \Delta H° = -393.5\ \text{kJ}$

(3) $\quad 2\,CO(g) + O_2(g) \longrightarrow 2\,CO_2(g) \qquad \Delta H° = -566.0\ \text{kJ}$

SOLUTION: We need to develop a strategy for combining the information in equations 2 and 3 to obtain the enthalpy change of equation 1. In equation 3, CO is a reactant. To produce a reaction with CO as a product, we need to reverse equation 3 and change the sign of $\Delta H°$:

(4) $\quad 2\,CO_2(g) \longrightarrow 2\,CO(g) + O_2(g) \qquad \Delta H° = +566.0\ \text{kJ}$

If we combined equations 2 and 4, we would get an equation with C on the left and CO on the right as in reaction 1, but the quantities of CO_2 in equations 2 and

4 are not the same and so would not cancel. To get them to cancel, we must first multiply equation 2 by two:

$$2\ C(s) + O_2(g) \longrightarrow CO_2(g) \qquad \Delta H° = -393.5 \text{ kJ}]$$

or

(5) $\qquad 2\ C(s) + 2\ O_2(g) \longrightarrow 2\ CO_2(g) \qquad \Delta H° = -787.0 \text{ kJ}$

Combining equations 4 and 5 gives us the desired equation and its enthalpy change:

$$2\ CO_2(g) + 2\ C(s) + 2\ O_2(g) \longrightarrow$$
$$2\ CO(g) + O_2(g) + 2\ CO_2(g)\ \Delta H° = (-787.0 + 566.0) \text{ kJ}$$

which simplifies to

$$2\ C(s) + O_2(g) \longrightarrow 2\ CO(g) \qquad \Delta H° = -221.0 \text{ kJ}$$

Thus, the incomplete combustion of carbon releases 221.0 kJ of heat, which works out to 110.5 kJ for each mole for carbon. This amount is less than a third of the energy (393.5 kJ) produced by the complete combustion of a mole of carbon to CO_2.

PRACTICE EXERCISE: At high temperatures, such a those in the combustion chambers of automobile engines, molecular nitrogen and oxygen combine in an endothermic reaction forming nitrogen monoxide:

$$N_2(g) + O_2(g) \longrightarrow 2\ NO(g) \qquad \Delta H° = 180 \text{ kJ}$$

When NO is released into the environment, it may undergo further oxidation to NO_2 in the following exothermic reaction:

$$2\ NO(g) + O_2(g) \longrightarrow 2\ NO_2(g) \qquad \Delta H° = -112 \text{ kJ}$$

Is the overall reaction between N_2 and O_2 forming NO_2

$$N_2(g) + 2\ O_2(g) \longrightarrow 2\ NO_2(g) \qquad \Delta H° = ?$$

exothermic or endothermic? What is $\Delta H°$ for this reaction? (See Problems 107–112.)

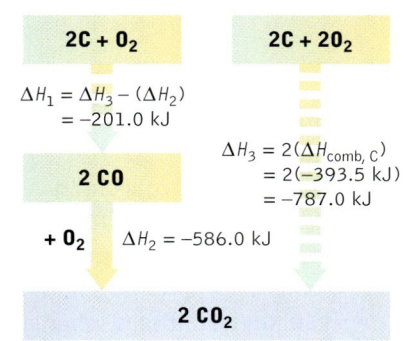

The enthalpy of formation of carbon monoxide can be calculated from the ΔH_{comb} of carbon and the ΔH_{comb} of carbon monoxide by using Hess's law.

CHAPTER REVIEW

Summary

SECTION 11.1

Fossil fuels (coal, crude oil, and natural gas) are today's major sources of energy, but they are nonrenewable.

SECTION 11.2

The law of conservation of energy states that energy is converted from one form into another but is not created or destroyed. Measurement of the energy changes resulting from the breaking and making of chemical bonds is called thermochemistry. Energy is defined as the capacity to do work and can take many forms. Chemical energy is a form of potential energy that is stored in the bonds of substances.

SECTION 11.3

Alkanes, or saturated hydrocarbons, have the general formula C_nH_{2n+2}. Alkanes with $n = 1$ to 4 are constituents of natural gas. Condensed structural formulas are simpler than Lewis structures in that the C–H bonds are not drawn. In straight-chain hydrocarbons, the C–C bonds also may not be shown.

SECTION 11.4

The combustion of hydrocarbons produces H_2O and CO_2 and releases energy. This energy may be in the form of heat (q) or work (w) or both. A reaction or process under study is called a system. Everything else is called its surroundings. The amount of energy gained or lost by the system during a chemical reaction or physical change of state equals the amount of heat lost or gained by its surroundings. The study of energy flow into and out of a system is called thermodynamics. Changes that release energy are exothermic; those that absorb energy are endothermic. The transfer of energy to a system changes its internal energy by an amount $\Delta E = q + w$.

SECTION 11.5

The flow of heat during a chemical or physical change at constant pressure equals a change in enthalpy (ΔH). Reversing a process changes the sign of its enthalpy change.

SECTION 11.6

The heat required to raise the temperature of a substance is related to the quantity of the substance, the molar heat capacity of the substance, and the difference between the initial and the final temperatures of the substance. The solid, liquid, and gaseous states of a substance have different molar heat capacities. The heat required to melt a quantity of a solid substance can be calculated from the quantity of the substance and its molar heat of fusion, ΔH_{fus}. The heat required to convert a liquid into vapor can be calculated from the quantity of the substance and its molar heat of vaporization, ΔH_{vap}.

SECTION 11.7

The energy required to break a chemical bond is its bond energy. The enthalpy change during a chemical reaction can be estimated from the difference in average bond energies of the products and reactants. The enthalpy change of a combustion reaction is called its heat of combustion, ΔH_{comb}.

SECTION 11.8

The enthalpy changes of chemical reactions such as combustion reactions can be measured with a bomb calorimeter, in which heat produced by the reaction warms its surroundings. Knowing the heat capacity and temperature change of the surroundings enables us to calculate the enthalpy change of the reaction.

SECTION 11.9

The standard enthalpy of formation of a compound, $\Delta H_f°$, is the enthalpy change that takes place when 1 mole of the compound is formed under standard conditions from its elements in their standard states. The $\Delta H_f°$ of the most stable form of a pure element is defined as zero. The difference in enthalpies of formation of products minus reactants can be used to calculate the enthalpy change of a reaction.

Chapter Review

SECTION 11.0

The fuel value of a substance is the enthalpy change that takes place when 1 gram of the substance undergoes combustion.

SECTION 11.11

According to Hess's law, the enthalpy change for an overall process is the sum of the enthalpy changes for the constituent steps in the process.

Key Terms

alkane (saturated hydrocarbon) (p. 516)
bond energy (p. 532)
calorimetry (p. 535)
chemical energy (p. 515)
endothermic (p. 519)
energy (p. 513)
enthalpy (p. 523)
exothermic (p. 519)
first law of thermodynamics (p. 518)
fuel value (p. 540)

heat capacity (p. 526)
Hess's law (p. 543)
hydrocarbon (p. 515)
internal energy (p. 520)
law of conservation of energy (p. 515)
molar heat capacity (p. 526)
molar heat of combustion (p. 534)
molar heat of fusion (p. 526)
molar heat of vaporization (p. 527)

potential energy (p. 514)
saturated hydrocarbon (alkane) (p. 516)
standard enthalpy of formation (p. 537)
standard state (p. 537)
state function (p. 514)
surroundings (p. 518)
system (p. 518)
thermochemistry (p. 513)
thermodynamics (p. 518)

Key Skills and Concepts

SECTION 11.1

Understand that the energy requirements of the world economy are mostly being met by the combustion of finite supplies of fossil fuels.

SECTION 11.2

Understand the meaning of *state function*.
Understand that heat and work are interchangeable forms of energy.
Be able to state the law of conservation of energy.

SECTION 11.3

Know the names, formulas, and structures of the C_1 to C_4 alkane hydrocarbons.
Be able to convert condensed structural formulas into Lewis structures.

SECTION 11.4

Be able to complete and balance equations for hydrocarbon combustion reactions.
Be able to identify a thermodynamic system and its surroundings.

Be able to state the first law of thermodynamics.
Understand the meaning of *exothermic* and *endothermic*.
Understand the relation between the energy change of a system, the heat flow to or from it, and the work done by or on it.

SECTION 11.5

Know the meaning of *enthalpy* and the relation between the change in enthalpy (ΔH) and heat flow at constant pressure.

SECTION 11.6

Understand the enthalpy changes associated with heating or cooling and with changes of state and be able to calculate these changes for given amounts of substances.

SECTION 11.7

Be able to estimate the enthalpy changes for balanced combustion reactions (heats of combustion) of specified reactants from bond energies.

SECTION 11.8

Be able to calculate enthalpy changes from data obtained from a bomb-calorimeter experiment.

SECTION 11.9

Understand the concept of standard heat of formation (ΔH_f°). Be able to calculate the enthalpy change of a reaction from the heats of formation of the reactants and products.

SECTION 11.10

Be able to calculate the fuel value of a substance from its heat of combustion and molar mass.

SECTION 11.11

Be able to use Hess's Law to manipulate reactions and their associated enthalpy changes to obtain ΔH for a particular reaction.

Key Equations and Relations

SECTION 11.2

Mechanical work (w) equals force (F) times distance (d):

$$w = F \times d \quad (11.1)$$

The kinetic energy (K.E.) of an object with mass m moving at velocity u is

$$K.E. = \tfrac{1}{2} m u^2 \quad (11.2)$$

SECTION 11.4

The change in internal energy of a system (ΔE) is the sum of the heat (q) flow into the system and the amount of work done on the system:

$$\Delta E = q + w \quad (11.3)$$

When the volume of a system increases at constant pressure as a result of a chemical reaction, the change in internal energy of a system, ΔE, is

$$\Delta E = q - P\Delta V \quad (11.4)$$

SECTION 11.5

The enthalpy of a system is defined as the sum of its internal energy and its pressure times its volume:

$$H = E + PV \quad (11.5)$$

The change in enthalpy (ΔH) of a system undergoing a change in volume at constant pressure is

$$\Delta H = \Delta E + P\Delta V \quad (11.6)$$

An enthalpy change at constant pressure equals the flow of heat to the system:

$$\Delta H = q_p \quad (11.7)$$

SECTION 11.6

The amount of heat added or removed from a substance as its temperature changes by ΔT is

$$q = n c_P \Delta T \quad (11.8)$$

where n is the number of moles of substance and c_P is the molar heat capacity of the substance.

The amount of heat needed to melt (fuse) n moles of a solid substance is

$$q = n\Delta H_{fus} \quad (11.9)$$

where ΔH_{fus} is the substance's molar heat of fusion.

The amount of heat needed to vaporize n moles of a liquid substance is

$$q = n\Delta H_{vap} \quad (11.10)$$

where ΔH_{vap} is the substance's molar heat of vaporization.

The enthalpy change in a chemical reaction (ΔH_{rxn}) can be estimated from the difference in the average bond energies of the products minus the average bond energies of the bonds in the reactants.

SECTION 11.8

In calorimetry, the amount of heat released to the surroundings by a system is

$$q = C\Delta T \quad (11.1)$$

where C is the heat capacity and ΔT is the increase in the temperature of the surroundings.

QUESTIONS AND PROBLEMS

Defining Energy

CONCEPT REVIEW

1. How are energy and work related?
2. Explain the difference between potential energy and kinetic energy.
3. Explain what is meant by a state function.
4. Are kinetic and potential energies both state functions?

PROBLEMS

5. Describe the nature of the potential energy in
   a. a brand new battery for your portable CD player.
   b. a gallon of gasoline.
   c. the crest of a wave before it crashes onto shore.
6. Describe the kinetic energy in a stationary ice cube.

Natural Gas

CONCEPT REVIEW

7. Do all alkanes have the same percent composition?
8. Do all alkanes have the same empirical formula?
9. Explain the difference between a *methylene* group and a *methyl* group.
10. The butane molecule in Table 11.1 is called a *linear alkane*. Are the carbon chains in butane truly linear?
11. Why do the boiling points of alkanes increase with increasing molar mass?
12. What kinds of intermolecular forces act between alkane molecules?

PROBLEMS

13. Which of the following compounds are alkanes? (a) CH_4; (b) C_2H_2; (c) $C_{16}H_{32}$; (d) C_4H_{10}.
14. Which of the following structures are those of alkanes?

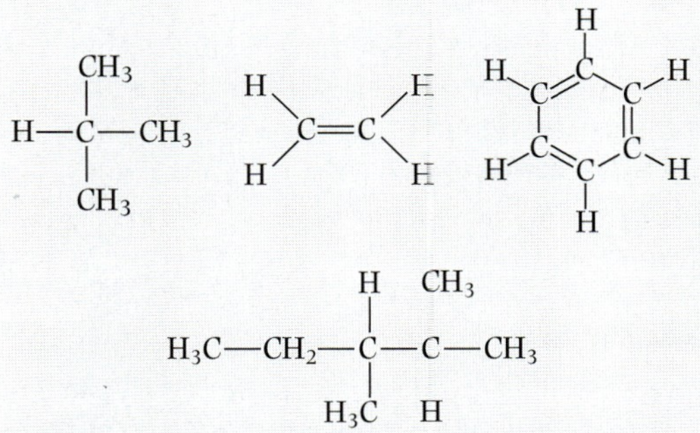

15. What is the value of the C–C–C bond angle in propane?
16. Write the molecular and structural formulas of butane.

Energy Transfer

CONCEPT REVIEW

17. What is meant by the terms *system* and *surroundings*?
18. What is the difference between an *exothermic* and an *endothermic* process?
19. What is meant by the term *internal energy*?
20. Give two ways of increasing the internal energy of a gaseous system.
21. In each of the following processes describe the system and give the sign of q_{system}, combustion of methane; freezing water to make ice; touching a hot stove.
22. In each of the following processes describe the system and give the sign of q_{system}, driving an automobile; applying ice to a sprained ankle; cooking a hot dog.

PROBLEMS

23. Which of the following processes are exothermic and which are endothermic?
    a. A match burns.
    b. A molten metal solidifies.
    c. Rubbing alcohol feels cold on the skin.
24. Which of the following processes are exothermic and which are endothermic?
    a. Ice cubes solidify in the freezer.
    b. Water evaporates from a glass left on a windowsill.
    c. Dew forms on grass overnight.
25. What happens to the internal energy of a liquid at its boiling point when it vaporizes?
26. What happens to the internal energy of a gas when it expands (with no heat flow)?
27. How much $P\Delta V$ work does a gas system do on its surroundings at a constant pressure of 1.00 atm if the volume of gas triples from 250. mL to 750. mL? Express your answer in L · atm and joules.
28. An expanding gas does 150. J of work on its surroundings at a constant pressure of 1.01 atm. If the gas initially occupied 68 mL, what is the final volume of the gas?

29. Calculate ΔE for the following situations:
    a. $q = 100$ J and $w = -50$ J
    b. $q = 6.2$ kJ and $w = 0.7$ J
    c. $q = -615$ J and $w = -325$ J
30. Calculate ΔE for a system that absorbs 726 kJ of energy from its surroundings and does 526 kJ of work on its surroundings.
31. Calculate ΔE for the combustion of a gas that releases 210 kJ of energy to its surroundings and does 65.5 kJ of work on its surroundings.
32. Calculate ΔE for a chemical reaction that produces 90.7 kJ of energy but does no work on its surroundings.
*33. The following reactions take place in a cylinder equipped with a movable piston at atmospheric pressure. Which reactions will result in work being done on the surroundings? Assume the system returns to an initial temperature of 110°C. Hint: The volume of a gas is proportional to n at constant T and P.
    a. $CH_4(g) + 2\,O_2(g) \rightarrow CO_2(g) + 2\,H_2O(g)$
    b. $C_3H_8(g) + 5\,O_2(g) \rightarrow 3\,CO_2(g) + 4\,H_2O(g)$
    c. $N_2(g) + 2\,O_2(g) \rightarrow 2\,NO_2(g)$

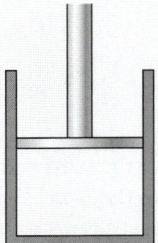

*34. In which direction will the piston described in Problem 33 have moved when the following reactions are carried out at atmospheric pressure inside the cylinder and after the system has returned to its initial temperature of 110°C?
    a. $N_2(g) + 3\,H_2(g) \rightarrow 2\,NH_3(g)$
    b. $C(s) + O_2(g) \rightarrow CO_2(g)$
    c. $C_2H_5OH(g) + 2\,O_2(g) \rightarrow 2\,CO_2(g) + 3\,H_2O(g)$

Enthalpy

CONCEPT REVIEW

35. What is meant by *enthalpy change*?
36. Describe the difference between an internal energy change ΔE and an enthalpy change ΔH.
37. Why is the sign of ΔH negative for an exothermic process?
38. What happens to the magnitude and sign of the enthalpy change when a process is reversed?

PROBLEMS

39. Adding Drano to a clogged sink causes the drain pipe to get warm. What is the sign of ΔH for this process?

40. Breaking a small pouch of water inside a larger bag containing ammonium nitrate activates chemical cold packs. What is the sign of ΔH for the process that takes place in the cold pack?

41. The stable form of oxygen under standard conditions is the diatomic molecule O_2. What is the sign of $\Delta H°$ for the following process?

$$O_2(g) \longrightarrow 2\,O(g)$$

42. Gypsum is the common name of calcium sulfate dihydrate ($CaSO_4 \cdot 2H_2O$). When gypsum is heated to 150°C, it loses most of the water in its formula and forms plaster of Paris ($CaSO_4 \cdot 0.5\,H_2O$).

$$2\,CaSO_4 \cdot 2H_2O(s) \longrightarrow 2\,CaSO_4 \cdot 0.5H_2O\,(s) + 3H_2O\,(g)$$

What is the sign of ΔH for making plaster of Paris?

43. A solid with metallic properties is formed when hydrogen gas is compressed under extremely high pressures. Predict the sign of the enthalpy change for the following reaction:

$$H_2(g) \longrightarrow H_2(s)$$

44. A simple "kitchen chemistry" experiment requires placing some vinegar in a soda bottle. A deflated balloon containing baking soda is stretched over the mouth of the bottle. Adding the baking soda to the vinegar starts the following reaction and inflates the balloon.

$$NaHCO_3(aq) + HCH_3CO_2(aq) \longrightarrow$$
$$NaCH_3CO_2(aq) + CO_2(g) + H_2O(l)$$

If the contents of the bottle are considered a system, is work done on the surroundings or on the system?

Heating Curves

CONCEPT REVIEW

45. What is the difference between *specific heat* and *heat capacity*?

46. What happens to the heat capacity of a material if its mass is doubled? Is the same true for the specific heat?

47. Are the heats of fusion and vaporization of a given substance usually the same?

48. An equal mount of heat is added to pieces of metal A and metal B having the same mass. Does the metal with the larger heat capacity reach the higher temperature?

*__49.__ Most automobile engines are cooled by water circulating through them and a radiator. However, the original Volkswagen Beetles had air-cooled engines. Why might car designers choose water cooling over air cooling?

*__50.__ The reactor-core cooling systems in some nuclear power plants use liquid sodium as the coolant. Sodium has a thermal conductivity of 1.42 J/cm · s · K compared with that of water (6.1×10^{-3} J/cm · s · K. The respective molar heat capacities are 28.28 J/mol · K and 75.31 J/mol · K. What advantage is there in using liquid sodium instead of water in this application?

PROBLEMS

51. How much heat is needed to raise the temperature of 100. g of water from 30°C to 100°C?

52. One hundred grams of water at 30°C absorbs 290 kJ of heat from a mountain climber's stove at an elevation where the boiling point of water is 93°C. Is this amount of energy sufficient to heat the water to its boiling point?

53. Use the data in the following table to sketch a heating curve for 1 mole of methanol. Start the curve at $-100°C$ and end it at 100°C.

Boiling point	65°C
Melting point	$-94°C$
ΔH_{vap}	37 kJ/mol
ΔH_{fus}	3.18 kJ/mol
Molar heat capacity	81.1 J/mol · C°

54. Use the data in the following table to sketch a heating curve for 1 mole of octane. Start the curve at $-75°C$ and end it at 150°C.

Boiling point	125.7°C
Melting point	$-56.8°C$
ΔH_{vap}	41.5 kJ/mol
ΔH_{fus}	20.7 kJ/mol
Molar heat capacity	254.6 J/mol · C°

55. During a strenuous workout, an athlete generates 2000. kJ of heat energy. What mass of water would have to evaporate from the athlete's skin to dissipate this much heat?

56. The same quantity of energy is added to 10.00-g pieces of gold, magnesium, and platinum, all initially at 25°C. The molar heat capacities of these three metals are 25.41 J/mol · C°, 24.79 J/mol · C°, and 25.95 J/mol · C°, respectively. Which piece of metal will have the highest final temperature?

*57. Exactly 10 mL of water at 25°C was added to a hot iron skillet. All of the water was converted into steam at 100°C. If the mass of the pan was 1.20 kg and the molar heat capacity of iron is 25.19 J/mol · C°, what was the temperature change of the skillet?

*58. A 20.0-g piece of iron and a 20.0-g piece of gold at 100°C were dropped into 1.00 L of water at 20°C. The molar heat capacities of iron and gold are 25.19 J/mol · C° and 25.41 J/mol · C°, respectively. What is the final temperature of the water and pieces of metal?

Estimating ΔH from Average Bond Energies

CONCEPT REVIEW

59. Why must the stoichiometry of a reaction be known in order to estimate the enthalpy change from bond energies?

60. Why must the structures of the reactants and products be known in order to estimate the enthalpy change of a reaction from bond energies?

61. Which fuel should have a higher fuel value: liquid butane or gaseous butane?

62. If the energy needed to break 2 moles of C=O double bonds is greater than the sum of the energies needed to break the O=O bonds in a mole of O_2 and vaporize a mole of carbon, why does the combustion of C release heat?

PROBLEMS

63. Use the average bond energies in Table 11.2 to estimate the enthalpy changes of the following reactions.
    a. $2 N_2(g) + 3 H_2(g) \rightarrow 2 NH_3(g)$
    b. $N_2(g) + 2 H_2(g) \rightarrow H_2NNH_2(g)$
    c. $2 N_2(g) + O_2(g) \rightarrow 2 N_2O(g)$

64. Use the average bond energies in Table 11.2 to estimate the enthalpy changes of the following reactions.
    a. $CO_2(g) + H_2(g) \rightarrow H_2O(g) + CO(g)$
    b. $N_2(g) + O_2(g) \rightarrow 2 NO(g)$
    c. $C(s) + CO_2(g) \rightarrow 2 CO(g)$
    (The heat of sublimation of graphite (C(s)) is 719 kJ/mol)

65. The combustion of carbon monoxide releases 283 kJ/mol of CO. If the O=O and C=O bond energies are 497 and 799 kJ/mol, respectively, what is the predicted bond energy of carbon monoxide?

$$2 CO(g) + O_2(g) \longrightarrow 2 CO_2(g) \quad \Delta H_{comb} = -566 \text{ kJ}$$

66. The reaction of H_2 with F_2 produces HF with $\Delta H = -269$ kJ/mol of HF. If the H–H and H–F bond energies are 436 and 565 kJ/mol, respectively, what is the F–F bond energy?

$$H_2(g) + F_2(g) \longrightarrow 2 HF(g)$$

67. Use the average bond energies in Table 11.2 to estimate how much less energy is released during the incomplete combustion of 1 mol of methane to carbon monoxide and water vapor instead of complete combustion to carbon dioxide and water vapor.

68. Use the average bond energies in Table 11.2 to estimate how much more energy is released by the reaction:

$$C(s) + O_2(g) \longrightarrow CO_2(g)$$

than by the reaction:

$$C(s) + \tfrac{1}{2}O_2(g) \longrightarrow CO(g)$$

*69. Estimate $\Delta H°$ for the following reaction, starting with the bond energies in Table 11.2:

$$4 NH_3(g) + 7 O_2(g) \longrightarrow 4 NO_2(g) + 6 H_2O(g)$$

*70. The value of $\Delta H°_{rxn}$ for the reaction:

$$2 H_2S(g) + 3 O_2(g) \longrightarrow 2 SO_2(g) + 2 H_2O(g)$$

is −1036 kJ. Estimate the energy of the bonds in SO_2 from the appropriate average bond energies in Table 11.2.

Calorimetry

CONCEPT REVIEW

71. Why is it necessary to know the heat capacity of a calorimeter?
72. Could an endothermic reaction be used to measure the heat capacity of a calorimeter?
73. If we replace the water in a bomb calorimeter with another liquid, do we need to redetermine the heat capacity of the calorimeter?
74. When measuring the heat of combustion of a very small amount of material, would you prefer to use a calorimeter having a small or a large heat capacity?

PROBLEMS

75. Calculate the heat capacity of a calorimeter if the combustion of 5.000 g of benzoic acid led to a temperature increase of 16.397 °C.
76. Calculate the heat capacity of a calorimeter if the combustion of 4.663 g of benzoic acid led to an increase in temperature of 7.149 °C.
77. The complete combustion of 1.200 g of cinnamaldeyde (C_9H_8O, one of the compounds in cinnamon) in a bomb calorimeter ($C_{calorimeter}$ = 3.640 kJ/°C) produced an increase in temperature of 12.79 Celsius degrees. Calculate the molar enthalpy of combustion of cinnamaldehyde, ΔH_{comb} (kilojoules per mole of cinnamaldehyde).
78. The aromatic hydrocarbon cymene ($C_{10}H_{14}$) is found in nearly 100 spices and fragrances including coriander, anise, and thyme. The complete combustion of 1.608 g of cymene in a bomb calorimeter ($C_{calorimeter}$ = 3.640 kJ/°C) produced an increase in temperature of 19.35 Celsius degrees. Calculate the molar enthalpy of combustion of cymene, ΔH_{comb} (kilojoules per mole of cymene).
79. Phthalates used as plasticizers in rubber and plastic products are believed to act as hormone mimics in humans. The ΔH_{comb} of dimethylphthalate ($C_{10}H_{10}O_4$) is 4685 kJ/mol. 1.00 g of dimethylphthalate is combusted in a calorimeter with $C_{calorimeter}$ = 7.854 kJ/C° at 20.215°C. What is the final temperature of the calorimeter?
80. The flavor of anise is due to anethole, a compound with the molecular formula $C_{10}H_{12}O$. The ΔH_{comb} of anethole is 5541 kJ/mol. If 0.950 g of anethole is combusted in a calorimeter with $C_{calorimeter}$ = 7.854 kJ/C° at 20.611°C, what is the final temperature of the calorimeter?

Enthalpies of Formation and Reaction

CONCEPT REVIEW

81. Oxygen and ozone are examples of allotropes. Are the standard heats of formation of allotropes such as oxygen and ozone the same?
82. Why are the standard heats of formation of elements in their standard states assigned a value of zero?

PROBLEMS

83. For which of the following reactions does ΔH_{rxn} represent an enthalpy of formation?
    a. $C(s) + O_2(g) \rightarrow CO_2(g)$
    b. $CO_2(g) + C(s) \rightarrow 2\ CO(g)$
    c. $CO_2(g) + H_2(g) \rightarrow H_2O(g) + CO(g)$
    d. $2\ H_2(g) + C(s) \rightarrow CH_4(g)$
84. For which of the following reactions does ΔH_{rxn} also represent an enthalpy of formation?
    a. $2\ N_2(g) + 3\ O_2(g) \rightarrow 2\ NO_2(g) + 2\ NO(g)$
    b. $N_2(g) + O_2(g) \rightarrow 2\ NO(g)$
    c. $2\ NO_2(g) \rightarrow N_2O_4(g)$
    d. $N_2(g) + 2\ O_2(g) \rightarrow 2\ NO_2(g)$
85. Use heats of formation to calculate the enthalpy of reaction for the following methane-generating reaction of methanogenic bacteria:

    $$4\ H_2(g) + CO_2(g) \longrightarrow CH_4(g) + 2\ H_2O(g)$$

86. Use heats of formation to calculate the enthalpy of reaction for the following methane-generating reaction of methanogenic bacteria:

    $$4\ CH_3NH_2(g) + 2\ H_2O(l) \longrightarrow 3\ CH_4(g) + CO_2(g) + 4\ NH_3(g)$$

87. Ammonium nitrate decomposes to N_2O and water at temperatures between 250 and 300°C. Write a balanced chemical reaction describing the decomposition of ammonium nitrate, and calculate the standard heat of reaction by using the appropriate heats of formation.

88. Above 300°C, ammonium nitrate decomposes to N_2, O_2, and H_2O. Write a balanced chemical reaction describing the decomposition of ammonium nitrate, and determine the heat of reaction by using the appropriate heats of formation.

89. The bomb that destroyed the Murrow Federal Office Building in Oklahoma City in April 1995 was constructed from ordinary materials: fertilizer (ammonium nitrate) and fuel oil (a mixture of long-chain hydrocarbons, similar to decane, $C_{10}H_{22}$). Determine the enthalpy change of the following explosive reaction by using the appropriate heats of formation ($\Delta H_{f,C_{10}H_{22}} = 249.7$ kJ/mol):

$$3\ NH_4NO_3(s) + C_{10}H_{22}(l) + 14\ O_2(g) \longrightarrow$$
$$3\ N_2(g) + 17\ H_2O(g) + 10\ CO_2(g)$$

*90. A highly explosive compound is trinitrotoluene, or TNT. The thermal decomposition of TNT is described by the following chemical equation:

$$2\ C_7H_5N_3O_6(s) \longrightarrow 12\ CO(g) + 5\ H_2(g) + 3\ N_2(g) + 2\ C(s)$$

If ΔH_f° for TNT is $-10{,}153$ kJ/mol, how much TNT is needed to equal the explosive power of 1 mol of ammonium nitrate in Problem 89?

Fuel Values

CONCEPT REVIEW

91. What is meant by *fuel value*?
92. What are the units of fuel values?
93. How are fuel values calculated from molar heats of combustion?
94. Is the fuel value of liquid propane the same as that of propane gas?

PROBLEMS

95. If all the energy obtained from burning 1.00 pound of propane were used to heat water, how many kilograms of water could be heated from 20°C to 45°C?
96. A 1995 article in *Discover* magazine on world-class sprinters contained the following statement: "In one race, a field of eight runners releases enough energy to boil a gallon jug of ice at 0°C in ten seconds!" How much "energy" do the runners release in 10 seconds? Assume that the ice weighs 128 ounces.
97. Lightweight camping stoves typically use "white gas," a mixture of C_5 and C_6 hydrocarbons.
    a. Calculate the fuel value of C_5H_{12}, given that $\Delta H_{comb} = 3535$ kJ/mol.
    b. How much heat is released during the combustion of 1.00 kg of C_5H_{12}?
    c. How many grams of C_5H_{12} must be combusted to heat 1.00 kg of water from 20°C to 90°C? Assume that all the heat released during combustion is used to heat the water.
98. The heavier hydrocarbons in "white gas" are hexanes (empirical formula C_6H_{14}).
    a. Calculate the fuel value of C_6H_{14}, given that $\Delta H_{comb} = 4163$ kJ/mol.
    b. How much heat is released during the combustion of 1.00 kg of C_6H_{14}?
    c. How many grams of C_6H_{14} are needed to heat 1.00 kg of water from 25°C to 85°C? Assume that all of the heat released during combustion is used to heat the water.
    *d. If "white gas" is 25% C_5 hydrocarbons and 75% C_6 hydrocarbons, how many grams of white gas are needed to heat 1.00 kg of water from 25°C to 85°C?
99. The combustion of methane produces less energy per mole than the combustion of hexane (see Problem 98), but more energy per gram. However, methane is a gas at room temperature with a density of 0.66 g/L, whereas hexane is a liquid with a density of 0.66 g/mL. Calculate the "energy densities" of both fuels expressed in kJ/L.
100. Calculate the densities of the hydrocarbon gases in Table 11.4 at STP (see Section 8.3) and then convert the fuel values in Table 11.4 into energy densities (kJ/L).

Hess's Law

CONCEPT REVIEW

101. How is Hess's law consistent with the law of conservation of energy?

102. Why is the heat of formation of $CO(g)$ difficult to measure experimentally?

103. Explain how the use of $\Delta H_f°$ to calculate $\Delta H_{rxn}°$ is an example of Hess's law.

104. Explain how the calculation of lattice energies in Chapter 9 is an example of Hess's law.

PROBLEMS

105. How can the first two of the following reactions be combined to obtain the third reaction?

$$CO(g) + \tfrac{1}{2}O_2(g) \longrightarrow CO_2(g)$$
$$C(s) + O_2(g) \longrightarrow CO_2(g)$$
$$C(s) + \tfrac{1}{2}O_2(g) \longrightarrow CO(g)$$

106. How can the heat of formation of $CO(g)$ be calculated from the heat of formation of $CO_2(g)$ and the heat of combustion of $CO(g)$?

107. Calculate the enthalpy of formation of $SO_2(g)$ from the enthalpy changes of the following reactions:

$2 SO_2(g) + O_2(g) \longrightarrow 2 SO_3(g)$ $\Delta H = -196$ kJ
$2 S(s) + 3 O_2(g) \longrightarrow 2 SO_3(g)$ $\Delta H = -790$ kJ
$S(s) + O_2(g) \longrightarrow SO_2(g)$ $\Delta H_f° = ?$

108. The destruction of ozone by chlorofluorocarbons can be described by the following reactions:

$ClO(g) + O_3(g) \longrightarrow Cl(g) + 2 O_2(g)$ $\Delta H° = -29.90$ kJ
$2 O_3(g) \longrightarrow 3 O_2(g)$ $\Delta H° = 24.18$ kJ

Determine the value of $\Delta H°$ for

$Cl(g) + O_3(g) \longrightarrow ClO(g) + O_2(g)$ $\Delta H° = ?$

109. The mineral spodumene ($LiAlSi_2O_6$) exists in two crystalline forms, called α and β. Use Hess's law and the following information to calculate $\Delta H_{rxn}°$ for the conversion of α-spodumene into β-spodumene.

$$Li_2O(s) + 2\, Al(s) + 4\, SiO_2(s) + \tfrac{3}{2}O_2(g) \longrightarrow$$
$$2\, \alpha-LiAlSi_2O_6(s)$$
$$\Delta H° = -1870.6 \text{ kJ}$$

$$Li_2O(s) + 2\, Al(s) + 4\, SiO_2(s) + \tfrac{3}{2}O_2(g) \longrightarrow$$
$$2\, \beta-LiAlSi_2O_6(s)$$
$$\Delta H° = -1814.6 \text{ kJ}$$

110. Use the following data to determine whether the conversion of diamond into graphite is exothermic or endothermic.

$C_{diamond}(s) + O_2(g) \longrightarrow CO_2(g)$ $\Delta H = -395.4$ kJ
$2\, CO_2(g) \longrightarrow 2\, CO(g) + O_2(g)$ $\Delta H = 566.0$ kJ
$2\, CO(g) \longrightarrow C_{graphite}(s) + CO_2(g)$ $\Delta H = -172.5$ kJ
$C_{diamond}(s) \longrightarrow C_{graphite}(s)$ $\Delta H = ?$ kJ

111. Determine ΔH for the decomposition of NOCl from the following data:

$\tfrac{1}{2}N_2(g) + \tfrac{1}{2}O_2(g) \longrightarrow NO(g)$ $\Delta H = 90.3$ kJ
$NO(g) + \tfrac{1}{2}Cl_2(g) \longrightarrow NOCl(g)$ $\Delta H = -38.6$ kJ
$2\, NOCl(g) \longrightarrow N_2(g) + O_2(g) + Cl_2(g)\; \Delta H = ?$

Which of the ΔH_{rxn} values represent heats of formation?

112. The heat of decomposition of NO_2Cl is -114 kJ. Use the following data to calculate the heat of formation of NO_2Cl from N_2, O_2, and Cl_2.

$NO_2Cl(g) \longrightarrow NO_2(g) + \tfrac{1}{2}Cl_2(g)$ $\Delta H = -114$ kJ
$\tfrac{1}{2}N_2(g) + O_2(g) \longrightarrow NO_2(g)$ $\Delta H_f = 33.2$ kJ
$\tfrac{1}{2}N_2(g) + O_2(g) + \tfrac{1}{2}Cl_2(g) \longrightarrow NO_2Cl(g)\; \Delta H_f = ?$

12 Energy
And Organic Chemistry

About 7% of the corn grown in the United States is fermented to make ethanol, most of which is added to gasoline to reduce pollution from combustion and to add fuel value. Unfortunately, it takes about two-thirds of the energy in ethanol to make it this way: to grow and harvest the corn, to ferment it, and to isolate ethanol from the fermentation mixture by distillation.

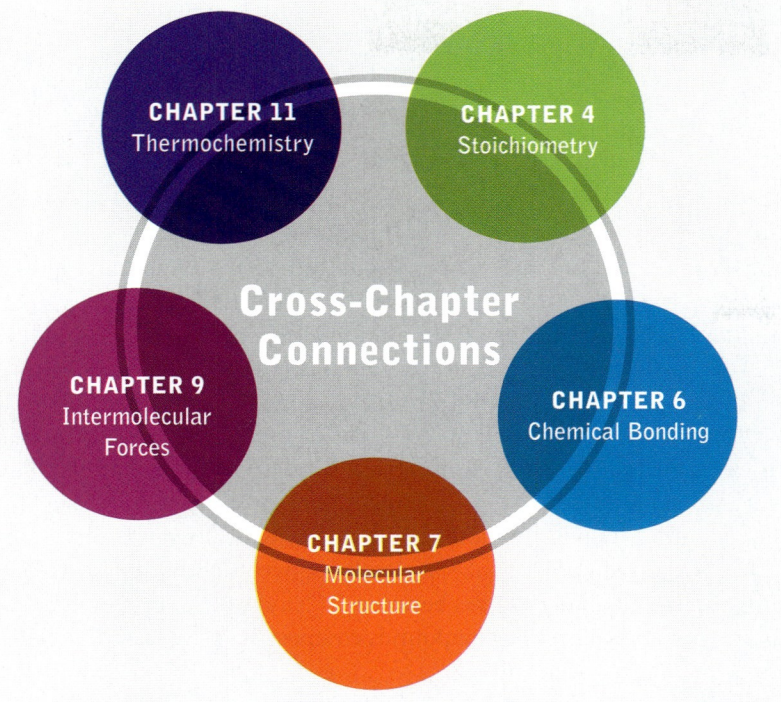

12.1	Petroleum refining: fractional distillation and Raoult's law
12.2	Alkanes in gasoline and structural isomerism
	Cycloalkanes
	Structural isomerism and octane ratings
	Rules for naming alkanes
12.3	Aromatic hydrocarbons
12.4	Alcohols, ethers, and reformulated gasoline
12.5	Carbohydrates
	Structures of sugar
	Condensation reactions
	Starch and cellulose
12.6	More fuels from biomass
	Carboxylic acids
	Amines
12.7	Coal
12.8	Hydrogen as fuel
12.9	Combustion analysis and elemental composition
12.10	Alkenes and alkynes

A Look Ahead

From an examination in Chapter 11 of how the chemical energy of volatile hydrocarbons in natural gas is converted into heat and light during combustion reactions, we turn to a discussion of the molecular structures and fuel values of more-complex fuels. The formulation of gasoline additives and extenders leads to a discussion of the structures of alcohols and ethers. Next, we examine the conversion of biomass into liquid and gaseous fuels, including ethanol and methane, in processes catalyzed by the biological activity of microorganisms. This examination will address the structure and reactivity of carbohydrates and other classes of organic compounds. We will also consider the use of combustion as an analytical tool for determining the composition of organic compounds.

Gasoline is a mixture of mostly C_5 to C_{12} hydrocarbons.

✓ **Fractional distillation** is a method of separating a mixture of compounds on the basis of their different boiling points.

 CONNECTION: Distillation as a method of separating mixtures is described in Section 1.1.

12.1 PETROLEUM REFINING: FRACTIONAL DISTILLATION AND RAOULT'S LAW

As noted in Chapter 11, carbon compounds have been accumulating in Earth's crust since the onset of biological activity. When decaying aquatic plant and animal matter is buried in rapidly accumulating sediments, its carbon and hydrogen content is shielded from oxidation to CO_2 and H_2O. The matter slowly decomposes. In the initial stages of decomposition, microorganisms that live under anaerobic (oxygen-free) conditions convert part of the material into methane gas and the rest into a complex mixture of insoluble organic compounds called *kerogen*. As kerogen is buried by more sediment, it is subjected to increasing pressure and temperature, which promotes additional chemical transformations. When these transformations take place at depths between about 1 and 4 km below Earth's surface and temperatures between 75° and 150°C, the products may include the complex mixture of compounds composed mostly of carbon and hydrogen that we know as crude oil.

Crude oil contains hydrocarbons with five or more carbon atoms in their molecular structures. (The C_1 to C_4 hydrocarbons are usually found in deposits of natural gas, though they are also found dissolved in deposits of crude oil—see the next Concept Test.) After crude oil has been extracted from the ground, it is separated at refineries into fractions, or *cuts*, on the basis of their volatilities in a process called **fractional distillation** (Figure 12.1). The most volatile fraction is used in gasoline. The next most volatile cut is used in cleaning solvents and as paint thinner; the next most volatile after that is used in kerosene and diesel fuel. Compounds with even higher boiling points are viscous liquids and are used as lubricants.

CONCEPT TEST: The nonpolar constituents of natural gas may dissolve in crude oil, which consists of mostly nonpolar liquid components. As crude oil is pumped to the surface from thousands of meters deep in Earth's crust, how will the solubility of natural gas be affected?

As noted in Chapter 9, the presence of nonvolatile solutes increases the boiling point of a pure solvent by lowering its vapor pressure. The relation between vapor pressure and concentration is described by Raoult's law. Does Raoult's law also apply to homogeneous mixtures of volatile compounds, such as crude oil? Indeed it does. The total vapor pressure of a solution of volatile compounds equals the sum of the vapor pressures of each pure component x (P_x) multiplied by the mole

> Does Raoult's law also apply to homogeneous mixtures of volatile compounds, such as crude oil?

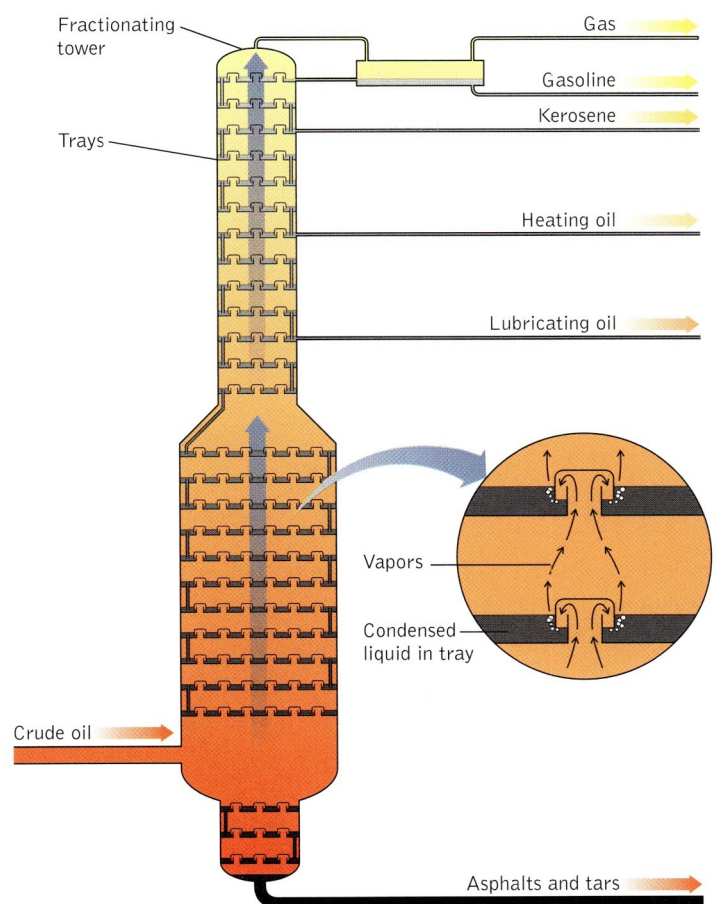

FIGURE 12.1 Refineries separate crude oil into different hydrocarbon fractions on the basis of their volatilities in a process known as fractional distillation. Vapors released from heated crude oil rise through a fractionating tower where they condense and collect in trays of progressively cooler temperatures. In this way, the most-volatile, lowest-boiling components reach the top of the tower and the less-volatile components condense in lower trays. The nonvolatile fraction of crude oil, including tar and asphaltic compounds, accumulates at the bottom of the tower.

fraction of that component in the solution (X_x) as shown in Equation 12.1 and Figure 12.2.

$$P_{\text{total}} = X_1 P_1^\circ + X_2 P_2^\circ + X_2 P_2^\circ + \cdots \quad (12.1)$$

CONNECTION: Vapor-pressure lowering due to the presence of nonvolatile solutes is described in Section 9.6.

SAMPLE EXERCISE 12.1: Calculate the vapor pressure of a solution prepared by dissolving 13 grams of heptane (C_7H_{16}) in 87 grams of octane (C_8H_{18})

FIGURE 12.2 The total vapor pressure of a mixture of volatile components equals the sum of the vapor pressures of its individual components. In this example, the total vapor pressure (purple line) of mixtures of n-octane (C_8H_{18}) and n-heptane (C_7H_{16}) at 98°C equals the sum of the vapor pressures of n-octane (blue line) and n-heptane (red line). According to Raoult's law, the vapor pressure of each component depends on the vapor pressure of the pure compound and its mole fraction (X) in the mixture (see Equation 12.1).

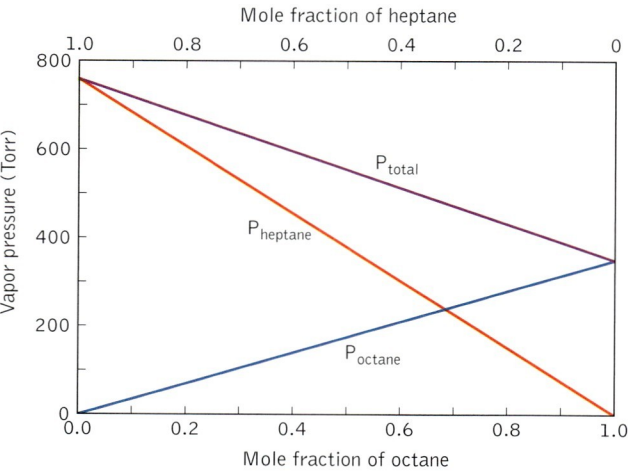

at 25°C. How enriched is the more-volatile component in the vapor phase compared with the liquid? The vapor pressures of octane and heptane at 25°C are 11 torr and 31 torr, respectively.

SOLUTION: We can use Equation 12.1 to determine the total vapor pressure of the solution from the vapor pressures of the ingredients after we have converted the composition of the solution into mole fractions:

$$87 \text{ g } C_8H_{18} \times \frac{1 \text{ mol } C_8H_{18}}{114 \text{ g } C_8H_{18}} = 0.76 \text{ mol } C_8H_{18}$$

$$13 \text{ g } C_7H_{14} \times \frac{1 \text{ mol } C_7H_{16}}{100 \text{ g } C_7H_{16}} = 0.13 \text{ mol } C_7H_{16}$$

The mole fraction of each component in the mixture is

$$X_{octane} = \frac{0.76 \text{ mol}}{(0.76 + 0.13) \text{ mol}} = 0.85$$

$$X_{heptane} = 1 - X_{octane} = 0.15$$

Using these mole fraction values and the vapor pressures of the two alkanes in Equation 12.1, we have

$$P_{total} = X_{heptane} P°_{heptane} + X_{octane} P°_{octane}$$

$$= 0.15(31 \text{ torr}) + 0.85(11 \text{ torr})$$

$$= 4.6 \text{ torr} + 9.4 \text{ torr} = 14 \text{ torr}$$

To calculate how enriched the vapor phase is in the more-volatile component (i.e., the component with the greater vapor pressure, heptane), we need to recall the message in Dalton's law of partial pressures (see Section 8.4): the partial pressure of a gas in a mixture of gases is proportional to its mole fraction in the mixture. In equation form, this principle becomes

$$P_x = X_x P_{total}$$

Therefore, the ratio of the mole fraction of heptane to that of octane in the vapor phase should be the ratio of their two vapor pressures:

$$\frac{X_{heptane,\ vapor}}{X_{octane,\ vapor}} = \frac{4.6\ \text{torr}}{9.4\ \text{torr}} = 0.49$$

The mole ratio of heptane to octane in the liquid mixture was

$$\frac{0.13\ \text{mol}}{0.76\ \text{mol}} = 0.17$$

Therefore, the vapor phase is enriched in heptane by a factor of

$$\frac{0.49}{0.17} = 2.9$$

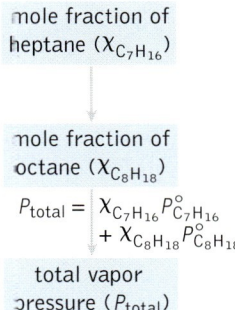

mole fraction of heptane ($X_{C_7H_{16}}$)

mole fraction of octane ($X_{C_8H_{18}}$)

$P_{total} = X_{C_7H_{16}} P°_{C_7H_{16}} + X_{C_8H_{18}} P°_{C_8H_{18}}$

total vapor pressure (P_{total})

DISCUSSION: This result illustrates how fractional distillation works. If the vapors of heptane and octane produced by a boiling mixture of the two are condensed and collected, they will be richer in heptane than the boiling mixture is. If this collected mixture is then redistilled, the products of that distillation will be even more enriched in heptane. In a fractional distillation apparatus, repeated distillation steps allow components with only slightly different volatilities to be separated from one another.

PRACTICE EXERCISE: Benzene is a trace component of gasoline. What is the mole ratio of benzene to octane in the vapor above a solution of 10% benzene and 90% octane by mass at 25°C? Given: The vapor pressures of octane and benzene at 25°C are 11 torr and 95 torr, respectively. (See Problems 11 and 12.)

Raoult's law works best when the strengths of solute–solvent, solvent–solvent, and solute–solute intermolecular interactions are similar. Under these conditions, a solution behaves as an **ideal solution** and Raoult's law is obeyed. However, strong intermolecular interactions between solute and solvent molecules tend to lower the vapor pressure of the solution and produce negative deviations from the vapor pressures predicted by Raoult's law, as shown in Figure 12.3A. On the other hand, relatively weak solute–solvent intermolecular interactions make it easier for molecules to vaporize, which leads to greater-than-predicted vapor pressures and positive deviations from Raoult's law, as shown in Figure 12.3B. A mixture of hydrocarbons such as octane and heptane is expected to behave like an ideal solution because intermolecular interactions between the components are all London forces acting on molecules of similar structure and size. However, the addition of methanol to gasoline, which is done to dissolve trace amounts of water that may be present in automobile gas tanks, creates a nonideal solution due to the strong hydrogen bonds between the methanol molecules (see Section 12.4).

 An **ideal solution** obeys Raoult's law.

 CONNECTION: Methanol molecules contain O—H bonds as described in Section 12.4; so they form hydrogen bonds between molecules (see Section 9.3).

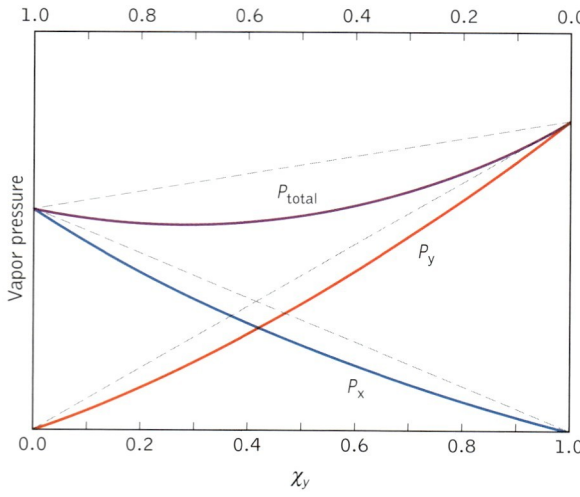 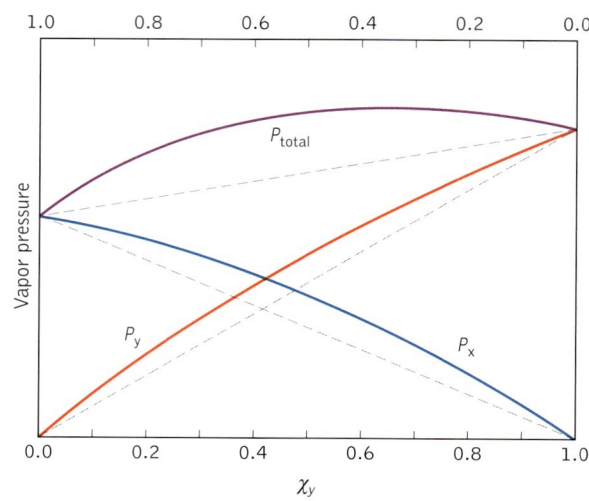

FIGURE 12.3 In a mixture of volatile substances x and y, P_x and P_y may deviate from the ideal behavior (the dashed lines in the two graphs) predicted by Raoult's law and Equation 12.1. A. In some mixtures, the total vapor pressure is lower than that predicted by Raoult's law. B. In other mixtures, the total vapor pressure is higher than that predicted by Raoult's law.

12.2 ALKANES IN GASOLINE AND STRUCTURAL ISOMERISM

Table 12.1 contains information about an important class of hydrocarbons found in crude oil. They are *normal* alkanes, abbreviated *n*-alkanes, in which each carbon atom is bound to no more than two other carbon atoms, forming a

TABLE 12.1 Some Properties of the C_5 to C_{12} *n*-Alkanes

Name	Condensed Structural Formula	Molar Mass (g/mol)	Melting Point (°C)	Boiling Point (°C)
Pentane	$CH_3CH_2CH_2CH_2CH_3$, or $CH_3(CH_2)_3CH_3$	72	−130	36
Hexane	$CH_3(CH_2)_4CH_3$	86	−95	69
Heptane	$CH_3(CH_2)_5CH_3$	100	−91	98
Octane	$CH_3(CH_2)_6CH_3$	114	−57	126
Nonane	$CH_3(CH_2)_7CH_3$	128	−51	151
Decane	$CH_3(CH_2)_8CH_3$	142	−30	174
Undecane	$CH_3(CH_2)_9CH_3$	156	−26	196
Dodecane	$CH_3(CH_2)_{10}CH_3$	170	−10	216

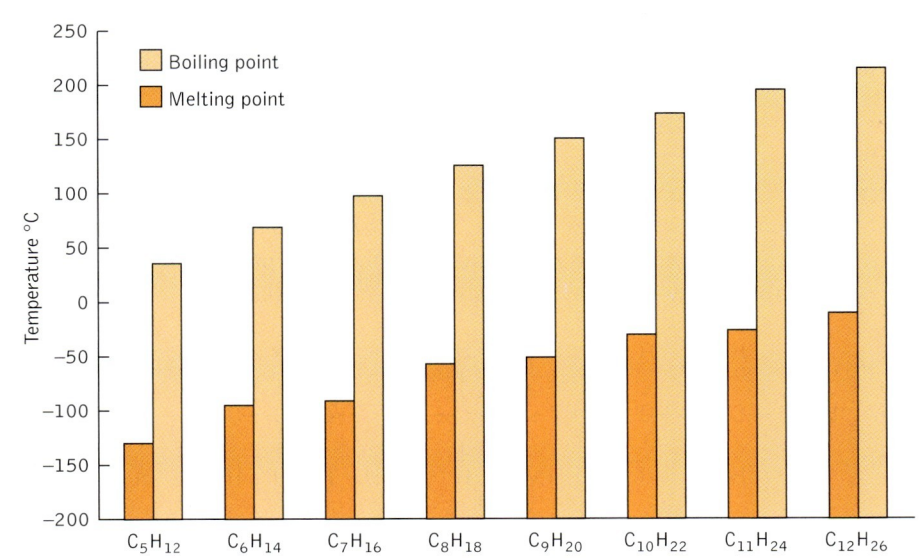

FIGURE 12.4 Melting points and boiling points of the C_5 to C_{12} n-alkanes increase with increasing molar mass.

chain of C–C bonds. The *n*-alkanes in Table 12.1 have boiling points that make them part of the crude-oil fraction known as gasoline. Note the similarities in their structures and the trends in their melting and boiling points. A group of compounds such as these with structures that differ by the number of methylene groups in their chains is called a homologous series. The physical properties of the members of a homologous series vary with changing molar mass in uniform and predictable ways. Certainly, the melting and boiling points of these compounds do so (Figure 12.4).

The names of the *n*-alkanes in Table 12.1 contain information about the number of carbon atoms in each of their molecules. The prefixes preceding the -*ane* ending may be familiar to you as the Greek prefixes that indicate, for example, the number of sides in a *penta*gon (5), or the number of field-and-track events in the *deca*thlon (10). These prefixes are also widely used for naming compounds with structures that include hydrocarbon chains, as we shall see throughout this chapter. Alkanes with chains of more than 10 carbon atoms often include combinations of prefixes in their name. For example, a molecule of octadecane, $CH_3(CH_2)_{16}CH_3$, has 18-carbon-atom chains, which corresponds to the double prefix denoting 8 + 10.

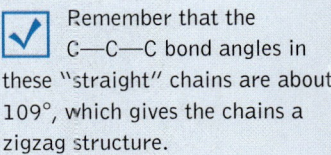

Remember that the C—C—C bond angles in these "straight" chains are about 109°, which gives the chains a zigzag structure.

CONNECTION: Trends in boiling points for the C_1 to C_4 alkanes in natural gas are listed in Table 11.1 (Section 11.3).

CONCEPT TEST: What kind of intermolecular interaction accounts for the trends in melting and boiling points shown in Figure 12.4?

SAMPLE EXERCISE 12.2: Draw the Lewis structure and condensed structural formulas of pentadecane.

SOLUTION: The Greek prefix *penta* means "five" and *deca* means "ten." Together, they indicate that the carbon chain in this normal alkane is 15 carbon atoms long. All alkanes have the generic formula C_nH_{2n+2}, which means that there must be $2n + 2$, or $2(15) + 2 = 32$, hydrogen atoms in each molecule of pentadecane. The Lewis structure for such a combination of carbon and hydrogen atoms with four single bonds to each carbon atom looks like this:

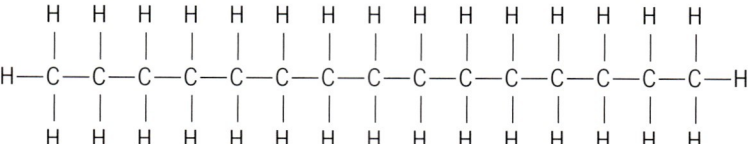

The condensed structural formula, which does not show the C—H bonds, looks like this:

$$CH_3-CH_2-CH_2-CH_2-CH_2-CH_2-CH_2$$
$$-CH_2-CH_2-CH_2-CH_2-CH_2-CH_2-CH_2-CH_3$$

The condensed structural formula can be further condensed by not showing the C—C bonds:

$$CH_3CH_2CH_2CH_2CH_2CH_2CH_2CH_2CH_2CH_2CH_2CH_2CH_2CH_2CH_3$$

or, simply,

$$CH_3(CH_2)_{13}CH_3$$

PRACTICE EXERCISE: Calculate the molar mass and determine the value of *x* in the condensed structural formula $(CH_3(CH_2)_xCH_3)$ of heptadecane. (See Problems 13 and 14.)

CONCEPT TEST: Octadecane, $CH_3(CH_2)_{16}CH_3$, is a solid at room temperature (its melting point is 28°C). On the basis of this information and that in Figure 12.4, predict whether pentadecane, $CH_3(CH_2)_{13}CH_3$, is a solid, liquid, or gas at room temperature (25°C).

Cycloalkanes

Gasolines are formulated from cuts of crude oil that boil below about 200°C. Their components are principally C_5 to C_{12} saturated hydrocarbons. Not all of these alkanes are straight-chain molecules. In some of their molecular struc-

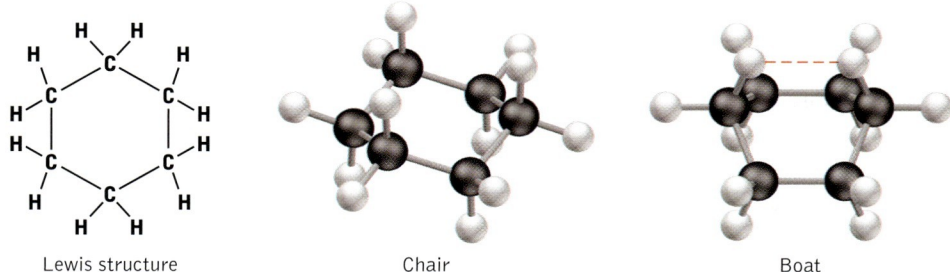

| Lewis structure | Chair | Boat |

FIGURE 12.5 In these three views of the molecular structure of cyclohexane, the Lewis structure shows the bonding pattern but not the true shape. The ball-and-stick models show that there are two different three-dimensional forms called *chair* and *boat*. Repulsion (shown by the dashed red line) between hydrogen atoms on opposite sides of the ring cause the boat form to be less stable than the chair form.

tures, the carbon atoms form rings rather than chains. These *cyclo*alkanes have two fewer hydrogen atoms per molecule than do the *n*-alkanes with the same number of carbon atoms (one from each methyl group at the two ends of the corresponding straight chain) and so have the general molecular formula C_nH_{2n}.

The most abundant cycloalkane in gasoline is cyclohexane, C_6H_{12}. If the shape of cyclohexane really were a planar ring, as suggested by its Lewis structure in Figure 12.5, the C–C–C bond angles would be 120 degrees. (Each inside angle of a regular polygon of *n* sides is

$$\left(\frac{n-2}{n} \times 180\right)°$$

The C–C–C bond angles in cyclohexane are actually much closer to the 109 degrees that we would predict for end-on overlap of sp^3 hybrid orbitals on adjacent carbon atoms. To achieve this geometry, the ring is puckered instead of flat. There are two possible pucker patterns, as shown in Figure 12.5: one is called the *chair* form, and the other is called the *boat* form. Note that, in the boat form, two hydrogen atoms across the ring from each other are pointed toward each other. Repulsion between these two atoms causes this form to be less stable than the chair form, and so the chair form is preferred.

 CONNECTION: Orbital hybridization schemes that account for the geometries of hydrocarbon molecules are described in Section 7.4.

CONCEPT TEST: The most stable cycloalkane ring structures are those that allow end-on overlap of sp^3 hybrid orbitals and C–C–C bond angles in the ring of about 109 degrees. On the basis of this geometry, rank the following cyclic alkanes in order of decreasing stability: cyclopropane (C_3H_6), cyclobutane (C_4H_8), and cyclopentane (C_5H_{10}).

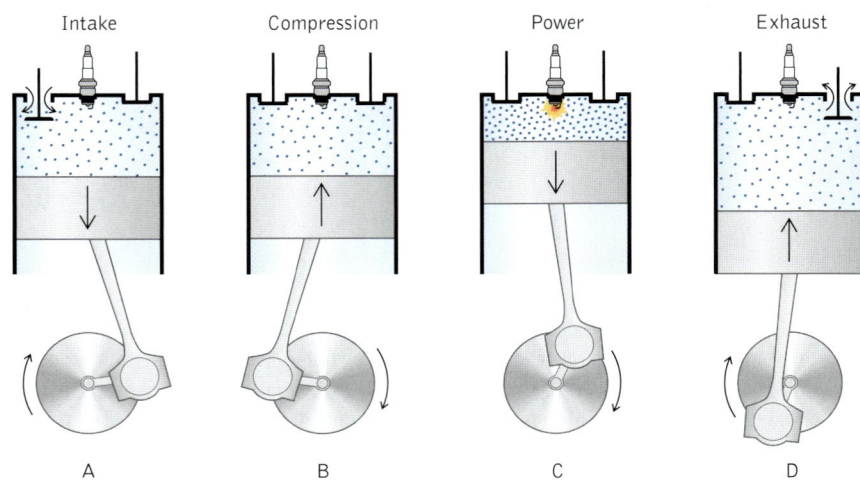

FIGURE 12.6 The combustion process in most internal combustion engines comprises four stages. A. A mixture of gasoline and air enters the combustion cylinder. B. Upward motion of a piston compresses the mixture. C. A spark ignites the mixture and the energy released during rapid combustion drives the piston down. D. Upward motion of the piston pushes exhaust gases out of the cylinder.

Structural isomerism and octane ratings

In the combustion process in a gasoline-fueled engine (Figure 12.6), motion of the engine's pistons compresses a mixture of air and gasoline vapor into a small volume just before a spark ignites the mixture. During this compression stage, the piston does $P\Delta V$ work on the fuel and air gases, and they heat up. If the mixture gets hot enough, it may ignite before the piston reaches the top of its stroke, creating a pinging or knocking sound as the piston is slammed backward.

CONNECTION: $P\Delta V$ work is described in Sections 11.5 and 11.6.

The octane number of gasoline is a measure of its ignition temperature and its ability to resist engine knock. A gasoline's octane rating is determined by comparing its antiknock properties with a compound sometimes called isooctane. The prefix *iso* means that not all the carbon atoms in the molecule form a straight chain. In isooctane, the longest straight chain in the molecule is only five carbon atoms long. The other three carbon atoms are in methyl groups, shown in red in the following Lewis structure, that are attached to the sides of the chain.

The structure of isooctane is not the same as that of *n*-octane, yet the two compounds have the same molecular formula: C_8H_{18}. It turns out that all alkanes with more than three carbon atoms per molecule can have more than one molecular structure. Molecules with the same formula but different structures are called **structural isomers.** Consider butane, the fuel used in disposable lighters. There are two ways to arrange the bonds between carbon atoms in butane. They can be linked together in a straight chain (the structure on the left) or arranged in a branched structure with one CH_3 substituent, or side group, on a three-carbon chain (the structure on the right):

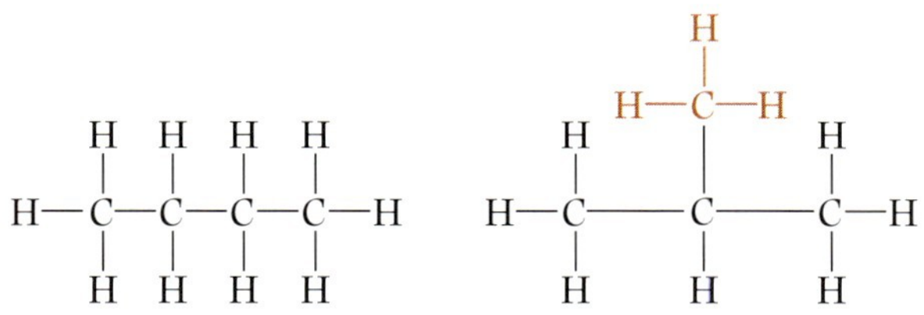

> ✓ **Structural isomers** are molecules having the same molecular formula but different bonding arrangements of their atoms.

Because both structures might be called butane, we need to add more information to that name to distinguish between them. The next subsection describes an approach for naming alkanes on the basis of their molecular structures. We will use similar approaches to name other classes of organic molecules.

Rules for naming alkanes

1. Locate the longest carbon chain, or *backbone*, of the molecule.
2. Assign a base name that corresponds to the number of carbon atoms in the longest chain. Thus, the name of the isomer of butane in the left-hand structure shown above should rightly be *butane* because there are four carbon atoms in its backbone. The isomer depicted in the right-hand structure has only a three-carbon backbone, and so it has the base name *propane*.
3. The name of a linear, or straight-chain, alkane is simply its base name, though the names of these *normal* alkanes may be preceded by *n-*. Thus, butane is sometimes referred to as *n*-butane.
4. The name of an isomer with substituents bonded to one or more middle carbon atoms includes the name of the backbone with a prefix that includes:
   (a) the number(s) of the carbon atom(s) to which the substituents are bonded;
   (b) the number of substituents (if there is more than one) of each type using the appropriate prefix—for example, *di* for 2, *tri* for 3, and *tetra* for 4; and
   (c) the names of the substituents.

Applying these rules to the right-hand isomer of butane, we find that one CH_3 (methyl) group is bonded to the second (2-) carbon atom in the propane backbone of the molecule. Therefore, the proper name for the right-hand structure is 2-methylpropane. Note that the middle carbon of the propane chain is number 2 no matter from which end we start numbering. This is not the case for isooctane (Figure 12.7). Applying the preceding rules to isooctane, we note that there are a total of three methyl-group substituents bonded to a five-carbon, or pentane,

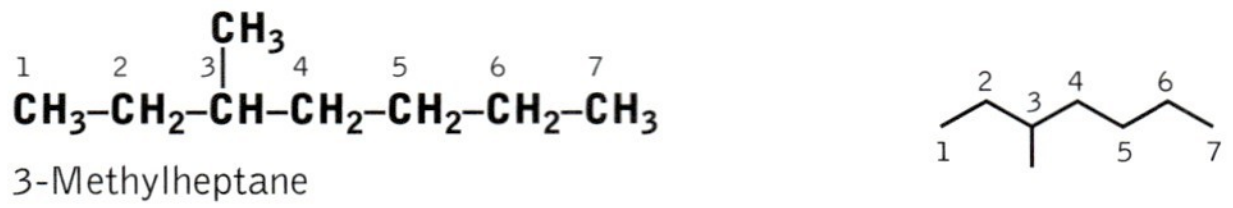

FIGURE 12.7 Octane, a main ingredient in gasoline, has many isomers including the three shown here. Note that the numbering pattern for the longest carbon chain in each compound starts at the end that provides the lowest numbers for the substituent bonding sites. The corresponding line structures in the column on the right provide a better representation of the orientation of the C–C bonds in these molecules.

$$\underset{\text{3-Methylheptane}}{\overset{\overset{\overset{CH_3}{|}}{}}{CH_3-CH_2-\underset{}{\overset{}{C}H}-CH_2-CH_2-CH_2-CH_3}}$$
1 2 3 4 5 6 7

$$\underset{\text{2,3-Dimethylhexane}}{CH_3-\overset{\overset{CH_3}{|}}{C}H-\overset{\overset{CH_3}{|}}{C}H-CH_2-CH_2-CH_3}$$
1 2 3 4 5 6

$$\underset{\text{2,2,4-Trimethylpentane (iso octane)}}{CH_3-\overset{\overset{CH_3}{|}}{\underset{\underset{CH_3}{|}}{C}}-CH_2-\overset{\overset{CH_3}{|}}{C}H-CH_3}$$
1 2 3 4 5

> An **alkyl group** is a saturated hydrocarbon substituent group in an organic molecule.

backbone. Two methyl groups are bonded to the second carbon atom from the left, and a third methyl group is bonded to the fourth carbon atom from the left. The name of this compound based on the naming rules is 2,2,4-trimethylpentane.

Had we numbered the carbon atoms starting from the right end of the pentane chain in the structure for isooctane, we would have ended up with the name 2,4,4-trimethylpentane. Is this name an acceptable alternative to 2,2,4-trimethylpentane? The answer is no, because we must follow another rule in naming organic compounds: number the longest chain by starting at the end that gives the greatest number of low values for substituent bonding sites.

A molecule of *methane*, CH_4, minus one hydrogen atom is called a *methyl*, CH_3, substituent group (see Section 11.3). This name change is another recurring part of chemical nomenclature: we replace the *-ane* ending of a parent alkane with *-yl* when the alkane loses a hydrogen atom. These substituent groups are really molecular fragments and are called **alkyl groups.** A few of the more common ones are shown in Table 12.2. The second four-carbon group is named *sec*-butyl because the connecting carbon is a *secondary* carbon, that is, it is bonded to two other carbon atoms. In the last of the butyl groups in Table 12.2, the connecting carbon atom is bonded to three other carbon atoms, making it a *tertiary* carbon and making the group *tert*-butyl. There are tertiary carbon atoms in two other groups in Table 12.2. Their names, isopropyl and isobutyl, carry the *iso* prefix to indicate branched isomers of octane and other alkanes.

CONCEPT TEST: There are more octane isomers than those shown in Figure 12.7. Draw Lewis structures for two more of them and name them.

TABLE 12.2 Some Common Alkyl Groups

Number of Carbon Atoms	Structure	Name
1	CH₃—	Methyl
2	CH₃CH₂—	Ethyl
3	CH₃CH₂CH₂—	Propyl
	CH₃CHCH₃ (with bond)	Isopropyl
4	CH₃(CH₂)₂CH₂—	Butyl
	CH₃CHCH₂CH₃ (with bond)	sec-Butyl
	(CH₃)CHCH₂—	Isobutyl
	(CH₃)₃C—	tert-Butyl

TABLE 12.3 Octane Ratings of Some Alkanes in Gasoline

Formula	Name	Octane Rating
C₅H₁₂	Pentane	66
	2-Methylbutane	104
C₆H₁₄	Hexane	22
	2-Methylpentane	82
C₇H₁₆	Heptane	0
	1-Methylhexane	22
	2,2-Dimethylpentane	89
	2,2,3-Trimethylbutane	112
C₈H₁₈	Octane	−16
	2,2,4-Trimethylpentane	100

Table 12.3 lists the octane numbers of isooctane and several other hydrocarbons found in gasoline. Note that branched alkanes have higher antiknock numbers than do straight-chain hydrocarbons. Unfortunately, they are not as abundant in the gasoline fraction of most crude oils as normal alkanes. Fortunately, branched alkanes can be synthesized from normal alkanes in chemical reactions called *isomerization* reactions. The costs of synthesizing these compounds contributes to the higher prices for high-octane gasoline.

SAMPLE EXERCISE 12.3: The line structure of cyclohexane, in which each short line represents a C—C single bond and each carbon atom is bonded to the maximum number of hydrogen atoms, may be drawn this way:

Which of the following line structures represent compounds that are structural isomers of cyclohexane?

a. b. c. d. e. f.

SOLUTION: Each of the six corners in the line structure of cyclohexane is occupied by a carbon atom bonded to two other carbon atoms and to two hydrogen atoms. Therefore, the molecular formula of cyclohexane is $(CH_2)_6$ or C_6H_{12}. To be a structural isomer of cyclohexane, another compound must have the same molecular formula, but have a different bonding pattern.

All six possibilities have different bonding patterns; so we need to determine which of them has the molecular formula C_6H_{12}. It turns out that all six compounds have six carbon atoms per molecule; so the key variable is the number of hydrogen atoms. To inventory the hydrogen atoms we need to keep in mind that each carbon atom in these structures is bonded to four other atoms; so those bonded to only one other carbon atom are in methyl groups (—CH_3), those bonded to two other carbon atoms are in methylene groups (—CH_2—) and those bonded to three other carbon atoms are bonded to only one hydrogen atom. Assigning the number (in red) of hydrogen atoms to each carbon atom in these structures we have:

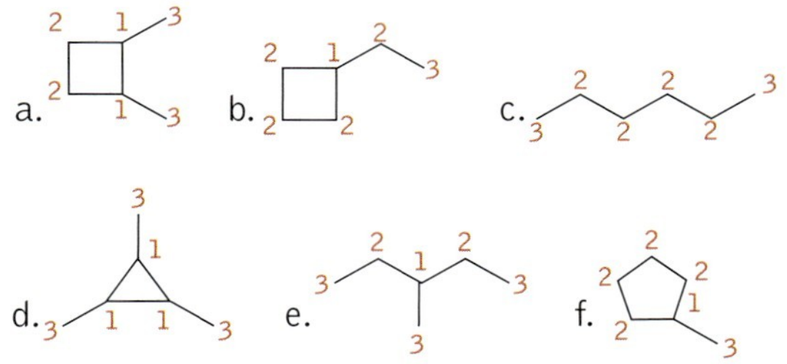

Summing the number of hydrogen atoms in each structure we find that structures a., b., d. and f. have 12 and so have the molecular formula C_6H_{12}. They are structural isomers of cyclohexane. Structures c. and e. (the two that are not cyclic alkanes) have 14 hydrogen atoms per molecule and the molecular formula C_6H_{14}. These compounds are not structural isomers of cyclohexane.

PRACTICE EXERCISE: Draw all the structural isomers of cyclobutane that contain only C—C single bonds. (See Problems 17 and 18.)

SAMPLE EXERCISE 12.4: Name the compounds represented by line structures a., b., and c., in Sample Exercise 12.3.

SOLUTION: Structure a. is based on a 4-carbon ring; so the base of its name is *cyclobutane*. There are two methyl groups bonded to the cyclobutane structure. We need to indicate in the name that they are bonded to adjacent carbon atoms and not to two at opposite corners of the four-carbon ring. We do this by numbering the carbon atoms in the ring. To keep the numbers of those bonded to methyl groups as small as possible we number the top right-hand carbon atom C-1; then the one below it is C-2:

and the name of the compound is 1,2 dimethylcyclobutane.

Structure b. is also based on a 4-carbon ring; so the base of its name is *cyclobutane*. There is a single 2-carbon alkyl group, called an ethyl group (see Table 12.2), attached to a carbon atom in the ring. Since all the carbon atoms in the ring are equivalent, there is no need to identify which one is the site of attachment; so the name of the compound is simply ethylcyclobutane.

Structure c. represents a straight-chain, or *n*-, alkane with six carbon atoms in the chain. Therefore, the name of the compound is simply *n*-hexane.

PRACTICE EXERCISE: Name the compounds represented by line structures d., e., and f., in Sample Exercise 12.3. (See Problems 19 and 20.)

12.3 AROMATIC HYDROCARBONS

Among the minor components of gasoline that play an important role in increasing octane ratings are a class of compounds called **aromatic hydrocarbons**. As their class name implies, aromatic hydrocarbons have distinctive odors. However, "aromaticity" from a chemist's perspective is associated with their molecular structures. The most distinctive features of these structures are planar, hexagonal rings in which six carbon atoms are joined by a combination of σ and π bonds, as shown in Figure 12.8. The simplest of these molecules, with only one such ring, is benzene, C_6H_6. The Lewis structures of benzene are two resonance forms that represent extreme views, not realistic ones, of a bonding system in which π electrons can move between pairs of atoms (Figure 12.8A). In reality, there are no alternating C—C and C=C bonds in a benzene ring. Instead, there are circular clouds of π electrons above and below the ring, as shown in Figure 12.8C. These electrons reside in sideways overlapping $2p_z$ orbitals on the carbon atoms in the ring (Figure 12.8B). They are completely spread out, or delocalized, over all the carbon atoms in the ring. Delocalization leads to considerable stability. Indeed, benzene rings are among the most stable molecular structures.

Aromatic rings are part of a large number of compounds in which one or more substituent groups take the place of hydrogen atoms around the ring. For example, when a methyl group replaces a hydrogen atom, we have methylbenzene, also known by its common name, toluene:

If we wish to attach two methyl groups to a benzene ring, we have three attachment options; so there are three structural isomers of dimethylbenzene, also known as xylene. We distinguish between these isomers by numbering the car-

An **aromatic hydrocarbon** is a compound composed of carbon and hydrogen with one or more rings of carbon atoms in which there are delocalized π electrons that are spread over all the carbon atoms in the ring(s).

CONNECTION: Resonance in molecules is described in Section 6.6.

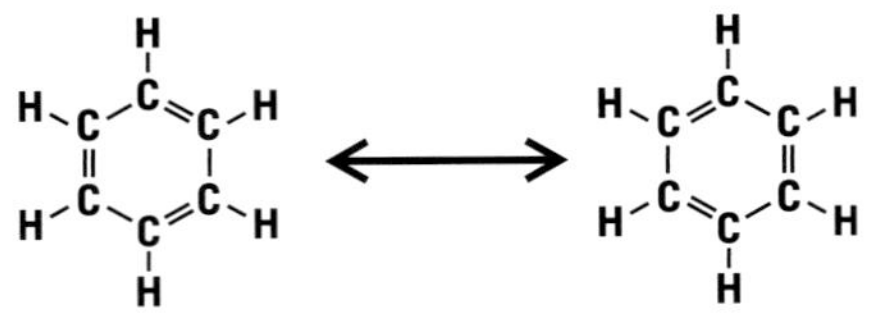

FIGURE 12.8 Aromatic compounds have planar, hexagonal ring structures. A. Lewis structures of the resonance forms of benzene feature alternating single and double C–C bonds. B. The σ bonds in benzene are formed by end-on overlap of the sp^2 hybrid orbitals of the carbon atoms with one another and with the $1s$ orbitals of hydrogen atoms. C. The three π bonds are spread evenly over all six carbon atoms, creating clouds of delocalized π electrons above and below the ring.

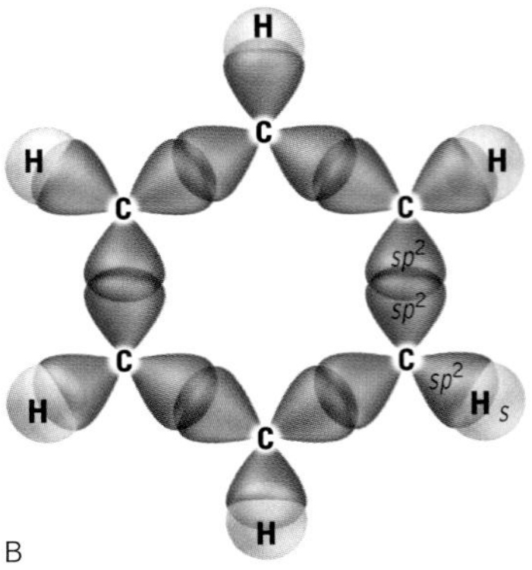

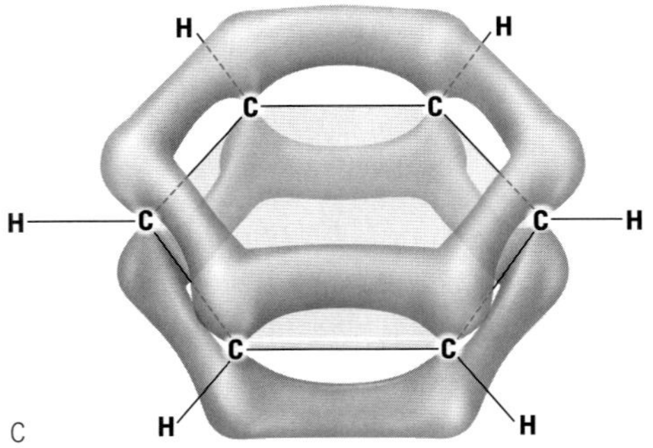

bon atoms in the ring consecutively, starting with the atom bonded to one of the methyl groups. Starting the numbering in this way results in the smallest values for the carbon atoms that are substituent bonding sites. This convention is analogous to numbering carbon atoms in a straight chain by starting from the end that minimizes the numbers on carbon atoms that are bonded to substituent groups. The three isomers of dimethylbenzene are

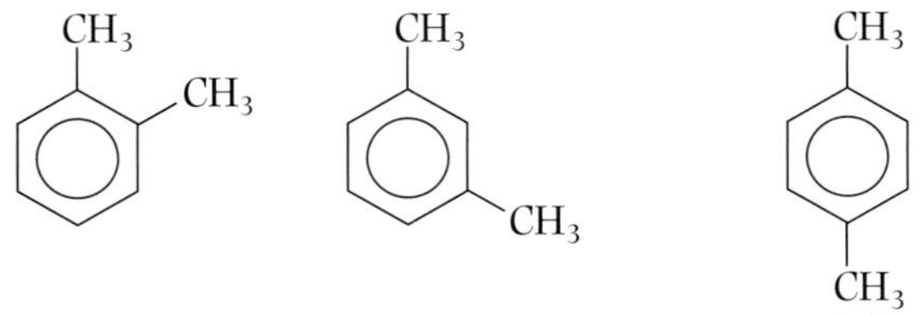

1,2-Dimethylbenzene (*o*-Xylene)   1,3-Dimethylbenzene (*m*-Xylene)   1,4-Dimethylbenzene (*p*-Xylene)

Sometimes chemists refer to these and similarly substituted aromatic compounds by using an older, but still widely used, naming system in which substituents in the 1,2- positions are said to be *ortho* (*o*-); those in the 1,3- positions are said to be *meta* (*m*-), and those in the 1,4- positions are said to be *para* (*p*-) to each other. The differences in these molecular structures are responsible for the sometimes small, but still measurable, differences in the physical properties of these isomers, as noted in Table 12.4.

12.4 Alcohols, Ethers, and Reformulated Gasoline

TABLE 12.4 Properties of the Isomers of Dimethylbenzene

Methyl-Group Bonding Sites	Molar Mass (g/mol)	Melting Point (°C)	Boiling Point (°C)	Density at 20°C (g/mL)
1,2	106.17	−25	144	0.8802
1,3	106.17	−48	139	0.8642
1,4	106.17	+13	138	0.8611

Petroleum also contains compounds with structures in which benzene rings are fused together by sharing one or more of their hexagonal sides.

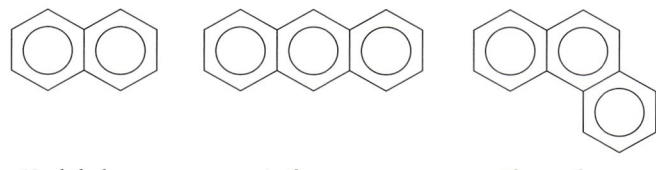

Naphthalene   Anthracene   Phenanthrene

These compounds with two or more aromatic rings belong to a class of compounds called *polycyclic aromatic hydrocarbons* (PAHs). Extensive delocalization of the π electrons among their rings make these structures particularly stable. They may be formed in the incomplete combustion of hydrocarbons and are found at particularly high concentrations in the soot from incinerators and diesel engines. When introduced into the environment, these compounds are among the most long-lived of hydrocarbons. Unfortunately, some of them are quite toxic and have been shown to cause cancer.

The incomplete combustion of diesel fuel produces unburned carbon soot that also contains polycyclic aromatic hydrocarbons.

12.4 ALCOHOLS, ETHERS, AND REFORMULATED GASOLINE

In January 1995, air-quality regulations went into effect in many U.S. cities that mandated reductions in atmospheric concentrations of pollutants associated with emissions from gasoline-fueled engines. The new regulations led to the widespread use of "reformulated" gasolines containing additives to promote complete combustion and boost octane ratings. These additives are not hydrocarbons, but rather other kinds of **organic compounds**. Like hydrocarbons, all organic compounds contain C–C or C–H bonds or both in their molecular structures. Unlike hydrocarbons, all other organic compounds contain additional elements.

Among the compounds added to gasolines to promote combustion are "oxygenated" additives that contain oxygen in their molecular structures in the form of either **alcohol** (–C–O–H) or **ether** (–C–O–C–) functional groups. A **functional group** is a part of a molecule's structure that is common to all members of its particular class of molecules. Functional groups are responsible for most of

> ✓ An **organic compound** is made of molecules in which carbon atoms are bonded to other carbon atoms or to hydrogen atoms or to both. These molecules may also contain atoms of other elements.
>
> An **alcohol** is an organic compound with a R–O–H functional group. In a phenol, the –O–H group is attached to an aromatic ring.
>
> An **ether** is an organic compound with a –R–O–R– functional group.
>
> A **functional group** is a combination of atoms bonded together in organic compounds that imparts particular properties to those compounds.

the chemical properties of organic compounds. Thus members of the same class of compounds have similar chemical properties and are used in similar ways.

CONCEPT TEST: Many alcohols and ethers are more soluble in polar solvents, such as water, than are hydrocarbons. Can you explain why?

One widely used gasoline additive in the 1990s was methyl *tert*-butyl ether (MTBE). As the name of this ether suggests, each molecule of MTBE consists of a methyl group and a *tert*-butyl group (see Table 12.3) joined by an oxygen atom as shown in Figure 12.9. Actually, organic chemists call this compound *tert*-butyl methyl ether because, according to the convention for naming ethers, the two groups joined by oxygen are named in alphabetical order, and so butyl should come before methyl (the *tert*- prefix does not count). Still, this compound is best known in environmental science by the abbreviation MTBE.

Unlike the nonpolar hydrocarbons in gasoline, MTBE is soluble in water. Consequently, gasoline spills, leakage from storage tanks, and releases from watercraft can produce extensive MTBE contamination of groundwater and drinking water supplies and have done so since the early 1990s. After toxicity tests showed MTBE to be a possible carcinogen (cancer-causing agent), several states, including California, where more than 25% of the world's production of MTBE was being used in gasoline, made plans to eliminate the use of MTBE as a gasoline additive. These plans raised the question of which additive(s) would replace MTBE. The most likely candidate is ethanol.

CONNECTION: The intermolecular forces responsible for the solubility of MTBE in water are discussed in Section 9.3.

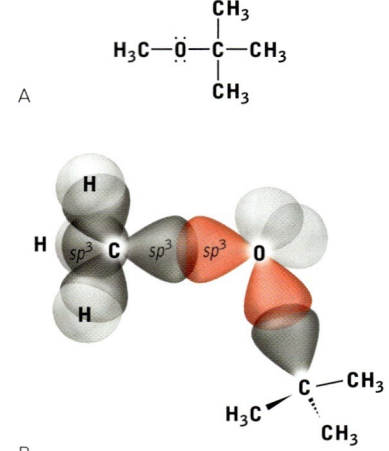

FIGURE 12.9 Two views of the molecular structure of MTBE. A. Each molecule of MTBE contains a methyl and a *tert*-butyl group linked by an ether (R–O–R) functional group. B. The bond angles in MTBE molecules are consistent with sp^3 hybridization of carbon and oxygen atoms.

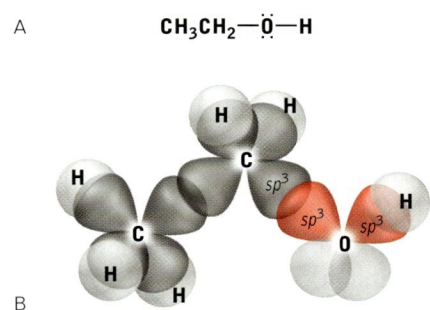

FIGURE 12.10 A. The molecular structure of ethanol contains an ethyl group bonded to a hydroxyl group, forming an alcohol (R–O–H) functional group. B. The bond angles in ethanol molecules are consistent with overlapping sp^3 orbitals of carbon and oxygen atoms.

12.4 Alcohols, Ethers, and Reformulated Gasoline

Ethanol (also called ethyl alcohol or grain alcohol) makes up about 1% of the gasoline used in the United States. In some regions, particularly the Midwest, a formulation of gasoline called gasohol contains 10% ethanol. The structure of ethanol resembles that of ethane, except that one of the hydrogen atoms has been replaced with an O–H group, as shown in Figure 12.10. Thus, the *-ane* ending on ethane (or the name of any alkane) changes to an *-ol* ending on the name of the corresponding alcoh*ol*.

Most of the ethanol used in gasoline is produced by fermentation of sugar derived from corn. Ethanol burns readily in air:

$$CH_3CH_2OH(l) + 3\ O_2(g) \longrightarrow 2\ CO_2(g) + 3\ H_2O(l) \qquad \Delta H° = -1367\text{ kJ}$$

Ethanol is added to gasoline, forming a mixture called gasohol, to promote complete combustion and provide additional fuel value.

CONCEPT TEST: Ethane is a gas at room temperature (b.p. = −89°C), but ethanol is a liquid (b.p. = 78.5°C). What intermolecular forces account for this large difference in the boiling points of these compounds?

Methanol, or methyl alcohol (CH_3OH), also is used as an additive to gasoline. It is manufactured by reacting carbon monoxide with hydrogen:

$$CO(g) + 2\ H_2(g) \longrightarrow CH_3OH(l)$$

Methanol burns according to the following thermochemical equation:

$$2\ CH_3OH(l) + 3\ O_2(g) \longrightarrow 2\ CO_2(g) + 4\ H_2O(l) \qquad \Delta H° = -1454\text{ kJ}$$

If we divide the absolute value of $\Delta H°$ by twice the molar mass of methanol (because the reaction consumes 2 moles of methanol), we get a fuel value for methanol of

$$\frac{1454\text{ kJ}}{(32.0\text{ g/mol} \times 2\text{ mol})} = 22.7\text{ kJ/g}$$

Let's compare the fuel value of methanol with that of octane. The thermochemical equation of the combustion of octane is

$$2\ C_8H_{18}(l) + 25\ O_2(g) \longrightarrow 16\ CO_2(g) + 18\ H_2O(l) \qquad \Delta H° = -1.091 \times 10^4\text{ kJ}$$

By dividing the absolute value of $\Delta H°$ by twice the molar mass of octane, we get a fuel value of

$$\frac{1.091 \times 10^4\text{ kJ}}{(114.2\text{ g/mol} \times 2\text{ mol})} = 47.8\text{ kJ/g}$$

Thus, the fuel value of methanol is a little less than half that of octane (and most of the other hydrocarbons in gasoline). Why is the fuel value of methanol much less than that of octane? The answer is in its molecular structure. The amount of energy released during combustion depends on the number of carbon atoms available for forming C=O bonds and the number of hydrogen atoms

Why is the fuel value of methanol much less than that of octane?

CONNECTION: The calculation of fuel values is described in Section 11.11.

available for forming O—H bonds as CO_2 and H_2O are produced. The presence of an oxygen atom in the structure of CH_3OH adds significantly to its mass (methanol is 50% oxygen by mass) but adds nothing to its fuel value. The oxygen content of a combustible substance essentially dilutes its energy value. The more "oxygenated" a fuel is, the lower its fuel value.

CONCEPT TEST: Predict, on the basis of their molecular formulas, which alcohol has the greater fuel value: methanol or ethanol.

The first internal combustion engines built in the 1870s burned pure ethanol, and early automobiles, including the Model T Ford, could be modified to run on either gasoline or pure ethanol. The use of alcohol as a gasoline additive and volume extender resulted in a sharp increase in ethanol production toward the end of the twentieth century, with annual production in the United States reaching 1.5 billion gallons. However, several challenges limit the wide use of ethanol as an automobile fuel. Like methanol, ethanol has less fuel value than that of a comparable mass or volume of gasoline owing to the presence of oxygen in its molecular structure. And considerable energy and money is consumed in its production: from growing and harvesting corn to converting corn starch (see Section 12.5) into sugar and then into alcohol and to distilling the alcohol from the fermentation mixture. It is estimated that more than two-thirds of the energy released in the combustion of ethanol derived from corn is consumed in its production. Thus, ethanol produced in this way is more expensive than gasoline, even at today's prices. Some alternative, more energy-efficient ways to produce ethanol are described in Section 12.6.

CONNECTION: The role of photosynthesis in the carbon cycle is described in Section 4.7.

12.5 CARBOHYDRATES

Section 12.4 discussed the use of ethanol as an oxygen-containing additive for reducing pollution from gasoline-fueled engines. The starting material for most of the 1.5 billion gallons of ethanol used for this purpose in the United States each year is corn, or, to be more precise, corn starch. Corn starch is an example of a **carbohydrate**, or **saccharide**. Starch belongs to the class of carbohydrates called **polysaccharides**. The giant molecules of polysaccharides are made of long chains of the most common **monosaccharide** (simple sugar), glucose. Glucose is formed (see Section 4.7) in green plants by the endothermic process driven by the energy of sunlight known as photosynthesis:

$$6\ CO_2(g) + 6\ H_2O(l) \longrightarrow C_6H_{12}O_6(s) + 6\ O_2(g) \qquad \Delta H° = 2801\ \text{kJ}$$

Carbohydrates are a class of organic compounds with the general molecular formula $C_x(H_2O)_y$. They include simple sugars (**monosaccharides**), such as glucose, $C_6H_{12}O_6$, and **polysaccharides**, including starch and cellulose, that are made of chains of glucose molecules.

The name *carbohydrate* comes from early attempts to determine their elemental composition. Glucose has the chemical formula $C_6H_{12}O_6$, which may also be written $C_6(H_2O)_6$. In fact, all carbohydrates have an elemental composition that fits the general formula $C_x(H_2O)_y$. When concentrated sulfuric acid is added to

FIGURE 12.11 Addition of concentrated H_2SO_4 causes sugar to dehydrate, leaving a residue of carbon.

sugar, the sugar rapidly and exothermically decomposes, releasing water molecules and leaving behind a cinder of carbon (Figure 12.11). This evidence led scientists to incorrectly assume that molecules of sugar consist of groups of carbon atoms surrounded by water molecules, and so sugar was given the chemical name *hydrate of carbon* or *carbohydrate*.

Molecular structures of glucose and other sugars

Glucose has several molecular structures, and none of them incorporates carbon atoms surrounded by water molecules. One of these structures is shown in Figure 12.12. The structure has a backbone of six carbon atoms. The number-one carbon atom (C-1) is part of a functional group that we have not considered before. Surrounding C-1 is a C—C bond, a C—H bond, and a C=O double bond. This combination of bonds constitutes an **aldehyde group** with the structure shown in Figure 12.13.

In glucose, each of the carbon atoms from C-2 to C-6 is bonded to a hydroxyl group. The base name of this six-carbon aldehyde is *hexanal*, in which the *-al* ending designates the presence of an aldehyde group. The five O—H, or *hydroxyl*,

> ✓ An **aldehyde** is an organic compound with a functional group containing a carbon atom double bonded to an oxygen atom and single bonded to a hydrogen atom:
>
>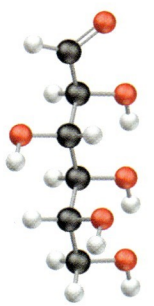

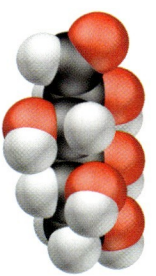

FIGURE 12.12 This open-chain structure of glucose features a backbone of six covalently bonded carbon atoms. The carbon atom labeled "1" is double bonded to the oxygen atom of an aldehyde group. Each of the other carbon atoms in this structure is bonded to an OH (alcohol) group.

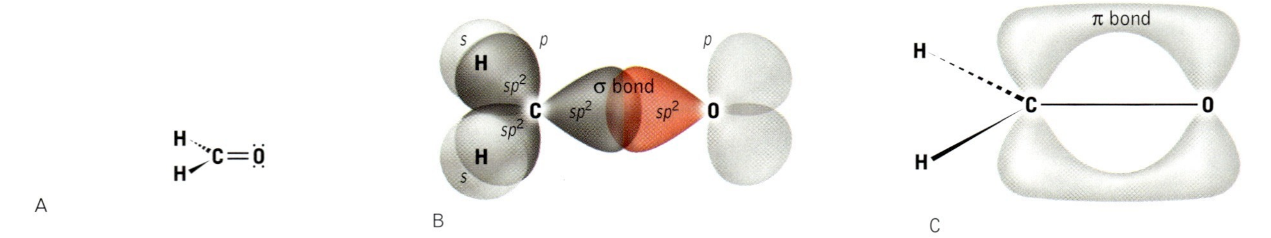

FIGURE 12.13 Formaldehyde is the simplest of aldehydes. A. It has a trigonal planar molecular shape. B. The bond angles are consistent with sp^2 hybridization of the orbitals of carbon. C. The π bond in the C=O double bond includes regions of high electron density above and below the plane of the σ bonds.

groups bonded at C-2 through C-6 make this compound a *pentahydroxyhexanal*. Unfortunately, the naming of biomolecules is sometimes complicated by the fact that chemists and biologists use different names for the same substance. To a biologist, glucose is an *aldohexose*. The logic behind this name is that glucose is an aldehyde (*aldo*) with six carbons (*hex*), and it's a sugar, which means that its name ends in *-ose*.

The C=O double bond exists in many organic compounds, including those of biological importance, as we shall see in this chapter and the next. Unequal sharing of the bonding electrons between the carbon and highly electronegative oxygen atom makes C=O bonds polar and can add to the permanent dipole moment of the molecule in which it exists. Thus, many of the organic compounds with C=O bonds are soluble in water, as discussed in Section 9.5. The C=O bond is so important in organic chemistry that it is considered a functional group all by itself: the **carbonyl group**.

Another structure of glucose is produced by a chemical reaction that occurs when the carbon backbone of the open chain wraps around so that the hydroxyl group at C-5 approaches and reacts with the aldehyde group at C-1, as shown in Figure 12.14. In the reaction, the O—H bond of the hydroxyl group at C-5 is broken. The hydrogen atom that is released as the bond breaks ends up attached to the oxygen atom of the aldehyde group, and the hydroxyl oxygen atom bonds with the C-1 carbon atoms. The product of this rearrangement is a cyclic form of glucose: a six-membered ring of five carbon atoms and the oxygen atom that was originally part of the hydroxyl group at C-5.

The cyclic product of this ring-closing reaction has the same chemical formula as that of open-chain glucose ($C_6H_{12}O_6$) and so is a structural isomer of it. However, the cyclic form is no longer an aldehyde. Instead, it is a **hemiacetal**—a compound in which a carbon atom (C-1 in this case) is bonded to both an –OH group and an –OR group, where R represents a group attached to an organic molecule by a carbon atom.

The ring-closure, or *cyclization*, reaction is an example of an internal, or *intramolecular*, reaction, because the same molecule supplied both the aldehyde and the hydroxyl functional groups for the reaction. Hemiacetals also form in reactions

✓ A **carbonyl group** is the C=O double bond in an organic compound. It is a feature of several organic functional groups and adds to the polarity of organic substances.

CONNECTION: The presence of carbonyl groups in the structures of organic compounds contributes to their solubilities in polar solvents, such as water (Section 9.5).

✓ A **hemiacetal** is a functional group that includes a carbon atom bonded to a hydroxyl group and to an oxygen atom that is also bonded to an alkyl group:

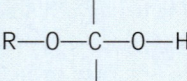

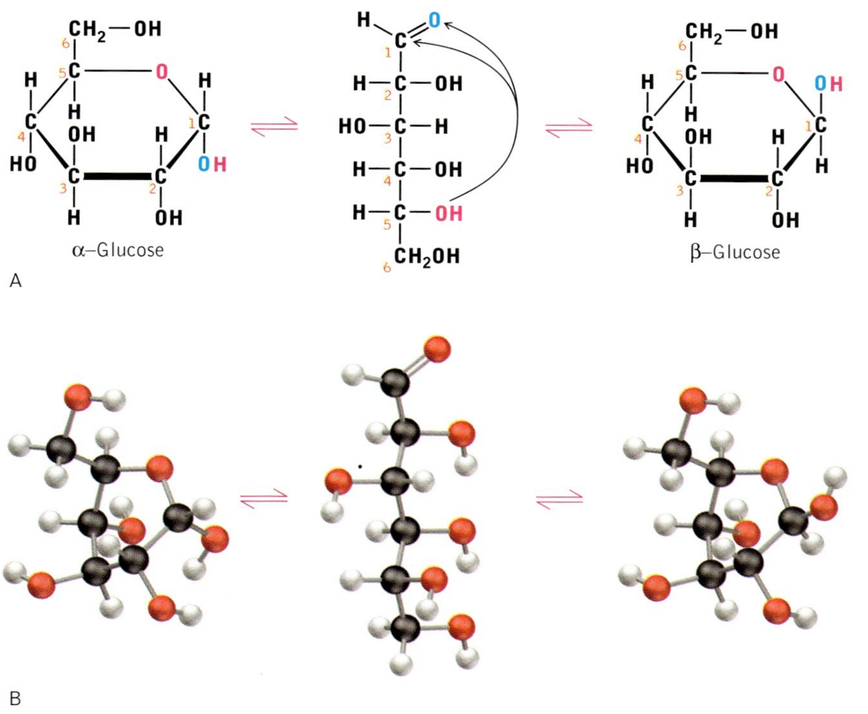

FIGURE 12.14 A molecule of glucose can have an open-chain or one of two cyclic hemiacetal structures. A. The open-chain form in the center is transformed into the cyclic forms on either side in a process in which the open-chain wraps around itself so that the hydroxyl group at C-5 (highlighted in red) approaches and reacts with the aldehyde group at C-1. A bond forms between the O atom of the hydroxyl group and the C-1 carbon atom, creating a six-atom ring. As the ring closes, the hydrogen atom of the C-5 hydroxyl group bonds to the O atom of the aldehyde group (in blue), forming a hydroxyl group at C-1. Depending on the orientation of the groups at C-1 and C-5, the process yields either α-glucose or β-glucose. The double arrows indicate that the cyclization process is reversible. B. Ball and stick models provide three-dimensional perspectives of the process and the structures of the products.

---

between different molecules, such as an alcohol and an aldehyde, resulting in the two reactants being chemically bonded together through a C–O–C link.

To understand how the open-chain structure of a sugar relates to that of the corresponding cyclic form, let's compare an open-chain form of glucose with the corresponding cyclic hemiacetal forms. Notice that the orientations of the OH groups on C-2, C-3, and C-4 in the chain are right, left, right. In the cyclic form, they are down, up, down when the CH₂OH group is on top of the ring. However, the orientation of the –OH group on C-1 may be up or down depending on how the aldehyde and hydroxyl groups of the chain are oriented when they come together during the ring-closing reaction. It turns out that β-glucose (–OH group up) forms 64% of the time and α-glucose (–OH group down) forms 36% of the time.

This same structural relation is observed for the open-chain and cyclic hemiacetal forms of other monosaccharides, as illustrated below for talose. The OH groups on C-2, C-3, and C-4 in the chain are all oriented to the left. In both cyclic forms, all three OH groups are oriented upward, as shown in the following structures. They have been simplified by leaving out the carbon atoms in the rings and the hydrogen atoms bonded to them.

CONNECTION: The double headed arrows indicate the cyclization reactions are reversible (see Section 6.6). Aqueous solutions of glucose are mixtures of 64% β-glucose and 36% α-glucose less than 0.1% is in the linear form.

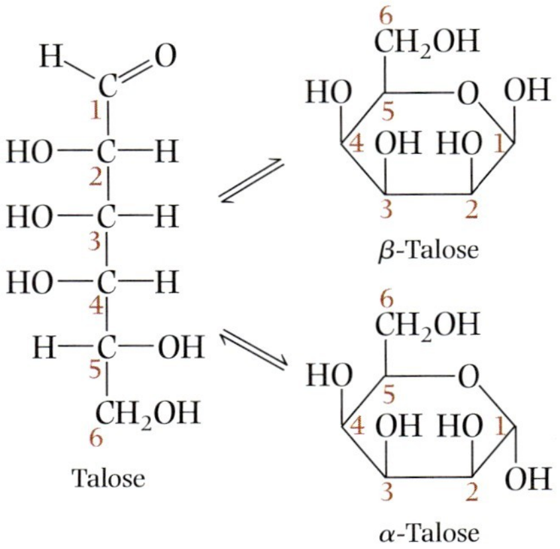

SAMPLE EXERCISE 12.5: The structure of α-mannose is shown below. Draw the structure of the aldohexose form and the structure of β-mannose.

SOLUTION: To form the aldohexose structure, the bond between the ring oxygen atom and C-1 breaks and the hydrogen atom on the –OH group bonded to C-1 migrates to the ring oxygen atom. These changes produce an aldehyde on C-1 and an alcohol (OH) group on C-5. The distribution of the OH groups on C-2, C-3 and C-4 is consistent with their locations on the carbon backbone: those on the same side of the ring as the CH_2OH (up) are on the left side of the aldohexose structure; those pointed down in α-mannose are on the right. When the aldohexose closes to form the β-mannose structure, the OH group on C-1 is on the same side of the ring as the CH_2OH group.

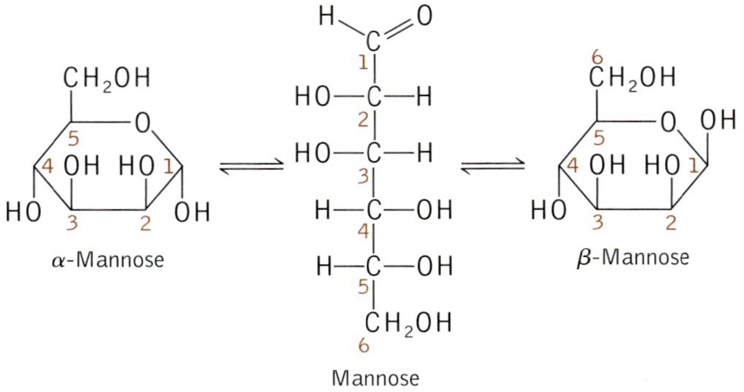

PRACTICE EXERCISE: Draw the α- and β-hemiacetal forms of the aldohexose shown below. (See Problems 57 and 58.)

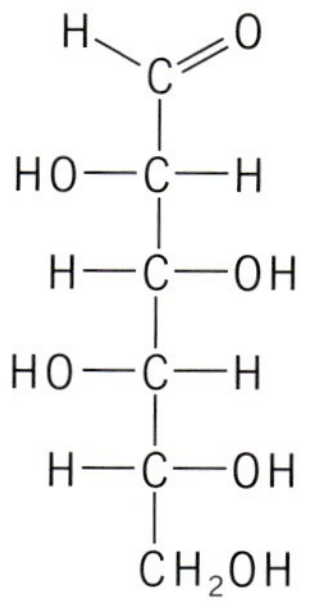

$$\begin{array}{c} H\diagdownC{=}O \\ HO-C-H \\ H-C-OH \\ HO-C-H \\ H-C-OH \\ CH_2OH \end{array}$$

Figure 12.15 shows the structures of open-chain and cyclic forms of another common sugar: fructose. Fructose and glucose have the same formula and so are structural isomers of one another. However, the C=O double bond in fructose is not on the end carbon in the open-chain form of the molecule. Instead, it is part of a functional group of the following form:

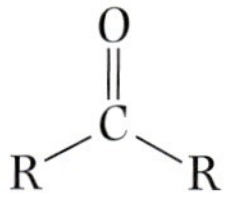

$$\begin{array}{c} O \\ \| \\ R-C-R \end{array}$$

This functional group is a **ketone group**, and its presence in the structure of fructose makes this sugar a *keto* sugar, or *ketohexose*. Fructose can cyclize to form six-atom rings if the ring closure reaction involves the hydroxyl group at C-6 and

✓ A **ketone** is a compound in which a carbonyl group is bonded to two carbon atoms:

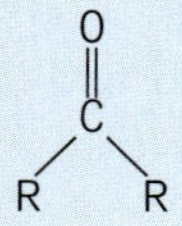

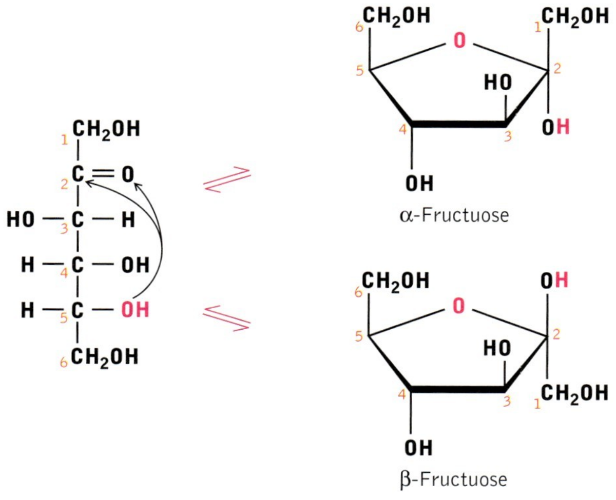

$$\begin{array}{c} O \\ \| \\ R-C-R \end{array}$$

FIGURE 12.15 Fructose is a ketohexose that undergoes cyclization when its carbon backbone bends so that the hydroxyl group (shown in red) at C-5 approaches and reacts with the aldehyde group at C-2. A bond forms between the O atom of the hydroxyl group and the C-2 carbon atom, creating a five-atom ring. As the ring closes, the hydrogen atom of the C-5 hydroxyl group bonds to the O atom of the aldehyde group (in blue), forming a hydroxyl group at C-2 that can be either below or above the ring, producing α-fructose or β-fructose, respectively.

the carbonyl group at C-2. However, fructose can also form five-atom rings as shown in Figure 12.15. In these cyclization reactions the hydroxyl group at C-5 reacts with the carbonyl group at C-2. The products are molecules with two —CH_2OH groups that are either on the same side of the ring (α-fructose) or on opposite sides (β-fructose).

Condensation reactions

All of the sugars we have discussed are examples of "simple" sugars or monosaccharides. In nature there are more complex carbohydrates made up of molecules of simple sugars bonded together. The smallest of this class of compounds are disaccharides Among them is sucrose, or table sugar, whose molecules consist of one molecule each of glucose and fructose, bonded together as shown in Figure 12.16.

Look closely at the structures of glucose and fructose and note how they react to make sucrose. Notice how the hydroxyl group of the glucose molecule has coupled with the hydroxyl group of fructose, producing a C–O–C (ether) link between the two sugars. One molecule of water is lost in this process. This coupling reaction is an example of a **condensation** reaction. The name makes sense if we think of the condensation of water vapor on a cold object as the *loss* of water vapor from the atmosphere. In a condensation reaction between the two sugar molecules, they collectively *lose* the ingredients of a water molecule.

Coupling reactions can take place at more than one hydroxyl group on the same sugar molecule. In this way, sugar molecules can join together to form a

> ✓ In a **condensation** reaction, functional groups containing oxygen and hydrogen react together and lose a molecule of water in the process.

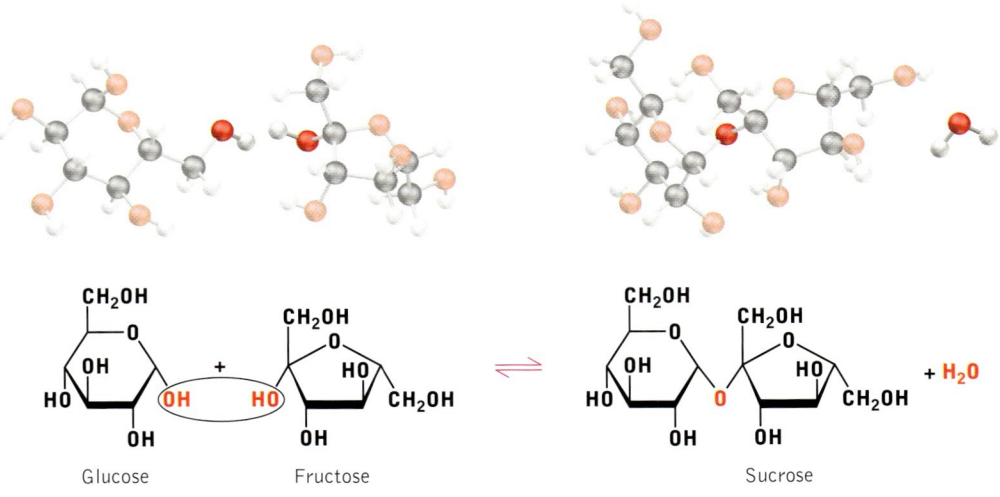

FIGURE 12.16 A condensation reaction between two monosaccharides—glucose and fructose—produces the disaccharide sucrose (table sugar) and water.

FIGURE 12.17 Starch is a long chain of glucose molecules that are connected to one another through bonds formed by condensation reactions.

long chain of many monosaccharides—that is, a polysaccharide. Condensation is part of the process by which a plant stores chemical energy from photosynthesis. Water-soluble glucose is converted into large-molecular-weight, insoluble starch with a structure such as that shown in Figure 12.17. The dotted lines at each end of the fragment in Figure 12.17 indicate that this biological polymer, or *biopolymer*, is considerably longer than the four-monomer (tetramer) segment shown.

The conversion of glucose into starch is an effective way to store energy only if starch can be converted back into glucose. Fortunately, the condensation process is reversible: a molecule of water can be added to each of the C–O–C links between the monosaccharide units in the polysaccharide chain. Such a reaction is an example of **hydrolysis,** which is any chemical reaction in which water is a reactant. Thus, the chemical process that converts corn starch into ethanol begins with the hydrolysis of corn starch into glucose. Then the glucose is converted into ethanol and carbon dioxide by yeast fermentation:

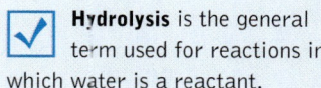 **Hydrolysis** is the general term used for reactions in which water is a reactant.

$$C_6H_{12}O_6(aq) \longrightarrow 2\ CH_3CH_2OH(l) + 2\ CO_2(g) \qquad \Delta H° = -66\ \text{kJ}$$

The slightly exothermic reaction provides energy to the yeast cells. It also produces ethanol, a compound with considerable fuel value.

Starch and cellulose

Starch is much less soluble in water than simple sugars and so can be stored as a future source of energy for the plant that produced it and for the animal that consumes it. Polysaccharides with a slightly different structure from that of starch provide the physical structures of plants. These materials are called *cellulose*.

Plants synthesize cellulose from glucose to build stems and other support structures. Unlike grazing animals, we humans cannot digest cellulose. Much of the dehydrated biomass of celery, for example, is cellulose, and so celery has little nutritional value.

Starch and cellulose are both polymers of glucose; so how is it that we can digest one and not the other? The answer lies in the orientation of the hydroxyl group at C-1 of cyclic glucose. If the —CH_2OH group at C-5 is above the ring and the OH group at C-1 is below the ring, we have α-glucose. If the OH group at C-1 is above the ring, we have β-glucose. When molecules of α-glucose form polysaccharides through condensation reactions, the chemical bond between C-1 of one α-glucose molecule and C-4 of another is called an *α-1,4 glycosidic bond* as shown in Figure 12.18. When molecules of β-glucose form polysaccharides, a similar condensation reaction leads to the formation of β-1,4 glycosidic bonds. Starch contains α-glycosidic bonds, whereas cellulose has only β-glycosidic bonds. Evolution has

> ✓ The linked glucose structures in Figure 12.18 are more-realistic representations of the hemiacetal forms of glucose than are the plane hexagons we use in other figures in that the structures below show the actual bond angles. Note the similarities between these structures and the chair form of cyclohexane in Figure 12.5.

> Starch and cellulose are both polymers of glucose; so how is it that we can digest one and not the other?

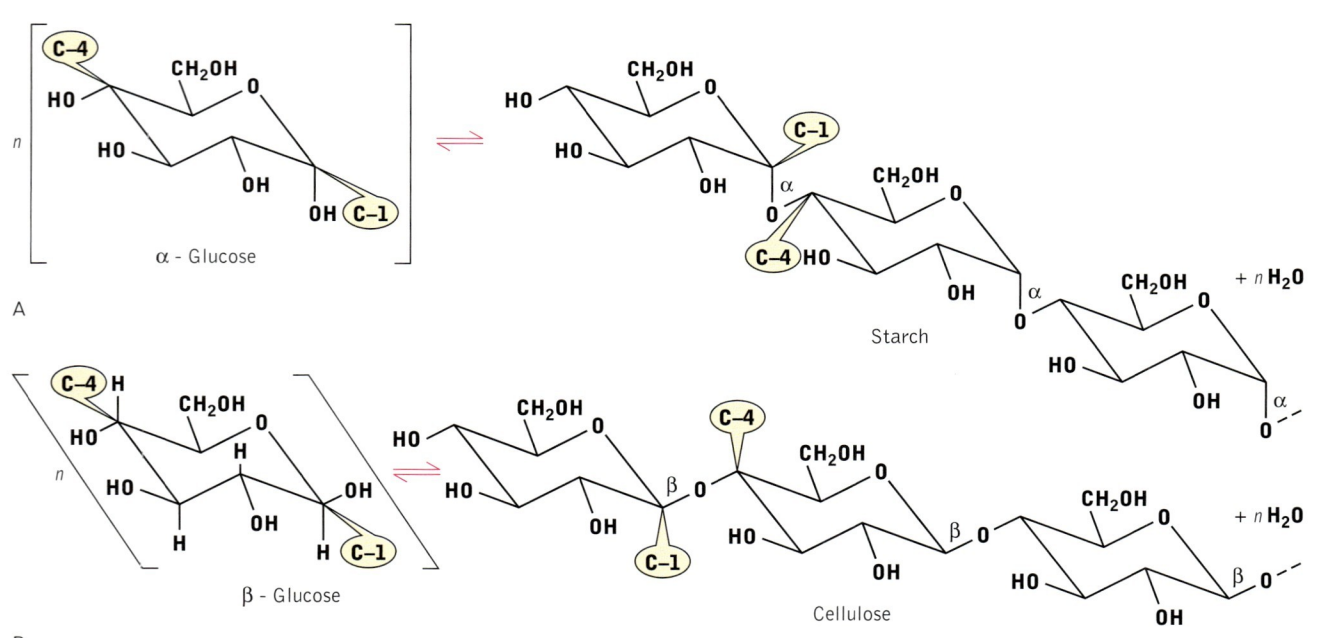

FIGURE 12.18 α-Glucose and β-glucose differ in the orientation of the OH group at C-1. A. The polysaccharide formed from α-glucose has α-glycosidic links between the monomer units. This structure is that of starch. B. The polysaccharide formed from β-glucose has β-glycosidic links between the monomer units. This structure is that of cellulose.

provided humans with digestion enzymes that can break α-glycosidic bonds but not β-glycosidic bonds.

SAMPLE EXERCISE 12.6: α-Lactose is a disaccharide formed by a condensation reaction between the C-1 hydroxyl group of β-galactose and the C-4 hydroxyl group of α-glucose. Draw the condensed molecular structure of α-lactose and identify the type of glycosidic bond in α-lactose. The structures of β-galactose and α-glucose are shown below.

SOLUTION: The linking hydroxyl groups are shown in red. When they link together, releasing a molecule of H_2O, they form a glycosidic bond that is designated 1,4 because it links C-1 and C-4 carbons and it is a β-glycosidic bond because the OH on C-1 of galactose is in the β-orientation. Therefore, the α-lactose drawn below includes a β-1,4 bond.

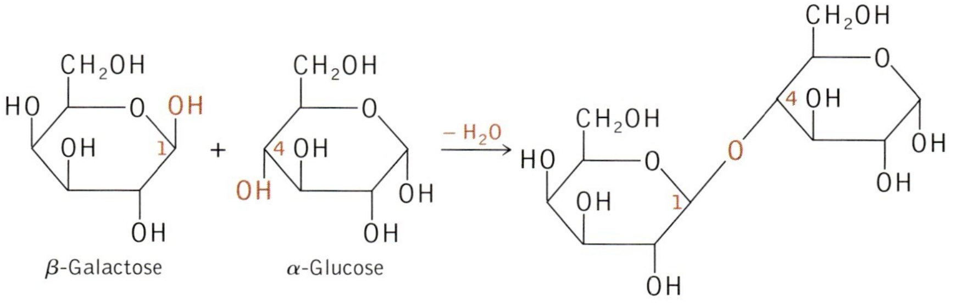

PRACTICE EXERCISE: Maltose (malt sugar) is a disaccharide used to sweeten prepared foods. A molecule of maltose consists of two molecules of α-glucose linked between C-1 in one glucose molecule and C-4 in the other. Draw the condensed molecular structure of maltose and identify the type of glycosidic bond in its structure. (See Problems 67–70.)

Wood fiber is composed mostly of three types of biological polymers: cellulose, hemicellulose, and lignin. As already noted, cellulose is a polysaccharide formed from monomeric units of glucose, $C_6H_{12}O_6$, linked together by β-glycosidic bonds. These bonds are formed during condensation reactions that result in the loss of 1 mole of H_2O for every mole of $C_6H_{12}O_6$, which leaves cellulose with monomeric units that have the formula:

$$C_6H_{12}O_6 - H_2O = C_6H_{10}O_5$$

Therefore the formula of a cellulose polymer composed of n monomeric units is $(C_6H_{10}O_5)_n$. The value of n represents the *degree of polymerization*. Hemicelluloses have the chemical composition of a polysaccharide, but lignins are more complex molecules with ether and alcohol functional groups and aromatic hydrocarbon structures (Figure 12.19).

FIGURE 12.19 Lignins are a component of wood fiber and of soft coal. They have complex structures that incorporate aromatic rings and ether and alcohol functional groups.

Methane gas is a renewable source of energy because it can be produced by the bacterial degradation of organic matter. Methane is produced in swamps and in the treatment of sewage and other wastes.

Like oxygenated additives for gasoline, wood is a heavily "oxygenated" fuel, particularly that part of its structure consisting of cellulose and hemicellulose. This information alone means that the fuel value of wood is less than those of hydrocarbons for the same reasons that alcohols have less fuel value than do hydrocarbons. In addition to a diminished fuel value, wood often contains some water; so some of the heat generated by combustion is wasted converting this water into steam.

Wood has been used as fuel to heat homes and cook food for thousands of years. It and other forms of cellulose, including municipal solid waste, straw, corn stalks, and other agricultural residues, are now being used as inexpensive sources of sugar for fermentation and alcohol production. Just as the α-glycosidic bonds that link the building blocks of starch together can be broken, so, too, can the β-glycosidic bonds in cellulose, releasing glucose. Until now, the destruction of these bonds has required either strong acids, which produce poor yields of glucose, or cellulase enzymes, which are in short supply for commercial alcohol production. Currently, scientists are trying to bioengineer microorganisms to increase the supply and drive down the cost of cellulase enzymes. It is estimated that the use of cellulose from forestry and agricultural residues instead of the starch and sugar in corn or grain as feedstocks for ethanol production will increase the net energy value of ethanol from its current 30% to as high as 80% of its theoretical fuel value.

12.6 MORE FUELS FROM BIOMASS

Although biomass in the form of starch and cellulose has less fuel value than do hydrocarbon fuels, biological processes can convert biomass into the hydrocarbon with the highest fuel value, methane. Since the sixteenth century, swamps have been known to produce a flammable gas called "swamp gas," now identified as methane. The digestive systems of many animals, including cows, produce methane through bacterial action. The conversion of biomass into methane is an attractive future source of hydrocarbons, provided the complexities of bacterial action can be adapted for large-scale production.

Carboxylic acids

The production of methane from plant residues that are mostly cellulose requires the sequential action of several types of bacteria. In the first stage, *hydrolytic* and *transitional* bacteria break up cellulose into mixtures of small molecules including hydrogen, carbon dioxide, and, depending on the bacterial strain, alcohols (methanol, ethanol, 2-propanol) or two **carboxylic acids:** formic and acetic acids (Figure 12.20). The carboxylic acid functional group consists of a carbon atom bonded to an OH group and double bonded to an oxygen atom. When these compounds dissolve in water, some of the O–H bonds ionize. Both of the electrons in the O–H bond remain on the oxygen atom, producing a negatively charged **carboxylate anion** (—COO⁻) and a positively charged hydrogen ion (H⁺), as shown in Figure 12.20D. The acidic properties of these compounds are related to the ionization process and will be discussed in more detail in Chapter 16.

> ✓ **Carboxylic acids** are organic compounds with
>
>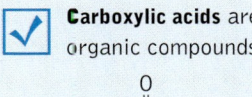
>
> functional groups. They partly ionize in water, forming **carboxylate anions** and protons.

CONNECTION: Carboxylic acids are weak electrolytes (Section 5.3) and weak acids (Section 16.2).

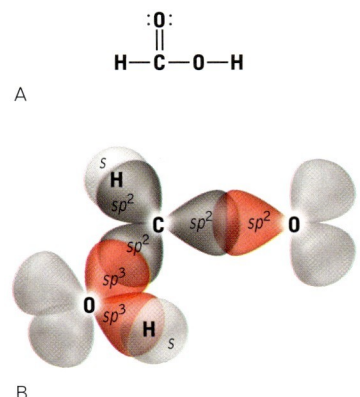

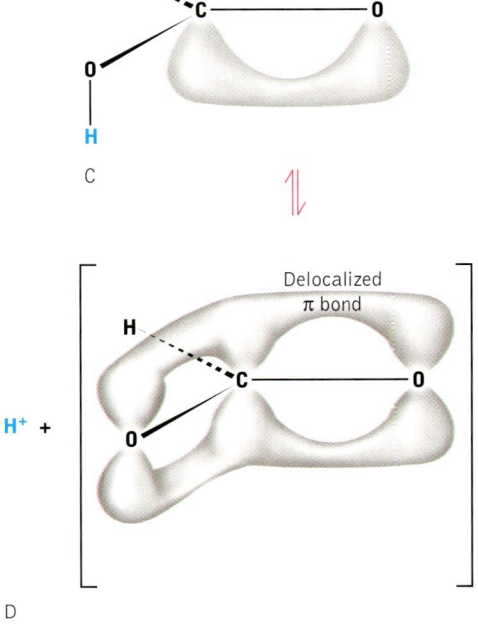

FIGURE 12.20 Formic acid is the simplest of carboxylic acids. A. The Lewis structure of formic acid features the —COOH carboxylic acid functional group. B. The bond angles around the carbon atom are consistent with sp^2 hybridization of the orbitals of carbon. C. The electrons in the C=O π bond are concentrated above and below the plane of the σ bonds around carbon. D. When formic acid ionizes into the formate ion HCOO⁻ and H⁺, π electrons are delocalized over both C–O bonds.

TABLE 12.5 Names of Some Primary Alcohols, Aldehydes, and Carboxylic Acids

Number of Carbon Atoms	Alkane	Alcohol	Aldehyde	Carboxylic Acid
1	Methane	Methanol (methyl alcohol)	Methanal (formaldehyde)	Methanoic acid (formic acid)
2	Ethane	Ethanol (ethyl alcohol)	Ethanal (acetaldehyde)	Ethanoic acid (acetic acid)
3	Propane	1-Propanol (n-propyl alcohol)	Propanal	Propanoic acid (propionic acid)
4	Butane	1-Butanol	Butanal	Butanoic acid (butyric acid)

Note: Common names are in parentheses.

In the final stage of methane production, the products of the first stage undergo several reactions promoted by the metabolism of *methanogenic* bacteria. Acetic acid is the reactant in the second reaction; formic acid is the reactant in the third. Methane is produced in all four of these reactions:

$$4\,H_2(g) + CO_2(g) \longrightarrow CH_4(g) + 2\,H_2O(l)$$

$$CH_3CO_2H(aq) \longrightarrow CH_4(g) + CO_2(g)$$

$$4\,HCO_2H(aq) \longrightarrow CH_4(g) + 3\,CO_2(g) + 2\,H_2O(l)$$

$$4\,CH_3OH(aq) \longrightarrow 3\,CH_4(g) + CO_2(g) + 2\,H_2O(l)$$

Acetic acid and formic acid have names that predate the current, more systematic approach to naming organic compounds. According to the current system (see Table 12.5), the names of the C_1 and C_2 acids should be methanoic acid and ethanoic acid. The names of the carboxylate anions these acids form when they ionize are obtained by replacing the -ic ending of the name of the acid with -ate. Thus, acetic (ethanoic) acid forms acetate (ethanoate) ions. Some of the old, but still widely used, common names appear in Table 12.5 in parentheses.

Amines

Some strains of methanogenic bacteria produce methane from a class of nitrogen-containing organic compounds called **amines.** These compounds are responsible for much of the foul odor of decaying biological tissue.

Amines have structures that resemble that of ammonia (NH_3). They have a trigonal pyramidal molecular geometry with three single bonds and a lone pair of electrons around the nitrogen atom. In an amine, one, two, or all three of the

✓ An **amine** is an organic compound containing a nitrogen atom bonded to at least one carbon atom.

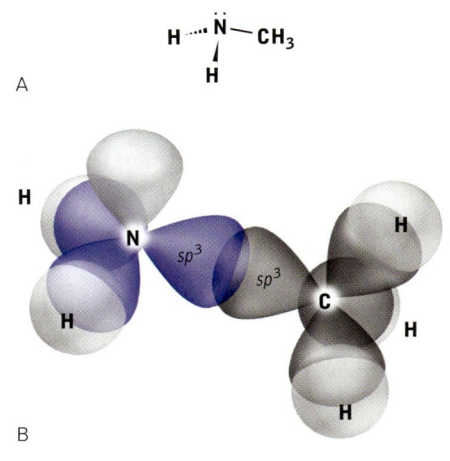

FIGURE 12.21 A. In a molecule of methylamine a nitrogen atom is bonded to two hydrogen atoms and a methyl group. B. All bonds to nitrogen are single bonds formed by overlap of three of the four sp^3 hybrid orbitals on nitrogen with an sp^3 hybrid orbital on carbon or s orbitals on hydrogen atoms.

hydrogen atoms in ammonia are replaced by R groups. The names of amines are derived from the names of the attached substituent groups in alphabetical order. A prefix is used to denote the number of groups of the same type: *di* for two and *tri* for three. Amines having one R group (Figure 12.21) are called *primary* amines; those with two or three R groups are called *secondary* or *tertiary* amines, respectively.

Biochemical processes in the cells of bacteria of the *Methanosarcina* genus hydrolyze primary, secondary, and tertiary methylamines to methane, carbon dioxide, and ammonia:

$$4\ CH_3NH_2(aq) + 2\ H_2O(l) \longrightarrow 3\ CH_4(g) + CO_2(g) + 4\ NH_3(aq)$$

$$2\ (CH_3)_2NH(aq) + 2\ H_2O(l) \longrightarrow 3\ CH_4(g) + CO_2(g) + 2\ NH_3(aq)$$

$$4\ (CH_3)_3N(aq) + 6\ H_2O(l) \longrightarrow 9\ CH_4(g) + 3\ CO_2(g) + 4\ NH_3(aq)$$

The chemical reactions in this section describe some of the pathways by which methane can be produced from the decay of biomass. The industrial development of these processes for fuel production has been slow because fossil fuels are still plentiful enough and not expensive enough to make such development cost effective. The biggest natural gas field ever discovered is being developed off the coast of Qatar in the Arabian Gulf. Therefore, the economic incentives to develop alternative fuel-production methods do not yet exist. One limitation on their development is that the reactions presented in this section are far from complete. A lot of plant matter remains undigested in a natural anaerobic environment. Lignin is particularly difficult to degrade to methane. In nature, incomplete conversion leads to the formation of lignin-rich sediments that turn into coal after millions of years. Current research in the biochemistry of methanogenic bacteria, coupled with the power of genetic engineering, may enable the conversion of a broad range of biological materials into CH_4 and other high-energy fuels at reasonable costs.

The odor of decaying fish is due to trimethylamine, $N(CH_3)_3$.

12.7 COAL

Coal is the most abundant fossil fuel in the United States and relatively easy to extract from the earth. It is not surprising that coal provides more than one-fourth of the energy consumed each year in the United States. That value would be even higher were it not for the significant concentrations of sulfur (see Table 12.6) and other impurities in coal that pollute the environment as coal burns.

Coal is plant tissue that has been subjected to high temperatures and pressure through millions of years of burial in geological formations. During that time, the plant residues lose much of their hydrogen and oxygen content, becoming mostly graphite carbon with lesser proportions of hydrocarbons and other compounds. The longer this transformation process goes on, the more carbon-rich the coal becomes. The most carbon rich and the hardest coal is called anthracite. Grades of coal with lower carbon content include soft (bituminous) coal and lignite, the grade with the least carbon content (see Table 12.6). Coal contains small quantities of alkanes, ranging from C_8 to C_{31}, and aromatic compounds (Figure 12.22). The graphite structure also contains ether (–C–O–C–) linkages and hydroxyl groups.

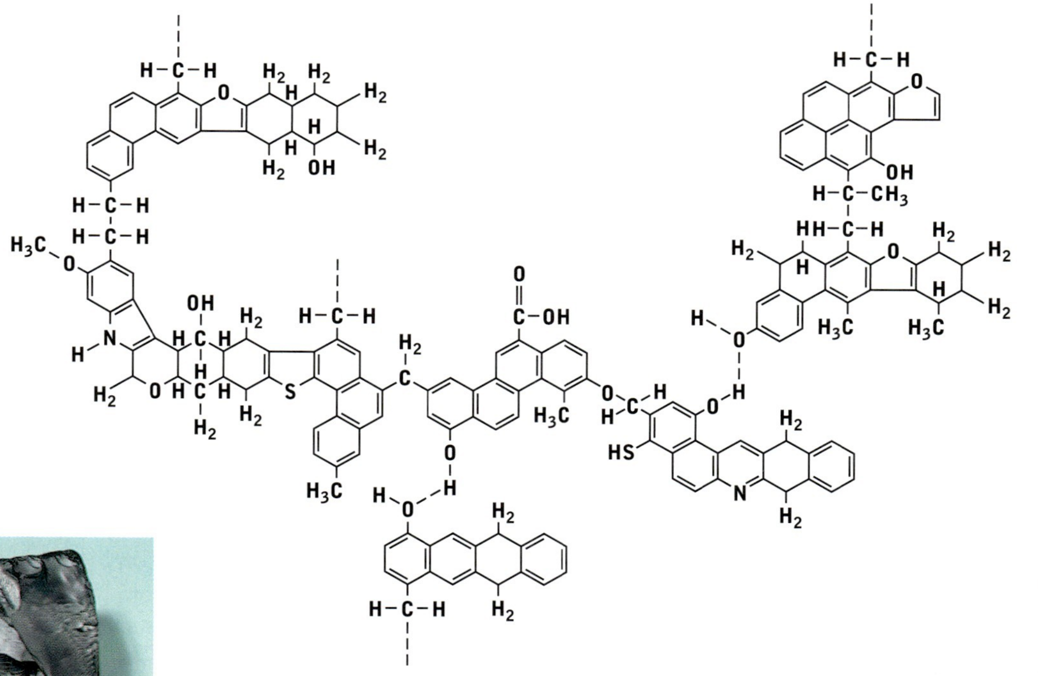

FIGURE 12.22 Coal is a complex mixture of large-molar-mass substances. The composition of coal varies, depending on its type and where it is mined. Various studies have documented the presence of aromatic hydrocarbons, ethers, alcohols, and compounds containing nitrogen and sulfur. The model shown here is one of several that have been proposed for anthracite coal found in western Pennsylvania.

TABLE 12.6 Typical Elemental Composition (Percent by Mass) of Several Grades of Coal

Grade	Carbon	Hydrogen	Oxygen	Sulfur	Nitrogen
Anthracite (hard coal)	90	4	4	1	1
Bituminous (soft coal)	80	6	10	2	1
Subbituminous	75	5	18	1	1
Lignite	67	6	25	1	1

CONCEPT TEST: On the basis of the elemental composition data in Table 12.6, predict which grade of coal has the highest fuel value.

12.8 HYDROGEN AS FUEL

Hydrogen is an attractive fuel for transporting people and cargo on several counts. Its fuel value is released in a simple combustion reaction in which the only product is water:

$$2\,H_2(g) + O_2(g) \longrightarrow 2\,H_2O(g) \qquad \Delta H^\circ_{rxn} = -483\text{ kJ} \qquad (12.2)$$

Although water vapor absorbs infrared radiation, most of the water vapor produced during combustion reactions eventually condenses and falls to Earth as rain or snow. Therefore, water vapor is not a significant contributor to global warming. In fact, some scientists believe that the increased evaporation of seawater that will accompany global warming will contribute to increased cloud cover, which will reflect sunlight and moderate the greenhouse effect.

Because 2 moles of water are formed in Equation 12.2, ΔH°_{rxn} should be twice the heat of formation of water vapor, $\Delta H^\circ_{f,H_2O}(g)$. A check of the data in Appendix 5 confirms that it is. Hydrogen's considerable heat of combustion coupled with its low molar mass results in a fuel value that is much higher than that of any other fuel. Hydrogen's high energy-to-mass ratio is one reason why hydrogen is the fuel of choice for the main engines of U.S. space shuttles and the booster rockets of other space vehicles. Despite its high fuel value, hydrogen is not a widely used fuel. The main reason is its low density. Even liquid hydrogen has a density of only 0.071 g/mL at its boiling point of 20 K. Multiplying its fuel value (120 kJ/g) by this density, we find that the combustion of liquid hydrogen yields only 8.5 kJ/mL. This value is much less than the 30.2 kJ/mL of octane. This low heat-to-volume ratio is one factor limiting the use of liquid hydrogen as an alternative fuel to gasoline for internal combustion engines. Another is hydrogen's extremely low boiling point, which makes it difficult (and expensive) to liquefy and store as a liquid.

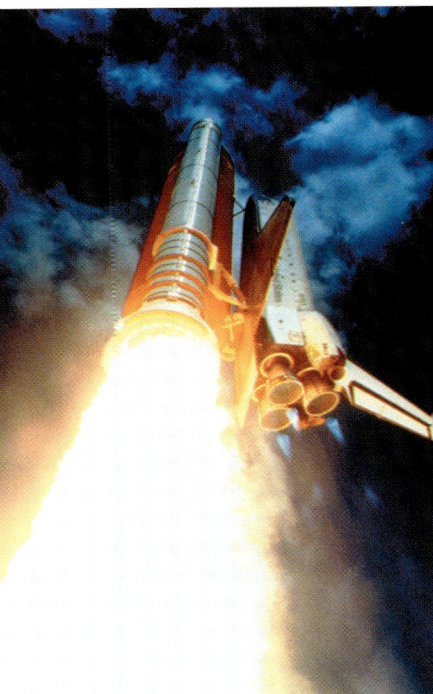

The three main engines of the U.S. space shuttles use hydrogen as fuel because of its high fuel value (122 kJ/g).

The problem of hydrogen storage has inspired scientists to investigate the ability of some materials to absorb hydrogen gas much as a sponge absorbs water. This research has focused on a class of compounds called metal *hydrides*. Some metals can absorb molecular hydrogen by breaking the H—H bond and forming two metal–hydrogen bonds:

$$M(s) + H_2(g) \longrightarrow H\text{—}M\text{—}H(s)$$

Hydrogen atoms are very small and can fit into the space of a cubic or hexagonal closest-packed arrangement of larger metal atoms, as described in Chapter 10. The structures of two hydrides of titanium, TiH and TiH_2, are shown in Figure 12.23. Hydrogen atoms reside in holes that are also called *interstices*, and so these metal hydrides are known as *interstitial* metal hydrides. Because hydrogen atoms are so small, interstitial metal hydrides have the same volumes as those of the metals alone. When metal hydrides are heated, the hydrogen absorption reaction is reversed, providing a steady stream of hydrogen fuel to an engine. A major advantage in using metal hydrides for hydrogen fuel storage is that they are less likely than hydrogen to explode.

Among the Group 4B metals, zirconium, hafnium, and titanium can chemically absorb as much as 1 mole of H_2 per mole of metal, forming hydrides with the formula MH_2. Among the metals in Group 5A, niobium forms a 1:1 compound with hydrogen, NbH. Palladium metal does not form stable metal hydrides but Pd metal can still absorb as much as 1000 times its volume of H_2 gas at standard temperature and pressure [273 K and 1 atm]. Still other metals form hydrides that are used in rechargeable batteries, which we will discuss in Chapter 17.

CONNECTION: The locations and sizes of holes in crystal structures are described in Sections 10.2 and 10.3.

CONNECTION: The use of hydrogen in fuel cells is described in Section 17.8.

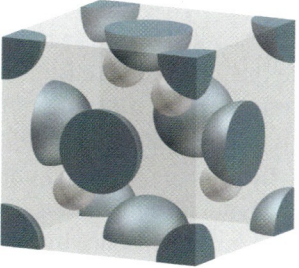

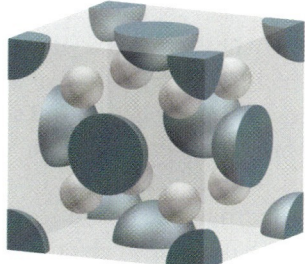

A Titanium hydride   B Titanium dihydride

FIGURE 12.23 Hydrogen can be stored in the crystal structures of metals, forming metal hydrides. A. In this expanded view of the unit cell of titanium hydride (TiH) hydrogen atoms occupy half the tetrahedral holes in the lattice of cubic closest-packed titanium atoms (blue spheres). B. Hydrogen atoms occupy all the tetrahedral holes formed by titanium atoms in titanium dihydride (TiH_2).

12.9 COMBUSTION ANALYSIS AND ELEMENTAL COMPOSITION

The combustion of organic compounds releases energy. It also gives products that enable chemists to determine the chemical composition of unknown organic compounds. In these determinations chemists exploit the fact that the complete combustion of organic compounds converts all of the carbon in them to CO_2 and all of the hydrogen to H_2O.

Consider the following scenario. A hydrocarbon of unknown composition is combusted in a chamber through which a stream of pure oxygen flows as shown in Figure 12.24. The gases produced flow through two tubes one after the other. One tube is packed with material that selectively absorbs all the water vapor produced during combustion; the second absorbs all of the carbon dioxide. The masses of these tubes are measured before and after the combustion reaction. The mass of the tube that traps CO_2 is found to have increased by 1.320 g, and the mass of the tube that traps water vapor has increased by 0.540 g. How can we use these results to determine the composition and molecular formula of the unknown compound? Let's establish what we know about the unknown compound:

1. It's a hydrocarbon, and so it contains only carbon and hydrogen.
2. Conversion of its carbon content into CO_2 yields 1.320 g of CO_2.
3. Conversion of its hydrogen content into H_2O produces 0.540 g of H_2O.

Our goal is to develop a molecular formula for the unknown compound; so we need information about the molar proportions of the elements in it. Our first task is to determine the number of moles of carbon in 1.320 g of carbon dioxide and the number of moles of hydrogen in 0.540 g of water vapor. By dividing

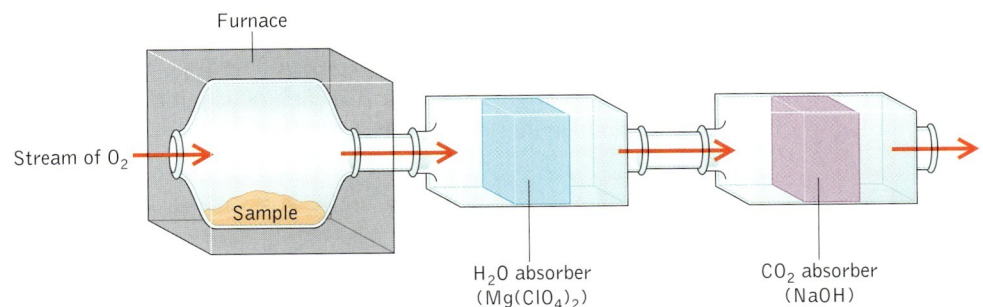

FIGURE 12.24 The combustion of organic compounds in excess oxygen produces water and carbon dioxide. Water and CO_2 are absorbed on $Mg(ClO_4)_2$ and NaOH, respectively. The empirical formula of the compound is calculated from the mass of H_2O and CO_2 absorbed.

these masses by the appropriate molar masses, we can calculate the number of moles of CO_2 and H_2O that they represent. Next, we convert these numbers of moles of CO_2 and H_2O into corresponding numbers of moles of the elements of interest: C and H. Because 1 mole of CO_2 contains 1 mole of C and because 1 mole of H_2O contains 2 moles of H, these conversions are easy to make, as shown in the following equations.

$$1.320 \text{ g CO}_2 \times \frac{1 \text{ mol CO}_2}{44.01 \text{ g CO}_2} \times \frac{1 \text{ mol C}}{1 \text{ mol CO}_2} = 0.0300 \text{ mol C}$$

$$0.540 \text{ g H}_2\text{O} \times \frac{1 \text{ mol H}_2\text{O}}{18.01 \text{ g H}_2\text{O}} \times \frac{2 \text{ mol H}}{1 \text{ mol H}_2\text{O}} = 0.0600 \text{ mol H}$$

Our next step is to convert these moles of C and H into simple whole numbers. The best way to do so is to divide both by the smaller of the two:

$$\frac{0.0300 \text{ mol C}}{0.0300 \text{ mol C}} = 1 \qquad \frac{0.0600 \text{ mol H}}{0.0300 \text{ mol C}} = 2$$

CONNECTION: The method of determining empirical formulas from percent composition values is described in Section 4.6.

CONNECTION: A carbon atom in CH_2 would have only 6 valence electrons—2 less than it needs for a complete octet.

The H:C mole ratio of 2:1 corresponds the formula CH_2. Because this formula was derived experimentally, it is called an *empirical* formula (see Section 4.5). It describes the simplest ratio of the atoms in a molecule. Is there actually a molecule with the molecular formula CH_2? On the basis of our understanding of bonding in Chapters 7 and 8, we can predict that a compound with the molecular formula CH_2 is highly unlikely, because a carbon atom would not have a complete octet with only two C—H bonds.

The actual molecular formula is probably not CH_2, but it could be C_2H_4 or C_3H_6 or C_4H_8 or any other formula of the form $(CH_2)_n$, where n is a positive integer. All of these molecules have an H:C ratio of 2:1. To determine the true molecular formula, we need additional information about the compound. If, for example, we know that the unknown is a gas at room temperature, then the value of n must be small, because hydrocarbons with large molar masses sizes are liquids or solids. We can determine the molecular formula of a compound exactly if we also know its molar mass. Suppose through some other analytical technique we determine that the molar mass of our compound is 28 g/mol. Then the molecular formula could not be CH_2, because CH_2 would have a molar mass of 14 g/mol. To have twice the molar mass, the compound would have to have *twice* the number of carbon and hydrogen atoms in each of its molecules. Therefore, $n = 2$ and the actual molecular formula is $(CH_2)_2$, or C_2H_4.

SAMPLE EXERCISE 12.7: A compound containing only carbon and hydrogen is combusted, producing 0.660 g of CO_2 and 0.410 g of H_2O. The molar mass of the compound is 30 g/mol. Calculate the empirical formula and the molecular formula of this compound.

SOLUTION: The moles of carbon and hydrogen can be determined from the masses of CO_2 and H_2O, respectively:

$$0.660 \text{ g CO}_2 \times \frac{1 \text{ mol CO}_2}{44.01 \text{ g CO}_2} \times \frac{1 \text{ mol C}}{1 \text{ mol CO}_2} = 0.0150 \text{ mol C}$$

$$0.410 \text{ g H}_2\text{O} \times \frac{1 \text{ mol H}_2\text{O}}{18.02 \text{ g H}_2\text{O}} \times \frac{2 \text{ mol H}}{1 \text{ mol H}_2\text{O}} = 0.0450 \text{ mol H}$$

The hydrogen-to-carbon ratio (0.0450 mol H/0.0150 mol C) simplifies to 3:1 when we divide both values by the smaller of the two, 0.0150. Thus the empirical formula is CH_3.

To determine the molecular formula, we first determine the mass of one of the CH_3 "building blocks" in the molecular formula. In whole mass units, this value is $[12 + (3 \times 1)] = 15$ g/building block. Dividing the molar mass by the mass of one building block, we determine the value of n in $(CH_3)_n$:

$$\frac{30 \text{ g/mol}}{15 \text{ g/building block}} = \frac{2 \text{ building blocks}}{\text{mol}} = n$$

Thus, the molecular formula is $(CH_3)_2$, or C_2H_6, which is the molecular formula of ethane.

PRACTICE EXERCISE: Cigarette smoke contains many compounds that contain only carbon and hydrogen. A 171-mg sample of one of these compounds, suspected of being carcinogenic, is subjected to combustion analysis and produces 594 mg of CO_2 and 81 mg of H_2O. Determination of the molar mass gives a value of 228 g/mol. What are the empirical and molecular formulas of the compound? (See Problems 91 and 92.)

Note that, in the preceding sample exercise, we did not need to know the initial mass of the sample to determine its empirical formula. The reason is that we knew the sample was a hydrocarbon and that it was completely converted into the CO_2 and H_2O that were collected and weighed. What if the composition of the compound were totally unknown? What if it were a compound isolated from a tropical plant in the hope that it had beneficial pharmacological properties? Many such compounds are made of carbon, hydrogen, oxygen, and sometimes nitrogen. To calculate the empirical formula of a compound made of C, H, and O, for example, we need to know the proportion of oxygen in it, but there is no simple way of measuring that directly when the compound is combusted in an atmosphere of pure oxygen.

The following sample exercise shows how you could directly determine the carbon and hydrogen content of a known mass of such a compound and then indirectly determine its oxygen content and its empirical formula.

SAMPLE EXERCISE 12.8: When 1.000 g of an unknown organic compound containing carbon, hydrogen, and oxygen is combusted, 2.36 g of CO_2 and 0.640 g of H_2O are produced. What is the empirical formula of the compound?

SOLUTION: First, let's calculate the numbers of moles of carbon and hydrogen from the masses of CO_2 and H_2O as in the preceding sample exercise:

$$2.36 \text{ g } CO_2 \times \frac{1 \text{ mol } CO_2}{44.01 \text{ g } CO_2} \times \frac{1 \text{ mol C}}{1 \text{ mol } CO_2} = 0.0536 \text{ mol C}$$

$$0.640 \text{ g } H_2O \times \frac{1 \text{ mol } H_2O}{18.0 \text{ g } H_2O} \times \frac{2 \text{ mol H}}{1 \text{ mol } H_2O} = 0.0714 \text{ mol H}$$

Next, we need to compare the masses of C and H in these moles of C and H with the mass of the initial sample. The difference is the mass of oxygen in the sample.

$$0.0536 \text{ mol C} \times \frac{12.01 \text{ g C}}{1 \text{ mol C}} = 0.643 \text{ g C}$$

$$0.0714 \text{ mol H} \times \frac{1.008 \text{ g H}}{1 \text{ mol H}} = 0.0718 \text{ g H}$$

The sum of these two masses (0.643 g + 0.0718 g = 0.715 g) is less than the mass of the starting sample (1.000 g), and the difference must be the mass of oxygen in the sample.

$$\text{total mass} = 1.000 \text{ g} = \text{mass of C} + \text{mass of H} + \text{mass of O}$$

$$\text{mass of O} = 1.000 - 0.715 = 0.285 \text{ g}$$

Now that the mass of oxygen is known, we can determine the number of moles of O in the sample.

$$0.285 \text{ g O} \times \frac{1 \text{ mol O}}{16.00 \text{ g O}} = 0.0178 \text{ mol O}$$

Thus, the mole ratio of the three elements in the sample is 0.0714 mol H : 0.0536 mol C : 0.0178 mol O. Dividing by the smallest value (0.0178 mol) we get an H:C:O mole ratio of 4:3:1. Therefore the empirical formula of the sample is C_3H_4O.

PRACTICE EXERCISE: Vanillin is the compound containing carbon, hydrogen, and oxygen that gives vanilla beans their distinctive flavor. The combustion of 30.4 mg of vanillin produces 70.4 mg CO_2 and 14.4 mg H_2O. Determine the empirical formula of vanillin. (See Problems 93 and 94.)

12.10 ALKENES AND ALKYNES

In Section 12.9, we derived the molecular formula C_2H_4 for a hydrocarbon. This formula does not fit our generic formula for an alkane: C_nH_{2n+2}. Because our hydrocarbon has fewer than the maximum number of hydrogen atoms, we say that it is *unsaturated* with respect to hydrogen. With two fewer hydrogen atoms than the maximum number allowed, the carbon atoms must form more than one

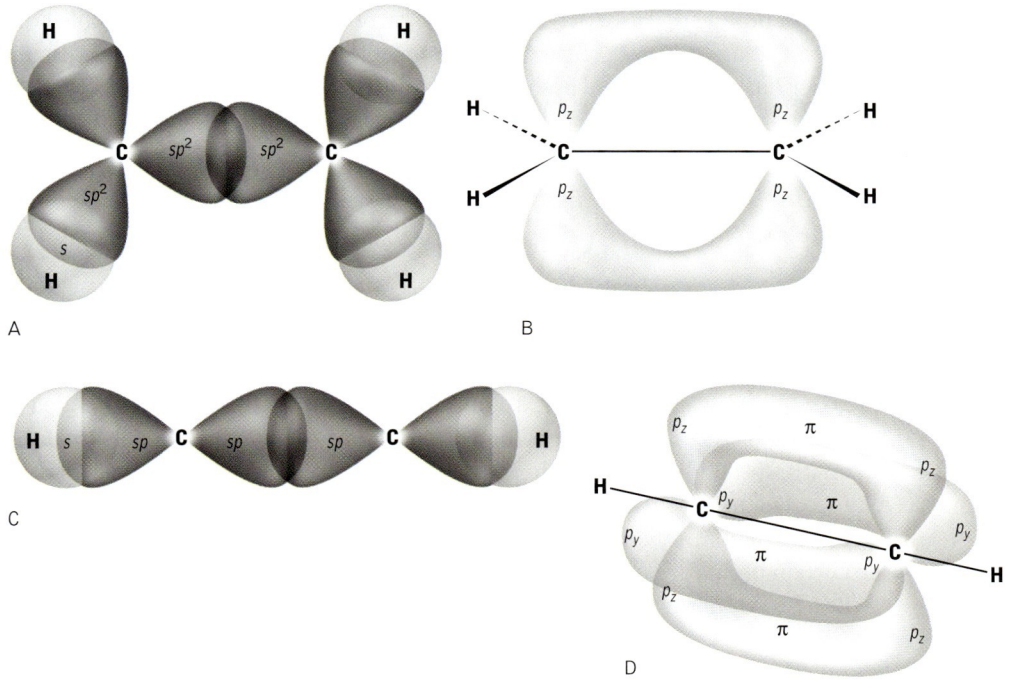

FIGURE 12.25 The bonds in ethylene and acetylene are π bonds formed by overlapping $2p$ orbitals of carbon atoms. The bonding in $CH_2=CH_2$ is consistent with (A) sp^2 hybridization. The C=C double bond includes (B) a π bond formed by overlapping $2p_z$ orbitals. The bonding in HC≡CH is consistent with (C) sp hybridization and (D) the formation of two π bonds by overlapping $2p_y$ and $2p_z$ orbitals of the two carbon atoms.

bond between themselves to fill their valence shells (Figure 12.25). A Lewis structure that satisfies the octet rule for the carbon atoms in C_2H_4 is:

$$\begin{array}{c} H \\ \diagdown \diagup H \\ C=C \\ \diagup \diagdown H \\ H \end{array}$$

The name of this compound is related to that of the corresponding alkane, ethane. We change the ending of ethane from -ane to -ene to indicate the presence of the C=C double bond. Thus the name of this compound is ethene. It also has a common, older name—ethylene. This naming convention applies to all hydrocarbons with a double bond—that is, to all **alkenes.** Ethylene is used to ripen fruit and as a starting material for a plastic material known as polyethylene. Propene (propylene) is used to manufacture polypropylene, which is used to make outdoor clothing and rope. We will explore the synthesis and properties of these polymeric materials in Chapter 18.

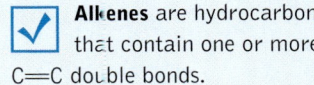

Alkenes are hydrocarbons that contain one or more C=C double bonds.

CONNECTION: The bonding in ethylene and other alkenes is discussed in Sections 6.6 and 7.5.

There is only one structural isomer of ethene, but there are two structural isomers of butene (C_4H_8), each with a four-carbon chain. One has a double bond between C-1 and C-2, which is shown on the left in the following structures, and the other has a double bond between C-2 and C-3:

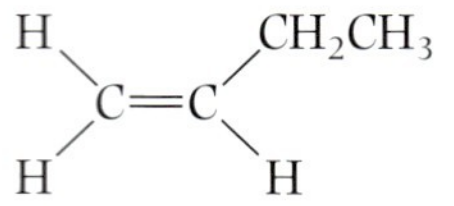

$$\underset{\text{1-butene}}{\overset{H}{\underset{H}{>}}C=C\overset{CH_2CH_3}{\underset{H}{<}}} \qquad \underset{\text{2-butene}}{\overset{H_3C}{\underset{H}{>}}C=C\overset{CH_3}{\underset{H}{<}}}$$

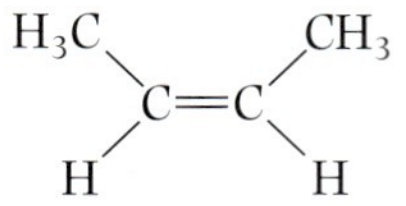

The different locations of the double bond impart different chemical and physical properties to these isomers. Therefore, we need to give them distinguishing names that indicate the location of the double bond. In the left-hand isomer, the first C–C bond from the end is a double bond, and so this isomer is called 1-butene. The right-hand isomer is named 2-butene because the double bond is between the second pair of carbon atoms. Can there be a 3-butene? As with alkanes, we number the carbon atoms (and the bonds between them) so that the lowest possible numbers are used. Therefore, 3-butene is really 1-butene when the carbon atoms are numbered correctly.

Besides, rotating the structure of 1-butene 180° around a vertical axis puts the double bond on the right end of the structure, but it's still the same compound! Therefore, there are only two structural isomers of butane: one with the double bond at one end, the other with the double bond in the middle.

Alkenes such as 2-butene also exhibit a different kind of isomerism that is based on how substituents are arranged around the C=C double bond. Consider these two structures of 2-butene:

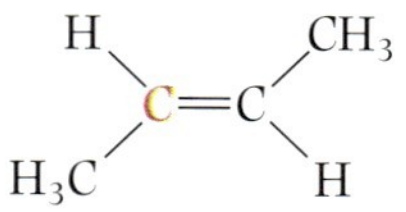

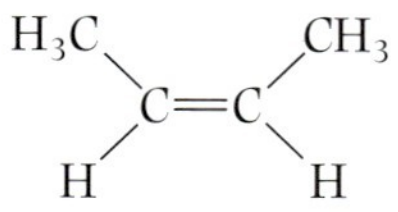

$$\overset{H}{\underset{H_3C}{>}}C=C\overset{CH_3}{\underset{H}{<}} \qquad \overset{H_3C}{\underset{H}{>}}C=C\overset{CH_3}{\underset{H}{<}}$$

If the carbon atom shown in red could rotate 180° around the axis of the double bond, the left-hand structure would become the right-hand structure. However, such rotation around a double bond cannot happen. The combination of σ and π bonds makes double bonds rigid. Therefore, we have two structures that represent two different compounds. These compounds are **geometric isomers** of each other. The one on the left, with two methyl groups on opposite sides of the double bond, is *trans*-2-butene; the one on the right with the methyl groups on the same side is *cis*-2-butene. In general, trans isomers have the same (or similar) substituent groups on opposite sides of a double bond, and cis isomers have the same (or similar) substituent groups on the same side of a double bond.

✓ **Geometric isomers** are compounds with the same atoms and bonds, but with different spatial arrangements of their bonds and atoms around rigid structures, such as double bonds.

CONCEPT TEST: Why doesn't 1-butene have cis and trans isomers?

SAMPLE EXERCISE 12.9: Name the compounds with the following condensed molecular structures:

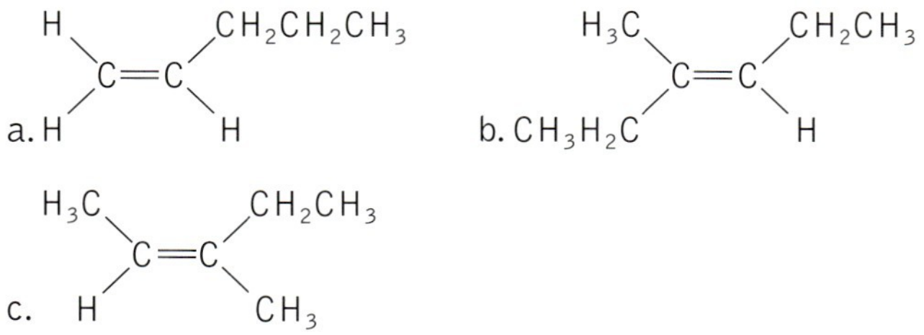

a.
$$\begin{array}{c} H \\ \diagdown \\ H \end{array} C=C \begin{array}{c} CH_2CH_2CH_3 \\ \diagup \\ H \end{array}$$

b.
$$\begin{array}{c} H_3C \\ \diagdown \\ CH_3H_2C \end{array} C=C \begin{array}{c} CH_2CH_3 \\ \diagup \\ H \end{array}$$

c.
$$\begin{array}{c} H_3C \\ \diagdown \\ H \end{array} C=C \begin{array}{c} CH_2CH_3 \\ \diagup \\ CH_3 \end{array}$$

SOLUTION: a. This structure has five carbon atoms and one double bond. This combination makes the compound a pentene. The double bond is between C-1 and C-2, so the compound is 1-pentene.

b. The longest chain of carbon atoms in the structure goes from the lower left end to the upper right. There are six carbon atoms in this chain, parts of which are on opposite sides of a double bond. That pattern makes this the structure of a trans isomer of hexene. The double bond is between C-3 and C-4 no matter which end of the chain we use to start numbering carbon atoms. If we start numbering from the left end, the methyl group is on C-3. (This numbering direction is better than right-to-left because it provides a lower carbon number for the site of the methyl group.) Therefore, the name of the compound is trans-3-methyl-3-hexene where the first 3 in the name indicates the location of the methyl group and the second 3 indicates the location of the double bond.

c. The longest chain of carbon atoms in the structure goes from the upper left to the upper right and so is on the same side of a double bond between C-2 and C-3. Therefore, this is the structure of a *cis*-2-pentene. The methyl group is attached at C-3; so the name of the compound is *cis*-3-methyl-2-pentene.

PRACTICE EXERCISE: Draw condensed molecular structures of:
  a. *trans*-2-pentene
  b. 2-methyl-2-butene
  c. *cis*-3-methyl-3-hexene
(See Problems 103 and 104.)

How do we name a hydrocarbon with a C≡C triple bond? Consider the compound with the following structure:

$$H-C\equiv C-H$$

To name a hydrocarbon with a C≡C triple bond, we change the ending of the parent alkane (ethane in this case) from -*ane* to -*yne*, which is the ending for **alkynes**. That makes the systematic name of this compound ethyne. Ethyne has a more familiar name, acetylene, and is a widely used fuel in the torches used by metalworkers to cut through steel and other metals.

Like alkenes, longer-chain alkynes, such as butyne, can have structural isomers:

$$H-C\equiv C-CH_2CH_3 \qquad CH_3-C\equiv C-CH_3$$
  1-butyne   2-butyne

✓ **Alkynes** are hydrocarbons that contain one or more C≡C triple bonds.

We again distinguish the two isomers by numbering the bonds in such a way that the triple bond has the lowest possible number. Thus the left-hand isomer is 1-butyne and the right-hand structure is 2-butyne.

CONCEPT TEST: Can alkynes ever have geometrical isomers?

SAMPLE EXERCISE 12.10: Name the compound with the following condensed molecular structure.

$$CH_3-C\equiv C-CH_2CH(CH_3)_2$$

SOLUTION: The name of the compound is based on the longest carbon chain in its structure. Sometimes it is easier to identify the longest chain by redrawing the structure showing all the carbon-carbon bonds. Doing so, we get:

$$CH_3-C\equiv C-CH_2-CH-CH_3$$
$$|$$
$$CH_3$$

The carbon backbone is six carbon atoms long. If we number the atoms from left to right to minimize the number of the location of the C≡C triple bond (between C-2 and C-3), then the methyl side chain is attached at C-5 and the name of the compound is 5-methyl-2-hexyne.

PRACTICE EXERCISE: Name the compound with the following condensed molecular structure:

$$CH_3CH_2-C\equiv C-CH(CH_3)_2$$

(See Problems 105 and 106.)

CHAPTER REVIEW

Summary

SECTION 12.1

Crude oil forms from decomposing plant and animal tissues under anaerobic conditions followed by additional transformations deep in the earth at high temperatures and pressures. Crude oil is separated into gasoline and other products by fractional distillation. The total vapor pressure of an ideal solution of volatile components is predicted by Raoult's law.

SECTION 12.2

Alkanes have the general formula C_nH_{2n+2}, where n is the number of carbon atoms per molecule. The boiling and melting points of alkanes generally increase with increasing n. Cycloalkanes have the formula C_nH_{2n}. The chair molecular shape of cyclohexane has lower energy than the boat form. Structural isomers have the same molecular formula but different bonding arrangements of their atoms and different properties.

SECTION 12.3

The octane grade of gasoline is boosted by the addition of aromatic compounds with planar, hexagonal rings and delocalized π electrons, as in benzene (C_6H_6). Aromatic compounds, particularly polycyclic aromatic hydrocarbons (PAHs), which consist of fused benzene rings, have very stable molecular structures.

SECTION 12.4

Organic compounds containing alcohol (R—O—H) and ether (R—O—R) functional groups are added to gasoline to promote combustion, though they have lower fuel values than those of hydrocarbons.

SECTION 12.5

Carbohydrates are a class of organic compounds with the general molecular formula $C_x(H_2O)_y$. Among the smallest carbohydrates are simple sugars, such as glucose ($C_6H_{12}O_6$). These sugars in their open-chain form contain either aldehyde or ketone functional groups. These chains undergo intramolecular reactions that result in the formation of ring structures that are the dominant form of sugars. Dissacharides and polysaccharides, including starch and cellulose, form from condensation reactions of simple sugar molecules. Hydrolysis is the reverse of a condensation reaction, converting insoluble polysaccharides such as starch back into soluble monosaccharides.

SECTION 12.6

The major constituents of wood are cellulose, hemicelluloses, and lignin. Efforts are underway to increase the efficiency of obtaining ethanol from biomass.

SECTION 12.7

Coal, the product of high-pressure and high-temperature plant decomposition over millions of years, is graded according to its carbon content and fuel value. Anthracite has the highest values of both.

SECTION 12.8

Hydrogen has a high fuel value, but its low density and difficulty of storage make it currently unsuitable for large-scale use. Storage media include metals, such as titanium, zirconium, and hafnium, that form metal hydrides with apparent formula MH_2.

SECTION 12.9

Methanogenic bacteria convert plant biomass into methane. Other products include ethanol and the simplest carboxylic acids: formic acid and acetic acid. Amines are organic compounds containing an —NR_2 functional group, where R can be a hydrogen atom or a hydrocarbon substituent. *Methanosarcina* bacteria convert methylamines into methane, carbon dioxide, and ammonia.

SECTION 12.10

The combustion of organic compounds gives CO_2 and H_2O, which can be trapped and weighed to give the atomic carbon and hydrogen content. The oxygen content is generally obtained as the remainder. The results lead to an empirical formula for the sample; the empirical formula can be converted into a molecular formula if the molar mass has been measured.

SECTION 12.11

Alkenes contain one or more double bonds, the rigidity of which leads to the formation of geometric isomers. Simple alkynes contain at least one triple bond.

Key Terms

alcohol (p. 573)
aldehyde (p. 577)
alkene (p. 597)
alkyl group (p. 568)
alkyne (p. 599)
amine (p. 588)
aromatic hydrocarbon (p. 571)
carbohydrate (saccharide) (p. 576)

carbonyl group (p. 578)
carboxylate anion (p. 587)
carboxylic acid (p. 587)
condensation (p. 581)
ether (p. 573)
fractional distillation (p. 558)
functional group (p. 573)
geometric isomer (p. 598)

hemiacetal (p. 576)
hydrolysis (p. 583)
ideal solution (p. 561)
ketone (p. 581)
monosaccharide (p. 576)
organic compound (p. 573)
polysaccharide (p. 576)
structural isomer (p. 567)

Key Skills and Concepts

SECTION 12.1

Know how to calculate the vapor pressure of an ideal solution of volatile components from Raoult's law.

SECTION 12.2

Be able to name and draw the Lewis structures of n-alkanes with as many as 19 carbon atoms per molecule.

Be able to sketch the boat and chair conformations of cyclohexane and explain why the chair form is the more stable conformation.

Be able to draw the Lewis structures and name branched alkane isomers.

SECTION 12.3

Be able to draw the structures of benzene and substituted benzene compounds.

Understand the meaning of the term *polycyclic* aromatic hydrocarbons.

SECTION 12.4

Understand the meaning of the term *functional group*.

Be able to recognize the *alcohol* functional group (R–O–H) and the *ether* functional group (R–O–R).

Understand why the fuel values of oxygenated compounds such as alcohols and ethers are lower than those of the corresponding alkanes.

SECTION 12.5

Recognize the *aldehyde* group and the *ketone* group and understand that each includes a *carbonyl* group.

Understand the cyclization reactions that result in the *hemiacetal* forms of aldohexoses.

Understand how monosaccharides are reversibly linked together through condensation reactions to make polysaccharides.

Know that fermentation converts starch into ethanol.

SECTION 12.6

Understand the structural differences between starch and cellulose.

SECTION 12.7

Know the characteristics of the different grades of coal.

SECTION 12.8

Be able to explain the advantages and disadvantages of hydrogen as a fuel.

SECTION 12.9

Be able to list the principal products of the bacterial conversion of cellulose.

Be able to draw the Lewis structures of simple carboxylic acids and the anions formed by their ionization in water.

Be able to draw Lewis structures and give the names of simple primary, secondary, and tertiary amines.

SECTION 12.10

Be able to use combustion analysis data to calculate the empirical formula of a substance containing carbon, hydrogen, and oxygen.

SECTION 12.11

Be able to name and draw molecular structures of simple alkenes and alkynes.

Key Equations and Relations

The total vapor pressure (P_{total}) of a solution of volatile compounds equals the sum of the vapor pressures of each pure component x (P_x°) multiplied by the mole fraction of that component in the solution (X_x):

$$P_{total} = X_1 P_1^\circ + X_2 P_2^\circ + X_2 P_2^\circ + \cdots \quad (12.1)$$

QUESTIONS AND PROBLEMS

Alkanes, Aromatic Compounds, and Fuel Values

CONCEPT REVIEW

1. What physical property of the components of crude oil is used to separate them?
2. Do alkanes and cyclic alkanes with the same number of carbon atoms per molecule also have the same number of hydrogen atoms per molecule?
3. What is the hybridization of carbon in alkanes?
4. Are hexane and cyclichexane structural isomers?
5. Why is cyclohexane not planar?
6. Are cycloalkanes saturated hydrocarbons?
7. Do structural isomers always have the same molecular formula?
8. Do structural isomers always have the same chemical properties?
9. Why is benzene a planar molecule?
10. Why are aromatic molecular structures stable?

PROBLEMS

11. At 20°C, the vapor pressure of ethanol is 45 torr and the vapor pressure of methanol is 92 torr. What is the vapor pressure at 20°C of a solution prepared by mixing 25. g of methanol and 75. g of ethanol?

12. A bottle is half-filled with a 50:50 (mole-to-mole) mixture of heptane and octane 25°C. What is the mole ratio of heptane vapor to octane vapor in the air space above the liquid in the bottle? The vapor pressures of octane and heptane at 25°C are 11 torr and 31 torr, respectively.

13. How many carbon atoms are there in a molecule of each of the following hydrocarbons? (a) methane; (b) pentane; (c) cycloheptane; (d) cyclodecane.

14. How many carbon atoms are there in a molecule of each of the following hydrocarbons? (a) octane; (b) nonane; (c) cyclobutane; (d) cyclopropane.

15. Draw and name all the structural isomers of C_5H_{12}.
16. Draw and name all the structural isomers of C_6H_{14}.
17. Which of the following molecules are structural isomers of n-octane?

18. Which of the following molecules are structural isomers of n-heptane?

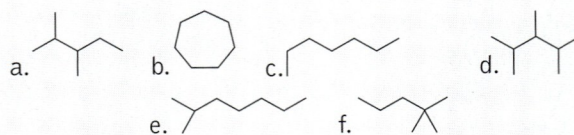

19. Name the hydrocarbons with the following line structures. Which of them are likely to improve the octane rating of gasoline?

20. Name the hydrocarbons with the following line structures and indicate which fractional distillation cut of petroleum is most likely to contain them.

21. Draw the Lewis structures of (a) 2-methylpentane, (b) 2,3-dimethylhexane, (c) 2,3-dimethylbutane, and (d) 2-methyl-3-ethylpentane.
22. Draw the Lewis structures of (a) 3,4,4,5-tetramethylheptane, (b) 3,3,4-trimethylhexane, (c) 2,2,4-trimethylpentane, and (d) 2,2,3,3-tetramethylpentane.
23. Correct the errors in the following names of alkanes and assign a correct name to each compound: (a) 4-methylpentane, (b) 2-ethylbutane, (c) 1-methylpropane.
24. The following names are incorrect: (a) 1-methyl-2-ethylbutane, (b) 1,1,1-trimethylpentane, and (c) 2-ethylhexane. Give the correct name of each compound.
25. Which of the following line structures are aromatic hydrocarbons?

26. Which of the following line structures are aromatic hydrocarbons?

27. Draw line structures of and name the three isomers of diethylbenzene.
28. Draw line structures of and name all of the isomers of trichlorobenzene.
29. Write a balanced chemical equation for the complete combustion of undecane.
30. Waxes contain hydrocarbons with high molar masses. Write a balanced chemical equation for the complete combustion of one such hydrocarbon, $C_{30}H_{62}$.
31. How many grams of methanol must be combusted to raise the temperature of 454 g of water from 20.0°C to 50.0°C? Assume that the transfer of heat from the methanol flame to the water is 100% efficient. How many grams of carbon dioxide are produced in this combustion reaction?
32. Some homes rely on furnaces fueled by heating oil for heat and hot water. How many kilograms of pentadecane ($C_{15}H_{32}$), a component of heating oil, are needed to raise the temperature of 10.0 kg of water from 15.0°C to 50.0°C if 90% of the heat released during combustion ($\Delta H_{comb,C_{15}H_{32}} = -9046$ kJ/mol) is transferred to the water?

Alcohols and Ethers

CONCEPT REVIEW

33. Explain why ethanol boils at a higher temperature than methanol.
34. Why is ethanol much more soluble in water than is either ethane or dimethyl ether?
35. Why are the fuel values of ethanol and dimethyl ether lower than that of ethane?
36. Diethyl ether and butanol have the same molecular formula, $C_4H_{10}O$. Do you think they also have the same heat of combustion?

PROBLEMS

37. Draw the Lewis structures of all the alcohols with the formula C_3H_8O.

38. Draw the Lewis structures of all the alcohols with the formula C$_4$H$_{10}$O.
39. Which of the following are not structural isomers of *n*-hexanol?

40. Which of the following compounds are not structural isomers of methyl *n*-pentyl ether?

41. Draw the condensed molecular structures of
    (a) 2,3-dimethyl-2-butanol, (b) 2-ethyl-1-pentanol, and (c) 5,5-diethyl-2-heptanol.
42. Draw the condensed molecular structures of
    (a) 3-ethyl-3-hexanol, (b) 3,3,5-trimethyl-1-octanol, and (c) 2-methylcyclopentanol.
43. Give the correct chemical name of each of the following alcohols.

44. Name each of the following alcohols.

45. Draw the line structures of (a) ethyl propyl ether, (b) diisopropyl ether, and (c) methyl cyclohexyl ether.
46. Draw the line structures of (a) *n*-butyl cyclopropyl ether, (b) *n*-butyl ethyl ether, and (c) *sec*-butyl ethyl ether.
47. Assign names to the following ethers.

48. Assign names to the following ethers.

Aldehydes, Ketones, and Polysaccharides

CONCEPT REVIEW

49. Name two classes of organic compounds that contain carbonyl functional groups.
50. Why is fructose called a ketohexose, whereas glucose and mannose are called aldohexoses?
51. Can an aldohexose and a ketohexose have the same molecular formula?
52. Give a reason why the hemiacetal of glucose forms between the aldehyde on C-1 and the OH group on C-5 and not the OH group on C-3.

PROBLEMS

53. Which of the following compounds are structural isomers of pentanal (C$_5$H$_{10}$O)?

54. Ethanal and 2-propanone (acetone) have the same molecular formula but different structures. Which compound is a ketone?

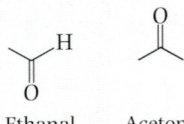
Ethanal Acetone

55. Which of the following compounds are aldehydes? Are any of the molecules structural isomers of one another?

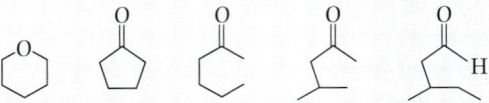

56. Which of the following compounds are ketones?

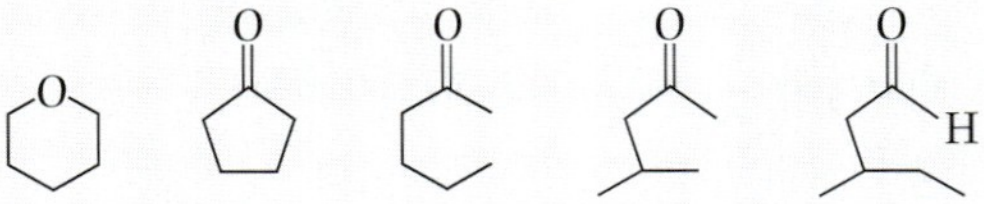

57. Draw a hemiacetal structure for each of the following aldohexoses.

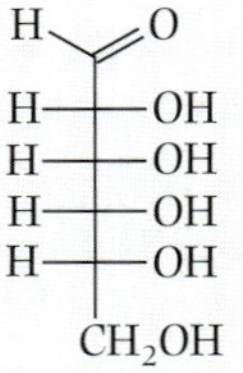

Allose 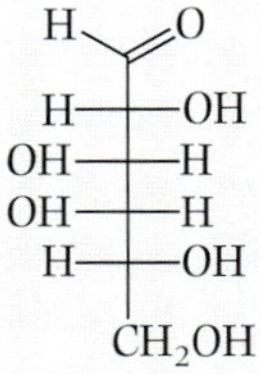 Galactose

58. Draw the hemiacetal structure for each of these monosaccharides.

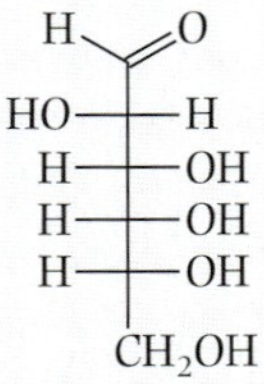

Altrose 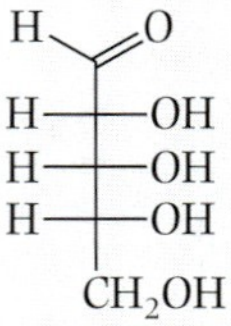 Ribose

59. Which of the following structures are α and β isomers of the same monosaccharide?

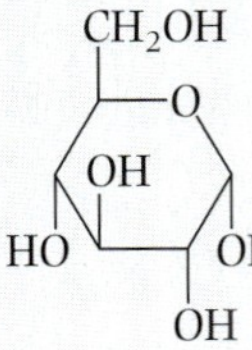

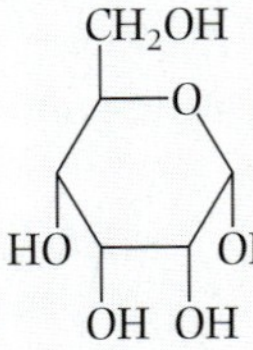

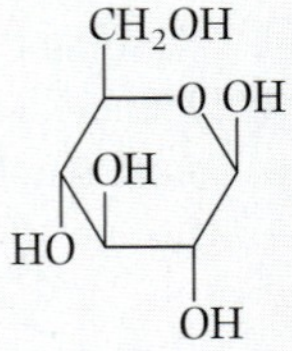

60. Identify the α and β isomeric pair from among these structures.

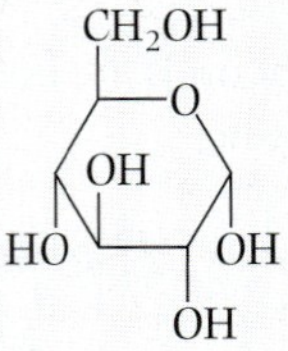

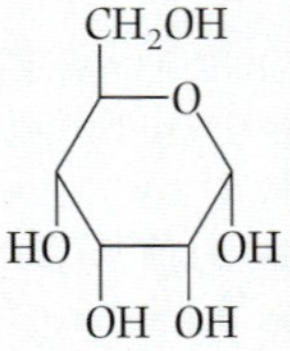

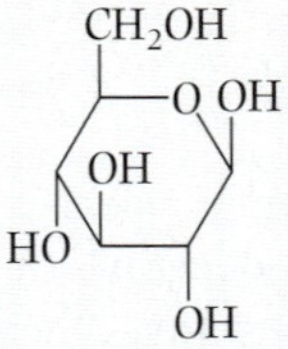

61. Is it possible to form both α- and β-glucose from cyclization of the following aldohexose?

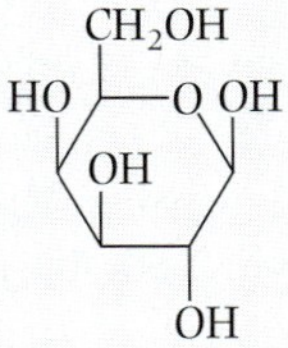

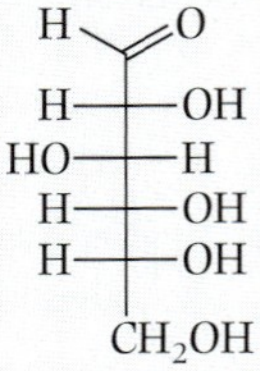

62. Which of the following hemiacetals cannot be obtained from the aldohexose form of glucose?

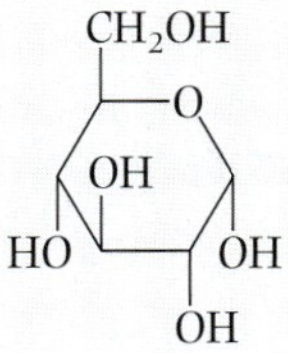

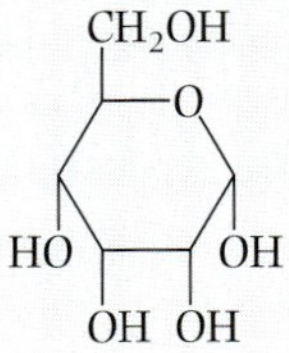

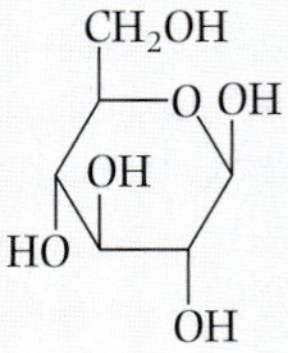

Cellulose

CONCEPT REVIEW

63. What is the molecular structural difference between starch and cellulose?
64. Why are the fuel values of the principal constituents of wood—cellulose, hemicellulose, and lignin—lower than those of alkanes?
65. Explain why the presence of moisture in wood reduces its fuel value.
66. Why is the discovery of enzymes that catalyze cellulose hydrolysis a worthwhile objective?

PROBLEMS

67. Draw the condensed molecular structures of the products of the following reactions:
    a. α-glucose + α-glucose →
    b. α-glucose + β-glucose →
    c. α-glucose + β-mannose →
68. Draw the condensed molecular structures of the products of the following reactions:
    a. α-mannose + β-talose →
    b. β-allose + β-glucose →
    c. β-allose + α-talose →

69. Identify the monosaccharides linked together in each di- or trisaccharide. Which of these saccharides is or are digestible by humans?

70. Draw the structure of a disaccharide that cannot be metabolized by humans.

Coal and Hydrogen As Fuels

CONCEPT REVIEW

71. Even though anthracite coal is often more expensive than bituminous coal, why might it still be a good value?
72. What are some advantages and disadvantages of H_2 as a fuel?
73. Coal deposits are often classified as lignite, subbituminous, bituminous, or anthracite. Predict the order of fuel values for these classes of coal using the data below and explain your reasoning.

Type	C (%)	H (%)
Lignite	70	4
Subbituminous	77	5
Bituminous	80	5
Anthracite	90	3

74. The following empirical formula has been proposed for coal: $C_{135}H_{96}O_9NS$. Describe some of the limitations on using this formula.

PROBLEMS

75. An equimolar mixture of H_2 and CO obtained from coal can be used directly as a fuel (reaction 1) or the mixture can first be converted into methanol (reaction 2a) and then burned (reaction 2b). Which approach, reaction 1 or reactions 2a and 2b, produces more heat per gram of the H_2 and CO mixture?
    (1) $H_2(g) + CO(g) + O_2(g) \rightarrow H_2O(g) + CO_2(g)$
    (2a) $2 H_2(g) + CO(g) \rightarrow CH_3OH(l)$
    (2b) $CH_3OH(l) + \frac{3}{2} O_2(g) \rightarrow 2 H_2O(g) + CO_2(g)$

76. One way of utilizing coal is to convert it into fuel gases in processes called *coal gasification*. Which of the following gasification reactions is exothermic?
    (1) $C(s) + 2 H_2(g) \rightarrow CH_4(g)$
    (2) $2 C(s) + O_2(g) \rightarrow CO(g)$
    (3) $C(s) + H_2O(g) \rightarrow H_2(g) + CO(g)$

Carboxylic Acids and Amines

CONCEPT REVIEW

77. Explain why carboxylic acids tend to be more soluble in water than are aldehydes with the same number of carbon atoms.
78. Explain why methylamine (CH_3NH_2) is more soluble in water than butylamine ($C_4H_9NH_2$).
79. Predict, on the basis of its structure, which compound has a higher fuel value: ethane (CH_3CH_3) or ethanol (CH_3CH_2OH).
80. Predict, on the basis of its structure, which compound has a higher fuel value: acetic acid (CH_3COOH) or ethanal (CH_3CHO).

PROBLEMS

81. Draw the Lewis structures, including all resonance structures, of propanoic acid and the propanoate anion.
82. Draw the Lewis structures of (a) ethyldimethylamine, (b) *t*-butyldimethylamine, and (c) dicyclohexylamine.
83. Assign chemical names to each of the following amines.

 a. b. c. d.

84. Assign chemical names to each of the following amines.

 a. b. c. d.

85. Use the heat-of-formation data in Appendix 4 to calculate $\Delta H°_{rxn}$ for the following reaction, which is catalyzed by methanogenic bacteria:

 $4\ CH_3NH_2(g) + 2\ H_2O(l) \longrightarrow 3\ CH_4(g) + CO_2(g) + 4\ NH_3(g)$

86. Use the heat-of-formation data in Appendix 5 to calculate $\Delta H°_{rxn}$ for the following reaction:

 $4\ CH_3NH_2(g) + 13\ O_2(g) \longrightarrow 10\ H_2O(l) + 4\ CO_2(g) + 4\ NO_2(g)$

Elemental Analyses from Combustion

CONCEPT REVIEW

87. Explain why it is important for combustion analysis to be carried out in an excess of oxygen.
88. Why isn't the quantity of CO_2 obtained in a combustion analysis a direct measure of the oxygen content of the starting compound?
89. Describe the difference between an empirical formula and a molecular formula.
90. What additional information is needed to determine a molecular formula from the results of an elemental analysis of an organic compound?

PROBLEMS

91. The combustion of 135 mg of a hydrocarbon produces 440 mg of CO_2 and 135 mg H_2O. The molar mass of the hydrocarbon is 270 g/mol. Determine the empirical and molecular formulas of this compound.
92. A sample of a carboxylic acid (mass = 0.1000 g) is burned in oxygen, producing 0.1783 g of CO_2 and 0.0734 g of H_2O. Determine the empirical formula of the carboxylic acid.
93. Methylheptenone contains the elements C, H, and O and has a citrus-lemon odor. The complete combustion of 192 mg of methylheptenone produces 528 mg of CO_2 and 216 mg of H_2O. What is the empirical formula of methylheptenone?
94. The combustion of 40.5 mg of a compound that is extracted from the bark of the sassafras tree and that contains C, H, and O, produces 110. mg of CO_2 and 22.5 mg of H_2O. The molar mass of the compound is 162. g/mol. Determine its empirical and molecular formulas.

Alkenes and Alkynes

CONCEPT REVIEW

95. How do structural isomers differ from geometric isomers?

96. Explain why alkanes don't have geometric isomers.

97. Can combustion analysis distinguish between an alkene and a cycloalkane containing the same number of carbon atoms?

98. Can combustion analysis data be used to distinguish between an alkyne and a cycloalkene containing the same number of carbon atoms?

99. Why don't 1-alkenes have cis and trans isomers?

100. Why don't 2-alkynes have cis and trans isomers?

PROBLEMS

101. Draw the Lewis structures of (a) propene and (b) propyne.

102. Draw the Lewis structures of (a) 2-pentene and (b) 2-butyne.

103. Draw the condensed molecular structures of *cis*-3-heptene, *cis*-1,2-dichloro-1-propene, and *trans*-4-methyl-2-hexene.

104. Draw the condensed molecular structures of *trans*-1,2-dichloro-1-butene, *cis*-4-methyl-2-pentene, and *trans*-3-methyl-3-hexene.

105. Assign names to the following alkene and alkyne.

106. Assign names to the following alkene and alkyne.

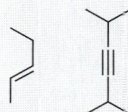

13 Entropy and Free Energy

And Fueling the Human Engine

This well-ordered array of young trees and other landscaping plants in a nursery is unlike the random distributions of plants we find in natural settings. On a microscopic scale, ordered arrays of atoms, ions, and molecules represent lower "entropy" states than those in which particles have more disorder and freedom of motion. A spontaneous reaction results in an increase in the entropy of a reaction system or its surroundings, or both.

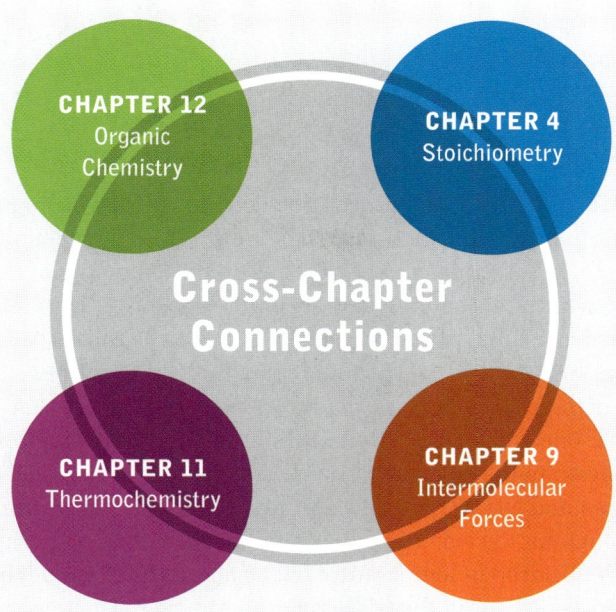

13.1 Enthalpies of solution

13.2 Entropy and why endothermic processes take place

13.3 Entropy calculations

13.4 Free energy
Connecting ΔH and ΔS
The meaning of free energy

13.5 Fueling the human engine
Carbohydrates revisited
Amino acids and proteins
Stereoisomerism
Lipids

13.6 The energy values of carbohydrates, fats, and proteins

13.7 Driving the human engine

13.8 DNA and making proteins

A Look Ahead

In this chapter, we address the question, "Why do some physical changes and chemical reactions happen spontaneously, but others do not?" The answer to this question comes from the second law of thermodynamics and the notion that processes that are exothermic ($\Delta H < 0$) or that create a more disordered system are more likely to occur than processes that are endothermic ($\Delta H > 0$) or that create a more ordered system. Disorder and freedom of motion are expressed by the thermodynamic parameter entropy (S). Processes that yield products with more freedom of motion than the starting materials have a positive ΔS. Processes with $\Delta H < 0$ or $\Delta S > 0$ or both are likely to result in a decrease in free energy ($\Delta G < 0$). These processes occur spontaneously. Many spontaneous biochemical processes consist of more than one step; those steps with negative ΔG values couple with and drive those with positive ΔG values.

13.1 ENTHALPIES OF SOLUTION

In Chapters 11 and 12, we examined the release of energy during the combustion of various fuels. Combustion reactions are exothermic reactions that occur in the gas phase. In this chapter, we will focus on the energetics of chemical reactions and physical changes that occur in solution. Some of them release energy, but some others consume it. We will start with a device that is perhaps familiar to you: the "instant" cold pack, a popular item among the coaches of youth sports teams. Many of these cold packs are made of a sealed plastic bag that contains two compartments: one is filled with water; the other contains ammonium nitrate, NH_4NO_3. To activate the cold pack, the membrane separating the two compartments is broken and NH_4NO_3 dissolves. Ammonium and nitrate ions are rapidly dispersed throughout the water and the temperature of the cold pack drops.

What does this drop in temperature tell us about the thermodynamics of the dissolution process? Clearly, the process is endothermic. The temperature of the solution (mostly water) falls because the dissolution of ammonium nitrate *absorbs* energy, causing a flow of heat from the surroundings (e.g., your injured ankle) into the system (the cold pack). The temperature change of the reaction is related to the enthalpy change that takes place as ammonium nitrate dissolves. This enthalpy change can be measured with a calorimeter of the design described in the following sample exercise. The degree to which a solute absorbs or releases heat as it dissolves is often expressed in terms of its **heat of solution** (ΔH_{soln}): the enthalpy change that takes place as 1 mole of solute dissolves.

CONNECTION: Calorimetry is used to determine heats of reaction (Section 11.9).

✓ **Heat of solution**(ΔH_{soln}) is the change in enthalpy when a mole of solute dissolves in a solvent, such as water. It is the net result of the enthalpy changes associated with the separation of solute and solvent particles from other solute and solvent particles and mixing with one another.

SAMPLE EXERCISE 13.1: Calculate the heat of solution (ΔH_{soln}) of ammonium nitrate in water from the results of the following experiment. An 0.80 g sample of NH_4NO_3 is dissolved in 100. mL of water initially at 20.00°C in an

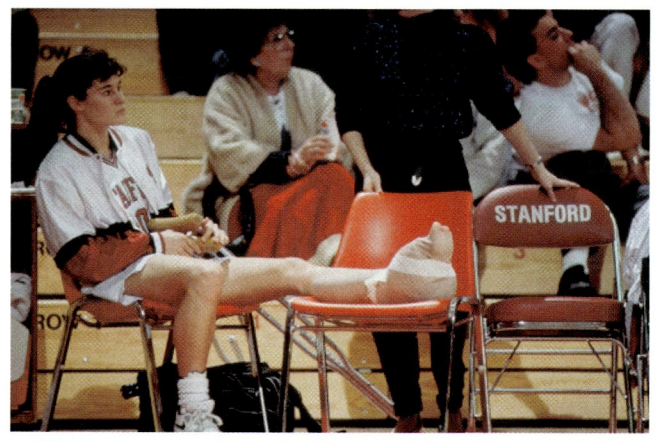

First-aid procedures for minor athletic injuries often include a bag of ice administered at the point of injury. When a supply of ice is unavailable, chemical cold packs based on endothermic dissolution process may be used. Many of these packs are "activated" by squeezing or punching them so that the solid NH_4NO_3 dissolves, forming a solution of ammonium (NH_4^+) and nitrate (NO_3^-) ions. The dissolution reaction is endothermic ($\Delta H_{soln} > 0$).

insulated container of negligible heat capacity. The temperature of the resulting solution is 18.63°C. The density of water is 1.00 g/mL, and the molar heat capacity of liquid water is 75.36 J/mol·C°.

SOLUTION: We can use Equation 11.9 and the decrease in temperature (ΔT) to calculate the quantity of heat lost by the water. We are given the molar heat capacity of water; so we must first convert the given volume of water into an equivalent number of moles.

$$100. \text{ mL } H_2O \left(\frac{1.00 \text{ g}}{\text{mL}}\right)\left(\frac{1 \text{ mol}}{18.02 \text{ g}}\right) = 5.55 \text{ mol } H_2O$$

To calculate the flow of heat (q), we multiply this quantity of water by the molar heat capacity (c) of water and by the temperature change ($\Delta T = T_{final} - T_{initial}$):

$$q = nc\Delta T$$

$$= 5.55 \text{ mol } H_2O\left(\frac{75.36 \text{ J}}{\text{mol·C°}}\right)(18.63 - 20.00)\text{C°} = -264 \text{ J} = -0.264 \text{ kJ}$$

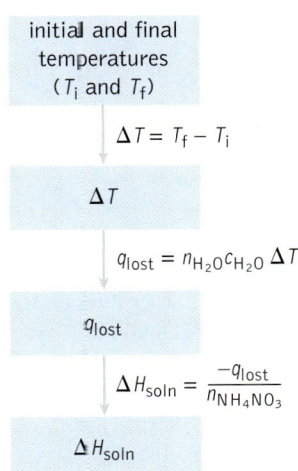

initial and final temperatures (T_i and T_f)

$\Delta T = T_f - T_i$

ΔT

$q_{lost} = n_{H_2O} c_{H_2O} \Delta T$

q_{lost}

$\Delta H_{soln} = \dfrac{-q_{lost}}{n_{NH_4NO_3}}$

ΔH_{soln}

The decrease in water temperature means that the dissolution process is endothermic. If 0.80 g of NH_4NO_3 absorbs 0.264 kJ of heat when it dissolved in water, then a mole of NH_4NO_3 ($M = 80.0$ g/mol) absorbs an amount of heat (ΔH_{soln}) that can be calculated as follows:

$$\Delta H_{soln} = \frac{0.264 \text{ kJ}}{0.80 \text{ g}}\left(\frac{80.0 \text{ g}}{\text{mol}}\right) = 26.4 \frac{\text{kJ}}{\text{mol}}$$

PRACTICE EXERCISE: Some commercially available drain cleaners contain sodium hydroxide. If $\Delta H_{soln, NaOH} = -44.5$ kJ/mol, what is the final temperature of 100 mL of water initially at 18.00°C after 1.00 g of NaOH has dissolved in it? (See Problems 9–14.)

Ammonium nitrate has a positive heat of solution, but ΔH_{soln} values for many other compounds are negative. For example, calcium chloride ($CaCl_2$) is widely used to melt ice on sidewalks in the winter because the presence of Ca^{2+} and Cl^- lowers the freezing point of water and because the heat released as solid $CaCl_2$ dissolves provides additional ice-melting capacity that other salts, such as NaCl, do not have. Sodium thiosulfate ($Na_2S_2O_3$) is another ionic compound with a large negative ΔH_{soln}. It is used in some chemical "hot packs."

A soluble ionic compound has a negative or positive ΔH_{soln}, depending on the relative strengths of the intermolecular intereactions (ion–ion, dipole–dipole, and ion–dipole) associated with the dissolution process. The enthalpy changes that take place in the dissolution of a mole of NH_4NO_3 are shown in Figure 13.1, along with a molecular view of the dissolution process. The ions in NH_4NO_3 (or any ionic solid) are packed together in a crystal lattice, and molecules of liquid water are held together by hydrogen bonds. To dissolve an ionic solid, the lattice energy that holds its cations and anions together must be overcome, which

CONNECTION: The ability of a solute to lower the freezing point of a solvent is described in Section 5.4.

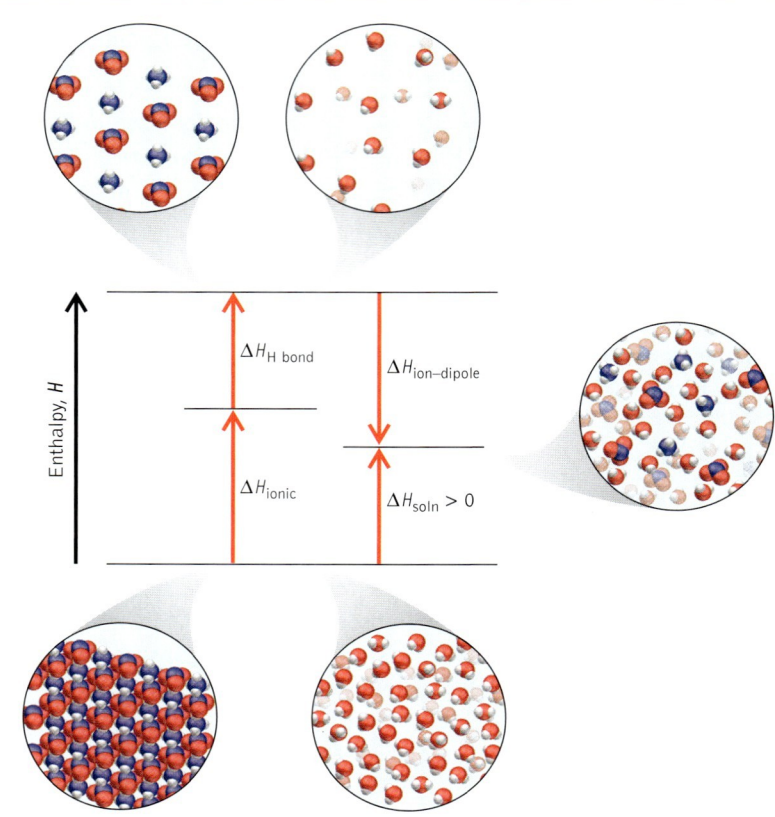

FIGURE 13.1 Enthalpy changes are associated with several processes that take place in the dissolution of a mole of NH_4NO_3. Ionic interactions between ammonium and nitrate ions in NH_4NO_3 must be overcome and the hydrogen bonds between water molecules must be broken. These processes are endothermic ($\Delta H_{\text{H bond}} > 0$ and $\Delta H_{\text{ionic}} > 0$). The formation of ion–dipole interactions between NH_4^+ and H_2O and between NO_3^- and H_2O are exothermic processes ($\Delta H_{\text{ion–dipole}} < 0$). According to Hess's law, ΔH_{soln} equals the sum of $\Delta H_{\text{ionic}} + \Delta H_{\text{H bond}} + \Delta H_{\text{ion–dipole}}$.

CONNECTION: The nature and strength of intermolecular interactions are described in Chapter 9.

requires energy. As the ions dissolve, molecules of water must realign themselves to accommodate the ions. This process entails breaking some hydrogen bonds between water molecules—another process that requires energy and has a positive ΔH. Forming new ion–dipole interactions between dissolved NH_4^+ and NO_3^- ions and polar water molecules is equivalent to forming weak bonds and so is an exothermic process with a negative ΔH. The overall enthalpy change associated with the dissolution process (ΔH_{soln}) will be, according to Hess's law, the sum of the endothermic separation of oppositely charged ions (ΔH_{ionic}) and breaking hydrogen bonds ($\Delta H_{\text{H bond}}$) and the exothermic formation of ion–dipole interactions ($\Delta H_{\text{ion–dipole}}$):

$$\Delta H_{\text{soln}} = \Delta H_{\text{ionic}} + \Delta H_{\text{H bond}} + \Delta H_{\text{ion–dipole}} \qquad (13.1)$$

If the sum of ΔH_{ionic} and $\Delta H_{\text{H bond}}$ more than offsets the negative value of $\Delta H_{\text{ion–dipole}}$, then the overall reaction will be endothermic, which is the case for the chemical cold pack.

The terms on the right-hand side of Equation 13.1 can be subdivided into two groups: one connected with interactions that take place in the solid phase (ΔH_{ionic}) and the other connected with interactions that take place in the liquid phase ($\Delta H_{\text{H bond}}$ and $\Delta H_{\text{ion–dipole}}$). The energy required to separate the anions and cations in a mole of an ionic compound has the same magnitude (but the opposite sign) as the compound's lattice energy (U), as described in Section 9.2. Thus,

$$\Delta H_{ionic} = -U$$

Breaking the hydrogen bonds between water molecules and forming ion–dipole interactions as ions of solute dissolves are processes that, taken together, are called *hydration*. The enthalpy change that accompanies hydration, called the **heat of hydration** ($\Delta H_{hydration}$), can be defined as follows:

$$\Delta H_{hydration} = \Delta H_{H\,bond} + \Delta H_{ion-dipole}$$

With these concepts in mind, we can redefine the enthalpy change that takes place as an ionic solute dissolves (Figure 13.2):

$$\Delta H_{soln} = \Delta H_{hydration} - U \qquad (13.2)$$

As noted in Chapter 9, the attraction between oppositely charged ions is the strongest kind of intermolecular force. Therefore, pulling apart a mole of ions, even those with charges of only $+1$ and -1, requires hundreds of kilojoules of energy. Thus, the dissolution process will be exothermic only if $\Delta H_{hydration}$ has an even larger negative value. It turns out that $\Delta H_{hydration}$ for the ions that we frequently encounter (Table 13.1) do indeed have large negative values. Just as the strengths of ion–ion interactions increase with charge and decrease with increasing distance between ions, so, too, do the strengths of ion–dipole interactions. Thus, $\Delta H_{hydration}$ values for the $+2$ ions in Table 13.1 are nearly four times as great as those of the $+1$ ions of comparable size; within a group of ions with the same charge, the smallest ions have the greatest $\Delta H_{hydration}$ values.

CONNECTION: Lattice energy is the change in enthalpy when gaseous ions combine to form a mole of a solid ionic compound (see Section 9.2).

✓ **hydration** is the process by which water molecules rearrange themselves around particles of solute in solution. The resulting enthalpy change is called the **heat of hydration**.

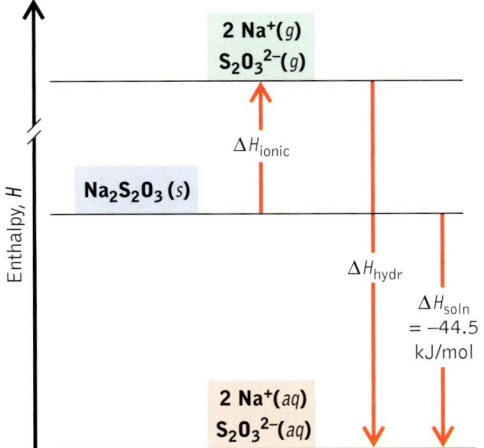

FIGURE 13.2 The heats of solution of ionic compounds are the result of small differences in their lattice energies and the heats of hydration of the ions. In this example these differences lead to a negative ΔH_{soln} for $Na_2S_2O_3$.

TABLE 13.1 Average Heats of Hydration of Some Common Ions

Ion	Ionic radius (pm)	$\Delta H_{hydration}$ (kJ/mol)
Li^+	76	-510
Na^+	102	-410
K^+	138	-336
Mg^{2+}	72	-1903
Ca^{2+}	100	-1591
F^-	133	-431
Cl^-	181	-313
Br^-	196	-284
I^-	220	-247

CONCEPT TEST: Use the trends in heats of hydration in Table 13.1 to develop an explanation for why the dissolution of ammonium nitrate is endothermic.

CONCEPT TEST: Using the concepts and relative sizes of ΔH_{ionic}, $\Delta H_{\text{H bond}}$, and $\Delta H_{\text{ion–dipole}}$, explain why some ionic compounds dissolve in water and others do not.

13.2 ENTROPY AND WHY ENDOTHERMIC PROCESSES TAKE PLACE

Many exothermic reactions are **spontaneous** reactions; that is, they *occur without outside intervention.* Sometimes spontaneous reactions need a little boost to get going, such as the spark from a spark plug igniting a mixture of gasoline and air inside an automobile engine. However, once a spontaneous reaction starts, it will keep going on its own. Although it is true that most exothermic reactions *are* spontaneous, our experience with the cold-pack reaction tells us that some endothermic reactions are spontaneous, too. Similarly, ice cubes melting in a cold drink on a summer day constitute a system undergoing a spontaneous endothermic phase change. Why are some endothermic chemical and physical processes spontaneous? To answer this question, we need to understand the concept of **entropy** (*S*).

> Why are some endothermic chemical and physical processes spontaneous?

Entropy is a measure of the degree of randomness or disorder in a system or in its surroundings. There is a tendency in nature toward randomness and disorder, which means that processes that result in an increase in entropy are favored over those that do not.

Imagine what an urban park would look like if maintenance workers did not pick up the twigs and leaves that fall from trees or the trash left by uncaring visitors. What if they didn't cut the grass, prune the trees, trim the hedges, or paint the benches and tables? In a few years, the ground would be covered with a mixture of natural and man-made litter scattered unevenly among grasses and weeds of random height and density. Trees and hedges would grow into an assortment of sizes and shapes. The parts of benches and tables made of iron or steel would rust away, and untreated wood would rot. In time, the park would begin to take on the appearance of a natural woodland, attractive in its own right but not of much use as a park. Thus, the park's maintenance staff must do work to inhibit or undo processes that increase entropy.

The tendency toward disorder is codified in the **second law of thermodynamics,** which states that *spontaneous reactions are accompanied by a net increase*

☑ In thermodynamics, a **spontaneous process** is one that occurs by itself, without the need for the addition of energy.

☑ **Entropy** (*S*) is a measure of the disorder, or freedom of motion, of a substance or system.

☑ The **second law of thermodynamics** states that a process occurs spontaneously if it results in an increase in the entropy of the universe.

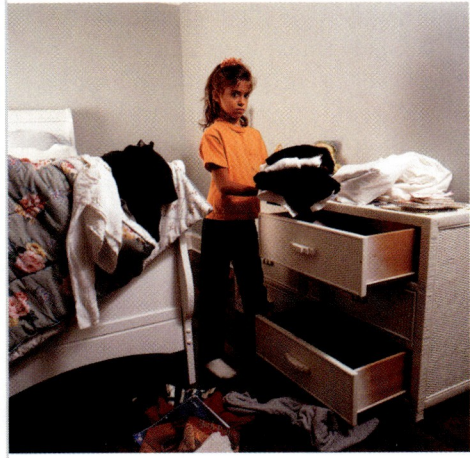 

Entropy is a measure of the disorder or randomness of a system. A neat room has a lower entropy than a messy room. The tendency toward disorder is evident by the seeming ease with which a neat room becomes a messy room.

in the entropy of the universe. By definition, the universe is the sum of a system and its surroundings. Therefore, we can write an equation that relates the change in entropy of the universe (ΔS_{univ}) to the change in entropy of the system (ΔS_{sys}) and of the system's surroundings (ΔS_{surr}), as follows:

$$\Delta S_{univ} = \Delta S_{sys} + \Delta S_{surr} \tag{13.3}$$

If a process is spontaneous, then the entropy of the universe must increase, or

$$\Delta S_{univ} = \Delta S_{sys} + \Delta S_{surr} > 0 \tag{13.4}$$

Let's consider the meaning of Equation 13.4 in the context of the dissolution of ammonium nitrate in a cold pack. The key point is that the dissolution of a solid solute in a solvent increases the freedom of motion of the atoms, ions, or molecules in the solid. The ordered arrangement of dipoles in water is disrupted by the presence of dissolved ions but then reordered as water molecules orient themselves around the ions. The net effect is an overall positive change in entropy of the system ($\Delta S_{sys} > 0$). Similarly, if a concentrated solution is allowed to mix with more solvent, particles of solute have more volume in which to move and $\Delta S_{sys} > 0$. Moreover, if the solute dissolves spontaneously, then, according to the second law, there must have been an overall increase in the entropy of the universe, that is, $\Delta S_{univ} > 0$.

The entropy of a substance is linked to its temperature. This relation is predicted by the kinetic molecular theory: higher temperatures mean higher kinetic energies and root-mean-square speeds of atoms, ions, and molecules. Greater speeds mean more random motion and so more entropy. Conversely,

 CONNECTION: The surroundings are everything in the universe except the system (see Section 11.5); so the system plus its surroundings equals the universe.

 CONNECTION: The relation between temperature and root-mean-square speed of gas molecules is discussed in Section 8.6.

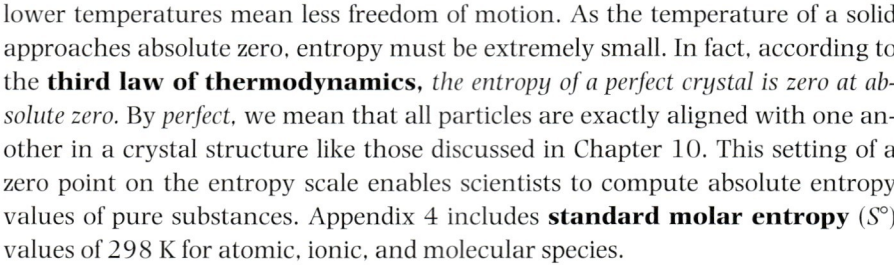

According to the **third law of thermodynamics,** the entropy of a perfect crystal at absolute zero is zero.

The **standard molar entropy** ($S°$) of a pure substance is its entropy at 1 bar of pressure and 298 K. For ions in solution, the standard conditions also include an ion concentration of 1 M.

CONNECTION: The crystal structures in Sections 10.2 and 10.3 represent perfect crystals. Real crystals have defects in their structures, as described in Chapter 18.

lower temperatures mean less freedom of motion. As the temperature of a solid approaches absolute zero, entropy must be extremely small. In fact, according to the **third law of thermodynamics,** *the entropy of a perfect crystal is zero at absolute zero.* By *perfect*, we mean that all particles are exactly aligned with one another in a crystal structure like those discussed in Chapter 10. This setting of a zero point on the entropy scale enables scientists to compute absolute entropy values of pure substances. Appendix 4 includes **standard molar entropy** ($S°$) values of 298 K for atomic, ionic, and molecular species.

Higher temperatures can also lead to changes of state that are linked to changes in entropy. Liquids and solids are condensed phases; so their particles have limited freedom of motion. Still, molecules of liquid water can tumble over and slide by each other. Even the particles in solids have some freedom of motion: although they may not be free to move past their neighbors, they can still vibrate within the space that they occupy, and, the higher their temperature, the faster they vibrate. Atoms, ions, and molecules in the gaseous state have the most freedom of motion and so the most entropy. The changes in entropy that occur as ice at a temperature below its melting point is heated are shown in Figure 13.3. Note the jump in entropy as the ice melts and the even bigger jump (almost five times bigger) when liquid water vaporizes.

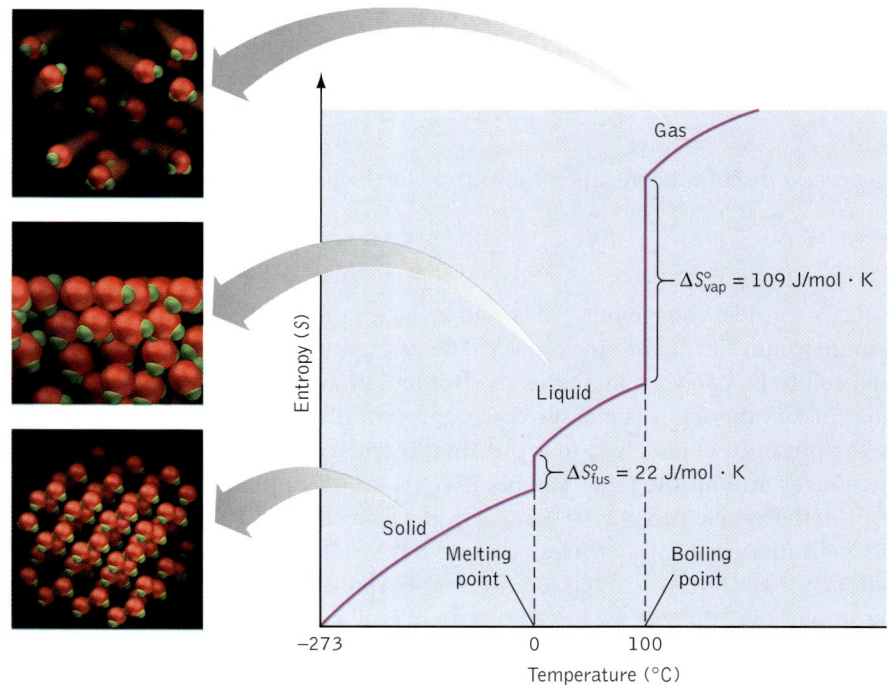

FIGURE 13.3 The entropy of a quantity of water is linked to the motion of its molecules, which increases with increasing temperature. Abrupt increases in entropy accompany changes of state, with the greatest increase during the transition from liquid to gas.

SAMPLE EXERCISE 13.2: Which of the following phenomena result in an increase in entropy ($\Delta S_{sys} > 0$)?
  a. A drinking glass shatters when it is dropped on the floor.
  b. Books scattered on tables and desks in a library are collected and reshelved.
  c. A glass of water evaporates.
  d. The air in a greenhouse is warmed by the sun on a cold winter day.
  e. A puddle of water freezes during a cold winter night.

SOLUTION: An increase in entropy results when the physical state of a substance becomes more disordered. This happens when its temperature increases or when it changes from solid to liquid, liquid to gas, or solid to gas. On the basis of these principles, processes c and d result in increases in entropy. So does process a because a single glass object has become many pieces of glass. Reshelving books (process b) reduces the disorder of their positions in the library and so results in a decrease in entropy.

Water freezing at cold temperatures also results in a decrease in entropy. Because it occurred spontaneously, there must have been a positive change in the entropy of the universe ($\Delta S_{univ} > 0$). We know that ΔS_{sys} is less than 0 as a substance freezes; so ΔS_{surr} must be greater than 0. To make the surroundings more disordered, heat must have flowed into them. That could happen only if freezing were an exothermic process, which it is.

PRACTICE EXERCISE: Describe a process in which (a) ΔS_{sys} and ΔS_{surr} are positive and another process in which (b) ΔS_{sys} and ΔS_{surr} are negative. (See Problems 31 and 32.)

One more important factor influences the absolute entropy of a substance: *the complexity of its structure.* Consider the following standard molar entropies of the C_1 to C_4 alkanes of natural gas.

Compound:	CH_4	C_2H_6	C_3H_8	C_4H_{10}
$S°$ (J/mol · K):	186	230	270	310

Note how entropy increases with increasing molecular size. We can explain this trend by considering the *internal motion* of the atoms in these molecules. Chemical bonds and molecules themselves can stretch and bend and twist and rotate (see Chapter 7). The more bonds there are in a molecule, the more opportunities there are for internal motion, and so the greater the absolute entropy the molecule has. Conversely, a more rigid molecule has less entropy than one that is less rigid. For this reason, carbon in the form of diamond, with its extended tetrahedral network of C–C bonds, has a smaller $S°$ (2.4 J/mol · K) than does carbon in the form of graphite ($S° = 5.7$ J/mol · K) with its networks of hexagons arranged in sheets that can easily slide by one another (Figure 13.4).

Let's reconsider the instant cold pack in the context of the second law of thermodynamics. As its ingredients mix and its temperature drops, the cold pack becomes colder than its surroundings and heat flows from its surroundings into

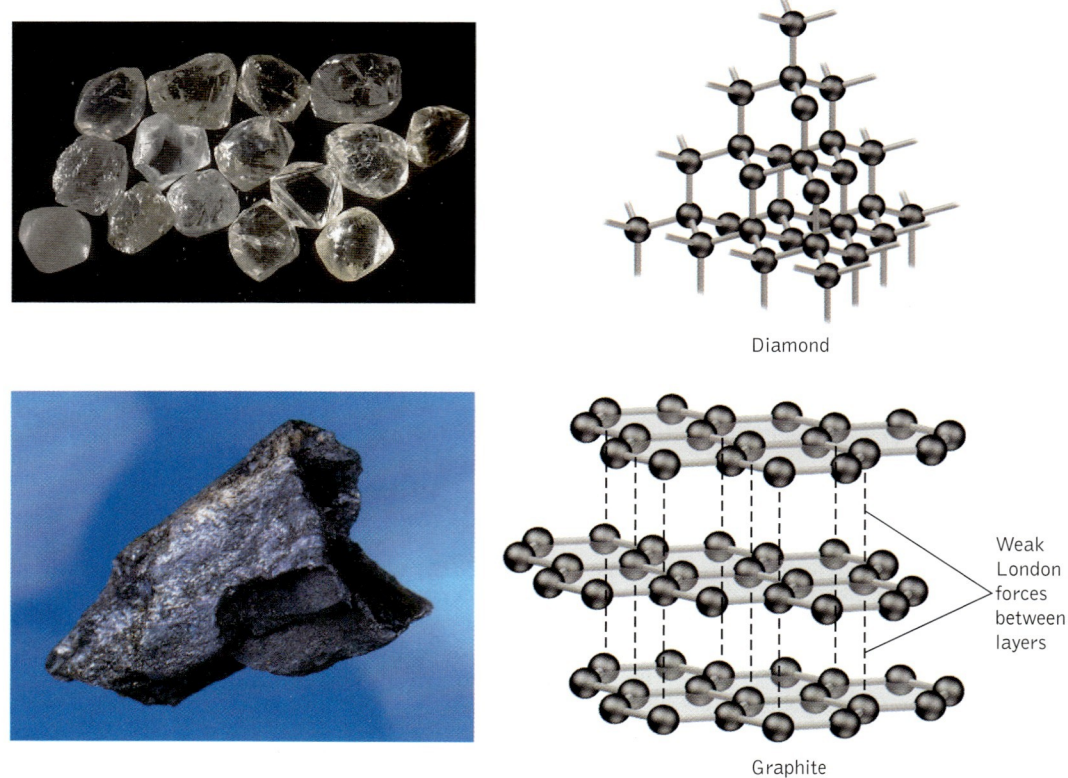

FIGURE 13.4 Graphite has a higher molar entropy than diamond. The four covalent bonds attached to each carbon atom in diamonds makes the three-dimensional array of atoms rigid. However, the sheets of carbon atoms in graphite can slide past each other because they are held together by relatively weak forces, not covalent bonds.

the cold pack. This transfer of heat energy should correspond to a drop in the entropy of its surroundings ($\Delta S_{surr} < 0$). The cold-pack process is spontaneous; so there must be an increase in the entropy of the universe. These observations and conclusions can be expressed in mathematical terms as follows:

1. $\Delta S_{univ} > 0$
2. $\Delta S_{sys} > 0$
3. $\Delta S_{surr} < 0$

It follows from these three inequalities and the second law of thermodynamics that the increase in the entropy of the system must be greater than the loss in the entropy of the surroundings and so, overall, there is an increase in the entropy of the universe.

SAMPLE EXERCISE 13.3: Predict whether the products of each of the following reactions have more or less entropy than the reactants.

a. $C_3H_8(g) + 5\,O_2(g) \rightarrow 3\,CO_2(g) + 4\,H_2O(g)$
b. $CaCO_3(s) + 2\,HCl(aq) \rightarrow CaCl_2(aq) + CO_2(g) + H_2O(l)$
c. $NH_3(g) + H_2O(l) \rightarrow NH_4^+(aq) + OH^-(aq)$

SOLUTIONS:
a. All of the reactants and products in the first reaction are gases, and so they all have relatively high entropies compared with liquids and solids. There are 6 moles of reactant gases and 7 moles of gaseous products. Thus, there is a net increase by one in the number of moles of (high-entropy) gas in the reaction, which means that there is likely to be an increase in entropy ($\Delta S_{sys} > 0$).
b. In this reaction, a total of 3 moles of condensed-phase reactants yield 2 moles of condensed-phase products and, more significantly, 1 mole of gaseous product. The high entropy of gases ensures that this reaction also has a positive ΔS_{sys}.
c. This reaction describes the chemical changes that take place when ammonia gas dissolves in water. The key factor here is the loss of 1 mole of gas to the condensed, aqueous phase. Therefore, there will be a loss in entropy in this reaction ($\Delta S_{sys} < 0$). Note: This example differs from the cold-pack reaction in that the process entails the dissolution of a high-entropy gas, not a low-entropy solid.

PRACTICE EXERCISE: Which of the following processes increase the entropy of the system?
a. Amorphous sulfur crystallizes.
b. Solid carbon dioxide sublimes at room temperature.
c. Iron rusts.

(See Problems 35 and 36.)

13.3 ENTROPY CALCULATIONS

We can calculate the change in entropy under standard conditions for any chemical reaction (ΔS°_{rxn}) from the difference in the weighted sums of the standard molar entropies of m moles of reactants and n moles of products:

$$\Delta S^\circ_{rxn} = \sum n S^\circ_{products} - \sum m S^\circ_{reactants} \qquad (13.5)$$

The symbol Σ represents the sum of all of the entropies of the reactants or products under standard conditions. Each standard entropy, S°, is multiplied by the appropriate coefficient for that substance in the balanced chemical reaction. Thus, for example, 2 moles of a reactant or product has twice the entropy of 1 mole. In other words, entropy is an extensive thermodynamic property that depends on the amount of a substance present or the amount consumed or produced in a reaction.

SAMPLE EXERCISE 13.4: Calculate $\Delta S°$ for the dissolution of ammonium nitrate, given the following standard entropy values:

Species	$NH_4NO_3(s)$	→	$NH_4^+(aq)$	+	$NO_3^-(aq)$
$S°$ (J/mol · K):	151.04		112.8		146.4

SOLUTION: We can calculate $\Delta S°_{rxn}$ by using the given $S°$ values and Equation 13.5:

$$\Delta S°_{rxn} = \sum n S°_{products} - \sum n S°_{reactants}$$

$$= \left(1 \text{ mol} \times 112.8 \frac{\text{J}}{\text{mol·K}} + 1 \text{ mol} \times 146.4 \frac{\text{J}}{\text{mol·K}}\right)$$

$$- \left(1 \text{ mol} \times 151.0 \frac{\text{J}}{\text{mol·K}}\right)$$

$$= 108.2 \text{ J/K}$$

The value of $\Delta S°$ is positive, as expected for the dissolution of a solid.

PRACTICE EXERCISE: Calculate the standard entropy change ($\Delta S°_{rxn}$) for the combustion of methane gas using the appropriate $S°$ values in Appendix 4. (See Problems 37–40.)

13.4 FREE ENERGY

So far, we have identified two driving forces that make chemical reactions happen:

1. The formation of low-energy products from high-energy reactants in exothermic reactions ($\Delta H < 0$).
2. The formation of products that have more entropy than the reactants ($\Delta S > 0$).

A reaction that is *both* exothermic *and* that has a positive ΔS_{sys} must be spontaneous because the flow of heat produced by the exothermic reaction leads to a positive ΔS_{surr}. If ΔS_{sys} and ΔS_{surr} are both positive, then ΔS_{univ} must be positive and, according to the second law of thermodynamics, the reaction must be spontaneous. This condition is shown in Figure 13.5A. On the other hand, a reaction that cools its surrounding ($\Delta S_{surr} < 0$) because it is endothermic ($\Delta H > 0$) and that also results in a loss of entropy ($\Delta S_{sys} < 0$) will *never* be spontaneous, because ΔS_{univ} has to be negative, as shown in Figure 13.5B. Left unanswered is whether an endothermic reaction with a positive ΔS or an exothermic reaction with a negative ΔS will be spontaneous. Either combination may or may not be spontaneous, as illustrated by the various combinations of ΔS_{surr} and ΔS_{sys} shown in Figure 13.5C through F.

13.4 Free Energy

It is not practical to measure changes in the entropy of the universe; so we need a parameter based on the system that enables us to predict when a process will produce a positive ΔS_{univ} and thus be spontaneous. Of particular interest are exothermic reactions that result in a loss of entropy ($\Delta H < 0$, $\Delta S < 0$) or endothermic reactions with a gain in entropy ($\Delta H > 0$, $\Delta S > 0$). As shown in Figure 13.5C through F, such reactions may or may not be spontaneous. Clearly, we need a mathematical equation that allows us to quantitatively balance the effects of ΔH and ΔS.

CONNECTION: The method of calculating $\Delta H°$ for a reaction from the standard heats of formation of its reactants and products is described in Section 11.10.

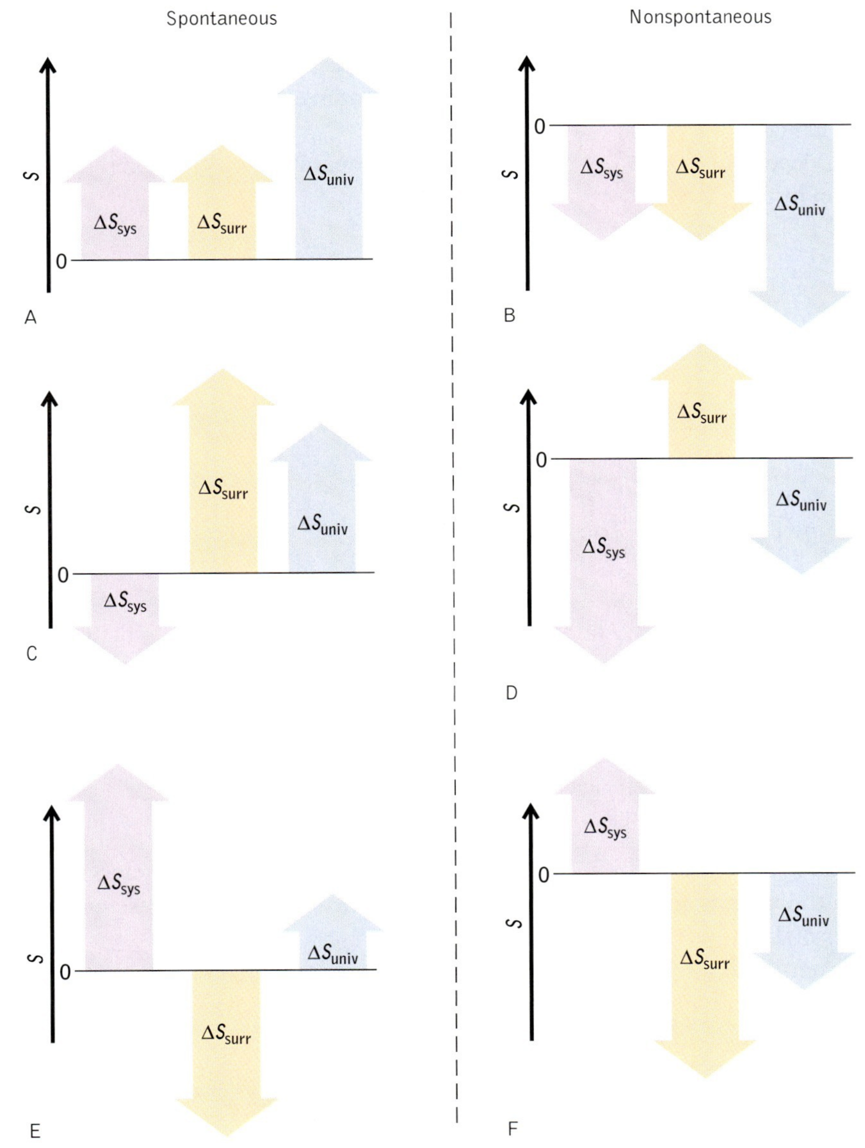

FIGURE 13.5 The magnitudes and signs of ΔS_{sys} and ΔS_{surr} determine the sign and magnitude of ΔS_{univ} and so determine whether a process (including a chemical reaction) is spontaneous. The spontaneous processes in the left-hand column include exothermic processes A and C. The heat they produce warms their surroundings, producing increases in ΔS_{surr}. In process A, ΔS_{sys} is also positive, and so ΔS_{univ} must be positive, too. In process C a decrease in ΔS_{sys} is offset by a greater increase in ΔS_{surr} and so ΔS_{univ} is still positive. Even an endothermic process may be spontaneous when, as in E, a positive ΔS_{sys} is large enough to offset a negative ΔS_{surr}. At the top of the right-hand column endothermic process B results in decreases in both ΔS_{sys} and ΔS_{surr} and so process B must be nonspontaneous. At the bottom of the column endothermic process F is nonspontaneous because a positive ΔS_{sys} is not large enough to offset a negative ΔS_{surr}. In the middle of the column exothermic process D, in which there is a loss in entropy, is nonspontaneous because it does not release enough heat to produce a ΔS_{surr} that is large enough to offset a negative ΔS_{sys}.

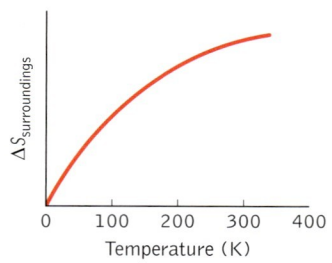

FIGURE 13.6 Entropy increases with increasing temperature, but the dependence of entropy on temperature is not linear. An increase of 100 K between 0 and 100 K will have a greater effect on S than will a 100-K increase between 300 and 400 K.

Connecting ΔH and ΔS

To develop an equation connecting ΔH and ΔS, let's start with the heat (q) produced by an exothermic reaction. This heat flows from the system to the surroundings, so that the heat lost by the system (q_{sys}) equals the heat gained by its surroundings (q_{surr}), or

$$q_{surr} = -q_{sys} \tag{13.6}$$

This flow of heat increases the motion of atoms, ions, and molecules in the surroundings and so increases their entropy ($\Delta S_{surr} > 0$). How large is ΔS_{surr}? The answer depends on the amount of heat released (q). It also depends on temperature. Why? Because the effect of q is greater if the temperature is lower. The addition of heat to surroundings that are cold has a greater relative effect on their entropy than does the addition of heat to surroundings that are already hot, as shown in Figure 13.6.

The dependence of ΔS_{surr} on q_{surr} and on temperature can be expressed in equation form:

$$\Delta S_{surr} = q_{surr}/T$$

or

$$\Delta S_{surr} = -q_{sys}/T \tag{13.7}$$

Now let's assume that the exothermic reaction takes place at constant pressure so that the heat generated is equal to the change in enthalpy of the reaction, or

$$\Delta H_{sys} = q_{sys} \tag{13.8}$$

Combining Equations 13.7 and 13.8, we have

$$\Delta S_{surr} = -\Delta H_{sys}/T \tag{13.9}$$

Combining Equations 13.9 and 13.4 produces a new equation for predicting reaction spontaneity:

$$\Delta S_{sys} - \Delta H_{sys}/T > 0 \tag{13.10}$$

If we multiply both sides by $-T$, we have:

$$\Delta H_{sys} - T\Delta S_{sys} < 0 \tag{13.11}$$

The beauty of Equation 13.11 is that all the thermodynamic parameters relate only to the system and so are easier to measure or calculate than are entropy changes of the surroundings or the universe.

The left-hand side of Equation 13.11 provides a mathematical description of the change in **free energy** of the system and is given the symbol ΔG in honor of American scientist and engineer J. Willard Gibbs (1839–1903). Gibbs coined the term free energy in 1877 and is credited with developing the mathematical foundation of modern thermodynamics. In equation form the change in free energy of a reaction that proceeds at constant temperature (T) and pressure is written:

✓ Gibbs **free energy** is a thermodynamic-state function related to temperature and two other state functions: enthalpy and entropy. Processes in which there is a decrease in free energy are spontaneous.

$$\Delta G = \Delta H - T\Delta S \qquad (13.12)$$

where the lack of subscripts means that all the terms refer to the *system*. If ΔG is negative, then the reaction is spontaneous. We describe reactions with a negative ΔG not as exothermic, though they may be, but as **exergonic**. Reactions with a positive ΔG are called **endergonic**.

Because many reactions, including those in living organisms, take place at nearly constant T and P, ΔG is an extremely useful predictor of their spontaneity. Note that ΔG will be negative (and a reaction will be spontaneous) if ΔH has a large negative value or $T\Delta S$ has a large positive value or both. Thus, Equation 13.12 mathematically expresses our understanding of the two thermodynamic forces that drive chemical reactions. Table 13.2 summarizes the effects of the signs of ΔH and ΔS on ΔG and on reaction spontaneity.

The influence of temperature on the spontaneity of ice melting is presented in terms of ΔH and $T\Delta S$ in Figure 13.7. At higher temperatures, the larger $T\Delta S$ term in Equation 13.12 for ice melting outweighs the unfavorable, positive ΔH term, and so ice melts spontaneously. At temperatures below 0°C, $T\Delta S$ is smaller than the positive ΔH term; so ΔG is positive and ice does not melt spontaneously. At temperatures below 0°C, ΔG for the opposite process (water freezing) is negative, and so, as we know, water freezes spontaneously. There is an important message here: the ΔG of a process or reaction is equal in magnitude but opposite in sign to the ΔG of the reverse process or reaction.

If water freezes spontaneously at $T < 0°C$ and if ice melts at $T > 0°C$, then what happens to ice and water when $T = 0°C$? At that temperature and 1 atm of pressure, the value of ΔG for ice melting or water freezing is zero. What does it mean when the free-energy change for a process equals zero? Theoretically, ice should not melt spontaneously, but water should not freeze spontaneously either. Therefore the composition of a mixture of ice and water should not change at 0°C, and the two phases should *coexist in equilibrium* with each other. A similar situation exists at 100°C and 1 atm, a

> An **exergonic reaction** is one in which reactants form products that have lower free energies, and so ΔG is negative. Exergonic reactions are spontaneous.
>
> An **endergonic reaction** is one in which reactants form products that have higher free energies, and so ΔG is positive. Endergonic reactions are not spontaneous.

What does it mean when the free-energy change for a process equals zero?

TABLE 13.2 Effects of ΔH, ΔS, and T on ΔG and Spontaneity

ΔH	ΔS	ΔG	Spontaneity
<0	>0	Always <0	Always spontaneous
<0	<0	<0 at low temperatures	Spontaneous at low temperatures
>0	>0	<0 at high temperatures	Spontaneous at high temperatures
>0	<0	Always >0	Never spontaneous

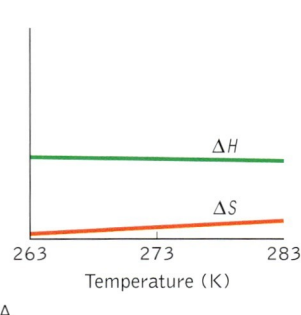 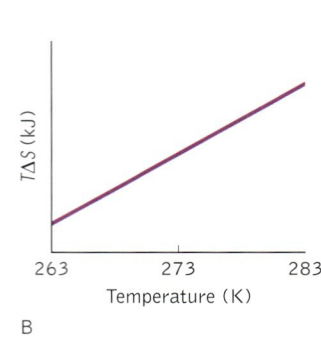 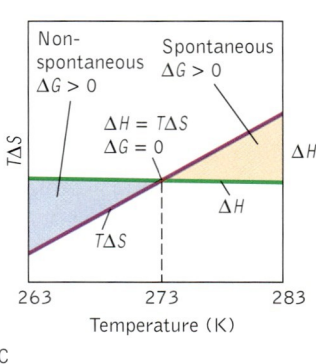

FIGURE 13.7 Changes in entropy (ΔS), enthalpy (ΔH), and free energy (ΔG) for ice melting (fusion) at temperatures ranging from $-10°C$ (263 K) to $+10°C$ (283 K). When ΔH and $T\Delta S$ are plotted on the same graph, we observe that, below 0°C, $T\Delta S$ is less than ΔH; so $\Delta G > 0$ and ice does not melt. Below 0°C, the reverse process is spontaneous and water freezes. Above 0°C, $T\Delta S$ exceeds ΔH; so $\Delta G < 0$ and ice melts spontaneously.

temperature and pressure at which the ΔG of the vaporization of water and the ΔG for the condensation of water vapor equal zero. At 100°C, water and water vapor can exist in equilibrium with each other.

The temperature at which the free-energy change for a process equals zero and the process achieves equilibrium can be calculated from Equation 13.12 if we know the values of ΔH and ΔS for the process. For example, the changes in standard enthalpy and entropy for the fusion of water (melting of ice) are

$$H_2O(s) \longrightarrow H_2O(l) \qquad \Delta H° = 6.03 \times 10^3 \text{ J}; \Delta S° = 22.1 \text{ J/K}$$

We can use this information and Equation 13.12 to calculate the temperature at which ΔG for ice melting will be zero. In doing this calculation, we assume that the values of $\Delta H°$ and $\Delta S°$ do not change significantly with small changes in absolute temperature. For this process and for most physical and chemical changes, such an assumption is acceptable. Therefore we can assume that

$$\Delta G = \Delta H - T\Delta S \cong \Delta H° - T\Delta S°$$

Inserting the values of $\Delta H°$ and $S°$ and the requirement that ΔG be zero, we have:

$$\Delta G = 6.03 \times 10^3 \text{ J} - T(22.1 \text{ J/K}) = 0$$

Solving for T:

$$T = \frac{6.03 \times 10^3 \text{ J}}{22.1 \text{ J/K}} = 273 \text{ K, or } 0°C$$

we obtain the reassuringly familiar value for the melting point of ice.

SAMPLE EXERCISE 13.5: Calculate the free energy change for the dissolution of ammonium nitrate at 298 K, given $\Delta H° = 26.4$ kJ and $\Delta S° = 108.2$ J·K.

SOLUTION: Substituting the values of $\Delta H°$ and $\Delta S°$ for ΔH and ΔS in Equation 13.12, we have

$$\Delta G = \Delta H - T = 26.4 \text{ kJ} - (298 \text{ K})(-0.108 \text{ kJ} \cdot \text{K}) = -5.8 \text{ kJ}$$

PRACTICE EXERCISE: The standard enthalpy and entropy changes for the combustion of methane are $\Delta H° = -801$ kJ and $\Delta S° = -5$ J·K, respectively. Is the reaction spontaneous under standard conditions?

$$CH_4(g) + 2\ O_2(g) \longrightarrow CO_2(g) + 2\ H_2O(g)$$

(See Problems 53 and 54.)

enthalpy change (ΔH), entropy change (ΔS), and temperature (T)

$\Delta G = \Delta H - T\Delta S$

free-energy change (ΔG)

Chemical reactions also can have zero free energies. As a spontaneous chemical reaction ($\Delta G < 0$) proceeds, reactants become products and ΔG becomes less negative (for reasons that will become clear in Chapter 15). In fact, ΔG may reach zero before all the reactants are consumed. In this case, no more products form and no more reactants are consumed. Rather, reactants and products coexist in equilibrium with each other. This state is called *chemical equilibrium* and is the subject of Chapters 15 and 16.

SAMPLE EXERCISE 13.6: A chemical reaction is spontaneous at low temperature but is not spontaneous at high temperature. What are the signs of ΔH and ΔS?

SOLUTION: A reaction is spontaneous at all temperatures if $\Delta H < 0$ and $\Delta S > 0$, and a reaction will never be spontaneous if $\Delta H > 0$ and $\Delta S < 0$. Thus we are left with two other possibilities for the reaction in this exercise: (1) both ΔH and ΔS are positive or (2) both ΔH and ΔS are negative. To decide which is correct, remember that the importance of ΔS increases with increasing temperature, because the product of the two parameters ($T\Delta S$) appears in Equation 13.12. If the reaction is not spontaneous at high temperature where the magnitude of $T\Delta S$ is more likely to overwhelm an unfavorable ΔH value, then entropy considerations work against the reaction being spontaneous. Therefore there must be a loss in entropy, or $\Delta S < 0$. For the reaction to be spontaneous at all, as it is at lower temperatures, ΔH must be negative.

PRACTICE EXERCISE: The synthesis of ammonia from N_2 and H_2 is described by the following chemical equation. The reaction is spontaneous under standard conditions but has a negative $\Delta S°$. Is the reaction exothermic or endothermic? Is the reaction spontaneous at all temperatures?

$$N_2(g) + 3\ H_2(g) \longrightarrow 2\ NH_3(g)$$

(See Problems 57 and 58.)

The meaning of free energy

Perhaps you are wondering why J. Willard Gibbs called his new thermodynamic parameter *free* energy. Let's explore the meaning of free energy by using as a model the combustion of gasoline in an automobile engine. The combustion reaction releases energy (ΔE) in the form of heat (q) and work (w) as defined by Equation 11.3:

$$\Delta E = q + w$$

Much of the heat is wasted energy that flows from the engine to its surroundings. The useful energy is contained in the rapid expansion of gaseous products of combustion. The product of the pressure (P) exerted by these gases on the engine's pistons times the volume change (ΔV) produced by expansion of the gases as the pistons are pushed down is useful ($P\Delta V$) work (see Section 12.3).

The concept of free energy is linked to our attempts to obtain useful work from chemical reactions. The free energy released by a spontaneous chemical reaction at constant T and P is a measure of the *maximum amount of energy that is **free** to do useful work*. To see how this amount compares with the total energy released, let's rearrange the terms in Equation 13.12:

$$\Delta H = \Delta G + T\Delta S \qquad (13.13)$$

Equation 13.13 tells us that the enthalpy change (ΔH) during a chemical process may be divided into two parts. One of them (ΔG) is the useful part that we can try to harvest with devices such as internal combustion engines, steam generators for electricity, or batteries and fuel cells. The other part ($T\Delta S$) is a temperature-dependent change in entropy that is not usable. It is energy that is dissipated as an increase in the entropy of the universe (see Figure 13.8).

Not only is all of the $T\Delta S$ part wasted, but the amount of ΔG that is actually used is never the maximum amount. No device for transforming chemical free

FIGURE 13.8 The combustion of methane can be used to power motor vehicles, but some of the enthalpy of combustion ($\Delta H_{comb} = -802$ kJ/mol) is lost as $T\Delta S$. The remaining free energy, ΔG, is the maximum energy available to do work—in this case, to propel a bus. Not all of this energy is converted into useful work, because the conversion of chemical free energy into mechanical energy is inevitably less than 100% efficient.

$$CH_4(g) + 2\ O_2(g) \longrightarrow CO_2(g) + H_2O(g)$$

energy into, for example, mechanical energy, is 100% efficient. Additional heat losses always flow into the surroundings without doing anything useful. Efficiency is usually inversely related to the rate at which a spontaneous reaction takes place. The slower the reaction, the better the chances of harnessing its change in free energy; the faster the reaction, the greater the amount of free energy wasted heating up the environment.

The free-energy change in a reaction can be calculated from Equation 13.12. It can also be calculated from the standard free energies of formation, ΔG_f°, of the products and reactants (see Appendix 4 for a list of ΔG_f° values). The superscript stands for values of the free energies of formation under standard conditions. We can use free energies of formation to calculate ΔG_{rxn}° by using Equation 13.14. Note the similarities between Equation 13.14 and similar expressions for ΔH_{rxn}° (Equation 11.14) and ΔS_{rxn}° (Equation 13.5). According to Equation 13.14, the free-energy change for a chemical reaction having m moles of products and producing n moles of reactants is the difference in the weighted sums of the free energies of formation of the products minus the reactants. By definition, the free energy of formation of an element in its most stable form under standard conditions is zero.

$$\Delta G_{rxn}^\circ = \sum n \Delta G_{f,\,products}^\circ - \sum m \Delta G_{f,\,reactants}^\circ \qquad (13.14)$$

SAMPLE EXERCISE 13.7: The physical properties of structural isomers of C_8 alkanes are significantly different. So, too, are their free energies of formation, as shown in the following data.

C_8 alkane	ΔG_f° (kJ/mol)
Octane	16.3
2-Methylheptane	11.7
3,3-Dimethylhexane	12.6

Each compound burns according to the following chemical equation:

$$2\ C_8H_{18}(l) + 25\ O_2(g) \longrightarrow 16\ CO_2(g) + 18\ H_2O(g)$$

Predict whether ΔG_{rxn}° of the combustion reactions of these isomers will be the same or different.

SOLUTION: The free energy of combustion can be calculated according to Equation 13.14; however, there is actually no need to do the calculation. The products of the reaction are the same for each isomer; so $\Delta G_{f,\,products}^\circ$ will be the same in each combustion reaction. Because ΔG° of oxygen is zero by definition and ΔG_f° is different for each isomer, $\Delta G_{f,\,reactants}^\circ$ and ΔG_{rxn}° for the combustion reaction are different for each isomer.

PRACTICE EXERCISE: Calculate ΔG_{rxn}° for each of the isomers in the preceding sample exercise. (See Problems 55 and 56.)

13.5 FUELING THE HUMAN ENGINE

Like an automobile engine, the human body has the ability to convert the chemical energy of fuel (food, that is) into heat and mechanical energy. The chemical reactions that convert food into energy in the body consist of many more steps than combustion does. The array of chemical reactions by which food supplies energy is called *catabolism*. The body also uses food as starting material from which it synthesizes the chemicals that make up body tissue in another sequence of chemical reactions known collectively as *anabolism*. In the remaining sections of this chapter, we will focus on the catabolism of the three major categories of energy-producing compounds: carbohydrates, fats, and proteins.

Carbohydrates revisited

The dietary proportions of the major food groups recommended by the U.S. Food and Drug Administration are shown in Figure 13.9. The base of this "food pyramid" consists of whole-grain cereals and breads. Fruits and vegetables are on the next layer. All of these foods are excellent sources of complex carbohydrates (polysaccharides) and are (or should be) the most important source of energy in your diet. In Chapter 12, we noted that carbohydrates come in a variety of molecular shapes and sizes. The simplest carbohydrates are monosaccharides or simple sugars, such as glucose ($C_6H_{12}O_6$), that can have open-chain or cyclic structures. These sugars undergo condensation reactions in which two of them combine to form disaccharides or in which many glucose molecules combine to

CONNECTION:
Carbohydrate structures are described in Section 12.3.

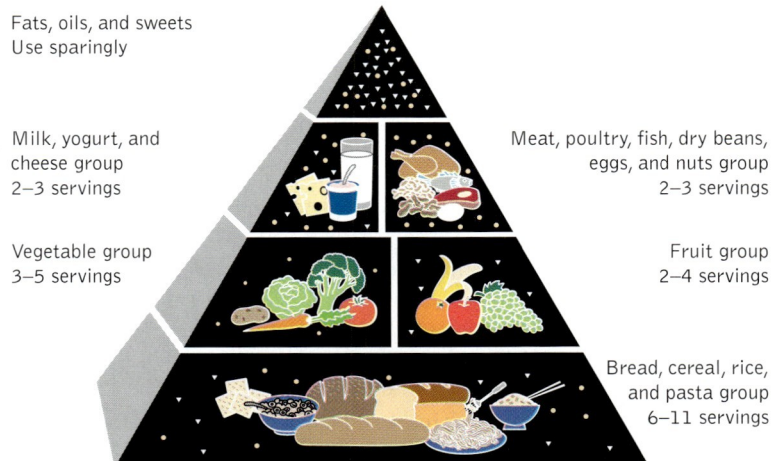

FIGURE 13.9 The U.S. Food and Drug Administration (FDA) recommends a diet consisting mostly of foods that are rich in complex carbohydrates: cereals, grains, fruits, and vegetables. The food pyramid shown here is a convenient way to illustrate these dietary guidelines.

form polysaccharides, such as starch or cellulose. We can digest plant starch because we can hydrolyze the α-glycosidic links between the sugar units in starch. Fortunately for us, plants store starch in easily harvested and rather tasty locations such as fruit, tubers (e.g., potatoes), or seeds, such as peas, beans, corn, and wheat. Unfortunately for us, most of the biomass of agricultural crops is in the form of a polysaccharide that we cannot digest—cellulose. Digesting cellulose means hydrolyzing β-glycosidic bonds in cellulose, which requires an enzyme, cellulase, that we do not have.

Let's use Equation 13.14 to calculate the change in free energy associated with the catabolism of 1.00 gram of glucose to CO_2 and H_2O by using the following standard free energies of formation:

Compound	ΔG_f° (kJ/mol)
$C_6H_{12}O_6(s)$	−910.1
$CO_2(g)$	−394.4
$H_2O(l)$	−237.2

The balanced chemical equation for the reaction is

$$C_6H_{12}O_6(s) + 6\ O_2(g) \longrightarrow 6\ CO_2(g) + 6\ H_2O(l)$$

The stoichiometry and listed ΔG_f° values can be used in Equation 13.14 to calculate the change in free energy as a result of catabolizing one mole of glucose:

$$\Delta G^\circ_{rxn} = [(6\ \text{mol} \times \Delta G^\circ_{f,\ CO_2}) + (6\ \Delta G^\circ_{f,\ H_2O})] - [\Delta G^\circ_{f,\ C_6H_{12}O_6} + (6 \times \Delta G^\circ_{f,\ O_2})]$$

$$= 6\ \text{mol} \times (-394.4\ \text{kJ/mol}) + 6\ \text{mol}(-237.2\ \text{kJ/mol})$$

$$- (-910.1\ \text{kJ} + 6\ \text{mol} \times 0\ \text{kJ/mol})$$

$$= -2366.4 - 1423.2 + 910.1 = -2879\ \text{kJ}$$

Clearly, the catabolism of glucose is spontaneous under standard conditions. Equally clearly, the reverse of the reaction:

$$6\ CO_2(g) + 6\ H_2O(l) \longrightarrow C_6H_{12}O_6(s) + 6\ O_2(g) \qquad \Delta G^\circ_{rxn} = +2879\ \text{kJ}$$

is not spontaneous and will not happen unless energy is added to a mixture of carbon dioxide and water, which occurs during photosynthesis.

To complete our calculation, we use ΔG_{rxn} to calculate the free-energy change in metabolizing 1.00 gram of glucose, the molar mass of which is 180.16 g/mol:

$$1.00\ \text{g}\left(\frac{1\ \text{mol}}{180.16\ \text{g}}\right)\left(\frac{-2879\ \text{kJ}}{\text{mol}}\right) = -16.0\ \text{kJ}$$

Amino acids and proteins

Proteins are biopolymers made of molecular building blocks called **α-amino acids.** These molecules consist of an amino (—NH_2) group and a carboxylic acid (—CO_2H) group bonded to a central carbon atom. One of the other two tetrahedral

> ✓ The prefix α defines the proximity of the two functional groups to each other. Groups that are α to each other are bonded to the same carbon; those that are β to each other are bonded to adjacent carbon atoms; those that are separated by three carbon atoms are γ to each other, and so on through the Greek alphabet.
>
> An **α-amino acid** is a compound in which a carboxylic acid group and an amino group are bonded to the same carbon atom.

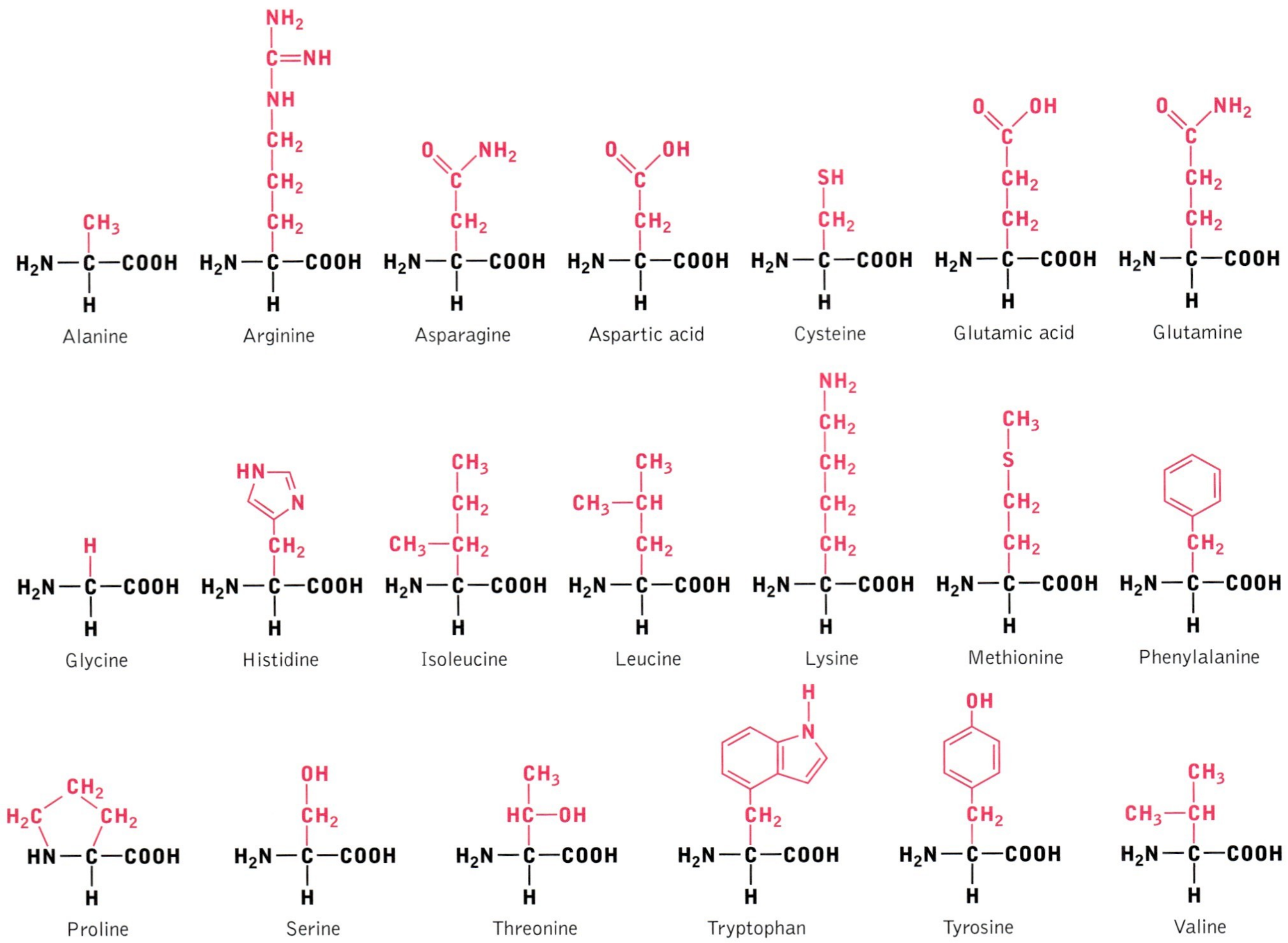

FIGURE 13.10 Proteins are combinations of 20 naturally occurring amino acids. Note that the differences between the amino acids are in the structures of their side chains (highlighted in red) extending from the central carbon atom.

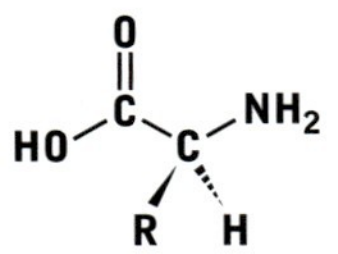

An amino acid contains an NH_2 (amine) group separated from a carboxylic acid (CO_2H) function by a single carbon atom. This carbon atom is covalently bonded to a hydrogen atom and 1 of 20 groups (shown in Figure 13.10) in naturally occurring amino acids.

bonds on this carbon atom is to a hydrogen atom; the second bond is to one of the 20 different R groups or H as shown in Figure 13.10. These groups range in size from the simple methyl group of alanine to the conjugated rings of tryptophan.

Of the 20 different amino acids found in nature and necessary to sustain life, our bodies can manufacture 10. The remaining 10, called the essential amino acids (Table 13.3), must be obtained from foods that are high in protein content, such as dairy products, meat, and some grains. Meat and dairy products contain all 10 essential amino acids, but grains do not. Most grains, for example, are high in methionine but low in lysine. Legumes, including peas and beans, are high in lysine but low in methionine. Therefore, vegetarians must eat foods with complementary amino acid content.

TABLE 13.3 The Essential Amino Acids

Histidine	Lysine	Threonine
Isoleucine	Methionine	Tryptophan
Leucine	Phenylalanine	Valine
Arginine*		

*Essential for children but can be synthesized by adults.

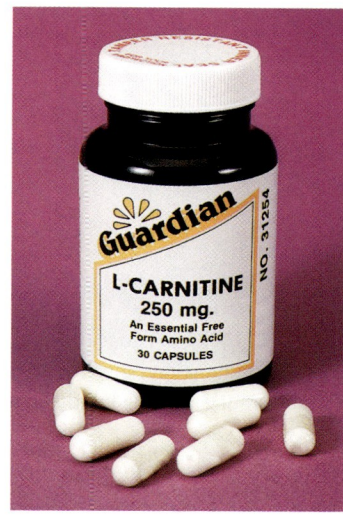

Amino acids are available in health food stores and pharmacies. Some people believe that taking supplemental amino acids is a key to good health—a belief not shared by all nutritionists.

SAMPLE EXERCISE 13.8: Identify which, if any, of the following line structures is that of an α-amino acid.

a. 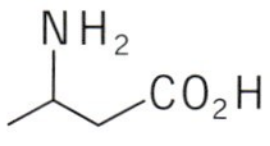 b. 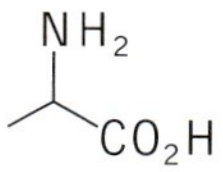 c.

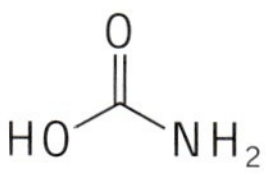

SOLUTION: An α-amino acid contains an amino group and a carboxylic acid group bonded to the same carbon atom. In structure a, there are two carbon atoms between the —NH_2 and —CO_2H groups. In structure c, there is no carbon between the two groups. Structure b is the only one in which the —NH_2 and —CO_2H groups are bonded to the same carbon atom, making this structure the only α-amino acid of the three.

PRACTICE EXERCISE: Which one of the two —CO_2H groups in aspartic acid makes it an α-amino acid (the one on the left or the one on the right in the following structure)?

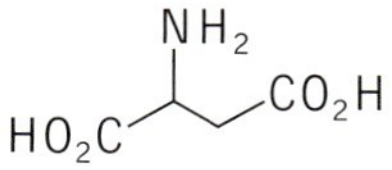

(See Problems 65 and 66.)

Amino acids are linked together in proteins by **peptide bonds.** These bonds form in condensation reactions between the amine group of one amino acid and the carboxylic acid group of another, as shown in Figure 13.11. As proteins from foods are digested, their peptide bonds are hydrolyzed, forming fragments called **peptides,** and eventually forming individual amino acids. This process is the reverse of the condensation reactions that form a protein by linking its component amino acids together. Our bodies use the amino acids released during digestion to synthesize the proteins that we need to build and repair tissue; their catabolism also supplies energy.

> ✓ Large proteins and smaller **peptides** consist of amino acids linked together by **peptide bonds.** A peptide bond is formed by a condensation reaction between the amino group on one amino acid and the carboxylic acid group on another.

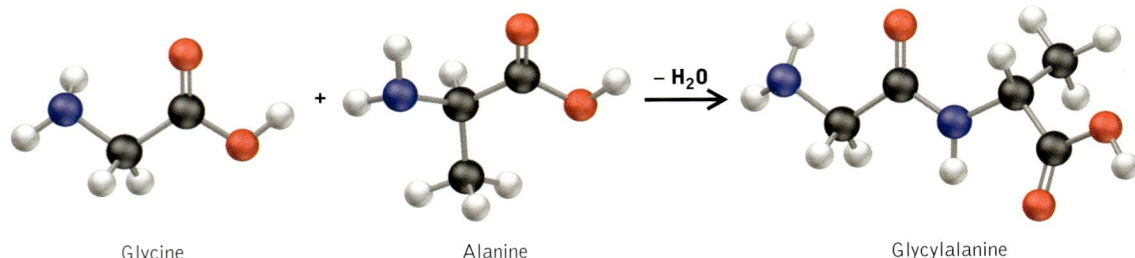

FIGURE 13.11 A condensation reaction between the —NH₂ group of alanine and the —CO₂H group of glycine results in the formation of a peptide bond and a dipeptide called glycylalanine. Large proteins contain hundreds of amino acids connected by peptide bonds.

SAMPLE EXERCISE 13.9: The artificial sweetener aspartame has the molecular structure shown here. The bond shown in red is similar to a peptide bond and undergoes hydrolysis in a similar way.

Aspartame

Draw the structures of the molecules produced by the hydrolysis of the red bond and try to name the products.

SOLUTION: Hydrolysis of the red bond results in addition of OH to the carbonyl carbon atom on the left and H to the nitrogen atom on the right end of the red bond. A check of the structures in Figure 13.10 reveals that the compound formed from the left side of the aspartame structure is aspartic acid. The right-hand side becomes a molecule that resembles phenylalanine, except that the hydrogen atom of the —CO₂H group has been replaced by a methyl group. This substitution makes this compound the methyl *ester* of phenylalanine. Ester groups are described later in this section.

PRACTICE EXERCISE: Draw the dipeptide formed by a condensation reaction between the —NH₂ group of methionine and the —CO₂H group of alanine. (See Problems 67–70.)

Stereoisomerism

We have seen how subtle differences in the orientation of the bonds in α- and β-glucose lead to dramatic differences in their biological activity. Similarly, the biological activity of amino acids strongly depends on the arrangements of bonds around their central carbon atoms. To understand why bond orientation plays such a key role in the behavior of molecules in biological processes, we need to understand the concept of **stereoisomerism**.

The effect of *structural* isomerism on the chemical properties of molecular substances was introduced in Chapter 12 in the context of the octane ratings of the alkanes in gasoline (see Table 12.3). Structural isomers have the same formulas, but their atoms are bonded to one another in different patterns. On the other hand, *stereoisomers* have the same formulas *and* the same chemical bonds. The difference between two stereoisomers is in the *orientation* of their bonds. We saw this difference in Chapter 12—in the cis and trans forms of 2-butene:

> **Stereoisomers** are compounds with the same formula and bonding arrangement that differ in the spatial orientation of the atoms in the molecules.

Both compounds have the same bonds on each carbon atom; the difference is in the orientation of these bonds. As noted in Chapter 12, this difference in bond orientation is called *geometric* isomerism. Geometric isomerism is one form of stereoisomerism.

There is another. In 19 of the 20 amino acids shown in Figure 13.10 (the exception is glycine), the central carbon atom is bonded to four *different* groups: NH_2, $-CO_2H$, $-H$, and $-R$.

Consider the simplest of these amino acids: alanine. There are two, spatially different ways to arrange the bonds around the central carbon atom in alanine, as shown in Figure 13.12. There is no way to turn or twist the form of alanine on the left to make it match its mirror image on the right. Because the two structures cannot be superimposed, they represent two *different compounds*. When stereoisomers cannot be superimposed on their mirror images, they are called **enantiomers**. A molecule with this kind of asymmetry is said to be **chiral**, from the Greek word *cheir*, meaning "hand" (the reflection of a left hand in a mirror looks like a right hand). The carbon atoms bonded to four different groups in chiral compounds are *chiral centers*.

> **Enantiomers** are stereoisomers whose mirror images are not superimposable. Enantiomeric molecules are **chiral** molecules.

CONCEPT TEST: The four bonds around the central carbon atom are arranged in a tetrahedral geometry corresponding to sp^3 hybridization of the carbon atom's valence-shell orbitals. This array makes most amino acids chiral molecules. Could four different groups arranged around a central atom in a square planar geometry also produce a chiral molecule?

CHAPTER 13 Entropy and Free Energy

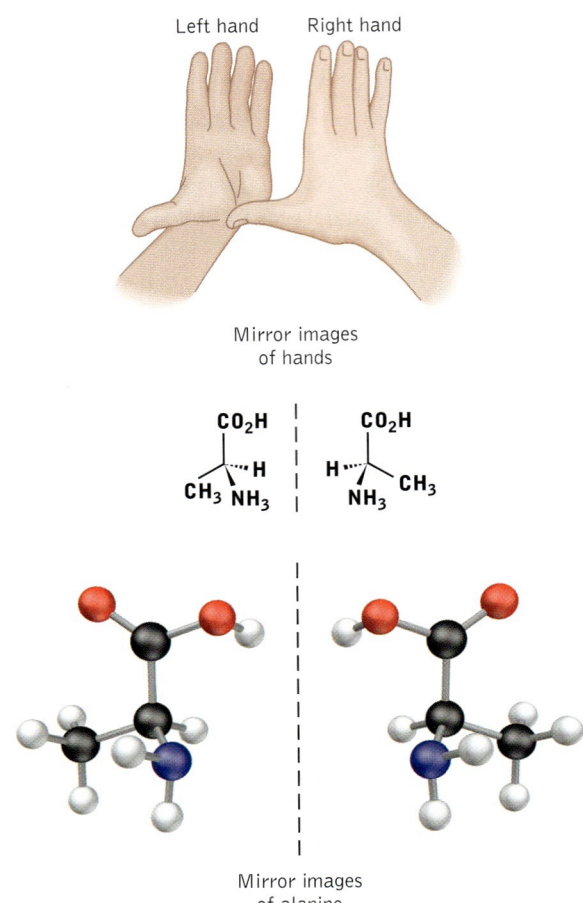

FIGURE 13.12 There are two spatially different ways of placing four different groups in a tetrahedral geometry around a carbon atom. The two isomers of alanine are mirror images of each other that cannot be superimposed on each other. Your hands are another example of two nonsuperimposable objects that are mirror images of each other. To prove it try putting your right hand in a left-hand glove.

The differences between the enantiomers of alanine may be clearer if we view the molecule along the carbon–hydrogen bond, as shown in Figure 13.13. This perspective of the orientation of the other three bonds is known as a Newman projection. We can simplify this perspective by labeling the other three groups "a" for the amino group, "b" for the carboxylic acid group, and "c" for the methyl side group (Figure 13.13). This sequence of labels is determined by the atomic numbers of the atoms of the groups bonded to carbon. Nitrogen ($Z = 7$) has a larger atomic number than that of carbon ($Z = 6$); so the amine group is labeled "a." The carboxylic acid and methyl groups are both connected by C–C bonds, but two bonds within the carboxylic acid group are C–O bonds; the methyl group contains only C–H bonds. Because oxygen has a larger atomic number than that of hydrogen, the carboxylic acid group is designated "b," leaving "c" for the methyl group. In the isomer on the right in Figure 13.13, the arrow marking the pathway from a to b to c is in the clockwise direction. An isomer with this clock-

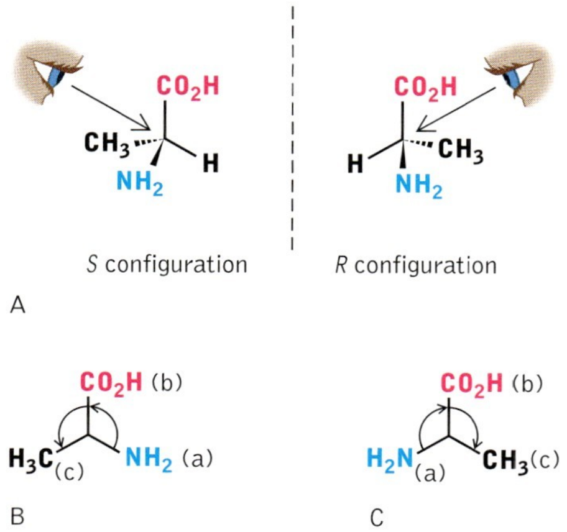

FIGURE 13.13 We can distinguish between the two stereoisomers of alanine by looking at the molecule down the C—H bond (carbon atom in front). The other three groups are labeled "a" (amine group), "b" (carboxylic acid group) and "c" (methyl group). A clockwise pathway connecting group a, b, and c is defined as the R configuration, whereas a counterclockwise pathway is defined as the S configuration.

wise array of substituent groups is designated an (R) isomer, after the Latin *rectus*, meaning "right." In an (S) isomer (from the Latin *sinister*, meaning "left") the arrow describing the a-to-b-to-c pathway is counterclockwise.

SAMPLE EXERCISE 13.10: The (R) form of the drug Prozac is the therapeutic form. Some patients suffer from negative side effects if administered high doses of the drug that contain some of the (S) form.
  a. Identify the chiral center in the structure of Prozac shown here.
  b. Which stereoisomer is it: (R) or (S)?

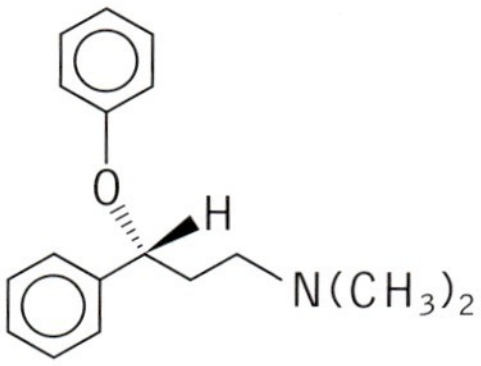

SOLUTION:
  a. The carbon atom in the center of the structure is bonded to four different atoms or groups of atoms: —H, —O—C_6H_5, —C_6H_5, and —CH_2—CH_2—$N(CH_3)_2$. Therefore, the center carbon atom is chiral.
  b. To view the molecule down the C—H bond you would have to be behind the page and looking up to the right. From this perspective the three

different groups are arranged: —O—C_6H_5 at the top, —C_6H_5 to the lower right, and —CH_2—CH_2—$N(CH_3)_2$ to the lower left. The —O—C_6H_5 group has an oxygen atom bonded to the chiral center and so is labeled "a" because oxygen has a larger atomic number than that of carbon. Both of the other groups are bonded to the chiral center by a carbon atom, but the carbon atom in the ring is also bonded to two other carbon atoms, whereas the connecting carbon atom in —CH_2—CH_2—$N(CH_3)_2$ is bonded to only one other carbon atom. Therefore, the —C_6H_5 group is "b." An arrow marking the path a–b–c would turn clockwise, making this structure the (R) configuration.

PRACTICE EXERCISE: Is the form of glyceraldehyde shown below the (R) or the (S) isomer?

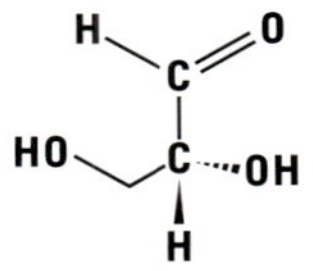

(See Problems 73 and 74.)

Enantiomers have the same composition and nearly the same structures. We might expect them to have similar, if not identical, properties. Are there differences in the physical and chemical properties of enantiomers? Indeed there are. Let's reconsider the structure of glucose:

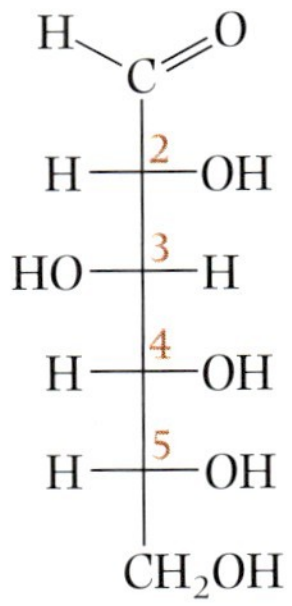

Note that the four carbon atoms from C-2 to C-5 are each bonded to four different substituent groups. This bonding pattern makes each one of them a chiral center. If any of the OH groups at C-2 to C-4 were on the opposite of the carbon atom chain, we would have a different sugar (see Figure 12.15). However, if the OH at C-5 were on the left side of the above structure instead of the right, we would have another stereoisomer of glucose. We distinguish between the two stereoisomers using the following naming convention: when an open-chain structure is drawn as above with the aldehyde (or ketone in

> Are there differences in the physical and chemical properties of enantiomers?

a ketose sugar) group at the top, the structure represents an (R) isomer if the OH group of the chiral carbon atom the farthest from the aldehyde (or ketone) group is pointed to the right. In an (S) isomer the OH group is pointed to the left as shown below.

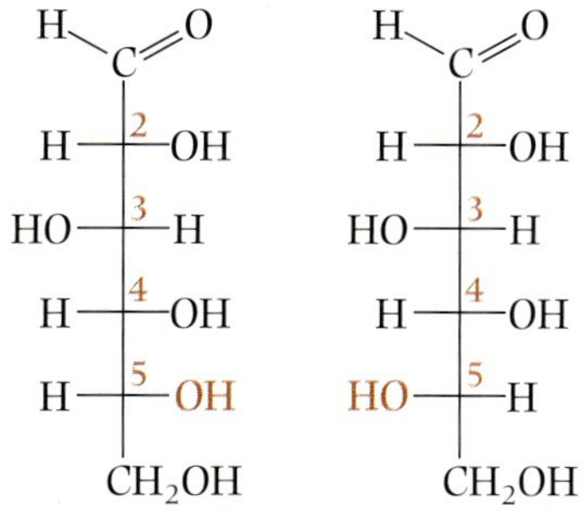

$$
\begin{array}{cc}
\text{H}\diagdown_{\text{C}}\diagup^{\text{O}} & \text{H}\diagdown_{\text{C}}\diagup^{\text{O}} \\
\text{H}\overset{2}{-}\text{OH} & \text{H}\overset{2}{-}\text{OH} \\
\text{HO}\overset{3}{-}\text{H} & \text{HO}\overset{3}{-}\text{H} \\
\text{H}\overset{4}{-}\text{OH} & \text{H}\overset{4}{-}\text{OH} \\
\text{H}\overset{5}{-}\text{OH} & \text{HO}\overset{5}{-}\text{H} \\
\text{CH}_2\text{OH} & \text{CH}_2\text{OH} \\
(R)\text{-glucose} & (S)\text{-glucose}
\end{array}
$$

There is a significance to the handedness of carbohydrates: most of those found in nature and all of those described in Chapter 12 are right-handed (R) isomers. On the other hand (so to speak), we can metabolize and make new proteins only from those amino acids that are S isomers.

Then there is the tragedy regarding the use of the drug thalidomide (Figure 13.14). In the 1950s, the drug was widely prescribed in Europe as a sedative and tranquilizer, particularly to pregnant women. The results of tests with laboratory animals had disclosed no significant side effects. However, its use in humans disclosed that the drug was a powerful *teratogen* (from the Latin *teras*, meaning

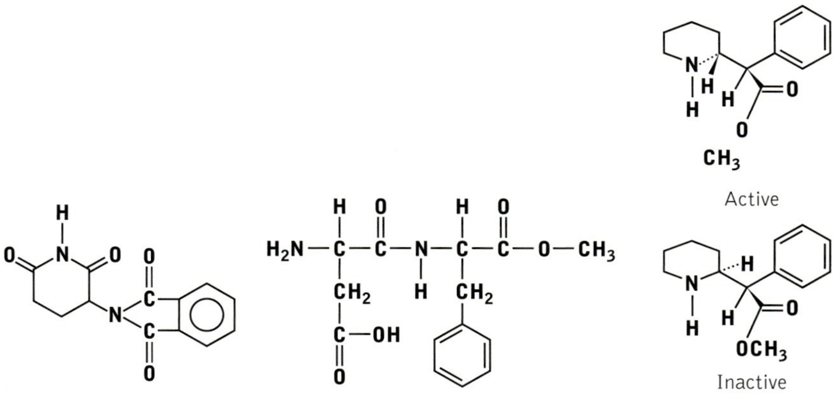

FIGURE 13.14 Chirality is an important property in compounds used in the pharmaceutical industry. The use of a mixture of enantiomers of thalidomide by pregnant women caused birth defects in their babies in the 1950s. The artificial sweetener Nutrasweet is a dipeptide prepared from two (S)-amino acids. The results of some studies suggest that one enantiomer of Ritalin is more potent than the other.

THE CHEMISTRY OF GROUP 5B

Many transition metals are essential to life. The biological functions of some of them are well known: the transport of oxygen by the iron-containing protein hemoglobin, for example, or the benefits to good health of the cobalt-containing enzyme known as vitamin B-12. Lesser-known transition elements, usually ingested in only microgram amounts per day, also may play important biological roles in humans and other organisms. Among them are several elements located in Groups 5B, 6B, and 7B of the periodic table.

Vanadium (V) was named after Vanadis, the Norse goddess of beauty and youth, because of its brightly colored minerals (see the photograph of vanadinite). Vanadium is the 20th most abundant element in Earth's crust. It is present in several minerals, such as vanadinite (lead chlorovanadate, $Pb_5(VO_4)_3Cl$), and as patronite (VS_4) in seams of coal. It is also found in crude-oil deposits, where it is sometimes the most abundant metal present. The ash formed by burning fuels derived from such oil can be more than 50% vanadium, making the ash an important industrial source of the metal. In in-

Vanadinite, a mineral containing lead, chlorine, vanadium, and oxygen, is a principal source of vanadium.

dustry, most of the vanadium produced is used with iron to make particularly strong and hard ferrovanadium steel. Vanadium compounds are widely used as catalysts in, for example, the production of sulfuric acid and synthetic polymers, such as nylon.

Vanadium has an interesting biochemistry. Marine creatures called tunicates accumulate vanadium in their

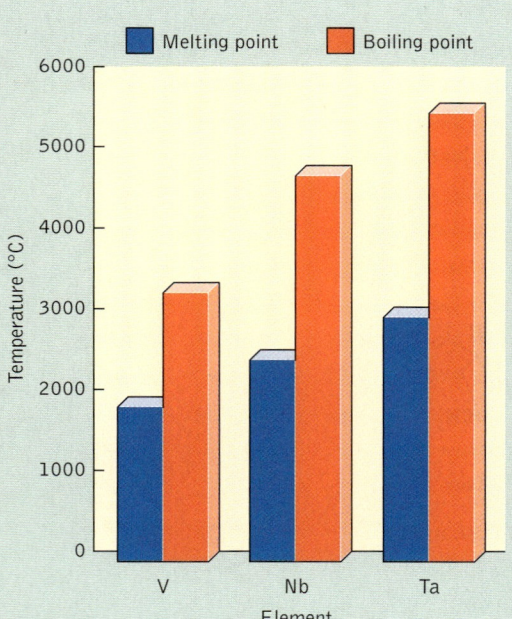

The Group 5B elements show a trend toward increasing melting and boiling points with increasing atomic number.

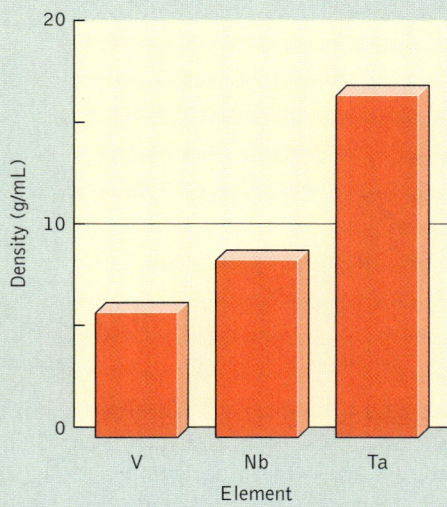

The densities of the Group 5B elements increase with increasing atomic number.

blood at concentrations 10^6 times that in seawater—for reasons that are as yet unknown but that may have something to do with growing the rubbery shell, or tunic, that protects tunicates from potential predators. In some mammals, vanadium may inhibit the biosynthesis of cholesterol. Vanadium has been shown to be an essential nutrient in the growth of rats and chicks, and there is some evidence that it, like chromium, plays a role in lessening the effects of complex metabolic diseases, such as diabetes. Although the VO^{2+} oxocation has been found in substances that mimic the action of insulin, the mechanisms of vanadium biochemistry are, as yet, largely undefined.

The heavier elements in Group 5B play important roles in medicine. Niobium (Nb) and tantalum (Ta) exist together in minerals such as niobite (also called columbite because niobium was once called columbium), which has the formula $(Fe,Mn)(Nb,Ta)_2O_6$, and niobite-tantalite, $(Fe,Mn)(Ta,Nb)_2O_6$. Treatment of these minerals with HF followed by extraction with organic solvents allows the separation of the two metals as their fluorides. After conversion into the oxides Nb_2O_5 and Ta_2O_5, the pure metals are isolated by reduction with Na metal or hot carbon. Niobium-containing metals are superconducting and are used to make the magnets in magnetic resonance imaging (MRI) systems. MRI is a noninvasive technique for diagnosing a range of soft-tissue injuries and illnesses. Tantalum is highly resistant to corrosion, making it ideal material for manufacturing artificial joints and implanted devices.

The atomic and ionic radii of the Group 5B elements increase slightly with increasing atomic number.

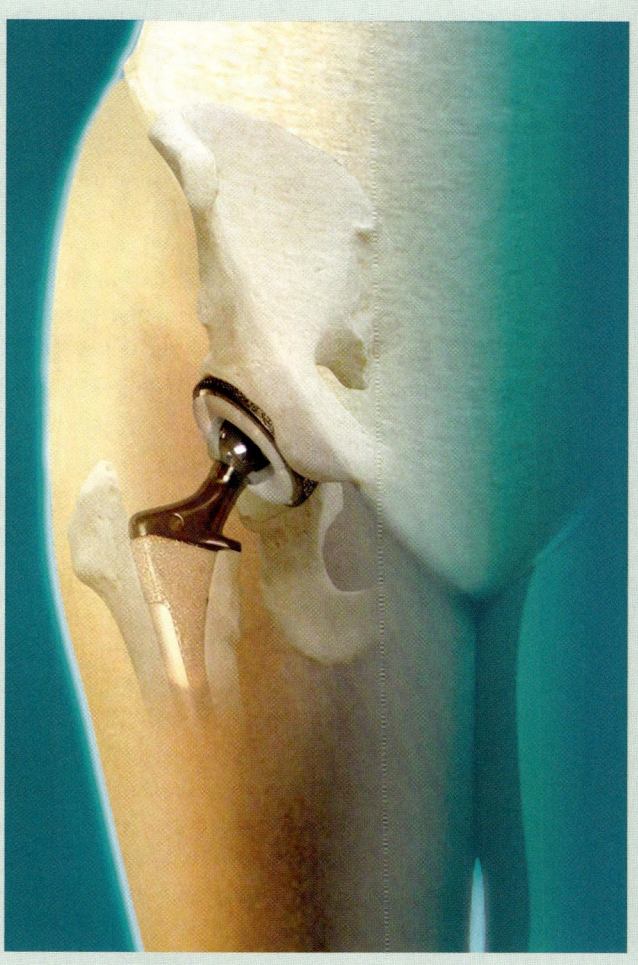

Tantalum is used to coat artificial hip joints.

"monster"); that is, it caused birth defects. Many women who took the drug during the first 12 weeks of pregnancy gave birth to babies suffering from phocomelia, a condition characterized by flipperlike or absent arms and legs and many other deformities. Many thalidomide babies would have been born to American mothers were it not for the refusal of Dr. Frances Oldham Kelsey of the FDA to approve the drug for sale to pregnant women. She recognized that thalidomide had not been tested enough, particularly because the compound had a chiral center and existed as two enantiomers. The drug given to pregnant women was really a 50:50 mixture (called a *racemic* mixture) of the two enantiomers. One of them was a sedative and tranquilizer; the other caused birth defects in thousands of babies.

Lipids

Lipids constitute the third major category of substances that fuel the human engine. They are also a major fuel storage medium in the body. These functions are related to their chemical structures and to their limited solubilities in water. Indeed, lipids are defined as biological molecules that are more soluble in nonpolar solvents than they are in water. This solubility pattern is linked to the large, hydrocarbon-like nonpolar parts of their molecular structures.

There are many categories of lipids, but we will focus on the one that supplies, for many people, a greater proportion of their dietary fuel than it should: fats and oils. These substances are formed from condensation reactions between glycerol (an alcohol with three —OH groups) and compounds known as fatty acids such as the one shown in Figure 13.15. In such a reaction, the —OH group of the carboxylic acid combines with the hydrogen atom of the —OH group of the alcohol, forming a molecule of water and a molecule of a compound called an **ester**:

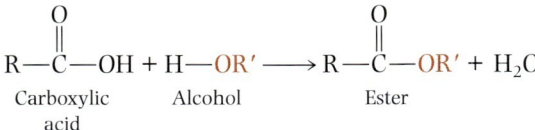

A fatty acid consists of a long hydrocarbon chain ending in a carboxylic acid (—CO_2H) group. When the hydrocarbon chain contains carbon atoms connected only by single C—C bonds, the hydrogen/carbon mole ratio in the chain is the highest that it can be, and the chain is said to be *saturated*, as described in Section 11.1. A fatty acid with such a hydrocarbon chain is called a *saturated fatty acid* (Figure 13.16). However, some fatty acids contain hydrocarbon chains with C=C double bonds. A double bond means that the chain contains two fewer C—H bonds and the hydrogen/carbon ratio is smaller than that of a saturated fatty acid. Such a molecule is called an *unsaturated* fatty acid. Fatty acids that contain only one C=C double bond are called *monounsaturated*; those that contain more than one C=C bond are commonly called *polyunsaturated*.

The greater the degree of saturation, the stronger the London forces between the hydrocarbon chains. These stronger intermolecular forces are the reason

> A **lipid** is a natural substance found in plants and animals that is soluble in nonpolar solvents.

> An **ester** is an organic compound in which a carbonyl group is bonded to an —O—R group. Ester groups are formed by reactions between carboxylic acids and alcohols in a process called esterification.

CONNECTION: The structures of saturated and unsaturated hydrocarbons are described in Sections 11.3 and 12.9, respectively.

why the saturated fatty acid stearic acid (see Figure 13.15) is a solid at room temperature (melting point = 70°C), whereas polyunsaturated linolenic acid (see Figure 13.16), which also has 18 carbon atoms per molecule, is a liquid (melting point = −5°C).

Carbon–carbon double bonds are more chemically reactive than C—C single bonds, which is why unsaturated fatty acids are more easily digested and metabolized by the body. It is a the reason why dietary guidelines encourage the consumption of unsaturated, easily digested fats rather than saturated fats.

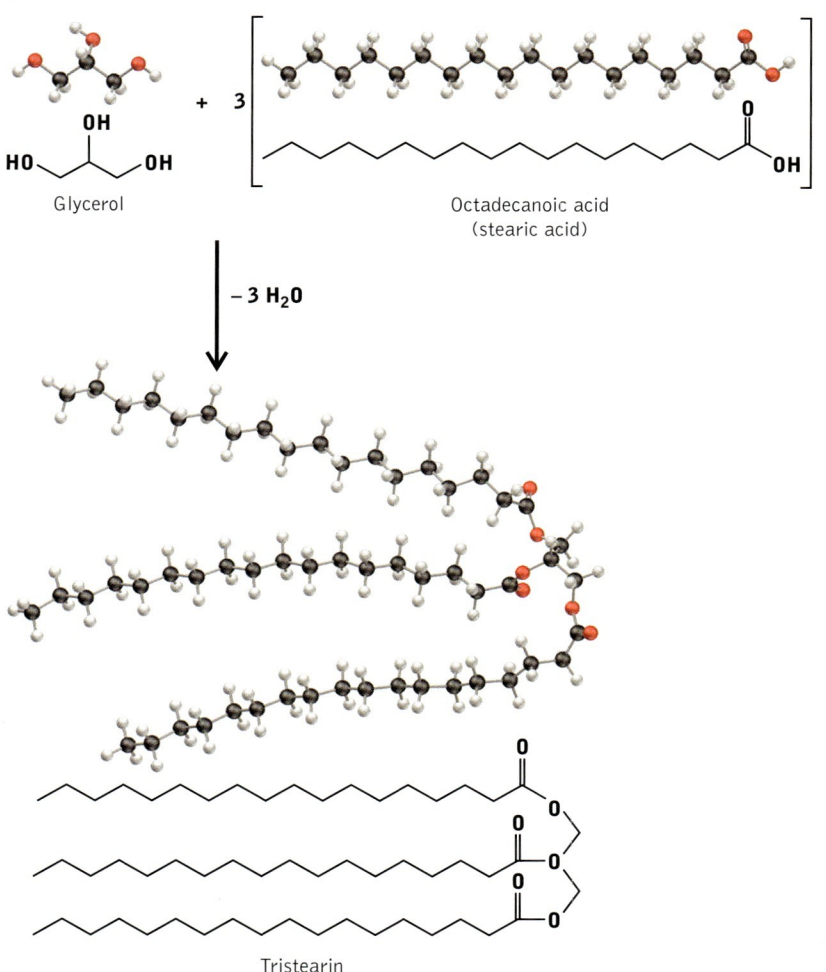

FIGURE 13.15 Condensation reactions between fatty acids and glycerol produce triglycerides, or fats. In this example, the reaction between three —OH groups on a molecule of glycerol and the —CO$_2$H groups on three molecules of octadecanoic (stearic) acid produces a common fat found in mammals.

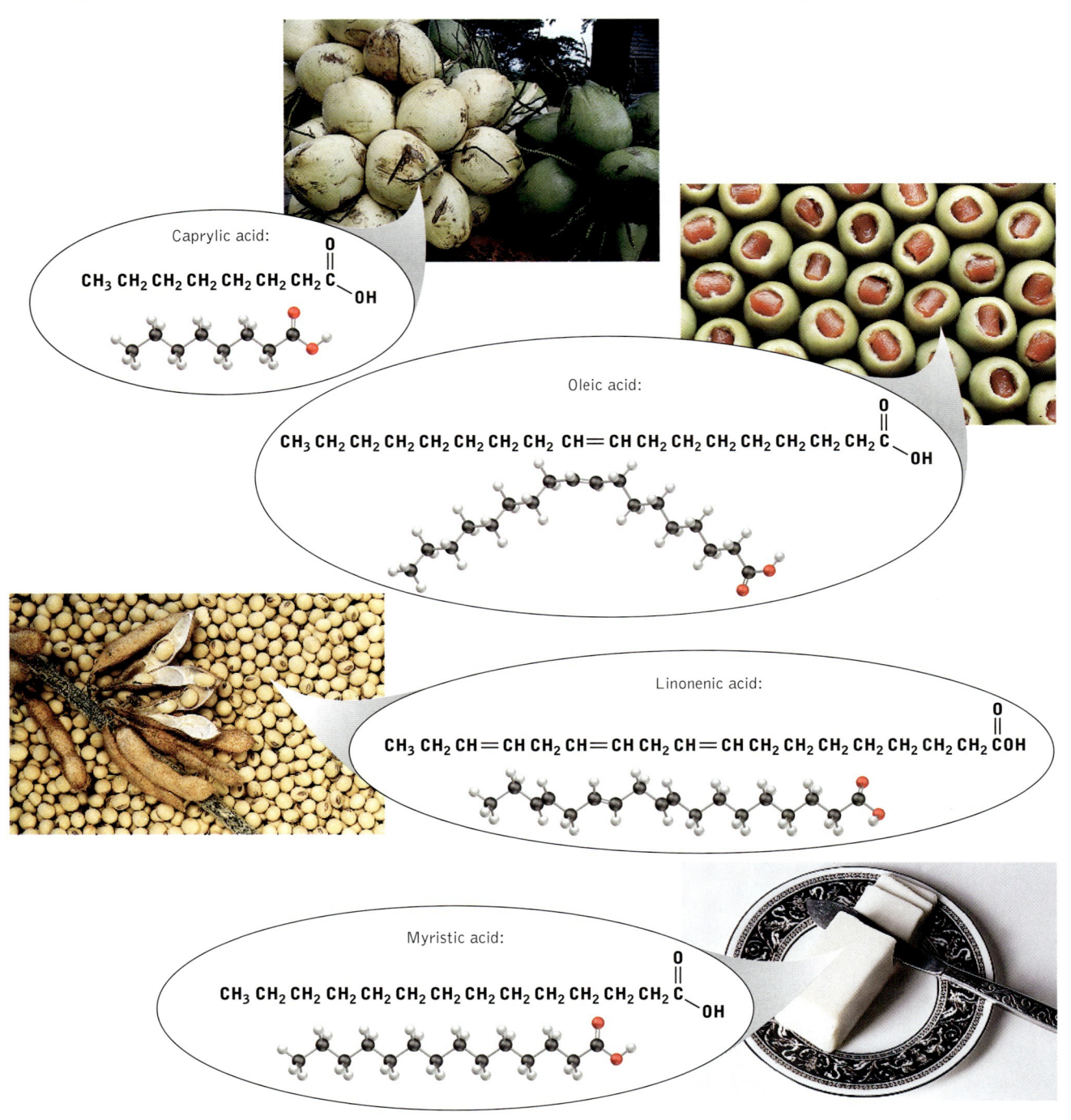

FIGURE 13.16 Fatty acids in fruits and vegetables, such as olives and soybeans, have hydrocarbon chains with one or more C=C double bonds and so are unsaturated. However coconut oil is high in saturated fats that contain, for example, caprylic acid and lauric acid ($CH_3(CH_2)_{10}CO_2H$). Animal products, such as butter fat, are also high in saturated fatty acids including myristic acid, palmitic acid ($CH_3(CH_2)_{14}CO_2H$), and stearic acid (see Figure 13.15).

SAMPLE EXERCISE 13.11: Draw the line structure of the triglyceride produced by the reaction between glycerol and octanoic acid.

SOLUTION: Octanoic acid is a carboxylic acid containing eight carbon atoms. The terminal carbon atom is in the CO_2H functional group, giving the condensed structural formula $CH_3(CH_2)_6CO_2H$. Thus, reaction with glycerol produces the following fatty acid:

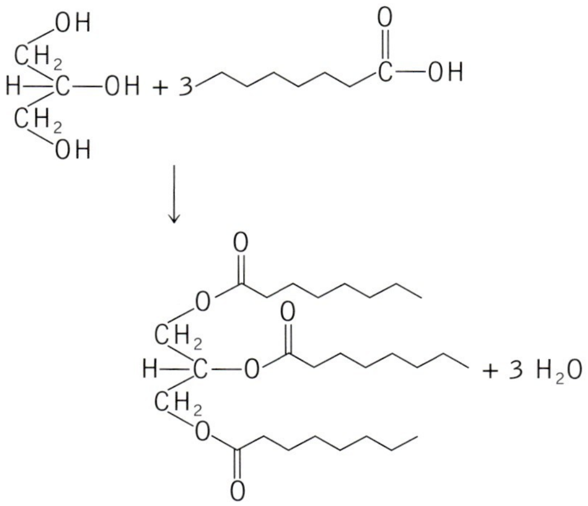

PRACTICE EXERCISE: Draw the line structure of an unsaturated triglyceride with eight carbon atoms in each acid side chain and with C=C double bonds between C-2 and C-3 and between C-5 and C-6. (See Problems 75 and 76.)

13.6 THE ENERGY VALUES OF CARBOHYDRATES, FATS, AND PROTEINS

Carbohydrates, fats, and proteins are important sources of food energy because they are rich in carbon and hydrogen. As noted in Chapters 11 and 12, the combustion reactions of compounds rich in carbon and hydrogen are highly exothermic. Although combustion and catabolism are quite different processes, the overall chemical equations that describe them are the same.

Let us examine how much energy is stored in glucose—a representative carbohydrate. Glucose is a good choice because (1) it is the most common of all sugars and (2) it is a major end product of the decomposition of more-complex carbohydrates in the digestive system; hence it is commonly called "blood sugar." The final products of glucose catabolism are carbon dioxide and water:

$$C_6H_{12}O_6(s) + 6\,O_2(g) \longrightarrow 6\,CO_2(g) + 6\,H_2O(l) \qquad \Delta H° = -2801 \text{ kJ}$$

Taking the absolute value of $\Delta H°$ and dividing by the molar mass of glucose, we obtain its fuel value:

$$\frac{2801 \text{ kJ}}{\text{mol}} \times \frac{1 \text{ mol}}{180.16 \text{ g}} = 15.55 \text{ kJ/g}$$

How does this release of energy compare with the energy obtained from proteins? Proteins are complex macromolecules with different proportions of 20 different amino acids (see Figure 13.10). To enhance our understanding of the range of energy contained within different amino acids, let's consider two of them: alanine, with a relatively small R group (methyl), and phenylalanine, with a relatively large one. Reference books, such as the *Handbook of Chemistry and Physics*, contain tables of thermochemical data including heats of combustion ($\Delta H°_{comb}$) of common organic compounds. These data assume that the C, H, and N atoms in amino acids are converted into CO_2, liquid H_2O, and N_2 during combustion. Such data for alanine and phenylalanine are presented in the following table, with the fuel values obtained by dividing $\Delta H°_{comb}$ by molar mass.

Amino acid	Formula	Molar mass (g/mol)	$\Delta H°_{comb}$ (kJ/mol)	Fuel value (kJ/g)
Alanine	$C_3H_7O_2N$	89.10	−1622	18.2
Phenylalanine	$C_9H_{11}O_2N$	165.19	−4650	28.1

Note that the amino acid with the greater carbon and hydrogen content (phenylalanine) has more value as a fuel. Most amino acids have less hydrocarbon content, and so the actual fuel value of a protein is closer to that of alanine.

Now let's consider the fuel values of fats. We need to select a model that reasonably mimics the fuel value of the large array of compounds that are fats. Let's work with a common triglyceride, tristearin ($C_{57}H_{110}O_6$). The balanced chemical equation describing the conversion of tristearin into CO_2 and liquid H_2O is

$$2 \text{ } C_{57}H_{110}O_6(s) + 163 \text{ } O_2(g) \longrightarrow 114 \text{ } CO_2(g) + 110 \text{ } H_2O(l)$$

The value of $\Delta H°_{rxn}$ for this reaction is $-75{,}520$ kJ. Because two moles of tristearin are consumed, the heat of combustion of tristearin must be half the value of $\Delta H°_{rxn}$ or $-37{,}760$ kJ. Dividing the absolute value of $\Delta H°_{comb}$ by the molar mass of tristearin, 891.51 g/mol, we get a fuel value of

$$\frac{37{,}760 \text{ kJ}}{\text{mol}} \times \frac{1 \text{ mol}}{891.51 \text{ g}} = 42.35 \text{ kJ/g}$$

The condensation reactions that lead to the formation of disaccharides and complex carbohydrates remove water molecules from simple sugars. Because these water molecules have no fuel value, the larger saccharides should have somewhat greater fuel value than that of glucose. In fact, the average fuel value for carbohydrates is about 17 kJ/g. Proteins have, on the average, about the same fuel value as that of carbohydrates. Fats have less oxygen and more carbon and hydrogen content than do carbohydrates and proteins and more than twice the fuel value, about 38 kJ/g on the average. Anyone who has been on a weight-reducing diet is familiar with this difference. One of the most effective ways to lose weight is to reduce the consumption of foods rich in fat. On the other hand, marathon runners and others who specialize in high-endurance athletic events perform better on diets in which nearly half the food energy comes from fats. They are better able to exercise longer at peak capacity, experience less muscle fatigue, use stored energy more efficiently, and avoid infection and inflammation than are high-endurance athletes on low-fat diets.

CONCEPT TEST: The average fuel value of dietary fat, which is a mixture of saturated and unsaturated fats, is about 38 kJ/g, or about 10% less than the fuel value of the saturated fat tristearin (42 kJ/g). Why do unsaturated fats with one or more C=C double bonds in their fatty acid side chains have less fuel value than saturated fats with no C=C double bonds?

Food labels contain information about the amount of carbohydrate, fat, and protein in the products. Information on the energy content of a serving also is provided. This information is related to the energy stored in the bonds of the constituent molecules.

Consider a 1-cup serving of low-fat yogurt. The label on the yogurt container provides the following nutrition facts: 46 g of carbohydrate, 9 g of protein, and 3 g of fat. Using the preceding average fuel values for these three classes of compounds, we can calculate the total fuel value of the yogurt and compare it with the number printed on the label: 240 Calories. Note that a dietary "Calorie" written with a capital "C" is the same as 1000 calories (lowercase "c"), which scientists have used in the past to express quantities of energy. The connection between Calories and kilojoules is

$$1 \text{ Calorie} = 1 \text{ kilocalorie} = 4.184 \text{ kJ}$$

Let's calculate a theoretical value for the energy content of the low-fat yogurt from its composition and the contributions to total caloric content made by the following three groups of nutrients. First, we will calculate the fuel value for each group:

1. Carbohydrates:

$$46 \text{ g} \times \frac{17 \text{ kJ}}{\text{g}} \times \frac{1 \text{ Cal}}{4.184 \text{ kJ}} = 187 \text{ Cal}$$

2. Proteins:

$$9 \text{ g} \times \frac{17 \text{ kJ}}{\text{g}} \times \frac{1 \text{ Cal}}{4.184 \text{ kJ}} = 36 \text{ Cal}$$

3. Fats:

$$3 \text{ g} \times \frac{38 \text{ kJ}}{\text{g}} \times \frac{1 \text{ Cal}}{4.184 \text{ kJ}} = 27 \text{ Cal}$$

Summing these values, we get a total fuel value of 250 Calories, a value close to that on the container label.

SAMPLE EXERCISE 13.12: A 1.0-oz (28. g) serving of tortilla chips contains 7 g of total fat and a total of 140 Calories of fuel value. From the fuel value of one serving and the mass of fat in it, calculate the number of Calories from fat in the serving and what fraction of the total caloric content comes from fat. Assume a fuel value for fat of 9 Calories/gram.

SOLUTION: The calories in 7 g of fat is $7 \text{ g} \left(\frac{9 \text{ Cal}}{\text{g}} \right) = 63$ Cal (the label lists 60 Cal). This value is $\frac{63 \text{ Cal}}{140 \text{ Cal}} = 0.45 = 45\%$ of the fuel value in the tortilla chips. The fat comes almost entirely from the vegetable oil in which the chips are deep fried. Evidence of this may be found on the label for baked (not fried) tortilla chips: a 1.0-oz serving has a fuel value of 110 Calories, of which only 5 Calories are derived from fat.

PRACTICE EXERCISE: Check the "Nutrition Facts" label on a package of your favorite snack food. From the fuel value of one serving and the mass of fat in it, calculate the number of Calories from fat in the serving and what fraction of the total caloric content comes from fat.

13.7 DRIVING THE HUMAN ENGINE

Young women have an average daily nutritional need of 2100 kilocalories; for teenage boys, the figure is 2900 kilocalories, or a little more than 12,000 kJ each day. How do our bodies regulate the production of this much energy without causing us to overheat or to run out of fuel or oxidants or both and die? Millions of years of evolution have resulted in a complex series of chemical reactions that mediate the conversion of blood sugar (glucose) into carbon dioxide and water in our cells. Several sequences of chemical reactions are required. In one sequence, known as **glycolysis,** each mole of glucose, $C_6H_{12}O_6$, is converted into 2 moles of 2-oxopropanoic acid ($CH_3COCOOH$), which is commonly called pyruvic acid. Pyruvic acid or, more precisely, the pyruvate ion (CH_3COCOO^-), feeds the Krebs cycle (citric acid cycle) in which carbon is converted into CO_2. The hydrogen components are eventually converted to H_2O.

These sequences of chemical reactions are not presented here to be memorized. Rather, we will examine a few of their reactions to gain an understanding of how changes in free energy allow spontaneous reactions to drive nonspontaneous reactions and so sustain life. We will first consider an early stage in glycolysis: the conversion of glucose into glucose 6-phosphate. This conversion is an example of a **phosphorylation** reaction (Figure 13.17) in which glucose reacts with hydrogenphosphate ion (HPO_4^{2-}), producing glucose 6-phosphate and

> ✓ In a series of enzyme-catalyzed reactions, **glycolysis** converts each mole of glucose into 2 moles of pyruvate.

> ✓ Reaction of glucose with HPO_4^{2-} gives glucose 6-phosphate in an example of a **phosphorylation** reaction.

water. This reaction is not spontaneous ($\Delta G° = +12.8$ kJ). The energy needed to make this reaction happen comes from a compound called adenosine triphosphate (ATP). Adenosine triphosphate functions in our cells both as a storehouse of energy and as an energy-transfer agent. Adenosine *tri*phosphate hydrolyzes to adenosine *di*phosphate (ADP) in a reaction (see Figure 13.17) that produces a dihydrogenphosphate ion and energy: $\Delta G° = -30.5$ kJ.

In the body, the hydrolysis of ATP and the phosphorylation of glucose consume and produce H_2O and HPO_4^{2-} in complementary ways. As a result, they are coupled together: the spontaneous ATP → ADP reaction supplies the energy that "drives" the nonspontaneous formation of glucose 6-phosphate, as shown in Figure 13.17. This pattern of reactions with negative free-energy changes coupling with and driving those with positive $\Delta G°$ values is present throughout the steps in glycolysis (see Figure 13.10), and in many other sequences of catabolic reactions.

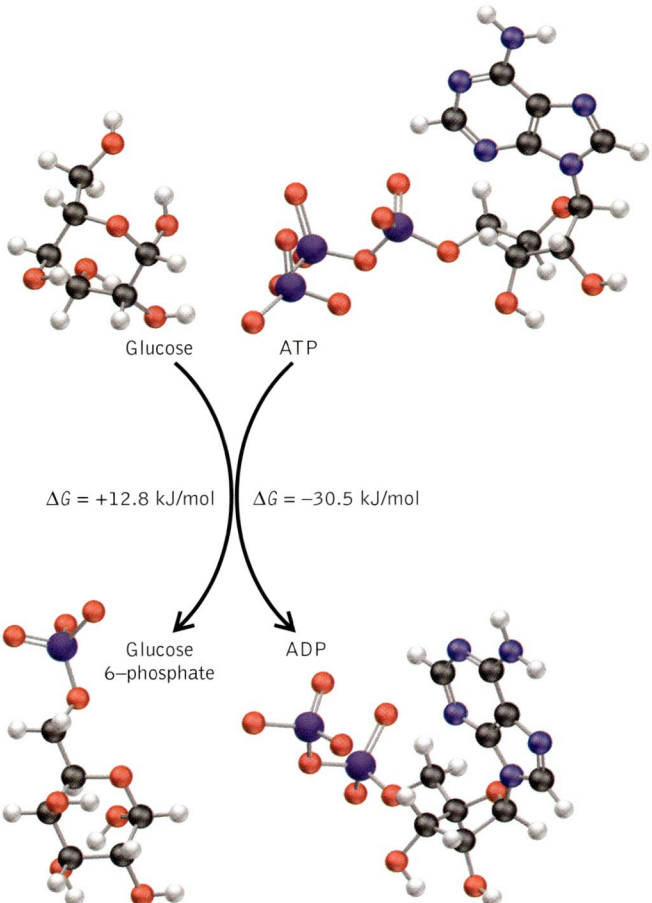

FIGURE 13.17 The conversion of glucose into glucose 6-phosphate, the first step in the catabolism of glucose, requires energy released by the hydrolysis of ATP to ADP. The free-energy change for the overall reaction between glucose and ATP to produce glucose 6-phosphate and ADP is the sum of the free-energy changes of the two reactions.

THE CHEMISTRY OF GROUP 7B

Manganese (Mn) is a relatively abundant element in Earth's crust, ranking third among the transition metals after iron and titanium. The most important Mn ores are pyrolusite (MnO_2) and hausmannite (Mn_3O_4). Manganese metal is produced by the reduction of manganese oxides with hot carbon. More than 90% of the manganese produced in this way is used in metallurgy. Manganese is used with copper or aluminum and in steel and cast iron to enhance their strength, hardness, and resistance to wear.

In biology, manganese is an essential micronutrient for plant growth, and manganese-containing enzymes play a key role in photosynthesis. Manganese is also an essential trace element in higher animals in which it participates in the action of many enzymes. In these biomolecules the manganese ions are usually surrounded by nitrogen atoms in amino acids that may share their lone pairs of electrons with the empty d orbitals of the metal ion. The products of such interactions are a class of proteins called *metalloproteins*.

There are no stable isotopes of technetium (Tc), and so this element is not found in nature. Technetium is formed in the fission of uranium and can be synthesized by neutron bombardment of ^{98}Mo:

$$^{98}_{42}Mo + ^{1}_{0}n \longrightarrow ^{99}_{42}Mo$$

which undergoes β-decay, giving metastable technetium-99:

$$^{99}_{42}Mo \longrightarrow ^{99m}_{43}Tc + ^{0}_{-1}\beta$$

The radioactive-decay process of ^{99m}Tc is described in Chapter 2. The isotope is injected into patients as the pertechnate ion, $^{99m}TcO_4^{2-}$, to radioimage their cardiovascular systems.

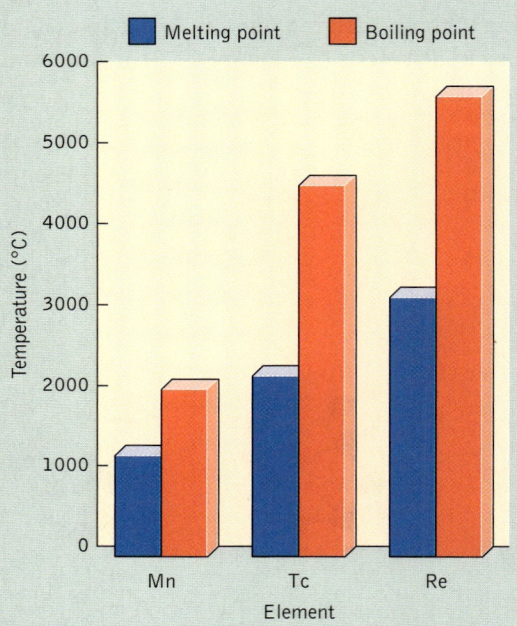

The Group 7B elements show a trend toward increasing melting and boiling points with increasing atomic number.

The free energies of the reactions in these biochemical pathways are additive. In other words, the coupling efficiency is 100%, and all of the energy produced by a favored reaction is available to drive another. Consider the first two steps in glycolysis:

Step 1 $ATP + H_2O \longrightarrow ADP + HPO_4^{2-}$ $\Delta G° = -30.5$ kJ

Step 2 $C_6H_{12}O_6 + HPO_4^{2-} \longrightarrow C_6H_{11}O_6PO_3H^{2-} + H_2O$ $\Delta G° = +12.8$ kJ

If we add the reactions in steps 1 and 2 together and add their free energies of reaction, we get

Rhenium (Re) is the rarest of the Group 5B, 6B, and 7B elements, present at only 0.0007 ppm in Earth's crust and usually as an impurity in molybdenite (MoS_2). It is separated from the molybdenum by conversion into the volatile oxide Re_2O_7, which is deposited from the vapor phase as a solid. Rhenium metal is produced by treating the oxide with hot H_2. Radioactive isotopes of rhenium have recently been used to treat colon-rectal cancer.

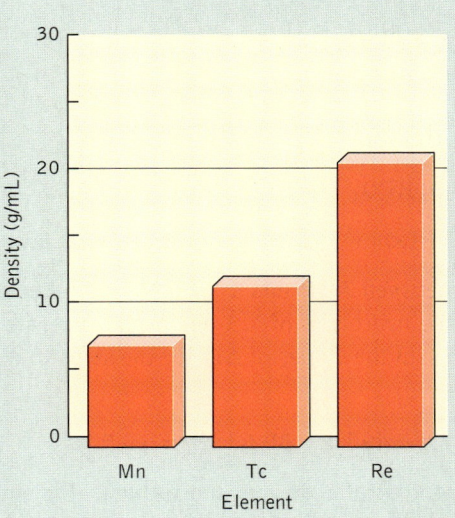

The densities of the Group 7B elements increase with increasing atomic number.

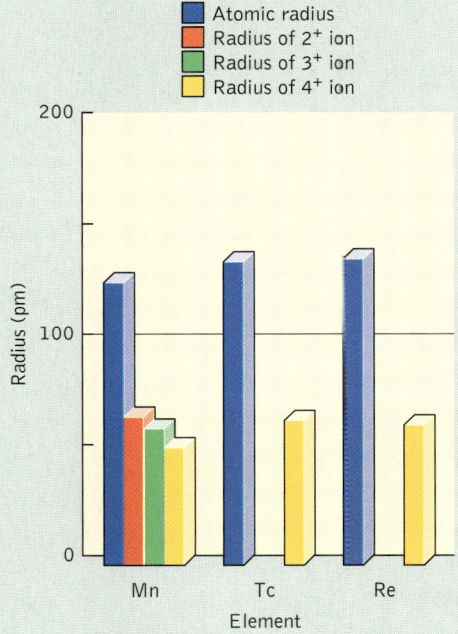

The atomic radii of the Group 7B elements increase with increasing atomic number. Compounds of Tc^{3+} or Re^{3+} have not been identified.

$$ATP + H_2O + C_6H_{12}O_6 + HPO_4^{2-} \longrightarrow$$
$$ADP + HPO_4^{2-} + C_6H_{11}O_6PO_3^{2-} + H_2O$$

$$\Delta G° = -30.5 + 12.8 = -17.7 \text{ kJ}$$

Because equal quantities of H_2O and HPO_4^{2-} appear on both sides of the combined equation, they cancel out to leave the net overall reaction:

$$C_6H_{12}O_6 + ATP \longrightarrow ADP + C_6H_{11}O_6PO_3^{2-} \qquad \Delta G° = -17.7 \text{ kJ}$$

SAMPLE EXERCISE 13.13: The hydrolysis of 1,3-diphosphoglycerate has a large negative $\Delta G°$ value and can drive the conversion of ADP into ATP. Calculate $\Delta G°$ for the reaction ADP + 1,3-diphosphoglycerate $\longrightarrow$ 3-phosphoglycerate + ATP:

[structural formulas: 1,3-diphosphoglycerate + ADP + H⁺ $\longrightarrow$ 3-phosphoglycerate + ATP]

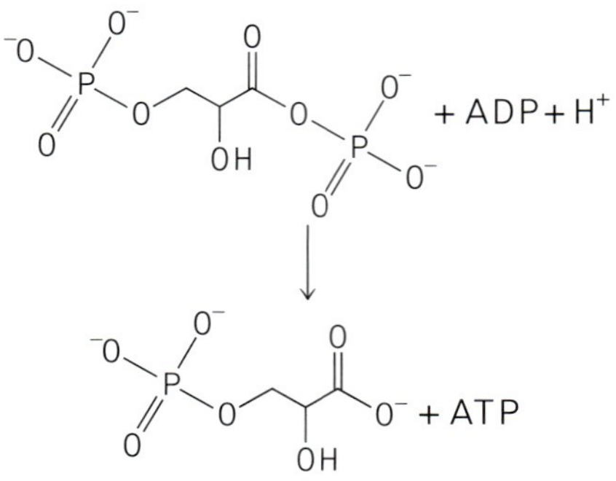

from the following changes in standard free energy:

(1) 1,3-diphosphoglycerate(aq) + H_2O(l) $\longrightarrow$

$\quad$ 3-phosphoglycerate(aq) + HPO_4^{2-}(aq) $\qquad \Delta G° = -49$ kJ

(2) ADP(aq) + HPO_4^{2-}(aq) $\longrightarrow$ ATP(aq) + H_2O(l) $\qquad \Delta G° = +30.5$ kJ

SOLUTION: The overall reaction between ADP and 1,3-diphosphoglycerate is the sum of the reactions describing the hydrolysis of 1,3-diphosphoglycerate and the phosphorylation of ADP. We can sum the values of $\Delta G°$ for the steps to determine $\Delta G°$ for the overall reaction:

$$\Delta G°_{overall} = \Delta G°_{step\ 1} + \Delta G°_{step\ 2} = -49\ kJ + 30.5\ kJ = -18.5\ kJ$$

Thus, the hydrolysis of 1,3-diphosphoglycerate provides more than sufficient energy for the phosphorylation of ADP.

PRACTICE EXERCISE: The conversion of glucose into lactic acid drives the phosphorylation of 2 moles of ADP to ATP:

$C_6H_{12}O_6$(aq) + 2 $H_2PO_4^-$(aq) + 2 ADP(aq) $\longrightarrow$
$\quad$ 2 $CH_3CH(OH)COOH$(aq) + 2 ATP(aq) + 2 H_2O(l) $\qquad \Delta G° = -135$ kJ/mol

What is $\Delta G°$ for the conversion of glucose into lactic acid?

$$C_6H_{12}O_6(aq) \longrightarrow 2\ CH_3CH(OH)COOH(aq)$$

(See Problems 93 and 94.)

Altogether, glycolysis consists of six nonspontaneous reactions with positive $\Delta G°$ values driven by two spontaneous reactions with negative $\Delta G°$ values. Two of the nonspontaneous reactions are driven by the release of energy from the hydrolysis of ATP to ADP. Excess energy from the two spontaneous reactions is used to convert ADP back into ATP. By summing the ΔG° values for the eight steps of the glycolysis pathway, we can calculate the free energy change for the conversion of glucose into pyruvate, as shown in Figure 13.18. That value is -110 kJ/mole of glucose.

The glycolysis pathway is anaerobic because it does not use oxygen. Thus, glycolysis does not produce carbon dioxide as we would expect from the combustion of glucose. However, another series of aerobic reactions, the Krebs cycle,

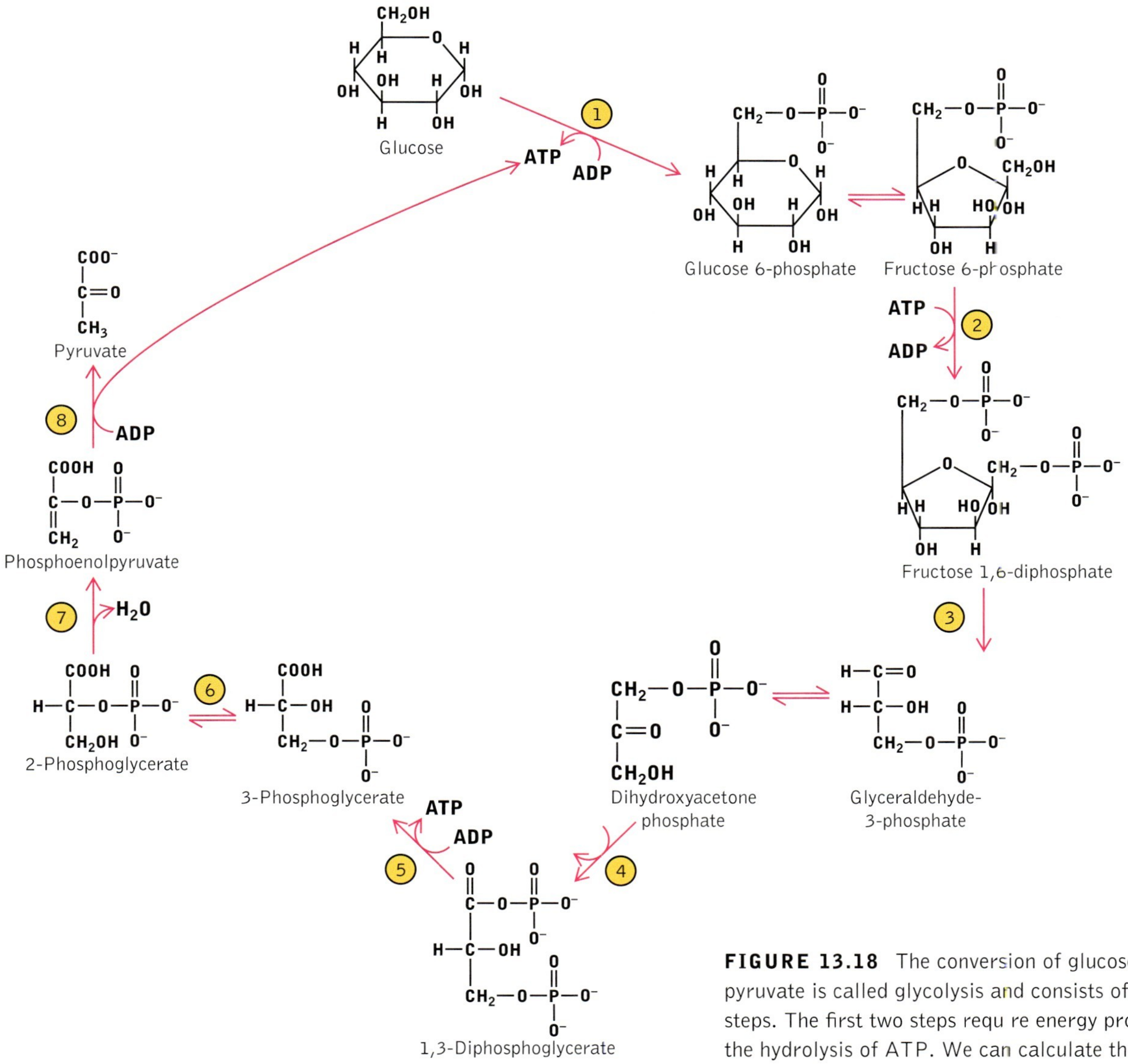

FIGURE 13.18 The conversion of glucose into pyruvate is called glycolysis and consists of eight steps. The first two steps require energy provided by the hydrolysis of ATP. We can calculate the free energy change in these steps and for the overall process by summing the values of ΔG for each step.

leads to the formation of carbon dioxide. By the end of the decomposition of pyruvate, we have an overall reaction in which 1 mole of pyruvate is converted into 3 moles of carbon dioxide and 2 moles of water. The energy released by this reaction is used to convert 30 moles of ADP into 30 moles of ATP, rather than being lost as heat. The ATP produced in the Krebs cycle can be used to drive other nonspontaneous cellular reactions.

If our muscle cells lack sufficient oxygen, as happens during highly strenuous exercise, then the glycolysis pathway takes a detour and pyruvate is converted into lactate until sufficient oxygen becomes available for the conversion of lactate into pyruvate. Then the Krebs cycle can continue. Under these conditions, oxygen is a limiting reactant. The buildup of lactate in muscle tissue causes the soreness that you feel in the aftermath of overexertion.

We have seen that our bodies can draw considerable energy from fats. In each case, the large triglyceride molecules are broken down into fatty acids and glycerol. Fatty acids are catabolized through the fatty acid cycle (Figure 13.19) in which they are converted into acetyl coenzyme A (acetyl CoA). Each fatty acid molecule loses a fragment containing two carbon atoms with each pass through the fatty acid cycle. Acetyl coenzyme A also connects glycolysis to the Krebs cycle in glucose catabolism. The glycerol produced in fat hydrolysis is converted into dihydroxyacetone phosphate, which enters the glycolysis pathway, as shown in Figure 13.18.

The food energy in proteins is released through their hydrolysis into constituent amino acids. The amino acids are stripped of their NH_2 groups and the products are converted into pyruvate or acetyl coenzyme A, again with the use of ATP as an energy storehouse and energy-transfer agent.

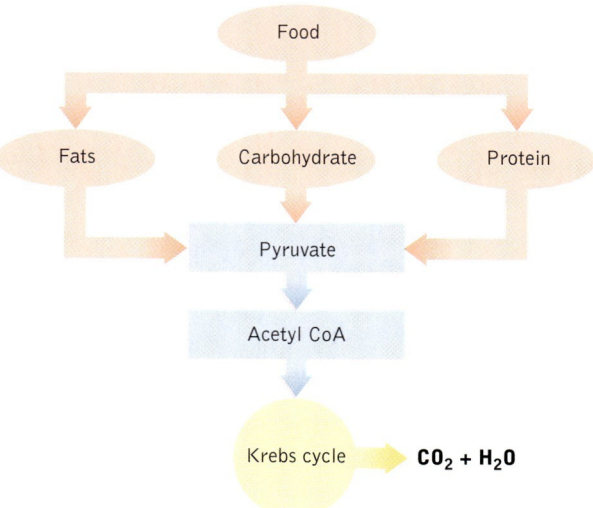

FIGURE 13.19 The conversion of food into energy (catabolism) by living organisms is linked to the formation and reactions of acetyl CoA. The chemical reactions of fats, carbohydrates, and proteins all result in the production of acetyl CoA.

13.8 DNA AND MAKING PROTEINS

As noted in Section 13.5, our bodies synthesize the proteins that we need by linking amino acids together in condensation reactions. There are thousands of different proteins in the body with as many different functions. How does the molecular machinery of protein synthesis guide the process so that the right protein is available in the appropriate cells and tissues when needed? The key to understanding protein synthesis is to understand the molecular structure and function of another class of biological compounds: nucleic acids.

By the 1940s, scientists knew that the nuclei of living cells contain substances different from proteins, fats, and carbohydrates. The enzymes that they had isolated that could degrade these other compounds did not degrade the substances found in the nucleus. When nucleic samples were finally subjected to chemical analysis, they were found to contain huge macromolecules consisting of building blocks (now called **nucleotides**) that could be further subdivided into three subunits: a phosphate group, a five-carbon cyclic sugar called deoxyribose with the molecular formula $C_5H_{10}O_4$, and a nitrogen-containing base (Figure 13.20).

Each phosphate group is bonded at C-5 of one sugar molecule and at C-3 of another. The result is formation of a *phosphate diester* linkage between the two sugar molecules:

$$[C–3]—O—\overset{\overset{\displaystyle O}{\|}}{\underset{\underset{\displaystyle O^-}{|}}{P}}—O—[C–5]$$

Each sugar molecule is also bonded at C-1 to one of four nitrogen-containing bases. These groups are bases because the lone pair of electrons on their nitrogen atoms are potential hydrogen ion acceptors, in much the way that nitrogen atoms in ammonia molecules are hydrogen ion acceptors. These four bases are frequently identified by the first letters of their names: A (adenine), C (cytosine), G (guanine), and T (thymine). They are weak bases, and so the more strongly acidic properties of the hydrogen atom on each phosphate group makes these nucleic substances, overall, acidic. Hence their name: **deoxyribonucleic acids**, or DNA.

In early studies of the structure of DNA, scientists discovered that, although the proportions of the four nitrogen-containing bases in DNA from different organisms are unique, the percentages of adenine and thymine in each type of DNA are always nearly the same, as are the percentages of guanine and cytosine. Thus, for every A there is a T, and for every C there is a G. The significance of these molecular equalities was revealed in 1952 when X-ray diffraction patterns produced by crystals of DNA indicated that the molecular subunits were arranged in a helix.

Further analysis of the diffraction patterns by British biophysicist Francis H. C. Crick (b. 1916) in collaboration with American scientist James D. Watson (b. 1928) disclosed the presence of a repeating structure that consisted of two

> ☑ **Deoxyribonucleic acid** (DNA) is a biological polymer consisting of units called **nucleotides**, each of which is made of three subunits: a nitrogen-containing base, deoxyribose sugar, and a phosphodiester. The sequence of bases carries genetic information.

CONNECTION: The structures and properties of acids and bases are described in Section 5.6.

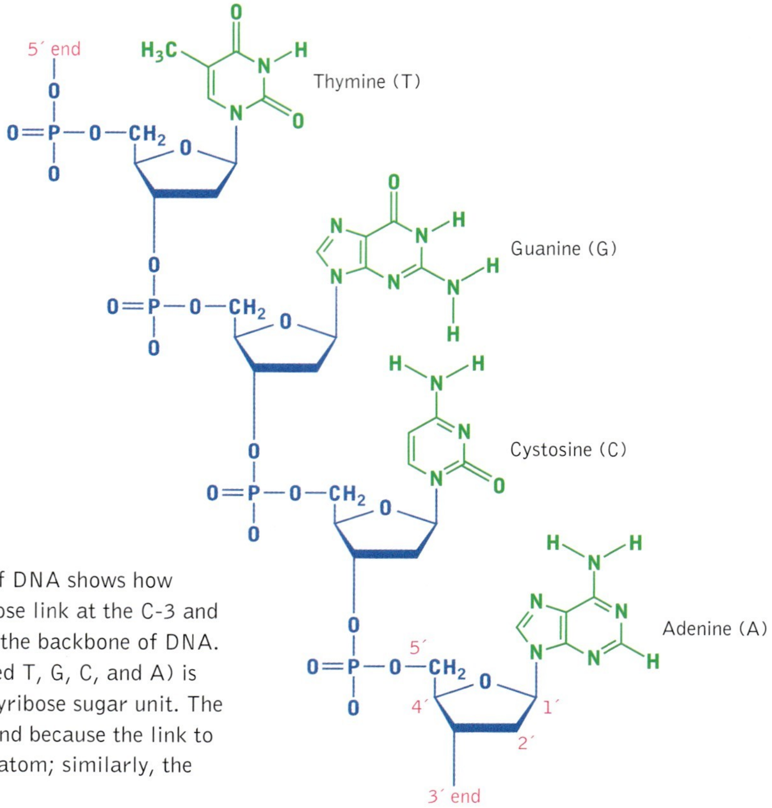

FIGURE 13.20 A part of a single strand of DNA shows how phosphate groups and molecules of deoxyribose link at the C-3 and C-5 carbon atoms of the sugar units to form the backbone of DNA. One of four nitrogen-containing bases (labeled T, G, C, and A) is bonded to the C-1 carbon atom of each deoxyribose sugar unit. The bottom end of the fragment is called the 3′ end because the link to the next phosphate unit is at the C-3 carbon atom; similarly, the top end is called the 5′ end.

matched helices, or a double helix, as shown in Figure 13.21A. The two strands of the double helix are held together by hydrogen bonding between groups on A and T base pairs and on C and G base pairs. This complementary structure is crucial in providing DNA with the ability to replicate itself. In the replication process, the two strands unwind and then serve as molecular templates for the synthesis of complementary strands, using individual nucleotides present in the cell (Figure 13.21B).

If the only distinguishing feature between one DNA nucleotide and the next is the identity of its base, then how does the sequencing of just four bases in DNA store enough information to control the synthesis of proteins from 20 amino acids? Clearly, the four-letter DNA alphabet must be used to write molecular "words" that identify particular proteins. On a molecular level, words are synonymous with the sequence of bases along a strand of DNA. How many bases does it take to specify a particular protein? Two-letter words would not be long enough, because there are only 16 two-letter combinations of four different letters (4^2). Therefore, there must be at least three letters in these molecular words.

> How does the sequencing of just four bases in DNA store enough information to control the synthesis of proteins from 20 amino acids?

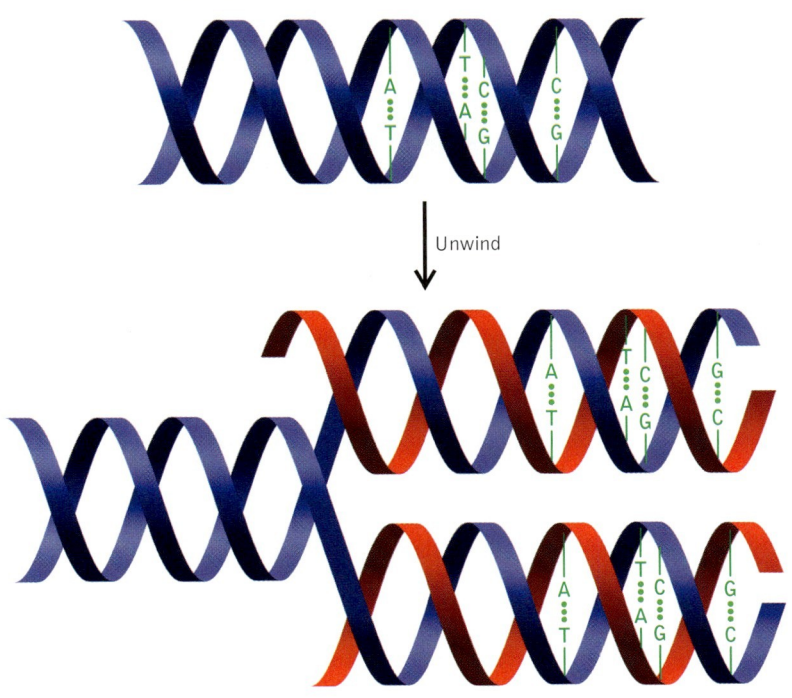

FIGURE 13.21 A. The two strands of the double helix of DNA are held together by hydrogen bonds (green dots) between adenosine and thymine (A–T) and between cytosine and guanine (C–G) base pairs. B. In replication these hydrogen bonds break and the double helix unwinds. Then each of the original strands serves as a molecular template for the synthesis of a new strand that has a complementary sequence of bases for every A in the original strand there is a T in the new strand, and so on.

SAMPLE EXERCISE 13.14: The sequence of bases in a segment of one strand of DNA is

A-T-T-G-A-C-T-G-G

What is the corresponding sequence of bases in the complementary strand?

SOLUTION: The base pairing of DNA nucleotides on complementary strands are A–T and G–C. Therefore, the sequence on the other strand must be

T-A-A-C-T-G-A-C-C

PRACTICE EXERCISE: What is the base sequence of the strand of DNA complementary to one in which the sequence is

C-A-C-G-T-T-A-G-C

(See Problem 101.)

Keep in mind that 16 different words are possible only if the order of the nucleotides is important—for example, only if AT is a different sequence from TA. Are AT and TA really different at the molecular level? The answer is yes. Words are read in only one direction along a strand of DNA or RNA: from the 5′ end to the 3′ end, as shown for DNA structure in Figure 13.20.

To put genetic codes to work synthesizing proteins, they must be transmitted from DNA in the nuclei of cells to the sites of protein synthesis at structures called

THE CHEMISTRY OF GROUP 6B

At the top of Group 6B is chromium (Cr), an element that is about as abundant as vanadium (see p. 640) in Earth's crust and that also owes its name to its colorful compounds. As noted in Chapter 10, rubies get their distinctive red color from the presence of trace concentrations of chromium(III) in crystals of aluminum oxide (Al_2O_3). The principal chromium-bearing mineral is chromite, or iron chromium oxide ($FeCr_2O_4$). It is the raw material from which chromium is produced for use mostly in making a variety of steels, including stainless steels, which are from 10% to 30% chromium. The presence of chromium metal in these alloys adds strength, hardness, luster, and corrosion resistance, as discussed in Chapter 18.

The biochemistry of chromium is linked to the very different chemical properties of the element in its two most common oxidation states: +3 and +6. Chromium(III) compounds play a role in regulating glucose levels in the blood by enhancing the function of insulin. The results of recent clinical studies indicate that taking chromium(III) supplements benefits patients suffering from some forms of diabetes. Results of other studies indicate that chromium(III) may help control blood-lipid levels in people suffering from cardiovascular disease.

On the other hand, chromium(VI) compounds, or *chromates*, are both toxic and carcinogenic (they cause cancer). The chromate ion (CrO_4^{2-}) is actively transported across cell membranes and into cells by the biochemical carriers that normally transport HPO_4^{2-} and SO_4^{2-}. Inside cells, CrO_4^{2-} is reduced to Cr^{3+} in processes that result in chromium ions becoming tightly attached to proteins, amino acids, and DNA. The products of attachment of one chemical species to another are called *adducts*. The presence of DNA adducts in cells is particularly sinister, because the altered DNA may be unable to provide the right genetic information to control, for example, the production of proteins or the growth of cells. A consequence of this kind of DNA alteration, or damage, is uncontrolled cell growth and the formation of cancerous cells. Fortunately, cells have the ability to repair at least some of the damage caused by the formation of DNA adducts as new DNA is replicated from old DNA. In addition, chromium–DNA adducts are sensitive to treatment with EDTA—a compound with the ability to complex with metal ions, as described in Chapter 16, and detoxify them.

Molybdenum (Mo) resides just below chromium in the periodic table, though its abundance is only about 1% that of chromium in Earth's crust and about the same as that of the element just below it, tungsten. The principal molybdenum-containing mineral is molybdenite (MoS_2). The molybdenum in MoS_2 can be converted to MoO_3 by roasting the mineral in air. Then the oxide is reduced with H_2 to free molybdenum metal at temperatures near 850°C. Molybdenum produced in this way is used mostly in the production of high-performance steels that keep their strength at high temperatures. Small additions of molybdenum also enhance the corrosion resis-

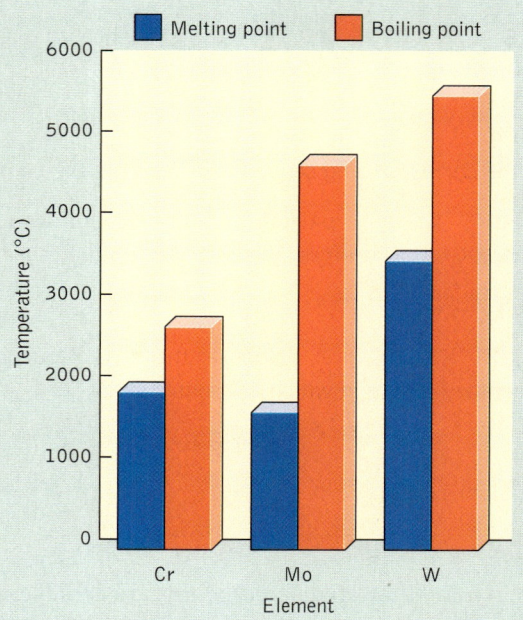

The Group 6B elements show a trend toward increasing melting and boiling points with increasing atomic number. The melting point of molybdenum is an exception to this trend, with a melting point slightly less than that of chromium.

tance of the stainless steels used to manufacture reaction vessels and other equipment for the pharmaceutical industry.

Molybdenum is an essential trace element in plants. In legumes, clusters of molybdenum, iron, and sulfur (see structure below) are essential to the ability of the enzyme molybdenum nitrogenase in bacteria to catalyze nitrogen fixation; that is, to convert N_2 from the air into water-soluble NH_3. The energy that drives this endergonic process comes from the exergonic conversion of ATP into ADP. As noted elsewhere in this chapter, high-energy ATP is generated from ADP in reactions that are coupled to the catabolism of glucose.

Recently, enzymes isolated from organisms growing near deep-sea vents were found to contain tungsten (W) atoms at the sites normally occupied by molybdenum atoms (W and Mo appear to have similar biochemical properties) in enzymes found in other marine organisms. The tungsten-based enzymes appear to maintain their biological function at higher temperatures than do the enzymes based on molybdenum.

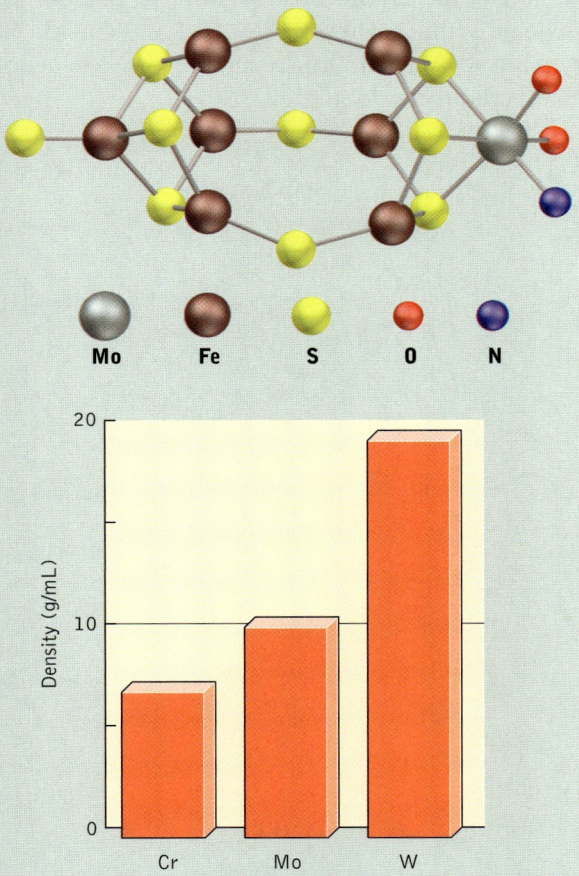

The densities of the Group 6B elements increase with increasing atomic number.

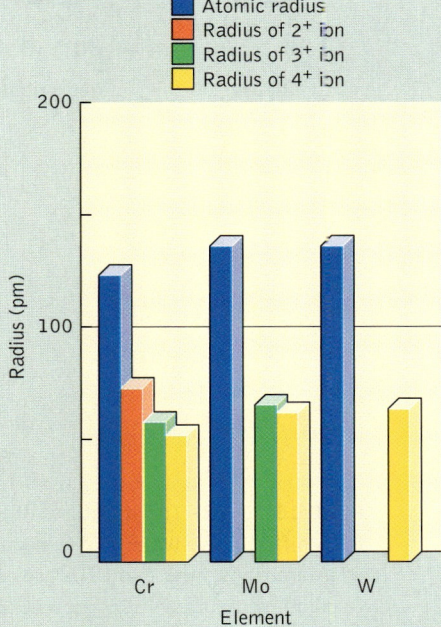

The atomic and ionic radii of the Group 6B elements increase with increasing atomic number.

660   **CHAPTER 13**   Entropy and Free Energy

> ✓ Nucleotides containing the monosaccharide ribose are called **ribonucleotides**. Ribonucleotides are the building blocks of **ribonucleic acid (RNA)**.
>
> In a cell, **messenger RNA (mRNA)** carries transcribed genetic messages from DNA to direct protein synthesis.
>
> The process by which information in DNA is used to synthesize RNA is called **transcription**.

ribosomes. The transmission process is handled by **messenger RNA** (mRNA). Messenger RNA is synthesized in accord with the genetic information contained in DNA in a process called **transcription**. Transcription resembles the replication of DNA but with some important differences. For one thing, the building blocks of mRNA are not deoxyribonucleotides but rather **ribonucleotides**. Ribonucleotides differ from deoxyribonucleotides in that they have a different sugar in their structures: ribose instead of deoxyribose (Figure 13.22). There is also a different kind of nitrogen-containing base in mRNA. It is called uracil. Its structure is very similar to that of thymine, the difference being one methyl group, as shown in Figure 13.22B. Like thymine, uracil pairs with adenine. Thus, the deoxyribonucleotide sequence AAGACCT in DNA becomes the ribonucleotide sequence UUCUGGA in the mRNA that is transcribed from it.

After RNA encounters and attaches itself to a ribosome, its genetic code is decoded into a sequence of amino acids that will eventually be linked together to form a protein. Each sequence of three bases forms a three-letter word, or *codon*, that identifies a particular amino acid. A total of 61 of the possible 64 sequences are used in this way. Thus most of the 20 amino acids have more than one codon. For example, all the sequences that begin with G-G, G-G-A, G-G-C, G-G-G, and G-G-U encode the same amino acid: glycine. The remaining three sequences of the 64 signal the protein synthesis process to stop.

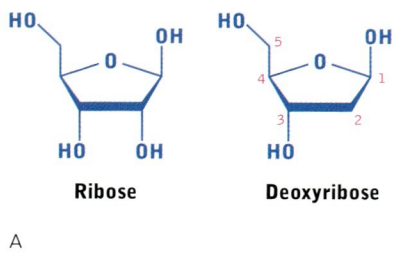

FIGURE 13.22 The structures of RNA and DNA differ in two ways. A. The sugar units in RNA are ribose; those in DNA are deoxyribose (a ribose without an —OH group C-2). B. Three of the four nitrogen bases in RNA and DNA are the same, but RNA has uracil (U) instead of thymine (T) hydrogen bonded to adenine on complementary strands.

In the decoding process, which is called **translation,** the genetic code contained in mRNA is passed on to another kind of nucleic acid, **transfer RNA** (tRNA). Each tRNA molecule binds to only one kind of amino acid and carries it to the site of protein synthesis on a strand of mRNA. This biomolecular synthesis proceeds at a rate of about 15 amino acids per second until a codon is encountered that signals the end of translation and causes the protein to be released. With about 3×10^5 molecules of mRNA in circulation, one of our cells synthesizes about 1×10^4 protein molecules every second, sustaining us and helping define who we are.

> **Translation** decodes genetic messages carried by RNA for use in protein synthesis.
>
> **Transfer RNA** (tRNA) transports and positions specific amino acids for protein synthesis.

SAMPLE EXERCISE 13.15: One sequence of mRNA bases that encodes the amino acid sequence arginine-glycine-proline is

C-G-C-G-G-U-C-C-A

What was the DNA sequence that produced this mRNA sequence?

SOLUTION: The base pairing between mRNA nucleotides and DNA nucleotides is A–T, G–C, and U–A. Therefore, the genetic code for this sequence of three amino acids in the DNA strand is

G-C-G-C-C-A-G-G-T

PRACTICE EXERCISE: One sequence of mRNA bases that encodes the amino acid sequence leucine-histidine-valine followed by the signal to terminate the synthesis process is

C-U-A-C-A-U-G-U-C-U-G-A

What was the DNA sequence that produced this mRNA sequence? (See Problem 102.)

CHAPTER REVIEW

Summary

SECTION 13.1

The dissolution of a solute in a solvent can be divided into several thermodynamic steps, the enthalpy changes of which add together to determine whether the process is endothermic or exothermic.

SECTION 13.2

Spontaneous processes occur without intervention when there is an increase in entropy of the universe. Forming a gas, converting a solid into a liquid, or dissolving a solid in a liquid result in an increase in entropy of the system. Pure crystals have zero entropy at absolute zero.

SECTION 13.3

The entropy change in a reaction under standard conditions can be calculated from the standard entropies of the products and reactants and their coefficients in the balanced chemical equation.

SECTION 13.4

Spontaneous processes have negative ΔG values. Reversing a process changes the sign of ΔG. For a system at equilibrium, $\Delta G = 0$. The temperature range over which a process is spontaneous depends on the signs and magnitudes of ΔH and ΔS. Free-energy change is the maximum useful work that the system can do on its surroundings.

SECTION 13.5

Carbohydrates, proteins, and fats are the principal energy sources in foods. Proteins are made of α-amino acids, which have an amino group and a carboxylic acid group attached to the same carbon atom in the molecule. Except for glycine, α-amino acids are chiral. Chiral molecules have a nonsuperimposable mirror image. The molecule and its mirror image are enantiomers. Fats are esters of glycerol and long-hydrocarbon-chain carboxylic acids. The chains can be saturated (all C–C bonds are single bonds) or unsaturated (includes one or more C=C double bonds).

SECTION 13.6

Fats have twice the fuel value of carbohydrates and proteins because fats have higher carbon and hydrogen content.

13.7

Many important biochemical processes are made possible by coupled spontaneous and nonspontaneous reactions. The free energy released in such processes (e.g., the hydrolysis of ATP to form ADP) is used in our bodies to drive nonspontaneous processes.

SECTION 13.8

The code for synthesizing proteins is contained in the sequences of nucleotides in the DNA in our cells. This code is transcribed into RNA, which translates the code of three-letter words in the synthesis of the array of proteins that define our physical characteristics.

Key Terms

α-amino acid (p. 631)
chiral molecule (p. 636)
deoxyribonucleic acid (DNA) (p. 657)
enantiomer (p. 636)
endergonic reaction (p. 625)
entropy (p. 616)
ester (p. 642)
exergonic reaction (p. 625)
free energy (p. 624)

glycolysis (p. 650)
heat of hydration (p. 615)
heat of solution (p. 612)
lipid (p. 642)
messenger RNA (mRNA) (p. 660)
nucleotide (p. 657)
peptide (p. 633)
peptide bond (p. 633)
phosphorylation (p. 650)

ribonucleic acid (RNA) (p. 660)
second law of thermodynamics (p. 616)
spontaneous process (p. 616)
standard molar entropy (p. 618)
stereoisomerism (p. 635)
third law of thermodynamics (p. 618)
transcription (p. 660)
transfer RNA (tRNA) (p. 661)
translation (p. 661)

Key Skills and Concepts

SECTION 13.1

Be able to describe at the molecular level how an ionic solid dissolves in water.

SECTION 13.2

Be able to describe the terms *entropy* and *spontaneity*. Understand the concept of *system*, *surroundings*, and *universe*.

Be able to predict the signs of entropy changes for common events and specified chemical processes.

SECTION 13.3

Be able to calculate the change in entropy of a reaction from the entropies of the reactants and products.

SECTION 13.4

Understand that a spontaneous process is one in which there is a decrease in free energy of a system.
Be able to predict whether a process is spontaneous at high or low temperatures from the signs of its enthalpy and entropy changes.
Understand the meaning of $\Delta G = 0$.
Be able to calculate free-energy changes from the free energies of formation of the reactants and products or from the corresponding enthalpy and entropy changes.

SECTION 13.5

Understand the molecular basis of chirality.
Know the structure and bonding of fats, peptides, proteins, and amino acids.
Be able to distinguish a saturated from an unsaturated fatty acid.

SECTION 13.6

Know that, the higher the carbon and hydrogen content of a food substance, the higher its energy (fuel) value.

SECTION 13.7

Understand how the free energy released in some biochemical processes (e.g., the hydrolysis of ATP to form ADP) is used to drive nonspontaneous processes in living systems.

SECTION 13.8

Understand how the genetic code stored in DNA can be used to control the synthesis of proteins

Key Equations and Relations

SECTION 13.1

The enthalpy change for dissolution of an ionic solute in water:

$$\Delta H_{soln} = \Delta H_{ionic} + \Delta H_{H\ bond} + \Delta H_{ion-dipole} \quad (13.1)$$

which simplifies to

$$\Delta H_{soln} = -U + \Delta H_{hydration} \quad (13.2)$$

where U is the lattice energy of the ionic solute.

SECTION 13.2

The total entropy change in a process is the sum of the entropy changes of the system and its surroundings.

$$\Delta S_{univ} = \Delta S_{sys} + \Delta S_{surr} \quad (13.3)$$

A spontaneous process results in an increase in the entropy of the universe.

$$\Delta S_{univ} = \Delta S_{sys} + \Delta S_{surr} > 0 \quad (13.4)$$

SECTION 13.3

The entropy change of a reaction under standard conditions is

$$\Delta S°_{rxn} = \sum n S°_{products} - \sum m S°_{reactants} \quad (13.5)$$

where n and m are the coefficients of the various products and reactants, respectively, in the balanced chemical equation for the process.

SECTION 13.4

The relation between the changes in free energy, enthalpy, and entropy of a system is given by:

$$\Delta G = \Delta H - T\Delta S \quad (13.12)$$

The free-energy change in a reaction at standard conditions:

$$\Delta G°_{rxn} = \sum n \Delta G°_{f,\ products} - \sum m \Delta G°_{f,\ reactants} \quad (13.14)$$

QUESTIONS AND PROBLEMS

Enthalpies of Solution

CONCEPT REVIEW

1. Why is calcium chloride more effective than sodium chloride in melting ice on a sidewalk?
2. What factors determine whether the heat of solution of a solute is positive or negative?
3. Which of the terms in Equation 13.1 for $\Delta H_{solution}$ are always positive?
4. What is the connection between a large positive $\Delta H_{solution}$ and the solubility of an ionic compound?
5. What is the effect of strongly exothermic ion–dipole interactions ($\Delta H_{ion-dipole} < 0$) on the heat of solution?
6. Would you expect the heat of hydration of Al^{3+} ions to be greater or less than the heat of hydration of Mg^{2+} ions?
7. Would you expect the heat of hydration of NO_3^- ions to be greater or less than the heat of hydration of Cl^- ions?
8. Why does the value of the heat of hydration of an ion depend on its size and its charge?

PROBLEMS

9. The heats of solution of NH_4NO_3 and $MgSO_4$ are $+26$ and -91 kJ/mol, respectively. If 0.30 mol of NH_4NO_3 and 0.10 mol of $MgSO_4$ are dissolved in a beaker of water, does the temperature of the water increase, decrease, or remain the same?
10. The heats of solution of NH_4NO_3 and NaOH are $+26$ and -44.5 kJ/mol, respectively. If 40 g of NH_4NO_3 and 10 g of NaOH are dissolved in a beaker of water, does the temperature of the water increase, decrease, or remain the same?
11. What is the temperature of a solution prepared by dissolving 16 g of $MgSO_4$ in 1.00 kg of water initially at 20.0°C? Given: $\Delta H_{MgSO_4} = -91$ kJ/mol. Assume that the heat capacity of the solution is the same as that of water.
12. The heat of solution of NaOH is -44.5 kJ/mol. What is the final temperature of a solution prepared by dissolving 40.0 g of NaOH in 1000 mL of water at 20°C? Assume that the heat capacity of the solution is the same as that of water.
13. The addition of 10.00 g of sodium acetate to 1.000 L of water at 20.00°C produces a solution with a temperature of 19.60°C. Calculate the molar heat of solution of sodium acetate. Assume that the heat capacity of the solution is the same as that of water.
14. When 8.0 g of sodium fluoride dissolve in 100. mL of water, initially at 20.00°C, the final temperature of the solution is 19.60°C. Calculate the molar heat of solution of NaF. Assume that the heat capacity of the solution is the same as that of water.
15. Which of the following relations is true if $\Delta H_{solution} > 0$?
    a. $\Delta H_{H\ bond} + \Delta H_{ionic} > \Delta H_{ion-dipole}$
    b. $\Delta H_{H\ bond} + \Delta H_{ionic} < \Delta H_{ion-dipole}$
    c. $\Delta H_{H\ bond} + \Delta H_{ionic} = \Delta H_{ion-dipole}$
16. Which of the following relations is true if a compound has a large negative $\Delta H_{solution}$?
    a. $\Delta H_{H\ bond} + \Delta H_{ionic} > \Delta H_{ion-dipole}$
    b. $\Delta H_{H\ bond} + \Delta H_{ionic} < \Delta H_{ion-dipole}$
    c. $\Delta H_{H\ bond} + \Delta H_{ionic} = \Delta H_{ion-dipole}$
17. Calculate the lattice energies of LiCl, NaCl, and KCl from the following data:

Compound	$\Delta H_{solution}$ (kJ/mol)	$\Delta H_{hydration}$ (kJ/mol)
LiCl	+17	−823
NaCl	+47	−723
KCl	+52	−649

18. Use the following enthalpies of solution and lattice energies to calculate the enthalpy of hydration ($\Delta H_{hydration}$) for the dissolution of each compound.

Compound	$\Delta H_{solution}$ (kJ/mol)	U (kJ/mol)
LiBr	−49	−803
NaBr	−0.60	−741
KBr	+20	−678

19. Which of the compounds in the following table dissolve in water with a release of heat?

Compound	$\Delta H_{hydration}$ (kJ/mol)	U (kJ/mol)
NaF	−841	−914
NaCl	−723	−770
NaBr	−694	−728

20. Which of the Group 2A chlorides in the following table dissolve in water with a release of heat?

Compound	$\Delta H_{hydration}$ (kJ/mol)	U (kJ/mol)
$MgCl_2$	−2529	−1616
$CaCl_2$	−2117	−2016
$SrCl_2$	−2050	−1988

Entropy

CONCEPT REVIEW

21. In living cells, small molecules combine to make much larger molecules. Are these processes accompanied by increases or decreases in entropy of the molecules?
22. Which physical state of a substance—solid, liquid, or gas—has the highest standard molar entropy?
23. Does the dissolution of a gas in a liquid result in an increase or a decrease in entropy?
24. Adding solid Drano to water causes the temperature of the water to increase. If Drano is the system, what are the signs of ΔS_{sys} and ΔS_{surr}?
25. Ice cubes melt in a glass of lemonade, cooling the lemonade from 10°C to 0°C. If the ice cubes are the system, what are the signs of ΔS_{sys} and ΔS_{surr}?
26. Diamond and fullerenes are two allotropes of carbon. On the basis of their different structures and properties, predict which has the higher standard molar entropy?
27. The 1996 Nobel Prize in physics was awarded to Douglas Osheroff, Robert Richardson, and David Lee for their discovery of "superfluidity" in ^{3}He. When ^{3}He is cooled to 2.7 mK, the liquid settles into an "ordered" superfluid state. What is the predicted sign of the entropy change for the conversion of liquid ^{3}He into its superfluid state?
28. What happens to the sign of the entropy change when a process is reversed?
29. Why is the standard molar entropy of gaseous H_2S greater than the standard molar entropy of gaseous H_2O?
30. Why is the standard molar entropy of gaseous S_8 ($S° = 430$ J/mol · K) greater than the standard molar entropy of gaseous S_2 ($S° = 228$ J/mol · K)?

PROBLEMS

31. Which of the following combinations of entropy changes for a process are possible?
    a. $\Delta S_{sys} < 0$, $\Delta S_{surr} > 0$, $\Delta S_{univ} > 0$
    b. $\Delta S_{sys} < 0$, $\Delta S_{surr} < 0$, $\Delta S_{univ} > 0$
    c. $\Delta S_{sys} < 0$, $\Delta S_{surr} > 0$, $\Delta S_{univ} < 0$
32. In each of the following pairs, which alternative has the greater entropy:
    a. a pound of ice cubes or a pint of water?
    b. a spoonful of sugar in a sugar bowl or a spoonfull of sugar dissolved in cup of coffee?
    c. a cup of hot water or a cup of cold water?
    d. a mole of cyclohexane (C_6H_{12}) or a mole of 1-hexene ($CH_3CH_2CH_2CH_2CH$=CH_2)?
33. Rank the compounds in each of the following groups in order of increasing standard molar entropy, $S°$.
    a. $CH_4(g)$, $CF_4(g)$, and $CCl_4(g)$.
    b. $CH_3OH(l)$, $C_2H_5OH(l)$, and $C_3H_7OH(l)$.
    c. $HF(g)$, $H_2O(g)$, and $NH_3(g)$.
34. Rank the compounds in each of the following groups in order of increasing standard molar entropy, $S°$.
    a. $CH_4(g)$, $C_2H_6(g)$, and $C_3H_8(g)$.
    b. $CCl_4(l)$, $CHCl_3(l)$, and $CH_2Cl_2(l)$.
    c. $CO_2(sl)$, $CO_2(g)$, and $CS_2(g)$.
35. Predict the sign of ΔS for each of the following processes.
    a. A bricklayer builds a wall out of a random pile of bricks.
    b. Rake a yard full of leaves into a single pile.
    c. $Ag^+(aq) + Cl^-(aq) \rightarrow AgCl(s)$.
    d. $Zn(s) + 2 HCl(aq) \rightarrow H_2(g) + ZnCl_2(aq)$.
36. Predict the sign of ΔS for each of the following processes.
    a. Sweat evaporates.
    b. Solid silver chloride dissolves in aqueous ammonia.
    c. $C_3H_8(g) + 5 O_2(g) \rightarrow 3 CO_2(g) + 4 H_2O(l)$.
    d. $N_2O_5(g) \rightarrow NO_2(g) + NO_3(g)$.
37. Use the standard molar entropies in Appendix 4 to calculate $\Delta S°$ values for each of the following atmospheric reactions, which contribute to the formation of photochemical smog (described in Chapter 14).
    a. $N_2(g) + O_2(g) \rightarrow 2 NO(g)$
    b. $2 NO(g) + O_2(g) \rightarrow 2 NO_2(g)$
    c. $NO(g) + \frac{1}{2} O_2(g) \rightarrow NO_2(g)$
    d. $2 NO_2(g) \rightarrow N_2O_4(g)$
38. Use the standard molar entropies in Appendix 4 to calculate the $\Delta S°$ value for each of the following reactions of sulfur compounds.
    a. $H_2S(g) + \frac{3}{2} O_2(g) \rightarrow H_2O(g) + SO_2(g)$
    b. $2 SO_2(g) + O_2(g) \rightarrow 2 SO_3(g)$
    c. $SO_3(g) + H_2O(l) \rightarrow H_2SO_4(aq)$
    d. $S(g) + O_2(g) \rightarrow SO_2(g)$
39. The following reaction plays a key role in the destruction of ozone in the atmosphere. The standard entropy change ($\Delta S°$) is 19.9 J/K. Use the standard molar entropies ($S°$) in Appendix 4 to calculate the $S°$ value for ClO.

    $Cl(g) + O_3(g) \longrightarrow ClO(g) + O_2(g) \quad \Delta S° = 19.9$ J/K

40. Calculate the $\Delta S°$ value for the conversion of ozone into oxygen

    $$O_3(g) \longrightarrow \tfrac{3}{2} O_2(g)$$

    in the absence of Cl atoms and compare it with the $\Delta S°$ value calculated in Problem 39.

Free-Energy Changes

CONCEPT REVIEW

41. A nineteenth-century scientist, Marcellin Berthoulet, stated that all exothermic reactions are spontaneous. Is this statement correct?
42. Under what conditions does an increase in temperature turn a nonspontaneous process into a spontaneous one?
43. Give an example of a process that is spontaneous at any temperature.
44. Give an example of a process that is never spontaneous no matter what the temperature is.

PROBLEMS

45. What are the signs of ΔS, ΔH, and ΔG for the sublimation of dry ice (solid CO_2) at 25°C?
46. What are the signs of ΔS, ΔH, and ΔG for the formation of dew on a cool night?
47. Indicate whether each of the following processes is spontaneous.
    a. The fragrance of a perfume spreads through a room.
    b. A broken clock is mended.
    c. An iron fence rusts.
    d. An ice cube melts in a glass of tap water.
48. Indicate whether each of the following processes is spontaneous.
    a. Charcoal is converted into carbon dioxide when ignited in air.
    b. Dry ice (solid CO_2) subliming at room temperature.
    c. Sugar dissolves in hot water.
    d. CH_4 and O_2 are formed from CO_2 and H_2O.
49. Calculate the free energy change for the dissolution of one mole of NaBr and mole mole of NaI at 298 K, given $\Delta H°_{soln} = -1$ and -7 kJ/mol, respectively, for NaBr and NaI. The corresponding values for $\Delta S°_{soln}$ are $+57$ and $+74$ J/mol · K.
50. The values of $\Delta H°_{rxn}$ and $\Delta S°_{rxn}$ for the reaction $2\ NO + O_2 \rightarrow 2NO_2$ are -112 kJ and -147 J/K, respectively. Calculate $\Delta G°$ at 298 K for this reaction. Why do you think the value of $\Delta S°$ is less than zero?
51. A mixture of CO and H_2 is produced by passing steam over charcoal. Calculate the $\Delta G°$ value for the reaction and predict the lowest temperature at which the reaction is spontaneous.

$$H_2O(g) + C(s) \longrightarrow H_2(g) + CO(g)$$

52. For each of the following combustion reactions,

    (1) $2\ CH_3OH(g) + 3\ O_2(g) \longrightarrow 2\ CO_2(g) + 4\ H_2O(g)$
    (2) $CH_4(g) + 3\ O_2(g) \longrightarrow CO_2(g) + 2\ H_2O(l)$
    (3) $2\ H_2(g) + O_2(g) \longrightarrow 2\ H_2O(g)$

    a. predict the sign of $\Delta S°$ before calculating its value.
    b. calculate $\Delta S°$ and $\Delta G°$ from the data in Appendix 4.

53. Deposits of elemental sulfur are often seen near active volcanoes. Their presence there may be because of the following reaction of SO_2 with H_2S:

$$SO_2(g) + 2\ H_2S(g) \longrightarrow 3\ S(s) + 2\ H_2O(g)$$

Calculate $\Delta H°$ and $\Delta S°$ for this reaction, and predict the temperature range over which the reaction is spontaneous.

54. Methanogenic bacteria convert acetic acid (CH_3COOH) to CO_2 and CH_4. Is this process endothermic or exothermic under standard conditions? Is the reaction spontaneous?
55. Use the data in Appendix 4 to calculate the $\Delta G°$ value for each of the following reactions.
    a. $N_2(g) + O_2(g) \rightarrow 2\ NO(g)$
    b. $2\ NO(g) + O_2(g) \rightarrow 2\ NO_2(g)$
    c. $NO(g) + \frac{1}{2}O_2(g) \rightarrow NO_2(g)$
    d. $2\ NO_2(g) \rightarrow N_2O_4(g)$
56. Use the data in Appendix 4 to calculate the $\Delta G°$ value for each of the following reactions.
    a. $H_2S(g) + \frac{3}{2}O_2(g) \rightarrow H_2O(g) + SO_2(g)$
    b. $2\ SO_2(g) + O_2(g) \rightarrow 2\ SO_3(g)$
    c. $SO_3(g) + H_2O(l) \rightarrow H_2SO_4(l)$
    d. $S(g) + O_2(g) \rightarrow SO_2(g)$
57. Which of the reactions in Problem 55 is spontaneous at
    a. high temperature?
    b. low temperature?
    c. all temperatures?
58. Which of the reactions in Problem 56 is spontaneous at
    a. high temperature?
    b. low temperature?
    c. all temperatures?

Human Fuels

CONCEPT REVIEW

59. What are *enantiomers*?
60. What is meant by the term *chiral*?
61. What is a *racemic* mixture?
62. What is the difference between a saturated and an unsaturated fatty acid?
63. Meteorites contain more (S)-amino acids than (R)-amino acids, which are the forms that make up the proteins in our bodies. What is meant by (S)-amino acids?
64. Of the 20 amino acids necessary to sustain life, which one is not chiral?
65. Which of the following compounds is not an α-amino acid?

a. 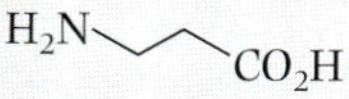 H_2N—CH$_2$CH$_2$—CO$_2$H   b. 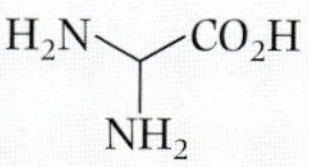 H_2N—CH(NH$_2$)—CO$_2$H   c. 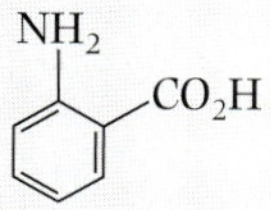 (ortho-aminobenzoic acid)

66. Which of the following compounds are α-amino acids?

a. 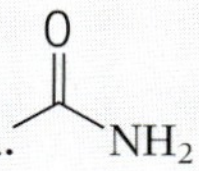 (acetamide structure)   b. 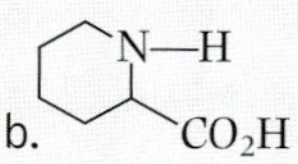 (piperidine-2-carboxylic acid)   c. 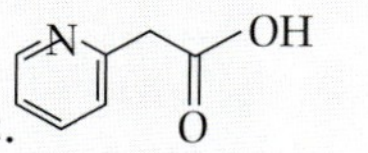 (pyridyl acetic acid)

PROBLEMS

67. Draw structures of the peptides produced from condensation reactions of the following (S)-amino acids (see Figure 13.10):
  a. alanine + serine
  b. alanine + phenylalanine
  c. alanine + valine
  d. methionine + alanine + glycine
  e. methionine + valine + alanine
  f. serine + glycine + tyrosine

68. Draw structures of the peptides produced from condensation reactions of the following (S)-amino acids (see Figure 13.10):
  a. glycine + alanine
  b. isoleucine + isoleucine
  c. tyrosine + phenylalanine
  d. proline + serine + cysteine
  e. serine + proline + serine
  f. valine + arginine + phenylalanine

69. The human brain produces polypeptides called *endorphins* that help in controlling pain. The following pentapeptide is called entephalin. Identify the five amino acids that make up entephalin.

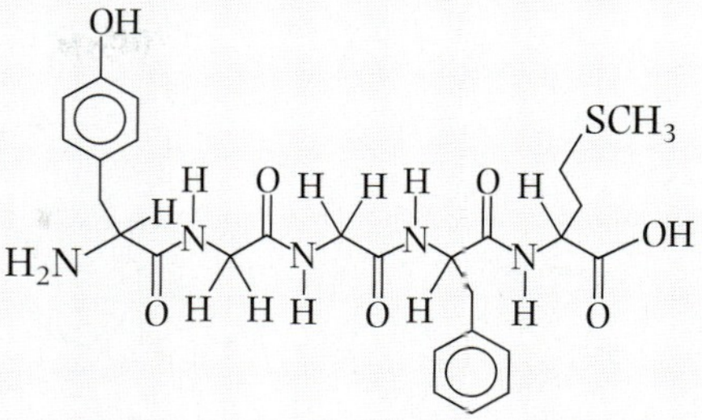

70. Angiotensin II is a polypeptide that regulates blood pressure. The structure of angiotensin II is shown here. Which amino acids are in the structure?

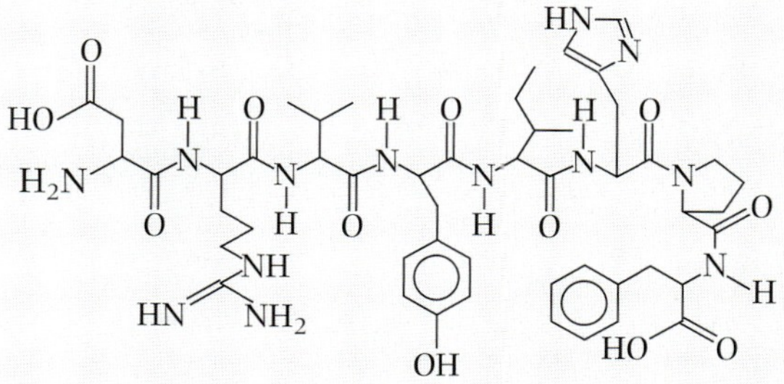

71. Which of the following objects are chiral? (a) a golf club; (b) a tennis racket; (c) a glove; (d) a shoe.

72. Which of the following objects are chiral? (a) a key; (b) a screwdriver; (c) a light bulb; (d) a baseball.

73. Which of the following amino acids are (R) and which are (S) stereoisomers?

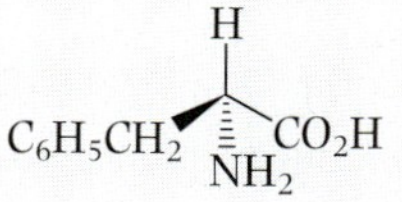

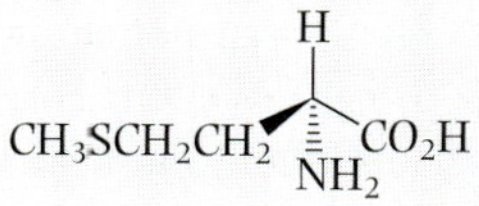

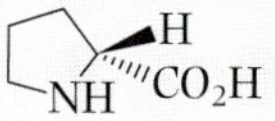

$C_6H_5CH_2$—C(H)(NH$_2$)—CO$_2$H $CH_3SCH_2CH_2$—C(H)(NH$_2$)—CO$_2$H

(proline structure)

74. Which of the following amino acids are (R) and which are (S) enantiomers?

a. 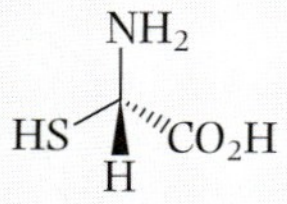 HS—C(H)(NH$_2$)—CO$_2$H   b. H_5C_6—C(CH$_3$)(NH$_2$)—CO$_2$H

c. H—C(NH$_2$)(OCH$_3$)—CO$_2$H   d. 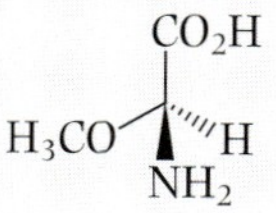 H_3CO—C(CO$_2$H)(H)—NH$_2$

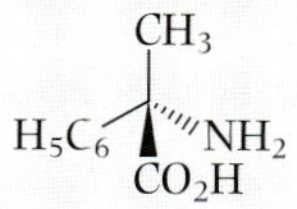

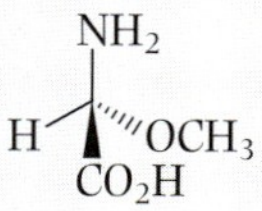

75. Which of the following compounds are unsaturated fats?

a.

b.

c.

76. Draw the structures of the fats formed by reaction of glycerol with octanoic acid, decanoic acid, and dodecanoic acid.

Fuel Values

CONCEPT REVIEW

77. Is the fuel value of leucine less than, greater than, or equal to that of isoleucine?
78. Why are the average fuel values of fats higher than those of carbohydrates and proteins?
79. Do glucose, mannose, and galactose all have the same fuel value? The structures of these monosaccharides are given in Figure 12.15.
80. Is the fuel value of glucose in the linear (aldohexose) form the same as that in the cyclic (hemiacetal) form?
81. Some Arctic explorers have eaten sticks of butter on their explorations. Give a nutritional reason for this unusual cuisine.
82. Which has a higher fuel value: glucose or starch?

PROBLEMS

83. To lose weight, the calories expended must exceed the calories consumed in food. The following table lists the energy expended during bicycling and running. How many "12-minute miles" would you have to run to consume 1 pound of fat? (Remember that a food Calorie equals 1 kcal.)

Activity	Calories/hour
Bicycling at 12 mph	480–600
Running at 5 mph	480–600

84. It has been said that pedaling a bicycle 3 miles is equivalent to running 1 mile. Use the data in Problem 83 to evaluate the validity of this statement.
85. Catabolism of 1.0 gram of fat (fuel value = 38kJ/g) can raise the temperature of how many grams of drinking water from 4.0 to 37.0 °C.
86. What is the final temperature of 125 g of water originally at 20°C if we heated the sample by the combustion of 1.0 gram of protein (fuel value = 17kJ/g)?

Driving the Human Engine

CONCEPT REVIEW

87. The second step in glycolysis converts glucose 6-phosphate into fructose 6-phosphate. Can you think of a reason why $\Delta G°$ for this reaction is close to zero?

88. Why is it important that at least some of the exergonic steps in glycolysis convert ADP to ATP?

89. How do we calculate the overall free-energy change of a process consisting of two steps?

90. For which of the steps in glycolysis would you predict an increase in entropy?

PROBLEMS

91. Identify the missing product in the following metabolic reaction.

$$H_2N-\text{CH}(\text{NH}_2)-\text{C}(=O)-\text{OH} + H_2O \longrightarrow HO-\text{CH}(\text{NH}_2)-\text{C}(=O)-\text{OH} + \,?$$

92. Identify the missing product in the following metabolic reaction.

(structure with NH$_2$, OH groups) $+ H_2O \longrightarrow$ (structure with NH$_2$, OH groups) $+\,?$

93. The hydrolysis of a mole of maltose ($\Delta G_f° = -2246.6$ kJ/mol) produces 2 moles of glucose ($\Delta G_f° = -1274.5$ kJ/mol):

$$\text{maltose} + H_2O \longrightarrow 2\ \text{glucose}$$

If the value of $\Delta G_f°$ of water is -285.8 kJ/mol, what is the change in free energy of the hydrolysis reaction?

94. The hydrolysis of fats to fatty acids and glycerol are exergonic reactions as are the hydrolysis reactions of other esters. For example, the standard free-energy change in the hydrolysis of ethyl acetate ($CH_3CO_2C_2H_5$) to ethanol ($\Delta G_f° = -487.0$ kJ/mol) and acetic acid is -19.7 kJ. From this value and the standard free energy of formation values of liquid ethanol, acetic acid, and water, calculate the value of $\Delta G_f°$ of ethyl acetate.

DNA and Making Proteins

CONCEPT REVIEW

95. What are the three kinds of molecular-building subunits in DNA? Which two form the "backbone" of DNA strands?

96. Why does a codon consist of a sequence of three, and not two, ribonucleotides?

97. What kind of intermolecular force holds together the strands of DNA in the double-helix configuration?

98. What is meant by "base pairing" in DNA? Which bases are paired up?

PROBLEMS

99. Draw the structure of adenosine 5'-monophosphate, one of the four ribonucleotides in a strand of RNA.

100. Draw the structure of deoxythymidine 5'-monophosphate, one of the four nucleotides in a strand of DNA.

101. In the replication of DNA, a segment of an original strand has the sequence T-C-G-G-T-A. What is the complementary sequence on the new strand?

102. In transcription, a segment of the strand of DNA that is transcribed has the sequence T-C-G-G-T-A. What is the corresponding sequence of nucleotides on the messenger RNA that is produced in transcription?

14 Chemical Kinetics
And Air Pollution

Brown Los Angeles haze (photochemical smog) is the product of many chemical reactions, starting with the formation of nitrogen monoxide inside automobile engines. The rates of these reactions and the intensity of the southern California sun determine when, and for how long, a smog event occurs.

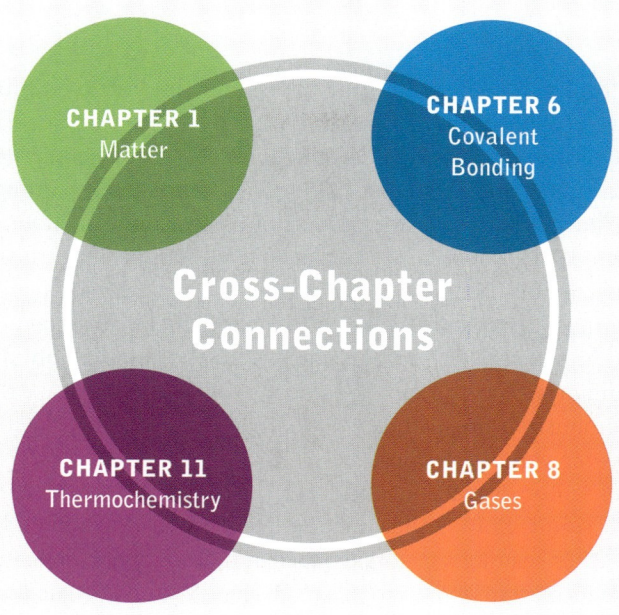

14.1 Photochemical smog
14.2 Reaction rates
14.3 Effect of concentration on reaction rate
  Reaction order and initial rates
  The single-experiment approach
  Second-order reactions
14.4 Reaction mechanisms
14.5 Reaction rates, temperature, and the Arrhenius equation
14.6 Catalysis

A Look Ahead

The chemistries of photochemical smog formation and stratospheric ozone depletion provide contexts for an examination of the rates of chemical reactions—that is, chemical kinetics. We will see that most chemical reactions take place at rates that depend on the concentrations of the reactants: rates decrease as reactants are consumed and their concentrations decrease. This concentration dependency is expressed in the rate law for each chemical reaction. Each rate-law expression includes a term called the rate constant whose value increases with increasing temperature. There are energy barriers, called activation energies, that must be overcome as chemical reactions proceed. Catalysts reduce the height of these barriers and speed up the rates of chemical reactions.

14.1 PHOTOCHEMICAL SMOG

The skies above many metropolitan areas are often obscured by layers of gray or brown haze, even on otherwise clear, sunny days. The haze that nearly obscures the skyline in the photograph on page 670 is caused by sunlight scattering off liquid and solid atmospheric aerosols suspended in the air. Some of these aerosols

✓ The term "smog" was first used to describe the combination of *smo*ke and *fog* that blanketed London in the 1950s.

CONNECTION: Combustion reactions of fossil fuels are described in Chapters 11 and 12.

✓ The carbon monoxide concentration in exhaust from a properly tuned automobile engine is about 0.05% (by volume).

CONNECTION: Ozone formation and depletion are described in Section 6.6.

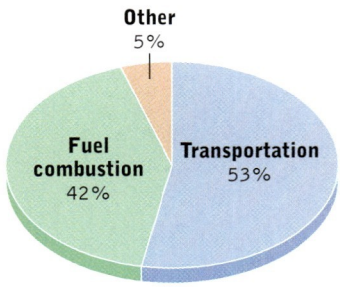

In 1998, 24.5 million tons of nitrogen oxides (NO_x) were released into the atmosphere in the United States. More than half of these nitrogen oxides came from motor vehicles.

are just tiny droplets of water that may have condensed on even tinier particles of dust because of high humidity and the mixing of warm and cool air masses. However, the atmosphere above a densely populated, industrialized urban area often contains high concentrations of particles produced by human activity. Many of these particles are the by-products of combustion reactions that generate the energy needed to run factories, produce electricity, and transport people and the products of our civilization. These particles and gaseous pollutants create a form of atmospheric pollution known as photochemical smog.

How does smog form? The process starts with the combustion of gasoline in automobile engines. The principal products of combustion are carbon dioxide and water vapor, but there is a long list of minor or trace-level products that result from either incomplete combustion (including carbon monoxide) or other reactions that take place at high temperatures. Among these reactions is the formation of nitric oxide (NO) from N_2 and O_2. The thermochemical equation for this reaction

$$N_2(g) + O_2(g) \longrightarrow 2\,NO(g) \qquad \Delta H° = +180.6 \text{ kJ} \qquad (14.1)$$

indicates that it is highly endothermic and so does not proceed except at very high temperatures. Nitric oxide released into the atmosphere may react with more oxygen, producing nitrogen dioxide:

$$2\,NO(g) + O_2(g) \longrightarrow 2\,NO_2(g) \qquad \Delta H° = -114.2 \text{ kJ} \qquad (14.2)$$

The energy ($h\nu$) of ultraviolet radiation from the sun can break the N–O bonds in NO_2, forming NO and oxygen atoms:

$$NO_2(g) \xrightarrow{h\nu} NO(g) + O(g) \qquad (14.3)$$

Photochemically generated oxygen atoms may react with molecular oxygen, producing ozone:

$$O_2(g) + O(g) \longrightarrow O_3(g) \qquad (14.4)$$

Ozone is an important ingredient in photochemical smog because of the risks that it poses to human health and because it is a reactant in chemical reactions that form other noxious compounds. For example, volatile hydrocarbons in the atmosphere react with ozone and nitrogen oxides, forming, in a series of reactions that will not be discussed in detail, aldehydes and a compound named peroxyacetyl nitrate (PAN):

A strong lachrymator, meaning that it brings tears to the eyes, PAN also causes some people to have difficulty breathing. Thus, a web of chemical reactions simultaneously leads to the formation of NO, NO_2, O_3, and a variety of other

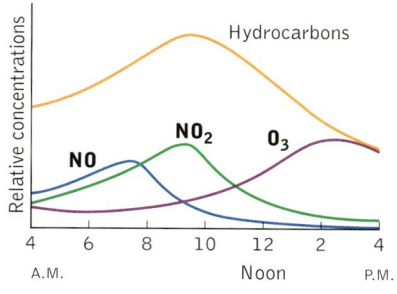

FIGURE 14.1 In a photochemical smog event, NO from vehicles builds up in the morning but reacts with O_2 in the atmosphere, forming NO_2. The photodecomposition of NO_2 leads to the formation of O_3.

unpleasant substances in polluted air. To understand how these processes link together and degrade the environment, we need to know the rates at which they take place. Because the products of some of these reactions are the reactants for others, the relative rates of these reactions will influence when these pollutants appear, how long they persist, and what their concentrations will be.

Figure 14.1 shows how concentrations of NO, NO_2, and O_3 rise and fall as photochemical smog forms. Note that there is a maximum in the concentration of NO just after the morning rush hour (though no comparable maximum occurs after the evening rush hour). Later in the morning, the concentration of NO_2 reaches a maximum. This sequence makes sense because NO is a precursor to NO_2 formation, meaning that it forms first and is then consumed as NO_2 is formed. The highest ozone concentrations are reached in the middle of the afternoon when the photochemical reactions of NO_2 (Equation 14.3) are in full swing, producing a supply of free oxygen atoms for the formation of ozone (Equation 14.4). Why isn't there an increase in NO after the evening rush hour? One reason is the presence of ozone, which reacts with NO, forming NO_2 and O_2:

> **Why isn't there an increase in NO after the evening rush hour?**

$$O_3(g) + NO(g) \longrightarrow O_2(g) + NO_2(g) \qquad (14.5)$$

Cracks in the surfaces of automobile tires are caused by chemical weathering due, in part, to atmospheric ozone. The molecular structures of both natural and many of the synthetic polymers used to make tires and other rubber products contain C=C double bonds. The bonds may be broken by chemical reactions with O_3, which weaken rubber and make it brittle. Synthetic rubber made of polymers without C=C double bonds is resistant to O_3 deterioration.

14.2 REACTION RATES

The amount of NO that forms inside an automobile engine depends on the rate of the reaction in Equation 14.1. We can express reaction rate as the increase in the concentration of NO ($\Delta[NO]$) that occurs in a tiny interval of time (Δt). The rate of change in [NO] is the difference between the *final* NO concentration ($[NO]_f$) and the *initial* concentration ($[NO]_i$) divided by the interval of time ($\Delta t = t_f - t_i$):

$$\text{rate of appearance of NO} = \frac{\Delta[NO]}{\Delta t} = \frac{[NO]_f - [NO]_i}{t_f - t_i} \qquad (14.6)$$

✓ Brackets surrounding the formula of a compound are used as a symbol for concentration in moles of the compound per liter (molarity).

The rate of the reaction between N_2 and O_2 can also be expressed as the rate of disappearance of either of the reactants. Because their concentrations decrease as the reaction proceeds, we need a negative sign in front of the concentration term of a reactant to obtain a positive value for the reaction rate:

$$\text{reaction rate} = \frac{-\Delta[N_2]}{\Delta t} = \frac{[N_2]_f - [N_2]_i}{t_f - t_i} \quad (14.7)$$

Similarly, the rate of disappearance of oxygen can be used to define the reaction rate:

$$\text{reaction rate} = \frac{-\Delta[O_2]}{\Delta t} = \frac{[O_2]_f - [O_2]_i}{t_f - t_i} \quad (14.8)$$

Because 2 moles of NO are formed by 1 mole of N_2 and 1 mole of O_2, the rate of appearance of NO should be *twice* the rate of disappearance of N_2 or O_2. This relation may be expressed in equation form as follows:

$$\frac{\Delta[NO]}{\Delta t} = -2\frac{\Delta[N_2]}{\Delta t} = -2\frac{\Delta[O_2]}{\Delta t} \quad (14.9)$$

Note that a balanced chemical equation enables us to predict the *relative* rates at which reactants are consumed and products are formed. However, an equation such as Equation 14.9 does not provide any information on the magnitude of these rates. Rate values can only be obtained experimentally, as we will see later in this section.

SAMPLE EXERCISE 14.1: Consider the synthesis of ammonia (NH_3) from the reaction of N_2 with H_2:

$$N_2(g) + 3 H_2(g) \longrightarrow 2 NH_3(g)$$

How is the rate of formation of NH_3 related to the rates of consumption of N_2 and H_2?

SOLUTION: The relative rates at which reactants are consumed and products are formed can be determined from the stoichiometry of the reaction. Since 2 moles of NH_3 are formed for every mole of N_2 consumed, the rate of formation of ammonia is twice as fast as the rate of disappearance of N_2. By a similar argument, the rate at which NH_3 is formed is 2/3 the rate at which H_2 is consumed because 2 moles of NH_3 are formed as 3 moles of H_2 are consumed. The relative rates of change in the concentrations of the three substances can be described by the following equation:

$$\frac{\Delta[NH_3]}{\Delta t} = -2\frac{\Delta[N_2]}{\Delta t} = -\frac{2}{3}\frac{\Delta[H_2]}{\Delta t}$$

where the negative signs in front of the N_2 and H_2 terms indicate that their concentrations decrease as the concentration of NH_3 increases.

PRACTICE EXERCISE: In the oxidation of carbon monoxide to carbon dioxide,

$$2\ CO(g) + O_2(g) \longrightarrow 2\ CO_2(g)$$

which reactant disappears at a higher rate: CO or O_2? Is the rate of appearance of CO_2 higher or lower than the rate of disappearance of O_2? (See Problems 5 and 6.)

SAMPLE EXERCISE 14.2: If, during the reaction between NO and O_2 (see Equation 14.2), $\Delta[O_2]/\Delta t = -0.033$ M/s, what is the rate of formation of NO_2?

SOLUTION: According to the stoichiometry of the chemical equation, the concentration of NO_2 increases at twice the rate at which the concentration of O_2 decreases; so:

$$\frac{\Delta[NO_2]}{\Delta t} = -2\frac{\Delta[O_2]}{\Delta t} = -2(-0.033\ M/s) = 0.066\ M/s$$

PRACTICE EXERCISE: NO reacts with H_2, forming N_2:

$$2\ NO(g) + 2\ H_2(g) \longrightarrow 2\ H_2O(g) + N_2(g)$$

If $\Delta[NO]/\Delta t = -1.5$ M/s, what are the rates of change of $[N_2]$ and $[H_2O]$? (See Problems 13 and 14.)

Average reaction rates and the formation of NO

We can track the rate of the endothermic reaction $N_2 + O_2 \rightarrow 2\ NO$ by measuring the changing concentrations of N_2, O_2 and NO in a high-temperature reaction vessel. The data in Table 14.1 and plotted in Figure 14.2 are based on such measurements. Note that the concentrations of N_2 and O_2 are the same throughout

TABLE 14.1 The Rate* of the Reaction $N_2 + O_2 \longrightarrow 2\ NO$

Time (μs)	$[N_2]$, $[O_2]$ (μM)	[NO] (μM)
0	17.0	0.0
5	13.1	7.8
10	9.6	14.8
15	7.6	18.6
20	5.8	22.2
25	4.5	24.8
30	3.6	26.7

*Reaction temperature = 3200 K

the experiment, as they should because the initial concentrations of these reactants are the same, and so are the rates at which they are consumed. The curve of [NO] rises steeply because NO forms twice as rapidly as N_2 or O_2 is consumed.

The data in Table 14.1 can be used to calculate the *average* reaction rate based on the change in $[N_2]$, $[O_2]$ or [NO] over a particular time interval. An average rate is the slope of a straight line connecting concentration values at the beginning and end of that interval as shown in Figure 14.3. For example, the average rate of change in $[N_2]$ (or $[O_2]$) between 5 and 10 μs is

$$\frac{\Delta[N_2]}{\Delta t} = \frac{[N_2]_f - [N_2]_i}{t_f - t_i} = \frac{(9.6 - 13.1)\mu M}{(10 - 5)\mu s}$$

$$= -0.70 M/s$$

During the same interval the average rate of formation of NO is:

$$\frac{\Delta[NO]}{\Delta t} = \frac{[NO]_f - [NO]_i}{t_f - t_i} = \frac{(14.8 - 7.8)\mu M}{(10 - 5)\mu s}$$

$$= 1.40\ M/s$$

Since the coefficients of N_2 or O_2 are one, let's use their rate of consumption as our reference for the rate of the overall reaction. Then the average rate of the reaction between 5 and 10 μs is:

$$\text{rate} = -\frac{\Delta[N_2]}{\Delta t} = -\frac{\Delta[O_2]}{\Delta t} = \frac{1}{2}\frac{\Delta[NO]}{\Delta t}$$

or 0.70 M/s.

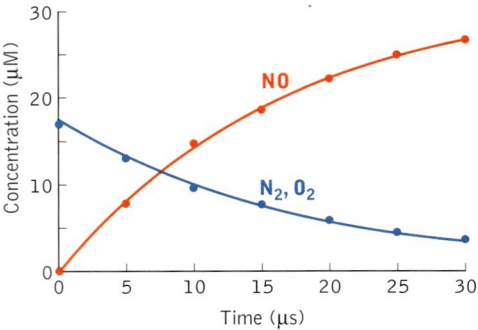

FIGURE 14.2 Changes in the concentrations of N_2, O_2, and NO over 50 μs of the reaction $N_2 + O_2 \longrightarrow 2\ NO$.

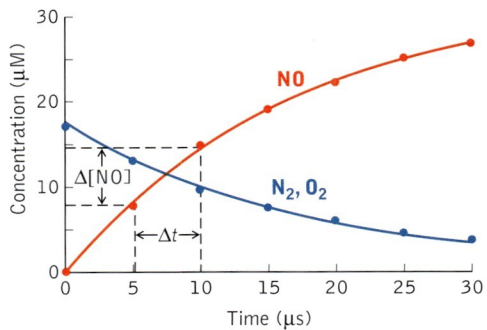

FIGURE 14.3 The average rate of the reaction between N_2 and O_2 can be determined between 5 and 10 μs from half the value of $\Delta[NO]/\Delta t$ for this time interval.

Instantaneous reaction rates and the formation of NO₂

Now let's consider the conversion of the NO from engine exhaust into NO_2 (the "brown" in "brown LA haze") by its reaction with oxygen in the air:

$$2\,NO(g) + O_2(g) \longrightarrow 2\,NO_2(g)$$

The rate of this reaction is described by the data in Table 14.2 and plotted in Figure 14.4. Note that the rate of disappearance of O_2 is half the rate of disappearance of NO, as predicted by the stoichiometry of the reaction. The *instantaneous* reaction rate at any moment in the course of the reaction can be determined

TABLE 14.2 The Rate of the Reaction between NO and O_2

Time (s)	[NO] (M)	[O₂] (M)	[NO₂] (M)
0	0.0100	0.0100	0.0000
285	0.0090	0.0095	0.0010
660	0.0080	0.0090	0.0020
1175	0.0070	0.0085	0.0030
1895	0.0060	0.0080	0.0040
2975	0.0050	0.0075	0.0050
4700	0.0040	0.0070	0.0060
7800	0.0030	0.0065	0.0070

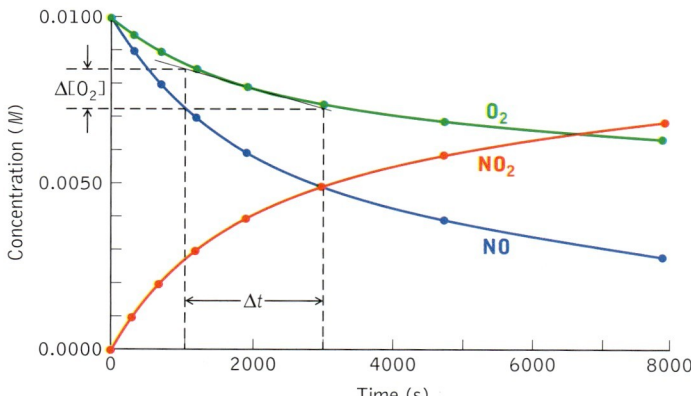

FIGURE 14.4 The instantaneous rate of change in $[O_2]$ in the reaction $2\,NO + O_2 \longrightarrow 2\,NO_2$ can be determined from the slope of the curve of $[O_2]$ versus t. For example, the instantaneous rate of change in $[O_2]$ at $t = 2000$ s is equal to the slope of the tangent to the curve at $t = 2000$ s.

graphically from the slope of a line tangent to one of the curves in Figure 14.4. For example, if we wish to use the instantaneous rate of change of $[O_2]$ at $t = 2000$ s, we first draw a line tangent to the curve for $[O_2]$ at $t = 2000$ s and then pick two convenient points (e.g., at $t = 1000$ and 3000 s) along that tangent line. From the differences in the x and in the y coordinates of these points, we calculate the slope of the line, which is a measure of $\Delta[O_2]/\Delta t$ and the instantaneous reaction rate at $t = 2000$ s:

$$\text{rate} = \frac{-\Delta[O_2]}{\Delta t} = \frac{-(0.0084 - 0.0072)\, M}{(1000 - 3000)\, s} = 6.\times 10^{-7}\, M/s$$

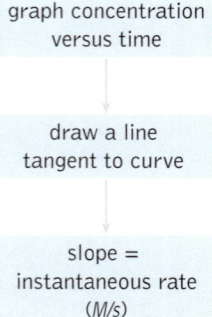

graph concentration versus time

draw a line tangent to curve

slope = instantaneous rate (M/s)

SAMPLE EXERCISE 14.3: Express the rate of the reaction between NO and O_2 in Figure 14.4 in terms of the instantaneous rate of change in [NO] at $t = 2000$ s.

SOLUTION: The instantaneous rate of change in [NO] at $t = 2000$ s can be determined from the slope of a line tangent to the curve of [NO] versus time at $t = 2000$ s. Choosing points along the tangent line at $t = 1000$ and 3000 s, we have

$$\text{rate} = \frac{-\Delta[NO]}{\Delta t} = \frac{-(0.0046 - 0.0070)\, M}{(3000 - 1000)\, s} = 1.2 \times 10^{-6}\, M/s$$

Note that this rate of change in [NO] is twice the reaction rate ($6.\times 10^{-7}$ M/s) based on the rate of change in $[O_2]$. This difference in rates is what we would expect from the stoichiometry of the reaction: 2 moles of NO are consumed for every mole of O_2.

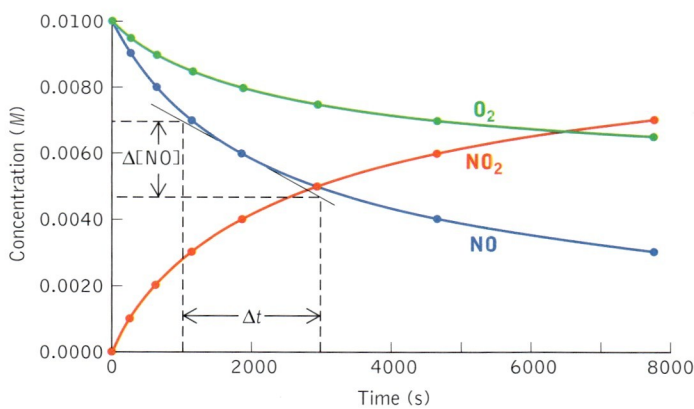

PRACTICE EXERCISE: Calculate the instantaneous rate of change in $[NO_2]$ in the reaction between NO and O_2 at $t = 3000$ s from the plot in Sample Exercise 14.3. (See Problems 17 and 18.)

14.3 EFFECT OF CONCENTRATION ON REACTION RATE

The curves in Figures 14.2 through 14.4 indicate that the most rapid changes in the concentrations of reactants and product take place early in the progress of reactions. This trend is common in chemical reactions: the rates of reactions tend to decrease as reactions proceed and reactants are consumed. The trend makes sense if we assume that reactions take place as a result of collisions between molecules of reactants. The more molecules there are, the more collisions take place per unit time. This molecular view is linked to the kinetic molecular theory of gases, which says that the pressure exerted by a gas is proportional to the frequency with which molecules of the gas collide with their container. The more molecules present in a given volume, the greater the number of collisions that take place per unit time and the greater the pressure. By a similar argument, the more collisions there are between reacting species per unit time, the higher the rate of a chemical reaction.

 CONNECTION: The pressure of a gas is proportional to the frequency and force of molecular collisions (Sections 8.2 and 8.6).

Reaction order and initial rates

The dependence of reaction rates on reactant concentrations can be used to find out *how* reactions take place. Let's consider the results of a study of the rate of the following reaction:

$$2\,NO(g) + O_2(g) \longrightarrow 2\,NO_2(g)$$

in which different concentrations of NO and O_2 are introduced into a reaction vessel. The values of the initial rates of the reaction can be determined from the slopes of lines tangent to plots of decreasing concentrations of NO (see Figure 14.5) or O_2

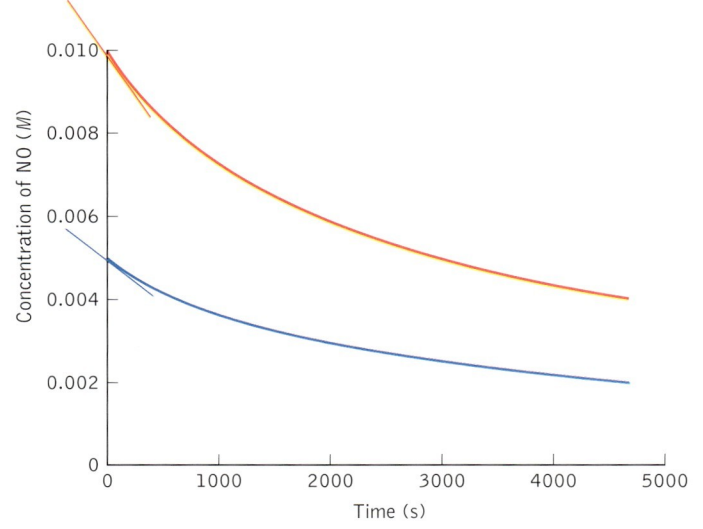

FIGURE 14.5 Initial reaction rates can be determined from the slopes of lines that are tangent to plots of reactant concentration versus time at $t = 0$. In this figure, initial rates for the reaction between NO and O_2 are determined for initial NO concentration ($[NO]_0$) values of 0.0100 (in blue) and 0.0050 M (in red). In both experiments $[O_2]_0 = 0.0100\,M$.

TABLE 14.3 Effect of Reactant Concentrations on Initial Rates of
$2 NO + O_2 \longrightarrow 2 NO_2$

Experiment	$[NO]_0$	$[O_2]_0$	Initial Reaction Rate $\left(\dfrac{-\Delta[NO]}{\Delta t}, \dfrac{M}{s}\right)$
1	0.0100	0.0100	2.0×10^{-6}
2	0.0100	0.0050	1.0×10^{-6}
3	0.0050	0.0100	5.0×10^{-7}
4	0.0050	0.0050	2.5×10^{-7}

or to increasing concentrations of NO_2 versus time at $t = 0$. Table 14.3 contains the results of several such determinations of initial reaction rates with the concentrations of NO and O_2 that produced them. The zero subscripts indicate that the concentrations of NO and O_2 are at $t = 0$.

To interpret the data in Table 14.3, we select pairs of experiments in which the concentration of one reactant is different, but the concentration of the other is the same. For example, in experiments 1 and 2, $[NO]_0$ is the same, but the value of $[O_2]_0$ in experiment 1 is twice the value of $[O_2]_0$ in experiment 2. Note that the reaction rate in experiment 1 is twice that in experiment 2. Thus, doubling the concentration of O_2 causes the reaction rate to double. We may conclude that the rate of the reaction is proportional to the concentration of O_2; that is:

$$\text{rate} \propto [O_2]$$

Now let's consider a pair of experiments in which $[NO]_0$ differs but $[O_2]_0$ is the same. For example, the concentration of NO in experiment 1 is twice that in experiment 3. Comparing the reaction rates for experiments 1 and 3, we find that the rate in experiment 1 is not twice but four times that in experiment 3 ($2.00 \times 10^{-6}/5.00 \times 10^{-7} = 4$). Clearly, reaction rate is not proportional to [NO]. However, it *is* proportional to $[NO]^2$. Expressing this result mathematically, we have

$$\text{rate} \propto [NO]^2$$

Now let's combine these two rate expressions by multiplying the two right sides together:

$$\text{rate} \propto [O_2][NO]^2$$

You may be wondering why we didn't simply add the concentration terms instead of multiplying them. The reason may be gleaned from the numbers of collisions that different numbers of O_2 and NO molecules may undergo, as shown in Figure 14.6. The possible collisions between different pairs of O_2 and NO molecules are represented by double-headed arrows. Note that the number of these arrows is proportional to the *product* of the concentrations of the two species.

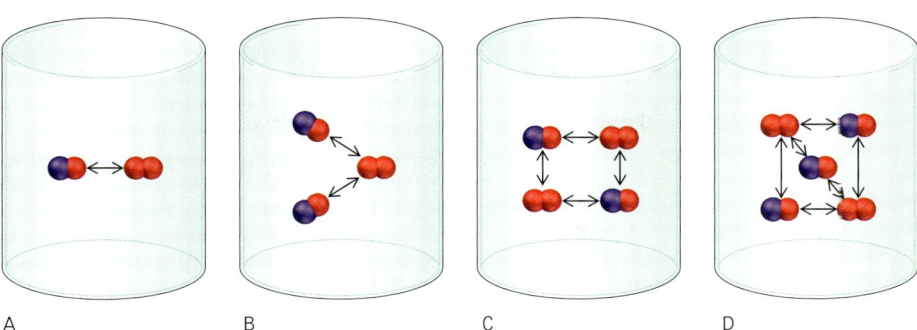

FIGURE 14.6 The rate of a chemical reaction increases with increasing concentrations of reactants because the reactants are more likely to collide and react with each other. Note that the number of possible collisions between pairs of reacting molecules (depicted by the double-headed arrows) is proportional to the product of the number of molecules of each type.

Therefore, it is logical that the expression for the rate of the reaction between O_2 and NO also will be based on the product of the concentrations of O_2 and NO.

We can make an equation out of the reaction-rate expression if we add a constant of proportionality (k):

$$\text{rate} = k[O_2][NO]^2 \qquad (14.10)$$

Equation 14.10 is called the **rate law** for the reaction between O_2 and NO. This equation tells us that the rate of the reaction is proportional to the concentration of O_2 to the first power and to the concentration of NO to the second power. The constant of proportionality k is called the **rate constant**. Its value changes with changing temperature. For a given temperature, we can calculate the value of k by using initial reaction rate data such as those in Table 14.3 and will do so shortly. But, first, let's consider the significance of the exponents of the concentration terms on the right-hand side of Equation 14.10. In chemical kinetics, we describe the dependence of reaction rate on concentration in terms of **reaction order.** The reaction between O_2 and NO is said to be *first order* in O_2 and *second order* in NO. The sum of the exponents of the concentration terms describes the reaction's *overall order*. Because $2 + 1 = 3$, this rate-law expression is *third order* overall. The significance of reaction order in describing how a reaction takes place will be addressed in Section 14.4.

First, let's calculate the value of k in the rate-law expression in Equation 14.10. Because the rate of the reaction is proportional to $[O_2]$ and to $[NO]^2$, we can use the rate and concentration data from one of the experiments in Table 14.3 to calculate the value of k. Which experiment we use doesn't matter; all of them should give the same value of k. Using the data from experiment 1 (the rate of formation of $NO_2 = 2.00 \times 10^{-6}$ M/s; [NO] = 0.0100 M, and $[O_2]$ = 0.0100 M in Equation 14.10), we have

$$\text{rate} = 2.00 \times 10^{-6} \text{ M/s} = k[O_2][NO]^2 = k(0.1)(0.01)^2$$

> ✓ The **rate law** of a chemical reaction is an equation relating the rate of the reaction to the concentrations of the reactants.
>
> The **rate constant** (k) is the constant of proportionality in a rate-law expression for a chemical reaction that connects its rate at a given temperature to the concentrations of reactants.
>
> **Reaction order** is the power to which the concentration of a reactant is raised in the rate-law expression for the reaction.

Solving for k, we find that

$$k = \frac{2.00 \times 10^{-6} \text{ M/s}}{(0.01 \text{ M})(0.01 \text{ M})^2} = 2.00 \text{ M}^{-2}/\text{s}$$

and substituting this value of k in the rate-law expression, we have

$$\text{rate} = 2.00 \text{ M}^{-2} \text{ s}^{-1} [O_2][NO]^2$$

We need to keep in mind that this value for the rate constant for this reaction is valid only for the temperature at which the experiments in Table 14.3 were carried out, though the rate-law expression with no specified value for the rate constant applies to any temperature.

CONCEPT TEST: Why doesn't it matter which experiment in Table 14.3 is used to calculate the value of k for the reaction $2 \text{ NO} + O_2 \rightarrow 2 \text{ NO}_2$?

Reaction order determines the units of the rate constant. The rate of a reaction is always expressed in terms of the rate of appearance of a product or disappearance of a reactant and so should have units of concentration change per unit time interval, such as molarity/second (M/s, or $M \text{ s}^{-1}$). For a first-order reaction in which concentration is expressed in molarity and time in seconds, the units of k must be per second (s^{-1}):

$$\text{rate} = k[X]^1$$

or

$$k = \frac{\text{rate}}{[X]}$$

$$= \frac{\frac{M}{s}}{M} = \text{s}^{-1}$$

The units of second-order rate constants can be derived in a similar way:

$$\text{rate} = k[X]^2$$

or

$$k = \frac{\text{rate}}{[X]^2}$$

$$= \frac{\frac{M}{s}}{M^2} = M^{-1} \text{ s}^{-1}$$

Similarly the units of a third-order rate constant, such as the one for the reaction between O_2 and NO, are $M^{-2} \text{ s}^{-1}$.

SAMPLE EXERCISE 14.4: The reaction $NO + O_3 \rightarrow NO_2 + O_2$ is first order in both reactants.
  a. Write the rate law for the reaction.
  b. Determine the overall reaction order.

SOLUTION: a. A general form of the rate law can be written as follows:

$$\text{rate} = k[NO]^m [O_3]^n$$

The reaction is first order in both reactants; so $m = n = 1$. The rate law becomes

$$\text{rate} = k[NO][O_3]$$

b. The overall reaction order is $m + n = 1 + 1 = 2$, or second order overall.

PRACTICE EXERCISE: Suppose the following reaction is second order overall and first order in both A and B.

$$A + B + C \longrightarrow D + 2E$$

  a. Determine the reaction order with respect to C.
  b. Write an expression for the rate law and determine the units of the rate constant for this reaction. (See Problems 27 and 28.)

SAMPLE EXERCISE 14.5: Write the rate law for the reaction of N_2 with O_2 at 3195 K on the basis of the following data, and determine the rate constant for the reaction.

$$N_2 + O_2 \longrightarrow 2 NO$$

Experiment	$[N_2]_0$	$[O_2]_0$	$\Delta[NO]/\Delta t$ (M/s)
1	0.040	0.040	1000
2	0.040	0.010	500
3	0.010	0.010	125

SOLUTION: The rate law for the reaction has the form

$$\text{rate} = k[N_2]^m [O_2]^n$$

To determine the value of m, we can use the data from experiments 2 and 3, in which the concentration of O_2 is the same but the concentration of N_2 is four times as great in experiment 1 as in experiment 3. The rate of the reaction is four times as fast in experiment 1, which leads us to conclude that the reaction is first order in N_2 and the value of m is 1.

The data in experiments 1 and 2 enable us to calculate the value of n. In these two experiments, the concentration of N_2 is the same but the concentration of O_2 is four times as great in experiment 1. However, the reaction rate in experiment

1 is only two times as great as that in experiment 2. Thus, we need to answer the question, "Four raised to what power equals two?" The answer is the 1/2 power; the reaction is only 1/2 order in O_2 and 3/2 order overall:

$$\text{rate} = k[\text{N}_2][\text{O}_2]^{1/2}$$

Substituting the data from experiment 1 gives

$$1000 \text{ M/s} = k(0.040)(0.040)^{1/2} = k(0.080)$$

$$k = 1.25 \times 10^4 \text{ M}^{-1/2} \text{ s}^{-1}$$

What does a reaction order of 1/2 mean? The meaning of a fractional reaction order will become clearer when we examine the mechanisms of reactions in Section 14.4. It turns out that many chemical reactions take place in two or more steps, only one of which controls the overall rate of the reaction. The concentrations of the reactants in this key step and their relations to the initial reactants determine the rate-law expression for the overall reaction.

PRACTICE EXERCISE: Nitric oxide reacts rapidly with unstable nitrogen trioxide (NO_3), forming NO_2:

$$\text{NO} + \text{NO}_3 \longrightarrow 2 \text{ NO}_2$$

The following data were collected at 298 K:

Experiment	[NO]	[NO_3]	Initial rate (M/s)
1	1.25×10^{-3}	1.25×10^{-3}	2.45×10^4
2	2.50×10^{-3}	1.25×10^{-3}	4.90×10^4
3	2.50×10^{-3}	2.50×10^{-3}	9.80×10^4

Determine the rate law for the reaction and calculate the value of the rate constant. (See Problems 39–42.)

The single-experiment approach

There are some reactions for which we can determine the rate law from the results of just one experiment. Some of these reactions have a single reactant and so their reaction rates depend on the concentration of only one substance. An important class of single-reactant reactions start with the absorption of photons of electromagnetic radiation. As noted in the discussion of the reactions that lead to photochemical smog formation, nitrogen dioxide may photochemically decompose to NO and atomic oxygen:

$$\text{NO}_2 \xrightarrow{h\nu} \text{NO} + \text{O}$$

Atomic oxygen produced by this reaction then combines with oxygen, forming ozone:

$$\text{O} + \text{O}_2 \longrightarrow \text{O}_3$$

Ozone itself undergoes photochemical decomposition. As each molecule of ozone absorbs a photon of UV radiation, one of the oxygen–oxygen bonds breaks, producing a free oxygen atom and a molecule of oxygen:

$$O_3 \xrightarrow{h\nu} O_2 + O$$

This decomposition reaction can be studied in the laboratory by using a high-intensity source of ultraviolet light, such as a mercury-vapor lamp. One such study yielded the results listed in Table 14.4 and plotted in Figure 14.7. Because ozone is the only reactant (except for photons, which we assume are available in abundance), the rate law for the decomposition reaction should depend only on the concentration of ozone. Does this dependence mean that the reaction is first order in O_3? If it is, then the rate-law expression should be

$$\text{rate} = \frac{-\Delta[O_3]}{\Delta t} = k[O_3]$$

CONNECTION: The atomic emission spectrum of mercury includes an intense line at 253.6 nm (see Section 3.2).

This rate-law expression can be transformed (with the use of calculus) into an expression that relates the concentration of ozone, $[O_3]$, at any time in the course of the reaction to its concentration at the beginning of the reaction, $[O_3]_0$.

$$\ln \frac{[O_3]}{[O_3]_0} = -kt \qquad (14.11)$$

The transformed version of the rate-law equation is called the **integrated rate law** (because integral calculus is used to derive it) for this first-order reaction. The general integrated rate law for any reaction that is first order in reactant X is

$$\ln \frac{[X]}{[X]_0} = -kt \qquad (14.12)$$

✓ The **integrated rate law** for a chemical reaction is a mathematical expression describing the change in concentration of a reactant with time.

To find out whether the photodecomposition of ozone is first order, we can rearrange the terms in this equation by splitting the quotient on the left side of Equation 14.11 in two:

$$\ln [O_3] - \ln [O_3]_0 = -kt$$

TABLE 14.4 The Rate of the Photochemical Decomposition of Ozone

Time (s)	$[O_3]$	$\ln [O_3]$
0	1.000×10^{-4}	−9.315
100	0.896×10^{-4}	−9.320
200	0.803×10^{-4}	−9.430
300	0.719×10^{-4}	−9.540
400	0.644×10^{-4}	−9.650
500	0.577×10^{-4}	−9.760
600	0.517×10^{-4}	−9.870

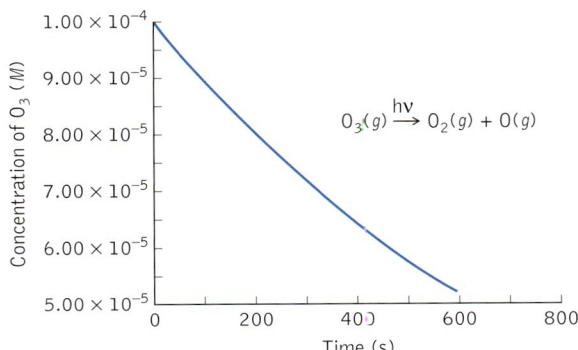

FIGURE 14.7 Ozone decomposes in the presence of intense UV light in a reaction that is first order in O_3.

Rearranging these terms, we get

$$\ln [O_3] = -kt + \ln [O_3]_0$$

which is the equation of a straight line ($y = mx + b$), where $\ln [O_3]$ is the y variable and time (t) is the x variable. The slope of the line (m) equals $-k$, and the y intercept (b) equals $\ln [O_3]_0$. The graph in Figure 14.8 is a plot of the $\ln [O_3]$ versus t data in Table 14.4, and it is indeed a straight line. Therefore, the reaction is first order in O_3. If the reaction were not first order, the line would be curved, not straight. Calculating the slope of the line, $\Delta(\ln [O_3])/\Delta t$, from two points on it yields the value of the reaction rate constant (k):

$$k = -\text{slope} = -(-1.1 \times 10^{-3}\,\text{s}^{-1}) = 1.1 \times 10^{-3}\,\text{s}^{-1}.$$

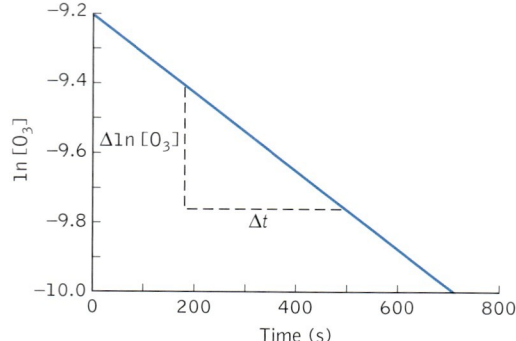

FIGURE 14.8 A plot of ln [O_3] versus time during the photodecomposition of ozone is a straight line, indicating that the decomposition reaction is first order in O_3. The rate constant of the reaction is the negative value of the slope of the line, or $1.1 \times 10^{-3}\,\text{s}^{-1}$.

SAMPLE EXERCISE 14.6: One of the less-common nitrogen oxides in the atmosphere is dinitrogen pentoxide (N_2O_5). One reason why the concentrations of N_2O_5 are small is that it is unstable and rapidly decomposes to N_2O_4 and O_2:

$$N_2O_5 \longrightarrow N_2O_4 + \tfrac{1}{2} O_2$$

Suppose a study of the decomposition of N_2O_5 yields the following data:

Time (s)	[N_2O_5]
0	0.1000
50	0.0707
100	0.0500
200	0.0250
300	0.0125
400	0.00625

Determine whether the decomposition of N_2O_5 is first order in N_2O_5, and determine the value of the rate constant.

SOLUTION: If the decomposition of N_2O_5 is first order, then the plot of ln $[N_2O_5]$ versus time will be a straight line. To find out if it is, we need to calculate ln $[N_2O_5]$ values from the data given. Adding these values to the table of data, we have:

Time (s)	$[N_2O_5]$	ln $[N_2O_5]$
0	0.1000	−2.303
50	0.0707	−2.649
100	0.0500	−2.996
200	0.0250	−3.689
300	0.0125	−4.382
400	0.0062	−5.075

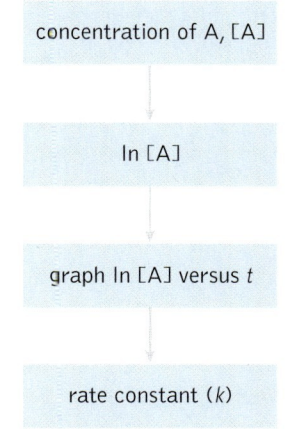

The plot of ln $[N_2O_5]$ versus t is shown in the adjoining illustration. It is indeed a straight line, indicating that the reaction is first order in N_2O_5. By selecting two convenient sets of data points along the line, at $t = 100$ and 300 s, we can calculate the slope of the line:

$$\text{slope} = \frac{\Delta y}{\Delta x} = \frac{-4.382 - (-2.996)}{300 - 100} = \frac{-1.386}{200} = -0.00693$$

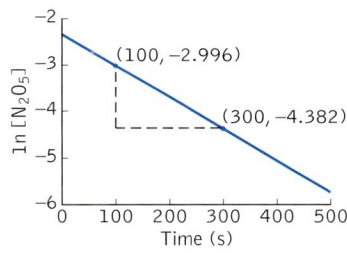

The slope of the line equals $-k$. Therefore, the rate constant $k = 0.00693$ s^{-1}. The y intercept (arrow on the graph) $= -2.30$, which is the natural log of the initial concentration.

PRACTICE EXERCISE: Hydrogren peroxide (H_2O_2) decomposes into water and oxygen:

$$H_2O_2(l) \longrightarrow H_2O(l) + \tfrac{1}{2} O_2(g)$$

The following data were collected for the decomposition of H_2O_2 at a constant temperature. Determine whether the decomposition of H_2O_2 is first order in H_2O_2, and calculate the value of the rate constant at the temperature of the experiments.

Time (s)	$[H_2O_2]$
0	0.500
100	0.460
200	0.424
500	0.330
1000	0.218
1500	0.144

(See Problems 43 and 44.)

> The **half-life** of a chemical reaction is the time required for half of the initial quantity of a reactant to be consumed.

A parameter related to the rate constant of a chemical reaction is the **half-life** ($t_{1/2}$) of the reaction, which is the interval during which the concentration of a reactant decreases by half. Let's consider this concept in the context of another nitrogen oxide found in the atmosphere: dinitrogen oxide (N_2O), also called nitrous oxide, the principal ingredient in laughing gas. Atmospheric concentrations of this potent greenhouse gas have been increasing in recent years, though the principal source of atmospheric N_2O is biological: bacterial degradation of nitrogen compounds in soil. Dinitrogen oxide is not a product of combustion reactions in internal combustion engines, because at typical engine temperatures any N_2O that might be formed rapidly decomposes into nitrogen and oxygen:

$$N_2O \longrightarrow N_2 + \tfrac{1}{2} O_2$$

The rate of this reaction, which is first order in N_2O, can be expressed as the half-life of N_2O at a given temperature. The effect of the passage of several half-lives on a population of N_2O molecules is shown in Figure 14.9. The shape of this graph resembles those presented in Chapter 1 to describe the rate of decay of radioactive isotopes. It should: radioactive decay also is a first-order process, and so a graph of the natural log of the quantity of radioactive isotope in a sample versus time is a straight line with a slope of $-k$, where k is the *radioactive decay constant*.

CONNECTION: Radioactive decay (Section 1.5) is an example of a first-order reaction.

CONCEPT TEST: Why is radioactive decay a first-order process?

We can derive a mathematical relation between the half-life and the rate constant of a chemical reaction. Let's start with the integrated rate law for a reaction that is first order in reactant X (Equation 14.12):

$$\ln \frac{[X]}{[X]_0} = -kt$$

After one half-life has passed ($t = t_{1/2}$), the concentration of X is half its original value; that is, $[X] = \tfrac{1}{2}[X]_0$. Inserting this information into Equation 14.12, we have

$$\ln \frac{\tfrac{1}{2}[X_0]}{[X_0]} = -kt_{1/2}$$

or

$$\ln \tfrac{1}{2} = -kt_{1/2}$$

The natural log of $\tfrac{1}{2} = -0.693$, and so:

$$-0.693 = -kt_{1/2}$$

or

$$t_{1/2} = 0.693/k \tag{14.13}$$

Thus, the value of the half-life of a first-order reaction is inversely proportional to the rate constant and is not dependent on concentration. Whether the concentration of reactant is high or low, half of it is consumed in one half-life.

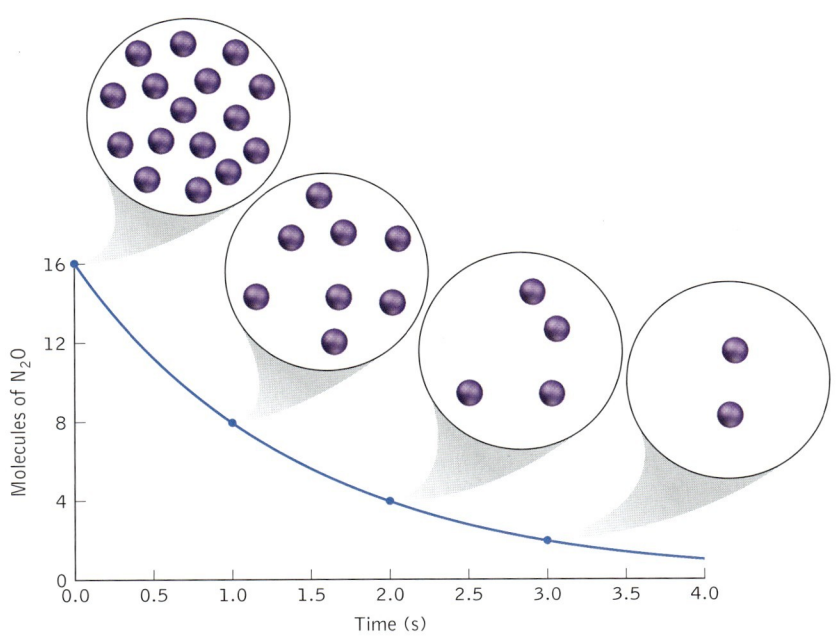

FIGURE 14.9 The decomposition reaction

$$N_2O \longrightarrow N_2 + \tfrac{1}{2}O_2$$

is first order in N_2O. If the half-life ($t_{1/2}$) of the reaction is 1.0 s, then, on average, half a population of 16 N_2O molecules (represented by purple spheres) decomposes in 1.0 s, half of the remaining 8 molecules decompose in the next 1.0 s, and so on.

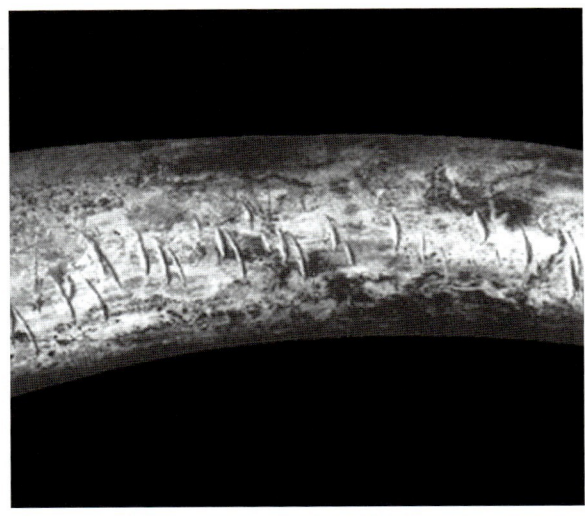

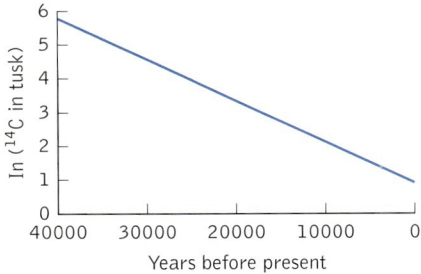

A mammoth tusk with grooves made by a sharp stone edge (indicating the presence of humans or Neanderthals) was uncovered at a prehistoric campsite in the Ural Mountains in 2001. Radiocarbon dating based on the decay of ^{14}C ($t_{1/2}$ = 5730 years) put the age of the tusk at nearly 40,000 years. Radioactive decay obeys first-order kinetics; so a graph of ln ^{14}C in the tusk versus time is a straight line with a slope of

$$-\frac{0.693}{5730 \text{ yr}} = -1.21 \times 10^{-4} \text{ yr}^{-1}.$$

SAMPLE EXERCISE 14.7: The rate constant (k) for the thermal decomposition of N_2O_5 at a particular temperature is 7.8×10^{-3} s^{-1}. What is the half-life of N_2O_5 at that temperature?

SOLUTION: Equation 14.13 relates the rate constant (k) of a first order reaction to its half-life ($t_{1/2}$). Inserting the value of k given into Equation 14.13, we have

$$t_{1/2} = \frac{0.693}{k} = \frac{0.693}{7.8 \times 10^{-3}\,\text{s}^{-1}} = 89\,\text{s}$$

PRACTICE EXERCISE: Environmental scientists studying the transport and fate of pollutants often calculate the half-life of these pollutants in the ecosystems into which they are released. In a study of the gasoline additive MTBE in Donner Lake, California, scientists from the University of California, Davis, found that, in the summer, the half-life of MTBE in the lake was 28 days. What was the transport rate constant of MTBE out of Donner Lake during the study? (See Problems 51 and 52.)

Second-order reactions

At high temperatures, nitrogen dioxide decomposes into NO and O_2:

$$2\,NO_2(g) \longrightarrow 2\,NO(g) + O_2(g)$$

The rate-law expression for this reaction should take the form

$$\text{rate} = \frac{-\Delta[NO_2]}{\Delta t} = k[NO_2]^m \quad (14.14)$$

Table 14.5 contains data on the rate of change in the concentration of NO_2 over time. Let's use it to evaluate the order (m) of the reaction and the rate constant (k). A plot of ln $[NO_2]$ versus time for the thermal decomposition of NO_2 is shown in Figure 14.10A. The curve is clearly not linear, which tells us that the thermal decomposition of NO_2 is *not* first order. What is the order of the reaction? The answer is hidden in how the reaction takes place. If the decomposition process were of single molecules of NO_2 simply falling apart, we would expect it to be first order. However, if the reaction is due to collisions between NO_2 molecules, then the reaction would be second order in NO_2.

> How can we determine if the thermal decomposition of NO_2 is second order?

How can we determine if the thermal decomposition of NO_2 is second order? One way to find out is to assume that it is and then test that assumption. The test entails transforming Equation 14.14 into the corresponding integrated-rate-law expression for a second-order reaction, again by using calculus. The result of the transformation is

TABLE 14.5 The Rate of Decomposition of NO_2

Time (s)	$[NO_2]$	ln $[NO_2]$	$1/[NO_2]$
0	1.00×10^{-2}	−4.605	100
100	6.48×10^{-3}	−5.039	154
200	4.79×10^{-3}	−5.341	209
300	3.80×10^{-3}	−5.573	263
400	3.15×10^{-3}	−5.760	317
500	2.69×10^{-3}	−5.918	372
600	2.35×10^{-3}	−6.057	426

$$\frac{1}{[NO_2]} = kt + \frac{1}{[NO_2]_0} \qquad (14.15)$$

Equation 14.15 is that of a straight line ($y = mx + b$) if we make $1/[NO_2]$ the y variable. Such a plot appears in Figure 14.10B. It is indeed linear. Therefore, the thermal decomposition of NO_2 is second order. The slope of the line provides a direct measure of k, the second-order rate constant, which, in this case, is 0.544 $M^{-1} s^{-1}$ A generic form of Equation 14.15 that applies to any reaction that is second order in substance X is

$$\frac{1}{[X]} = kt + \frac{1}{[X]_0} \qquad (14.16)$$

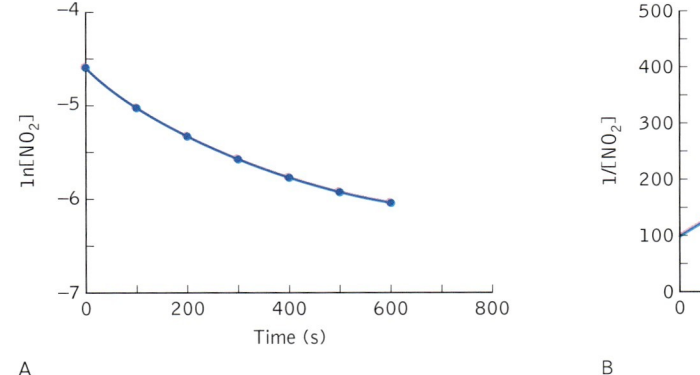
A

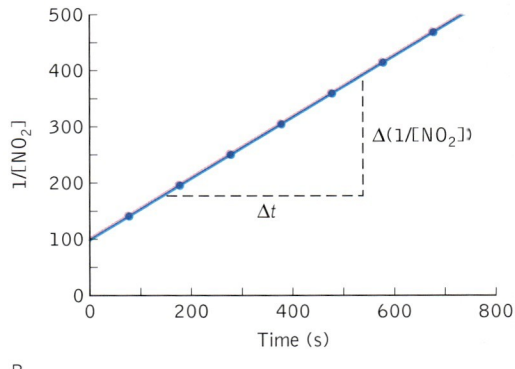
B

FIGURE 14.10 Pure NO_2 slowly decomposes at high temperatures to NO and O_2. A. A plot of ln $[NO_2]$ versus time is not linear, indicating that the reaction is not first order. B. A plot of $1/[NO_2]$ versus time is linear, indicating that the reaction is second order in NO_2. The slope of the line equals the second-order rate constant (k), which is 0.544 $M^{-1} s^{-1}$ at the temperature of the experiment.

SAMPLE EXERCISE 14.8: Chlorine monoxide (ClO) accumulates in the stratosphere above Antarctica in the winter and plays a key role in the formation of the ozone hole above the South Pole in the spring of each year. Eventually, ClO decomposes according to the following equation:

$$2\ ClO \longrightarrow Cl_2 + O_2$$

The kinetics of this reaction were studied in a laboratory experiment at 298 K, and the following data were obtained. Determine the order of the decomposition reaction, the rate law, and the value of the rate constant at 298 K.

Time (ms)	[ClO] (M)
0	1.50×10^{-8}
10	7.19×10^{-9}
20	4.74×10^{-9}
30	3.52×10^{-9}
40	2.81×10^{-9}
100	1.27×10^{-9}
200	0.66×10^{-9}

SOLUTION: To distinguish between a first- and second-order reaction in which there is a single reactant, we need to plot ln [ClO] versus time and 1/[ClO] versus time. If the plot of ln [ClO] versus time is linear, then the reaction is first order; but, if it is not and a plot of 1/[ClO] versus time *is* linear, then the reaction is second order. To evaluate these possibilities, we need to calculate the natural log and reciprocal values of the tabulated concentrations of ClO:

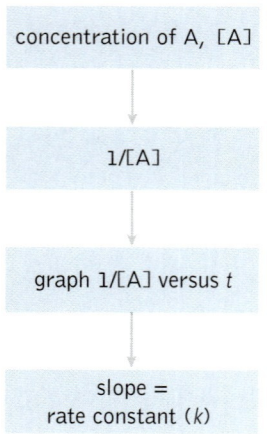

Time (ms)	[ClO]	ln [ClO]	1/[ClO]
0	1.50×10^{-8}	-18.015	6.67×10^7
10	7.19×10^{-9}	-18.751	1.39×10^8
20	4.74×10^{-9}	-19.167	2.11×10^8
30	3.52×10^{-9}	-19.465	2.84×10^8
40	2.81×10^{-9}	-19.690	3.56×10^8
100	1.27×10^{-9}	-20.484	7.89×10^8
200	0.66×10^{-9}	-21.139	1.51×10^9

and plot ln [ClO] and 1/[ClO] versus time, as shown on page 693.

As you can see, the plot of ln [ClO] versus time is not linear, but the plot of 1/[ClO] versus time is linear. Therefore, the reaction is second order in ClO and so has the following rate-law expression:

$$\text{rate} = k[ClO]^2$$

According to the integrated rate law for this second-order reaction,

$$\frac{1}{[ClO]} = kt + \frac{1}{[ClO]_0}$$

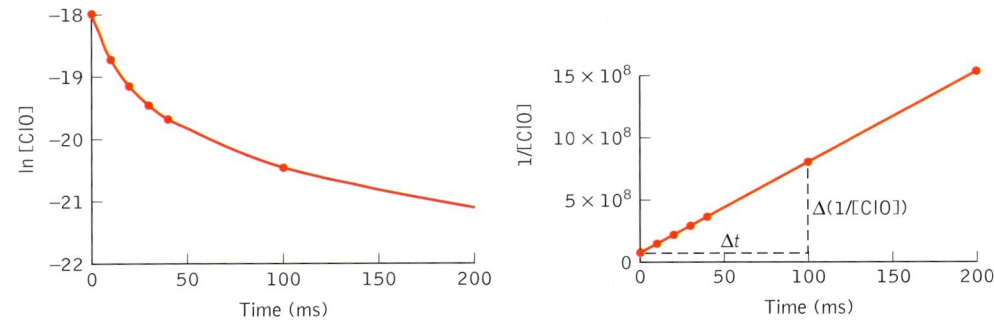

The slope of the plot of $1/[\text{ClO}]$ versus t equals the rate constant. By choosing two convenient pairs of data (0 and 100 ms), we can calculate the slope (and k) as follows:

$$k = \text{slope} = \frac{\Delta y}{\Delta x} = \frac{\Delta\left(\frac{1}{[\text{ClO}]}\right)}{\Delta t} = \frac{(7.89 - 0.67) \times 10^8 \, M^{-1}}{(100 - 0) \times 10^{-3} \, s} = 7.22 \times 10^9 \, M^{-1} \, s^{-1}$$

PRACTICE EXERCISE: The reaction of NO_2 with CO produces NO and CO_2:

$$NO_2 + CO \rightarrow NO + CO_2$$

The reaction rate is independent of the CO concentration (it is *zero* order in CO — see Section 14.4). Determine whether the reaction is first or second order in NO_2, and calculate the rate law from the following data at 498 K.

Time (s)	$[NO_2]$
0	0.250
1000	0.238
2000	0.226
3000	0.216
5000	0.198
7000	0.183
9000	0.170

(See Problems 45–50.)

The concept of half-life can also be applied to second-order reactions. The relation between the second-order rate constant (k) and the half-life ($t_{1/2}$) for the thermal decomposition of NO_2 can be derived from Equation 14.15 if we first rearrange the terms to solve for t:

$$kt = \frac{1}{[NO_2]} - \frac{1}{[NO_2]_0}$$

After one half-life has elapsed ($t = t_{1/2}$), $[NO_2]$ has decreased to half the initial concentration: $[NO_2] = \frac{1}{2}[NO_2]_0$. Substituting these terms into the preceding equation, we have

$$kt_{1/2} = \frac{1}{\frac{1}{2}[NO_2]_0} - \frac{1}{[NO_2]_0}$$

$$= \frac{2}{[NO_2]_0} - \frac{1}{[NO_2]_0} = \frac{1}{[NO_2]_0}$$

or

$$t_{1/2} = \frac{1}{k[NO_2]_0}$$

An equation of similar form can be written for any reaction that is second order in substance X:

$$t_{1/2} = \frac{1}{k[X]_0} \tag{14.17}$$

SAMPLE EXERCISE 14.9: Calculate the half-life of the second-order thermal decomposition of NO_2 if the rate constant is $0.543 \times 10^{-3}\ M^{-1}\ s^{-1}$ and the initial concentration of NO_2 is 0.0100 M.

SOLUTION: Substituting the values for k and $[NO_2]_0$ into Equation 14.17, we obtain

$$t_{1/2} = \frac{1}{k[NO_2]_0} = \frac{1}{(0.543\ M^{-1}\ s^{-1})(1.00 \times 10^{-2}\ M)} = 184\ s$$

PRACTICE EXERCISE: The second-order rate constant for the decomposition of ClO (see Sample Exercise 14.8) is $7.22 \times 10^9\ M^{-1}\ s^{-1}$. Determine the half-life of ClO when its initial concentration is $1.50 \times 10^{-8}\ M$. (See Problems 53–56.)

---

second-order rate constant, k
initial concentration, $[A]_0$

$$t_{1/2} = \frac{1}{k[A]_0}$$

half-life, $t_{1/2}$ (s)

Equation 14.17 applies only to reactions that are second order *in a single reactant*. It does not apply to the following reaction between ozone and NO:

$$O_3 + NO \longrightarrow O_2 + NO_2$$

or to the reaction between ozone and NO_2:

$$O_3 + 2\ NO_2 \longrightarrow 2\ NO_3 + O_2$$

because both of these reactions are second order *overall*, but only *first order in each of the reactants*.

The integrated rate law for a reaction that is first order in two reactants is quite complicated. However, a simple rate-law expression for such a reaction can be derived when one of the reactants is present at much higher concentration than the other. For example, the concentrations of ozone in polluted urban air are often from hundreds to thousands of times as great as the concentrations of NO. With such a large excess of ozone, the concentration of ozone remains more or less constant in the course of the reaction:

$$O_3 + NO \longrightarrow O_2 + NO_2$$

The rate law for this reaction:

$$\text{rate} = k[\text{NO}][\text{O}_3]$$

may, in this case, be simplified to:

$$\text{rate} = k'[\text{NO}] \qquad (14.18)$$

where $k' = k[\text{O}_3]_0$ and where $[\text{O}_3]_0$ is the initial concentration of ozone that remains relatively constant throughout the reaction.

Equation 14.18 looks like the rate law for a first-order reaction. It is considered a *pseudo-first-order* rate law because it *appears* to obey first-order kinetics. The pseudo-first-order rate constant, k', can be determined from a plot of ln [NO] versus time. The value of k can then be determined by dividing k' by $[\text{O}_3]_0$ as shown in the following sample exercise.

SAMPLE EXERCISE 14.10: The following data were obtained in a study of the oxidation of trace levels NO (note that NO concentrations are expressed in numbers of molecules per cubic centimeter) by a large excess of ozone:

Time (μs)	Concentration of NO (molecules/cm^3)
0	1.00×10^9
100	8.36×10^8
200	6.98×10^8
300	5.83×10^8
400	4.87×10^8
500	4.07×10^8
1000	1.65×10^8

a. Determine the pseudo-first-order rate constant, k', for the reaction.
b. If the initial concentration of ozone is 100 times the initial concentration of NO, or 1.00×10^{11} molecules/cm^3, what is the second-order rate constant k for the reaction?

SOLUTION: a. To evaluate a pseudo-first-order rate constant for the reaction, we need to calculate the natural logarithms of the concentration values

Time (μs)	[NO] (molecules/cm^3)	ln[NO]
0	1.00×10^9	20.723
100	8.36×10^8	20.544
200	6.98×10^8	20.364
300	5.83×10^8	20.184
400	4.87×10^8	20.004
500	4.07×10^8	19.824
1000	1.65×10^8	18.915

and then plot them against time as shown in Figure 14.11.

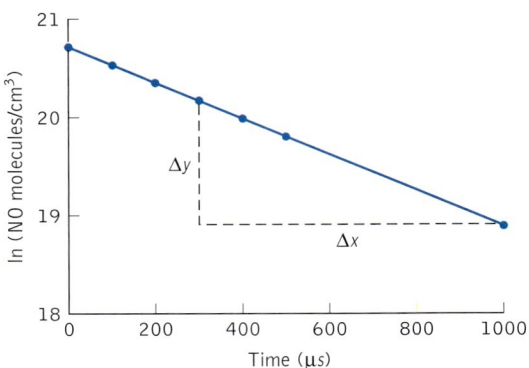

FIGURE 14.11 The reaction NO + O$_3$ ⟶ NO$_2$ + O$_2$ is first order in both NO and O$_3$ and second order overall. However, in the presence of a large excess of O$_3$, the reaction becomes pseudo-first order in NO, as indicated by the linear plot of ln [NO] versus time. The slope of the line, $\Delta(\ln[\text{NO}])/\Delta t = -0.00181\ \text{s}^{-1}$, is the negative of the pseudo-first-order rate constant k'; so $k' = 0.00181\ \text{s}^{-1}$.

The slope of the line ($\Delta y/\Delta x$) is calculated in the usual way from a convenient pair of data points, such as (300, 20.184) and (1000, 18.915):

$$\text{slope} = \frac{\Delta y}{\Delta x} = \frac{(20.184 - 18.915)}{(300 - 1000)\ \text{s}} = -0.00181\ \text{s}^{-1}$$

Therefore, the value of the pseudo-first-order rate constant $k' = k\,[\text{O}_3]$ is $0.00181\ \text{s}^{-1}$.

 b. To calculate the second-order rate constant, k, we divide k' by the concentration of ozone (which does not change in the course of the reaction.

or

$$k = \frac{k'}{[\text{O}_3]} = \frac{0.00181\ \text{s}^{-1}}{1.00 \times 10^{11}\ \text{molecules/cm}^3} = \frac{1.81 \times 10^{-14}\ \text{cm}^3\ \text{s}^{-1}}{\text{molecules}}$$

To obtain a value of k with the conventional units (M^{-1} s^{-1}), we need to convert the reciprocal concentration units of cm^3/molecule into M^{-1}:

$$k = \frac{1.81 \times 10^{-14}\ \cancel{\text{cm}^3}\ \text{s}^{-1}}{\cancel{\text{molecules}}} \times \frac{6.022 \times 10^{23}\ \cancel{\text{molecules}}}{\text{mol}} \times \frac{1\ \text{L}}{1000\ \cancel{\text{cm}^3}}$$

$$= 1.09 \times 10^7\ \frac{\text{L} \cdot \text{s}^{-1}}{\text{mol}} = 1.09 \times 10^7\ \text{M}^{-1}\ \text{s}^{-1}$$

PRACTICE EXERCISE: Determine the pseudo-first-order and the second-order rate constants for the reaction between chlorine atoms and ozone

$$Cl + O_3 \longrightarrow ClO + O_2$$

from the following results of an experiment in which the ozone concentration was a constant 8.5×10^{-11} M:

Time (μs)	[Cl] (M)
0	5.60×10^{-14}
100	5.27×10^{-14}
600	3.89×10^{-14}
1200	2.69×10^{-14}
1850	1.81×10^{-14}

(See Problems 57 and 58.)

14.4 REACTION MECHANISMS

Let's revisit the chemical equation describing the thermal decomposition of NO_2:

$$2\ NO_2(g) \longrightarrow 2\ NO(g) + O_2(g)$$

The reaction is second order in NO_2, which indicates that it takes place as a result of the collision of two molecules of NO_2. But, how? How do the bonds in two colliding molecules of NO_2 rearrange into the bonds that hold together a molecule of NO and a molecule of O_2? The answer to this question is contained in the **mechanism** of the reaction.

One such mechanism for the thermal decomposition of NO_2 is shown in Figure 14.12. In the first stage of the reaction, a collision between a pair of NO_2 molecules produces a very short lived **activated complex** in which the two molecules share an oxygen atom. Activated complexes such as this one represent midway points in chemical reactions. They have extremely brief lifetimes—in the femto (10^{-15}) second range or less—and rapidly fall apart, either forming products or re-forming reactants. To form products, the shared oxygen atom is transferred from one NO_2 molecule to the other, forming a molecule of NO and a molecule of NO_3 that rapidly decomposes by forming another activated complex. The bonds in this second complex rearrange so that two oxygen atoms become bonded together, forming a molecule of O_2 and leaving behind a molecule of NO.

> How do the bonds in two colliding molecules of NO_2 rearrange into the bonds that hold together a molecule of NO and a molecule of O_2?

> ✓ The **mechanism** of a chemical reaction is the sequence of bonds being broken and new ones being formed in the transformation of reactants into products.

> An **activated complex** is a high-energy, unstable species that exists momentarily in the course of a chemical reaction and that quickly falls apart, either forming products or re-forming reactants.

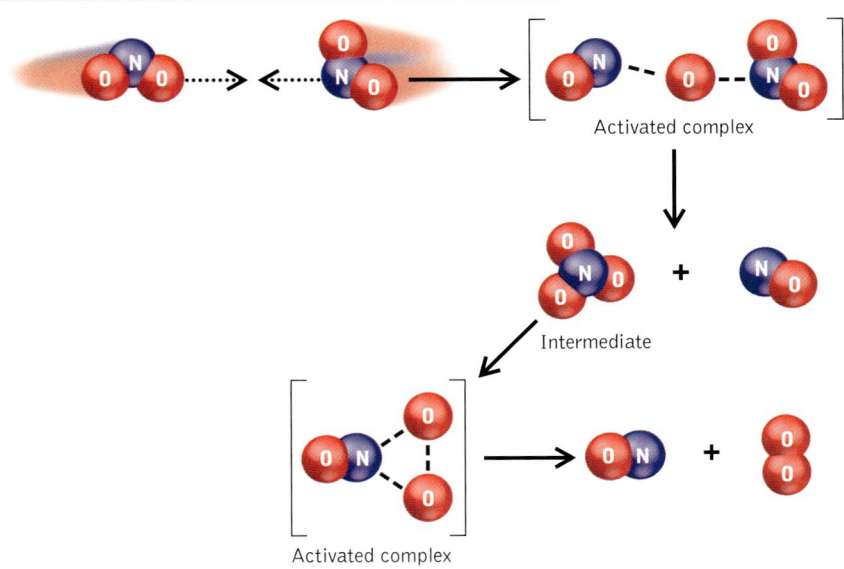

FIGURE 14.12 The decomposition of NO_2 begins when two NO_2 molecules collide, producing NO and NO_3. NO_3 rapidly decomposes into NO and O_2. The second step is faster than the first. Thus the first step is the rate-determining step, and the overall reaction is second order in NO_2.

✓ A reaction mechanism consists of one or more **elementary steps** that provide a molecular view of the reaction process.

✓ The **molecularity** of an elementary step is the number of reacting particles (atoms, ions, or molecules) in the step.

NO_3 is considered an *intermediate* in this mechanism, meaning that it is produced in one step and then consumed in the next. Intermediates are not considered reactants or products and so do not appear in the overall equation describing a reaction.

The reaction mechanism in Figure 14.12 is a combination of two **elementary steps,** which provide detailed molecular views of the reaction. Elementary steps may involve one or more molecules. Those that involve a single molecule are called *unimolecular*. If more than one molecule takes part, then all of the molecules must react simultaneously. Elementary steps based on collisions between two molecules are called *bimolecular*. They are much more common than *termolecular* (three-molecule) steps, because the chances of three molecules colliding simultaneously are much less than those of bimolecular collisions. These italicized terms are used by chemists to describe the **molecularity** of an elementary step, referring to the number of atomic-scale particles (free atoms, ions, or molecules) that collide with one another in the step. Note that the molecularity of an elementary step and its reaction order are the same and that the rate law for an elementary step can be written directly from its stoichiometry. The same cannot be said for an overall chemical reaction that includes two or more elementary steps.

Is there experimental evidence that supports the mechanism in Figure 14.12? A valid mechanism must be consistent with the stoichiometry of the overall reaction and its observed rate law. In this case, the sum of the two steps *does* match the stoichiometry:

Step 1: $2 NO_2 \longrightarrow NO + NO_3$

Step 2: $NO_3 \longrightarrow NO + O_2$

Summing steps 1 and 2,

$$2\,NO_2 + \cancel{NO_3} \longrightarrow 2\,NO + \cancel{NO_3} + O_2$$

and simplifying, we get the overall reaction:

$$2\,NO_2 \longrightarrow 2\,NO + O_2$$

The observed rate law for the decomposition of NO_2 is second order in NO_2. This rate law is predicted for step 1 but not for step 2, which means that the rate of the overall process must be controlled by step 1 rather than step 2. The first step is the controlling step *if it is the slower of the two*. Then the first step acts as a throttle on the overall reaction. It is said to be the **rate-determining step** in the reaction. Its rate law:

$$\text{rate of step 1} = \frac{-\Delta[NO_2]}{\Delta t} = k_1[NO_2]^2$$

if its rate constant (k_1) is much less than the rate constant for step 2 (k_2):

$$\text{rate of step 2} = \frac{-\Delta[NO_3]}{\Delta t} = k_2[NO_3]$$

Is there any evidence that $k_1 \ll k_2$? Actually, there is. Chemists have extensively studied the kinetics of the formation and decomposition of NO_3 and have determined that the rate constant for step 1 is only about $1 \times 10^{-10}\,M^{-1}\,s^{-1}$ at 300 K. The rate constant for step 2 is enormous by comparison: $6.3 \times 10^4\,s^{-1}$. Therefore, as soon as any NO_3 is formed in step 1, it rapidly falls apart, forming NO and O_2 in step 2.

Now let's consider the reverse of the thermal decomposition of NO_2—namely, the formation of NO_2 by the reaction of NO and O_2, a key step in the formation of photochemical smog:

$$2\,NO + O_2 \longrightarrow 2\,NO_2$$

The observed rate law for this reaction is second order in NO and first order in O_2:

$$\text{rate} = k[NO]^2[O_2]$$

One proposed mechanism includes two elementary steps, as shown in Figure 14.13.

Step 1: $\qquad\qquad NO + O_2 \longrightarrow NO_3$

Step 2: $\qquad\qquad NO_3 + NO \longrightarrow 2\,NO_2$

If the first step were the rate-determining step, then the reaction would be first order in NO and O_2, but that is not what is observed. If the second step were the rate-determining step, then the reaction would be first order in NO and NO_3, but that is not what is observed either. Are there any other options? It turns out there is at least one. Consider what happens if the first step is fast *and reversible*; then, during the reaction, NO_3 is rapidly produced from NO and O_2 but decomposes just as rapidly back into NO and O_2. In other words, the rate of step 1 in the

✓ The overall balanced equation for a reaction is the sum of its elementary steps.

The **rate-determining step** in a chemical reaction is the slowest elementary step.

The "break" at the beginning of game of pool produces multiple collisions of pairs of balls (analogous to bimolecular collisions of two molecules), but rarely do three balls collide simultaneously (analogous to a termolecular collision). It is reasonable, then, that reactions that rely on the simultaneous collision of three molecules are much less likely to take place (and will proceed much more slowly) than those requiring the collision of only two molecules.

forward direction matches the rate of step 1 in reverse. Expressed mathematically with the rate constants for the forward and reverse reactions given the symbols k_f and k_r, respectively, we have

$$\text{rate forward reaction} = k_f[NO][O_2] = \text{fast}$$

$$\text{rate of reverse reaction} = k_r[NO_3] = \text{also fast}$$

Combining the two rate-law expressions, we have

$$k_f[NO][O_2] = k_r[NO_3]$$

Solving for $[NO_3]$, we have

$$[NO_3] = \frac{k_f}{k_r}[NO][O_2] \qquad (14.19)$$

Now let's focus on the rate-law expression for step 2. Doing so is appropriate because, if step 1 is fast and step 2 is slow, then step 2 will be the rate-determining step. The rate law for step 2 is

$$\text{rate} = k_2[NO_3][NO] \qquad (14.20)$$

Now let's combine Equations 14.19 and 14.20 by replacing the $[NO_3]$ term in Equation 14.20 with the right-hand side of Equation 14.19:

$$\text{rate} = \frac{k_2 k_f}{k_r}[NO]^2[O_2]$$

The three rate constants can be combined into one overall rate constant:

$$k_{overall} = \frac{k_2 k_f}{k_r}$$

and so the rate-law expression for the reaction becomes

$$\text{rate} = k_{overall}[NO]^2[O_2]$$

This expression matches the observed order of the reaction, and so the proposed mechanism may indeed be valid.

For these airline passengers, the "rate-determining" step in getting through the airport to their planes is passing through this security checkpoint.

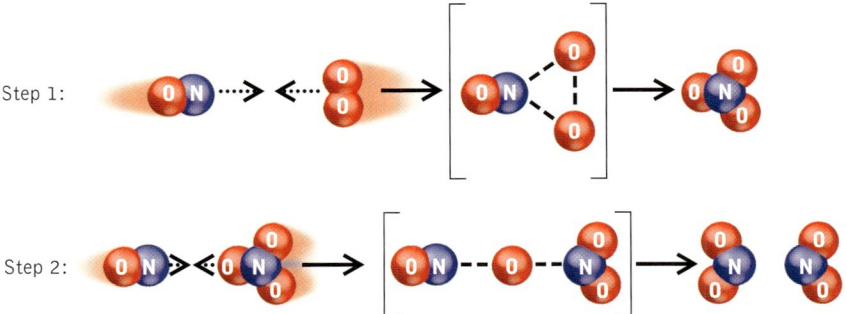

FIGURE 14.13 There are two steps in this mechanism for the formation of NO_2 from NO and O_2: (1) in a fast bimolecular reaction, NO and O_2 form NO_3, and (2) in a slower bimolecular reaction, NO_3 reacts with a second molecule of NO, forming two molecules of NO_2. The structures of the activated complexes formed in the two steps are shown in brackets.

THE CHEMISTRY OF THE GROUP 8B METALS

There are nine transition metals (Fe, Ru, Os, Co, Rh, Ir, Ni, Pd, and Pt) in three adjacent columns in the periodic table that are all considered to be in one group: Group 8B. This unusual multicolumn grouping of elements is based on the similar chemistries of iron (Fe), cobalt (Co), and nickel (Ni) coupled with the expected similarities among elements in each of the columns. The heavier six elements—ruthenium (Ru), osmium (Os), rhodium (Rh), iridium (Ir), palladium (Pd), and platinum (Pt)—are sometimes called the "platinum group metals."

Iron is by far the most abundant of these elements, making up most of our planet's core, much of its mantle, and 6.2% of its crust. The other Group 8B elements are much less abundant. Earth's crust contains only 99 μg/g Ni and 29 μg/g Co. The six platinum-group metals are quite rare, with abundances ranging from 2×10^{-2} to 1×10^{-4} μg/g.

As described in Chapter 4, iron occurs in Earth's crust primarily as oxide ores, including the minerals hematite (Fe_2O_3) and magnetite (Fe_3O_4). Rich deposits of iron ore are located in a region of northern Minnesota known as the Iron Range. Iron oxides are reduced to molten iron with red-hot carbon, also known as coke:

$$2\ Fe_2O_3(s) + 3\ C(s) \longrightarrow 4\ Fe(l) + 3\ CO_2(g)$$

The principal cobalt and nickel minerals include linnaeite (Co_3S_4), cobaltite (CoAsS), smaltite ($CoAs_2$),

The Iron Range in northern Minnesota has one of the richest deposits of iron ore in the world.

niccolite (NiAs), and millerite (NiS). They are usually found in ores containing other elements. The largest deposits of nickel-containing ores are found in Sudbury, Ontario. These ores are typically mixtures of nickel, copper, iron, and cobalt sulfides and arsenides. The similar sizes and identical charges of Cu^{2+}, Co^{2+}, Ni^{2+}, and Fe^{2+} allow these cations to substitute for one another in crystals formed with As^{3-} and S^{2-} anions.

To isolate individual Group 8B elements from such a mixture requires a series of steps that separate the various elements based on their volatilities or solubilities. Some steps require redox chemistry to change the oxidation states of the elements in ways that affect their solubilities, or to obtain the elements as free metals.

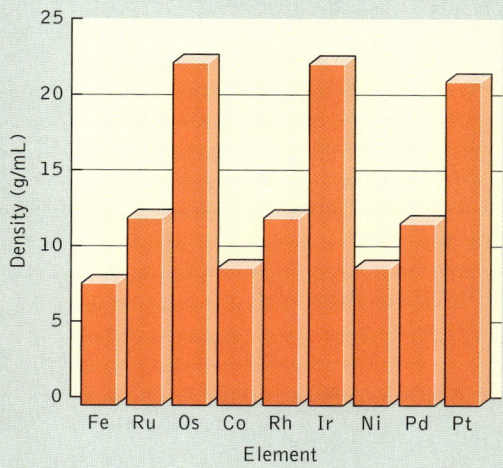

The densities of the Group 8B elements increase with increasing atomic number down the group. Osmium, iridium, and platinum are among the densest metals.

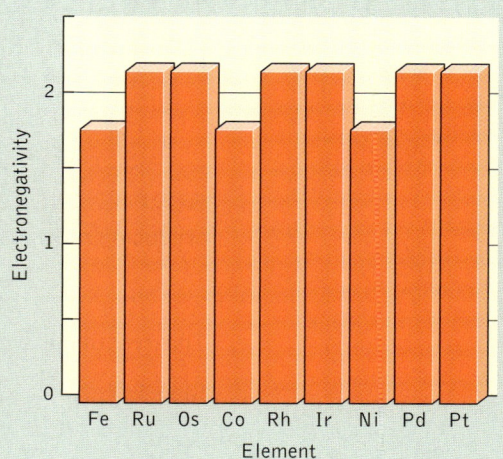

Only small variations in the electronegativity of the Group 8B elements are observed down the group and across the rows in the group.

The refining process starts by heating ore samples to melt the metallic compounds in them. Silica is added to convert iron to solid $FeSiO_3$. Any ruthenium and osmium in the samples is distilled as their volatile tetraoxides: RuO_4 and OsO_4.

After Fe, Ru, and Os have been removed, the ore sample is again heated in air. Any cobalt present is converted into Co_3O_4, any sulfur into SO_2, and any arsenic into As_4O_{10}. Nickel oxidizes slowly under these conditions and so remains as the sulfide, NiS. To recover cobalt, the mixture is treated with sulfuric acid, which dissolves Co_3O_4. Then the solution is treated with NaOCl, which causes Co to precipitate as $Co(OH)_3$. The precipitate is converted back to pure Co_3O_4, and then the Co_3O_4 is reacted with hot hydrogen gas, producing free cobalt metal:

$$Co_3O_4(s) + 4\ H_2(g) \longrightarrow 3\ Co(s) + 4\ H_2O(g)$$

The remaining sample residue may still contain nickel sulfides along with mixtures of reduced metals. The nickel sulfides can be reacted with oxygen to obtain NiO and reduced with coke (hot carbon):

$$NiO(s) + C(s) \longrightarrow Ni(s) + CO(g)$$

Mixtures of all six platinum metals can be separated by precipitation reactions or solvent extraction.

The reactivity of the Group 8B elements with oxygen, water, and halogens decreases with increasing atomic number both across and down the group. Nickel is an important constitutent of steel, including many of the forms of stainless steel. Cobalt is also used in iron-based alloys, mostly for those used in permanent magnets and at high temperatures.

The platinum-group metals are widely used catalysts. About 140 tons of platinum, 70 tons of palladium, and

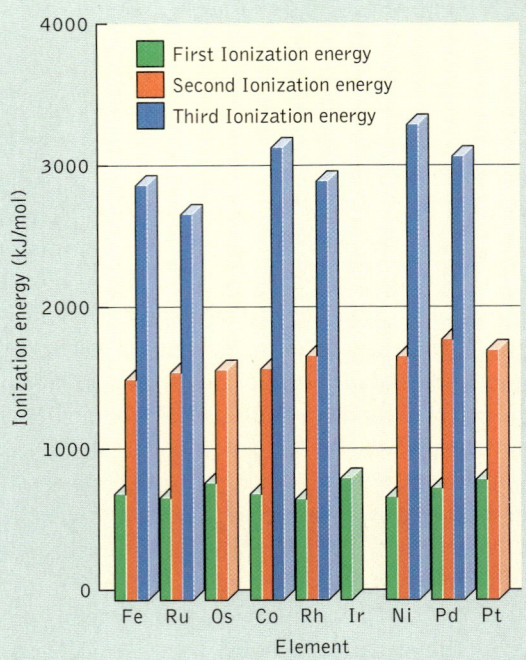

The third ionization energies of the Group 8B elements decrease down the group but increase from left to right. Both trends are consistent with those observed for the ionization energies of other groups and periods of elements. However, the first and second ionization energies of the 8B elements do not consistently follow these trends.

SAMPLE EXERCISE 14.11: At high temperature, NO reacts with hydrogen, producing nitrogen and water vapor:

$$2\ NO + 2\ H_2 \longrightarrow 2\ H_2O + N_2$$

This reaction is believed to take place in two elementary steps:

Step 1: $\qquad H_2 + 2\ NO \longrightarrow N_2O + H_2O$

Step 2: $\qquad N_2O + H_2 \longrightarrow N_2 + H_2O$

Identify the molecularity of each elementary step and write the rate law for each step.

SOLUTION: The molecularity of an elementary step equals the number of particles that collide. In this case, the first step is a termolecular reaction and the

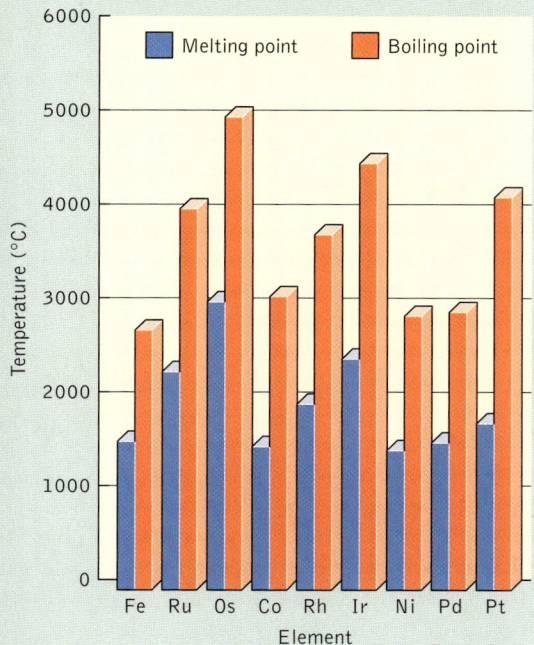

All the Group 8B metals have high melting and extremely high boiling points that increase with increasing atomic number down the group, but decrease from left to right across the same row.

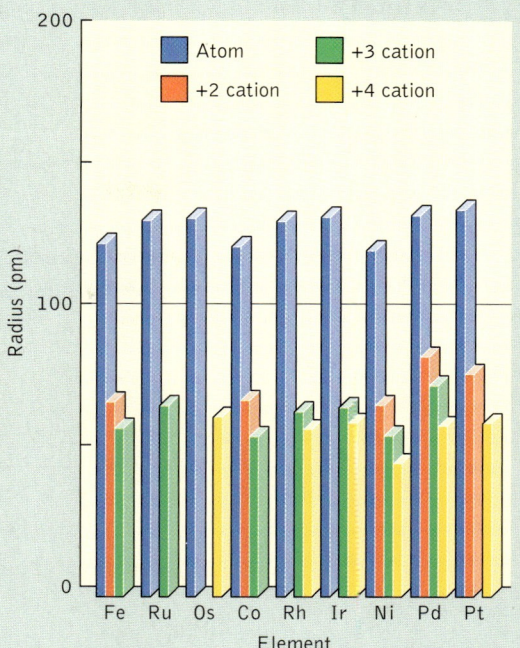

The atomic radii of the Group 8B elements increase with increasing atomic number within a column; there are only small changes in atomic radius across the three columns. Ionic radii generally increase down a group and across the series in cases where accurate data are available. Data for some cations are missing because of the absence of compounds in which those cations are unambiguously present.

15 tons of rhodium are used worldwide each year in the manufacture of the catalytic converters used in automobiles and light trucks. Numerous industrial processes also require precious-metal catalysts either as pure metals or as metal compounds.

second a bimolecular reaction. The rate laws for elementary steps can be written directly from their stoichiometries:

Step 1: $\qquad \text{rate} = k_1 [H_2][NO]^2$

Step 2: $\qquad \text{rate} = k_2 [N_2O][H_2]$

PRACTICE EXERCISE: Here is another possible mechanism for the reduction of NO by H_2:

Step 1: $\qquad H_2 + NO \longrightarrow N + H_2O$

Step 2: $\qquad N + NO \longrightarrow N_2 + O$

Step 3: $\qquad H_2 + O \longrightarrow H_2O$

Determine the molecularity and the rate law for each of the three elementary steps. (See Problems 61–64.)

CONCEPT TEST: The experimentally determined rate law for the reaction between NO and H_2 is:

$$\text{rate} = k[NO]^2[H_2]$$

Which of the proposed mechanisms in the preceding sample exercise and practice exercise appears to be correct?

Before we leave this section on reaction mechanisms, let's consider one more reaction of pollutants in automobile exhaust: the reaction between NO_2 and CO, which produces NO and CO_2:

$$NO_2 + CO \longrightarrow NO(g) + CO_2$$

This reaction is believed to take place in two elementary steps:

Step 1: $\quad 2\,NO_2 \longrightarrow NO_3 + NO$

Step 2: $\quad NO_3 + CO \longrightarrow NO_2 + CO_2$

The experimentally determined rate law for the reaction is

$$\text{rate} = k[NO_2]^2$$

This experimental evidence indicates that step 1 must be the slow, rate-determining step. There is no [CO] term in the rate law for this reaction. This independence of reaction rate on the concentration of a reactant makes the reaction *zero* order in that reactant. One interpretation of zero order is that the rate law contains a [CO] term to the power of zero. Because any value to the zero power is unity, the rate expression is

$$\text{rate} = k[NO_2]^2[CO]^0 = k[NO_2]^2(1) = k[NO_2]^2$$

The fact that the reaction is zero order in CO is evidence that (1) the reaction is a multistep reaction and (2) CO is consumed in a step that is not the rate-determining step.

14.5 REACTION RATES, TEMPERATURE, AND THE ARRHENIUS EQUATION

In Section 14.3, we introduced the idea that chemical reactions take place in the gas phase when molecules collide with sufficient energy to break some of the chemical bonds in the reactants while forming bonds in the products. This energy is called **activation energy** (E_a), and every chemical reaction has a characteristic activation-energy value, usually expressed in kilojoules. Activation energy is an energy barrier that must be overcome if a reaction is to proceed—like the hill that must be climbed by the bicyclist in Figure 14.14.

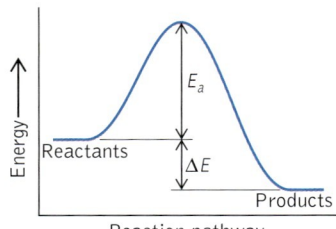

A

B

FIGURE 14.14 The energy profile (A) of an exothermic reaction includes an activation energy barrier (E_a) that is analogous to (B) the hill that a bicyclist must climb en route from a starting point on a high plateau to a destination on a lower plain.

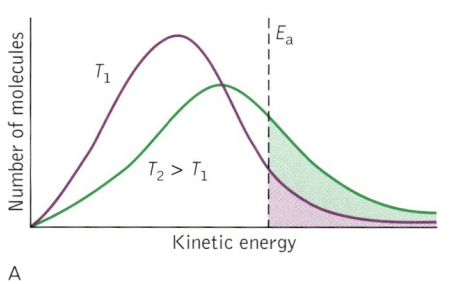
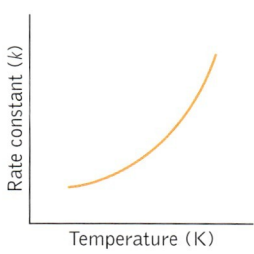

FIGURE 14.15 According to kinetic molecular theory, some fraction (the shaded areas in A) of a population of reactant molecules will have kinetic energies equal to or greater than the activation energy (E_a) of the reaction in which they engage. As temperature increases from T_1 to T_2, the number of molecules with energies exceeding E_a increases, leading to an increase in reaction rate. B. Thus, the rate constant for any reaction increases with increasing temperature.

According to kinetic molecular theory, the fraction of molecules with kinetic energies greater than a given activation energy should increase with increasing temperature, as shown in Figure 14.15A. Therefore, the rates of chemical reactions should increase with increasing temperature, and indeed they do (Figure 14.15B).

The mathematical connections between temperature, the value of the rate constant (k) of a reaction, and its activation energy (E_a) are given by the **Arrhenius equation:**

$$k = A\, e^{-E_a/RT} \tag{14.21}$$

Parameter A is called the **frequency factor.** The frequency factor is the product of collision frequency and a factor that corrects for the fact that not every collision results in a chemical reaction. Some collisions do not lead to products because the colliding molecules are not oriented toward each other in the right way. To examine the importance of molecular orientation during collisions, let's revisit a reaction that maintains the level of NO_2 in polluted urban air after the evening rush hour:

$$O_3 + NO \longrightarrow O_2 + NO_2$$

Two of the ways in which ozone and nitric oxide molecules might approach each other are shown in Figure 14.16. Only one of these orientations, the one in which an oxygen atom on an O_3 molecule approaches the nitrogen atom on NO, leads to a chemical reaction between the two molecules.

A collision between O_3 and NO molecules with the correct orientation and enough kinetic energy may result in the formation of an activated complex in which some of the bonds that hold these molecules together break and the bonds that hold O_2 and NO_2 together form. Activated complexes have high internal energies and are therefore high-energy states, called **transition states.** These states are of higher energy than the energies of the starting materials or products. In fact, the energies of these transition states define the heights of the activation-energy barriers of chemical reactions.

> ✓ **Activation energy** is the energy barrier that colliding species must overcome if they are to react with each other.
>
> The **Arrhenius equation** relates the rate constant of a chemical reaction to absolute temperature (T), the reaction's activation energy (E_a), and a collisional frequency factor (A).
>
> The collisional **frequency factor** (A) is the product of the frequency of collisions between reacting species and the orientation factor required for them to react.
>
> ✓ A **transition state** is the high-energy state associated with the formation of an activated complex.

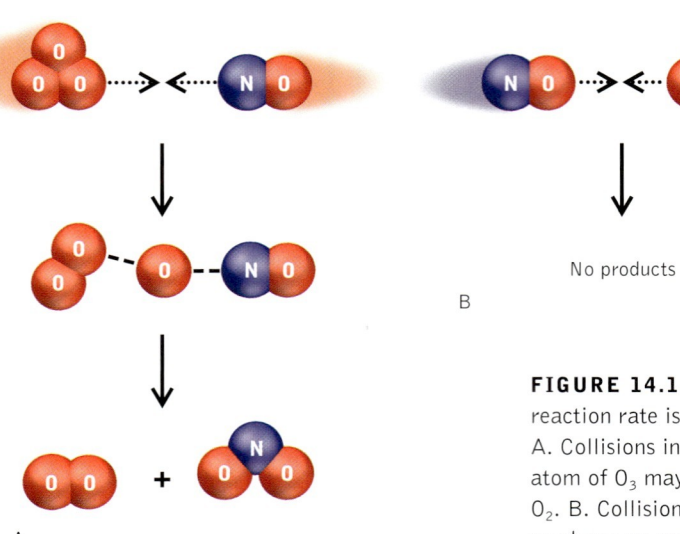

FIGURE 14.16 The importance of molecular orientation on reaction rate is illustrated in the reaction between NO and O_3. A. Collisions in which the nitrogen on NO collides with an oxygen atom of O_3 may lead to an activated complex that yields NO_2 and O_2. B. Collisions between oxygen atoms on NO and O_3 molecules produces no such activated complex and no reaction.

CONNECTION: Entropy, free energy, and their relation to enthalpy are described in Section 13.4.

Consider the energy profile for the reaction between nitric oxide and ozone in Figure 14.17. The x-axis of the profile represents the progress of the reaction. Distances along the y-axis represent changes in chemical energy. The size of the activation-energy barrier depends on the direction from which it is approached. If a reaction is exothermic in the forward direction, such as the one illustrated in Figure 14.17, then E_a is smaller in the forward direction than in the reverse direction. A smaller activation-energy barrier means that the exothermic reaction proceeds at a faster rate than the reverse, endothermic reaction.

One of the many uses of the Arrhenius equation is in calculating the value of E_a for a chemical reaction. To see how this procedure works, let's first rewrite Equation 14.21 by taking the natural logarithm of both sides:

$$\ln k = -\frac{E_a}{R}\left(\frac{1}{T}\right) + \ln A \tag{14.22}$$

Equation 14.22 fits the general equation of a straight line ($y = mx + b$) if we make ($\ln k$) the y variable and ($1/T$) the x variable. We can calculate E_a by determining the rate constants for a reaction at several temperatures. Plotting $\ln k$ versus $1/T$ should yield a straight line, the slope of which is $-E_a/R$. Table 14.6 and Figure 14.18 show the results of measurements of the rate of the reaction between NO and O_3 at six different temperatures. The straight line that best fits the points in Figure 14.18 has a slope of -1257 K. There are no units for the y values (they are logarithms), and the units of the x values are 1/K. Therefore, the units for the slope $\Delta y/\Delta x$ will be $1/(1/K)$, or simply K. The slope equals $-E_a/R$; so:

$$E_a = -\text{slope} \times R$$
$$= -(-1257 \text{ K})\left(\frac{8.314 \text{ J}}{\text{mol} \cdot \text{K}}\right) = 1.04 \times 10^4 \text{ J/mol} = 10.4 \text{ kJ/mol}$$

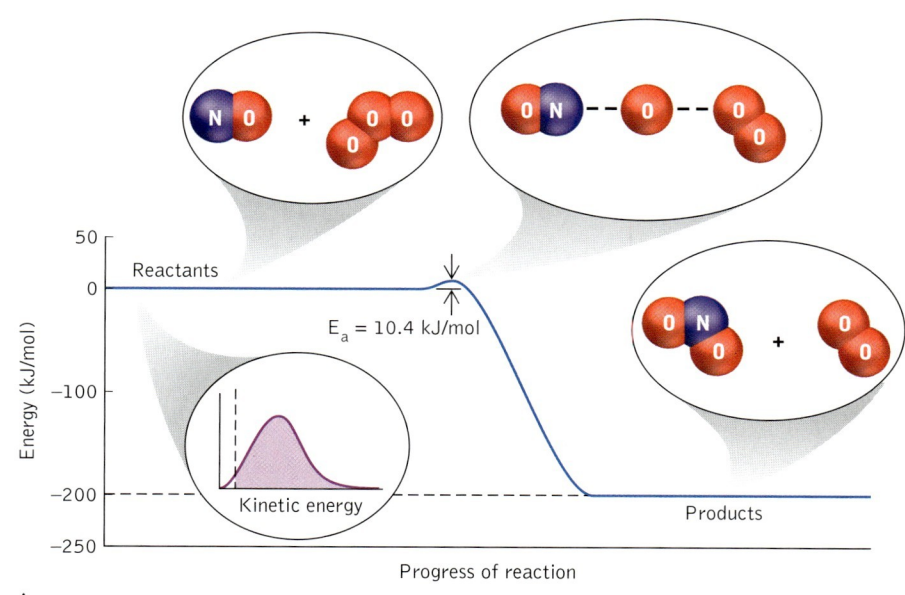

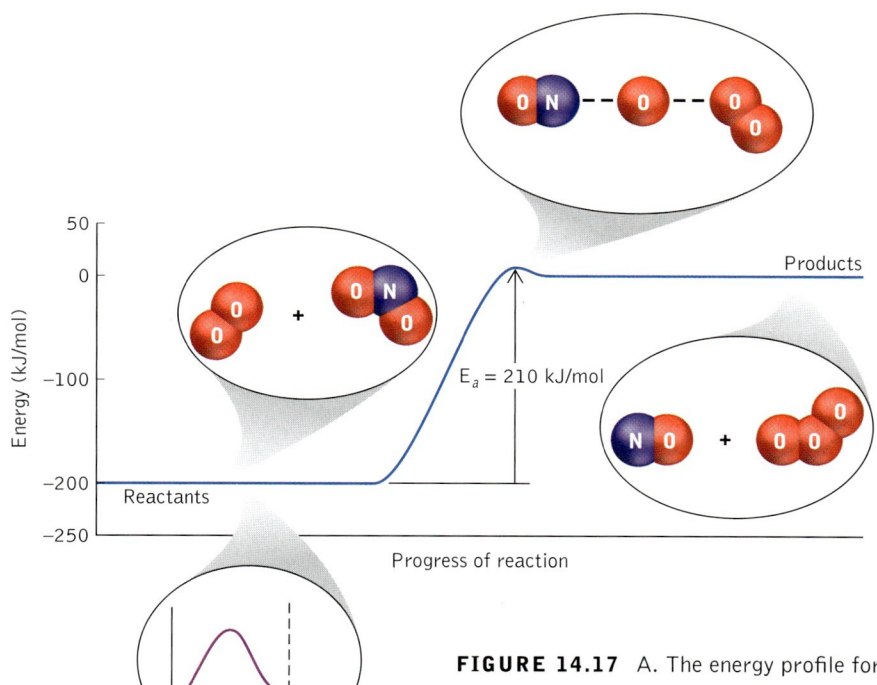

FIGURE 14.17 A. The energy profile for the exothermic reaction O_3 + NO ⟶ O_2 + NO_2 includes an activation-energy barrier of 10.4 kJ/mol. B. The reverse reaction has an activation energy of 210 kJ/mol. Fewer molecules have the kinetic energy needed to overcome the larger activation energy of the reverse reaction as shown by the different shaded areas under the kinetic energy curves. Therefore, the reverse reaction proceeds much more slowly than the forward reaction between O_3 and NO.

TABLE 14.6 Temperature Dependence of the Rate of the Reaction
$NO + O_3 \longrightarrow NO_2 + O_2$

T (K)	k (M^{-1} s^{-1})	ln k	1/T (K^{-1})
300	1.21 × 10^{10}	23.216	3.33 × 10^{-3}
325	1.67 × 10^{10}	23.539	3.08 × 10^{-3}
350	2.20 × 10^{10}	23.814	2.86 × 10^{-3}
375	2.79 × 10^{10}	24.052	2.67 × 10^{-3}
400	3.45 × 10^{10}	24.264	2.50 × 10^{-3}
425	4.15 × 10^{10}	24.449	2.35 × 10^{-3}

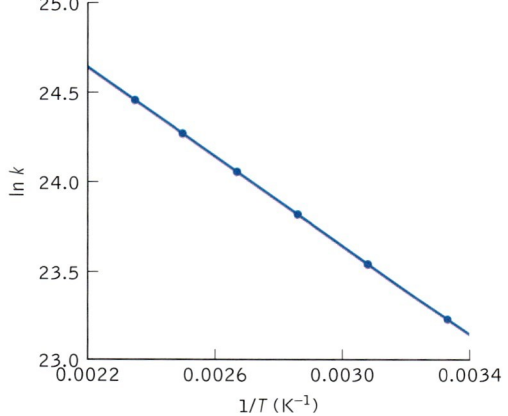

FIGURE 14.18 A graph of the natural logarithm of the rate constant (ln k) versus 1/T yields a straight line with a slope equal to $-E_a/R$ and a y intercept equal to the natural logarithm of the frequency factor (ln A).

The y intercept of the line that best fits the data is 27.4. This value represents ln A. The value of A is

$$A = e^{27.4} = 7.9 \times 10^{11}$$

We can use the values of E_a and A to calculate k at any temperature. For example, at T = 250 K:

$$k = Ae^{-E_a/RT} = A \exp\left(-\frac{E_a}{RT}\right)$$

$$= 7.9 \times 10^{11} \exp\left[-\frac{1.04 \times 10^4 \text{ J/mol}}{(8.314 \text{ J/mol} \cdot \text{K})(250 \text{ K})}\right]$$

$$= 5.2 \times 10^9 \; M^{-1} \, s^{-1}$$

SAMPLE EXERCISE 14.12: The following data were collected in a study of the effect of temperature on the rate of decomposition of ClO to Cl$_2$ and O$_2$. Determine the activation energy for the reaction.

$$2 \, \text{ClO} \longrightarrow \text{Cl}_2 + \text{O}_2$$

$k\,(M^{-1}\,s^{-1})$	$T\,(K)$
1.9×10^9	238
3.1×10^9	258
4.9×10^9	278
7.2×10^9	298

SOLUTION: The value of the activation energy of a reaction can be determined from the slope of a plot of ln k versus $1/T$. Thus, the first step is to expand the data table to include ln k and $1/T$:

$k\,(M^{-1}\,s^{-1})$	ln k	$T\,(K)$	$1/T\,(K^{-1})$
1.9×10^9	21.365	238	4.20×10^{-3}
3.1×10^9	21.855	256	3.90×10^{-3}
4.9×10^9	22.313	278	3.60×10^{-3}
7.2×10^9	22.697	298	3.36×10^{-3}

A plot of ln k versus $1/T$ yields a straight line:

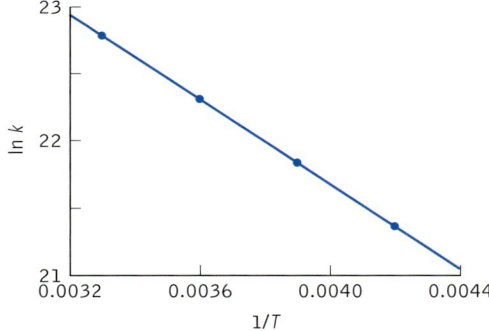

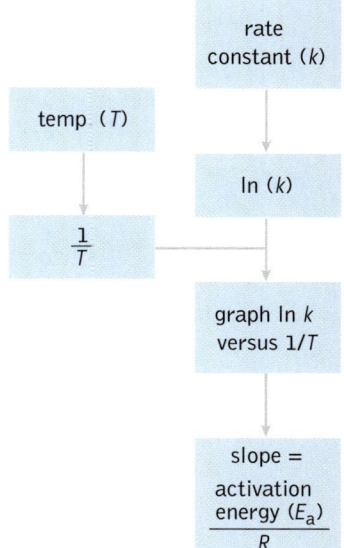

the slope of which is -1580 K. According to the Arrhenius equation, the slope equals $-E_a/R$. Solving for E_a, we have

$$E_a = -\text{slope} \times R$$

$$= -(-1580\,\cancel{K}) \times \left(\frac{8.314\,J}{\text{mol}\cdot\cancel{K}}\right) = 1.314 \times 10^4\,J/\text{mol}$$

$$= 13.1\,kJ/\text{mol}$$

PRACTICE EXERCISE: Atomic bromine reacts with ozone, forming BrO and O_2:

$$Br + O_3 \longrightarrow BrO + O_2$$

The rate constant for the reaction was determined at four temperatures ranging from 238 K to 298 K. Calculate the activation energy for this reaction.

T (K)	k (cm³/molecule · s)
238	5.9×10^{-13}
258	7.7×10^{-13}
278	9.6×10^{-13}
298	1.2×10^{-12}

(See Problems 81–86.)

How does the concept of activation energy apply to multi–step reactions? To answer this question, let's revisit the thermal decomposition of NO_2:

$$2\ NO_2(g) \longrightarrow 2\ NO(g) + O_2(g)$$

This reaction takes place in two steps (see Figure 14.12) with the formation of a reactive, short-lived intermediate, NO_3. These two elementary steps produce a two-hill energy profile for the reaction as shown in Figure 14.19. In step 1, collisions between pairs of NO_2 molecules result in the formation of an activated complex associated with the first transition state in Figure 14.19. As this activated complex transforms into NO and NO_3, the energy of the system drops, reaching

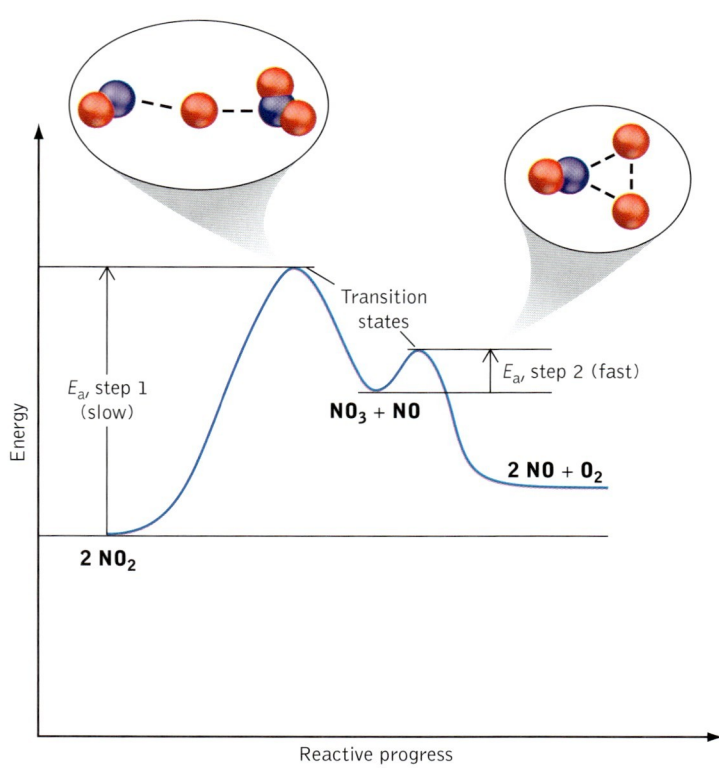

FIGURE 14.19 The reaction-energy profile for the decomposition of NO_2 to NO and O_2 features activation-energy barriers for both elementary steps in the overall reaction mechanism. The activation energy of the first step is larger than that of the second step; so the first step is the slower of the two and the rate-determining step.

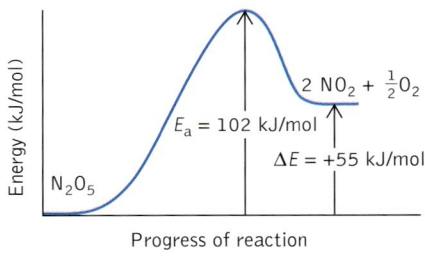
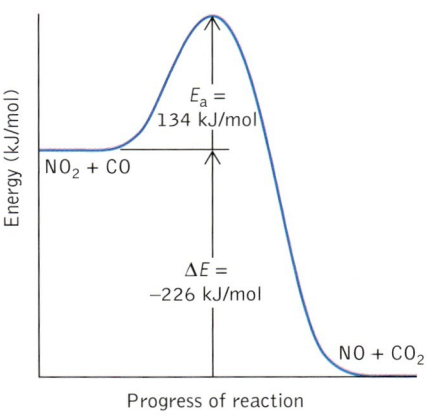

FIGURE 14.20 Spontaneous reactions are not necessarily rapid reactions. Here are two reaction-energy profiles for (A) the endergonic reaction $N_2O_5 \longrightarrow 2\ NO_2 + \frac{1}{2} O_2$ with an activation energy of 102 kJ/mol and (B) the exergonic reaction $NO_2 + CO \longrightarrow NO + CO_2$ with an activation energy of 134 kJ/mol. The nonspontaneous endergonic reaction is likely to be faster than the exergonic reaction.

the bottom of the valley between the two hills. The height of the second energy barrier represents the transition state associated with a second activated complex in which the bonds in NO_3 rearrange. Two N–O bonds break and an O–O double bond forms to give the final products, NO and O_2.

In the forward direction, the activation energy for step 1 is much greater than that for step 2. This difference is consistent with the relative rates of the two steps. The first one is the slower, rate-determining step and that is the one with the higher activation-energy barrier. The profile approached from the reverse direction (in which NO and O_2 react, forming NO_2) has a smaller initial activation-energy barrier, and so we would expect a faster first step in the two-step process. That is consistent with the mechanism described in Section 14.4 in which we assumed a rapid and reversible first step followed by the slower, rate-determining reaction between NO_3 and NO to form NO_2.

Earlier in this section, we noted that activation-energy barriers tend to be lower when approached from the reaction direction that has a negative ΔE. This observation raises a question: Do exothermic reactions always proceed more rapidly than endothermic reactions? The answer is, not necessarily. Consider the reaction-energy profiles for the highly exergonic reaction between CO and NO_2 and the decidedly endergonic decomposition of N_2O_5 in Figure 14.20. The decomposition of N_2O_5 to NO_2 and O_2 has an activation energy of 102 kJ. However, the exergonic reaction of NO_2 with CO has an even larger activation-energy barrier (134 kJ/mol) and so may be the slower of the two at any given temperature. Thus, a reaction with a large decrease in free energy is spontaneous from a thermodynamics point of view—but spontaneous does not necessarily mean rapid. The rates of some spontaneous reactions are so slow, they appear to be nonspontaneous: they have favorable thermodynamics, but unfavorable kinetics.

14.6 CATALYSIS

Let's consider the photochemical decomposition of ozone into oxygen:

$$2 O_3 \xrightarrow{h\nu} 3 O_2$$

This reaction is important both in the lower atmosphere, where we would like to reduce ozone levels associated with smog formation, and in the stratosphere, where the loss of ozone may pose a threat to us and other organisms by allowing more ultraviolet radiation to penetrate through the atmosphere. The process begins with the absorption of a photon of UV radiation and the generation of atomic oxygen:

Step 1: $$O_3(g) \xrightarrow{h\nu} O_2(g) + O(g)$$

Oxygen atoms generated in this fashion can react with additional ozone molecules to form two more molecules of oxygen:

Step 2: $$O_3 + O \longrightarrow 2 O_2$$

The rate of the second step is slowed by an activation-energy barrier of 17.7 kJ/mol of O_3 Figure 14.21. However, gases in the atmosphere can reduce the size of this barrier. In Section 6.6, we considered the role of CFCs and atomic chlorine in the decomposition of stratospheric ozone. The photodecomposition of chlorofluorocarbons results in the release of chlorine atoms that react with ozone, forming ClO and O_2 (Equation 6.16):

Step 1: $$Cl + O_3 \longrightarrow ClO + O_2$$

The activation energy for this reaction is only 2.2 kJ/mol. Thus, the presence of chlorine atoms greatly accelerates ozone destruction. In addition, the chlorine monoxide formed in this reaction can react with more ozone, forming O_2 and regenerating chlorine atoms (Equation 6.17):

Step 2: $$ClO + O_3 \longrightarrow Cl + 2 O_2$$

This reaction is so fast that an accurate value for its activation energy has not been determined. We do know that the rate-limiting step in the reaction of ozone

> **CONNECTION:** The catalytic role of chlorine in ozone depletion is described in Section 6.6. Recall that a catalyst is a substance that increases the rate of a reaction but is not consumed by it.

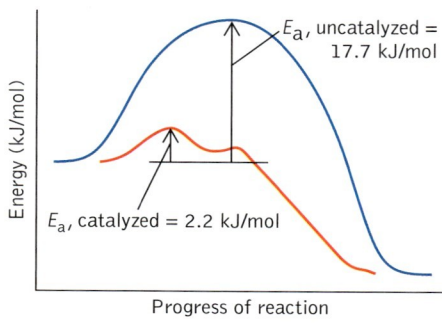

FIGURE 14.21 The photodecomposition of O_3 to O_2 has a larger activation energy (17.7 kJ/mol) than does the decomposition of O_3 in the presence of chlorine atoms (2.2 kJ/mol). The catalytic effect of chlorine is key to the depletion of stratospheric ozone and the formation of ozone holes over the South Pole.

with chlorine is step 1. Chlorine atoms act as a catalyst in the destruction of ozone by providing a different reaction pathway for the conversion of O_3 to O_2, a pathway that lowers the activation energy and so speeds up the process. Because the catalyst (Cl) and reacting species (O_3) are both gases and so exist in the same physical phase, chlorine is considered a *homogeneous* catalyst.

SAMPLE EXERCISE 14.13: The following reactions have been proposed for the decomposition of ozone in the presence of NO at high temperatures. If the rate of the overall reaction is more rapid than the uncatalyzed decomposition of ozone to oxygen, is NO a catalyst in the reaction?

Step 1: $\quad\quad\quad\quad\quad\quad\quad\quad O_3 + NO \longrightarrow O_2 + NO_2$

Step 2: $\quad\quad\quad\quad\quad\quad\quad\quad NO_2 \longrightarrow NO + O$

Step 3: $\quad\quad\quad\quad\quad\quad\quad\quad O + O_3 \longrightarrow 2\,O_2$

SOLUTION: The definition of a catalyst requires that it is neither produced nor consumed in the overall reaction. If we sum the reactions in steps 1 through 3, we obtain the overall reaction

$$2\,O_3 \longrightarrow 3\,O_2$$

which does not include NO. Because NO is consumed in the first step and regenerated in the second, it serves as a catalyst.

PRACTICE EXERCISE: The combustion of fossil fuels results in the release of SO_2 into the atmosphere. When there, SO_2 may react with NO_2, forming SO_3 and NO.

$$NO_2 + SO_2 \longrightarrow NO + SO_3$$

If the NO that is formed is then oxidized to NO_2,

$$2\,NO + O_2 \longrightarrow 2\,NO_2$$

is NO_2 a catalyst for the following reaction?

$$2\,SO_2 + O_2 \longrightarrow 2\,SO_3$$

(See Problems 93 and 94.)

Significant progress has been made in the past 30 years in reducing the concentrations of nitrogen oxides and carbon monoxide in automobile exhaust. One of the most important advances was the development of *catalytic converters.* Figure 14.22 provides a molecular view of how an automobile catalytic converter removes NO from engine exhaust. Hot exhaust gases flow through the converter, passing over ceramic structures coated with thin layers of metals and metal oxides. Molecules of NO are adsorbed onto these surfaces. As a result of the adsorption process, N–O bonds are weakened and break, forming free atoms of N and O. Nitrogen atoms combine with other nitrogen atoms forming N_2 and

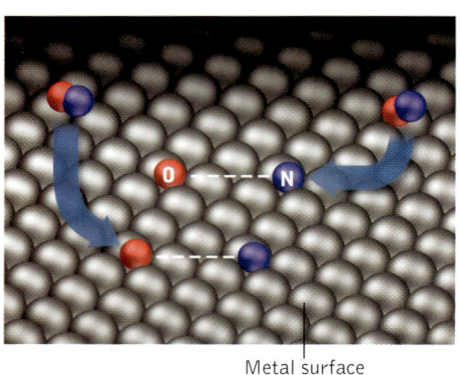
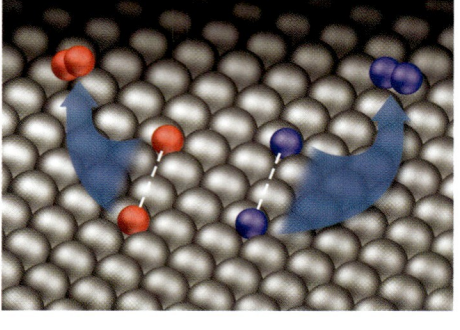

A   Metal surface   B

FIGURE 14.22 Catalytic converters in automobiles reduce emissions of NO by lowering the activation energy of its decomposition into N_2 and O_2. (A) NO molecules are adsorbed onto the metal surface where their NO bonds are broken and (B) pairs of O atoms and N atoms form O_2 and N_2, respectively. The O_2 and N_2 desorb from the surface and are released to the atmosphere.

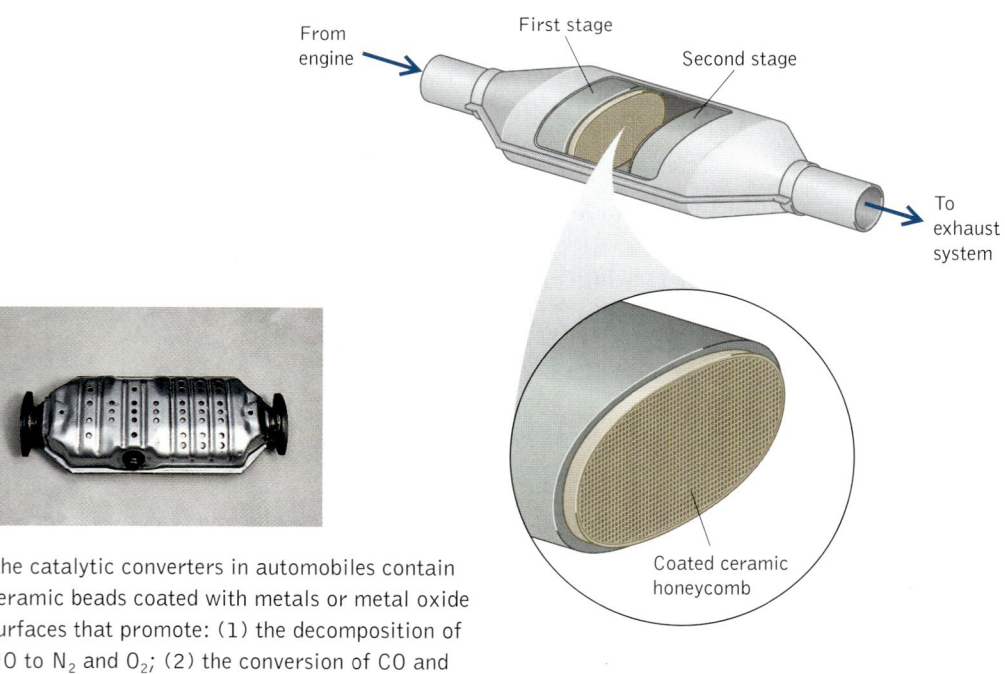

The catalytic converters in automobiles contain ceramic beads coated with metals or metal oxide surfaces that promote: (1) the decomposition of NO to N_2 and O_2; (2) the conversion of CO and hydrocarbons to CO_2 and H_2O.

pairs of oxygen atoms form O_2. In this way catalytic converters increase the rates of the reaction:

$$2\ NO \longrightarrow N_2 + O_2$$

Catalytic converters also promote the complete combustion of any unburned or partly oxidized hydrocarbon fuel to CO_2 and H_2O and the oxidation of CO to CO_2.

Automotive catalysts are *heterogeneous* catalysts because the reactants and catalyst are in different phases. The active surfaces of the first generation of catalytic converters developed in the 1970s were made of platinum (from 1 to 2 g) and palladium (from 0.25 to 1.0 g) and reduced nitrogen oxide emissions to an average of 3.1 g/mile. Later amendments to the U.S. Clean Air Act mandated further reductions in emissions of NO_x, CO, and hydrocarbons (Table 14.7 and Figure 14.23) and led to the development of other catalysts. Rhodium (Rh) proved

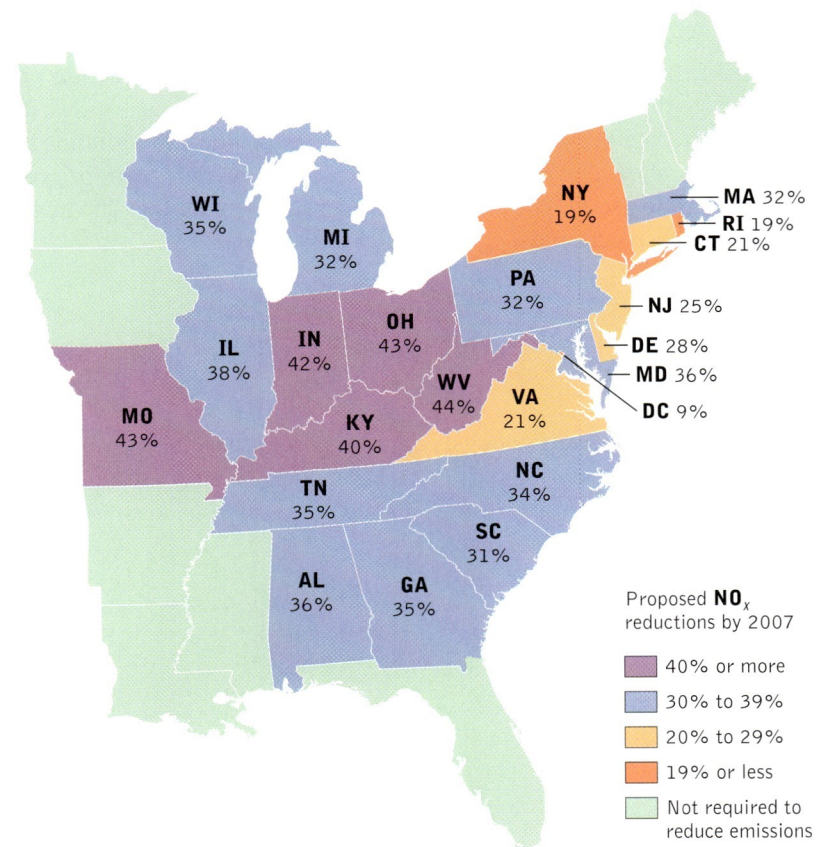

FIGURE 14.23 Proposed EPA (Environmental Protection Agency) regulations for NO_x levels in 22 states in the eastern United States will require substantial cuts in estimated levels for the year 2007.

TABLE 14.7 Federal Emissions Standards for Automobile Exhaust

	NO_x (g/mile)	CO (g/mile)	Hydrocarbons (g/mile)
Before 1976	3.5–7.0	83–90	13–16
1976	3.1	15.0	1.5
1991	1.0	3.4	0.41
2004	0.2	1.7	0.125

Source: U.S. EPA.

to be an excellent catalyst for NO_x reduction and had replaced palladium in catalytic converters by 1986. Cerium(IV) oxide (CeO_2) also was added to catalyze the oxidation of CO.

What is the molecularity of decomposition reactions on heterogeneous catalysts such as those used in automobile exhaust systems? It turns out that many of them appear to be *zero* order in the reactants, which means that reaction rates do not depend on the concentrations of pollutants in the exhaust gases. It also means that concentrations of these pollutants decrease linearly with time in contact with the catalyst. The slope of a plot of reactant concentration (or pressure) versus time (Figure 14.24) equals the zero-order rate constant, k. The integrated rate law for such a reaction is

$$[A] = -kt + [A]_0$$

and the half life is

$$t_{1/2} = [A]_0/2k$$

Reactions on heterogeneous catalysts are zero order when the number of available reactive sites on a catalyst's surface is limited. When all of the active sites are covered with reactant molecules, the rate of the reaction is constant and independent of the concentration of reactant.

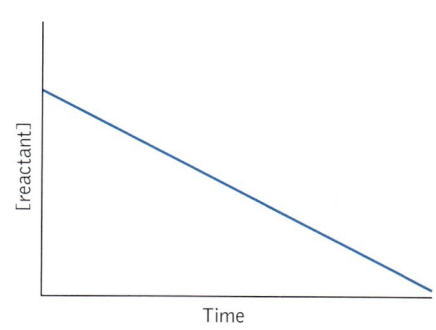

FIGURE 14.24 The change in concentration of a reactant in a zero-order reaction is constant over time.

CHAPTER REVIEW

Summary

SECTION 14.1

The study of reaction rates is called chemical kinetics. Familiarity with chemical kinetics is important in understanding how natural processes and those caused by human activity proceed. For example, the concentrations of various pollutants in a smog event are related to the rates of a series of chemical reactions.

SECTION 14.2

The rate of a reaction can be expressed as the rate of change in the concentrations of reactants or products. The rates of disappearance of reactants and appearance of products are related by the stoichiometry of the reaction and are typically expressed in molarity per unit time. Reaction rates can be known only from experimental measurements. In most reactions, the reaction rate decreases with increasing reaction time.

SECTION 14.3

The rate of a reaction, such as $a\text{A} + b\text{B} \longrightarrow c\text{C}$, depends on the concentrations of A and B as determined experimentally and expressed in the rate law for the reaction:

$$\text{rate} = k[\text{A}]^m[\text{B}]^n$$

where the powers m and n are the orders of the reaction with respect to A and B, respectively, and k is the rate constant. The units of a rate constant depend on the overall reaction order, which is the sum of the reaction orders with respect to individual reactants.

The order of a reaction can be determined from differences in the initial rates of reaction in reaction mixtures with different concentrations of reactants. Reaction order can also be determined from the results of single kinetics experiments and plots of reactant concentration versus time. The half-life ($t_{1/2}$) of a reaction is the time required for the concentration of a reactant to decrease to one-half of its starting concentration; $t_{1/2}$ is inversely proportional to k. The half-life of a first-order reaction is independent of reactant concentration, but $t_{1/2}$ of a second-order reaction is inversely proportional to the initial reactant concentration.

SECTION 14.4

The mechanism of a chemical reaction consists of one or more elementary steps that describe on a molecular level how the reaction takes place. The balanced overall reaction is the sum of these elementary steps. Elementary steps that involve one, two, or three molecules are said to be unimolecular, bimolecular, and termolecular, respectively. The rate law for a reaction applies to the slowest elementary step, which is called the rate-determining step.

SECTION 14.5

A reaction's activation energy is a barrier that separates the sum of the internal energies of the reactants from the energies of the products. Reactions with large activation energies are usually slow. The top of the energy barrier is the transition state related to the internal energy of a short-lived activated complex.

SECTION 14.6

Catalysts increase the rates of reactions. Homogeneous catalysts are in the same gas or liquid phase as the reactants. Solid-state catalysts that are in a separate phase from that of the reactants, such as the catalytic converters in automobile exhaust systems, are examples of heterogeneous catalysts. The bonds in reacting species are more easily broken when they are adsorbed on the surface of a solid catalyst than when in the gas phase.

Key Terms

activated complex (p. 697)
activation energy (p. 704)
Arrhenius equation (p. 705)
elementary step (p. 698)
frequency factor (p. 705)

half-life of a reaction (p. 688)
integrated rate law (p. 685)
mechanism (p. 697)
molecularity (p. 698)
rate constant (p. 681)

rate-determining step (p. 699)
rate law (p. 681)
reaction order (p. 681)
transition state (p. 705)

Key Skills and Concepts

SECTION 14.2

Be able to relate the rates of change in the concentrations of reactants and products to the stoichiometry of a chemical reaction.

Be able to calculate average and instantaneous rates from tabulated kinetics data or graphs.

SECTION 14.3

Be able to determine the order of a reaction with respect to its reactants and derive the rate law for a reaction and its rate constant (with proper units) from initial rate and initial concentration data.

Be able to relate the overall order of a rate law to the units of its rate constant.

Be able to tell whether a reaction is first order or second order from appropriate plots of reactant concentration vs. time.

Understand the relation between k and $t_{1/2}$ for first- and second-order reactions.

SECTION 14.4

Understand the concept of the rate-determining step and the relation of this slowest step to the overall rate law.

SECTION 14.5

Be able to calculate the activation energy and frequency factor for a reaction from rate constant and temperature data and the logarithm form of the Arrhenius equation.

Understand the concept of an activated complex and the energy level known as a transition state in the progress of a reaction.

SECTION 14.6

Understand the role of a catalyst in altering the mechanism of a reaction and increasing its rate.

Understand the difference between homogeneous and heterogeneous catalysis.

Key Equations and Relations

SECTION 14.2

In the progress of a reaction such as $a\,A + b\,B \longrightarrow c\,C$, the average rate of change of [C],

$$\frac{\Delta[C]}{\Delta t} = \frac{[C]_f - [C]_i}{t_f - t_i}$$

where the subscripts "f" and "i" refer to values at some final and initial times, respectively, is related by the stoichiometry of the reaction to the rates of change in [A] and [B]:

$$\frac{\Delta[C]}{c\Delta t} = \frac{-\Delta[A]}{a\Delta t} = \frac{-\Delta[B]}{b\Delta t}$$

SECTION 14.3

The rate of a reaction, such as $a\,A + b\,B \longrightarrow c\,C$, depends on the concentrations of A and B as determined experimentally and expressed in the rate law for the reaction:

$$\text{rate} = k[A]^m[B]^n$$

The reaction $a\,A \longrightarrow b\,B$ is first order in A when a plot of ln [A] versus time is linear. The slope of the line is the negative of the first-order rate constant k. The half-life ($t_{1/2}$) for a first-order reaction with rate constant k is given by

$$t_{1/2} = \frac{0.693}{k} \qquad (14.13)$$

The reaction $aA \longrightarrow bB$ is second order in A when a plot of ln [A] versus time is not linear but when a plot of 1/[A] versus time is linear. The slope of the plot is the second-order rate constant k. The half-life of a second-order reaction

$$t_{1/2} = \frac{1}{k[A]_0} \quad (14.17)$$

depends on the initial concentration of A, $[A]_0$.
If the reaction $aA + bB \longrightarrow cC$ is first order in both A and B (rate = $k[A][B]$) and $[B] \gg [A]$, then the reaction is pseudo-first order in A with pseudo-first-order rate constant, k', where

$$k' = k[B]$$

SECTION 14.5

The Arrhenius equation

$$k = A\,e^{-E_a/RT} \quad (14.21)$$

relates the rate constant, k, of a reaction to the absolute temperature, T, the frequency factor, A, and the activation energy, E_a. The natural logarithm of the Arrhenius equation

$$\ln k = -\frac{E_a}{R}\left(\frac{1}{T}\right) + \ln A \quad (14.22)$$

is the equation of a straight line; so plotting $\ln k$ versus $1/T$ gives a line whose slope is $-E_a/R$ where R is the ideal-gas constant (8.314 J/mol·K).

QUESTIONS AND PROBLEMS

Reaction Rates

CONCEPT REVIEW

1. Explain the difference between the average rate and the instantaneous rate of a chemical reaction.
2. Can the average rate and instantaneous rate of a chemical reaction ever be the same?
3. Why do the average rates of most reactions change with time?
4. Does the instantaneous rate of a chemical reaction change with time?

PROBLEMS

5. *Nitrosomononas* bacteria convert ammonia into nitrite in the presence of oxygen by the following reaction:

$$2\,NH_3(aq) + 3\,O_2(g) \longrightarrow 2\,H^+(aq) + 2\,NO_2^-(aq) + 2\,H_2O(l)$$

   a. How are the rates of appearance of H^+ and NO_2^- related to the rate of disappearance of NH_3?
   b. How is the rate of appearance of NO_2^- related to the rate of disappearance of O_2?
   c. How is the rate of disappearance of NH_3 related to the rate of disappearance of oxygen?

6. Catalytic converters in automobiles combat air pollution by converting NO and CO into N_2 and CO_2:

$$2\,CO + 2\,NO \longrightarrow N_2 + 2\,CO_2$$

   a. How is the rate of appearance of N_2 related to the rate of disappearance of CO?
   b. How is the rate of appearance of CO_2 related to the rate of disappearance of NO?
   c. How is the rate of disappearance of CO related to the rate of disappearance of NO?

7. Write expressions for the rate of appearance of products and the rate of disappearance of reactants in each of the following reactions:
   a. $H_2O_2 \longrightarrow 2\,OH$
   b. $ClO + O_2 \longrightarrow ClO_3$
   c. $N_2O_5 + H_2O \longrightarrow 2\,HONO_2$

8. Write expressions for the rate of appearance of products and the rate of disappearance of reactants in each of the following reactions:
   a. $Cl_2O_2 \longrightarrow 2\,ClO$
   b. $N_2O_5 \longrightarrow NO_2 + NO_3$
   c. $2\,INO \longrightarrow I_2 + 2\,NO$

9. Nitrous oxide decomposes to nitrogen and oxygen by the following reaction:

$$2\,N_2O \longrightarrow 2\,N_2 + O_2$$

In the following graph, which curve represents $[N_2O]$ and which curve represents $[O_2]$?

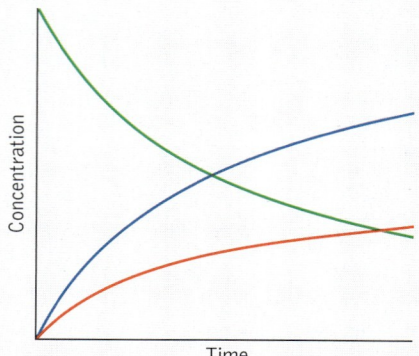

10. Sulfur trioxide is formed by the reaction: $SO_2 + 1/2\,O_2 \longrightarrow SO_3$. In the graph, which curve represents $[SO_2]$ and which curve represents $[O_2]$? All three gases are present initially.

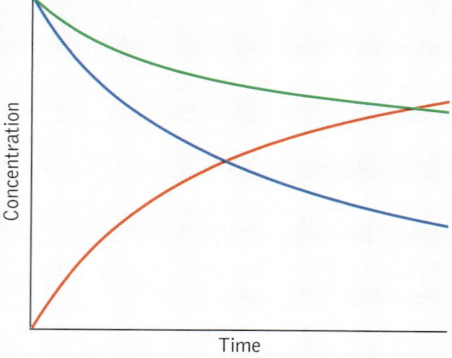

11. Sulfur dioxide emissions in power plant stack gases may react with carbon monoxide:

$$SO_2 + 3\,CO \longrightarrow 2\,CO_2 + COS$$

Write an equation relating
a. the rate of appearance of CO_2 to the rate of disappearance of CO.
b. the rate of appearance of COS to the rate of disappearance of SO_2.
c. the rate of disappearance of CO to the rate of disappearance of SO_2.

12. Nitric oxide (NO) can be removed from gas-fired power-plant emissions by reaction with methane:

$$CH_4 + 4\,NO \longrightarrow 2\,N_2 + CO_2 + 2\,H_2O$$

Write an equation relating
a. the rate of appearance of N_2 to the rate of appearance of CO_2.
b. the rate of appearance of CO_2 to the rate of disappearance of NO.
c. the rate of disappearance of CH_4 to the rate of appearance of H_2O.

13. Chlorine monoxide (ClO) plays a major role in the creation of the ozone holes in the stratosphere over Earth's polar regions.
a. If $\Delta[ClO]/\Delta t$ at 298 K is -2.95×10^6 M/s, what is the rate of change in $[Cl_2]$ and $[O_2]$ in the following reaction?

$$2\,ClO \longrightarrow Cl_2 + O_2$$

b. If $\Delta[ClO]/\Delta t$ is -9.03×10^3 M/s, what is the rate of appearance of oxygen and ClO_2 in the following reaction?

$$ClO + O_3 \longrightarrow O_2 + ClO_2$$

14. The chemistry of smog formation includes NO_3 as an intermediate in several reactions.
a. If $\Delta[NO_3]/\Delta t$ is -1.54×10^4 μM/min in the following reaction, what is the rate of appearance of NO_2?

$$NO_3 + NO \longrightarrow 2\,NO_2$$

b. What is the rate of change in $[NO_2]$ in the following reaction if $\Delta[NO_3]/\Delta t$ is -3.2×10^{-1} μM/min?

$$2\,NO_3 \longrightarrow 2\,NO_2 + O_2$$

15. Nitrite ion reacts with ozone in aqueous solution, producing nitrate ion and oxygen:

$$NO_2^- + O_3 \longrightarrow NO_3^- + O_2$$

The following data were collected for this reaction at 298 K. Calculate the average reaction rate between 0 and 100 μs and between 200 and 300 μs.

Time (μs)	$[O_3]$ (M)
0	1.13×10^{-2}
100	9.93×10^{-3}
200	8.70×10^{-3}
300	8.15×10^{-3}

16. Dinitrogen pentoxide (N_2O_5) decomposes to nitrogen dioxide and nitrogen trioxide:

$$N_2O_5 \longrightarrow NO_2 + NO_3$$

Calculate the average rate of this reaction between consecutive measurement times in the following table.

Time (s)	[N_2O_5] (molecules/cm^3)
0	1.500×10^{12}
1.45	1.357×10^{12}
2.90	1.228×10^{12}
4.35	1.111×10^{12}
5.80	1.005×10^{12}

17. The following data were collected for the dimerization of ClO to Cl_2O_2 at 298 K.

Time (s)	[ClO] (molecules/cm^3)
0	2.60×10^{11}
1	1.08×10^{11}
2	6.83×10^{10}
3	4.99×10^{10}
4	3.93×10^{10}
5	3.24×10^{10}
6	2.76×10^{10}

Plot [ClO] and [Cl_2O_2] as a function of time and determine the instantaneous rates of change in both at 1 second.

18. Tropospheric ozone is rapidly consumed in many reactions, including:

$$O_3 + NO \longrightarrow NO_2 + O_2$$

Use the following data to calculate the instantaneous rate of the preceding reaction at $t = 0.000$ s and $t = 0.052$ s.

Time (s)	[NO] (M)
0.000	2.0×10^{-3}
0.011	1.8×10^{-3}
0.027	1.6×10^{-3}
0.052	1.4×10^{-3}
0.102	1.2×10^{-3}

Effect of Concentration on Reaction Rates: Reaction Order

CONCEPT REVIEW

19. Can two different chemical reactions have the same rate-law expression?
20. Why are the units of the rate constants different for reactions of different order?
21. Does the half-life of a second-order reaction have the same units as the half-life for a first-order reaction?
22. Does the half-life of a first-order reaction depend on the concentration of the reactants?
23. What effect does doubling the initial concentration of a reactant have on the half-life of a reaction that is second order in the reactant?
24. Two first-order decomposition reactions of the form $A \longrightarrow B + C$ have the same rate constant at a given temperature. Do the reactants in the two reactions have the same half-lives at this temperature?

PROBLEMS: WRITING RATE-LAW EXPRESSIONS

25. For each of the following reactions, determine the order with respect to reactants and the overall reaction order.
    a. Rate = $k[A][B]$
    b. Rate = $k[A]^2[B]$
    c. Rate = $k[A][B]^3$
26. Determine the overall order of the following rate laws and the order with respect to reactants.
    a. Rate = $k[A]^2[B]$
    b. Rate = $k[A]^2[B][C]$
    c. Rate = $k[A][B]^3[C]^{1/2}$
27. Write rate laws and determine the units of the rate constant (by using M for concentration and s for time) for the following reactions:
    a. The reaction of oxygen atoms with NO_2 is first order in both reactants.

b. The reaction between NO and Cl_2 is second order in NO and first order in chlorine.
c. The reaction between Cl_2 and chloroform ($CHCl_3$) is first order in $CHCl_3$ and one-half order in Cl_2.
d. The decomposition of ozone (O_3) to O_2 is second order in ozone.

28. Compounds A and B react to give a single product, C. Write the rate law for each of the following cases and determine the units of the rate constant using M for concentration and s for time:
    a. The reaction is first order in A and second order in B.
    b. The reaction is first order in A and second order overall.
    c. The reaction is independent of the concentration of A and second order overall.
    d. The reaction is second order in both A and B.

29. Predict the rate law for the reaction $2\ BrO \longrightarrow Br_2 + O_2$ if
    a. the rate doubles when the concentration of BrO doubles.
    b. the rate quadruples when the concentration of BrO doubles.
    c. the rate is halved when the concentration of BrO is halved.
    d. the rate is unchanged when the concentration of BrO is doubled.

30. Predict the rate law for the reaction $NO + Br_2 \longrightarrow NOBr_2$ if
    a. the rate doubles when the concentration of NO is doubled and [Br_2] remains constant.
    b. the rate doubles when the concentration of Br_2 is doubled and [NO] remains constant.
    c. the rate increases by 1.56 times when [NO] is increased 1.25 times and [Br_2] remains constant.
    d. the rate is halved when [NO] is doubled and [Br_2] remains constant.

31. In the reaction of NO with ClO, the initial rate of reaction quadruples when the concentrations of both reactants are doubled. What additional information do we need to determine whether the reaction is first order in each reactant?

    $NO + ClO \longrightarrow NO_2 + Cl$

32. The reaction between chlorine monoxide and nitrogen dioxide produces chlorine nitrate ($ClONO_2$):

    $ClO + NO_2 + M \longrightarrow ClONO_2 + M$

    A third molecule (M) takes part in the reaction but is unchanged by it. The reaction is first order in NO and in ClO.

a. Write the rate law for this reaction.
b. What is the reaction order with respect to M?

33. The reaction of NO_2 with ozone produces NO_3 in a second-order reaction overall.

    $NO_2 + O_3 \longrightarrow NO_3 + O_2$

    a. Write the rate law for the reaction if the reaction is first order in each reactant.
    b. The rate constant for the reaction is $1.93 \times 10^4\ M^{-1}s^{-1}$ at 298 K. What is the rate of the reaction when [NO_2] $= 1.8 \times 10^{-8}\ M$ and [O_3] $= 1.4 \times 10^{-7}\ M$?
    c. What is the rate of the appearance of NO_3 under these conditions?
    d. What happens to the rate of the reaction if the concentration of O_3 is doubled?

34. The reaction between N_2O_5 and water is a source of nitric acid in the atmosphere.

    $N_2O_5 + H_2O \longrightarrow 2\ HNO_3$

    a. The reaction is first order in each reactant. Write the rate law for the reaction.
    b. When [N_2O_5] is $0.132\ \mu M$ and [H_2O] $= 230\ \mu M$, the rate of the reaction is $4.55 \times 10^{-4}\ \mu M^{-1}\ min^{-1}$. What is the rate constant for the reaction?

35. Each of the following reactions is first order in the reactants and second order overall. Which reaction is fastest if the initial concentrations of the reactants are the same? All reactions are at 298 K.
    a. $ClO_2 + O_3 \longrightarrow ClO_3 + O_2$ $k = 3.0 \times 10^{-19}$ cm^3/molecule·s
    b. $ClO_2 + NO \longrightarrow NO_2 + ClO$ $k = 3.4 \times 10^{-13}$ cm^3/molecule·s
    c. $ClO + NO \longrightarrow Cl + NO_2$ $k = 1.7 \times 10^{-11}$ cm^3/molecule·s
    d. $ClO + O_3 \longrightarrow ClO_2 + O_2$ $k = 1.5 \times 10^{-17}$ cm^3/molecule·s

36. Two reactions in which there is a single reactant have nearly the same rate constant. One is first order; the other is second order.
    a. If the initial concentrations of the reactants are both $1.0\ mM$, which reaction will proceed at the higher rate?
    b. If the initial concentrations of the reactants are both $2.0\ M$, which reaction will proceed at the higher rate?

PROBLEMS: METHOD OF INITIAL RATES

37. In the presence of water, NO and NO_2 react to form nitrous acid (HNO_2) by the following reaction:

$$NO + NO_2 + H_2O \longrightarrow 2\ HNO_2$$

When the concentration of NO or NO_2 is doubled, the initial rate of reaction doubles. If the rate of the reaction does not depend on $[H_2O]$, what is the rate law for this reaction?

38. The hydroperoxyl radical, HO_2, is one of the highly reactive substances present in trace concentrations in the atmosphere during a smog event. The rate constant for the reaction $2\ HO_2 + SO_3 \longrightarrow H_2SO_3 + 2\ O_2$ at 298 K is $2.6 \times 10^{11}\ M^{-1}\ s^{-1}$. The initial rate of the reaction doubles when the concentration of SO_3 or HO_2 is doubled. What is the rate law for the reaction?

39. Chlorine dioxide (ClO_2) is a disinfectant used in municipal water-treatment plants. It dissolves in basic solution, producing ClO_3^- and ClO_2^-:

$$2\ ClO_2(g) + 2\ OH^-(aq) \longrightarrow ClO_3^-(aq) + ClO_2^-(aq) + H_2O(l)$$

The following kinetic data were obtained for the reaction:

Experiment	$[ClO_2]_0$	$[OH^-]_0$	Initial rate (M/s)
1	0.060	0.030	0.0248
2	0.020	0.030	0.00827
3	0.020	0.090	0.0247

Determine the rate law and the rate constant for this reaction.

40. The following kinetic data were collected at 298 K for the reaction of ozone with nitrite ion, producing nitrate and oxygen.

$$NO_2^-(aq) + O_3(g) \longrightarrow NO_3^-(aq) + O_2(g)$$

Experiment	$[NO_2^-]_0$	$[O_3]_0$	Initial rate (M/s)
1	0.0100	0.0050	25
2	0.0150	0.0050	37.5
3	0.0200	0.0050	50.0
4	0.0200	0.0200	200.0

Determine the rate law for the reaction and the value of the rate constant.

41. Hydrogen gas reduces NO to N_2 by the following reaction:

$$H_2 + 2\ NO \longrightarrow 2\ H_2O + N_2$$

The initial reaction rates of four mixtures of H_2 and NO were measured at 900°C with the following results:

Experiment	$[H_2]_0$ (M)	$[NO]_0$ (M)	Initial rate (M/s)
1	0.212	0.136	0.0248
2	0.212	0.272	0.0991
3	0.424	0.544	0.793
4	0.848	0.544	1.59

Determine the rate law and the rate constant for the reaction at 900°C.

42. The rate of the reaction $NO_2 + CO \longrightarrow NO + CO_2$ was determined in three experiments at 225°C. The results are given in the following table.

$$NO_2 + CO \longrightarrow NO + CO_2$$

Experiment	$[NO_2]_0$ (M)	$[CO]_0$ (M)	Initial rate $-\Delta[NO_2]/\Delta t$ (M/s)
1	0.263	0.826	1.44×10^{-5}
2	0.263	0.413	1.44×10^{-5}
3	0.526	0.413	5.76×10^{-5}

a. Determine the rate law for the reaction.
b. Calculate the value of the rate constant at 225°C.
c. Calculate the rate of appearance of CO_2 when $[NO_2] = [CO] = 0.500\ M$.

PROBLEMS: INTEGRATED RATE LAWS

43. Kinetic data for the reaction $Cl_2O_2 \longrightarrow 2\ ClO$ are summarized in the following table. Determine the value of the first-order rate constant.

Time (μs)	$[Cl_2O_2]$ (M)
0	6.60×10^{-8}
172	5.68×10^{-8}
345	4.89×10^{-8}
517	4.21×10^{-8}
690	3.62×10^{-8}
862	3.12×10^{-8}

44. Radioactive isotopes such as ^{32}P are used to follow biological processes. The following radioactivity data (in picocuries) were collected for a sample containing ^{32}P:

Time (days)	Radioactivity (pCi)
0	10.0
1	9.53
2	9.08
5	7.85
10	6.16
20	3.79

a. Write the rate law for the decay of ^{32}P.
b. Determine the value of the first-order rate constant.

45. The dimerization of ClO

$$2\,\text{ClO}(g) \longrightarrow \text{Cl}_2\text{O}_2$$

is second order in ClO. Use the following data to determine the value of k at 298 K.

Time (s)	[ClO] (molecules/cm^3)
0	2.60×10^{11}
1	1.08×10^{11}
2	6.83×10^{10}
3	4.99×10^{10}
4	3.93×10^{10}

46. Nitrogen trioxide decomposes to NO$_2$ and O$_2$ by the following reaction:

$$2\,\text{NO}_3 \longrightarrow 2\,\text{NO}_2 + \text{O}_2$$

The following data were collected at 298 K. Determine the rate law for the reaction, and calculate the value of the rate constant at 298 K.

Time (min)	[NO$_3$] (μM)
0	1.470×10^{-3}
10	1.463×10^{-3}
100	1.404×10^{-3}
200	1.344×10^{-3}
300	1.288×10^{-3}
400	1.237×10^{-3}
500	1.190×10^{-3}

47. Nitrous acid (HNO$_2$) slowly decomposes to NO, NO$_2$, and water by the following reaction:

$$2\,\text{HNO}_2(aq) \longrightarrow \text{NO}(g) + \text{NO}_2(g) + \text{H}_2\text{O}(l)$$

Use the data below to determine the rate law and the rate constant for this reaction.

Time (min)	[HNO$_2$] (μM)
0	0.1560
1000	0.1466
1500	0.1424
2000	0.1383
2500	0.1345
3000	0.1309

48. There are two structural isomers of ClO$_2$:

$$\text{O}-\overset{\text{Cl}}{\text{O}} \qquad \text{Cl}-\overset{\text{O}}{\text{O}}$$

The isomer with the Cl–O–O skeletal arrangement is unstable and rapidly decomposes according to the reaction

$$2\,\text{ClOO} \longrightarrow \text{Cl}_2 + 2\,\text{O}_2.$$ The following data were collected for the decomposition of ClOO at 298 K:

Time (μs)	[ClOO] M
0.00	1.76×10^{-6}
0.67	2.36×10^{-7}
1.3	3.56×10^{-8}
2.1	3.23×10^{-9}
2.8	3.96×10^{-10}

Determine the rate law for the reaction and the value of the rate constant at 298 K.

49. At high temperatures, ammonia spontaneously decomposes into N$_2$ and H$_2$. The following data were collected at one such temperature:

Time (s)	[NH$_3$] (M)
0	2.56×10^{-2}
12	2.47×10^{-2}
56	2.16×10^{-2}
224	1.31×10^{-2}
532	5.19×10^{-3}
746	2.73×10^{-3}

Determine the rate law for the decomposition of ammonia and the value of the rate constant at the temperature of the experiment.

50. Atmospheric chemistry involves highly reactive, odd-electron molecules such as the hydroperoxyl radical HO$_2$, which decomposes into H$_2$O$_2$ and O$_2$. From the following data obtained at 298 K, determine the rate law for the reaction and the value of the rate constant at 298 K:

Time (μs)	[HO$_2$] (μM)
0	8.5
0.6	5.1
1.0	3.6
1.4	2.6
1.8	1.8
2.4	1.1

PROBLEMS: HALF-LIFE

51. Nitrous oxide (N$_2$O) is used as an anesthetic (laughing gas) and in aerosol cans to produce whipped cream. It is a potent greenhouse gas and decomposes slowly to N$_2$ and O$_2$:

$$2\,\text{N}_2\text{O} \longrightarrow 2\,\text{N}_2 + \text{O}_2$$

a. If the plot of ln [N_2O] as a function of time is linear, what is the rate law for the reaction?
b. How many half-lives will it take for the concentration of the N_2O to reach 6.25% of its original concentration?

52. Determine the half-life of ^{32}P from the data in Problem 44.

53. The unsaturated hydrocarbon butadiene (C_4H_6) dimerizes to cyclooctadiene (C_8H_{12}). When data collected in studies of the kinetics of this reaction were plotted against reaction time, plots of [C_4H_6] or ln [C_4H_6] versus time produced curved lines, but the plot of 1/[C_4H_6] versus time was linear.
a. What is the rate law for the reaction?
b. How many half-lives will it take for the [C_4H_6] to decrease to 3.1% of its original concentration?

54. Determine the half-life for the decomposition of HNO_2 in Problem 47.

55. Determine the half-life for the dimerization of ClO in Problem 45.

56. Determine the half-life for the decomposition of Cl_2O_2 in Problem 43.

PROBLEMS: PSEUDO-FIRST-ORDER REACTIONS

57. The catabolism of table sugar (sucrose, $C_{12}H_{22}O_{11}$) begins with the hydrolysis of this disaccharide to glucose:

$$C_{12}H_{22}O_{11}(aq) + H_2O(l) \longrightarrow 2\ C_6H_{12}O_6(aq)$$

The kinetics of the reaction were studied at 24°C in a reaction system with a large excess of water; so the reaction was pseudo-first order in sucrose. Determine the rate law and the pseudo-first-order rate constant for the reaction from the following data:

Time (s)	[$C_{12}H_{22}O_{11}$] (M)
0	0.562
612	0.541
1600	0.509
2420	0.484
3160	0.462
4800	0.417

58. Hydroperoxyl radicals (HO_2) react rapidly with ozone, producing oxygen and OH radicals:

$$HO_2 + O_3 \longrightarrow OH + 2\ O_2$$

The rate of this reaction was studied in the presence of a large excess of ozone. Determine the pseudo-first-order rate constant and the second-order rate constant for the reaction from the following data:

Time (ms)	[HO_2] (M)	[O_3] (M)
0	3.2×10^{-6}	1.0×10^{-3}
10	2.9×10^{-6}	1.0×10^{-3}
20	2.6×10^{-6}	1.0×10^{-3}
30	2.4×10^{-6}	1.0×10^{-3}
80	1.4×10^{-6}	1.0×10^{-3}

Reaction Mechanisms

CONCEPT REVIEW

59. The rate law for the reaction between NO and H_2 is second order in NO and third order overall, whereas the reaction of NO with Cl_2 is first order in each reactant and second order overall. Why can't these reactions proceed by similar mechanisms?

60. The rate law for the reaction of NO with Cl_2 (rate = k[NO][Cl_2]) is the same as that for the reaction of NO_2 with F_2 (rate = k[NO_2][F_2]). Is it possible that these reactions have similar mechanisms?

PROBLEMS

61. Write the rate laws for the following elementary steps and identify them as uni-, bi-, or termolecular steps:
a. $SO_2Cl_2 \longrightarrow SO_2 + Cl_2$
b. $NO_2 + CO \longrightarrow NO + CO_2$
c. $2\ NO_2 \longrightarrow NO_3 + NO$

62. Write the rate laws for the following elementary steps and identify them as uni-, bi-, or termolecular steps:
a. $Cl + O_3 \longrightarrow ClO + O_2$
b. $2\ NO_2 \longrightarrow N_2O_4$
c. $^{14}_{6}C \longrightarrow ^{14}_{7}N + ^{0}_{-1}\beta$

63. Write the overall reaction that consists of the following elementary steps:
$$N_2O_5 \longrightarrow NO_3 + NO_2$$
$$NO_3 \longrightarrow NO_2 + O$$
$$2\,O \longrightarrow O_2$$

64. What overall reaction consists of the following elementary steps?
$$ClO^-(aq) + H_2O(aq) \longrightarrow HClO(aq) + OH^-(aq)$$
$$I^-(aq) + HClO(aq) \longrightarrow HIO(aq) + Cl^-(aq)$$
$$OH^-(aq) + HIO(aq) \longrightarrow H_2O(l) + IO^-(aq)$$

*65. In the following mechanism for NO formation, oxygen atoms are produced by breaking O=O bonds at high temperature in a fast reversible reaction. If $\Delta[NO]/\Delta t = k[N_2][O_2]^{1/2}$, which step in the mechanism is the rate-determining step?

Step 1: $O_2 \rightleftharpoons 2O$
Step 2: $O + N_2 \longrightarrow NO + N$
Step 3: $N + O \longrightarrow NO$
Overall: $N_2 + O_2 \longrightarrow 2\,NO$

66. A proposed mechanism for the decomposition of hydrogen peroxide consists of three elementary steps:
$$H_2O_2 \longrightarrow 2\,OH$$
$$H_2O_2 + OH \longrightarrow H_2O + HO_2$$
$$HO_2 + OH \longrightarrow H_2O + O_2$$
If the rate law for the reaction is first order in H_2O_2, which step in the mechanism is the rate-determining step?

67. At a given temperature, the rate of the reaction between NO and Cl_2 is proportional to the product of the concentrations of the two gases: $[NO][Cl_2]$. The following two-step mechanism was proposed for the reaction:

Step 1: $NO + Cl_2 \longrightarrow NOCl_2$
Step 2: $NOCl_2 + NO \longrightarrow 2\,NOCl$
Overall: $2\,NO + Cl_2 \longrightarrow NOCl_2$

Which step must be the rate-determining step if this mechanism is correct?

68. Ozone decomposes thermally to oxygen by the following reaction:
$$2\,O_3 \longrightarrow 3\,O_2$$
The following mechanism has been proposed:
$$O_3 \longrightarrow O + O_2$$
$$O + O_3 \longrightarrow 2\,O_2$$
The reaction is second order in ozone. What properties of the two elementary steps (particularly, relative rate and reversibility) are consistent with this mechanism?

69. The rate laws for the thermal and photochemical decomposition of NO_2 are different. Which of the following mechanisms are possible for the thermal decomposition of NO_2 and which are possible for the photochemical decomposition of NO_2? Given:

rate of thermal decompostion = $k[NO_2]^2$
rate of photochemical decomposition = $k[NO_2]$

a. $NO_2 \xrightarrow{\text{slow}} NO + O$
   $O + NO_2 \xrightarrow{\text{fast}} NO + O_2$

b. $NO_2 + NO_2 \xrightarrow{\text{fast}} N_2O_4$
   $N_2O_4 \xrightarrow{\text{slow}} NO + NO_3$
   $NO_3 \xrightarrow{\text{fast}} NO + O_2$

c. $NO_2 + NO_2 \xrightarrow{\text{slow}} NO + NO_3$
   $NO_3 \xrightarrow{\text{fast}} NO + O_2$

70. The rate laws for the thermal and photochemical decomposition of NO_2 are different. Which of the following mechanisms are possible for the thermal decomposition of NO_2 and which are possible for the photochemical decomposition of NO_2? Given:

Rate of thermal decomposition = $k[NO_2]^2$
Rate of photochemical decomposition = $k[NO_2]$

a. $NO_2 + NO_2 \xrightarrow{\text{slow}} N_2O_4$
   $N_2O_4 \xrightarrow{\text{fast}} N_2O_3 + O$
   $N_2O_3 + O \xrightarrow{\text{fast}} N_2O_2 + O_2$
   $N_2O_2 \xrightarrow{\text{fast}} 2\,NO$

b. $NO_2 + NO_2 \xrightarrow{\text{slow}} NO + NO_3$
   $NO_3 \xrightarrow{\text{fast}} NO + O_2$

c. $NO_2 \xrightarrow{\text{slow}} N + O_2$
   $N + NO_2 \xrightarrow{\text{fast}} N_2O_2$
   $N_2O_2 \xrightarrow{\text{slow}} 2\,NO$

Reaction Rates and Temperature

CONCEPT REVIEW

71. Which, if any, of the following statements is true?
   a. Exothermic reactions are always fast.
   b. Reactions with $\Delta G > 0$ are slow.
   c. Endothermic reactions are always slow.
   d. Reactions accompanied by an increase in entropy are fast.

72. Which, if any, of the following statements is true?
   a. Reactions with $\Delta G < 0$ are always fast.
   b. Reactions with $\Delta H > 0$ are always fast.
   c. Reactions with $\Delta S < 0$ are always slow.
   d. Reactions with $\Delta H < 0$ are fast only at low temperature.

73. The order of a reaction is independent of temperature, but the value of the rate constant varies with temperature. Why?

74. Why is the value of E_a for an exergonic reaction less than the E_a value for the same reaction running in reverse?

*75. Two first-order reactions have activation energies of 15 and 150 kJ/mol. Which reaction will show the larger increase in rate as temperature is increased?

*76. The rate of a chemical reaction is too slow to measure at room temperature. We could either raise the temperature or add a catalyst. Which would be a better solution for making an accurate determination of the rate constant?

PROBLEMS

77. The hypothetical reaction $A \longrightarrow B$ has an activation energy of 50.0 kJ/mol. Draw a reaction profile for each of the following mechanisms:
   a. a single elementary step.
   b. a two-step reaction in which the activation energy of the second step is 15 kJ/mol.
   c. a two-step reaction in which the activation energy of the second step is the rate determining barrier.

78. Draw a reaction profile for the following reaction:

$$A + B \longrightarrow C \longrightarrow D + E$$

in which:
   a. C is an activated complex.
   b. the reaction has two elementary steps; the first step is rate-determining and C is an intermediate.
   c. the reaction has two elementary steps; the second step is rate-determining and C is an intermediate.

79. Which of the following mechanisms is consistent with the following reaction profile?

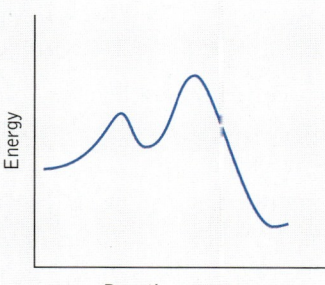

   a. $2\,A \xrightarrow{\text{slow}} B$
      $B \xrightarrow{\text{fast}} C$
   b. $A + B \longrightarrow C$
   c. $2\,A \xrightarrow{\text{fast}} B$
      $B \xrightarrow{\text{slow}} C$

80. Which of the following mechanisms is consistent with the following reaction profile?

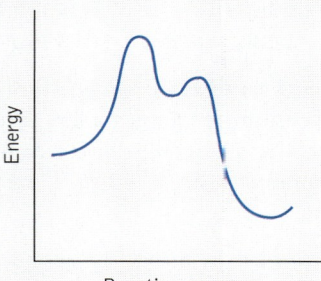

   a. $A + B \xrightarrow{\text{slow}} C$
      $C \xrightarrow{\text{fast}} D$
   b. $A + B \longrightarrow C$
   c. $2\,A \xrightarrow{\text{fast}} C$
      $B + C \xrightarrow{\text{slow}} D$

81. The rate constant for the reaction of ozone with oxygen atoms was determined at four temperatures. Calculate the activation energy and frequency factor A for the reaction $O + O_3 \longrightarrow 2\, O_2$, given the following data:

T (K)	k (cm³/molecule · s)
250	2.64×10^{-4}
275	5.58×10^{-4}
300	1.04×10^{-3}
325	1.77×10^{-3}

82. The rate constant for the reaction $NO_2 + O_3 \longrightarrow NO_3 + O_2$ was determined over a 40-Celsius-degree range with the following results:

T (K)	k (M^{-1} s^{-1})
203	4.14×10^{5}
213	7.30×10^{5}
223	1.22×10^{6}
233	1.96×10^{6}
243	3.02×10^{6}

a. Determine the activation energy for the reaction.
b. Calculate the rate constant of the reaction at 300 K.

83. Values of the rate constant of the reaction $N_2 + O_2 \longrightarrow 2\, NO$ are as follows at the temperatures indicated:

T (K)	k (M$^{-1/2}$ s^{-1})
2000	318
2100	782
2200	1770
2300	3733
2400	7396

a. Calculate the activation energy of the reaction.
b. Calculate the frequency factor for the reaction.
c. Calculate the value of the rate constant at ambient temperature, $T = 300$ K.

84. Values of the rate constant for the decomposition of N_2O_5 at four temperatures are as follows:

T (K)	k (s^{-1})
658	2.14×10^{5}
673	3.23×10^{5}
688	4.81×10^{5}
703	7.03×10^{5}

a. Determine the activation energy of the decomposition reaction.
b. Calculate the value of the rate constant at 300 K.

85. The value of the rate constant for the reaction between chlorine dioxide and ozone was measured at four temperatures between 193 and 208 K. The results are as follows:

T (K)	k (M^{-1} s^{-1})
193	34.0
198	62.8
203	112.8
208	196.7

Calculate the values of the activation energy and the frequency factor for the reaction.

86. Chlorine atoms react with methane, forming HCl and CH_3. The rate constant for the reaction is 6.0×10^7 M^{-1} s^{-1} at 298 K. When the experiment was repeated at three other temperatures, the following data were collected:

T (K)	k (M^{-1} s^{-1})
303	6.5×10^{7}
308	7.0×10^{7}
313	7.5×10^{7}

Calculate the values of the activation energy and the frequency factor for the reaction.

Catalysis and Zero-Order Reactions

CONCEPT REVIEW

87. Does a catalyst affect both the rate and the rate constant of a reaction?
88. Is the rate law for a catalyzed reaction the same as that for the uncatalyzed reaction?
89. Does a substance that increases the rate of a reaction also increase the rate of the reverse reaction?
90. The rate of the reaction between NO_2 and CO is independent of [CO]. Does this mean that CO is a catalyst for the reaction?
91. Under what reaction conditions does a bimolecular reaction obey pseudo-first-order reaction kinetics?
92. Why doesn't the concentration of a homogeneous catalyst appear in the rate law for the reaction it catalyses?

PROBLEMS

93. Is NO a catalyst for the decomposition of N_2O in the following two-step reaction mechanism or is N_2O a catalyst for the conversion of NO to NO_2?

Step 1: $NO(g) + N_2O(g) \longrightarrow N_2(g) + NO_2(g)$
Step 2: $2\ NO_2(g) \longrightarrow 2\ NO(g) + O_2(g)$

94. Explain why NO is a catalyst in the following two-step process that results in the depletion of ozone in the stratosphere:

Step 1: $NO(g) + O_3(g) \longrightarrow NO_2(g) + O_2(g)$
Step 2: $\dfrac{O(g) + NO_2(g) \longrightarrow NO(g) + O_2(g)}{O(g) + O_3(g) \longrightarrow 2O_2(g)}$

95. Overall, on the basis of the frequency factors and activation-energy values of the following two reactions, determine which one will have the larger rate constant at room temperature (298 K).

$O_3 + O \longrightarrow O_2 + O_2 \qquad A = 8.0 \times 10^{-12}$ cm^3/molecule·s
$E_a = 17.1$ kJ/mol

$O_3 + Cl \longrightarrow ClO + O_2 \qquad A = 2.9 \times 10^{-11}$ cm^3/molecule·s
$E_a = 2.16$ kJ/mol

96. On the basis of the frequency factors and activation-energy values of the following two reactions, determine which one will have the larger rate constant at room temperature (298 K).

$O_3 + Cl \longrightarrow ClO + O_2 \qquad A = 2.9 \times 10^{-11}$ cm^3/molecule·s
$E_a = 2.16$ kJ/mol

$O_3 + NO \longrightarrow NO_2 + O_2 \qquad A = 2.0 \times 10^{-12}$ cm^3/molecule·s
$E_a = 11.6$ kJ/mol

15 Chemical Equilibrium
And Why Smog Persists

The town of Mexican Hat in southeastern Utah owes its name to a nearby geological formation, the top of which is a 2500-ton rock, nearly 20 m wide, that looks like an upside-down sombrero. The forces of physical and chemical erosion have carved a mass of rock that seems to be precariously balanced but has remained so for thousands of years. This chapter examines how chemical reactions reach a state of balance as products and reactants coexist in unchanging concentrations over time.

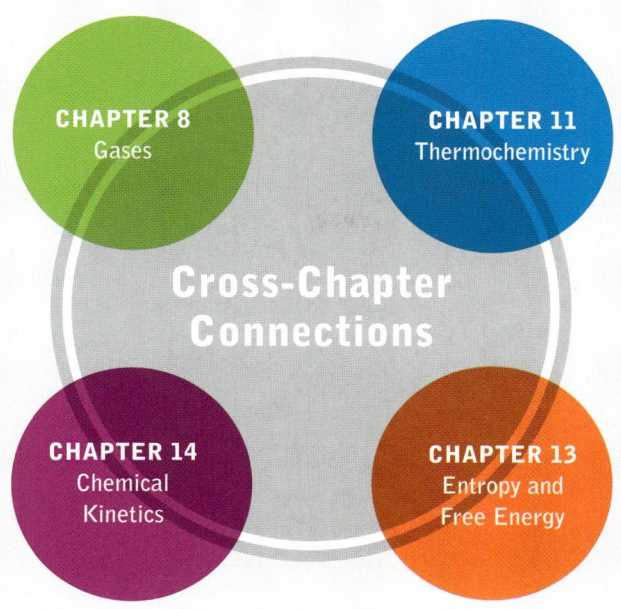

15.1 Achieving equilibrium

15.2 Equilibrium constants and reaction quotients
- Reactions in reverse
- K and Q for combined equations
- Multiplying a chemical equation by a constant

15.3 Equilibrium in the gas phase and K_p

15.4 K, Q, and ΔG

15.5 Le Châtelier's principle

15.6 The role of catalysts

15.7 Calculations based on K

15.8 Changing K with changing temperature

15.9 Heterogeneous equilibria

A Look Ahead

Many chemical reactions do not go to completion. Instead, they proceed until they reach chemical equilibrium, at which time the rate at which reactants form products is matched by the rate at which products form reactants. In this chapter, chemical reactions of compounds implicated in atmospheric pollution provide the context for introducing the concept of chemical equilibrium. We derive an expression called an equilibrium constant (K) that relates the concentrations (or partial pressures) of reactants and products in a reaction at equilibrium. We examine the connections between K and ΔG, the effect of temperature on K, the differences between K values based on concentrations and on partial pressures in gas-phase reactions, and how Le Châtelier's principle predicts how systems at equilibrium respond to stress.

CONNECTION: The kinetics of the formation of NO are discussed in Section 14.1.

☑ Chemical equilibrium is the state in which the rate of a reaction in the forward direction matches its rate in the reverse direction and in which the concentrations of reactant(s) and product(s) do not change.

15.1 ACHIEVING EQUILIBRIUM

In Chapter 14, we examined some of the many chemical reactions that lead to the formation of photochemical smog. Smog formation begins inside internal combustion engines where high temperatures (~2300 K) promote the endothermic reaction:

$$N_2(g) + O_2(g) \longrightarrow 2\,NO(g) \qquad \Delta H° = +180.6\ \text{kJ}$$

An important feature of this reaction is that it appears to stop before all of the O_2 (the limiting reactant in this case) is consumed. Why is this so? One explanation has to do with the brief time spent by N_2 and O_2 at high temperatures before they are pushed out of the engine and into a cooler exhaust system and the atmosphere. This explanation is based on reaction kinetics. However, there is another explanation—one based on the concept of *chemical equilibrium*.

CONCEPT TEST: Why should O_2, and not N_2, be the limiting reactant in the formation of NO in the combustion chambers of a car's engine?

To gain an understanding of what is meant by chemical equilibrium, let's consider a hypothetical experiment that takes place inside a gas-tight furnace at 2300 K. Suppose we introduce a mixture of N_2 and O_2 into the furnace. The initial concentrations of these gases are 0.1000 and 0.0005 M, respectively. Then we monitor the concentrations of N_2, O_2, and NO in the furnace over time. The resulting concentration profiles might look like those plotted in Figure 15.1. At first, the concentrations of N_2 and O_2 decrease and the concentration of NO increases. Eventually, the changes in concentration slow down and finally stop altogether; the concentration of O_2 (the limiting reactant in this experiment) never decreases to zero. The reaction has reached chemical equilibrium.

It is important to note that, though the concentrations of reactants and products have stopped changing, the *chemical reaction has not stopped.* Chemical equi-

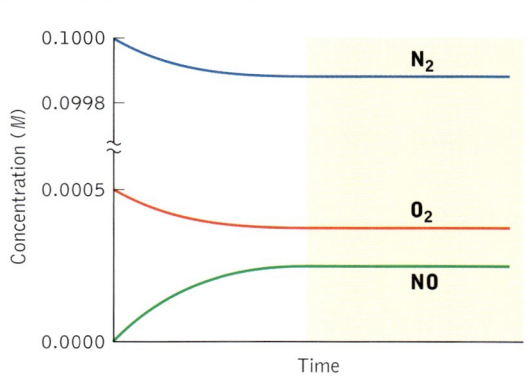

FIGURE 15.1 The concentrations of N_2, O_2, and NO change as they approach equilibrium in the reaction $N_2 + O_2 \rightleftharpoons 2\,NO$. Initially $[N_2] = 0.1000\ M$; $[O_2] = 0.0005\ M$; $[NO] = 0.0000\ M$. (There is a break in the *y*-axis so that small changes in the concentrations of all three gases can be seen.) Concentrations of the gases do not change in the yellow zone and thereafter, indicating that equilibrium has been achieved.

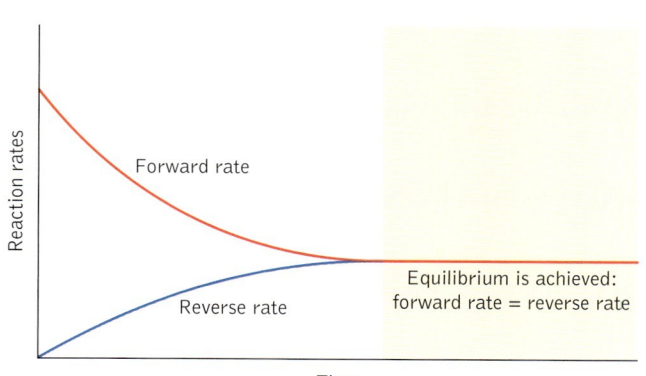

FIGURE 15.2 The rates of the forward and reverse reactions for $N_2 + O_2 \rightleftharpoons 2\,NO$ (or any reaction) are the same when chemical equilibrium has been achieved. Initially, only reactants are present and the rate of the forward reaction is high. The reverse reaction rate is zero because there is no reactant for the reverse reaction. As reactants are consumed, the forward reaction slows while the rate of the reverse reaction increases. After a time, the two rates become equal and chemical equilibrium is achieved.

librium means the rate at which NO forms as the reaction proceeds from left to right equals the rate at which NO dissociates, forming N_2 and O_2 as the reaction proceeds *in reverse* (see Figure 15.2). The concentrations of reactants and products, once they have reached their equilibrium values, will not change over time unless something is done to perturb the equilibrium. We will examine the effects of such perturbations later in this chapter.

Another important feature of the equilibrium between N_2, O_2 and NO is that equilibrium may be approached from either the forward or the reverse direction. For example, if we were to inject a sample of NO into a furnace at 2300 K and monitor the concentrations of NO, N_2, and O_2, we would observe a decrease in [NO] and increases in [N_2] and [O_2] as the reaction ran in reverse. In this case, the rate of the reverse reaction, $2\,NO \longrightarrow N_2 + O_2$, is high initially and decreases with time, whereas the rate of the forward reaction is zero initially and then increases until it matches the rate of the reverse reaction. At that time, we once again have a system in chemical equilibrium.

The chemistry of another ingredient of photochemical smog, NO_2, provides us visual images of chemical equilibrium. A sample of pure NO_2 gas in a container at high pressure—say 10 atm—has a distinctive deep brown color (see Figure 15.3). Now suppose the gas is slowly released from the container. You might expect the intensity of the color to decrease as the pressure in the container decreases so that a plot of intensity vs. pressure produces a straight line. However, that is not what is observed. Instead, the color decreases less than the pressure decreases as shown by the results plotted in Figure 15.3.

We can explain these results by invoking chemical equilibrium. Molecules of NO_2 at high pressure are known to combine with other NO_2 molecules, forming molecules of a colorless dimer, N_2O_4:

$$2\,NO_2(g) \longrightarrow N_2O_4(g)$$

Not all of the NO_2 turns into N_2O_4 because NO_2 gas at high pressure has a dark brown color. Therefore we may assume that NO_2 and N_2O_4 reach a state of chemical equilibrium:

$$2\,NO_2(g) \rightleftharpoons N_2O_4(g)$$

The numbers of passengers on a subway train and on the platform next to it illustrate the concept of dynamic equilibrium. If an equal number of passengers board and leave the train, the "equilibrium concentrations" of passengers on the train and on the platform are constant.

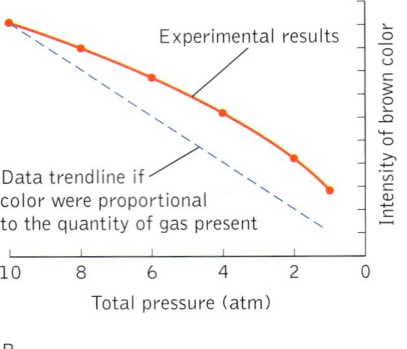

FIGURE 15.3 An equilibrium mixture of NO_2, which has a deep brown color, and N_2O_4, which is colorless, is present at high-pressure in a transparent container. A. The brown color of the mixture indicates a high concentration of NO_2. B. As gas is vented from the container, the intensity of the brown color decreases; but it does not decrease as rapidly as the pressure inside the container drops because the chemical equilibrium $2NO_2(g) \rightleftharpoons N_2O_4(g)$ has shifted to the left. C. At lower pressures there is proportionately more NO_2 and less N_2O_4 in the container.

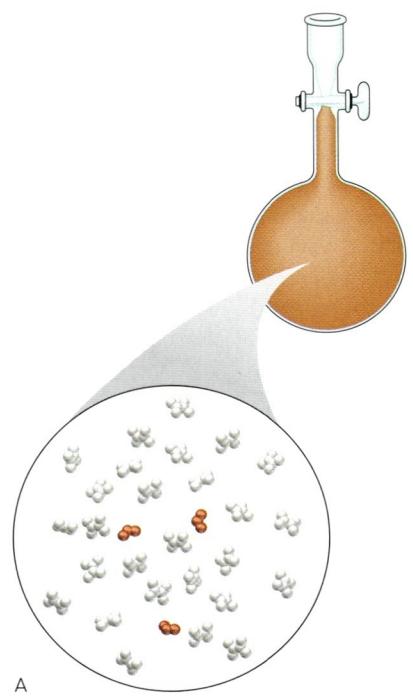

A

B

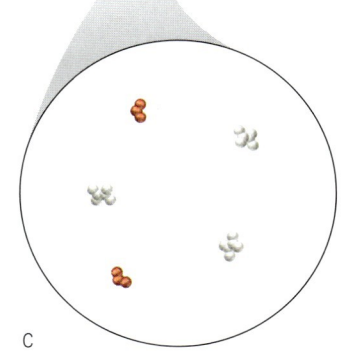

C

CONNECTION: The molecular structures of NO_2 and N_2O_4 are shown in Section 7.4.

The presence of some N_2O_4 in a sample of NO_2 at high pressure accounts for the experimental results plotted in Figure 15.3. As the pressure in the container decreases, fewer collisions between molecules of NO_2 lead to a reduction in the rate of the forward reaction so that, overall, molecules of N_2O_4 dissociate more rapidly than molecules of NO_2 combine to form them. Thus, there is proportionately more NO_2 present as pressure decreases, which gives the gas mixture a darker brown color that we might have expected. (We will address the response of gases at chemical equilibrium to changes in pressure in more detail in Section 15.5.) It turns out that the reaction is second order in NO_2 (see Section 14.3) and so the rate law of the forward reaction is

$$\text{rate}_f = k_f[NO_2]^2$$

The rate of the reverse reaction (rate_r) depends on $[N_2O_4]$:

$$\text{rate}_r = k_r[N_2O_4]$$

When the rates of the forward and reverse reactions become the same,

$$\text{rate}_f = \text{rate}_r$$

the following equality also is true:

$$k_f[NO_2]^2 = k_r[N_2O_4]$$

Rearranging the terms in this equation, we have

$$\frac{k_f}{k_r} = \frac{[N_2O_4]}{[NO_2]^2}$$

The left side of this expression is the ratio of two constants and so is itself a constant. It is called an **equilibrium constant** and is given the symbol K:

$$K = \frac{k_f}{k_r} = \frac{[N_2O_4]}{[NO_2]^2}$$

The right side of this equation is called the **equilibrium constant expression** or the **mass action expression** for the dimerization reaction. The term *mass action* comes from research conducted in the mid-nineteenth century by two Norwegian chemists, Cato Guldberg (1836–1902) and Peter Waage (1833–1900). They discovered that any reversible reaction eventually reaches a state in which the ratio of the concentration terms of the products to the reactants, each raised to a power corresponding to the coefficient for that substance in the balanced chemical equation, has a characteristic value at a given temperature. They called this phenomenon the **law of mass action.** It applies to any system of reactants and products that have reached equilibrium or that are on their way to doing so.

To see how the law of mass action works, let's consider a general chemical reaction in which a moles of substance A react with b moles of substance B, forming c moles of substance C and d moles of substance D:

$$aA + bB \rightleftharpoons cC + dD$$

The double arrows indicate that the reaction is reversible and that it will reach a state of chemical equilibrium. According to the law of mass action, the equilibrium concentrations of A, B, C, and D should, at a particular temperature, be such that the following ratio is a constant:

$$K_c = \frac{[C]^c[D]^d}{[A]^a[B]^b] \quad (15.1)$$

The subscript "c" is used to indicate that the value of K is based on a ratio of *concentrations.* As we will see, there are other ways to express the proportions of reactants and products in a system. For example, equilibrium constants for reactions taking place in the gas phase may be based on partial pressures and so are given the symbol K_p. If substances A, B, C, and D in Equation 15.1 were all gases, then the chemical equilibrium they achieve could be expressed with their partial pressures:

$$K_p = \frac{(P_C)^c(P_D)^d}{(P_A)^a(P_B)^b} \quad (15.2)$$

Note that each term in Equations 15.1 and 15.2 is raised to a power corresponding to the coefficient for that substance in the balanced chemical equation. A key concept here is that the mass action expression for any reaction can be written directly from the balanced chemical equation—that is, the stoichiometry of the reaction. Thus mass action expressions are not necessarily linked to the rate laws for the forward and reverse reactions, even though equilibrium is achieved only when the rates of the forward and reverse reaction are equal. Why is this so? One way to rationalize this independence of equilibrium states from reaction rates is that the time required to achieve equilibrium is not what counts.

> ✓ The **equilibrium constant expression,** or **mass action expression,** for a chemical reaction is the ratio of concentration terms for products divided by that for reactants in accordance with the balanced equation of the reaction.
>
> The **equilibrium constant** (K) is the numerical value for the equilibrium constant expression for a chemical reaction.
>
> The **law of mass action** states that the mass action expression for a chemical system at equilibrium will have a characteristic value at a given temperature.

What is important is the composition of the system after equilibrium has been achieved.

SAMPLE EXERCISE 15.1: Write the equilibrium constant expressions, K_c and K_p, for the synthesis of ammonia from nitrogen and hydrogen:

$$N_2(g) + 3\,H_2(g) \longrightarrow 2\,NH_3(g).$$

SOLUTION: Equilibrium constant expressions are ratios of product (NH_3 in this case) to reactant (N_2 and H_2) terms based on the coefficients of the reactants and product in the balanced chemical equation. These coefficients become exponents in the concentration terms of the K_c expression and the partial-pressure terms of the K_p expression. The two expressions are therefore

$$K_c = \frac{[NH_3]^2}{[N_2][H_2]^3} \quad \text{and} \quad K_p = \frac{P_{NH_3}^2}{P_{N_2}\,P_{H_2}^3}$$

PRACTICE EXERCISE: Write the K_c and K_p mass action expressions for the reaction: $2\,NO(g) + O_2(g) \longrightarrow 2\,NO_2(g)$. (See Problems 13 and 14.)

15.2 EQUILIBRIUM CONSTANTS AND REACTION QUOTIENTS

Let's develop the mass action expression for the formation of NO in a car's engine according to the reaction $N_2 + O_2 \rightleftharpoons 2\,NO$. The numerator will consist of an [NO] term raised to the second power, and the denominator will be the product of the concentrations of N_2 and O_2 each raised to the first power:

$$\frac{[NO]^2}{[N_2][O_2]}$$

According to the law of mass action, this fraction will have a particular value at equilibrium at a particular temperature—say, 2300 K. We could determine that value if we had concentration data for the three gases at equilibrium at that temperature. Let's use the concentration profiles in Figure 15.1 for this purpose. At equilibrium, we have the following molar concentrations:

$[N_2]$	$[O_2]$	$[NO]$
0.09988	0.00038	0.00024

Using these values in the chemical equilibrium expression, we have

$$K_c = \frac{[NO]^2}{[N_2][O_2]} = \frac{(0.00024)^2}{(0.09988)(0.00038)} = 0.0015$$

The significance of this value of K_c is that it applies to any reaction mixture of these three gases no matter what their initial concentrations may have been, as long as they are are equilibrium at 2300 K. As will be discussed in Section 15.8, the value of K_c for this endothermic reaction is much smaller at lower temperatures, such as those encountered in the atmosphere.

SAMPLE EXERCISE 15.2: A sealed chamber contains an equilibrium mixture of NO_2 and N_2O_4. The partial pressures of the two gases are P_{NO_2} = 0.101 atm and $P_{N_2O_4}$ = 0.074 atm. What is the value of K_p at this temperature for the following reaction?

$$2\ NO_2(g) \rightleftharpoons N_2O_4(g)$$

SOLUTION: The mass action expression for the reaction based on partial pressures is

$$K_p = \frac{P_{N_2O_4}}{P_{NO_2}^2}$$

Inserting the given partial pressures and solving for K_p, we have

$$K_p = \frac{0.074}{0.101^2} = 7.2$$

Even though the partial pressure of N_2O_4 is slightly less than that of NO_2, the equilibrium constant is greater than 1 because the partial pressures of both gases are less than 1 and the partial pressure of NO_2 is squared in the denominator. If at equilibrium P_{NO_2} had been 1.0 atm, then $P_{N_2O_4}$ would have been 7.2 atm, or seven times as high. In general, as the partial pressure of NO_2 increases, the proportion of it that is converted into N_2O_4 increases. We will discuss why this happens in Section 15.5.

PRACTICE EXERCISE: A reaction vessel contains an equilibrium mixture of SO_2, O_2, and SO_3. Given the following partial pressures of these gases — P_{SO_2} = 0.0018 atm, P_{O_2} = 0.0032 atm, and P_{SO_3} = 0.0166 atm — calculate the value of K_p for the following reaction:

$$2\ SO_2(g) + O_2(g) \rightleftharpoons 2\ SO_3(g)$$

(See Problems 15–18.)

Let's reconsider the reaction profile for the formation of NO in Figure 15.1. If we sampled the reaction mixture at a time before equilibrium was achieved, the concentration of NO would be less than its concentration at equilibrium. Therefore the numerator in the mass action expression:

$$\frac{[NO]^2}{[N_2][O_2]}$$

> The **reaction quotient** (Q) is the numerical value of the mass action expression when values for the concentrations (or partial pressures) of reactants and products are inserted in it.
>
> At equilibrium, the value of the reaction quotient (Q) equals that of the equilibrium constant (K).

would be less than its equilibrium value, given the quantities of reactants present. Because not all of the reactants that would eventually react had done so, the concentrations in the denominator would be greater than their equilibrium values. Thus, the overall value of the mass action expression would be less than that of the equilibrium constant. Still, it has some significance: it provides a kind of status report on the progress of a reaction that is on its way to equilibrium but hasn't gotten there yet. It is called a **reaction quotient** (Q). In our example, there is too much reactant and not enough product in the reaction mixture for the system to be at equilibrium; so the corresponding value of Q_c is less than that of K_c. As the reaction continues, the value of Q_c approaches K_c. When the reaction has reached equilibrium, $Q_c = K_c$.

Can the value of Q ever be larger than that of K? Definitely. For the reaction $N_2 + O_2 \rightleftharpoons 2\ NO$, Q would be larger than K if we had started with pure NO in the reaction chamber or if the reaction had come to equilibrium and then more NO had been added. Under those conditions, the numerator in the equilibrium constant equation would be temporarily too large and the reaction would run in reverse, reducing the concentration of NO and increasing the concentrations of N_2 and O_2 until equilibrium was achieved or restored. These changes would have the effect of reducing the size of the numerator and increasing the size of the denominator of the mass action expression until equilibrium was restored. The relative values of Q and K and their consequences are summarized in Table 15.1 and Figure 15.4.

> **Can the value of Q ever be larger than that of K?**

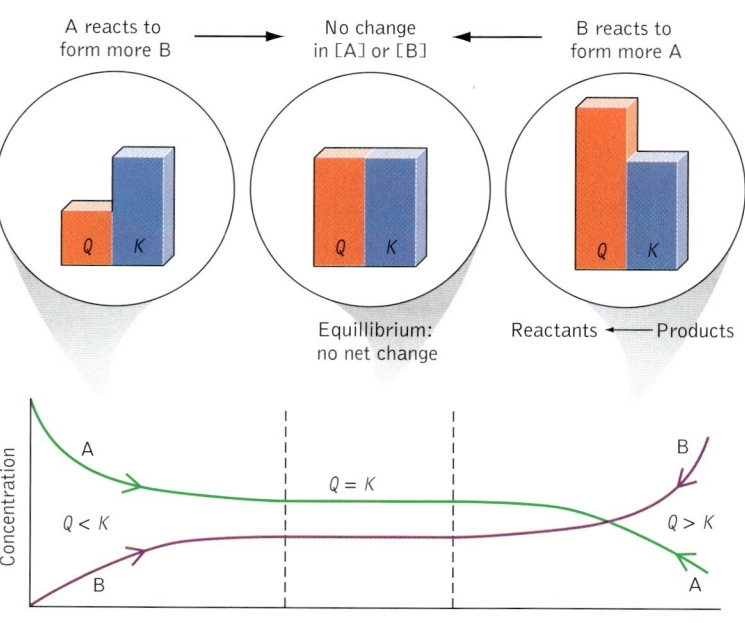

FIGURE 15.4 The value of the reaction quotient, Q, relative to the equilibrium constant, K, for the reaction $A \longrightarrow B$ predicts the direction in which a reaction proceeds. If $Q < K$, the rate of the forward reaction is greater than that of the reverse reaction, leading to the formation of more product ($A \longrightarrow B$). If $Q > K$, the reaction is faster in the reverse direction toward the formation of reactant ($A \longleftarrow B$). In both cases, the reaction proceeds until $Q = K$, a state in which the forward and reverse reactions proceed at the same rate and the concentrations of A and B do not change.

15.2 Equilibrium Constants and Reaction Quotients

TABLE 15.1 Comparison of Q and K Values

Value of Q	What It Means
$Q < K$	Reaction proceeds in the forward direction ($\rightarrow$)
$Q = K$	Reaction is at equilibrium ($\rightleftharpoons$)
$Q > K$	Reaction proceeds in the reverse direction ($\leftarrow$)

SAMPLE EXERCISE 15.3: A reaction vessel at 2300 K contains the following molar concentrations of gases:

$[N_2]$	$[O_2]$	$[NO]$
0.50	0.25	0.0042

If $K_c = 1.5 \times 10^{-3}$ for the reaction $N_2(g) + O_2(g) \rightleftharpoons 2\, NO(g)$ at 2300 K, is this reaction mixture at equilibrium? If not, in what direction will the reaction proceed to reach equilibrium?

SOLUTION: The mass action expression for this reaction based on concentrations is

$$K_c = \frac{[NO]^2}{[N_2][O_2]}$$

If we insert the given concentration values into the mass action expression, we get the following reaction quotient:

$$Q_c = \frac{[0.0042]^2}{[0.50][0.25]} = 1.4 \times 10^{-4}$$

This value of Q_c is less than the value of K_c (1.5×10^{-3}), and so the reaction mixture is not at equilibrium. To achieve equilibrium, there must be more of the product and less of the reactants, which can happen if the reaction proceeds in the forward direction.

PRACTICE EXERCISE: The value of K_c for the reaction $2\, NO_2 \rightleftharpoons N_2O_4$ is 0.21 at 398 K. Is a mixture of the two gases in which $[NO_2] = 0.025\, M$ and $[N_2O_4] = 0.0014\, M$ in chemical equilibrium? If not, in which direction will the reaction proceed so that equilibrium is achieved? (See Problems 21 and 22.)

concentrations of N_2, O_2, and NO

$$Q_c = \frac{[NO]^2}{[N_2][O_2]}$$

Q_c

compare Q_c with K_c

$Q_c < K_c$

reaction proceeds in forward direction

Reactions in reverse

In Section 15.1, we derived the equilibrium constant expression for the reaction $2\, NO_2 \longrightarrow N_2O_4$:

$$K_c = \frac{[N_2O_4]}{[NO_2]^2} \tag{15.3}$$

Now let's write the expression for K_c for the reverse reaction in which NO_2 forms from N_2O_4:

$$N_2O_4(g) \longrightarrow 2\,NO_2(g)$$

The equilibrium constant expression for the dimerization reaction is

$$K_c = \frac{[NO_2]^2}{[N_2O_4]} \qquad (15.4)$$

Now compare the expressions in Equations 15.3 and 15.4. Note that one is the reciprocal of the other. As a general rule, *the equilibrium constant for a reaction is the reciprocal of the equilibrium constant for the reaction written in the reverse direction.* Expressing these relations mathematically gives

$$K_f = \frac{1}{K_r} \qquad (15.5)$$

where the subscripts "f" and "r" indicate the *forward* and *reverse* reaction directions. This relation makes sense: if a reaction has a large equilibrium constant, then there will be mostly product and little reactant at equilibrium. In the reverse reaction, there should be little product and mostly reactant at equilibrium and a correspondingly small equilibrium constant. For example, the equilibrium constant (K_c) for the reaction $2\,NO_2 \rightleftharpoons N_2O_4$ is 0.21 at 100°C, so the value of the equilibrium constant for the reverse reaction, $N_2O_4 \rightleftharpoons 2\,NO_2$, at 100°C is

$$K_r = \frac{1}{K_f} = \frac{1}{0.21} = 4.8$$

✓ The equilibrium constant for the reverse of a reaction is the reciprocal of the equilibrium constant for the reaction in the forward direction.

K and Q for combined equations

The chemistry of photochemical smog formation includes many reactions in which the products of some are the reactants in others. For example, the NO produced in the combustion chambers of car and truck engines may be further oxidized to NO_2. If we sum the two reactions to obtain an overall reaction for the formation of NO_2, we get

(1) $N_2(g) + O_2(g) \rightleftharpoons 2\,NO(g)$

(2) $2\,NO(g) + O_2(g) \rightleftharpoons 2\,NO_2(g)$

Overall: $N_2(g) + 2\,O_2(g) + \cancel{2\,NO(g)} \rightleftharpoons \cancel{2\,NO(g)} + 2\,NO_2(g)$

or

$$N_2(g) + 2\,O_2(g) \rightleftharpoons 2\,NO_2(g)$$

The mass action expression for the overall reaction is

$$K_c = \frac{[NO_2]^2}{[N_2][O_2]^2} \qquad (15.6)$$

We can derive this expression from the equilibrium constant expressions for reactions 1 and 2, which are

$$K_1 = \frac{[NO]^2}{[N_2][O_2]} \quad \text{and} \quad K_2 = \frac{[NO_2]^2}{[NO]^2[O_2]}$$

If we multiply K_1 by K_2, we get the mass action expression for the overall reaction:

$$K_1 \times K_2 = \frac{[\cancel{NO}]^2}{[N_2][O_2]} \times \frac{[NO_2]^2}{[\cancel{NO}]^2[O_2]} = \frac{[NO_2]^2}{[N_2][O_2]^2} = K_{overall}$$

This approach works for all series of reactions, and so, as a general rule,

$$K_{overall} = K_1 \times K_2 \times K_3 \times K_4 \times \cdots \times K_n \qquad (15.7)$$

To apply this principle, let's calculate the value of K_c for the overall reaction $N_2 + 2\,O_2 \rightleftharpoons 2\,NO_2$ in the exhaust system of an automobile at 1000 K. At that temperature, the equilibrium constants for reactions 1 and 2 are

$$K_1 = \frac{[NO]^2}{[N_2][O_2]} = 7.2 \times 10^{-9}$$

$$K_2 = \frac{[NO_2]^2}{[NO]^2[O_2]} = 0.020$$

The equilibrium constant for the overall reaction should be the product of $K_1 \times K_2$, or

$$K_{overall} = K_1 \times K_2 = 7.2 \times 10^{-9} \times 0.020 = 1.4 \times 10^{-10}$$

✓ The overall equilibrium constant for a combination of two or more reactions is the product of the equilibrium constants of the two or more reactions; that is, $K_{overall} = K_1 \times K_2 \times K_3 \times \cdots \times K_n$.

SAMPLE EXERCISE 15.4: At 1000 K, the equilibrium constant of the reaction $2\,NO_2 \rightleftharpoons N_2O_4$ is $K_c = 6.6 \times 10^{-7}$. On the basis of this value and the value of the equilibrium constant for $N_2 + 2\,O_2 \rightleftharpoons 2\,NO_2$ at 1000 K ($K_c = 1.4 \times 10^{-10}$), calculate the value of the equilibrium constant at 1000 K for the reaction $N_2 + 2\,O_2 \rightleftharpoons N_2O_4$.

SOLUTION: We need to determine how the two reactions for which we know K_c values combine to give the overall reaction. Inspection of the three suggests that the overall reaction is the sum of the two.

(1) $N_2 + 2\,O_2 \rightleftharpoons 2\,NO_2$

+ (2) $2\,NO_2 \rightleftharpoons N_2O_4$

Overall: $N_2 + 2\,O_2 + \cancel{2\,NO_2} \rightleftharpoons \cancel{2\,NO_2} + N_2O_4$

or

$$N_2 + 2\,O_2 \rightleftharpoons N_2O_4$$

When an overall reaction is the sum of two other reactions (1 and 2), then $K_{overall} = K_1 \times K_2$. Using the equilibrium constant values provided, we have

$$K_{overall} = 1.4 \times 10^{-10} \times 6.6 \times 10^{-7} = 9.2 \times 10^{-17}$$

PRACTICE EXERCISE: What is the value of the equilibrium constant (K_c) for the reaction $N_2O_4 \rightleftharpoons N_2 + 2\,O_2$ at $T = 1000$ K? (See Problems 23–26.)

Multiplying a chemical equation by a constant

There is often more than one acceptable way to write a chemical equation. For example, the oxidation of NO to NO_2 may be written

(1) $$2\ NO(g) + O_2(g) \rightleftharpoons 2\ NO_2(g)$$

or, multiplying this equation by one-half, we have

(2) $$NO(g) + \tfrac{1}{2} O_2(g) \rightleftharpoons NO_2(g)$$

How does multiplying equation 1 by one-half affect the value of its equilibrium constant? To find out, let's write the equilibrium constant expressions for the two equations:

$$K_1 = \frac{[NO_2]^2}{[NO]^2[O_2]} \quad \text{and} \quad K_2 = \frac{[NO_2]}{[NO][O_2]^{1/2}}$$

All of the exponents in the second expression are half the corresponding exponents in the first one. Expressing this difference in equation form, we have

$$K_2 = (K_1)^{1/2}$$

These equations may leave you wondering how the same reaction having the same reactants and product can have two equilibrium constant values. Surely the same ingredients should be present in the same proportions at equilibrium no matter how we choose to write a balanced equation for their reaction. And they are. The difference in K values is not chemical, it is mathematical. It is related to how we use the equilibrium concentrations in calculating their ratios. It is, for example, about choosing to use [NO] and not $[NO]^2$ in such a calculation. That choice does not affect the actual concentration of NO at equilibrium.

> ✓ If the coefficients in a chemical equation are doubled, the value of the equilibrium constant for the reaction is squared.

SAMPLE EXERCISE 15.5: An important reaction in the formation of atmospheric aerosols of sulfuric acid (and acid rain) is the oxidation of SO_2 to SO_3, which combines with H_2O to form H_2SO_4. One way to write a chemical equation for the oxidation reaction is

$$SO_2(g) + \tfrac{1}{2} O_2(g) \rightleftharpoons SO_3(g)$$

The value of K_c for this reaction at 298 K is 2.8×10^{12}. What is the value of K_c at 298 K for the following reaction?

$$2\ SO_2(g) + O_2(g) \rightleftharpoons 2\ SO_3(g)$$

SOLUTION: The second chemical equation is the first equation with all coefficients multiplied by two. Thus the equilibrium constant for the second reaction should be the square of the equilibrium constant of the first:

$$K_2 = (K_1)^2$$
$$= (2.8 \times 10^{12})^2 = 7.8 \times 10^{24}$$

PRACTICE EXERCISE: At 1000 K, the reaction

$$N_2(g) + 3 H_2(g) \rightleftharpoons 2 NH_3(g)$$

has a K_c value of 2.4×10^{-3}. What is the value of K_c at 1000 K for the following reaction?

$$\tfrac{1}{3} N_2(g) + H_2(g) \rightleftharpoons \tfrac{2}{3} NH_3(g)$$

(See Problems 27–30.)

15.3 EQUILIBRIUM IN THE GAS PHASE AND K_p

Let's consider the decomposition of nitrogen dioxide into nitric oxide and oxygen:

$$2 NO_2(g) \rightleftharpoons 2 NO(g) + O_2(g)$$

The equilibrium constant expression, based on the concentrations of reactant and products, is

$$K_c = \frac{[NO]^2 [O_2]}{[NO_2]^2} \quad (15.8)$$

As noted in Section 15.1, it is sometimes more convenient to describe gaseous reactants and products in terms of their *partial pressures* instead of their molar concentrations. To relate an equilibrium constant based on partial pressures (K_p) to the corresponding K_c, we need to relate the partial pressure of each gas to its molar concentration. We start with the ideal-gas law:

$$PV = nRT$$

Solving for P, we get

$$P = \frac{n}{V} RT$$

Because n is the number of moles of gas and V is the volume of the gas in liters, the fraction n/V represents molarity (M):

$$P = MRT$$

Rearranging the terms in this equation to solve for M, we have

$$M = P \frac{1}{RT} \quad (15.9)$$

CONNECTION: The ideal-gas law is derived in Section 8.2; partial pressures of gases are defined in Section 8.9.

Equation 15.9 relates the molar concentration of a gaseous substance to its partial pressure. Let's use this relation in Equation 15.8 by replacing the molar-concentration terms with the right-hand side of Equation 15.9. We get the following equilibrium constant based on partial pressures:

$$K_c = \frac{[NO]^2[O_2]}{[NO_2]^2} = \frac{\left(\dfrac{P_{NO}}{RT}\right)^2 \left(\dfrac{P_{O_2}}{RT}\right)}{\left(\dfrac{P_{NO_2}}{RT}\right)^2}$$

Combining all the $1/RT$ terms (there are three in the numerator and two in the denominator because there are 3 moles of products and 2 moles of reactant), we get

$$K_c = \frac{(P_{NO})^2(P_{O_2})}{(P_{NO_2})^2} \frac{1}{RT} \tag{15.10}$$

The ratio of the partial pressures of the gases in this reaction, each raised to a power in keeping with the stoichiometry of the reaction, is given the symbol K_p:

$$K_p = \frac{(P_{NO})^2(P_{O_2})}{(P_{NO_2})^2} \tag{15.11}$$

Combining Equations 15.10 and 15.11, we get

$$K_c = K_p \frac{1}{RT}$$

or

$$K_p = K_c(RT) \tag{15.12}$$

Equation 15.12 relates K_p and K_c for the decomposition of NO_2. It also relates K_p and K_c for any reaction in which one more mole of gas is produced than is consumed in the balanced chemical equation. For other reactions of gases, the relation between K_p and K_c can be written as

$$K_p = K_c(RT)^{\Delta n} \tag{15.13}$$

where Δn is the number of moles of gaseous products minus the number of moles of gaseous reactants in the balanced chemical equation. If these numbers of moles happen to be the same, then $\Delta n = 0$ and $K_p = K_c$.

SAMPLE EXERCISE 15.6: Express K_p in terms of K_c for the following gas-phase reactions.

    a. $O(g) + O_2(g) \rightleftharpoons O_3(g)$
    b. $N_2(g) + O_2(g) \rightleftharpoons 2\ NO(g)$

SOLUTION: The relation between K_p and K_c is based on the difference in the numbers of moles of gaseous products and reactants; so we need to calculate this value (Δn) for reactions a and b.

a. In the ozone-formation reaction, 2 moles of reactants form 1 mole of product. Therefore:

$$\Delta n = 1 - 2 = -1$$

substituting the value of Δn into Equation 15.13 gives

$$K_p = K_c(RT)^{\Delta n} = K_c(RT)^{-1}$$

b. In the formation of nitric oxide, 2 moles of reactants form 2 moles of products. In this example, $\Delta n = 0$ and so

$$K_p = K_c(RT)^0 = K_c$$

PRACTICE EXERCISE: What is the K_p/K_c ratio for the synthesis of ammonia?

$$N_2(g) + 3\,H_2(g) \rightleftharpoons 2\,NH_3(g)$$

(See Problems 33–36.)

Let's calculate the values of K_c and K_p for the oxidation of sulfur dioxide to sulfur trioxide, a key reaction in the formation of acid rain in the eastern United States and in Europe, from experimental concentration data. The overall reaction is described by the following equilibrium:

$$2\,SO_2(g) + O_2(g) \rightleftharpoons 2\,SO_3(g)$$

Suppose a test mixture of SO_2, SO_3, and O_2 at equilibrium at 1000 K has the following composition:

$$[SO_2] = 3.77 \times 10^{-3}\,M$$

$$[SO_3] = 4.13 \times 10^{-3}\,M$$

$$[O_2] = 4.30 \times 10^{-3}\,M$$

The concentration-based equilibrium constant expression for the oxidation of SO_2 to SO_3 is

$$K_c = \frac{[SO_3]^2}{[SO_2]^2[O_2]}$$

Inserting the concentration values given, we have

$$K_c = \frac{(4.13 \times 10^{-3}\,M)^2}{(3.77 \times 10^{-3}\,M)^2(4.30 \times 10^{-3}\,M)}$$

$$= 279\,M^{-1}$$

The units of the K_c value are M^{-1} (or L/mol) because 3 moles of reactants produce only 2 moles of products. Therefore, the units of the denominator of the equilibrium constant expression are M^3, the units of the numerator are M^2, and so the ratio of the two has units of M^{-1}. Because there is one less mole of gaseous product than reactants, the value of Δn in Equation 15.13 as it applies to this reaction is -1. Thus, we can convert the value of K_c into an equivalent value of K_p in the following way:

$$K_p = K_c(RT)^{\Delta n} = K_c(RT)^{-1} = \frac{K_c}{RT}$$

$$= \frac{279.\frac{\text{L}}{\text{mol}}}{\left(0.0821 \frac{\text{L} \cdot \text{atm}}{\text{mol} \cdot \text{K}}\right)(1000\ K)}$$

$$= 3.4\ \text{atm}^{-1}$$

> ✓ Equilibrium constant values are usually not written with units. We left them on these values of K_c and K_p to show how they cancel as K_c is converted to K_p.

Note that the value of K_c is nearly two orders of magnitude larger than that of K_p (actually $0.0821 \times 1000. = 82.1$ times as large). The presence of the RT term in Equation 15.13 means that reactions in which $\Delta n \neq 0$ will have K_c and K_p values that often differ by an order of magnitude or more, depending on temperature and the value of Δn. These differences are the result of the different values

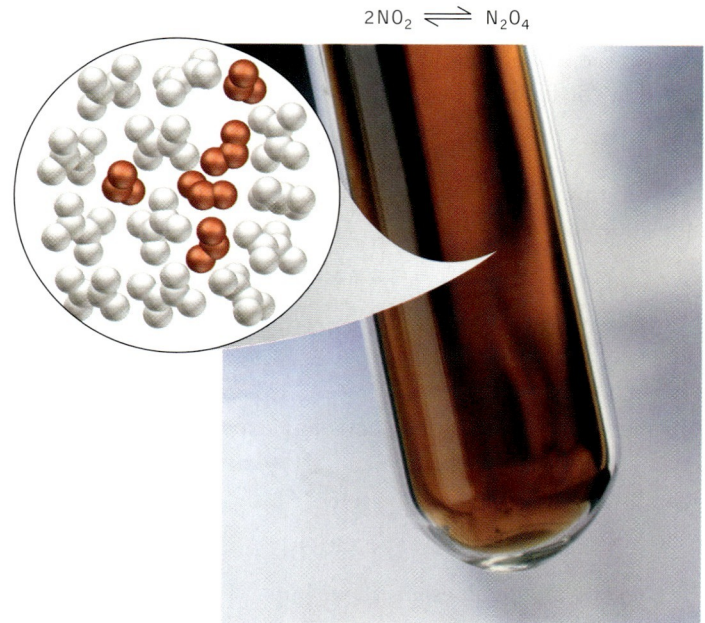

$2NO_2 \rightleftharpoons N_2O_4$

Another view of the meaning of K_p and K_c values is provided by the equilibrium $2\ NO_2 \rightleftharpoons N_2O_4$. At 25°C the value of K_c is 216 M^{-1}, but the value of K_p is only:

$$K_p = K_c(RT)^{\Delta n} = \left(216\frac{\text{L}}{\text{mol}}\right)\left[\left(0.0821\frac{\text{L} \cdot \text{atm}}{\text{mol} \cdot \text{K}}\right)(298\ K)\right]^{-1}$$

$$= \frac{216\frac{\text{L}}{\text{mol}}}{24.5\frac{\text{L} \cdot \text{atm}}{\text{mol}}} = 8.8\ \text{atm}^{-1}$$

An equilibrium mixture of these gases in a test tube at a total pressure of one atmosphere has about 2.5 times as many molecules of colorless N_2O_4 as brown NO_2. Still, the concentration of NO_2 is high enough to give the mixture the dark brown color you see in the test tube above.

that we encounter when expressing the composition of a gas in molarity instead of partial pressure. Remember that, at STP (0°C, 1 atm), a mole of an ideal gas occupies a volume of 22.4 L. Thus, the molar concentration of a pure gas at STP is 1.00 mol/22.4 L, or only 0.0446 M.

CONNECTION: The value of molar volume of an ideal gas at STP (22.4 L) is derived in Section 8.2.

15.4   K, Q, AND ΔG

We have seen that the value of the equilibrium constant for a reaction does not depend on the kinetics of the reaction. However, it is directly related to the thermodynamics of the reaction and, in particular, to the change in free energy, ΔG. If ΔG is negative, then a reaction is spontaneous and should proceed in the forward direction (see Chapter 13, Table 13.2). If ΔG is positive, then a reaction is considered nonspontaneous. Instead, the reverse of the reaction will have a negative ΔG value and will proceed, converting the products of the forward reaction into reactants. Either way, a reaction should proceed in the spontaneous direction until it achieves equilibrium. At that point, there is no longer any free energy left to drive the reaction; ΔG has gone to zero, and the reaction has achieved equilibrium.

This thermodynamic view of equilibrium is described mathematically in the following equation:

$$\Delta G = \Delta G° + RT \ln Q \qquad (15.14)$$

A spontaneous reaction with a negative $\Delta G°$ value proceeds until

1. the concentrations of products are large, and
2. the concentrations of reactants are small, making
3. the value of Q so large that
4. the value of $RT \ln Q$ is large enough to offset the negative value of $\Delta G°$.

Equilibrium is achieved when ΔG becomes zero. At that point, the value of the reaction quotient Q matches that of the equilibrium constant K and Equation 15.14 becomes

$$\Delta G = \Delta G° + RT \ln K = 0$$

or

$$\Delta G° = -RT \ln K \qquad (15.15)$$

Equation 15.15 can be rearranged to allow us to calculate the value of the equilibrium constant for a reaction from its change in standard free energy. First, we rearrange the terms:

$$\ln K = -\Delta G°/RT$$

and then take the antilogs of both sides:

$$K = e^{-\Delta G°/RT} \qquad (15.16)$$

Equation 15.16 provides the following interpretation of reaction spontaneity: if $\Delta G°$ is less than zero, then $(-\Delta G°/RT)$ is positive, then $e^{-\Delta G°/RT}$ is greater than one, making K greater than 1. Thus, a nonspontaneous reaction is one for which $\Delta G°$ is greater than 0, and K is less than 1.

To illustrate the connection between $\Delta G°$ and K, let's revisit the formation of NO from molecular nitrogen and oxygen but this time at room temperature. A balanced equation for the reaction may be written

$$\tfrac{1}{2} N_2(g) + \tfrac{1}{2} O_2(g) \rightleftharpoons NO(g)$$

The value of $\Delta G°$ for the reaction is easily calculated from the standard free energy of formation of NO, because the reaction describes the formation of NO from its elemental components. Therefore the value of $\Delta G°$ at 298 K is the same as the value of $\Delta G_f°$ for NO, which is +86.6 kJ. We can use Equation 15.16 and this value of $\Delta G°$ to calculate the equilibrium constant for the reaction at 298 K. First, let's calculate the value of the exponent in Equation 15.6:

$$\frac{-\Delta G°}{RT} = \frac{-\left(\frac{86.6 \text{ kJ}}{\text{mol}}\right)\left(\frac{1000 \text{ J}}{\text{kJ}}\right)}{\left(\frac{8.314 \text{ J}}{\text{mol} \cdot \text{K}}\right)(298 \text{ K})} = -35.0$$

Inserting this value into Equation 15.16 gives

$$K = e^{-\Delta G°/RT} = e^{-35.0} = 6.6 \times 10^{-16}$$

Thus, the large, positive value of $\Delta G°$ for the reaction (+86.6 kJ/mol) can be interpreted two ways:

1. The reaction is not spontaneous or
2. it proceeds in the forward direction to such a small degree (because K is much less than 1) that only a very minute quantity of NO forms.

On the other hand, the reverse reaction in which NO decomposes into N_2 and O_2

$$NO(g) \rightleftharpoons \tfrac{1}{2} N_2(g) + \tfrac{1}{2} O_2(g)$$

has a very negative $\Delta G°$ value at 298 K: -86.6 kJ/mol. This value also can be interpreted in one of two ways:

1. The reaction is spontaneous or
2. it proceeds in the forward direction to such a large degree that only a tiny concentration of NO is present when equilibrium is achieved ($K \gg 1$).

Visual and graphical images of chemical equilibria in three reactions with $\Delta G°$ values that are greater than zero, less than zero, and equal to zero are shown in Figure 15.5.

CONNECTION: The method for calculating free-energy changes from standard free energies of formation is described in Section 13.5.

SAMPLE EXERCISE 15.7: Using the appropriate $\Delta G_f°$ values, calculate (a) $\Delta G°$ and (b) the equilibrium constant for the formation of NO_2 from NO and O_2 at 298 K.

$$NO(g) + \tfrac{1}{2} O_2(g) \longrightarrow NO_2(g)$$

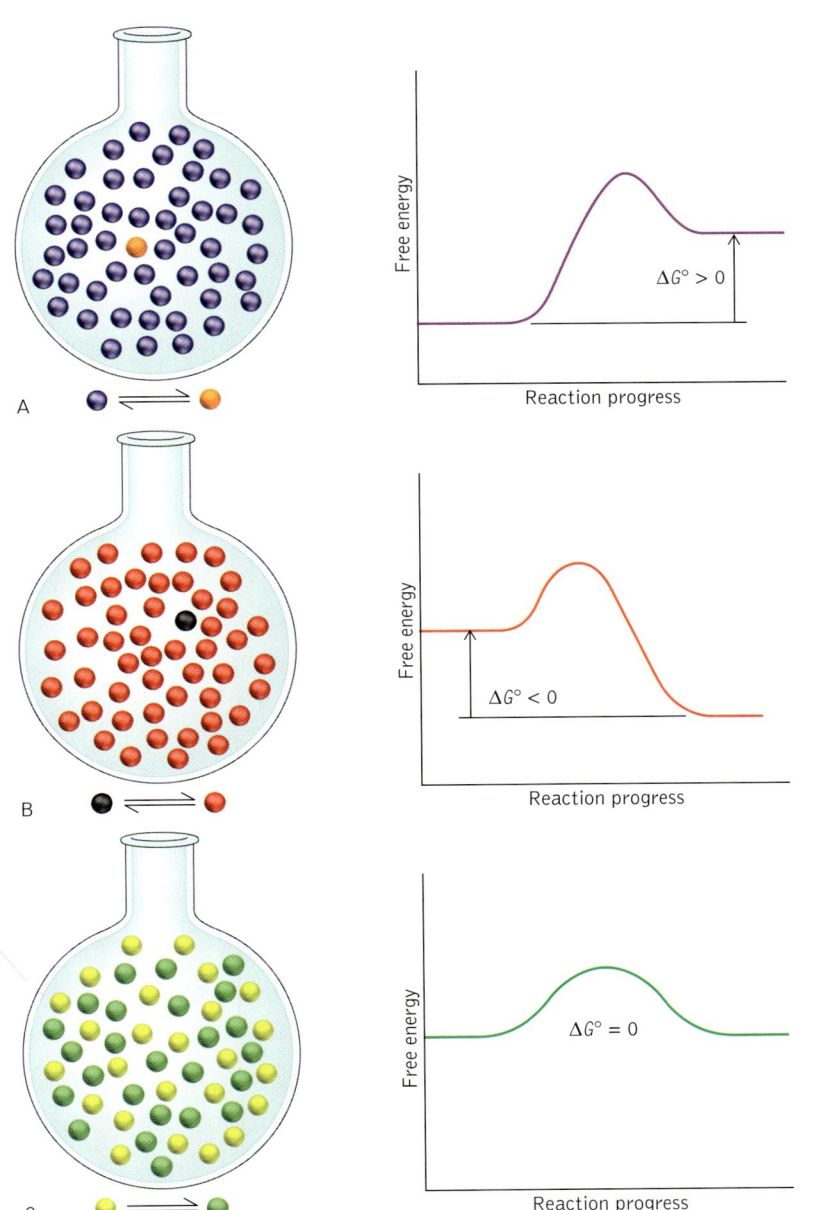

FIGURE 15.5 The value of the equilibrium constant for a reaction is related to the change in standard free energy in the course of the reaction: $\Delta G° = -RT \ln K$. A. If $K < 1$, then $\Delta G° > 0$. B. A value of $K > 1$ corresponds to $\Delta G° < 0$. C. $K = 1$ when $\Delta G° = C$.

SOLUTION: a. Calculating the free-energy change for a chemical reaction involves taking the difference in the weighted sum of the $\Delta G_f°$ values of its product(s) minus reactant(s) (Equation 13.14):

$$\Delta G°_{rxn} = \sum n \Delta G°_{f,prod} - \sum m \Delta G°_{f,react}$$

Inserting the appropriate values of ΔG_f° gives

$$\Delta G_{rxn}^\circ = [\Delta G_f^\circ (NO_2)] - [\Delta G_f^\circ (NO) + \tfrac{1}{2} \Delta G_f^\circ (O_2)]$$

$$= [1 \text{ mol } (51.3 \text{ kJ/mol})] - [1 \text{ mol } (86.6 \text{ kJ/mol}) + \tfrac{1}{2} \text{ mol } (0.0 \text{ kJ/mol})]$$

$$= (51.3 - 86.6) \text{ kJ}$$

$$= -35.3 \text{ kJ}$$

or $-35{,}300$ J per mole of NO_2 produced.

b. Using this value of ΔG° in Equation 15.16 to calculate the equilibrium constant (K), we have

$$\frac{-\Delta G^\circ}{RT} = -\frac{\left(\dfrac{-35{,}300 \text{ J}}{\text{mol}}\right)}{\left(\dfrac{8.314 \text{ J}}{\text{mol} \cdot \text{K}}\right)(298 \text{ K})} = 14.25$$

$$K = e^{-\Delta G^\circ / RT} = e^{14.25} = 1.5 \times 10^6$$

PRACTICE EXERCISE: The ΔG_f° value for ammonia gas is -16.5 kJ/mol at 298 K. What is the value of the equilibrium constant for the following reaction at 298 K? Note: The equation as written produces 2 moles of NH_3.

$$N_2(g) + 3 H_2(g) \longrightarrow 2 NH_3(g)$$

(See Problems 47–52.)

What type of equilibrium constant, K_c or K_p, is linked to ΔG° by Equation 15.15? The answer is: It depends. The symbol ΔG° represents a change in free energy under standard conditions. For a gaseous reactant or product, standard conditions mean that its *partial pressure* is 10^5 Pa. Thus, the ΔG° of a reaction in the gas phase is linked by Equation 15.15 to its K_p value. However, standard conditions for reactions in solution, which is the focus of Chapter 16, mean that all dissolved reactants and products are present at a *concentration* of 1.00 M. Thus, the ΔG° of a reaction in solution is linked by Equation 15.15 to its K_c value.

> What type of equilibrium constant, K_c or K_p, is linked to ΔG° by Equation 15.15?

15.5 LE CHÂTELIER'S PRINCIPLE

In Section 15.2, we described what happens when a chemical reaction at equilibrium is "perturbed" by adding more product. Increasing the concentration of a product causes the reaction to run in reverse, consuming some of the added

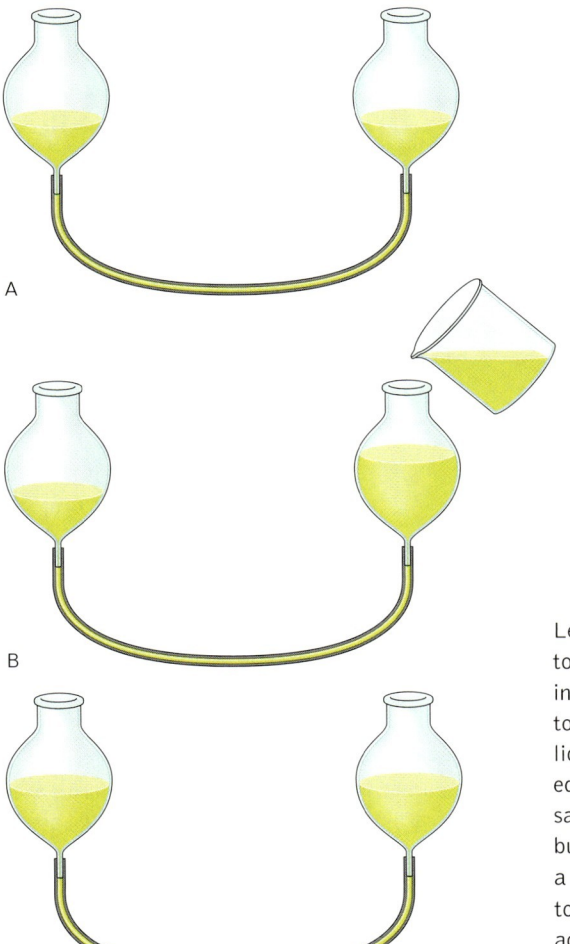

Le Châtelier's principle states that applying stress to a system at equilibrium will force the equilibrium in the direction that will relieve the stress. One way to illustrate this principle is with the levels of liquid in two bulbs connected by a thin tube. At equilibrium (A), liquid levels in both bulbs are the same. When (B) more liquid is added to one of the bulbs (equivalent to adding a reactant or product to a mixture at chemical equilibrium), liquid flows toward the other bulb (equivalent to consuming the added reactant or product) until (C) the levels are again equal.

product and converting the consumed part into reactants. This response to perturbation, or "stress," is described in a principle named after French chemist Henri Louis Le Châtelier (1850–1936). **Le Châtelier's principle** states that *if a system at equilibrium is subjected to a stress, the position of the equilibrium will shift in the direction that will relieve that stress.*

Let's examine some stresses to chemical systems at equilibrium and the responses that those stresses induce. Figure 15.6 illustrates one system's response. This one is based on the equilibrium between NO_2 and its dimer, N_2O_4, discussed in Section 15.1:

$$2\,NO_2(g) \rightleftharpoons N_2O_4(g)$$

The images in Figure 15.6 show a reaction mixture of the two gases at equilibrium. The brown color indicates the presence of some NO_2, but it is not the dark brown of pure NO_2. Therefore, there must be a significant amount of colorless N_2O_4 present, too. Then the mixture is compressed. Increasing the pressure on

> **Le Châtelier's principle** states that a system at equilibrium responds to a stress in such a way that it relieves that stress.

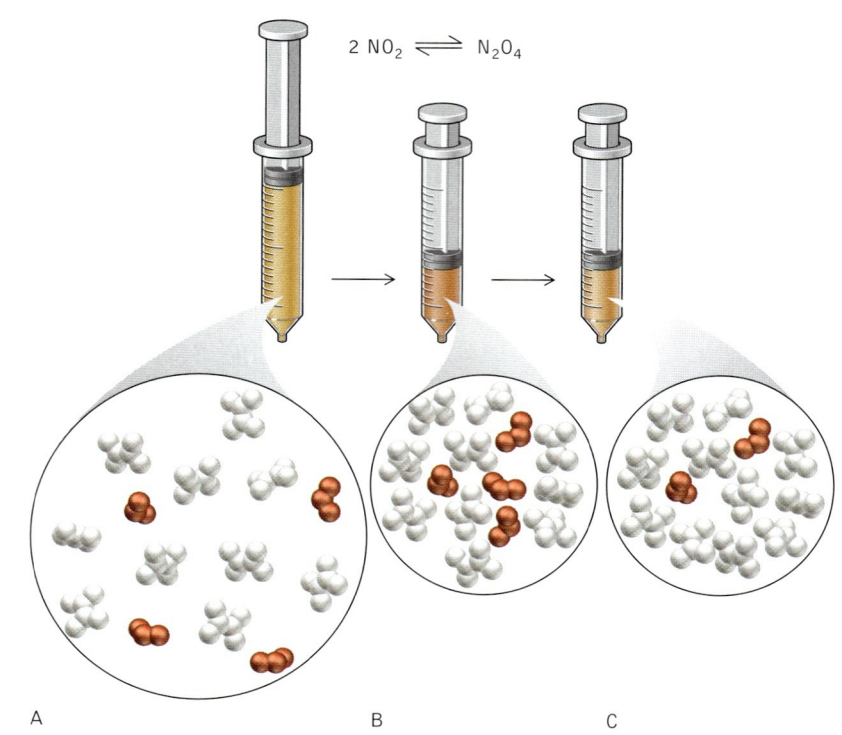

FIGURE 15.6 A. A gas-tight syringe contains an equilibrium mixture of NO_2 and N_2O_4. B. When the plunger is pushed in, increasing the pressure inside the syringe, the color of the gas mixture is temporarily darker as the molecules of NO_2 are compressed into a smaller volume. C. With the passage of time, the color of the mixture fades as NO_2 forms more N_2O_4. Two moles of gas are consumed for every mole that is formed, so the total number of moles of gas in the syringe is reduced, partly relieving the increase in pressure.

the equilibrium gas mixture puts a stress on the equilibrium that will be relieved by reducing the pressure. According to the ideal-gas law, the pressure of a gas at constant temperature and volume is proportional to the number of moles of gas in the mixture:

$$PV = nRT$$

or

$$P = \left(\frac{RT}{V}\right)n$$

Therefore, one way to reduce the total pressure of a mixture of gases is to reduce the total number of moles of gas in the mixture. The number of moles of gas in a mixture of NO_2 and N_2O_4 can be reduced by shifting the equilibrium in the direction of making more N_2O_4, because 2 moles of NO_2 combine to form only 1 mole of N_2O_4. Thus, compressing the mixture produces an instantaneous darkening of its color (Figure 15.6B) as more molecules of NO_2 are squeezed into a smaller volume, thereby increasing their concentration. But then the color lightens, as shown in Figure 15.6C, as some of the NO_2 dimerizes, forming N_2O_4 and partly relieving the pressure.

As you might expect, changing pressure shifts the equilibrium of any reaction in which there is a change in the number of moles of gas as the reaction proceeds. *Increased* pressure shifts the equilibrium toward the side of the reaction with *fewer* moles of gas. A *decrease* in pressure shifts the equilibrium toward the side of the reaction with *more* moles of gas.

CONCEPT TEST: Describe the perturbations caused by removing or adding a reactant to an equilibrium mixture. Indicate how the composition of the mixture would change as the system relieves the stresses caused by these changes.

SAMPLE EXERCISE 15.8: In which of the following equilibria would an increase in pressure promote the formation of more product(s)?

a. $N_2(g) + O_2(g) \rightleftharpoons 2\ NO(g)$
b. $2\ NO(g) + O_2(g) \rightleftharpoons 2\ NO_2(g)$
c. $N_2O_4(g) \rightleftharpoons 2\ NO_2(g)$
d. $H_2O(l) + CO_2(g) \rightleftharpoons H_2CO_3(aq)$
e. $CaCO_3(s) \rightleftharpoons CaO(s) + CO_2(g)$

SOLUTION: The only two forward reactions favored by an increase in pressure are *b* and *d* because they are the only two in which there are more moles of gaseous reactants than products. Note that reactions *d* and *e* involve substances in more than one phase. Such heterogeneous equilibria are discussed in Section 15.9.

PRACTICE EXERCISE: In which of the equilibria in the preceding sample exercise would an increase in pressure lead to the formation of less product? (See Problems 59–64.)

Industrial chemists frequently exploit Le Châtelier's principle to obtain more of a desired product. Consider the reaction between nitrogen and hydrogen to form ammonia:

$$N_2(g) + 3\ H_2(g) \rightleftharpoons 2\ NH_3(g)$$

This reaction is at the heart of an industrial process for producing ammonia (see The Chemistry of Ammonia on page 754). The process uses reaction vessels that operate at 200 atm of pressure. Why run the ammonia reaction at high pressure? To understand why, note that a total of 4 moles of reactant gases combine to make only 2 moles of product. When the reaction is carried out in a rigid, sealed chamber, the pressure inside the chamber decreases as the reaction proceeds because the total number of moles of gas in the chamber decreases. Running the reaction at high pressure puts a stress on the system that is relieved by making more ammonia.

THE CHEMISTRY OF AMMONIA

Ammonia (NH_3) is a colorless gas that irritates mucous membranes and is toxic at high concentrations. More than 18 million tons of NH_3 is produced in the United States every year; worldwide production exceeds 120 million tons. About 85% of the ammonia produced is used as fertilizer either directly as NH_3 or as ammonium nitrate, ammonium phosphate, ammonium sulfate, or urea:

$$(NH_4)_3PO_4 \xleftarrow{H_3PO_4} NH_3 \xrightarrow{HNO_3} NH_4NO_3$$
$$(NH_4)_2SO_4 \xleftarrow{H_2SO_4} \quad \xrightarrow{CO_2} H_2N-\underset{\underset{O}{\|}}{C}-NH_2$$

Most of the ammonia produced for industry and agriculture is synthesized by the reaction

$$N_2(g) + 3\,H_2(g) \rightleftharpoons 2\,NH_3(g)$$

The industrial process for synthesizing ammonia based on this reaction was developed by two German scientists, Fritz Haber (1868–1935) and Karl Bosch (1874–1940), just before World War I. The reaction is spontaneous ($\Delta G° = -33$ kJ/mol) and exothermic ($\Delta H° = -92$ kJ/mol) under standard conditions. However, the reaction is extremely slow at room temperature. A key step in Haber's research was his discovery of catalysts that could speed up the reaction. The best catalysts are made of iron metal and metal oxides such as MgO and Al_2O_3. These catalysts provide acceptable reaction rates at temperatures between 380° and 450°C.

The ammonia reaction has an equilibrium constant K_p of 6×10^5 at 298K. Increasing the temperature increases the reaction rate; however, it dramatically reduces the value of K_p and the reaction yield. Le Châtelier's principle can be used to offset the low values of K_p at high tem-

Fritz Haber.

peratures by running the reactions at high pressure. Because only 2 moles of ammonia gas are produced for each 4 moles of reactants consumed, high pressure (about 200 atm is typically used) shifts the equilibrium toward the side of the reaction with fewer moles of gas and the formation of more ammonia.

The Haber-Bosch process incorporates another step to increase yield: when a hot reaction mixture reaches chemical equilibrium, it flows through a system of chilled condensers. There the ammonia in the mixture, which has a higher boiling point than those of either hydrogen or nitrogen, condenses as a liquid and is thus removed from the gaseous reaction mixture. This removal produces an-

✓ An increase in pressure shifts the position of a chemical equilibrium toward the side of the reaction that has fewer moles of gas.

To make even more ammonia, the hot (400°C) reaction mixture is pumped through a system of chilled condensers where ammonia, which has a much higher boiling point than either hydrogen or nitrogen, condenses as a liquid and so is lost from the gaseous reaction mixture. The removal of ammonia constitutes another stress on the equilibrium that is relieved by producing more of it.

other stress on the equilibrium, one that is relieved by the formation of more ammonia when the ammonia-depleted mixture returns to the high-temperature reaction chamber.

The Haber-Bosch process has had an enormous effect on history and the human condition. Some historians believe that World War I would have ended more than 2 years earlier were it not for the Haber-Bosch process. Had it ended earlier, there would have been millions fewer casualties—an outcome that might have dramatically changed the terms of Germany's surrender and the course of events in Germany in the 1920s and 1930s. How could one chemical reaction affect the course of human events so dramatically? During the war, Germany was cut off from its South American supplies of $NaNO_3$, which it needed to make gunpowder and other explosives. However, the Haber-Bosch process provided a way to take nitrogen from the air and turn it into ammonia. Ammonia can be converted into the nitrates needed for explosives by a series of three reactions known as the Ostwald process. In the first step, ammonia reacts with oxygen to produce NO:

(1) $2 NH_3(g) + 5 O_2(g) \longrightarrow 4 NO(g) + 3 H_2O(g)$
$\Delta H° = -905$ kJ

This reaction is another of those "spontaneous" reactions that proceeds very slowly. However, the use of a catalyst made of an alloy of platinum and rhodium and a reaction temperature of about 900°C leads to yields of NO of about 97%. Next, NO is further oxidized to NO_2:

(2) $2 NO(g) + O_2(g) \longrightarrow 2 NO_2(g)$
$\Delta H° = -113$ kJ

In the third step, NO_2 is dissolved in water, producing nitric acid and NO that is recycled back to step 2:

(3) $3 NO_2(g) + H_2O(l) \longrightarrow 2 HNO_3(aq) + 2 NO(g)$
$\Delta H° = -139$ kJ

Reacting HNO_3 with NH_3 produces NH_4NO_3, a key ingredient in explosives.

Manufacturing explosives is not the principal legacy of the Haber-Bosch process; nor the reason why Haber won the Nobel Prize in chemistry in 1919. Rather, the most enduring legacy is the manufacture of nitrogen fertilizers. As previously noted, most of the ammonia produced by the chemical industry is used in agriculture. In most plants, including the major food crops, more than 10 times as much nitrogen is needed for plant growth as phosphorus or potassium, the other key micronutrients. Thus nitrogen-containing fertilizers are one of the reasons why agricultural productivity is dramatically greater today than it was a century ago and why much of the world has an adequate food supply.

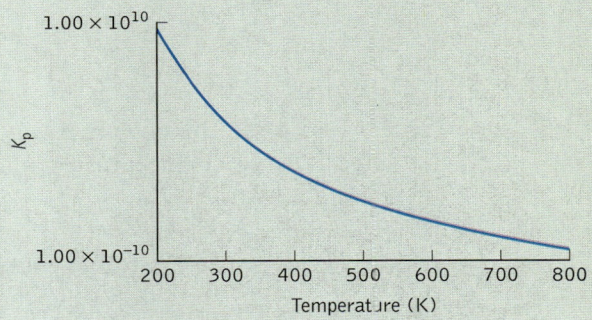

The equilibrium constant K_p for the reaction $N_2(g) + 3 H_2(g) \rightleftharpoons 2 NH_3(g)$ decreases with increasing temperature. Note that the values of K_p are plotted logarithmically: the value of K_p at 200 K is more than 10^{18} times as large as that at 800 K. However, high temperatures and catalysts are needed to increase the *rate* of the reaction.

Removing product from a reaction mixture perturbs the system and promotes the formation of more product. Removing product also yields a Q expression that has a smaller numerator (and so a smaller value) than K. According to Table 15.1, the reaction will proceed from left to right and more product will be made. Similarly, the addition of product to an equilibrium mixture perturbs the

> ✓ Adding a reactant or removing a product shifts the position of a chemical equilibrium in the forward direction, favoring the formation of more product.

equilibrium and yields a Q expression that has a larger value than K. According to Le Châtelier's principle, the stress of this perturbation is relieved by the conversion of some of the added product into reactant(s), which is just what we would predict when Q is greater than K.

CONCEPT TEST: Explain why the equilibrium constant for a highly endothermic reaction is likely to decrease with decreasing temperature.

Hb(aq) + 4 O$_2$(g) ⇌ Hb(O$_2$)$_4$(aq)

Mountain climbers can acclimate to the thin air and low oxygen concentrations encountered at high altitudes. Oxygen is transported in human blood bound to a protein called hemoglobin (Hb). As [O$_2$] in the air decreases with increasing elevation, LeChâtelier's principle predicts a shift in the equilibrium of O$_2$ + Hb ⇌ Hb · O$_2$ to the left, decreasing [Hb · O$_2$], which means that less O$_2$ is delivered to the cells. With time, the body responds to this stress by manufacturing more hemoglobin, increasing [Hb] in the blood and pushing the equilibrium back to the right. In this way, [Hb · O$_2$] is restored to near-normal levels. There is a limit to how much hemoglobin the body can produce and the blood can carry. At the very high altitudes of Earth's tallest mountains, this compensation mechanism is not sufficient and so most climbers breathe bottled oxygen.

15.6 THE ROLE OF CATALYSTS

This chapter began with a discussion of the formation of NO in the combustion chambers of car and truck engines:

$$N_2(g) + O_2(g) \rightleftharpoons 2\,NO(g)$$

To control NO emissions, these vehicles may be equipped with catalytic converters that speed up the rate of the decomposition of NO:

$$2\,NO(g) \rightleftharpoons N_2(g) + O_2(g)$$

As discussed in Chapter 14, a catalyst promotes the rate of a chemical reaction by lowering its activation energy. This role raises a question: If a catalyst increases the rate of a reaction does the addition of a catalyst affect the equilibrium constant of a reaction?

To answer this question, consider the catalyzed and uncatalyzed energy profiles of the exothermic reaction in Figure 14.21. Note that the height of the activation-energy barrier that reactants must overcome is reduced by the same amount whether the reaction pro-

> Does the addition of a catalyst affect the equilibrium constant of a reaction?

ceeds in the forward direction or in reverse. As a result, the increases in reaction rates produced by the catalyst will be the same in both directions. This leads to the conclusion, which is supported by experimental data, that a catalyst does not change the equilibrium constant of a reaction or the composition of a equilibrium reaction mixture. On the other hand, a catalyst should (and does) decrease the time that it takes for a reaction to reach equilibrium.

> Catalysts decrease the time needed to achieve chemical equilibrium; they do not affect the equilibrium state or the value of the equilibrium constant.

CONNECTION: A catalyst simultaneously increases the rate of a reaction in both the forward and the reverse direction (see Section 14.6) and so does not alter the value of the equilibrium constant.

15.7 CALCULATIONS BASED ON K

Reference books and the tables in Appendix 5 of this book contain lists of equilibrium constants for chemical reactions. These values are used in several kinds of calculations, including those in which:

1. equilibrium concentrations of the reactants and products are known and we wish to calculate the equilibrium constant, as we did for K_c in the formation of nitric oxide (Section 15.2), and
2. we know the value of K and the starting quantities of reactants, and we want to calculate the concentrations of reactants and products in the reaction at equilibrium.

Let's calculate the quantity of product formed in a reaction at equilibrium for which we know the initial concentrations of the reactants and the value of K. In our first example, we will calculate how much nitric oxide (NO) forms in a sample of air heated to a temperature at which K_p for the formation of NO from N_2 and O_2 is 1.00×10^{-5}. The initial partial pressures of N_2 and O_2 are 0.79 and 0.21 atmospheres, respectively, and we assume there is no NO present.

The equilibrium constant expression for the formation of NO from N_2 and O_2 is

$$K_p = \frac{(P_{NO})^2}{(P_{N_2})(P_{O_2})}$$

Let's put the initial partial pressures of the reactants and product in the first row of a data table known as an ICE table. This acronym means that ICE tables contain the **i**nitial partial pressures (or concentrations) of reactants and products, the **c**hanges in their pressures or concentrations as the reaction proceeds to equilibrium, and their pressures or concentrations after **e**quilibrium has been achieved. Filling in the first row of the following ICE table with the information given in the example, we have

	P_{N_2}	P_{O_2}	P_{NO}
Initial	0.79	0.21	0
Change			
Equilibrium			

To fill in the remaining rows, we need to do some basic algebra. We do not know how much N_2 or O_2 will be consumed during the reaction or how much nitric oxide will be produced. Let's define the change in P_{N_2} as $-x$ (it has to be a negative value because N_2 is consumed during the reaction). The mole ratio of N_2 to O_2 in the balanced chemical equation is 1:1; so the change in P_{O_2} also will be $-x$. Two moles of NO are produced from each mole of N_2 and O_2; so the change in P_{NO} will be $+2x$. Inserting these algebraic values in the second row of the ICE table, we have

	P_{N_2}	P_{O_2}	P_{NO}
Initial	0.79	0.21	0
Change	$-x$	$-x$	$+2x$
Equilibrium			

Combining the *initial* and *change* rows of the ICE table, we obtain the following algebraic terms for the partial pressures for all three gases at equilibrium:

	P_{N_2}	P_{O_2}	P_{NO}
Initial	0.79	0.21	0
Change	$-x$	$-x$	$+2x$
Equilibrium	$0.79-x$	$0.21-x$	$2x$

The last row of partial-pressure terms can be substituted into the K_p equation for the reaction:

$$K_p = \frac{(P_{NO})^2}{(P_{N_2})(P_{O_2})} = \frac{(2x)^2}{(0.79-x)(0.21-x)} = 1.00 \times 10^{-5}$$

Multiplication of the binomial terms in the denominator gives

$$K_p = 1.00 \times 10^{-5} = \frac{4x^2}{0.1659 - 1.00x + x^2}$$

Cross-multiplying, we get

$$1.659 \times 10^{-6} - 1.00 \times 10^{-5}x + 1.00 \times 10^{-5}x^2 = 4x^2$$

Combining the x^2 terms and rearranging, we have

$$3.99999x^2 + 1.00 \times 10^{-5}x - 1.659 \times 10^{-6} = 0$$

You may recognize this equation as one that fits the general form of a quadratic equation:

$$ax^2 + bx + c = 0$$

which can be solved for x by using the following formula:

$$x = \frac{-b \pm \sqrt{b^2 - 4ac}}{2a}$$

Inserting the values for a, b, and c into the quadratic formula, we have

$$x = \frac{-1.00 \times 10^{-5} \pm \sqrt{(-1.00 \times 10^{-5})^2 - 4(3.99999)(-1.659 \times 10^{-6})}}{2(4.00)}$$

$$= \frac{-1.00 \times 10^{-5} \pm \sqrt{1.00 \times 10^{-10} + 2.6544 \times 10^{-5}}}{8.000}$$

$$= \frac{-1.00 \times 10^{-5} \pm 5.152 \times 10^{-3}}{8.000}$$

There are two possible values of x, but only one of them is positive: 6.428×10^{-4}. A gas can't have a negative partial pressure, so we will use only the positive value to calculate equilibrium partial pressures:

$P_{O_2} = 0.21 - x = 0.21 - (6.428 \times 10^{-4}) = 0.21$ atm (to two significant figures)

$P_{N_2} = 0.79 - x/2 = 0.79 - (6.428 \times 10^{-4}) = 0.79$ atm

$P_{NO} = 2x = 2(6.428 \times 10^{-4}) = 1.2855 \times 10^{-3} = 0.0013$ atm

Note that the equilibrium partial pressures of the reactants decrease by insignificant amounts from their initial partial pressures. These insignificant decreases make sense given the small value (1.0×10^{-5}) of the equilibrium constant. Because the x terms in the denominator of the equilibrium constant equation are small compared with the initial partial pressures, there is a much simpler approach to calculating P_{NO}: we can ignore the x terms in the denominator and use the initial values for P_{N_2} and P_{O_2} instead:

$$K_p = 1.0 \times 10^{-5} = \frac{(P_{NO})^2}{(P_{N_2})(P_{O_2})} = \frac{4x^2}{(0.79 - x)(0.21 - x)} \cong \frac{4x^2}{(0.79)(0.21)} = \frac{4x^2}{0.1659}$$

$$4x^2 = (0.79)(0.21)(1.0 \times 10^{-5})$$
$$= 1.659 \times 10^{-6}$$
$$x^2 = 4.148 \times 10^{-7}$$
$$x = 6.440 \times 10^{-4}$$

Note that the difference between this value of x and that obtained from the solution to the quadratic equation (6.428×10^{-4}) is

$$(6.440 \times 10^{-4}) - (6.428 \times 10^{-4}) = 0.0012 \times 10^{-4}$$

Dividing this difference by the original value of x, we have a relative difference of only

$$\frac{0.012 \times 10^{-4} \, M}{6.428 \times 10^{-4} \, M} \times 100\% = 0.19\%$$

In almost all equilibrium calculations, such a small difference is unimportant. Certainly, it is insignificant when we know the initial partial pressures to only two significant figures. As a general rule, we can ignore the x component of the concentration or partial-pressure terms for reactants if, after doing a simplified calculation, the value of x is less than 5% of the smallest initial value.

SAMPLE EXERCISE 15.9: Much of the H_2 used in the Haber process (see the Chemistry of Ammonia on page 754) is produced by reacting methane from natural gas with high-temperature steam. In the first stage of a two-stage reaction, CO and H_2 form. In the second stage, the CO formed in the first stage is reacted with more steam in a reaction known as the *water-gas shift reaction*:

$$CO(g) + H_2O(g) \rightleftharpoons CO_2(g) + H_2(g)$$

If a reaction vessel at 400°C is charged with an equimolar mixture of CO and steam such that $P_{CO} = P_{H_2O} = 2.00$ atm, what will be the partial pressure of H_2 at equilibrium? Given: $K_p = 10$ at 400°C.

SOLUTION: Let's set up an ICE table for this reaction. The initial partial pressures are $P_{CO} = P_{H_2O} = 2.00$ atm and $P_{CO_2} = P_{H_2} = 0.00$ atm. Let x be the increase in partial pressure of H_2 as a result of the reaction. According to the stoichiometry of the reaction, 1 mole of CO_2 forms for each mole of H_2 that is formed, and 1 mole of CO and 1 mole of H_2O are consumed. Therefore, the changes in P_{CO_2} and P_{H_2} will both be x, and the changes in P_{CO} and P_{H_2O} will both be $-x$. Inserting these numbers into the ICE table and using them to develop equilibrium values for the partial pressures of the reactants and products, we have

	P_{CO}	P_{H_2O}	P_{CO_2}	P_{H_2}
Initial	2.00	2.00	0.00	0.00
Change	$-x$	$-x$	$+x$	$+x$
Equilibrium	$2.00-x$	$2.00-x$	x	x

Inserting the equilibrium partial-pressure terms into the equilibrium-constant expression for the water-gas shift reaction, we have

✓ **Note:** There is a shortcut to solving this K_p expression for x that avoids using the quadratic equation. Can you find it?

$$K_p = \frac{(P_{CO_2})(P_{H_2})}{(P_{CO})(P_{H_2O})} = \frac{(x)(x)}{(2.00-x)(2.00-x)} = 10$$

Multiplication of the binomial terms in the denominator gives

$$K_p = \frac{x^2}{4.00 - 4.00x + x^2} = 10$$

Cross-multiplying, we get

$$40.0 - 40.0x + 10x^2 = x^2$$

Combining the x^2 terms and rearranging, we have

$$9x^2 - 40.0x + 40.0 = 0$$

Using these coefficients in the quadratic formula gives

$$x = \frac{-(-40) \pm \sqrt{(-40)^2 - 4(9)(40)}}{2(9.00)}$$

$$= \frac{+40 \pm \sqrt{1600 + 1440}}{18.00} = \frac{+40 \pm 12.65}{18.00}$$

$$= \frac{27.35}{18.00} = 1.52 \text{ or } \frac{52.65}{18.00} = 2.92$$

Of the two solutions to the quadratic equation, one of them (2.92) has no meaning for us, because inserting it into the equilibrium-constant expression yields negative partial pressures of CO and H_2O. Therefore, the partial pressure of H_2 (and CO_2) at equilibrium is 1.52 atm.

PRACTICE EXERCISE: The balanced chemical equation for the format on of hydrogen iodide from H_2 and I_2 is $H_2(g) + I_2(g) \rightleftharpoons 2\,HI(g)$. The value of K_p for the reaction is 50. at 450°C. What will be the partial pressure of HI in a sealed reaction vessel at 450°C if the initial values of P_{H_2} and P_{I_2} are both 0.100 atm and there is no HI present? (See Problems 73–78.)

SAMPLE EXERCISE 15.10: Let's revisit the chemical equilibrium associated with the second step in the industrial synthesis of H_2:

$$CO(g) + H_2O(g) \rightleftharpoons CO_2(g) + H_2(g)$$

Suppose a reaction vessel at 400°C contains a mixture of CO, steam, and some H_2 formed in the first step in the synthesis. As a result, the reactants and products have the following initial partial pressures: $P_{CO} = 2.00$ atm, $P_{H_2O} = 2.00$ atm, $P_{H_2} = 0.15$ atm, and $P_{CO_2} = 0.00$ atm. What will be the partial pressure of H_2 at equilibrium? Given: $K_p = 10$ at 400°C.

SOLUTION: Let's set up the ICE table for this reaction with the initial partial pressures specified. Let x be the increase in partial pressure of H_2 as a result of the reaction. The change in P_{CO_2} will also be x, and the change in P_{CO} and in P_{H_2O} will be $-x$. Inserting these numbers into the ICE table and using them to develop equilibrium terms for the partial pressures of the reactants and products, we have

	P_{CO}	P_{H_2O}	P_{CO_2}	P_{H_2}
Initial	2.00	2.00	0.00	0.15
Change	$-x$	$-x$	$+x$	$+x$
Equilibrium	$2.00-x$	$2.00-x$	x	$0.15+x$

Inserting these equilibrium terms into the equilibrium constant expression for the reaction gives

$$K_p = \frac{(P_{CO_2})(P_{H_2})}{(P_{CO})(P_{H_2O})} = \frac{(x)(0.15 + x)}{(2.00 - x)(2.00 - x)} = 10$$

Multiplication of the binomial terms in the denominator gives us

$$K_p = \frac{x^2 + 0.15x}{4.00 - 4.00x + x^2} = 10$$

Cross-multiplying gives

$$40.0 - 40.0x + 10x^2 = x^2 + 0.15x$$

Combining the x^2 and x terms and rearranging, we have

$$9x^2 - 40.15x + 40.0 = 0$$

Using these coefficients in the quadratic formula gives

$$x = \frac{-(-40.15) \pm \sqrt{(-40.15)^2 - 4(9)(40)}}{2(9.00)}$$

$$= \frac{+40.15 \pm \sqrt{1612 - 1440}}{18.00} = \frac{+40.15 \pm 13.15}{18.00}$$

$$= \frac{27.00}{18.00} = 1.50 \text{ or } \frac{53.30}{18.00} = 2.96$$

Of the two solutions to the quadratic equation, one of them (2.96) is not physically possible, because inserting it into the equilibrium constant expression gives us negative partial pressures of CO and H_2O. Therefore, $x = 1.50$ and the partial pressure of H_2 at equilibrium is $1.50 + 0.15 = 1.65$ atm.

PRACTICE EXERCISE: The value of K_c for the reaction of $2\ NO_2 \rightleftharpoons N_2O_4$ is 0.21 at 398 K. If a reaction vessel at that temperature initially contains 0.030 M concentrations of both NO_2 and N_2O_4, what will be the concentrations of the two gases when equilibrium is achieved? (See Problems 79 and 80.)

Let's compare the values of $[H_2]$ from the two preceding sample exercises. Note that the presence of a small quantity of H_2 at the beginning of the reaction in the second exercise resulted in slightly less of it being formed during the reaction. This result make sense because the product terms in the mass action expression achieved equilibrium values without the need to produce as much H_2.

On the other hand, the total partial pressure of H_2 at equilibrium is slightly higher for the second case in which some H_2 was present initially. This outcome also makes sense. Less additional H_2 was needed to reach the equilibrium partial pressure of H_2 achieved in the first action. At that stage in the second reaction, slightly more of the reactants were left unreacted. As they proceeded to react, more H_2 formed and a higher equilibrium partial pressure of H_2 was reached.

15.8 CHANGING K WITH CHANGING TEMPERATURE

Earlier in this chapter, we derived the relation between the change in standard free energy of a chemical reaction, $\Delta G°$, and its equilibrium constant, K under standard conditions (Equation 15.15):

$$\Delta G° = -RT \ln K$$

In Chapter 13, we derived the following equation relating the changes in standard free energy, enthalpy, and entropy of a chemical reaction (Equation 13.12):

$$\Delta G° = \Delta H° - T\Delta S°$$

If we rearrange the terms in Equation 15.15 to solve for $\ln K$, we get

$$\ln K = \frac{-\Delta G°}{RT}$$

Let's combine this equation with Equation 13.12:

$$\ln K = -\frac{\Delta H°}{RT} + \frac{\cancel{T}\Delta S°}{R\cancel{T}}$$

$$= -\frac{\Delta H°}{RT} + \frac{\Delta S°}{R}$$

$$\ln K = -\frac{\Delta H°}{R}\left(\frac{1}{T}\right) + \frac{\Delta S°}{R} \qquad (15.17)$$

CONNECTION: The relation between free energy, enthalpy, entropy, and temperature is discussed in Section 13.4.

Several messages are embedded in Equation 15.17. One of them is that a large gain in entropy (positive $\Delta S°$) helps make $\ln K$ positive, which helps create equilibrium states with large concentrations of products and small concentrations of reactants. In addition, the influence of $\Delta S°$ on K does not depend on temperature. Also, exothermic reactions (with negative $\Delta H°$ values) tend to have positive $\ln K$ values because of the negative sign in front of the $\Delta H°/R$ term. However, the influence of negative $\Delta H°$ values depends on temperature: the higher the temperature, the smaller the $1/T$ term and the smaller the influence of a negative $\Delta H°$ on the value of K.

This temperature dependence makes sense if we think of heat as a product in exothermic reactions. Adding heat to an exothermic reaction affects the composition of the reaction mixture in much the way that adding more product to a reaction mixture affects it. According to Le Châtelier's principle, adding a product shifts the equilibrium to the left, forming more reactants and reducing the quantities of the other products in the system. So it is with heat in exothermic reactions: when such a reaction is run at higher temperature, proportionately smaller quantities of products are formed from given quantities of reactants, reflecting a smaller value of the equilibrium constant.

CONNECTION: Exothermic reactions release energy, endothermic reactions consume it (Section 11.4).

N_2O_4 + heat ⇌ $2NO_2$

FIGURE 15.7 The value of nearly all equilibrium constants changes with changing temperature. For example, the dimerization of NO_2 is an exothermic process, and so the reaction is not favored at high temperature such as in a bath of boiling water at 100°C. At that temperature there is little N_2O_4 dimer (which is colorless) present, and the mixture is mostly brown NO_2. As the contents of the flask are cooled, more dimer forms and the brown color of NO_2 fades. At 0°C NO_2 exists as a pale yellow liquid (which is mostly N_2O_4) and at even colder temperatures it becomes a colorless solid (essentially all N_2O_4).

On the other hand, an endothermic reaction requires heat to proceed. Adding heat promotes the formation of more products from a given quantity of reactants, and so the values of the equilibrium constants for these reactions increase with increasing temperature as shown in Figure 15.7.

Let's examine the effect of temperature on an exothermic reaction by revisiting the decomposition of NO into N_2 and O_2. Even at the high temperatures at which automobile exhaust catalysts normally operate, this equilibrium favors the decomposition, rather than the formation, of NO. We can calculate the value of the equilibrium constant of this reaction (or any reaction) at an elevated temperature (or any temperature) if we know the value of K for a reference temperature. We start with Equation 15.17:

$$\ln K = -\frac{\Delta H°}{R}\left(\frac{1}{T}\right) + \frac{\Delta S°}{R}$$

This equation fits the format of a straight line ($y = mx + b$) when the y variable is $\ln K$ and the x variable is $1/T$. The slope of the line is $-\Delta H°/R$ and the intercept is $\Delta S°/R$. Now let's select two points along such a line: temperatures, T_1 and T_2, at which the values of the equilibrium constant are K_1 and K_2, respectively:

$$\ln K_1 = -\frac{\Delta H°}{R}\left(\frac{1}{T_1}\right) + \frac{\Delta S°}{R}$$

$$\ln K_2 = -\frac{\Delta H°}{R}\left(\frac{1}{T_2}\right) + \frac{\Delta S°}{R}$$

The difference between $\ln K_1$ and $\ln K_2$ is related to the difference between T_1 and T_2 as follows:

$$\ln K_1 - \ln K_2 = -\frac{\Delta H°}{R}\left(\frac{1}{T_1}\right) + \frac{\Delta H°}{R}\left(\frac{1}{T_2}\right)$$

Rearranging the terms, we have

$$\ln\left(\frac{K_1}{K_2}\right) = -\frac{\Delta H°}{R}\left(\frac{1}{T_1} - \frac{1}{T_2}\right)$$

or

$$\ln\left(\frac{K_2}{K_1}\right) = -\frac{\Delta H°}{R}\left(\frac{1}{T_2} - \frac{1}{T_1}\right) \tag{15.18}$$

Equation 15.18 is called the *van't Hoff* equation. Let's use it to calculate K_p for the decomposition of NO:

$$NO(g) \rightleftharpoons \tfrac{1}{2}N_2(g) + \tfrac{1}{2}O_2(g)$$

at 500°C, which is a typical temperature in the catalytic converter of an automobile (see Section 14.9). To use Equation 15.18 we first need a reference value for K_p. We also need the standard enthalpy change ($\Delta H°$) of the reaction. In Section 15.4 we calculated the concentration-based (K_c) equilibrium constant for the formation of NO at 298 K:

$$\tfrac{1}{2}N_2 + \tfrac{1}{2}O_2(g) \rightleftharpoons NO \qquad K_c = 6.6 \times 10^{-16}$$

This reaction is the reverse of the decomposition reaction, so its equilibrium constant is the reciprocal of the value we need. In addition, the total number of moles of gaseous reactants and products is the same, so $K_c = K_p$. Therefore, the value of K_p for the decomposition reaction is

$$K_p = \frac{1}{6.6 \times 10^{-16}} = 1.5 \times 10^{15}$$

The value of $\Delta H°$ can also be derived from the reverse reaction, which represents the formation of NO from its elemental components in their standard states. Thus, $\Delta H°$ for the reverse reaction is $\Delta H°_f$ for NO. This value (see Appendix 4) is 90.3 kJ, which means that the value of $\Delta H°$ for the decomposition reaction is -90.3 kJ per mole of NO.

Using Equation 15.18 to calculate K_c for the reaction at 500°C (773 K), we let $T_1 = 298$ K and $T_2 = 773$ K:

$$\ln\left(\frac{K_2}{K_1}\right) = -\frac{\Delta H°}{R}\left(\frac{1}{T_2} - \frac{1}{T_1}\right)$$

$$\ln\left(\frac{K_2}{1.5 \times 10^{15}}\right) = -\frac{-90.3 \frac{\text{kJ}}{\text{mol}} \times \frac{1000 \text{ J}}{\text{kJ}}}{8.314 \text{ J/mol} \cdot \text{K}}\left(\frac{1}{773 \text{ K}} - \frac{1}{298 \text{ K}}\right)$$

$$\ln\left(\frac{K_2}{1.5 \times 10^{15}}\right) = -22.40$$

$$\left(\frac{K_2}{1.5 \times 10^{15}}\right) = e^{-22.40} = 1.87 \times 10^{-10}$$

$$K_2 = 2.8 \times 10^5$$

This large value of K indicates that little NO should remain in engine exhaust *if chemical equilibrium is achieved*. Recall that the job of the catalytic converter is to help achieve chemical equilibrium by increasing the rates of reactions among exhaust gases.

SAMPLE EXERCISE 15.11: Ammonia is synthesized from nitrogen and hydrogen gas in an exothermic chemical reaction that requires high temperatures to increase its rate. Use the data in Appendix 4 to calculate

a. the equilibrium constant (K_p) for the reaction $N_2(g) + 3 H_2(g) \rightleftharpoons 2 NH_3(g)$ at 298 K.
b. the equilibrium constant (K_p) for the reaction $N_2(g) + 3 H_2(g) \rightleftharpoons 2 NH_3(g)$ at 773 K.

SOLUTION: a. We can calculate the value of K_p for this reaction at 298 K by using Equation 15.16 and the value of $\Delta G_f°$ for NH_3 from Appendix 4. We need to be mindful that the $\Delta G_f°$ value (-16.5 kJ/mol) is for the formation of 1 mole of NH_3, but 2 moles of NH_3 are formed in the reaction. Therefore, $\Delta G°$ for the reaction is

$$2 \times -16.5 \text{ kJ/mol} = -33 \text{ kJ/mol, or } -33,000 \text{ J/mol}$$

Using this value in Equation 15.16, we have

$$K_p = e^{-\Delta G°/RT} = \exp\left(\frac{-\Delta G°}{RT}\right)$$

$$= \exp\left(\frac{(-)-33000 \text{ J/mol}}{8.314 \text{ J/mol} \cdot \text{K}) \times 298 \text{ K}}\right)$$

$$= 6.1 \times 10^5$$

b. We can calculate K_p for the reaction at 773 K by using Equation 15.18. Let $T_1 = 298$ and $T_2 = 773$. The value of $\Delta H°$ for the reaction is twice the $\Delta H_f°$ of NH_3 (46.1 kJ/mol):

$$2 \times -46.1 \text{ kJ/mol} = -98.2 \text{ kJ/mol, or } -98,200 \text{ J/mol}$$

$$\ln\left(\frac{K_2}{K_1}\right) = -\frac{\Delta H°}{R}\left(\frac{1}{T_2} - \frac{1}{T_1}\right)$$

$$\ln\left(\frac{K_2}{6.1 \times 10^5}\right) = -\frac{-(-98{,}200 \text{ J/mol})}{8.314 \text{ J/mol} \cdot \text{K}} \left(\frac{1}{773 \text{ K}} - \frac{1}{298 \text{ K}}\right)$$

$$\ln\left(\frac{K_2}{6.1 \times 10^5}\right) = -24.36$$

$$\left(\frac{K_2}{6.1 \times 10^5}\right) = e^{-24.36} = 2.6 \times 10^{-11}$$

$$K_2 = 1.6 \times 10^{-5}$$

Thus, the synthesis of ammonia has an equilibrium constant (K_p) that is greater than 10^5 at room temperature, but only about 10^{-5} at 773 K. As noted in page 754, ammonia is usually synthesized from N_2 and H_2 at temperatures near 400°C. These high temperatures are used to increase the *rate* of the reaction, even though raising the temperature decreases the value of K and so decreases how much ammonia can be made when chemical equilibrium is achieved. In this case, a more-favorable reaction rate outweighs the less-favorable thermodynamics of running the reaction at high temperature.

PRACTICE EXERCISE: Use the data in Appendix 4 to calculate the equilibrium constant (K_p) for the reaction $2 N_2(g) + O_2(g) \rightleftharpoons 2 N_2O(g)$ at 298 K and 2000 K. (See Problems 85–88.)

15.9 HETEROGENEOUS EQUILIBRIA

So far in this chapter we have focused on reactions in the gas phase. However, the principles of chemical equilibrium also apply to reactions in the liquid phase, particularly reactions in solution, which are the focus of Chapter 16. There are also reactions in which reactants and products are in more than one phase. In Section 15.3, we considered the equilibrium associated with the oxidation of SO_2 to SO_3, a key step in the formation of H_2SO_4 both in the atmosphere and in acid rain. One way to prevent this reaction from happening is to "scrub" SO_2 from the exhaust gases from power plants and factories where sulfur-containing fuels are burned. Lime, CaO, is a widely used scrubbing agent. It is sprayed into the exhaust gases and combines with SO_2:

> ✓ Heterogeneous reactions involve reactants and products in more than one phase.

$$CaO(s) + SO_2(g) \longrightarrow CaSO_3(s)$$

The large quantities of lime needed for this reaction and for many other industrial uses are obtained by heating pulverized limestone, $CaCO_3$, to temperatures at which the equilibrium of its highly endothermic decomposition reaction:

$$CaCO_3(s) \rightleftharpoons CaO(s) + CO_2(g) \qquad \Delta H° = 178.1 \text{ kJ} \qquad (15.19)$$

is shifted to the right, favoring the formation of products. At room temperature, the equilibrium position of this reaction must be far to the left; otherwise $CaCO_3$ would not be the abundant component of Earth's crust that it is. How high a temperature is needed to form CaO from $CaCO_3$? To answer this question, let's first write the mass action expression for the decomposition reaction. Doing so in the usual way gives us

> **How high a temperature is needed to form CaO from $CaCO_3$?**

$$K_c = \frac{[CaO][CO_2]}{[CaCO_3]}$$

The terms in the formula raise an interesting question: What are the concentrations of solid CaO and $CaCO_3$? One way to answer this question requires calculating the number of moles of CaO in a liter of CaO from its density (2.71 g/mL) and molar mass (56.08 g/mol):

$$\frac{2.71 \text{ g}}{\text{ml}} \times \frac{1000 \text{ mL}}{\text{L}} \times \frac{1 \text{ mol}}{56.08 \text{ g}} = 48.8 \text{ M}$$

We could do a similar calculation for $CaCO_3$, but there is no point to it, because the concentrations of pure solid substances do not really change during a reaction unless they disappear altogether. Solid substances either are present or they are not. If they are present, then their concentrations are constant. Because the concentration terms of the solids are constant, we can remove them from the right-hand side of the mass action expression and incorporate them into the value of the equilibrium constant on the left-hand side, creating a new equilibrium constant, K'_c:

$$K_c \frac{[CaCO_3]}{[CaO]} = [CO_2]$$

or

$$K'_c = [CO_2]$$

By convention, the new equilibrium constant is the only one that is ever used. Therefore, there is no need to label it K'_c; K_c is perfectly acceptable, and so, the mass action expression for the decomposition of $CaCO_3$ is simply

$$K_c = [CO_2]$$

The convention of ignoring the concentrations of pure solids also applies to pure liquids. Thus, mass action expressions of most reactions in which water is a reactant or product do not include an $[H_2O]$ term. However, the concentrations of solutes dissolved in water (or other solvents) *are included* in mass action expressions, because the values of these concentrations are likely to change as the reactions proceed.

The kilns used to convert limestone into lime operate at about 900°C. At this temperature, the value of K_c for the decomposition of $CaCO_3$ is about 0.011.

Because $K_c = [CO_2]$, the equilibrium concentration of CO_2 is 0.011 M. This concentration may not seem like much, but 0.011 M is actually greater than the concentration of a pure ideal gas at 900°C. If enough heat is available to produce an atmosphere of pure CO_2 from solid limestone, then 900°C is hot enough to shift the equilibrium in Equation 15.19 to the right in favor of the formation of products.

SAMPLE EXERCISE 15.12: Write K_c expressions for the following reactions:

a. $CaO(s) + SO_2(g) \rightleftharpoons CaSO_3(s)$
b. $CO_2(g) + H_2O(l) \rightleftharpoons H_2CO_3(aq)$

SOLUTION: a. If we apply the conventional approach to writing a K_c expression for the reaction, we have

$$K_c = \frac{[CaSO_3]}{[CaO][SO_2]}$$

However, the concentrations of the two solids are constants and so do not appear in the expression, leaving only the term for $[SO_2]$:

$$K_c = \frac{1}{[SO_2]}$$

b. In the equilibrium constant expression for this reaction, $[H_2O]$ is a contant and so is not included, leaving

$$K_c = \frac{[H_2CO_3]}{[CO_2]}$$

PRACTICE EXERCISE: Write K_p expressions for the following reactions:

a. $C(s) + CO_2(g) \rightleftharpoons 2\ CO(g)$
b. $CO_2(g) + H_2(g) \rightleftharpoons CO(g) + H_2O(l)$

(See Problems 89–92.)

CHAPTER REVIEW

Summary

SECTION 15.1

Chemical equilibrium can be approached from either reaction direction and is achieved when the forward and reverse reaction rates are the same. The expression for the equilibrium constant K_c of a chemical reaction is based on concentration terms for the reactants and products, each raised to the corresponding stoichiometric coefficient in the balanced equation.

SECTION 15.2

The reaction quotient Q is the value of the mass action expression for concentrations (or partial pressures) of reactants and products that may or may not be in equilibrium. If $Q < K$, the reaction proceeds in the forward direction. If $Q > K$, it runs in reverse. At equilibrium, $Q = K$. Reversing a reaction inverts its equilibrium constant. If reactions are summed to give an overall reaction, their equilibrium constants are multiplied together to obtain an overall equilibrium constant. Doubling the coefficients of a reaction equation means that the equilibrium constant is squared.

SECTION 15.3

The equilibrium constant K_p for a reaction of gases is written in terms of the equilibrium partial pressures of the gases.

SECTION 15.4

The equilibrium constant is a thermodynamic property of a chemical reaction: the greater the decrease in free energy, the larger the value of the equilibrium constant.

SECTION 15.5

Systems, including chemical reactions, at equilibrium respond to stress by shifting position to relieve the stress. Adding or removing a reactant or product, or applying pressure to a reaction mixture that includes gases creates stress that shifts the position of a chemical equilibrium.

SECTION 15.6

A catalyst decreases the time that it takes a system to achieve equilibrium, but it does not change the value of the equilibrium constant.

SECTION 15.7

Equilibrium constants K_c or K_p can be calculated from known equilibrium concentrations or from partial pressures. Equilibrium concentrations or pressures of reactants and products can be calculated from initial concentrations or pressures, the reaction stoichiometry, and the value of the equilibrium constant.

SECTION 15.8

Higher reaction temperatures increase equilibrium constant values of endothermic reactions but decrease equilibrium constant values of exothermic reactions.

SECTION 15.9

The concentrations of pure liquids and solids do not change during a reaction and so are omitted from equilibrium constant expressions.

Key Terms

equilibrium constant (p. 735)
equilibrium constant expression
  (mass action expression) (p.735)
law of mass action (p. 735)
Le Châtelier's principle (p. 751)
reaction quotient (p. 738)

Key Skills and Concepts

SECTION 15.1

Be able to write the K_c or K_p expression for a balanced chemical reaction.

SECTION 15.2

Be able to calculate the reaction quotient Q and compare it to the value of K to discover whether a reaction is proceeding in the forward direction or in reverse, or is at equilibrium.

Be able to combine K values for the individual steps in an overall reaction to calculate $K_{overall}$.

Be able to calculate the new equilibrium constant for a system when the coefficients of the equilibrium constant expression are multiplied by a constant.

Be able to calculate the equilibrium constants K_c and K_p from known equilibrium concentrations and partial pressures.

SECTION 15.3

Be able to relate K_p and K_c for an equilibrium involving gaseous reactants or products.

SECTION 15.4

Be able to calculate the value of K from the value of $\Delta G°$.

SECTION 15.5

Be able to predict how an equilibrium mixture that includes gaseous components will respond to a change in pressure.

SECTION 15.6

Understand that catalysts affect the rates of reactions but not the final equilibrium state.

SECTION 15.7

Be able to calculate the concentrations or partial pressures of the components of an equilibrium reaction mixture.

SECTION 15.8

Be able to calculate the value of K at one temperature from its value at another temperature and $\Delta H°$.

SECTION 15.9

Be able to write the equilibrium constant expression for a heterogeneous reaction.

Key Equations and Relations

SECTION 15.1

The mass action expression for the reaction $aA + bB \rightleftharpoons cC + dD$ is

$$K_c = \frac{[C]^c[D]^d}{[A]^a[B]^b} \quad (15.1)$$

SECTION 15.3

The relation between K_p and K_c for an equilibrium system involving gases is

$$K_p = K_c(RT)^{\Delta n} \quad (15.13)$$

where Δn is the number of moles of gaseous products minus the number of moles of gaseous reactants in the balanced chemical equation.

SECTION 15.4

The relation between the change in free energy of a reaction and the reaction quotient is

$$\Delta G = \Delta G° + RT \ln Q \quad (15.14)$$

The relation between the standard free-energy change of a reaction and its equilibrium constant is

$$\Delta G° = -RT \ln K \quad (15.15)$$

or

$$K = e^{-\Delta G°/RT} \quad (15.16)$$

SECTION 15.8

The relation between the equilibrium constant of a chemical reaction and the changes in standard enthalpy and entropy is

$$\ln K = -\frac{\Delta H°}{R}\left(\frac{1}{T}\right) + \frac{\Delta S°}{R} \quad (15.17)$$

Calculating the value of an equilibrium constant K_2 at a different temperature T_2 from a known value K_1 at temperature T_1 and the standard enthalpy change.

$$\ln\left(\frac{K_2}{K_1}\right) = -\frac{\Delta H°}{R}\left(\frac{1}{T_2} - \frac{1}{T_1}\right) \quad (15.18)$$

QUESTION AND PROBLEMS

Achieving Equilibrium

CONCEPT REVIEW

1. How are reaction rates related to the state of chemical equilibrium?
2. Describe an example of dynamic equilibrium that you experienced today.
3. Does the following graph of concentration versus time indicate that the reaction A ⟶ 2B reached equilibrium at 20 μs?

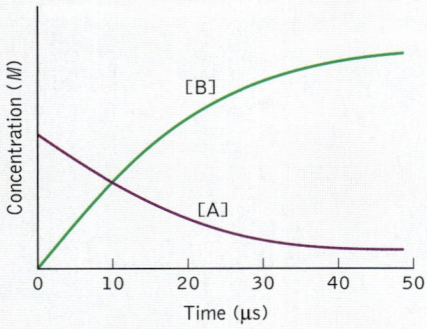

4. At equilibrium, is the sum of the concentrations of all the reactants always equal to the sum of the concentrations of all of the products?

PROBLEMS

5. In a study of the reaction $2\,N_2O \rightleftharpoons 2\,N_2 + O_2$, quantities of all three gases were injected into a reaction vessel. The N_2O consisted entirely of isotopically labeled $^{15}N_2O$. Analysis of the reaction mixture after 1 day revealed the presence of compounds with molar masses 28, 29, 30, 32, 44, and 46 g/mol. Identify the compounds and account for their appearance.
6. A mixture of ^{13}CO, $^{12}CO_2$, and O_2 in a sealed reaction vessel was used to follow the reaction $2\,CO + O_2 \rightleftharpoons 2\,CO_2$. Analysis of the reaction mixture after 1 day revealed the presence of compounds with molar masses 28, 29, 32, 44, and 45 g/mol. Identify the compounds and account for their appearance.

Equilibrium Constants and Reaction Quotients

CONCEPT REVIEW

7. Suppose the forward rate constant of the reaction $A \rightleftharpoons B$ is greater than the rate constant of the reverse reaction at a given temperature. Is the value of the equilibrium constant less than, greater than, or equal to 1?
8. The activation energy of the reaction $2\,NO + O_2 \longrightarrow 2\,NO_2$ is significantly less than the activation energy for the reaction $2\,NO_2 \longrightarrow 2\,NO + O_2$. Is the equilibrium constant for the reaction $2\,NO + O_2 \rightleftharpoons 2\,NO_2$ likely to be less than, greater than, or equal to 1?
9. In which direction will a reaction proceed if $Q < K$?
10. Explain how it is possible for a reaction to have a large equilibrium constant but a small rate constant.

PROBLEMS

11. Suppose the reaction $A \rightleftharpoons B$ in the forward direction is first order in A and the rate constant is $1.50 \times 10^{-2}\,s^{-1}$. The reverse reaction is first order in B and the rate constant is $4.50 \times 10^{-2}\,s^{-1}$ at the same temperature. What is the value of the equilibrium constant for this reaction at this temperature?
12. The equilibrium constant, K_c, for the reaction between NO and O_2 at 700 K is 8.7×10^6. The rate constant for the reverse reaction at this temperature is $0.54\,M^{-1}s^{-1}$. What is the rate constant for the forward reaction at 700 K?
13. Nitrogen oxides play important roles in air pollution. Write mass action expressions for the following reactions involving nitrogen oxides.
    a. $N_2(g) + 2\,O_2(g) \rightleftharpoons N_2O_4(g)$
    b. $3\,NO(g) \rightleftharpoons NO_2(g) + N_2O(g)$
    c. $2\,N_2O(g) \rightleftharpoons 2\,N_2(g) + O_2(g)$
14. Write equilibrium constant expressions for the following reactions, which contribute to the destruction of stratospheric ozone:

a. $Cl(g) + O_3(g) \rightleftharpoons ClO(g) + O_2(g)$
b. $2\ ClO(g) \rightleftharpoons Cl_2(g) + O_2(g)$
c. $2\ O_3(g) \rightleftharpoons 3\ O_2(g)$

15. Use the following graph to estimate the value of the equilibrium constant (K_c) for the reaction $N_2O \rightleftharpoons N_2 + \frac{1}{2} O_2$.

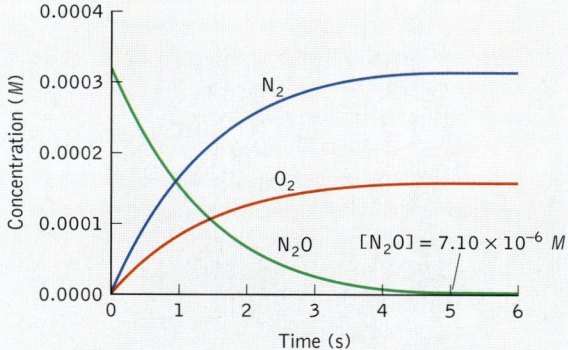

16. Estimate the value the equilibrium constant (K_c) for the reaction $2\ NO + O_2 \rightleftharpoons 2\ NO_2$ from the following graph.

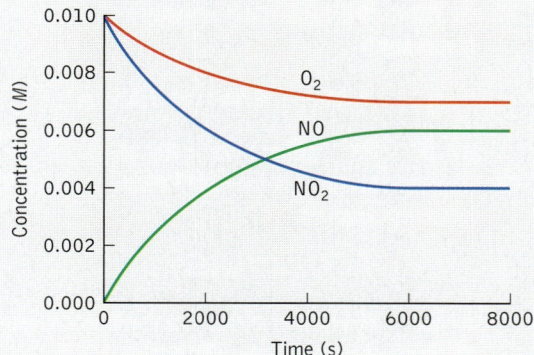

17. At equilibrium, the concentrations of N_2, O_2, and NO in a sealed reaction vessel are $[N_2] = 3.3 \times 10^{-3}$ M, $[O_2] = 5.8 \times 10^{-3}$ M, and $[NO] = 3.1 \times 10^{-3}$ M. What is the value of K_c for the reaction $N_2 + O_2 \rightleftharpoons 2\ NO$ at the temperature of the reaction mixture?

18. Analyses of an equilibrium mixture of N_2O_4 and NO_2 yielded the following results: $[NO_2] = 4.2 \times 10^{-3}$ M and $[N_2O_4] = 2.9 \times 10^{-3}$ M. What is the value of the equilibrium constant K_c for the following reaction at the temperature of the mixture?

$$2\ NO_2(g) \rightleftharpoons N_2O_4(g)$$

19. A sealed reaction vessel initially contains 1.50×10^{-2} moles of water vapor and 1.50×10^{-2} moles of CO. After the following reaction

$$H_2O(g) + CO(g) \rightleftharpoons H_2(g) + CO_2(g)$$

has come to equilibrium, the vessel contains 8.3×10^{-3} moles of CO_2. What is the value of the equilibrium constant, K_c, of the reaction at the temperature of the vessel?

*20. A 100-mL reaction vessel initially contains 2.60×10^{-2} moles of NO and 1.30×10^{-2} moles of H_2. At equilibrium, the concentration of NO in the vessel is 0.161 M. At equilibrium the vessel also contains N_2, H_2O, and H_2. What is the value of the equilibrium constant, K_c, for the following reaction?

$$2\ H_2(g) + 2\ NO(g) \rightleftharpoons 2\ H_2O(g) + N_2(g)$$

21. If the equilibrium constant (K_c) for the reaction $A \rightleftharpoons B$ is 22 at a given temperature and if $[A] = 0.10$ M and $[B] = 2.0$ M in a reaction mixture at that temperature, is the reaction at chemical equilibrium? If not, in which direction will the reaction proceed to reach equilibrium?

22. The equilibrium constant (K_c) for the reaction $2C \rightleftharpoons D + E$ is 3×10^{-3}. At a particular time, the composition of a reaction mixture is $[C] = [D] = [E] = 5 \times 10^{-4}$ M. In which direction will the reaction proceed to reach equilibrium?

23. The following reaction is one of the elementary steps in the oxidation of NO:

$$NO(g) + NO_3(g) \rightleftharpoons 2\ NO_2(g)$$

Write an expression for equilibrium constant, K_c, of this reaction and for the reverse reaction:

$$2\ NO_2(g) \rightleftharpoons NO(g) + NO_3(g)$$

How are the two K_c expressions related?

24. The value of the equilibrium constant (K_p) for the formation of ammonia,

$$N_2(g) + 3\ H_2(g) \rightleftharpoons 2\ NH_3(g)$$

is 4.5×10^{-5} at 450°C. What is the value of K_p for the following reaction?

$$2\ NH_3(g) \rightleftharpoons N_2(g) + 3\ H_2(g)$$

25. Calculate the value of the equilibrium constant (K_c) for the reaction

$$2\ D \rightleftharpoons A + 2\ B$$

from the following information:

$A + 2\ B \rightleftharpoons C$ $K_c = 3.3$
$C \rightleftharpoons 2\ D$ $K_c = 0.041$

26. Calculate the value of the equilibrium constant (K_c) for the reaction

$$Q + X \rightleftharpoons M$$

from the following information:

$2\,M \rightleftharpoons Z$ $\quad K_c = 6.2 \times 10^{-4}$
$Z \rightleftharpoons 2\,Q + 2\,X$ $\quad K_c = 5.6 \times 10^{-2}$

27. The reaction between sulfur dioxide and oxygen can be written in two ways:

$$SO_2(g) + \tfrac{1}{2}O_2(g) \rightleftharpoons SO_3(g)$$

and

$$2\,SO_2(g) + O_2(g) \rightleftharpoons 2\,SO_3(g)$$

Write expressions for the equilibrium constants for both reactions. How are they related?

28. At a given temperature, the equilibrium constant, K_c, for the reaction

$$2\,NO(g) + 2\,H_2(g) \rightleftharpoons N_2(g) + 2\,H_2O(g)$$

is 0.11. What is the equilibrium constant for the following reaction?

$$NO(g) + H_2(g) \rightleftharpoons \tfrac{1}{2}N_2(g) + H_2O(g)$$

29. At a given temperature, the equilibrium constant (K_c) for the reaction

$$2\,SO_2(g) + O_2(g) \rightleftharpoons 2\,SO_3(g)$$

is 2.4×10^{-3}. What is the equilibrium constant for the following reactions at that temperature?
a. $SO_2(g) + \tfrac{1}{2}O_2(g) \rightleftharpoons SO_3(g)$
b. $2\,SO_3(g) \rightleftharpoons 2\,SO_2(g) + O_2(g)$
c. $SO_3(g) \rightleftharpoons SO_2(g) + \tfrac{1}{2}O_2(g)$

30. If the equilibrium constant (K_c) for the reaction $2\,NO + O_2 \rightleftharpoons 2\,NO_2$ is 5×10^{12}, what is the value of K_c for each of the following reactions?
a. $NO + \tfrac{1}{2}O_2 \rightleftharpoons NO_2$
b. $2\,NO_2 \rightleftharpoons 2\,NO + O_2$
c. $NO_2 \rightleftharpoons NO + \tfrac{1}{2}O_2$

K_c and K_p

CONCEPT REVIEW

31. Under what conditions are the values of K_c and K_p equal?
32. At 298 K, is K_p greater than or less than K_c if there is a net increase in the number of moles of gas in a reaction and if $K_p > 1$.

PROBLEMS

33. For which of the following reactions will the values of K_c and K_p be equal?
a. $2\,SO_2(g) + O_2(g) \rightleftharpoons 2\,SO_3(g)$
b. $Fe(s) + CO_2(g) \rightleftharpoons FeO(s) + CO(g)$
c. $H_2O(g) + CO(g) \rightleftharpoons H_2(g) + CO_2(g)$

34. For which of the following reactions will the values of K_c and K_p be different?
a. $2\,NO_2(g) \rightleftharpoons N_2O_4(g)$
b. $2\,NO(g) + O_2(g) \rightleftharpoons 2\,NO_2(g)$
c. $2\,O_3(g) \rightleftharpoons 3\,O_2(g)$

35. Phosgene ($COCl_2$) is used in the manufacture of foam rubber and bulletproof glass. It is formed from carbon monoxide and chlorine in the following reaction:

$$Cl_2(g) + CO(g) \rightleftharpoons COCl_2(g)$$

The value of K_c for the reaction is 5.0 at 325°C. What is the value of K_p at 325°C?

36. At 500°C, the equilibrium constant K_p for the synthesis of ammonia

$$N_2(g) + 3\,H_2(g) \rightleftharpoons 2\,NH_3(g)$$

is 1.45×10^{-5}. What is the value of K_c?

37. Suppose the value of the equilibrium constant K_p of the following reaction

$$A + B \rightleftharpoons C$$

is 1.00 at 300 K. Are either of the following reaction mixtures at chemical equilibrium at 300 K?
a. $P_A = P_B = P_C = 1.0$ atm
b. $[A] = [B] = [C] = 1.0$ M

38. In which direction will the following reaction proceed to reach equilibrium under the conditions given?

$$A + B \rightleftharpoons C \qquad K_p = 1.00 \text{ at } 300 \text{ K}$$

a. $P_A = P_C = 1.0$ atm; $P_B = 0.50$ atm
b. $[A] = [B] = [C] = 1.0$ M

39. If the value of the equilibrium constant, K_c, for the following reaction is 5×10^5 at 298 K, what is the value of K_p at 298 K?

$$2\,CO(g) + O_2(g) \rightleftharpoons 2\,CO_2(g)$$

40. If the value of K_p for the following reaction

$$SO_2(g) + NO_2(g) \rightleftharpoons NO(g) + SO_3(g)$$

is 3.45 at 298 K, what is the value of K_c for the reverse reaction?

41. If the equilibrium constant K_c for the reaction

$$N_2(g) + O_2(g) \rightleftharpoons 2\,NO(g)$$

is 1.5×10^{-3}, in which direction will the reaction proceed if the partial pressures of the three gases are all 1.00×10^{-3} atm?

42. At 650 K, the value of the equilibrium constant K_p for the ammonia synthesis reaction

$$N_2(g) + 3\,H_2(g) \rightleftharpoons 2\,NH_3(g)$$

is 4.3×10^{-4}. If a vessel contains a reaction mixture in which $[N_2] = 0.010\ M$, $[H_2] = 0.030\ M$, and $[NH_3] = 0.00020\ M$, will more ammonia form?

K, Q, and ΔG

CONCEPT REVIEW

43. Do all reactions with equilibrium constants < 1 have values of $\Delta G° > 0$?

*44. The equation $\Delta G° = RT \ln K$ relates the value of K_p, not K_c, to the change in standard free energy for a reaction in the gas phase. Explain why.

PROBLEMS

45. The standard free energies of formation for the three interhalogen compounds containing Cl are $\Delta G°_{f,\,ClF} = -57.7$ kJ, $\Delta G°_{f,\,ClBr} = -1.0$ kJ, $\Delta G°_{f,\,ICl} = -13.95$ kJ. Which of the following reactions will have the largest equilibrium constant?

$$Cl_2(g) + F_2(g) \rightleftharpoons 2\,ClF(g)$$
$$Cl_2(g) + Br_2(g) \rightleftharpoons 2\,ClBr(g)$$
$$Cl_2(g) + I_2(g) \rightleftharpoons 2\,ICl(g)$$

46. Three of the steps in glycolysis (described in Chapter 13) are shown here along with the corresponding change in standard free energy. Which reaction has the largest equilibrium constant?

fructose 1,6-diphosphate $\rightleftharpoons$ 2 glyceraldehyde 3-phosphate $\Delta G° = 24$ kJ
3-phosphoglycerate $\rightleftharpoons$ 2-phosphoglycerate $\Delta G° = 4.4$ kJ
2-phosphoglycerate $\rightleftharpoons$ phosphoenolpyruvate $\Delta G° = 1.8$ kJ

47. The value of $\Delta G°$ for the phosphorylation of glucose in glycolysis is $+13.8$ kJ/mol. What is the equilibrium constant for the reaction?

48. In glycolysis, the hydrolysis of ATP to ADP is used to drive the phosphorylation of glucose (see Chapter 13):

glucose + ATP $\rightleftharpoons$ ADP + glucose 6-phosphate $\Delta G° = -16.7$ kJ

What is K_c for the reaction at 298 K?

49. The value of the equilibrium constant K_p for the reaction:

$$H_2(g) + CO_2(g) \rightleftharpoons H_2O(g) + CO(g)$$

is 0.534 at 700°C.
a. Calculate the value of $\Delta G°$ for the reaction using Equation 15.15.
b. Using $\Delta G°_f$ values from Appendix 4, calculate $\Delta G°$ for the reaction and compare the result with that obtained in part a.

50. Sucrose enters the series of reactions in glycolysis after its hydrolysis into glucose and fructose:

sucrose + H_2O $\rightleftharpoons$ glucose + fructose
$K_c = 5.3 \times 10^{12}$ at 298 K

What is the value of $\Delta G°$ for this process?

51. Use the following data to calculate the value of K_p at 298 K for the reaction $N_2(g) + 2\,O_2(g) \rightleftharpoons 2\,NO_2(g)$:

$N_2(g) + O_2(g) \rightleftharpoons 2\,NO(g)$ $\Delta G° = 173.2$ kJ
$2\,NO(g) + O_2(g) \rightleftharpoons 2\,NO_2(g)$ $\Delta G° = -69.7$ kJ

52. Under the appropriate conditions, NO forms N_2O and NO_2:

$$3\,NO(g) \rightleftharpoons N_2O(g) + NO_2(g)$$

Use the values for $\Delta G°$ for the following reactions to calculate the value of K_p for the above reaction at 500°C.

$2\,NO(g) + O_2(g) \longrightarrow 2\,NO_2(g)$ $\Delta G° = -69.7$ kJ
$2\,N_2O(g) \longrightarrow 2\,NO(g) + N_2(g)$ $\Delta G° = -33.8$ kJ

Le Châtelier's Principle

CONCEPT REVIEW

53. Does adding reactants to a system at equilibrium increase the value of the equilibrium constant?
54. Increasing the concentration of a reactant shifts the position of chemical equilibrium toward the formation of more products. What effect does adding a reactant have on the rates of the forward and reverse reactions?
55. Patients suffering from carbon monoxide poisoning are treated with pure oxygen to remove CO from the hemoglobin (Hb) in their blood. The two relevant equilibria are

$$Hb + 4\ CO(g) \rightleftharpoons Hb(CO)_4$$
$$Hb + 4\ O_2(g) \rightleftharpoons Hb(O_2)_4$$

The value of the equilibrium constant for CO binding to Hb is greater than that for O_2. How, then, does this treatment work?

56. Is the equilibrium constant (K_p) for the reaction $2\ NO_2 \rightleftharpoons N_2O_4$ in air the same in Los Angeles as in Denver, if the atmospheric pressure in Denver is lower but the temperature is the same?
57. Henry's law (Chapter 8) predicts that the solubility of a gas increases with its partial pressure. Explain Henry's law in relation to LeChâtelier's principle.
58. Why does adding an inert gas such as argon to an equilibrium mixture of CO, O_2, and CO_2 in a sealed vessel increase the total pressure of the system but not affect the position of the equilibrium $2\ CO + O_2 \rightleftharpoons 2\ CO_2$?

PROBLEMS

59. Which of the following equilibria will shift toward the formation of more products if an equilibrium mixture is compressed into half its volume?
    a. $2\ N_2O \rightleftharpoons 2\ N_2 + O_2(g)$
    b. $2\ CO + O_2 \rightleftharpoons 2\ CO_2(g)$
    c. $N_2 + O_2 \rightleftharpoons 2\ NO(g)$
    d. $2\ NO + O_2 \rightleftharpoons 2\ NO_2(g)$

60. Which of the following equilibria will shift toward the formation of more products if the volume of a reaction mixture at equilibrium expands by a factor of 2?
    a. $2\ SO_2 + O_2 \rightleftharpoons 2\ SO_3(g)$
    b. $NO + O_3 \rightleftharpoons NO_2 + O_2(g)$
    c. $2\ N_2O_5 \rightleftharpoons 2\ NO_2 + O_2(g)$
    d. $N_2O_4 \rightleftharpoons 2\ NO_2(g)$

61. What would be the effect of the following changes on the equilibrium concentrations of reactants and products in the reaction

$$2\ O_3(g) \rightleftharpoons 3\ O_2(g)$$

    a. O_3 is added to the system.
    b. O_2 is added to the system.
    c. The mixture is compressed to one tenth its initial volume.

62. How will the following changes affect the position of the equilibrium:

$$2\ NO_2(g) \rightleftharpoons NO(g) + NO_3(g)$$

    a. The concentration of NO is increased.
    b. The concentration of NO_2 is increased.
    c. The volume of the system is allowed to expand to 5 times its initial value.

63. In the reaction $2\ SO_2 + O_2 \rightleftharpoons 2\ SO_3$, how would reducing the partial pressure of O_2 affect the position of the equilibrium?

*64. Ammonia is added to a reaction mixture containing H_2, Cl_2, and HCl that is at chemical equilibrium. How will the addition of ammonia affect the relative concentrations of H_2, Cl_2, and HCl if the equilibrium constant of reaction 2 is much greater than the equilibrium constant of reaction 1?

    (1) $H_2(g) + Cl_2 \rightleftharpoons 2\ HCl(g)$
    (2) $HCl(g) + NH_3(g) \rightleftharpoons NH_4Cl(s)$

Calculations Based on K

PROBLEMS

65. With regard to the following reaction,

$$PCl_5(g) \rightleftharpoons PCl_3(g) + Cl_2(g) \qquad K_p = 23.6 \text{ at } 500\ K$$

    a. calculate the equilibrium partial pressures of the reactants and products if the initial pressures are: $P_{PCl_5} = 0.560$ atm and $P_{PCl_3} = P_{Cl_2} = 0$ atm.
    b. if more chlorine is added after equilibrium is reached, how will the concentrations of PCl_5 and PCl_3 change?

66. Enough NO_2 is injected into a cylindrical vessel to produce a partial pressure, P_{NO_2}, of 0.900 atm. Calculate the equilibrium partial pressures of NO_2 and N_2O_4, given:

$$2\,NO_2(g) \rightleftharpoons N_2O_4(g) \quad\quad K_p = 4$$

67. The value of K_c for the reaction between water vapor and dichlorine monoxide:

$$H_2O(g) + Cl_2O(g) \rightleftharpoons 2\,HOCl(g)$$

is 0.0900 at 25°C. Determine the equilibrium concentrations of all three compounds if the starting concentrations of both reactants are 0.00432 M.

68. The value of K_p for the reaction:

$$3\,H_2(g) + N_2(g) \rightleftharpoons 2\,NH_3(g)$$

is 4.3×10^{-4} at 648 K. Determine the equilibrium partial pressure of NH_3 in a reaction vessel that initially contained 0.900 atm N_2 and 0.500 atm H_2.

69. The value of K_p for the reaction $NO + \frac{1}{2}O_2 \longrightarrow NO_2$ is 2×10^6 at 25°C. At equilibrium, what is the ratio of P_{NO_2} to P_{NO} in air at 25°C? Assume that $P_{O_2} = 0.21$ atm and does not change.

70. During volcanic eruptions, gases as hot as 700°C and rich in SO_2 are released into the atmosphere. As air mixes with these gases, the following reaction converts some of this SO_2 into SO_3:

$$2\,SO_2(g) + O_2(g) \rightleftharpoons 2\,SO_3(g)$$

If the value of K_p for this reaction is 4.5 at 700°C, what is the ratio of P_{SO_3} to P_{SO_2} in equilibrium with $P_{O_2} = 0.21$ atm?

71. A flask containing pure NO_2 was heated to 1000 K, a temperature at which the value of K_p for the decomposition of NO_2

$$2\,NO_2(g) \rightleftharpoons 2\,NO(g) + O_2(g)$$

is 158. The partial pressure of O_2 at equilibrium is 0.136 atm.
a. Calculate the partial pressures of NO and NO_2.
b. Calculate the total pressure in the flask at equilibrium.

72. The equilibrium constant K_p for the reaction $2\,SO_3 \rightleftharpoons 2\,SO_2 + O_2$ is 7.69 at 830°C. If a vessel at this temperature initially contains pure SO_3, what is the partial pressure of O_2 in the flask at equilibrium if the partial pressure of SO_3 at equilibrium is 0.100 atm?

73. In a study of the formation of NO_x air pollution, a chamber heated to 2200°C was filled with air (0.79 atm N_2; 0.21 atm O_2). What are the equilibrium partial pressures of N_2, O_2, and NO if $K_p = 0.050$ for the following reaction?

$$N_2(g) + O_2(g) \rightleftharpoons 2\,NO(g)$$

74. The equilibrium constant K_p for the thermal decomposition of NO_2

$$2\,NO_2(g) \rightleftharpoons 2\,NO(g) + O_2(g)$$

is 6.5×10^{-6} at 450°C. If a reaction vessel at this temperature initially contains 0.500 atm NO_2, what will be the partial pressures of NO_2, NO, and O_2 in the vessel when equilibrium has been reached?

75. The value of K_c for the thermal decomposition of hydrogen sulfide

$$2\,H_2S(g) \rightleftharpoons 2\,H_2(g) + S_2(g)$$

is 2.2×10^{-4} at 1400 K. A sample of gas in which $[H_2S] = 0.600$ M is heated to 1400 K in a sealed vessel. After chemical equilibrium has been achieved, what is the value of $[H_2S]$? Assume no H_2 or S_2 was present in the original sample.

76. On a particularly smoggy day, the concentration of NO_2 in the air over an urban area reaches 2.2×10^{-7} M. If the temperature of the air is 25°C, what is the concentration of the NO_2 dimer N_2O_4 in the air? Given:

$$N_2O_4(g) \rightleftharpoons 2\,NO_2(g) \quad\quad K_c = 6.1 \times 10^{-3}$$

77. Phosgene ($COCl_2$) is produced by the reaction of carbon monoxide and chlorine:

$$CO(g) + Cl_2(g) \rightleftharpoons COCl_2(g)$$

The value of K_c for this reaction is 5.0 at 600 K. What are the equilibrium partial pressures of the three gases if a reaction vessel initially contains a mixture of the reactants in which $P_{CO} = P_{Cl_2} = 0.265$ atm and $P_{COCl_2} = 0.000$ atm?

78. At 2000°C, the value of K_c for the reaction $2\,CO + O_2 \rightleftharpoons CO_2$ is 1.0. What is the ratio of [CO] to $[CO_2]$ at 2000°C in an atmosphere in which $[O_2] = 0.0045$ M?

*79. The water-gas shift reaction is an important source of hydrogen. The value of K_c for the reaction

$$CO(g) + H_2O(g) \rightleftharpoons CO_2(g) + H_2(g)$$

at 700 K is 5.1. Calculate the equilibrium concentrations of the four gases if the initial concentration of each of them is 0.050 M.

*80. Sulfur dioxide reacts with NO_2, forming SO_3 and NO:

$$SO_2(g) + NO_2(g) \rightleftharpoons SO_3(g) + NO(g)$$

If the value of K_c for the reaction is 2.50, what are the equilibrium concentrations of the products if the reaction mixture was initially 0.50 M SO_2, 0.50 M NO_2, 0.0050 M SO_3 and 0.0050 M NO?

Changing K with Changing Temperature

CONCEPT REVIEW

81. The value of the equilibrium constant for a reaction decreases with increasing temperature. Is this reaction endothermic or exothermic?
82. The reaction $2\ CO + O_2 \rightleftharpoons 2\ CO_2$ is exothermic. Does the value of K_p increase or decrease with increasing temperature?
83. The value of K_p for the water-gas shift reaction (see Problem 79)

$$CO(g) + H_2O(g) \rightleftharpoons H_2(g) + CO_2(g)$$

increases as the temperature decreases. Is the reaction exothermic or endothermic?
84. Does the value of K_p for the following reaction

$$CH_4(g) + H_2O(g) \rightleftharpoons 3\ H_2(g) + CO(g) \qquad \Delta H° = 206\ kJ$$

increase, decrease, or remain unchanged as temperature increases?

PROBLEMS

85. At 2000°C, the value of K_c for the reaction

$$N_2(g) + O_2(g) \rightleftharpoons 2\ NO(g) \qquad \Delta H° = 180.6\ kJ$$

is 4.10×10^{-4}. What is the value of K_c at 25°C?
86. The value of K_p for the reaction

$$N_2(g) + 3\ H_2(g) \rightleftharpoons 2\ NH_3(g) \qquad \Delta H° = -92.2\ kJ$$

is 41 at 400 K. What is the value of K_p at 700 K?
87. The equilibrium constant for the reaction

$$2\ NO(g) + O_2(g) \rightleftharpoons 2\ NO_2(g)$$

decreases from 1.5×10^5 at 430°C to 23 at 1000°C. From these data, calculate the value of $\Delta H°$ for the reaction.
88. The value of K_c for the reaction $A \rightleftharpoons B$ is 0.455 at 50°C and 0.655 at 100°C. Calculate $\Delta H°$ for the reaction.

Heterogeneous Equilibria

CONCEPT REVIEW

89. Write the equilibrium constant (K_p) expression for the following reaction:

$$N_2(g) + 3\ H_2(g) \rightleftharpoons 2\ NH_3(l)$$

90. Write the equilibrium constant (K_p) expression for the following reaction, which is linked to the formation of acid rain:

$$SO_3(g) + H_2O(g) \rightleftharpoons H_2SO_4(l)$$

91. Why does the K_c expression for the reaction

$$CaCO_3(s) \rightleftharpoons CaO(s) + CO_2(g)$$

not contain terms for the concentrations of $CaCO_3$ and CaO?
92. Use the equations $\Delta G° = \Delta H° - T\Delta S°$ and $\Delta G° = -RT \ln K$ to explain why the value of K_p for the reaction

$$CaCO_3(s) \rightleftharpoons CaO(s) + CO_2(g)$$

increases with increasing temperature.

PROBLEMS

93. Passing steam over hot carbon produces a mixture of carbon monoxide and hydrogen:

$$H_2O(g) + C(s) \rightleftharpoons CO(g) + H_2(g)$$

The value of K_c for the reaction at 1000°C is 3.0×10^{-2}.
a. Calculate the equilibrium partial pressures of the products and reactants if $P_{H_2O} = 0.442$ atm and $P_{CO} = P_{H_2} = 0$ atm at the start of the reaction. Assume that the carbon is in excess.
b. Determine the equilibrium partial pressures of the reactants and products after CO and H_2 at 0.075 atm are added to the equilibrium mixture in part a.

94. The value of K_p for the reaction $CO_2(g) + C(s) \rightleftharpoons 2\ CO(g)$ is 1.5 at 700°C. Calculate the equilibrium partial pressure of CO and CO_2 if initially $P_{CO_2} = 0.368$ atm and $P_{CO} = 0$ atm. Pure graphite is also present initially and when equilibrium is achieved.

95. Ammonium hydrogen sulfide (NH_4SH) was detected in the atmosphere of Jupiter subsequent to its collision with the comet Shoemaker-Levy. The equilibrium between ammonia, hydrogen sulfide, and NH_4SH is described by the following equation:

$$NH_4SH(s) \rightleftharpoons NH_3(g) + H_2S(g)$$

The value of K_p for the reaction at 24°C is 0.126. Suppose a sealed flask contains an equilibrium mixture of NH_4SH, NH_3, and H_2S. At equilibrium, the partial pressure of H_2S is 0.355 atm, what is the partial pressure of NH_3?

96. Calcium oxide is used to remove SO_2 from exhaust gases. The $\Delta G°$ of the overall reaction

$$CaO(s) + SO_2(g) + \tfrac{1}{2}O_2(g) \rightleftharpoons CaSO_4(s)$$

is -417.6 kJ. What is P_{SO_2} in equilibrium with air ($P_{O_2} = 0.21$ atm) and solid CaO?

16 Equilibrium in the Aqueous Phase

And Acid Rain

Acid precipitation can alter the chemical balance of soils. One result is the death of trees, such as these spruce in the Green Mountains of Vermont, and other vegetation.

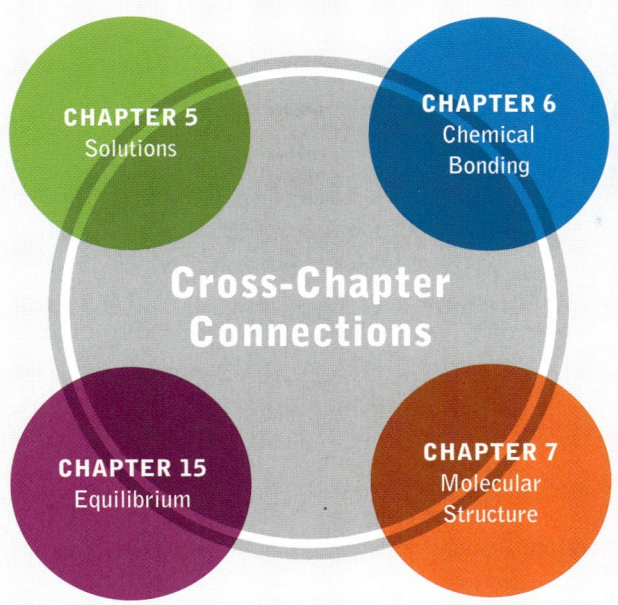

Weak and strong acids
Diprotic acids

16.2 **Acids and bases: a molecular view**
Acids in water
Bases in water
Lewis acids and bases
Conjugate pairs

16.3 **Acid strength and molecular structure**

16.4 **The concept of pH**
The pH scale
The pH of "natural" and acid rain
pOH

16.5 **The pH of solutions of acidic and basic salts**

16.6 **Buffer solutions and the pH of natural waters**

16.7 **Acid–base indicators**

16.8 **Acid–base titrations**

16.9 **Solubilities of minerals and other compounds**

16.10 **Complex ions**
Complexation and solubility
Metal complexes in biomolecules

A Look Ahead

In this chapter, the concepts of chemical equilibrium introduced in Chapter 15 are applied to reactions in solution. We begin with the nitrogen and sulfur oxides that contribute to atmospheric pollution. When these compounds dissolve in condensing water vapor, the weak and strong acids that they form fall to Earth as acid rain. The acidities of rain and of other solutions are calculated from their concentrations and acid-ionization equilibrium constants. Reactions between acidic precipitation and basic minerals initiate a detailed examination of neutralization reactions, acid–base titrations, and how pH affects the solubilities of some ionic compounds.

16.1 ACID RAIN AND ACID STRENGTH

Preceding chapters described how pollutants released in combustion processes can increase atmospheric concentrations of nitrogen and sulfur oxides, including NO_2 and SO_3. When these oxides of nitrogen and sulfur dissolve in drops of rain, they form acid rain, as shown in Figure 16.1. For example, when NO_2 dissolves in water, it may form nitric acid, HNO_3, and nitrous acid, HNO_2:

$$2\ NO_2(g) + H_2O(l) \longrightarrow HNO_3(aq) + HNO_2(aq) \tag{16.1}$$

CONNECTION: See Section 5.6 for a review of the acidic and basic properties of the nonmetal and metal oxides.

CONCEPT TEST: In addition to forming two acids, the reaction in Equation 16.1 is an oxidation–reduction reaction. Which element is oxidized and which is reduced? What are the changes in oxidation numbers? (Hint: The method for determining the oxidation states of elements in compounds is described in Section 5.4.)

Weak and strong acids

Nitric acid is a strong acid, which means that it ionizes completely in water (see Section 5.6), forming H^+ and NO_3^-:

$$HNO_3(aq) \longrightarrow H^+(aq) + NO_3^-(aq) \tag{16.2}$$

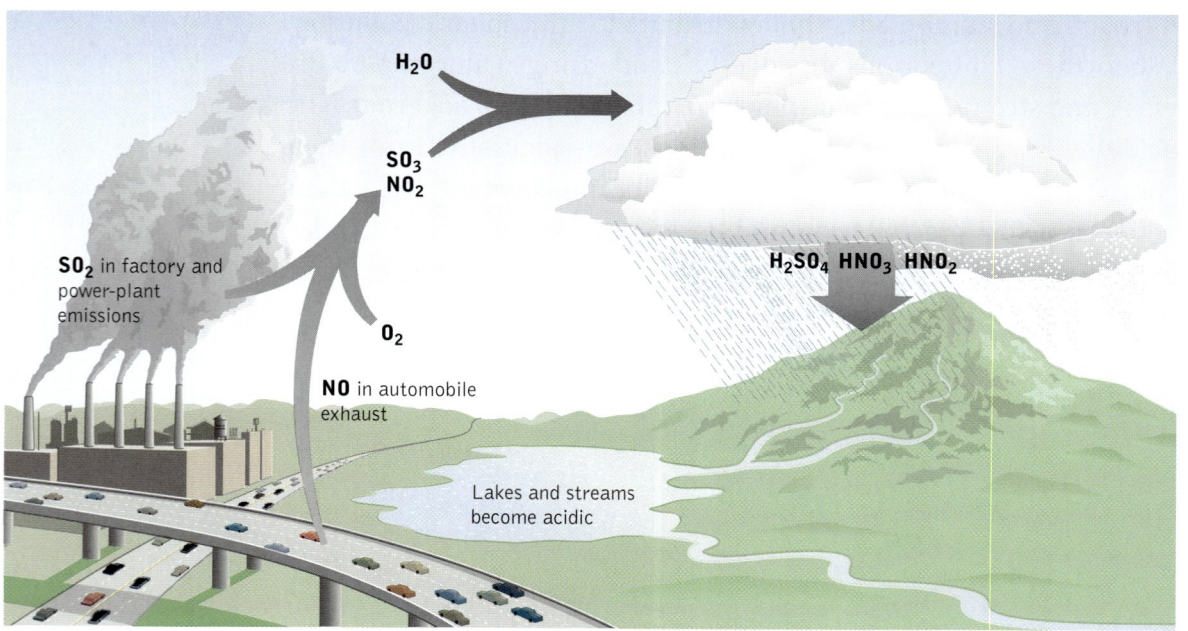

FIGURE 16.1 Acid rain forms when volatile, nonmetal oxides such as NO_2 and SO_3 dissolve in rain, forming nitric (HNO_3) and sulfuric (H_2SO_4) acids.

However, nitrous acid is a weak acid. It is only partly ionized and reaches a state of equilibrium in which the rate at which HNO_2 ionizes is matched by the rate at which H^+ and NO_2^- combine to form molecules of HNO_2. The equilibrium state is described by the value of its equilibrium constant (K_a) at 25°C:

$$HNO_2(aq) \rightleftharpoons H^+(aq) + NO_2^-(aq) \qquad K_a = 4.0 \times 10^{-4} \qquad (16.3)$$

We use the subscript "a" to indicate that the equilibrium constant K is an **a**cid-ionization equilibrium constant. The equilibrium constant expression is

$$K_a = \frac{[NO_2^-][H^+]}{[HNO_2]} = 4.0 \times 10^{-4}$$

The relatively small value of K_a tells us that at equilibrium $[HNO_2]$ will usually be greater than $[H^+]$ and $[NO_2^-]$. In other words, the equilibrium lies to the left, favoring the formation of the reactant. We have the opposite situation with a strong acid, such as HNO_3, for which the ionization equilibrium (Equation 16.2) lies far to the right. For example, in a 0.100 M solution of nitric acid, HNO_3 is completely ionized, and so $[H^+] = [NO_3^-] = 0.100\ M$ and $[HNO_3]$ is essentially zero. Figure 16.2 provides graphical and molecular views of the different degrees of ionization of weak and strong acids.

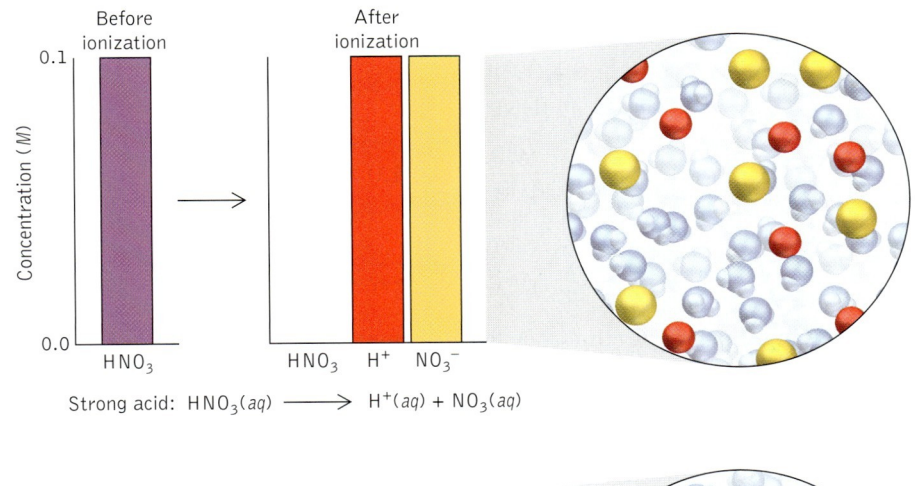

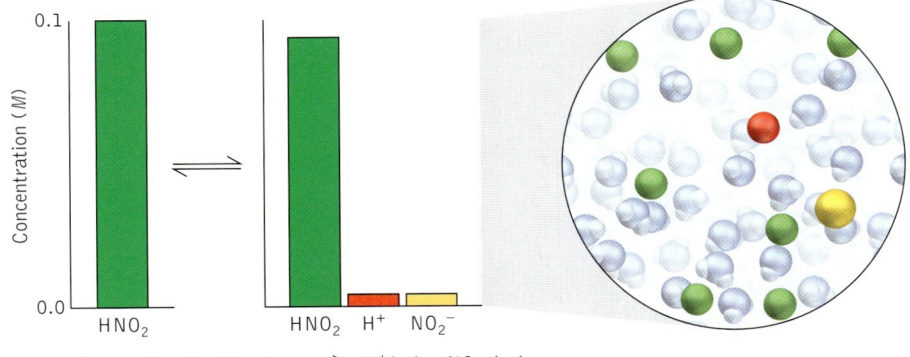

FIGURE 16.2 These graphs and molecular views illustrate the differences in the degree of ionization of 0.100 M solutions of a strong (HNO_3) and a weak (HNO_2) acid.

To obtain a quantitative view of the degree to which HNO_2 ionizes in aqueous solution, let's calculate the value of $[H^+]$ in 0.100 M HNO_2. We will use the mathematical approach developed in Chapter 15 for calculating concentrations of reactants and products at equilibrium. Recall that we use ICE tables to derive terms for the concentrations of reactants and products initially (I) and after equilibrium (E) has been achieved; the difference between the two values represents the change (C) that takes place as the reaction proceeds (see Section 15.7).

To set up an ICE table for Equation 16.3, we need to make an important assumption: the ionization of HNO_2 is the only significant source of H^+ ions in the solution. Therefore, the initial concentration of H^+ will be tiny (essentially zero) compared with $[H^+]$ at equilibrium. It turns out that this assumption is valid. If we let x be $[H^+]$ at equilibrium, then the change in $[H^+]$ will be $+x$. The change in $[NO_2^-]$ also will be $+x$ and the change in $[HNO_2]$ will be $-x$. Using these values to complete the ICE table, we have

	$[HNO_2]$	$[H^+]$	$[NO_2^-]$
Initial	0.100	0.000	0.000
Change	$-x$	$+x$	$+x$
Equilibrium	$0.100 - x$	x	x

Inserting the equilibrium concentration terms in the equilibrium constant expression for the ionization of HNO_2, we have

$$K_a = \frac{[NO_2^-][H^+]}{[HNO_2]}$$

$$= \frac{(x)(x)}{(0.100 - x)} = 4.0 \times 10^{-4}$$

Cross multiplying gives

$$4.0 \times 10^{-5} - 4.0 \times 10^{-4}x = x^2$$

Rearranging the terms so that they fit the format of a quadratic equation ($ax^2 + bx + c = 0$)

$$x^2 + 4.0 \times 10^{-4}x - 4.0 \times 10^{-5} = 0$$

and solving the quadratic, we have

$$x = \frac{-b \pm \sqrt{b^2 - 4ac}}{2a} = \frac{-4.0 \times 10^{-4} \pm \sqrt{(4.0 \times 10^{-4})^2 - 4(1)(-4.0 \times 10^{-5})}}{2(1)}$$

$$= \frac{-4.0 \times 10^{-4} \pm \sqrt{1.60 \times 10^{-7} + 1.60 \times 10^{-4}}}{2(1)}$$

$$= \frac{-4.0 \times 10^{-4} \pm 1.26 \times 10^{-2}}{2(1)} = -6.5 \times 10^{-3}, \text{ or } +6.1 \times 10^{-3}$$

The negative value of x has no meaning (there is no such thing as a negative concentration); so, at equilibrium,

$$[H^+] = 6.1 \times 10^{-3} M$$

To put this result in perspective, the value of [H⁺] in a solution of a strong, completely ionized acid, such as nitric acid, is usually the same as the concentration of the acid itself. Thus, [H⁺] in a 0.100 M solution of HNO₃ is 0.100 M. Taking the ratio of these two [H⁺] values,

$$\frac{0.100\ M}{6.1 \times 10^{-3}\ M} = 16$$

we find that [H⁺] in HNO₃ is 16 times as great as in HNO₂. If we divide the initial concentration of HNO₂ (0.100 M) into the part of it that ionized (6.1×10^{-3}), we have

$$\frac{6.1 \times 10^{-3}\ M}{0.100\ M} = 0.061 = 6.1\%$$

This value represents the **degree of ionization** (sometimes called the **degree of dissociation**) of HNO₂ in a 0.100 M solution of the acid. As you might expect, the degree of ionization of an acid depends on the value of its K_a: the larger the value of K_a, the larger the degree of ionization. The degree of ionization also depends on the concentration of a weak acid, as we will discover later in this section.

> **Degree of ionization** is the ratio of the quantity of a substance that is ionized to the total quantity of the substance present. The degree of ionization is also called the **degree of dissociation**.

SAMPLE EXERCISE 16.1: Calculate the value of [H⁺] of 0.100 M acetic acid, which is the principal ingredient in vinegar, and has the condensed structural formula:

$$\text{H}_3\text{C}-\overset{\overset{\displaystyle O}{\|}}{\text{C}}-\text{O}-\text{H}$$

Given:

$$\text{HC}_2\text{H}_3\text{O}_2(aq) \rightleftharpoons \text{C}_2\text{H}_3\text{O}_2^{-}(aq) + \text{H}^+(aq) \qquad K_a = 1.76 \times 10^{-5}$$

SOLUTION: The method used to calculate the value of [H⁺] of a solution of nitrous acid can be used for a solution of any weak acid. Let's set up an ICE table for the reaction. Initially, [HC₂H₃O₂] = 0.100 M, and [C₂H₃O₂⁻] = 0.000 M. We assume that the initial [H⁺] is insignificant (essentially zero) compared with the equilibrium concentration, which we let be x. Given the 1:1 stoichiometry between H⁺ and C₂H₃O₂⁻, the changes in the concentrations of both H⁺ and C₂H₃O₂⁻ are $+x$, and the change in [HC₂H₃O₂] is $-x$. The resulting ICE table is

	[HC₂H₃O₂]	[H⁺]	[C₂H₃O₂⁻]
Initial	0.100	0.000	0.000
Change	$-x$	$+x$	$+x$
Equilibrium	$0.100 - x$	x	x

Now we make an important simplifying assumption. The value of K_a is quite small (1.76×10^{-5}), and the concentration of the acid is relatively large (0.100 M). Under these conditions, the value of x will probably be small compared with 0.100 M. If so, then $(0.100 - x)$ is approximately the same as 0.100, which means that the denominator of the chemical equilibrium expression

CONNECTION: The use of ICE tables in solving equilibrium problems is described in Section 15.7.

$$K_a = \frac{[C_2H_3O_2^-][H^+]}{[HC_2H_3O_2]} = 1.76 \times 10^{-5}$$

can be simplified from

$$\frac{(x)(x)}{(0.100-x)} = 1.76 \times 10^{-5}$$

to

$$\frac{(x)(x)}{(0.100)} = 1.76 \times 10^{-5}$$

Cross multiplying

$$x^2 = 1.76 \times 10^{-6}$$

and taking the square root, we get

$$x = [H^+] = 1.33 \times 10^{-3} \; M$$

We should check the validity of our assumption that x is small compared with 0.100 M. Dividing x by 0.100 M and expressing the result as a percentage, we have

$$\frac{1.33 \times 10^{-3} \; M}{0.100 \; M} = 0.0133 = 1.33\%$$

As a rule, simplifying assumptions that produce x values that are less than 5% of initial concentrations are acceptable, and so is our assumption.

PRACTICE EXERCISE: Calculate the value of $[H^+]$ in 0.050 M benzoic acid,

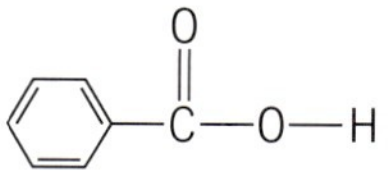

which is a widely used preservative in soft drinks and other foods. Given:

$$HC_7H_5O_2(aq) \rightleftharpoons C_7H_5O_2^-(aq) + H^+(aq) \qquad K_a = 6.5 \times 10^{-5}$$

(See Problems 9 and 10.)

Diprotic acids

In many regions of North America and Europe, the principal cause of acid precipitation is the presence of atmospheric aerosols of sulfuric acid that are produced when SO_3 combines with water vapor in the atmosphere:

$$SO_3(g) + H_2O(g) \longrightarrow H_2SO_4(l)$$

Sulfuric acid that has dissolved in water undergoes acid ionization in two steps.

In the first step, H_2SO_4 completely ionizes into HSO_4^- and H^+:

$$H_2SO_4(aq) \longrightarrow HSO_4^-(aq) + H^+(aq) \qquad K_{a1} \gg 1 \qquad (16.4)$$

However, the second step may not be complete, because the equilibrium constant for it is less than 1 at 25°C:

$$HSO_4^-(aq) \rightleftharpoons SO_4^{2-}(aq) + H^+(aq) \qquad K_{a2} = 1.2 \times 10^{-2} \qquad (16.5)$$

The symbols K_{a1} and K_{a2} are used here to identify which step in a two-step ionization process each equilibrium constant describes.

Because the first ionization step is complete but the second step may not be, most solutions of sulfuric acid contain between 1 and 2 moles of H^+ for every mole of H_2SO_4. Let's determine the value of $[H^+]$ in 0.100 M H_2SO_4. Because the first ionization step is complete, the starting point in this calculation is a solution in which all the H_2SO_4 has ionized to HSO_4^- and H^+, as described in Equation 16.4. Therefore, before the second step begins, $[HSO_4^-] = [H^+] = 0.100$ M. Ionization of HSO_4^- (Equation 16.5) produces additional H^+. Let's set up an ICE table for the second ionization step in which x is the change in $[H^+]$.

	$[HSO_4^-]$	$[H^+]$	$[SO_4^{2-}]$
Initial	0.100	0.100	0.000
Change	$-x$	$+x$	$+x$
Equilibrium	$0.100 - x$	$0.100 + x$	x

Inserting the equilibrium concentration terms in the equilibrium constant expression (K_{a2}), we have

$$K_{a2} = \frac{[H^+][SO_4^{2-}]}{[HSO_4^-]}$$

$$= \frac{(0.100 + x)(x)}{(0.100 - x)} = 1.2 \times 10^{-2}$$

Cross multiplying, we have

$$1.2 \times 10^{-3} - 1.2 \times 10^{-2}x = 0.100x + x^2$$

Rearranging the terms so that they fit the format of a quadratic equation ($ax^2 + bx + c = 0$) gives

$$x^2 + 0.112x - 1.2 \times 10^{-3} = 0$$

Solving the quadratic, we have

$$x = \frac{-b \pm \sqrt{b^2 - 4ac}}{2a} = \frac{-0.112 \pm \sqrt{(0.112)^2 - 4(1)(-1.2 \times 10^{-3})}}{2(1)}$$

$$= \frac{-0.112 \pm \sqrt{0.01254 + 0.0048}}{2}$$

$$= \frac{-0.112 \pm 0.132}{2} = +0.010, \text{ or } -0.122$$

The negative value for x has no meaning, because it corresponds to a negative $[SO_4^{2-}]$ concentration value; so we will use only the +0.010 value. At equilibrium,

$$[H^+] = (0.100 + x)\,M = (0.100 + 0.010)\,M = 0.110\,M$$

As predicted, the value of $[H^+]$ is between one and two times the initial concentration of H_2SO_4 (0.100 M). The degree of ionization of HSO_4^- is

$$\frac{[SO_4^{2-}]_{\text{equilibrium}}}{[HSO_4^-]_{\text{initial}}} = \frac{0.010\,M}{0.100\,M} = 0.10 = 10\%$$

SAMPLE EXERCISE 16.2: Calculate the the value of $[H^+]$ in a 100-mL sample of melted snow that contains 2.30 mg of H_2SO_4.

SOLUTION: Because a molarity concentration is required, we should begin by converting the given concentration of sulfuric acid into molarity:

$$\frac{2.30\;\cancel{\text{mg }H_2SO_4}}{100\text{ mL}} \times \frac{1\text{ mmol }H_2SO_4}{98.0\;\cancel{\text{mg }H_2SO_4}} = 2.35 \times 10^{-4}\text{ mol }H_2SO_4/L = 2.35 \times 10^{-4}\,M$$

Note that the molar mass used in this calculation has the units of milligrams per millimole, which is equivalent to the more conventional grams-per-mole units but handier to use in this case. Similarly, molarity is understood to mean moles per liter, but it also means millimoles per milliliter. After all, if a liter of a solution contains a mole of solute, then a milliliter of the same solution must contain a millimole of solute.

Next, we calculate $[H^+]$ in equilibrium with this concentration of sulfuric acid. Sulfuric acid is a strong, diprotic acid, which means that it ionizes in two steps as shown in Table 16.1. Given the values of K_{a1} and K_{a2} we may assume that

1. *All* the H_2SO_4 is ionized, forming HSO_4^- and H^+, according to the reaction

$$H_2SO_4(aq) \longrightarrow HSO_4^-(aq) + H^+(aq) \qquad K_{a1} \gg 1$$

2. *Some* of the HSO_4^- forms SO_4^{2-} and more H^+, according to the reaction

$$HSO_4^-(aq) \rightleftharpoons SO_4^{2-}(aq) + H^+(aq) \qquad K_{a2} = 1.2 \times 10^{-2}$$

As a result of the first acid-ionization step, all of the sulfuric acid is in the form of H^+ and HSO_4^-. Therefore, at the beginning of the second step, the initial concentrations of these ions is the same as the original concentration of H_2SO_4, 2.35×10^{-4} M. Let's set up an ICE table based on these initial concentration values, in which we let the increase in $[H^+]$ from the second step (dissociation of HSO_4^-) be x:

	$[HSO_4^-]$	$[H^+]$	$[SO_4^{2-}]$
Initial	2.35×10^{-4}	2.35×10^{-4}	0
Change	$-x$	$+x$	$+x$
Equilibrium	$2.35 \times 10^{-4} - x$	$2.35 \times 10^{-4} + x$	x

Using the equilibrium constant K_{a2} for HSO_4^-, we have

TABLE 16.1 Ionization Equilibria for Three Diprotic Acids

Acid		Ionization Equilibria	K_{a1}	K_{a2}
Carbonic acid	Step 1:	$H_2CO_3 + H_2O \rightleftharpoons H_3O^+ + HCO_3^-$	4.3×10^{-7}	4.7×10^{-11}
	Step 2:	$HCO_3^- + H_2O \rightleftharpoons H_3O^+ + CO_3^{2-}$		
Sulfurous acid	Step 1:	$H_2SO_3 + H_2O \rightleftharpoons H_3O^+ + HSO_3^-$	1.3×10^{-3}	6.2×10^{-8}
	Step 2:	$HSO_3^- + H_2O \rightleftharpoons H_3O^+ + SO_3^{2-}$		
Sulfuric acid	Step 1:	$H_2SO_4 + H_2O \rightleftharpoons H_3O^+ + HSO_4^-$	Strong acid	1.2×10^{-2}
	Step 2:	$HSO_4^- + H_2O \rightleftharpoons H_3O^+ + SO_4^{2-}$	($K_{a1} \gg 1$)	

$$K_{a2} = \frac{[H^+][SO_4^{2-}]}{[HSO_4^-]} = \frac{(2.35 \times 10^{-4} + x)(x)}{(2.35 \times 10^{-4} - x)} = 1.2 \times 10^{-2}$$

Cross multiplying gives

$$2.82 \times 10^{-6} - 1.2 \times 10^{-2}x = 2.35 \times 10^{-4}x + x^2$$

Rearranging the terms into the standard quadratic format

$$x^2 + 1.2235 \times 10^{-2}x - 2.82 \times 10^{-6} = 0$$

and solving for x, we have

$$x = \frac{-1.2235 \times 10^{-2} \pm \sqrt{(1.2235 \times 10^{-2})^2 - (4 \times 1 \times (-2.82 \times 10^{-6}))}}{2}$$

$$= \frac{(-1.2235 \pm 1.2688) \times 10^{-2}}{2} = -1.252 \times 10^{-2}, \text{ or } 2.26 \times 10^{-4} \text{ M}$$

Ignoring the negative solution, we obtain the equilibrium value of $[H^+]$ by summing the positive value of x and 2.35×10^{-4} M from the first ionization step:

$$[H^+] = 2.35 \times 10^{-4} \text{ M} \quad (\text{Step 1})$$
$$+\ 2.26 \times 10^{-4} \text{ M} \quad (\text{Step 2})$$
$$4.61 \times 10^{-4} \text{ M} \quad (\text{Total})$$

PRACTICE EXERCISE: Calculate the value of $[H^+]$ and the degree of ionization of HSO_4^- in 0.00100 M H_2SO_4. (See Problems 19 and 20.)

The results of the preceding calculation and sample and practice exercises illustrate a pattern observed for all weak acids: the degree to which they ionize increases with decreasing concentration. The graph in Figure 16.3, which shows the dependence of the degree of ionization of HNO_2 on $[HNO_2]$, illustrates this trend. A difference between weak and strong acids is that weak acids are completely ionized only at low concentrations, whereas strong acids are completely ionized at much higher concentrations.

✓ The degree of ionization of a weak acid increases with decreasing acid concentration.

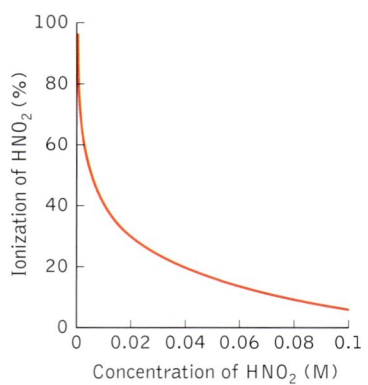

FIGURE 16.3 The degree of ionization of weakly acidic HSO_4^- in solutions of H_2SO_4 decreases with increasing concentration of H_2SO_4.

Table 16.1 lists the K_{a1} and K_{a2} values for three diprotic acids found in acid rain. Carbonic acid is present because (1) the atmosphere is about 0.036% (by volume) CO_2 and (2) CO_2 is slightly soluble in water. When CO_2 dissolves in water, carbonic acid, H_2CO_3, forms:

$$CO_2(g) + H_2O(l) \rightleftharpoons H_2CO_3(aq) \qquad (16.6)$$

Note that, in each pair of K_{a1} and K_{a2} values in Table 16.1, the value of K_{a2} is smaller than that of K_{a1}. Why is the value of K_{a2} for a diprotic acid much less than that of K_{a1}? We can rationalize the difference on the basis of electrostatic attractions between oppositely charged ions. The first ionization step produces a negatively charged oxoanion. The second ionization step requires that a positive ion (H^+) leave a negative ion, producing an even more negatively charged oxoanion. Separating oppositely charged ions that are naturally attracted to each other is not a process that we would expect to be favored, and the smaller values for K_{a2} confirm our expectations. For this reason, the K_{a2} of any diprotic acid is less (often much less) than K_{a1}.

> Why is the value of K_{a2} for a diprotic acid much less than that of K_{a1}?

The large differences between K_{a1} and K_{a2} values for the two weak diprotic acids in Table 16.1 have an important consequence: the second ionization step has little (usually an insignificant) effect on $[H^+]$ in solutions of these acids. Essentially all of the strength of these acids is due to the first ionization step.

16.2 ACIDS AND BASES: A MOLECULAR VIEW

In Section 5.6, an acid was defined as a substance that donates hydrogen ions, and a base was defined as a substance that accepts hydrogen ions. These descriptions of acids and bases were developed independently by Danish chemist Johannes Brønsted (1879–1947) and English chemist Thomas Lowry (1874–1936) and published in the same year, 1923. Today, these descriptions are known as the *Brønsted-Lowry* model of acids and bases.

Acids in water

The Brønsted-Lowry model applies to acids and bases in aqueous solutions. If an acid is a substance that donates hydrogen ions in aqueous solution, then another substance must be the recipient of these donated hydrogen ions. This other substance, acting as a hydrogen-ion acceptor, is behaving as a base. In an aqueous solution of hydrochloric acid, the only other substance present is water; so we must conclude that water acts as a base in a solution of HCl. This donation and

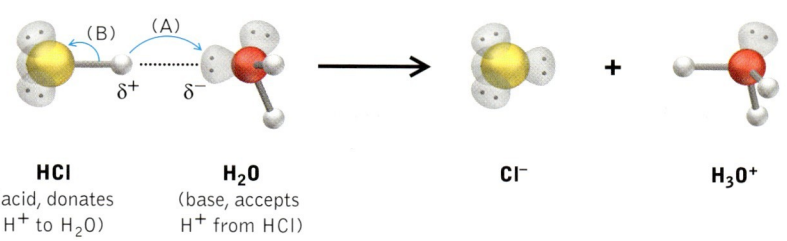

FIGURE 16.4 When hydrogen chloride dissolves in water, HCl ionizes into H^+, which bonds (arrow A) to H_2O, forming H_3O^+. The bonding pair of electrons in each molecule of HCl moves to the Cl atom (arrow B), forming Cl^-.

acceptance of hydrogen ions can be written in equation form as follows (note that pure HCl, or hydrogen chloride, is a gas at room temperature):

$$HCl(aq) + H_2O(l) \longrightarrow Cl^-(aq) + H_3O^+(aq) \qquad (16.7)$$

This reaction may be viewed on a molecular basis (Figure 16.4) in the following way. Strong intermolecular interactions between O atoms of H_2O molecules and H atoms of HCl molecules result in ionization of HCl, forming H^+ and Cl^- ions, and, at the same time, an O—H bond forms as an O atom shares one of its lone pairs of electrons with the H^+ ion. The product of the interaction is a species called a *hydronium ion*, which has the formula H_3O^+.

Equation 16.7 is a source of insight into the meaning of the symbol $H^+(aq)$ in chemical equations such as Equations 16.2 through 16.5. Although it is convenient to describe acidic and basic properties in terms of the donation and acceptance of hydrogen ions, which we will routinely do, keep in mind that there are no free protons (H^+) in acidic solutions. Moreover, strong attraction between the electronegative oxygen atoms in water molecules and H_3O^+ ions leads to the formation of clusters of molecules of H_2O around H_3O^+ ions as shown in Figure 16.5).

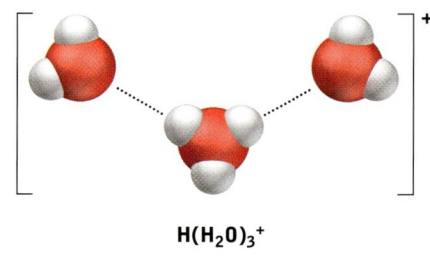
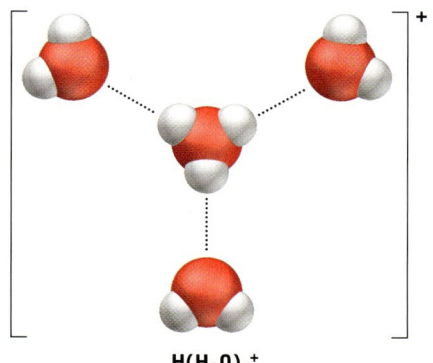

FIGURE 16.5 Water molecules cluster around hydronium (H_3O^+) ions, forming species with the general formula $H(H_2O)_n^+$. In the examples shown, the clusters have the formulas $H(H_2O)_3^+$ and $H(H_2O)_4^+$.

Bases in water

Let's consider what happens when basic compounds dissolve in water. Among the most basic compounds are the hydroxides of the Group 1A elements and three of the Group 2A hydroxides: $Ca(OH)_2$, $Ba(OH)_2$, and $Sr(OH)_2$. These Group 1A and 2A hydroxides are strong bases because all of them completely dissociate in water, releasing their OH^- ions. For reasons that we will soon explore, the OH^- ion is the strongest base possible in aqueous solution. In fact, an early theory of acids and bases developed in 1887 by Swedish chemist Svante August Arrhenius (1859–1927) defined a base as a substance that donates OH^- ions. Therefore, we can relate the strength of a base to the concentration of OH^- ions it produces when it dissolves in water.

SAMPLE EXERCISE 16.3: Calcium hydroxide, $Ca(OH)_2$, also known as *slaked lime*, is widely used to treat drinking water. Calculate the value of $[OH^-]$ in 1.50×10^{-3} M $Ca(OH)_2$.

SOLUTION: Calcium hydroxide is a strong base that dissociates completely in water. The balanced equation for the dissociation of $Ca(OH)_2$ is

$$Ca(OH)_2(aq) \longrightarrow Ca^{2+}(aq) + 2\,OH^-(aq)$$

According to the stoichiometry of the dissociation process, 2 moles of OH^- are produced from each mole of $Ca(OH)_2$; so $[OH^-]$ is

$$1.50 \times 10^{-3} \text{ M Ca(OH)}_2 \times \frac{2 \text{ mol OH}^-}{1 \text{ mol Ca(OH)}_2} = 3.00 \times 10^{-3} \text{ M OH}^-$$

PRACTICE EXERCISE: Calculate the value of $[OH^-]$ in 4.2×10^{-2} M $Ba(OH)_2$. (See Problems 23 and 24.)

CONNECTION: Amines are organic compounds containing R–NH_2 groups in their molecular structures as described in Section 13.7.

Now let's consider what happens when a weakly basic compound, such as ammonia or an amine, dissolves in water. If ammonia is to act as a Brønsted-Lowry base and accept hydrogen ions, then water must act as the acid that supplies those hydrogen ions. The transfer of hydrogen ions from H_2O to NH_3 can be written in equation form as follows:

$$NH_3(g) + H_2O(l) \rightleftharpoons NH_4^+(aq) + OH^-(aq) \tag{16.8}$$

Note that the reaction results in the production of OH^- ions; so NH_3 also fits the Arrhenius definition of a base.

The reaction of ammonia with water entails intermolecular forces and covalent bonds breaking and forming (Figure 16.6). Strong intermolecular interactions between the N atoms of NH_3 molecules and the H atoms of H_2O molecules result in ionization of O–H bonds in H_2O, forming H^+ and OH^- ions and in the transfer of H^+ to NH_3, forming NH_4^+. The lone pair of electrons on each N atom is shared with an H^+ ion to make a fourth N–H bond in each NH_4^+ ion.

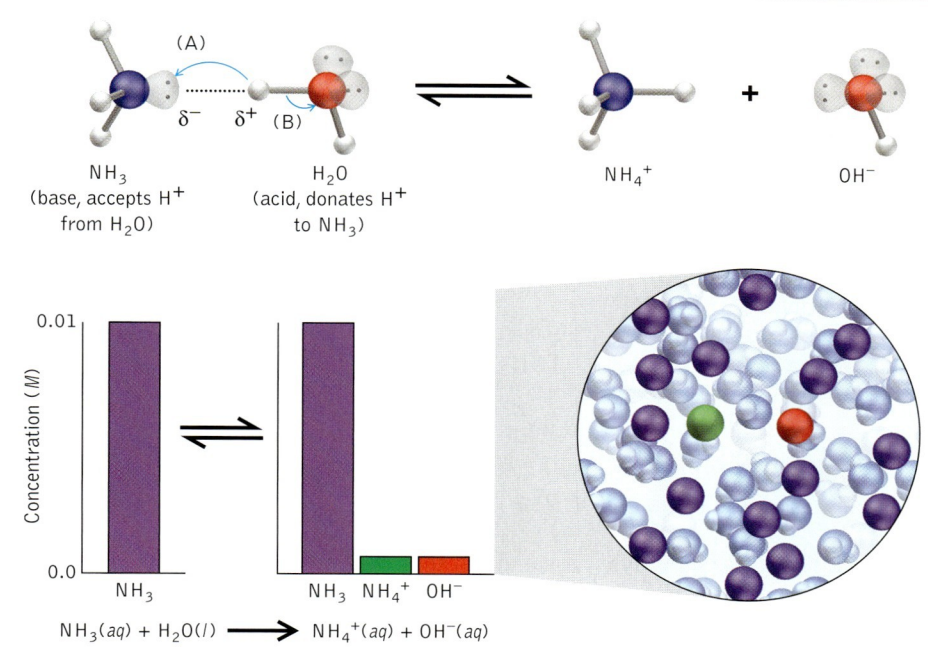

FIGURE 16.6 In the reaction of NH_3 with H_2O, hydrogen bonding between a H atom in H_2O and the N atom in NH_3 leads to the ionization of H_2O and the formation of a fourth N–H bond (arrow A), producing NH_4^+. A bonding pair of electrons in each molecule of H_2O moves to the O end of an O–H bond (arrow B), forming hydroxide ion, OH^-. The graphs and molecular view show that only a small fraction of the NH_3 molecules react with H_2O to form NH_4^+ and OH^-.

In a solution of ammonia not all of the molecules of NH_3 accept hydrogen ions, forming NH_4^+. Instead, the reaction described by Equation 16.8 reaches an equilibrium in which much of the ammonia is likely to be present as NH_3. The limited strength of ammonia as a base is indicated by the small value of the equilibrium constant for Equation 16.8 at 25°C:

$$K_b = \frac{[NH_4^+][OH^-]}{[NH_3]} = 1.76 \times 10^{-5}$$

Here we use the subscript "b" to indicate that the reaction is a **b**ase-ionization reaction that results in the acceptance of hydrogen ions from, in this case, H_2O. Note that there is no concentration term for H_2O in the denominator. It is left out because the concentration of water in an aqueous reaction is not likely to change significantly in the course of the reaction. To simplify equilibrium constant expressions and calculations, the molar concentration of water is a constant that is incorporated into the values of equilibrium constants.

✓ Equilibrium constant expressions for reactions in aqueous solutions do not include $[H_2O]$ terms, because $[H_2O]$ does not change significantly in the course of the reaction.

CONCEPT TEST: What is the molar "concentration" of water in pure water? Assume that the density of water is 1.00 g/mL.

SAMPLE EXERCISE 16.4: Calculate the value of $[OH^-]$ in 0.250 M NH_3.

SOLUTION: The basic properties of NH_3 are described by the reaction

$$NH_3(g) + H_2O(l) \rightleftharpoons NH_4^+(aq) + OH^-(aq)$$

and its equilibrium constant

$$K_b = \frac{[NH_4^+][OH^-]}{[NH_3]} = 1.76 \times 10^{-5}$$

Initially, $[NH_3] = 0.250$ and $[NH_4^+] = 0.000$. We assume that the initial $[OH^-]$ is insignificant (essentially zero) compared with the equilibrium concentration, which we let be x. Given the 1:1 stoichiometry between NH_4^+ and OH^-, the changes in the concentrations of both are $+x$, and the decrease in $[NH_3]$ is $-x$. The resulting ICE table is

	$[NH_3]$	$[OH^-]$	$[NH_4^+]$
Initial	0.250	0.000	0.000
Change	$-x$	$+x$	$+x$
Equilibrium	$0.250 - x$	x	x

Now we make the simplifying assumption that the value of x will be small compared with 0.250 M, and so $0.250 - x$ is approximately 0.250. Inserting the appropriate values in the equilibrium constant reaction, we have

$$K_b = \frac{[NH_4^+][OH^-]}{[NH_3]} = \frac{(x)(x)}{(0.250)} = 1.76 \times 10^{-5}$$

Cross multiplying

$$x^2 = 4.4 \times 10^{-6}$$

and taking the square root, we get

$$x = [OH^-] = 2.1 \times 10^{-3} \, M$$

This $[OH^-]$ is less than 1% of the initial concentration of ammonia, making our assumptions and the results of our calculation valid.

PRACTICE EXERCISE: Methylamine is a weak base ($K_b = 4.4 \times 10^{-4}$). Calculate the value of $[OH^-]$ in 0.200 M CH_3NH_2. (See Problems 37 and 38.)

Lewis acids and bases

There is another way to view the behavior of ammonia as a base. Think of the reaction in Equation 16.8 as the donation and acceptance of *electrons* rather than *protons*. A new N–H bond forms as a molecule of NH_3 *donates a pair of electrons* and as each acidic compound (water in this case) *accepts a pair of electrons*. This transfer of electron pairs is another basis for defining acids and bases:

- A *base* is a substance that *donates* pairs of electrons.
- An *acid* is a substance that *accepts* pairs of electrons.

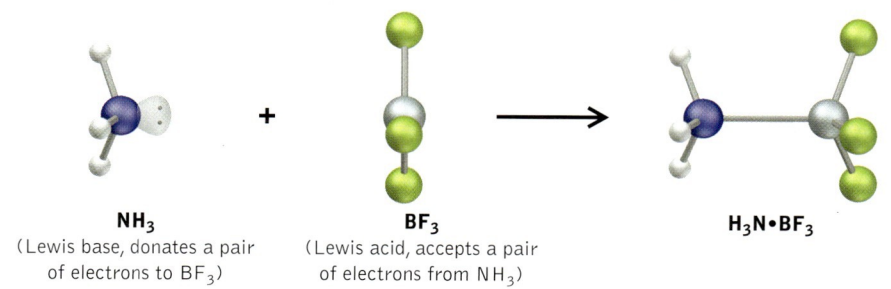

FIGURE 16.7 In the reaction between NH_3 and BF_3, the nitrogen atom in NH_3 donates a pair of electrons (and so acts as a Lewis base), and the boron atom in BF_3 accepts the donated pair of electrons and so acts as a Lewis acid.

These definitions were developed by Gilbert N. Lewis as he probed the nature of the chemical bond (see Section 6.2). Substances that fit these definitions are called *Lewis acids* or *Lewis bases*. The Lewis definition of a base is consistent with the concept of a Brønsted-Lowry base (hydrogen-ion acceptor) because a substance must be able to donate a pair of electrons if it is to form a bond to a hydrogen ion and thereby accept the H^+ ion (Figure 16.4). Thus, all species that can accept hydrogen ions must also be able to donate pairs of electrons. However, the Lewis definition of an acid encompasses an array of species that have no hydrogen ions to donate but can still accept electrons. One such compound is boron trifluoride, BF_3. There are only six valence electrons around the boron atom in BF_3, giving it the capacity to accept another pair to complete its octet. A suitable electron-pair donor is NH_3, as may be seen in the reaction and molecular models in Figure 16.7.

CONNECTION: Gilbert Lewis's pioneering theories of the nature of covalent bonding are described in Section 6.2.

Conjugate pairs

The reactants and products in Equation 16.8 are at equilibrium when the rates of the forward and reverse reactions are the same. In the forward direction, ammonia acts as a base, and water acts as an acid. However, in the reverse direction, ammonium ion is the hydrogen-ion donor (Brønsted-Lowry acid), and hydroxide ion is the hydrogen-ion acceptor (Brønsted-Lowry base). This behavior makes NH_4^+ and NH_3 an acid–base **conjugate** pair. The ammonium ion is the conjugate acid of ammonia. Similarly, H_2O and OH^- are an acid–base conjugate pair: OH^- is the conjugate base of water. Note that, in each conjugate pair, the structural difference between the acid and the base is a H^+ ion that the acid has and that the base does not.

 Adding a H^+ ion to a base produces its **conjugate** acid.

$$NH_3\,(g) + H_2O\,(l) \rightleftharpoons NH_4^+\,(aq) + OH^-\,(aq)$$
$$\text{(base)} \quad\quad \text{(acid)} \quad\quad\quad \text{(conjugate acid)} \quad \text{(conjugate base)}$$

The small value (1.76×10^{-5} at 25°C) of the equilibrium constant for this reaction tells us that the equilibrium lies far to the left. We may conclude that OH^-

is a stronger base than NH_3 because OH^- is better at taking hydrogen ions away from NH_4^+ than NH_3 is at taking hydrogen ions away from H_2O. In fact, OH^- is a very strong base, as are those substances, including NaOH and the other hydroxides of the Group 1A and 2A elements that dissociate completely, producing OH^-, when they dissolve in water.

Let's reexamine the acid-ionization reaction of HCl—this time identifying the acid–base conjugate pairs:

$$HCl\,(g) + H_2O\,(l) \longrightarrow Cl^-\,(aq) + H_3O^+\,(aq)$$
$$\text{(acid)} \quad \text{(base)} \quad\quad \text{(conjugate base)} \quad \text{(conjugate acid)}$$

This reaction goes to completion, which means that the reverse reaction essentially does not happen. Therefore, HCl must be a very strong acid, and Cl^- must be a very weak base. This contrast in acid–base behavior is common for all conjugate pairs: strong acids have very weak conjugate bases, and strong bases have very weak conjugate acids. Moderately weak acids (such as HNO_2) have moderately weak conjugate bases (such as NO_2^-), and moderately weak bases (such as NH_3) have moderately weak conjugate acids (such as NH_4^+). The complementary strengths of a number of the common acid–base conjugate pairs are shown in Figure 16.8.

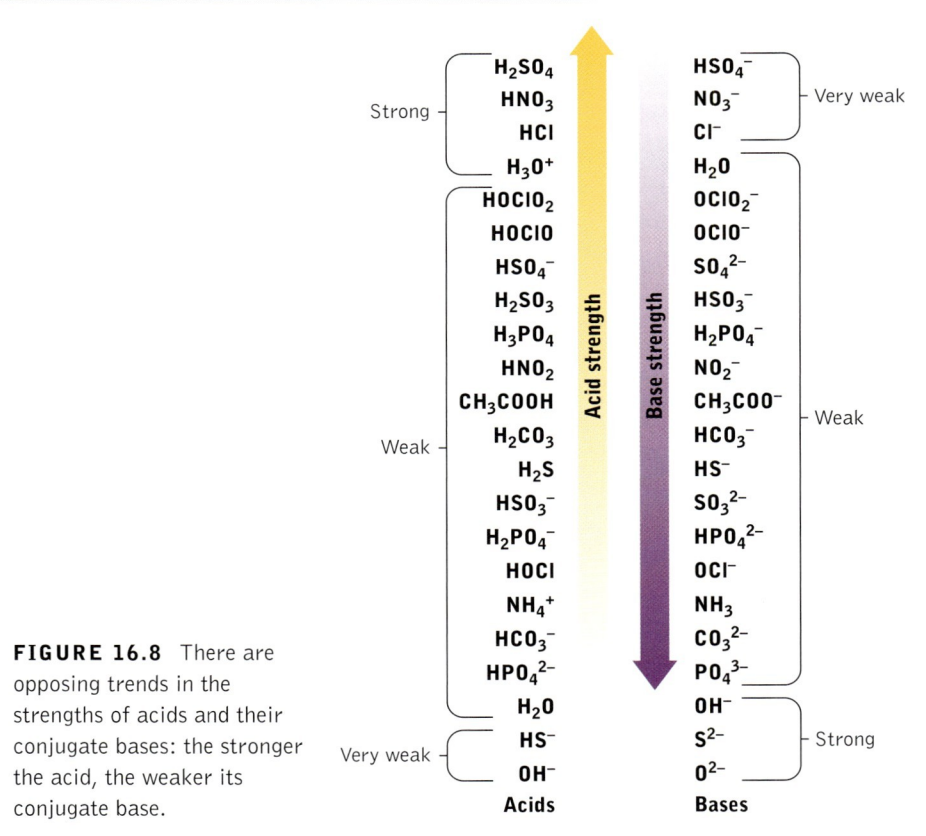

FIGURE 16.8 There are opposing trends in the strengths of acids and their conjugate bases: the stronger the acid, the weaker its conjugate base.

SAMPLE EXERCISE 16.5: Identify the conjugate base of each of the following acids: (a) H_2SO_4; (b) H_2S; (c) H_3PO_4; (d) HCO_3^-; (e) HPO_4^{2-}.

SOLUTION: Conjugate bases have structures and formulas similar to those of their acids, but they have one less hydrogen ion. Therefore, the conjugate bases of the acids in parts *a* through *e* are: (a) HSO_4^-; (b) HS^-; (c) $H_2PO_4^-$; (d) CO_3^{2-}; (e) PO_4^{3-}.

PRACTICE EXERCISE: Identify the conjugate acid of each of the following bases: (a) SO_4^{2-}; (b) CH_3NH_2; (c) HCO_3^-; (d) OH^-; (e) HPO_4^{2-}.
(See Problems 39 and 40.)

We need to keep in mind that, no matter how strong an acid is, it can't be more than 100% ionized, forming H_3O^+, in an aqueous solution. Therefore, the hydronium ion is the strongest acid that can exist in aqueous solutions. Similarly, OH^- is the strongest base that can exist in water. When an even stronger base, such as the sulfide ion, S^{2-}, is added to water, it is completely converted to SH^- and OH^-.

$$S^{2-}(aq) + H_2O(l) \longrightarrow SH^-(aq) + OH^-(aq)$$

Thus, water acts as a controller, or leveler, of the strengths of very strong acids and bases that dissolve in it. Those acidic substances that ionize completely, forming H_3O^+, are strong acids, and those ionic hydroxide compounds that dissociate completely are strong bases.

16.3 ACID STRENGTH AND MOLECULAR STRUCTURE

As stated at the beginning of this chapter, nitric acid (HNO_3) is a strong acid, whereas nitrous acid (HNO_2) is a weak acid. Similarly, sulfurous acid (H_2SO_3), which is formed when sulfur dioxide dissolves in water,

$$SO_2(g) + H_2O(l) \rightleftharpoons H_2SO_3(aq)$$

> Why are these "... ic" acids so much stronger than the corresponding "... ous" acids?

is a weak acid, whereas sulfuric acid (H_2SO_4) is a strong acid. Why are these "... ic" acids so much stronger than the corresponding "... ous" acids? The answer lies in their molecular structures. Consider the structural differences between H_2SO_4 and H_2SO_3 shown in Figure 16.9. Note that the ionizable hydrogen atoms in both molecules are bonded to oxygen atoms that are also bonded to the central sulfur atoms. In each case, the central sulfur atom is also double bonded to one

THE CHEMISTRY OF TWO STRONG ACIDS: SULFURIC AND NITRIC ACIDS

Two strong acids, nitric acid (HNO_3) and sulfuric acid (H_2SO_4), are among the top 12 industrial chemicals produced in the world. Sulfuric acid ranks first, with a worldwide production of more than 150 million tons (40 million tons in the United States alone) each year. Nitric acid production is about 11th on the industrial chemicals list (U.S. production was 8.0 million tons in 2001). About 70% of the H_2SO_4 and 75% of the HNO_3 produced in the United States is used to make fertilizer. The rest is used in a variety of chemical manufacturing processes, including the preparation of synthetic fibers described in Chapter 18.

Pure sulfuric acid is a dense, colorless, oily liquid. It fumes when heated, partly decomposing into H_2O and SO_3 and producing a solution that is 98.3% H_2SO_4 and 1.7% H_2O. This solution, which is 18 M H_2SO_4, is the liquid sold as "concentrated" sulfuric acid. Concentrated sulfuric acid is very hygroscopic (absorbs water) and is used as a drying agent and to remove water from many compounds. It can dehydrate sugar, turning this carbohydrate into carbon. Sulfuric acid dissolves in water in a process so exothermic that the solution may boil, which is why concentrated sulfuric acid must be diluted by slowly adding it to cold water. Never add water to concentrated sulfuric acid.

The synthesis of sulfuric acid starts with the combustion of sulfur or sulfide minerals to sulfur dioxide followed by the oxidation of SO_2 to SO_3:

$$S(s) + O_2(g) \longrightarrow SO_2(g)$$
$$+SO_2(g) + \tfrac{1}{2}O_2(g) \rightleftharpoons SO_3(g)$$

Overall: $S(s) + \tfrac{3}{2}O_2(g) \rightleftharpoons SO_3(g)$

Both reactions are equilibrium processes, but the equilibrium constant for the formation of SO_2 is large and the reaction essentially goes to completion. The equilibrium between SO_2, O_2, and SO_3 in the second step is strongly affected by Le Châtelier's principle. The reac-

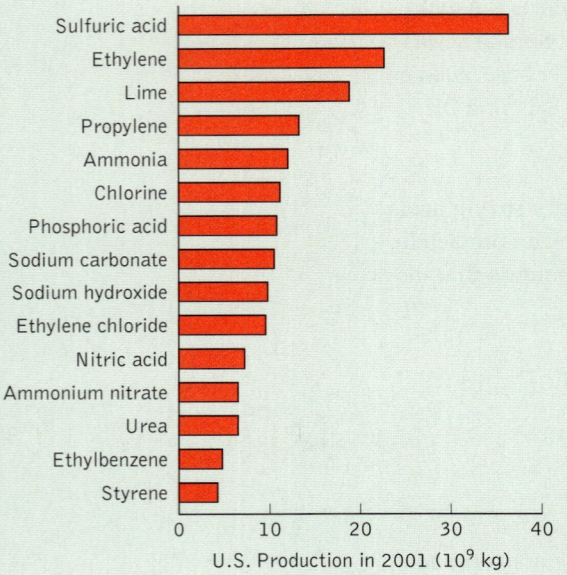

The top 15 industrial chemicals produced in the United States in 2001 included sulfuric acid (1st) and nitric acid (11th).

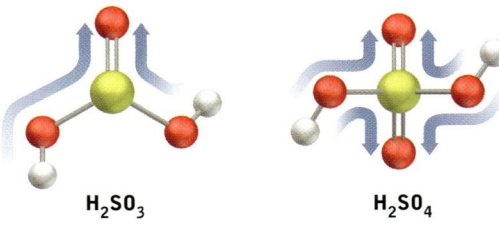

FIGURE 16.9 Sulfuric acid (H_2SO_4) is a stronger acid than sulfurous acid (H_2SO_3) because of the greater electron-withdrawing power (shown by the curved arrows) of an additional double-bonded oxygen atom in a molecule of H_2SO_4. When more electron density is drawn away from the hydrogen atoms of an acid, they are more easily ionized.

tion is exothermic but slow at ambient temperature. Higher temperatures speed the rate of the reaction but decrease the equilibrium concentration of product. The yield is improved by (1) increasing the pressure of the reactants (Do you understand why?), (2) using an excess of O_2, and (3) harvesting SO_3 during the reaction. Vanadium(V) oxide, V_2O_5, is used as a catalyst for the reaction, allowing it to proceed at an acceptable rate at moderate temperatures. Sulfuric acid itself is produced by reaction between sulfur trioxide and water:

$$SO_3(g) + H_2O(l) \longrightarrow H_2SO_4(l)$$

Note that the steps in the production of sulfuric acid are the same as those that lead to acid precipitation in the environment.

The production of nitric acid is linked to the production of ammonia because NH_3 is a reactant in the synthesis of HNO_3. The controlled, selective oxidation of ammonia to NO and subsequent conversion into nitric acid is known as the Ostwald process. Developing it earned Wilhelm Ostwald (1853–1932) the Nobel Prize in chemistry in 1909. The three steps in the process are as follows:

(1) $4\ NH_3(g) + 5\ O_2(g) \longrightarrow 4\ NO(g) + 6\ H_2O(g)$

(2) $2\ NO(g) + O_2(g) \longrightarrow 2\ NO_2(g)$

(3) $3\ NO_2(g) + H_2O(l) \longrightarrow 2\ HNO_3(l) + NO(g)$

Note that the oxidation number of nitrogen increases from -3 (in NH_3) to $+2$ (in NO) to $+4$ (in NO_2) to $+5$ (in HNO_3) during the process. A catalyst composed of platinum or platinum and rhodium and a reaction temperature of 850°C are needed to achieve a rapid-rate conversion of ammonia into nitric acid.

Wilhelm Ostwald was a professor of physical chemistry at Leipzig University in Germany from 1887 until 1906. During that time, he mentored several students whose names should be familiar to you: Jacobus Henricus van't Hoff, who won the Nobel Prize in chemistry in 1901, and Svante August Arrhenius, who won the same prize two years later. In Chapter 17, we will encounter the name of another of Ostwald's students, Walther Hermann Nernst, whose pioneering work in thermochemistry led to his receiving the Nobel Prize in chemistry in 1920 (though we will encounter his name in the context of electrochemical reactions and yet another equation named after one of Ostwald's brilliant students).

(in H_2SO_3) or two (in H_2SO_4) other oxygen atoms. Oxygen is one of the most electronegative (see Section 6.5) elements ($X = 3.5$). Oxygen atoms double bonded to the central atom of oxoacids, including H_2SO_3 and H_2SO_4, attract electron density toward themselves and away from the other ends of the molecules. These other ends are occupied by the hydrogen atoms of O–H groups. The more electron density that is drawn away from these O–H groups, the easier it is to break their O–H bonds, releasing hydrogen ions. Thus, sulfuric acid, with two electron-withdrawing S=O bonds, is a stronger acid than sulfurous acid, with only one S=O bond.

Acid	Structure	Oxidation state of Cl*	K_a
Hypochlorous	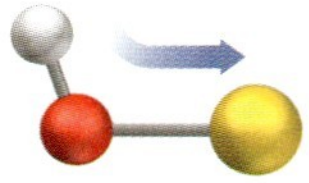	+1	2.9×10^{-8}
Chlorous	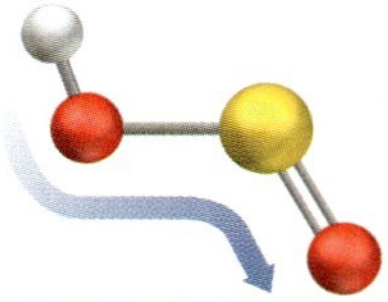	+3	1.1×10^{-2}
Chloric	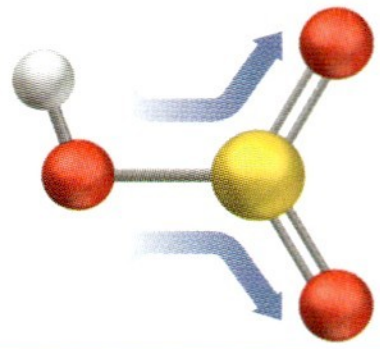	+5	~1
Perchloric	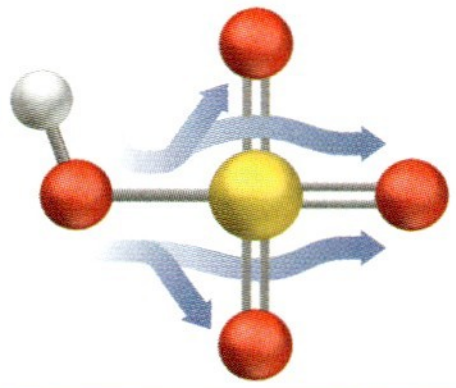	+7	~1×10^8

FIGURE 16.10 The strengths of the oxoacids of chlorine increase with increasing oxidation number of chlorine and the corresponding greater withdrawal of electron density (shown by the number of the arrows) from the hydrogen atoms on the molecules, making them more easily ionized.

* Recall that we calculate oxidation numbers in polyatomic compounds by first assigning a value or −2 to oxygen atoms and +1 to hydrogen atoms (see Section 5.5.).

CONNECTION: The rules for assigning oxidation numbers are described in Section 5.5.

CONNECTION: The periodic trend in electronegativities of the elements is described in Section 6.5.

This trend of increasing acid strength with increasing numbers of oxygen atoms bonded to the central atom (i.e., with increasing oxidation number of the central atom) is true for all oxoacids. It is exemplified in the strongly acidic properties of HNO_3 and the weak acidity of HNO_2. It is also illustrated in the strengths of the oxoacids of chlorine, as shown in Figure 16.10.

The connection between the electron-withdrawing power of the central atom in an oxoacid and the strength of the acid is also evident when we compare the strengths of acids having different central atoms but otherwise similar structures. Consider, for example, the relative strength of the three hypohalous acids in Figure 16.11. The differences between them are the identities (and electronegativities) of the halogen atoms in their structures. The most electronegative of the three (Cl) has the greatest attraction for the pair of electrons that it shares with oxygen. This attraction draws electron density toward chlorine, which in turn pulls electron density away from hydrogen and toward the oxygen end of the already polar O–H bond. These shifts in electron density make it easier for hypochlorous acid (HOCl) to ionize, forming hypochlorite (OCl⁻) ions and H^+. Thus, HOCl is the strongest of the three acids, followed by hypobromous acid (HOBr) and hypoiodous acid (HOI).

Acid	Structure	Electronegativity of central atom	K_a
Hypochlorous	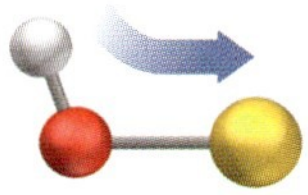	3.0	2.9×10^{-8}
Hypobromus	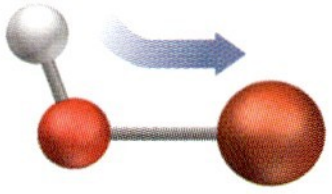	2.8	2.3×10^{-9}
Hypoiodous	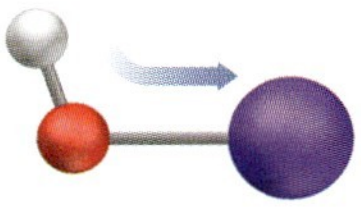	2.5	2.3×10^{-11}

FIGURE 16.11 The strengths of these three hypohalous acids are related to the electronegativities of their halogen atoms and the greater shifts in electron density (depicted by the thickness of the arrows) away from the hydrogen end of the molecule, which makes ionization easier.

16.4 THE CONCEPT OF pH

The relative acidities of solutions may be expressed in terms of their concentrations of H_3O^+ but there is another way. It makes use of the **pH** scale. To understand pH, we need to understand the *autoionization* of water. As the term implies, water molecules have the ability to ionize each other. The process resembles the ionization of water by ammonia described earlier in this chapter. On a molecular scale, the process is driven by hydrogen bonds between hydrogen and oxygen atoms of different water molecules. These hydrogen bonds are strong enough to ionize some water molecules, as shown in Figure 16.12. We can write a chemical equation for this autoionization reaction as follows:

$$\underset{\text{(acid)}}{H_2O\,(l)} + \underset{\text{(base)}}{H_2O\,(l)} \rightleftharpoons \underset{\substack{\text{(conjugate} \\ \text{base)}}}{OH^-\,(aq)} + \underset{\substack{\text{(conjugate} \\ \text{acid)}}}{H_3O^+\,(aq)} \quad (16.9)$$

In this reaction, one water molecule (acting as an acid) donates a hydrogen ion to another water molecule (acting as a base). The first molecule forms its conjugate base (OH^-), and the second forms its conjugate acid (H_3O^+). We encountered this dual acid–base nature of water in Section 16.3: water can act as a proton acceptor (base) in solutions of hydrochloric acid but as an acid in a solution of ammonia. When a substance can take on both acidic and basic characteristics—donating hydrogen ions in basic solutions and accepting hydrogen ions in acidic solutions—the substance is said to be **amphoteric**. The autoionization of water is an example of amphoteric behavior.

The equilibrium constant for the autoionization of water can be written as follows:

$$K_c = \frac{[H_3O^+][OH^-]}{[H_2O][H_2O]}$$

✓ **pH** is a scale for expressing the acidity or basicity of a substance.

✓ An **amphoteric** substance is capable of behaving both as an acid and as a base.

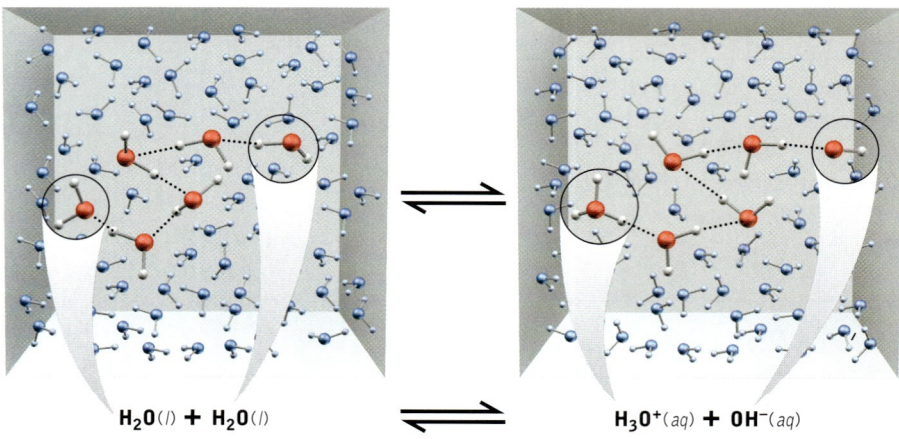

FIGURE 16.12 A computer simulation of the autoionization of water shows how a network of hydrogen bonds can pull apart two different water molecules, transforming one molecule into an OH⁻ ion and the other into H₃O⁺. The two frames represent "snapshots" of the same array of water molecules taken 150 femto (10⁻¹⁵) seconds apart. Hydrogen bonds temporarily link the six highlighted water molecules in the left-hand frame. The formation of these bonds leads (right-hand frame) to the loss of a H⁺ ion by the molecule at the upper-right end, transforming it into an OH⁻ ion. Through a series of intermolecular interactions hydrogen bonds become O–H covalent bonds and vice versa, and the H⁺ ion is, in effect, passed down the chain until it ends up on the water molecule at the lower-left end, forming H₃O⁺.

As we have done before, we will incorporate the [H$_2$O] terms into the value of the equilibrium constant because the concentration of water is essentially a constant. Doing so simplifies the equilibrium constant expression, which we label K_w, to

$$K_w = [\text{H}_3\text{O}^+][\text{OH}^-] \tag{16.10}$$

If we were to measure [H$_3$O$^+$] and [OH$^-$] in pure water at 25°C, we would find that both have the same value: 1.0×10^{-7} M. By using this concentration value in Equation 16.10, we can determine the value of K_w:

$$K_w = [\text{H}_3\text{O}^+][\text{OH}^-] = (1.0 \times 10^{-7})(1.0 \times 10^{-7}) = 1.0 \times 10^{-14} \tag{16.11}$$

Such a tiny value for K_w indicates that autoionization does not happen to a great extent. Rather, the reverse of autoionization—namely, the reaction between [H$_3$O$^+$] and [OH$^-$] to produce H$_2$O—has an equilibrium constant of $1/K_w$ or 1.0×10^{14} and so essentially goes to completion:

$$\text{H}_3\text{O}^+(aq) + \text{OH}^-(aq) \longrightarrow 2\,\text{H}_2\text{O}(l) \qquad K_c = 1/K_w = 1.0 \times 10^{14}$$

Equation 16.11 applies to *all aqueous solutions*, not just to pure water. Therefore, in any aqueous sample, there will be an inverse relation between [H$_3$O$^+$] and [OH$^-$]: as the value of one increases, the value of the other must decrease so that the product of the two is always 1.0×10^{-14}. A solution in which [H$_3$O$^+$] is greater than [OH$^-$] is considered acidic; a solution in which [H$_3$O$^+$] is less than [OH$^-$] is basic; and a solution in which the concentrations of the two are equal (1.0×10^{-7} M) is neither acidic nor basic, but rather is *neutral*.

The tiny value of K_w means that autoionization of water does not contribute significantly to [H⁺] in solutions of acids or to [OH⁻] in solutions of bases, and so we can ignore the contribution of autoionization in most calculations of acidity or basicity. Just keep in mind that autoionization means that initial values of [H⁺] and [OH⁻] are never truly zero. They are just too small (1×10^{-7} M) to be significant. However, if acids or bases are extremely weak or if their concentrations are extremely dilute, the autoionization of H_2O may need to be taken into account. See, for example, Problem 54 at the end of the chapter.

The pH scale

In the early 1900s, scientists developed an electrochemical device called the hydrogen electrode to directly determine the acidity, or [H_3O^+], of solutions. The electrical voltage, or *potential*, produced by the hydrogen electrode is not a linear function of [H_3O^+], but rather a linear function of the *logarithm* of [H_3O^+]. This relation led Danish biochemist Søren Sørenson (1868–1939) to propose a scale for expressing acidity (and basicity) based on what he termed "the potential of the hydrogen ion," abbreviated **pH**. Mathematically, we define pH as the *negative logarithm to the base 10 of* [H_3O^+], or

$$\text{pH} = -\log [H_3O^+] \qquad (16.12)$$

Sørenson's pH scale has several attractive features. Because it is a logarithmic scale, there are no exponents, as are commonly found in values of [H_3O^+]. This feature also means that a change of one pH unit corresponds to a 10-fold change in [H_3O^+]; so a seemingly small pH change corresponds to a large change in [H_3O^+]. The negative sign in front of the logarithmic term means that almost all pH values are positive. That's the good news. The bad news is that *large* pH values correspond to *small* values [H_3O^+], which can be confusing at first. Keep in mind that acidic solutions have pH values less than 7 ([H_3O^+] > 1.0×10^{-7} M) and basic solutions have pH values greater than 7 ([H_3O^+] < 1.0×10^{-7} M). The pH values for some common materials are shown in Figure 16.13.

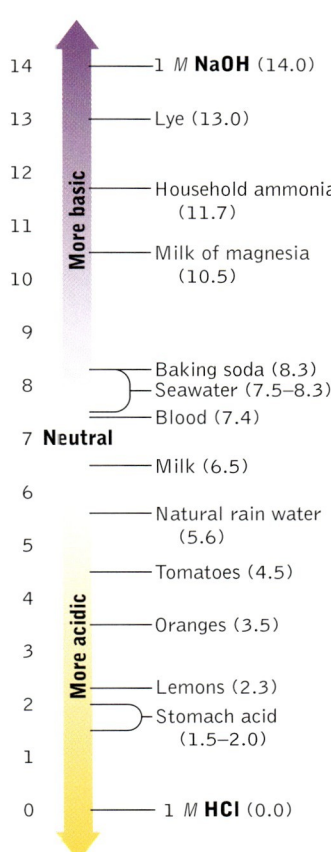

FIGURE 16.13 The pH scale is a convenient way to express the range of acidic or basic properties of some common materials.

The pH of "natural" and acid rain

Let's calculate the pH of "natural" rain—that is, the pH of rain that does not contain quantities of strongly acidic nonmetal oxides produced by human activities (volcanic activity and other natural processes also can add these gases to the atmosphere). As noted in Section 16.3, all rain contains some carbonic acid as it forms in our atmosphere, which is 0.036% CO_2. The limited solubility of CO_2 produces a dilute solution of carbonic acid (Equation 16.6):

$$CO_2(g) + H_2O(l) \rightleftharpoons H_2CO_3(aq)$$

The resulting [H_2CO_3] is about 1.2×10^{-5} M at 25°C. We will assume, as we usually do, that the only significant source of H_3O^+ in solution is carbonic acid. Carbonic acid ionizes in two steps (see Table 16.1):

Step 1: $H_2CO_3(aq) + H_2O(l) \rightleftharpoons HCO_3^-(aq) + H_3O^+(aq)$ $K_{a1} = 4.3 \times 10^{-7}$

Step 2: $HCO_3^-(aq) + H_2O(l) \rightleftharpoons CO_3^{2-}(aq) + H_3O^+(aq)$ $K_{a2} = 4.7 \times 10^{-11}$

The value of K_{a1} is about 10^4 times that of K_{a2}; so let's proceed on the assumption that the first ionization step is the one that controls pH. In setting up an ICE table based on step 1, we will assume that the concentration of carbonic acid is a constant 1.2×10^{-5} M. This is a reasonable assumption because $[H_2CO_3]$ is controlled by the concentration of carbon dioxide in the atmosphere and the equilibrium in Equation 16.6. If there is a decrease in $[H_2CO_3]$ due to its ionization to HCO_3^- and H_3O^+, then, according to Le Châtelier's principle, more CO_2 from the atmosphere will dissolve in the rain and $[H_2CO_3]$ will be restored to 1.2×10^{-5} M. Therefore, $[H_2CO_3]$ does not change in the course of the reaction.

Let $x = [H_3O^+]$ at equilibrium. Because 1 mole of HCO_3^- is formed for every mole of H_3O^+, $[HCO_3^-]$ at equilibrium also will be x. Thus, both changes in the concentrations of the two products are $+x$. The appropriate ICE table looks like this:

	$[H_2CO_3]$	$[H_3O^+]$	$[HCO_3^-]$
Initial	1.2×10^{-5}	0	0
Change	0	$+x$	$+x$
Equilibrium	1.2×10^{-5}	x	x

Using the equilibrium terms from the ICE table in the equilibrium constant expression for K_{a1}, we have

$$K_{a1} = \frac{[H_3O^+][HCO_3^-]}{[H_2CO_3]} = \frac{(x)(x)}{1.2 \times 10^{-5}} = 4.3 \times 10^{-7}$$

Cross multiplying gives

$$x^2 = (1.2 \times 10^{-5})(4.3 \times 10^{-7}) = 5.2 \times 10^{-12}$$

Solving for x, we have

$$x = [H_3O^+] = 2.3 \times 10^{-6}\ M$$

Taking the negative logarithm of $[H_3O^+]$ to calculate pH gives

$$pH = -\log[H_3O^+] = -\log(2.3 \times 10^{-6}) = -(-5.64) = 5.64$$

SAMPLE EXERCISE 16.6: Calculate the pH of the melted-snow sample ($[H^+] = 4.61 \times 10^{-4}$ M) described in Sample Exercise 16.2 on page 788.

SOLUTION: To calculate pH, we take the negative logarithm of $[H^+]$:

$$pH = -\log[H^+] = -\log 4.60 \times 10^{-4} = 3.34$$

This pH is quite acidic, indicating that the original snow sample fell from an atmosphere polluted with SO_3, the gas that forms H_2SO_4 when it combines with water. Note also that, in this rather dilute solution of sulfuric acid, nearly all the HSO_4^- dissociates into SO_4^{2-} and H_3O^+.

PRACTICE EXERCISE: Calculate the pH of a 2.73×10^{-3} M solution of HCl. (See Problems 49 and 50.)

pOH

The use of negative logarithms permeates the chemical literature on acid–base and other equilibria. Thus, numerous expressions of solution composition and chemistry start with a lowercase "p" symbolizing the $-\log$ math function. For example, just as every solution has a pH value, it also has a *pOH* value. The latter is particularly useful in describing the strength of a basic solution. We define pOH mathematically by the equation

$$\text{pOH} = -\log [\text{OH}^-] \qquad (16.13)$$

A solution's pH and pOH values are connected as its $[\text{H}_3\text{O}^+]$ and $[\text{OH}^-]$ are connected. To see how, let's take the negative logarithm of each term in Equation 16.11:

$$-\log K_w = -\log \{[\text{H}_3\text{O}^+][\text{OH}^-]\} = -\log (1.00 \times 10^{-14})$$

Keeping in mind the mathematical rule that the logarithm of (AB) = log A + log B, we get

$$\text{p}K_w = \text{pH} + \text{pOH} = 14.00 \qquad (16.14)$$

Note that we have also taken the negative logarithm of the equilibrium constant in Equation 16.14. Many tables of equilibrium constants list pK, rather than K, values because doing so does not require the use of exponential notation and so is more convenient.

SAMPLE EXERCISE 16.7: Calculate the pOH and pH of 0.0100 M NaOH.

SOLUTION: Let's begin with the ionization reaction:

$$\text{NaOH}(aq) \longrightarrow \text{Na}^+(aq) + \text{OH}^-(aq)$$

which tells us that 1 mole of NaOH dissociates completely in water, producing 1 mole of OH^-. Therefore, a 0.0100 M solution of NaOH will be 0.0100 M in OH^-. We take the negative logarithm of this value to calculate pOH:

$$\text{pOH} = -\log [\text{OH}^-] = -\log 0.0100 = 2.00$$

The connection between pH and pOH is

$$\text{pH} + \text{pOH} = \text{p}K_w = 14.00$$

Therefore,

$$\text{pH} = 14.00 - \text{pOH} = 14.00 - 2.00 = 12.00$$

(See Problems 51 and 52.)

16.5 THE pH OF SOLUTIONS OF ACIDIC AND BASIC SALTS

Seawater and the fresh water in many rivers and lakes have pH values that range from neutral to weakly alkaline. How can they have such high pH values when the rain that serves, directly or indirectly, as their water supply has a pH of 5.6 or, as we have seen, even less? The answer is that, when rain falls to Earth and becomes surface or groundwater, its pH changes as it flows over rocks and through soils that contain basic components. One such component is calcium carbonate, $CaCO_3$, the principal constituent of limestone, marble, and shellfish shells. (It is also the active ingredient in many antacid tablets.) If acid rain containing dilute sulfuric acid soaks into soil containing $CaCO_3$, the acid is converted into either environmentally benign carbonic acid

$$CaCO_3(s) + H_2SO_4(aq) \rightleftharpoons CaSO_4(s) + H_2CO_3(aq) \quad (16.15)$$

or, if enough $CaCO_3$ is available, into calcium sulfate and soluble calcium bicarbonate:

$$2\ CaCO_3(s) + H_2SO_4(aq) \rightleftharpoons CaSO_4(s) + Ca(HCO_3)_2(aq) \quad (16.16)$$

As long as carbonates and other basic substances are present in soils and in the sediments of rivers and lakes, nature has the capacity to neutralize the acids in acid rain and maintain pH in a range that supports aquatic life.

CONNECTION: The reaction of acidic groundwater with calcium carbonate and its connection to the formation of limestone caves is described in Section 5.7.

Equations 16.15 and 16.16 are examples of acid–base neutralization reactions. Sulfuric acid is the H^+ donor; and the carbonate ion in $CaCO_3$ is the H^+ ion acceptor (base) in both reactions. The effectiveness of carbonate as a base is illustrated by the chemical reactions that take place when a soluble carbonate salt, such as Na_2CO_3, is added to water. Sodium carbonate, also known as *soda ash*, is the active ingredient in the products used to raise the pH of swimming pools. This adjustment is necessary after a rainstorm has added enough acid rain to a pool to drop its pH below the range (7.2–7.4) comfortable for swimming.

CONNECTION: Neutralization reactions were introduced in Section 5.6.

Like ammonia (see Section 16.2) and other bases, carbonate reacts with water when it dissolves, splitting water molecules into H^+ and OH^-. The CO_3^{2-} ion acquires the H^+ ion, forming HCO_3^- and releasing an OH^- ion:

$$CO_3^{2-}(aq) + H_2O(l) \rightleftharpoons HCO_3^-(aq) + OH^-(aq) \quad (16.17)$$

The corresponding K_b expression is

$$K_b = \frac{[HCO_3^{2-}][OH^-]}{[CO_3^{2-}]}$$

Appendix 5 contains K_b values for a number of bases but not for the CO_3^{2-} ion. The reason for its omission is that you should be able to calculate K_b from K_{a2} for H_2CO_3. How are the basic properties of CO_3^{2-} connected to the acidic properties of H_2CO_3? To find the answer, consider the second step in the ionization of carbonic acid:

$$HCO_3^-(aq) + H_2O(l) \rightleftharpoons CO_3^{2-}(aq) + H_3O^+(aq)$$

Its equilibrium constant expression, K_{a2}, is

$$K_{a2} = \frac{[CO_3^{2-}][H_3O^+]}{[HCO_3^-]} = 4.7 \times 10^{-11}$$

Look closely at the K_b expression for carbonate and the K_{a2} expression for carbonic acid. Each resembles the reciprocal of the other. The only difference is the presence of a term for $[H_3O^+]$ or $[OH^-]$ in one that is not in the reciprocal of the other.

> How are the basic properties of CO_3^{2-} connected to the acidic properties of H_2CO_3?

Ground limestone, sometimes applied from the air, may be used to neutralize the effects of acid rain on lakes and ponds.

However, we have seen that $[H_3O^+]$ and $[OH^-]$ are connected by K_w. Consider what happens when we multiply K_w by the reciprocal of K_{a2} for carbonic acid:

$$K_w \cdot \left(\frac{1}{K_{a2}}\right) = [OH^-][H_3O^+] \cdot \frac{[HCO_3^-]}{[H_3O^+][CO_3^{2-}]} = \frac{[OH^-][HCO_3^-]}{[CO_3^{2-}]}$$

The resulting expression matches the K_{b1} expression for the carbonate ion. From the values of K_w and K_{a2}, we can calculate the value of K_b:

$$K_{b1} = \frac{K_w}{K_{a2}} = \frac{1.0 \times 10^{-14}}{4.7 \times 10^{-11}} = 2.1 \times 10^{-4}$$

The inverse relation between the K_a of an acid (HCO_3^-, in this case) and the K_b of its conjugate base (CO_3^{2-}), holds for all conjugate pairs. The relation expressed in equation form is

$$K_b = K_w/K_a \qquad (16.17)$$

This equation reinforces the complementary nature of an acid and its conjugate base: as the strength (K_a) of the acid increases, the strength (K_b) of its conjugate base decreases, and vice versa. In the case under discussion, the bicarbonate ion is very weakly acidic ($K_a = 4.7 \times 10^{-11}$), whereas the carbonate ion is a moderately strong base ($K_b = 2.1 \times 10^{-4}$).

A similar calculation based on the first ionization step of carbonic acid

$$H_2CO_3(aq) + H_2O(l) \rightleftharpoons HCO_3^-(aq) + H_3O^+(aq) \qquad K_{a1} = 4.3 \times 10^{-7}$$

allows us to calculate the K_{b2} value for CO_3^{2-} and the reaction:

$$HCO_3^-(aq) + H_2O(l) \rightleftharpoons H_2CO_3(aq) + OH^-(aq) \qquad (16.18)$$

$$K_{b2} = \frac{K_w}{K_{a1}} = \frac{1.0 \times 10^{-14}}{4.3 \times 10^{-7}} = 2.3 \times 10^{-8}$$

In much the way the sequential acid ionization reactions of carbonic acid:

$$H_2CO_3 \xrightarrow[K_{a1}]{-H^+} HCO_3^- \xrightarrow[K_{a2}]{-H^+} CO_3^{2-}$$

are described by acid ionization equilibrium constants K_{a1} and K_{a2}; the ability of CO_3^{2-} to accept H^+ ions is related to the values of K_{b1} and K_{b2}:

$$CO_3^{2-} \xrightarrow[K_{b1}]{+H^+} HCO_3^- \xrightarrow[K_{b2}]{+H^+} H_2CO_3$$

where

$$K_{b1} = \frac{K_w}{K_{a2}} \quad \text{and} \quad K_{b2} = \frac{K_w}{K_{a1}}$$

Let's use K_{b1} to calculate the pH of $0.100\ M$ Na_2CO_3. In setting up an ICE table, we assume that the only important source of OH^- is the carbonate reaction with water and that the autoionization of water will not contribute significantly to $[OH^-]$ at equilibrium. Let x be the equilibrium value of $[OH^-]$. Then $[HCO_3^-]$ also is x. The changes in the two must both be $+x$, and the change in $[CO_3^{2-}]$ must be $-x$. Completing the ICE table, we have

	$[CO_3^{2-}]$	$[HCO_3^-]$	$[OH^-]$
Initial	0.100	0	0
Change	$-x$	$+x$	$+x$
Equilibrium	$0.100 - x$	x	x

Using the equilibrium terms in the K_{b1} expression for carbonate, we have

$$K_{b1} = 2.1 \times 10^{-4} = \frac{[HCO_3^{2-}][OH^-]}{[CO_3^{2-}]} = \frac{(x)(x)}{0.100 - x}$$

Let's simplify the rest of the calculation by assuming that the value of x will be much less than $0.100\ M$. If it is, then

$$\frac{x^2}{0.100} = 2.1 \times 10^{-4}$$

$$x^2 = 2.1 \times 10^{-5}$$

$$x = 4.6 \times 10^{-3}\ M = [OH^-]$$

Now is a good time to check on the validity of our two assumptions. Is the calculated $[OH^-]$ much greater than $[OH^-]$ in pure water ($1.0 \times 10^{-7}\ M$)? Indeed it is; so the assumption that the autoionization of water could be ignored was valid. As for the assumption that there was no significant decrease in $[CO_3^{2-}]$ as the reaction proceeded to equilibrium, there was in fact a 4.6% decrease:

$$\frac{\text{change in }[CO_3^{2-}]}{\text{initial }[CO_3^{2-}]} = \frac{-4.6 \times 10^{-3}\ \cancel{M}}{0.100\ \cancel{M}}(100\%) = -4.6\%$$

This value is just below our 5% acceptable error guideline (see Section 16.1). To assess its effect, let's redo the calculation, leaving $-x$ in the denominator:

$$\frac{x^2}{0.100 - x} = 2.1 \times 10^{-4}$$

By cross multiplying

$$x^2 = 2.1 \times 10^{-5} - 2.1 \times 10^{-4} x$$

and rearranging the terms

$$x^2 + 2.1 \times 10^{-4} x - 2.1 \times 10^{-5} = 0$$

we have a quadratic equation whose positive solution is 4.5×10^{-3}. This value is quite close to the initial result of 4.6×10^{-3}, and our simplification is acceptable for most purposes.

To calculate pH from [OH⁻] let's start by calculating pOH:

$$\text{pOH} = -\log[\text{OH}^-] = -\log[4.6 \times 10^{-3}] = 2.34$$

Equation 16.21 links pH and pOH. Rearranging the terms and inserting the pOH value just calculated, we have

$$\text{pH} = pK_w - \text{pOH} = 14.00 - 2.34 = 12.66$$

A pH value of 12.66 is quite basic. If you swam in pool water at that pH you would experience skin irritation and painful burning in your eyes. To make sure the right amount of soda ash is used in swimming pools, small quantities are usually added and the pH of the water is tested. Usually these tests rely on color indicators of pH, which are described in the next section.

If the salt of a weak acid is basic, then it makes sense that the salt of a weak base should be acidic. Consider ammonium chloride. It is the salt that is produced when NH_3 (weak base) is neutralized with a strong acid (HCl):

$$NH_3(aq) + HCl(aq) \rightleftharpoons NH_4Cl(aq)$$

Given this weak-base, strong-acid pedigree, we might expect NH_4Cl to be, on balance, an acidic salt and the pH values of solutions of NH_4Cl to be less than 7.00. These expectations are correct, as shown in the following sample exercise.

SAMPLE EXERCISE 16.8: Calculate the pH of 0.25 M NH_4Cl.

SOLUTION: As noted, NH_4Cl is the salt of a weak base and a strong acid, and so the result of this calculation should be a pH value less than 7.00. The acidic properties of NH_4Cl are derived from the ability of the ammonium ion, NH_4^+, to act as a H^+ donor:

$$NH_4^+(aq) + H_2O(l) \rightleftharpoons NH_3(aq) + H_3O^+(aq)$$

The equilibrium constant expression for this reaction

$$K_a = \frac{[NH_3][H_3O^+]}{[NH_4^+]}$$

is connected to the equilibrium constant expression for the conjugate base of NH_4^+, namely, NH_3

> ✓ The Cl⁻ ion is a pathetically weak H^+ ion acceptor (see Figure 16.8), and so it does not contribute to the acid-base properties of NH_4Cl.

$$K_b = \frac{[NH_4^+][OH^-]}{[NH_3]} = 1.76 \times 10^{-5}$$

by Equation 16.17:

$$K_b = K_w/K_a$$

Rearranging the terms in Equation 16.17 to solve for K_a and then inserting the numerical value of K_b for NH_3, we have

$$K_a = K_w/K_b = \frac{1.0 \times 10^{-14}}{1.76 \times 10^{-5}} = 5.6 \times 10^{-10} = \frac{[NH_3][H_3O^+]}{[NH_4^+]}$$

Now let's set up the ICE table for the reaction in which we make the usual assumptions that the reaction is the only significant source of H_3O^+ and that $x = [H_3O^+]$ at equilibrium:

	$[NH_4^+]$	$[NH_3]$	$[H_3O^+]$
Initial	0.25	0	0
Change	$-x$	$+x$	$+x$
Equilibrium	$0.25 - x$	x	x

Inserting the equilibrium concentration terms into the equilibrium constant expression for K_a gives

$$K_a = 5.6 \times 10^{-10} = \frac{[NH_3][H_3O^+]}{[NH_4^+]} = \frac{(x)(x)}{0.25 - x}$$

Let's simplify the rest of the calculation by assuming that the value of x will be much less than 0.25. If it is, then

$$\frac{x^2}{0.25} = 5.6 \times 10^{-10}$$

$$x^2 = 1.4 \times 10^{-10}$$

$$x = 1.2 \times 10^{-5} = [H_3O^+]$$

Taking the negative logarithm of $[H_3O^+]$ to obtain the pH of the solution, we have

$$pH = -\log[H_3O^+] = -\log[1.2 \times 10^{-5}] = 4.9$$

This result is reasonable: a solution of a very weak acid ($K_a = 5.6 \times 10^{-10}$) should be weakly acidic. The calculated $[H_3O^+]$ is more than 100 times that produced by the autoionization of water (1.0×10^{-7} M); so ignoring the contribution of the latter process is permitted. So, too, is assuming that x is small compared with 0.25 M.

PRACTICE EXERCISE: The acetate ion is the conjugate base of acetic acid ($K_a = 1.76 \times 10^{-5}$) and reacts with water according to the following reaction equation: $C_2H_3O_2^-(aq) + H_2O(l) \rightleftharpoons HC_2H_3O_2(aq) + OH^-(aq)$. Calculate the pH of 0.25 M sodium acetate, $NaC_2H_3O_2$. (See Problems 73 and 74.)

16.6 BUFFER SOLUTIONS AND THE pH OF NATURAL WATERS

Suppose that a sample of river water contains 1.2×10^{-5} M carbonic acid as a result of atmospheric carbon dioxide dissolving in it. Let's also suppose that the sediments of the river contain limestone or other carbonate minerals. We have seen that the carbonate ion is an effective base. This property allows an acid–base reaction to take place in which solid calcium carbonate reacts with carbonic acid, forming soluble calcium bicarbonate:

$$CaCO_3(s) + H_2CO_3(aq) \rightleftharpoons Ca(HCO_3)_2(aq)$$

Other acidic substances in the river water also would be neutralized by carbonate minerals (see, for example, Equation 16.15). All of these neutralization reactions produce bicarbonate ions. Suppose that the total concentration of bicarbonate ion produced by these reactions is 1.0×10^{-4} M. What would the pH of the water be?

We know the concentrations of both carbonic acid and bicarbonate ion. These substances are related by the ionization reaction of carbonic acid

$$H_2CO_3(aq) + H_2O(l) \rightleftharpoons HCO_3^-(aq) + H_3O^+(aq) \quad (16.19)$$

and by the following mass action expression:

$$K_{a1} = \frac{[HCO_3^-][H_3O^+]}{[H_2CO_3]} = 4.3 \times 10^{-7}$$

The difference between this river water and a solution containing only carbonic acid is that the only HCO_3^- in a dilute solution of H_2CO_3 is due to the ionization of H_2CO_3. The river water, however, has an additional source of HCO_3^-. According to Le Châtelier's principle, added HCO_3^- should shift the chemical equilibrium in Equation 16.19 to the left. This would lower $[H_3O^+]$ and so raise pH to values above the 5.6 calculated earlier for natural rainwater.

This phenomenon illustrates a principle known as the **common-ion effect**. In any ionic equilibrium, the reaction that produces an ion is suppressed when another source of the same ion is added to the system. In the river-water sample, the ionization of H_2CO_3 is suppressed by additional HCO_3^- from neutralization reactions in which carbonate minerals take part. To assess the effect of the added bicarbonate, we will insert the given values for $[H_2CO_3]$ and $[HCO_3^-]$ into the K_{a1} expression (letting $x = [H_3O^+]$):

$$K_{a1} = \frac{[HCO_3^-][H_3O^+]}{[H_2CO_3]} = \frac{(1.0 \times 10^{-4})x}{1.2 \times 10^{-5}} = 4.3 \times 10^{-7}$$

Solving for x, we obtain a hydronium-ion concentration of 5.2×10^{-8} M. Taking the negative logarithm of this value, we find that the pH of the river water is 7.29. Thus, the presence of bicarbonate results in a solution with a pH nearly two units above that of carbonic acid alone.

In addition to having a slightly alkaline pH, our river water also has a capacity to neutralize any acid rain that falls into it. Let's examine the effect of adding a

> The **common-ion effect** is the shift in the position of an equilibrium caused by the addition of an ion taking part in the reaction. We encountered this effect in the second ionization step of sulfuric acid (Section 16.1).

small amount of a strong acid on the river water's pH and on the pH of the same volume of pure (pH 7.00) water at 25°C. We will start with either 1.00 L of pure water or 1.00 L of pH 7.29 river water containing 1.0×10^{-4} M bicarbonate ion. To these samples, we will add 10. mL of 1.0×10^{-3} M HNO_3.

Adding acid to pure water is an exercise in dilution: 10 mL of acid diluted to a final volume of 1.01 L. The number of moles of acid is calculated as follows:

$$(10 \text{ mL}) \left(\frac{1.0 \times 10^{-3} \text{ mol}}{\text{L}} \right) \left(\frac{1 \text{ L}}{1000 \text{ mL}} \right) = 1.0 \times 10^{-5} \text{ mol}$$

and is not changed by the dilution. In other words:

Number of moles of acid present initially =
number of moles of acid present finally (after dilution)

Restating this principle in equation form and using the subscripts "i" for the initial acid solution and "f" for the final diluted solution, we have

$$V_i \times M_i = V_f \times M_f$$

Expressing 10 mL as 0.010 L and substituting the given information, we have

$$(0.010 \text{ L})(1.0 \times 10^{-3} \text{ M}) = (1.01 \text{ L})(x)$$

$$x = 9.9 \times 10^{-6} \text{ M}$$

CONNECTION: Calculations involving dilutions and the equation $V_i \times M_i = V_f \times M_f$ were introduced in Section 5.2.

If we assume that nitric acid is completely dissociated, the final concentration of H_3O^+ also must be 9.9×10^{-6} M. The corresponding pH value ($-\log 9.9 \times 10^{-6}$ M) is 5.00. Thus, adding 10 mL of 1.0×10^{-3} M nitric acid to a 1-liter sample of pure water lowers the pH of the sample from 7.00 to 5.00.

To assess the effect of adding the same amount of nitric acid to a liter of our river water, we need to reexamine the carbonic acid–bicarbonate equilibrium:

$$H_2CO_3(aq) + H_2O(l) \rightleftharpoons HCO_3^-(aq) + H_3O^+(aq)$$

If strong acid (H_3O^+) is added to this mixture, the reaction's equilibrium position shifts to the left, consuming some of the bicarbonate and producing carbonic acid. If the river water is saturated with respect to atmospheric carbon dioxide, then the concentration of carbonic acid will not increase permanently; rather the excess will decompose into carbon dioxide and water:

$$H_2CO_3(aq) \rightleftharpoons CO_2(g) + H_2O(l)$$

Because the hydronium ion is a much stronger acid than carbonic acid, H_3O^+ ions will react with HCO_3^- ions:

$$H_3O^+(aq) + HCO_3^-(aq) \longrightarrow H_2CO_3(aq) + H_2O(l)$$

until essentially all of the 1.0×10^{-5} moles of H_3O^+ that was added has been consumed. Therefore, 1.0×10^{-5} moles of HCO_3^- is consumed. The number of moles of bicarbonate present initially was

$$(1.0 \times 10^{-4} \text{ M}) \times (1.00 \text{ L}) = 1.0 \times 10^{-4} \text{ mol}$$

Subtracting the number of moles consumed, we have 9.0×10^{-5} moles of HCO_3^-

in a final volume of 1.01 L. The resulting pH can be predicted from the K_{a1} for carbonic acid. Letting $x = [H_3O^+]$ we have:

$$K_{a1} = \frac{[HCO_3^-][H_3O^+]}{[H_2CO_3]} = \frac{\left(\frac{9.0 \times 10^{-5} \text{ mol}}{1.01 \text{ L}}\right)x}{1.2 \times 10^{-5} \text{ M}} = 4.3 \times 10^{-7}$$

Cross multiplying and solving for x, gives

$$x = [H_3O^+] = 5.7 \times 10^{-8} \text{ M}$$

Taking $-\log [H_3O^+]$:

$$pH = -\log 5.7 \times 10^{-8} = 7.24$$

Recall that the original pH was 7.29. Thus the addition of nitric acid lowered the pH of the river water by only 0.05 pH units.

Not only does bicarbonate in the river water minimize the changes in pH when small amounts of acid are added, the carbonic acid that is also present provides the capacity to neutralize small additions of strong base because of the following neutralization reaction:

$$H_2CO_3(aq) + OH^-(aq) \longrightarrow HCO_3^-(aq) + H_2O(l)$$

The resulting increase in $[HCO_3^-]$ will increase pH but only slightly. The beauty of this base-neutralization reaction is that, as H_2CO_3 is consumed, more carbon dioxide from the atmosphere dissolves in the water, restoring $[H_2CO_3]$:

$$CO_2(g) + H_2O(l) \rightleftharpoons H_2CO_3(aq)$$

In this way, the atmosphere provides an enormous reservoir of base-neutralization capacity.

A solution that has the capacity to resist pH change by neutralizing small additions of acid or base is called a **pH buffer.** Buffers are an important component of natural water systems; indeed, they are also essential to life itself. The internal pH of most living cells is regulated by buffer systems, including the carbonic acid–bicarbonate system just described. When pH stabilization provided by these buffers is disturbed, proteins do not function optimally and the health of individual cells and whole organisms is in jeopardy. For example, people on crash diets may suffer from *acidosis* as the consumption of body fat for energy produces organic acid by-products, such as acetoacetic acid and β-hydroxybutyric acid. These compounds may cause blood pH to fall below its normal range of 7.2 to 7.4. A result of such a drop is impaired O_2 transport by hemoglobin (see Section 16.10), which can lead to lethargy and depression.

The capacity of a buffer to resist pH change is a function of the concentrations of its acid and base components: the greater their concentrations, the greater the **buffer capacity** (Figure 16.14).

Optimum buffering action is achieved when the concentrations of the acid and base components of the buffer system are equal. To examine the significance of this equality, let's consider a generic form of the acid-ionization process:

$$acid \rightleftharpoons base + H^+$$

> A **pH buffer** is a solution of acidic and basic solutes that resists changes in its pH when acids or bases are added to it.

> **Buffer capacity** is the quantity of acid or base that can be added to a pH buffer without significantly changing the pH of the buffer.

FIGURE 16.14 The changes in pH of buffer solutions caused by adding strong acid (red curve) or strong base (blue curve) increase as the concentrations of the buffers decrease. In this illustration, five 100-mL solutions containing 1.00, 0.30, 0.10, 0.03, and 0.015 M acetic acid and sodium acetate all have an initial pH of 4.75. The graph shows the pH values of these solutions after adding 1.00 mL of 1.00 M HCl or 1.00 mL of 1.00 M NaOH.

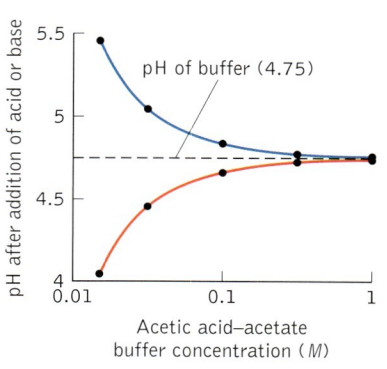

which has the equilibrium constant expression

$$K_a = \frac{[\text{H}^+][\text{base}]}{[\text{acid}]}$$

Now let's take $-\log$ of both sides of the mass action expression. Doing so transforms K_a into pK_a and $[\text{H}^+]$ into pH:

$$pK_a = \text{pH} - \log\frac{[\text{base}]}{[\text{acid}]}$$

Rearranging the terms to solve for pH, we have

$$\text{pH} = pK_a + \log\frac{[\text{base}]}{[\text{acid}]} \tag{16.20}$$

Equation 16.20 is a particularly useful one and is called the **Henderson-Hasselbalch equation.** It enables us to calculate the pH of a solution in which we know the concentration of an acid and its conjugate base (or a base and its conjugate acid). However, do not try to use it to calculate the pH of a solution of

> ✓ The **Henderson-Hasselbalch equation** is used to calculate the pH of a solution in which the concentrations of acid and conjugate base are known:
>
> $\text{pH} = pK_a + \log\frac{[\text{base}]}{[\text{acid}]}$.

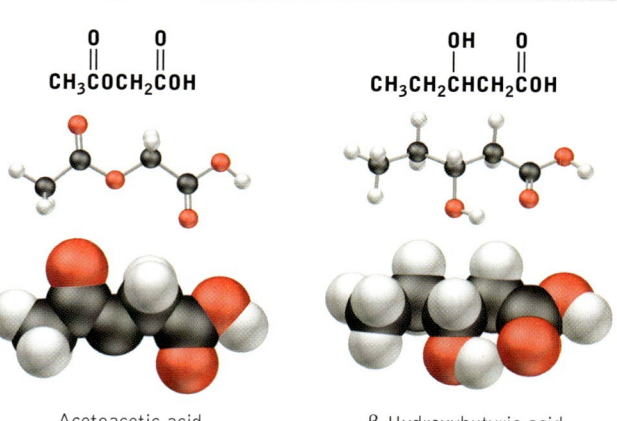

High concentrations of acetoacetic and β-hydroxybutyric acids, by-products of crash diets, can cause blood pH to fall.

Acetoacetic acid   β-Hydroxybutyric acid

only a weak acid or weak base. That approach does not work, because both pH and either the numerator or denominator would be unknown, and you cannot solve for two unknowns with only one equation. Instead, use the approaches described earlier to calculate the pH values of solutions containing only weak acids or only weak bases.

Consider what happens to the logarithmic term in the Henderson-Hasselbalch equation when the concentration of the acid is the same as that of its conjugate base. Then the numerator and denominator in the log term are equal, and the value of the fraction is 1. The log of 1 is 0, and so $pH = pK_a$. This equality serves as a handy reference point in an acid–conjugate base system. If the concentration of the basic component is greater than that of the acid, then the logarithmic term is greater than zero; so $pH > pK_a$. If the concentration of the basic component is less than that of the acid, then the logarithmic term is less than zero; so $pH < pK_a$. Consider the case in which the concentration of the base is ten times the concentration of the acid, or $[base] = 10[acid]$. Substituting this ratio into the Henderson-Hasselbalch equation, we have

$$pH = pK_a + \log \frac{[base]}{[acid]}$$

$$= pK_a + \log \frac{10 \cdot [acid]}{[acid]}$$

$$= pK_a + \log 10 = pK_a + 1$$

Therefore, a 10-fold higher concentration of base than acid produces a pH one unit above the pK_a value. Similarly, if the concentration of the acid component is 10 times that of the base, pH is one unit below pK_a. A buffer has the most buffer capacity within these pH values—that is, when $pH = pK_a \pm 1$. In this range, a buffer solution will resist pH change from additions of either acid or base. As you might guess, buffer capacity is maximum when $pH = pK_a$.

SAMPLE EXERCISE 16.9: Calculate the pH of a buffer that is 0.200 M in NH_3 and 0.300 M in NH_4Cl.

SOLUTION: The Henderson-Hasselbalch equation can be used to calculate the pH of a solution of an acid and its conjugate base. The ingredients in this buffer include such an acid–conjugate-base pair in NH_4^+ (NH_4Cl dissociates completely, forming NH_4^+ and Cl^-) and NH_3. The strength of NH_3 as a base is given by the value of its K_b: 1.76×10^{-5}. The corresponding pK_b value is

$$pK_b = -\log K_b = -\log 1.76 \times 10^{-5} = 4.75$$

This pK_b value is related to the pK_a value of the conjugate acid (NH_4^+) by the following equation:

$$pK_b + pK_a = 14.00$$

Thus,

$$pK_a = 14.00 - pK_b = 14.00 - 4.75 = 9.25$$

Inserting this value and the given concentrations of NH_3 and NH_4^+ into the Henderson-Hasselbalch equation, we have

$$pH = pK_a + \log \frac{[\text{base}]}{[\text{acid}]}$$

$$= 9.24 + \log \frac{0.200}{0.300} = 9.07$$

The K_b value of NH_3 is greater than the K_a value of NH_4^+. In other words, ammonia is a stronger base than the ammonium ion is an acid. Therefore, a solution containing similar concentrations of both should be slightly basic.

PRACTICE EXERCISE: Calculate the pH of a buffer that is 0.150 M in benzoic acid and 0.100 M in sodium benzoate. (See Problems 77 and 78.)

16.7 ACID–BASE INDICATORS

As noted in Section 16.5, pool owners routinely check the pH of their pools by using colored pH indicators. These indicators are substances that change color with changing pH. One such compound is phenol red (Figure 16.15). It functions as a *colorimetric* (based on color) **pH indicator** because (1) it is a weak acid (pK_a = 7.6) and (2) the un-ionized form (which, for convenience, we assign the generic formula HIn) of phenol red is actually yellow, but the ionized (In$^-$) form is violet. At a pH one unit above the pK_a, or 8.6, the ratio of In$^-$ to HIn is, according to the Henderson-Hasselbalch equation, 10:1, and a phenol-red solution is violet. At pH less than 6.6, phenol red is largely un-ionized, and so a solution of the indicator is yellow. In the pH range from about 6.8 to 8.4, the color of a

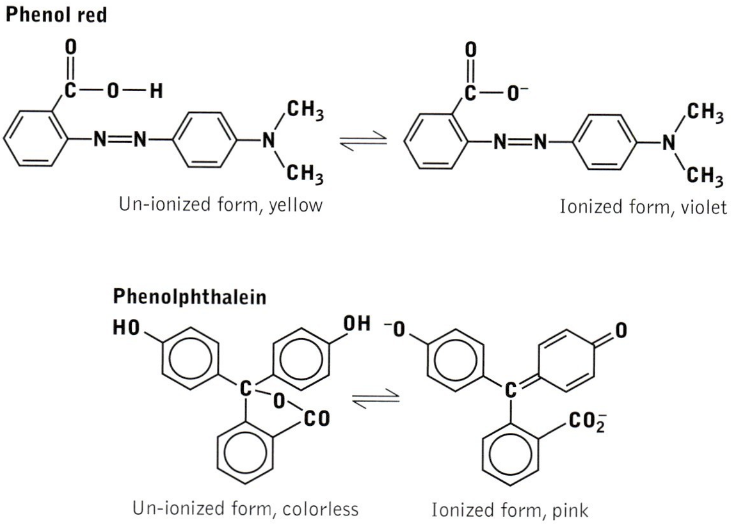

> A **pH indicator** is a water-soluble weak organic acid (HIn) that changes color when it ionizes: HIn → H$^+$ + In$^-$.

FIGURE 16.15 Molecular structures of phenol red and phenolphthalein.

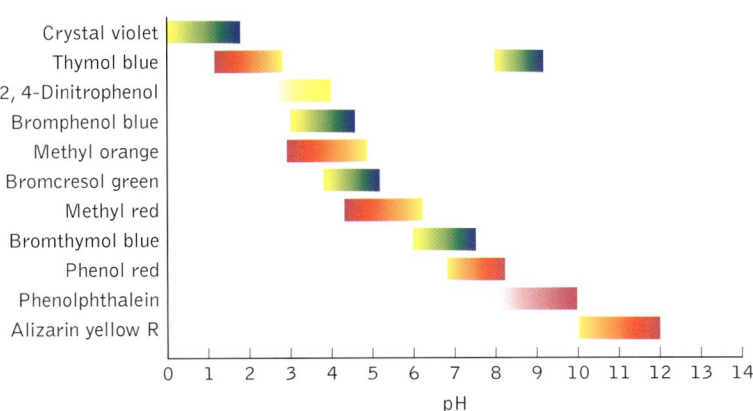

FIGURE 16.16 pH indicators are used to determine the pH of solutions and pH changes in titrations. An indicator is useful within a range of 1 pH unit above and below the pK_a value for the acid group of the indicator.

phenol-red solution changes from yellow to orange to red to violet with increasing pH. These pH-dependent color changes allow the pH of a pool to be adjusted to the optimal pH for a swimmer's comfort (~7.4) to within about 0.2 pH units of the desired value. The analysis consists of adding a few drops of concentrated indicator to about 5 to 10 mL of sample and comparing the color of the sample with a pH color chart. If the pH of the water is too low (the usual problem), Na_2CO_3 is added and the pH test is repeated.

Phenol red is suitable for testing pool water because its pK_a is close to the desired pH, and so the pH range over which it changes color is centered on the desired value. Other color indicators are used to detect changes in pH over other pH ranges, as shown in Figure 16.16. All of them have a useful pH range from 1 pH unit below to 1 pH unit above their individual pK_a values; that is,

$$\text{pH range} = pK_a \pm 1.0 \qquad (16.21)$$

Indicators may be used to determine the pH of solutions, and, as we are about to see, they may also be used to detect the large changes in pH that occur in determinations of the concentrations of acids and bases in acid–base titrations.

16.8 ACID–BASE TITRATIONS

The term *alkalinity* is sometimes used to indicate the buffer capacity of a water sample with respect to resisting pH lowering due to additions of strong acid. Although the buffer mixtures found in natural waters can be more complex than in our model river water, the carbonic acid–bicarbonate buffer system is often

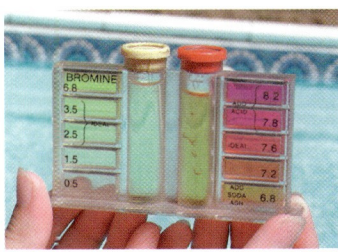

A

B

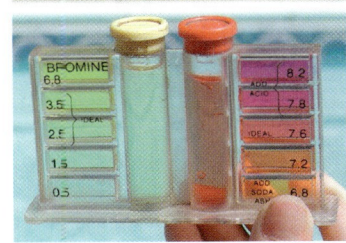

C

Many pool test kits include the pH indicator phenol red. A few drops are added to a sample of pool water collected in the tube with the red cap. A. After a rainstorm, the pH of a pool sample is 6.8 (or less), as indicated by the yellow color of the sample. B. Soda ash (Na_2CO_3) is added to the pool to raise pH. C. A follow-up test produces a red-orange color, indicating that the pH of the pool has been properly adjusted.

the most important one. Acid rain poses a threat to an aquatic ecosystem when the alkalinity of the water—that is, its buffer capacity—is exceeded. Most aquatic ecosystems tolerate slightly acidic pH, but many fish cannot tolerate pH values less than 5.5, and most aquatic life disappears at pH < 5.0.

How is alkalinity determined? The technique commonly used is acid–base titration. The mechanics of titrations were introduced in Section 5.9 (see Figure 5.33): a known volume of sample undergoes reaction with a known volume of a standard solution (of known concentration). These three "knowns" allow us to determine the concentration of the substance of interest in the sample and, depending on our skill, to make this determination with considerable precision and accuracy. It is not unusual to obtain titration results that are repeatable (precise) to three or even four significant figures.

How is alkalinity determined?

CONNECTION: The apparatus used in titration measurements is described in Section 5.9.

To determine the alkalinity of a sample, a few drops of a pH color indicator are added to a known volume of the sample before a strongly acidic standard solution (titrant) is slowly added to the sample from a buret. If carbonate is present in the sample, then the first additions of strong acid will convert carbonate into bicarbonate:

$$\underset{\text{(titrant)}}{H_3O^+ (aq)} + \underset{\text{(sample)}}{CO_3^{2-} (aq)} \longrightarrow HCO_3^- (aq) + H_2O(l)$$

Until the carbonate in the sample is almost completely consumed, the pH of the sample does not change much (see Figure 16.17) as titrant is added. However, when just enough titrant has been added to consume all the carbonate present in the sample, the pH of the sample drops sharply at the first equivalence point—a change that is detected by a change in the color of the pH indicator. The sharpest drop in pH occurs when the number of moles of acid added exactly matches the number of moles of carbonate initially in the sample. At this point, essentially all of the CO_3^{2-} present initially has been converted into HCO_3^-.

The conversion of carbonate into bicarbonate is the first stage of our alkalinity titration. In the second stage, the bicarbonate formed in the first stage plus any bicarbonate present in the original sample reacts with additional acid titrant,

$$\underset{\text{(titrant)}}{H_3O^+ (aq)} + \underset{\text{(sample)}}{HCO_3^- (aq)} \longrightarrow H_2CO_3(aq) + H_2O(l)$$

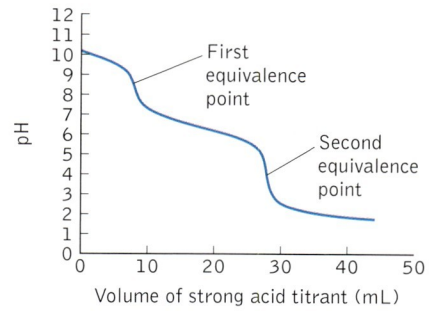

FIGURE 16.17 These results of an alkalinity titration show two equivalence points: the first marks the complete conversion of carbonate into bicarbonate, and the second marks the conversion of bicarbonate into carbonic acid.

forming carbonic acid. If more carbonic acid is produced than is soluble, then the carbonic acid leaves the solution in the form of bubbles of carbon dioxide:

$$H_2CO_3(aq) \longrightarrow H_2O(l) + CO_2(g)$$

In the second stage, the titration curve (see Figure 16.17) has a second region in which added acid has little effect on pH. Then, as the bicarbonate in the sample is completely consumed, pH again drops sharply, producing a second equivalence point that is detected by the change in color of a second indicator.

Note that the initial pH of the sample is slightly above 10, which is quite alkaline and above the pH range tolerated by many species of aquatic life. However, such alkaline water may be found in arid regions, such as the U.S. Southwest, where there are deposits of limestone and other basic minerals. The pH of most seawater and fresh water is 8.2 or less, which corresponds to points after the first equivalence point in Figure 16.17. Therefore, there can be little carbonate present in these waters, and most of their alkalinity is due to the presence of bicarbonate. The alkalinity titration curves for these samples will have only one equivalence point at the pH of the second equivalence point in Figure 16.17.

Lakes and springs in the desert southwest of the United States can have highly alkaline water owing to the presence of basic minerals containing, for example, carbonate (CO_3^{2-}) and tetraborate ($B_4O_7^{2-}$) ions.

CONCEPT TEST: In the alkalinity titration of a sample that initially contains both CO_3^{2-} and HCO_3^-, the volume of titrant required to reach the first equivalence point is less than that required to go from the first to the second. Why?

Which color indicators could be used to detect the equivalence points in an alkalinity titration? The phenol-red indicator used to test swimming-pool water would not be a good candidate, because it changes color between pH 6.8 and 8.4. This range is just below the pH of the first equivalence point and well above the second. To detect the first equivalence point, we need a color indicator with a pK_a near the pH of the first equivalence point (8.5). One candidate is phenolphthalein ($pK_a = 8.8$), which is pink in its basic form and colorless at low pH. To detect the second equivalence point, we could add bromcresol green ($pK_a = 4.6$) after the first equivalence point has been reached. We would not add it before the first equivalence point, because its blue green color in basic solutions would obscure the pink-to-colorless transition of phenolphathalein. On the other hand, phenolphthalein would not obscure the color change obtained by using bromcresol green because phenolphthalein is colorless in acidic solutions.

> Which color indicators could be used to detect the equivalence points in an alkalinity titration?

Alkalinity determinations are only one of many applications of acid–base titrations. Some entail the titration of a base that can accept only 1 mole of H_3O^+ ions per mole of base, such as NaOH. These titrations have only a single equivalence point, as shown in Figure 16.18. The size of the pH change at the equivalence point decreases with decreasing concentration of base in the sample, but even millimolar concentrations of strong bases produce titration curves with sharp breaks that can be easily detected.

FIGURE 16.18 The titration curves for 20.0 mL of 0.100 M NH₃ (green) and 20.0 mL of 0.100 M NaOH (blue) with 0.100 M HCl differ in the region before the equivalence point but not in the region after it. After the equivalence point, the bases are completely consumed and the pH of the titration solution is controlled only by the increasing amount of excess HCl that has been added.

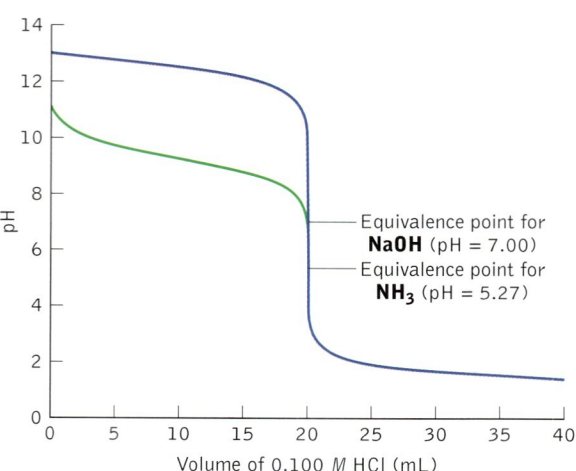

The pH change near the equivalence point in the titration of a weak base, such as NH_3, is not as great as that of a strong base, as shown in the titration curves in Figure 16.18. The starting pH of a weak base is not as high as that of a strong base, and so the decrease in pH at the equivalence point is not as great. Note that, at the equivalence point in the ammonia titration, pH = 5.27, not 7.00. Why is the pH not 7.00? To answer this question, we need to think about what ions are in solution at the equivalence point in the titration. The titration reaction is:

$$HCl(aq) + NH_3(aq) \longrightarrow NH_4^+(aq) + Cl^-(aq)$$

Therefore, when 0.100 M HCl is added to an equal volume of 0.100 M NH_3, as in the titration in Figure 16.18, we have at the equivalence point a solution that is 0.0500 M in NH_4Cl (do you see why?). As stated in Section 16.5, NH_4Cl is an acidic salt because it is the salt of a strong acid (HCl) and a weak base (NH_3). Thus, solutions of NH_4Cl are expected to have pH values less than 7.00.

CONCEPT TEST: How does the pH of a solution of NH_4Cl change with increasing concentration of this salt?

Just as the concentrations of basic compounds can be determined by titrating them with known quantities of strong acids, the concentrations of acidic compounds can be determined by titrating them with standard solutions of strong base. Figure 16.19 illustrates two such applications: in one of them, 20.0 mL of 0.100 M HCl is titrated with 0.100 M NaOH; in the other, the sample is 20.0 mL of 0.100 M acetic acid ($HC_2H_3O_2$). As in Figure 16.18, the two titration curves differ until their equivalence points are reached (pH values are consistently

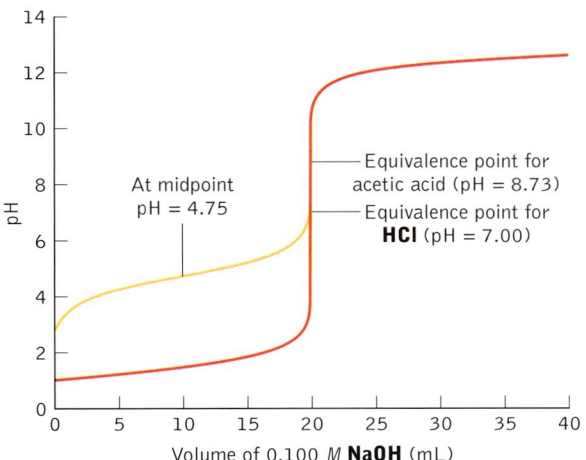

FIGURE 16.19 The titration curves for 20.0 mL of 0.100 M acetic acid ($HC_2H_3O_2$, purple) and 20.0 mL of 0.100 M HCl (red) with 0.100 M NaOH differ in the region before the equivalence point. The titration reaction with the weaker acid has higher pH values until the end point is reached at 20.0 mL of NaOH, where the pH = 8.73.

higher for the weak acid), but they overlap beyond their equivalence points. In the region after the equivalence points, pH is controlled by the increasing concentration of the same NaOH titrant used in both experiments.

Two pH values along the titration curve of acetic acid deserve our attention. The first is at the equivalence point (pH = 8.7). This value is well above pH 7.0, even though it is the point in the titration at which just enough NaOH has been added to exactly neutralize the acetic acid in the sample. The pH of the equivalence point is slightly basic because the product of the titration reaction

$$HC_2H_3O_2(aq) + NaOH(aq) \longrightarrow NaC_2H_3O_2(aq) + H_2O(l)$$

is a 0.0500 M solution of sodium acetate ($NaC_2H_3O_2$). Sodium acetate is the salt of weak acid and strong base, and so its aqueous solutions should be slightly basic (see Sample Exercise 16.8 on page 809 and its paired practice exercise).

The other important milestone in the titration of acetic acid is halfway to the equivalence point. At that point, half of the acetic acid initially in the sample has been converted into acetate ion. Therefore the concentrations of the acetic acid still in solution and the acetate that has been produced are the same. This equality means that the logarithmic term in the Henderson-Hasselbalch equation (Equation 16.20)

$$pH = pK_a + \log \frac{[base]}{[acid]}$$

is zero, and so

$$pH = pK_a$$

The pK_a of acetic acid is

$$-\log(1.76 \times 10^{-5}) = 4.75$$

which is indeed the pH halfway to the equivalence point of the acetic acid titration.

To see how the math works in acid–base titrations, consider the following titration of a sample of vinegar to determine its acetic acid content. We'll use a pipet to add 10.00 mL of vinegar to a 125-mL Erlenmeyer flask and then add distilled water to bring the total volume to about 50 mL. We then add a few drops of pH indicator solution and titrate the diluted vinegar with 0.1050 M NaOH. We repeat this procedure two more times and calculate the average volume of NaOH solution needed to reach the equivalence point. If that volume is 16.24 mL, what was the concentration of acetic acid in the vinegar?

The key pieces of information in this titration are the volume of sample (10.00 mL), the volume of titrant (16.24 mL), and the concentration of titrant (0.1050 M). We also need the balanced equation for the titration reaction, which is

$$HC_2H_3O_2(aq) + NaOH(aq) \longrightarrow NaC_2H_3O_2(aq) + H_2O(l)$$

This stoichiometry tells us that 1 mole of acetic acid is consumed for every mole of NaOH. Because we are working with milliliter volumes, a more useful quantity is the millimole and the following equality:

$$\text{mmol } HC_2H_3O_2 = \text{mmol NaOH}$$

The number of millimoles of a solute in solution is equal to the volume of the solution in milliliters times the molarity of the solute. Thus, the preceding equality can be expressed as

$$(V_{HC_2H_3O_2})(M_{HC_2H_3O_2}) = (V_{NaOH})(M_{NaOH}) \tag{16.22}$$

or

$$M_{HC_2H_3O_2} = \frac{(V_{NaOH})(M_{NaOH})}{V_{HC_2H_3O_2}}$$

Inserting the experimental data, we have

$$M_{HC_2H_3O_2} = \frac{(16.24 \text{ mL})(0.1050 \text{ } M)}{10.00 \text{ mL}}$$
$$= 0.1705 \text{ } M$$

We can write an equation for calculating the concentration of any acidic or basic solute from the results of a titration by using a general version of Equation 16.22:

$$V_A M_A = \frac{n_A}{n_B} V_B M_B \tag{16.23}$$

in which V_A and M_A represent the volume and molarity of the acid, V_B and M_B represent the volume and molarity of the base, and n_A/n_B is the ratio of the moles of acid to moles of base in the balanced chemical equation describing the reaction. For example, in a titration of a solution of sulfuric acid with NaOH,

$$H_2SO_4(aq) + 2 \text{ NaOH}(aq) \longrightarrow Na_2SO_4(aq) + 2 H_2O(l)$$

the number of moles of NaOH consumed is twice the number of moles of H_2SO_4 consumed. Therefore,

16.8 Acid–Base Titrations

and Equation 16.23 becomes

$$\frac{n_A}{n_B} = \frac{1}{2}$$

$$(V_{H_2SO_4})(M_{H_2SO_4}) = \tfrac{1}{2}(V_{NaOH})(M_{NaOH})$$

SAMPLE EXERCISE 16.10: Suppose a 50.00-mL sample of water from an alkaline spring in the desert of southern Nevada is titrated with 0.02075 M HCl. A few drops of phenolphthalein are added at the beginning of the titration, and the solution turns pink. It takes 11.21 mL of titrant to reach the pink-to-clear equivalence point. Then a few drops of bromcresol green are added, and it takes 32.28 mL of titrant before its blue green color changes to yellow. What were the initial concentrations of carbonate and bicarbonate in the spring-water sample?

SOLUTION: The titration takes place in two stages. In the first stage (to the phenolphthalein color change), carbonate is titrated to bicarbonate. The stoichiometry of the titration reaction

$$HCl(aq) + CO_3^{2-}(aq) \longrightarrow HCO_3^{-}(aq) + Cl^{-}(aq)$$

indicates that the acid and carbonate react in a 1:1 mole ratio; so, at the first equivalence point,

$$\text{mmol HCl added} = \text{mmol CO}_3^{2-} \text{ consumed}$$

and so

$$(V_{HCl})(M_{HCl}) = (V_{CO_3^{2-}})(M_{CO_3^{2-}})$$

Solving for $M_{CO_3^{2-}}$ and inserting the values given, we have

$$M_{CO_3^{2-}} = \frac{(V_{HCl})(M_{HCl})}{V_{CO_3^{2-}}}$$

$$= \frac{(11.21 \text{ mL})(0.02075 \text{ } M)}{50.00 \text{ mL}}$$

$$= 4.652 \times 10^{-3} \text{ } M$$

At the first equivalence point, all CO_3^{2-} initially present has been converted into HCO_3^{-}. In the second stage in the titration, this bicarbonate plus all of that present initially reacts with more HCl titrant, forming carbonic acid:

$$HCl(aq) + HCO_3^{-}(aq) \longrightarrow H_2CO_3(aq) + Cl^{-}(aq)$$

If there were no HCO_3^{-} present initially in the sample, another 11.21 mL of titrant would be needed just to titrate the HCO_3^{-} produced in the first stage. Subtracting this volume from that required to reach the second equivalence point

$$32.28 \text{ mL} - 11.21 \text{ mL} = 21.07 \text{ mL}$$

we find that 21.07 mL of HCl was needed to titrate the HCO_3^{-} initially present in the sample. Using this volume to calculate the initial $[HCO_3^-]$, we have

$$M_{HCO_3^-} = \frac{(V_{HCl})(M_{HCl})}{V_{HCO_3^-}}$$

$$= \frac{(21.07 \text{ mL})(0.02075 \text{ M})}{50.00 \text{ mL}}$$

$$= 8.744 \times 10^{-3} \text{ M}$$

PRACTICE EXERCISE: Your job is to determine the concentration of ammonia in a cleaning solution. In the titration of a 25.00-mL sample of the solution, the equivalence point is reached after 10.49 mL of 0.01155 M HCl has been added.

a. What is the concentration of ammonia in the solution?
b. Which of the pH indicators in Figure 16.16 would be suitable for detecting the equivalence point in this titration?

(See Problems 99 and 100.)

CONCEPT TEST: Which of the pH indicators in Figure 16.16 would be useful in detecting the equivalence point in the titration of acetic acid with NaOH?

16.9 SOLUBILITIES OF MINERALS AND OTHER COMPOUNDS

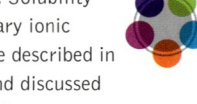

CONNECTION: Solubility trends for binary ionic compounds are described in Section 5.7 and discussed further in Section 9.2.

As water that has seeped into the ground flows through soil that contains calcium carbonate ($CaCO_3$), small concentrations of the $CaCO_3$ dissolve. We can write a balanced equation describing the dissolution process:

$$CaCO_3(s) \rightleftharpoons Ca^{2+}(aq) + CO_3^{2-}(aq)$$

The corresponding equilibrium constant expression is

$$K = \frac{[Ca^{2+}][CO_3^{2-}]}{[CaCO_3]}$$

Because calcium carbonate is a solid, its effective concentration does not change (until it is all gone); so, by convention, $[CaCO_3]$ is assigned a value of unity, and the $[CaCO_3]$ term drops out of the equilibrium constant expression, leaving

$$K_{sp} = [Ca^{2+}][CO_3^{2-}]$$

where K_{sp} represents an equilibrium constant called the **solubility-product constant** or, simply, the **solubility product.**

☑ The **solubility-product constant** (K_{sp}) (also called the **solubility product**) is an equilibrium constant relating the dissolved concentrations of the ions produced by the dissolution of a slightly soluble compound.

16.9 Solubilities of Minerals and other Compounds

The K_{sp} values of $CaCO_3$ and some other slightly soluble compounds are listed in Appendix 5. These values can be used to calculate the solubility of such compounds in water. Consider, for example, calcium fluoride, CaF_2. It is found in nature in the mineral *fluorite*, which is the principal source of fluoride for industry, including the manufacture of glass and semiconductors. At 25°C, the K_{sp} of CaF_2 is 3.9×10^{-11}. Let's use this value to calculate how many moles of CaF_2 dissolve in a liter of water and how many grams of CaF_2 dissolve in 50.0 mL of water at 25°C.

The dissolution of CaF_2 is described in the following balanced equation:

$$CaF_2(s) \rightleftharpoons Ca^{2+}(aq) + 2F^-(aq)$$

The corresponding solubility-product expression is

$$K_{sp} = [Ca^{2+}][F^-]^2 = 3.9 \times 10^{-11}$$

If we let x be the number of moles of CaF_2 that dissolves in 1.000 L of water at 25°C, then x moles of Ca^{2+} and $2x$ moles of F^- are produced. Because these quantities are dissolved in 1.000 L of water, $[Ca^{2+}] = x$ and $[F^-] = 2x$. Substituting the algebraic terms into the K_{sp} expression, we have

$$K_{sp} = [Ca^{2+}][F^-]^2 = (x)(2x)^2 = x(4x^2) = 4x^3$$

$$4x^3 = 3.9 \times 10^{-11}$$

$$x = 2.1 \times 10^{-4} \, M$$

Note that the entire algebraic expression for $[F^-]$ was squared in this calculation. Forgetting to square the coefficient is a mistake that students sometimes make. Also note that the molar solubility of CaF_2 is much greater than its solubility product. This difference is true for all sparingly soluble ionic compounds because each K_{sp} is the product of small concentration values multiplied together, producing an even smaller overall K_{sp} value.

If the solubility of CaF_2 is 2.1×10^{-4} M, how many grams of CaF_2 will dissolve in 50.0 mL of solution? To answer this question, we convert the moles of solute that dissolve in 1 liter of solution into an equivalent number of grams (the molar mass of CaF_2 is 78.08 g/mol) and then into the mass that dissolves in 50.0 mL of solution:

$$\frac{2.1 \times 10^{-4} \, \cancel{\text{mol}}}{1.000 \, \cancel{\text{L}}} \times \frac{78.08 \text{ g}}{\cancel{\text{mol}}} \times \frac{1.000 \, \cancel{\text{L}}}{1000 \, \cancel{\text{mL}}} \times 50.0 \, \cancel{\text{mL}} = 8.3 \times 10^{-4} \text{ g}$$

Let's derive a general equation relating the solubility (s) of any ionic compound M_mX_x to its K_{sp} value. We start with the dissolution process:

$$M_mX_x \rightleftharpoons mM(aq) + xX(aq)$$

If s moles of M_mX_x dissolve in 1 liter of solution, then $(s \cdot m)$ moles of M and $(s \cdot x)$ moles of X are produced. Inserting these values into the corresponding equilibrium constant expression for M_mX_x, we have

$$K_{sp} = [M]^m[X]^x = (s \cdot m)^m (s \cdot x)^x$$

or

$$K_{sp} = (m^m x^x)s^{(m+x)} \tag{16.24}$$

Let's use Equation 16.24 to calculate the solubility of calcium phosphate, $Ca_3(PO_4)_2$, the principal ingredient in the mineral apatite:

$$K_{sp} = (m^m x^x)s^{(m+x)} = (3^3 2^2)s^{(3+2)} = 36s^5$$

Inserting the value of the K_{sp} of $Ca_3(PO_4)_2$ listed in Appendix 5 and solving for s, we have

$$K_{sp} = 2.1 \times 10^{-33} = 36s^5$$

$$s = \sqrt[5]{\frac{2.1 \times 10^{-33}}{36}} = 1.4 \times 10^{-7} M$$

SAMPLE EXERCISE 16.11: The mineral barite is mostly barium sulfate ($BaSO_4$), and is widely used in industry and in medical imaging of the digestive system. Calculate the solubility (in molarity) of $BaSO_4$ in pure water and in seawater that contains 2.8 g/L of SO_4^{2-}. Given: $K_{sp} = 9.1 \times 10^{-11}$.

SOLUTION: The stoichiometry of the dissolution of $BaSO_4$ is simple: 1 mole of barium ion and 1 mole of sulfate ion from each mole of $BaSO_4$ that dissolves:

$$BaSO_4(s) \rightleftharpoons Ba^{2+}(aq) + SO_4^{2-}(aq)$$

Therefore, if x moles of $BaSO_4$ dissolve per liter of pure water, then $[Ba^{2+}] = x$ and $[SO_4^{2-}] = x$. Inserting these values into the K_{sp} expression gives

$$K_{sp} = [Ba^{2+}][SO_4^{2-}] = (x)(x) = 9.1 \times 10^{-11}$$

$$x^2 = 9.1 \times 10^{-11}$$

$$x = 9.5 \times 10^{-6} M$$

Before calculating the solubility of $BaSO_4$ in a solution that already contains sulfate ions, we should think about what the outcome is likely to be. Is $BaSO_4$ more soluble in seawater or in pure water? The answer lies in the dissolution equilibrium:

> Is $BaSO_4$ more soluble in seawater or in pure water?

$$BaSO_4(s) \rightleftharpoons Ba^{2+}(aq) + SO_4^{2-}(aq)$$

According to Le Châtelier's principle (and the common-ion effect), additional sulfate ion will shift the equilibrium to the left, and less $BaSO_4$ should dissolve in seawater. We can account for the sulfate that is already present by calculating the background value of $[SO_4^{2-}]$

$$\text{background } [SO_4^{2-}] = \frac{2.8 \text{ g}}{L} \times \frac{1 \text{ mol}}{96.06 \text{ g}} = \frac{0.029 \text{ mol}}{L}$$

and adding this value to the $[SO_4^{2-}]$ term in the K_{sp} expression:

$$K_{sp} = [Ba^{2+}][SO_4^{2-}] = (x)(0.029 + x) = 9.1 \times 10^{-11}$$

Solving for x is simplified if we ignore its contribution to the total concentration of sulfate. Doing so is reasonable because x is likely to be much less than 0.029 M. Thus:

$$(x)(0.029) \cong 9.1 \times 10^{-11}$$

$$x = 3.1 \times 10^{-9} \text{ M}$$

The concentration of sulfate added by the dissolution of $BaSO_4$ is much smaller than the background concentration, and so our simplifying assumption is justified.

This calculation is another illustration of the common-ion effect. In this example, the dissolution process is suppressed by the additional sulfate present, which causes an equilibrium shift from right to left.

PRACTICE EXERCISE: What is the solubility in pure water of $MgCO_3$, the principal component of the mineral dolomite? Given: $K_{sp} = 6.8 \times 10^{-6}$. (See Problems 121–124.)

16.10 COMPLEX IONS

In Chapter 9, we described how water molecules are oriented around dissolved anions and cations in such a way that their hydrogen atoms and (+) dipoles are directed toward dissolved anions and their oxygen atoms and (−) dipoles are directed toward dissolved cations. For some of these hydrated cations, ion-dipole interactions lead to the sharing of lone pairs of electrons on oxygen atoms with empty orbitals on the cations. This sharing of electron pairs meets our definition of chemical-bond formation. In forming these bonds, the metal ion acts as an acceptor of electron pairs—that is, as a Lewis acid—and water acts as a Lewis base. A cluster of water molecules bonded to a metal ion in this way is an example of a **complex ion**. The water molecules in such a complex ion are called **ligands**. For example, four water molecules may form a tetrahedral complex with a dissolved zinc (Zn^{2+}) ion, as shown in Figure 16.20. Similar reactions can result in the formation of complexes with other geometries including octahedral—as in $Fe(H_2O)_6^{3+}$ and $Al(H_2O)_6^{3+}$.

> In a **complex ion**, ions and molecules called **ligands** surround and bond to a metal ion.

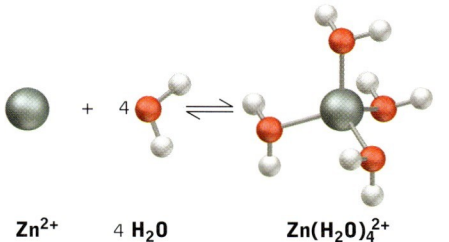

Zn^{2+} + 4 H_2O ⇌ $Zn(H_2O)_4^{2+}$

FIGURE 16.20 Dissolved metal ions can act as Lewis acids (electron-pair acceptors). Here the oxygen atoms of four water molecules donate pairs of electrons to empty orbitals on a zinc (Zn^{2+}) ion as it dissolves.

Complexation and solubility

There is mutual attraction between positive metal ions and the bonding electrons of the surrounding water molecules. Metal ions draw these shared electrons toward the center of the complex ions and away from the hydrogen atoms of the water molecules. This shift in electron density makes it easier for the water molecules to ionize, donating H^+ ions to nearby molecules of water. This ionization reaction produces free H_3O^+ ions, and leaves OH^- ions attached to the metal ions, as shown in Figure 16.21. Thus, metal cations complexed with H_2O may act as Brønsted-Lowry acids. Cations with relatively large positive charges, such as $+2$ and $+3$, are particularly effective at attracting shared pairs of electrons, as indicated by the acid strengths (K_a values) of several of these complex ions in Table 16.2.

In Table 16.2 and elsewhere in this section, we will not routinely include the water molecules surrounding a dissolved ion in its chemical formula. For example, "$Fe^{3+}(aq)$" represents $Fe(H_2O)_6^{3+}(aq)$ and "$Fe(OH)^{2+}(aq)$" represents $Fe(H_2O)_5(OH)^{2+}(aq)$. Because many hydrated metal ions are acidic, they do not exist in most natural waters simply as $M^{n+}(aq)$. Even in lakes acidified by acid rain or streams contaminated by acidic mining wastes, the principal iron(III) and aluminum(III) species are more likely to be $Fe(OH)^{2+}(aq)$ and $Al(OH)^{2+}(aq)$ than $Fe^{3+}(aq)$ and $Al^{3+}(aq)$. In alkaline waters iron(III) and aluminum(III) complexes incorporate even more hydroxide ions, forming two-hydroxide (*dihydroxo*) complexes such as $Fe(OH)_2^+(aq)$ and $Al(OH)_2^+(aq)$, or insoluble neutral species such as $Fe(OH)_3(s)$ and $Al(OH)_3(s)$, or even the anionic *tetrahydroxo* complex $Al(OH)_4^-(aq)$.

The dissolved concentrations of $Fe(OH)_3$ and $Al(OH)_3$ cannot be very large because of the limited solubilities of these compounds: their K_{sp} values are only 1.1×10^{-36} and 1.9×10^{-33}, respectively. However, we need to keep in mind that these tiny K_{sp} values apply to equilibrium constant expressions that are based on the concentrations of *free metal ions*—for example,

$$K_{sp} = [Al^{3+}][OH^-]^3 = 1.9 \times 10^{-33}$$

We can use this expression to calculate the solubility of Al^{3+} at pH = 7.0:

$$[Al^{3+}] = \frac{K_{sp}}{[OH^-]^3} = \frac{1.9 \times 10^{-33}}{(1.0 \times 10^{-7})^3} = 1.9 \times 10^{-11} \, M$$

This result may suggest to you that aluminum(III) compounds are essentially insoluble at neutral pH, but what it really means is that $[Al^{3+}]$ must be very small

TABLE 16.2 Acid-Ionization Equilibrium Constants for Some Hydrated Metal Ions

Ion	K_a
$Fe^{3+}(aq)$	3×10^{-3}
$Cr^{3+}(aq)$	1×10^{-4}
$Al^{3+}(aq)$	1×10^{-5}
$Cu^{2+}(aq)$	3×10^{-8}
$Pb^{2+}(aq)$	3×10^{-8}
$Zn^{2+}(aq)$	1×10^{-9}
$Co^{2+}(aq)$	2×10^{-10}
$Ni^{2+}(aq)$	1×10^{-10}

↑ Acid strength

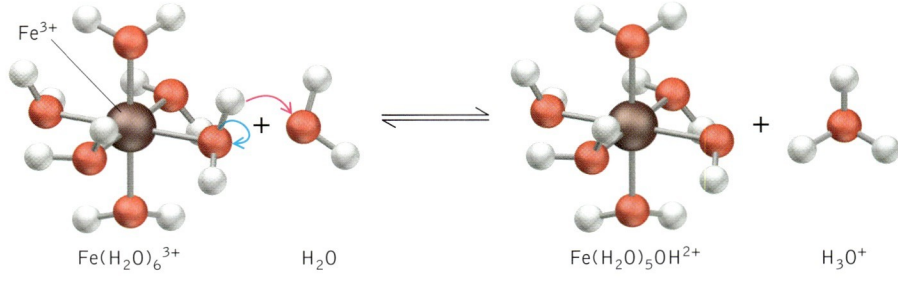

FIGURE 16.21 Hydrated metal ions, such as $Fe(H_2O)_6^{3+}$, are moderately strong Brønsted-Lowry acids. The complex ion donates an H^+ ion to a nearby water molecule, forming an H_3O^+ ion and a second complex, $Fe(H_2O)_5(OH)^{2+}$ in which one of the ligands is an OH^- ion instead of an H_2O molecule.

at neutral pH. However, we have seen that most of the aluminum(III) in solution at pH 7.0 is not $Al^{3+}(aq)$, but rather a mixture of hydroxo complexes. The formation of these (and other) complexes allows the overall solubility of aluminum(III) in neutral solutions to be much higher than 1.9×10^{-11} M. Moreover, the solubility of aluminum(III) compounds increase at high pH, where charged (soluble) $Al(OH)_4^-(aq)$ is the dominant species in solution, and at low pH, where $Al(OH)^{2+}(aq)$ and $Al(OH)_2^+(aq)$ predominate.

The greater solubility of $Al(OH)_3$ (and the hydroxides and oxides of some other metals) at both high and low pH is a result of the capacity of these compounds to behave as both acids and bases. In a strongly basic solution $Al(OH)_3$ is a hydrogen-ion donor, a behavior that is easy to follow if we add the waters of hydration to the formulas of the complex ions:

$$Al(H_2O)_3(OH)_3(s) + OH^-(aq) \rightleftharpoons Al(H_2O)_2(OH)_4^-(aq) + H_2O(l)$$

In an acidic solution, $Al(OH)_3$ becomes a hydrogen-ion acceptor:

$$Al(H_2O)_3(OH)_3(s) + H_3O^+(aq) \rightleftharpoons Al(H_2O)_4(OH)_2^+(aq) + H_2O(l)$$

This exhibition of both acidic and basic properties makes $Al(OH)_3$ an amphoteric compound (see page 801). Zinc hydroxide, $Zn(OH)_2$, and chromium(III) hydroxide, $Cr(OH)_3$, also exhibit this behavior, but calcium hydroxide, $Ca(OH)_2$, and iron(III) hydroxide, $Fe(OH)_3$, do not. These differences in behavior are exploited in the purification of aluminum ore, as described in Chapter 18.

Other Lewis bases form stable complexes with metal ions. Some of the common inorganic ligands include Cl^-, OH^-, CN^- (cyanide), and NH_3. Appendix 5 contains a table of some of the complex ions formed by these ligands. If, for example, an excess of ammonia is added to a solution of copper (II) sulfate, $CuSO_4$ tetraamine copper(II), $Cu(NH_3)_4^{2+}$, ions form. The complexation reaction can be followed visually because solutions of Cu^{2+}, like those of most transition-metal ions with unfilled d orbitals, have distinctive colors. A solution of $CuSO_4$ has a characteristic robin's egg blue color due to the presence of $Cu^{2+}(aq)$. However, when ammonia is added to a solution of $CuSO_4$, the solution turns a deep navy blue, as shown in Figure 16.22. Deep blue is the color of $Cu(NH_3)_4^{2+}$.

FIGURE 16.22 A solution containing $Cu(H_2O)_4^{2+}$ has a characteristic robin's-egg blue color. After a clear solution of ammonia has been added (middle test tube), the resulting solution is the deep navy blue of $Cu(NH_3)_4^{2+}$.

FIGURE 16.23 Many metal ions, including Cu^{2+}, form complexes with NH_3. The formation constant for the reaction in which 4 molecules of NH_3 displace 6 molecules of H_2O, producing $Cu(NH_3)_4^{2+}$, the square planar $Cu(NH_3)_4^{2+}$ complex ion, is 5.0×10^{13}.

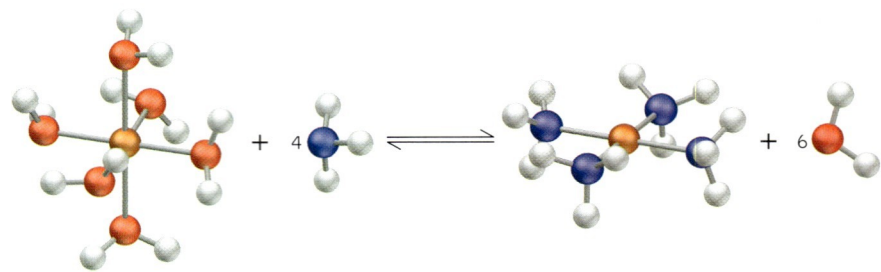

A tetraamine copper(II) ion forms from $Cu^{2+}(aq)$ when four molecules of ammonia displace six molecules of water, and form the square planar complex ion, as shown in Figure 16.23. The color changes because the different ligands (ammonia nitrogen compared with the oxygen atoms in molecules of water) affect the energies of the d orbitals in Cu^{2+} differently. These energy differences become wavelength differences as electron transitions between partly filled d orbitals result in the absorption and emission of distinctive colors of light.

The change in color when NH_3 is added to a solution of $Cu(aq)^{2+}$ indicates that the equilibrium position for the reaction

$$Cu^{2+}(aq) + 4\, NH_3(aq) \rightleftharpoons Cu(NH_3)_4^{2+}(aq) \qquad K_f = 5.0 \times 10^{13}$$

is far to the right, favoring the formation of the complex. This position is also indicated by the large value of the equilibrium constant. The subscript "f" is used to indicate that this equilibrium constant describes the **f**ormation of a complex ion. In fact, K_f values are often called simply **formation constants**. The equilibrium constant expression for the formation of $Cu(NH_3)_4^{2+}$ is

$$K_f = \frac{[Cu(NH_3)_4^{2+}]}{[Cu^{2+}][NH_3]^4}$$

For the general case in which a complex ion MX_n forms from n moles of ligand X for every 1 mole of metal-ion M, the formation constant expression is

$$K_f = \frac{[MX_n]}{[M][X]^n} \qquad (16.25)$$

Formation constants are sometimes used to calculate the concentration of a complex ion that forms from particular initial concentrations of a metal ion and a ligand. They can also be used to calculate the concentration of a free, uncomplexed metal ion in equilibrium with a given (often much larger) concentration of ligand, as in the following sample exercise.

CONNECTION: The colors of transition-metal solutions arise from differences in the energies of valence-shell d orbitals, the phenomenon that gives gemstones their distinctive colors (Section 10.7).

✓ The **formation constant** (K_f) is an equilibrium constant relating the concentration of a complex ion to the concentrations of the metal ion and ligand(s) that form the complex.

SAMPLE EXERCISE 16.12: Ammonia gas is dissolved in a 1.00×10^{-4} M solution of $CuSO_4$ so that $[NH_3] = 2.00 \times 10^{-3}$ M. Calculate the concentration of $Cu^{2+}(aq)$ ions in the solution after the addition of ammonia.

SOLUTION: The concentrations of $NH_3(aq)$, $Cu^{2+}(aq)$, and $Cu(NH_3)_4^{2+}(aq)$ are related by the formation constant expression:

$$K_f = \frac{[Cu(NH_3)_4^{2+}]}{[Cu^{2+}][NH_3]^4} = 5.0 \times 10^{13}$$

We may assume that all the copper(II) in the solution of $CuSO_4$ was present as $Cu^{2+}(aq)$ before NH_3 was added. After NH_3 was added, but before any complexation occurred, the initial concentration of ammonia (2.00×10^{-3} M) was 20 times that of $Cu^{2+}(aq)$: more than enough to convert all of the $Cu^{2+}(aq)$ into $Cu(NH_3)_4^{2+}$. Because the value of K_f is large, we can assume that nearly all the $Cu^{2+}(aq)$ is converted into $Cu(NH_3)_4^{2+}$, and so only a tiny concentration (call it x) of free $Cu^{2+}(aq)$ ion remains uncomplexed at equilibrium. If all but x of the initial concentration of $Cu^{2+}(aq)$ (1.00×10^{-4} M) is converted into $Cu(NH_3)_4^{2+}$, then, at equilibrium, $[Cu(NH_3)_4^{2+}] = 1.00 \times 10^{-4} - x$.

According to the stoichiometry of the reaction, 4 moles of NH_3 are consumed for every 1 mole of $Cu(NH_3)_4^{2+}$ produced. If the change in $[Cu(NH_3)_4^{2+}]$ is $1.00 \times 10^{-4} - x$, then the change in $[NH_3]$ is -4 times that. Using these values to complete the ICE table, we have

	$[Cu^{2+}]$	$[NH_3]$	$[Cu(NH_3)_4^{2+}]$
Initial	1.00×10^{-4}	2.00×10^{-3}	0
Change	$-(1.00 \times 10^{-4} - x)$	$-4(1.00 \times 10^{-4} - x)$	$1.00 \times 10^{-4} - x$
Equilibrium	x	$1.60 \times 10^{-3} + 4x$	$1.00 \times 10^{-4} - x$

Now we make the simplifying assumption that the value of x is small compared with 1.00×10^{-4}. If it is, then $4x$ also must be small compared with 1.60×10^{-3}. Inserting the simpler values in the formation constant expression gives

$$K_f = \frac{[Cu(NH_3)_4^{2+}]}{[Cu^{2+}][NH_3]^4}$$

$$= \frac{1.00 \times 10^{-4}}{x \cdot (1.60 \times 10^{-3})^4} = 5.0 \times 10^{13}$$

Solving for x, we have

$$x = \frac{1.00 \times 10^{-4}}{(1.60 \times 10^{-3})^4 (5.0 \times 10^{13})}$$

$$= 3.0 \times 10^{-7} \text{ M} = [Cu^{2+}]$$

This result confirms our assumption that the concentration of Cu^{2+} is much less than that of the complex. In fact, more than 99% of the Cu(II) in the solution is present as $Cu(NH_3)_4^{2+}$.

PRACTICE EXERCISE: Calculate the concentration of $Ag^+(aq)$ in a solution that is 0.100 M $AgNO_3$ and 0.800 M NH_3. Given: $Ag^+(aq) + 2\ NH_3(aq) \rightleftharpoons Ag(NH_3)_2^+(aq)$; $K_f = 1.7 \times 10^7$. (See Problems 121 and 122.)

Metal complexes in biomolecules

In living systems, the biological availability of essential micronutrients, such as iron(II), iron(III), chromium(III), and other transition metals, is enhanced by the formation of complexes with biomolecular ligands. Indeed, many of the biological functions of these metals are based on their forming complexes with biomolecules.

Consider, for example, the foundation of our food chain and the process of photosynthesis. How do green plants harvest the energy of the sun? A key ingredient in the process is a metal complex. Unlike the inorganic ligands that donate single pairs of electrons per molecule, the structure responsible for harvesting

> How do green plants harvest the energy of the sun?

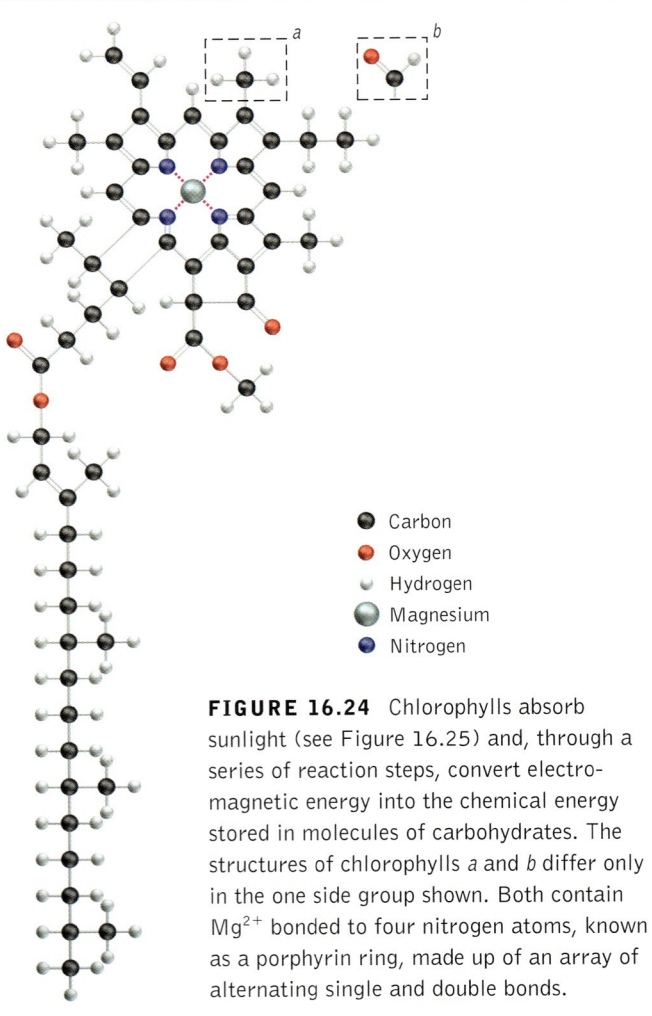

- Carbon
- Oxygen
- Hydrogen
- Magnesium
- Nitrogen

FIGURE 16.24 Chlorophylls absorb sunlight (see Figure 16.25) and, through a series of reaction steps, convert electromagnetic energy into the chemical energy stored in molecules of carbohydrates. The structures of chlorophylls *a* and *b* differ only in the one side group shown. Both contain Mg^{2+} bonded to four nitrogen atoms, known as a porphyrin ring, made up of an array of alternating single and double bonds.

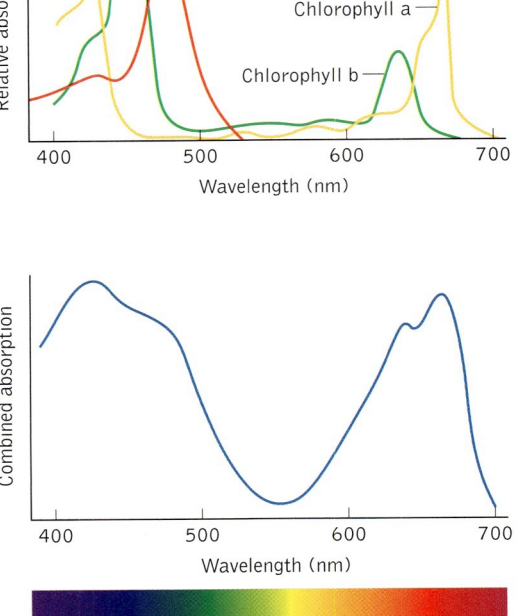

FIGURE 16.25 Green plants contain 3 pigments with similar absorption spectra: chlorophylls *a* and *b* and caratenoid. Taken together, their absorption spectrum absorbs most of the visible radiation emitted by the sun (except the yellow green color of leaves).

the sun's energy has four pairs of electrons on four different nitrogen atoms directed toward a single Mg^{2+} ion (Figure 16.24). These four nitrogen atoms are part of a circular structure of alternating, or *conjugated*, single and double bonds known as a porphyrin ring. Delocalization of the π electrons of the conjugated double bonds of the ring gives it stability and accounts for chlorophyll's ability to absorb wavelengths of red and blue-violet radiation (Figure 16.25).

Another important class of biological molecules is based on iron–porphyrin complexes contained within proteins. These complexes are called *heme* groups (see Figure 16.26). The four nitrogen atoms in the porphyrin ring occupy equatorial positions in an octahedral complex of iron(II). Below them a fifth bond is formed between Fe^{2+} and a lone pair of electrons on the ring nitrogen atom of the amino acid histidine. The nature of the sixth bond, located above the porphyrin ring, depends on the function of the protein. In the protein hemoglobin, this site is occupied either by a molecule of oxygen (as in the oxygenated hemoglobin in blood leaving the lungs) or by a molecule of water (in the *deoxyhemoglobin* returning through a vein to the lungs to be reoxygenated). In other proteins, another amino acid, such as methionine, is located in the top axial position. This structure with a second bound amino acid does not have the capacity to transport or store O_2, which are the functions of hemoglobin and the related protein myoglobin. Instead, these other proteins, called *cytochromes*, mediate oxidation and reduction processes connected with the catabolism of nutrients, such as carbohydrates. In these proteins, the heme group serves as a biomolecular conveyer of electrons as the half reaction

$$Fe^{3+} + e^- \rightleftharpoons Fe^{2+}$$

rapidly and reversibly consumes or releases electrons as needed to ensure the efficient conversion of the chemical energy that sustains life.

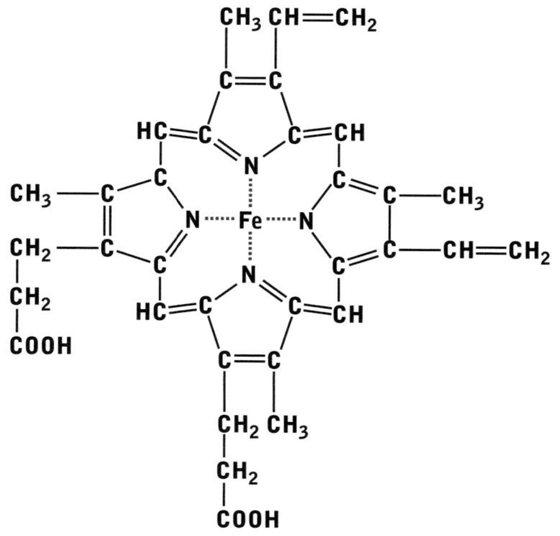

FIGURE 16.26 A molecule of hemoglobin contains four heme groups, each one of which contains an Fe^{2+} ion surrounded by 4 nitrogen atoms of a porphyrin ring. A molecule of O_2 can reversibly donate a pair of electrons to an empty d orbital of Fe^{2+} that is oriented above the porphyrin ring, forming a complex that is the principal oxygen carrier in the blood.

CHAPTER REVIEW

Summary

SECTION 16.1

Some acids donate all their ionizable protons and are strong acids, but most donate only a fraction of their ionizable protons and are weak acids. The strength of an acid that is only partly ionized is described by the value of its acid-ionization constant, K_a. The degree of ionization of a weak acid increases with decreasing acid concentration.

SECTION 16.2

A Brønsted-Lowry acid donates H^+ ions to other substances; a Brønsted-Lowry base accepts H^+ ions from other substances. A strong base is an effective H^+-ion acceptor, whereas a weak base is a poor H^+-ion acceptor. Most bases are weak bases. The ability of a base to accept H^+ ions is expressed by the value of its base equilibrium constant, K_b. An acid that donates H^+ ions becomes its conjugate base, and a base that accepts H^+ ions becomes its conjugate acid. Strong acids have extremely weak conjugate bases. Very weak acids have very strong conjugate bases. Lewis acids are electron-pair acceptors, and Lewis bases are electron-pair donors.

SECTION 16.3

The acidity of oxoacids increases as the number of X=O double bonds and the oxidation number of the central X atom increase.

SECTION 16.4

pH is a logarithmic scale for expressing the acidities of substances. Water can act as an acid or a base and so is amphoteric. Hydrogen ions in aqueous solutions are bonded to water molecules, forming hydronium ions, H_3O^+. In aqueous solutions $[H_3O^+]$ is inversely proportional to $[OH^-]$.

SECTION 16.5

The value of K_b of a base is inversely proportional to the value of K_a of its conjugate acid. The product of the two constants also is a constant, K_w, and so K_b can be calculated from K_a and vice versa.

SECTION 16.6

The common-ion effect is the shift in the position of an equilibrium caused by the addition of an ion that takes part in the reaction. A pH buffer is a solution of weak acid and its conjugate base that resists change in its pH when acids or bases are added to it. Buffer capacity is the quantity of acid or base that can be added to a pH buffer without significantly changing the pH of the buffer.

SECTIONS 16.7 AND 16.8

pH color indicators are weakly acidic or basic substances whose acid and base forms have different colors. Color indicators are used to estimate the pH of solutions and to monitor the progress of acid–base titrations in which the concentration of an acidic or basic solute is determined.

SECTION 16.9

The solubility of a slightly soluble ionic compound is described by an equilibrium constant called the solubility product, K_{sp}. The solubilities of many ionic compounds are reduced by the common-ion effect. The solubilities of amphoteric compounds are increased by the addition of strong acids or bases.

SECTION 16.10

A complex ion consists of a metal ion bonded to surrounding ligands, which are ions or molecules or both. Complex ions have characteristic formation constants.

Key Terms

amphoteric substance (p. 801)
buffer capacity (p. 813)
common-ion effect (p. 811)
complex ion (p. 827)
conjugate (p. 795)
degree of dissociation (p. 785)
degree of ionization (p. 785)
formation constant (p. 830)
Henderson-Hasselbalch equation (p. 814)
ligand (p. 827)
pH (p. 801)
pH buffer (p. 813)
pH indicator (p. 816)
solubility-product constant (solubility product) (p. 824)

Key Skills and Concepts

SECTION 16.1

Be able to describe the properties of a weak acid, a strong acid, a weak base, and a strong base.

Be able to write the acid-ionization constant expression (K_a) of a weak acid.

Be able to write the base-ionization constant expression (K_b) of a weak base.

Be able to calculate [H_3O^+] in a solution of a weak acid of specified concentration.

Be able to calculate the degree of ionization of a weak acid of specified concentration.

SECTION 16.2

Understand how weak bases react with water, forming OH^-.

Be able to distinguish between a Brønsted-Lowry acid and base and a Lewis acid and base.

Be able to give the conjugate acid of a base and the conjugate base of an acid.

SECTION 16.3

Be able to relate the strengths of oxoacids to the oxidation state and electronegativity of their central atoms.

Be able to write K_{a1} and K_{a2} equilibrium constant expressions for diprotic acids.

SECTION 16.4

Understand the relation between [H_3O^+] and [OH^-] in an aqueous solution.

Know the definitions of pH, pOH, and pK.

Be able to calculate the pH of a solution of a strong acid, weak acid, strong base, or weak base.

SECTION 16.5

Be able to calculate the pH of a solution of an acidic or basic salt.

Understand how the common-ion effect influences acid–base equilibria.

SECTION 16.6

Be able to calculate the pH of a buffer solution by using the Henderson-Hasselbalch equation.

Be able to explain how a buffer solution responds to small additions of strong acid or strong base.

SECTION 16.7

Explain how pH indicators work and how they are used.

SECTION 16.8

Be able to select the appropriate pH indicator to detect the equivalence point in an acid–base titration.

Be able to calculate the concentration of an acidic or basic substance from the results of a titration.

SECTION 16.9

Be able to write the solubility product (K_{sp}) expression for the dissolution of an ionic compound in water.

Be able to calculate the molar solubility of a compound from its K_{sp} value and vice versa.

Explain how the solubility of a compound is affected by the common-ion effect, and be able to calculate the solubility of a compound in the presence of a common dissolved ion and its K_{sp} value.

SECTION 16.10

Be able to write the formation constant (K_f) expression for the formation of a complex ion.

Understand how the formation of complex ions can increase the solubility of metal ions.

Key Equations and Relations

SECTION 16.4

pH and pOH:

$$K_w = [H_3O^+][OH^-] = 1.00 \times 10^{-14} \quad (16.11)$$
$$pH = -\log[H_3O^+] \quad (16.12)$$
$$pK_w = pH + pOH = 14.00 \quad (16.14)$$

SECTION 16.6

Calculation of the pH of a buffer solution or any solution in which the concentrations of an acid and its conjugate base are known:

$$pH = pK_a + \log\frac{[\text{base}]}{[\text{acid}]} \quad (16.20)$$

SECTION 16.7

Useful range of a colorimetric pH indicator, HIn:

$$pH = pK_a \pm 1.0$$

where pK_a is the negative logarithm of the acid-ionization constant of the indicator (HIn $\rightleftharpoons$ H$^+$ + In$^-$).

SECTION 16.8

In a titration in which n_B moles of base (B) are consumed for every n_A moles of acid (A):

$$(V_A)(M_A) = \frac{n_B}{n_A}(V_B)(M_B) \quad (16.23)$$

SECTION 16.9

The solubility, s, of an ionic compound of M_mX_x is related to its solubility product, K_{sp}:

$$K_{sp} = (m^m x^x)s^{(m+x)} \quad (16.24)$$

SECTION 16.10

The concentration of complex ion MX_n is related to the concentrations of the metal ion (M) and ligand (X) and to the formation constant (K_f) of the complex ion:

$$K_f = \frac{[MX_n]}{[M][X]^n} \quad (16.25)$$

QUESTIONS AND PROBLEMS

Acid Strength: K_a and Degree of Ionization

CONCEPT REVIEW

1. Does the degree of dissociation of a weak acid increase or decrease as the concentration of the acid decreases?
2. One-molar solutions of the following acids are prepared. On the basis of their acid-ionization equilibrium constants, predict which solution will have the largest concentration of hydrogen ions.

Acid	K_a
CH_3CO_2H	1.8×10^{-5}
HNO_2	4.5×10^{-4}
$HOCl$	3.5×10^{-8}
$HOCN$	3.5×10^{-4}

3. On the basis of the following degree-of-ionization data for 0.100 M solutions, select which acid has the smallest K_a.

Acid	Degree of ionization (%)
$HC_7H_5O_2$	2.5
HF	8.5
HN_3	1.4
$HC_2H_3O_2$	1.3

4. The K_a values of amino acids depend on the solvent. For example, the K_a of proline is 2.5×10^{-11} in water, 2.8×10^{-11} in an aqueous solution that is 28% ethanol, and 1.66×10^{-8} in aqueous formaldehyde at 25°C. In which solvent is proline the strongest acid?
5. The K_a of alanine in aqueous ethanol is less than its K_a in water. In which solvent does alanine ionize to the largest degree?
*6. Why is the K_a value of alanine lower in ethanol than in water?
7. Why is the value of K_{a2} for a diprotic acid usually much smaller than the value of K_{a1}?

8. There are two hydrogen-ion acceptor sites (lone pairs on nitrogen atoms) in one molecule of ethylene diamine ($H_2NCH_2CH_2NH_2$), and so this base has K_{b1} and K_{b2} equilibrium constants. Which one do you think is larger? Explain why.

PROBLEMS

9. If the K_a of acetic acid is 1.76×10^{-5}, what is the value of $[H^+]$ in 0.500 M acetic acid?
10. If the K_{a1} of alanine is 2.82×10^{-4}, what is the value of $[H^+]$ in a 0.100 M solution of alanine?
11. The muscle fatigue felt during strenuous exercise is caused by the buildup of lactic acid. In an aqueous 1.00 M solution, 2.94% of lactic acid is ionized. What is the value of K_a?
12. The odor of spoiled butter is due in part to butanoic acid, which results from the chemical breakdown of butter fat. A 0.100 M solution of butanoic acid is 1.23% ionized. Calculate the value of K_a for butanoic acid.
13. The active ingredient in aspirin is acetylsalicyclic acid, $HC_8H_7O_4$. Determine the degree of ionization of 0.100 M acetylsalicyclic acid ($K_a = 3.0 \times 10^{-4}$).
14. Determine the degree of ionization of 0.00100 M acetic acid ($K_a = 1.76 \times 10^{-5}$).
15. At equilibrium, the value of $[H^+]$ in a 0.250 M solution of an unknown acid is 4.07×10^{-3} M. Determine the degree of ionization and the K_a of this acid.
16. Nitric acid (HNO_3) is a strong acid that is completely ionized in aqueous solutions of concentrations ranging from 1% to 10% (1.5 M). However, in more concentrated solutions, part of the nitric acid is present as un-ionized molecules of HNO_3. For example, in a 50% solution (7.5 M) at 25°C, only 33% of the molecules of HNO_3 dissociate into H^+ and NO_3^-. What is the value of K_a of HNO_3?
17. A solution is 0.100 M in HCl and HF. Calculate the concentration of hydrogen ions in the solution.
18. Calculate the value of $[H^+]$ in 0.200 M H_2SO_4.
19. Ascorbic acid (vitamin C) is a diprotic acid with $K_{a1} = 7.94 \times 10^{-5}$ and $K_{a2} = 1.62 \times 10^{-12}$ at 25°C. Calculate the value of $[H^+]$ and the degree of ionization of 0.250 M ascorbic acid. Does the second ionization contribute significantly to the overall percent ionization?
*20. Citric acid has three ionizable hydrogen atoms with $K_{a1} = 7.10 \times 10^{-4}$, $K_{a2} = 1.68 \times 10^{-5}$, and $K_{a3} = 6.4 \times 10^{-6}$ at 25°C. Calculate the value of $[H^+]$ and the degree of ionization of 0.0330 M citric acid.
21. A 1.00-L solution of nitric acid is prepared by diluting 10.0 mL of 15.0 M HNO_3 with water. What is the acid strength, $[H^+]$, of the dilute solution?
22. What is the hydrogen-ion concentration in a solution of hydrochloric acid that was prepared by diluting 20.0 mL of concentrated (11.6 M) HCl to a final volume of 500 mL?
23. Calculate the value of $[OH^-]$ in 0.0800 M $Sr(OH)_2$.
24. Calculate the value of $[OH^-]$ in a solution produced when 5.0 grams of KOH (the active ingredient in some drain cleaners) is dissolved in enough water to make 250 mL of solution.
25. Describe how you would prepare 2.50 L of a NaOH solution in which $[OH^-] = 0.70$ M, starting with solid NaOH.
26. How many milliliters of 1.00 M NaOH do you need to prepare 250 mL of a solution in which $[OH^-] = 0.0200$ M?

Acids and Bases: A Molecular View

CONCEPT REVIEW

27. A 1.0 M aqueous solution of $NaNO_2$ is a much better conductor of electricity than is a 1.0 M solution of HNO_2. Explain why.
28. Hydrogen chloride and water are molecular compounds, yet a solution of HCl dissolved in H_2O is an excellent conductor of electricity. Explain why.
29. Describe the intermolecular forces and changes in bonding that lead to the formation of a basic solution when methylamine (CH_3NH_2) dissolves in water.
30. Describe the chemical reactions of sulfur that begin with the burning of high-sulfur fossil fuel and that end with the reaction between acid rain and building exteriors made of marble ($CaCO_3$).
31. Which of the following compounds are Brønsted-Lowry acids? HCl, NH_3, BF_3, HNO_3.
32. Why is BF_3 a Lewis acid but not a Brønsted-Lowry acid?
33. Explain why the K_{a1} of H_2SO_4 is much greater than the K_{a1} of H_2SeO_4.
34. Explain why the K_{a1} of H_2SO_4 is much greater than the K_{a1} of H_2SO_3.

PROBLEMS

35. Identify the acids and bases in the following reactions:
    a. $HNO_3(aq) + NaOH(aq) \rightarrow H_2O(l) + NaNO_3(aq)$
    b. $CaCO_3(s) + 2HCl(aq) \rightarrow CaCl_2(aq) + CO_2(aq) + H_2O(l)$
    c. $NH_3(aq) + HCN(aq) \rightarrow NH_4CN(aq)$

36. Identify the acids and bases in the following reactions:
    a. $NH_2^-(aq) + H_2O(l) \rightleftharpoons NH_3(aq) + OH^-(aq)$
    b. $HClO_4(aq) + H_2O(l) \rightleftharpoons ClO_4^-(aq) + H_3O^+(aq)$
    c. $HSO_4^-(aq) + CO_3^{2-}(aq) \rightleftharpoons SO_4^{2-}(aq) + HCO_3^-(aq)$
37. The K_b of dimethylamine $((CH_3)_2NH)$ is 5.9×10^{-4} at 25°C. Calculate the value of $[OH^-]$ in 1.20×10^{-3} M dimethylamine. Given:

$$(CH_3)_2NH(aq) + H_2O(l) \rightleftharpoons (CH_3)_2NH_2^+(aq) + OH^-(aq)$$

38. Trimethylamine is a contributor to the odor of rotting fish. Calculate the value of $[OH^-]$ in 0.0200 M $(CH_3)_3N$ ($K_b = 6.5 \times 10^{-5}$ at 25°C).

39. Identify the conjugate base of each of the following acids: HNO_2, $HOCl$, H_3PO_4, and NH_3.
40. Identify the conjugate acid of each of the following bases: NH_3, ClO_2^-, SO_4^{2-}, and OH^-.
41. Predict which acid in the following pairs of acids is the stronger acid. (a) H_2SO_3 or H_2SeO_3; (b) H_2SeO_4 or H_2SeO_3.
42. Predict which acid in the following pairs of acids is the stronger acid. (a) HOBr or HOBrO; (b) HOCl or HOBr.

pH

CONCEPT REVIEW

43. Explain why pH values decrease as acidity increases.
44. Solution A is 100 times as acidic as solution B. What is the difference in the pH values of solution A and solution B?
45. Under what conditions is the pH of a solution negative?
*46. In principle, a polar liquid such as ethanol (C_2H_5OH) can undergo autoionization. Explain why the value of the equilibrium constant for the autoionization of ethanol is much less than that of water.

PROBLEMS

47. Calculate the pH and pOH of the solutions with the following hydrogen-ion or hydroxide-ion concentrations. Indicate which solutions are acidic, basic, or neutral.
    a. $[H^+] = 3.45 \times 10^{-8}$ M.
    b. $[H^+] = 2.0 \times 10^{-5}$ M.
    c. $[H^+] = 7.0 \times 10^{-8}$ M.
    d. $[OH^-] = 8.56 \times 10^{-4}$ M.
48. Calculate the pH and pOH of the solutions with the following hydrogen-ion or hydroxide-ion concentrations. Indicate which solutions are acidic, basic, or neutral.
    a. $[OH^-] = 7.69 \times 10^{-3}$ M.
    b. $[OH^-] = 2.18 \times 10^{-9}$ M.
    c. $[H^+] = 4.0 \times 10^{-8}$ M.
    d. $[H^+] = 3.56 \times 10^{-4}$ M.
49. Calculate the pH of stomach acid in which $[HCl] = 0.155$ M.
50. Calculate the pH of 0.00500 M HNO_3.
51. Calculate the pH and pOH of 0.0450 M NaOH.
52. Calculate the pH and pOH of 0.160 M KOH.
53. Calculate the pH of 1.33 M HNO_3.
*54. Calculate the pH of 6.9×10^{-8} M HBr.
55. The venom of biting ants contains formic acid, HCO_2H ($K_a = 1.8 \times 10^{-4}$ at 25°C). Calculate the pH of a 0.100 M solution of formic acid.
56. Uric acid can collect in joints, giving rise to a medical condition known as gout. If the pK_a of uric acid is 3.89, what is the pH of 0.150 M uric acid?
57. The average pH of precipitation in Minnesota is about 5.2; in northern Maine, it is 4.5. Calculate the value of $[H^+]$ in solutions with these pH values.
58. A newspaper reported that the "level of acidity" in a sample taken from an extensively studied watershed in New Hampshire in February, 1998, was "an astounding 200 times lower than the worst measurement" taken in the preceding 23 years. What is this difference expressed in units of pH?

Diprotic Acids

PROBLEMS

59. Adipic acid, $HO_2C(CH_2)_4CO_2H$, is used in the manufacture of certain types of nylon. It also imparts the distinctive taste of beets. Its acid-ionization constants are $K_{a1} = 3.71 \times 10^{-5}$ and $K_{a2} = 3.87 \times 10^{-6}$ at 25°C. Determine the pH of a 0.175 M solution of adipic acid.
60. The leaves of the rhubarb plant contain high concentrations of diprotic oxalic acid (HO_2CCO_2H) and must be removed before the stems are used to make rhubarb pie. If $pK_{a1} = 1.23$ and $pK_{a2} = 4.19$, what is the pH of a 0.0288 M solution of oxalic acid?

Acid Rain and Basic Minerals

CONCEPT REVIEW

61. The pK_a values of the conjugate acids of pyridine derivatives shown below increase as more methyl groups are added. Do more methyl groups increase or decrease the strength of these pyridine bases?

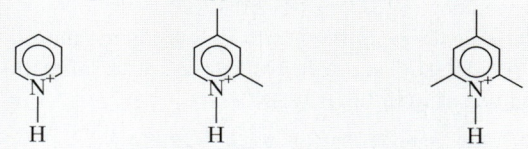

Pyridine   2,4-Dimethylpyridine   2,4,6-Trimethylpyridine
5.18   6.99   7.43

62. Why is it unnecessary to include a table of K_b values of the conjugate bases of weak acids whose K_a values are known?

63. Which of the following salts produces an acidic solution in water: ammonium acetate, ammonium nitrate, or sodium formate?

64. Which of the following salts produces a basic solution in water: NaF, KCl, NH$_4$Cl?

PROBLEMS

65. Morphine is an effective pain killer but is also highly addictive. Calculate the pH of a 0.115 M solution of morphine if $pK_b = 5.79$.

66. Codeine is a popular prescription pain killer because it is much less addictive than morphine. Codeine contains a basic nitrogen atom that can be protonated to give the conjugate acid of codeine. Calculate the pH of a 3.42×10^{-4} M solution of codeine if the pK_a of the conjugate acid is 8.21.

67. Quinine occurs naturally in the bark of the cinchona tree. For centuries it was the only treatment for malaria. Quinine contains two weakly basic nitrogen atoms, with $K_{b1} = 3.31 \times 10^{-6}$ and $K_{b2} = 1.35 \times 10^{-9}$ at 25°C. Calculate the pH of a 0.01050 M solution of quinine in water.

68. Pyridine (C$_5$H$_5$N) is a particularly foul-smelling substance used in manufacturing pesticides and plastic resins. Calculate the pH of 0.125 M pyridine ($K_b = 1.4 \times 10^{-9}$).

69. Nicotine is responsible for the addictive properties of tobacco. If $K_{b1} = 1.05 \times 10^{-6}$ and $K_{b2} = 1.32 \times 10^{-11}$, determine the pH of a 1.00×10^{-3} M solution of nicotine.

70. Ethylenediamine, H$_2$NCH$_2$CH$_2$NH$_2$, is used extensively in the synthesis of compounds containing transition-metal complexes in water. If $pK_{b1} = 3.29$ and $pK_{b2} = 6.44$, what is the pH of 2.50×10^{-4} M H$_2$NCH$_2$CH$_2$NH$_2$?

71. If the K_a of the conjugate acid of the artificial sweetener saccharin is 2.1×10^{-11}, what is pK_b for saccharin?

72. If the K_{a1} and K_{a2} values of chromic acid (H$_2$CrO$_4$) are 0.16 and 3.2×10^{-7}, respectively, what is the value of K_b of the CrO$_4^{2-}$ anion?

73. Sodium fluoride is added to many municipal water supplies to reduce tooth decay. Calculate the pH of a 0.00339 M solution of NaF. Given: the K_a of HF = 6.8×10^{-4} at 25°C.

74. Calculate the pH of a 1.25×10^{-2} M solution of the decongestant ephedrine hydrochloride if the pK_b of ephedrine (its conjugate base) is 3.86.

Controlling pH: Buffers

CONCEPT REVIEW

75. Why is a solution of sodium acetate and acetic acid a better pH buffer than is a solution of sodium chloride and hydrochloric acid?

76. Why does a solution of a weak base and its conjugate acid act as a better buffer than does a solution of the weak base alone?

PROBLEMS

77. Calculate the pH of a buffer that contains 0.244 M acetic acid and 0.122 M sodium acetate. The K_a of acetic acid is 1.76×10^{-5} at 25°C. What is the pH of this mixture at 0°C ($K_a = 1.64 \times 10^{-5}$)?

78. Calculate the pH of a buffer that contains a 0.100 M pyridine ($pK_b = 5.25$) and 0.275 M pyridinium chloride.

79. Calculate the pH and pOH of 500.0 mL of a phosphate buffer containing 0.225 M HPO$_4^{2-}$ and 0.225 M PO$_4^{3-}$; K_a for HPO$_4^{2-}$ = 4.2×10^{-13} at 25°C.

80. Determine the pH and pOH of 0.250 L of a buffer containing 0.0200 M boric acid and 0.0250 M sodium borate; pK_a for B(OH)$_3$ = 9.00 at 25°C.

81. What is the ratio of acetate ion to acetic acid ($K_a = 1.76 \times 10^{-5}$) in a buffer containing these compounds at pH = 3.56?
82. What is the ratio of lactic acid ($K_a = 1.37 \times 10^{-4}$) to lactate in a solution with pH = 4.00?
*83. How much 10 M HNO_3 must be added to 1.00 L of a buffer containing 0.010 M acetic acid and 0.10 M sodium acetate to reduce the pH to 5.00?
*84. How much 6.0 M NaOH must be added to 0.500 L of a 0.0200 M acetic acid and 0.0250 M sodium acetate buffer to raise the pH to 5.75?
85. What is the pH of a solution that results from mixing together equal volumes of 0.05 M ammonia and 0.025 M hydrochloric acid?
86. What is the pH of a solution that results from mixing together equal volumes of 0.05 M acetic acid and 0.025 M sodium hydroxide?
*87. Calculate the pH of 1.00 L of a buffer containing 0.120 M HNO_2 and 0.150 M $NaNO_2$ before and after the addition of 1.00 mL of 12.0 M HCl.
*88. Calculate the pH of 100.0 mL of a buffer containing 0.100 M NH_4Cl and 0.100 M NH_3 before and after the addition of 1.0 mL of 6 M HNO_3.

Titrations and Acid–Base Indicators

CONCEPT REVIEW

89. What are the differences between the titration curve of a strong acid titrated with a strong base and that of a weak acid titrated with a strong base?
90. Do all titrations of a strong base with a strong acid have the same pH at the equivalence point?
91. Do all titrations of a weak acid with a strong base have the same pH at the equivalence point?
92. What properties must a compound have to serve as an acid-base indicator?

PROBLEMS

93. A 25.0-mL sample of 0.100 M acetic acid is titrated with 0.125 M NaOH. Calculate the pH of the titration mixture after 10.0, 20.0, and 30.0 mL of base have been added.
94. A 25.0-mL sample of 0.100 M aqueous trimethylamine is titrated with 0.125 M HCl. Calculate the pH of the solution after 10.0, 20.0, and 30.0 mL of acid have been added; pK_b of $(CH_3)_3N$ = 4.19 at 25°C.
95. Sketch a titration curve for the titration of 50.0 mL of 0.250 M HNO_2 with 1.00 M NaOH. What is the pH at the equivalence point?
96. Red cabbage juice is a sensitive acid-base indicator with colors ranging from red at acidic pH to yellow in alkaline solutions. What color would red cabbage juice have when 25 mL of a 0.10 M solution of acetic acid is titrated with 0.10 M NaOH to its end point?
97. Sketch a titration curve for the titration of the malaria drug quinine if 40.0 mL of 0.100 M quinine is titrated with 0.100 M HCl. Given: K_{b1} for quinine = 3.31×10^{-6} and $K_{b2} = 1.35 \times 10^{-9}$.
98. Sketch a titration curve for the titration of 100 mL of 1.25×10^{-2} M adipic acid, $HO_2C(CH_2)_4CO_2H$, with 1.00×10^{-2} M NaOH. Given: K_{a1} of adipic acid = 3.71×10^{-5} and $K_{a2} = 3.87 \times 10^{-6}$. What color indicators could be used to detect the equivalence points?
99. What is the concentration of ammonia in a solution if 22.35 mL of 0.1145 M HCl are needed to titrate a 100.0-mL sample of the solution?
100. In an alkalinity titration of a 100.0-mL sample of water from a hot spring, 2.56 mL of 0.0355 M HCl is needed to reach the first equivalence point (pH = 8.3) and another 10.42 mL is needed to reach the second equivalence point (pH = 4.0). If the alkalinity of the spring water is due only to the presence of carbonate and bicarbonate, what are the concentrations of each of them?

Solubility Products

CONCEPT REVIEW

101. What is the difference between *solubility* and *solubility product*?
102. Describe how the common-ion effect limits the dissolution of a sparingly soluble ionic compound.
103. Which of the following cations will precipitate first as a carbonate mineral from an equimolar solution of Mg^{2+}, Ca^{2+}, and Sr^{2+}?
104. If the solubility of a compound increases with increasing temperature, does K_{sp} increase or decrease?

105. The K_{sp} of strontium sulfate increases from 2.77×10^{-7} at 3°C to 3.81×10^{-7} at 17°C. Is the dissolution of strontium sulfate endothermic or exothermic?
106. How will adding concentrated NaOH (aq) affect the solubility of an Al(III) salt?
*107. Fluoride ions in drinking water and in toothpaste convert hydroxyapatite in tooth enamel into fluoroapatite:

$$Ca_5(PO_4)_3OH(s) + F^-(aq) \rightleftharpoons Ca_5(PO_4)_3F(s) + OH^-(aq)$$

Why is fluoroapatite less susceptible than hydroxyapatite to erosion by acids?

*108. Tooth enamel is composed of a mineral known as hydroxyapatite with the formula $Ca_5(PO_4)_3OH$. Explain why tooth enamel is eroded by acidic substances.

PROBLEMS

109. At equilibrium, a saturated solution of barium sulfate contains 1.04×10^{-5} M Ba^{2+}. Starting with this information, calculate the K_{sp} of barium sulfate.
110. A saturated solution of barium fluoride contains 1.5×10^{-2} M F^-. Starting with this information, calculate the K_{sp} of BaF_2.
111. Calculate the equilibrium concentrations of Cu^+ and Cl^- in a saturated solution of copper(I) chloride, given $K_{sp} = 1.02 \times 10^{-6}$.
112. Calculate the equilibrium concentration of Pb^{2+} and F^- in a saturated solution of lead fluoride if $K_{sp} = 3.2 \times 10^{-8}$.
113. Calculate the solubility of calcite ($CaCO_3$) in grams per milliliter, given $K_{sp} = 9.9 \times 10^{-9}$ at 15° C.
114. Calculate the solubility of silver iodide in grams per milliliter if $K_{sp} = 1.5 \times 10^{-16}$ at 25°C.
115. Calculate the pH of a saturated solution of silver hydroxide; $K_{sp} = 1.52 \times 10^{-8}$.
116. Calculate the pH of a saturated solution of zinc hydroxide; $K_{sp} = 4 \times 10^{-17}$.
117. You have 100 mL of each of the following solutions. In which one will the most $CaCO_3$ dissolve? (a) 0.1 M NaCl; (b) 0.1 M Na_2CO_3; (c) 0.1 M NaOH; (d) 0.1 M HCl.
118. In which of the following solutions will CaF_2 be most soluble? (a) 0.010 M $Ca(NO_3)_2$; (b) 0.01 M NaF; (c) 0.001 M NaF; (d) 0.10 M $Ca(NO_3)_2$.
119. Calculate the solubility of calcite ($CaCO_3$) in 0.00100 M Na_2CO_3 ($K_{sp} = 4.5 \times 10^{-9}$ at 25°C).
120. Calculate the solubility of silver chloride in seawater with a chloride concentration of 0.547 M.
121. During the processing of black-and-white photographic film, excess silver(I) is removed by washing the film in a bath containing sodium thiosulfate. This treatment is based on the following complexation reaction:

$$Ag^+(aq) + 2\,S_2O_3^{2-}(aq) \longrightarrow Ag(S_2O_3)_2^{3-}(aq) \quad K_f = 5 \times 10^{13}$$

What is the ratio of $[Ag^+]$ to $[Ag(S_2O_3)_2^{3-}]$ in a bath in which $[S_2O_3^{2-}]$ is 0.233 M?

122. Lead poisoning in children and some cardiovascular diseases in adults are treated with intravenous injections of a complexing agent called EDTA. If the concentration of EDTA in the blood is 2.5×10^{-8} M and the formation constant for the Pb^{2+}-EDTA complex is 2.0×10^{18}, what is the concentration ratio of the free (and potentially toxic) $Pb^{2+}(aq)$ in the blood to the much less toxic Pb^{2+}-EDTA complex?
*123. Calculate the pH of a saturated solution of lithium carbonate (Li_2CO_3), for which $K_{sp} = 1.7 \times 10^{-3}$.
*124. Calculate the pH of 0.100 M $Fe(NO_3)_3$. Hint: the acidic properties of hydrated metal ions are described on page 828.

17 Electrochemistry
And Electrical Energy

The fuel for this four-passenger Honda is H_2 gas. Power comes from a stack of fuel cells that can generate 60 kilowatts of electrical power—enough to give the car a top speed of 150 km/h (93 mph)—as they extract electrical energy from the reaction $H_2 + 2\,O_2 \longrightarrow 2\,H_2O$.

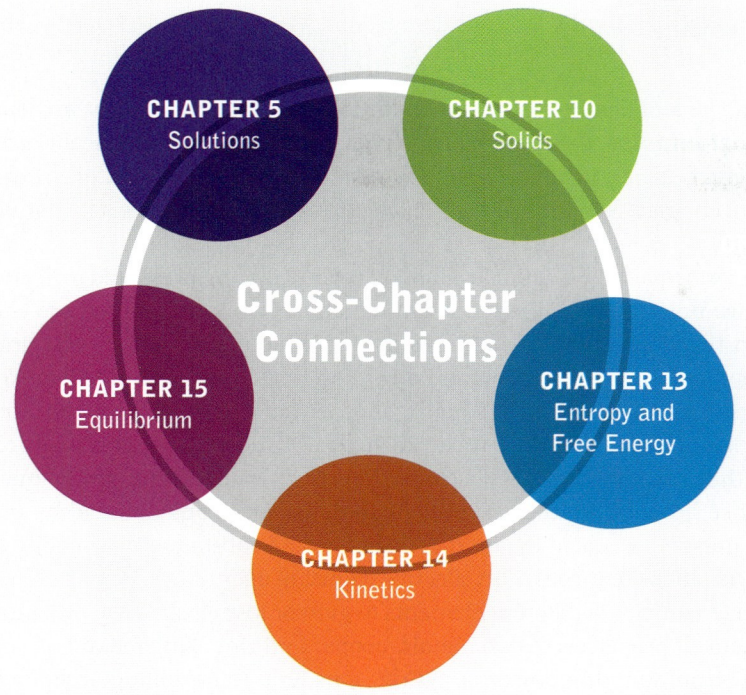

17.1 Voltaic cells

17.2 Voltage and free energy

17.3 The chemistries of some common batteries
  Dry cells
  Alkaline batteries
  Nickel–cadmium batteries

17.4 Standard potentials and batteries for laptops
  Cell potentials
  A reference point: the standard hydrogen electrode
  Nickel–metal hydride batteries
  Lithium-ion batteries

17.5 The effect of concentration on potential

17.6 Quantities of reactants and battery power

17.7 Electrolytic cells and recharging batteries

17.8 "Low emission" vehicles and more voltaic devices
  Hybrid vehicles
  Fuel cells
  Photochemical cells
  Biochemical fuel cells

A Look Ahead

In earlier chapters, we addressed atmospheric pollution problems caused by burning fossil fuels to generate electricity and to propel vehicles. In this chapter, we explore alternative energy conversion processes in which chemical energy is transformed into electrical energy in voltaic cells such as batteries. We use standard potentials to predict the electromotive force of batteries and the Nernst equation to predict how their energy content declines as reactants become products. To recharge a battery, electrical energy is converted into chemical energy as the reaction in the voltaic cell is reversed. During recharging, the battery's reactants are reformed from products as the voltaic cell becomes an electrolytic cell. We examine several applications of electrochemical cells including fuel cells, photovoltaic cells, and hybrid propulsion systems for motor vehicles.

17.1 VOLTAIC CELLS

In Chapters 14 through 16 we discussed some of the environmental pollution problems associated with the combustion of fossil fuels. In this chapter, we will examine new technologies for powering vehicles and producing electrical energy

Electrochemistry is the branch of chemistry that examines the transformations between chemical and electrical energy.

CONNECTION: In a redox reaction, the reducing agent is oxidized and the oxidizing agent is reduced (see Section 5.5).

that can dramatically reduce air pollution. These technologies are based on **electrochemistry,** the branch of chemistry that connects chemical reactions to the production or consumption of electrical energy. At the heart of electrochemistry are chemical reactions in which electrons are gained or lost—that is, redox chemistry.

The principles of oxidation–reduction reactions were presented in Section 5.5. There we noted that half-reactions in which electrons are gained are called reduction reactions and that half-reactions in which electrons are lost are called oxidation reactions. Reduction and oxidation half-reactions must happen simultaneously: as one reactant is reduced, another must be oxidized so that the number of electrons gained equals the number lost. A substance that is easily oxidized is one that readily gives up electrons. This electron-donating power makes the substance an effective reducing agent (because electrons donated by one substance must be accepted by another). Similarly, a substance that readily accepts electrons (and is reduced) is an effective oxidizing agent.

The tendencies of different elements to gain or lose electrons are illustrated by the chemical properties of two metals, copper and zinc. Zinc metal has considerable electron-donating power, and Cu^{2+} ions tend to be willing acceptors of donated electrons. These chemical properties are captured in the images in Figure 17.1. When Zn metal is placed in a solution of Cu^{2+} ions, Zn atoms spontaneously donate electrons to Cu^{2+} ions, forming Zn^{2+} ions and Cu atoms. The progress of this reaction can be monitored visually: shiny metallic zinc turns dark as a highly textured coating of copper accumulates, and the distinctive blue color of Cu^{2+} ions in solution fades as these ions acquire electrons and become atoms of copper metal.

Writing a chemical equation that describes the spontaneous reaction between Zn and Cu^{2+} is easily done because the metals react in a simple 1 : 1 stoichiometry:

$$Zn(s) + Cu^{2+}(aq) \longrightarrow Cu(s) + Zn^{2+}(aq) \qquad (17.1)$$

FIGURE 17.1 A strip of zinc immersed in a solution of copper sulfate becomes encrusted with a dark layer of copper, and the blue color of the solution fades as Cu^{2+} ions are reduced to Cu metal while Zn metal is oxidized to (colorless) Zn^{2+} ions.

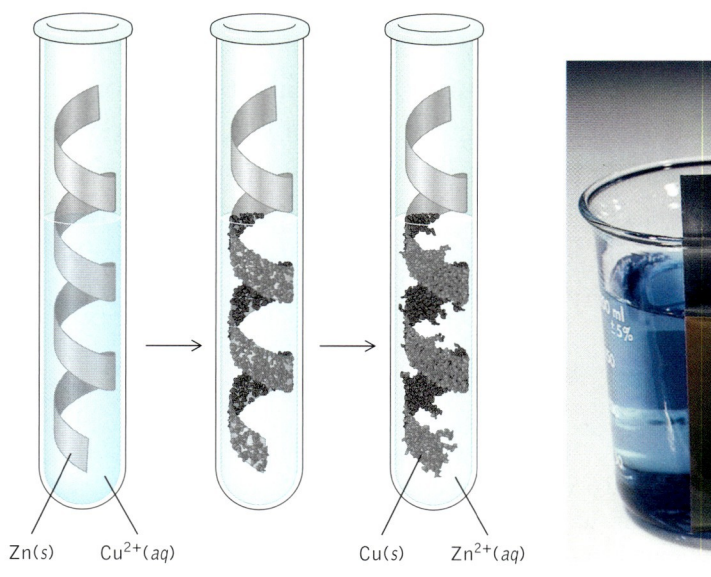

In the reaction, every mole of Zn loses 2 moles of electrons, and every mole of Cu^{2+} ions gains 2 moles of electrons.

Let's divide Equation 17.1 into the two half-reactions that describe the chemical changes undergone by the two metals individually as Zn atoms lose electrons:

$$Zn(s) \longrightarrow Zn^{2+}(aq) + 2\ e^- \qquad (17.2)$$

and Cu^{2+} ions gain them:

$$Cu^{2+}(aq) + 2\ e^- \longrightarrow Cu(s) \qquad (17.3)$$

Equation 17.2 is an example of an oxidation half-reaction; Equation 17.3 is an example of a reduction half-reaction. When we add them together (and cancel out the electron terms common to both sides of the reaction arrow), we get the original, complete reaction (Equation 17.1).

Having separated the oxidation and reduction half-reactions on paper, let's try to separate them physically by isolating the oxidation of Zn in one reaction chamber and the reduction of Cu^{2+} in another. An apparatus that achieves this goal is shown in Figure 17.2. One chamber contains a strip of zinc in a 1 M solution of $Zn(NO_3)_2$, and the other contains a strip of copper immersed in a 1 M solution of $Cu(NO_3)_2$. We need to connect the two chambers together with a structure called a salt bridge, essentially an upside-down U-tube with porous plugs at each end and filled with a solution of a strong electrolyte, such as sodium nitrate. The bridge provides an electrical connection between the two chambers even though the reactants and products of the two half-reactions are physically separated.

CONNECTION: A half-reaction is an equation describing the gain of electrons during the reduction of a substance or the loss of electrons during the oxidation of a substance (see Section 5.5).

✓ In many electrochemical cells, the salt bridge is replaced by a permeable membrane that separates the two compartments but allows ions to migrate from one compartment to the other.

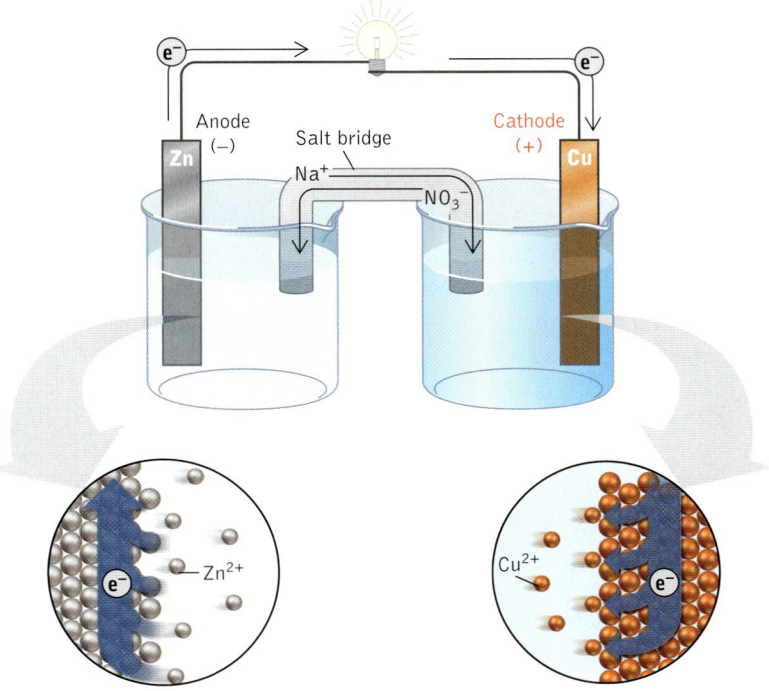

FIGURE 17.2 This voltaic cell consists of two compartments: one contains a zinc anode immersed in a 1 M solution of Zn^{2+}; the other contains a copper cathode immersed in a 1 M solution of Cu^{2+}. The salt bridge connection between them also contains a high concentration of $NaNO_3$. Migration of Na^+ and NO_3^- ions through the salt bridge allows electricity to flow between the compartments and through the external circuit with enough power to light a small flashlight bulb. The power comes from the oxidation of Zn metal at the anode and the reduction of Cu^{2+} ions at the cathode.

The two half-cell reactions proceed if we connect the Cu and Zn strips to an external electrical conductor, such as a metal wire. Then the electrons lost by Zn atoms flow from the strip of zinc through the wire to the copper strip where they are acquired by Cu^{2+} ions near the surface of the strip, forming more Cu atoms. With the passage of time, the mass of the Zn strip decreases as the mass of the Cu strip increases, and the blue color of $Cu^{2+}(aq)$ fades.

CONCEPT TEST: Will the increase in mass of the copper strip match the decrease in mass of the zinc strip during the reaction just described?

As the two half-reactions take place, Cu^{2+} ions are consumed in one compartment and Zn^{2+} ions are produced in the other, which should result in a buildup of positive charge on the Zn side and the creation of excess negative charge on the Cu side. In reality, there is no such charging of the compartments, because Na^+ ions in the salt bridge migrate toward the Cu compartment, replacing the Cu^{2+} ions that were converted into Cu atoms. At the same time NO_3^- ions migrate toward the Zn compartment, neutralizing the positive charges of the Zn^{2+} ions being produced there. This migration of ions completes the electrical circuit and allows the apparatus in Figure 17.2 to function as a battery.

At this point we need to address nomenclature. The apparatus in Figure 17.2, which includes the Cu and Zn strips, the compartments containing solutions of reactants and products, and the salt bridge, constitutes an *electrochemical cell*. This particular cell in which a spontaneous chemical reaction produces a current in an external circuit is called a *voltaic cell*. The Cu and Zn strips that provide electrical connections to the external circuit serve as electrodes. The electrode at which the oxidation half-reaction takes place (the Zn electrode in this case) is called an **anode**, and the electrode at which reduction takes place (the Cu electrode) is called a **cathode**.

> In a *voltaic cell,* chemical energy is transformed into electrical energy by a spontaneous redox reaction.
>
> An **anode** is an electrode at which an oxidation half-reaction takes place; a **cathode** is an electrode at which a reduction half-reaction takes place.

CONNECTION: Electrodes are defined in Section 5.3 as structures that make electrical connections between electrolytes and external electrical circuits.

SAMPLE EXERCISE 17.1: Write the voltaic-cell reaction, and sketch a cell in which nickel metal is oxidized to Ni^{2+} ions and Ag^+ ions are reduced to silver metal. In the sketch, include the reactants and products of the cell reaction; identify the anode and the cathode, and show the direction of electron flow in a wire connecting the two electrodes.

SOLUTION: We start by drawing two compartments connected by a salt bridge, as in Figure 17.2. A strip of nickel metal in a solution of Ni^{2+} ions is in one compartment, and a strip of silver immersed in a solution of Ag^+ ions is in the other. The oxidation of Ni

$$Ni \longrightarrow Ni^{2+} + 2\ e^-$$

produces a supply of electrons that travel through the wire to the silver electrode where they are consumed by the reduction of silver ions:

$$Ag^+ + e^- \longrightarrow Ag$$

Oxidation takes place at the nickel electrode; so it is the anode. Reduction takes place at the silver electrode, making it the cathode. The cell reaction is the sum of the preceding half-reactions. However, before adding them together, we need to multiply the silver half-reaction by two so that the number of electrons produced at the anode equals the number of electrons consumed at the cathode:

$$Ni \longrightarrow Ni^{2+} + 2\ e^-$$
$$+ 2\ (Ag^+ + e^- \longrightarrow Ag)$$
$$\overline{Ni + 2\ Ag^+ + \cancel{2\ e^-} \longrightarrow Ni^{2+} + 2\ Ag + \cancel{2\ e^-}}$$

Canceling out the electrons, we have the overall cell reaction:

$$Ni + 2\ Ag^+ \longrightarrow Ni^{2+} + 2\ Ag$$

A sketch of the cell might look like this:

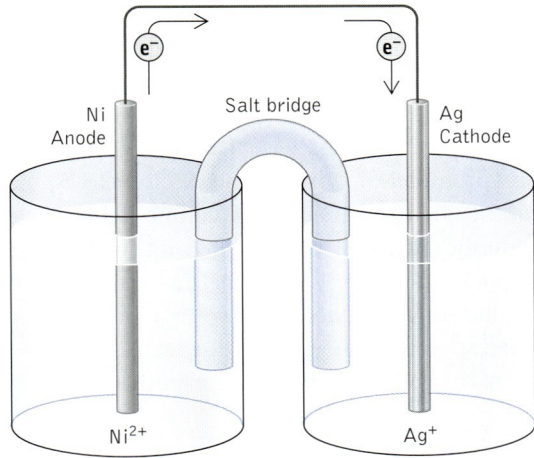

PRACTICE EXERCISE: Write the balanced equation for the cell reaction and sketch a voltaic cell in which zinc metal is oxidized to Zn^{2+} and Ni^{2+} ions are reduced to nickel metal. Identify the anode and the cathode. (See Problems 5–8.)

17.2 VOLTAGE AND FREE ENERGY

A voltaic cell is an electrochemical cell that produces a **cell potential (E_{cell})**, or **electromotive force**, that can force an electrical current through an external circuit. For example, the cell in Figure 17.2 produces a cell potential of 1.10 V. When such a cell potential is used to push electrons (and their charges) through an electrical circuit, the cell is doing work, as defined in Section 11.1. The SI unit for electrical charge is the coulomb, C. In metallic conductors, the charge carriers are the mobile valence electrons that form metallic bonds (see Section 10.6). Each electron has a charge of -1.602×10^{-19} C (Section 1.5); so the quantity of electrical charge in a mole of electrons (e^-) is

☑ **Cell potential (E_{cell}) (electromotive force)** is the voltage between the electrodes of a voltaic cell.

There are 9.65×10^4 C of electrical charge in a mole of positively charged particles. The charge on a mole of electrons is -9.65×10^4 C.

$$\frac{-1.602 \times 10^{-19} \text{ C}}{\text{e}^-} \times \frac{6.022 \times 10^{23} \text{ e}^-}{\text{mol e}^-} = \frac{-9.65 \times 10^4 \text{ C}}{\text{mol e}^-}$$

The corresponding quantity of positive charge, 9.65×10^4 C/mol, is called the Faraday constant (F) after Michael Faraday, the British scientist whose discoveries helped scientists understand that redox reactions take place when electrons are transferred between reacting species. The quantity of charge (C) flowing through an electrical circuit connected to a voltaic cell equals the product of the number of moles of electrons (n) flowing through the circuit times the Faraday constant:

$$C = nF \tag{17.4}$$

CONNECTION: Michael Faraday found that passing an electrical current through solutions caused chemical reactions to take place (see Section 6.2).

The electrical work (w_{elec}) done on the circuit equals the product of this charge times the electromotive force that is pushing it:

$$w_{elec} = CE_{cell} = nFE_{cell} \tag{17.5}$$

There must be a connection between this work done and the loss of chemical energy inside the cell. They aren't equal, because the conversion of chemical energy into electrical energy is not 100% efficient, just as the chemical energy in gasoline and oxygen is not converted with 100% efficiency into the kinetic energy that propels an automobile down the road (Section 13.5). As in the automobile engine, some of the chemical energy of a voltaic cell is wasted, and only the change in "free" energy (ΔG) will be available to do work moving electrical charge through an external circuit. To express this connection in equation form, we need to keep in mind that work done *by* a voltaic cell on its surroundings (the electrical circuit) corresponds to energy *lost* during the cell reaction. Therefore, from the perspective of the cell, this work has a negative sign, and so

$$\Delta G = -nFE_{cell} \tag{17.6}$$

CONNECTION: *Free* energy is the maximum amount of energy in a reaction that is available to do useful work (see Section 13.4).

You may be wondering about the different units on the quantities on the left and right sides of Equation 17.6. We usually express energy in joules, but the units on the right are:

$$(\text{mol})(\text{coulomb}/\text{mol})(\text{volt})$$

or

$$\text{coulomb} \cdot \text{volt}$$

It turns out that an important conversion factor for units of energy is the following relation:

$$1 \text{ coulomb} \cdot \text{volt} = 1 \text{ joule } (1 \text{ C} \cdot \text{V} = 1 \text{ J})$$

and so the left and right sides of Equation 17.6 are indeed equivalent.

The negative sign on the right side of Equation 17.6 makes sense for another reason. Voltaic cells must produce a voltage if they are to do work in an external circuit. Thus E in Equation 17.6 must be a positive value. However, the reaction in a voltaic cell must also be spontaneous, which requires that ΔG be negative. Because n and F have positive values, we need a negative sign in Equation 17.6 so that a positive E_{cell} corresponds to a negative ΔG_{cell}.

✓ In a voltaic cell, a spontaneous cell reaction ($\Delta G_{cell} < 0$) produces a positive cell potential, or electromotive force ($E_{cell} > 0$).

Let's calculate ΔG for the Zn–Cu cell reaction in Figure 17.2 from its cell potential of 1.10 V. The half-reactions of the cell tell us that 2 moles of electrons are transferred for each mole of reactant in the cell reaction (Equation 17.1). Therefore, the value of n in Equation 17.6 is 2 and the calculation proceeds as follows:

$$\Delta G = -nFE_{cell}$$
$$= -2 \text{ mol} \times \frac{9.65 \times 10^4 \text{ C}}{\text{mol}} \times 1.10 \text{ V} = -2.51 \times 10^5 \text{ V} \cdot \text{C}$$
$$= -2.51 \times 10^5 \text{ J}$$
$$= -251 \text{ kJ}$$

As expected, the spontaneous cell reaction has a sizeable, negative ΔG. To put this value in perspective, it is about the same as $\Delta G°$ for the combustion of 1 mole of hydrogen gas:

$$H_2(g) + \tfrac{1}{2} O_2(g) \longrightarrow H_2O(l) \qquad \Delta G° = -237 \text{ kJ}$$

SAMPLE EXERCISE 17.2: Calculate the value of ΔG for the cell reaction in the lead–acid batteries (see Figure 17.3) used to start car and truck engines if each cell in the battery produces 2.04 V. The cell reaction is

$$Pb(s) + PbO_2(s) + 2 H_2SO_4(aq) \longrightarrow 2 PbSO_4(s) + 2 H_2O(l)$$

SOLUTION: To calculate the value of ΔG, we need to determine the number of moles of electrons (n) transferred in the cell reaction. One way to do so is to determine the changes in the oxidation states of lead at the two electrodes. Lead metal is a pure element and so has an oxidation number of 0. Lead in PbO_2 must

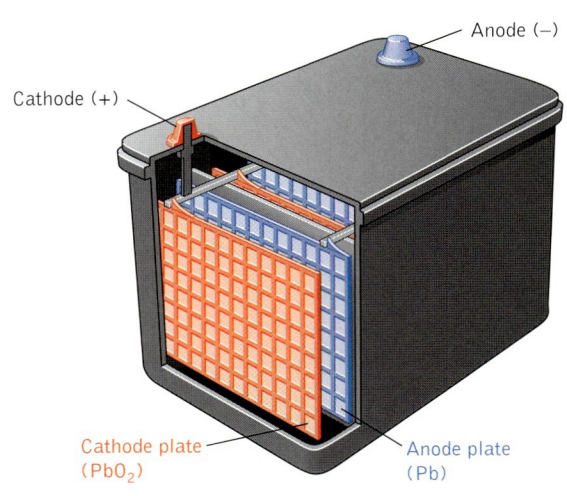

FIGURE 17.3 The lead–acid battery that provides power to start most motor vehicles contains six cells, each with an anode made of spongy lead and a cathode made of PbO_2 immersed in an electrolyte of 4.5 M H_2SO_4. The electrodes are formed into plates and held in place by grids made of a lead alloy. The grids connect the cells (each providing 2.0 V) together in series so that the operating voltage of the battery is the sum of the six cell potentials, or 12.0 V.

> ✓ In a working lead storage battery, $PbO_2(s)$ is reduced to $PbSO_4(s)$, and $Pb(s)$ is oxidized to $PbSO_4(s)$. This $PbSO_4$ is converted back into PbO_2 and Pb metal when the battery is recharged.

have an oxidation number of +4 to balance the −2 of two oxide ions. The charge on the sulfate ion is −2; so the oxidation number of lead in $PbSO_4$ must be +2. Thus, the oxidation number of lead metal at the anode increases by 2 during the battery reaction, and the oxidation number of lead in PbO_2 decreases by 2. These changes mean that 2 moles of electrons are transferred for every mole of Pb metal that is oxidized and every mole PbO_2 that is reduced. Inserting this value of n and the voltage of the cell into Equation 17.6, we have

$$\Delta G = -nFE_{cell}$$

$$\Delta G = -2 \; \text{mol} \left(\frac{9.65 \times 10^4 \; C}{\text{mol}} \right)(2.041 \; V) = -3.94 \times 10^5 \; V \cdot C$$

$$= -3.94 \times 10^5 \; J = -394 \; kJ$$

PRACTICE EXERCISE: Fuel cells (described in Section 17.8) are voltaic cells in which the energy released by chemical reactions, such as the following one between hydrogen and oxygen, is converted into electrical energy:

$$H_2(g) + \tfrac{1}{2} O_2(g) \longrightarrow H_2O(l) \qquad \Delta G° = -237 \; kJ$$

If all the free-energy change in the cell reaction were converted into electrical energy, what would be the potential ($E°$) of the cell? (See Problems 15 and 16.)

17.3 THE CHEMISTRIES OF SOME COMMON BATTERIES

Table 17.1 lists a few of the many types of voltaic cells commonly used as energy sources. Some are familiar, such as the lead–acid battery used to start automobiles and the batteries used to power flashlights and portable electronic devices. In this section and those that follow, we will examine the chemical processes that supply the electrical power in several of these battery designs.

Dry cells

For many years, batteries called "dry cells" were widely used to power flashlights, portable radios, and other electronic devices. The name of these batteries comes from the fact the electrolyte inside them is solid rather than liquid, even though the solid is really a moist paste. The dry-cell battery was invented by French engineer Georges Leclanché (1839–1882) in 1866. His design featured a carbon-rod electrode inserted in a paste of NH_4Cl and MnO_2 and surrounded by a zinc container as shown in Figure 17.4. The container serves as the anode of the cell; so the oxidation half-reaction is

$$Zn(s) \longrightarrow Zn^{2+}(aq) + 2 \; e^-$$

17.3 The Chemistries of Some Common Batteries

TABLE 17.1 Electrochemical Energy Sources

Application	Voltaic Cell Design
Automobiles	Lead–acid battery
	Nickel–metal hydride battery
	Polymer membrane fuel cell
Portable electronics	Nickel–cadmium battery
Flashlights	Dry cell battery
	Alkaline battery
Computers	Nickel–metal hydride battery
	Lithium-ion battery
Watches	Mercury button battery
Spacecraft	Alkaline fuel cell
Power generation	Phosphoric acid fuel cell
	Molten carbonate fuel cell
	Solid oxide fuel cell

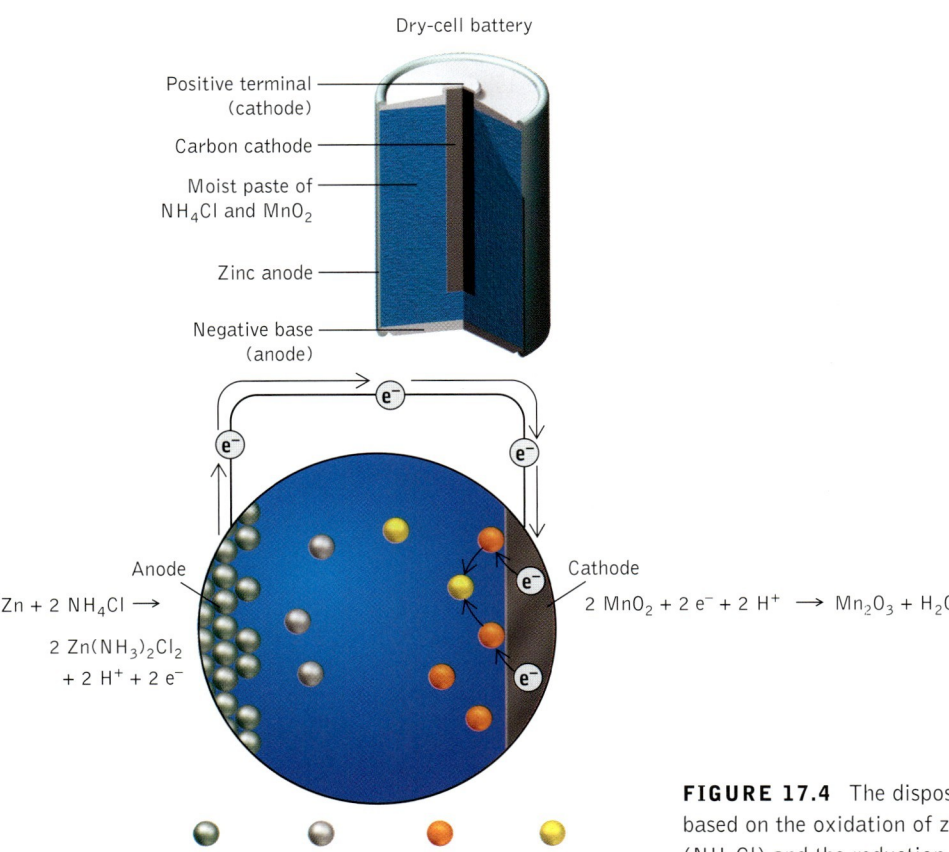

FIGURE 17.4 The disposable "dry cell" battery is based on the oxidation of zinc in an acid medium (NH_4Cl) and the reduction of MnO_2 to Mn_2O_3. The overall cell potential is 1.5 V. The anode is consumed as zinc oxidizes to $Zn(NH_3)_2Cl_2$.

The reduction of ammonium ions to ammonia and hydrogen takes place at the carbon cathode (the carbon itself does not change oxidation state) as described by the half-reaction

$$2\,NH_4^+(aq) + 2\,e^- \longrightarrow 2\,NH_3(g) + H_2(g)$$

Production of NH_3 and H_2 would cause the dry cell to explode if it were not for the presence of $ZnCl_2$ and MnO_2. Zinc(II) combines with ammonia and chloride to form solid diamminozinc chloride:

$$Zn^{2+}(aq) + 2\,NH_3(g) + 2\,Cl^-(aq) \longrightarrow Zn(NH_3)_2Cl_2(s)$$

while manganese(IV) oxide (commonly called manganese dioxide) oxidizes hydrogen gas to water:

$$2\,MnO_2(s) + H_2(g) \longrightarrow Mn_2O_3(s) + H_2O(l)$$

All of these reactions added together result in a rather complicated cell reaction:

$$Zn(s) + 2\,NH_4Cl(s) + 2\,MnO_2(s) \longrightarrow Zn(NH_3)_2Cl_2(s) + Mn_2O_3(s) + H_2O(l)$$

Alkaline batteries

Today, alkaline batteries (Figure 17.5) have largely replaced dry-cell batteries in many applications. Alkaline batteries produce the same cell potential (1.5 V) as a dry cell, but they have considerably more capacity to produce electric current at that potential. Like dry cells, alkaline batteries have zinc anodes, but the electrolyte is strongly alkaline potassium hydroxide instead of ammonium chloride, which is a weakly acidic compound. Under alkaline conditions, the anodic half-reaction produces insoluble $Zn(OH)_2$:

$$Zn(s) + 2\,OH^-(aq) \longrightarrow Zn(OH)_2(s) + 2\,e^-$$

while at the cathode solid manganese(IV) oxide is reduced to solid manganese(III) oxide:

$$2\,MnO_2(s) + H_2O(l) + 2\,e^- \longrightarrow Mn_2O_3(s) + 2\,OH^-(aq)$$

Combining these two equations, we get the following voltaic cell reaction:

$$Zn(s) + 2\,MnO_2(s) \longrightarrow Mn_2O_3(s) + Zn(OH)_2(s)$$

Alkaline batteries have simpler cell chemistry than dry cells, and the cell reaction does not result in the formation of gases. The latter provides an important operational advantage: alkaline batteries produce steadier voltages at higher currents. High current loads in a dry cell may result in such rapid production of hydrogen and ammonia that these gases are not totally consumed by reactions with Zn^{2+} and MnO_2. The resulting buildup of gases impedes ion transport within the battery, causing a high internal resistance and a drop in battery voltage.

Dry cells also have shorter shelf lives than alkaline batteries because of zinc metal in contact with ammonium salts slowly oxidizes. Alkaline batteries also last longer in use because they have greater quantities of reactants than do dry

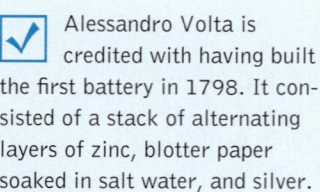

CONNECTION: The formation of complex ions between transition-metal ions and Lewis bases such as NH_3 is described in Section 16.10.

✓ Alessandro Volta is credited with having built the first battery in 1798. It consisted of a stack of alternating layers of zinc, blotter paper soaked in salt water, and silver.

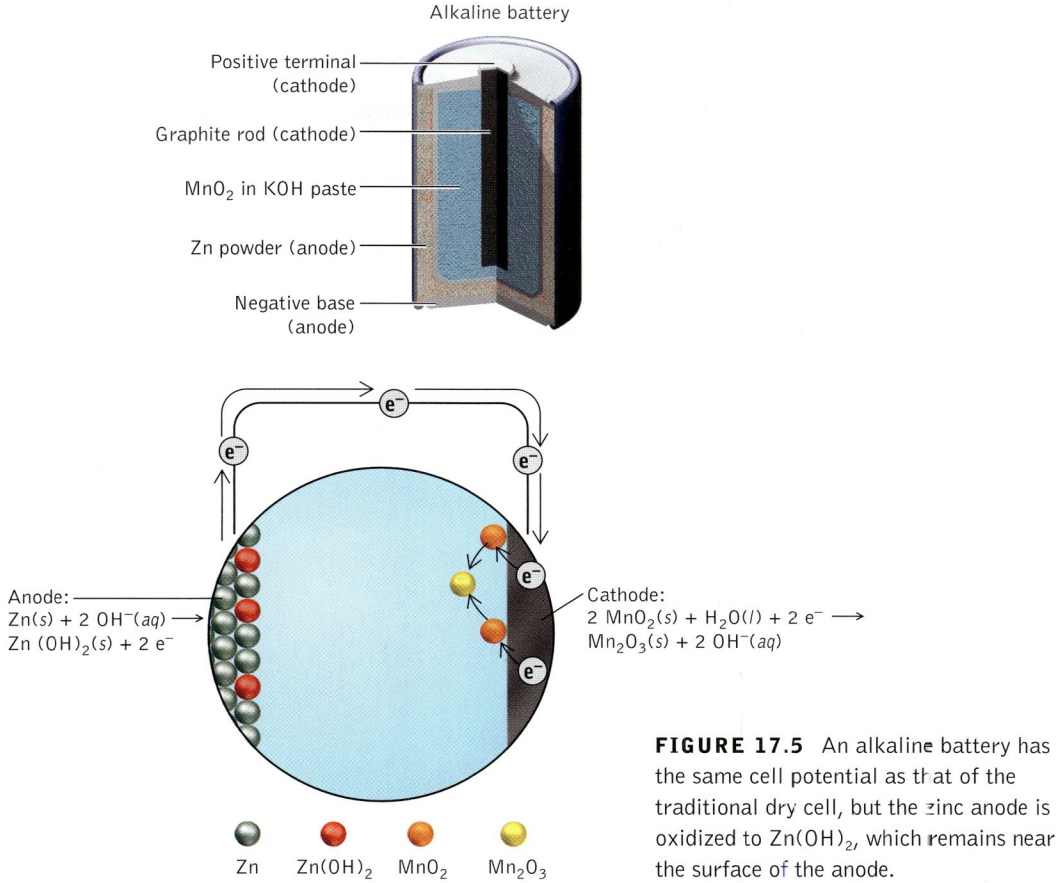

FIGURE 17.5 An alkaline battery has the same cell potential as that of the traditional dry cell, but the zinc anode is oxidized to Zn(OH)$_2$, which remains near the surface of the anode.

cells of the same dimensions. Electrolyte (ammonium chloride) is consumed in the standard dry-cell reaction but not in the alkaline zinc–manganese batteries, which allows more space for cathode material (MnO$_2$) in the alkaline batteries. In addition, powdered zinc is used in place of a zinc can to increase the amount of oxidizable material. With more oxidizable and more reducible starting materials, alkaline batteries do indeed outlast dry cells of comparable size—as suggested by TV ads featuring the Energizer Bunny.

Nickel–cadmium batteries

Many of the rechargeable batteries used in flashlights and cordless tools and telephones are nickel–cadmium (Ni-Cd, pronounced "nicad") batteries (Figure 17.6). The electrical power of Ni-Cd batteries comes from two chemical reactions in a strongly basic electrolyte at two electrodes: one made of cadmium metal and the other of a solid nickel compound, NiO(OH). Note that there are two different

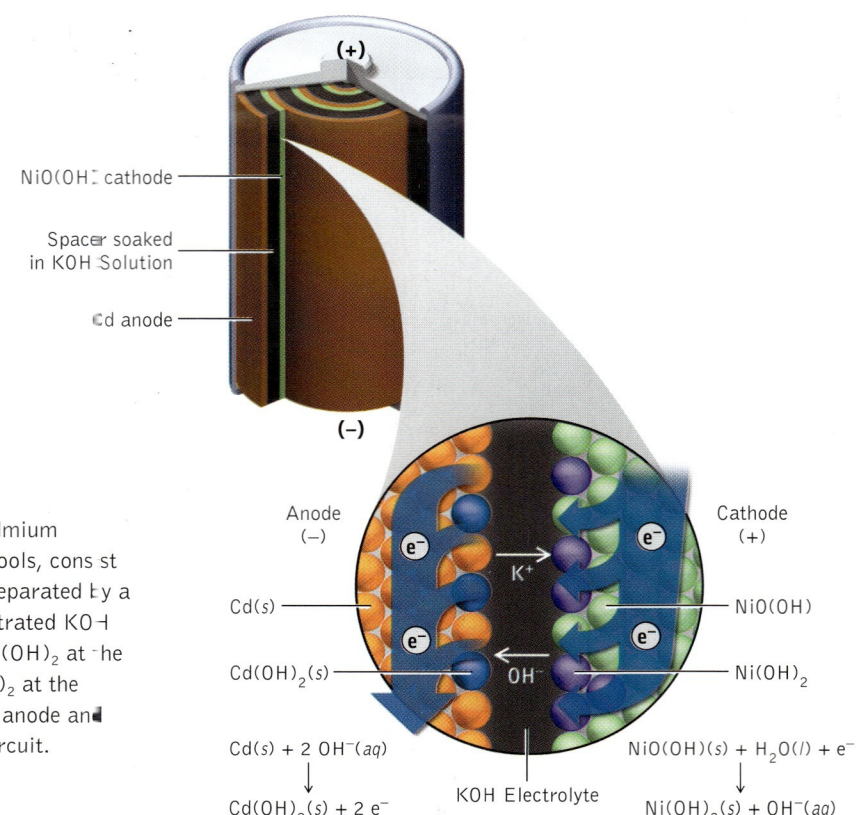

FIGURE 17.6 Rechargeable nickel–cadmium batteries, such as those used in cordless tools, consist of thin plates of Cd metal and NiO(OH) separated by a porous plastic spacer soaked with concentrated KOH solution. Cadmium is oxidized to solid $Cd(OH)_2$ at the anode, and NiO(OH) is reduced to $Ni(OH)_2$ at the cathode. The migration of K^+ toward the anode and OH^- toward the cathode completes the circuit.

anions bonded to nickel in this electrode material. This is not an unusual occurrence in transition-metal chemistry: there are many such *oxohydroxides*. In nature, they form when transition-metal oxides on rocks and minerals come in contact with water.

As a Ni-Cd battery generates electricity, cadmium metal is oxidized, forming solid $Cd(OH)_2$ at the battery's anode, and solid NiO(OH) is reduced at the cathode. The oxidation half-reaction in strongly basic electrolyte (typically 22 M KOH) can be written as follows:

$$Cd(s) + 2\ OH^-(aq) \longrightarrow Cd(OH)_2(s) + 2\ e^- \qquad (17.7)$$

At the cathode, each mole of NiO(OH) gains 1 mole of electrons according to the half-reaction

$$NiO(OH)(s) + H_2O(l) + e^- \longrightarrow Ni(OH)_2(s) + OH^-(aq) \qquad (17.8)$$

Note the participation of hydroxide ions in both half-reactions. High electrolyte pH is key to the Ni-Cd cell reaction because $Cd(OH)_2$ and $Ni(OH)_2$, like most transition-metal hydroxides, are not very soluble. Thus a solid coating of $Cd(OH)_2$ forms around the anode and a solid coating of $Ni(OH)_2$ forms around the cathode as the

✓ In a working Ni-Cd battery, NiO(OH)(s) is reduced to $Ni(OH)_2(s)$, and Cd(s) is oxidized to $Cd(OH)_2(s)$. The products adhere to their respective electrodes, enabling the battery to be recharged.

battery's cell reaction proceeds. As we shall see in Section 17.7, it is important that these products adhere to the electrode so that they can be converted back to the starting electrode materials as the battery is recharged.

The overall reaction of a Ni-Cd battery is the sum of the two half-reactions shown in Equations 17.7 and 17.8. Before we can add them together, we need to balance the number of electrons gained and lost by multiplying Equation 17.8 by two

$$2\ NiO(OH)(s) + 2\ H_2O(l) + 2\ e^- \longrightarrow 2\ Ni(OH)_2(s) + 2\ OH^-(aq)$$

Adding this equation to Equation 17.7, we have

$$Cd(s) + 2\ NiO(OH)(s) + \cancel{2\ OH^-(aq)} + 2\ H_2O(l)$$
$$\longrightarrow Cd(OH)_2(s) + 2\ Ni(OH)_2(s) + \cancel{2\ OH^-(aq)} + \cancel{2e^-}$$

After canceling out the hydroxide ions and free electrons that appear on both sides of the reaction arrow, we obtain the overall equation for the Ni-Cd cell reaction:

$$Cd(s) + 2\ NiO(OH)(s) + 2\ H_2O(l) \longrightarrow Cd(OH)_2(s) + 2\ Ni(OH)_2(s)$$

CONNECTION: To write balanced equations for cell reactions, be sure that the gains and losses of electrons in the reduction and oxidation half-reactions are equal (see Section 5.5).

SAMPLE EXERCISE 17.3: Some of the small, button-shaped batteries used in watches and calculators consist of a moist KOH electrolyte between an anode made of zinc and a cathode at which HgO is reduced to elemental mercury as the battery discharges. Write the anode and cathode half-reactions and the overall cell reaction for these batteries.

SOLUTION: At the cathode, 2 moles of electrons are needed for each mole of HgO that is reduced to elemental mercury, which leads to the following incomplete half-reaction as a starting point:

$$HgO(s) + 2\ e^- \longrightarrow Hg(l)$$

To balance this half-reaction, we need an oxygen-containing species among the products. We also need 2 moles of negative charges on the product side to balance the charges on the 2 moles of electrons on the reactant side. The electrolyte is moist KOH; so adding 2 moles of 2 OH^- ions is a logical way to address both needs:

$$HgO(s) + 2\ e^- \longrightarrow Hg(l) + 2\ OH^-(aq)$$

Now we have 1 more mole of O and 2 more moles of H on the right-hand side. This problem is solved by adding 1 mole of H_2O to the left-hand side, which produces a complete and balanced reduction half-reaction:

(1) $$HgO(s) + H_2O(l) + 2\ e^- \longrightarrow Hg(l) + 2\ OH^-(aq)$$

The half-reaction at the anode is the same as that in alkaline batteries:

(2) $$Zn(l) + 2\ OH^-(aq) \longrightarrow Zn(OH)_2(s) + 2\ e^-$$

The battery cell reaction is the sum of half-reactions 1 and 2:

$$Zn(s) + HgO(s) + \cancel{2\,OH^-(aq)} + H_2O(l) + \cancel{2\,e^-} \longrightarrow$$
$$Zn(OH)_2(s) + \cancel{2\,OH^-(aq)} + Hg(l) + \cancel{2\,e^-}$$

or

$$Zn(s) + HgO(s) + H_2O(l) \longrightarrow Zn(OH)_2(s) + Hg(l)$$

PRACTICE EXERCISE: Other button batteries are based on the reduction of Ag_2O instead of HgO at their cathodes. They also have Zn anodes, with moist KOH as an electrolyte. Write the anode and cathode half-reactions and the overall cell reaction for silver button batteries. (See Problems 17–20.)

17.4 STANDARD POTENTIALS AND BATTERIES FOR LAPTOPS

How much electrical energy can a battery produce? Electrical energy is defined by the electromotive force (voltage) of the battery and the quantity of charge that it can deliver at that voltage. In this section, we will focus on the voltages that batteries produce. These voltages depend on the half-reactions that take place inside them and on the **standard potentials ($E°$)** of those half-reactions. Appendix 6 lists the standard potentials of some of the more commonly encountered electrochemical half-reactions. By convention, these half-reactions are written as reductions (and so are sometime called *standard reduction potentials*) and have the general form:

> **How much electrical energy can a battery produce?**

$$\text{oxidized form} + \text{electron(s)} \longrightarrow \text{reduced form}$$

There is no need for a separate table of standard oxidation potentials, because reduction half-reactions can always be written backward, making them oxidation half-reactions. Reversing the reduction half-reaction does not affect the magnitude of its standard potential, but it does change the sign. In this respect, standard potentials are like other thermodynamic parameters of reactions, such as $\Delta G°$, $\Delta H°$, and $\Delta S°$. Recall that reversing a chemical reaction changes the sign of these parameters, but their magnitudes stay the same.

> ✓ A **standard potential ($E°$)** is the electromotive force of a half-reaction written as a reduction in which all reactants and products are in their standard states, which means that the concentrations of all dissolved substances are 1 molar, and the partial pressures of all gases are 1 bar.

Cell potentials

The **standard cell potential ($E°_{cell}$)**, of a battery is calculated from the difference in the standard potentials of its cathode and anode:

$$E°_{cell} = E°_{cathode} - E°_{anode} \qquad (17.9)$$

For example, the value of $E°_{cell}$ for the Ni-Cd battery cell reaction can be calculated from the standard potential of the cathode half-reaction:

> ✓ **Standard cell potential ($E°_{cell}$)** is the potential of a cel when all reactants and products are in their standard states.

$$\text{NiO(OH)}(s) + \text{H}_2\text{O}(l) + e^- \longrightarrow \text{Ni(OH)}_2 + \text{OH}^-(aq) \quad E°_{\text{cathode}} = +1.32 \text{ V}$$

minus the standard potential of the anode half-reaction:

$$\text{Cd(OH)}_2(s) + 2\,e^- \longrightarrow \text{Cd}(s) + 2\,\text{OH}^-(aq) \quad\quad E°_{\text{anode}} = 0.019 \text{ V}$$

and so

$$E°_{\text{cell}} = E°_{\text{cathode}} - E°_{\text{anode}}$$
$$= 1.32 \text{ V} - 0.019 \text{ V} = 1.30 \text{ V}$$

Note that we used the standard (reduction) electrode potentials of both half-reactions in this calculation, even though the anode half-reaction is actually an oxidation half-reaction. Doing so may seem odd, but when we *subtract* $E°_{\text{anode}}$ from $E°_{\text{cathode}}$ (rather than adding it), we are compensating for the fact that the $E°_{\text{anode}}$ is the standard potential of an *oxidation* half reaction.

The superscript symbol has its usual meaning: all reactants and products are in their standard states, which means that the concentrations of all dissolved substances are 1 molar, the pressure of all gases is 1 bar (10^5 Pa). We also assume a temperature of 25°C. Equation 17.9 is a special case of a general equation that applies to the potential of any cell:

$$E_{\text{cell}} = E_{\text{cathode}} - E_{\text{anode}} \quad\quad (17.10)$$

Voltaic cells produce positive cell potentials; that is, $E_{\text{cell}} > 0$. A *positive* electromotive force, like a *negative* change in free energy (ΔG), means that the cell reaction is *spontaneous*.

A reference point: the standard hydrogen electrode

As already noted, oxidation and reduction must take place simultaneously. Consequently, we can directly measure only the *net result* of simultaneous oxidation and reduction half-reactions; that is, E_{cell}. How can we know the individual values of E_{cathode} and E_{anode}? One way to define these values is to arbitrarily assign a value to the potential produced by a particular half-reaction and then reference the potentials of all other half-reactions to it. The designated reference point is the standard potential for the reduction of hydrogen ions to hydrogen gas (see Equation 17.11). The $E°$ for this half-reaction is defined as 0.000 V. An electrode that generates this reference potential, called the **standard hydrogen electrode** (SHE), consists of a platinum electrode in contact with a solution of strong acid ($[\text{H}^+] = 1.00\ M$) and hydrogen gas, as shown in Figure 17.7. Platinum is not changed by the electrode reaction. Rather, it serves as a chemically inert conveyor of electrons and sensor of the potential generated by the reduction of hydrogen ions to hydrogen gas:

> How can we know the individual values of E_{cathode} and E_{anode}?

> ✓ The potential of the standard hydrogen electrode (SHE) is defined as 0.000 V, and so the SHE serves as reference electrode against which other electrode half-reactions can be measured.

$$2\,\text{H}^+(aq) + 2\,e^- \longrightarrow \text{H}_2(g) \quad\quad E° = 0.000 \text{ V} \quad\quad (17.11)$$

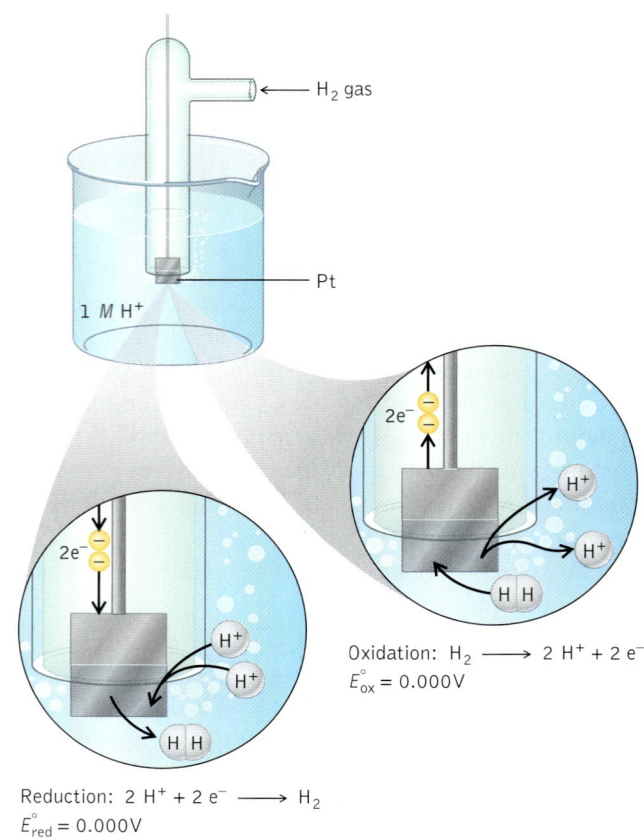

FIGURE 17.7 The standard hydrogen electrode (SHE) consists of a platinum electrode in a solution of 1.0 M HCl and bathed in a stream of pure H_2 gas.

Because the electrode potential of an SHE is 0.000 V, the value of E_{cell} for any cell in which the SHE is one of the two components will be the potential produced by the other component. Suppose, for example, that a cell consists of a strip of zinc metal immersed in a 1.00 M solution of Zn^{2+} ions and connected by a salt bridge to an SHE. Suppose further that a meter measures the electromotive force pushing electrons from the zinc electrode to the SHE, as shown in Figure 17.8A. This direction of current flow means that the zinc electrode is the anode because Zn atoms are being oxidized to Zn^{2+} ions, thereby losing electrons to the electrode, and that the SHE is the cathode. At 25°C, the meter reads 0.76 V. Inserting this value into Equation 17.9 (which is appropriate because all the reactants and products are in their standard states), we have

$$E^\circ_{cell} = E^\circ_{cathode} - E^\circ_{anode}$$

$$0.76\ V = E^\circ_{SHE} - E^\circ_{Zn} = 0.00\ V - E^\circ_{Zn}$$

Therefore

$$E^\circ_{Zn} = -0.76\ V$$

Sometimes the reference electrode is the anode of an electrochemical cell. Consider the voltaic cell in Figure 17.8B. An SHE is coupled to a copper electrode immersed in a 1 M solution of Cu^{2+} ions. In this cell, current flows from the SHE

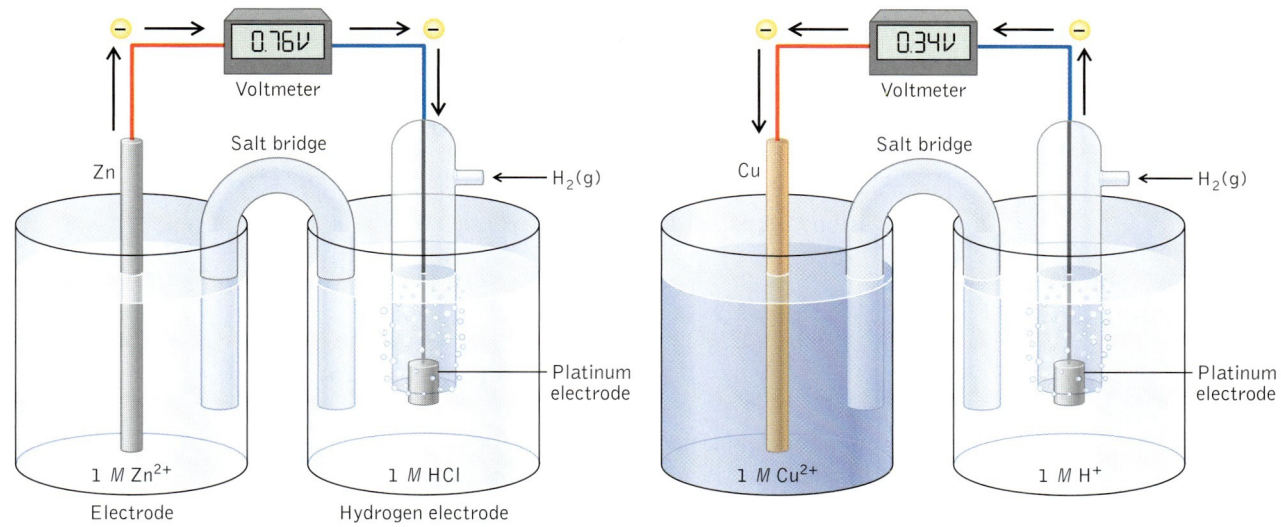

FIGURE 17.8 The standard hydrogen electrode when used as a reference electrode allows the determination of the electrode potential of another half-reaction. A. When coupled to a Zn electrode under standard conditions, the SHE is the cathode; the Zn electrode is the anode, and $E°_{cell} = 0.000\ V - E°_{Zn}$. B. When coupled to a Cu electrode under standard conditions, the SHE is the anode, the Cu electrode is the cathode, and $E°_{cell} = E°_{Cu} - 0.000\ V = E°_{Cu}$.

through an external circuit to the copper electrode at a cell potential of 0.34 V at 25°C. The SHE is the anode, copper is the cathode, and $E°_{Cu}$ is calculated as follows:

$$E°_{cell} = E°_{cathode} - E°_{anode}$$
$$= E°_{Cu} - E°_{SHE}$$
$$= E°_{Cu} - 0.000\ V = 0.34\ V$$

Therefore

$$E°_{Cu} = 0.34$$

SAMPLE EXERCISE 17.4: The half-reactions in a lead–acid battery are the oxidation of Pb to PbSO$_4$ at the anode and the reduction of PbO$_2$ to PbSO$_4$ at the cathode. Calculate the value of $E°_{cell}$ for a lead–acid battery.

SOLUTION: To calculate $E°_{cell}$, we need to find the appropriate standard half-reactions in Appendix 6. One of them must be the reverse of the anode reaction because all the half-reactions in Appendix 6 are written as reductions. A search uncovers the following entries:

(1) $PbSO_4(s) + 2\ e^- \longrightarrow Pb(s) + SO_4^{2-}(aq)$ $E° = -0.356\ V$

(2) $PbO_2(s) + 4\ H^+(aq) + SO_4^{2-}(aq) + 2\ e^- \longrightarrow PbSO_4(s) + 2\ H_2O(l)$
     $E° = +1.685\ V$

Half-reaction (1) is the reverse of the anode half-reaction because Pb is oxidized to $PbSO_4$ in the cell reaction. Therefore, the cell reaction is the sum of half-reaction (1) written in reverse and half-reaction (2):

$$Pb(s) + SO_4^{2-}(aq) \rightarrow PbSO_4(s) + 2e^-$$
$$+ PbO_2(s) + 4H^+(aq) + SO_4^{2-}(aq) + 2e^- \rightarrow PbSO_4(s) + 2H_2O(l)$$
$$\overline{Pb(s) + PbO_2(s) + 4H^+(aq) + 2SO_4^{2-} \rightarrow 2PbSO_4(s) + 2H_2O(l)}$$

The standard potential of this cell reaction is

$$E°_{cell} = E°_{cathode} - E°_{anode}$$
$$= E°_{(2)} - E°_{(1)}$$
$$= 1.685 \text{ V} - (-0.356 \text{ V}) = 2.041 \text{ V}$$

Even without knowing the cell reaction, we can predict from the $E°$ values of half-reactions (1) and (2) which is the cathode and which is the anode half-reaction. Remember that a voltaic cell has a positive E_{cell} value. To obtain a positive E_{cell} value, we must subtract the negative value from the positive one. In general, the half-reaction with the smaller $E°$ value will be the one that takes place (in reverse) at the anode.

PRACTICE EXERCISE: The first operational battery may have been Allesandro Volta's stack of zinc and silver foil separated by blotter paper soaked with $NaNO_3$ solution. If we attempted to reproduce one cell of this battery by immersing a silver electrode in a 1 M solution of Ag^+ ions separated by a salt bridge from a zinc electrode immersed in a 1 M solution of Zn^{2+} ions, what voltage would be generated? (See Problems 35 and 36.)

Nickel–metal hydride batteries

The first laptop computers were powered by Ni-Cd batteries but, by the mid-1990s, most new laptops were powered by nickel–metal hydride (NiMH) batteries (Figure 17.9). These batteries derive their electrical power from the reduction of NiO(OH) at the cathode (as in Ni-Cd batteries) and the oxidation of hydrogen at an anode made of a transition-metal alloy, such as $LaNi_5$. The two electrodes are separated by a porous spacer containing an aqueous solution of KOH.

The cathode half-reaction in a NiMH battery is the same as in a Ni-Cd—namely, the reduction of NiO(OH) to $Ni(OH)_2$:

$$NiO(OH)(s) + H_2O(l) + e^- \longrightarrow Ni(OH)_2(s) + OH^-(aq) \qquad E° = 1.32 \text{ V} \quad (17.12)$$

The anode half-reaction is based on hydrogen, but it is not simply the SHE half-reaction written in reverse:

$$H_2(g) \longrightarrow 2H^+(aq) + 2e^-$$

The SHE reaction is not appropriate here for two reasons. First, the reaction takes place in a strongly alkaline medium, in which OH^- ions far outnumber H^+ ions.

CONNECTION: Atoms of hydrogen in transition-metal hydrides are located in holes, or *interstices*, formed by three-dimensional arrays of metal atoms, as described in Section 12.8. The location and sizes of the holes in close-packed structures are described in Sections 10.6 and 10.7.

CONNECTION: Alloys are solid solutions of two or more metals (see Section 18.2).

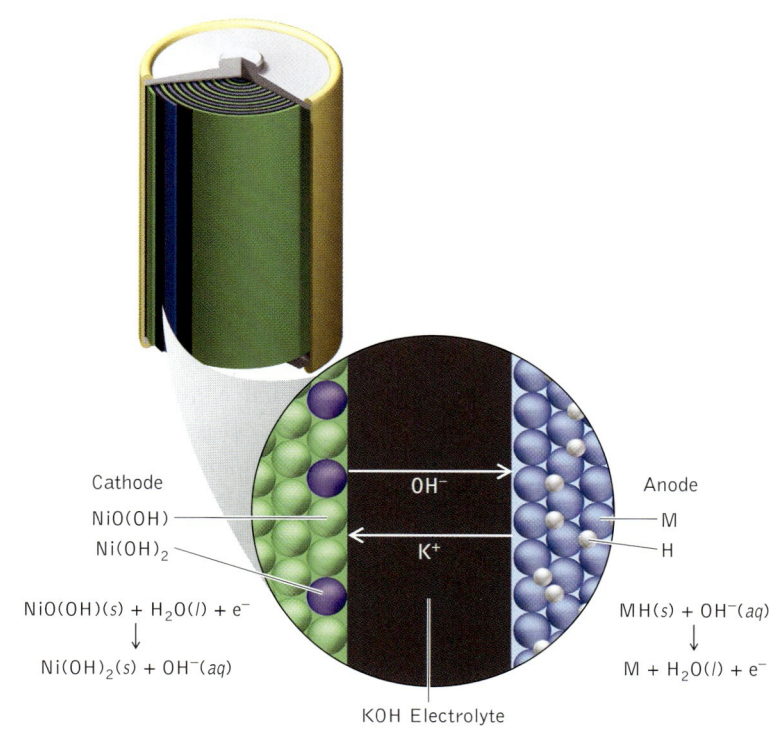

FIGURE 17.9 In the nickel–metal hydride (NiMH) battery, hydrogen atoms are stored in the interstices of a metal matrix from which they may migrate, participate in the anode half-reaction, and be oxidized at the anode while NiO(OH) is reduced at the cathode.

A search of Appendix 6 for a reduction half-reaction in which H_2 is a product in a basic solution uncovers the following half-reaction:

$$2\,H_2O(l) + 2\,e^- \longrightarrow H_2(g) + 2\,OH^-(aq) \qquad E° = -0.83\text{ V} \quad (17.13)$$

Note the difference between this $E°$ value and that of the SHE (0.000 V). The negative $E°$ value for the alkaline half-reaction indicates that hydrogen is less easily formed from the reduction of water than it is from the reduction of H^+ ions. On the other hand, if the half-reaction ran in reverse (as an oxidation), then it would have a positive standard potential of $+0.83$ V. This value means that H_2 is more readily oxidized in strong base than in strong acid.

Unfortunately, the half-reaction in Equation 17.13 is still not the best representation of the NiMH anode reaction. Hydrogen is stored in the metal hydride as free atoms, not as H_2 molecules. The partial pressure of H_2 gas in one of these metal hydrides at 25°C is extremely small. It turns out that the best way to write the anode half-reaction is based on the oxidation of hydrogen atoms in a metal hydride (MH):

$$MH(s) + OH^-(aq) \longrightarrow M(s) + H_2O(l) + e^- \qquad (17.14)$$

where M stands for a transition metal or metal alloy that forms a hydride. The standard reduction potential for the reverse of this half-reaction depends on the chemical properties of MH, but generally the values are near 0 V.

To obtain the overall cell reaction, we simply combine the anode and cathode half-reactions (Equations 17.12 and 17.14) because one electron is gained in one and lost in the other:

$$\text{MH}(s) + \text{NiO(OH)}(s) + \cancel{\text{OH}^-(aq)} + \cancel{\text{H}_2\text{O}(l)} + e^- \longrightarrow$$
$$\text{M}(s) + \text{Ni(OH)}_2(s) + \cancel{\text{H}_2\text{O}(l)} + \text{OH}^-(aq) + e^-$$

Canceling all the terms that are common to both sides of the reaction arrow gives the nickel–metal hydride cell reaction:

$$\text{MH}(s) + \text{NiO(OH)}(s) \longrightarrow \text{M}(s) + \text{Ni(OH)}_2(s)$$

The $E°_{cell}$ for the nickel–hydrogen battery cannot be determined without an accurate value of $E°_{anode}$. It turns out that most NiMH batteries deliver about 1.2 V during most of their discharge cycle. Thus, Ni-Cd and NiMH batteries produce similar cell potentials and are used in many of the same applications. However, NiMH batteries are lighter than Ni-Cd batteries of comparable current capacity, and NiMH batteries are made of less-hazardous components (cadmium compounds are quite toxic).

Lithium-ion batteries

By the late 1990s, the higher power demands of high-speed processors, large video displays, and other components of modern laptop computers created a demand for a new generation of batteries with greater current capacity and higher voltage than provided by NiMH batteries. This demand was met by the development of batteries based, not on redox chemistry, but on the concentration-driven migration of lithium ions. In these batteries, Li^+ ions are stored in anodes made of highly pure graphite: each ring of six carbon atoms in the structure of graphite can store one Li^+ ion. Highly porous cathodes made of transition-metal oxides, such as MnO_2, form stable complexes with Li^+ ions. In a fully charged lithium-ion battery, there is a high concentration of Li^+ ions in the anode and a low concentration of Li^+ ions in the cathode. This difference means that there is a *concentration gradient* between the two electrodes that is large enough to produce a cell potential of about 4 V. During discharge, the migration of Li^+ ions (and their positive charges) down the concentration gradient from anode to cathode is

☑ The voltage of a lithium-ion battery is produced by a higher concentration of Li^+ ions in the anode than in the cathode. The dependence of electrode potentials on concentration is described in Section 17.5

Lithium batteries are popular in laptop computers, where power demands are high.

accompanied by a flow of electrons in an external circuit from the anode to the cathode. This parallel flow of positive and negative charges means that no electrical charge accumulates on either electrode.

The electrodes, particularly the anodes, of lithium-ion batteries react with oxygen and water; so these batteries have to be completely solid state devices or they must use nonaqueous electrolytes. Typically, the electrolytes are solutions of lithium salts, such as lithium hexafluorophosphate, $LiPF_6$, or lithium hexafluoroarsenate, $LiAsF_6$, dissolved in nonaqueous solvents such as tetrahydrofuran, ethylene carbonate, or propylene carbonate:

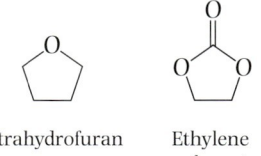

Tetrahydrofuran   Ethylene carbonate   Propylene carbonate

CONNECTION: Graphite consists of two-dimensional arrays of aromatic rings (see Section 10.8). These layers are held together only by van der Waals forces, which makes the insertion of atoms or ions between them relatively easy.

17.5 THE EFFECT OF CONCENTRATION ON POTENTIAL

Reactions stop when one of the reactants is consumed. However, a commercial battery usually stops supplying its rated voltage at a useful current value just before its reactants are completely consumed. Why does the cell potential of a battery decrease when it is nearly discharged? The answer is that the electromotive force produced by a voltaic cell is determined by the concentrations of the reactants and products.

> Why does the cell potential of a battery decrease when it is nearly discharged?

Standard reduction potentials are useful for approximating the E_{cell} generated by voltaic cells, but standard potentials are only valid for 1 molar aqueous solutions at 25°C. The potential of any half-reaction changes with changing temperature and with changing concentrations of reactants and products. German scientist Walther Nernst (1864–1941) studied the dependence of cell potential on temperature and concentration and developed an expression (the Nernst equation) that describes these dependencies:

$$E_{cell} = E°_{cell} - \frac{RT}{nF} \ln Q \qquad (17.15)$$

where R is the gas constant (8.3145 J/K · mol), n is the number of moles of electrons transferred in the balanced cell reaction, F is the Faraday constant, and Q is the reaction quotient.

A handy form of the Nernst equation is obtained when we replace the symbols of the constants with their actual values. If we assume that $T = 25°C$ (298 K) and we switch from natural logarithms to logarithms to the base 10 (which is a

CONNECTION: This reaction quotient Q is the same parameter used to track the progress of a reaction as it approaches equilibrium (see Section 15.2).

THE CHEMISTRY OF THE GROUP 2B ELEMENTS

The last group in the transition metals (Zn, Cd, Hg) contains elements found in several of the batteries described in this chapter. None of the three metals is particularly abundant in Earth's crust. Zinc is the most abundant of the three (78 mg/kg), and mercury is about as abundant as silver. All three are found in sulfide ores, including sphalerite (ZnS), greenockite (CdS), and cinnabar (HgS). Zinc is also found in carbonates and silicates. Most zinc ores contain cadmium impurities.

Zinc metal is recovered from zinc ore by heating the ore in air to produce zinc oxide:

$$2\ ZnS(s) + 3\ O_2(g) \longrightarrow 2\ ZnO(s) + 2\ SO_2(g)$$

which is used in antibiotic creams, in sunscreens, and on the inside of fluorescent light bulbs. Zinc oxide is reduced to molten metal with hot carbon:

$$2\ ZnO(s) + C(s) \longrightarrow 2\ Zn(l) + CO_2(g)$$

Cadmium is produced along with zinc from impurities in the ore and is separated by distillation of the molten metals.

Mercury is produced directly from reactions of HgS with oxygen:

$$HgS(s) + O_2(g) \longrightarrow Hg(l) + SO_2(g)$$

Mercury is the only metal and one of only two elements (bromine is the other) that is a liquid at room temperature. Zinc and cadmium are low-melting, silvery solids that react rapidly with moist air to form the corresponding oxides. The oxides and sulfides of zinc and cadmium are semiconductors (see Chapter 18).

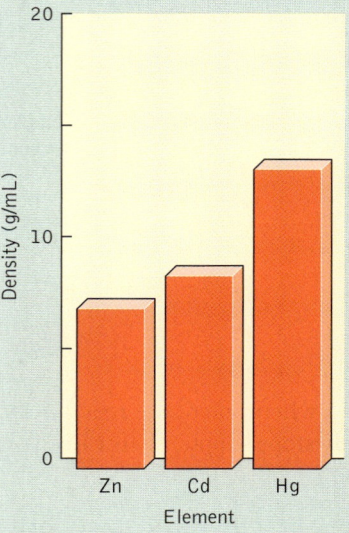

The density of the Group 2B elements increases with increasing atomic number.

Zinc oxide is used to coat the inside of fluorescent light tubes and to make topical creams that also serve as effective sunscreens.

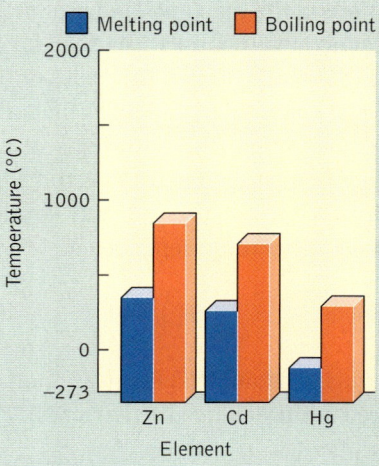

The Group 2B elements are low-melting metals (Hg is a liquid at room temperature) and relatively low-boiling metals compared with other metals.

The Group 2B metals have $(n-1)d^{10}ns^2$ electron configurations in their valence shells and, like the metals in Group 2A, tend to lose two valence electrons and form M^{2+} cations. Mercury also forms a Hg_2^{2+} cation, which contains a covalent Hg—Hg bond. The Group 2B elements are more likely than the alkaline earth (Group 2A) metals to form covalent bonds.

Zinc is an essential element in living organisms, but cadmium and mercury have no known health benefits and are quite toxic. The toxicity of these metals is related to their attachment to sulfur groups on amino acids. These interactions allow the metals to be transported throughout the body attached to the peptides and proteins in blood. If they reach the brain, interactions with the proteins found there can lead to mental retardation and other neurological disorders. The toxicity of mercury is increased by the formation of covalent alkyl Hg compounds such as methyl mercuric chloride, CH_3HgCl, and dimethyl mercury, $(CH_3)_2Hg$. Mercury in these chemical forms can more easily penetrate hydrophobic cell membranes than can inorganic Hg^{2+} and so can more severely disrupt internal cell functions. Microbial methylation of inorganic mercury wastes dumped in Minamata Bay, Japan, in the 1950s led to 52 deaths and recognition of the environmental hazards of improper disposal of mercury and other metals.

The disposal of mercury wastes in Minamata Bay, Japan, in the 1950s resulted in such high concentrations of methyl mercury compounds in the shellfish harvested from the bay that more than 50 people died from methyl mercury poisoning and many others suffered neurological damage.

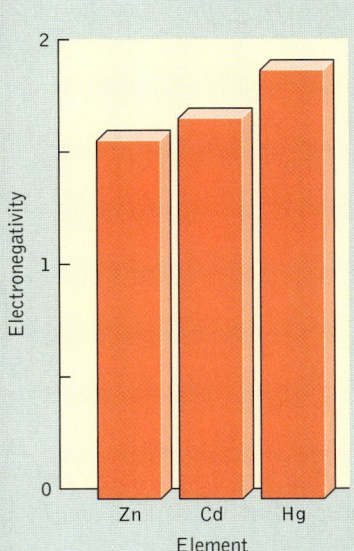

The electronegativities of the Group 2B elements increase slightly with increasing atomic number.

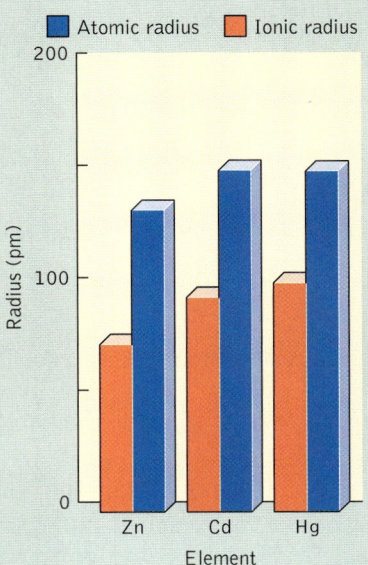

The radii of the +2 cations of the Group 2B elements increase with increasing atomic number. The atomic radii are less dependent on atomic number.

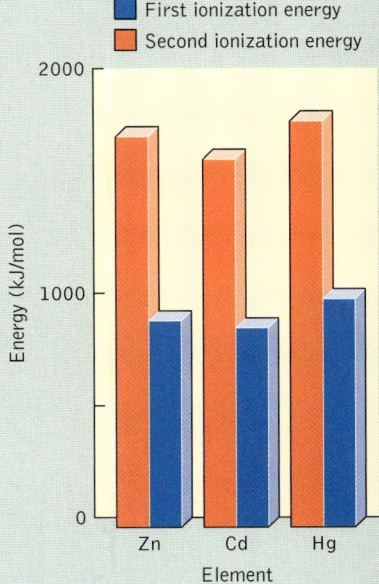

The first and second ionization energies of the Group 2B elements show little dependence on atomic number.

matter of multiplying by 2.303), then the coefficient of the logarithmic term has the value 0.0592 and the Nernst equation becomes

$$E_{cell} = E°_{cell} - \frac{0.0592}{n} \log Q$$

In a cell reaction in which a moles of A is oxidized to c moles of C while b moles of B is reduced to d moles of D

$$aA + bB \longrightarrow cC + dD$$

the Nernst equation becomes

$$E_{cell} = E°_{cell} - \frac{0.0592}{n} \log \frac{[C]^c[D]^d}{[A]^a[B]^b} \quad (17.16)$$

SAMPLE EXERCISE 17.5: In Section 17.1, we considered a voltaic cell consisting of a strip of copper immersed in a solution of $CuSO_4$ and a strip of zinc immersed in a solution of $ZnSO_4$. If the concentration of $ZnSO_4$ is 10 times the concentration of $CuSO_4$, what is the value of E_{cell} at 25°C? Given:

$$Cu^{2+}(aq) + Zn(s) \longrightarrow Cu(s) + Zn^{2+}(aq) \qquad E° = +1.10 \text{ V}$$

SOLUTION: The Nernst equation can be used to calculate the value of E_{cell} for different concentrations of reactants and products. In this reaction, solid copper and zinc are product and reactant, respectively. We assume that both metals have an effective constant "concentration" of unity and so do not appear in the reaction quotient, which leaves just $[Zn^{2+}]$ and $[Cu^{2+}]$ in the logarithmic term of the Nernst equation. The value of n in this cell reaction is 2: Cu^{2+} gains two electrons and Zn loses two in forming Zn^{2+}. Thus, the Nernst equation for the Cu–Zn cell reaction is

$$E_{cell} = 1.10 \text{ V} - \frac{0.0592}{2} \log \frac{[Zn^{2+}]}{[Cu^{2+}]}$$

If $[Zn^{2+}] = 10 [Cu^{2+}]$, then the logarithmic term is

$$\log \frac{10 \, [\cancel{Cu^{2+}}]}{[\cancel{Cu^{2+}}]} = \log 10 = 1$$

Inserting this value into the Nernst equation gives

$$E_{cell} = 1.10 \text{ V} - \frac{0.0592}{2} \text{V} = 1.07 \text{ V}$$

PRACTICE EXERCISE: Calculate the cell potential of a Cu–Zn battery if the concentration of copper ion is 100 times the concentration of the zinc ion. (See Problems 41 and 42.)

CONCEPT TEST: Appendix 6 lists the following potentials for two half-cell reactions in which O_2 is the oxidizing agent.

In acid solutions:

$$O_2(g) + 4\,H^+(aq) + 4\,e^- \longrightarrow 2\,H_2O(l) \qquad E° = 1.229\,V$$

In basic solutions:

$$O_2(g) + 2\,H_2O(l) + 4\,e^- \longrightarrow 4\,OH^-(aq) \qquad E° = 0.401\,V$$

Which of these half-reactions could be used to calculate cell potentials of redox reactions including O_2 at pH = 7.0?

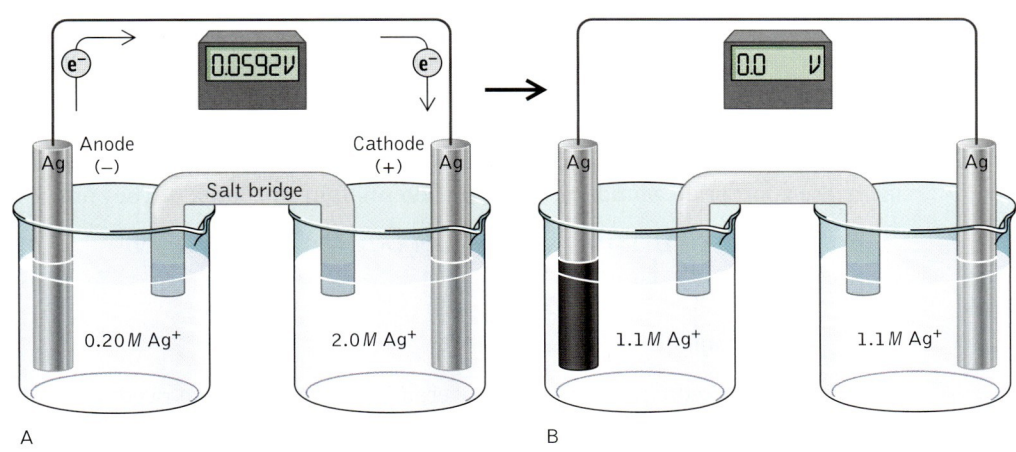

$E_{anode} = E°_{Ag} - \dfrac{0.0592}{1}\log\left[\dfrac{1}{0.20}\right]$

$E_{cathode} = E°_{Ag} - \dfrac{0.0592}{1}\log\left[\dfrac{1}{2.0}\right]$

$E_{cell} = E_{cathode} - E_{anode}$

$= -0.0592\log\left[\dfrac{0.20}{2.0}\right]$

$= 0.0592\,V$

The Nernst equation predicts that a voltaic cell can be constructed by using two identical half-reactions but different concentrations of reactants. (A) If silver wires are immersed in 2.0 M and 0.20 M solutions of Ag^+ in an electrochemical cell such as that shown here, a cell potential of 0.0592 V is generated. (B) As the reaction proceeds and current flows through the external circuit, [Ag^+] in the anode compartment increases while Ag^+ is consumed at the cathode. With the passage of time, the concentrations of Ag^+ in both compartments approach the same value (1.1 M) and the value of E_{cell} approaches zero.

As a battery's cell reaction proceeds, the concentrations of the reactants ([A] and [B] in the denominator of Equation 17.16) decrease and the concentrations of the products ([C] and [D] in the numerator) increase. Therefore the value of the logarithmic term increases but, because there is a negative sign in front of the logarithmic term, the value of E_{cell} decreases. Thus, the voltage produced by a battery decreases as the concentrations of its reactants decrease, as shown in Figure 17.10. Eventually the value of E_{cell} approaches zero. As it does, the cell reaction stalls or at least there is no further change in the concentrations of reactants or products. This condition meets our definition of chemical equilibrium. It means that the value of the reaction quotient (Q) is equal to the equilibrium constant (K) for the cell reaction. At equilibrium, Equation 17.16 becomes

$$0 = E°_{cell} - \dfrac{0.0592}{n}\log K$$

or

$$\log K = \frac{nE°_{cell}}{0.0592} \quad (17.17)$$

Equation 17.17 makes a mathematical connection between the equilibrium constant of a cell reaction and the standard cell potential. It applies to any redox reaction, not just those in voltaic cells, as demonstrated in the following example.

A method used by ecologists to determine whether a natural area should be classified a "wetland" (aside from its being noticeably wet) is to examine the color of the soil. If the soil has a hue ranging from yellow to reddish brown, the iron in it is present as Fe^{3+}. If the soil is dark gray to black, the iron is present as Fe^{2+}, and the area is probably a wetland. The basis for this conclusion is the fact that atmospheric oxygen cannot easily penetrate water-logged soils, and the concentration of O_2 determines whether iron exists as Fe^{2+} or Fe^{3+}. The connection between the oxygen concentration and the oxidation state of iron can be made by using the appropriate standard reduction potentials in Appendix 6. They are:

$$O_2(g) + 4\,H^+(aq) + 4\,e^- \longrightarrow 2\,H_2O(l) \qquad E° = 1.229 \text{ V}$$

and

$$Fe^{3+}(aq) + e^- \longrightarrow Fe^{2+}(aq) \qquad E° = 0.770 \text{ V}$$

To write a balanced equation for the overall redox reaction, we need to decide which of these half-reactions to reverse. Because we want to assess the effect of O_2 as a reactant, it makes sense to leave O_2 as a reactant and reverse the iron half-reaction:

$$Fe^{2+}(aq) \longrightarrow Fe^{3+}(aq) + e^-$$

Next, we need to multiply the iron half-reaction by four

$$4\,Fe^{2+}(aq) \longrightarrow 4\,Fe^{3+}(aq) + 4\,e^-$$

✓ Note that multiplying this half-reaction by four does not change its $E°$ value. Standard electrode potentials are intensive properties of half-reactions.

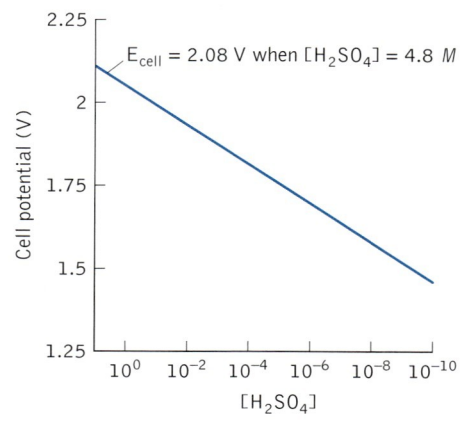

FIGURE 17.10 The potential of a voltaic cell decreases as reactants are converted into products. The value of $E°_{cell}$ of the lead–acid battery cell reaction

$$Pb(s) + PbO_2(s) + 2\,H_2SO_4(aq) \longrightarrow 2\,PbSO_4(s) + 2\,H_2O(l)$$

is 2.04 V. As the concentration of sulfuric acid (the only dissolved component) decreases, the E_{cell} value decreases. The relation between the E_{cell} value and $[H_2SO_4]$ is given by the Nernst equation for the cell reaction

$$E_{cell} = 2.04 \text{ V} - \frac{0.0592}{2} \log \frac{1}{[H_2SO_4]^2}$$

In a fully charged automobile battery, $[H_2SO_4] = 4.8\ M$ and $E_{cell} = 2.08$ V.

to balance its loss of electrons with the electrons gained (four) in the oxygen half-reaction. Now we can combine the two half-reactions to get the overall redox reaction:

$$O_2(g) + 4\,H^+(aq) + 4\,Fe^{2+}(aq) \longrightarrow 4\,Fe^{3+}(aq) + 2\,H_2O(l)$$

To calculate the value of $E°$ for the overall reaction, we subtract the standard reduction potential of the oxidation half-reaction ($E°_{ox}$) from the standard potential of the reduction half-reaction ($E°_{red}$). This calculation is equivalent to subtracting $E°_{anode}$ from $E°_{cathode}$ as in Equation 17.9. During the reaction, Fe^{2+} is oxidized (so $E°_{ox} = E°_{Fe}$) and O_2 is reduced (so $E°_{red} = E°_{O_2}$). Thus the standard potential of the overall reaction is

$$E°_{rxn} = E°_{red} - E°_{ox}$$
$$= E°_{O_2} - E°_{Fe}$$
$$E°_{rxn} = (1.229 - 0.770) = 0.459\,V$$

Using Equation 17.17 to calculate the corresponding equilibrium constant, we have

$$\log K = \frac{nE°_{cell}}{0.0592} = \frac{4(0.459)}{0.0592} = 31.0$$

and

$$K = 1.0 \times 10^{31}$$

This enormous value for K may suggest that finding Fe^{2+} in the presence of O_2 is very unlikely. However, we need to keep in mind that the standard conditions for these calculations included $[H^+] = 1.00\,M$, or pH = 0.00 and an atmosphere of pure O_2. Wetland soils are never that acidic, and the partial pressure of O_2 in the atmosphere is only 0.21 atm. To obtain a more realistic view of the reduction potential of O_2 (and so the strength of O_2 as an oxidizing agent) in environmental waters, let's use the Nernst equation to calculate the value of E_{O_2} at pH = 6.00 and $P_{O_2} = 0.21$ atm:

$$E_{O_2} = E°_{O_2} - \frac{0.0592}{n} \log \frac{1}{[H^+]^4 P_{O_2}}$$

$$= 1.229 - \frac{0.0592}{4} \log \frac{1}{(1.0 \times 10^{-6})^4(0.21)}$$

$$= 1.229 - \frac{0.0592}{4} \log \frac{1}{2.1 \times 10^{-25}}$$

$$= 0.864\,V$$

This potential provides a more reasonable measure of the strength of O_2 as an oxidizing agent in weakly acidic environmental waters or water-logged soils. Environmental scientists use such values to describe the *oxidation potential* of an aquatic sample. Any increase in pH or decrease in P_{O_2} will decrease the value of the oxidation potential. If it should fall to 0.770 V, the standard potential of the

$E°_{cell}$

$\log K = \dfrac{nE°_{cell}}{0.0592}$

$\log K$

$K = 10^{\log K}$

K

reduction of Fe^{3+} to Fe^{2+}, then, according to the Nernst equation for the iron reaction, the ratio of $[Fe^{2+}]$ to $[Fe^{3+}]$ would be unity:

$$E_{Fe} = E_{Fe}^\circ - \frac{0.0592}{n} \log \frac{[Fe^{2+}]}{[Fe^{3+}]}$$

$$0.770 = 0.770 - \frac{0.0592}{1} \log \frac{[Fe^{2+}]}{[Fe^{3+}]}$$

$$0.000 = \log \frac{[Fe^{2+}]}{[Fe^{3+}]}$$

$$\frac{[Fe^{2+}]}{[Fe^{3+}]} = 1$$

At even lower P_{O_2}, the oxidation potential will be less than 0.770 V and, at equilibrium, $[Fe^{2+}]$ will be greater than $[Fe^{3+}]$.

> ✓ The potential of any redox reaction or half-reaction that includes H^+ or OH^- depends on pH.

SAMPLE EXERCISE 17.6: Many of the methods used to determine the levels of mercury contamination in environmental samples include the conversion of all the mercury in the samples to Hg^{2+} and then reduction to elemental Hg with Sn^{2+}. Use the appropriate standard potentials in Appendix 6 to calculate the equilibrium constant for this reduction reaction at 25°C.

SOLUTION: Equation 17.17 relates the standard potential of an electrochemical cell to the equilibrium constant of the cell reaction. As we have seen, Equation 17.17 may be used to calculate the equilibrium constant for any redox reaction from the standard potential of the reaction. So, first we need to calculate E_{rxn}°. According to the data in Appendix 6, the standard potential for the Hg^{2+} to Hg half-reaction is

$$Hg^{2+}(aq) + 2\,e^- \longrightarrow Hg(l) \qquad E^\circ = 0.851 \text{ V}$$

Because Sn^{2+} is the reducing agent in the reaction, it must be oxidized. Therefore, we need to search for a reduction half-reaction in which tin in a higher oxidation state is reduced to Sn^{2+}. The following half-reaction appears in the Appendix 6:

$$Sn^{4+}(aq) + 2\,e^- \longrightarrow Sn^{2+}(aq) \qquad E^\circ = 0.154 \text{ V}$$

We can combine these two half-reactions if we first reverse the tin reaction to make Sn^{2+} a reactant:

$$Hg^{2+}(aq) + 2\,e^- \longrightarrow Hg(l)$$
$$+\ Sn^{2+}(aq) \longrightarrow Sn^{4+}(aq) + 2\,e^-$$
$$Hg^{2+}(aq) + Sn^{2+}(aq) \longrightarrow Hg^0(l) + Sn^{4+}(aq)$$

To calculate E_{rxn}°, we subtract E_{Sn}° (Sn^{2+} is oxidized) from E_{Hg}° (Hg^{2+} is reduced):

$$E_{rxn}^\circ = E_{Hg}^\circ - E_{Sn}^\circ$$
$$= 0.851 - (0.154) = 0.697 \text{ V}$$

Inserting this value and $n = 2$ into Equation 17.17, we have

$$\log K = \frac{nE°_{cell}}{0.0592} = \frac{2(0.697)}{0.0592} = 23.5$$

$$K = 10^{23.5} = 3 \times 10^{23}$$

PRACTICE EXERCISE: Using the standard reduction potentials in Appendix 6, calculate the value of the equilibrium constant at 25°C at for the reaction:

$$5\ Fe^{2+}(aq) + MnO_4^-(aq) + 8\ H^+(aq) \longrightarrow 5\ Fe^{3+}(aq) + Mn^{2+}(aq) + 4\ H_2O(l)$$

(See Problems 43 and 44.)

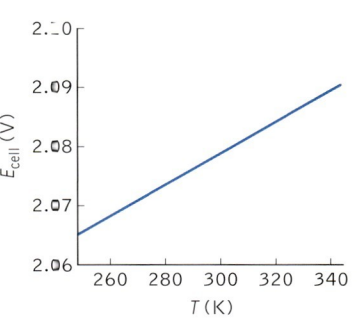

FIGURE 17.11 The Nernst equation predicts the effect of temperature on cell potential. A plot of E_{cell} for the lead–acid battery reaction as a function of temperature shows that the value of E_{cell} decreases only slightly with decreasing temperature. Other temperature-dependent effects, such as the viscosity of the electrolyte, have a more dramatic effect on battery performance.

The Nernst equation predicts that the value of E_{cell} changes slightly with changing temperature, as shown for the lead–acid battery reaction in Figure 17.11. Frequently, this change is overwhelmed by other temperature effects. For example, temperatures below 0°C can significantly increase the viscosity of the sulfuric acid electrolyte in automobile batteries. Increased viscosity means greater resistance to ion migration. The result is decreased battery performance under heavy current loads, such as when an automobile battery is called on to start an engine on a cold winter morning.

17.6 QUANTITIES OF REACTANTS AND BATTERY POWER

An important performance characteristic of a battery is its capacity to do work, which means to deliver electrical charge (C) at its design cell potential (Equation 17.5):

$$w_{elec} = C\ E_{cell}$$

Thus, we could express battery capacity in coulomb · volts (which equal joules), but, as we are about to see, there are other ways to do so. The first way is based on the definition of an ampere as the rate of flow of electrical charge equal to 1 coulomb per second:

$$1\ ampere = 1\ coulomb/second$$

or

$$1\ coulomb = 1\ ampere \cdot second$$

Therefore, w_{elec} could be expressed in volt · ampere · seconds. The ability to do work at a specified rate is defined as power: the faster a voltaic device can do a given quantity of work is done, the more powerful the device is. A widely used unit of electrical power is the watt, where

$$1\ watt = 1\ volt \cdot ampere = 1\ joule/second$$

 CONNECTION: The ampere (A) is the SI base unit of electrical current (see Chapter 1).

Thus a cell with a cell potential of 1 volt that produces 1 ampere of current produces 1 watt of power. Watts of power, like joules of energy, are rather small units, and so the power of practical voltaic devices is often expressed in kilowatts rather than watts.

The capacity of a lead–acid battery in an automobile to deliver electrical charge fast enough to start its engine is described by the parameter "cold cranking amps," or CCA. Heavy-duty car batteries have CCA ratings of more than 700, meaning that they can produce more than 700 amperes at a potential of at least 7.2 volts for 30 seconds (considered to be an adequate interval in which to start a car's engine) at 0°C. Inexpensive lead–acid batteries have CCA ratings of 500 or less. As you might guess, the CCA rating of a battery is directly related to the quantities of reactants (Pb and PbO_2—see Sample Exercise 17.4 on page 859) in its electrodes that can be oxidized and reduced quickly.

Let's calculate how many grams of Pb and PbO_2 would be consumed in a lead–acid battery if it delivered 600 amperes of current in a period of 30 seconds. Because an ampere is the same as a coulomb/second (C/s), the quantity of electrical charge that is delivered is

$$30\,s\left(\frac{600\ C}{s}\right) = 1.80 \times 10^4\ C$$

A mole of charge is equivalent to 9.65×10^4 coulombs; so the moles of charge (and electrons) that flow over the 30-s interval is

$$1.80 \times 10^4\ \cancel{C}\left(\frac{1\ mol\ e^-}{9.65 \times 10^4\ \cancel{C}}\right) = 0.187\ mol\ e^-$$

To relate moles of electrons to mole of reactants, we need to look again at the anode and cathode half-reactions for the lead–acid battery. In the oxidation of lead at the anode,

$$Pb(s) + SO_4^{2-}(aq) \longrightarrow PbSO_4(s) + 2\ e^-$$

2 moles of electrons are lost by each mole of Pb metal. The reduction of a mole of PbO_2 at the cathode

$$PbO_2(s) + 4\ H^+(aq) + 2\ SO_4^{2-}(aq) + 2\ e^- \longrightarrow PbSO_4(s) + 2\ H_2O(l)$$

results in the consumption of 2 moles of electrons. The stoichiometry of the anode half-reaction leads to the following approach to calculating the mass of lead consumed:

$$0.187\ \cancel{e^-}\left(\frac{1\ \cancel{mol\ Pb}}{2\ \cancel{e^-}}\right)\left(\frac{207.2\ g\ Pb}{1\ \cancel{mol\ Pb}}\right) = 19.3\ g\ Pb$$

A similar calculation of the amount of lead(IV) oxide consumed gives the following result:

$$0.187\ \cancel{e^-}\left(\frac{1\ \cancel{mol\ PbO_2}}{2\ \cancel{e^-}}\right)\left(\frac{239.2\ g\ PbO_2}{1\ \cancel{mol\ PbO_2}}\right) = 22.3\ g\ PbO_2$$

☑ The quantities of reactants consumed as a battery discharges depend on the quantity of electrical charge that it delivers.

SAMPLE EXERCISE 17.7: A lithium-ion battery is capable of producing 4.00 ampere · hours of charge at 3.6 V. How many grams of Li^+ ions must migrate from the anode to the cathode to produce this much electricity?

SOLUTION: The connection between quantities of electricity and quantities of substances is the Faraday constant: 1 mole of electrons contains 9.65×10^4 coulombs of charge. Let's first calculate the total quantity of charge delivered by the battery. We will use the fact that a coulomb (C) of charge is transferred by a current of 1 ampere (A) flowing for 1 second (1 C = 1 A · s).

$$4.00 \text{ A} \cdot \text{hr} \left(\frac{3600 \text{ s}}{\text{hr}}\right)\left(\frac{1 \text{ C}}{\text{A} \cdot \text{s}}\right) = 14,400 \text{ C}$$

Dividing this quantity by the Faraday constant gives the moles of electrons:

$$14,400 \text{ C} \left(\frac{1 \text{ mol e}^-}{9.65 \times 10^4 \text{ C}}\right) = 0.149 \text{ mol e}^-$$

The flow of 0.149 mole of electrons in an external circuit corresponds to a flow of 0.149 mole of Li^+ ions in the battery. The mass of this quantity of Li^+ ions is

$$0.149 \text{ mol e}^- \left(\frac{1 \text{ mol Li}}{\text{mol e}^-}\right)\left(\frac{6.941 \text{ g Li}}{\text{mol Li}}\right) = 1.04 \text{ g Li}$$

PRACTICE EXERCISE: Magnesium metal is produced by passing an electrical current through molten $MgCl_2$. The reaction at the cathode is

$$Mg^{2+}(l) + 2 \text{ e}^- \longrightarrow Mg(l)$$

How many grams of magnesium metal are produced if an average current of 63.7 amperes flows for 4.50 hours? Assume all of the current was consumed by the preceding half-reaction. (See Problems 65 and 66.)

17.7 ELECTROLYTIC CELLS AND RECHARGING BATTERIES

Several of the batteries described so far in this chapter are rechargeable, which means that the cell reactions in which chemical energy is converted into electrical energy can be reversed. The application of an opposing electrical potential greater than the cell potential forces the cell reaction to run backward, converting electrical energy into chemical energy. Recharging a battery is an example of electrolysis as the battery is transformed from a voltaic cell into an **electrolytic cell** (Figure 17.12).

The chemical changes that take place in a lead–acid battery in the spontaneous discharge process and then during recharging are shown in Figure 17.13. During recharging, current flows in the opposite direction through the battery.

> ✓ An **electrolytic cell** is one in which an external source of electrical energy does work on a chemical system, turning low-energy reactant(s) into high-energy product(s).

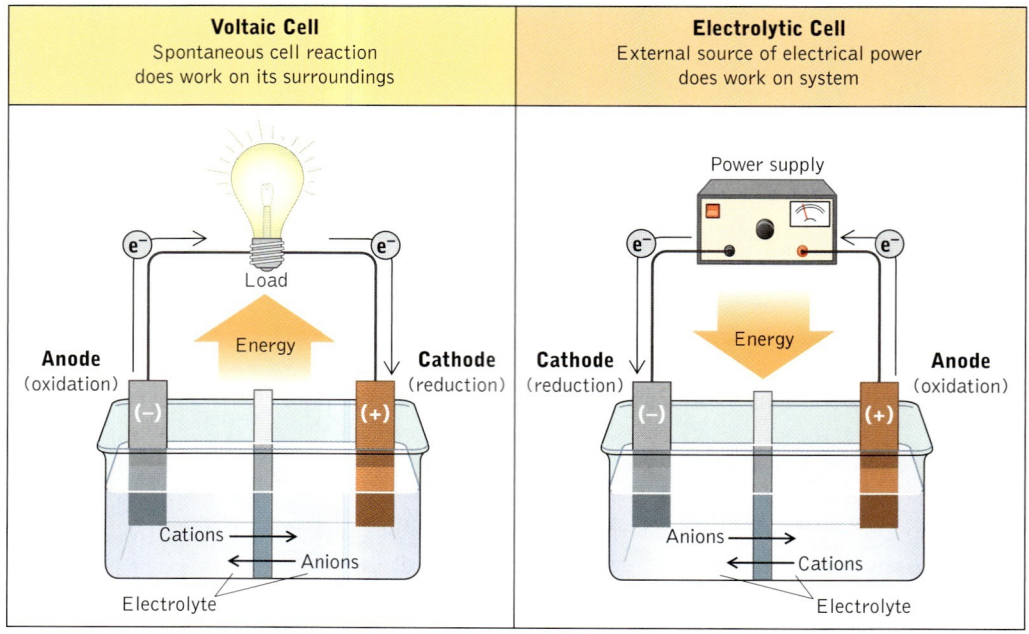

FIGURE 17.12 In a voltaic cell, a spontaneous (exergonic) chemical reaction produces electrical energy and does work, such as lighting a light bulb, on its surroundings. In an electrolytic cell, an external supply of electrical energy does work on the chemical system in the cell, driving a nonspontaneous (endergonic) reaction.

CONNECTION: Electrolysis is a chemical reaction caused by the passage of an electric current through a cell (see Section 5.5).

Thus the lead electrodes that serve as anodes during the discharge reaction are turned into cathodes during recharge as any $PbSO_4$ that formed on them during discharge is reduced back to Pb:

$$PbSO_4(s) + 2\,e^- \longrightarrow Pb(s) + SO_4^{2-}(aq) \qquad E° = -0.356\text{ V}$$

Similarly, the PbO_2 electrodes that serve as cathodes during discharge become anodes during recharge as any $PbSO_4$ that formed on them during discharge is oxidized back to PbO_2:

$$PbSO_4(s) + 2\,H_2O(l) \longrightarrow PbO_2(s) + 4\,H^+(aq) + 2\,SO_4^{2-}(aq) + 2\,e^- \qquad E° = -1.685\text{ V}$$

Recharging a lead–acid battery or any battery with an aqueous electrolyte can result in some undesirable and even dangerous side reactions. If too high a voltage forces too great a current through a lead–acid battery, the reduction of lead sulfate at the lead electrode may not keep up with the flow of electrons to the electrode surface. In that event, some other reducible species, such as the hydrogen ions in a concentrated solution of sulfuric acid, may accept the influx of electrons as described by Equation 17.11:

$$2\,H^+(aq) + 2\,e^- \longrightarrow H_2(g) \qquad E° = 0.000\text{ V}$$

If both H^+ ions and $PbSO_4$ are present at the surface of a lead electrode, which of them is preferentially reduced? Note that the $E°$ for reducing $PbSO_4$ (-0.356 V)

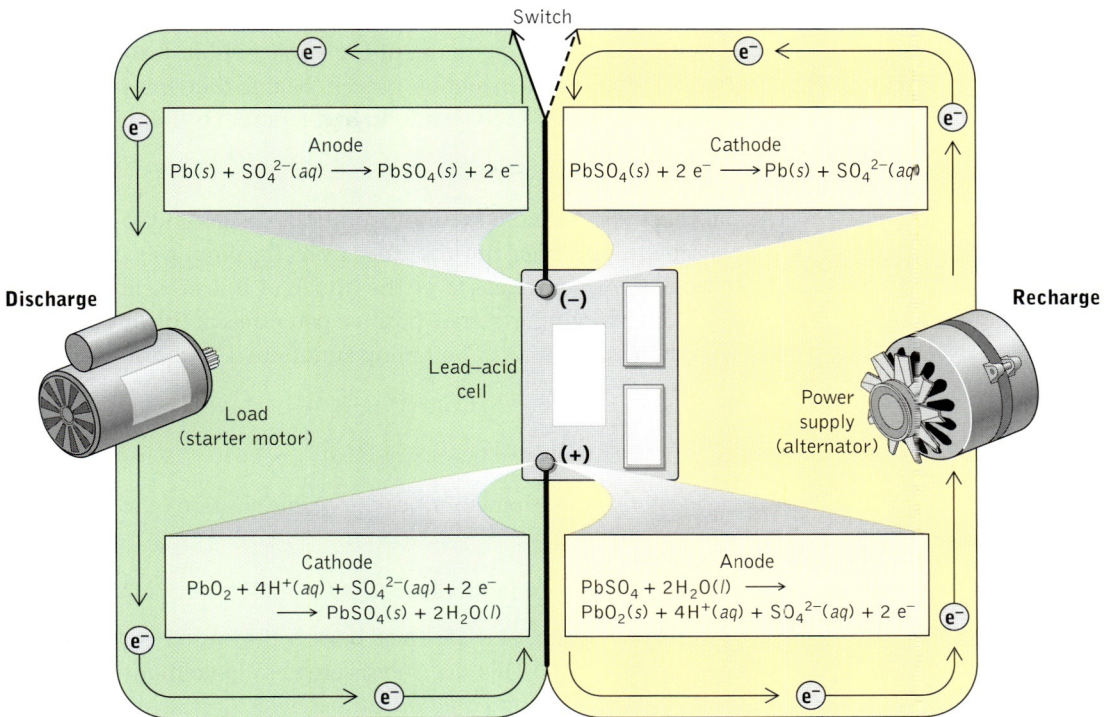

FIGURE 17.13 The lead–acid batteries used in many vehicles are based on the oxidation of Pb and the reduction of PbO_2. As the battery discharges (green circuit), Pb is oxidized, and PbO_2 is reduced. When the engine is running, an alternator generates electrical energy that flows into the battery, recharging it as both electrode reactions are reversed: $PbSO_4$ is oxidized to PbO_2, and $PbSO_4$ is reduced to Pb.

is more negative than the standard potential for reducing hydrogen ions (0.000 V). A species requiring a less-negative potential should be more easily reduced than another requiring a more-negative potential. On the basis of this analysis, you might conclude that hydrogen ions would be reduced to hydrogen gas before lead sulfate is reduced to lead metal. Your logic would be sound, but your conclusion would be wrong. Another factor affects the potential at which an electrode reaction begins. For many half-reactions, particularly those in which a gas evolves, the electron transfer process at the electrode surface may be extremely slow. The reaction may not proceed at a significant rate until a large **overpotential** (an even greater potential) has been applied. The reduction of H^+ to H_2 is such a reaction. The overpotential for hydrogen-ion reduction at the platinum electrode used in the standard hydrogen electrode is small, but it is 1 volt or more at a lead electrode, depending on current density (current per unit area of electrode surface). Thus more negative potentials are required to reduce hydrogen ions at a lead cathode than to reduce $PbSO_4$ to Pb, and so the latter reduction is more likely to happen.

A similar situation exists during recharging at the anode of a lead–acid battery. There the desired oxidation process is the conversion of $PbSO_4$ back into

☑ An **overpotential** is the additional potential above that calculated from standard potential values that must be applied to carry out an electrolysis cell reaction. Overpotentials often accompany the formation of gases.

PbO$_2$. However, during rapid recharging, PbSO$_4$ may not react quickly enough to keep up with the electric current flowing from the charging device to the battery. If that happens, the potential of the anode will drift toward more-positive values at which another oxidation process begins that can meet the unmet demand for electrons. One such oxidation half-reaction is that of water:

$$2\,H_2O(l) \longrightarrow O_2(g) + 4\,H^+(aq) + 4\,e^- \qquad E° = -1.229\text{ V} \qquad (17.18)$$

This half-reaction also includes the generation of a gas and so it also has a significant overpotential. Under normal rates of recharging, this overpotential limits the amount of oxygen released at the anode of a lead–acid battery. However, during rapid recharging, strongly positive potentials at the anode plates promote the oxidation of water to O$_2$. Combining Equations 17.11 and 17.18

$$2\,\{2\,H^+(aq) + 2\,e^- \longrightarrow H_2(g)\} \qquad E° = 0.000\text{ V}$$
$$2\,H_2O(l) \longrightarrow O_2(g) + 4\,H^+(aq) + 4\,e^- \qquad E° = -1.229\text{ V}$$

we have a cell reaction that describes the decomposition of water into molecular hydrogen and oxygen:

$$2\,H_2O(l) \longrightarrow 2\,H_2(g) + O_2(g) \qquad E° = -1.229\text{ V} \qquad (17.19)$$

Thus, the recharging of a lead–acid battery may also result in the electrolysis of water (Figure 17.14). This electrolysis creates a potentially explosive gas mixture above the battery that could be ignited by a spark when jumper cables are used to start an automobile with a dead (discharged) battery. To minimize the chances of such an explosion, the last jumper cable connection should be made to a point on the engine or body of the automobile away from the battery instead of the negative (grounded) electrode of the battery.

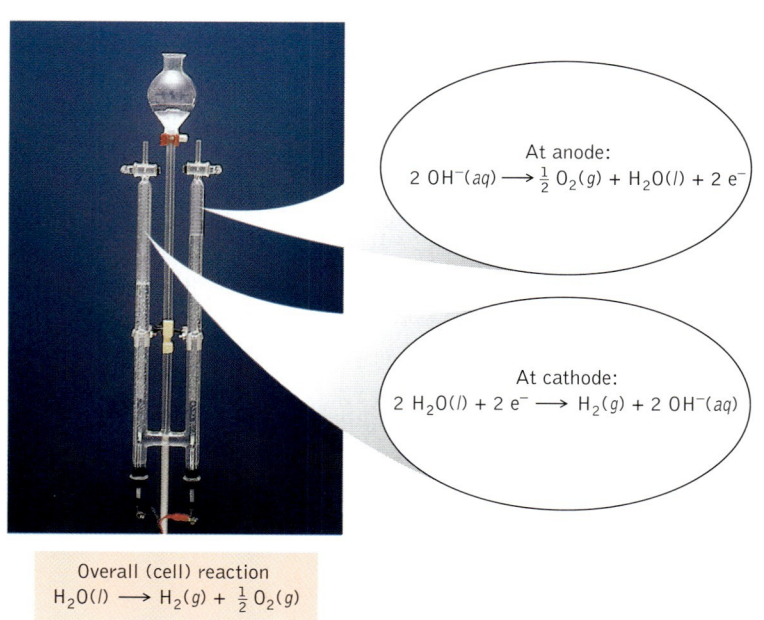

FIGURE 17.14 The electrolysis of water produces a mixture of two volumes of H$_2$ (at the cathode) to one volume of O$_2$ (at the anode). To promote the reaction, an electrolyte, such as sodium carbonate, is added to make the water a better conductor. Solutions of Na$_2$CO$_3$ are basic; so the electrode half-reactions are not the same as those that occur when H$_2$O is electrolyzed in acidic media. Although the two sets of half-reactions differ, the overall cell reactions are the same, as explained in Section 17.8.

At anode:
$2\,OH^-(aq) \longrightarrow \tfrac{1}{2}O_2(g) + H_2O(l) + 2\,e^-$

At cathode:
$2\,H_2O(l) + 2\,e^- \longrightarrow H_2(g) + 2\,OH^-(aq)$

Overall (cell) reaction
$H_2O(l) \longrightarrow H_2(g) + \tfrac{1}{2}O_2(g)$

17.7 Electrolytic Cells and Recharging Batteries

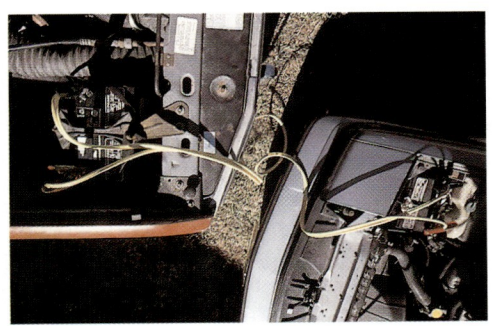

If a working lead–acid battery of one vehicle is connected to the discharged battery of another, enough energy can be transferred to start the second vehicle. However, care must be taken because a rapidly recharging battery can generate H_2 and O_2 gas. To prevent an explosion, make the last connection to the body of the second car rather than to the (−) terminal of its battery. Any sparks are thus formed well away from the battery. (In most automobile electrical systems, the negative terminal of the battery is "grounded" to the body of the vehicle.)

SAMPLE EXERCISE 17.8: How many minutes would it take to recharge a Ni-Cd battery at a charging current of 1.00 A if, during the discharge reaction, 1.12 g of Cd had been converted to $Cd(OH)_2$?

SOLUTION: The problem provides the mass of one of the reactants (Cd) in an electrochemical reaction. Because we need to relate this mass to a quantity of electrical current, we need to convert the mass into an equivalent number of moles of Cd:

$$n_{Cd} = 1.12 \text{ g Cd } \frac{(1 \text{ mol})}{112.41 \text{ g}} = 0.0100 \text{ mol Cd}$$

As the battery discharged, Cd was oxidized to $Cd(OH)_2$ according to the following half-reaction:

$$Cd(s) + 2\ OH^-(aq) \longrightarrow Cd(OH)_2(s) + 2\ e^-$$

The stoichiometry of the half-reaction indicates that 2 moles of electrons are released for every mole of Cd metal oxidized, for a total of

$$0.0100 \text{ mol Cd} \left(\frac{2 \text{ mol } e^-}{\text{mol Cd}}\right) = 0.0200 \text{ mol } e^-$$

The Faraday constant (9.65×10^4 C/mol) provides the connection between moles of electrons and coulombs of charge. Additionally, 1 ampere is a flow of electrical current equal to 1 coulomb per second; so $1 \text{ C} = 1 \text{ A} \cdot \text{s}$. Putting these facts into equation form, we have:

$$0.0200 \text{ mol } e^- \left(\frac{9.65 \times 10^4 \text{ C}}{\text{mol } e^-}\right)\left(\frac{1 \text{ A} \cdot \text{s}}{\text{C}}\right) = 1.93 \times 10^3 \text{ A} \cdot \text{s}$$

Therefore, the charger must deliver 1.00 A of current for 1.93×10^3 s, or

$$(1.93 \times 10^3 \text{ s}) \left(\frac{1 \text{ min}}{60 \text{ s}}\right) = 32.2 \text{ min}$$

mol Cd (n_{Cd})

↓ $n_{e^-} = n_{Cd}\left(\frac{2 \text{ mol } e^-}{\text{mol Cd}}\right)$

mol e^- (n_{e^-})

↓ (n_{e^-}) $\left(\frac{9.65 \times 10^4 \text{ C}}{\text{mol } e^-}\right)$

coulombs (C)

↓ $1 \text{ C} = 1 \text{ A} \cdot \text{s}$

current · time
(A · s)

↓ ÷ current (A)

time (s)

PRACTICE EXERCISE: Suppose a car's starter motor draws 230 A of current for 6.0 seconds to start the car's engine on a cold winter morning. What mass of Pb would have been oxidized to $PbSO_4$ in the car's battery to supply this much electricity? (See Problems 65–72.)

Electrolysis is used in many processes other than the recharging of batteries. Electrolytic cells are used to electroplate thin layers of silver (Figure 17.15), gold, and other metals onto objects, giving these objects the appearance, conductivity, resistance to corrosion, and other properties of the electroplated metal, but at a fraction of the cost of fabricating the entire object out of the metal.

In the chemical industry, the electrolysis of molten salts is used to produce high-energy, highly reactive substances, such as chlorine and fluorine, alkali and alkaline earth metals, and aluminum. When a salt such as NaCl is heated to just above its melting point, it becomes an ionic liquid with an enormous capacity to conduct electricity. If a sufficiently large voltage is applied to carbon electrodes immersed in molten NaCl, sodium ions will be attracted to the negative electrode and will be reduced, producing sodium metal:

> The electrolysis of molten NaCl gives Na metal at the cathode and Cl_2 at the anode.

$$Na^+(l) + e^- \longrightarrow Na(l)$$

Chloride ions will be attracted to the positive electrode and oxidized:

$$2\ Cl^-(l) \longrightarrow Cl_2(g) + 2\ e^-$$

Multiplying the first reaction by two and combining it with the second, we have the overall electrolytic cell reaction:

$$2\ Na^+(l) + 2\ Cl^-(l) \longrightarrow 2\ Na(l) + Cl_2(g)$$

Note that the anode in an electrolytic cell is the positive electrode and the cathode is the negative electrode, because electrons are supplied to the cathode from an external power supply, giving the cathode a negative charge and making the anode the positive electrode. However, in a voltaic cell, the oxidation half of the spontaneous cell reaction releases electrons to the anode and from there to an external circuit, making the anode the negative electrode. Thus, the anodes of voltaic and electrolytic cells have opposite electrical signs, and so do the cathodes.

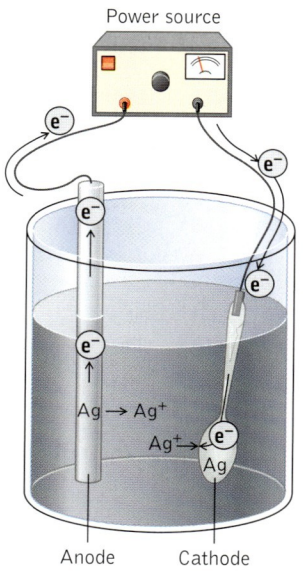

FIGURE 17.15 Items of silverware have only thin coatings of silver, which are applied by a type of electrolysis known as electroplating. In the process, the positive pole of an electric power supply is attached to a piece of pure silver, and the negative pole is connected to the cutlery to be electroplated. The oxidation half-reaction at the silver electrode (anode), $Ag \rightarrow Ag^+ + e^-$, runs in reverse at the cathodes (cutlery) as Ag^+ is reduced to the free metal at the surfaces of the cutlery.

CONCEPT TEST: The electrolysis of molten NaCl produces Na metal at the cathode. However, the electrolysis of an aqueous solution of NaCl results in the formation of gases at both the cathode and the anode. On the basis of the standard potentials in Appendix 6, predict which gases are formed at each electrode.

17.8 "LOW EMISSION" VEHICLES AND MORE VOLTAIC DEVICES

Electric cars outnumbered gasoline-powered cars in the early days of the automobile. In 1911, there were many more charging stations (360) for electric vehicles in Boston than there were gasoline stations. The limited range of electric vehicles coupled with the rapid development and refinement of the internal combustion engine led to the commercial demise of electric cars. A century later, concerns about pollutants emitted by internal combustion engines and poor air quality in some urban areas have renewed interest in electrically powered vehicles.

Hybrid vehicles

The development of electric automobiles has been limited by the availability of batteries with sufficient energy-storage capacity to compete with the driving range and performance of gasoline-powered cars. More than a ton of lead–acid batteries is required for an electric passenger car with a driving range of 100 miles (160 km). NiMH batteries offer increased driving range and less weight, but they are more expensive.

These limitations have inspired automobile makers to develop "hybrid" vehicles. These vehicles use highly efficient propulsion systems that include gasoline engines about half the size of those in cars of comparable size. To make up for the small engines, additional power comes from electric motors powered by NiMH batteries. Efficiency extends to deceleration and a process called "regenerative braking" in which the kinetic energy of a car is converted into electrical energy as the electric motor functions as an electric generator and recharges its batteries. This design circumvents the range limitations imposed by battery-powered, all-electric propulsion systems. Hybrid vehicles get about 50% better gas mileage and so emit about two-thirds of the pollutants and greenhouse gases produced by conventional cars of comparable size.

Hybrid gasoline–electric vehicles get better mileage and produce less air pollution than do gasoline-only vehicles. This 2003 Honda Civic is powered by a small (1.3-liter) gasoline engine and a 13-horsepower electric motor and gets more than 50 miles per gallon of gasoline.

Fuel cells

Fuel cells are promising alternative energy sources for electric vehicles. Like batteries, fuel cells rely on redox reactions to generate electrical energy. Unlike batteries, fuel cells are continuously supplied with reactants for the cell reaction. A popular reaction is based on the oxidation of H_2 to H_2O:

$$2\ H_2(g) + O_2(g) \longrightarrow 2\ H_2O(g) \qquad E° = 1.229\ V \qquad (17.20)$$

A **fuel cell** is a voltaic device in which there is a flow of reactants to the anode and cathode.

Note that Equation 17.20 is the reverse of the electrolysis of water (Equation 17.19). If Equation 17.20 is to be the basis for a voltaic cell, then it must be divisible into anode and cathode half-reactions. In acidic electrolytes, they are

At the anode: $\quad H_2(g) \longrightarrow 2\ H^+(aq) + 2\ e^- \qquad E° = 0.00\ V \qquad (17.21)$

At the cathode: $\ O_2(g) + 4\ H^+(aq) + 4\ e^- \longrightarrow 2\ H_2O(l) \quad E° = 1.229\ V \quad (17.22)$

In fuel cells with alkaline material between the electrodes, such as molten sodium carbonate, the high-pH versions of Equations 17.21 and 17.22 take place at the anode:

$$H_2(g) + 2\ OH^-(aq) \longrightarrow 2\ H_2O(l) + 2\ e^- \qquad E° = -0.828\ V \quad (17.23)$$

and at the cathode:

$$O_2(g) + 2\ H_2O(l) + 4\ e^- \longrightarrow 4\ OH^-(aq) \qquad E° = 0.401\ V \quad (17.24)$$

Note that, if we combine the two pairs of standard potentials by subtracting $E°_{anode}$ from $E°_{cathode}$, we get the same $E°_{cell}$ value (1.229 V). This equality is logical: the energy released under standard conditions ($\Delta G°$) by the oxidation of hydrogen (Equation 17.20) should have but one value, and so $E°_{cell}$ should have only one value.

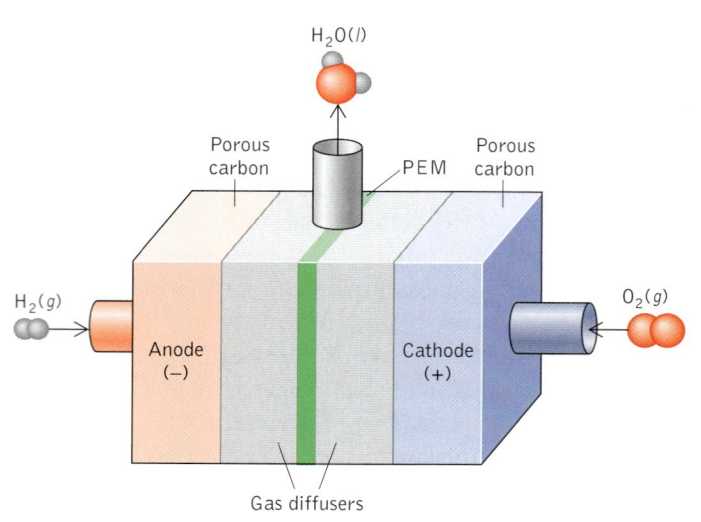

FIGURE 17.16 Most of the fuel cells used in vehicles incorporate a proton-exchange membrane (PEM) between the two halves of the cell. Gas diffusers that incorporate catalysts to speed up the fuel-cell reaction bring H_2 and O_2 to opposite sides of the membrane. On the anode side, H_2 is oxidized to H^+ ions. Through a series of ion-exchange reactions with sites in the PEM membrane, H^+ ions are transported to the cathode side, where they participate in the reduction of O_2 to H_2O. In other fuel cells with alkaline media between the electrodes, the oxidation of H_2 consumes OH^- ions, which are produced in the reduction of O_2 at the cathode.

FIGURE 17.17 In 2002 the four-passenger Honda FCX became the first commercial fuel-cell powered car in the U.S. and Japan. The FCX has a top speed of 150 km/h and two tanks that can store 156 liters of H_2 gas at 350 atm give the car a driving range of about 350 km.

The same free-energy change could be used to power a vehicle by burning hydrogen in an internal combustion engine. However, typically less than a third of the energy from combustion is converted into mechanical energy to power the car. Most of it is lost to the surroundings as heat. In contrast, fuel-cell technologies can convert most of the energy released in a fuel-cell reaction to electrical energy. In addition, fuel-cell-powered vehicles emit only water vapor; they produce no oxides of nitrogen that can contribute to smog formation and no carbon monoxide or carbon dioxide is released.

How does a fuel cell work? As in a battery, the half-reactions in a fuel cell are separated from each other by a barrier through which electrolyte ions can pass. In one design, porous graphite electrodes containing nickel and nickel(II) oxide are used for the cathode and anode respectively. Hydrogen gas and oxygen gas are supplied to the anode and cathode, respectively. Hydroxide ions formed by the reaction in Equation 17.24 migrate through the electrolyte to the cathode, where they combine with hydrogen as it is oxidized to water (Equation 17.23). In a commercial fuel cell developed for use in automobiles, a thin polymer membrane acts as the electrolyte and the ion-transport medium (see Figure 17.16).

> **How does a fuel cell work?**

The prototype fuel-cell-powered car shown in Figure 17.17 has two tanks that can hold 130 liters of H_2 at pressures as high as 300 atm. Even at these high pressures, the car has an operating range of only about 300 km. There is also the problem of where to refuel. It is likely that most fuel-cell-powered vehicles will be buses and fleet vehicles operating from a central location. For example, fuel-cell-powered buses (with very large fuel tanks) in Chicago that can carry 60 passengers have an operating range of 400 km.

Even longer ranges would be possible if better methods for storing hydrogen gas were available. Currently under development are mobile chemical-processing plants that can extract hydrogen from liquid fuels such as gasoline or methanol as needed by the vehicle. Scientists are also developing new materials that can be used to store large quantities of hydrogen safely. In one promising technology, a carbon material, related in its structure to bucky balls, has been used to store many times the mass of hydrogen that could be compressed into the same volume. Moreover, if an automobile were in an accident, hydrogen stored in this way would not be released and explode, as might happen if a high-pressure tank filled with H_2 ruptured.

Photochemical cells

Another approach to hydrogen production is to harness the energy of the sun to split water into H_2 and O_2. A **photochemical cell** with this capability is shown in Figure 17.18. Its anode is a thin piece of platinum metal. The cathode consists of a "sandwich" made from a layer of gallium arsenide, GaAs, between two thin films of platinum. One of the platinum layers is coated with gallium indium phosphide, GaInP. This coating absorbs light and converts it into electricity, which passes through an external circuit where it can be used to electrolyze water.

Photochemical cells directly convert solar energy into electricity.

How much hydrogen can be produced photochemically from sunlight? The sun's energy reaches Earth at a rate of 8.3 joules per square centimeter per minute. If a cell such as that shown in Figure 17.18 is in direct sunlight for 8 hours (480 min), the solar energy falling on each square centimeter of its surface will be

> How much hydrogen can be produced photochemically from sunlight?

$$\frac{(8.3 \text{ J})}{\text{cm}^2 \cdot \text{min}} (1.0 \text{ cm}^2)(480 \text{ min}) = 4.0 \times 10^3 \text{ J} = 4.0 \text{ kJ}$$

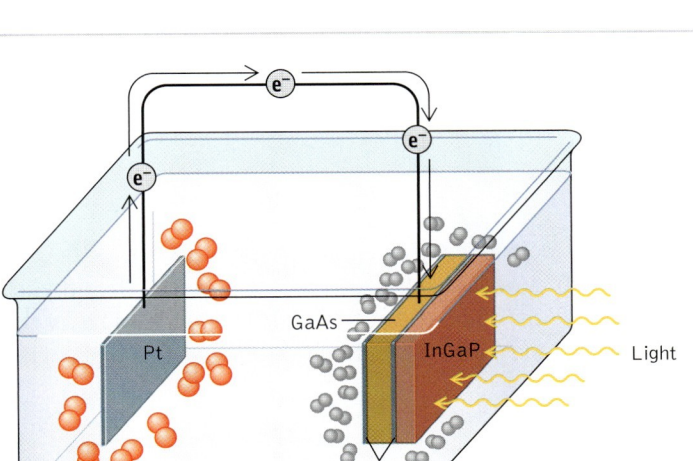

FIGURE 17.18 This photochemical cell uses the energy of sunlight to electrolyze H_2O into H_2 at the cathode and O_2 at the anode. The anode is a thin piece of platinum metal. The cathode consists of a "sandwich" made from a layer of gallium arsenide (GaAs) between two thin films of platinum (Pt). One of the platinum layers is coated with gallium indium phosphide (GaInP). This coating absorbs sunlight and converts it into a potential sufficiently large that water can be electrolyzed.

Let's compare this value with the heat of formation of liquid water under standard conditions:

$$H_2(g) + \tfrac{1}{2} O_2(g) \longrightarrow H_2O(l) \qquad \Delta H° = -286 \text{ kJ}$$

This quantity of energy would have to be added to separate a mole of H_2O into a mole of H_2 and a half mole of O_2:

$$H_2O(l) \longrightarrow H_2(g) + \tfrac{1}{2} O_2(g) \qquad \Delta H° = 286 \text{ kJ}$$

If we divide the 4.0 kJ of sun energy by the 286 kJ/mol, we get

$$\frac{4.0 \text{ kJ}}{286 \frac{\text{kJ}}{\text{mol}}} = 0.014 \text{ mol } H_2$$

Unfortunately, the energy of the sun can never be captured and converted into electrical energy with 100% efficiency. Modern photocells are about 12% to 18% efficient. Therefore, current technology limits H_2 production from solar-powered photochemical cells to about 2×10^{-3} mol (0.004 g) of H_2 per square centimeter of cell-surface area per day. The cost of the materials for these cells has so far limited their commercial development.

Biochemical fuel cells

Scientists are currently developing fuel cells that can harvest the energy released by the conversion of glucose to carbon dioxide and water:

$$C_6H_{12}O_6(s) + 6 O_2(g) \longrightarrow 6 CO_2(g) + 6 H_2O(l)$$

This reaction is a redox reaction and so has the potential for producing electricity if the sites for releasing and taking in electrons can be physically isolated. The standard cell potential for the reaction can be calculated from its $\Delta G°$ value with Equation 17.6:

$$\Delta G° = -nFE°$$

but first we need to determine how many electrons are transferred. The easiest way to determine the value of n is to consider the reduction of O_2 half-reaction (Equation 17.22):

$$O_2(g) + 4 H^+(aq) + 4 e^- \longrightarrow 2 H_2O(l) \qquad E° = 1.229 \text{ V}$$

Four electrons are needed per mole of oxygen. There are 6 moles of O_2 in the glucose reaction, and so a total of 24 moles of electrons are transferred. The number of electrons gained by oxygen must equal the number lost by glucose; so n for the cell reaction must be 24. Thus, the cathode reaction for the process may be written as follows:

$$6 H_2O(l) + C_6H_{12}O_6(aq) \longrightarrow 6 CO_2(g) + 24 H^+(aq) + 24 e^-$$

Rearranging Equation 17.6 and inserting the appropriate values and conversion factors,

$$E° = \frac{-\Delta G°}{nF} = \left(\frac{-(-2816 \text{ kJ})}{(24 \text{ mol e}^-)\left(\frac{9.65 \times 10^3 \text{C}}{\text{mol e}^-}\right)}\right)\left(\frac{1000 \text{ J}}{\text{kJ}}\right)\left(\frac{\text{V} \cdot \text{C}}{\text{J}}\right) = 1.22 \text{ V}$$

we find that the standard potential of the oxidation of glucose is about the same as that of the oxidation of hydrogen gas. The challenge is to design a fuel cell that can harvest this energy.

In 1995, researchers at King's College in London successfully prepared an electrochemical cell that used the bacterium *E. coli* to metabolize a broth of carbohydrates (Figure 17.19). It produced enough electricity to run a digital clock for one day. The bacteria reproduce themselves, and plant carbohydrates are readily replenished. In the years ahead, scientists may develop other, more-powerful biochemical fuel cells. They would be less polluting, renewable sources of electrical energy.

✓ In principle, the energy released in any spontaneous redox reaction, including biochemical processes, can be converted into electrical energy.

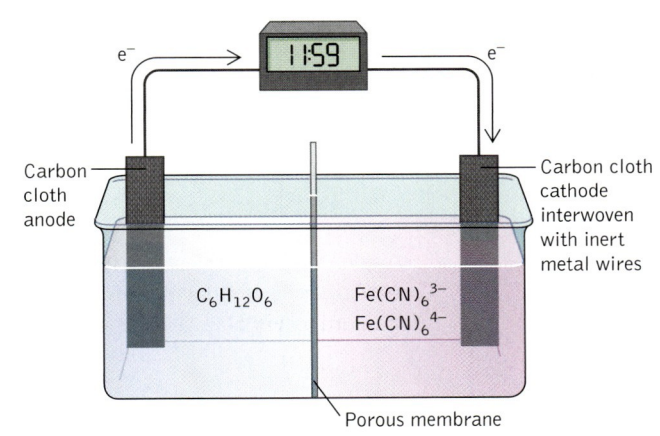

FIGURE 17.19 The "biological" battery shown here uses bacteria to oxidize glucose and an electrode made of woven carbon fibers to collect the electrons released by the process. The cathode reaction that completes the electrical circuit is based on the reduction of the cyano complex of Fe^{3+} to the cyano complex of Fe^{2+}: $Fe(CN)_6^{3-}(aq) + e^- \rightarrow Fe(CN)_6^{4-}(aq)$.

CHAPTER REVIEW

Summary

SECTION 17.1

A voltaic cell produces an electrical current in an external circuit as a result of oxidation at the anode and reduction at the cathode. Electrons flow from the anode through an external circuit to the cathode. A salt bridge between the two half-reactions completes the electrical circuit.

SECTION 17.2

In voltaic cells spontaneous cell reactions ($\Delta G < 0$) produce positive cell potentials ($E_{cell} > 0$).

SECTION 17.3

In a Ni-Cd battery, $NiO(OH)(s)$ is reduced to $Ni(OH)_2(s)$, and $Cd(s)$ is oxidized to $Cd(OH)_2(s)$. Dry cells and alkaline batteries have cell reactions based on oxidation of Zn and reduction of MnO_2.

SECTION 17.4

Every half-reaction has a characteristic standard electrode potential. The scale used to express standard potential has as its zero point the potential of the standard hydrogen electrode. The standard potential of a cell can be calculated from the standard potentials of its anode and cathode half-reactions.

SECTION 17.5

The potential of a voltaic cell approaches zero as the cell reaction reaches equilibrium. The Nernst equation predicts how the potential of a cell or half-reaction changes with temperature and the concentrations of the cell reactants and products.

SECTION 17.6

A watt is a unit of electrical power. One watt is equal to 1 volt · ampere, or 1 joule/s. The quantities of reactants consumed in a voltaic cell reaction are directly proportional to the coulombs of electrical charge that it delivers.

SECTION 17.7

Reversing a spontaneous reaction by passing electric current through a system is called electrolysis. Electrode reactions that produce gases such as H_2 often have high overpotentials.

SECTION 17.8

Fuel cells directly convert stored chemical energy into electrical energy. Photochemical cells directly convert solar energy into electricity. In principle, the energy released in any spontaneous redox process, including biochemical processes, can be converted into electrical energy.

Key Terms

anode (p. 846)
cathode (p. 846)
cell potential (p. 847)
electrochemistry (p. 844)

electromotive force (p. 847)
fuel cell (p. 879)
overpotential (p. 875)
photochemical cell (p. 881)

standard hydrogen electrode (p. 857)
standard cell potential (p. 856)
standard potential (p. 856)

Key Skills and Concepts

SECTION 17.1

Be able to specify the components and sketch a voltaic cell based on the half-reactions of a redox reaction.

SECTION 17.2

Know that the free-energy change of a voltaic-cell reaction is related to the electrical potential (voltage) of the cell, and be able to calculate one from the other.

SECTION 17.3

Be able to explain how common batteries work.

SECTION 17.4

Be able to use the standard potentials of a cell's half reactions to calculate $E°_{cell}$.

SECTION 17.5

Be able to use the Nernst equation to calculate the potentials of half-reactions and cell reactions.
Be able to calculate the equilibrium constant of a cell reaction from its standard potential.

SECTION 17.6

Be able to calculate the changes in the quantities of cell reactants needed to deliver a given quantity of electrical charge.

SECTION 17.7

Be able to calculate the amounts of materials obtained by electrolysis from the quantity of charge passing through an electrolytic cell.

Be able to calculate the time required to electrolytically produce a specified quantity of product from the cell reaction and the electrical current flowing through the cell.

SECTION 17.8

Be able to explain how a fuel cell works.
Be able to explain the concept of a photochemical cell.

KEY EQUATIONS AND RELATIONS

SECTION 17.2

The electrical potential (E) of a voltaic cell is related to the change in free energy (ΔG) of the cell reaction:

$$\Delta G = -nFE \qquad (17.6)$$

where F is the Faraday constant (9.65×10^4 C/mol) and n is the number of electrons transferred in the balanced cell reaction.

SECTION 17.4

The standard potential of an electrochemical cell:

$$E°_{cell} = E°_{cathode} - E°_{anode} \qquad (17.9)$$

SECTION 17.5

The potential of a half-reaction or cell reaction at 25°C is given by the Nernst equation:

$$E = E° - \frac{0.0592}{n} \log Q$$

where Q is the reaction quotient for the half-reaction or cell reaction.

The equilibrium constant (K) for a voltaic-cell reaction at 25°C is related to its standard potential:

$$\log K = \frac{nE°_{cell}}{0.0592} \qquad (17.17)$$

SECTION 17.6

Electrical current (in amperes), electrical charge (in coulombs), electrical energy (in joules), and electrical power (in watts) are related as follows:

1 coulomb = 1 ampere · second
1 watt = 1 volt · ampere = 1 joule/second

QUESTIONS AND PROBLEMS

Voltaic Cells

CONCEPT REVIEW

1. What is the role of the salt bridge in an electrochemical cell?
2. Why can't a wire perform the same function as a salt bridge in an electrochemical cell?
3. In a voltaic cell, why is the cathode labeled the negative terminal and the anode the positive terminal?
4. In the voltaic cell shown here, the greater density of a concentrated solution of CuSO$_4$ allows a less concentrated solution of ZnSO$_4$ solution to be (carefully) layered on top of it. Why is a salt bridge not needed in this cell?

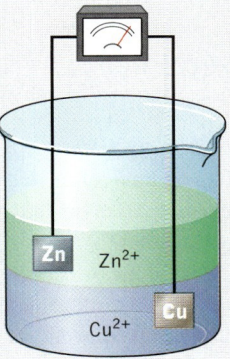

PROBLEMS

5. A voltaic cell with an aqueous electrolyte is based on the reaction between $Zn^{2+}(aq)$ and $Pb(s)$, producing $Zn(s)$ and $Pb^{2+}(aq)$.
   a. Write half-reactions for the anode and cathode.
   b. Write a balanced cell reaction.
   c. Sketch a diagram of the cell. Indicate the composition of the electrodes and the dissolved species that take part in the cell reaction and the direction of electron flow in an external circuit.

6. A voltaic cell is based on the reaction between $Ag^+(aq)$ and $Ni(s)$, producing $Ag(s)$ and $Ni^{2+}(aq)$.
   a. Write the anode and cathode half-reactions.
   b. Write a balanced cell reaction.
   c. Sketch a diagram of the cell. Indicate the composition of the electrodes and the dissolved species that take part in the cell reaction and the direction of electron flow in an external circuit.

7. A voltaic cell with a basic aqueous electrolyte is based on the oxidation of $Cd(s)$ to $Cd(OH)_2(s)$ and the reduction of $MnO_4^-(aq)$ to $MnO_2(s)$.
   a. Write half-reactions for the cell's anode and cathode.
   b. Write a balanced cell reaction.
   c. Sketch a diagram of the cell. Indicate the composition of the electrodes and the dissolved species that take part in the cell reaction and the direction of electron flow in an external circuit.

8. A voltaic cell based on the reduction of $Ag^+(aq)$ to $Ag(s)$ and the oxidation of $Sn(s)$ to $Sn^{2+}(aq)$.
   a. Write half-reactions for the cell's anode and cathode.
   b. Write a balanced cell reaction.
   c. Sketch a diagram of the cell. Indicate the composition of the electrodes and the dissolved species that take part in the cell reaction and the direction of electron flow in an external circuit.

9. In 1999, scientists in Israel developed a battery based on the following cell reaction with iron(VI), nicknamed "super iron":

   $$2\,K_2FeO_4(aq) + 3\,Zn(s) \longrightarrow Fe_2O_3(s) + ZnO(s) + 2\,K_2ZnO_2(aq)$$

   a. Determine the number of electrons transferred in the cell reaction.
   b. Determine the oxidation states of the transition metals in the reaction.
   c. Identify the substance of which the anode is made.

10. The Hawaiian bobtail squid produces hypochlorite by reacting hydrogen peroxide with Cl^- ions from seawater:

    $$H_2O_2(aq) + Cl^-(aq) \longrightarrow OCl^-(aq) + H_2O(l)$$

    Determine the oxidation states of Cl and O in the reactants and products.

11. Identify which elements, if any, undergo oxidation or reduction in each of the following reactions.
    a. $ClO_3^-(aq) + Br^-(aq) + 6\,H^+(aq) \to Cl^-(aq) + 3\,Br_2(g) + 3\,H_2O(l)$
    b. $Cl_2(g) + 2\,OH^-(aq) \to Cl^-(aq) + OCl^-(aq) + H_2O(l)$
    c. $AgNO_3(aq) + NaIO_3(aq) \to AgIO_3(s) + NaNO_3(aq)$

12. Identify which elements, if any, undergo oxidation or reduction in each of the following reactions:
    a. $2\,HIO_3(aq) + 5\,H_2O_2(aq) \to 5\,O_2(g) + I_2(s) + 6\,H_2O(l)$
    b. $Ba(BrO_3)_2(aq) + H_2SO_4(aq) \to 2\,HBrO_3(aq) + BaSO_4(s)$
    c. $4\,ClO_3^-(aq) \to Cl^-(aq) + 3\,ClO_4^-(aq)$

Voltage and Free Energy

CONCEPT REVIEW

13. The potential of one voltaic cell is twice that of another. How is the value of the ΔG of the first cell reaction related to the ΔG of the other?

*14. In physics, work (w) is done by exerting a force (F) to move an object through a distance (d): $w = F \times d$. Explain how this definition of work relates to electrical work ($w_{elec} = C \times E$).

PROBLEMS

15. Starting with the appropriable standard free energies of formation in Appendix 5, calculate the value of $\Delta G°$ and $E°_{cell}$ of the following reactions.
    a. $2\,Cu^+(aq) \to Cu^{2+}(aq) + Cu(s)$
    b. $Ag(s) + Fe^{3+}(aq) \to Ag^+(aq) + Fe^{2+}(aq)$

16. Starting with the appropriate standard free energies of formation in Appendix 5, calculate the values of $\Delta G°$ and $E°_{cell}$ of the following reactions.

a. $FeO(s) + H_2(g) \rightarrow Fe(s) + H_2O(l)$
b. $2\ Pb(s) + O_2(g) + 2\ H_2SO_4(aq) \rightarrow 2\ PbSO_4(s) + 2\ H_2O(l)$

17. In the aluminum-air battery Al metal is oxidized to $Al(OH)_3$ and O_2 from the air is reduced in an alkaline electrolyte. Write the appropriate cathode and anode half-reactions and the overall cell reaction for the aluminum-air battery.

18. Sometimes the cell reaction of NiCd batteries is written with Cd metal as the anode and solid NiO_2 as the cathode. Assuming the products of the electrode reactions are solid hydroxides of Cd(II) and Ni(II), respectively, write a balanced chemical equation for the cell reaction.

19. Manganese oxides are used in many types of batteries, including the dry cell. Balance the following redox reactions of manganese oxides in acid solutions.
a. $MnO_4^-(aq) + H_2O_2(aq) \rightarrow Mn^{2+}(aq) + O_2(g)$
b. $MnO_2(s) + Cl^-(aq) \rightarrow Mn^{2+}(aq) + Cl_2(g)$
c. $K_2MnO_4(aq) \rightarrow KMnO_4(aq) + MnO_2(s)$

20. Sodium–nickel chloride and zinc–air batteries have been proposed as cheaper alternatives to the nickel–metal-hydride battery. Balance the following redox reactions and identify the oxidizing and reducing agents.
a. $Ni^{2+}(aq) + H_2(g) \rightarrow Ni(s) + 2\ H^+(aq)$
b. $Na(s) + NiCl_2(l) \rightarrow NaCl(l) + Ni(s)$
c. $Zn(s) + O_2(g) \rightarrow ZnO(s)$

Half-Reactions and Standard Cell Potentials

CONCEPT REVIEW

21. What is the function of platinum in the standard hydrogen electrode?

22. Is it possible to build a battery in which the anode chemistry is based on a half-reaction such as:

$$Fe^{2+}(aq) \rightarrow Fe^{3+}(aq) + e^-$$

in which none of the species is a solid conductor?

23. If the container (Zn electrode) in a dry cell were replaced by another with twice the thickness of the original, how would the cell potential be affected?

24. A voltaic cell similar to the Cu-Zn cell in Figure 17.2 has a cathode made of Ni immersed in a 1 M solution of $NiSO_4$. Will the potential of this Ni-Zn cell be greater than, the same as, or less than the Cu-Zn cell in Figure 17.2?

25. To calculate the standard potential of an electrochemical cell, why don't we sum the standard potentials of the anode and cathode instead of subtracting $E°_{anode}$ from $E°_{cathode}$ (as in Equation 17.11)?

26. Suppose there were a scale for expressing electrode potentials in which the standard potential for the reduction of water in base

$$2\ H_2O(l) + 2\ e^- \rightarrow H_2(g) + 2\ OH^-(aq)$$

is assigned an $E°$ value of 0.000 V. On this new scale,
a. what would be the $E°$ value of the following half-reaction?

$$2\ H^+(aq) + 2\ e^- \rightarrow H_2(g)$$

b. what would be the $E°$ value of the following half-reaction?

$$Fe^{3+}(aq) + e^- \rightarrow Fe^{2+}(aq)$$

c. what would be the $E°_{cell}$ value of the lead–acid battery reaction?

$$Pb(s) + PbO_2(s) + 2\ H_2SO_4(aq) \rightarrow 2\ PbSO_4(s) + 2\ H_2O(l)$$

27. If a piece of silver is placed in a solution in which $[Ag^+] = [Cu^{2+}] = 1.00\ M$, will the following reaction proceed spontaneously?

$$2\ Ag(s) + Cu^{2+}(aq) \rightarrow 2\ Ag^+(aq) + Cu(s)$$

28. A piece of cadmium is placed in a solution in which $[Cd^{2+}] = [Sn^{2+}] = 1.00\ M$. Will the following reaction proceed spontaneously?

$$Cd(s) + Sn^{2+}(aq) \rightarrow Cd^{2+}(aq) + Sn(s)$$

PROBLEMS

29. Starting with the standard potentials in Appendix 6, calculate the values of $E°_{cell}$ and $\Delta G°$ of the following reactions.
a. $Cu(s) + Sn^{2+}(aq) \rightarrow Cu^{2+}(aq) + Sn(s)$
b. $Zn(s) + Ni^{2+}(aq) \rightarrow Zn^{2+}(aq) + Ni(s)$

30. Starting with the standard potentials in Appendix 6, calculate the values of $E°_{cell}$ and $\Delta G°$ of the following reactions.
a. $Fe(s) + Cu^{2+}(aq) \rightarrow Fe^{2+}(aq) + Cu(s)$
b. $Ag(s) + Fe^{3+}(aq) \rightarrow Ag^+(aq) + Fe^{2+}(aq)$

31. Voltaic cells based on the following pairs of half-reactions are prepared so that all reactants and products are in their standard states. For each pair, write a balanced equation for the cell reaction, and identify which half-reaction takes place at the anode and at the cathode.
a. $ZnO(s) + H_2O(l) + 2\ e^- \rightarrow Zn(s) + 2\ OH^-(aq)$
$MnO_2(s) + 2\ H_2O(l) + 2\ e^- \rightarrow Mn(OH)_2(s) + 2\ OH^-$

b. $S(l) + 2\,e^- \rightarrow S^{2-}(l)$
   $Na^+(l) + e^- \rightarrow Na(l)$
c. $Al(OH)_4^-(aq) + 2\,e^- \rightarrow Al(s) + 4\,OH^-(aq)$
   $O_2(g) + 2\,H_2O(l) + 2\,e^- \rightarrow 4\,OH^-(aq)$

32. Voltaic cells based on the following pairs of half-reactions are constructed. For each pair, write a balanced equation for the cell reaction, and identify which half-reaction takes place at each anode and cathode.
    a. $Cd^{2+}(aq) + 2\,e^- \rightarrow Cd(s)$
       $Ag^{2+}(aq) + e^- \rightarrow Ag(s)$
    b. $AgBr(s) + e^- \rightarrow Ag(s) + Br^-(aq)$
       $MnO_2(s) + 4\,H^+(aq) + 2\,e^- \rightarrow Mn^{2+}(aq) + 4\,H_2O(l)$
    c. $PtCl_4^{2-}(aq) + 2\,e^- \rightarrow Pt(s) + 4\,Cl^-(aq)$
       $AgCl(s) + e^- \rightarrow Ag(s) + Cl^-(aq)$

33. In the voltaic cell shown here, the concentrations of Cu^{2+} and Cd^{2+} are 1.00 M. Identify, on the basis of the standard potentials in Appendix 6, which electrode is the anode and which is the cathode. Indicate the direction of electron flow.

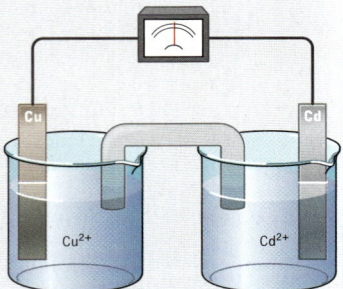

34. In the voltaic cell shown here, $[Ag^+] = [H^+] = 1.00M$. Identify, on the basis of the standard potentials in Appendix 6, which electrode is the anode and which is the cathode. Indicate the direction of electron flow.

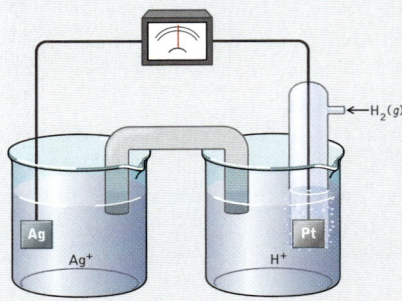

35. The half-reactions and standard potentials for a NiMH battery with a titanium–zirconium anode are as follows:

    Cathode: $NiOOH(s) + H_2O(l) + e^- \rightarrow$
             $Ni(OH)_2(s) + OH^-(aq)$ $E° = 1.32\ V$
    Anode: $TiZr_2H(s) + OH^-(aq) \rightarrow$
             $TiZr_2(s) + H_2O(l) + e^-$ $E° = 0.00\ V$

    a. Write the overall cell reaction for this battery.
    b. Calculate the standard cell potential.

36. A zinc–air battery contains a zinc anode, a carbon cathode, and KOH electrolyte. The anode half-reaction is

    $Zn(s) + 2\,OH^-(aq) \rightarrow Zn(OH)_2(s) + 2e^-$

    a. If the active ingredient at the cathode is "air," predict the cathode half-reaction.
    b. Calculate the standard cell potential on the basis of your prediction in part a and $E° = -0.34V$ for the reduction of $Zn(OH)_2(s)$ to $Zn(s)$.

Effect of Concentration on Potential

CONCEPT REVIEW

37. The standard potential of the Cu–Zn cell reaction

    $Zn(s) + Cu^{2+}(aq) \rightarrow Zn^{2+}(aq) + Cu(s)$

    is 1.10 V. Would the potential of the Cu–Zn cell differ from 1.10 V if the concentrations of both Cu^{2+} and Zn^{2+} were 0.25 M?

38. Would you expect the potential of a battery based on the following reaction

    $Zn(s) + Cu^{2+}(aq) \rightarrow Zn^{2+}(aq) + Cu(s)$

    to change as the cell reaction proceeds and the battery is discharged? Explain your answer.

*39. The cell potential of a lead–acid battery changes as the concentration of sulfuric acid in the battery's electrolyte changes (see Figure 17.8). Does the potential of a nickel–metal-hydride battery change if the concentration of KOH in its electrolyte is changed?

*40. The cell potential of a Ni-Cd battery remains fairly constant as it discharges. Why?

PROBLEMS

41. Calculate the E_{cell} value at 298 K for the cell based on the following reaction:

$$Fe^{3+}(aq) + Cr^{2+}(aq) \rightarrow Fe^{2+}(aq) + Cr^{3+}(aq)$$

when $[Fe^{3+}] = [Cr^{2+}] = 1.50 \times 10^{-3}$ M and $[Fe^{2+}] = [Cr^{3+}] = 2.5 \times 10^{-4}$ M at 298 K.

42. Calculate the E_{cell} value at 298 K for the cell based on the following reaction:

$$Cu(s) + 2 Ag^+(aq) \rightarrow Cu^{2+}(aq) + 2 Ag(s)$$

when $[Ag^+] = 2.56 \times 10^{-3}$ M and $[Cu^{2+}] = 8.25 \times 10^{-4}$ M.

43. Determine the equilibrium constant for the following reaction at 298 K. Use the appropriate standard potentials in Appendix 6.

$$Fe^{3+}(aq) + Cr^{2+}(aq) \rightarrow Fe^{2+}(aq) + Cr^{3+}(aq)$$

44. Determine the equilibrium constant at 298 K for the following reaction between MnO_2 and Fe^{2+} in acid solution. Use the appropriate standard potentials in Appendix 6.

$$4 H^+(aq) + MnO_2(s) + 2 Fe^{2+}(aq) \rightarrow Mn^{2+}(aq) + 2 Fe^{3+}(aq) + 2 H_2O(l)$$

45. If the potential of a hydrogen electrode based on the following half-reaction

$$2 H^+(aq) + 2 e^- \longrightarrow H_2(g)$$

is 0.000 V at pH = 0.00, what is the potential of the same electrode at pH = 7.00?

46. Given the following standard potentials for the reduction of nicotinamide adenine dinucleotide (NAD^+) and oxaloacetate (an intermediate in the Krebs cycle),

$$NAD^+(aq) + 2 H^+(aq) + 2 e^- \longrightarrow NADH(aq) + H^+(aq) \quad E° = -0.320 \text{ V}$$

$$\text{oxaloacetate}(aq) + 2 H^+(aq) + 2 e^- \longrightarrow \text{malate}(aq) \quad E° = -0.166 \text{ V}$$

a. calculate the standard cell potential for the following reaction:

$$\text{oxaloacetate}(aq) + NADH(aq) + H^+(aq) \longrightarrow \text{malate}(aq) + NAD(aq)$$

b. calculate the equilibrium constant for the cell reaction at 298 K.

47. Permanganate ion can oxidize sulfite to sulfate in basic solution:

$$2 MnO_4^-(aq) + 3 SO_3^{2-}(aq) + H_2O(l) \longrightarrow 2 MnO_2(s) + 3 SO_4^{2-}(aq) + 2 OH^-(aq)$$

Determine the potential for the reaction at 298 K when the concentrations of the reactants and products are as follows: $[MnO_4^-] = 0.150$ M $[SO_3^{2-}] = 0.256$ M, $[SO_4^{2-}] = 0.178$ M, and $[OH^-] = 0.0100$ M. Will the value of E_{cell} increase or decrease as the reaction proceeds?

48. Manganese dioxide is reduced by iodide ion in acid solution:

$$MnO_2(s) + 2 I^-(aq) + 4 H^+(aq) \longrightarrow Mn^{2+}(aq) + I_2(aq) + 2 H_2O(l)$$

Determine the electrical potential of the reaction at 298 K when the initial concentrations of the components are as follows: $[I^-] = 0.225$ M, $[H^+] = 0.900$ M, $[Mn^{2+}] = 0.100$ M, and $[I_2] = 0.00114$ M. Will the value of E_{cell} increase or decrease as the reaction proceeds if the solubility of iodine in water is approximately 0.114 M?

49. A copper penny dropped into a solution of nitric acid produces a mixture of nitrogen oxides. The following reaction describes the formation of one of them, NO:

$$3 Cu(s) + 8 H^+(aq) + 2 NO_3^-(aq) \longrightarrow 2 NO(g) + 3 Cu^{2+}(aq) + 4 H_2O(l)$$

a. Calculate $E°$ for this reaction.
b. Calculate E at 298 K when $[H^+] = 0.100$ M, $[NO_3^-] = 0.0250$ M, $[Cu^{2+}] = 0.0375$ M, and the partial pressure of NO = 0.00150 atm.

50. Chlorine dioxide (ClO_2) is produced by the reaction of chlorate (ClO_3^-) with Cl^- in acid solution:

$$2 ClO_3^-(aq) + 2 Cl^-(aq) + 4 H^+(aq) \longrightarrow 2 ClO_2(g) + Cl_2(g) + 2 H_2O(l)$$

a. Determine $E°$ for the reaction.
b. The reaction produces an atmosphere in the reaction vessel in which $P_{ClO_2} = 2.0$; $P_{Cl_2} = 1.00$ atm. Calculate $[ClO_3^-]$ if, at equilibrium ($T = 298$ K), $[H^+] = [Cl^-] = 10.0$ M.

51. The oxidation of NH_4^+ to NO_3^- in acid solution is described by the following equation:

$$NH_4^+(aq) + 2 O_2(g) + H_2O(l) \longrightarrow NO_3^-(aq) + 2 H^+(aq)$$

a. Calculate $E°$ for the reaction.
b. If the reaction is in equilibrium with air ($P_{O_2} = 0.21$ atm) at pH 5.60, what is the ratio of $[NO_3^-]$ to $[NH_4^+]$ at 298 K?

52. Recalculate the answers to Problem 51 for the oxidation of NH_4^+ at an altitude where $P_{O_2} = 0.18$ atm.

Battery Power

CONCEPT REVIEW

53. One 12-volt lead–acid battery has a CCA rating of 550; another has a CCA rating of 800. In which of the following parameters are the two likely to be different?
a. Individual cell potentials
b. Anode half-reactions
c. Total masses of electrode material
d. Number of cells
e. Composition of their electrolytes
f. Combined surface areas of their electrodes

54. In a voltaic cell based on the Cu-Zn cell reaction

$$Zn(s) + Cu^{2+}(aq) \longrightarrow Cu(s) + Zn^{2+}(aq)$$

there is exactly 1 mole of each reactant and product. A second cell based on the Cu-Cd cell reaction:

$$Cd(s) + Cu^{2+}(aq) \longrightarrow Cu(s) + Cd^{2+}(aq)$$

also has exactly 1 mole of each reactant and product. Which of the following statements about these two cells is true?
a. Their cell potentials are the same.
b. The masses of their electrodes are the same.
c. The quantities of electrical charge that they can produce are the same.
d. The quantities of electrical energy that they can produce are the same.

PROBLEMS

55. Which of the following voltaic cells will produce the greatest quantity of electrical charge per gram of anode material?

$$Cd(s) + 2\,NiO(OH)(s) + 2\,H_2O(l) \longrightarrow 2\,Ni(OH)_2(s) + Cd(OH)_2(s)$$

or

$$4\,Al(s) + 3\,O_2(g) + 6\,H_2O(l) + 4\,OH^-(aq) \longrightarrow 4\,Al(OH)_4^-(aq)$$

56. Which of the following voltaic cells will produce the greatest quantity of electrical charge per gram of anode material?

$$Zn(s) + MnO_2(s) + H_2O(l) \longrightarrow ZnO(s) + Mn(OH)_2(s)$$

or

$$Li(s) + MnO_2(s) \longrightarrow LiMnO_2(s)$$

57. Which of the following voltaic cell reactions delivers the most power per gram of anode material?

$$Cd(s) + 2\,NiO(OH)(s) + 2\,H_2O(l) \longrightarrow 2\,Ni(OH)_2(s) + Cd(OH)_2(s)$$

or

$$Li(s) + MnO_2(s) \longrightarrow LiMnO_2(s) \quad E° = 2.69\text{ V}$$

58. Which of the following voltaic cell reactions delivers the most power per gram of anode material?

$$Zn(s) + MnO_2(s) + H_2O(l) \longrightarrow ZnO(s) + Mn(OH)_2(s)$$

or

$$2\,Na(s) + S(l) \longrightarrow Na_2S(l)$$

Electrolytic Cells and Recharging Batteries

CONCEPT REVIEW

59. The electrolysis of molten $MgCl_2$ obtained from evaporated seawater is used to produce magnesium metal. Do we need to remove the principal ingredient in sea salt (NaCl) before electrolysis?

60. The salts obtained from the evaporation of seawater can act as a source of halogens, principally Cl_2 and Br_2, through the electrolysis of the molten alkali metal halides. As the potential of the anode in an electrolytic cell is increased, which of these two halogens forms first?

61. In the electrolysis described in Question 59, why is it necessary to use molten salts rather than seawater itself?

62. In the electrolysis of H_2O to H_2 and O_2, a strong electrolyte, such as Na_2CO_3, is added to help speed up the electrolysis process. How does Na_2CO_3 speed it up?

63. In the electrolysis of H_2O to H_2 and O_2, why should Na_2CO_3, but not NaCl, be added to speed up electrolysis?

64. Electrolysis can be used to determine the concentration of Cu^{2+} in a given volume of solution by electrolyzing the solution in a cell equipped with a platinum cathode. If all of the Cu^{2+} is reduced to Cu metal at the cathode, the increase in mass of the electrode provides a measure of the concentration of Cu^{2+} present in the original solution. To ensure the complete (99.99%) removal of the Cu^{2+} from a solution in which $[Cu^{2+}]$ is initially about 1.0 M, will the potential of the cathode (versus SHE) have to be more negative or less negative than 0.34 V (the standard potential for $Cu^{2+} + 2\,e^- \rightarrow Cu$).

PROBLEMS

65. Suppose the current flowing from a battery is used to electroplate an object with silver. Calculate the mass of silver that would be deposited by a battery that delivers 1.7 ampere · hours of charge.
66. A battery charger used to recharge the NiMH batteries used in a digital camera can deliver as much as 0.50 A of current to each battery. If it takes 100. minutes to recharge one battery, how much $Ni(OH)_2$ (in grams) is oxidized to NiO(OH)?
67. A Ni-Cd battery containing 5.00 g of Cd and 4.10 g of NiO(OH) was 50% discharged when it was connected to a charger with an output of 2.00 A at 1.3 V. How long does it take to recharge the battery?
68. How long does it take to deposit a 1.00-μm-thick coating of gold on a disk-shaped medallion 4.0 cm in diameter and 2.0 mm thick at a constant current of 85. A? The density of gold is 19.3 g/cm^3.
69. Nuclear submarines can stay under water nearly indefinitely because they can produce their own oxygen by the electrolysis of water.
    a. How many liters of O_2 at STP (273 K and 1.00 atm) are produced in 1 hour in an electrolysis cell operating at a current of 0.025 A?
    b. Could seawater be used as the source of oxygen in the electrolysis process? Explain why or why not.
70. In the electrolysis of water, how long will it take to produce 100. L of H_2 at STP (273 K and 1.00 atm) by using an electrolysis cell through which flows a current of 50. mA?
71. Calculate the minimum (least-negative) cathode potential (versus SHE) needed to begin electroplating nickel from 0.35 M Ni^{2+} onto a piece of iron. Assume that the overpotential for the reduction of Ni^{2+} to Ni metal is negligible.
72. What is the minimum (least-negative) cathode potential (versus SHE) needed to electroplate silver onto cutlery in a solution of Ag^+ and NH_3 in which, because most of the silver ions are present as the diammino complex, $Ag(NH_3)_2^+$, the concentration of free silver ions, $[Ag^+]$, is only 3.50×10^{-5} M?

Low-Emission Vehicles

CONCEPT REVIEW

73. What are the advantages of hybrid (gasoline engine–electric motor) power systems over all-electric systems? What are the disadvantages?
74. To make the refueling of fuel cells easier, several manufacturers offer converters that turn readily available hydrocarbon fuels—such as natural gas, propane, and even gasoline—into H_2 for the fuel cells and CO_2. Although vehicles with such power systems are not truly "zero emission," they still offer significant environmental benefits over vehicles powered by internal combustion engines. Name a few of them.

PROBLEMS

75. Fuel cells with molten alkali metal carbonates as electrolytes can use methane as a fuel. The methane is first converted into hydrogen in a two-step process:

    $$CH_4(g) + H_2O(g) \longrightarrow CO(g) + 3\,H_2(g)$$
    $$CO(g) + H_2O(g) \longrightarrow H_2(g) + CO_2(g)$$

    a. Assign oxidation numbers to carbon and hydrogen in the reactants and products.
    b. Using the standard free energy of formation values in Appendix 5, calculate the standard free-energy changes in the two reactions and the overall $\Delta G°$ for the formation of $H_2 + CO_2$ from methane and steam.
76. Molten carbonate fuel cells fueled with H_2 convert as much as 60% of the free energy released by the formation of water from H_2 and O_2 into electrical energy. Determine the quantity of electrical energy obtained from converting 1 mole of H_2 into $H_2O(l)$ in such a fuel cell.

18 Materials Chemistry
Past, Present, and Future

The Leonard P. Zakim Bunker Hill Bridge across the Charles River in Boston was the widest (56 m and ten lanes of traffic) cable-stayed bridge in the world when it opened for traffic in 2003. Each cable can support over 5 million kilograms. This strength is derived from the crystal structures formed by iron and other metals in the steel strands from which the cables are made.

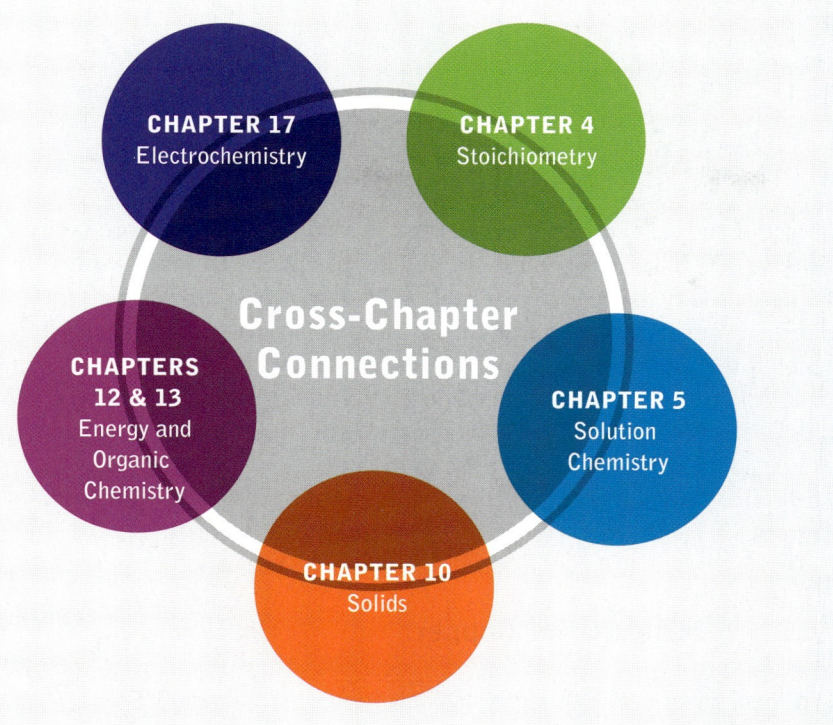

- 18.1 **Metals**
  - The Age of Copper
  - The Bronze Age
  - The Iron Revolution
  - Aluminum alloys: lightweight and high performance
- 18.2 **Ceramics**
  - Made of clay
  - Making ceramics
  - Superconducting ceramics
- 18.3 **Semiconductors**
- 18.4 **Fibers for Clothing and Other Uses**
  - Natural fibers
  - Synthetic polymers from condensation reactions
  - Synthetic polymers from addition reactions
- 18.5 **The Scientific Method—Revisited**

A Look Ahead

This chapter draws on many of the chemical concepts discussed in earlier chapters of the text as we explore the historical development of examples from three classes of materials: metals, ceramics, and fibers. We begin with the first metal tools, the rise of smelting metal ores, and production of the first alloys during the Bronze Age. We relate the properties of these and other alloys to their atomic structures. Then we track the development of ceramics, including the chemical changes that occur when materials made of clay are fired in kilns, and we examine why most ceramics are insulators and a few are superconductors. We compare the molecular structures and properties of fibers made from natural polymers (cellulose and proteins) with those made from totally synthetic polymers, and end with a discussion of the discovery of Teflon and a lesson about the dynamics of scientific inquiry.

18.1 METALS

Most of the elements in the periodic table are metals. They are typically hard, shiny, malleable (easily shaped), ductile (easily drawn out) conductors of heat and electricity. However, only gold, silver, and copper are found as free metals in Earth's crust. Of these three metals only gold is found principally as the free metal; the principal silver ore is argentite (Ag_2S) and copper occurs mostly in chalcopyrite, $CuFeS_2$, chalcocite, Cu_2S, and malachite, $Cu_2(CO_3)(OH)_2$. Most metals have relatively low ionization energies and they form stable complex ions and ionic compounds. Thus, they are more likely to occur in Earth's crust and hydrosphere as cations surrounded by anions and neutral ligands, and not as free atoms.

CONNECTION: Malleable materials can be hammered or rolled into thin sheets without breaking; ductile materials can be drawn out into thin wires (see Section 1.1).

The Age of Copper

Archeological evidence suggests that a major technological advance occurred around 6000 years ago (or, as archaeologists would say, 6000 bp *(before present)*) when artisans in Mesopotamia and the Indus River valley learned how to convert copper ores, such as $CuFeS_2$, to copper metal (Figure 18.1). The ores were crushed and then heated in air, producing copper(II) oxide. The CuO was reduced to metallic Cu in a process called **smelting** by reacting it with carbon monoxide formed by burning wood or charcoal in a furnace with an insufficient supply of air.

$$CuO + CO \longrightarrow Cu + CO_2 \tag{18.1}$$

Smelting is the high temperature reduction of metallic oxides to form free metals.

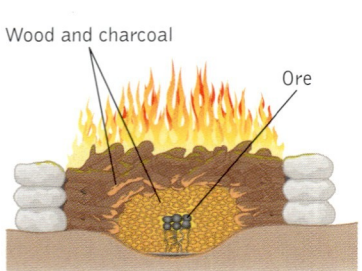

A A primitive furnace

B Bronze Age furnace

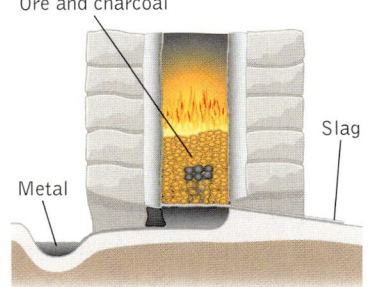

C A Roman furnace

The discovery that copper could be extracted from copper minerals about 6000 years ago led to the development of furnaces designed for this purpose. The key to their operation was incomplete combustion of wood and charcoal, producing CO that reacted with CuO, forming CO_2 and Cu. Bellows were used to control the ratio of fuel to air in Bronze Age furnaces. In a Roman design dense, molten copper metal flowed from the bottom of the furnace and was separated from less dense, unreacted minerals and solid by-products of the smelting process, called *slag*.

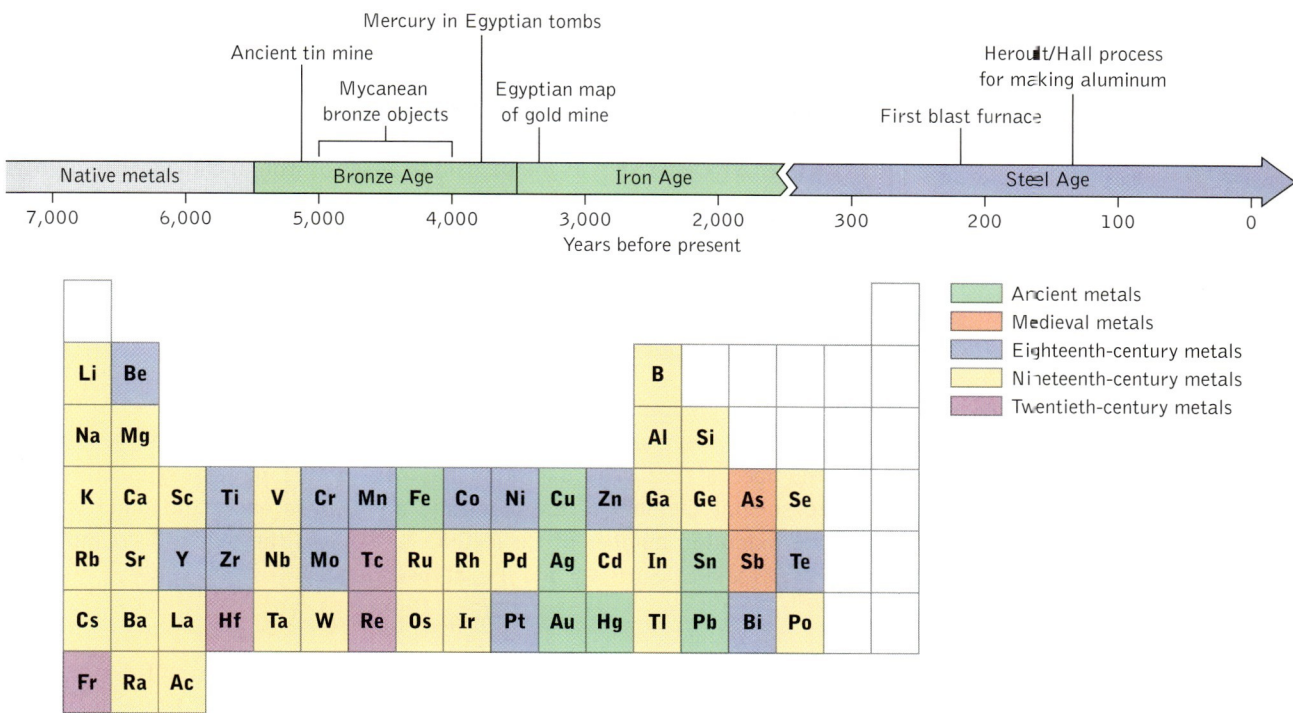

FIGURE 18.1 For thousands of years humans have searched for new materials from which to craft tools and weapons and to build shelters. This search has led to the discovery and processing of many of the metals in the periodic table. Copper was the first metal to be refined from a mineral and the first to be blended with other metals into alloys. Iron- and aluminum-based alloys are the dominant industrial metals today.

SAMPLE EXERCISE 18.1: The principal ores of copper are chalcopyrite, $CuFeS_2$, chalcocite, Cu_2S, and malachite, $Cu_2(CO_3)(OH)_2$. Which ore is richest, that is, which contains the greatest mass percent of copper?

SOLUTION: We need to calculate the mass of copper in a given mass of each of the minerals. Starting with chalcopyrite, 1 mole of $CuFeS_2$, which has a molar mass of 183.44 g/mol, contains 1 mole of Cu (63.456 g). The mass percent of copper is the molar mass of Cu divided by the molar mass of $CuFeS_2$ and multiplied by 100:

$$\left(\frac{63.46 \text{ g Cu}}{183.44 \text{ g CuFeS}_2}\right)(100\%) = 34.46\% \text{ Cu in CuFeS}_2$$

Chalcocite and malachite contain 2 moles of copper per mole of mineral, so their mass percent Cu values are:

$$(2 \text{ mol Cu})\left(\frac{63.46 \text{ g Cu/mol Cu}}{159.16 \text{ g Cu}_2\text{S}}\right)(100\%) = 79.74\% \text{ Cu in Cu}_2\text{S}$$

Copper artifacts, including these axe heads from an excavation in southern Jordan were made in a metal factory that operated about 4500 years ago.

CONNECTION: See Section 4.5 for a review of how to calculate percent composition from a chemical formula.

and

$$(2 \; \cancel{\text{mol Cu}}) \left(\frac{63.46 \text{ g Cu}/\cancel{\text{mol Cu}}}{220.94 \text{ g Cu}_2(\text{CO}_3)(\text{OH})_2} \right) (100\%) = 57.44\% \text{ Cu in Cu}_2(\text{CO}_3)(\text{OH})_2$$

Thus, Cu_2S has the greatest mass percent of Cu.

PRACTICE EXERCISE: Is the percent by mass of iron in chalcopyrite greater than or less than the percent by mass of copper? See Problems 5 and 6.

SAMPLE EXERCISE 18.2: Assign oxidation numbers to the elements in the reactants and products in Equation 18.1 and identify the oxidizing and reducing agents. How many electrons are transferred in the process?

SOLUTION: Any free element, such as Cu, has an oxidation number of 0. Oxygen is the more electronegative element in CuO and CO_2, and is assigned an oxidation number of -2. The oxidation numbers of the remaining elements are assigned so that the sum of the oxidation numbers in each compound is zero. Thus, the oxidation number of Cu in CuO is $+2$, and the oxidation numbers of carbon in CO and CO_2 are $+2$ and $+4$, respectively:

CONNECTION: The rules for calculating oxidation numbers are discussed in Section 5.5. Keep in mind that the oxidation number of oxygen in most of its compounds (peroxides are notable exceptions) is -2.

$$\overset{+2 \; -2}{\text{CuO}} + \overset{+2 \; -2}{\text{CO}} \longrightarrow \overset{0}{\text{Cu}} + \overset{+4 \; -2}{\text{CO}_2}$$

The oxidation number of Cu in CuO is reduced during the reaction from $+2$ to 0 as the carbon in CO is oxidized. Because carbon is oxidized, carbon monoxide is the reducing agent in this reaction, and in other reactions (see Equation 18.2) in which metals are formed from ores of their oxides.

PRACTICE EXERCISES: Roasting Cu_2S in air to form CuO is also a redox reaction. Write a balanced chemical equation for the reaction (you will need to predict the source of O and the final form of S) and assign oxidation numbers to the elements in all the reactants and products (see Problems 10 and 11).

CONNECTION: Remember that the oxidizing agent is reduced in a redox reaction, and the reducing agent is oxidized (see Section 17.4).

The copper produced in ancient kilns formed at temperatures above its melting point. As this molten copper cooled it crystallized in a face-centered cubic (fcc) array of atoms (Figure 18.2). During the Age of Copper, molten Cu was poured into molds to form cast Cu tools and weapons. Because of copper's considerable malleability, these casts could be further shaped and their edges sharpened by hammering.

CONNECTION: All the Group 1B metals (Cu, Ag, and Au) form fcc crystal structures (Section 10.6).

Why are copper and other metals malleable? One explanation is derived from the model of metallic bonding in which

> **Why are copper and other metals malleable?**

atoms of metals are bonded to each other by a "sea" of electrons (see Section 10.6). The atoms are able to slide by each other as shown in Figure 18.3. Isolated defects or dislocations in the array of atoms (Figure 18.4) simply migrate through the crystal when it is bent or stretched. One way to strengthen a metal against bending is to deliberately increase the number of dislocations in the crystal by a process called **work-hardening.** Work-hardening involves repetitive

✓ **Work-hardening** produces imperfections and dislocations in metallic structures that reduce the ease with which atoms can slide by each other, making the metal stiffer and harder.

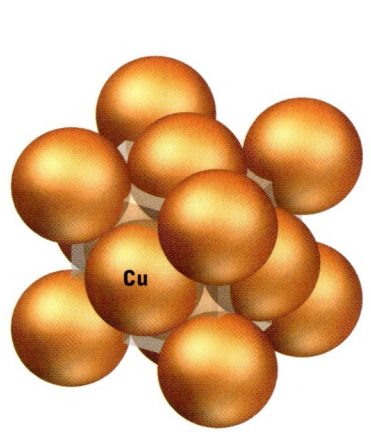

FIGURE 18.2 Copper and the other Group 1B metals form face-centered cubic crystal structures when they solidify from the molten state.

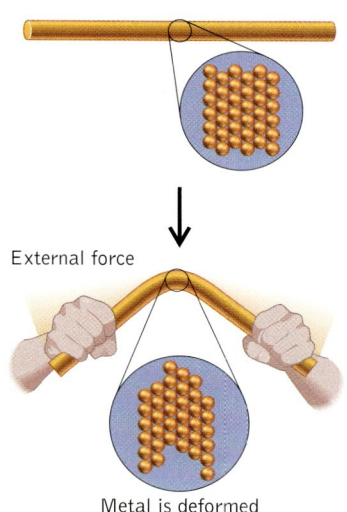

FIGURE 18.3 The "sea of electrons" model of metallic bonding provides an atom-scale view of why metals are malleable and ductile: rows of their atoms are able to slide by each other when subjected to an external force.

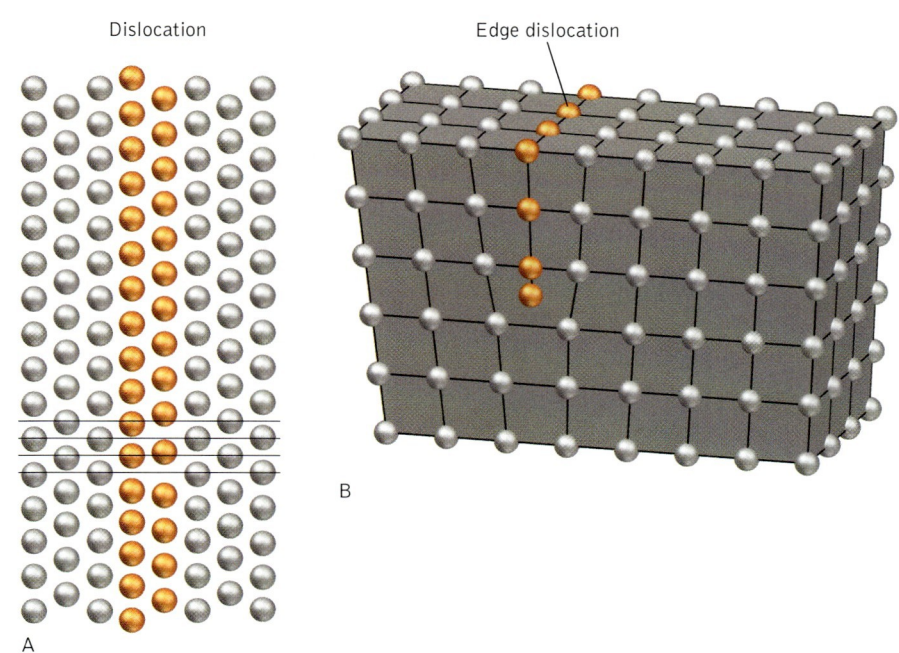

FIGURE 18.4 Real crystals are never perfect. Among the imperfections are dislocations illustrated by the misaligned atoms (gold spheres) shown here. If these imperfections are isolated, they can migrate as the metal is bent or drawn out.

THE CHEMISTRY OF THE GROUP 1B ELEMENTS

The Group 1B elements, copper, silver, and gold, collectively called the coinage metals, are probably the most familiar group of metallic elements in the periodic table. They are, of course, widely used in the manufacture of jewelry, coins, and as repositories of wealth. The search for gold (and wealth) inspired the unsuccessful efforts of medieval alchemists to change cheaper metals into gold—an obsession that contributed to the development of the chemical sciences. The Group 1B metals are the only three metals found as free elements in Earth's crust. Copper (as the free metal and in ores) is the most abundant of the group, with a crustal abundance of 68 ppm (about the same as nickel). Silver and gold are much more rare with abundances of only 0.08 and 0.004 ppm, respectively.

The process for smelting copper from copper minerals is described on page 894. It is based on conversion of copper ore to CuO and then reduction of CuO to Cu metal with carbon monoxide. Silver often occurs as a minor ingredient in the ores of Cu, Pb, and Zn, and is recovered when these ores are smelted and refined. Elemental silver and gold may be extracted from crushed rock by reaction with cyanide, CN^-, followed by reduction with zinc. The reactions for gold may be written:

$$4\ Au(s) + 8\ CN^-(aq) + O_2(g) + 2\ H_2O(l) \longrightarrow 4\ Au(CN)_2^-(aq) + 4\ OH^-(aq)$$

$$2\ Au(CN)_2^-(aq) + Zn(s) \longrightarrow 2\ Au(s) + Zn(CN)_4^{2-}(aq)$$

The valence shells of atoms of Cu, Ag, and Au have the electron configuration $d^{10}s^1$. Atoms of all three metals form ions with a charge of +1 by losing their single s electrons. In this respect the metals of Group 1B resemble the metals of Group 1A. However, the Group 1B metals do not react rapidly with air, water, or mineral acids (HCl, HBr). Copper and silver do dissolve in oxidizing acids, such as nitric acid:

$$Cu(s) + 4\ HNO_3(aq) \longrightarrow Cu(NO_3)_2(aq) + 2\ H_2O(l) + 2\ NO_2(g)$$

Gold dissolves in HNO_3 only when three volumes of concentrated HCl are added to each volume of concentrated HNO_3, forming a highly corrosive solution known as *aqua regia*:

$$Au(s) + NO_3^-(aq) + 4\ H^+(aq) + 4\ Cl^-(aq) \longrightarrow AuCl_4^-(aq) + 2\ H_2O(l) + NO(g)$$

All the +1 ions of the Group 1B metals form unsoluble halides. When Cu^+ ions are present in solution, they readily undergo disproportionation, simultaneously un-

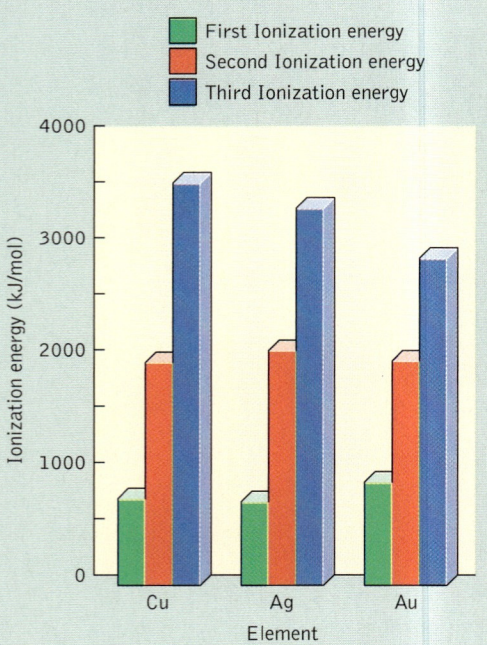

The third ionization energies of the Group 1B elements decrease with increasing atomic number, but the first and second ionization energies do not.

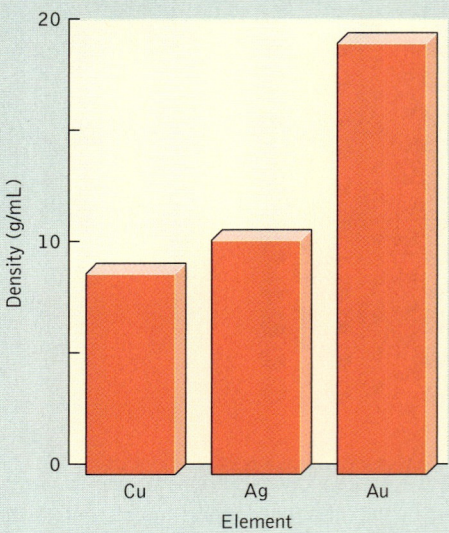

The densities of the Group 1B elements increase with increasing atomic number.

dergoing oxidation and reduction to form Cu^{2+} and elemental Cu so that, at equilibrium:

$$2\,Cu^+(aq) \rightleftharpoons Cu^{2+}(aq) + Cu(s) \quad K = 1.2 \times 10^6$$

Based on the large equilibrium constant for this reaction, we might expect that the principal oxidation state of Cu in nature should be +2. That expectation is supported by the composition of the principal Cu-containing minerals: chalcopyrite ($CuFeS_2$) and malachite ($Cu_2(CO_3)(OH)_2$).

However, copper minerals formed under strongly reducing conditions (which tend to be rich in sulfide) contain Cu(I) as in chalcocite, (Cu_2S).

The Group 1B elements are widely used in industry. About 55% of the copper mined each year is used to make wire and other electrical components and about 15% is used to make pipes and other plumbing fixtures. Silver metal has many uses. In addition to making jewelry it is also used in electrical components and in dental amalgams. Silver halides, particularly AgBr, are widely used in photography. Light oxidizes the Br^- ion in the AgBr in photographic film and reduces Ag^+ to tiny particles of Ag metal. When the film is developed, additional AgBr near the particles of Ag metal are reduced to Ag, forming the darkest image where the most light reached the film. Unreacted AgBr in the film is dissolved in sodium thiosulfate solution:

$$AgBr(s) + 2\,S_2O_3^{2-}(aq) \longrightarrow Ag(S_2O_3)_2^{3-}(aq) - Br^-(aq)$$

Removing unreacted AgBr "fixes" the original negative image on the film. The fixing process works because AgBr, and the other silver halides, which are only slightly soluble in water, readily dissolve in solutions of sodium thiosulfate. This solubility is linked to the considerable stability of the $Ag(S_2O_3)_2^{3-}$ complex:

$$Ag^+(aq) + 2\,S_2O_3^{2-}(aq) \longrightarrow Ag(S_2O_3)_2^{3-}(aq) \quad K_c = 5 \times 10^{13}$$

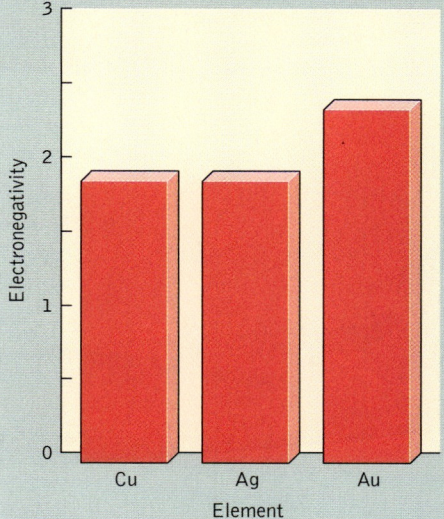

The electronegativities of the Group 1B elements are high for metallic elements.

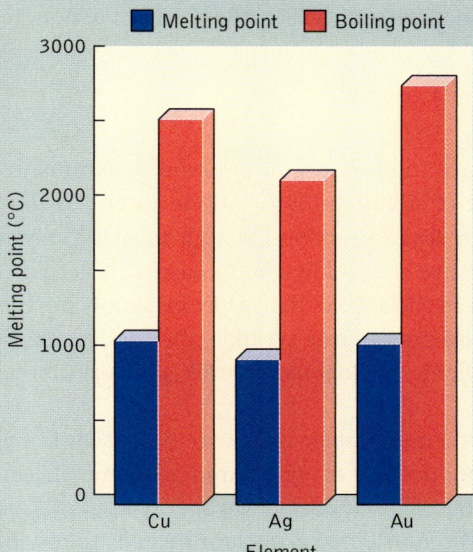

Silver has the lowest melting and boiling points of all the Group 1B elements.

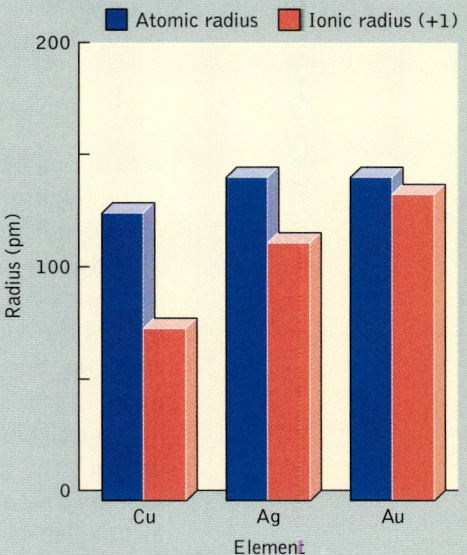

The radii of the atoms and +1 ions of the Group 1B elements increase with increasing atomic number.

A Single dislocation

B Intersecting (pinned) dislocations

FIGURE 18.5 Work-hardening is a process that increases the number of dislocations in a metal, rendering it more resistant to bending. A. The effect of work-hardening can be illustrated by trying to remove wrinkles from a carpet. A single wrinkle can be easily moved to one side of the carpet and eliminated. B. Two or more intersecting wrinkles are much more difficult to move, just as many intersecting dislocations make it difficult to bend a work-hardened metal.

bending or hammering to introduce intersecting or *pinned* dislocations in the metal (Figure 18.5). The presence of additional dislocations entangles the line defects making it more difficult for them to move through the bulk material. This situation is like trying to remove a wrinkle from a carpet. It is relatively easy to move a single wrinkle toward the edge of the carpet by simply pushing it toward the edge. If there is a second intersecting wrinkle, the task of removing either one becomes much more difficult, analogous to bending a work-hardened metal.

Copper has long been valued as a material for cookware because it is a good conductor of heat. The thermal conductivity of copper and other metals can be explained on an atomic level in terms of bands of electron energy levels discussed in Chapter 10. When atoms on the outside of a pot are heated, they vibrate faster. Electrons absorb this thermal energy and populate empty or partially

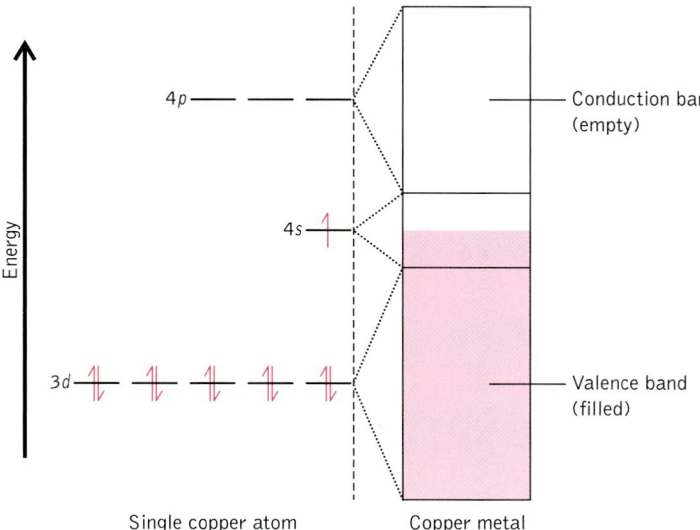

FIGURE 18.6 The high thermal conductivity of copper (and other metals) results from filled valence and unfilled conduction with energies that are close to each other or even overlap.

filled, higher energy orbitals in conduction bands (Figure 18.6). Thermal energy in the form of electrons in high-energy conduction bands can move rapidly through a copper pot from the source of heat to the contents of the pot.

The Bronze Age

One of the disadvantages of making tools and weapons from copper is that the metal is easily bent out of shape. However, about 5500 years ago people living around the Aegean Sea discovered, probably by accident, that mixing tin with copper produced a new material that was both easier to cast and much stronger. This discovery ushered in the Bronze Age, which lasted some 3000 years. Copper/tin bronzes are examples of **homogeneous alloys**: uniform mixtures of two or more elements with metallic properties. Tin is miscible with copper in all proportions, forming homogeneous solid solutions of the two elements. At the concentrations of tin used in Bronze Age tools (< 30% by mass), the tin atoms substitute for copper atoms in the face-centered cubic structure of copper metal creating a **substitutional alloy** (Figure 18.7). When the proportion of tin is greater than 40%, the structure of the alloy changes to a hexagonal close-packed arrangement of the larger tin atoms with the smaller copper atoms occupying octahedral holes.

Substitutional alloys of the general formula $A_x B_{(1-x)}$, where x can have a value from 0 to 1, may occur when metals have the same crystal structure and atomic radii that are within 15% of each other. The radii of copper and tin atoms are 128 and 140 pm, respectively, a difference of about 9%. The larger size of the tin atom makes it more difficult for the planes of copper atoms to slip past one another

CONNECTION: The mobility of electrons in partially filled energy bands also accounts for the high electrical conductivity of copper (see Section 10.6).

 A **homogeneous alloy** is a solid solution of two or more metals. The solution also has metallic properties.

A **substitutional alloy** is a homogeneous alloy of metals with similar atomic radii (within 15%). Atoms of the minor components substitute for atoms of the major component in the crystal structure of the major component.

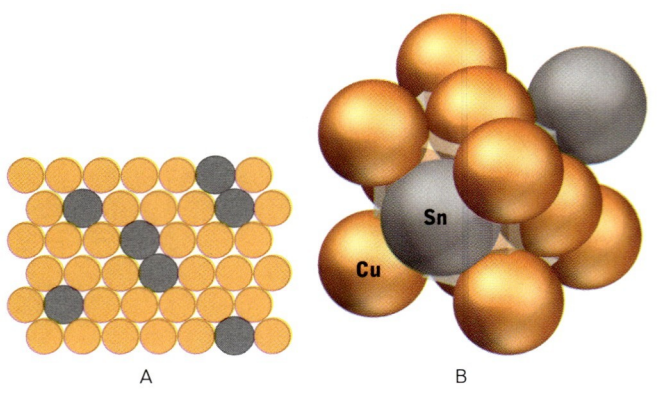

FIGURE 18.7 Substitutional alloys are formed when the atoms of one metal replace those of a host metal while retaining the crystal structure of the host: (A) a two-dimensional and a (B) three-dimensional view of a bronze alloy made of copper (Cu) and tin (Sn).

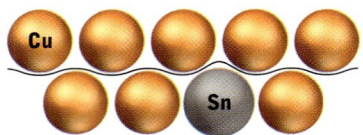

FIGURE 18.8 The larger Sn atoms in bronze create dislocations in the crystal structure of Cu and ripples in the slip plane (wavy line), making it more difficult for Cu atoms to slide by each other when an external force is applied. This is an atomic view of why bronze tools and weapons are less easily bent than those made of copper.

CONNECTION: The sizes, numbers, and locations of octahedral holes in a face-centered cubic crystal structure are described in Section 10.6.

CONNECTION: The photosynthetic incorporation of $^{14}CO_2$ into living plants and the decay of ^{14}C in plant tissue is described in Section 2.8.

(Figure 18.8). Thus bronze tools, utensils and weapons are less malleable than those made of pure copper—less malleable, but harder and stronger.

Ancient metalworkers needed tin to make bronze objects. The source was usually the mineral cassiterite, SnO_2, though the locations of the ancient cassiterite mines long remained a mystery. However, in 1989 archaeologists discovered an ancient mine in the Taurus Mountains of Turkey that they suspected dated from the Bronze Age. To test their hypothesis they analyzed charred wood in ashes from the mine using radiocarbon dating techniques. They found that the ^{14}C radioactivity of the sample was 53.6% of that in modern charred wood. These values can be used to determine the age of the ashes with Equation 1.7:

$$A_t/A_0 = (0.5)^n$$

where A_t is the activity of the ^{14}C in the ash, A_0 is the initial (assumed to be the same as modern) activity of ^{14}C in ash, and n is the number of half-lives. Taking the natural logarithm of both sides we obtain:

$$\ln (A_t/A_0) = n \ln (0.5)$$

and substituting the analytical results:

$$\ln (0.536) = -0.693\, n$$

$$n = 0.901 \text{ half-lives}$$

The age of the ashes in years equals the product of the number of half-lives times the years per half-life or:

$$n \times t_{1/2} = (0.901 \text{ half-lives})(5.73 \times 10^3 \text{ yr/half-life}) = 5.16 \times 10^3 \text{ yr}$$

This age is near the beginning of the period known as the Bronze Age (about 5500 bp).

Bronze alloys are relatively easy to cast because the melting points of Cu/Sn alloys are lower than the melting point of pure copper. For example, the melting point of a Cu/Sn alloy that is 25% by weight Sn is 798 °C while that of pure Cu is 1085 °C.

Why is the melting point of the Cu alloy lower than that of pure Cu? The answer is related to the perturbations in the copper lattice produced by presence of the slightly larger tin atoms—perturbations that can weaken the metallic bonds that hold together the atoms in the lattice. Weakened interaction between atoms in the solid phase make it easier to convert the solid to a liquid. As a result, the temperatures at which many alloys melt are lower than the melting points of the principal elements.

> Why is the melting point of a Cu alloy lower than that of pure Cu?

SAMPLE EXERCISE 18.3: Sterling silver is an alloy that is 93% Ag and 7% Cu. The presence of Cu inhibits tarnishing, but also affects the temperature at which the alloy softens and begins to melt. The melting points of pure Ag and pure Cu are 962 and 1085 °C, respectively. Does sterling silver melt:
 (a) above 1085 °C, (b) between 962 and 1085 °C, or (c) below 962 °C?

SOLUTION: Atoms of Cu (atomic radius = 128 pm—see Appendix 3) fit into the locations normally occupied by Ag atom (atomic radius = 144 pm) with room to spare. The resulting lack of perfect fit of atoms in the crystal structure will mean slightly weaker metallic bonds than in pure Ag and so the alloy should (and does) melt below 962 °C.

DISCUSSION: For the same reason that copper alloys of silver melt at lower temperatures than pure silver, adding silver to a copper reduces the melting point of copper. As you might suspect, there is a particular alloy of silver and copper that has the lowest melting point of all copper/silver alloys. It is 28% Cu, 72% Ag, and melts at 779 °C.

PRACTICE EXERCISE: A soft solder that is 70.0% Sn and 30.0% Pb melts:
 a. Below the melting point of pure Sn (232 °C).
 b. Above the melting point of pure Sn (232 °C) but below that of pure Pb (327 °C).
 c. Above the melting point of pure Pb (327 °C). (See Concept Review 18.)

Other substitutional alloys based on tin and copper had been developed by the time of the Roman Empire and remain important materials today. For example, zinc alloyed with copper forms brass. Pewter was originally a tin/lead alloy. However, the pewter used in dishes today contains about 90% tin alloyed with copper and antimony. Copper and tin can also form **intermetallic compounds,** materials with constant compositions and structures different from those of the

> ✓ An **intermetallic compound** is a homogeneous alloy of fixed proportions and characteristic physical and chemical properties.

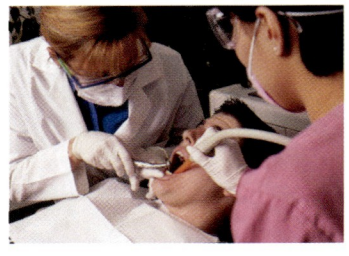

Dental amalgams are mercury-based alloys that may release minute quantities of mercury. The health effects of these releases became the source of considerable debate in the 1990s. The position of the U.S. Food and Drug Administration (FDA) and the American Dental Association is that health risks are not significant for most people, though the FDA was considering new regulations on the use of dental amalgams in late 2002.

parent elements. For example, Cu_3Sn (38.4% Sn) is a crystallene solid hexagonal closest-packed arrangement of copper and tin atoms.

CONCEPT TEST: The material dentists use to fill most cavities is an alloy called an amalgam. It contains approximately 50% mercury, 35% silver, 9% tin, 6% copper, and a trace of zinc. Based on the atomic radii of these elements (see Appendix 3) predict whether an alloy containing these proportions of these metals could be a substitutional alloy.

The Iron Revolution

The next major advance in metallurgy occurred around 3100 years ago with the discovery that iron oxides could be reduced to iron metal in wood or charcoal fires with limited air supplies. As in the smelting of CuO, the reducing agent is carbon monoxide:

$$Fe_2O_3(s) + 3\ CO(g) \rightarrow 2\ Fe(s) + 3\ CO_2(s) \tag{18.2}$$

Smelting iron required higher temperatures than those needed to reduce copper or tin compounds, but once ancient metalworkers had developed the iron smelting process, iron quickly replaced bronze as the metal of choice for fabricating tools and weapons. There were two reasons for this: (1) iron is much more abundant than copper and tin in Earth's crust (see Section 4.5), and (2) tools and weapons made of iron and iron-containing alloys are much stronger than those made of bronze.

The high melting point of iron (1538 °C) prevented early metalworkers from casting iron objects as they could copper and its alloys. However, they could hammer hot pieces of crude iron into useful shapes. Implements fashioned in this way are said to be made of *wrought* iron.

The temperatures in the first iron forges probably did not exceed 1300 °C. However, tinkering with the design of the forges enabled early ironworkers to increase these temperatures. Above 1300 °C, carbon begins to dissolve in iron producing a lower melting iron alloy that could be cast into shapes using technologies developed in the Bronze Age. This material is called *cast* iron. Cast iron is a brittle material that cracks easily. Still, the ability to cast iron made it an attractive material for manufacturing a range of goods from cannonballs to stoves.

Today the reduction of iron ore is done in blast furnaces: enormous reaction vessels (some more than 50 m tall) that operate continuously at about 1600 °C. Iron ore, hot carbon (coke), and limestone are added to the top of the vessel and molten iron and solid by-products (called slag) are harvested from the bottom. Blast furnaces get their name from the blasts of preheated air (about 1000 °C) that are injected through nozzles near the bottom and that suspend the reactants until the iron reduction process is complete. It may take as long as 8 hrs for the reactants to fall to the bottom of a blast furnace. On their way down, O_2 in the hot air partially oxidizes coke to carbon monoxide, and then CO reduces the iron in iron ore as described in Equation 18.2. At the temperatures in the furnace, limestone, $CaCO_3$, decomposes to CaO:

$$CaCO_3(s) \longrightarrow CaO(s) + CO_2(s)$$

Calcium oxide reacts with silica impurities in the ore forming calcium silicate:

$$CaO(s) + SiO_2(s) \longrightarrow CaSiO_3(l)$$

This calcium silicate becomes part of the slag that floats on the more dense molten iron at the bottom of the furnace and is drawn off after the iron has been harvested. The largest blast furnaces operating continuously produce 10,000–15,000 tons of iron a week.

Molten iron produced in a blast furnace may contain up to 5% carbon. To remove the carbon, the molten iron is transferred to a second furnace into which hot O_2 and additional CaO are injected. In this *basic oxygen process* the carbon is oxidized to CO_2, and any remaining silicon impurities form a slag of $CaSiO_3$. Sulfur and phosphorus impurities in the fuels used in the blast furnace also end up in the slag. If a chemical analysis indicates that it is warranted, the process is repeated, and then the vessel is tipped on its side to tap the refined molten iron below the slag level. Then the vessel is tilted the other way and the slag is removed. Worldwide, about 60% of the iron used to make steel is refined using the basic oxygen process.

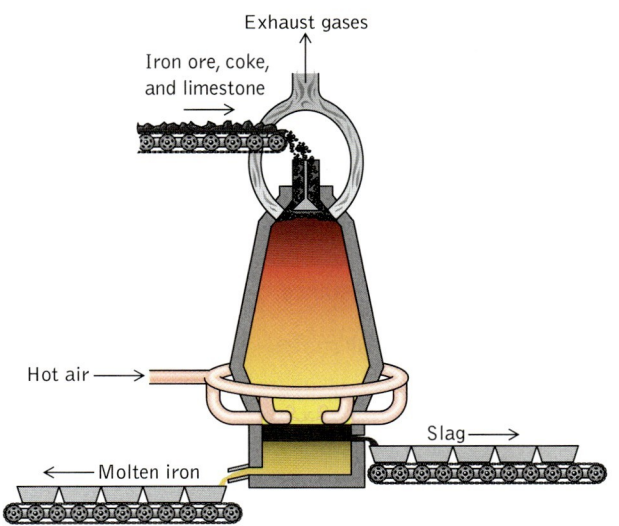

A Blast furnace

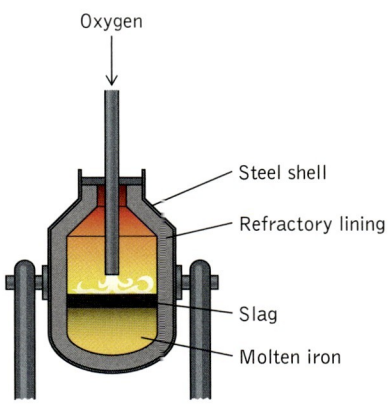

B Basic oxygen process

A. Blast furnaces operate continuously at temperatures near 1600 °C, converting iron ore into iron at rates as high as 2000 tons of iron per day. B. Impurities in molten iron are removed in the basic oxygen process. Injected O_2 converts C to CO_2 and lime is added to react with Si and P impurities, which become part of the slag layer. The reaction vessel pivots on its side arms allowing the purified iron to be tapped off below the slag layer.

> **Interstitial alloys** are homogeneous alloys in which the atoms of a much smaller element fit into the holes (interstices) of the crystal structure of the host metal. In steel, carbon atoms are located in the holes of the crystal structure of iron.

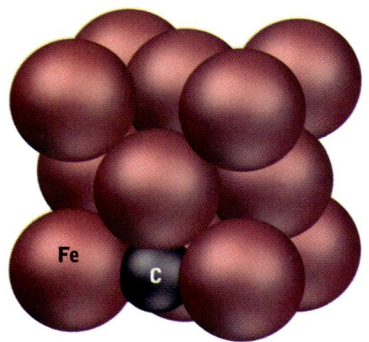

FIGURE 18.9 Steel is an interstitial alloy of carbon in iron. The face-centered cubic form of iron (austentite) that forms at high temperatures can accommodate carbon atoms in its octahedral holes. The ratio of the atomic radii of carbon to iron (77 pm/124 pm = 0.62) is within the range (0.44–0.73—see Table 10.2) in which a smaller atom fits into the octahedral holes of the face-centered cubic crystal of a larger atom.

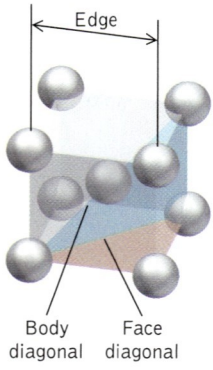

Bcc unit cell

When molten iron cools to its melting point of 1538 °C, it crystallizes in a face-centered cubic structure called *austentite*. With further cooling the structure changes into a body-centered cubic structure called *ferrite*. The holes between the iron atoms in either crystal structure can accommodate carbon atoms. Iron containing carbon atoms in such holes is an example of an **interstitial alloy** because the carbon atoms occupy spaces or *interstices* between the iron atoms (Figure 18.9). Austentite can dissolve up to 2% carbon at 1150 °C, but ferrite can dissolve only a few hundredths of a percent carbon. As austentite cools and turns into ferrite, carbon that was soluble in austentite, but not in ferrite, precipitates as either elemental carbon, or it forms iron carbide, Fe_3C. Excess carbon and Fe_3C create dislocations in the crystal structure of Fe, pinning the Fe atoms around their precipitation points. This pinning hardens the material, but also distorts the iron lattice, making it brittle.

SAMPLE EXERCISE 18.4: Compare the sizes of the holes in body-centered cubic (bcc) and face-centered cubic (fcc) crystalline structures of Fe to the size of carbon atoms. Use these comparisons to explain the greater solubility of carbon in austentite (fcc iron). The atomic radii of iron and carbon are 124 and 77 pm, respectively.

SOLUTION: If atoms of one element are to fit into the octahedral holes of an fcc crystal of another element, the ratio of the radii of the intercalating to host atoms must be between 0.414 and 0.732 (see Table 10.2). The corresponding range for fitting into tetrahedral holes is 0.225–0.414. In the case of carbon atoms in iron structures, r_C/r_{Fe} = 77 pm/124 pm = 0.62. This ratio indicates that carbon should fit into octahedral holes but not into tetrahedral holes.

A bcc unit cell of Fe is shown below. The unit cell contains distorted octahedral holes located in the center of each face. The central atom from an adjacent unit cell completes the octahedral hole.

The dimensions of the distorted octahedral hole are different in different directions. The gap between the center atoms in adjacent unit cells is one of the dimensions. The gap distance equals the edge length of the bcc unit cell minus twice the radius of an iron atom. Let's call this distance "(hole dimension)$_1$":

$$\text{(hole dimension)}_1 = \text{edge} - 2r_{Fe} \quad (1)$$

The other key dimension is the length of a diagonal line across a face of the unit cell minus twice the radius of the iron atoms at the ends of the diagonal. We'll label it "(hole dimension)$_2$":

$$\text{(hole dimension)}_2 = \text{face diagonal} - 2r_{Fe} \quad (2)$$

The edge and face diagonal distances are related to each other by the Pythagorean theorem because two adjacent edges and a diagonal form a right triangle. According to Pythagoras:

$$(\text{face diagonal})^2 = 2\,(\text{edge})^2 \quad (3)$$

Equations (1), (2), and (3) together tell us that (hole dimension)$_1$ is related to the edge distance in the same way (hole dimension)$_2$ is related to the face diagonal. Also, the face diagonal is longer than the edge distance. Therefore, (hole dimension)$_2$ is greater than (hole dimension)$_1$. This makes (hole dimension)$_1$ the limiting factor in controlling the size of an interstitial atom or ion. To determine the actual length of (hole dimension)$_1$ we need to express the edge distance in terms of a parameter we know, namely, r_{Fe}. This requires some more geometry. In a bcc unit cell, the Fe atoms touch along the body diagonal, which means that this diagonal distance is equal to twice the diameter of an Fe atom, or $4r_{Fe}$. Consider a right triangle composed of this diagonal, a face diagonal, and an edge.

Applying the Pythagorean theorem we have:

$$(4r_{Fe})^2 = (edge)^2 + (face\ diagonal)^2 \quad (4)$$

Substituting Equation 4 into Equation 3:

$$(4r_{Fe})^2 = (edge)^2 + (face\ diagonal)^2$$
$$= (edge)^2 + 2\ (edge)^2 + 3\ (edge)^2$$

and so:

$$edge = \frac{4r_{Fe}}{\sqrt{3}} = 2.3094\ r_{Fe}$$

Therefore, the size of (hole dimension)$_1$ is:

$$(hole\ dimension)_1 = edge - 2r_{Fe}$$
$$= (2.3094 - 2.0000)r_{Fe} = 0.3094\ r_{Fe}$$

To fit into such a hole the radius of an atom would have to be less than half this dimension, or:

$$r < \left(\frac{1}{2} \times 0.3094\ r_{Fe}\right) = \left(\frac{1}{2} \times 0.3094 \times 126\ pm\right)$$

$$< 19.5\ pm$$

However, the radius of carbon atoms is 77 pm, and so they will not fit into the octahedral holes of the bcc structure of Fe.

PRACTICE EXERCISE: Materials made of iron and steel are sometimes treated using a process known as "plasma nitriding" in which nitrogen atoms are incorporated into the interstices between iron atoms. In which type of hole—tetrahedral, octahedral, or bcc—would you expect the nitrogen in Fe$_4$N to be found? Given: $r_N = 75$ pm. (See Problems 25 and 26.)

Since ancient times metalworkers have known that repeated heating followed by hammering and rapid quenching in water dramatically increases the hardness and strength of wrought iron. We now know that the improved mechanical properties are due to the concentration of carbon at the iron surface. Also, implements manufactured from certain iron ores that contained small amounts of manganese, aluminum, and other elements give superior performance compared to those made of higher-purity iron. These observations led to the development of the iron alloys known collectively as steel. Steel is the principal ingredient in commercial and industrial construction and in building automobiles, ships, trains, and the rails trains run on. Worldwide production of steel rose dramatically during the twentieth century (Figure 18.10).

There are three major categories of modern steels:

- carbon steel (up to 3% C; 90% of all steels are carbon steels)
- low-alloy steel (up to 8% alloying elements)
- high-alloy steel (>8% alloying elements)

Carbon steels contain from about 0.05% to 3% carbon and up to 2% other alloying elements. Small differences in carbon content produce large differences in the properties of the steel as indicated by the information in Table 18.1.

In general, the higher the carbon concentration, the more iron carbide, Fe_3C, forms during cooling and the stronger the steel. For example, a steel containing 0.8% carbon can withstand three times more pressure than a steel with 0.05%

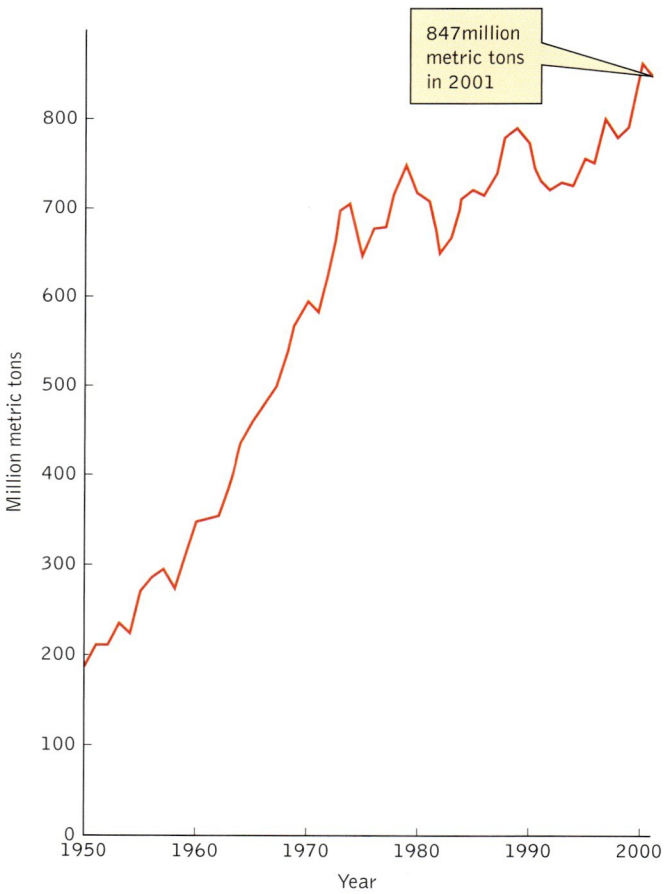

FIGURE 18.10 Worldwide steel production increased from less than 200 million metric tons to nearly 850 million metric tons over the last half of the twentieth century. *Source:* International Iron and Steel Institute.

TABLE 18.1

Carbon Content (%)	Designation	Properties	Used to Make
0.05–0.19	Low-carbon	Malleable, ductile	Nails, cables
0.20–0.49	Medium-carbon	High-strength	Construction girders
0.5–3.0	High-carbon	Hard but brittle	Cutting tools

carbon. This increased strength and hardness comes at the cost of increased brittleness. The same 0.05% carbon steel can be stretched or bent over three times more than 0.8% carbon steel.

Iron alloys that contain chromium (up to 25%) and nickel (up to 20%) are corrosion-resistant **stainless steels.** A common stainless steel, called "302," contains 0.08% carbon, 8% nickel, and 18% chromium. Stainless steel is less likely to corrode because the added chromium and nickel oxidize more easily than iron. Surface atoms of chromium and nickel form a protective coating of chromium and nickel oxides that inhibit further oxidation of the material.

> **Stainless steel** is a class of alloys of iron that are corrosion-resistant because of the presence of at least 12% chromium in their structures. Many stainless steels also contain other alloy elements, such as nickel, to enhance corrosion resistance or to provide other desired properties.

Aluminum alloys: lightweight and high-performance

The need for strong, lightweight materials—particularly for the facades of buildings, airplanes, and other vehicles—has led to the development of aluminum alloys. Aluminum is the most abundant metallic element in Earth's crust, and one of the most reactive. Aluminum occurs in nature mostly in the compounds it forms with oxygen, silicon, and hydrogen. Chemical weathering of aluminosilicate rocks may result in dissolution of the silicon oxide portion, leaving behind mixtures of several Al(III) compounds. The three most common of them are:

Mineral	Formula
Diaspore	AlO(OH)
Boehmite	AlO(OH)
Gibbsite	Al(OH)$_3$

Diaspore and boehmite have the same chemical formula (that of aluminum oxohydroxide), but they have different crystal structures. Diaspore, which is the more stable form at higher temperatures, consists of hexagonal close-packed oxide ions in which Al^{3+} ions occupy octahedral holes. The boehmite structure consists of layers of cubic close-packed (fcc) O^{2-} and OH^- ions with Al^{3+} ions in half the octahedral holes. Mixtures of all three minerals are found in aluminum ore, called *bauxite*.

THE CHEMISTRY OF GROUP 3A ELEMENTS: BORON, ALUMINUM, GALLIUM, INDIUM, AND THALLIUM

Group 3A contains one of the most abundant elements in Earth's crust, aluminum, and four others that occur in much smaller proportions. Boron has a relatively low crustal abundance (9 ppm), but there are some major deposits of boron-containing minerals in reasonably accessible places, such as the salt flats of the southwestern United States. Gallium (with a crustal abundance of only 19 ppm) and indium (0.24 ppm) occur mostly as impurities in zinc sulfide. Their ions replace Zn^{2+} in the sphaelerite structure described in Section 10.3. Thallium (0.7 ppm) occurs most often as an impurity in galena, PbS. Commercially important quantities of all three elements are recovered during the processing of ores containing ZnS and PbS.

The atoms of the Group 3A elements each have one p and two s electrons in their valence shells. Despite their common outer shell electron configurations, their chemical properties are quite different. The first, second, and third ionization energies of boron are much higher than for the other elements in the group—so high that boron does not form ions with a +3 charge, as the other elements do. Instead, boron forms only covalent bonds and molecular compounds, for example BCl_3, in which it has less than an octet of electrons in its outermost shell.

With increasing atomic number, the elements in the group are more likely to form ions with charges of +1 rather than +3 as each of their atoms loses its valence shell p electron, but not the two s electrons. This trend is the result of similar ionization energies, but much greater ionic radii with increasing atomic number in the group. This combination means that the energy required to form +3 ions from the free atoms is about the same, but the lattice energies of the ionic compounds formed by the larger ions +3 ions in the group, particularly In^{3+} and Tl^{3+}, are much less, as predicted by Coulomb's law. This trend in lattice energies means that, depending on reaction conditions, forming InCl and TlCl is favored over forming $InCl_3$ and $TlCl_3$.

All Group 3A elements react at high temperature with O_2, forming compounds with the formula M_2O_3, but none of the elements react with water at room temperature. All except boron dissolve in HCl with the production of hydrogen gas.

The methods for producing aluminum from aluminum ore are described on pages 912 and 913. Elemental boron is produced by first converting borate minerals such as sodium tetraborate decahydrate, $Na_2B_4O_7 \cdot 10\ H_2O$ (better known by its common name *borax*), to boric acid, H_3BO_3, and then dehydrating the boric acid at high temperature:

$$2\ H_3BO_3(s) \xrightarrow{\Delta} B_2O_3(s) + 3\ H_2O(g)$$

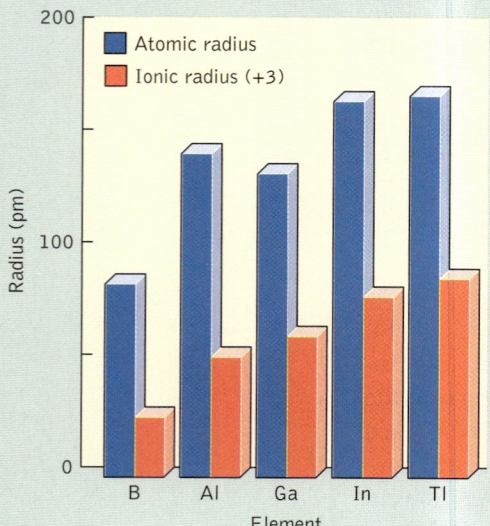

The radii of the atoms and +3 ions of the Group 3A elements generally increase with increasing atomic number.

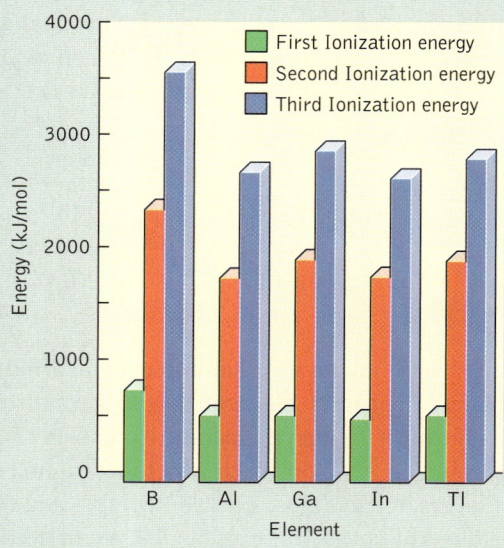

The first three ionization energies of the Group 3A elements tend to decrease with increasing atomic number.

Boron trioxide is then reduced to elemental boron by reduction with a Group 1A or 2A metal, such as Mg:

$$B_2O_3(s) + 3\ Mg(s) \longrightarrow 2\ B(s) + 3\ MgO(s)$$

This reduction process was independently developed nearly 200 years ago by two pioneers in chemistry, the Frenchman Joseph-Louis Gay-Lussac (1778–1850) and the Englishman Sir Humphrey Davy (1778–1829). Small additions (<0.005%) of elemental boron are widely used to strengthen steel alloys. Boron reacts with various metals to form *borides*. These compounds are hard, relatively inert, and have higher melting points than the corresponding metals. Some, such as aluminum boride, are used as abrasives for grinding and polishing. Some of the many uses of other boron compounds are listed below.

Compound	Formula	Uses
Sodium tetraborate (borax)	$Na_2B_4O_7 \cdot 10\ H_2O$	Cleanser, mild antiseptic, soldering flux (cleaner)
Boric acid	H_3BO_3	Fire retardant for fabrics, mild antiseptic, eye wash, metal plating bath
Boron trioxide	B_2O_3	Borosilicate (heat resistant) glass, e.g., Pyrex, fiberglass
Diborane	B_2H_6	Synthesis of organic compounds
Boron nitride	BN	Abrasive

Gallium, indium, and thallium are used in electronic devices. They form semiconducting materials with the Group 5A elements, e.g., InP. Gallium arsenide, GaAs, is used in the lasers in scanners, CD players, and in light-emitting diodes (LEDs). A ceramic oxide of thallium and three other metals, $Tl_2Ba_2Ca_2Cu_3O_{10}$, is a superconductor with one of the highest critical temperatures (125 K) yet discovered.

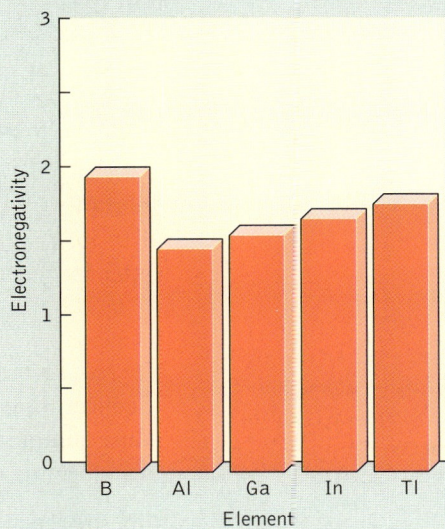

Boron is the most electronegative of the Group 3A elements, and forms only covalent bonds with nonmetals. The other elements in the group have electronegativities above 1.5 and so may form molecular as well as ionic compounds.

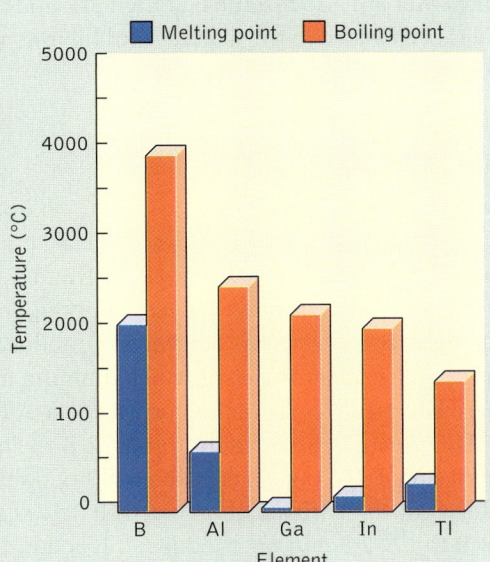

The melting and boiling points of all of the Group 3A elements tend to decrease with increasing atomic number.

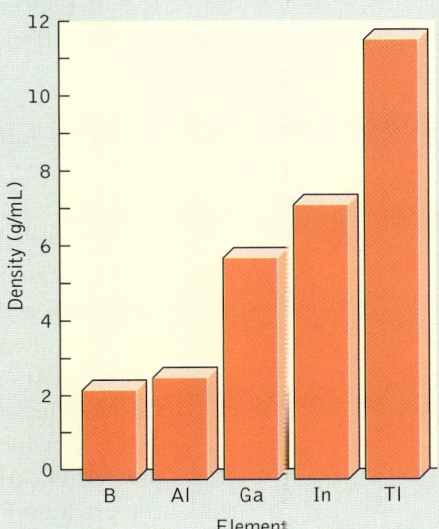

The densities of the Group 3A elements increase with increasing atomic number.

CONCEPT TEST: The following chemical equilibria influence the chemical composition of bauxite deposits:

$$Al(OH)_3(s) \rightleftharpoons AlO(OH)(s) + H_2O(g)$$

$$2\, AlO(OH)(s) \rightleftharpoons Al_2O_3(s) + H_2O(g)$$

Describe how increasing temperature is likely to influence which form of Al(III) is favored.

Hint: Remember that a "favored" reaction has a negative ΔG where

$$\Delta G = \Delta H - T\Delta S$$

CONCEPT TEST: There are no molecules of water in the crystal structure of solid AlO(OH); still, some older texts represent AlO(OH) with the formula $Al_2O_3 \cdot H_2O$. How might experimental observations have led scientists to believe that the latter formula was correct?

> The method of extracting and purifying aluminum oxide using strong base is called the **Bayer process**.

To convert bauxite to aluminum metal, the ore is crushed and extracted with a concentrated solution of NaOH in a reaction called the **Bayer process**. Because aluminum hydroxide is amphoteric (see Section 16.10), it dissolves in base, forming the tetrahydroxo complex, $Al(OH)_4^-$:

Gibbsite reaction: $\quad Al(OH)_3(s) + OH^-(aq) \longrightarrow Al(OH)_4^-(aq)$

Diaspore reaction: $\quad AlO(OH)(s) + OH^-(aq) + H_2O(l) \longrightarrow Al(OH)_4^-(aq)$

Impurities such as $CaCO_3$, Fe_2O_3, and TiO_2 are insoluble at high pH and can be removed by filtration. Any silicon present in the ore combines with some of the aluminum, forming aluminosilicates that are also insoluble at high pH, and they too are removed by filtration. Acid is then added to the filtered solution and purified $Al(OH)_3$ precipitates:

$$Al(OH)_4^-(aq) + H^+(aq) \longrightarrow Al(OH)_3(s) + H_2O(l)$$

The precipitate is converted to pure Al_2O_3 by heating:

$$2\, Al(OH)_3(s) \xrightarrow{\Delta} Al_2O_3(s) + 3\, H_2O(g)$$

CONNECTION: Solid $Al(OH)_3$ is amphoteric because it is soluble in base (forming $Al(OH)_4^-$) and also soluble in acid, forming $Al(OH)_2^+$ (aq), $Al(OH)^{2+}$ (aq), and Al^{3+} (aq) with decreasing pH (see Section 16.10).

In 1886 two 23-year-old chemists, Charles Hall (1863–1914) and Paul-Louis-Toussaint Héroult (1863–1914) (contemporaries, indeed), working independently in the United States and France, respectively, developed an industrial process (called the Hall-Héroult process) for preparing elemental aluminum from alumina (Al_2O_3). The process is based on electrolysis. Molten cryolite, Na_3AlF_6 is the electrolyte because: (1) solid Al_2O_3 readily dissolves in it and (2) it melts at 1021 °C, which is certainly hot, but about 1000 C° below the melting point of pure Al_2O_3. Aluminum(III) is reduced at the cathode of an electrolytic cell and the graphite anode is itself oxidized to CO_2, giving an overall cell reaction that can be written:

$$2\, Al_2O_3 \text{ (in molten } Na_3AlF_6) + 3\, C(s) \longrightarrow 4\, Al(l) + 3\, CO_2(g) \quad (18.3)$$

This sketch of the Hall's first aluminum-production electrolytic cells shows large trays made of cast iron that held the molten reaction mixture. Carbon anodes were attached by copper rods to copper plates suspended over the trays. Molten aluminum collected at carbon cathodes at the bottom of the trays and was poured into the inglot molds that are on the

CONCEPT TEST: Before the Hall-Héroult process was developed, other scientists had tried to produce aluminum by electrolyzing the basic solutions produced by the Bayer process. That approach did not work. Explain why.

SAMPLE EXERCISE 18.5: How much time would it take to produce 10.00 kg of Al in an electrolytic cell such as the ones developed by Hall and Héroult if the current through it is 1.25×10^5 A? Assume that the only reduction process is the reduction of Al(III) to Al^0.

SOLUTION: The Faraday constant, 9.65×10^4 C/mol, links the quantity of electrical charge (coulombs) in an electrochemical reaction to the quantity (moles) of electrons lost and gained by the reactants. Let's use the Faraday constant to calculate the number of coulombs of electrical charge needed to produce 10.00 kg of Al, keeping in mind that 3 moles of electrons are needed to reduce 1 mole of Al(III) to the free metal:

$$10.00 \text{ kg Al} \times \frac{1000 \text{ g}}{\text{kg}} \times \frac{1 \text{ mol Al}}{26.98 \text{ g Al}} \times \frac{3 \text{ mol e}^-}{\text{mol Al}}$$
$$\times \frac{9.65 \times 10^4 \text{ C}}{1 \text{ mol e}^-} = 1.073 \times 10^8 \text{ C}$$

An ampere of electrical current is the flow of coulomb of charge per second. In equation form:

$$1 \text{ A} = 1 \text{ C/s}$$

or

$$1 \text{ C} = 1 \text{ A} \cdot \text{s}$$

Charles M. Hall and Paul-Louis-Toussaint Héroult independently developed the same electrolytic process for producing aluminum metal from alumina. Hall's sister, Julia, who was also a chemistry major at Oberlin College, assisted her brother in the lab, and her business skills helped make their aluminum production company in Pittsburgh, Pennsylvania a financial success. The company eventually became the Aluminum Company of America, a name that was shortened in the 1990s to Alcoa, Inc.

The time for passage of 1.073×10^8 C of charge at a current of 1.25×10^5 A is:

$$\frac{1.073 \times 10^8 \cancel{C}}{1.25 \times 10^5 \cancel{A}} \times \frac{A \cdot s}{\cancel{C}} = 858 \text{ s} = 14.3 \text{ min}$$

PRACTICE EXERCISE: What mass of nickel plate could be applied to metal objects using an electrolytic plating bath containing dissolved Ni(II) if a current of 7.5×10^4 A passes through the bath for 20.0 minutes? (See Problems 28 and 29.)

Let's consider the energy requirements of producing Al from Al_2O_3 or recycling products made of Al metal. The principal energy cost of the Hall-Héroult process is the electricity needed to reduce Al_2O_3. The major cost in recycling is the energy required to melt aluminum metal. We can estimate the energy required for the electrolytic reduction of aluminum by calculating the standard enthalpy change for the reaction:

$$2\ Al_2O_3(s) + 3\ C(s) \rightarrow 4\ Al(s) + 3\ CO_2(g)$$

(Note that in this calculation we do not account for the energy required to melt the cryolite electrolyte nor the fact that aluminum produced in molten cryolite is itself molten.) Starting with standard enthalpies of formation, we have:

$$\Delta H^0_{rxn} = [3(\Delta H^0_{f,\ CO_2}) + 4(\Delta H^0_{f,\ Al})] - [2(\Delta H^0_{f,\ Al_2O_3}) + 3(\Delta H^0_{f,\ C})]$$

$$= [3 \text{ mol } CO_2 (-393.5 \text{ kJ/mol } CO_2) + 1 \text{ mol Al } (0.0 \text{ kJ/mol Al})]$$

$$- [2 \text{ mol } Al_2O_3 (-1675.7 \text{ kJ/mol } Al_2O_3) + 3 \text{ mol C } (0.0) \text{ kJ/mol C}]$$

$$= 2170.9 \text{ kJ}$$

Dividing this value by the 4 moles of Al produced, we get 542.7 kJ per mole of Al. Now let's calculate the energy required to recycle 1 mole of Al by heating it from 25 °C to its melting point (660 °C) and then melting it. The energy needed to heat 1 mole of aluminum from 25 °C to 660 °C can be calculated from its molar heat capacity (24.31 J/mol · C°) and Equation 11.9:

$$q = n\ c_p\ \Delta T$$

$$= 1.00 \text{ mol} \times \frac{24.31 \text{ kJ}}{\text{mol} \cdot \text{C}°} \times (660 - 25)\text{C}°$$

$$= 15.4 \text{ kJ}$$

The heat that must be added to melt a mole of Al at its melting point is its heat of fusion ($\Delta H_{fus} = 10.7$ kJ/mol):

$$\frac{10.7 \text{ kJ}}{\text{mol}} \times 1.00 \text{ mol} = 10.7 \text{ kJ}$$

Thus, the total energy needed to heat and melt 1.00 mole of Al is:

$$15.4 \text{ kJ} + 10.7 \text{ kJ} = 26.1 \text{ kJ}$$

Aluminum recycling symbol. Aluminum recycling is a billion dollar a year industry in the United States because recycling saves manufacturers 95% of the energy needed to produce aluminum from aluminum ore.

This value represents

$$\frac{26.1 \text{ kJ}}{542.7 \text{ kJ}} \times 100\% = 4.8\%$$

of the energy needed to electrochemically reduce Al(III) to Al⁰. There are other energy costs associated with aluminum production and recycling, but, overall, recycling saves aluminum manufacturers about 95% of the energy required to produce the metal from aluminum ore. This energy savings has inspired the rapid growth of a global aluminum recycling industry (aluminum recycling in the United States is a \$1 billion/year business). Currently junkyards in the U.S. recycle 85% of the aluminum in cars (5×10^8 kg/year). Over 50% of the aluminum in food and beverage containers has been recycled. The value of scrap aluminum is about 10 times that of scrap steel because of the above energy considerations and because it takes less energy to melt aluminum than it does to melt the same mass of scrap steel.

CONNECTION: Methods for calculating the quantities of heat needed to raise the temperature of a given mass of a material and to either melt or boil it are described in Section 11.7.

CONCEPT TEST: List several of the energy costs not included in the above calculations that might be associated with recycling aluminum products.

Aluminum forms alloys with many other elements. Alloys with magnesium, silicon, copper, and zinc are the most common. For example, an alloy containing 0.8% Si, 4.4% Cu, 0.4% Mg, and 0.8% Mn and another containing 1.6% Cu, 2.5% Mg, 5.6% Zn, and 0.25% Cr are widely used in aircraft construction. The ends of beverage cans are manufactured from an aluminum alloy containing 4.5% Mg and 0.35% Mn. Alloys containing lithium are even more attractive for applications where light weight is important because lithium has the smallest molar mass (6.941 g/mol) and density (0.534 g/mL) of all the metallic elements. Small amounts of Cu, Mg, and Zr are added to improve the ductility and strength of Al-Li alloys.

Aluminum is a reactive metal that should be readily oxidized by O_2 in the air in a very exothermic reaction:

$$4 \text{ Al}(s) + 3 O_2(g) \rightarrow 2 \text{ Al}_2O_3(s) \quad \Delta H° = -3351 \text{ kJ}$$

However, aluminum and aluminum alloys are widely used as building materials and as fuselages of airplanes because they are *corrosion resistant*. Why do materials made of aluminum and aluminum alloys resist corrosion? The answer is that aluminum surfaces do indeed oxidize to Al_2O_3, but the thin layer of surface oxide acts as a shield, preventing further reaction between the metal beneath it and air in much the way that coatings of Cr and Ni oxides inhibit the corrosion of stainless steel.

> Why do materials made of aluminum and aluminum alloys resist corrosion?

THE CHEMISTRY OF GROUP 4B METALS: Ti, Zr, AND Hf

Titanium, zirconium, and hafnium are Group 4B elements. This designation suggests that the properties of these elements are at least somewhat related to those of the Group 4A elements: C, Si, Ge, Sn, and Pb. The outer shell electron configurations of the two groups are similar: Group 4A elements have two s and two p electrons in their valence shells and Group 4B elements have two s and two d electrons in their valence shells. Still, there are relatively few similarities in the chemical properties of the nonmetals and metalloids at the top of Group 4A and the metals in Group 4B.

Ti, Zr, and Hf are all shiny metals with high melting points and malleabilities typical of metals. None are found as pure elements in Earth's crust, but rather combined (usually with oxygen) in minerals. Titanium is the ninth most abundant element (6.3%) in Earth's crust, about $\frac{1}{3}$ as abundant as sodium. Zirconium (162 ppm) and hafnium (2.8 ppm) are much less common. The principal minerals containing these elements are ilmenite, $FeTiO_3$, rutile, TiO_2, baddeleyite, ZrO_2, and alvite, $MSiO_4$ (M = Zr, Hf, or thorium, Th). Crystals of $ZrSiO_4$ are called zircons; they have the appearance of diamonds (to the nondiscriminating eye) but not the price.

To produce all three elements, their oxides are treated with halogens and carbon at high temperatures, producing the corresponding tetrahalides:

$$2\ FeTiO_3(s) + 7\ Cl_2(g) + 6\ C(s) \xrightarrow{\Delta}$$
$$2\ TiCl_4(g) + 2\ FeCl_3(s) + 6\ CO(g)$$

$$ZrO_2(s) + 2\ I_2(g) + 4\ C(s) \xrightarrow{\Delta} ZrI_4(g) + 4\ CO(g)$$

$$(Zr, Hf)SiO_4(s) + 2\ I_2(g) + 2\ C(s) \xrightarrow{\Delta}$$
$$(Zr, Hf)I_4(g) + 2\ CO(g) + SiO_2(s)$$

Direct reduction of their oxides with carbon (as in the smelting of iron ore) does not work for the Group 4B metals because they would form hard, ceramic carbides, TiC, ZrC, and HfC. The tetrahalides, such as $TiCl_4$, are reduced to the pure metal with elemental magnesium:

$$TiCl_4(g) + 2\ Mg(l) \xrightarrow{900\ °C} Ti(s) + 2\ MgCl_2(l)$$

Titanium(IV) chloride in combination with aluminum compounds is used as a catalyst in the polymerization of propylene described in this chapter.

Titanium metal is used in alloys with aluminum in the aerospace industry where strong, lightweight, chemically unreactive materials are required. Zirconium alloys tend to be even more resistant to corrosion than those containing titanium. Hafnium is a strong neutron absorber and is used in control rods of nuclear submarines.

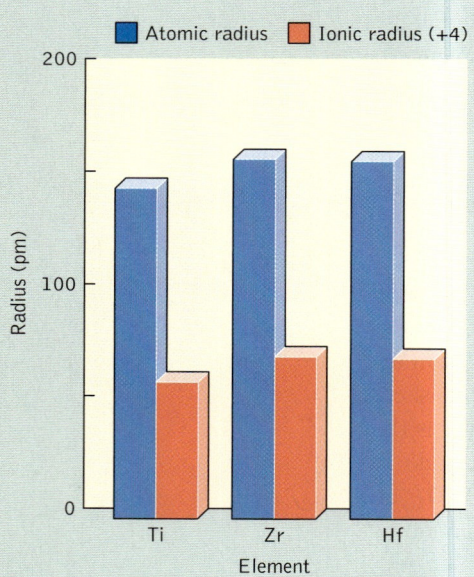

The atomic and ionic (+4) radii of the Group 4B elements increase with increasing atomic number.

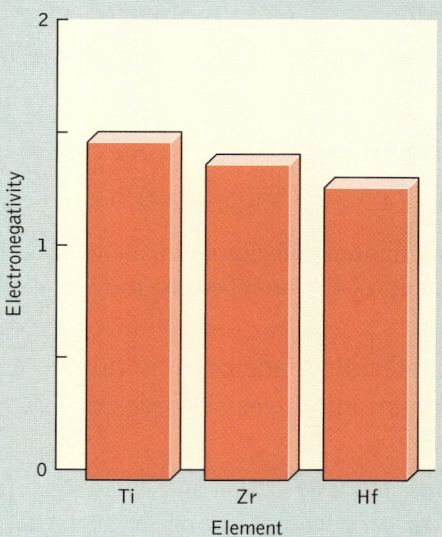

The electronegativities of the Group 4B elements decrease slightly with increasing atomic number.

In general, the reactions of Group 4B elements with oxygen or strong acids become significant only at elevated temperature. All three metals react with hydrogen gas, H_2, forming interstitial hydrides. Titanium dihydride, TiH_2, has a structure containing cubic close-packed titanium atoms with hydrogen occupying all of the tetrahedral holes.

Titanium dioxide, TiO_2, is the most widely used pigment in white paint. Although titanium dioxide can be mined directly it usually contains colored impurities. These are removed by conversion of titanium ore to titanium(IV) chloride followed by reaction with O_2:

$$TiCl_4(g) + 2\ O_2(g) \xrightarrow{1000-1400\ °C} TiO_2(s) + 2\ Cl_2(g)$$

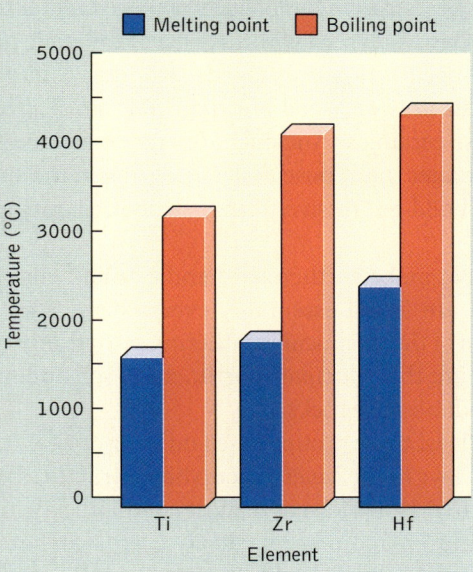

A trend toward increasing melting and boiling points with increasing atomic number is observed for the Group 4B elements.

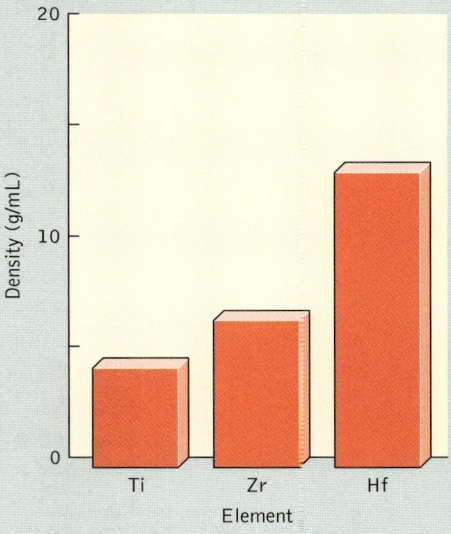

The densities of the Group 4B elements increase with increasing atomic number.

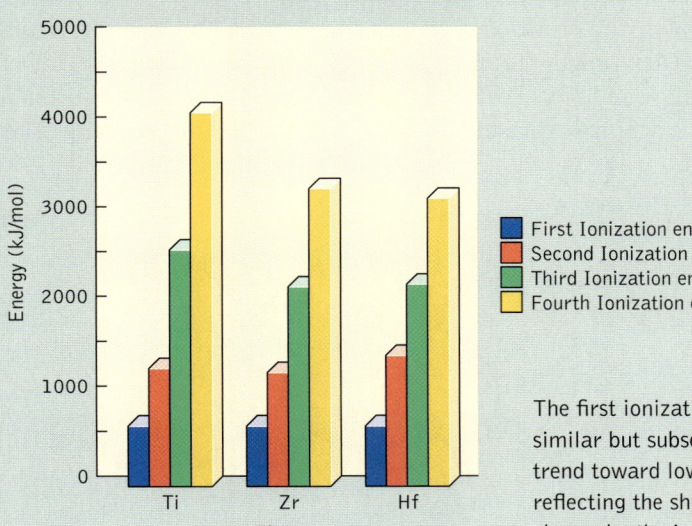

The first ionization energies of the Group 4B elements are similar but subsequent ionization energies show the expected trend toward lower values with increasing atomic number, reflecting the shielding of the outer electrons from the nuclear charge by the inner electrons.

18.2 CERAMICS

Made of Clay

A **ceramic** may be defined as a compound or mixture of compounds that has been heated to transform it to a harder, more heat- and chemical-resistant material. Humans recognized the benefits of heating objects made of clay minerals at least 7000 years ago. Indeed, the use of ceramic materials preceded metal technology in many ancient cultures. Clay reinforced with sticks or grass made primitive bricks, and clay could be shaped into sturdy bowls, pitchers, and other cookware. Heating earthenware in fires or wood-burning kilns (in a process known as *firing*) produced the first ceramics. Ancient Romans discovered that bricks fired in kilns were much stronger than those made by drying in the sun—a discovery that enabled them to build impressively large structures, including the Coliseum in Rome.

Clay minerals have been used to make earthenware and ceramic materials since ancient times because they are found essentially everywhere. They are among the most common products of the chemical weathering or hydrothermal alterations of igneous rocks. They are the principal constituent of soil and make up about 40% of the minerals in sedimentary rocks.

Geologists classify clays into at least three groups based on their crystal structures and chemical compositions. Clays in the *kandite* group have structures similar to that of kaolinite (see Figure 18.11), which is the most common clay in the group (and the world). Some kaolinite is found in practically every soil, but rich deposits of nearly pure kaolinite are found in highly weathered soils, such as those in the southeastern United States and in tropical rainforests. Kaolinite has a relatively simple (by clay standards) chemical composition, $Al_2Si_3O_5(OH)_4$. Other metal ions, such as Na, K, Ca, Mg, and Fe that are found in igneous rocks are largely absent in kaolinite. Their absence indicates that kaolinite deposits form under *acidic* weathering conditions. Under these conditions H^+ ions displace Na^+, K^+, Ca^{2+}, and Mg^{2+} from ion exchange sites in the aluminosilicate structure (Figure 18.11). Other major groups, including *smectites* and *illites* form under alkaline weathering conditions. These clays are likely to contain an assortment of alkali, alkaline earth, and transition metal cations at the ion exchange sites in their structures.

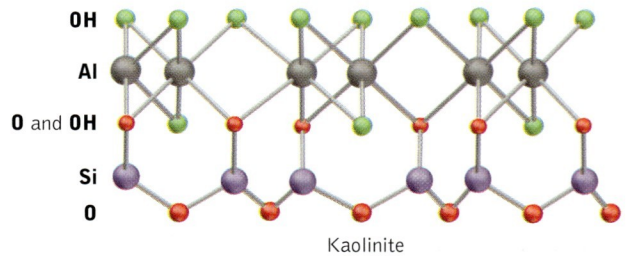

FIGURE 18.11 Kaolinite is an example of a 1:1 clay mineral in that its structure consists of pairs of one silica tetrahedral layer and one aluminum-containing octahedral layer. These pairs of layers are held together by hydrogen bonding between the hydrogen atoms in the octahedral layer and the oxygen atoms of the tetrahedral layer directly above it.

In the crystal structure of kaolinite each SiO_4 tetrahedron shares two corners with adjacent SiO_4 tetrahedra and one oxygen atom with an aluminum atom directly above. Each aluminum atom resides in an octahedron of six oxygen atoms; two of them are shared with SiO_4 tetrahedra, and four are part of OH groups that are each shared with another aluminum atom. Alternating layers of silicon-containing tetrahedra and aluminum-containing octahedra make kaolinite an example of a 1 : 1 clay mineral. The pairs of layers are held to the pairs directly above and below by hydrogen bonding between the hydrogen atoms of the OH groups in octahedral layers and the O atoms of the tetrahedral layer directly above. The strengths of these bonds are such that the layers in kaolinite are not easily separated and, unlike most clays, molecules of water cannot penetrate between them.

Because water cannot penetrate between its layers, kaolinite does not expand when it is added to water. Nor does it dissolve. However, kaolinite and other clays are easily suspended in water because particles of clay tend to be small. Much of the turbidity of bodies of natural water, such as the Missouri River, is due to suspended clay minerals. Actually, geologists use the term "clay" to describe the smallest particles found in soils and sediments—those less than 2 μm in diameter. Some clay particles are much smaller, as little as 0.001 μm, and form colloidal suspensions that are stable indefinitely.

Why don't clay particles in a colloid settle out of suspension? Under a microscope, colloidal particles are seen to be constantly moving in random directions because of collisions with water molecules. On a molecular scale these collisions are really ion-dipole interactions between the edges of clay particles, which tend to be negatively charged in mildly acidic to basic aqueous solutions, with the dipoles of water molecules. As long as the particles are small enough, these collisions and the random *Brownian* motion they produce keep the particles in suspension. Repulsion between similarly charged clay particles prevents them from adhering to each other when they collide, and so they do not form conglomerates that are large enough to settle out of suspension.

> Why don't clay particles in a colloid settle out of suspension?

Clay pottery, such as these Portuguese bowls and jars, is often coated with ceramic glazes that allow it to hold liquids and that add to its beauty. The coating process includes firing at temperatures as high as 1300°C. The appearance of a glaze depends on its ingredients (brown colors may be produced by adding iron(III) oxide), its firing temperature, and whether firing takes place in an oxidizing or reducing atmosphere.

CONNECTION: Colloidal suspensions appear cloudy because particles too small to see can still scatter light; short wavelengths are scattered more than longer ones (see Section 5.12).

Making ceramics

For the same reason that kaolinite does not swell when it is suspended in water, it does not shrink as much as most clays when it is dehydrated at high temperatures. This property is one of those that makes kaolinite a good starting material for making ceramics. Another is that high-purity kaolinite is bright white, making it the material of choice for the manufacture of the porcelain used in bathroom fixtures and fine china. Finally, moist kaolinite is *plastic*, meaning that it can be molded into a shape that it keeps after heating and cooling.

Creating ceramic objects from kaolinite and other clays takes several steps. First, moist clay is formed into pots, bricks, and other objects on a potter's wheel or in molds or presses. Drying at just above 100 °C removes much of the water

 The word "plastic" is widely used to describe materials made from synthetic polymers (which are discussed at the end of this chapter), but it comes from the Greek *plastikos* meaning simply "to form."

This large kaolinite deposit is located at the Hilltop pit in Lancaster County, North Carolina. (The pine tree in the foreground is about 2 m tall.)

that made the clay plastic. Further heating to about 450 °C removes water that was adsorbed onto the surfaces of the clay particles or that remained between the layers of non-kaolinite clays. Between 450 and 650 °C the hydroxyl groups of the octahedral layer decompose with loss of water, leaving behind half their number of oxygen atoms:

$$Al_2Si_2O_5(OH)_4(s) \xrightarrow{450-650\,°C} 2\,H_2O\,(g) + Al_2Si_2O_7(s)$$

The first major structural change occurs just below 1000 °C when $Al_2Si_2O_7$ is converted into another aluminosilicate, $Al_4Si_3O_{12}$:

$$2\,Al_2Si_2O_7(s) \xrightarrow{\sim 950\,°C} Al_4Si_3O_{12}(s) + SiO_2(s)$$

In this transformation, SiO_2 is separated from the aluminosilicate structure, forming a second solid phase. At even higher temperatures $Al_4Si_3O_{12}$ continues to lose SiO_2:

$$Al_4Si_3O_{12}(s) \xrightarrow{>950\,°C} 2\,Al_2SiO_5(s) + SiO_2(s)$$

$$3\,Al_2SiO_5(s) \xrightarrow{1350\,°C} Al_6Si_2O_{13}(s) + SiO_2(s)$$

The last product is called *mullite*. Sometimes its formula is written $3\,Al_2O_3 \cdot 2\,SiO_2$ indicating that it is a blend of alumina and silica structures. Mullite is a widely used ceramic material in the manufacture of furnaces, boilers, ladles, and kilns that must tolerate temperatures as high as 1700 °C. These products are used as containers of molten metals and in the glass, chemical, and cement industries. Mullite is very hard and is also used as an abrasive.

Other clay-based ceramics are used in high-temperature applications: from cookware to fireplace bricks to the tiles on the bottom of the U.S. space shuttles. Ceramics are well suited to these uses because of their high melting points and because they are good thermal and electrical insulators. For example, the thermal conductivity of aluminum metal at 100 °C is over eight times the thermal conductivity of its ceramic oxide, Al_2O_3.

The insulating properties of ceramics can be understood at the atomic level in terms of the same band theory we used to explain the conductive properties of copper and other metals. Unlike metals, ceramics have fully occupied valence bands with significant energy gaps between their valence and conduction bands (Figure 18.12). As a result, their valence electrons have very limited mobilities.

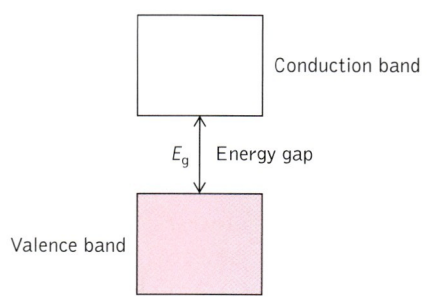

FIGURE 18.12 Most ceramics have band structures with low-energy, fully filled valence bands and much higher-energy, empty conduction bands. The energy difference between the two bands is called the band gap, E_g. The magnitude of E_g is such that few valence electrons reach the conduction band, and so most ceramics are poor thermal and electrical conductors.

Superconducting ceramics

Some ceramics composed of copper, barium, oxygen, and a lanthanide element, such as yttrium, have very unusual electrical conductivities. Unlike simple binary ceramics, such as the SiO_2 used as insulators in microelectronic devices, the electrical resistance of a piece of $YBa_2Cu_3O_7$ *drops to zero* when it is cooled to 95 K. At that temperature $YBa_2Cu_3O_7$ is no longer an insulator; it is not even a conventional conductor of electricity. Instead, it is a **superconductor.** Superconductivity is an attractive property in many electrical and electronic applications. Imagine the savings in energy if none of the electrical energy in an electrical transmission line or other device were lost as heat to the surroundings.

> A **superconductor** is a material that conducts electricity with zero resistance when the material is cooled below its critical temperature.

The crystalline structure of $YBa_2Cu_3O_7$, which materials scientists call a *perovskite* structure, is shown in Figure 18.13. In a perfect perovskite structure, Cu ions occupy the corners of a cube that has either a Y or Ba ion at its center and 12 oxide ions along the edges. Stacking three perovskite unit cells (Figure 18.13B) gives a material with the chemical formula $YBa_2Cu_3O_9$. Material with this composition is not superconducting, but it becomes superconducting with the loss of eight oxygen atoms from edge sites, forming $YBa_2Cu_3O_7$ as shown in Figure 18.13C.

Yttrium-barium-copper oxides and related materials behave as superconductors because of the formation of electron pairs called *Cooper pairs* in their crystals (Figure 18.14). When an electron is introduced into a superconductor, electrostatic forces produce a slight displacement in the positively charged cations in the crystal. The attraction between electrons and these positively charged cations can bring two electrons into close proximity and result in formation of a Cooper pair. The movement of Cooper pairs rather than individual electrons through the material accounts for its superconductivity. To understand the superconductivity of a Cooper pair think of a pair of horses harnessed together to pull a wagon. The team of horses will travel forward, avoiding obstacles in their

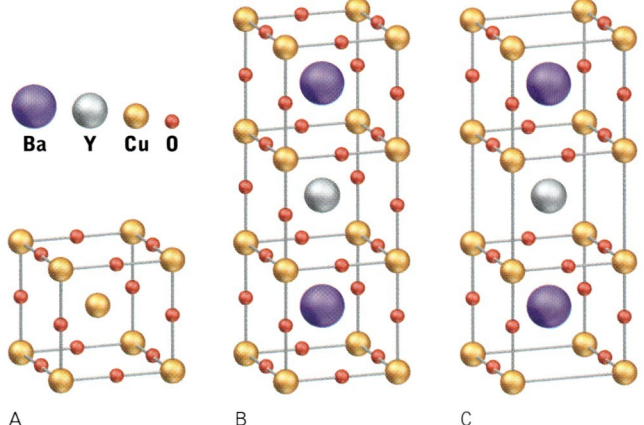

FIGURE 18.13 The crystal structure of the high-temperature superconductor $YBa_2Cu_3O_7$ is based on a stack of the unit cells shown in A. In B, there is a barium ion in the center of the top and bottom cells, and a yttrium ion in the center of the middle cell. Assigning fractional values to the ions that are shared with other unit cells, the array of three cells in B has 9 oxide ions ($\frac{1}{4}$ of the 20 that occupy edges of the three-cell array plus $\frac{1}{2}$ of the 8 that occupy faces—9 in all) and 3 copper ions ($\frac{1}{8}$ of the 8 on the top and bottom surfaces plus $\frac{1}{4}$ of the 8 around the middle or 3 in all). Structure C has 8 fewer oxide ions in edge positions, which reduces the number assignable to the structure by 2. The structure is that of a superconductor with the formula $YBa_2Cu_3O_7$.

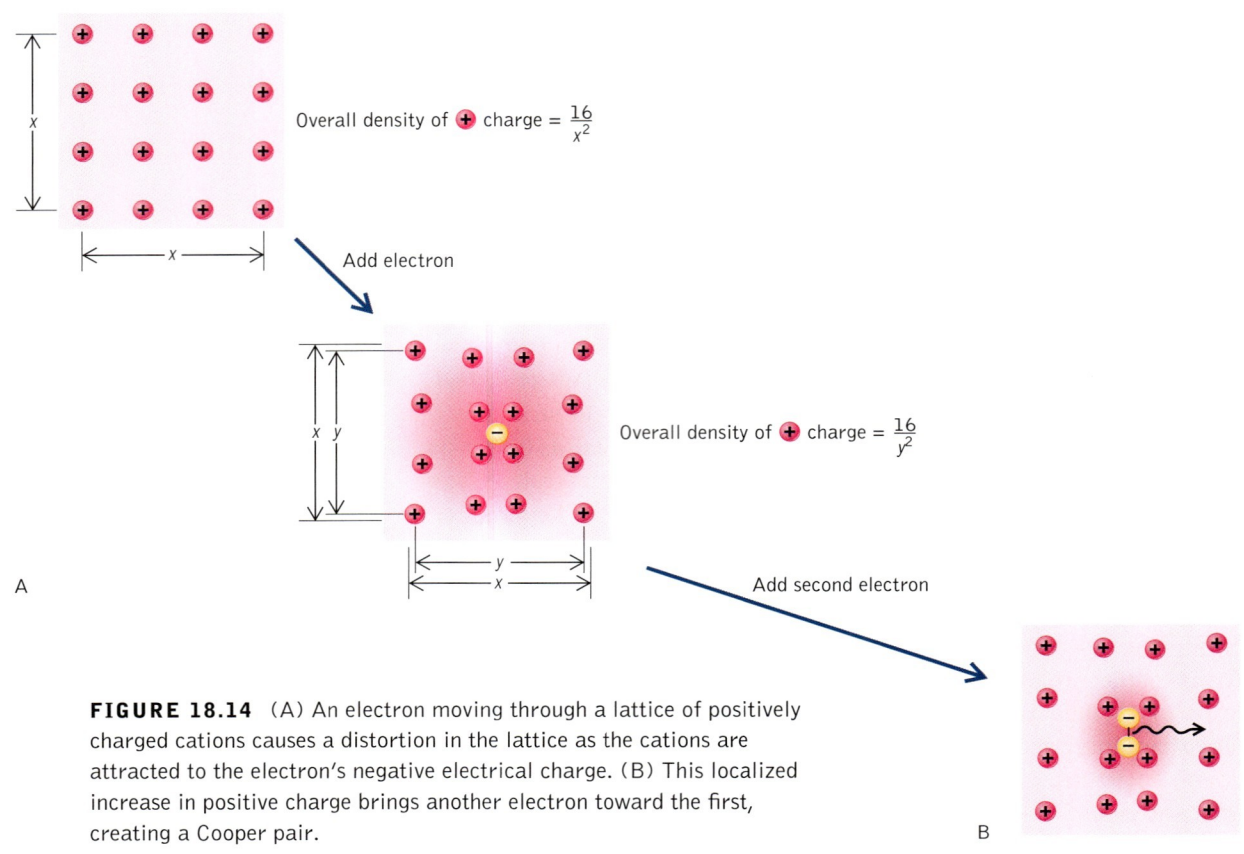

FIGURE 18.14 (A) An electron moving through a lattice of positively charged cations causes a distortion in the lattice as the cations are attracted to the electron's negative electrical charge. (B) This localized increase in positive charge brings another electron toward the first, creating a Cooper pair.

path but never separating from each other. Single animals, however, will take divergent paths, scattering over a wide area. Similarly, single electrons are easily deflected by the atoms or ions in a solid, but electrons "harnessed" together in a Cooper pair are scattered much less than single electrons. Less scattering leads to less (actually zero) electrical resistance.

Superconductors are of technological interest because, in principle, high currents can flow through them with no resistance. In conventional metallic conductors the vibration of metal atoms in their crystalline lattices interferes with the flow of free electrons: the higher the temperature, the more the vibration and the greater the resistance to electron flow. However, at temperatures approaching absolute zero these vibrations become very small. In the early twentieth century scientists discovered that the lattice vibrations in mercury and some metal alloys are so small at temperatures below about 20 K (called *critical temperatures* (T_c)) that electrons can pass freely through these materials, and they become superconductors. Unfortunately, it is difficult and expensive to chill materials to these ultracold temperatures.

THE CHEMISTRY OF GROUP 3B AND THE LANTHANIDES

The Group 3B elements, scandium, (Sc), yttrium, (Y), lanthanum, (La), and actinium (Ac) are just to the right of the alkaline earth elements in the periodic table. The locations of La and Ac serve as detour points for locating the 14 elements that comprise the *lanthanide* (also called *rare earth*) elements ($Z = 58–71$) and the *actinide* elements ($Z = 90–103$) that appear at the bottom of the table. Most of the Group 3B and lanthanide elements are widely distributed in Earth's crust. For example, scandium and yttrium are similar in abundance to the essential (to human health) element cobalt. However, the significant natural abundances of these metals is obscured by the fact that they often occur as impurities in minerals of other metals. As a result, elaborate separation schemes are needed to extract them from the ores of these other metals.

Pure Group 3B and lanthanide elements are produced by electrolysis of mixtures of their chlorides in molten sodium chloride or reduction of halides with calcium metal. Most of the elements have a silvery luster although europium (Eu) and ytterbium (Yb) are yellow. Reaction with oxygen produces oxides with the formula M_2O_3. The oxides react with HX (X = F, Cl, Br, I) producing the corresponding halides, MX_3.

The Group 3B elements have chemical properties that resemble those of the Group 3A elements as much as they do the transition metals. Scandium, yttrium, lanthanum, and actinium readily lose their three valence shell electrons (two in an s orbital and one in a d) to form cations with +3 charges. The ground state electron configurations of most lanthanide elements (except Ce, Gd, and Lu) is $[Xe]4f^n5d^06s^2$. Gadolinium ($[Xe]4f^75d^16s^2$) and lutetium ($[Xe]4f^{14}5d^16s^2$) deviate from this pattern because of: (1) the similar energies of electrons in 4f and 5d orbitals and (2) the stability of half-filled or completely filled sets of 4f orbitals. Note how these two electron configurations imitate the s^1d^5 and s^1d^{10} valence shell electron configurations of the Group 6B and 1B transition metals, though there is a difference: the Group 6B metals acquire more stable electron configurations by promoting a valence shell s electron to a d orbital to half-fill their outermost d orbitals, whereas Gd achieves greater stability by promoting an f electron to a d orbital to retain a configuration with a half-filled set of f orbitals.

Cerium is unusual among the lanthanides in that it form cations with a charge of +4 as well as +3. This behavior is linked to its electron configuration,

The radii of the atoms and +3 ions of the Group 3B elements increase with increasing atomic number. (The atomic radius of radioactive actinium is not precisely known.)

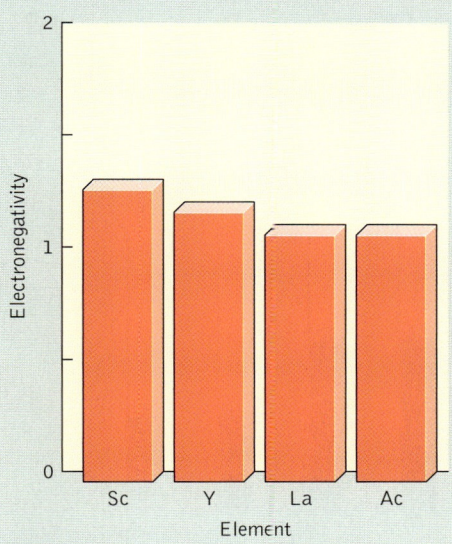

The electronegativities of the group 3B elements decrease with increasing atomic number.

[Xe]$4f^1 5d^1 6s^2$, and the small energy differences between its $4f$, $5d$, and $6s$ orbitals. All lanthanide atoms form cations with a charge of +3 by losing the their two $6s$ electrons and either a $5d$ or $4f$ electron, leaving ions with the generic electron configuration [Xe]$4f^n$. Europium forms a stable +2 cation because doing so produces an ion with the electron configuration [Xe]$4f^7$ and a half-filled set of f orbitals.

The Group 3B and lanthanide elements are used to produce alloys and their oxides are used as catalysts. For example, self-cleaning ovens contain some CeO_2 on the interior walls. The cerium (IV) oxide acts as a cata-

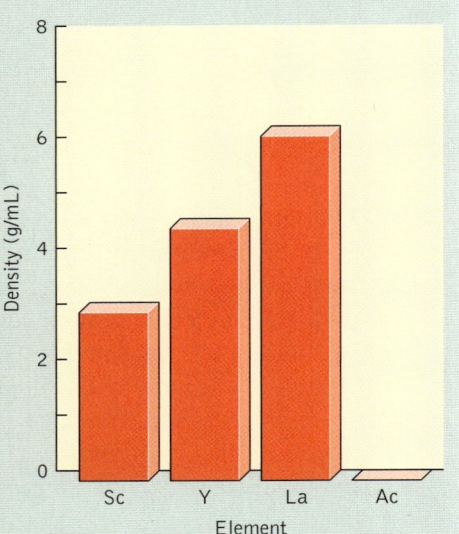

The densities of the Group 3B elements increase with increasing atomic number. The density of actinium has not been determined but is predicted to be 10.1 g/mL.

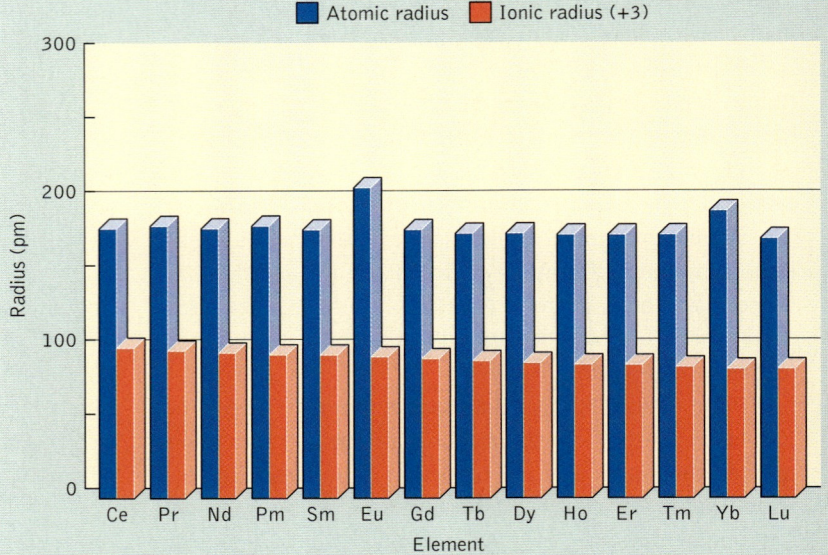

The atomic and ionic radii of the lanthanide elements generally decrease with increasing atomic number.

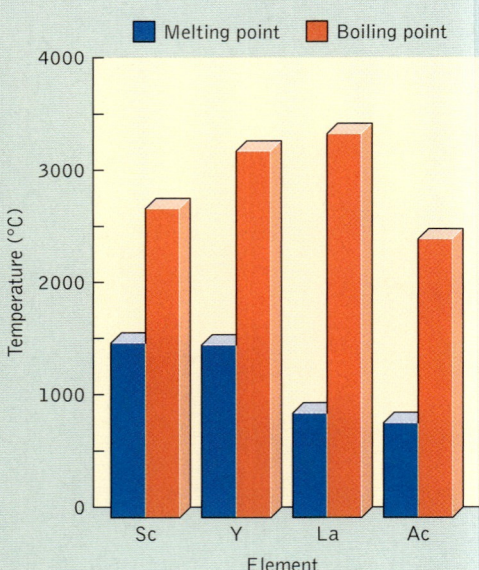

The melting points of the Group 3B elements decrease with increasing atomic number, but the boiling points increase from Sc to La.

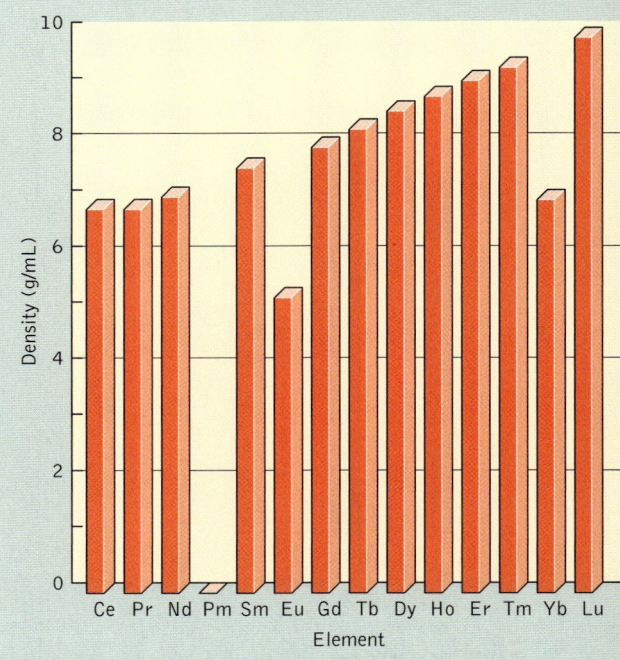

The densities of the lanthanide elements generally increase with increasing atomic number.

lyst for combustion during self-cleaning cycles that turn baked-on carbon deposits into CO_2 and flakes of ash. Yttrium and other lanthanide oxides are the materials that emit the colors produced by televisions and computer monitors that contain cathode ray tubes (CRTs). Excited states of the lanthanide ions are produced by absorbing the energy that is contained in three beams of high-speed electrons emitted by cathodes in the back of the television. The energy of the electrons in each beam corresponds to the energy needed to excite different lanthanide ions in the television screen. For example, the energy of the "red" beam of electrons, when absorbed by a **phosphor** containing Eu^{2+} or Y^{3+} ions, excites these ions. When they return to their ground states they emit photons of red light.

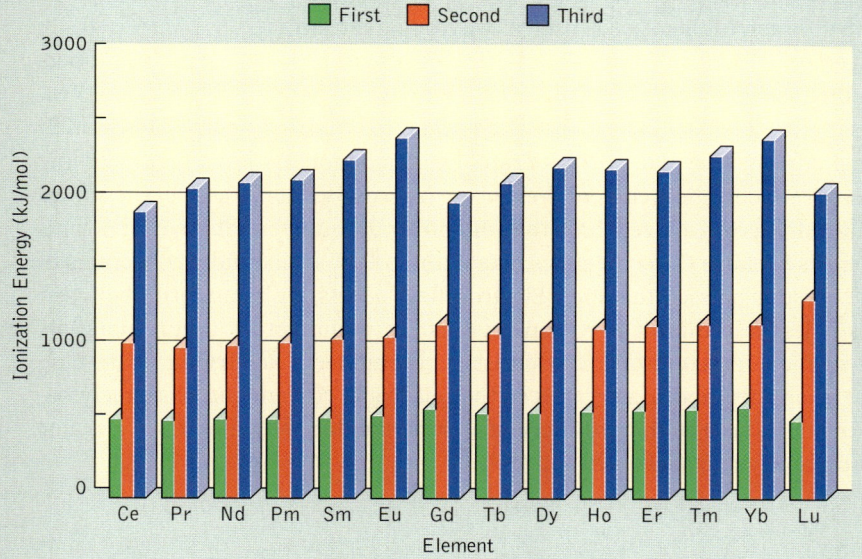

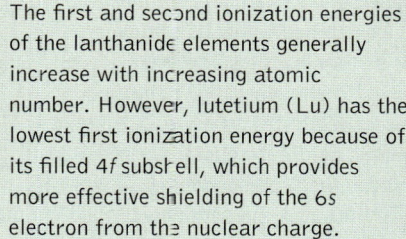

A **phosphor** is a material that emits visible light when exposed to a beam of electrons or electromagnetic radiation. In a color CRT, three phosphors that emit red, green, and blue light are arranged as dots or stripes on the inside of the CRT screen.

The first and second ionization energies of the lanthanide elements generally increase with increasing atomic number. However, lutetium (Lu) has the lowest first ionization energy because of its filled $4f$ subshell, which provides more effective shielding of the $6s$ electron from the nuclear charge.

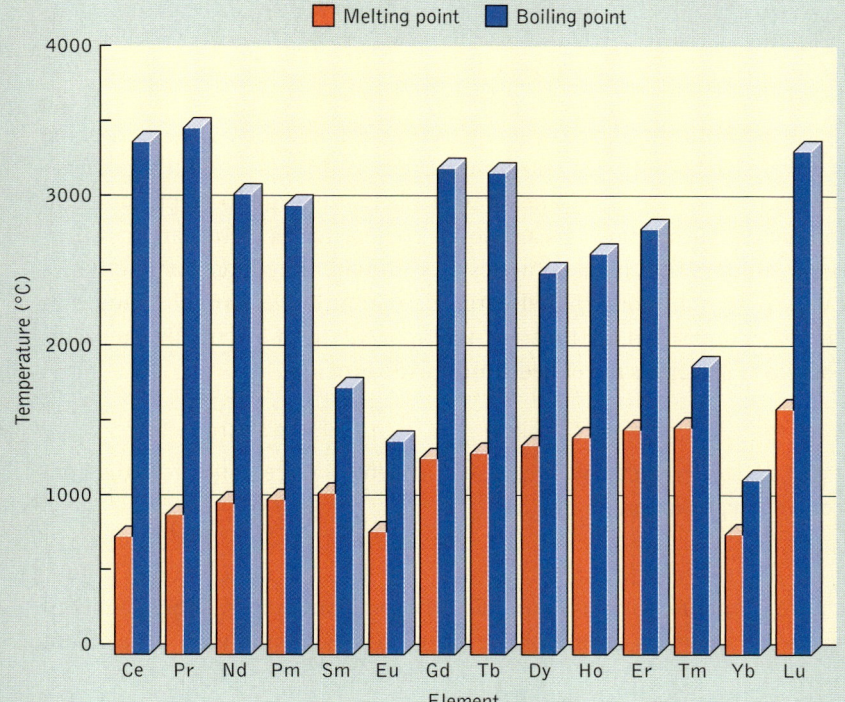

The melting points, though not the boiling points, of the lanthanide elements generally increase with increasing atomic number.

However, in 1986 scientists discovered that ceramic materials become superconductors at temperatures above the boiling point of liquid nitrogen (77 K). Part of the interest in superconductors is related to their ability to exclude magnetic fields—a property known as the *Meissner effect* and illustrated in Figure 18.15. A small cylinder of superconductive material cooled below its critical temperature floats in a magnetic field because magnetic lines of force are excluded from it. Because of this exclusion, the magnet below the superconductor repels it, suspending it in air. This magnetic levitation can be used to float a train above electromagnetic tracks and to propel it along the tracks at relatively low cost and at much higher speeds than a conventional train.

Since 1986 scientists from around the world have produced a variety of superconducting ceramic materials with critical temperatures up to 133 K ($-140°C$). At present the applications of these new materials are limited by how much electrical current they can carry. Ceramic superconductors have relatively low critical current densities (current divided by the cross-sectional area of the superconductive material) above which they are no longer superconductive. This limitation exists because these ceramics, like many solids, are actually collections of discrete crystals pressed together. Electron flow across the boundaries between crystals is less free than it is within crystals. To manufacture useful materials from ceramic superconductors, scientists and engineers are trying to develop ways to orient neighboring crystals so that electrical current can pass freely between them. If these efforts are successful, superconductivity may have a major impact on how you travel in the future.

FIGURE 18.15 A small cylinder of superconductive material that has been chilled below its critical temperature levitates above a magnet. The magnet repels the superconductor because the magnetic field it produces cannot penetrate the superconductor. As a result, the superconductor floats above the magnet—an example of the Meissner effect.

A **semiconductor** is material with a conductivity that can be varied over several orders of magnitude by altering its chemical composition or by applying electrical potentials.

CONNECTION: The proportion of the electrons in a system that are in high-energy states decreases exponentially with increasing energies of the states, as described by the Boltzmann equation (Section 3.4).

18.3 SEMICONDUCTORS

Most ceramics have a large gap between the energies of their valence and conduction bands (see Figure 18.12)—a gap that keeps electrons immobilized in the valence band and makes these ceramics electrical insulators. However, some materials have a small band gap, as shown in Figure 18.16A, and a small fraction of the electrons in valence bands have sufficient energy to be promoted to conduction bands. These materials are not as conductive as metals are, but they are not the insulators that most ceramics are. Instead, they have conductivities that can vary over many orders of magnitude with relatively small changes in their chemical composition, or through the application of electrical potentials. These in-between materials are called **semiconductors**.

CONCEPT TEST: Predict how increasing its temperature will affect the conductivity of a semiconductor.

Some pure elements, such as Si and Ge, are semiconductors. Other semiconductors are solid solutions of one of the Group 4A elements, with an element

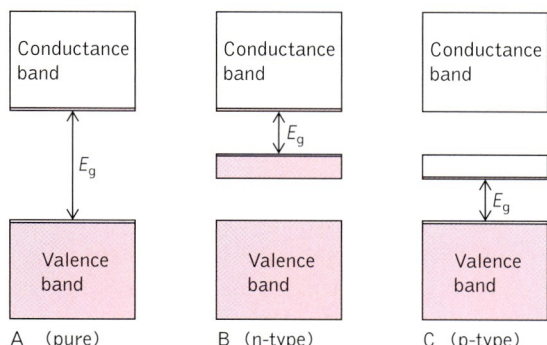

FIGURE 18.16 Defects in the crystals of semiconductors can enhance the electrical conductivity of the material. (A) A pure semiconductor element, such as Si or Ge, has a band gap E_g. (B) Doping with an element that has more valence electrons per atom than the host creates a filled band in the band gap of an n-type semiconductor. (C) Doping with an element with fewer valence electrons than the host material creates an empty conductance band in the band gap of a p-type semiconductor. The transistors used in solid state electronics are combinations of n- and p-type semiconductors.

with a similar atomic radius but from another group in the periodic table. If the alloying element is from Group 5A, each of its atoms adds one extra valence electron to the alloy. These electrons populate a band within the silicon band gap, as shown in Figure 18.16B. This arrangement makes it easier for valence electrons to reach the conduction band, thereby enhancing the conductivity of the semiconductive Group 4A element. This type of semiconductor, in which atoms with more valence electrons occupy sites in a crystal structure, essentially adding extra **n**egatively charged particles (electrons) to the structure, is called an *n-type* semiconductor.

In another type of semiconductor alloy, called *p-type*, atoms of a Group 3A element, such as gallium, with one fewer valence electrons replace some of the atoms of a Group 4A element. The addition of the Group 3A element, which is an example of a substitution process called *doping*, produces an empty band in the band gap as shown in Figure 18.16C. In this case, less energy is needed to promote an electron from the filled valence band of silicon to the empty (conduction) band of gallium, which also increases the electrical conductivity of the material.

SAMPLE EXERCISE 18.6: Does the addition of arsenic to a crystal of germanium create an n- or p-type semiconductor?

SOLUTION: Since arsenic has five valence electrons and germanium has four, an n-type semiconductor is formed.

PRACTICE EXERCISE: Gallium arsenide, GaAs, is a semiconductor used in the optical scanners in retail stores. GaAs can be made an n-type or p-type semiconductor by replacing some of the As with another element. Which of these elements would work: Se, Sb, or Sn? (See Concept Review 33 and 34.)

Materials scientists make extensive use of doping to manipulate the properties of materials, often by changing the size of the energy gap between valence and conduction bands. For example, GaAs is a semiconductor that emits red ($\lambda = 874$ nm) light when connected to an electrical circuit. This emission is used in devices such as red laser pointers, the bar code readers in stores, and in CD players. The energy of each photon of red light corresponds to the energy gap between the valence and conduction bands in GaAs. When electrical energy is applied to the material, some of it raises electrons to the conduction band. When they fall back to the valence band they emit light. If aluminum is substituted for gallium in GaAs, the band gap increases, and, predictably, the wavelength of emitted light decreases. For example, a material with the composition $Al_{0.5}Ga_{0.5}As$ emits yellow-orange light ($\lambda = 620$ nm). Many of the multicolored indicator lights in electronic devices such as sound system components rely on light emitting diodes, or LEDs, based on Al_xGa_xAs. Similarly, substitution of phosphorus for arsenic in GaAs creates LED materials with the generic formula GaP_xAs_{1-x}. The electron energy band gap in these materials varies from 1.42 eV ($x = 0$) to 2.25 eV for pure GaP. LEDs made of pure GaP emit green light. Those with little or no phosphorus emit red or infrared radiation. Other LEDs made of zinc selenide, ZnSe, have a band gap of 2.75 eV and emit blue light ($\lambda = 450$ nm).

SAMPLE EXERCISE 18.7: Differences in the energies of electronic states (and band gaps) are sometimes expressed in electron-volts (eV) where 1 eV = 1.602×10^{-19} J. What is the band gap eV in GaAs that corresponds to $\lambda = 874$ nm?

SOLUTION: A change in the electronic energy, ΔE, of an atom or molecule is related to the wavelength of light absorbed or emitted by the atom or molecule by the formula $\Delta E = hc/\lambda$, where h is Planck's constant (6.626×10^{-34} J·s) and λ is wavelength. Thus, the ΔE of the band gap in GaAs is:

$$\Delta E = hc/\lambda = \frac{(6.626 \times 10^{-34} \text{ J·s})(3.00 \times 10^8 \text{ m/s})}{(874 \text{ nm})(10^{-9} \text{ m/nm})}$$

$$= 2.27 \times 10^{-19} \text{ J}$$

This energy difference in electron-volts is:

$$\Delta E = -2.27 \times 10^{-19} \text{ J} \frac{(1 \text{ eV})}{1.602 \times 10^{-19} \text{ J}} = 1.42 \text{ eV}$$

PRACTICE EXERCISE: Determine the wavelength of light emitted by an aluminum arsenide laser with a band gap of 2.15 eV. (See Problems 43 and 44.)

18.4 FIBERS FOR CLOTHING AND OTHER USES

Paleontologists believe that humans began twisting together handfuls of plant fibers to make string and thread as early as 30,000 years ago. These threads and string were knotted and laced together to make primitive woven fabrics. The results of radiocarbon dating of Early Neolithic artifacts indicate that humans began to use looms to weave fabric as early as 8000 years ago. Ancient Egyptian weavers used fibers from the flax plant to weave linen cloth at least 6500 years ago, and Chinese weavers crafted intricate looms for weaving silk as long as 3700 years ago—about the same time that weavers in India learned how to make cloth with cotton fibers. From then until well into the last century the fabrics used to make clothing were manufactured exclusively from natural fibers, that is, from biopolymers. Then, beginning in the mid-twentieth century, a series of discoveries, many by scientists working for one company, led to a host of synthetic polymers for making fibers and other materials. In this section we examine the molecular structures of both natural and synthetic fibers, and we relate the properties of these fibers to their molecular structures.

Natural fibers

Fibers obtained from plants are made principally of cellulose, the polymeric carbohydrate that gives plant tissues their structural integrity. Some fibers are obtained from the seed pods (cotton, for example) and others come from the stems or inner bark (for instance, the flax from which linen is made). The molecules of cellulose in cotton have around 10,000 monomeric units of glucose molecules linked together by β-glycosidic bonds as shown in Figure 18.17. Adjacent chains are also linked together by hydrogen bonds. These intermolecular hydrogen bonds give a multidimensional ordering to the cellulose polymers that enhances their strength and imparts a degree of stiffness or inelasticity to cotton fibers. Linen fibers have about the same degree of polymerization as cotton fibers, and their cellulose chains are even more ordered, making linen fibers and cloth woven from them even less elastic than those made of cotton.

The presence of many —OH groups in the cellulose fibers means that fabrics woven from them have a high affinity for H_2O. Water molecules are able to penetrate the spaces between cellulose chains forming hydrogen bonds to these —OH groups. These interactions make cotton and linen the materials of choice for towels and tropical clothing. Cotton and linen are among the few fibers that are stronger wet than dry. However, moist cotton fibers are digested by the microorganisms that produce mildew and rot. Linen is more resistant to weathering and rotting.

Fibers derived from silk and wool consist of proteins. In particular, they are members of the class of proteins called *fibrous proteins*. Unlike soluble proteins, such as hemoglobin, insoluble fibrous proteins consist of long, thin strands that

 CONNECTION: Cotton and linen fibers are made of cellulose, a polysaccharide consisting of glucose monomers linked together by β-glycosidic bonds (see Sections 12.3–12.4).

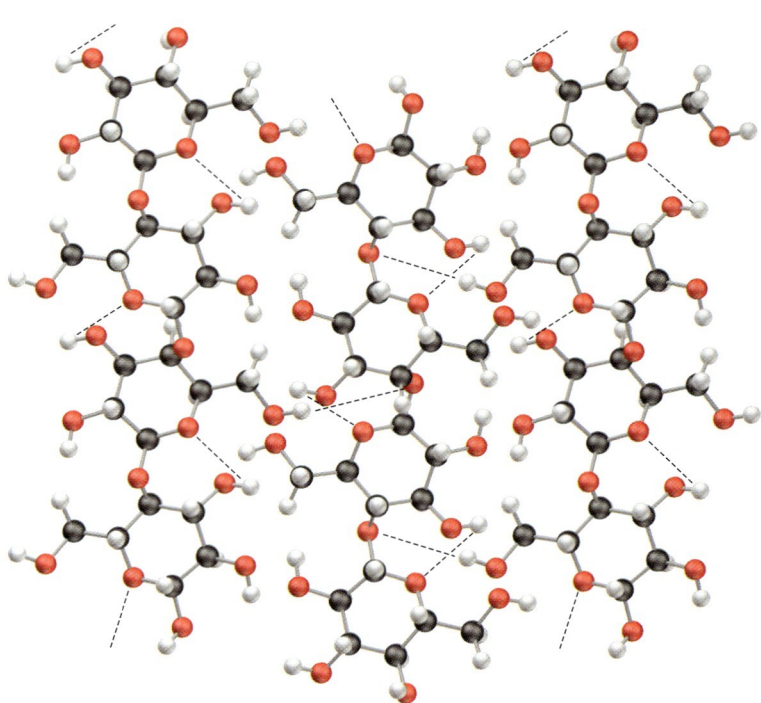

FIGURE 18.17 Both intra- and interchain hydrogen bonding leads to a rigid three-dimensional structure that makes cotton fiber strong but less elastic than those made from other natural and synthetic polymers.

> The **primary structure** of a protein is the sequence of amino acids in the polypeptide chain(s) of the protein.
>
> The **secondary structure** of a protein is the geometric arrangement of the amino acids in the polypeptide chain(s) of the protein that results from hydrogen bonds and other interactions between amino acids.

wrap around each other, forming even longer fibers. Wool fibers are only about 15 cm long, but silk fibers, when stretched to their maximum lengths, are up to 1.5 km long! Wool and silk proteins also differ in the sequence of amino acids in their polypeptide chains, that is, in their **primary structures.** The proteins in the silk produced by the mulberry silkworm contain ten units with the six–amino acid sequence: serine-glycine-alanine-glycine-alanine-glycine. Neighboring chains of silk proteins, or different sections of the same chain that have folded back on themselves, are held together by hydrogen bonds in a pleated array called a β-pleated sheet as shown in Figure 18.18. This arrangement of amino acids along a protein that is stabilized by *intramolecular* hydrogen bonds, is an example of a **secondary structure** of a protein. When sections of a protein (or any polymeric chain) have this sort of multisegment or multichain ordering, they are said to be *crystalline*. As we shall see, the degree of crystallinity of a polymeric material strongly influences its physical properties.

> How can silk be both strong and elastic?

Crystalline regions in polymers generally stiffen as well as strengthen the material, so fibers made from them tend to be less elastic. However, natural silk can be stretched by 20–25% of its normal length. How can silk be both strong and elastic? The answer lies in additional structural features of the proteins of which silk is made. The crystalline regions of β-pleated sheets are separated from each

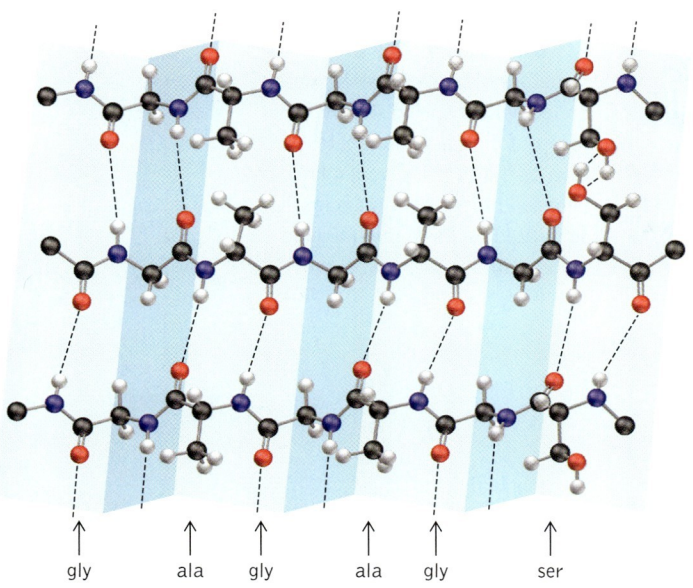

gly　ala　gly　ala　gly　ser

FIGURE 18.18 After feeding on mulberry leaves for several weeks these worms secrete filaments of silk up to 1.5 km long to make their cocoons. Silk is composed of fibrous proteins, regions of which are held together by hydrogen bonds in two-dimensional arrays called β-pleated sheets.

other by 33 other amino acid units that take on a secondary structure called a *random coil*. As the name suggests, a random coil is a structure that has no definable structure. The only limitation on the orientation of this segment is that it not collide with other segments. The resulting freedom of motion allows the random coil to stretch when subjected to an external force. When the force is removed, the random coil returns to its less elongated, more random conformation. Thus, the combination of β-pleated sheet and random coil regions impart both strength and elasticity to silk fibers.

The combined effect of a material's resistance to stretching and its ability to stretch a little without breaking is expressed in a fiber's stiffness, or *modulus of elasticity*. The modulus of elasticity is obtained from the initial slope of a plot of **tensile strength** as a function of elongation. Values for the modulus of elasticity of silk are the range of 7–10 gigapascals (1 GPa = 10^9 Pa). On a molecular scale, the combined effect of the β-pleated sheet and random coil regions in silk proteins is an overall three-dimensional shape that is called the **tertiary structure** of the protein. Tertiary structures in biological polymers are strongly influenced by intramolecular hydrogen bond formation, but other factor also play a role.

Wool fibers aren't as strong as silk, but they can be stretched considerably more: up to 30–40% of their original length. Wool fibers are elastic because the protein chains in wool coil up like springs in structures called α-helices. Proteins form these structures because of intramolecular hydrogen bonding between—NH hydrogen atoms and C=O oxygen atoms on every fourth amino acid along the chain. These groups approach each other and hydrogen bonds form when the protein forms a coil with 3.6 amino acids per turn (see Figure 18.19).

> ✓ **Tensile strength** is the pulling force needed to break a fiber or other material divided by its cross-sectional area.
>
> The **tertiary structure** of a protein is the overall three-dimensional shape of the protein.

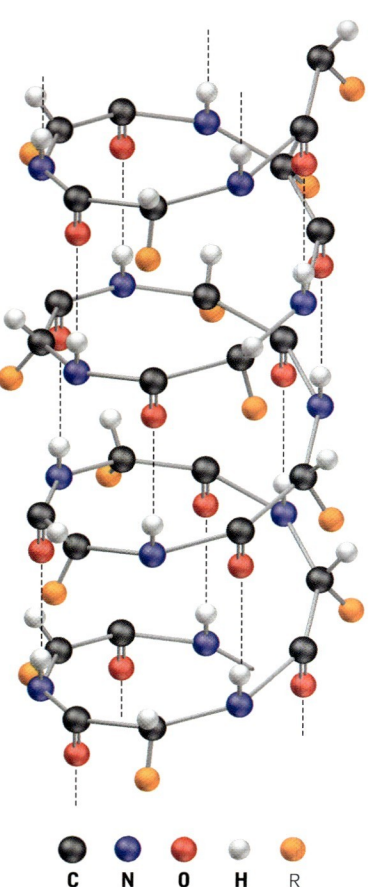

FIGURE 18.19 The α-helix structure of proteins resembles a coiled helical spring. Every fourth amino acid completes one turn of the coil. Hydrogen bonds between pairs of amino acids separated by three other amino acids stabilize the structure.

C N O H R

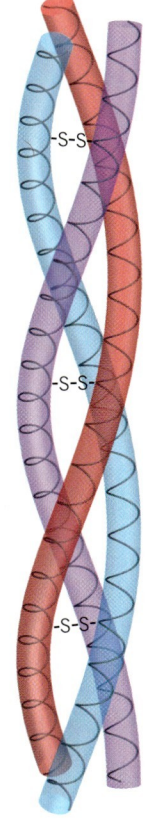

FIGURE 18.20 A triple helix forms when α-helical regions of three proteins wrap around each other, and are held in place by hydrogen bonds, —S—S— covalent bonds, and other intermolecular forces.

Sometimes the α-helical regions of three protein molecules in wool wind around each other forming a *triple helix* (Figure 18.20). Triple helices are stabilized by hydrogen bonding and by covalent bonding between the peptide chains. Wool proteins contain relatively high proportions of the sulfur-containing amino acid cysteine. When the side groups (—CH$_2$SH) of two cysteine molecules approach each other the —SH (sulfhydryl) groups on the ends may react with each other forming a —S—S— (disulfide) bond that covalently links the two peptide chains together.

The elasticity of wool fiber can be partially explained by changes in its secondary structure under strain. Stretching an α-helix breaks some of the *intra*molecular hydrogen bonds, but new *inter*molecular hydrogen bonds may form. Thus, the helical structure of wool may be replaced by a sheet structure like that of silk. Intermolecular disulfide bonds keep the polypeptide chains aligned so that the α-helical structure is restored when the elongating force on the wool fiber is released.

Synthetic polymers from condensation reactions

The early twentieth century saw a flurry of advances in the synthesis of new fibers based initially on chemical modifications of natural fibers. The first truly synthetic textile fiber was invented in 1935 by Wallace Hume Carothers (1896–1937) at the DuPont Experimental Station near Wilmington, Delaware. Carother's discovery was initially given a number for a name, 66, because it was prepared from the condensation reaction between 1,6-dihexanoic acid and 1,6-diaminohexane:

$$n \; \underset{HO}{\overset{O}{\text{C}}}{-}(CH_2)_4{-}\underset{OH}{\overset{O}{\text{C}}} + n \; \underset{H}{\overset{H}{\text{N}}}{-}(CH_2)_6{-}\underset{H}{\overset{H}{\text{N}}} \longrightarrow$$

$$\left[\underset{\underset{H}{N}}{\overset{O}{\text{C}}}{-}(CH_2)_4{-}\underset{\underset{H}{N}}{\overset{O}{\text{C}}}{-}(CH_2)_6 \right]_n + 2n \; H_2O$$

During the reaction the atoms shown in red: the OH atoms in the carboxylic acid groups and one of the H atoms bonded to each nitrogen atom in 1,6-diaminohexane—combine to form a molecule of water. At the same time, covalent bonds form at both ends of each reacting molecule, forming amide bonds and members of an important class of condensation polymers called *polyamides*. Today this particular polyamide is best known by its DuPont trade name, Nylon.

How do the properties of Nylon compare with those of natural fibers? Nylon is stronger than either silk, wool, or cotton and has an elasticity comparable to that of silk. The polymer chains in Nylon and other polyamides are linked together by hydrogen bonds forming a secondary structure such as that shown in Figure 18.21. These intermolecular hydrogen bonds provide strength while the —(CH₂)₅— regions give Nylon polymers their ability to stretch easily.

> **How do the properties of nylon compare with those of natural fiber?**

CONCEPT TEST: Fabric made of Nylon rapidly dissolves if acid is spilled on it. Where in the structure of nylon might reactions, induced by strong acid, occur that result in the destruction of this polyamide?

CONNECTION: Condensation reactions are so named because water is one of their products. Condensation reactions are described in Section 12.5 in the context of linking sugar molecules together to form polysaccharides.

CONNECTION: Note that the amide bonds that link Nylon monomeric units together have the same structure as the peptide bonds that link amino acids together in proteins (see Section 13.7).

The invention of Nylon was followed by the development of many other synthetic fibers based on condensation polymerization reactions. In 1962 a highly elastic polymer, or *elastomer*, called Lycra (spandex) was commercialized by DuPont. Lycra has a molecular structure with two different subunits (Figure 18.22). The (—O—CH₂CH₂—)$_x$ portion is flexible and stretches easily (Lycra fibers can stretched up to seven times their normal length) while the segment containing aromatic rings adds strength.

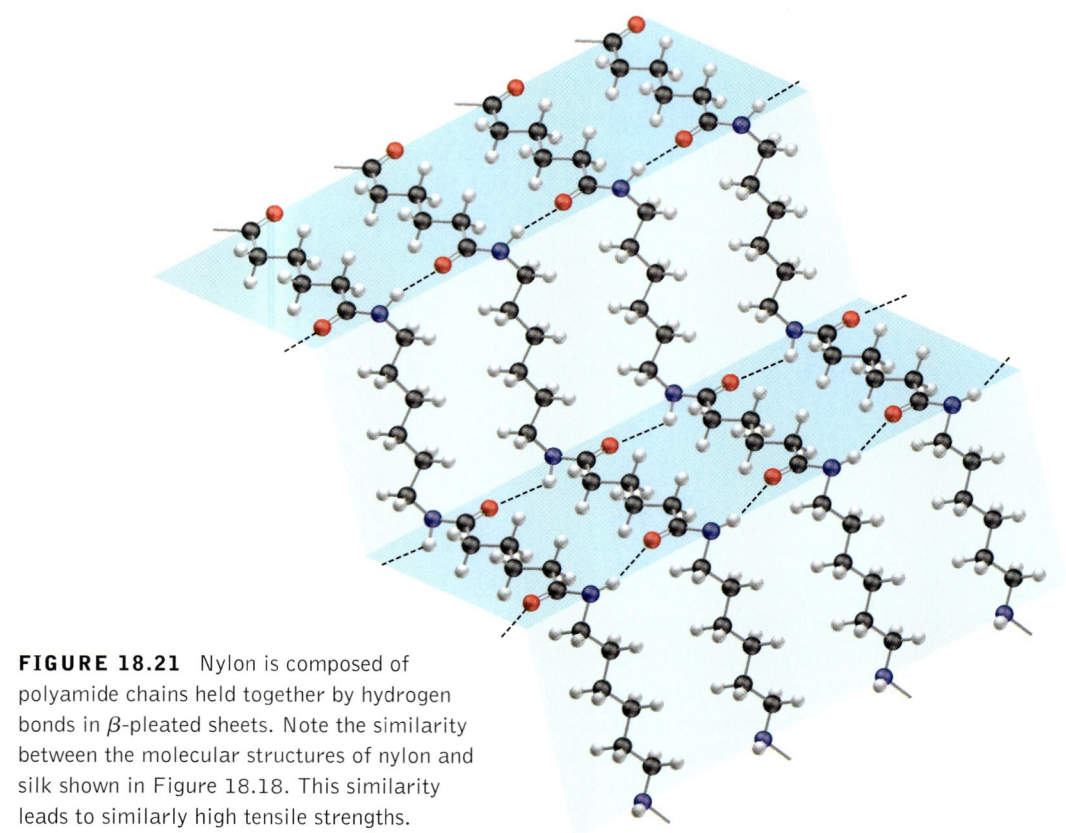

FIGURE 18.21 Nylon is composed of polyamide chains held together by hydrogen bonds in β-pleated sheets. Note the similarity between the molecular structures of nylon and silk shown in Figure 18.18. This similarity leads to similarly high tensile strengths.

The structures of two more polyamides, with the trade names Kevlar and Nomex, are shown below:

Kevlar

Nomex

Both are used to make extremely strong fibers. Note that the only difference in these structures is in the attachment points to the aromatic rings: they are para (C-1 and C-4) in Kevlar and meta (C-1 and C-3) in Nomex. There is extensive hydrogen bonding between amide groups on adjacent polymeric chains (shown for Kevlar in Figure 18.23), which adds to the strengths of both polymeric materials. Kevlar is used to make ropes, cables, bullet-resistant vests, and other clothing. Garments made of Nomex are flame-resistant as well as strong and so are used to make protective clothing for firefighters, rescue workers, and race-car drivers.

18.4 Fibers for Clothing and Other Uses

[Structure of Lycra polymer showing elastic segment —(O—CH₂—CH₂)ₓ—O—C(=O)—NH— connected to rigid segment containing aromatic rings and amide groups]

Elastic segment | Rigid segment

FIGURE 18.22 Lycra is a condensation polymer used to make strong, elastic fabrics such as those worn by Tour de France winner Lance Armstrong. The strength of the fibers is derived mostly from the rigid segment shown on the right in the structure above. This segment includes aromatic rings and amide functional groups that can form hydrogen bonds with those on adjacent polymers (as in nylon—see Figure 18.21). Elasticity comes from the segment on the left that has only single bonds in building blocks linked by either (C—O—C) groups.

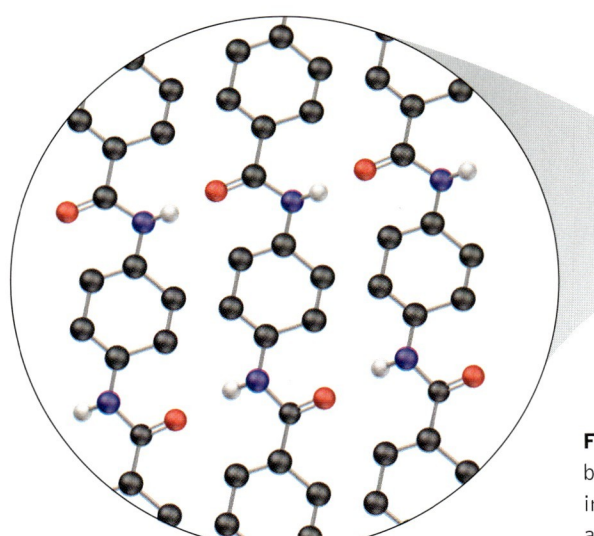

FIGURE 18.23 The strength of the Kevlar fibers used to make bullet-resistant vests relates to the presence of rigid aromatic rings in its structure and hydrogen bonding between —NH hydrogen atoms and C=O oxygen atoms of amide groups on adjacent chains.

SAMPLE EXERCISE 18.8: Polyesters are an important class of condensation polymers that have been widely used to make fibers for clothing. One of them, best known by the trade name Dacron, is the product of the reaction between 1,2 ethanediol (ethylene glycol):

$$HO-CH_2CH_2-OH$$

and an organic acid (called terephthalic acid) made of molecules that have two carboxylic acid groups at the para (1,4) positions of a benzene ring:

[Structure: HO—C(=O)—C₆H₄—C(=O)—OH]

Draw the condensed molecular structure of the repeating units in Dacron.

CONNECTION: Esterification reactions were introduced in Chapter 13 in the context of forming fats (triglycerides) from the reaction of fatty acids with triglycerol, an alcohol with three —OH groups per molecule.

SOLUTION: Esters are formed in condensation reactions between carboxylic acids and alcohols. In these reactions the —OH group (shown in red below) of the carboxylic acid and the hydrogen atom (also shown in red) of the —OH group of the alcohol combine, forming water, as the acid and alcohol covalently bond together:

$$\underset{\text{Carboxylic acid}}{-\overset{O}{\underset{\|}{C}}-OH} + \underset{\text{Alcohol}}{HO-CH_2-} \rightarrow \underset{\text{Ester}}{-\overset{O}{\underset{\|}{C}}-O-CH_2-} + \underset{\text{Water}}{H_2O}$$

To make Dacron, both of the carboxylic acid groups of terephthalic acid combine with both of the —OH groups of 1,2-ethanediol forming a polyester chain. A chemical equation based on the condensed molecular structures of the reactants and products and involving n molecules of reactants to make a polymer with n monomeric units looks like this:

$$n\,HO-CH_2CH_2-OH + n\,HO-\overset{O}{\underset{\|}{C}}-\underset{\text{Terephthalic acid}}{\bigcirc}-\overset{O}{\underset{\|}{C}}-OH \rightarrow$$

$$\underset{\text{Dacron}}{\left[O-CH_2CH_2-O-\overset{O}{\underset{\|}{C}}-\bigcirc-\overset{O}{\underset{\|}{C}} \right]_n} + 2n\,H_2O$$

PRACTICE EXERCISE: Draw the condensed molecular structure of the polymer produced when 6-aminohexanoic acid, $H_2N(CH_2)_5COOH$, polymerizes with itself through amide bond formation. (Fibers made from this polymer are sold under the trade name Perlon.) (See Problems 59 and 60.)

☑ An **addition polymer** is formed when monomers containing double bonds combine through an **addition reaction**. In this reaction the π bonds in the monomers' double bonds are broken and σ bonds that link the monomers together are formed.

Synthetic polymers from addition reactions

There are many commercially important synthetic polymers, and we have space here to discuss only a few of them. A summary of the structures and physical properties of the some of the more widely used condensation polymers appears in Table 18.2. Also in the table are several polymers produced not by condensation reactions, but by **addition reactions.** To understand how these **addition polymers** are made, let's consider the one that is the most widely used to make fibers for clothing and home furnishings: polypropylene. Polypropylene is prepared from the polymerization of propene (or propylene), $CH_2\!=\!CHCH_3$. There are many ways to carry out this reaction, and various catalysts may be used to promote and manipulate it. We will focus on one of the more common methods. It begins with a peroxide initiator R—O—O—R that decomposes into odd-electron radicals (R—O ·). These radicals attack the π bonds of propene as shown in Figure 18.24. During the reaction the π bond breaks and a σ bond

TABLE 18.2

Condensation Polymers

Name and Formula	Uses
Poly(ethylene terephthalate) (PET) $\left(\text{CH}_2-\text{CH}_2-\text{O}-\overset{\text{O}}{\underset{\|}{\text{C}}}-\text{C}_6\text{H}_4-\overset{\text{O}}{\underset{\|}{\text{C}}}-\text{O}\right)_n$	Beverage bottles, fibers (e.g., Dacron), food packaging, photographic film, home furnishings
Nylon $\left(\text{N}(H)-(\text{CH}_2)_6-\text{N}(H)-\overset{\text{O}}{\underset{\|}{\text{C}}}-(\text{CH}_2)_4-\overset{\text{O}}{\underset{\|}{\text{C}}}\right)_n$	Fibers for carpets, clothing, upholstery, book bags, soft luggage
Kevlar $\left(\overset{\text{O}}{\underset{\|}{\text{C}}}-\text{C}_6\text{H}_4-\overset{\text{O}}{\underset{\|}{\text{C}}}-\text{N}(H)-\text{C}_6\text{H}_4-\text{N}(H)\right)_n$	Fibers for strengthening clothing, structures, and other materials
Nomex $\left(\overset{\text{O}}{\underset{\|}{\text{C}}}-\text{C}_6\text{H}_4-\overset{\text{O}}{\underset{\|}{\text{C}}}-\text{N}(H)-\text{C}_6\text{H}_4-\text{N}(H)\right)_n$	Fibers for flame-resistant fabrics
Poly(ethylene naphthalate) (PEN) $\left(\text{O}-\overset{\text{O}}{\underset{\|}{\text{C}}}-\text{C}_{10}\text{H}_6-\overset{\text{O}}{\underset{\|}{\text{C}}}-\text{O}-\text{CH}_2-\text{CH}_2\right)_n$	Bar code labels

Addition Polymers

Name and Formula	Uses
Polyethylene $(\text{CH}_2-\text{CH}_2)_n$	Bags, bottles, sheeting, toys
Polypropylene $\left(\underset{\text{CH}_3}{\text{CH}}-\text{CH}_2\right)_n$	Fibers for home furnishings and clothing, lab ware
Poly(vinyl chloride) (PVC) $\left(\text{CH}_2-\underset{\text{Cl}}{\text{CH}}\right)_n$	Pipes and plumbing fixtures, construction materials
Polyacrylonitrile (PAN) $\left(\text{CH}_2-\underset{\text{CN}}{\text{CH}}\right)_n$	Carpets, knitwear
Polytetrafluoroethylene (Teflon) $(\text{CF}_2-\text{CF}_2)_n$	Non-stick cooking surfaces, sealants, bearings

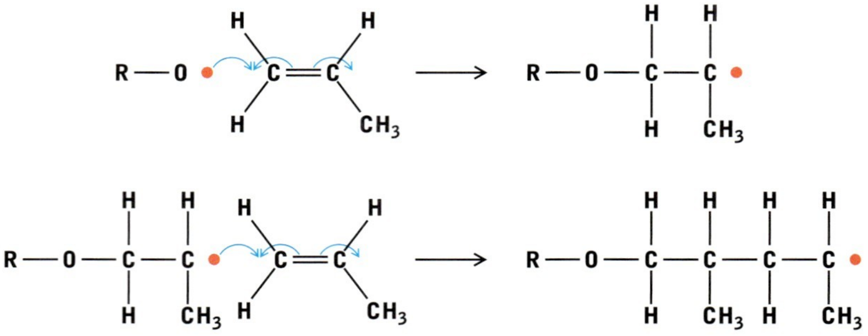

FIGURE 18.24 Polypropylene is formed in an addition reaction initiated by an R—O• radical. Attack of the radical breaks the C=C π bond in a propene molecule. One of the two π bond electrons combines with the odd electron on the radical, forming an O—C bond; the other remains unpaired. This newly formed radical can attack another propene molecule where another addition reaction takes place, and another, and another. . . .

> The **dalton (D)** is a unit of mass equal to 1 amu. Scientists usually express the mass of biological and synthetic polymers in kilodaltons (kD).

> An **atactic polymer** has a random orientation of side groups attached to chiral centers along the polymer chain; an **isotactic polymer** has groups oriented mostly in one direction.

A. Atactic

B. Isotactic

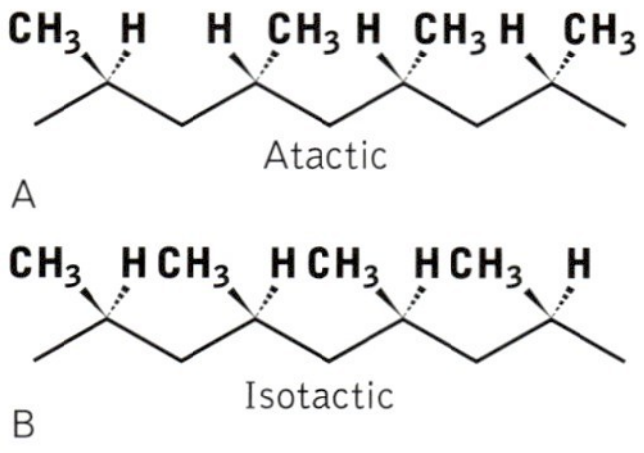

FIGURE 18.25 A. In atactic polypropylene, methyl groups are oriented randomly along the polymer backbone. B. In isotactic polypropylene (the commercially more important form), the methyl groups are aligned in the same direction.

---

forms between R—O and propene. The product of the reaction a new highly reactive radical. This new radical attacks another propene molecule, forming another C—C σ bond as shown in Figure 18.24. Similar addition reactions continue and the chain grows until there is so little propene left that the polymeric radical is more likely to react with another radical instead of adding another propene building block. When this happens, the odd electrons of the two radicals pair up, forming a C—C bond, and we have a molecule of polypropylene with a mass of about 50,000–200,000 amu [50–200 kilodaltons (kD)].

Polypropylene can be manufactured with the CH_3 group concentrated on one side of the zigzag polymer chain formed by C—C bonds, or they may be randomly distributed on both sides. Their distribution depends on the reaction conditions, and particularly on which catalyst was used to promote the reaction. Polypropylene with a random distribution of methyl (—CH_3) groups is called **atactic**; polypropylene with significantly more methyl groups on one side of the chain than the other is called **isotactic** (see Figure 18.25).

> Does tacticity affect the properties of polypropylene?

Does tacticity affect the properties of polypropylene? Yes indeed, because isotactic polymer chains can align themselves in a more systematic and orderly fashion than atactic polymers. This leads to more crystallinity in the secondary structure of the isotactic polymer and more stiffness and strength in material made from it. Materials made from atactic polypropylene tend to be softer, weaker, and more flexible.

The properties of fibers made of polypropylene depend on the secondary structure of the polymer and to the absence of hydrogen bonds. The tetrahedral arrangement of groups around each carbon atom allows the chain to assume a helical secondary structure as shown in Figure 18.26. Like a spring, the flexibility of the coiled helix accounts for the elasticity of polypropylene fibers. Stretching absorbs a significant amount of tension that might otherwise break a less elastic material. Polypropylene breaks only when enough force has been applied to stretch the helix into an extended zigzag structure and then additional force ruptures C—C bonds. The effect of the helix on the properties of polypropylene and other polymers formed from alkenes becomes evident when the strength and elasticity along their chains are compared with strength and elasticity perpendicular to the chains. For example, both the strength and modulus of elasticity along

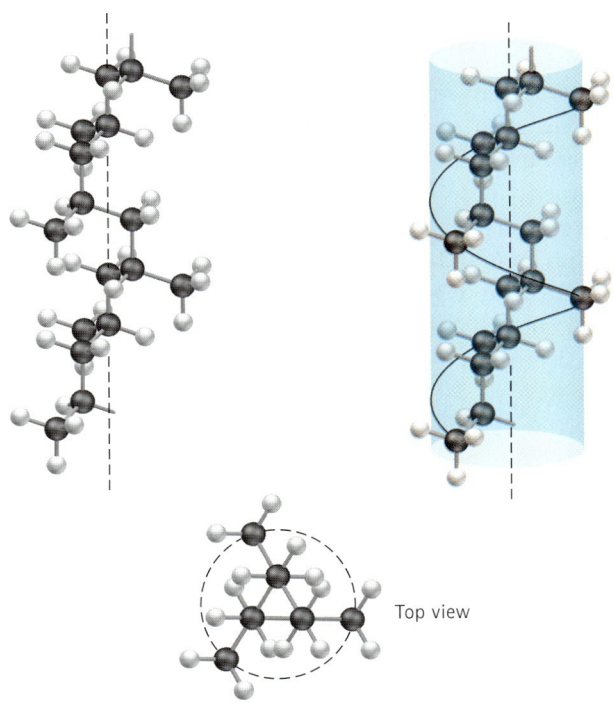

FIGURE 18.26 The strength and stiffness of rope and other materials made of isotactic polypropylene are related to the helical molecular structures of this polymer. In the structure shown here, the helix turns counterclockwise from bottom to top. There are three monomeric units per turn, as can be seen if we view the helix from the top. In isotactic polypropylene, clockwise- and counterclockwise-wound helices are equally distributed.

the chains of polyethylene (described below) are about 100 times greater than perpendicular to them.

The properties of fibers made of polypropylene are also due to the absence of hydrogen bonds between the chains of these nonpolar polymers. Weak interchain interactions allow these materials to be melted and drawn into fibers at relatively low temperatures. This kind of processing is not possible for hydrogen-bonded natural fibers such as wool or cotton. A lack of hydrogen bonding also accounts for the water-repellent properties of garments made of densely woven polypropylene fibers. Knitted garments made of polypropylene fibers, including swimsuits, socks, and underwear, allow moisture to pass through with little being absorbed by the fibers. Therefore, they dry more quickly than comparable items made of cotton.

Other hydrocarbons with C=C double bonds also undergo addition polymerization. Among the most commonly encountered products is polystyrene,

$$+CH_2-CH+_n$$
          |
          C_6H_5

which is used to make Styrofoam cups and building materials, and polyethylene:

$$+CH_2-CH_2+_n$$

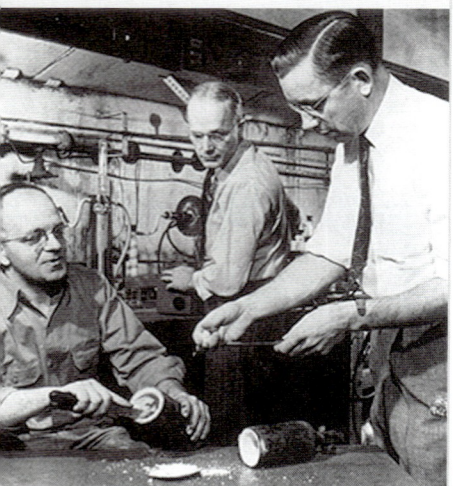

In this photograph, Roy Plunkett (on the right) joins assistants Jack Rebok (left) and Robert McHarness in reenacting their discovery of Teflon.

CONNECTION: According to Graham's law the rate of effusion of a gas is inversely proportional to the square root of the mass of its molecules (see Section 8.6). The difference in the rates of effusion of $^{235}UF_6$ vs. $^{235}UF_6$ is only 0.3%—not much, but enough for the brilliant scientists and engineers of the Manhattan Project to enrich ^{235}U from 0.7% (its natural abundance) to over 50%.

which is widely used in plastic bags and containers for household liquids. Yet another polymer was discovered by accident by Roy J. Plunkett (1910–1994), a young chemist working at a DuPont research laboratory in Deepwater, New Jersey. Plunkett's research team was working on the development of new chlorofluorohydrocarbons (the same CFCs now implicated in stratospheric ozone depletion) when, in April 1938, he directed an assistant to fill metal canisters with tetrafluoroethylene (TFE) gas (a starting material for making CFCs). The canisters were chilled in dry ice to liquefy the TFE and later warmed to room temperature. At that point the canisters should have been full of TFE gas again, but they were not. Instead of gaseous TFE, all the canisters contained white greasy flakes. Plunkett correctly concluded that the white flakes were polymerized TFE, or poly(tetrfluoroethylene)—a substance that would become better known by its trade name, Teflon.

Plunkett, being a good scientist, realized the significance of this discovery and set about investigating the properties of the white flakes. Not only were they very slippery, but they were also the most chemically inert, corrosion-resistant material in the world, and they possessed these properties over an enormous temperature range: from -250 °C to above $+250$ °C.

In the late 1930s, the cost of producing Teflon was so high that it had little commercial value. However, World War II and the Manhattan Project changed that. To build an atomic bomb, scientists in the Manhattan Project needed to concentrate the fissionable isotope of uranium (^{235}U) from the naturally occurring mixture of uranium isotopes—mostly nonfissionable ^{238}U. To do so they converted uranium ore to volatile uranium hexafluoride. The hexafluoride of the fissionable isotope ($^{235}UF_6$) has a slightly smaller mass than the more abundant $^{238}UF_6$. According to Graham's law (Section 8.6), the gas with the smaller mass should effuse faster through a porous membrane. Before scientists could exploit this property to seperate the two isomers, they needed vessels in which uranium hexafluoride, which is extremely corrosive, could be synthesized and processed. That's where Teflon came in. The wartime demand for Teflon led to the development of new, less costly methods for making it. Further cost reductions in the last half of the twentieth century led to the large-scale production of a variety of fluorinated polymers—most with Teflon in their names—that are used to make many industrial and consumer products, including lubricants, sealants, and the nonstick cooking surfaces in pots and pans.

18.5 THE SCIENTIFIC METHOD REVISITED

The story of the discovery of Teflon allows us to revisit an important aspect of scientific inquiry: that major advances sometimes happen when scientists discover things they weren't even looking for. In Chapter 1, we discussed the discovery of the cosmic microwave background in the 1960s by Wilson and Penzias at Bell Labs. These scientists were trying to develop the world's first satellite-based communication system; they weren't searching for an echo of the Big Bang. The significance of their finding became known because they and other scientists understand the importance of networking and sharing ideas. Roy Plunkett was a brilliant scientist because he realized the significance of his accidental discovery of Teflon and pursued it, following a line of research that had little to do with making new refrigerants. This pursuit of knowledge in uncharted directions is part of the dynamic of basic research. Successful scientific inquiry is about being prepared for the unexpected and responding creatively to it.

CHAPTER REVIEW

Summary

SECTION 18.1

Processing ores involves roasting them at temperatures at which metals in them are converted to their oxides, followed by reducing the oxides to free metals in processes called smelting.

Most metals are malleable and ductile. These properties are consistent with the model of atoms in metals slipping by each other in "seas of electrons." Work-hardening a metal by repetitive hammering or bending results in intersecting or pinned dislocations in the metal that hinder the slipping process.

Alloys are blends of a host metal and one or more other metals added to enhance the properties of the host including strength, hardness, and corrosion resistance. In substitutional alloys atoms of the added metals replace atoms of the host metal in its crystal structure. In interstitial alloys atoms of alloying elements are located in the spaces between atoms of the host metal. Intermetallic compounds have constant compositions and structures that are different from those of their constituent metals.

SECTION 18.2

Ceramics are electrical insulators due to the large energy separation between their filled valence and empty conduction bands. Many ceramics are made by heating clays at high temperature in a process called firing, which alters the chemical composition of the clay and makes the material harder, denser, and stronger. Some ceramics become superconductors at low (liquid N_2) temperatures because of the formation of pairs of electrons called Cooper pairs.

SECTION 18.3

Substitution of Group 5A elements for Group 4A elements such as silicon gives n-type semiconductors. Substituting a Group 3A element such as gallium creates a p-type semiconductor. These substitutions, called doping, enhance the conductivities of the Group 4A elements.

SECTION 18.4

Natural fibers are made of cellulose (cotton and linen) or protein (silk and wool). The primary structure of a protein is the sequence of the amino acids in the protein chain. The arrangement in space of the chain defines the secondary structure of a protein and is determined by interactions (for example, hydrogen bonds and covalent bonds) within and between protein chains. Secondary structures common to all proteins include pleated sheets (which impart rigidity and strength) and helical coils (which add elasticity). The overall three-dimensional shape of a protein is called its tertiary structure.

Synthetic polymers are produced by condensation reactions (analogous to the reactions that produce polysaccharides and proteins in living systems) and addition reactions involving reactants with π bonds in their structures. Some synthetic polymers, such as polypropylene, have chiral centers in the polymer backbone.

Random orientation of side groups at these chiral centers produces atactic polymers; nonrandom orientations produce isotactic polymers that are stiffer and stronger.

SECTION 18.5

The discovery of Teflon is another example of the serendipity, first discussed in Chapter 1, that sometimes surrounds major scientific discoveries.

Key Terms

addition polymer (p. 936)
addition reaction (p. 936)
atactic polymer (p. 938)
Bayer process (p. 914)
homogeneous alloy (p. 901)
intermetallic compound (p. 903)
interstitial alloy (p. 906)

isotactic polymer (p. 938)
phosphor (p. 925)
primary structure (p. 930)
secondary structure (p. 930)
semiconductor (p. 926)
smelting (p. 894)

stainless steel (p. 909)
substitutional alloy (p. 901)
superconductor (p. 921)
tensile strength (p. 931)
tertiary structure (p. 931)
work-hardening (p. 896)

QUESTIONS AND PROBLEMS

Refining Ores

CONCEPT REVIEW

1. Ancient metalworkers recovered copper from its ore chalcopyrite, $CuFeS_2$, by first roasting the ore in air and then smelting the product of the roasting process in a reaction that was deprived of air. How did these reaction conditions alter the oxidation state of Cu?
2. Among the products of roasting Cu_2S in air are CuO and SO_2. Which of the three elements Cu, S, and O are oxidized and which are reduced in this reaction?
3. Metal smelting using charcoal (carbon) requires high temperatures. Give some reasons why high temperatures would tend to promote the smelting process.
4. In some texts the formula of the soluble form of Al(III) at high pH is written, AlO_2^-, instead of $Al(OH)_4^-$. Are these two formulas of dissolved Al(III) species equivalent? Explain your answer.

PROBLEMS

5. Which of the following iron ores contains the greatest percent iron by mass: $FeCO_3$ (siderite), Fe_2O_3 (hematite), Fe_3O_4 (magnetite), or FeS_2 (pyrite)?
6. Cobalt was called a "strategic mineral" during much of the twentieth century. Large deposits of the major cobalt minerals, smaltite ($CoAs_2$), cobaltite (CoAsS), and linnaeite (Co_3O_4), are found in Africa and Australia. Which of these minerals contains the greatest percent cobalt by mass?
7. The iron in copper-containing ores such as chalcopyrite is removed during the roasting and smelting process in a reaction with silica that forms iron silicate:

$$FeCuS_2(s) + O_2(g) \longrightarrow FeO(s) + SO_2(g) + CuS(s)$$
$$FeO(s) + SiO_2(s) \longrightarrow FeSiO_3(s)$$

Balance these chemical equations and determine the oxidation state of Fe in each Fe-containing substance.

8. Extensive mining in the nineteenth and twentieth centuries consumed many of the richest deposits of gold in the world. However, it is still profitable to extract gold from rocks with Au concentrations as low as 5 ppm by mass by using a process in which: (1) the crushed ore is treated with sodium cyanide to produce soluble $Au(CN)_2^-$; (2) this resulting solution is then reacted with zinc dust to produce gold. One of the by-products of the

second reaction is $Zn(CN)_4^{2-}$. Write balanced chemical equations for the two reactions and identify the oxidation states of Au and Zn in their cyano complexes.

9. Magnesium can be produced at high pressure by reaction of the magnesium/calcium oxide (approximate formula $CaMg(CO_3)_2$) with an iron/silicon alloy (FeSi). Identify the oxidizing and reducing agents in the equation:
$$2\ CaMg(CO_3)_2 + FeSi(s) \rightarrow 2\ Mg(s) + Ca_2SiO_4(s) + Fe(s) + 2\ CO_2(g)$$

10. Elemental boron may be prepared by reacting B_2O_3 with magnesium metal. Magnesium oxide is also produced. Write a balanced chemical equation describing the reaction and identify the oxidizing and reducing agents.

11. Reacting Ga_2O_3 with Mg metal and O_2 produces $MgGa_2O_4$. Write a balanced chemical equation describing the reaction and identify the oxidation states of Ga and Mg in the reactants and product.

12. Molybdenite (MoS_2) ore is the most important industrial source of molybdenum metal. The ore is heated in air to a temperature at which molybdenite is converted to MoO_3, which is then reacted with H_2 to form Mo metal.
    a. Write balanced chemical equations describing these two reactions (you will need to predict the likely final chemical forms of H and S).
    b. What are the highest and lowest oxidation states of Mo during the overall smelting process?

13. Use the following heats of formation to calculate the standard enthalpy of reaction (ΔH^0) of the reduction of the following copper ores to copper metal. Assume the reducing agent is CO (g) and the products include Cu(s), $H_2O(g)$ and $CO_2(g)$.

Ore	Formula	ΔH_f^0 (kJ/mol)
Cuprite	Cu_2O	−168.6
Malachite	$CuCO_3 \cdot Cu(CH)_2$	−1051.4
Azurite	$(CuCO_3)_2 \cdot Cu(OH)_2$	−1063.6

14. Use the following heats of formation to calculate the standard enthalpy of reaction (ΔH^0) of the reduction of three iron ores to iron metal using carbon (graphite) as the reducing agent. The products are Fe(s) and $CO_2(g)$.

Ore	Formula	ΔH°_f (kJ/mol)
Siderite	$FeCO_3$	−741
Hematite	Fe_2O_3	−824
Magnetite	Fe_3O_4	−1,118

Alloys

CONCEPT REVIEW

15. White gold was originally developed to give the appearance of platinum. One formulation of white gold contains 25% nickel and 75% gold. Which is more malleable, white gold or pure gold?
16. Is there a difference between a solid solution and a homogeneous alloy?
17. The abundances of copper and zinc in Earth's crust are nearly the same. There are some deposits of copper metal in the crust, but zinc occurs only in ores of zinc compounds. Describe the property (or properties) of these metals that account for this difference in their chemical forms in nature.
18. Explain why an alloy that is 28% Cu and 72% Ag melts at a temperature lower than that of the melting points of either Cu or Ag.

PROBLEMS

19. An interstitial alloy is prepared from two metals A and B where B has the smaller radius. Metal A crystallizes in a face-centered cubic structure. What is the formula of the alloy if B occupies:
    a. all of the octahedral holes,
    b. half the octahedral holes,
    c. $\frac{1}{4}$ of the octahedral holes.
20. An interstitial alloy was prepared from two metals. Metal A with the larger atomic radius has a hexagonal close-packed structure. What is the formula of the alloy if atoms of metal B occupy:
    a. all of the tetrahedral holes,
    b. $\frac{1}{2}$ of the tetrahedral holes,
    c. $\frac{1}{2}$ of the octahedral holes.

21. An interstitial alloy contains one atom of B for every five atoms of host element A, which has a face-centered cubic structure. What fraction of the octahedral holes are occupied in this alloy?
22. If the B atoms in the alloy described in the previous problem occupied tetrahedral holes in A, what percent of the holes would they occupy?
23. A highly ordered substitutional alloy between two metals X and Y has a face-centered cubic unit cell. If all of the X atoms occupy the faces of the unit cell and the Y atoms occupy all of the corner positions, what is the empirical formula of the alloy?
24. Metal M has a body-centered cubic unit cell. In an alloy of M, atoms of X occupy 25% of the sites at the centers of the unit cells. What is the empirical formula of the alloy?
25. Vanadium reacts with carbon to form vanadium carbide, an interstitial alloy. Given the atomic radii of V (135 pm) and C (77 pm), are the carbon atoms more likely to occupy octahedral or tetrahedral holes in a cubic close-packed array of vanadium atoms?
26. What is the minimum atomic radius required for a cubic close-packed metal to accommodate hydrogen atoms in its octahedral holes?
27. Magnesium metal is produced by electrolysis of molten $MgCl_2$ in what is called the Dow process. How much magnesium is produced using a current of 1.00×10^5 A for 4.00 hours? What volume of chlorine is also produced at 1.00 atm of pressure and 800 °C?
28. The principal industrial process for producing sodium metal is based on the electrolysis of molten NaCl. How much sodium is produced using a current of 4.50×10^6 A for 90.0 minutes? What volume of chlorine is produced at 1 atm pressure and 725 °C?

Ceramics

CONCEPT REVIEW

29. Describe the chemical changes that occur when an object made of kaolinite is fired in a kiln to 1500 °C.
30. What properties of kaolinite make it a useful starting material for making porcelain fixtures?
31. Why do metal oxides tend to have higher melting points than the free metals?
32. Predict which compound in each of the following pairs has the higher melting point.
    a. CaF_2 or ZrO_2
    b. UO_2 or CeO_2
    c. MgO or CaO
33. Does the addition of Al_2O_3 to ZnO create an n-type or p-type semiconductor?
34. Does doping ZnO with Li_2O create an n-type or p-type semiconductor?

PROBLEMS

35. Aluminum nitride and silicon nitride are ceramic materials that can be produced from the corresponding oxides at high temperature.

    $Al_2O_3(s) + 3\ C(s) + N_2(g) \longrightarrow 2\ AlN(s) + 3\ CO(g)$
    $3\ SiO_2(s) + 6\ C(s) + 2\ N_2(s) \longrightarrow Si_3N_4(s) + 6\ CO(g)$

    Identify the elements that are oxidized or reduced in these reactions.

36. If nitrogen is left out of the reactions in Problem 35, the products of the two reactions are aluminum carbide, Al_4C_3, and silicon carbide, SiC, respectively. Write balanced chemical equations for these reactions and determine whether they are redox processes.
37. X-ray diffraction patterns are obtained for two samples of mica, a layered aluminosilicate mineral. One of them is muscovite, $KAl_2(OH)_2(AlSi_3O_{10})$, and the other is paragonite, $NaAl_2(OH)_2(AlSi_3O_{10})$. The diffraction patterns indicate the space between successive pairs of Al- and Si- containing layers are 193.2 and 200.3 pm, respectively. Which dimension corresponds to the muscovite sample?
38. If the wavelength of the x-rays used in Problem 37 is 154 pm, what are the values of 2θ for first-order ($n = 1$) diffraction of x-rays by muscovite and paragonite?
39. Grossular garnets have the general formula $Ca_3Al_2(SiO_4)_3$. Within the silicate crystal structure the two cations may occupy tetrahedral, octahedral, and distorted cubic holes.
    a. Without consulting a table of ionic radii, predict which cation in these garnets is the larger.
    b. Predict the types of holes in which the cations are likely to be found in the garnet structure.
40. The ceramic material barium titanate ($BaTiO_3$) is used in devices that measure pressure. The radii of Ba^{2+}, Ti^{4+}, and O^{2-} are 149, 56, and 126 pm, respectively. If the

O^{2-} ions are in a closest-packed structure, which hole(s) can accommodate the metal cations?

41. Decomposition of 1.00 kg of kaolinite ($Al_2Si_2O_5(OH)_4$) fired in a kiln at 950 °C produces a mixture of mullite ($Al_6Si_2O_{13}$), silica, and water vapor. Calculate the percent weight loss during firing.

42. Talc, ($Mg_3(OH)_2Si_4O_{10}$,) is a soft, white mineral used in cosmetics. It is formed by heating the mineral antigorite, $Mg_6(OH)_8Si_4O_{10}$, in a decomposition that also produces magnesium oxide and water vapor. Write a balanced chemical equation describing this reaction and calculate the weight lost during the decomposition of a 150.0-g sample of antigorite.

43. The series of nitride ceramics, AlN, GaN, and LnN are all semiconductors used in the microelectronics industry. Their band gaps are 580.6, 322.1, and 192.9 kJ/mol, respectively. Which, if any, of these energies correspond to radiation in the visible region of the spectrum?

44. Calculate the wavelengths of light emitted by the semiconducting phosphides AlP, GaP, and InP, which have band gaps of 241.1, 216.0, and 122.5 kJ/mol, respectively.

*45. At 1400 K, zinc oxide is in equilibrium with Zn and O_2:

$$2\ ZnO(s) \rightleftharpoons 2\ Zn(g) + O_2(g)$$

If ΔG_{rxn} for the reaction is 183 kJ and $P_{O_2} = 0.21$ atm, what is P_{Zn} in a mixture at equilibrium at 1400 K?

*46. Magnesium oxide decomposes at 1000 K to magnesium gas and O_2. If $\Delta G_{f, MgO} = -493$ kJ/mol at 1000 K, determine the partial pressures of O_2 and Mg at 1000 K for the equilibrium:

$$2\ MgO(s) \rightleftharpoons 2\ Mg(g) + O_2(g)$$

Fibers

CONCEPT REVIEW

47. Predict the signs of ΔS and ΔH for the spontaneous polymerization of propene (propylene) to polypropylene.
48. Describe the functions of the catalysts used in the polymerization of propene.
49. According to valence bond theory, what is the hybridization of the carbon atoms in the $(CH_2)_n$ backbone of flexible, elastic polymers?
50. Explain the difference between the primary, secondary, and tertiary structures of proteins.
51. Explain the contribution of the primary, secondary, and tertiary structures of silk to its strength and elasticity.
52. Describe the differences in the forces responsible for the helical secondary structure of proteins and polypropylene.
53. Are the π-electrons in Kevlar likely to be delocalized over the entire polymer chain?
54. Are the π-electrons in Nomex likely to be delocalized over the entire polymer chain?
55. In the structure of silk, each amino acid can form at least two hydrogen bonds with N—H and C=O groups on neighboring strands. Which amino acids can form more than two hydrogen bonds?
56. A biotechnology firm produces a series of synthetic polypeptides containing glycine, alanine, and serine during the development of synthetic silk. How does a higher percentage of serine affect the elasticity of the polymer?

PROBLEMS

57. Using the average bond energies in Table 11.2, estimate ΔH^0 for the polymerization of propene.

$$n CH_2{=}CH{-}CH_3 \rightarrow {-}(CH_2{-}CH(CH_3)){-}_n$$

58. Using the average bond energies in Table 11.2, estimate ΔH^0 for the formation of a peptide bond between two glycine molecules.

[Structural diagram: two glycine molecules $H_2N{-}CH_2{-}C(=O){-}OH$ combining to form dipeptide $H_2N{-}CH_2{-}C(=O){-}N(H){-}CH_2{-}C(=O){-}OH + H_2O$]

59. Silly Putty is a condensation polymer of dihydroxydimethylsilane:

$$HO{-}Si(CH_3)(CH_3){-}OH$$

Draw the condensed molecular structure of the repeating monomeric unit in Silly Putty.

60. Kevlar can be formed by a condensation reaction between the two compounds shown below.

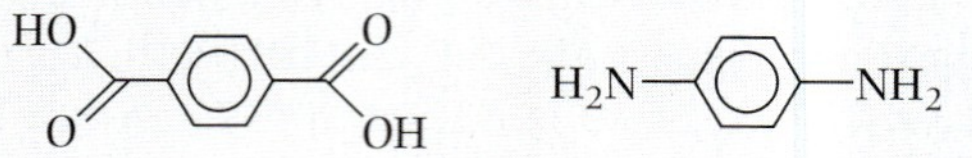

Draw the condensed molecular structure of the repeating monomeric unit of the polymer that would be formed if the diamino compound on the right (above) were replaced with the following compound in the polymerization reaction:

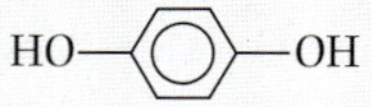

61. Silicones are polymeric materials with the formula $\{R_2SiO\}_n$ where R is CH_3 or C_6H_5 (phenyl).

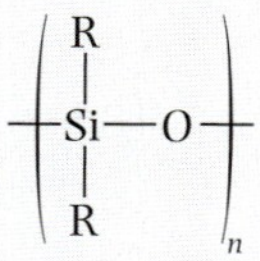

Why are silicones water repellent?

62. The following three polymers are widely used in the plastics industry. In which of them are the intermolecular forces per mole of monomer the strongest?

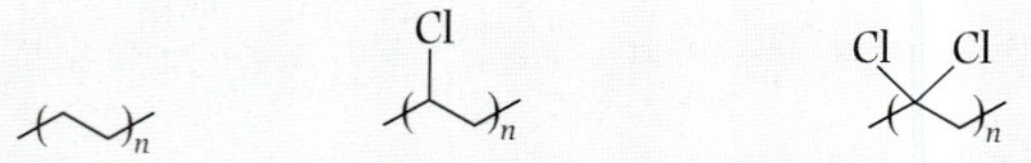

Polyethylene   Poly(vinylchloride)   Poly(vinylidenechloride)

63. Which of the following polymers would you expect to be more elastic?

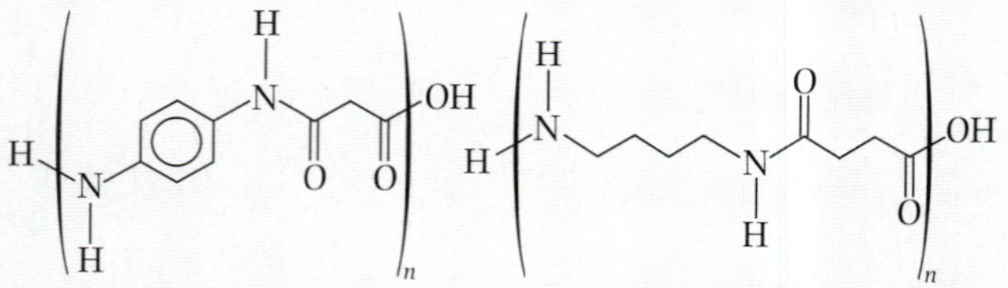

64. Synthetic rubber can be prepared by addition polymerization of either butadiene or isoprene.

Butadiene   Isoprene

Draw line structures of the repeating monomeric units in poly(butadiene) and poly(isoprene).

65. Polymerization of which of the following alkenes could produce an isotactic addition polymer?

a   b   c

66. Draw the line structure of a three-monomer segment of an isotactic polymer formed by addition polymerization of one of the compounds in Problem 65.

67. Poly(thiophene) and poly(pyrrole) change color when they are reduced electrochemically. Identify the reactants in the polymerization reactions that produce each of them. Are these polymers addition or condensation polymers?

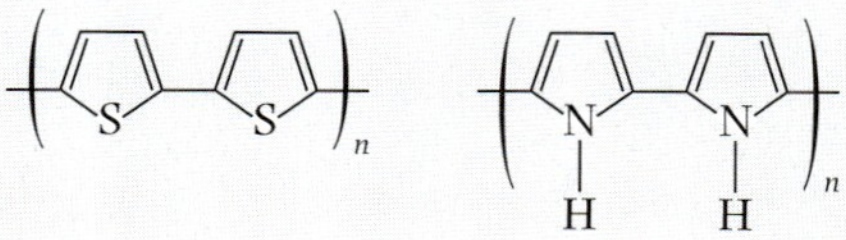

68. The condensed molecular structure of Poly(acrylonitrile), which is marketed as Orlon™ and Acrilon™, is shown below. Identify the monomeric reactant that produces this polymer. Is poly(acrylonitrile) an addition or a condensation polymers?

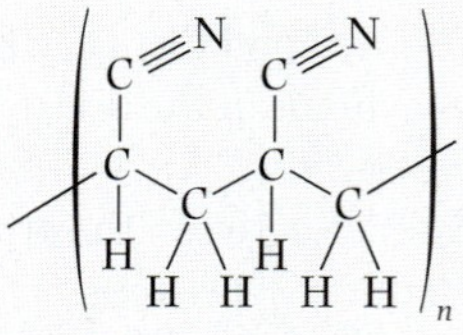

Poly(acrylonitrile)

69. Identify the reactants in the polymerization reactions that produce each of the following polymers. Are they addition or condensation polymers?

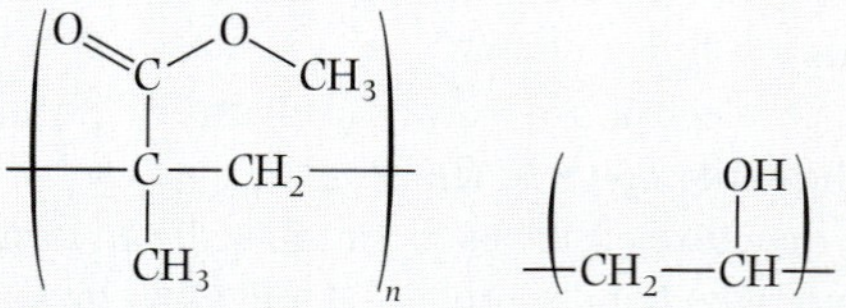

70. Identify the reactants in the polymerization reactions that produce each of the following polymers. Which of

them is an addition and which is a condensation polymer?

71. Computer diskettes are made of poly(vinyl acetate). Draw the condensed molecular structure of this polymer. The structure of its monomer is:

72. The 2000 Nobel Prize in Chemistry was awarded for research on the electrically conductive polymer poly(acetylene).
    a. Draw the molecular structure of three monomeric units of the addition polymer that results from polymerization of acetylene, HC≡CH.
    b. Draw the molecular structure of two isomers of poly(acetylene).
    c. Identify the feature of the structure of the polymer that makes it conductive.

APPENDIX 1
Mathematical Procedures

WORKING WITH EXPONENTIAL NOTATION

Quantities that scientists work with often are very large, such as Earth's mass, or very small, such as the mass of an electron. It is easier to work with these numbers if they are expressed in exponential notation.

The general form of standard exponential notation is a value between 1 and 10 multiplied by 10 raised to an integral power. According to this definition, 598×10^{22} kg (Earth's mass) is not in standard exponential notation, but 5.98×10^{24} kg is. It is good practice to use and report data or parameters in standard exponential notation.

1. **To convert an "ordinary" number to standard exponential notation** move the decimal point to the left for a large number, or to right for a small one, so that the decimal point is located after the first nonzero digit.
    A. For example, to express Earth's average density ($5,517$ kg/m^3) in exponential notation requires moving the decimal three places to the left. Doing so is the same as dividing the number by 1000, or 10^3. To keep the value the same we add an exponent to multiply it by 10^3. So, Earth's density in standard exponential notation is 5.517×10^3 kg/m^3.
    B. If you move the decimal point of a value less than one to the right to express it in exponential notation, then the exponent is a negative integer equal to the number of places you moved the decimal point to the right. For example, the value of R used in solving ideal gas law problems is 0.08206 L $\cdot$ atom/mol $\cdot$ K. Moving the decimal point two places to the right, converts the value of R to exponential notation: 8.206×10^{-2} L $\cdot$ atom/mol $\cdot$ K.
    C. Another value of R, 8.314 J/mol $\cdot$ K, does not need an exponent, though it could be written 8.314×10^0 J/mol $\cdot$ K.

2. **To add or subtract numbers in exponential notation,** their exponents must be the same. This may require that you to change the exponents of some values. (Note: This step is not necessary when using a scientific calculator because the calculator will make all necessary conversions.)

Sample Exercise 1 Calculate the sum of the masses of the subatomic particles in an atom of lithium-7. Given the following masses in grams:

Particle	Mass (g)
proton	1.67263×10^{-24}
neutron	1.67494×10^{-24}
electron	9.10939×10^{-28}

Solution An atom of ^{7}Li has three protons, three electrons, and four neutrons. Therefore, the total mass of the subatomic particles in an atom is the sum of three times the masses of a proton and electron and four times the mass of a neutron. Doing the multiplication steps first:

mass of three protons = $3(1.67263 \times 10^{-24}\text{ g}) = 5.01789 \times 10^{-24}$ g

mass of four protons = $4(1.67494 \times 10^{-24}\text{ g}) = 6.69976 \times 10^{-24}$ g

mass of three electrons = $3(9.10939 \times 10^{-28}\text{ g}) = 2.732817 \times 10^{-27}$ g

Before adding these masses together, we must express them all with the same exponent. The most convenient is 10^{-24}. To express the combined mass of three electrons using this exponent requires shifting the decimal point in 2.732817×10^{-27} g three places to the left, which makes the number before the exponent 1000 times smaller and compensates for the exponent becoming 1000 times larger. Adding the resulting mass of the three electrons, $0.002732817 \times 10^{-24}$ g, to the other two masses, we have:

$$5.01789 \times 10^{-24}\text{ g}$$
$$+\ 6.69976 \times 10^{-24}\text{ g}$$
$$\underline{+\ 0.002732817 \times 10^{-24}\text{ g}}$$
$$11.720382817 \times 10^{-24}\text{ g}$$

We need to round off this sum so that we have only five digits to the right of the decimal place, because that is the smallest number of digits to the right in two of the three values being summed. Therefore, the final answer is

$$11.72038 \times 10^{-24}\text{ g}$$

3. **To multiply values with exponents,** the values in front of the exponents are multiplied together, but the exponents are added (these steps happen automatically with scientific calculators).

Sample Exercise 2 American Steve Fossett circumnavigated the globe in early summer, 2002, in the *Spirit of Freedom* balloon, which was partially filled with 5.5×10^5 ft^3 of helium. What is this volume in liters? Given: 1 ft^3 = 2.83×10^4 L

Solution We convert the starting value by multiplying 5.5 by 2.83 and adding the exponents (4 + 5):

$$(5.5 \times 10^5\text{ ft}^3)\left(\frac{2.83 \times 10^4\text{ L}}{\text{ft}^3}\right) = 15.5 \times 10^9\text{ L or } 1.55 \times 10^{10}\text{ L}$$

4. **To divide values with exponents,** the values in front of the exponents are divided, but the exponents are subtracted (again, these steps happen automatically with scientific calculators).

Sample Exercise 3 The speed of light is 2.998×10^8 m/s. What is the equivalent speed in miles per second? Given: 1 mile = 1.609×10^3 m

Solution Expressing the speed of light in miles per second requires dividing the speed of light in meters per second by the conversion factor given in the exercise. Dividing the values in front of the exponents (2.998/1.609) and subtracting their exponents (8 − 3), we have:

$$\frac{2.998 \times 10^8\text{ m/s}}{1.609 \times 10^3\text{ m/mi}} = 1.863 \times 10^5\text{ mi/s}$$

Working with Logarithms

A logarithm to the base 10 has the following form:

$$\log_{10} x = \log x = p, \text{ where } x = 10^p$$

We usually abbreviate the logarithm function "log" if the logarithm is to the base 10, which means the scale in which the log 10 = 1.

A logarithm to the base e, called a *natural* logarithm, has the following form:

$$\log_e x = \ln x = q, \text{ where } x = e^q$$

Scientific calculators have "log" and "ln" buttons, so it is easy to convert a number into its log or ln form. The directions below apply to most non-graphing calculators.

Sample Exercise 4 Find the logarithm to the base 10 of 4.5 (log 4.5).

Solution Enter 4.5 into your calculator and press the "LOG" button.* The answer should be 0.6532 (to four significant figures).

Sample Exercise 5 Find the logarithm to the base 10 of 100 (log 100).

Solution Enter 100 into your calculator and press the "LOG" button. The answer should be 2. This answer is as expected, because 10 (the base) raised to the power of the log (2) = 10^2 = 100.

Sample Exercise 6 Find the natural logarithm of 4.5.

Solution Enter 4.5 into your calculator and press the "LN" button. The answer should be 1.504.

Sample Exercise 7 Find the natural logarithm of 100.

Solution Enter 100 into your calculator and press the "ln" button. The answer should be 4.61.

Let's compare the results of the four previous exercises. In both pairs of ln and log values, the ln value is 2.303 times the log value. These examples fit the general equation:

$$\ln x = 2.30 \log x$$

It is reasonable that the ln of a value is greater that the log of the same value because ln is based on *e* (2.718) while log is to the base 10. The smaller base of ln units means that there are more of them than log units in a given value.

---

*If you have a graphing calculator such as a TI 83, press the "LOG" button, enter 4.5, and then press the "ENTER" button.

Sample Exercise 8 Calculate $\ln 1.2 \times 10^{-3}$.

Solution Enter 1.2 into the calculator and then press the "EXP" button, and then enter 3, and "+/−". Finally press "LN". The corresponding keystrokes with a graphing calculator are "LN", 1.2, "x", "^", "(−)", 3, "ENTER". The result should be −6.725, which is negative because the original number is less than 1. Keep in mind that values greater than zero have positive logarithm values; those less than zero have negative logarithm values.

5. **Combining logs:** The following equations summarize how logarithms of the products or quotients of two or more values are related to the individual logs of those values.

$$\text{logarithm } ab = \text{logarithm } a + \text{logarithm } b$$

and

$$\text{logarithm } a/b = \text{logarithm } a - \text{logarithm } b$$

Converting Logarithms into Numbers

If we know the value of log x, what is the value of x? This question frequently arises when working with pH (see Chapter 16), which is the negative log of the concentration of hydrogen ions, [H$^+$], in solution:

$$\text{pH} = -\log [\text{H}^+]$$

Suppose the pH of a solution of a weak acid is 2.50. The concentration of H$^+$ is related to this pH value as follows:

$$2.50 = -\log[\text{H}^+]$$

or

$$-2.50 = \log[\text{H}^+]$$

To find the value of [H$^+$], enter 2.5 in your calculator and press the "+/−" button to change its sign to −2.5. The next step depends on the type of calculator you have. If yours has a "10^x" button, push it to find the value of $10^{-2.5}$, which is the number we are looking for. The corresponding keystrokes with a graphing calculator are "10^x", "(−)", 2.5, "ENTER". On some calculators there is no "10^x" key, but there is an inverse function, or "INV" key that is used to invert other function keys. Hitting the "INV" key followed by the "log" key takes the inverse of a log, called an *antilog*, which is the same as raising 10 to the power (−2.5 in this case) that was entered. Some calculators, including the virtual one in many Windows operating systems, has an "x^y" key. To use it you enter 10, push the "x^y" key, enter 2.5, and then push the "+/−" key followed by the equals sign. All of these approaches do the same calculation: taking 10 to the −2.50 power, and give the same answer: [H$^+$] = 3.2×10^{-3}.

Sample Exercise 9 Calculate the hydrogen ion concentration in rainwater in which pH = 5.62.

Solution Using one of the methods described above you should find that the value of $10^{-5.62}$ is 2.4×10^{-6}.

Solving Quadratic Equations

If the terms in an equation can be rearranged so that they take the form:

$$ax^2 + bx + c = 0$$

they have the form of a quadratic equation. The value(s) of x can be determined from the values of the coefficients a, b, and c, by using the equation:

$$x = \frac{-b \pm \sqrt{b^2 - 4ac}}{2a}$$

For example, if the solution to a problem yields the following expression where x is the concentration of a solute:

$$x^2 + 0.112x - 1.2 \times 10^{-3} = 0$$

the value of x can be determined as follows:

$$x = \frac{-b \pm \sqrt{b^2 - 4ac}}{2a}$$
$$= \frac{-0.112 \pm \sqrt{(0.112)^2 - 4(1)(-1.2 \times 10^{-3})}}{2(1)}$$
$$= \frac{-0.112 \pm \sqrt{0.01254 + 0.0048}}{2}$$
$$= \frac{-0.112 \pm 0.132}{2} = +0.010 \text{ or } -0.122$$

In this example, the negative value for x satisfies the equation, but it has no meaning since we cannot have negative concentration values; so, we use only the +0.010 value.

Expressing Data in Graphical Form

Fitting curves to plots of experimental data is a powerful tool in determining the relations between variables. Many natural phenomena obey exponential functions. For example, the rate constant (k) of a chemical reaction increases exponentially with increasing absolute temperature (T). This relation is described by the Arrhenius equation (see Chapter 14):

$$k = Ae^{\frac{-E_a}{RT}}$$

Table 1 Rate Constant k as a Function of Temperature T

Temperature T, (K)	$1/T$, (K^{-1})	Rate constant k	$\ln k$
500	0.0020	0.030	−3.5
550	0.0018	0.38	−0.97
600	0.0017	2.9	1.1
650	0.0015	17	2.8
700	0.0014	75	4.3

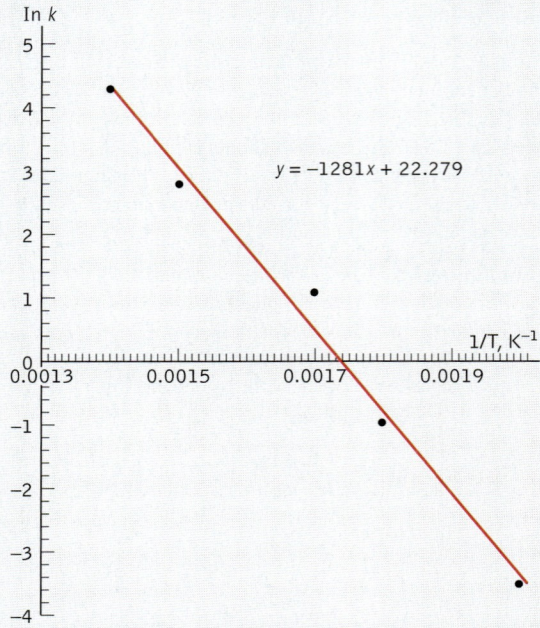

FIGURE 1

where A is a constant for a particular reaction (called the frequency factor), E_a is the activation energy of the reaction, and R is the ideal gas constant. Taking the natural logarithms of both sides of the Arrhenius equation gives:

$$\ln k = \ln A - \left(\frac{E_a}{RT}\right)$$

This equation fits the general equation of a straight line ($y = mx + b$) if ($\ln k$) is the y-variable and ($1/T$) is the x-variable. Plotting ($\ln k$) versus ($1/T$) should give a straight line with a slope equal to $-E_a/R$. The slopes of these plots are negative because the activation energies, E_a, of chemical reactions are positive. The data for a reaction given in columns 2 and 4 of Table 1 are plotted in Figure 1. The slope of the straight line (-1281K) is used to calculate the value of E_a:

$$-1281\text{K} = -\frac{E_a}{R}$$

$$E_a = -(-1281\text{K})(8.31 \text{ J/K mol})$$
$$= 10{,}645 \text{ J/mol} = 10.6 \text{ kJ/mol}$$

APPENDIX 2
SI Units and Conversion Factors

Table A2.1 Some SI Base Units

SI Base Quantity	Unit	Symbol
length	meter	m
mass	kilogram	kg
time	second	s
amount of substance	mole	mol
temperature	kelvin	K
electric current	ampere	A

Table A2.2 Some SI Derived Units

SI Derived Quantity	Unit	Symbol	Dimensions
electric charge	coulomb	C	$A \cdot s$
electric potential	volt	V	J/C
force	newton	N	$kg \cdot m/s^2$
frequency	hertz	Hz	s^{-1}
momentum	newton-second	—	$kg \cdot m/s$
power	watt	W	J/s
pressure	pascal	Pa	N/m
radioactivity	becquerel	Bq	s^{-1}
speed or velocity	meter per second	—	m/s
work, energy, or heat	joule (newton-meter)	J	$kg \cdot m^2/s^2$

Table A2.3 SI Prefixes

Prefix	Symbol	Multiplier	Prefix	Symbol	Multiplier
deci	d	10^{-1}	deka	da	10^1
centi	c	10^{-2}	hecto	h	10^2
milli	m	10^{-3}	kilo	k	10^3
micro	μ	10^{-6}	mega	M	10^6
nano	n	10^{-9}	giga	G	10^9
pico	p	10^{-12}	tera	T	10^{12}
femto	f	10^{-15}	peta	P	10^{15}
atto	a	10^{-18}	exa	E	10^{18}

Table A2.4 Special Units and Conversion Factors

Quantity	Unit	Symbol	Conversion
energy	electron-volt	eV	1 eV = 1.60217733 × 10^{-19} J
force	pound	lb	1 lb = 0.453592 kg (Earth)
heat	calorie	cal	1 cal = 4.184 J
length	angstrom	Å	1 Å = 10^{-8} cm = 10^{-10} m
length	inch	in	1 in = 2.54 cm
mass	atomic mass unit	u	1 u = 1.6605402 × 10^{-27} kg
pressure	atmosphere	atm	1 atm = 1.01325 × 10^5 Pa
pressure	torr	torr	1 torr = 1 atm/760
temperature	Celsius scale	°C	°C = K − 273.15
temperature	Fahrenheit scale	°F	°F = $\frac{9}{5}$(°C) + 32
time	minute	min	1 min = 60 s
time	hour	h	1 h = 60 min = 3600 s
time	day	d	1 d = 24 hr = 86,400 s
time	year	y	1 y = 365.25 d
volume	liter	L	1 L = 1 dm^3 = 10^{-3} m^3
volume	cubic centimeter	cm^3, cc	1 cm^3 = 1 mL = 10^{-3} L

Table A2.5 Physical Constants

Quantity	Symbol	Value
acceleration due to gravity (Earth)	g	9.80665 m/s^2 (exact)
Avogadro's number	N_0	6.0221367 × 10^{23} mol^{-1}
Bohr radius	a_0	0.529177249 × 10^{-10} m
Boltzmann's constant	k_B	1.380658 × 10^{-23} J/K
electronic charge-to-mass ratio	$-e/m_e$	1.75881962 × 10^{11} C/kg
elementary charge	e	1.60217733 × 10^{-19} C
Faraday constant	$\mathscr{F}$	9.6485309 × 10^4 C/mol
mass of an electron	m_e	9.10939 × 10^{-31} kg
mass of a neutron	m_n	1.67494 × 10^{-27} kg
mass of a proton	m_p	1.67263 × 10^{-27} kg
molar volume of ideal gas at STP	V_m	22.41410 L/mol
permittivity of vacuum	ϵ_0	8.854187817 × 10^{-12} C^2/N·m^2
Planck's constant	h	6.6260755 × 10^{-34} J·s
speed of light in vacuum	c	2.99792458 × 10^8 m/s (exact)
universal gas constant	R	8.3145 J/mol·K
		0.082058 L·atm/mol·K

APPENDIX 3
The Elements and Their Properties

Table A3.1 Ground-State Electron Configurations, Atomic Radii, and First Ionization Energies of the Elements

Element	Symbol	Atomic Number Z	Ground-State Configuration	Atomic Radius (Å)	Ionization Energy (kJ/mol)
hydrogen	H	1	$1s^1$	0.37	1312.0
helium	He	2	$1s^2$	0.32	2372.3
lithium	Li	3	$[He]2s^1$	1.52	520.2
beryllium	Be	4	$[He]2s^2$	1.13	899.5
boron	B	5	$[He]2s^2 2p^1$	0.88	800.6
carbon	C	6	$[He]2s^2 2p^2$	0.77	1086.5
nitrogen	N	7	$[He]2s^2 2p^3$	0.70	1402.3
oxygen	O	8	$[He]2s^2 2p^4$	0.66	1313.9
fluorine	F	9	$[He]2s^2 2p^5$	0.64	1681.0
neon	Ne	10	$[He]2s^2 2p^6$	0.69	2080.7
sodium	Na	11	$[Ne]3s^1$	1.86	495.3
magnesium	Mg	12	$[Ne]3s^2$	1.60	737.7
aluminum	Al	13	$[Ne]3s^2 3p^1$	1.43	577.5
silicon	Si	14	$[Ne]3s^2 3p^2$	1.17	786.5
phosphorus	P	15	$[Ne]3s^2 3p^3$	1.10	1011.8
sulfur	S	16	$[Ne]3s^2 3p^4$	1.04	999.6
chlorine	Cl	17	$[Ne]3s^2 3p^5$	0.99	1251.2
argon	Ar	18	$[Ne]3s^2 3p^6$	0.97	1520.6
potassium	K	19	$[Ar]4s^1$	2.27	418.8
calcium	Ca	20	$[Ar]4s^2$	1.97	589.8
scandium	Sc	21	$[Ar]4s^2 3d^1$	1.61	633.1
titanium	Ti	22	$[Ar]4s^2 3d^2$	1.45	658.8
vanadium	V	23	$[Ar]4s^2 3d^3$	1.31	650.9
chromium	Cr	24	$[Ar]4s^1 3d^5$	1.25	652.9
manganese	Mn	25	$[Ar]4s^2 3d^5$	1.37	717.3
iron	Fe	26	$[Ar]4s^2 3d^6$	1.24	762.5
cobalt	Co	27	$[Ar]4s^2 3d^7$	1.25	760.4
nickel	Ni	28	$[Ar]4s^2 3d^8$	1.25	737.1
copper	Cu	29	$[Ar]4s^1 3d^{10}$	1.28	745.5
zinc	Zn	30	$[Ar]4s^2 3d^{10}$	1.34	906.4
gallium	Ga	31	$[Ar]4s^2 3d^{10} 4p^1$	1.22	578.8
germanium	Ge	32	$[Ar]4s^2 3d^{10} 4p^2$	1.22	762.2

Continued on next page

Table A3.1 Ground-State Electron Configurations, Atomic Radii, and First Ionization Energies of the Elements (*Continued*)

Element	Symbol	Atomic Number Z	Ground-State Configuration	Atomic Radius (Å)	Ionization Energy (kJ/mol)
arsenic	As	33	$[Ar]4s^2 3d^{10} 4p^3$	1.21	947.0
selenium	Se	34	$[Ar]4s^2 3d^{10} 4p^4$	1.17	941.0
bromine	Br	35	$[Ar]4s^2 3d^{10} 4p^5$	1.14	1139.9
krypton	Kr	36	$[Ar]4s^2 3d^{10} 4p^6$	1.10	1350.8
rubidium	Rb	37	$[Kr]5s^1$	2.47	403.0
strontium	Sr	38	$[Kr]5s^2$	2.15	549.5
yttrium	Y	39	$[Kr]5s^2 4d^1$	1.78	599.8
zirconium	Zr	40	$[Kr]5s^2 4d^2$	1.59	640.1
niobium	Nb	41	$[Kr]5s^1 4d^4$	1.43	652.1
molybdenum	Mo	42	$[Kr]5s^1 4d^5$	1.36	684.3
technetium	Tc	43	$[Kr]5s^2 4d^5$	1.35	702.4
ruthenium	Ru	44	$[Kr]5s^1 4d^7$	1.32	710.2
rhodium	Rh	45	$[Kr]5s^1 4d^8$	1.34	719.7
palladium	Pd	46	$[Kr]4d^{10}$	1.38	804.4
silver	Ag	47	$[Kr]5s^1 4d^{10}$	1.44	731.0
cadmium	Cd	48	$[Kr]5s^2 4d^{10}$	1.49	867.8
indium	In	49	$[Kr]5s^2 4d^{10} 5p^1$	1.63	558.3
tin	Sn	50	$[Kr]5s^2 4d^{10} 5p^2$	1.40	708.6
antimony	Sb	51	$[Kr]5s^2 4d^{10} 5p^3$	1.41	833.6
tellurium	Te	52	$[Kr]5s^2 4d^{10} 5p^4$	1.43	869.3
iodine	I	53	$[Kr]5s^2 4d^{10} 5p^5$	1.33	1008.4
xenon	Xe	54	$[Kr]5s^2 4d^{10} 5p^6$	1.30	1170.4
cesium	Cs	55	$[Xe]6s^1$	2.65	375.7
barium	Ba	56	$[Xe]6s^2$	2.17	502.9
lanthanum	La	57	$[Xe]6s^2 5d^1$	1.87	538.1
cerium	Ce	58	$[Xe]6s^2 4f^1 5d^1$	1.82	534.4
praseodymium	Pr	59	$[Xe]6s^2 4f^3$	1.82	527.2
neodymium	Nd	60	$[Xe]6s^2 4f^4$	1.81	533.1
promethium	Pm	61	$[Xe]6s^2 4f^5$	1.81	535.5
samarium	Sm	62	$[Xe]6s^2 4f^6$	1.80	544.5
europium	Eu	63	$[Xe]6s^2 4f^7$	2.00	547.1
gadolinium	Gd	64	$[Xe]6s^2 4f^7 5d^1$	1.79	593.4
terbium	Tb	65	$[Xe]6s^2 4f^9$	1.76	565.8
dysprosium	Dy	66	$[Xe]6s^2 4f^{10}$	1.75	573.0
holmium	Ho	67	$[Xe]6s^2 4f^{11}$	1.74	581.0
erbium	Er	68	$[Xe]6s^2 4f^{12}$	1.73	589.3
thulium	Tm	69	$[Xe]6s^2 4f^{13}$	1.72	596.7
ytterbium	Yb	70	$[Xe]6s^2 4f^{14}$	1.94	603.4
lutetium	Lu	71	$[Xe]6s^2 4f^{14} 5d^1$	1.72	523.5
hafnium	Hf	72	$[Xe]6s^2 4f^{14} 5d^2$	1.56	658.5
tantalum	Ta	73	$[Xe]6s^2 4f^{14} 5d^3$	1.43	761.3
tungsten	W	74	$[Xe]6s^2 4f^{14} 5d^4$	1.37	770.0
rhenium	Re	75	$[Xe]6s^2 4f^{14} 5d^5$	1.34	760.3
osmium	Os	76	$[Xe]6s^2 4f^{14} 5d^6$	1.34	839.4
iridium	Ir	77	$[Xe]6s^2 4f^{14} 5d^7$	1.36	878.0
platinum	Pt	78	$[Xe]6s^1 4f^{14} 5d^9$	1.37	868.4

Continued on next page

Table A3.1 Ground-State Electron Configurations, Atomic Radii, and First Ionization Energies of the Elements (*Continued*)

Element	Symbol	Atomic Number Z	Ground-State Configuration	Atomic Radius (Å)	Ionization Energy (kJ/mol)
gold	Au	79	$[Xe]6s^1 4f^{14} 5d^{10}$	1.44	890.1
mercury	Hg	80	$[Xe]6s^2 4f^{14} 5d^{10}$	1.50	1007.1
thallium	Tl	81	$[Xe]6s^2 4f^{14} 5d^{10} 6p^1$	1.70	589.4
lead	Pb	82	$[Xe]6s^2 4f^{14} 5d^{10} 6p^2$	1.75	715.6
bismuth	Bi	83	$[Xe]6s^2 4f^{14} 5d^{10} 6p^3$	1.55	703.3
polonium	Po	84	$[Xe]6s^2 4f^{14} 5d^{10} 6p^4$	1.67	812.1
astatine	At	85	$[Xe]6s^2 4f^{14} 5d^{10} 6p^5$	1.40	924.6
radon	Rn	86	$[Xe]6s^2 4f^{14} 5d^{10} 6p^6$	1.45	1037.1
francium	Fr	87	$[Rn]7s^1$	2.7	380
radium	Ra	88	$[Rn]7s^2$	2.23	509.3
actinium	Ac	89	$[Rn]7s^2 6d^1$	1.88	499
thorium	Th	90	$[Rn]7s^2 6d^2$	1.80	587
protactinium	Pa	91	$[Rn]7s^2 5f^2 6d^1$	1.61	568
uranium	U	92	$[Rn]7s^2 5f^3 6d^1$	1.38	587
neptunium	Np	93	$[Rn]7s^2 5f^4 6d^1$	1.30	597
plutonium	Pu	94	$[Rn]7s^2 5f^6$	1.51	585
americium	Am	95	$[Rn]7s^2 5f^7$	1.84	578
curium	Cm	96	$[Rn]7s^2 5f^7 6d^1$	—	581
berkelium	Bk	97	$[Rn]7s^2 5f^9$	—	601
californium	Cf	98	$[Rn]7s^2 5f^{10}$	—	608
einsteinium	Es	99	$[Rn]7s^2 5f^{11}$	—	619
fermium	Fm	100	$[Rn]7s^2 5f^{12}$	—	627
mendelevium	Md	101	$[Rn]7s^2 5f^{13}$	—	635
nobelium	No	102	$[Rn]7s^2 5f^{14}$	—	642
lawrencium	Lr	103	$[Rn]7s^2 5f^{14} 6d^1$	—	—
rutherfordium	Rf	104	$[Rn]7s^2 5f^{14} 6d^2$	—	—
dubnium	Db	105	$[Rn]7s^2 5f^{14} 6d^3$	—	—
seaborgium	Sg	106	$[Rn]7s^2 5f^{14} 6d^4$	—	—
bohrium	Bh	107	$[Rn]7s^2 5f^{14} 6d^5$	—	—
hassium	Hs	108	$[Rn]7s^2 5f^{14} 6d^6$	—	—
meitnerium	Mt	109	$[Rn]7s^2 5f^{14} 6d^7$	—	—
element 110	[110]	110	$[Rn]7s^2 5f^{14} 6d^8$	—	—
element 111	[111]	111	$[Rn]7s^2 5f^{14} 6d^9$	—	—
element 112	[112]	112	$[Rn]7s^2 5f^{14} 6d^{10}$	—	—

Table A3.2 Miscellaneous Physical Properties of the Elements[a]

Element	Symbol	Atomic Number	Physical State[b,c]	Density[d] (g/mL)	Melting Point (°C)	Boiling Point (°C)
hydrogen	H	1	gas	0.000090	−259.14	−252.87
helium	He	2	gas	0.000179	<−272.2	−268.93
lithium	Li	3	solid	0.534	180.5	1347
beryllium	Be	4	solid	1.848	1283	2484
boron	B	5	solid	2.34	2300	3650
carbon	C	6	solid (gr)	1.9–2.3	≈3350	sublimes
nitrogen	N	7	gas	0.00125	−210.00	−195.8
oxygen	O	8	gas	0.00143	−218.8	−182.95
fluorine	F	9	gas	0.00170	−219.62	−188.12
neon	Ne	10	gas	0.00090	−248.59	−246.08
sodium	Na	11	solid	0.971	97.72	883
magnesium	Mg	12	solid	1.738	650	1090
aluminum	Al	13	solid	2.6989	660.32	2467
silicon	Si	14	solid	2.33	1414	2355
phosphorus	P	15	solid (wh)	1.82	44.15	280
sulfur	S	16	solid	2.07	115.21	444.60
chlorine	Cl	17	gas	0.00321	−101.5	−34.04
argon	Ar	18	gas	0.00178	−189.3	−185.9
potassium	K	19	solid	0.862	63.28	759
calcium	Ca	20	solid	1.55	842	1484
scandium	Sc	21	solid	2.989	1541	2380
titanium	Ti	22	solid	4.54	1668	3287
vanadium	V	23	solid	6.11	1910	3407
chromium	Cr	24	solid	7.19	1857	2671
manganese	Mn	25	solid	7.3	1246	1962
iron	Fe	26	solid	7.874	1538	2750
cobalt	Co	27	solid	8.9	1495	2870
nickel	Ni	28	solid	8.902	1455	2730
copper	Cu	29	solid	8.96	1084.6	2562
zinc	Zn	30	solid	7.133	419.53	907
gallium	Ga	31	solid	5.904	29.76	2403
germanium	Ge	32	solid	5.323	938.25	2833
arsenic	As	33	solid (gy)	5.727	614	sublimes
selenium	Se	34	solid (gy)	4.79	221	685
bromine	Br	35	liquid	3.12	−7.2	58.78
krypton	Kr	36	gas	0.00373	−157.36	−153.22
rubidium	Rb	37	solid	1.532	39.31	688
strontium	Sr	38	solid	2.54	777	1382
yttrium	Y	39	solid	4.469	1526	3336
zirconium	Zr	40	solid	6.506	1855	4409
niobium	Nb	41	solid	8.57	2477	4744
molybdenum	Mo	42	solid	10.22	2623	4639
technetium	Tc	43	solid	11.50	2157	4538
ruthenium	Ru	44	solid	12.41	2334	3900
rhodium	Rh	45	solid	12.41	1964	3695
palladium	Pd	46	solid	12.02	1555	2963
silver	Ag	47	solid	10.50	961.78	2212

Continued on next page

Table A3.2 Miscellaneous Physical Properties of the Elements (*Continued*)

Element	Symbol	Atomic Number	Physical State[b,c]	Density[d] (g/mL)	Melting Point (°C)	Boiling Point (°C)
cadmium	Cd	48	solid	8.65	321.07	767
indium	In	49	solid	7.31	156.60	2072
tin	Sn	50	solid (wh)	7.31	231.9	2270
antimony	Sb	51	solid	6.691	630.63	1750
tellurium	Te	52	solid	6.24	449.5	998
iodine	I	53	solid	4.93	113.7	184.4
xenon	Xe	54	gas	0.00589	−111.75	−108.0
cesium	Cs	55	solid	1.873	28.44	671
barium	Ba	56	solid	3.5	727	1640
lanthanum	La	57	solid	6.145	920	3455
cerium	Ce	58	solid	6.770	799	3424
praseodymium	Pr	59	solid	6.773	931	3510
neodymium	Nd	60	solid	7.008	1016	3066
promethium	Pm	61	solid	7.264	1042	≈3000
samarium	Sm	62	solid	7.520	1072	1790
europium	Eu	63	solid	5.244	822	1596
gadolinium	Gd	64	solid	7.901	1314	3264
terbium	Tb	65	solid	8.230	1359	3221
dysprosium	Dy	66	solid	8.551	1411	2561
holmium	Ho	67	solid	8.795	1472	2694
erbium	Er	68	solid	9.066	1529	2862
thulium	Tm	69	solid	9.321	1545	1946
ytterbium	Yb	70	solid	6.966	824	1194
lutetium	Lu	71	solid	9.841	1663	3393
hafnium	Hf	72	solid	13.31	2233	4603
tantalum	Ta	73	solid	16.654	3017	5458
tungsten	W	74	solid	19.3	3422	5660
rhenium	Re	75	solid	21.02	3186	5596
osmium	Os	76	solid	22.57	3033	5012
iridium	Ir	77	solid	22.42	2446	4130
platinum	Pt	78	solid	21.45	1768.4	3825
gold	Au	79	solid	19.3	1064.18	2856
mercury	Hg	80	liquid	13.546	−38.83	356.73
thallium	Tl	81	solid	11.85	304	1473
lead	Pb	82	solid	11.35	327.46	1749
bismuth	Bi	83	solid	9.747	271.4	1564
polonium	Po	84	solid	9.32	254	962
astatine	At	85	solid	unknown	302	337
radon	Rn	86	gas	0.00973	−71	−61.7
francium	Fr	87	solid	unknown	27	677
radium	Ra	88	solid	5	700	1737
actinium	Ac	89	solid	10.07	1051	≈3200
thorium	Th	90	solid	11.72	1750	4788
protactinium	Pa	91	solid	15.37	1572	unknown
uranium	U	92	solid	18.95	1132	3818

[a]For relative atomic masses and alphabetical listing of the elements, see the flyleaf at the front of this volume.
[b]Normal state at 25°C and 1 atm.
[c]Allotropes: gr = graphite, gy = gray, wh = white.
[d]Liquids and solids at 25°C and 1 atm; gases at 0°C and 1 atm (STP).

Table A3.3 A Selection of Stable Isotopes[a]

Isotope ^AX	Natural Abundance (%)	Atomic Number Z	Neutron Number N	Mass Number A	Atomic Mass (amu)	Binding Energy per Nucleon (MeV)[b]
^{1}H	99.985	1	0	1	1.007825	—
^{2}H	0.015	1	1	2	2.014000	1.160
^{3}He	0.000137	2	1	3	3.016030	2.572
^{4}He	99.999863	2	2	4	4.002603	7.075
^{6}Li	7.5	3	3	6	6.015121	5.333
^{7}Li	92.5	3	4	7	7.016003	5.606
^{9}Be	100.0	4	5	9	9.012182	6.463
^{10}B	19.9	5	5	10	10.012937	6.475
^{11}B	80.1	5	6	11	11.009305	6.928
^{12}C	98.90	6	6	12	12.000000	7.680
^{13}C	1.10	6	7	13	13.003355	7.470
^{14}N	99.634	7	7	14	14.003074	7.476
^{15}N	0.366	7	8	15	15.000108	7.699
^{16}O	99.762	8	8	16	15.994915	7.976
^{17}O	0.038	8	9	17	16.999131	7.751
^{18}O	0.200	8	10	18	17.999160	7.767
^{19}F	100.0	9	10	19	18.998403	7.779
^{20}Ne	90.48	10	10	20	19.992435	8.032
^{21}Ne	0.27	10	11	21	20.993843	7.972
^{22}Ne	9.25	10	12	22	21.991383	8.081
^{23}Na	100.0	11	12	23	22.989770	8.112
^{24}Mg	78.99	12	12	24	23.985042	8.261
^{25}Mg	10.00	12	13	25	24.985837	8.223
^{26}Mg	11.01	12	14	26	25.982593	8.334
^{27}Al	100.0	13	14	27	26.981538	8.331
^{28}Si	92.23	14	14	28	27.976927	8.448
^{29}Si	4.67	14	15	29	28.976495	8.449
^{30}Si	3.10	14	16	30	29.973770	8.521
^{31}P	100.0	15	16	31	30.973761	8.481
^{32}S	95.02	16	16	32	31.972070	8.493
^{33}S	0.75	16	17	33	32.971456	8.498
^{34}S	4.21	16	18	34	33.967866	8.584
^{36}S	0.02	16	20	36	35.967080	8.575
^{35}Cl	75.77	17	18	35	34.968852	8.520
^{37}Cl	24.23	17	20	37	36.965903	8.570
^{36}Ar	0.337	18	18	36	35.967545	8.520
^{38}Ar	0.063	18	20	38	37.962732	8.614
^{40}Ar	99.600	18	22	40	39.962384	8.595
^{39}K	93.258	19	20	39	38.963707	8.557
^{41}K	6.730	19	22	41	40.961825	8.576
^{40}Ca	96.941	20	20	40	39.962591	8.551
^{42}Ca	0.647	20	22	42	41.958618	8.617
^{43}Ca	0.135	20	23	43	42.958766	8.601
^{44}Ca	2.086	20	24	44	43.955480	8.658
^{46}Ca	0.004	20	26	46	45.953689	8.669
^{48}Ca	0.187	20	28	48	47.952533	8.666
^{45}Sc	100.0	21	24	45	44.955910	8.619

Continued on next page

Table A3.3 A Selection of Stable Isotopes (*Continued*)

Isotope ^AX	Natural Abundance (%)	Atomic Number Z	Neutron Number N	Mass Number A	Atomic Mass (amu)	Binding Energy per Nucleon (MeV)[b]
^{46}Ti	8.0	22	24	46	45.952629	8.656
^{47}Ti	7.3	22	25	47	46.951764	8.661
^{48}Ti	73.8	22	26	48	47.947947	8.723
^{49}Ti	5.5	22	27	49	48.947871	8.711
^{50}Ti	5.4	22	28	50	49.944792	8.756
^{51}V	99.750	23	28	51	50.943962	8.742
^{50}Cr	4.345	24	26	50	49.946046	8.701
^{52}Cr	83.789	24	28	52	51.940509	8.776
^{53}Cr	9.501	24	29	53	52.940651	8.760
^{54}Cr	2.365	24	30	54	53.938882	8.778
^{55}Mn	100.0	25	30	55	54.938049	8.765
^{54}Fe	5.9	26	28	54	53.939612	8.736
^{56}Fe	91.72	26	30	56	55.934939	8.790
^{57}Fe	2.1	26	31	57	56.935396	8.770
^{58}Fe	0.28	26	32	58	57.933277	8.792
^{59}Co	100.0	27	32	59	58.933200	8.768
^{204}Pb	1.4	82	122	204	203.973020	7.880
^{206}Pb	24.1	82	124	206	205.974440	7.875
^{207}Pb	22.1	82	125	207	206.975872	7.870
^{208}Pb	52.4	82	126	208	207.976627	7.868
^{209}Bi	100.0	83	126	209	208.980380	7.848

[a] Selection is complete through cobalt-59. Where natural abundances do not add to 100%, the differences are made up by radioactive isotopes with exceedingly long half-lives: potassium-40 (0.0117%, $t_{1/2} = 1.3 \times 10^9$ y); vanadium-50 (0.250%, $t_{1/2} > 1.4 \times 10^{17}$ y).

[b] 1 MeV (million electron-volts) = 1.602189×10^{-13} J.

Appendix 3 The Elements and Their Properties

Table A3.4 A Selection of Radioactive Isotopes

Isotope ^{A}X	Decay Mode[a]	Half-Life $t_{1/2}$	Atomic Number Z	Neutron Number N	Mass Number A	Atomic Mass (amu)	Binding Energy per Nucleon (MeV)
^{3}H	β^-	12.3 y	1	2	3	3.01605	2.827
^{8}Be	α	$\approx 7 \times 10^{-17}$ s	4	4	8	8.005305	7.062
^{14}C	β^-	5.7×10^3 y	6	8	14	14.003241	7.520
^{22}Na	β^+	2.6 y	11	11	22	21.994434	7.916
^{24}Na	β^-	15.0 h	11	13	24	23.990961	8.064
^{32}P	β^-	14.3 d	15	17	32	31.973907	8.464
^{35}S	β^-	87.2 d	16	19	35	34.969031	8.538
^{59}Fe	β^-	44.5 d	26	33	59	58.934877	8.755
^{60}Co	β^-	5.3 y	27	33	60	59.933819	8.747
^{90}Sr	β^-	29.1 y	38	52	90	89.907738	8.696
^{99}Tc	β^-	2.1×10^5 y	43	56	99	98.906524	8.611
^{109}Cd	EC	462 d	48	61	109	108.904953	8.539
^{125}I	EC	59.4 d	53	72	125	124.904620	8.450
^{131}I	β^-	8.04 d	53	78	131	130.906114	8.422
^{137}Cs	β^-	30.3 y	55	82	137	136.907073	8.389
^{222}Rn	α	3.82 d	86	136	222	222.017570	7.695
^{226}Ra	α	1600 y	88	138	226	226.025402	7.662
^{232}Th	α	1.4×10^{10} y	90	142	232	232.038054	7.615
^{235}U	α	7.0×10^8 y	92	143	235	235.043924	7.591
^{238}U	α	4.5×10^9 y	92	146	238	238.050784	7.570
^{239}Pu	α	2.4×10^4 y	94	145	239	239.052157	7.560

[a]Modes of decay include alpha emission (α), beta emission (β^-), positron emission (β^+), electron capture (EC).

APPENDIX 4
Chemical Bonds and Thermodynamic Data

Table A4.1 Average Lengths and Strengths of Covalent Bonds

Atom	Bond	Bond Length (nm)	Bond Strength (kJ/mol)
H	H—H	75	432
	H—F	92	565
	H—Cl	127	431
	H—Br	141	363
	H—I	161	295
C	C—C	154	348
	C=C	134	614
	C≡C	120	839
	C—H	112	411
	C—N	143	293
	C=N	138	615
	C≡N	116	891
	C—O	143	358
	C=O[a]	123	748
	C≡O	113	1072
	C—F	133	453
	C—Cl	177	339
	C—Br	179	276
	C—I	215	238
N	N—N	147	163
	N=N	124	418
	N≡N	110	941
	N—H	104	386
	N—O	136	201
	N=O	122	607
	N≡O	106	678
O	O—O	148	146
	O=O	121	495
	O—H	96	458
S	S—O	151	265
	S=O	143	523
	S—S	204	266
	S—H	134	347
F	F—F	143	159
Cl	Cl—Cl	200	243
Br	Br—Br	228	193
I	I—I	266	157

[a] The bond strength of C=O in CO_2 is 799 kJ/mol.

Table A4.2 Critical Temperatures (T_c) and van der Waals Parameters (a, b) of Real Gases

Gas[a]	Molar Mass (g/mol)	T_c (K)	a (L²·atm/mol²)	b (L/mol)
H_2O	18.015	647.14	5.537	0.0305
Br_2	159.808	588	9.75	0.0591
CCl_3F	137.367	471.2	14.68	0.1111
Cl_2	70.906	416.9	6.343	0.0542
CO_2	44.010	304.14	3.658	0.0429
Kr	83.80	209.41	2.325	0.0396
CH_4	16.043	190.53	2.300	0.0430
O_2	31.999	154.59	1.382	0.0319
Ar	39.948	150.87	1.355	0.0320
F_2	37.997	144.13	1.171	0.0290
CO	28.010	132.91	1.472	0.0395
N_2	28.013	126.21	1.370	0.0387
H_2	2.016	32.97	0.245	0.0265
He	4.003	5.19	0.035	0.0238

[a]Listed in descending order of critical temperature.

Table A4.3 Thermodynamic Properties at 25°C

Substance[a,b]	$\mathcal{M}$ (g/mol)	$\Delta H_f°$ (kJ/mol)	$S°$ (J/mol·K)	$\Delta G_f°$ (kJ/mol)
Elements and Monatomic Ions				
Ag^+(aq)	107.868	105.6	72.7	77.1
Ag(g)	107.868	284.9	173.0	246.0
Ag(s)	107.868	0.0	42.6	0.0
Al^{3+}(aq)	26.982	−531	−321.7	−485
Al(g)	26.982	330.0	164.6	289.4
Al(s)	26.982	0.0	28.3	0.0
Ar(g)	39.948	0.0	154.8	0.0
Au(g)	196.967	366.1	180.5	326.3
Au(s)	196.967	0.0	47.4	0.0
B(g)	10.811	565.0	153.4	521.0
B(s)	10.811	0.0	5.9	0.0
Ba^{2+}(aq)	137.327	−537.6	9.6	−560.8
Ba(g)	137.327	180.0	170.2	146.0
Ba(s)	137.327	0.0	62.8	0.0
Be(g)	9.012	324.0	136.3	286.6
Be(s)	9.012	0.0	9.5	0.0
Br^-(aq)	79.904	−121.6	82.4	−104.0
Br(g)	79.904	111.9	175.0	82.4
Br_2(g)	159.808	30.9	245.5	3.1
$Br_2(\ell)$	159.808	0.0	152.2	0.0
C(g)	12.011	716.7	158.1	671.3
C(s, diamond)	12.011	1.9	2.4	2.9
C(s, graphite)	12.011	0.0	5.7	0.0

Continued on next page

Table A4.3 Thermodynamic Properties at 25°C (*Continued*)

Substance[a,b]	M (g/mol)	$\Delta H_f°$ (kJ/mol)	$S°$ (J/mol·K)	$\Delta G_f°$ (kJ/mol)
$Ca^{2+}(aq)$	40.078	−542.8	−55.3	−553.6
$Ca(g)$	40.078	177.8	154.9	144.0
$Ca(s)$	40.078	0.0	41.6	0.0
$Cl^-(aq)$	35.453	−167.2	56.5	−131.2
$Cl(g)$	35.453	121.3	165.2	105.3
$Cl_2(g)$	70.906	0.0	223.0	0.0
$Co^{2+}(aq)$	58.933	−58.2	−113	−54.4
$Co^{3+}(aq)$	58.933	92	−305	134
$Co(g)$	58.933	424.7	179.5	380.3
$Co(s)$	58.933	0.0	30.0	0.0
$Cr(g)$	51.996	396.6	174.5	351.8
$Cr(s)$	51.996	0.0	23.8	0.0
$Cs^+(aq)$	132.905	−258.3	133.1	−292.0
$Cs(g)$	132.905	76.5	175.6	49.6
$Cs(s)$	132.905	0.0	85.2	0.0
$Cu^+(aq)$	63.546	71.7	40.6	50.0
$Cu^{2+}(aq)$	63.546	64.8	−99.6	65.5
$Cu(g)$	63.546	337.4	166.4	297.7
$Cu(s)$	63.546	0.0	33.2	0.0
$F^-(aq)$	18.998	−332.6	−13.8	−278.8
$F(g)$	18.998	79.4	158.8	62.3
$F_2(g)$	37.996	0.0	202.8	0.0
$Fe^{2+}(aq)$	55.845	−89.1	−137.7	−78.9
$Fe^{3+}(aq)$	55.845	−48.5	−315.9	−4.7
$Fe(g)$	55.845	416.3	180.5	370.7
$Fe(s)$	55.845	0.0	27.3	0.0
$H^+(aq)$	1.0079	0.0	0.0	0.0
$H(g)$	1.0079	218.0	114.7	203.3
$H_2(g)$	2.0158	0.0	130.6	0.0
$He(g)$	4.0026	0.0	126.2	0.0
$Hg_2^{2+}(aq)$	401.18	172.4	84.5	153.5
$Hg^{2+}(aq)$	200.59	171.1	−32.2	164.4
$Hg(g)$	200.59	61.4	175.0	31.8
$Hg(\ell)$	200.59	0.0	75.9	0.0
$I^-(aq)$	126.904	−55.2	111.3	−51.6
$I(g)$	126.904	106.8	180.8	70.2
$I_2(g)$	253.808	62.4	260.7	19.3
$I_2(s)$	253.808	0.0	116.1	0.0
$K^+(aq)$	39.098	−252.4	102.5	−283.3
$K(g)$	39.098	89.0	160.3	60.5
$K(s)$	39.098	0.0	64.7	0.0
$Li^+(aq)$	6.941	−278.5	13.4	−293.3
$Li(g)$	6.941	159.3	138.8	126.6
$Li^+(g)$	6.941	685.7	133.0	648.5
$Li(s)$	6.941	0.0	29.1	0.0
$Mg^{2+}(aq)$	24.305	−466.9	−138.1	−454.8
$Mg(g)$	24.305	147.1	148.6	112.5
$Mg(s)$	24.305	0.0	32.7	0.0

Continued on next page

Table A4.3 Thermodynamic Properties at 25°C (Continued)

Substance[a,b]	$\mathcal{M}$ (g/mol)	ΔH_f° (kJ/mol)	S° (J/mol·K)	ΔG_f° (kJ/mol)
$Mn^{2+}(aq)$	54.938	−220.8	−73.6	−228.1
$Mn(g)$	54.938	280.7	173.7	238.5
$Mn(s)$	54.938	0.0	32.0	0.0
$N(g)$	14.0067	472.7	153.3	455.5
$N_2(g)$	28.0134	0.0	191.5	0.0
$Na^+(aq)$	22.990	−240.1	59.0	−261.9
$Na(g)$	22.990	107.5	153.7	77.0
$Na^+(g)$	22.990	609.3	148.0	574.3
$Na(s)$	22.990	0.0	51.3	0.0
$Ne(g)$	20.180	0.0	146.3	0.0
$Ni^{2+}(aq)$	58.693	−54.0	−128.9	−45.6
$Ni(g)$	58.693	429.7	182.2	384.5
$Ni(s)$	58.693	0.0	29.9	0.0
$O(g)$	15.999	249.2	161.1	231.7
$O_2(g)$	31.998	0.0	205.0	0.0
$P(g)$	30.974	314.6	163.1	278.3
$P_4(s, red)$	123.896	−17.6	22.8	−12.1
$P_4(s, white)$	123.896	0.0	41.1	0.0
$Pb^{2+}(aq)$	207.2	−1.7	10.5	−24.4
$Pb(g)$	207.2	195.2	162.2	175.4
$Pb(s)$	207.2	0.0	64.8	0.0
$Rb^+(aq)$	85.468	−251.2	121.5	−284.0
$Rb(g)$	85.468	80.9	170.1	53.1
$Rb(s)$	85.468	0.0	76.8	0.0
$S(g)$	32.006	277.2	167.8	236.7
$S(s)$	32.006	0.0	32.1	0.0
$Sc(g)$	44.956	377.8	174.8	336.0
$Si(g)$	28.086	450.0	168.0	405.5
$Si(s)$	28.086	0.0	18.8	0.0
$Sn(g)$	118.710	301.2	168.5	266.2
$Sn(s, gray)$	118.710	−2.1	44.1	0.1
$Sn(s, white)$	118.710	0.0	51.2	0.0
$Sr^{2+}(aq)$	87.62	−545.8	−32.6	−559.5
$Sr(g)$	87.62	164.4	164.6	130.9
$Sr(s)$	87.62	0.0	52.3	0.0
$Ti(g)$	47.867	473.0	180.3	428.4
$Ti(s)$	47.867	0.0	30.7	0.0
$V(g)$	50.942	514.2	182.2	468.5
$V(s)$	50.942	0.0	28.9	0.0
$Zn^{2+}(aq)$	65.39	−153.9	−112.1	−147.1
$Zn(g)$	65.39	130.4	161.0	94.8
$Zn(s)$	65.39	0.0	41.6	0.0
Polyatomic Ions				
$CH_3COO^-(aq)$	59.045	−486.0	86.6	−369.3
$CO_3^{2-}(aq)$	60.009	−677.1	−56.9	−527.8
$C_2O_4^{2-}(aq)$	88.020	−825.1	45.6	−673.9
$CrO_4^{2-}(aq)$	115.994	−881.2	50.2	−727.8

Continued on next page

Table A4.3 Thermodynamic Properties at 25°C (*Continued*)

Substance[a,b]	$\mathcal{M}$ (g/mol)	$\Delta H_f°$ (kJ/mol)	$S°$ (J/mol·K)	$\Delta G_f°$ (kJ/mol)
$Cr_2O_7^{2-}(aq)$	215.988	−1490.3	261.9	−1301.1
$HCOO^-(aq)$	45.018	−425.6	92	−351.0
$HCO_3^-(aq)$	61.017	−692.0	91.2	−586.8
$HSO_4^-(aq)$	97.072	−887.3	131.8	−755.9
$MnO_4^-(aq)$	118.936	−541.4	191.2	−447.2
$NH_4^+(aq)$	18.038	−132.5	113.4	−79.3
$NO_3^-(aq)$	62.005	−205.0	146.4	−108.7
$OH^-(aq)$	17.007	−230.0	−10.8	−157.2
$PO_4^{3-}(aq)$	94.971	−1277.4	−222	−1018.7
$SO_4^{2-}(aq)$	96.064	−909.3	20.1	−744.5
Inorganic Compounds				
$AgCl(s)$	143.321	−127.1	96.2	−109.8
$AgI(s)$	234.773	−61.8	115.5	−66.2
$AgNO_3(s)$	169.873	−124.4	140.9	−33.4
$Al_2O_3(s)$	101.961	−1675.7	50.9	−1582.3
$B_2H_6(g)$	27.669	35.0	232.0	86.6
$B_2O_3(s)$	69.622	−1263.6	54.0	−1184.1
$BaCO_3(s)$	197.34	−1216.3	112.1	−1137.6
$BaSO_4(s)$	233.39	−1473.2	132.2	−1362.2
$CaCO_3(s)$	100.087	−1206.9	92.9	−1128.8
$CaCl_2(s)$	110.984	−795.4	108.4	−748.8
$CaF_2(s)$	78.075	−1228.0	68.5	−1175.6
$CaO(s)$	56.077	−634.9	38.1	−603.3
$Ca(OH)_2(s)$	74.093	−985.2	83.4	−897.5
$CaSO_4(s)$	136.142	−1434.5	106.5	−1322.0
$CO(g)$	28.010	−110.5	197.7	−137.2
$CO_2(g)$	44.010	−393.5	213.6	−394.4
$CO_2(aq)$	44.010	−412.9	121.3	−386.2
$CS_2(g)$	76.143	115.3	237.8	65.1
$CS_2(\ell)$	76.143	87.9	151.0	63.6
$CsCl(s)$	168.358	−443.0	101.2	−414.6
$CuSO_4(s)$	159.610	−771.4	109.2	−662.2
$FeCl_2(s)$	126.750	−341.8	118.0	−302.3
$FeCl_3(s)$	162.203	−399.5	142.3	−334.0
$FeO(s)$	71.844	−271.9	60.8	−255.2
$Fe_2O_3(s)$	159.688	−824.2	87.4	−742.2
$HBr(g)$	80.912	−36.3	198.7	−53.4
$HCl(g)$	36.461	−92.3	186.9	−95.3
$HF(g)$	20.006	−273.3	173.8	−275.4
$HI(g)$	127.912	26.5	206.6	1.7
$HNO_3(g)$	63.013	−135.1	266.4	−74.7
$HNO_3(\ell)$	63.013	−174.1	155.6	−80.7
$HNO_3(aq)$	63.013	−206.6	146.0	−110.5
$HgCl_2(s)$	271.50	−224.3	146.0	−178.6
$Hg_2Cl_2(s)$	472.09	−265.4	191.6	−210.7
$H_2O(g)$	18.015	−241.8	188.7	−228.6
$H_2O(\ell)$	18.015	−285.8	70.0	−237.2

Continued on next page

Table A4.3 Thermodynamic Properties at 25°C (*Continued*)

Substance[a,b]	$\mathcal{M}$ (g/mol)	ΔH_f° (kJ/mol)	S° (J/mol·K)	ΔG_f° (kJ/mol)
$H_2S(g)$	34.082	−20.17	205.6	−33.01
$H_2O_2(g)$	34.015	−136.3	232.7	−105.6
$H_2O_2(\ell)$	34.015	−187.8	109.6	−120.4
$H_2SO_4(\ell)$	98.079	−814.0	156.9	−690.0
$H_2SO_4(aq)$		−909.2	20.1	−744.5
$KBr(s)$	119.002	−393.8	95.9	−380.7
$KCl(s)$	74.551	−436.5	82.6	−408.5
$LiBr(s)$	86.845	−351.2	74.3	−342.0
$LiCl(s)$	42.394	−408.6	59.3	−384.4
$Li_2CO_3(s)$	73.891	−1215.9	90.4	−1132.1
$MgCl_2(s)$	95.211	−641.3	89.6	591.8
$Mg(OH)_2(s)$	58.320	−924.5	63.2	−833.5
$MgSO_4(s)$	120.369	−1284.9	91.6	−1170.6
$MnO_2(s)$	86.937	−520.0	53.1	−465.1
$NaC_2H_3O_2(s)$	82.034	−708.8	123.0	−607.2
$NaBr(s)$	102.894	−361.1	8.8	−349.0
$NaBr(s)$	102.894	−361.4	86.82	−349.3
$NaCl(s)$	58.443	−411.2	72.1	−384.2
$NaCl(g)$	58.443	−181.4	229.8	−201.3
$Na_2CO_3(s)$	105.989	−1130.7	135.0	−1044.4
$NaHCO_3(s)$	84.007	−950.8	101.7	−851.0
$NaNO_3(s)$	84.995	−467.9	116.5	−367.0
$NaOH(s)$	39.997	−425.6	64.5	−379.5
$Na_2SO_4(s)$	142.043	−1387.1	149.6	−1270.2
$NF_3(g)$	71.002	−132.1	260.8	−90.6
$NH_3(aq)$	17.031	−80.29	111.3	−26.50
$NH_3(g)$	17.031	−46.1	192.3	−16.5
$NH_4Cl(s)$	53.491	−314.4	94.6	−203.0
$NH_4NO_3(s)$	80.043	−365.6	151.1	−183.9
$N_2H_4(g)$	32.045	95.40	238.5	159.4
$NiCl_2(s)$	129.60	−305.3	97.7	−259.0
$NiO(s)$	74.60	−239.7	38.0	−211.7
$NO(g)$	30.006	90.3	210.7	86.6
$NO_2(g)$	46.006	33.2	240.0	51.3
$N_2O(g)$	44.013	82.1	219.9	104.2
$N_2O_4(g)$	92.011	9.2	304.2	97.8
$NOCl(g)$	65.459	51.7	261.7	66.1
$O_3(g)$	47.998	142.7	238.8	163.2
$PCl_3(g)$	137.33	−288.07	311.7	−269.6
$PCl_3(\ell)$	137.33	−319.6	217	−272.4
$PF_5(g)$	125.96	−1594.4	300.8	−1520.7
$PH_3(g)$	33.998	5.4	210.2	13.4
$PbCl_2(s)$	278.1	−359.4	136.0	−314.1
$PbSO_4(s)$	303.3	−920.0	148.5	−813.0
$SO_2(g)$	64.065	−296.8	248.2	−300.1

Continued on next page

Table A4.3 Thermodynamic Properties at 25°C (*Continued*)

Substance[a,b]	$\mathcal{M}$ (g/mol)	ΔH_f° (kJ/mol)	S° (J/mol·K)	ΔG_f° (kJ/mol)
$SO_3(g)$	80.064	−395.7	256.8	−371.1
$ZnCl_2(s)$	136.30	−415.1	111.5	−369.4
$ZnO(s)$	81.37	−348.0	43.9	−318.2
$ZnSO_4(s)$	161.45	−982.8	110.5	−871.5
Organic Molecules				
$CCl_4(g)$	153.823	−102.9	309.7	−60.6
$CCl_4(\ell)$	153.823	−135.4	216.4	−65.3
$CH_4(g)$	16.043	−74.8	186.2	−50.8
$HC_2H_3O_2(g)$	60.053	−432.8	282.5	−374.5
$HC_2H_3O_2(\ell)$	60.053	−484.5	159.8	−389.9
$CH_3OH(g)$	32.042	−200.7	239.7	−162.0
$CH_3OH(\ell)$	32.042	−238.7	126.8	−166.4
$C_2H_2(g)$	26.038	226.7	200.8	209.2
$C_2H_4(g)$	28.054	52.3	219.5	68.1
$C_2H_6(g)$	30.070	−84.7	229.5	−32.9
$C_2H_5OH(g)$	46.069	−235.1	282.6	−168.6
$C_2H_5OH(\ell)$	46.069	−277.7	160.7	−174.9
$CH_3CHO(g)$	44.05	−166	266	−133.7
$C_3H_8(g)$	44.097	−103.9	269.9	−23.5
$n\text{-}C_4H_{10}(g)^c$	58.123	−124.7	310.0	−15.7
$n\text{-}C_4H_{10}(\ell)^c$	58.123	−147.6	231.0	−15.0
$CH_3COCH_3(\ell)$	46.07	−248.4	199.8	
$CH_3COCH_3(g)$	46.07	−217.1	295.3	−152.7
$C_4H_9OH(\ell)$	74.12	−327.3	225.8	
$(C_2H_5)_2O(\ell)$	74.12	−279.6	172.4	
$(C_2H_5)_2O(g)$	74.12	−252.1	342.7	
$(CH_3)_2C{=}C(CH_3)_2(\ell)$	84.16	66.6	362.6	−69.2
$(CH_3)_2NH(\ell)$	45.09	−43.9	182.3	
$(CH_3)_2NH(g)$	45.09	−18.5	273.1	
$(C_2H_5)_2NH(\ell)$	73.14	−103.3		
$(C_2H_5)_2NH(g)$	73.14	−71.4		
$(CH_3)_3N(\ell)$	59.11	−46.0	208.5	
$(CH_3)_3N(g)$	59.11	−23.6	287.1	
$(CH_3CH_2)_3N(\ell)$	101.19	−134.3		
$(CH_3CH_2)_3N(g)$	101.19	−95.8		
$C_6H_6(g)$	78.114	82.9	269.2	129.7
$C_6H_6(\ell)$	78.114	49.0	172.8	124.5
$C_6H_{12}O_6(s)$	180.158	−1274.4	212.1	−910.1
$n\text{-}C_8H_{18}(\ell)^c$	114.231	−249.9	361.1	6.4
$C_{12}H_{22}O_{11}(s)$	342.300	−2221.7	360.2	−1543.8
$HCOOH(\ell)$	46.026	−424.7	129.0	−361.4

[a]Substances are arranged alphabetically by chemical formula within each class: (1) elements and monatomic ions; (2) polyatomic ions; (3) inorganic compounds (including CO and CO_2); (4) organic molecules (hydrocarbon-based).
[b]Symbols denote standard enthalpy of formation (ΔH_f°), standard third-law entropy (S°), and standard Gibbs free energy of formation (ΔG_f°). Entropies in aqueous solution are referred to $S^\circ[H^+(aq)] = 0$, not to absolute zero.
[c]The symbol *n* denotes the "normal" unbranched alkane.

APPENDIX 5
Equilibrium Constants

Table A5.1 Ionization Constants of Selected Acids at 25°C

Acid	Step	Aqueous Equilibrium^a	K_a	pK_a
acetic	1	$CH_3COOH \rightleftharpoons H^+ + CH_3COO^-$	1.76×10^{-5}	4.75
ammonium ion	1	$NH_4^+ \rightleftharpoons H^+ + NH_3$	5.7×10^{-10}	9.25
arsenic	1	$H_3AsO_4 \rightleftharpoons H^+ + H_2AsO_4^-$	5.5×10^{-3}	2.26
	2	$H_2AsO_4^- \rightleftharpoons H^+ + AsO_4^{2-}$	1.7×10^{-7}	6.77
	3	$HAsO_4^{2-} \rightleftharpoons H^+ + AsO_4^{3-}$	5.1×10^{-12}	11.29
benzoic	1	$C_6H_5COOH \rightleftharpoons H^+ + C_6H_5COO^-$	6.46×10^{-5}	4.19
boric	1	$H_3BO_3 \rightleftharpoons H^+ + H_2BO_3^-$	5.4×10^{-10}	9.27
	2	$H_2BO_3^- \rightleftharpoons H^+ + HBO_3^{2-}$	$<10^{-14}$	>14
bromoacetic	1	$CH_2BrCOOH \rightleftharpoons H^+ + CH_2BrCOO^-$	2.0×10^{-3}	2.70
butanoic	1	$C_3H_7COOH \rightleftharpoons H^+ + C_3H_7COO^-$	1.5×10^{-5}	4.82
carbonic	1	$H_2CO_3 \rightleftharpoons H^+ + HCO_3^-$	4.3×10^{-7}	6.37
	2	$HCO_3^- \rightleftharpoons H^+ + CO_3^{2-}$	4.7×10^{-11}	10.33
chloric	1	$HClO_3 \rightleftharpoons H^+ + ClO_3^-$	~ 1	~ 0
chlorous	1	$HClO_2 \rightleftharpoons H^+ + ClO_2^-$	1.1×10^{-2}	1.96
chloroacetic	1	$CH_2ClCOOH \rightleftharpoons H^+ + CH_2ClCOO^-$	1.4×10^{-3}	2.85
dichloroacetic	1	$CHCl_2COOH \rightleftharpoons H^+ + CHCl_2COO^-$	5.5×10^{-2}	1.26
ethanol	1	$CH_3CH_2OH \rightleftharpoons H^+ + CH_3CH_2O^-$	1.3×10^{-16}	15.9
fluoroacetic	1	$CH_2FCOOH \rightleftharpoons H^+ + CH_2FCOO^-$	2.6×10^{-3}	2.59
formic	1	$HCOOH \rightleftharpoons H^+ + HCOO^-$	1.77×10^{-4}	3.75
germanic	1	$H_2GeO_3 \rightleftharpoons H^+ + HGeO_3^-$	9.8×10^{-10}	9.01
	2	$HGeO_3^- \rightleftharpoons H^+ + GeO_3^{2-}$	5×10^{-13}	12.3
hydr(o)azoic	1	$HN_3 \rightleftharpoons H^+ + N_3^-$	1.9×10^{-5}	4.72
hydr(o)iodic	1	$HI \rightleftharpoons H^+ + I^-$	$\gg 1$ (strong)	<0
hydrobromic	1	$HBr \rightleftharpoons H^+ + Br^-$	$\gg 1$ (strong)	<0
hydrochloric	1	$HCl \rightleftharpoons H^+ + Cl^-$	$\gg 1$ (strong)	<0
hydrocyanic	1	$HCN \rightleftharpoons H^+ + CN^-$	6.2×10^{-10}	9.21
hydrofluoric	1	$HF \rightleftharpoons H^+ + F^-$	6.8×10^{-4}	3.17
hydrosulfuric	1	$H_2S \rightleftharpoons H^+ + HS^-$	8.9×10^{-8}	7.05
	2	$HS^- \rightleftharpoons H^+ + S^{2-}$	$\approx 10^{-19}$	≈ 19
hypobromous	1	$HBrO \rightleftharpoons H^+ + BrO^-$	2.3×10^{-9}	8.55
hypochlorous	1	$HClO \rightleftharpoons H^+ + ClO^-$	2.9×10^{-8}	7.54
hypoiodous	1	$HIO \rightleftharpoons H^+ + IO^-$	2.3×10^{-11}	10.5
iodic	1	$HIO_3 \rightleftharpoons H^+ + IO_3^-$	1.7×10^{-1}	0.77
iodoacetic	1	$CH_2ICOOH \rightleftharpoons H^+ + CH_2ICOO^-$	7.6×10^{-4}	3.12

Continued on next page

Table A5.1 Ionization Constants of Selected Acids at 25°C (*Continued*)

Acid	Step	Aqueous Equilibrium[a]	K_a	pK_a
lactic	1	$CH_3CHOHCOOH \rightleftharpoons H^+ + CH_3CHOHCOO^-$	1.4×10^{-4}	3.85
malonic	1	$HOOCCH_2COOH \rightleftharpoons H^+ + HOOCCH_2COO^-$	1.5×10^{-3}	2.82
	2	$HOOCCH_2COO^- \rightleftharpoons H^+ + {}^-OOCCH_2COO^-$	2.0×10^{-6}	5.70
nitric	1	$HNO_3 \rightleftharpoons H^+ + NO_3^-$	$\gg 1$ (strong)	<0
nitrous	1	$HNO_2 \rightleftharpoons H^+ + NO_2^-$	4.0×10^{-4}	3.40
oxalic	1	$HOOCCOOH \rightleftharpoons H^+ + HOOCCOO^-$	5.9×10^{-2}	1.23
	2	$HOOCCOO^- \rightleftharpoons H^+ + {}^-OOCCOO^-$	6.4×10^{-5}	4.19
perchloric	1	$HClO_4 \rightleftharpoons H^+ + ClO_4^-$	$\gg 1$ (strong)	<0
periodic	1	$HIO_4 \rightleftharpoons H^+ + IO_4^-$	2.3×10^{-2}	1.64
phenol	1	$C_6H_5OH \rightleftharpoons H^+ + C_6H_5O^-$	1.3×10^{-10}	9.89
phosphoric	1	$H_3PO_4 \rightleftharpoons H^+ + H_2PO_4^-$	7.52×10^{-3}	2.12
	2	$H_2PO_4^- \rightleftharpoons H^+ + HPO_4^{2-}$	6.2×10^{-8}	7.21
	3	$HPO_4^{2-} \rightleftharpoons H^+ + PO_4^{3-}$	2.2×10^{-13}	12.66
propionic	1	$CH_3CH_2COOH \rightleftharpoons H^+ + CH_3CH_2COO^-$	1.4×10^{-5}	4.86
sulfuric	1	$H_2SO_4 \rightleftharpoons H^+ + HSO_4^-$	$\gg 1$ (strong)	<0
	2	$HSO_4^- \rightleftharpoons H^+ + SO_4^{2-}$	1.2×10^{-2}	1.92
sulfurous	1	$H_2SO_3 \rightleftharpoons H^+ + HSO_3^-$	1.3×10^{-2}	1.9
	2	$HSO_3^- \rightleftharpoons H^+ + SO_3^{2-}$	6.2×10^{-8}	7.1
thiocyanic	1	$HSCN \rightleftharpoons H^+ + SCN^-$	$\gg 1$ (strong)	<0
trichloroacetic	1	$CCl_3COOH \rightleftharpoons H^+ + CCl_3COO^-$	2.3×10^{-1}	0.64
trifluoroacetic	1	$CF_3COOH \rightleftharpoons H^+ + CF_3COO^-$	5.9×10^{-1}	0.23
water	1	$H_2O \rightleftharpoons H^+ + OH^-$	1.0×10^{-14}	14.00

[a] The formulas of the carboxylic acids are written using an RCOOH format to highlight their molecular structures.

Table A5.2 Strengths of Selected Bases at 25°C

Base	Aqueous Equilibrium	K_b	pK_b
Ammonia	$NH_3 + H_2O \rightleftharpoons NH_4^+ + OH^-$	1.76×10^{-5}	4.75
Aniline	$C_6H_5NH_2 + H_2O \rightleftharpoons C_6H_5NH_3^+ + OH^-$	4.0×10^{-10}	9.4
Diethylamine	$(C_2H_5)_2NH + H_2O \rightleftharpoons (C_2H_5)_2NH_2^+ + OH^-$	8.6×10^{-4}	3.1
Dimethylamine	$(CH_3)_2NH + H_2O \rightleftharpoons (CH_3)_2NH_2^+ + OH^-$	5.9×10^{-4}	3.2
Methylamine	$CH_3NH_2 + H_2O \rightleftharpoons CH_3NH_3^+ + OH^-$	4.4×10^{-4}	3.4
Nicotine (1)	[structure]	1.0×10^{-6}	6.0
(2)	[structure]	1.3×10^{-11}	10.9
Pyridine	$C_5H_5N + H_2O \rightleftharpoons C_5H_5NH^+ + OH^-$	1.7×10^{-9}	8.8
Quinine (1)	[structure]	3.3×10^{-6}	5.5
(2)	[structure]	1.4×10^{-10}	5.5
Urea	$H_2NCONH_2 + H_2O \rightleftharpoons > H_2NCONH_3^+ + OH^-$	1.3×10^{-14}	13.9

Table A5.3 Solubility-Product Constants at 25°C

Cation	Anion	Heterogeneous Equilibrium[a]	K_{sp}
aluminum	hydroxide	$Al(OH)_3 \rightleftharpoons Al^{3+} + 3OH^-$	1.9×10^{-33}
	phosphate	$AlPO_4 \rightleftharpoons Al^{3+} + PO_4^{3-}$	9.8×10^{-21}
barium	carbonate	$BaCO_3 \rightleftharpoons Ba^{2+} + CO_3^{2-}$	2.6×10^{-9}
	fluoride	$BaF_2 \rightleftharpoons Ba^{2+} + 2F^-$	1.0×10^{-6}
	sulfate	$BaSO_4 \rightleftharpoons Ba^{2+} + SO_4^{2-}$	1.1×10^{-10}
calcium	carbonate	$CaCO_3 \rightleftharpoons Ca^{2+} + CO_3^{2-}$	5.0×10^{-9}
	fluoride	$CaF \rightleftharpoons Ca^{2+} + 2F^-$	3.9×10^{-11}
	hydroxide	$Ca(OH)_2 \rightleftharpoons Ca^{2+} + 2OH^-$	4.7×10^{-6}
	phosphate	$Ca_3(PO_4)_2 \rightleftharpoons 3Ca^{2+} + 2PO_4^{3-}$	2.1×10^{-33}
	sulfate	$CaSO_4 \rightleftharpoons Ca^{2+} + SO_4^{2-}$	7.1×10^{-5}
copper(I)	bromide	$CuBr \rightleftharpoons Cu^+ + Br^-$	6.3×10^{-9}
	chloride	$CuCl \rightleftharpoons Cu^+ + Cl^-$	1.0×10^{-6}
	iodide	$CuI \rightleftharpoons Cu^+ + I^-$	1.3×10^{-12}
copper(II)	phosphate	$Cu_3(PO_4)_2 \rightleftharpoons 3Cu^{2+} + 2PO_4^{3-}$	1.4×10^{-37}
	hydroxide	$Cu(OH)_2 \rightleftharpoons Cu^{2+} + 2OH^-$	4.8×10^{-20}
iron(II)	carbonate	$FeCO_3 \rightleftharpoons Fe^{2+} + CO_3^{2-}$	3.1×10^{-11}
	fluoride	$FeF_2 \rightleftharpoons Fe^{2+} + 2F^-$	2.4×10^{-6}
	hydroxide	$Fe(OH)_2 \rightleftharpoons Fe^{2+} + 2OH^-$	4.9×10^{-17}
lead	bromide	$PbBr_2 \rightleftharpoons Pb^{2+} + 2Br^-$	6.6×10^{-6}
	carbonate	$PbCO_3 \rightleftharpoons Pb^{2+} + CO_3^{2-}$	1.5×10^{-13}
	chloride	$PbCl_2 \rightleftharpoons Pb^{2+} + 2Cl^-$	1.6×10^{-5}
	fluoride	$PbF_2 \rightleftharpoons Pb^{2+} + 2F^-$	7.1×10^{-7}
	iodide	$PbI_2 \rightleftharpoons Pb^{2+} + 2I^-$	8.5×10^{-9}
	sulfate	$PbSO_4 \rightleftharpoons Pb^{2+} + 2SO_4^{2+}$	1.8×10^{-8}
lithium	carbonate	$Li_2CO_3 \rightleftharpoons 2Li^+ + CO_3^{2-}$	8.2×10^{-4}
magnesium	carbonate	$MgCO_3 \rightleftharpoons Mg^{2+} + CO_3^{2-}$	6.8×10^{-6}
	fluoride	$MgF_2 \rightleftharpoons Mg^{2+} + 2F^-$	6.5×10^{-9}
	hydroxide	$Mg(OH)_2 \rightleftharpoons Mg^{2+} + 2OH^-$	5.6×10^{-12}
manganese(II)	carbonate	$MnCO_3 \rightleftharpoons Mn^{2+} + CO_3^{2-}$	2.2×10^{-11}
	hydroxide	$Mn(OH)_2 \rightleftharpoons Mn^{2+} + 2OH^-$	2.1×10^{-13}
mercury(I)	bromide	$Hg_2Br_2 \rightleftharpoons Hg_2^{2+} + 2Br^-$	6.4×10^{-23}
	carbonate	$Hg_2CO_3 \rightleftharpoons Hg_2^{2+} + CO_3^{2-}$	3.7×10^{-17}
	chloride	$Hg_2Cl_2 \rightleftharpoons Hg_2^{2+} + 2Cl^-$	1.5×10^{-18}
	iodide	$Hg_2I_2 \rightleftharpoons Hg_2^{2+} + 2I^-$	5.3×10^{-29}
	sulfate	$Hg_2SO_4 \rightleftharpoons Hg_2^{2+} + SO_4^{2-}$	8.0×10^{-7}
mercury(II)	hydroxide	$Hg(OH)_2 \rightleftharpoons Hg^{2+} + 2OH^-$	3.1×10^{-26}
	iodide	$HgI_2 \rightleftharpoons Hg^{2+} + 2I^-$	2.8×10^{-29}
silver	bromide	$AgBr \rightleftharpoons Ag^+ + Br^-$	5.4×10^{-13}
	carbonate	$Ag_2CO_3 \rightleftharpoons 2Ag^+ + CO_3^{2-}$	8.5×10^{-12}
	chloride	$AgCl \rightleftharpoons Ag^+ + Cl^-$	1.8×10^{-10}
	chromate	$Ag_2CrO_4 \rightleftharpoons 2Ag^+ + CrO_4^{2-}$	1.1×10^{-12}
	hydroxide	$AgOH \rightleftharpoons Ag^+ + OH^-$	1.8×10^{-8}
	iodide	$AgI \rightleftharpoons Ag^+ + I^-$	8.3×10^{-17}
	phosphate	$Ag_3PO_4 \rightleftharpoons 3Ag^+ + PO_4^{3-}$	8.9×10^{-17}
	sulfate	$Ag_2SO_4 \rightleftharpoons 2Ag^+ + SO_4^{2-}$	1.2×10^{-5}
strontium	carbonate	$SrCO_3 \rightleftharpoons Sr^{2+} + CO_3^{2-}$	5.6×10^{-10}
	fluoride	$SrF_2 \rightleftharpoons Sr^{2+} + 2F^-$	4.3×10^{-9}
	sulfate	$SrSO_4 \rightleftharpoons Sr^{2+} + SO_4^{2-}$	3.4×10^{-7}
zinc	carbonate	$ZnCO_3 \rightleftharpoons Zn^{2+} + CO_3^{2-}$	1.2×10^{-10}
	hydroxide	$Zn(OH)_2 \rightleftharpoons Zn^{2+} + 2OH^-$	3.0×10^{-16}

[a]Equilibrium is between solid phase and aqueous solution.

Table A5.4 Solubility of Ionic compounds in Water[a]

	Br^-	CH_3COO^-	CO_3^{2-}	Cl^-	F^-	I^-	NO_3^-	OH^-	PO_4^{3-}	S^{2-}	SO_4^{2-}
Ag^+	i	s	i	i	s	i	s	i	i	i	ss
Ba^{2+}	s	s	i	s	ss	s	s	s	i	d	i
Ca^{2+}	s	s	i	s	i	s	s	ss	i	ss	ss
Cs^+	s	s	s	s	s	s	s	s	—	s	s
Hg_2^{2+}	i	s	i	i	d	i	s	—	i	i	ss
Hg^{2+}	ss	s	i	s	d	i	s	i	i	i	d
Li^+	s	s	s	s	ss	s	s	s	ss	s	s
K^+	s	s	s	s	s	s	s	s	s	s	s
Mg^{2+}	s	s	i	s	i	s	s	i	i	d	s
NH_4^+	s	s	s	s	s	s	s	s	s	s	s
Na^+	s	s	s	s	s	s	s	s	s	s	s
Pb^{2+}	ss	s	i	ss	ss	i	s	ss	i	i	i
Rb^+	s	s	s	s	s	s	s	s	—	s	s
Sr^{2+}	s	s	i	s	ss	s	s	ss	i	i	i

[a] Compounds of cation and anion are understood to form in proper stoichiometric ratios.
s: soluble (solubility greater than 10 g L^{-1} at room temperature)
ss: slightly soluble (solubility between 0.1 g L^{-1} and 10 g L^{-1})
i: insoluble (solubility less than 0.1 g L^{-1})
d: decomposes

Table A5.5 Formation Constants of Complexes at 25°C

Complex Ion	Aqueous Equilibrium	K_f
$[Ag(NH_3)_2]^+$	$Ag^+ + 2NH_3 \rightleftharpoons Ag(NH_3)_2^+$	1.7×10^7
$[AgCl_2]^-$	$Ag^+ + 2Cl^- \rightleftharpoons AgCl_2^-$	2.5×10^5
$[Ag(CN)_2]^-$	$Ag^+ + 2CN^- \rightleftharpoons Ag(CN)_2^-$	1.0×10^{21}
$[Al(OH)_4]^-$	$Al^{3+} + 4OH^- \rightleftharpoons Al(OH)_4^-$	7.7×10^{33}
$[Au(CN)_2]^-$	$Au^+ + 2CN^- \rightleftharpoons Au(CN)_2^-$	2.0×10^{38}
$[Co(NH_3)_6]^{2+}$	$Co^{2+} + 6NH_3 \rightleftharpoons Co(NH_3)_6^{2+}$	7.7×10^4
$[Co(NH_3)_6]^{3+}$	$Co^{3+} + 6NH_3 \rightleftharpoons Co(NH_3)_6^{3+}$	5.0×10^{31}
$[Cu(NH_3)_4]^{2+}$	$Cu^{2+} + 4NH_3 \rightleftharpoons Cu(NH_4)_4^{2+}$	5.0×10^{13}
$[Cu(CN)_4]^{2-}$	$Cu^{2+} + 4CN^- \rightleftharpoons CU(CN)_4^{2-}$	1.0×10^{25}
$[HgCl_4]^{2-}$	$Hg^{2+} + 4Cl^- \rightleftharpoons HgCl_4^{2-}$	1.2×10^{15}
$[Ni(NH_3)_6]^{2+}$	$Ni^{2+} + 6NH_3 \rightleftharpoons Ni(NH_3)_6^{2+}$	5.5×10^8
$[PbCl_4]^{2-}$	$Pb^{2+} + 4Cl^- \rightleftharpoons PbCl_4^{2-}$	2.5×10^1
$[Zn(NH_3)_4]^{2+}$	$Zn^{2+} + 4NH_3 \rightleftharpoons Zn(NH_3)_4^{2+}$	2.9×10^9
$[Zn(OH)_4]^{2-}$	$Zn^{2+} + 4OH^- \rightleftharpoons Zn(OH)_4^{2-}$	2.8×10^{15}

APPENDIX 6
Standard Reduction Potentials

Table A6.1 Standard Reduction Potentials at 25°C

Half-reaction	n	$E°$ (V)
$F_2 + 2e^- \longrightarrow 2F^-$	2	2.866
$H_2N_2O_2 + 2H^+ + 2e^- \longrightarrow N_2 + 2H_2O$	2	2.65
$O(g) + 2H^+ + 2e^- \longrightarrow H_2O$	2	2.421
$Cu^{3+} + e^- \longrightarrow Cu^{2+}$	1	2.4
$XeO_3 + 6H^+ + 6e^- \longrightarrow Xe + 3H_2O$	6	2.10
$O_3 + 2H^+ + 2e^- \longrightarrow O_2 + H_2O$	2	2.076
$OH + e^- \longrightarrow OH^-$	1	2.02
$Co^{3+} + e^- \longrightarrow Co^{2+}$	1	1.92
$H_2O_2 + 2H^+ + 2e^- \longrightarrow 2H_2O$	2	1.776
$N_2O + 2H^+ + 2e^- \longrightarrow N_2 + H_2O$	2	1.766
$Au^+ + e^- \longrightarrow Au$	1	1.692
$PbO_2 + SO_4^{2-} + 4H^+ + 2e^- \longrightarrow PbSO_4 + 2H_2O$	2	1.6913
$MnO_4^- + 4H^+ + 3e^- \longrightarrow MnO_2 + 2H_2O$	3	1.673
$NiO_2 + 4H^+ + 2e^- \longrightarrow Ni^{2+} + 2H_2O$	2	1.678
$HClO + H^+ + e^- \longrightarrow \frac{1}{2}Cl_2 + H_2O$	1	1.63
$Ce^{4+} + e^- \longrightarrow Ce^{3+}$	1	1.61
$Mn^{3+} + e^- \longrightarrow Mn^{2+}$	1	1.542
$MnO_4^- + 8H^+ + 5e^- \longrightarrow Mn^{2+} + 4H_2O$	5	1.507
$BrO_3^- + 6H^+ + 5e^- \longrightarrow \frac{1}{2}Br_2 + 3H_2O$	5	1.52
$ClO_3^- + 6H^+ + 5e^- \longrightarrow \frac{1}{2}Cl_2 + 3H_2O$	5	1.47
$PbO_2 + 4H^+ + 2e^- \longrightarrow Pb^{2+} + 2H_2O$	2	1.455
$Au^{3+} + 3e^- \longrightarrow Au$	3	1.40
$Cl_2 + 2e^- \longrightarrow 2Cl^-$	2	1.3583
$Cr_2O_7^{2-} + 14H^+ + 6e^- \longrightarrow 2Cr^{3+} + 7H_2O$	6	1.33
$MnO_2 + 4H^+ + 2e^- \longrightarrow Mn^{2+} + 2H_2O$	2	1.23
$O_2 + 4H^+ + 4e^- \longrightarrow 2H_2O$	4	1.229
$IO_3^- + 6H^+ + 5e^- \longrightarrow \frac{1}{2}I_2(s) + 3H_2O$	5	1.195
$IO_3^- + 6H^+ + 6e^- \longrightarrow I^- + 3H_2O$	6	1.085
$Br_2(\ell) + 2e^- \longrightarrow 2Br^-$	2	1.066
$HNO_2 + H^+ + e^- \longrightarrow NO + H_2O$	1	1.00
$VO_2 + 2H^+ + e^- \longrightarrow VO^{2+} + H_2O$	1	1.00
$NO_3 + 4H^+ + 3e^- \longrightarrow NO + 2H_2O$	3	0.96
$2Hg^{2+} + 2e \longrightarrow Hg_2^{2+}$	2	0.92
$ClO^- + H_2O + 2e^- \longrightarrow Cl^- + 2OH^-$	2	0.89

Continued on next page

Table A6.1 Standard Reduction Potentials at 25°C (*Continued*)

Half-reaction	n	$E°$ (V)
$HO_2^- + H_2O + 2e^- \longrightarrow 3OH^-$	2	0.88
$Hg^{2+} + 2e^- \longrightarrow Hg$	2	0.851
$Ag^+ + e^- \longrightarrow Ag$	1	0.7996
$Hg_2^{2+} + 2e^- \longrightarrow 2Hg$	2	0.7973
$Fe^{3+} + e^- \longrightarrow Fe^{2+}$	1	0.770
$PtCl_4^{2-} + 2e^- \longrightarrow Pt + 4Cl^-$	2	0.73
$O_2 + 2H^+ + 2e^- \longrightarrow H_2O_2$	2	0.68
$MnO_4^- + 2H_2O + 3e^- \longrightarrow MnO_2 + 4OH^-$	3	0.59
$H_3A_3O_4 + 2H^+ + 2e \longrightarrow H_3A_5O_3 + H_2O$	2	0.559
$I_2(s) + 2e^- \longrightarrow 2I^-$	2	0.5355
$Cu^+ + 2e^- \longrightarrow Cu$	1	0.521
$H_2SO_3 + 4H^+ + 4e^- \longrightarrow S + 3H_2O$	4	0.449
$Ag_2CrO_4 + 2e^- \longrightarrow 2Ag + CrO_4^{2-}$	2	0.4470
$O_2 + 2H_2O + 4e^- \longrightarrow 4OH^-$	4	0.401
$Fe(CN)_6^{3-} + e^- \longrightarrow Fe(CN)_6^{4-}$	1	0.36
$Ag_2O + H_2O + 2e^- \longrightarrow 2Ag + 2OH^-$	2	0.342
$Cu^{2+} + 2e^- \longrightarrow Cu$	2	0.3419
$BrO^+ + 2H^+ + 3e^- \longrightarrow Br + H_2O$	3	0.32
$AgCl + e^- \longrightarrow Ag + Cl^-$	1	0.2223
$HSO_4^- + 3H^+ + 2e^- \longrightarrow H_2SO_3 + H_2O$	2	0.17
$Sn^{4+} + 2e^- \longrightarrow Sn^{2+}$	2	0.154
$Cu^{2+} + e^- \longrightarrow Cu^+$	1	0.153
$S + 2H^+ + 2e^- \longrightarrow H_2S$	2	0.141
$HgO + H_2O + 2e^- \longrightarrow Hg + 2OH^-$	2	0.0977
$AgBr + e^- \longrightarrow Ag + Br^-$	1	0.095
$Ag(S_2O_3)^{3-} + e^- \longrightarrow Ag + 2S_2O_3^{2-}$	2	0.01
$2H^+ + 2e^- \longrightarrow H_2$	2	0.0000
$Pb^{2+} + 2e^- \longrightarrow Pb$	2	−0.126
$CrO^{2-} + 4H_2O + 3e^- \longrightarrow Cr(OH)_3 + 5OH^-$	3	−0.13
$Sn^{2+} + 2e^- \longrightarrow Sn(s)$	2	−0.136
$AgI + e^- \longrightarrow Ag + I^-$	1	−0.1522
$CuI + e^- \longrightarrow Cu + I^-$	1	−0.185
$N_2 + 5H^+ + 4e^- \longrightarrow N_2H_5^+$	4	−0.23
$Ni^{2+} + 2e^- \longrightarrow Ni$	2	−0.257
$PbSO_4 + H^+ + 2e^- \longrightarrow Pb + HSO_4$	2	−0.356
$Co^{2+} + 2e^- \longrightarrow Co$	2	−0.277
$Ag(CN)^- + e^- \longrightarrow Ag + 2CN^-$	1	−0.31
$Cd^{2+} + 2e^- \longrightarrow Cd$	2	−0.403
$Cr^{3+} + e^- \longrightarrow Cr^{2+}$	1	−0.41
$Fe^{2+} + 2e^- \longrightarrow Fe$	2	−0.447
$2CO_2 + 2H^+ + 2e^- \longrightarrow H_2C_2O_4$	2	−0.49
$Ni(OH)_2 + 2e^- \longrightarrow Ni + 2OH^-$	2	−0.72
$Cr^{3+} + 3e^- \longrightarrow Cr$	3	−0.74
$Zn^{2+} + 2e^- \longrightarrow Zn$	2	−0.7618
$2H_2O + 2e^- \longrightarrow H_2 + 2OH^-$	2	−0.8277
$SO_4^{2-} + H_2O + 2e^- \longrightarrow SO_3^{2-} + 2OH^-$	2	−0.92
$N_2 + 4H_2O + 4e^- \longrightarrow 4OH^- + N_2H_4$	4	−1.16

Continued on next page

Table A6.1 Standard Reduction Potentials at 25°C (*Continued*)

Half-reaction	n	$E°$ (V)
$Mn^{2+} + 2e^- \longrightarrow Mn$	2	-1.185
$Zn(OH)_2 + 2e^- \longrightarrow Zn + 2OH^-$	2	-1.249
$Al^{3+} + 3e^- \longrightarrow Al$	3	-1.662
$Mg^{2} + 2e^- \longrightarrow Mg$	2	-2.37
$Na^+ + e^- \longrightarrow Na$	1	-2.71
$Ca^{2+} + 2e^- \longrightarrow Ca$	2	-2.868
$Ba^{2+} + 2e^- \longrightarrow Ba$	2	-2.912
$K^+ + e^- \longrightarrow K$	1	-2.95
$Li^+ + e^- \longrightarrow Li$	1	-3.05

Photo Credits

NOTE: "un" denotes an unnumbered figure.

Chapter 1

1.0: © Roger Ressmeyer/Corbis ; un1.1a: Bettmann/Corbis ; un1.1b: © Corbis; un1.1c: Bettmann/Corbis; un1.2: © Ecoscene/Corbis; un1.6: Bettmann/Corbis; 1.1: © IBM/SPL/Photo Researchers, Inc.; 1.3: © Michael S. Yamashita/Corbis; 1.4: NASA; 1.9: © Marilyn "Angel" Wynn Nativestock.com; 1.16: © Richard Hamilton Smith/Corbis; 1.17: © Tecmap Corporation; Eric Curry/Corbis; 1.20: NASA; 1.29: Created by NASA Goddard Space Flight Center under the guidance of the COBE Science Working Group; Provided by the NSSDC; Problem 53: NON SEQUITUR ©1996 Wiley Miller. Dist. By UNIVERSAL PRESS SYNDI-CATE. Reprinted with permission. All rights reserved.

Chapter 2

2.0: NASA/CXC/SAO; un2.1: Roger Ressmeyer/Corbis; un2.2: Hulton-Deutsch Collection/CORBIS; un2.3a: © AKG; un2.4: Archivo Iconografico, S.A./Corbis; un2.5: © David Lees/Corbis; box2.1: Photographers Blais/Turnbull; Parks Canada; 2.1: R. Thompson, M. Rieke, and G. Schneider (Univ. Arizona), NASA; 2.4: NASA/HST/ J. Morse/K. Davidson; 2.10: Hulton Archive/Getty Images; 2.11: J. Hester (ASU), NASA; 2.15: © Yann Arthus-Bertrand/Corbis; 2.18: Bettmann/Corbis; 2.23: Wellcome Dept. of Cognitive Neurology/ SPL/Photo Researchers, Inc.; 2.25: © Ric Ergenbright/Corbis; 2.26: © Kevin Schafer/Corbis

Chapter 3

3.0: © 1996 Wendy Carlos and Fred Espenak; un3.1: Mark Newman/ Visuals Unlimited; un3.3: IBM Almaden Research Center Visualization; box3.1: © Kevin R. Morris/Corbis; 3.12: © Dr. E. R. Degginger/COLOR-PIC, INC.; 3.23a: Bettmann/Corbis; 3.23b: Corbis

Chapter 4

4.0: © Science Visuals Unlimited; un4.1: NASA/JPL/Caltech/Johns Hopkins University Applied Physics Laboratory; un4.2: © Cathy Crawford/Corbis; un4.3: © Ric Ergenbright/Corbis; un4.5: courtesy of CLOROX® Liquid Bleach; un4.6a: Brad Mogen/Visuals Unlimited; un4.6b: © John Elk III; un4.7a, b © Tom Pantages; box4.1: Tim Fisher/The Military Picture Library/Corbis; 4.1a, b: NASA/JPL/Caltech; 4.4: D. Roddy (LPI); 4.5: Austin Post, USGS/CVO/Glaciology Project; 4.6: Brad Mogen/Visuals Unlimited; 4.15a: © Dr. E. R. Degginger/COLOR-PIC, Inc.; 4.15b1: © Carl & Ann Purcell/Corbis; 4.15b2: Frank Lane Picture Agency/Corbis

Chapter 5

5.0: © Rick Doyle/Corbis; box5.1: © Bryan Pickering; Eye Ubiquitous/ Corbis; un5.1: © David Wrobel/Visuals Unlimited; un5.2: © Raymond Gehman/Corbis; un5.2b: © David Muench/Corbis; un5.3: © José Manuel Sanchis Calvete/Corbis; un5.4: © Dr. E. R. Degginger/COLOR-PIC, INC.; un5.5: © David Fleetham/Visuals Unlimited; un5.6: © Dr. John Cunningham/Visuals Unlimited; un5.6: © Tom Pantages; 5.1a: © Owen Franken/Corbis; 5.1b: NASA; 5.2a, b: © Dr. E. R. Degginger/COLOR-PIC, INC.; 5.3: © W. Wayne Lockwood, M.D./Corbis; 5.5a, b: © Tom Pantages; 5.6: © Tom Pantages; 5.8: © Tom Pantages; 5.10: © David M. Phillips/Visuals Unlimited; 5.11: © Richard T. Nowitz/Corbis; 5.24: © Richard Thom/Visuals Unlimited; 5.25: © Dr. E. R. Degginger/COLOR-PIC, INC.; 5.26: © Dr. E. R. Degginger/COLOR-PIC, INC.; 5.27: Thomas R. Gilbert; 5.28: © Tom Pantages; 5.32: © Dr. E. R. Degginger/COLOR-PIC, INC.; 5.33a: © Jonathan Blair/Corbis; 5.33b: © Craig Lovell/Corbis

Chapter 6

6.0: © Dr. E. R. Degginger/COLOR-PIC, INC.; un6.1: Bettmann/Corbis; un6.2: © W. Perry Conway/Corbis; un6.3: © Kevin R. Morris/Corbis; box6.1: © Paul Souders; box6.2: © Joseph Sohm; ChromoSohm

Photo Credits

Inc./Corbis; box6.3: © Dr. E. R. Degginger/COLOR-PIC, INC; 6.2: AIP Emilio Segrè Visual Archives; 6.12: © Geir Ole Braathen; 6.14: © Dr. E. R. Degginger/COLOR-PIC, INC.; 6.18: © Yoav Levy/Phototake; 6.21: G. Schneider (UofA), K. L. Luhman (CfA), et al., NICMOS IDT, NASA WFPC2 data: C. O'Dell and S. Wong (Rice)

Chapter 7

7.0: NASA; un7.4: © Patrick Bennett/Corbis; un7.5: John Sohlden/Visuals Unlimited; box7.1: © Dr. E. R. Degginger/COLOR-PIC, INC.; 7.1: NASA/NSSDC; 7.2a: © Dr. E. R. Degginger/COLOR-PIC, INC.; 7.2b: NASA/NSSDC; 7.10: © Tom Pantages; 7.15: © Tom Pantages

Chapter 8

8.0: NASA/Corbis; un8.1: Bettman/Corbis; un8.2: NOAA; un8.4: Romilly Lockyer/The Image Bank/Getty Images; un8.6b: © Paul Souders; un8.7a, b: © Penni Zivian; box8.1: Custom Medical Stock Photo; 8.4a: © Galen Rowell/Corbis; 8.4b: © Hubert Stadler/Corbis; 8.4c: © Bill Ross/Corbis; 8.14: © Phil Degginger/COLOR-PIC, INC.

Chapter 9

9.0: © Richard Hamilton Smith/Corbis; un9.1: © Owen Franken/Corbis; un9.2: David Wrobel/Visuals Unlimited; un9.3: Larry Stepanowicz/Visuals Unlimited; 9.1a–c: © Johann Schumacher Design; 9.4a: © Dwight Kuhn; 9.7a, b: © Tom Pantages; 9.8: © Tom Pantages; 9.9: Ken Lucas/Visuals Unlimited; 9.14a: © Dr. E. R. Degginger/COLOR-PIC, INC.; 9.14b: Tom Pantages/ © 1998 MBARI; 9.15: Jeff J. Daly/Visuals Unlimted; 9.16: © Dr. E. R. Degginger/COLOR-PIC, INC.; 9.17a: © Tom Pantages; 9.17b: © Phil Degginger/COLOR-PIC, INC.; 9.19: © Paul Souders

Chapter 10

10.0: © Jose Manuel Sanchis Calvente/Corbis; un10.1: Historical Pictures Archive/Corbis; un10.2a: © Paul Souders; un10.2b: Karlene/Lowell Schwartz; un10.16: © Dr. E.R. Degginger/COLOR-PIC, INC.; un10.17a, b: courtesy of David Loope, University of Nebraska; un10.17c: John Betts Fine Minerals; 10.1: © Dr. E. R. Degginger/COLOR-PIC, INC.; 10.2a: © Jose Manuel Sanchis Calvete/Corbis; 10.2b: © Ric Erginbright/Corbis; 10.4: © Tom Pantages; 10.10: © Tom Pantages; 10.11: © Tom Pantages; 10.12: Fundamental Photographs; 10.14: courtesy John Holfert, Mineral Market; 10.15: Adrienne Hart-Davis/Science Photo Library; 10.16a: © Martin Miller/Visuals Unlimited; 10.16b: © Ken Lucas/Visuals Unlimited; 10.16c: © Tom Pantages; 10.16d: © A. J. Copley/Visuals Unlimited; 10.26: © Tom Pantages; 10.27: © Adam Woolfitt/Corbis; 10.28: Vaughan Fleming/Science Photo Library

Chapter 11

11.0: © Lowell Georgia/Corbis; un11.2: © Science Visuals Unlimited; un11.3a: © RDF/Visuals Unlimited; un11.3b: Gary W. Carter/Visuals Unlimited; un11.4: John Gerlach/Visuals Unlimited; un11.7a: D. Yeske/Visuals Unlimited; un11.5: Susan M. Davis/Caribbean Design & Marketing Consultants.; un11.7b: © Warren Stone/Visuals Unlimited; 11.2a, b: Raymond Malace; 11.3a, b: © Phil Degginger/COLOR-PIC, INC.; 11.3c, d: © Dr. E. R. Degginger/COLOR-PIC, INC.; 11.8: © AFP/Corbis; 11.9: © Jon Gnass

Chapter 12

12.0: © Philip Gould/Corbis; un12.1: Owaki-Kulla/Corbis; un12.2: © Dr. E. R. Degginger/COLOR-PIC, INC.; un12.3: Bettmann/Corbis; un12.4: NASA; un12.5: Visuals Unlimited; un12.6: © Phil Degginger/COLOR-PIC, INC.; 12.11: © Dr. E. R. Degginger/COLOR-PIC, INC.; 12.17: Dr. E. R. Degginger/COLOR-PIC, INC.; 12.20: © Dr. E. R. Degginger/COLOR-PIC, INC.

Chapter 13

13.0: © Steve Terrill/Corbis; un13.1: © Michael Topolovac/David Madison Sports Images, Inc.; un13.3a: Andreas von Einsiedel; Elizabeth Whiting & Associates/Corbis; un13.3b © Richard Hutchings/Corbis; un13.4: © Phil Degginger/COLOR-PIC, INC.; un13.6: © Phil Degginger/COLOR-PIC, INC.; un13.7: © Dr. E. R. Degginger/COLOR-PIC, INC.; un13.8a: © Phil Degginger/COLOR-PIC, INC.; 13.4: © Charles O'Rear/Corbis; 13.5: © Lester V. Bergman/Corbis; 13.8: Cleanairbus; 13.17a: © Francis E. Caldwell/Visuals Unlimited; 13.17b: © Gary Randall/Visuals Unlimited; 13.17c: © Inga Spence/Visuals Unlimited; 13.17d: © Bud Nielsen

Chapter 14

14.0: © Robert Landau/Corbis; un14.2: © E. R. Degginger/COLOR-PIC, INC.; un14.7: © William Schick/Corbis; un14.8: AFP/Corbis; un14.10: © Dr. E. R. Degginger/COLOR-PIC, INC.; 14.14: © David Stoecklein/Corbis

Chapter 15

15.0: Wilderland Images; un15.1: © Jim Pickerell/Stock Connection/PictureQuest; un15.2: © 1999 Richard Megna/Fundamental Photographs; un15.4: © Tim Hauf Photography/Visuals Unlimited; box15.1: © Austrian Archives/Corbis; 15.6a–c: © 1993 Richard Megna/Fundamental Photographs.; 15.7a, b: © Dr. E. R. Degginger/COLOR-PIC, INC.

Chapter 16

16.0: © D. Like/Visuals Unlimited; un16.2: © E. R. Degginger/COLOR-PIC, INC.; un16.4: © Larry Stepanowicz/Visuals Unlimited; un16.5: © Dr. E. R. Degginger/COLOR-PIC, INC.; un16.6: © Gerald A. Corsi/Visuals Unlimited; un16.7: © Dr. E. R. Degginger/COLOR-PIC, INC.; un16.9: © Hulton-Deutsch Collection/Corbis; 16.22: © E. R. Degginger/COLOR-PIC, INC.

Chapter 17

17.0: © Chuck PLACE; un17.2: © Dr. E. R. Degginger/COLOR-PIC, INC.; un17.3: © Kim Fennema/Visuals Unlimited; un17.4: courtesy of Honda; 17.1: © Dr. E. R. Degginger/COLOR-PIC, INC.; 17.7: © Tom Pantages; 17.10: © Tom Pantages; 17.15: © Tom Pantages; 17.17: © Shamim Ferdaus/Corbis Sygma; 17.18: © AFP/Corbis

Chapter 18

18.0: © Andy Ryan Photography; un18.1: John D. Cunningham/Visuals Unlimited; un18.5a: courtesy of The Edgar Fahs Smith Collection, University of Pennsylvania; un18.5b: courtesy of Alcoa; un18.7: Jeff Greenberg/Visuals Unlimited; un18.8: courtesy of Nora Foley Ayuso, U.S. Geological Survey; un18.9: Hagley Museum and Library; 18.15: © Tom Pantages; 18.18: © Inga Spence/Visuals Unlimited; 18.22: © Dr. Jeremy Burgess/Photo Researchers, Inc.; 18.23: © Dr. E. R. Degginger/COLOR-PIC, INC.; 18.26: © Tom Pantages

Glossary

absolute zero (0 K) the lowest temperature possible.

accuracy the agreement between an experimental value and the true value.

acid 1. a compound that has a formula that begins with hydrogen and ends with the formula of an oxoanion. 2. a substance that can donate H^+ to water and other substances.

acid rain highly acidic rain owing to pollutant nonmetal oxides and other acidic compounds dissolved in it.

actinide element one of the elements 90 through 103 in the periodic table.

activated complex a high-energy, unstable species that exists momentarily in the course of a chemical reaction and that quickly falls apart, either forming products or re-forming reactants.

activation energy the energy barrier that colliding species must overcome if they are to react with each other.

addition polymer a polymer formed when monomers containing double bonds (usually C–C bonds) combine through an addition reaction.

addition reaction a reaction that forms polymers, when the π bonds in the monomers are broken and Σ bonds that link the monomers together are formed.

alcohol an organic compound with a R—O—H functional group. In a phenol, the —O—H group is attached to an aromatic ring.

aldehyde an organic compound with a functional group containing a carbon atom double-bonded to an oxygen atom and single-bonded to a hydrogen atom.

alkali metal one of the elements that belongs to Group 1A of the periodic table and exists in nature as a cation with a charge of 11.

alkaline earth metal one of the elements that belongs to Group 2A of the periodic table and exists in nature as a cation with a charge of 12.

alkane a molecule in which all the C—C bonds are single bonds. Also called saturated hydrocarbon.

alkene a hydrocarbon that contain one or more C—C double bonds.

alkyl group a saturated hydrocarbon substituent group in an organic molecule.

alkyne a hydrocarbon that contains one or more C—C triple bonds.

allotrope a form of the same element with different structures and different physical and chemical properties.

alpha (α) decay the loss of a ^4_2H particle by the nucleus of a heavy element.

alpha(α)-amino acid a compound in which a carboxylic acid group and an amino group are bonded to the same carbon atom.

amine an organic compound containing a R—N— functional group.

amorphous solid a solid that has no long-range ordering in its structures.

amphoteric a term describing a substance that is capable of behaving both as an acid and as a base.

amplitude the height of a wave's crests.

angular (bent) molecular geometry an orientation of groups of valence electrons on the central atom of a molecule that forms a tetrahedron with two lone pairs and two bonding pairs.

angular molecular geometry an orientation of electron groups on the central atom of a molecule with pairs of electrons grouped together by double and triple bonds with the top corner occupied by a lone pair and a triangular shape that produces a bond angle of 116.5 degrees.

angular momentum quantum number (1) an integer from zero to $n - 1$ that defines the shape of an orbital.

anion a negatively charged atom that has gained electrons; a negative ion.

anode an electrode at which an oxidation half-reaction takes place.

antibonding molecular orbital one of the two orbitals formed by interaction between two atomic orbitals. The antibonding orbital is higher-energy, and has two separate, teardrop-shaped lobes of electron density and a region of no electron density (a node) between the two hydrogen atoms.

antimatter a group of subatomic particles that are the charge opposites of those typically found in atoms.

aqueous solution a solution in which water is the solvent.

aromatic hydrocarbon a compound composed of carbon and hydrogen with one or more rings of carbon atoms in which there are delocalized π electrons that are spread over all the carbon atoms in the ring(s).

Arrhenius equation an equation that relates the rate constant of a chemical reaction to absolute temperature (T), the reaction's activation energy (E_a), and a collisional frequency factor (A).

atactic polymer a polymer with a random orientation of side groups attached to chiral centers along the polymer chain.

atom the smallest particle of an element that retains the characteristics of the element.

atomic mass unit (amu) unit widely used to express the relative masses of atoms and subatomic particles.

atomic number the number of protons in each of an element's atoms.

Avogadro's number ($N_A = 6.022 \times 10^{23}$) the number of particles equal to the number of atoms in exactly 12 grams of carbon-12, which is a very large number of atoms. To four significant figures, it equals 6.022×10^{23} atoms. Avogadro's number represents the number of particles that make up one mole.

base a substance that accepts H^+ and so can produce an excess of OH^2 in aqueous solutions.

basic forces of nature natural phenomena, including gravity, electromagnetism, strong nuclear force, and weak nuclear force.

Bayer process the method of extracting and purifying aluminum oxide using strong base.

belt of stability a plot of neutrons versus protons in the nuclei of stable and radioactive nuclides.

beta (β) decay the process by which a neutron disintegrates into a proton and a high-energy electron called a β particle.

binary ionic compound a compound that consists of crystals (three-dimensional arrays with repeating patterns) of alternating cations (ions with positive charges) and anions (ions with negative charges).

binary molecular compound a compound that consists of molecules containing atoms of only two elements.

binding energy (E) the energy that binds nucleons together. It can be determined by using the equation $E = (\Delta m)c^2$, where Δm is the mass defect of the nucleus.

body-centered cubic unit cell (bcc) a unit cell that has lattice points at the eight corners of a cube and at the center of the cell.

boiling-point elevation the increase in the boiling point of a solvent when particles of solute are dissolved in it.

bond dipole a phenomenon in which two bonded atoms have partial electrical charges of equal but opposite sign.

bond energy (ΔH_{rxn}) the enthalpy change required to break 1 mole of bonds in the gas phase. Average bond energies can be used to estimate the enthalpy change in chemical reactions.

bond order the number of bonds between atoms. Single, double, and triple bonds have bond orders of 1, 2, and 3, respectively.

bonding molecular orbital one of the two molecular orbitals formed by interaction between two atomic orbitals. It is a lower-energy MO, is oval in shape, and spans the two atomic centers.

Boyle's law the product of the pressure times the volume of a gas is constant at constant temperature. In other words, volume is inversely proportional to pressure.

Bragg equation an equation that relates the angle of diffraction (θ) of X-rays to the spacing (d) of layers of ions or atoms in a crystal.

breeder reactor a nuclear reactor in which fissionable material is produced during normal reactor operation.

buffer capacity the quantity of acid or base that can be added to a pH buffer without significantly changing the pH of the buffer.

calorimetry the measurement of the change in heat that occurs during a physical or chemical change.

capillary action the rise of a liquid up a narrow tube as a result of adhesive forces between the liquid and the tube and cohesive forces within the liquid.

carbohydrate a class of organic compounds with the general molecular formula $C_x(H_2O)_y$. They include simple sugars, such as glucose, $C_6H_{12}O_6$, and polysaccharides, including starch and cellulose, formed from condensation reactions of sugar molecules.

carbonyl group the C—O double bond in an organic compound. It is a feature of several organic functional groups and adds to the polarity of organic substances.

carboxylate anion an ion formed from carboxylic acids and water.

carboxylic acids organic compounds with —C—OH functional groups. They partly ionize in water, forming carboxylate anions and protons.

carrier a stable isotope used along with a radioactive isotope in studies on the movement of radionuclides in the body and their accumulation in certain organs.

catalyst a substance that increases the rate of a reaction without being consumed by it.

cathode an electrode at which a reduction half-reaction takes place.

cation a positively charged atom that has lost electrons; a positive ion.

cell potential see **electromotive force**.

chain reaction self-sustaining fission due to the production of neutrons that split other nuclei.

Charles's law the volume of a gas at constant pressure is proportional to its absolute temperature.

chemical energy a kind of potential energy that is a function of the *composition*, not the *position*, of a substance.

chemical equation an equation that describes the proportions of the substances called reactants that are consumed during a chemical reaction and the substances (called products) that are formed.

chemical equilibrium 1. a point at which the rate of formation of products from reactants is equal to the rate of formation of reactants from products. 2. the state in which the rate of a reaction in the forward direction matches its rate in the reverse direction and in which the concentrations of reactant(s) and product(s) do not change.

chemical formula a formula that uses atomic symbols and subscripts to identify the constituent elements in a substance and their proportions.

chemical property a characteristic of a substance that can be observed only through a chemical reaction that includes the substance.

chemical weathering a chemical reaction of constituents of the atmosphere (often dissolved in rain) with rocks and minerals. As a result parts of the rocks and minerals dissolve.

chemistry the science of matter: its composition, structure, and properties.

chiral molecule an enantiomeric molecule.

closest packing the most efficient stacking of particles in a crystalline solid.

colligative property a property of a solution that depends on the concentration but not the identity of the particles dissolved in it.

colloid a stable suspension of particles with diameters ranging from 1 to 100 nm in liquids or gases.

combustion a reaction in which an element or a compound burns in an atmosphere containing oxygen.

common-ion effect the shift in the position of an equilibrium caused by the addition of an ion taking part in the reaction.

complex ion a phenomenon in which ions and molecules called ligands surround and bond to a metal ion.

compound a substance composed of two or more elements combined together in fixed proportions.

concentration the ratio of the quantity of solute to either the volume or the mass of the solution or solvent in which the solute is dissolved.

condensation a reaction in which functional groups containing oxygen and hydrogen react together and lose a molecule of water in the process.

conjugate adding an H^+ ion to a base.

constructive interference a phenomenon that occurs when the crests and troughs of two waves coincide.

continuum radiation electromagnetic radiation of all wavelengths within a wavelength range.

coordination number the number of surrounding particles that touch a particle in a crystalline structure.

cosmic background microwave radiation found throughout the universe and believed to be an echo of the Big Bang. It was predicted to exist but was later discovered by accident in the 1960s.

Coulomb's law the energy (E) of the interaction between two ions is proportional to the product of the charges of two ions (Q_1 and Q_2) and inversely proportional to the distance (d) between them.

covalent bond a bond formed from electron sharing by atoms in molecules.

critical mass the mass of a fissionable substance needed to produce a self-sustaining chain reaction.

critical point the point of temperature and pressure at which the liquid and gaseous states of a substance are indistinguishable.

critical pressure the pressure above which the liquid and gaseous state of a substance are indistinguishable.

critical temperature the temperature above which the liquid and gaseous state of a substance are indistinguishable.

crystal field splitting the splitting of a set of d orbitals into two subsets with different energies because of interaction with pairs of electrons in orbitals on surrounding atoms.

crystal field splitting energy (Δ) the difference in energy between two subsets created by crystal field splitting.

crystal field theory the colors and magnetic properties of species containing transition-metal ions are due to the splitting of the energies of the ions' d orbitals caused by interactions with electrons on surrounding atoms.

crystalline solid a solid made of ordered arrays of atoms, ions, or molecules that gives characteristic X-ray diffraction patterns.

cubic closest packing (ccp) a crystal structure in which the composition of layers of particles in face-centered cubic unit cells has an *abcabcabc* . . . pattern.

cyclotron a device used to accelerate particles to very high velocities that uses a combination of magnetic and electrical fields to swirl positive particles in spiral pathways until they exit the device and smash into target nuclei.

Dalton's law the total pressure of a mixture of gases equals the sum of the partial pressures of each of the gases in the mixture.

degenerate orbital an orbital in a shell that has the same energy as another orbital.

degree of dissociation see **degree of ionization**.

degree of ionization the ratio of the quantity of a substance that is ionized to the total quantity of the substance present. The degree of ionization of a weak acid increases with decreasing acid concentration. The degree of ionization is also called the degree of dissociation.

delocalization the process in which molecular orbitals are spread out.

density the ratio of the mass of an object to its volume.

deoxyribonucleic acid (DNA) a biological polymer consisting of units called nucleotides, each of which is made of three subunits: a nitrogen-containing base, deoxyribose sugar, and a phosphodiester. The sequence of bases carries genetic information.

destructive interference a phenomenon that occurs when the crests and troughs of two waves do not coincide.

deuteron the combination of one proton and one neutron.

diamagnetic having no unpaired electrons.

diffraction the bending of electromagnetic rays around the edge of an object or as they pass through minute openings, forming circular waves.

diffusion the spread of one substance through another.

dipole–dipole interaction the attractive force between polar molecules.

dipole-induced dipole interaction the ability of polar solvent molecules to induce dipoles in nonpolar solutes.

dispersion force the relatively weak intermolecular force caused by temporary dipoles induced by neighboring atoms or molecules. The strength of the force depends on how easily the atoms or molecules can be polarized.

distillation a separation technique in which the more volatile components of a mixture are vaporized and then condensed, separating them from the less volatile components.

Doppler effect the shift in perceived frequency of waves caused by movement of their source away from, or toward, the observer.

effective nuclear charge (Z_{eff}) the attractive force toward the nucleus felt by an electron in an atom.

effusion the process by which a gas escapes through tiny holes into a space of lower pressure.

electrochemical cell a reaction system in which oxidation and reduction reactions at separate electrodes either consume (in an electrolytic cell) or produce (in a voltaic cell) electrical energy.

electrochemistry the branch of chemistry that examines the transformations between chemical and electrical energy.

electrode a conducting solid used to make electrical contact with a liquid or a solution.

electrolysis a chemical reaction caused by the passage of an electrical current through the reactant.

electrolyte a substance that dissociates into ions when it dissolves, enhancing the conductivity of water.

electrolytic cell a cell in which an external source of electrical energy does work on a chemical system, turning low-energy reactant(s) into high-energy product(s).

electromagnetic radiation radiant (electromagnetic) energy including light and ultraviolet, infrared, X-ray, γ ray, and other forms of radiation.

electromagnetism the force of attraction between oppositely charged particles or between the north and south poles of magnets.

electromotive force cell potential; the voltage between the electrodes of a voltaic cell.

electron affinity the energy change that occurs when a mole of an element's gaseous atoms combine with a mole of electrons.

electron capture the process by which the proton : neutron ratio in a proton-rich nucleus is reduced, by the nucleus "capturing" one of its own electrons.

electron configuration the distribution of electrons among the orbitals of an atom or ion.

electronegativity a measure of an element's ability to attract bonding electrons.

electron-spin resonance (ESR) techniques used to determine the location of an unpaired electrons. Molecules and ions with an odd number of electrons exhibit electron-spin resonance. Electron-spin resonance spectra enable scientists to predict where unpaired electrons are likely to be in a molecule.

element a pure substance composed of only one kind of atom.

elementary step a simple reaction that represents the progress of the overall reaction. Elementary steps provide a molecular view of the reaction progress.

empirical agreeing with experimental measurements but not necessarily having a theoretical basis.

empirical formula the simplest whole-number ratio of the elements in a compound. An empirical formula of a molecular compound is not necessarily the same as its true molecular formula.

enantiomer a stereoisomer whose mirror images are not superimposable.

endergonic reaction a reaction in which reactants form products that have higher free energies, and so ΔG is positive. Endergonic reactions are not spontaneous.

endothermic a term describing the process by which heat is absorbed by the system from its surroundings.

energy the capacity to move an object through a distance or to transfer heat.

enthalpy (H) the sum of a system's internal energy and the product of the pressure and volume of the system.

entropy (S) a measure of the disorder, or freedom of motion, of a substance or system.

equilibrium constant (K) the value of the reaction quotient (Q) at equilibrium.

equilibrium constant expression the ratio of concentration terms for products divided by that for reactants in accordance with the balanced equation of the reaction. The equilibrium constant (K) is the numerical value for the equilibrium constant expression for a chemical reaction.

equivalence point a point in titration that is reached when just enough standard solution has been added to completely react with a solute in the sample. The concentration of the solute is calculated from the volume and concentration of the standard solution and the volume of the sample.

ester an organic compound in which a carbonyl group is bonded to an —O—R group. Ester groups are formed by reactions between carboxylic acids and alcohols in a process called esterification.

ether an organic compound with a —R—O—R— functional group.

excited state any state of an atom with energy higher than that of the ground state.

exergonic reaction a reaction in which reactants form products that have lower free energies, and so ΔG is negative. Exergonic reactions are spontaneous.

exothermic a term describing the process by which heat from the system flows to its surroundings.

face-centered cubic unit cell (fcc) a unit cell that has the same kind of particle at the eight corners of a cube and at the center of each face.

filtration a process for separating particles suspended in a liquid from that liquid by passing the mixture through a medium that retains the particles.

first law of thermodynamics the energy gained or lost by a system must equal the energy lost or gained by its surroundings.

formal charge the number of valence electrons on the free atom minus the number assigned to the atom in the molecular structure. Formal charges are used to assess the relative stability of resonance structures.

formation constant (Kf) an equilibrium constant relating the concentration of a complex ion to the concentrations of the metal ion and ligand(s) in the complex.

fractional distillation a method of separating a mixture of compounds on the basis of their different boiling points.

free energy a thermodynamic-state function related to temperature and two other state functions: enthalpy and entropy. Processes in which there is a decrease in free energy are spontaneous.

freezing-point depression the lowering of the freezing point of a solvent when particles of solute are dissolved in it.

frequency (ν) the rate at which wave crests pass a point per unit time.

frequency factor (A) the product of the frequency of collisions between reacting species and the orientation factor required for them to react.

fuel cell a voltaic device in which there is a flow of reactants to the anode and cathode.

fuel value the quantity of energy released when 1 gram of a substance undergoes combustion.

functional group a combination of atoms bonded together in an organic compound that imparts particular properties to that class of compounds.

gas constant (R) the constant of proportionality in the equation $PV = nRT$, the ideal-gas equation.

geometric isomer a compound with the same atoms and bonds as another compound, but with different spatial arrangements of their bonds and atoms around rigid structures, such as double bonds.

glycolysis a process that converts each mole of glucose in a series of enzyme-catalyzed reactions into 2 moles of pyruvate.

Graham's law of effusion the rate of effusion of a gas is inversely proportional to the square root of its molar mass.

gravity the attraction between bodies based on their masses.

ground state the lowest-energy, or most stable, state of an atom.

half-life the time interval in which half of a quantity of a radioactive substance decays.

half-life of a reaction the time required for half of the initial quantity of a reactant to be consumed.

half-reaction the oxidation or reduction component of an oxidation-reduction reaction.

heat capacity (C) the quantity of heat required to increase the temperature of an object (or group of objects) by 1 Celsius degree.

heat of hydration the enthalpy change that results from hydration.

heat of solution (ΔH_{soln}) the change in enthalpy when a mole of solute dissolves in a solvent, such as water. It is the net result of the enthalpy changes associated with the separation of solute and solvent particles from other solute and solvent particles and mixing with one another.

Heisenberg's uncertainty principle the principle that says that you cannot at the same time determine the position and momentum of an electron in an atom.

hemiacetal a functional group that includes a carbon atom bonded to a hydroxyl group and to an oxygen atom that is also bonded to another carbon atom (included in the R group).

Henderson-Hasselbalch equation ($pH = pKa + \log$) an equation used to calculate the pH of a solution in which the concentrations of acid and conjugate base are known.

Henry's law the solubility of a gas in a liquid is proportional to the partial pressure of the gas.

Hess's law the enthalpy change of a reaction that is the sum of two or more other reactions is equal to the sum of the enthalpy changes of the constituent reactions.

hexagonal closest packing (hcp) a crystal structure based on a hexagonal unit cell in which the composition of layers of particles has an *ababab* . . . pattern.

homogeneous alloy a solid solution of two or more metals. The solution also has metallic properties.

Hund's rule the lowest energy state of an atom or ion contains the maximum number of unpaired electrons with parallel spins.

hybrid atomic orbital an orbital associated only with a particular atom in a molecule.

hydration 1. the process by which a dissolved ion or molecule is surrounded by an ordered array of water molecules. 2. the process by which water molecules rearrange themselves around particles of solute in solution.

hydrocarbon an organic compound composed of only carbon and hydrogen.

hydrogen bond a particularly strong kind of dipole–dipole interaction between a hydrogen atom bonded to a highly electronegative element and an atom of one of these elements in another molecule.

hydrolysis the general term used for reactions in which water is a reactant.

hydrophilic ("water-loving") a term describing substances that are attracted to water and dissolve in it.

hydrophobic ("water-fearing") a term describing substances are repelled by water and are insoluble in it.

hypothesis a tentative explanation for a series of observations.

ideal gas a hypothetical gas whose behavior at the temperatures and pressures encountered in nature can be accounted for by the ideal gas equation.

ideal-gas equation ($PV = nRT$) an equation that expresses the ideal-gas law by relating the pressure, volume, and temperature of a given quantity of gas.

ideal-gas law the relation between pressure, volume, and temperature for any gas, provided it behaves as an ideal gas.

ideal solution a solution that obeys Raoult's law.

immiscible liquid a liquid that is not completely soluble in another liquid.

infrared radiation a "warm" region of the electromagnetic spectrum. Infrared waves have longer wavelengths than visible light.

infrared spectrum the spectrum of infrared radiation. A compound's infrared spectrum can serve as a molecular fingerprint, enabling chemists to detect the compound's presence in a sample and to determine its molecular structure.

integrated rate law a mathematical expression describing the change in concentration of a reactant in a chemical reaction with time.

intermediate a substance formed in one step of a multistep process and consumed in a later step.

intermetallic compound a homogeneous alloy of fixed proportions and characteristic physical and chemical properties.

internal energy (E) the sum of the potential and kinetic energies of the constituents of a system. The internal energy of a system is increased by heating the system or by doing work on it. Internal energy decreases as a system loses heat or does work on its surroundings.

interstitial alloy a homogeneous alloy in which the atoms of a much smaller element fit into the holes (interstices) of the crystal structure of the host metal. In steel, carbon atoms are located in the holes of the crystal structure of iron.

ion an electrically charged atom or group of atoms.

ion–dipole interaction the intermolecular force between an ion and the end of a polar molecule with the opposite partial charge.

ion-exchange reaction a reaction in which ions on a solid phase are replaced by ions in solution.

phase diagram a graphical presentation of the dependence of the stabilities of the physical states of a substance on temperature and pressure.

phosphorylation a reaction in which a phosphate group is added to an organic molecule. The reaction of glucose with dihydrogenphosphate that gives glucose 6-phosphate is an example.

photochemical cell a cell that directly converts solar energy into electricity.

photoelectric effect the release of electrons, called photoelectrons, from a metal as a result of the absorption of electromagnetic radiation. The minimum frequency of light for releasing photoelectrons depends on the metal. Higher frequencies give electrons with more kinetic energy.

photoelectron an electron emitted during the photoelectric effect.

photoionization the process by which the energy needed to separate an electron comes from electromagnetic radiation.

photon the smallest "building block" of electromagnetic radiation, such as that emitted or absorbed by a single atom.

physical property a characteristic of a substance that can be observed without changing it into another substance.

pi (π) bond a bond formed by electrons occupying π orbitals. In a covalent π bond, electron density is greatest above and below or in front of and behind the bonding axis.

pi (π) molecular orbital an orbital formed by the mixing of atomic orbitals that are not oriented along the bonding axis in a molecule.

pi* (π*) molecular orbital an orbital formed by the mixing of atomic orbitals that are not oriented along the bonding axis in a molecule.

plasma hot, extensively ionized gas.

polar covalent bond a bond that results from unequal electron sharing of bonding pairs of electrons between atoms.

polarization the process by which the electron cloud surrounding a molecule, ion, or atom is perturbed.

polyatomic ion an ion that contains more than one atom.

polymorph a compound with the same molecular formula but with different structures and different properties.

polysaccharide a carbohydrate formed from condensation reactions of sugar molecules, including starch and cellulose.

positron a particle with the mass of an electron but with a positive charge.

positron emission a process by which the proton : neutron ratio in a proton-rich nucleus is reduced.

potential energy the energy that an object has because of its position.

precipitate 1. (v.) to be separated from a solution as a solid. 2. (n.) a solid separated from a solution.

precipitation reaction a reaction in which a solid product called a precipitate is formed from a reaction in solution.

precision the repeatability of a measurement.

primary structure the sequence of amino acids in the polypeptide chain(s) of a protein.

principal quantum number (n) a positive integer that indicates the relative size of an orbital or a group of orbitals in an atom.

product a substance that is formed during a chemical reaction.

pure substance a particular kind of matter with well-defined properties and a fixed chemical composition.

qualitative chemical analysis analysis that tells us what substances are present in a sample.

quantitative analysis analysis that tells us how much of one or more substances is present.

quantum a quantity.

quantum mechanics the formulation and solution of wave equations; the description of behavior of particles as waves. The wavelike properties of electrons can be observed in the interference patterns produced by beams of electrons.

quantum number a solution to a wave equation, a unique combination of three integers.

radial distribution plot a graphical representation of the probability of finding an electron near the nucleus of an atom.

radioactive decay the spontaneous disintegration of unstable particles accompanied by the release of radiation.

radiocarbon dating the process of determining the age of an artifact based on the presence of trace amounts of radioactive ^{14}C found in the carbon dioxide that plants incorporate into their structures during photosynthesis.

Raoult's law the vapor pressure of a solution containing nonvolatile solutes is proportional to the mole fraction of the solvent.

rare earth element see **lanthanide element**.

rate constant (k) a proportion of the reaction rate to the concentration of reactant.

rate-determining step the slowest elementary step in a chemical reaction.

rate law the relationship of the rate of a reaction to the rate constant and the concentrations of reactants.

reactant a substance that is consumed during a chemical reaction.

reaction mechanism a mechanism that consists of one or more elementary steps that provide a molecular view of the reaction process.

reaction quotient (Q) the numerical value of the mass action expression when values for the concentrations (or partial pressures) of reactants and products are inserted in it.

redox reaction an oxidation–reduction reaction.

reducing agent an ion that provides the electrons in a redox reaction. In the reaction, the reducing agent is oxidized.

reduction a chemical change accompanied by a gain of electrons.

refraction the change in direction of a beam of electromagnetic radiation as it passes from one medium to another.

resonance structure one of two or more Lewis structures with the same skeletal structure, but different bonding arrangements.

rest mass the mass of a particle of light with zero velocity.

reverse osmosis a technique used in the water-purification process in which water is pumped through semipermeable membranes, leaving dissolved impurities behind.

ribonucleotide a nucleotide that contains the monosaccharide ribose.

root-mean-square speed (u_{rms}) the square root of the average of the squares of the velocities of a collection of a gas's molecules.

saturated containing the maximum amount of solute that can be dissolved in a given volume or mass of solvent.

saturated hydrocarbons see **alkanes**.

scientific law a statement, often expressed in a mathematical equation, that summarizes a broad range of observations and experimental results.

scientific method an approach to acquiring knowledge based on careful observation of phenomena.

second law of thermodynamics a process occurs spontaneously if it results in an increase in the entropy of the universe.

secondary structure the geometric arrangement of the amino acids in the polypeptide chain(s) of a protein that results from hydrogen bonds and other interactions between amino acids.

semiconductor a material with a conductivity that can be varied over several orders of magnitude by altering its chemical composition or by applying electrical potentials.

semipermeable slightly porous.

semipermeable membrane a membrane that blocks the passage of all solution components except the solvent.

sigma (Σ) bond a covalent bond in which electron density is greatest between the bonded atoms along the bonding axis.

sigma (Σ) molecular orbital an orbital formed by the mixing of atomic orbitals oriented along the bonding axis.

simple cubic unit cell a unit cell that has lattice points only at the eight corners of a cube.

smelting the roasting of metal ores to form metal oxides that are subsequently reduced to the free metal.

solubility product see **solubility-product constant**.

solubility-product constant (Ksp) an equilibrium constant relating the dissolved concentrations of the ions produced by the dissolution of a slightly soluble compound (also called the solubility product).

solute an ingredient in a solution aside from the solvent.

solution a homogeneous mixture of two or more substances.

solvent the substance present in a solution in the greatest proportion (in number of moles).

sp hybridization the process by which a linear orientation of valence electrons and the formation of as many as two s bonds and two p bonds are achieved.

sp^2 hybridization the process by which the trigonal planar orientation of valence electrons is achieved, which can result in the formation of as many as three s bonds and one p bond.

sp^3 hybridization the process by which the tetrahedral orientation of valence electrons is achieved, which can result in the formation of as many as four s bonds.

spectator ion an ion that does not appear in the net ionic equation, even though it was present in solution, because it was not affected in the reaction.

spectrum the distribution of the radiant energy that an object or a substance emits or absorbs over different wavelengths.

sphere of hydration an ordered array of water molecules.

spin magnetic quantum number (m_s) a quantum number to account for two spin orientations. It is $+\frac{1}{2}$ or $-\frac{1}{2}$, indicating the electron-spin orientation, up or down.

spontaneous process a process in thermodynamics that occurs by itself, without the need for the addition of energy.

stainless steel a class of alloys of iron that are corrosion-resistant because of the presence of at least 12% chromium in their structures. Many stainless steels also contain other alloy elements, such as nickel, to enhance corrosion resistance or to provide other desired properties.

standard enthalpy of formation the enthalpy change that takes place when 1 mole of a substance in its standard state is formed from its elements in their standard states.

standard hydrogen electrode (SHE) an electrode that consists of a platinum plate immersed in an acid solution in which [H^+] is 1 molar and bathed in bubbles of H_2 gas at 1 atmosphere pressure. By definition, the standard electrode potential for the SHE is 0.000 V at 298 K.

standard molar entropy ($S°$) the entropy of an atom or molecule at 1 atm of pressure and 298 K. For ions in solution, the standard conditions also include an ion concentration of $1\ M$.

standard potential the electromotive force of a half-reaction written as a reduction in which all reactants and products are in their standard states, which means that the concentration of all dissolved substances is 1 molar.

standard solution a solution with accurately measured volume containing a known concentration of a substance that reacts with the solution of unknown concentration.

standard state the most stable physical form of a substance at standard temperature and pressure.

standard temperature and pressure (STP) 0°C and 1 atmosphere of pressure. The molar volume at STP is 22.4 L.

standing wave a wave that is confined to a given space and has a wavelength that is some multiple of the dimensions of the space.

state function a property based solely on a system's chemical or physical state or both and not on how it acquired those states.

stereoisomer a compound with the same formula and bonding arrangement as another compound, but that differs in the spatial orientation of the atoms in the molecules.

stoichiometry the relation between the quantities of reactants and the quantities of products in a chemical reaction.

strong nuclear force a force of nature that is extremely strong but operates over only very small distances.

structural isomer a molecule that has the same molecular formula but different bonding arrangements of its atoms.

substitutional alloy a homogeneous alloy of metals with similar atomic radii (within 15%). Atoms of the minor components substitute for atoms of the major component in the crystal structure of the major component.

superconductor a material that conducts electricity with zero resistance when the material is cooled below its critical temperature.

supercritical fluid a substance at temperatures above its critical temperature and critical pressure, in which the liquid and gaseous states of that substance are indistinguishable.

supersaturated temporarily containing more than the maximum amount of solute that can be dissolved in a given volume or mass of solvent.

surface tension the energy needed to separate the molecules of a unit area of the surface of a liquid.

surroundings everything that is not part of a system.

system that part of the universe that is the focus of study. In thermochemistry, the system includes the reaction vessel and its contents.

tensile strength the pulling force needed to break a fiber or other material divided by its cross-sectional area.

tertiary structure the overall three-dimensional shape of a protein.

tetrahedral molecular geometry an orientation of groups of valence electrons on the central atom of a molecule that forms a true tetrahedron with 4 bonding pairs.

theoretical yield the mass of a product that should be formed by a given mass of reactant. In nature or in the laboratory, the actual yield is often less than the theoretical yield.

theory a general explanation of a widely observed phenomenon that has been extensively tested.

thermochemistry the study of the heat produced or consumed in chemical reactions.

thermodynamics the study of energy and its transformations.

third law of thermodynamics the entropy of a perfect crystal at absolute zero temperature is zero.

titration an approach for determining the concentration of halides in a solution in which a carefully measured volume of a standard solution of known concentration is added to a sample solution. The concentration of the solute in the sample is calculated from the volume of the sample, the concentration of the standard solution, and the volume of the standard solution needed to react with all the solute.

tracer a species that has been labeled with a radioactive isotope of a constituent element.

transcription the process by which information in DNA is used to synthesize RNA.

transfer RNA (tRNA) nucleic acid that transports and positions specific amino acids for protein synthesis.

transition metal an element in Groups 3 to 12. Atoms of the transition metals typically have partly filled d orbitals.

transition state the high-energy state associated with the formation of an activated complex.

translation a process that decodes genetic messages carried by RNA for use in protein synthesis.

transmutation the conversion of the nucleus of an element into the nucleus of another.

trigonal bipyramidal orientation an orientation of electron groups on the central atom of a molecule with five pairs of valence electrons. Three of the five pairs are 120 degrees apart in the middle, or *equatorial*, plane, and the other two pairs are oriented 90 degrees, or *axial*, to the plane.

trigonal planar molecular geometry an orientation of electron groups on the central atom of a molecule with pairs of electrons grouped together by double and triple bonds with three groups of electrons on the central atom and a triangular shape that produces bond angles of about 120 degrees.

trigonal pyramidal molecular geometry an orientation of groups of valence electrons on the central atom of a molecule that forms a tetrahedron with one lone pair and three bonding pairs.

triple point a particular temperature and pressure at which all three states of a substance are stable.

turbidity the tendency of liquids containing suspended solids to scatter light.

Tyndall effect a light-scattering phenomenon caused by colloidal particles.

ultraviolet radiation the region of electromagnetic spectrum in which components are refracted even more than violet light. Ultraviolet (UV) rays have shorter wavelengths than visible light.

unit cell the basic repeating unit of the arrangement of atoms, ions, or molecules in a crystalline solid.

unsaturated containing less than the maximum amount of solute that can be dissolved in a given volume or mass of solvent.

valence-bond theory the theory that atomic orbitals can be mixed together to create hybrid atomic orbitals that yield observed molecular geometries.

valence electron the outermost electron of an atom that forms chemical bonds.

van der Waals equation $[(P + n^2a/V^2)(V - nb) = nRT]$ the modification of the ideal gas equation to accommodate the behavior of a real gas. It includes corrections for intermolecular interactions and the volume occupied by particles of gas.

van't Hoff factor (i) the ratio of the moles of particles actually in a solution and the moles of solute that dissolved.

vapor pressure the pressure exerted by a vapor in equilibrium when a liquid is at a constant temperature. The evaporation of a liquid substance produces gas molecules that are responsible for the vapor pressure of the liquid.

viscosity a measure of the resistance to flow of a fluid (liquid or gas).

volatile a term describing a gas or an easily vaporized liquid or solid.

voltaic cell a cell in which chemical energy is transformed into electrical energy by a spontaneous redox reaction.

wave equation Schrödinger's mathematical description of electron waves. It is a second-order differential equation.

wave function a solution of a wave equation.

wave mechanics see **quantum mechanics**.

wavelength (λ) the distance from one wave crest to the next.

weak electrolyte a substance that only partly dissociates into ions when it dissolves and only slightly increases the conductivity of water.

weak nuclear force a force of nature that is stronger than gravity but not as strong as electromagnetism.

work function (ϕ) the amount of energy needed to release a photoelectron from the surface of a metal.

work-hardening a process that produces imperfections and dislocations in metallic structures that reduce the ease with which atoms can slide by each other, making the metal stiffer and harder.

X-ray diffraction a process to determine the structures of crystalline solids. X-rays are diffracted when they bounce off the nuclei and electrons of atoms.

Answers to In-Chapter Questions and Problems

CHAPTER 1

Concept Tests

p. 4. Physical properties can be observed without changing the composition of a substance, but its chemical properties are based on chemical reactions. This means that *b* and *c*, the melting point and malleability of gold, are physical properties, and reactions of gold with *aqua regia* or cyanide (*a* and *d*) are chemical properties.

p. 8. Air and water were considered elements because they are common substances that could not, until recently, be separated into pure elements as we know them today. The components of air can be separated by cryogenic cooling and water can be separated by electrolysis.

p. 11. Density is defined as the mass divided by the volume. If the volume of the universe is increasing as the mass remains constant, its density will decrease.

p. 23. We can develop a single equation for the calculation of time by first rearranging the equation $d = v \times t$ to read $t = d/v$. Earlier we developed a relation between v, v', v and c: $(v - v'/v) = v/c$. Solving for v, we find: $v = (v - v'/v)c$. Substituting for v in the equation $t = d/v$ yields a single equation that relates t, d, c and the red-shift:

$$t = d/(v - v'/v)c$$

p. 24. When an object made of iron rusts, oxygen combines with the iron resulting in an increase in the mass of the object.

p. 26. The error in the measurement makes it impossible to determine whether or not the object is pure gold. If the calculation were accurate to 3 significant figures, one could conclude that the object is not pure gold because its density is too low.

p. 45. The nucleus of a helium atom contains two protons, and twice the amount of positive charge in the nucleus of a hydrogen atom. The greater positive charge exerts a stronger pull on the electrons of helium making it more difficult to ionize helium. We will return to this question in Chapter 3.

Practice Exercises

1.1. $3.82 \times 10^{14}\ s^{-1}$
1.2. $1.99 \times 10^{-9}\ m = 1.99\ nm$
1.3. The pitch will be 16% higher.
1.4. No answer required.
1.5. 1.2×10^4 mph
1.6. 1.14
1.7. 8.4×10^5
1.8. $°F = 9/5°C + 32$
1.9. 51.5% ^{107}Ag, 48.5% ^{109}Ag

CHAPTER 2

Concept Tests

p. 92. Even though limestone contains carbon, the half-life of ^{14}C is too short to provide an accurate date for a limestone formation, because the ^{14}C content after only 1 million years (approximately 175 half-lives) would be too small to determine accurately.

p. 94. These elements are absent because all of their isotopes are radioactive with relatively short half lives (on a geological time scale).

Practice Tests

2.1. $E = 9.034 \times 10^{-12}$ J

2.2. $^{17}_{8}O$

2.3. $^{18}_{9}F \longrightarrow ^{18}_{8}O$ + positron
The same nuclide ($^{18}_{8}O$) could be formed by electron capture.

2.4. rubidium-37 (^{37}Rb)

2.5. 12.9% or about $\frac{1}{8}$ remains

2.6. 4930 yr

CHAPTER 3

Concept Tests

p. 106. Helium arises from α-decay of uranium minerals. Alpha particles are helium nuclei, which acquire electrons readily.

p. 110. The volume of water that evaporates from a lake and the time required to get to class vary continuously. The number of eggs and red traffic lights vary by discrete values.

p. 144. Cobalt and nickel would be interchanged in Mendeleev's periodic table as would tellurium and iodine.

p. 146. The outermost electrons in Ar are easier to remove (ionize) than those in Ne or He because they are shielded from the nucleus by inner shell electrons.

Practice Exercises

3.1. 9.11×10^{-8} m = 91.1 nm

3.2. 486, 434, and 410 nm

3.3. $E = 2.42 \times 10^{-19}$ J

3.4. 1.02×10^{-34} m

3.5. four orbitals total, one 2s and three 2p

3.6. [orbital diagrams for O^{2-}, Mg^{2+}, Al^{3+}, Cl^- showing 1s, 2s, 2p, 3s, 3p fillings]

3.7. all of them

3.8. $[Kr]5s^2 4d^7$

CHAPTER 4

Concept Tests

p. 167. In arid regions, the absence of precipitation prevents the highly soluble NaCl deposits from being washed away by rain or snowmelt. (In certain areas such as the Great Divide Basin in Utah, rivers flow to the middle and evaporate, leaving salt deposits and a very salty lake.)

Practice Exercises

4.1. The ratio of O in the two compounds is 2 : 3.

4.2. magnesium oxide, carbon monoxide, arsenic(III) oxide, phosphorus(V) chloride

4.3. (a) $SrCl_2$; (b) MgO; (c) NaF; (d) $CaBr_2$

4.4. $MnCl_2$, MnO_2. These two compounds represent the most common oxidation states of manganese, and so might also be called manganous chloride and manganic oxide (or manganese dioxide). However, manganese has more than two oxidation states, so it's better to use Roman numerals to indicate oxidation state.

4.5. (a) $Sr(NO_3)_2$; (b) K_2SO_3; (c) BaI_2; (d) K_2O

4.6. (a) ammonium nitrate; (b) magnesium chlorate; (c) lithium nitrite; (d) sodium hypochlorite; (e) potassium selenate

4.7. (a) hypochlorous acid; (b) chlorous acid; (c) carbonic acid

4.8. 3.3×10^{-23} mol DNA

4.9. 1.5×10^{10} atoms Au

4.10. 0.0765 mol Cu

4.11. 49.3 g

4.12. (a) CO_2: 44.01 g/mol; (b) H_2CO_3: 62.02 g/mol

4.13. 26.6 g N_2O_5; 0.492 mol HNO_3

4.14. $P_4O_{10} + 6 H_2O \longrightarrow 4 H_3PO_4$

4.15. $C_3H_8 + 5 O_2 \longrightarrow 3 CO_2 + 4 H_2O$

4.16. $MgSiO_3$ contains 47.81% O, 27.98% Si, 24.21% Mg

4.17. $FeCr_2O_4$

4.18. P_2O_5

4.19. $2 C_4H_{10} + 13 O_2 \longrightarrow 8 CO_2 + 10 H_2O$
3.03 g CO_2

4.20. 6.09×10^{26} g

4.21. 89.8% yield

CHAPTER 5

Concept Tests

p. 214. A homogeneous mixture has a uniform composition throughout. Filtered air and cough syrup (*c* and *d*) are examples of homogeneous solutions. Suspended particles in muddy river water (*a*) are likely to settle out, making it a heterogeneous mixture. Helium (*b*) is a pure element and not a mixture at all.

p. 217. a. The salinity of oceans depends on the flow of fresh water from melting polar ice and coastal rivers, which reduce salinity, and evaporation, which increases it.

p.217. b. Dividing the concentration of chloride ion in mol/L (0.558) by the concentration in mol/kg (0.546) leads to units of kg/L. The numerical answer, 1.02 kg/L, tells us that one liter of sea water weighs slightly more than 1 kg.

p. 224. Strong electrolytes are better conductors of electricity than weak electrolytes or non-electrolytes. This means that ethanol (a non-electrolyte) and acetic acid (a weak electrolyte) will be poor conductors of electricity. The three ionic compounds are strong electrolytes but $CaCl_2$ will be the best conductor because it forms three moles of ions per mole in solution, whereas NaCl and KBr form only two, and each mole of Ca^{2+} has twice the charge-carrying capacity as a mole of Na^+ or K^+.

p. 225. Osmosis forces water to flow from cucumber cells to the brine because the concentration of salts is greater in the brine than in the cells. The loss of water leads to the shriveled "pickle."

The opposite situation exists for the flower. In this case, the concentration of solutes is greater inside the cells than in the surrounding water, so water flows into the flower cells.

p. 233. Molarity and molality of dilute aqueous solutions are similar because one liter of a dilute solution contains nearly one kilogram of water. This will not necessarily be the case for other solvents, or for concentrated aqueous solutions.

p. 242. Hydrogen (H_2) is flammable but chlorine (Cl_2) is not. Inserting a match into a volume of gas collected from the top of the device in Figure 5.21 should produce a rapid reaction and loud pop. In the presence of chlorine there is no reaction. Of course, this experiment should be performed with great care in a fume hood, as chlorine gas is toxic.

p. 250. The five acids are assigned the following names: sulfurous acid (H_2SO_3), sulfuric acid (H_2SO_4), nitrous acid (HNO_2), nitric acid (HNO_3), and carbonic acid (H_2CO_3).

p. 253. Water is the acid and ammonia is the base, since water donates a proton to ammonia (the proton acceptor).

p. 266. Distillation requires an apparatus that consumes considerable energy to heat water to its boiling point (100°C) and additional water to cool the condenser. An ion-exchange column operates at ambient temperature.

Practice Exercises

5.1. Water is the solvent.
5.2. $3.3 \times 10^{-2} \, \mu M$ barium
5.3. $0.0199 \, M \, Na^+$
5.4. $2.76 \, g \, NH_4NO_3$
5.5. $8.7 \times 10^{-4} \, M$ or $0.87 \, mM$
5.6. 13.8 mL
5.7. 20 atm
5.8. 4.24 atm
5.9. 0.84 m
5.10. 2.2 m NaCl
5.11. 109.3°C
5.12. −0.12°C
5.13. 180 g/mol
5.14. 6.40×10^4 g/mol
5.15. (a) +4; (b) +1; (c) +5
5.16. $S_8 + 4 \, O_2 + 8 \, OH^- \longrightarrow 4 \, S_2O_3^{2-} + 4 \, H_2O$
5.17. $2 \, HNO_3 + Mg(OH)_2 \longrightarrow 2 \, H_2O + Mg(NO_3)_2$
5.18. $Ca(OH)_2(s) + 2 \, H^+(aq) \longrightarrow 2 \, H_2O(l) + Ca^{2+}(aq)$
5.19. **a.** $BaSO_4$; **b.** $Ba^{2+}(aq) + SO_4^{2-}(aq) \longrightarrow BaSO_4(s)$
5.20. $2.42 \, mM \, Br^-$
5.21. $1.31 \times 10^{-4} \, M$

CHAPTER 6

Concept Tests

p. 296. Noble gases should have small negative, or even positive, values of electron affinity, since noble gases have a completed octet. Adding an electron to such an atom requires populating (partially filling) a higher energy orbital.

p. 321. Mixing 2s and 2p atomic orbitals raises the energy of the σ_{2p} molecular orbital of N_2 above the energy of the π_{2p}. The same pattern appears true for Li_2, Be_2,

B_2, and C_2 (although Be_2 is not stable). There is considerably less s-p mixing in O_2, F_2, and Ne_2 where σ_{2p} is lower in energy and π_{2p}.

Practice Exercises

6.1. SO should have the same structure and Lewis symbol as O_2, since S and O are isoelectronic. The Lewis structure $\ddot{S}=\ddot{O}$ shows a double bond between S and O.

6.2. The bromine end is more electron-rich.

6.3. The bonds Be—Cl and C—F are the most polar. They have the same difference in electronegativity between the two elements.

6.4.

$:\ddot{Cl}-\underset{\underset{:\ddot{Cl}:}{|}}{\overset{\overset{H}{|}}{C}}-\ddot{Cl}:$ $H-C\equiv N:$

6.5. $[:\ddot{O}-H]^-$

6.6.

:O: :Ö: :Ö:
 ‖ | |
 S   ↔   S   ↔   S
:Ö: :Ö: :Ö: :Ö: :Ö: :Ö:

6.7. $[:\ddot{N}-N\equiv N:]^- \leftrightarrow [\ddot{N}=N=\ddot{N}]^- \leftrightarrow [:N\equiv N-\ddot{N}:]^-$

6.8.
$\underset{best}{\overset{0\ \ 0\ \ 0}{\ddot{S}=C=\ddot{O}}} \leftrightarrow \overset{-1\ \ 0\ +1}{:\ddot{S}-C\equiv O:} \leftrightarrow \overset{+1\ \ 0\ -1}{:S\equiv C-\ddot{O}:}$

6.9.
$\overset{-2\ \ +1\ \ 0}{[:\ddot{N}-N\equiv N:]^-} \leftrightarrow \underset{best}{\overset{-1\ +1\ -1}{[\ddot{N}=N=\ddot{N}]^-}} \leftrightarrow \overset{0\ \ +1\ -2}{[:N\equiv N-\ddot{N}:]^-}$

$\overset{-1\ +1\ +1}{[:\ddot{O}-N=O:]^+} \leftrightarrow \underset{best}{\overset{0\ +1\ 0}{[\ddot{O}=N=\ddot{O}]^+}} \leftrightarrow \overset{+1\ +1\ -1}{[:O\equiv N-\ddot{O}:]^+}$

6.10 H_2^+ has a single electron in the σ_{1s} (bonding) orbital; so H_2^+ could exist.

6.11.

σ^*_{2p}	☐		σ^*_{2p}	☐
π^*_{2p}	↑ ☐		π^*_{2p}	↑ ↑
π_{2p}	↑↓ ↑↓		π_{2p}	↑↓ ↑
σ_{2p}	↑↓		σ_{2p}	↑↓
σ^*_{2s}	↑↓		σ^*_{2s}	↑↓
σ_{2s}	↑↓		σ_{2s}	↑↓

Bond order for O_2^+ = 2.5; bond order for the excited state, O_2^{+*} = 1.5.

CHAPTER 7

Concept Tests

p. 358. The lone pair of valence electrons on the central oxygen atom is relatively close to the atom, so it strongly repels the two sets of bonding pairs, forcing them together and reducing the angle between them.

p. 359. In CO_2, there are only two independent groups of valence electrons (each with four electrons) around carbon and these are oriented to opposite sides of the carbon atom. However, there are three independent groups of valence electrons around the sulfur atom in SO_2, including a single lone pair. Repulsion between this pair and the two sets of bonding pairs results in an O—S—O bond angle of less than 180 degrees.

p. 366. The central atoms in CCl_4 and NH_3 have four independent pairs of valence electrons and sp^3 hybridization.

p. 368. The following resonance forms of CO_2

$:O\equiv C-\ddot{O}: \leftrightarrow \ddot{O}=C=\ddot{O} \leftrightarrow :\ddot{O}-C\equiv O:$

each contain two σ and two π bonds, and so the hybridization of the carbon atom in all three is sp.

p. 376. Asymmetric stretching of the C—H bonds in CH_4 is one of the vibration modes producing an infrared absorption band.

Practice Exercises

7.1 The cyanate ion, OCN^-, has the same number of valence electrons as the thiocyanate ion, SCN^- (Sample Exercise 7.1) and so has three resonance forms with similar bonding patterns:

(1)   (2)   (3)
$[:O\equiv C-\ddot{N}:]^- \leftrightarrow [\ddot{O}=C=\ddot{N}]^- \leftrightarrow [:\ddot{O}-C\equiv N:]^-$

Among the infrared spectral features that could be used to distinguish between these resonance forms are those characteristic of C≡O triple bond and C—N bond stretching vibrations in structure (1), C=O and N=O double bond stretching vibrations. All three resonance forms have different bonding patterns. If resonance form (1) predominates, we would expect to see absorption at frequencies corresponding to stretching of C≡O and C—N bonds. If resonance form (2) predominates, we would expect frequencies for C=O and C=N. If resonance form (3) predominates we would expect frequencies for C—O and C≡N stretching vibrations.

7.2. The Lewis structures of PCl_4^+, PCl_5 and PCl_6^- are

[Lewis structures shown]

Phosphorus has an expanded octet in PCl_5 (10 valence electrons) and PCl_6^- (12 valence electrons).

7.3. A Lewis structure that puts the negative formal charges on the more electronegative (O) atoms is:

[Lewis structure shown]

7.4. The structure on the left minimizes interactions between bonding pairs of electrons and is the preferred structure.

7.5. The Lewis structure of NOF is Ö=N—F: and it has an angular molecular geometry. The Lewis structure of SO_2Cl_2 is

[Lewis structure shown]

It has four independent groups of valence electrons, and so has a tetrahedral molecular geometry.

7.6. The carbon atoms in C_2H_6 each form four σ bonds and so have sp^3 hybridization. There are two σ and two π bonds in CS_2, and so the hybridization of the carbon atom is sp.

7.7. Molecules of CF_4 are symmetrical, and so are nonpolar.

7.8. $Q = 0.177$ e

CHAPTER 8

Concept Tests

p. 390. Yes, the height of the column would be the same. A larger force (F) would be required to push up the contents of a tube with a larger cross-sectional area (A), but, because force is proportional to area at constant pressure ($F = P \times A$), the atmosphere would exert this larger force and the height of the column would be the same.

p. 404. a. The densest gas is the one with the largest molar mass, i.e., Kr.

p. 404. b. The greater decrease in density is produced by decreasing pressure from 2.00 atm to 1.00 atm.

p. 406. The values are the same because at constant temperature and pressure the volume of a gas is proportional to the number of moles of the gas.

p. 421. The number of molecules in a given volume decreases with increasing altitude, so the distance between intermolecular collisions increases.

Practice Exercises

8.1. 62 N
8.3. a. 922 kPa; b. 0.910 atm; c. 692 torr
8.4. 10 L
8.5. 2.1×10^3 L
8.6. 32 psi
8.7. 1.8×10^3 L
8.8. 106 g
8.9. The O_2-filled balloon will sink because O_2 (M = 32 g/mol) is denser than air.
8.10. M = 44 g/mol (CO_2)
8.11. 50% He, 50% O_2
8.12. 0.042
8.13. 0.0022 g H_2
8.14. 1.0×10^{-4} mol/L
8.15. 1.37×10^3 m/s; 2.65 times the root-mean-square speed of N_2.
8.16. argon
8.17. The five gases have similar molar masses, but only HF is polar, and so it has the strongest intermolecular interactions, which lead to non-ideal behavior.

CHAPTER 9

Concept Tests

p. 446. Acetone has stronger intermolecular interactions and a higher boiling point.

p. 447. The two compounds have similar molecular structures, but CCl_4 has a larger molar mass and its molecules have larger clouds of electrons that are more easily polarized, leading to stronger dispersion forces and a higher boiling point for CCl_4 than for CF_4.

p. 450. a. Atoms of radon have the largest, most polarizable clouds of electrons and so Rn is the noble gas most soluble in a polar solvent such as water.

p. 450. b. The charges on ions can perturb the electrons clouds of molecules of a nonpolar substance, creating ion-induced dipole interactions.

p. 455. Longer cooking times are required at higher altitudes because water boils at lower temperatures.

p. 456. a. Because CO_2 is nonpolar, supercritical CO_2 is a better solvent for nonpolar, rather than polar, solutes.

p. 456. b. The propane in a propane tank can be in the liquid state because it is under high pressure.

p. 457. At higher temperatures water molecules move more rapidly, making it easier for a needle to slip between them.

p. 459. The viscosity of seawater should be higher than distilled water because there are relatively strong ion-dipole interactions between dissolved sea salts and water molecules in seawater.

Practice Exercises

9.1. $BaO > CaCl_2 > NaCl$
9.2. TiO_2
9.3. 3831 kJ/mol. This value is nearly four times that of LiF.
9.4. $O_2 < CO < H_2O$
9.5. 19.1 torr
9.6. Gaseous CO_2 turns into solid CO_2.

CHAPTER 10

Concept Tests

p. 476. Yes, there is only one hole in the unit cell of a simple cubic structure of Cl^- ions and a Cs^+ ion occupies it.

p. 484. abcabcabc . . .

p. 488. The olivine rich in iron should be denser because the molar mass of Fe is greater than the molar mass of Mg.

p. 490. The region labeled a is diamond. The region labeled c includes high temperature and low pressure and so represents carbon vapor. The liquid phase is stable at the high temperatures and pressures of region b. The remaining two regions represent graphite and diamond. Because the layers of carbon atoms in graphite are relatively far apart, graphite is less dense than diamond and so is stable at the lower pressures of the region labeled d.

p. 498. Fe(II)/Fe(III) = $\frac{1}{2}$ because the average oxidation number is 2.67.

Practice Exercises

10.1. 410 pm; 30.2°; 40.6°
10.2. 288 pm
10.3. 656 pm; 2.80 g/mL
10.4. 36%
10.5. It appears red because it absorbs green radiation ($\lambda = 516$ nm).
10.6. Only the low-spin state of Co^{3+} is diamagnetic.

CHAPTER 11

Concept Tests

p. 514. If the skiers' masses m_1 and m_2 are different, then the force of gravity on the skiers is different and the skier with the greater mass has the greater potential energy.

p. 518. The boiling points increase with molecular size because the intermolecular forces here are London (dispersion) forces and these increase with increasing molecular size.

p. 520. Warm water evaporates to make water vapor (heat flows from water to water vapor; q is positive). Lower air temperature results in vapor condensation (heat flows from water vapor to air; q is negative) and tiny water droplets form that hang in the air as fog.

p. 522. The units of work according to the two definitions are the same. For example, using English units we would have: work = force × distance = pounds × feet; pressure × volume = (pounds/feet^2) × feet^3 = pounds × feet.

p. 530. The molar heat capacity of aluminum metal (24.4 J/mol · °C) is much smaller than the molar heat capacity of liquid water (75.3 J/mol · °C).

p. 541. CH_4 has a much higher molar heat of combustion than H_2, so methane wins on a molar basis. However, H_2 has only one eighth the molar mass of CH_4, so the fuel value (kJ/g) of H_2 is larger than that of CH_4.

Practice Exercises

11.1 Boil water, $q > 0$; condense steam, $q < 0$.
11.2 1.6×10^6 J
11.3 970 L
11.4 261.5 kJ released
11.5 0°C
11.6 -1152 kJ/mol
11.7 24.61 kJ released; 49.21 kJ released
11.8 -41.2 kJ
11.10 6.6×10^9 metric tons
11.11 endothermic; $+68$ kJ

CHAPTER 12

Concept Tests

p. 558. As crude oil is withdrawn from an underground deposit where pressures are much higher than at the surface, the solubilities of the gases dissolved in the oil will decrease.

p. 562. The intermolecular forces between a polar solute (methanol) and a nonpolar solvent (gasoline) should be weaker than between molecules of pure solute or pure solvent, so positive deviations from Raoult's law are expected.

p. 564. a. The melting point of pentadecane should be below 25°C and its boiling point should be over 100°C, making it a liquid at room temperature.

p. 564. b. The principal intermolecular forces in nonpolar substances are London dispersion forces.

p. 565. cyclopentane > cyclobutane > cyclopropane

p. 569. For example:

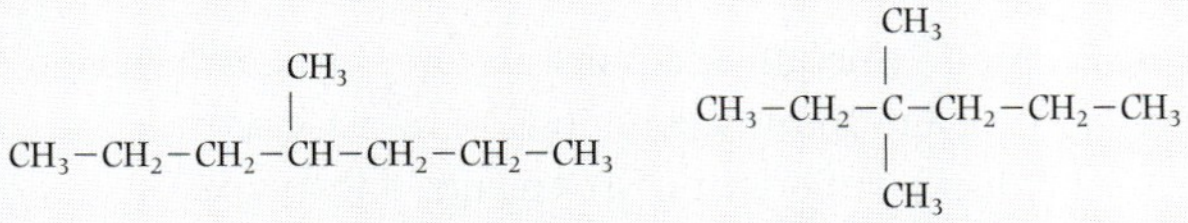

CH$_3$–CH$_2$–CH$_2$–CH–CH$_2$–CH$_2$–CH$_3$
                              |
                              CH$_3$
4-methylheptane

          CH$_3$
          |
CH$_3$–CH$_2$–C–CH$_2$–CH$_2$–CH$_3$
          |
          CH$_3$
3,3-dimethylhexane

p. 574. Unequal sharing of the bonding electrons in alcohol (C—O—H) and ether (C—O—C) functional groups imparts permanent dipole moments to alcohols and ethers, making many of them more soluble in polar solvents than in nonpolar solvents.

p. 576. Oxygen (with no fuel value) accounts for a greater fraction of the molar mass of methanol than ethanol; so methanol has a smaller fuel value.

p. 591. Anthracite has the highest carbon and lowest oxygen content and so has the largest fuel value.

p. 598. There are two hydrogen atoms bonded to C-1 in 1-butene, so there is only one possible structure.

p. 600. No, to minimize the interactions between a triple and a single bond, they must be on opposite sides of the carbon atom, producing a bond angle of 180°.

Practice Exercises

12.1. $n_{\text{benzene}}/n_{\text{octane}} = 58/42 = 1.4$
12.2. $\mathcal{M} = 240.46$ g/mol; $x = 15$
12.3. □ △ 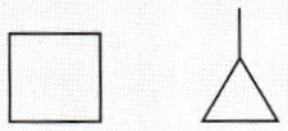
12.4. d. 1,2,3-trimethylcyclopentane; e. 3-methylpentane; f. methylcycolopentane
12.5. [α-form and β-form structures of glucose shown]
12.6. The structure of α-maltose is: [structure shown]

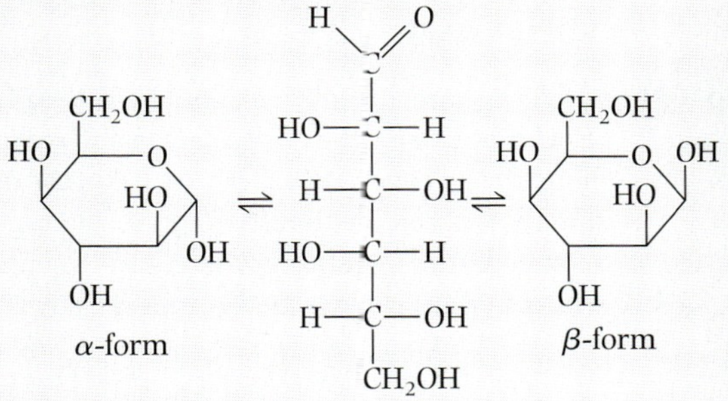

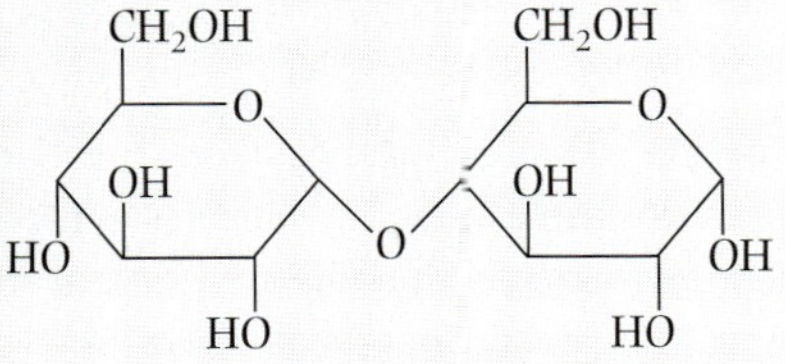

The two rings are connected by an α-1,4-glycosidic bond.

12.7. The empirical formula is C$_3$H$_2$; the molecular formula is C$_{18}$H$_{12}$.
12.8. The empirical formula is C$_8$H$_8$O$_3$.
12.9. a. [cis-2-pentene structure]   b. [trans-2-butene structure]   c. [structure shown]

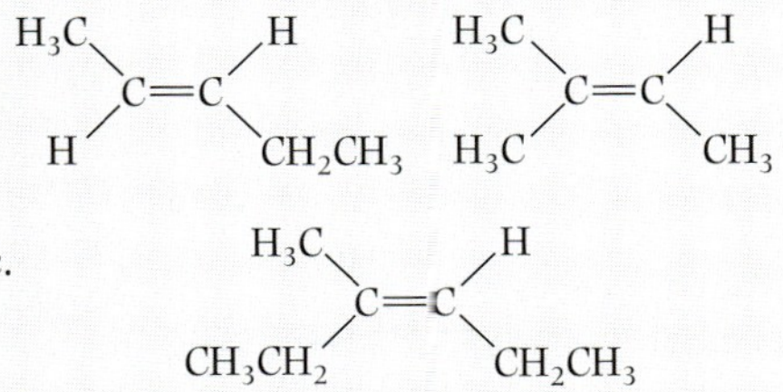

12.10 2-methyl-3-hexyne

CHAPTER 13

Concepts Tests

p. 616. a. Heats of hydration become less negative as cations and anions increase in size. The heats of hydration of the relatively large NH_4^+ cation and $CH_3CO_2^-$ anion are not sufficiently negative to give an overall negative heat of solution, so the dissolution of ammonium acetate is endothermic.

p. 616. b. Exothermic ion–dipole forces between the solvent and the ions do not compensate for the strong forces between ions that have to be broken in ionic compounds such as CaO in order to make a solution. This makes dissolution very endothermic and disfavors high solubility of such compounds.

p. 636. Draw a square and label the corners a, b, c, d (clockwise). Draw the mirror image and rotate it along the ab edge to fit exactly on top of the original square. You can see that such square planar geometries are not chiral, whereas sp^3 geometries with four different attached groups are chiral.

p. 644. Hydrogen atoms contribute strongly to fuel values because forming water by hydrocarbon combustion is very exothermic. In addition, H atoms have very low relative mass. Unsaturated fats have fewer H atoms than saturated fats of nearly the same mass, so the fuel values of the unsaturated fats are lower.

Practice Exercises

13.1. 71°C
13.2. a. evaporation of a liquid; b. condensation of a vapor
13.3. b and c
13.4. −5 J/mol·K
13.5. yes
13.6. exothermic; yes
13.7. n-octane, −10458 kJ; 2-methylheptane, −10449 kJ; 2,3-dimethylhexane, −10450 kJ
13.8. the one on the left
13.9. [structure]
13.10. [structures S and R]
13.11. [structure]
13.12. For one serving of Tostitos®: total fuel value 130 Calories, Calories from fat 50, total fat 6g. Calculated Calories from fat 54, 42% fuel value from fat.
13.13. −196 kJ
13.14. G-T-G-C-A-A-T-C-G
13.15. G-A-U-G-U-A-C-A-G-A-C-U

CHAPTER 14

Concept Tests

p. 682. For an uncatalyzed reaction the rate constant (i) is the same for any combination of reactant concentrations and (ii) depends only on the temperature. So any data set in Table 14.3 can be used to determine the rate constant.

p. 688. Radioactive decay results from processes in an atom's nucleus. Since the nuclei of two radioactive atoms are not in direct contact, it makes sense that radioactive decay is invariably first-order.

p. 764. Step 1 of the sample exercise is termolecular, with a rate law that matches the experimental rate law.

The two-step mechanism of the sample exercise in which Step 1 is the rate-determining step would be consistent with the experimental rate law for the reaction of NO with H_2.

Practice Exercises

- 14.1. CO; higher
- 14.2. 1.5 M/s and 0.75 M/s, respectively
- 14.3. 6.7×10^{-7} M/s
- 14.4. a. zero; b. rate = [A][B], units of k: $M^{-1}s^{-1}$
- 14.5. rate = k[NO][NO$_3$]; $k = 1.6 \times 10^{10} M^{-1}s^{-1}$
- 14.6. first-order; $k = 8 \times 10^{-4}\,s^{-1}$
- 14.7. $2.5 \times 10^{-2}\,d^{-1}$
- 14.8. first-order; $k = 4 \times 10^{-5}\,s^{-1}$
- 14.9. 9.23×10^{-3} s
- 14.10. $k' = 611\,s^{-1}$; $k = 7.2 \times 10^{12}\,M^{-1}\,s^{-1}$
- 14.11. All three steps are bimolecular; rate = k[A][B].
- 14.12. $E_a = 6.9$ kJ/mol
- 14.13. yes

CHAPTER 15

Concept Tests

p. 732. Equal numbers of N_2 and O_2 molecules react to make NO in automobile engines. They are present at 79% and 20% v/v, respectively, in the air where the reaction takes place. Since the molar masses of N_2 and O_2 are nearly the same, O_2 is the limiting reactant. Also most of the O_2 is consumed during the combustion process.

p. 753. Any equilibrium has a fixed equilibrium constant at constant temperature. Adding a reactant makes $Q < K$, and the system shifts from left to right to make more products and restore $Q = K$. Removing a reactant makes $Q > K$, and the system shifts from right to left to make more reactants and restore $Q = K$.

p. 756. Lower temperatures are unfavorable for endothermic equilibrium reactions because heat acts as a reactant. A lower temperature corresponds to less available heat, which makes the left-to-right reaction less likely to occur. So the equilibrium constants for endothermic equilibrium reactions are lower at lower temperatures.

Practice Exercises

- 15.1. $K_c = \dfrac{[NO]^2}{[N_2][O_2]}$; $K_p = \dfrac{P_{NO_2}^2}{P_{N_2}P_{O_2}}$
- 15.2. $K_p = 2.66 \times 10^4$
- 15.3. No; reaction will move to the left to reach equilibrium.
- 15.4. $K_c = 1.1 \times 10^{16}$ (Sample Exercise gives $1/K_c$)
  $N_2O_4 \rightleftharpoons N_2 + 2\,O_2$
- 15.5. $K_c = 1.3 \times 10^{-1}$
- 15.6. $K_p/K_c = (RT)^{-2}$
- 15.7. $K_p = 2.7 \times 10^5$
- 15.8. c and e
- 15.9. $P_{HI} = 0.155$ atm
- 15.10. $[NO_2] = 3.58\,M$; $[N_2O_4] = 2.71\,M$
- 15.11. $K_p = 4.8 \times 10^{-37}$ (298 K); $K_p = 3.8 \times 10^{-6}$ (2000 K)
- 15.12. $K_p = P_{CO_2}/P_{CO_2}$; $K_p = \dfrac{P_{CO}}{P_{CO_2}P_{H_2}}$

CHAPTER 16

Concept Tests

p. 782. The oxidation number of N in the reactant NO_2 is +4, while in the products HNO_3 and HNO_2 it is +5 and +3, respectively. Thus, the N in one NO_2 has been oxidized from +4 to +5, and the N in the other NO_2 has been reduced from +4 to +3. No other elements are oxidized or reduced in this "disproportionation" reaction.

p. 793. A liter of water weighs 1 kg, which contains 1000 g ÷ 18.01 g/mol = 55.5 mol of water. Thus the "concentration" of water is 55.5 mol L^{-1}, or 55.5 M. Quotes are used for "concentration" because pure water is not a solution.

p. 819. In the titration with strong acid, the CO_3^{2-} in the sample (say, x) is converted to HCO_3^- (x), which adds to the HCO_3^- (y) already in the sample. All the HCO_3^- ($x + y$) is converted to CO_2 ($x + y$) in the second titration step. This explains why more strong acid is needed in the second step of the titration than in the first step.

p. 820. The acidity of an NH_4Cl solution is due to NH_4^+ acting as a weak acid. Increasing $[NH_4^+]$ in the solution shifts the weak acid equilibrium to the right, so that $[H_3O^+]$ increases and the pH of the solution decreases.

p. 824. The titration product is a solution of sodium acetate. Acetate is a weak base and the pH of the product solution is 8–9, depending on the acetate concentration. Phenolphthalein is often used as an indicator for the titration of a weak acid with a strong base.

17.4. $E_{cell}° = 1.56$ V
17.5. $E_{cell}° = 1.16$ V
17.6. $K = 3.2 \times 10^{62}$
17.7. 130 g Mg
17.8. 1.48 g Pb

Practice Exercises

16.1. $[H^+] = 1.8 \times 10^{-3}$ M
16.2. $[H^+] = 0.00185 \times 10^{-4}$ M; H_2SO_4 is 93% ionized.
16.3. $[OH^-] = 8.4 \times 10^{-2}$ M
16.4. $[OH^-] = 9.4 \times 10^{-3}$ M
16.5. a. HSO_4^-; b. $CH_3NH_3^+$; c. H_2CO_3; d. H_2O; e. $H_2PO_4^-$
16.6. pH = 2.56
16.7. No answer required.
16.8. pH 11.3
16.9. pH 4.0
16.10. a. $[NH_3] = 4.85 \times 10^{-3}$ M; b. methyl red
16.11. 1.15×10^{-7} M
16.12. $[Ag^+] = 9.8 \times 10^{-9}$ M

CHAPTER 17

Concept Tests

p. 846. No. Although the number of moles of zinc metal dissolved is the same as the number of moles of copper metal deposited, the masses are different because zinc and copper have different atomic weights.

p. 866. Either reaction gives the same $E°$ at pH 7.0.

p. 878. We predict the products to be $H_2(g)$ at the cathode and $Cl_2(g)$ at the anode.

Practice Exercises

17.1. $Zn(s) + Ni(aq)^{2+} \longrightarrow Zn(aq)^{2+} + Ni(s)$; Zn is the anode, Ni is the cathode.
17.2. $E° = 1.23$ V
17.3. anode: $Zn(s) + 2\ OH^-(aq) \longrightarrow ZnO(s) + H_2O(l) + 2\ e^-$
cathode: $Ag_2O(s) + H_2O(l) + 2\ e^- \longrightarrow 2\ Ag(s) + 2\ OH^-$

CHAPTER 18

Concept Tests

p. 904. Yes, the radii are sufficiently alike that the elements could form a substitutional alloy.

p. 912. High temperatures favor Al_2O_3. One mole of steam is released when Al_2O_3 forms from AlO(OH); a total of three moles of steam are released when it forms from $Al(OH)_3$. Therefore, in both cases, its formation is accompanied by large increases in entropy and so is favored at high temperatures.

p. 912. Heating a sample of AlO(OH) may convert it to Al_2O_3. In the process, steam is given off.

p. 913. In basic aqueous solutions the reduction of water

$$2\ H_2O + 2\ e^- \longrightarrow H_2 + 2\ OH^- \quad E° = -0.8277\ V$$

has a less negative standard reduction potential than the reduction of aluminum ions. Therefore, hydrogen gas, not aluminum metal, forms at the cathodes of electrolytic cells.

p. 915. Energy is needed to collect and transport aluminum materials to processing centers, to fabricate the containers and other consumable materials used to recycle aluminum, and to operate recycling and reprocessing centers.

p. 926. Higher temperatures should increase the number of electrons that populate higher conduction bands, and so conductivity should increase.

p. 933. Strong acid could catalyze hydrolysis of the amide bonds, reversing the polymerization reaction shown on page 933. This reaction is analogous to the acid-catalyzed hydrolysis of peptide bonds in proteins by stomach acid.

Practice Exercises

18.1. Copper has a larger molar mass than iron and so makes up a larger percent by mass of the substance.

18.2. Oxidation numbers appear over the elements in the chemical equation:
$$\overset{+1\ -2}{Cu_2S}(s) + 2\,\overset{0}{O_2}(g) \rightarrow 2\,\overset{+2\ -2}{CuO}(s) + \overset{+4\ -2}{SO_2}(g)$$

18.3. a. below 232°C

18.4. The nitrogen would be found in an octahedral hole because the ratio of the radii of N and Fe atoms is 0.56, which is between 0.44 and 0.73 (see Table 10.2).

18.5. 5.5×10^4 g

18.6. Sn

18.7. 577 nm

18.8.

$$\left[\begin{array}{c} O \\ \parallel \\ -C-N-(CH_2)_5- \\ H \end{array} \right]_n$$

Perlon

Answers to Selected End-of-Chapter Questions and Problems

CHAPTER 1

1. a. anything that has mass and occupies space, such as this textbook
    b. matter made up of only a single type of atoms, such as carbon, hydrogen, or iron
    c. a pure substance made up of two or more elements chemically combined, such as water (H_2O), glucose ($C_6H_{12}O_6$), and sodium chloride (NaCl)
    d. an element that has metallic luster, is malleable and ductile and a good conductor of electricity, such as aluminum, copper, nickel, and gold
    e. an element that is neither malleable nor ductile and is a poor conductor of heat and electricity. Examples are chlorine, oxygen, sulfur, and nitrogen.
    f. a positively charged particle, such as Na^+, Fe^{2+}, and Al^{3+}
    g. a negatively charged particle. Some examples are Cl^-, O^{2-}, and N^{3-}.

2. A substance (or pure substance) is a particular kind of matter with well-defined properties and fixed chemical composition.
    A compound is a pure substance made up of two or more elements chemically combined. An element is the simplest form of a pure substance.

3. luster, malleability, ductility, good conductor of heat and electricity

4. Salt is soluble in water. If you add water to the mixture of salt and sand, the salt will dissolve and the sand can be separated by filtration. The salt can then be recovered by evaporating the water from the salt solution.

5. Salt has a very high boiling point (1413°C) compared with water (100°C). Seawater can be desalinated by distillation in an apparatus like that shown in Figure 1.7. When heat is applied to the boiling flask, the water is vaporized to steam while the salt remains in the flask. The pure steam is then cooled by the condenser to give pure liquid water.

6. (b) combustion

7. (a) iron and (d) sodium chloride

9. a. physical
    b. physical
    c. physical
    d. physical
    e. shiny is physical, tarnishing is chemical
    f. chemical

11. c and f

13. You can separate water and ethanol by distillation using the apparatus in Figure 1.7.

15. results of observations and/or experiments

16. A hypothesis becomes a theory if it stands up to repeated tests.

17. yes

18. The theory that matter consists of atoms is universally accepted because it explains a vast array of observations and chemical laws such as the laws of conservation of mass and definite proportions. Predictions based on this theory have led to innumerable advances in science and technology. Moreover, despite the passage of 200 years since the modern atomic theory was formulated, no experimental evidence has been found that is inconsistent with this theory.

19. Hubble discovered galaxies outside our Milky Way and found that these galaxies were moving away from our own and that the speeds with which these galaxies were receding were proportional to the distance from Earth to them. If other galaxies are moving away from the Earth, then the universe must be expanding.

20. Today, we can prove that the Earth is round by taking a picture of the Earth from space, or you could travel in the same direction, e.g., due west, and eventually end up where you started.

21. X-rays are able to penetrate skin and can cause damage to the human body.

22. The longer the wavelength, the lower the frequency. From Figure 1.14, one can see that the top picture (longer wavelength) has fewer complete waves than the bottom picture (shorter wavelength).

23. Ultraviolet radiation has a greater frequency and greater energy than does infrared radiation. The higher energy associated with ultraviolet radiation is responsible for the increased chances of cancer associated with it.

24. When a wave strikes a surface with a small opening, it can bend around the edge of the opening forming circular waves. If there are two small openings, the circular waves interact producing patterns that can either reinforce or cancel out each other (constructive and destructive interference).

25. $4.87 \times 10^{14}\,\text{s}^{-1}$

27. a. 2.882 m
    b. 2.950 m
    c. 2.979 m

29. radio waves

31. 6.42×10^7 m/s or about 1.43×10^8 mph

33. mass—kilogram; length—meter; temperature—kelvin; time—second; amount of substance—mole

34. 1.00 mg

35. 6.7034×10^8 mph

37. 93.2%

39. United States

41. 1.33×10^3 Calories

43. 1.3×10^4 ft
    2.5 miles

45. 23 g

47. $5.1\,\text{g/cm}^3$

49. yes

51. $0.28\,\text{cm}^3$

53. a. Accuracy is a measure of how close a measurement (or the average value) is to the known value; precision indicates how close the results of repeated measurements are to each other.
    b. No, the scale could be accurate to the nearest pound even if it is not accurate to the nearest ounce.
    c. Yes. It is possible to be *both* precise *and* accurate.
    d. Measurements that are accurate to the nearest pound have an uncertainty of ±0.5 pounds.

55. b, c, and d

57. a. 17.4
    b. 1×10^{-13}
    c. 5.70×10^{-23}
    d. 0.160

59. a. $_{-1}^{0}e$
    b. $_{1}^{1}p$
    c. $_{0}^{1}n$
    d. $_{1}^{2}d$
    e. $_{2}^{4}\alpha$

60. 3.125 g

61. 94.7%

63. Each quantity does not contribute equally in computing the average.

64. A nuclide is an atom with a particular combination of neutrons and protons in its nucleus. The atoms of an isotope of a particular element all have the same number of neutrons in their nuclei.

65. The average atomic mass is a weighted average that takes into account the mass of each isotope and its natural abundance.

66. a and b

67. 63.55 amu

69. 32.06 amu

71. 39.96 amu

73. The hotter the solid the shorter the wavelengths of radiation it emits.

74. The zero point of an absolute temperature scale is absolute zero—the coldest temperature possible.

75. $-269°C$

77. $39.2°C$

79. $-89.2°C$, 184.0 K

81. On the Celsius scale the three values are $-181°C$, $-141°C$, and $-250°C$. Therefore, $HgBa_2CaCu_2O_6$ has the highest critical temperature.

83. a. 9×10^3 nm
    b. 8×10^3 nm
    c. 1.5×10^3 nm
    d. 3×10^{-1} nm

CHAPTER 2

1. electron $\quad {}_{-1}^{0}e$ $\qquad$ β particle $\quad {}_{-1}^{0}e$ or ${}_{-1}^{0}\beta$
   positron $\quad {}_{1}^{0}e$ $\qquad$ proton $\quad {}_{1}^{1}H$ or ${}_{1}^{1}p$
   neutron $\quad {}_{0}^{1}n$ $\qquad$ α particle $\quad {}_{2}^{4}\alpha$ or ${}_{2}^{4}He$
   deuteron $\quad {}_{1}^{2}H$

2. Negatively charged: electron, β particle
   Positively charged: positron, proton, α particle, and deuteron
   Neutral: neutron

3. electrons = β particles = positrons < protons < neutrons < deuterons < α particles

4. the γ ray region

5. A hydrogen atom is made up of a proton and an electron. Anti-hydrogen is made up of an anti-proton (negatively charged proton) and a positron (a positively charged electron).

6. no

7. The mass defect is the difference between the mass of the particles (nucleons) in a nucleus and the actual mass of the nucleus. The binding energy is the energy released when neutrons and protons come together to form a nucleus, or the energy required to separate the particles in a nucleus into individual nucleons.

8. Matter is destroyed and converted into energy when the nuclei of the light elements fuse and form the nuclei of heavier elements with up to 26 protons in their nuclei. However, above $Z = 26$ there is no loss in mass during fusion.

9. 2.425×10^{-12} m

11. 8.42×10^{-11} J

13. a. -4.36×10^{-12} J
    b. -6.80×10^{-12} J
    c. -2.867×10^{-12} J
    d. -1.601×10^{-12} J

15. 1.1358×10^{-12} J/nucleon

17. Proton capture requires a very high temperature ($> 4 \times 10^7$ K) to produce the necessary velocity to overcome the positive-positive repulsion that is expected when a proton approaches a positively-charged nucleus.

    Neutron capture can occur with a much lower velocity because the neutrons have no charge; so there is no repulsive force to overcome.

18. We can consider a neutron to be a particle made of a proton fused with an electron. If the electron part of the neutron is expelled from the nucleus (as a β particle), a proton remains. This increases the atomic number (or nuclear charge) by 1.

19. The belt of stability is a band running diagonally through a plot of neutrons versus protons in the nuclei of stable and radioactive nuclides. Nuclei that fall above the belt are likely to undergo β decay; those below the belt are likely to undergo positron emission or electron capture.

20. The net result for both electron capture and positron emission is that the neutron-to-proton ratio increases. In electron capture, a proton and an electron combine to form a neutron. This increases the number of neutrons by one, while decreasing the number of protons by one. In positron emission, a positively charged particle is lost from the nucleus. This means that the nucleus will contain one more neutron and one less proton, just as with electron capture.

21. no

22. c

23. a and b

25. greater than

27. ${}_{13}^{26}Al \rightarrow {}_{12}^{26}Mg + {}_{+1}^{0}e$

29. a. β decay
    b. electron capture or positron emission
    c. β decay
    d. electron capture or positron emission
    e. electron capture or positron emission

31. decay by electron capture or positron emission

33. In fission the nucleus of a heavier atom is split into two lighter ones. In fusion two lighter nuclei are fused to make the nucleus of a heavier atom.

34. A fissionable nucleus is one that can be split into two or more nuclei by neutron bombardment.

35. Fission of ^{235}U is initiated by absorption of a neutron by the nucleus.

36. For every neutron absorbed, 3 neutrons are created. A critical mass is needed so that on average at least one of these neutrons splits another nucleus, sustaining the chain reaction.

37. In a nuclear reactor, rods (made of cadmium or boron) absorb neutrons and so control the rate of fission of the fuel.

38. A breeder reactor is a nuclear reactor in which fissionable material is produced during normal reactor operation.

39. The nuclides produced by nuclear fission all have smaller neutron-to-proton ratios than the heavy nuclides that produced them. Therefore, neutrons must be released in the fission process.

40. Possible nuclear reactions are:

    i) For ^{12}C $\quad ^{12}_{6}\text{C} + ^{1}_{1}\text{H} \rightarrow ^{13}_{7}\text{N} \rightarrow ^{10}_{5}\text{B} + 2\,^{1}_{1}\text{H} + ^{1}_{0}\text{n}$
    $\quad\quad\quad\quad\quad\quad ^{12}_{6}\text{C} + ^{1}_{1}\text{H} \rightarrow ^{13}_{7}\text{N} \rightarrow ^{10}_{5}\text{B} + 2\,^{0}_{1}\text{e} + 3\,^{1}_{0}\text{n}$

    ii) For ^{14}N $\quad ^{14}_{7}\text{N} + ^{1}_{1}\text{H} \rightarrow ^{15}_{8}\text{O} \rightarrow ^{10}_{5}\text{B} + ^{1}_{1}\text{H} + ^{4}_{2}\alpha$
    $\quad\quad\quad\quad\quad\quad ^{14}_{7}\text{N} + ^{1}_{1}\text{H} \rightarrow ^{15}_{8}\text{O} \rightarrow ^{10}_{5}\text{B} + ^{0}_{1}\text{e} + ^{4}_{2}\alpha + ^{1}_{0}\text{n}$

41. a. 52 protons and 86 neutrons; $^{138}_{52}\text{Te}$
    b. 51 protons and 82 neutrons; $^{133}_{51}\text{Sb}$
    c. 55 protons and 88 neutrons; $^{143}_{55}\text{Cs}$

43. a. $^{4}_{2}\text{He}$ or $^{4}_{2}\alpha$
    b. $^{0}_{-1}\text{e}$ or $^{0}_{-1}\beta$
    c. $^{0}_{1}\text{e}$
    d. $^{0}_{-1}\text{e}$ or $^{0}_{-1}\beta$

45. Particle accelerators (linear accelerator or cyclotrons) are devices used to smash high-velocity protons, α particles, and other ions into target nuclei, forming the nuclei of heavier elements not found in nature.

46. To make "superheavy" elements, scientists bombard heavy isotopes with the nuclei of medium-weight elements. To do this, the medium-weight nuclei needs to be moving fast enough to overcome the electrostatic repulsion with the target nuclei; however, if the medium-weight nuclei are moving too fast, the "superheavy" nucleus that is formed will be unstable and undergo fission.

47. a. $^{32}_{15}\text{P}$
    b. $4\,^{1}_{0}\text{n}$
    c. $2\,^{1}_{1}\text{H}$
    d. $^{125}_{54}\text{Xe}$, $^{0}_{1}\text{e}$

49. a. $^{122}_{53}\text{I}$
    b. $^{10}_{5}\text{B}$
    c. $^{58}_{26}\text{Fe}$
    d. $^{68}_{30}\text{Zn}$

51. $^{208}_{82}\text{Pb} + ^{86}_{36}\text{Kr} \rightarrow ^{293}_{118}\text{X} + ^{1}_{0}\text{n}$

    $^{244}_{94}\text{Pu} + ^{48}_{20}\text{Ca} \rightarrow ^{289}_{114}\text{XX} + 3\,^{1}_{0}\text{n}$

53. Photographic film (as in a radiation dosimeter) or a Geiger counter

54. The level of radioactivity is a measure of the radioactivity present at a particular time. It is usually measured in units of radioactive decay events per unit time. Dose of radioactivity is the quantity of ionizing radiation absorbed by a unit mass of matter (a cumulative effect).

55. Exposure to radiation may damage the genes in living cells or destroy the cells completely. The symptoms of radiation sickness include vomiting, intestinal bleeding, infection, and death.

56. Radon-222 is a gas. If you inhale it and it undergoes decay in your lungs, it gives off an α particle. The resulting radioactive decay product, ^{218}Po, is a solid that lodges in body tissue, where it gives off an additional α particle producing ^{214}Pb. These alpha particles can damage the genes in the cells of lung tissues, leading to lung cancer.

57. a. A short half-life is desired so that the radioactivity does not remain in the body for a long period of time.
    b. Beta emitters or γ ray emitters are more desirable than α emitters because of the damage that the greater ionizing ability of the α particles can cause.
    c. The decay products (daughter nuclei) should be non-radioactive so they cannot cause biological damage.

58. Carbon-11 labeled glucose is injected into a patient and becomes concentrated in organs with high rates of energy consumption, such as the brain. Positrons emitted as ^{11}C decays are detected and used to create 3-D images of the brain.

59. After about 10 half-lives (57,000 years) most of the radioactive carbon-14 has decomposed to nitrogen-14. This makes it very difficult to accurately determine the level of carbon-14 left in the object.

60. c

61. The half-life for ^{40}K is 1.28×10^9 years. This is a very long half-life. It will take at least 300,000 years for there to be a measurable decrease in the activity levels of the potassium-40 in a sample.

62. The author of the article could have noted that the changes to molecular structure can lead to slowing down the growth or killing bacteria cells.

63. 5×10^{-4} rads; 2.5×10^{-4} J

65. a. 11.1 disintegrations per second per liter
    b. 5.3×10^6 radon atoms per liter

67. bismuth-213

69. a. $^{10}_{5}B + ^{1}_{0}n \rightarrow ^{11}_{5}B \rightarrow ^{7}_{3}Li + ^{4}_{2}He$
    b. 4.484×10^{-13} J
    c. Alpha emitters have a much higher relative biological effectiveness (RBE) than other forms of radiation, and so destroy tumors more effectively.

71. 25%

73. 492 days

75. a. Iridium-92 could not possibly exist because it would have only 15 neutrons. The known nuclides of iridium contain between 105 and 121 neutrons.
    b. $^{192}_{77}Ir \rightarrow ^{0}_{-1}e + ^{192}_{78}Pt$
    c. The reader is correct.

77. a. electron capture
    b. electron capture
    c. non-radioactive
    d. positron emission

79. 74.3 days

81. yes

83. 2.15 ng

85. 8.1 days

87. 135 minutes

89. no atoms of A
    2.5×10^5 atoms of B
    7.5×10^5 atoms of C

91. 11%

93. a. $^{188}_{74}W \rightarrow ^{188}_{75}Re + ^{0}_{-1}e$
    b. Since ^{188}W has more neutrons than most W atoms, we can deduce that this isotope must be above the belt of stability. Electron capture and positron emission is a typical decay scheme for isotopes below the belt of stability.

95. 34.9% of the original ^{14}C remains in the sample.

97. 0.850

99. 36,600 years

CHAPTER 3

1. A photon is a particle of electromagnetic radiation.

2. An emission spectrum appears as a series of colored lines on a black background and is produced when electrons in atoms lose energy and drop from a higher energy level to a lower one.

   An absorption spectrum appears as a series of dark lines in the continuous rainbow of colors of the spectrum. It is produced when electrons gain energy and move from a lower energy level to a higher one and light of specific wavelength is absorbed.

3. Several of the Fraunhofer lines corresponded to the lines produced by the emission spectra of known elements.

4. atomic absorption

5. The Bohr atom model and quantum theory account for the electromagnetic spectrum of hydrogen. Also, the photoelectric effect links the energy of radiation with electron energy levels in atoms, and the spectra of ions are different from those of their parent atoms.

6. Energies that bind electrons to metals are different for different elements.

7. The diameter of an atom is about 10,000 times greater that the diameter of its nucleus.

8. Aluminum atoms have far fewer protons in their nuclei than do gold atoms. An α particle would undergo fewer deflections upon striking aluminum foil compared to the same thickness of gold foil.

9. a, c, and e are quantities with discrete values (there can't be $6\frac{1}{2}$ eggs left in a carton).

11. no

13. 6.96×10^{-19} J

15. potassium, 8.03×10^5 m/s

17. Hydrogen should have the simplest atomic spectrum because it has only one electron and so there are no electron-electron interactions that perturb electron energy levels.

18. The principal quantum number n can be used to calculate the energy of the individual orbits and to gauge how far the electron is from the nucleus.

19. The energy of light emitted by a hydrogen atom depends only on the difference between two allowed energy states n_1 and n_2.

20. In the ground state of a hydrogen atom, its single electron is in the lowest energy level ($n = 1$). In an excited state, the single electron is in a higher energy level ($n > 1$).

21. In multielectron elements and ions, electrons interact with each other. Bohr's model does not account for these interactions.

22. Lyman lines are in the ultraviolet region.

23. a

24. a

25. The emission spectra of H and He^+ contain similar patterns of lines, but the wavelengths of the He^+ lines are shorter than the wavelengths of the corresponding H lines (see problem 29).

26. no

27. No, these transitions result in absorption of infrared radiation. The Fraunhofer lines are in the visible spectrum of the sun.

28. No, these transitions produce UV emission lines.

29. a. decreases
    b. no

31. 3.03×10^{-19} J

33. 486.1 nm (no significant redshift)

35. λ is the wavelength of particle.
    h is Planck's constant (6.626×10^{-34} J·s).
    m is the mass of the particle.
    v is the velocity of the particle.

    This equation indicates that the characteristic wavelength of a particle, especially a small one moving at near the speed of light, is inversely proportional to its mass.

    Diffraction is the result of interference patterns produced by multiple waves. Because electrons undergo diffraction, they must have wave-like properties.

37. 2.21×10^{-34} m

39. b and c

41. a. 9.70×10^2 m/s
    b. 5.28×10^{-1} m/s
    c. 5.27×10^{-1} m/s
    d. 1.33×10^{-1} m/s

43. A Bohr orbit is a circular pathway in which an electron of constant energy revolves around a nucleus. Orbitals are three-dimensional regions of space within which an atom's electrons are likely to be found.

44. n—the principal quantum number defines the average distance of the orbital from the nucleus; l—the angular momentum quantum number defines the shape of the orbital; m_l—the magnetic quantum number defines the orientation of the orbital.

45. three

46. four

47. n^2 or 1, 4, 9, 16, and 25, respectively

49. l = 3, 2, 1, or 0

51. a. $2s$
    b. $3p$
    c. $4d$
    d. $1s$

53. a. 2
    b. 2
    c. 10
    d. 2

55. b

57. c < a < d < b

59. An unoccupied orbital *exists* to the extent that it *could* be occupied by an electron. Also, it represents a real solution to Schrödinger's wave equation.

60. s orbitals are spherical.

    p orbitals have two balloon-shaped lobes and resemble three-dimensional figure eights.

Four of the five d orbitals have cloverleaf arrays of four lobes that resemble two overlapping figure eights. The remaining d orbital consists of two lobes resembling a p orbital, with a doughnut shape (taurus) in the plane dividing the lobes.

61. Effective nuclear charge is the charge that attracts outer-shell electrons to the nucleus of an atom based on the number of protons in the nucleus less the shielding effect from inner-shell electrons.

62. Degenerate orbitals are equal in energy.

63. yes

64. The column an element occupies in the periodic table can be used to predict its valence configuration (i.e., the configuration of the highest energy, outer-shell electrons).

65. The elements in Groups 1A and 2A have valence electron configurations ns^1 and ns^2 respectively. The B group elements that follow 1A and 2A have partially filled d orbitals in their n-1 shells.

66. The 4s orbitals fill before the 3d orbitals because the 4s orbital is lower in energy.

67. Li atom has 3 electrons $\quad 1s^2\,2s^1$
    Li$^+$ ion has 2 electrons $\quad 1s^2$
    He atom has 2 electrons $\quad 1s^2$
    F$^-$ ion has 10 electrons $\quad 1s^2\,2s^2\,2p^6$
    Ne atom has 10 electrons $\quad 1s^2\,2s^2\,2p^6$
    Na$^+$ ion has 10 electrons $\quad 1s^2\,2s^2\,2p^6$
    Mg^{2+} ion has 10 electrons $\quad 1s^2\,2s^2\,2p^6$
    Al^{3+} ion has 10 electrons $\quad 1s^2\,2s^2\,2p^6$

    The only species with the configuration [He] $2s^1$ is the Li atom.

69. Neon (Ne) has 10 electrons, while sodium (Na) has 11 electrons, so they differ by the presence of the eleventh (3s) electron:

    Ne $1s^2\,2s^2\,2p^6$
    Na $1s^2\,2s^2\,2p^6\,3s^1$ or [Ne] $3s^1$

71. a. 3
    b. 2
    c. 0
    d. 0

73. The ions K$^+$, S^{2-}, and I$^-$ all have the electron configurations of noble gases.

75. a. $1s^2\,2s^2\,2p^6\,3s^1$ or [Ne] $3s^1$
    b. $1s^2\,2s^2\,2p^6\,3s^2\,3p^5$ or [Ne] $3s^2\,3p^5$
    c. $1s^2\,2s^2\,2p^2\,3s^2\,3p^6\,4s^2\,3d^5$ or [Ar] $4s^2\,3d^5$
    d. $1s^2\,2s^2\,2p^6\,3s^2\,3p^6\,3d^5$ or [Ar] $3d^5$

77. a and d

79. two (the first configuration is an excited state, the last two are the ground state at the same element, scandium)

81. $1s^2\,2s^2\,2p^6\,3s^2\,3p^3$

83. As, 3
    Te, 2
    Sn, 2
    Ge, 2

85. Ti, 2
    Cr, 6
    Cu, 1
    Zn, 0

87. The highest energy electrons in ^{131}I are the five 5p electrons. For these 5p electrons:

    $n = 5$ (from the 5 in 5p)
    $l = 1$ (l always equals 1 for p electrons)
    $m_l = 1, 0,$ or -1
    $m_s = +\frac{1}{2}$ or $-\frac{1}{2}$

    All the isotopes of an element have the same electron configuration, and so their electrons have the same sets of quantum numbers.

89. Ionization energy is the energy required to remove an electron from a gaseous atom or ion.

90. a. As one goes down a group of elements in the periodic table the first ionization energies are smaller.
    b. As one moves from left to right across the periodic table the first ionization energies generally increase.

91. a. Helium has a filled shell, while lithium does not, so helium should have a higher first ionization energy than lithium. Also the outermost electron in lithium is farther from the nucleus than the outermost electron in helium, and this also makes it easier to remove the outermost electron from lithium.
    b. Beryllium should have a greater first ionization energy than lithium because of a greater effective nuclear charge felt by the outermost electron in beryllium.
    c. Beryllium has a greater first ionization energy than boron because, in the boron atom, the filled 2s orbital shields the more distant 2p electron from the nucleus. The 2p electron in the boron atom should be easier to remove than the 2s electron in beryllium.
    d. The half-filled 2p subshell in the nitrogen atom is a more stable configuration than the $2p^4$ configuration in oxygen. It is more difficult to remove the outermost electron in the nitrogen atom.

93. Einstein's statement reflects his idea that everything in the universe should have a definite explanation or value. Dealing in probabilities is not acceptable because it does not give a definite value, only the chances of an event occurring.

    Bohr's comment implies that anything that occurs must happen for a reason and only God knows what that reason is.

CHAPTER 4

1. Fe (*l*)
2. Al_2O_3
3. He^+
4. If the Earth had developed a core before Eros was formed, we should expect that the elements in the Earth's core (Fe and Ni) would be less abundant in Eros.
5. The laws of definite proportions and multiple proportions are compatible because the law of definite proportions concerns different samples of a single compound, whereas the law of multiple proportions concerns different compounds made of the same elements.
6. The law of definite proportions is a brief statement or mathematical equation that can be used to summarize scientific data. Dalton's atomic theory explains the chemical properties of elements including the law of definite proportions.
7. 1 : 1.5 or 2 : 3
9. XO_2^{2-}
10. No, because there are only two oxoanions and they are distinguishable by using only the suffixes *ite* and *ate*.
11. The Roman numeral in the name of a compound that contains a transition metal corresponds to the charge on the transition metal atom.
12. All alkali metals form ions with a +1 charge. All alkaline earth metals form ions with a +2 charge. Since there is no ambiguity in the charge on alkali or alkaline earth metals, a Roman numeral designation is unnecessary.

13. a. nitrogen trioxide
    b. dinitrogen pentoxide
    c. dinitrogen tetroxide
    d. nitrogen dioxide
    e. dinitrogen trioxide
    f. nitrogen monoxide
    g. dinitrogen monoxide
    h. tetranitrogen monoxide

15. a. Na_2S sodium sulfide
    b. $SrCl_2$ strontium chloride
    c. Al_2O_3 aluminum oxide
    d. LiH   lithium hydride

17. CoO, cobalt(II) oxide; Co_2O_3, cobalt(III) oxide; CoO_2, cobalt(IV) oxide

19. NaCl, sodium chloride; Na_2SO_4, sodium sulfate $MgCl_2$, magnesium chloride; $MgSO_4$, magnesium sulfate; $CaCl_2$, calcium chloride; $CaSO_4$, calcium sulfate KCl, potassium chloride; K_2SO_4, potassium sulfate $SrCl_2$, strontium chloride; $SrSO_4$, strontium sulfate

21. a. BrO^-
    b. SO_4^{2-}
    c. IO_3^-
    d. NO_2^-

23. a. nickel(II) carbonate
    b. sodium cyanide
    c. lithium hydrogencarbonate or lithium bicarbonate
    d. calcium hypochlorite

25. a. hydrofluoric acid
    b. bromic acid
    c. H_3PO_4
    d. HNO_2

27. The disadvantage of using "dozen" in place of "mole" is that we would have to use very large numbers for measurable quantities of ions, atoms, etc.

28. The molar mass of a molecule or compound is the mass of 1 mole of the compound or 6.022×10^{23} of its molecules. The mass of one molecule is $1/6.022 \times 10^{23}$ the mass of a mole of that substance.

29. no

30. NO_2

31. a. 7.3×10^{-10} mol Ne
    b. 7.0×10^{-11} mol CH_4
    c. 4.2×10^{-12} mol O_3
    d. 8.1×10^{-15} mol NO_2

33. a. 1.7×10^{-8} μmol
    b. 1.7×10^{-10} μmol
    c. 2.3×10^{-15} μmol

35. a. 7.528×10^{22} Ti atoms
    b. 7.528×10^{22} Ti atoms
    c. 1.506×10^{23} Ti atoms
    d. 2.258×10^{23} Ti atoms

37. a. Both samples contain the same number of moles of oxygen.
    b. 1 mol N_2O_4
    c. 2 mol CO_2

39. a. 3.0 mol Al
    b. 4.5 mol Al
    c. 1.5 mol Al

41. 41.7 mol C

43. 0.25 mol Ca^{2+} ions; 10. g Ca^{2+} ions

45. a. 64.07 g/mol
    b. 48.00 g/mol
    c. 44.01 g/mol
    d. 108.02 g/mol

47. a. 152.16 g/mol
    b. 164.24 g/mol
    c. 148.24 g/mol
    d. 132.17 g/mol

49. a. To three significant figures, the molar masses of N_2O and CO_2 are the same (44.0 g/mol), so the two balloons contain the same number of molecules.
    b. CO_2
    c. O_2

51. 0.752 mol SiO_2

53. 10.3 g $MgCO_3$

55. the cube of diamond

57. no

58. no

59. No, unless all of the reactants and products are gases.

60. no

61. a. $CH_4(g) + H_2O(g) \rightarrow CO(g) + 3 H_2(g)$
    b. $2 NH_3(g) \rightarrow N_2(g) + 3 H_2(g)$
    c. $CO(g) + H_2O(g) \rightarrow CO_2(g) + H_2(g)$

63. a. $3 FeSiO_3(s) + 4 H_2O(l) \rightarrow Fe_3Si_2O_5(OH)_4(s) + H_4SiO_4(aq)$
    b. $Fe_3Si_2O_5(OH)_4(s) + 3 CO_2(g) + 2 H_2O(l) \rightarrow 3 FeCO_3(s) + 2 H_4SiO_4(aq)$
    c. $Fe_2SiO_4(s) + 2 CO_2(g) + 2 H_2O(l) \rightarrow 2 FeCO_3(s) + H_4SiO_4(aq)$

65. a. $N_2(g) + O_2(g) \rightarrow 2 NO(g)$
    b. $2 NO(g) + O_2(g) \rightarrow 2 NO_2(g)$
    c. $NO(g) + NO_3(g) \rightarrow 2 NO_2(g)$
    d. $2 N_2(g) + O_2(g) \rightarrow 2 N_2O(g)$

67. a. $N_2O_5(g) + Na(s) \rightarrow NaNO_3(s) + NO_2(g)$
    b. $N_2O_4(g) + H_2O(l) \rightarrow HNO_3(aq) + HNO_2(aq)$
    c. $3 NO(g) \rightarrow N_2O(g) + NO_2(g)$

69. $2 C_2H_2(g) + 5 O_2(g) \rightarrow 4 CO_2(g) + 2 H_2O(l)$

71. An empirical formula is the smallest whole number ratio of the different atoms in a substance. The molecular formula gives us the actual number of atoms of each element in a molecule of a substance.

72. yes

73. no

74. the same

75. a. 74.2% Na; 25.8% O
    b. 57.4% Na; 40.0% O; 2.5% H
    c. 27.4% Na; 57.1% O; 1.2% H; 14.3% C
    d. 43.4% Na; 45.3% O; 11.3% C

77. c

79. mole basis: N_2O_3 and NO_2
    weight percent: NO, N_2O_3, and NO_2

81. no

83. $ZrSiO_4$

85. a. MgO
    b. $2 Mg(s) + O_2(g) \rightarrow 2 MgO(s)$

87. a. $Mg_3Si_2H_3O_8$
    b. $Mg_6Si_4H_6O_{16}$

89. no

90. no

91. a. 4.5×10^{11} mol C
    b. 1.98×10^{10} kg CO_2

93. a. $2 NaHCO_3(s) \rightarrow Na_2CO_3(s) + CO_2(g) + H_2O(g)$
    b. 13.1 g CO_2

95. 1.17 kg NaAlO$_2$
97. 1.5 ton SO$_2$
99. a. 1.48 kg HF
    b. 1.11 × 10^3 g UF$_6$
101. a. 346 g Cu
     b. 353 g CuFeS$_2$
103. c
104. No, not all of the oxygen was consumed in the reaction.
105. The theoretical yield is the maximum amount of a substance that can be produced from the given amounts of reactants. The percent yield is the percentage of the theoretical yield that is actually formed in a reaction.
106. no
107. The same reactants may undergo different chemical reactions, yielding different sets of products. Sometimes the rate of reaction is so slow that reactants remain unreacted even after an extended period of time. Other reactions do not go to completion no matter how long they are allowed to run, yielding a mixture of reactants and products whose composition does not change with time.
108. No, for the reasons given in 107.
109. 3 cups of Hollandaise sauce
111. 0.844
113. NH$_3$ + HCl → NH$_4$Cl
     0.7 g NH$_3$ is left unreacted
115. 59%
117. a. C$_6$H$_{12}$O$_6$(aq) → 2 C$_2$H$_5$OH(l) + 2 CO$_2$(g)
     b. 77.1%

CHAPTER 5

1. The component that is present in the greater quantity expressed in moles is considered to be the solvent.
2. water
3. yes
4. No, some solutes are only slightly soluble; so their saturated solutions are not concentrated.
5. Molarity is defined as moles of solute per liter of solution.
6. 1.00 M
7. 1.00 M
8. a. homogeneous
   b. heterogeneous
   c. homogeneous
9. a. 3.7 M
   b. 0.027 M
   c. 3.25 M
   d. 5.92 M
11. a. 0.14 M Na$^+$
    b. 0.11 M Cl$^-$
    c. 0.096 M SO$_4^{2-}$
    d. 0.20 M Ca^{2+}
13. a. 5.4 g NaCl
    b. 5.98 g CuSO$_4$
    c. 3.2 g CH$_3$OH
15. 0.570 g
17. a. 9.6 × 10^{-3} mol Lindane
    b. 7.80 × 10^{-4} mol Dieldrin
    c. 8.77 × 10^{-5} mol DDT
    d. 4.22 × 10^{-3} mol Aldrin
19. orchard: 3.4 × 10^{-10} M DDT
    residential: 5.7 × 10^{-11} M DDT
    residential after a storm: 3.2 × 10^{-8} M DDT
21. 1.9 × 10^{-9} M C$_6$Cl$_6$
    2 × 10^{-10} M C$_{12}$H$_6$Cl$_6$O
    3.5 × 10^{-9} M C$_6$H$_6$Cl$_6$
23. 4.4 × 10^{-5} M Cu^{2+}
25. 2 × 10^{-14} M
27. a. 24 L
    b. 96 mL
    c. 1.85 mL
29. 1.125 mL
31. table salt and formic acid
32. 1.0 M MgCl$_2$
33. a. 0.025 M
    b. 0.050 M
    c. 0.075 M

35. a. The greater the solute concentration, the greater the osmotic pressure.
    b. Dissolving a solute in a solvent results in a solution that freezes at a lower temperature than the pure solvent.
    c. Dissolving a solute in a solvent results in a solution that boils at a higher temperature than the pure solvent.

36. A semipermeable membrane allows certain substances (e.g., water) to pass through but not others.

37. The solvent flows across the membrane from the side with a higher solvent concentration to the side with lower solvent concentration, or in this case from the pure solvent side of the membrane to the solution side. This occurs because of the tendency to equalize the concentration of solute on both sides of the membrane. Because the membrane is semipermeable, equalizing the concentrations is due to movement of solvent rather than solute.

38. The solvent molecules flow from the side with lower solute concentration to the side with higher solute concentration. The solvent molecules move so as to equalize the concentration of solute on both sides of the membrane.

39. Osmotic pressure is directly proportional to both the molarity of the solute and the temperature of the solution ($\pi = MRT$).

40. Reverse osmosis is a process that reverses the natural direction of solvent flow. Solvent molecules flow from the region of higher solute concentration to the region of lower solute concentration.

41. Colligative properties depend on the number of solute particles. The van't Hoff factor allows one to determine the actual number of solute particles present per formula unit and so the total concentration of solute particles.

42. In each of these techniques one can use experimentally measured data to find the concentration of the unknown solute. From this information and either the volume of solution or the mass of solvent used, one can find the moles of solute in the sample. If the mass of solute was measured at the beginning of the experiment, we also know this quantity. If we know both the mass of solute and moles of solute, we can determine the molar mass (molar mass = grams of solute/mole of solute).

It is important to know whether the solute is an electrolyte or a nonelectrolyte. If the solute is a nonelectrolyte, then the measured colligative property can be used to determine solute concentration. However, if the solute is an electrolyte, then the solute concentration determined by measuring a colligative property will be too high because dissociation of the solute increases the number of particles in solution.

43. false

44. The NaCl solution has the highest osmotic pressure and the glucose solution has the lowest.

45. Molality is defined as moles of solute divided by kilograms of solvent. Molarity is defined as moles of solute divided by liters of solution.

46. increase

47. $0.5\ m\ CaCl_2$

48. $0.5\ m\ CaCl_2$

49. Seawater contains dissolved salts. This lowers the freezing point of the water.

50. The total solute (mostly sugar) concentration in the regular, nondiet soft drink is greater than those of the diet drinks.

51. a. Solvent (water) molecules flow from the NaCl solution to the KCl solution.
    b. Solvent (water) molecules flow from the NaBr solution to the $CaCl_2$ solution.
    c. Solvent (water) molecules flow from the dextrose solution to the NaCl solution.

53. a. 57.5 atm
    b. 0.682 atm
    c. 54.2 atm
    d. 46. atm

55. a. $i = 1, 0.0275\ M$
    b. $i = 1, 0.00111\ M$
    c. $i = 2.47, 0.0100\ M$

57. 2.29 atm

59. a. $0.58\ m$
    b. $0.18\ m$
    c. $1.12\ m$

61. a. 307 g
    b. 86.8 g
    c. 28.8 g

63. NH_3, 6.5×10^{-5} m
    NO_2^-, 8.7×10^{-6} m
    NO_3^-, 0.0220 m

65. The boiling point is increased by 3.81°C above its normal boiling point of 76.8°C.

67. 0.252 m

69. -1.90°C

71. 0.0100 m $Ca(NO_3)_2$

73. 0.06 m $FeCl_3 < 0.20$ m $KCl < 0.10$ m $MgCl_2$

75. yes

77. ammonium chloride, 1.85
    ammonium sulfate, 2.46

79. 94.0 g/mol

81. 164 g/mol, $C_{10}H_{12}O_2$

83. A half-reaction describes only the oxidation or reduction part of a redox reaction.

84. reduction (gain of electrons); oxidation (loss of electrons)

85. The greater the change in oxidation numbers of an element, the more electrons it gains or loses.

86. In this process, sodium ions are converted to sodium atoms (reduction) and chloride ions are converted to chlorine gas (oxidation).

87. 0

88. OH^- -1
    NH_4^+ $+1$
    SO_4^{2-} -2
    PO_4^{3-} -3

89. a. $+1$
    b. $+5$
    c. $+7$

91. a. $Br_2(l) + 2 e^- \rightarrow 2 Br^-(aq)$ (reduction)
    b. $Pb(s) + 2 Cl^-(aq) \rightarrow PbCl_2(s) + 2 e^-$ (oxidation)
    c. $O_3(g) + 2 H^+(aq) + 2 e^- \rightarrow O_2(g) + H_2O(l)$ (reduction)
    d. $H_2S(g) \rightarrow S(s) + 2 H^+(aq) + 2 e^-$ (oxidation)

93. $2 Fe_3O_4(s) + H_2O(l) \rightarrow 3 Fe_2O_3(s) + 2 H^+(aq) + 2 e^-$

95. a. SiO_2: Si = $+4$, O = -2
    Fe_3O_4: Fe = $8/3$, O = -2
    Fe_2SiO_4: Fe = $+2$, Si = $+4$, O = -2
    Fe is reduced and O is oxidized.

b. SiO_2: Si = $+4$, O = -2
    Fe(s): Fe = 0
    $O_2(g)$: O = 0
    Fe_2SiO_4: Fe = $+2$, Si = $+4$, O = -2
    Fe(s) is oxidized and $O_2(g)$ is reduced.
c. FeO: Fe = $+2$, O = -2
    H_2O: H = $+1$, O = -2
    O_2: O = 0
    $Fe(OH)_3$: Fe = $+3$, O = -2, H = $+1$
    FeO is oxidized and O_2 is reduced.

97. a. $4 FeCO_3 + O_2 \rightarrow 2 Fe_2O_3 + 4 CO_2$
    b. $6 FeCO_3 + O_2 \rightarrow 2 Fe_3O_4 + 6 CO_2$
    c. $4 Fe_3O_4 + O_2 \rightarrow 6 Fe_2O_3$

99. $NH_4^+(aq) + 2 O_2(g) \rightarrow NO_3^-(aq) + 2 H^+(aq) + H_2O(l)$

101. a. $4 OH^- + 2 Mn^+ + O_2 \rightarrow 2 MnO_2 + 2 H_2O$
     b. $4 H^+ + MnO_2 + 2 I^- \rightarrow Mn^{2+} + I_2 + 2 H_2O$
     c. $I_2 + 2 S_2O_3^{2-} \rightarrow S_4O_6^{2-} + 2 I^-$

103. a. $8 H_2O + 2 MnO_4^-(aq) + 7 S^{2-}(aq) \rightarrow 2 MnS(s) + 5 S(s) + 16 OH^-(aq)$
     b. $H_2O + 2 MnO_4^-(aq) + 3 CN^-(aq) \rightarrow 3 CNO^-(aq) + 2 MnO_2(s) + 2 OH^-(aq)$
     c. $H_2O + 2 MnO_4^- + 3 SO_3^{2-}(aq) \rightarrow 3 SO_4^{2-}(aq) + 2 MnO_2(s) + 2 OH^-(aq)$

105. An acid is defined as a proton donor.

106. A strong acid completely dissociates, forming H^+ ions, in water. A weak acid only partially ionizes in water.

107. Strong acids:
     hydrochloric acid   HCl   nitric acid HNO_3
     hydrobromic acid   HBr   perchloric acid $HClO_4$
     hydroiodic acid   HI   sulfuric acid H_2SO_4
     Weak acids:
     acetic acid $HC_2H_3O_2$ citric acid $H_3C_6H_5O_7$
     carbonic acid H_2CO_3 phosphoric acid H_3PO_4
     nitrous acid HNO_2 formic acid $HCHO_2$

108. H_2SO_4 donates a proton more easily than does HSO_4^- because ionizing HSO_4^- requires removing a positive ion (H^+) from a negative ion (HSO_4^-).

109. A proton acceptor is called a base.

110. A strong base completely dissociates, or reacts with water, forming OH^- ions, while a weak base only partially dissociates or reacts to form OH^- ions.

111. Strong bases include the hydroxides of the Group 1A and Group 2A elements:
    LiOH   NaOH   KOH   RbOH   CsOH
    Ca(OH)$_2$ Sr(OH)$_2$ Ba(OH)$_2$
    Weak bases:
    NH$_3$ ammonia   C$_5$H$_5$N   pyridine
    Na$_2$CO$_3$ sodium carbonate

112. H$^+$(aq) + OH$^-$(aq) → H$_2$O(l)

113. a. acid, H$_2$SO$_4$; base, Ca(OH)$_2$

    H$^+$(aq) + OH$^-$(aq) → H$_2$O(l)

    b. acid, H$_2$SO$_4$; base, PbCO$_3$

    PbCO$_3$(s) + 2 H$^+$(aq) + SO$_4^{2-}$(aq) → PbSO$_4$(s) + CO$_2$(g) + H$_2$O(l)

    c. acid, HC$_2$H$_3$O$_2$; base, Ca(OH)$_2$

    Ca(OH)$_2$(s) + 2 HC$_2$H$_3$O$_2$(aq) → Ca^{2+}(aq) + 2 C$_2$H$_3$O$_2^-$(aq) + 2 H$_2$O(l)

115. A saturated solution contains the maximum amount of solute that can be dissolved in a given quantity of solvent. In a saturated solution the dissolved solute is in equilibrium with undissolved solute. A supersaturated solution contains more dissolved solute than a saturated solution and is unstable; the excess dissolved solute may precipitate with minimal disturbance.

116. grams of dissolved solute per mL of water (g/mL)
    grams of dissolved solute per L of water (g/L)
    moles of dissolved solute per liter of solution (M, molarity)

117. In a precipitation reaction solutions containing soluble substances are mixed to form at least one insoluble solid product (a precipitate).

118. One solution contained cations and the other contained anions which combined together to form an insoluble ionic compound.

119. barium sulfate, lead hydroxide, calcium phosphate

121. For lead(II) carbonate the reaction is

    PbCO$_3$(s) + 2 H$^+$(aq) → Pb^{2+}(aq) + CO$_2$(g) + H$_2$O(l)

    For lead(II) hydroxide the reaction is

    Pb(OH)$_2$(s) + 2 H$^+$(aq) → Pb^{2+}(aq) + 2 H$_2$O(l)

123. Cr^{3+}(aq) + 3 OH$^-$(aq) → Cr(OH)$_3$(s)
    Cd^{2+}(aq) + 2 OH$^-$(aq) → Cd(OH)$_2$(s)

125. 7.2 × 10^{-3} g or 7.2 mg

127. The cations in the water are replaced by H$^+$ ions and the anions are replaced by OH$^-$ ions. The H$^+$ and OH$^-$ released by the resin do not remain as free ions but combine to form additional water molecules.

129. a. 5.00 mL
    b. 29.7 mL
    c. 68.0 mL

131. 400 mL

133. 2.4 g

135. 1.5 × 10^3 L

137. 0.134 g

139. Colloidal particles scatter light (the Tyndall effect).

140. 1 nm to 100 nm

141. hydrophobic (water fearing) and hydrophilic (water loving)

142. adding electrolytes to a hydrophobic colloid; heating a hydrophilic colloid

CHAPTER 6

1. the electrons that are in the highest-energy shell

2. yes, in atoms of H and He

3. yes

4. yes

5. Li·   ·Mg·   ·Al·

7. [Na]$^+$ [:Äs:]$^{3-}$ [Ca]$^{2+}$ [:S:]$^{2-}$

9. I$^-$

11. a. Group 1A
    b. Group 4A, Group 4B
    c. Group 6A, Group 6B

13. a. any +1 ions of Group 2A elements, e.g., [Mg·]$^+$
    b. any +3 ions of Group 3A or Group 3B elements, e.g., [Al]$^{3+}$ [Sc]$^{3+}$

15. [Mg]$^{2+}$

17. Covalent bonds are formed when atoms share electrons. Ionic compounds consist of cations and anions that are held together by electrostatic forces.

18. The basis for the criticism is the notion that when two atoms share a pair of electrons, both atoms get credit for both electrons as they fill their octets.

19. no

20. No, though a diatomic molecule could have 8 valence electrons (see problem 23).

21. In a polar covalent bond, the electrons in the bond are not shared equally.

22. Every element has a particular ability to attract electrons involved in a covalent bond. If atoms of two different elements form a covalent bond, one will attract the shared electrons more than the other.

23. a. 8
    b. 8
    c. 8
    d. 10

25. a. :C≡O:
    b. Ö=Ö
    c. [:Cl̈—Ö:]⁻
    d. [:C≡N:]⁻

27. a. 3
    b. 2
    c. 1
    d. 3

29. ionization energy and electron affinity

30. When the difference in electronegativities (Δ_χ) of the two atoms involved in a bond is greater than 2.0, the bond is considered to be ionic in nature.

31. As you move down a group, electronegativity values generally decrease. As you move from left to right across a period, electronegativity values generally increase.

32. This is a direct result of Z_{eff} increasing across a row and decreasing down a group. It makes sense that an electron feeling a larger Z_{eff} will require more energy to remove it from an atom.

33. Essentially they do follow the same trend, but it is the sign of the electron affinity values that make the trend seem different. If an element and an electron combine to form a more stable system than when they were apart, energy will be released and ΔH will be negative. The more stable the anion formed the more negative the value of the electron affinity.

34. Z_{eff} increases across a row, and decreases down a group. It makes sense that electrons feeling a larger Z_{eff} will be pulled closer to the nucleus and result in a smaller atomic size.

35. Energy must be added to force the electron and the atom to combine to form an anion.

36. no

37. C—S, C—O, N—H, C—H; the element underlined in each bond has higher electronegativity (S, O, N, and C).

39. polar covalent, a and b; ionic, c and d

41. A hydrogen atom is capable of sharing a maximum of 2 electrons. In H—H—O, the central hydrogen atom would have to share 4 electrons.

42. no

43. a. CF_2Cl_2

    :F̈:
    |
    :F̈—C—C̈l:
    |
    :C̈l:

    b. Cl_2FCCF_2Cl

    :F̈: :F̈:
    | |
    :C̈l—C—C—C̈l:
    | |
    :C̈l: :F̈:

    c. C_2Cl_3F

    :C̈l: :C̈l:
       \\C=C//
    :C̈l: :F̈:

45. $CH_3CH_2CH_2CH_2SH$

    H H H H
    | | | |
    H—C—C—C—C—S̈—H
    | | | |
    H H H H

    H_2S

    H—S̈—H

47. Cl_2O

    :C̈l—Ö—C̈l:

    $[ClO_3]^-$

    [:Ö=Cl(—Ö:)(—Ö:)]⁻

49. Resonance refers to the movement of bonded and lone pairs of electrons in a molecule or polyatomic ion. Resonance is indicated when more than one valid Lewis structure can be drawn for a particular array of atoms.

The actual electronic structure of the molecule is a combination of the valid Lewis structures.

50. yes

51. Only compounds with one or more multiple bonds, such as SO_2, can undergo resonance. The two bonds in H_2S are both S—H single bonds.

52. All resonance forms must have the same skeletal arrangement of atoms and have the same number of valence electrons.

53. The concept of resonance helps us go beyond the limitations of single Lewis structures in understanding chemical bonding.

54. Resonance forms are needed when it is possible to draw more than one Lewis structure for a given skeletal structure of a molecule.

55. No, the skeletal structures are different.

56. The two structures are not resonance forms of S_2O because the connectivities are different (S—S—O and S—O—S).

57. C_6H_6

[Two resonance structures of benzene showing alternating double bonds with H atoms attached to each C]

59. S_2O_2

$\ddot{O}=\ddot{S}-\ddot{S}-\ddot{O}: \leftrightarrow :\ddot{O}-\ddot{S}=\ddot{S}-\ddot{O}:$
$\leftrightarrow \ddot{O}=\ddot{S}-\ddot{S}=\ddot{O}$

S_2O_3

[Three resonance structures of S_2O_3]

61. CO_3^{2-}

$\left[:\ddot{O}-C=\ddot{O}\atop|\atop:\ddot{O}:\right]^{2-} \leftrightarrow \left[\ddot{O}=C-\ddot{O}:\atop|\atop:\ddot{O}:\right]^{2-} \leftrightarrow \left[:\ddot{O}-C-\ddot{O}:\atop\|\atop:O:\right]^{2-}$

63. In addition to helping predict which resonance forms contribute the most to molecular bonding, formal charges can also help predict how atoms are arranged in molecules, according to the following rules:

   i. Forms with zero formal charges or those closest to zero are more stable.

   ii. Forms with negative formal charges on the most electronegative elements are more stable. It follows that forms with positive formal charges on the least electronegative elements are more stable.

64. Resonance forms with negative formal charges on the most electronegative elements are more stable. It follows that forms with positive formal charges on the least electronegative elements are more stable.

65. The structure with the negative formal charge on the O, the more electronegative element, will contribute more to the bonding in the molecule.

66. The structure with the positive formal charge on the N, the less electronegative element, will contribute more to the bonding in the molecule.

67. N_2O

$:\ddot{N}=O=\ddot{N}: \leftrightarrow :N\equiv O-\ddot{N}: \leftrightarrow :\ddot{N}-O\equiv N:$
$-1 \; +2 \; -1 \quad\quad 0 \; +2 \; -2 \quad\quad -2 \; +2 \; 0$

With the NON arrangement, the oxygen atom has a $+2$ formal charge in each resonance form, yet it is the most electronegative element in this molecule. Stable forms generally have a negative formal charge on the more electronegative element. Since there are no resonance forms of the NON arrangement where this is the case, the NON structure is not likely to be stable.

69. HCN   HNC
H—C≡N: H—N≡C
0   0   0   0 +1 −1

For HCN, the formal charge of each atom is 0 in the only valid Lewis structure.

For HNC, the formal charges are as follows: H = 0, N = +1, and C = −1

71. H_2NCN

[Two resonance structures with formal charges shown: 0 for all atoms in first structure; +1, 0, −1 in second structure]

The structure with the C≡N bond is preferred because all the formal charges are 0.

73. a. CH₃NO₂

$$H-\underset{\underset{H}{|}}{\overset{H}{\underset{|}{C}}}-N-\overset{:\ddot{O}:}{\underset{|}{\overset{\|}{O}}}\ddot{:} \longleftrightarrow H-\underset{\underset{H}{|}}{\overset{H}{\underset{|}{C}}}-N=\ddot{O}$$

with :Ö: group

b. CNNO₂

C—N—N(=O)(—O)

:C≡N—N(=O)(—O:)   ↔   :C≡N—N(—Ö:)(=O:)
   0 +1 +1 0   0 +1 +1 -1 0

N—C—N(=O)(—O)

:N≡C—N(=O)(—O:)   ↔   :N≡C—N(—Ö:)(=O:)
  0 0 +1 0   0 0 +1 -1 0

The NCNO₂ arrangement is preferred because its resonance forms have fewer formal charges.

c. No. Resonance forms must have the same skeletal arrangement of atoms.

75. CNO⁻

$[\ddot{C}=N=\ddot{O}]^{-1} \longleftrightarrow [:C\equiv N-\ddot{O}:]^{-1} \longleftrightarrow [:\ddot{C}-N\equiv O:]^{-1}$
  -2 +1 0   -1 +1 -1   -3 +1 +1

NCO⁻

$[\ddot{N}=C=\ddot{O}]^{-1} \longleftrightarrow [:N\equiv C-\ddot{O}:]^{-1} \longleftrightarrow [:\ddot{N}-C\equiv O:]^{-1}$
  -1 0 0   0 0 -1   -2 0 +1

CON⁻

$[\ddot{C}=O=\ddot{N}]^{-1} \longleftrightarrow [:C\equiv O-\ddot{N}:]^{-1} \longleftrightarrow [:\ddot{C}-O\equiv N:]^{-1}$
  -2 +2 -1   -1 +2 -2   -3 +2 0

NCO⁻ has the smallest formal charges and the most atoms with a charge of 0, compared to the other two structures.

77. No, the resonance structures of NO₂ and NO₃ indicate that the N—O bonds in NO₂ have more double bond character and so should be slightly shorter.

78. The resonance structures of O₃ indicate that the O—O bonds in O₃ are between single and double bonds. The O—O bond in O₂ is a true double bond and so will be shorter.

79. In the four resonance forms of N₂O₄, each N has one N—O single bond and one N—O double bond. In the two most important resonance forms of N₂O, one has an N—O single bond and the other has an N—O double bond. Thus, the N—O bonds in both molecules have similar proportions of single bond and double bond character and should have similar lengths.

80. Yes, if all the atoms have complete octets, then the S—O bonds in SO₃²⁻ and in SO₄²⁻ are all single bonds and should have about the same lengths.

81. NO⁺ < NO₂⁻ < NO₃⁻

83. C₂O₃

$$:O\equiv C-\overset{:O:}{\underset{|}{C}}-\ddot{O}: \longleftrightarrow :O\equiv C-\overset{:\ddot{O}:}{\underset{\|}{C}}=\ddot{O}$$

We expect two different bond lengths based on the Lewis structures above. One CO bond will be shorter (a triple bond), while the other two will be longer (between a single and a double bond because of resonance).

85. no

86. no

87. no

88. no

89.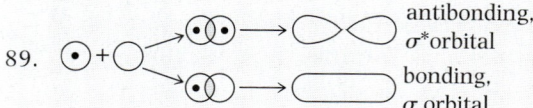

91. Bond order is half the difference in the number of valence electrons in bonding and antibonding orbitals. The MO diagrams below allow us to determine these numbers

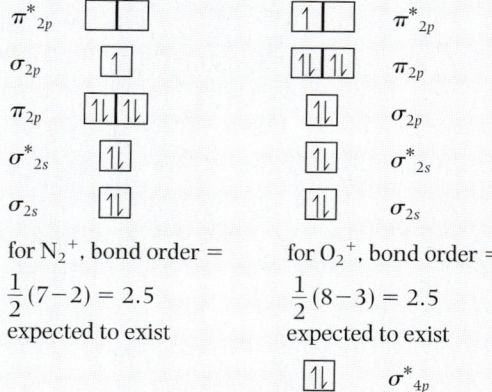

for N₂⁺, bond order = $\frac{1}{2}(7-2) = 2.5$
expected to exist

for O₂⁺, bond order = $\frac{1}{2}(8-3) = 2.5$
expected to exist

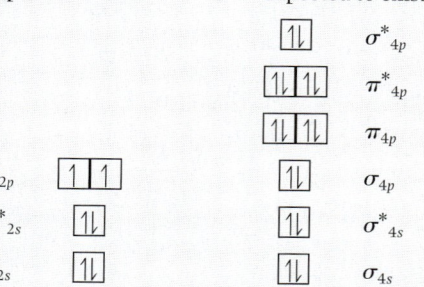

for C_2^{2+}, bond order = $\frac{1}{2}(4-2) = 1$ expected to exist

All the valence shell bonding and antibonding MOs of Br_2^{2-} are filled, giving it a bond order of zero. Therefore Br_2^{2-} is not expected to exist

93. N_2^+, O_2^+, C_2^{2+}
95. $N_2^{2-}, O_2^{2-}, Br_2^{2-}$
97. a and b
99. no

CHAPTER 7

1. After obtaining the spectrum of the compound, one should be able to determine if the compound has a C—N or a C=N bond by comparing the absorption bands in the unknown compound to those in compounds known to have C—N and C=N bonds.
2. yes
3. Photons of infrared radiation don't have enough energy.
4. There are no covalent bonds that could absorb IR radiation in a monoatomic gas such as Ar.
5. yes
6. If the structure were S—S—O, we would expect to see IR absorption bands characteristic of an S—S bond. We would expect no such absorption bands if the connectivity were S—O—S.
7. SO < SO_2 < SO_3
   highest   lowest
   (assuming sulfur obeys the octet rule)
9. NO^+ will have the highest N—O stretching frequency and NO_4^{3-} will have the lowest stretching frequency.
11. All of these elements are in the second period of the periodic table. Atoms in the second row are not able to expand their octets. The $3s$ orbitals are too high in energy compared to the $2p$ orbitals to be used for bonding in second period elements.
12. Yes, any odd electron species must have a minimum of one atom with an odd number of electrons, which violates the octet rule.

13. a, b, and c
15. b
17. a. 12
    b. 8
    c. 8
    d. 10
19. a. 2
    b. 4
    c. 1
    d. 5
    e. 4
21. In order for OF_4 to exist, the central oxygen must have an expanded octet. Since oxygen is in the second period, it cannot expand its octet. In OF_4, the fluorines cannot expand their octets either.
23. Since the central P of POF_3 can have an expanded octet, we can draw a Lewis structure with a P=O double bond. All formal charges are 0. The N in NOF_3 cannot expand its octet and the best Lewis structure we can draw for the molecule has an N—O single bond, an N atom with a +1 formal charge, and an O atom with a −1 formal charge.
25. both
27. yes
31. c, d, and e
33. a. S
    b. N
    c. C
    d. O
35. d
37. Cl_2O_6 with Cl—Cl bond

:O: :O:
 ‖   ‖
Ö=Cl—Cl=Ö
 ‖   ‖
:O: :O:

Cl_2O_6 with Cl—O—Cl linkage

:O: :O:
 ‖   ‖
Ö=Cl—Ö—Cl=Ö
 ‖
:O:

ClO_2

Ö=Cl=Ö

39. NCCN

41.
:N≡C—S̈—F̈:
         |
        :F̈:
       (with another F)

The formal charges for each atom are: N(0), C(0), S(0), F(0).

43.
$$\left[\begin{array}{c} :\ddot{F}: \\ :\ddot{F}-Te-\ddot{F}: \\ :\ddot{F}: \:\ddot{F}: \end{array}\right]^{2-}$$
(with F and O substituents)

To minimize interactions between bonding pairs of electrons, five of the atoms bonded to Te are at the corners of a pentagon; the other two atoms are directly above and below the center of the pentagon, making a pentagonal bipyramid.

45. With increasing nuclear charge, the energy of the atomic orbitals within a series of atoms decreases. So, the $2p$ orbitals of oxygen are lower in energy than those of nitrogen.

46. The MO diagram for CO should be more distorted because the energies of the atomic orbitals in the C and O atoms are further apart than those in the N and O atoms.

47.
σ_{2p} [↑]
π_{2p} [↑↓][↑↓]
σ^*_{2s} [↑↓]
σ_{2s} [↑↓]

The bond order for CN = $\frac{1}{2}(7-2) = 2.5$

49.
π^*_p [↑][ ]
π_p [↑↓][↑↓]
σ_p [↑↓]
σ^*_s [↑↓]
σ_s [↑↓]

There are five more electrons in bonding MOs than in antibonding MOs, giving a bond order of 2.5.

51. Xe_2^+ has a total of 15 valence electrons. The MO diagram below yields a bond order of 0.5.

σ^*_{5p} [↑]
π^*_{5p} [↑↓][↑↓]
π_{5p} [↑↓][↑↓]
σ_{5p} [↑↓]
σ^*_{5s} [↑↓]
σ_{5s} [↑↓]

53. They must have unpaired electrons, the spins of which produce ESR signals.

54. The splitting in the spectrum allows one to determine where the unpaired electron resides. Splittings in ESR spectra arise from interactions between the spins of unpaired electrons and the spins of the protons in the nucleus. Oxygen has no unpaired proton spins, and a one-line spectrum would result if the unpaired electron was on the oxygen. Nitrogen has one unpaired proton spin, and a three-line spectrum is expected if the unpaired electron resides on the nitrogen. A three-line spectrum is observed, and hence ESR leads us to believe that the unpaired electron is on the N.

55. Since a three-line spectrum is observed, the free electron should on the nitrogen. A possible Lewis structure is

$$\left[\begin{array}{c} :O: \:\ddot{O}: \:O: \\ \| \quad \quad \| \\ \ddot{O}=S-N-S=\ddot{O} \\ \| \quad \quad \| \\ :O: \quad :O: \end{array}\right]^{2-}$$

57. electron pairs

58. Electron pairs involved in bonding are attracted by two nuclei while lone pair electrons are only attracted by one nucleus. Therefore, the lone pair on N is closer to the bonding pairs than another bonding pair would be, leading to more repulsion.

59. The Lewis structure is needed to determine the number of lone pairs and bonded groups around the atoms in a molecule or ion.

60. yes

61. The bond angle in a triatomic species depends on the number of lone pairs of electrons on the central atom. For example, if the central atom has an octet (4 pairs) of valence electrons, it may have 0, 1, or 2 lone pairs. If it has no lone pairs the molecule will be linear (180° bond angle). If the central atom has one lone pair, the molecule will have an angular geometry, with a bond angle between 109.5° and 120°. If there are two lone pairs, the molecule will have an angular geometry, with a bond angle between 90° and 109.5°.

62. In each of these molecules, the central atom has three groups of shared pairs of electrons and no lone pairs.

63. a. tetrahedral
     b. trigonal pyramidal
     c. angular
     d. tetrahedral

65. a. tetrahedral
    b. trigonal planar
    c. angular
    d. square pyramidal

67. a. tetrahedral
    b. tetrahedral
    c. trigonal planar
    d. linear

69. O_3 and SO_2

71. SCN^- and CNO^-

73. S_2O is angular; S_2O_2 is trigonal planar

75. XeF_4 is square planar. There are seven pairs of valence electrons around XeF_5^- arranged in a pentagonal bipyramid. The five bonding pairs form a pentagon. The two lone pairs are above and below the plane of the pentagon.

77. The geometry about the P is tetrahedral.

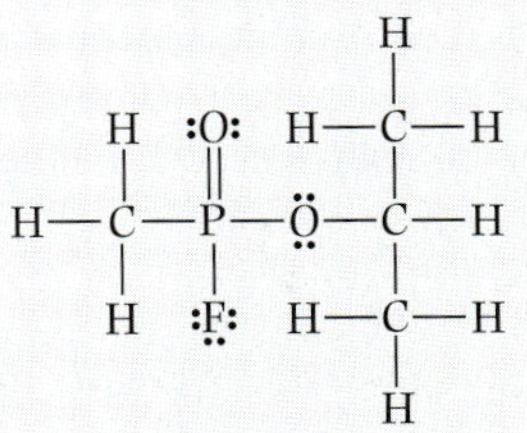

    (All formal charges are zero.)

79. Orbitals that are sp^3 hybridized are not capable of forming π bonds because of their shape. The maximum electron density is along the bond axis, not above and below the axis, which would be necessary to form a π bond.

80. valence bond theory

81. molecular orbital theory

82. d^2sp^3 hybrid orbitals

83. Free atoms are generally regarded as not having hybrid orbitals. Normally, the central atom mixes its orbitals into hybrids (this requires energy) in order to bond with other atoms (this releases energy). Without the benefit of bonding, a free atom has no need to hybridize and will remain in its ground state.

84. yes

85. a. sp
    b. sp^2
    c. sp
    d. sp^2
    3. sp^2

87. N_3^- is linear; in N_3F, one N is sp hybridized and the other two are sp^2 hybridized.

89. The S in SF_2 is sp^3 hybridized. The S in SF_4 is dsp^3 hybridized. The S in SF_6 is d^2sp^3 hybridized.

91. In the central SOS linkage, sp^3 hybridized orbitals on the two S atoms overlap with sp^3 hybridized orbitals on oxygen.

93. no

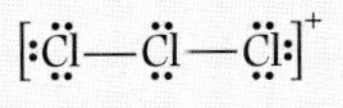

    $[:\ddot{\text{Cl}}-\ddot{\text{Cl}}-\ddot{\text{Cl}}:]^+$

95. Cl_3^+ is angular and the central Cl atom is sp^3 hybridized.

97. no

99. benzene, all carbons sp^2 hybridized; cyclobutane, all carbons sp^3 hybridized; cyclobutene, two carbons sp^3 hybridized and two carbons sp^2 hybridized.

101. A polar bond is a bond with an uneven or unsymmetrical electron density between the two bonding atoms. A polar molecule is one with an overall uneven charge distribution which results in a permanent dipole moment.

102. yes

103. yes

104. A dipole moment is a measure of the magnitude of a molecule's polarity.

105. CCl_4 and CO_2 are nonpolar. SO_2, $CHCl_3$ and H_2S are polar.

107. All are polar.

109. a. $CBrF_3$
     b. CHF_2Cl
     c. Cl_2FCCF_2Cl

111. $COI_2 < COCl_2 < COBr_2$

113. The asymmetry of the structure on the left should result in a larger dipole moment.

115. Partial charge on C = +0.0206 e.

CHAPTER 8

1. Pressure is force per unit area.

2. There are two opposing forces at work: gravity pulling the mercury down the glass tube and air pressure pushing the mercury up the tube. Thus, the height of the mercury in the glass tube is proportional to atmospheric pressure.

3. 1 atmosphere = 760 torr

4. 1 millibar = 100 pascals

5. the one containing ethanol

6. The denser the liquid, the shorter the column height needed.

7. Sharp blades have a smaller area than dull blades, which causes the force/area ratio to increase. Since this ratio defines the pressure, it must also increase and the sharp blades exert a greater pressure.

8. The increased area of the snowshoe results in a lower pressure being exerted on the snow.

9. 3.92×10^3 Pa

11. a. 0.020 atm
    b. 0.739 atm
    c. 4.14 atm

13. Pressure is related to the total mass of molecules of gas above a given altitude. As altitude increases, the mass of the gases still above that altitude must decrease. Less mass means less pressure.

14. STP is 0°C and 1 atmosphere of pressure. The volume of an ideal gas at STP is 22.4 L/mol.

15. increase

16. The new pressure is 10 times the original pressure.

17. 2.0 atm

19. 2.33 atm, 13.3 m (42.9 ft)

21. yes

23. 323°C

25. 4.27 L

27. plot 1

29. raising the temperature

31. a. The volume does not change.
    b. The volume decreases to 25% of its original value.
    c. The volume increases by a factor of 1.17 or about 17%.

33. line 2

35. 1.73×10^3 L

37. 1.7 L

39. 0.67 mol of air

41. 12.1 atm

43. 812 g of H_2

45. a. 0.042 mol of C_2H_2
    b. 11 g

47. 6.2×10^2 g of $KClO_3$

49. 715 g of sodium peroxide, Na_2O_2

51. no

52. Warm air rises because it is less dense than colder air.

53. a. Increasing the pressure of a gas at constant temperature will increase the density of the gas.
    b. Decreasing the temperature at constant pressure will increase the density of a gas.

54. One method of determining the density of a gas is by using a special type of glass bulb of known volume. The bulb is evacuated and weighed empty. It is then filled with a gas. The bulb is then re-weighed. The difference between the two mass measurements is the mass of the gas in the glass bulb. The mass of the gas is divided by the known volume of the bulb to determine the density of the gas.

55. a. 9.07 g/L
    b. The density of radon gas is much greater than the density of air, so radon concentrations are likely to be greater in the basement.

57. line 1

59. SO_2

61. CO

63. the pressure of one gas in a mixture of gases

64. No, a barometer measures *total* atmospheric pressure.

65. c

66. Both balloons contain the same number of particles.

67. 0.2

69. $P_{tot} = 2.5$ atm; $P_{N_2} = 1.7$ atm; $P_{H_2} = 0.49$ atm; $P_{CH_4} = 0.25$ atm

71. 0.0190 mol of O_2

73. a. greater
    b. less
    c. greater

75. There is 28 times as much O_2.

77. 680 mm Hg

79. 25%

81. As gas pressure increases, gas molecules collide with and pass through the surface of a liquid more frequently, leading to a greater number dissolved in the liquid.

82. O_2 is more likely to be consumed by the body's biochemical processes.

83. No, the Henry's law constant for air would be a weighted average of the individual constants for the atmospheric gases.

84. k_H

85. A small amount of CO_2 reacts with H_2O forming HCO_3^-, CO_3^{2-} and H^+ ions. This results in more CO_2 being dissolved in the water, so k_H is larger.

86. The bubbles are dissolved air, which is expelled as the water is heated because air is less soluble at higher temperatures.

87. 7.8×10^{-3} mol/L or 7.8×10^{-3} M

89. a. 0.088 g/L
    b. 0.75 g/L

91. at 10°C, 1.8×10^{-3} M/atm
    at 20°C, 1.3×10^{-3} M/atm
    at 30°C, 1.0×10^{-3} M/atm

93. Root-mean-square speed is the speed of a gas molecule that has the average kinetic energy of the gas.

94. In the many collisions that occur between gas particles, the kinetic energy of each particle in the collision can change. This results in particles with many different speeds.

95. a. The root-mean-square speed is inversely proportional to the square root of the molar mass.
    b. The root-mean-square speed is directly proportional to the square root of the Kelvin temperature.

96. Yes. The pressure of a gas can affect the temperature of the gas. If the temperature of the gas changes then its root-mean-square speed would also change.

97. Graham's law of effusion relates the rate of effusion and the molar mass, $\dfrac{\text{rate}_A}{\text{rate}_B} = \sqrt{\dfrac{M_B}{M_A}}$. If one has an unknown gas and a known gas, the relative rates of effusion can be experimentally measured. Since we would know the rate of effusion of both the known and unknown gas and the molar mass of the known gas, we can use Graham's law to find the molar mass of the unknown gas.

98. yes

99. Diffusion is the spreading out of particles through another substance, such as a solution. Effusion is the escape of gas particles through a small opening.

100. yes

101. $SO_2 < NO_2 < CO_2$

103. C

105. 716 m/s

107. D_2 molecules move 0.707 times as fast as H_2 molecules.

109. SO_2 should have a higher more defined curve, while the CO_2 curve should be flatter. C_3H_8 should have a curve very similar to CO_2 because C_3H_8 and CO_2 both have a molar mass of 44.0 g/mol.

111. 32.0 g/mol

113. 18.0 g/mol

115. $^{12}CO_2$ effuses 1.01 times faster than $^{13}CO_2$.

117. The hydrogen balloon has a smaller volume after 24 hours.

119. Gas particles are in constant motion. As H_2S travels, there will be several collisions between the gas particles that change their direction and/or slow them down. The net result is that gas molecules take a longer time to reach a destination point than is expected based only on the root-mean-square velocity.

121. i. Molecules of real gases occupy space.
     ii. There are forces of attraction between gas particles.

122. Water is a polar molecule. Nitrogen, N_2, is a nonpolar substance. The polarity of water molecules causes the molecules to interact more strongly with each other. The net result is that there are fewer individual gas particles in water vapor than in the same quantity of N_2. This decrease in the proportion of "free" molecules is reflected in the larger "a" value for water.

123. The pressures of real gases will be lower at low temperatures because their molecules are more likely to interact with each other at lower temperature.

124. Since the volume of atoms increases with increasing atomic weight, we should expect the van der Waals "b" value to increase with atomic weight.

125. The "a" value is associated with the attractive forces between gas particles. As molar mass increases, the attractive forces between particles get stronger. As attractive forces become stronger, the "a" value becomes larger.

126. The more polar the molecules, the more strongly they interact with each other, leading to fewer collisions with their surroundings and lower pressure. The greater the deviation from ideality, the larger the "a" value.

127. Molecules of CH_4 are much larger than atoms of helium; so the volume they occupy will be more important.

129. 916 atm

CHAPTER 9

1. gaseous
2. sublimation
3. London forces
4. no
5. First, methane is a nonpolar compound. Secondly, methane has no N, O, or F atoms bonded to a hydrogen atom.
6. Ion–dipole interactions tend to be stronger than dipole–dipole interactions because one of the interacting species has a full positive or negative charge. According to Coulomb's law, more charge means stronger interaction.
7. Hydrogen bonds are a type of dipole–dipole interaction. The reason why they are considered a separate type of interaction is because of their strength relative to most other dipole–dipole interactions.
8. polar
9. As molecular size increases, the number of electrons increases, and the size of the electron cloud generally increases. London forces tend to increase with increasing size of the electron cloud because larger clouds are more easily distorted, creating larger temporary dipoles.
10. The difference in solubility comes from the fact that the ions in $CaSO_4$ are more difficult to separate relative to those in NaCl. $CaSO_4$ is made up Ca^{2+} and SO_4^{2-} and the magnitudes of the charges on these ions are double those found in NaCl.
11. Ar has a larger electron cloud than He. The larger electron cloud of Ar is more polarizable and hence Ar has stronger London forces than He. The van der Waals constant a accounts for intermolecular attractions, so we would expect Ar to have a larger van der Waals constant a because it has stronger interparticle (in this case, London) forces.

12. A molecule of CS_2 has a larger, more easily distorted cloud of electrons than a molecule of CO_2. Therefore London forces and the value of a are larger in CS_2.

13. NH_3 has the ability to form hydrogen bonds and PH_3 does not.

14. Water has a larger dipole moment than methanol, and each molecule of H_2O can form up to two hydrogen bonds, while molecules of CH_3OH can form only one.

15. This is due to the greater London forces in CH_2Cl_2 relative to CH_2F_2. The larger electron clouds of the Cl atoms are more polarizable and this leads to stronger London forces.

16. This is due to the much larger electron cloud on HBr. This larger electron cloud results in stronger London forces and a higher boiling point.

17. The London forces will be similar since the molar masses are similar. The difference is that Br_2 is nonpolar and ICl is polar. Therefore, molecules of I-Cl also experience dipole-dipole forces. This additional interaction in ICl results in a higher boiling point.

18. The interactions that need to be overcome in boiling pure NaCl are ion–ion interactions. Ion–ion interactions are much stronger than the ion–dipole and hydrogen-bonding interactions that must be overcome in the boiling water in a solution of NaCl.

19. $CsBr < KBr < SrBr_2$

21. The melting points should decrease with increasing atomic number of the halide because an increase in ionic size means, according to Coulomb's law, weaker ion–ion interactions.

23. -723 kJ

25. a. CCl_4
    b. C_3H_8

27. CO_2

29. a. $CHCl_3$
    b. CH_3OH
    c. NaCl

31. CaO

33. As temperature increases, the average kinetic energy of the molecules in a liquid increases. The increased average kinetic energy allows more of the molecules to overcome the attractive forces keeping them in the liquid state.

34. The liquid boils.
35. $CH_3CH_2OH < CH_3OCH_3 < CH_3CH_2CH_3$
37. mole fraction of water = 0.70
    vapor pressure = 17 torr
39. a. solid
    b. gas
40. In order for a substance to be considered a supercritical fluid, it must exist at a temperature above the critical temperature at a pressure above the critical pressure.
41. yes
42. no
43. from liquid to gas
45. $-56.6°C$
47. a. liquid
    b. gas
    c. solid
49. decrease
53. expand
55. Water forms stronger hydrogen bonds, which result in higher surface tension.
56. In capillary action adhesive and cohesive forces combine to overcome the force of gravity and pull water upward. The smaller mass of water in a narrow tube means less gravitational force must be overcome than in a wider tube.
57. Different liquids have different adhesive and cohesive forces at work as the liquids rise up the tubes, as well as different densities. The liquids will rise until the adhesive and cohesive forces balance the force of gravity; so different liquids will rise to different heights in a capillary tube.
58. Ice floats on water because ice is less dense than water.
59. Below 0°C, the liquid water may freeze. Since ice is less dense than water, it will expand to a larger volume and possibly break the pipes while trying to expand.
60. As temperature increases, the average kinetic energy of a gas increases. More particles will have the energy necessary to break the adhesive and cohesive forces responsible for capillary action. We expect a hot liquid to rise to a lower height than a cold liquid.
61. no
62. yes
63. Mercury is nonpolar and will not have strong adhesive forces with the glass, SiO_2. In addition, the cohesive forces in mercury will be maximized if it assumes a spherical shape, which minimizes the surface area. This produces a convex meniscus.
64. The mercury resists flowing into the tube because the cohesive forces that keep Hg atoms together outside the tube are much stronger than the adhesive forces between a polar (glass) tube and nonpolar Hg.
65. water

CHAPTER 10

1. In order for an X-ray pattern to contain distinct peaks, the sample studied must have a crystalline lattice.
2. The atoms, molecules, and/or ions present in a solution are in constant random motion. X-ray diffraction requires that these particles be locked in a repeating array, or they will not be able to act like a diffraction grating.
3. X-ray diffraction occurs when repeating arrays of atoms, molecules, or ions make up a diffraction grating. Diffraction tends to work best when the spaces in the grating and the wavelength of the light are close in size. Microwaves have much longer wavelengths than the distances between atoms, ions, and molecules in the solid state.
4. A crystallographer may change the wavelength of the X-ray source if the crystal to be examined has a larger or smaller spacing between its layers than the wavelength of the X-ray source in use.
5. NaCl
7. If $n = 2$ and 3, then the results are consistent with a d spacing of 586 pm.
9. $2.38°$
11. Species at the corners, faces, and edges of a unit cell are shared with other cells that touch the unit cell at those locations. Particles on faces are split between two cells and hence we count $\frac{1}{2}$ of the particle per unit cell. Particles on the edges of cubic structures are split between four cells and hence we count $\frac{1}{4}$ of the particle per unit cell. Particles at the corners of cubic structures are split between eight cells and hence we count $\frac{1}{8}$ of the particle per unit cell.

12. body-centered cubic
13. face-centered cubic
14. no
15. In unit cells, smaller ions tend to fit into the holes made by larger ions. Using the radius of the larger spheres, the size of various types of holes in various packing schemes can be obtained. The cation-to-anion ionic radius ratio can be used to predict the types of holes into which the smaller ions can fit.
16. The difference in the two types of packing arises in how layers are stacked. In cubic close packing, layers are arranged in an *abcabcabc* . . . pattern. In hexagonal close packing, the layers are arranged in an *ababab* . . . pattern.
17. A sphere would have to be much smaller to fit directly between three larger spheres than in spaces above or below them.
18. The arrangement of ions in which Cs^+ ions are in the holes of a simple cubic array of Cl^- ions is the same as a body-centered cubic array of Cl^- ions in which the center Cl^- ions are replaced with Cs^+ ions.
19. In the pattern on the left, there are eight light-colored squares and one dark square per unit cell.

    In the pattern on the right, there are two light and two dark squares per unit cell.
21. $\frac{1}{2}A$, $\frac{1}{2}B$, $1X$
23. A_8B_8X
25. The formula of this unit cell is $Mg_{1/2}B$ and the formula unit of this ionic compound would be MgB_2.
27. face-centered
29. 5.18 g/cm^3
31. 104.2 pm
33. 52.36%
35. BaF_2
37. $MgAl_2O_4$
39. radius = 499 pm
    density = 1.71 g/cm^3
41. yes, octahedral
43. tetrahedral. To have three times as many metal atoms as C_{60} spheres, all the tetrahedral and all the larger octahedral holes would have to be occupied.
45. A flat, symmetrical octagon with sides of equal length should have interior angles of 135°, but the S—S—S bond angles in S_8 are expected to be smaller than 109.5° because of lone-pair repulsion. To accommodate optimal S—S—S bond angles, the octagon must be puckered.
46. We expect any elemental form of phosphorus to have at least three bonds to other phosphorus atoms. A phosphorus atom with three bonds would also have a lone pair of electrons and a trigonal pyramidal arrangement of P atoms that would not lie within the same plane.
47. allotropes
48. no
49. 1.38 g/cm^3
51. There are 2 oxide ions and 4 hydrogen ions in one unit cell of ionic ice.
    density = 2.44 g/cm^3

    $[\text{H}]_2^+ [\ddot{\text{Ö}}\!:\!]^{2-}$

53. $Mg_3(OH)_4(Si_2O_5)$
55. In the electron-sea model, copper nuclei are assumed to be dispersed in a shared sea of electrons that are free to move throughout the bulk sample of copper. The band theory is based on molecular orbital (MO) theory and the notion that the atomic orbitals of, for example, copper atoms overlap with each other forming a vast array of molecular orbitals that extend over the entire piece of copper.
56. The ionic interactions in NaCl are stronger than the metallic bonds in sodium metal. Stronger intermolecular forces result in higher melting points.
57. The lower-energy band is filled.
59. Systems with partially filled *d* orbitals are colored because the energies absorbed correspond to wavelengths in the visible spectrum. The wavelengths that are not absorbed are reflected back and that is what we see.
60. The orbitals of d_{z^2} and $d_{x^2-y^2}$ in an octahedral field are pointed directly toward the atoms or ions defining the octahedron. The repulsions between these corner atoms or ions and the lobes of the d_{z^2} and $d_{x^2-y^2}$ orbitals of the transition metal raise the energy of these two orbitals relative to the other three *d* orbitals (d_{xy}, d_{xz}, and d_{yz}).

61. The lobes of $d_{x^2-y^2}$ orbitals in a square planar field point directly toward the atoms or ions in the square plane. The repulsions between these corner atoms or ions and the lobes of the $d_{x^2-y^2}$ orbitals of the transition metal raise the energy of this orbital relative to the other d orbitals.

62. The value of Δ_o helps determine if a system is high spin or low spin. Systems with large values of Δ_o usually fill their lower-energy d orbitals first. This arrangement promotes the pairing up of electrons in the lower-energy d orbitals. Systems with small values of Δ_o have partially filled low and high energy d orbitals. This often results in many unpaired electrons.

63. Zn^{2+} ions have the configuration [Ar] $4s^0\ 3d^{10}$. Since their $3d$ subshell is completely filled, this does not allow for transitions that correspond to wavelengths within the visible spectrum.

64. Ti^{4+} is isoelectronic with Ar and has no partially filled d orbitals that give the compounds of other transition metals their distinctive colors.

65. Fe^{2+}, Fe^{3+}, Cr^{3+}

67. $\lambda = 450$ nm

69. For $\lambda = 580$ nm, $\Delta_o = 3.4 \times 10^{-19}$ J

71. Fe^{3+} (d^5) has 5 unpaired electrons in a high-spin configuration.
    Rh^+ (d^8) has 2 unpaired electrons in a high-spin configuration.
    V^{3+} (d^2) has 2 unpaired electrons in a high-spin configuration.
    Mn^{3+} (d^4) has 2 unpaired electrons in a low-spin configuration.

73. The complexes in this problem are Co(II) (d^7) and Co(III) (d^6). The number of unpaired electrons is the clue to figuring out the relative magnitudes of Δ_o. The Co(II) complex must be high spin (small Δ_o) to have 3 unpaired electrons. The Co(III) complex must be low spin (large Δ_o) in order to have 0 unpaired electrons.

CHAPTER 11

1. Energy is the ability to do work.

2. Kinetic energy is energy associated with moving objects. Potential energy is the energy of position.

3. A state function depends on the initial and final conditions, but not on the path taken to get from the initial to the final set of conditions.

4. only potential energy

5. a. A battery has stored (potential) energy (in the form of chemical energy).
   b. The potential energy here is stored in the chemical bonds present in gasoline. As the gasoline molecules react old bonds are broken and new ones are formed. The result is that chemical energy is transformed into heat and kinetic energy. The kinetic energy is used to move your car.
   c. The potential energy in a wave depends on the height of the wave. The higher the wave the greater the potential energy present.

7. no

8. no

9. A methylene group consists of a carbon atom that is bonded to 2 hydrogen atoms and 2 other atoms. A methyl group is bonded to 3 hydrogen atoms and only 1 other atom.

10. no

11. The larger the molecule the more polarizable is its charge cloud of bonding electrons and the more energy is required to overcome the attractive (London) forces. The result is that the boiling points of alkanes tend to increase with molar mass.

12. The only kind of intermolecular force present is London (dispersion) forces.

13. a and d

15. about 109°

17. The system includes the chemical and physical changes being studied. The surroundings are everything else except the system.

18. An exothermic process gives off heat. An endothermic process absorbs heat.

19. The internal energy is the sum of all the kinetic and potential energies of the components of the system.

20. i. Doing work on the system by, for example, compressing it.

    ii. Adding heat to the system from the surroundings.

21. In the combustion of methane the flame produced by the combustion reaction and the reactants and products are the system, and $\Delta q_{system} < 0$; assuming water and ice are the system, $\Delta q_{system} < 0$ when water freezes; assuming your hand is the system when you touch a hot stove, then $\Delta q_{system} > 0$.

22. Let the combustion reaction mixture in the car engine be the system and $\Delta q_{system} < 0$. If the ankle is the system, then $\Delta q_{system} < 0$. If the hot dog is the system, then cooking a hot dog has a positive Δq_{system}.

23. a. exothermic
    b. exothermic
    c. endothermic

25. At the boiling point, heat is absorbed from the surroundings. In this case q (heat) is a positive quantity. The internal energy increases.

27. $0.500 \text{ L} \cdot \text{atm}$ or -50.6 J

29. a. 50 J
    b. 6.2 kJ
    c. −940 J

31. 276 kJ

33. b

35. the heat change at constant pressure

36. Enthalpy change (ΔH) is the sum of the change in internal energy (ΔE) of a system and the work ($P\Delta V$) done by the system.

37. In an exothermic process, the system gives off heat to the surroundings. If the system loses heat, then the energy content of the system must be lower after the change and ΔH is negative.

38. When a process is reversed, the sign of ΔH changes, but not the numerical value of ΔH.

39. negative

41. positive

43. negative

44. the surroundings

45. Specific heat is the heat required to increase the temperature of *1 g of a substance* by 1°C (or 1 K). Heat capacity is the heat required to raise the temperature of *an object* by 1°C (or 1 K).

46. The heat capacity doubles if the mass of the material being studied is doubled. The specific heat is unchanged.

47. no

48. no

49. Water has a higher molar heat capacity than air, and should therefore keep the engine cooler.

50. The larger thermal conductivity of Na(*l*) means that it absorbs and releases heat more quickly as it passes through the heat transfer system of a reactor. This capacity offsets the lower specific heat of Na(*l*). It requires less energy to heat 1 mole of sodium compared to 1 mole of water, but the much greater thermal conductivity of sodium means it absorbs heat much more effectively than water.

51. 29.3 kJ

55. 885 g

57. −47.5°C

59. The enthalpy change depends on the number and types of bonds broken and formed. The only way we can determine these quantities is from the balanced equation.

60. The structures must be known because the bonds between different atoms have different strengths and the strength of a multiple bond is greater than the strength of a single bond between the same two atoms.

61. gaseous butane

62. Because the high C=O bond energy is *released* during the reaction when CO_2 is produced.

63. a. −74 kJ
    b. +103 kJ
    c. +183 kJ

65. 1166.5 kJ/mol

67. 384 kJ

69. −569 kJ

71. The heat capacity of a calorimeter is needed to relate a change in its temperature to the quantity of heat released or absorbed.

72. yes

73. yes

74. small

75. 8.044 kJ/°C

77. 5121 kJ/mol

79. 23.08°C
81. no
82. The standard heat of formation of a substance is defined as the value of ΔH when a mole of the substance forms from its component elements in their standard states. Therefore, the heat of formation of any element in its standard state must be zero since no energy is involved in forming it from itself.
83. b and c
85. -165 kJ
87. $NH_4NO_3(s) \rightarrow N_2O(g) + 2\ H_2O(g)$
    -35.9 kJ
89. -6691 kJ
91. a heat of combustion
92. kJ/g
93. To convert from a molar heat of combustion to a fuel value, divide the heat of combustion by the molar mass of the fuel.
94. no
95. 202 kg
97. a. 49.1 kJ/g
    b. 4.91×10^4 kJ
    c. 5.97 g
99. octane, 3.11×10^4 kJ/L; gaseous methane, 33.3 kJ/L
101. If a certain quantity of energy is released when A forms B and a different quantity of energy is released when B forms C, then it makes sense, based on the law of conservation of energy, that the sum of these two quantities is released when A forms C directly.
102. It is difficult to directly measure the heat of formation of carbon monoxide because burning carbon in a limited supply of oxygen is more likely to produce carbon dioxide and soot (unburned carbon) than pure carbon monoxide.
103. If one writes a formation reaction for each reactant and product and then combines them to obtain an overall reaction, the sum of the ΔH values for the individual reactions equals ΔH for the overall reaction. This summation is an application of Hess's law.
104. The lattice energy (U) is released in the last step in forming an ionic crystal from the elements that make it up. Summing the ΔH values of the other steps and subtracting this sum from the ΔH of the overall reaction is an application of Hess's law.
105. Leave the second reaction as written and reverse the first reaction.
107. -297 kJ/mol
109. 28.0 kJ
111. -103.4 kJ

CHAPTER 12

1. The physical property used to separate the components of crude oil is volatility or boiling point.
2. no
3. Increasing chain-length and molar mass lead to increasing London forces between molecules that result in more energy (and higher temperatures) to melt and boil these compounds.
4. no
5. Cyclohexane is not planar because each carbon atom is bonded to four other atoms. This results in a bond angle of about 109 degrees that are not compatible with the bond angles in flat geometric shapes such as a hexagon.
6. yes
7. yes
8. no
9. Benzene is planar because the sp^2 hybridized carbon atoms in its ring form bonds with 120° bond angles. This value matches the angle that the sides of a planar hexagon make.
10. Aromatic structures are stable because of the delocalization of the π bonds found in these substances.
11. 60. torr (or mm Hg)
13. a. 1
    b. 5
    c. 7
    d. 10

15.
CH₃—CH(CH₃)—CH₂—CH₃   (2-methylbutane)

and

CH₃—C(CH₃)(CH₃)—CH₃   (2,2-dimethylpropane)

17. a and d

19. a. 2-methylpentane
    b. 2,3-dimethylbutane
    c. 3-methylpentane
    d. 2-methylbutane

    All are more likely to be found in gasoline.

21. (four structural formulas shown)

23. 2-methylpentane, 3-methylpentane, butane

25. c and d

27. 1,2-diethylbenzene (o-diethylbenzene),

    1,3-diethylbenzene (m-diethylbenzene), and

    1,4-diethylbenzene (p-diethylbenzene).

29. $C_{11}H_{24} + 17\,O_2 \rightarrow 11\,CO_2 + 12\,H_2O$

31. 2.84 g of methanol; 3.90 g of CO_2

33. The molar mass of ethanol is larger than the molar mass of methanol. Therefore, the London forces in ethanol are stronger and so its boiling point is higher.

34. Ethanol and water are polar substances that form hydrogen bonds. Therefore they interact strongly with one another and are miscible in all proportions. Ethane is a nonpolar substance and dimethylether does not form hydrogen bonds. Therefore both should be less soluble in water than ethanol.

35. Ethane contains only carbon and hydrogen. However, ethanol and dimethylether also contain oxygen, which has no fuel value. The more "oxygenated" a fuel is, the lower its fuel value.

36. no

37.
H—C(H)(H)—C(H)(H)—C(H)(H)—Ö—H   H—C(H)(H)—C(H)(:Ö:H)(H)—C(H)(H)—H

39. b and e

43. a. 4-methyl-2-pentanol
    b. 2-methyl-3-pentanol
    c. 5-ethyl-6-methyl-2-heptanol
    d. 1-hexanol

47. a. ethyl propyl ether
    b. sec-butyl methyl ether
    c. sec-butyl ethyl ether

49. aldehydes and ketones

50. In fructose, the carbonyl group is not on an end carbon. In the open-chain form, fructose contains a ketone functional group and so is called a ketohexose. In glucose and mannose, the carbonyl is on the end carbon atom. These sugars are both aldehydes and are called aldohexoses.

51. yes

52. A reaction between a carbonyl group on C-1 (of an aldohexose) and the OH group bonded to C-3 would result in a 4-membered ring that would be much less stable than a 6-membered ring.

53. all of them

55. Only the last structure (structure e) is an aldehyde. Structures a and b are structural isomers of one another; c, d, and e are structural isomers of one another.

57.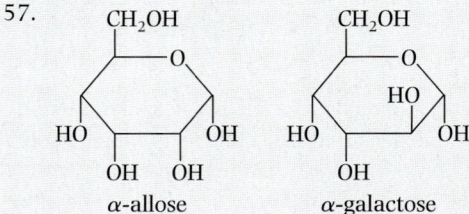

α-allose   α-galactose

59. Structures a and c differ only in the orientation of the OH group bonded to C-1. They are α and β forms of the same monosaccharide.

61. yes

63. Starch contains only α-glycosidic linkages, whereas cellulose has only β-glycosidic linkages.

64. All the components of wood contain several oxygen atoms. Oxygen adds mass to a substance but no fuel value. Therefore the energy produced per gram is lower for these oxygenated substances than for alkanes (which contain no oxygen).

65. Wood contains a significant amount of water. Some of the heat generated by the combustion process is lost in converting liquid water to steam (water vapor).

66. These enzymes can convert cellulose into sugar and ethanol. Discovering such an enzyme would make it possible to make sugar and ethanol cheaply. Lowering the price of ethanol would make it much more competitive as an energy source and a good alternative to fossil fuels.

69. a. β-glucose and α-galactose
    b. α-glucose and α-mannose
    c. α-mannose, β-mannose, and α-talose

71. Anthracite has the greatest percentage of carbon and the smallest percentage of oxygen in its composition. As a result anthracite has the greatest fuel value of the four types of coal.

72. Advantages: Complete combustion produces only water. Hydrogen has a higher fuel value than any other fuel. Disadvantages: Low density results in lower yields of energy per milliliter of liquid. Its low heat-to-volume ratio is one factor limiting the use of liquid hydrogen for internal combustion engines. It has a very low boiling point, which makes it difficult and expensive both to liquefy and to store as a liquid. It is highly flammable.

73. anthracite > bituminous > subbituminous > lignite

74. Coal is not a compound. It is a mixture of solid carbon-based compounds and impurities and, therefore, does not have a definite composition.

75. 2a and 2b

77. The OH group of a carboxylic acid is capable of forming hydrogen bonds with water molecules. Aldehydes tend to be less polar than carboxylic acids of similar molar mass also.

78. Methylamine contains only a single carbon atom. Butylamione contains a 4-carbon hydrocarbon chain. Since a larger proportion of butylamine is non-polar we would expect it to be much less soluble in water than methylamine.

79. ethane

80. ethanal

81. (structure)

83. a. ethylmethylisopropylamine, or N-ethyl-N-methylisopropylamine
    b. dimethylisopropylamine, or N,N-dimethylisopropylamine
    c. methyl-sec-butylamine, or N-methyl-sec-butylamine
    d. methylisobutylamine or N-methylisobutylamine

85. -138.7 kJ

87. The assumed products of a combustion reaction of a hydrocarbon are carbon dioxide (CO_2) and water (H_2O). If an excess of oxygen is not present, not all of the original sample may be completely converted to these products.

88. There is always another source of oxygen in the combustion reaction.

89. A molecular formula tells us the actual number of atoms of each element in a substance. An empirical formula is the smallest whole number ratio of the atoms of each element in a substance.

90. the molar mass

91. C_2H_3

93. $C_8H_{16}O$

95. Structural isomers are substances that have the same molecular formula but a different linkage of atoms. Geometric isomers have an identical structure in terms of which atoms are bonded to each other, but some of the atoms are oriented in different fixed positions.

96. In open-chain alkanes all of the carbon–carbon bonds are single bonds and are free to rotate relative to each other. Geometric isomers are not possible because the atoms do not have fixed orientations.

97. no

98. no

99. Both groups bonded to C-1 are hydrogen atoms. A requirement for *cis-trans* isomerism is that each carbon that forms the double bond must have two different groups bonded to it.

100. A triple bond between C-2 and C-3 means that there is a single bond between C-1 and C-2 and a C—C—C bond angle of 180°. Structural isomerism is not possible in such a linear molecule.

101.
```
    H  H  H           H
    |  |  |           |
H—C—C=C         H—C—C≡C—H
    |     |           |
    H     H           H
  (a) propene      (b) propyne
```

103.
CH₃CH₂\   /CH₂CH₂CH₃
       C=C
      H/   \H

Cl\   /Cl   CH₃CH₂\   /H
   C=C   C=C
  H/   \CH₃   H₃C/   \CH₂CH₃

105. 2-hexene; 4-methyl-2-hexyne

CHAPTER 13

1. The dissolution of $CaCl_2$ is exothermic, releasing heat that helps melts the ice.

2. If the amount of energy required to overcome the attractive forces between solute particles and between solvent particles is less than the amount of energy released when solute and solvent particle interact, the process is exothermic.

3. $\Delta H_{\text{H Bond}}$ and ΔH_{ionic}.

4. The more positive the heat of solution, the less soluble a compound is likely to be.

5. If the energy released when ions and water dipoles interact is large, the dissolution is likely to be exothermic.

6. greater than

7. less than

8. The larger the charge, or the smaller the size of an ion, the stronger the ion-dipole interactions with water molecules when it dissolves and the more exothermic the hydration process.

9. increase

11. 22.8°C

13. 13.9 kJ/mol

15. a

17. $U = \Delta H_{\text{hydration}} - \Delta H_{\text{solution}}$
    For LiCl: $U = -823 - 17 = -840$ kJ/mol
    For NaCl: $U = -723 - 47 = -770$ kJ/mol
    For KCl: $U = -649 - 52 = -701$ kJ/mol

19. They all have a positive heat of solution, so none of these salts releases heat on dissolving.

21. decreases

22. gas

23. decrease

24. ΔS_{sys}, positive; ΔS_{surr}, positive

25. ΔS_{sys}, positive; ΔS_{surr}, negative

26. fullerenes

27. negative

28. Reversing a process changes the sign of the entropy change (ΔS).

29. A molecule of H_2S is larger than a molecule of H_2O with more electrons that give rise to more internal energy states.

30. S_8 contains 8 atoms per molecule. This allows more internal motions leading to a greater absolute molar entropy than in S_2.

31. a and c

33. a. $CH_4(g) < CF_4(g) < CCl_4(g)$
    b. $CH_3OH(l) < C_2H_5OH(l) < C_3H_7OH(l)$
    c. $HF(g) < H_2O(g) < NH_3(g)$

35. a. decreases
    b. decreases
    c. decreases
    d. increases

37. a. $\Delta S° = 24.7$ J/K, or 24.7 J/°C
    b. $\Delta S° = -146.4$ J/K, or -146 J/°C
    c. $\Delta S° = -73.2$ J/K, or -73.2 J/°C
    d. $\Delta S° = -167.6$ J/K, or -167.6 J/°C

39. $S°_{\text{ClO}} = +218.8$ J/mol·K or 218.8 J/mol·°C

41. no

42. when $\Delta H > 0$ and $\Delta S > 0$.

43. The combustion of coal (and many other fuels) is both exothermic (negative $\Delta H°$) and results in an increase in entropy ($\Delta S°$ is positive), and so will be spontaneous at all temperatures.

44. Going up to the top of a mountain is a process that is never spontaneous. It requires some form of energy to be used for this to occur.

45. ΔS is positive; ΔG is negative; ΔH is positive

47. a. spontaneous
    b. nonspontaneous
    c. spontaneous
    d. spontaneous

49. For NaBr: $\Delta G° = -18.0$ kJ/mol
    For NaI: $\Delta G° = -29.1$ kJ/mol

51. $+91.4$ kJ; 981 K

53. $\Delta H° = -145.6$ kJ
    $\Delta S° = -186.7$ J/K
    This reaction is spontaneous at temperatures below 780 K.

55. a. $\Delta G° = +173.2$ kJ
    b. $\Delta G° = -70.6$ kJ
    c. $\Delta G° = -35.3$ kJ
    d. $\Delta G° = -4.8$ kJ

57. a. 55a and 55c
    b. 55b and 55d
    c. none of them

59. Enantiomers are isomers with structures that are mirror images of each other but that cannot be superimposed.

60. The term *chiral* refers to a molecule that can exist as enantiomers. It usually contains an atom that has four different groups bonded to the central atom.

61. A racemic mixture is one that contains a 50:50 ratio of two enantiomers.

62. In a saturated fatty acid all the carbons are linked by single bonds. An unsaturated fatty acid contains one or more carbon-carbon double bonds.

63. (S)-amino acids are structures where the priority order of the groups bonded to the chiral atom traces out a counter-clockwise direction; that is, the groups a, b, and c are ordered counter-clockwise.

64. glycine

65. a and c

66. a

69. tyrosine, glycine, glycine, phenylalanine, and methionine

71. a, c, and d

73. They are all (S) stereoisomers.

75. a. saturated
    b. unsaturated
    c. unsaturated

77. same

78. Fats contain a greater percentage of carbon and hydrogen and less oxygen when compared to carbohydrates and proteins. This greater carbon and hydrogen contents increases the fuel value of fats.

79. Glucose, mannose, and galactose all have the same molecular formula. However, there are slight differences in structure, so the fuel values are similar but not identical.

80. greater than

81. Butter is made up of a greater percentage of fat than many other foods. Fats have more than twice the fuel value (caloric content) of carbohydrates or proteins.

82. starch

83. 34–43 miles

85. 121 g H_2O

87. The reaction is essentially an isomerization reaction producing a substance with the same chemical composition as the reactant

88. Because the exergonic conversion of ATP to ADP drives the endergonic steps.

89. The reactions, along with their corresponding $\Delta G°$ values, are additive, that is, we can use Hess's law.

90. Steps 3 and 7

91. The missing product is NH_3.

93. -16.6 kJ

95. The three molecular subunits in DNA are a phosphate group, a five-carbon cyclic sugar called deoxyribose, and a nitrogen-containing base The two components that form the "backbone" are the phosphate group and the sugar.

96. A codon is a sequence of three nucleotides on an mRNA unit that codes for a specific amino acid. There are 20 amino acids. If only 2 nucleotides were used to make a codon for a specific amino acid there would be only 16 possible combinations—not enough to produce a unique code for each of the 20 amino acids.

97. hydrogen bonds

98. Base pairing refers to the hydrogen bonding that occurs between two specific pairs of bases that hold the two strands of a DNA molecule together. In DNA, complementary base pairing links adenine to thymine, and guanine to cytosine.

101. A-G-C-C-A-T

CHAPTER 14

1. The average rate is the rate of a reaction averaged over a specific time interval. The instantaneous rate is the rate at any particular moment during a chemical reaction.

2. yes

3. Because the reactants are used up.

4. yes

5. a. $\dfrac{\Delta[H^+]}{\Delta t} = \dfrac{\Delta[NO_2^-]}{\Delta t} = -\dfrac{\Delta[NH_3]}{\Delta t}$
   b. $\dfrac{\Delta[NO_2^-]}{\Delta t} = -\dfrac{2}{3}\dfrac{\Delta[O_2]}{\Delta t}$
   c. $\dfrac{\Delta[NO_2^-]}{\Delta t} = -\dfrac{2}{3}\dfrac{\Delta[O_2]}{\Delta t}$

7. a. $\dfrac{\Delta[OH]}{\Delta t} = -2\dfrac{\Delta[H_2O_2]}{\Delta t}$
   b. $\dfrac{\Delta[ClO_3]}{\Delta t} = -\dfrac{\Delta[ClO]}{\Delta t} = -\dfrac{\Delta[O_2]}{\Delta t}$
   c. $\dfrac{\Delta[HONO_2]}{\Delta t} = -2\dfrac{\Delta[N_2O_5]}{\Delta t}$

9. The green curve represents [N_2O], the red curve represents [O_2].

11. a. $\dfrac{\Delta[CO_2]}{\Delta t} = -\dfrac{2}{3}\dfrac{\Delta[CO]}{\Delta t}$
    b. $\dfrac{\Delta[COS]}{\Delta t} = \dfrac{-\Delta[SO_2]}{\Delta t}$
    c. $\dfrac{\Delta[CO]}{\Delta t} = 3\dfrac{\Delta[SO_2]}{\Delta t}$

13. a. $\dfrac{\Delta[Cl_2]}{\Delta t} = \dfrac{\Delta[O_2]}{\Delta t} = 1.48 \times 10^6$ M/s
    b. $\dfrac{\Delta[O_2]}{\Delta t} = -\dfrac{\Delta[ClO]}{\Delta t} = 9.03 \times 10^3$ M/s

15. Between 0 and 100 μs,
    $\dfrac{\Delta[O_3]}{\Delta t} = -13.7$ M/s

    Between 200 and 200 μs,
    $\dfrac{\Delta[O_3]}{\Delta t} = -5.5$ M/s

19. no

20. Since the units of "rate" in a rate law expression are conventionally M/s, reactions of various orders must have rate constants with different units in order to ensure that the units of both sides of the expression agree.

21. yes

22. no

23. Doubling the initial concentration of a second-order reactant will decrease the half-life by a factor one half.

24. yes

25. a. first order with respect to A, first order with respect to B, and second order overall
    b. second order with respect to A, first order with respect to B, and third order overall
    c. first order with respect to A, third order with respect to B, and fourth order overall

27. a. rate = $k[O][NO_2]$, $M^{-1}\,s^{-1}$
    b. rate = $k[NO]^2[Cl_2]$, $M^{-2}\,s^{-1}$
    c. rate = $k[CHCl_3][Cl_2]^{1/2}$, $M^{-1/2}\,s^{-1}$
    d. rate = $k[O_3]^2$, $M^{-1}\,s^{-1}$

29. a. rate = $k[BrO]$
    b. rate = $k[BrO]^2$
    c. rate = $k[BrO]$
    d. rate = $k[BrO]^0 = k$

31. the order for one of the individual reactants

33. a. rate = $k[NO_2][O_3]$
    b. rate = $k[NO_2][O_3]$
       rate = 4.9×10^5 M/s
    c. The rate of appearance of NO_2 will be 4.9×10^5 M/s.
    d. Since the reaction is first order with respect to O_3, if the concentration of O_3 is doubled, the rate of the reaction will be doubled as long as all other conditions are kept constant.

35. c

37. rate = k [NO][NO$_2$]

39. rate = k [ClO$_2$][OH$^-$]
    $k = 1.4 \times 10^1 \, M^{-1} \, s^{-1}$

41. rate = k [H$_2$][NO]^2
    $k = 6.32 \, M^{-2} \, s^{-1}$

43. $8.69 \times 10^{-4} \, \mu s^{-1}$

45. $5.40 \times 10^{-12} \, M^{-1} \, s^{-1}$

47. rate = k [HNO$_2$]^2
    $k = 4.097 \times 10^{-4} \, \mu M^{-1} \, \text{min}^{-1}$

49. rate = k [NH$_3$]
    $k = 3.00 \times 10^{-3} \, s^{-1}$

51. rate = k [N$_2$O], 4 half-lives

53. a. rate = k [C$_4$H$_6$]^2
    b. 29.3

55. 7.12×10^{-1} s

57. rate = k [C$_{12}$H$_{22}$O$_{11}$][H$_2$O]
    $k = 3.46 \times 10^{-3} \, M^{-1} \, s^{-1}$

59. They involve different reactants.

60. yes

61. a. rate = k[SO$_2$Cl$_2$]; unimolecular
    b. rate = k[NO$_2$][CO]; bimolecular
    c. rate = k[NO$_2$]^2; bimolecular

63. 2 N$_2$O$_5 \rightarrow$ 4 NO$_2$ + O$_2$

65. step 2

67. step 1

69. a. photochemical
    b. thermal
    c. thermal

71. a. False, they may have large activation energies.
    b. False, they may be fast but the reverse reaction would be even faster.
    c. False, they could be rapid at high temperatures.
    d. False, they may or may not be favored thermodynamically, but they could still be slow.

72. a. false
    b. false
    c. false
    d. False, in all cases the thermodynamic property is not directly linked to reaction kinetics.

73. Rate is temperature dependent since it involves how often molecules collide with the correct orientation in an effective collision. The rate order relates rate to the concentration of reactants and the mechanism of the reaction, which is not changed by changing temperature. As the temperature of a reaction is varied, both the translational and rotational movements of the molecules change. These changes affect their ability to react in a given time period.

74. The value of E_a of an exergonic reaction is lower than the E_a of the reverse reaction, by an amount equal to ΔG.

75. the second one

76. changing temperature

79. c

81. $E_a = 2.48 \times 10^2$ J/mol

83. a. 4.54×10^3 J/mol
    b. 5.06×10^{10}
    c. $1.05 \times 10^{-44} \, M^{-1/2} \, s^{-1}$

85. $E_a = 5.64 \times 10^2$ J/mol
    $A = 3.31 \times 10^{12}$

87. yes

88. yes

89. yes

90. no

91. A bimolecular reaction will obey pseudo-first-order kinetics when the concentration of one reactant is high enough to remain essentially constant during the course of the reaction.

93. NO is the catalyst, not N$_2$O

95. the second one

CHAPTER 15

1. The rate of the forward reaction is equal to the rate of the reverse reaction.

2. One example: the rates at which water flows from a shower head and down the drain is the same, keeping the volume of water in the shower constant.

3. no

4. no

5. $^{14}N_2$: 28 g/mol; original reactant
   $^{14}N^{15}N$: 29 g/mol; possibly formed by $^{15}N_2O$ reacting with $^{14}N^{15}NO$
   $^{16}O_2$: 32 g/mol; original reactant
   $^{14}N^{15}NO$: 45 g/mol; possibly formed by the reaction of $^{14}N^{15}N$ and O_2
   $^{15}N_2O$: 46 g/mol; original reactant

7. $K > 1$

8. $K > 1$

9. towards products

10. K is the ratio of the rate constants for the forward and reverse reactions. K can be large when the rate constant is small, as long as the rate constant for the reverse reaction is even smaller.

11. $K = 3.33 \times 10^{-1}$

13. a. $K = \dfrac{[N_2O_4]}{[N_2][O_2]^2}$

    b. $K = \dfrac{[NO_2][N_2O]}{[NO]^3}$

    c. $K = \dfrac{[N_2]^2[O_2]}{[N_2O]^2}$

15. $K_c \approx 0.55$

17. $K_c = 5.0 \times 10^{-1}$

19. $K_c = 1.53$

21. no; towards products

23. $K_{c(\text{forward})} = \dfrac{[NO_2]^2}{[NO][NO_3]}$ $\quad K_{c(\text{reverse})} = \dfrac{[NO][NO_3]}{[NO_2]^2}$

25. $K_c = 7.4$

27. $K_c = \dfrac{[SO_3]}{[SO_2][O_2]^{1/2}}$ $\quad K'_c = \dfrac{[SO_3]^2}{[SO_2]^2[O_2]}$
    $K'_c = K_c^2$

29. a. 4.9×10^{-2}
    b. 4.2×10^2
    c. 2.0×10^1

31. when $\Delta n = 0$

32. $K_p = K_c(RT)^{\Delta n}$

33. b and c

35. $K_p = 1.01 \times 10^{-1}$

37. a. yes
    b. no

39. $K_p = 2.0 \times 10^4$

41. toward reactants

43. yes

44. Both can be used depending upon the reaction. K_p is used when all species are gases and K_c is used in the other cases.

45. the formation of $ClF(g)$

47. $K = 3.81 \times 10^{-3}$

49. a. $\Delta G° = 7.80 \times 10^4$ J

51. a. $\Delta G° = 3.01 \times 10^4$ J/mol
    b. $\Delta G° = 2.86 \times 10^4$ J

53. no

54. The rate of the forward reaction is increased and the rate of the reverse reaction is decreased.

55. A large increase in P_{O_2} drives the second reaction toward products.

56. no

57. Dissolving a gas in a liquid is an equilibrium process. Increasing the partial pressure of a soluble gas drives the reaction towards products (in this case more dissolved gas).

58. The partial pressures of the gases involved in the mass action expression do not change with the addition of an inert gas.

59. b and d

61. towards reactants

63. a. towards reactants
    b. towards products
    c. no change

65. a. $P_{Cl_2} = P_{PCl_3} = 0.547$ atm; $P_{PCl_5} = 0.013$ atm
    b. P_{PCl_5} will increase and P_{PCl_3} will decrease.

67. $[H_2O] = [Cl_2O] = 0.00376\ M$; $[HOCl] = 1.13 \times 10^{-3}\ M$

69. $P_{NO_2}/P_{NO} = 9 \times 10^5$

71. a. $P_{NO} = 0.272$ atm; $P_{NO_2} = 7.98 \times 10^{-3}$ atm
    b. $P_{\text{total}} = 0.416$ atm

73. $P_{N_2} = 0.75$ atm; $P_{O_2} = 0.17$ atm; $P_{NO} = 0.08$ atm

75. $[H_2S] = 0.546\ M$

77. $P_{CO} = P_{Cl_2} = 0.258$ atm; $P_{COCl_2} = 6.67 \times 10^{-3}$ atm
79. $[CO] = [H_2O] = 0.031\ M$; $[CO_2] = [H_2] = 0.069\ M$
81. exothermic
82. K_p decreases
83. exothermic
84. K_p increases
85. $K_c = 1.28 \times 10^{-31}$
87. $\Delta H° = -1.1 \times 10^5$ J/mol
89. $K_p = \dfrac{1}{(P_{N_2})(P_{H_2})^3}$
90. $K_p = \dfrac{1}{(P_{SO_3})(P_{H_2O})}$
91. They are not included because the concentrations of solids are considered to be constant.
92. This reaction has a positive value for $\Delta H°$ and a positive value of $\Delta S°$. These types of reactions are spontaneous at high temperatures. The value of ΔG will become more negative as the temperature increases. This will lead to an increase in K.
93. a. $P_{H_2O} = 0.049$ atm; $P_{CO} = P_{H_2} = 0.393$ atm
    b. $P_{H_2O} = 0.065$ atm; $P_{CO} = P_{H_2} = 0.452$ atm
95. $P_{NH_3} = 0.355$ atm

CHAPTER 16

1. The percent dissociation increases as concentration decreases.
2. HNO_2
3. $HC_2H_3O_2$
4. formaldehyde solution
5. water
6. Alanine ionizes more readily in water than in ethanol. Alanine is a stronger acid in water than it is in ethanol, because water is a better base than ethanol.
7. It's more difficult to remove an H^+ from the anion produced in the first ionization because of coulombic attraction.
8. K_{b1}. Addition of the first H^+ ion produces a cation that will repel a second H^+ ion.
9. $[H^+] = 2.96 \times 10^{-3}\ M$
11. $K_a = 8.91 \times 10^{-4}$
13. 5.3%
15. 1.63% ionized; $K_a = 6.7 \times 10^{-5}$
17. $[H^+] = 0.100\ M$
19. $[H^+] = 4.42 \times 10^{-3}\ M$; 1.76% ionized. K_{a2} does not contribute significantly.
21. $[H^+] = 1.50 \times 10^{-1}\ M$
23. $[OH^-] = 0.160\ M$
25. Slowly combine 70.0 g of NaOH with enough water to make 2.50 L of solution.
27. 1.0 M $NaNO_2$ is a stronger electrolyte than HNO_2, because HNO_2 is a weak acid.
28. The HCl and the H_2O react to form H_3O^+ and Cl^-. Having these ions in solution facilitates the conduction of electricity.
29. CH_3NH_2 and H_2O are attracted by dipole–dipole forces. The lone pair on the nitrogen atom in CH_3NH_2 forms a coordinate covalent bond with one hydrogen ion donated by a molecule of water. This results in the formation of $CH_3NH_3^+$ and OH^-.
30. Sulfur reacts with oxygen to form sulfur trioxide. Sulfur trioxide reacts with water to form sulfuric acid. Sulfuric acid reacts with calcium carbonate to form calcium sulfate, water, and carbon dioxide.
31. HCl and HNO_3
32. Because BF_3 can accept a pair of electrons, it is a Lewis acid. BF_3 is not capable of being a Brønsted-Lowry acid since it has no ionizable hydrogen atoms.
33. S is more electronegative than Se.
34. The O—H bond of H_2SO_4 is weaker than that of H_2SO_3 because of less electron density due to the greater electron-withdrawing power of one more oxygen atom bonded to sulfur in H_2SO_4.
35. a. acid: HNO_3; base: NaOH
    b. acid: HCl; base: $CaCO_3$
    c. acid: HCN; base: NH_3
37. $[OH^-] = 3.89 \times 10^{-9}\ M$; 3.24×10^{-4} % ionized
39. $HNO_2 : NO_2^-$; $HOCl : OCl^-$; $H_3PO_4 : H_2PO_4^-$; $NH_3 : NH_2^-$

41. a. H_2SO_3
    b. H_2SeO_4

43. because pH is the *negative* log of $[H^+]$

44. 2 pH units

45. when $[H^+]$ is greater than 1 M

46. Water is a better base and acid unto itself than ethanol is unto itself.

47. a. pH = 7.462; pOH = 6.538; basic
    b. pH = 4.70;   pOH = 9.30;   acidic
    c. pH = 7.15;   pOH = 6.85;   basic
    d. pH = 10.932; pOH = 3.068; basic

49. pH = 0.810

51. pH = 12.653; pOH = 1.347

53. pH = −0.124

55. pH = 2.38

57. $[H^+]_{Minnesota} = 6 \times 10^{-6}$ M; $[H^+]_{Maine} = 3 \times 10^{-5}$ M

59. pH = 2.597

61. Base strength increases as the number of methyl groups increases.

62. They are easy to calculate by using $K_a K_b = K_w$.

63. NH_4NO_3

64. NaF

65. pH = 10.63

67. pH = 10.27

69. 3.19% ionized

71. $pK_b = 3.32$

73. pH = 7.36

75. Buffers have the ability to neutralize both acids and bases. An NaCl/HCl buffer has no ability to neutralize acids.

76. A good buffer has the ability to neutralize both acids and bases. A buffer containing only a weak base without its conjugate acid would not be able to neutralize bases.

77. pH = 4.48

79. pH = 12.45; pOH = 1.55

81. $[C_2H_3O_2^-]/[HC_2H_3O_2] = 6.54 \times 10^{-3}$

83. 2.93 mL

85. pH = 9.26

87. pH = 3.44 before addition of acid; pH = 3.37 after addition of acid

89. Strong acid-strong base titration curves have gradual pH changes except in the region near the equivalence point. Near the equivalence point the pH change is quite rapid.

90. yes

91. no

92. A pH indicator should have a clearly visible color change over a narrow pH range, and this range should coincide with the equivalence point of the titration.

93. pH = 4.74; pH = 8.75; pH = 12.36

95. 7.60

99. $[NH_3] = 0.0256$ M

101. Solubility is the concentration of a compound that dissolves in solution. The solubility product is the product of the molar concentrations of all of the ions from a given compound that dissolve in solution each raised to a power based on the number of moles of that ion in a mole of the compound.

102. It is a consequence of Le Châtelier's principle.

103. Inspection of K_{sp} values indicates that $SrCO_3$ will precipitate first.

104. K_{sp} increases.

105. endothermic

106. The fluoride ion (F^-) in fluoroapatite is a weaker base than the OH^- ion in hydroxyapatite, so acidic solutions react less readily with the F^- in fluoroapatite than with the OH^- ion in hydroxyapatite.

107. reactants

109. $K_{sp} = 1.08 \times 10^{-10}$

111. $[Cu^+] = [Cl^-] = 1.01 \times 10^{-3}$ M

113. 9.9×10^{-6} g $CaCO_3$/mL

115. pH = 10.09

117. d

119. 9.8×10^{-6} M

121. $[Ag^+]/[Ag(S_2O_3)_2^{3-}] = 4 \times 10^{-13}$

123. pH = 11.56

CHAPTER 17

1. To keep the charge balanced between the two half-cells and to complete the circuit.

2. A wire would not be able to emit ions without being part of the electrochemical reaction itself.

3. Oxidation occurs at the anode and reduction occurs at the cathode.

4. Ions can pass freely between the two aqueous solutions.

5. a. Anode: $Pb(s) \rightarrow Pb^{2+}(aq) + 2\ e^-$
   Cathode: $Zn^{2+}(aq) + 2\ e^- \rightarrow Zn(s)$
   b. $Pb(s) + Zn^{2+}(aq) \rightarrow Pb^{2+}(aq) + Zn(s)$

7. a. Anode: $Cd(s) + 2\ OH^-(aq) \rightarrow Cd(OH)_2(s) + 2\ e^-$
   Cathode: $2\ H_2O(l) + MnO_4^- + 3\ e^- \rightarrow MnO_2(s) + 4\ OH^-$
   b. $3\ Cd(s) + 4\ H_2O(l) + 2\ MnO_4^-(aq) \rightarrow 2\ MnO_2(s) + 2\ OH^- + 3\ Cd(OH)_2(s)$

9. a. $n = 6$ for this reaction.
   b. In K_2FeO_4 the oxidation number of iron is $+6$.
   In Fe_2O_3 the oxidation number of iron is $+3$.
   In Zn the oxidation number of zinc is 0.
   In ZnO the oxidation number of zinc is $+2$.
   In K_2ZnO_2 the oxidation number of zinc is $+2$.
   c. The anode is made of $Zn(s)$.

11. a. Chlorine is being oxidized and bromine is being reduced.
    b. Chlorine is both oxidized and reduced.
    c. This is not a redox reaction.

13. The cell reaction with the more negative ΔG value has two times the E_{cell} value of the other.

14. In electrochemical work, electric charge is being moved through an external circuit.

15. a. $\Delta G° = -34.5$ kJ
    $E°_{cell} = +0.358$ V
    b. $\Delta G° = +2.9$ kJ
    $E°_{cell} = -0.030$ V

17. a. $4\ Na(s) + O_2(g) \rightarrow 2\ Na_2O(s)$
    b. $Cl_2(g) + 2\ NaBr(s) \rightarrow 2\ NaCl(s) + Br_2(g)$
    c. $3\ PbO(s) + 2\ NH_3(g) \rightarrow N_2(g) + 3\ H_2O(l) + 3\ Pb(s)$

19. a. $6\ H^+(aq) + 2\ MnO_4^-(aq) + 5\ H_2O_2(aq) \rightarrow 2\ Mn^{2+}(aq) + 8\ H_2O(l) + 5\ O_2(g)$
    b. $4\ H^+(aq) + MnO_2(s) + 2\ Cl^-(aq) \rightarrow Mn^{2+}(aq) + 2\ H_2O(l) + Cl_2(g)$
    c. $4\ H^+(aq) + 3\ MnO_4^{2-}(aq) \rightarrow 2\ MnO_4^-(aq) + 2\ H_2O(l) + MnO_2(s)$

21. The Pt serves only as an inert conductor.

22. yes

23. It should not be affected.

24. The potential will be smaller for the Ni-Zn cell.

25. The half-reaction at the anode is an oxidation half-reaction and we must change the sign of the anode potential since the table contains only reduction potentials. If we subtract $E°_{anode}$ from $E°_{cathode}$, this takes care of the change of sign.

26. a. $E°_{red} = +0.8277$ V
    b. $E°_{red} = +1.598$ V
    c. $E°_{cell} = +1.817$ V

27. No, $E°_{cell}$ is negative.

28. Yes, $E°_{cell}$ is positive.

29. a. $E°_{cell} = -0.478$ V
    $\Delta G° = +92.2$ kJ
    b. $E°_{cell} = +0.505$ V
    $\Delta G° = -97.4$ kJ

31. a. $Zn(s) + MnO_2(s) + H_2O(l) \rightarrow Mn(OH)_2(s) + ZnO(s)$
    Manganese is reduced at the cathode and zinc is oxidized at the anode.
    b. $S_8(s) + 16\ Na(s) \rightarrow 16\ Na^+(aq) + 8\ S^{2-}(aq)$
    S_8 is reduced at the cathode and Na is oxidized at the anode.
    c. $4\ Al(s) + 4\ OH^-(aq) + 3\ O_2(g) + 6\ H_2O(l) \rightarrow 4\ Al(OH)_4^-(aq)$
    O_2 is reduced at the cathode and Al is oxidized at the anode.

33. Electron flow is from anode to cathode. $Cu^{2+}(aq)$ will be reduced at the cathode and $Cd(s)$ will be oxidized at the anode.

35. a. $NiOOH(s) + 2\ H_2O(l) + TiZr_2(s) \rightarrow Ni(OH)_2(s) + 2\ OH^-(aq) + TiZr_2H(s)$
    b. $E°_{cell} = +1.32$ V

37. In this case, no. However, cell potentials are concentration dependent, so the effect of concentrations must be determined from the Nernst equation.

38. Yes, cell potentials are concentration dependent. The concentration of Zn^{2+} will increase and $[Cu^{2+}]$ will decrease as the reaction occurs.

39. No. All species involved in the cell reaction are solids.

40. All species involved in the cell reaction are solids and their concentration does not change as the reaction proceeds.

41. $E_{cell} = +1.27\ V$

43. $K = 8.56 \times 10^{19}$

45. $E_{red} = -0.414\ V$

47. $E_{cell} = +1.49\ V$ and decreases as the reaction proceeds.

49. $E°_{cell} = +0.62\ V$
    $E°_{cell} = +0.61\ V$

51. a. $E°_{cell} = +273\ V$
    b. $E_{cell} = 0$ and $[NO_3^-]/[NH_4^+] = 631$

53. c.

54. c.

55. The reaction involving Al(s).

57. The reaction with the Li(s) anode will have more power per gram.

59. No, Mg^{2+} is easier to reduce than Na^+.

60. Br_2 is formed first since Br^- is easier to oxidize.

61. If water is not removed, it will be oxidized and/or reduced, especially in the presence of salts, which make the solution conductive.

62. It increases the conductivity of the solution, promoting electrolysis.

63. Cl^- could be reduced to Cl_2.

64. The potential should be more negative.

65. 6.8 g Ag(s)

67. 36 min

69. a. $0.0052\ L\ O_2$
    b. No. The negative ions in seawater are oxidized preferentially during electrolysis.

71. The least negative potential necessary would be $-0.244\ V$.

73. Advantages include increased driving range, increased gas mileage, decreased emissions.

74. Vehicles powered by fuel cells are "cleaner" because only water is emitted and less energy is wasted as heat when converting fuels into mechanical energy.

75. a. CH_4:   C (−4)   H (+1)
    H_2O: H (+1)
    CO: C (+2)
    H_2:   H (0)
    CO_2:   C (+4)
    b. reaction 1: $\Delta G° = +142.2\ kJ$
       reaction 2: $\Delta G° = -28.6\ kJ$
       overall: $\Delta G° = 113.6\ kJ$

CHAPTER 18

1. The copper-containing mineral is oxidized to produce copper(II) oxide. The copper(II) oxide then reacts with CO to produce copper metal. Overall, an ionic form of copper is reduced to elemental copper.

2. Cu^+ is oxidized to Cu^{2+}.
   O_2 is reduced to O^{2-}.
   S^{2-} is oxidized to S^{+4}.

3. At higher temperatures, the C(s) can be converted to CO(g). The gaseous CO will react with the solid mineral more readily than the solid carbon.

4. Yes, they differ only by how much water is assumed to be in the structure of the ions:
   $AlO_2^- + 2\ H_2O \rightarrow Al(OH)_4^-$

5. magnetite (Fe_3O_4), 72.4%

6. linaeite (Co_3O_4), 73.4%

7. $2\ FeCuS_2(s) + 3\ O_2(g) \rightarrow 2\ FeO(s) + 2\ SO_2(g) + 2\ CuS(s)$
   $FeO(s) + SiO_2(s) \rightarrow FeSiO_3(s)$
   $FeCuS_2$:   Fe (+3)
   FeO: Fe (+2)
   $FeSiO_3$:   Fe (+2)

8. $4\ Au + 8\ NaCN + O_2 + 2\ H_2O \rightarrow 4\ Na[Au(CN)_2] + 4\ NaOH$
   $2\ Na[Au(CN)_2] + Zn \rightarrow 2\ Au + Na_2Zn(CN)_4$
   $Na[Au(CN)_2]$: Au (+1)
   $Na_2Zn(CN)_4$: Zn (+2)

9. FeSi is the oxidizing agent and $MgCaO_2$ is the reducing agent.

10. $B_2O_3 + 3\ Mg \rightarrow 2\ B + 3\ MgO$
    Mg is the reducing agent and B_2O_3 is the oxidizing agent.

11. $2\ Ga_2O_3(s) + 2\ Mg(s) + O_2(g) \rightarrow 2\ MgGa_2O_4(s)$
    Ga_2O_3: Ga (+3)
    Mg: Mg (0)
    $MgGa_2O_4$: Mg (+2), Ga (+3)

12. a. $2\ MoS_2 + 7\ O_2 \rightarrow 2\ MoO_3 + 4\ SO_2$
       $MoO_3 + 3\ H_2 \rightarrow Mo + 3\ H_2O$
    b. The oxidation states of Mo in the two equations from part a range from 0 (Mo) to +6 (MoO_3).

15. pure gold

16. no

17. Zn is more readily oxidized than Cu.

18. The mixture of atoms of different sizes in the solid phase weakens the metallic bonds that hold the lattice together. This results in a lower melting point for the alloy relative to the pure metal.

19. a. AB
    b. A_2B
    c. A_4B

21. One fifth of the octahedral holes are occupied.

23. X_3Y

25. octahedral holes

27. a. 1.81×10^5 g Mg
    b. 2.16×10^5 L $Cl_2(g)$

28. 5.79×10^6 g Na
    1.03×10^7 L $Cl_2(g)$

29. First, water that was adsorbed on the clay particles is driven off. Next, some hydroxide ions are decomposed to water and oxide ions. The water formed is driven off and the oxides remain. As the temperature continues to rise, units of silica (SiO_2) are split off sequentially. The process converts $Al_2Si_2O_5(OH)_4$ to 3 $Al_2O_3 \cdot 2\ SiO_2$ (mullite).

31. There are greater attractive forces between the ions in the oxides than between the metal atoms.

33. p-type

35. In both reactions carbon is oxidized and nitrogen is reduced.

37. The larger spacing corresponds to the potassium-containing compound.

39. a. Ca^{2+} is larger than Al^{3+}.
    b. Al^{3+} will likely fill tetrahedral holes and Ca^{2+} will likely fill octahedral holes.

41. 3 $Al_2Si_2O_5(OH)_4 \rightarrow Al_6Si_2O_{13} + 4\ SiO_2 + 6\ H_2O$
    45.% of the mass is lost in the conversion of kaolinite to mullite.

43. The band gap for InN corresponds to 620 nm light, which is in the visible region.

45. $P_{Zn} = 8.4 \times 10^{-4}$ atm

47. $\Delta S < 0$
    $\Delta H < 0$

48. to promote the reaction and to manipulate the properties of the resultant polymer

49. The carbon atoms in $(CH_2)_n$ are sp^3 hybridized.

50. The primary structure tells us the order of amino acids in the protein. The secondary structure refers to interactions with the protein chain to form β-sheets, β-turns, etc.

    The tertiary structure tells us how various parts of the protein fold over each other to form its overall three-dimensional shape.

51. See pages 930–931 of the text.

52. Hydrogen bonding is responsible for the helical structure in proteins, while polypropylene secondary structure depends on London forces.

53. No, there are only single bonds to nitrogen atoms in the amide groups.

55. Tyrosine, tryptophan, serine, threonine, lysine, arginine, histidine, aspartate, glutamate, asparagine, and glutamine can all form more than two hydrogen bonds.

57. $\Delta H = -82$ kJ

61. The side chain groups of silicones are hydrophobic.

63. The polymer with more sp^3 hybridized atoms in its backbone.

65. Both a and b could produce atactic polymers.

67. Poly(thiophene) and poly(pyrrole) are both condensation polymers. Their respective starting materials are thiophene and pyrrole.

69. Poly(vinylalcohol) is an addition polymer made from $H_2C=C-OH$

    Poly(methylmethacrylate) is an addition polymer of

    $$\begin{array}{c} H_3C \\ \\ H \end{array} \!\!\! \diagdown C = CH - \overset{\overset{\displaystyle O}{\|}}{C}OCH_3$$

71. $[-CH_2-CH=CH-CH_2-]_n$
    This polymer contains carbon-carbon double bonds.

Index

Page numbers in *italics* indicate figures, the letters *mn* after a page number indicate a marginal note reference, and the letter *t* after a page number indicates a table reference.

absolute zero, 37, 399, 435, 618, 922
absorption spectroscopy, 323–324
    by atoms, 106, *107*, 109
    by molecules, 309, 316, 323–324, 334–337, 341–342, 376, 684, 712
    by transition metal ions, 495–498, 829, 832
accelerators, particle, 74, *75*
accuracy, 30–31
acetate ion, 172
acetic acid, 172, *222*, 223, 587, 588
    titration of, 821–822
acetoacetic acid, *814*
acetone, 448–449
acetyl coenzyme A (acetyl CoA), 654
acetylene, 188, *189*, 492, 599
acid(s), 171–172, 250
    Brønsted-Lowry, 790
    hydrated metal ions as, 828
    in water, 790–791, 797
    ionization of, 783–785, 789
    Lewis, 795, 827
    molecular view of, 790–797
    and pH buffers, 813–815
    and pH values, 812–813
    strong, 250–251, 782–785, 789, 796, 797, 798–799, 812–813
    titration of, 820–821
    weak, 251, 252, 782–785, 789, 796, 797
acid-base conjugate pairs, 795–796
acid-base indicators, 816–817
acid-base neutralization reactions, 250–256, 806
acid-base titrations, 817–823
acidic solutions, 802
    pH values of, 803, 806–809
acid-ionization equilibrium constants, 783, 828*t*
acid-ionization reaction, 813–814, 828
acid rain, 252–256, 745, *780*, 782, *783*, 786, 790, 806, 811, 817, 828

acid strength, 782–785
    and molecular structure, 797–800
    see also acids: strong; acids: weak
actinide elements, 142, 923
actinium, 923, *924*
actinium-227, 178
activated complexes, 697, 705
activation energy, 704–711
addition polymers, 936–938
addition reactions, 936–940
adenine, 655, *656*, 657
adenosine diphosphate (ADP), 649–654, 659
adenosine triphosphate (ATP), 649–654, 659
adhesive forces, 457–458
agriculture
    and fertilizers, 755
air
    composition of, 7–8
    see also atmosphere
air bags, *402*
alanine, 635–637
    fuel value of, 646
alcohol
    as gasoline additive, 575–576
alcohols, 573, 588*t*
aldehydes, 577–578, 588*t*, 672
alkali metals, 167, 178–179
alkaline batteries, 851*t*, 852–853
alkaline earth metals, 167, 254–255
alkaline fuel cell, 851*t*
alkaline waters
    and metal ions, 828
alkalinity, 817–818
    see also acid-base titrations; buffer capacity
alkalinity titration, 817–818
alkanes, 516–517, 562–563, 564–565, 566–567, 569
    and coal, 590
    cyclic, 565
    naming of, 567, 568
    octane ratings of, 569*t*
alkenes, 597–598
alkyl groups, 568, 569*t*, 865
alkynes, 599, 600

allotropes, 298, 304, 348, 350, 414, 489–490, 492–493
alloys
    homogeneous, 901, 903
    interstitial, 906
    substitutional, 901–902, 903
alpha (α) decay, 64, 69–70
alpha (α) functional groups, 631*m*
alpha (α)-helix, 931–932
alpha particles, 35, 40, 56, 57, 61, 84
    and radiation dosage, 81, *82*
    and radon, *85*
    scattering of, *108*
    and transmuting elements, 74
alumina, 497, 912, 914
aluminosilicates, 263, 489, 912, 919–920
aluminum, 254, 909, 910, 912, *913*
    alloys of, *895*, 909–915
    chemistry of, 910, 828, 829
    ionization energy of, 146
    recycling of, 914–915
aluminum boride, 911
aluminum hydroxide, 828–829
aluminum oxide, *see* alumina
aluminum oxohydroxide, 909
amines, 588–589
    as bases in water, 792
amino acids, 631–633, 634, 833
    codons of, 660
    energy value of, 646
    essential, 632, 633*t*
    in wool and silk proteins, 930, 931–932
ammonia
    as a base, 792–793, 794, 820
    chemistry of, 754–755
    dipole moment of, *372*
    molecular geometry of, *356*
    and nitric acid, 799
    and nitrogen compounds, 414
    orbitals of, 365
    production of, 753–755
ammonium chloride, 809, 820
ammonium nitrate, 414
    and cold packs, 612–614
ammonium phosphate, 415
amorphous solids, 469
ampere, 19*t*, 871–872

Index

amphoteric substances, 801, 826, 912
amplitude (of a wave), *14*, 15
anabolism, 630
angle of diffraction, 470–471
angle of incidence, 470
angular (bent) geometries, 355, *356*, *357mn*, 358, *363t*
angular momentum quantum number, 124, *125*
anions, 6, 167, 287
    Lewis structures of polyatomic, 304
    hydration of dissolved, *445*
    sizes of, *440*
anodes, 846
    in lead-acid battery, *849*
anoxia, 411
antacids, *252*, 256
Antarctica, *338*
    and ozone layer, 305, *306*
anthracene, 573
anthracite, 590, *591t*
antibonding molecular orbitals, 317, 318–320
antifluorite structures, 480
antimatter, 57
antimony, 415, 903
antimony trioxide, 415
apatite, 414, 415, 826
aquamarine, 500–501
aqua regia, 4
aquatic ecosystems
    and pH, 817–818
aquatic organisms
    and oxygen solubility, 450
aqueous equilibria, 782–833
    acid-base indicators, 816–817
    acid-base titrations, 817–824
    acids and bases, 782–805
    acidic and basic salts, 806–810
    buffers, 811–816
    common-ion effect, 811, 826
    complexation reactions, 827–831
    solubility equilibria, 824–827
aqueous reactions
    acid-base, 250
    ion-exchange, 263
    precipitation, 257
    redox, 244
    titrations, 266
    net ionic equations, 243
aqueous solutions, 210–212
    boiling points of, 230–233
    concentrations of, 214–220
    freezing points of, 234–236
    osmotic pressures of, 224–230
    vapor pressures of, *452*
Arabian Gulf, 589
argentite, 894
argon, 134
    concentration of in air, *343t*
    electron configuration of, 137, 139
Aristotle, *8*
aromatic compounds in coal, 590
aromatic hydrocarbons, 571–573

aromatic rings, 571
Arrhenius, Svante August, 792, 799
Arrhenius equation, 705, 706–708
arsenic, 415
arsenic acid, 415
arsine, 415
artificial isotopes and elements, 74–78
astatine, 307, 308, *309*
asteroids, *162*
    and collisions with Earth, 163, *164*
atactic polymers, 938
atmosphere, 392–400
atmosphere (atm), 390
atmospheric pressure, 388–391, 392–394, 406
    see also partial pressures
atom, central, 299
atomic bombs, 74, 75, *78*
atomic mass unit, 32, 175
atomic number, 38
atomic orbitals, 316, 318, *319*
    hybrid, 363, 364–370
    and valence-bond theory, 363, 364
atomic structure, 106–109
    orbital model of, 147
    plum-pudding model of, 107–108
atomic theory of matter, Dalton's, 165
atoms, 2
    diamagnetic, 321
    excited states of, 116, 117–118
    formation of, 37–41
    ground states of, 116, 117–118
    isoelectronic, 136
    mass of, 38, 40
    orbital energies in, *140*
    paramagnetic, 321
    Rutherford's model of, 108, *109*
    Thomson's plum-pudding model of, *287*
auroras, *284*, 285–286, 322–324
    origins of colors in, *322t*
austenite, 906
autoionization
    of water, 801–803
automobile exhaust, 713–716
    federal emission standards, *716t*
    pollutants in, 704
automobiles, 576
    cooling systems, 234, 453
    combustion in, 628
    electric, 878–879
    recharging batteries in, 876, *877*
    tires and ozone, *673*
    see also gasoline; gasoline additives
autumnal turnover, 460–461
Avogadro, Amedeo, 174
Avogadro's law, *394*
Avogadro's number, 173, 174

background-radiation dosage, 83
Bacon, Francis, 9, *10*
bacteria
    and electricity, 883
    hydrolytic and transitional, 587
    methanogenic, 588, 589

baddelelyte, 916
baking powder, 179
baking soda, 179
ball-and-stick models, 472, *473*, 517
balloons, hot-air, 521–522
Balmer, Johann, 113–115, 129
Balmer series of spectral lines, 114–115, 117
band theory, 491
    and ceramics, 920
barite (barium sulfate), 254
barium, 254, 255
barium-141, *72*
barium hydroxide, 792
barometer, 389, 390
bases, 250
    acid-base conjugate pairs, 795–796
    acid-base indicators, 816–817
    acid-base titrations, 817–823
    Brønsted-Lowry model of, 790
    conjugate, 795–796
    Lewis bases, 794–795, 827
    molecular view of, 790–797
    and pH buffers, 813–815
    titration of, 819–820
    and water, 792–793, 797
base-ionization reaction, 793
basic oxygen process, 905
basic solutions, 802
    pH values of, 803, 806–809
batteries
    alkaline, *851t*, 852–853
    biological, 883
    chemistries of, 850–855
    dry cell, 850–852
    lead-acid, 849, 850, *851t*, 871, 872, 873–876, *877*, 879
    lithium-ion, 862–863
    nickel-cadmium (nicad), *851t*, 853–855
    nickel-metal hydride (NiMH), *851t*, 860–862, 879
    recharging, 873–878
    standard potentials of, 856–857, 863
battery capacity, 871
battery power, 871–872
bauxite, 909, 912, *913*
Bayer process, 912
becquerel, 79, *80t*
Becquerel, Henri, 78, 79
belt of stability, 65, 66
bending vibrations, 336
bends, the, 417
bent geometries, *see* angular (bent) geometries
benzene, 183, 571
    freezing and boiling points, *240t*
benzene ring, 571–572
benzoic acid, 536
beryl, 488, 494–495
beryllium, 254, 255
    electron configuration of, 136
    hybrid orbitals for, 368
    ionization energy of, 144
    isotopes of, 58
    in universe, 93

beryllium chloride, 368
beryllium-8, 58, 61
beta decay, 62–64, 67t
   and supernovae, 67, 68, 69, 70
beta particles, 62, 63, 121
   and radiation dosage, 81, 82
bicarbonate
   and alkalinity titration, 817–818
   ions of, 811–812
Big Bang, 10–11, 33, 35, 45
Bikini Island atomic bomb test, 78
bimolecular element steps, 698, 699, 700
binary ionic compounds, 167–168
binary molecular compounds, 166
binding energy, 58-62
biochemical fuel cells, 882–883
biological batteries, 883
biomass, 512
   and methane, 586–587
biomolecules
   metal complexes in, 831–833
biopolymers, 929
bismite, 415
bismuth, 415
   crystal structure of, 494
bismuth-209, 64
bituminous coal, 590, 591t
blast furnaces, 904–905
bleach, household, 171
blimps, 133
blood, see hemoglobin
blood oxygen, 410–412
blood pH, 814
blood sugar, see glucose
boat form (of molecular structure), 565
body-centered cubic structures, 906
body-centered cubic (bcc) unit cell, 473, 474
boehmite, 909
Bohr, Niels, 112–114, 150
Bohr model of the hydrogen atom, 112–118, 120, 121
bohrium, 78t
boiling-point elevation, 232, 235, 236
   and measuring molar mass, 239
   for selected solvents, 240t
boiling points, 235
   and phase diagrams, 454
Boltzmann, Ludwig, 117–118
Boltzmann's constant, 118
bomb calorimeter, 535, 536
bond angles, 301, 336, 355, 355–361, 363t
bond dipole, 371–372
bond energy, 532–534
   average, 532–533, 535
   and bond order, 534
   of covalent bonds, 533t
bond formation, 315t, 532–534
bonding, chemical, 284–324
   and electronegativity, 293
   and valence-bond theory, 363–370
   see also metallic bonding
bonding molecular orbital, 316
bond order, 320
   and bond energy, 534

bonds, chemical, 286–287, 827
   see also covalent bonds; metallic bonds; polar covalent bonds
bond length, 315t, 316, 345, 375,
   and bond strength (energy), 315t
bond vibration modes, 334–447
bone marrow, 81
Bonneville salt flats, 167
borax, 910, 911
boric acid, 910, 911
borides, 911
Born, Max, 123, 150
boron, 65, 136
   chemistry of, 910–911
   electron configuration of, 136
   hybrid orbitals for, 368
   ionization energy of, 144–145
   in universe, 93
boron compounds, 911
boron-11, 65, 93
boron nitride, 911
boron trifluoride, 368, 795
boron trioxide, 911
Bosch, Karl, 754
boundary-surface representation, 126–127, 130
Bowen, Ira, 109
Boyle, Robert, 394
Boyle's law, 392–394, 411
Bragg equation, 471–472
brass, 903
breathing, 411–412
breeder reactor, 75
bristlecone pine
   dating of, 92
Broglie, Louis de, 119–120, 121, 122
Broglie, de, equation, 119
bromcresol green, 819
bromine, 307, 308, 864
Brønsted, Johannes, 790
Brønsted-Lowry acids and bases, 790, 795, 828
bronze, 901–903
Bronze Age, 901–902
Brownian motion, 919
brucite, 254
Buckminsterfullerenes (bucky balls, fullerenes), 490
buffer capacity, 813–815
   see also alkalinity
buffer solutions, 813–815
   and pH, 813–815
Bunsen, Robert Wilhelm, 106
buret, 267, 268
butane, 516, 517t, 567
2-butanone, 449
butene, 598

cadmium, 63, 864, 865
cadmium-110, 63
calcite, 484
calcium, 254, 255
   electron configuration of, 139
   ionization of, 118, 146
   line spectra of atoms and ions of, 109

calcium carbide, 492
calcium carbonate, 170, 254, 256, 484, 806, 811
   solubility of, 824–825
calcium chloride, 613
calcium fluoride, 825
calcium hydrogen carbonate (calcium bicarbonate), 252, 806, 811
calcium hydroxide, 829
calcium hypochlorite, 307
calcium oxide (lime), 252, 254, 297, 349, 767–769, 905
calcium phosphate, 254
   solubility of, 826
calcium silicate, 905
calcium sulfate, 349, 806
calcium sulfide, 168
calcium sulfite, 349
California
   and gasoline additives, 574
californium-254, 78
calorimetry, 535–536
cancer
   and radiation therapy, 86–87
candela, 19t
capillary action, 458–459
caprylic acid, 644
caratenoid, 832
carbohydrates, 576–577, 630–631
   energy values of, 645
   fuel value of, 646
   handedness of, 639
carbon
   allotropes of, 489–490, 492
      entropies of, 619, 620
   chemistry of, 492
   electron configuration of, 136–137
   in fossil fuels, 511–512
   and fuel value, 540–541
   hybrid orbitals of, 364, 366, 367
   ionization energy of, 145
   isotopes of, 60, 61, 62, 63, 64–65, 67t, 93
   phase diagram of, 490
   and reduction of iron ore, 904–906
   in steel, 908–909
   see also Buckminsterfullerenes; diamonds; graphite; hydrocarbons
carbonates
   and alkalinity, 817–818
   and ionization, 806–808
   and neutralization reactions, 806–807
carbon cycle, 190–192
carbon dioxide, 492, 543–544
   and absorption of infrared radiation, 334–337
   bond dipole of, 371
   bond vibration modes of, 376
   concentration of in air, 343t
   and Earth's temperature, 337, 339–340
   and global warming, 456
   infrared spectrum of, 335
   molar mass and density of, 403
   on Venus, 337
   phase diagram of, 456
   and radiocarbon dating, 90–92

structure of, 311–313, 358–359, 367
  sublimation of, 492
carbon-11, 64–65, 67t
carbon-14, 62, 63, 67t
carbonic acid, 252–253, 790, 803–804,
      806, 811–813
  ionization of, 806–808
  ionization equilibrium for, 789t
carbonic acid–bicarbonate system, 817–819
carbon isotopes, 67t
carbon monoxide, 182, 543–544
  and auto exhaust, 713–716
  Lewis structure of, 340–341
  and nitrogen dioxide, 704
  and reduction of iron ore, 904, 905
  and smog, 672
  spectrum of, 324
carbon-oxygen bond, 334–336
carbon tetrachloride
  freezing and boiling points of, 240t
carbonyl group, 578
carboxylate anion, 587, 588
carboxylate group, 265
carboxylic acids, 587–588, 631
Carothers, Wallace Hume, 933
casserite, 902
cast iron, 904
catabolism, 630, 631, 654
  of glucose, 645–646
catalysis, 712–716
catalysts, 311, 754
  in chemical equilibrium, 756–757
  heterogeneous, 715, 716
  homogeneous, 713
catalytic converters, 703, 713–716, 756,
      765, 766
cathode ray tubes (CRTs), 925
cathodes, 846
  in lead-acid battery, 849
cations, 6, 167, 287
  electron configurations of, 142
  and transition metals, 169
caves
  limestone, 252
cell potential, see electromotive force;
      standard cell potentials
cellulase enzymes, 586
cellulose, 583–586, 631
  fibers, 929
Celsius scale, 35, 36
central atom, 299
ceramics, 918–926
  superconducting, 921–926
cerium, 923–925
cerium(IV) oxide, 716
cesium, 106, 178
cesium chloride, 481
chain reaction, 71, 72, 73
chair form (of cyclohexane), 565
chalcocite, 894, 899
chalcogens, see Group 6A elements
chalcopyrite, 894, 899
charcoal, 512, 541
Charles, Jacques, 399–400
Charles's law, 400

chemical bonding, see bonding, chemical
chemical energy, 515
chemical equations, 173
  completing and balancing, 181–189
  and equilibrium constants, 742
chemical equilibrium, see equilibrium,
      chemical
chemical formulas, 166
chemical industry, 539
chemical properties, 2
chemical reactions, see reactions, chemical;
      and entries under reaction
chemical weathering, 250
  of feldspars, 258–259
chemistry
  defined, 2, 3
Chernobyl explosion (1986), 80, 82–83
chiral centers, 635, 638
chiral molecules, 635
chlorates, 308
chloride
  molarity of in seawater, 232–233
  oxidation of, 241–242
chlorine, 307–309
  oxoacids of, 800
  oxoanions of, 171
  and ozone, 310–311, 712–713
chlorine monoxide, 310, 311
chlorofluorocarbons, 712, 940
  and Lewis structures, 306
  and ozone, 305–310
chlorophyll, 832
chromium, 141, 641, 658
  and stainless steel, 909
chromium(III) hydroxide, 829
cinnabar, 864
cis isomers, 598
citric acid cycle, see Krebs cycle
clays, 918
Clean Air Act, 715
cleaning solvents, 558
climate
  of Earth, 337–340
closest-packed structure (of a crystal), 477
clothing fibers, 929–940
coal, 267, 511, 512, 590–591
  elemental composition of, 591t
coastal zones
  fresh water and seawater, 219–220
cobalt, 701, 702
codons, 660
cohesive forces, 457–458
coinage metals
  chemistry of, 898–899
coke, 492, 701, 904–905
cold cranking amps (CCA), 872
cold packs, 612, 617, 619–620
colligative properties, 228–236
collisions, elastic, 421
collisions, molecular
  of gases, 420–422
colloidal suspensions, 919
colloids, 269–270
color, 2
colorimetric pH indicators, 816–817

columbite, see niobite
combustion, 183
  and energy transfer, 518–522
  enthalpy of, 528
  measuring heat of, 535–536
  molar heat of, 534
combustion analysis, 593–595
combustion reactions
  energy of, 531–534
  in gasoline engines, 566
common-ion effect, 811
communications satellites, 42
compact disk (CD), 18
complex ions, 827–832
compounds, 2
  binary ionic, 167–168
  binary molecular, 166
  composition of, 164–165
  empirical formulas of, 186–188, 189
  molar mass of, 175
  naming of, 166–172
  and oxidation number, 244
concentration, 214–219, 221
  effect on standard potentials, 863–870
  calculations of, 216–220
  in lithium-ion battery, 862–863
  and physiological saline, 225
  and reaction rates, 679–695
condensation polymers, 937t
condensation reactions, 582–583
  synthetic polymers from, 933–936, 937t
conduction bands, 491, 494, 901, 920
  and semiconductors, 926–928
conductivity
  of ceramics, 920, 921–926
  of copper, 900
  of semiconductors, 926–928
conjugated double bonds, 632
  in porphyrin rings, 832, 833
conjugate pairs, 795–796, 807
conservation of energy, law of, 515
conservation of mass, law of, 180
constructive interference, 16, 17
continuum radiation, 12, 43–45
Cooper pairs, 921–922
coordination number, 473–474
Copernicus, Nicolaus, 9, 10
copper, 894–896, 897, 903–904
  and aluminum alloys, 915
  and bronze, 901–902
  chemistry of, 898–899
  conductivity of, 900
  electron configuration of, 141
  melting point of, 903
  reaction with zinc, 844–846
  smelting of, 898
copper compounds, 169
copper/tin bronzes, 901–902
copper(II) oxide, 894
copper(II) sulfate, 829
corn starch
  ethanol from, 576, 583
corona, 104
coronium, 109
corrosion-resistant materials, 915

cosmic microwave background radiation, 42, 45, 46
cosmic radiation, 83, 84
cotton, 929
coulomb, 32t, 373, 848, 871
Coulomb's law, 439
covalent bonds, 288, 292, 533t
  and bond lengths and strengths, 315t
Crick, Francis H. C., 655
cristobalite, 485, 486
critical mass, 71
critical pressure and temperature, 455, 922
crude oil, 511, 512, 558, 559
  and vanadium, 640
cryolite, 912
crystal field splitting, 495–500
crystal field splitting energy, 495–500
crystal field theory, 495–500
crystal lattices, 467–472
crystalline solids, 467–471
crystals
  dislocations in, 897
crystal structures, 474–484
cubic closest packing (ccp), 477–480, 482–483
cubic holes, 480t, 481
curie, 79, 80t
Curie, Marie, 78, 79, 80, 350
Curie, Pierre, 79, 350
cyanide, 829
  and gold mining, 4, 898
cyclic alkanes, 565
cyclization reaction, 578–579
cyclohexane, 565
cyclo-octasulfur, 490
cyclotron, 74, 76
Cygnus Loop nebula, 69
cysteine, 932
cytochromes, 833
cytosine, 655, 656, 657

D5W, 226
Dalton, John, 165
Dalton's law (of partial pressures), 406, 408
dating
  radiocarbon, 90–93, 689
  radiochemical, 89–93
Davisson, Clinton J., 121, 122
Davy, Sir Humphrey, 911
debye, 373
decane, 562t
decomposition (of matter), 558
definite proportions, law of, 165
degenerate orbitals, 129
degree of dissociation, see degree of ionization
degree of ionization, 785
deionized water, 266
delocalization (of molecular orbitals), 316
density, 2
dental amalgams, 904mn
deoxyribonucleic acid, see DNA
deoxyribose, 655, 656
desalinization, 229, 230–231
destructive interference, 18
detritus, 190

deuterium, 38, 56
deuteron, 35
diamagnetism, 321
diamonds, 489–490, 492
  artificial, 490
  entropy of, 619, 620
diaspore, 909, 912
diborane, 911
Dicke, Robert, 42
diesel fuel, 558, 573
diffraction, 16–18
  see also angle of diffraction; X-ray diffraction
diffusion (of a gas), 419, 421
dilution, 220–221
  of acid and pure water, 812
dimethylbenzene (xylene), 571, 573t
dimethyl mercury, 865
dinitrogen oxide, see nitrous oxide
dinitrogen pentoxide, 181
dipole-dipole interactions, 445–446
dipole-induced dipole interactions, 449, 450
dipole moments, 372–373, 376
dipoles, temporary, 446
diprotic acids, 789t, 790
disaccharides, 582, 630
dispersion forces (London forces), 135, 446–447
dissolution process, 612–615
distillation, 6–7
  fractional, 558, 559
  of seawater, 7, 230–231
DNA, 655–660
DNA adducts, 658
Dobson units, 306
dodecane, 562t
dolomite, 254
Dolomite Range (Italy), 254
doping (of semiconductors), 927, 928
Doppler effect, 20
double helix, 657
Dover, white cliffs of, 170
drain cleaner, 179
Drake, Edwin L., 512
drilling muds, 254
dry ice, 8
dubnium, 77
ductility, 2
DuPont, 933, 940
dynamic equilibrium, 733

E. coli
  and electricity, 883
Earth
  and asteroid collisions, 163, 164
  atmosphere of, 392–400
  atmospheric pressure of, 388–391
  carbon dioxide absorption band of, 337
  and climatic conditions, 337–340
  composition of, 162–164
  dating age of, 92, 93
  early atmosphere of, 160, 164, 173, 181, 182, 183, 190, 209
  emission spectra of surface of, 336–337

  and oceans and life, 208
  primordial, 173, 181, 182, 183, 209
  and water, 209–211, 212
Earth's crust
  and alkali metals, 178
  carbon in, 492
  common elements in, 258t
  Group 8B metals in, 701
  nitrogen in, 414
  phosphorus in, 414
  silicon in, 492
earthenware, 918
eclipses, solar, 104
effective nuclear charge, 127, 294–295
effusion (of a gas), 419–420
eicosene, 239
Einstein, Albert, 31, 110, 111, 118–119, 121, 149, 150
elastic collisions, 421
elasticity, modulus of, 931
elastomers, 933
electrical energy (of batteries), 856
electricity
  conduction of, 223
  salt and conduction of, 222–223
  water as conductor of, 221–223
electric vehicles, 878–881
electrochemical cells, 243, 846, 883
  see also voltaic cells
electrochemical energy sources, 851t
electrochemistry, 844
electrodes, 223, 846
  in lead-acid battery, 849
  and oxidation-reduction reactions, 243
  see also anodes; cathodes
electrolysis, 243, 878
  and battery charging, 873
  of molten salts, 878
  of water, 876
electrolytes, 222, 223
  and van't Hoff factor, 237, 239
  weak, 223
electrolytic cells, 243, 873–878
electromagnetic radiation, 12–21
  energies of, 18
  and line spectra, 106, 109
  photons of, 110
  wavelike and particle-like properties of, 19
electromagnetic spectrum, 12–18
electromagnetism, 32
electromotive force, 847–848, 856–857
electron affinity, 295–296
electron capture, 64, 65, 69
electron configurations, 132–143
  abbreviated, 136
  and ionization energies, 144–148
  and periodic table organization, 140
electron density, 126, 127
electron diffraction, 121–122, 315
electron-distribution profiles, 128
electronegativity, 293–297
electron-electron repulsion, 439, 440
electron pairs
  and acids and bases, 794–795
  sharing of, 286–293

Index

electrons, 32
  and atomic orbits, 126–131, 132
  and atomic structure, 106–109
  and the Big Bang, 33, 35
  ejection of, 146–147
  and Heisenberg's uncertainty principle, 150
  in hydrogen atoms, 112–113
  orbits of, 113, 118
  paired, 321
  properties of, 32t
  spin of, 118, 131–132, 501–502
  unpaired, 352–354
  see also orbitals; photoelectrons; valence electrons
electron spin, 501–502
  flip of, 353
  in odd-electron molecules, 352, 353
electron-spin resonance, 352–354
electron-spin resonance spectrometers, 353
electron transfer, 241–242, 287
electron waves, 120–122
electroplating, 878
elementary steps (in reaction mechanisms), 698
elements, 2, 3
  allotropes of, 298
  artificial, 74–78
  classes of, 2–4
  and electron affinity, 295–296
  and electron configuration, 132–133
  electronegativity of, 293–294
  emission and absorption line spectra, 106
  families of, 142
  main-group (representative), 293–294, 295, 296
  and oxidation number, 244
  physical and chemical properties of, 2, 3
  standard state of, 537
  superheavy, 77–78
  transmutation of, 74
  transuranic, 74
elements, theory of the four, 8
emeralds, 494–498
emission lines, 106, 109
emission spectra
  of free atoms and ions, 109
empirical equations, 110
empirical formulas, 186–188, 189, 594
empirical information, 186
enantiomers, 635, 638–642
endergonic reactions, 625
endothermic processes, 519, 524, 614, 616, 622, 623, 706, 711, 764
energy
  activation, 704–711
  of combustion reactions, 531–534
  defined, 513–514
  electrochemical sources of, 851t
  and fossil fuels, 512
  and frequency, 18
  and matter, 31–35
  of X rays, 146–147
  see also binding energy; chemical energy; internal energy; ionization energies; kinetic energy; potential energy

energy, law of conservation of, 515
energy transfer, 518–522
  see also thermodynamics
energy units
  conversion factors for, 848
engine exhaust, 677
engine knock, 566
enthalpy, 523–524, 537
  of solution, 612–615
enthalpy changes, 523–527, 532–540, 542–543, 628, 765
enthalpy values, 537
entropy, 616–620, 622
  at absolute zero, 618
  calculations of, 621
  standard molar, 618
entropy change, 617, 628
  of the universe, 617
Environmental Protection Agency, 715
Epsom salts, 254
equations
  empirical, 110
  molecular, 256
  net ionic, 243, 256
  thermonuclear, 542–543
equations, chemical, *see* chemical equations
equilibrium, chemical, 251, 627, 732–736, 747
  and battery cell reactions, 867
  catalysts in, 756–757
  and free energy, 747–750
  in the gas phase, 743–747
  and reaction quotient, 738
  and stress, 751–756
equilibrium, dynamic, 733
equilibrium calculations, 757–760
equilibrium constant expressions, 735, 739, 743–747
  for reactions in solutions, 793
  for reactions in the gas phase, 757–760, 762
equilibrium constants, 735, 736–742, 746mn, 747–750, 757–760, 768
  acid-ionization, 783, 790
  and autoionization of water, 801–802
  of cell reactions, 868–869
  and formation constants, 830
  and free energies, 747–750
  overall, 740–741
  and reaction quotients, 739t
  relating K_c and K_p, 744–745
  and solubility products, 824–825
  and temperature changes, 763–766
equivalence point, 267
essential amino acids, 632, 633t
esters, 642
ethane, 517, 421, 588, 595
  fuel for hot-air balloons, 521–522
ethanol, 222, 223, 574–575
  as auto fuel, 576
  energy value of, 586
  freezing and boiling points, 240t
  molecular structure of, 574
  production of, 583
  van't Hoff factor for, 237
ethene, 597, 598

ethers, 573
ethyl alcohol, *see* ethanol
ethylene, 189, 597
ethyne, *see* acetylene
Europa, 334
europium, 923, 924
excited-state electron configurations, 138, 322, 323
excited states, 116, 117–118, 286, 290, 296
exergonic reactions, 625
exosphere, 286
exothermic processes, 519, 521, 524, 532, 538, 614, 616, 622–623, 706, 707, 711, 763
expanded-octet molecules, 359–361
explosives, 414, 755

fabrics, 929
face-centered cubic (fcc) unit cell, 473, 475, 479, 480, 483, 906
  of copper, 896, 897
Fahrenheit scale, 35, 36
Fajans, Kasimir, 288
Faraday, Michael, 286, 848
Faraday constant, 848
fats, 642–643, 644, 646–647, 654
fatty acids, 642–643
feedstocks, 512
feldspar, 264
  chemical weathering of, 258–259
Fermi, Enrico, 77
ferrite, 906
fertilizers, 414, 415, 755, 798
fibers for clothing, 929–940
  natural, 929–932
  synthetic, 933
fibrous proteins, 929–930
filtration, 5–6
  of seawater, 212
fireworks, 308
firing ceramics, 918
first law of thermodynamics, 518
first-order reactions, 685–688, 689, 694–695
flame retardants, 415
flares, 139
flax, 929
fluoride, 825
fluorinated polymers, 940
fluorine, 135, 307
  electron affinity of, 295
fluorite, 479, 480, 825
fog
  particles in, 269
food, 630–632
Food and Drug Administration, 630
food labels, 647
fool's gold, *see* iron pyrite; pyrite
forces (of nature), 32
formal charges, 312–313
formaldehyde, 578
  infrared spectrum of, 341
  Lewis structure of, 340–341, 357, 358
formation constants, 830
formic acid, 587–588

formulas
  chemical, 166
  empirical, 186–188, *189*, 594
formula unit, 177
fossil fuels, *191*, 511–512
  see also coal; crude oil; natural gas; oil
433 Eros (asteroid), *162*
Fowler, Joanna S., 88
fractional distillation, 558, *559*
francium, 178
Fraunhofer, Joseph von, 20, 105
Fraunhofer lines, 20, 105–106, 108–109, 117–118, 134
free energy, 624–625, 626–627, 628–629
  and electrical charge, 848–849
  and chemical equilibrium, 627, 747–750
  of formation, 629
  and the meaning of "free," 628
  and temperature, 763
freezing-point depression, 234, 236, 239
  for selected solvents, 240*t*
freezing points, 234–235
frequency, 14–15
  and energy, 18
frequency factor (of molecular collisions), 705
fresh water
  major ions in, 214–216
  mixed with seawater, 219–220
  pH of, 806, 819
fructose, 581–582
fuel cells, 851*t*
  biochemical, 882–883
  and electric vehicles, 879–881
fuel values, 540–541
Fuller, R. Buckminster, 490
fullerenes, *see* Buckminsterfullerenes
functional groups, 573–574, 631*mn*
fusion (melting process), 525, 526
  see also helium fusion; hydrogen fusion; nuclear fusion

gadolinium, 923
Galilei, Galileo, 9–10, *390*
Galileo spacecraft, *332*
gallium, *927*, 928
  chemistry of, 910, 911
  crystal structure of, *494*
gallium arsenide, 881, 911
gallium indium phosphide, 881
gamma rays, 83
  and radiation dosage, 81, *82*
gas (as state of matter), 438–439
gas constant, 400–401
  values for, 401*t*
gas density, 403–405
gas giants (planets), *162*
gases, 519
  effusion and diffusion rates of, 419–420, 421
  and elastic collisions, 421
  and chemical equilibrium, 743–747
  and Henry's law, 411–412
  kinetic molecular theory of, 417–418, 421
  and molecular collisions, 420–422
  partial pressure of, 411–412
  in phase diagrams, 454–455

pressure of, 392–395
properties of, 386–425
real, 422–424
root-mean-square speed of, 417–419, *420*, 421
solubility of, 411–412
total pressure of and equilibrium, 752–753
van der Waals constants of, 424*t*
  see also ideal gases; natural gas; noble gases
gasohol, 575
gasoline, 558, 561, 563, 564
  additives, 561, 573–576
  combustion of, 628
  high-octane, 569
  octane number of, 566
  reformulated, 573
  and viscosity, 459
Gay-Lussac, Joseph-Louis, 911
Geiger, Hans, 78–79, 108
Geiger counter, 78–79
gemstones, 494–501
geometric isomerism, 598, 635
geothermal vents, 246, *247*
Gerlach, Walther, 132
germanium, 493
Gesellschaft für Schwerionenforschung (Society for Heavy Ion Research), 77, 78*t*
Gibbs, J. Willard, 624
Gibbs free energy, *see* free energy
gibbsite, 251, 909, 912
glass, 469
global warming, 337–340, *456*, 591
glucose, 223, 576
  catabolism of, 631, 645–646
  and cellulose, 586
  conversion of to pyruvate, 653–654
  conversion of to starch, 583
  cyclic form of, 578–580, 584
  energy value of, 645–646
  and ethanol production, 583
  open-chain, 578–580
  and phosphorylation, 648–649
  and photosynthesis, 193–194
  structure of, 577–580, 638
  see also glycolysis
glycerol, *643*
glycine, 660
glycolysis, 648, 649–654
glycosidic bonds, 584–585, 586, 929
gold, 2, 3, 4, 28, 108, 894, 898
Goudsmit, Samuel, 131
Graham, Thomas, 420
Graham's law of effusion, 420, 940
grain alcohol, *see* ethanol
gram, 24
Grand Canyon, *210*
granite
  weathering of, *259*
graphite, 489, 490, 492
  entropy of, 619, *620*
gravity, 32
Great Salt Lake, *167*
green plants
  and photosynthesis, 832

greenhouse effect, 334, *335*, 396
greenhouse gases, 334, 337, 338, *339*, 340, 688
greenockite, 864
ground states, 116, 117–118, 286, 298, 322–323, 327, 496
Group 1A elements, 439
Group 1B elements, 898–899
Group 2A elements, 439
Group 2B elements, 864–865
Group 3A elements, 910–911
Group 3B elements, 923–925
Group 4A elements, 492–493, 916
Group 4B elements, 916–917
Group 5A elements, 414–416
Group 5B elements, 640–641
Group 6A elements, 348–350
Group 6B elements, 658–659
Group 7A elements, 307–309
Group 7B elements, 650–651
Group 8B elements, 701–703
guanine, 655, *656*, 657
Guldberg, Cato, 735
gunpowder, 414
gypsum, 254, 349

Haber, Fritz, 414, 754, 755
Haber-Bosch process, 414, 754–755
hafnium, 592, 916–917
half-life, 33, 34, 688, *689*
  and radiochemical dating, 89–93
  of first-order reaction, 688–689
  of second-order reactions, 693–694
half-reactions, 242, 844, 845, 846, 855*mn*, 856
  potentials of, 870*mn*
  standard potentials of, 856–857
halide ions, 260
halite, 167
Hall, Charles, 912, *913*
Hall, Julia, *913*
Hall-Héroult process, 912, *913*, 914
halogens
  chemistry of, 307–309
  melting and boiling points of, 447*t*
Handbook of Chemistry and Physics, 257
hardness, 2
hard water, 265
hassium, 78*t*
hausmannite, 650
heat, *see* combustion; specific heat
heat capacity, 526, 535–536
heat flow
  in changes of states of matter, 519–520
  and enthalpy, 523–524
  and melting snow, 525–526
  into water, 524–530
  and work on a system, 521
heat of formation, 537
heat of hydration, 615
heat of solution, 612
heat of reaction, 539
heat transfer, 528–531
heavy water, *73*
heavy-water reactors, *73*
Heisenberg, Werner, 150
Heisenberg's uncertainty principle, 150

helium, 5, 106, 134
　atomic mass of, 41
　buoyancy of, 133
　concentration of in air, 343t
　density of, 403
　effective nuclear charge in, 129
　electron configuration of, 133
　formation of, 46–47, 58
　in hot-air balloons, 521
　molecular-orbital diagram of, 317
　in universe, 40–41, 93
helium fusion, 56
helium-3 nuclei, 56
hematite, 184, 186, 701
heme groups, 833
hemiacetal, 578–579
hemicellulose, 585
hemoglobin, 411–412, 502, 833
　oxygen-binding sites on, 412
　and oxygen transport, 756
Henderson-Hasselbalch equation, 814–815
Henry's law, 411–412
Henry's-law constant, 411
heptane, 560, 562t
herbicides, 415
Héroult, Paul-Louis-Toussaint, 912, 913
Herschel, William, 12
Hess's law, 543
heterogeneous catalysts, 715, 716
heterogeneous equilibria, 767–769
heterogeneous mixtures, 3, 5, 211, 212
hexagonal closest packing (hcp), 477, 480t, 482–484
hexagonal unit cells, 483
hexanal, 577
hexane, 562t
high-alloy steel, 908
high spin electron configurations, 501–502
Hindenburg explosion (1937), 133
Hiroshima bombing (1945), 71
histidine, 833
homogeneous alloys, 901, 903
homogeneous catalysts, 713
homogeneous mixtures, 3, 5, 210, 211, 212
homologous series, 563
hot packs, 613
hot springs, 304
　and radium, 86
Hubble, Edwin Powell, 11
Hubble Space Telescope, 515
Hund, Friedrich, 316
Hund's rule, 137, 495
Hurricane Andrew, 392
hybrid atomic orbitals, 363, 364–370
hybridization, 368–370
hybrid vehicles, 879
hydration, 615
　heat of, 615
　of ions, 445, 615t
　sphere of, 445
hydride ions, 245
hydrocarbons, 515–516
　aromatic, 571–573
　in crude oil, 558
　and emission reduction, 715
　of natural gas, 541t

　polycyclic aromatic, 573
　see also alkanes; alkenes; alkynes; cycloalkanes
hydrochloric acid, 251, 790–791
　acid-ionization reaction of, 796
hydrogen
　atomic mass of, 41
　atomic orbitals of, 317
　compounds of, 4
　in fossil fuels, 511
　as fuel, 591–592
　and fuel values, 540–541
　molecular orbitals of, 316–317
　oxidation number of, 245
　physical and chemical properties of, 5
　production of, 881
　spectral lines of, 117
　in universe, 40–41, 93
　see also water
hydrogen, liquid, 591
hydrogen atom, 592
　Bohr model of, 112–118, 120, 121
　energy states in, 116
　orbitals, 127–129
hydrogen bonds, 446, 458, 460
　and cotton fibers, 929, 930
　and Nylon, 933, 934
　and water, 457
　and wool fiber, 931–932
hydrogen cyanide, 289
hydrogen fluoride
　dipole moment of, 372
hydrogen fusion, 55–57, 58
hydrogen gas
　storage of, 592, 880–881
　synthesis of, 542–543
hydrogen ions, 256
　and acid-base neutralization reactions, 250–251
　reduction of, 875
hydrogen isotopes, 38, 40
hydrogen spectrum, 114–115, 131
hydrogen sulfide, 246–248, 348
hydrolysis, 583
hydrolytic bacteria, 587
hydronium ion, 791, 797, 812
hydrophilic interactions, 449
hydrophilic substances, 269–270
hydrophobic interactions, 449
hydrophobic substances, 269–270
hydroxide ions, 256
hydroxides, 260, 792
β-hydroxybutyric acid, 814
hydroxyl, 577–578
hypobromous acid, 800, 801
hypochlorous acid, 800, 801
hypohalous acids, 800, 801
hypoiodous acid, 800, 801
hypothesis, 9, 10

ice, 460
　melting of, 618, 625, 626
　molar heat capacity of, 526–527
Ice Age, 337
ICE tables, 757–758, 784, 787, 804, 808
ideal gases, 401
　behavior of, 422, 423

　density of, 403–405
　molar volume of, 403
ideal-gas law, 401, 402, 404, 409, 423, 743
ideal solutions, 561
illites, 918
ilmenite, 916
imaging, medical
　with radionuclides, 87–89
immiscible liquids, 449
incidence, angle of, 470
indigo, 216
indium, 494, 910
indium-115, 63
induced-dipole–induced-dipole interactions, 447
infrared radiation, 13, 340–342, 376
　and carbon dioxide, 334–337
　and greenhouse gases, 337
　and water, 336
infrared spectrophotometer, 340, 341
infrared spectroscopy, 323, 340–342
integrated rate law, 685, 688
intermediates (in reaction mechanisms), 46, 698
intermetallic compounds, 903–904
intermolecular interactions, 423–424, 438–461
　and dipole–dipole interactions, 445–446
　and ideal solutions, 561
　and phase diagrams, 454–456
　and water, 457
internal combustion engines, 576
internal energy, 520–521
International Union of Pure and Applied Chemistry, 77
interstices in crystals, 592, 906
interstitial alloys, 906
interstitial metal hydrides, 592
intramolecular reactions, 578
iodine, 307, 308
ion-dipole interactions, 445–446
ion-dipole intermolecular force, 445
ion exchange, 263–266
　and water purification, 265
ionic clusters, 236
ionic compounds, 6, 287, 439–443
　and conduction of electricity, 222–223
　lattice energies of, 440t
　solubility of, 257, 259–261, 445, 825–826
ionic equilibrium
　and common-ion effect, 811
ionic solids, 441–442, 445
　dissolution of, 613–614
ion-ion interactions, 439–441
ionization, 84
　of acids, 789–790, 791
　base-ionization reaction, 793
　of bases, 792–793
　degree of, 785
　of sulfuric acid, 786–788
ionization energies, 144–148, 295
ionization equilibria
　for diprotic acids, 789t
ionizing radiation, 80–82, 83–84
　effects of, 82t
　in U.S., 84
ion pairs, 236, 237

ions, 6
    common-ion effect, 811
    complex, 827–832
    and electrolytes, 223
    heats of hydration of, 615t
    hydrated and solvated, 445
    isoelectronic, 136
    molecular, 290–291
    names and charges of, 170t
    packing efficiency of, 476–484
    polyatomic, 169–173, 244, 484
    spectator, 243, 256
iridium, 701
iron, 512, 701–702
    melting point of, 904
    see also cast iron; wrought iron
iron alloys, 895
    stainless steel, 909
    see also steel
iron carbide, 906, 908
iron ore, 184, 186, 701
    reduction of, 904–906
iron oxides, 701
iron-porphyrin complexes, 833
iron pyrite, 348–349
Iron Range (Minnesota), 184, 701
Iron Revolution, 904
iron smelting, 904–906
iron sulfide, 267
iron(III) compounds, 828, 829
isobars, 391
isobutyl, 568
isoelectronic atoms and ions, 136
isomerization reactions, 569
isomers
    cis, 598
    geometric, 598, 635
    right (R) and left (S), 637, 639
    structural, 567, 635
    trans, 598
isooctane, 566, 567–568
isopropyl, 568
isotactic polymers, 938–939
isotonic solutions, 228
isotopes, 38
    artificial, 74–78

Joliot-Curie, Frédéric, 74
Joliot-Curie, Irène, 74, 81
joule (J), 18, 19t, 848, 871

kaolinite, 259, 263, 264, 489, 918–919
Kelvin scale, 19t, 36, 37
kerogen, 558
kerosene, 558
ketohexose, 581
ketones, 578, 581, 638–639
keto sugar, 581
Kevlar, 934, 935, 937t
kinetic energy, 146–147, 514, 520, 521
    of a gas, 417, 421
kinetic molecular theory, 417–418, 617, 679
    and activation energies, 705
Kirchoff, Gustav Robert, 106
Krebs cycle, 648, 653–654

krypton, 134, 135, 137
    concentration of in air, 343t
    electron configuration of, 137
krypton-91, 72

lactate/lactic acid, 652, 840
    in muscles, 637, 654
lanthanide (rare earth) elements, 142
    chemistries of, 923–925
    densities of, 924
    ionization energies of, 925
    melting and boiling points of, 925
lanthanum, 923
laptop computers, 860
lattice energy, 440–444, 615
lattice points (of a crystalline array), 472
lauric acid, 644
Lavoisier, Antoine, 348
law of conservation of energy, 515
law of conservation of mass, 180
law of definite proportions, 165
law of mass action, 735, 736
law of multiple proportions, 165
Le Châtelier, Henri Louis, 751
Le Châtelier's principle, 751–756
lead, 493
lead acetate, 260
lead-acid batteries, 849, 850, 851t, 871, 872
    electric cars, 879
    recharging, 873–876, 877
lead nitrate, 260
lead oxides, 493
lead sulfide, 260
lead-214, 70
Leclanché, Georges, 850
Lemaître, Georges, 10
length
    equivalent units of, 28t
leukemia, 81
leukocytes, 81
Lewis, Gilbert N., 287–288, 292, 795
Lewis acids, 795, 827
Lewis bases, 795, 827
Lewis structures, 290–292, 298–311
    and chlorofluorocarbons, 306
    and formal charge, 312
    of ozone, 299–301, 304
Lewis symbols, 289–290
Libby, Willard, 90, 91
life
    dependence on water, 210
ligands, 827
light
    line sources of, 12
    Newton's particle view of, 19
    particles of, 110–112
    speed of, 24, 118–119
    wavelike properties of, 11–21, 14
light emitting diodes (LEDs), 928
lightning, 304
light scattering
    and particles, 269
    and turbidity, 211
light-year, 24
lignins, 585, 586, 589

lignite, 590, 591t
lime, see calcium oxide
limestone, 252, 767–769, see also calcium carbonate
    and smelting iron, 904, 905
    and neutralization of natural waters, 806, 807
limiting reactant, 194–196
linear accelerator, 74, 75
linear molecular geometries, 358–359, 361, 363t, 367
linen, 929
line spectra, 106, 109
    see also emission lines; Fraunhofer lines
linolenic acid, 643, 644
lipids, 642–643, 644
liquid(s), 438–439, 519
    immiscible, 449
    meniscus of, 457
    miscible, 448
    and phase diagrams, 454–455
    surface tension of, 457, 458
    viscosity of, 459
liter, 24
lithium, 129, 178
    electron configuration of, 136
    ionization energy of, 144
    in universe, 93
lithium alloys, 915
lithium-ion batteries, 851t, 862–863
lithium salts, 863
Lockyer, Joseph, 106
London, Fritz, 446
London dispersion forces, see van der Waals forces
London forces, see dispersion forces
lone pairs (of valence electrons), 290
low-alloy steel, 908
Lowry, Thomas, 790
low-spin electron configurations, 501–502
lubricants, 558
lung cancer, 85
lungs
    and breathing, 411–412
luster, 2
lutetium, 923
Lycra (spandex), 933, 935
Lyman series of spectral lines, 114–115
lysine, 632

magnesium, 254, 255
    and aluminum alloys, 915
    electron configuration of, 139
    ionization energy of, 146
magnesium fluoride, 168
magnesium-24, 61
magnetic quantum number, 124, 125
magnetic resonance imaging (MRI), 641
magnetite, 186, 498, 701
main-group elements, 293–294, 295, 296
    electron affinity values of, 296
    electronegativities of, 295
    ionic radii of, 439, 440
malachite, 894, 899

Index

malleability, 2
  of metals, 896, *897*
manganese
  electron configuration of, 142
  chemistry of, 650
Manhattan Project, 940
Mars
  life on, 334
  polar caps of, 324
  and water, 210
Marsden, Ernest, 108
mass
  equivalent units of, 28*t*
  and moles, 177
mass, law of conservation of, 180
mass action, law of, 735, 737–738
mass defect, 58
mass numbers, 32
matter
  classes and properties of, 2–8
  creation theories of, 9–11
  Dalton's atomic theory of, 165
  and energy, 31–35
  states of, 438, 519–520
matter waves, 118–122
Maxwell, James Clerk, 13, *14*
McHarness, Robert, *940*
mean free path, 421
measurements, 23–31
  uncertainty of, 25–26
  see also SI units
mechanisms of reactions, *see* reaction mechanisms
Meissner effect, 926
meitnerium, 78*t*
melting points
  and phase diagrams, 454–455
melting process (fusion), 525, 526
Mendeleev, Dimitri, 143
Mendeleev's periodic table, 143
  see also periodic table of the elements
meniscus, 457, *458*
Mercury (planet)
  atmosphere of, *162*
mercury
  chemistry of, 864, 865
  crystal structure of, *494*
  and dental amalgams, 904*mn*
mercury button battery, 851*t*
mesosphere, *286*
messenger RNA (mRNA), 660–661
metal(s), 894
  band theory of, 491
  coinage, 898–899
  crystal structures of, *494*
  malleability of, 896, *897*
  properties of, 2, 3–4
  and smelting, 894
  work-hardening of, 896–900
  see also transition metals
metal complexes, 827–833
  in biomolecules, 831–833
metal hydrides, 592
metal ions
  acid-ionization equilibrium constants for, 827, 828*t*

metallic bonding, 491, 896, *897*
metalloids
  properties of, 2, 3–4
metalloproteins, 650
metasilicates, 488–489
meteorites, *27*
meter, 19*t*, 23, 24
methane, 182–183, 516, 517*t*, 568, 586–587
  combustion of, 538, 540
  concentration of in air, 343*t*
  dipole moment of, *372*
  hybrid orbitals of, 364–365
  molecular geometry of, 355, *356*
  molecular shape of, 363
  production of, 587–588, 589
methanogenic bacteria, 588, 589
methanol
  fuel value of, 575
  and gasoline, 561, 575
methionine, 632, 833
methyl alcohol, *see* methanol
methylamines, 589
methylbenzene, *see* toluene
methylene groups, 516
methyl groups, 516, 568
methyl mercuric chloride, 865
methyl mercury poisoning, *865*
methyl *tert*-butyl ether (MTBE), 574
metric system, 24
Mexican Hat, Utah, *730*
microcurie, 79–80
microelectronics industry, 415, 492
microscope, scanning electron, *121*
microwave radiation, 324, 353
  see also cosmic microwave background radiation
microwaves, 42
milk of magnesia, 252, 254
Milky Way galaxy, *46*
millerite, 701
millibars, 391
millicurie, 79
Milliken, Robert S., 316
Minamata Bay, Japan, 865
minerals, 486
  solubilities of, 824–826
mining, 492
miscible liquids, 448
Missouri River, 5, *919*
mixtures, 5
  and filtration, 6
  heterogeneous, 211, *212*
  homogeneous, 210, 211, *212*
modulus of elasticity, 931
molality, 232–233
molar concentrations, 215
molar heat capacity, 526–527
molar heat of combustion, 534
molar heat of fusion, 526
molar heat of vaporization, 527
molarity, 215–219, 232–233
molar mass, 175–177, 403–405, 420
  measuring, 238–240
molar volume
  of an ideal gas, 403

molasses
  viscosity of, *459*
mole (mol), 19*t*, 173–180
  chemical reactions and, 173–180
  and mass, 177
molecular compounds, 4
molecular equations, 256
molecular formulas, 4
molecular geometry, 354–362, 363*t*, 367, 370
molecular interactions, *see* intermolecular interactions
molecular ions, 290–291
molecularity (of an elementary step), 698
molecular motion, 392–393, 417–422
molecular-orbital diagrams, 318–321, 322
molecular orbitals (MOs), 316–323, 364, 491–492
  antibonding, 318–320
  of hydrogen, 316–317
molecular-orbital theory, 364, 491–492
  and spectroscopic properties, 322–324
molecular orientation
  and reaction rate, 705, *706*
molecular shape, *see* molecular geometry
molecular vibration, 376
molecules, 4
  expanded-octet, 359–361
  gas-phase, 315
  linear, 361
  nonpolar, 447
  odd-electron, 347, 352, 353
  polar, 292, 371–373, 423, 445–446
  polyatomic, 299
  spectral signatures of, 323
mole fraction, 406–408
mole ratio, 180
Molina, Mario, 305
molten carbonate fuel cell, 851*t*
molybdenite, 651, 658
molybdenum, 658–659
momentum, 119
monosaccharides, 576, 580, 582, 630
monounsaturated fatty acids, 642
Montreal Protocol, 305
motion, molecular, 420–421
motor vehicles
  and lead-acid batteries, 849
  see also automobiles
mountain climbing
  and oxygen, *756*
mullite, 920
multiple proportions, law of, 165
muscalure, *189*
myoglobin, 833
myristic acid, 644

NaCl structure, *see* rock-salt structure
Nagasaki bombing (1945), 74
naming systems
  for alkanes, 567, 568
  for aromatic compounds, 572
  for binary compounds, 166–169
  for biomolecules, 578
  for ethers, 574
  for hydrocarbon chains, 563

naming systems (continued)
  for organic compounds, 588
  for oxoacids, 172
  for polyatomic ions, 169–170
naphthalene, 573
natural gas, 511, 512, 515–517, 589
  alkanes of, 516, 517t, 619
    fuel values of, 541t
  combustion of, 518
natural waters
  buffers in, 817
  pH of, 817
Near Earth Asteroid Rendezvous (NEAR)
    Shoemaker spacecraft, 162
neon, 134
  concentration of in air, 343t
  electron configuration of, 137
  ionization energy of, 145
neon-20, 61
neptunium-237, 92
neptunium-239, 74
Nernst, Walther Hermann, 799, 863
Nernst equation, 863–866, 867, 871
net ionic equations, 243, 256
network solid, 484
neutralization reactions, 806–807
neutral solutions, 802
neutrino, 32–33
neutron bombardment, 74
neutron capture, 62–64, 69, 74
neutrons, 32
  and Big Bang, 33
  decay of, 33–34
  thermal, 62
Newman projection, 636
Newton, Sir Isaac, 11–12, 14, 16, 19
NGC 2264 star cluster, 56
niccolite, 701
Ni-Cd (nicad) batteries, see nickel-cadmium
    batteries
nickel, 701, 702
  and stainless steel, 909
nickel-cadmium (nicad) batteries, 851t, 853–855
nickel-metal hydride (NiMH) batteries, 851t, 860–862
  in electric cars, 879
niobite, 641
niobite-tantalite, 641
niobium, 592, 641
nitrates, 414
nitric acid, 181, 414, 782–783, 812–813
  chemistry of, 798–799
  production of, 798, 799
nitric oxide (nitrogen monoxide), 347
  decomposition of, 764–766
  molecular orbital diagram of, 351–352
  reaction rate with oxygen, 677t
  reaction rates and formation of, 675–676
  reaction with ozone, 706–708
  and smog, 672–674
nitrides, 254
nitrogen
  in blood, 417
  chemistry of, 414–416
  concentration of in air, 343t
  electron configuration of, 137

ionization energy of, 145
Lewis structure of, 290–291
molecular-orbital diagram of, 318–321, 322, 323
oxides of, 164, 165
nitrogen compounds
  and ammonia, 414
nitrogen dioxide, 347, 414
  and carbon monoxide, 704
  concentration of in air, 343t
  decomposition of, 690–691, 743
  dimerization of, 352, 733–734, 764
  electron spin resonance spectrum of, 354
  formation of, 699–700
  reaction rates and formation of, 677–678
  reactions of, 352
  and smog, 672–674
  thermal decomposition of, 710–711
nitrogen fertilizers, 755
nitrogen-14, 63
nitrogen monoxide, see nitric oxide
nitrogen oxides
  and auto exhaust, 713–716
  U.S. production of, 672mn
nitrous acid, 414, 783–784
  ionization of, 784
nitrous oxide (dinitrogen oxide), 688
  formation of, 736–737, 748
  infrared spectrum of, 341–342
  Lewis structure of, 342
  production of, 740–741
noble gases, 446–447
  chemistry of, 134–135
  electron configurations of, 137, 138
  melting and boiling points of, 447t
nodes (in atomic orbitals), 127
Nomex, 934, 937t
nonane, 562t
nonelectrolytes, 222, 223
  van't Hoff factor of, 239
nonmetals
  properties of, 2, 3–4
nonpolar compounds
  viscosity of, 459
nonpolar molecules, 447
nonpolar solutes, 448–449
normal alkanes (n-alkanes), 562–563
northern lights (aurora borealis), 285
n-type semiconductor, 927
nuclear bombs, 78
nuclear decay, 149
nuclear fission, 71, 72, 73
nuclear fusion, 60–61, 62
nuclear reactors, 71, 73, 74–75
nuclei
  formation of, 34–35
nucleic acids, 655
nucleons, 35
nucleotides, 655–657
nucleus
  mass number of, 35
nuclides, 38, 64–65, 66
  artificial, 74
  radioactive, and cancer therapy, 86–87
  and supernovae, 78–79
  synthesis of, 77

Nutrasweet, 639
Nylon, 933, 934, 937t

obsidian, 469
oceans, 208
ocean waves, 436, 437
octadecane, 563
octahedral geometry, 359, 361, 363t
octahedral holes, 478, 480
octane
  fuel value of, 575
  isomers of, 568
  properties of, 562t
  vapor pressure of, 560
octane number, 566
octet rule, 287
  exceptions to, 343, 344–345, 347
octets
  expansion of, 345, 346mn, 347
odd-electron molecules, 347, 351–352, 353
O-H stretching vibration, 336
oil, see crude oil; petroleum
Oil Creek, Pennsylvania, 512
oil prices, 512
oil-producing and -exporting countries (OPEC), 512
oil wells, 512
oleic acid, 644
olivine, 486–488
orbital degeneracy, 136
orbital diagrams, 137, 138
orbital model of atomic structure, 147
orbitals, 123–125, 136, 144–148
  degenerate, 129
  and electron configuration, 132–143
  energies of in multielectron atoms, 140
  shapes and sizes of, 126–131, 132
  see also antibonding molecular orbitals;
    atomic orbitals; bonding molecular
    orbital; hybrid atomic orbitals;
    molecular orbitals
organic compounds, 573–574
  combustion of, 593–595
orthosilicates, 486–488
osmium, 701, 702
osmosis, 224
osmosis, reverse, 228, 229
osmotic pressure, 224, 226–228, 236
  and determining molar mass, 239, 240
Ostwald, Wilhelm, 799
Ostwald process, 755, 799
overpotential, 875, 876
oxidation half-reactions, 844, 845, 855mn, 856, 857
oxidation numbers, 244–245, 800
oxidation potential, 869–870
oxidation-reduction reactions, 241–249, 844
oxidation state, see oxidation numbers
oxides
  of metals as bases, 250
  of nonmetals as acids, 250
oxidizing agents, 245
oxoacids, 344, 799–800
  strengths of, 800–801
oxoanions, 170–172
oxohydroxides, 854

oxygen
  allotropes of, 298
  in blood, 410–412, 756
  chemistry of, 5, 348–350
  concentration of in air, 343t
  ionization energy of, 145
  Lewis structure of, 290–291
  molecular-orbital diagram of, 318–321
  oxidation states of, 244–245
  partial pressure of, 408
  rates of reaction with nitrogen and nitric oxide, 675–676, 677t
oxygen-binding sites
  on hemoglobin, 412
oxygen-16, 60, 61
ozone, 298–299
  concentration of in air, 343t
  and energy profile with nitric oxide, 706–708
  formation of, 684–685
  Lewis structures of, 299–301, 357–358
  loss of, 712
  photochemical decomposition of, 685–686, 712–713
  and resonance structures, 303, 304, 305
  and smog, 672–674
ozone layer
  depletion of, 305–311

packing efficiency (of ions), 476–484
paint thinner, 558
paired electrons, 321
palladium, 592, 701, 702–703, 715, 716
palmitic acid, 644
paper industry, 307, 308
papyrus
  dating of, 91
paramagnetic electron configurations, 321, 501
partial pressures, 406–408, 743–747, 750, 757–760, 762
particle acceleration, 74, 75
pascal, 388
Pascal, Blaise, 388
Paschen series of spectral lines, 115
Pauli, Wolfgang, 132
Pauli exclusion principle, 132
Pauling, Linus, 293, 363–364
pentahydroxyhexanal structure, 578
pentane, 562t
Penzias, Arno A., 42, 45, 324, 940
peptide bonds, 633, 634
peptides, 633
percent composition, 40–41, 184–186, 187
percent yield, 197
perchlorates, 308
periodic table of the elements, 4, 38–40
  and electron affinity, 295–296
  and electronegativity values, 293–294, 295, 296–297
  and ion charges, 167, 168
  and ionization energies, 144–146, 295
  and Lewis symbols, 289
  Mendeleev's, 143
  and naming compounds, 166
  and orbitals, 132–143

  organization of, 140
  and valence electrons, 293–294
periods (of periodic table), 136
permanent dipole moments, 372–373
perovskite structure, 921
peroxides, 178–179
  as oxidizing agents, 244–245
peroxyacetyl nitrate (PAN), 672
Persian Gulf, 230–231
perturbation (of chemical equilibria), see stress
PET imaging, see positron emission tomography
petroleum, 512
  benzene rings of, 573
  refining of, 558–561
pewter, 903
phase change, 525
phase diagrams, 454–456
pH buffers, 813–815
phenanthrene, 573
phenolphthalein, 267, 268, 816, 819
phenol red, 816–817, 819
phenylalanine
  fuel value of, 646
pH, 805
  and acid-base titrations, 819–821
  of solutions of acidic and basic salts, 806–809
  of common substances, 803
  defined, 803
  and Henderson-Hasselbalch equation, 814–815
  lowering of, 817–823
  and pOH, 805
  and oxidation potential, 870mn
  of river water, 811–813
  scale 801, 803
pH indicators, 816–817, 819
phosphate diester, 655
phosphate groups, 655, 656
phosphine, 415
phosphor, 925
phosphoric acid, 415, 851t
phosphorus
  chemistry of, 414–416
  see also red phosphorus; white phosphorus
phosphorus pentachloride, 368–370
phosphorus pentafluoride
  molecular geometry of, 359–360, 361
phosphorylation, 648–649
photochemical cells, 881–882
photochemical smog, see smog, photochemical
photodissociation, 286
photoelectric effect, 111–112
photoelectrons, 111–112
photoionization, 44
photons, 19, 110
  energy of, 117, 119
  and photoelectric effect, 111–112
  spontaneous emissions of, 149
  and zero rest mass, 119
photosynthesis, 190, 193–194, 461, 576, 832
phototube, 111
physical properties (of elements), 2

physiological saline, 225
phytoplankton, 211, 212, 340
pi (π) bonds, 318, 364, 366, 367
pi (π and π^*) molecular orbitals, 318–320
pitchblende, 86
Planck, Max, 18, 110
Planck constant, 18–19
planetesimals, 162
planets
  formation of, 162
plants
  and photosynthesis, 832
  physical structure of, 583–584
plasma, 35, 44, 919
platinum, 701, 702–703, 715
  isotopes of, 40
  in photochemical cells, 881
platinum group metals, 701, 702–703
Plexiglas, 937t
plum-pudding model of atomic structure, 107–108
Plunkett, Roy J., 940, 941
plutonium, 75
plutonium-239, 74, 75
pOH values, 805
polar covalent bonds, 292, 371
polarity
  and solubility, 447–450
polarization (of molecules, ions, and atoms), 446–447
polar molecules, 292, 371–373, 423
  interactions of, 445–446
polar solutes, 448
pollution
  atmospheric, 672–673
  and auto exhaust, 704
  see also acid rain; smog, photochemical
polonium, 84, 348, 350
polonium-218, 70
polyacrylonitrile (PAN), 937t
polyamides, 933–934
polyatomic ions, 169–173, 484
  and oxidation number, 244
polyatomic molecules, 299
polybutadiene, 937t
polycyclic aromatic hydrocarbons (PAHs), 573
polyethylene, 937t, 939–940
polyethylene naphthalate (PEN), 937t
polyethylene terephthalate (PET), 937t
polyethylene vinyl acetate (EVA), 937t
polymerization
  degree of, 585
polymer membrane fuel cell, 851t
polymers
  addition 936–938
  atactic and isotactic, 938–939
  condensation, 937t
  crystalline regions in, 930
  fluorinated, 940
  synthetic, 929, 933–940
polymethyl methacrylate (Plexiglas), 937t
polymorphs, 485
polypropylene, 597, 936–939
polysaccharides, 576, 583, 631
polystyrene, 937t, 939
polytetrafluoroethylene (Teflon), 937t, 940

polyunsaturated fatty acids, 642–643
polyvinyl chloride (PVC), 937t
porphyrin ring, 832, 833
positron emission, 64–65, 67t, 69
positron emission tomography (PET), 88–89
positrons, 56, 57, 58
potassium, 178
   electron configuration of, 139
   ionization of, 146
   XPS spectrum of, 148
potassium hydroxide, 256
potassium nitrate, 179, 414
potential, *see* electromotive force (cell potential); overpotential; oxidation potential; standard cell potentials
potential energy, 513, 514, 515, 520, 521
   see also chemical energy
power
   and work, 871
precipitation reactions, 257–261
precision (in measurement), 30, 31
pressure
   critical, 455
   of mixture of gases, 752–753
   and phase diagrams, 454
   and temperature, 396–398, 399–400
   see also atmospheric pressure; osmotic pressure
pressures, partial, *see* partial pressures
principal quantum number, 123
products (of chemical reactions), 173
promethium, 66
propane, 516, 517
   combustion of, 531–534, 540, 541
   and hot-air balloons, 521–522
propanone, *see* acetone
propene (propylene), 597, 936–938
protein chains
   in silk, 930
   in wool, 931–932
proteins, 631–632
   and DNA, 655–656
   energy values of, 646
   fibrous, 929–930
   food energy in, 654
   fuel value of, 646
   see also peptides
   structures of
      crystalline sections of, 930
      α-helix, 932
      β-pleated sheets, 930
      primary, 930
      random coil, 931
      secondary, 930
      tertiary, 931
protein synthesis, 655–661
proton-exchange membrane (PEM), 880
protons, 32
   and Big Bang, 33, 35
   properties of, 32t
Proust, Joseph Louis, 165
pseudo-first-order rate law and rate constant, 695
p-type semiconductor, 927
pure substances, 2, 3, 5

pyrite, 248
pyroxenes, 487, 488
pyruvate
   conversion of glucose to, 653–654
pyruvic acid, 648

qualitative chemical analysis, 323
quantitative chemical analysis, 323
quantum (of energy), 110
quantum mechanics, 122, 150
quantum numbers, 123–125
   and hydrogen atoms, 113, 115
quantum theory, 149
quarks, 31, 33
quartz, 469, 484–485
   XRD scan of, 470
   see also silicon

racemic mixture, 642
rad (radiation absorbed dose), 80t, 81
radial distribution plot, 126
radiation
   assessing risks of, 83–85
   background, 83
   biological effects of, 80–85
   continuum, 12, 43–45
   cosmic, 42, 46, 83
   particle description of, 110
   units for expressing quantities of, 80t
   see also cosmic microwave background radiation; electromagnetic radiation; infrared radiation; ionizing radiation; ultraviolet radiation
radiation absorbed dose (rad), 80t, 81
radiation dosage, 81–83
radiation dosimeters, 78
radiation sickness, 83
radiation therapy, 86–87
   radionuclides used in, 87t
radicals, 347
radioactive decay, 31–33, 688
   rates of, 33–35
radioactive decay constant, 688
radioactive tracers, 87–88
radioactivity, 78
   detecting, 78–79
   measuring, 79–80
radiocarbon dating, 90–93, 689
radiochemical dating, 89–93
radionuclides, 64, 76
   artificial, 74
   medical imaging with, 87–89
radium, 86, 254, 255
radium-226, 70, 86
radon, 83–85, 134
   chemistry of, 86
   electron configuration of, 137
radon-222, 70, 83, 84, 86
rain
   pH of, 803
rainbows, 11–12
Ramsay, Sir William, 106, 134
Raoult, François Marie, 452
Raoult's law, 452, 558–561, 562
rare earth elements, *see* lanthanide elements

rate constant, 681–682, 688, 689, 705
   pseudo-first-order, 695
   and temperature, 706–708
rate-determining step, 699
rate law, 681–682, 684–685
   integrated, 685
   pseudo-first-order, 695
RBE (relative biological effectiveness), 81, 82
reactants, 173
   limiting, 194
   battery power and quantities of, 872
reaction mechanisms, 697–704
reaction order, 681
   and reaction rates, 679–682
reaction quotient, 738
   for combined equations, 740–741
   and equilibrium constant, 739t
reaction rates, 673–678
   effect of reactant concentration on, 679–695
   and formation of nitrogen dioxide, 677–678
   forward and reverse, 733–734
   instantaneous, 677–678
   and reaction order, 679–682
   and temperature, 705, 708t
reactions, chemical
   addition, 936–938
   electron loss and gain in, 844
   elementary steps in, 698
   endergonic, 625
   endothermic, 519, 614, 706, 711
   enthalpy changes of, 539
   and equilibrium, 732–733, 747
   and equilibrium constants, 736–742
   exergonic, 625
   exothermic, 519, 521, 532, 538, 616, 622–623, 624, 706, 707, 711
   first-order, 685–688, 689, 694–695
   half-life of, 688, 689
   intramolecular, 528
   ion-exchange, 263–266
   and the mole, 173–180
   neutralization, 806–807
   precipitation, 257–261
   rate-determining steps in, 699
   reverse, and equilibrium, 739–740
   second-order, 690–695
   zero-order, 716
   see also combustion reactions; condensation reactions; cyclization reaction; spontaneous reactions
reactors
   heavy-water, 73
   nuclear, 71, 73, 74–75
real gases, 422–424
Rebok, Jack, 940
recession speed, 21
recycling
   aluminum, 914–915
red blood cells, 224–225
redox reactions (oxidation-reduction reactions), 241–249, 844, 846mn, 848, 868, 883mn
   balancing of, 246–249
   potential of, 870mn

red phosphorus, 414–415
redshift, 20–21
reducing agents, 245
reduction half-reactions, 844, 845, 855mn, 856
refraction, 12
regenerative braking, 879
relative biological effectiveness (RBE), 81, 82
relativity, special theory of, 118–119
rem (roentgen equivalent man), 80t, 81
resonance structures, 303, 305, 312–313
respiration
  and carbon cycle, 190, 191
rest mass, 119
reverse osmosis, 228, 229
rhenium, 651
rhodium, 701, 703, 715–716
ribonucleic acid (RNA), 657, 660–661
  see also messenger RNA; transfer RNA
ribosomes, 660
ring-closure reaction, see cyclization reaction
(R) isomers, 637, 639
Ritalin, 639
Ritter, Johann Wilhelm, 13
road salt, 261
rock-salt structure (of crystals), 478
roentgen, 80t, 81
roentgen equivalent man (rem), 80t, 81
root-mean-square speed (of gas molecules), 417–419, 420, 421
Rowland, Sherwood, 305
rubidium, 106, 178
rubies, 497–498, 658
rule of eight, see octet rule
ruthenium, 701, 702
Rutherford, Ernest, 74, 108, 143
rutherfordium, 77
rutile, 916

saccharides, 576
  fuel value of, 646
saline
  physiological, 225
salt, 167, 179
  as conductor of electricity, 222–223
  and freezing, 234, 235
  see also sea salt; sodium chloride
salt bridge, 845, 846
saltpeter, see potassium nitrate
salt water, see seawater
saturated fatty acids, 642–643, 644
saturated hydrocarbons, see alkanes
saturated solutions, 257, 258
scandium, 923
  electron configuration of, 139, 142
Schrödinger, Erwin, 122–123
Schrödinger equation, 131
scientific law, 165
scientific method, 9, 10, 940–941
scuba divers, 413–417
scuba tanks, 521
Seaborg, Glenn T., 76, 77
seaborgium, 77
sea of electrons model of metallic bonding, 896, 897

Searles Lake, 476
sea salt, 7, 436, 437–438
sea spray, 436, 437–438
seawater, 210–211, 212, 257, 258, 437–438
  boiling point of, 231
  desalinization of, 229, 230–231
  distillation of, 7, 230–231
  as electricity conductor, 221–223
  and freezing point, 235
  major ions in, 214–216
  mixed with fresh water, 219–220
  particles suspended in, 269
  pH values of, 806, 819
  vapor pressure of, 450–451
sec-butyl group, 568
Secchi disk, 213
secondary structure, 930
second law of thermodynamics, 616–617, 619–620
second-order reactions, 690–695
seesaw geometries, 361, 363t
selenic acid, 350
selenides, 350
selenium, 348, 350
semiconductor alloys, 927
semiconductors, 926–928
semipermeable membrane, 224
Shroud of Turin, 90
sigma (σ) bonds, 316–317, 364, 366, 367
sigma (σ and σ^*) molecular orbitals, 316–317
significant figures, 25–29
silicates, 486–489
silicates, condensed, see metasilicates
silicon, 2
  in aluminum alloys, 915
  chemistry of, 492
silk, 929–930
silk proteins, 930–931
silver, 894
  chemistry of, 864, 898
  uses of, 899
silver halides, 899
silverware, 878
simple cubic structure, 474, 480t, 481
simple sugars, see monosaccharides
(S) isomers, 637, 639
SI units, 16t, 19t, 23–24
  conversion factors for, 28t
slag, 894, 904, 905
smectites, 918
smelting, 894
  of iron, 904–906
smog, photochemical, 671–673
  formation of, 732, 740
smoke
  particles in, 269, 269
snow
  melting of, 525–526, 527
soda ash, see sodium carbonate
sodium, 178
  electron configuration of, 137–138
  ionization energy of, 145–146
  spectrum of, 131

sodium acetate, 259, 821
sodium atoms, 148–149
  absorption and emission lines of, 106, 107
  ground-state, 118
sodium bicarbonate, 179
sodium carbonate (soda ash), 179, 256, 806
sodium chloride
  crystal structure of, 467–468, 472–475, 476
  density of, 475
  and electricity conduction, 222–223
  freezing point of, 237
  ions of, 223
  lattice energy of, 441, 443
  Lewis diagram of, 292–293
  packing efficiency of, 476–478
  and reduction reaction, 241, 243
  as road salt, 261
  saturated solution of, 258
  van't Hoff factor for, 236
  see also halite; rock-salt structure; salt
sodium fluoride, 168
sodium ions, 243
sodium nitrate, 414
sodium tetraborate decahydrate, see borax
sodium thiosulfate, 613
soft water, 265
solar eclipse, 104
solar energy
  and photochemical cells, 881–882
solar nebula, 162
solar spectrum, 12–13, 108–109, 110
  and Fraunhofer lines, 20, 105–106
solar wind, 162, 286
solids, 438–439, 519
  amorphous, 469
  in chemical equilibria, 768
  crystalline, 467–471, 472–476
  network, 484
  and phase diagrams, 454–455
solid solutions, 210
solubility
  of ionic compounds, 259–261
  and polarity, 447–450
solubility-product constant, 824–825
solute concentration, 214, 266–267
  and vapor pressure, 452
solutes, 210, 211, 224
  nonpolar, 448–449
solution, heat of, 612
solution concentration, 214–219, 221
solutions, 210–211, 212
  acidic, 802
  aqueous, 210
  basic, 802
  buffer, 813–815
  chemical reactions in and life, 208
  colligative properties of, 228–236
  concentrations in, 214, 266–267
  enthalpies of, 612–615
  freezing and boiling points of, 235
  ideal, 561
  isotonic, 228
  neutral, 802
  pH values of, 805, 814–815
    and pOH values, 805

solutions (continued)
   saturated, 257, *258*
   solid, 210
   standard, 266
   supersaturated, 257, *259*
   and titration, 266–267
   and van't Hoff factor, 236
   vapor pressure of, 451
Solvay process, 179
solvents, 210, *211*, 224
Sørenson, Søren, 803
space-filling models, 472, *473*, 517
space shuttle, *4*, *332*, 591
spandex, *see* Lycra
specific heat, 526
spectator ions, 243, 256
spectroscopy
   UV-visible absorption, 323–324
   infrared absorption, 323, 340–342
   X-ray photoelectron, 146–148
spectrum, 12
   continuous, 110
   electromagnetic, 13
sphalerite, 479, 483, 864, 910, *see also* zinc sulfide
sphere of hydration, 445
sp hybrid orbitals, 367, 368
sp^2 hybrid orbitals, 366, 368, *370*
sp^3 hybrid orbitals, 364–365, *370*
spin (of electrons), 118, 131–132, 501–502
spin magnetic quantum number, 131, *132*
spontaneous processes, 616–617, 622–623, 625, 627, 711
   and free energy, 628, 747–748
   (voltaic) cell reactions, 857
spring bloom, 460
square planar geometries, *360*, 363t, 500
square pyramidal geometries, *360*, 363t
stainless steels, 909
stalactites, 252, 257
stalagmites, 252, 257
standard cell potentials, 856–857, 867–869
   of batteries, 856–857, 863
   effect of concentration on, 863–870
   and temperature, 871
standard electrode potentials, 856, 863, 868mn
standard enthalpy, 537–539
standard heat of formation, 537
standard hydrogen electrode (SHE), 857–859
standard molar entropy, 618
standard state (of an element), 537
standard temperature and pressure (STP), 402–403
standing waves, 120
starch, 631
   hydrolysis of, 583–585
   structure of, 583
starlight
   redshift in, 20–21
stars
   formation of, 55, 56
   see also supernovae
state function, 513, 514, 515
steam-reforming process, 539
stearic acid, 643, 644

steel, 658, 702, 907–909, 911
   worldwide production of, 908
steelmaking, 905, 906mn
stereoisomerism, 635–642
Stern, Otto, 132
stibnite, 415
stilbine, 415
stoichiometry, 180, *193*
stomach acid, *252*, 256
stratosphere, *286*, 298
   CFCs in, 310
   and ozone, 305
stress
   and equilibrium, 751–756
strong nuclear force, 32
strontium, 255
   emission spectrum of, *139*
strontium-90, 71
structural isomers, 567, 635
Styrofoam, 939
subbituminous coal, 591t
sublimation, 454
sublimation points, 454
substitutional alloys, 901–902, 903
sucrose, 582
sugar, *see* disaccharides; fructose; glucose; monosaccharides; sucrose
sulfur, 308
   chemistry of, 348–350
   deposits of, 348
   oxidation number of, 248
   oxidation state of, 246
   oxides of, 164, 165
   structure of, 490
sulfur dioxide, 349
   concentration of in air, 343t, 344
   oxidation of, 745, 767
sulfur hexafluoride, 346mn, 368, 369
   molecular geometry of, *359*, *360*
sulfuric acid, 173, 253–256, 267, *268*, 349–350, 786–788, 797–799
   chemistry of, 798–799
   ionization of, 786–788, 789t
   production of, 798–799
   structure of, 344–345
   synthesis of, 798–799
sulfurous acid, 797–799
   ionization of, 789t
sulfur tetrafluoride
   molecular geometry, 360–361
sulfur trioxide, 745
sun, *see* entries under solar
sunlight
   and photochemical cells, 881–882
   and rainbows, 11–12
sunscreens, 864
superconductors, 911, 921
   ceramic, 921–926
supercritical fluid, 455
superheavy elements, 77–78
supernovae, 67–71, 78–79
superoxides, 179
supersaturated solutions, 257, *259*, 476
surface tension, 457, *458*
surroundings (of a thermodynamic system), 518
swamp gas, *see* methane

swimming pools
   and pH, 806, 809, 816–817
symmetric-stretch vibration, 334–336, 376
synthetic polymers, 933–940
Système International d'Unités, *see* SI units
systems, thermodynamic, 518, 520

table salt, 467–468
tacticity
   and polypropylene, 938
talose, 580
tantalum, 641
Taurus Mountains, Turkey, 902
technetium, 66, 650
technetium-99m, 88
Teflon, 937t, 940
television screens, 925
tellurium, *348*, 350
Telstar, 42
temperature
   and altitude, 396
   critical, 455, 922
   and equilibrium constants, 763–766
   and molecular motion, 396
   and phase diagrams, 454
   and pressure, 396–398, 399–400
   and reaction rate, 708t
temperature scales, 35–37
temporary dipole, 376
tensile strength, 931
teratogen, 639
termolecular elementary steps, 698, *699*
tert-butyl group, 568, 574
tert-butyl methyl ether, 574
tetraamine copper(II), 829
tetrafluoroethylene (TFE) gas, 940
tetrahedral holes, 477, 478, 479, 480
tetrahedral molecular geometry, 355–357, 363t, *499*
thalidomide, 639–642
thallium, 910, 911
theoretical yield, 197
theory, scientific, 9
thermal neutrons, 62
thermochemical equations, 542–543
thermochemistry, 513
thermocline, 460–461
thermodynamic systems, 518, 520
thermodynamics, first law of, 518
thermodynamics, second law of, 616–617
thermodynamics, third law of, 618
thermosphere, *285–286*, 290
thiosulfate
   oxidation of, 248–249
third law of thermodynamics, 618
Thomson, George, 121–122
Thomson, J. J., 287
thorium, 916
thorium-230, 70
thorium-234, 69–70
"302" stainless steel, 909
thymine, 655, *656*, 657
tin
   and bronze, 901–902
   chemistry of, 493
   oxides of, 493

tin-116, 63
titanium, 254, 592
   chemistry of, 916–917
   electron configuration of, 141, 142
titanium dihydride, 917
titanium dioxide, 917
titanium(IV) chloride, 916, 917
titanium hydrides, 592
titration, 266–267, 268
   acid-base, 817–823
toluene (methylbenzene), 571
torr, 390
Torricelli, Evangelista, 389, 390
tracer, radioactive, 87–88
transcription, 660
transfer RNA, 661
trans isomers, 598
transitional bacteria, 587
transition metals, 139–141, 494, 701–703
   binary compounds of, 169
   in biomolecules, 831
transition states, 705
translation, 661
transmutation, 65
transuranic elements, 74
tridymite, 485
triglycerides, 643
trigonal bipyramidal molecular geometry, 359*mn*, 360, 361, 363*t*, 368–370
trigonal planar molecular geometry, 357, 358, 363*t*, 366, 368
trigonal pyramidal molecular geometry, 356, 363*t*
trimethylamine, 589
triple helix, 932
triple point, 455
tristearin, 646
tritium, 38
troposophere, 286
   constituents of, 343
   changing composition of, 311
   and temperatures, 396
T-shaped molecular geometry, 361, 363*t*
tungsten, 659
tunicates, 640–641
turbidity, 211, 213
turquoise, 499–500
Tyndall effect, 269

Uhlenbeck, George, 131
ultraviolet (UV) radiation, 13
   and ozone, 298, 305
ultraviolet spectra, 322–324
undecane, 562*t*
unimolecular elementary steps, 698
unit cell, 472–475
United States
   carbon dioxide production in, 339
   ethanol production in, 576
   ionizing radiation levels, 84
   nitrogen oxide emissions in, 672, 715
   production of chemicals in, 798
   radon concentrations in, 85
U.S. Food and Drug Administration, 630
universe
   atomic abundances of elements in, 94
   composition of, 93–94, 162–163

   theories of creation of, 9–11, 162
   early, 40–46
   entropy of the, 617
   expansion of, 11, 21, 35
unpaired electrons, 352–354
unsaturated fatty acids, 643, 644
unsaturated solutions, 257
uracil, 660
Ural Mountains, 689
uraninite (uranium dioxide), 86
uranium, 307
   chemistry of, 86
uranium hexafluoride, 940
uranium isotopes, 69–70
uranium mines
   and lung cancer, 85
uranium-235, 86, 92
uranium-235 fission, 71, 72, 73
uranium-238, 69–70, 74, 75, 86
Uranus
   atmosphere of, 162

valence bands
   of insulators, 920
   of metals, 901
   of semiconductors, 926–928
valence-bond theory, 363–370
valence electrons, 288–289, 292, 312
   and Lewis structures, 299–300
   lone pairs, 290
   and the periodic table, 293–294
valence shell electron pair repulsion (VSEPR) model, 355, 367
valence-shell electrons, 439, 440
valency, 288
vanadinite, 640
vanadium, 640–641
   electron configuration of, 141
van der Waals constants, 424
van der Waals equation, 423–424
van der Waals forces, 135
van't Hoff, Jacobus Henricus, 799
van't Hoff equation, 765
van't Hoff factor, 236, 237, 239
vaporization, molar heat of, 527
vapor pressure, 409, 450–453
   and solute concentration, 452
   of solutions, 451
   and temperature, 452–453
   of volatile compounds, 558–559, 560, 562
   of water, 409*t*
vehicles
   electric, 878–881
   hybrid, 879
Venus
   atmosphere of, 324, 334
   and carbon dioxide, 337
vibration, molecular, 341, 342, 376, 521
viscosity, 459
visible spectra, 322–324
volatile compounds, 7
   vapor pressure of, 558–559, 560, 562
volcanic activity, 164
   and Earth's early atmosphere, 160
   on primordial Earth, 173
volcanic glass, *see* obsidian

Volta, Alessandro, 852*mn*
voltages
   of batteries, 856
voltaic cells, 243, 845, 846, 847–848, 851*t*, 857, 867, 874, 878
volts, 848, 871
volume
   equivalent units of, 28*t*
   SI unit for, 24
VSEPR model *see* valence shell electron pair repulsion model

Waage, Peter 735
water, 4
   acids in, 790–791, 797, 828
   alkaline, 819, 828
   autoionization of, 801–803
   bases in, 792–793, 797
   conversion to steam, 527
   deionized, 266
   distillation of, 6–7
   and Earth, 209–211, 212
   as conductor of electricity, 221–223
   filtration of, 5–6
   freezing of, 460, 626
   freezing and boiling points of, 240*t*
   and freezing-point depression, 234
   hard, 265
   heat flow and transfer in, 524–530
   and ionic compounds solubility in, 445
   on Mars, 210
   molecular geometry of, 356, 365
   phase diagram of, 454–455
   as polar molecule, 371–374
   properties of, 5, 457–461
   purification of, 228, 265
   soft, 265
   triple point of, 454, 455
   vapor pressure of, 409*t*, 452–453
   vibration modes of, 376
   viscosity of, 459
   see also aqueous solutions; fresh water; natural waters; seawater
water gas, 539
water supply
   and salt contamination, 261
water vapor
   and absorption of infrared radiation, 336
   in Earth's atmosphere, 337–338
   and pressure, 451–452
Watson, James D., 655
watt, 871–872
wave functions, 122–123
wavelength, 14–15
wave mechanics, 122
   see also quantum mechanics
waves
   electron, 120–122
   standing, 120
weak electrolytes, 223
weak-link principle, 26–28
weak nuclear force, 32
weathering, chemical, 250
   of feldspars, 258–259
weather maps, 391
weaving, 929
West Caicos Island, 476

wetlands, 868, 869
white-hot radiation, *43*
white-hot temperature, 44
white light, 18
white phosphorus, 414–415
Wilson, Robert W., 42, 45, 324
Wolf, Alfred P., 88
Wollaston, William Hyde, 20
wood
    components of, 585
    as fuel, 512, 586
wood preservatives, 415
wool, 929–930, 931–932
work, 513–514, 871
    on a thermodynamic system, 521–522
work function, 112
work-hardening, 896–900
wrought iron, 904, 907
würtzite, 483–484, *see also* zinc sulfide
wüstite, 184

xenon, 134, 135
    electron configuration of, 137
X-ray diffraction, *121*, 469*mn*, 470–471
X-ray diffractometer, *470*, 471
X-ray imaging, 13
X-ray photoelectron spectroscopy (XPS), 146–148
X rays
    energy of, 146–147
    and radiation dosage, 81
    wavelike properties of, *121*
xylene (dimethyl benzene), 571, 573*t*

yeast fermentation, 583
yield
    percent and theoretical, 197
yogurt
    fuel value of, 647
Young, Thomas, 16
ytterbium, 923

yttrium, 921, 923
yttrium-barium-copper oxides, 921

zero-order reactions, 716
zero rest mass, 119
zinc
    and aluminum alloys, 915
    chemistry of, 864, 865
    reaction with copper(II) compounds, 844–846
    reduction reactions, 244, *245*
zinc blende, *see* sphalerite
zinc chloride, 244
zinc hydroxide, 829
zinc oxide, 864
zinc selenide, 928
zinc sulfide, 479, 483–484, 864, 910
zirconium, 592
    chemistry of, 916–917
zircons, 916

SI Base Units

SI Base Quantity	Unit	Symbol
length	meter	m
mass	kilogram	kg
time	second	s
amount of substance	mole	mol
temperature	kelvin	K
electric current	ampere	A
luminous intensity	candela	cd

SI Derived Units

SI Derived Quantity	Unit	Symbol	Dimensions
acceleratiobn	meter per second per second	—	$m\ s^{-2}$
Celsius temperature	degree Celsius	°C	K
electric charge	coulomb	C	$A\ s$
electric field	volt per meter	—	$V\ m^{-1}$ or $N\ C^{-1}$
electric potential	volt	V	$J\ C^{-1}$
force	newton	N	$kg\ m\ s^{-2}$
frequency	hertz	Hz	s^{-1}
momentum (impulse)	newton second	—	$kg\ m\ s^{-1}$
power	watt	W	$J\ s^{-1}$
pressure	pascal	Pa	$N\ m^{-2}$
radioactivity	becquerel	Bq	s^{-1}
speed or velocity	meter per second	—	$m\ s^{-1}$
work, energy, or heat	joule (newton meter)	J	$kg\ m^2\ s^{-2}$

SI Prefixes

Prefix	Symbol	Multiplier	Prefix	Symbol	Multiplier
deci	d	10^{-1}	deka	da	10^1
centi	c	10^{-2}	hecto	h	10^2
milli	m	10^{-3}	kilo	k	10^3
micro	μ	10^{-6}	mega	M	10^6
nano	n	10^{-9}	giga	G	10^9
pico	p	10^{-12}	tera	T	10^{12}
femto	f	10^{-15}	peta	P	10^{15}
atto	a	10^{-18}	exa	E	10^{18}

Special Units and Conversion Factors

Quantity	Unit	Symbol	Conversion
energy	electron volt	eV	$1\text{ eV} = 1.60217733 \times 10^{-19}$ J
force	pound	lb	$1\text{ lb} = 0.453592$ kg (earth)
heat	calorie	cal	$1\text{ cal} = 4.184$ J
length	angstrom	Å	$1\text{ Å} = 10^{-8}\text{ cm} = 10^{-10}$ m
length	inch	in	$1\text{ in} = 2.54$ cm
mass	atomic mass unit	u	$1\text{ u} = 1.6605402 \times 10^{-27}$ kg
pressure	atmosphere	atm	$1\text{ atm} = 1.01325 \times 10^5$ Pa
pressure	torr	torr	$1\text{ torr} = 1\text{ atm}/760$
temperature	Celsius scale	°C	$°C = K - 273.15$
temperature	Fahrenheit scale	°F	$°F = \frac{9}{5}(°C) + 32$
time	minute	min	$1\text{ min} = 60$ s
time	hour	h	$1\text{ h} = 60\text{ min} = 3500$ s
time	day	d	$1\text{ d} = 24\text{ h} = 86{,}400$ s
time	year	y	$1\text{ y} = 365.25$ d
volume	liter	L	$1\text{ L} = 1\text{ dm}^3 = 10^{-3}\text{ m}^3$
volume	cubic centimeter	cm^3, cc	$1\text{ cm}^3 = 1\text{ mL} = 10^{-3}$ L

Physical Constants

Quantity	Symbol	Conversion
acceleration due to gravity (earth)	g	9.80665 m s^{-2} (exact)
Avogadro's number	N_0	$6.0221367 \times 10^{23}\text{ mol}^{-1}$
Bohr radius	a_0	$0.529177249 \times 10^{-10}$ m
Boltzmann's constant	k_B	$1.380658 \times 10^{-23}\text{ J K}^{-1}$
electronic charge-to-mass ratio	$-e/m_e$	$1.75881962 \times 10^{11}\text{ C kg}^{-1}$
elementary charge	e	$1.60217733 \times 10^{-19}$ C
Faraday constant	$\mathscr{F}$	$9.6485309 \times 10^4\text{ C mol}^{-1}$
gravitational constant (universal)	G	$6.67259 \times 10^{-11}\text{ m}^3\text{ kg}^{-1}\text{ s}^{-2}$
mass of an electron	m_e	$9.1093897 \times 10^{-31}$ kg
mass of a neutron	m_n	$1.6749286 \times 10^{-27}$ kg
mass of a proton	m_p	$1.6726231 \times 10^{-27}$ kg
molar volume of ideal gas at STP	V_m	$22.41410\text{ L mol}^{-1}$
permittivity of vacuum	ϵ_0	$8.854187817 \times 10^{-12}\text{ C}^2\text{ N}^{-1}\text{ m}^{-2}$
Planck's constant	h	$6.6260755 \times 10^{-34}$ J s
Rydberg constant	R_∞	$2.1798741 \times 10^{-18}$ J
speed of light in vacuum	c	$2.99792458 \times 10^8\text{ m s}^{-1}$ (exact)
universal gas constant	R	$8.314510\text{ J mol}^{-1}\text{ K}^{-1}$
		$0.08205783\text{ atm L mol}^{-1}\text{ K}^{-1}$